Biology

seventh edition

Sylvia S. Mader

Boston Burr Ridge, IL Dubuque, IA Madison, WI New York San Francisco St. Louis
Bangkok Bogotá Caracas Lisbon London Madrid
Mexico City Milan New Delhi Seoul Singapore Sydney Taipei Toronto

McGraw-Hill Higher Education

A Division of The ***McGraw-Hill*** *Companies*

BIOLOGY, SEVENTH EDITION

Published by McGraw-Hill, an imprint of The McGraw-Hill Companies, Inc., 1221 Avenue of the Americas, New York, NY 10020.

Some ancillaries, including electronic and print components, may not be available to customers the United States.

This book is printed on recycled, acid-free paper containing 10% postconsumer waste.

1 2 3 4 5 6 7 8 9 0 VNH/VNH 0 9 8 7 6 5 4 3 2 1 0

ISBN 0–07–013657–2
ISBN 0–07–118080–X (ISE)

Vice president and editor-in-chief: *Kevin T. Kane*
Publisher: *Michael D. Lange*
Sponsoring editor: *Patrick E. Reidy*
Developmental editor: *Anne Melde*
Editorial assistant: *Dianne Love*
Senior marketing manager: *Lisa L. Gottschalk*
Senior project manager: *Marilyn M. Sulzer*
Production supervisor: *Sandy Ludovissy*
Designer: *K. Wayne Harms*
Cover/interior designer: *Christopher Reese*
Cover image: *Mitch Reardon/Tony Stone Images*
Senior photo research coordinator: *Lori Hancock*
Photo research: *Connie Mueller*
Senior supplement coordinator: *Audrey A. Reiter*
Compositor: *GTS Graphics, Inc.*
Typeface: *10/12 Palatino*
Printer: *Von Hoffman Press, Inc.*

The credits section for this book begins on page C–1 and is considered an extension of the copyright page.

Library of Congress Cataloging-in-Publication Data

Mader, Sylvia S.
Biology / Sylvia S. Mader. — 7th ed.
p. cm.
Includes bibliographical references and index.
ISBN 0-07-013657-2
1. Biology. I. Title.

QH308.2 .M23 2001
570—dc21 00–035477
CIP

INTERNATIONAL EDITION ISBN 0–07–118080–X

www.mhhe.com

Brief Contents

Table of Contents

Preface

When I write *Biology* I always feel that I am speaking directly to you, the reader. This occurs, no doubt, because I taught biology before I began to write about it. During those years, I began to develop techniques for making science-shy students successful in their study of biology. The first edition of *Biology* had a learning system that has continued to improve through feedback from both students and instructors. The end result is the seventh edition you hold in your hands. The major topics in *Biology* are now numbered. This allows instructors to assign just certain portions of the chapter and permits study in terms of the concepts listed in the chapter outline. With the advent of the computer it became possible for me to page the text so that each illustration is on the same or facing page as the reference.

Most illustrations in this edition are new or revised to improve their student appeal and educational design. Each type of molecule (fat, protein, nucleic acids) is colored the same, not just in the chemistry chapters but throughout the book. Each type organelle is colored the same in the cell chapter and also in all chapters. The addition of icons relates the part to the whole and the addition of micrographs relates art to an actual structure. Icons occur in the cell, ecology, diversity, and systems chapters. A cell icon shows the usual location of organelles. A cellular respiration icon depicts the connections between various pathways. Phylogenetic tree icons remind students of the evolutionary relationship among groups of organisms.

The emphasis on the scientific process has been strengthened in this edition of *Biology*. The first chapter has an improved description of the scientific process using a more appealing experiment from literature. The end matter of each chapter now contains *Thinking Scientifically* questions and these give students an opportunity to participate in the scientific process and learn to think critically. The *Science Focus* readings are more varied this edition. Some of these are written by contemporary biologists who tell how they go about doing their research and how these findings can be applied to human beings.

A *Bioethical Issue* section has been added to the end of most chapters. These issues pertain to a wide range of topics that require thoughtful consideration by society at this time. The issues end with appropriate questions to help students center their thoughts and arrive at an opinion. The myriad of topics considered include genetic disease testing, human cloning, AIDS vaccine trials, animal rights, responsibility for one's health, and fetal research.

New Chapters and Sections

Biology is an introductory text that covers the concepts and principles of biology from the structure and function of the cell to the organization of the biosphere. It draws upon the entire world of living things to bring out an evolutionary theme that is introduced from the start. The concept of evolution is necessary to understanding the unity and diversity of life and serves as a background for the study of ecological principles and modern ecological problems.

Every chapter of *Biology* has been revised and skillfully updated so that this edition has not grown in page length. These chapters are new:

Chapter 9, Cellular Reproduction and the Cell Cycle, has been reorganized and now ends with a discussion of the relationship between the cell cycle and cancer.
Chapter 27, Conservation Biology, is a new chapter which discusses the current biodiversity crisis including why we should care, the root causes, and how to preserve species and prevent extinctions.
Chapter 30, The Protists, was thoroughly updated to better reflect modern concepts regarding protist diversity.
Chapter 37, Nutrition and Transport in Plants, now begins with a section that includes a discussion of soil formation and its nutritional functions. In this chapter, new figures support an improved discussion of xylem and phloem transport.
Chapter 46, Neurons and Nervous Systems, has been reorganized and rewritten for better flow. The more thorough discussion of the central nervous system includes a new section on learning and memory.
Chapter 49, Hormones and Endocrine Systems, better emphasizes the role of hormones in homeostasis and the effects of imbalance on the human phenotype.

A more complete list of updates for the seventh edition is available on the *Biology* website.

Readings

The readings are organized into four types. The *Science Focus* readings are often written by contemporary scientists who tell us about their particular type of research and how they became interested in this field of biology. The *Health Focus* readings usually give practical information concerning some particular topic of interest, such as proper nutrition and how to prevent cancer. The *Ecology Focus* readings draw attention to some particular environmental problem such as the need to preserve tropical rain forests and the relationship between ozone holes and skin cancer. The *Closer Look* readings are designed to expand, in an interesting way, on the core information presented in each chapter.

Science Focus

Ecology Focus

Health Focus

A Closer Look

The Learning System

Each chapter in *Biology* is comprised of basic features that serve as the pedagogical framework for the chapter. Before you begin reading the text, spend a little time looking over these pages. They provide a quick guide to the learning tools found throughout the text that have been designed to enhance your understanding of biology.

The Essential Study Partner CD-ROM

An icon has been placed throughout the text to alert you to topics that are also discussed in detail on the Essential Study Partner CD-ROM that accompanies their text.

Chapter Concepts

The chapter begins with an integrated outline that numbers the major topics of the chapter and lists the concepts for each topic.

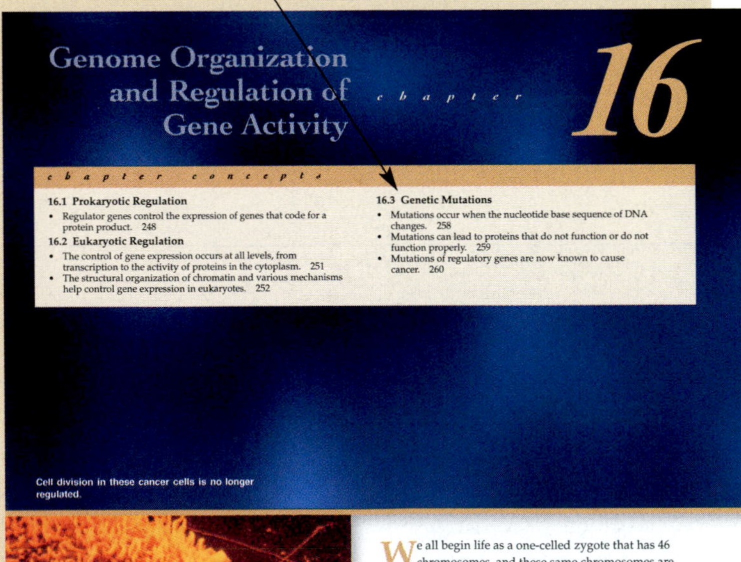

We all begin life as a one-celled zygote that has 46 chromosomes, and these same chromosomes are passed to all the daughter cells during mitosis. Yet cells become specialized in structure and function. The genes for digestive enzymes are expressed only in digestive tract cells and the genes for muscle proteins are expressed only in muscle cells. This shows it is possible to control transcription and translation of genes in each particular type cell.

In this chapter we will see that regulation of gene activity extends from the nucleus (whether a gene is transcribed) to the cytoplasm (whether an enzyme is functioning). Complete knowledge of how gene expression is controlled has far-ranging applications. Only then will we understand how development progresses and what causes the occurrence of various disorders, including cancer. Cancer develops when genes that promote cell division are expressed and when genes that suppress cell division are not expressed. It has become increasingly apparent that the regulation of gene expression is of major importance to the health of all species, including human beings.

chapter 16

chapter 16 Genome Organization and Regulation of Gene Activity 16-5 251

16.2 Eukaryotic Regulation

The **genome** is all of a cell's DNA including all of its genes that code for a protein product. As we have seen in prokaryotes, the genome is organized into regulators and operons that control the expression of structural genes. In eukaryotes, there is no universal regulatory mechanism that controls the activity of coding genes. Instead regulation is possible at any point in the pathway from gene to functional protein. These four levels of gene expression are commonly seen in eukaryotes (Fig. 16.4):

1. **Transcriptional control:** In the nucleus a number of mechanisms serve to control which structural genes are transcribed or the rate at which transcription of the genes occurs. These include the organization of chromatin and the use of transcription factors that initiate transcription, the first step in the process of gene expression.
2. **Posttranscriptional control:** Posttranscriptional control occurs in the nucleus after DNA is transcribed and primary mRNA is formed. Differential processing of preliminary mRNA before it leaves the nucleus, and also the speed with which mature mRNA leaves the nucleus, can affect the ultimate amount of gene product.
3. **Translational control:** Translational control occurs in the cytoplasm after mRNA leaves the nucleus and before there is a protein product. The life expectancy of mRNA molecules (how long they exist in the cytoplasm) can vary, as can their ability to bind ribosomes. It is also possible that some mRNAs may need additional changes before they are translated at all.
4. **Posttranslational control:** Posttranslational control, which also takes place in the cytoplasm, occurs after protein synthesis. The polypeptide product may have to undergo additional changes before it is biologically functional. Also, a functional enzyme is subject to feedback control—the binding of an end product can change the shape of an enzyme so that it is no longer able to carry out its reaction.

We will see that regulation of gene expression in eukaryotes means there is cellular control of the amount and activity of a gene product.

Control of gene expression occurs at four levels in eukaryotes. In the nucleus there is transcriptional and posttranscriptional control; in the cytoplasm there is translational and posttranslational control.

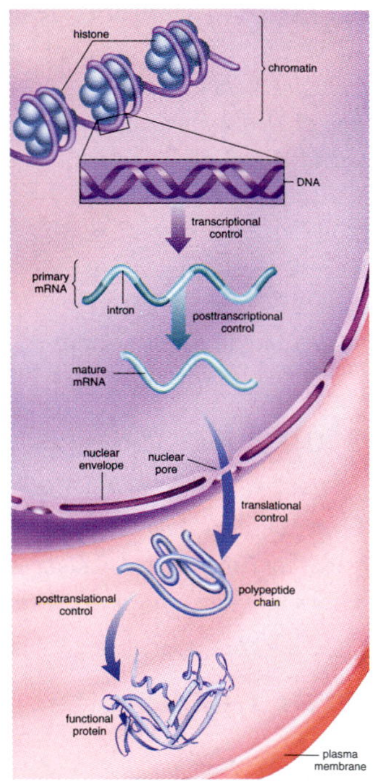

Figure 16.4 Levels of gene expression control in eukaryotic cells.
Transcriptional and posttranscriptional control occur in the nucleus. Translational and posttranslational control occur in the cytoplasm.

Internal Summary Statements

Internal summaries stress the chapter's key concepts. These appear at the ends of major sections and help you focus their study efforts on the basics.

Connecting Concepts

Basic research into the nature and organization of genes in various organisms allowed geneticists to produce recombinant DNA molecules. A knowledge of transcription and translation made it possible for scientists to manipulate the expression of genes in foreign organisms. These breakthroughs have spurred a biotechnology revolution. One result is that bacteria and eukaryotic cells are now used to produce vaccines, hormones, and growth factors for use in humans. Today, plants and animals are also engineered to produce a product or have characteristics desired by humans. In addition, biotechnology offers the promise of treating and even someday curing human genetic disorders such as muscular dystrophy, cystic fibrosis, hemophilia, and many others. Mapping the human genome relies on both traditional linkage data and biotechnology techniques to discover the locus of many other genetic disorders. This is the first step toward the gene therapy that may result in a cure of various genetic disorders.

Some people are concerned that transgenic organisms may spread out of control and wreak ecological havoc, much as plants and animals new to an area sometimes do. Or they worry that transgenic plants may pass on their traits, such as resistance to pesticides or herbicides, to weeds that will then prosper as never before. Thus far a "superweed" has not occurred but that does not mean it could not happen in the future. The alteration of organisms as a result of genetic changes is one of the definitions of evolution, a topic that will be discussed in detail in the next section of the text.

Summary

17.1 Cloning of a Gene

Two methods are currently available for making copies of DNA. Recombinant DNA contains DNA from two different sources. A restriction enzyme is used to cleave plasmid DNA and to cleave foreign DNA. The "sticky ends" produced facilitate the insertion of foreign DNA into vector DNA. The foreign gene is sealed into the vector DNA by DNA ligase. Both bacterial plasmids and viruses can be used as vectors to carry foreign genes into bacterial host cells. A genomic library can be used as a source of genes to be cloned. A radioactive or fluorescent probe is used to identify the location of a single gene among the cloned fragments of an organism's DNA.

If just a portion of the genome is of interest or available, PCR uses the enzyme DNA polymerase to make multiple copies of target DNA. Analysis of DNA segments following PCR can involve gel electrophoresis to identify the DNA as belonging to a particular organism or can involve determining the base sequence of the DNA segment. The entire genome is used for fingerprinting. First, restriction enzymes are used to fragment DNA. Then the use of probes is necessary in order to see a pattern following gel electrophoresis.

17.2 Biotechnology Products

Transgenic organisms are ones that have had a foreign gene inserted into them. Genetically engineered bacteria, agricultural plants, and farm animals now produce commercial products of interest to humans, such as hormones and vaccines. Bacteria secrete the product. The seeds of plants and the milk of animals contain the product.

Transgenic bacteria have been produced to promote the health of plants, perform bioremediation, extract minerals, and

Connecting Concepts

These appear at the close of the text portion of the chapter, and they stimulate critical thinking by showing how the concepts of the chapter are related to others in the text.

Summary

The major headings are repeated in the summary because it is organized according to the major sections of the chapter. The summary helps you review the important concepts and topics discussed in each chapter.

Reviewing the Chapter

These are a series of page-referenced study questions that follow the sequence of the chapter.

Thinking Scientifically

New this edition, these critical thinking questions give you an opportunity to reason as a scientist. Detailed answers to these questions are found on the Online Learning Center.

produce chemicals. Transgenic agricultural plants which have been engineered to resist herbicides and pests are commercially available. Transgenic animals have been given various genes, in particular the one for bovine growth hormone (bGH). Cloning of animals is now possible.

17.3 The Human Genome Project

The Human Genome Project has two goals. The first goal is to construct a genetic map which will show the sequence of the genes on each chromosome. Although the known location of genes is expanding, the genetic map still contains many RFLPs (restriction fragment length polymorphisms) instead as the first step toward finding more genes. The second goal of the project is to sequence the bases on all the chromosomes. To do this researchers rely on PCR and automatic sequencing instruments. As the sequencers have improved so has the speed with which DNA is sequenced. The hope is that knowing the location and sequence of bases in a gene will promote the possibility of gene therapy.

17.4 Gene Therapy

Gene therapy is used to correct the genotype of humans and to cure various human ills. There are ex vivo and in vivo methods of gene therapy. Gene therapy has apparently helped children with SCID to lead a normal life; the treatment of cystic fibrosis has been less successful. A number of imaginative therapies are being employed in the war against cancer and other human illness such as cardiovascular disease.

Reviewing the Chapter

1. What is the methodology for producing recombinant DNA to be used in gene cloning? 266
2. Bacteria can be used to clone a gene or produce a product. Explain. 266–67
3. What is a genomic library, and how do you locate a gene of interest in the library? 267
4. What is the polymerase chain reaction (PCR), and how is it carried out to produce multiple copies of a DNA segment? 268
5. What is DNA fingerprinting, a process that utilizes the entire genome? 269
6. What are some practical applications of DNA segments analysis following PCR? 268–69
7. For what purposes have bacteria, plants, and animals been genetically altered? 270–73
8. Explain how and why transgenic animals that secrete a product are often cloned. 272–73
9. Explain the two primary goals of the Human Genome Project. What are the possible benefits of the project? 274
10. Explain and give examples of ex vivo and in vivo gene therapies in humans. 275

Testing Yourself

Choose the best answer for each question.

1. Which of these is a true statement?
   a. Both plasmids and viruses can serve as vectors.
   b. Plasmids can carry recombinant DNA but viruses cannot.
   c. Vectors carry only the foreign gene into the host cell.
   d. Only gene therapy uses vectors.
   e. Both a and d are correct.
2. Which of these is a benefit to having insulin produced by biotechnology?
   a. It is just as effective.
   b. It can be mass-produced.
   c. It is nonallergenic.
   d. It is less expensive.
   e. All of these are correct.
3. Restriction fragment length polymorphisms (RFLPs)
   a. are achieved by using restriction enzymes.
   b. identify individuals genetically.
   c. are the basis for DNA fingerprints.
   d. can be subjected to gel electrophoresis.
   e. All of these are correct.
4. Which of these would you not expect to be a biotechnology product?
   a. vaccine
   b. modified enzyme
   c. DNA probe
   d. protein hormone
   e. steroid hormone
5. What is the benefit of using a retrovirus as a vector in gene therapy?
   a. It is not able to enter cells.
   b. It incorporates the foreign gene into the host chromosome.
   c. It eliminates a lot of unnecessary steps.
   d. It prevents infection by other viruses.
   e. Both b and c are correct.
6. Gel electrophoresis
   a. cannot be used on nucleotides.
   b. measures the size of plasmids.
   c. tells whether viruses are infectious.
   d. measures the charge and size of proteins and DNA fragments.
   e. All of these are correct.
7. Put these phrases in the correct order to form a plasmid carrying recombinant DNA.

   1 use restriction enzymes
   2 use DNA ligase
   3 remove plasmid from parent bacterium
   4 introduce plasmid into new host bacterium

   a. 1, 2, 3, 4
   b. 4, 3, 2, 1
   c. 3, 1, 2, 4
   d. 2, 3, 1, 4
8. Which of these is incorrectly matched?
   a. xenotransplantation—source of organs
   b. protoplast—plant cell engineering
   c. RFLPs—DNA fingerprinting
   d. DNA polymerase—PCR
   e. DNA ligase—mapping human chromosomes
9. The restriction enzyme called EcoRI has cut double-stranded DNA in this manner. The piece of foreign DNA to be inserted has what bases from left to the right?

10. The following drawings pertain to gene therapy. Label the drawings, using these terms: retrovirus, recombinant RNA (twice), defective gene, recombinant DNA, reverse transcription, and human genome.

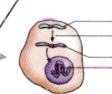

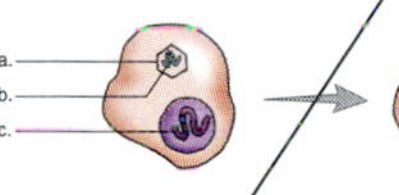

Thinking Scientifically

1. cDNA libraries contain only expressed DNA sequences. Therefore a cDNA library produced for a liver cell will contain genes unique to the cell. How could a CDNA library for a liver cell be used to acquire a complete copy of a liver-cell gene from a DNA library that contains the entire genome? A complete gene contains the promoter and introns.
2. There has been much popular interest in recreating extinct animals from DNA obtained from various types of fossils. However, such DNA is always badly degraded, consisting of extremely short pieces. Even if one hypothetically had 10 intact genes from a dinosaur, why might it still be impossible to create a dinosaur?

Bioethical Issue

Somatic gene therapy attempts to treat or prevent human illnesses. Some day, for example, it may be possible to give children who have cystic fibrosis, Huntington disease, or any other genetic disorder a normal gene to make up for the inheritance of a faulty gene. And gene therapy is even more likely for treatment of diseases like cancer, AIDS, and heart disease.

Germ-line gene therapy is the expression now being used to mean the use of gene therapy solely to improve the traits of an individual. It would be the genetic equivalent of procedures like body building, liposuction, or hair transplants. If genetic interference occurred early—that is, on the eggs, sperm, or embryo—it's possible that it would indeed affect the germ line—that is, all the future descendants of the individual.

How might germ-line gene therapy become routine? Consider this scenario. Presently, a gene for VEGF (vascular endothelial growth factor), a protein produced by cells to grow new blood vessels, is being used to treat atherosclerosis. Improved arterial circulation in the legs of some recent patients did away with the threat of possible amputation. This same gene is now being considered for the treatment of blocked coronary arteries. There may be instances, though, in which healthy people want to grow new blood vessels for enhancement purposes. Runners might want to improve their circulation in order to win races, and parents might think that increased circulation to the brain would increase the intelligence of their children. Bioethicist Eric Juengst of Case Western thinks that as a society, there is nothing we can now do to prevent us from crossing the line between the use of gene therapy for therapeutic purposes and its use for enhancement reasons. Do you have any concerns about gene therapy for enhancement reasons, such as its cost and general availability to everyone?

Understanding the Terms

bacteriophage 267
clone 266
complementary DNA (cDNA) 267
DNA fingerprinting 269
DNA ligase 266
gene therapy 275
genetic engineering 270
genome 267
genomic library 267
plasmid 266
polymerase chain reaction (PCR) 268
probe 267
recombinant DNA (rDNA) 266
restriction enzyme 266
restriction fragment length polymorphism (RFLP) 269
tissue engineering 271
transgenic organism 270
vector 266
xenotransplantation 271

Match the terms to these definitions:

a. ______________ Bacterial enzyme that stops viral reproduction by cleaving viral DNA; used to cut DNA at specific points during production of recombinant DNA.
b. ______________ Free-living organisms in the environment that have had a foreign gene inserted into them.
c. ______________ Known sequences of DNA that are used to find complementary DNA strands; can be used diagnostically to determine the presence of particular genes.
d. ______________ Production of identical copies; in genetic engineering, the production of many identical copies of a gene.
e. ______________ Self-duplicating ring of accessory DNA in the cytoplasm of bacteria.

Web Connections

Exploring the Internet

http://www.mhhe.com/biosci/genbio/mader
(click on *Biology 7/e*)

The *Biology 7/e* Online Learning Center provides many resources for studying the material in this chapter including links to the following sites:

The Howard Hughes Medical Institute's site, Blazing a Genetic Trail, focuses on the gene mutations that cause disease as they review recent genetic research and its applications.

http://www.hhmi.org/GeneticTrail/

The Human Genome Project describes the history of the project and progress made to date.

http://www.nhgri.nih.gov/HGP/

National Centre for Biotechnology Education. An overview of the origins and nature of biotechnology and enzymes used in food production can be found in Food Biotechnology: past, present, and future. This well-written, comprehensive website includes examples of current plant biotechnology, and plants that are currently undergoing field trials. Some of the concerns for environmental safety are explained.

http://www.ncbe.reading.ac.uk/

Biotechnology and Scientific Services Home Page. Information is found here on subjects ranging from GM (genetically modified) foods to regulation in the industry.

http://www.aphis.usda.gov/biotech/index.html

Testing Yourself

This section consists of objective questions that allow you to test your ability to answer recall-based questions. Answers to *Testing Yourself* questions are given in Appendix A.

Bioethical Issue

A *Bioethical Issue* has been added to the end of many chapters this edition. These short readings discuss a variety of controversial topics that our society continues to confront. The reading ends with appropriate questions to help you fully consider the issue and arrive at an opinion.

Web Connections

This section directs you to the Online Learning Center, and informs you of links that are available for further study of the topics discussed in the chapter.

Essential Study Partner CD-ROM

A free study partner that engages, investigates, and reinforces what you are learning from your textbook. You'll find the **Essential Study Partner** for *Biology* to be a complete, interactive student study tool packed with hundreds of animations and learning activities. From quizzes to interactive diagrams, you'll find that there has never been a better study partner to ensure the mastery of core concepts. Best of all, it's **FREE** with your new textbook purchase.

The unit pop-up menu is accessible at any time within the program. Clicking on the current unit will bring up a menu of other units available in the program.

The topic menu contains an interactive list of the available topics. Clicking on any of the listings within this menu will open your selection and will show the specific concepts presented within this topic. Clicking any of the concepts will move you to your selection. You can use the UP and DOWN arrow keys to move through the topics.

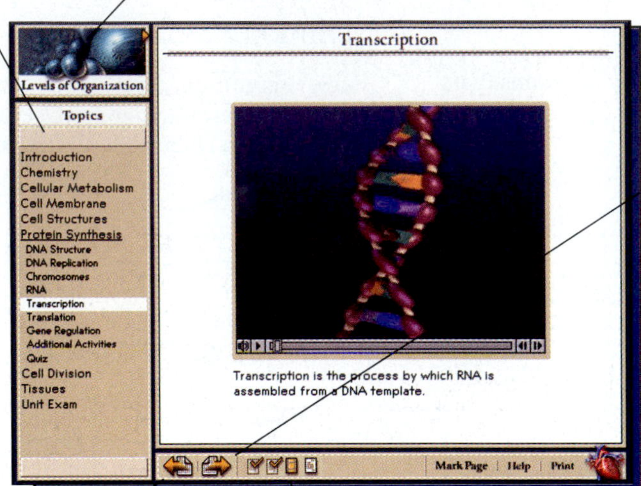

To the right of the arrows is a row of icons that represent the number of screens in a concept. There are three different icons, each representing different functions that a screen in that section will serve. The screen that is currently displayed will highlight yellow and visited ones will be checked.

The film icon represents an animation screen.

Along the bottom of the screen you will find various navigational aids. At the left are arrows that allow you to page forward and backward through text screens or interactive exercise screens. You can also use the LEFT and RIGHT arrows on your keyboard to perform the same function.

The activity icon represents an interactive learning activity.

The page icon represents a page of informational text.

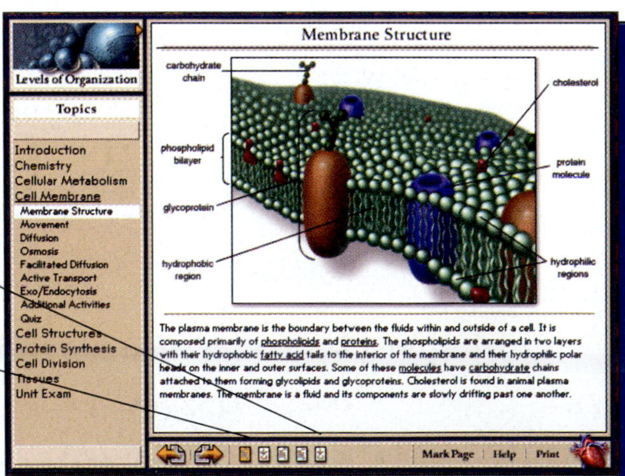

Contact your McGraw-Hill sales representative for more information or visit *www.mhhe.com*.

The Online Learning Center

Your Password to Success

www.mhhe.com/biosci/genbio/maderbiology7

This text-specific website allows students and instructors from all over the world to communicate. Instructors can create a more interactive course with the integration of this site, and students will find tools that help them improve their grades and learn that biology can be fun.

Student Resources

Study questions
Quizzing with immediate feedback
Links to chapter-related websites
Case studies
Interactive art labeling exercises
Critical thinking exercises

Instructor Resources

Instructor's Manual
Activities that can be assigned as coursework
Links to related websites to expand on particular topics
Classroom activities
Lecture outlines
Case studies

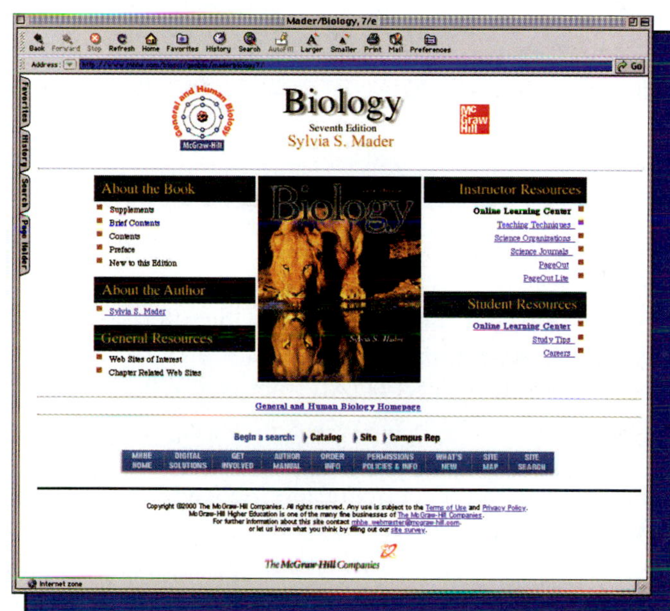

Imagine the advantages of having so many learning and teaching tools all in one place—all at your fingertips—FREE.

Contact your McGraw-Hill sales representative for more information or visit *www.mhhe.com.*

McGraw-Hill Higher Education Introduces:

Coming Soon

INTERACTIVE e-SOURCE

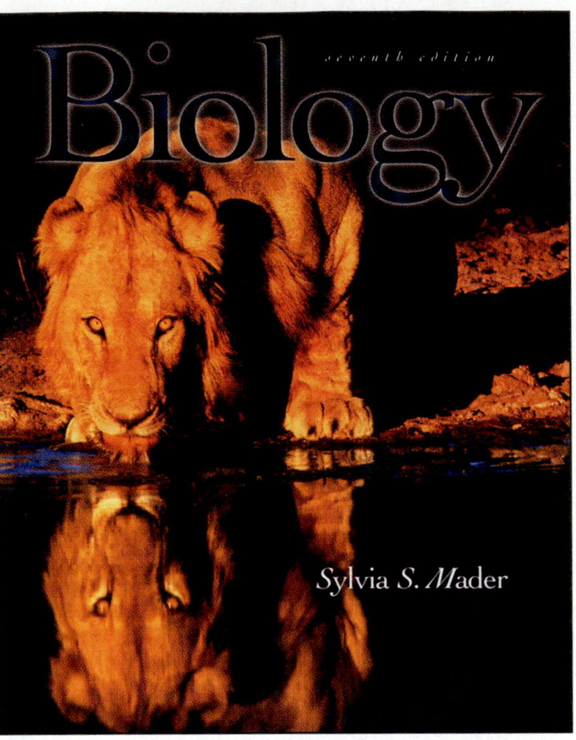

Biology, Seventh Edition

by Sylvia S. Mader

2001 • 0-07-240998-3
www.mhhe.com/biosci/genbio/maderbiology7

INTERACTIVE e-SOURCE is an exciting student resource that combines McGraw-Hill print, media, study, and web-based materials into one easy-to-use CD-ROM. This CD-ROM provides cutting-edge technology that accommodates all learning styles, and works as a complement to the printed text. This CD provides a truly non-linear experience by utilizing video, art, web-based, and other course materials to help students organize their study materials. Best of all, INTERACTIVE e-SOURCE will be free with your new textbook purchase.

The following features illustrate, in depth, the benefits of INTERACTIVE e-SOURCE.

- All **narrative, art and photos**, PLUS expertly crafted **animations and video** are included in the interlinked HTML files of the full textbook.
- Targeted Web links encourage **focused Web research**.
- A **powerful Search Engine** enables students to improve studying by locating targeted content quickly and easily.
- An **annotation function** allows students to enhance their studying by personalizing e-SOURCE material with their own specific notes or comments.
- By using the **highlighting function** (and their choice of highlight color), students can quickly identify important concepts as they review their materials.
- The **Media Gallery** organizes all non-text content on the CD and presents it as a clear menu. This allows students to perform non-linear searches as well as retrieval of any media content relevant to their specific studying needs.
- A **Customized Binder** function allows students to extract important information and customize their materials for future reference.
- A special **Read** feature converts written text to audio, enabling students to listen to important concepts and hear the proper pronunciations of key terms.

Contact your McGraw-Hill sales representative for more information or visit *www.mhhe.com*.

Visual Resource Library CD-ROMs

These CD-ROMs are electronic libraries of educational presentation resources that instructors can use to enhance their lectures. View, sort, search, and print catalog images, play chapter-specific slideshows using PowerPoint, or create customized presentations when you:

- Find and sort thumbnail image records by name, type, location, and user-defined keywords
- Search using keywords or terms
- View images at the same time with the Small Gallery View
- Select and view images at full size
- Display all the important file information for easy file identification
- Drag and place or copy and paste into virtually any graphics, desktop publishing, presentation, or multimedia application

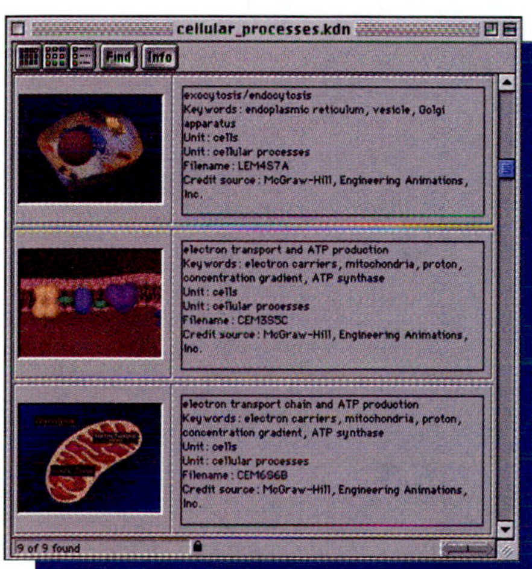

Life Science Animations Visual Resource Library CD-ROM

This instructor's tool, containing more than 125 animations of important biological concepts and processes—found in the *Essential Study Partner* and *Dynamic Human CD-ROMs*—is perfect to support your lecture. The animations contained in this library are not limited to subjects covered in the text, but include an expansion of general life science topics.

Biology Visual Resource Library CD-ROM

This helpful CD-ROM contains ALL 1,500 photographs and illustrations from *Biology*. You'll be able to create interesting multimedia presentations with the use of these images, and students will have the ability to easily access the same images in their texts to later review the content covered in class.

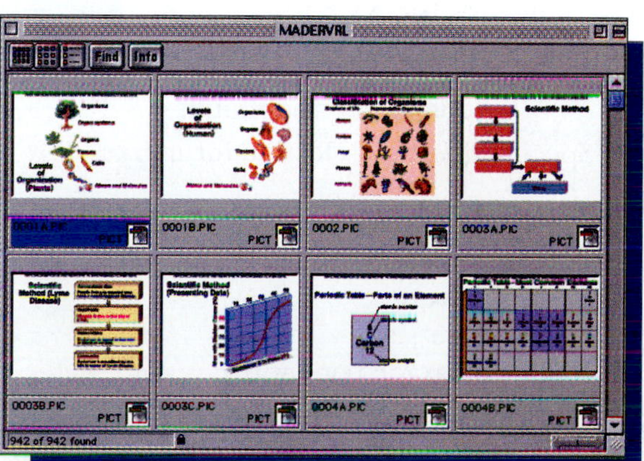

Contact your McGraw-Hill sales representative for more information or visit *www.mhhe.com*.

PageOut

Proven. Reliable. Class-tested.

More than 6,000 professors have chosen **PageOut** to create course websites. And for good reason: **PageOut** offers powerful features, yet is incredibly easy to use.

Now you can be the first to use an even better version of **PageOut**. Through class-testing and customer feedback, we have made key improvements to the grade book, as well as the quizzing and discussion areas. Best of all, **PageOut** is still free with every McGraw-Hill textbook. And students needn't bother with any special tokens or fees to access your **PageOut** website.

Customize the site to coincide with your lectures.

Complete the **PageOut** templates with your course information and you will have an interactive syllabus online. This feature lets you post content to coincide with your lectures. When students visit your **PageOut** website, your syllabus will direct them to components of McGraw-Hill web content germane to your text, or specific material of your own.

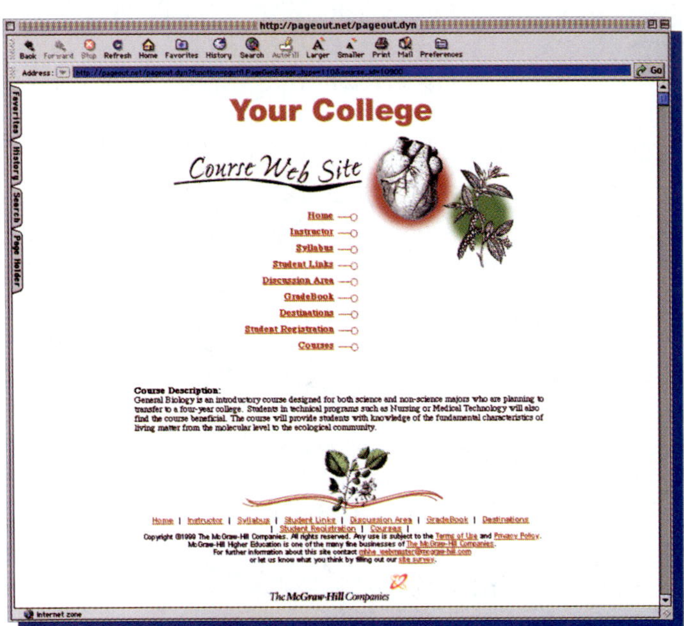

New Features based on customer feedback:

- Specific question selection for quizzes
- Ability to copy your course and share it with colleagues or use as a foundation for a new semester
- Enhanced grade book with reporting features
- Ability to use the **PageOut** discussion area, or add your own third party discussion tool
- Password protected courses

Short on time? Let us do the work.

Send your course materials to our McGraw-Hill service team. They will call you for a 30 minute consultation. A team member will then create your **PageOut** website and provide training to get you up and running. Contact your McGraw-Hill representative for details.

Contact your local McGraw-Hill sales representative for more information or visit *www.mhhe.com*.

Course Solutions

Simplify your life.

Course Solutions 2000 is the answer to your teaching needs. With a full range of multimedia products and special services married to our market-respected, time-tested textbooks, the *Course Solutions 2000* program is designed to make your life easier. Everything you need for effective, interactive teaching is at your fingertips.

WHY USE COURSE SOLUTIONS?

- **McGraw-Hill Learning Architecture** Each *McGraw-Hill Online Learning Center* is ready to be imported into our *McGraw-Hill Learning Architecture*—a full course management software system for *Local Area Networks* and *Distance Learning Classes.* Developed in conjunction with Top Class software, *McGraw-Hill Learning Architecture* is a powerful course management system available upon special request.

- **Course Consultant Service** In addition to the *Course Integration Guide,* instructors using Course Solutions textbooks can access a special curriculum-based Course Consultant Service via a web-based threaded discussion list within each *Online Learning Center.*

- **Instructor Syllabus Service** For *new* adopters of *Course Solutions* textbooks, McGraw-Hill will help correlate all text, supplementary, and appropriate materials and services to your course syllabus. Simply call your McGraw-Hill representative for assistance.

- **PageOut** Use this intuitive software to create your own course website.

- **Other Delivery Options** *Online Learning Centers* are also compatible with a number of full-service online course delivery systems or outside educational service providers.

- **Student Tutorial Service** Within each *Online Learning Center* resides a free student tutorial service.

- **e-BOOKS** Each *Course Solutions* title will feature its own e-book—an on-screen version of the actual textbook with numerous in-text links to other valuable digital resources. E-books—which are extremely valuable and very affordable—are available in one of three formats, depending on the textbook:
  - Interactive e-Source CD-ROM
  - e-Text CD-ROM
  - Online Learning Center

- **Web CT Linkage** Specially prepared *Online Learning Centers,* available with *Course Solutions* textbooks, load easily into any *Web CT Course Management* system.

- **Enhanced New Media Integration Guide** Each *Course Solutions* title features a valuable *Integration Guide* for instructors, significantly improved over past Integration Guides. Each guide indicates where and how to use available media resources with the text. Each now includes detailed descriptions of the content of each relevant new media module or exercise.

- **Online Animations** Selected *Course Solutions* titles now include high-quality animations in their *Online Learning Centers.*

Contact your local McGraw-Hill sales representative for more information or visit *www.mhhe.com.*

New Technology

Essential Study Partner CD-ROM

This interactive student study tool is packed with over 100 animations and more than 200 learning activities. From quizzes to interactive diagrams, your students will find that there has never been a more exciting way to study biology. A self-quizzing feature allows students to test their knowledge of a topic before moving on to a new module. Additional unit exams give students the opportunity to review an entire subject area. The quizzes and unit exams hyperlink students back to tutorial sections so they can easily review coverage for a more complete understanding. The text-specific *Essential Study Partner CD-ROM* supports and enhances the material presented in ***Biology*** and is correlated with the text.

Biology Online Learning Center

McGraw-Hill text-specific websites allow students and instructors from all over the world to communicate. By visiting this site, students can access additional study aids—including quizzes—explore links to other relevant biology sites, and catch up on current information. Log on today!

www.mhhe.com/biosci/genbio/mader

McGraw-Hill Course Solutions

Designed specifically to help you with your individual course needs, *Course Solutions* will assist you in integrating your syllabus with ***Biology***, and state-of-the-art new media tools.

At the heart of *Course Solutions* you'll find integrated multimedia, a full-scale Online Learning Center, and an enhanced Integration Guide. These unparalleled services are also available as a part of *Course Solutions:* e-Books, Web CT linkage, Online animations, McGraw-Hill Learning Architecture, McGraw-Hill Course Consultation Service, Visual Resource Library Image Licensing, McGraw-Hill Student Tutorial Service, McGraw-Hill Instructor Syllabus Service, PageOut Lite, PageOut: The Course Website Development Center, and other delivery options.

PageOut™ Put together your own customized website with the use of PageOut, a program designed specifically for instructors wanting to put course information on the web. No experience in web publishing is necessary, just choose from a collection of templates to create your class website.

Visual Resource Library CD-ROM

This helpful CD-ROM contains approximately 1,500 images and select animations that can be easily imported into PowerPoint to create multimedia presentations. Or, you may use the already prepared PowerPoint presentations.

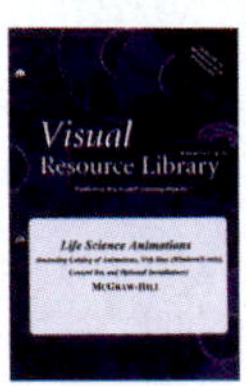

Life Science Animations CD-ROM

This two CD-ROM set contains more than 125 animations of important biological concepts and processes.

Classroom Testing Software (MicroTest)

This helpful testing software—available in either Macintosh or Windows format—provides well-written and researched book-specific questions featured in the Test Item File.

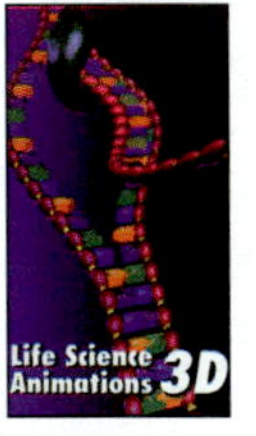

Life Science Animations Videotape Series

Animations of key biological processes are available on seven videotapes. The animations bring visual movement to biological processes that are difficult to understand on the text page.

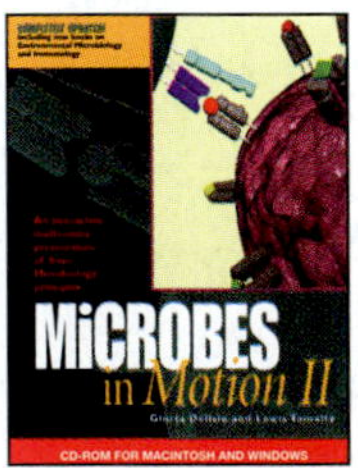

Microbes in Motion CD-ROM, Version 2.0

This interactive CD-ROM allows students to actively explore microbial structure and function. Great for self-study, preparation for class or exams, or for classroom presentations.

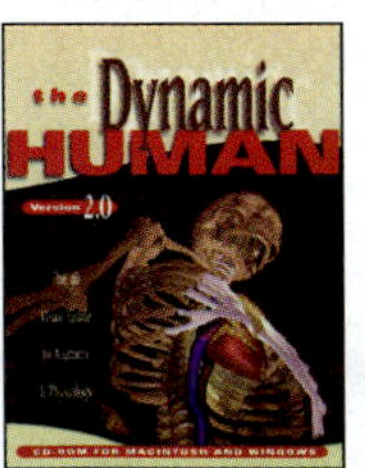

The Dynamic Human CD-ROM, Version 2.0

This guide to anatomy and physiology interactively illustrates the complex relationships between anatomical structures and their functions in the human body. Realistic, three-dimensional visuals are the premier feature of this exciting learning tool.

Dynamic Human Videodisc

Enhance your classroom presentations with movement, sound, and motion of internal organs, cells, and systems. More than 80 premier 3-D animations covering all body systems from the outstanding *Dynamic Human CD-ROM* are included.

Other Available Supplements

Instructor's Manual
The *Instructor's Manual* is designed to assist instructors as they plan and prepare for classes using ***Biology.*** The *Instructor's Manual* contains both an extended lecture outline and lecture enrichment ideas, which together review in detail the contents of the text chapter. The technology section lists videos and computer software items that are available from outside sources and also those that are available from McGraw-Hill.

Student Study Guide
To ensure close coordination with the text, the author Dr. Sylvia S. Mader has written the *Student Study Guide* that accompanies the text. Each text chapter has a corresponding study guide chapter that includes a listing of objectives, study questions, and a chapter test. Answers to the study questions and the chapter tests are provided to give students immediate feedback.

The concepts in the study guide are the same as those in the text, and the study questions in the study guide are sequenced according to these concepts. Instructors who make their choice of concepts known to the students can thereby direct student learning in an efficient manner. Instructors and students who make use of the *Student Study Guide* should find that student performance increases dramatically.

Laboratory Manual
Dr. Mader has also written the ***Laboratory Manual*** to accompany ***Biology.*** With few exceptions, each chapter in the text has an accompanying laboratory exercise in the manual (some chapters have more than one accompanying exercise). In this way, instructors are better able to emphasize particular portions of the curriculum, if they wish. The 35 laboratory sessions in the manual are designed to further help students appreciate the scientific method and to learn the fundamental concepts of biology and the specific content of each chapter. All exercises have been tested for student interest, preparation time, and feasibility. This lab manual can be customized to fit your lab course, contact your McGraw-Hill representative for details.

Transparencies
This set of transparency acetates to accompany the text, has been expanded to *650* full-color acetates.

Micrograph Slides
This ancillary provides a boxed set of 100 color slides of photomicrographs and electron micrographs in the text.

HealthQuest CD-ROM
ISBN 0-697-29723-3 (Windows)
ISBN 0-07-039335-4 (Macintosh)

Virtual Biology Laboratory CD-ROM
ISBN 0-697-37991-4

Life Science Living Lexicon CD-ROM
ISBN 0-697-37993-0

How to Study Science, Third Edition
ISBN 0-697-36051-2

Basic Chemistry for Biology, Second Edition
ISBN 0-697-36087-3

Schaum's Outlines: Biology
ISBN 0-07-022405-6

Understanding Evolution, by Volpe and Rosenbaum
ISBN 0-697-05137-4

AIDS Booklet
ISBN 0-697-29428-5

The Internet Primer
ISBN 0-07-303203-4

Critical Thinking Case Study Workbook
ISBN 0-697-14556-5

How Scientists Think, by George Johnson
ISBN 0-697-27875-1

To order any of these great study tools, contact your bookstore manager or call our Customer Service Department at 800-338-3987.

Acknowledgments

Biology has always been a team effort and it is fitting to acknowledge all those who worked so diligently on this edition. My publisher Michael Lange was always there to offer advice and my editor Patrick Reidy stepped in when needed to encourage us all. The clear, steady directives of my developmental editor, Anne Melde, were very much appreciated. She also contributed heavily to the success of the multimedia that accompanies the text.

Those in production also worked unfailingly toward the success of this edition. Marilyn Sulzer was pleasant, efficient, and cooperative while balancing a multitude of tasks. Marilyn doubled as the art coordinator and Lori Hancock was the photo research coordinator. And I especially want to thank Wayne Harms for the beautiful book he designed for all of us to enjoy.

In my office, Evelyn Jo Hebert has consistently provided steadfast and proficient support through several editions of the text. Everyone on the book team remained cheerful and helpful while going beyond the call of duty. My thanks to them all.

The Reviewers

Many instructors have contributed not only to this edition of *Biology* but also to previous editions. I am extremely thankful to each one, for they have all worked diligently to remain true to our calling and provide a product that will be the most useful to our students.

In particular, it is appropriate to acknowledge the help of the following individuals for the seventh edition:

Christine Tachibana
University of Washington

Larry Frederick
Southeastern Louisiana University

Dennis Bakewicz
New York City Technical College

Linda Smith-Staton
Pellissippi State Technical Community College

Shirley Porteous-Gafford
Fresno City College

Jerald Dosch
College of Visual Arts

Todd Christian Yetter
Cumberland College

Rita Hoots
Woodland College

Gary Greer
West Virginia State College

Bernard Wood
George Washington University

Robert Grammer
Belmont University

Beth Montelone
Kansas State University

Howard Duncan
Norfolk State University

Peter I. Ekechukwu
Horry-Georgetown Technical College

Lynne Lohmeier
Mississippi Gulf Coast Community College

Brian Hagenbuch
Holyoke Community College

Thomas J. Montagno
Simmons College

Margaret Nordlie
University of Mary

Brenda Zink
Northeastern Junior College

Jun Tsuji
Siena Heights University

Ralph C. Goff
Mansfield University of Pennsylvania

Judy Bluemer
Morton College

Tammy J. Liles
Lexington Community College

Marjorie Miller
Greenville Technical College

Angela Mason
Beaufort County Community College

Lyndell Robinson
Lincoln Land Community College

Andy Neill
Joliet Junior College

Debbie Firak
McHenry County College

Richard Bounds
Mount Olive College

John Mathwig
College of Lake County

Terry F. Werner
Harris-Stowe State College

Suzanne Kempke
Armstrong Atlantic State University

Thurman Wilson
Prairie State College

Stephen A. Miller
College of the Ozarks

R. Douglas Lyng
Indiana University–Purdue University at Fort Wayne

Andrew Goliszek
North Carolina Agricultural and Technical State University

James Enderby Bidlack
University of Central Oklahoma

Billy Williams
Dyersburg State Community College

Timothy Metz
Campbell University

George Williams, Jr.
Southern University and Agricultural and Mechanical College of Baton Rouge

James Fitch
Jones Junior College

Suzzette F. Chopin
Texas A&M University–Corpus Christi

Sue Trammell
Rend Lake College

Anton Lawson
Arizona State University–Main

Philip Snider
University of Houston

Robert David Allen
University of Nevada–Las Vegas

Jane Aloi
Saddleback College

Carol St. Angelo
Hofstra University

George W. Cox
San Diego State University

A View of Life

chapter 1

chapter concepts

1.1 How to Define Life

- There are various levels of biological organization. At each higher level properties emerge that cannot be explained by the sum of the parts. 2
- Although life is quite diverse, it can be defined by certain common characteristics. 2
- All living things have characteristics in common because they are descended from a common ancestor; but they are diverse because they are adapted to different environments. 4

1.2 How the Biosphere Is Organized

- The biosphere is made up of ecosystems where living things interact with each other and the physical environment. 5
- Tropical rain forests, which are noted for their diversity, are endangered due to human activities. 7

1.3 How Living Things Are Classified

- Living things are classified into categories according to their evolutionary relationships. 8

1.4 The Process of Science

- The scientific process is a way to gather information and to come to conclusions about the natural world. 9
- Various conclusions pertaining to the same area of interest are sometimes used to arrive at a theory, a general concept about the natural world. 12

Snakes and humans, like all living things, have many characteristics in common.

From bacteria to bats, toadstools to trees, whippoorwills to whales—the diversity of the living world boggles the mind. Yet, all organisms, over all of time, are united by a common bond. Just as you are descended from your parents, grandparents, and so forth, going back for many generations, all forms of life that ever lived are tied together by an unbroken lineage that can be traced back through time to the infancy of our planet. Humans and snakes share many similarities because they have a recent common ancestor. Both have a heart, a liver, intestines, and backbone. Why? Because they are both descended from the first vertebrates that ever existed.

Through adaptation to particular environments, groups of organisms diversified. Now the diversity of life including tigers, lions, gorillas, and elephants is threatened by human activities. What evolution has accomplished over billions of years can be destroyed in a much shorter length of time. What do we want to do—preserve or destroy diversity? We are in the driver's seat, but which road will we choose to take? The future of life is in our hands.

Courtesy of J. William Schopf, Director, UCLA Center for the Study of Evolution and the Origin of Life.

1.1 How to Define Life

You intuitively know that the cat and not the couch where the cat snuggles is alive. But it would most likely be difficult for you to come up with a one-line definition of life. The same is true for biologists; they can't easily define life either. Therefore, they rely on listing certain characteristics shared by all living things.

Living Things Are Organized

The complex organization of living things begins with the cell, the basic unit of life. Cells are made up of molecules that contain atoms, which are the smallest units of matter that can enter into chemical combination. In multicellular organisms, similar cells combine to form a tissue—nerve cells form nerve tissue, for example. Tissues make up organs, as when various tissues combine to form the brain. Organs, in turn, work together in systems; for example, the brain works with the spinal cord and a network of nerves to form the nervous system. A multicellular individual has several other organ systems also.

There are levels of biological organization that extend beyond the individual organism. All organisms of one type in a particular area belong to a population. In a temperate deciduous forest, there is a population of gray squirrels and a population of oak trees. The populations of various animals and plants in the forest make up a community (Fig. 1.1). The populations interact among themselves and with the physical environment (soil, atmosphere, etc.) forming an ecosystem.

Summing the Parts

In the living world, the whole is more than the sum of its parts. Each new level of biological organization has emergent properties that are due to interactions between the parts making up the whole. For example, when cells are broken down into bits of membrane and oozing liquids, these parts themselves cannot carry out the business of living. Slice a frog and arrange the slices, and the frog cannot flick out its tongue and catch flies.

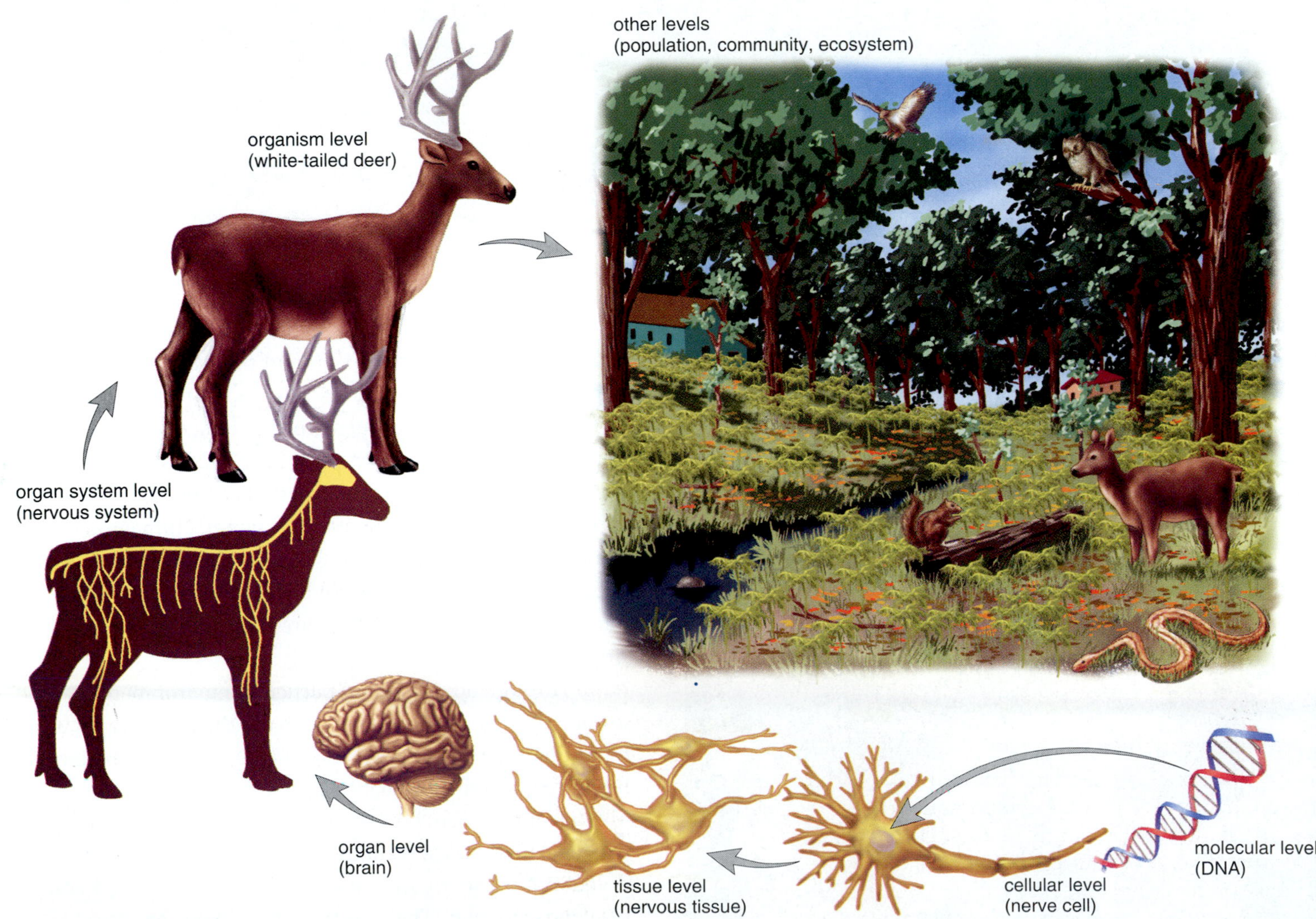

Figure 1.1 **Levels of biological organization.**

a. b.

Figure 1.2 Acquiring nutrient materials and energy. Plants and other types of photosynthesizers are able to produce organic nutrients; animals feed directly on photosynthesizers or other animals. **a.** A lioness charges her prey, a group of zebras. **b.** One zebra of the herd is taken and becomes a meal for hungry lionesses.

Living Things Acquire Materials and Energy

Living things cannot maintain their organization or carry on life's activities without an outside source of materials and energy (Fig. 1.2). Food provides nutrient molecules, which are used as building blocks or for energy. **Energy** is the capacity to do work, and it takes work to maintain the organization of the cell and the organism. When cells use nutrient molecules to make their parts and products, they carry out a sequence of chemical reactions. **Metabolism** [Gk. *meta*, implying change] is all the chemical reactions that occur in a cell.

The ultimate source of energy for nearly all life on earth is the sun. Plants and certain other organisms are able to capture solar energy and carry on **photosynthesis,** a process that transforms solar energy into chemical energy in the bonds of organic nutrient molecules. Animals and plants get energy by metabolizing the nutrient molecules made by photosynthesizers.

Remaining Homeostatic

For metabolic processes to continue, living things need to keep themselves stable in temperature, moisture level, acidity, and other physiological factors. This is **homeostasis** [Gk. *homoios*, like, resembling, and *stasis*, standing]—the maintenance of internal conditions within certain boundaries.

Many organisms depend on behavior to regulate their internal environment. A chilly lizard may raise its internal temperature by basking in the sun on a hot rock. When it starts to overheat, it scurries for cool shade. Other organisms have control mechanisms that do not require any conscious activity. When a student is so engrossed in her textbook that she forgets to eat lunch, her liver releases stored sugar to keep the blood sugar level within normal limits. Hormones regulate sugar storage and release, but in other instances the nervous system is involved in maintaining homeostasis.

Living Things Respond

Living things find energy and/or nutrients by interacting with their surroundings. Even unicellular organisms can respond to their environment. In some, the beating of microscopic hairs, and in others, the snapping of whiplike tails, moves them toward or away from light or chemicals. Multicellular organisms can manage more complex responses. A vulture can smell meat a mile away and soar toward dinner. A monarch butterfly can sense the approach of fall and begin its flight south where resources are still abundant.

The ability to respond often results in movement: the leaves of a plant turn toward the sun and animals dart toward safety. Appropriate responses help ensure survival of the organism and allow it to carry on its daily activities. All together, we call these activities the behavior of the organism.

Figure 1.3 Adaptations of rockhopper penguins, *Eudyptes*.
Male and female rockhoppers with their offspring. The stubby forelimbs of penguins are modified as flippers for fast swimming. Only a little over a half meter tall, rockhopper penguins are named for their skill in leaping from rock to rock.

Living Things Have Adaptations

Adaptations [L. *ad*, toward, and *aptus*, fit, suitable] are modifications that make organisms suited to their way of life. For example, penguins are adapted to an aquatic existence in the Antarctic (Fig. 1.3). Most birds have forelimbs proportioned for flying, but a penguin has stubby, flattened wings suitable for swimming. Their feet and tails serve as rudders in the water, but the flat feet also allow them to walk on land. Rockhopper penguins have a bill adapted to eating small shellfish. Their eggs—one, or at most two—are carried on their feet, where they are protected by a pouch of skin. This allows the birds to huddle together for warmth while standing erect and incubating eggs.

Organisms become modified over time by a process called **natural selection.** Certain members of a **species** [L. *species*, model, kind], defined as a group of interbreeding individuals, may inherit a genetic change that causes them to be better suited to a particular environment. These members can be expected to produce more surviving offspring who also have the favorable characteristic. In this way, the attributes of the species' members change over time.

Living Things Reproduce and Develop

Life comes only from life. Every type of living thing can **reproduce,** or make another organism like itself (Fig. 1.3). Bacteria, protozoa, and other unicellular organisms simply split in two. In most multicellular organisms, the reproductive process begins with the pairing of a sperm from one partner and an egg from the other partner. The union of sperm and egg, followed by many cell divisions, results in an immature individual, which grows and develops through various stages to become the adult.

An embryo develops into a sperm whale or a yellow daffodil because of a blueprint inherited from its parents. The blueprint or instructions for their organization and metabolism are encoded in the genes. The **genes,** which contain specific information for how the organism is to be ordered, are made of long molecules of DNA (deoxyribonucleic acid). All cells have a copy of the hereditary material, DNA, whose shape resembles a spiral staircase with millions of steps.

Descent with Modification

All living things share the same basic characteristics discussed in this chapter. They are all composed of cells organized in a similar manner. Their genes are composed of DNA, and they carry out the same metabolic reactions to acquire energy and maintain their organization. This unity suggests that all living things are descended from a common ancestor—the first cell or cells. However, **evolution** [L. *evolutio*, an unrolling] is descent with modification. One species can give rise to several species, each adapted to a particular set of environmental conditions. Specific adaptations allow species to play particular roles in an ecosystem. The diversity of life-forms is best understood in terms of the many different ways in which organisms carry on their life functions within an ecosystem where they live, acquire energy, and reproduce.

Descent from a common ancestor explains the unity of life. Adaptations to different ways of life account for the great diversity of life-forms.

1.2 How the Biosphere Is Organized

Individual organisms belong to a **population,** all the members of a species within a **community.** All communities taken together make up the **biosphere** [Gk. *bios,* life, and L. *sphaera,* ball], a thin layer of life that encircles earth. One of the most interesting discoveries about communities is that they are highly dynamic. The number of species, kinds of species, and size of populations within most communities is constantly changing due to disturbances and climatic variability. Because species composition is so variable, ecologists often concentrate on studying the movement of energy and nutrients through communities. Such a study involves the physical environment. The populations within a community interact among themselves and with the physical environment (soil, atmosphere, etc.), forming an **ecosystem** [Gk. *oikos,* home, house, and *systema,* ordered arrangement].

A major feature of the interactions between populations of a community pertains to who eats whom. Plants produce organic nutrients for themselves, and the animals that eat plants are food for other animals. Such a sequence of organisms is called a food chain (Fig. 1.4).

Both plants and animals interact with the physical environment. Plants take in inorganic nutrients, like carbon dioxide and water; both plants and animals return carbon dioxide to the atmosphere when they respire. When organisms die and decay, inorganic nutrients are made available to plants once more. The blue arrows in Figure 1.4 show how chemicals cycle through the various populations of an ecosystem.

In contrast, the yellow to red arrows in Figure 1.4 show how energy flows through an ecosystem: solar energy used by plants to produce organic nutrients is eventually converted to heat when organisms, including plants, use organic nutrients as an energy source. Therefore, a constant supply of solar energy is required for an ecosystem and for life to exist.

The ecosystems that make up the biosphere are classified first as terrestrial or aquatic. Among the terrestrial ecosystems, tropical rain forests are of extreme interest at present because of their great diversity and the threat of their imminent destruction as discussed in the reading on page 7.

In ecosystems, the same nutrients keep cycling through populations, but energy flows because it is eventually converted to heat.

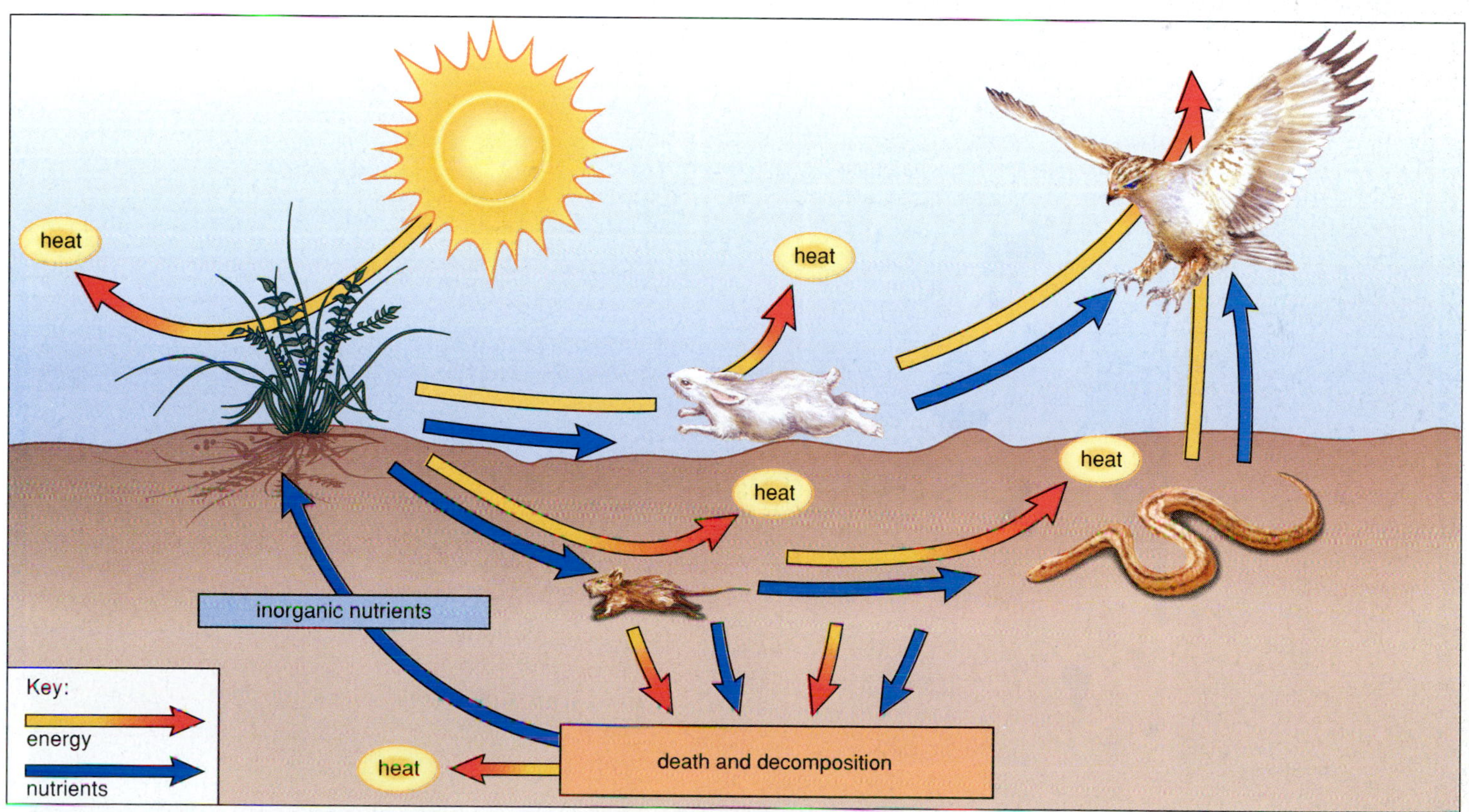

Figure 1.4 **Ecosystem organization.**
Within an ecosystem, nutrients cycle (see blue arrows): from plants, which use solar energy and inorganic nutrients to produce organic nutrients, to decomposers, which make inorganic nutrients available to plants once more. Energy flows (see yellow to red arrows) from the sun through all populations and eventually is converted to heat which dissipates.

Tropical Rain Forest, a Terrestrial Ecosystem

Of all the ecosystems, the greatest diversity of species occurs in tropical rain forests. They are found near the equator where there is plentiful sun and rainfall the entire year. Major rain forests are located in South America (Fig. 1.5), central and west Africa, and Southeast Asia. Rain forests have a multilayered canopy which consists of broad-leaved evergreen trees of different heights. Most animal populations live in the canopy where they interact with each other. Typically, brightly colored birds, such as toucans and macaws, fly around eating fruits, buds, and pollen. Other birds, such as long-billed hummingbirds feed on nectar often taken from small plants that grow independently on the trees. Tree sloths and spider monkeys are mammals that live in the canopy and are preyed upon by jaguars. Other canopy animals include butterflies, tree frogs, and dart-poison frogs. Many canopy animals, such as bats, are active only at night. Snakes, spiders, and ants are animals that live on or near the ground and not in the canopy.

Figure 1.5 Tropical rain forest populations interact in the manner described in Figure 1.4.

Ecology Focus

Tropical Rain Forests: Can We Live Without Them?

So far, nearly 2 million species of organisms have been discovered and named. Two-thirds of the plant species, 90% of the nonhuman primates, 40% of birds of prey, and 90% of the insects live in the tropics. Many more species of organisms (perhaps as many as 30 million) are estimated to live in the tropical rain forests but have not yet been discovered.

Ten years ago tropical forests spanned the planet on both sides of the equator and covered 6–7% of the total land surface of the earth—an area roughly equivalent to our contiguous forty-eight states. Every year humans destroy an area of forest equivalent to the size of Oklahoma. At this rate, these forests and the species they contain will disappear completely in just a few more decades. If the forest areas now legally protected survive, 56–72% of all tropical forest species would still be lost.

The loss of tropical rain forests results from an interplay of social, economic, and political pressures. Many people already live in the forest, and as their numbers increase, more of the land is cleared for farming. People move to the forests because internationally financed projects build roads and open the forests up for exploitation. Small-scale farming accounts for about 60% of tropical deforestation, and this is followed by commercial logging, cattle ranching, and mining. International demand for timber promotes destructive logging of rain forests in Southeast Asia and South America. A market for low-grade beef encourages the conversion of tropical rain forests to pastures for cattle. The lure of gold draws miners to rain forests in Costa Rica and Brazil.

The destruction of tropical rain forests produces only short-term benefits but is expected to cause long-term problems. The forests soak up rainfall during the wet season and release it during the dry season. Without them, a regional yearly regime of flooding followed by drought is expected to destroy property and reduce agricultural harvests. Worldwide, there could be changes in climate that would affect the entire human race. On the other hand, the preservation of tropical rain forests offers benefits. For example, the rich diversity of plants and animals would continue to exist for scientific and pharmacological study. One-fourth of the medicines we currently use come from tropical rain forests. The rosy periwinkle from Madagascar has produced two potent drugs for use against Hodgkin's disease, leukemia, and other blood cancers. It is hoped that many of the still-unknown plants will provide medicines for other human ills.

Studies show that if the forests were used as a sustainable source of nonwood products, such as nuts, fruits, and latex rubber, they would generate as much or more revenue while continuing to perform their various ecological functions. And biodiversity could still be preserved. Brazil is exploring the concept of "extractive reserves," in which plant and animal products are harvested but the forest itself is not cleared. Ecologists have also proposed "forest farming" systems, which mimic the natural forest as much as possible while providing abundant yields. But for such plans to work maximally, the human population size and the resource consumption per person must be stabilized.

Preserving tropical rain forests is a wise investment. Such action promotes the survival of most of the world's species—indeed, the human species, too.

The Human Population

The human population tends to modify existing ecosystems for its own purposes. As more and more ecosystems are converted to towns and cities, fewer of the natural cycles are able to function adequately to sustain an ever growing human population. It is important to do all we can to preserve ecosystems, because only then can we be assured that we will continue to exist. The recognition that natural ecosystems need to be conserved is one of the most important developments of our new ecological awareness.

We now know that tropical rain forests perform many services for us. For example, they act like a giant sponge and absorb carbon dioxide, a pollutant that pours into the atmosphere from the burning of fossil fuels such as oil and coal. If rain forests continue to be depleted as they are now, an increased amount of carbon dioxide in the atmosphere is expected to cause an increase in the average daily temperature. Problems with acid rain are also expected to increase, since carbon dioxide combines with water to form carbonic acid, a component of acid rain.

The present **biodiversity** (number and size of populations in a community) of our planet is being threatened. It has been estimated that the number of species in the biosphere may be as high as 80 million species, but only about 2 million have been identified and named. We may never have a chance to identify all there are because we may be presently losing from 24 to even 100 species a day due to human activities. The existence of the species featured in Figure 1.5 is threatened because tropical rain forests are being reduced in size. Most biologists are alarmed over the present rate of extinction and believe the rate may eventually rival the mass extinctions that have periodically occurred during our planet's history.

Because all living things are dependent upon the normal functioning of the biosphere, the wisest approach is to conserve ecosystems as much as possible.

1.3 How Living Things Are Classified

Since life is so diverse it is helpful to have a classification system to group organisms according to their similarities (see Appendix B). **Taxonomy** [Gk. *tasso,* arrange, classify, and *nomos,* usage, law] is the discipline of identifying and classifying organisms according to certain rules. Biologists give each living thing a binomial [L. *bis,* two, and *nomen,* name] or two-part name. For example, the scientific name for the garden pea is *Pisum sativum.* The first word is the genus and the second word is the specific epithet of a species within a genus. Species are grouped into ever more inclusive categories: genera, families, orders, classes, phyla, kingdoms, and finally domains.

It has been common practice the past few years to recognize five kingdoms (called Monera, Protista, Fungi, Plantae, and Animalia) but biochemical evidence now suggests that kingdoms should be organized into three higher categories called domains. Further, both domain Bacteria and domain Archaea contain unicellular prokaryotes which are structurally simple but metabolically complex. They lack a membrane bounded nucleus found in the eukaryotes of domain Eukarya. Domain Eukarya contains the kingdoms most familiar to most (Fig. 1.6). Protists (kingdom Protista) range from unicellular to multicellular organisms and include the algae and protozoans. Among the fungi (kingdom Fungi) are the familiar molds and mushrooms. Plants (kingdom Plantae) are well known as multicellular photosynthesizers of the world while animals (kingdom Animalia) are multicellular and ingest their food.

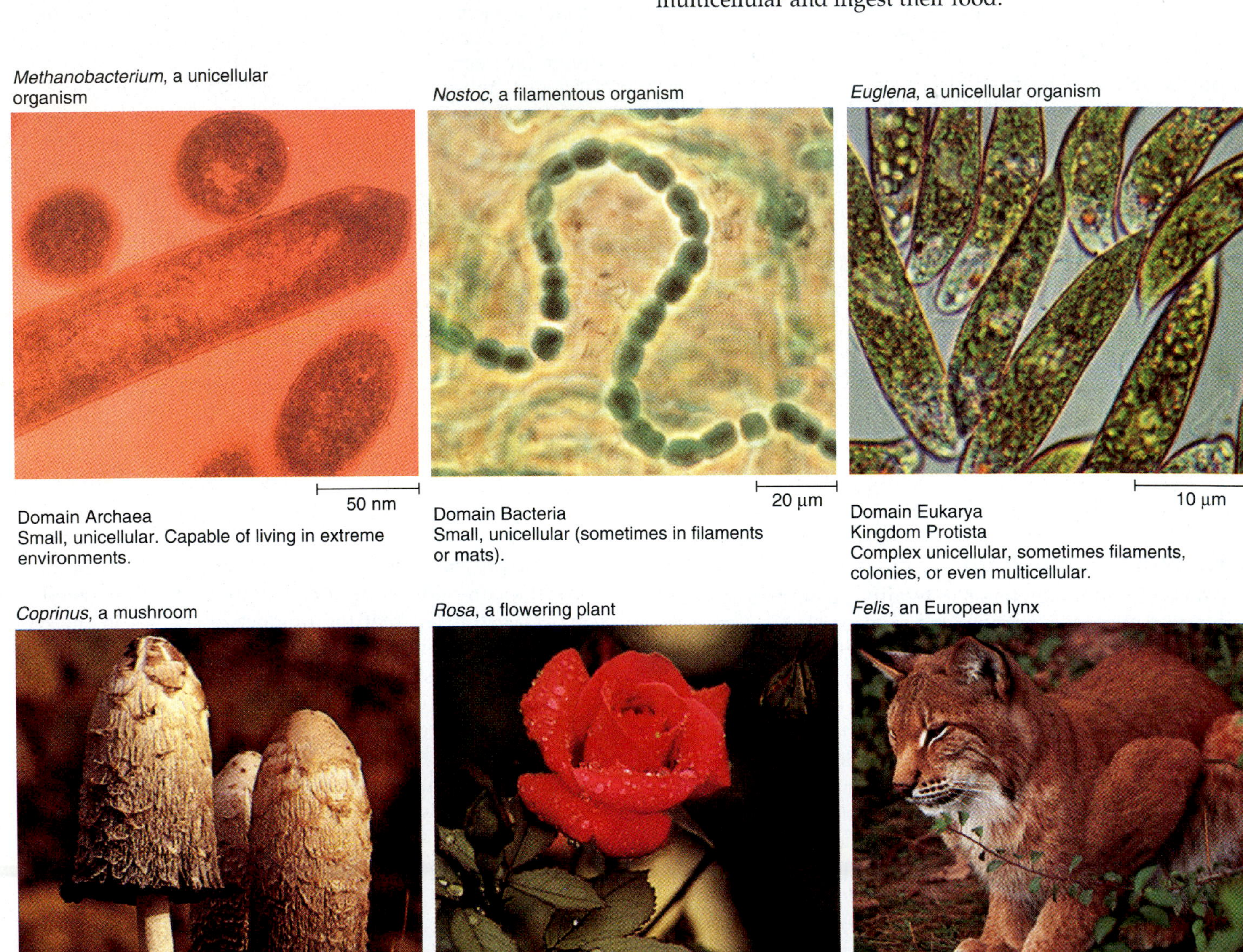

Methanobacterium, a unicellular organism
50 nm
Domain Archaea
Small, unicellular. Capable of living in extreme environments.
Nostoc, a filamentous organism
20 μm
Domain Bacteria
Small, unicellular (sometimes in filaments or mats).
Euglena, a unicellular organism
10 μm
Domain Eukarya
Kingdom Protista
Complex unicellular, sometimes filaments, colonies, or even multicellular.
Coprinus, a mushroom
Domain Eukarya
Kingdom Fungi
Multicellular and absorb food.
Rosa, a flowering plant
Domain Eukarya
Kingdom Plantae
Multicellular and photosynthesize.
Felis, an European lynx
Domain Eukarya
Kingdom Animalia
Multicellular and ingest food.

Figure 1.6 **Pictorial representation of the three domains of life.**

1.4 The Process of Science

Biology is the scientific study of life. Religion, aesthetics, ethics, and, science, are all ways that human beings have of finding order in the natural world. Science differs from other human ways of knowing and learning by its process. **Science** considers only what is observable by the senses or by instruments that extend the ability of the senses as a way to understand the natural world. Microscopes extend the ability of sight way beyond what can be seen by the naked eye, for example. Observations can be made by any one of our five senses; we can observe with our noses that dinner is almost ready; observe with our fingertips that a surface is smooth and cold; and observe with our ears that a piano needs tuning. Observations help scientists conduct various types of investigations.

Once a scientist becomes interested in an observable event, called a **phenomenon,** he or she will most likely study up on it. The goal will be to find any past studies by using the Internet, the library, or other scientists to discover what ideas there may be about this phenomenon. Then a scientist uses inductive reasoning to formulate a hypothesis. **Inductive reasoning** occurs when a person uses isolated facts and creative thinking to come up with a possible explanation. In this case, the possible explanation is called a **hypothesis.** All past experiences of the individual no matter what they be, will most likely influence the formulation of a hypothesis. Einstein said that aesthetic qualities like beauty and simplicity helped him to formulate his hypotheses about the workings of the universe. Indeed, most every scientist looks favorably upon the simplest hypothesis for a phenomenon and tests that one first.

Science only considers hypotheses that can be tested. Moral and religious beliefs while very important to our lives differ between cultures and through time and they are not always testable by further observations and/or experimentation. Hypotheses are tested either in the laboratory setting or in a natural setting. The laboratory is a place that lends itself to carrying on **experiments,** which are artificial situations devised to test hypotheses. Experiments can also be carried out in a natural setting, called the field.

A Field Investigation

David P. Barash, who was observing the mating behavior of mountain bluebirds, decided to perform an experiment to test the hypothesis that aggression of the male varies during the reproductive cycle (Fig. 1.7). Once the hypothesis has been stated, deductive reasoning comes into play. **Deductive reasoning** begins with a general statement that infers a specific conclusion. It often takes the form of an "if, then" statement. In this case, if male aggression varies during the reproductive cycle, then aggression may change after the nest is built, after the first egg is laid, and after the hatching of the eggs.

a. Scientist makes observations, studies previous data, and formulates a hypothesis.

b. Scientist performs experiment and collects objective data.

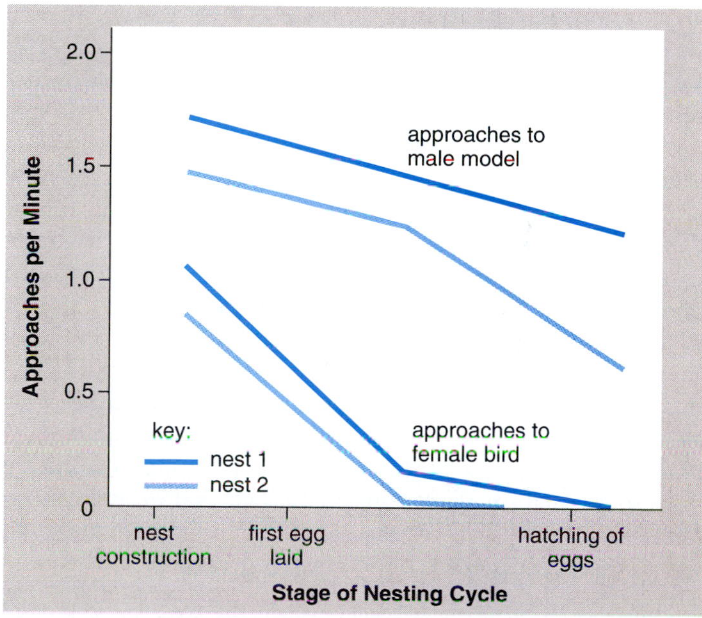

c. Scientist studies results and comes to a conclusion.

Figure 1.7 An experiment in the field.
a. Observation of male bluebird behavior allowed David Barash to formulate a testable hypothesis. Then, he **(b)** collected data and **(c)** displayed in a graph. Finally, he came to a conclusion.

Barash posted a male bluebird model near nests while the resident male was out foraging. The behavior of the male toward the model and toward his female mate were noted during the first ten minutes of the male's return. Aggression was most severe when the male model was presented before the first egg was laid, less severe when the model was presented after an egg was laid, and least severe after the eggs had hatched (see Fig. 1.7).

Experiments are considered more rigorous when they include a control group. Control in this context does not refer to the power of the experimenter—rather it refers to the validity of the experimental results. A **control group** experiences all the steps in the experiment but does not contain the variable being tested. Barash used a control group. He posted a male robin instead of a male bluebird near certain nests. (The male bluebirds did not respond at all to the robin model.) When an experiment has a control group, we know that the results are due to the variable being tested and not due to some nonidentifiable chance event that occurred during the experiment.

The Experimental Results

The results of an experiment are referred to as the **data.** Data often take the form of a table or a graph that allows one to see the results in an organized manner. Barash's data are shown in Figure 1.7. Data should be objective rather than subjective, and mathematical data are conducive to objectivity. For example, Barash didn't report that it seemed to him that the male bluebirds were more aggressive before the eggs were laid (subjective); rather, he defined aggression as the "number of approaches per minute" (objective) and reported that number. When Mendel, the father of genetics, did a pea cross concerning height, he didn't report that he could tell at a glance there were more tall than short plants; he reported that the ratio was three tall to every short plant. In fact, Mendel repeated one particular experiment so many times that he counted 7,324 peas!

To ensure that the results are not due to chance or some unknown variable, experiments are often repeated not only by the original scientist but by others in the same area of expertise. Therefore, scientists must keep careful records of how they perform their experiments so that others can repeat them. If the same results are not obtained over and over again, the experiment is not considered valid.

The Conclusion

After studying the results, the experimenter comes to a **conclusion** whether the results support or **falsify** (show to be untrue) the hypothesis. The data allowed Barash to conclude that aggression in male bluebirds is related to their reproductive cycle. Therefore, his hypothesis was supported. If male bluebirds were always aggressive even toward male robin models, his hypothesis would have been proven false.

Science progresses and any hypothesis may have to be modified in the future. There is always the possibility that a more sophisticated experiment using perhaps more advanced technology might falsify the hypothesis. Therefore, a scientist never says that the data "prove" the hypothesis to be "true." Because of this feature, some think of science as what is left after alternative hypotheses have been rejected.

Scientists report their findings in scientific journals so that their methodology and results are available to the scientific community. Barash reported his experiment in the *American Naturalist.*[1] The reporting of experiments results in accumulated data that will help other scientists formulate hypotheses. Also, it results in a body of information that is made known to the general public through the publishing of books, such as this biology textbook.

A Laboratory Investigation

When scientists are studying a phenomenon, they often perform experiments in a laboratory where conditions can be kept constant. A **variable** in a controlled experiment is a factor that can cause an observable change during the experiment. The component in an experiment being tested is called the **experimental variable;** in other words, the investigator deliberately manipulates this step of the experiment. Then the investigator observes the effects of the experiment. In other words, he or she observes the **dependent variable:**

Experimental Variable	Dependent Variable
Component of the experiment being tested	Result or change that occurs due to the experimental variable

Suppose, for example, physiologists want to determine if sweetener S is a safe food additive. On the basis of available information, they formulate a hypothesis that sweetener S is a safe food additive even when 50% of diet is sweetener S. They decide on testing these two groups:

Test group: 50% of diet is sweetener S

Control group: diet contains no sweetener S

[1] Barash, D. P. 1976. The male responds to apparent female adultery in the mountain bluebird, *Sialia currucoides:* An evolutionary interpretation. *American Naturalist* 110: 1097–1101.

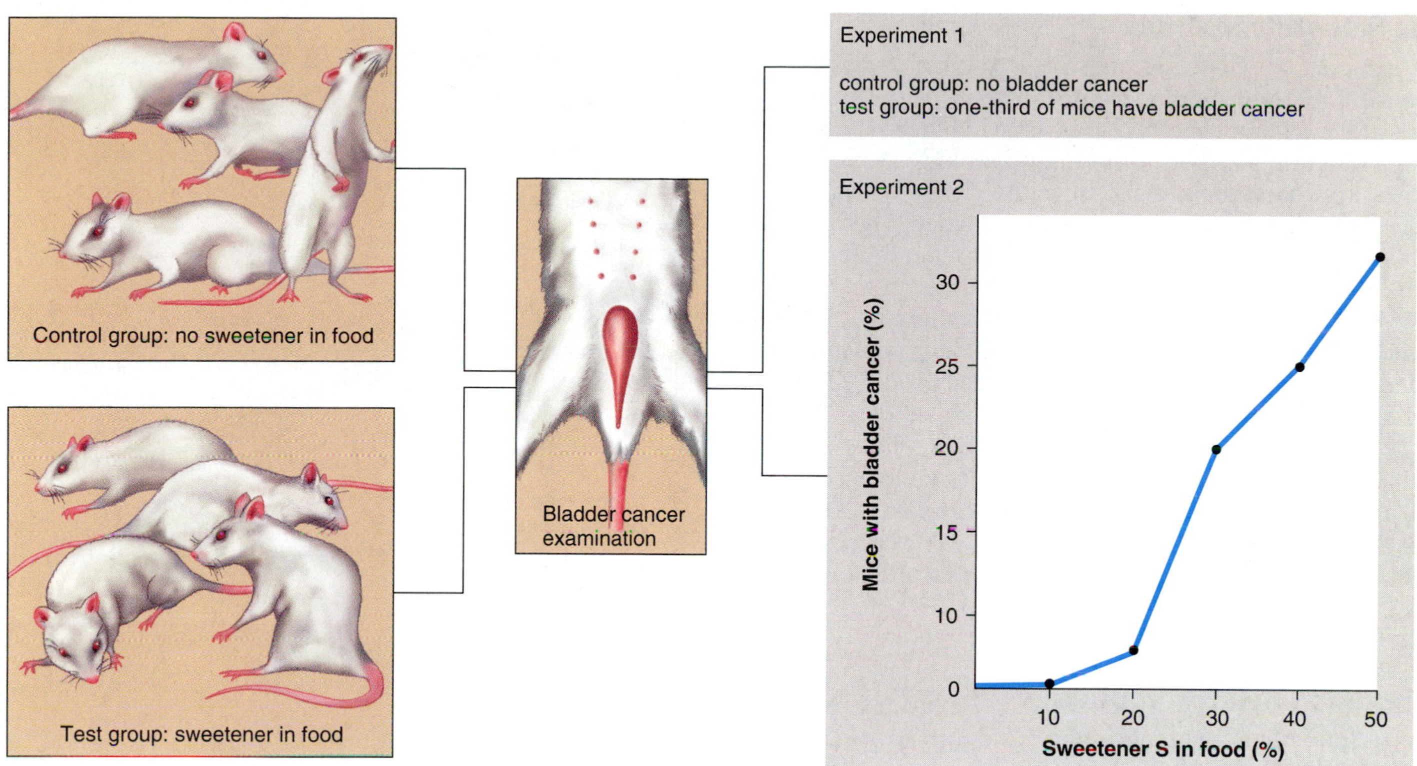

Figure 1.8 A controlled experiment.
Genetically similar mice are randomly divided into a control group and (a) test group(s) that contain 100 mice each. All groups are exposed to same conditions, such as housing, temperature, and water supply. The control group is not subjected to sweetener S in the food. At the end of an experiment all mice are examined for bladder cancer. The results of experiment 1 and experiment 2, which are described in the text, are shown on the far right.

To help ensure that the two groups are identical, the researchers place a certain number of randomly chosen inbred (genetically identical) mice into the various groups—say, 100 mice per group. If any of the mice are different from the others, it is hoped random selection has distributed them evenly among the groups. The researchers also make sure that all conditions, such as availability of water, cage setup, and temperature of the surroundings, are the same for both groups. The food for each group is exactly the same except for the amount of sweetener S.

At the end of the experiment, both groups of mice are to be examined for bladder cancer. Let's suppose that one-third of the mice in the test group are found to have bladder cancer, while none in the control group have bladder cancer. The results of this experiment do not support the hypothesis that sweetener S is a safe food additive when 50% of diet is sweetener S.

Continuing the Experiment

Science is ongoing, and one experiment frequently leads to another. Physiologists might now wish to hypothesize that sweetener S is safe if the diet contains a limited amount of sweetener S. They feed sweetener S to groups of mice at ever-greater concentrations:

Group 1: diet contains no sweetener S (the control)

Group 2: 5% of diet is sweetener S

Group 3: 10% of diet is sweetener S

↓

Group 11: 50% of diet is sweetener S

Again the data is presented in the form of a table or a graph (Fig. 1.8). Researchers might run a statistical test to determine if the difference in the number of cases of bladder cancer between the various groups is significant. After all, if a significant number of mice in the control group develop cancer, the results are invalid. Scientists prefer mathematical data because such data lends itself to objectivity.

On the basis of the data, the experimenters try to develop a recommendation concerning the safety of sweetener S in the food of humans. They might caution, for example, that the intake of sweetener S beyond 10% of the diet is associated with too great a risk of bladder cancer.

Laboratory experiments have a control group, which is not exposed to the experimental variable.

The Scientific Method

The process of science is often described in terms of the **scientific method.** Some scientists object to outlining the steps of the scientific method as is done in Figure 1.9 because such a diagram suggests a rigid methodology. Actually, scientists approach their work in many different ways and even sometimes make discoveries by chance. The most famous case pertains to penicillin. When examining a petri dish in 1928, Alexander Fleming noticed an area around a mold that was free of bacteria. Upon investigating, Fleming found that the mold produced an antibacterial substance he called penicillin. Penicillin was later mass produced and is still a successfully used antibiotic in humans today.

A discussion of the scientific process is not complete without an admission that scientists do have to accept certain assumptions. They have to believe, for example, that nature is real and understandable and knowable by observing it; that nature is orderly and uniform; that measurements yield knowledge of the thing measured; and that natural laws are not affected by time.

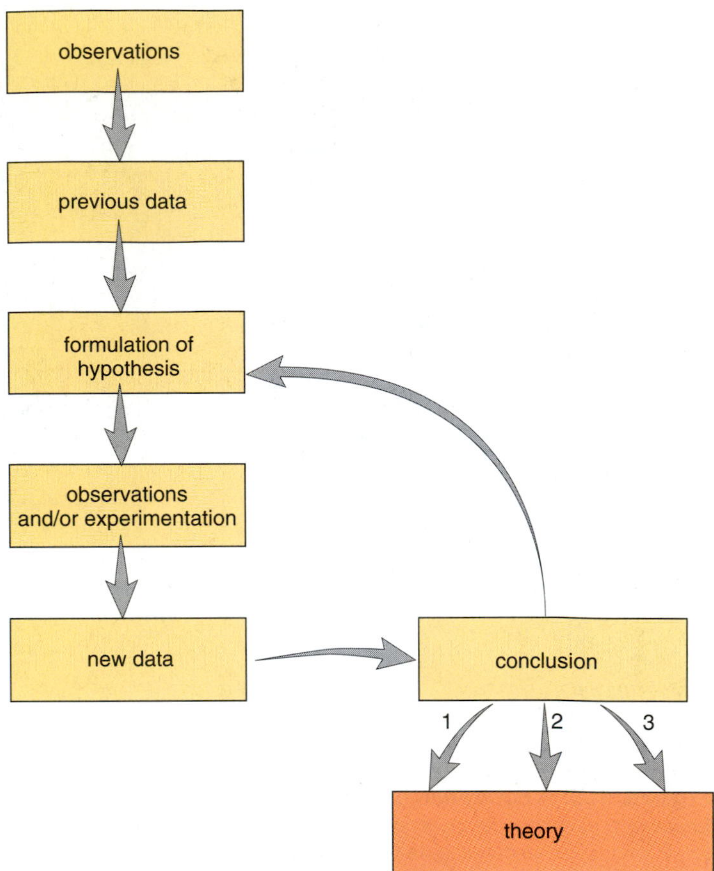

Figure 1.9 Flow diagram for the scientific method. On the basis of observations and previous data, a scientist formulates a hypothesis. The hypothesis is tested by further observations or a controlled experiment, and new data either support or falsify the hypothesis. The return arrow indicates that a scientist often chooses to retest the same hypothesis or to test a related hypothesis. Conclusions from many different but related experiments may lead to the development of a scientific theory. For example, studies pertaining to development, anatomy, and fossil remains all support the theory of evolution.

Scientific Theories in Biology

The ultimate goal of science is to understand the natural world in terms of **scientific theories,** which are concepts that join together well-supported and related hypotheses. In a movie, a detective may claim to have a theory about the crime. Or you may say that you have a theory about the won-lost record of your favorite baseball team. But in science, the word *theory* is reserved for a conceptual scheme that is supported by a broad range of observations, experiments, and data.

Some of the basic theories of biology are:

Name of Theory	Explanation
Cell	All organisms are composed of cells.
Biogenesis	Life comes only from life.
Evolution	All living things have a common ancestor and are adapted to a particular way of life.
Gene	Organisms contain coded information that dictates their form, function, and behavior.

Evolution is the unifying concept of biology because it pertains to many different aspects of living things. For example, the theory of evolution enables scientists to understand the history of life, the variety of living things, and the anatomy, physiology, and development of organisms—even their behavior.

Barash gave an evolutionary interpretation to his results. It was adaptive, he said, for male bluebirds to be less aggressive after the first egg is laid because by that time the male bird is "sure the offspring is his own" and maladaptive for the male bird to waste time and energy being too aggressive toward a rival and his mate after hatching because his offspring are already present. (When an organism is adapted to its environment, it is better able to survive and produce offspring.)

Fruitful theories are ones that help scientists generate new hypotheses, and the theory of evolution has been a very fruitful theory. In fact, it probably helped Barash develop the hypothesis he chose to test. Because the theory of evolution has been supported by so many observations and experiments for over 100 years, some biologists refer to the **principle** of evolution, suggesting that this is the appropriate term for theories that are generally accepted by an overwhelming number of scientists. The term **law** instead of principle is preferred by some. In another chapter concerning energy relationships, for example, we will examine the laws of thermodynamics.

The scientific process eventually enables biologists to arrive at a theory that is generally accepted by all.

Connecting Concepts

What we know about biology and what we'll learn in the future results from objective observation and testing of the natural world. The ultimate goal of science is to understand the natural world in terms of theories—conceptual schemes supported by abundant research and not yet found lacking. Evolution is a theory that accounts for the differences that divide and the unity that joins all living things. All living things have the same levels of organization and function similarly because they are related—even back to the first living cells on earth.

Scientific creationism, which states that God created all species as they are today, cannot be considered science because creationism upholds a supernatural cause rather than a natural cause for events. When faith is involved, a hypothesis cannot be tested in a purely objective way. Just as science does not test religious beliefs, it does not make ethical or moral decisions. The general public may want scientists to label certain research as "good" or "bad" and to predict whether any resulting technology will primarily benefit or harm. Yet science, by its very nature, is impartial and simply attempts to study natural phenomena.

Scientists should provide the public with as much information as possible when such issues as recombinant DNA technology or various methods to preserve the environment are being debated. Then, they, along with other citizens, can help make decisions about what is most likely best for society. All men and women have a responsibility to decide how to use scientific knowledge so that it benefits all living things, including the human species.

This textbook was written to help you understand the scientific process and learn the basic concepts of general biology so that you will be better informed. This chapter introduced you to the levels of biological organization from the cell to the biosphere. The cell, the simplest of living things, is composed of nonliving molecules. Therefore, we must begin our study of biology with a brief look at cellular chemistry. In the next two chapters, you will study some important inorganic and organic molecules in the cell. Then, you will learn how the cell makes use of energy and materials in order to maintain itself and reproduce.

Summary

1.1 How to Define Life

Although living things are diverse, they share certain characteristics in common. Living things (a) are organized, and there are levels of organization from the cell to ecosystems, (b) need an outside source of materials and energy, (c) respond to external stimuli, (d) reproduce, passing on genes to their offspring, and (e) have adaptations suitable to their way of life in a particular environment.

The process of evolution explains both the unity and the diversity of life. Descent from a common ancestor explains why all organisms share the same characteristics, and adaptation to various ways of life explains the diversity of life-forms.

1.2 How the Biosphere Is Organized

Within an ecosystem, populations interact with one another and with the physical environment. Nutrients cycle within and between ecosystems, but energy flows unidirectionally, and eventually becomes heat. Adaptations of organisms allow them to play particular roles within an ecosystem, as when the trees within a tropical rain forest photosynthesize and provide a home for other organisms.

1.3 How Living Things Are Classified

Each living thing is given an italicized binomial name that consists of the genus and specific epithet. For example, *Pisum sativum* is the name of the garden pea. Each species belongs to genus, family, order, class, phylum, kingdom, and finally domain from the least inclusive to the most inclusive category. The three domains of life are Archaea, Bacteria, and Eukarya. The first two domains contain unicellular organisms which are structurally simple but metabolically complex. Domain Eukarya contains the kingdoms Protista, Fungi, Plantae, and Animalia. Protists range from unicellular to multicellular organisms and include the protozoans and algae. Among the fungi are the familiar molds and mushrooms. Plants are well known as multicellular photosynthesizers of the world while animals are multicellular and ingest their food.

1.4 The Process of Science

When studying the world of living things, biologists and other scientists use the scientific process. Observations along with previous data are used to formulate a hypothesis. New observations and/or experiments are carried out in order to test the hypothesis.

Scientists often do controlled experiments. In a controlled experiment, the experimental variable is that portion of the experiment being manipulated and the dependent variable is the change due to the experimental variable. The control group is not exposed to the experimental variable.

New data may support a hypothesis or they may prove it false. Hypotheses cannot be proven true. Several conclusions in a particular area may allow scientists to arrive at a theory—such as the cell theory, gene theory, or the theory of evolution. The theory of evolution is a unifying theory of biology.

Reviewing the Chapter

1. What evidence can you cite to show that living things are organized? 2
2. Why do living things require an outside source of materials and energy? 3
3. What are the common characteristics of life listed in the text? 3–4
4. What is passed from generation to generation when organisms reproduce? What has to happen to the hereditary material DNA in order for evolution to occur? 4
5. How does evolution explain both the unity and the diversity of life? 4
6. What is an ecosystem, and why should human beings preserve ecosystems? 5, 7
7. Explain the scientific name of an organism. What are the categories of classification? What four kingdoms are in the domain Eukarya? 8
8. Describe the series of steps involved in the scientific process. Which of the steps requires the use of inductive reasoning? deductive reasoning? 9
9. Give an example of a laboratory experiment. Name the experimental variable and the dependent variable. 10
10. What is the ultimate goal of science? Give an example that supports your answer. 12

Testing Yourself

Choose the best answer for each question. For questions 1–4, match the statements in the key with the sentences below.

Key:

a. Living things are organized.
b. Living things metabolize.
c. Living things respond.
d. Living things reproduce.
e. Living things evolve.

1. Genes made up of DNA are passed from parent to child.
2. Zebras run away from approaching lions.
3. Cells use materials and energy for growth and repair.
4. There are many different kinds of living things.
5. Evolution from the first cell(s) best explains why
   a. ecosystems have populations of organisms.
   b. photosynthesizers produce food.
   c. diverse organisms share common characteristics.
   d. human activities are threatening the biosphere.
   e. All of these are correct.
6. Adaptation to a way of life best explains why living things
   a. display homeostasis.
   b. are diverse.
   c. began as single cells.
   d. are classified into three domains.
   e. mate with their own kind.
7. Into which kingdom would you place a multicellular land organism that carries on photosynthesis?
   a. Protista
   b. Fungi
   c. Plantae
   d. Animalia
   e. Archaea
8. Which is the experimental variable in the experiment concerning sweetener S?
   a. Conditions like temperature and housing are the same for all groups.
   b. The amount of sweetener S in food.
   c. Two percent of the group fed food that was 10% sweetener S got bladder cancer, and 90% of the group fed food that was 50% sweetener S got bladder cancer.
   d. The data were presented as a graph.
   e. The use of a control versus a test group.
9. Which is the control group in this same experiment?
   a. All mice in this group died because a lab assistant forgot to give them water.
   b. All mice in this group received food that was 10% sweetener S.
   c. All mice in this group received no sweetener S in food.
   d. Some mice in all groups got bladder cancer; therefore, there was no control group.
   e. All of these are correct.
10. With which of these steps in the scientific method do you associate inductive reasoning?
    a. experimentation
    b. conclusion
    c. researching the scientific literature
    d. formulating the hypothesis
    e. Both b and d are correct.
11. An investigator spills dye on a culture plate and then notices that the bacteria live despite exposure to sunlight. He hypothesizes that the dye protects bacteria against death by ultraviolet (UV) light. To test this hypothesis, he decides to expose two hundred culture plates to UV light. One hundred plates contain bacteria and dye; the other hundred plates contain only bacteria. Result: after exposure to UV light, the bacteria on both plates die. Fill in the right-hand portion of this diagram.

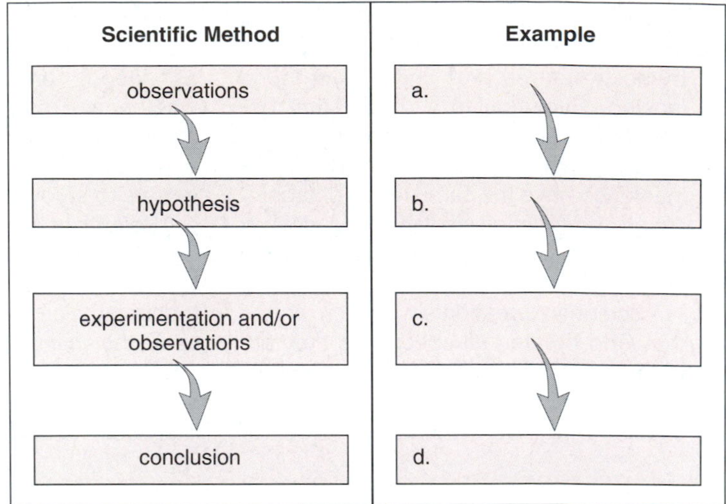

Thinking Scientifically

1. You were part of a study testing a new drug to treat migraine headaches. Some 25% of the control group, which had been given a placebo[2], and 46% of the experimental group reported an improvement in their symptoms. Would you conclude that this is an effective drug? What variables in study participants might have influenced which of them benefited from the new drug?
2. Viruses are small, infectious particles. They have a genome of either DNA or RNA, but no organelles characteristic of cells. They do not divide by mitosis but are duplicated by having their constituent parts produced and assembled by the infected cell. Viruses can evolve in the same way as living things. Genome changes due to random mutations is followed by selection. Would a virus be considered alive? What are the criteria on which you base your answer?

Bioethical Issue

The Endangered Species Act requires the federal government to identify endangered and threatened species and to protect their habitats, even to the extent of purchasing the land where they live. Developers feel that the act protects wildlife at the expense of jobs for U.S. citizens. In an effort to allow development in sensitive areas, it is now possible to move forward after a Habitat Conservation Plan (HCP) is approved. An HCP permits, for instance, new home construction or logging on a part of the land if wildlife habitat is conserved on another part. Habitat conservation can also mean helping the government buy suitable land someplace else.

The nation's first HCP was approved in 1980. It permitted housing construction on San Bruno Mountain, near San

[2] A placebo is an inert substance that looks like the medication taken by the experimental group.

Francisco, if 97% of the living space for the endangered mission blue butterfly was preserved. That sounds pretty good, but over time, hundreds of HCPs have been approved, and the conservation requirement has slipped a bit. Tim Cullinan, a director of the National Audubon Society, recently found that logging companies in the Pacific Northwest are proposing the exchange of habitat on public land for the right to log privately owned, old forests. In other words, nothing has been given up. Today, there are so many HCPs presented, they are rubber-stamped by government officials with no public review at all.

Do you favor development over preservation of habitats or vice versa? Do you think the federal government should be in the business of trying to preserve endangered species? Do you think that public review of HCPs should be allowed, even if it slows down the approval process? Is it the public's responsibility to remain vigilant or is governmental review of HCPs sufficient?

Understanding the Terms

adaptation 4
biodiversity 7
biology 9
biosphere 5
community 5
conclusion 10
control group 10
data 10
deductive reasoning 9
dependent variable 10
ecosystem 5
energy 3
evolution 4
experiment 9
experimental variable 10
falsify 10
gene 4
homeostasis 3
hypothesis 9
inductive reasoning 9
law 12
metabolism 3
natural selection 4
phenomenon 9
photosynthesis 3
population 5
principle 12
reproduce 4
science 9
scientific method 12
scientific theory 12
species 4
taxonomy 8
variable 10

Match the terms to these definitions:

a. ______________ All of the chemical reactions that occur in a cell during growth and repair.
b. ______________ Changes that occur among members of a species with the passage of time, often resulting in increased adaptation to the prevailing environment.
c. ______________ Component in an experiment that is manipulated as a means of testing it.
d. ______________ Process by which plants utilize solar energy to make their own organic food.
e. ______________ Sample that goes through all the steps of an experiment except the one being tested; a standard against which results of an experiment are checked.

Web Connections

Exploring the Internet

http://www.mhhe.com/biosci/genbio/mader
(click on *Biology 7/e*)

The *Biology 7/e* Online Learning Center provides many resources for studying the material in this chapter including links to the following sites:

Links to Many Specific Career Descriptions. At least 200 links to websites can be found through this site which is updated frequently. An alphabetical listing of occupations in biology allows the user to see websites under many of the listings that include detailed descriptions of careers.

http://www.furman.edu/~snyder/careers/careerlist.html

The Tree of Life explores the phylogenetic relationships between great numbers of organisms and is continually updated.

http://phylogeny.arizona.edu/tree/phylogeny.html

Microscopy Society of America with current information about the society and information on many kinds of microscopy.

http://www.msa.microscopy.com/

National Biological Information Infrastructure. This is a gateway site to biological information from a myriad of sources, both governmental and private. Links abound.

http://www.nbii.gov/index.html

New Scientist. Science and technology news is the focus of this website. Internet Magazine *called this the best website in 1995. A must-see site for amateur and professional scientists.*

http://www.newscientist.com/

Further Readings for Chapter One

Balick, M. J., and Cox, P. A. 1996. *Plants, people, and culture: The science of ethnobotany.* New York: Scientific American Library. Discusses the importance of rain forest conservation.

Carey, S. S. 1997. *A beginner's guide to scientific method.* 2d ed. Belmont, Calif.: Wadsworth Publishing. The basics of scientific method are explained.

Cox, G. W. 1997. *Conservation biology.* Dubuque, Ia.: Wm. C. Brown Publishers. This text examines the field of conservation, surveys basic principles of ecology, and considers steps to preserve biodiversity.

Dobson, A. P. 1996. *Conservation and biodiversity.* New York: Scientific American Library. Attempts to manage endangered species and preserve biodiversity are presented.

Drewes, F. 1997. *How to study science.* 2d ed. Dubuque, Ia.: Wm. C. Brown Publishers. This supplement shows students how to study, take notes, and interpret text figures.

Johnson, G. B. 1996. *How scientists think.* Dubuque, Ia.: Wm. C. Brown Publishers. Presents 21 fundamental experiments in genetics and molecular biology.

Kellert, S. R. 1996. *The value of life: Biological diversity and human society*. Washington D.C.: Island Press/Shearwater Books. The importance of biological diversity to the well-being of humanity is explored.

Marchuk, W. N. 1992. *A life science lexicon.* Dubuque, Ia.: Wm. C. Brown Publishers. Helps students master life science terms.

Minkoff, E. C., and Baker, P. J. 1996. *Biology today: An issues approach.* New York: The McGraw-Hill Companies, Inc. This introductory text emphasizes understanding of selected biological issues, and discusses each issue's social context.

Primak, R. B. 1995. *A primer of conservation biology.* Sunderland, Mass.: Sinauer Associates. Addresses the loss of biological diversity throughout the world, and suggests remedies.

part

i The Cell

A cell is the basic unit of life, and all living things are composed of cells. Therefore, our knowledge of the structure and function of a cell can be applied directly to the organism. Our study of the cell begins with the atoms and molecules that make up the structure and carry on the functions of cells. Cellular organization requires an ongoing input of matter and energy. Metabolic pathways carry out the transformations needed to keep the cell operational. In most ecosystems, the energy of the sun maintains life. Plant cells, and other types of photosynthetic cells, capture solar energy and store it in molecules that later are a source of matter and energy for all living things.

Cells come from preexisting cells. A better understanding of the regulation of cellular reproduction has contributed greatly to cancer research. Knowledge of chemistry, energy transformations, metabolic pathways, and cellular reproduction increases our ability to keep ourselves healthy and our world fit to live in.

Basic Chemistry

chapter 2

chapter concepts

2.1 Chemical Elements

2.2 Compounds and Molecules

2.3 Chemistry of Water

Eastern gray kangaroo, *Macropus giganteus*

A hundred years ago, scientists believed that only nonliving things, like rocks and metals, consisted of chemicals. They thought that living things like kangaroos and sunflowers had a special force, called a vital force, which was necessary for life. Scientific investigation, however, has repeatedly shown that both nonliving and living things have the same physical and chemical bases. Although living things contain molecules not found in inanimate objects, such molecules must still be understood by studying basic chemical properties.

Suppose you have a special interest in kangaroos, and you read books about them and even go to Australia to watch kangaroos in the wild. Still, it would be necessary for you to study chemistry in order to fully understand kangaroos. We perceive the world in terms of whole objects like kangaroos, sunflowers, mushrooms, and humans, but in this chapter you will discover that all organisms consist of atoms and molecules linked together in specific ways that give them properties different from nonliving things. In order to understand how kangaroos jump, the kangaroo expert must study the physical and chemical nature of the kangaroo's muscular system.

2.1 Chemical Elementse

Matter refers to anything that takes up space and has mass. It is helpful to remember that matter can exist as a solid, a liquid, or a gas. Then we can realize that not only are we matter but so too are the water we drink and the air we breathe.

All matter, both nonliving and living, is composed of certain basic substances called **elements.** It is quite remarkable that there are only 92 naturally occurring elements. We know these are elements because they cannot be broken down to substances with different properties (a property is a chemical or physical characteristic, such as density, solubility, melting point, and reactivity).

Both the earth's crust and organisms are made up of elements, but they differ as to which ones are predominant (Fig. 2.1). Only six elements—carbon, hydrogen, nitrogen, oxygen, phosphorus, and sulfur—make up most (about 98%) of the body weight of most organisms. The acronym CHNOPS helps us remember these six elements. The properties of these elements are essential to the uniqueness of living things from cells to organisms.

All living and nonliving things are matter composed of elements. Six elements in particular are commonly found in living things.

Figure 2.1 Elements of earth's crust and organisms. The earth's crust primarily contains the elements oxygen, silicon, and aluminum. Organisms primarily contain the elements hydrogen, oxygen, carbon, and nitrogen. Along with phosphorus and sulfur, these elements make up most biological molecules.

Atomic Structure

In the early 1800s, the English scientist John Dalton proposed that elements actually contain tiny particles called **atoms** [Gk. *atomos*, uncut, indivisible]. He also deduced that there is only one type of atom in each type of element. You can see why, then, the name assigned to each element is the same as the name assigned to the type of atom it contains. Some of the names we use for the elements (atoms) are derived from English and some are derived from Latin. One or two letters create the **atomic symbol,** which stands for this name. For example, the symbol H stands for a hydrogen atom, and the symbol Na (for *natrium* in Latin) stands for a sodium atom. Table 2.1 gives the atomic symbols for the other elements (atoms) commonly found in living things.

From our discussion of elements, we would expect each atom to have a certain mass. The mass of an atom is in turn dependent upon the presence of certain subatomic particles. Although physicists have identified a number of subatomic particles, we will consider only the most stable of these: **protons, neutrons,** and **electrons** [Gk. *elektron*, amber, electricity]. Protons and neutrons are located within the nucleus of an atom, and electrons move about the nucleus. Figure 2.2 shows the arrangement of the subatomic particles in helium, an atom that has only two electrons. In Figure 2.2*a*, the stippling shows the probable location of electrons, and in Figure 2.2*b*, the circle represents the average location of electrons.

Table 2.1

Common Elements in Living Things

Element*	Atomic Symbol	Atomic Number	Atomic Mass**
Hydrogen	H	1	1
Carbon	C	6	12
Nitrogen	N	7	14
Oxygen	O	8	16
Sodium	Na	11	23
Magnesium	Mg	12	24
Phosphorus	P	15	31
Sulfur	S	16	32
Chlorine	Cl	17	35
Potassium	K	19	39
Calcium	Ca	20	40

The atomic number gives the number of protons (and electrons in electrically neutral atoms). The number of neutrons is equal to the atomic mass minus the atomic number.

* *In the Periodic Table of the Elements (see inside back cover) and here, elements are arranged in order of ascending atomic number and mass.*

** *Average of most common isotopes.*

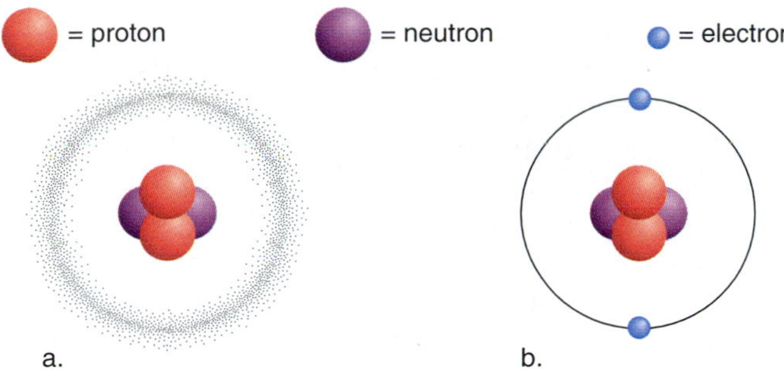

Figure 2.2 Model of helium (He).
Atoms contain subatomic particles, which are located as shown. Protons and neutrons are found within the nucleus, and electrons are outside the nucleus. **a.** The stippling shows the probable location of the electrons in the helium atom. **b.** The average location of electrons is sometimes represented by a circle.

Our concept of an atom has changed greatly since Dalton's day. If we could draw an atom the size of a football field, the nucleus would be like a gumball in the center of the field and the electrons would be tiny specks whirling about in the upper stands. Most of an atom is empty space. We should also realize that we can only indicate where the electrons are expected to be most of the time. In our analogy, the electrons may even stray outside the stadium at times.

In effect, the **atomic mass** of an atom is just about equal to the sum of its protons and neutrons. Protons and neutrons are assigned one atomic mass unit each. Electrons are so small that their mass is assumed to be zero in most calculations (Table 2.2). The term atomic mass is preferred over the term atomic weight because mass is constant while weight changes according to the gravitational force of a planet. The gravitational force of the earth is greater than that of the moon, therefore substances weigh less on the moon even though their mass has not changed.

All atoms of an element have the same number of protons. This is called the atom's **atomic number.** In Table 2.1, atoms are listed according to increasing atomic number, as they are in the periodic table of the elements found inside the back cover. This indicates that it is the number of protons (i.e., the atomic number) that makes an atom unique. The atomic number is often written as a subscript to the lower left of the atomic symbol. The atomic mass is often written as a superscript to the upper left of the atomic symbol. For example, the carbon atom can be noted in this way:

atomic mass —— $^{12}_{6}C$ —— atomic symbol
atomic number ——

Atoms have an atomic symbol, mass, and number. The subatomic particles (protons, neutrons, and electrons) determine the characteristics of atoms.

Table 2.2

Subatomic Particles

Particle	Electric Charge	Atomic Mass	Location
Proton	+1	1	Nucleus
Neutron	0	1	Nucleus
Electron	−1	0	Electron shells

Isotopes

The atomic mass given in Table 2.1 is the average mass of each type of atoms listed. This is because atoms of the same element may differ in the number of neutrons. Atoms of the same element that have the same number of protons, and differ only in the number of neutrons, are called **isotopes** [Gk. *isos,* equal, and *topos,* place]. Three isotopes of carbon can be written in the following manner:

$$^{12}_{6}C \quad ^{13}_{6}C \quad ^{14}_{6}C$$

Carbon 12 has six neutrons, carbon 13 has seven neutrons, and carbon 14 has eight neutrons. Unlike the other two isotopes, carbon 14 is unstable; it breaks down into elements with lower atomic numbers. When it decays, it emits radiation in the form of radioactive particles or radiant energy. Therefore carbon 14 (^{14}C) is called a **radioactive isotope.**

Isotopes have many uses. Because proportions of isotopes in various food sources are known, biologists can now determine the proportion of isotopes in mummified or fossilized human tissues to know what ancient peoples ate. Radioactive isotopes are used as tracers in biochemical experiments. For example, ^{14}C was used to detect the sequential biochemical steps that occur during photosynthesis. And because ^{14}C decays at a known rate, the amount of ^{14}C remaining is often used to determine the age of fossils.

Radioactive isotopes are also used in medicine. If a patient is injected with radioactive iodine, the thyroid gland will take it up, and a scan of the thyroid will indicate if any abnormality is present. Glucose labeled with a radioactive isotope can be injected into the body and will be taken up by metabolically active tissues. In PET (positron emission tomography), the radiation given off is used by a computer to generate cross-sectional images that indicate the metabolic activity of various tissues.

Atoms of the same element that have the same number of protons but a different number of neutrons and a different mass are called isotopes.

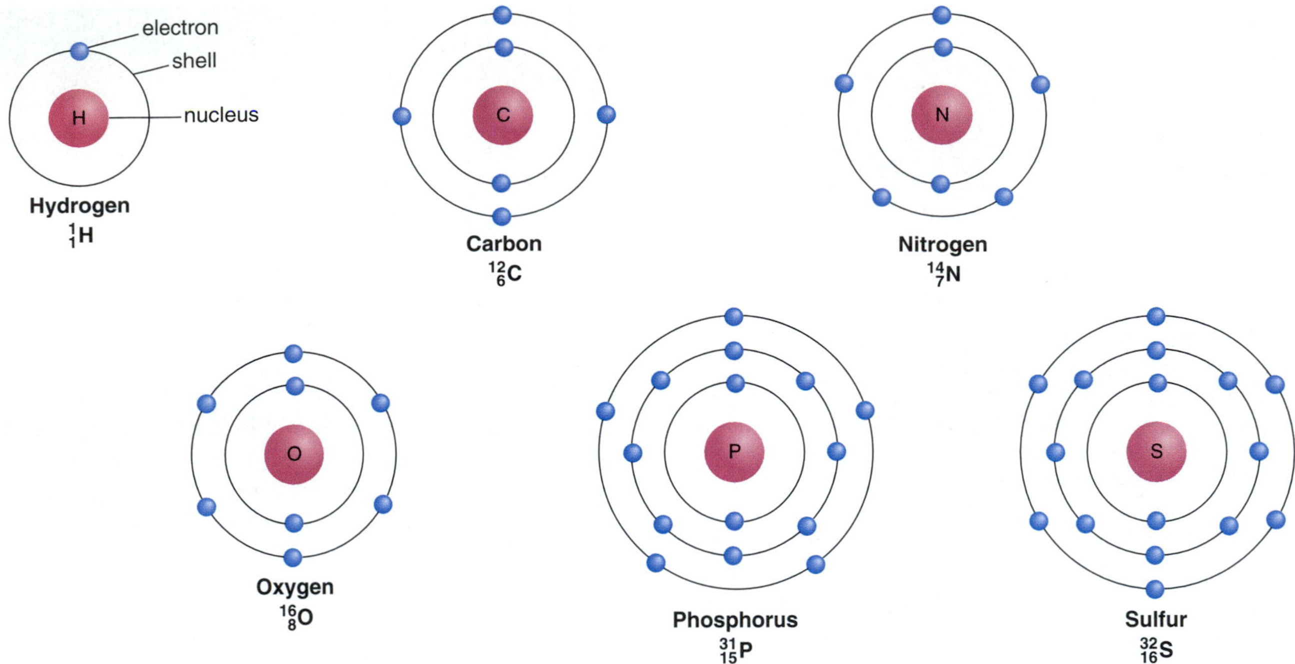

Figure 2.3 Bohr models of atoms.
Electrons are placed in energy levels (electron shells) according to certain rules: the first shell can contain up to two electrons, and each shell thereafter can contain up to eight electrons as long as we consider only atoms with an atomic number of 20 or below. Each shell is to be filled before electrons are placed in the next shell.

Energy Levels

Protons and electrons carry a charge; protons have a positive (+) electrical charge, and electrons have a negative (−) electrical charge. When an atom is electrically neutral, the number of protons equals the number of electrons. Therefore, in electrically neutral atoms, the atomic number tells you the number of protons and the number of electrons. For example, a carbon atom has six protons, and when electrically neutral, it also has six electrons.

In 1913, the Danish physicist Niels Bohr proposed that electrons orbit in concentric energy levels (called **electron shells**) about the nucleus. He based his model on a previous discovery that although electrons have the same mass and charge, they vary in energy content. **Energy** is defined as the ability to do work. Electrons differ in the amount of their potential energy, that is, stored energy ready to do work. The electron shells indicate the relative amounts of stored energy electrons have. Electrons with the least amount of potential energy are located in the shell closest to the nucleus, called the *K* shell. Electrons in the next higher shell, called the *L* shell, have more energy, and so forth, as we proceed from shell to shell outside the nucleus.

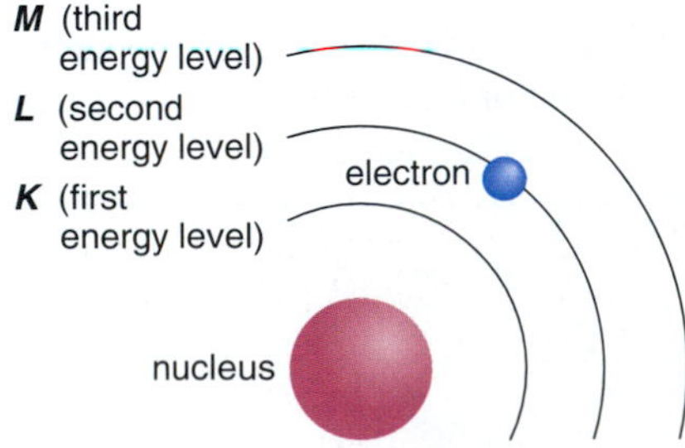

An analogy may help you appreciate that the farther electrons are from the nucleus, the more potential energy they possess. Falling water has energy, as witnessed by how it can turn a waterwheel connected to a shaft that transfers the energy to machinery for grinding grain, cutting wood, or weaving cloth. The higher the waterfall, the greater the amount of energy released per unit amount of water. The potential energy possessed by electrons is not due to gravity; it is due to the attraction between the positively charged protons and the negatively charged electrons. It takes energy to keep an electron farther away from the nucleus as opposed to closer to the nucleus. You have often heard that sunlight provides the energy for photosynthesis in green plants, but it may come as a surprise to learn that when a pigment such as chlorophyll absorbs the energy of the sun, electrons move to higher energy levels about nuclei.

Figure 2.3 shows you how the Bohr model helps you determine how many electrons are in the outer shell of an atom. For atoms up to an atomic number of 20 (i.e., calcium), the first shell can contain as many as two electrons; thereafter, each additional shell can contain eight electrons. For these atoms, each lower shell is filled with electrons before the next higher shell contains any electrons. The sulfur atom, with an atomic number of 16, has three shells (two electrons in the first shell, eight electrons in the second shell, and six electrons in the third, or outer, shell).

In the periodic table (see inside back cover), elements are arranged in rows according to the number of electrons in the outer shell. If an atom has only one shell, the outer shell

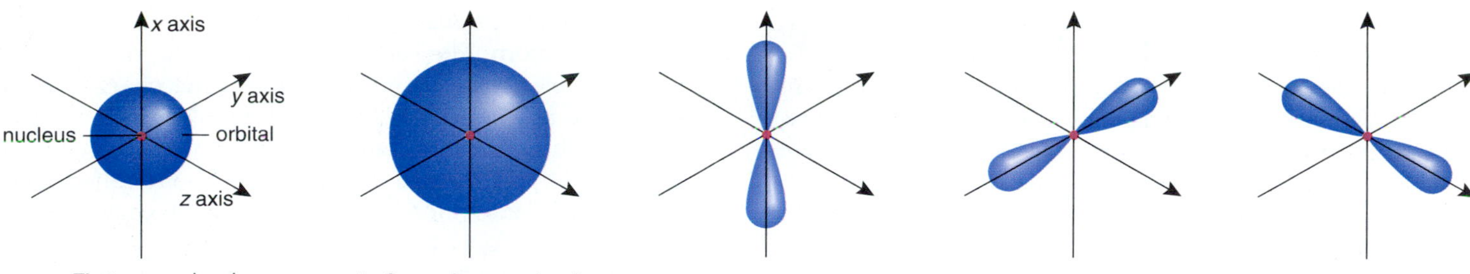

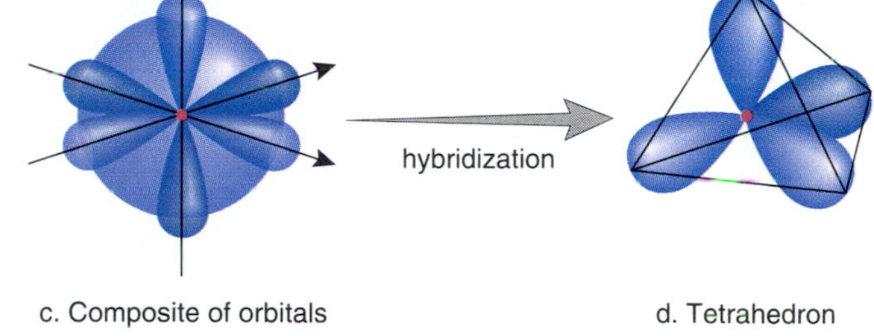

Figure 2.4 Electron orbitals.
Each electron energy level (see Fig. 2.3) has one or more orbitals, a volume of space in which the rapidly moving electrons are most likely found. The nucleus is at the intersection of *x, y,* and *z* axes. **a.** The first energy level (electron shell) has only one orbital, which has a spherical shape. Two electrons can occupy this orbital. **b.** The second energy level has one spherically-shaped orbital and three dumbbell-shaped orbitals at right angles to each other. Since two electrons can occupy each orbital, there are a total of eight electrons in the second electron shell. **c.** The orbitals of the second shell superimposed on one another. **d.** When the orbitals of the second shell hybridize, teardrop-shaped orbitals point toward the corners of a tetrahedron (a triangular pyramid).

is complete when it has two electrons. Otherwise, the **octet rule,** which states that the outer shell is most stable when it has eight electrons, holds. Atoms with eight electrons in the outer shell do not ordinarily react; they are said to be inert. Atoms with fewer than eight electrons in the outer shell react with other atoms in such a way that after the reaction, each has a stable outer shell. Atoms can give up, accept, or share electrons in order to have a stable outer shell.

Electron orbitals. The Bohr model was later modified because it is not actually possible to pinpoint the location of an electron. Rather than an orbit, an electron occupies an **orbital,** which is a volume of space where a rapidly moving electron is statistically predicted to be found. An orbital has a characteristic energy state and a characteristic shape. At the first energy level, there is only a single spherical orbital, where at most two electrons are found about the nucleus (Fig. 2.4*a*). The space is spherical shaped because the most likely location for each electron is a fixed distance in all directions from the nucleus. At the second energy level, there are four orbitals; one of these is spherical shaped but the other three are dumbbell shaped (Fig. 2.4*b*). This is the shape that allows the electrons to be most distant from one another. Since each orbital can hold two electrons, there are a maximum of eight electrons in the L shell (Fig. 2.4*c*). When bonding occurs, the orbitals of this shell sometimes hybridize, forming teardrop-shaped orbitals that point to the corners of a triangular pyramid called a tetrahedron (Fig. 2.4*d*).

The number of electrons in the outer shell determines the manner in which atoms react with one another.

Chemical Formulas and Equations

When writing chemical formulas, atomic symbols are used to represent the atoms, and subscripts are used to indicate how many atoms of each type there are in a substance. For example, the chemical formula H_2O (read as H-two-O) indicates that a water molecule contains two hydrogen atoms and one oxygen atom. The chemical formula for glucose contains many atoms:

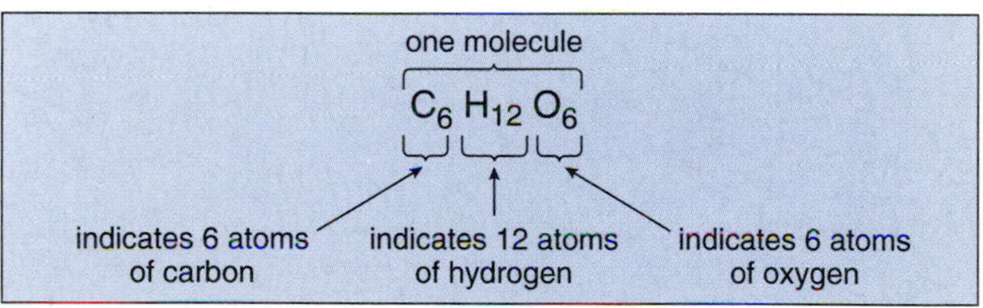

Formulas are used in chemical equations to represent chemical reactions that occur between atoms and molecules:

$$6CO_2 + 6H_2O \longrightarrow C_6H_{12}O_6 + 6O_2$$

carbon dioxide + water → glucose + oxygen

In this equation, which is often used to represent photosynthesis, six molecules of carbon dioxide react with six molecules of water to yield one glucose molecule and six molecules of oxygen. The reactants (molecules that participate in the reaction) are shown on the left of the arrow, and the products (molecules formed by the reaction) are shown on the right. Notice that the equation is "balanced," that is, there are the same number of each type of atom on both sides of the arrow.

2.2 Compounds and Molecules

Molecules can form when two or more atoms of the same element react with one another. For example, oxygen does not exist in nature as a single atom, O; instead, two oxygen atoms are joined as O_2. When atoms of two or more different elements react or bond together, a **compound** results. Water (H_2O) is a compound that contains the elements hydrogen and oxygen. We can also speak of H_2O as a **molecule** [L. *moles,* mass], because a molecule is the smallest part of a compound that still has the properties of that compound.

Electrons possess energy and the bonds that exist between atoms contain energy. Organisms are dependent upon chemical-bond energy to maintain their organization. When a chemical reaction occurs, electrons shift in their relationship to one another and energy may be given off. This is the energy that we make use of to carry on our daily lives.

Ionic Bonding

Ionic bonds form when electrons are transferred from one atom to another. For example, sodium (Na), with only one electron in its third shell, tends to be an electron donor (Fig. 2.5*a*). Once it gives up this electron, the second shell, with eight electrons, becomes its outer shell. Chlorine (Cl), on the other hand, tends to be an electron acceptor. Its outer shell has seven electrons, so it needs only one more electron to have a completed outer shell. When a sodium atom and a chlorine atom come together, an electron is transferred from the sodium atom to the chlorine atom. Now both atoms have eight electrons in their outer shells.

This electron transfer, however, causes a charge imbalance in each atom. The sodium atom has one more proton than it has electrons; therefore, it has a net charge of +1 (symbolized by Na^+). The chlorine atom has one more electron than it has protons; therefore, it has a net charge of −1 (symbolized by Cl^-). Such charged particles are called **ions.** Sodium (Na^+) and chlorine (Cl^-) are not the only biologically important ions. Some, such as potassium (K^+), are formed by the transfer of a single electron to another atom; others, such as calcium (Ca^{2+}) and magnesium (Mg^{2+}), are formed by the transfer of two electrons.

Ionic compounds are held together by an attraction between the charged ions called an **ionic bond.** When sodium reacts with chlorine, an ionic compound called sodium chloride (NaCl) results, and the reaction is called an ionic reaction. Sodium chloride is a salt commonly known as table salt because it is used to season our food (Fig. 2.5*b*). Salts can exist as dry solids, but when such a compound is placed in water, the ions separate as the salt dissolves, as when NaCl separates into Na^+ and Cl^-. Ionic compounds are most commonly found in this dissociated (ionized) form in biological systems because these systems are 70–90% water.

The transfer of electron(s) between atoms results in ions that are held together by an ionic bond, the attraction of negative and positive charges.

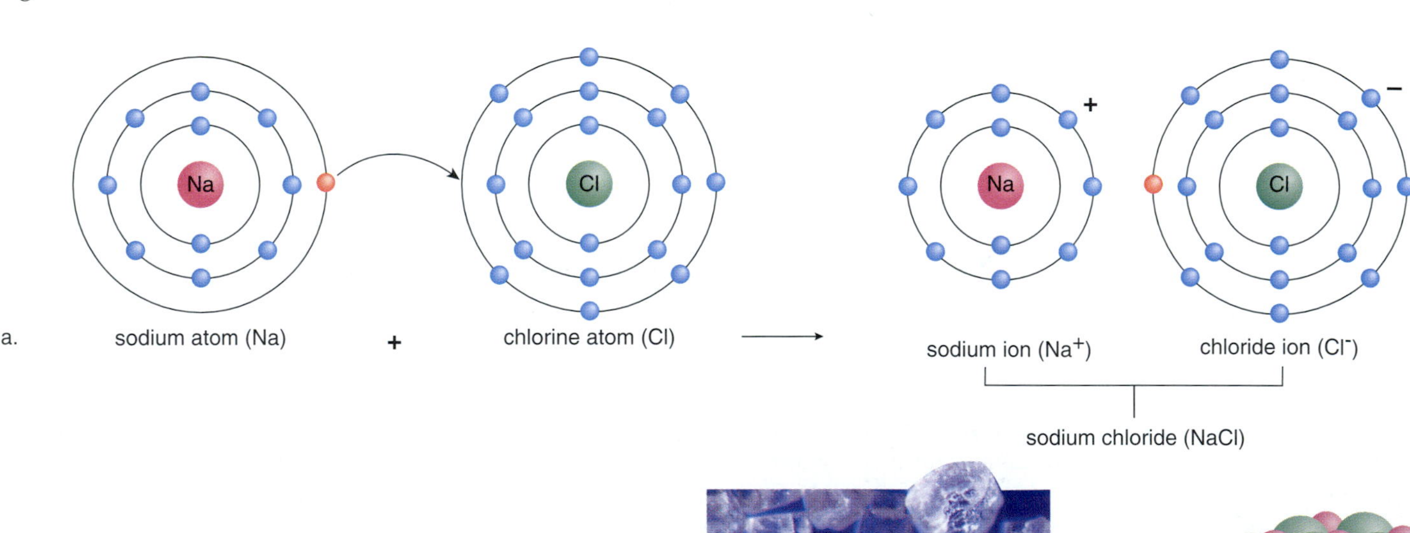

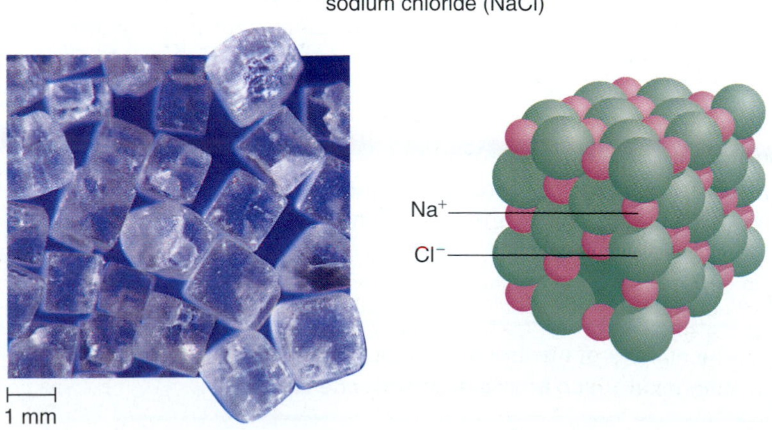

Figure 2.5 Ionic reaction.
a. During the formation of sodium chloride, an electron is transferred from the sodium atom to the chlorine atom. At the completion of the reaction, each atom has eight electrons in the outer shell, but each also carries a charge as shown. **b.** In a sodium chloride crystal, ionic bonding between Na^+ and Cl^- causes the atoms to assume a three-dimensional lattice in which each sodium ion is surrounded by six chlorine ions, and each chlorine ion is surrounded by six sodium ions.

Covalent Bonding

A **covalent bond** [L. *co,* together, with, and *valens,* strength] results when two atoms share electrons in such a way that each atom has an octet of electrons in the outer shell (or two electrons, in the case of hydrogen). In a hydrogen atom the outer shell is complete when it contains two electrons. If hydrogen is in the presence of a strong electron acceptor, it gives up its electron to become a hydrogen ion (H^+). But if this is not possible, hydrogen can share with another atom and thereby have a completed outer shell. For example, a hydrogen atom can share with another hydrogen atom. In this case, the two orbitals overlap and the electrons are shared between them (Fig. 2.6*a*). Because they share the electron pair, each atom has a completed outer shell. When a reaction results in a covalent molecule, it is called a covalent reaction.

A more common way to symbolize that atoms are sharing electrons is to draw a line between the two atoms, as in the structural formula H—H. In a molecular formula, the line is omitted and the molecule is simply written as H_2.

Electron Model	Structural Formula	Molecular Formula
a. H H	H−H	H_2
b. O O	O=O	O_2
c. H H C H H	H–C(H)(H)–H	CH_4

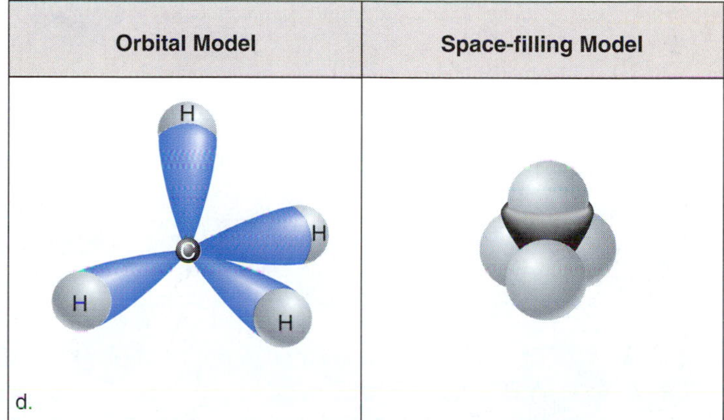

Figure 2.6 Covalently bonded molecules.
In a covalent bond, atoms share electrons and each atom has a completed outer shell. **a.** A molecule of hydrogen (H_2) contains two hydrogen atoms sharing a pair of electrons. This single covalent bond can be represented in any of the three ways shown. **b.** A molecule of oxygen (O_2) contains two oxygen atoms sharing two pairs of electrons. This results in a double covalent bond. **c.** A molecule of methane (CH_4) contains one carbon atom bonded to four hydrogen atoms. By sharing pairs of electrons, each has a completed outer shell. **d.** When carbon binds to four other atoms as in methane, the orbitals of the outer shell hybridize forming a tetrahedron. The space-filling model is also a three-dimensional representation of the molecule.

Like a single bond between two hydrogen atoms, a double bond can also allow two atoms to complete their octets. In a double covalent bond, two atoms share two pairs of electrons (Fig. 2.6*b*). In order to show that oxygen gas (O_2) contains a double bond, the molecule can be written as O=O.

It is even possible for atoms to form triple covalent bonds as in nitrogen gas (N_2), which can be written as N≡N. Single covalent bonds between atoms are quite strong, but double and triple bonds are even stronger.

Shape of Molecules

Structural formulas make it seem as if molecules are one dimensional but actually molecules have a three dimensional shape that often determines their biological function. Molecules consisting of only two atoms are always linear but a molecule like methane with five atoms (Fig. 2.6*c*) has a tetrahedral shape. Why? Because each of carbon's four hybrid orbitals is sharing with a hydrogen atom. The space-filling model comes closest to the actual shape of the molecule (Fig. 2.6*d*).

The shapes of molecules are necessary to the structural and functional roles they play in living things. For example, hormones have shapes that allow them to be recognized by the cells in the body. One form of diabetes occurs when the receptors of cells fail to recognize the hormone insulin. On the other hand, AIDS occurs when certain blood cells have receptors that bind to the HIV virus, allowing it to enter, multiply, and destroy the cell.

In a covalent molecule, atoms share electrons; the final shape of the molecule often determines the role it plays in cells and organisms.

Nonpolar and Polar Covalent Bonds

When the sharing of electrons between two atoms is fairly equal, the covalent bond is said to be a **nonpolar covalent bond.** All the molecules in Figure 2.6, including methane (CH_4), are nonpolar. In the case of water (H_2O), however, the sharing of electrons between oxygen and each hydrogen is not completely equal. The larger oxygen atom, with the greater number of protons, dominates the H_2O association. The attraction of an atom for the electrons of a covalent bond is called its **electronegativity.** The oxygen atom is more electronegative than the hydrogen atom, and it can attract the electron pair to a greater extent. In a water molecule, this causes the oxygen atom to assume a slightly negative charge (δ^-), and it causes the hydrogen atoms to assume a slightly positive charge (δ^+). The unequal sharing of electrons in a covalent bond creates a **polar covalent bond,** and, in the case of water, the molecule itself is a polar molecule (Fig. 2.7*a*).

The water molecule is a polar molecule and has a symmetric distribution of charge: one end of the molecule (the oxygen atom) carries a slightly negative charge, and the other ends of the molecule (the hydrogen atoms) carry slightly positive charges.

Hydrogen Bonding

Polarity within a water molecule causes the hydrogen atoms in one molecule to be attracted to the oxygen atoms in other polar molecules (Fig. 2.7*b*). This attractive force creates a weak bond called a **hydrogen bond.** This bond is often represented by a dotted line because a hydrogen bond is easily broken. Hydrogen bonding is not unique to water. A biological molecule can contain many polar covalent bonds involving hydrogen and usually oxygen or nitrogen. An electropositive hydrogen atom can be attracted to an electronegative oxygen or nitrogen atom within the same molecule or in another molecule.

Although a hydrogen bond is more easily broken than a covalent bond, many hydrogen bonds taken together are quite strong. Hydrogen bonds between parts of cellular molecules help maintain their proper structure and function. We will see that some of the important properties of water are also due to hydrogen bonding.

A hydrogen bond occurs between a slightly positive hydrogen atom of one molecule and a slightly negative atom of another molecule or between parts of the same molecule.

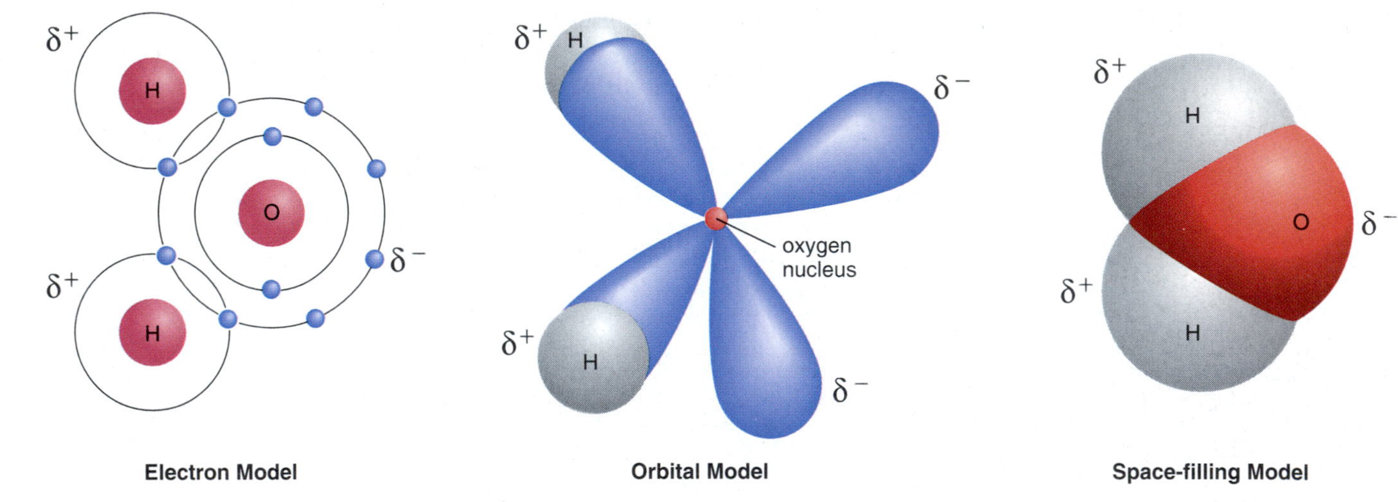

Figure 2.7 Water molecule.
a. Three models for the structure of water. The electron model does not indicate the shape of the molecule. The orbital model shows that the four orbitals of the outer shell (see Fig. 2.4*b*) have hybridized so that they point toward the corners of a tetrahedron. In water only two of these orbitals are utilized in covalent bonding. The space-filling model shows the V shape of a water molecule. Water is a polar molecule; the oxygen attracts the electrons more strongly than do the hydrogens, and there is a slightly positive charge on each hydrogen and a slightly negative charge on the oxygen. **b.** Hydrogen bonding between water molecules. A hydrogen bond is the attraction of a slightly positive hydrogen to a slightly negative atom in the vicinity. Each water molecule can bond to four other molecules in this manner. When water is in its liquid state, some hydrogen bonds are forming and others are breaking at all times.

2.3 Chemistry of Water

The first cell(s) evolved in water, and all living things are 70–90% water. What are the unique properties of water that make water essential to the continuance of life? Water is a polar molecule and water molecules are hydrogen bonded to one another (Fig. 2.7). A hydrogen bond is much weaker than a covalent bond within a water molecule, but taken together, hydrogen bonds cause water molecules to cling together. Without hydrogen bonding between molecules, water would boil at −80°C and freeze at −100°C, making life as we know it impossible. But because of hydrogen bonding, water is a liquid at temperatures suitable for life. It boils at 100°C and freezes at 0°C.

Properties of Water

The temperature of water rises and falls more slowly than other liquids under the same conditions. A calorie is the amount of heat energy needed to raise the temperature of one gram of water 1°C. In comparison, other covalently bonded liquids require only about half this amount of energy to rise in temperature 1°C. The many hydrogen bonds that link water molecules help water absorb heat without a great change in temperature.

When water cools down, heat is released. Converting one gram of the coldest liquid water to ice requires the loss of 80 calories of heat energy (Fig. 2.8). However, water holds heat, and its temperature falls more slowly than other liquids. This property of water is important not only for aquatic organisms but also for all living things. Water protects organisms from rapid temperature changes and helps them maintain their normal internal temperatures.

Water has a high heat of vaporization. Converting one gram of the hottest water to steam requires an input of 540 calories of heat energy. This high heat of vaporization means that water has a high boiling point. Because water boils at 100°C, it is in a liquid state at temperatures suitable to living things (Fig. 2.8).

Hydrogen bonds must be broken to change water to steam; this accounts for the very large amount of heat needed for evaporation. This property of water helps moderate the earth's temperature and keep it suitable to living things. It also gives animals in a hot environment an efficient way to release excess body heat. When an animal sweats, body heat is used to vaporize the sweat, thus cooling the animal.

a.

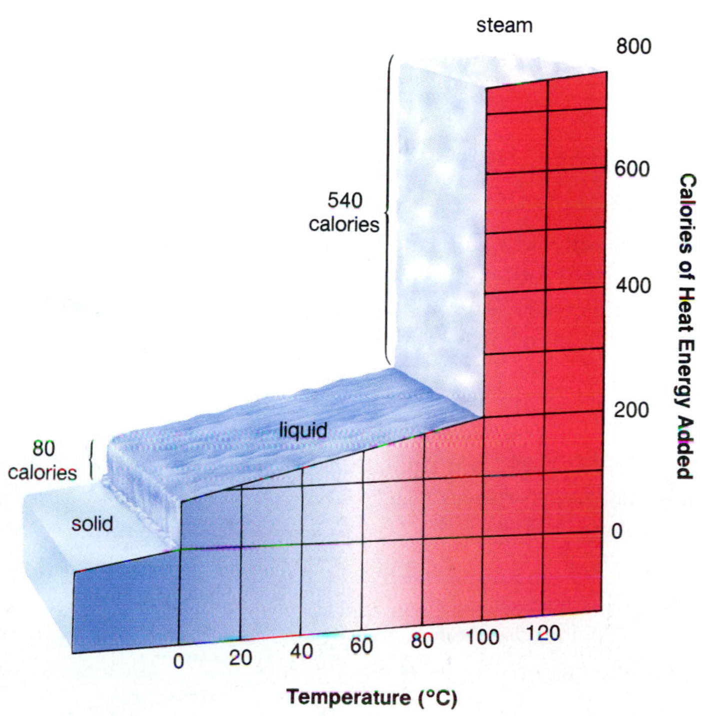

b.

Figure 2.8 Temperature and water.
a. Water has a high heat of vaporization; therefore, splashing water on the body when temperatures rise keeps animals cool. b. Water is the only common molecule that can be a solid, a liquid, or a gas according to naturally occurring environmental temperatures. At ordinary temperatures and pressure, water is a liquid, and it takes a large input of heat to change it to steam. In contrast, water gives off heat when it freezes, and this heat will keep the environmental temperature higher than expected. Can you see why there are less severe changes in temperature along the coasts?

Water facilitates chemical reactions both outside of and within living systems. Water dissolves a great number of solutes because it is a polar molecule. When a salt, such as sodium chloride (NaCl), is put into water, the negative ends of the water molecules are attracted to the sodium ions, and the positive ends of the water molecules are attracted to the chloride ions. This causes the sodium ions and the chloride ions to separate and to dissociate in water:

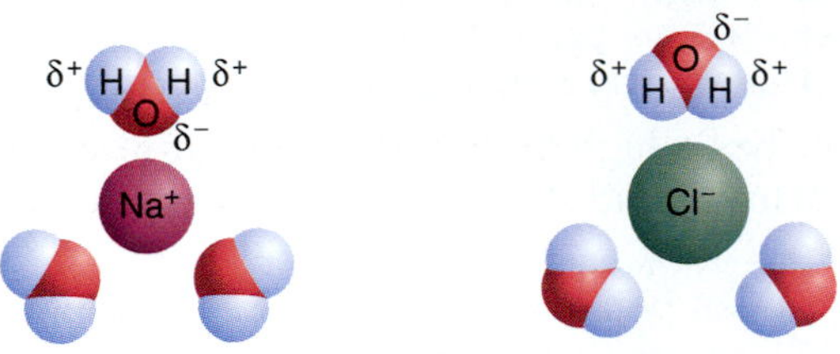

The salt NaCl dissolves in water

Water is also a solvent for larger molecules that contain ionized atoms or are polar molecules:

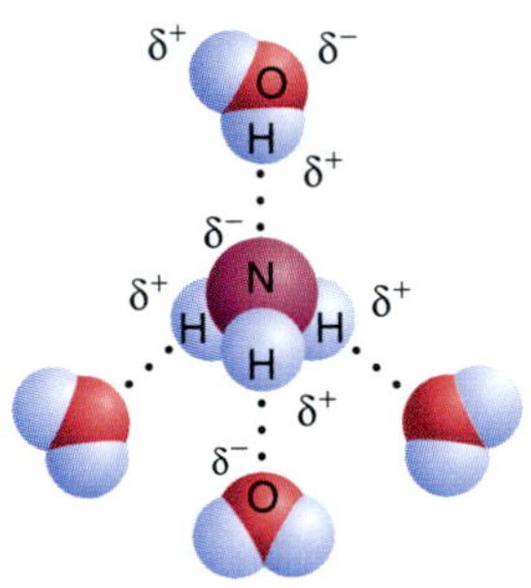

A polar molecule dissolves in water

When ions and molecules disperse in water, they move about and collide, allowing reactions to occur. Those molecules that can attract water are said to be **hydrophilic** [Gk. *hydrias*, of water, and *phileo*, love]. Nonionized and nonpolar molecules that cannot attract water are said to be **hydrophobic** [Gk. *hydrias*, of water, and *phobos*, fear].

Water molecules are cohesive and adhesive. Cohesion is apparent because water flows freely, yet water molecules do not separate from each other. They cling together because of hydrogen bonding. Because water molecules have a positive and negative pole, they adhere to surfaces, particularly polar surfaces; therefore, water exhibits adhesion. Because water can fill a tubular vessel and still flow, dissolved and suspended molecules are evenly distributed throughout a system. For these reasons, water is an excellent transport system both outside of and within living organisms. One-celled organisms rely on external water to transport nutrient and waste molecules, but multicellular organisms often contain internal vessels in which water serves to transport nutrients and wastes. For example, the liquid portion of our blood is 90% water that contains dissolved and suspended substances.

Cohesion and adhesion both contribute to the transport of water in plants. Plants have their roots anchored in the soil where they absorb water, but the leaves are uplifted and exposed to solar energy. How is it possible for water to rise to the top of even very tall trees (Fig. 2.9)? The properties of water account for the transport of water in plants. A plant contains a system of vessels that reaches from the roots to the leaves. Water evaporating from the leaves is immediately replaced with water molecules from the vessels. Because water molecules are cohesive, a tension is created that pulls water up from the roots. Adhesion of water to the walls of the vessels also helps prevent the water column from breaking apart.

Water has a high surface tension. It's possible to skip rocks on water because it has a high surface tension. Surface tension is measured by determining how difficult it is to break the surface of a liquid. As with cohesion, hydrogen bonding causes water to have a high surface tension. A water strider can even walk on the surface of a pond without breaking the water's surface. Table 2.3 summarizes the properties of water.

Table 2.3

Water

Properties	Chemical Reason	Effect
Resists change of state (from liquid to ice and from liquid to steam)	Hydrogen bonding	Moderates earth's temperature
Resists changes in temperature	Hydrogen bonding	Helps keep body temperatures constant
Universal solvent	Polarity	Facilitates chemical reactions
Is cohesive and adhesive	Hydrogen bonding; polarity	Serves as transport medium
Has a high surface tension	Hydrogen bonding	Difficult to break surface
Less dense as ice than as liquid water	Hydrogen bonding changes	Ice floats on water

Unlike most substances, frozen water is less dense than liquid water. As water cools, the molecules come closer together. They are densest at 4°C, but they are still moving about (Fig. 2.10). At temperatures below 4°C, there is only vibrational movement, and hydrogen bonding becomes more rigid but also more open. This means that water expands as it freezes, which is why cans of soda burst when placed in a freezer or why frost heaves make northern roads bumpy in the winter. It also means that ice is less dense than liquid water, and therefore ice floats on liquid water. If ice did not float on water, ice would sink, and once it began to accumulate at the bottom of ponds, lakes, and perhaps even the ocean, they would freeze solid and never melt, making life impossible in the water and also on land.

Instead, bodies of water always freeze from the top down. When a body of water freezes on the surface, the ice acts as an insulator to prevent the water below it from freezing. This protects many aquatic organisms so that they can survive the winter. As ice melts in the spring, it draws heat from the environment, helping to prevent a sudden change in temperature that might be harmful to life.

Water has unique properties that allow cellular activities to occur and that make life on earth possible.

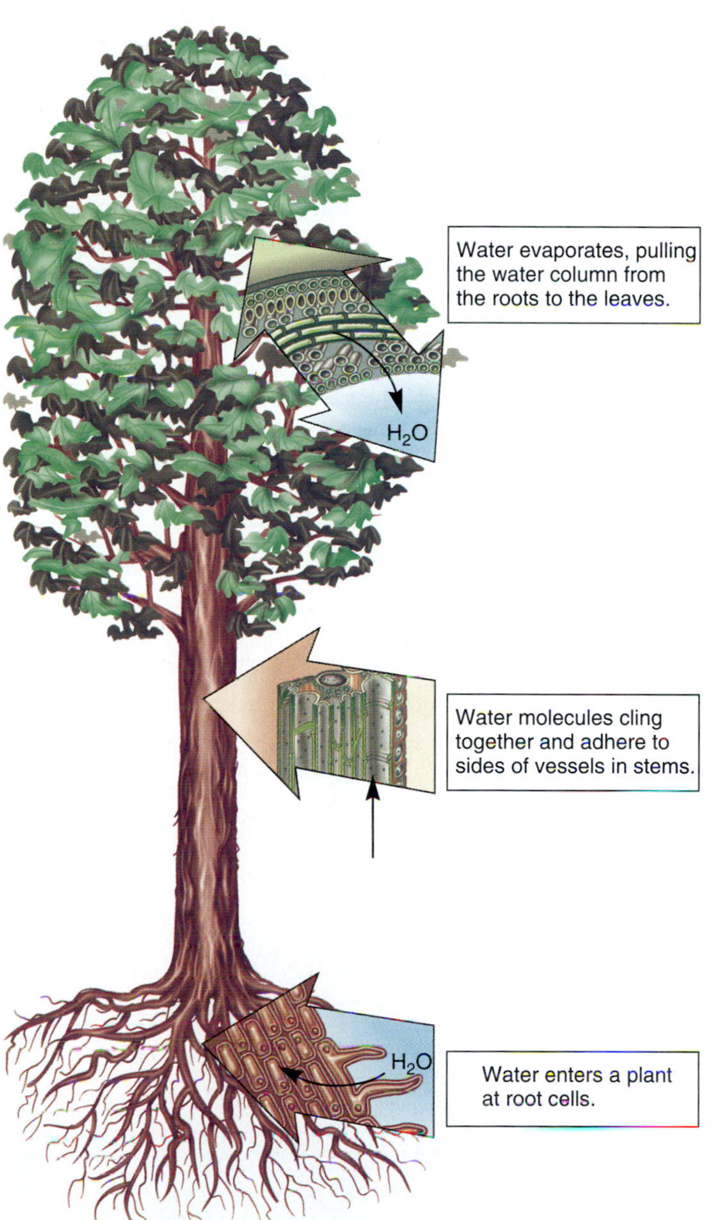
Water evaporates, pulling the water column from the roots to the leaves.
H_2O
Water molecules cling together and adhere to sides of vessels in stems.
H_2O
Water enters a plant at root cells.

Figure 2.9 **Water as a transport medium.**
How does water rise to the top of tall trees? Vessels are water-filled open pipelines from the roots to the leaves. When water evaporates from the leaves, this water column is pulled upward due to the cohesion of water molecules with one another and the adhesion of water molecules to the sides of vessels.

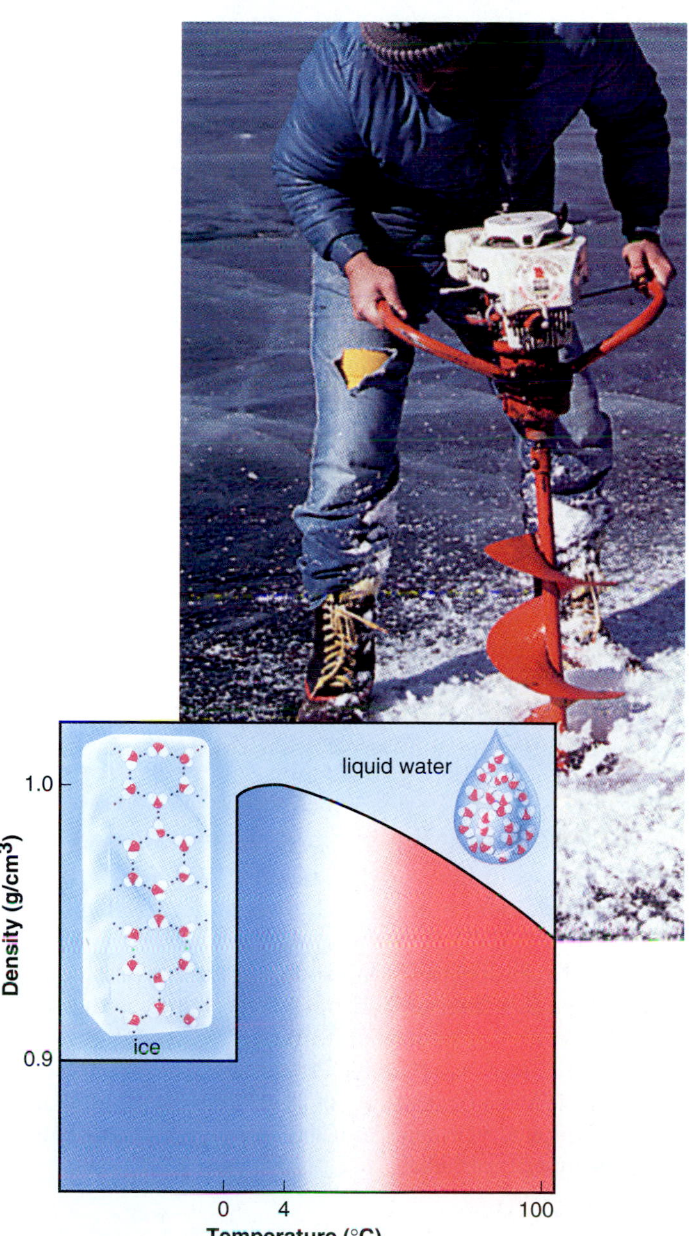
liquid water
ice
1.0
0.9
Density (g/cm^3)
0
4
100
Temperature (°C)

Figure 2.10 **Water as ice.**
Most substances contract when they solidify, but water expands because the water molecules in ice form a lattice in which the hydrogen bonds are farther apart than in liquid water. Water is more dense at 4°C than at 0°C; therefore, water freezes from the top down, making it necessary to drill a hole in order to go ice fishing.

Acids and Bases

When water ionizes, it releases an equal number of **hydrogen ions (H^+)** and **hydroxide ions (OH^-):**

$$\text{H–O–H} \rightleftharpoons H^+ + OH^-$$

water	hydrogen ion	hydroxide ion

Only a few water molecules at a time are dissociated, and the actual number of these ions is very small (10^{-7} moles/liter).[1]

Acids (High H^+ Concentration)

Lemon juice, vinegar, tomatoes, and coffee are all familiar acids. What do they have in common? **Acids** are molecules that dissociate in water, releasing hydrogen ions (H^+).[2] For example, an important inorganic acid is hydrochloric acid (HCl), which dissociates in this manner:

$$HCl \longrightarrow H^+ + Cl^-$$

Dissociation is almost complete; therefore, this is called a strong acid. If hydrochloric acid is added to a beaker of water, the number of hydrogen ions (H^+) increases greatly.

Bases (Low H^+ Concentration)

Milk of magnesia and ammonia are common bases that most people have heard of. **Bases** are molecules that either take up hydrogen ions (H^+) or release hydroxide ions (OH^-). For example, an important inorganic base is sodium hydroxide (NaOH), which dissociates in this manner:

$$NaOH \longrightarrow Na^+ + OH^-$$

Dissociation is almost complete; therefore, sodium hydroxide is called a strong base. If sodium hydroxide is added to a beaker of water, the number of hydroxide ions increases.

pH Scale

The **pH scale**[3] is used to indicate the acidity and basicity (alkalinity) of a solution. A pH of exactly 7 is neutral pH. Pure water has an equal number of hydrogen ions (H^+) and hydroxide ions (OH^-), and therefore, one of each is released when water dissociates. One mole of pure water contains only 10^{-7} moles/liter of hydrogen ions, which is the source of the pH value for neutral solutions.

The pH scale was devised to simplify discussion of the hydrogen ion concentration $[H^+]$ and consequently of the hydroxide ion concentration $[OH^-]$; it eliminates the use of cumbersome numbers. For example,

$[H^+]$ (moles per liter)	pH
0.000001 = 1×10^{-6}	6
0.0000001 = 1×10^{-7}	7
0.00000001 = 1×10^{-8}	8

Each pH unit has 10 times the amount of hydrogen ions (H^+) as the next higher unit.

In order to understand the relationship between hydrogen ion concentration and pH, consider the following question. Of the two other values listed above, which indicates a higher hydrogen ion concentration than pH 7 (neutral pH) and therefore refers to an acidic solution? A number with a smaller negative exponent indicates a greater quantity of hydrogen ions (H^+) than one with a larger negative exponent. Therefore, pH 6 is an acidic solution.

Bases add hydroxide ions (OH^-) to solutions and increase the hydroxide ion concentration $[OH^-]$ of water. Basic (also called alkaline) solutions, then, have fewer hydrogen ions (H^+) compared to hydroxide ions. The pH 8 refers to a basic solution because it indicates a lower hydrogen ion concentration $[H^+]$ (greater hydroxide ion concentration) than pH 7.

The pH scale (Fig. 2.11) ranges from 0 to 14. It is a logarithmic scale as opposed to an exponential scale. As we move toward a lower pH, each unit has 10 times the acidity of the previous unit, and as we move toward a higher pH,

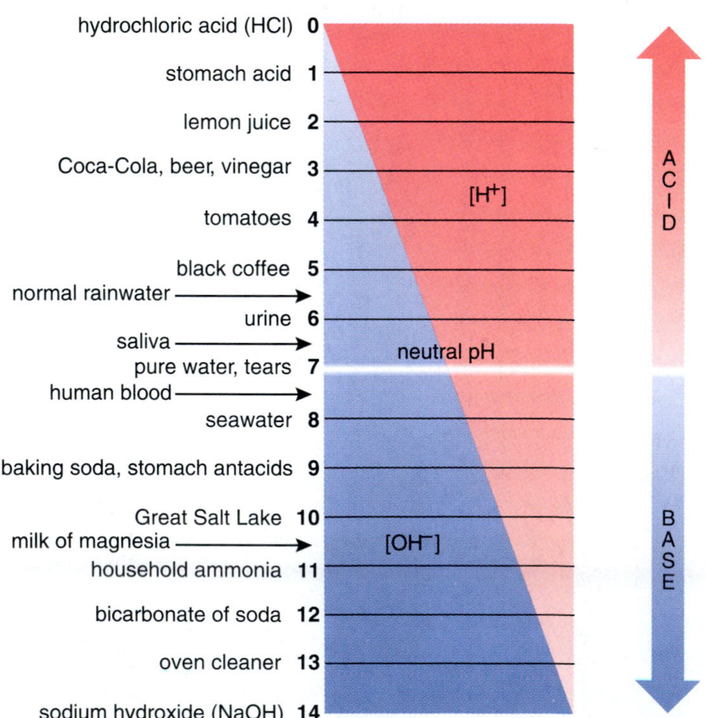

Figure 2.11 The pH scale.
The proportionate amount of hydrogen ions to hydroxide ions is indicated by the diagonal line. Any pH above 7 is basic, while any pH below 7 is acidic.

[1] In chemistry, a mole is defined as the amount of matter that contains as many objects (atoms, molecules, ions) as the number of atoms in exactly 12 grams of ^{12}C.

[2] A hydrogen atom contains one electron and one proton. A hydrogen ion has only one proton, so is often called a proton.

[3] pH is defined as the negative log of the hydrogen ion concentration $[H^+]$. A log is the power to which 10 must be raised to produce a given number.

Ecology Focus

The Harm Done by Acid Deposition

Normally, rainwater has a pH of about 5.6 because the carbon dioxide in the air combines with water to give a weak solution of carbonic acid. Rain falling in northeastern United States and southeastern Canada now has a pH between 5.0 and 4.0. We have to remember that a pH of 4 is ten times more acidic than a pH of 5 to comprehend the increase in acidity this represents.

There is very strong evidence that this observed increase in rainwater acidity is a result of the burning of fossil fuels, like coal and oil, as well as gasoline derived from oil. When fossil fuels are burned, sulfur dioxide and nitrogen oxides are produced, and they combine with water vapor in the atmosphere to form the acids sulfuric acid and nitric acid. These acids return to earth contained in rain or snow, a process properly called wet deposition, but more often called acid rain. Dry particles of sulfate and nitrate salts descend from the atmosphere during dry deposition.

Unfortunately, regulations that require the use of tall smokestacks to reduce local air pollution only cause pollutants to be carried far from their place of origin. Acid deposition in southeastern Canada is due to the burning of fossil fuels in factories and power plants in the Midwest.

Acid deposition adversely affects lakes, particularly in areas where the soil is thin and lacks limestone (calcium carbonate, $CaCO_3$), a buffer to acid deposition. It leaches aluminum from the soil, carries aluminum into the lakes, and converts mercury deposits in lake bottom sediments to soluble and toxic methyl mercury. Lakes not only become more acidic, but they also show accumulation of toxic substances. The increasing deterioration of thousands of lakes and rivers in southern Norway and Sweden during the past two decades has been attributed to acid deposition. Some lakes contain no fish and others have decreasing amounts of fish. The same phenomenon has been observed in Canada and the United States (mostly in the Northeast and upper Midwest).

In forests, acid deposition weakens trees because it leaches away nutrients and releases aluminum. By 1988, most spruce, fir, and other conifers atop North Carolina's Mt. Mitchell were dead from being bathed in ozone and acid fog for years. The soil was so acidic, new seedlings could not survive. Many countries in northern Europe have also reported woodland and forest damage most likely due to acid deposition (see photograph).

Lake and forest deterioration aren't the only effects of acid deposition. Reduction of agricultural yields, damage to marble and limestone monuments and buildings, and even illnesses in humans have been reported. Acid deposition has been implicated in the increased incidence of lung cancer and possibly colon cancer in residents of the East Coast. Tom McMillan, former Canada Minister of the Environment, says that acid rain is "destroying our lakes, killing our fish, undermining our tourism, retarding our forests, harming our agriculture, devastating our heritage, and threatening our health."

each unit has 10 times the basicity of the previous unit. A pH of 7 has an equal concentration of hydrogen ions (H^+) and hydroxide ions (OH^-). Above pH 7 there are more hydroxide ions than hydrogen ions, and below pH 7 there are more hydrogen ions than hydroxide ions.

Buffers

In living things, pH needs to be maintained within a narrow range or there are health consequences. The pH of our blood when we are healthy is always about 7.4, that is, just slightly basic (alkaline). Normally, pH stability is possible because the body has built-in mechanisms to prevent pH changes. **Buffers** are the most important of these mechanisms. Buffers help keep the pH within normal limits because they are chemicals or combinations of chemicals that take up excess hydrogen ions (H^+) or hydroxide ions (OH^-). For example, carbonic acid (H_2CO_3) is a weak acid that minimally dissociates and then reforms in the following manner:

$$\underset{\text{carbonic acid}}{H_2CO_3} \underset{\text{re-forms}}{\overset{\text{dissociates}}{\rightleftharpoons}} \underset{\text{bicarbonate ion}}{H^+ + HCO_3^-}$$

Blood always contains a combination of some carbonic acid and some bicarbonate ions. When hydrogen ions (H^+) are added to blood, the following reaction occurs:

$$H^+ + HCO_3^- \longrightarrow H_2CO_3$$

When hydroxide ions (OH^-) are added to blood, this reaction occurs:

$$OH^- + H_2CO_3 \longrightarrow HCO_3^- + H_2O$$

These reactions prevent any significant change in blood pH.

Acids have a pH that is less than 7, and bases have a pH that is greater than 7. Buffers, which can combine with both hydrogen ions and hydroxide ions, help to keep the pH of internal body fluids near pH 7 (neutral).

Connecting Concepts

The methods of science applied to the structure of matter have revealed that all substances consist of various combinations of the same set of 92 elements. Living things consist primarily of just six of these elements—carbon, hydrogen, nitrogen, oxygen, phosphorus, and sulfur (CHNOPS for short). These elements combine to form the unique types of molecules found in living cells. Many other elements exist in smaller amounts in organisms as ions, and their functions are dependent on their charged nature. Cells consist largely of water, a molecule that contains only hydrogen and oxygen. Covalent bonding between the atoms and hydrogen bonding between the molecules give water properties that sustain life. No other planet we know of has liquid water.

In the next chapter, we will learn that carbon combines covalently with the other five elements (i.e., HNOPS) to form the organic molecules of cells. It is these unique molecules that set living forms apart from nonliving objects. Carbon-containing molecules can be modified in numerous ways and this accounts for the diversity of life. In a cheetah, for example, such molecules are modified to become its black and beige fur. Another molecule called chlorophyll, which is necessary for photosynthesis, gives plants their green color.

It is difficult for us to visualize that a kangaroo, cheetah, or a pine tree is a combination of molecules and ions, but later in this book we will learn that even our thoughts of these organisms are simply the result of molecules flowing from one brain cell to another. An atomic, ionic, and molecular understanding of the variety of processes unique to life provides a deeper understanding of the definition of life and offers tools for the improvement of its quality, preservation of its diversity, and appreciation of its beauty.

Summary

2.1 Chemical Elements

Both living and nonliving things are composed of matter consisting of elements. Each element contains atoms of just one type. The acronym CHNOPS tells us the most common elements (atoms) found in living things: carbon, hydrogen, nitrogen, oxygen, phosphorus, and sulfur. Atoms contain subatomic particles. Protons and neutrons in the nucleus determine the atomic mass of an atom. The atomic number indicates the number of protons and the number of electrons in electrically neutral atoms. Protons have a positive charge and electrons have a negative charge. Isotopes are atoms of the same type that differ in their number of neutrons. Radioactive isotopes are used as tracers in biological experiments and medical procedures.

Electrons occupy energy levels (electron shells) at discrete distances from the nucleus. The number of electrons in the outer shell determines the reactivity of an atom. The first shell is complete when it has two electrons. In atoms up through calcium, number 20, every shell thereafter is complete with eight electrons. The octet rule states that atoms react with one another in order to have a stable outer shell that contains eight electrons. Most atoms, including those common to living things, do not have stable outer shells. This causes them to react with one another to form compounds and/or molecules. Following the reaction, the atoms have stable outer shells.

Electron shells have orbitals. The first shell has a single spherical-shaped orbital. The second shell has four orbitals: the first is spherical shaped and the others are dumbbell shaped. When hybridization occurs, the orbitals of the second shell are teardrop shaped and point to the corners of a tetrahedron. Atoms have a three-dimensional shape which determines the shape of molecules.

2.2 Compounds and Molecules

Ionization is the loss or addition of one or more electrons from an atom's outer shell and a more stable atomic configuration results. An ionic bond is an attraction between oppositely charged ions. When a covalent reaction occurs, atoms share electrons. A covalent bond is the sharing of these electrons. There are single, double, and triple covalent bonds.

When carbon combines with four other atoms, the resulting molecule has a tetrahedral shape because the orbitals of carbon's outer shell have hybridized. The shape of a molecule is important to its biological role. Hormones and other molecules, for example, are recognized by a cell's receptors because of their specific shapes.

In polar covalent bonds, the sharing of electrons is not equal; one of the atoms exerts greater attraction for the electrons than the other and a slight charge results on each atom. A hydrogen bond is a weak attraction between a slightly positive hydrogen atom of one molecule and a slightly negative oxygen or nitrogen atom within the same or a different molecule. Hydrogen bonds help maintain the structure and function of cellular molecules.

2.3 Chemistry of Water

Water is a polar molecule. The polarity of water molecules causes hydrogen bonding to occur between water molecules. These two features account for the unique properties of water, which are summarized in Table 2.3. These features allow cellular activities to occur and life to exist on Earth.

A small fraction of water dissociates to produce an equal number of hydrogen ions and hydroxide ions. This is termed neutral pH. In acidic solutions, there are more hydrogen ions than hydroxide ions; these solutions have a pH less than 7. In basic solutions, there are more hydroxide ions than hydrogen ions; these solutions have a pH greater than 7. Cells are sensitive to pH changes. Biological systems often contain buffers that help keep the pH within a normal range.

Reviewing the Chapter

1. Name the kinds of subatomic particles studied; list their mass, charge, and location in an atom. Which of these varies in isotopes? 18–19
2. Using the Bohr model draw an atomic structure for a carbon atom that has six protons and six neutrons. 20
3. Draw an atomic representation for the molecule $MgCl_2$. Using the octet rule, explain the structure of the compound. 21
4. State the number of orbitals per energy level. With reference to the second energy level, explain how a tetrahedral shape comes about. 21

5. Explain whether CO_2 (O=C=O) is an ionic or a covalent compound. Why does this arrangement satisfy all atoms involved? 22–23
6. Show that the shape of methane is dependent on the shape of carbon's orbitals. Of what significance is the shape of molecules in organisms? 23
7. Explain why water is a polar molecule. What is the relationship between the polarity of the molecule and the hydrogen bonding between water molecules? 24
8. Name six properties of water and relate them to the structure of water, including its polarity and hydrogen bonding between molecules. 25–27
9. Define an acid and a base. On the pH scale, which numbers indicate an acid, a base, and a neutral pH? 28
10. What are buffers, and why are they important to life? 29

Testing Yourself

Choose the best answer for each question.

1. An atom, such as calcium, which has two electrons in the outer shell, would most likely
   a. share to acquire a completed outer shell.
   b. lose these two electrons and become a negatively charged ion.
   c. lose these two electrons and become a positively charged ion.
   d. bind with carbon by way of hydrogen bonds.
   e. bind with another calcium to satisfy its energy needs.
2. The atomic number tells you the
   a. number of neutrons in the nucleus.
   b. number of protons in the atom.
   c. mass of the atom.
   d. the number of its electrons when neutral.
   e. Both b and d are correct.
3. Orbitals
   a. have the same volume and charge as an electron shell.
   b. always occupy the same space and have the same energy content.
   c. are the volume of space most likely occupied by an electron.
   d. influence the shape of atoms.
   e. Both c and d are correct.
4. A covalent bond is indicated by
   a. plus and minus charges attached to atoms.
   b. dotted lines between hydrogen atoms.
   c. concentric circles about a nucleus.
   d. overlapping electron shells or a straight line between atomic symbols.
   e. the touching of atomic nuclei.
5. The shape of a molecule
   a. is dependent in part on the shape of orbitals.
   b. influences their biological function.
   c. is dependent on its electronegativity.
   d. is dependent on its place in the periodic table.
   e. Both a and b are correct.
6. In which of these are the electrons shared unequally?
   a. double covalent bond
   b. triple covalent bond
   c. hydrogen bond
   d. polar covalent bond
   e. ionic and covalent bonds
7. In the molecule

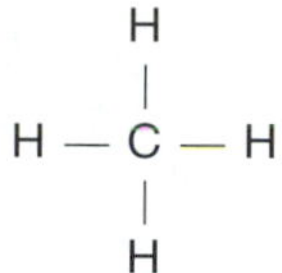

   a. all atoms have eight electrons in the outer shell.
   b. all atoms are sharing electrons.
   c. carbon could accept more hydrogen atoms.
   d. the orbitals point to the corners of a square.
   e. All of these are correct.
8. Which of these properties of water is not due to hydrogen bonding between water molecules?
   a. stabilizes temperature inside and outside cell
   b. molecules are cohesive
   c. is a solvent for many molecules
   d. ice floats on water
   e. Both b and c are correct.
9. Acids
   a. release hydrogen ions.
   b. have a pH value above 7.
   c. take up hydroxide ions and become neutral.
   d. increase the number of water molecules.
   e. Both a and b are correct.
10. Complete this diagram of a nitrogen atom by placing the correct number of protons and neutrons in the nucleus and electrons in the shells. Explain why the correct formula for ammonia is NH_3 and not NH_4.

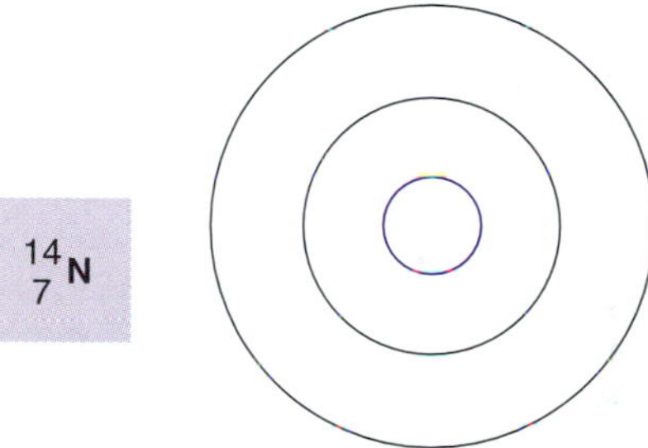

Thinking Scientifically

1. Natural antifreeze molecules allow many animals to exist in conditions cold enough to freeze the blood (or equivalent fluid) of animals without these additives. The usual explanation about how these molecules work is that they bind to tiny ice crystals in the blood and prevent them from getting larger. However, the exact mechanism is still not completely understood. Knowing the role of hydrogen bonding in the transition from liquid water to ice, how might these "natural antifreeze" molecules interact with ice/water to prevent crystal growth?
2. DNA is a large molecule that is made of two long, thin strands. All of the atoms within each strand are held together by covalent bonds, but the two strands are held together by hydrogen bonds only. Knowing that individual hydrogen bonds are weaker than covalent bonds, but that many hydrogen bonds together make a strong connection, what would you predict about the stability of the DNA molecule?

Bioethical Issue

Like a well-tuned system, your body constantly balances acidic and basic chemicals to keep your pH level at a steady state. Normally, this system works fine.

But when a person gets sick or endures physical stress—as in, for example, childbirth—pH levels may dip or rise too far, endangering that person's life. In most American hospitals, doctors routinely administer IVs, or intravenous infusions, of certain fluids to maintain a patient's pH level.

However, some people oppose IVs for philosophical reasons. For example, women preferring a natural childbirth may refuse an IV. That's relatively safe, as long as the woman is healthy.

Problems arise when hospital policy dictates an IV, even though a patient does not want one. Should a patient be allowed to refuse an IV? Or does a hospital have the right to insist, for health reasons, that patients accept IV fluids? And what role should doctors play—patient advocates or hospital representatives?

Understanding the Terms

acid 28
atom 18
atomic mass 19
atomic number 19
atomic symbol 18
base 28
buffer 29
compound 22
covalent bond 23
electron 18
electron shells 20
electronegativity 24
element 18
energy 20
hydrogen bond 24
hydrogen ion 28
hydrophilic 26
hydrophobic 26
hydroxide ion 28
ion 22
ionic bond 22
isotope 19
matter 18
molecule 22
neutron 18
nonpolar covalent bond 24
octet rule 21
orbital 21
pH scale 28
polar covalent bond 24
proton 18
radioactive isotope 19

Match the terms to these definitions:

a. ______________ Bond in which the sharing of electrons between atoms is unequal.

b. ______________ Charged particle that carries a negative or positive charge(s).

c. ______________ Molecules tending to raise the hydrogen ion concentration in a solution and to lower its pH numerically.

d. ______________ Union of two or more atoms of the same element; also the smallest part of a compound that retains the properties of the compound.

e. ______________ Measurement scale for the hydrogen ion concentration $[H^+]$ of a solution.

Web Connections

Exploring the Internet

http://www.mhhe.com/biosci/genbio/mader
(click on *Biology 7/e*)

The *Biology 7/e* Online Learning Center provides many resources for studying the material in this chapter including links to the following sites:

Chemist's Art Gallery. This site consists of a series of visualizations and animations of various chemicals and chemical processes. This is a huge site with many links.

http://www.csc.fi/lul/chem/graphics.html

Water and Ice Module is a site with online activities for students to learn more about the structure of water and ice.

http://cwis.nyu.edu/pages/mathmol/modules/water/water_teacher.html

Learn more about the elements, their structures, and properties in Chemicool, a periodic table of the elements produced by MIT.

http://www-tech.mit.edu/Chemicool/

ChemCenter, a site sponsored by the American Chemical Society, includes information on graduate programs, chemistry related news, and events.

http://www.chemcenter.org/

Visual Elements is a site designed by the Chemistry Society and has much text and visual images for all elements in the periodic table.

http://www.chemsoc.org/viselements/pages/periodic_table.html

The Chemistry of Organic Molecules

chapter 3

chapter concepts

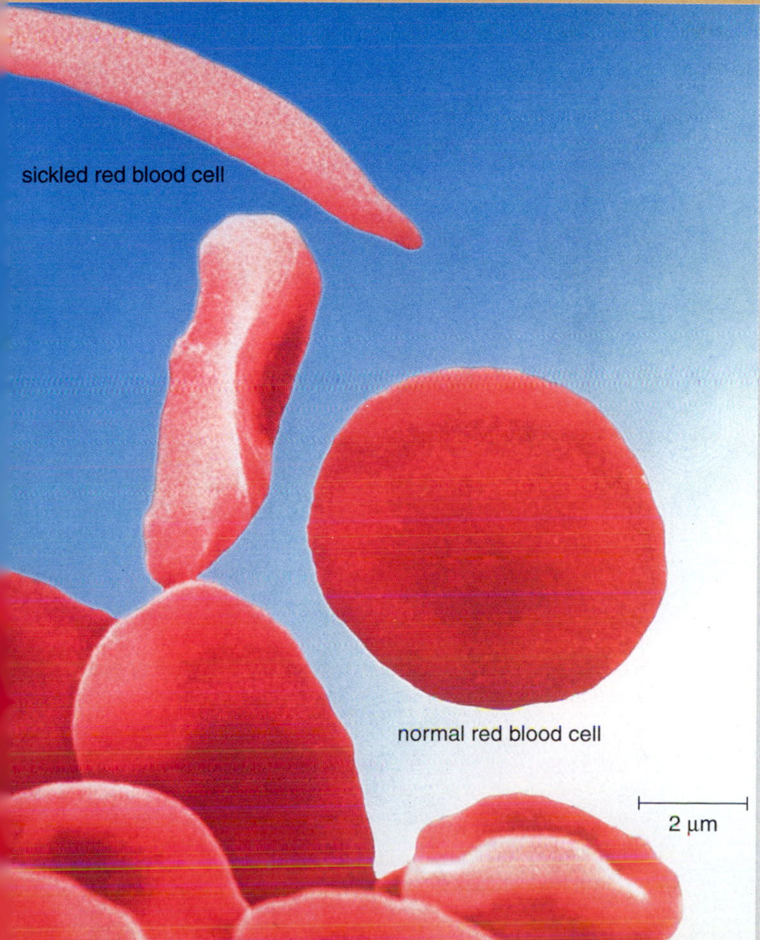

Normal red blood cell compared to a sickled red blood cell.

Normal red blood cells are biconcave disks that easily change shape as they squeeze along tiny blood vessels. A person with sickle-cell disease has sickle-shaped red blood cells that aren't as flexible. Their inflexibility causes them to either break down or clog small blood vessels. Typically, the individual suffers from anemia, poor circulation, and lack of resistance to infection. Internal hemorrhaging leads to jaundice, episodic pain of the abdomen and joints, and damage to internal organs.

Red blood cells contain the respiratory pigment hemoglobin. The chemistry of sickle-cell hemoglobin is different from that of normal hemoglobin. Sickle-cell hemoglobin isn't as soluble as normal hemoglobin. It stacks up into long semirigid rods that distort the red blood cells into a sickle shape. The molecule acts this way because a different chemical (amino acid) is found at one tiny spot compared to normal hemoglobin. This one difference accounts for abnormally shaped red blood cells and the symptoms of sickle-cell disease. Sickle-cell disease reminds us that the structure and shape of organic chemicals making up our bodies can influence our health.

3.1 Organic Molecules

The most common elements in living things are carbon, hydrogen, nitrogen, and oxygen, which constitute about 95% of your body weight. But we will see that both the sameness and the diversity of life are dependent upon the chemical characteristics of carbon, an atom whose chemistry is essential to living things.

The bonding of hydrogen, oxygen, nitrogen, and other atoms to carbon creates the molecules known as **organic molecules.** Laypeople sometimes use the word organic to mean wholesome. To biologists, organic molecules are those that always contain carbon and hydrogen. It is the organic molecules that characterize the structure and the function of living things like the primrose, the blueshell crab, and the bacterium in Figure 3.1. **Inorganic molecules** constitute nonliving matter, but even so, inorganic molecules like salts (e.g., NaCl) also play important roles in living things. Table 3.1 contrasts inorganic and organic molecules.

Living things are highly organized; therefore, it is easy for us to differentiate between the plant, the animal, and the bacterium in Figure 3.1. Despite their diversity all living things contain the same classes of primary molecules: carbohydrates, proteins, lipids, and nucleic acids. But within these classes there is molecular diversity. For example, the plant, the animal, and the bacterium all utilize a carbohydrate molecule for structural purposes, but the exact carbohydrate is different in each of them.

Table 3.1

Inorganic Versus Organic Molecules

Inorganic Molecules	Organic Molecules
Usually contain positive and negative ions	Always contain carbon and hydrogen
Usually ionic bonding	Always covalent bonding
Always contain a small number of atoms	May be quite large with many atoms
Often associated with nonliving matter	Usually associated with living organisms

Of the elements most common to living things, the chemistry of carbon allows the formation of varied organic molecules. This accounts for both the sameness and the diversity of living things.

Figure 3.1 Carbohydrates as structural materials.
a. Plants are held erect partly by the incorporation of the polysaccharide cellulose into the cell wall, which surrounds each cell. **b.** The shell of crabs contains chitin, a different type of polysaccharide. **c.** Most bacterial cells, like plant cells, are enclosed by a cell wall, but in this case the wall is strengthened by another type of polysaccharide known as peptidoglycan.

Carbon Skeletons and Functional Groups

Carbon has four electrons in its outer shell, and this allows it to bond with as many as four other atoms. Moreover, because these bonds are covalent, they are quite strong. Usually carbon bonds to hydrogen, oxygen, nitrogen, or another carbon atom. The ability of carbon to bond to itself results in carbon chains of various lengths and shapes. Long chains containing 50 or more carbon atoms are not unusual in living systems. Carbon can also share two pairs of electrons with another atom, giving a double covalent bond. Carbon-to-carbon bonding sometimes results in ring compounds of biological significance.

Carbon chains make up the skeleton or backbone of organic molecules. **Functional groups,** which are clusters of certain atoms that always behave in a certain way, can be attached to the carbon chain. The organic molecules in living things—sugars, fatty acids, amino acids, and nucleotides—all have a carbon backbone, and in addition, they often have one or more of the functional groups listed in Figure 3.2.

Molecules composed only of carbon and hydrogen, are **hydrophobic** (not attracted to water). But the addition of a functional group like —OH, or —C=O makes the molecule polar and able to interact with other polar molecules. In particular, polar molecules are **hydrophilic** (attracted to water). Since cells are 70–90% water, the ability to interact with and be soluble in water profoundly affects the function of organic molecules in cells.

Organic molecules containing carboxyl groups (—COOH) are both polar and acidic. They tend to ionize and release hydrogen ions in solution:

$$\text{—COOH} \longrightarrow \text{—COO}^- + H^+$$

Functional groups determine the polarity of an organic molecule and also the types of reactions it will undergo. We will see that alcohols react with carboxyl groups when a fat forms, and amino groups react with carboxyl groups during protein formation.

Isomers

Isomers also contribute to the diversity of organic molecules. **Isomers** [Gk. *isos,* equal, and *meros,* part, portion] are molecules that have identical molecular formulas, but they are different molecules because the atoms in each are arranged differently. For example, the compounds in Figure 3.3 are isomers.

Functional Groups

Name	Structure	Found in	Chemical characteristics
hydroxyl (alcohol)	*R*—OH	some amino acids, nucleotides, sugars	polar, forms hydrogen bonds
carboxyl (acid)	*R*—C(=O)—OH	fats amino acids	polar, acidic
ketone	*R*—C(=O)—*R*	some sugars	polar
aldehyde	*R*—C(=O)—H	some sugars	polar
amino	*R*—N(H)—H	amino acids proteins	polar, basic, forms hydrogen bonds
sulfhydryl	*R*—SH	some amino acids proteins	forms disulfide bonds
phosphate	*R*—O—P(=O)(OH)—OH	phospholipids nucleotides nucleic acids	polar, acidic

R = remainder of molecule

Figure 3.2 Functional groups.
Molecules with the same type of backbone can still differ according to the type of functional group attached to the backbone. Many of these functional groups are polar, helping to make the molecule soluble in water. In this illustration, the remainder of the molecule (aside from the functional groups) is represented by an *R*.

a. Glyceraldehyde: H—C(H)(OH)—C(H)(OH)—C(=O)—H

b. Dihydroxyacetone: H—C(H)(OH)—C(=O)—C(H)(OH)—H

Figure 3.3 Isomers.
Isomers have the same molecular formula but different configurations. Both of these compounds have the formula $C_3H_6O_3$. **a.** In glyceraldehyde, oxygen is double-bonded to an end carbon. **b.** In dihydroxyacetone, oxygen is double-bonded to the middle carbon.

Organic molecules are diverse. Carbon skeletons vary in size and shape; functional groups have various chemical characteristics, and isomers occur.

Building Polymers

The four classes of organic compounds in living things are carbohydrates, lipids, proteins, and nucleic acids. Among the carbohydrates, polysaccharides are **polymers:** they are long chains of unit molecules called **monomers.** Simple sugars (monosaccharides) are the monomers within polysaccharides, amino acids join to form the polypeptides of proteins, and nucleotides are the monomers of nucleic acids:

Polymer	Monomer
Polysaccharide	Monosaccharide
Polypeptide	Amino acid
Nucleic acid	Nucleotide

Each type of monomer comes in different varieties because functional groups attached to monomers and/or the placement of functional groups can vary. Also the bonding between monomers can differ according to the particular polymer. For these reasons even polymers that contain the same type monomer can have different chemical characteristics and functions. In the pages that follow, note the names of unit molecules and how they join to form large molecules. You will also want to know the functions of each type of organic molecule discussed.

Polymers are long chains of unit molecules called monomers. Even polymers that contain the same type monomer can vary because monomers come in modified form and bonding between monomers is characteristic of the polymer.

Condensation and Hydrolysis

Polymers can be so large that they are called macromolecules. A polypeptide can contain hundreds of amino acids, and a nucleic acid can contain millions of nucleotides. How do polymers get so large? Cells use the modular approach when constructing polymers. Just as a train increases in length as box cars are hitched together one by one, so a polymer gets longer as monomers bond to one another.

Regardless of the type of bond that holds monomers join by an identical mechanism. During **condensation synthesis,** when monomers join, a hydroxyl (—OH) group is removed from one monomer and a hydrogen (—H) is removed from the other. Notice in Figure 3.4*a* that water is given off during condensation synthesis. Water is removed (*condensation*) and a bond is made (*synthesis*). Condensation synthesis does not take place unless the proper enzyme (a molecule that speeds a chemical reaction in cells) is present and the monomers are in an activated energy-rich form.

Polymers are broken down by **hydrolysis,** which is essentially the reverse of condensation synthesis. Notice in Figure 3.4*b* that water is added during hydrolysis: an —OH group from water attaches to one monomer and an —H from water attaches to the other monomer. In other words, during hydrolysis, water is used to break a bond.

Organic molecules are routinely built up in cells by the removal of water (H_2O) during condensation synthesis. They are broken down in cells by the addition of water during hydrolysis.

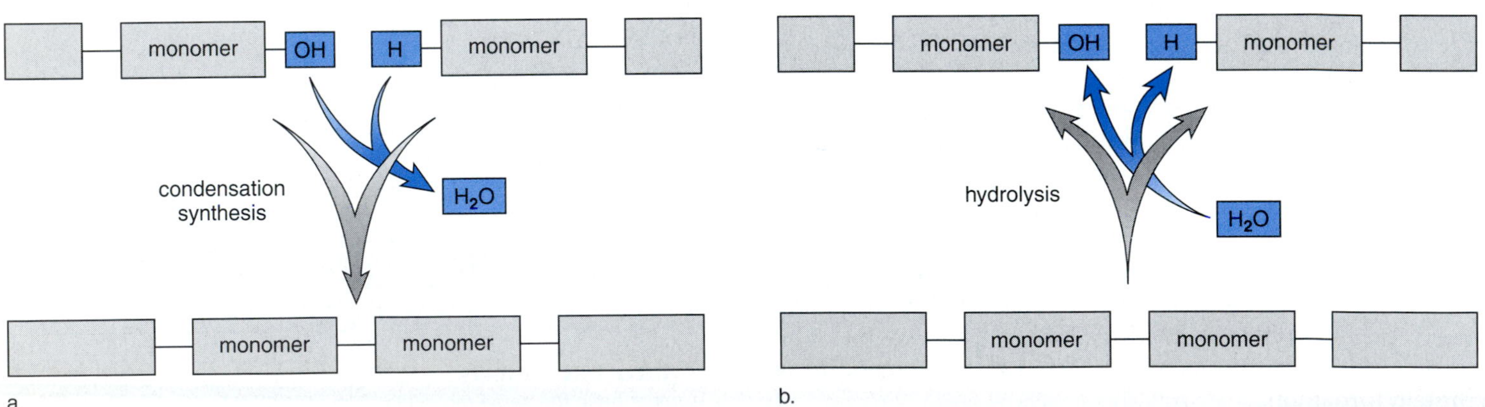

Figure 3.4 Condensation synthesis and hydrolysis of polymers.
a. In cells, synthesis often occurs when monomers bond following condensation (removal of H_2O). **b.** Hydrolysis occurs when the monomers in a polymer separate after the addition of H_2O.

3.2 Carbohydrates

The formula CH_2O is sometimes used to represent **carbohydrates** [L. *carbo,* charcoal, and Gk. *hydatos,* water], molecules that bear many hydroxyl groups (Fig. 3.5). Carbohydrates include monosaccharides (one sugar), disaccharides (two sugars bonded together), and polysaccharides (many sugars bonded together). Sugars and some polysaccharides (starches and glycogen) serve as energy storage compounds. Cellulose, a polysaccharide, is the most abundant organic compound on Earth because it supports plant cell walls. The carbohydrates attached to the surfaces of animal cells are specific to the individual and antigenic to certain other individuals.

Monosaccharides are simple sugars with a carbon backbone of three to seven carbon atoms. The best known sugars are those that have six carbons **(hexoses). Glucose** [Gk. *glykys,* sweet, and *osus,* full of] is in the blood of animals, and **fructose** is frequently found in fruits. These sugars are isomers of each other. They both have the molecular formula $C_6H_{12}O_6$, but they differ in structure (Fig. 3.5). Structural differences cause molecules to vary in shape, which is best seen by using space-filling models. Shape is very important in determining how molecules interact with one another.

Ribose and **deoxyribose** are two five-carbon sugars **(pentoses)** of significance because they are found respectively in the nucleic acids RNA and DNA. RNA and DNA are discussed later in the chapter.

A **disaccharide** contains two monosaccharides that have joined by condensation. **Lactose** is a disaccharide that contains galactose and glucose and is found in milk. **Maltose** (composed of two glucose molecules) is a disaccharide of interest because it is found in our digestive tract as a result of starch digestion (Fig. 3.6).

Sucrose is a disaccharide that contains glucose and fructose. Sugar is transported within the body of a plant in the form of sucrose, and this is the sugar we use at the table to sweeten our food. We acquire this sugar from plants, such as sugarcane and sugar beets.

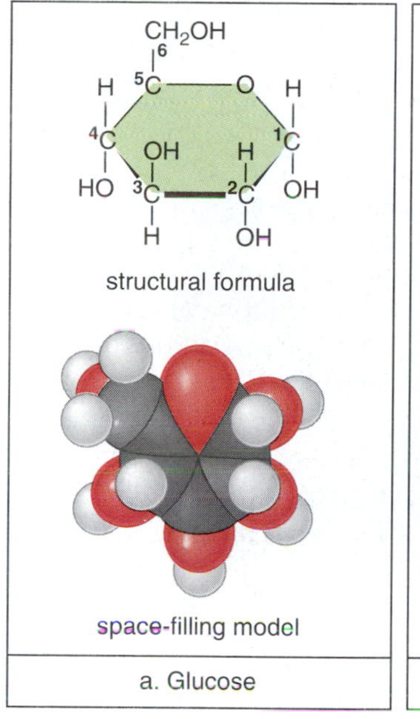

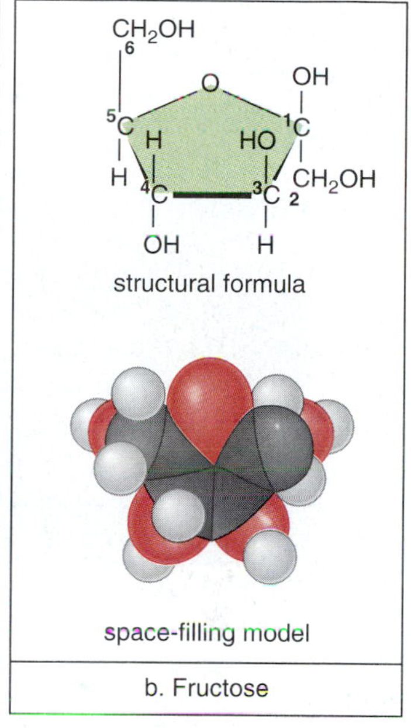

Figure 3.5 Two common six-carbon sugars.
a. Glucose is a six-carbon sugar that usually exists as a ring compound in cells. The shape of glucose is indicated in the space-filling model.
b. Fructose is an isomer of glucose. It too has the molecular formula $C_6H_{12}O_6$. Notice, though, that the atoms are bonded in a slightly different manner. The carbon atoms in glucose and fructose are numbered.

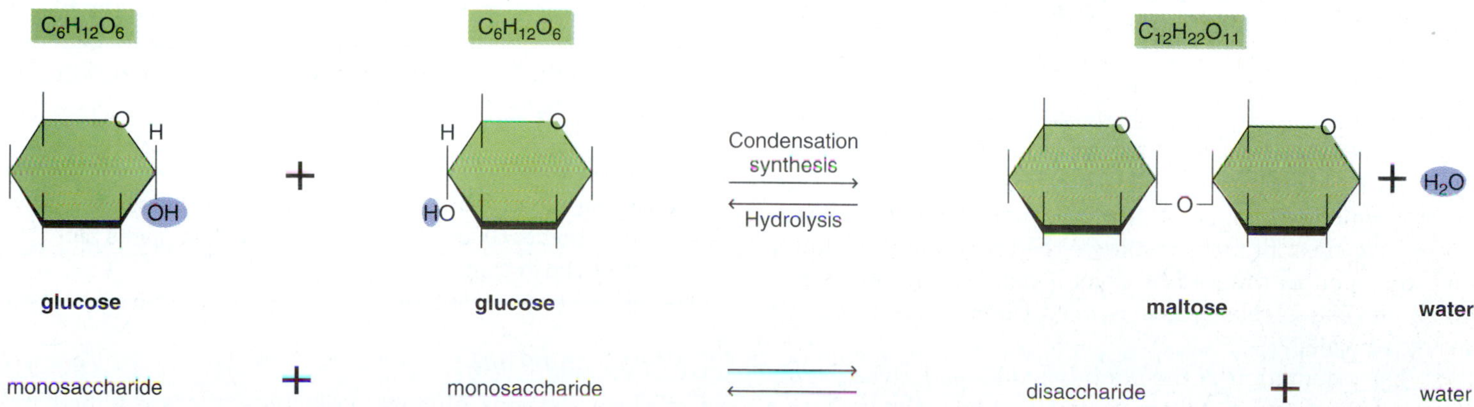

Figure 3.6 Condensation synthesis and hydrolysis of maltose, a disaccharide.
During condensation synthesis of maltose, a bond forms between two glucose molecules and the components of water are removed. During hydrolysis, the components of water are added and the bond is broken. This illustration uses a simplified way to represent glucose, only the bonds between the carbon atoms and bonds between attached groups are indicated.

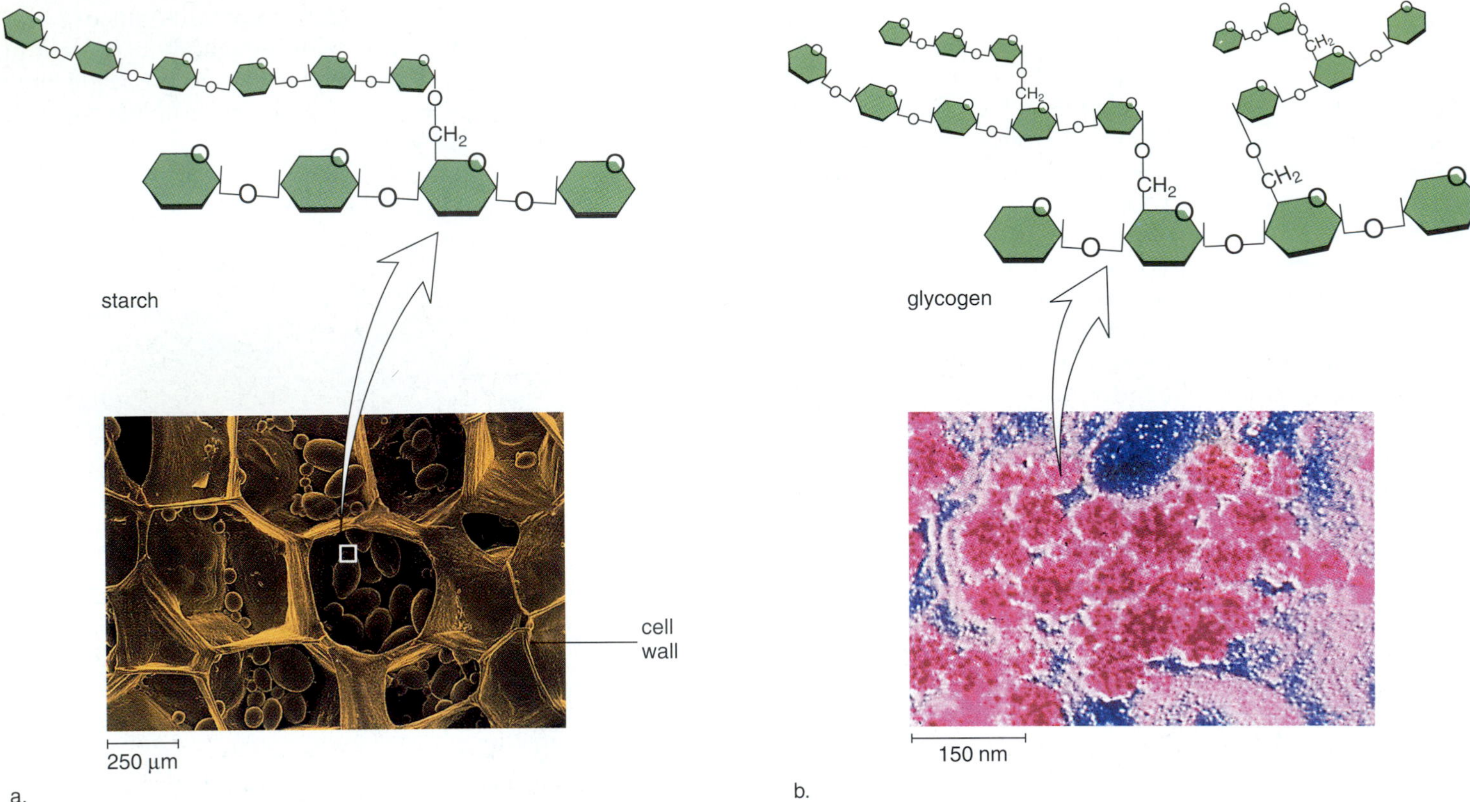

Figure 3.7 Starch and glycogen structure and function.
a. Starch is a chain of glucose molecules that branches as shown. The electron micrograph shows the location of starch in plant cells. Glucose is stored in plants as starch. **b.** Glycogen is a highly-branched polymer of glucose molecules. The electron micrograph shows glycogen deposits in a portion of a liver cell. Glucose is stored in animals as glycogen.

Polysaccharides

Polysaccharides are long polymers of monosaccharides formed by condensation synthesis. The most common polysaccharides in living things are starch, glycogen, and cellulose, which are chains of glucose molecules. Chitin is a polysaccharide that contains a modified glucose molecule.

Starch and Glycogen

The structures of starch and glycogen differ only slightly (Fig. 3.7). **Glycogen** [Gk. *glykys*, sweet, and *genitus*, producing] is characterized by many branches, which are side chains of glucose that go off from the main chain. **Starch** has fewer branches. Branching allows the breakdown of starch and glycogen to proceed at several points simultaneously. Once glucose is released, it is used directly as an energy source for cells.

The orientation of the bond in starch and glycogen allows these polymers to form compact spirals, making these polymers suitable as storage compounds. When leaf cells are actively producing sugar by photosynthesis, they store some of this sugar within the cell as starch (Fig. 3.7*a*). Otherwise, roots store sugar as starch, and so do seeds. During seed germination, starch is broken down into maltose and then glucose, which provides energy for growth.

Animal cells store extra carbohydrates as glycogen, sometimes called "animal starch." After we eat, the muscles and liver join glucose molecules to form glycogen (Fig. 3.7*b*). Between meals, the liver releases glucose to keep the blood concentration of glucose near the normal 0.1%. Glucose moves from the blood into cells where it participates in cellular metabolism.

Cells use sugars, especially glucose, as an immediate energy source. Glucose is stored as starch in plants and as glycogen in animals.

Cellulose and Chitin

Cellulose contains glucose molecules that are joined together differently than they are in starch and glycogen. The orientation of the bond between glucose molecules in cellulose causes the polymers to be straight and fibrous, making cellulose suitable as a structural compound.

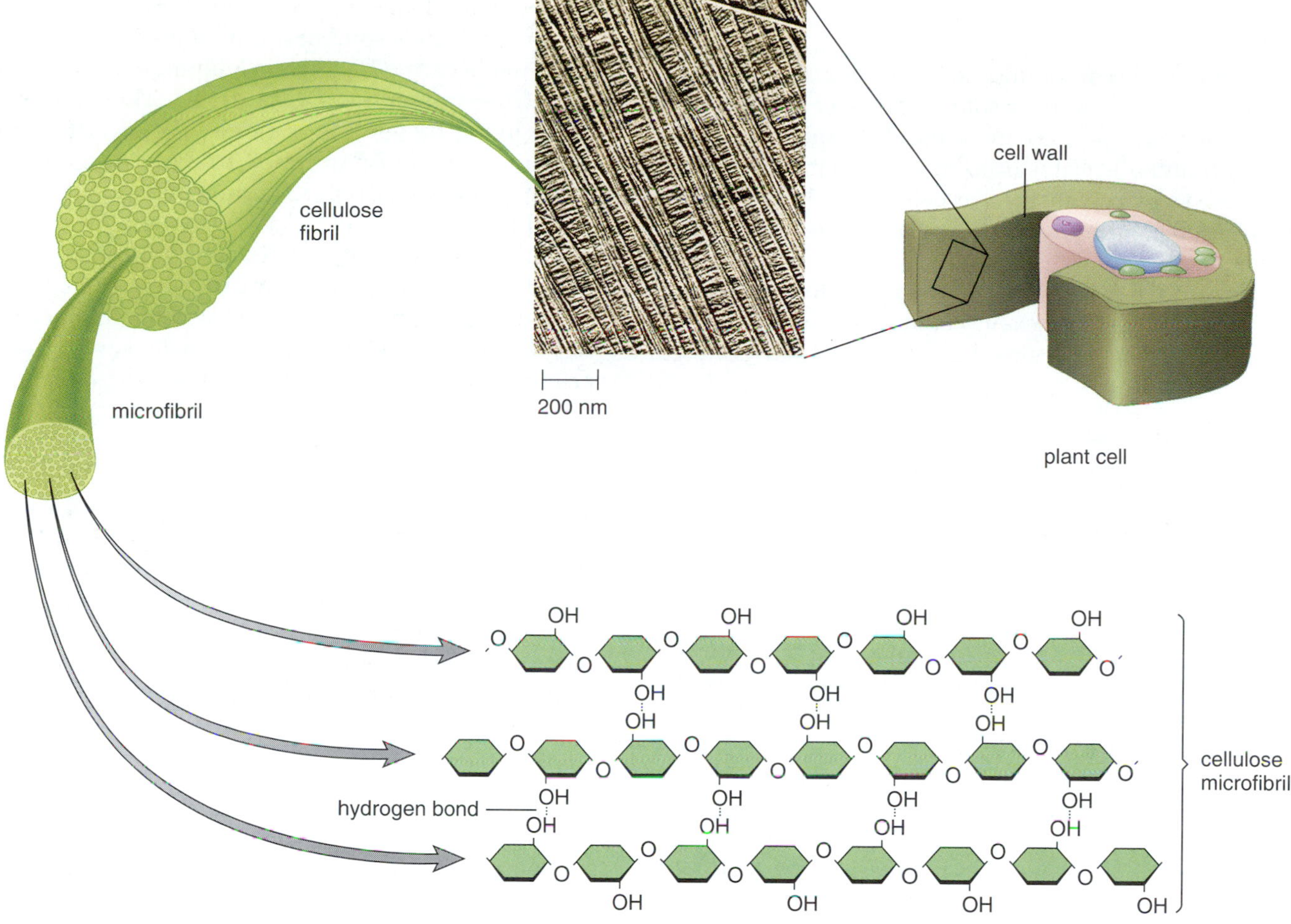

Figure 3.8 **Cellulose fibrils.**
In plant cell walls, each cellulose fibril contains several microfibrils. Each microfibril contains many polymers of glucose hydrogen-bonded together. Three such polymers are shown.

As shown in Figure 3.8, the long, unbranched polymers of glucose in cellulose are held together by hydrogen bonding within microfibrils. Several microfibrils, in turn, make up a cellulose fibril. Layers of cellulose fibrils occur in plant cell walls; fibrils are parallel in each layer, and the layers lie at angles to one another. This arrangement makes plant cell walls stronger. Humans have found many uses for cellulose and cotton fibers which are almost pure cellulose. We often wear cotton clothing. Furniture and buildings are made from wood, which contains a high percentage of cellulose.

Most animals, including humans, digest little cellulose because digestive enzymes are unable to break the linkage between the glucose molecules in cellulose. Even so, cellulose is a recommended part of our diet because it provides bulk (also called fiber or roughage) that helps the body maintain regularity of elimination.

Cattle and sheep receive nutrients from grass because they have a special stomach chamber, the rumen, where bacteria live that can digest cellulose. Unfortunately, cattle are often kept in feedlots, where they are fed grains instead of grass. This practice wastes fossil fuel energy because it takes energy to grow the grain, to process it, to transport it, and finally to feed it to the cattle. Also, range-fed cattle move about and produce a leaner meat than do cattle raised in a feedlot. There is growing evidence that lean meat is healthier for humans than fatty meat.

Chitin [Gk. *chiton*, tunic], which is found in the exoskeleton of crabs and related animals like lobsters and insects, is also a polymer of glucose. Each glucose unit, however, has an amino group attached to it. The linkage between the glucose molecules is like that found in cellulose; therefore, chitin is not digestible by humans. Recently, scientists have discovered how to turn chitin into thread that can be used as suture material. They hope to find other uses for treated chitin. If so, all those discarded crab shells that pile up beside crabmeat processing plants will not go to waste.

Plant cell walls contain cellulose. The shells of crabs and related animals contain chitin.

3.3 Lipids 

A variety of organic compounds are classified as **lipids** [Gk. *lipos*, fat]. Many of these are insoluble in water because they lack any polar groups. The most familiar lipids are those found in fats and oils. Fat is utilized for both insulation and energy reserves by organisms. Fat below the skin of whales is called blubber; in humans it is given slang expressions such as "spare tire" and "love handles."

Phospholipids and steroids are also important lipids found in living things. For example, they are components of the plasma membrane, a sheetlike structure that encloses the cell.

Lipids are quite varied in structure, but they tend to be insoluble in water.

Fats and Oils

Fats and **oils** contain two types of unit molecules: fatty acids and glycerol. Each **fatty acid** consists of a long hydrocarbon chain with a carboxyl (acid) group at one end. Because the carboxyl group is a polar group, fatty acids are soluble in water. Most of the fatty acids in cells contain 16 to 18 carbon atoms per molecule, although smaller ones are also found. Fatty acids are either saturated or unsaturated. **Saturated** fatty acids have no double bonds between the carbon atoms. The carbon chain is saturated, so to speak, with all the hydrogens that can be held. **Unsaturated** fatty acids have double bonds in the carbon chain wherever the number of hydrogens is less than two per carbon atom:

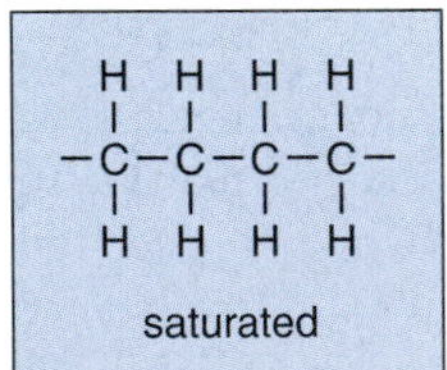


```
 H  H  H  H
 |  |  |  |
–C–C=C–C–
 |        |
 H        H
```

unsaturated

Glycerol is a compound with three hydroxyl (—OH) groups. Hydroxyl groups are polar; therefore, glycerol is soluble in water. When a fat is formed, the acid portions of three fatty acids react with these hydroxyl groups. Aside from a fat molecule, three molecules of water result. This is condensation synthesis reaction and a fat can be hydrolyzed to its components (Fig. 3.9*a*). Because there are three fatty acids per glycerol molecule, fats and oils are sometimes called a **triglyceride.** Notice that unlike fatty acids, triglycerides lack polar groups; therefore they do not mix with water. Despite the liquid nature of both cooking oils and water, cooking oils separate out of water even after shaking.

Triglycerides containing fatty acids with unsaturated bonds melt at a lower temperature than those containing fatty acids with saturated bonds. This is because a double bond makes a kink that prevents close packing between the hydrocarbon chains (Fig. 3.9*b* and *c*). We can reason that butter, a fat, which is solid at room temperature, must be saturated while corn oil, which is a liquid even when placed in the refrigerator, must be unsaturated. This difference is useful to living things. For example, the feet of reindeer and penguins contain unsaturated triglycerides, and this helps protect these exposed parts from freezing.

In general, however, fats, which are most often of animal origin, are solid at room temperature and oils, which are liquid at room temperature, are of plant origin. Diets high in animal fat have been associated with circulatory disorders. Replacement of fat whenever possible with oils such as peanut oil and sunflower oil has been suggested.

Nearly all animals use fat in preference to glycogen for long-term energy storage. Gram per gram, fat stores more energy than glycogen. The C—H bonds of fatty acids makes them a richer source of chemical energy than glycogen, which has many C—OH bonds. Also, fat droplets, being nonpolar, do not contain water. Small birds, like the broad-tailed hummingbird, store a great deal of fat before they start their long spring and fall migratory flights. About 0.15 gram of fat per gram of body weight is accumulated each day. If the same amount of energy were stored as glycogen, a bird would be so heavy it would not be able to fly.

Fats and oils are triglycerides (one glycerol plus three fatty acids). They are used as long-term energy-storage compounds in plants and animals.

Waxes

In **waxes,** a long-chain fatty acid bonds with a long-chain alcohol.

fatty acid

$$\begin{array}{l} O{=}C-CH_2-CH_2-CH_2-CH_2-CH_2-CH_2-CH_2-CH_2-CH_2-CH_3 \\ \quad | \\ O \\ \quad | \\ \;\; C-CH_2-CH_2-CH_2-CH_2-CH_2-CH_2-CH_2-CH_2-CH_3 \end{array}$$

long-chain alcohol

Waxes are solid at normal temperatures because they have a high melting point. Being hydrophobic, they are also waterproof and resistant to degradation. In many plants, waxes form a protective cuticle (covering) that retards the loss of water for all exposed parts. In animals, waxes are involved in skin and fur maintenance. In humans, wax is produced by glands in the outer ear canal. Here its function is to trap dust and dirt particles, preventing them from reaching the eardrum.

A honeybee produces wax in glands on the underside of its abdomen. The wax is used to make the six-sided cells of the comb where honey is stored. Honey contains the sugars fructose and glucose, breakdown products of the sugar sucrose.

glycerol + 3 fatty acids ⇌ (condensation synthesis / hydrolysis) fat + 3 H_2O (3 waters)

a. Formation of a fat.

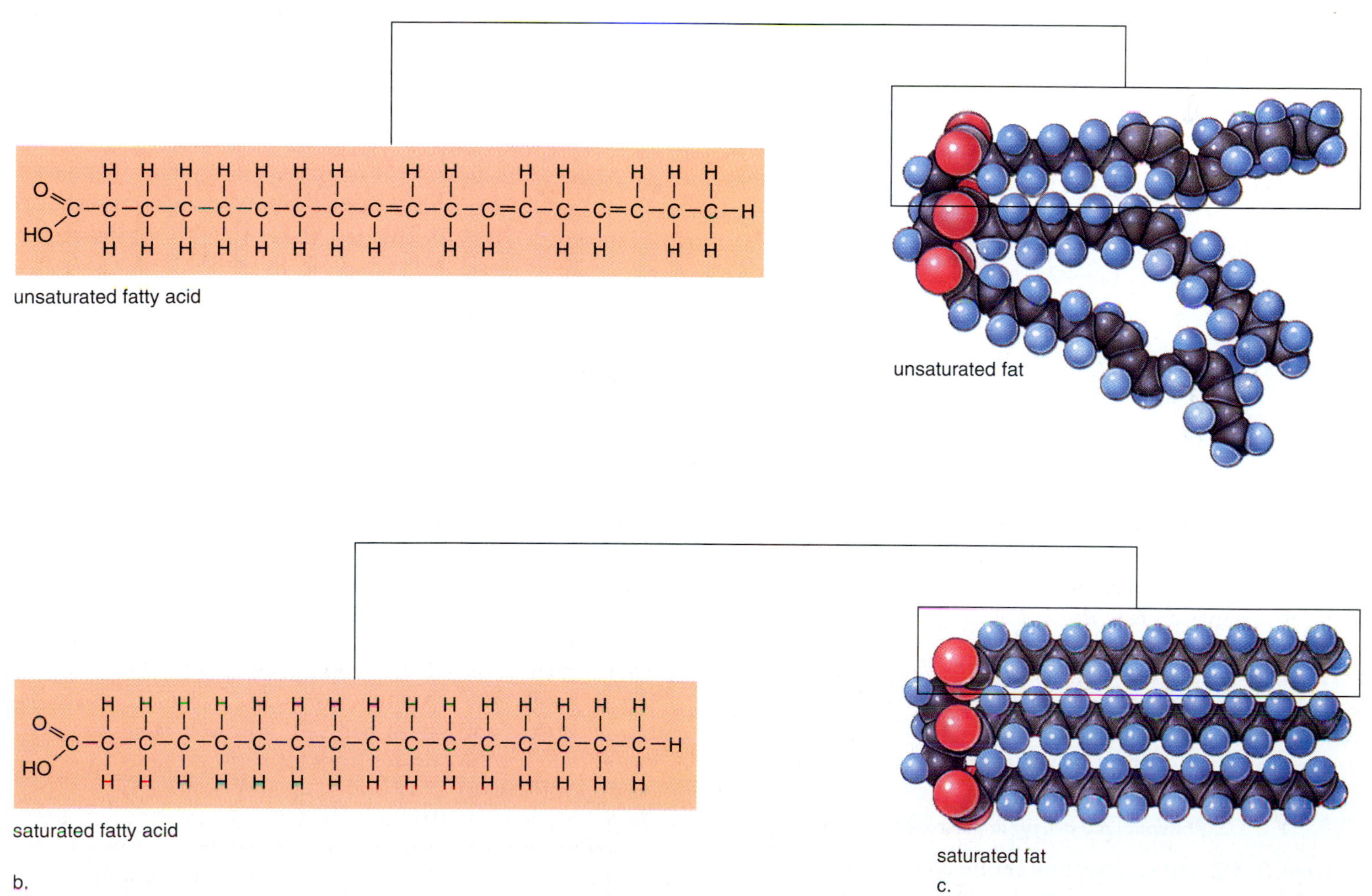

Figure 3.9 Fat and fatty acids.
a. During condensation synthesis of a fat, glycerol combines with three fatty acid molecules. During hydrolysis, the components of water are added, and the bonds are broken. b. A fatty acid has a carboxyl group attached to a long hydrocarbon chain. If there are double bonds between some of the carbons in the chain, the fatty acid is unsaturated. If there are no double bonds, the fatty acid is saturated. c. Space-filling models of an unsaturated fat and a saturated fat.

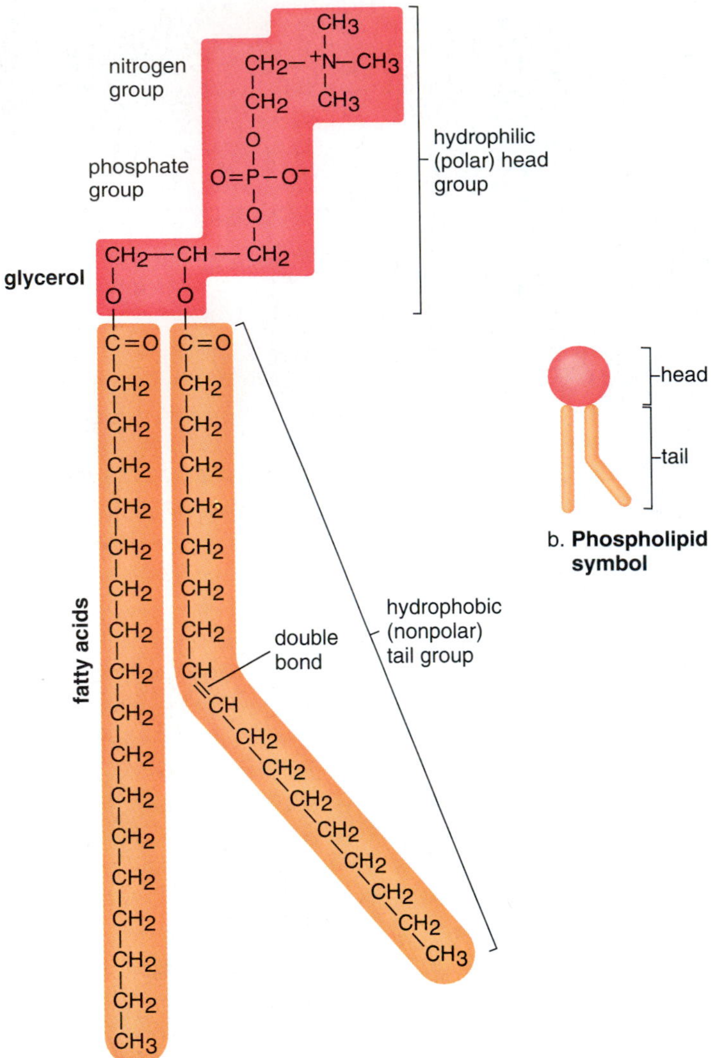

a. **Lecithin, a phospholipid**

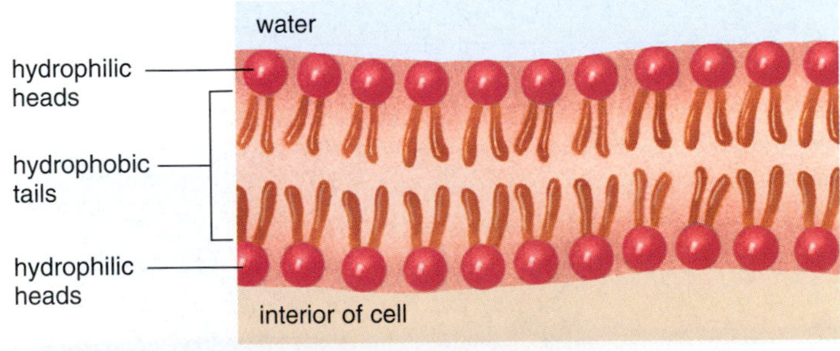

c. **Phospholipid bilayer in plasma membrane**

Figure 3.10 Phospholipid structure and shape.
a. Phospholipids are constructed like fats, except that they contain an ionized phosphate group. Lecithin also contains an ionized nitrogen-containing group. **b.** The hydrophilic (polar) head group is soluble in water, whereas the two hydrophobic (nonpolar) tail groups are not. **c.** This causes the molecules to arrange themselves as a bilayer in the plasma membrane which surrounds cells.

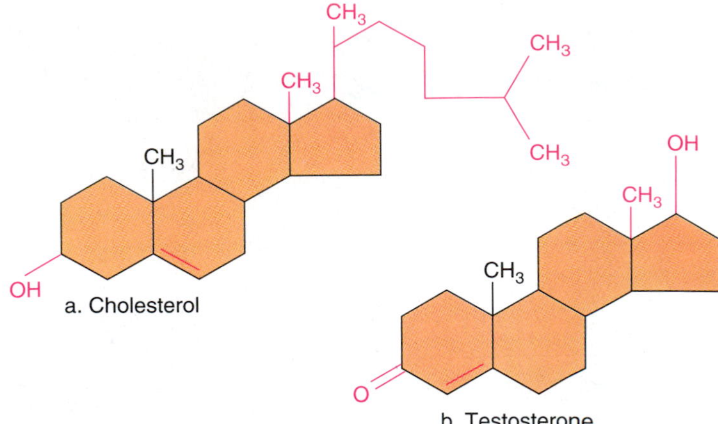

Figure 3.11 Steroid diversity.
a. Cholesterol, like all steroid molecules, has four adjacent rings, but their effects on the body largely depend on the attached groups indicated in red. **b.** Testosterone is the male sex hormone.

Phospholipids

Phospholipids [Gk. *phos*, light, and *lipos*, fat], as implied by their name, contain a phosphate group. Essentially, phospholipids are constructed like neutral fats, except that in place of the third fatty acid, there is a phosphate group or a grouping that contains both phosphate and nitrogen. This group becomes the polar head of the molecule, while the hydrocarbon chains of the fatty acids become the nonpolar tails (Fig. 3.10). The plasma membrane which surrounds cells is a phospholipid bilayer in which the polar heads face outward into a watery medium and the tails face each other because they are water repelling.

Steroids

Steroids are lipids that have an entirely different structure from that of fats. Steroid molecules have a backbone of four fused carbon rings, but each one differs primarily by the type of functional groups attached to the rings. Cholesterol is a component of an animal cell's plasma membrane and is the precursor of several other steroids, such as the sex hormones estrogen and testosterone (Fig. 3.11).

We know that a diet high in saturated fats and cholesterol can lead to circulatory disorders. This type of diet causes fatty material to accumulate inside the lining of blood vessels and blood flow is reduced. As discussed in the Health Focus reading on page 43, nutrition labels are now required to list the calories from fat per serving and the percent daily value from saturated fat and cholesterol.

Phospholipids, unlike other lipids, are soluble in water because they have a hydrophilic group. Steroids are ring compounds that have a similar backbone but vary according to the attached functional groups.

Nutrition Labels

Since May 1994, packaged foods have been labeled as depicted in Figure 3A. The nutrition information given here is based on the serving size (i.e., 1¼ cup, 57 grams) of the cereal. A Calorie* is a measurement of energy. One serving of the cereal provides 220 Calories, of which 20 are from fat. It's also of interest that the cereal provides no cholesterol nor saturated fat. The suggestion that we study nutrition labels to determine how much cholesterol and fat (whether saturated or nonsaturated) they contain is based on innumerable statistical and clinical studies of three types:

(1) Clinical trials that show an elevated blood cholesterol level is a risk factor for coronary heart disease (CHD). The Framingham Heart Study conducted in Framingham, Massachusetts, concludes that as the blood cholesterol level in over 5,000 men and women rises so does the risk of CHD. Elevated blood cholesterol appears to be one of the three major CHD risk factors along with smoking and high blood pressure. Other studies have shown the same. The Multiple Risk Factor Intervention Trial followed more than 360,000 men and found also that there was a direct relationship between an increasing blood cholesterol level and the risk of a heart attack.

(2) Clinical trials indicate that lowering high blood cholesterol levels will reduce the risk of CHD. For example, the Coronary Primary Prevention Trial found that 9% reduction in total blood cholesterol levels produced a 19% reduction in CHD deaths and nonfatal heart attacks. The Cholesterol Lowering Atherosclerosis Study collected X-ray evidence that substantial cholesterol lowering produces slowed progression and regression of plaque in coronary arteries. Plaque is a buildup of soft fatty material including cholesterol beneath the inner linings of arteries. Plaque can accumulate to the point that blood can no longer reach the heart and a heart attack occurs.

(3) Clinical studies show that there is a relationship between diet and blood cholesterol levels. The national Research Council reviewed all sorts of scientific studies before concluding that the intake of saturated fatty acids raises blood cholesterol levels, while the substitution of unsaturated fats and carbohydrates in the diet lowers the blood cholesterol level. An Oslo Study, a Los Angeles Veterans Administration study, and a Finnish Mental Hospital Study showed that diet alone can produce 10–15% reductions in blood cholesterol level.

The desire of consumers to follow dietary recommendations led to the development of the type of nutrition label shown in 3A, which shows the amount of total fat, saturated fat, and cholesterol in a serving of food. The carbohydrate content in a food is of interest because carbohydrates aren't usually associated with health problems. In fact, carbohydrates should compose the largest proportion of the diet. Breads and cereals containing complex carbohydrates are preferable to candy and ice cream containing simple carbohydrates because they are likely to contain dietary fiber (nondigestible plant material). Insoluble fiber has a laxative effect and may reduce the risk of colon cancer; soluble fiber combines with the cholesterol in food and prevents the cholesterol from entering the body proper.

The amount of dietary sodium (as in table salt) is of concern because excessive sodium intake has been linked to high blood pressure in some people. It is recommended that the intake of sodium be no more than 2,400 mg per day. A serving of this cereal provides what percent of this maximum amount?

Vitamins are essential requirements needed in small amounts in the diet. Each vitamin has a recommended daily intake, and the food label tells what percent of the recommended amount is provided by a serving of this cereal.

AQUA PUFFS

Nutrition Facts

Serving Size: 1¼ cup (57 g)
Servings per container: 8

Amount per Serving	Cereal
Calories	220
Calories from Fat	20
	% Daily Value
Total fat: 2 g	3%
Saturated fat: 0g	0%
Cholesterol: 0 mg	0%
Sodium: 320 mg	13%
Total Carbohydrate: 46 g	15%
Soluble fiber: less than 1 g	
Insoluble fiber: 6 g	
Sugars: 11 g	
Other carbohydrates: 28 g	
Protein: 5 g	

Vitamin A — 0% • Vitamin C — 10%
Calcium — 0% • Iron — 80%

		2,000 Calories	2,500 Calories
Total fat	Less than	65 g	80 g
Saturated fat	Less than	20 g	25 g
Cholesterol	Less than	300 mg	300 mg
Sodium	Less than	2,400 mg	2,400 mg
Total carbohydrate		300 mg	375 mg
Dietary fiber		25 g	30 g

Calories per gram:
Fat 9 • carbohydrate 4 • protein 4

AQUA PUFFS

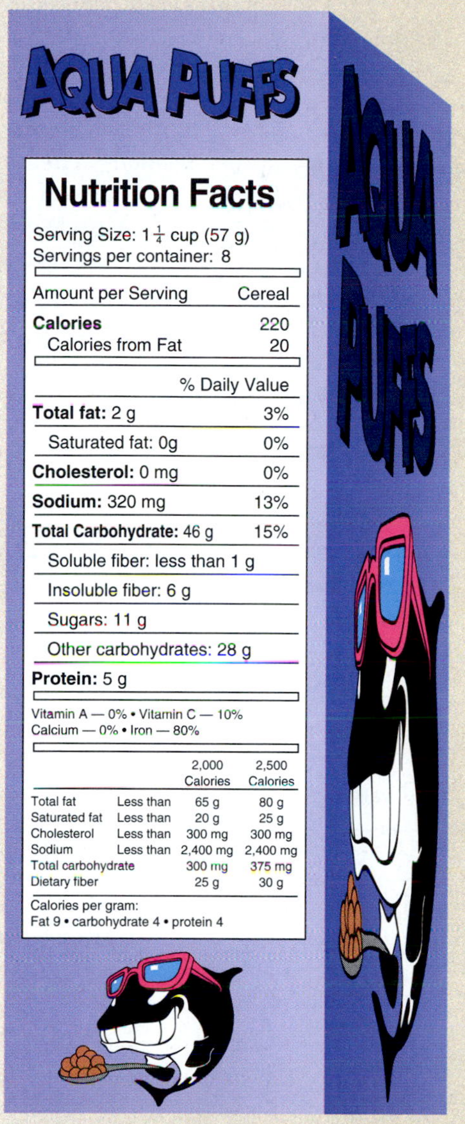

Figure 3A Nutrition label on side panel of cereal box.

*A calorie is the amount of heat required to raise the temperature of 1 gram of water 1°C. A Calorie (capital C), which is used to measure food energy, is equal to 1,000 calories.

3.4 Proteins

A **protein** is composed of one or more polypeptides, which are polymers of amino acids. Proteins perform many functions, and only a few functions can be discussed here.

Support Some proteins such as keratin, which makes up hair and nails, and collagen, which lends support to ligaments, tendons, and skin are structural proteins.

Enzymes Enzymes bring reactants together and thereby speed chemical reactions in cells. They are specific for one particular type of reaction and can function at body temperature.

Transport Channel proteins in the plasma membrane allow substances to enter and exit cells. Carrier proteins in the membrane actually transport molecules into and out of the cell. Some other proteins transport molecules in the blood of animals; hemoglobin is a complex protein that transports oxygen.

Defense Antibodies are proteins. They combine with foreign substances, called antigens. In this way they prevent antigens from destroying cells and upsetting homeostasis.

Hormones Hormones are regulatory proteins. They serve as intercellular messengers that influence the metabolism of cells. Insulin regulates the content of glucose in the blood and in cells; the presence of growth hormone determines the height of the individual.

Motion The contractile proteins actin and myosin allow parts of cells to move and cause muscles to contract. Muscle contraction accounts for the movement of animals from place to place.

Vertebrate cells function differently according to the type of protein they contain; muscle cells contain actin and myosin; red blood cells contain hemoglobin; support cells contain collagen, and so forth.

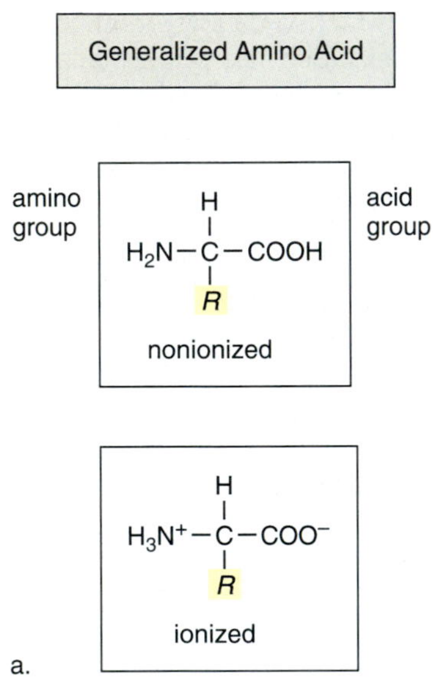

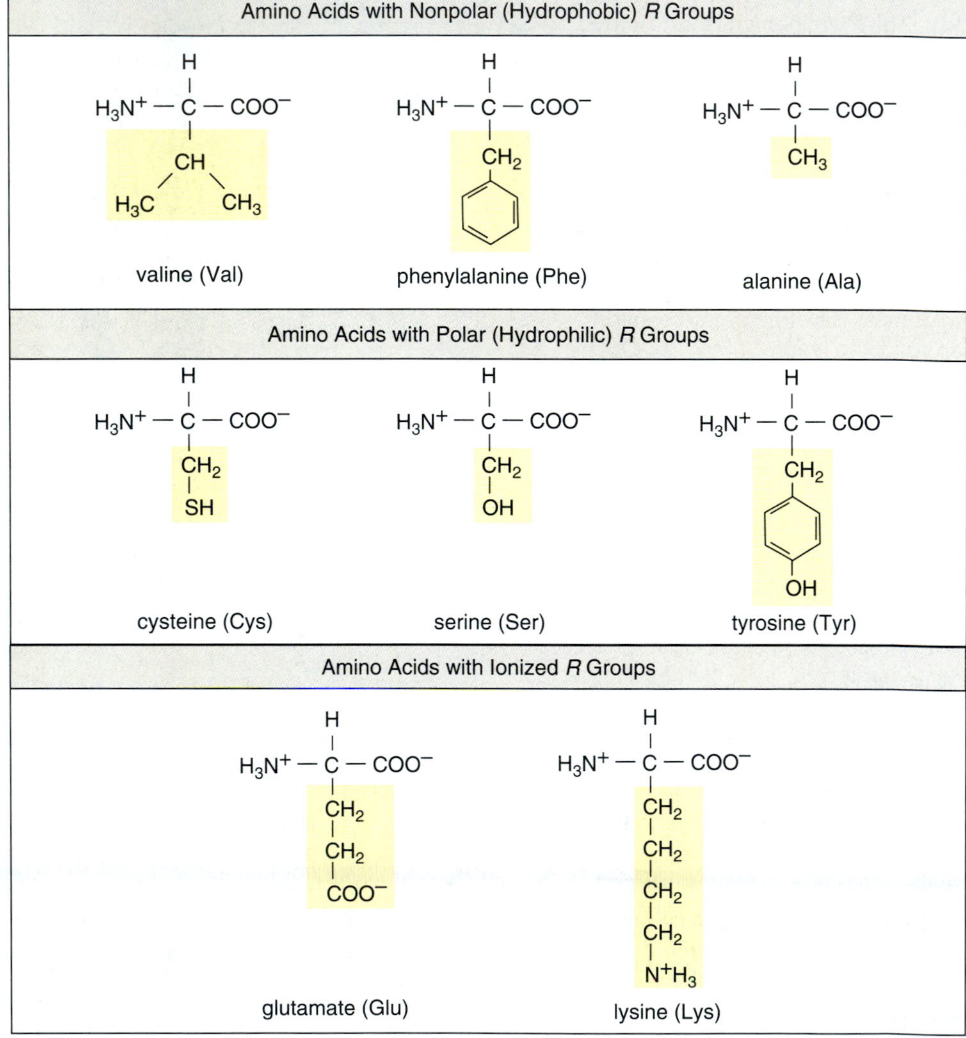

Figure 3.12 Amino acids.
a. Generalized amino acid showing the amino group, the acid (carboxyl) group and the *R* group. Both the nonionized and ionized form are shown.
b. Polypeptides contain 20 different kinds of amino acids, of which only eight are shown here. Amino acids differ by the particular *R* group (yellow) attached to the central carbon. Some *R* groups are nonpolar and hydrophobic, some are polar and hydrophilic, and some are ionized and hydrophilic.

Amino Acids

An **amino acid** has a carbon atom bonded to one hydrogen atom and in addition three groups of atoms. The name of the molecule is appropriate because one of these groups is an amino group (—NH_2) and another is an acidic group (—COOH). The last group is called an *R* group because it is the *R*emainder of the molecule (Fig. 3.12).

Amino acids differ according to their particular *R* group. The *R* groups range in complexity from a single hydrogen atom to a complicated ring compound. The unique chemical properties of an amino acid depend on those of the *R* group. For example, some *R* groups are polar and some are not. Also, the amino acid cysteine has an *R* group that ends with a sulfhydryl (—SH) group that often serves to connect one chain of amino acids to another by a disulfide bond, —S—S—. There are 20 different amino acids commonly found in cells, of which Figure 3.12 gives several examples.

Peptides

Figure 3.13 shows how two amino acids join by a condensation reaction between the carboxyl group of one and the amino group of another. The resulting covalent bond between two amino acids is called a **peptide bond.** The atoms associated with the peptide bond share the electrons unevenly because oxygen is more electronegative than nitrogen. Therefore, the hydrogen attached to the nitrogen has a slightly positive charge, while the oxygen has a slightly negative charge.

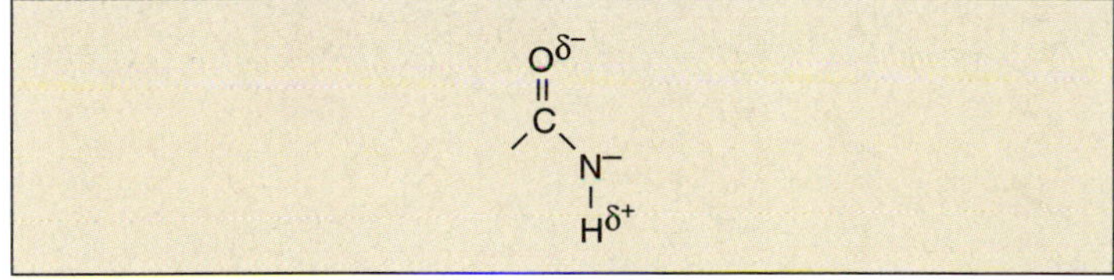

The polarity of the peptide bond means that hydrogen bonding is possible between the —CO of one amino acid and the —NH of another amino acid in a polypeptide.

A **peptide** is two or more amino acids bonded together, and a **polypeptide** is a chain of many amino acids joined by peptide bonds. A protein may contain more than one polypeptide chain; therefore, you can see why a protein could have a very large number of amino acids.

Levels of Protein Structure

The final shape of a protein, in large measure, determines the function of a protein in the cell and body of an organism. A protein can have up to four levels of structure but not all proteins have all four levels. Fibrous proteins have no specific tertiary structure and many proteins have no quaternary structure.

Primary Structure

The primary structure of a protein is the sequence of the amino acids joined by peptide bonds. When amino acids join together a polypeptide results. In other words, the primary structure is a polypeptide with its own particular sequence of amino acids. Therefore, polypeptides differ by the number of amino acids they contain and the sequence of their *R* groups. The chemical characteristics of the *R* groups are important to other levels of structure.

In 1953, Frederick Sanger determined the amino acid sequence of the hormone insulin, the first protein to be sequenced. He did it by first breaking insulin into fragments and then determining the amino acid sequence of the fragments before determining the sequence of the fragments themselves. It was a laborious ten-year task, but it established for the first time that each polypeptide has a particular sequence of amino acids. Today, there are automated sequencers that tell scientists the sequence of amino acids in a polypeptide within a few hours.

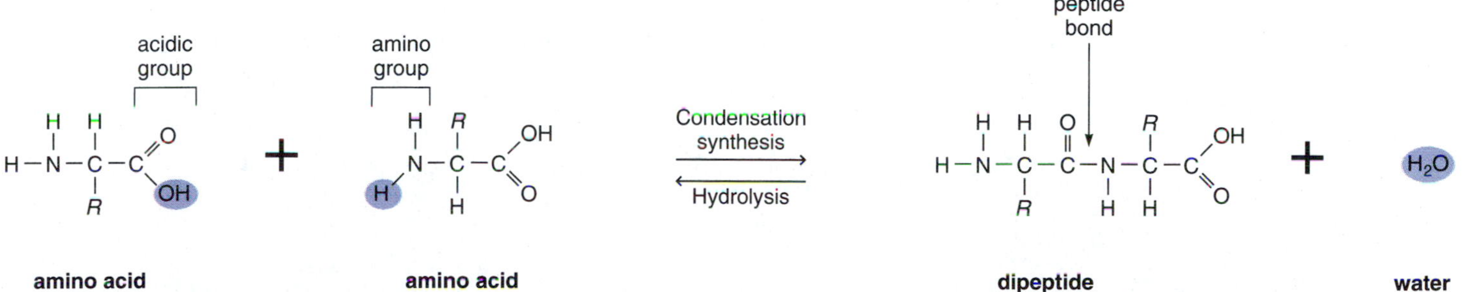

Figure 3.13 Condensation synthesis and hydrolysis of a dipeptide.
As two amino acids join, a peptide bond forms and a water molecule is given off. During hydrolysis, water is added and the peptide bond is broken.

Secondary Structure

The secondary structure of a protein occurs when segments of a polypeptide coil or fold in a particular way (Fig. 3.14).

Linus Pauling and Robert Corey, who began studying the structure of amino acids in the late 1930s, concluded that a coiling, they called an α helix, and a pleated sheet, they

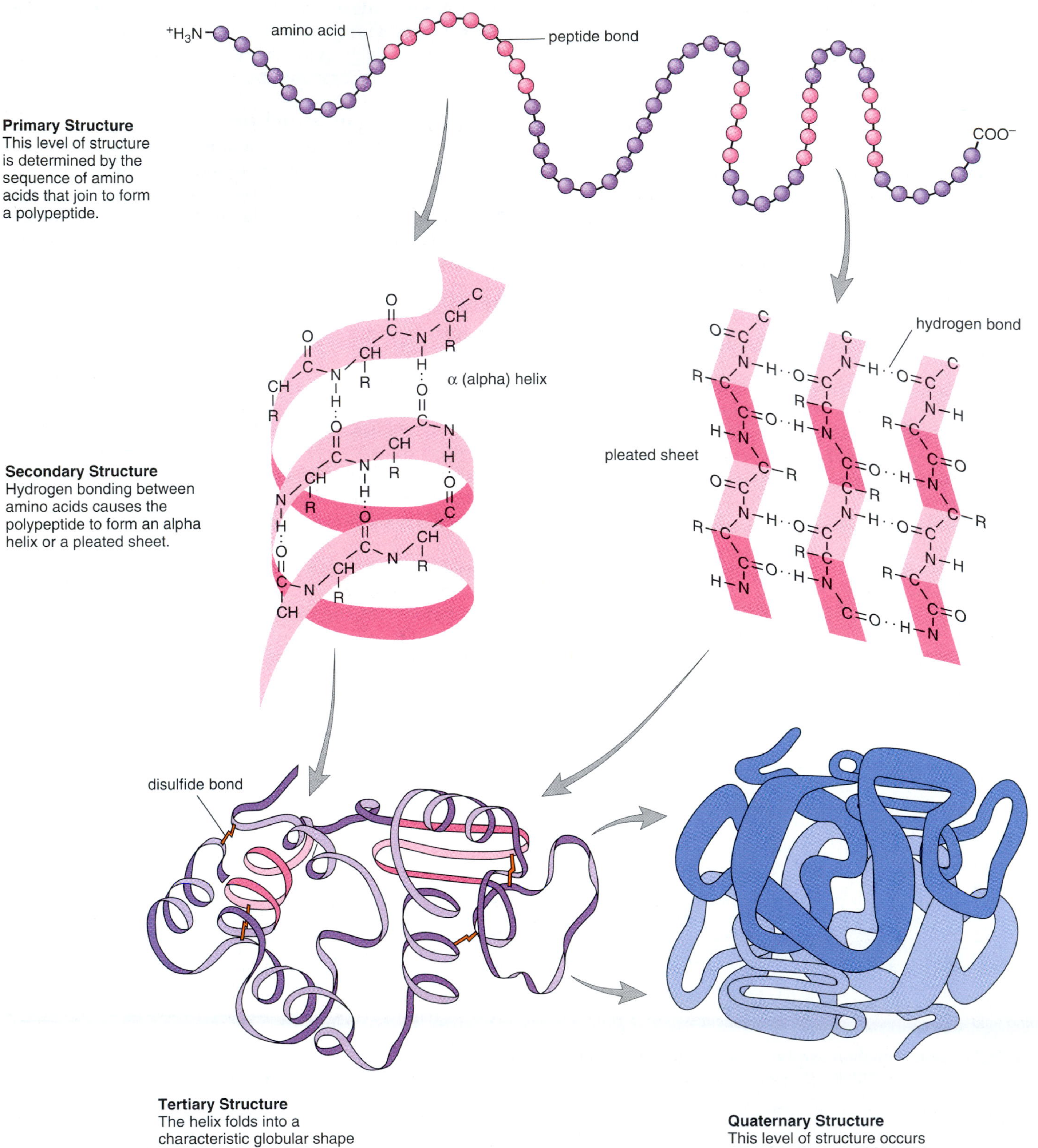

^+H_3N
amino acid
peptide bond
COO^-
Primary Structure
This level of structure is determined by the sequence of amino acids that join to form a polypeptide.
α (alpha) helix
hydrogen bond
pleated sheet
Secondary Structure
Hydrogen bonding between amino acids causes the polypeptide to form an alpha helix or a pleated sheet.
disulfide bond
Tertiary Structure
The helix folds into a characteristic globular shape due in part to covalent bonding between *R* groups.
Quaternary Structure
This level of structure occurs when two or more polypeptides join to form a single protein.

Figure 3.14 Levels of protein organization.

called the β (beta) sheet, were two basic patterns of amino acids within a polypeptide. They called it the α helix because it was the first pattern they discovered, and the β sheet because it was the second pattern they discovered.

Hydrogen bonding often holds the secondary structure of a polypeptide in place. Hydrogen bonding between every fourth amino acid accounts for the spiral shape of the helix. In a β sheet, the polypeptide turns back upon itself and hydrogen bonding occurs between extended lengths of the polypeptide. In keratin, α helices are covalently bonded to one another by disulfide linkages (—S—S—) between two cysteine amino acids. When you get a permanent, the hair is rolled and a reducing agent is used to break the present disulfide bonds. Each —S—S— bond becomes two —SH groups instead. When the reducing agent is removed, newly formed disulfide bonds make the hair curly.

Tertiary Structure

The tertiary structure is a folding and twisting that results in the final three-dimensional shape of a polypeptide.

Various types of bonding between the *R* groups of the amino acids bring about the tertiary structure. Hydrogen bonds, ionic bonds, and covalent bonds all contribute to the tertiary structure of a polypeptide. Strong disulfide linkages in particular help maintain the tertiary shape. Hydrophobic *R* groups do not bond with other *R* groups, and they tend to collect in a common region where they are not exposed to water. These are called hydrophobic interactions. Although hydrophobic interactions are not bonds, they are very important in creating and stabilizing the tertiary structure.

Quaternary Structure

Some proteins have a quaternary structure because they consist of more than one polypeptide. Hemoglobin is a much-studied globular protein that consists of four polypeptides and therefore has a quaternary structure. Each polypeptide in hemoglobin has a primary, secondary, and tertiary structure.

Denaturation of Proteins

Both temperature and pH can bring about a change in polypeptide shape. For example, we are all aware that the addition of acid to milk causes curdling; heating causes egg white, which consists mainly of a protein called albumin, to congeal, or coagulate. When a protein loses its normal configuration, it is said to be **denatured.** Denaturation occurs because the normal bonding patterns between parts of a molecule have been disturbed. Once a protein loses its normal shape, it is no longer able to perform its usual function.

Each type of enzyme is specific to the reaction it speeds. The shape of an enzyme is suitable for receiving its substrate(s) so that an enzyme-substrate complex forms. When an enzyme is denatured its shape changes and it can no longer bring the substrates, which are the reactants, together for the reaction to occur. Enzymes work best at body temperature and each one also has an optimal pH at which the rate of the reaction is highest. At this pH value, the enzyme has its normal shape. A change in pH can change the polarity of *R* groups and disrupt the normal interactions that maintain the normal shape of the enzyme.

If the conditions which caused denaturation were not too severe, and if these are removed, some proteins regain their normal shape and biological activity. This shows that it is simply the sequence of amino acids that cause the levels of structure and the protein's final shape (Fig. 3.15).

The sequence of amino acids in a polypeptide determines its final shape because various *R* groups interact differently. The function of a protein is dependent on its shape.

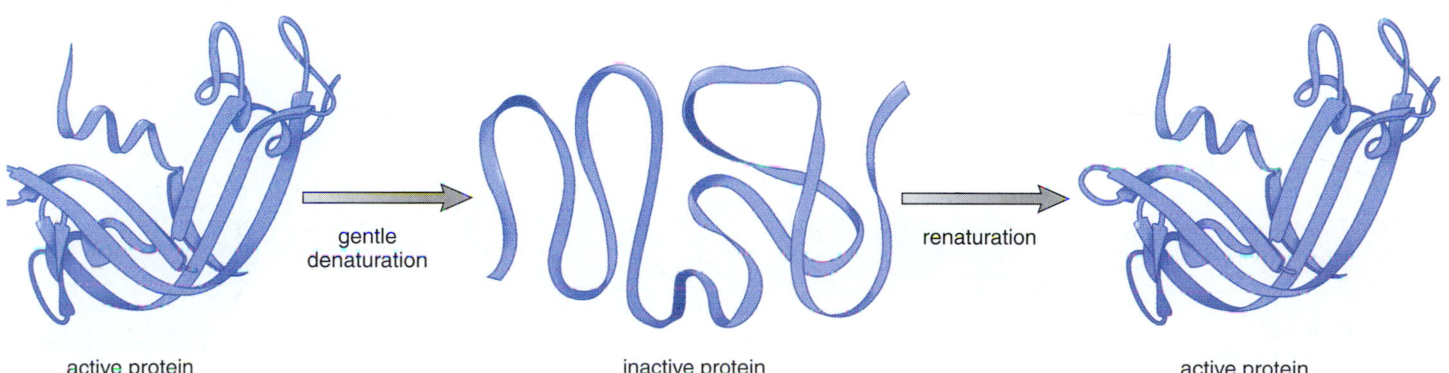

Figure 3.15 Denaturation and renaturation of a protein.
When a protein is denatured, it loses its normal shape and activity. If denaturation is gentle and if the original conditions are restored, some proteins regain their normal shape. This shows that the normal conformation of the molecule is due to the various interactions among a set sequence of amino acids. Each type of protein has a particular sequence of amino acids.

3.5 Nucleic Acids

Nucleic acids are polymers of nucleotides with very specific functions in cells. **DNA (deoxyribonucleic acid)** is the genetic material that stores information regarding its own replication and the order in which amino acids are to be joined to make a protein. **RNA (ribonucleic acid)** is another type of nucleic acid. An RNA molecule called messenger RNA (mRNA) is an intermediary in the process of protein synthesis, conveying information from DNA regarding the amino acid sequence in a protein.

Some nucleotides have independent metabolic functions in cells. For example, some are components of coenzymes, which facilitate enzymatic reactions. **ATP (adenosine triphosphate)** is a nucleotide that supplies energy for synthetic reactions and for various other energy-requiring processes in cells.

Structure of DNA and RNA

Every **nucleotide** is a molecular complex of three types of molecules: phosphate (phosphoric acid), a pentose sugar, and a nitrogen-containing base (Fig. 3.16*a*). In DNA the pentose sugar is deoxyribose, and in RNA the pentose sugar is ribose. A difference in the structure of these 5-carbon sugars accounts for their respective names because deoxyribose lacks an oxygen atom found in ribose (Fig. 3.16*b*).

There are four types of nucleotides in DNA and four types of nucleotides in RNA (Fig. 3.16*c*). The base of a nucleotide can be a pyrimidine with a single ring or a purine with a double ring. In DNA, the pyrimidine bases are cytosine and thymine; in RNA, the pyrimidine bases are cytosine and uracil. In both DNA and RNA, the purine bases are adenine or guanine. These molecules are called bases because their presence raises the pH of a solution.

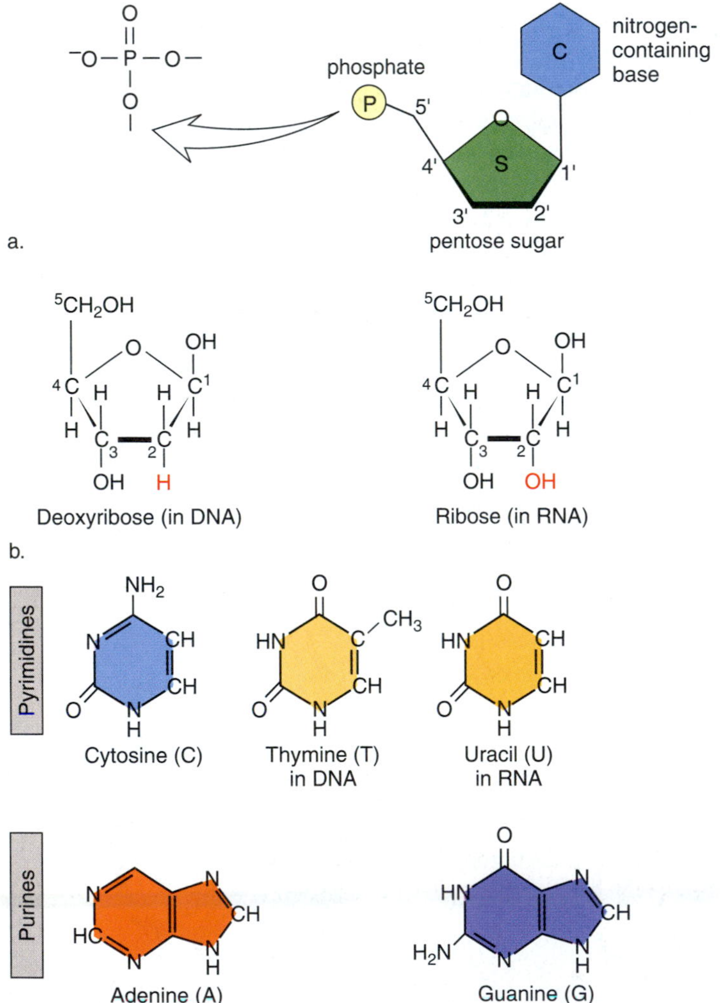

Figure 3.16 Nucleotides.
a. A nucleotide consists of a phosphate molecule, a pentose sugar, and a nitrogen-containing base. **b.** RNA contains the sugar ribose and DNA contains the sugar deoxyribose. **c.** RNA contains the bases C, U, A, and G. DNA contains the bases C, T, A, and G.

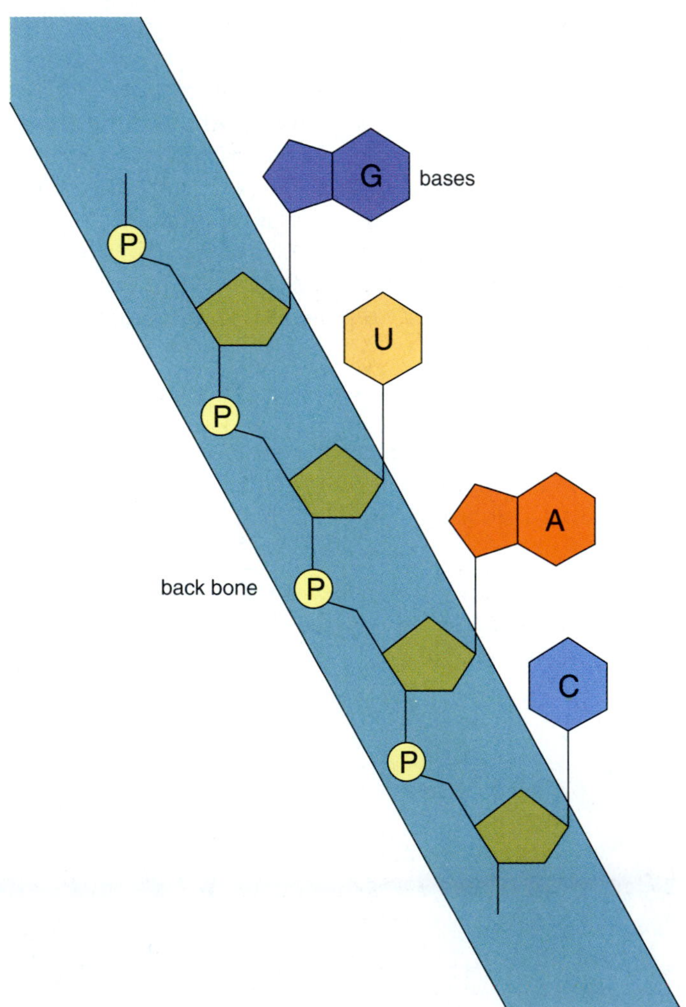

Figure 3.17 RNA structure.
RNA is a single-stranded polymer of nucleotides. When the nucleotides join, the phosphate group of one is bonded to the sugar of the next. The bases project out to the side of the resulting sugar-phosphate backbone.

Nucleotides join in a definite sequence when DNA and RNA form by condensation synthesis. The polynucleotide is a linear molecule called a strand in which the backbone is made up of a series of sugar-phosphate-sugar-phosphate molecules. The bases project to one side of the backbone. Since the nucleotides occur in a definite order, so do the bases. RNA is single stranded (Fig. 3.17).

DNA is double stranded, with the two strands usually twisted about each other in the form of a double helix. The two strands are held together by hydrogen bonds between pyrimidine and purine bases. The bases can be in any order within a strand, but between strands, thymine (T) is always paired with adenine (A), and guanine (G) is always paired with cytosine (C). This is called **complementary base pairing.** Therefore, regardless of the order or the quantity of any particular base pair, the number of purine bases (A + G) always equals the number of pyrimidine bases (T + C) (Fig. 3.18). Table 3.2 summarizes the differences between RNA and DNA.

Table 3.2

DNA Structure Compared to RNA Structure

	DNA	RNA
Sugar	Deoxyribose	Ribose
Bases	Adenine, guanine, thymine, cytosine	Adenine, guanine, uracil, cytosine
Strands	Double stranded with base pairing	Single stranded
Helix	Yes	No

The nucleic acids DNA and RNA are polymers of nucleotides. DNA is the genetic material and RNA is an intermediary during the process of protein synthesis.

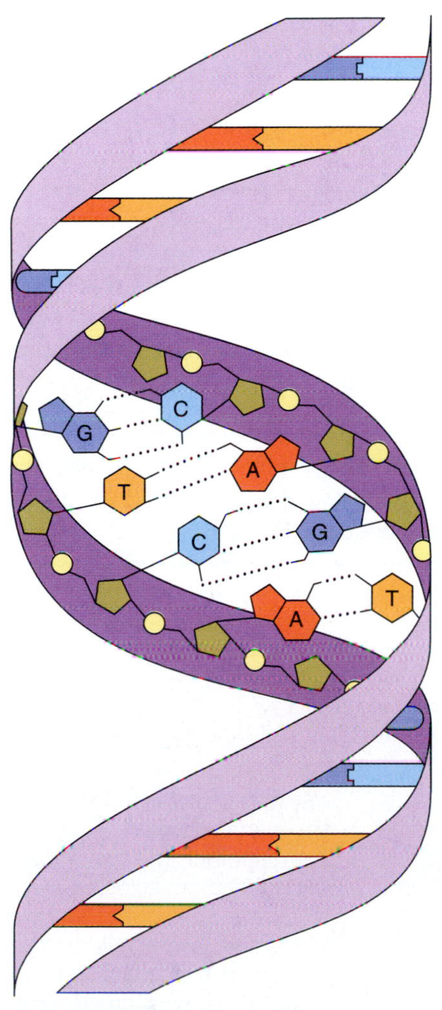

Figure 3.18 **DNA structure.**
DNA is a double helix in which the two nucleotide strands twist about each other. **a.** There are hydrogen bonds (dotted lines) between the complementarily paired bases: C is always paired with G and A is always paired with T. **b.** Space-filling model.

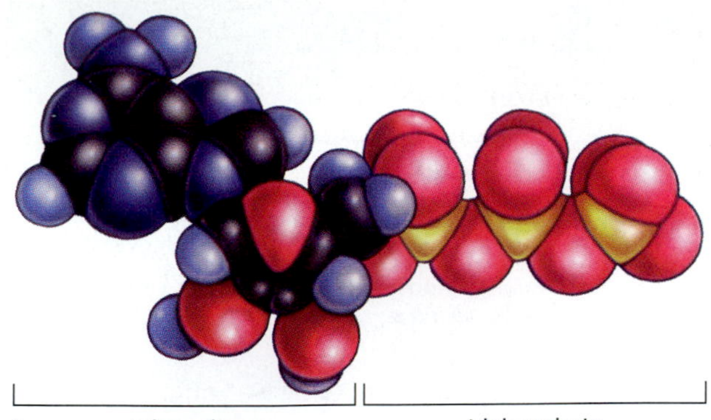

Figure 3.19 ATP
ATP, the universal energy currency of cells, is composed of adenosine and three phosphate groups. **a.** Space-filling model of ATP. **b.** When cells require energy, ATP usually becomes ADP + Ⓟ and energy is released.

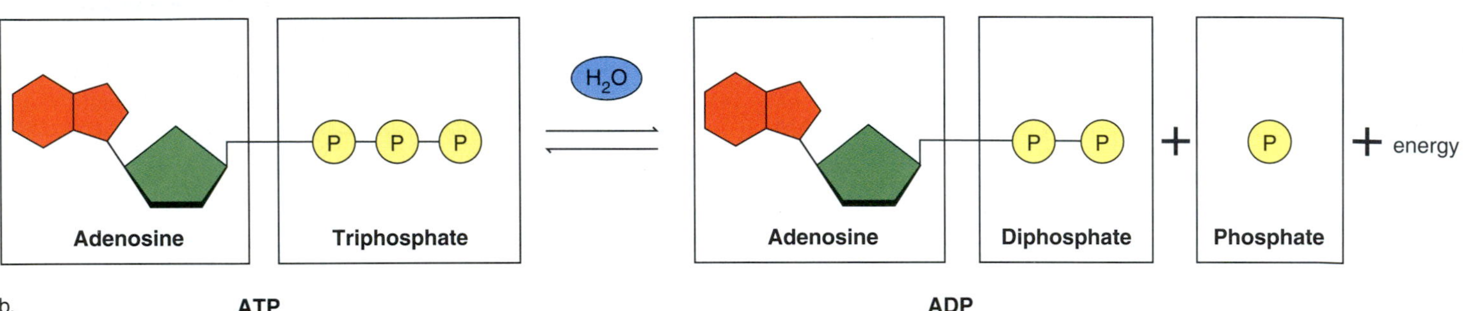

ATP (Adenosine Triphosphate)

ATP is a nucleotide in which **adenosine** is composed of adenine and ribose. Triphosphate stands for the three phosphate groups that are attached to ribose, the pentose sugar (Fig. 3.19). ATP is a high-energy molecule because the last two phosphate bonds are unstable and are easily broken. Usually in cells the terminal phosphate bond is hydrolyzed to give the molecule **ADP (adenosine diphosphate)** and a molecule of inorganic phosphate Ⓟ.

The energy that is released by ATP breakdown is coupled to energy-requiring processes in cells, such as the synthesis of macromolecules like carbohydrates and proteins. In muscle cells, the energy is used for muscle contraction, and in nerve cells, it is used for the conduction of nerve impulses. Just as you spend money when you pay for a product or a service, cells "spend" ATP in the same way. Therefore ATP is called the energy currency of cells.

Because energy is released when the last phosphate bonds of ATP are hydrolyzed, it is sometimes called a high-energy bond, symbolized by a wavy line. But this terminology is misleading—the breakdown of ATP releases energy because the products of hydrolysis (ADP and Ⓟ) are more stable than ATP. It is the entire molecule that releases energy and not a particular bond.

ATP is a high-energy molecule. ATP breaks down to ADP + Ⓟ, releasing energy, which is used for all metabolic work done in a cell.

Connecting Concepts

What does the term organic mean? For some, organic means that food products have been grown without the use of chemicals or have been minimally processed. But you now know that biochemically speaking, organic refers to molecules containing carbon and hydrogen. In biology, organic also refers to living things or anything that has been alive in the past. Therefore, the food we eat and the wood we burn are organic substances. Fossil fuels (coal and oil) formed over 300 million years ago from plant and animal life that, by chance, did not decompose, are also organic. When burned, they release carbon dioxide into the atmosphere just as we do when we breathe!

Although living things are very complex, their macromolecules are simply polymers of small organic molecules. Simple sugars are the monomers of complex carbohydrates; amino acids are the monomers of proteins; nucleotides are the monomers of nucleic acids. Fats are composed of fatty acids and glycerol.

This system of forming macromolecules still allows for diversity. Monomers exist in modified forms and can combine in slightly different ways; therefore, a variety of macromolecules can come about. In cellulose, a plant product, glucose monomers are linked in a slightly different way from glucose monomers in glycogen, an animal product. One protein differs from another by the sequence of the same 20 amino acids.

There is no doubt that the chemistry of carbon is the chemistry of life. The groups of molecules discussed in this chapter, as well as other small molecules and ions, are assembled into structures that make up cells. As discussed in the next chapter, each structure has a specific function necessary to the life of a cell.

Summary

3.1 Organic Molecules

The chemistry of carbon accounts for the diversity of organic molecules found in living things. Carbon can bond with as many as four other atoms. It can also bond with itself to form both chains and rings. Differences in the carbon backbone and attached functional groups cause organic molecules to have different chemical properties. The chemical properties of a molecule determine how it interacts with other molecules and the role the molecule plays in the cell. Some functional groups are hydrophobic and some are hydrophilic.

There are four classes of organic compounds in cells: carbohydrates, lipids, proteins, and nucleic acids (Table 3.3). Polysaccharides, the largest of the carbohydrates, are polymers of simple sugars called monosaccharides. The polypeptides of proteins are polymers of amino acids and nucleic acids are polymers of nucleotides. Polymers are formed by the joining together of monomers. For each bond formed during condensation synthesis, a molecule of water is removed, and for each bond broken during hydrolysis, a molecule of water is added.

3.2 Carbohydrates

Monosaccharides, disaccharides, and polysaccharides are all carbohydrates. Therefore, the term carbohydrate includes both the monomers (e.g., glucose) and the polymers (e.g., starch, glycogen, and cellulose). Glucose is the immediate energy source of cells. Polysaccharides like starch, glycogen, and cellulose are polymers of glucose. Starch in plants and glycogen in animals are energy-storage compounds, but cellulose in plants and chitin in arthropods have structural roles. Chitin's monomer is glucose with an attached amino group.

3.3 Lipids

Lipids include a wide variety of compounds that are insoluble in water. Fats and oils, which allow long-term energy storage, contain one glycerol and three fatty acids. Both glycerol and fatty acids have polar groups but fats and oils do not have polar groups and this accounts for the insolubility of fats and oils. Fats tend to contain saturated fatty acids, and oils tend to contain unsaturated fatty acids. Saturated fatty acids do not have double bonds but unsaturated fatty acids do have double bonds in their hydrocarbon chain.

In a phospholipid, one of the fatty acids is replaced by a phosphate group. In the presence of water, phospholipids form a double layer because the head of each molecule is ionized and the tails are not.

Waxes are composed of a long fatty acid bonded to an alcohol with a long hydrocarbon chain. Steroids have the same four-ring structure as cholesterol but each differs by the groups attached to these rings.

3.4 Proteins

Proteins are polymers of amino acids. Proteins carry out many diverse functions in cells and organisms. Various ones play a support, enzymatic, defensive, transport, motion, and regulatory function.

Table 3.3

Organic Molecules in Cells

	Categories	Elements	Examples	Functions
Carbohydrates	Monosaccharides	C, H, O		
	6-carbon sugar		Glucose	Immediate energy source
	5-carbon sugar		Deoxyribose, ribose	Structure of DNA, RNA
	Disaccharides 12-carbon sugar	C, H, O	Sucrose	Transport sugar in plants
	Polysaccharides polymer of glucose	C, H, O	Starch, glycogen Cellulose	Energy storage in plants, animals Plant cell wall structure
Lipids	Triglycerides 1 glycerol + 3 fatty acids	C, H, O	Fats, oils	Long-term energy storage
	Waxes fatty acid + alcohol	C, H, O	Cuticle Ear wax	Protective covering in plants Protective wax in ears
	Phospholipids like triglyceride except the head group contains phosphate	C, H, O, P	Lecithin	Plasma membrane component
	Steroids backbone of 4 fused rings	C, H, O	Cholesterol Testosterone	Plasma membrane component Male sex hormone
Proteins	Polypeptides polymer of amino acids	C, H, O, N, S	Enzymes Myosin and actin Insulin Hemoglobin Collagen	Speed up cellular reactions Muscle cell components Regulates sugar content of blood Oxygen carrier in blood Fibrous support of body parts
Nucleic Acids	Nucleic acids polymer of nucleotides	C, H, O, N, P	DNA RNA	Genetic material Protein synthesis
	Nucleotides		ATP Coenzymes	Energy carrier Assist enzymes

A polypeptide is a long chain of amino acids joined by peptide bonds. There are 20 different amino acids in cells that differ only by their *R* groups. Polarity and nonpolarity are important aspects of the *R* groups. A protein has three levels of structure: the primary level is the sequence of the amino acids; the secondary level contains α helices and β (pleated) sheets held in place by hydrogen bonding between amino acids along the polypeptide chain; and the tertiary level is the final folding and twisting of the polypeptide that is held in place by bonding and hydrophobic interactions between *R* groups. Proteins that contain more than one polypeptide have a quaternary level of structure, as well.

A protein can be denatured and lose its normal shape and activity, but if renaturation occurs, it assumes both again. This shows that the final shape of a protein is dependent upon the primary sequence of amino acids.

3.5 Nucleic Acids

The nucleic acids DNA and RNA are polymers of nucleotides. Each nucleotide has three components: a phosphate (phosphoric acid), a 5-carbon sugar, and a nitrogen-containing base.

DNA, which contains the sugar deoxyribose, is the genetic material that stores information for its own replication and for the order in which amino acids are to be sequenced in proteins. DNA, with the help of mRNA, specifies protein synthesis. DNA contains phosphate, the sugar deoxyribose, the bases A, T, C, and G and is a double-stranded helix. RNA contains phosphate, the sugar ribose, the bases A, U, C, and G and is single stranded.

ATP, with its unstable phosphate bonds, is the energy currency of cells. Hydrolysis of ATP to ADP + Ⓟ releases energy that is used by the cell to make a product or do any other type of metabolic work.

Reviewing the Chapter

1. How do the chemical characteristics of carbon affect the characteristics of organic molecules? 34–35
2. Give examples of functional groups, and discuss the importance of their being hydrophobic or hydrophilic. 35
3. What molecules are monomers of the polymers studied in this chapter? How do monomers join to produce polymers, and how are polymers broken down to monomers? 36
4. Name several monosaccharides, disaccharides, and polysaccharides, and give a function of each. How are these molecules structurally distinguishable? 37–38
5. What is the difference between a saturated and unsaturated fatty acid? Explain the structure of a fat molecule by stating its components and how they join together. 40–41
6. How does the structure of a phospholipid differ from that of a fat? How do phospholipids form a bilayer in the presence of water? 42
7. Describe the structure of a generalized steroid. How does one steroid differ from another? 42
8. Draw the structure of an amino acid and a dipeptide, pointing out the peptide bond. 44
9. Discuss the four possible levels of protein structure and relate each level to particular bonding patterns. 46
10. How is the tertiary structure of a polypeptide related to its primary structure? Mention denaturation as evidence of this relationship. 44–46
11. How do nucleotides bond to form nucleic acids? State and explain several differences between the structure of DNA and that of RNA. 48–49
12. Discuss the structure and function of ATP. 49

Testing Yourself

Choose the best answer for each question.

1. Which of these is not a characteristic of carbon?
   a. forms four covalent bonds
   b. bonds with itself
   c. is sometimes ionic
   d. forms long chains
   e. sometimes shares two pairs of electrons with another atom
2. The functional group —COOH is
   a. acidic.
   b. basic.
   c. never ionized.
   d. found only in nucleotides.
   e. All of these are correct.
3. A hydrophilic group is
   a. attracted to water.
   b. a polar and/or ionized group.
   c. found in fatty acids.
   d. the opposite of a hydrophobic group.
   e. All of these are correct.
4. Which of these is an example of hydrolysis?
   a. amino acid + amino acid $\longrightarrow$ dipeptide + H_2O
   b. dipeptide + H_2O $\longrightarrow$ amino acid + amino acid
   c. denaturation of a polypeptide
   d. Both a and b are correct.
   e. Both b and c are correct.
5. Which of these makes cellulose nondigestible in humans?
   a. a polymer of glucose subunits
   b. a fibrous protein
   c. the linkage between the glucose molecules
   d. the peptide linkage between the amino acid molecules
   e. The carboxyl groups ionize.
6. A fatty acid is unsaturated if it
   a. contains hydrogen.
   b. contains double bonds.
   c. contains an acidic group.
   d. bonds to glycogen.
   e. bonds to a nucleotide.
7. Which of these is not a lipid?
   a. steroid
   b. fat
   c. polysaccharide
   d. wax
   e. phospholipid

8. The difference between one amino acid and another is found in the
   a. amino group.
   b. carboxyl group.
   c. *R* group.
   d. peptide bond.
   e. carbon atoms.
9. The shape of a polypeptide is
   a. maintained by bonding between parts of the polypeptide.
   b. important to its function.
   c. ultimately dependent upon the primary structure.
   d. necessary to its function.
   e. All of these are correct.
10. Which of these contains a peptide bond?

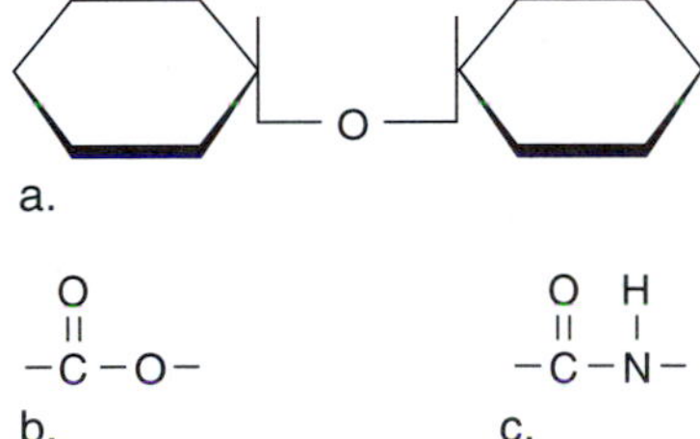

11. Nucleotides
    a. contain a sugar, a nitrogen-containing base, and a phosphate molecule.
    b. are the monomers for fats and polysaccharides.
    c. join together by covalent bonding between the bases.
    d. are present in both DNA and RNA.
    e. Both a and d are correct.
12. ATP
    a. is an amino acid.
    b. has a helical structure.
    c. is a high-energy molecule that can break down to ADP and phosphate.
    d. provides enzymes for metabolism.
    e. is most energetic when in the ADP state.
13. Label the following diagram using the terms monomer, hydrolysis, condensation, and polymer, and explain the diagram:

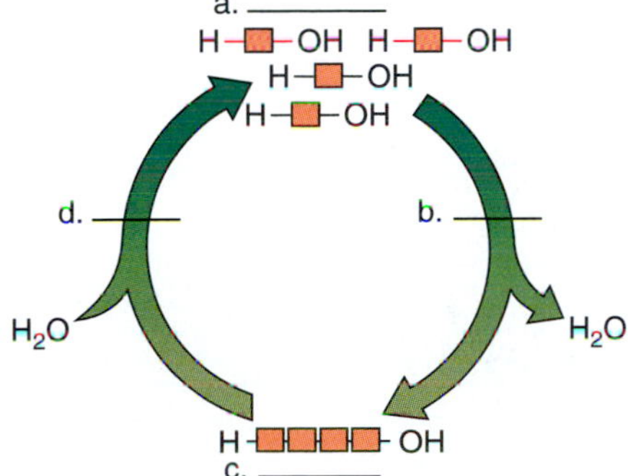

Thinking Scientifically

1. You are studying the fat content of different types of seeds. You have discovered that some types of seeds have a much higher percentage of saturated fatty acids than others. You know the property difference (solid versus liquid) and the structural difference (more hydrogen versus less) between saturated and unsaturated fatty acids. How might these fatty acid differences correlate with climate (tropical compared to temperate), the size of seeds (small compared to large) and environmental conditions for germination (favorable compared to unfavorable)?
2. You are investigating molecules that inhibit a bacterial enzyme. You discovered that the addition of several phosphate groups to an inhibitor improves its effectiveness. Why would knowledge of the three-dimensional structure of the bacterial enzyme help you understand why the phosphate groups improve the inhibitor's effectiveness?

Bioethical Issue

Eric Stevenson and more than 110,000 veterans of the Gulf War are mysteriously ill. They complain of conditions like skin rashes, breathing difficulties, fatigue, diarrhea, muscle and joint pain, headaches, and loss of memory. While the Pentagon doesn't recognize what is called the Gulf War Syndrome (GWS), it does admit that soldiers were exposed to at least four categories of organic chemicals:

- Petroleum products such as kerosene, diesel fuel, leaded gasoline, and smoke from oil-well fires.
- Pesticides and insect repellents. These were applied to clothing and sprayed into the air.
- Drugs and vaccines. Pyridostigmine bromide was given to protect against nerve gas. Vaccines against anthrax and botulism were given in case of biological warfare.
- Biological and chemical weapons. Reluctantly, the Pentagon estimates that hundreds of thousands of soldiers may have been exposed to nerve gas released into the air when Iraqi ammunition depots were bombed.

While the military says that exposure can make you sick, they cling to the idea that toxic chemicals either kill you outright or you recover completely. Therefore, they suggest that the veterans are suffering from post-traumatic stress disorder. Lingering symptoms of stress are to be expected in a certain number of soldiers that come home from war.

Epidemiologist Robert Haley, however, has studied the effects of these chemicals on chickens, and found that their combination does produce Gulf War Syndrome symptoms. He says, "The Defense Department should agree that this is a physical injury, a brain injury, aided and abetted by acute stress."

For what reasons might the Pentagon be reluctant to accept the possibility that GWS is due to exposure of troops to organic chemicals? In so doing are they shirking their responsibility to Gulf War veterans?

Understanding the Terms

adenosine 50
ADP (adenosine diphosphate) 50
amino acid 45
ATP (adenosine triphosphate) 48, 50
carbohydrate 37
cellulose 38
chitin 39
complementary base pairing 49
condensation synthesis 36
denatured 47
deoxyribose 37
disaccharide 37
DNA (deoxyribonucleic acid) 48
fat 40
fatty acid 40
fructose 37
functional group 35
glucose 37
glycerol 40
glycogen 38
hexose 37
hydrolysis 36
hydrophilic 35
hydrophobic 35
inorganic molecule 34
isomer 35
lactose 37
lipid 40
maltose 37
monomer 36
monosaccharide 37
nucleic acid 48
nucleotide 48
oil 40
organic molecule 34
pentose 37
peptide 45
peptide bond 45
phospholipid 42
polymer 36
polypeptide 45
polysaccharide 38
protein 44
ribose 37
RNA (ribonucleic acid) 48
saturated 40
starch 38
steroid 42
sucrose 37
triglyceride 40
unsaturated 40
wax 40

Match the terms to these definitions:

a. ______________ Class of organic compounds that includes monosaccharides, disaccharides, and polysaccharides.
b. ______________ Class of organic compounds that tend to be soluble in nonpolar solvents such as alcohol.
c. ______________ Macromolecule consisting of covalently bonded monomers.
d. ______________ Molecules that have the same molecular formula but a different structure and, therefore, shape.
e. ______________ Two or more amino acids joined together by covalent bonding.

Web Connections

Exploring the Internet

http://www.mhhe.com/biosci/genbio/mader
(click on *Biology 7/e*)

The *Biology 7/e* Online Learning Center provides many resources for studying the material in this chapter including links to the following sites:

Amino Acids. This site lists the 20 different amino acids, presents their structure, and discusses their abbreviations.

http://www.chemie.fu-berlin.de/chemistry/bio/amino-acids.html

Large Molecule Review. The chapters that follow are from a hypertextbook. They cover lipids, sugars, and nucleic and amino acids.

http://esg-www.mit.edu:8001/esgbio/lm/lmdir.html

and

http://esg-www.mit.edu:8001/esgbio/lm/lipids/lipids.html
http://esg-www.mit.edu:8001/esgbio/lm/sugars/sugars.html
http://esg-www.mit.edu:8001/esgbio/lm/proteins/aa/aminoacids.html
http://esg-www.mit.edu:8001/esgbio/lm/nucleicacids/nucleicacids.html

Library of 3-D Molecular Structures. Both organic and inorganic molecular structures are seen.

http://www.nyu.edu/pages/mathmol/library/

Cell Structure and Function

chapter

chapter concepts

4.1 Cellular Level of Organization

- All organisms are composed of cells, which arise from preexisting cells. 56
- A microscope is usually needed to see a cell because most cells are quite small. 57
- Cell-volume-to-cell-surface relationships explain why cells are so very small. 57

4.2 Bacterial Cells

- Prokaryotic cells do not have a membrane-bounded nucleus nor other organelles of eukaryotic cells. 60

4.3 Eukaryotic Cells

- Organelles are membrane-bounded compartments specialized to carry out specific functions. 61
- The endosymbiotic hypothesis states that certain eukaryotic organelles were originally prokaryotes taken up by a larger cell. 64
- Eukaryotic cells have a membrane-bounded nucleus that contains DNA within chromosomes. 65
- Chloroplasts use solar energy to produce organic molecules that are broken down, releasing energy in mitochondria. 70
- The cytoskeleton, a complex system of filaments and tubules, gives the cell its shape and accounts for the movement of the cell and its organelles. 72

Micrograph of a plant cell. *Arabidopsis thaliana.*

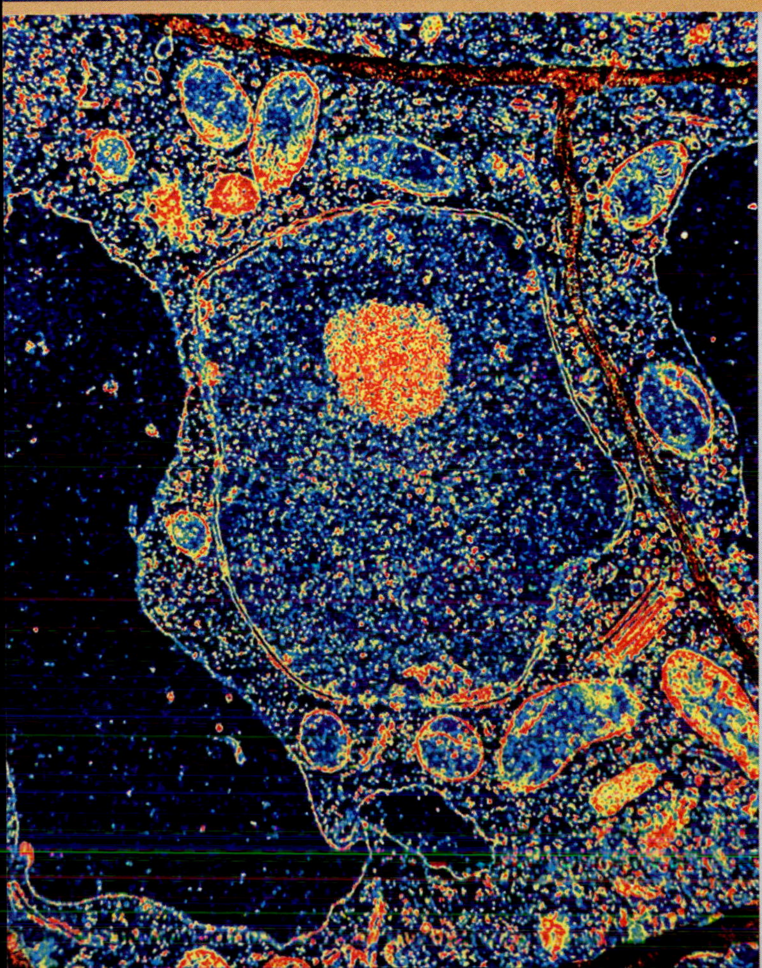

Today we are accustomed to thinking of living things as being constructed of cells. But the word cell didn't come into use until the seventeenth century. Antonie van Leeuwenhoek of Holland is famous for observing tiny, unicellular living things that no one had seen before. Leeuwenhoek sent his findings to an organization of scientists called the Royal Society in London. Robert Hooke, an Englishman, confirmed Leeuwenhoek's observations and was the first to use the term cell. The tiny chambers he observed in the honeycomb structure of cork reminded him of the rooms, or cells, in a monastery. Naturally, he referred to the boundaries of these chambers as walls.

Even today, a light micrograph of a cell is not overly impressive. You can make out the nucleus and its nucleolus near the cell's center but that's about it. The rest of the cell seems to contain only an amorphous matrix. Through electron microscopy and biochemical analysis it was discovered that the cell actually contains organelles, tiny specialized structures performing specific cellular functions. Further, the nucleus contains numerous chromosomes and thousands of genes!

4.1 Cellular Level of Organization

In the 1830s, Matthias Schleiden stated that all plants are composed of cells and Theodor Schwann stated that all animals are composed of cells. These Germans based their ideas not only on their own work but on the work of all who had studied tissues under microscopes. Today we recognize that all the organisms we see about us are made up of cells. Figure 4.1 illustrates that in our daily lives we observe the whole organism, but if it were possible to view them internally with a microscope, we would then see their cellular nature. A **cell** is the smallest unit of living matter.

There are unicellular organisms but most, including ourselves, are multicellular. A cell is not only the structural unit, it is also the functional unit of organs and, therefore, organisms. This is very evident when you consider certain illnesses of the human body such as diabetes or prostate cancer. It is the cells of the pancreas or the prostate that are malfunctioning, rather than the organ itself.

All organisms are made up of cells, and a cell is the structural and functional unit of organs and, ultimately, organisms.

Cells reproduce. Once a growing cell gets to a certain size, it divides. Unicellular organisms reproduce themselves when they divide. Multicellular organisms grow when their cells divide. Cells are also involved in the reproduction of multicellular organisms. Therefore, there is always a continuity of cells from generation to generation. Soon after Schleiden and Schwann had presented their findings, another German scientist, Rudolf Virchow, used a microscope to study the life of cells. He came to the conclusion that cells do not arise on their own accord; rather, "every cell comes from a preexisting cell." By the middle of the nineteenth century, biologists clearly recognized that all organisms are composed of self-reproducing cells.

Cells are capable of self-reproduction, and cells come only from preexisting cells.

The previous two highlighted statements are often called the **cell theory.** The cell theory is one of the unifying concepts of biology. There are extensive data to support its two basic principles and it is universally accepted by biologists.

Figure 4.1 Organisms and cells. All organisms, whether plants or animals, are composed of cells. This is not readily apparent because a microscope is usually needed to see the cells. **a.** Corn. **b.** Light micrograph of corn leaf showing many individual cells. **c.** Rabbit. **d.** Light micrograph of a rabbit's intestinal lining showing that it, too, is composed of cells. The dark-staining bodies are nuclei.

a. Corn, *Zea mays*

c. Rabbit, *Oryctolagus* sp.

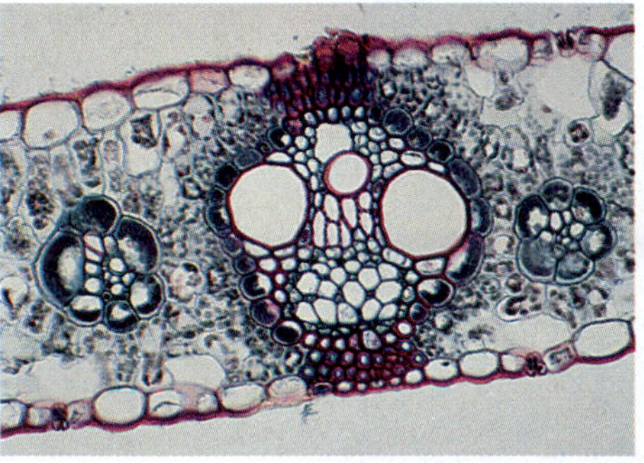

b.

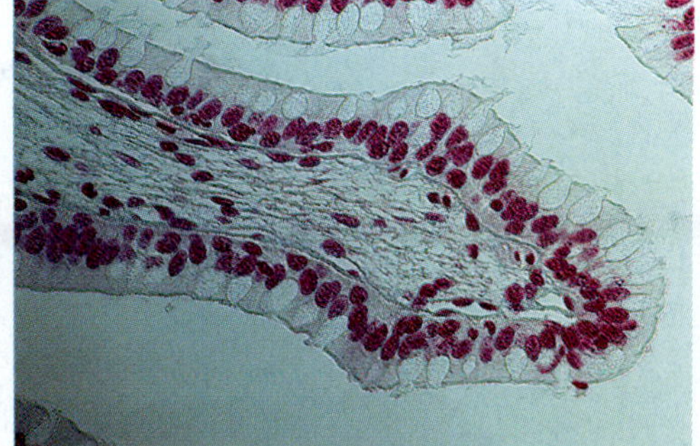

d.

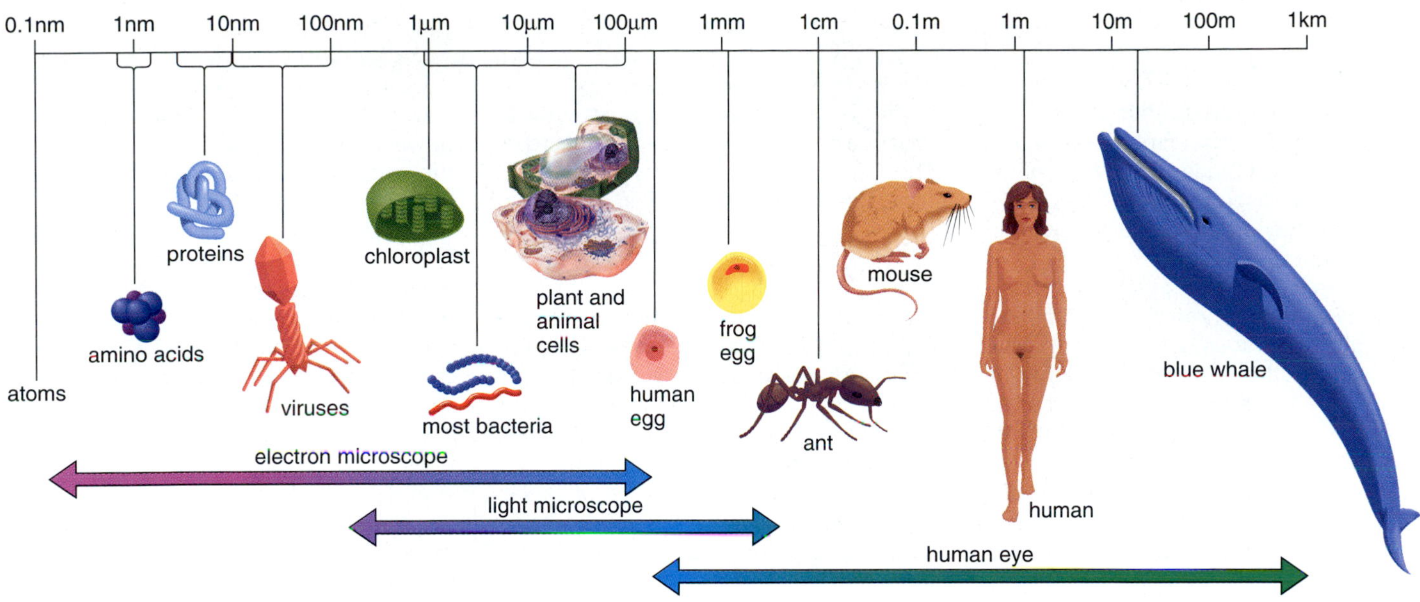

Figure 4.2 The sizes of living things and their components.
Whales', humans', mice's, and even frogs' eggs can be seen by the human eye. It takes a microscope to see most cells and lower levels of biological organization. Cells are visible with the light microscope, but not in much detail. It takes an electron microscope to see organelles in detail and to make out viruses and molecules. Notice that in this illustration each higher unit is 10 times greater than the lower unit. (In the metric system, see inside back cover, 1 meter = 10^2 cm = 10^3 mm = 10^6 mm = 10^9 nm.)

Cell Size

Cells are quite small. A frog's egg, at about one millimeter (mm) in diameter, is large enough to be seen by the human eye. Most cells are far smaller than one millimeter; some are even as small as one micrometer (µm)—one thousandth of a millimeter. Cell inclusions and macromolecules are even smaller still than a micrometer and are measured in terms of nanometers (nm).

Why are cells so small? To answer this question consider that a cell needs a surface area large enough to allow adequate nutrients to enter and to rid itself of wastes. A small cube that is 1 mm tall has a surface area of 6 mm^2 and a volume of 1 mm^3. This is a ratio of surface area to volume of 6:1. But a cube that is 2 mm tall has a surface area of 24 mm^2 and a volume of 8 mm^3. This is a ratio of only 3:1. Therefore a small cell has more surface area per volume than does a large cell:

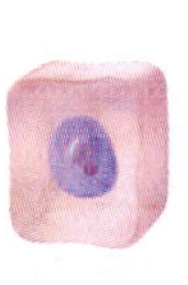

small cell—
more surface area
per volume

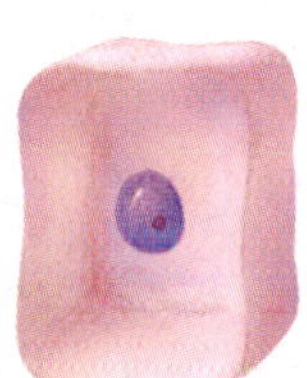

large cell—
less surface area
per volume

Small cells, not large cells, are likely to have an adequate surface area for exchange of wastes for nutrients. We would expect, then, a size limitation for an actively metabolizing cell. A chicken's egg is several centimeters in diameter, but the egg is not actively metabolizing. Once the egg is incubated and metabolic activity begins, the egg divides repeatedly without growth. Cell division restores the amount of surface area needed for adequate exchange of materials. Further, cells that specialize in absorption have modifications that greatly increase the surface area per volume of the cell. The columnar cells along the surface of the intestinal wall have surface foldings called microvilli (sing., microvillus), which increase their surface area.

A cell needs a surface area that can adequately exchange materials with the environment. Surface-area-to-volume considerations require that cells stay small.

Figure 4.2 outlines the visual range of the eye, light microscope, and electron microscope. The discussion of microscopy in the reading on pages 58–59 explains why the electron microscope allows us to "see" so much more detail than the light microscope does.

Microscopy of Today

Cells were not discovered until the invention of the microscope in the seventeenth century. Since that time, various types of microscopes have been developed for the study of cells and their components.

In the *bright-field light microscope,* light rays passing through a specimen are brought into focus by a set of glass lenses, and the resulting image is then viewed by the human eye. In the *transmission electron microscope,* electrons passing through a specimen are brought into focus by a set of magnetic lenses, and the resulting image is projected onto a fluorescent screen or photographic film. In the *scanning electron microscope* (*SEM*), a narrow beam of electrons is scanned over the surface of the specimen, which is coated with a thin metal layer. The metal gives off secondary electrons that are collected by a detector to produce an image on a television screen. The SEM permits the development of three-dimensional images (Fig. 4A).

Magnification, Resolution, and Contrast

Almost everyone knows that the magnifying capability of a transmission electron microscope is greater than that of a light microscope. A light microscope can magnify objects a few thousand times, but an electron microscope can magnify them hundreds of thousands of times. The difference lies in the means of illumination. The path of light rays and electrons moving through space is wavelike, but the wavelength of electrons is much shorter than the wavelength of light. This difference in wavelength accounts for the electron microscope's greater magnifying capability and its greater resolving power. The greater the resolving power, the greater the detail eventually seen. Resolution is the minimum distance between two objects at which they can still be seen, or resolved, as two separate objects. If oil is placed between the sample and the objective lens of the light microscope, the resolving power is increased, and if ultraviolet light is used instead of visible light, it is also increased. But typically, a light microscope can resolve down to 0.2 µm, while the electron microscope can resolve down to 0.0002 µm. If the resolving power of the average human eye is set at one, then that of the typical light microscope is about 500, and that of the electron microscope is 100,000. This means that the electron microscope distinguishes much greater detail (Fig. 4A).

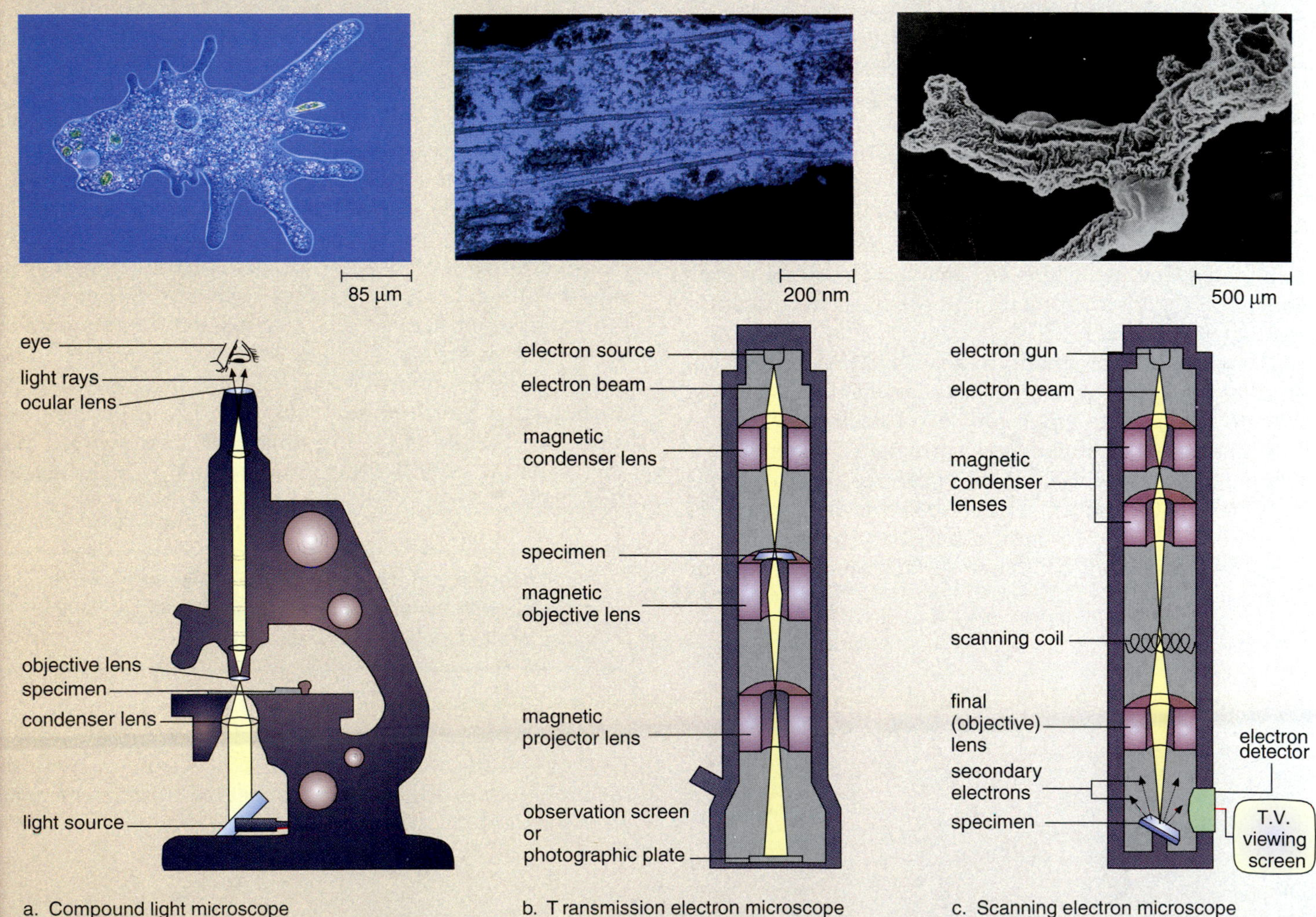

Figure 4A Diagram of microscopes with accompanying micrographs of *Amoeba proteus.*

Some microscopes view living specimens, but often specimens are treated prior to observation. Cells are killed, fixed so they do not decompose, and embedded into a matrix. The matrix strengthens the specimen so that it can be thinly sliced. These sections are often stained with colored dyes (light microscopy) or with electron-dense metals (electron microscopy) to provide contrast. Another way to increase contrast is to use optical methods such as phase contrast and differential interference contrast (Fig. 4B). In addition to optical and electronic methods for contrasting transparent cells, a third very prominent research tool is called *immunofluorescence microscopy,* because it uses fluorescent antibodies to reveal the location of a protein in the cell (see Fig. 4.14). The importance of this method is that the cellular distribution of a single type of protein can be examined.

Illumination, Viewing, and Recording

Light rays can be bent (refracted) and brought to a focus as they pass through glass lenses, but electrons do not pass through glass. Electrons have a charge that allows them to be brought into a focus by magnetic lenses. The human eye utilizes light to see an object but cannot utilize electrons for the same purpose. Therefore, electrons leaving the specimen in the electron microscope are directed toward a screen or a photograph plate that is sensitive to their presence. Humans can view the image on the screen or photograph.

A major advancement in illumination has been the introduction of *confocal microscopy,* which uses a laser beam scanned across the specimen to focus on a single shallow plane within the cell. The microscopist can "optically section" the specimen by focusing up and down, and a series of optical sections can be combined in a computer to create a three-dimensional image, which can be displayed and rotated on the computer screen.

An image from a microscope may be recorded by replacing the human eye with a television camera. The television camera converts the light image into an electronic image, which can be entered into a computer. In *video-enhanced contrast microscopy,* the computer makes the darkest areas of the original image much darker and the lightest areas of the original much lighter. The result is a high-contrast image with deep blacks and bright whites. Even more contrast can be introduced by the computer if shades of gray are replaced by colors.

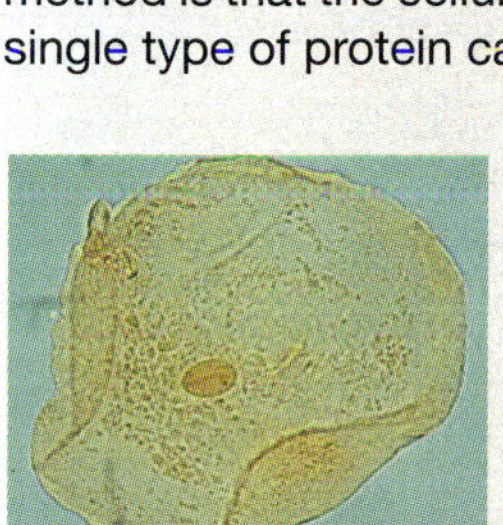

Bright-field. Light passing through the specimen is brought directly into focus. Usually, the low level of contrast within the specimen interferes with viewing all but its largest components.

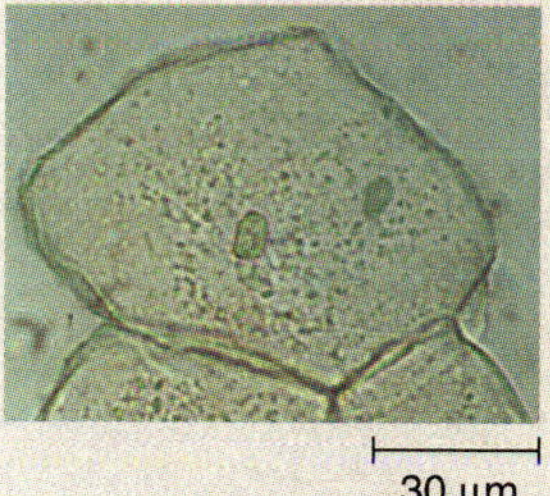

Bright-field (stained). Dyes are used to stain the specimen. Certain components take up the dye more than other components, and therefore contrast is enhanced.

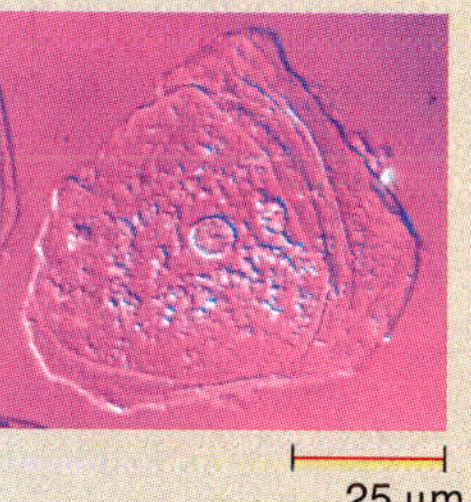

Differential interference contrast. Optical methods are used to enhance density differences within the specimen so that certain regions appear brighter than others. This technique is used to view living cells, chromosomes, and organelle masses.

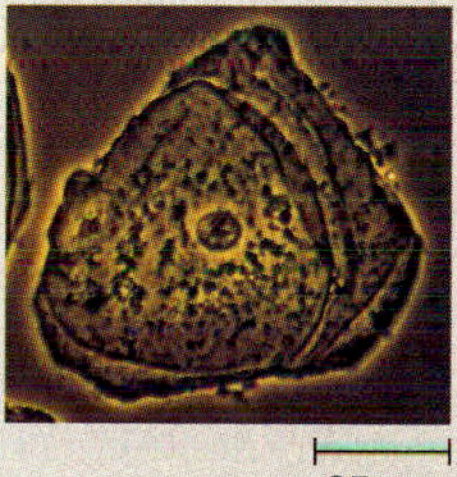

Phase contrast. Density differences in the specimen cause light rays to come out of "phase." The microscope enhances these phase differences so that some regions of the specimen appear brighter or darker than others. This technique is widely used to observe living cells and organelles.

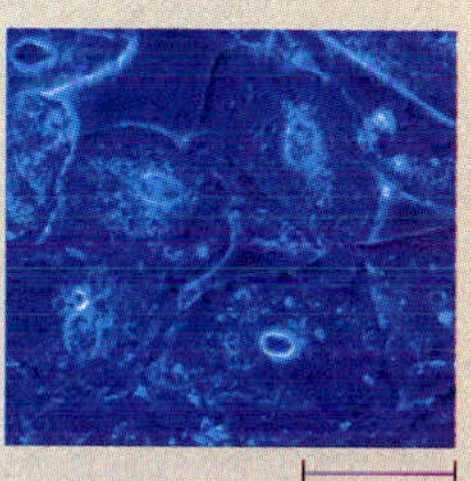

Darkfield. Light is passed through the specimen at an oblique angle so that the objective lens receives only light diffracted and scattered by the object. This technique is used to view organelles, which appear quite bright against a dark field.

Figure 4B Cheek cells. Photomicrographs of cheek cells, illustrating different types of light microscopy.

4.2 Bacterial Cells

Bacteria [Gk. *bacterion*, rod] are **prokaryotic cells**[1] [Gk. *pro*, before, and *karyon*, kernel, nucleus] in the domain Bacteria. Most bacteria are between 1–10 μm in size; therefore, they are just visible with the light microscope.

Figure 4.3 illustrates the main features of bacterial anatomy. The **cell wall** contains peptidoglycan, a complex molecule with chains of a unique amino disaccharide joined by peptide chains. In some bacteria, the cell wall is further surrounded by a **capsule** and/or gelatinous sheath called a **slime layer.** Motile bacteria usually have long, very thin appendages called flagella (sing., **flagellum**) that are composed of subunits of the protein called flagellin. The flagella, which rotate like propellers, rapidly move the bacterium in a fluid medium. Some bacteria also have fimbriae, which are short appendages that help them attach to an appropriate surface.

A membrane called the **plasma membrane** regulates the movement of molecules into and out of the cytoplasm, the interior of the cell. **Cytoplasm** in a prokaryotic cell consists of **cytosol,** a semifluid medium, and thousands of **ribosomes,** small bodies that coordinate the synthesis of proteins. In prokaryotes, most genes are found within a single chromosome (loop of DNA) located within the **nucleoid** [L. *nucleus*, nucleus, kernel and Gk. *-eides*, like], but they may also have small accessory rings of DNA called **plasmids.** In addition, the photosynthetic cyanobacteria have light-sensitive pigments, usually within the membranes of flattened disks called **thylakoids.**

Although bacteria seem structurally simple, they are actually metabolically diverse. Bacteria are adapted to living in almost any kind of environment and are diversified to the extent that almost any type of organic matter can be used as a nutrient for some particular bacterium. Given an energy source, most bacteria are able to synthesize any kind of molecule they may need. Therefore, the cytoplasm is the site of thousands of chemical reactions and bacteria are more metabolically competent than are human beings. Indeed, the metabolic capability of bacteria is exploited by humans who use them to produce a wide variety of chemicals and products for human use.

Bacteria are prokaryotic cells. Prokaryotic cells have these constant features.

Outer boundary	cell wall plasma membrane
Cytoplasm	ribosomes thylakoids (cyanobacteria)
Nucleoid	innumerable enzymes chromosome (loop of DNA)

[1] Archaea (domain Archaea) are another type of prokaryote. Chapter 29 discusses archaea, and also bacteria in more detail.

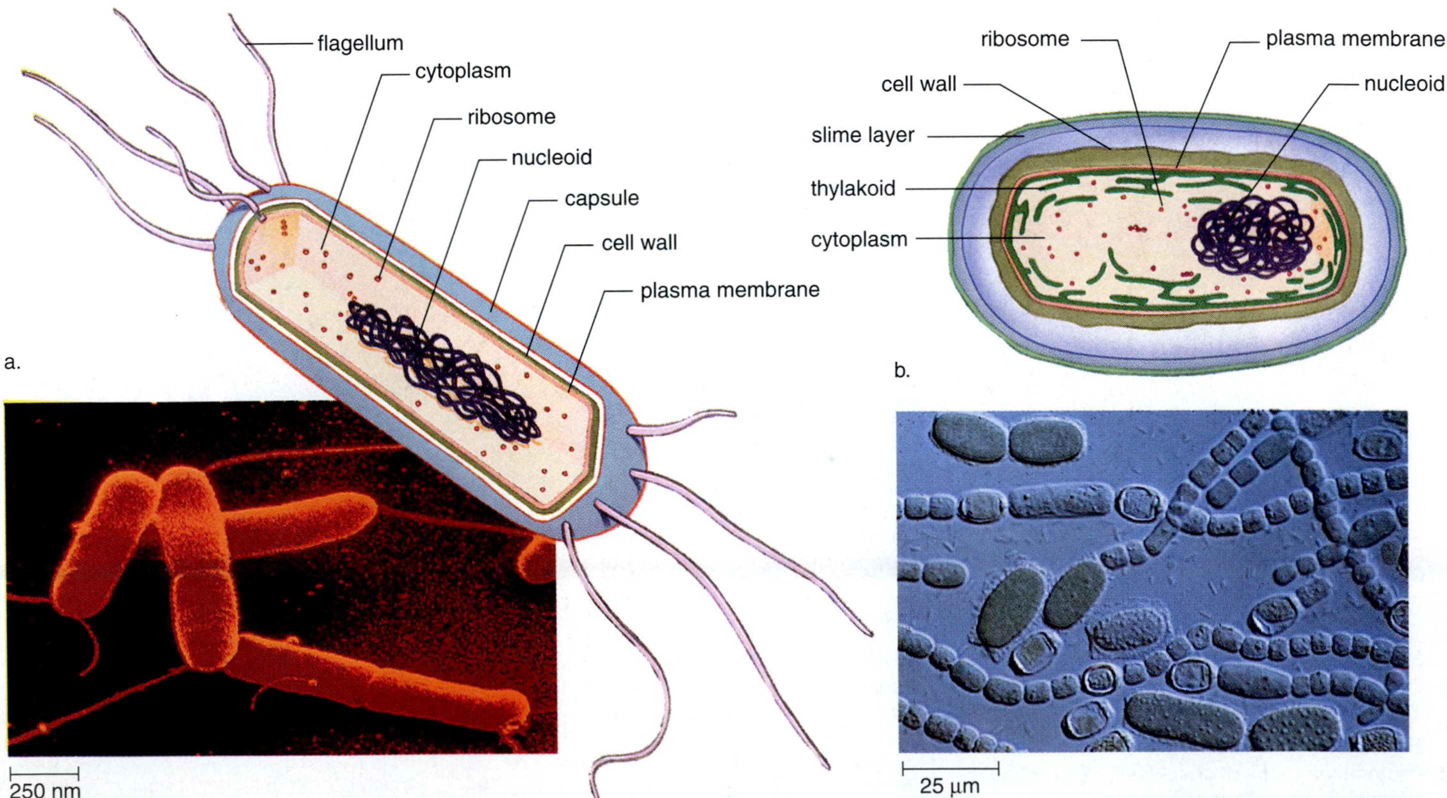

Figure 4.3 Bacterial (prokaryotic) cells.
a. Nonphotosynthetic bacterium. **b.** Cyanobacterium, a photosynthetic bacterium, formerly called a blue-green alga.

4.3 Eukaryotic Cells

In general, eukaryotic cells are larger than prokaryotic cells. In contrast to prokaryotic cells, **eukaryotic cells** [Gk. *eu*, true, and *karyon*, kernel, nucleus] have a true nucleus. A nucleus is a membrane-bounded structure where DNA is housed within threadlike structures called chromatin.

Eukaryotic cells have a membrane-bounded nucleus, and prokaryotic cells lack a nucleus.

A membrane is a phospholipid bilayer with embedded proteins.

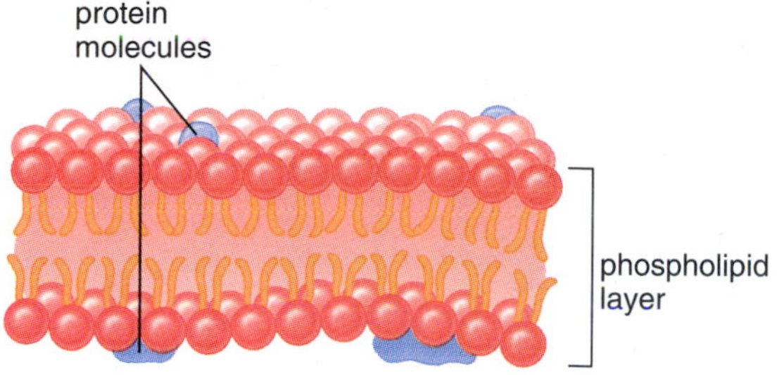

In eukaryotic cells, **organelles** are small bodies, each with a specific structure and function (Table 4.1). Many organelles are membrane-bounded. The cytosol, which is a semifluid medium outside the nucleus, is divided up and compartmentalized by the organelles. Compartmentalization keeps the cell organized and keeps its various functions separate from one another. The cytosol in eukaryotic cells has an organized lattice of protein filaments called the cytoskeleton.

Eukaryotic cells, like prokaryotic cells, have a plasma membrane that separates the contents of the cell from the environment and regulates the passage of molecules into and out of the cell. Some eukaryotic cells, notably plant cells, also have an outer boundary called a **cell wall.** A plant cell wall contains cellulose fibrils and therefore has a different composition than the cell wall of bacteria. A cell wall supports and protects the cell but does not interfere with the movement of molecules across the plasma membrane.

Notice in Figures 4.4 and 4.5, both animal and plant cells contain mitochondria while only plant cells have chloroplasts. Animal cells, but not plant cells, have centrioles. The color chosen to represent each structure in the plant and animal cell is used for that structure throughout the chapters of this part and throughout the text.

Eukaryotic cells have a number of organelles and prokaryotic cells lack membrane-bounded organelles.

Table 4.1

Structures in Eukaryotic Cells

Name	Composition	Function
Cell wall	Contains cellulose fibrils	Support and protection
Plasma membrane	Phospholipid bilayer with embedded proteins	Define cell boundary; regulation of molecule passage into and out of cell
Nucleus	Nuclear envelope surrounding nucleoplasm, chromatin, and nucleoli	Storage of genetic information; synthesis of DNA and RNA
Nucleolus	Concentrated area of chromatin, RNA, and proteins	Ribosomal formation
Ribosome	Protein and RNA in two subunits	Protein synthesis
Endoplasmic reticulum (ER)	Membranous flattened channels and tubular canals	Synthesis and/or modification of proteins and other substances, and transport by vesicle formation
Rough ER	Studded with ribosomes	Protein synthesis
Smooth ER	Having no ribosomes	Various; lipid synthesis in some cells
Golgi apparatus	Stack of membranous saccules	Processing, packaging, and distribution of proteins and lipids
Vacuole and vesicle	Membranous sacs	Storage of substances
Lysosome	Membranous vesicle containing digestive enzymes	Intracellular digestion
Peroxisomes	Membranous vesicle containing specific enzymes	Various metabolic tasks
Mitochondrion	Membranous cristae bounded by an outer membrane	Cellular respiration
Chloroplast	Membranous grana bounded by two membranes	Photosynthesis
Cytoskeleton	Microtubules, intermediate filaments, actin filaments	Shape of cell and movement of its parts
Cilia and flagella	9 + 2 pattern of microtubules	Movement of cell
Centriole	9 + 0 pattern of microtubules	Formation of basal bodies

Figure 4.4 **Animal cell anatomy.**
a. Generalized drawing. **b.** Transmission electron micrograph. See Table 4.1 for a description of these structures, along with a listing of their functions.

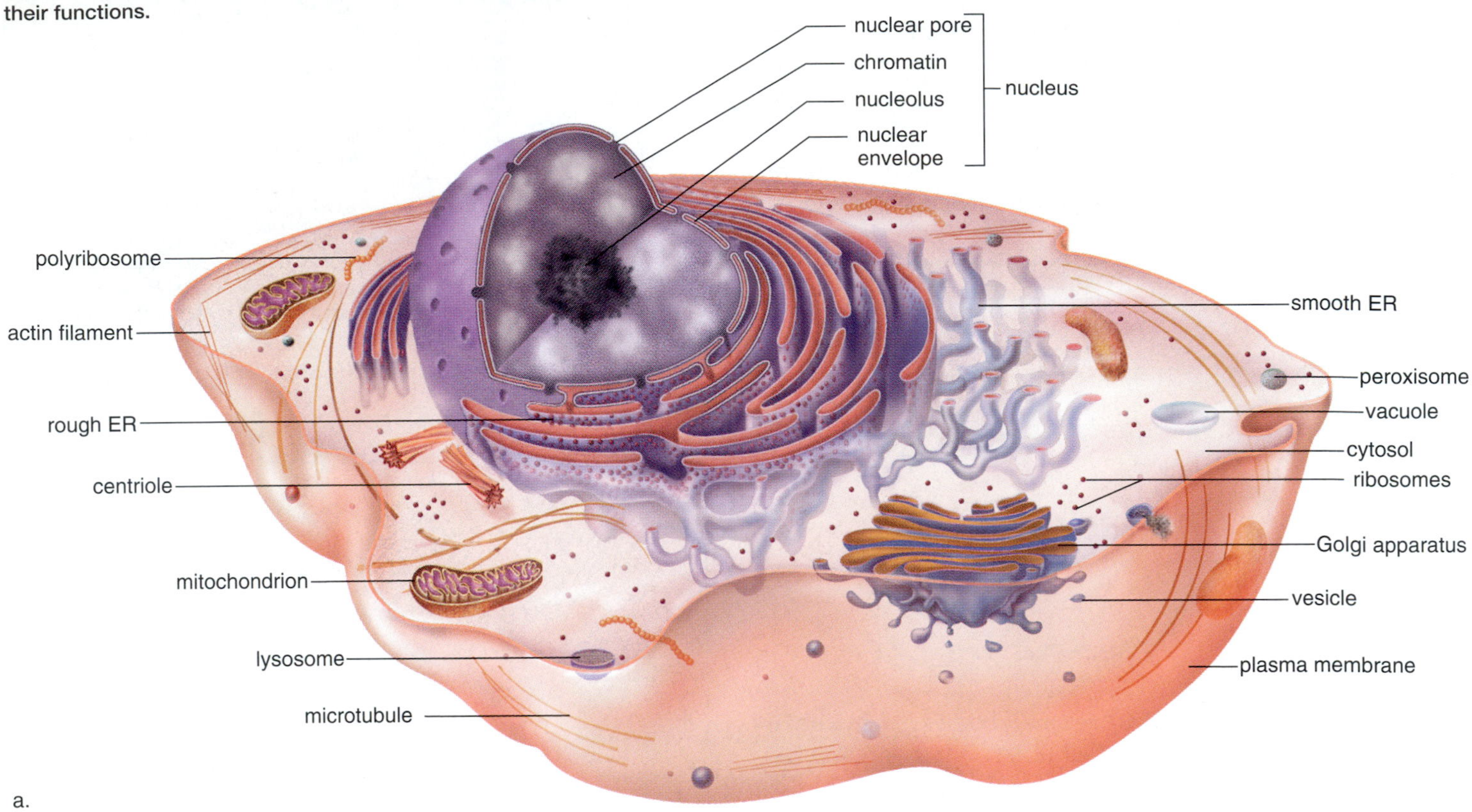

nuclear pore
chromatin
nucleolus
nuclear envelope
nucleus
polyribosome
actin filament
rough ER
centriole
mitochondrion
lysosome
microtubule
smooth ER
peroxisome
vacuole
cytosol
ribosomes
Golgi apparatus
vesicle
plasma membrane

a.

plasma membrane
nuclear envelope
chromatin
nucleolus
endoplasmic reticulum
50 nm

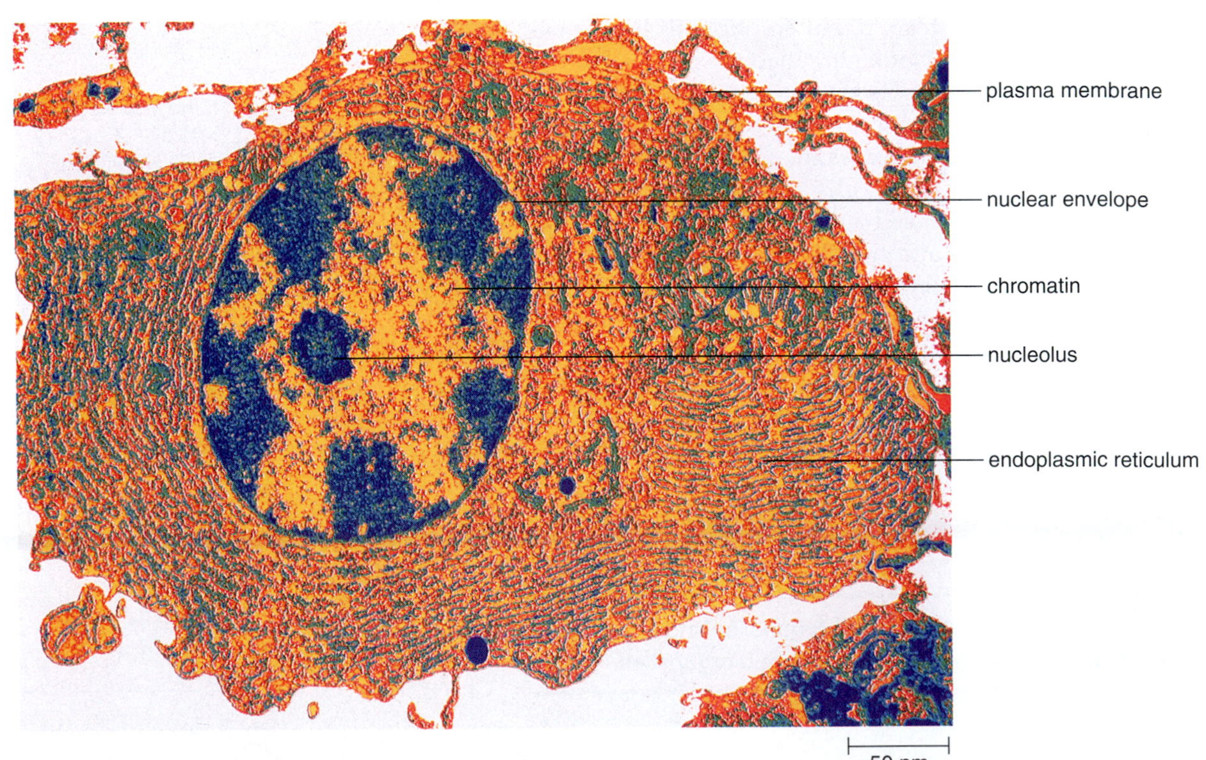

b.

Figure 4.5 Plant cell anatomy.
a. Generalized drawing. **b.** Transmission electron micrograph of young leaf cell. See Table 4.1 for a description of these structures, along with a listing of their functions.

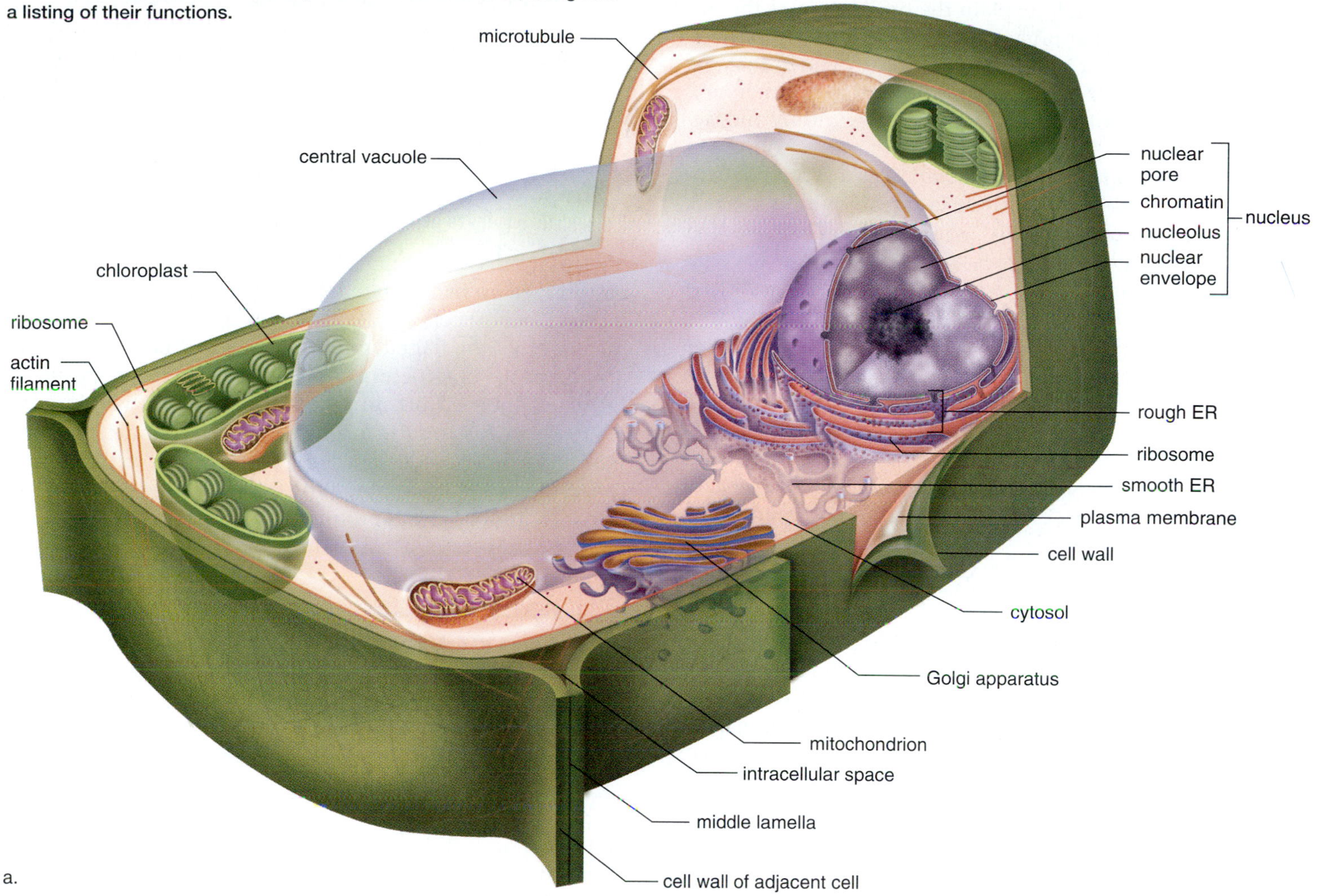

microtubule
central vacuole
chloroplast
ribosome
actin filament
nuclear pore
chromatin
nucleolus
nuclear envelope
nucleus
rough ER
ribosome
smooth ER
plasma membrane
cell wall
cytosol
Golgi apparatus
mitochondrion
intracellular space
middle lamella
cell wall of adjacent cell

a.

mitochondrion
nucleus
ribosomes
plasma membrane
central vacuole
chloroplast
cell wall
1 µm

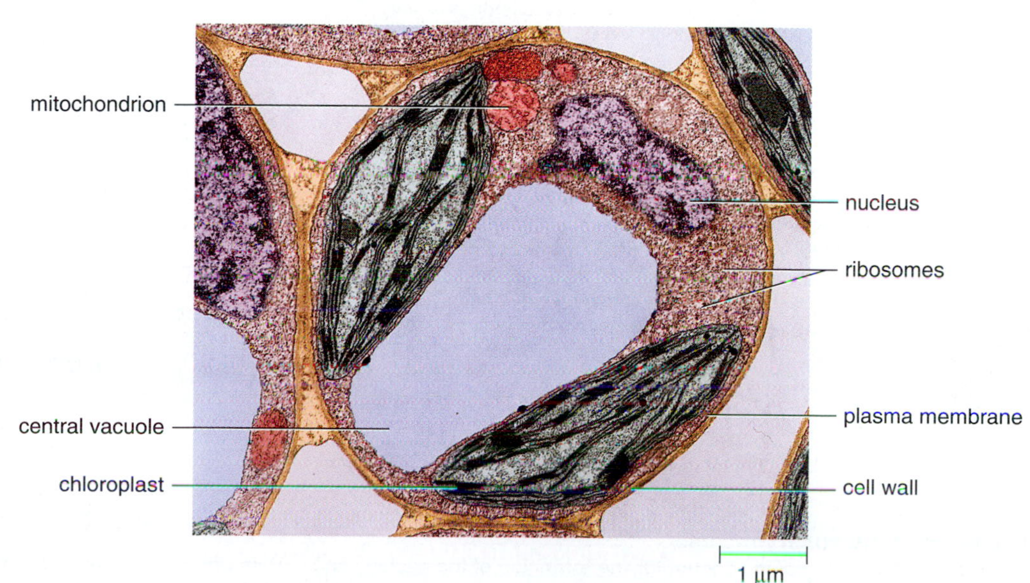

b.

The Evolution of the Eukaryotic Cell

How did the eukaryotic cell arise? Invagination of the plasma membrane might explain the origination of the nucleus and certain other organelles, such as the endoplasmic reticulum and the Golgi apparatus. Some believe that the rest of the organelles could also have arisen in this manner.

Another hypothesis has gained recognition and acceptance by many. In the laboratory, it has been observed that an amoeba infected with bacteria can become dependent upon them so that the bacteria remain within the cell. Some investigators, especially Lynn Margulis, believe that mitochondria and chloroplasts are derived from bacteria that were taken up by a much larger cell that already had a nucleus (Fig. 4.6). Perhaps mitochondria were originally aerobic bacteria that take in oxygen and chloroplasts were originally cyanobacteria that photosynthesize. The host cell would have benefited from an ability to utilize oxygen and/or make its own organic food if by chance the bacteria were not destroyed. In other words, after these prokaryotes entered by *endocytosis,* a *symbiotic* relationship was established. Some of the evidence for the **endosymbiotic hypothesis** is as follows:

1. Mitochondria and chloroplasts are similar to bacteria in size and in structure.
2. Both organelles are bounded by a double membrane—the outer membrane may be derived from the engulfing vesicle, and the inner one may be derived from the plasma membrane of the original prokaryote.
3. Mitochondria and chloroplasts contain a limited amount of genetic material and divide by splitting. Their DNA is a circular loop like that of bacteria.
4. Although most of the proteins within mitochondria and chloroplasts are now produced by the eukaryotic host, they do have their own ribosomes and they do produce some proteins. Their ribosomes resemble those of bacteria.
5. The RNA (ribonucleic acid) composition of their ribosomes suggests a eubacterial origin for chloroplasts and mitochondria.

Margulis even suggests that the flagella of eukaryotes are derived from a spirochete prokaryote that became attached to a host cell (Fig. 4.6). However, it is important to remember that the flagella of eukaryotes but not prokaryotes are composed of structures called microtubules. In any case, the acquisition of microtubules by eukaryotes would have led to the ability to form a spindle which separates the chromosomes during cell division. The type of cell division called meiosis is associated with sexual reproduction of eukaryotes and sexual reproduction contributes to variation among members of a population and the evolution of new species.

According to the endosymbiotic hypothesis, aerobic bacteria became mitochondria and cyanobacteria became chloroplasts after being taken up by precursors to modern-day eukaryotic cells.

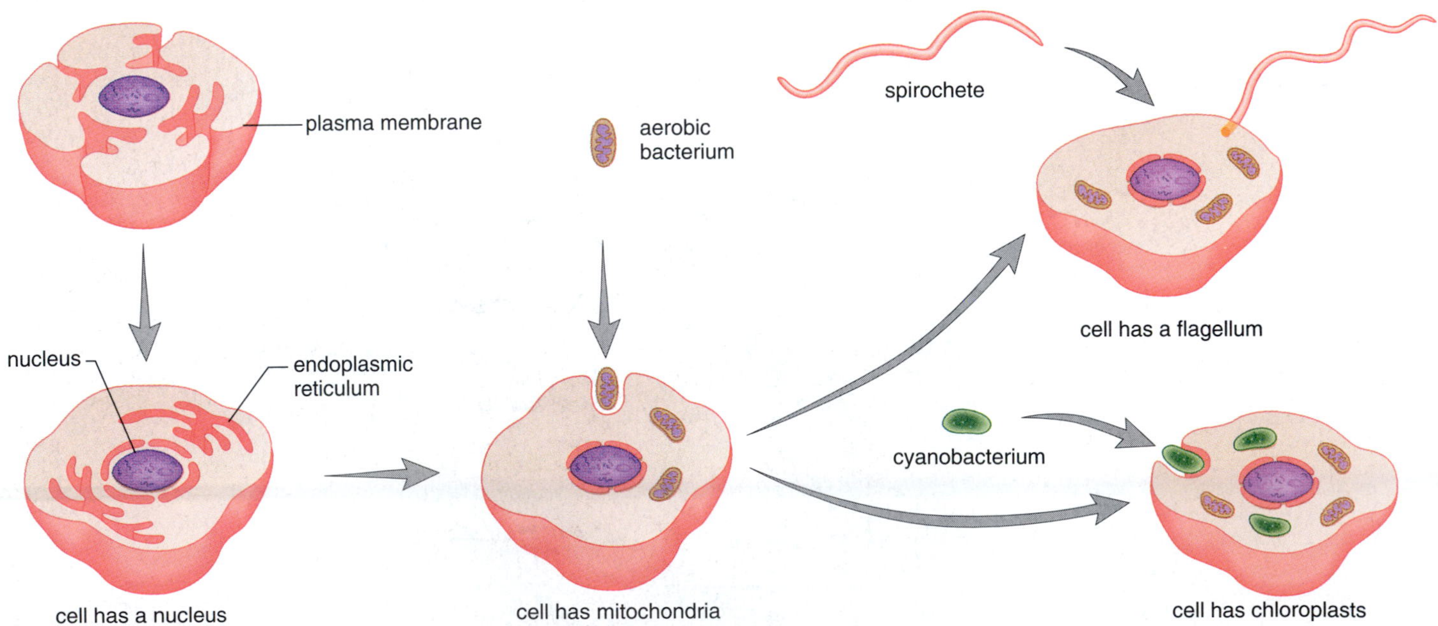

Figure 4.6 Evolution of the eukaryotic cell.
Invagination of the plasma membrane could account for the formation of the nucleus and certain other organelles. The endosymbiotic hypothesis suggests that mitochondria, chloroplasts, and flagella are derived from prokaryotes that were taken up by a much larger eukaryotic cell.

The Nucleus

The **nucleus** [L. *nucleus,* kernel], which has a diameter of about 5 µm, is a prominent structure in the eukaryotic cell (Fig. 4.7). The nucleus contains **chromatin** [Gk. *chroma,* color, and *teino,* stretch] in a semifluid medium called the **nucleoplasm.** Chromatin looks grainy, but actually it is a network of strands that undergoes coiling into rodlike structures called **chromosomes** [Gk. *chroma,* color, and *soma,* body], just before the cell divides. Chemical analysis shows that chromatin, and therefore chromosomes, contains DNA and much protein, and some RNA (ribonucleic acid).

All forms of RNA, of which there are several, are produced in the nucleus. Some dark regions of chromatin are nucleoli (sing., **nucleolus**) where a type of RNA, called ribosomal RNA (rRNA), is produced and where rRNA joins with proteins to form the subunits of ribosomes. (Ribosomes are small bodies in the cytoplasm where protein synthesis occurs.)

The nucleus is separated from the cytoplasm by a double membrane known as the **nuclear envelope.** The nuclear envelope has **nuclear pores** of sufficient size (100 nm) to permit the passage of proteins into the nucleus and ribosomal subunits out of the nucleus. High-power electron micrographs show that the pores have nonmembranous components associated with them that form a nuclear pore complex.

The nucleus is of primary importance because it contains DNA, the molecule which stores genetic information and determines the characteristics of a cell and its metabolic functioning. Activated DNA, with the help of messenger RNA (mRNA) acting as an intermediary, specifies the sequencing of amino acids during protein synthesis. The proteins of a cell determine its structure and the functions it can perform.

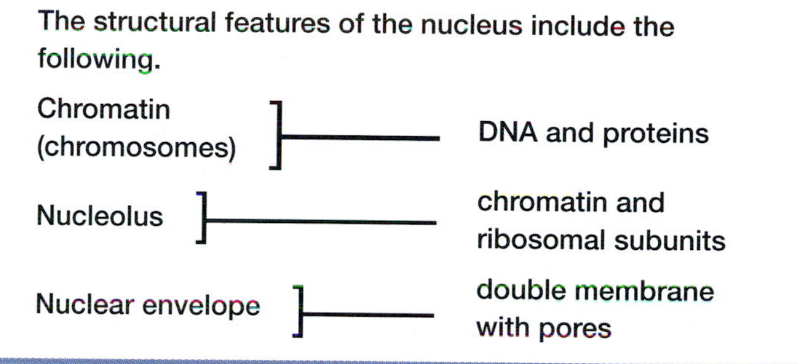
The structural features of the nucleus include the following.

Feature	Composition
Chromatin (chromosomes)	DNA and proteins
Nucleolus	chromatin and ribosomal subunits
Nuclear envelope	double membrane with pores

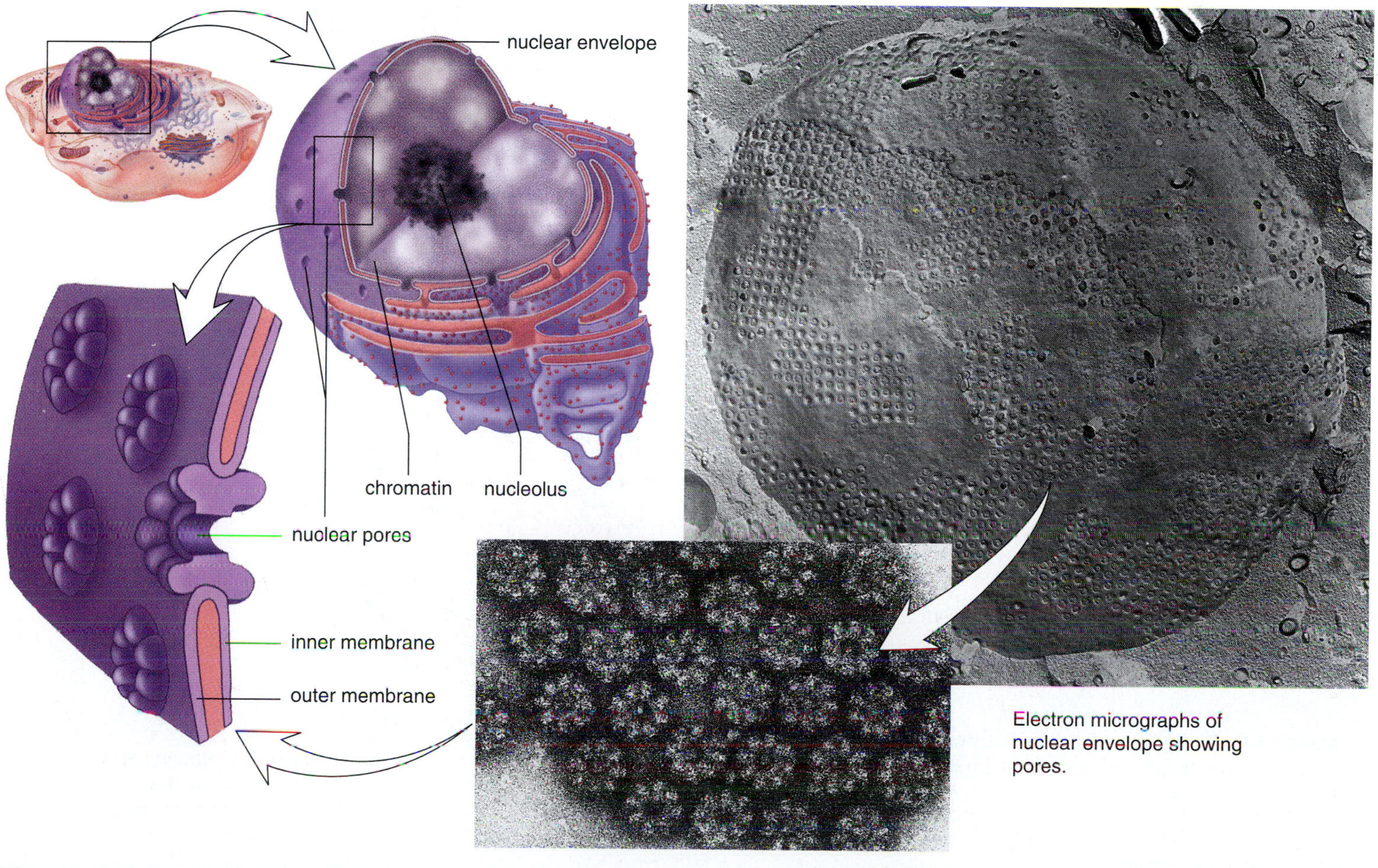

Electron micrographs of nuclear envelope showing pores.

Figure 4.7 Anatomy of the nucleus.
The nucleus contains chromatin. Chromatin has a special region called the nucleolus, which is where rRNA is produced and ribosomal subunits are assembled. The nuclear envelope contains pores, as is shown in this micrograph of a freeze-fractured nuclear envelope. Each pore is lined by a complex of eight proteins.

Ribosomes

Unlike many of the organelles discussed in this chapter, **ribosomes** are found in both prokaryotes and eukaryotes. In eukaryotes, ribosomes are 20 nm by 30 nm, and in prokaryotes they are slightly smaller. In both types of cells, ribosomes are composed of two subunits, one large and one small (Fig. 4.8*c*). Each subunit has its own mix of proteins and rRNA.

In eukaryotic cells some ribosomes occur free within the cytosol either singly or in groups, called **polyribosomes,** and others are attached to the endoplasmic reticulum (ER), a membranous system of saccules and channels discussed in the next section.

Ribosomes, as mentioned, are sites of protein synthesis. They receive mRNA from the nucleus and this nucleic acid carries a coded message from DNA indicating the correct sequence of amino acids in a protein. Proteins synthesized by cytoplasmic ribosomes are used in the cytosol and those synthesized by attached ribosomes end up in a cellular membrane, stored in a **vesicle** (tiny membranous sacs), or secreted out of the cell.

What causes a ribosome to bind to endoplasmic reticulum? Binding occurs only if the protein being synthesized by a ribosome has an ER signal sequence. An ER signal sequence enables a ribosome to bind to a receptor protein on the endoplasmic reticulum (Fig. 4.8*d*). Receptor proteins are sometimes called "docking proteins" because they form a docking site for a particular molecule.

Ribosomes are small organelles where protein synthesis occurs. Ribosomes occur in the cytosol, both singly and in groups (i.e., polyribosomes). Numerous ribosomes are attached to the endoplasmic reticulum.

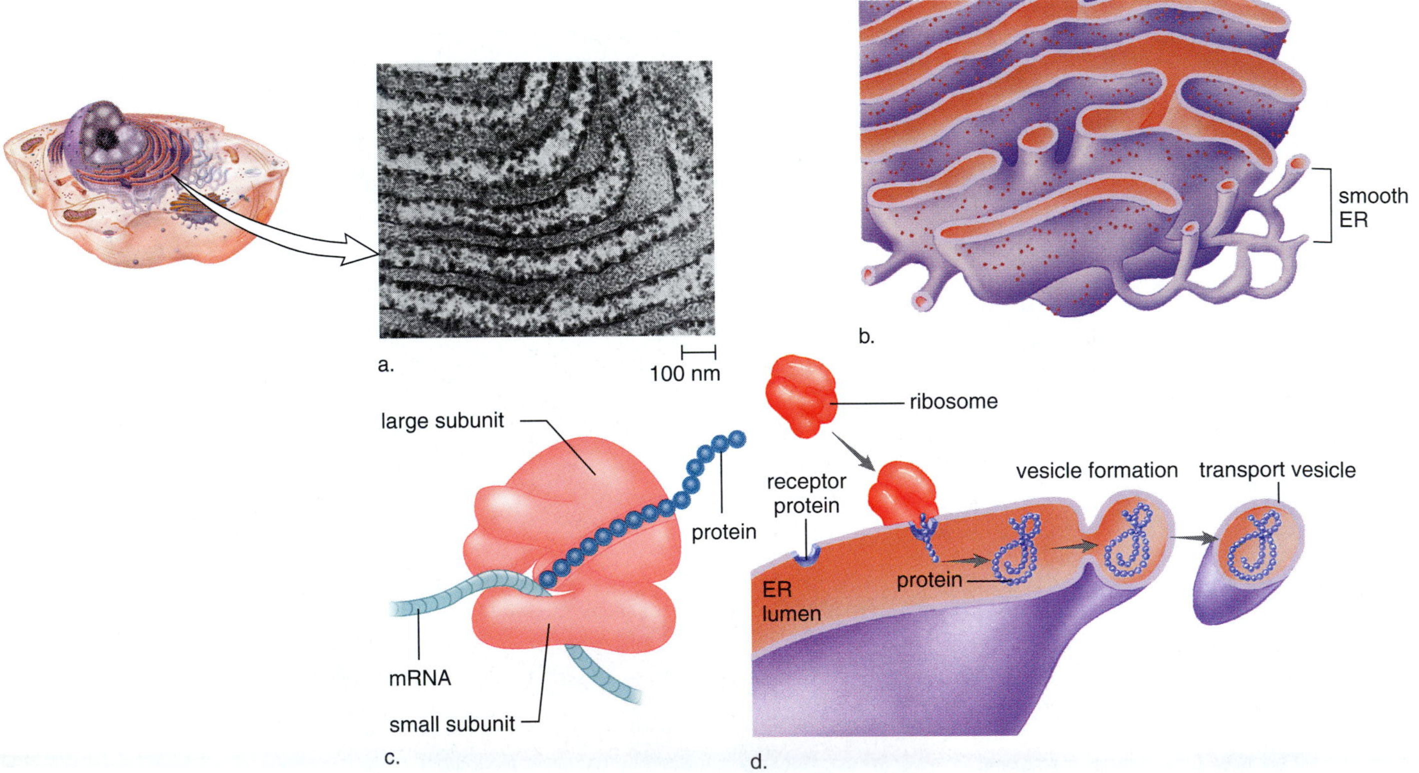

Figure 4.8 Rough endoplasmic reticulum (ER).
a. In this mouse liver cell micrograph, rough ER consists of flattened vesicles studded on the outer surface with ribosomes. **b.** Rough ER is continuous with smooth ER as illustrated. **c.** A ribosome carries out protein synthesis. The messenger RNA (mRNA) molecule, situated between the large subunit and small subunit, contains coded information which specifies the sequence of amino acids in a protein. **d.** When a ribosome attaches to the endoplasmic reticulum the growing protein enters the lumen of the ER. Transport vesicles, which bud from smooth ER, carry protein molecules to other locations in the cell.

The Endomembrane System

The endomembrane system consists of the nuclear envelope, the endoplasmic reticulum, the Golgi apparatus, and several vesicles. This system compartmentalizes the cell so that particular enzymatic reactions are restricted to specific regions. Membranes that make up the endomembrane system are connected by direct physical contact and/or by the transfer of vesicles from one part to the other.

Endoplasmic Reticulum

The **endoplasmic reticulum (ER)** [Gk. *endon,* within, *plasma,* something molded, and L. *reticulum,* net], consisting of a complicated system of membranous channels and saccules (flattened vesicles), is physically continuous with the outer membrane of the nuclear envelope. **Rough ER** is studded with ribosomes on the side of the membrane that faces the cytoplasm; therefore it is correct to say that rough ER synthesizes proteins. It also modifies proteins after they have entered the ER lumen (Fig. 4.8). Certain ER enzymes add carbohydrate (sugar) chains to proteins and then they are called glycoproteins. Others assist the folding process that results in higher levels of protein organization.

Smooth ER, which is continuous with rough ER, does not have attached ribosomes. Smooth ER synthesizes phospholipids, steroids, and fatty acids. It also has various other functions depending on the particular cell. In the testes, it produces testosterone and, in the liver, it helps detoxify drugs. Regardless of any specialized function, smooth ER also forms vesicles in which large molecules are transported to other parts of the cell. Often these vesicles are on their way to the plasma membrane or the Golgi apparatus.

Rough ER is involved in protein synthesis and modification. Smooth ER is generally a site for lipid metabolism. Molecules that are produced or modified in the ER are eventually enclosed in transport vesicles.

Research Methods Modern microscopic techniques can be counted on to reveal the structure of organelles like rough and smooth ER. But researchers have to turn to biochemical analysis in order to discover the function of organelles. Figure 4.9 shows how organelles can be separated out of the cell by a process called cell fractionation. Cell fractionation allows researchers to concentrate on finding the function of particular parts of cell. Microscopy and biochemical analysis has allowed researchers to distinguish the functions of all the different components of the endomembrane system.

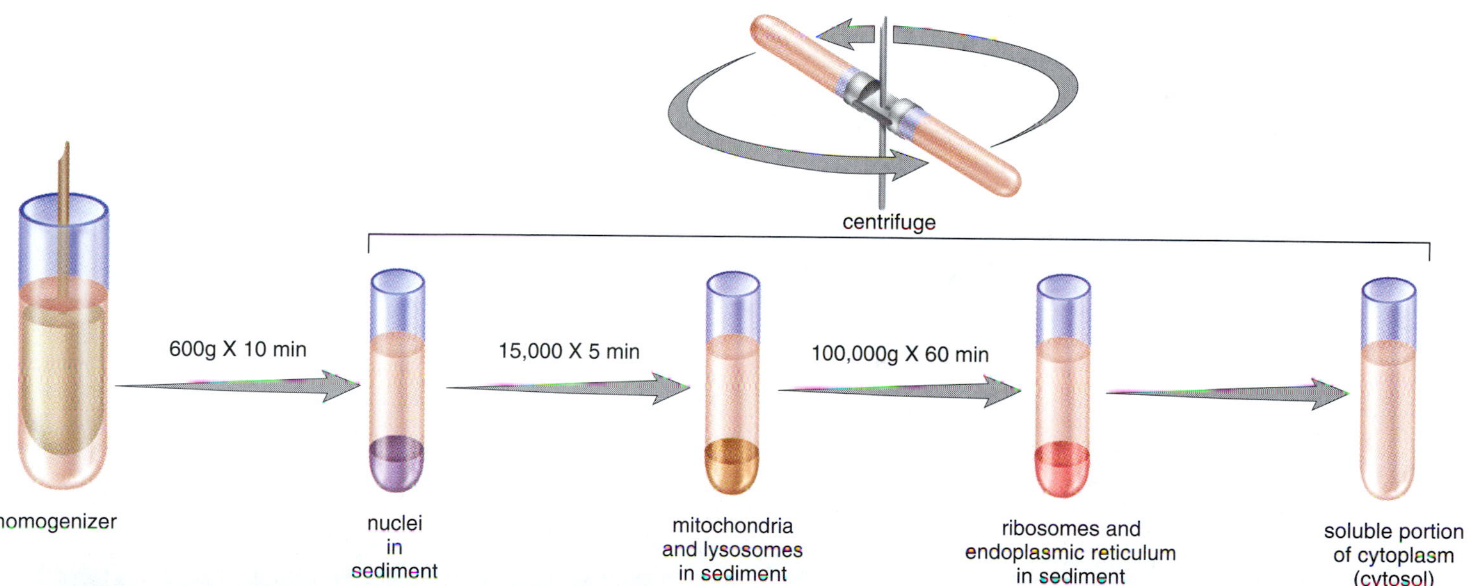

Figure 4.9 Cell fractionation.
Cells are broken open mechanically by the action of the pestle against the side of a centrifuge tube. Then, a centrifuge spins the tubes and this action separates out the contents of the cell. The notations above the arrows indicate the speed and length of time necessary to separate out the structures listed. With ever increasing speed, first the larger and then smaller components of the cell separate out. Once a purified sample of an organelle has been obtained, its proteins can be identified. Also, under the proper conditions, an organelle will continue to work in isolation so that its functions can be determined.

The Golgi Apparatus

The **Golgi apparatus** is named for Camillo Golgi, who discovered its presence in cells in 1898. The Golgi apparatus consists of a stack of three to twenty slightly curved saccules whose appearance can be compared to a stack of pancakes (Fig. 4.10). In animal cells, one side of the stack (the inner face) is directed toward the ER, and the other side of the stack (the outer face) is directed toward the plasma membrane. Vesicles can frequently be seen at the edges of the saccules.

Protein or lipid-filled vesicles that bud from the smooth ER are received by the Golgi apparatus at its inner face. Thereafter, the apparatus carries out its functions as these substances move through its saccules. The Golgi apparatus contains enzymes that alter the carbohydrate chains first attached to proteins in the rough ER. For example, one monomer could be substituted for another or a phosphate might be added to the carbohydrate chain. In some cases the modified carbohydrate chain serves as a signal molecule that determines the protein's final destination in the cell.

The Golgi apparatus packages its products in vesicles that depart the Golgi apparatus at the outer face. Some of these are vesicles called lysosomes, which are discussed next. Some vesicles are on their way to other organelles that have docking sites for them. In that case a glycoprotein with a signaling sequence is a part of the vesicle's membrane. Some vesicles proceed to the plasma membrane, where they discharge their contents as secretion occurs. Secretion is termed exocytosis because the substance exits the cytosol.

The Golgi apparatus receives transport vesicles from the smooth ER. It then modifies, sorts, and packages proteins for use in the cell or secretion from the cell.

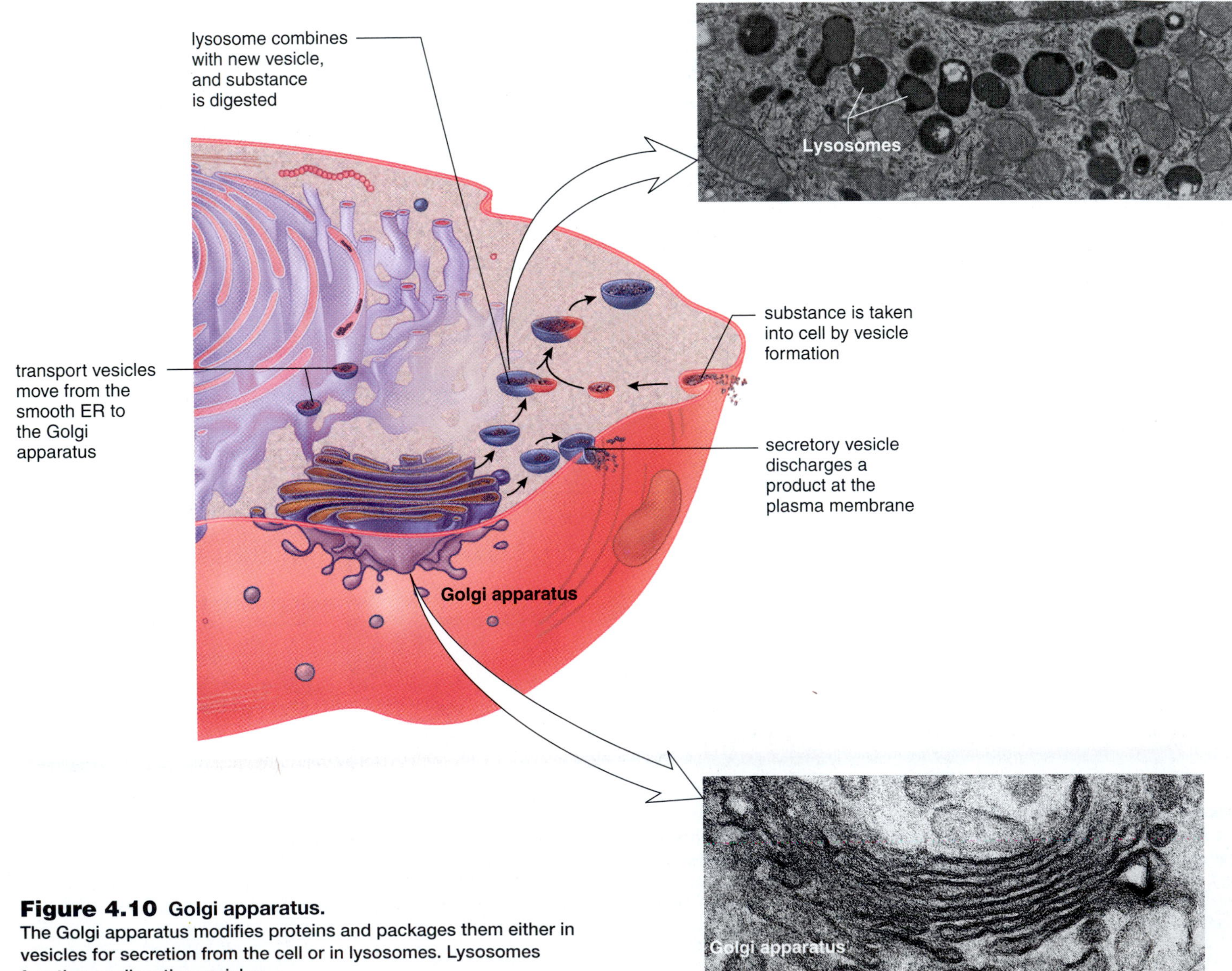

Figure 4.10 **Golgi apparatus.**
The Golgi apparatus modifies proteins and packages them either in vesicles for secretion from the cell or in lysosomes. Lysosomes function as digestive vesicles.

Lysosomes

Lysosomes [Gk. *lyo,* loose, and *soma,* body] are membrane-bounded vesicles produced by the Golgi apparatus. They have a very low pH and contain powerful hydrolytic digestive enzymes.

Sometimes macromolecules are brought into a cell by vesicle formation at the plasma membrane (Fig. 4.10). When a lysosome fuses with such a vesicle, its contents are digested by lysosomal enzymes into simpler subunits that then enter the cytoplasm. Some white blood cells defend the body by engulfing bacteria that are then enclosed within vesicles. When lysosomes fuse with these vesicles, the bacteria are digested. It should come as no surprise, then, that even parts of a cell are digested by its own lysosomes (called autodigestion). Normal cell rejuvenation most likely takes place in this manner, but autodigestion is also important during development. Lysosomes participate in programmed cell death technically termed **apoptosis** which is a normal part of development. When a tadpole becomes a frog, lysosomes digest away the cells of the tail. The fingers of a human embryo are at first webbed, but they are freed from one another as a result of lysosomal action.

Occasionally, a child is born with a metabolic disorder involving a missing or inactive lysosomal enzyme. In these cases, the lysosomes fill to capacity with macromolecules that cannot be broken down. The cells become so full of these lysosomes that the child dies. Someday soon it may be possible to provide the missing enzyme for these children.

Lysosomes are produced by the Golgi apparatus, and their hydrolytic enzymes digest macromolecules from various sources.

Peroxisomes are membrane-bounded vesicles that contain specific enzymes imported from the cytosol (Fig. 4.11). Peroxisomes have enzymes for oxidizing small organic molecules with the formation of hydrogen peroxide (H_2O_2):

$$RH_2 + O_2 \longrightarrow R + H_2O_2$$

Hydrogen peroxide, a toxic molecule, is immediately broken down to water and oxygen by another peroxisomal enzyme called catalase. Peroxisomes are abundant in cells that metabolize lipids; in liver and yeast cells they detoxify alcohol.

Peroxisomes have additional roles in plants. In germinating seeds, peroxisomes, called glyoxysomes, oxidize fatty acids into molecules that can be converted to sugars needed by the growing plant. Peroxisomes also carry out a reaction in leaves that uses up oxygen and releases carbon dioxide that can be used for photosynthesis.

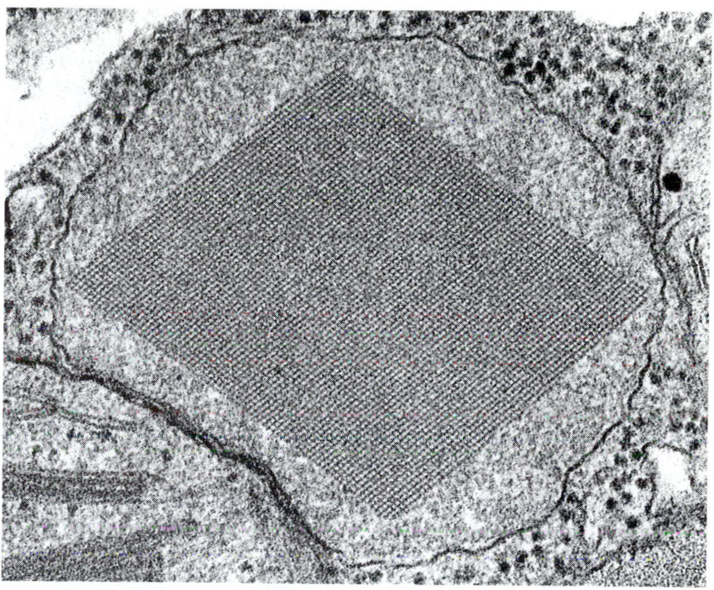

Figure 4.11 Peroxisome in a tobacco leaf.
Peroxisomes are vesicles that oxidize organic substances with a resulting build-up of hydrogen peroxide. The crystalline, squarelike core of a peroxisome contains the enzyme catalase, which breaks down hydrogen peroxide (H_2O_2) to water and oxygen.

Vacuoles

A **vacuole** is a large membranous sac. A vesicle is smaller than a vacuole. Typically, plant cells have a large central vacuole so filled with a watery fluid that it gives added support to the cell (see Fig. 4.5).

Vacuoles store substances. Plant vacuoles contain not only water, sugars, and salts but also pigments and toxic molecules. The pigments are responsible for many of the red, blue, or purple colors of flowers and some leaves. The toxic substances help protect a plant from herbivorous animals. The vacuoles present in some protists are quite specialized, and they include contractile vacuoles for ridding the cell of excess water and digestive vacuoles for breaking down nutrients.

The organelles of the endomembrane system are as follows.

Endoplasmic reticulum (ER): synthesis and modification and transport of proteins and other substances
- Rough ER: protein synthesis
- Smooth ER: lipid synthesis, in particular

Golgi apparatus: modifies, sorts, and packages protein molecules

Lysosomes: intracellular digestion

Peroxisomes: metabolize small organic molecules

Vacuoles: storage areas

Energy-Related Organelles

Life is possible only because of a constant input of energy used to maintain the structure of cells. Chloroplasts and mitochondria are the two eukaryotic membranous organelles that specialize in converting energy to a form that can be used by the cell. **Chloroplasts** use solar energy to synthesize carbohydrates, and carbohydrate-derived products are broken down in mitochondria (sing., **mitochondrion**) to produce ATP molecules.

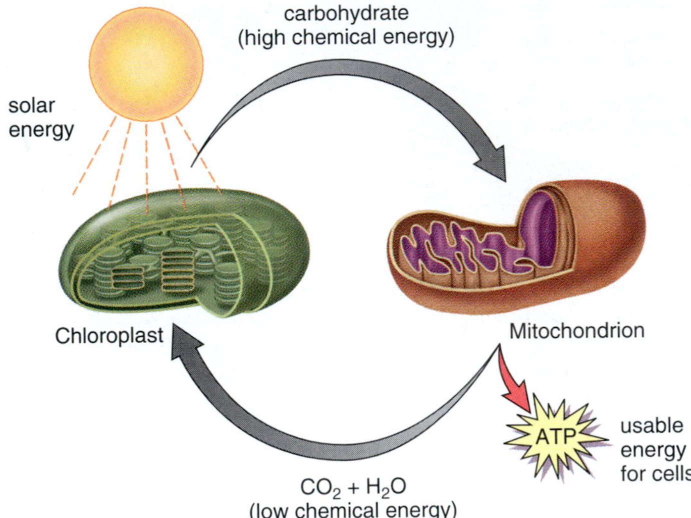

Photosynthesis, which occurs in chloroplasts [Gk. *chloros*, green, and *plastos*, formed, molded], is the process by which solar energy is converted to chemical energy within carbohydrates. Photosynthesis can be represented by this equation:

solar energy + carbon dioxide + water ⟶ carbohydrate + oxygen

Here the word *energy* stands for solar energy, the ultimate source of energy for cellular organization. Only plants, algae, and certain bacteria are capable of carrying on photosynthesis in this manner.

Cellular respiration is the process by which the chemical energy of carbohydrates is converted to that of ATP (adenosine triphosphate). Cellular respiration can be represented by this equation:

carbohydrate + oxygen ⟶ carbon dioxide + water + energy

Here the word *energy* stands for ATP molecules. When a cell needs energy, ATP supplies it. The energy of ATP is used for synthetic reactions, active transport, and all energy-requiring processes in cells. In eukaryotes mitochondria are necessary to the process of cellular respiration, which produces ATP.

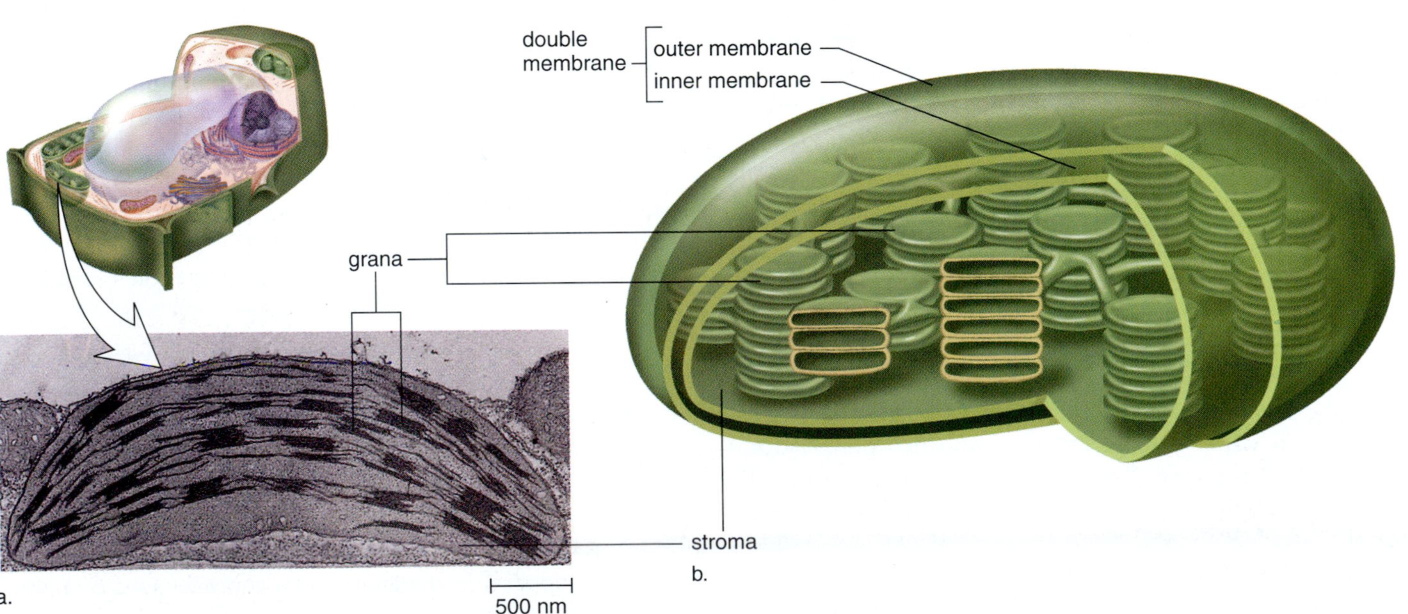

Figure 4.12 Chloroplast structure.
Chloroplasts carry out photosynthesis. **a.** Electron micrograph. **b.** Generalized drawing in which the outer and inner membranes have been cut away to reveal the grana.

Chloroplasts

The photosynthetic cells of algae and plants contain chloroplasts. Chloroplasts are about 4–6 μm in diameter and 1–5 μm in length; they belong to a group of organelles known as plastids. Among the plastids are also the amyloplasts, which store starch, and the chromoplasts, which contain red and orange pigments.

A chloroplast is bounded by two membranes that enclose a fluid called the **stroma.** The stroma contains enzymes that synthesize carbohydrates. It also contains DNA and ribosomes. Chloroplasts are able to synthesize some of the proteins necessary to their function; others are imported into the organelle from the cytosol.

A membrane system within the stroma is organized into interconnected flattened sacs called **thylakoids** [Gk. *thylakos*, sack, and *eides*, like, resembling]. In certain regions, the thylakoids are stacked up in structures called grana (sing., **granum**). There can be hundreds of grana within a single chloroplast (Fig. 4.12). Chlorophyll, a green pigment that is located within the thylakoid membranes of grana, captures solar energy. In chapter 7, you will learn how the organization of a chloroplast allows it to not only capture solar energy but also produce carbohydrates.

Mitochondria

Most eukaryotic cells, whether protists, fungi, plant or animal cells, contain mitochondria. This means that algal and plant cells contain both chloroplasts and mitochondria. The shape and size of mitochondria is variable according to the species but they are usually 0.5–1.0 μm in diameter and 2–5 μm in length.

Mitochondria, like chloroplasts, are bounded by two membranes. The inner membrane invaginates to form **cristae.** Cristae provide a much greater surface area to accommodate the protein complexes and other participants in cellular respiration. The cristae project into the **matrix,** an inner space filled with semifluid medium that contains enzymes. These enzymes break down carbohydrate products, releasing energy that is used for ATP production on the cristae (Fig. 4.13). The matrix also contains DNA and ribosomes. Mitochondria are able to synthesize some of the proteins necessary to their function; others are imported into the organelle from the cytosol.

Chloroplasts and mitochondria are membranous organelles whose structure lends itself to the processes that occur within them.

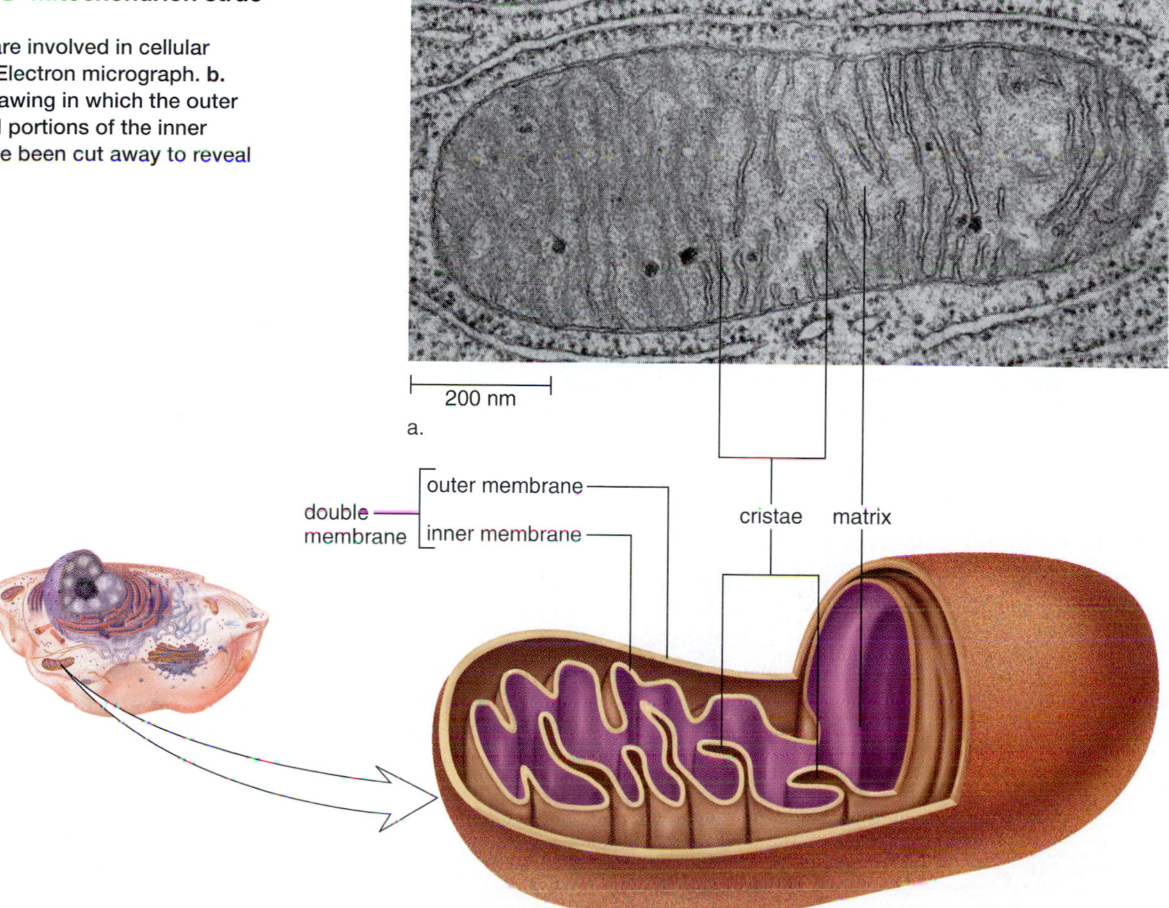

Figure 4.13 Mitochondrion structure.
Mitochondria are involved in cellular respiration. **a.** Electron micrograph. **b.** Generalized drawing in which the outer membrane and portions of the inner membrane have been cut away to reveal

The Cytoskeleton

The **cytoskeleton** [Gk. *kytos*, cell, and *skeleton*, dried body] is a network of interconnected filaments and tubules that extends from the nucleus to the plasma membrane in eukaryotic cells. Prior to the 1970s, it was believed that the cytosol was an unorganized mixture of organic molecules. Then, high-voltage electron microscopes, which can penetrate thicker specimens, showed that the cytosol was instead highly organized. The technique of immunofluorescence microscopy identified the makeup of specific protein fibers within the cytoskeletal network (Fig. 4.14).

The name *cytoskeleton* is convenient in that it allows us to compare the cytoskeleton to the bones and muscles of an animal. Bones and muscles give an animal structure and produce movement. Similarly, we will see that the elements of the cytoskeleton maintain cell shape and cause the cell and its organelles to move. The cytoskeleton is dynamic especially because elements can undergo rapid assembly and disassembly. Such changes occur at rates that are measured in seconds and minutes. The entire cytoskeletal network can even disappear and reappear at various times in the life of a cell. Before a cell divides, for instance, the elements disassemble and then reassemble into a structure called a spindle that distributes chromosomes in an orderly manner. At the end of cell division, the spindle disassembles and the elements reassemble once again into their former array.

The cytoskeleton contains three types of elements: actin filaments, intermediate filaments, and microtubules, which are responsible for cell shape and movement.

Actin Filaments

Actin filaments (formerly called microfilaments) are long, extremely thin fibers (about 7 nm in diameter) that occur in bundles or meshlike networks. The actin filament contains two chains of globular actin monomers twisted about one another in a helical manner.

Actin filaments play a structural role when they form a dense complex web just under the plasma membrane, to which they are anchored by special proteins. They are also seen in the microvilli that project from intestinal cells, and their presence most likely accounts for the ability of microvilli to alternately shorten and extend into the intestine. In plant cells, actin filaments apparently form the tracks along which chloroplasts circulate or stream in a particular direction. Also, the presence of a network of actin filaments lying beneath the plasma membrane accounts for the formation of pseudopods, extensions that allow certain cells to move in an amoeboid fashion.

How are actin filaments involved in the movement of the cell and its organelles? They interact with **motor molecules,** which are proteins that can attach, detach, and reattach farther along an actin filament. In the presence of ATP, myosin pulls actin filaments along in this way. Myosin has both a head and a tail. In muscle cells, the tails of several muscle myosin molecules are joined to form a thick filament. In nonmuscle cells, cytoplasmic myosin tails are bound to membranes but the heads still interact with actin:

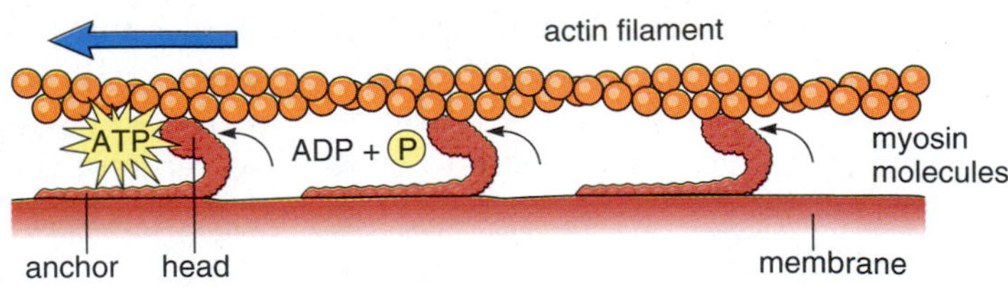

During animal cell division, the two new cells form when actin, in conjunction with myosin, pinches off the cells from one another.

Intermediate Filaments

Intermediate filaments (8–11 nm in diameter) are intermediate in size between actin filaments and microtubules. They are a ropelike assembly of fibrous polypeptides, but the specific type varies according to the tissue. Some intermediate filaments support the nuclear envelope, whereas others support the plasma membrane and take part in the formation of cell-to-cell junctions. In the skin, the filaments, which are made of the protein keratin, give great mechanical strength to skin cells. Recent work has shown intermediate filaments to be highly dynamic. They also assemble and disassemble but need to have phosphate added first by soluble enzymes.

Microtubules

Microtubules [Gk. *mikros*, small, little, and L. *tubus*, pipe] are small hollow cylinders about 25 nm in diameter and from 0.2–25 μm in length.

Microtubules are made of a globular protein called tubulin. There is a slightly different amino acid sequence in α tubulin compared to β tubulin. When assembly occurs, α and β tubulin molecules come together as dimers and the dimers arrange themselves in rows. Microtubules have 13 rows of tubulin dimers surrounding what appears in electron micrographs to be an empty central core.

The regulation of microtubule assembly is under the control of a microtubule organizing center (MTOC). In most eukaryotic cells the main MTOC is in a structure called the **centrosome** [Gk. *centrum*, center, and *soma*, body], which lies near the nucleus. Microtubules radiate from the MTOC, helping to maintain the shape of the cell and acting as tracks along which organelles can move. Whereas the motor

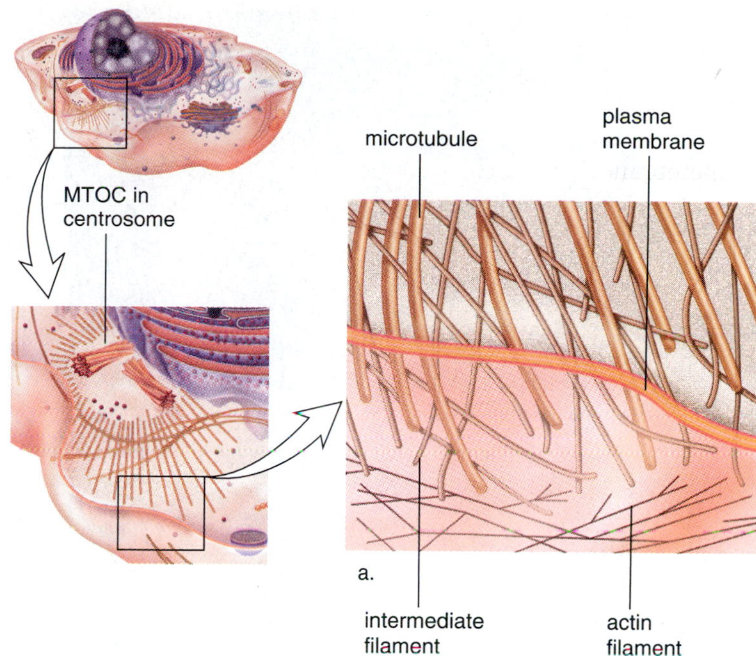

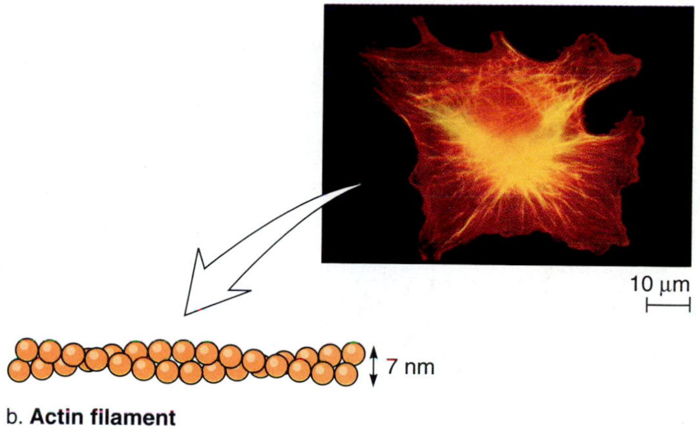

b. **Actin filament**

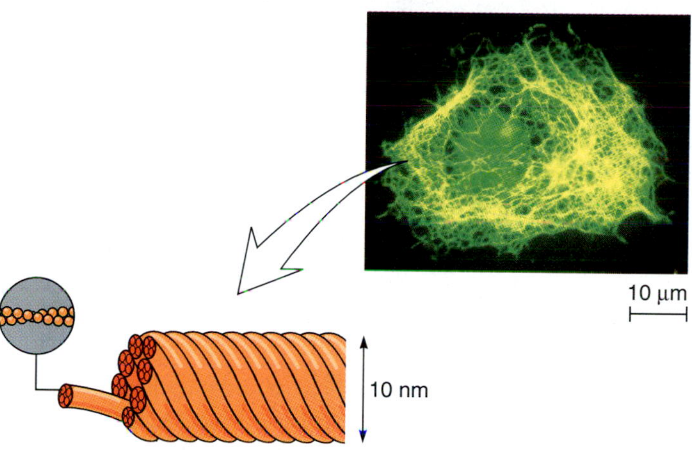

c. **Intermediate filament**

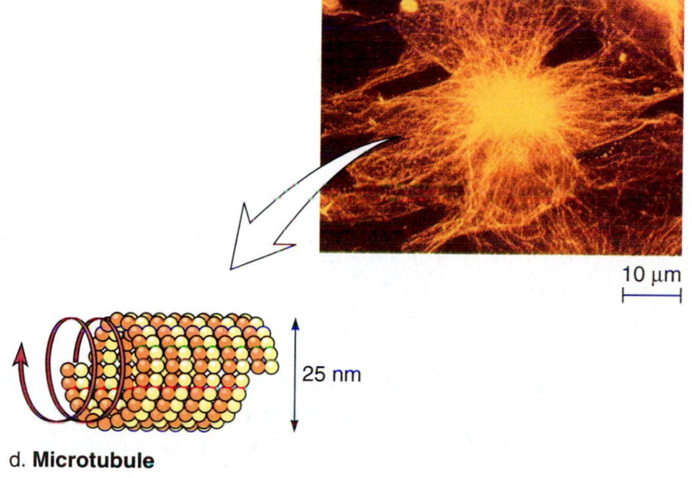

d. **Microtubule**

Figure 4.14 The cytoskeleton.
a. Diagram comparing the size relationship of actin filaments, intermediate filaments, and microtubules. Immunofluorescence, a technique based on the binding of fluorescent antibodies to specific proteins, is used to detect the location of (**b**) actin filaments, (**c**) intermediate filaments, and (**d**) microtubules in the cell.

molecule myosin is associated with actin filaments, the motor molecules kinesin and dynein are associated with microtubules:

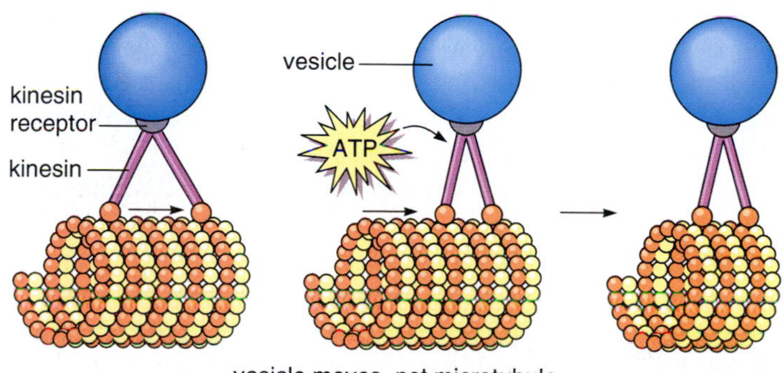

There are different types of kinesin proteins, each specialized to move one kind of vesicle or cellular organelle. Kinesin moves vesicles or organelles in an opposite direction from dynein. Cytoplasmic dynein is closely related to the molecule dynein found in flagella.

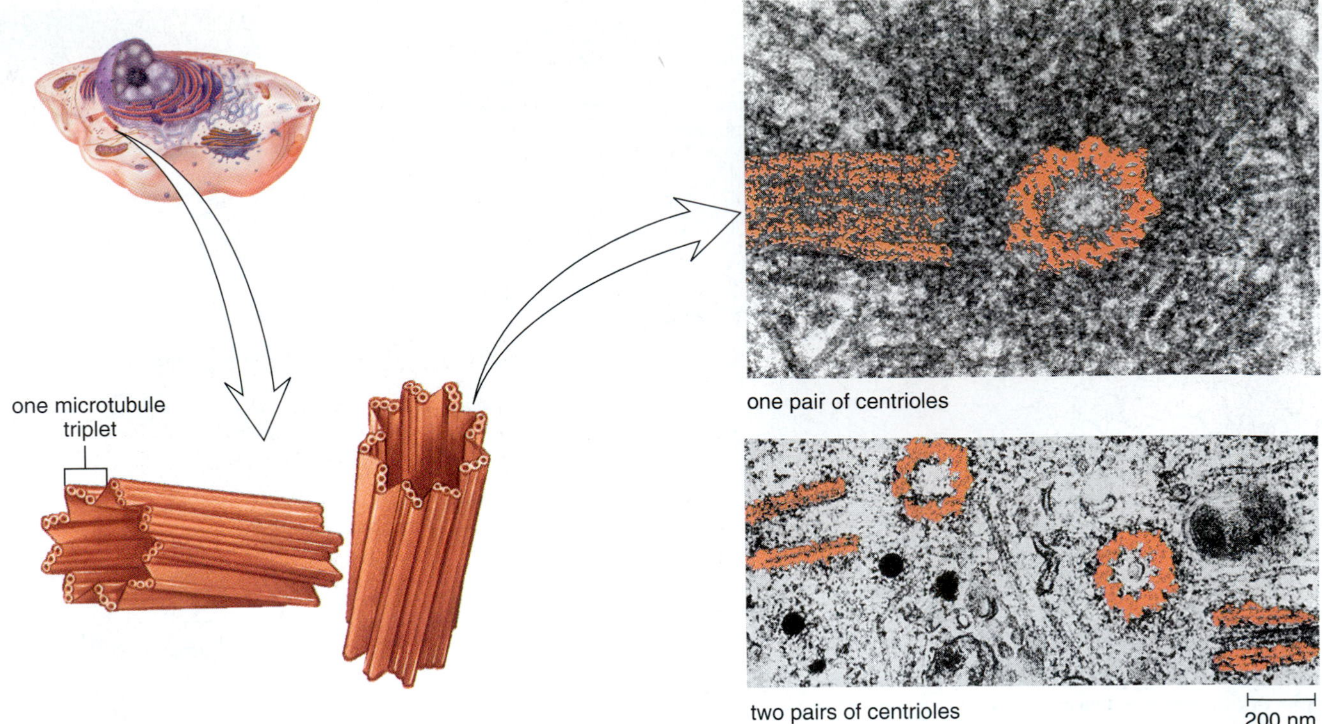

Figure 4.15 Centrioles.
In a nondividing cell there is a pair of centrioles in a centrosome outside the nucleus. Just before a cell divides, the centrioles replicate and there are two pairs of centrioles. During cell division, centrioles in two centrosomes separate so that each new cell has one centrosome containing one pair of centrioles.

Centrioles

Centrioles [Gk. *centrum,* center] are short cylinders with a 9 + 0 pattern of microtubule triplets—that is, a ring having nine sets of triplets with none in the middle. In animal cells and most protists, a centrosome contains two centrioles lying at right angles to each other. A centrosome, you will recall, is the major microtubule organizing center for the cell. Therefore, it is possible that centrioles are also involved in the process by which microtubules assemble and disassemble.

Before an animal cell divides, the centrioles replicate and the members of each pair are at right angles to one another (Fig. 4.15). Then each pair becomes part of a separate centrosome. During cell division the centrosomes move apart and most likely function to organize the mitotic spindle. In any case, each new cell has its own centrosome and pair of centrioles. Plant and fungal cells have the equivalent of a centrosome but it does not contain centrioles, suggesting that centrioles are not necessary to the assembly of cytoplasmic microtubules.

In cells with cilia and flagella, centrioles are believed to give rise to basal bodies that direct the organization of microtubules within these structures. In other words, a basal body may do for a cilium (or flagellum) what the centrosome does for the cell.

Centrioles, which are short cylinders with a 9 + 0 pattern of microtubule triplets, may give rise to the basal bodies of cilia and flagella.

Cilia and Flagella

Cilia [L. *cilium,* eyelash, hair] and **flagella** [L. *flagello,* whip] are hairlike projections that can move either in an undulating fashion, like a whip, or stiffly, like an oar. Cells that have these organelles are capable of movement. For example, unicellular paramecia move by means of cilia, whereas sperm cells move by means of flagella. The cells that line our upper respiratory tract have cilia that sweep debris trapped within mucus back up into the throat, where it can be swallowed. This action helps keep the lungs clean.

In eukaryotic cells, cilia are much shorter than flagella, but they have a similar construction. Both are membrane-bounded cylinders enclosing a matrix area. In the matrix are nine microtubule doublets arranged in a circle around two central microtubules. This is called the 9 + 2 pattern of microtubules. Cilia and flagella move when the microtubule doublets slide past one another (Fig. 4.16).

As mentioned, each cilium and flagellum has a basal body lying in the cytoplasm at its base. Basal bodies have the same circular arrangement of microtubule triplets as centrioles and are believed to be derived from them. It is possible that basal bodies organize the microtubules within cilia and flagella, but this is not supported by the observation that cilia and flagella grow by the addition of tubulin dimers to their tips.

Cilia and flagella, which have a 9 + 2 pattern of microtubules, are involved in the movement of cells.

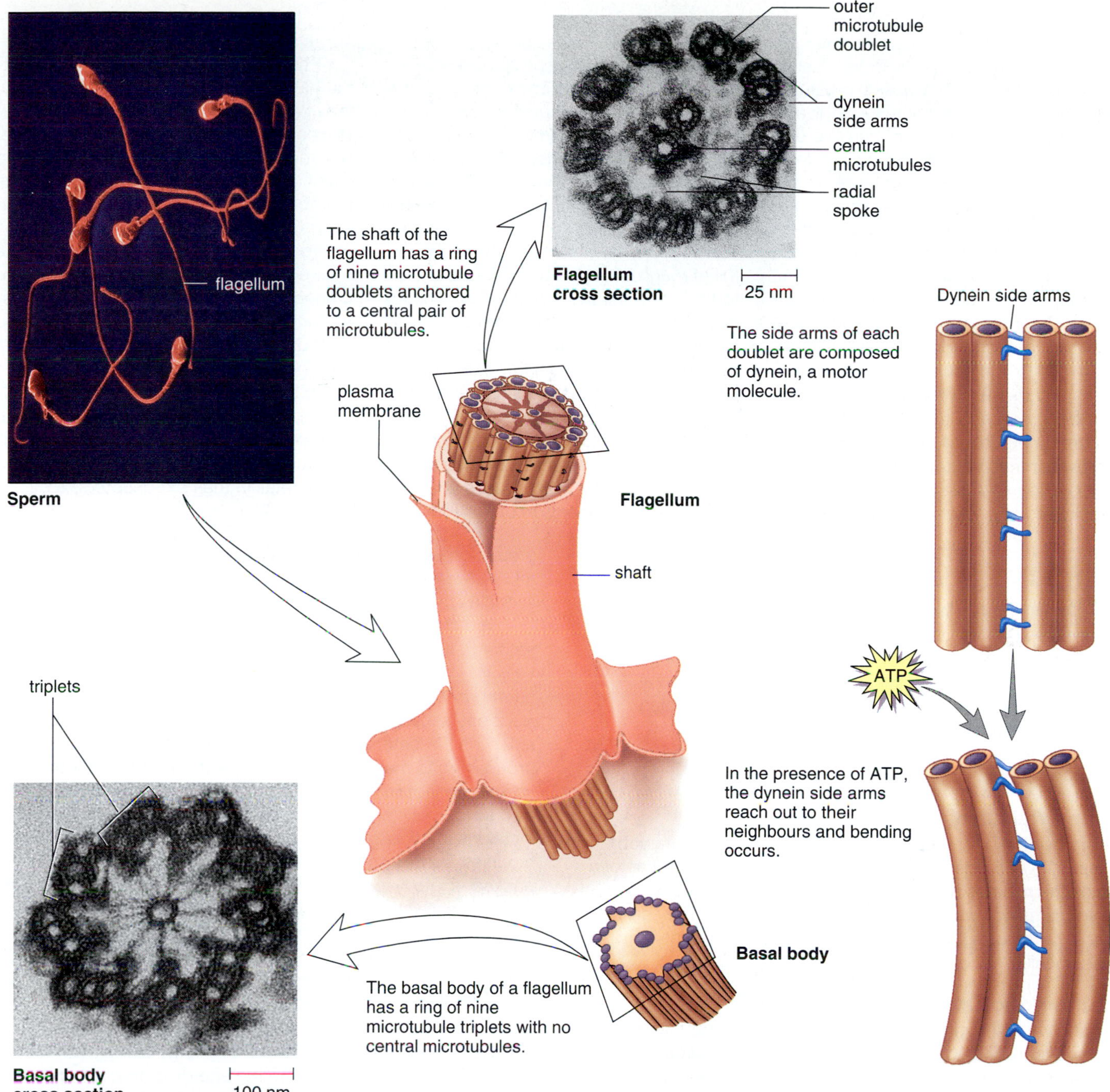
flagellum
Sperm
outer microtubule doublet
dynein side arms
central microtubules
radial spoke
Flagellum cross section
25 nm
The shaft of the flagellum has a ring of nine microtubule doublets anchored to a central pair of microtubules.
plasma membrane
Flagellum
shaft
Dynein side arms
The side arms of each doublet are composed of dynein, a motor molecule.
ATP
In the presence of ATP, the dynein side arms reach out to their neighbours and bending occurs.
triplets
Basal body
The basal body of a flagellum has a ring of nine microtubule triplets with no central microtubules.
Basal body cross section
100 nm

Figure 4.16 **Structure of cilium or flagellum.**
A flagellum has a basal body with a 9 + 0 pattern of microtubule triplets. (Notice the ring of nine triplets, with no central microtubules.) The shaft of the flagellum has a 9 + 2 pattern (a ring of nine microtubule doublets surrounds a central pair of single microtubules). Compare the cross section of the basal body to the cross section of the flagellum shaft and note that in place of the third microtubule, the outer doublets have side arms of dynein, a motor molecule. In the presence of ATP, the dynein side arms reach out and attempt to move along their neighboring doublet. Because of the radial spokes connecting the doublets to the central microtubules, bending occurs.

Connecting Concepts

When the Dutchman Antonie van Leeuwenhoek observed one-celled creatures under his homemade microscope in the 1600s, he probably had no idea the unicellular organisms he found so fascinating had the same basic structure as the cells in his own body. Eukaryotic cells are compartmentalized and therefore can be likened to a miniature factory, where each department has a specific function. The nucleus, which contains the chromosomes, determines the characteristics of the cell. Mitochondria produce high-energy ATP molecules from carbohydrates and oxygen. Proteins are made on the ribosomes and the Golgi apparatus prepares materials to be secreted by the cell. Lysosomes degrade chemical refuse for recycling of molecular subunits. The cytoskeleton is highly organized with protein fibers of different dimension contributing to cell shape and motility as well as movement of structures within the cell.

Unicellular organisms are generalists, while cells from multicellular organisms are specialized for specific functions. Muscle cells, which depend on a ready supply of ATP, are full of mitochondria; pancreatic cells, which secrete insulin, are full of rough ER, a Golgi apparatus and so forth. "Form follows function," a concept developed by architects from designing buildings, applies equally well to cells and even to their organelles. For example, we will see that both chloroplasts and mitochondria rely on membrane compartmentalization to produce ATP. In chloroplasts this ATP is used to produce glucose and in mitochondria the ATP is made available to the cell as an energy source.

Prokaryotic cells lack the compartmentalization seen in eukaryotic cells, but they carry out all the functions of eukaryotic cells. They have a plasma membrane, but their single chromosome is located within a nucleoid region. They lack a true membrane-bounded nucleus. Like eukaryotes, prokaryotes have ribosomes where proteins are synthesized. Despite their simple organization, prokaryotes are among the most adaptable and successful forms on earth. This is largely due to their diverse metabolic capabilities; most any organic molecule can be broken down by some type of prokaryote.

Most cellular processes are more complex than presented here and we will investigate certain cellular processes in more detail in the next chapters.

Summary

4.1 Cellular Level of Organization

All organisms are composed of cells, the smallest units of living matter. Cells are capable of self-reproduction, and existing cells come only from preexisting cells. Cells are very small and are measured in micrometers. The plasma membrane regulates exchange of materials between the cell and the external environment. Cells must remain small in order to have an adequate amount of surface area per cell volume.

4.2 Bacterial Cells

There are two major groups of cells: prokaryotic and eukaryotic. Both types have a plasma membrane and cytoplasm. Prokaryotic cells lack the nucleus of eukaryotic cells. Prokaryotic cells have a nucleoid that is not bounded by a nuclear envelope. They also lack most of the other organelles that compartmentalize eukaryotic cells.

4.3 Eukaryotic Cells

Eukaryotic cells, like prokaryotic cells, have an outer plasma membrane surrounding a cytosol. In addition, the cytosol of eukaryotic cells contains various organelles, each with a specific structure and function. The possible evolution of the eukaryotic cell is of interest. The nuclear envelope most likely evolved through invagination of the plasma membrane, but mitochondria and chloroplasts may have arisen through endosymbiotic events.

The nucleus of eukaryotic cells is bounded by a nuclear envelope containing pores. These pores serve as passageways between the cytoplasm and the nucleoplasm. Within the nucleus, chromatin, which contains DNA, undergoes coiling into chromosomes at the time of cell division. The nucleolus is a special region of the chromatin where rRNA is produced and where proteins from the cytoplasm gather to form ribosomal subunits. These subunits are joined in the cytoplasm.

Ribosomes are organelles that function in protein synthesis. They can be bound to ER or exist within the cytosol singly or in groups called polyribosomes. When protein synthesis occurs a mRNA leaves the nucleus with a coded message from DNA that specifies the sequence of amino acids in that protein.

The endomembrane system includes the ER (both rough and smooth), the Golgi apparatus, the lysosomes, and other types of vesicles and vacuoles. The endomembrane system serves to compartmentalize the cell and keep the various biochemical reactions separate from one another. Newly produced proteins enter the ER lumen, where they may be modified before proceeding to the interior of the smooth ER. The smooth ER has various metabolic functions depending on the cell type, but it also forms vesicles that carry proteins and lipids to different locations, particularly to the Golgi apparatus. The Golgi apparatus modifies, sorts, and repackages proteins. Some proteins are packaged into lysosomes, which carry out intracellular digestion, or into vesicles that fuse with the plasma membrane. Following fusion, secretion occurs. The endomembrane system also includes peroxisomes. When these organelles oxidize molecules they produce hydrogen peroxide that is subsequently broken down. The large single plant cell vacuole not only stores substances but lends support to the plant cell.

Cells require a constant input of energy to maintain their structure. Chloroplasts capture the energy of the sun and carry on photosynthesis, which produces carbohydrate. Carbohydrate-derived products are broken down in mitochondria as ATP is produced. This is an oxygen-requiring process called cellular respiration.

The cytoskeleton contains actin filaments, intermediate filaments, and microtubules. These maintain cell shape and allow it and the organelles to move. Actin filaments, the thinnest filaments, interact with the motor molecule myosin in muscle cells to bring about contraction; in other cells, they pinch off daughter cells and have other dynamic functions. Intermediate filaments support the nuclear envelope and the plasma membrane and probably participate in cell-to-cell junctions. Microtubules radiate out from the centrosome and are present in centrioles, cilia, and flagella. They serve as tracks, along which vesicles and other organelles move, due to the action of specific motor molecules.

Reviewing the Chapter

1. What are the two basic tenets of the cell theory? 56
2. Why is it advantageous for cells to be small? 57
3. Roughly sketch a bacterial (prokaryotic) cell, label its parts, and state a function for each of these. 60
4. What similar features do prokaryotic cells and eukaryotic cells have? What is their major difference? 60–61
5. Describe the endosymbiotic hypothesis and the evidence to support it. 64
6. Describe the structure and the function of the nuclear envelope and the nuclear pores. 65
7. Distinguish between the nucleolus, rRNA, and ribosomes. 65–66
8. Trace the path of a protein from rough ER to the plasma membrane. 66
9. Give the overall equations for photosynthesis and cellular respiration, contrast the two, and tell how they are related. 70
10. What are the three components of the cytoskeleton? What are their structures and functions? 72–73
11. Relate the structure of flagella (and cilia) to centrioles, and discuss the function of both. 74–75

Testing Yourself

Choose the best answer for each question.

1. The small size of cells best correlates with
   a. the fact they are self-reproducing.
   b. their prokaryotic versus eukaryotic nature.
   c. an adequate surface area for exchange of materials.
   d. the fact they come in multiple sizes.
   e. All of these are correct.
2. Which of these is not a true comparison of the light microscope and the transmission electron microscope?

	Light	**Electron**
a.	uses light to "view" object	uses electrons to "view" object
b.	uses glass lenses for focusing	uses magnetic lenses for focusing
c.	specimen must be killed and stained	specimen may be alive and nonstained
d.	magnification is not as great	magnification is greater
e.	resolution is not as great	resolution is greater

3. Which of these best distinguishes a prokaryotic cell from a eukaryotic cell?
   a. Prokaryotic cells have a cell wall, but eukaryotic cells never do.
   b. Prokaryotic cells are much larger than eukaryotic cells.
   c. Prokaryotic cells have flagella, but eukaryotic cells do not.
   d. Prokaryotic cells do not have a membrane-bounded nucleus, but eukaryotic cells do have such a nucleus.
   e. Prokaryotic cells have ribosomes but eukaryotic cells do not have ribosomes.
4. Which organelle would not have originated by endosymbiosis?
   a. mitochondria
   b. flagella
   c. nucleus
   d. chloroplasts
   e. Both b and c are correct.
5. Which of these is not found in the nucleus?
   a. functioning ribosomes
   b. chromatin that condenses to chromosomes
   c. nucleolus that produces rRNA
   d. nucleoplasm instead of cytoplasm
   e. all forms of RNA
6. Vesicles from the smooth ER most likely are on their way to the
   a. rough ER.
   b. lysosomes.
   c. Golgi apparatus.
   d. plant cell vacuole only.
   e. location suitable to their size.
7. Lysosomes function in
   a. protein synthesis.
   b. processing and packaging.
   c. intracellular digestion.
   d. lipid synthesis.
   e. production of hydrogen peroxide.
8. Mitochondria
   a. are involved in cellular respiration.
   b. break down ATP to release energy for cells.
   c. contain grana and cristae.
   d. are present in animal but not plant cells.
   e. All of these are correct.
9. Which organelle releases oxygen?
   a. ribosome
   b. Golgi apparatus
   c. mitochondrion
   d. chloroplast
   e. smooth ER
10. Which of these is not true?
    a. Actin filaments are found in muscle cells.
    b. Microtubules radiate out from the ER.
    c. Intermediate filaments sometimes contain keratin.
    d. Motor molecules use microtubules as tracts.
    e. Eukaryotic cells without centrosomes still produce a spindle.
11. Cilia and flagella
    a. bend when microtubules try to slide past one another.
    b. contain myosin that pulls on actin filaments.
    c. are organized by basal bodies derived from centrioles.
    d. are constructed similarly in prokaryotes and eukaryotes.
    e. Both a and c are correct.

12. Study the example given in (a) below. Then for each other organelle listed, state another that is structurally and functionally related. Tell why you paired these two organelles.
    a. The nucleus can be paired with nucleoli because nucleoli are found in the nucleus. Nucleoli occur where chromatin is producing rRNA.
    b. mitochondria
    c. centrioles
    d. ER
13. Label these parts of the cell that are involved in protein synthesis and modification. Give a function for each structure.

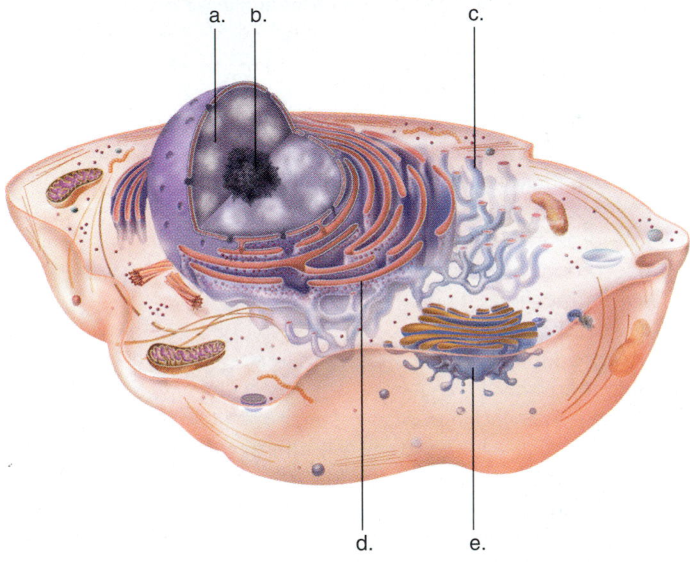

Thinking Scientifically

1. Vesicles that leave the Golgi apparatus as lysosomes move to different locations including the plasma membrane and the cytoplasm. What type of system could direct each budding vesicle to its proper location? What role could be played by the cytoskeleton?
2. In a cell biology laboratory students are examining sections of plant cells. Some students report seeing the nucleus, others do not. Some students see a large vacuole, others see a small vacuole. Some see evidence of an extensive endoplasmic reticulum, others see almost none. How can these observations be explained if the students are looking at the same cells?

Understanding the Terms

Match the terms to these definitions:

a. ______________ Organelle consisting of saccules and vesicles that processes, packages, and distributes molecules about or from the cell.
b. ______________ Network of strands consisting of DNA and associated proteins observed within a nucleus that is not dividing.
c. ______________ Dark-staining, spherical body in the cell nucleus that produces ribosomal subunits.
d. ______________ Internal framework of the cell, consisting of microtubules, actin filaments, and intermediate filaments.
e. ______________ Membrane-bounded vesicle that contains hydrolytic enzymes for digesting macromolecules.

Web Connections

Exploring the Internet

http://www.mhhe.com/biosci/genbio/mader/
(click on *Biology 7/e*)

The *Biology 7/e* Online Learning Center provides many resources for studying the material in this chapter including links to the following sites:

Cells Alive! is a fun website with downloadable videos and animations of a variety of cell topics, including the cytoskeleton, bacterial mobility, comparative sizes of cells, and much more.

http://www.cellsalive.com/

View a wide variety of different cell types using Histoweb.

http://www.kumc.edu/instruction/medicine/anatomy/histoweb/

Organelles: an overview of cellular organelle structure and function.

http://www.dcn.davis.ca.us/~carl/organell.htm

Prokaryotes, Eukaryotes, & Viruses Tutorial. A discussion of the 6 kingdoms, and the basic functions of the eukaryotic cell organelles.

http://www.biology.arizona.edu/cell_bio/tutorials/pev/main.html

MIT site showing and describing cellular parts. Quite detailed with many links.

http://esg-www.mit.edu:8001/esgbio/cb/cbdir.html

Membrane Structure and Function

chapter

chapter concepts

5.1 Membrane Models

- The present-day fluid-mosaic model proposes that membrane is less rigid and more dynamic than suggested by the previous sandwich model. 80

5.2 Plasma Membrane Structure and Function

- The plasma membrane regulates the passage of molecules and ions into and out of the cell. 81
- The membrane contains lipids and proteins, each with specific functions. 83

5.3 Permeability of the Plasma Membrane

- Small, noncharged molecules tend to pass freely across the plasma membrane. 84
- Some molecules diffuse (move from an area of higher concentration to an area of lower concentration) across a plasma membrane. 85
- Water diffuses across the plasma membrane and this can affect cell size and shape. 86
- Carrier proteins assist the transport of some ions and molecules across the plasma membrane. 88
- Vesicle formation takes other substances into the cell, and vesicle fusion with the plasma membrane discharges substances from the cell. 90

5.4 Modification of Cell Surfaces

- The cell wall of a plant cell supports the cell. The extracellular matrix of animal cells influences their shape, movement, and function. 92
- The activities of cells within a tissue are coordinated in part because cells are linked and directly communicate with one another. 93

A needle is poised to puncture the plasma membrane of a cell.

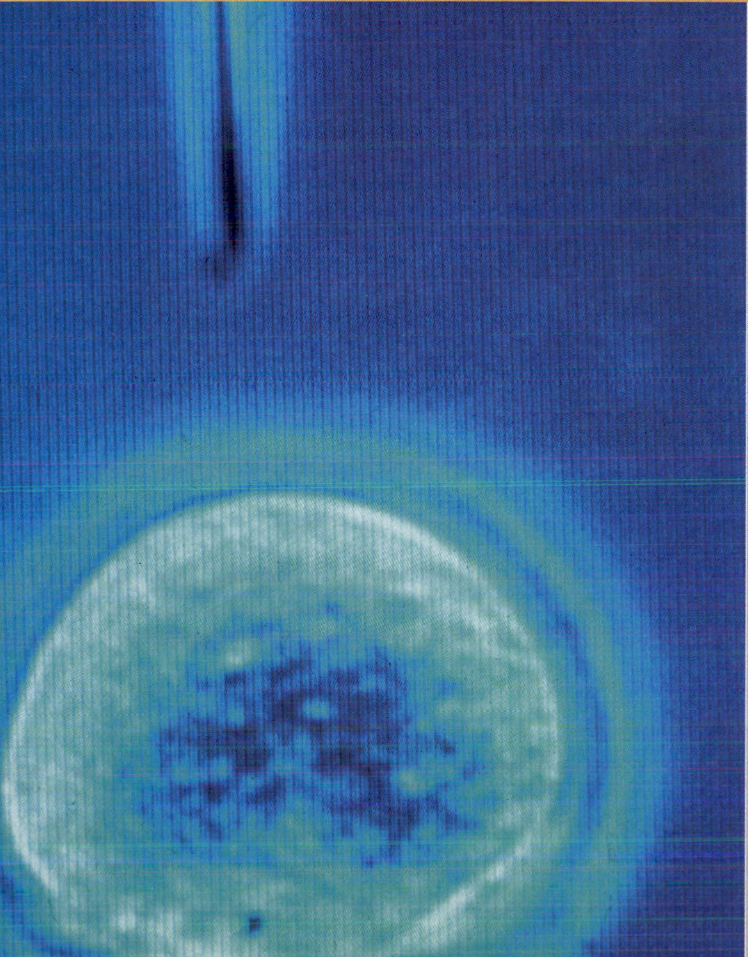

At first glance, a short pygmy, an overweight diabetic, and a young child with a high cholesterol level seem to have little in common. In reality, however, each suffers from a defect of their cells' plasma membrane. The pygmy's cells won't allow growth hormone to enter; the diabetic's cells do not bind to insulin, and the young child's cells prevent the entrance of a lipoprotein. A plasma membrane was essential to the origin of the first cell(s) and its proper functioning is essential to our good health today.

A plasma membrane encloses every cell, whether the cell is a unicellular amoeba or one of many from the body of a cockroach, peony, mushroom, or human. Universally, a plasma membrane protects a cell by acting as a barrier between its living contents and the surrounding environment. It regulates what goes into and out of the cell and marks the cell as being unique to the organism. In multicellular organisms, cell junctions requiring specialized features of the plasma membrane connect cells together in specific ways and pass on information to neighboring cells so that the activities of tissues and organs are coordinated.

5.1 Membrane Models

At the turn of the century, investigators noted that lipid-soluble molecules entered cells more rapidly than water-soluble molecules. This prompted them to suggest that lipids are a component of the plasma membrane. Later, chemical analysis disclosed that the plasma membrane contains phospholipids. In 1925, E. Gorter and G. Grendel measured the amount of phospholipid extracted from red blood cells and determined that there is just enough to form a bilayer around the cells. They further suggested that the nonpolar (hydrophobic) tails are directed inward and the polar (hydrophilic) heads are directed outward, forming a phospholipid bilayer:

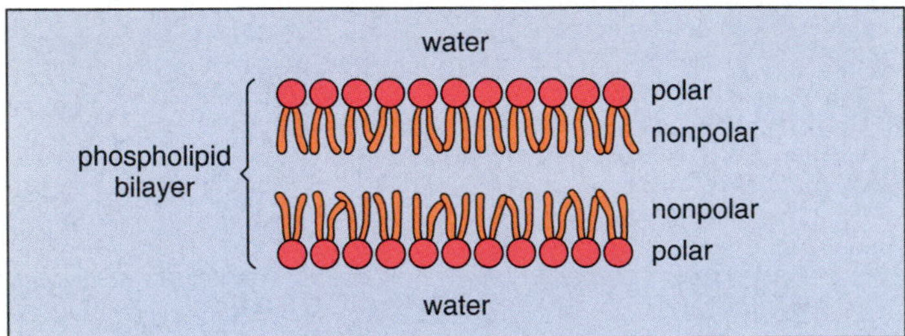

The presence of lipids cannot account for all the properties of the plasma membrane, such as its permeability to certain nonlipid substances. Such observations prompted J. Danielli and H. Davson to suggest in the 1940s that proteins are also a part of the membrane. They proposed a sandwich model in which the phospholipid bilayer is a filling between two continuous layers of proteins. By the late 1950s, electron microscopy had advanced to allow viewing of the plasma membrane and indeed it does have a sandwich appearance (Fig. 5.1*a*). At that time, J. D. Robertson modified the sandwich model and proposed that the outer dark layer (stained with heavy metals) contained protein *plus* the hydrophilic heads of the phospholipids. The interior was assumed to be the hydrophobic tails of these molecules. Robertson went on to suggest that all membranes in various cells have basically the same composition. Therefore his proposal was called the unit membrane model (Fig. 5.1*b*).

After some years, investigators began to doubt the accuracy of the unit membrane model. Not all membranes have the same appearance in electron micrographs and they do not have the same function. For example, the inner membrane of a mitochondrion, which is coated with rows of particles, functions in cellular respiration and it has a far different appearance from the plasma membrane. Finally, in 1972, S. Singer and G. Nicolson introduced the **fluid-mosaic model** of membrane structure, which proposes, in part, that the membrane is a fluid phospholipid bilayer in which protein molecules are either partially or wholly embedded. The proteins are scattered throughout the membrane in an irregular pattern that can vary from membrane to membrane. The mosaic distribution of proteins is supported especially by electron micrographs of freeze-fractured membranes (Fig. 5.1*c* and *d*).

a. Electron micrograph of red blood cell plasma membrane

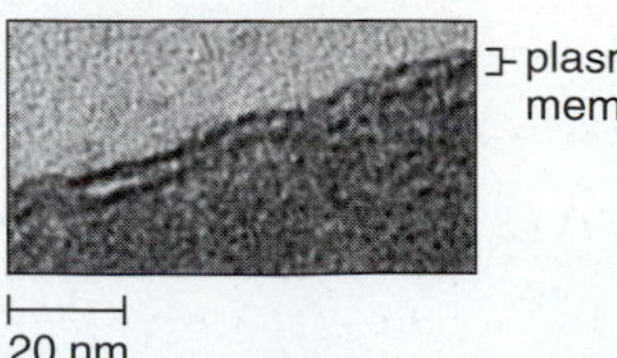

b. Two possible models

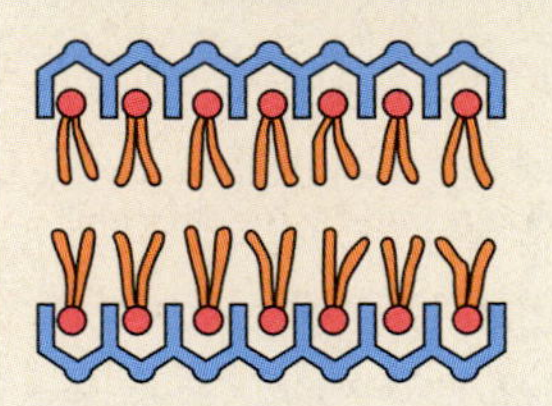

Robertson unit membrane

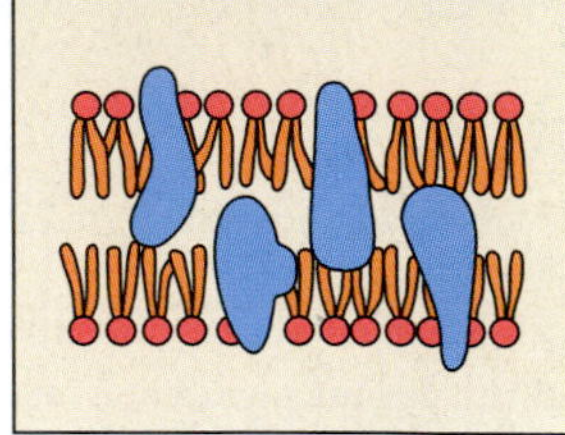

Singer and Nicolson fluid-mosaic model

c. Freeze-fracture of membrane

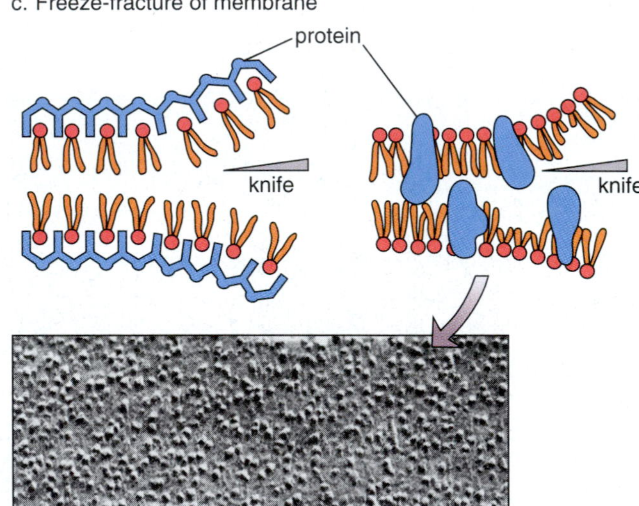

d. Electron micrograph of freeze-fractured membrane shows presence of particles

Figure 5.1 Membrane structure.
a. The red blood cell plasma membrane typically has a three-layered appearance in electron micrographs. **b.** Robertson's unit membrane model proposed that the outer dark layers in electron micrographs were made up of protein and polar heads of phospholipid molecules, while the inner light layer was composed of the nonpolar tails. The Singer and Nicolson fluid-mosaic model put protein molecules within the lipid bilayer. **c.** A technique called freeze-fracture allows an investigator to view the interior of the membrane. Cells are rapidly frozen in liquid nitrogen and then fractured with a special knife. Platinum and carbon are applied to the fractured surface to produce a faithful replica that is observed by electron microscopy. **d.** A fracture in the middle of the bilayer shows the presence of particles, consistent with the fluid-mosaic model.

The fluid-mosaic model of membrane structure is widely accepted at this time.

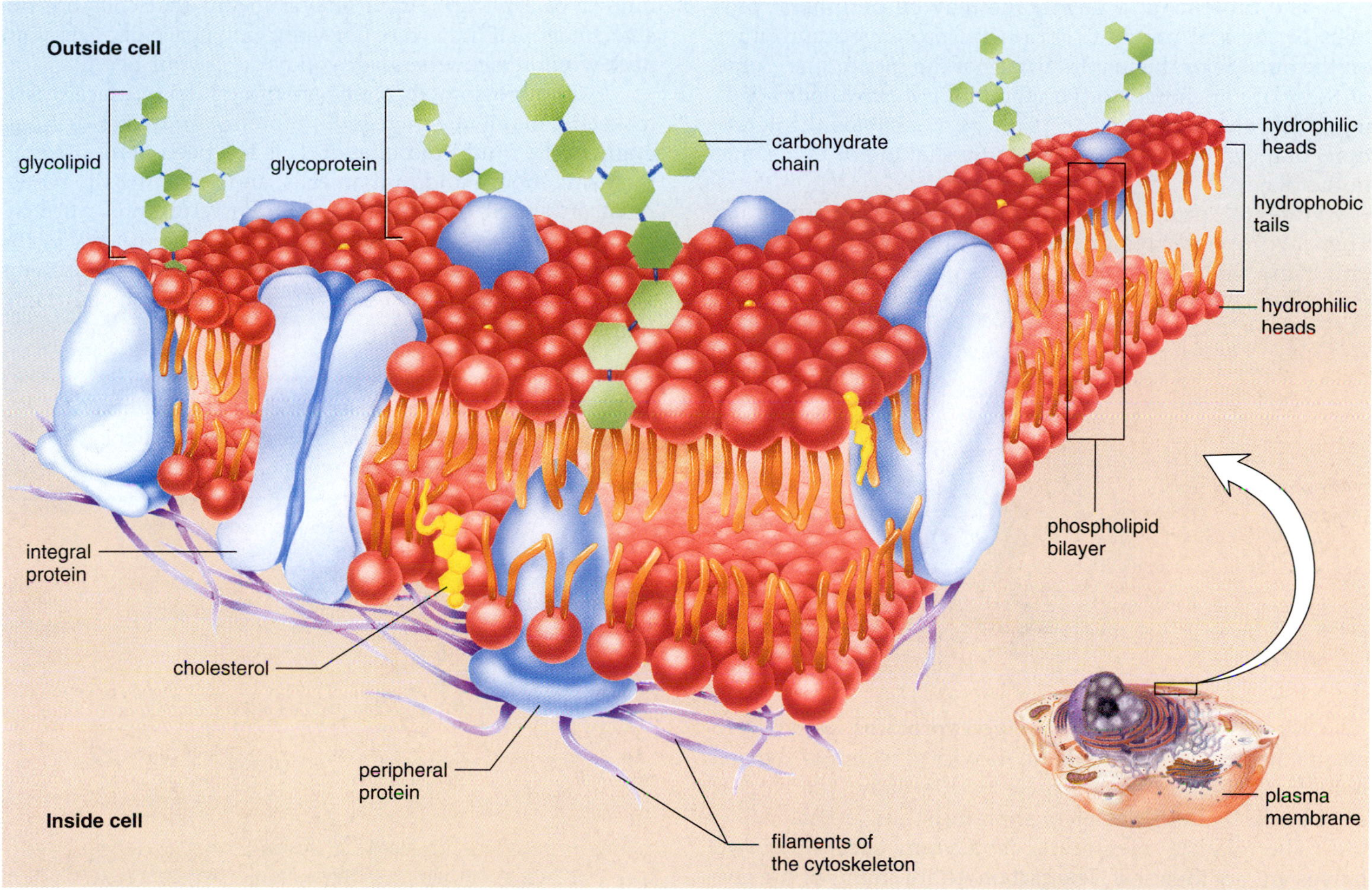

Figure 5.2 Fluid-mosaic model of plasma membrane structure.
The membrane is composed of a phospholipid bilayer in which proteins are embedded. The hydrophilic heads of phospholipids are a part of the outside surface and the inside surface of the membrane. The hydrophobic tails make up the interior of the membrane. Note the plasma membrane's asymmetry—carbohydrate chains are attached to the outside surface and cytoskeleton filaments are attached to the inside surface.

5.2 Plasma Membrane Structure and Function

According to the fluid-mosaic model of membrane structure, a membrane has two components: lipids and proteins. Most of the lipids in the plasma membrane are **phospholipids** which are known to arrange themselves spontaneously into a bilayer. The hydrophilic (polar) heads of the phospholipid molecules face the intracellular and extracellular fluids. The hydrophobic (nonpolar) tails face each other (Fig. 5.2).

In addition to phospholipids, there are two other types of lipids in the plasma membrane. **Glycolipids** have a structure similar to phospholipids except that the hydrophilic head is a variety of sugars joined to form a straight or branching carbohydrate chain. Glycolipids have a protective function and also have various other functions to be discussed in this chapter.

Cholesterol is a lipid that is found in animal plasma membranes; related steroids are found in the plasma membrane of plants. Cholesterol reduces the permeability of the membrane to most biological molecules.

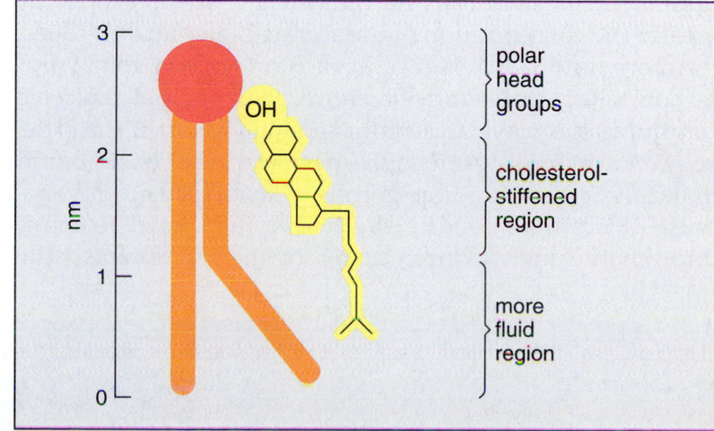

The proteins in a membrane may be peripheral proteins or integral proteins. Peripheral proteins occur either on the outside or the inside surface of the membrane. Some of these are anchored to the membrane by covalent bonding. Still others are held in place by noncovalent interactions that can be disrupted by gentle shaking or by change in the pH.

Integral proteins are found within the membrane. Their hydrophobic regions are embedded within the membrane and their hydrophilic regions project from both surfaces of the bilayer:

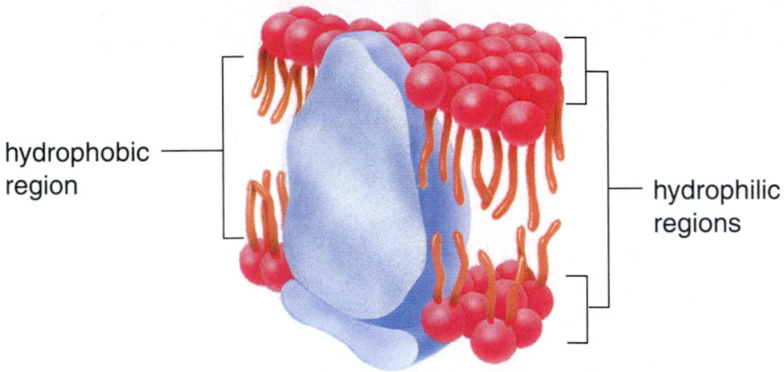

Many integral proteins are **glycoproteins,** which have an attached carbohydrate chain. As with glycolipids, the carbohydrate chain of sugars projects externally. Therefore it can be said that the plasma membrane is "sugar-coated."

The plasma membrane is asymmetrical: the two halves are not identical. The carbohydrate chains of the glycolipids and proteins occur only on the outside surface and the cytoskeletal filaments attach to proteins only on the inside surface.

The plasma membrane consists of a phospholipid bilayer. Peripheral proteins are found on the outside and inside surface of the membrane. Integral proteins span the lipid bilayer and often have attached carbohydrate chains.

Fluidity of the Plasma Membrane

At body temperature, the phospholipid bilayer of the plasma membrane has the consistency of olive oil. The greater the concentration of unsaturated fatty acid residues, the more fluid is the bilayer. In each monolayer, the hydrocarbon tails wiggle, and the entire phospholipid molecule can move sideways at a rate averaging about 2 μm—the length of a prokaryotic cell—per second. (Phospholipid molecules rarely flip-flop from one layer to the other, because this would require the hydrophilic head to move through the hydrophobic center of the membrane.) The fluidity of a phospholipid bilayer means that cells are pliable. Imagine if they were not—the long nerve fibers in your neck would crack whenever you nodded your head!

Although some proteins are often held in place by cytoskeletal filaments, in general, proteins are free to drift laterally in the fluid lipid bilayer. This has been demonstrated by fusing mouse and human cells, and watching the movement of tagged proteins (Fig. 5.3). Forty minutes after fusion, the proteins are completely mixed. The fluidity of the membrane is needed for the functioning of some proteins such as enzymes which become inactive when the membrane solidifies.

The fluidity of the membrane, which is dependent on its lipid components, is critical to the proper functioning of the membrane's proteins.

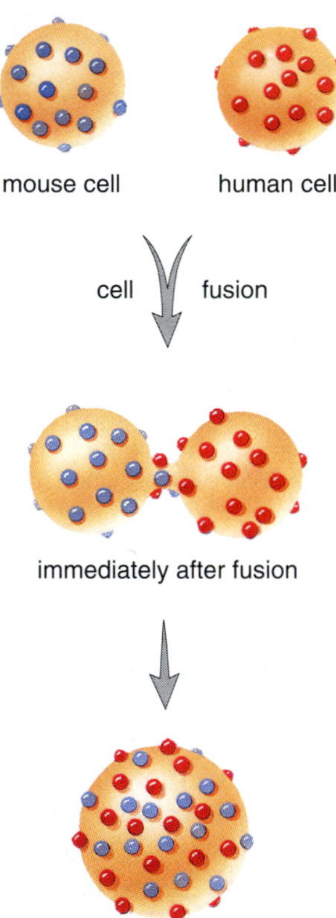

Figure 5.3 Experiment to demonstrate lateral drifting of plasma membrane proteins.
After human and mouse cells fuse, the plasma membrane proteins of the mouse (blue circles) and of the human cell (red circles) mix within a short time.

Mosaic Quality of the Membrane

The plasma membranes of various cells and the membranes of various organelles each have their own unique collections of proteins. The proteins form different patterns according to the particular membrane and also within the same membrane at different times. When you consider that the plasma membrane of a red blood cell contains over 50 different types of proteins, you can see why the membrane is said to be a mosaic.

The integral proteins largely determine a membrane's specific functions. As we will discuss in more detail, certain plasma membrane proteins are involved in the passage of molecules through the membrane. Some of these are **channel proteins** through which a substance can simply move across the membrane; others are **carrier proteins** that combine with a substance and help it to move across the membrane. Still others are receptors; each type of **receptor protein** has a shape that allows a specific molecule to bind to it. The binding of a molecule, such as a hormone (or other signal molecule), can cause the protein to change its shape and bring about a cellular response. Some plasma membrane proteins are **enzymatic proteins** that carry out metabolic reactions directly. The peripheral proteins associated with the membrane often have a structural role in that they help stabilize and shape the plasma membrane.

Figure 5.4 depicts the various functions of membrane proteins.

The mosaic pattern of a membrane is dependent on the proteins, which vary in structure and function.

Cell-Cell Recognition

The carbohydrate chains of glycolipids and glycoproteins serve as the "fingerprints" of the cell. The possible diversity of the chain is enormous; it can vary by the number of sugars (15 is usual, but there can be several hundred), by whether the chain is branched, and by the sequence of the particular sugars.

Glycolipids and glycoproteins vary from species to species, from individual to individual of the same species, and even from cell to cell in the same individual. Therefore, they make cell-cell recognition possible. Researchers working with mouse embryos have shown that as development proceeds, the different type cells of the embryo develop their own carbohydrate chains and that these chains allow the tissues and cells of the embryo to sort themselves out.

As you probably know, transplanted tissues are often rejected by the body. This is because the immune system is able to recognize that the foreign tissue's cells do not have the same glycolipids and glycoproteins as the rest of the body's cells. We also now know that a person's particular blood type is due to the presence of particular glycoproteins in the membrane of red blood cells.

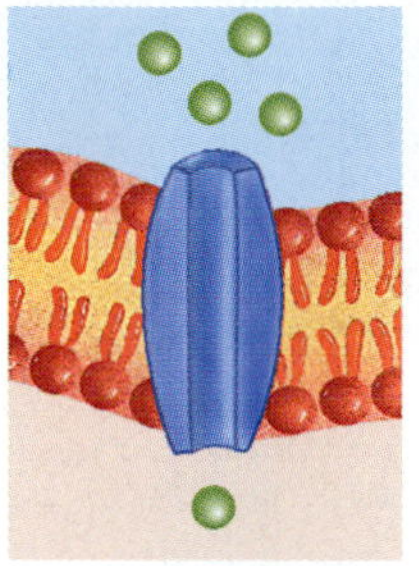

Channel Protein
Allows a particular molecule or ion to cross the plasma membrane freely. Cystic fibrosis, an inherited disorder, is caused by a faulty chloride (Cl^-) channel; a thick mucus collects in airways and in pancreatic and liver ducts.

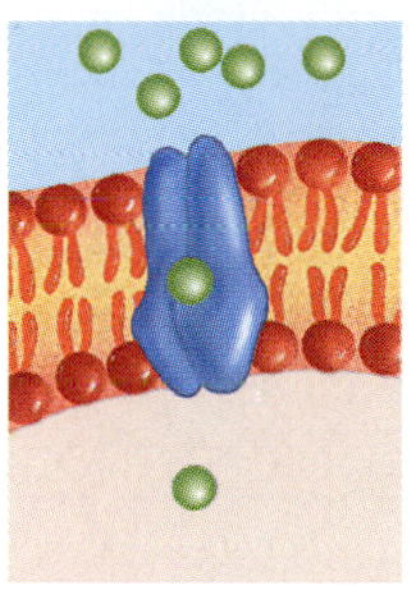

Carrier Protein
Selectively interacts with a specific molecule or ion so that it can cross the plasma membrane. The inability of some persons to use energy for sodium-potassium (Na^+–K^+) transport has been suggested as the cause of their obesity.

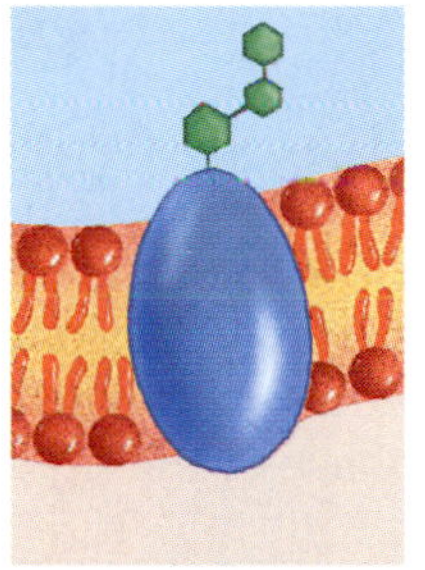

Cell Recognition Protein
The MHC (major histocompatibility complex) glycoproteins are different for each person, so organ transplants are difficult to achieve. Cells with foreign MHC glycoproteins are attacked by blood cells responsible for immunity.

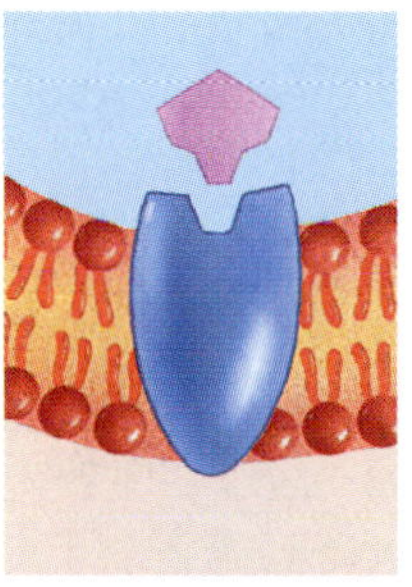

Receptor Protein
Is shaped in such a way that a specific molecule can bind to it. Pygmies are short, not because they do not produce enough growth hormone, but because their plasma membrane growth hormone receptors are faulty and cannot interact with growth hormone.

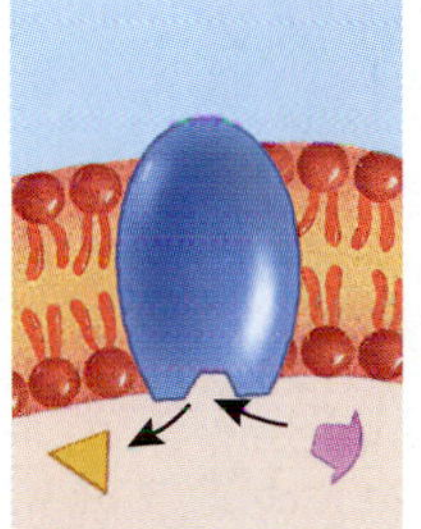

Enzymatic Protein
Catalyzes a specific reaction. The membrane protein, adenylate cyclase, is involved in ATP metabolism. Cholera bacteria release a toxin that interferes with the proper functioning of adenylate cyclase; sodium ions and water leave intestinal cells and the individual dies from severe diarrhea.

Figure 5.4 Membrane protein diversity.
These are some of the functions performed by proteins found in the plasma membrane.

Table 5.1

Passage of Molecules into and out of Cells

	Name	Direction	Requirement	Examples
Passive Transport Means	Diffusion	Toward lower concentration	Concentration gradient only	Lipid-soluble molecules, water, and gases
	Facilitated transport	Toward lower concentration	Carrier and concentration gradient	Some sugars and amino acids
Active Transport Means	Active transport	Toward greater concentration	Carrier plus cellular energy	Other sugars, amino acids, and ions
	Exocytosis	Toward outside	Vesicle fuses with plasma membrane	Macromolecules
	Endocytosis			
	Phagocytosis	Toward inside	Vacuole formation	Cells and subcellular material
	Pinocytosis (includes receptor-mediated endocytosis)	Toward inside	Vesicle formation	Macromolecules

5.3 Permeability of the Plasma Membrane

The plasma membrane is **differentially** (selectively) **permeable.** Some substances can move across the membrane and some cannot (Fig. 5.5). Macromolecules cannot diffuse across the membrane because they are too large. Ions and charged molecules cannot cross the membrane because they are unable to enter the hydrophobic phase of the lipid bilayer.

Noncharged, lipid-soluble molecules such as alcohols and oxygen can cross the membrane with ease. They are able to slip between the hydrophilic heads of the phospholipids and pass through the hydrophobic tails of the membrane.

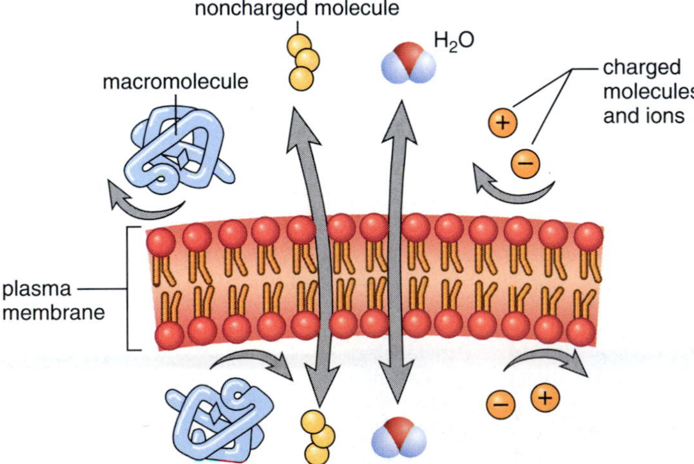

Figure 5.5 How molecules cross the plasma membrane. The curved arrows indicate that these substances cannot cross the plasma membrane and the back and forth arrows indicate that these substances can cross the plasma membrane.

Small polar molecules such as carbon dioxide and water also have no difficulty crossing through the membrane. These molecules follow their **concentration gradient** which is a gradual decrease in concentration over distance. To take an example, oxygen is more concentrated outside the cell than inside the cell because a cell uses oxygen during cellular respiration. Therefore oxygen follows its concentration gradient as it enters a cell. Carbon dioxide, on the other hand, which is produced when a cell carries on cellular respiration, is more concentrated inside the cell than outside the cell, and therefore it moves down its concentration gradient as it exits a cell.

Special means are sometimes used to get ions and charged molecules into and out of cells. Macromolecules can cross a membrane when they are taken in or out by vesicle formation (Table 5.1). Ions and molecules, like amino acids and sugars, are assisted across by one of two classes of transport proteins. Carrier proteins combine with an ion or molecule before transporting it across the membrane. Channel proteins form a channel that allows an ion or charged molecule to pass through. Our discussion in this chapter is largely restricted to carrier proteins. Carrier proteins are specific for the substances they transport across the plasma membrane.

Ways of crossing a plasma membrane are classified as passive or active (Table 5.1). Passive ways, which do not use chemical energy, involve diffusion or facilitated transport. These passive ways depend on the motion energy of ions and molecules. Active ways, which do require chemical energy, include active transport, endocytosis, and exocytosis.

The plasma membrane is differentially permeable. Certain substances can freely pass through the membrane and others must be transported across either by carrier proteins or by vacuole formation.

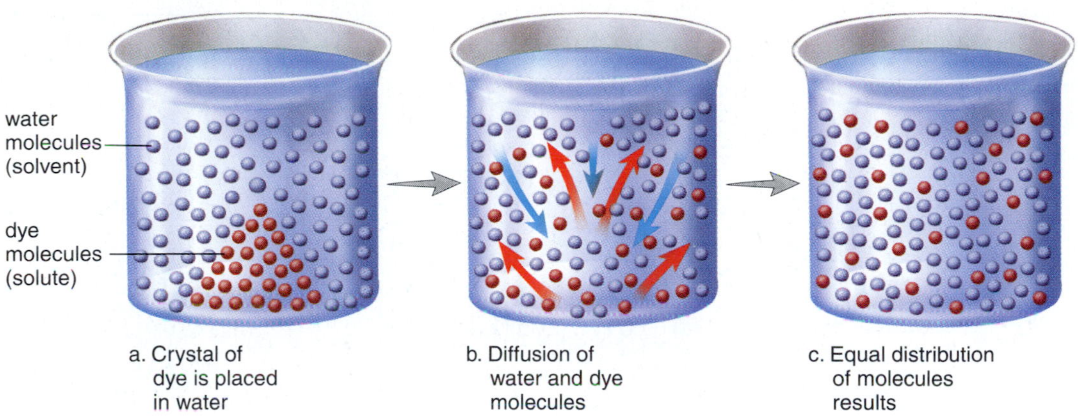

Figure 5.6 Process of diffusion.
Diffusion is spontaneous, and no chemical energy is required to bring it about. **a.** When a dye crystal is placed in water, it is concentrated in one area. **b.** The dye dissolves in the water, and there is a net movement of dye molecules from higher to lower concentration. There is also a net movement of water molecules from a higher to a lower concentration. **c.** Eventually, the water and the dye molecules are equally distributed throughout the container.

Diffusion and Osmosis

Diffusion is the movement of molecules from a higher to a lower concentration—that is, down their concentration gradient—until equilibrium is achieved and they are distributed equally. Diffusion is a physical process that can be observed with any type of molecule. For example, when a dye crystal is placed in water (Fig. 5.6), the dye and water molecules move in various directions, but their net movement, which is the sum of their motion, is toward the region of lower concentration. Therefore, the dye is eventually dissolved in the water, resulting in a colored solution. A **solution** contains both a solute, usually a solid, and a solvent, usually a liquid. In this case, the **solute** is the dye and the solvent is the water molecules. Once the solute and **solvent** are evenly distributed, they continue to move about, but there is no net movement of either one in any direction.

As discussed, the chemical and physical properties of the plasma membrane allow only a few types of molecules to enter and exit a cell simply by diffusion. Gases can diffuse through the lipid bilayer; this is the mechanism by which oxygen enters cells and carbon dioxide exits cells. Also, consider the movement of oxygen from the alveoli (air sacs) of the lungs to blood in the lung capillaries (Fig. 5.7). After inhalation (breathing in), the concentration of oxygen in the alveoli is higher than that in the blood; therefore, oxygen diffuses into the blood.

Molecules diffuse down their concentration gradients. A few types of small molecules can simply diffuse through the plasma membrane.

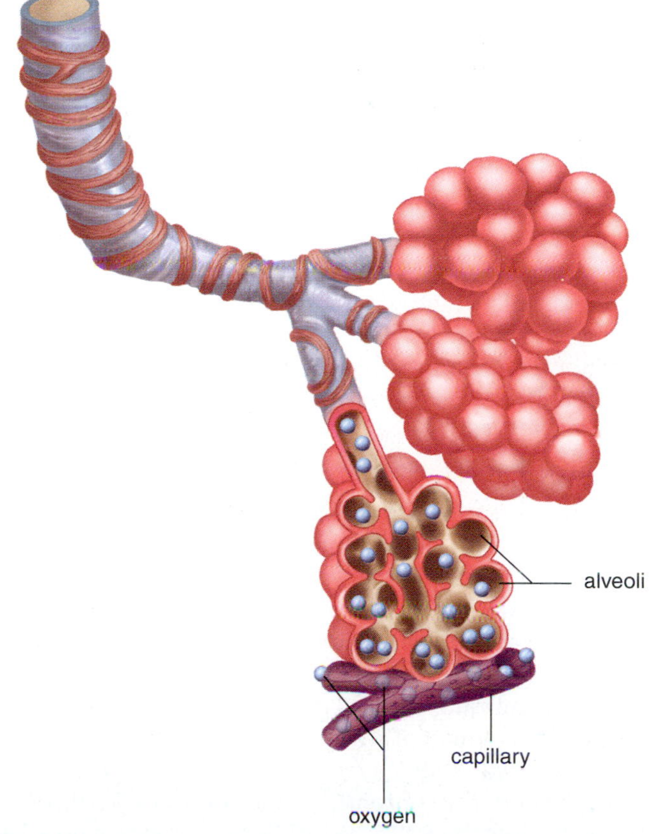

Figure 5.7 Gas exchange in lungs.
Oxygen (O_2) diffuses into the capillaries of the lungs because there is a higher concentration of oxygen in the alveoli (air sacs) than in the capillaries.

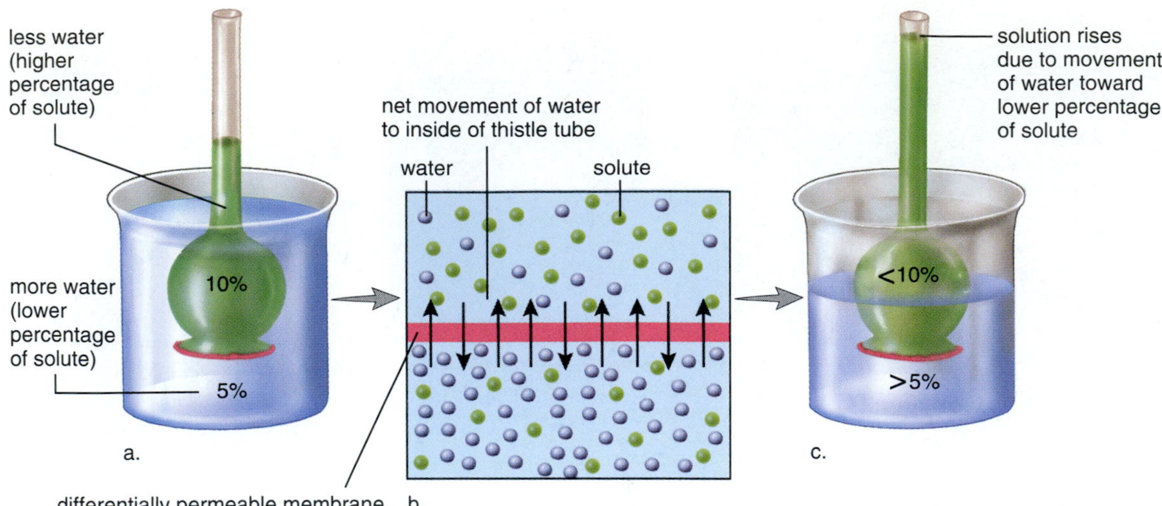

Figure 5.8 Osmosis demonstration.
a. A thistle tube, covered at the broad end by a differentially permeable membrane, contains a 10% sugar solution. The beaker contains a 5% sugar solution. **b.** The solute (green circles) is unable to pass through the membrane, but the water (blue circles) passes through in both directions. There is a net movement of water toward the inside of the thistle tube, where there are fewer water molecules per volume. **c.** Due to the incoming water molecules, the level of the solution rises in the thistle tube.

Osmosis

Osmosis [Gk. *osmos,* a pushing] is the diffusion of water into and out of cells. To illustrate osmosis, a thistle tube containing a 10% sugar solution[1] is covered at one end by a differentially permeable membrane and is then placed in a beaker containing a 5% sugar solution (Fig. 5.8). The beaker contains more water molecules (lower percentage of solute) per volume, and the thistle tube contains fewer water molecules (higher percentage of solute) per volume (Fig. 5.8*a*). Under these conditions, there is a net movement of water from the beaker to the inside of the thistle tube across the membrane (Fig. 5.8*b*). The solute is unable to pass through the membrane; therefore, the level of the solution within the thistle tube rises (Fig. 5.8*c*).

Notice the following in this illustration of osmosis:

1. A differentially permeable membrane separates two solutions. The membrane does not permit passage of the solute.
2. The beaker has more water (lower percentage of solute), and the thistle tube has less water (higher percentage of solute) per volume.
3. The membrane permits passage of water, and there is a net movement of water from the beaker to the inside of the thistle tube.
4. In the end, the concentration of solute in the thistle tube is less than 10%. Why? Because there is now less solute per volume. And the concentration of solute in the beaker is greater than 5%. Why? Because there is now more solute per volume.

Water enters the thistle tube due to the osmotic pressure of the solution within the thistle tube. **Osmotic pressure** is the pressure that develops in a system due to osmosis.[2] In other words, the greater the possible osmotic pressure the more likely water will diffuse in that direction. Due to osmotic pressure, water is absorbed from the human large intestine, is retained by the kidneys, and is taken up by capillaries from tissue fluid.

Tonicity

Tonicity is the degree to which a solution's concentration of solute versus water causes water to move into or out of cells. In the laboratory, cells are normally placed in **isotonic solutions;** that is, the solute concentration is the same on both sides of the membrane, and therefore there is no net gain or loss of water (Fig. 5.9). The prefix *iso* means *the same as,* and the term *tonicity* refers to the strength of the solution. A 0.9% solution of the salt sodium chloride (NaCl) is known to be isotonic to red blood cells. Therefore, intravenous solutions medically administered usually have this tonicity.

Solutions that cause cells to swell, or even to burst, due to an intake of water are said to be **hypotonic solutions.** The prefix *hypo* means *less than,* and refers to a solution with a lower percentage of solute (more water) than the cell. If a cell is placed in a hypotonic solution, water enters the cell; the net movement of water is from the outside to the inside of the cell.

Any concentration of a salt solution lower than 0.9% is hypotonic to red blood cells. Animal cells placed in such a solution expand and sometimes burst due to the buildup of

[1] Percent solutions are grams of solute per 100 ml of solvent. Therefore, a 10% solution is 10 g of sugar with water added to make up 100 ml of solution.

[2] Osmotic pressure is measured by placing a solution in an osmometer and then immersing the osmometer in pure water. The pressure that develops is the osmotic pressure of a solution.

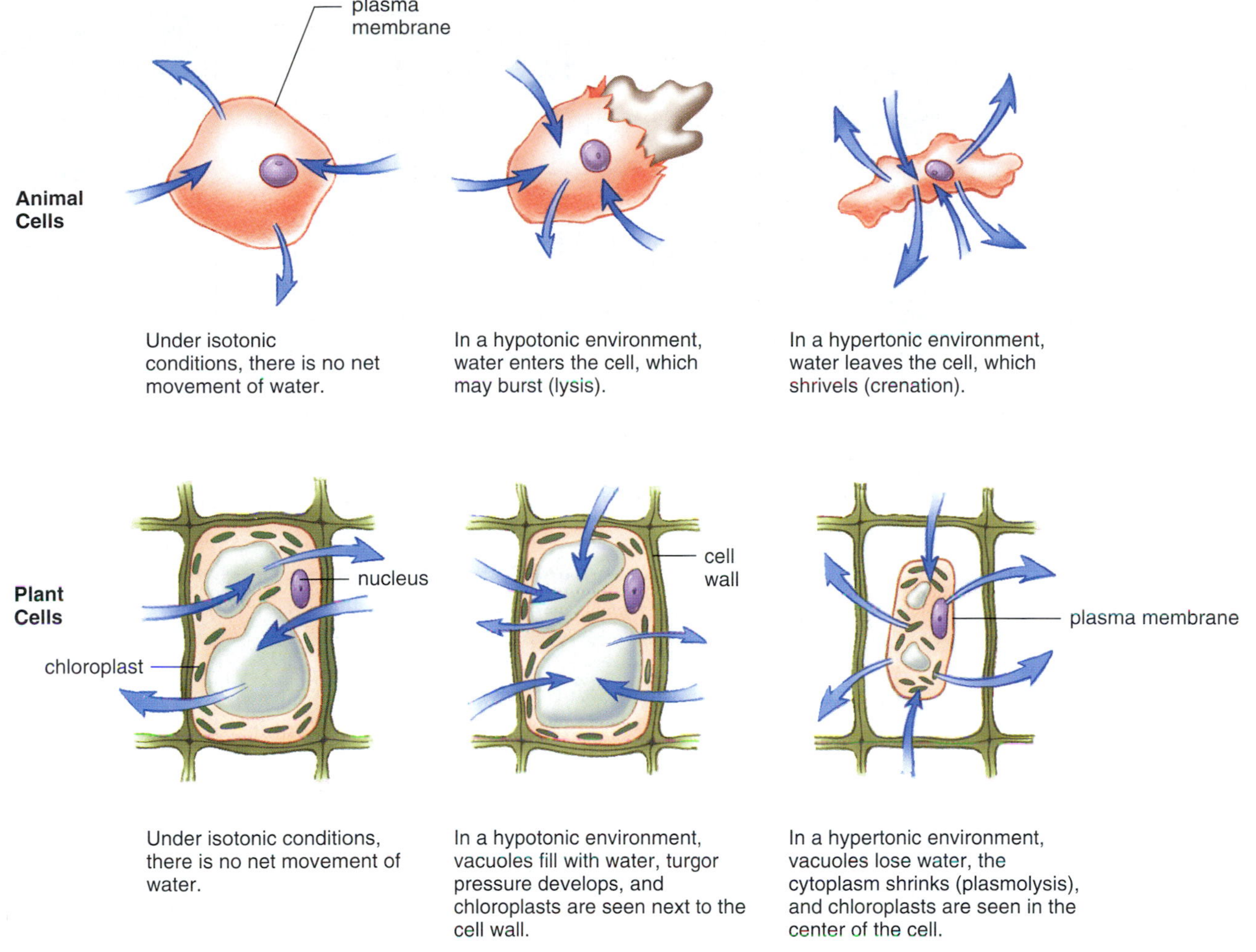

Figure 5.9 Osmosis in animal and plant cells.
The arrows indicate the net movement of water. In an isotonic solution, a cell neither gains nor loses water; in a hypotonic solution, a cell gains water; and in a hypertonic solution, a cell loses water.

pressure. The term *lysis* is used to refer to disrupted cells; hemolysis, then, is disrupted red blood cells.

The swelling of a plant cell in a hypotonic solution creates **turgor pressure** [L. *turgor,* swelling]. When a plant cell is placed in a hypotonic solution, we observe expansion of the cytoplasm because the large central vacuole gains water and the plasma membrane pushes against the rigid cell wall. The plant cell does not burst because the cell wall does not give way. Turgor pressure in plant cells is extremely important to the maintenance of the plant's erect position. If you forget to water your plants they wilt due to decreased turgor pressure.

Solutions that cause cells to shrink or to shrivel due to a loss of water are said to be **hypertonic solutions.** The prefix *hyper* means *more than,* and refers to a solution with a higher percentage of solute (less water) than the cell. If a cell is placed in a hypertonic solution, water leaves the cell; the net movement of water is from the inside to the outside of the cell.

Any solution with a concentration higher than 0.9% sodium chloride is hypertonic to red blood cells. If animal cells are placed in this solution, they shrink. The term **crenation** [L. *cranatus,* notched, wrinkled] refers to red blood cells in this condition. Meats are sometimes preserved by salting them. The bacteria are not killed by the salt but by the lack of water in the meat.

When a plant cell is placed in a hypertonic solution, the plasma membrane pulls away from the cell wall as the large central vacuole loses water. This is an example of **plasmolysis,** a shrinking of the cytoplasm due to osmosis. Dead plants you see along a salted roadside after the winter died because they were exposed to a hypertonic solution.

In an isotonic solution, a cell neither gains nor loses water. In a hypotonic solution, a cell gains water. In a hypertonic solution, a cell loses water and the cytoplasm shrinks.

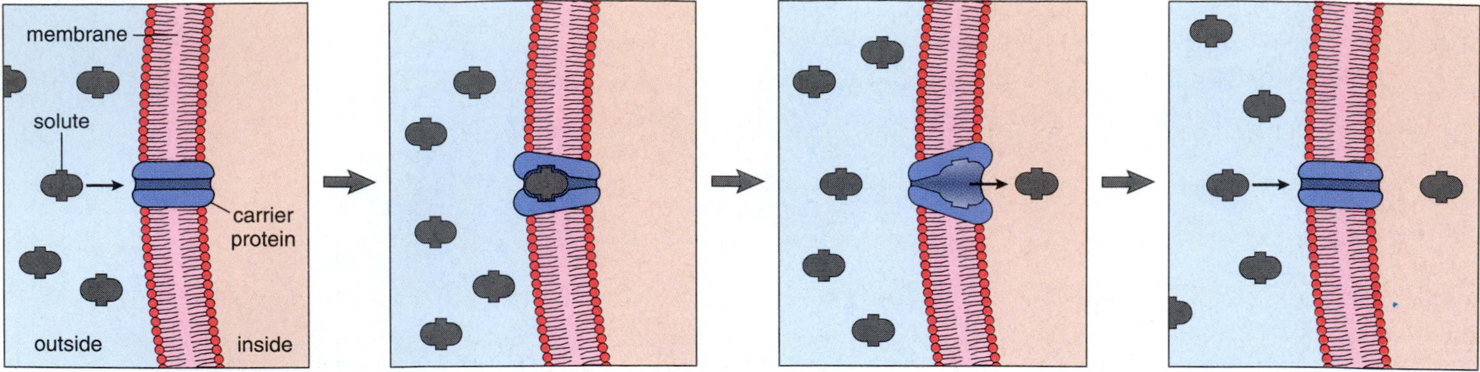

Figure 5.10 Facilitated transport.
During facilitated transport, a carrier protein speeds the rate at which the solute crosses the plasma membrane toward a lower concentration. Note that the carrier protein undergoes a change in shape as it moves a solute across the membrane.

Transport by Carrier Proteins

The plasma membrane impedes the passage of all but a few substances. Yet, biologically useful molecules are able to enter and exit the cell at a rapid rate because there are carrier proteins in the membrane. **Carrier proteins** are specific; each can combine with only a certain type of molecule or ion, which is then transported through the membrane. It is not completely understood how carrier proteins function, but after a carrier combines with a molecule, the carrier is believed to undergo a change in shape that moves the molecule across the membrane. Carrier proteins are required for facilitated transport and active transport (see Table 5.1).

Some of the proteins in the plasma membrane are carriers. They transport biologically useful molecules into and out of the cell.

Facilitated Transport

Facilitated transport explains the passage of such molecules as glucose and amino acids across the plasma membrane even though they are not lipid-soluble. The passage of glucose and amino acids is facilitated by their reversible combination with carrier proteins, which in some manner transport them through the plasma membrane. These carrier proteins are specific. For example, various sugar molecules of identical size might be present inside or outside the cell, but glucose can cross the membrane hundreds of times faster than the other sugars. As stated earlier, this is the reason that the membrane can be called differentially permeable.

A model for facilitated transport (Fig. 5.10) shows that after a carrier has assisted the movement of a molecule to the other side of the membrane, it is free to assist the passage of other similar molecules. Neither diffusion, explained previously, nor facilitated transport requires an expenditure of energy because the molecules are moving down their concentration gradient in the same direction they tend to move anyway.

Active Transport

During **active transport,** molecules or ions move through the plasma membrane, accumulating either inside or outside the cell. For example, iodine collects in the cells of the thyroid gland; glucose is completely absorbed from the gut by the cells lining the digestive tract; and sodium can be almost completely withdrawn from urine by cells lining the kidney tubules. In these instances, molecules have moved to the region of higher concentration, exactly opposite to the process of diffusion.

Both carrier proteins and an expenditure of energy are needed to transport molecules against their concentration gradient. In this case, chemical energy (ATP molecules usually) is required for the carrier to combine with the substance to be transported. Therefore, it is not surprising that cells involved primarily in active transport, such as kidney cells, have a large number of mitochondria near a membrane where active transport is occurring.

Proteins involved in active transport often are called pumps because, just as a water pump uses energy to move water against the force of gravity, proteins use energy to move a substance against its concentration gradient. One type of pump that is active in all animal cells, but is especially associated with nerve and muscle cells, moves sodium ions (Na^+) to the outside of the cell and potassium ions (K^+) to the inside of the cell. These two events are linked, and the carrier protein is called a **sodium-potassium pump.** A change in carrier shape after the attachment and again after the detachment of a phosphate group allows it to combine alternately with sodium ions and potassium ions (Fig. 5.11). The phosphate group is donated by ATP when it is broken down enzymatically by the carrier.

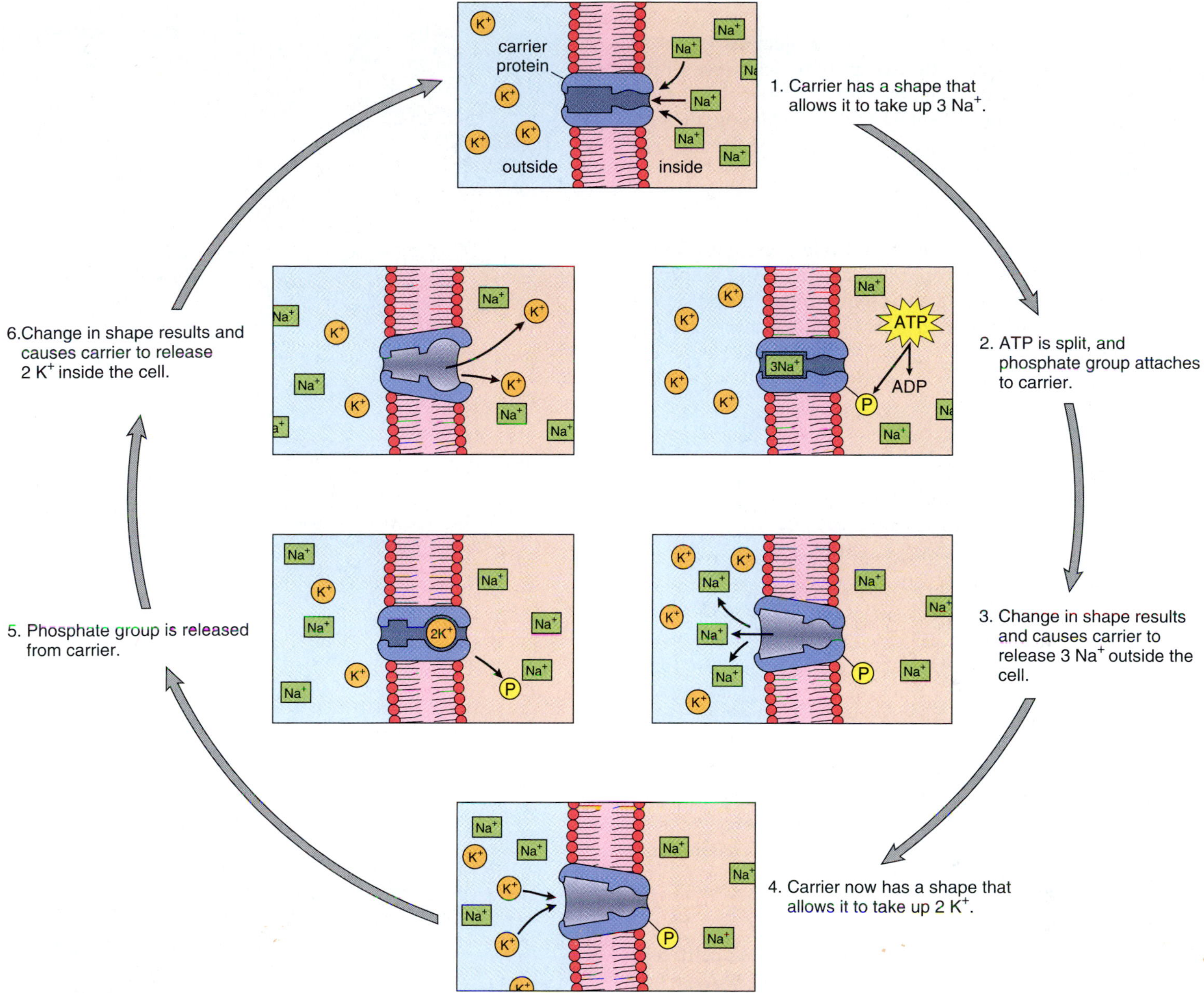

Figure 5.11 The sodium-potassium pump.
The same carrier protein transports sodium ions (Na^+) to the outside of the cell and potassium ions (K^+) to the inside of the cell because it undergoes an ATP-dependent change in shape. Three sodium ions are carried outward for every two potassium ions carried inward; therefore, the inside of the cell is negatively charged compared to the outside.

The sodium-potassium pump results in both a concentration gradient and an electrical gradient for these ions across the plasma membrane.

The passage of salt (NaCl) across a plasma membrane is of primary importance in cells. The chloride ion (Cl^-) usually crosses the plasma membrane because it is attracted by positively charged sodium ions (Na^+). First sodium ions are pumped across a membrane and then chloride ions simply diffuse through channels that allow their passage. As noted in Figure 5.4, the chloride ion channels malfunction in persons with cystic fibrosis, leading to the symptoms of this inherited (genetic) disorder.

During facilitated transport, small molecules follow their concentration gradient. During active transport, small molecules and ions move against their concentration gradient.

Membrane-Assisted Transport

What about the transport of macromolecules such as polypeptides, polysaccharides, or polynucleotides, which are too large to be transported by carrier proteins? They are transported in or out of the cell by vesicle formation, thereby keeping the macromolecules contained so that they do not mix with those in the cytoplasm.

Exocytosis

During **exocytosis** [Gk. *ex*, out of, and *kytos*, cell], vesicles often formed by the Golgi apparatus and carrying a specific molecule, fuse with the plasma membrane as secretion occurs. This is the way that insulin leaves insulin-secreting cells, for instance.

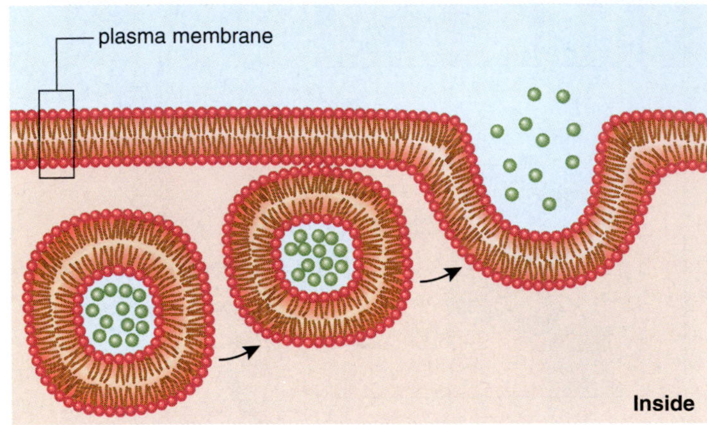

Notice that the membrane of the vesicle becomes a part of the plasma membrane. During cell growth, exocytosis is probably used as a means to enlarge the plasma membrane, whether or not secretion is also taking place.

Endocytosis

During **endocytosis** [Gk. *endon*, within, and *kytos*, cell], cells take in substances by vesicle formation (Fig. 5.12). A portion of the plasma membrane invaginates to envelop the substance, and then the membrane pinches off to form an intracellular vesicle.

When the material taken in by endocytosis is large, such as a food particle or another cell, the process is called **phagocytosis** [Gk. *phagein*, to eat, and *kytos*, cell]. Phagocytosis is common in unicellular organisms like amoebas and in ameboid cells like macrophages, which are large cells that engulf bacteria and worn-out red blood cells in mammals. When the endocytic vesicle fuses with a lysosome, digestion occurs.

Pinocytosis [Gk. *pino*, drink, and *kytos*, cell] occurs when vesicles form around a liquid or very small particles. Blood cells, cells that line the kidney tubules or intestinal wall, and plant root cells all use this method of ingesting substances (solutes). Whereas phagocytosis can be seen with the light microscope, the electron microscope must be used to observe pinocytic vesicles, which are no larger than 1–2 μm.

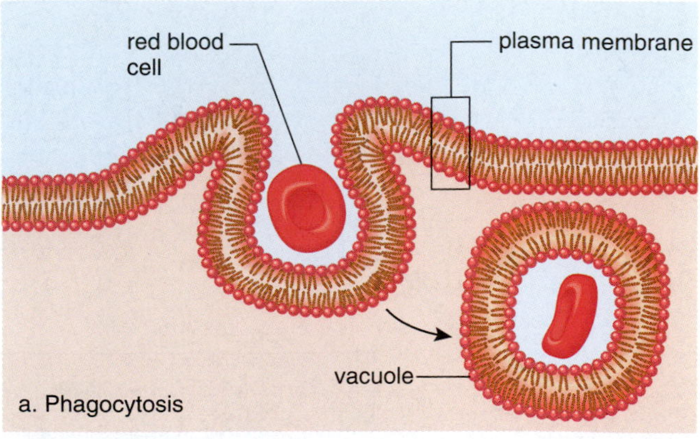

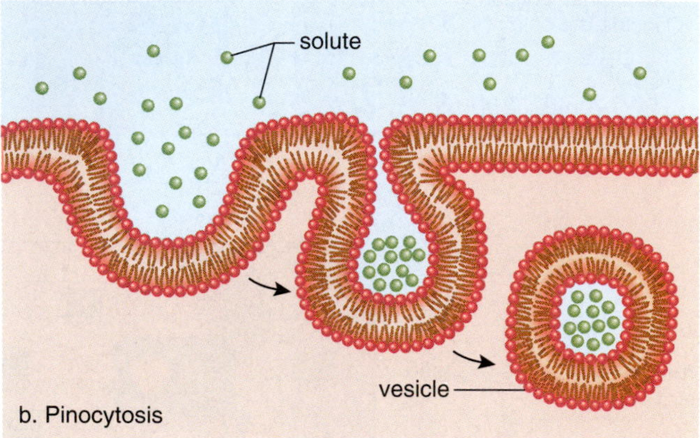

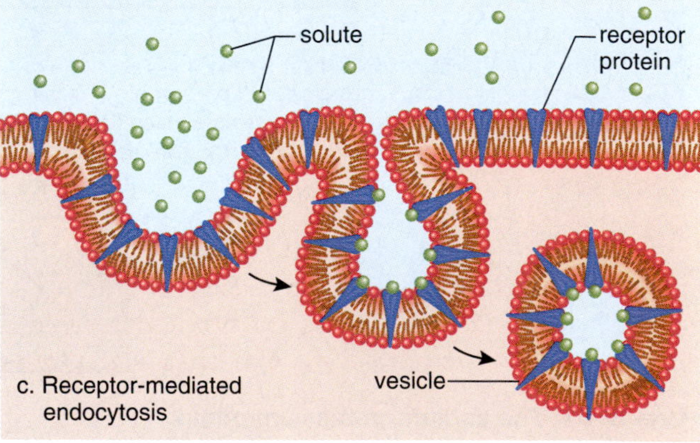

Figure 5.12 Three methods of endocytosis.
a. Phagocytosis occurs when the substance to be transported into the cell is large: white blood cells can engulf bacteria by phagocytosis. Digestion occurs when the resulting vacuole fuses with a lysosome.
b. Pinocytosis occurs when a solute such as a polypeptide is to be transported into the cell. The result is a small vacuole or vesicle.
c. Receptor-mediated endocytosis is a form of pinocytosis. The solute to be taken in first binds to specific receptor proteins which migrate to a pit or are already in a pit. The vesicle that forms contains the solute and its receptors. Sometimes the receptors are recycled, as shown in Figure 5.13.

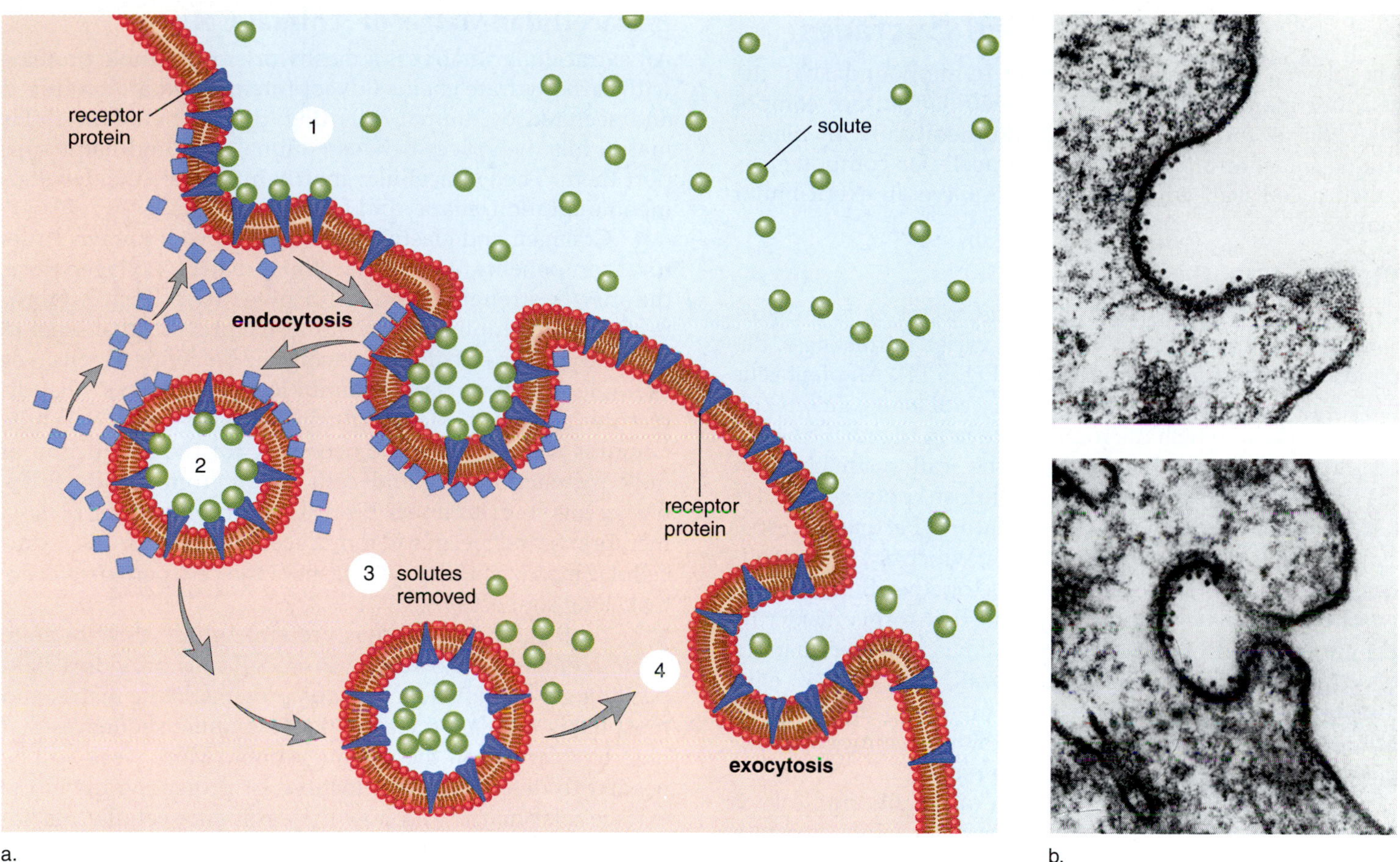

Figure 5.13 Receptor-mediated endocytosis.
a. (1) The receptor proteins (blue triangles) in a pit combine only with a specific solute (green circles). (2) The vesicle that forms is at first coated with a fibrous protein (blue squares), but soon the vesicle loses its coat. (3) Solutes sometimes leave the vesicle undigested. (4) When exocytosis occurs, membrane and therefore receptors are returned to the plasma membrane. **b.** Electron micrographs of a coated pit in the process of forming a vesicle.

Receptor-mediated endocytosis is a form of pinocytosis that is quite specific because it involves the use of a receptor protein shaped in such a way that a specific substance (called a ligand) such as vitamins, peptide hormones, and lipoproteins can bind to it. The binding of a substance to receptor proteins causes the receptors to gather at one location. This location is called a coated pit because there is a layer of fibrous protein on the cytoplasmic side (see step 1, Fig. 5.13). Once the vesicle is formed, the fibrous coat is released and the vesicle appears uncoated (see step 2). The fate of the vesicle and its contents depends on the kind of substance it contains. Sometimes the substance simply enters the cytoplasm (step 3). A spent hormone, on the other hand, may be digested when the vesicle fuses with a lysosome. The membrane of the vesicle and, therefore, the receptor proteins are returned to the plasma membrane (step 4), or the vesicle can go to other membranous locations.

Aside from simply allowing substances to enter cells selectively from an extracellular fluid, coated pits are also involved in the transfer and exchange of substances between cells. Such exchanges take place when the substances move from maternal blood into fetal blood at the placenta, for example.

The importance of receptor-mediated endocytosis is demonstrated by a genetic disorder called familial hypercholesterolemia. Cholesterol is transported in blood by a complex of lipids and proteins called low-density lipoprotein (LDL). Individuals with familial hypercholesterolemia have inherited a gene that causes them to have a reduced number and/or defective receptors for LDL in their plasma membranes. Instead of cholesterol entering cells, it accumulates in the walls of arterial blood vessels, leading to high blood pressure, occluded (blocked) arteries, and heart attacks.

Substances are secreted from a cell by exocytosis. Substances enter a cell by endocytosis. Receptor-mediated endocytosis allows cells to take up specific kinds of molecules and then release them within the cell.

5.4 Modification of Cell Surfaces

The plasma membrane is the outer living boundary of the cell, but many cells have an extracellular surface component that is formed exterior to the membrane. In plants, fungi, algae, and bacteria, the extracellular component is called a cell wall while animal cells have an extracellular matrix.

Plant Cell Walls

In addition to a plasma membrane, plant cells are surrounded by a porous **cell wall** that varies in thickness, depending on the function of the cell (Fig. 5.14). All plant cells have a primary cell wall. The primary cell wall contains cellulose fibrils in which microfibrils are held together by noncellulose substances. Pectins allow the wall to stretch when the cell is growing, and noncellulose polysaccharides harden the wall when the cell is mature. Pectins are especially abundant in the middle lamella, which is a layer of adhesive substances that holds the cells together. Some cells in woody plants have a secondary wall that forms inside the primary cell wall. The secondary wall has a greater quantity of cellulose fibrils than the primary wall, and layers of cellulose fibrils are laid down at right angles to one another. Lignin, a substance that adds strength, is a common ingredient of secondary cell walls in woody plants.

In plant tissues, the cytoplasm of neighboring cells is sometimes connected by **plasmodesmata** (sing., plasmodesma), numerous narrow membrane-lined channels that pass through the cell wall. Cytoplasmic strands within these channels allow direct exchange of materials between neighboring plant cells. Coordination of cellular activities within a tissue occurs because of these linkages.

Extracellular Matrix of Animal Cells

An extracellular matrix is a meshwork of insoluble proteins with carbohydrate chains (glycoproteins) that are produced and secreted by animal cells (Fig. 5.15). The extracellular matrix fills the spaces between animal cells and helps support them. The extracellular matrix influences the development, migration, shape, and function of cells.

Collagen and elastin fibers are two well-known structural components of the extracellular matrix. Collagen gives the matrix strength and elastin gives it resilience. Fibronectins and laminins are two adhesive proteins that seem to play a dynamic role in influencing the behavior of cells. For example, fibronectin and laminin form "highways" that direct the migration of cells during development. Recently, laminins were found to be necessary for the production of milk by mammary gland cells taken from a mouse. Fibronectins and laminins bind to receptors in the plasma membrane and permit communication between the extracellular matrix and the cytoplasm of the cell, perhaps via cytoskeletal connections.

Proteoglycans are glycoproteins whose carbohydrate chains contain amino sugars. Proteoglycans provide a rigid packing gel that joins the various proteins in the matrix and most likely regulate the activity of signaling sequences that bind to receptors in the plasma protein. More work will be needed to determine the functions of proteoglycans in the extracellular matrix and how they influence cellular metabolism via signaling proteins like hormones.

Cells have extracellular structures. Cellulose and noncellulose substances occur in plant cell walls; the extracellular matrix of animal cells is rich in glycoproteins. Certain of these have functions that affect cell behavior.

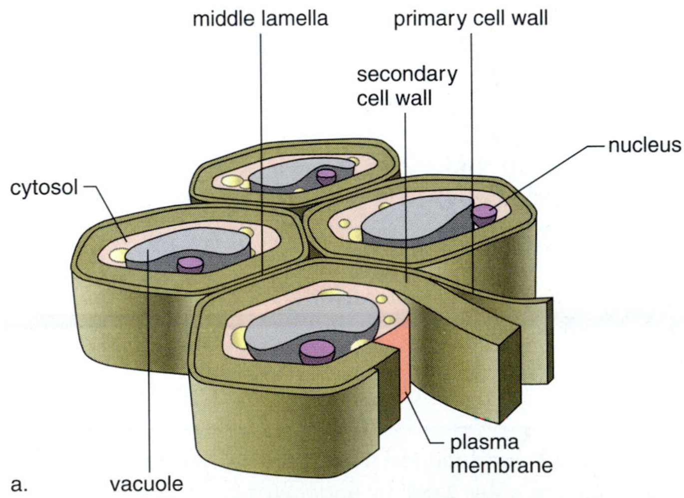

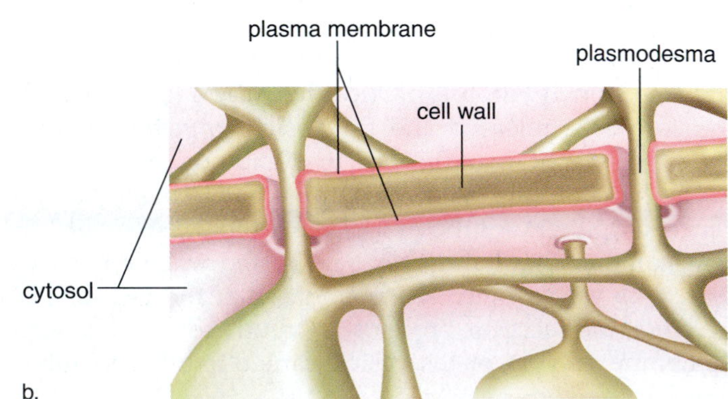

Figure 5.14 Plant cell wall and plasmodesmata.
a. All plant cells have a primary cell wall and some have a secondary cell wall. The cell wall, which lends support to the cell, is freely permeable.
b. Plasmodesmata are strands of cytoplasm that run in membrane-lined channels connecting certain plant cells.

Junctions Between Cells

For the cells of a tissue to act in a coordinated manner, it is beneficial for the plasma membranes of adjoining cells to interact. The plasmodesmata that link adjacent plant cells and the junctions that occur between animal cells are examples of such cellular interactions (Fig. 5.16).

In **adhesion junctions,** internal cytoplasmic plaques, firmly attached to the cytoskeleton within each cell, are joined by intercellular filaments. In some organs—like the heart, stomach, and bladder, where tissues get stretched—adhesion junctions hold the cells together.

Adjacent cells are even more closely joined by **tight junctions,** in which plasma membrane proteins actually attach to each other, producing a zipperlike fastening. The cells of tissues that serve as barriers are held together by tight junctions; in the intestine the digestive juices stay out of the body, and in the kidneys the urine stays within kidney tubules, because the cells are joined by tight junctions.

A **gap junction** allows cells to communicate. A gap junction is formed when two identical plasma membrane channels join. The channel of each cell is lined by six plasma membrane proteins. A gap junction lends strength to the cells, but it also allows small molecules and ions to pass between them. Gap junctions are important in heart muscle and smooth muscle because they permit a flow of ions that is required for the cells to contract as a unit.

Three types of junctions are seen between animal cells: adhesion junctions (desmosomes), tight junctions, and gap junctions.

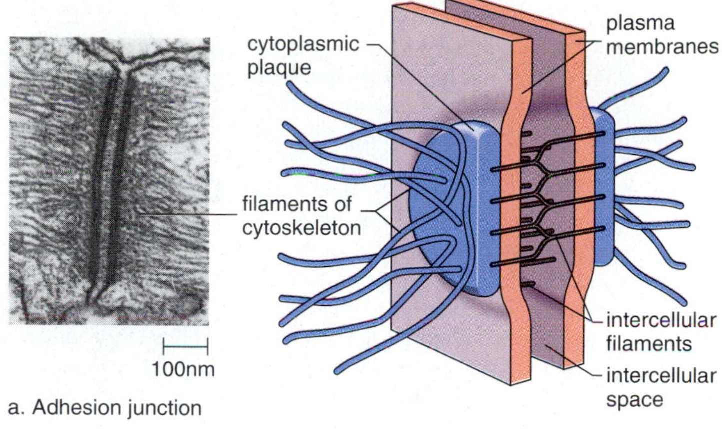

a. Adhesion junction

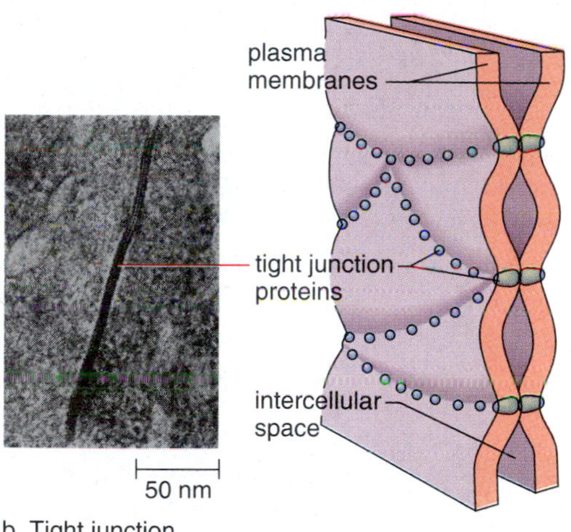

b. Tight junction

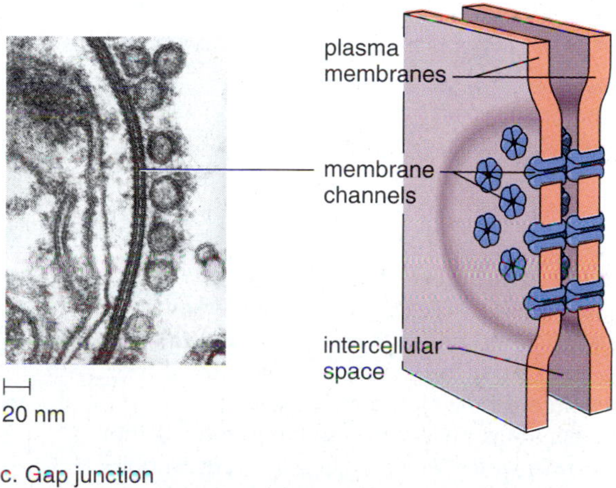

c. Gap junction

Figure 5.16 Junctions between cells of the intestinal wall.
a. In adhesion junctions (desmosomes), intercellular filaments run between two cells. **b.** Tight junctions between cells form an impermeable barrier because their adjacent plasma membranes are joined. **c.** Gap junctions allow communication between two cells because adjacent plasma membrane channels are joined.

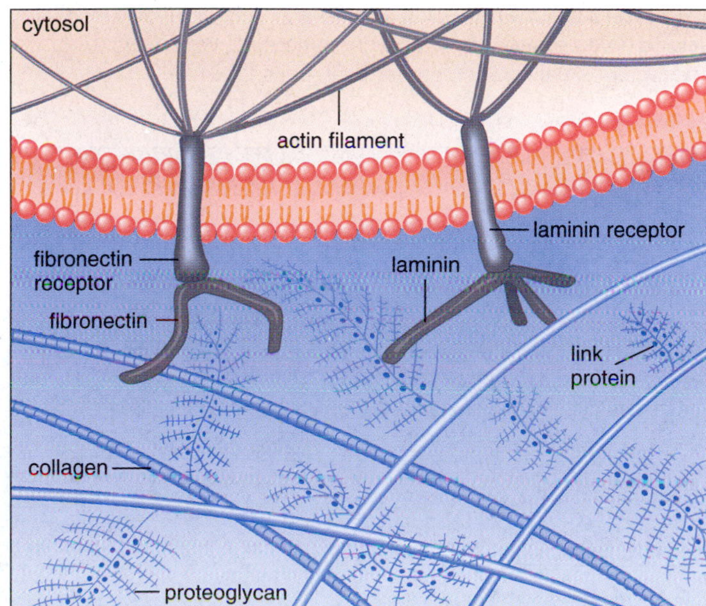

Figure 5.15 Animal cell extracellular matrix.
The glycoproteins in the extracellular matrix support an animal cell and also affect its behavior. Collagen and elastin also have a support function, while fibronectins and laminins bind to receptors in the plasma membrane and most likely assist cell communication processes.

Connecting Concepts

The plasma membrane is quite appropriately called the gatekeeper of the cell because it maintains the integrity of the cell and stands guard over what enters and leaves the cell. But we have seen that the plasma membrane does so much more than this. Its glycoproteins and glycolipids mark the cell as belonging to the organism. Its numerous proteins allow communication between cells and allow tissues to function as a whole. Now it appears that the extracellular material secreted by cells assists the plasma membrane in its numerous functions.

The progression in our knowledge about the plasma membrane illustrates how science works. The concepts and techniques of science evolve and change, and the knowledge we have today is amended and expanded by new investigative work. Basic science has applications that promote the well-being and health of human beings. To know that the plasma membrane is misfunctioning in a person who is diabetic, or those with a high cholesterol count is a first step toward curing these conditions. Even cancer is sometimes due to receptor proteins that signal the cell to divide even when no growth factor is present.

Our ability to understand the functioning of the plasma membrane is dependent on a thorough understanding of the molecules and ions that make up the cell. Today, it is impossible to deny the premise that biology and medicine have a biochemical basis.

Summary

5.1 Membrane Models

The fluid-mosaic model of membrane structure developed by Singer and Nicolson was preceded by several other models. Electron micrographs of freeze-fractured membranes support the fluid-mosaic model, and not Robertson's unit membrane concept based on the Danielli and Davson sandwich model.

5.2 Plasma Membrane Structure and Function

There are two components of the plasma membrane, lipids and proteins. In the lipid bilayer, phospholipids are arranged with their hydrophilic (polar) heads at the surfaces and their hydrophobic (nonpolar) tails in the interior. The lipid bilayer has the consistency of oil but acts as a barrier to the entrance and exit of most biological molecules. Membrane glycolipids and glycoproteins are involved in marking the cell as belonging to a particular individual and tissue.

The hydrophobic portion of a transmembrane protein lies in the lipid bilayer of the plasma membrane, and the hydrophilic portion lies at the surfaces. Proteins act as receptors, carry on enzymatic reactions, join cells together, form channels, or act as carriers to move substances across the membrane.

5.3 Permeability of the Plasma Membrane

Some molecules (lipid-soluble compounds, water, and gases) simply diffuse across the membrane from the area of higher concentration to the area of lower concentration. No metabolic energy is required for diffusion to occur.

The diffusion of water across a differentially permeable membrane is called osmosis. Water moves across the membrane into the area of higher solute (less water) content per volume. When cells are in an isotonic solution, they neither gain nor lose water. When cells are in a hypotonic solution, they gain water, and when they are in a hypertonic solution, they lose water (Table 5.2).

Other molecules are transported across the membrane by carrier proteins that span the membrane. During facilitated transport, a carrier protein assists the movement of a molecule down its concentration gradient. No energy is required.

During active transport, a carrier protein acts as a pump that causes a substance to move against its concentration gradient. The sodium-potassium pump carries Na^+ to the outside of the cell and K^+ to the inside of the cell. Energy in the form of ATP molecules is required for active transport to occur.

Larger substances can enter and exit a membrane by exocytosis and endocytosis. Exocytosis involves secretion. Endocytosis includes phagocytosis, pinocytosis, and receptor-mediated endocytosis. Receptor-mediated endocytosis makes use of receptor proteins in the plasma membrane. Once a specific solute (e.g., ligand) binds to receptors, a coated pit becomes a coated vesicle. After losing the coat, the vesicle can join with the lysosome, or after discharging the substance, the receptor-containing vesicle can fuse with the plasma membrane.

5.4 Modification of Cell Surfaces

Plant cells have a freely permeable cell wall, with cellulose as its main component. Plant cells are joined by small membrane-lined channels called plasmodesmata that span the cell wall and contain strands of cytoplasm that allow materials to pass from one cell to another.

Animal cells have an extracellular matrix that determines their shape and influences their behavior. Junctions between animal cells include adhesion junctions and tight junctions, which help to hold cells together, and gap junctions, which allow passage of small molecules between cells.

Table 5.2

Effect of Osmosis on a Cell

Tonicity of Solution	Concentrations: Solute	Concentrations: Water	Net Movement of Water	Effect on Cell
Isotonic	Same as cell	Same as cell	None	None
Hypotonic	Less than cell	More than cell	Cell gains water	Swells, turgor pressure
Hypertonic	More than cell	Less than cell	Cell loses water	Shrinks, plasmolysis

Reviewing the Chapter

1. Describe the fluid-mosaic model of membrane structure as well as the models that preceded it. Cite the evidence that either disproves or supports these models. 80–81
2. Tell how the phospholipids are arranged in the plasma membrane. What other lipids are present in the membrane, and what functions do they serve? 81–82
3. Describe how proteins are arranged in the plasma membrane. What are their various functions? Describe an experiment indicating that proteins can laterally drift in the membrane. 82–83
4. Define diffusion. What substances can diffuse through a differentially permeable membrane? 85
5. Define osmosis. Describe verbally and with drawings what happens to an animal cell when placed in isotonic, hypotonic, and hypertonic solutions. 86–87
6. Describe verbally and with drawings what happens to a plant cell when placed in these solutions. 86–87
7. Why do most substances have to be assisted through the plasma membrane? Contrast movement by facilitated transport with movement by active transport. 88
8. Draw and explain a diagram that shows how the sodium-potassium pump works. 88–89
9. Describe and contrast three methods of endocytosis. 90–91
10. Give examples to show that cell surface modifications help plant and animal cells communicate. 92–93

Testing Yourself

Choose the best answer for each question.

1. Write hypotonic solution or hypertonic solution beneath each cell. Justify your conclusions.

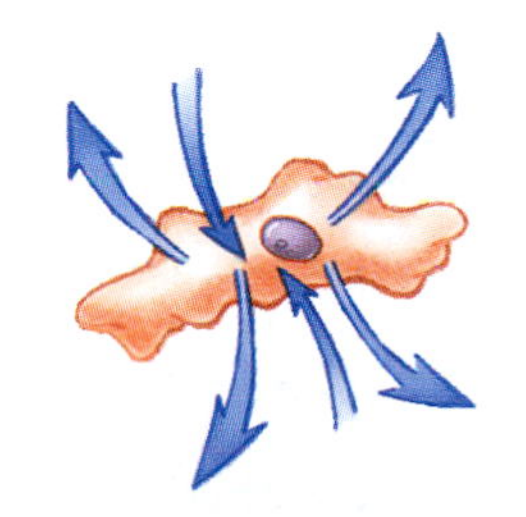

a. ______________________

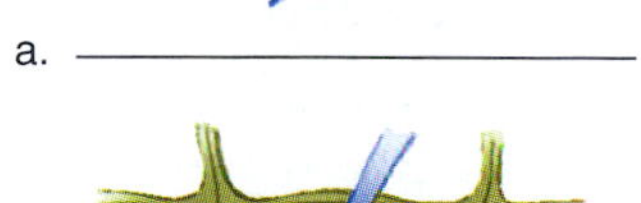

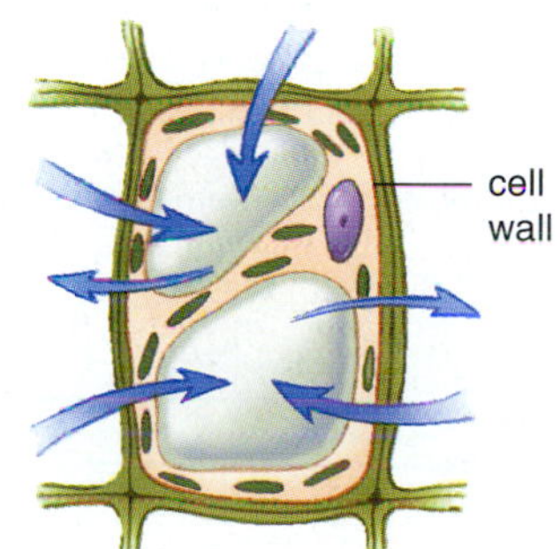

b. ______________________

2. Electron micrographs following freeze-fracture of the plasma membrane indicate that
   a. the membrane is a phospholipid bilayer.
   b. some proteins span the membrane.
   c. protein is found only on the surfaces of the membrane.
   d. glycolipids and glycoproteins are antigenic.
   e. there are receptors in the membrane.
3. A phospholipid molecule has a head and two tails. The tails are found
   a. at the surfaces of the membrane.
   b. in the interior of the membrane.
   c. spanning the membrane.
   d. where the environment is hydrophilic.
   e. Both a and b are correct.
4. During diffusion,
   a. solvents move from the area of higher to lower concentration but not solutes.
   b. there is a net movement of molecules from the area of higher to lower concentration.
   c. a cell must be present for any movement of molecules to occur.
   d. molecules move against their concentration gradient if they are small and charged.
   e. All of these are correct.
5. When a cell is placed in a hypotonic solution,
   a. solute exits the cell to equalize the concentration on both sides of the membrane.
   b. water exits the cell toward the area of lower solute concentration.
   c. water enters the cell toward the area of higher solute concentration.
   d. solute exits and water enters the cell.
   e. Both c and d are correct.
6. When a cell is placed in a hypertonic solution,
   a. solute exits the cell to equalize the concentration on both sides of the membrane.
   b. water exits the cell toward the area of lower solute concentration.
   c. water exits the cell toward the area of higher solute concentration.
   d. solute exits and water enters the cell.
   e. Both a and c are correct.
7. Active transport
   a. requires a carrier protein.
   b. moves a molecule against its concentration gradient.
   c. requires a supply of chemical energy.
   d. does not occur during facilitated transport.
   e. All of these are correct.
8. The sodium-potassium pump
   a. helps establish an electrochemical gradient across the membrane.
   b. concentrates sodium on the outside of the membrane.
   c. utilizes a carrier protein and chemical energy.
   d. is present in the plasma membrane.
   e. All of these are correct.
9. Receptor-mediated endocytosis
   a. is no different from phagocytosis.
   b. brings specific solutes into the cell.
   c. helps to concentrate proteins in vesicles.
   d. results in high osmotic pressure.
   e. All of these are correct.

10. Plant cells
    a. always have a secondary cell wall, even though the primary one may disappear.
    b. have channels between cells that allow strands of cytoplasm to pass from cell to cell.
    c. develop turgor pressure when water enters the nucleus.
    d. do not have cell-to-cell junctions like animal cells.
    e. All of these are correct.

Thinking Scientifically

1. The mucus in bronchial tubes must be thin enough for cilia to move bacteria and viruses up into the throat away from the lungs. Which way would Cl^- normally cross the plasma membrane (see Fig. 5.4, top) of bronchial tube cells in order for mucus to be thin? Use the concept of osmosis to explain your answer.
2. Winter wheat is planted in the early fall, grows over the winter when the weather is colder, and is harvested in the spring. As the temperature drops, the makeup of the plasma membrane of winter wheat changes. Unsaturated fatty acids replace saturated fatty acids in the phospholipids of the membrane. Why is this a suitable adaptation?

Understanding the Terms

active transport 88
adhesion junction 93
carrier protein 83, 88
cell wall 92
channel protein 83
cholesterol 81
concentration gradient 84
crenation 87
differentially permeable 84
diffusion 85
endocytosis 90
enzymatic protein 83
exocytosis 90
facilitated transport 88
fluid-mosaic model 80
gap junction 93
glycolipid 81
glycoprotein 82
hypertonic solution 87
hypotonic solution 86
isotonic solution 86
osmosis 86
osmotic pressure 86
phagocytosis 90
phospholipid 81
pinocytosis 90
plasmodesmata 92
plasmolysis 87
receptor-mediated endocytosis 91
receptor protein 83
sodium-potassium pump 88
solute 85
solution 85
solvent 85
tight junction 93
tonicity 86
turgor pressure 87

Match the terms to these definitions:

a. ______________ Type of plasma membranes that regulates the passage of substances into and out of the cell, allowing some to pass through and preventing the passage of others.
b. ______________ Diffusion of water through the plasma membrane of cells.
c. ______________ Higher solute concentration (less water) than the cytosol of a cell; causes cell to lose water by osmosis.
d. ______________ Causes an increase in liquid on the side of the membrane with higher solute concentration.
e. ______________ One of the major lipids found in animal plasma membranes; makes the membrane impermeable to many molecules.

Web Connections

Exploring the Internet

http://www.mhhe.com/biosci/genbio/mader/
(Click on *Biology 7/e*)

The *Biology 7/e* Online Learning Center provides many resources for studying the material in this chapter including links to the following sites:

Membrane Structure contains a thorough, well-illustrated discussion of membrane structure.

http://cellbio.utmb.edu/cellbio/membrane.html

Cell Membrane Structure and Composition. This site describes the fluid-mosaic model of membrane structure.

http://esg-www.MIT.EDU:8001/esgbio/cb/membranes/structure.html

Cell Membranes. This site discusses the structure and function of proteins in cell membranes.

http://esg-www.MIT.EDU:8001/esgbio/cb/membranes/intro.html

Membrane Transport Mechanisms. This site describes the mechanisms of cellular transport. Examples include diffusion, active transport, and the sodium-potassium pump.

http://esg-www.MIT.EDU:8001/esgbio/cb/membranes/transport.html

Electron Microscopic View of Membranes. A large number of electron microscopic photos of cell membranes, showing gap junctions, integral and peripheral proteins, and text to describe.

http://cellbio.utmb.edu/cellbio/membrane3.htm#polarized cells

Metabolism: Energy and Enzymes

chapter 6

chapter concepts

Painted lady butterfly, *Vanessa cardui*, on purple coneflower, *Echinacea purpurea*

Living things cannot maintain their organization nor carry on life's other activities without a source of organic food. Green plants utilize solar energy, carbon dioxide, and water to make organic food for themselves and all living things. Animals, such as butterflies and human beings, feed on plants or other animals that have eaten plants.

Food provides nutrient molecules, which are used as a source of energy and building blocks. Energy is the capacity to do work, and it takes work to maintain the organization of a cell and the organism, including the beautiful wings of this butterfly. When nutrients are broken down, they provide the necessary energy to make ATP (adenosine triphosphate). ATP fuels chemical reactions in cells such as the synthetic reactions that produce cell parts and products. The use of ATP in cells is substantial evidence of the relatedness of all life-forms.

Metabolism is all the chemical reactions that occur in a cell. Enzymes are protein molecules that speed metabolic reactions in cells at a relatively low temperature. This chapter deals with energy and enzymes, two essential requirements for cellular metabolism.

6.1 Energy

Living things can't grow, reproduce, or exhibit any of the characteristics of life without a ready supply of energy. **Energy,** which is the capacity to do work, occurs in many forms. Light energy comes from the sun; electrical energy powers kitchen appliances; and heat energy warms our houses. **Kinetic energy** is the energy of motion. All moving objects have kinetic energy. Thrown baseballs, falling water, and contracting muscles have kinetic energy. **Potential energy** is stored energy. Water behind a dam, or a rock at the top of a hill, or ATP, has potential energy that can be converted to kinetic energy. **Chemical energy** is in the interactions of atoms, one to the other, in a molecule.

Not only do chemicals have potential energy, they also have varying amounts of potential energy. Glucose has much more energy than its breakdown products, carbon dioxide and water.

Two Laws of Thermodynamics

Early researchers who first studied energy and its relationships and exchanges formulated two **laws of thermodynamics.** *The first law, also called the "law of conservation of energy," says that energy cannot be created or destroyed but can only be changed from one form to another.* Think of the conversions that occur when coal is used to power a locomotive. First, the chemical energy of coal is converted to heat energy and then heat energy is converted to kinetic energy in a steam engine. Similarly, the potential energy of coal or gas is converted to electrical energy by power plants. Do energy transformations occur in the human body? Yes, indeed. As an example, consider that the chemical energy in the food we eat is changed to the chemical energy of ATP, and then this form of potential energy is used for, say, muscle contraction (Fig. 6.1).

The second law of thermodynamics says that energy cannot be changed from one form to another without a loss of usable energy. Only about 25% of the chemical energy of gasoline is converted to the motion of a car; the rest is lost as heat. Heat, of course, is a form of energy, but heat is the most random form of energy and quickly dissipates into the environment. When muscles use the chemical energy within ATP to bring about contraction, some of this energy becomes heat right away. With conversion upon conversion, eventually all usable forms of energy become heat that is lost by the organism to the environment. And because heat dissipates, it can never be converted back to a form of potential energy. Organisms abide by laws of thermodynamics but so do larger systems. The reading on the next page discusses how ecosystems also obey the second law of thermodynamics.

Entropy

Entropy [Gk. *entrope,* a turning inward] is a measure of randomness or disorder. An organized, usable form of energy has low entropy, whereas an unorganized, less stable form of energy such as heat has high entropy. A neat room has a much lower entropy than a messy room. We know that a neat room always tends toward messiness. In the same way, energy conversions eventually result in heat, and therefore the entropy of the universe is always increasing.

How does an ordered system such as a neat room or an organism come about? You know very well that it takes an input of usable energy to keep your room neat. In the same way, it takes a constant input of usable energy from the food you eat to keep you organized. This input of energy goes through many energy conversions, and the output is finally heat, which increases the entropy of the universe.

A civilized society such as ours requires a great deal of low entropy energy. Fossil fuel energy, such as coal and oil, is used to grow and process your food and is used to keep your environment ordered. Building houses and schools and roads all require an input of usable energy, which is finally converted to heat. Our civilized society is increasing the entropy of the universe at a much higher rate than any society in the past.

Figure 6.1 Energy for life.
All of the energy needed to move this athlete is provided by the food he has eaten. Once food has been processed in the digestive tract, nutrients are transported about the body, including to the muscles. The energy of nutrient molecules is converted to that of ATP molecules which power muscle contraction.

The laws of thermodynamics explain why the entropy of the universe spontaneously increases and why organisms need a constant input of usable energy to maintain their organization.

Ecosystems and the Second Law of Thermodynamics

A cell converts the energy of one chemical molecule into another but ecosystems also convert energy. In an ecosystem, the energy stored in a prey population is used by a predator population. Because of this, energy flows through an ecosystem (Fig. 6A). As transformations of energy occur, useful energy is lost to the environment in the form of heat, until finally useful energy is completely used up. Since energy cannot recycle, there is a need for an ultimate source of energy. This source, which continually supplies almost all living things with energy, is the sun. The entire universe is tending toward disorder, but in the meantime, solar energy is sustaining living things.

Human beings also feed on other organisms. We feed directly on plants, such as corn, or on animals like poultry and cattle that have fed on corn. In the United States, however, much supplemental energy in addition to solar energy is used to produce food. Even before planting time, there is an input of fossil fuel energy for the processing of seeds, and the making of tools, fertilizers, and pesticides. Then, fossil fuel energy is used to transport these materials to the farm. At the farm, fuel is needed to plant the seeds, to apply fertilizers, and pesticides, and to irrigate, harvest, and dry crops. After harvesting, still more fuel is used to process crops to make the neatly packaged products we buy in the supermarket. Most of the food we eat today has been processed in some way. Even farm families now buy at least some of their food from supermarkets in nearby towns.

Since 1940 the amount of supplemental fuel used in the American food system has greatly increased. By now the amount of supplemental energy is at least three or four times that of the caloric content of the food produced! This is partially caused by the trend to produce more food on less land by growing high-yielding hybrid wheat and corn plants. These plants require more care and about twice as much supplemental energy as the traditional varieties of wheat and corn. Cattle confined to feedlots and fed grain that has gone through the whole production process require about twenty times the amount of supplemental energy as do range-fed cattle. Our food system has been labeled energy-intensive because it requires such a large input of supplemental energy.

Our energy-intensive food system is a matter for concern because it increases the cost of food and the burning of fossil fuels adds pollutants to the atmosphere. What can be done? First of all, we could grow crops that do not require so much supplemental energy. And second, we could eat primarily vegetables and grains. It is estimated that only about 10% of the energy contained in one population is actually taken up by the next population. (About 90% is lost as heat.) This means that about ten times the number of people can be sustained on a diet of vegetables and grain rather than a diet of meat. And when we do eat meat we could depend more on range-fed cattle. Cattle kept close to farmland supply manure that can substitute, in part, for chemical fertilizer. Biological control, the use of natural enemies to control pests, would cut down on pesticide use. Solar and wind energy could be used instead of fossil fuel energy, particularly on the farm. For example, wind-driven irrigation pumps are feasible.

Finally, of course, consumers could help matters. We could overcome our prejudice against vegetables that have slight blemishes. We could avoid processed foods and buy cheaper cuts of beef, which have come from range-fed cattle. And we could avoid using electrically powered gadgets when preparing food at home.

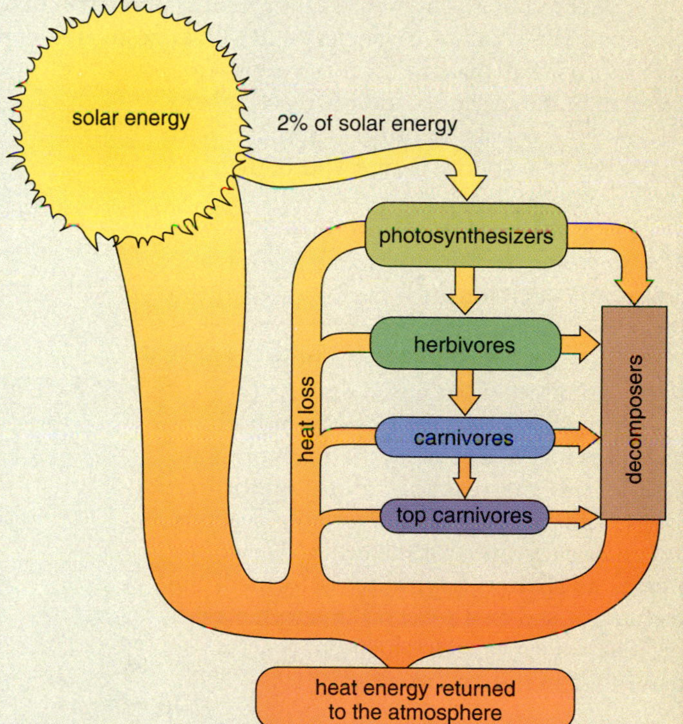

Figure 6A **Energy loss in an ecosystem.** Ordinarily about 2% of the solar energy reaching the earth is taken up by photosynthesizers (plants and algae). This is the energy that allows them to make their own food. Herbivores obtain their food by eating plants, and carnivores obtain food by eating other animals. Whenever the energy content of food is used by organisms, it is eventually converted to heat. With death and decay by decomposers, all the energy temporarily stored in organisms returns as heat to the atmosphere. In order to support a very large population, human beings supplement solar energy with fossil fuel energy to grow crops. Usually, humans feed on crops directly or on animals (herbivores) that have been fed on crops.

6.2 Metabolic Reactions and Energy Transformations

Metabolism is the sum of all the reactions that occur in a cell. **Reactants** are substances that participate in a reaction, while **products** are substances that form as a result of a reaction. In the reaction A + B → C + D, A and B are the reactants while C and D are the products. How would you know that this reaction will occur spontaneously—that is, without an input of energy? Using the concept of entropy, it is possible to state that a reaction will occur spontaneously if it increases the entropy of the universe. But this is not very helpful because in cell biology we don't wish to consider the entire universe. We simply want to consider this reaction. In such instances, cell biologists use the concept of free energy. **Free energy** is the amount of energy available—that is, energy that is still "free" to do work after a chemical reaction has occurred. Free energy is denoted by the symbol G after Josaih Gibbs who first developed the concept. A negative ΔG (change in free energy) means that the products have less free energy than the reactants and the reaction will occur spontaneously. In our reaction, if C and D have less free energy than A and B, then the reaction will "go."

Exergonic reactions are ones in which ΔG is negative and energy is released, while **endergonic reactions** are ones in which the products have more free energy than the reactants. Endergonic reactions can only occur if there is an input of energy.

If the change in free energy in both directions is just about zero, the reaction is reversible and the reaction is at equilibrium. How could you make a reversible reaction "go" in one direction or the other? Very often in cells, as soon as a product is formed, the product is used as a reactant in another reaction. Such occurrences cause the reaction to go in the direction of the product.

Coupled Reactions

Can the energy released by an exergonic reaction be used to "drive" an endergonic reaction? In the body many reactions such as protein synthesis, nerve conduction, or muscle contraction are endergonic: they require an input of energy. On the other hand, the breakdown of ATP to ADP + Ⓟ is exergonic and energy is released (Fig. 6.2).

In **coupled reactions,** the energy released by an exergonic reaction is used to drive an endergonic reaction. ATP breakdown is often coupled to cellular reactions that require an input of energy. Coupling, which requires that the exergonic reaction and the endergonic reaction be closely tied, can be symbolized like this:

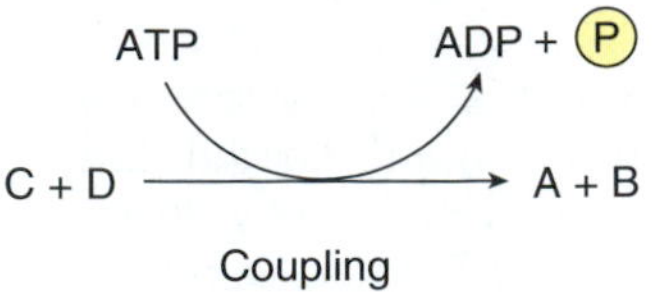

When ATP breakdown is coupled to an endergonic reaction, the overall reaction becomes exergonic (Fig. 6.2). Not all the energy of ATP breakdown is captured and some is immediately lost as heat. Eventually it is all lost as heat.

How is a cell assured of a supply of ATP? Recall that glucose breakdown during cellular respiration provides the energy for the buildup of ATP in mitochondria. Only 39% of the free energy of glucose is transformed to ATP; the rest is lost as heat.

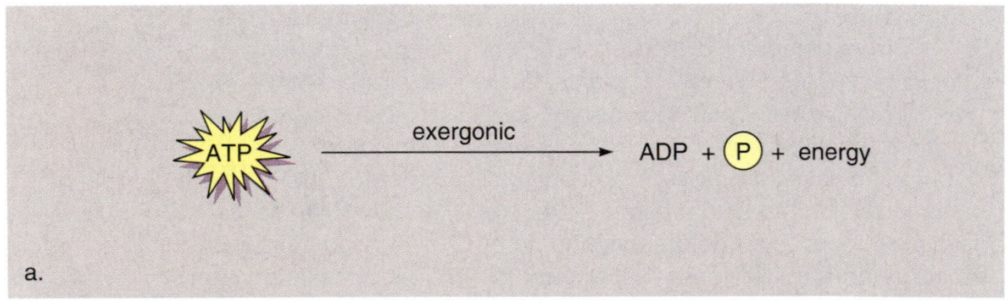

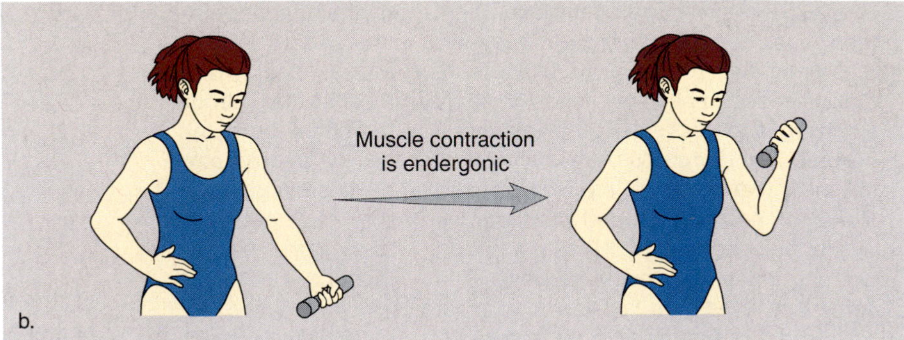

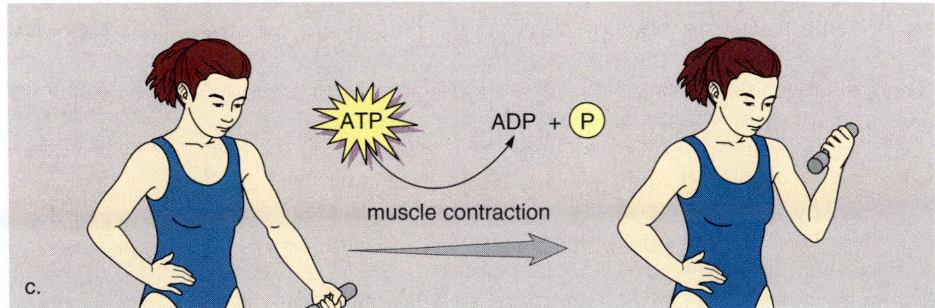

Figure 6.2 Coupled reactions.
a. The breakdown of ATP is exergonic. **b.** Muscle contraction is endergonic and therefore cannot occur without an input of energy. **c.** Muscle contraction is coupled to ATP breakdown, making the overall process exergonic. Now muscle contraction can occur.

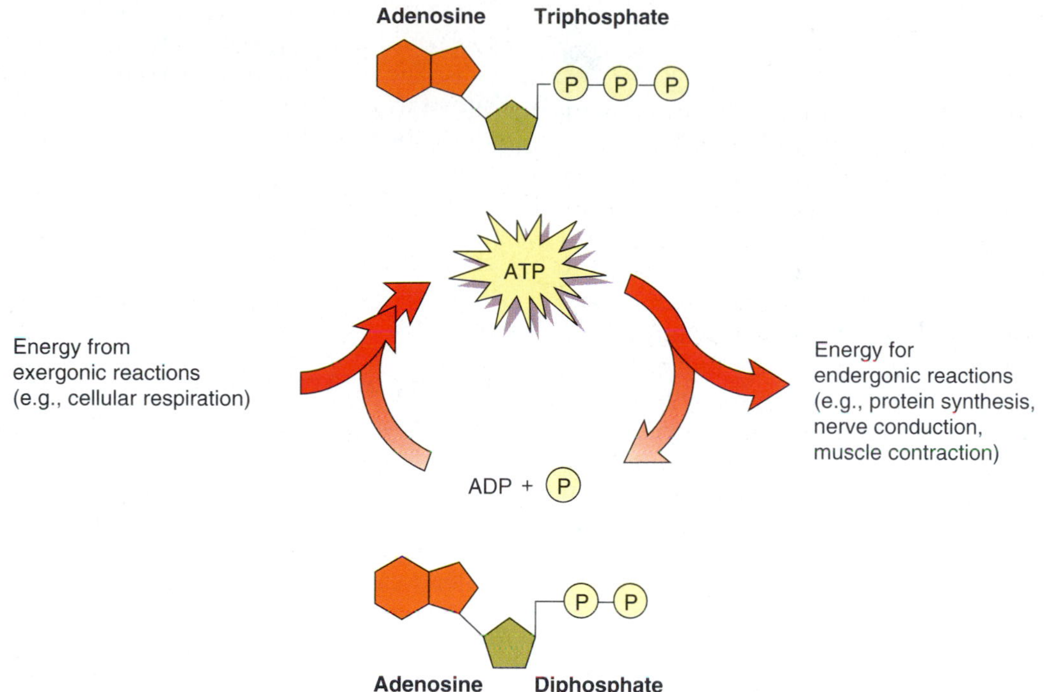

Figure 6.3 **The ATP cycle.**
In cells, the exergonic breakdown of glucose during cellular respiration is coupled to the buildup of ATP, and then the exergonic breakdown of ATP is coupled to various endergonic reactions in cells. When a phosphate group is removed by hydrolysis, ATP releases the appropriate amount of energy for most metabolic reactions. The high-energy content of ATP comes from the complex interaction of the atoms within the molecule.

ATP: Energy for Cells

ATP (adenosine triphosphate) is the common energy currency of cells: when cells require energy, they "spend" ATP. You may think that this causes our bodies to produce a lot of ATP, and it does; however, the amount stored in cells is minimal because ATP is constantly being broken down to **ADP (adenosine diphosphate)** and Ⓟ (Fig. 6.3).

The use of ATP as a carrier of energy has some advantages: (1) It provides a common energy currency that can be used in many different types of reactions. (2) When ATP becomes ADP + Ⓟ, the amount of energy released is just about enough for the biological purposes mentioned in the following section, and so little energy is wasted. (3) ATP breakdown is coupled to endergonic reactions in such a way that it minimizes energy loss.

Function of ATP

Recall that at various times we have mentioned at least three uses for ATP.

Chemical work Supplies the energy needed to synthesize macromolecules that make up the cell.

Transport work Supplies the energy needed to pump substances across the plasma membrane.

Mechanical work Supplies the energy needed to permit muscles to contract, cilia and flagella to beat, chromosomes to move, and so forth.

Structure of ATP

ATP is a nucleotide composed of the base adenine and the sugar ribose (together called adenosine) and three phosphate groups. ATP is called a "high-energy" compound because a phosphate group is easily removed. Under cellular conditions, the amount of energy released when ATP is hydrolyzed to ADP + Ⓟ is about 7.3 kcal per mole.[1]

ATP is a carrier of energy in cells. It is the common energy currency because it supplies energy for many different types of reactions.

[1] A mole is the number of molecules present in the molecular weight of a substance (in grams).

6.3 Metabolic Pathways and Enzymes

Reactions do not occur haphazardly in cells; they are usually a part of a **metabolic pathway,** a series of linked reactions. Metabolic pathways begin with a particular reactant and terminate with an end product. While it is possible to write an overall equation for a pathway as if the beginning reactant went to the end product in one step, there are actually many specific steps in between. In the pathway, one reaction leads to the next reaction, which leads to the next reaction, and so forth in an organized, highly structured manner. This arrangement makes it possible for one pathway to lead to several others, because various pathways have several molecules in common. Also, metabolic energy is captured and utilized more easily if it is released in small increments rather than all at once.

A metabolic pathway can be represented by the following diagram:

$$\begin{array}{ccccccccccccc} & E_1 & & E_2 & & E_3 & & E_4 & & E_5 & & E_6 & \\ A & \rightarrow & B & \rightarrow & C & \rightarrow & D & \rightarrow & E & \rightarrow & F & \rightarrow & G \end{array}$$

In this diagram, the letters A–F are reactants and letters B–G are products in the various reactions. The letters E_1–E_6 are enzymes.

An **enzyme** is a protein molecule[2] that functions as an organic catalyst to speed a chemical reaction. In a crowded ballroom, a mutual friend can cause particular people to interact. In the cell, an enzyme brings together particular molecules and causes them to react with one another.

The reactants in an enzymatic reaction are called the **substrates** for that enzyme. In the first reaction, A is the substrate for E_1 and B is the product. Now B becomes the substrate for E_2, and C is the product. This process continues until the final product G forms.

Any one of the molecules (A–G) in this linear pathway could also be a substrate for an enzyme in another pathway. A diagram showing all the possibilities would be highly branched.

Energy of Activation

Molecules frequently do not react with one another unless they are activated in some way. In the absence of an enzyme, activation is very often achieved by heating the reaction flask to increase the number of effective collisions between molecules. The energy that must be added to cause molecules to react with one another is called the **energy of activation** (E_a). Figure 6.4 compares E_a when an enzyme is not present to when an enzyme is present, illustrating that enzymes lower the amount of energy required for activation to occur.

In baseball, a home-run hitter must not only hit the ball to the fence, but over the fence. When enzymes lower the energy of activation, it is like removing the fence; then it is possible to get a home run by simply hitting the ball as far as the fence was.

[2] Catalytic RNA molecules are called ribozymes and are not enzymes.

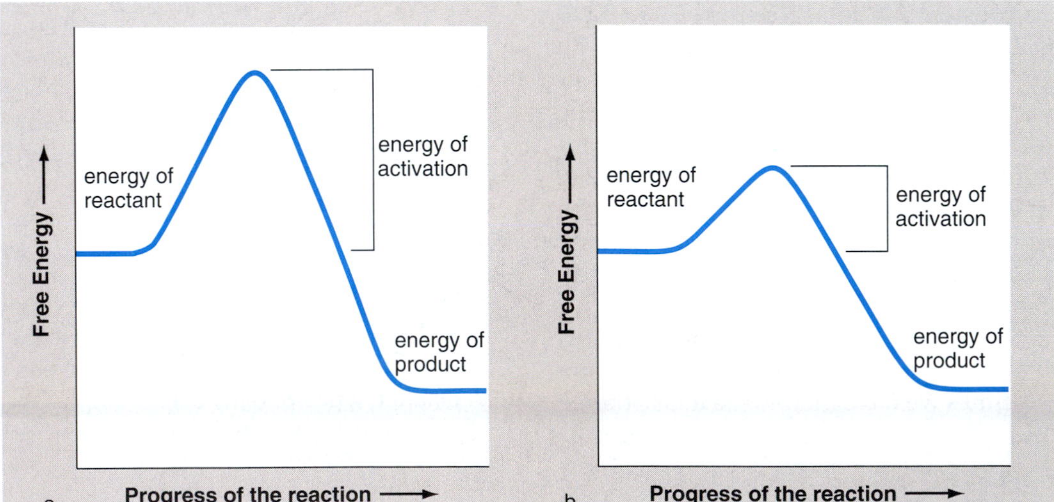

Figure 6.4 **Energy of activation (E_a).**
Enzymes speed the rate of chemical reactions because they lower the amount of energy required to activate the reactants. **a.** Energy of activation when an enzyme is not present. **b.** Energy of activation when an enzyme is present. Even spontaneous reactions like this one speed up when an enzyme is present.

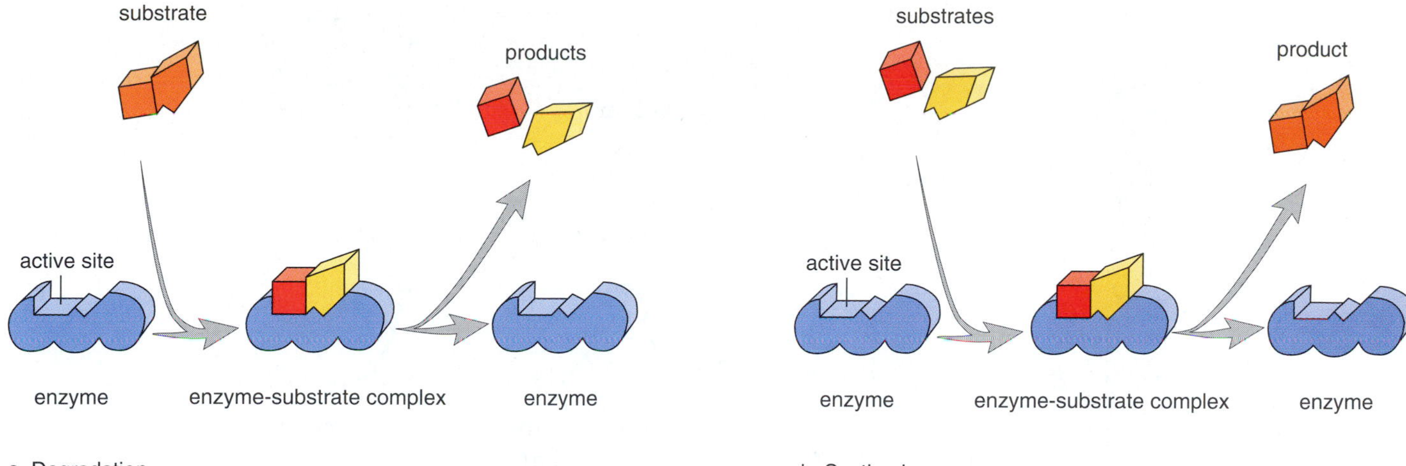

Figure 6.5 Enzymatic action.
An enzyme has an active site, which is where the substrates and enzyme fit together in such a way that the substrates are oriented to react. Following the reaction, the products are released and the enzyme is free to act again. **a.** Some enzymes carry out degradation; the substrate is broken down to smaller products. **b.** Other enzymes carry out synthesis; substrates are combined to produce a larger product.

Enzyme-Substrate Complexes

The following equation, which is pictorially shown in Figure 6.5, is often used to indicate that an enzyme forms a complex with its substrate:

E	+	S	→	ES	→	E	+	P
enzyme		substrate		enzyme-substrate complex				product

In most instances only one small part of the enzyme, called the **active site,** complexes with the substrate(s). It is here that the enzyme and substrate fit together, seemingly like a key fits a lock; however, it is now known that the active site undergoes a slight change in shape in order to accommodate the substrate(s). This is called the **induced-fit model** because the enzyme is induced to undergo a slight alteration to achieve optimum fit (Fig. 6.6).

The change in shape of the active site facilitates the reaction that now occurs. After the reaction has been completed, the product(s) is released, and the active site returns to its original state, ready to bind to another substrate molecule. Only a small amount of enzyme is actually needed in a cell because enzymes are not used up by the reaction.

Some enzymes do more than simply complex with their substrate(s); they actually participate in the reaction. Trypsin digests protein by breaking peptide bonds. The active site of trypsin contains three amino acids with *R* groups that actually interact with members of the peptide bond—first to break the bond and then to introduce the components of water. This illustrates that the formation of the enzyme-substrate complex is very important in speeding up the reaction.

Sometimes it is possible for (a) particular reactant(s) to produce more than one type of product(s). The presence or absence of an enzyme determines which reaction takes place. If a substance can react to form more than one product, then the enzyme that is present and active determines which product is produced.

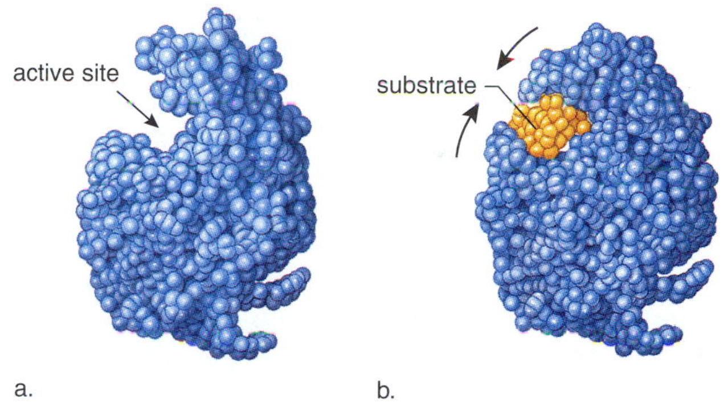

Figure 6.6 Induced fit model.
Computer-generated images of an enzyme called lysozyme that hydrolyzes its substrate, a polysaccharide that makes up bacterial cell walls. **a.** Configuration of enzyme when no substrate is bound to it. **b.** After the substrate binds, the configuration of the enzyme changes so that hydrolysis can better proceed.

Every reaction in a cell requires its specific enzyme. Because enzymes only complex with their substrates, they are named for their substrates. For example, the enzyme that digests lipid molecules is called lipase and the enzyme that digests urea is called urease. Notice that the name of an enzyme ends in "ase."

> Enzymes are protein molecules that speed chemical reactions by lowering the energy of activation. They do this by forming an enzyme-substrate complex.

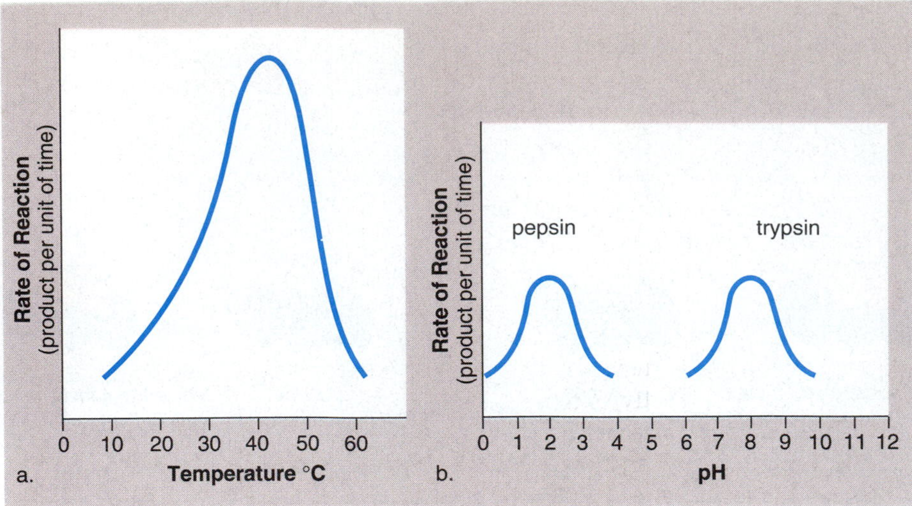

Figure 6.7 **Rate of an enzymatic reaction as a function of temperature and pH.**
a. At first, as with most chemical reactions, the rate of an enzymatic reaction doubles with every 10°C rise in temperature. In this graph, the rate of reaction is maximum at about 40°C; then it decreases until the reaction stops altogether, because the enzyme has become denatured. **b.** Pepsin, an enzyme found in the stomach, acts best at a pH of about 2, while trypsin, an enzyme found in the small intestine, performs optimally at a pH of about 8. The shape that enables these proteins to bind with their substrates is not properly maintained at other pHs.

Factors Affecting Enzymatic Speed

Enzymatic reactions proceed quite rapidly. Consider, for example, the breakdown of hydrogen peroxide (H_2O_2) as catalyzed by the enzyme catalase: $2\ H_2O_2 \rightarrow 2\ H_2O + O_2$. The breakdown of hydrogen peroxide can occur 600,000 times a second when catalase is present. To achieve maximum product per unit time, there should be enough substrate to fill active sites most of the time. Temperature and optimal pH also increase the rate of an enzymatic reaction.

Substrate Concentration

Generally, enzyme activity increases as substrate concentration increases because there are more collisions between substrate molecules and the enzyme. As more substrate molecules fill active sites, more product results per unit time. But when the enzyme's active sites are filled almost continuously with substrate, the enzyme's rate of activity cannot increase anymore. Maximum rate has been reached.

Temperature and pH

As the temperature rises, enzyme activity increases (Fig. 6.7*a*). This occurs because as the temperature rises there are more effective collisions between enzyme and substrate. However, if the temperature rises beyond a certain point, enzyme activity eventually levels out and then declines rapidly because the enzyme is **denatured.** An enzyme's shape changes during denaturation, and then it can no longer bind its substrate(s) efficiently.

Each enzyme also has an optimal pH at which the rate of the reaction is highest. Figure 6.7*b* shows the optimal pH for the enzymes pepsin and trypsin. At this pH value, these enzymes have their normal configurations. The globular shape of an enzyme is dependent on interactions, such as hydrogen bonding, between *R* groups. A change in pH can alter the ionization of these side chains and disrupt normal interactions, and under extreme conditions of pH, denaturation eventually occurs. Again, the enzyme has an altered shape and is then unable to combine efficiently with its substrate.

Enzyme Concentration

Since enzymes are specific, a cell regulates which enzymes are present and/or active at any one time. Otherwise enzymes may be present that are not needed, or one pathway may negate the work of another pathway.

One way to control enzyme activity after it is present is to activate or deactivate the enzyme. Phosphorylation is one way to activate an enzyme. Signal molecules received by membrane receptors often turn on kinases, which then activate enzymes by phosphorylating them:

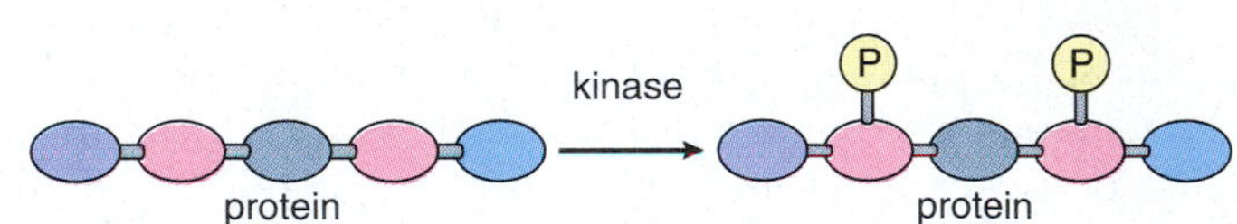

Enzyme Inhibition

Enzyme inhibition occurs when an active enzyme is prevented from combining with its substrate. In *competitive inhibition,* another molecule is so close in shape to the enzyme's substrate that it can compete with the true substrate for the enzyme's active site. This molecule inhibits the reaction because only the binding of the true substrate results in a product. In *noncompetitive inhibition,* a molecule binds to an enzyme, but not at the active site. The other binding site is called the **allosteric site** [Gk. *allo,* other, and *steric,* space, structure]. In this instance, inhibition occurs when binding of a molecule causes a shift in the three-dimensional structure so that the substrate cannot bind to the active site.

The activity of almost every enzyme in a cell can be regulated by its product. When a product is in abundance, it binds competitively with its enzyme's active site; as the product is used up, inhibition is reduced and more product can be produced. In this way, the concentration of the product is always kept within a certain range. Most metabolic pathways are regulated by **feedback inhibition,** but the end product of the pathway binds at an allosteric site on the first enzyme of the pathway (Fig. 6.8). This binding shuts down the pathway, and no more product is produced.

In inhibition, a metabolite binds to the active site or binds to an allosteric site on an enzyme.

Poisons are often enzyme inhibitors. Cyanide is an inhibitor for an essential enzyme (cytochrome *c* oxidase) in all cells, which accounts for its lethal effect on humans. Penicillin blocks the active site of an enzyme unique to bacteria. When penicillin is administered, bacteria die but humans are unaffected.

Enzyme Cofactors

Many enzymes require an inorganic ion or organic, but nonprotein, molecule to function properly. These necessary ions or molecules are called **cofactors.** The inorganic ions are metals such as copper, zinc, or iron. The organic, nonprotein molecules are called **coenzymes.** Coenzymes assist the enzyme and may even accept or contribute atoms to a reaction.

It is interesting that vitamins are often components of coenzymes. **Vitamins** are relatively small organic molecules that are required in trace amounts in our diet and in the diet of other animals for synthesis of coenzymes. The vitamin becomes a part of the coenzyme's molecular structure. For example, vitamin niacin is part of the coenzyme NAD^+ and B_{12} is a part of the coenzyme FAD.

A deficiency of any vitamin results in a lack of a coenzyme and therefore enzymatic actions. In humans, this eventually results in vitamin-deficiency symptoms: niacin deficiency results in a skin disease called pellagra, and riboflavin deficiency results in cracks at the corners of the mouth.

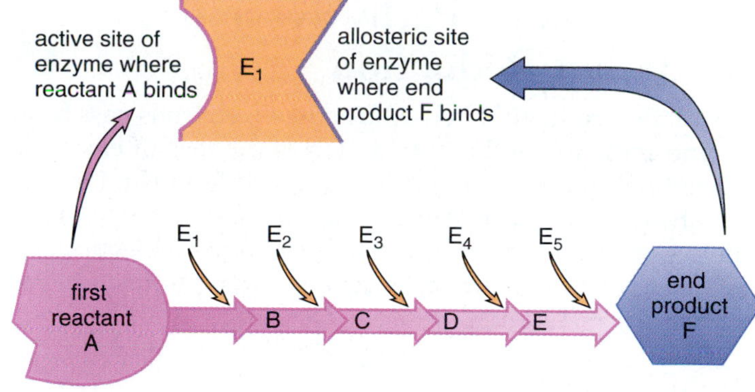

a. Overall view of pathway

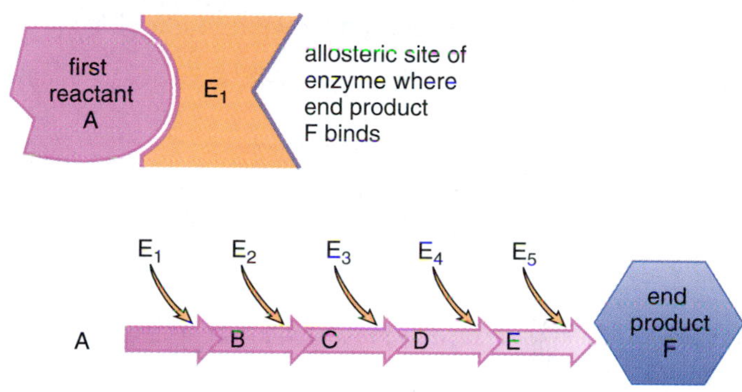

b. View of active pathway

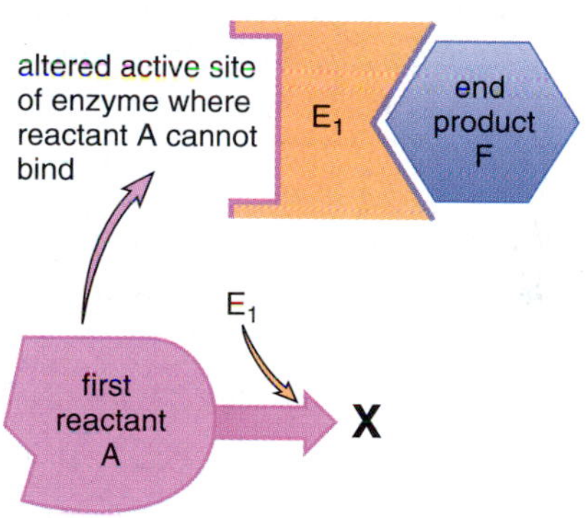

c. View of inhibited pathway

Figure 6.8 Feedback inhibition.
a. This hypothetical metabolic pathway is regulated by feedback inhibition. **b.** When reactant A binds to the active site of E_1, the pathway is active and the end product is produced. **c.** Once there is sufficient end product, some binds to the allosteric site of E_1. Now a change of shape prevents reactant A from binding to the active site of E_1 and the end product is no longer produced.

Enzymes speed a reaction by forming a complex with the substrate. Various factors affect enzymatic speed, including substrate concentration, temperature, pH, enzyme concentration, the presence of inhibitors or necessary cofactors.

6.4 Metabolic Pathways and Oxidation-Reduction

In oxidation-reduction (redox) reactions, electrons pass from one molecule to another. **Oxidation** is the loss of electrons and **reduction** is the gain of electrons. Oxidation and reduction always take place at the same time because one molecule accepts the electrons given up by another molecule. Oxidation-reduction reactions occur during photosynthesis and cellular respiration.

Photosynthesis

In living things, hydrogen ions often accompany electrons, and oxidation is a loss of hydrogen atoms ($e^- + H^+$). Reduction is a gain of hydrogen atoms. For example, the overall reaction for photosynthesis can be written like this:

$$\underset{\text{Carbon dioxide}}{6\,CO_2} + \underset{\text{Water}}{6\,H_2O} + \text{Energy} \rightarrow \underset{\text{Glucose}}{C_6H_{12}O_6} + \underset{\text{Oxygen}}{6\,O_2}$$

This equation shows that when hydrogen atoms are transferred from water to carbon dioxide, glucose is formed. Water has been oxidized and carbon dioxide has been reduced. Since glucose is a high-energy molecule, an input of energy is needed to make the reaction go. Chloroplasts are able to capture solar energy and convert it by way of an electron transport system (discussed in the next section) to the chemical energy of ATP molecules. ATP is then used along with hydrogen atoms to reduce glucose.

A coenzyme of oxidation-reduction called **$NADP^+$ (nicotinamide adenine dinucleotide phosphate)** is active during photosynthesis. This molecule carries a positive charge, and therefore is written as $NADP^+$. During photosynthesis, $NADP^+$ accepts electrons and a hydrogen ion derived from water and later passes them by way of a metabolic pathway to carbon dioxide, forming glucose. The reaction that reduces $NADP^+$ is:

$$NADP^+ + 2\,e^- + H^+ \rightarrow NADPH$$

Cellular Respiration

The overall equation for cellular respiration is opposite that of photosynthesis:

$$\underset{\text{Glucose}}{C_6H_{12}O_6} + \underset{\text{Oxygen}}{6\,O_2} \rightarrow \underset{\text{Carbon dioxide}}{6\,CO_2} + \underset{\text{Water}}{6\,H_2O} + \text{Energy}$$

In this reaction, glucose has lost hydrogen atoms (been oxidized) and oxygen has gained hydrogen atoms (been reduced). When oxygen gains hydrogen atoms it becomes water. Since glucose is a high-energy molecule and water is a low-energy molecule, energy has been released. The organelles called mitochondria use the energy released from glucose breakdown to build ATP molecules by way of an electron transport system that passes electrons to oxygen. Oxygen then becomes water.

In metabolic pathways, most oxidations such as those that occur during cellular respiration involve a coenzyme called **NAD^+ (nicotinamide adenine dinucleotide).** This molecule carries a positive charge, and therefore is represented as NAD^+. During oxidation reactions, NAD^+ accepts two electrons but only one hydrogen ion. The reaction that reduces NAD^+ is:

$$NAD^+ + 2\,e^- + H^+ \rightarrow NADH$$

Electron Transport System

As mentioned above, chloroplasts use solar energy to generate ATP and mitochondria use glucose energy to generate ATP by way of an electron transport system. An **electron transport system** is a series of membrane-bound carriers that pass electrons from one carrier to another. High-energy electrons are delivered to the system and low-energy electrons leave it. Every time electrons are transferred to a new carrier, energy is released; this energy is ultimately used to produce ATP molecules (Fig. 6.9).

In certain redox reactions, the result is release of energy, and in others, energy is required. In an electron transport system, each carrier is reduced and then oxidized in turn. The overall effect of oxidation/reduction as electrons are passed from carrier to carrier of the electron transport system is the release of energy for ATP production.

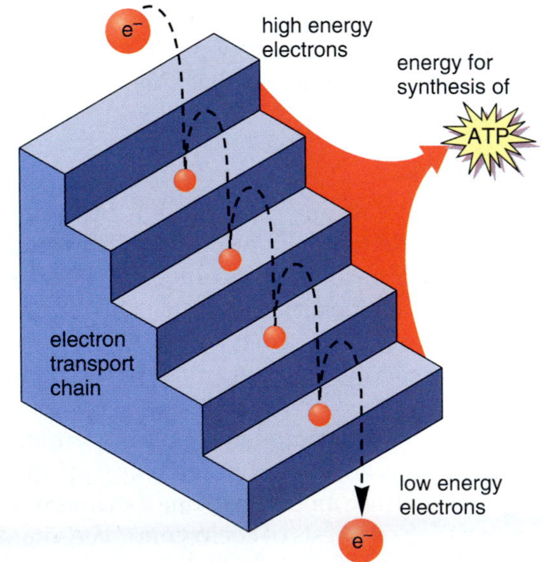

Figure 6.9 Electron transport system.
High-energy electrons enter the system and with each step, as they pass from carrier to carrier, energy is released and used for ATP production.

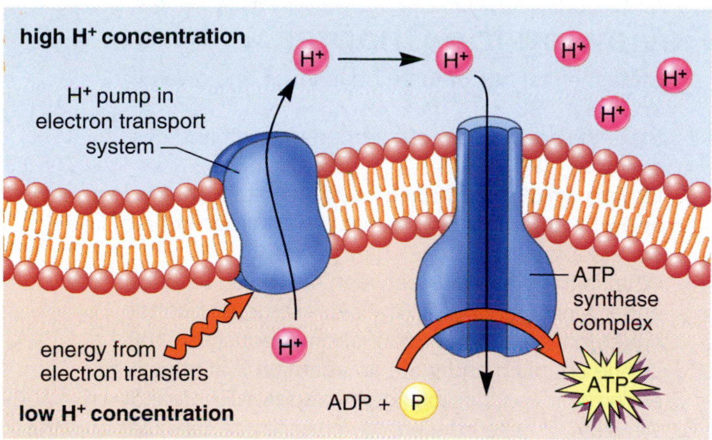

Figure 6.10 **Chemiosmosis.**
Carriers in the electron transport system pump hydrogen ions (H^+) across a membrane. When the hydrogen ions flow back across the membrane through a protein complex, ATP is synthesized by an enzyme called ATP synthase. Chemiosmosis occurs in mitochondria and chloroplasts.

ATP Production

For many years, it was known that ATP synthesis was somehow coupled to the electron transport system, but the exact mechanism could not be determined. Peter Mitchell, a British biochemist, received a Nobel Prize in 1978 for his chemiosmotic theory of ATP production in both mitochondria and chloroplasts.

In mitochondria and chloroplasts, the carriers of the electron transport system are located within a membrane. Hydrogen ions (H^+), which are often referred to as protons in this context, tend to collect on one side of the membrane because they are pumped there by certain carriers of the electron transport system. This establishes an electrochemical gradient across the membrane that can be used to provide energy for ATP production. Particles, called **ATP synthase complexes,** span the membrane. Each complex contains a channel that allows hydrogen ions to flow down their electrochemical gradient. The flow of hydrogen ions through the channel provides the energy for the ATP synthase enzyme to produce ATP from ADP + Ⓟ (Fig. 6.10). **Chemiosmosis** [Gk. *osmos,* push] is the production of ATP due to a hydrogen ion gradient across a membrane.

Consider this analogy to understand chemiosmosis. The sun's rays evaporate water from the seas and help create the winds that blow clouds to the mountains, where water falls in the form of rain and snow. The water in a mountain reservoir has a higher potential energy than that of water in the ocean. The potential energy is converted to electrical energy when water is released and used to turn turbines in an electrochemical dam before it makes its way to the ocean. The continual release of water results in a continual production of electricity.

Similarly, during photosynthesis, solar energy collected by chloroplasts continually leads to ATP production. ATP is produced because thylakoid membrane acts like a dam to maintain an energy gradient in the form of a concentration gradient of hydrogen ions. The hydrogen ions flow through membrane channels that couple the flow of hydrogen ions to the formation of ATP, like the turbines in a hydroelectric dam system couple the flow of water to the formation of electricity.

Similarly, during cellular respiration, glucose breakdown provides the energy to establish a hydrogen ion gradient across the inner membrane of mitochondria. And again hydrogen ions flow through membrane channels that couple the flow of hydrogen ions to the formation of ATP.

As oxidation-reduction occurs, the electron transport system deposits hydrogen ions (H^+) on one side of a membrane. When the ions flow down an electrochemical gradient through an ATPase complex, ATP is formed.

Connecting Concepts

All cells use energy. Energy is the ability to do work, to bring about change, to make things happen, whether it's a leaf growing or a human running. The metabolic pathways inside cells utilize the chemical bond energy of ATP to synthesize molecules, cause muscle contraction, and even allow us to read these words.

A metabolic pathway consists of a series of individual chemical reactions, each with its own enzyme. The cell can regulate the activity of the very many hundreds of different enzymes taking part in cellular metabolism. Enzymes are proteins and as such they are sensitive to environmental conditions, including pH, temperature, and even certain pollutants as will be discussed later in this text.

ATP is called the universal energy "currency" of life. This is an apt analogy—before we can spend currency (i.e., money), we must first make some money. Similarly, before the cell can spend ATP molecules, it must make them. Cellular respiration in mitochondria transforms the chemical bond energy of carbohydrates to that of ATP molecules. An ATP is spent when it is hydrolyzed and the resulting energy is coupled to an endergonic metabolic reaction. All cells are continually making and breaking down ATPs. If ATP is lacking, the organism dies.

What is the ultimate source of energy for ATP production? Except for a few deep ocean vents and also certain cave communities, the answer is the sun. Photosynthesis inside chloroplasts transforms solar energy into the chemical bond energy of carbohydrates. And then carbohydrate products are broken down in mitochondria as ATP is built up. Chloroplasts and mitochondria are the cellular organelles that permit a flow of energy from the sun through all living things.

Summary

6.1 Energy

There are two energy laws that are basic to understanding energy-use patterns at all levels of biological organization. The first law states that energy cannot be created or destroyed, but can only be transferred or transformed. The second law states that one usable form of energy cannot be completely converted into another usable form. As a result of these laws, we know that the entropy of the universe is increasing and that only a constant input of energy maintains the organization of living things.

6.2 Metabolic Reactions and Energy Transformations

Metabolism is a term that encompasses all the chemical reactions occurring in a cell. Considering individual reactions, only those that result in a negative free energy difference—that is, the products have less usable energy than the reactants—occur spontaneously. Such reactions, called exergonic reactions, release energy. Endergonic reactions, which require an input of energy, occur only in cells because it is possible to couple an exergonic process with an endergonic process. For example, glucose breakdown is an exergonic metabolic pathway that drives the buildup of many ATP molecules. These ATP molecules then supply energy for cellular work. Thus, ATP goes through a cycle in which it is constantly being built up from, and then broken down, to ADP + Ⓟ.

6.3 Metabolic Pathways and Enzymes

A metabolic pathway is a series of reactions that proceed in an orderly, step-by-step manner. Each reaction requires a specific enzyme. Reaction rates increase when enzymes form a complex with their substrates. Generally, enzyme activity increases as substrate concentration increases; once all active sites are filled, maximum rate has been achieved.

Any environmental factor, such as temperature and pH, affects the shape of a protein and, therefore, also affects the ability of an enzyme to do its job. Cellular mechanisms regulate enzyme quantity and activity. The activity of most metabolic pathways is regulated by feedback inhibition. Many enzymes have cofactors or coenzymes that help them carry out a reaction.

6.4 Metabolic Pathways and Oxidation-Reduction

Photosynthesis is a metabolic pathway in chloroplasts that transforms solar energy to the chemical energy within carbohydrates (e.g., glucose). Cellular respiration is a metabolic pathway completed in mitochondria that transforms this energy into that of ATP molecules.

The overall equation for photosynthesis is the opposite of that for cellular respiration. During photosynthesis, $NADP^+$ is a coenzyme that reduces substrates, and during cellular respiration, NAD^+ is a coenzyme that oxidizes substrates. Redox reactions are a major way in which energy transformation occurs in cells.

Both processes make use of an electron transport system in which electrons are transferred from one carrier to the next one with the release of energy that is ultimately used to produce ATP molecules. Chemiosmosis explains how the electron transport system produces ATP. The carriers of this system deposit hydrogen ions (H^+) on one side of a membrane. When the ions flow down an electrochemical gradient through an ATPase complex, an enzyme utilizes the release of energy to make ATP from ADP and Ⓟ.

Reviewing the Chapter

1. State the first law of thermodynamics and give an example. 98
2. State the second law of thermodynamics and give an example. 98
3. Explain why the entropy of the universe is always increasing and why an organized system like an organism requires a constant input of useful energy. 98
4. What is the difference between exergonic reactions and endergonic reactions? Why can exergonic but not endergonic reactions occur spontaneously? 100
5. Define coupling and write an equation that shows an endergonic reaction being coupled to ATP breakdown. 100
6. Why is ATP called the energy currency of cells? What is the ATP cycle? 101
7. Diagram a metabolic pathway. Label the reactants, products, and enzymes. 102
8. Why is less energy needed for a reaction to occur when an enzyme is present? 102
9. Why are enzymes specific, and why can't each one speed up many different reactions? 103
10. Name and explain the manner in which at least three factors can influence the speed of an enzymatic reaction. How do cells regulate the activity of enzymes? 104–05
11. What are cofactors and coenzymes? 105
12. Describe how oxidation-reduction occurs in cells and discuss the overall equations for photosynthesis and cellular respiration in terms of oxidation-reduction. 106
13. Describe an electron transport system. 106
14. Tell how cells form ATP during chemiosmosis. 107

Testing Yourself

Choose the best answer for each question.

1. Consider this reaction: $A + B \rightarrow C + D +$ energy.
   a. This reaction is exergonic.
   b. An enzyme could still speed the reaction.
   c. ATP is not needed to make the reaction go.
   d. A and B are reactants; C and D are products.
   e. All of these are correct.
2. The active site of an enzyme
   a. is similar to that of any other enzyme.
   b. is the part of the enzyme where its substrate can fit.
   c. can be used over and over again.
   d. is not affected by environmental factors like pH and temperature.
   e. Both b and c are correct.
3. If you wanted to increase the amount of product per unit time of an enzymatic reaction, do not increase
   a. the amount of substrate.
   b. the amount of enzyme.
   c. the temperature somewhat.
   d. the pH.
   e. All of these are correct.
4. An allosteric site on an enzyme is
   a. the same as the active site.
   b. nonprotein in nature.
   c. where ATP attaches and gives up its energy.
   d. often involved in feedback inhibition.
   e. All of these are correct.

5. During photosynthesis, carbon dioxide
   a. is oxidized to oxygen.
   b. is reduced to glucose.
   c. gives up water to the environment.
   d. is a coenzyme of oxidation-reduction.
   e. All of these are correct.
6. Electron transport systems
   a. are found in both mitochondria and chloroplasts.
   b. release energy as electrons are transferred.
   c. are involved in the production of ATP.
   d. are located in a membrane.
   e. All of these are correct.
7. The difference between NAD^+ and $NADP^+$ is that
   a. only NAD^+ production requires niacin in the diet.
   b. one is an organic molecule and the other is inorganic because it contains phosphate.
   c. one carries electrons to the electron transport system and the other carries them to synthetic reactions.
   d. one is involved in cellular respiration and the other is involved in photosynthesis.
   e. Both c and d are correct.
8. Chemiosmosis is dependent upon
   a. the diffusion of water across a differentially permeable membrane.
   b. an outside supply of phosphate and other chemicals.
   c. the establishment of an electrochemical hydrogen ion (H^+) gradient.
   d. the ability of ADP to join with Ⓟ even in the absence of a supply of energy.
   e. All of these are correct.
9. Use these terms to label this diagram: substrates, enzyme (used twice), active site, product, and enzyme-substrate complex. Explain the importance of an enzyme's shape to its activity.

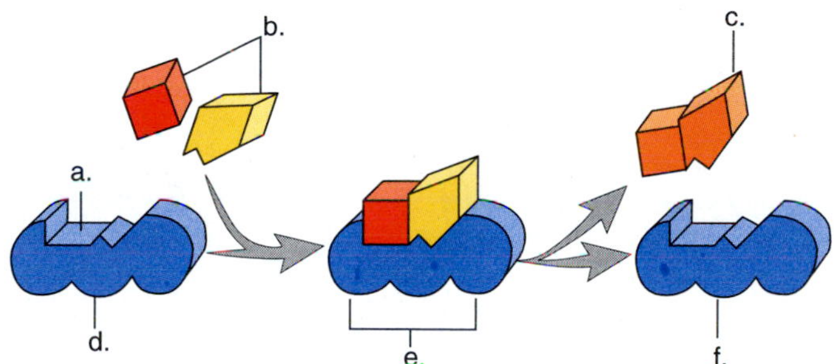

10. Label this diagram describing chemiosmosis.

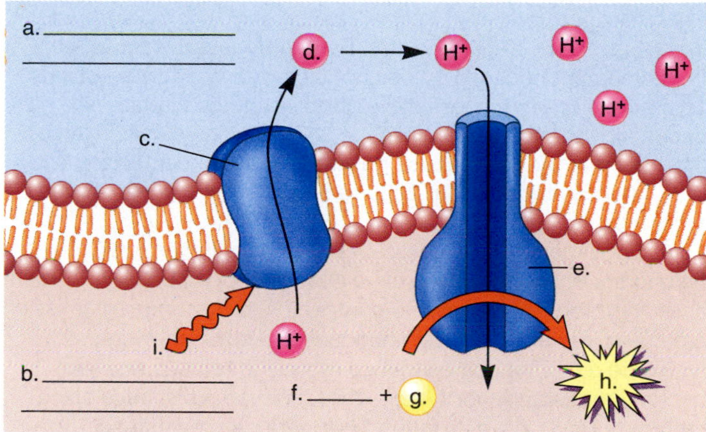

Thinking Scientifically

1. A certain flower generates heat. This heat attracts pollinating insects to the flower. While the evolutionary benefit of attracting insects is obvious, the metabolic cost of this particular adaptation is high. What metabolic mechanism(s) might a plant use to generate heat and under what circumstances would the metabolic cost be high?
2. The free energy of carbon dioxide and water is considerably less than the free energy of sucrose (table sugar). However, the conversion of sucrose to carbon dioxide and water is never spontaneous under normal conditions. How would you explain this observation?

Bioethical Issue

Today, we are very much concerned about emerging diseases caused by parasites. Emerging diseases are ones like AIDS and Ebola, which emerge from their natural host to cause illness in humans. In 1993, the hantavirus strain emerged from the common deer mouse and killed about 60 young people in the Southwest. In the case of hantavirus, we know that climate was involved. An unusually mild winter and wet spring caused piñon trees to bloom well and provide pine nuts to the mice. The increasing deer mouse population came into contact with humans, and the hantavirus leaped easily from mice to humans.

The prediction is that global warming, caused in large part by the burning of fossil fuels, will upset normal weather cycles and result in outbreaks of hantavirus as well as malaria, dengue and yellow fevers, filariasis, encephalitis, schistosomiasis, and cholera. Clearly any connection between global warming and emerging diseases offers another reason why greenhouse gases should be curtailed when fossil fuels like gasoline are consumed. Greenhouse gases are those like carbon dioxide and methane, which allow the sun's rays to pass through but then trap the heat from escaping.

In December of 1997, 159 countries met in Kyoto, Japan, to work out a protocol that would reduce greenhouse gases worldwide. It is believed that the emission of greenhouse gases, especially from power plants, will cause earth's temperature to rise 1.5°–4.5° by 2060. The U.S. Senate does not want to ratify the agreement because it does not include a binding emissions commitment from the less-developed countries that are only now becoming industrialized. While the U.S. presently emits a large proportion of the greenhouse gases, China is expected to pass the United States in about 2020 to become the biggest source of greenhouse emissions.

Negotiations with the less-developed countries is still going on and some creative ideas have been put forward. Why not have a trading program that allows companies to buy and sell emission credits across international boundaries? Accompanying that would be a market in greenhouse reduction techniques. If it became monetarily worth their while, companies in developing countries would have an incentive to reduce greenhouse emissions. If you were a CEO, would you be willing to reduce greenhouse emissions simply because they cause a deterioration of the environment and probably cause human illness? Why or why not? Instead, do you approve of giving companies monetary incentives to reduce greenhouse emissions? Why or why not?

Understanding the Terms

active site 103
ADP (adenosine diphosphate) 101
allosteric site 105
ATP (adenosine triphosphate) 101
ATP synthase complex 107
chemical energy 98
chemiosmosis 107
coenzyme 105
cofactor 105
coupled reactions 100
denatured 104
electron transport system 106
endergonic reaction 100
energy 98
energy of activation 102
entropy 98
enzyme 102
enzyme inhibition 105
exergonic reaction 100
feedback inhibition 105
free energy 100
induced-fit model 103
kinetic energy 98
laws of thermodynamics 98
metabolic pathway 102
metabolism 100
NAD^+ (nicotinamide adenine dinucleotide) 106
$NADP^+$ (nicotinamide adenine dinucleotide phosphate) 106
oxidation 106
potential energy 98
product 100
reactant 100
reduction 106
substrate 102
vitamin 105

Match the terms to these definitions:

a. ______________ All of the chemical reactions that occur in a cell during growth and repair.
b. ______________ Stored energy as a result of location or spatial arrangement.
c. ______________ Essential requirement in the diet, needed in small amounts. They are often part of coenzymes.
d. ______________ Measure of disorder or randomness.
e. ______________ Nonprotein organic molecule that aids the action of the enzyme to which it is loosely bound.

Web Connections

Exploring the Internet

http://www.mhhe.com/biosci/genbio/mader
(click on *Biology 7/e*)

The *Biology 7/e* Online Learning Center provides many resources for studying the material in this chapter including links to the following sites:

Enzymes Tutorials. Answer questions as you learn more about enzymes, how scientists study enzyme-catalyzed reactions, and enzyme kinetics.

http://biog-101-104.bio.cornell.edu/BioG101_104/tutorials/enzymes.html

Animated GIFs and Protein Chemistry. In a short visit to this site, you can encounter the three-dimensional structure of an enzyme with its substrate bound to the active site.

http://www2.ucsc.edu/people/straycat/cpa.html

Enzymes in the Manufacture of Fruit Juices. Information on the practical uses of enzymes utilized in the commercial manufacture of fruit juices.

http://134.225.167.114/NCBE/FEATURES/juice.html

Enzyme Mechanisms. This chapter is from an MIT hypertextbook. Chemical energetics, enzyme mechanics, enzyme kinetics, and feedback inhibition are covered, and there are feedback regulation problems included.

http://esg-www.mit.edu:8001/esgbio/eb/ebdir.html

Whole Food Formula—Enzymes. Visit this rather scientifically incorrect site to see how many people pay lots of money due to lack of knowledge of enzymes. As a scientist, why can you see that this is just a waste of money? What statements in this ad are correct facts and which are not?

http://www.wimall.com/friend/enzyme.html

Photosynthesis

chapter 7

chapter concepts

7.1 Solar Energy

- Plants make use of solar energy in the visible light range when they carry on photosynthesis. 112

7.2 Structure and Function of Chloroplasts

- Photosynthesis takes place in chloroplasts, organelles that contain membranous thylakoids surrounded by a fluid called stroma. 114
- Photosynthesis has two sets of reactions: solar energy is captured by the pigments in thylakoids, and carbon dioxide is reduced by enzymes in the stroma. 114

7.3 Solar Energy Capture

- Solar energy energizes electrons and permits a buildup of ATP. 116

7.4 Carbohydrate Synthesis

- Carbon dioxide reduction requires energized electrons and ATP. 119
- Plants use C_3, or C_4, or CAM photosynthesis, which are distinguishable by the manner in which CO_2 is fixed. 122

Giant panda, *Ailuropoda melanoleuca*, feeding on bamboo, *Bambusa arundinacea*.

Why would most life on earth perish without photosynthesis? Through photosynthesis plants and algae produce food for themselves and all other living things. Animals, like giant pandas, feed directly on photosynthesizers while other animals feed indirectly on photosynthesizers.

When photosynthesis occurs, carbon dioxide is absorbed and oxygen is released. Oxygen is required by organisms when they carry on cellular respiration. Oxygen also rises high in the atmosphere and forms the ozone shield that protects terrestrial organisms from the damaging effects of the ultraviolet rays of the sun. Planting trees and saving forests helps purify the air of carbon dioxide which is poisonous to many organisms and contributes to global warming.

The human population feeds on plants, but they are also a source of fabrics, rope, paper, lumber, fuel, and pharmaceuticals. In short, plants make modern society possible. And while we are thanking green plants for human survival, let's not forget the simple beauty of a magnolia bloom or the majesty of an old-growth forest.

7.1 Solar Energy

Solar energy makes our modern way of life possible not only because it can be used directly to heat buildings, but also because the bodies of plants became the fossil fuel coal which we still use today to generate electricity. Photosynthetic organisms, like plants and algae, produce organic food for the biosphere. Organic nutrients serve as an energy source and the building blocks that permit growth and repair. Photosynthesis is carried out in this manner:

solar energy + carbon dioxide + water → carbohydrate + oxygen

Because **photosynthesis** is an energy transformation in which solar energy is converted to the chemical bond energy of carbohydrate molecules, we will examine solar radiation in a little more detail.

Electromagnetic Spectrum

Solar radiation can be described in terms of its energy content and its wavelength. The energy comes in discrete packets called **photons** [Gk. *phos*, light]. So, in other words, you can think of radiation as photons that travel in waves:

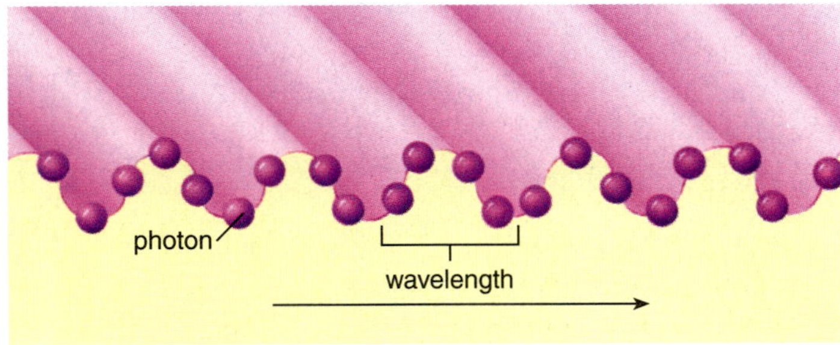

Figure 7.1*a* illustrates that solar radiation, or the **electromagnetic spectrum,** can be divided on the basis of wavelength—gamma rays have the shortest wavelength and radio waves have the longest wavelength. The energy content of photons is inversely proportional to the wavelength of the particular type of radiation; that is, short-wavelength radiation has photons of a higher energy content than long-wavelength radiation. High-energy photons, such as those of short-wavelength ultraviolet radiation, are dangerous to cells because they can break down organic molecules. Low-energy photons, such as those of infrared radiation, do not damage cells. They only increase the vibrational or rotational energy of molecules; and do not break bonds. Photosynthesis utilizes only the portion of the electromagnetic spectrum known as **visible light.** (It is called visible light because it is the part of the spectrum that the eye can see.) Photons of visible, or white light, have just the right amount of energy to promote electrons to a higher electron shell in atoms without harming cells. White light is actually made up of a number of different wavelengths of radiation; when it is passed through a prism (or through raindrops), we see these wavelengths as different colors of light.

Energy Balance Sheet

Only 42% of solar energy directed toward earth actually reaches the earth's surface; the rest is absorbed by or reflected into the atmosphere and becomes heat.

Of this usable portion, only about 2% is eventually utilized by plants; the rest becomes heat.

Of this, only 0.1–1.6% is ever incorporated into plant material; the rest becomes heat.*

Of this, only 20% is eaten by herbivores; a large proportion of the remainder becomes heat.*

Of this, only 30% is ever eaten by carnivores; a large proportion becomes heat.*

Conclusion: Most of the available energy is never utilized by living things.

** Plant and animal remains became the fossil fuels that we burn today to provide energy. So eventually this energy becomes heat.*

Energy Balance Sheet

Only about 42% of solar radiation passes through the earth's atmosphere and reaches its surface. Most of this radiation is within the visible-light range. Higher energy wavelengths are screened out by the ozone layer in the atmosphere, and lower energy wavelengths are screened out by water vapor and carbon dioxide before they reach the earth's surface. The conclusion is, then, that organic molecules and processes within organisms, like vision and photosynthesis, are chemically adapted to the radiation that is most prevalent in the environment—that is, visible light.

The energy balance sheet given on this page shows that most of the available energy reaching the earth is never utilized by living things. Photosynthetic pigments capture less than 2% of the solar energy that reaches the earth. Of this only a small percentage ever becomes a part of living things.

Photosynthesis utilizes the portion of the electromagnetic spectrum (solar radiation) known as visible light.

Photosynthetic Pigments

The pigments found in photosynthesizing cells are capable of absorbing various portions of visible light. This is called their absorption spectrum. Photosynthetic organisms differ by the type of chlorophyll they utilize. In plants, chlorophyll *a* and chlorophyll *b* play a prominent role in photosynthesis. The absorption spectra for chlorophylls *a* and *b* are shown in Figure 7.1*b*. Both chlorophylls *a* and *b* absorb violet, blue, and red light better than the light of other colors. Because green light is only minimally absorbed, plant leaves appear green to us. Other plant pigments such as the carotenoids

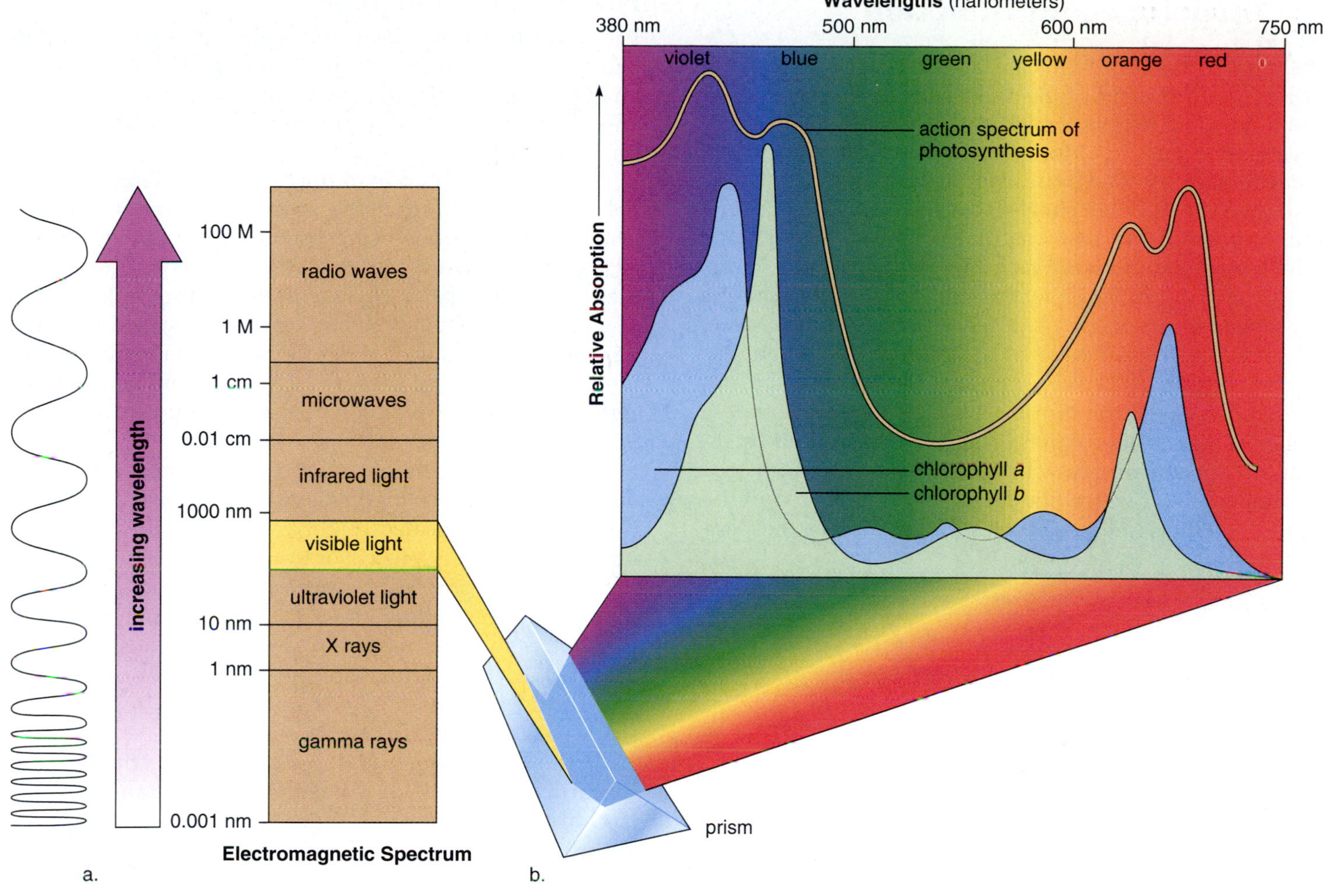

Figure 7.1 The electromagnetic spectrum and chlorophylls *a* and *b*.
a. The electromagnetic spectrum contains forms of energy that differ according to wavelength. Visible light is only a small portion of the electromagnetic spectrum. b. Chlorophylls *a* and *b*, significant pigments in plants, absorb certain wavelengths within visible light. This is their absorption spectrum. The action spectrum for photosynthesis in plants—the wavelengths that are used when photosynthesis is taking place—matches well the sum of the absorption spectrums for chlorophylls *a* and *b*.

are shades of yellow and orange and are able to absorb light in the violet-blue-green range. These pigments become noticeable in the fall when chlorophyll breaks down.

How do you determine the absorption spectrum of pigments? To identify the absorption spectrum of a particular pigment, a purified sample is exposed to different wavelengths of light inside an instrument called a spectrophotometer. A spectrophotometer measures the amount of light that passes through the sample, and from this it can be calculated how much was absorbed. The amount of light absorbed at each wavelength is plotted on a graph, and the result is a record of the pigment's absorption spectrum (Fig. 7.1*b*). How do we know that the peaks shown in the absorption spectrum for chlorophyll *a* and *b* indicate which wavelengths plants use for photosynthesis? Because photosynthesis gives off oxygen, we can use the production rate of oxygen as a means to measure the rate of photosynthesis at each wavelength of light. When such data are plotted, the resulting graph is a record of the action spectrum for photosynthesis in plants. An action spectrum is those portions of the electromagnetic spectrum that are used to perform a function, in this case, photosynthesis. Because the sum of the absorption spectrum for chlorophylls *a* and *b* matches the action spectrum for plant photosynthesis, we are confident that the light absorbed by the chlorophylls does contribute extensively to photosynthesis in plants.

The sum of the absorption spectrums for chlorophylls *a* and *b* matches well the action spectrum for photosynthesis.

7.2 Structure and Function of Chloroplasts

It wasn't until the end of the nineteenth century that scientists understood the photosynthetic process and that, in eukaryotes, it occurs in **chloroplasts** [Gk. *chloros*, green, and *plastos*, formed, molded]. The overall equation for photosynthesis given on page 112 suggests that in the presence of sunlight, carbon dioxide, and water are combined to produce a high-energy food molecule such as carbohydrate.

In 1930, C. B. van Niel of Stanford University found that oxygen given off by photosynthesis comes from water and not from carbon dioxide as had been originally thought. This was proven by two separate experiments. When plants were exposed to carbon dioxide that contained an isotope of oxygen, called heavy oxygen (^{18}O), the O_2 given off by the plant did not contain heavy oxygen. Only when heavy oxygen was a part of water (indicated by the color red in this equation) did this isotope appear in O_2 given off by the plant:

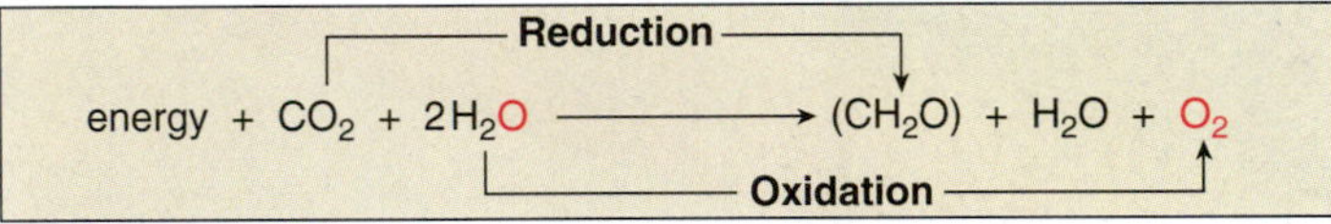

Notice that carbon dioxide is reduced and water is oxidized during photosynthesis. This equation is preferred by some because it has the advantage of keeping the chemical arithmetic correct. In this equation CH_2O stands for carbohydrate.

Structure of Chloroplasts

In a chloroplast, a double membrane surrounds a fluid called the **stroma** [Gk. *stroma*, bed, mattress]. The stroma is an enzyme-rich solution where CO_2 is first attached to an organic compound and then is reduced. A membrane system within the stroma forms flattened sacs called **thylakoids** [Gk. *thylakos*, sack, and *eides*, like, resembling], which in some places are stacked to form grana (sing., **granum**), so called because they looked like piles of seeds to early microscopists. The space within each thylakoid is thought to be connected to the space within every other thylakoid, thereby forming an inner compartment within chloroplasts called the thylakoid space. **Chlorophyll** and other pigments are found within the membranes of the thylakoids. These pigments absorb the solar energy, which will energize electrons prior to reduction of CO_2 in the stroma (Fig. 7.2).

Because chlorophyll, within thylakoids, absorbs solar energy, electrons are available for reduction of CO_2 in the stroma.

Function of Chloroplasts

An overall equation tells us the beginning reactants and the end products of a metabolic pathway. What about what goes on in between the first reactants (carbon dioxide and water) and the final products (carbohydrate and water) during photosynthesis?

In 1905, F. F. Blackman suggested that there were probably two sets of reactions involved in photosynthesis because when light is being maximally absorbed, a rise in temperature still increases the rate of photosynthesis. (Enzymatic reactions speed up when temperature is increased.) The first set of photosynthetic reactions is called the **light-dependent reactions** because they cannot take place unless light is present. The second set of reactions is called the **light-independent reactions** because they can take place whether light is present or not.

The light-dependent reactions occur in the thylakoid membrane where the pigments chlorophyll *a* and *b*, plus the carotenoids, are located. Light can be absorbed by (taken up), reflected by (given off), or transmitted (passed through) a pigment. Chlorophyll *a* appears blue-green and chlorophyll *b* appears yellow-green to us because these are the very colors they do not absorb and are instead reflected to our eyes or transmitted through these pigments (see Fig. 7.1). For the same reason, the carotenoids appear as various shades of yellow and orange to our eyes.

The light-dependent reactions are the energy-capturing reactions—low-energy electrons removed from H_2O are energized when thylakoid membrane pigments absorb solar energy. These electrons move from chlorophyll *a* down an electron transport system, which produces ATP from ADP and Ⓟ. Energized electrons are also taken up by $NADP^+$. After $NADP^+$ accepts electrons, it becomes NADPH. This molecule temporarily holds energy, in the form of energized electrons, that will be used to reduce CO_2.

The light-dependent reactions capture solar energy.

The second set of reactions involved in photosynthesis occurs in the stroma of a chloroplast. They are called the light-independent reactions because they can take place in either the light or the dark. The light-independent reactions are the synthesis reactions, which use the ATP and NADPH formed in the thylakoids to reduce CO_2.

The light-independent reactions synthesize carbohydrate.

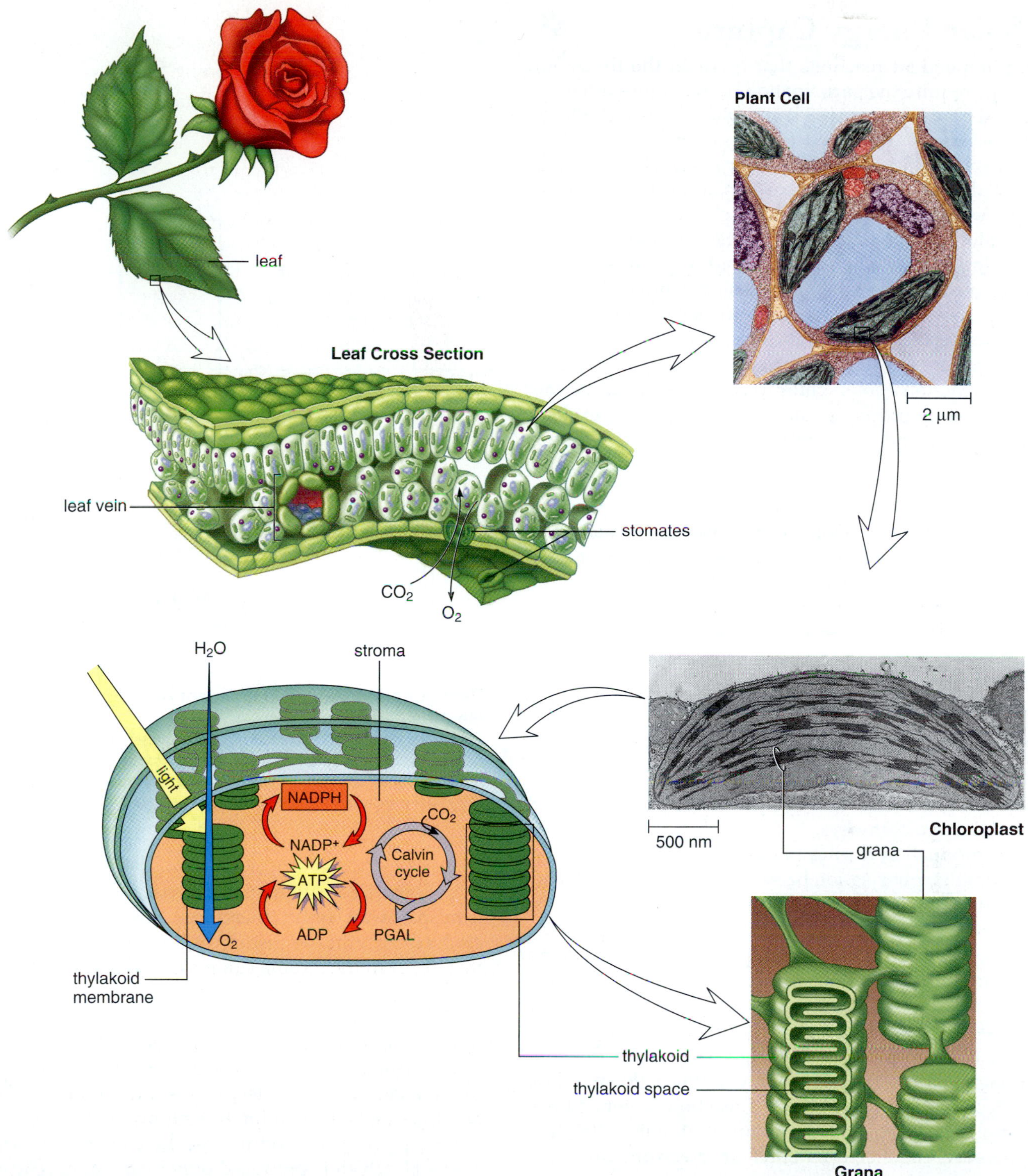

Figure 7.2 Chloroplast structure and function.
The cells that form the bulk of a leaf contain many chloroplasts. Each chloroplast is bounded by a double membrane and contains a fluid called stroma. The membranous thylakoids stacked in grana are interconnected so that there is a single thylakoid space. The light-dependent reactions, which produce ATP and NADPH, occur in the thylakoid membrane. These molecules are used by the light-independent reactions to reduce carbon dioxide during a cyclical series of enzymatic reactions called the Calvin cycle. The Calvin cycle occurs in the stroma.

7.3 Solar Energy Capture

The light-dependent reactions that occur in the thylakoid membranes require the participation of two light-gathering units called photosystem I (PS I) and photosystem II (PS II). The photosystems are named for the order in which they were discovered and not for the order in which they occur in the thylakoid membrane or participate in the photosynthetic process. Each **photosystem** has closely packed molecules of chlorophyll *a* and chlorophyll *b* molecules and accessory pigments, such as carotenoid pigments. These pigment molecules are called an antenna complex because they serve as an "antenna" for gathering solar energy. Solar energy is passed from one pigment to the other until it is concentrated into one of two particular chlorophyll *a* molecules, the reaction-center chlorophylls. Electrons in the reaction-center chlorophyll *a* molecules become so excited that they escape and move to a nearby electron-acceptor molecule.

In a photosystem, the light-gathering antenna complex absorbs solar energy and funnels it to a reaction-center chlorophyll *a* molecule, which then sends energized electrons to an electron-acceptor molecule.

Electron Pathways

Electrons can follow a cyclic electron pathway or a noncyclic electron pathway during the first phase of photosynthesis. The cyclic electron pathway generates only ATP, while the noncyclic pathway results in both NADPH and ATP. The cell regulates the proportion of ATP to NADPH by the relative activity of the two pathways.

ATP production during photosynthesis is sometimes called photophosphorylation because light is involved. The production of ATP during the cyclic electron pathway is called cyclic photophosphorylation whereas ATP production during the noncyclic electron pathway is called noncyclic photophosphorylation.

Cyclic Electron Pathway

The **cyclic electron pathway** (Fig. 7.3) begins after the PS I antenna complex absorbs solar energy. In this pathway, high-energy electrons (e^-) leave the PS I reaction-center chlorophyll *a* molecule but eventually return to it. Before they return, however, the electrons enter an **electron transport system,** a series of carriers that pass electrons from one to the other. As the electrons pass from one carrier to the next, energy is released and stored in the form of a hydrogen (H^+) gradient. When these hydrogen ions flow down their electrochemical gradient through ATP synthase complexes, ATP production occurs (see page 118).

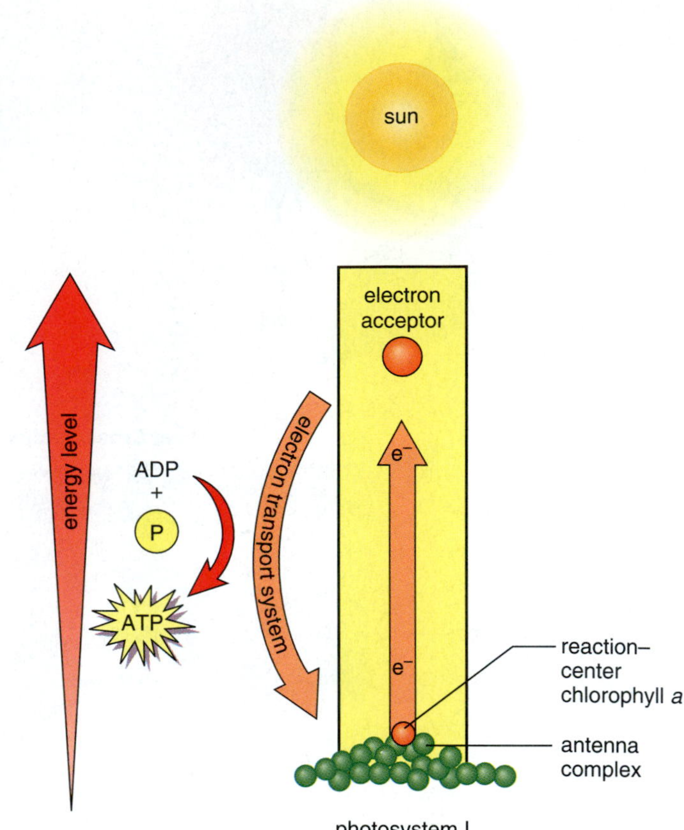

Figure 7.3 The light-dependent reactions: the cyclic electron pathway.
Energized electrons (e^-) leave the photosystem I (PS I) reaction-center chlorophyll *a* and are taken up by an electron acceptor, which passes them down an electron transport system before they return to PS I. Only ATP production results from this pathway.

Some photosynthetic bacteria utilize the cyclic electron pathway only; therefore, this pathway probably evolved early in the history of life. In plants the reactions that occur in the stroma can make use of this extra ATP for the Calvin cycle because this cycle requires a larger number of ATP than NADPH. Also, there are other enzymatic reactions aside from those involving photosynthesis occurring in the stroma that can make use of this ATP. It's possible that the cyclic flow of electrons is utilized alone when carbon dioxide is in such limited supply that carbohydrate is not being produced. At this time there would be no need for NADPH, which is produced only by the noncyclic electron pathway.

The cyclic electron pathway, from PS I back to PS I, has only one effect: production of ATP.

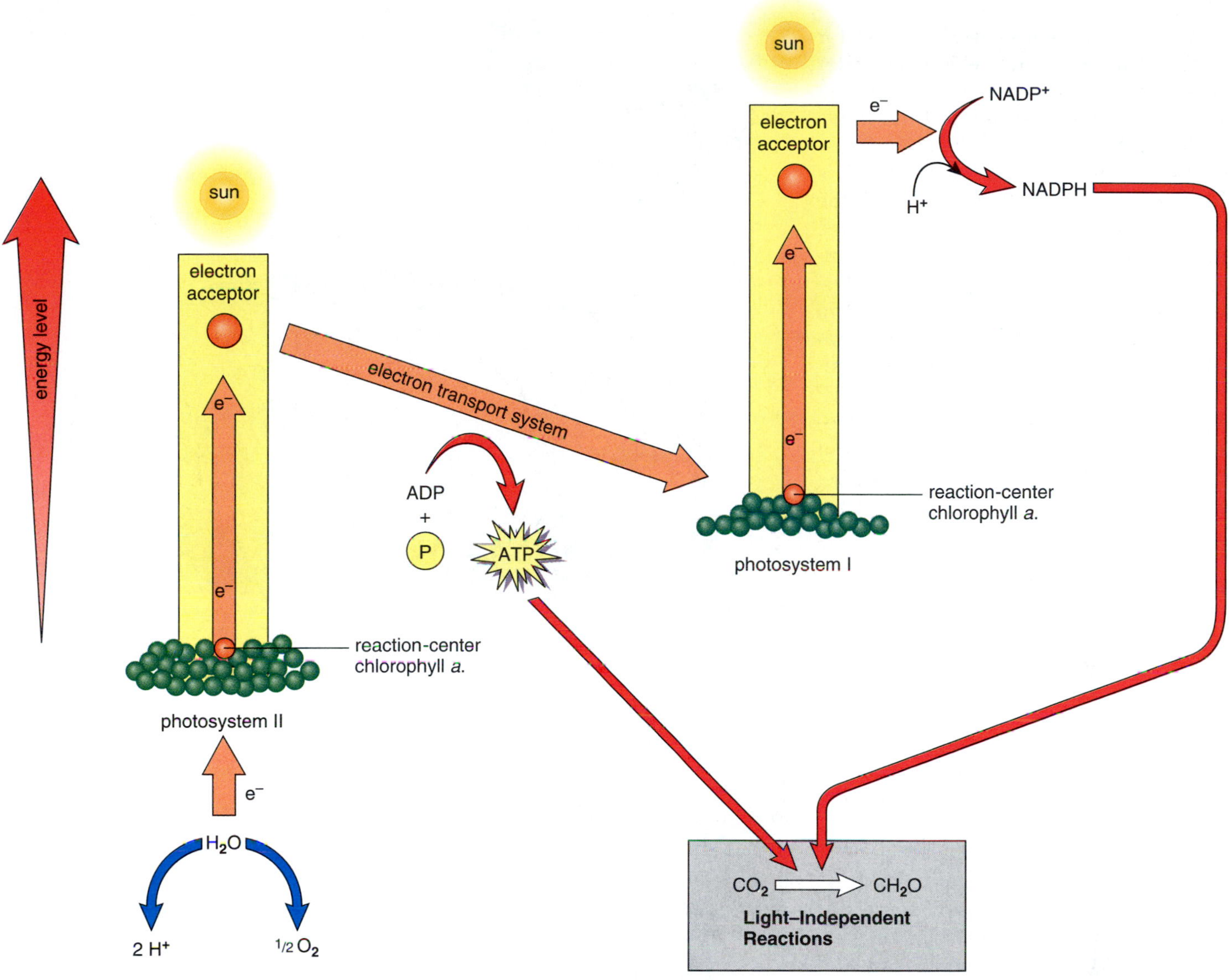

Figure 7.4 The light-dependent reactions: the noncyclic electron pathway.
Electrons, taken from water, move from photosystem II (PS II) to photosystem I (PS I) to $NADP^+$. The ATP and NADPH produced will be used by the light-independent reactions to reduce carbon dioxide (CO_2) to a carbohydrate (CH_2O).

Noncyclic Electron Pathway

During the **noncyclic electron pathway,** electrons move from water (H_2O) through PS II to PS I and then on to $NADP^+$ (Fig. 7.4). This pathway begins when the PS II antenna complex absorbs solar energy and high-energy electrons (e^-) leave the reaction-center chlorophyll *a* molecule. PS II takes replacement electrons from water, which splits, releasing oxygen. This oxygen is released from the chloroplast and the plant as oxygen gas. The hydrogen ions (H^+) stay in the thylakoid space and contribute to the formation of a hydrogen ion gradient.

The high-energy electrons that leave PS II are captured by an electron acceptor, which sends them to an electron transport system. As the electrons pass from one carrier to the next, energy is released and stored in the form of a hydrogen ion (H^+) gradient. When these hydrogen ions flow down their electrochemical gradient through ATP synthase complexes, ATP production occurs (see page 118).

Low-energy electrons leaving the electron transport system enter PS I. When the PS I antenna complex absorbs solar energy, high-energy electrons leave the reaction-center chlorophyll *a* and are captured by an electron acceptor. This time, the electron acceptor passes the electrons on to $NADP^+$. $NADP^+$ now takes on an H^+ and becomes NADPH. The NADPH and ATP produced by the noncyclic flow of electrons in the thylakoid membrane are used by enzymes in the stroma during the light-independent reactions.

Results of noncyclic electron flow: water is oxidized (split), yielding H^+, e^-, and O_2; ATP is produced; and $NADP^+$ becomes NADPH.

ATP Production

The thylakoid space acts as a reservoir for hydrogen ions (H^+). First, each time water is oxidized, two H^+ remain in the thylakoid space. Second, as the electrons move from carrier to carrier in the electron transport system, they give up energy, which is used to pump H^+ from the stroma into the thylakoid space. Therefore, there is a large number of H^+ in the thylakoid space compared to the number in the stroma. The flow of H^+ (often referred to as protons in this context) from high to low concentration across the thylakoid membrane provides the energy that allows an **ATP synthase** enzyme to enzymatically produce ATP from ADP + Ⓟ. This method of producing ATP is called **chemiosmosis** because ATP production is tied to an electrochemical gradient.

The Thylakoid Membrane

Both biochemical and structural techniques have been used to determine that there are intact complexes (particles) in the thylakoid membrane (Fig. 7.5):

PS II consists of a protein complex and a light-gathering antenna complex shown to one side. PS II oxidizes water and produces oxygen.

The cytochrome complex acts as the transporter of electrons between PS II and PS I. The pumping of H^+ occurs during electron transport.

PS I consists of a protein complex and a light-gathering antenna complex to one side. Notice that PS I is associated with the enzyme that reduces $NADP^+$ to NADPH.

ATP synthase complex has an H^+ channel and a protruding ATP synthase. As H^+ flows down its concentration gradient through this channel from the thylakoid space into the stroma, ATP is produced from ADP + Ⓟ.

The light-dependent reactions that occur in the thylakoid membrane produce ATP and NADPH and oxidize water, giving off oxygen.

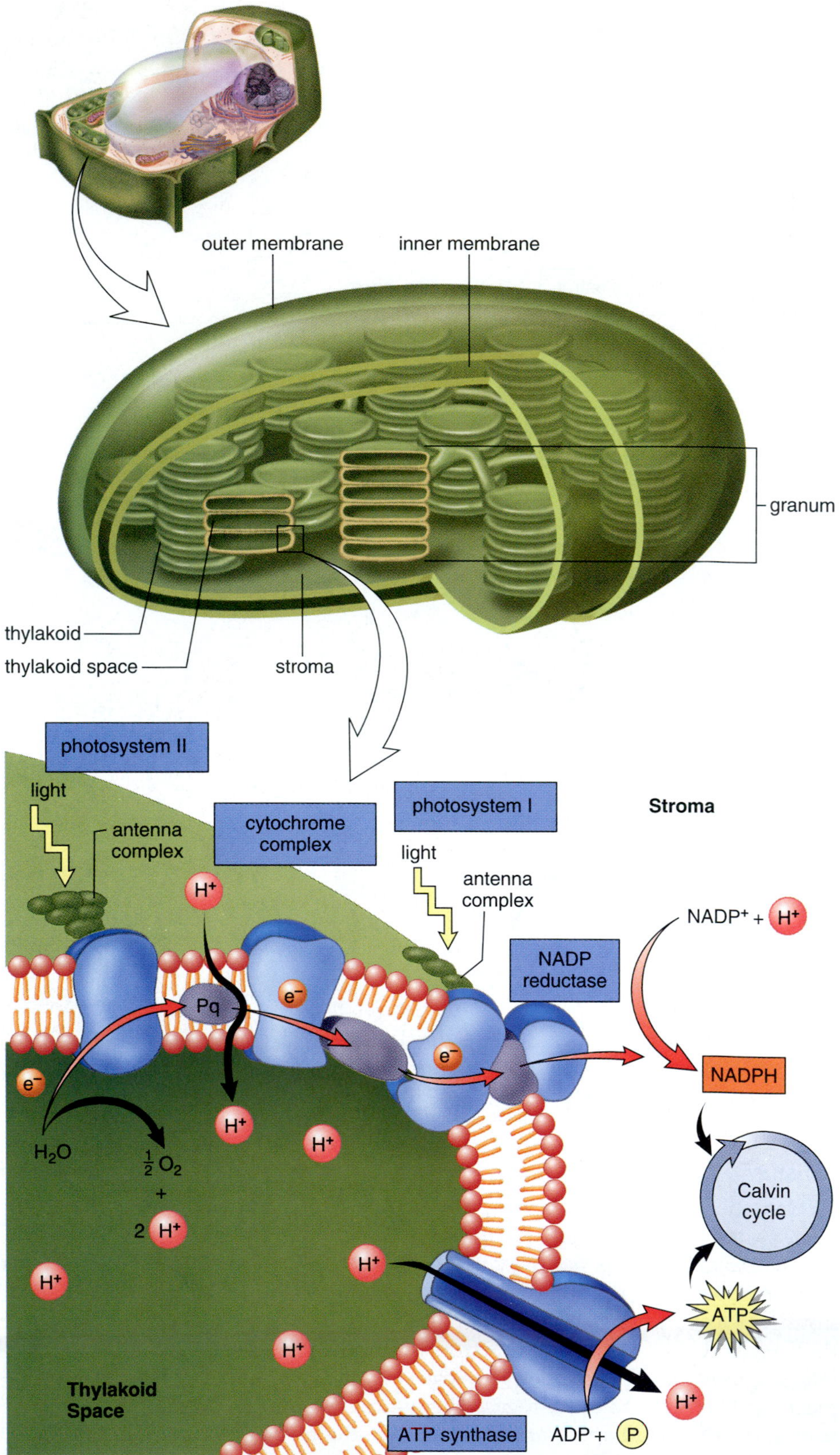

Figure 7.5 Organization of the thylakoid.
Protein complexes within the thylakoid membrane pump hydrogen ions from the stroma into the thylakoid space. *Pq* is a mobile carrier that also transfers hydrogen ions (H^+) from the stroma to the thylakoid space. When hydrogen ions flow back out of the space into the stroma through the ATP synthase complex, ATP is produced from ADP + Ⓟ.

7.4 Carbohydrate Synthesis

The light-independent reactions are the second stage of photosynthesis. They take their name from the fact that light is not directly required for these reactions to proceed. These reactions occur when CO_2 has entered the leaf and ATP and NADPH have been produced during the light-dependent reactions. PS I sets in motion a regulatory mechanism by which the enzymes of the light-independent reactions are turned on.

In this stage of photosynthesis, NADPH and ATP are used to reduce carbon dioxide: CO_2 becomes CH_2O within a carbohydrate molecule. Electrons and energy needed for this reduction synthesis are supplied by NADPH and ATP.

The reduction of carbon dioxide occurs in the stroma of a chloroplast by a series of reactions known as the **Calvin cycle.** In a cyclic series of reactions, the final product will also be the first reactant when the cycle of reactions begins again.

The Calvin cycle is named for Melvin Calvin, one of the individuals who was instrumental in identifying the reactions that make up the cycle (Fig. 7.6). Calvin received a Nobel Prize in 1961 for his part in determining these reactions.

The light-independent reactions use the ATP and NADPH from the light-dependent reactions to reduce carbon dioxide.

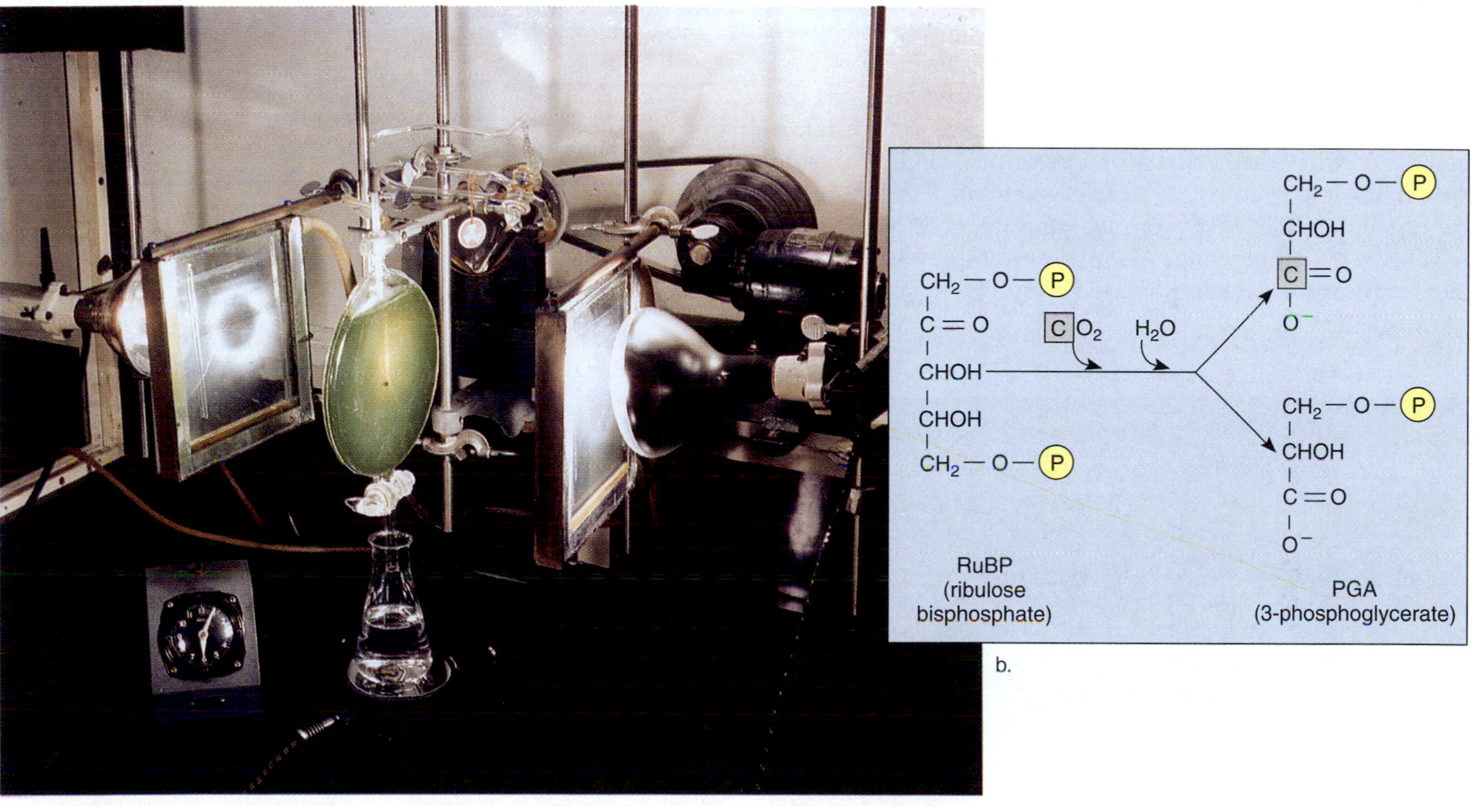

Figure 7.6 Identifying the reactions of the Calvin cycle.
a. Calvin and his colleagues used the apparatus shown; algae (*Chlorella*) were placed in the flat flask, which was illuminated by two lamps. Radioactive carbon dioxide ($^{14}CO_2$) was added, and the algae were killed a short time later by transferring them into boiling alcohol (in the beaker below the flask). This treatment instantly stopped all chemical reactions in the cells. Carbon-based compounds were then extracted from the cells and analyzed. **b.** By tracing the radioactive carbon, Calvin and his colleagues were able to trace the pathway by which CO_2 was metabolized. They found that the radioactive carbon (gray box) is first added to RuBP and that the resulting molecule splits to form two molecules of PGA. By gradually increasing the time (2 seconds, 5 seconds, and so on) after initial exposure of the algae to $^{14}CO_2$ before killing, they were able to identify the rest of the molecules in the cycle that is now called the Calvin cycle (see Fig. 7.8).

The Importance of PGAL

PGAL (glyceraldehyde-3-phosphate) is the product of the Calvin cycle that can be converted to all sorts of organic molecules. Compared to animal cells, algae and plants have enormous biochemical capabilities. They use PGAL for the purposes described in Figure 7.7.

Notice that glucose phosphate is among the organic molecules that result from PGAL metabolism. This is of interest to us because glucose is the molecule that plants and animals most often metabolize to produce the ATP molecules they require for their energy needs. Glucose is blood sugar in human beings.

Glucose phosphate can be combined with fructose (and the phosphate removed) to form sucrose, the molecule that plants use to transport carbohydrates from one part of the body to the other. Glucose phosphate is also the starting point for the synthesis of starch and cellulose. Starch is the storage form of glucose. Some starch is stored in chloroplasts, but most starch is stored in amyloplasts in roots. Cellulose is a structural component of plant cell walls and becomes fiber in our diet because we are unable to digest it. A plant can utilize the hydrocarbon skeleton of PGAL to form fatty acids and glycerol, which are combined in plant oils. We are all familiar with corn oil, sunflower oil, or olive oil, which we use in cooking. Also, when nitrogen is added to the hydrocarbon skeleton derived from PGAL, amino acids are formed.

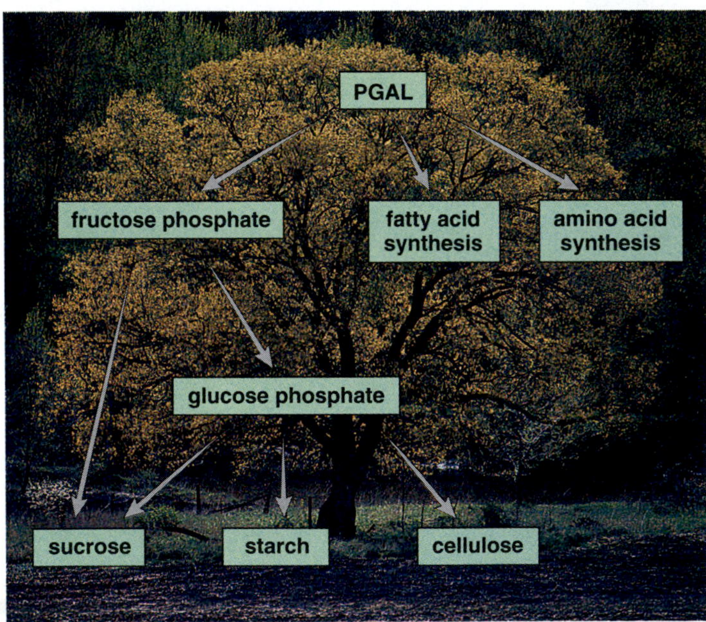

Figure 7.7 Fate of PGAL.
PGAL is the first reactant in a number of plant cell metabolic pathways. Two PGALs are needed to form glucose phosphate: glucose is often considered the end product of photosynthesis. Sucrose is the transport sugar in plants; starch is the storage form of glucose, and cellulose is a major constituent of plant cell walls.

The Calvin Cycle

The reactions of the Calvin cycle are the light-independent reactions that synthesize carbohydrate. The Calvin cycle includes (1) carbon dioxide fixation, (2) carbon dioxide reduction, and (3) regeneration of RuBP (ribulose bisphosphate).

Fixation of Carbon Dioxide

Carbon dioxide (CO_2) fixation, the attachment of carbon dioxide to an organic compound, is the first event in the Calvin cycle. **RuBP (ribulose bisphosphate),** a five-carbon molecule, combines with carbon dioxide (Fig. 7.8). The enzyme that speeds up this reaction, called RuBP carboxylase, is a protein that makes up about 20–50% of the protein content in chloroplasts. The reason for its abundance may be that it is unusually slow (it processes only a few molecules of substrate per second compared to thousands per second for a typical enzyme), and so there has to be a lot of it to keep the Calvin cycle going.

Carbon dioxide fixation occurs when carbon dioxide combines with RuBP.

Reduction of Carbon Dioxide

The six-carbon molecule resulting from carbon dioxide fixation immediately breaks down to form two PGA (3-phosphoglycerate) three-carbon molecules. Each of the two PGA molecules undergoes reduction to PGAL in two steps:

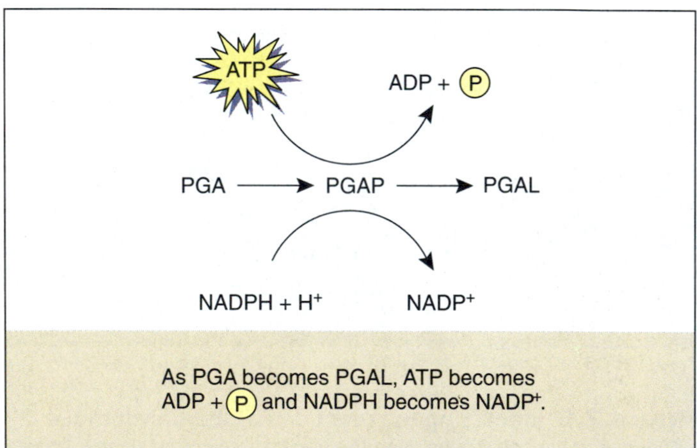

This is the sequence of reactions that uses NADPH from the light-dependent reactions and one of the two sequences that uses ATP from the same source. This sequence signifies the reduction of carbon dioxide (CO_2) to a carbohydrate (CH_2O). Electrons and energy are needed for this reduction reaction, and these are supplied by NADPH and ATP, respectively.

Figure 7.8 Light-independent reactions.
The Calvin cycle is divided into three portions: CO_2 fixation, CO_2 reduction, which requires NADPH and ATP, and regeneration of RuBP. Because five PGAL are needed to re-form three RuBP, it takes three turns of the cycle to have a net gain of one PGAL. Two PGAL molecules are needed to form one glucose ($C_6H_{12}O_6$) molecule.

Metabolites of the Calvin Cycle	
RuBP	ribulose bisphosphate
PGA	3-phosphoglycerate
PGAP	1,3-bisphosphoglycerate
PGAL	glyceraldehyde-3-phosphate

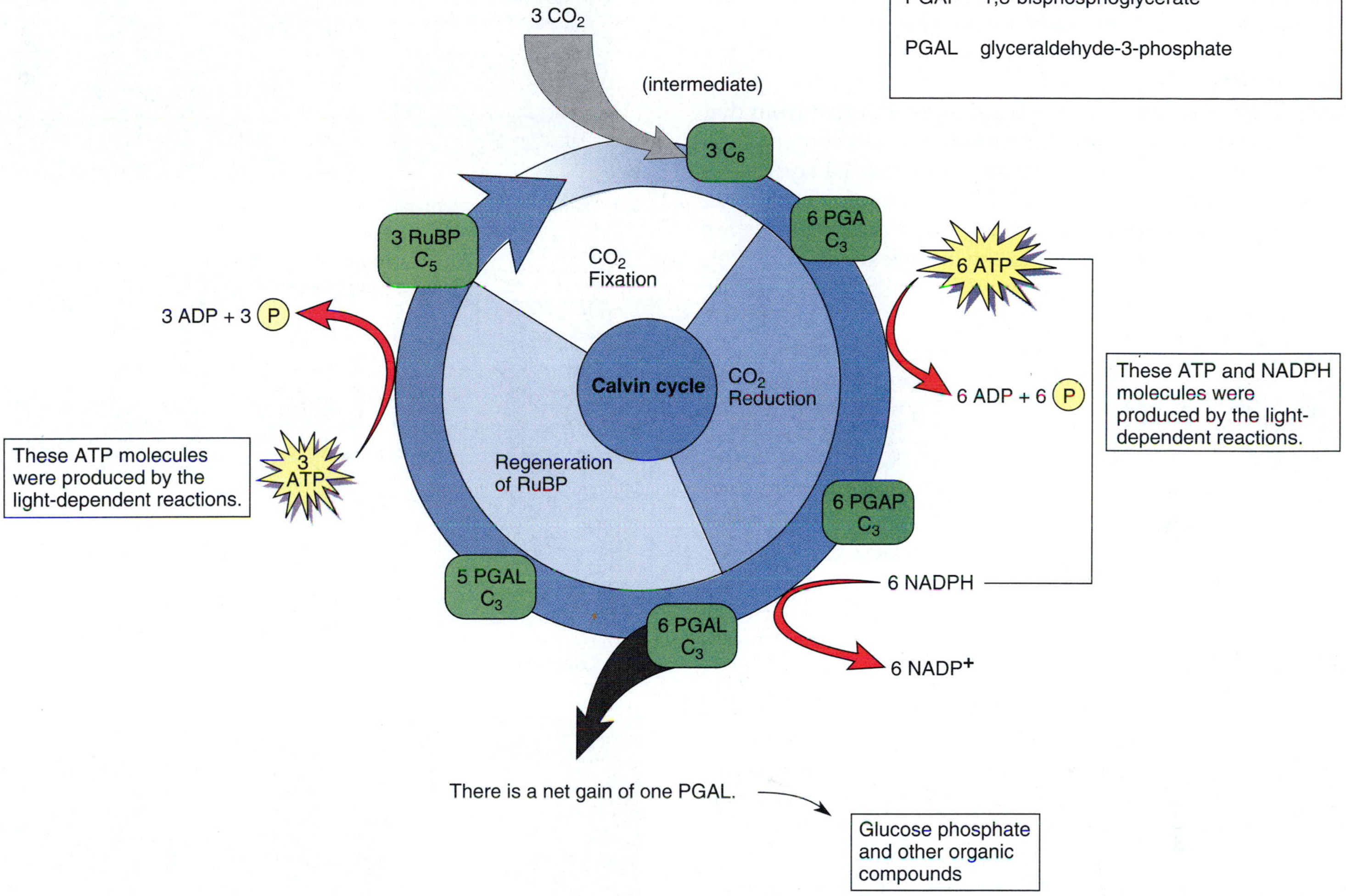

Regeneration of RuBP

For every three turns of the Calvin cycle, five molecules of PGAL are used to re-form three molecules of RuBP so that the cycle can continue. Notice that 5 × 3 (carbons in PGAL) = 3 × 5 (carbons in RuBP):

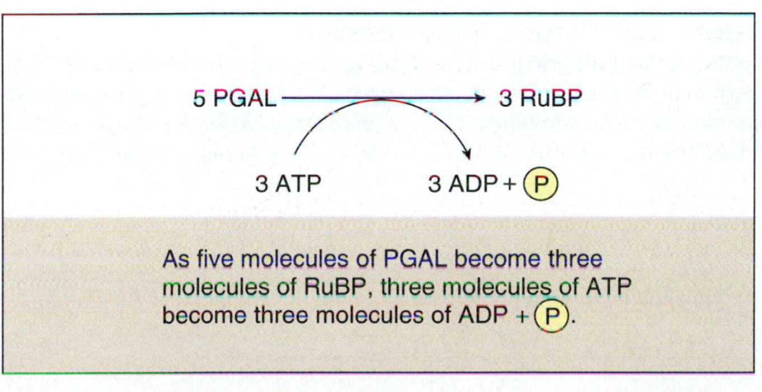

As five molecules of PGAL become three molecules of RuBP, three molecules of ATP become three molecules of ADP + (P).

The net gain of three turns of the Calvin cycle is one PGAL molecule. This sequence of reactions also utilizes some of the ATP produced by the light-dependent reactions.

Five out of every six PGAL molecules of the Calvin cycle are used to regenerate three RuBP molecules, so the cycle can continue.

The first detectable molecule identified by Melvin Calvin—namely PGA—has three carbons (Fig. 7.8). Therefore the Calvin cycle is also known as the C_3 cycle. Since Melvin Calvin did his research work, it's been discovered that plant species differ in the way they fix carbon dioxide and the first detectable molecule following carbon dioxide fixation is not always a C_3 molecule.

Modes of Photosynthesis

Three modes of photosynthesis are known (Fig. 7.9). In **C_3 plants,** the Calvin cycle fixes carbon dioxide (CO_2) directly, and the first detectable molecule following fixation is PGA, a C_3 molecule. **C_4 plants** fix CO_2 by forming a C_4 molecule prior to the involvement of the Calvin cycle. **CAM plants** fix CO_2 by forming a C_4 molecule at night when stomates can open without much loss of water.

C_4 Photosynthesis

The structure of a leaf from a C_3 plant is different from that of a C_4 plant. In a C_3 plant, the mesophyll cells contain well-formed chloroplasts and are arranged in parallel layers. In a C_4 leaf, the bundle sheath cells, as well as the mesophyll cells, contain chloroplasts. Further, the mesophyll cells are arranged concentrically around the bundle sheath cells:

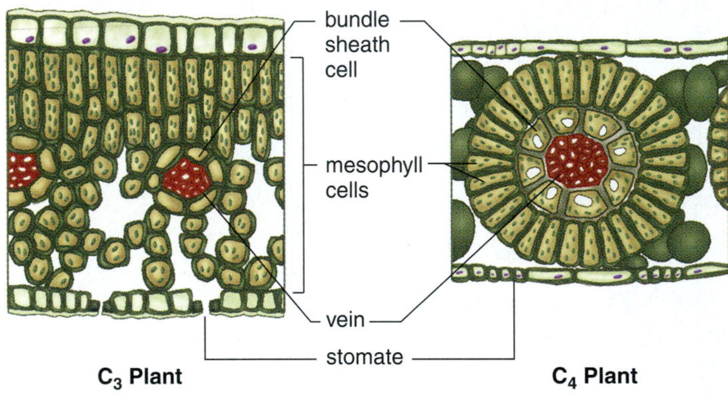

C_3 plants use the enzyme RuBP carboxylase to fix CO_2 to RuBP in mesophyll cells. The first detected molecule following fixation is PGA. C_4 plants use the enzyme PEP carboxylase (PEPCase) to fix CO_2 to PEP (phosphoenolpyruvate, a C_3 molecule). The result is oxaloacetate, a C_4 molecule.

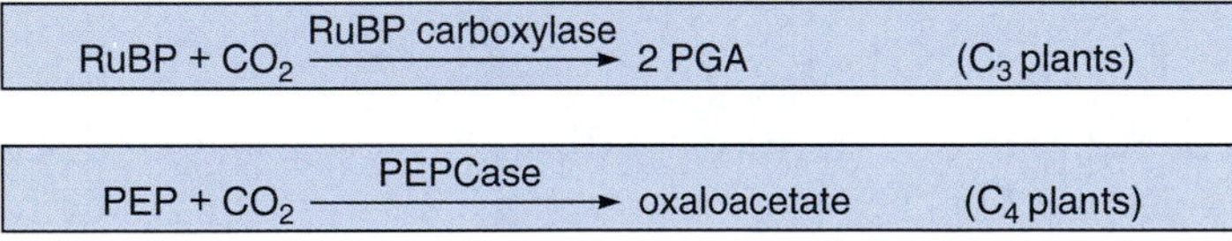

$$\text{RuBP} + CO_2 \xrightarrow{\text{RuBP carboxylase}} 2\ \text{PGA} \quad (C_3\ \text{plants})$$

$$\text{PEP} + CO_2 \xrightarrow{\text{PEPCase}} \text{oxaloacetate} \quad (C_4\ \text{plants})$$

In a C_4 plant, CO_2 is taken up in mesophyll cells and then malate, a reduced form of oxaloacetate, is pumped into the bundle sheath cells (Fig. 7.9). Here, and only here, CO_2 enters the Calvin cycle. It takes energy to pump molecules, and you would think that the C_4 pathway would be disadvantageous. Yet in hot, dry climates, the net photosynthetic rate of C_4 plants such as sugarcane, corn, and Bermuda grass is about two to three times that of C_3 plants such as wheat, rice, and oats. Why do C_4 plants enjoy such an advantage? The answer is they can avoid photorespiration, discussed next.

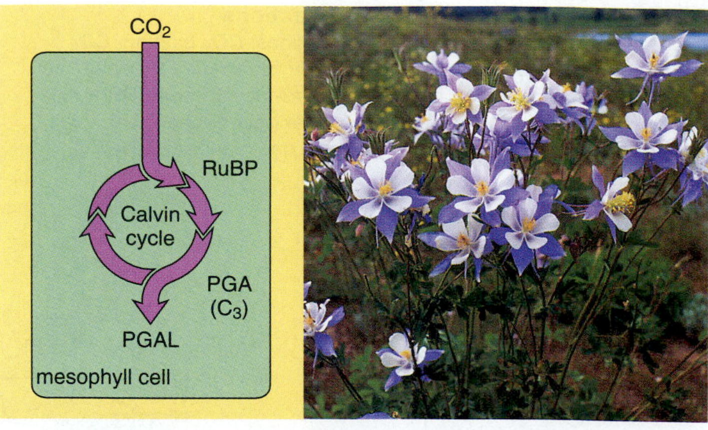

CO_2 fixation in a C_3 plant, blue columbine, *Aquilegia caerulea*

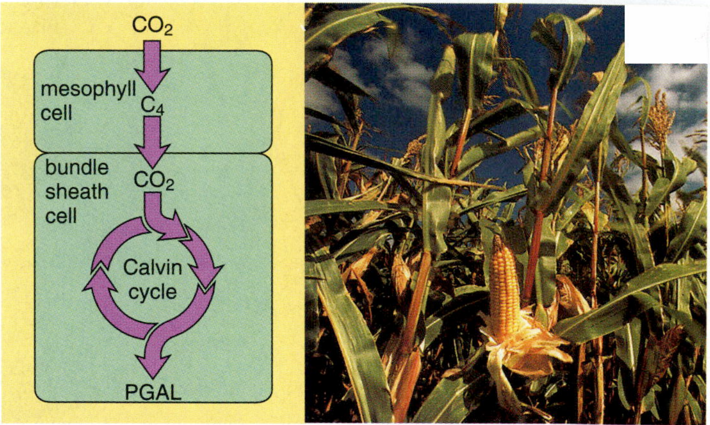

CO_2 fixation in a C_4 plant, corn, *Zea mays*

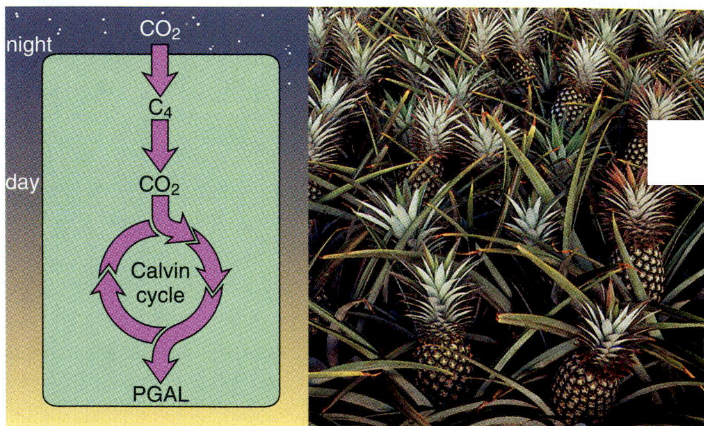

CO_2 fixation in a CAM plant, pineapple, *Ananas comosus*

Figure 7.9 Carbon dioxide fixation.
Plants can be categorized according to the way carbon dioxide (CO_2) is fixed. **a.** In C_3 plants, CO_2 is taken up by the Calvin cycle directly. **b.** C_4 plants utilize a C_4 molecule prior to releasing CO_2 to the Calvin cycle. **c.** CAM plants use this same C_4 molecule only at night to take up CO_2.

Photorespiration

Notice in the diagrams of leaves on the previous page, there are little openings called stomates (sing., stomate) in the surfaces of leaves through which water can leave and carbon dioxide (CO_2) can enter. If the weather is hot and dry, the stomates close, conserving water. (Water loss might cause the plant to wilt and die.) Now the concentration of CO_2 decreases in leaves, while oxygen, a by-product of photosynthesis, increases. When oxygen rises in C_3 plants, it combines with RuBP carboxylase instead of CO_2. The result is one molecule of PGA and eventually the release of CO_2. This is called **photorespiration** because in the presence of light (*photo*), oxygen is taken up and CO_2 is released (*respiration*).

Photorespiration does not occur in C_4 leaves because PEPCase does not combine with O_2. Even when stomates are closed CO_2 is delivered to the Calvin cycle in the bundle sheath cells. When the weather is moderate, C_3 plants have the advantage, but when the weather becomes hot and dry, C_4 plants have the advantage, and we can expect them to predominate. In the early summer, C_3 plants such as Kentucky bluegrass and creeping bent grass predominate in lawns in the cooler parts of the United States, but by midsummer, crabgrass, a C_4 plant, begins to take over.

When the weather is hot and dry, photorespiration occurs in C_3 plants but not C_4 plants. This gives C_4 plants an advantage under these conditions.

CAM Photosynthesis

CAM stands for crassulacean-acid metabolism; the Crassulaceae is a family of flowering succulent (water-containing) plants that live in warm, arid regions of the world. CAM was first discovered in these plants, but now it is known to be prevalent among most succulent plants that grow in desert environments, including the cacti.

Whereas a C_4 plant represents partitioning in space—carbon dioxide fixation occurs in mesophyll cells and the Calvin cycle occurs in bundle sheath cells—CAM is partitioning by the use of time. During the night, CAM plants use PEPCase to fix some CO_2 forming C_4 molecules, which are stored in large vacuoles in mesophyll cells. During the day, C_4 molecules (malate) release CO_2 to the Calvin cycle when NADPH and ATP are available from the light-dependent reactions. The primary reason for this partitioning again has to do with the conservation of water. CAM plants open their stomates only at night, and therefore only at that time is atmospheric CO_2 available. During the day, the stomates close to conserve water and CO_2 cannot enter the plant.

Photosynthesis in a CAM plant is minimal because of the limited amount of CO_2 fixed at night, but it does allow CAM plants to live under stressful conditions.

CAM plants use PEPCase to fix carbon dioxide at night when the stomates are open. During the day, the stored carbon dioxide can enter the Calvin cycle.

Connecting Concepts

"Have You Thanked a Green Plant Today?" is a bumper sticker that you may have puzzled over until now. Plants, you now know, capture solar energy and store it in carbon-based organic nutrients that are passed to other organisms when they feed on plants and/or on other organisms. In this context, plants are called autotrophs because they make their own organic food. Heterotrophs are organisms that take in preformed organic food.

Both autotrophs and heterotrophs contribute to the global carbon cycle. In the carbon cycle, organisms in both terrestrial and aquatic ecosystems exchange carbon dioxide with the atmosphere. Autotrophs, like plants, take in carbon dioxide when they photosynthesize. Carbon dioxide is returned to the atmosphere when autotrophs and heterotrophs carry on cellular respiration. In this way, the very same carbon atoms cycle from the atmosphere to autotrophs, then to heterotrophs, and then back to autotrophs again.

Living and dead organisms contain organic carbon and serve as a reservoir of carbon in the carbon cycle. Some 300 million years ago, a host of plants died and did not decompose. These plants were compressed to form the coal that we mine and burn today. (Oil has a similar origin, but most likely formed in marine sedimentary rocks that included animal remains.)

The amount of carbon dioxide in the atmosphere is increasing steadily because we humans burn fossil fuels to run our modern industrial society. Many fear that this buildup of carbon dioxide will contribute to global warming, because carbon dioxide is a greenhouse gas. Like the glass of a greenhouse, it allows sunlight to pass through, but then traps the resulting heat. Yet while we are pumping carbon dioxide into the atmosphere due to fossil fuel burning, we are destroying vast tracts of tropical rain forests, which soak up carbon dioxide like a sponge. Clearly, these actions are not in our self-interest and it would behoove us to reduce the burning of fossil fuels and preserve tropical rain forests.

Summary

7.1 Solar Energy

Photosynthesis produces carbohydrates and releases oxygen, both of which are utilized by the majority of living things. Photosynthesis uses solar energy in the visible-light range; the photons of this range contain the right amount of energy to energize the electrons of chlorophyll molecules. Specifically, chlorophylls *a* and *b* absorb violet, blue, and red wavelengths best. This causes chlorophyll to appear green to us.

7.2 Structure and Function of Chloroplasts

A chloroplast is bounded by a double membrane and contains two main components: the liquid stroma and the membranous grana made up of membranous thylakoids. The light-dependent reactions take place in the thylakoids, and the light-independent reactions take place in the stroma.

7.3 Solar Energy Capture

In the cyclic electron pathway, electrons energized by the sun leave PS I. They pass down an electron transport system and back to PS I again. Chemiosmosis follows and ATP is generated. In keeping with its name, PS I probably evolved first and is the only photosynthetic pathway present in certain bacteria today. Most photosynthetic cells regulate the activity of the cyclic and noncyclic pathways to suit the needs of the cell.

The noncyclic electron pathway of the light-dependent reactions begins when solar energy enters PS II. PS II energized electrons are picked up by an electron acceptor. The oxidation (splitting) of water replaces these electrons in the reaction-center chlorophyll *a* molecule. Oxygen is released to the atmosphere and hydrogen ions (H^+) remain in the thylakoid space. The acceptor molecule passes electrons to PS I by way of an electron transport (cytochrome) system. When solar energy is absorbed by PS I, energized electrons leave and are ultimately received by $NADP^+$, which also combines with H^+ from the stroma to become NADPH.

The energy made available by the passage of electrons down the electron transport system allows carriers to pump H^+ into the thylakoid space. The buildup of H^+ establishes an electrochemical gradient. When H^+ flows down this gradient through the channel present in ATP synthase complexes, ATP is synthesized from ADP and Ⓟ by ATP synthase. This method of producing ATP is called chemiosmosis.

Chemiosmosis requires an organized membrane. The thylakoid membrane is highly organized: PS II functions to oxidize (split) water, the cytochrome complex transports electrons and pumps H^+, PS I is associated with an enzyme that reduces $NADP^+$, and ATP synthase produces ATP.

7.4 Carbohydrate Synthesis

The energy yield of the light-dependent reactions is stored in ATP and NADPH. These molecules are used by the light-independent reactions to reduce carbon dioxide (CO_2) to carbohydrate, namely PGAL, which is then converted to all the organic molecules a plant needs.

During the first stage of the Calvin cycle, the enzyme RuBP carboxylase fixes CO_2 to RuBP, producing a six-carbon molecule that immediately breaks down to two C_3 molecules. During the second stage, CO_2 is incorporated into an organic molecule and is reduced to carbohydrate (CH_2O). This step requires the NADPH and some of the ATP from the light-dependent reactions. For every three turns of the Calvin cycle, the net gain is one PGAL molecule; the other five PGAL molecules are used to re-form three molecules of RuBP. This step also requires ATP for energy. It required two PGAL molecules to make one glucose molecule.

In C_4 plants, as opposed to the C_3 plants just described, the enzyme PEPCase fixes carbon dioxide to PEP to form a four-carbon molecule, oxaloacetate, within mesophyll cells. A reduced form of this molecule is pumped into bundle sheath cells where CO_2 is released to the Calvin cycle. C_4 plants avoid photorespiration by a partitioning of pathways in space: carbon dioxide fixation occurs in mesophyll cells and the Calvin cycle occurs in bundle sheath cells.

During CAM photosynthesis, PEPCase fixes CO_2 to PEP at night. The next day, CO_2 is released and enters the Calvin cycle within the same cells. This represents a partitioning of pathways in time: carbon dioxide fixation occurs at night and the Calvin cycle occurs during the day. The plants that carry on CAM are desert plants, in which the stomates only open at night in order to conserve water.

Reviewing the Chapter

1. Why is it proper to say that almost all living things are dependent on solar energy? 112
2. Discuss the electromagnetic spectrum and the absorption spectrum of chlorophyll. Why is chlorophyll a green pigment? 112–13
3. Name the two major components of chloroplasts and associate each portion with the two sets of reactions that occur during photosynthesis. How are the two pathways related? 114
4. What roles do PS I and PS II play during the light-dependent reactions? 116
5. Trace the cyclic electron pathway, naming and explaining the main events that occur as electrons cycle. 116
6. Trace the noncyclic electron pathway, naming and explaining all the events that occur as the electrons move from water to $NADP^+$. 117
7. Explain what is meant by chemiosmosis, and relate this process to the electron transport system present in the thylakoid membrane. 118
8. How is the thylakoid membrane organized? Name the main complexes in the membrane and give a function for each. 118
9. Describe the three stages of the Calvin cycle. Which stage utilizes the ATP and NADPH from the light-dependent reactions? 120–21
10. Explain C_4 photosynthesis, contrasting the actions of RuBP carboxylase and PEPCase. 122–23
11. Explain CAM photosynthesis, contrasting it to C_4 photosynthesis in terms of partitioning a pathway. 123

Testing Yourself

Choose the best answer for each question.

1. The absorption spectrum of chlorophyll
   a. is not the same as that of carotenoids.
   b. approximates the action spectrum of photosynthesis.
   c. explains why chlorophyll is a green pigment.
   d. shows that some colors of light are absorbed more than others.
   e. All of these are correct.
2. The final acceptor of electrons during the noncyclic electron pathway is
   a. PS I.
   b. PS II.
   c. ATP.
   d. $NADP^+$.
   e. water.
3. A photosystem contains
   a. pigments, a reaction center, and an electron acceptor.
   b. ADP, Ⓟ, and hydrogen ions (H^+).
   c. protons, photons, and pigments.
   d. cytochromes only.
   e. Both b and c are correct.
4. Which of these should not be associated with the electron transport system?
   a chloroplasts
   b. cytochromes
   c. movement of H^+ into the thylakoid space
   d. formation of ATP
   e. absorption of solar energy
5. PEPCase has an advantage compared to RuBP carboxylase. The advantage is that
   a. PEPCase is present in both mesophyll and bundle sheath cells, but RuBP carboxylase is not.
   b. RuBP carboxylase fixes carbon dioxide (CO_2) only in C_4 plants, but PEPCase does it in both C_3 and C_4 plants.
   c. RuBP carboxylase combines with O_2, but PEPCase does not.
   d. PEPCase conserves energy but RuBP carboxylase does not.
   e. Both b and c are correct.
6. The NADPH and ATP from the light-dependent reactions are used to
   a. split water.
   b. cause RuBP carboxylase to fix CO_2.
   c. re-form the photosystems.
   d. cause electrons to move along their pathways.
   e. convert PGA to PGAL.
7. CAM photosynthesis
   a. is the same as C_4 photosynthesis.
   b. is an adaptation to cold environments in the southern hemisphere.
   c. is prevalent in desert plants that close their stomates during the day.
   d. occurs in plants that live in marshy areas.
   e. stands for chloroplasts and mitochondria.
8. Chemiosmosis
   a. depends on protein complexes in the thylakoid membrane.
   b. depends on an electrochemical gradient.
   c. depends on a difference in H^+ concentration between the thylakoid space and the stroma.
   d. results in ATP formation.
   e. All of these are correct.
9. Label this diagram of a chloroplast.

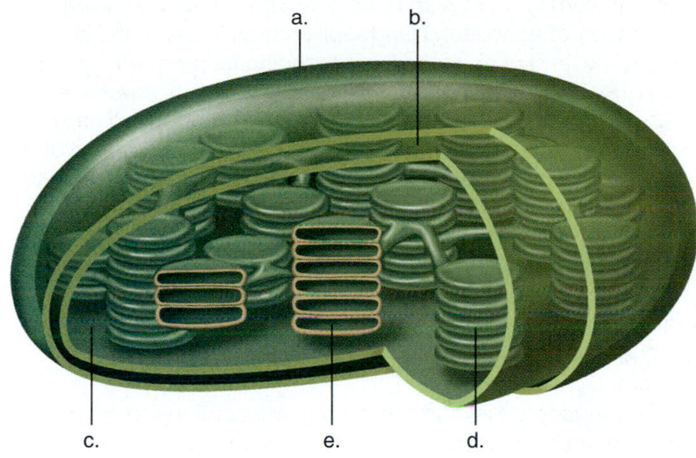

   f. The light-dependent reactions occur in which part of a chloroplast?
   g. The light-independent reactions occur in which part of a chloroplast?
10. Label this diagram using these labels: water, carbohydrate, carbon dioxide, oxygen, ATP, ADP + Ⓟ, NADPH, and $NADP^+$.

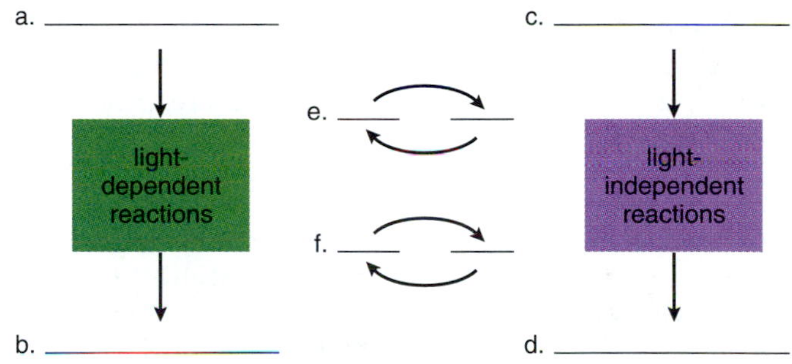

Thinking Scientifically

1. In the fall of the year the leaves of many trees change from green to red or yellow. Two hypotheses can explain this color change: (1) In the fall, chlorophyll degenerates and red or yellow pigments that were earlier masked by chlorophyll become apparent. (2) In the fall, red or yellow pigments are synthesized and they mask the color of chlorophyll. How could you test these two hypotheses?
2. You have discovered a type of bacteria that seems to be capable of photosynthesis but has no chloroplasts. However, there are invaginations of the plasma membrane inside the bacteria. Would photosynthesis be possible with such an arrangement? What would you expect to find in the invaginated membrane if these bacteria were photosynthetic?

Bioethical Issue

Whether there will be enough food to feed the increase in population expected by the middle of the twenty-first century is unknown. Over the past 40 years, the world's food supply has expanded faster than the population, due to the development of high-yielding plants, and the increased use of irrigation, pesticides, and fertilizers. Unfortunately, modern farming techniques result in pollution of the air, water, and land. One of the most worrisome threats to food security is an increasing degradation of agricultural land. Soil erosion in particular is robbing the land of its topsoil and reducing its productivity.

In 1986, it was estimated that humans already use nearly 40% of the Earth's terrestrial photosynthetic production and therefore we should reach maximum capacity in the middle of the twenty-first century when the population is projected to double its 1986 size. Most population growth will occur in the less developed countries in Africa, Asia, and Latin America that are only now becoming industrialized. Even the United States will face an increased drain on its economic resources and increased pollution problems due to population growth. In the developing countries the technical gains needed to prevent a disaster will be enormous.

Some people feel that technology will continue to make great strides for many years to come. They maintain technology hasn't begun to reach the limits of performance and therefore, will be able to solve the problems of increased population growth. Others feel that technology's successes are self-defeating. The newly developed hybrid crops that led to enormous increases in yield per acre also cause pollution problems that degrade the environment. These scientists are in favor of calling a halt to an increasing human population by all possible measures and as quickly as possible. Which of these approaches do you favor?

Understanding the Terms

ATP synthase 118
C_3 plant 122
C_4 plant 122
Calvin cycle 119
CAM plant 122
carbon dioxide (CO_2) fixation 120
chemiosmosis 118
chlorophyll 114
chloroplast 114
cyclic electron pathway 116
electromagnetic spectrum 112
electron transport system 116
granum 114
light-dependent reactions 114
light-independent reactions 114
noncyclic electron pathway 117
photon 112
photorespiration 123
photosynthesis 112
photosystem 116
RuBP (ribulose bisphosphate) 120
stroma 114
thylakoid 114
visible light 112

Match the terms to these definitions:

a. ______________ Energy-capturing portion of photosynthesis that takes place in thylakoid membranes of chloroplasts and cannot proceed without solar energy; it produces ATP and NADPH.

b. ______________ Photosynthetic unit where solar energy is absorbed and high-energy electrons are generated; contains an antenna complex and an electron acceptor.

c. ______________ Passage of electrons along a series of carrier molecules from a higher to a lower energy level; the energy released is used for the synthesis of ATP.

d. ______________ Process usually occurring within chloroplasts whereby chlorophyll traps solar energy and carbon dioxide is reduced to a carbohydrate.

e. ______________ Series of photosynthetic reactions in which carbon dioxide is fixed and reduced to PGAL.

Web Connections

Exploring the Internet

http://www.mhhe.com/biosci/genbio/mader
(click on *Biology 7/e*)

The *Biology 7/e* Online Learning Center provides many resources for studying the material in this chapter including links to the following sites:

Introduction to Photosynthesis contains a thorough treatment of the topic and includes the evolution and discovery of photosynthesis, the light and dark reactions, and chloroplast structure, illustrated with diagrams and formulas.

http://esg-www.mit.edu:8001/esgbio/ps/intro.html

For a more detailed account of photosynthesis, choose Photosynthesis—the Light Reactions, which includes sections on the physics of light and the photosystem, the light and dark reactions, and alternate pathways for photosynthesis.

http://esg-www.mit.edu:8001/esgbio/ps/light.html

Photosynthetic Molecules Page. Three-dimensional representations of chlorophyll and other molecules involved in photosynthesis.

http://www.nyu.edu:80/pages/mathmol/library/photo/index.html

Photosynthesis—Other Approaches. This MIT hypertextbook chapter covers C_4 mechanisms as well as chemosynthesis, with both text and diagrams.

http://esg-www.mit.edu:8001/esgbio/ps/other.html

Photosynthesis. This page has links to lectures on photosynthesis and problem sets.

http://mss.scbe.on.ca/DSPHOTOS.HTM

Cellular Respiration

chapter

chapter concepts

Effort requires energy.

When you go snow boarding, take an aerobics class, or just sit around, ATP molecules allow your muscles to contract. ATP molecules are produced during cellular respiration, a process that requires the participation of mitochondria. There are numerous mitochondria in muscle cells.

The glucose and oxygen for cellular respiration are delivered to cells by the circulatory system, and the end products—water and carbon dioxide—are removed by the circulatory system. Cellular respiration consists of many small steps mediated by specific enzymes. High-energy electrons are removed from glucose breakdown products and are passed down an electron transport system located on the cristae of mitochondria. As the electrons move from one carrier to the next, energy is released and captured for the production of ATP molecules.

One form of chemical energy (glucose) cannot be transformed completely into another (ATP molecules) without the loss of usable energy in the form of heat. When ATP is produced, heat is given off. Since exercise requires plentiful ATP, your body's internal temperature rises and you begin sweating.

8.1 Cellular Respiration

Cellular respiration includes all the various metabolic pathways which break down carbohydrates and other metabolites with the concomitant buildup of ATP. **Cellular respiration,** as implied by its name, is a cellular process that requires oxygen and gives off carbon dioxide (CO_2). Most often it involves the complete breakdown of glucose to carbon dioxide and water (H_2O):

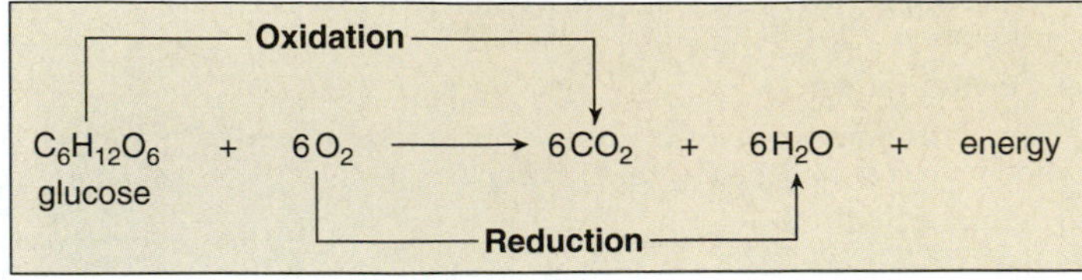

Glucose is a high-energy molecule, and its breakdown products, CO_2 and H_2O, are low-energy molecules. Therefore, we expect the process to be exergonic and release energy. As breakdown occurs, electrons are removed from substrates and eventually are received by oxygen atoms, which then combine with H^+ to become H_2O.

The equation shows changes in regard to hydrogen atom (H) distribution. But remember that a hydrogen atom consists of a hydrogen ion plus an electron ($H^+ + e^-$). Therefore when hydrogen atoms are removed from glucose so are electrons. Since oxidation is the loss of electrons, and reduction is the gain of electrons, glucose breakdown is an oxidation-reduction reaction. Glucose is oxidized and O_2 is reduced.

On the other hand, the buildup of ATP is an endergonic reaction that requires energy. The pathways of cellular respiration allow the energy within a glucose molecule to be released slowly so that ATP can be produced gradually. Cells would lose a tremendous amount of energy if glucose breakdown occurred all at once—much energy would become nonusable heat. The step-by-step breakdown of glucose to carbon dioxide and water usually realizes a maximum yield of 36 or 38 ATP molecules dependent on conditions to be discussed later. The energy in these ATP molecules is equivalent to about 39% of the energy that was available in glucose. This conversion is more efficient than many others; for example, only about 25% of the energy within gasoline is converted to the motion of a car.

NAD^+ and FAD

Cellular respiration involves many individual metabolic reactions, each one catalyzed by its own enzyme. Enzymes of particular significance are the redox enzymes that utilize the coenzyme **NAD^+.** When a metabolite is oxidized, NAD^+ accepts two electrons plus a hydrogen ion (H^+) and NADH results. The electrons received by NAD^+ are high-energy electrons that are usually carried to the electron transport system. Figure 8.1 illustrates how NAD^+ carries electrons.

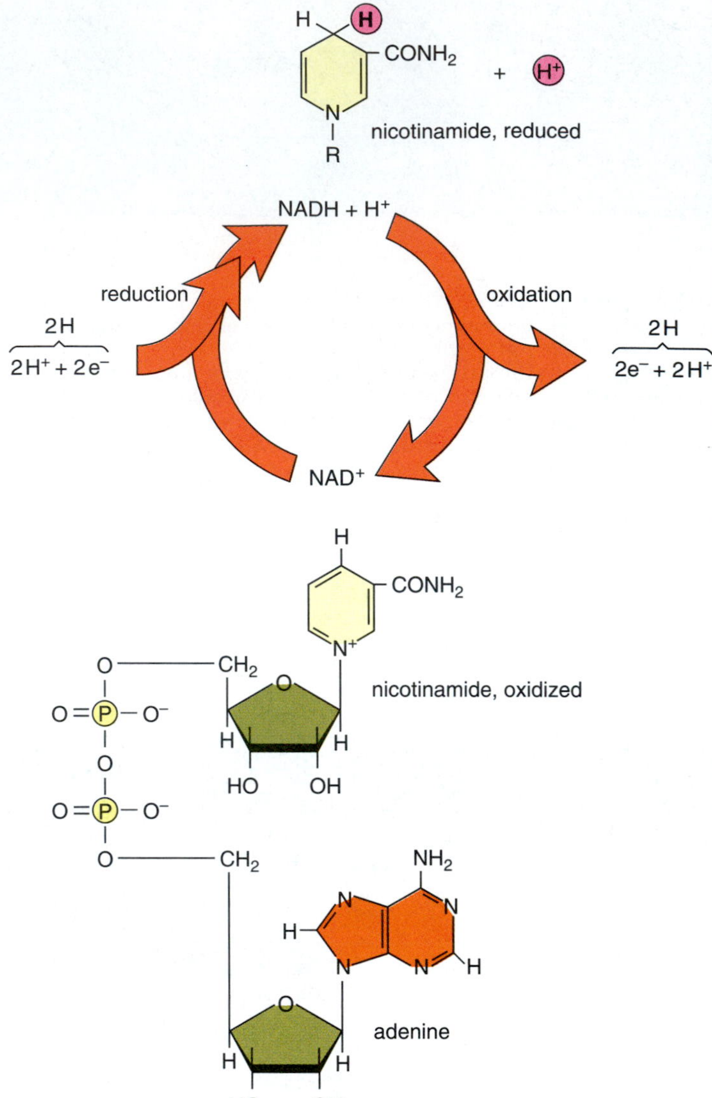

Figure 8.1 The NAD^+ cycle.
The coenzyme NAD^+ is a dinucleotide (two nucleotides joined by bonding between their phosphates). NAD^+ accepts two electrons (e^-) plus a hydrogen ion (H^+) and NADH + H^+ result. When NADH passes the electrons to another substrate or carrier, NAD^+ is formed.

NAD^+ is called a coenzyme of oxidation-reduction because it can oxidize a metabolite by accepting electrons and can reduce a metabolite by giving up electrons. Only a small amount of NAD^+ need be present in a cell, because each NAD^+ molecule is used over and over again. **FAD** is another coenzyme of oxidation-reduction which is sometimes used instead of NAD^+. FAD accepts two electrons and two hydrogen ions (H^+) to become $FADH_2$.

NAD^+ and FAD are two coenzymes of oxidation-reduction that are active during cellular respiration.

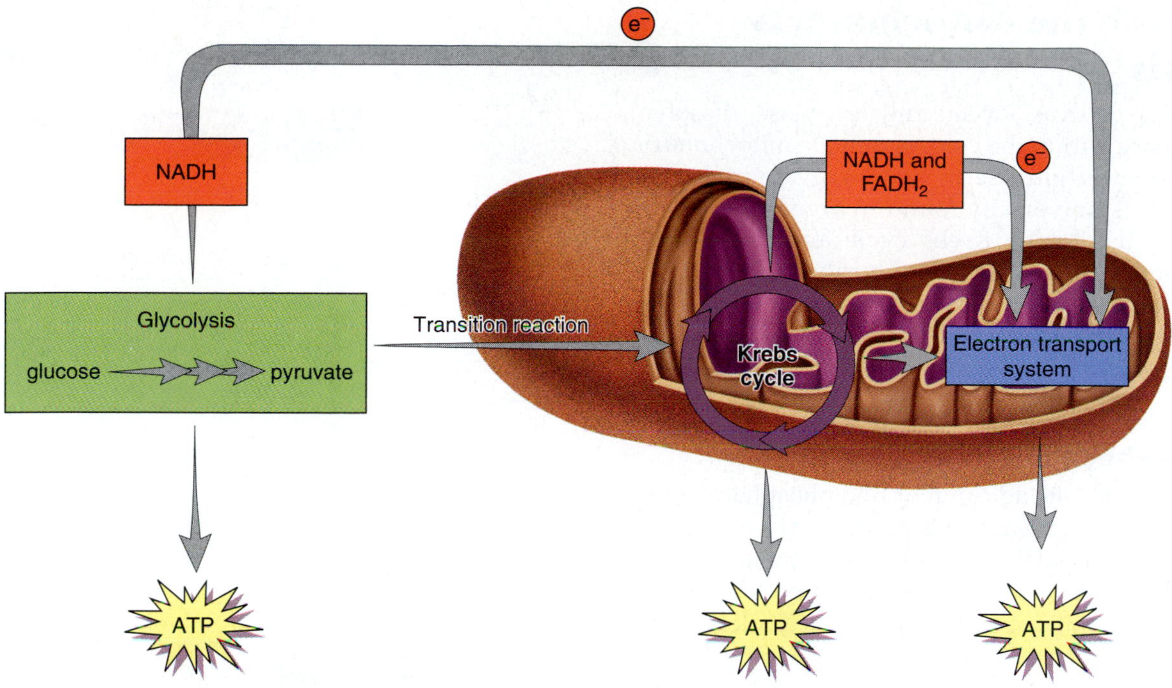

Figure 8.2 The four phases of complete glucose breakdown.
The complete breakdown of glucose consists of four phases. Glycolysis in the cytosol produces pyruvate which enters mitochondria if oxygen is available. The transition reaction and the Krebs cycle which follow occur inside the mitochondria. Also, inside mitochondria, the electron transport system receives the electrons that were removed from glucose breakdown products. The end result of glucose breakdown is 36 to 38 ATP depending on the particular cell.

Phases of Complete Glucose Breakdown

Glucose breakdown involves four phases: three metabolic pathways plus one individual reaction. These phases will be color coded throughout this chapter in the same manner as shown in Figure 8.2. Glycolysis takes place outside the mitochondria and does not utilize oxygen. The other phases of cellular respiration take place inside the mitochondria where oxygen is utilized.

- **Glycolysis** is the breakdown of glucose to two molecules of pyruvate. Enough energy is released for the immediate buildup of two ATP.
- During the **transition reaction,** pyruvate is oxidized to an acetyl group, and CO_2 is removed. Since glycolysis ends with two molecules of pyruvate, the transition reaction occurs twice per glucose molecule.
- The **Krebs cycle** is a cyclical series of reactions that release CO_2 and produce ATP. The Krebs cycle turns twice because two acetyl-CoA molecules enter the cycle per glucose molecule. Altogether, the Krebs cycle accounts for two immediate ATP molecules per glucose molecule.
- The **electron transport system** is a series of carriers that accept the electrons removed from glucose during glycolysis, the transition reaction, and the Krebs cycle. Usually the coenzyme NAD^+ carries these electrons from these phases to the electron transport system. The system passes the electrons along from one carrier to the next until they are finally received by oxygen. As the electrons pass from a higher energy to a lower energy state, energy is released and stored for ATP buildup. The electron transport system accounts for 32 or 34 ATP, depending on the particular cell.

Pyruvate is a pivotal metabolite in cellular respiration. If oxygen is not available to the cell, **fermentation,** an anaerobic process, occurs in the cytosol. During fermentation, glucose is incompletely metabolized to lactate or to carbon dioxide and alcohol, depending on the organism. As we shall see on page 137, fermentation results in a net gain of only two ATP per glucose molecule.

The complete breakdown of glucose to carbon dioxide and water is aerobic and requires oxygen, which is utilized in mitochondria.

8.2 Outside the Mitochondria: Glycolysis

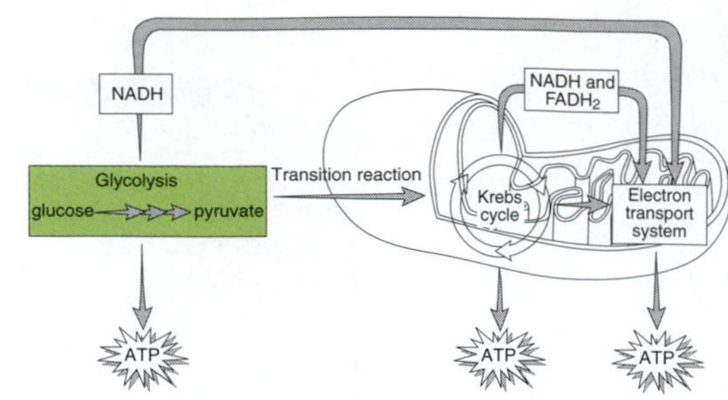

Glycolysis [Gk. *glykeros,* sweet, and *lyo,* loose, dissolve], which takes place within the cytosol outside mitochondria, is the breakdown of glucose to two pyruvate molecules. Since glycolysis is universally found in organisms, it most likely evolved before the Krebs cycle and the electron transport system. This may be why glycolysis occurs in the cytosol and does not require oxygen. Bacteria evolved before other organisms, and there are some bacteria today that are anaerobic; and indeed they die in the presence of oxygen.

Energy Investment Steps

As glycolysis begins, the addition of two phosphate groups activates glucose (C_6) to react. This requires two separate reactions and uses two ATP. The resulting C_6 molecule splits into two C_3 molecules, each of which carries a phosphate group. From this point on, each C_3 molecule undergoes the same series of reactions.

Energy Harvesting Steps

During glycolysis, oxidation occurs by the removal of electrons which are accepted by NAD^+. Each NAD^+ accepts two electrons and one hydrogen ion (H^+); altogether two NADH molecules result. Enough energy is released to generate four ATP molecules by **substrate-level phosphorylation** [Gk. *phos,* light, and *phoreus,* carrier]. During substrate-level phosphorylation, a phosphate is enzymatically transferred from a high-energy metabolite to ADP. For example, in Figure 8.3, the molecule designated as PGAP gives up a phosphate to ADP, and ATP results.

Subtracting the two ATP that were used to get started, there is a net gain of two ATP from glycolysis. Much chemical energy still remains in a glucose molecule and this energy becomes available when glycolysis is followed by the breakdown of pyruvate inside mitochondria. Pyruvate enters mitochondria if oxygen is available. If oxygen is not available, glycolysis becomes a part of fermentation (see page 137).

Figure 8.4 gives the main reactions of glycolysis. After glucose is split, each succeeding reaction occurs twice. Therefore, these reactions are preceded by a 2×. Altogether the inputs and outputs of glycolysis are as follows:

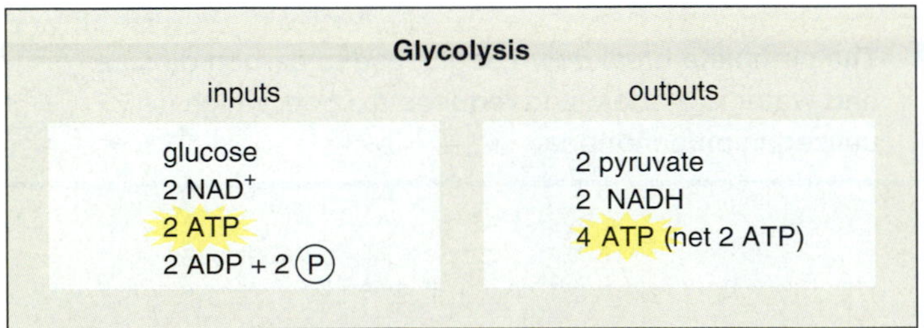

Glycolysis

inputs	outputs
glucose	2 pyruvate
2 NAD^+	2 NADH
2 ATP	4 ATP (net 2 ATP)
2 ADP + 2 Ⓟ	

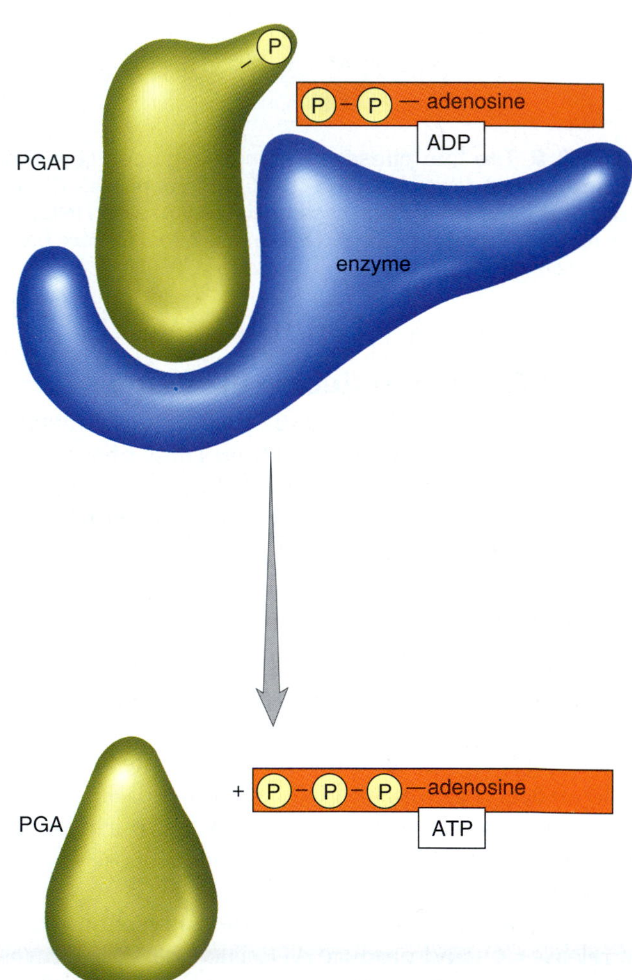

Figure 8.3 Substrate level phosphorylation.
During substrate level phosphorylation, a phosphate group is transferred from a high-energy substrate in a metabolic pathway to ADP. The result is an immediate gain of one ATP molecule. When PGAP, a substrate in the glycolytic pathway shown in Figure 8.4, transfers a phosphate group to ADP, it becomes PGA.

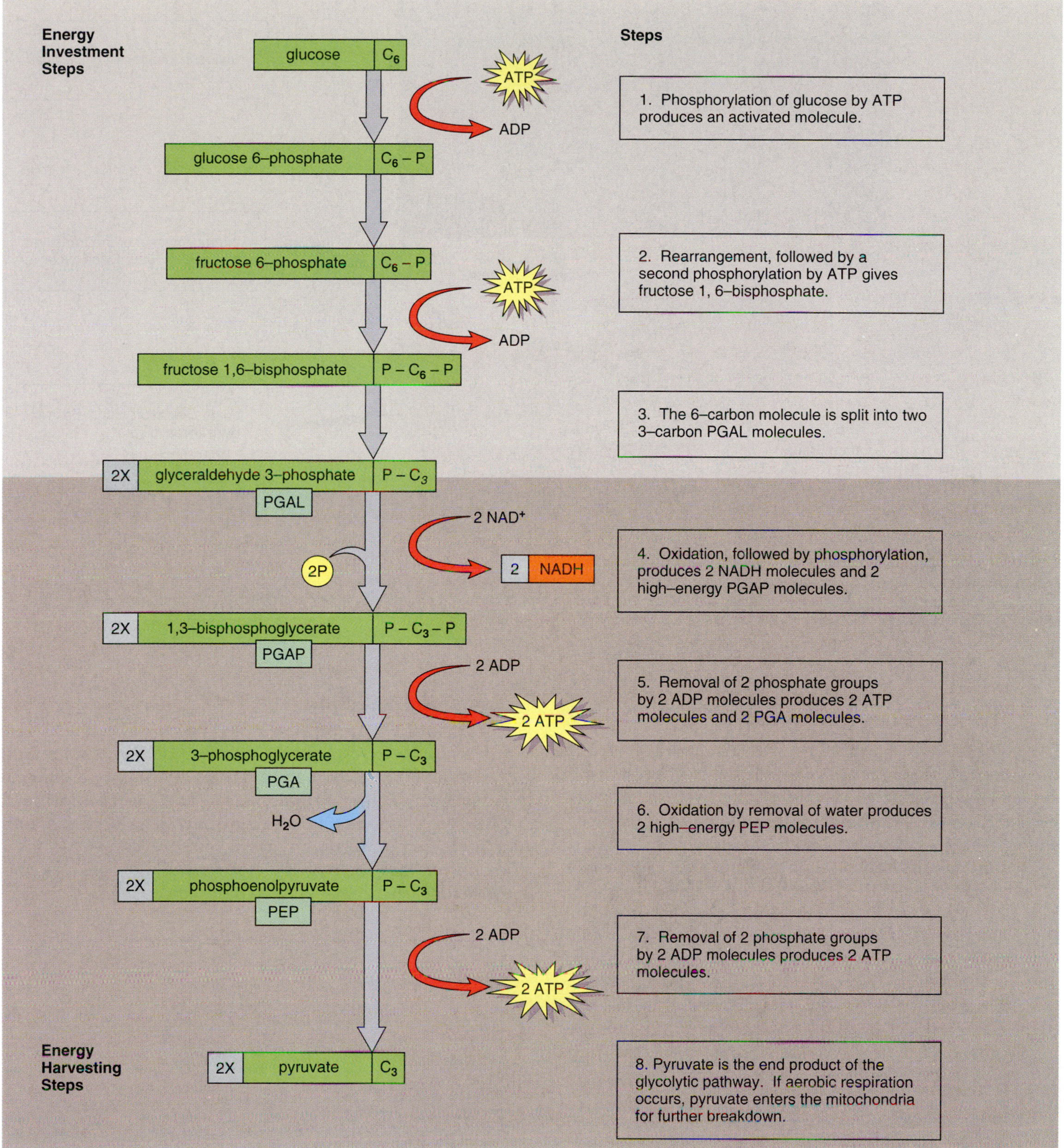

Figure 8.4 **Glycolysis.**
This metabolic pathway begins with glucose and ends with pyruvate. Net gain of two ATP molecules can be calculated by subtracting those expended during the energy investment steps from those produced during the energy harvesting steps. Text in boxes to the far right explains the reactions.

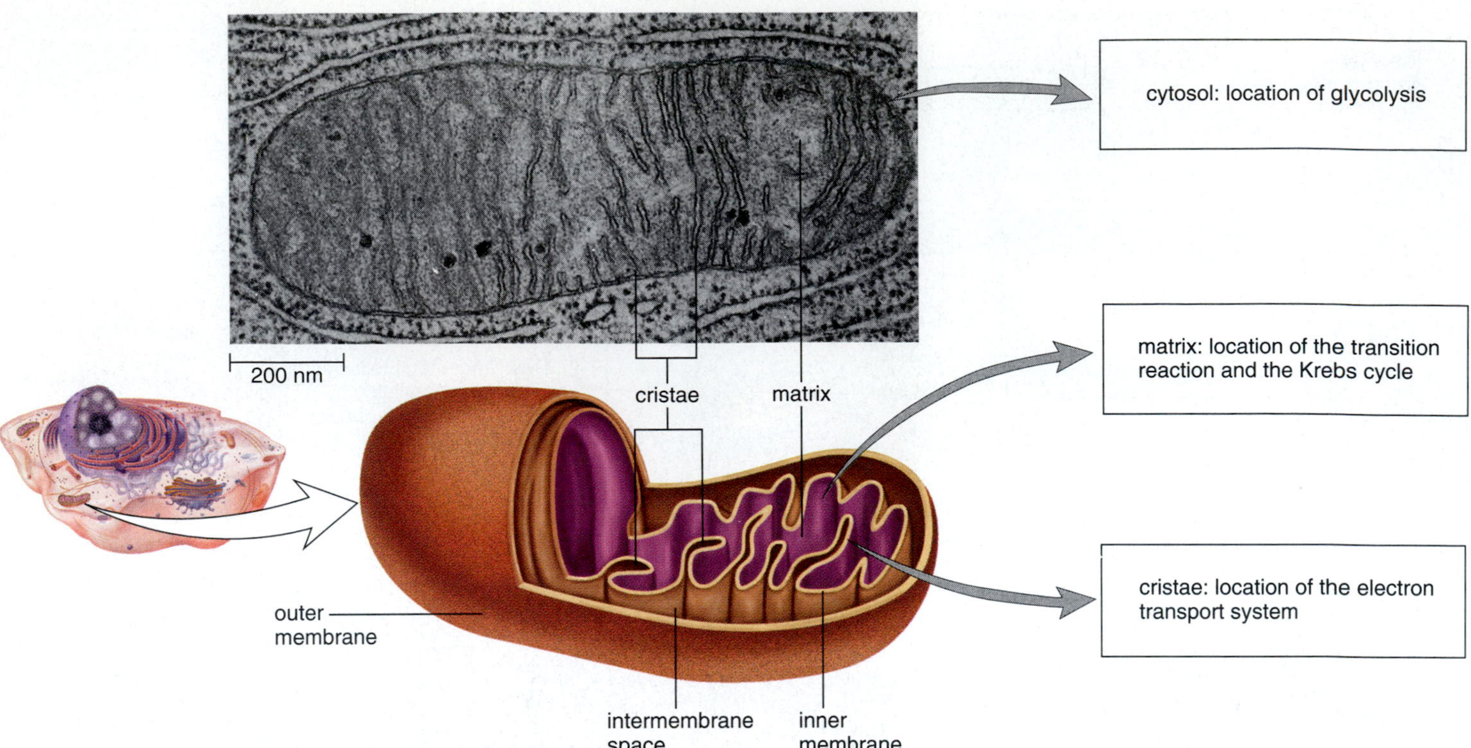

Figure 8.5 **Mitochondrion structure and function.**
A mitochondrion is bounded by a double membrane with an intermembrane space. The inner membrane invaginates to form the shelf-like cristae. Glycolysis takes place in the cytosol outside mitochondria. The transition reaction and the Krebs cycle occur within the mitochondrial matrix. The electron transport system is located on the cristae of a mitochondrion.

8.3 Inside the Mitochondria

Glucose breakdown continues in mitochondria where the transition reaction, Krebs cycle, and the electron transport system are located (Fig. 8.5). Eventually, pyruvate from glycolysis is broken down completely to carbon dioxide and water.

A **mitochondrion** has a double membrane, with an intermembrane space (between the outer and inner membrane). Cristae are folds of inner membrane that jut out into the mitochondrial matrix, the innermost compartment, which is filled with a gel-like fluid. The transition reaction and the Krebs cycle enzymes are located in the matrix, and the electron transport system is located in the cristae. Most of the ATP produced during cellular respiration is produced in mitochondria; therefore, mitochondria are often called the powerhouses of the cell.

Transition Reaction

The transition reaction is so called because it connects glycolysis to the Krebs cycle. In this reaction, pyruvate is converted to a two-carbon acetyl group attached to coenzyme A, collectively termed **acetyl-CoA,** and carbon dioxide is given off. This is an redox reaction in which electrons are removed from pyruvate by an enzyme that uses NAD^+ as a coenzyme that receives the electrons. The reaction occurs twice for each original glucose molecule.

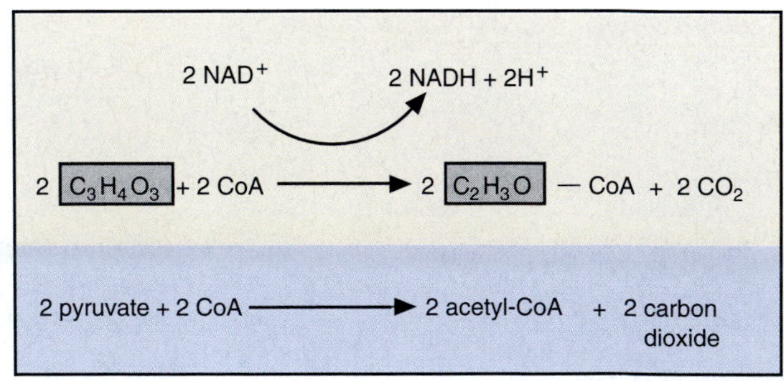

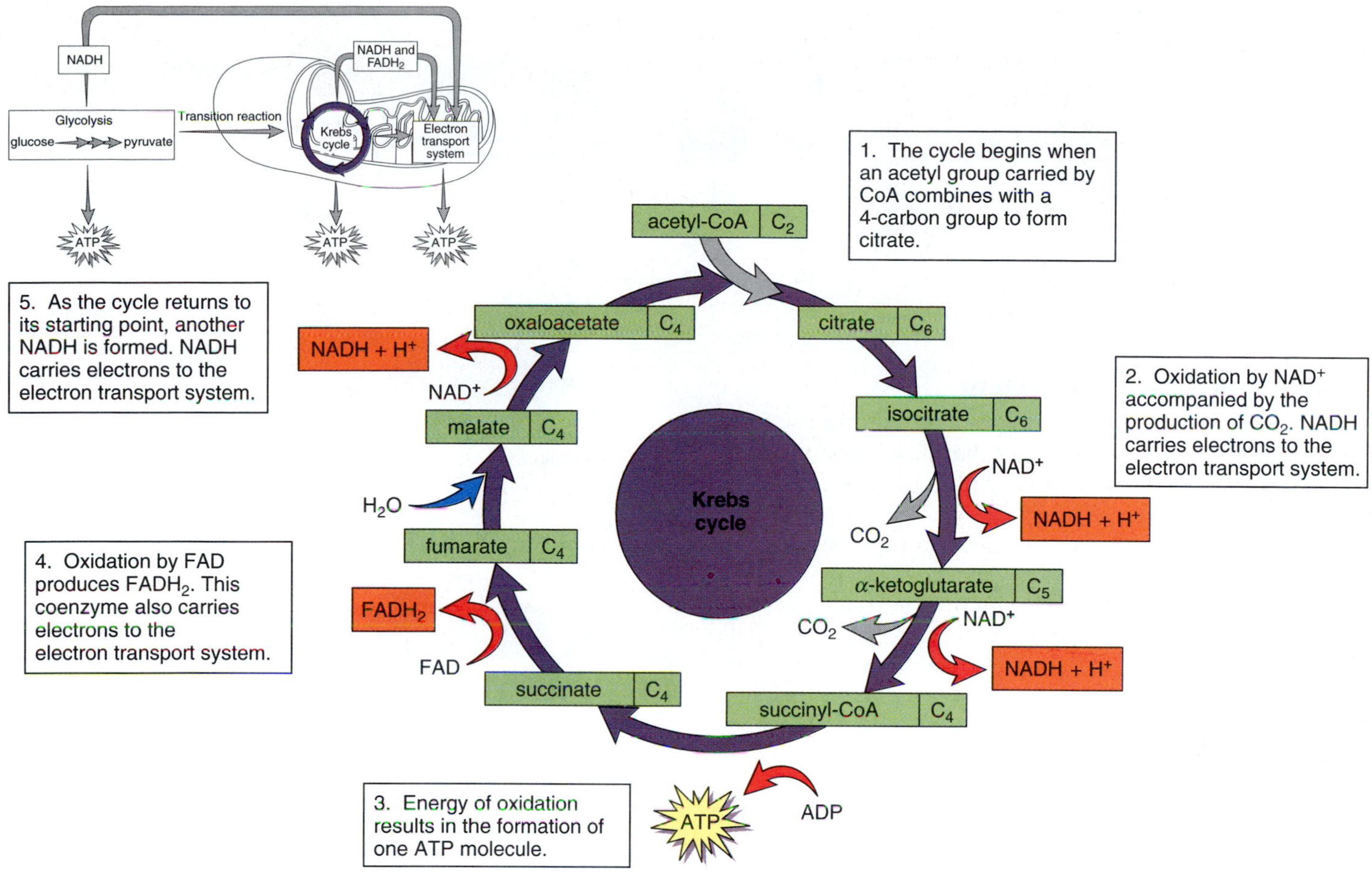

NADH
NADH and FADH2
Glycolysis
glucose pyruvate
Transition reaction
Krebs cycle
Electron transport system
ATP
ATP
ATP
1. The cycle begins when an acetyl group carried by CoA combines with a 4-carbon group to form citrate.
acetyl-CoA C2
citrate C6
isocitrate C6
2. Oxidation by NAD+ accompanied by the production of CO2. NADH carries electrons to the electron transport system.
NAD+
NADH + H+
CO2
α-ketoglutarate C5
CO2
NAD+
NADH + H+
succinyl-CoA C4
ADP
ATP
3. Energy of oxidation results in the formation of one ATP molecule.
succinate C4
FAD
FADH2
4. Oxidation by FAD produces FADH2. This coenzyme also carries electrons to the electron transport system.
fumarate C4
H2O
malate C4
NAD+
NADH + H+
oxaloacetate C4
5. As the cycle returns to its starting point, another NADH is formed. NADH carries electrons to the electron transport system.
Krebs cycle

Figure 8.6 **Krebs cycle.**
The net result of this cycle is the removal of two molecules of carbon dioxide (CO_2) as oxidation occurs. There is a gain of one ATP during the cycle but much energy still resides in the electrons that were received by three molecules of NAD^+ and one molecule of FAD. NADH and $FADH_2$ take these electrons to the electron transport system. (Keep in mind that the Krebs cycle turns twice per glucose molecule.)

The Krebs Cycle

The Krebs cycle is a cyclical metabolic pathway located in the mitochondrial matrix. The Krebs cycle is named for Sir Hans Krebs, a German-born British scientist who received the Nobel Prize for his part in identifying these reactions in the 1930s.

At the start of the Krebs cycle, the C_2 acetyl group produced by the transition reaction joins with a C_4 molecule, forming citrate, a C_6 molecule. For this reason, the cycle is also known as the citric acid cycle (Fig. 8.6). As the substrates undergo oxidation by removal of electrons, two molecules of carbon dioxide (CO_2) form. (Carbon dioxide is one of the end products of the complete breakdown of glucose.)

During the oxidation process most of the electrons are accepted by NAD^+ and NADH then forms. But in one instance, electrons are taken by FAD, and $FADH_2$ forms. NADH and $FADH_2$ carry these electrons to the electron transport system. Some of the energy released when oxidation occurs is used immediately to form ATP by substrate-level phosphorylation, as in glycolysis. A high-energy metabolite accepts a phosphate group and subsequently passes it on to ADP so that ATP forms.

The Krebs cycle turns twice for each original glucose molecule. Therefore, the input and output of the Krebs cycle per glucose molecule are as follows:

Krebs Cycle

inputs	outputs
2 acetyl groups	4 CO_2
2 ADP + 2 Ⓟ	2 ATP
6 NAD^+	6 NADH
2 FAD	2 $FADH_2$

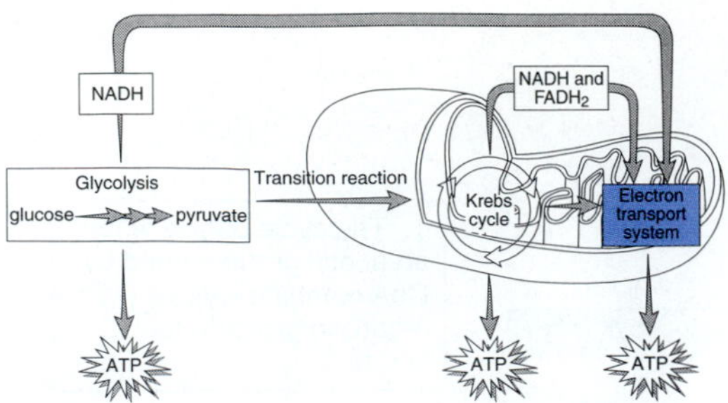

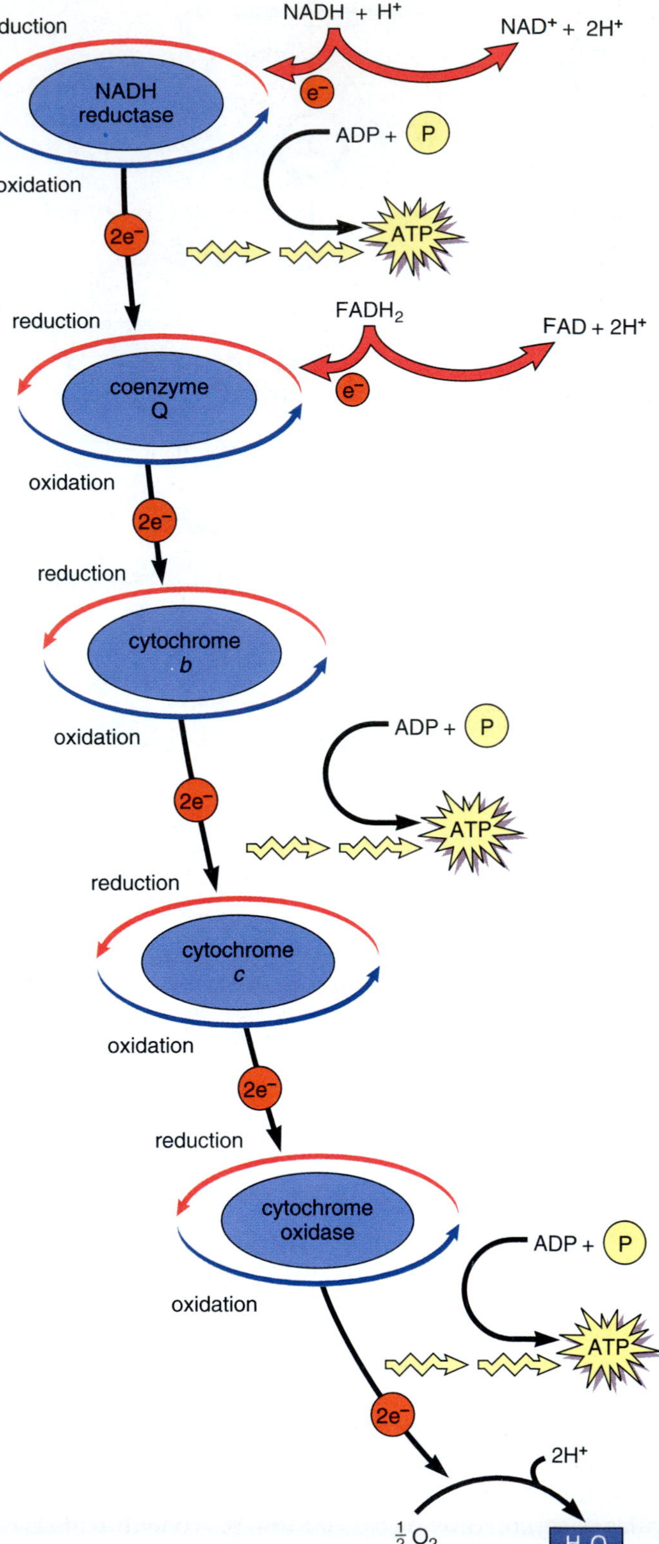

Figure 8.7 The electron transport system.
NADH and $FADH_2$ bring electrons to the electron transport system. As the electrons move down the system, energy is released and used to form ATP. For every pair of electrons that enters by way of NADH, three ATP result. For every pair of electrons that enters by way of $FADH_2$, two ATP result. Oxygen, the final acceptor of the electrons, becomes a part of water.

The Electron Transport System

The electron transport system located in the cristae of the mitochondria is a series of carriers that pass electrons from one to the other. Notice that some of the protein carriers of the system are cytochrome molecules.

The electrons that enter the electron transport system are carried by NADH and $FADH_2$. Figure 8.7 is arranged to show that high-energy electrons enter the system, and low-energy electrons leave the system. When NADH gives up its electrons, it becomes NAD^+ and the next carrier gains the electrons and is reduced. This oxidation-reduction reaction starts the process, and each of the carriers in turn becomes reduced and then oxidized as the electrons move down the system. As the pair of electrons is passed from carrier to carrier, energy is released and stored in the form a hydrogen ion (H^+) gradient. When these ions flow down their electrochemical gradient through an ATP synthase complex, ATP is formed (Fig. 8.8).

Oxygen receives the energy-spent electrons from the last of the carriers. After receiving electrons, oxygen combines with hydrogen ions and water forms. The term **oxidative phosphorylation** refers to the production of ATP as a result of energy released by the electron transport system.

When NADH delivers electrons to the first carrier of the electron transport system, enough energy is released by the time the electrons are received by O_2 to permit the production of three ATP molecules. When $FADH_2$ delivers electrons to the electron transport system, only two ATP are produced.

The cell needs only a limited supply of the coenzymes NAD^+ and FAD because they are constantly being recycled and reused. Therefore, once NADH has delivered electrons to the electron transport system, it is "free" to return and pick up more hydrogens. In the same manner, the components of ATP are recycled in cells. Energy is required to join ADP + Ⓟ, and then when ATP is used to do cellular work, ADP and Ⓟ are made available once more. The recycling of coenzymes and ADP increases cellular efficiency since it does away with the necessity to synthesize large quantities of NAD^+, FAD, and ADP anew.

As electrons pass down the electron transport system, energy is released and ATP is produced.

The Cristae of a Mitochondrion

The carriers of the electron transport system and the proteins concerned with ATP synthesis are spatially arranged in a particular manner in the cristae of mitochondria. Essentially the electron transport system consists of three protein complexes and two protein mobile carriers. The mobile carriers transport electrons between the complexes (Fig. 8.8).

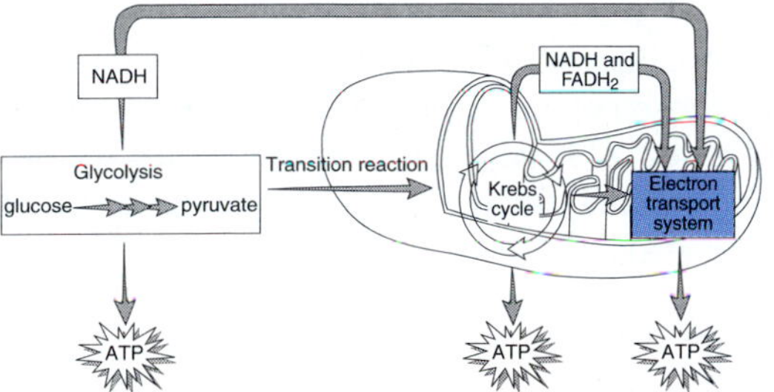

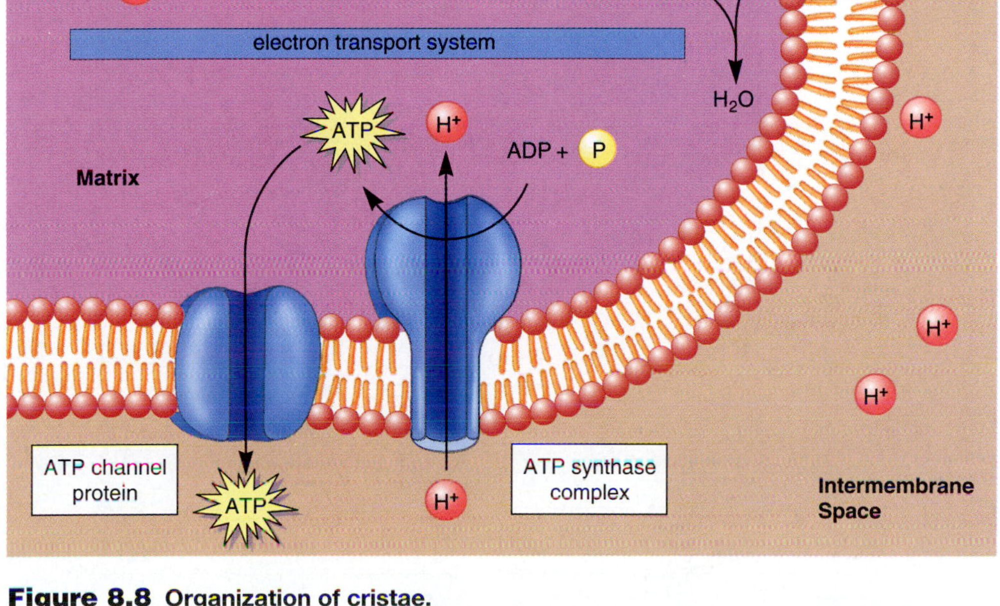

Figure 8.8 Organization of cristae.
The electron transport system is located in the cristae. As electrons move from one complex to the other, hydrogen ions (H^+) are pumped from the matrix into the intermembrane space. As hydrogen ions flow down a concentration gradient from the intermembrane space into the mitochondrial matrix, ATP is synthesized by the enzyme ATP synthase, which is a part of the ATP synthase complex. ATP leaves the matrix by way of a channel protein.

The carriers of the electron transport system accept electrons, which they pass from one to the other. What happens to the hydrogen ions (H^+) carried by NADH and $FADH_2$? The carriers of the electron transport system use the released energy to pump these hydrogen ions from the matrix into the intermembrane space of a mitochondrion. The vertical arrows in Figure 8.8 show that the NADH dehydrogenase complex, the cytochrome *b-c* complex, and the cytochrome oxidase complex all pump H^+ into the intermembrane space. This establishes a strong electrochemical gradient; there are about ten times as many hydrogen ions in the intermembrane space than there are in the matrix.

ATP Production The cristae contain an ATP synthase complex through which hydrogen ions flow down a gradient from the intermembrane space into the matrix. As hydrogen ions flow from high to low concentration, the enzyme ATP synthase synthesizes ATP from ADP + Ⓟ. Just how H^+ flow drives ATP synthesis is not known; perhaps the hydrogen ions participate in the reaction, or perhaps they cause a change in the shape of ATP synthase and this brings about ATP synthesis. Mitochondria produce ATP by **chemiosmosis,** so called because ATP production is tied to an electrochemical gradient that is an H^+ gradient.

The finding of respiratory poisons has lent support to the chemiosmotic model of ATP synthesis. When one poison inhibits ATP synthesis, the H^+ gradient becomes larger than usual, and when another type makes the membrane leaky to H^+, no ATP is made.

Once formed, ATP molecules diffuse out of the mitochondrial matrix by way of a channel protein.

Mitochondria synthesize ATP by chemiosmosis. ATP production is dependent upon an electrochemical gradient established by the pumping of H^+ into the intermembrane space.

Energy Yield from Glucose Breakdown

Figure 8.9 calculates the ATP yield for the complete breakdown of glucose to CO_2 and H_2O in eukaryotes. The diagram includes the number of ATP produced by substrate phosphorylation and the number that is produced by oxidative phosphorylation as a result of electrons passing down the electron transport system.

Substrate-Level Phosphorylation

Per glucose molecule, there is a net gain of two ATP from glycolysis in the cytosol and two more from the Krebs cycle in the matrix of mitochondria. Altogether, four ATP are formed outside the electron transport system.

Oxidative Phosphorylation

Most of the ATP produced by cellular respiration is due to oxidative phosphorylation and energy released by the electron transport system. Per glucose molecule, ten NADH and two $FADH_2$ take electrons to the electron transport system. For each NADH formed *inside* the mitochondria by the Krebs cycle, three ATP result, but for each $FADH_2$, only two ATP are produced. Figure 8.7 explains the reason for this difference: $FADH_2$ delivers its electrons to the transport system after NADH and therefore these electrons cannot account for as much ATP production.

What about the ATP yield of NADH generated *outside* the mitochondria by the glycolytic pathway? NADH cannot cross mitochondrial membranes, but there is a "shuttle" mechanism that allows its electrons to be delivered to the electron transport system inside the mitochondria. The shuttle consists of an organic molecule, which can cross the outer membrane, accept the electrons, and in most cells, deliver them to a FAD molecule in the inner membrane. If FAD is utilized, only two ATP result.

Heart and liver cells, which have high metabolic rates, are exceptions; in these cells cytoplasmic NADH results in the production of three ATP. Also, since prokaryotes (bacteria) do not have mitochondria, each NADH produces three ATP for a total of 38 ATP.

Efficiency of Complete Glucose Breakdown

It is interesting to calculate how much of the energy in a glucose molecule eventually becomes available to the cell. The difference in energy content between the reactants (glucose and O_2) and the products (CO_2 and H_2O) is 686 kcal. An ATP phosphate bond has an energy content of 7.3 kcal, and 36 to 38 of these are usually produced during glucose breakdown. Taking the lower number, 36 phosphates are equivalent to a total of 263 kcal. Therefore, at least 263 kcal/686 kcal × 100, or about 39% of the available energy is transferred from glucose to ATP as a result of glucose breakdown.

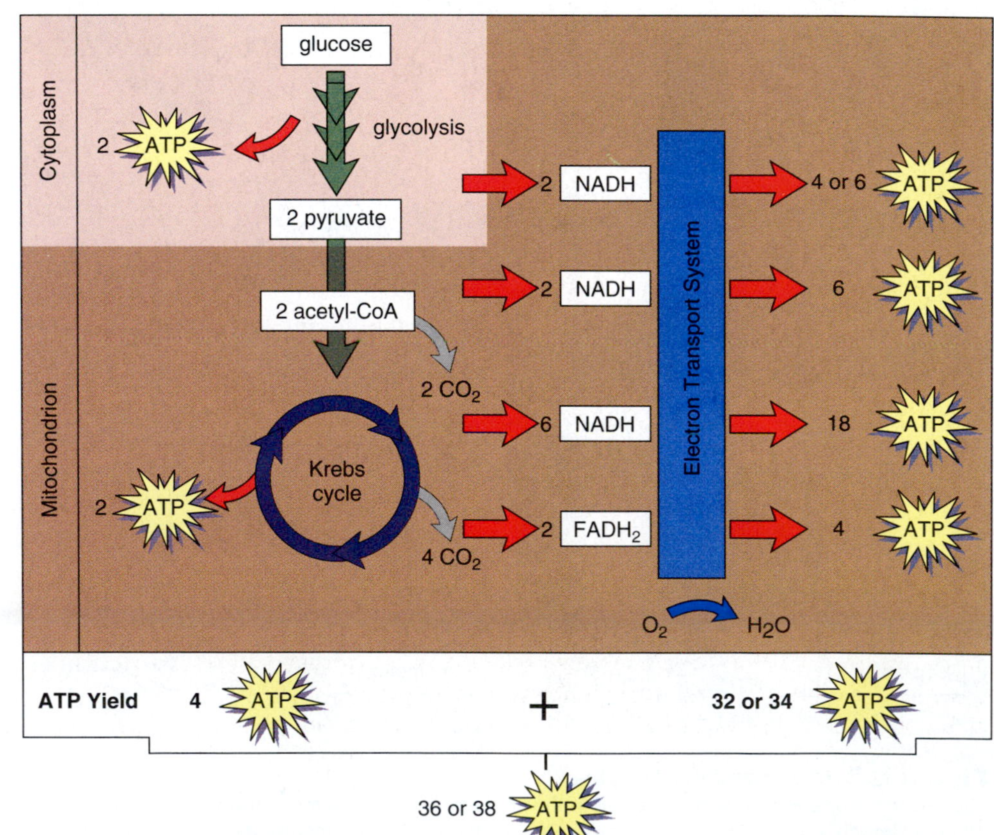

Figure 8.9 Energy yield per glucose molecule.
Substrate-level phosphorylation during glycolysis and the Krebs cycle results in the production of four ATP molecules per glucose molecule. Oxidative phosphorylation accounts for 32 or 34 ATP, and the grand total of ATP is therefore 36 or 38 ATP. The delivery of electrons from NADH generated outside the mitochondrion explains this variation. If the electrons from NADH are delivered by a shuttle mechanism at the start of the electron transport system, six ATP result per NADH; otherwise, four ATP result.

8.4 Fermentation

Cellular respiration also includes fermentation. **Fermentation** consists of glycolysis followed by the reduction of pyruvate by NADH to either lactate or alcohol and CO_2 (Fig. 8.10). The pathway operates anaerobically because after NADH transfers its electrons to pyruvate, it is "free" to return and pick up more electrons during the earlier reactions of glycolysis.

Certain anaerobic bacteria such as lactic-acid bacteria, which help us manufacture cheese, consistently produce lactate in this manner. Other bacteria produce chemicals of industrial importance anaerobically: isopropanol, butyric acid, propionic acid, and acetic acid. Yeasts are good examples of organisms that generate alcohol and CO_2. Yeast is used to leaven bread; the CO_2 produced makes bread rise. On the other hand, yeast is used to ferment sugar when wine is made; in that case, it is the ethyl alcohol that is desired. Eventually yeasts are killed by the very alcohol they produce.

Animal, including human, cells are similar to lactic-acid bacteria in that pyruvate, when produced faster than it can be oxidized through the Krebs cycle, is reduced to lactate.

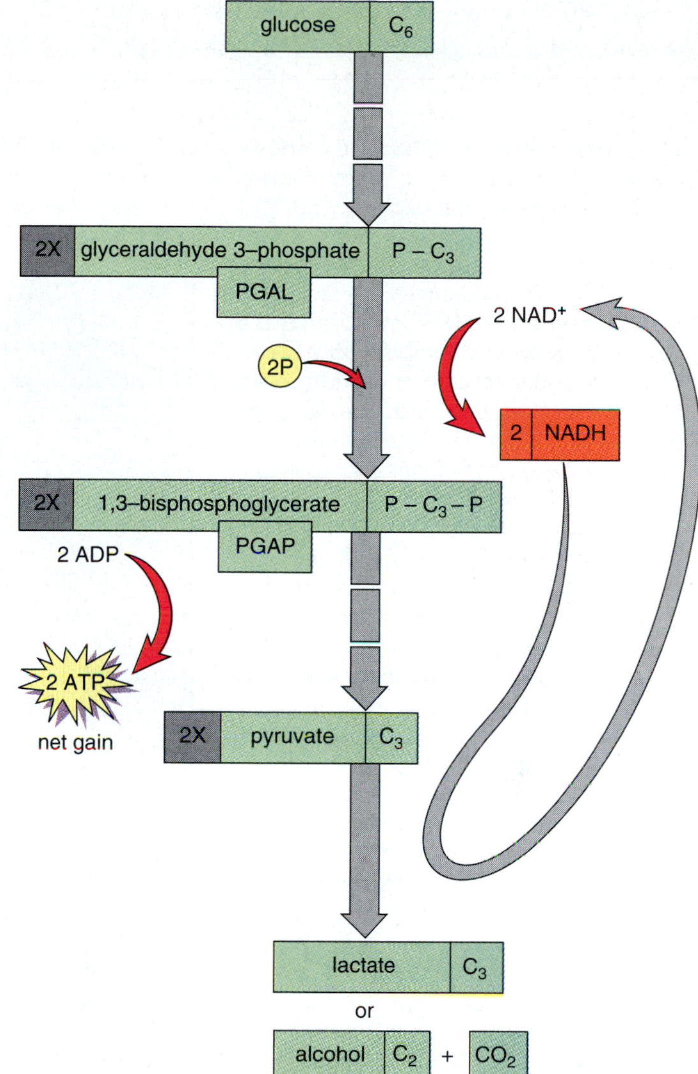

Figure 8.10 Fermentation.
Fermentation consists of glycolysis followed by a reduction of pyruvate. This "frees" NAD^+ and it returns to the glycolytic pathway to pick up more electrons.

Advantages and Disadvantages of Fermentation

Despite its low yield, fermentation is essential to humans because it can provide a rapid burst of ATP; muscle cells more than other cells are apt to carry on fermentation. When our muscles are working vigorously over a short period of time, fermentation is a way to produce ATP even though oxygen is temporarily in limited supply.

Lactate, however, is toxic to cells. At first, blood carries away all the lactate formed in muscles. Eventually, however, lactate begins to build up, changing the pH and causing the muscles to fatigue so that they no longer contract. When we stop running, our bodies are in **oxygen debt,** a term that refers to the amount of oxygen needed to restore ATP to its former level and rid the body of lactate. Oxygen debt is signified by the fact that we continue to breathe very heavily for a time. Recovery involves transporting lactate to the liver, where it is converted back to pyruvate. Some of the pyruvate is respired completely, and the rest is converted back to glucose.

Efficiency of Fermentation?

The two ATP produced per glucose during fermentation is equivalent to 14.6 kcal. Complete glucose breakdown to CO_2 and H_2O represents a possible energy yield of 686 kcal per molecule. Therefore, the efficiency for fermentation is only 14.6 kcal/686 kcal × 100, or 2.1%. This is much less efficient than the complete breakdown of glucose. The inputs and outputs of fermentation are as shown next.

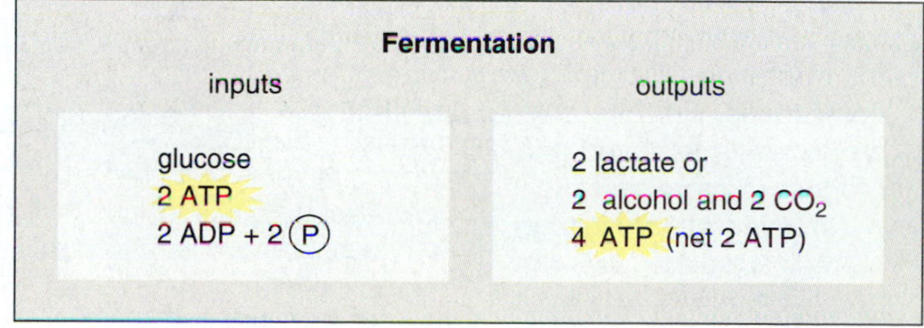

Fermentation

inputs	outputs
glucose	2 lactate or
2 ATP	2 alcohol and 2 CO_2
2 ADP + 2 Ⓟ	4 ATP (net 2 ATP)

Exercise: A Test of Homeostatic Control

Exercise is a dramatic test of the body's homeostatic control systems—there is a large increase in muscle oxygen (O_2) requirement, and a large amount of carbon dioxide (CO_2) is produced. These changes must be countered by increases in breathing and blood flow to increase oxygen delivery and removal of the metabolically produced carbon dioxide. Also, heavy exercise can produce a large amount of lactic acid due to the utilization of fermentation, an anaerobic process. Both the accumulation of carbon dioxide and lactic acid can lead to an increase in intracellular and extracellular acidity. Further, during heavy exercise, working muscles produce large amounts of heat that must be removed to prevent overheating. In a strict sense, the body rarely maintains true homeostasis while performing intense exercise or during prolonged exercise in a hot or humid environment. However, a better maintenance of homeostasis is observed in those who have had endurance training.

The number of mitochondria increases in the muscles of persons who train; and, therefore, there is greater reliance on the Krebs cycle and the electron transport system to generate energy. Muscle cells with few mitochondria must have a high ADP concentration to stimulate the limited number of mitochondria to start consuming oxygen. After an endurance training program, the large number of mitochondria start consuming oxygen as soon as the ADP concentration starts rising due to muscle contraction and subsequent breakdown of ATP. Therefore, a steady state of oxygen intake by mitochondria is achieved earlier in the athlete. This faster rise in oxygen uptake at the onset of work means that the oxygen deficit is less, and the formation of lactate due to fermentation is less. Further, any lactate that is produced is removed and processed more quickly.

Training also results in greater reliance on the Krebs cycle and increased fatty acid metabolism, because fatty acids are broken down to acetyl-CoA, which enters the Krebs cycle. This preserves plasma glucose concentration and also helps the body maintain homeostasis.

Source: *Scott K. Powers and Edward T. Howley,* Exercise Physiology, *2d ed., 1994. McGraw-Hill Companies, Dubuque, IA.*

In athletes, there is:

- a smaller oxygen deficit due to a more rapid increase in oxygen uptake at the onset of work;
- an increase in fat metabolism that spares blood glucose;
- a reduction in lactate and hydrogen ion (H^+) formation;
- an increase in ability to remove and process lactate.

8.5 Metabolic Pool

Degradative reactions, collectively called **catabolism,** break down molecules and tend to be exergonic. Synthetic reactions, collectively called **anabolism,** tend to be endergonic. Is it correct to say then, that catabolism drives anabolism?

Catabolism

We already know that glucose is broken down during cellular respiration. However, other molecules can also undergo catabolism. When a fat molecule is used as an energy source, it breaks down to glycerol and three fatty acids. As Figure 8.11 indicates, glycerol is converted to PGAL, a metabolite in glycolysis. The fatty acids are converted to acetyl-CoA and the acetyl groups enter the Krebs cycle. An 18-carbon fatty acid results in nine acetyl-CoA molecules. In the human body, oxidation of these acetyl-CoA molecules can produce a total of 216 ATP molecules. For this reason, fats are an efficient form of stored energy—there are, after all, three long fatty acid chains per fat molecule.

The carbon skeleton of amino acids can also be broken down. In humans, the carbon skeleton is produced in the liver when an amino acid undergoes **deamination,** or the removal of the amino group. The amino group becomes ammonia (NH_3), which enters the urea cycle and becomes part of urea, the primary excretory product of humans. Just where the carbon skeleton begins degradation is dependent on the length of the *R* group, since this determines the number of carbons left after deamination.

Anabolism

We have already mentioned that the ATP produced during catabolism drives anabolism. But there is another way catabolism is related to anabolism. The substrates making up the pathways in Figure 8.11 can be used as starting materials for synthetic reactions. In other words, compounds that enter the pathways are oxidized to substrates that can be used for biosynthesis. This is the cell's **metabolic pool** [Gk. *meta,* implying change] in which one type of molecule can be converted to another. In this way, carbohydrate intake can result in the formation of fat. PGAL can be converted to glycerol, and acetyl groups can be joined to form fatty acids. Fat synthesis follows. This explains why you gain weight from eating too much candy, ice cream, and cake.

Some metabolites of the Krebs cycle can be converted to amino acids through transamination, the transfer of an amino group to an organic acid, forming a different amino acid. Plants are able to synthesize all of the amino acids they need. Animals, however, lack some of the enzymes necessary for synthesis of all amino acids. Adult humans, for example, can synthesize eleven of the common amino acids, but they cannot synthesize the other nine. The amino acids that cannot be synthesized must be supplied by the diet; they are called the essential amino acids. The nonessential amino acids can be synthesized. It is quite possible for animals to suffer from protein deficiency if their diet does not contain adequate quantities of all the essential amino acids.

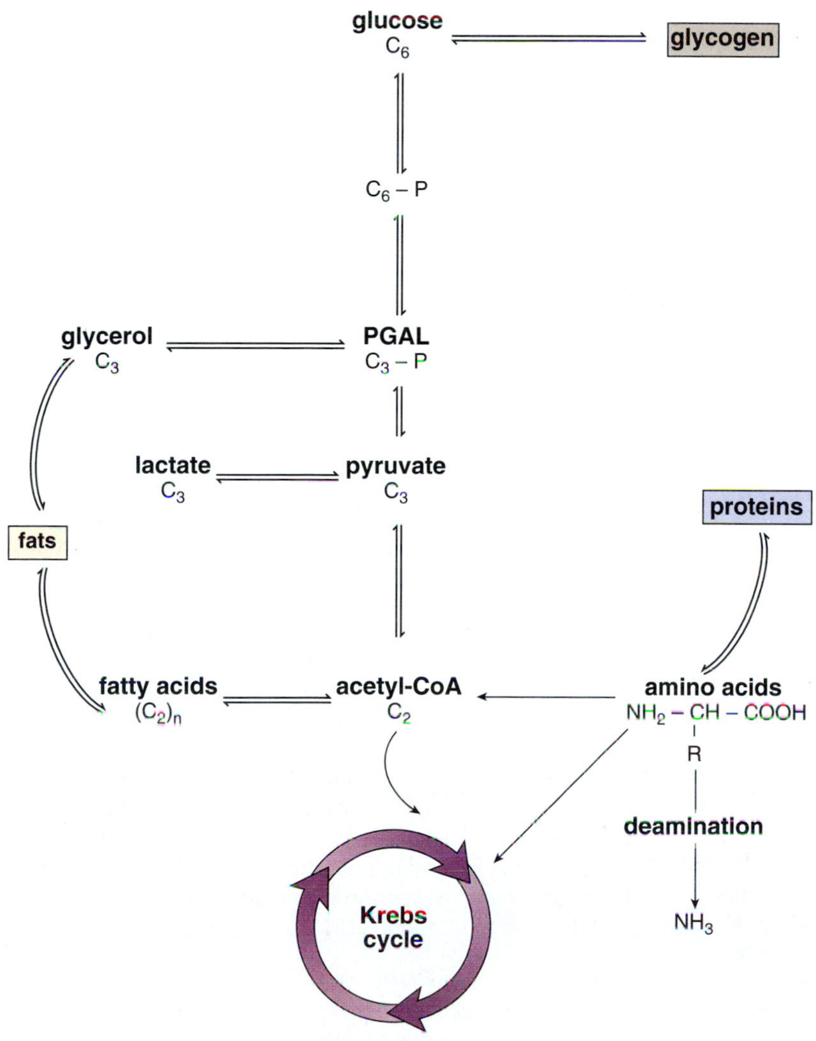

Figure 8.11 The metabolic pool concept.
When they are used as energy sources, carbohydrates, fats, and proteins enter degradative pathways at specific points. Catabolism produces metabolites that can also be used for anabolism of other compounds.

All the reactions involved in cellular respiration are a part of a metabolic pool; the metabolites from the pool can be used for catabolism or for anabolism.

Connecting Concepts

The organelles of eukaryotic cells have a structure that suits their function. Cells didn't arise until they had a membranous covering, and membrane is also absolutely essential to the organization of chloroplasts and mitochondria. In a chloroplast, membrane forms the grana, which are stacks of interconnected, flattened membranous sacs called thylakoids. The inner membrane of a mitochondrion invaginates to form the convoluted cristae.

The detailed structure of chloroplasts and mitochondria is different, but essentially they operate similarly: an assembly line of particles in the thylakoid membrane and cristae carry out functions necessary to photosynthesis and cellular respiration, respectively. In both organelles, electron transport system carriers pump hydrogen ions into an enclosed space, establishing an electrochemical gradient. When hydrogen ions flow down this gradient through ATP synthase complex, the energy released is used to produce ATP. In chloroplasts the electrons sent to the electron transport system are energized by the sun; in mitochondria energized electrons are removed from the substrates of the glycolytic and Krebs cycle.

In chloroplasts, the gel-like fluid of the stroma contains the enzymes of the Calvin cycle, which reduce carbon dioxide to a carbohydrate, and in mitochondria, the gel-like fluid of the matrix contains the enzymes of the Krebs cycle, which oxidize carbohydrate products received from the cytosol. In chloroplasts reduction of carbon dioxide requires ATP produced by chemiosmosis, while in mitochondria oxidation of substrates releases carbon dioxide; the ATP produced by chemiosmosis is made available to the cell.

According to the endosymbiotic hypothesis, chloroplasts and mitochondria were independent prokaryotic organisms at one time. Indeed, each contains genes (DNA) not found in the eukaryotic nucleus. Through evolution all organisms are related, and therefore these organelles may be related also. The similar organization of these organelles suggests that this is the case.

Summary

8.1 Cellular Respiration

The oxidation of glucose to CO_2 and H_2O is an exergonic reaction that drives ATP synthesis, an endergonic reaction. Four phases are required: glycolysis, the transition reaction, the Krebs cycle, and passage of electrons along the electron transport system. Oxidation of substrates involves the removal of hydrogen atoms ($H^+ + e^-$) from substrate molecules, usually by the coenzyme NAD^+, but in one case by FAD.

8.2 Outside the Mitochondria: Glycolysis

Glycolysis, the breakdown of glucose to two molecules of pyruvate, is a series of enzymatic reactions that occur in the cytosol. Breakdown releases enough energy to immediately give a net gain of two ATP by substrate-level phosphorylation. Two NADH are formed.

8.3 Inside the Mitochondria

When oxygen is available, pyruvate from glycolysis enters the mitochondrion, where the transition reaction takes place. During this reaction, oxidation occurs as CO_2 is removed from pyruvate. NAD^+ is reduced, and CoA receives the C_2 acetyl group that remains. Since the reaction must take place twice per glucose molecule, two NADH result.

The acetyl group enters the Krebs cycle, a cyclical series of reactions located in the mitochondrial matrix. Complete oxidation follows, as two CO_2 molecules, three NADH molecules, and one $FADH_2$ molecule are formed. The cycle also produces one ATP molecule. The entire cycle must turn twice per glucose molecule.

The final stage of glucose breakdown involves the electron transport system located in the cristae of the mitochondria. The electrons received from NADH and $FADH_2$ are passed down a chain of carriers until they are finally received by oxygen, which combines with H^+ to produce water. As the electrons pass down the system, energy is released and stored for ATP production. The term oxidative phosphorylation is sometimes used for ATP production associated with the electron transport system.

The cristae of mitochondria contain protein complexes of the electron transport system that pass electrons from one to the other and pump H^+ into the intermembrane space, setting up an electrochemical gradient. When H^+ flows down this gradient through an ATP synthase complex, energy is released and used to form ATP molecules from ADP and Ⓟ. This is ATP synthesis by chemiosmosis.

Of the 36 to 38 ATP formed by complete glucose breakdown, four are the result of substrate-level phosphorylation and the rest are produced by oxidative phosphorylation. The energy for the latter comes from the electron transport system. For most NADH molecules that donate electrons to the electron transport system, three ATP molecules are produced. In some cells each NADH formed in the cytosol, however, results in only two ATP molecules. This occurs when the hydrogen atoms are shuttled across the mitochondrial membrane by a carrier that passes them to FAD. $FADH_2$ results in the formation of only two ATP because its electrons enter the electron transport system at a lower energy level.

8.4 Fermentation

Fermentation involves glycolysis followed by the reduction of pyruvate by NADH either to lactate or to alcohol and carbon dioxide (CO_2). The reduction process "frees" NAD^+ so that it can accept more hydrogen atoms from glycolysis.

Although fermentation results in only two ATP molecules, it still serves a purpose. In vertebrates, it provides a quick burst of ATP energy for short-term, strenuous muscular activity. The accumulation of lactate puts the individual in oxygen debt because oxygen is needed when lactate is completely metabolized to CO_2 and H_2O.

8.5 Metabolic Pool

Carbohydrate, protein, and fat can be catabolized by entering the degradative pathways at different locations. These pathways also provide metabolites needed for the anabolism of various important substances. Catabolism and anabolism, therefore, both utilize the same pools of metabolites.

Reviewing the Chapter

1. What is the overall chemical equation for the complete breakdown of glucose to CO_2 and H_2O? Explain how this is an oxidation-reduction reaction. Why is the reaction able to drive ATP buildup? 128
2. What are NAD^+ and FAD? What are their functions? 128
3. What are the three pathways involved in the complete breakdown of glucose to carbon dioxide (CO_2) and water (H_2O)? What reaction is needed to join two of these pathways? 129
4. What are the main events of glycolysis? How is ATP formed? 130
5. Give the substrates and products of the transition reaction. Where does it take place? 132
6. What are the main events during the Krebs cycle? 133
7. What is the electron transport system, and what are its functions? 134
8. Describe the organization of protein complexes within the cristae. Explain how the complexes are involved in ATP production. 135
9. Calculate the energy yield of glycolysis and complete glucose breakdown. Distinguish between substrate-level phosphorylation and oxidative phosphorylation. 136
10. What is fermentation and how does it differ from glycolysis? Mention the benefit of pyruvate reduction during fermentation. What types of organisms carry out lactate fermentation, and what types carry out alcoholic fermentation? 137
11. Give examples to support the concept of the metabolic pool. 139

Testing Yourself

Choose the best answer for each question. For questions 1–8, identify the pathway involved by matching them to the terms in the key.

Key:

a. glycolysis
b. Krebs cycle
c. electron transport system

1. carbon dioxide (CO_2) given off
2. water (H_2O) formed
3. PGAL
4. NADH becomes NAD^+
5. oxidative phosphorylation
6. cytochrome carriers
7. pyruvate
8. FAD becomes $FADH_2$
9. The transition reaction
   a. connects glycolysis to the Krebs cycle.
   b. gives off CO_2.
   c. utilizes NAD^+.
   d. results in an acetyl group.
   e. All of these are correct.
10. The greatest contributor of electrons to the electron transport system is
    a. oxygen.
    b. glycolysis.
    c. the Krebs cycle.
    d. the transition reaction.
    e. fermentation.
11. Substrate-level phosphorylation takes place in
    a. glycolysis and the Krebs cycle.
    b. the electron transport system and the transition reaction.
    c. glycolysis and the electron transport system.
    d. the Krebs cycle and the transition reaction.
    e. Both b and d are correct.
12. Which of these is not true of fermentation?
    a. net gain of only two ATP
    b. occurs in cytosol
    c. NADH donates electrons to electron transport system
    d. begins with glucose
    e. carried on by yeast
13. Fatty acids are broken down to
    a. pyruvate molecules, which take electrons to the electron transport system.
    b. acetyl groups, which enter the Krebs cycle.
    c. amino acids, which excrete ammonia.
    d. glycerol, which is found in fats.
    e. All of these are correct.
14. Label this diagram of a mitochondrion and state a function for each portion named.

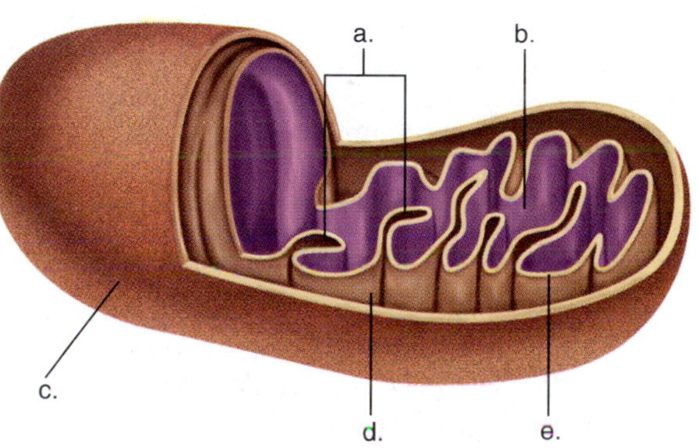

Thinking Scientifically

1. In mitochondria, NADH supplies electrons to the electron transport system. Molecular oxygen (with the highest affinity for electrons) is the final receiver. In photosynthesis, however, the oxygen in water (H_2O) is the source of electrons and the final receiver is $NADP^+$. How can oxygen act as both a donor and receiver of electrons?
2. You are working with pyruvate molecules that contain radioactive carbon. They are incubated with all the components of the Krebs cycle long enough for one turn of the cycle. Carbon dioxide is produced, but only one third of it is radioactive. How can this observation be explained?

Bioethical Issue

Feeling tired and run-down? Want to jump start your mitochondria? If you seem to have no specific ailment, you might be tempted to turn to what is now called alternative medicine. Alternative medicine includes such nonconventional therapies as herbal supplements, acupuncture, chiropractic therapy, homeopathy, osteopathy, and therapeutic touch (e.g., laying on of hands).

In 1992, Congress established what is now called the National Center for Complementary and Alternative Medicine (NCCAM), whose budget has grown from $2 million to $50 million for 1999. In 1994, the Dietary Supplement Health and Education Act allowed the marketing of vitamins, minerals, and herbs without the requirement that they be approved by the Food and Drug Administration (FDA).

Many scientists feel that alternative medicine should be subjected to the same rigorous clinical testing as traditional medicine. They approve of a control study that is now testing the efficacy of the herb Saint-John's-wort. The results are expected to tell whether Saint-John's-wort, a relatively inexpensive remedy, works just as well as a relatively expensive drug produced by pharmaceutical firms. Would it be satisfactory to you if alternative medical practices succeed simply for psychological reasons rather than for their physical benefit?

Understanding the Terms

acetyl-CoA 132
anabolism 139
catabolism 139
cellular respiration 128
chemiosmosis 135
deamination 139
electron transport system 129
FAD 128
fermentation 129, 137
glycolysis 129, 130
Krebs cycle 129
metabolic pool 139
mitochondrion 132
NAD^+ 128
oxidative phosphorylation 134
oxygen debt 137
pyruvate 129
substrate-level phosphorylation 130
transition reaction 129

Match the terms to these definitions:

a. ______________ Anaerobic breakdown of glucose that results in a gain of two ATP and end products such as alcohol and lactate.
b. ______________ Cycle of reactions in mitochondria that begins with citric acid; it produces CO_2, ATP, NADH, and $FADH_2$; also called the citric acid cycle.
c. ______________ Metabolites that are the products of and/or the substrates for key reactions in cells allowing one type of molecule to be changed into another type, such as the conversion of carbohydrates to fats.
d. ______________ Molecule made up of a two-carbon group attached to coenzyme A. During cellular respiration, the two-carbon group enters the Krebs cycle for further breakdown.
e. ______________ Series of membrane-bounded carrier molecules that pass electrons from one to the other, releasing energy that is used for the synthesis of ATP.

Web Connections

Exploring the Internet

http://www.mhhe.com/biosci/genbio/mader
(click on *Biology 7/e*)

The *Biology 7/e* Online Learning Center provides many resources for studying the material in this chapter including links to the following sites:

Learn more about the basics of cell carbohydrate metabolism in Basic Concepts of Metabolism: The Totality of Chemical Reactions of Living Matter. At this site, the user can find information about catabolism and anabolism, autotrophic and heterotrophic pathways, and redox reactions.

http://www.blc.arizona.edu/interactive/metabolism2.95/metabolism.html

Lipid Metabolism. This site describes in great detail the pathways by which lipids are metabolized in the body.

http://web.indstate.edu/thcme/mwking/lipsynth.html

Glycolysis and the Krebs Cycle. This MIT hypertextbook has terrific graphics and text describing these processes.

http://esg-www.mit.edu:8001/esgbio/glycolysis/dir.html

Fermentations: Changing the Course of Human History. An interesting description of fermentation processes used by humans.

http://www.accessexcellence.org/TSN/SS/ferm_background.html

Selected References on Fermentation, including intriguing publications and links, including "How to Brew Your First Beer."

http://www.accessexcellence.org/TSN/SS/ferm_references.html

Cellular Reproduction and the Cell Cycle

chapter

chapter concepts

From one cell, two cells.

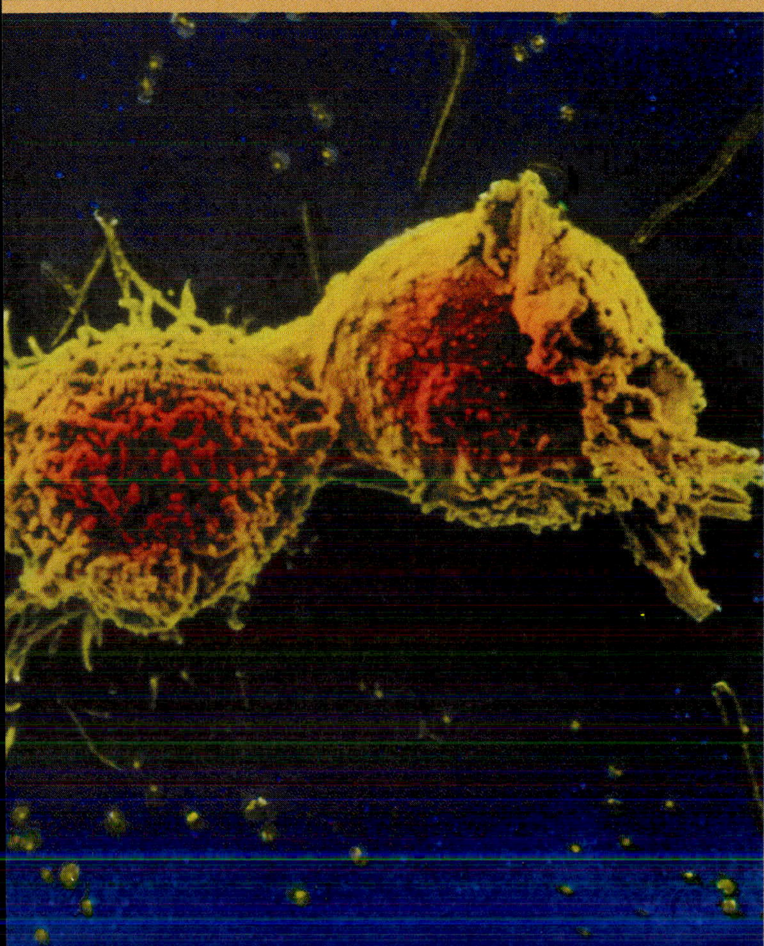

Consider the development of a human being. We begin life as one cell—an egg fertilized by a sperm. Yet in nine short months, we become complex organisms consisting of trillions of cells. How is such a feat possible? Cell division enables a single cell to produce many cells, allowing an organism to grow in size and to replace worn-out tissues.

The instructions for cell division lie in the genes. During the first part of an organism's life, the genes instruct all cells to divide. When adulthood is reached, however, only specific cells—human blood and skin cells, for example—continue to divide daily. Other tissues, such as nervous tissue, no longer routinely produce new cells.

Why don't all types of adult cells reproduce routinely? They contain the full complement of genetic material in their nuclei. Such questions are being intensely studied. Cell biologists have recently discovered that specific enzymes regulate the cell cycle, the period that extends from the time a new cell is produced until it too completes division. Proper function of these enzymes ensures that cells divide normally. Cancer may result when the enzymes regulating the cell cycle go awry.

9.1 How Prokaryotic Cells Divide

In all organisms, the genes consist of DNA located in one or more chromosomes. The reproduction of cells usually consists of duplicating the chromosome(s) and distributing a complete set to the daughter cells. In this way the daughter cells are genetically identical to the parent cell.

Bacteria and protists, such as amoeboids and paramecia, are unicellular. Cell division in unicellular organisms produces two new individuals. This is **asexual reproduction** because one parent has produced identical offspring.

Plants and animals are multicellular. Cell division in multicellular organisms is part of a lifelong developmental process. Growth occurs as the multicellular form develops into the adult organism and it occurs during renewal and repair.

Cell division allows unicellular organisms to reproduce and is necessary to the growth and repair of multicellular organisms.

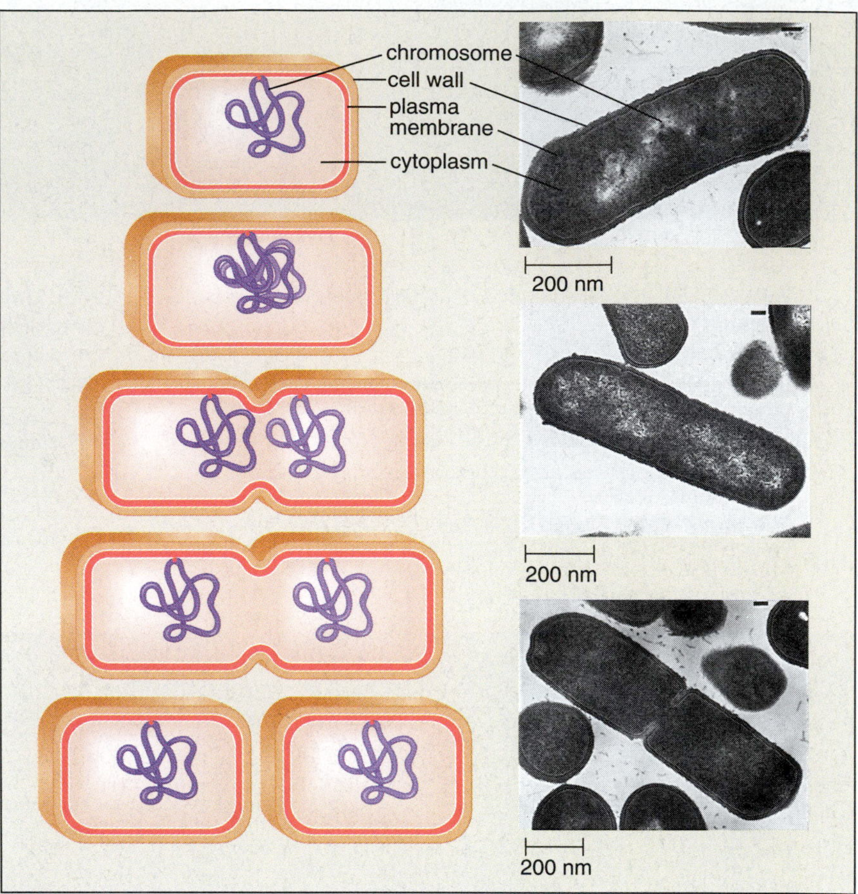

Figure 9.1 Binary fission.
In electron micrographs, it is possible to observe a bacterium dividing to become two bacteria. The diagrams depict chromosomal duplication and distribution. First DNA replicates, and as the plasma membrane lengthens, the two chromosomes separate. Upon fission, each bacterium has its own chromosome.

The Prokaryotic Chromosome

Prokaryotes (unicellular bacteria and archaea) lack a nucleus and other membranous organelles found in eukaryotic cells. They do have a **chromosome** which is composed of DNA associated with proteins. Prokaryotes have a single chromosome that contains just a few proteins. A eukaryotic chromosome has many more proteins than a prokaryotic chromosome.

In electron micrographs the bacterial chromosome appears as an electron-dense, irregularly shaped region called the **nucleoid** [L. *nucleus,* nucleus, kernel, and Gk. *-eides,* like], which is not enclosed by a membrane. When stretched out, the chromosome is seen to be a circular loop attached to the inside of the plasma membrane. Its length is up to about 1,000 times the length of the cell, which is why it needs to be folded inside the cell.

The prokaryotic chromosome is largely a single loop of DNA that is tightly folded into a region called the nucleoid.

Binary Fission

Prokaryotes reproduce asexually by binary fission. The process is termed **binary fission** because division (fission) produces two (binary) daughter cells that are identical to the original parent cell. Before division takes place, DNA is replicated so that there are two chromosomes attached to the inside of the plasma membrane. Following replication, the two chromosomes separate by an elongation of the cell that pulls the chromosomes apart. When the cell is approximately twice its original length, the plasma membrane grows inward and the new cell wall forms, dividing the cell into two approximately equal portions (Fig. 9.1).

Escherichia coli, which lives in our intestines, has a generation time (time it takes the cell to divide) of about 20 minutes under favorable conditions. In about seven hours a single cell can increase to over one million cells! Most bacteria, however, have a generation time of from one to three hours and others may require even more than 24 hours.

Asexual reproduction in prokaryotes is by binary fission. DNA replicates and the two resulting chromosomes separate as the cell elongates.

9.2 How Eukaryotic Cells Divide

Some eukaryotes are unicellular but most are multicellular. In multicellular forms cell division is necessary to growth and repair of tissues.

Eukaryotic Chromosomes

The DNA in the chromosomes of eukaryotes is associated with various proteins including *histone proteins* that are especially involved in organizing chromosomes. When a eukaryotic cell is not undergoing division, the DNA (and associated proteins) within a nucleus is a tangled mass of thin threads called **chromatin** [Gk. *chroma,* color, and *teino,* stretch]. At the time of division, chromatin becomes highly coiled and condensed, and it is easy to see the individual chromosomes.

When the chromosomes are visible it is possible to photograph and count them. Each species has a characteristic chromosome number (Table 9.1); for instance, human cells contain 46 chromosomes, corn has 20 chromosomes, and a crayfish has 200! This is called the full or **diploid (2n) number** [Gk. *diplos,* twofold, and *-eides,* like] of chromosomes that is found in all cells of the body. The diploid number includes two chromosomes of each kind. Half the diploid number, called the **haploid (n) number** [Gk. *haplos,* single, and *-eides,* like] of chromosomes, contains only one of each kind of chromosome. In the life cycle of many animals, only sperm and eggs have the haploid number of chromosomes.

Cell division in eukaryotes involves nuclear division (karyokinesis) and **cytokinesis,** which is division of the cytoplasm. The nuclei of somatic, or body, cells undergo **mitosis**—nuclear division in which the chromosome number stays constant. A 2n nucleus divides to produce daughter nuclei that are also 2n. Before nuclear division takes place, DNA replicates, duplicating the chromosomes. Each chromosome now has two identical parts called **sister chromatids** (Fig. 9.2). Sister chromatids are genetically identical; that is, they contain exactly the same genes. Sister chromatids are constricted and attached to each other at a region called the **centromere** [Gk. *centrum,* center, and *meros,* part]. During nuclear division the two sister chromatids separate at the centromeres, and in this way each duplicated chromosome gives rise to two daughter chromosomes. These chromosomes, which consist of only one chromatid, are distributed equally to the daughter cells. In this way, each daughter cell gets a copy of each chromosome.

Each type of eukaryote has a characteristic number of chromosomes in the nucleus of each cell. The chromosomes duplicate prior to mitosis, so that despite nuclear division the chromosome number stays constant.

Table 9.1

Diploid Chromosome Number of Some Eukaryotes

Type of Organism	Name of Organism	Chromosome Number
Fungi	*Aspergillus nidulans* (mold)	8
	Neurospora crassa (mold)	14
	Saccharomyces cerevisiae (yeast)	32
Plants	*Vicia faba* (broad bean)	12
	Zea mays (corn)	20
	Solanum tuberosum (potato)	48
	Nicotiana tabacum (tobacco)	48
Animals	*Musca domestica* (housefly)	12
	Rana pipiens (frog)	26
	Felis domesticus (cat)	38
	Homo sapiens (human)	46
	Pan troglodytes (chimp)	48
	Equus caballus (horse)	64
	Gallus gallus (chicken)	78
	Canis familiaris (dog)	78

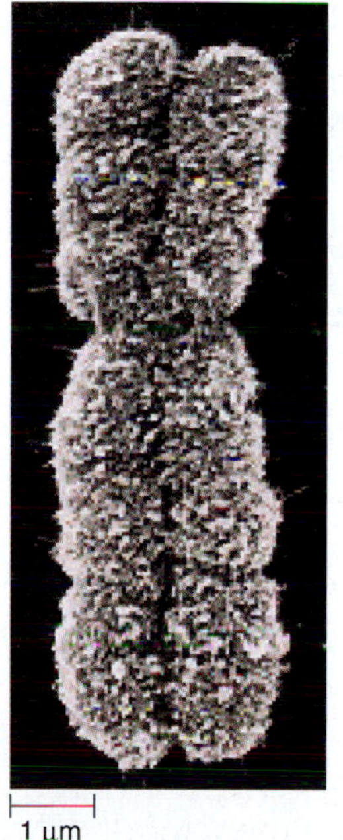

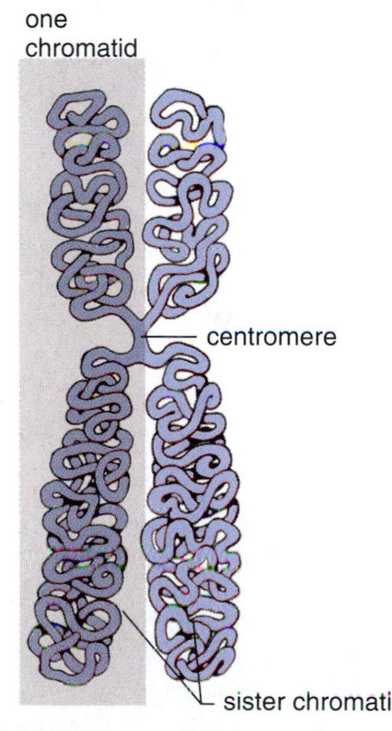

Figure 9.2 Duplicated chromosomes.
A duplicated chromosome contains two sister chromatids, each with copies of the same genes. **a.** Electron micrograph of a highly coiled and compacted chromosome, typical of a nucleus about to divide. **b.** Diagrammatic drawing of a compacted chromosome. One chromatid is screened in blue. The chromatids are held together at a region called the centromere.

The Mitotic Spindle

The **centrosome** [Gk. *centrum,* center, and *soma,* body], the main microtubule organizing center of the cell, divides before mitosis begins. Centrosomes are believed to be responsible for organizing the spindle, a structure that brings about chromosome movement during cell division. Each centrosome contains a pair of barrel-shaped organelles called **centrioles;** however, the fact that plant cells lack centrioles suggests that centrioles are not required for spindle formation.

The **spindle** contains many fibers, each composed of a bundle of microtubules. Microtubules are hollow cylinders found in the cytoplasm and other structures such as flagella and centrioles. Microtubules are made up of the protein tubulin. They assemble when tubulin subunits join and disassemble when tubulin subunits become free once more. The cytoskeleton, which is a network of interconnected filaments and tubules, begins to disassemble when spindle fibers begin forming. Most likely, the cytoskeleton provides material for spindle formation.

Mitosis in Animal Cells

Mitosis is a continuous process that is arbitrarily divided into five phases for convenience of description (Fig. 9.3).

Prophase

It is apparent during *prophase* that nuclear division is about to occur because chromatin has condensed and the chromosomes are now visible. As the chromosomes continue to compact, the nucleolus disappears and the nuclear envelope fragments.

The already duplicated chromosomes are composed of two sister chromatids held together at a centromere. Counting the number of centromeres in diagrammatic drawings gives the number of chromosomes for the cell depicted. During prophase, the chromosomes have no apparent orientation within the cell. However, specialized protein complexes called *kinetochores* develop on either side of each centromere, and these are important to future chromosome orientation.

The spindle begins to assemble as pairs of centrosomes migrate away from one another. Short microtubules radiate out in a starlike **aster** [Gk. *aster,* star] from the pair of centrioles located in each centrosome.

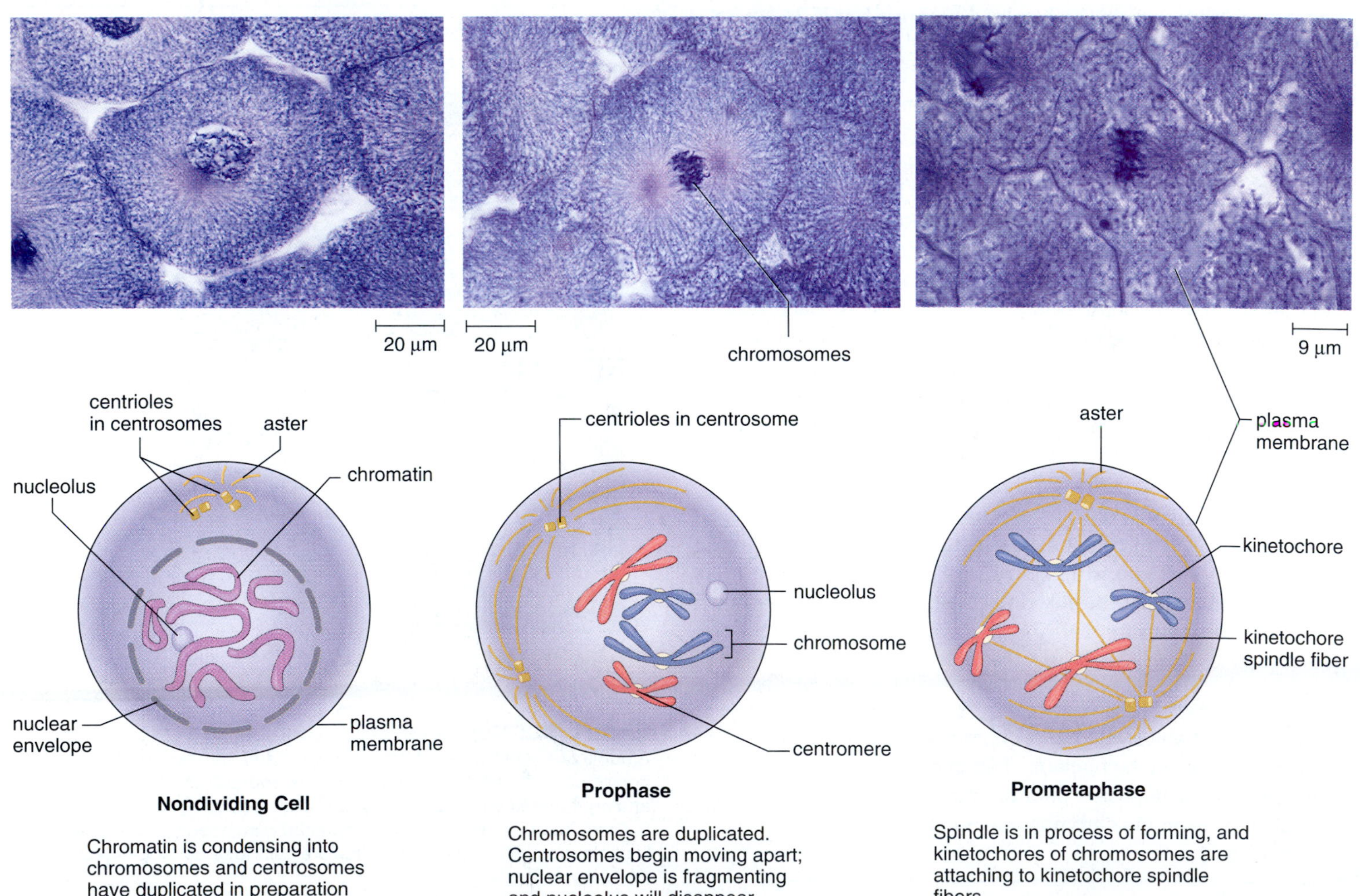
20 µm
20 µm
chromosomes
9 µm
centrioles in centrosomes
aster
nucleolus
chromatin
nuclear envelope
plasma membrane
Nondividing Cell
Chromatin is condensing into chromosomes and centrosomes have duplicated in preparation for mitosis.
centrioles in centrosome
nucleolus
chromosome
centromere
Prophase
Chromosomes are duplicated. Centrosomes begin moving apart; nuclear envelope is fragmenting and nucleolus will disappear.
aster
plasma membrane
kinetochore
kinetochore spindle fiber
Prometaphase
Spindle is in process of forming, and kinetochores of chromosomes are attaching to kinetochore spindle fibers.

Figure 9.3 Phases of mitosis in animal cells.

Prometaphase

As *prometaphase* begins, the spindle consists of poles, asters, and fibers, which are bundles of parallel microtubules. An important event during prometaphase is the attachment of the chromosomes to the spindle and their movement as they align at the metaphase plate (equator) of the spindle. The kinetochores of sister chromatids capture spindle fibers coming from opposite poles. Such spindle fibers are called *kinetochore spindle fibers.* In response to attachment by first one kinetochore and then the other, a chromosome moves first toward one pole and then toward the other until the chromosome is aligned at the metaphase plate of the spindle.

Metaphase

During *metaphase,* the chromosomes, attached to kinetochore fibers, are aligned at the metaphase plate. There are many nonattached spindle fibers called *polar spindle fibers,* some of which reach beyond the metaphase plate and overlap.

Anaphase

At the start of *anaphase,* the two sister chromatids of each duplicated chromosome separate at the centromere, giving rise to two daughter chromosomes. Daughter chromosomes, each with a centromere and single chromatid, begin to move toward opposite poles. What accounts for the movement of the daughter chromosomes? First, the polar spindle fibers lengthen as they slide past one another. Second, the kinetochore spindle fibers disassemble at the region of the kinetochores, and this pulls the daughter chromosomes to the poles.

Telophase

During *telophase,* the spindle disappears as new nuclear envelopes form around the daughter chromosomes. Each daughter nucleus contains the same number and kinds of chromosomes as the original parent cell. Remnants of the polar spindle fibers are still visible between the two nuclei.

The chromosomes become more diffuse chromatin once again, and a nucleolus appears in each daughter nucleus. Cytokinesis is nearly complete, and soon there will be two individual daughter cells, each with a nucleus that contains the diploid number of chromosomes.

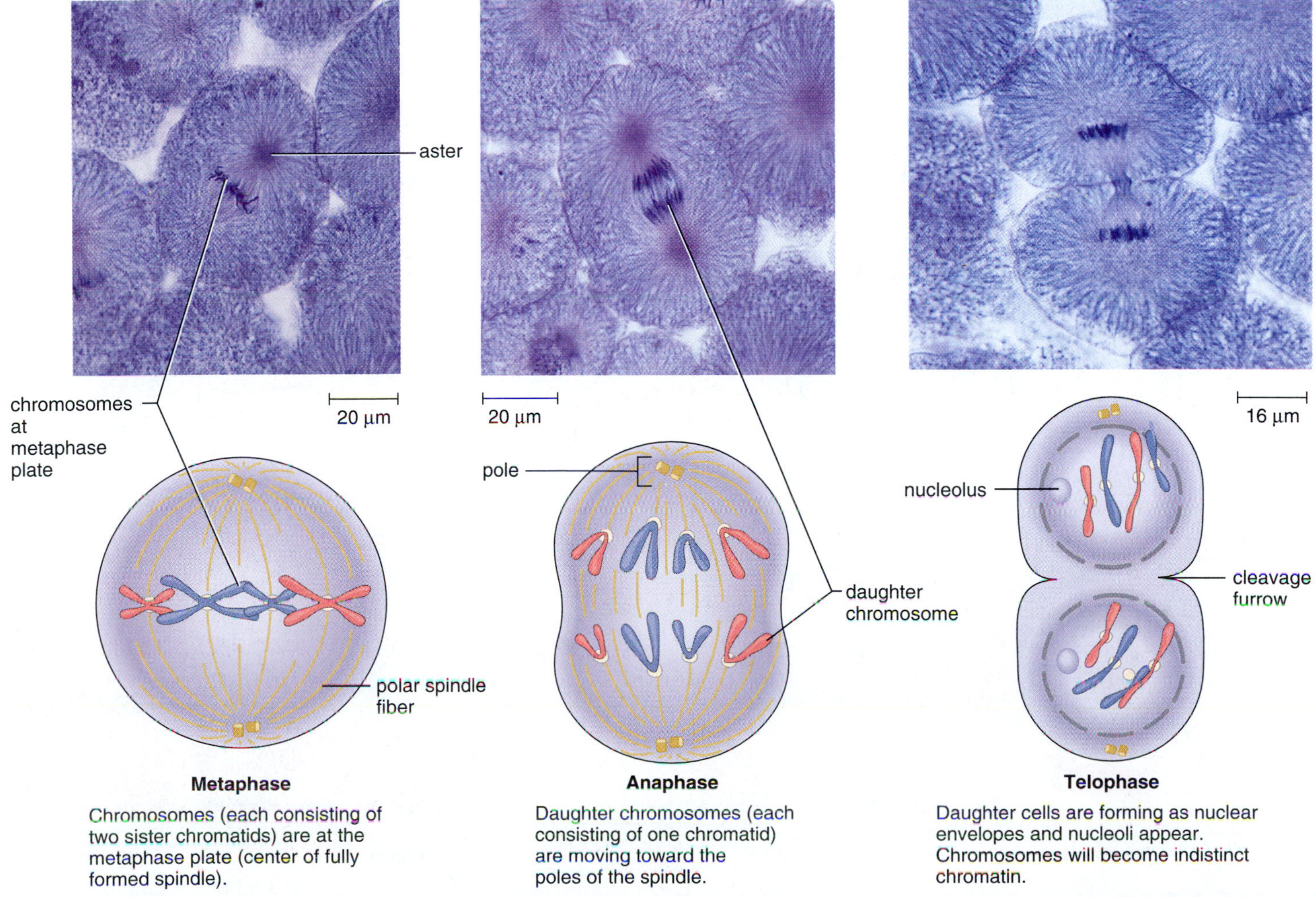

Metaphase
Chromosomes (each consisting of two sister chromatids) are at the metaphase plate (center of fully formed spindle).

Anaphase
Daughter chromosomes (each consisting of one chromatid) are moving toward the poles of the spindle.

Telophase
Daughter cells are forming as nuclear envelopes and nucleoli appear. Chromosomes will become indistinct chromatin.

Mitosis in Plant Cells

As with animal cells, mitosis in plant cells permits growth and repair. A certain type of plant tissue, called meristematic tissue, retains the ability to divide throughout the life of a plant. Root tip and shoot tip meristematic tissue allows these structures to increase in length. Lateral meristem accounts for the ability of trees to widen their girth.

Figure 9.4 illustrates mitosis in plant cells. Note that there are exactly the same stages in plant cells as in animal cells. During prophase, the chromatin condenses into scattered previously duplicated chromosomes and the spindle forms; during prometaphase (not illustrated), chromosomes attach to spindle fibers; during metaphase, the chromosomes are at the metaphase plate of the spindle; during anaphase, the daughter chromosomes move to the poles of the spindle; and during telophase, cytokinesis begins. Although plant cells have a centrosome and spindle, there are no centrioles nor asters during cell division.

Mitosis in plant and animal cells ensures the daughter cells receive the same number and kinds of chromosomes as the parent cell.

Cytokinesis in Plant and Animal Cells

Cytokinesis, or cytoplasmic cleavage, usually accompanies mitosis. By the end of mitosis each newly forming cell has received a share of the cytoplasmic organelles which duplicated during interphase. Division of the cytoplasm begins in anaphase, continues in telophase, but does not reach completion until the following interphase begins.

Cytokinesis in Plant Cells

Cytokinesis in plant cells occurs by a process different from that seen in animal cells (Fig. 9.5). The rigid cell wall that surrounds plant cells does not permit cytokinesis by furrowing. Instead, the Golgi apparatus, a membranous organelle in cells, produces membranous sacs called vesicles, which move along microtubules to the midpoint between the two daughter nuclei. These vesicles fuse, forming a **cell plate.** Their membrane completes the plasma membrane for both cells. They also release molecules that signal the formation of plant cell walls, which are strengthened by the addition of cellulose fibrils.

A spindle forms during mitosis in plant cells, but there are no centrioles or asters. Cytokinesis in plant cells involves the formation of a cell plate.

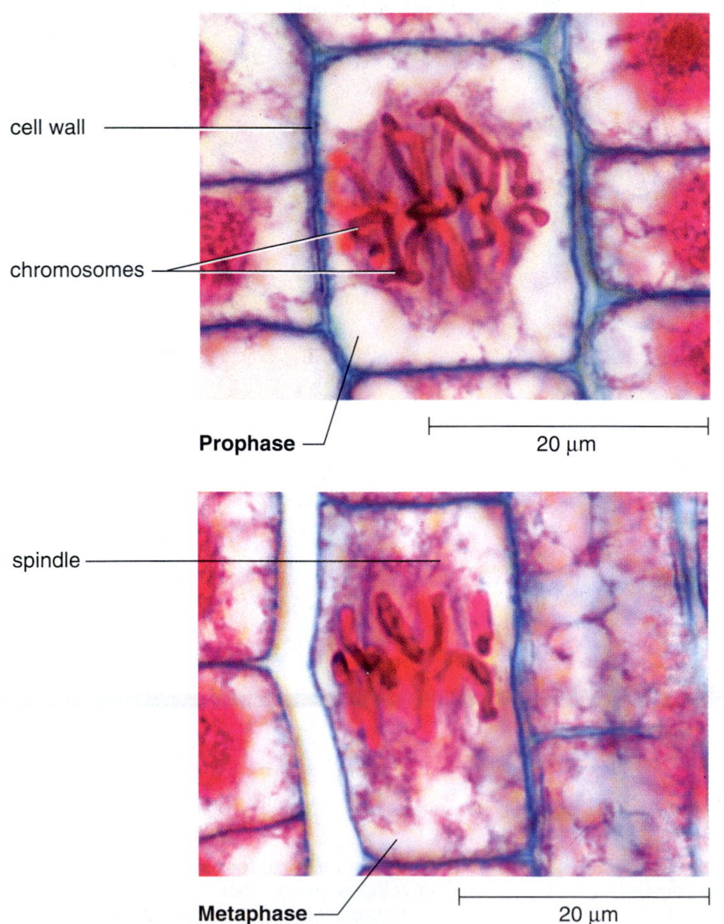

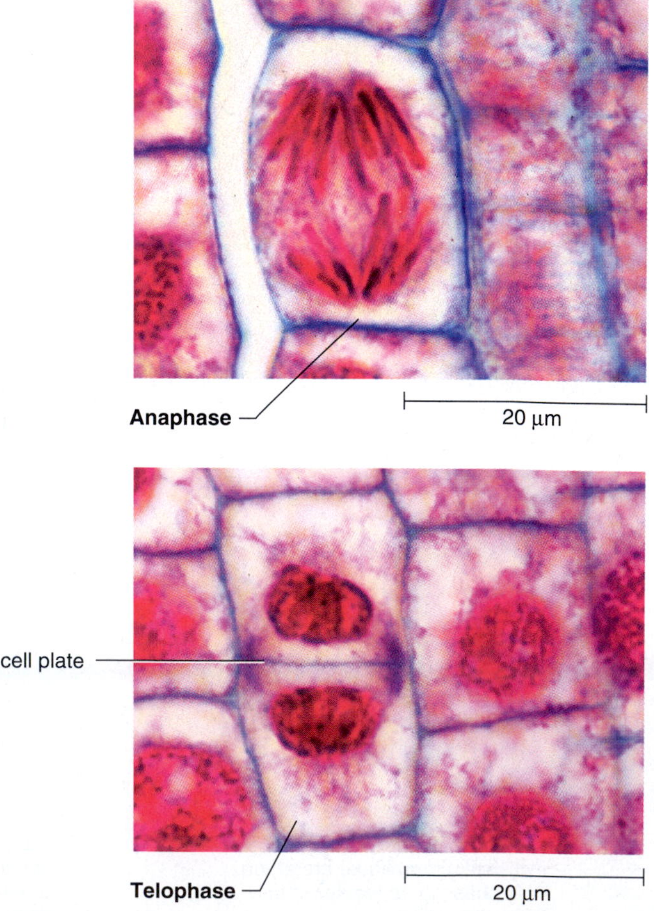

Figure 9.4 Phases of mitosis in plant cells.
Note the absence of centrioles and asters and the presence of the cell wall. In telophase, a cell plate develops between the two daughter cells. The cell plate marks the boundary of the new daughter cells, where new plasma membrane and a new cell wall will form for each cell.

Cytokinesis in Animal Cells

In animal cells, a cleavage furrow, which is an indentation of the membrane between the two daughter nuclei, begins as anaphase draws to a close. The cleavage furrow deepens when a band of actin filaments, called the contractile ring, slowly forms a constriction between the two daughter cells. The action of the contractile ring can be likened to pulling a drawstring ever tighter about the middle of a balloon. As the drawstring is pulled tight, the balloon constricts in the middle.

A narrow bridge between the two cells can be seen during telophase, and then the contractile ring continues to separate the cytoplasm until there are two independent daughter cells (Fig. 9.6).

> Cytokinesis in animal cells is accomplished by a furrowing process.

Cell Division in Other Eukaryotic Organisms

Protists and fungi also undergo mitosis and cytokinesis. In fungi and some groups of protists the nuclear envelope does not fragment. It's possible that the first role of microtubules was to support the nuclear envelope during mitosis, and only later did microtubules become attached to the chromosomes themselves. When mitosis is complete in these protists and fungi, the nuclear envelope divides and one nucleus goes to each daughter cell.

In plants and animals, as we have seen, the nuclear envelope fragments and plays no role in mitosis and cytokinesis.

> The cells of all organisms divide and new cells only come from preexisting cells. This is a tenet of the cell theory.

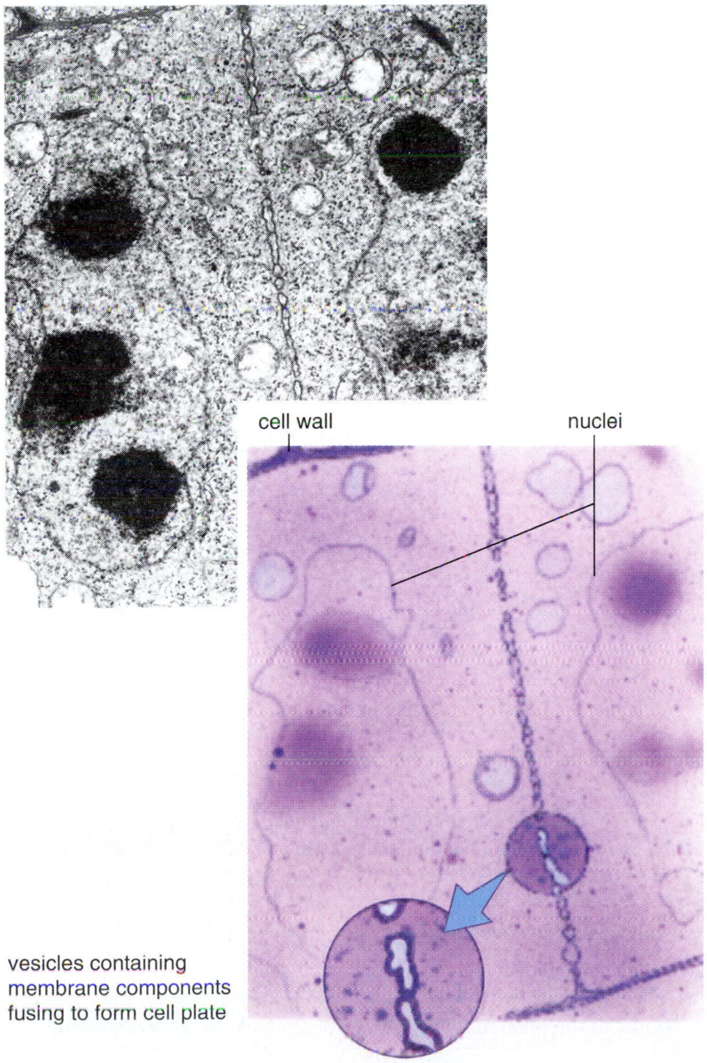

cell wall

nuclei

vesicles containing membrane components fusing to form cell plate

Figure 9.5 Cytokinesis in plant cells.
During cytokinesis in a plant cell, a cell plate forms midway between two daughter nuclei and extends to the plasma membrane.

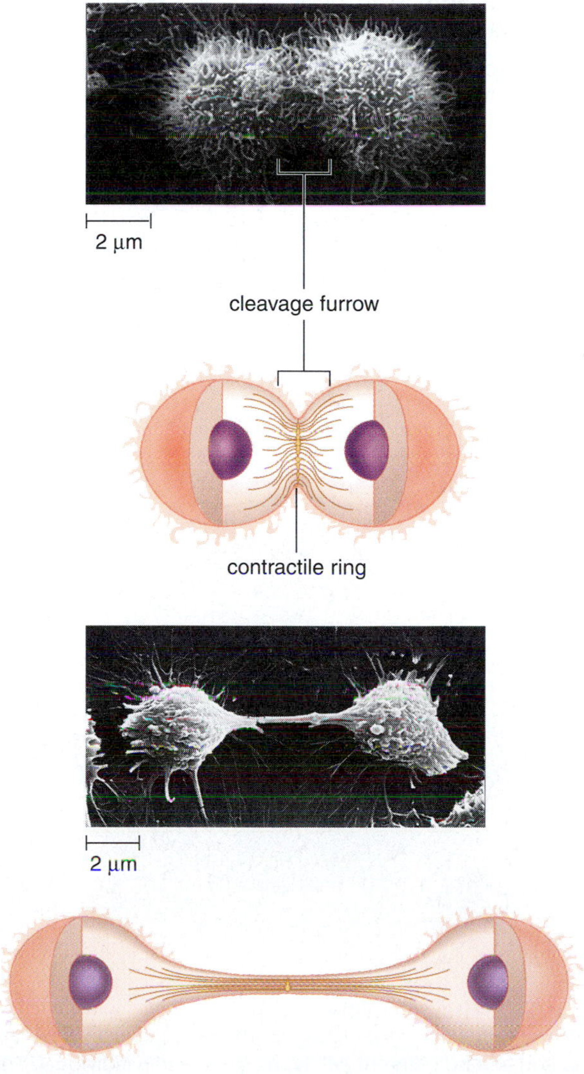

Figure 9.6 Cytokinesis in animal cells.
A single cell becomes two cells by a furrowing process. A contractile ring composed of actin filaments gradually gets smaller, and the cleavage furrow pinches the cell into two cells.

9.3 How Eukaryotic Cells Cycle 

By the 1870s, microscopy could provide detailed and accurate descriptions of chromosomal movements during mitosis, but there was no knowledge of cellular events between divisions. Because there was little visible activity between divisions, this period of time was dismissed as a resting state termed **interphase** [L. *inter*, between, and Gk. *phasis*, appearance]. When it was discovered in the 1950s that DNA replication occurs during interphase, the **cell cycle** concept was proposed.

Cells grow and divide during a cycle that has four stages (Fig. 9.7). The entire cell division stage, including both mitosis and cytokinesis, is termed the *M stage* (M = mitosis). The period of DNA synthesis when replication occurs is termed the *S stage* (S = synthesis) of the cycle. The proteins associated with DNA in eukaryotic chromosomes are also synthesized during this stage. There are two other stages of the cycle. The period of time prior to the S stage is termed the *G_1 stage*, and the period of time prior to the M stage is termed the *G_2 stage*. At first, not much was known about these stages, and they were thought of as G = gap stages. Now we know that during the G_1 stage, the cell grows in size and the cellular organelles increase in number. During the G_2 stage, various metabolic events occur in preparation for mitosis. Some biologists today prefer the designation G = growth for these two G stages. In any case, interphase consists of G_1, S, and G_2 stages.

Cells undergo a cycle that includes the G_1, S, G_2, and M stages.

The Cell-Cycle Clock

Some cells, such as skin cells, divide continuously throughout the life of the organism. Other cells, such as skeletal muscle cells and nerve cells, are arrested in the G_1 stage. If the nucleus from one of these cells is placed in the cytoplasm of an S-stage cell, it finishes the cell cycle. Cardiac muscle cells are arrested in the G_2 stage. If an arrested cell is fused with a cell undergoing mitosis, it too starts to undergo mitosis. It appears, then, that there are stimulatory substances that cause the cell to proceed through two critical checkpoints:

$$G_1 \text{ stage} \rightarrow \text{S stage}$$
$$G_2 \text{ stage} \rightarrow \text{M stage}$$

Over the past few years biologists have made remarkable progress in identifying the proteins that cause a cell to move from the G_1 stage to the S stage and the proteins that cause a cell to move from the G_2 stage to the M stage. Some of these biologists worked with frog eggs, others with yeast cells, and still others used cell cultures as their experimental material. The researchers identified two types of proteins of interest—kinases and cyclins. A **kinase** is an enzyme that removes a phosphate group from ATP (the form of chemical energy used by cells) and adds it to a protein. Activation by a kinase is a common way for cells to turn on a cellular process, but it turned out that the kinases involved in the cell cycle are themselves activated when they combine with a protein called cyclin. **Cyclins** are so named because their quantity is not constant in the cell.

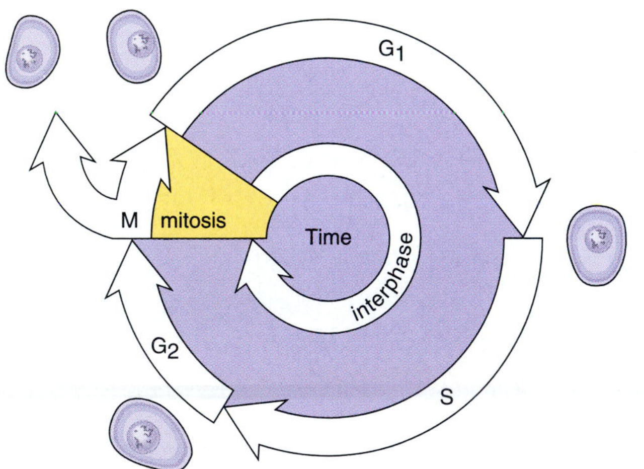

	Stage	Main Events	Length of time (hours)	
			Vicia faba	*Homo sapiens* (cultured fibroblasts)
Interphase	G_1	Organelles begin to double in number	4.9	6.3
	S	Replication of DNA	7.5	7.0
	G_2	Synthesis of proteins	4.9	2.0
	M	Mitosis	2.0	0.7
		Total:	19.3	16.0

Figure 9.7 **The cell cycle.**
Cells go through a cycle that consists of four stages (G_1, S for synthesis, G_2, and M for mitosis). The length of the different stages varies both among species and among different cell types in the same individual. The approximate lengths of the phases for the broad bean (*Vicia faba*) and humans (*Homo sapiens*) are given.

Figure 9.8 is an illustration that shows the process in a clockwise diagram. After S-kinase combines with S-cyclin, the kinase phosphorylates a protein that causes the cell to move from the G_1 stage to the S stage when DNA is synthesized (replicated). Then, S-cyclin is destroyed, and S-kinase is no longer active.

Similarly, after M-kinase combines with M-cyclin, the kinase phosphorylates a protein that causes the cell to move from the G_2 stage to the M stage when mitosis occurs. Three things occur: (1) chromosomes condense, (2) the nuclear envelope disassembles, and (3) the spindle forms. Now M-cyclin is destroyed.

Until recently, the mechanics of the cell cycle and the causes of cancer were thought to be distantly related. Now they are known to be intimately related. For example, oncogenes are cancer-causing genes, and it is possible that they control genes that code for cyclins that no longer function as they should. Growth factors are molecules that attach to plasma membrane receptors and thereby bring about cell growth. Ordinarily, a cyclin might combine with its kinase only when a growth factor is present. But a cyclin that has gone awry might combine with its kinase even when a growth factor is not present. The result would be uncontrolled cell growth as seen in cancer. On the other hand, tumor-suppressor genes usually function to prevent cancer from occurring. It has been shown that the product of one major tumor-suppressor gene (*p53*) brings about the production of a protein that can combine with a cyclin so that a kinase never attaches to it. In this way *p53* is instrumental in stopping the cell cycle. The next section discusses in greater detail the role of *p53* in preventing the development of cancer.

Kinases activated by combination with a cyclin at two critical checkpoints in the cell cycle phosphorylate a protein, and this triggers the beginning of the S stage and the M stage.

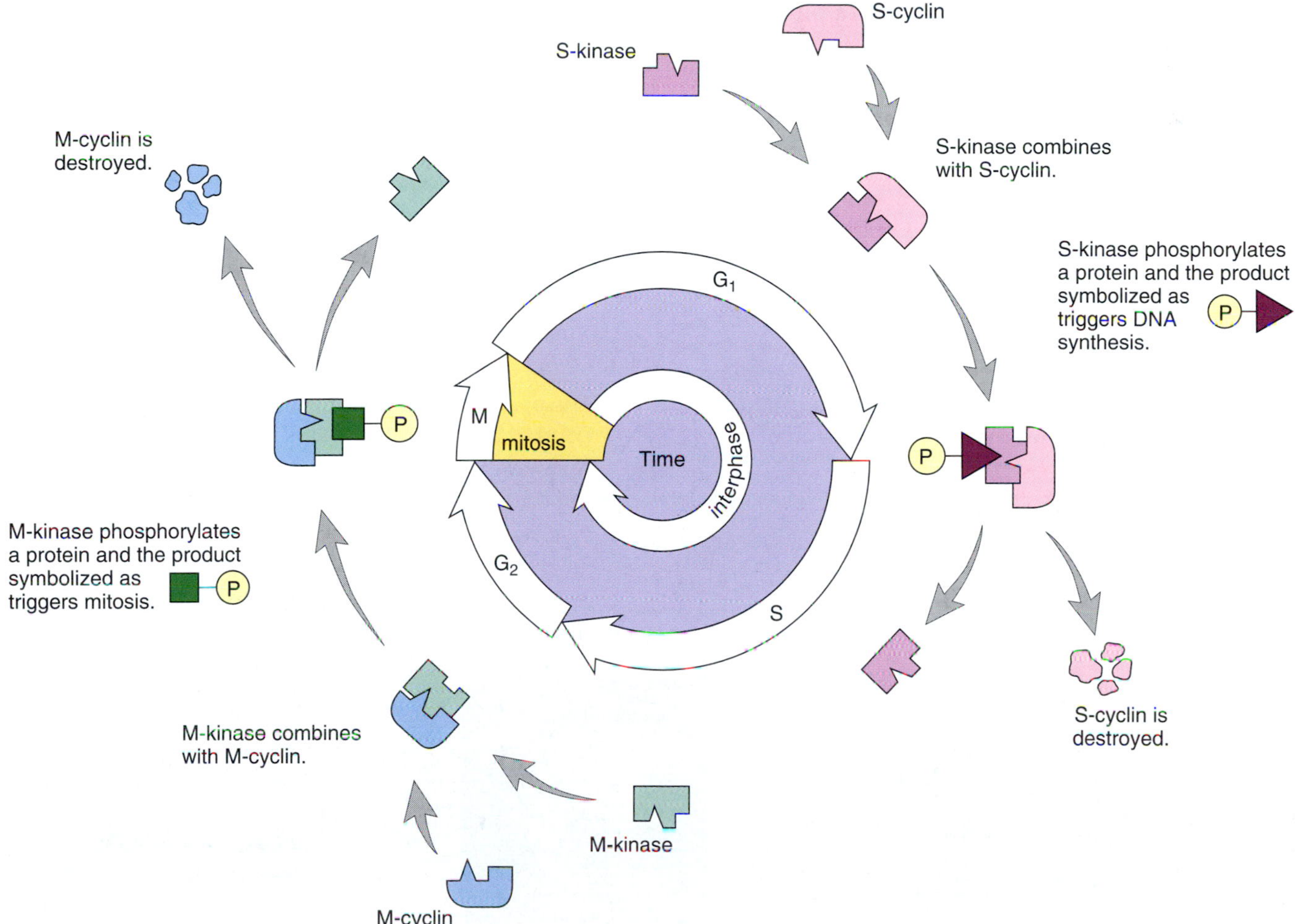

Figure 9.8 Control of the cell cycle.
At two critical checkpoints a kinase combines with a cyclin, and this moves the cell cycle forward. When the kinases phosphorylate proteins, the proteins are activated and produce effects appropriate to the particular stage. At the start of the S stage, S-kinase combines with S-cyclin, then synthesis (replication) of DNA takes place. At the start of the M stage, M-kinase combines with M-cyclin. Then mitosis occurs.

9.4 How Cancer Develops

Carcinogenesis is the development of cancer. **Cancer** is a genetic disease requiring a series of mutations, each propelling cells toward the development of a **tumor** [L. *tumor*, swelling], an abnormal mass of cells. The presence of a tumor indicates a failure in cell division control—a failure that is quite often due to a faulty *p53* gene. The ***p53* gene** ordinarily halts the cell cycle when DNA mutates and is in need of repair (Fig. 9.9). Why might DNA be in need of repair? DNA molecules are constantly acted on by physical and chemical agents that can cause mutations that lead to cancer. The best known mutagenic carcinogens are radiation, organic chemicals, and certain viruses. Radiation includes ultraviolet light, radon gas, X rays, and accidental emissions from nuclear power plants. Organic chemicals in tobacco smoke, certain foods, and pesticides and herbicides also bring about harmful changes in DNA molecules. A cell constantly monitors its DNA and if any agent has caused changes, repair enzymes set to work to correct the harm done. The p53 protein (the product of the *p53* gene) is involved in mobilizing repair enzymes and stopping the cell cycle while repair is going on. Only if repair is possible does the cell cycle start up again. If repair of DNA is not possible, the p53 protein promotes cell suicide, also called apoptosis. While ordinarily we think of cell death as a bad thing, it is obviously the best solution if cancer will develop otherwise.

The pivotal role of the *p53* gene is substantiated first by the observation that many different types of human cancers contain no or a faulty *p53* gene. Also, scientists have added the p53 protein to rapidly dividing cancer cells in a petri dish. The cells stopped dividing and died.

The two cellular responses to activation of *p53* are either arrest of the cell cycle while DNA repair is going on or apoptosis.

Apoptosis [Gk. *apo*, off, and *ptosis*, fall], defined as programmed cell death, is a set sequence of cellular changes involving shattering the nucleus, chopping up the chromosomes, digesting the cytoskeleton, and packaging the cellular remains into membrane-enclosed vesicles (apoptotic bodies) that can be engulfed by macrophages. The appearance of membrane blisters is a tip-off that a cell is undergoing apoptosis (Fig. 9.9).

A remarkable finding of the past few years is that cells routinely harbor the enzymes, now called caspases, that bring about apoptosis. The enzymes are ordinarily held in check by inhibitors but they can be unleashed in two ways. During human development, exterior signals cause certain cells to die, allowing paddles to be converted into fingers and toes, for example. In adults, cells that have DNA damaged beyond repair usually go ahead and kill themselves.

There are two sets of caspases. The first set are the "initiators" that receive the message to activate the "executioners," which then activate the enzymes that dismantle the cell. For example, there is an executioner that frees an endonuclease to enter the nucleus and start chopping DNA. Initiators, executioners, and dismantling enzymes all begin to work when they are clipped to make them shorter and ready for action.

Knowledge about apoptosis can possibly lead to new therapy regimens. Tumor cells, but not normal adult cells, contain high levels of a protein called survivin, which blocks apoptosis. If researchers can find a way to inactivate survivin, cancer cells might be more susceptible to radiation and chemotherapy. In Parkinson's disease and stroke, excess apoptosis may kill off brain cells. If so, inhibitors of apoptosis could be administered to keep brain cells alive.

The occurrence of apoptosis is usually essential for the good of the organism. A lack of apoptosis allows cancer to develop.

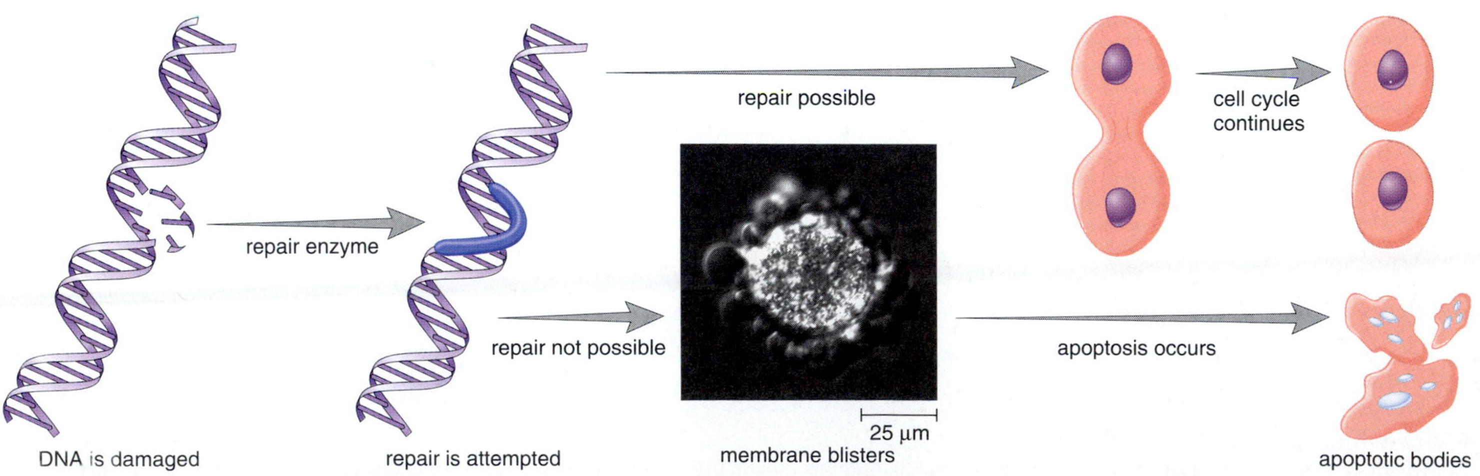

Figure 9.9 Functions of *p53*.
If DNA is damaged by a mutagen, *p53* is instrumental in stopping the cell cycle and activating repair enzymes. If repair is impossible, the p53 protein promotes apoptosis.

Characteristics of Cancer Cells

Cancer cells exhibit characteristics that indicate they have experienced a severe failure in regulating the cell cycle.

Cancer Cells Lack Differentiation

Most cells are specialized; they have a specific form and function that suits them to the role they play in the body. Cancer cells are nonspecialized and do not contribute to the functioning of a body part. A cancer cell does not look like a differentiated muscle, nervous, or connective tissue cell and instead has a shape and form that is distinctly abnormal (Fig. 9.10). Normal cells can enter the cell cycle for about 50 times, and then they die. Cancer cells can enter the cell cycle repeatedly, and in this way they are immortal. In cell tissue culture, they die only because they run out of nutrients or are killed by their own toxic waste products.

Cancer Cells Have Abnormal Nuclei

The nuclei of cancer cells are enlarged, and there may be an abnormal number of chromosomes. The chromosomes have mutated; some parts may be duplicated and some may be deleted. In addition, gene amplification (extra copies of specific genes) is seen much more frequently than in normal cells.

Cancer Cells Form Tumors

Normal cells anchor themselves to a substratum or adhere to their neighbors. They exhibit contact inhibition—when they come in contact with a neighbor, they stop dividing. In culture, normal cells form a single layer that covers the bottom of a petri dish. Cancer cells have lost all restraint; they pile on top of one another and grow in multiple layers. They have a reduced need for the growth factors that are needed by normal cells.

In the body, a cancer cell divides to produce a tumor which invades and destroys neighboring tissue. This new growth, termed neoplasia, is made of cells that are disorganized, a condition termed anaplasia. A benign tumor is a disorganized, usually encapsulated, mass that does not invade adjacent tissue.

Cancer Cells Undergo Angiogenesis and Metastasis

Angiogenesis, the formation of new blood vessels, is required to bring nutrients and oxygen to a cancerous tumor whose growth is not contained within a capsule. Cancer cells release a growth factor that causes neighboring blood vessels to branch into the cancerous tissue. Some modes of cancer treatment are aimed at preventing angiogenesis from occurring.

Cancer in situ is found in its place of origin without any invasion of normal tissue. Malignancy is present when **metastasis** [Gk. *meta*, between, and L. *stasis*, standing, a position] establishes new tumors that are distant from the primary tumor. To accomplish metastasis, cancer cells must first make their way across the extracellular matrix (substances including fibers secreted by the cell) and into a blood vessel or lymphatic vessel. It has been discovered that cancer cells have receptors that allow them to adhere to a component of the extracellular matrix; they also produce enzymes that degrade the matrix and allow them to invade underlying tissues. Cancer cells tend to be motile, have a disorganized internal cytoskeleton, and lack intact actin filament bundles. After traveling through the blood or lymph, cancer cells may then start tumors elsewhere in the body.

The patient's prognosis (probable outcome) is dependent on the degree to which the cancer has progressed: (1) whether the tumor has invaded surrounding tissues, (2) if so, whether there is any lymph node involvement, and (3) whether there are metastatic tumors in distant parts of the body. With each progressive step of the cancerous condition, the prognosis becomes less favorable.

Cancer cells are nonspecialized, have abnormal chromosomes, and divide uncontrollably. Because they are not constrained by their neighbors, they form a tumor. Then they metastasize, forming new tumors wherever they relocate.

Normal Cells

Controlled growth

Contact inhibition

One organized layer

Differentiated cells

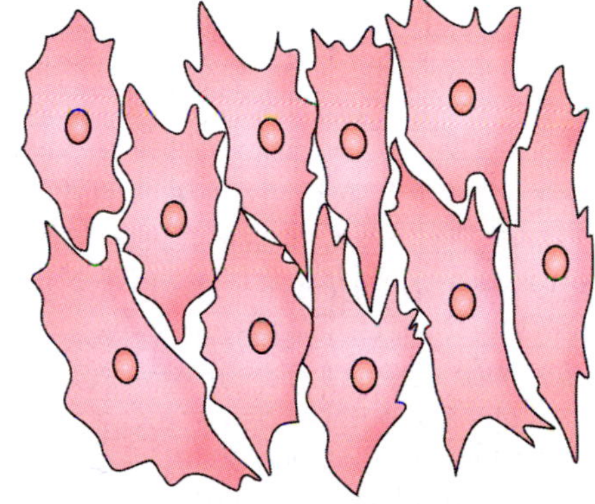

Cancer Cells

Uncontrolled growth

No contact inhibition

Disorganized, multilayered

Nondifferentiated cells

Abnormal nuclei

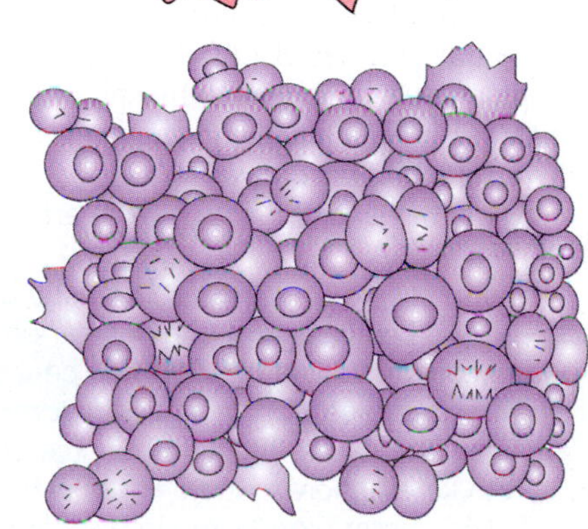

Figure 9.10 Cancer cells.
Cancer cells differ from normal cells in the ways noted.

Prevention of Cancer

As mentioned, certain environmental agents, such as radiation and certain organic chemicals and certain viruses are known to be carcinogenic. Prevention of cancer includes avoiding these agents whenever possible. Early detection is also important. For women, routine screening tests for cancer include: Pap smear, mammogram, and breast self-exam. Men should do a self-exam for testicular cancer. For colon cancer detection, there is a stool blood test that can be initiated at home and a physician can examine the colon for growths in his or her office.

Of interest to most is the discovery that diet can protect us from developing cancer. Fresh fruits and vegetables in the cabbage family have been found to be especially protective.

Connecting Concepts

Cells have an elaborate internal structure that is not revealed by use of the light microscope and complex enzymatic pathways that can only be discovered by biochemical analyses. Similarly, the process of cell division is more complicated than it first appears. We now know that cell division is a part of a tightly regulated cell cycle and there are negative consequences if the cell cycle should come out of sync. In humans, overproduction of skin cells due to an overstimulated cell cycle produces a chronic inflammatory condition known as psoriasis. In a rare condition called progeria, the reproductive capacity of all the body's cells is severely diminished and young people grow old and die at an early age. Many different types of cancer can result when the signals that keep the cell cycle in check are not transmitted or received properly. The hope is that if we learn to control the cell cycle, these conditions will be curable one day.

The newfound ability to clone farm animals from single adult cells shows that we are now able to initiate and control the process of cell division even in mammals. Tissue engineering during which organs are formed in the laboratory, as discussed in chapter 17, is an even better example that we can now manipulate the cell cycle.

The end product of ordinary cell division (i.e., mitosis) is two new cells, each with the same number and kinds of chromosomes as the parent cell. But what about the special type of cell division (i.e., meiosis) that produces the sex cells—sperm and egg? As we'll see in the next chapter, following meiosis, the sex cells have only half the chromosome number.

Summary

9.1 How Prokaryotic Cells Divide

A prokaryotic chromosome has a few proteins and a single, long loop of DNA. The chromosome is attached to the inside of the plasma membrane. Binary fission involves replication of DNA, followed by an elongation of the cell that pulls the chromosomes apart. Inward growth of the plasma membrane and formation of new cell wall material divide the cell in two.

9.2 How Eukaryotic Cells Divide

Between nuclear divisions, the chromosomes are not distinct and are collectively called chromatin. Each eukaryotic species has a characteristic number of chromosomes. The total number is called the diploid number, and half this number is the haploid number.

Among eukaryotes, cell division involves nuclear division (karyokinesis) and division of the cytoplasm (cytokinesis). Mitosis is nuclear division in which the chromosome number stays constant because each chromosome is duplicated and gives rise to two daughter chromosomes.

Mitosis has five phases, which are described here for animal cells.

Prophase—duplicated chromosomes are distinct, the nucleolus is disappearing, the nuclear envelope is fragmenting, and the spindle is forming between centrosomes. Asters radiate from the centrioles within the centrosomes.

Prometaphase—the kinetochores of sister chromatids attach to kinetochore spindle fibers extending from opposite poles. The chromosomes move back and forth until they are aligned at the metaphase plate.

Metaphase—the spindle is fully formed and the duplicated chromosomes are aligned at the metaphase plate. The spindle consists of polar spindle fibers that overlap at the metaphase plate and kinetochore spindle fibers that are attached to chromosomes.

Anaphase—daughter chromosomes move toward the poles. The polar spindle fibers slide past one another and the kinetochore spindle fibers disassemble. Cytokinesis by furrowing begins.

Telophase—nuclear envelopes re-form, chromosomes begin changing back to chromatin, the nucleoli reappear, and the spindle disappears.

Plant cells lack centrioles and, therefore, asters. Even so, the mitotic spindle forms and the same five mitotic phases are observed.

Cytokinesis in plant cells involves the formation of a cell plate from which the plasma membrane and cell wall are completed. Cytokinesis in animal cells is a furrowing process that divides the cytoplasm.

9.3 How Eukaryotic Cells Cycle

The cell cycle has four stages. During the G_1 stage the organelles increase in number; during the S stage, DNA replication occurs; during the G_2 stage, various proteins are synthesized; and during the M stage, mitosis occurs. Interphase consists of G_1, S, and G_2 stages.

Biologists have made great strides in understanding how the cell cycle is controlled. Kinases activated when they combine with cyclins phosphorylate proteins that trigger passage from the G_1 to the S stage, and the passage from the G_2 stage to the M stage.

9.4 How Cancer Develops

The gene *p53* ordinarily stops the cell cycle whenever DNA repair is needed to prevent harmful mutations. If repair is impossible, *p53* directs apoptosis, programmed cell death, to occur. Quite often when cancer occurs, *p53* is absent or nonfunctional.

The cell routinely contains the enzymes, called caspases, which bring about apoptosis. The initiator caspases are usually held in check by inhibitors, but when set in motion they activate executioner caspases which activate the cellular enzymes that destroy the cell. These enzymes shatter the nucleus, chop up the chromosomes, digest the cytoskeleton, cause membrane blisters, and package the cell in vesicles for absorption by macrophages.

Cancer cells are nondifferentiated, divide repeatedly, have abnormal nuclei, do not require growth factors, and are not constrained by their neighbors. After forming a tumor, cancer cells metastasize and start new tumors elsewhere in the body. Certain behaviors such as avoiding unnecessary radiation (e.g., solar radiation), exposure to certain organic chemicals (e.g., tobacco smoke), and adopting a diet rich in fruits and vegetables is protective against the development of cancer.

Reviewing the Chapter

1. Describe the prokaryotic chromosome and the process of binary fission. 144
2. Describe the eukaryotic chromosome and the significance of the duplicated chromosome. 145
3. Define the following words: chromosome, chromatin, chromatid, centriole, cytokinesis, centromere, and kinetochore. 144–46
4. Describe the events that occur during the phases of mitosis. 146–47
5. How does plant cell mitosis differ from animal cell mitosis? 148
6. Contrast cytokinesis in plant cells and animal cells. 148–49
7. Describe the cell cycle, including a description of interphase. 150
8. How are the two critical checkpoints in the cell cycle controlled? 150
9. What role does the *p53* gene play in the development of cancer? 152
10. Describe the process of apoptosis and how it is controlled. 152
11. List and discuss four characteristics of cancer cells that distinguish them from normal cells. 153

Testing Yourself

Choose the best answer for each question.

1. What feature in prokaryotes substitutes for the spindle action in eukaryotes?
   a. centrioles with asters
   b. fission instead of cytokinesis
   c. elongation of plasma membrane
   d. looped DNA
   e. presence of one chromosome
2. How does a prokaryotic chromosome differ from a eukaryotic chromosome? A prokaryotic chromosome
   a. is shorter and fatter.
   b. has a single loop of DNA.
   c. never replicates.
   d. contains many histones.
   e. All of these are correct.
3. The diploid number of chromosomes
   a. is the 2n number.
   b. is in a parent cell and therefore in the two daughter cells following mitosis.
   c. varies according to the particular organism.
   d. is in every somatic cell.
   e. All of these are correct.

For questions 4–6, match the descriptions that follow to the terms in the key.

Key:

a. centriole
b. chromatid
c. chromosome
d. centromere
e. cyclin

4. point of attachment for sister chromatids
5. found at a pole in the center of an aster
6. coiled and condensed chromatin
7. If a parent cell has fourteen chromosomes prior to mitosis, how many chromosomes will each daughter cell have?
   a. twenty-eight because each chromatid is a chromosome
   b. fourteen because the chromatids separate
   c. seven only after mitosis is finished
   d. any number between seven and twenty-eight
   e. seven in the nucleus and seven in the cytoplasm, for a total of fourteen
8. In which phase of mitosis are the chromosomes moving toward the poles?
   a. prophase
   b prometaphase
   c. metaphase
   d. anaphase
   e. telophase
9. Interphase
   a. is the same as prophase, metaphase, anaphase, and telophase.
   b. includes stages G_1, S, and G_2.
   c. requires the use of polar spindle fibers and kinetochore spindle fibers.
   d. is a stage in the cell cycle.
   e. Both b and d are correct.
10. Cytokinesis
   a. is mitosis in plants.
   b. requires the formation of a cell plate in plant cells.
   c. is the longest part of the cell cycle.
   d. is half a chromosome.
   e. is a form of apoptosis.
11. When cancer occurs,
   a. mutations have occurred.
   b. the gene *p53* is operational.
   c. apoptosis has occurred.
   d. the cells can no longer enter the cell cycle.
   e. All of these are correct.
12. Label this diagram of a cell in prophase of mitosis.

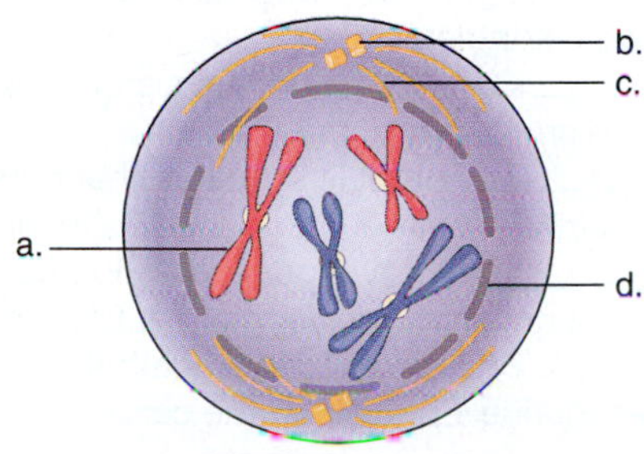

Thinking Scientifically

1. After DNA is duplicated in eukaryotes it must be bound to histone proteins. This requires the synthesis of hundreds of millions of new protein molecules, a process that the cell most likely regulates. With reference to Figure 9.8, when in the cell cycle would histones be made? At what point might histone synthesis be switched on and off?
2. When animal cells are grown in dishes in the laboratory most divide a few dozen times and then die. It would seem likely that the cell cycle is suddenly not functioning properly. With the help of Figure 9.8, hypothesize what could be wrong with the regulation of the cell cycle to turn it off.

Bioethical Issue

Turn on the television news or talk shows, or read a paper, and you're bound to hear about the latest research on breast cancer. Long overlooked as a disease, breast cancer is now winning national attention as a major factor in women's health. That, undoubtedly, is good. But attention tends to focus on one thing, at the expense of others. Some people suggest breast cancer has stolen the spotlight, leaving out prostate, liver, or stomach cancer, which also are deadly forms of cancer.

How should the nation prioritize cancer research? Is it fair that one form of the disease tends to dominate the news? Should attention focus on diseases that affect the highest number of people—or diseases that are most deadly? And how should research funding be allocated among cancers that affect only women or only men?

Understanding the Terms

apoptosis 152
asexual reproduction 144
aster 146
binary fission 144
cancer 152
cell cycle 150
cell plate 148
centriole 146
centromere 145
centrosome 144, 146
chromatin 145
chromosome 144
cyclin 150
cytokinesis 145
diploid (2n) number 145
haploid (n) number 145
interphase 150
kinase 150
metastasis 153
mitosis 145
nucleoid 144
p53 gene 152
sister chromatid 145
spindle 146
tumor 152

Match the terms to these definitions:

a. ______________ Central microtubule organizing center of cells consisting of granular material. In animal cells, it contains two centrioles.
b. ______________ Constriction where sister chromatids of a chromosome are held together.
c. ______________ Microtubule structure that brings about chromosome movement during nuclear division.
d. ______________ One of two genetically identical chromosome units that are the result of DNA replication.
e. ______________ Programmed cell death that is carried out by enzymes routinely present in the cell.

Web Connections

Exploring the Internet

http://www.mhhe.com/biosci/genbio/mader
(click on *Biology 7/e*)

The *Biology 7/e* Online Learning Center provides many resources for studying the material in this chapter including links to the following sites:

Cell Division: Binary Fission and Mitosis contains an excellent review of these topics, including color diagrams of the cell cycle, prokaryotic and eukaryotic modes of cell division, and the phases of mitosis. Great links to other sites.

http://gened.emc.maricopa.edu/bio/bio181/BIOBK/BioBookmito.html

The McGill University Mitosis Page is another great site that describes mitosis and includes full color micrographs and downloadable video and diagrams.

http://www.mcgill.ca/nrs/mitosis.htm

Mitosis. Photomicrographs and text description of mitosis in an animal cell.

http://clab.cecil.cc.md.us/faculty/biology1/mitosis.htm

Onion Root Tip Mitosis

http://biog-101-104.bio.cornell.edu/BioG101_104/tutorials/cell_division/onion_review.html

and

Whitefish Blastula Mitosis; nice photomicrographs of all phases of mitosis.

http://biog-101-104.bio.cornell.edu/BioG101_104/tutorials/cell_division/wf_review.html

And for a fun review of the subject:

Mitosis and Meiosis; an interactive review. Click on a cell and identify the phase of mitosis seen.

http://www.mssc.edu/biology/B101/biolab6.htm

and

Word Search Puzzle: find terms related to mitosis in the puzzle, and unscramble other words.

http://copland.udel.edu/~april/mitosis.html

Further Readings Part I

Alberts, B., et al. 1998. *Essential cell biology.* New York: Garland Publishing, Inc. This is an introductory molecular biology text.

Armbruster, P., and Hessberger, F. P. September 1998. Making new elements. *Scientific American* 279(3):72. The process of creating new artificial elements is examined.

Bayley, H. September 1997. Building doors into cells. *Scientific American* 277(3):62. Protein engineers are designing artificial pores for drug delivery.

Braun, V. December 18, 1998. Pumping iron through cell membranes. *Science* 282(5397):2202. Gated protein channels allow ions and small molecules to flow across a cell membrane in a regulated way.

Caret, R. L., et al. 1997. *Principles and applications of organic and biological chemistry.* 2d ed. Dubuque, Ia.: Wm. C. Brown Publishers. For undergraduates, this text emphasizes material unique to health-related studies.

Chang, R. 1998. *Chemistry.* 6th ed. Dubuque, Ia.: McGraw-Hill. This general chemistry text provides a foundation in chemical concepts and principles, and presents topics clearly.

Chapman, C. 1999. *Basic chemistry for biology.* 2d ed. Dubuque, Ia.: WCB/McGraw-Hill. The goal of this workbook is to provide a review of basic principles for biology students.

Cooper, G. M. 1997. *The cell: A molecular approach.* Sunderland, Mass.: Sinauer Associates, Inc. This text is for those beginning coursework in cell and molecular biology.

Ferber, D. January 8, 1999. Immortalized cells seem cancer-free so far. *Science* 283(5399):154. The gene for telomerase has been added to cells to make them "immortal"; perhaps these immortal cells might be used for cell replacement.

Flannery, M. February 1999. Proteins: The unfolding- and folding-picture. *American Biology Teacher* 61(2):150. Various proteins and protein-folding are discussed in depth.

Ford, B. J. April 1998. The earliest views. *Scientific American* 278(4):50. Presents experiments of early microscopists.

Foyer, C., and Noctor, G. April 23, 1999. Leaves in the dark see light. *Science* 284(5414):599. Excess light may be dangerous to plants because it can cause persistent decreases in rates of photosynthesis.

Frank, J. September/October 1998. How the ribosome works. *American Scientist* 86(5):428. New imaging techniques using cryo-electron microscopy allows researchers to study a three-dimensional map of the ribosome.

Gerstein, M., and Levitt, M. November 1998. Simulating water and the molecules of life. *Scientific American* 279(5):100. Computer models show how water affects the structure and movement of proteins and other biological molecules.

Gibson, A. November 1998. Photosynthetic organs of desert plants. *BioScience* 48(11):911. The structure of nonsucculent desert plants casts doubt on the view that saving water is their key strategy.

Ingber, D. E. January 1998. The architecture of life. *Scientific American* 278(1):48. Simple mechanical rules may govern cell movements, tissue organization, and organ development.

Krauskopf, S. January 1999. Doing the meiosis shuffle. *American Biology Teacher* 61(1):60. A playing card demonstration walks students through the stages of meiosis.

Lang, F., and Waldegger, S. September/October 1997. Regulating cell volume. *American Scientist* 85(5):456. Changes in cell volume may threaten organ or tissue function.

Nemecek, S. October 1997. Gotta know when to fold 'em. *Scientific American* 277(4):28. Details about how proteins fold are discussed.

Nurse, P., et al. October 1998. Understanding the cell cycle. *Nature Medicine* 4(1):1103. Article discusses the relevance of cell-cycle research.

Ojcius, D. M., et al. January/February 1998. Pore-forming proteins. *Science & Medicine* 5(1):44. Peptide molecules that cause pore formation in membranes are similar in structure and function.

Palazzo, R. E. March/April 1999. The centrosome. *Science & Medicine* 6(2):32. Article discusses centrosomal research.

Pennisi, E. February 5, 1999. Trigger for centrosome replication found. *Science* 283(5403):770. An enzyme called Cdk2 helps tell the dividing cell to copy its centrosome.

Pennisi, E. May 21, 1999. Nuclear transport protein does double duty in mitosis. *Science* 284(5418):1260. A protein called Ran, which plays a key role in nuclear transport, also triggers cell division.

Ross, F. C. 1997. *Foundation of allied health sciences: An introduction to chemistry and cell biology.* Dubuque, Ia.: Wm. C. Brown Publishers. This introductory text provides the background necessary for students in allied health sciences.

Scerri, E. R. November/December 1997. The periodic table and the electron. *American Scientist* 85(6):546. Electron configurations may only give an approximate explanation of the periodic table.

Scerri, E. R. September 1998. The evolution of the periodic system. *Scientific American* 279(3):78. Article discusses the history and evolution of the periodic table.

Schwartz, A. T., et al. 1997. *Chemistry in context: Applying chemistry to society.* 2d ed. Dubuque, Ia.: Wm. C. Brown Publishers. This introductory text is designed for students in the allied health fields.

Science 283(5407):1475. March 5, 1999. An entire section is devoted to topics involving mitochondria.

Sperelakis, N., editor. 1998. *Cell physiology source book.* 2d ed. San Diego: Academic Press. For advanced biology students, this is a comprehensive and authoritative text covering topics in cell physiology written by experts in the field.

Szalai, V. A., and Bridvig, G. W. November/December 1998. How plants produce dioxygen. *American Scientist* 86(6):542. Oxygen production through photosynthesis occurs in a manganese-containing complex in chloroplast membranes.

Zubay, G. L. 1998. *Biochemistry.* 4th ed. Dubuque, Ia.: Wm. C. Brown Publishers. This text for chemistry majors relates biochemistry to cell biology, physiology, and genetics.

p a r t

ii

Genetic Basis of Life

Hereditary information is stored in DNA, molecules that compose the genes located within chromosomes. A special form of cell division causes sex cells to have half the usual number of chromosomes. Fertilization gives the zygote the full number of chromosomes.

Principles of inheritance include those that allow us to predict the chances that an offspring will inherit a particular characteristic from one of the parents. These principles have been applied to the breeding of plants and animals and the study of human genetic disorders. But to go further, and to control an organism's characteristics, it is necessary to understand how DNA and RNA function in protein synthesis. The human endeavor known as biotechnology is based on our newfound knowledge of nucleic acid structure and function.

The principles of inheritance are central to understanding many other topics in biology—from the evolution and diversity of life to the reproduction and development of organisms.

Meiosis and Sexual Reproduction

chapter

chapter concepts

One sperm out of many fertilizes an egg.

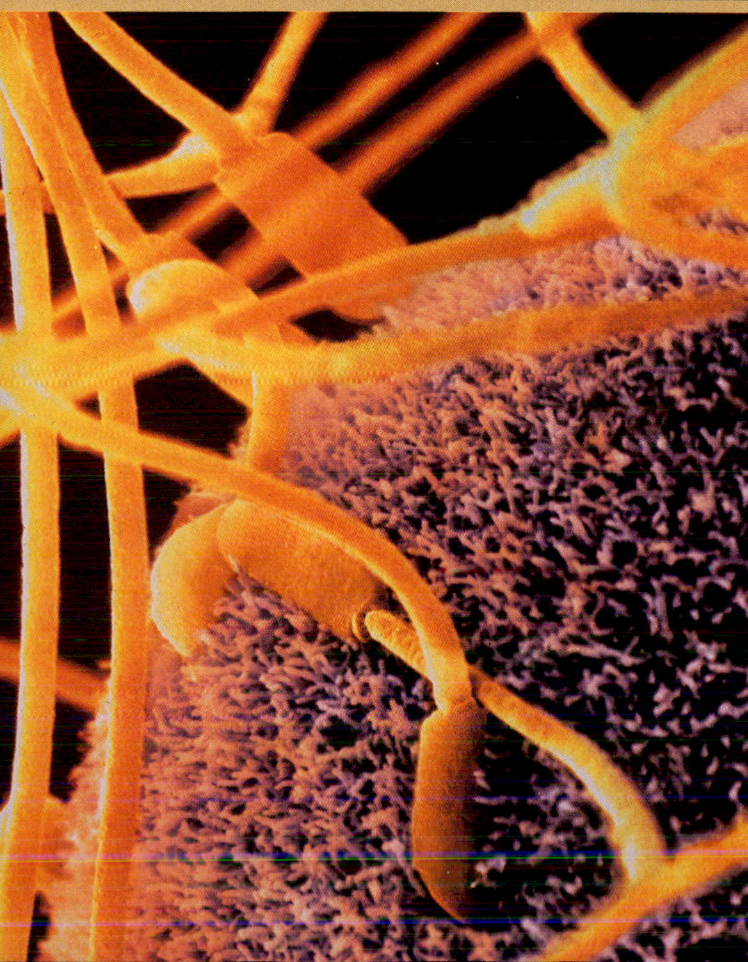

Think about what the sex act accomplishes—a sperm fertilizes an egg and a new individual begins life. The egg and sperm have to have half the number of chromosomes or else the chromosome number would double with each new generation. The production of reproductive cells also ensures that the next generation is populated with offspring that are genetically different from their parents and from each other.

During meiosis, the type of nuclear division involved in reproductive cell production, chromosome pairs come together. They often swap segments with each other producing different combination of genes on the chromosomes. Then, the reproductive cells receive only one of each kind of chromosome in all possible combinations. Finally, the sperm and egg that join during fertilization are usually from two different individuals and, if so, genetic recombination is bound to occur.

While asexual reproduction produces offspring that are identical to the single parent; sexual reproduction introduces genetic variability. In this way, sexual reproduction contributes to the process of evolution, especially if the environment is changing.

10.1 Halving the Chromosome Number

In sexually reproducing organisms, **meiosis** [Gk. *mio*, less, and *-sis*, act or process of] is the type of nuclear division that reduces the chromosome number from the diploid (2n) number [Gk. *diplos*, twofold, and *-eides*, like] to the haploid (n) number. The **haploid (n) number** [Gk. *haplos*, single, and *-eides*, like] of chromosomes is half the diploid number. **Gametes** (reproductive cells, often the sperm and egg) always have the haploid number of chromosomes. Gamete formation and then fusion of gametes to form a cell called a zygote are integral parts of **sexual reproduction.** A **zygote** always has the full or **diploid (2n) number** of chromosomes. In animals the zygote undergoes development to become the adult animal.

Obviously, if the gametes contained the same number of chromosomes as the body cells, the number of chromosomes would double with each new generation. Within a few generations, the cells of an animal would be nothing but chromosomes! The early cytologists (biologists who study cells) realized this, and Pierre-Joseph van Beneden, a Belgian, was gratified to find in 1883 that the sperm and the egg of the worm *Ascaris* each contained only two chromosomes, while the zygote and subsequent embryonic cells always have four chromosomes.

Homologous Pairs of Chromosomes

In diploid body cells, the chromosomes occur in pairs. Notice in Figure 10.1*a*, a pictorial display of human chromosomes, the chromosomes have been arranged according to pairs. The members of each pair are called **homologous chromosomes** or **homologues** [Gk. *homologos*, agreeing, corresponding]. The homologues look alike; they have the same length and centromere position. When stained, homologues have a similar banding pattern because they contain genes for the same traits. Although homologous chromosomes may have genes for the same traits such as finger length; the gene on one homologue may be for short fingers and the gene at the same location on the other homologue may be for long fingers.

The chromosomes in Figure 10.1*a* are duplicated as they would be just before nuclear division. Recall that during the S stage of the cell cycle, DNA replicates and the chromosomes become duplicated. The results of the duplication process are depicted in Figure 10.1*b*. When duplicated, a chromosome is composed of two identical parts called sister chromatids. The sister chromatids are held together at a region called the centromere.

Why does the zygote have two chromosomes of each kind? One member of a homologous pair was inherited from the male parent and the other was inherited from the female parent by way of the gametes. In Figure 10.1*b*, and throughout the chapter, the paternal chromosome is colored blue and the maternal chromosome is colored red. *Therefore, you should use size and centromere location to recognize homologues and not color*. We will see that while body cells contain two chromosomes of each kind, the gametes contain only one chromosome of each kind—derived from either the paternal or maternal homologue.

The zygote, which is always diploid, contains homologous chromosomes. Gametes are haploid due to meiosis, the type of cell division that reduces the chromosome number.

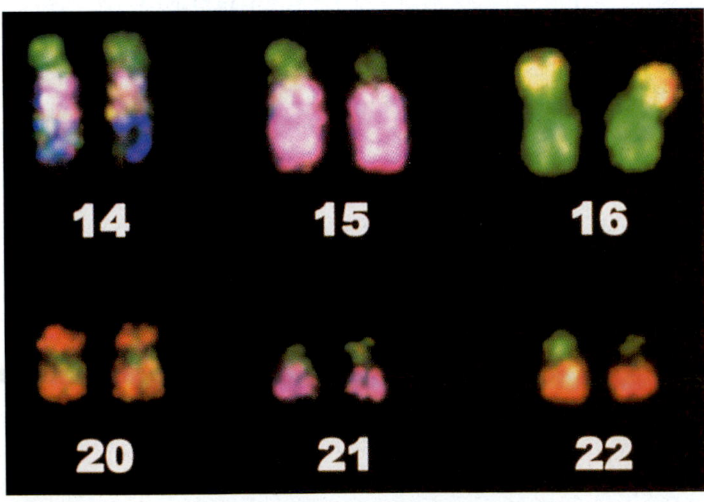

a.

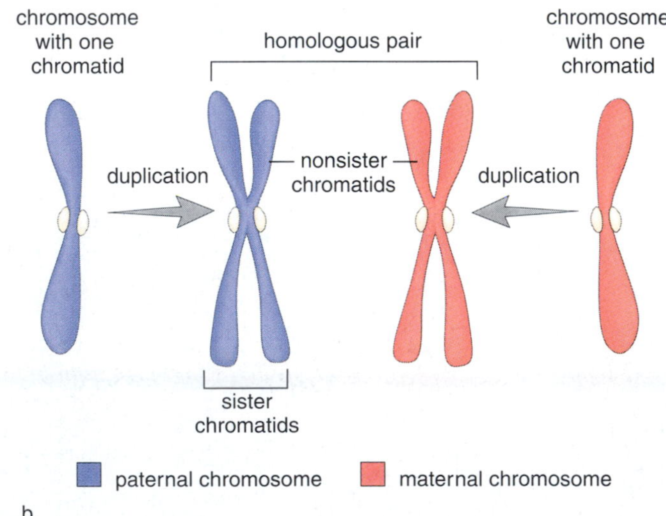

b.

Figure 10.1 Homologous chromosomes.
In a diploid body cell, the chromosomes occur in pairs called homologous chromosomes. **a.** In this micrograph of stained chromosomes from a human cell, the pairs have been numbered. **b.** These chromosomes are duplicated and each one is composed of two chromatids. The sister chromatids contain the exact same genes; the nonsister chromatids contain only genes for the same traits, like type of hair, color of eyes.

Overview of Meiosis

Meiosis requires two nuclear divisions and produces four haploid daughter cells, each having one of each kind of chromosome. Therefore, the daughter cells have half the total number of chromosomes present in the diploid parent nucleus. In this and other ways to be discussed, meiosis ensures that offspring will have a different combination of chromosomes and genes than either parent.

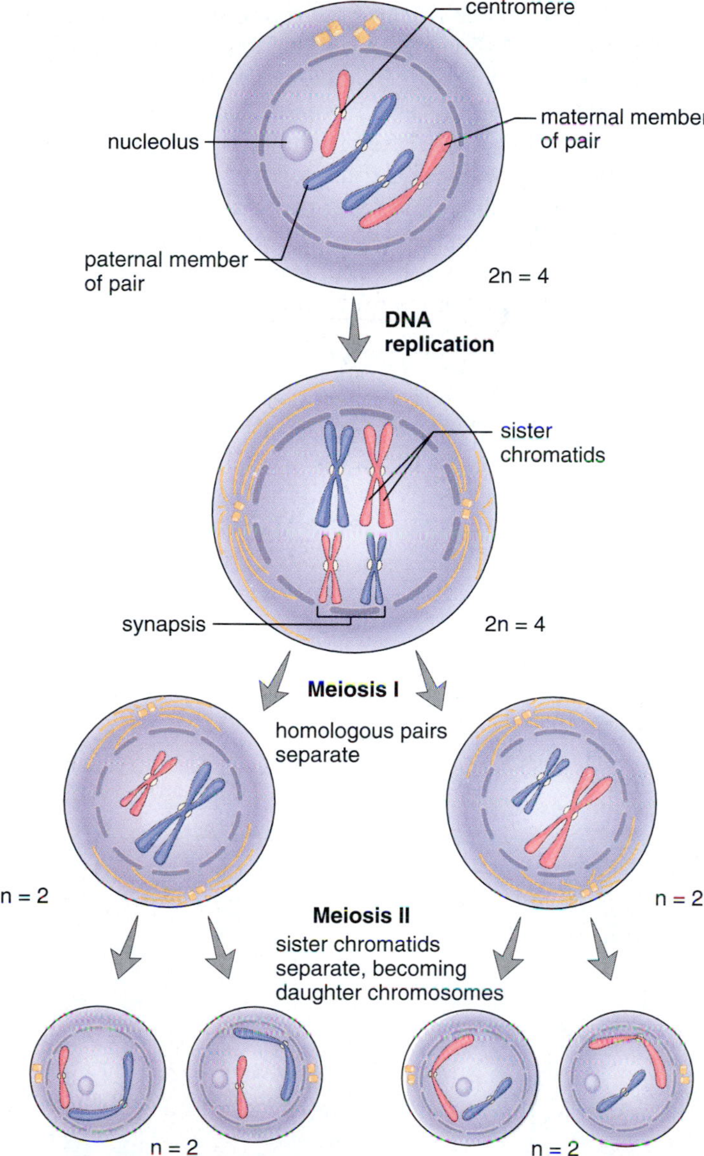

Figure 10.2 **Overview of meiosis.**
Following DNA replication, each chromosome is duplicated and consists of two chromatids. During meiosis I, the chromosome pairs separate, and during meiosis II, the sister chromatids of each duplicated chromosome separate. At the completion of meiosis, there are four haploid daughter cells.

Figure 10.2 presents an overview of meiosis, indicating the two cell divisions, meiosis I and meiosis II. Prior to meiosis I, DNA (deoxyribonucleic acid) replication has occurred; therefore, each chromosome has two sister chromatids. During meiosis I, something new happens that does not occur in mitosis. The homologous chromosomes come together and line up side by side due to a means of attraction still unknown. This so-called **synapsis** [Gk. *synaptos*, united, joined together] results in a **bivalent** [L. *bis*, two, and *valens*, strength]—that is, two homologous chromosomes that stay in close association during the first two phases of meiosis I. Sometimes the term tetrad [Gk. *tetra*, four] is used instead of bivalent because, as you can see, a bivalent contains four chromatids.

Following synapsis, the members of a homologous pair separate. This separation means that only one duplicated chromosome from each homologous pair reaches a daughter nucleus. It is important for daughter nuclei to have a member from each pair of homologous chromosomes because only in that way can there be a copy of each kind of chromosome in the daughter nuclei. The members of the homologous pairs separate in such a way that any particular kind of chromosome can be with any other kind in the daughter nuclei. Therefore, all possible combinations of chromosomes can occur within the gametes that result after meiosis is complete.

During meiosis I, homologous chromosomes separate and the daughter cells have one copy of each kind of chromosome.

No replication of DNA is needed between meiosis I and meiosis II because the chromosomes are already duplicated; they already have two sister chromatids. During meiosis II, the daughter chromosomes derived from sister chromatids move to opposite poles. Therefore, the chromosomes in the four daughter cells have only one chromatid. You can count the number of centromeres to verify that the parent cell has the diploid number of chromosomes and each daughter cell has the haploid number.

Following meiosis II, there are four haploid daughter cells and each chromosome consists of one chromatid.

In the animal life cycle, the daughter cells mature into gametes that fuse during fertilization. Fertilization restores the diploid number of chromosomes so that the body cells of an animal contains the diploid number of chromosomes.

10.2 Genetic Recombination

We have seen that meiosis provides a way to keep the chromosome number constant generation after generation. Without meiosis, the chromosome number of the next generation would continually increase. Meiosis also helps ensure that genetic recombination occurs with each generation. Due to genetic recombination, offspring have a different combination of genes than their parents. Asexually reproducing organisms, like prokaryotes, depend primarily on mutations to generate variation among offspring. This is sufficient because they produce great numbers of offspring within a limited amount of time. Mutation also occurs among sexually reproducing organisms, but the shuffling of genetic material during the process of reproduction contributes greatly to the possibility of offspring that may be better adapted to the environment. Meiosis brings about genetic recombination in two key ways: crossing-over and independent assortment of homologous chromosomes.

Crossing-Over of Nonsister Chromatids

It's often said that we inherit half our chromosomes from our mother and half from our father, but this is not strictly correct, because of crossing-over. **Crossing-over** is an exchange of genetic material between nonsister chromatids of a bivalent during meiosis I. At synapsis, homologues line up side by side, and a nucleoprotein lattice (called the synaptonemal complex) appears between them (Fig. 10.3). This lattice holds the bivalent together in such a way that the DNA of the nonsister chromatids is aligned. Now crossing-over may occur. As the lattice breaks down, homologues are temporarily held together by *chiasmata* (sing., chiasma), regions where the nonsister chromatids are attached due to crossing-over (Fig. 10.4). Then homologues separate and are distributed to different daughter cells.

In order to appreciate the significance of crossing-over, it is necessary to remember that the members of a homologous pair can carry slightly different instructions for the same genetic traits. For example, one homologue may carry

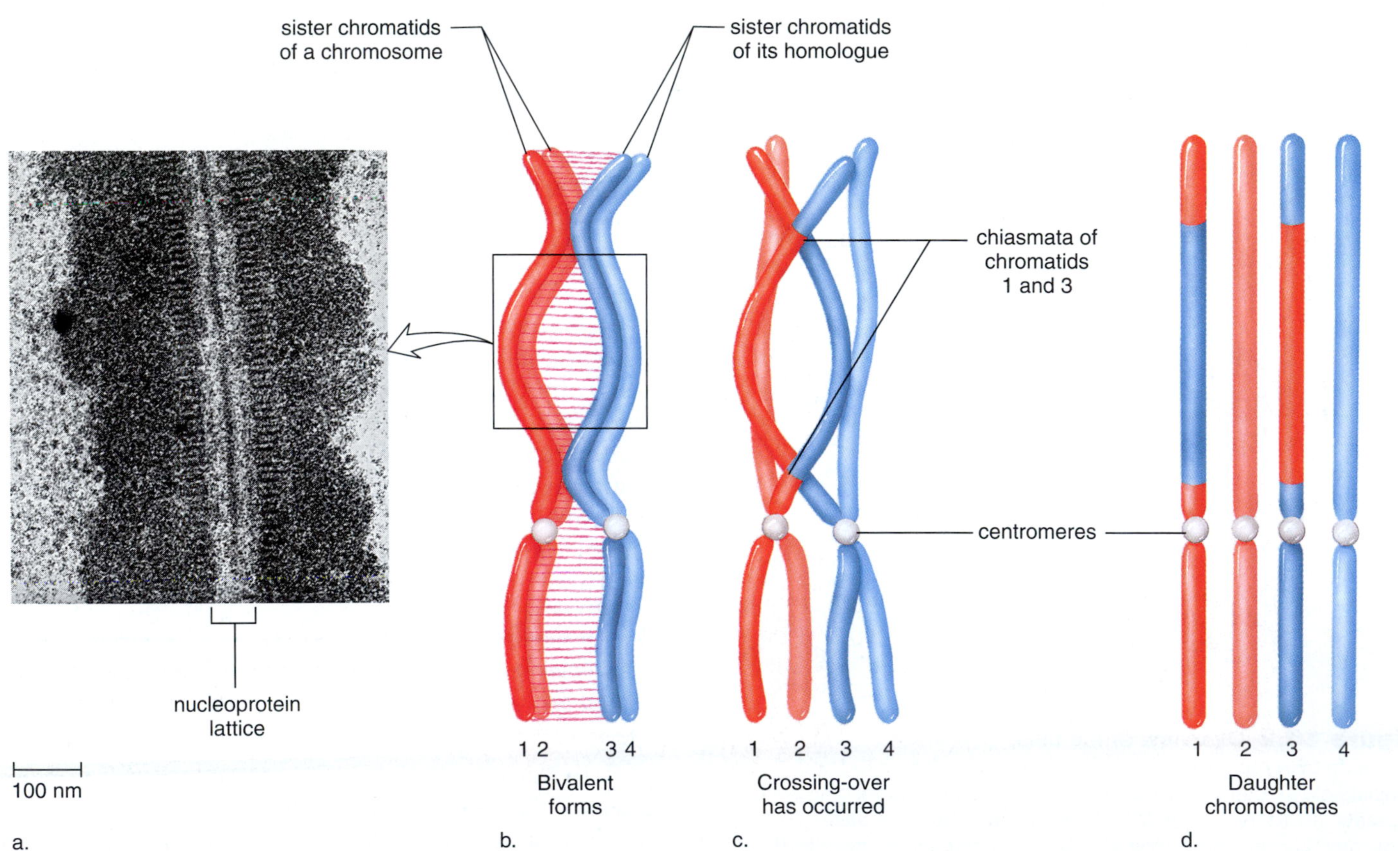

Figure 10.3 Crossing-over occurs during meiosis I.
a. The homologous chromosomes pair up, and a nucleoprotein lattice, called the synaptonemal complex, develops between them. This is an electron micrograph of the complex, which zippers the members of the bivalent together so that corresponding genes are in alignment. **b.** This diagrammatic representation shows only two places where nonsister chromatids 1 and 3 have come into contact. Actually, the other two nonsister chromatids most likely are also crossing over. **c.** Chiasmata indicate where crossing-over has occurred. The exchange of color represents the exchange of genetic material. **d.** Following meiosis II, daughter chromosomes have a new combination of genetic material due to crossing-over, which occurred between nonsister chromatids during meiosis I.

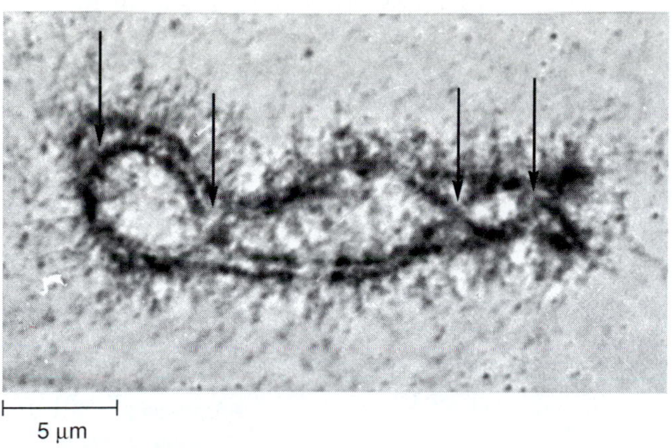

Figure 10.4 **Chiasmata**
Chiasmata (arrows) of a chromosome bivalent, from a testis cell of a grasshopper. The chiasmata mark the places where crossing-over between nonsister chromosomes of the bivalent has occurred. The chiasmata hold the members of the bivalent together until separation of homologues occurs.

instructions for brown eyes while the corresponding homologue may carry instructions for blue eyes. The same is true for all the genes carried on all the homologues. In the end, crossing-over means that the genetic instructions from a mother and father are mixed and the chromatids held together by a centromere are no longer identical. Therefore, when the chromatids separate during meiosis I, the daughter cells receive chromosomes with recombined genes.

Independent Assortment of Homologous Chromosomes

Due to **independent assortment** the homologous chromosomes separate independently or in a random manner. When homologues align at the metaphase plate, the maternal or paternal homologue may be orientated toward either pole. Figure 10.5 shows four possible orientations for a cell that contains only three pairs of chromosomes. Each orientation results in gametes that have a different combination of maternal and paternal chromosomes. Once all possible orientations are considered, the result will be 2^3 or eight combinations of maternal and paternal chromosomes in the resulting gametes from this cell.

In humans, where there are 23 pairs of chromosomes, the possible chromosomal combinations in the gametes is a staggering 2^{23}, or 8,388,608. And this does not even consider the genetic variations that are introduced due to crossing-over.

Fertilization

The variation that results from meiosis is enhanced by fertilization. Each person has genes for the same traits, but again, each gene's specific instructions can vary. Therefore, the gametes produced by one person are expected to be genetically different from the gametes produced by another person. When *the gametes fuse at fertilization*, the chromosomes donated by the parents are combined, and in humans, this means that $(2^{23})^2$, or 70,368,744,000,000, chromosomally different zygotes are possible, even assuming no crossing-over. If crossing-over occurs once, then $(4^{23})^2$, or 4,951,760,200,000,000,000,000,000,000, genetically different zygotes are possible for every couple.

> There are three ways by which genetic recombination comes about during sexual reproduction:
>
> 1. Independent alignment of bivalents at the metaphase plate means that gametes have different combinations of chromosomes.
> 2. Crossing-over means that the chromosomes in one gamete have a different combination of genes than chromosomes in another gamete.
> 3. Upon fertilization, combining of chromosomes from genetically different gametes occurs.

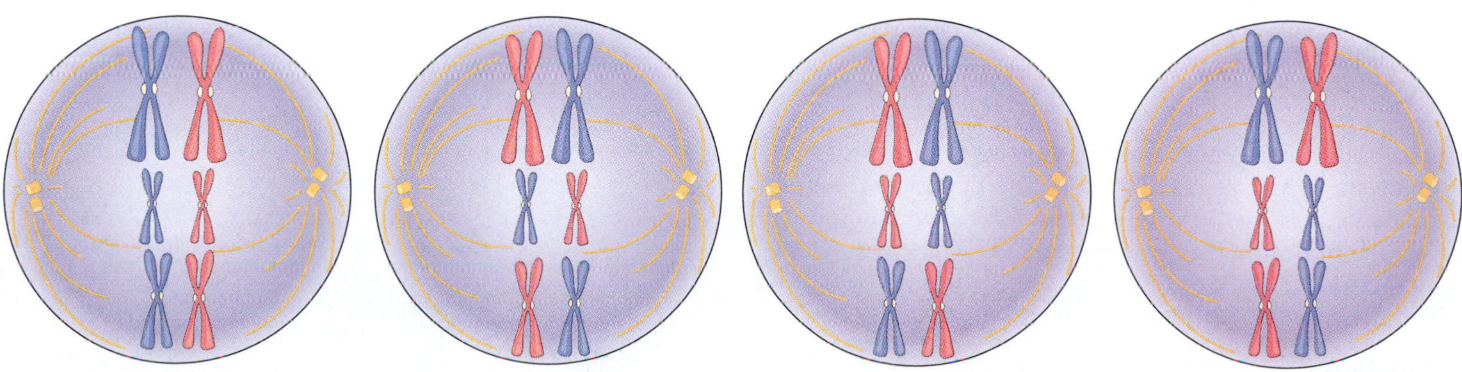

Figure 10.5 **Independent assortment.**
Four possible orientations of homologue pairs at the metaphase plate are shown. Each of these will result in daughter nuclei with a different combination of parental chromosomes. When a cell has three pairs of homologous chromosomes, there are 23 possible combinations of parental chromosomes in the daughter nuclei.

10.3 The Phases of Meiosis 

Both meiosis I and meiosis II have these phases: prophase, metaphase (preceded by prometaphase), anaphase, and telophase.

Prophase I

It is apparent during prophase I that nuclear division is about to occur because a spindle forms as the centrosomes migrate away from one another. The nuclear envelope fragments, and the nucleolus disappears.

The homologous chromosomes, each having two sister chromatids, undergo synapsis to form bivalents. As depicted in Figure 10.3 by the exchange of color, crossing-over between the nonsister chromatids may occur at this time. After crossing-over, the sister chromatids of a duplicated chromosome are no longer identical.

Throughout prophase I, the chromosomes have been condensing so that by now they have the appearance of metaphase chromosomes.

Metaphase I

During prometaphase I, the bivalents held together by chiasmata (see Fig. 10.4) have moved toward the metaphase plate (equator of the spindle). Metaphase I is characterized by a fully formed spindle and alignment of the bivalents at the metaphase plate. *Kinetochores,* protein complexes just outside the centromeres, are seen, and these are attached to spindle fibers called kinetochore spindle fibers.

Bivalents independently align themselves at the metaphase plate of the spindle. The maternal homologue of each bivalent may be orientated toward either pole, and the paternal homologue of each bivalent may be aligned toward either pole. This means that all possible combinations of chromosomes can occur in the daughter cells.

Anaphase I

During anaphase I, the homologues of each bivalent separate and move to opposite poles. Notice that each chromosome still has two chromatids.

Telophase I

In some species, there is a telophase I stage at the end of meiosis I. If so, the nuclear envelopes re-form and nucleoli appear. This phase may or may not be accompanied by cytokinesis, which is separation of the cytoplasm. Figure 10.6 shows only two of the four possible combinations of haploid chromosomes when the parent cell has two homologous pairs of chromosomes. Can you determine what the other two possible combinations of chromosomes are?

Interkinesis

Interkinesis is similar to interphase between mitotic divisions except that DNA replication does not occur—the chromosomes are already duplicated.

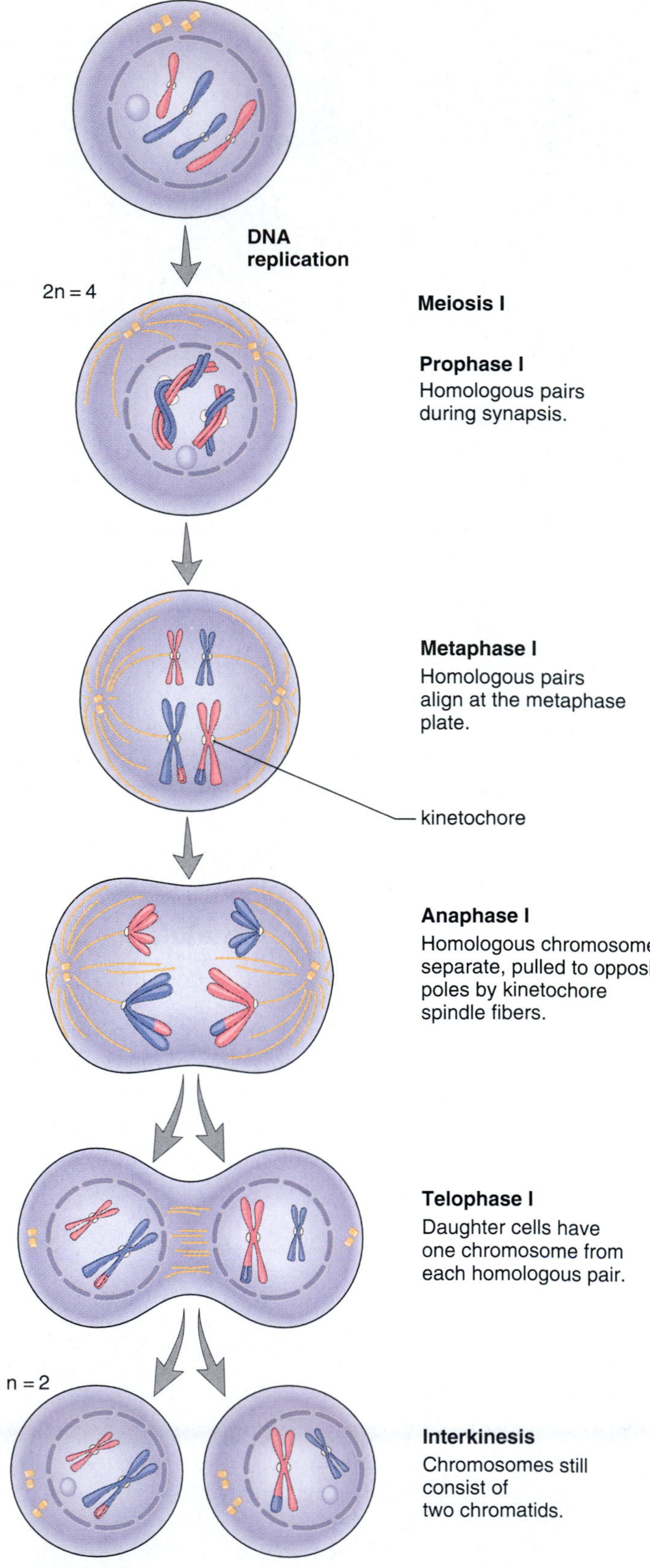
DNA replication
2n = 4
Meiosis I
Prophase I
Homologous pairs during synapsis.
Metaphase I
Homologous pairs align at the metaphase plate.
kinetochore
Anaphase I
Homologous chromosomes separate, pulled to opposite poles by kinetochore spindle fibers.
Telophase I
Daughter cells have one chromosome from each homologous pair.
n = 2
Interkinesis
Chromosomes still consist of two chromatids.

Figure 10.6 Meiosis I.
The exchange of color between nonsister chromatids represents crossing-over.

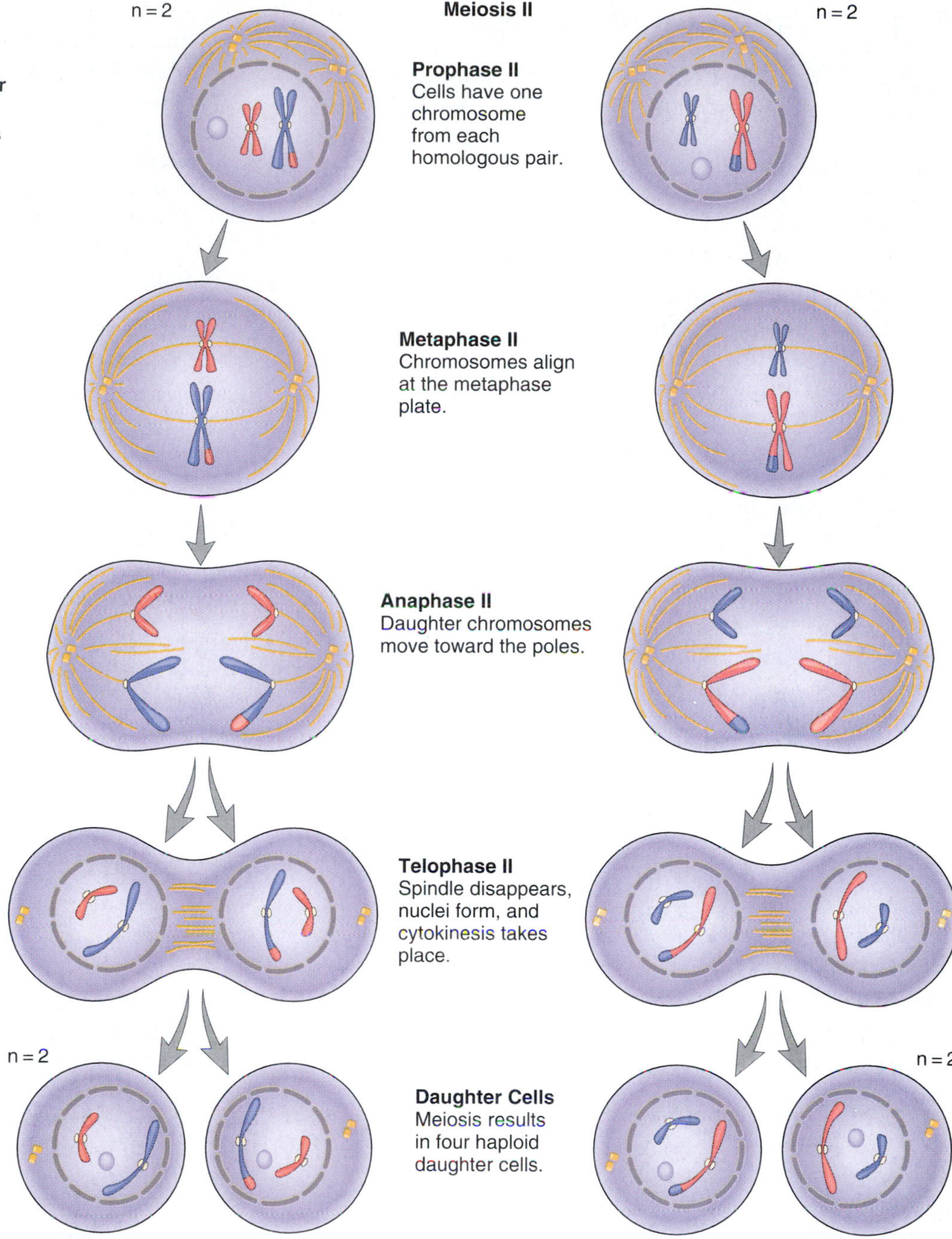
n = 2
Meiosis II
n = 2
Prophase II
Cells have one chromosome from each homologous pair.
Metaphase II
Chromosomes align at the metaphase plate.
Anaphase II
Daughter chromosomes move toward the poles.
Telophase II
Spindle disappears, nuclei form, and cytokinesis takes place.
n = 2
Daughter Cells
Meiosis results in four haploid daughter cells.
n = 2

Figure 10.7 Meiosis II.
During meiosis II, daughter chromosomes consisting of one chromatid each move to the poles. Following meiosis II, there are four haploid daughter cells. Comparing the number of centromeres in the daughter cells with the number in the parental cell at the start of meiosis I verifies that the daughter cells are haploid.

Meiosis II

During metaphase II, the haploid number of chromosomes, which are still duplicated, align at the metaphase plate (Fig. 10.7). During anaphase II, the two sister chromatids separate at the centromere, giving rise to two daughter chromosomes. These daughter chromosomes move toward the poles. At the end of telophase II and cytokinesis, there are four haploid cells. Due to crossing-over of chromatids, each gamete will most likely contain chromosomes with varied genes.

Following meiosis II, the haploid cells mature and become gametes in animals. In plants, they become spores that divide to produce a haploid adult generation. This haploid generation produces gametes that fuse to become the diploid generation. Thus, plants have both a haploid and diploid generation in their life cycle. In some fungi and some algae, the zygote that results from gamete fusion immediately undergoes meiosis, and therefore the adult is always haploid.

10.4 Comparison of Meiosis with Mitosis

Figure 10.8 compares mitosis to meiosis. The differences between these cellular divisions can be categorized according to process and occurrence.

Process

The following are differences between meiosis and mitosis:

1. DNA replication takes place only once during both meiosis and mitosis; but there are two nuclear divisions during meiosis and only one nuclear division during mitosis.
2. Four daughter cells are produced by meiosis; mitosis results in two daughter cells.
3. The four daughter cells formed by meiosis are haploid; the daughter cells produced by mitosis have the same chromosome number as the parental cell.
4. The daughter cells from meiosis are not genetically identical to each other or to the parental cell. The daughter cells from mitosis are genetically identical to each other and to the parental cell.

Comparison of Meiosis I to Mitosis

The following are distinctive differences between the processes of meiosis I and mitosis:

Meiosis I	Mitosis
Prophase I Pairing of chromosomes	Prophase No pairing of chromosomes
Metaphase I Homologous chromosomes at metaphase plate	Metaphase Duplicated chromosomes at metaphase plate
Anaphase I Homologous chromosomes separate	Anaphase Sister chromatids separate, becoming daughter chromosomes that move to the poles
Telophase I Daughter cells are haploid	Telophase Daughter cells are diploid

These events distinguish meiosis I from mitosis:

1. During prophase I of meiosis, homologous chromosomes pair and undergo crossing-over but this does not occur during mitosis.
2. During metaphase I of meiosis, paired homologous chromosomes align independently at the metaphase plate; during metaphase of mitosis, individual (duplicated) chromosomes align at the metaphase plate.
3. During anaphase I in meiosis, homologous chromosomes (with centromeres intact) separate and move to opposite poles; during anaphase of mitosis, sister chromatids separate, becoming daughter chromosomes that move to opposite poles.

Comparison of Meiosis II to Mitosis

The events of meiosis II are just like those of mitosis except that in meiosis II, the nuclei contain the haploid number of chromosomes. The following listing compares meiosis II to mitosis:

Meiosis II	Mitosis
Prophase II No pairing of chromosomes	Prophase No pairing of chromosomes
Metaphase II Haploid number of duplicated chromosomes at metaphase plate	Metaphase Diploid number of duplicated chromosomes at metaphase plate
Anaphase II Sister chromatids separate, becoming daughter chromosomes that move to the poles	Anaphase Sister chromatids separate, becoming daughter chromosomes that move to the poles
Telophase II Four daughter cells	Telophase Two daughter cells

Occurrence

Meiosis occurs only at certain times in the life cycle of sexually reproducing organisms. In humans, meiosis occurs only in the reproductive organs and produces the gametes. Mitosis is more common because it occurs in all tissues during growth and repair.

Meiosis is a specialized process that reduces the chromosome number and occurs only during the production of gametes. Mitosis is a process that occurs during growth and repair of all tissues.

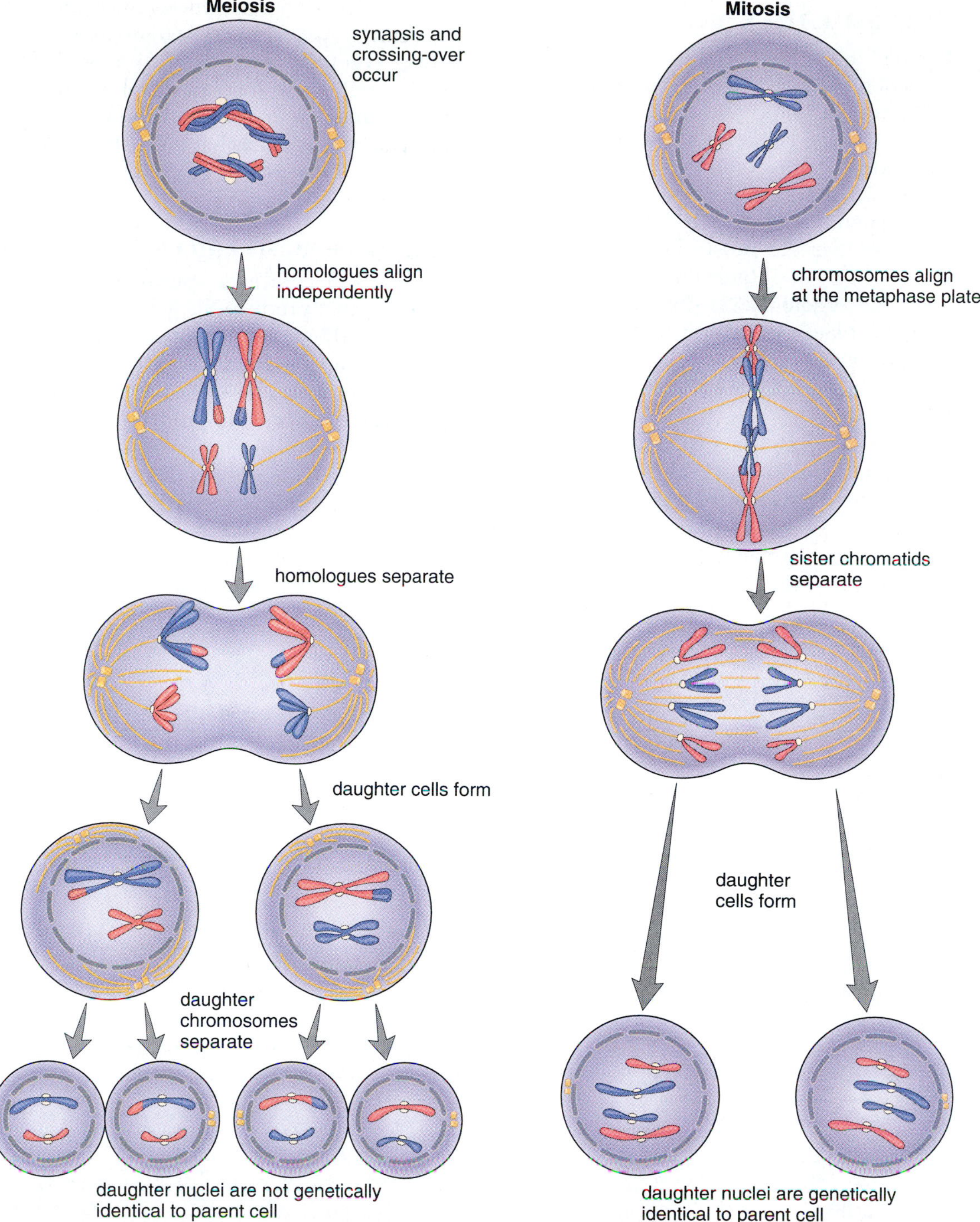

Figure 10.8 Comparison of meiosis and mitosis.
Why does meiosis produce haploid daughter cells while mitosis produces diploid daughter cells? Compare metaphase I of meiosis to metaphase of mitosis. Only in metaphase I are the homologous chromosomes paired at the metaphase plate. Members of the homologous chromosomes separate during anaphase I and therefore the daughter cells are haploid. The exchange of color between nonsister chromatids represents the crossing-over that occurred during meiosis I.

10.5 The Human Life Cycle

The term **life cycle** refers to all the reproductive events that occur from one generation to the next similar generation. In animals, the adult is always diploid and meiosis produces the gametes, the only haploid stage of the life cycle. In plants, there are two adult stages, one is diploid and the other is haploid. The haploid stage may be larger or smaller than the diploid stage depending on the species. The mosses growing on bare rock are haploid most of their life cycle while oak trees are diploid most of their life cycle. In fungi (and some algae) only the zygote is ever diploid and it undergoes meiosis. So the black mold you sometimes see growing on bread and the green coating you see growing on a pond are haploid. It is these haploid individuals that produce gamete nuclei.

In animals, such as humans, meiosis occurs during the production of gametes. In males, meiosis is a part of **spermatogenesis** [Gk. *sperma,* seed, and L. *genitus,* producing], which occurs in the testes and produces sperm. In females, meiosis is a part of **oogenesis** [Gk. *oon,* egg, and L. *genitus,* producing], which occurs in the ovaries and produces eggs. A sperm and egg join at fertilization and the resulting zygote undergoes mitosis during development of the fetus, which is the stage of development before birth. After birth, mitosis is involved in the continued growth of the child and repair of tissues at any time (Fig. 10.9). As a result of mitosis, each somatic cell in the body has the same number of chromosomes.

Spermatogenesis and Oogenesis in Humans

Figure 10.10 contrasts spermatogenesis with oogenesis, processes that produce the gametes in mammals, including humans. In the testes of males, primary spermatocytes with 46 chromosomes divide to form two secondary spermatocytes, each with 23 duplicated chromosomes. Secondary spermatocytes divide to produce four spermatids, also with 23 daughter chromosomes. Spermatids then differentiate into sperm (spermatozoa). The process of meiosis in males always results in four cells that become sperm.

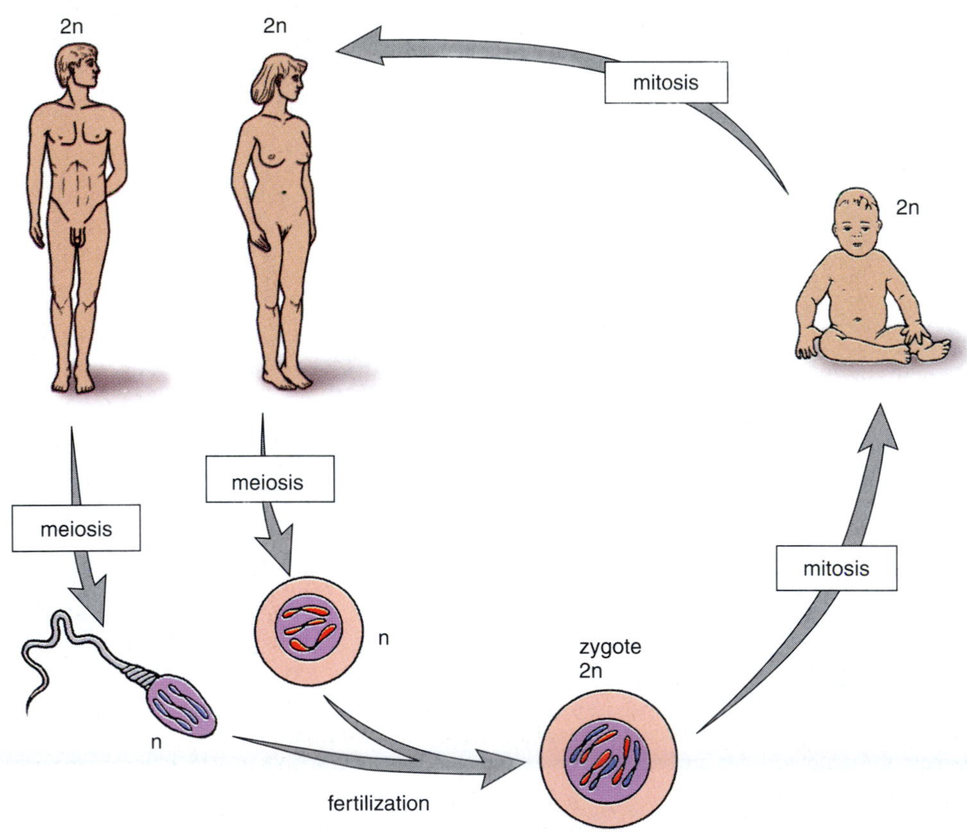

Figure 10.9 **Life cycle of humans.**
Meiosis in males is a part of sperm production, and meiosis in females is a part of egg production. When a haploid sperm fertilizes a haploid egg, the zygote is diploid. The zygote undergoes mitosis as it develops into a newborn child. Mitosis continues after birth until the individual reaches maturity; then the life cycle begins again.

In the ovaries of females, a primary oocyte has 46 chromosomes and divides meiotically into two cells, each having 23 chromosomes. One of these cells, termed the **secondary oocyte** [Gk. *oon*, egg, and *kytos*, cell], receives almost all the cytoplasm (Fig. 10.10). The other is a **polar body** that may disintegrate or may divide again. The secondary oocyte begins meiosis II and then stops at metaphase II. Then at ovulation, it leaves the ovary and enters an oviduct where it may be approached by a sperm. If a sperm does enter the oocyte, the oocyte is activated to continue meiosis II to completion. Therefore meiosis does not go to completion in females unless fertilization occurs. The mature egg has 23 chromosomes. Meiosis in human females produces only one egg and two polar bodies. The polar bodies are a way to discard unnecessary chromosomes while retaining much of the cytoplasm in the egg. The cytoplasm serves as a source of nutrients for the developing embryo.

In the animal life cycle, such as the human life cycle, the diploid adult produces the gametes by meiosis. Some other organisms are haploid as adults and meiosis occurs at some other juncture in the life cycle.

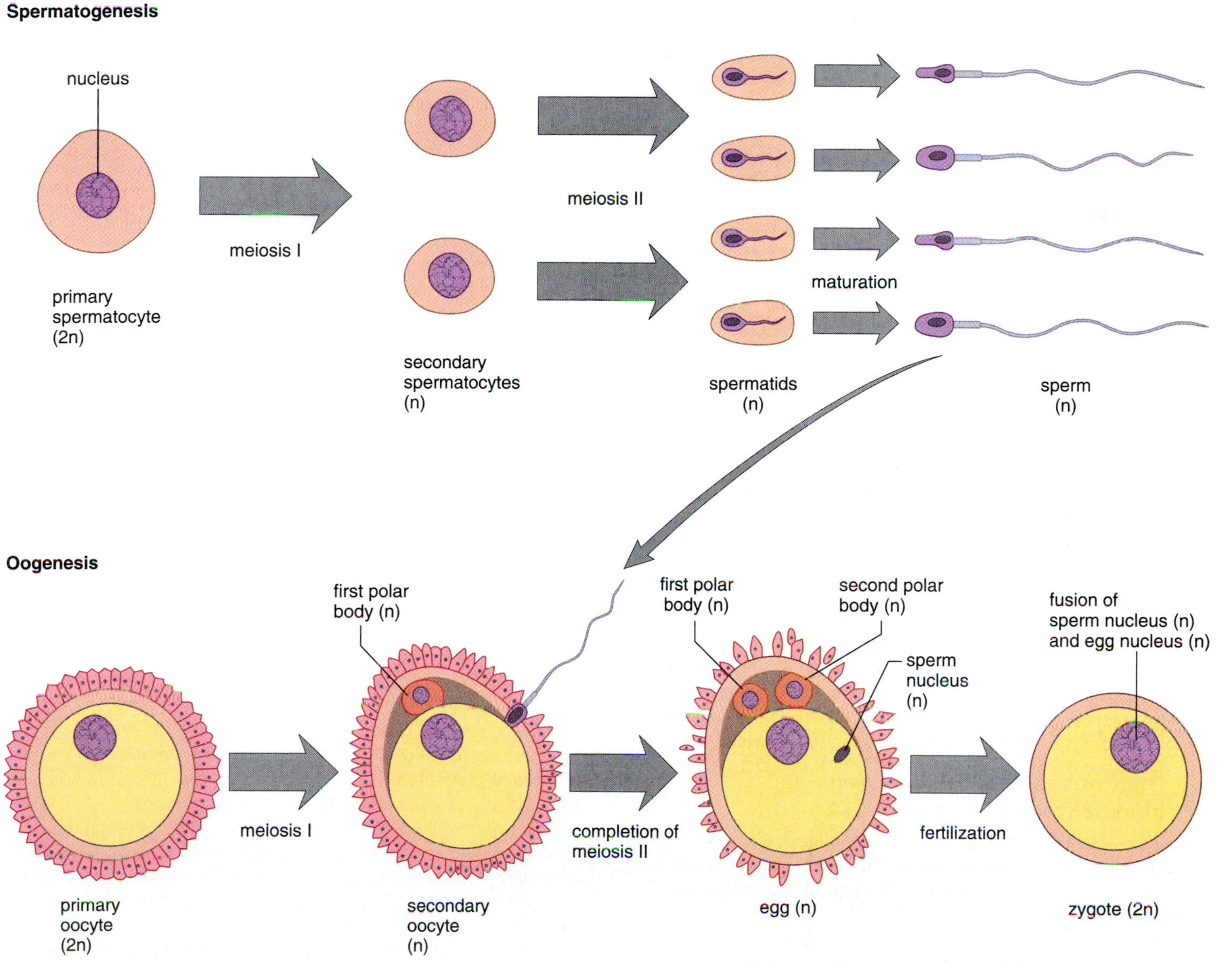

Figure 10.10 Spermatogenesis and oogenesis in mammals.
Spermatogenesis produces four viable sperm, whereas oogenesis produces one egg and at least two polar bodies. In humans, both sperm and egg have 23 chromosomes each; therefore, following fertilization, the zygote has 46 chromosomes.

Connecting Concepts

Meiosis is similar to the process of mitosis except that meiosis is a more elaborate process. During meiosis, controlling mechanisms ensure that homologous chromosomes first pair and then separate during the first division and that sister chromatids do not separate until the second division. In addition, signaling molecules ensure that only certain types of cells undergo meiosis during a restricted period of an organism's life span.

There is an evolutionary cost to sexual reproduction: the increased number of genes controlling the process can lead to an increased chance of mutations and the possibility of faulty or inviable gametes. But there is also an evolutionary advantage: sexually reproducing species have a greater likelihood of genetic diversity among offspring than do asexually reproducing species.

Understanding the behavior of chromosomes during meiosis is critical to understanding the manner in which genes segregate during gamete formation. The next chapter reviews the fundamental laws of genetics established by Gregor Mendel. Although Mendel had no knowledge of chromosome behavior, modern students have the advantage of applying their knowledge of meiosis to increase their understanding of Mendel's laws.

Summary

10.1 Halving the Chromosome Number

Meiosis ensures that the chromosome number in offspring stays constant generation after generation. The nucleus contains pairs of chromosomes, called homologous chromosomes (homologues).

Meiosis requires two cell divisions and results in four daughter cells. Replication of DNA takes place before meiosis begins. During meiosis I, the homologues undergo synapsis (resulting in a bivalent) and align independently at the metaphase plate. The daughter cells receive one member of each pair of homologous chromosomes. There is no replication of DNA during interkinesis. During meiosis II, the sister chromatids separate at the centromeres, giving rise to daughter chromosomes that move to opposite poles as they do in mitosis. The four daughter cells contain the haploid number of chromosomes and only one of each kind.

10.2 Genetic Recombination

Sexual reproduction ensures that the offspring have a different genetic makeup than the parents. Meiosis contributes to genetic variability in two ways: crossing-over and independent assortment of the homologous chromosomes. When homologous chromosomes lie side by side during synapsis, nonsister chromatids may exchange genetic material. Due to crossing-over the chromatids that separate during meiosis II have a different combination of genes.

When the homologous chromosomes align at the metaphase plate during metaphase I either the maternal or the paternal chromosome can be facing either pole. Therefore there will be all possible combinations of chromosomes in the gametes.

10.3 The Phases of sMeiosis

Meiosis I is divided into four phases:

Prophase I—Bivalents form, and crossing-over occurs as chromosomes condense; the nuclear envelope fragments.
Metaphase I—Bivalents independently align at the metaphase plate.
Anaphase I—Homologous chromosomes separate.
Telophase I—Nuclei become haploid, having received one duplicated chromosome from each homologous pair.

Meiosis II is divided into four phases:

Prophase II—Chromosomes condense and the nuclear envelope fragments.
Metaphase II—The haploid number of still duplicated chromosomes align at the metaphase plate.
Anaphase II—Daughter chromosomes move to the poles.
Telophase II—Four haploid daughter cells are genetically different from the parent cell.

10.4 Comparison of Meiosis with Mitosis

Mitosis and meiosis can be compared in this manner:

Meiosis I	Mitosis
Prophase	
Pairing of homologous chromosomes	No pairing of chromosomes
Metaphase	
Bivalents at metaphase plate	Duplicated chromosomes at metaphase plate
Anaphase	
Homologous chromosomes separate	Sister chromatids separate
Telophase	
Daughter nuclei are always haploid number	Daughter nuclei have the parent cell chromosome number

Meiosis II is like mitosis except the nuclei are haploid.

10.5 The Human Life Cycle

Meiosis occurs in any life cycle that involves sexual reproduction. In the animal life cycle, only the gametes are haploid; in plants and fungi, meiosis produces spores that develop into a multicellular haploid adult that produces the gametes. In unicellular protists, the zygote undergoes meiosis and spores become a haploid adult that gives rise to gametes.

During the life cycle of humans, and many other animals, meiosis is involved in spermatogenesis and oogenesis. Whereas spermatogenesis produces four sperm per meiosis, oogenesis produces one egg and two nonfunctional polar bodies. Spermatogenesis occurs in males and oogenesis occurs in females. When a sperm fertilizes an egg, the zygote has the diploid number of chromosomes.

Reviewing the Chapter

1. Why did early investigators predict that there must be a reduction division in the sexual reproduction process? 160
2. What are homologous chromosomes? Contrast the genetic makeup of sister chromatids with nonsister chromatids. 160
3. What is synapsis? Contrast in general meiosis I with meiosis II. 161
4. Draw and explain a diagram that illustrates synapsis and another that shows the significance of independent assortment of homologous pairs. How do these events ensure genetic variability of the gametes? 161–62
5. Draw and explain a series of diagrams that illustrate the stages of meiosis I. 164
6. Draw and explain a series of diagrams that illustrate the stages of meiosis II. 165
7. Construct a chart to describe the many differences between meiosis and mitosis. 166
8. What accounts for (1) the genetic similarity between daughter cells and the parent cell following mitosis and (2) the genetic dissimilarity between daughter cells and the parent cell following meiosis? 166
9. Explain the human (animal) life cycle and compare it to that of protists and plants. 168
10. Compare spermatogenesis of males to oogenesis in females. 168–69

Testing Yourself

Choose the best answer for each question.

1. A bivalent (tetrad) is
   a. a homologous chromosome.
   b. the paired homologous chromosomes.
   c. a duplicated chromosome composed of sister chromatids.
   d. the two daughter cells after meiosis I.
   e. the two centrioles in a centrosome.
2. If a parent cell has twelve chromosomes, then each of the daughter cells following meiosis will have
   a. forty-eight chromosomes.
   b. twenty-four chromosomes.
   c. twelve chromosomes.
   d. six chromosomes.
   e. Any one of these could be correct.
3. At the metaphase plate during metaphase I of meiosis, there are
   a. chromosomes consisting of one chromatid.
   b. unpaired duplicated chromosomes.
   c. bivalents (tetrads).
   d. homologous pairs of chromosomes.
   e. Both c and d are correct.
4. At the metaphase plate during metaphase II of meiosis, there are
   a. chromosomes consisting of one chromatid.
   b. unpaired duplicated chromosomes.
   c. bivalents (tetrads).
   d. homologous pairs of chromosomes.
   e. Both c and d are correct.
5. Gametes contain one of each kind of chromosome because
   a. the homologous chromosomes separate during meiosis.
   b. the chromatids separate during meiosis.
   c. only one replication of DNA occurs during meiosis.
   d. crossing-over occurs during prophase I.
   e. the parental cell contains only one of each kind of chromosome.
6. Crossing-over occurs between
   a. sister chromatids of the same chromosome.
   b. two different kinds of bivalents.
   c. two different kinds of chromosomes.
   d. nonsister chromatids of a bivalent.
   e. two daughter nuclei.
7. During which phase of meiosis do homologous chromosomes separate?
   a. prophase II
   b. telophase I
   c. metaphase I
   d. anaphase I
   e. anaphase II
8. Fertilization
   a. is a source of variation during sexual reproduction.
   b. is fusion of the gametes.
   c. occurs in both animal and plant life cycles.
   d. restores the diploid number of chromosomes.
   e. All of these are correct.
9. Which of these is not a difference between spermatogenesis and oogenesis in humans?

	Spermatogenesis	**Oogenesis**
a.	occurs in males	occurs in females
b.	produces four sperm per meiosis	produces one egg per meiosis
c.	produces haploid cells	produces diploid cells
d.	always goes to completion	does not always go to completion

10. Which of these drawings represents metaphase I? How do you know?

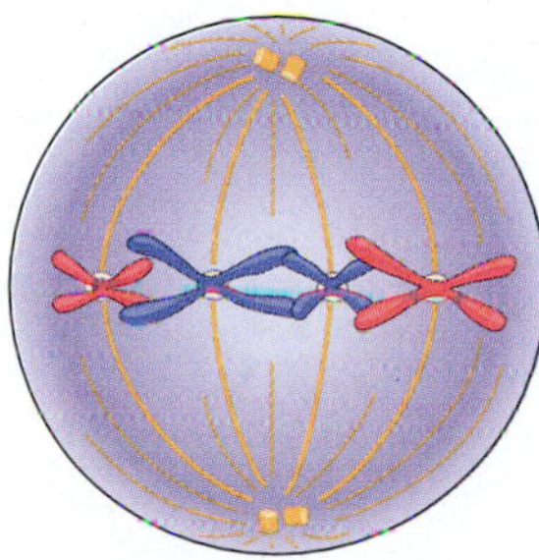

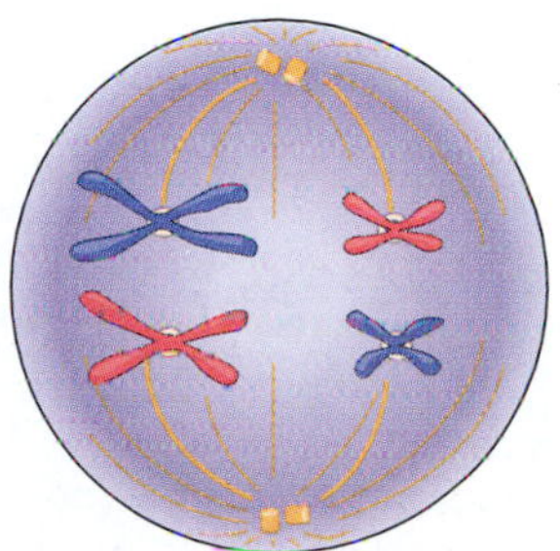

Thinking Scientifically

1. A population of lizards does not appear to contain any males, yet young are being born. If analysis shows that the offspring are diploid and have the same genetic traits as their mothers, what is the most likely source of maternal DNA to fertilize the eggs? How would you test this hypothesis?
2. During anaphase I of meiosis, spindle fibers pull the homologues of a bivalent in opposite directions. Explain with reference to attachment of spindle fibers to kinetochores. (see Figure 10.6)

Bioethical Issue

Cloning is making exact multiple copies of DNA or a cell or an organism. The first two procedures have been around for some time. Through biotechnology, bacteria produce cloned copies of human DNA. When a single bacterium reproduces asexually in a petri dish, a colony results. Each member of the colony is a clone of the original cell. Now, for the first time in our history, it is possible to produce a clone of an organism. No sperm and egg are required. The DNA of an adult cell is placed in an enucleated egg and the egg undergoes development to become an exact copy of the organism that donated the DNA. Some people fear that billionaires and celebrities will hasten to make multiple copies of themselves. Others feel that this is unlikely. Rather, they fear a different type of cloning.

Suppose it were possible to use the DNA of a burn victim to produce embryonic cells that are cajoled into becoming skin cells. These cells could be used to provide grafts of brand new skin. Would this be a proper use of cloning in humans?

Or suppose parents want to produce a child free of a genetic disease. Scientists produce a zygote through in vitro fertilization, and then they clone the zygote to produce any number of cells. Genetic engineering to correct the defect doesn't work on all the cells—only a few. They implant just those few in the uterus where development continues to term. Would this be a proper use of cloning in humans?

Well, what if science progressed to producing children with increased intelligence or athletic prowess in the same way? Would this be an acceptable use of cloning in humans? Presently, research in the cloning of humans is banned. Should it be? Why or why not?

Understanding the Terms

bivalent 161	life cycle 168
crossing-over 162	meiosis 160
diploid (2n) number 160	oogenesis 168
gamete 160	polar body 169
haploid (n) number 160	secondary oocyte 169
homologous chromosome 160	sexual reproduction 160
homologue 160	spermatogenesis 168
independent assortment 163	synapsis 161
	zygote 160

Match the terms to these definitions:

a. ______________ Production of sperm in males by the process of meiosis and maturation.

b. ______________ Pair of homologous chromosomes at the metaphase plate during meiosis I.

c. ______________ A nonfunctional product of oogenesis.

d. ______________ The functional product of meiosis I in oogenesis becomes the egg.

e. ______________ Member of a pair of chromosomes that carry genes for the same traits.

Web Connections

Exploring the Internet

http://www.mhhe.com/biosci/genbio/mader
(click on *Biology 7/e*)

The *Biology 7/e* Online Learning Center provides many resources for studying the material in this chapter including links to the following sites:

Meiosis and Genetic Recombination. This site is a section from the Mendelian genetics chapter of the MIT hypertextbook. Good discussion of crossing-over and recombination during meiosis.

http://esg-www.mit.edu:8001/esgbio/mg/meiosis.html

Meiosis Tutorial. This exercise shows the events of meiosis with both text and illustrations.

http://www.biology.arizona.edu/cell_bio/tutorials/meiosis/main.html

Meiosis—an Access Excellence short review of meiosis.

http://www.accessexcellence.org/AB/GG/meiosis.html

Meiosis. The first part of this laboratory exercise has a number of questions on meiosis that would serve as a good review.

http://www.hiline.net/~siremba/worksheets/meiosis.html

Lillium Meiosis. Labeled photomicrographs of the stages of meiosis in a lily.

http://www.dmacc.cc.ia.us/instructors/lillium.htm

Mendelian Patterns of Inheritance

chapter 11

chapter concepts

11.1 Gregor Mendel

- Mendel discovered certain laws of heredity after doing experiments with garden peas during the mid-1800s. 174

11.2 Monohybrid Inheritance

- The monohybrid cross told Mendel that each organism contains two factors for each trait and the factors segregate during formation of gametes. 176
- Today it is known that alleles located on chromosomes control the traits of individuals and homologous pairs of chromosomes separate during meiosis I. 177
- A testcross can be used to determine the genotype of an individual with the dominant phenotype. 180

11.3 Dihybrid Inheritance

- The dihybrid cross told Mendel that every possible combination of factors is present in the gametes. 181
- Today it is known that homologous pairs of chromosomes separate independently during meiosis I and this produces all possible combinations of alleles in the gametes. 182
- A testcross can also be used to determine the genotype of an individual that is dominant in two traits. 183

Like begets like.

The science of genetics explains why young giraffes resemble and have a combination of their parents' characteristics. An understanding of genetics has been acquired from studying a varied collection of organisms—peas, fruit flies, bread molds, bacteria, and humans.

Gregor Mendel, the father of genetics, chose the garden pea as his experimental material. He proposed that each parent donates particulate hereditary factors to offspring. Mendel worked in the nineteenth century, and his work was largely ignored until the twentieth century, when genetics took a modern turn. It wasn't until then that the term "gene" was coined and it was reasoned that the genes are on the chromosomes.

But what do genes do? Red bread mold experiments allowed Beadle and Tatum to conclude that genes in some way control the synthesis of enzymes. Not until the early 1950s did scientists know that genes are composed of DNA, and it was 1958 before Watson and Crick deduced the structure of DNA. From then on, work with the bacterium *Escherichia coli* brought us into the era of modern genetics and the ability to transform the genes of all organisms.

11.1 Gregor Mendel

Zebras always produce zebras, never bluebirds, and poppies always produce seeds for poppies, never dandelions. Almost everyone who observes such phenomena reasons that parents must pass hereditary information to their offspring. Many also observe, however, that offspring rarely resemble either parent exactly. After all, black-coated mice occasionally produce white-coated mice. The laws of heredity must explain not only the stability but also the variation that is observed between generations of organisms.

Gregor Mendel was an Austrian monk who formulated two fundamental laws of heredity in the early 1860s (Fig. 11.1). Previously, he had studied science and mathematics at the University of Vienna, and at the time of his genetic research he was a substitute natural science teacher at a local technical high school. Various hypotheses about heredity had been proposed before Mendel began his experiments. In particular, investigators were trying to support a blending concept of inheritance at this time.

Figure 11.1 Mendel working in his garden.
Mendel grew and tended the pea plants he used for his experiments. For each experiment, he observed as many offspring as possible. For a cross that required him to count the number of round seeds to wrinkled seeds, he observed and counted a total of 7,324 peas!

Blending Concept of Inheritance

When Mendel began his work, most plant and animal breeders acknowledged that both sexes contribute equally to a new individual. They felt that parents of contrasting appearance always produce offspring of intermediate appearance. Therefore, according to this concept, a cross between plants with red flowers and plants with white flowers would yield only plants with pink flowers. When red and white flowers reappeared in future generations, the breeders mistakenly attributed this to an instability in the genetic material.

A blending concept of inheritance offered little help to Charles Darwin, the father of evolution who wanted to give his ideas a genetic basis. If populations contained only intermediate individuals and normally lacked variations, how could diverse forms evolve? Only a particulate theory of inheritance, as proposed by Mendel, can account for the presence of discrete variations (differences) among the members of a population generation after generation. Although Darwin was a contemporary of Mendel, Darwin never learned of Mendel's work because it went unrecognized until 1900. Therefore, Darwin was never able to make use of the particulate theory of inheritance to support his theory of evolution.

At the time Mendel began his study of heredity, the blending concept of inheritance was popular.

Mendel's Experimental Procedure

Most likely his background in mathematics prompted Mendel to give a statistical basis to his breeding experiments. He prepared for his experiments carefully and conducted preliminary studies with various animals and plants. He then chose to work with the garden pea, *Pisum sativum* (Fig. 11.2*a*).

The garden pea was a good choice. The plants were easy to cultivate and had a short generation time. And although peas normally self-pollinate (pollen only goes to the same flower), they could be cross-pollinated by hand. Many varieties of peas were available, and Mendel chose twenty-two for his experiments. When these varieties self-pollinated, they were *true-breeding*—the offspring were like the parent plants and like each other. In contrast to his predecessors, Mendel studied the inheritance of relatively simple and distinguishable traits—seed shape, seed color, and flower color (Fig. 11.2*b*).

As Mendel followed the inheritance of individual traits, he kept careful records of the numbers of offspring that expressed each characteristic. And he used his understanding of the mathematical laws of probability to interpret the results. In other words, Mendel simply observed facts objectively, and if he had personal beliefs, he set them aside for the sake of the experiment. This is one of the qualities that make his experiments as applicable today as they were in 1860.

Mendel carefully designed his experiments and gathered mathematical data.

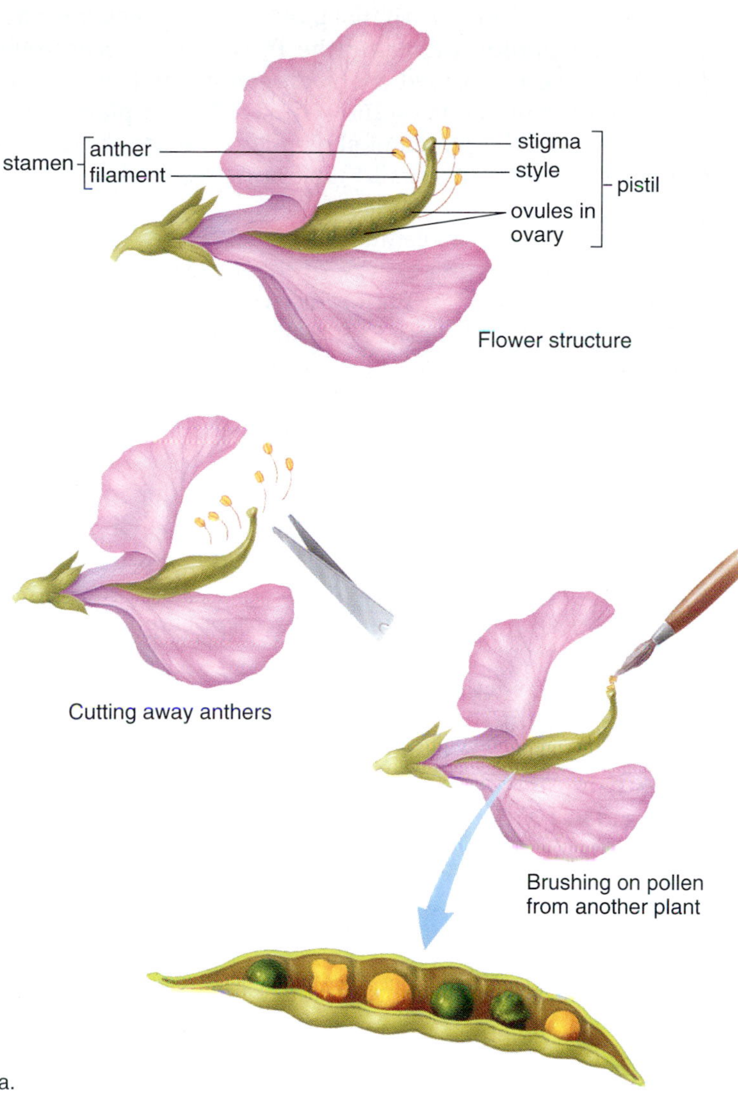

Trait	Characteristics		F_2 Results*	
			Dominant	Recessive
Stem length	Tall	Short	787	277
Pod shape	Inflated	Constricted	882	299
Seed shape	Round	Wrinkled	5,474	1,850
Seed color	Yellow	Green	7,022	2,001
Flower position	Axial	Terminal	651	207
Flower color	Purple	White	705	224
Pod color	Green	Yellow	428	152

*All of these produce approximately a 3:1 ratio. For example,

$$\frac{787}{277} \approx \frac{3}{1}$$

b.

Figure 11.2 Garden pea anatomy and traits.
a. In the garden pea, *Pisum sativum,* pollen grains produced in the anther contain sperm, and ovules in the ovary contain eggs. When Mendel did crosses he brushed pollen from one plant on the stigma of another plant. After sperm fertilized eggs, the ovules developed into seeds (peas). The open pod shows the results of a cross between plants with round, yellow seeds and plants with wrinkled, green seeds. **b.** Mendel selected these traits for study. He made sure his parental (P generation) plants bred true and then he cross-pollinated the plants. The offspring called F_1 (first filial) generation always resembled the parent with the dominant characteristic on the left. Mendel then allowed the F_1 plants to self-pollinate. In the F_2 (second filial) generation he always achieved a 3:1 (dominant to recessive) ratio. The text explains how Mendel went on to interpret these results.

11.2 Monohybrid Inheritance 

After ensuring that his pea plants were true-breeding, Mendel was then ready to perform a cross-pollination experiment between two strains. For these initial experiments, Mendel chose varieties that differed in only one trait. Crosses that involve only one trait are called monohybrid crosses. If the blending theory of inheritance were correct, then the cross should yield offspring with an intermediate appearance compared to the parents. For example, the offspring of a cross between a tall plant and short plant should be intermediate in height.

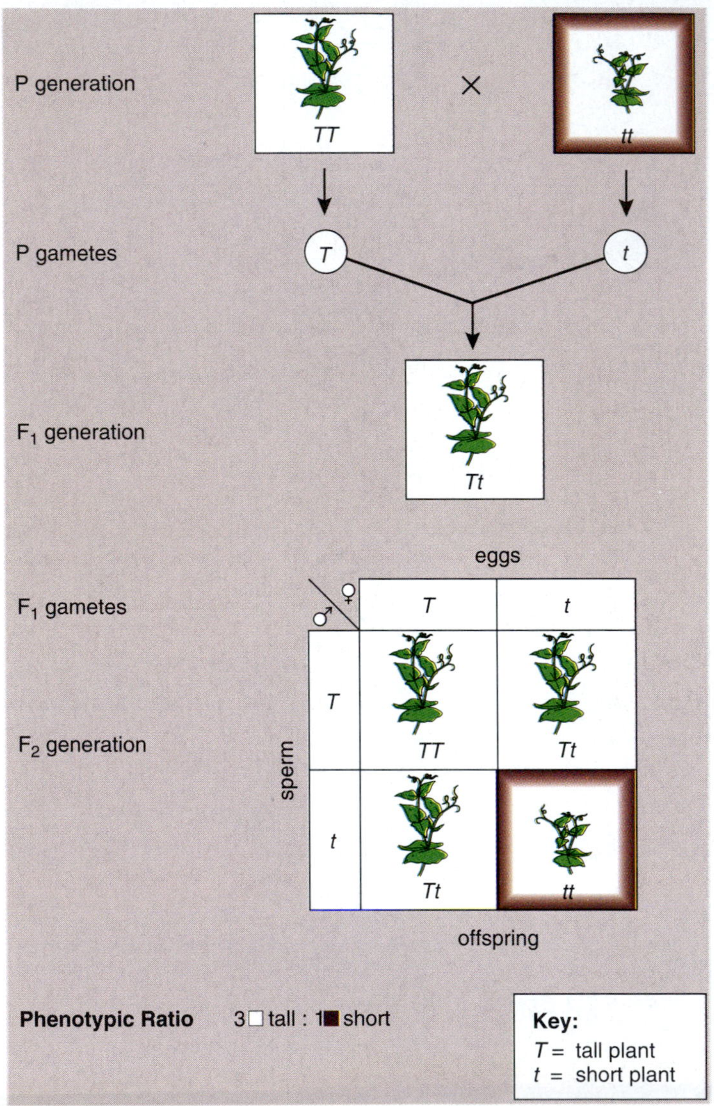

Figure 11.3 Monohybrid cross done by Mendel.
The P generation plants differ in one regard—length of the stem. The F_1 generation are all tall, but the factor for short has not disappeared because ¼ of the F_2 generation are short. The 3:1 ratio allowed Mendel to deduce that individuals have two discrete and separate genetic factors for each trait.

Mendel called the original parents the *P generation* and the first-generation offspring the *F_1* (for filial) *generation* (Fig. 11.3). He performed *reciprocal crosses:* first he dusted the pollen of tall plants on the stigmas of short plants, and then he dusted the pollen of short plants on the stigmas of tall plants. In both cases, all F_1 offspring resembled the tall parent.

Certainly, these results were contrary to those predicted by the blending theory of inheritance. Rather than being intermediate, the F_1 offspring were tall and resembled only one parent. Did these results mean that the other characteristic (i.e., shortness) had disappeared permanently? Apparently not, because when Mendel allowed the F_1 plants to self-pollinate, ¾ of the *F_2 generation* were tall and ¼ were short, a 3:1 ratio (Fig. 11.3).

Mendel counted many plants. For this particular cross, he counted a total of 1,064 plants, of which 787 were tall and 277 were short. In all crosses that he performed, he found a 3:1 ratio in the F_2 generation. The characteristic that had disappeared in the F_1 generation reappeared in ¼ of the F_2 offspring.

His mathematical approach led Mendel to interpret his results differently from previous breeders. He knew that the same ratio was obtained among the F_2 generation time and time again for the same type cross despite the particular trait, and he sought an explanation. A 3:1 ratio among the F_2 offspring was possible if:

1. the F_1 parents contained two separate copies of each hereditary factor, one of these being dominant and one being recessive;
2. the factors separated when the gametes were formed, and each gamete carried only one copy of each factor; and
3. random fusion of all possible gametes occurred upon fertilization.

In this way, Mendel arrived at the first of his laws of inheritance—the law of segregation.

Mendel's law of segregation:
Each organism contains two factors for each trait, and the factors segregate during the formation of gametes so that each gamete contains only one factor for each trait.

Mendel's law of segregation is in keeping with a particulate theory of inheritance because many individual factors are passed on from generation to generation. It is the reshuffling of these factors that explains how variations come about and why offspring differ from their parents.

As Viewed by Modern Genetics

Figure 11.3 also shows how we now interpret the results of Mendel's experiments on inheritance of stem length in peas. Each trait in a pea plant is controlled by two **alleles** [Gk. *allelon,* reciprocal, parallel], alternate forms of a gene that in this case control the length of the stem. In genetic notation, the alleles are identified by letters, the **dominant allele** (so named because of its ability to mask the expression of the other allele) with an uppercase (capital) letter and the **recessive allele** with the same but lowercase (small) letter. With reference to the cross being discussed, there is an allele for tallness (*T*) and an allele for shortness (*t*). Alleles occur on a homologous pair of chromosomes at a particular location that is called the **gene locus** (Fig. 11.4).

During meiosis, the type of cell division that reduces the chromosome number, the homologous chromosomes of a bivalent separate during meiosis I. Therefore, the process of meiosis gives an explanation for Mendel's law of segregation and for the reason there is only one allele for each trait in the gametes.

In Mendel's cross, the original parents (P generation) were true-breeding; therefore, the tall plants had two alleles for tallness (*TT*) and the short plants had two alleles for shortness (*tt*). When an organism has two identical alleles, as these had, we say it is **homozygous** [Gk. *homo,* same, and *zygos,* balance, yoke]. Because the parents were homozygous, all gametes produced by the tall plant contained the allele for tallness (*T*), and all gametes produced by the short plant contained an allele for shortness (*t*).

After cross-pollination, all the individuals of the resulting F_1 generation had one allele for tallness and one for shortness (*Tt*). When an organism has two different alleles at a gene locus, we say that it is **heterozygous** [Gk. *hetero,* different, and *zygos,* balance, yoke]. Although the plants of the F_1 generation had one of each type of allele, they were all tall. The allele that is expressed in a heterozygous individual is the dominant allele. The allele that is not expressed in a heterozygote is a recessive allele.

Genotype Versus Phenotype

It is obvious from our discussion that two organisms with different allelic combinations for a trait can have the same outward appearance (*TT* and *Tt* pea plants are both tall). For this reason, it is necessary to distinguish between the alleles present in an organism and the appearance of that organism.

The word **genotype** [Gk. *genos,* birth, origin, race, and *typos,* image, shape] refers to the alleles an individual receives at fertilization. Genotype may be indicated by letters or by short, descriptive phrases. Genotype *TT* is called homozygous dominant, and genotype *tt* is called homozygous recessive. Genotype *Tt* is called heterozygous.

The word **phenotype** [Gk. *phaino,* appear, and *typos,* image, shape] refers to the physical appearance of the individual. The homozygous dominant individual and the heterozygous individual both show the dominant phenotype and are tall, while the homozygous recessive individual shows the recessive phenotype and is short (Table 11.1).

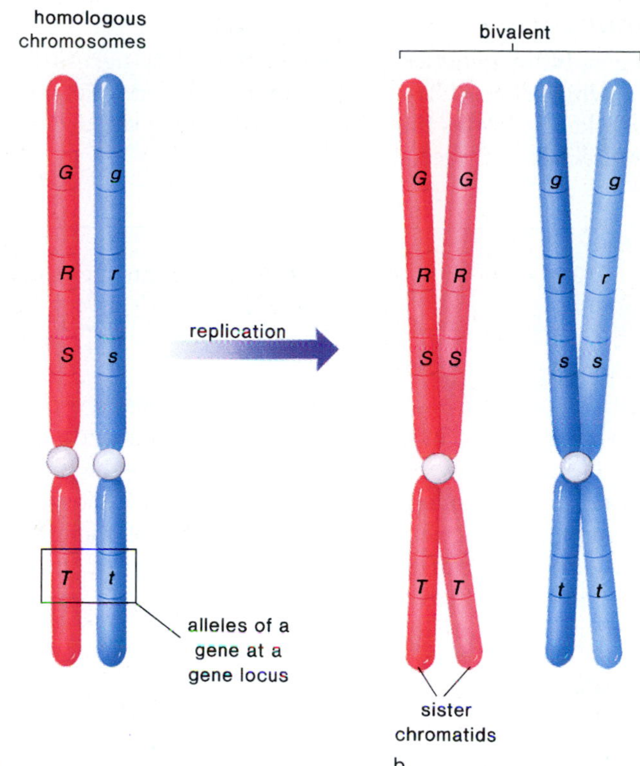

Figure 11.4 Homologous chromosomes.
a. The letters represent alleles; that is, alternate forms of a gene. Each allelic pair, such as *Gg* or *Tt,* is located on homologous chromosomes at a particular gene locus. **b.** Following DNA replication, each sister chromatid carries the same alleles in the same order.

Table 11.1

Genotype Versus Phenotype

Genotype	Genotype	Phenotype
TT	Homozygous dominant	Tall plant
Tt	Heterozygous	Tall plant
tt	Homozygous recessive	Short plant

Monohybrid Genetics Problems

When solving genetics problems, it is first necessary to know which characteristic is dominant. For example, the following key indicates that unattached earlobes are dominant over attached earlobes:

Key: *E* = unattached earlobes
e = attached earlobes

If a man, homozygous for unattached earlobes, reproduces with a woman who has attached earlobes, what type of earlobe will the child have? In row 1, P represents the parental generation, and the letters in this row are the genotypes of the parents. Row 2 shows that the gametes of each parent have only one type of allele for earlobes; therefore, the child (F generation) will have a heterozygous genotype and unattached earlobes:

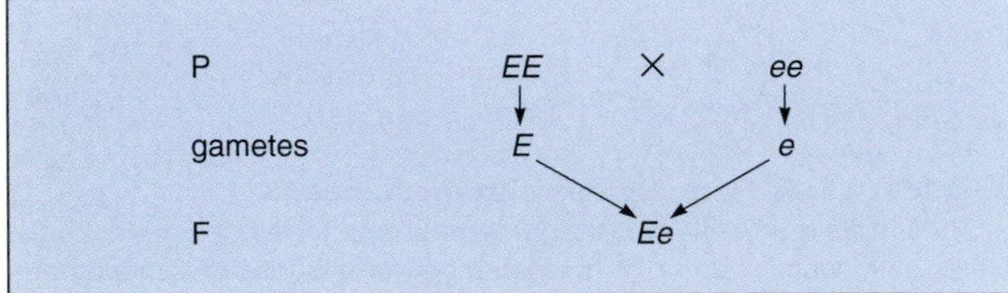

If two heterozygotes reproduce with one another, will the child have unattached or attached earlobes? In the P generation there was only one possible type of gamete for each parent because they were homozygous. Heterozygotes, however, can produce two types of gametes; ½ of the gametes will contain an *E*, and ½ will contain an *e*.

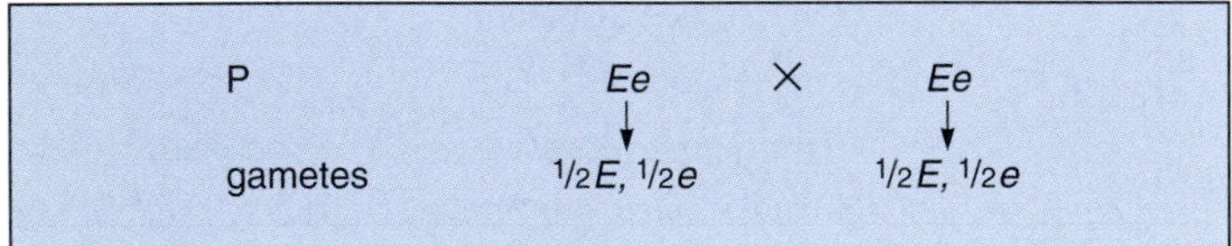

When determining the gametes, it is necessary to keep in mind that although an individual is diploid—that is, has two alleles for each trait—*each gamete is haploid—that is, has only one allele for each trait.* This is true of single-trait crosses as well as multiple-trait crosses.

When doing genetics problems, first decide on the appropriate key, and then determine the genotype of both parents and the various types of gametes for both parents.

Practice Problems 1 will help you learn to designate the gametes.

Practice Problems 1*

1. For each of the following genotypes, give all genetically different gametes, noting the proportion of each for the individual.
   a. *WW*
   b. *Ww*
   c. *Tt*
   d. *TT*
2. For each of the following, state whether a genotype (genetic makeup of an organism) or a type of gamete is represented.
   a. *D*
   b. *GG*
   c. *P*

** Answers to Practice Problems appear in Appendix A.*

Laws of Probability

When we calculate the expected results of the cross under consideration, or any genetic cross, we utilize the laws of probability. Imagine flipping a coin; *each time* you flip the coin there is a 50% chance of heads and a 50% chance of tails. In like manner, if the parent has the genotype *Ee*, what is the chance of any child inheriting either an *E* or *e* from that parent?

The chance of $E = 1/2$
The chance of $e = 1/2$

But in the cross *Ee* × *Ee*, a child will inherit an allele from both parents. The multiplicative law of probability states that *the chance, or probability, of two or more independent events occurring together is the product (multiplication) of their chance of occurring separately.* Therefore, the probability of receiving these genotypes is as follows:

1. The chance of $EE = 1/2 \times 1/2 = 1/4$
2. The chance of $Ee = 1/2 \times 1/2 = 1/4$
3. The chance of $eE = 1/2 \times 1/2 = 1/4$
4. The chance of $ee = 1/2 \times 1/2 = 1/4$

Now we have to consider the additive law of probability: *the chance of an event that can occur in two or more independent ways is the sum (addition) of the individual chances.* Therefore,

The chance of a child with unattached earlobes (*EE*, *Ee*, or *eE*) is ¾ (add **1, 2,** and **3**), or 75%.
The chance of a child with attached earlobes (*ee*) is ¼ (only **4**), or 25%.

The Punnett Square

The **Punnett square** was introduced by a prominent poultry geneticist, R. C. Punnett, in the early 1900s as a simple method to figure the probable results of a genetic cross. Figure 11.5 shows how the results of the cross under consideration (*Ee* × *Ee*) can be determined using a Punnett square. In a Punnett square all possible types of sperm are lined up vertically and all possible types of eggs are lined up horizontally (or vice versa), and every possible combination of alleles is placed within the squares. In our cross each parent has two possible types of gametes (*E* or *e*), so these two types of gametes are lined up vertically and horizontally.

The results of the Punnett square calculations show that the expected genotypes of offspring are ¼ *EE*, ½ *Ee*, and ¼ *ee*, giving a 1:2:1 genotypic ratio. Since ¾ have unattached earlobes and ¼ have attached earlobes, this is a 3:1 phenotypic ratio.

The gametes combine at random, and in practice it is usually necessary to observe a very large number of offspring before a 3:1 ratio can be verified. Only if many offspring are counted can it be assured that all genetically different sperm have had a chance to fertilize all genetically different eggs. If a number of heterozygotes produced 200 offspring, approximately 150 of these would have unattached earlobes and approximately 50 would have attached earlobes (a 3:1 ratio). In terms of genotypes, approximately 50 would be *EE*, about 100 would be *Ee*, and the remaining 50 would be *ee*.

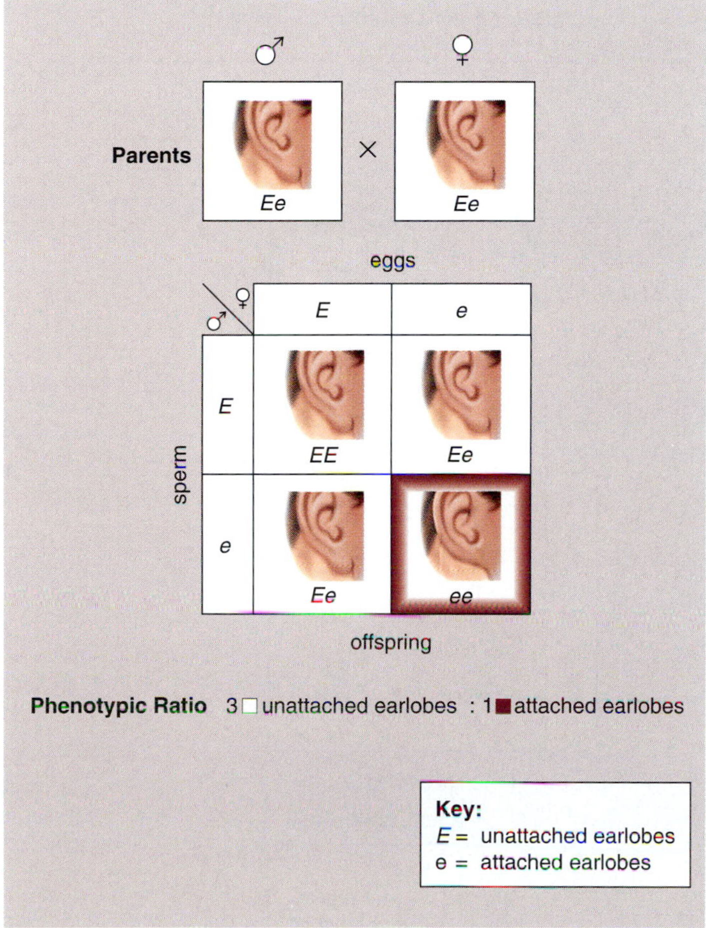

Figure 11.5 Genetic inheritance in humans.
When the parents are heterozygous, each child has a 75% chance of having the dominant phenotype and a 25% chance of having the recessive phenotype.

We cannot arrange crosses between humans in order to count a large number of offspring. Therefore, in humans the phenotypic ratio is used to estimate the chances any child has for a particular characteristic. In the cross under consideration, each child has a ¾ (or 75%) chance of having unattached earlobes and ¼ (or 25%) chance of having attached earlobes. And we must remember that *chance has no memory:* if two heterozygous parents already have a child with attached earlobes, the next child still has a 25% chance of having attached earlobes.

You will note that the Punnett square makes use of the laws of probability we mentioned in the previous section. First, we recognize there is a ½ chance of a child getting the *E* allele or the *e* allele from a parent. In this way we are able to designate the possible gametes. What operation is in keeping with the multiplicative law of probability? The operation of combining the alleles and placing them in the squares. What operation is in keeping with the additive law of probability? The operation of adding the results, and arriving at the phenotypic ratio.

Mendel was quite familiar with the laws of probability and was able to make use of them to see that inheritance of traits depended upon the passage of discrete factors from generation to generation.

The laws of probability allow one to calculate the probable results of one-trait genetic crosses.

Practice Problems 2*

1. In rabbits, if *B* = dominant black allele and *b* = recessive white allele, which of these genotypes (*Bb*, *BB*, *bb*) could a white rabbit have?
2. In pea plants, yellow seed color is dominant over green seed color. When two heterozygote plants are crossed, what percentage of plants would have yellow seeds? green seeds?
3. In humans, freckles is dominant over no freckles. A man with freckles reproduces with a woman with freckles, but the children have no freckles. What chance did each child have for freckles?
4. In horses, trotter (*T*) is dominant over pacer (*t*). A trotter is mated to a pacer, and the offspring is a pacer. Give the genotype of all horses.

* *Answers to Practice Problems appear in Appendix A.*

One-Trait Testcross

To test his idea about the segregation of alleles—in modern terms, that the F_1 was heterozygous—Mendel crossed his F_1 generation tall plants with true-breeding, short (homozygous recessive) plants. He reasoned that half the offspring should be tall and half should be short, producing a 1:1 phenotypic ratio (Fig. 11.6*a*). He obtained these results; therefore, his hypothesis that alleles segregate when gametes are formed was supported. It was Mendel's experimental use of simple dominant and recessive traits that allowed him to formulate and to test the law of segregation.

In Figure 11.6, the homozygous recessive parent can produce only one type of gamete—*t*—and so the same results would be obtained if the Punnett square had only one column. This is logical because the *t* means all the gametes that carry a *t*.

Today, a monohybrid **testcross** is used to determine if an individual with the dominant phenotype is homozygous dominant or heterozygous for a particular trait. Since both of these genotypes produce the dominant phenotype, it is not possible to determine the genotype by inspection. Figure 11.6*b* shows that if the F_1 in Mendel's cross had been homozygous dominant, then all the offspring would have been tall.

The results of a testcross indicate whether an individual with the dominant phenotype is heterozygous or homozygous dominant.

Practice Problems 3*

1. In horses, *B* = black coat and *b* = brown coat. What type of cross should be done to best determine whether a black-coated horse is homozygous dominant or heterozygous?
2. In fruit flies, *L* = long wings and *l* = short wings. The offspring exhibit a 1:1 ratio when a long-winged fly is crossed with a short-winged fly. What is the genotype of the parental flies?
3. In the garden pea, round seeds are dominant over wrinkled seeds. An investigator crosses a plant having round seeds with a plant having wrinkled seeds. He counts 400 offspring. How many of the offspring have wrinkled seeds if the plant having round seeds is a heterozygote?

** Answers to Practice Problems appear in Appendix A.*

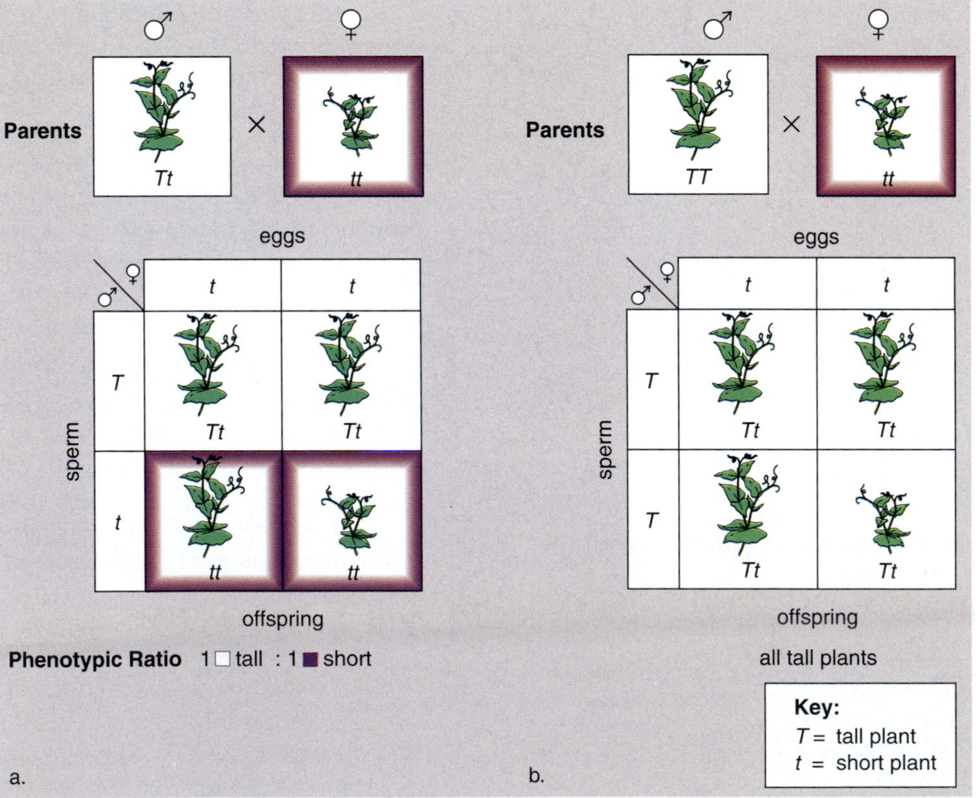

Figure 11.6 **Testcross.**
Crossing an individual with the dominant phenotype with a recessive individual indicates the genotype. **a.** If a parent with the dominant phenotype is heterozygous, the phenotypic ratio among the offspring is 1:1. **b.** If a parent with the dominant phenotype is homozygous, all offspring have the dominant phenotype.

11.3 Dihybrid Inheritance

Mendel performed a second series of crosses in which he crossed true-breeding plants that differed in two traits; such crosses are appropriately called dihybrid crosses. For example, he crossed tall plants having green pods with short plants having yellow pods (Fig. 11.7). The F_1 plants showed both dominant characteristics. As before, Mendel then allowed the F_1 plants to self-pollinate. Two possible results could occur in the F_2 generation:

1. If the dominant factors (*TG*) always segregate into the F_1 gametes together, and the recessive factors (*tg*) always stay together, then there would be two phenotypes among the F_2 plants—tall plants with green pods and short plants with yellow pods.
2. If the four factors segregate into the F_1 gametes independently, then there would be four phenotypes among the F_2 plants—tall plants with green pods, tall plants with yellow pods, short plants with green pods, and short plants with yellow pods.

Figure 11.7 shows that Mendel observed four phenotypes among the F_2 plants, supporting the second hypothesis. Therefore, Mendel formulated his second law of heredity—the law of independent assortment.

Mendel's law of independent assortment: Members of one pair of factors separate (assort) independently of members of another pair of factors. Therefore, all possible combinations of factors can occur in the gametes.

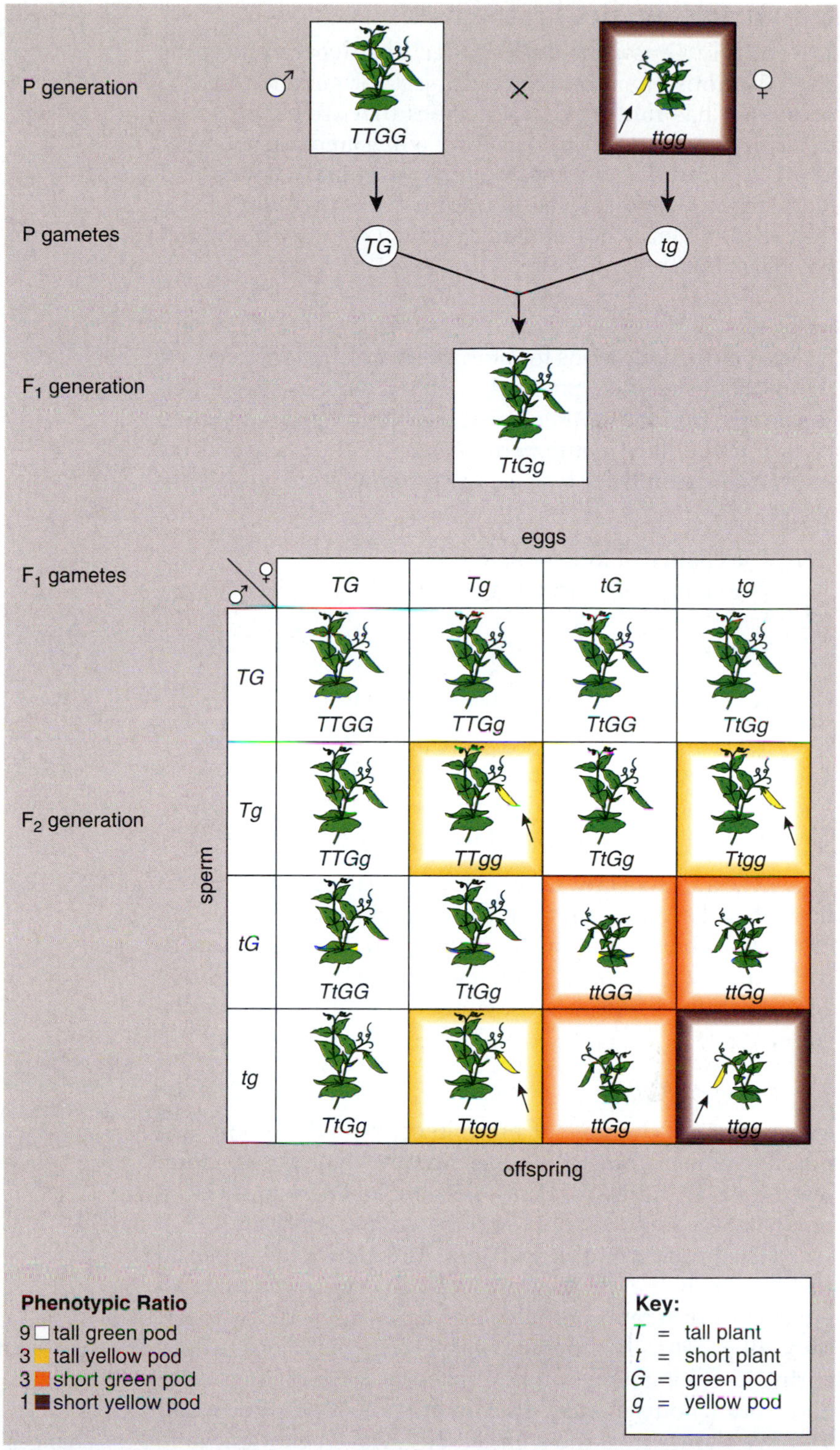

Practice Problems 4*

1. For each of the following genotypes, give all possible gametes, noting the proportion of each gamete for the individual.
   a. *TtGG*
   b. *TtGg*
   c. *TTGg*
2. For each of the following, state whether a genotype (genetic makeup of an organism) or a type of gamete is represented.
   a. *Tg*
   b. *WwCC*
   c. *TW*

* *Answers to Practice Problems appear in Appendix A.*

Phenotypic Ratio
9 tall green pod
3 tall yellow pod
3 short green pod
1 short yellow pod

Key:
T = tall plant
t = short plant
G = green pod
g = yellow pod

Figure 11.7 Dihybrid cross done by Mendel.
P generation plants differ in two regards—length of the stem and color of the pod. The F_1 generation shows only the dominant traits, but all possible phenotypes appear among the F_2 generation. The 9:3:3:1 ratio allowed Mendel to deduce that factors segregate into gametes independently of other factors.

Dihybrid Genetics Problems

The fruit fly, *Drosophila melanogaster*, less than one-fifth the size of a housefly, is a favorite subject for genetic research because it has mutant characteristics that are easily determined. A "wild-type" fly has long wings and a gray body. There are mutant flies with short (vestigial) wings and black (ebony) bodies. The key for a cross involving these traits is L = long wing, l = short wing, G = gray body, and g = black body.

Laws of Probability

If two flies heterozygous for both traits are crossed, what are the probable results? Since each characteristic is inherited separately from any other, it is possible to apply again the laws of probability mentioned on page 178. For example, we know the F_2 results for two separate monohybrid crosses are as listed here:

1. The chance of long wings = $3/4$
   The chance of short wings = $1/4$
2. The chance of gray body = $3/4$
   The chance of black body = $1/4$

Using the multiplicative law, we know that:

The chance of long wings and gray body = $3/4 \times 3/4 = 9/16$
The chance of long wings and black body = $3/4 \times 1/4 = 3/16$
The chance of short wings and gray body = $1/4 \times 3/4 = 3/16$
The chance of short wings and black body = $1/4 \times 1/4 = 1/16$

Using the additive law, we conclude that the phenotypic ratio is 9:3:3:1. Again, since all genetically different male gametes must have an equal opportunity to fertilize all genetically different female gametes to even approximately achieve these results, a large number of offspring must be counted.

The Punnett Square

In Figure 11.8, the flies of the P generation have only one possible type of gamete because they are homozygous dominant for both traits. All the F_1 flies are heterozygous (*LlGg*) and have the same phenotype (long wings, gray body).

The Punnett square in Figure 11.8 shows the expected results when the F_1 flies are crossed, assuming that all genetically different sperm have an equal opportunity to fertilize all genetically different eggs. Notice that $9/16$ of the offspring have long wings and a gray body, $9/16$ have long wings and a black body, $3/16$ have short wings and a gray body, and $1/16$ have short wings and a black body. This phenotypic ratio of 9:3:3:1 is expected whenever a heterozygote for two traits is crossed with another heterozygote for two traits and simple dominance is present in both genes.

A Punnett square can also be used to predict the chances of an offspring having a particular phenotype. What are the chances of an offspring with long wings and gray body? The chances are $9/16$. What are the chances of an offspring with short wings and gray body? The chances are $3/16$, and so forth.

Figure 11.8 Inheritance in fruit flies.
Each F_1 fly (*LlGg*) produces four types of gametes because all possible combinations of alleles can occur in the gametes. Therefore, all possible phenotypes appear among the F_2 offspring.

Today, we know that these results are obtained and the law of independent assortment holds because of the events of meiosis. The gametes contain one allele for each trait and in all possible combinations because homologues separate independently during meiosis I.

Two-Trait Testcross

A dihybrid testcross is used to determine if an individual is homozygous dominant or heterozygous for either of the two traits. Since it is not possible to determine the genotype of a long-winged, gray-bodied fly by inspection, the genotype may be represented as *L*__[1] *G*__.

When doing a dihybrid testcross, an individual with the dominant phenotype is crossed with an individual with the recessive phenotype. For example, a long-winged, gray-bodied fly is crossed with a short-winged, black-bodied fly. A long-winged, gray-bodied fly heterozygous for both traits will form four different types of gametes. The homozygous fly with short wings and a black body can form only one kind of gamete:

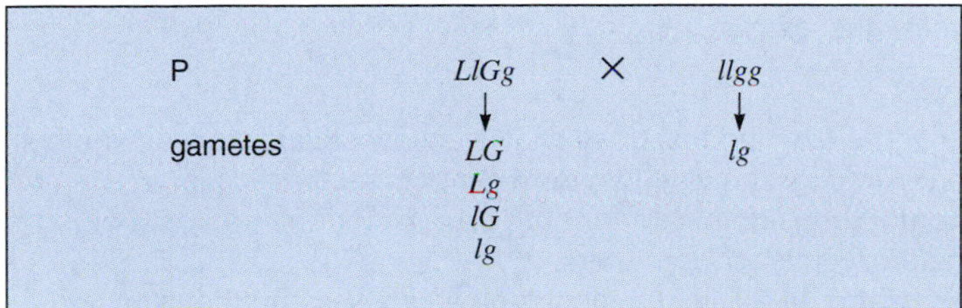

Figure 11.9 shows that ¼ of the expected offspring have long wings and a gray body; ¼ have long wings and a black body; ¼ have short wings and a gray body; and ¼ have short wings and a black body. This is a 1:1:1:1 phenotypic ratio. The presence of offspring with short wings and a black body shows that the *L*__*G*__ fly is heterozygous for both traits and has the genotype *LlGg*.

If the *L*__*G*__ fly is homozygous for both traits, then no offspring will have short wings or a black body when this fly is crossed with one that has the recessive phenotype for both traits. If the *L*__*G*__ fly is heterozygous for one trait but not the other, what is the expected phenotypic ratio among the offspring when this fly is crossed with one that has the recessive phenotype for both traits?

In dihybrid genetics problems, the individual has four alleles, two for each trait.

P generation ♂ *LlGg* × *llgg* ♀

eggs: *lg*; sperm: *LG*, *Lg*, *lG*, *lg*; offspring: *LlGg*, *Llgg*, *llGg*, *llgg*

Phenotypic Ratio
1 long wings, gray body
1 long wings, black body
1 short wings, gray body
1 short wings, black body

Key:
L = long wings
l = short wings
G = gray body
g = black body

Figure 11.9 Testcross.
Testcross to determine if a fly is heterozygous for both traits. If a fly heterozygous for both traits is crossed with a fly that is recessive for both traits, the expected ratio of phenotypes is 1:1:1:1. What would the result be if the test individual were homozygous dominant for both traits? homozygous dominant for one trait but heterozygous for the other?

[1] The blank means that a dominant or recessive allele can be present.

Practice Problems 5*

1. In horses, *B* = black coat, *b* = brown coat, *T* = trotter, and *t* = pacer. A black pacer mated to a brown trotter produces a black trotter offspring. Give all possible genotypes for this offspring.
2. In fruit flies, long wings (*L*) is dominant over short wings (*l*), and gray body (*G*) is dominant over black body (*g*). In each instance, what are the most likely genotypes of the previous generation if a student gets the following phenotypic results?
   a. 1:1:1:1 (all possible combinations in equal number)
   b. 9:3:3:1 (9 dominant; 3 mixed; 3 mixed; 1 recessive)
3. In humans, short fingers and widow's peak are dominant over long fingers and straight hairline. A heterozygote in both regards reproduces with a similar heterozygote. What is the chance of any one child having the same phenotype as the parents?

* *Answers to Practice Problems appear in Appendix A.*

Connecting Concepts

Humans have practiced selective plant and animal breeding since agriculture and animal domestication began thousands of years ago. Many early breeders came to realize that their programs were successful because they were manipulating inherited factors for particular characteristics. A good experimental design and a bit of luck allowed Mendel, and not one of these breeders, to work out his rules of inheritance.

The use of a large number of pea crosses and a statistical analysis of the large number of resulting offspring contributed to the success of Mendel's work. Although humans do not usually produce a large number of offspring it has been possible to come to the conclusion that Mendel's laws do apply to humans in many instances. Inheritance in large families that keep good historical records, such as Mormon families, has allowed researchers to show that a number of human genetic disorders are indeed controlled by a single allelic pair. Such disorders include cystic fibrosis, sickle-cell disease, albinism, achondroplasia (a form of dwarfism), neurofibromatosis, and Huntington disease.

Mendel was lucky in that he chose to study an organism, namely the garden pea, whose observable traits are often determined by a single allelic pair. In the common garden bean on the other hand, plant height, for example, is determined by several genes. Therefore, crosses of tall and short plants result in F_1 plants of intermediate height rather than all tall plants. This type of inheritance pattern, called polygenic, controls *many* important traits in farm animals, such as weight and milk production in cattle. You will learn more about polygenic and other types of inheritance patterns that differ from simple Mendelian inheritance in chapter 12.

Summary

11.1 Gregor Mendel

At the time Mendel began his hybridization experiments, the blending concept of inheritance was popular. This concept stated that whenever the parents were distinctly different, the offspring are intermediate between them. In contrast to preceding plant breeders, Mendel decided to do a statistical study, and he chose to study just seven distinctive traits of the garden pea.

11.2 Monohybrid Inheritance

When Mendel did his monohybrid crosses, he found that the recessive phenotype reappeared in about 1/4 of the F_2 plants. This allowed Mendel to deduce his law of segregation, which states that the individual has two factors for each trait and the factors segregate into the gametes.

The laws of probability can be used to calculate the expected phenotypic ratio of a cross. In practice, a large number of offspring must be counted in order to observe the expected results, because only in that way can it be ensured that all possible types of sperm have fertilized all possible types of eggs.

Because humans do not produce a large number of offspring, it is best to use a predicted ratio as a means of estimating the chances of an individual inheriting a particular characteristic.

Mendel also crossed the F_1 plants having the dominant phenotype with homozygous recessive plants. The results indicated that the recessive factor was present in these F_1 plants (i.e., that they were heterozygous). Today, we call this a testcross, because it is used to test whether an individual showing the dominant characteristic is homozygous dominant or heterozygous.

11.3 Dihybrid Inheritance

Mendel did dihybrid crosses, in which the F_1 individuals showed both dominant characteristics, but there were four phenotypes among the F_2 offspring. This allowed Mendel to deduce the law of independent assortment, which states that the members of one pair of factors separate independently of those from another pair. Therefore, all possible combinations of factors can occur in the gametes.

With regard to Mendel's dihybrid crosses, the laws of probability show that 9/16 of the F_2 offspring have the two dominant traits, 3/16 have one dominant trait with one recessive trait, 3/16 have the other dominant trait with the other recessive trait, and 1/16 have both recessive traits, for a 9:3:3:1 ratio.

The dihybrid testcross allows an investigator to test whether an individual showing two dominant characteristics is homozygous dominant for both traits or for one trait only, or is heterozygous for both traits.

Table 11.2 lists the most typical crosses and the expected results. Learning these saves the trouble of working out the results repeatedly for each of these crosses.

Table 11.2

Common Crosses Involving Simple Dominance

Examples	Phenotypic Ratios
Tt × *Tt*	3:1 (dominant to recessive)
Tt × *tt*	1:1 (dominant to recessive)
TtYy × *TtYy*	9:3:3:1 (9 dominant; 3 mixed; 3 mixed; 1 recessive)
TtYy × *ttyy*	1:1:1:1 (all possible combinations in equal number)

Reviewing the Chapter

1. How did Mendel's procedure differ from that of his predecessors? What mechanism did he use to set aside any personal beliefs he may have had? 174
2. How did the monohybrid crosses performed by Mendel refute the blending concept of inheritance? 174
3. Using Mendel's monohybrid cross as an example, trace his reasoning to arrive at the law of segregation. 176
4. Define these terms: allele, dominant allele, recessive allele, genotype, homozygous, heterozygous, and phenotype. 177
5. Use a Punnett square and the laws of probability to show the results of a cross between two individuals who are heterozygous for one trait. What are the chances of an offspring having the dominant phenotype? the recessive phenotype? 179
6. In what way does a monohybrid testcross support Mendel's law of segregation? 180
7. How is a monohybrid testcross used today? 180
8. Using Mendel's dihybrid cross as an example, trace his reasoning to arrive at the law of independent assortment. 181

9. Use a Punnett square and the laws of probability to show the results of a cross between two individuals who are heterozygous for two traits. What are the chances of an offspring having the dominant phenotype for both traits? having the recessive phenotype for both traits? 182
10. What would be the results of a dihybrid testcross if an individual was heterozygous for two traits? heterozygous for one trait and homozygous dominant for the other? homozygous dominant for both traits? 183

Testing Yourself

Choose the best answer for each question. For questions 1–4, match the cross with the results in the key:

Key:

a. 3:1
b. 9:3:3:1
c. 1:1
d. 1:1:1:1
e. 3:1:3:1

1. *TtYy* × *TtYy*
2. *Tt* × *Tt*
3. *Tt* × *tt*
4. *TtYy* × *ttyy*
5. Which of these could be a normal gamete?
   a. *GgRr*
   b. *GRr*
   c. *Gr*
   d. *GgR*
   e. None of these are correct.
6. Which of these properly describes a cross between an individual who is homozygous dominant for hairline but heterozygous for finger length and an individual who is recessive for both characteristics? (*W* = widow's peak, *w* = straight hairline, *S* = short fingers, *s* = long fingers)
   a. *WwSs* × *WwSs*
   b. *WWSs* × *wwSs*
   c. *Ws* × *ws*
   d. *WWSs* × *wwss*
7. In peas, yellow seed (*Y*) is dominant over green seed (*y*). In the F_2 generation of a monohybrid cross that begins when a dominant homozygote is crossed with a recessive homozygote, you would expect
   a. plants that produce three yellow seeds to every green seed.
   b. plants with one yellow seed for every green seed.
   c. only plants with the genotype *Yy*.
   d. only plants that produce yellow seeds.
   e. Both c and d are correct.
8. In humans, pointed eyebrows (*B*) are dominant over smooth eyebrows (*b*). Mary's father has pointed eyebrows, but she and her mother have smooth. What is the genotype of the father?
   a. *BB*
   b. *Bb*
   c. *bb*
   d. *BbBb*
   e. Any one of these is correct.
9. In guinea pigs, smooth coat (*S*) is dominant over rough coat (*s*) and black coat (*B*) is dominant over white coat (*b*). In the cross *SsBb* × *SsBb*, how many of the offspring will have a smooth black coat on average?
   a. 9 only
   b. about 9/16
   c. 1/16
   d. 6/16
   e. 2/6
10. In horses, *B* = black coat, *b* = brown coat, *T* = trotter, and *t* = pacer. A black trotter that has a brown pacer offspring is
    a. *BT*.
    b. *BbTt*.
    c. *bbtt*.
    d. *BBtt*.
    e. *BBTT*.
11. In tomatoes, red fruit (*R*) is dominant over yellow fruit (*r*) and tallness (*T*) is dominant over shortness (*t*). A plant that is *RrTT* is crossed with a plant that is *rrTt*. What are the chances of an offspring being heterozygous for both traits?
    a. none
    b. 1/2
    c. 1/4
    d. 9/16
    e. 1/3
12. In the cross *RrTt* × *rrtt*,
    a. all the offspring will be tall with red fruit.
    b. 75% (3/4) will be tall with red fruit.
    c. 50% (1/2) will be tall with red fruit.
    d. 25% (1/4) will be tall with red fruit.

Additional Genetics Problems*

1. If a man homozygous for widow's peak (dominant) reproduces with a woman homozygous for straight hairline (recessive), what are the chances of their children having a widow's peak? a straight hairline?
2. John has unattached earlobes (recessive) like his father, but his mother has attached earlobes (dominant). What is John's genotype?
3. In humans, the allele for short fingers is dominant over that for long fingers. If a person with short fingers who had one parent with long fingers reproduces with a person having long fingers, what are the chances of each child having short fingers?
4. In a fruit fly experiment (see key on page 182), two gray-bodied fruit flies produce mostly gray-bodied offspring, but some offspring have black bodies. If there are 280 offspring, how many do you predict will have gray bodies and how many will have black bodies? How many of the 280 offspring do you predict will be heterozygous? If you wanted to test whether a particular gray-bodied fly was homozygous dominant or heterozygous, what cross would you do?
5. Using the term 2n in which n = the number of heterozygous gene pairs, determine the number of possible gametes for the genotypes *BBHH*, *BbHh*, and *BBHh*.

* Answers to Additional Genetics Problems appear in Appendix A.

6. In rabbits, black color (*B*) is dominant over brown (*b*) and short hair (*S*) is dominant over long (*s*). In a cross between a homozygous black, long-haired rabbit and a brown, homozygous short-haired one, what would the F_1 generation look like? the F_2 generation? If one of the F_1 rabbits reproduced with a brown, long-haired rabbit, what phenotypes and in what ratio would you expect?
7. In horses, black coat (*B*) is dominant over brown coat (*b*) and being a trotter (*T*) is dominant over being a pacer (*t*). A black pacer is crossed with a brown trotter. The offspring is a brown pacer. Give the genotypes of all these horses.
8. The complete genotype of a long-winged, gray-bodied fruit fly is unknown (see key on page 182). When this fly is crossed with a short-winged, black-bodied fruit fly, the offspring all have a gray body but about half of them have short wings. What is the genotype of the long-winged, gray-bodied fly?
9. In humans, widow's peak hairline is dominant over straight hairline, and short fingers are dominant over long fingers. If an individual who is heterozygous for both traits reproduces with an individual who is recessive for both traits, what are the chances of their child also being recessive for both traits?

Thinking Scientifically

1. *Drosophila* that are homozygous recessive for the apterous (*apt*) mutation have no wings. Flies that are homozygous recessive for the vestigial (*vg*) mutation have small, nonfunctional wings. If these traits are alternative alleles of the same gene, what phenotype would you predict in the offspring from a cross between vestigial and apterous flies? If instead you obtain all flies with normal wings, what would you hypothesize about the number of genes that control wing development?
2. Mendel's paper reporting his results and conclusions included very detailed records of all of the crosses and offspring. The first parts of his paper show that the ratios he measured were not exactly 3:1 or 9:3:3:1. However, the data he shows for the last section is almost exactly what he had predicted. What would be a possible interpretation for this curious phenomenon?

Understanding the Terms

allele 177	homozygous 177
dominant allele 177	phenotype 177
gene locus 177	Punnett square 179
genotype 177	recessive allele 177
heterozygous 177	testcross 180

Match the terms to these definitions:

a. ______________ Allele that exerts its phenotypic effect only in the homozygote; its expression is masked by a dominant allele.

b. ______________ Alternative forms of a gene—which occur at the same locus on homologous chromosomes.

c. ______________ Allele that exerts its phenotypic effect in the heterozygote; it masks the expression of the recessive allele.

d. ______________ Cross between an individual with the dominant phenotype and an individual with the recessive phenotype to see if the individual with the dominant phenotype is homozygous or heterozygous.

e. ______________ Genes of an organism for a particular trait or traits; for example, *BB* or *Aa*.

Web Connections

Exploring the Internet

http://www.mhhe.com/biosci/genbio/mader
(click on *Biology 7/e*)

The *Biology 7/e* Online Learning Center provides many resources for studying the material in this chapter including links to the following sites:

Mendelian Genetics explores Mendel's First and Second Laws of Genetics, and independent assortment along with a number of more advanced topics (i.e., gene interactions, modifier genes, and variation in gene expression). An excellent site with a wealth of information.

http://www.cc.ndsu.nodak.edu/instruct/mcclean/plsc431/mendel/

Genetics Virtual Library is a website catalog of hundreds of genetics sites. Here you can find sites discussing every organism from bacteria to cattle to slime molds to mosquitoes. Transgenic and genome resource sites are also listed.

http://www.ornl.gov/TechResources/Human_Genome/vl.html

Online Mendelian Inheritance in Man. This site supported by the National Center for Biotechnology Information includes a catalog of human genes and genetic disorders. It provides text, images, references, and a map of human chromosomes.

http://www3.ncbi.nlm.nih.gov/Omim/

The Human Transcript Map. The genes of the human genome, circa 1996 with a link to the GeneMap'98. Text material describes disease conditions associated with individual genes. Clickable by chromosome number.

http://www.ncbi.nlm.nih.gov/SCIENCE96/

Mendel Web. This site is a resource for teachers and students interested in classical genetics, data analysis, and the history and literature of science. It includes Mendel's original paper (written in German) and an English translation of the paper.

http://www.netspace.org/MendelWeb/

Chromosomes and Genes

chapter

chapter concepts

12.1 Mendelism and the Genotype

- The genotype must be considered an integrated whole of all the genes because genes often work together to control the phenotype. 188
- There are forms of inheritance that involve degrees of dominance, interactions, multiple alleles, and polygenes. 188
- Environmental conditions can influence gene expression. 191

12.2 Mendelism and the Chromosomes

- The genes are on the chromosomes, and therefore they behave similarly during mitosis and meiosis. 192
- Certain chromosomes are called the sex chromosomes because they determine the sex of the individual. 192
- Certain traits, unrelated to sex, are controlled by genes located on the sex chromosomes. 192
- Alleles on the same chromosome are said to be linked, and crossing-over makes it possible to determine the order of genes on the chromosome. 194

12.3 Chromosomal Mutations

- Chromosomal mutations are caused by a change in chromosome number and by a change in chromosomal structure. 196

A human chromosome

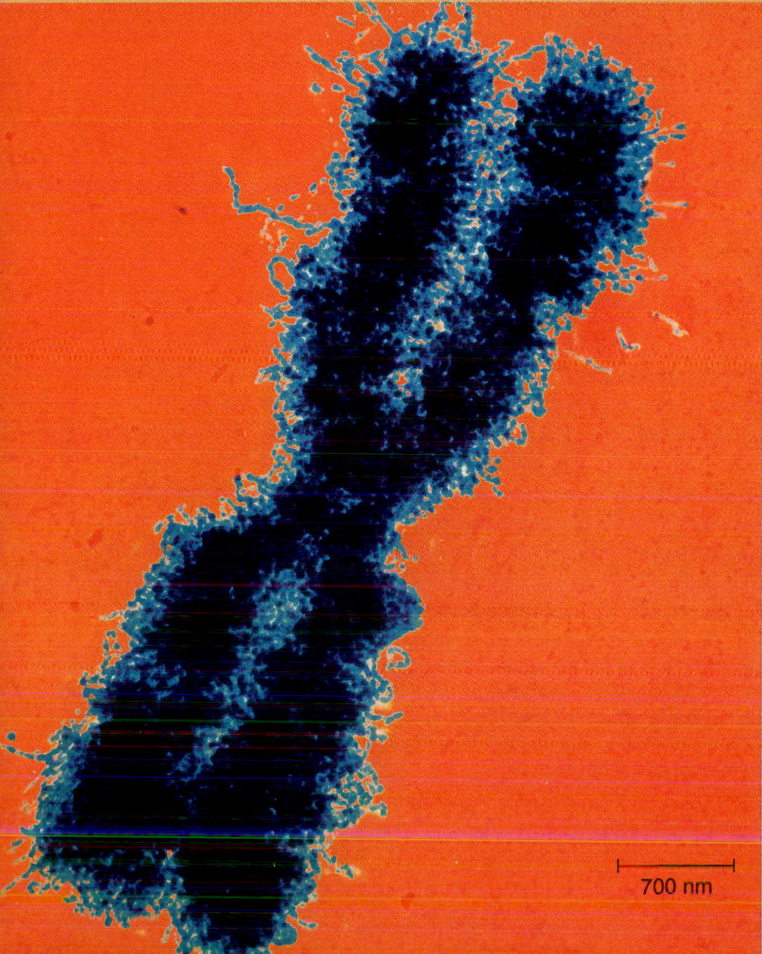

Considering that Mendel knew nothing about chromosomes, genes, or DNA, it is astounding that his laws about the transfer of genetic information from parent to offspring do apply to all organisms and complex patterns of inheritance. As brilliant and groundbreaking as Mendel's work was, it was only the first step toward a more thorough understanding of how the genotype functions.

Researchers did not know until the early 1900s that the genes are on the chromosomes, and the behavior of chromosomes explains Mendel's law of segregation and independent assortment as long as the genes are on separate chromosomes. Mendel would not have been able to formulate his two laws if he had been working with genes on the same chromosome! Determining the sequence of genes on the chromosomes began with *Drosophila*. Now researchers are sequencing the genes for all sorts of organisms, including humans. Chromosomes can undergo mutations, permanent changes in number and structure, and you can well imagine that such alterations would have a profound effect on the phenotype.

12.1 Mendelism and the Genotype

Since Mendel's day, variations in the dominant/recessive relationship he described have been discovered. These variations make it clear that the concept of the genotype should be expanded to include all the genes. The idea of just two alleles per trait, one dominant and one recessive, is too restrictive because various genes work together to bring about a phenotype which is also influenced by the environment.

Incomplete Dominance and Codominance

Incomplete dominance describes a situation in which a dominant allele is unable to make up for the presence of a recessive allele. In a cross between a true-breeding, red-flowered four-o'clock strain and a true-breeding, white-flowered strain, the offspring have pink flowers (Fig. 12.1). While this may appear to be an example of the blending theory of inheritance, it is not, because if these F_1 plants self-pollinate, the F_2 generation has a phenotypic ratio of 1 red-flowered : 2 pink-flowered : 1 white-flowered plant. The reappearance of the parental phenotypes in the F_2 generation makes it clear we are still dealing with an example of particulate inheritance of the type described by Mendel.

When doing an incomplete dominance problem, use the same letter to indicate the alleles but do not use a capital and lowercase letter because one allele is not completely dominant over the other. Therefore, Figure 12.1 uses the notations R_1 and R_2 for the alleles. The genotype R_1R_2 for the pink-flowered F_1 plant makes it clear that R_1 is not dominant over R_2. There is a biochemical explanation for incomplete dominance. The pink-flowered plants are producing one-half the amount of pigment as the red-flowered plants. The white-flowered plants lack pigment.

Similarly, in many instances of simple dominance the dominant allele is coding for a product (e.g., enzyme), while the recessive allele is not coding for a product. But in a case of simple dominance, the F_1 appears to have the same phenotype as the homozygous dominant parent. In the case of incomplete dominance, it is apparent to the most casual observer that the F_1 does not have the appearance of the parent with the dominant phenotype.

In **codominance,** another example that differs from Mendel's findings, both alleles are fully expressed. When an individual has blood type A, A type of glycoprotein appears on the red blood cells. When an individual has blood type B, B type of glycoprotein appears on the red blood cells. But when an individual has the blood type AB, both A and B types of glycoproteins appear on the red blood cells. The two different capital letters signify that both alleles are coding for an effective product.

In incomplete dominance, the F_1 does not resemble either parent because one functioning allele is unable to produce the dominant phenotype. In codominance, both alleles are functioning and coding for a product.

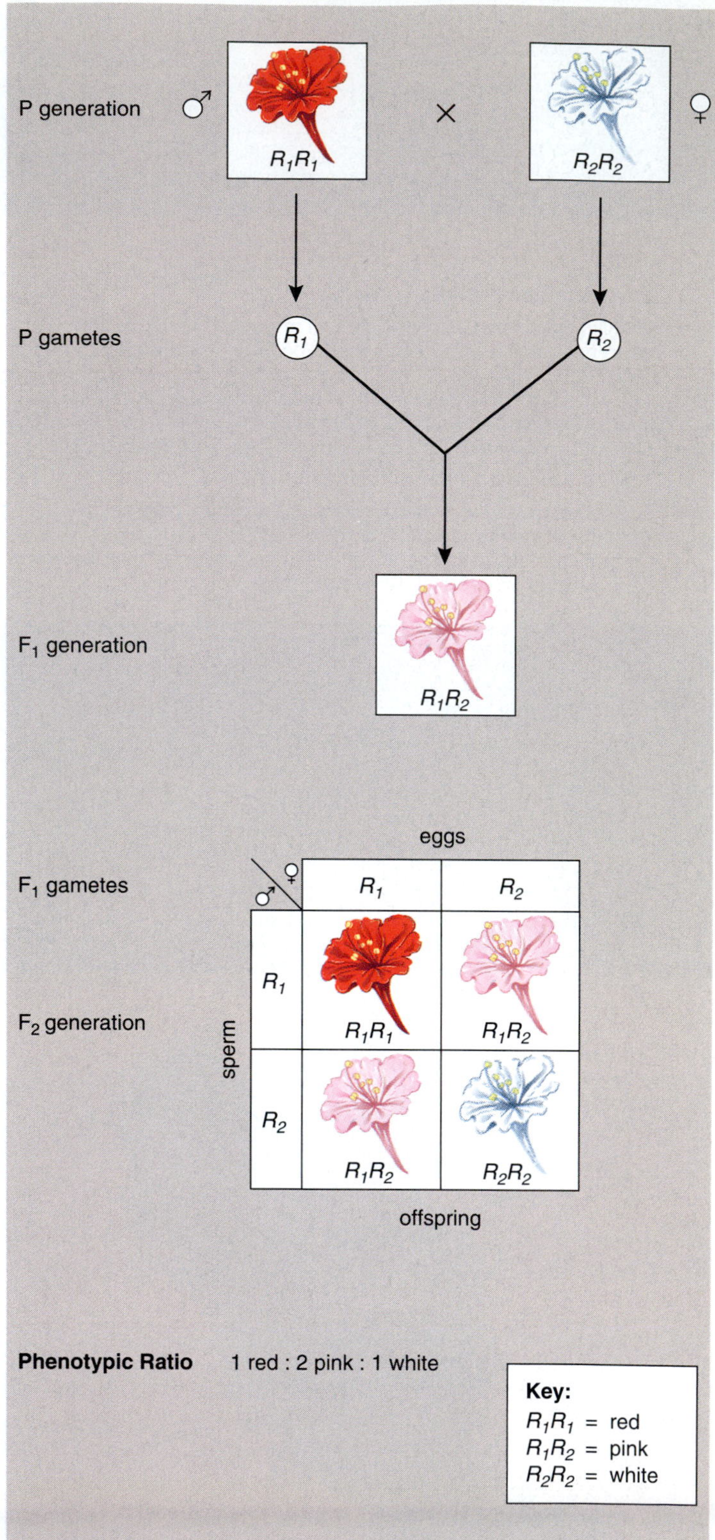

Figure 12.1 Incomplete dominance.
Incomplete dominance is illustrated by a cross between red- and white-flowered four-o'clocks. The F_1 plants are pink, a phenotype intermediate between those of the P generation; however, the F_2 results show that neither the red nor the white allele has disappeared. Therefore, this is still an example of particulate inheritance, though the heterozygote R_1R_2 is pink.

Epistasis

In **epistasis** [Gk. *epi,* on, and *stasis,* a standing], a gene at one locus interferes with the phenotypic expression of a gene at a different locus. In most instances, a recessive pair of alleles at one locus prevents the expression of a dominant allele at another locus. Both loci must have a functioning gene for the dominant phenotype to appear.

By studying the results of crosses in sweet peas, it has been deduced that there are two genes affecting pigmentation and that being homozygous recessive in either gene results in a lack of color. If the two varieties of white-flowered plants are crossed, the F_1 have purple flowers instead of the white flowers as expected. Among the F_2, following self-pollination of the F_1 plants, 9/16 of the plants have purple flowers and 7/16 have white flowers. Since the ratio is in sixteenths, the F_1 plants must have been heterozygous for both genes as shown in Figure 12.2*a.* Why do the F_1 plants have purple flowers? Because they have both a dominant *A* allele and a dominant *B* allele. Why is there a 9:7 ratio instead of a 9:3:3:1 ratio among the F_2 generation of plants? Whenever a plant is homozygous recessive for one of the genes, no pigment is produced. Both loci need at least one dominant allele to produce the dominant phenotype.

Although the exact details of pigmentation synthesis are known, the metabolic pathway shown in Figure 12.2*b* is hypothesized. If the *A* allele codes for the first enzyme and the *B* allele codes for the second enzyme, then being homozygous recessive for either gene would result in white instead of purple flowers.

A similar situation occurs in mammalian animals. If individuals inherit any one of several defects in the metabolic pathway for the synthesis of melanin, the individual is an albino.

In epistasis, genes at two different loci interact to control a single trait.

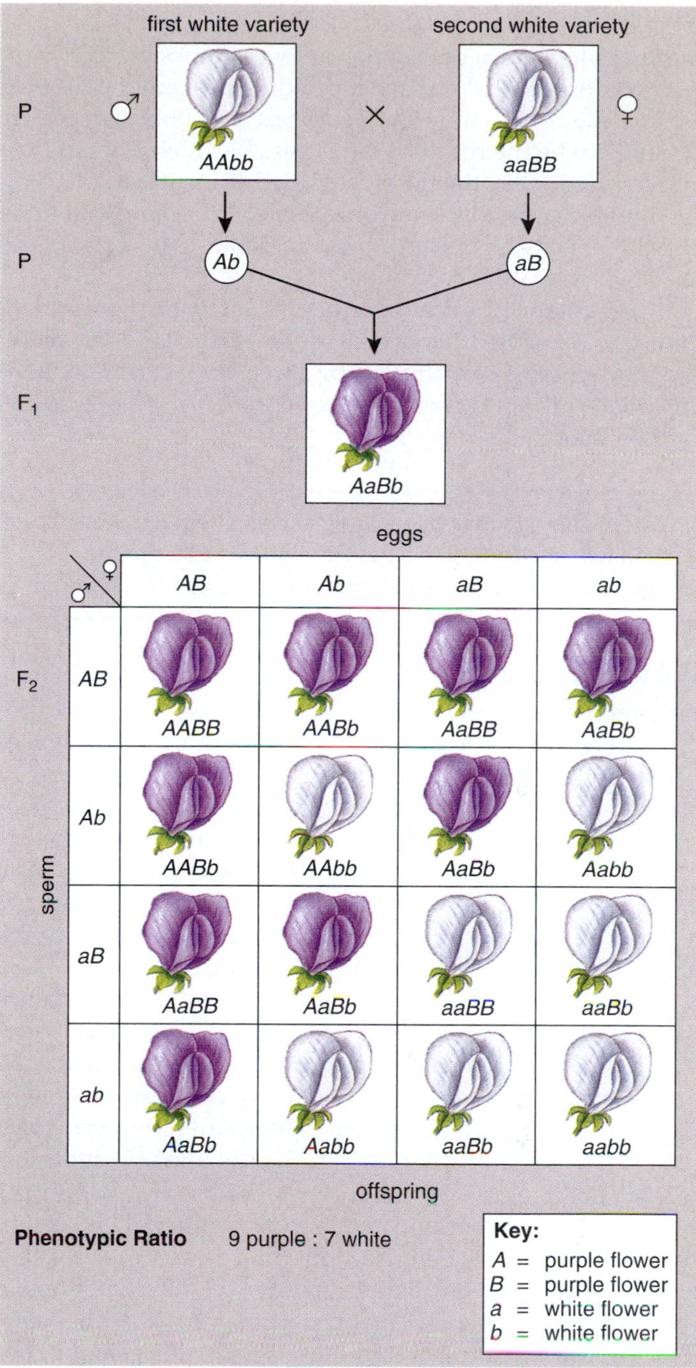

a.

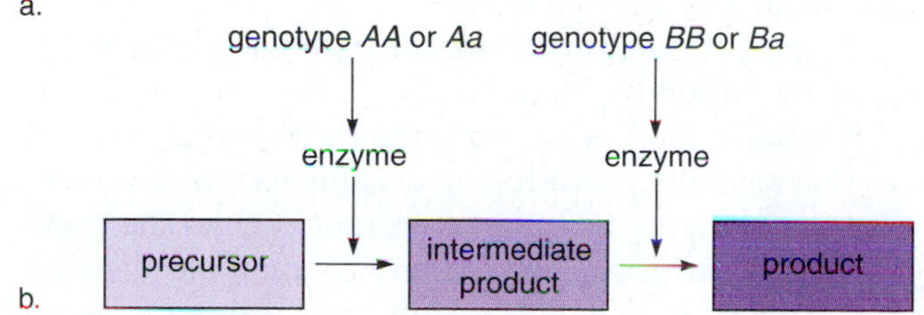

b.

Figure 12.2 Epistasis.
a. Two pairs of genes are believed to be involved in color production in sweet peas. Being homozygous recessive for either gene results in lack of color. **b.** It is hypothesized that each gene codes for an enzyme in a metabolic pathway that results in pigmentation. If either enzyme is lacking, an absence of color results.

Pleiotropy

The term **pleiotropy** [Gk. *pleion,* more, and *tropos,* turning] is used to describe a gene that affects more than one characteristic of the individual. Individuals with Marfan syndrome tend to be tall and thin with long legs, arms, and fingers. They are nearsighted, and the wall of the aorta is weak, causing it to enlarge and, eventually, to split. All of these effects are caused by an inability to produce a normal extracellular matrix protein called fibrillin. Fibrillin strengthens elastic fibers and allows extracellular components to adhere to one another.

There are some who suggest that Abraham Lincoln had Marfan syndrome. Certainly it would account for his lanky frame and the symptoms he described for posterity. If he had not been shot, it's possible he still would have died shortly thereafter from circulatory failure.

In pleiotropy, a single gene affects more than one trait.

Multiple Alleles

In **multiple alleles,** one gene has several possible alleles for its chromosomal locus. Each individual has only two of possible alleles. Three different peppered moth phenotypes are believed to be the result of three possible alleles: typical *(m)* is recessive to the other two; and mottled insularia *(M′)* is recessive to nearly black melanic allele *(M)*. Therefore, there are three possible phenotypes: melanic *(M—)*, insularia (*M′M′* or *M′m*) and typical *(mm)*.

Inheritance by multiple alleles often results in more than two possible phenotypes for a particular trait. Blood type in humans is controlled by three alleles (*A, B,* and *O*) and there are four possible phenotypes: A, B, AB, and O types of blood (see page 210).

When inheritance is by multiple alleles, there are more than two alleles for a single gene locus and there are more than two possible phenotypes among the offspring.

Polygenic Inheritance

Polygenic inheritance [Gk. *polys,* many, and L. *genitus,* producing] occurs when one trait is governed by several genes occupying different loci on the same homologous pair of chromosomes or on different homologous pairs of chromosomes. Each gene has two alleles. The contributing allele is represented by a capital letter, and the noncontributing allele is represented by a small letter. Each contributing allele has a quantitative effect on the phenotype, and therefore the allelic effects of all contributing alleles are additive.

For example, H. Nilsson-Ehle studied the inheritance of seed color in wheat. Genes at three different loci determine seed color. After he crossed plants that produced white and dark-red seeds, the F_1 plants were allowed to self-pollinate. Each of the F_2 plants produced seeds having one of seven degrees of color (Fig. 12.3). The alleles responsible for the color of the seeds are represented by dots in the squares. The number of capitals (contributing allele) is represented by coloring in the dots. Since each contributing allele has a small but equal quantitative effect, there is a range of F_2 phenotypes. The proportions of the various phenotypes can be graphed and these proportions can be connected to form a bell-shaped curve.

In chapter 10, we learned that sexual reproduction contributes to evolution by increasing variation among the offspring. Similarly, polygenic inheritance also increases the chances that sexually reproducing organisms will produce a phenotype more suited to the environment than the previous generation, particularly if the environment is changing. Polygenic inheritance which results in a range of phenotypes is observed quite often in complex organisms. If we were observing the phenotypes in nature, we would probably arrive directly at a bell-shaped curve because environmental

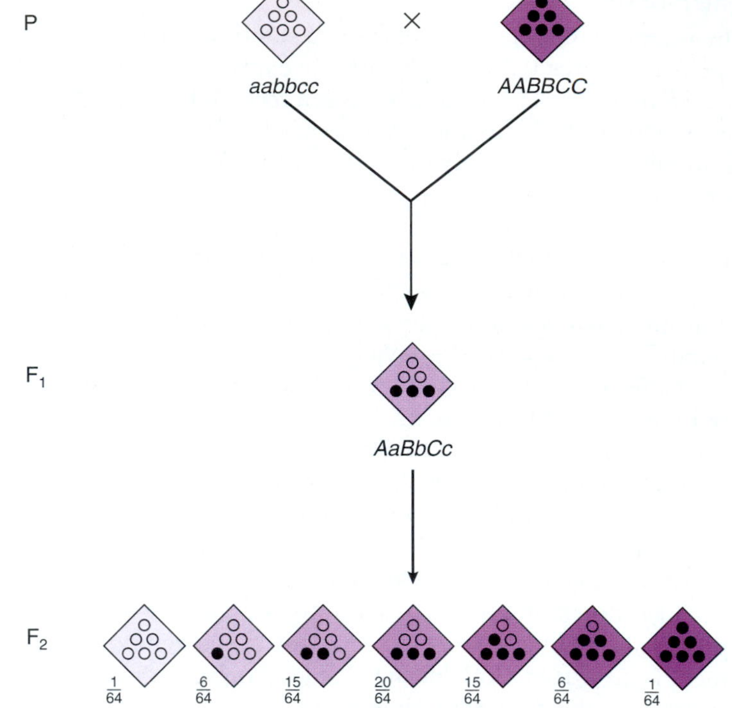

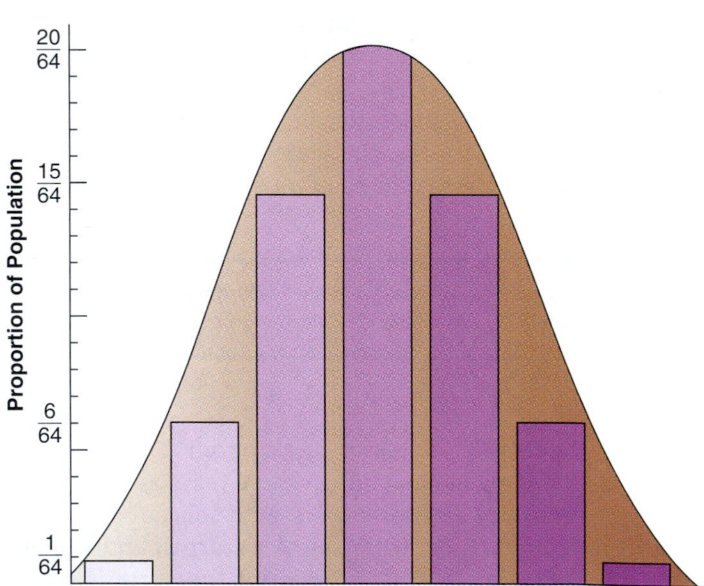

Figure 12.3 Polygenic inheritance.
In this example, a trait is controlled by three genes; only the alleles represented by a capital letter are contributing alleles. When two intermediate phenotypes are crossed, seven phenotypes are seen among the F_2 generation. The phenotype proportions are graphed. If environmental effects are considered, a bell-shaped curve results because of many intervening phenotypes.

effects cause many intervening phenotypes. Consider the inheritance of skin color or height in humans, which are believed to be examples of polygenic inheritance. In the first instance, exposure to sun can affect skin color and increase the number of phenotypic variations. In the second instance, nutrition can affect height, resulting in many more phenotypes than those expected.

In polygenic inheritance, each contributing allele adds to the phenotype. A bell-shaped curve of phenotypic variations is due to the number of alleles involved and is also due to environmental effects.

The Environment and the Phenotype

There is no doubt that both the genotype and the environment affect the phenotype. The relative importance of genetic and environmental influences on the phenotype can vary, but in some instances the environment seems to have an extreme effect. In the water buttercup, *Ranunculus peltatus*, the submerged part of the plant has a different appearance from the part above water. Apparently, the presence or absence of an aquatic environment dramatically influences the phenotype.

Temperature can also have a dramatic effect on the phenotypes of plants. Primroses have white flowers when grown above 32°C and red flowers when grown at 24°C. The coats of Siamese cats and Himalayan rabbits are darker in color at the ears, nose, paws, and tail. Himalayan rabbits are known to be homozygous for the allele *ch*, which is involved in production of melanin. Experimental evidence suggests that the enzyme encoded by this gene is active only at a low temperature and that, therefore, black fur only occurs at the extremities where body heat is lost to the environment (Fig. 12.4). When the animal is placed in a warmer environment, new fur on these body parts is light in color.

So, we come to the conclusion that matters are a little more complex than Mendel had envisioned. Not only do gene interactions contribute to the phenotype so does the environment.

An organism's environmental history in addition to the genotype can influence the phenotype.

a.

Figure 12.4 Coat color in Himalayan rabbits.
a. A Himalayan rabbit which has black extremities is usually homozygous recessive for the allele *ch*, which codes for an enzyme involved in the production of melanin. **b.** The influence of the environment on the phenotype has been demonstrated by plucking out the fur from one area and applying an ice pack. **c.** The new fur which grew in was black instead of white, showing that the enzyme is active only at low temperatures.

Practice Problems 1*

1. What genotypes and phenotypes are possible among the offspring if a moth with the genotype *Mm* is mated to a moth with the genotype *M′m*?
2. What are the results of a testcross involving incomplete dominance when pink-flowered four-o'clocks are crossed with white-flowered ones?
3. Breeders of dogs note various colors among offspring that range from white to black. If the coat color is controlled by three pairs of alleles, how many different shades are possible if all dominant genes have the same quantitative effect?
4. Investigators note that albino tigers usually have crossed eyes. What two inheritance patterns can account for a phenotype such as this?

* *Answers to Practice Problems appear in Appendix A.*

12.2 Mendelism and the Chromosomes 

The behavior of chromosomes during mitosis was described by 1875, and for meiosis in the 1890s. By 1902, both Theodor Boveri, a German, and Walter S. Sutton, an American, had independently noted the parallel behavior of genes and chromosomes and had proposed the **chromosomal theory of inheritance,** which states that the genes are located on the chromosomes. This theory is supported by the following observations:

1. Both chromosomes and factors (now called alleles) are paired in diploid cells.
2. In keeping with Mendel's law of segregation, both homologous chromosomes and alleles of each pair separate during meiosis so that the gametes have one-half the total number.
3. In keeping with Mendel's law of independent assortment, both homologous chromosomes and alleles of each pair separate independently so that the gametes contain all possible combinations.
4. Fertilization restores both the diploid chromosome number and the paired condition for alleles in the zygote.

The genes are on the chromosomes; therefore, they behave similarly during meiosis and fertilization.

Sex Chromosomes

In animal species, the **autosomes,** or nonsex chromosomes, are the same between the sexes. One special pair of chromosomes is called the **sex chromosomes** because this pair determines the sex of the individual. The sex chromosomes in the human female are XX and those in the male are XY. Because human males can produce two different types of gametes—those that contain an X and those that contain a Y—normally males determine the sex of the new individual.

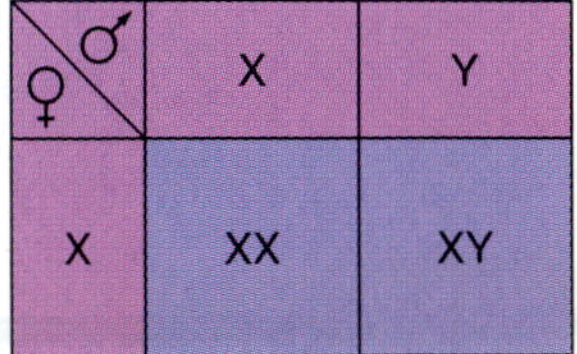

In addition to genes that determine sex, the sex chromosomes carry genes for traits that have nothing to do with the sex of the individual. By tradition, the term sex-linked or **X-linked** is used for genes carried on the X chromosome. The Y chromosome does not carry these genes and indeed carries very few genes.

X-Linked Alleles

The hypothesis that genes are on the chromosomes was substantiated by experiments performed by a Columbia University group of *Drosophila* geneticists, headed by Thomas Hunt Morgan. Fruit flies are even better subjects for genetic studies than garden peas: they can be easily and inexpensively raised in simple laboratory glassware; females mate only once and then lay hundreds of eggs during their lifetimes; and the generation time is short, taking only about ten days when conditions are favorable (Fig. 12.5).

Drosophila flies have the same sex chromosome pattern as humans, and this facilitates our understanding of a cross performed by Morgan. Morgan took a newly discovered mutant male with white eyes and crossed it with a red-eyed female:

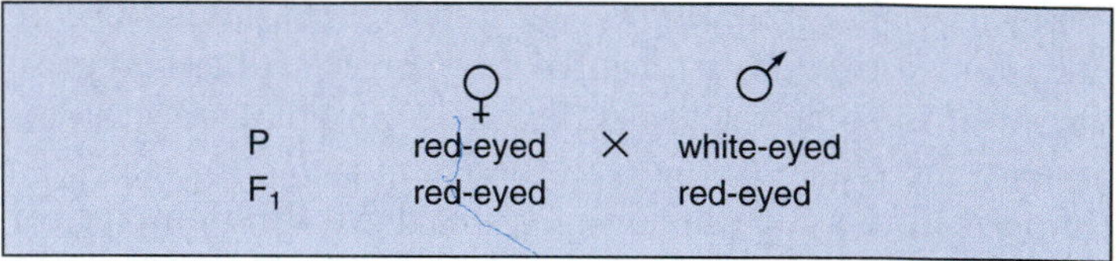

From these results, he knew that red eyes are dominant over white eyes. He then crossed the F_1 flies. In the F_2 generation, there was the expected 3 red-eyed : 1 white-eyed ratio, but it struck him as odd that all of the white-eyed flies were males:

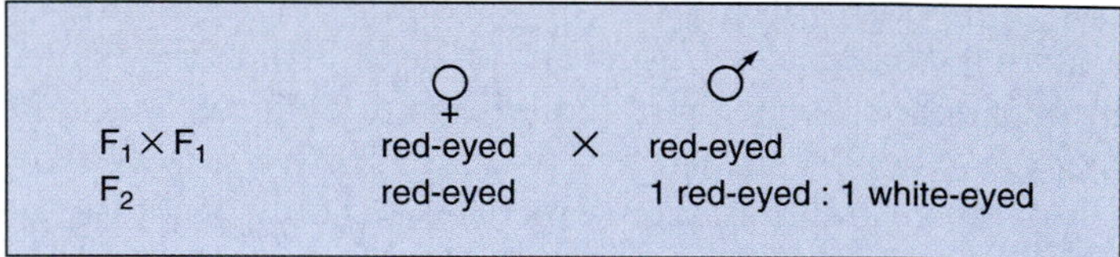

Obviously, a major difference between the male flies and the female flies was their sex chromosomes. Could it be possible that an allele for eye color was on the Y chromosome but not on the X? This idea could be quickly discarded because normal females have red eyes, and they have no Y chromosome. Perhaps an allele for eye color was on the X, but not on the Y, chromosome. Figure 12.5*b* indicates that this explanation would match the results obtained in the experiment. These results support the chromosomal theory of inheritance by showing that the behavior of a specific allele corresponds exactly with that of a specific chromosome—the X chromosome in *Drosophila*.

As discussed next, genes that are X-linked have a different pattern of inheritance than genes that are on the autosomes because the Y chromosome is blank for these genes and cannot offset the inheritance of a recessive allele

X-linked alleles are on the X chromosome, and the Y chromosome is blank for these alleles.

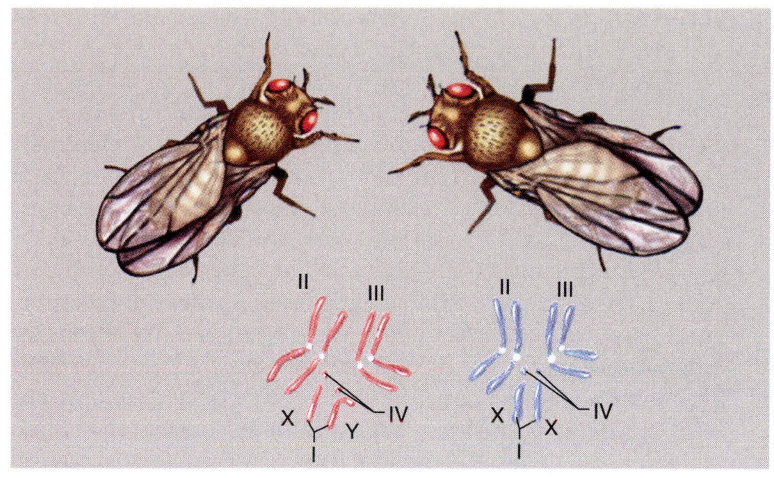

a.

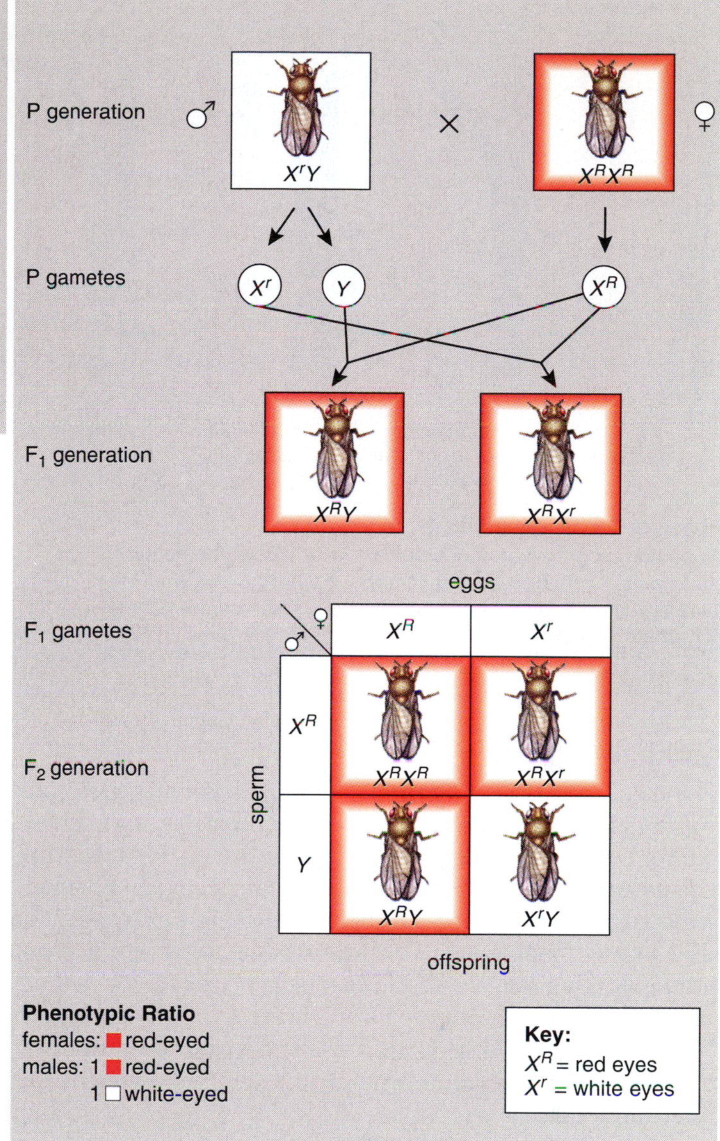

b.

Figure 12.5 X-linked inheritance.
a. Both males and females have eight chromosomes, three pairs of autosomes (II–IV), and one pair of sex chromosomes (I). Males are XY and females are XX. **b.** Once researchers deduced that the alleles for red/white eye color are on the X chromosome in *Drosophila*, they were able to explain their experimental results. Males with white eyes in the F_2 inherited the recessive allele only from the female parent; they receive a blank Y from the male parent.

X-Linked Problems

Recall that when solving autosomal genetics problems, the key and genotypes are represented as follows:

Key:	**Genotypes:**
L = long wing	*LL, Ll, ll*
l = short wings	

As noted in Figure 12.5, however, the key for an X-linked gene shows an allele attached to the X:

Key:

X^R = red eyes
X^r = white eyes

The possible genotypes in both males and females are as follows:

X^RX^R = red-eyed female
X^RX^r = red-eyed female
X^rX^r = white-eyed female
X^RY = red-eyed male
X^rY = white-eyed male

Notice that there are three possible genotypes for females, but only two for males. Females can be heterozygous X^RX^r, in which case they are carriers. Carriers usually do not show a recessive abnormality, but they are capable of passing on a recessive allele for an abnormality. Males cannot be carriers; if the dominant allele is on the single X chromosome, they have the dominant phenotype, and if the recessive allele is on the single X chromosome, they show the recessive phenotype.

Practice Problems 2*

1. Using the key X^B = bar eyes and X^b = normal eyes, give a. three possible genotypes for females and b. two possible genotypes for males. Also give the possible gametes for these males.
2. a. Which *Drosophila* cross would produce white-eyed males: (1) $X^RX^R \times X^rY$ or (2) $X^RX^r \times X^RY$? b. In what ratio?
3. A woman is color blind (X-linked recessive). a. What are the chances of her sons being color blind if she reproduces with a man having normal vision? b. What are the chances of her daughters being color blind? c. Being carriers?

* *Answers to Practice Problems appear in Appendix A.*

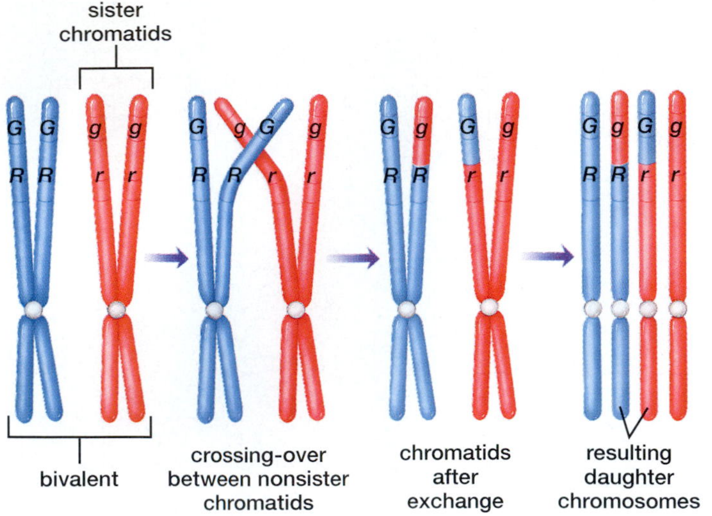

Figure 12.6 **Crossing-over.**
When homologous chromosomes are in synapsis, the nonsister chromatids exchange genetic material. Following crossing-over, recombinant chromosomes occur. Recombinant chromosomes contribute to recombinant gametes.

Linkage Groups

Drosophila probably has thousands of different genes controlling all aspects of its structure, biochemistry, and behavior. Yet it has only four pairs of chromosomes. This paradox led investigators like Sutton to conclude that each chromosome must carry a large number of genes. For example, it is now known that genes controlling eye color, wing type, body color, leg length, and antennae type are all located on the chromosomes numbered II in Figure 12.5. The alleles for these genes are said to form a **linkage group** because they are found on the same chromosome and they tend to be inherited together.

Crossing-over, you recall, occurs between nonsister chromatids when homologous pairs of chromosomes pair prior to separation during meiosis. During crossing-over the nonsister chromatids exchange genetic material and therefore genes. If crossing-over occurs between the two linked alleles of interest, a dihybrid produces four types of gametes instead of two (Fig. 12.6). Recombinant means a new combination of alleles. The recombinant gametes occur in reduced number because crossing-over is infrequent. Still, all possible phenotypes will occur among the offspring.

To take an actual example, suppose you are doing a cross between a gray-bodied, red-eyed heterozygote and a black-bodied, purple-eyed fly. Since the alleles governing these traits are both on chromosome II, you predict that the results will be 1:1 instead of 1:1:1:1, as is the case for unlinked genes. The reason, of course, is that linked alleles tend to stay together and do not separate independently, as predicted by Mendel's laws. Under these circumstances, the heterozygote forms only two types of gametes and produces offspring with only two phenotypes (Fig. 12.7*a*).

When you do the cross, however, you find that a very small number of offspring show recombinant phenotypes (i.e., those that are different from the original parents). Specifically, you find that 47% of the offspring have black bodies and purple eyes, 47% have gray bodies and red eyes, 3% have black bodies and red eyes, and 3% have gray bodies and purple eyes (Fig. 12.7*b*). What happened? In this instance, crossing-over led to a very small number of recombinant gametes, and when these were fertilized, recombinant phenotypes were observed in the offspring. An examination of chromosome II shows that these two sets of alleles are very close together. Doesn't it stand to reason that the closer together two genes are, the less likely they are to cross over? This is exactly what various crosses have repeatedly shown.

All the genes on one chromosome form a linkage group that tends to stay together, except when crossing-over occurs.

Chromosome Mapping

You can use the percentage of recombinant phenotypes to map the chromosomes, because there is a direct relationship between the frequency of crossing-over and the percentage of recombinant phenotypes. In our example (Fig. 12.7), a total of 6% of the offspring are recombinants, and for the sake of mapping the chromosomes, it is assumed that 1% of crossing-over equals one map unit. Therefore, the allele for black body and the allele for purple eyes are six map units apart.

Suppose you want to determine the order of any three genes on the chromosomes. To do so, you can perform crosses that tell you the map distance between all three pairs of alleles. If you know, for instance, that:

1. the distance between the black-body and purple-eye alleles = 6 map units,
2. the distance between the purple-eye and vestigial-wing alleles = 12.5 units, and
3. the distance between the black-body and vestigial-wing alleles = 18.5 units, then the order of the alleles must be as shown here:

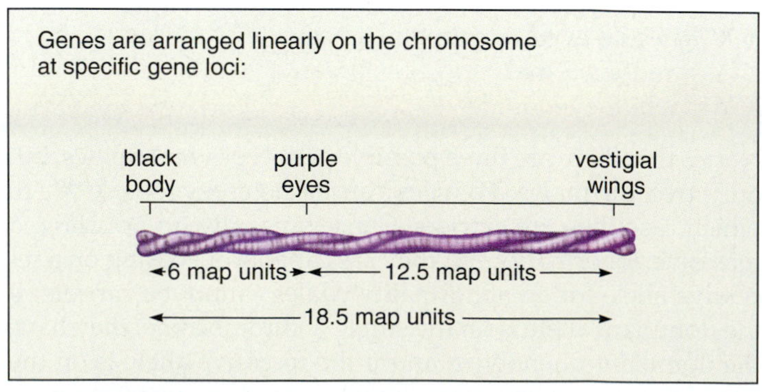

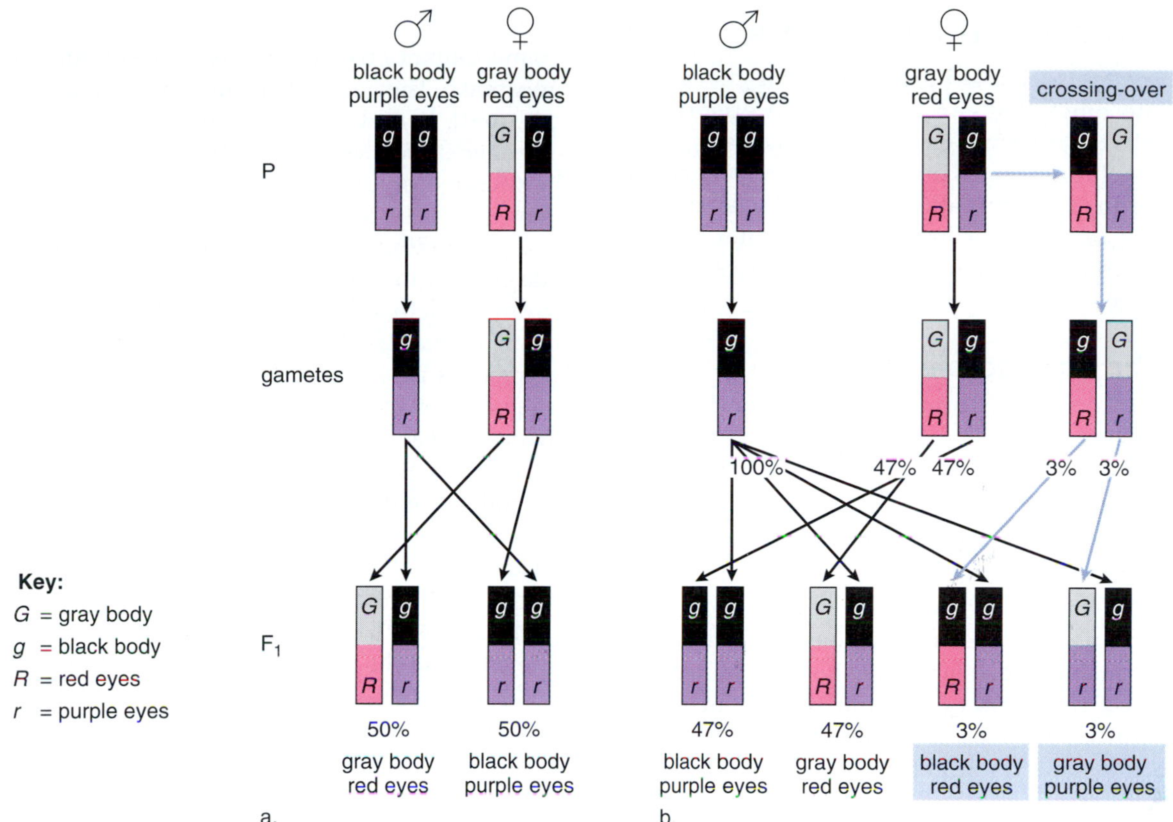

Figure 12.7 Complete linkage versus incomplete linkage.
a. Hypothetical cross in which genes for body and eye color on chromosome II of *Drosophila* are completely linked. Instead of the expected 1:1:1:1 ratio among the offspring, there would be a 1:1 ratio. b. Crossing-over occurs, and 6% of the offspring show recombinant phenotypes. The percentage of recombinant phenotypes is used to map the chromosomes (1% = 1 map unit).

Of the three possible orders (any one of the three genes can be in the middle), only the order given shows the proper distance between all three allelic pairs. Because it is possible to map the chromosomes, we conclude that genes are indeed located on chromosomes, they have a definite location, and they occur in a definite order.

Linkage data have been used to map the chromosomes of *Drosophila,* but the possibility of using linkage data to map human chromosomes is limited because we can only work with matings that have occurred by chance. This, coupled with the fact that humans tend not to have numerous offspring, means that additional methods are used to sequence the genes on human chromosomes. Today, it is customary to also rely on biochemical methods to map the human chromosomes.

The frequency of recombinant gametes that occurs due to the process of crossing-over has been used to map the chromosomes.

Practice Problems 3*

1. When *AaBb* individuals are allowed to self-breed, the phenotypic ratio is just about 3:1. a. What ratio was expected? b. What may have caused the observed ratio?
2. In two sweet pea strains, *B* = blue flowers, *b* = red flowers, *L* = long pollen grains, and *l* = round pollen grains. In a cross between a heterozygous plant with blue flowers and long pollen grains and a plant with red flowers and round pollen grains, 44% of offspring are blue, long; 44% are red, round; 6% are blue, round; and 6% are red, long. How many map units separate these two sets of alleles?
3. Investigators performed crosses that indicated bar-eye and garnet-eye alleles are 13 map units apart, scallop-wing and bar-eye alleles are 6 units apart, and garnet-eye and scallop-wing alleles are 7 units apart. What is the order of these alleles on the chromosome?

* *Answers to Practice Problems appear in Appendix A.*

12.3 Chromosomal Mutations

Mutations [L. *mutatus,* change] are permanent changes in genes or chromosomes that can be passed to offspring if they occur in cells that become gametes. Like crossing-over, recombination of chromosomes during meiosis, and gamete fusion during fertilization, mutations increase the amount of variation among offspring. Chromosomal mutations include changes in chromosome number and changes in chromosomal structure.

Changes in the Chromosome Number

Changes in the chromosome number include monosomies, trisomies, and polyploidy.

Monosomy and Trisomy

Monosomy occurs when an individual has only one of a particular type of chromosome (2n − 1), and *trisomy* occurs when an individual has three of a particular type of chromosome (2n + 1). The usual cause of monosomy and trisomy is nondisjunction during meiosis. *Nondisjunction* can occur during meiosis I if members of a homologous pair fail to separate, and during meiosis II if the daughter chromosomes go into the same daughter cell (see page 203). Although nondisjunction can also occur during mitosis, it is more common during meiosis.

Monosomy and trisomy occur in both plants and animals. In animals, autosomal monosomies and trisomies are generally lethal, although a trisomic individual is more likely to survive than a monosomic one. The survivors are characterized by a distinctive set of physical and mental abnormalities called a syndrome. In humans, Turner syndrome is a monosomy involving the sex chromosomes—the individual inherits a single X chromosome. The most common trisomy among humans is Down syndrome, which involves chromosome 21. Individuals with Turner syndrome are expected to live a full life span, and although those with Down syndrome are subject to various medical conditions, they generally do well also.

Polyploidy

Some mutant eukaryotes have more than two sets of chromosomes. They are called **polyploids** [Gk. *polys,* many, and *plo,* fold]. Polyploid organisms are named according to the number of sets of chromosomes they have. Triploids (3n) have three of each kind of chromosome, tetraploids (4n) have four sets, pentaploids (5n) have five sets, and so on.

Polyploidy is not seen in animals most likely because judging from trisomies, multiple copies of chromosomes would be lethal. Polyploidy is a major evolutionary mechanism in plants. It is estimated that 47% of all flowering plants are polyploids. Among these are many of our most important crops, such as wheat, corn, cotton, sugarcane, and fruits such as watermelons, bananas, and apples (Fig. 12.8). Also many attractive flowers, such as chrysanthemums and daylilies, are polyploids.

Polyploidy generally arises following hybridization. When two different species reproduce, the resulting organism, called a hybrid, may have an odd number of chromosomes. If so, the chromosomes cannot pair evenly during meiosis, and the organism will not be fertile. (Hybridization in animals can also lead to sterility—as when a horse and donkey produce a mule, which is unable to produce offspring.) In plants, hybridization, followed by doubling of the chromosome number, will result in an even number of chromosomes, and the chromosomes will be able to undergo synapsis during meiosis.

	Species		Hybrid	chromosome number doubles	Polyploid
	AA × *BB*	→	*AB*	→	*AABB*
chromosome number	(24) (30)		(27) sterile		(54) fertile

Monosomies (2n − 1) and trisomies (2n + 1) cause abnormalities. In plants, hybridization followed by polyploidy is a positive force for change.

a.

b.

c.

Figure 12.8 Polyploid plants.
Many food sources are specially developed polyploid plants. **a, b.** Among these are seedless watermelons and bananas. They are infertile (seeds poorly developed) triploids, but they can be propagated by asexual means. **c.** Polyploidy makes plants and their fruits larger, such as these jumbo McIntosh apples.

Changes in Chromosomal Structure

Some changes in chromosomal structure can be detected microscopically, and some cannot be detected. Various agents in the environment, such as radiation, certain organic chemicals, or even viruses, can cause chromosomes to break. Ordinarily, when breaks occur in chromosomes, the two broken ends reunite to give the same sequence of genes. Sometimes, however, the broken ends of one or more chromosomes do not rejoin in the same pattern as before, and the result is various types of **chromosomal mutations.** An inversion, a translocation, a deletion, and a duplication of chromosome segments are illustrated in Figure 12.9.

An *inversion* occurs when a segment of a chromosome is turned around 180°. You might think this is not a problem because the same genes are present, but the reversed sequence of genes can lead to altered gene activity.

A *translocation* is the movement of a chromosome segment from one chromosome to another, nonhomologous chromosome. In 5% of cases, a translocation that occurred in a previous generation between chromosomes 21 and 14 is the cause of Down syndrome. In these cases, Down syndrome is not related to the age of the mother, but instead tends to run in the family of either the father or the mother.

A *deletion* occurs when an end of a chromosome breaks off or when two simultaneous breaks lead to the loss of an internal segment. Even when only one member of a pair of chromosomes is affected, a deletion often causes abnormalities. An example is cri du chat (cat's cry) syndrome. The affected individual has a small head, is mentally retarded, and has facial abnormalities. Abnormal development of the glottis and larynx results in the most characteristic symptom—the infant's cry resembles that of a cat.

A *duplication* is the presence of a chromosomal segment more than once in the same chromosome. There are several ways a duplication can occur. A broken segment from one chromosome can simply attach to its homologue, or unequal crossing-over may occur, leading to a duplication and a deletion:

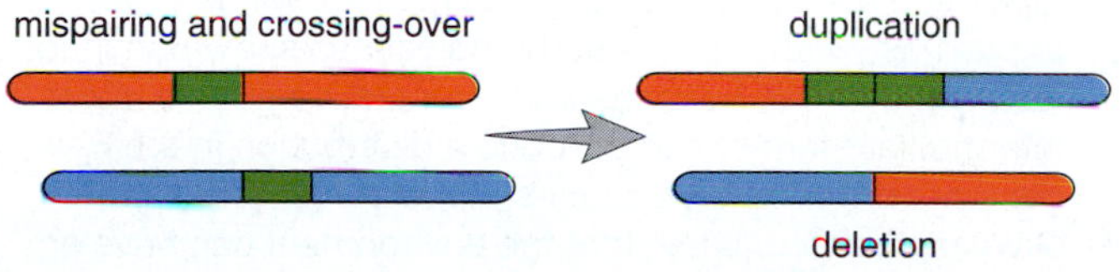

Multiple copies of genes can mutate differently and thereby provide additional genetic variation for the species. For example, there are several closely linked genes for human globin (globin is a part of hemoglobin, which is present in red blood cells and carries oxygen). This may have arisen by a process of duplication followed by different mutations in each gene.

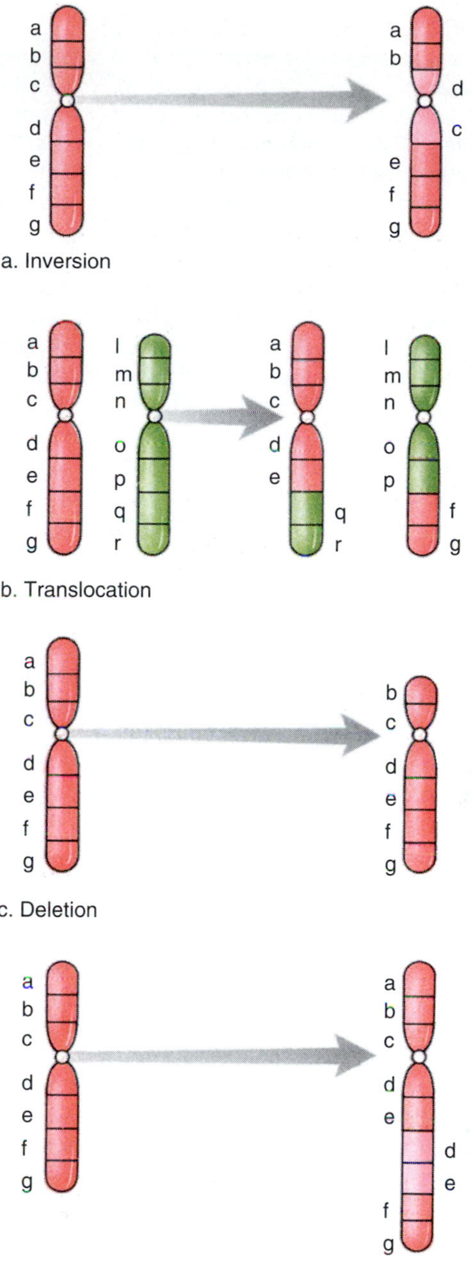

Figure 12.9 **Types of chromosomal mutations.**
a. Inversion occurs when a piece of chromosome breaks loose and then rejoins in the reversed direction. **b.** Translocation is the exchange of chromosome pieces between nonhomologous pairs. **c.** Deletion is the loss of a chromosome piece. **d.** Duplication occurs when the same piece is repeated within the chromosome.

Chromosomal mutations include observable changes in chromosome structure. Often chromosomal mutations lead to abnormal gametes and offspring. In rare instances, the increase in variation among offspring is beneficial to the species.

Connecting Concepts

By analyzing the outcome of his crosses, Mendel deduced that inheritable factors (now called alleles) contributed by each parent segregate from one another during the production of gametes. Early cytogeneticists (scientists who microscopically observe and study chromosomes) knew nothing of Mendel's work but they saw how homologous chromosome pairs separate during meiosis. Boveri and Sutton were the first to realize that Mendel's work and the work of cytogeneticists must be interrelated—in other words, the genes are on the chromosomes! The development of the chromosomal theory of inheritance is a fine example of how different avenues of research can result in a major scientific breakthrough. It also exemplifies why communication between scientists through publications and scientific conferences is so important.

Once it was understood that genes are carried on chromosomes, previously confusing phenomena such as linked genes, sex-linkage, and the consequences of chromosomal abnormalities became much easier to interpret correctly. The ability to map genes along chromosomes was an important step toward understanding the nature of genomes. It is no accident that the first eukaryotic genomes to be completely base sequenced are those of model organisms such as the yeast *Saccharomyces cerevisiae* and the worm *Caenorhabditis elegans* which already had excellent genetic maps. By contrast, constructing a base sequence map of the human genome was hampered by the lack of a conventional genetic map based on recombination distances. Humans cannot be asked to procreate in order to construct a genetic map! Chapter 17 discusses how new laboratory techniques now permit the mapping and sequencing of the human genome.

Summary

12.1 Mendelism and the Genotype

There are patterns of inheritance discovered since Mendel's original contribution that make it clear that the genotype should be thought of as involving all the genes. For example, different degrees of dominance have been observed. With incomplete dominance, the F_1 individuals are intermediate between the parental types; this does not support the blending theory because the parental phenotypes reappear in F_2. In epistasis, genes interact as when homozygous recessive alleles at one locus mask the effect of a dominant allele at another locus. In pleiotropy, a dominant allele has an effect that drastically alters the phenotype in many regards. Some genes have multiple alleles, although each individual organism has only two alleles. Coat color in rabbits is controlled by multiple alleles, and four different type coats are possible.

Polygenic traits are controlled by genes that have an additive effect on the phenotype, resulting in quantitative variations. A bell-shaped curve is seen because environmental influences bring about many intervening phenotypes. The relative influence of the genotype and the environment on the phenotype can vary—there are examples of extreme environmental influence.

12.2 Mendelism and the Chromosomes

The chromosomal theory of inheritance says that Mendel's factors (the genes) are located on the chromosomes, which accounts for the similarity of their behavior during meiosis and fertilization.

Sex determination in animals is dependent upon the chromosomes. Usually females are XX and males are XY. Solid experimental support for the chromosomal theory of inheritance came when Morgan and his group were able to determine that the white-eye allele in *Drosophila* is on the X chromosome.

Alleles on the X chromosome are called X-linked alleles. Therefore, when doing X-linked genetics problems, it is the custom to indicate the sexes by using sex chromosomes and to indicate the alleles by superscripts attached to the X. The Y is blank because it does not carry these genes.

All the alleles on a chromosome form a linkage group. Linked alleles do not obey Mendel's laws because they tend to go into the same gamete together. Crossing-over can cause recombinant gametes and recombinant phenotypes to occur. The percentage of recombinant phenotypes is used to measure the distance between genes and to map the chromosomes.

12.3 Chromosomal Mutations

Chromosomal mutations fall into two categories: changes in chromosome number and changes in chromosomal structure.

Monosomy occurs when an individual has only one of a particular type of chromosome (2n − 1); trisomy occurs when an individual has three of a particular type of chromosome (2n + 1). Polyploidy occurs when the eukaryotic individual has more than two complete sets of chromosomes.

Changes in chromosomal structure include inversions, translocations, deletions, and duplications.

Reviewing the Chapter

1. Compare the F_2 phenotypic ratio for a cross involving a dominant allele, an incompletely dominant allele, and codominant alleles. 188
2. Show that if a dihybrid cross between two heterozygotes involves an epistatic gene, you do not get a ratio of 9:3:3:1. 189
3. If you were studying a particular population of organisms, how would you recognize that several traits are being affected by a pleiotropic gene? 189
4. Explain inheritance by multiple alleles. List the human blood types and give the possible genotypes for each. 190
5. Explain why traits controlled by polygenes show continuous variation that can be measured quantitatively, have many intermediate forms, and produce a distribution in the F_2 generation that follows a bell-shaped curve. 190
6. Give examples to show that the environment can have an extreme effect on the phenotype despite the genotype. 191
7. What is the chromosomal theory of inheritance? List the ways in which genes and chromosomes behave similarly during meiosis and fertilization. 192
8. How is sex determined in humans? Which sex determines the sex of the offspring? 192
9. How did a *Drosophila* cross involving an X-linked gene help investigators show that certain genes are carried on certain chromosomes? 192

10. Show that a dihybrid cross between a heterozygote and a recessive homozygote involving linked genes does not produce the expected 1:1:1:1 ratio. What is the significance of the small percentage of recombinants that occurs among the offspring? 194–95
11. What are the two types of chromosomal mutations? What is a monosomy? A trisomy? Why would you expect a monosomy to be more lethal than a trisomy? Why is polyploidy not generally found in animals? In what way may polyploidy have assisted plant evolution? 196
12. What four types of changes in chromosomal structure were discussed? How might it be possible for a duplication and a deletion to occur at the same time? 197

Testing Yourself

Choose the best answer for each question. For questions 1–4, match the statements that follow to the items in the key.

Key:

a. multiple alleles
b. incomplete dominance
c. polygenes
d. epistatic gene
e. pleiotropic gene

1. A cross between oblong and round squash produced oval squash.
2. Although most people have an IQ of about 100, IQ generally ranges from about 50 to 150.
3. Investigators noted that whenever a particular species of plant had narrow instead of broad leaves, it was also short and yellow, instead of tall and green.
4. In humans, there are three alleles possible at the chromosomal locus that determine blood type.
5. When a white-eyed *Drosophila* female occurs,
   a. both parents could have red eyes.
   b. the female parent could have red eyes, but the male parent has to have white eyes.
   c. both parents must have white eyes.
   d. the female also has a gray body.
   e. Both a and b are correct.
6. Investigators found that a cross involving the mutant genes *a* and *b* produced 30% recombinants, a cross involving *a* and *c* produced 5% recombinants, and a cross involving *c* and *b* produced 25% recombinants. Which is the correct order of the genes?
   a. *a, b, c*
   b. *a, c, b*
   c. *b, a, c*
   d. Both a and b are correct.
   e. Both b and c are correct.
7. A boy is color blind (X-linked recessive) and has a straight hairline (autosomal recessive). Which could be the genotype of his mother?
   a. *bbww*
   b. X^bYWw
   c. bbX^wX^w
   d. X^BX^bWw
   e. X^WX^wBb
8. Which two of these chromosomal mutations are most likely to occur when homologous chromosomes are undergoing synapsis?
   a. inversion and translocation
   b. deletion and duplication
   c. deletion and inversion
   d. duplication and translocation
   e. inversion and duplication
9. Investigators do a dihybrid cross between two heterozygotes and get about a 3:1 ratio among the offspring. The reason must be due to
   a. polygenes.
   b. pleiotropic genes.
   c. linked genes.
   d. epistatic genes.
   e. multiple alleles.
10. In snakes, the recessive genotype *cc* causes the animal to be albino despite the inheritance of the dominant allele (*B* = black). What would be the results of a cross between two snakes with the genotype *CcBB*?
    a. all black snakes
    b. all white snakes
    c. 9:3:3:1
    d. 3 black : 1 albino
    e. 1 black : 1 albino

Additional Genetics Problems*

1. Chickens that are homozygous for the frizzled trait have feathers that are weak and stringy. When raised at low temperatures, the birds have circulatory, digestive, and hormonal problems. What inheritance pattern explains these results?
2. In radish plants, the shape of the radish may be long *(LL)*, round *(ll)*, or oval *(Ll)*. If oval is crossed with oval, what proportion of offspring will also be oval?
3. If blood type in cats were controlled by three codominant and multiple alleles, how many genotypes and phenotypes would occur? Could the cross $AC \times BC$ produce a cat with *AB* blood type?
4. In *Drosophila*, S = normal, s = sable body; W = normal, w = miniature wing. In a cross between a heterozygous normal fly and a sable-bodied, miniature-winged fly, the results were 99 normal flies, 99 with a sable body and miniature wings, 11 with a normal body and miniature wings, and 11 with a sable body and normal wings. What inheritance pattern explains these results? How many map units separate the genes for sable body and miniature wing?
5. In *Drosophila*, the gene that controls red eye color (dominant) versus white eye color is on the X chromosome. What are the expected phenotypic results if a heterozygous female is crossed with a white-eyed male?
6. Bar eyes in *Drosophila* is dominant and X-linked. What phenotypic ratio is expected for reciprocal crosses between true-breeding flies?
7. In *Drosophila*, a male with bar eyes and miniature wings is crossed with a female who has normal eyes and normal wings. Half the female offspring have bar eyes and miniature wings and half have bar eyes and normal wings. What is the genotype of the female parent?

* Answers to Additional Genetics Problems appear in Appendix A.

Thinking Scientifically

1. A man known to be a carrier of a genetic disease affecting a lysosomal enzyme has no symptoms of the disease. However, when the enzyme activity is actually measured, it is seen that the man's cells have half the activity as those from someone who is homozygous dominant. Can this allele be considered to be completely dominant? What would such a situation suggest for the treatment of patients with this genetic disease?
2. A flowering shrub grown in a large garden has produced offspring identical to the parent for several generations. When the owner of the garden sends seeds to a friend in another county, the friend reports a very different flower color in the plant. Two hypotheses are possible: a new mutation has occurred; or there is an environmental effect on flower color. How would you test these two hypotheses?

Bioethical Issue

Do you approve of choosing a baby's gender even before it is conceived? As you know, the sex of a child is dependent upon whether an X-bearing sperm or a Y-bearing sperm enters the egg. A new technique has been developed that can separate X-bearing sperm from Y-bearing sperm. First, the sperm are dosed with a DNA-staining chemical. Because the X chromosome has slightly more DNA than the Y chromosome, it takes up more dye. When a laser beam shines on the sperm, the X-bearing sperm shine a little more brightly than the Y-bearing sperm. A machine sorts the sperm into two groups on this basis. The results are not perfect. Following artificial insemination, there's about an 85% success rate for a girl and about a 65% success rate for a boy.

Some might argue that it goes against nature to choose gender. But what if the mother is a carrier of an X-linked genetic disorder such as hemophilia or Duchenne muscular dystrophy? Is it acceptable to bring a child into the world with a genetic disorder that may cause an early death? Wouldn't it be better to select sperm for a girl, who at worst would be a carrier like her mother? Some authorities do not find gender selection acceptable even for this reason. Once you separate reproduction from the sex act, they say, it opens the door to children that have been genetically designed in every way. As a society, should we accept certain ways of interfering with nature and not accept other ways? Why or why not?

Understanding the Terms

autosome 192
chromosomal mutation 197
chromosomal theory of inheritance 192
codominance 188
epistasis 189
incomplete dominance 188
linkage group 194
multiple allele 190
mutation 196
pleiotropy 189
polygenic inheritance 190
polyploid (polyploidy) 196
sex chromosome 192
X-linked 192

Match the terms to these definitions:

a. ______________ Any chromosome other than a sex chromosome.
b. ______________ Condition in which an organism has more than two complete sets of chromosomes.
c. ______________ Allele located on the X chromosomes that controls a trait unrelated to sex.
d. ______________ Pattern of inheritance in which a trait is controlled by several allelic pairs; each dominant allele contributes in an additive and like manner.
e. ______________ Condition in which an organism has more than two complete sets of chromosomes.

Web Connections

Exploring the Internet

http://www.mhhe.com/biosci/genbio/mader
(click on *Biology 7/e*)

The *Biology 7/e* Online Learning Center provides many resources for studying the material in this chapter including links to the following sites:

Genetic Linkage describes how linked genes yield distorted ratios in experimental crosses. This site is a thorough treatment of the subject, and includes how to determine the distance between linked genes, and detecting gene linkages in humans.

http://www.cc.ndsu.nodak.edu/instruct/mcclean/plsc431/linkage/

Variation in Chromosome Structure explores variations in chromosome number in plants and animals, including how variations in structure are related to human disorders.

http://www.cc.ndsu.nodak.edu/instruct/mcclean/plsc431/chromstruct/

and

Variation in Chromosome Number explores variations in chromosome number in plants and animals, including how variations in number are related to human disorders.

http://www.cc.ndsu.nodak.edu/instruct/mcclean/plsc431/chromnumber/

Pedigrees. A lengthy example of how pedigrees are used to determine patterns of inheritance in humans, from the MIT hypertextbook.

http://esg-www.mit.edu:8001/esgbio/mg/pedigrees.html

The Turner Center. A site with many links to other sources of information on aneuploidies.

http://www.aaa.dk/turner/engelsk/Download.htm#orientation

Self Quiz. This site has a number of questions in a quiz directly related to this chapter.

http://brooks.pvt.k12.ma.us/~bheun/apsq13.html

Human Genetics

chapter

chapter concepts

Is thrill-seeking behavior genetic?

Scarcely a week goes by without a newspaper or television report of another discovery in the exploding field of human genetics. Some researchers believe they have found genes which predispose a person toward alcoholism or a need for thrill-seeking behavior. While genes undoubtedly play a role in almost all human behavior, they act in conjunction *with* the environment. Few aspects of human behavior are controlled only by genes or only by environment. The question of nature versus nurture—what percentage of the distribution of a trait is controlled by genes and what percentage is controlled by environment—cannot now and may never be answered for most traits; both are important.

Genetic counseling is more extensive today because of the availability of so many prenatal tests to detect chromosomal and gene mutations. Amniocentesis followed by karyotyping can indicate whether a developing child has Down syndrome or one of the other chromosomal abnormalities. Specific chromosomes can be tested for the presence of a mutant allele for cystic fibrosis, neurofibromatosis, and sickle-cell disease, among others.

13.1 Inheritance of Chromosomes

Somatic (body) cells in humans have 46 chromosomes. To view the chromosomes, a cell can be photographed just prior to division so that a picture of the chromosomes is obtained. The picture may be entered into a computer and the chromosomes electronically arranged by pairs (Fig. 13.1). The members of a pair have the same size, shape, and constriction (location of the centromere), and they also have the same characteristic banding pattern upon staining. The resulting display of pairs of chromosomes is called a **karyotype.** Both males and females normally have 23 pairs of chromosomes, but one of these pairs is of unequal length in males. The larger chromosome of this pair is the **X chromosome** and the smaller is the **Y chromosome.** These are called the **sex chromosomes** because they contain the genes that determine sex. The other chromosomes, known as **autosomes,** include all the pairs of chromosomes except the X and Y chromosomes. In a karyotype, autosomes are usually ordered by size and numbered from the largest to smallest; the sex chromosomes are identified separately.

Sometimes there are medical reasons for doing a karyotype on an unborn child. As discussed on page 212 chorionic villi sampling and amniocentesis allow a physician to obtain a sample of cells in order to do prenatal testing including a karyotype. The chorionic villi project from an extraembryonic membrane in the region of the placenta where maternal blood exchange materials with embryonic blood. During **chorionic villi sampling,** a suction tube is used to remove a sample of chorionic villi cells for diagnostic purposes. The amniotic cavity contains amniotic fluid that protects the embryo from drying out and from mechanical

1. Blood is centrifuged to separate out blood cells.

2. Only white blood cells are transferred and treated to stop cell division.

3. Sample is fixed, stained, and spread on a microscope slide.

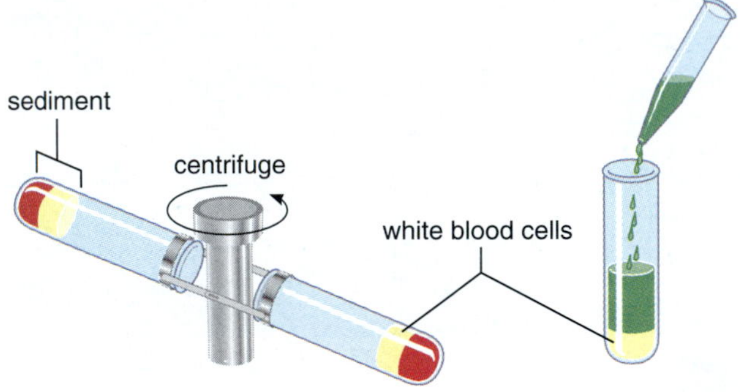

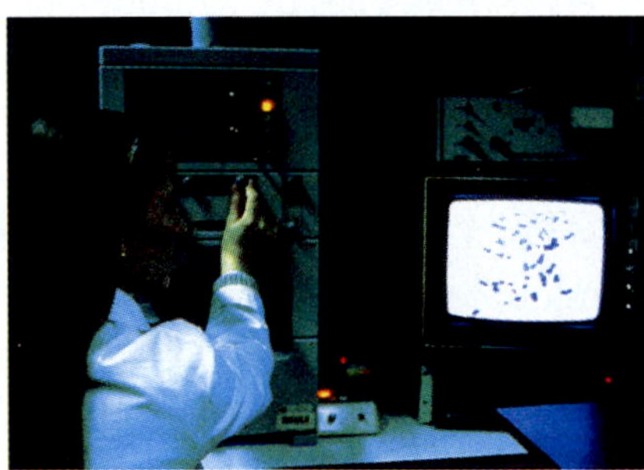

4. Slide is examined microscopically, and the chromosomes are photographed. Computer arranges the chromosomes into pairs.

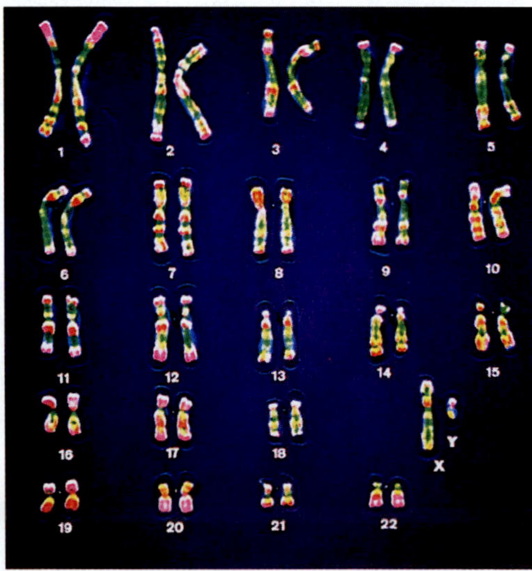

5. Karyotype: Chromosomes are paired by size, centromere location, and banding patterns.

Figure 13.1 Human karyotype preparation.
A karyotype displays the chromosomes of an individual from the largest to the smallest autosomal pairs. The sex chromosomes are either XX (females) or XY (males). As illustrated here, the stain used on the chromosomes can result in a banded appearance. The bands help researchers identify and analyze the chromosomes.

shock. During **amniocentesis,** a long needle is inserted into the amniotic cavity and a small sample of fluid and sloughed-off fetal cells are removed. The chromosomes of these cells can be karyotyped to see if the unborn child has a normal karyotype.

Nondisjunction during oogenesis or spermatogenesis can lead to the birth of a child with an abnormal chromosome number.

Nondisjunction

Gamete formation in humans involves meiosis, the type of cell division that reduces the chromosome number by one-half. **Nondisjunction** is the failure of homologous chromosomes to separate during meiosis I or daughter chromosomes to separate during meiosis II (Fig. 13.2*a*). Nondisjunction leads to gametes that have too few or too many chromosomes.

When these abnormal gametes fuse with normal gametes at fertilization, a monosomy or a trisomy can result. In a monosomy, instead of a pair of some particular chromosome, there is only one of that kind. In a trisomy there are three of a particular kind of chromosome instead of just two. Figure 13.2*b* lists various syndromes due to the inheritance of abnormal chromosome numbers. A *syndrome* is a group of symptoms that appear together when there is a particular medical condition.

The karyotype of a human being usually has 22 pairs of autosomes and one pair of sex chromosomes. Nondisjunction during meiosis leads to the inheritance of an abnormal chromosome number.

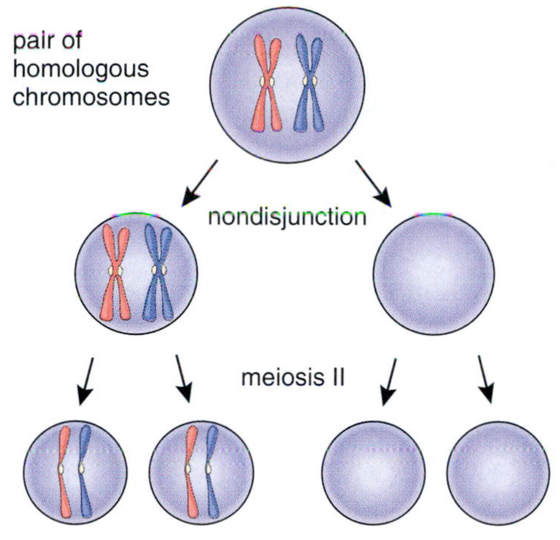

a.

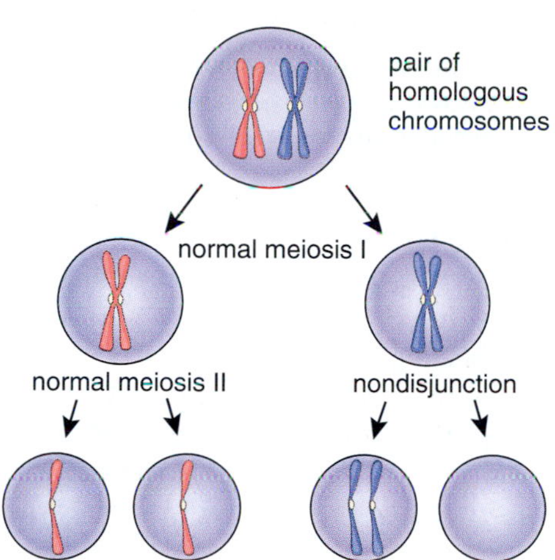

Syndrome	Description	Chromosomes	Incidences (newborns)
Down	Mental retardation; wide, flat face with upper eyelid fold, short stature; abnormal palm creases	Trisomy 21	1/800
Patau	Malformed internal organs, face, and head; extra digits; mental retardation	Trisomy 13	1/15,000
Edward	Malformed internal organs, face, and head; extreme muscle tone	Trisomy 18	1/6,000
Turner	Short stature; webbed neck; broad chest; no sexual maturity	XO	1/6,000
Klinefelter	Breast development possible; testes underdeveloped; no facial hair	XXY (or XXXY)	1/1,500
Triplo-X	Tall and thin with menstrual irregularities	XXX (or XXXX)	1/1,500
Jacob	Taller than average; persistent acne; speech and learning problems possible	XYY	1/1,000

b. From Robert F. Weaver and Philip W. Hedrick, *Genetics, 2d ed.* Copyright 1992 WCB.

Figure 13.2 Nondisjunction of autosomes during meiosis.
a. Nondisjunction can occur during meiosis I if homologous chromosomes fail to separate, and during meiosis II if the daughter chromosome fails to separate. In either case, abnormal gametes carry an extra chromosome or lack a chromosome. **b.** Frequency of syndromes caused by abnormal chromosomal numbers.

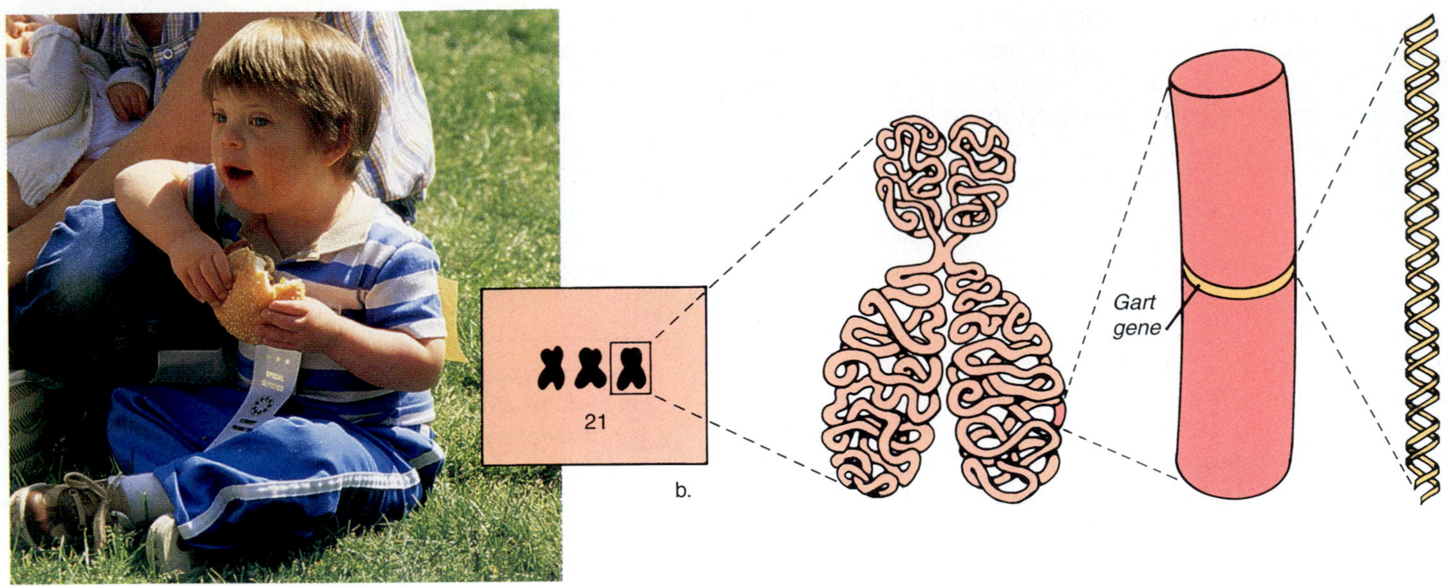

Figure 13.3 Down syndrome.
a. Characteristics of Down syndrome include a wide, flat face and narrow, slanting eyelids. b. The karyotype of an individual with Down syndrome often shows an extra chromosome 21. An extra *Gart* gene may account for the mental retardation that often accompanies the disorder.

Down Syndrome

Persons with Down syndrome, described in Figures 13.2*b* and 13.3, usually have three copies of chromosome 21 in their karyotype because the egg had two copies instead of one. The chances of a woman having a Down syndrome child increase rapidly with age, starting at about age 40, and the reasons for this are still being determined. In 23% of the cases of Down syndrome studied, the sperm, rather than the egg, had the extra chromosome 21. In about 5% of cases in either the sperm or the egg, the extra chromosome is attached to another chromosome, often chromosome 14. This abnormal chromosome arose because a translocation occurred between chromosomes 14 and 21 in one of the parents or even in a relative who lived generations earlier. Therefore, when Down syndrome is due to a translocation, it is not age related and instead tends to run in the family of either the father or the mother.

It is known that the genes that cause Down syndrome are located on the bottom third of chromosome 21 (Fig. 13.3*b*), and extensive investigative work has focused on discovering the specific genes responsible for the characteristics of the syndrome. Thus far, investigators have discovered several genes that may account for various conditions seen in persons with Down syndrome. For example, they have located genes most likely responsible for the increased tendency toward leukemia, cataracts, accelerated rate of aging, and mental retardation. The gene for mental retardation, dubbed the *Gart* gene, causes an increased level of purines in blood, a finding associated with mental retardation. It is hoped that someday it will be possible to control the expression of the *Gart* gene even before birth so that at least this symptom of Down syndrome can be suppressed.

Abnormal Sex Chromosomal Inheritance

Abnormal sex chromosomal inheritance can be due to receiving an abnormal number of sex chromosomes or an abnormal X chromosome.

Abnormal Number Syndromes

From birth, an XO individual with *Turner syndrome* has only one sex chromosome, an X; the O signifies the absence of a second sex chromosome. Turner females are short, have a broad chest, and folds of skin on the back of the neck. Even so, there is little reason to suspect a chromosomal abnormality until puberty. Turner females do not undergo puberty or menstruate, and there is a lack of breast development (Fig. 13.4*a*). The ovaries, oviducts, and uterus are very small and nonfunctional. They are usually of normal intelligence and can lead fairly normal lives, but they are also infertile even if they receive hormone supplements.

A male with *Klinefelter syndrome* has two or more X chromosomes in addition to a Y chromosome and is sterile. The testes and prostate gland are underdeveloped, and there is no facial hair. There may be some breast development (Fig. 13.4*b*). Affected individuals have large hands and feet and very long arms and legs. They are usually slow to learn but not mentally retarded unless they inherit more than two X chromosomes.

A *triplo-X* individual has more than two X chromosomes. It might be supposed that the XXX female is especially feminine, but this is not the case. Although in some cases there is a tendency toward learning disabilities, most triplo-X females have no apparent physical abnormalities except there may be menstrual irregularities, including

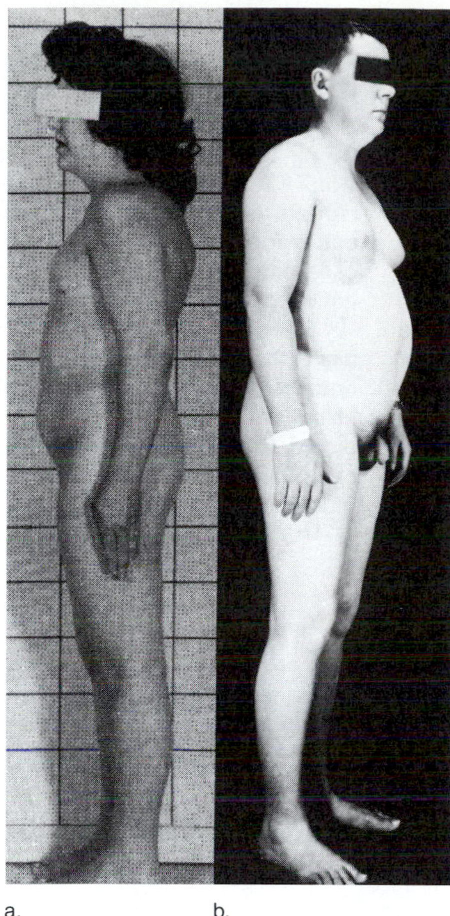

Figure 13.4 Abnormal sex chromosome inheritance.
a. A female with Turner syndrome (XO) is distinguished by a thick neck, short stature, and immature sexual features. **b.** A male with Klinefelter syndrome (XXY) has immature sex organs and shows breast development.

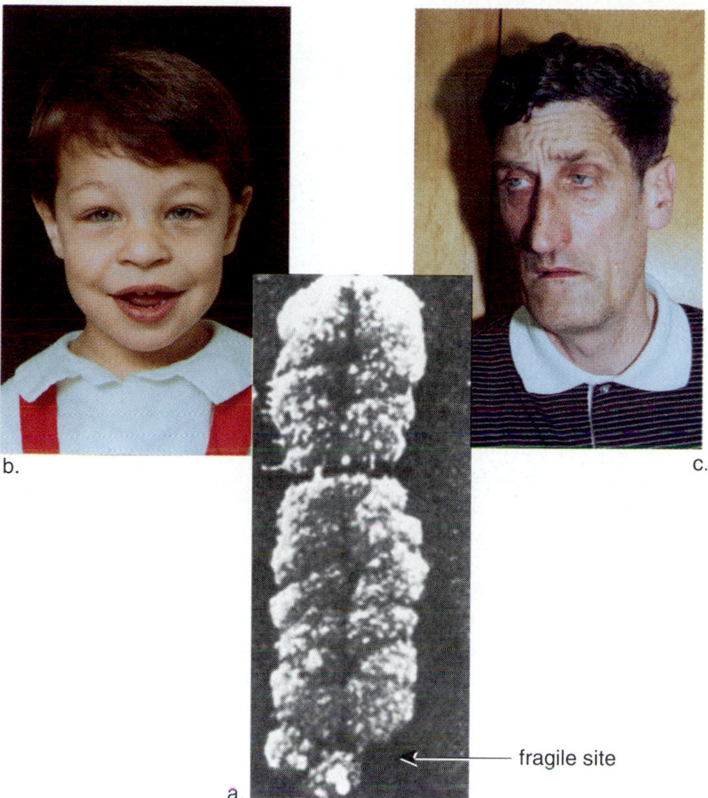

Figure 13.5 Fragile X syndrome.
a. An arrow points out the fragile site of this fragile X chromosome. **b.** A young person with the syndrome appears normal but **(c)** with age, the elongated face has a prominent jaw, and the ears noticeably protrude.

early onset of menopause. Triplo-X individuals have an increased risk of having daughters who are also triplo-X or XXY sons.

XYY males with Jacob syndrome result from nondisjunction during meiosis II (see Fig. 13.2*a*). Affected males usually are taller than average, suffer from persistent acne, and tend to have speech and reading problems. At one time it was suggested that these men were likely to be criminally aggressive, but it has since been shown that the incidence of such behavior among them may be no greater than among XY males.

Fragile X Syndrome

Males outnumber females by about 25% in institutions for the mentally retarded. In some of these males, the X chromosome is nearly broken, leaving the tip hanging by a flimsy thread. These males are said to have fragile X syndrome (Fig. 13.5).

As children, fragile X syndrome individuals may be hyperactive or autistic; their speech is delayed in development and often repetitive in nature. As adults, they have large testes and big, usually protruding ears. They are short in stature, but the jaw is prominent and the face is long and narrow. Stubby hands, lax joints, and a heart defect may also occur. Fragile X chromosomes also occurs in females, but when symptoms do appear in females, they tend to be less severe.

Fragile X syndrome is inherited in an unusual pattern; it passes from symptomless male carrier to a severely affected grandson, as discussed in the reading on page 206. The DNA sequence at the fragile site was isolated and found to have trinucleotide repeats. The base triplet CGG (cytosine, guanine, guanine) was repeated over and over again. There are about 6 to 50 copies of this repeat in normal persons but over 230 copies in persons with fragile X syndrome.

The most common autosomal chromosome abnormality is trisomy 21 (Down syndrome). Individuals also sometimes inherit an abnormal number of sex chromosomes or an abnormal X chromosome, as in fragile X syndrome.

Fragile X Syndrome

Fragile X syndrome is one of the most common genetic causes of mental retardation, second only to Down syndrome. It affects about one in 1,500 males and one in 2,500 females and is seen in all ethnic groups. It is called fragile X syndrome because its diagnosis used to be dependent upon observing an X chromosome whose tip is attached to the rest of the chromosome by a thin thread.

The inheritance pattern of fragile X syndrome is not like any other pattern we have studied (Fig. 13A). The chance of being affected increases in successive generations almost as if the pattern of inheritance switches from being a recessive one to a dominant one. Then, too, an unaffected grandfather can have grandchildren with the disorder; in other words, he is a carrier for an X-linked disease. This is contrary to what occurs with other mutant alleles on the X chromosome; in those cases, the male always shows the disorder.

In 1991, the DNA sequence at the fragile site was isolated and found to have trinucleotide repeats. The base triplet, CGG, was repeated over and over again. There are about 6 to 50 copies of this repeat in normal persons but over 230 copies in persons with fragile X syndrome. Carrier males have what is now termed a premutation; they have between 50 and 230 copies of the repeat and no symptoms. Both daughters and sons receive the premutation but only the daughters pass on the full mutation—that is, over 230 copies of the repeat. Any male, even those with fragile X syndrome and over 230 repeats, passes on at most the premutation number of repeats. It is unknown what causes the difference between males and females.

This type of mutation—called by some a dynamic mutation, because it changes, and by others an expanded trinucleotide repeat, because the number of triplet copies increases—is now known to characterize other conditions, such as certain forms of dystrophy, spinocerebellar ataxia type 1, and Huntington disease. With Huntington disease, the age of onset of the disorder is roughly correlated with the number of repeats, and the disorder is more likely to have been inherited from the father. For autosomal conditions, we would expect the sex of the parent to play no role in inheritance. The present exceptions have led to the genomic imprinting hypothesis—that the sperm and egg carry chromosomes that have been "imprinted" differently. Imprinting is believed to occur during gamete formation, and thereafter, the genes are expressed one way if donated by the father and another way if donated by the mother. Perhaps when we discover why more repeats are passed on by one parent than the other, we will discover the cause of so-called genomic imprinting.

What might cause repeats to occur in the first place? Something must go wrong during DNA replication prior to cell division. The difficulty that causes triplet repeats is not known. But when DNA codes for cellular proteins, the presence of repeats undoubtedly leads to nonfunctioning or malfunctioning proteins.

Scientists have developed a new technique that can identify repeats in DNA, and they expect that this technique will help them find the genes for other human disorders. They expect expanded trinucleotide repeats to be a very common mutation indeed.

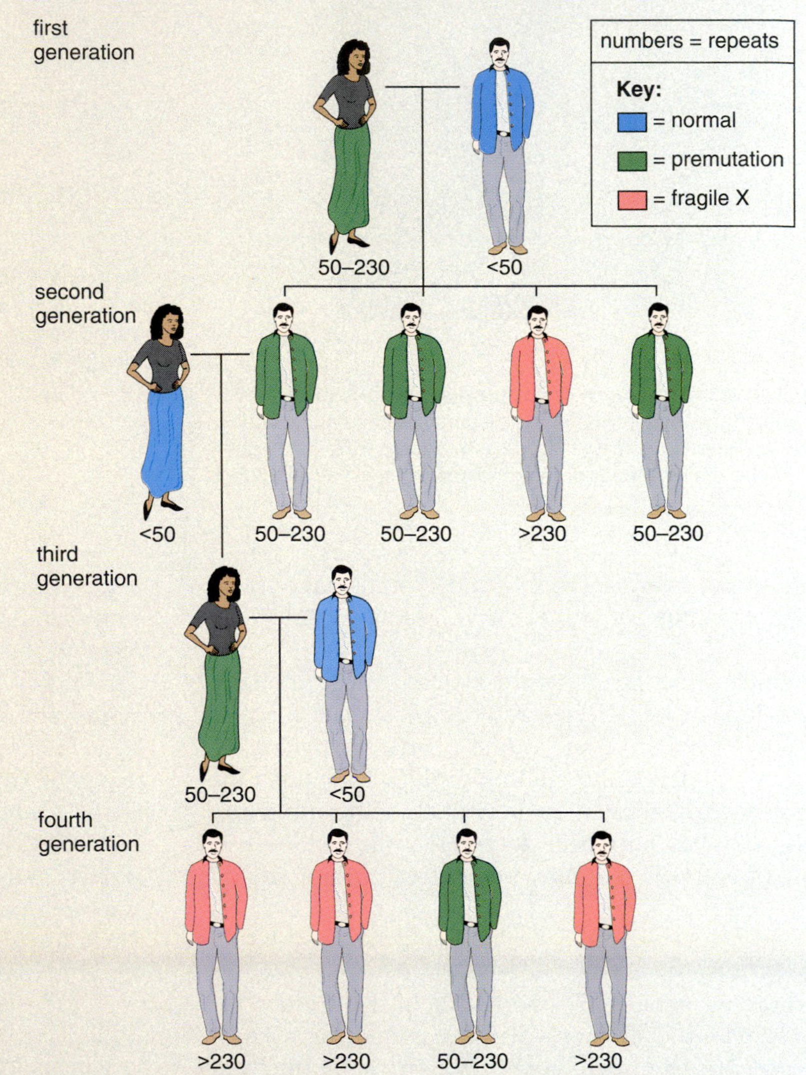

Figure 13A **Fragile X syndrome.**
Fragile X syndrome is a disorder caused by the presence of triplet repeats at a particular locus. Affected persons have over 230 repeats, a premutation is 50–230 repeats, and normal persons have fewer than 50 repeats. In each successive generation there are more affected individuals, usually males. The mutation is passed on by women with only a premutation number of repeats; men, even if affected, do not pass on the mutation.

13.2 Autosomal Genetic Disorders

Genetic disorders are medical conditions caused by alleles inherited from the parents. Some of these conditions are controlled by autosomal dominant or recessive alleles.

When a genetic disorder is a simple autosomal dominant, an individual with the alleles *AA* or *Aa* will have the disorder. When a genetic disorder is a simple autosomal recessive, only individuals with the alleles *aa* will have the disorder. Genetic counselors often construct pedigree charts to determine whether a condition is dominant or recessive. A **pedigree chart** shows the pattern of inheritance for a particular condition. Consider these two possible patterns of inheritance:

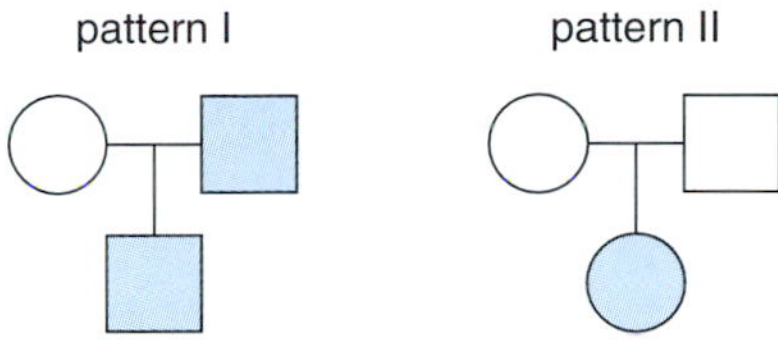

In both patterns, males are designated by squares and females by circles. Shaded circles and squares are affected individuals. A line between a square and a circle represents a union. A vertical line going downward leads, in these patterns, to a single child. (If there are more children, they are placed off a horizontal line.) Which pattern of inheritance do you suppose represents an autosomal dominant characteristic, and which represents an autosomal recessive characteristic?

In pattern I, the child is affected, as is one of the parents. When a disorder is dominant, an affected child usually has at least one affected parent. Of the two patterns, this one shows a dominant pattern of inheritance. Figure 13.6 shows a typical pedigree chart for an autosomal dominant disorder. Other ways to recognize an autosomal dominant pattern of inheritance are also given.

In pattern II, the child is affected but neither parent is; this can happen if the condition is recessive and the parents are *Aa*. Notice that the parents are **carriers** because they appear to be normal but are capable of having a child with a genetic disorder. Figure 13.7 shows a typical pedigree chart for an autosomal recessive genetic disorder. Other ways to recognize a recessive pattern of inheritance are also given in the figure.

It is important to realize that "chance has no memory," therefore, each child born to heterozygous parents has a 25% chance of having the disorder. In other words, it is possible that if a heterozygous couple has four children, each child might have the condition.

Dominant and recessive alleles have different patterns of inheritance.

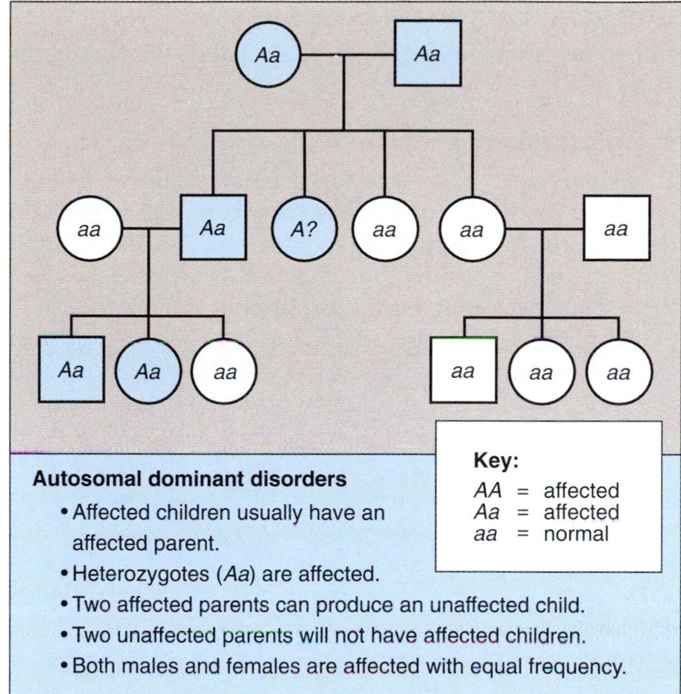

Figure 13.6 Autosomal dominant pedigree chart.
The list gives ways to recognize an autosomal dominant disorder.

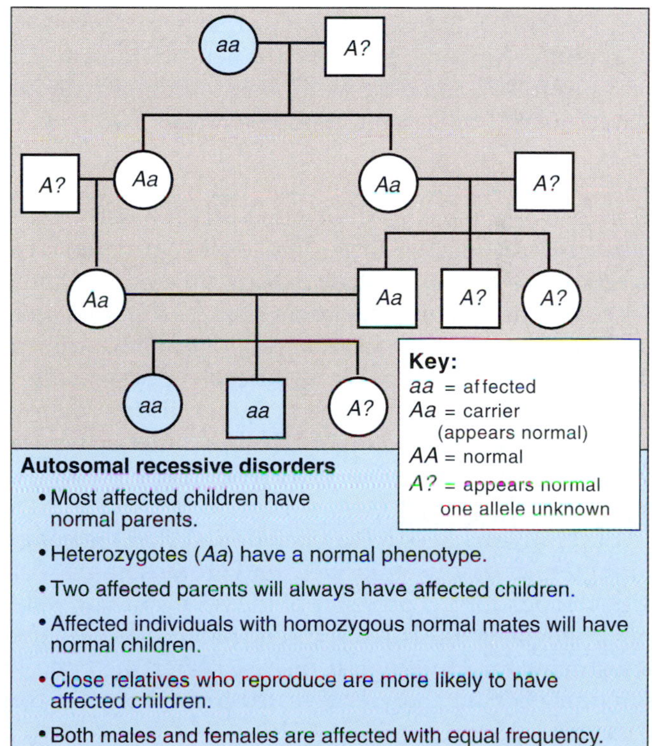

Figure 13.7 Autosomal recessive pedigree chart.
The list gives ways to recognize an autosomal recessive disorder. Only those affected with the recessive genetic disorder are shaded.

Autosomal Dominant Disorders

Of the many autosomal dominant disorders, we will discuss only two.

Neurofibromatosis

Neurofibromatosis,[1] sometimes mistakenly called von Recklinghausen disease, is one of the most common genetic disorders. It affects roughly one in 3,000 people, including an estimated 100,000 in the United States. It is seen equally in every racial and ethnic group throughout the world.

At birth or later, the affected individual may have six or more large, tan spots (known as cafe-au-lait) on the skin. Such spots may increase in size and number and may get darker. Small benign tumors (lumps) called neurofibromas may occur under the skin or in various organs. Neurofibromas are made up of nerve cells and other cell types.

This genetic disorder shows *variable expressivity.* In most cases, symptoms are mild and patients live a normal life. In some cases, however, the effects are severe. Skeletal deformities, including a large head, are seen, and eye and ear tumors can lead to blindness and hearing loss. Many children with neurofibromatosis have learning disabilities and are hyperactive.

In 1990, researchers isolated the gene for neurofibromatosis, which was known to be on chromosome 17. By analyzing the DNA (deoxyribonucleic acid), they determined that the gene was huge and actually included three smaller genes. This was only the second time that nested genes have been found in humans. The gene for neurofibromatosis is a tumor-suppressor gene active in controlling cell division. When it mutates, a benign tumor develops.

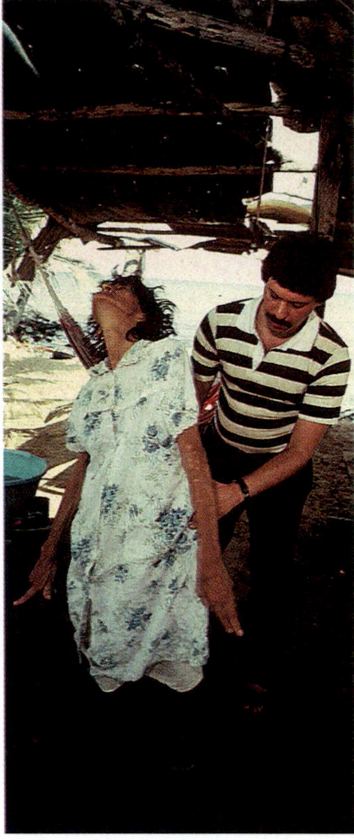

Figure 13.8 Huntington disease.
Persons with this condition gradually lose psychomotor control of the body. At first there are only minor disturbances, but the symptoms become worse over time. Unfortunately there is no known cure for Huntington disease.

Huntington Disease

One in 20,000 persons in the United States has *Huntington disease,* a neurological disorder that leads to progressive degeneration of brain cells, which in turn causes severe muscle spasms and personality disorders (Fig. 13.8). Most people appear normal until they are of middle age and have already had children who might also be stricken. Occasionally, the first signs of the disease are seen in these children when they are teenagers or even younger. There is no effective treatment, and death comes ten to 15 years after the onset of symptoms.

Several years ago, researchers found that the gene for Huntington disease was located on chromosome 4. A test was developed for the presence of the gene, but few want to know if they have inherited the gene because as yet there is no treatment for Huntington disease. After the gene was isolated in 1993, an analysis revealed that it contains many repeats of the base triplet CAG (cytosine, adenine, and guanine). Normal persons have 11 to 34 copies of the triplet, but affected persons tend to have 42 to more than 120 copies. The more repeats present, the earlier the onset of Huntington disease and the more severe the symptoms. It also appears that persons most at risk have inherited the disorder from their fathers. The latter observation is consistent with a new hypothesis called **genomic imprinting.** The genes are imprinted differently during formation of sperm and egg, and therefore the sex of the parent passing on the disorder becomes important.

It is now known that there are a number of other genetic diseases whose severity and time of onset vary according to the number of triplet repeats present within the gene. Moreover, the genes for these disorders are also subject to genomic imprinting.

There are many autosomal dominant disorders in humans. Among these are neurofibromatosis and Huntington disease, as listed in Table 13A on page 213.

[1] Although neurofibromatosis is commonly associated with Joseph Merrick, the severely deformed nineteenth-century Londoner depicted in *The Elephant Man,* researchers today believe Merrick actually suffered from a much rarer disorder called Proteus syndrome.

Autosomal Recessive Disorders

Of the many autosomal recessive disorders, we will discuss only three.

Cystic Fibrosis

Cystic fibrosis is the most common lethal genetic disease among Caucasians in the United States. About one in 20 Caucasians is a carrier, and about one in 2,500 children born to this group has the disorder. In these children, the mucus in the bronchial tubes and pancreatic ducts is particularly thick and viscous, interfering with the function of the lungs and pancreas. To ease breathing, the thick mucus in the lungs has to be manually loosened periodically, but still the lungs become infected frequently. The clogged pancreatic ducts prevent digestive enzymes from reaching the small intestine, and to improve digestion patients take digestive enzymes mixed with applesauce before every meal.

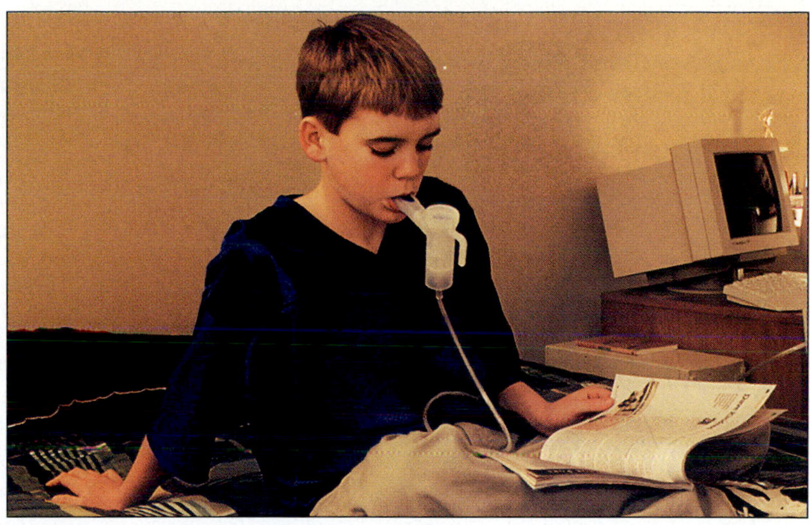

Figure 13.9 Cystic fibrosis therapy.
Antibiotic therapy is used to control lung infections of cystic fibrosis patients. The antibiotic tobramycin can be aerosolized and administered using a nebulizer. It is inhaled twice daily for about fifteen minutes.

In the past few years, much progress has been made in our understanding of cystic fibrosis, and new treatments have raised the average life expectancy to 17 to 28 years of age (Fig. 13.9). Research has demonstrated that chloride ions (Cl^-) fail to pass through plasma membrane channel proteins in these patients. Ordinarily, after chloride ions have passed through the membrane, water follows. It is believed that lack of water is the cause of abnormally thick mucus in bronchial tubes and pancreatic ducts. The cystic fibrosis gene, which is located on chromosome 7, has been isolated, and placed in a nasal spray. The spray restores about 25% of the ability of chloride ions to cross the membrane. Researchers are hopeful that gene therapy will one day be a realistic approach to curing the disease.

Genetic testing for the gene in adult carriers and in fetuses is possible; if a disease-causing gene is present, couples have to consider that gene therapy may be possible some day.

Tay-Sachs Disease

Tay-Sachs disease is a well-known genetic disease that usually occurs among Jews in the United States, most of whom are of central and eastern European descent. At first, it is not apparent that a baby has Tay-Sachs disease. However, development begins to slow down between four months and eight months of age, and neurological impairment and psychomotor difficulties then become apparent. The child gradually becomes blind and helpless, develops uncontrollable seizures, and eventually becomes paralyzed. There is no treatment or cure for Tay-Sachs disease, and most affected individuals die by the age of three or four.

Tay-Sachs disease results from a lack of the enzyme hexosaminidase A (Hex A) and the subsequent storage of its substrate, a fatty substance known as glycosphingolipid, in lysosomes. Although more and more lysosomes build up in many body cells, the primary sites of storage are the cells of the brain, which accounts for the onset, and the progressive deterioration of psychomotor functions.

Carriers of Tay-Sachs have about half the level of Hex A activity found in normal individuals. Prenatal diagnosis of the disease also is possible following either amniocentesis or chorionic villi sampling.

Phenylketonuria (PKU)

Phenylketonuria (PKU) occurs once in 5,000 births, so it is not as frequent as the disorders previously discussed. However, it is the most commonly inherited metabolic disorder to affect nervous system development. First cousins who marry are more apt to have a PKU child.

Affected individuals lack an enzyme that is needed for the normal metabolism of the amino acid phenylalanine, and an abnormal breakdown product, a phenylketone, accumulates in the urine. The PKU gene is located on chromosome 12, and there is a prenatal DNA test for the presence of this allele. Years ago, the urine of newborns was tested at home for phenylketone in order to detect PKU. Presently, newborns are routinely tested in the hospital for elevated levels of phenylalanine in the blood. If necessary, newborns are placed on a diet low in phenylalanine, which must be continued until the brain is fully developed, around age 7, or else severe mental retardation develops.

There are many autosomal recessive disorders in humans. Among these are Tay-Sachs disease, cystic fibrosis, and phenylketonuria (PKU) (see Table 13A, page 213).

Beyond Simple Mendelian Inheritance

There are patterns of inheritance other than simple dominant and recessive traits.

Polygenic Inheritance

Polygenic inheritance occurs when one trait is governed by two or more loci. Each dominant allele has a quantitative effect on the phenotype, and these effects are additive. The result is a continuous variation of phenotypes, resulting in a distribution of these phenotypes that resembles a bell-shaped curve. The more genes involved, the more continuous the variation and the distribution of the phenotypes. Also, environmental effects cause many intervening phenotypes; in the case of height, differences in nutrition ensure a bell-shaped curve.

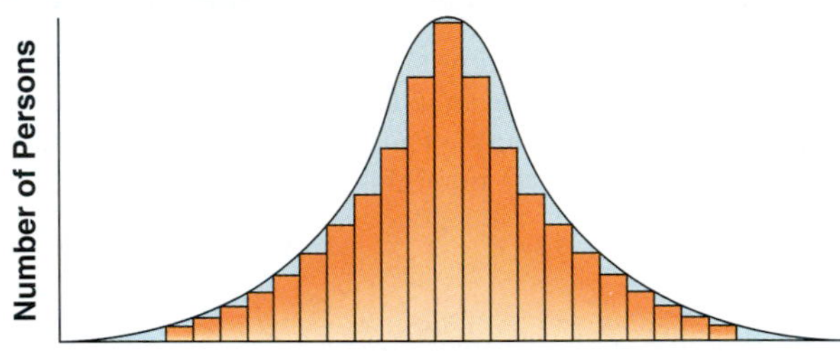

Just how many pairs of alleles control skin color is not known, but a range in colors can be explained on the basis of two pairs. When a very dark person reproduces with a very light person, the children have medium brown skin; and when two people with medium brown skin reproduce with one another, the children range in skin color from very dark to very light. This can be explained by assuming that skin color is controlled by two pairs of alleles and that each capital letter contributes to the color of the skin:

Phenotype	Genotypes
Very dark	*AABB*
Dark	*AABb* or *AaBB*
Medium brown	*AaBb* or *AAbb* or *aaBB*
Light	*Aabb* or *aaBb*
Very light	*aabb*

Notice again that there is a range in phenotypes and that there are several possible phenotypes in between the two extremes. Therefore, the distribution of these phenotypes is expected to follow a bell-shaped curve—few people have the extreme phenotypes, and most people have the phenotype that lies in the middle between the extremes.

Many human disorders, such as cleft lip and/or palate, clubfoot, congenital dislocations of the hip, hypertension, diabetes, schizophrenia, and even allergies and cancers, are most likely controlled by polygenes and subject to environmental influences. Therefore, many investigators are in the process of considering the nature versus nurture question; that is, what percentage of the distribution of the trait is controlled by genes and what percentage is controlled by the environment? Thus far, it has not been possible to come to precise, generally accepted percentages for any particular trait.

In recent years, reports have surfaced that all sorts of behavioral traits such as alcoholism, homosexuality, phobias, and even suicide can be associated with particular genes. No doubt behavioral traits are to a degree controlled by genes, but again, it is impossible at this time to determine to what degree.

Many human traits most likely controlled by polygenes are subject to environmental influences. The frequency of the phenotypes of such traits follows a bell-shaped curve.

Multiple Alleles

Some traits are controlled by **multiple alleles;** that is, more than two alleles. However, each person inherits only two of the total possible number of alleles. Three alleles for the same gene control the inheritance of ABO blood types: A = A antigen on red blood cells; B = B antigen on red blood cells; O = no antigens on red blood cells. Both A and B are dominant over O; therefore, there are two possible genotypes for type A blood and two possible genotypes for type B blood. If a person inherits one of each of these alleles, the blood type will be AB. Type O can only result from the inheritance of two O alleles:

Phenotype	Possible Genotype
A	*AA, AO*
B	*BB, BO*
AB	*AB*
O	*OO*

An examination of possible matings between different blood types sometimes produces surprising results. For example, if the cross is $AO \times BO$, the possible genotypes of children are *AB, OO, AO,* and *BO*.

Blood typing can sometimes aid in paternity suits. A man with type A blood (having genotype *AO*) could possibly be the father of a child with type O blood. On the other hand, a man with type AB blood cannot possibly be the father of a child with type O blood. Therefore, blood tests are legally used only to exclude a man from possible paternity.

The Rh factor is inherited separately from A, B, AB, or O blood types. In each instance, it is possible to be Rh positive (Rh^+) or Rh negative (Rh^-). It can be assumed that the inheritance of this antigen on red blood cells is controlled by a single allelic pair in which simple dominance prevails: the Rh-positive allele is dominant over the Rh-negative allele.

Inheritance by multiple alleles occurs when a gene exists in more than two allelic forms. However, each individual usually inherits only two alleles for these genes.

Degrees of Dominance

The field of human genetics also has examples of codominance and incomplete dominance. *Codominance* occurs when alleles are equally expressed in a heterozygote. We have already mentioned that the multiple alleles controlling blood type are codominant. An individual with the genotype *AB* has type AB blood.

Sickle-Cell Disease Sickle-cell disease is an example of a human disorder that is controlled by incompletely dominant alleles. Individuals with the Hb^AHb^A genotype are normal, those with the Hb^SHb^S genotype have sickle-cell disease, and those with the Hb^AHb^S genotype have the sickle-cell trait. Two individuals with sickle-cell trait can produce children with all three phenotypes, as indicated in Figure 13.10.

In persons with sickle-cell disease, the red blood cells aren't biconcave disks like normal red blood cells; they are irregular. In fact, many are sickle shaped. The defect is caused by an abnormal hemoglobin that accumulates inside the cells. Because the sickle-shaped cells can't pass along narrow capillary passageways like disk-shaped cells, they clog the vessels and break down. This is why persons with sickle-cell disease suffer from poor circulation, anemia, and poor resistance to infection. Internal hemorrhaging leads to further complications, such as jaundice, episodic pain of the abdomen and joints, and damage to internal organs.

Persons with sickle-cell trait do not usually have any sickle-shaped cells unless they experience dehydration or mild oxygen deprivation. Although a recent study found that army recruits with sickle-cell trait are more likely to die when subjected to extreme exercise, previous studies of athletes do not substantiate these findings. At present, most investigators believe that no restrictions on physical activity are needed for persons with the sickle-cell trait.

Among regions of malaria-infested Africa, infants with sickle-cell disease die, but infants with sickle-cell trait have a better chance of survival than the normal homozygote. When the parasite infects their red blood cells, the cells become sickle shaped and thereafter the cells lose potassium. This causes the parasite to die. The protection afforded by the sickle-cell trait keeps the allele for sickle-cell prevalent in populations exposed to malaria. As many as 60% of the population in malaria-infected regions of Africa have the allele. In the United States, about 10% of the African-American population carries the allele.

In a recent study, 22 children around the world with sickle-cell disease were given bone marrow transplants from healthy siblings and 16 were completely cured of the disease. Optimism over these results is dampened by the knowledge that only 6% of patients met the medical criteria for receiving treatment and the drugs used in the procedure result in infertility. Also, if unsuccessful, health could worsen instead of improve and there is a 10% risk of death from the treatment.

Innovative therapies are still being explored. For example, persons with sickle-cell disease produce normal fetal hemoglobin during development, and drugs that turn on the genes for fetal hemoglobin in adults are being developed. Mice have been genetically engineered to produce sickled red blood cells in order to test new antisickling drugs and various genetic therapies.

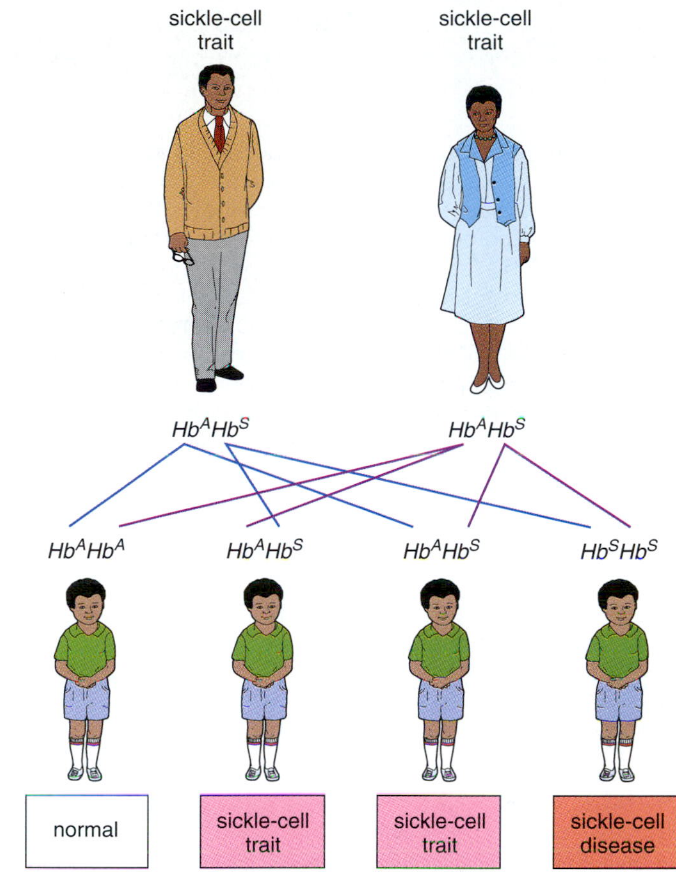

a.

b.

Figure 13.10 Inheritance of sickle-cell disease.
a. In this example, both parents have the sickle-cell trait. Therefore, each child has a 25% chance of having sickle-cell disease, a 25% chance of being perfectly normal and a 50% chance of having the sickle-cell trait. **b.** Sickled cells. Individuals with sickle-cell disease have sickled red blood cells that tend to clump, as illustrated here.

Sickle-cell disease is an inherited lifelong disorder that is being investigated on many fronts.

Genetic Counseling

Now that prospective parents are becoming aware that many illnesses are caused by faulty genes, more couples are seeking genetic counseling. The counselor studies the background of the couple and constructs a pedigree chart for the family. As much as possible, laboratory tests are performed on all persons involved. Blood tests can identify carriers of thalassemia and sickle-cell disease. By measuring enzyme levels in blood, ears, or skin cells, carriers of alleles coding for faulty enzymes can also be identified for certain inborn metabolic errors, such as Tay-Sachs disease. From this information, the counselor can sometimes predict the chances of a couple having a child with a genetic disorder.

If the woman is pregnant, chorionic villi sampling can be done early, and amniocentesis can be done later in the pregnancy. Women are only encouraged to undergo such procedures when there is reason for concern because the risk of complications resulting from these tests can sometimes be greater than the risk of having a child with a disorder. These procedures which are illustrated in Figure 13B, allow the testing of embryonic and fetal cells, respectively, to determine if the unborn child has a genetic disorder. Chromosomal tests are available for cystic fibrosis, neurofibromatosis, and Huntington disease. Appropriate biochemical tests identify other types of genetic disorders (Table 13A).

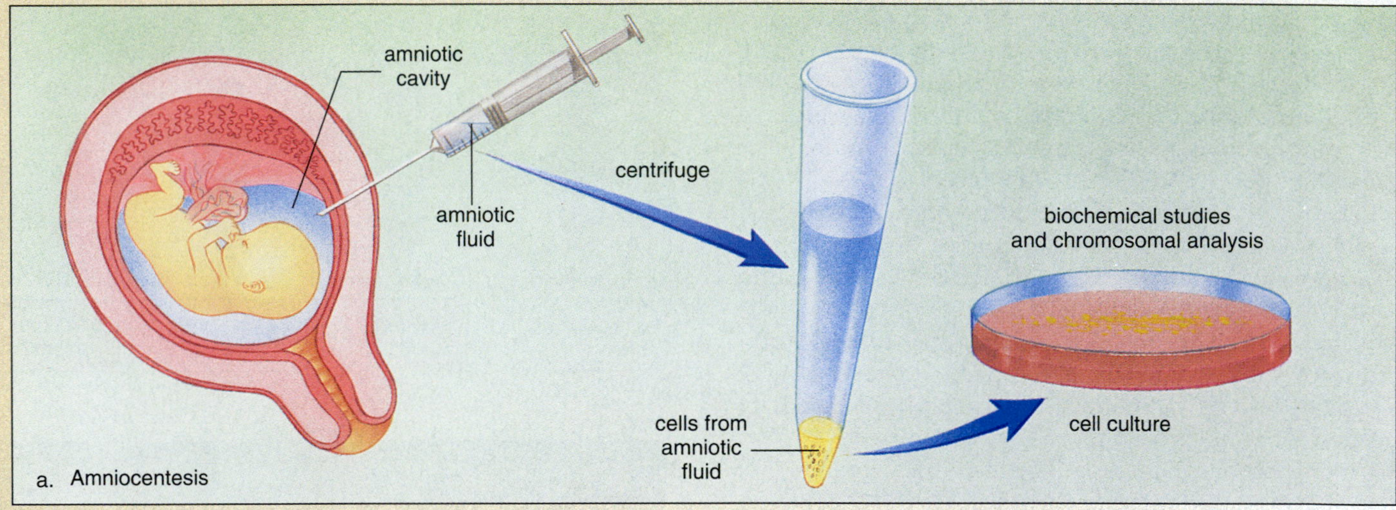

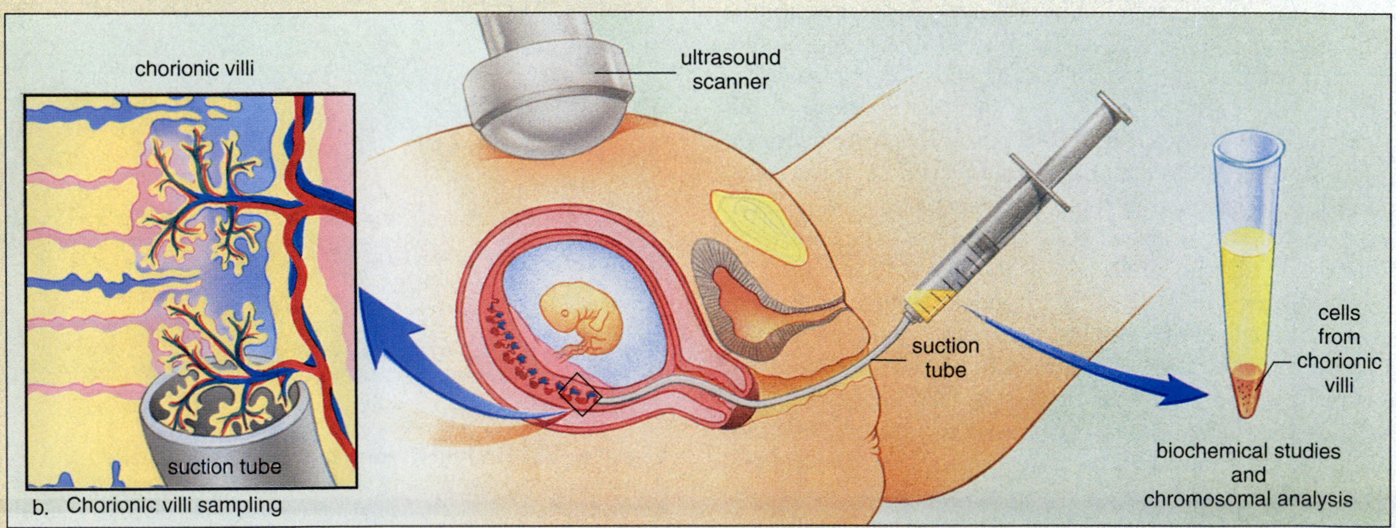

Figure 13B **Genetic defect testing before birth.**
a. Amniocentesis is usually performed from the fifteenth to the seventeenth week of pregnancy. A long needle is passed through the abdominal wall to withdraw a small amount of amniotic fluid, along with fetal cells. Since there are only a few cells in the amniotic fluid, testing may be delayed as long as four weeks until cell culture produces enough cells for testing purposes. About 40 tests are available for different defects.
b. Chorionic villi sampling can be done as early as the fifth week of pregnancy. The doctor inserts a long, thin tube through the vagina into the uterus. With the help of ultrasound, which gives a picture of the uterine contents, the tube is placed between the lining of the uterus and the chorion. Then a sampling of the chorionic villi cells is obtained by suction. Chromosome analysis and biochemical tests for genetic defects can be done immediately on these cells.

Traditionally, it might take four weeks following chorionic villi sampling and amniocentesis for the cells to grow and multiply in cell culture in adequate number for test purposes. This is a distinct disadvantage to those who may wish the option to abort the pregnancy if an abnormality is found. There is a new technique, however, that can be used to detect many chromosomal abnormalities just one or two days after cells are obtained. Fluorescent probes specific for certain regions of chromosomes can be applied to amniotic cells and they will bind to and light up those regions. For example, a probe for the X chromosome can tell how many X chromosomes the child inherited.

Every year more and more tests become available for genetic mutations associated with disorders ranging from psychological conditions to colon cancers. Maybe genetic testing shouldn't be done if the presence of a mutation only indicates a possible risk of developing a condition. As a society we have not yet decided how many and what types of genetic tests should be done and how to interpret some of the results once they are done. As an example, suppose one prenatal twin is XY and the other is XYY—do you think the parents should be informed? It might cause them to treat the XYY child differently when actually many XYY individuals live a normal life.

Table 13A

Test and Treatment for Some Human Genetic Disorders

Name	Description	Chromosome	Incidence (Newborns)	Status
Autosomal Dominant Disorders				
Neurofibromatosis	Benign tumors occur under the skin or deeper.	17	One in 3,000	Allele located; chromosome test now available*
Huntington disease	Minor disturbances in balance and coordination develop in middle age and progress towards severe neurological disturbances leading to death.	4	One in 20,000	Allele located; chromosome test now available*
Autosomal Recessive Disorders				
Cystic fibrosis	Mucus in the lungs and digestive tract is thick and viscous, making breathing and digestion difficult.	7	One in 2,500 Caucasians	Allele located; chromosome test now available;* treatment being investigated
Tay-Sachs disease	Neurological impairment and psychomotor difficulties develop early, followed by blindness and uncontrollable seizures before death occurs, usually before age 5.	15	One in 3,600 eastern European Jews	Biochemical test now available*
Phenylketonuria	Inability to metabolize phenylalanine, and if a special diet is not begun, mental retardation develops.	12	One in 5,000	Biochemical test now available; treatment available
Incomplete Dominance				
Sickle-cell disease	Poor circulation, anemia, internal hemorrhaging, due to sickle-shaped red blood cells.	11	One in 500 African Americans	Chromosome test now available*
X-Linked Recessive				
Duchenne muscular dystrophy	Muscle weakness develops early and progressively intensifies until death occurs, usually before age 20.	X	One in 5,000 male births	Allele located; biochemical tests of muscle tissue available; treatment being investigated
Hemophilia A	Propensity for bleeding, often internally, due to the lack of a blood clotting factor.	X	One in 15,000 male births	Treatment available

** Prenatal testing is done.*

13.3 Sex-Linked Genetic Disorders

The alleles for **sex-linked** genetic disorders are on the sex chromosomes. Most of these alleles are on the X chromosome, and therefore are said to be **X-linked.** Sex-linked alleles code for characteristics that have nothing to do with the sex of the individual.

X-Linked Recessive Disorders

Figure 13.11 gives a pedigree chart for an X-linked recessive condition. It also lists ways to recognize this pattern of inheritance. X-linked conditions can be dominant or recessive, but most known are recessive. More males than females have the trait because recessive alleles on the X chromosome are always expressed in males since the Y chromosome does not have a corresponding allele. If a male has an X-linked recessive condition, his daughters are often carriers; therefore, the condition passes from grandfather to grandson. Females who have the condition inherited the allele from both their mother and their father; and all the sons of such a female will have the condition.

Three well-known X-linked recessive disorders are color blindness, muscular dystrophy, and hemophilia.

Color Blindness

In humans, there are three different classes of cone cells, the receptors for color vision in the retina of the eyes. Only one pigment protein is present in each type of cone cell; there are blue-sensitive, red-sensitive, and green-sensitive cone cells. The gene for the blue-sensitive protein is autosomal, but the genes for the red- and green-sensitive proteins are on the X chromosome. About 8% of Caucasian men have red/green color blindness. Most of these see brighter greens as tans, olive greens as browns, and reds as reddish-browns. A few cannot tell reds from greens at all. They see only yellows, blues, blacks, whites, and grays. Opticians have special charts by which they detect those who are color blind.

Duchenne Muscular Dystrophy

Muscular dystrophy, as the name implies, is characterized by a wasting away of the muscles. The most common form, *Duchenne muscular dystrophy,* is X-linked and occurs in about one out of every 3,600 male births. Symptoms, such as waddling gait, toe walking, frequent falls, and difficulty in rising, may appear as soon as the child starts to walk. Muscle weakness intensifies until the individual is confined to a wheelchair. Death usually occurs by age 20; therefore, affected males are rarely fathers. The recessive allele remains in the population by passage from carrier mother to carrier daughter.

Once the gene for muscular dystrophy was isolated, it was discovered that the absence of a protein called dystrophin is the cause of the disorder. Much investigative work determined that dystrophin is involved in the release of calcium from the calcium-storage sites in muscle fibers. The lack of dystrophin causes calcium to leak into the cell, which promotes the action of an enzyme that dissolves muscle fibers. When the body attempts to repair the tissue, fibrous tissue forms, and this cuts off the blood supply so that more and more cells die.

A test is now available to detect carriers for Duchenne muscular dystrophy. Also, various treatments are being attempted. Immature muscle cells can be injected into muscles, and for every 100,000 cells injected, dystrophin production occurs in 30–40% of the patient's muscle fibers. The gene for dystrophin has been inserted into the thigh muscle cells of mice, and about 1% of these cells then produced dystrophin.

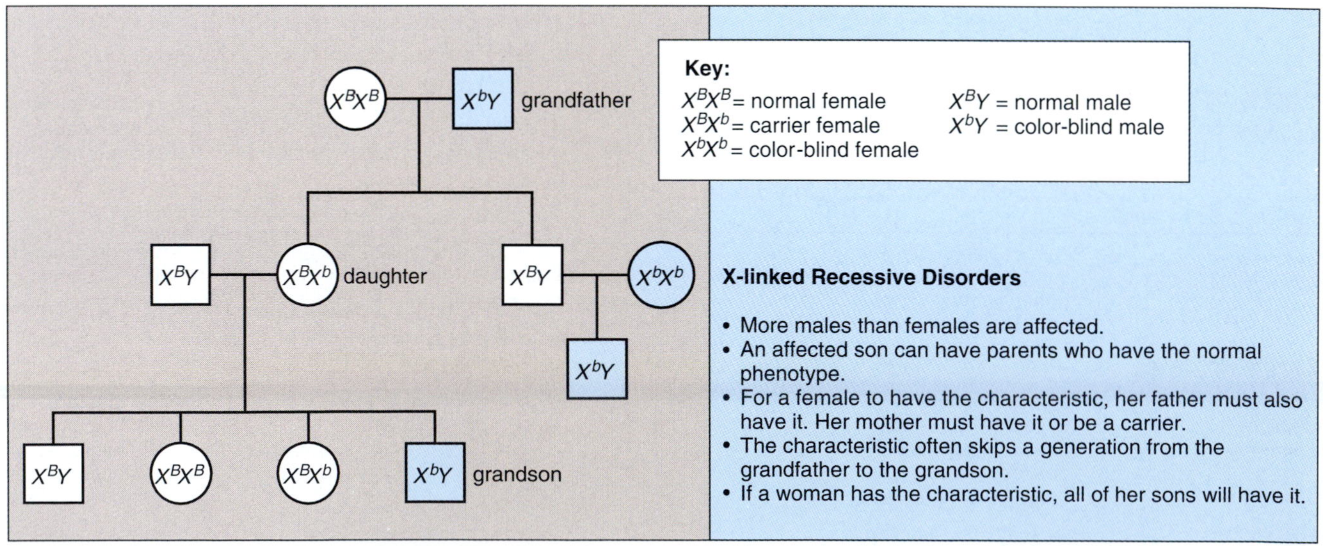

Figure 13.11 X-linked recessive pedigree chart.
The list gives ways of recognizing an X-linked recessive disorder. Only those affected with the recessive genetic disorder are shaded.

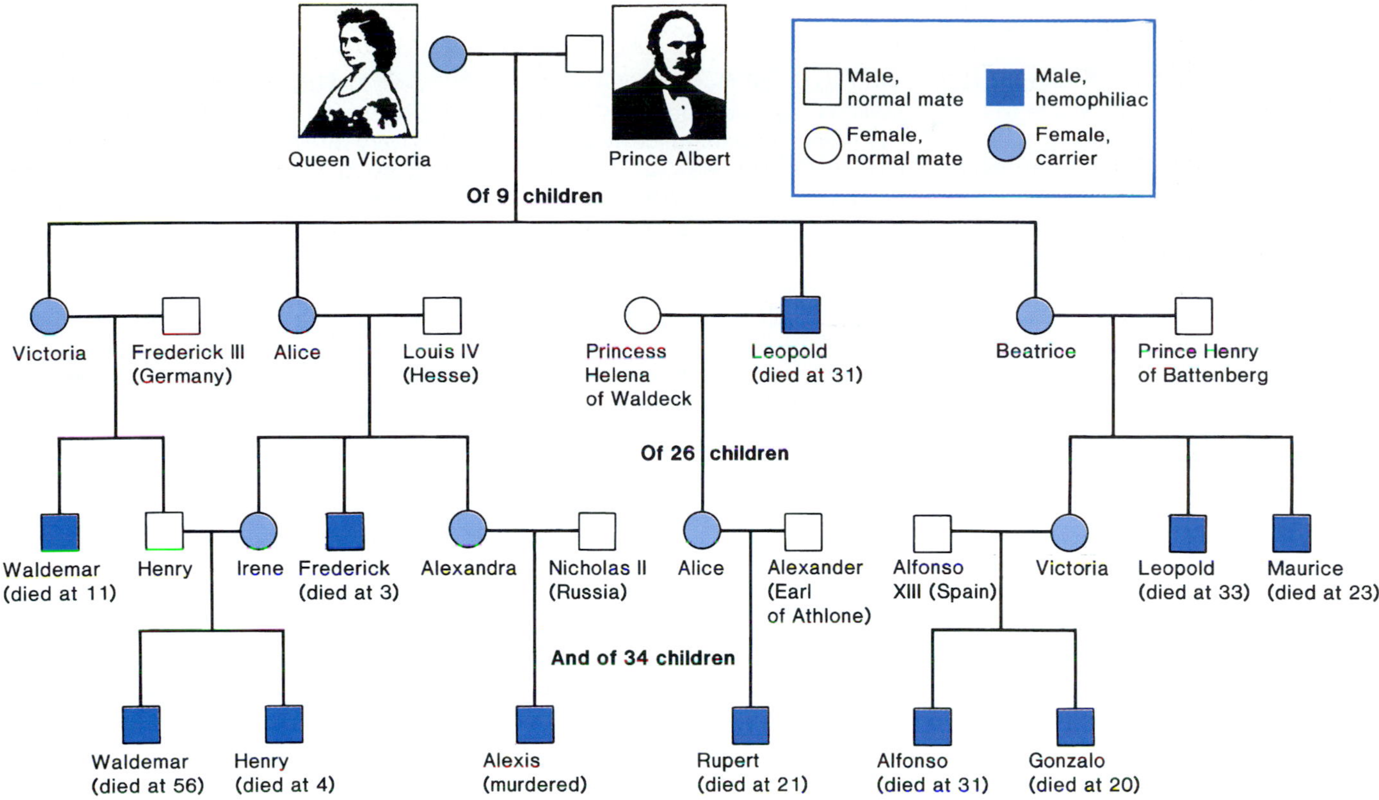

Figure 13.12 Hemophilia in European royal families.
Because Queen Victoria was a carrier, each of her sons had a 50% chance of having the disease and each of her daughters had a 50% chance of being a carrier. This pedigree shows only the affected individuals. Many others are unaffected, such as the members of the present British royal family.

Hemophilia

About one in 10,000 males is a hemophiliac. There are two common types of hemophilia: hemophilia A is due to the absence or minimal presence of a clotting factor known as factor IX, and hemophilia B is due to the absence of clotting factor VIII.

Hemophilia is called the bleeder's disease because the affected person's blood does not clot. Although hemophiliacs bleed externally after an injury, they also suffer from internal bleeding, particularly around joints. Hemorrhages can be checked with transfusions of fresh blood (or plasma) or concentrates of the clotting protein. Unfortunately, some hemophiliacs have contracted AIDS after receiving blood or using a blood concentrate, but donors are now screened more closely, and donated blood is now tested for HIV. Also, factor VIII is now available as a genetic engineering product.

At the turn of the century, hemophilia was prevalent among the royal families of Europe, and all of the affected males could trace their ancestry to Queen Victoria of England (Fig. 13.12). Of Queen Victoria's 26 grandchildren, five grandsons had hemophilia and four granddaughters were carriers. Because none of Queen Victoria's forebears or relatives was affected, it seems that the faulty allele she carried arose by mutation either in Victoria or in one of her parents. Her carrier daughters, Alice and Beatrice, introduced the gene into the ruling houses of Europe. Alexis, the last heir to the Russian throne before the Russian Revolution, was a hemophiliac. There are no hemophiliacs in the present British royal family because Victoria's eldest son, King Edward VII, did not receive the gene and therefore could not pass it on to any of his descendants.

Certain genetic disorders are caused by X-linked recessive alleles (see Table 13A, page 213). Males have only one X chromosome, and therefore X-linked recessive alleles are always expressed.

Certain traits common in males, like pattern baldness, are controlled by an autosomal allele that acts as if it were dominant in males but recessive in females. Most likely the activity of such genes is influenced by the sex hormones. Therefore they are called **sex-influenced traits.**

Connecting Concepts

It is apparent from a study of this chapter that Mendel's laws are applicable to human genetics: some conditions, such as cystic fibrosis, have a recessive inheritance pattern while others like Huntington disease have a simple Mendelian inheritance pattern. We also now know that sex linkage, incomplete dominance, and polygenic inheritance explain other types of human inheritance patterns. Therefore, the principles of inheritance determined by doing specific crosses with peas, fruit flies, or other model organisms (as described in Chapters 11 and 12) are relevant to the study of inherited traits in humans.

From a study of this chapter, it is also apparent that human conditions such as cystic fibrosis, muscular dystrophy, sickle-cell anemia, and color blindness, are due to alleles that code for abnormal forms of proteins. In other words, we began to look at the molecular basis for genetic inheritance. We will continue this examination in the next few chapters. Chapter 14 traces the history of the discovery of DNA, the genetic material, explains its chemistry and how it is normally copied. Chapter 15 tells how the expression of genes (the genotype) results in the characteristics (phenotype) of the organism. Chapter 16 discusses the many ways in which gene expression is regulated. Chapter 17 tells how the genes of all organisms can be manipulated.

Summary

13.1 Inheritance of Chromosomes

It is possible to treat and photograph the chromosomes of a cell so that they can be sorted and arranged in pairs. The resulting karyotype can be used to diagnose chromosomal abnormalities.

Down syndrome (trisomy 21) is the most common autosomal abnormality. The occurrence of this syndrome, which is often related to the mother's age, can be detected by amniocentesis. Most often, Down syndrome is due to nondisjunction during gamete formation, but in a small percentage of cases, there has been a translocation between chromosomes 14 and 21, in which chromosome 21 is attached to chromosome 14.

Turner syndrome (XO) is a monosomy for the X chromosome. There are several trisomies: Klinefelter syndrome, which is XXY; triplo-X, which is XXX; and Jacob syndrome. Fragile X syndrome, which has an unusual pattern of inheritance, is due to the presence of an abnormal X chromosome.

13.2 Autosomal Genetic Disorders

When studying human genes, biologists often construct pedigree charts to show the pattern of inheritance of a characteristic within a family. The particular pattern indicates the manner in which a characteristic is inherited.

Certain genetic disorders are inherited in a simple Mendelian manner. Neurofibromatosis is an autosomal dominant disorder that is inherited in this manner. We now know that Huntington disease, long classified as an autosomal dominant disease, is due to expanded trinucleotide repeats, and the greater the number of repeats, the earlier the onset of the disease. Also, the disease is more likely to have been inherited from the father for reasons that are still being explored. Cystic fibrosis, Tay-Sachs disease, and PKU are autosomal recessive disorders that have been studied in detail.

Polygenic traits are controlled by more than one gene loci and the individual inherits an allelic pair for each loci. Skin color illustrates polygenic inheritance. Traits controlled by polygenes are subject to environmental effects and show continuous variations whose frequency distribution forms a bell-shaped curve.

ABO blood type is an example of a human trait controlled by multiple alleles. Sickle-cell disease is a human disorder that is controlled by incompletely dominant alleles.

13.3 Sex-Linked Genetic Disorders

X-linked genes are on the X chromosome. Since the Y chromosome is blank for X-linked alleles, males are more apt to exhibit the recessive phenotype, which is inherited from their mothers. Color blindness, Duchenne muscular dystrophy, and hemophilia are X-linked recessive disorders. Table 13A on page 213 summarizes the genetic disorders discussed in this chapter.

Some traits, like pattern baldness, are sex-influenced traits controlled by autosomal alleles.

Reviewing the Chapter

1. What does the normal human karyotype look like? 202
2. Diagram how nondisjunction can occur during meiosis I and meiosis II. 203
3. What are the characteristics of Down syndrome? What is the most frequent cause of this condition? 204
4. What is the only known sex chromosome monosomy in humans? Name and describe three sex chromosome trisomies. 204–5
5. How might you distinguish an autosomal dominant trait from an autosomal recessive trait when viewing a pedigree chart? 207
6. Describe the symptoms of neurofibromatosis and Huntington disease. How does neurofibromatosis illustrate variable expressivity? For most autosomal dominant disorders, what are the chances of a heterozygote and a normal individual having an affected child? 208
7. Describe the symptoms of cystic fibrosis, Tay-Sachs disease, and PKU. For any one of these, what are the chances of two carriers having an affected child? 209
8. What observational data would allow you to hypothesize that a human trait is controlled by polygenes? 210
9. Why is ABO blood type an example of inheritance by multiple alleles? 210
10. Explain how sickle-cell disease is inherited. What are the symptoms of sickle-cell trait and sickle-cell disease? If two persons with the sickle-cell trait reproduce, what are the chances of having a child with the sickle-cell trait? with sickle-cell disease? What ethnic group is more likely to have this condition? Why? 211

11. Explain how color blindness, Duchenne muscular dystrophy, and hemophilia are inherited. What are the symptoms of these conditions? What are the chances of having an affected male offspring if the mother is a carrier and the father is normal? 214–15
12. How does the mode of inheritance for a sex-influenced trait differ from that of an X-linked recessive trait? 215

Testing Yourself

Choose the best answer for each question. For questions 1–3, match the conditions in the key with the descriptions below:

Key:

a. Down syndrome
b. Turner syndrome
c. Klinefelter syndrome
d. XYY
e. triplo-X individual

1. male with underdeveloped testes and some breast development
2. trisomy 21
3. XO female
4. Down syndrome
    a. is always caused by nondisjunction of chromosome 21.
    b. occurs less frequently today because nutrition is better now.
    c. is more often seen in children of mothers past the age of thirty-five.
    d. can be cured by removing one of the X chromosomes from the cells of a fetus.
    e. All of these are correct.
5. A person has a genetic disorder. Which of these is inconsistent with autosomal recessive inheritance?
    a. Both parents have the disorder.
    b. Both parents do not have the disorder.
    c. All the children (males and females) have the disorder.
    d. All of these are consistent.
    e. All of these are inconsistent.
6. A male has a genetic disorder. Which one of these is inconsistent with X-linked recessive inheritance?
    a. Both parents do not have the disorder.
    b. Only males in a pedigree chart have the disorder.
    c. Only females in previous generations have the disorder.
    d. The sons of a female with the disorder will all have the disorder.
    e. Both a and c are inconsistent.

For questions 7–10, match the conditions in the key with the following descriptions:

Key:

a. cystic fibrosis
b. Huntington disease
c. hemophilia
d. Tay-Sachs disease
e. sickle-cell disease

7. autosomal dominant
8. most often seen among eastern European Jews
9. X-linked recessive
10. thick mucus in lungs and pancreatic ducts
11. Determine if the characteristic possessed by the shaded squares (males) and circles (females) below is an autosomal dominant, an autosomal recessive, or an X-linked recessive disorder.

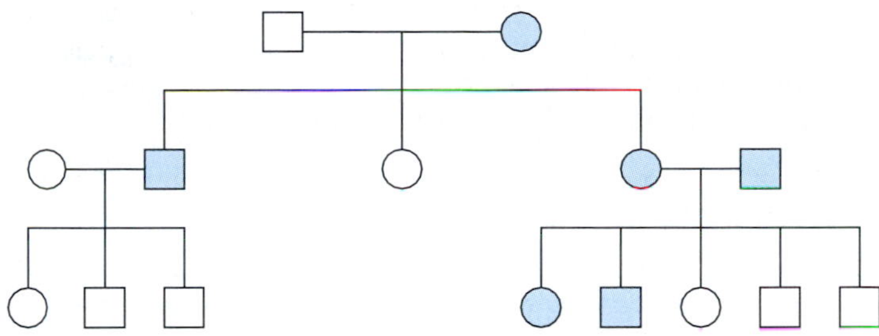

Additional Genetics Problems*

1. A hemophilic (X-linked recessive) man reproduces with a homozygous normal woman. What are the chances that their sons will be hemophiliacs? that their daughters will be hemophiliacs? that their daughters will be carriers?
2. A son with cystic fibrosis (autosomal recessive) is born to a couple who appear to be normal. What are the chances that any child born to this couple will have cystic fibrosis?
3. A man has type AB blood. What is his genotype? Could this man be the father of a child with type B blood? If so, what blood types could the child's mother have?
4. What is the genotype of a man who is color blind (X-linked recessive) and has a continuous hairline (autosomal recessive)? If this man has children by a woman who is homozygous dominant for normal color vision and widow's peak, what will be the genotype and phenotype of the children?
5. Determine if the characteristic possessed by the shaded squares (males) is dominant, recessive, or X-linked recessive. Write in the genotypes for the starred individual.

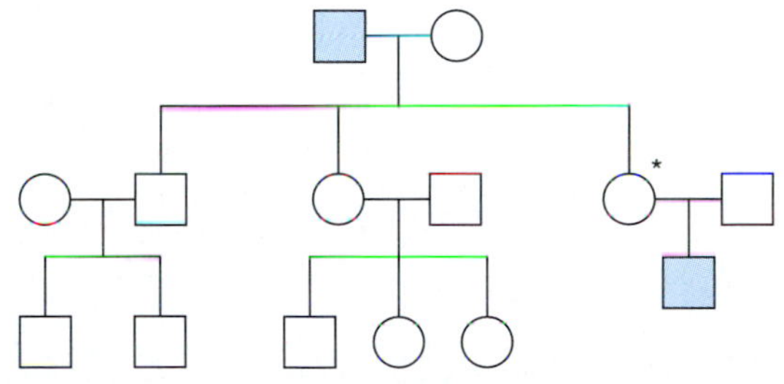

* Answers to Additional Genetics Problems appear in Appendix A.

Thinking Scientifically

1. Not all individuals with Down syndrome have equally severe mental retardation or other physical abnormalities. Explain on the basis of a difference in the expression of alleles.
2. The particular recessive allele that causes cystic fibrosis (CF) can vary. Seventy percent of patients with CF are homozygous recessive for any number of different alleles that also cause CF. The screening test for heterozygotes detects only the most common allele for CF. If a person with CF relatives tests negative for the common allele, how might s/he determine that s/he need not worry about the possibility of having a rarer form of the CF allele?

Bioethical Issue

Sickle-cell disease is a prevalent type of debilitating anemia in the African-American community because African Americans are more likely to have sickle-cell trait. They are not sick, but if two persons with sickle-cell trait reproduce with one another, each child has a 25% chance of having sickle-cell disease, which can lead to an early death.

In the 1970s, the federal government funded a screening program to detect those with sickle-cell trait because they thought prospective parents might want to have this information. The procedure is inexpensive and requires the testing of just one drop of blood. At first, members of the black community were in favor of testing; ministers conducted tests on their congregations and the black Panthers offered the test door-to-door in black communities.

Due to a lack of understanding, however, the results of testing were not treated as expected. The government, employers, and those with sickle-cell trait themselves thought they were sick. The Massachusetts legislature required that schools, for health reasons, identify all those with sickle-cell trait as well as students with sickle-cell disease. Some insurance companies began to deny coverage to those with sickle-cell trait on the grounds that they had a preexisting medical condition. The U.S. Air Force Academy rejected black applicants who had sickle-cell trait. Some commercial airlines refused to hire those with sickle-cell trait, thinking that they might faint at high altitudes. Some prominent scientists suggested that those with sickle-cell trait should forego having children.

Although this example suggests that genetic testing has many drawbacks, do you feel there are some benefits to testing if the rights of the individual are safeguarded?

Understanding the Terms

amniocentesis 203
autosome 202
carrier 207
chorionic villi sampling 212
genetic disorder 207
genomic imprinting 208
karyotype 202
multiple allele 210
nondisjunction 203
pedigree chart 207
sex chromosome 202
sex-influenced trait 215
sex-linked 214
X chromosome 202
Y chromosome 202
X-linked 214

Match the terms to these definitions:

a. ______________ Any chromosome other than a sex chromosome.
b. ______________ Autosomal phenotype controlled by an allele that is expressed differently in the two sexes; for example, the possibility of pattern baldness is increased by the presence of testosterone in males.
c. ______________ Chromosomes arranged by pairs according to their size, shape, and general appearance in mitotic metaphase.
d. ______________ Failure of homologous chromosomes or daughter chromosomes to separate during meiosis I and meiosis II, respectively.
e. ______________ Heterozygous individual who has no apparent abnormality but can pass on an allele for a recessively inherited genetic disorder.

Web Connections

Exploring the Internet

http://www.mhhe.com/biosci/genbio/mader
(click on *Biology 7/e*)

The *Biology 7/e* Online Learning Center provides many resources for studying the material in this chapter including links to the following sites:

Online Mendelian Inheritance in Man is a database of human genes and genetic disorders. It is a site designed for professionals, but can also be used by advanced students or those interested in taking a peek further into the field of genetics.

http://www3.ncbi.nlm.nih.gov/Omim/

Mendelian Genetics Chapter. A hypertextbook from MIT on genetics, including problems in linkage and pedigrees.

http://esg-www.mit.edu:8001/esgbio/mg/mgdir.html

Genomes Guide: Homo sapiens. *Clickable chromosomes, learn about diseases associated with particular chromosomes.*

http://www.ncbi.nlm.nih.gov/genome/guide/

Genes and Disease. A nifty site to look for a variety of genetically linked diseases.

http://www.ncbi.nlm.nih.gov/disease/

What is Fragile X? A site maintained by the Fragile X Association describes this condition.

http://www.fraxsocal.org/

DNA Structure and Functions

c h a p t e r

c h a p t e r   c o n c e p t s

14.1 The Genetic Material

- DNA is the genetic material, and therefore its structure and functions constitute the molecular basis of inheritance. 222
- DNA stores information regarding the development and metabolism of a cell. 222

14.2 The Structure of DNA

- DNA is a double helix; each of the two strands is a polymer of four different nucleotides. 225
- Four nucleotides permit a great deal of variability, and the sequence of base pairs in DNA varies from gene to gene. 226
- Hydrogen bonding between complementary bases joins the two strands. 226

14.3 Replication of DNA

- DNA is able to replicate, and in this way genetic information is passed from one cell generation to the next. 228
- Mutations occur when there are errors during the replication process. 231

Test tube containing DNA whose structure resembles a spiral staircase (digital composite).

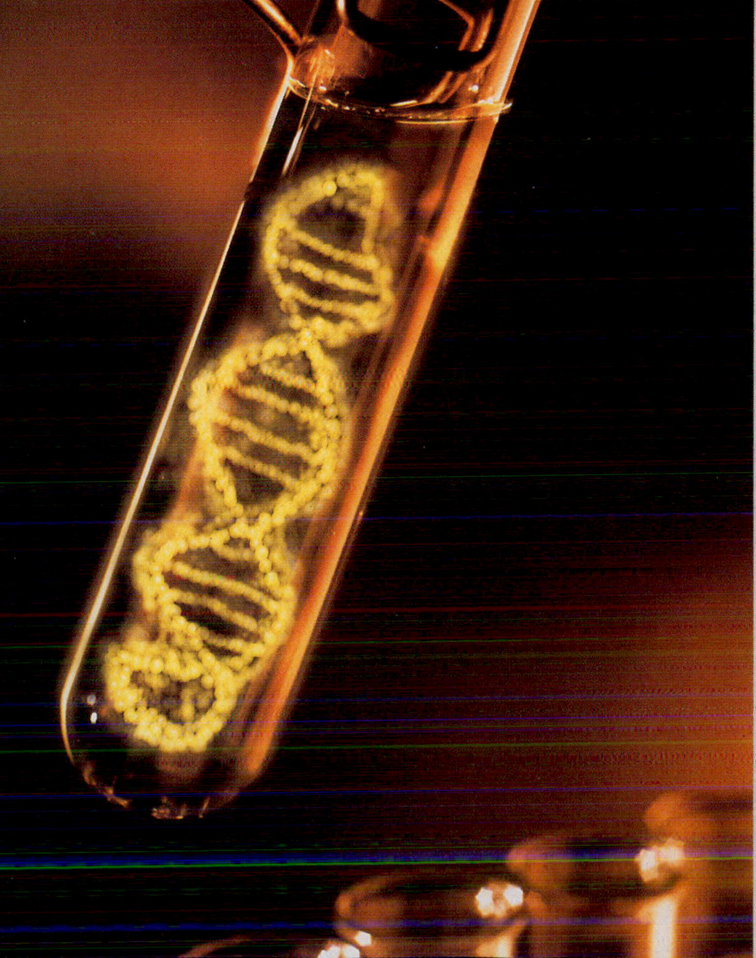

One of the most exciting periods of scientific activity in history occurred during the thirty short years between the 1930s and 1960s. Geneticists knew that chromosomes contain protein and DNA (deoxyribonucleic acid). Of these two organic molecules, proteins are seemingly more complicated; they consist of countless sequences of 20 amino acids, which can coil and fold into complex shapes. DNA, on the other hand, contains only four different nucleotides. Surely, the diversity of life-forms on earth must be the result of the endless varieties of proteins.

Due to several elegantly executed experiments, by the mid-1950s researchers realized that DNA, not protein, is the genetic material. But this finding only led to another fundamental question—*what exactly is the structure of DNA?* The biological community at the time knew that whoever determined the structure of DNA would get a Nobel Prize, and would go down in history. Consequently, researchers were racing against time and each other. The story of the discovery of DNA resembles a mystery, with each clue adding to the total picture until the breathtaking design of DNA—a double helix—was finally unraveled.

14.1 The Genetic Material

Even though previous investigators were able to confirm that the genes are on the chromosomes and were even able to map the *Drosophila* chromosomes, they still didn't know just what genes consisted of. You can well imagine, then, that the search for the genetic material was of utmost importance to biologists at the beginning of the twentieth century. They knew that this material must be:

1. able to *store information* that pertains to both the development and the metabolic activities of the cell or organism;
2. stable so that it *can be replicated* with high fidelity during cell division and be transmitted from generation to generation;
3. able to *undergo rare changes* called mutations [L. *muta*, change] that provide the genetic variability required for evolution to occur.

The genetic material must be able to store information, be replicated, and undergo mutations.

Knowledge about the chemistry of DNA was absolutely essential in order to come to the conclusion that DNA is the genetic material. In 1869, the Swiss chemist Friedrich Miescher removed nuclei from pus cells (these cells have little cytoplasm) and found that they contained a chemical he called *nuclein.* Nuclein, he said, was rich in phosphorus and had no sulfur, properties that distinguished it from protein. Later, other chemists did further work with nuclein and said that it contained an acidic substance they called **nucleic acid.** Soon it was realized that there are two types of nucleic acids: **DNA (deoxyribonucleic acid)** and **RNA (ribonucleic acid).**

Early in the twentieth century, it was discovered that nucleic acids contain only four types of **nucleotides,** molecules that are composed of a nitrogen-containing base, a phosphate, and a pentose (5-carbon sugar). Perhaps DNA was composed of repeating units, and each unit always had just one of each of the four different nucleotides. In that case, DNA could not vary between genes and could not be the genetic material! Everyone thought that the protein component of chromosomes must be the genetic material because proteins contain 20 different amino acids that can be sequenced in any particular way.

Transformation of Bacteria

In 1931, the bacteriologist Frederick Griffith performed an experiment with a bacterium (*Streptococcus pneumoniae,* or pneumococcus for short) that causes pneumonia in mammals. He noticed that when these bacteria are grown on culture plates, some, called S strain bacteria, produce shiny, smooth colonies and others, called R strain bacteria, produce colonies that have a rough appearance. Under the microscope, S strain bacteria have a capsule (mucous coat) but R strain bacteria do not. When Griffith injected mice with the S strain of bacteria, the mice died, and when he injected mice with the R strain, the mice did not die (Fig. 14.1). In an effort to determine if the capsule alone was responsible for the virulence (ability to kill) of the S strain bacteria, he injected mice with heat-killed S strain bacteria. The mice did not die.

Finally, Griffith injected the mice with a mixture of heat-killed S strain and live R strain bacteria. Most unexpectedly,

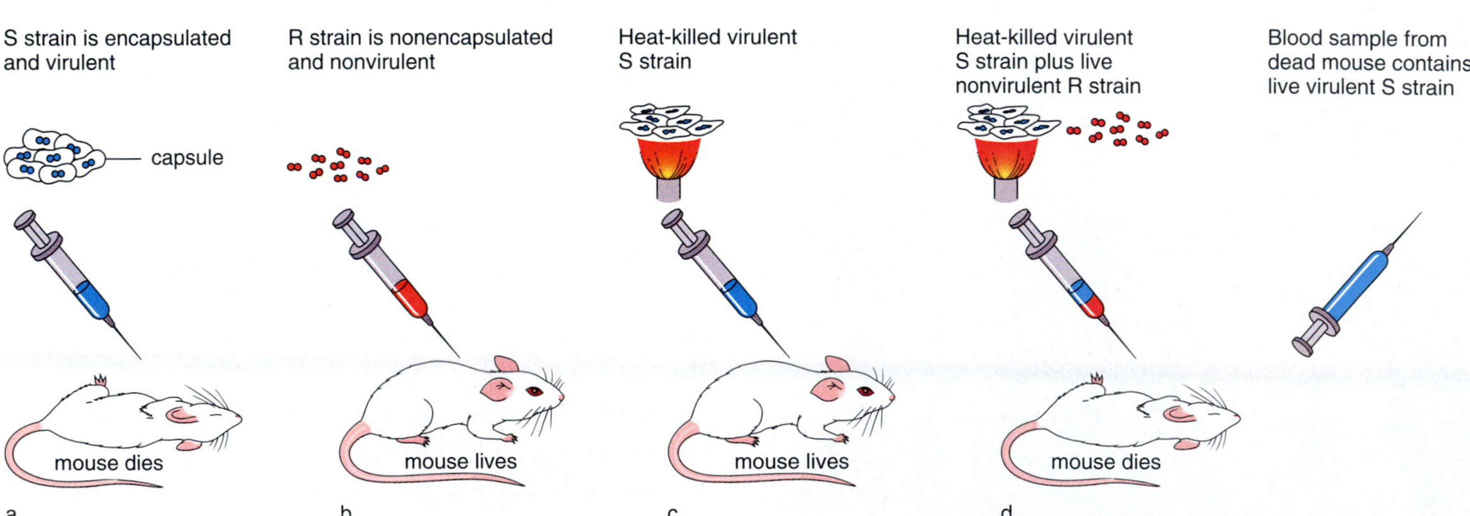

Figure 14.1 Griffith's transformation experiment.
a. Encapsulated S strain is virulent and kills the mouse. **b.** Nonencapsulated R strain is not virulent and does not kill the mouse. **c.** Heat-killed S strain bacteria do not kill the mouse. **d.** If heat-killed S strain and R strain are both injected into a mouse, it dies because the R strain bacteria have been transformed into the virulent S strain.

the mice died and living S strain bacteria were recovered from the bodies! Griffith concluded that some substance necessary to the synthesis of a capsule and, therefore, virulence must have passed from the dead S strain bacteria to the living R strain bacteria so that the R strain bacteria were *transformed* (Fig. 14.1*d*). This change in the phenotype of the R strain bacteria must be due to a change in their genotype. Indeed, couldn't the transforming substance that passed from S strain to R strain be genetic material? Reasoning such as this prompted investigators at the time to begin looking for the transforming substance to determine the chemical nature of the genetic material.

DNA: The Transforming Substance

Obviously, it is not convenient to look for the transforming substance in mice. It is not surprising, then, that the next group of investigators, led by Oswald Avery, isolated the genetic material in vitro (in laboratory glassware). After 16 years of research, this group published a paper demonstrating that the transforming substance is DNA. Their evidence included the following observations:

1. DNA from S strain bacteria causes R strain bacteria to be transformed.
2. Enzymes that degrade proteins cannot prevent transformation, nor did RNase, an enzyme that digests RNA.
3. Enzymatic digestion of the transforming substance with DNase, an enzyme that digests DNA, does prevent transformation.
4. The molecular weight of the transforming substance is so great that it must contain about 1,600 nucleotides! Certainly this is enough for some genetic variability.

These experiments certainly showed that DNA is the transforming substance. Some still remained skeptical that DNA was the genetic material, however. Perhaps, they said, DNA merely served to activate protein-based genes.

Transformation Experiments Today

Transformation experiments are often done today in various laboratories around the world from high schools to the most sophisticated research facilities. Transformation occurs when organisms receive foreign DNA and thereby acquire a new characteristic. One typical rudimentary experiment is to expose bacteria to DNA that allows them to grow in the presence of penicillin. Before they are transformed, the bacteria are unable to grow in the presence of penicillin but after they take up DNA from the medium they can grow in the presence of penicillin. The DNA contains a gene that makes bacteria resistant to penicillin!

Reproduction of Viruses

Soon after Avery published his results, many geneticists who were interested in determining the chemical nature of the genetic material began to work with bacteriophages [Gk. *bacterion*, rod, and *phagein*, to eat]. **Bacteriophages** are viruses that infect bacteria such as the bacterium *Escherichia coli* (*E. coli*), which normally lives within the human gut.

Viruses consist only of a protein coat called a capsid surrounding a nucleic acid core (Fig. 14.2). We now know that when a bacteriophage, or simply phage, latches onto a bacterium, only the nucleic acid (i.e., DNA) enters the cell and the capsid is left behind. Later the bacterium releases many hundreds of new viruses. Why? Because phage DNA contains the genetic information necessary to cause the bacterium to produce new phages. But we are getting ahead of our story. Back

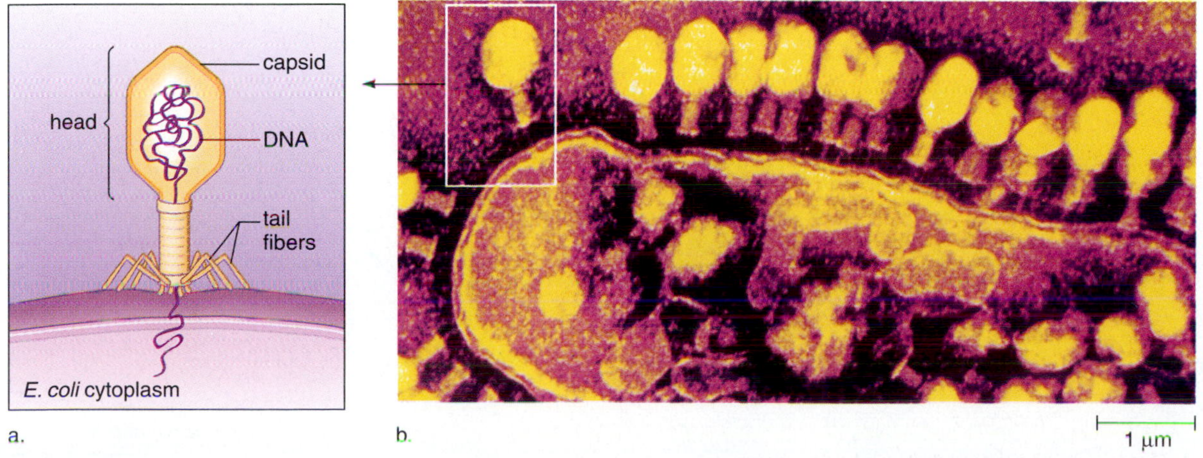

Figure 14.2 Bacteria and bacteriophages.
a. A T virus consists of a capsid (protein coat) surrounding a DNA core. When a T virus attacks a bacterium, only DNA enters the cell and later many fully formed viruses emerge from the cell. **b.** T viruses are pictured attacking an *E. coli* cell.

in the 1950s, biologists were still performing experiments to determine which one—DNA or protein—enters a virus. Whichever one does this has to be the genetic material.

In 1952, two experimenters, Alfred D. Hershey and Martha Chase, chose a bacteriophage known as T2 to determine which of the phage components—protein or DNA—entered bacterial cells and directed reproduction of the virus. Hershey and Chase relied on a chemical difference between DNA and protein.

Phosphorus (P) is present (but not sulfur) in DNA.
Sulfur (S) is present (but not phosphorus) in protein.

They used radioactive ^{32}P to label the DNA core of the phage and radioactive ^{35}S to label the protein in the capsid of the phage. Recall that radioactive isotopes serve as labels (i.e., tracers) in biological experiments because it is possible to detect the presence of radioactivity by standard laboratory procedures.

Hershey and Chase did two separate experiments (Fig. 14.3). In the first experiment phage DNA was labeled with radioactive ^{32}P. The phages were allowed to attach to and inject their genetic material into *E. coli* cells. Then the culture was agitated in a kitchen blender to remove whatever remained of the phages on the outside of the bacterial cells. Finally, the culture was centrifuged (spun at high speed) so that the bacterial cells collected as a pellet at the bottom of the centrifuge tube. In this experiment, as you would predict, they found most of the ^{32}P-labeled DNA in the cells and not in the liquid medium. Why? Because the DNA had entered the cells.

In the second experiment, phage protein in capsids was labeled with radioactive ^{35}S. The phages were allowed to attach to and inject their genetic material into *E. coli* bacterial cells. Then the culture was agitated in a kitchen blender to remove whatever remained of the phages on the outside of the bacterial cells. Finally the culture was centrifuged so that the bacterial cells collected as a pellet at the bottom of the centrifuge tube. In this experiment, as you would predict, they found ^{35}S-labeled protein in the liquid medium and not in the cells. Why? Because the radioactive capsids remained on the outside of the cells and were removed by the blender.

These results indicated that the DNA of a virus (and not protein) enters the host, where viral reproduction takes place. Therefore, DNA is the genetic material. It transmits all the necessary genetic information needed to produce new viruses.

The Hershey and Chase experiment showed that DNA and not protein is the genetic material.

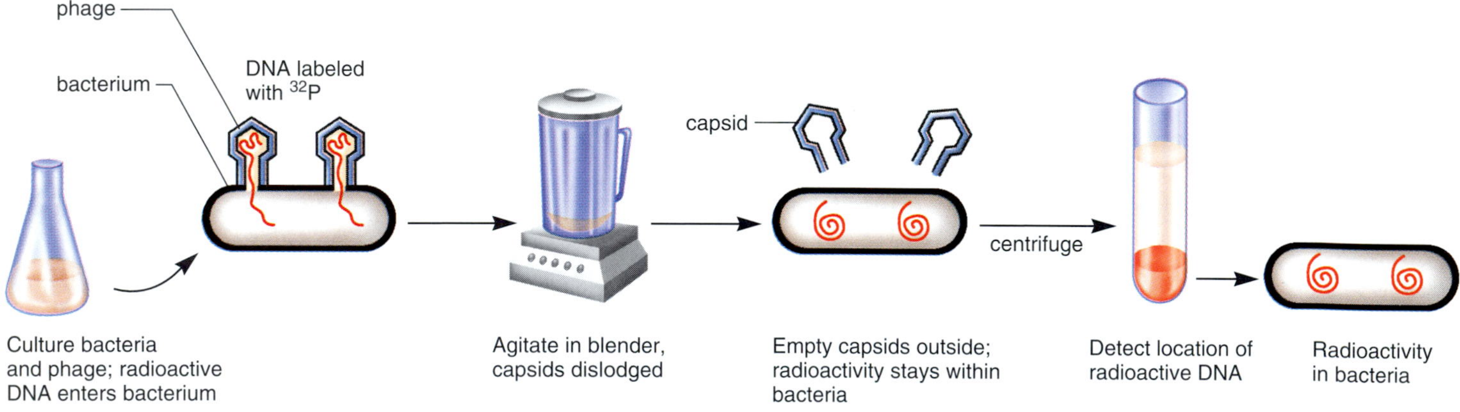

a. Viral DNA is labeled (red).

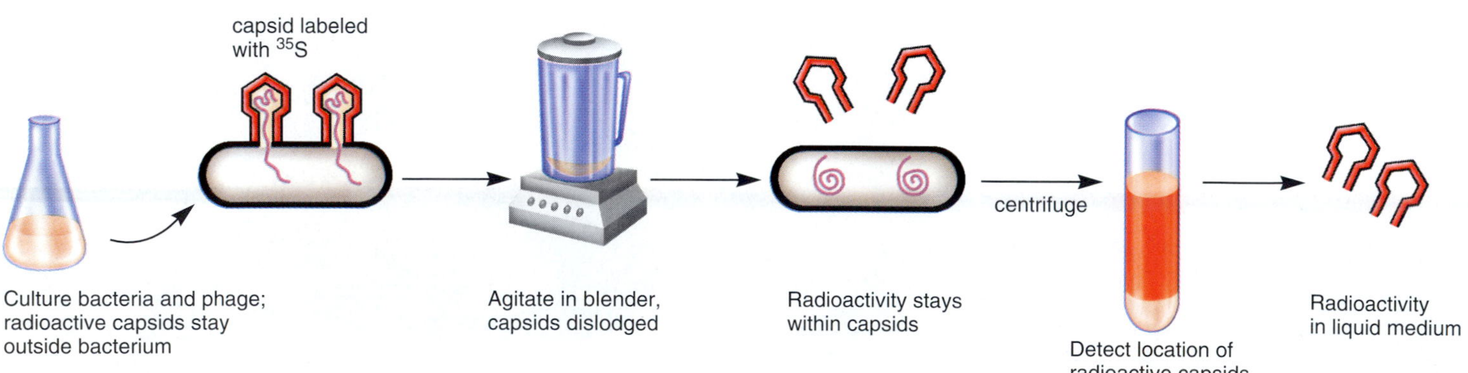

b. Viral capsid is labeled (red).

Figure 14.3 **Hershey and Chase experiment.**

14.2 The Structure of DNA 

During the same period of time that biologists were using viruses to show that DNA is the genetic material, biochemists were busy trying to determine its structure. The structure of DNA shows how DNA stores information, as is required of the genetic material.

Nucleotide Data

With the development of new chemical techniques in the 1940s, it was possible for Erwin Chargaff to analyze in detail the base content of DNA. It was known that DNA contains four different types of nucleotides: two with **purine** bases, **adenine (A)** and **guanine (G),** which have a double ring, and two with **pyrimidine** bases, **thymine (T)** and **cytosine (C),** which have a single ring (Fig. 14.4*a* and *b*). Did each species contain 25% of each kind of nucleotide, or did the amounts vary?

A sample of Chargaff's data is seen in Figure 14.4*c*. You can see that while some species—*E. coli* and *Zea mays* (corn), for example—do have approximately 25% of each type of nucleotide, most do not. Further, the percentage of each type of nucleotide differs from species to species. Therefore, DNA does have the *variability* between species required of the genetic material.

Within each species, however, DNA has the *constancy* required of the genetic material. Further, the percentage of A equals the percentage of T, and the percentage of G equals the percentage of C. The percentage of A + G equals 50%, and the percentage of T + C equals 50%. These relationships are called Chargaff's rules.

Chargaff's rules:

1. The amount of A, T, G, and C in DNA varies from species to species.
2. In each species, the amount of A = T and the amount of G = C.

The so-called tetranucleotide hypothesis, which said that DNA has repeating units, each with one of the four bases, was not supported by these data. Each species had its own constant base composition.

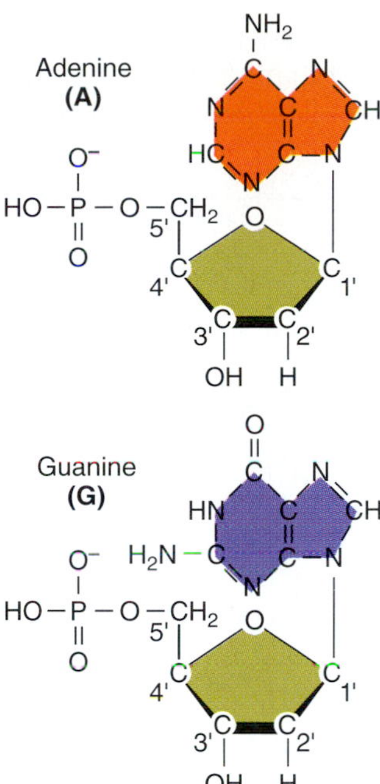

a. **Purine nucleotides**

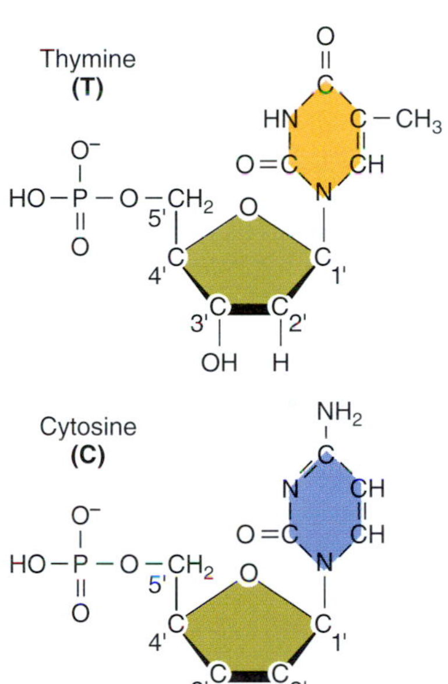

b. **Pyrimidine nucleotides**

Chargaff's DNA Database Composition in Various Species (%)

Species	A	T	G	C
Homo sapiens	31.0	31.5	19.1	18.4
Drosophila melanogaster	27.3	27.6	22.5	22.5
Zea mays	25.6	25.3	24.5	24.6
Neurospora crassa	23.0	23.3	27.1	26.6
Escherichia coli	24.6	24.3	25.5	25.6
Bacillus subtilis	28.4	29.0	21.0	21.6

c.

Figure 14.4 Nucleotide composition of DNA.
All nucleotides contain phosphate, a five-carbon sugar, and a nitrogen-containing base. In DNA, the sugar is deoxyribose—there is an absence of oxygen in the 2′ position—and the nitrogen-containing bases are **(a)** the purines adenine and guanine, which have a double ring, or **(b)** the pyrimidines thymine and cytosine, which have a single ring. **c.** Chargaff's data show that the DNA of various species differs. For example, in humans the A and T percentages are about 31%, but in fruit flies these percentages are about 27%.

Variation in Base Sequence

Chargaff's data suggest that A is always paired with T and G is always paired with C. The paired bases occur in any order:

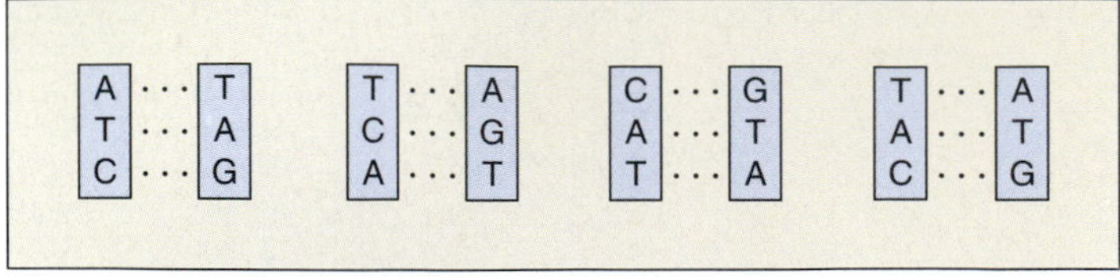

The variability that can be obtained is overwhelming. For example, it has been calculated that the human chromosome contains on the average about 140 million base pairs. Since any of the four possible nucleotides can be present at each nucleotide position, the total number of possible nucleotide sequences is $4^{140 \times 10^6}$ or $4^{140,000,000}$. No wonder each species has its own base percentages!

Diffraction Data

Rosalind Franklin, a student of M. H. F. Wilkins at King's College in London, studied the structure of DNA using X rays. She found that if a concentrated, viscous solution of DNA is made, it can be separated into fibers. Under the right conditions, the fibers are enough like a crystal (a solid substance whose atoms are arranged in a definite manner) that an X-ray pattern forms on a photographic film. Franklin's picture of DNA showed that DNA is a helix (Fig. 14.5). The helical shape is indicated by the crossed (X) pattern in the center of the photograph. The dark portions at the top and bottom of the photograph indicate that some portion of the helix is repeated.

The Watson and Crick Model

James Watson, an American, was on a postdoctoral fellowship at Cavendish Laboratories in Cambridge, England, and while there he began to work with the biophysicist Francis H. C. Crick. Using the data we have just presented, they constructed a model of DNA for which they received a Nobel Prize in 1962 (Fig. 14.6).

Watson and Crick knew, of course, that DNA is a polymer of nucleotides, but they did not know how the nucleotides were arranged within the molecule. This is what they hypothesized.

1. The Watson and Crick model shows that DNA is a double helix with sugar-phosphate backbones on the outside and paired bases on the inside. This arrangement fits the mathematical measurements provided by the X-ray diffraction data for the spacing between the base pairs (0.34 nm) and for a complete turn of the double helix (3.4 nm).
2. Chargaff's rules said that A = T and G = C. The model shows that A is hydrogen-bonded to T and G is hydrogen-bonded to C. This so-called **complementary base pairing** means that a purine is always bonded to a pyrimidine. Only in this way will the molecule have the width (2 nm) dictated by its X-ray diffraction pattern, since two pyrimidines together are too narrow and two purines together are too wide.

The double-helix model of DNA is like a twisted ladder; sugar-phosphate backbones make up the sides, and hydrogen-bonded bases make up the rungs, or steps, of the ladder.

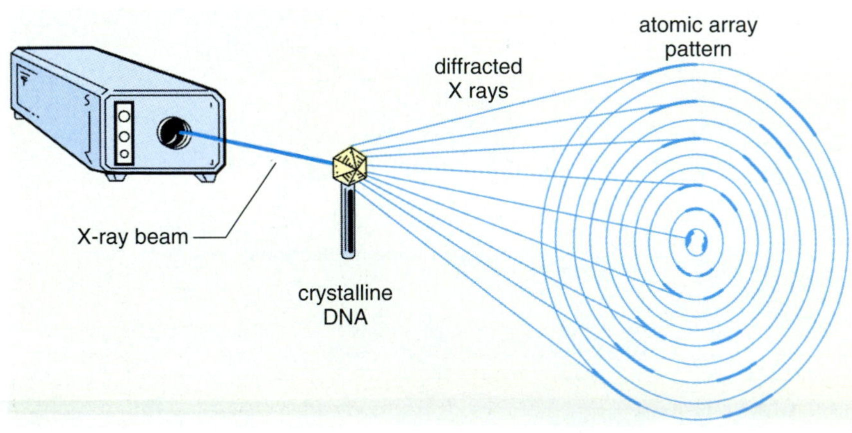

a.

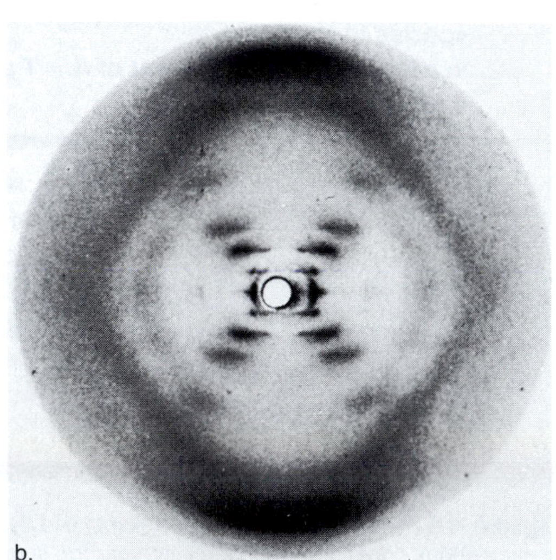

b.

Figure 14.5 X-ray diffraction of DNA.
a. When a crystal is X-rayed, the way in which the beam is diffracted reflects the pattern of the molecules in the crystal. The closer together two repeating structures are in the crystal, the farther from the center the beam is diffracted. **b.** The diffraction pattern of DNA produced by Rosalind Franklin. The crossed (X) pattern in the center told investigators that DNA is a helix, and the dark portions at the top and the bottom told them that some feature is repeated over and over. Watson and Crick determined that this feature was the hydrogen-bonded bases.

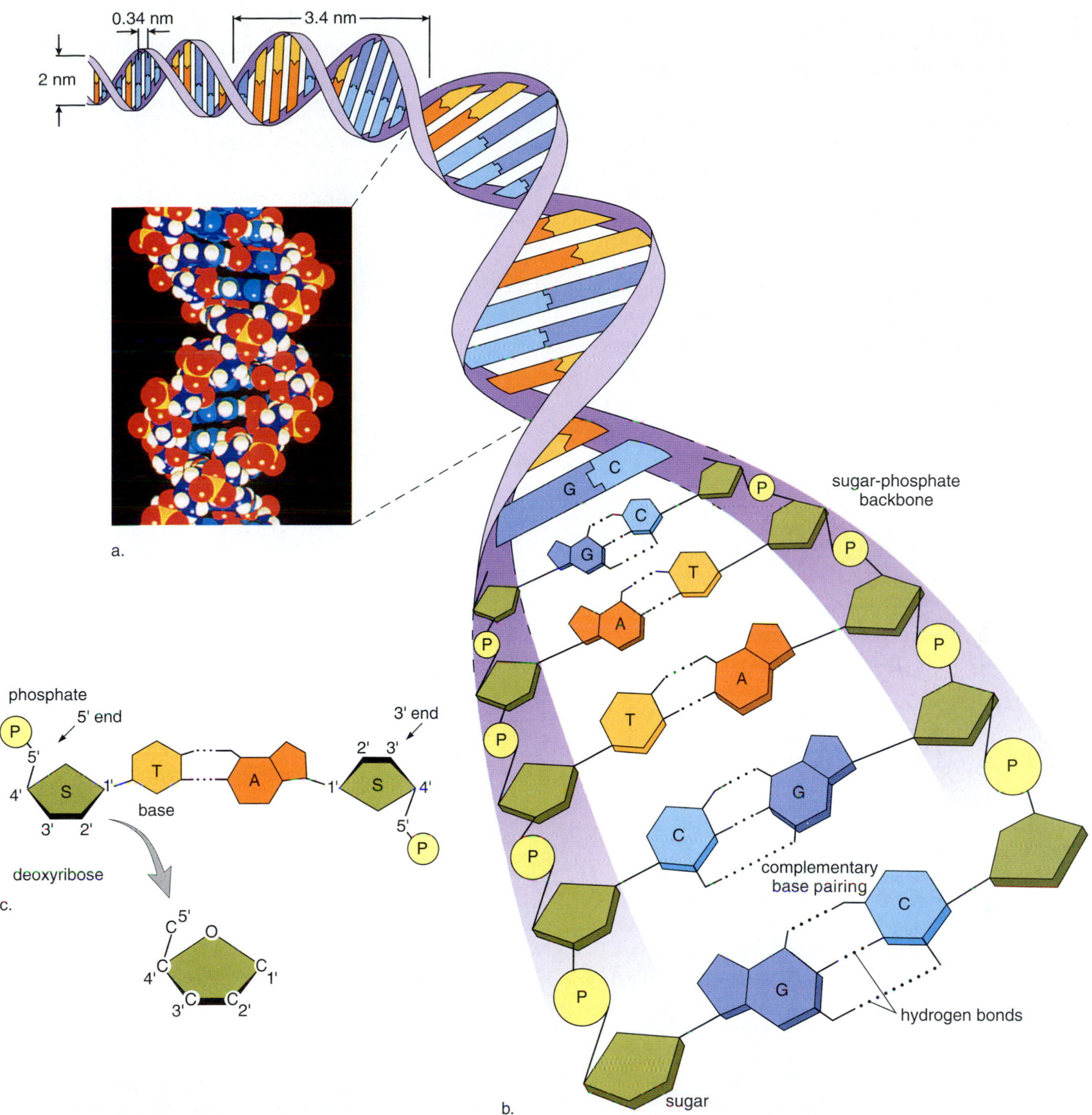

Figure 14.6 Watson and Crick model of DNA.
a. A space-filling model of DNA. Notice the close stacking of the paired bases, as determined by the X-ray diffraction pattern of DNA. (Color code for atoms: yellow = phosphate, dark blue = carbon, red = oxygen, turquoise = nitrogen, and white = hydrogen.) **b.** Diagram of DNA double helix shows that the molecule resembles a twisted ladder. Sugar-phosphate backbones make up the sides of the ladder, and hydrogen-bonded bases make up the rungs of the ladder. Complementary base pairing dictates that A is bonded to T and G is bonded to C. **c.** The two strands of the molecule are antiparallel; that is, the sugar-phosphate groups are oriented in different directions: the 5′ end of one strand is opposite the 3′ end of the other strand.

14.3 Replication of DNA

The term **DNA replication** refers to the process of copying a DNA molecule. Following replication, there is usually an exact copy of the DNA double helix. As soon as Watson and Crick developed their double-helix model, they commented, "It has not escaped our notice that the specific pairing we have postulated immediately suggests a possible copying mechanism for the genetic material."

During DNA replication, each old DNA strand of the parent molecule serves as a template for a new strand in a daughter molecule (Fig. 14.7). A **template** is most often a mold used to produce a shape complementary to itself. DNA replication is termed **semiconservative replication** because one of the old strands is conserved, or present, in each daughter molecule.

Replication requires the following steps:

1. *Unwinding.* The old strands that make up the parent DNA molecule are unwound and "unzipped" (i.e., the weak hydrogen bonds between the paired bases are broken). There is a special enzyme called helicase that unwinds the molecule.
2. *Complementary base pairing.* New complementary nucleotides, always present in the nucleus, are positioned by the process of complementary base pairing.
3. *Joining.* The complementary nucleotides join to form new strands. Each daughter DNA molecule contains an old strand and a new strand.

Steps 2 and 3 are carried out by an enzyme complex called **DNA polymerase.**[1] DNA polymerase works in the test tube as well as in cells.

In Figure 14.7, the backbones of the parent molecule (original double strand) are bluish, and each base is given a particular color. Following replication, the daughter molecules each have a pinkish backbone (new strand) and a bluish backbone (old strand). A daughter DNA double helix has the same sequence of bases as the parent DNA double helix had originally. Although DNA replication can be explained easily in this manner, it is actually a complicated process. Some of the more precise molecular events are discussed in the reading on page 228.

DNA replication must occur before a cell can divide. Cancer, which is characterized by rapidly dividing cells, is sometimes treated with chemotherapeutic drugs that are analogs (have a similar, but not identical structure) to one of the four nucleotides in DNA. When these are mistakenly used by the cancer cells to synthesize DNA, replication stops and the cells die off.

During DNA replication, the parent DNA molecule unwinds and unzips. Then each old strand serves as a template for a new strand.

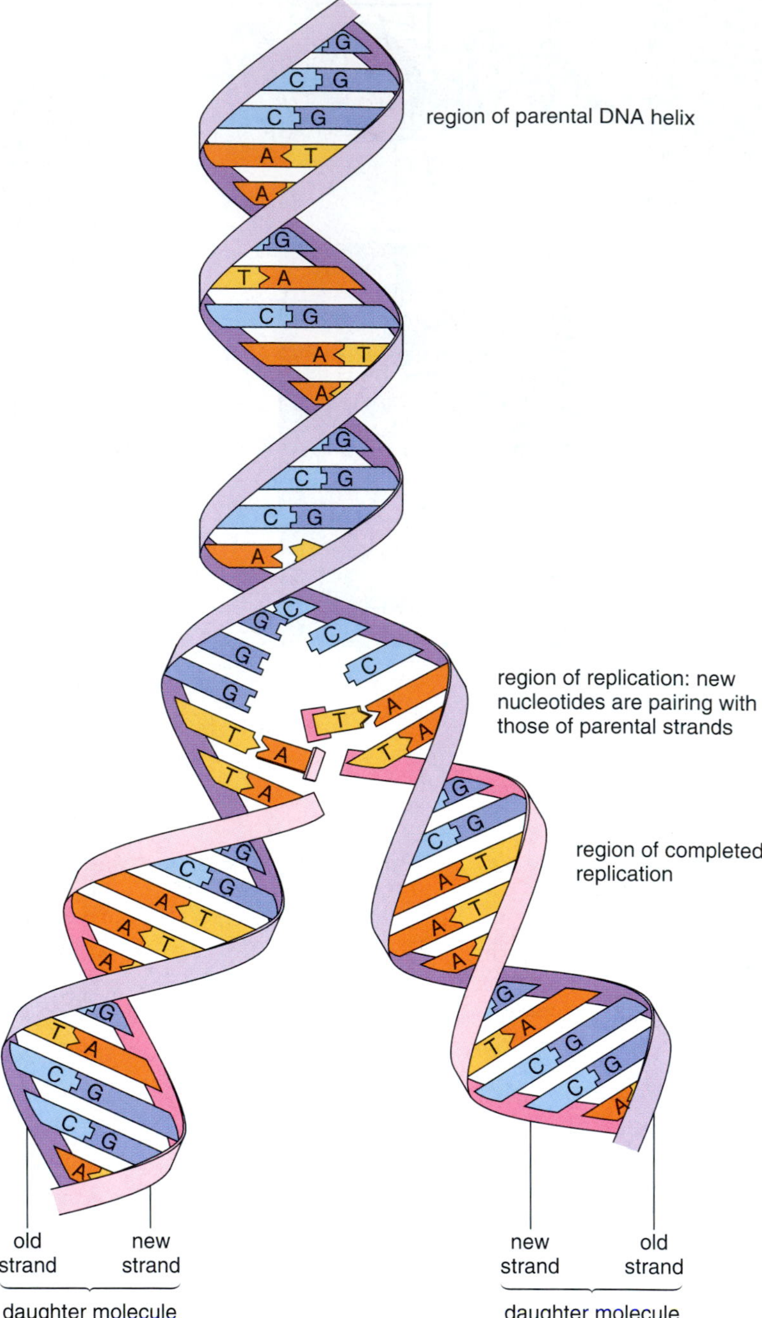

Figure 14.7 Semiconservative replication (simplified). After the DNA molecule unwinds, each old strand serves as a template for the formation of the new strand. Complementary nucleotides available in the cell pair with those of the old strand and then are joined together to form a strand. After replication is complete, there are two daughter DNA molecules. Each is composed of an old strand and a new strand. Each daughter molecule has the same sequence of base pairs as the parent molecule had before unwinding occurred.

[1] The complex most likely contains a number of different DNA polymerases with specific functions.

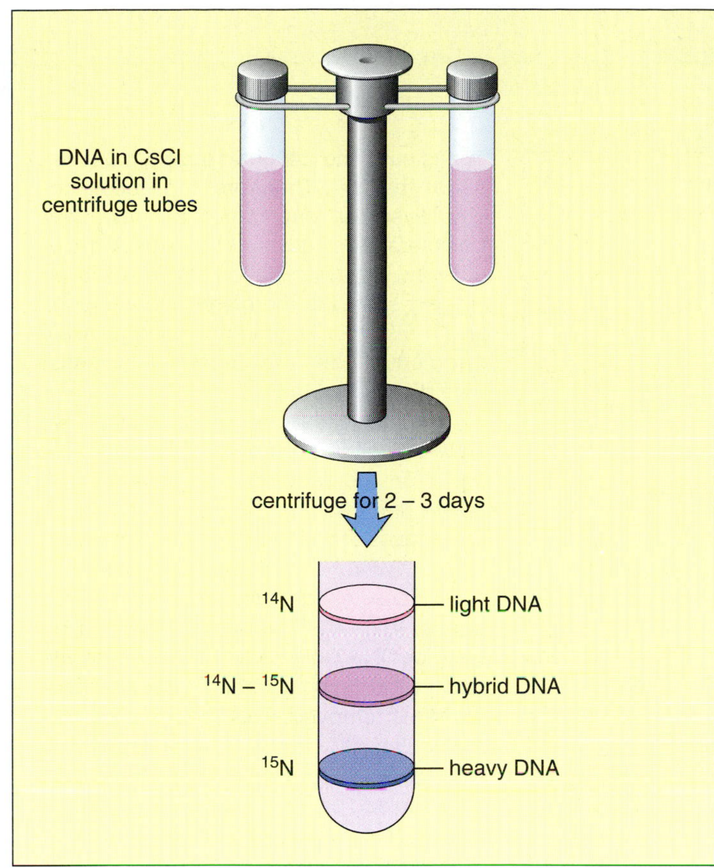

a. Possible results when DNA is centrifuged in CsCl

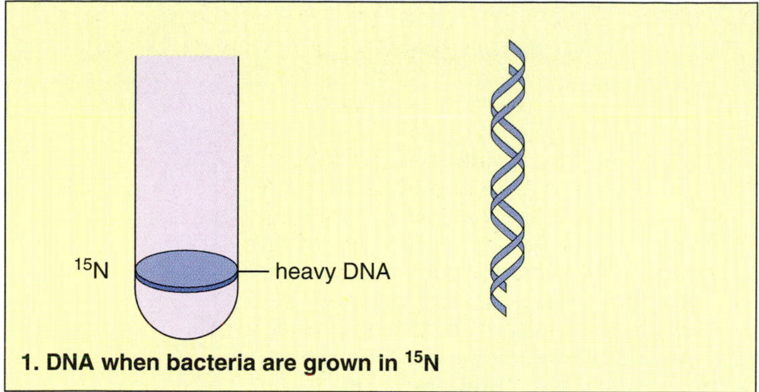

1. DNA when bacteria are grown in ^{15}N

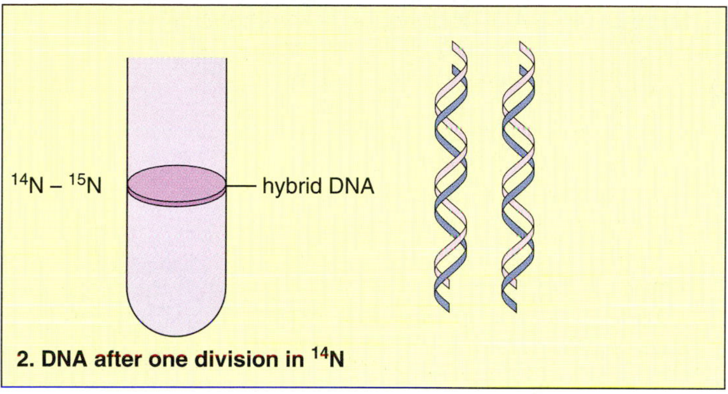

2. DNA after one division in ^{14}N

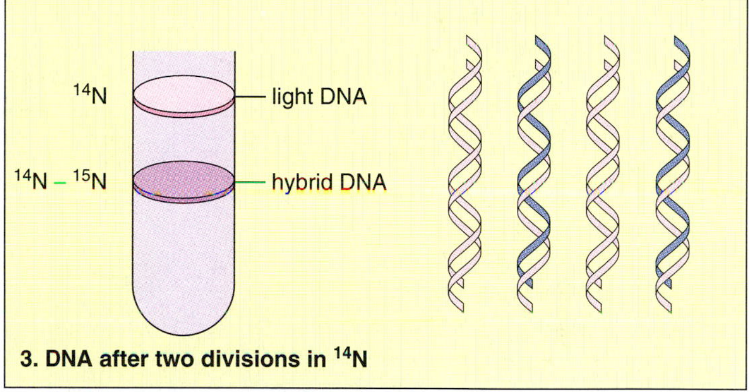

3. DNA after two divisions in ^{14}N

b. Steps in Meselson and Stahl experiment

Figure 14.8 Meselson and Stahl's DNA replication experiment.
a. When DNA molecules are centrifuged in a CsCl density gradient, they separate on the basis of density. **b.** Cells grown in heavy nitrogen (^{15}N) have dense (heavy DNA) strands. After one division in light nitrogen (^{14}N), DNA molecules are hybrid and have intermediate density. After two divisions, DNA molecules separate into two bands—one for light DNA and one for hybrid DNA.

Replication Is Semiconservative

DNA replication is termed semiconservative because each daughter double helix contains an old strand and a new strand. Semiconservative replication was experimentally confirmed by Matthew Meselson and Franklin Stahl in 1958.

Centrifuges spin tubes and in that way separate particles from the suspending fluid. Meselson and Stahl knew that it would be possible to centrifuge DNA molecules in a suspending fluid that would separate them on the basis of their different densities (Fig. 14.8*a*). A DNA molecule in which both strands contained heavy nitrogen (^{15}N, with an atomic weight of 15) is most dense. A DNA molecule in which both strands contained light nitrogen (^{14}N, with an atomic weight of 14) is least dense. A hybrid DNA molecule in which one strand is heavy and one is light has an intermediate density.

Meselson and Stahl first grew bacteria in a medium containing ^{15}N so that only heavy DNA molecules were extracted from the cells. Then they switched the bacteria to a medium containing ^{14}N. After one division, only hybrid DNA molecules were in the cells. After two divisions, half of the DNA molecules were light and half were hybrid. These were exactly the results to be expected if DNA replication is semiconservative.

DNA replication is semiconservative. Each new strand contains an old and a new strand.

Aspects of DNA Replication

Watson and Crick realized that the strands in DNA had to be antiparallel to allow for complementary base pairing. This opposite polarity of the strands introduces complications for DNA replication, as we will now see. First, it is important to take a look at a deoxyribose molecule in which the carbon atoms are numbered (Fig. 14A*a*). Use the structure to see that one of the strands of DNA (Fig. 14A*b*) runs from 3′ → 5′ in one direction and the other strand runs from 3′ → 5′ in the opposite direction.

During replication DNA polymerase has to synthesize the daughter strand in from 5′ → 3′. Why? Because DNA polymerase can only join a nucleotide to the free 3′ end of another nucleotide as shown in Figure 14A*b*. This also means that *DNA polymerase cannot start the synthesis of a DNA chain.* Therefore, an RNA polymerase lays down a short amount of RNA, called an RNA primer, that is complementary to the parental strand being replicated. After that, DNA polymerase can join DNA nucleotides to the 3′ end of the growing daughter (new) strand.

As the helicase enzyme unwinds DNA, one parental strand runs in the 3′ → 5′ direction toward the fork. This works out perfect for this daughter strand because it must be synthesized from the 5′ → 3′ direction. Therefore, this so-called *leading strand* is synthesized continuously toward the region of the fork (Fig. 14Ac.) What about the situation for replication of the other parental strand which runs in the 3′→ 5′ direction away from the fork? In this case the daughter strand must begin at the fork. Therefore, replication begins over and over again by new DNA polymerase molecules as the DNA molecule unwinds. Replication of this so-called *lagging strand* is therefore discontinuous and it results in segments called Okazaki fragments, after the Japanese scientist who discovered them.

Replication is only complete when the RNA primers are removed. This works out well for the lagging daughter strand. While proofreading, DNA polymerase removes the RNA primers and replaces them with complementary DNA nucleotides. Another enzyme, called DNA ligase, joins the fragments. There is no way for DNA polymerase to replicate the end of the leading strand after the RNA primer is removed. This means that DNA molecules get shorter as one replication follows another. The end of eukaryotic DNA molecules have a special nucleotide sequence called *telomeres.* Telomeres do not code for proteins and instead are repeats of a short nucleotide sequence like TTAGGG.

Mammalian cells grown in culture divide about 50 times and then they stop. After this number of divisions the continuous loss of telomeres apparently signals the cell to stop dividing. Ordinarily, telomeres are only added to chromosomes during gamete formation by an enzyme called telomerase. This enzyme unfortunately is often mistakenly turned on in cancer cells, an event that contributes to the ability of cancer cells to keep on dividing without limit.

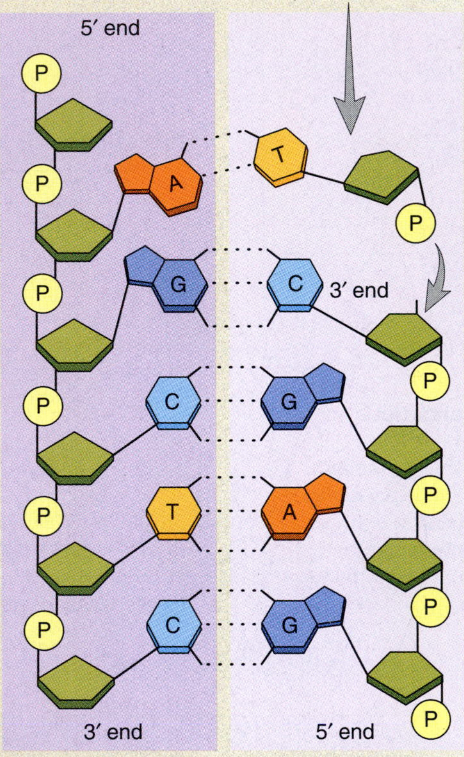

Figure 14A **DNA replication (in depth).**
a. Structure of deoxyribose, showing where nitrogen base and phosphate groups are attached. **b.** Template strand and replicated strand, the latter of which always grows from the 5′ end toward the 3′ end. **c.** Both parental strands are templates for a daughter strand. Replication is a continuous process for one daughter strand and a discontinuous process for the other daughter strand. The enzymes are much larger than shown.

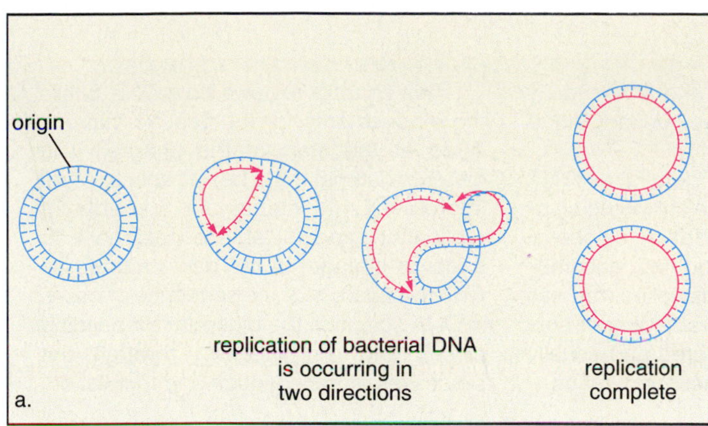

Figure 14.9 Prokaryotic versus eukaryotic replication. a. In prokaryotes, replication can occur in two directions at once because the DNA molecule is circular. b. In eukaryotes, replication occurs at numerous replication forks. The bubbles thereby created spread out until they meet.

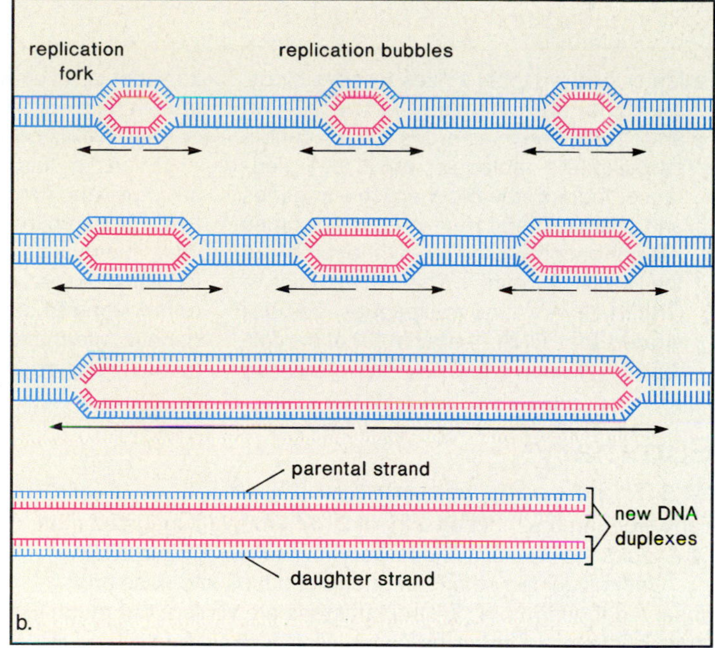

Prokaryotic Versus Eukaryotic Replication

Prokaryotes differ from eukaryotes in a number of ways, including how DNA replication takes place (Fig. 14.9).

Prokaryotic Replication

Bacteria have a single circular loop of DNA that must be replicated before the cell divides. In some circular DNA molecules, replication moves around the DNA molecule in one direction only. In others, as shown here, replication occurs in two directions. The process always occurs in the 5′ to 3′ direction.

Bacterial cells are able to replicate their DNA at a rate of about 10^6 base pairs per minute, and about 40 minutes are required to replicate the complete chromosome. Because bacterial cells are able to divide as often as once every 20 minutes, it is possible for a new round of DNA replication to begin even before the previous round is completed!

Eukaryotic Replication

In eukaryotes, DNA replication begins at numerous origins of replication along the length of the chromosome, and the so-called replication bubbles spread bidirectionally until they meet. Notice that there is a V shape wherever DNA is being replicated. This is called a **replication fork.**

Although eukaryotes replicate their DNA at a slower rate—500 to 5,000 base pairs per minute—there are many individual origins of replication. Therefore, eukaryotic cells complete the replication of the diploid amount of DNA (in humans over 6 billion base pairs) in a matter of hours!

Replication Errors

In molecular genetics, a gene is a section of DNA with a particular sequence of bases. A **genetic mutation** is a permanent change in this sequence of bases.

Some mutations are due to errors in DNA replication. During the replication process, DNA polymerase chooses complementary nucleotide triphosphates from the cellular pool. Then the nucleotide triphosphate is converted to a nucleotide monophosphate and aligned with the template nucleotide. A mismatched nucleotide slips through this selection process only once per 100,000 base pairs at most. The mismatched nucleotide causes a pause in replication, during which time it is excised from the daughter strand and replaced with the correct nucleotide. After this so-called **proofreading** has occurred, the error rate is only one mistake per one billion base pairs.

Some mutations are due to DNA damage. Many of these are caused by a variety of environmental factors. Antioxidants present in cells, the UV radiation in sunlight, organic chemicals in tobacco smoke, pesticides, and pollutants can all damage DNA. **DNA repair enzymes** are usually available and able to reverse most of these insults to DNA but some may escape notice.

The effect of a genetic mutation need not necessarily harm an organism. Some do a great deal of harm, as when they cause cancer in the individual and a genetic disease in offspring. Still, it is important to keep in mind that regardless of their effects, genetic variations are the raw material for the evolutionary process. Without changes in the genetic material, evolution would not be possible.

Mutations due to replication errors and environmental factors are rare due to the proofreading function of DNA polymerase and the presence of DNA repair enzymes in cells.

Connecting Concepts

Many of the Nobel Prizes in physiology, medicine, and chemistry between 1950 and 1990 were awarded to scientists studying the molecular basis of inheritance. Indeed, the biochemistry of genes and the basic structure of chromosomes did not become known until the middle of the twentieth century. The early studies of Griffith, Avery, and colleagues demonstrated that DNA contains genetic information; Watson and Crick later presented a model of its structure; and finally many researchers contributed to our knowledge of how DNA is copied.

Even to this date, details of DNA structure and function are still being clarified. Research in such fields as biochemistry, biophysics, bacteriology, and molecular biology, a new discipline, are still contributing to our understanding of how genetic information is stored and usually faithfully copied for the next generation.

This chapter reviews how DNA fulfills the requirements for the genetic material listed at the start of the chapter. The genetic material must be: (1) able to store information, (2) stable and capable of replication, and (3) able occasionally to undergo change. In the next chapter we will investigate how the sequence of bases in DNA specifies the blueprint for building a cell and an organism through the processes of transcription and translation.

Summary

14.1 The Genetic Material

Early work on the biochemistry of DNA wrongly suggested that DNA lacks the variability necessary for the genetic material.

Griffith injected strains of pneumococcus into mice and observed that smooth (S) strain bacteria are virulent but rough (R) strain bacteria are not. When heat-killed S strain bacteria were injected along with live R strain bacteria, however, virulent S strain bacteria were recovered from the dead mice. Griffith said that the R strain had been transformed by some substance passing from the dead S strain to the live R strain. Twenty years later, Avery and his colleagues reported that the transforming substance is DNA. Hershey and Chase turned to bacteriophage T2 as their experimental material. In two separate experiments, they labeled the protein coat with ^{35}S and the DNA with ^{32}P. They then showed that the radioactive P alone is largely taken up by the bacterial host and that reproduction of viruses proceeds normally. This convinced most researchers that DNA is the genetic material.

14.2 The Structure of DNA

Chargaff did a chemical analysis of DNA and found that A = T and G = C, and that the amount of purine equals the amount of pyrimidine. Franklin prepared an X-ray photograph of DNA that showed it is helical, has repeating structural features, and has certain dimensions. Watson and Crick built a model of DNA in which the sugar-phosphate molecules made up the sides of a twisted ladder and the complementary-paired bases were the rungs of the ladder.

14.3 Replication of DNA

The Watson and Crick model immediately suggested a method by which DNA could be replicated. The two strands unwind and unzip, and each parental strand acts as a template for a new (daughter) strand. In the end, each new duplex is like the other and like the parental duplex.

Meselson and Stahl demonstrated that replication is semiconservative by the following experiment: bacteria were grown in heavy nitrogen (^{15}N) and then switched to light nitrogen (^{14}N). The density of the DNA following replication was intermediate between these two, as measured by centrifugation of the molecules through a salt gradient.

The enzyme DNA polymerase joins the nucleotides together and proofreads them to make sure the bases have been paired correctly. Incorrect base pairs that survive the process are a mutation. Replication in prokaryotes proceeds from one point of origin until there are two copies of the circular chromosome. Replication in eukaryotes has many points of origin and many bubbles (places where the DNA strands are separating and replication is occurring). Replication occurs at the ends of the bubbles—at replication forks.

Reviewing the Chapter

1. List and discuss the requirements for genetic material. 220
2. Describe Griffith's experiments with pneumococcus, his surprising results, and his conclusion. 220–21
3. How did Avery and his colleagues demonstrate that the transforming substance is DNA? 221
4. Describe the experiment of Hershey and Chase, and explain how it shows that DNA is the genetic material. 222
5. What are Chargaff's rules? 223
6. Describe the Watson and Crick model of DNA structure. How did it fit the data provided by Chargaff and the X-ray diffraction pattern? 224–25
7. Explain how DNA replicates semiconservatively. What role does DNA polymerase play? What role does helicase play? 226–27
8. How did Meselson and Stahl demonstrate semiconservative replication? 227
9. List and discuss differences between prokaryotic and eukaryotic replication of DNA. 229
10. Explain why the replication process is a source of few mutations. 229

Testing Yourself

Choose the best answer for each question. For questions 1–4, match the names in the key to the statements below:

Key:

a. Griffith
b. Chargaff
c. Meselson and Stahl
d. Hershey and Chase
e. Avery

1. A = T and G = C.
2. Only the DNA from T2 enters the bacteria.
3. R strain bacteria became an S strain through transformation.
4. DNA replication is semiconservative.
5. If 30% of an organism's DNA is thymine, then
   a. 70% is purine.
   b. 20% is guanine.
   c. 30% is adenine.
   d. 70% is pyrimidine.
   e. Both c and d are correct.
6. If you grew bacteria in heavy nitrogen and then switched them to light nitrogen, how many generations after switching would you have some light/light DNA?
   a. never, because replication is semiconservative
   b. the first generation
   c. the second generation
   d. only the third generation
   e. Both b and c are correct.
7. The double-helix model of DNA resembles a twisted ladder in which the rungs of the ladder are
   a. a purine paired with a pyrimidine.
   b. A paired with G and C paired with T.
   c. sugar-phosphate paired with sugar-phosphate.
   d. 5′ end paired with a 3′ end.
   e. Both a and b are correct.
8. Cell division requires that the genetic material be able to
   a. store information.
   b. undergo replication.
   c. undergo rare mutations.
   d. condense into spindle fibers.
   e. All of these are correct.
9. In a DNA molecule, the
   a. bases are covalently bonded to the sugars.
   b. sugars are covalently bonded to the phosphates.
   c. bases are hydrogen-bonded to one another.
   d. nucleotides are covalently bonded to one another.
   e. All of these are correct.
10. In the following diagram, blue stands for heavy DNA (contains ^{15}N) and pink stands for light DNA (does not contain ^{15}N). Label each strand of all three DNA molecules as heavy or light DNA, and explain why the diagram is in keeping with the semiconservative replication of DNA.

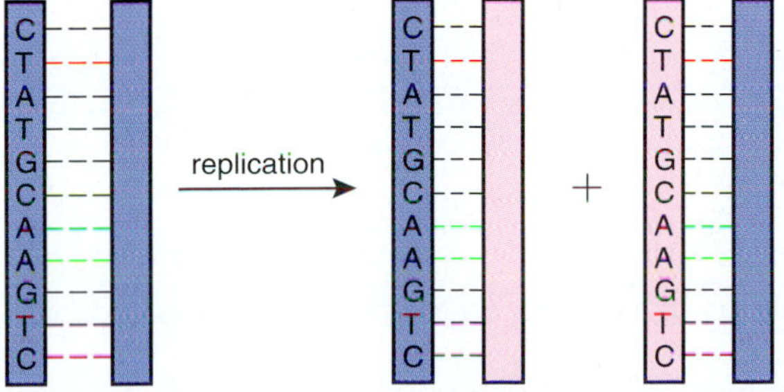

Thinking Scientifically

1. Replication of DNA in some viruses requires the use of proteins that have no enzymatic activity. Hypothesize how a protein might possibly act as a primer for DNA replication. What would be the advantage of using a protein rather than a RNA to get replication started? (See page 228.)
2. Skin cancer is much more common than brain cancer. Why might the frequency of cancer be related to rate of cell division in skin as opposed to rate of cell division in the brain? How might DNA sequencing help you test your hypothesis?

Bioethical Issue

If the results were absolutely private, would you want to know if your DNA harbors a genetic disease whose symptoms may not appear for several decades? If the disease were curable, it might be a good idea. Testing before symptoms appear encourages increased vigilance for the onset of breast and colon cancers. If gene therapy is the only way out, you might be first in line if such therapy is developed in the future.

Huntington disease is a particularly frightening genetic disorder, especially if you have seen a loved one suffer the mental and musculoskeletal deterioration that finally results in death. Although there is no cure for Huntington disease, you would be able to warn your children if you test positive, or it might help you decide whether to have a child. If you have this dominant allele, you will develop the condition and your children have a 50% chance of developing the condition. Children usually begin to have symptoms earlier than their parents exhibit symptoms.

Jason Brandt of the Johns Hopkins University School of Medicine has developed a program of pretest and posttest counseling for those who decide they want to be tested for Huntington disease. So far, about 200 have been tested, and almost twice that number have dropped out of pretest counseling, signifying no doubt they really didn't want to be tested.

Surprisingly, Brandt finds that people rarely make major life changes as a result of knowing they have the allele. Among 63 who tested positive, 10 got married to persons who knew they would come down with Huntington disease, and 10 others went on to have more children. One patient made a radical change after she knew she was negative for Huntington disease. She divorced her present husband, remarried, had another child and took up a career as a physical therapist.

Still, Brandt refuses to perform the test on persons he feels could not psychologically handle the results if they were positive. Do you think it is ever ethical to withhold medical information from patients?

Understanding the Terms

adenine (A) 223
bacteriophage 221
complementary base pairing 224
cytosine (C) 223
DNA (deoxyribonucleic acid) 220
DNA polymerase 226
DNA repair enzyme 229
DNA replication 226
genetic mutation 229
guanine (G) 223
nucleic acid 220
nucleotide 220
proofreading 229
purine 223
pyrimidine 223
replication fork 229
RNA (ribonucleic acid) 220
semiconservative replication 226
template 226
thymine (T) 223

Match the terms to these definitions:

a. _______________ Permanent change in DNA base sequence.
b. _______________ Bonding between particular purines and pyrimidines in DNA.
c. _______________ During replication, an enzyme that joins the nucleotides complementary to a DNA template.
d. _______________ Type of nitrogen-containing base, such as adenine and guanine, having a double-ring structure.
e. _______________ Virus that infects bacteria.

Web Connections

Exploring the Internet

http://www.mhhe.com/biosci/genbio/mader
(click on *Biology 7/e*)

The *Biology 7/e* Online Learning Center provides many resources for studying the material in this chapter including links to the following sites:

Legacies–Transformation and DNA explores how genetic material can be transformed from one bacterium to another, as discovered during Griffith's experiments in the late 1920s. This website also describes the follow-up works of Avery, MacLeod, and MacCarthy that showed that DNA was the genetic material of all organisms.

http://www.accessexcellence.org/AB/BC/Transformation_and_DNA.html

The One Gene/One Enzyme Hypothesis discusses Beadle and Tatum's 1941 breakthrough in finding the connection between genes and enzymes. How understanding the genetic code will advance the field of biotechnology is also considered.

http://www.accessexcellence.org/AB/BC/One_Gene_One_Enzyme.html

In Experiments that Inspire, the Hershey-Chase experiments are described, along with information on when viruses were discovered, and why there was the need to know what was the genetic material.

http://www.accessexcellence.org/AB/BC/Experiments_that_Inspire.html

Genetics Society of America. This site provides links to genetics societies concerned with human and medical genetics. It also provides links to educational sites and sites describing careers in genetics.

http://www.faseb.org/genetics/

Amino Acids Section. 3-D models of all amino acids.

http://www.nyu.edu:80/pages/mathmol/library/life/life1.html

Gene Activity: How Genes Work

chapter

chapter concepts

From four bases, many different organisms

Nearly two million species have so far been discovered and named. Yet one gene differs from another only by the sequence of the nucleotide bases in DNA. How does a difference in base sequence cause the uniqueness of the species—that is, for example, whether you are a tiger or a human? For that matter, whether a human has blue, brown, or hazel eye pigments?

By studying the activity of genes in cells, geneticists have confirmed that proteins are the link between the genotype and the phenotype. Mendel's peas are smooth or wrinkled according to the presence or absence of a starch-forming enzyme. The allele *S* in peas dictates the presence of the starch-forming enzyme, whereas the allele *s* does not.

Through its ability to specify proteins, DNA brings about the development of the unique structures that make up a particular type of organism. It follows that you have blue or brown or hazel eye pigments because of the type of enzymes contained within your cells. When studying gene expression in this chapter, keep in mind this flow diagram: DNA's sequence of nucleotides → sequences of amino acids → specific enzymes → structures in organism.

15.1 The Function of Genes

In the early 1900s, the English physician Sir Archibald Garrod suggested that there is a relationship between inheritance and metabolic diseases. He introduced the phrase *inborn error of metabolism* to dramatize this relationship. Garrod observed that family members often had the same disorder, and he said this inherited defect could be caused by the lack of a particular enzyme in a metabolic pathway. Since it was known at the time that enzymes are proteins, Garrod was among the first to hypothesize a link between genes and proteins.

Genes Specify Enzymes

Many years later, in 1940, George Beadle and Edward Tatum performed a series of experiments on *Neurospora crassa,* the red bread mold fungus, which reproduces by means of spores. Normally, the spores become a mold capable of growing on minimal medium (containing only a sugar, mineral salts, and the vitamin biotin) because mold can produce all the enzymes it needs. In their experiments, Beadle and Tatum induced mutations in asexually produced haploid spores by the use of X-rays. Some of the X-rayed spores could no longer become a mold capable of growing on minimal medium; however, growth was possible on medium enriched by certain metabolites. In the example given in Figure 15.1, a mold grows only when supplied with enriched medium that includes all metabolites or C and D alone. Since C and D are a part of this hypothetical pathway

$$A \xrightarrow{1} B \xrightarrow{2} C \xrightarrow{3} D$$

in which the numbers are enzymes and the letters are metabolites, it is concluded that the mold lacks enzyme 2. Beadle and Tatum further found that each of the mutant strains has only one defective gene leading to one defective enzyme and one additional growth requirement. Therefore, they proposed that each gene specifies the synthesis of one enzyme. This is called the *one gene–one enzyme hypothesis.*

Genes Specify a Polypeptide

The one gene–one enzyme hypothesis suggests that a genetic mutation causes a change in the structure of a protein. To test this idea, Linus Pauling and Harvey Itano decided to see if the hemoglobin in the red blood cells of persons with sickle-cell disease has a structure different from that of the red blood cells of normal individuals (Fig. 15.2). Recall that proteins are polymers of amino acids, some of which carry a charge. These investigators decided to see if there was a charge difference between normal hemoglobin (Hb^A) and

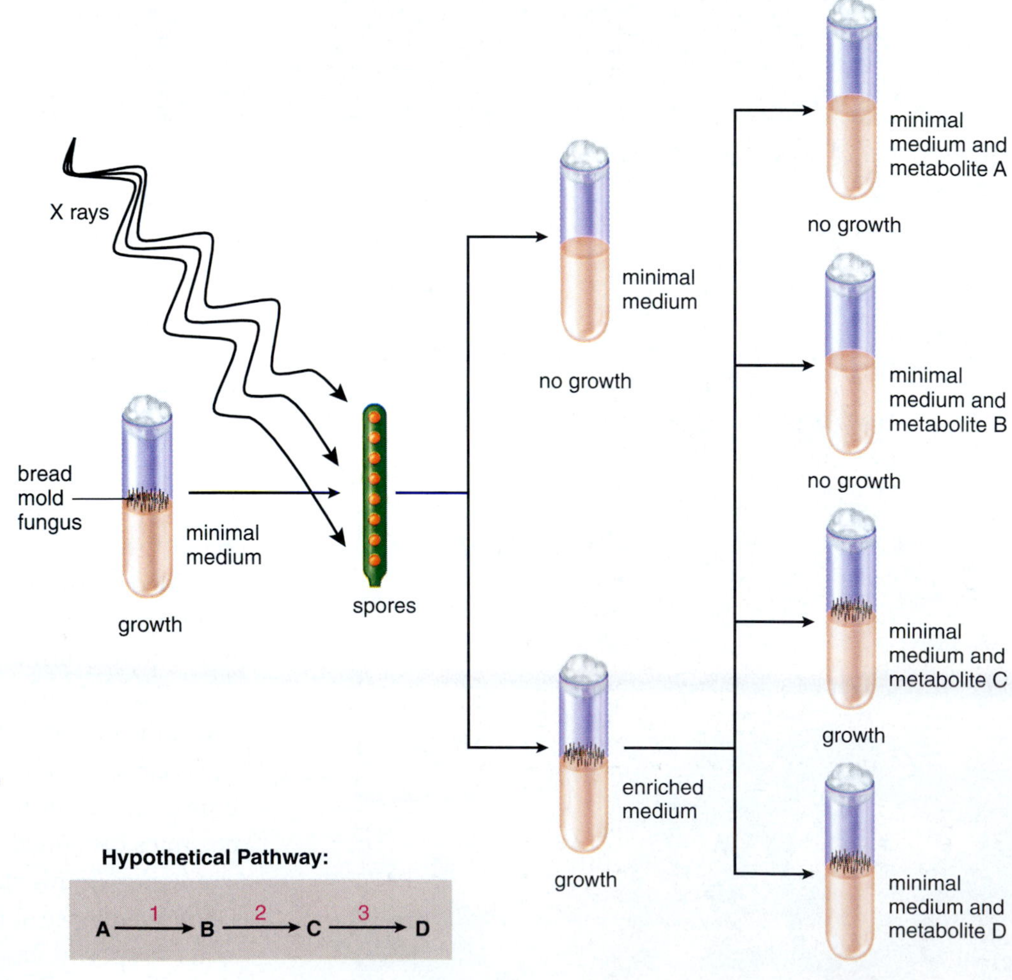

Figure 15.1 **Beadle and Tatum experiment.**
When haploid spores of *Neurospora crassa*, a bread mold fungus, are X-rayed, some are no longer able to germinate on minimal medium; however, they can germinate on enriched medium. In this example, the mycelia produced do not grow on minimal medium plus metabolite A or B, but they do grow on minimal medium plus metabolite C or D. This shows that enzyme 2 is missing from the hypothetical pathway.

sickle-cell hemoglobin (Hb^S). To determine this, they subjected hemoglobin collected from normal individuals, sickle-cell trait individuals, and sickle-cell disease individuals to electrophoresis, a procedure that separates molecules according to their size and charge, whether (+) or (−). Here is what they found:

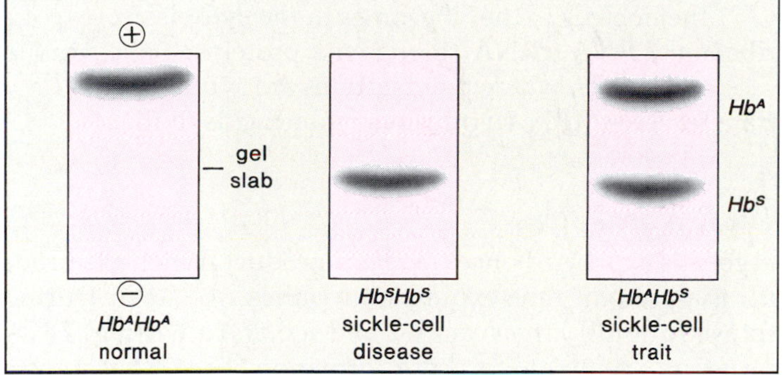

As you can see, there is a difference in migration rate toward the positive pole between normal hemoglobin and sickle-cell hemoglobin. Further, hemoglobin from those with sickle-cell trait separates into two distinct bands, one corresponding to that for Hb^A hemoglobin and the other corresponding to that for Hb^S hemoglobin. Pauling and Itano therefore demonstrated that a mutation leads to a change in the structure of a protein.

Several years later, Vernon Ingram was able to determine the structural difference between Hb^A and Hb^S. Normal hemoglobin Hb^A contains negatively charged glutamate; in sickle-cell hemoglobin the glutamate is replaced by nonpolar valine (Fig. 15.2*c*). This causes Hb^S to be less soluble and to precipitate out of solution, especially when environmental oxygen is low. At these times, the Hb^S molecules stack up into long, semirigid rods that push against the plasma membrane and distort the red blood cell into the sickle shape.

Hemoglobin contains two types of polypeptide chains, designated α (alpha) and β (beta). Only the β chain is affected in persons with sickle-cell trait and sickle-cell disease; therefore, there must be a gene for each type of chain. A refinement of the one gene–one enzyme hypothesis was needed, and it was replaced by the *one gene–one polypeptide hypothesis.*

A gene is a segment of DNA that specifies the sequence of amino acids in a polypeptide of a protein.

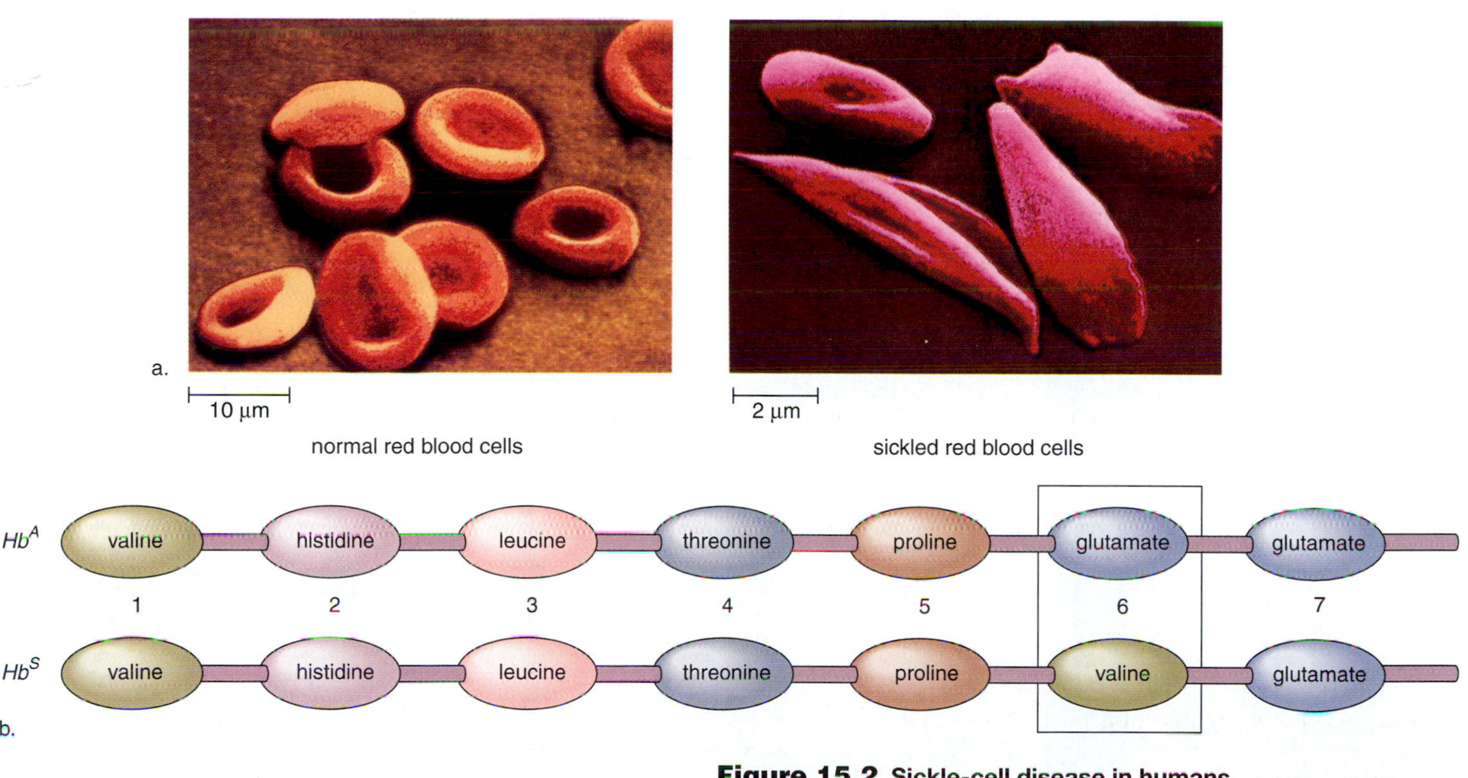

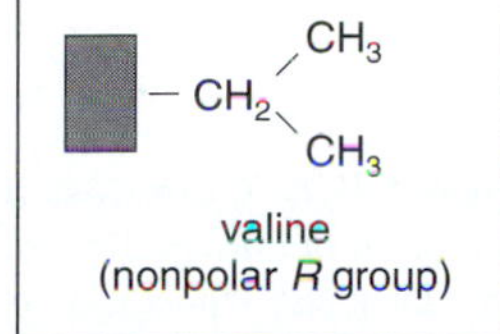

Figure 15.2 Sickle-cell disease in humans.
a. Scanning electron micrograph of normal *(left)* and sickled *(right)* red blood cells. **b.** Portion of the chain in normal hemoglobin Hb^A and in sickle-cell hemoglobin Hb^S. Although the chain is 146 amino acids long, the one change from glutamate to valine in the sixth position results in sickle-cell disease. **c.** Glutamate has a polar *R* group, while valine has a nonpolar *R* group, and this causes Hb^S to be less soluble and to precipitate out of solution, distorting the red blood cell into the sickle shape.

From DNA to RNA to Protein

Classical geneticists thought of a gene as a particle on a chromosome. To molecular geneticists, the complete definition of a gene is a sequence of DNA nucleotide bases that codes for a product. Therefore, a gene does not affect the phenotype directly; rather, the gene product affects the phenotype.

DNA specifies the production of proteins, even though in eukaryotes it is located in the nucleus, and proteins are synthesized at the ribosomes in the cytoplasm. **RNA (ribonucleic acid),** however, is not confined to the nucleus, it occurs in both the nucleus and the cytoplasm.

Types of RNA

Like DNA, RNA is a polymer of nucleotides. The nucleotides in RNA, however, contain the sugar ribose and the bases adenine (A), cytosine (C), guanine (G), and uracil (U). In other words, the base **uracil** replaces thymine found in DNA. Finally, RNA is single stranded and does not form a double helix in the same manner as DNA (Table 15.1 and Fig. 15.3).

There are three major classes of RNA, each with specific functions in protein synthesis:

messenger RNA (mRNA): takes a message from DNA in the nucleus to the ribosomes in the cytoplasm.
ribosomal RNA (rRNA): along with proteins, makes up the ribosomes, where polypeptides are synthesized.
transfer RNA (tRNA): transfers amino acids to the ribosomes.

Table 15.1

RNA Structure Compared to DNA Structure

	RNA	DNA
Sugar	Ribose	Deoxyribose
Bases	Adenine, guanine, uracil, cytosine	Adenine, guanine, thymine, cytosine
Strands	Single stranded	Double stranded with base pairing
Helix	No	Yes

The Required Steps

A gene is expressed once there is a product. When the product is a protein, gene expression requires two steps. During **transcription** [L. *trans*, across, and *scriptio*, a writing] DNA serves as a template for RNA formation. Then RNA moves into the cytoplasm. There are micrographs showing radioactively labeled RNA moving from the nucleus to the cytoplasm, where protein synthesis occurs. During **translation** [L. *trans*, across, and *latus*, carry, bear] an mRNA transcript directs the sequence of amino acids in a polypeptide (Fig. 15.4).

During transcription, DNA serves as a template for the formation of RNA. During translation, mRNA is involved in polypeptide synthesis.

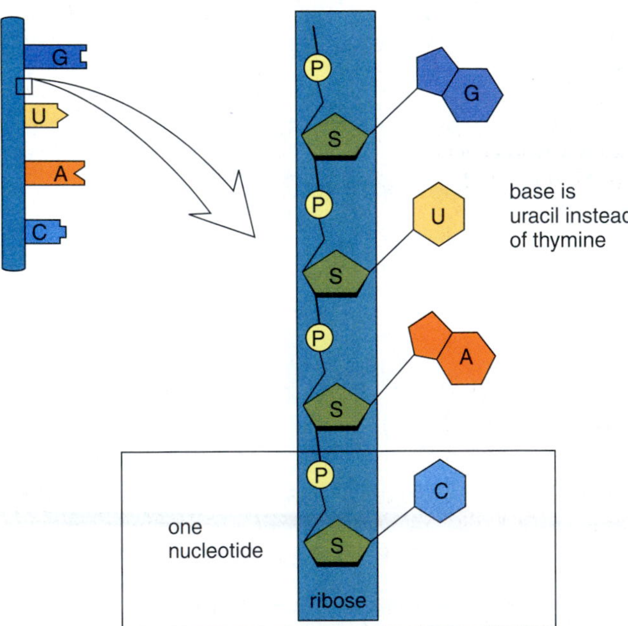

Figure 15.3 Structure of RNA.
Like DNA, RNA is a polymer of nucleotides. RNA, however, is single stranded, the pentose sugar is ribose, and uracil replaces thymine as one of the pyrimidine bases.

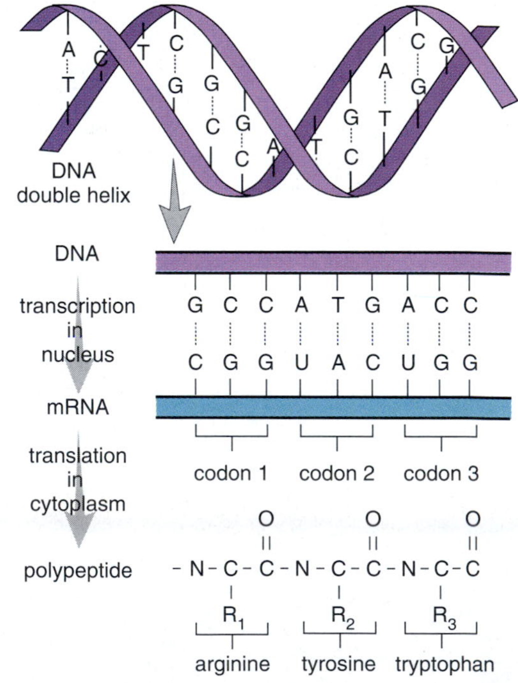

Figure 15.4 Overview of gene expression.
Transcription occurs in the nucleus when DNA acts as a template for mRNA synthesis. Translation occurs in the cytoplasm where the sequence of bases in mRNA determines the sequence of amino acids in a polypeptide.

15.2 The Genetic Code

The central dogma of molecular biology states that the sequence of nucleotides in DNA specifies (and a copy of this sequence in mRNA directs) the order of amino acids in a polypeptide. It would seem then that there must be a **genetic code** for each of the 20 amino acids found in proteins. But can four nucleotides provide enough combinations to code for 20 amino acids? If each code word, called a **codon,** were made up of two bases, such as AG, there could be only 16 codons (4×4)—not enough to code for 20 amino acids. But if each codon were made up of three bases, such as AGC, there would be 64 codons ($4 \times 4 \times 4$)—more than enough to code for 20 different amino acids:

Number of Bases in Genetic Code	Number of Different Amino Acids That Can Be Specified
1	4
2	16
3	64

It is expected, then, that the genetic code is a **triplet code** and that each codon, therefore, consists of three nucleotide bases.

Finding the Genetic Code

In 1961, Marshall Nirenberg and J. Heinrich Matthei performed an experiment that laid the groundwork for cracking the genetic code. First, they found a cellular enzyme could be used to construct a synthetic RNA (one that does not occur in cells), and then they found that the synthetic polymer could be translated in a cell-free system (a test tube that contains "freed" cytoplasmic contents of a cell). Their first synthetic RNA was composed only of uracil, and the protein that resulted was composed only of the amino acid phenylalanine. Therefore, the codon for phenylalanine was known to be UUU. Later, they were able to translate just three nucleotides at a time; in that way it was possible to assign an amino acid to each of the RNA codons (Fig. 15.5).

A number of important properties of the genetic code can be seen by careful inspection of Figure 15.5:

1. The genetic code is degenerate. This means that most amino acids have more than one codon; leucine, serine, and arginine have six different codons, for example. The degeneracy of the code is protective against potentially harmful effects of mutations.
2. The genetic code is unambiguous. Each triplet codon has only one meaning.
3. The code has start and stop signals. There is only one start signal but three stop signals.

First Base	Second Base: U	Second Base: C	Second Base: A	Second Base: G	Third Base
U	UUU phenylalanine	UCU serine	UAU tyrosine	UGU cysteine	U
U	UUC phenylalanine	UCC serine	UAC tyrosine	UGC cysteine	C
U	UUA leucine	UCA serine	UAA *stop*	UGA *stop*	A
U	UUG leucine	UCG serine	UAG *stop*	UGG tryptophan	G
C	CUU leucine	CCU proline	CAU histidine	CGU arginine	U
C	CUC leucine	CCC proline	CAC histidine	CGC arginine	C
C	CUA leucine	CCA proline	CAA glutamine	CGA arginine	A
C	CUG leucine	CCG proline	CAG glutamine	CGG arginine	G
A	AUU isoleucine	ACU threonine	AAU asparagine	AGU serine	U
A	AUC isoleucine	ACC threonine	AAC asparagine	AGC serine	C
A	AUA isoleucine	ACA threonine	AAA lysine	AGA arginine	A
A	AUG (*start*) methionine	ACG threonine	AAG lysine	AGG arginine	G
G	GUU valine	GCU alanine	GAU aspartate	GGU glycine	U
G	GUC valine	GCC alanine	GAC aspartate	GGC glycine	C
G	GUA valine	GCA alanine	GAA glutamate	GGA glycine	A
G	GUG valine	GCG alanine	GAG glutamate	GGG glycine	G

Figure 15.5 Messenger RNA codons.
Notice that in this chart, each of the codons (in squares) is composed of three letters representing the first base, second base, and third base. For example, find the blue square where C for the first base and A for the second base intersect. You will see that U, C, A, or G can be the third base. The three bases CAU and CAC are codons for histidine; the three bases CAA and CAG are codons for glutamine.

The Code Is Universal

The genetic code in Figure 15.5 is just about universally used by living things. This suggests that the code dates back to the very first organisms on earth and that all living things are related. Once the code was established, changes in it would have been very disruptive, and we know that it has remained largely unchanged over eons.

The genetic code is a triplet code. Each codon consists of three DNA nucleotides. Except for the stop codons, all the codons code for amino acids.

15.3 The First Step: Transcription

Transcription, which takes place in the nucleus of eukaryotic cells, is the first step required for gene expression—the making of a gene product. The formation of mRNA leads to a polypeptide as the gene product. The molecules tRNA and rRNA are transcribed from DNA templates, and these are products in and of themselves. Enzymes called **RNA polymerases** are involved in transcription.

Messenger RNA Is Formed

During transcription, an mRNA molecule is formed that has a sequence of bases complementary to a portion of one DNA strand; wherever A, T, G, or C is present in the DNA template, U, A, C, or G, respectively, is incorporated into the mRNA molecule (Fig. 15.6). A segment of the DNA helix unwinds and unzips, and complementary RNA nucleotides pair with DNA nucleotides of the strand that is being transcribed. An RNA polymerase joins the nucleotides together in the 5′→ 3′ direction. In other words, an RNA polymerase only adds a nucleotide to the 3′ end of the polymer under construction.

Transcription begins when RNA polymerase attaches to a region of DNA called a promoter. A **promoter** defines the start of a gene, the direction of transcription, and the strand to be transcribed. The RNA-DNA association is not as stable as the DNA helix. Therefore, only the newest portion of an RNA molecule that is associated with RNA polymerase is bound to the DNA, and the rest dangles off to the side. Elongation of the mRNA molecule continues until RNA polymerase comes to a DNA sequence called a **terminator.** The terminator causes RNA polymerase to stop transcribing the DNA and to release the mRNA molecule, now called an **RNA transcript.**

Many RNA polymerase molecules can be working to produce mRNA transcripts at the same time (Fig. 15.7). This allows the cell to produce many thousands of copies of the same mRNA molecule and eventually many copies of the same protein within a shorter period of time than if the single copy of DNA were used to direct protein synthesis.

As a result of transcription, there will be many mRNA molecules directing protein synthesis. Many more copies of a protein are produced within a given time versus using DNA directly.

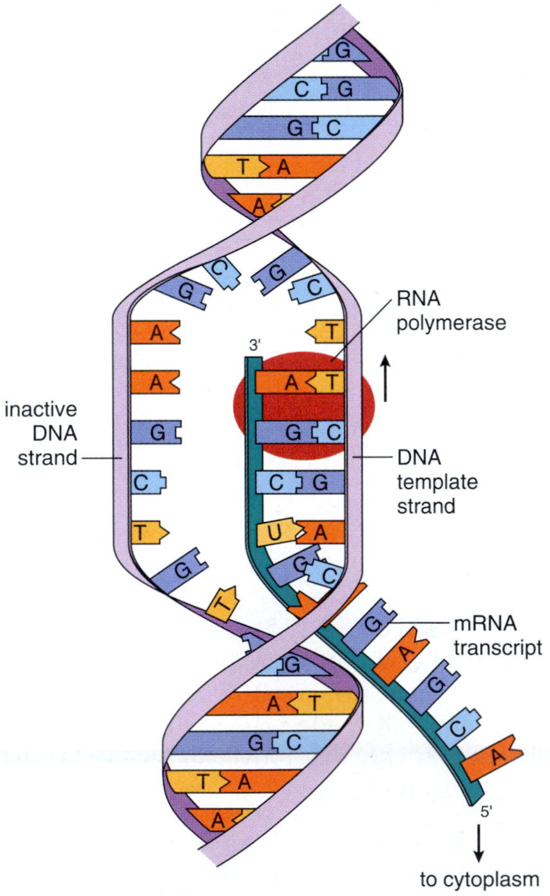

Figure 15.6 **Transcription.**
During transcription, complementary RNA is made from a DNA template. At the point of attachment of RNA polymerase, the DNA helix unwinds and unzips, and complementary RNA nucleotides are joined together. After RNA polymerase has passed by, the DNA strands rejoin and the RNA transcript dangles to the side.

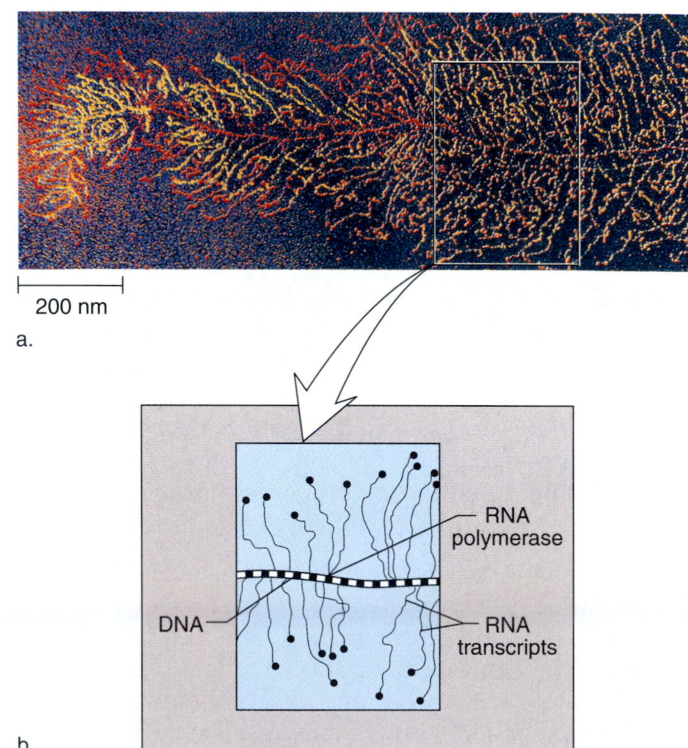

Figure 15.7 **RNA polymerase.**
a. Numerous RNA transcripts extend from a horizontal gene in an amphibian egg cell. **b.** The strands get progressively longer because transcription begins to the left. The dark dots along the DNA are RNA polymerase molecules. The dots at the end of the strands are spliceosomes involved in RNA processing (see Fig. 15.8).

Messenger RNA Is Processed

The newly formed mRNA molecule, called the primary *mRNA* transcript, is modified before it leaves the eukaryotic nucleus. First, the ends of the mRNA molecule are altered: a cap is put onto the 5′ end and a poly-A tail is put onto the 3′ end (Fig. 15.8). The "cap" is a modified guanine (G) nucleotide that helps tell a ribosome where to attach when translation begins. The "poly-A tail" consists of a chain of 150 to 200 adenine (A) nucleotides. The tail facilitates the transport of the mRNA out of the nucleus and also inhibits degradation of mRNA by hydrolytic enzymes.

Second, the primary mRNA transcript itself is changed. Most genes in plants and animals are interrupted by **introns,** segments of DNA that are never expressed. The introns occur between the **exons,** the segments of DNA that are expressed. During mRNA processing the introns are removed during an mRNA processing event called mRNA splicing (Fig. 15.8). The mRNA that has now been processed and leaves the nucleus is called mature mRNA.

How is mRNA processing carried out? And what is the function of introns in the first place? In most cases mRNA splicing is often done by *spliceosomes,* a complex that contains several kinds of ribonucleoproteins. A spliceosome cuts the primary mRNA and then rejoins the adjacent exons. There has been much speculation about the possible role of introns in the eukaryotic genome. It's possible that introns divide a gene into regions that can be joined in different combinations to produce various products in different cells. For instance, it's been shown that the thyroid gland and pituitary gland process the same primary mRNA transcript differently and therefore produce related but different hormones.

Some researchers are trying to determine whether introns exist in all organisms. They have found that the more simple the eukaryote, the less the likelihood of introns in the genes. An intron has been discovered in the gene for a tRNA molecule in *Anabaena,* a cyanobacterium. This particular intron is of interest because it is "self-splicing," that is, it has the capability of splicing itself out of an RNA transcript. RNAs with an enzymatic function are called **ribozymes.** Ribozymes are restricted in their function since each one cleaves RNA only at specific locations. Still, the discovery of ribozymes in prokaryotes supports the belief that RNA could have served as both the genetic material and as the first enzymes in the earliest living organisms.

Particularly in eukaryotes, the primary mRNA transcript is processed before it becomes a mature mRNA transcript.

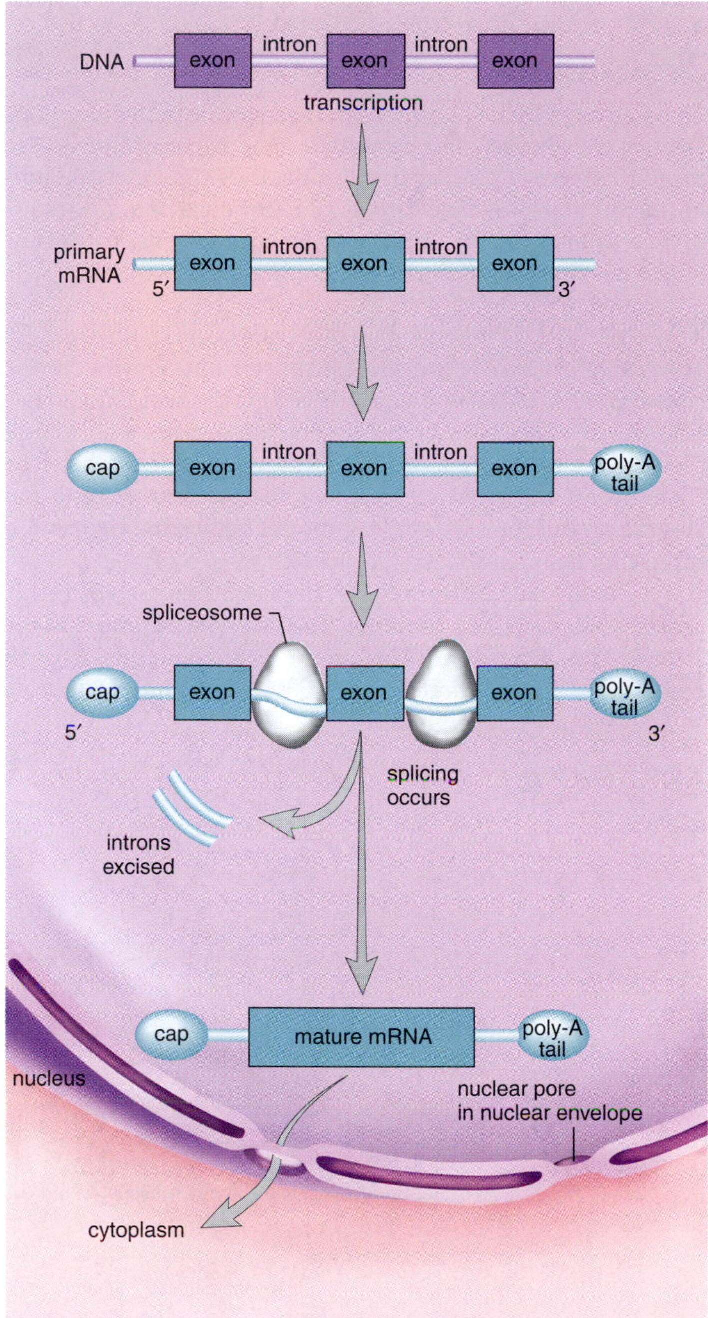

Figure 15.8 Messenger RNA (mRNA) processing in eukaryotes.
DNA contains both exons (coding sequences) and introns (noncoding sequences). Both of these are transcribed and are present in primary mRNA. During processing, a cap and a poly-A tail (a series of adenine nucleotides) are added to the molecule. Also, there is excision of the introns and a splicing together of the exons. This is accomplished by complexes called spliceosomes. Then the mature mRNA molecule is ready to leave the nucleus.

15.4 The Second Step: Translation

Translation, which takes place in the cytoplasm of eukaryotic cells, is the second step by which gene expression leads to protein synthesis. During translation, the sequence of codons in mRNA at a ribosome directs the sequence of amino acids in a polypeptide. In other words, one language (nucleic acids) gets translated into another language (protein).

The Role of Transfer RNA

Transfer RNA (tRNA) molecules transfer amino acids to the ribosomes. A tRNA molecule is a single-stranded nucleic acid that doubles back on itself to create regions where complementary bases are hydrogen-bonded to one another. The structure of a tRNA molecule is generally drawn as a flat cloverleaf, but the space-filling model shows the molecule's three-dimensional shape (Fig. 15.9).

There is at least one tRNA molecule for each of the 20 amino acids found in proteins. The amino acid binds to the 3′ end. The opposite end of the molecule contains an **anticodon,** a group of three bases that is complementary to a specific codon of mRNA. For example, a tRNA that has the anticodon GAA binds to the codon CUU and carries the amino acid leucine. (There are fewer tRNAs than there are codons because some tRNAs can pair with more than one codon. If the anticodon contains a U in the third position, it will pair with either an A or G in the third position of a codon. This is called the *wobble effect*.)

How does the correct amino acid become attached to the correct tRNA molecule? This task is carried out by amino acid-activating enzymes, called tRNA synthetases. Just as a key fits a lock, each enzyme has a recognition site for the amino acid to be joined to a particular tRNA. This is an energy-requiring process that utilizes ATP. Once the amino acid–tRNA complex is formed, it travels through the cytoplasm to a ribosome, where protein synthesis is occurring.

Transfer RNA (tRNA) molecules bind with one particular amino acid, and they bear an anticodon that is complementary to a particular codon—the codon for that amino acid.

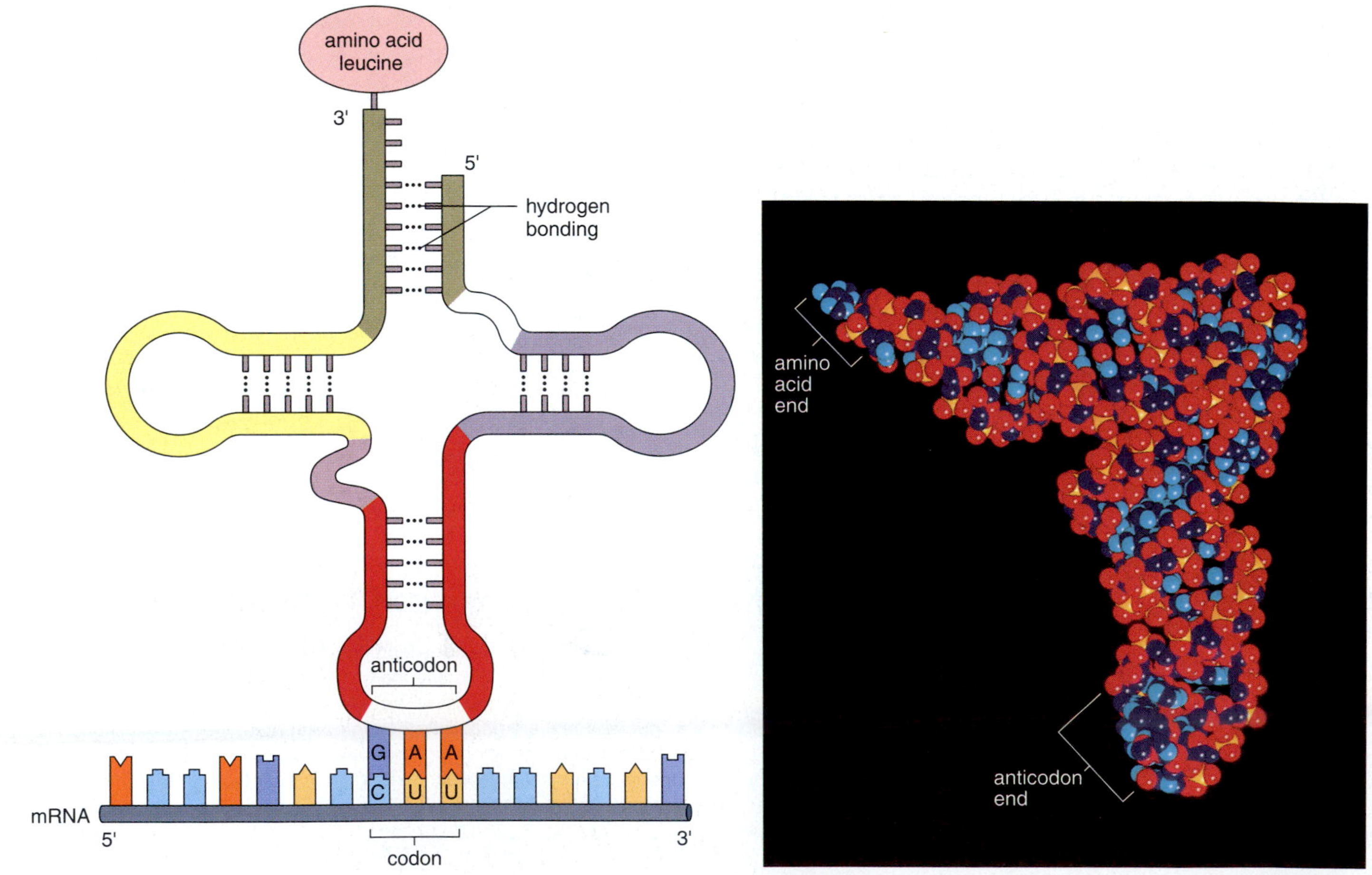

Figure 15.9 Structure of a transfer RNA (tRNA) molecule.
a. Complementary base pairing indicated by hydrogen bonding occurs between nucleotides of the molecule, and this causes it to form its characteristic loops. The anticodon that base-pairs with a particular messenger RNA (mRNA) codon occurs at one end of the folded molecule; the other two loops help hold the molecule at the ribosome. An appropriate amino acid is attached at the 3′ end of the molecule. For this mRNA codon and tRNA anticodon, the appropriate amino acid is leucine. **b.** Space-filling model of tRNA molecule.

The Role of Ribosomal RNA

Ribosomal RNA (rRNA) is produced off a DNA template in the nucleolus of a nucleus. Then the rRNA is packaged with a variety of proteins into ribosomal subunits, one of which is larger than the other. The subunits move separately through nuclear envelope pores into the cytoplasm where they combine when translation begins. Ribosomes can remain in the cytosol or they can become attached to endoplasmic reticulum.

Prokaryotic cells contain about 10,000 ribosomes, but eukaryotic cells contain many times that number. Ribosomes play a significant role in protein synthesis (Fig. 15.10). Ribosomes have a binding site for mRNA and binding sites for two transfer RNA (tRNA) molecules at a time. These binding sites facilitate complementary base pairing between tRNA anticodons and mRNA codons. Ribosomal RNA (i.e., a ribozyme) is also believed to join amino acids together, in a manner to be described, so that a polypeptide results. As the ribosome moves down the mRNA molecule, new tRNAs arrive; amino acids are joined, and the polypeptide forms and grows longer. Translation terminates once the polypeptide is fully formed; the ribosome dissociates into its two subunits and falls off the mRNA molecule.

As soon as the initial portion of mRNA has been translated by one ribosome, and the ribosome has begun to move down the mRNA, another ribosome attaches to the mRNA. Therefore, several ribosomes are often attached to and translating the same mRNA. The entire complex is called a **polyribosome** (Fig. 15.10).

What determines whether a ribosome remains in the cytosol or binds to endoplasmic reticulum (ER)? All ribosomes are at first free within the cytosol. But once synthesis begins, some polypeptides have a series of amino acids called a signal sequence, which enables a ribosome to bind to rough endoplasmic reticulum (rough ER). Then the polypeptide enters the rough ER lumen and is eventually secreted from the cell. Nearly all polypeptides have a signal sequence that targets them for their final location in the cell.

Ribosomal RNA (rRNA) is found in the ribosomes, structures that facilitate complementary base pairing between mRNA and tRNAs and join amino acids together.

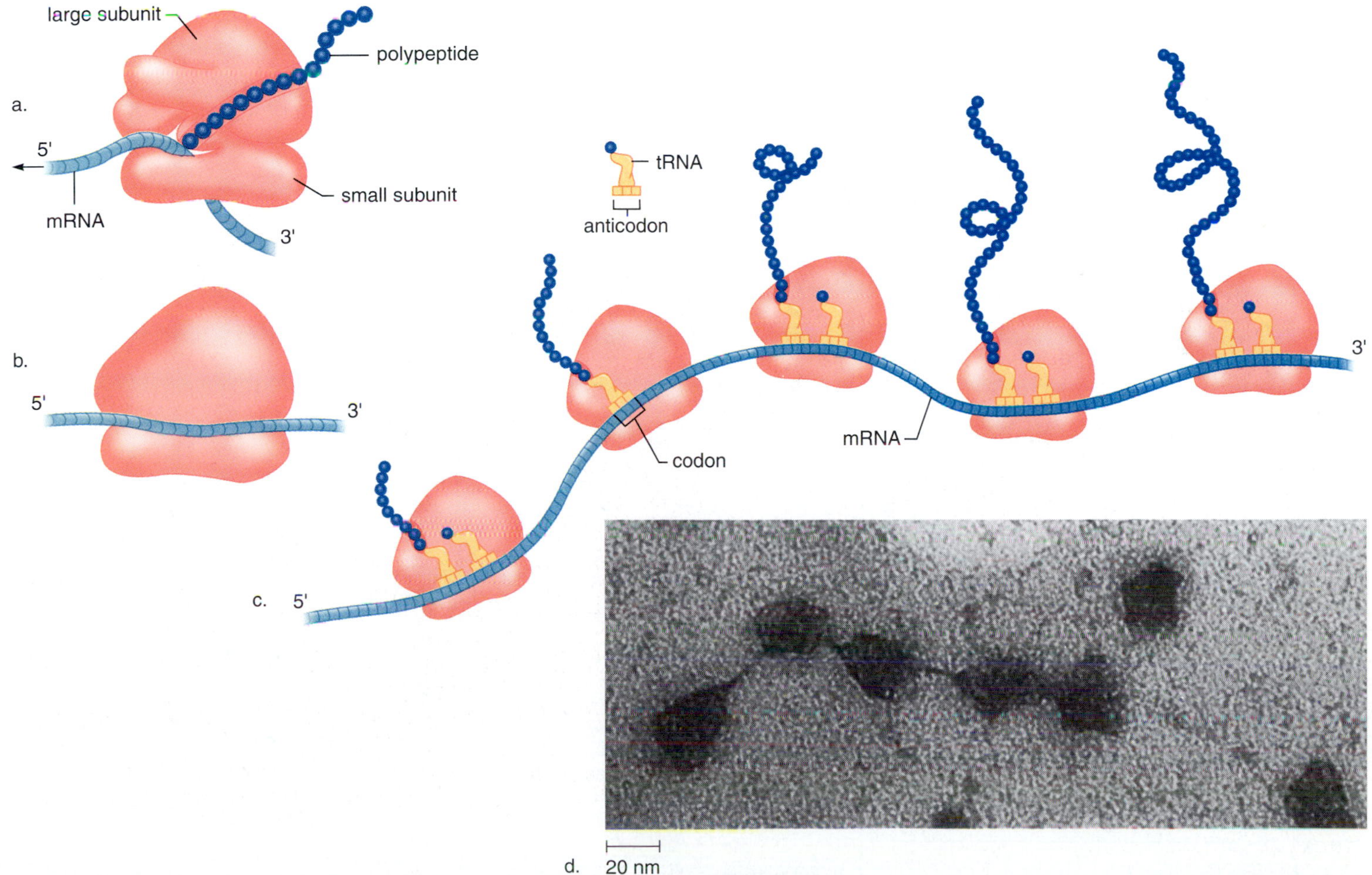

Figure 15.10 Polyribosome structure and function.
a. Side view of ribosome shows positioning of mRNA and growing polypeptide. b. A frontal view of a ribosome. c. Several ribosomes, collectively called a polyribosome, move along an mRNA at one time. Therefore, several polypeptides can be made at the same time. d. Electron micrograph of a polyribosome.

Translation Requires Three Steps

During translation, the codons of an mRNA base-pair with the anticodons of tRNA molecules carrying specific amino acids. The order of the codons determines the order of the tRNA molecules and the sequence of amino acids in a polypeptide. The process of translation must be extremely orderly so that the amino acids of a polypeptide are sequenced correctly.

Protein synthesis involves three steps: chain initiation, chain elongation, and chain termination. It should be kept in mind that enzymes are required for each of the steps to occur, and energy is needed for the first two steps also.

1. Chain initiation: In prokaryotes, a small ribosomal subunit attaches to the mRNA in the vicinity of the *start codon* (AUG). The first or initiator tRNA pairs with this codon. Then a large ribosomal subunit joins to the small subunit (Fig. 15.11*a*).

   Notice that a ribosome has two binding sites for tRNAs. One of these is called the P (for peptide) site, and the other is called the A (for amino acid) site. The initiator tRNA that is capable of binding to the P site carries the amino acid, methionine. The A site is for the next tRNA.

   Proteins called initiation factors are required to bring the necessary translation components (small ribosomal subunit, mRNA, initiator tRNA, and large ribosomal subunit) together. Energy is also expended.

Following chain initiation, mRNA and the first tRNA are now at a ribosome.

2. Chain elongation: During elongation, a tRNA with an attached peptide is already at the P site, and a tRNA carrying its appropriate amino acid is just arriving at the A site. Proteins called elongation factors facilitate complementary base pairing between the tRNA anticodon and the mRNA codon (Fig. 15.11*b*).

   The polypeptide is transferred and attached by peptide bond formation to the newly arrived amino acid. A ribozyme, which is a part of the larger ribosomal subunit, and energy are needed to bring about this transfer. Now the tRNA molecule at the P site leaves.

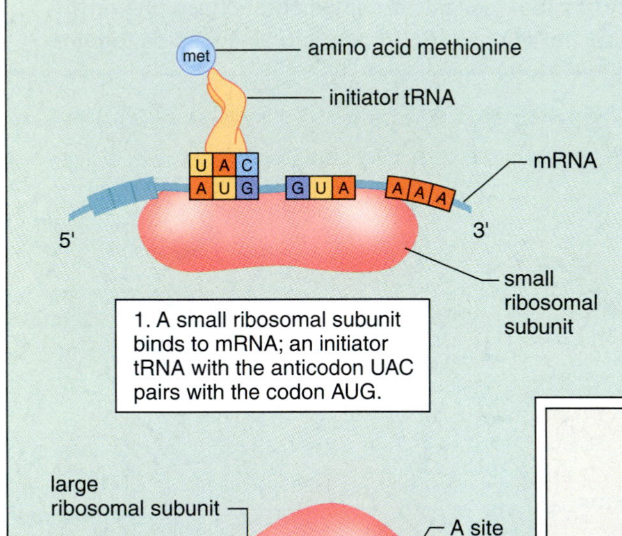

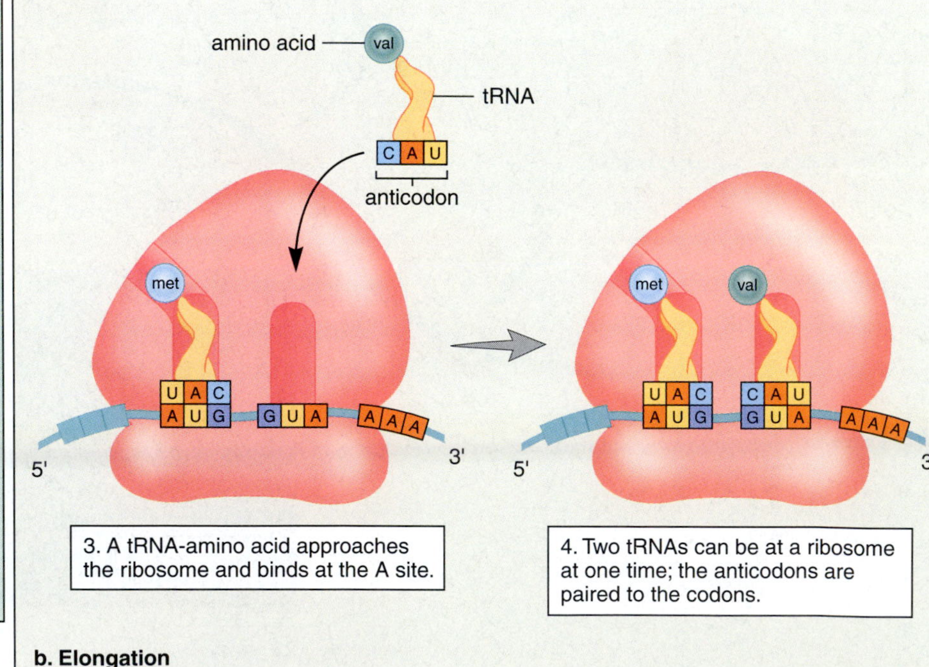

Figure 15.11 **Protein synthesis.**

Next *translocation* occurs: the mRNA, along with the peptide-bearing tRNA, moves from the A site to the empty P site. Since the ribosome has moved forward three nucleotides, there is a new codon now located at the empty A site.

The complete cycle—complementary base pairing of new tRNA, transfer of peptide chain, translocation—is repeated at a rapid rate (about 15 times each second in *Escherichia coli*).

During chain elongation, amino acids are added one at a time to the growing polypeptide.

3. Chain termination: Termination of polypeptide synthesis occurs at a stop codon, which does not code for an amino acid (Fig. 15.11*c*). The polypeptide is enzymatically cleaved from the last tRNA by a release factor. The tRNA and polypeptide leave the ribosome, which dissociates into its two subunits (see also Fig. 15.10).

During chain termination, the ribosome separates into its two subunits and the polypeptide is released.

A newly synthesized polypeptide may function alone or it may become a part of a protein that has more than one polypeptide. Proteins play a role in the anatomy and physiology of cells, as discussed earlier in this text. The plasma membrane of all cells contains proteins that carry out various functions and many proteins are enzymes that participate in cellular metabolism. The tissues of multicellular animals are distinguished by the uniqueness of their proteins. Properly functioning proteins are of paramount importance to the cell and to the organism. Organisms inherit genes that code for their own particular mix of proteins in their cells. If an organism inherits a mutated gene, the result can be a genetic disorder or a propensity toward cancer, in which proteins are not fulfilling their usual functions. This topic is explored further in chapter 16.

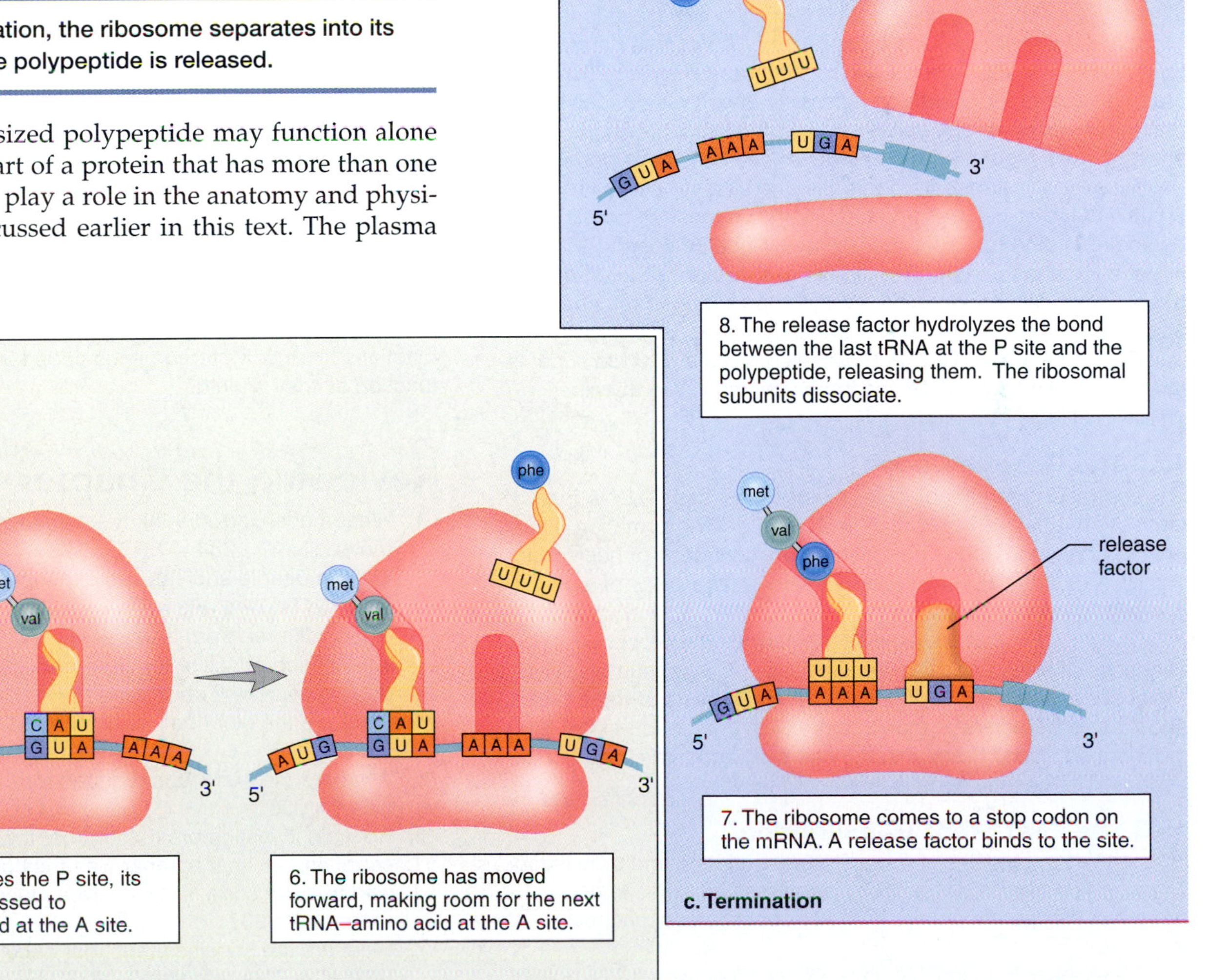
met
val
phe
UUU
GUA
AAA
UGA
5'
3'
8. The release factor hydrolyzes the bond between the last tRNA at the P site and the polypeptide, releasing them. The ribosomal subunits dissociate.
release factor
7. The ribosome comes to a stop codon on the mRNA. A release factor binds to the site.
c. Termination
UAC
CAU
AUG
5. As tRNA leaves the P site, its amino acid is passed to tRNA–amino acid at the A site.
6. The ribosome has moved forward, making room for the next tRNA–amino acid at the A site.

Figure 15.11 (cont.)

Connecting Concepts

Early geneticists were amazed to discover that a single nucleotide change in a huge DNA molecule can have profound phenotypic consequences. A well-known example in humans is the single nucleotide change that leads to an amino acid change at one position of the β chain of hemoglobin. The result is the often fatal condition known as sickle-cell disease because the red blood cells are sickle-shaped.

The genetic information stored by DNA is realized through the steps of transcription and translation. During transcription, DNA is used as a template for the production of a mRNA molecule. The actions of RNA polymerase, the enzyme that carries out transcription of DNA and DNA polymerase, the enzyme required for DNA replication, are similar enough to suggest that both enzymes evolved from a common ancestral enzyme. If RNA was the original genetic material, as has been proposed, then it must have specified the synthesis of an enzyme capable of replicating RNA. RNA replicase enzymes are present in RNA viruses still today. Such an enzyme could have been the common ancestor of the DNA and RNA polymerases known today.

Both RNA molecules and proteins are needed for the translation. During translation, protein synthesis is carried out according to the genetic information now borne by mRNA. Since a single gene can be transcribed many times and an mRNA molecule may be repeatedly translated, many copies of a protein may result. In the next chapter we will examine the ways cells can regulate the quantity of protein produced per gene.

Summary

15.1 The Function of Genes

Several investigators contributed to our knowledge of what genes do. Garrod is associated with the phrase "inborn error of metabolism" because he suggested that some of his patients had inherited an inability to carry out certain enzymatic reactions.

Beadle and Tatum X-rayed spores of *Neurospora crassa* and found that some of the subsequent cultures lacked a particular enzyme needed for growth on minimal medium. Since they found that the mutation of one gene results in the lack of a single enzyme, they suggested the one gene–one enzyme hypothesis.

Pauling and Itano found that the chemical properties of the β chain of sickle-cell hemoglobin differ from those of normal hemoglobin, and therefore the one gene–one polypeptide hypothesis was formulated instead. Later, Ingram showed that the biochemical change is due to the substitution of the amino acid valine (nonpolar) for glutamate (polar).

RNA differs from DNA in these ways: (1) the pentose sugar is ribose, not deoxyribose; (2) the base uracil replaces thymine; (3) RNA is single stranded.

15.2 The Genetic Code

The central dogma of molecular biology says that (1) DNA is a template for its own replication and also for RNA formation during transcription, and (2) the complementary sequence of nucleotides in mRNA directs the correct sequence of amino acids of a polypeptide during translation.

The sequence of bases in DNA specifies the proper sequence of amino acids in a polypeptide. The genetic code is a triplet code, and each codon (code word) consists of three bases. The code is degenerate; that is, more than one codon exists for most amino acids. There are also one start and three stop codons. The genetic code is just about universal.

15.3 The First Step: Transcription

Figure 15.12 provides a summary of transcription and translation. Transcription begins when RNA polymerase attaches to a promoter. Elongation occurs until RNA polymerase reaches a terminator.

Messenger RNA (mRNA) is processed following transcription, a cap is put onto the 5′ end and a poly-A tail is put onto the 3′ end; introns are removed in eukaryotes by spliceosomes.

15.4 The Second Step: Translation

Translation requires mRNA, ribosomal RNA (rRNA), and transfer RNA (tRNA). Each tRNA has an anticodon at one end and an amino acid at the other; there are amino acid-activating enzymes that ensure that the correct amino acid is attached to the correct tRNA. When tRNAs bind with their codon at a ribosome, the amino acids are correctly sequenced in a polypeptide according to the order predetermined by DNA.

Many ribosomes move along the same mRNA at a time. Collectively, these are called a polyribosome.

Translation requires these steps. During initiation, mRNA, the first (initiator) tRNA, and the ribosome all come together in the proper orientation at a start codon. During elongation, as the tRNAs bind to their codons, the growing peptide chain is transferred by peptide bonding to the next amino acid in the chain. During termination at a stop codon, the polypeptide is cleaved from the last tRNA. The ribosome now dissociates.

Polypeptides are a part of protein molecules with various functions in cells. Mutated genes code for proteins that may not function as they should.

Reviewing the Chapter

1. What made Garrod think there were inborn errors of metabolism? 234
2. Explain Beadle and Tatum's experimental procedure. 234
3. How did Pauling and Itano know that the chemical properties of Hb^S differed from those of Hb^A? What change does a substitution of valine for glutamate cause in the chemical properties of Hb^S compared to Hb^A? 234–35
4. What are the biochemical differences between RNA and DNA? 236
5. What are the two steps required for the expression of a gene? 236
6. How did investigators reason that the code must be a triplet code, and in what manner was the code cracked? Why is it said that the code is degenerate, unambiguous, and almost universal? 237
7. What are the specific steps that occur during transcription of RNA off a DNA template? 238–39
8. How is messenger RNA (mRNA) processed before leaving the eukaryotic nucleus? 239

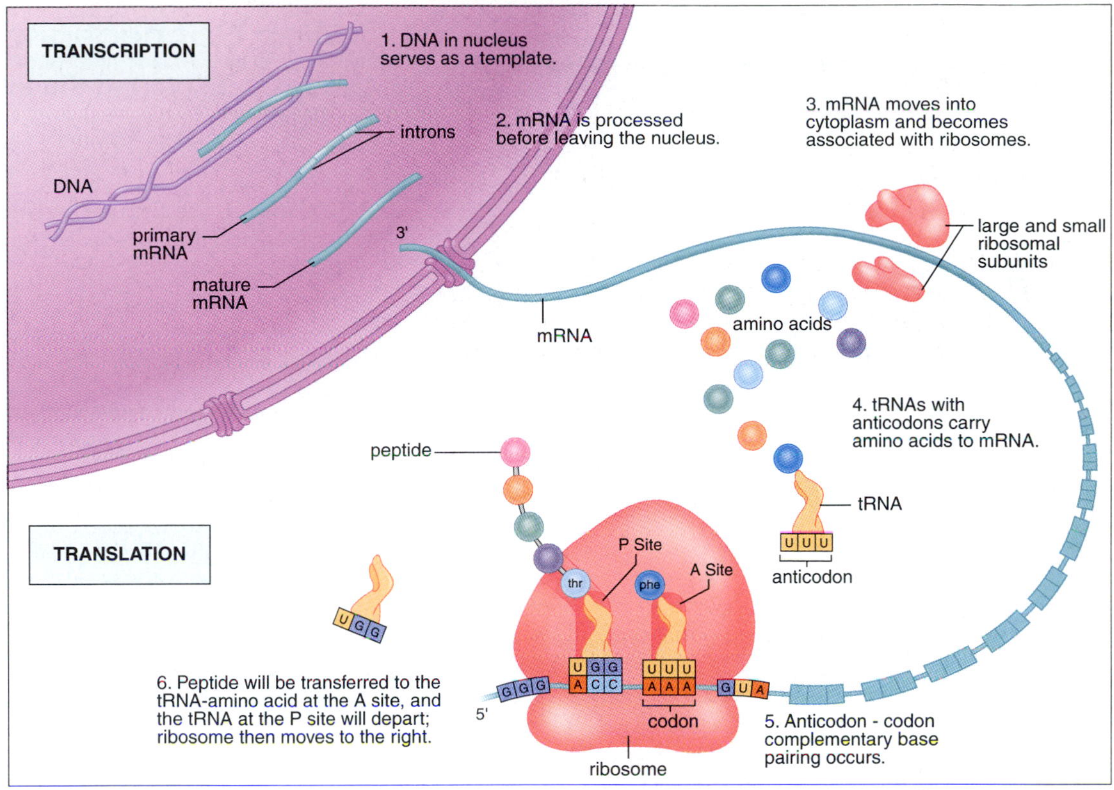
TRANSCRIPTION
1. DNA in nucleus serves as a template.
introns
2. mRNA is processed before leaving the nucleus.
3. mRNA moves into cytoplasm and becomes associated with ribosomes.
DNA
primary mRNA
3'
mature mRNA
large and small ribosomal subunits
mRNA
amino acids
4. tRNAs with anticodons carry amino acids to mRNA.
peptide
tRNA
TRANSLATION
P Site
A Site
thr
phe
UUU
anticodon
UGG
6. Peptide will be transferred to the tRNA-amino acid at the A site, and the tRNA at the P site will depart; ribosome then moves to the right.
UGG
ACC
UUU
AAA
GGG
GUA
5'
codon
5. Anticodon - codon complementary base pairing occurs.
ribosome

Figure 15.12 Summary of gene expression.

9. Compare the functions of mRNA, ribosomal RNA (rRNA), and transfer RNA (tRNA) during protein synthesis. What are the specific events of translation? 240–43
10. Genes specify the amino acid sequence in polypeptides. Explain with reference to Figure 15.12.

Testing Yourself

Choose the best answer for each question. For questions 1–4, match the investigator to the phrase in the key:

Key:

a. one gene–one enzyme hypothesis
b. inborn error of metabolism
c. one gene–one polypeptide hypothesis
d. one gene–one mRNA hypothesis

1. Sir Archibald Garrod
2. George Beadle and Edward Tatum
3. Linus Pauling and Harvey Itano
4. Considering the following pathway, if Beadle and Tatum found that *Neurospora* cannot grow if metabolite A is provided, but can grow if B, C, or D is provided, what molecule would be missing?

$$A \xrightarrow{1} B \xrightarrow{2} C \xrightarrow{3} D$$

   a. enzyme 1
   b. enzyme 2
   c. enzyme 3
   d. the substrate in addition to the product
   e. All of these are correct.
5. The central dogma of molecular biology
   a. states that DNA is a template for all RNA production.
   b. states that DNA is a template only for DNA replication.
   c. states that translation precedes transcription.
   d. pertains only to prokaryotes because humans are unique.
   e. All of these are correct.
6. If the sequence of bases in DNA is TAGC, then the sequence of bases in RNA will be
   a. ATCG.
   b. TAGC.
   c. AUCG.
   d. GCTA.
   e. Both a and b are correct.
7. RNA processing
   a. is the same as transcription.
   b. is an event that occurs after RNA is transcribed.
   c. is the rejection of old, worn-out RNA.
   d. pertains to function of transfer RNA during protein synthesis.
   e. Both b and d are correct.
8. During protein synthesis, an anticodon on transfer RNA (tRNA) pairs with
   a. DNA nucleotide bases.
   b. ribosomal RNA (rRNA) nucleotide bases.
   c. messenger RNA (mRNA) nucleotide bases.
   d. other tRNA nucleotide bases.
   e. Any one of these can occur.
9. Which statement regarding translation is out of order first?
   a. Ribosomal subunits bind to mRNA.
   b. The ribosome comes to a stop codon on the mRNA.
   c. As a tRNA leaves the ribosome it passes its amino acid to the growing polypeptide.
   d. A new tRNA-amino acid approaches the ribosome.
   e. The messenger RNA codon binds to the tRNA anticodon.

10. This is a segment of a DNA molecule. (Remember that the template strand only is transcribed.) What are (a) the RNA codons, (b) the tRNA anticodons, and (c) the sequence of amino acids in a protein?

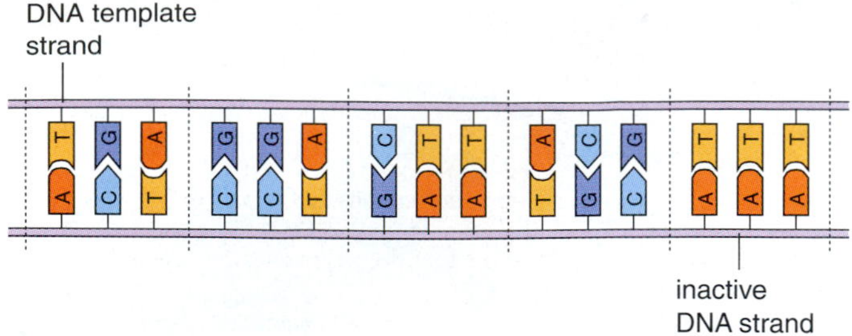

Thinking Scientifically

1. In one strain of bacteria you observe an x amount of some enzyme. In a mutant derived from that strain there is 1.5× the amount of the enzyme. A comparison of the sequence of the two strains shows that there is no change in the coding region of the gene. But there is a substitution of two A—T base pairs in the mutant for two G—C base pairs where DNA polymease attaches. What hypothesis would explain how such a mutation could produce this change in protein level? What change would be predicted in mRNA level?
2. Changes in DNA sequence, that is, mutations, account for evolutionary changes. Sometimes evolution seems to occur quite rapidly. What type of base change in DNA might bring about the greatest change in a protein and the greatest phenotypic change immediately?

Understanding the Terms

anticodon 240
codon 237
exon 239
genetic code 237
intron 239
messenger RNA (mRNA) 236
polyribosome 241
promoter 238
ribosomal RNA (rRNA) 236
ribozyme 239
RNA (ribonucleic acid) 236
RNA polymerase 238
RNA transcript 238
terminator 238
transcription 236
transfer RNA (tRNA) 236
translation 236
triplet code 237
uracil 236

Match the terms to these definitions:

a. _______________ Enzyme that speeds the formation of mRNA from a DNA template.
b. _______________ Noncoding segment of DNA that is transcribed but removed before mRNA leaves the nucleus.
c. _______________ Process whereby the sequence of codons in mRNA determines (is translated into) the sequence of amino acids in a polypeptide.
d. _______________ String of ribosomes, simultaneously translating different regions of the same mRNA strand during protein synthesis.
e. _______________ Three nucleotides of DNA or mRNA; it codes for a particular amino acid or termination of translation.

Web Connections

Exploring the Internet

http://www.mhhe.com/biosci/genbio/mader
(click on *Biology 7/e*)

The *Biology 7/e* Online Learning Center provides many resources for studying the material in this chapter including links to the following sites:

DNA Structure, Replication and Eukaryotic Chromatin is an excellent, thorough treatment of DNA structure, DNA replication, and eukaryotic chromosome structure.

http://www.cc.ndsu.nodak.edu/instruct/mcclean//plsc431/eukarychrom/index.htm

An easy-to-remember analogy for protein synthesis is presented in Protein Synthesis Activities. Although designed for a teacher, this site holds the key for understanding protein synthesis. Substantial background information on protein synthesis is also presented.

http://www.accessexcellence.org/AE/AEPC/WWC/1994/protein_synthesis.html

Gene Cards: human genes, proteins, and diseases. A database of human genes—this is a very large database.

http://bioinformatics.weizmann.ac.il/cards/

National Organization for Rare Disorders, Inc. This site discusses and contains links to more information concerning the prevention, treatment, and cure of rare "orphan" diseases, such as phenylketonuria. Although rare, there are over 1,000 diseases in their databank.

http://www.rarediseases.org/

Central Dogma Chapter. This MIT hypertextbook chapter covers transcription, translation, and includes practice problems.

http://esg-www.mit.edu:8001/esgbio/dogma/dogmadir.html

Genome Organization and Regulation of Gene Activity

chapter **16**

chapter concepts

Cell division in these cancer cells is no longer regulated.

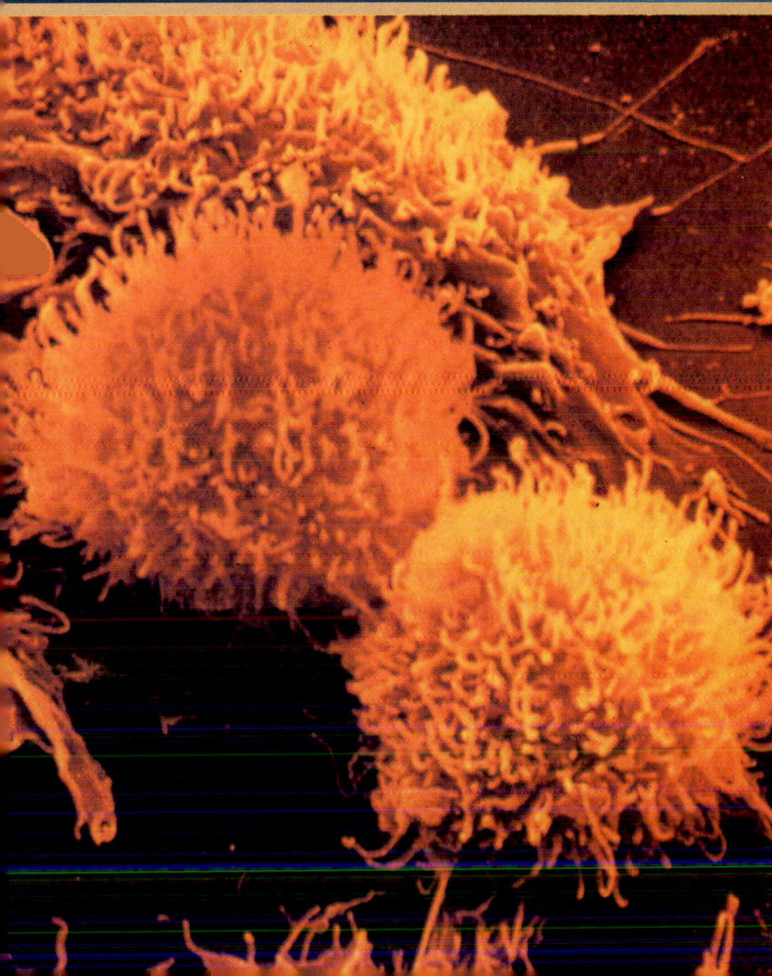

We all begin life as a one-celled zygote that has 46 chromosomes, and these same chromosomes are passed to all the daughter cells during mitosis. Yet cells become specialized in structure and function. The genes for digestive enzymes are expressed only in digestive tract cells and the genes for muscle proteins are expressed only in muscle cells. This shows it is possible to control transcription and translation of genes in each particular type of cell.

In this chapter we will see that regulation of gene activity extends from the nucleus (whether a gene is transcribed) to the cytoplasm (whether an enzyme is functioning). Complete knowledge of how gene expression is controlled has far-ranging applications. Only then will we understand how development progresses and what causes the occurrence of various disorders, including cancer. Cancer develops when genes that promote cell division are expressed and when genes that suppress cell division are not expressed. It has become increasingly apparent that the regulation of gene expression is of major importance to the health of all species, including human beings.

16.1 Prokaryotic Regulation 

Bacteria don't need the same enzymes (and possibly other proteins) all the time. Suppose, for example, the available nutrients vary—wouldn't it be advantageous to produce just the enzymes needed to produce whatever metabolites are missing and/or break down whichever organic nutrients are present?

In 1961, French microbiologists François Jacob and Jacques Monod showed that *Escherichia coli* is capable of regulating the expression of its genes. They proposed what is called the **operon** [L. *opera,* works] model to explain gene regulation in prokaryotes and later received a Nobel Prize for their investigations. A **regulator gene,** located outside the operon, codes for a repressor that controls whether the operon is active or not.

An operon includes the following elements:

Promoter—a short sequence of DNA where RNA polymerase first attaches when a gene is to be transcribed.

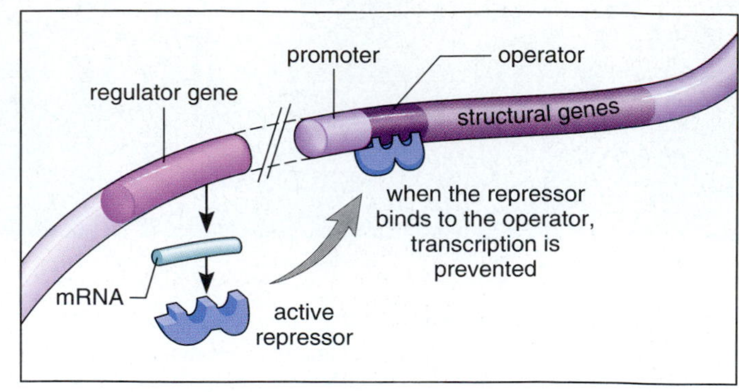

Operator—a short portion of DNA where an active repressor binds. When an active repressor is bound to the operator, RNA polymerase cannot attach to the promoter and transcription does not occur.

Structural genes—one to several genes coding for enzymes of a metabolic pathway that are transcribed as a unit.

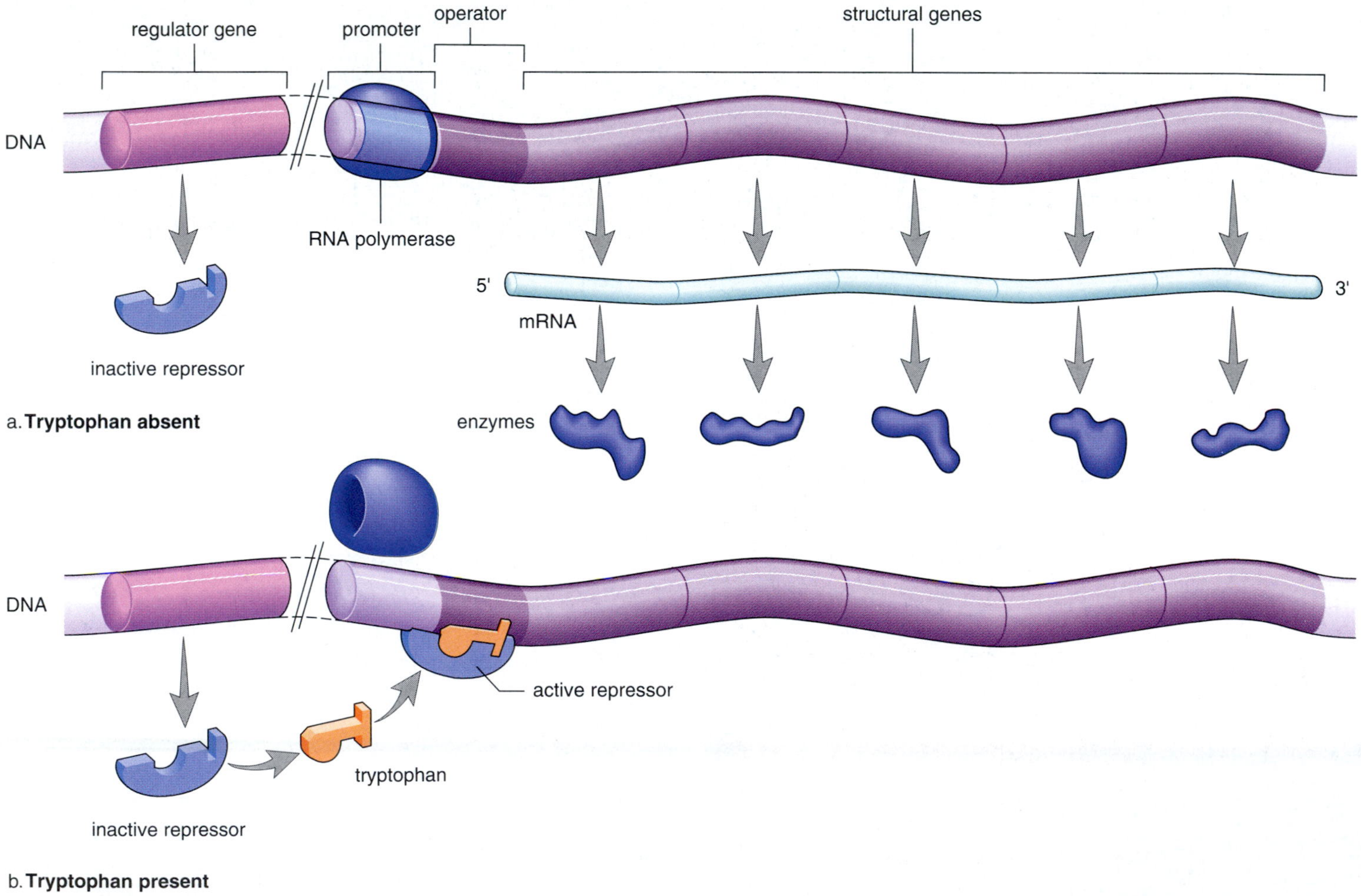

Figure 16.1 The *trp* operon.

a. The regulator gene codes for a repressor protein that is normally inactive. RNA polymerase attaches to the promoter and the structural genes are expressed. **b.** When the nutrient tryptophan is present, it binds to the repressor changing its shape. Now the repressor is active and can bind to the operator. RNA polymerase cannot attach to the promoter and the structural genes are not expressed.

The *trp* Operon

Jacob and Monod found that some operons in *E. coli* usually exist in the on rather than off condition. For example, in the *trp* operon, the regulator codes for a repressor that ordinarily is unable to attach to the operator. Therefore, RNA polymerase is able to bind to the promoter and the structural genes of the operon are ordinarily expressed (Fig. 16.1). Their products, five different enzymes, are a part of an anabolic pathway for the synthesis of the amino acid tryptophan.

If tryptophan happens to be already present in the medium, these enzymes are not needed by the cell and the operon is turned off by the following method. Tryptophan binds to the repressor. A change in shape now allows the repressor to bind to the operator and the structural genes are not expressed. The enzymes are said to be repressible, and the entire unit is called a **repressible operon.** Tryptophan is called the **corepressor.** Repressible operons are usually involved in anabolic pathways that synthesize a substance needed by the cell.

The *lac* Operon

When *E. coli* is denied glucose and is given the milk sugar lactose instead, it immediately begins to make the three enzymes needed for the metabolism of lactose. These enzymes are encoded by three genes: one gene is for an enzyme called β-galactosidase, which breaks down the disaccharide lactose to glucose and galactose; a second gene codes for a permease that facilitates the entry of lactose into the cell; and a third gene codes for an enzyme called transacetylase, which has an accessory function in lactose metabolism.

The three structural genes are adjacent to one another on the chromosome and are under the control of a single promoter and a single operator (Fig. 16.2). The regulator gene codes for a *lac* operon repressor that ordinarily binds to the operator and prevents transcription of the three genes. But when glucose is absent and lactose (or more correctly, allolactose, an isomer formed from lactose) is present, lactose binds to the **repressor,** and the repressor undergoes a change in shape that prevents it from binding to the operator.

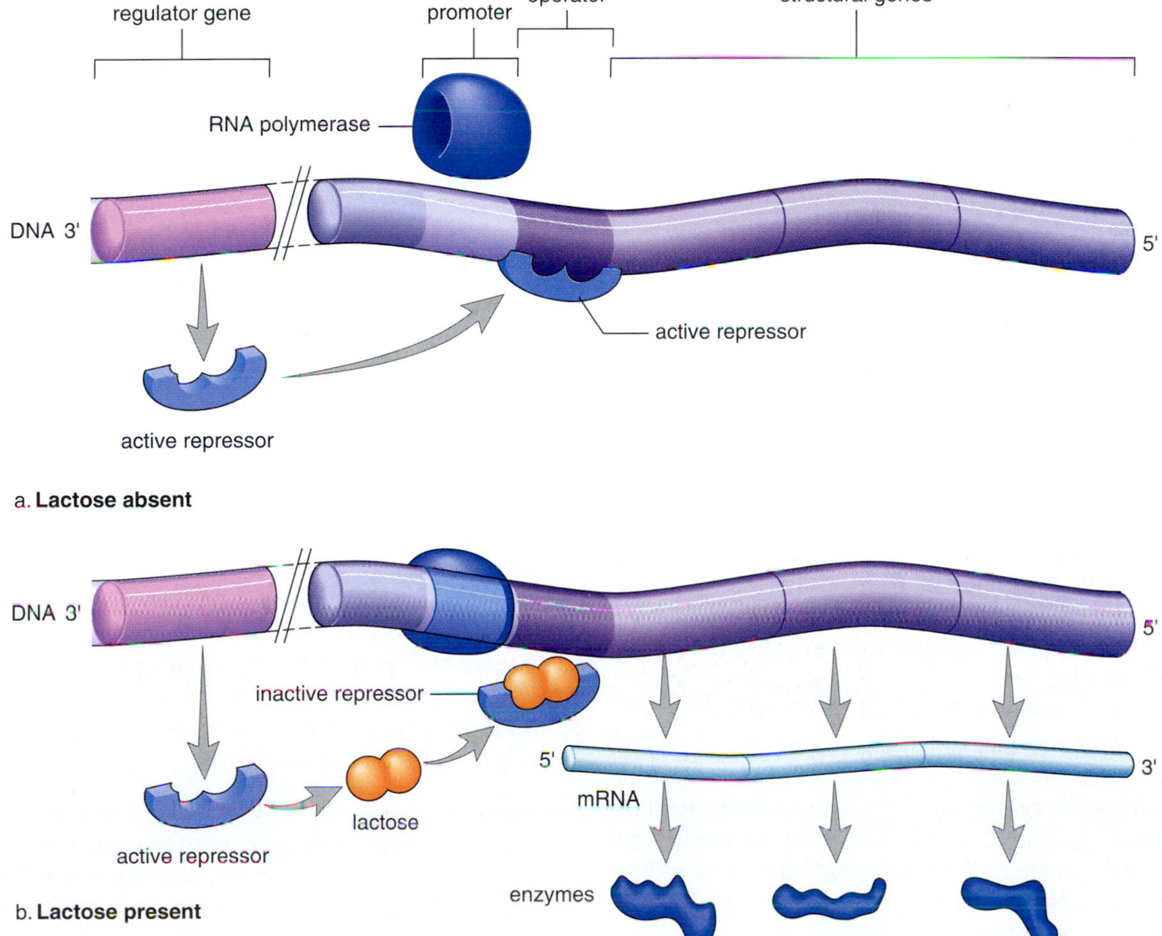

Figure 16.2 The *lac* operon.
a. The regulator gene codes for a repressor that is normally active. When it binds to the operator, RNA polymerase cannot attach to the promoter and structural genes are not expressed. **b.** When lactose is present, it joins to the repressor, changing its shape so that it is inactive and cannot bind to the operator. Now, RNA polymerase binds to the promoter and the structural genes are expressed.

Because the repressor is unable to bind to the operator, RNA polymerase is better able to bind to the promoter. After RNA polymerase carries out transcription, the three enzymes of lactose metabolism are synthesized.

Because the presence of lactose brings about expression of genes, it is called an **inducer** of the *lac* operon: the enzymes are said to be inducible enzymes, and the entire unit is called an **inducible operon.** Inducible operons are usually necessary to metabolic pathways that break down a nutrient. Why is that beneficial? Because these enzymes need only be active when the nutrient is present.

The *trp* operon is an example of a repressible operon because it is ordinarily turned on while the *lac* operon is an example of an inducible operon because it is ordinarily turned off.

Further Control of the lac *Operon*

E. coli preferentially breaks down glucose, and the bacterium has a way to ensure that the lactose operon is maximally turned on only when glucose is absent. When glucose is absent, a molecule called *cyclic AMP (cAMP)* accumulates. Cyclic AMP, which is derived from ATP, has only one phosphate group, which is attached to ribose at two locations.

adenine
5′ CH_2 O
P
3′
OH

Cyclic AMP
(cAMP)

Cyclic AMP binds to a molecule called a *catabolite activator protein (CAP)* and the complex attaches to a CAP binding site next to the *lac* promoter. When CAP binds to DNA, DNA bends exposing the promoter to RNA polymerase. RNA polymerase is now better able to bind to the promoter so that the *lac* operon structural genes are transcribed leading to their expression (Fig. 16.3).

When glucose is present there is little cAMP in the cell; CAP is inactive and the lactose operon does not function maximally. CAP affects other operons as well and takes its name for activating the catabolism of various other metabolites when glucose is absent. The cells' ability to encourage the metabolism of lactose and other metabolites when glucose is absent provides a backup system for survival when the preferred metabolite glucose is absent.

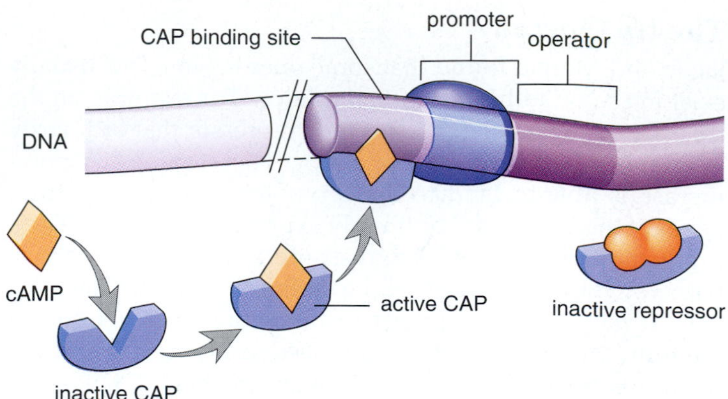

a. **Lactose present, glucose absent (cAMP level high)**

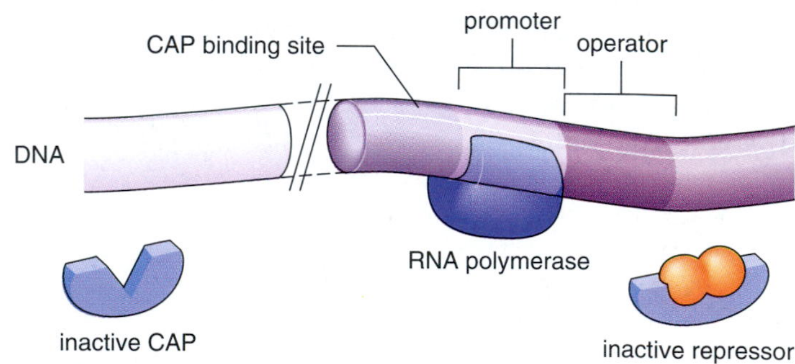

b. **Lactose present, glucose present (cAMP level low)**

Figure 16.3 **Action of CAP.**
When active CAP binds to its site on DNA, the RNA polymerase is better able to bind to the promoter so that the structural genes of the *lac* operon are expressed. **a.** CAP becomes active in the presence of cAMP, a molecule that is prevalent when glucose is absent. **b.** If glucose is present, CAP is inactive and RNA polymerase does not completely bind to the promoter.

Negative Versus Positive Control

Negative control of an operon is exemplified by the involvement of repressors. Why? Because active repressors shut down the activity of an operon. On the other hand, the CAP molecule is an example of positive control. Why? Because when this molecule is active it promotes the activity of an operon. The use of both a negative and a positive control mechanism to control an operon allows the cell to fine-tune its response. In the case of the *lac* operon, the operon is only maximally active when glucose is absent and lactose is present. If both glucose and lactose are present, the cell preferentially metabolizes glucose.

Bacterial DNA contains control regions whose sole purpose is to regulate the transcription of structural genes that code for enzymes or other products.

16.2 Eukaryotic Regulation 

The **genome** is all of a cell's DNA including all of its genes that code for a protein product. As we have seen in prokaryotes, the genome is organized into regulators and operons that control the expression of structural genes. In eukaryotes, there is no universal regulatory mechanism that controls the activity of coding genes. Instead regulation is possible at any point in the pathway from gene to functional protein. These four levels of gene expression are commonly seen in eukaryotes (Fig. 16.4):

1. **Transcriptional control:** In the nucleus a number of mechanisms serve to control which structural genes are transcribed or the rate at which transcription of the genes occurs. These include the organization of chromatin and the use of transcription factors that initiate transcription, the first step in the process of gene expression.
2. **Posttranscriptional control:** Posttranscriptional control occurs in the nucleus after DNA is transcribed and primary mRNA is formed. Differential processing of preliminary mRNA before it leaves the nucleus, and also the speed with which mature mRNA leaves the nucleus, can affect the ultimate amount of gene product.
3. **Translational control:** Translational control occurs in the cytoplasm after mRNA leaves the nucleus and before there is a protein product. The life expectancy of mRNA molecules (how long they exist in the cytoplasm) can vary, as can their ability to bind ribosomes. It is also possible that some mRNAs may need additional changes before they are translated at all.
4. **Posttranslational control:** Posttranslational control, which also takes place in the cytoplasm, occurs after protein synthesis. The polypeptide product may have to undergo additional changes before it is biologically functional. Also, a functional enzyme is subject to feedback control—the binding of an end product can change the shape of an enzyme so that it is no longer able to carry out its reaction.

We will see that regulation of gene expression in eukaryotes means there is cellular control of the amount and activity of a gene product.

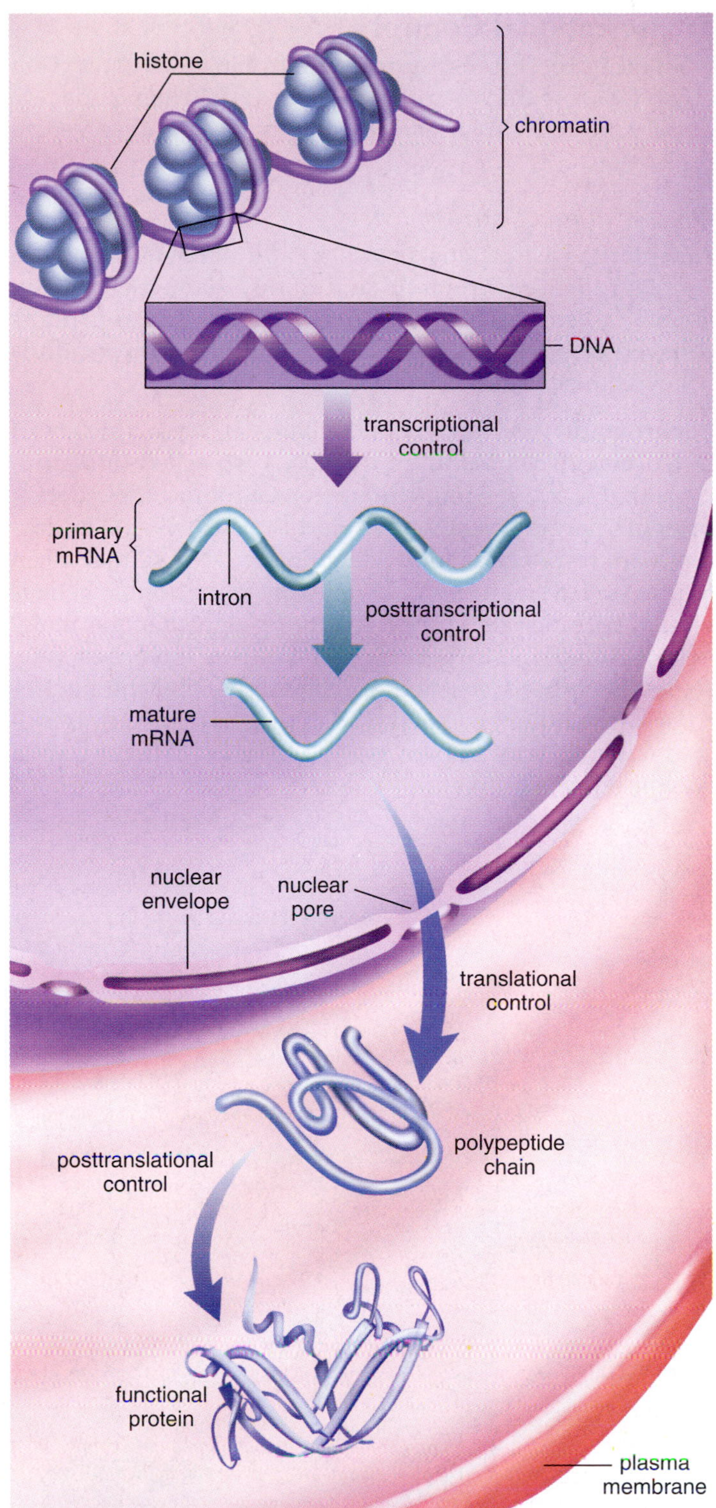

Figure 16.4 Levels of gene expression control in eukaryotic cells.
Transcriptional and posttranscriptional control occur in the nucleus. Translational and posttranslational control occur in the cytoplasm.

Control of gene expression occurs at four levels in eukaryotes. In the nucleus there is transcriptional and posttranscriptional control; in the cytoplasm there is translational and posttranslational control.

Transcriptional Control

We might expect transcriptional control in eukaryotes (1) to involve the organization of chromatin and (2) to include regulatory proteins such as those we have just observed in prokaryotes.

Organization of Chromatin

An interphase nucleus shows lightly stained chromatin regions called **euchromatin** and darkly stained chromatin regions called **heterochromatin** (Fig. 16.5). Euchromatin is believed to be genetically active, while heterochromatin is believed to be genetically inactive.

Euchromatin A human cell contains at least 2 meters of DNA (Fig. 16.6*a*). Yet all of this DNA is packed into a nucleus that is about 5 μm in diameter. Chromosomes contain proteins, largely histones, in addition to DNA. The histones are responsible for packaging the DNA so that it can fit into such a small space. Usually the DNA double helix is wound at intervals around a core of eight histone molecules, giving the appearance of a string of beads (Fig. 16.6*b*). Each bead is called a nucleosome, and the nucleosomes are said to be joined by "linker" DNA that is associated with a histone known as H1. This level of organization is typical of euchromatin.

In nucleosomes, the DNA double helix is wound around a core of eight histone proteins (Fig. 16.7*a*). There are four primary types of histone molecules in a nucleosome designated H2A, H2B, H3, and H4. Remarkably, the amino acid sequences of H3 and H4 vary little between organisms. For example, the H4 of peas is only two amino acids different from the H4 of cattle. This similarity suggests that there have been few mutations in the histone proteins during the course of evolution and that the histones therefore have very important functions.

Euchromatin can readily become genetically active. When active the nucleosomes allow access to the DNA, so that a gene can be turned on and be expressed in the cell (Fig. 16.7*b*). The histone molecules labeled H2 to H4 may regulate this change in chromosomal structure. Recent studies seem to suggest that histones can both repress and activate genes.

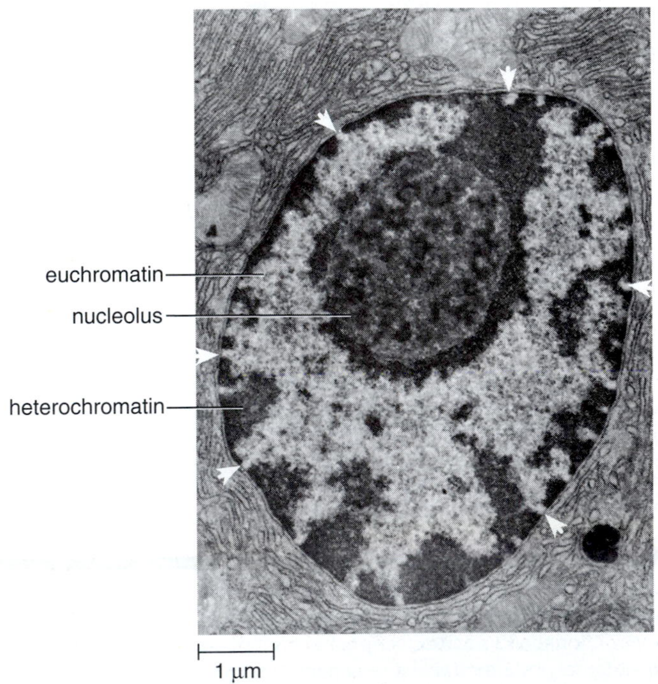

Figure 16.5 Eukaryotic nucleus.
The nucleus contains chromatin, DNA at two different levels of coiling and condensation. Euchromatin is at the level of coiled nucleosomes (30 nm), and heterochromatin is at the level of condensed chromatin (700 nm) in Figure 16.6. Arrows indicate nuclear pores.

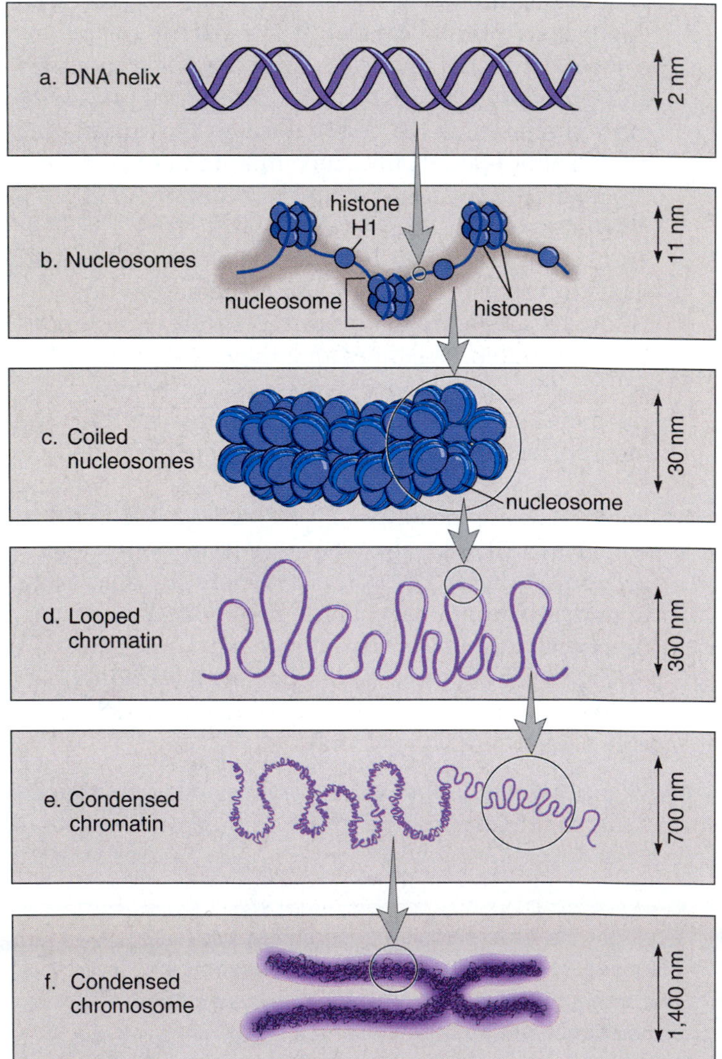

Figure 16.6 Levels of chromosomal structure.
Each drawing has a scale giving a measurement of length for that drawing. Notice that each measurement represents an ever-increasing length; therefore, it would take a much higher magnification to see the structure in **(a)** compared to that in **(f)**.

Heterochromatin As you know, chromatin condenses to become compact chromosomes at the time of cell division. With the aid of histone H1, the beaded string is first coiled tightly into a fiber that has six nucleosomes per turn. This fiber is about 30 nm in thickness (Fig. 16.6*c*). Then the fiber loops back and forth (Fig. 16.6*d*) and can condense still further to produce a highly compacted form (Fig. 16.6*e* and *f*) characteristic of metaphase chromosomes.

Though heterochromatin in an interphase nucleus does not achieve the same degree of compacting as a metaphase chromosome, there are similarities. In heterochromatin, the beaded string is compacted into a 30 nm fiber that is folded into looped regions attached to the inside of the nuclear envelope.

Heterochromatin Is Inactive To demonstrate that heterochromatin is genetically inactive, we can refer to the **Barr body,** a highly condensed structure. In mammalian females, one of the X chromosomes is found in the Barr body; chance alone determines which of the X chromosomes is condensed. For a given gene, if a female is heterozygous, 50% of her cells have Barr bodies containing one allele and the other 50% have the other allele. This causes the body of heterozygous females to be mosaic, with patches of genetically different cells (Fig. 16.8). The mosaic is exhibited in various ways: some females show columns of both normal and abnormal dental enamel; some have patches of pigmented and nonpigmented cells at the back of the eye; some have patches of normal muscle tissue and degenerative muscle tissue.

Barr bodies are not found in the cells of the female gonads, where the genes of both X chromosomes appear to be needed for maturation of the egg. But one X chromosome is inactive in the female zygote, lowering the amount of gene product during development to that seen in males. Most likely, development has become adjusted to this lower dosage, and a higher dosage would bring about abnormalities. We certainly know that, in regard to other chromosomes, any imbalance causes a greatly altered phenotype.

It appears that transcription is not occurring when chromatin is in the heterochromatin state.

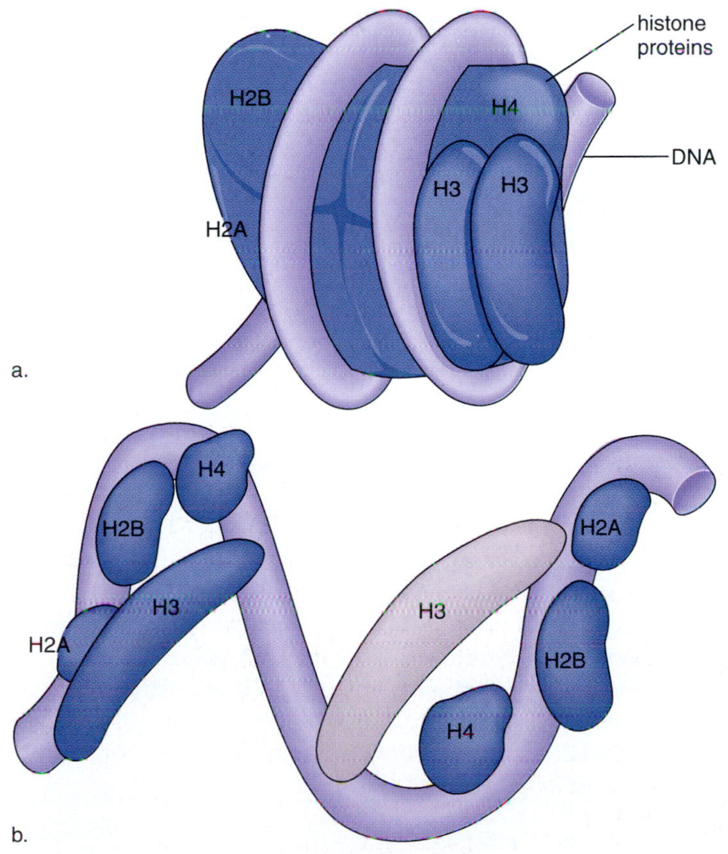

Figure 16.7 Nucleosome structure.
a. In a nucleosome, the DNA double helix winds twice around a group of histone proteins. Such tight packing is characteristic of heterochromatin. **b.** This model shows how it would be possible for the histones to loosen up, allowing transformation from heterochromatin to euchromatin as necessary for transcription to occur.

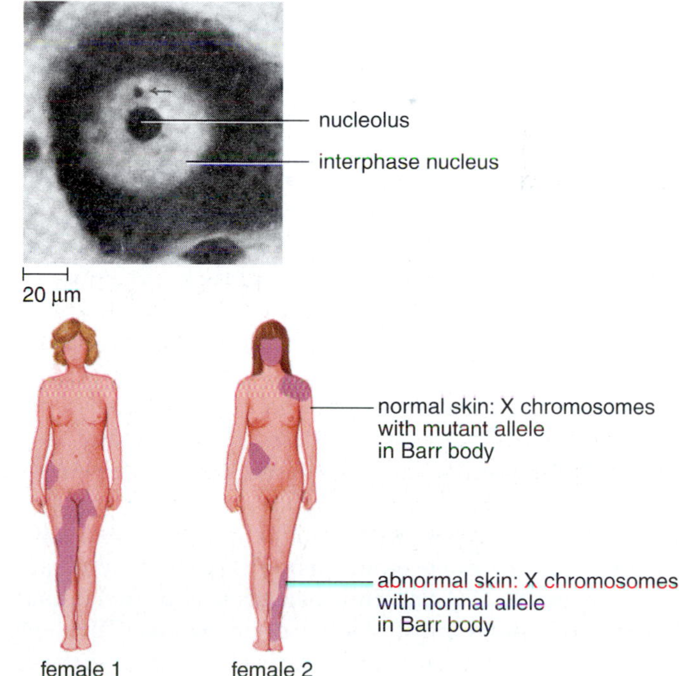

Figure 16.8 Barr body.
In the body of females, the nucleus of each cell contains a Barr body (at arrow) which is a condensed, inactive X chromosome. Chance alone dictates which of the two X chromosomes becomes a Barr body.

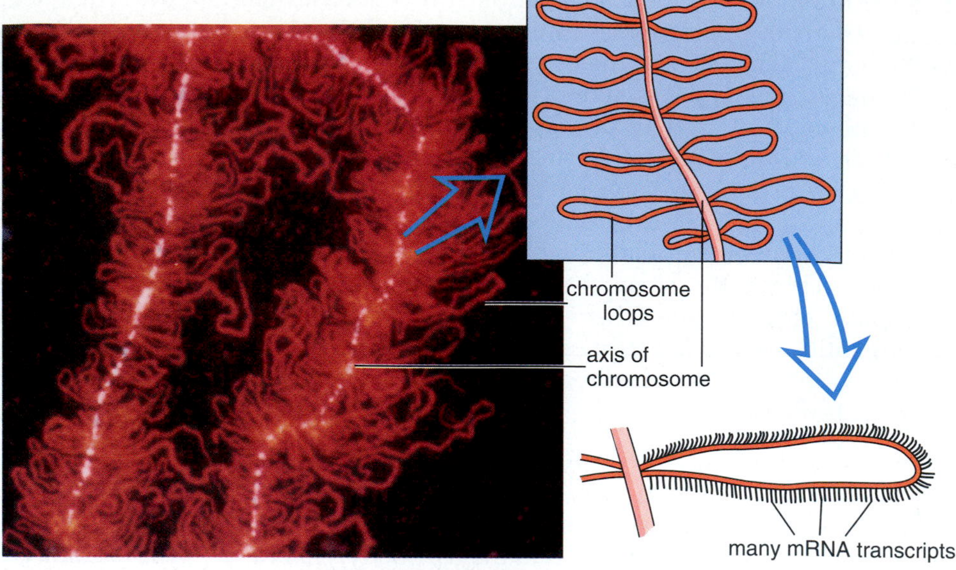

Figure 16.9 Lampbrush chromosomes.
These chromosomes are present in maturing amphibian egg cells and give evidence that when mRNA is being synthesized, chromosomes most likely decondense. Each chromosome has many loops extended from its axis (white). Many mRNA transcripts are being made off of these DNA loops (red).

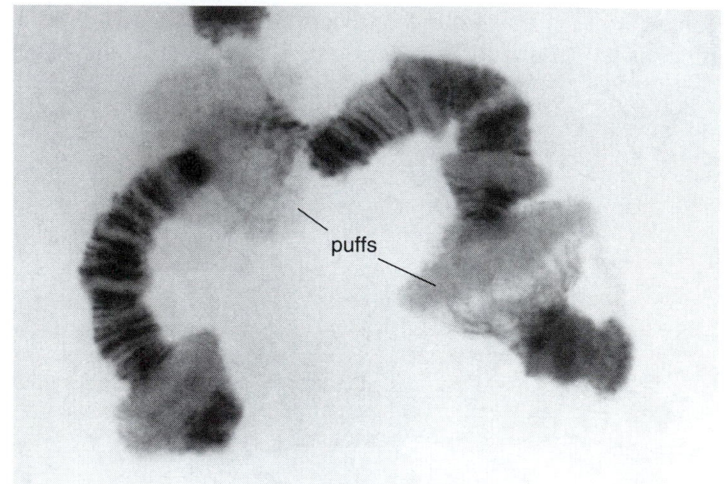

Figure 16.10 Polytene chromosome.
Polytene chromosomes, such as this one from an insect, *Trichosia pubescens,* have about 1,000 parallel chromatids. As development occurs, puffs appear at specific sites, perhaps indicating where DNA is being transcribed. When ecdysone, a hormone that stimulates molting, is added, a pattern of puffs characteristic of natural molting appears.

Euchromatin Is Active In contrast to heterochromatin, euchromatin is likely to be genetically active—it can be transcribed. The chromosomes within the developing egg cells of many vertebrates are called lampbrush chromosomes because they have many loops that appear to be bristles (Fig. 16.9). Here mRNA is being synthesized in great quantity as development proceeds.

In the salivary glands and other tissues of larval flies, the chromosomes duplicate and reduplicate many times without dividing mitotically. The homologues, each consisting of about 1,000 sister chromatids, synapse together to form giant chromosomes called polytene chromosomes [Gk. *poly,* many, and L. *taenia,* band, ribbon]. It is observed that as a larva develops, first one and then another of the chromosomal regions bulge out, forming chromosomal puffs (Fig. 16.10). The use of radioactive uridine, a label specific for RNA, indicates that DNA is being actively transcribed at these chromosomal puffs. It appears that the chromosome is decondensing at the puffs, allowing RNA polymerase to attach to a section of DNA.

Varying the amount of chromatin as in polytene chromosomes is another form of transcriptional control. When the gene product is either transfer RNA (tRNA) or ribosomal RNA (rRNA), only an increased amount of transcription permits an increase in the amount of gene product. In *Xenopus* (a frog) germ cells that are producing eggs, the number of nucleoli, and therefore copies of RNA genes, actually increases to nearly 1,000. This ensures an immense number of ribosomes in the egg cytoplasm to support protein synthesis during the early stages of development. This form of control is called *gene amplification.*

Transcription Factors

Although no operons like those of prokaryotic cells have been found in eukaryotic cells, transcription is controlled by DNA-binding proteins called **transcription factors.** Every cell contains many different types of transcription factors, and a different combination is believed to regulate the activity of any particular gene. A group of transcription factors binds to a promoter adjacent to a gene and then the complex

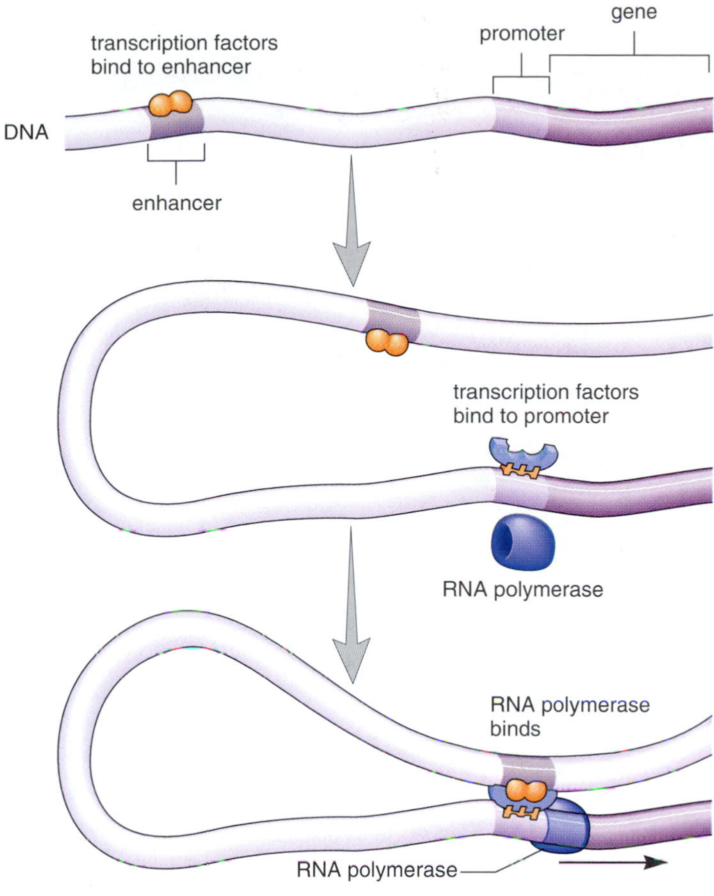

Figure 16.11 **Transcription factors.**
In this model of how transcription factors in eukaryotic cells work, transcription begins after transcription factors bind to a promoter and to an enhancer. The enhancer is far from the promoter, but physical contact is possible when the DNA loops bringing all factors into contact. Only then does transcription begin.

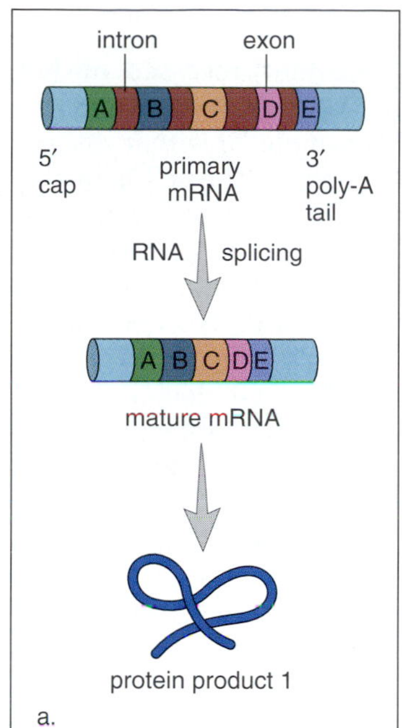

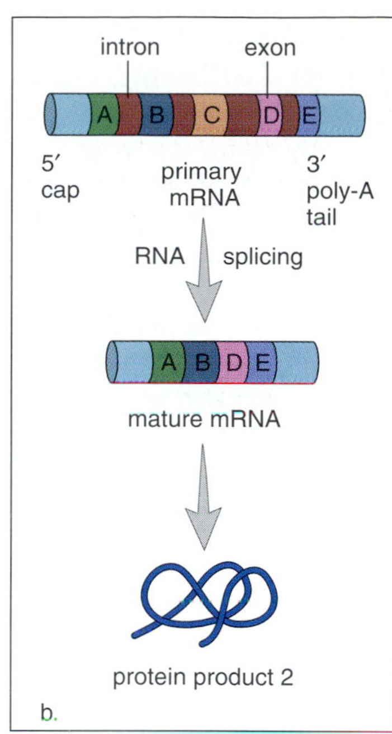

Figure 16.12 **Processing of mRNA transcripts.**
Because the primary mRNAs are processed differently in these two cells (a and b), distinct proteins result.

attracts and binds RNA polymerase, but transcription may still not begin. Quite often, enhancers are also involved in promoting transcription. Enhancers are regions where factors that help regulate transcription of the gene can also bind. *Enhancers* can be quite a distance from the promoter, but a hairpin loop in the DNA can bring the transcription factors attached to the enhancer into contact with the transcription factors at the promoter. Now RNA polymerase begins the transcription process (Fig. 16.11).

Transcription factors are always present in a cell and most likely they have to be activated in some way before they will bind to DNA. Activation probably occurs when they are phosphorylated by a kinase. Kinases, which add a phosphate group to molecules, and phosphatases, which remove a phosphate group, are known to be signaling proteins involved in a growth regulatory network that reaches from receptors in the plasma membrane to the nucleus. The growth regulatory network is discussed in more detail later in this chapter.

Posttranscriptional Control

Posttranscriptional control begins once there is an mRNA transcript. In eukaryotes, genes have both exons (coding regions) and introns (noncoding regions). Messenger RNA (mRNA) molecules (and other RNAs) are processed before they leave the nucleus and pass into the cytoplasm. Differential excision of introns and splicing of mRNA can vary the type of mRNA that leaves the nucleus. For example, both the hypothalamus and the thyroid gland produce a hormone called calcitonin, but the mRNA that leaves the nucleus is not the same in both types of cells. Radioactive labeling studies show that they vary because of a difference in mRNA splicing (Fig. 16.12). Evidence of different patterns of mRNA splicing is found in other cells, such as those that produce neurotransmitters, muscle regulatory proteins, and antibodies.

The speed of transport of mRNA from the nucleus into the cytoplasm can ultimately affect the amount of gene product realized per unit time following transcription. There is evidence that there is a difference in the length of time it takes various mRNA molecules to pass through a nuclear pore.

Transcriptional control in eukaryotes involves organization of chromatin and use of transcription factors. Posttranscriptional involves differential mRNA processing and factors that affect the length of time it takes mRNA to travel to the cytoplasm.

Translational Control

Translational control begins when the processed mRNA molecule reaches the cytoplasm. The eggs of frogs and certain other animals contain mRNA molecules that are not translated at all until fertilization occurs. The cell is unable to recognize these *masked messengers* as available for translation. As soon as fertilization occurs, unmasking takes place, and there is a rapid burst of specific gene product synthesis.

Obviously, the longer an active mRNA molecule remains in the cytoplasm, the more product there will eventually be. During maturation, mammalian red blood cells eject their nucleus, and yet they continue to synthesize hemoglobin for several months thereafter. This means that the necessary mRNAs are able to persist all this time. Differences in the guanine nucleotide cap at the 5′ end and the poly-A tail at the 3′ end of a mature mRNA transcript (see Fig. 15.8) might determine how long a particular transcript remains active before it is destroyed by a ribonuclease associated with ribosomes.

Hormones seem to cause the stabilization of certain mRNA transcripts. For example, the hormone prolactin promotes milk production in mammary glands by primarily affecting the length of time the mRNA for casein, a major protein in milk, persists and is translated. Similarly, an mRNA for phosphoprotein, called vitellin, persists for three weeks instead of 16 hours in amphibian cells if they are treated with estrogen. This hormone apparently interferes with the action of ribonuclease.

Posttranslational Control

Posttranslational control begins once a protein has been synthesized. Some proteins are not active immediately after synthesis. For example, bovine proinsulin is at first biologically inactive. After the single, long polypeptide folds into a three-dimensional structure, a sequence of about 30 amino acids is enzymatically removed from the middle of the molecule. This leaves two polypeptide chains that are bonded together by disulfide (S—S) bonds and results in an active protein.

Finally, most metabolic pathways are regulated by feedback inhibition (Fig. 16.13). When the end product of the pathway is present and binds at an allosteric site on the first enzyme of the pathway, the enzyme changes shape. Now the first reactant of the pathway cannot bind to the active site and the pathway is shut down.

Translational control determines the degree to which an mRNA is translated into a protein product. Posttranslational control affects the activity of a protein product, whether or not it is functional, and the length of time it is functional.

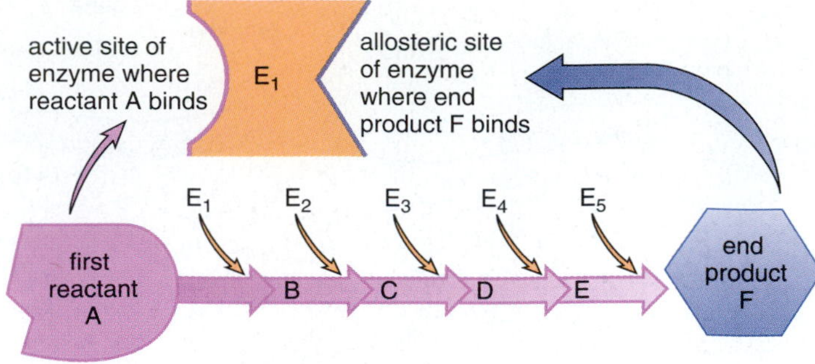

Overall view of pathway

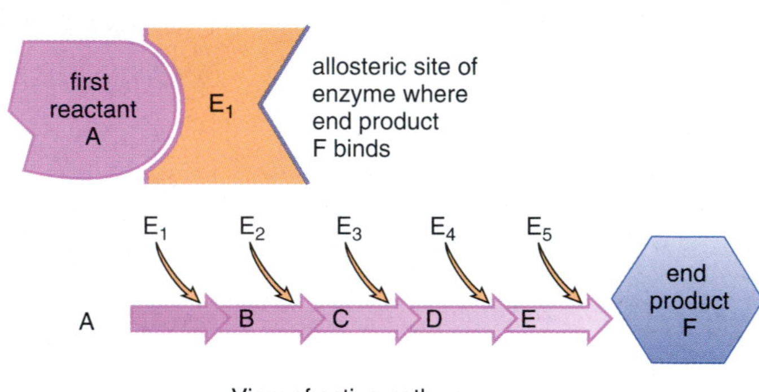

View of active pathway

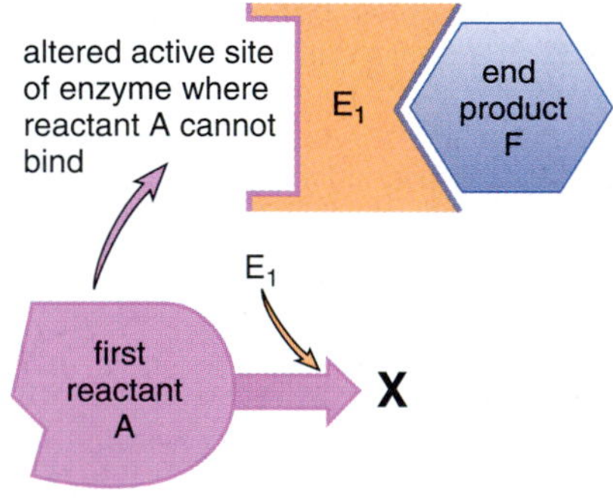

View of inhibited pathway

Figure 16.13 Feedback inhibition.
This hypothetical metabolic pathway is regulated by feedback inhibition. When reactant A binds to the active site of E_1, the pathway is active and the end product is produced. Once there is sufficient end product, some bind to the allosteric site of E_1. A change of shape prevents reactant A from binding to the active site of E_1 and the end product is no longer produced.

Science Focus

Barbara McClintock and the Discovery of Jumping Genes

When Barbara McClintock (Fig. 16A) first began studying inheritance in corn (maize) plants, geneticists believed that each gene had a fixed locus on a chromosome. Thomas Morgan and his colleagues at Columbia University were busy mapping the chromosomes of *Drosophila,* but McClintock preferred to work with corn. In the course of her studies, she came to the conclusion that "controlling elements" could move from one location to another on the chromosome. If a controlling element landed in the middle of a gene, it prevented the expression of that gene. Today, McClintock's controlling elements are called moveable genetic elements, transposons, or (in slang), "jumping genes."

Based on her experiments with maize, McClintock showed that because transposons are capable of suppressing gene expression, they could account for the pigment pattern of the corn strain popularly known as Indian corn. A colorless corn kernel results when cells are unable to produce a purple pigment due to the presence of a transposon within a particular gene needed to synthesize the pigment. While mutations are usually stable, a transposition is very unstable. When the transposon jumps to another chromosomal location, some cells regain the ability to produce the purple pigment and the result is a corn kernel with a speckled pattern (Fig. 16B).

When McClintock first published her results in the 1950s, the scientific community ignored them. Years later, when molecular genetics was well established, transposons were also discovered in bacteria, yeasts, plants, flies, and humans.

Geneticists now believe that transposons:

1. can cause localized mutations; that is, mutations that occur in certain cells and not others.
2. can carry a copy of certain host genes with them when they jump. Therefore, they can be a source of chromosome mutations such as translocations, deletions, and inversions.
3. can leave copies of themselves and certain host genes before jumping. Therefore, they can be a source of a duplication, another type of chromosome mutation.
4. can contain one or more genes that make a bacterium resistant to antibiotics.

Considering that transposition has a powerful effect on genotype and phenotype (Fig. 16C), it most likely has played an important role in evolution. For her discovery of transposons, McClintock was, in 1983, finally awarded the Nobel Prize in Physiology or Medicine. In her Nobel Prize acceptance speech, the eighty-one-year-old scientist proclaimed that "it might seem unfair to reward a person for having so much pleasure over the years, asking the maize plant to solve specific problems and then watching its responses."

Figure 16A Barbara McClintock.

Figure 16B Corn kernels.
Some kernels are purple, some colorless, and some speckled.

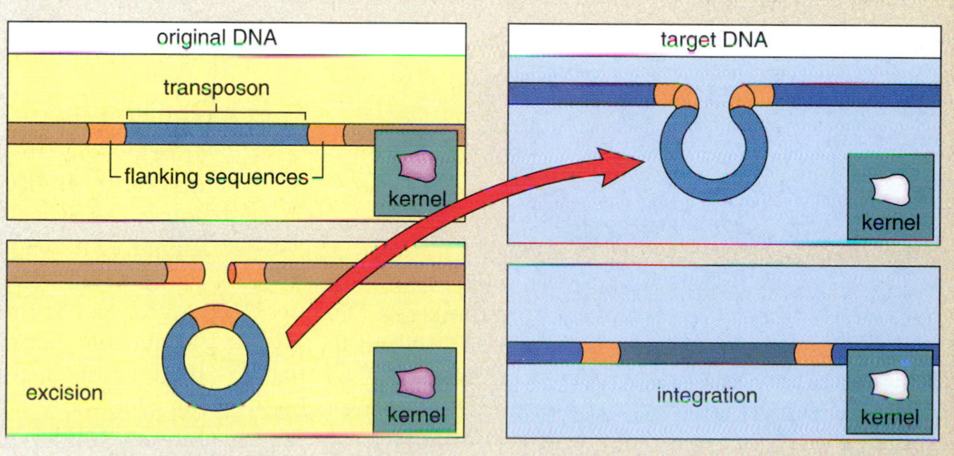

Figure 16C Transposon.
In its original location, a transposon is interrupting a gene (brown) that is not involved in kernel pigmentation. When the transposon moves to another chromosome position, it blocks the action of a gene (blue) required for synthesis of a purple pigment, and the kernel is now colorless.

16.3 Genetic Mutations

A **genetic mutation** is a permanent change in the sequence of bases in DNA. Genetic mutations vary in their phenotypic consequences depending on how protein activity is affected. Germ-line mutations are passed onto offspring while somatic mutations are not passed from one generation to the next. Germ-line mutations are the raw material for evolution; they produce the genetic variations that recombine to give novel genetic makeups. If one of these results in an adaptive phenotype, the individual will have more offspring than others of its kind. In this way certain genes and not others are preserved for future generations. Germ-line mutations coupled with genetic recombinations have resulted in the many species we see about us.

Cause of Mutations

Some mutations are spontaneous. They just happen for no apparent reason while others are due to environmental mutagens. Spontaneous mutations due to DNA replication errors are very rare. DNA polymerase, the enzyme that carries out replication, proofreads the new strand against the old strand and detects any mismatched nucleotides, and each is usually replaced with a correct nucleotide. In the end, there is only about one mistake for every one billion nucleotide pairs replicated.

Environmental Mutagens

A mutagen is an environmental agent that increases the chances of a mutation. Among the best-known mutagens are radiation and organic chemicals. **Cancer** is a genetic disorder caused by a failure in regulation of gene activity. Mutagens that increase the chances of cancer are called **carcinogens.**

Certain forms of radiation such as X rays and gamma rays are called ionizing radiation because they create free radicals, ionized atoms with unpaired electrons. Free radicals react with and alter the structure of other molecules, including DNA. Ultraviolet (UV) radiation is easily absorbed by the pyrimidines in DNA. Wherever there are two thymine molecules next to one another, ultraviolet radiation may cause them to bond together, forming thymine dimers.

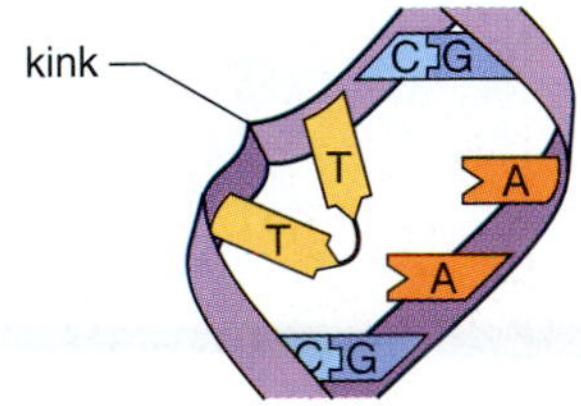

Usually, these dimers are removed from damaged DNA by repair enzymes which constantly monitor DNA and fix any irregularities. One enzyme excises a portion of DNA that contains the dimer; another makes a new section by using the other strand as a template; and still another seals the new section in place. The importance of repair enzymes is exemplified by individuals with the condition known as xeroderma pigmentosum. They lack some of the repair enzymes, and as a consequence these individuals have a high incidence of skin cancer (Fig. 16.14).

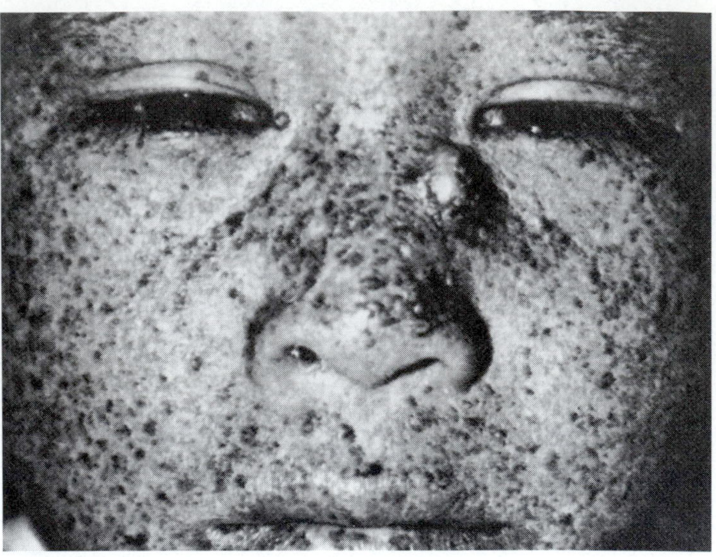

Figure 16.14 Xeroderma pigmentosum.
Deficient DNA repair enzymes leave the skin cells vulnerable to the mutagenic effects of ultraviolet light, and hundreds of skin cancers (small dark spots) appear on the skin exposed to the sun. This individual also has a tumor on the bridge of the nose.

Some organic chemicals act directly on DNA. The chemical 5-bromouracil pairs with thymine but then rearranges to a form that pairs with cytosine the next time DNA is replicated. What was once an A—T pair becomes a G—C pair. Chemicals that add hydrocarbon groups or remove amino groups from DNA bases cause mispairings. Tobacco smoke contains a number or organic chemicals that are known carcinogens, and it is estimated that one-third of all cancer deaths can be attributed to smoking. Lung cancer is the most frequent lethal cancer in the United States, and smoking is also implicated in the development of cancers of the mouth, larynx, bladder, kidney, and pancreas. The greater the number of cigarettes smoked per day, the earlier the habit starts, and the higher the tar content, the greater the possibility of cancer. When smoking is combined with drinking alcohol, the risk of these cancers increases even more.

Transposons

Transposons are specific DNA sequences that have the remarkable ability to move within and between chromosomes. As discussed in the reading on page 257, their movement to a new location sometimes alters neighboring genes, particularly increasing or decreasing their expression. This can happen if the transposon is a regulator gene. Although "moveable elements" in corn were described 40 years ago, their significance was only realized recently. So-called *jumping genes* now have been discovered in bacteria, fruit flies, humans, and many other organisms.

The white-eye mutant in *Drosophila,* first observed by Morgan, is due to a transposon located within a gene coding for a pigment-producing enzyme. In a rare human neurological disorder, called Charcot-Marie-Tooth disease, the muscles and nerves of the legs and feet gradually wither away. This syndrome is due to a transposon called Mariner, which is also found in *Drosophila!*

Effect on Protein Activity

The effect of a DNA base sequence change on protein activity can range from no effect to complete inactivity. To substantiate this statement we will consider point mutations and frameshift mutations.

Point mutations involve a change in a single DNA nucleotide, and therefore a change in a specific codon. Figure 16.15 gives an example in which a single base change could have no effect or a drastic effect, depending on the particular base change that occurs. You already know that sickle-cell disease is due to a single base change in DNA. Because the β chain of hemoglobin now contains valine instead of glutamate at one location, hemoglobin molecules form semirigid rods. The resulting sickle-shaped cells clog blood vessels and die off more quickly than normal-shaped cells.

The term *reading frame* applies to the sequence of codons because they are read from some specific starting point as in this sentence: THE CAT ATE THE RAT. If the letter C is deleted from this sentence and the reading frame is shifted, we read THE ATA TET HER AT—something that doesn't make sense. **Frameshift mutations** occur most often because one or more nucleotides are either inserted or deleted from DNA. The result of a frameshift mutation can be a completely new sequence of codons and nonfunctional proteins.

Nonfunctional Proteins

A single nonfunctioning protein can have a dramatic effect on the phenotype. For example, one particular metabolic pathway in cells is as follows:

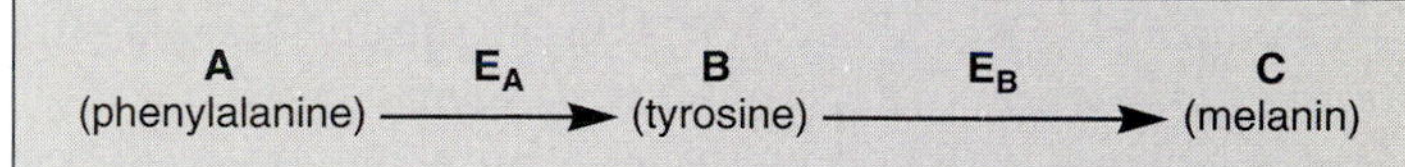

If a faulty code for enzyme E_A is inherited, a person is unable to convert the molecule A to B. Phenylalanine builds up in the system and the excess causes mental retardation and the other symptoms of the genetic disorder called phenylketonuria (PKU). In the same pathway, if a person inherits a faulty code for enzyme E_B, then B cannot be converted to C and the individual is an albino. The human transposon called *Alu* is responsible for hemophilia when *Alu* inserts into the gene for clotting factor IX and places a premature stop codon there.

Recall that cystic fibrosis is due to the inheritance of a faulty code for a chloride channel protein in the plasma membrane. A rare condition called androgen insensitivity is due to a faulty receptor for male sex hormones androgens, such as testosterone. Although there is plenty of testosterone in the blood, the cells are unable to respond to it. Female instead of male external genitals form and female instead of male secondary sexual characteristics occur. The individual who appears to be a normal female may be prompted to seek medical advice when menstruation never occurs. The karyotype is that of a male and not a female and the individual does not have the internal sexual organs of a female (Fig. 16.16).

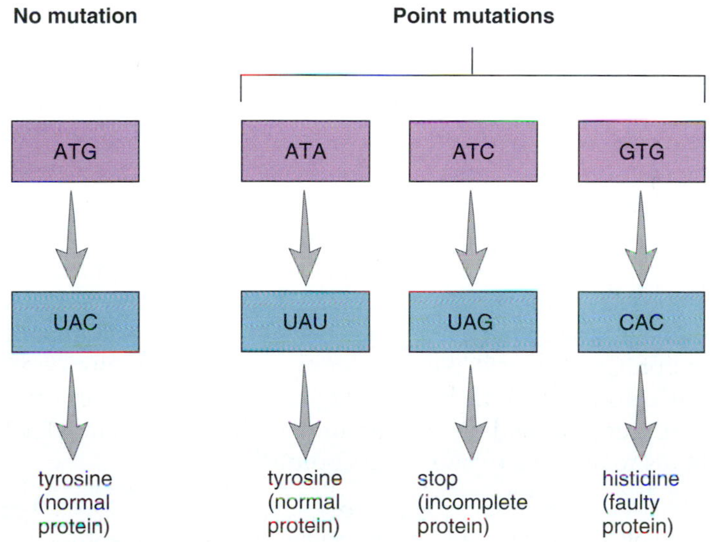

No mutation | Point mutations

No mutation	Point mutations		
ATG	ATA	ATC	GTG
UAC	UAU	UAG	CAC
tyrosine (normal protein)	tyrosine (normal protein)	stop (incomplete protein)	histidine (faulty protein)

Figure 16.15 Point mutation.
The effect of a DNA base alteration can vary. Starting at the left, if after the base change the DNA codes for the same amino acid, there is no noticeable effect; if it now codes for a stop codon, the resulting protein will be incomplete; and if it now codes for a different amino acid, a faulty protein is possible.

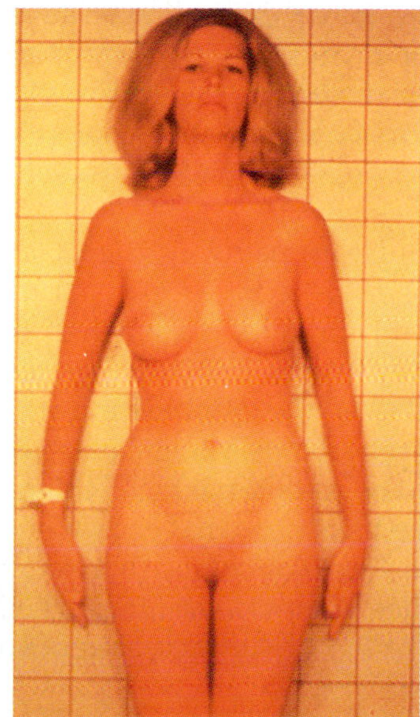

Figure 16.16 Androgen insensitivity.
This individual has a female appearance but the karyotype of a male, and there are testes instead of ovaries and a uterus in the abdominal cavity. Her cells are unable to respond to the male sex hormones because of a mutation that makes the androgen receptor ineffective.

Carcinogens

Factors that contribute to carcinogenesis (the origination of cancer) are listed in Figure 16.17. It's possible to inherit a gene that tends to cause cancer. The recently isolated *BRCA1* and *BRCA2* genes seem to account for about 20% of premenopausal breast cancer that runs in families and a substantial proportion of such ovarian cancers as well.

Tobacco is well known to cause lung and other cancers because it contains carcinogenic organic chemicals. Similarly, UV radiation is associated with skin cancer, and papillomavirus is implicated in cancer of the cervix. Even our diet influences whether we eventually develop cancer or not. Obesity is associated with uterine, postmenopausal breast cancer, as well as cancers of the colon, kidney, and gallbladder.

Regulation of the Cell Cycle

Carcinogens promote mutations in genes that control the cell cycle. An irregular mass of cells called a **tumor** is derived from an ancestral cell whose genes have undergone a series of mutations. These mutations cause the cell to repeatedly enter the cell cycle.

The cell cycle is a series of stages that occur in this sequence: G_1 stage—organelles begin to double in number; S stage—replication of DNA occurs and duplication of chromosomes occurs; G_2 stage—synthesis of proteins occurs that prepares the cell for mitosis; M stage—mitosis occurs. The passage of a cell from G_1 to the S stage is tightly regulated, as is the passage of a cell from G_2 to mitosis.

Two classes of genes known as proto-oncogenes and tumor-suppressor genes control the passage of cells through the cell cycle. **Proto-oncogenes** promote the cell cycle and **tumor-suppressor genes** inhibit the cell cycle. Each class of genes is a part of a regulatory pathway that involves extracellular signaling molecules called **growth factors,** plasma membrane growth factor receptors, internal signaling proteins within the cytoplasm, and a number of genes within the nucleus.

In the stimulatory pathway, a growth factor released by a neighboring cell is received by a plasma membrane receptor, and this sets in motion a whole series of enzymatic reactions that ends when stimulatory proteins enter the nucleus (Fig. 16.17). In the inhibitory pathway, a growth-inhibitory factor released by a neighboring cell is received by a plasma membrane receptor, which sets in motion a whole series of enzymatic reactions that ends when inhibitory proteins enter the nucleus. The balance between stimulatory signals and inhibitory signals determines whether proto-oncogenes are active or tumor-suppressor genes are active.

Oncogenes

Proto-oncogenes are so called because a mutation can cause them to become **oncogenes** (cancer-causing genes). An oncogene may code for a faulty receptor in the stimulatory pathway. A faulty receptor may be able to start the stimulatory process even when no growth factor is present! Or an oncogene may produce an abnormal protein product or else abnormally high levels of a normal product that stimulates the cell cycle to begin or go to completion. In either case, uncontrolled growth threatens.

Researchers have identified perhaps one hundred oncogenes that can cause increased growth and lead to tumors. The oncogenes most frequently involved in human cancers belong to the *ras* gene family. An alteration of only a single nucleotide pair is sufficient to convert a normally functioning *ras* proto-oncogene to an oncogene. The *ras*K oncogene is found in about 25% of lung cancers, 50% of colon cancers, and 90% of pancreatic cancers. The *ras*N oncogene is associated with leukemias (cancer of blood-forming cells) and lymphomas (cancers of lymphoid tissue), and both *ras* oncogenes are frequently found in thyroid cancers.

Tumor-Suppressor Genes

When a tumor-suppressor gene undergoes a mutation, the regulatory balance shifts in favor of cell cycle stimulation. Researchers have identified about a half-dozen tumor-suppressor genes. The *RB* tumor-suppressor gene was discovered when the inherited condition retinoblastoma was being studied. If a child receives only one normal *RB* gene and that gene mutates, eye tumors develop in the retina by the age of three. The *RB* gene has now been found to malfunction in cancers of the breast, prostate, and bladder, among others. Loss of the *RB* gene through chromosome deletion is particularly frequent in a type of lung cancer called small cell lung carcinoma. How the RB protein fits into the inhibitory pathway is known. When a particular growth-inhibitory factor attaches to a receptor, the RB protein is activated. An active RB protein turns off the expression of a proto-oncogene, whose product stimulates the cell cycle.

Another major tumor-suppressor gene is called *p53,* a gene that is more frequently mutated in human cancers than any other known gene. It has been found that the p53 protein acts as a transcription factor and as such is involved in turning on the expression of genes whose products are cell cycle inhibitors. The p53 protein can also stimulate **apoptosis,** programmed cell death. If DNA is damaged in any way, the p53 protein inhibits the cell cycle, and this gives cellular enzymes the opportunity to repair the damage. But if DNA damage should persist, the *p53* gene goes on to bring about apoptosis of the cell. No wonder many types of tumors contain cells that lack an active *p53* gene. Some cancer cells have a *p53* gene but make a large amount of bcl-2, a protein which binds to and inactivates the p53 protein.

Each cell contains regulatory pathways involving proto-oncogenes and tumor-suppressor genes that code for components active in the pathways.

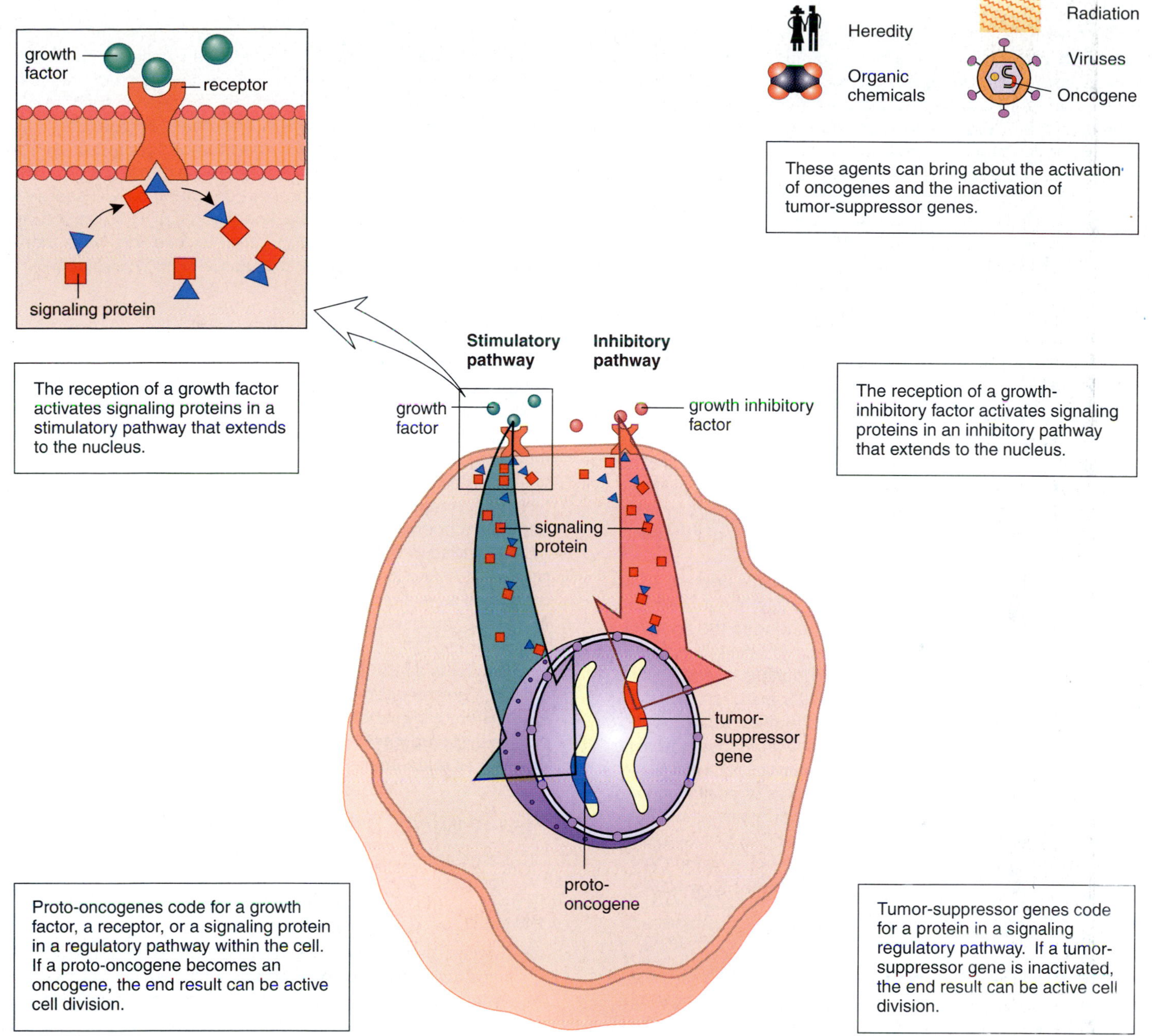

growth factor
receptor
signaling protein
Heredity
Radiation
Organic chemicals
Viruses
Oncogene
These agents can bring about the activation of oncogenes and the inactivation of tumor-suppressor genes.
Stimulatory pathway
Inhibitory pathway
The reception of a growth factor activates signaling proteins in a stimulatory pathway that extends to the nucleus.
growth factor
growth inhibitory factor
The reception of a growth-inhibitory factor activates signaling proteins in an inhibitory pathway that extends to the nucleus.
signaling protein
tumor-suppressor gene
proto-oncogene
Proto-oncogenes code for a growth factor, a receptor, or a signaling protein in a regulatory pathway within the cell. If a proto-oncogene becomes an oncogene, the end result can be active cell division.
Tumor-suppressor genes code for a protein in a signaling regulatory pathway. If a tumor-suppressor gene is inactivated, the end result can be active cell division.

Figure 16.17 **Causes of cancer.**
Two types of regulatory pathways extend from the plasma membrane to the nucleus. In the stimulatory pathway, plasma membrane receptors receive growth-stimulatory factors. Then, proteins within the cytoplasm and proto-oncogenes within the nucleus stimulate the cell cycle. In the inhibitory pathway, plasma membrane receptors receive growth-inhibitory factors. Then, proteins within the cytoplasm and tumor-suppressor genes within the nucleus inhibit the cell cycle from occurring. Whether cell division occurs or not depends on the balance of stimulatory and inhibitory signals received. Hereditary and environmental factors cause mutations of proto-oncogenes and tumor-suppressor genes. These mutations can cause uncontrolled growth and a tumor.

Connecting Concepts

The basic mechanisms by which genetic information in DNA is converted first to an mRNA transcript and then to a protein by means of translation were explained in the previous chapter. These processes occur in all cells but not all genes are expressed in all cells of a multicellular organism. Gene regulation determines whether genes are expressed and/or the degree to which a gene is expressed. In differentiated cells like nerve cells, muscle fibers, reproductive cells, and so forth, only certain genes are actively expressed. The collective work of many researchers was required to determine that the regulation of genes can occur at several different stages during the processes of transcription and translation.

Mutations of regulatory genes may cause structural genes to be abnormally turned off or on, or expressed in abnormal quantity. Such mutations can seriously affect a developing embryo or, as we saw in this chapter, can lead to development of cancer. Cancers can arise when proto-oncogenes, which are normally not expressed, become oncogenes that are expressed. On the other hand cancers can also arise when tumor-suppressor genes fail to be adequately expressed.

In the next chapter, you will see how the basic knowledge of DNA structure, replication, and expression have contributed to a biotechnology revolution.

Summary

16.1 Prokaryotic Regulation

Regulation in prokaryotes usually occurs at the level of transcription. The operon model developed by Jacob and Monod says that a regulator gene codes for a repressor, which sometimes binds to the operator. When it does, RNA polymerase is unable to bind to the promoter, and transcription of structural genes cannot take place.

The *trp* operon is an example of a repressible operon because when tryptophan, the corepressor, is present, it binds to the repressor. The repressor is then able to bind to the operator, and transcription of structural genes does not take place.

The *lac* operon is an example of an inducible operon because when lactose, the inducer, is present, it binds to the repressor. The repressor is unable to bind to the operator, and transcription of structural genes takes place if glucose is absent.

These are two examples of negative control because a repressor is involved. There are also examples of positive control. The structural genes in the *lac* operon are not maximally expressed unless glucose is absent in addition to lactose being present. At that time cAMP attaches to a molecule called CAP and then CAP binds to a site next to the promoter. Now RNA polymerase is better able to bind to the promoter and transcription occurs.

16.2 Eukaryotic Regulation

The genome is all the DNA in a cell, including all coding genes. The following levels of control of gene expression are possible in eukaryotes: transcriptional control, posttranscriptional control, translational control, and posttranslational control.

Chromatin organization helps regulate transcription. Highly compacted heterochromatin is genetically inactive, as exemplified by Barr bodies. Less compacted euchromatin is genetically active, as exemplified by lampbrush chromosomes in vertebrates and polytene chromosomal puffs in insects. Gene amplification is the replication of a gene such that there are more copies than were originally present in the zygotic nucleus. Regulatory proteins called transcription factors, as well as DNA sequences called enhancers, play a role in controlling transcription in eukaryotes.

Posttranscriptional control refers to variations in messenger RNA (mRNA) processing and the speed with which a particular mRNA molecule leaves the nucleus.

Translational control affects mRNA translation and the length of time it is translated. Posttranslational control affects whether or not an enzyme is active and how long it is active.

16.3 Genetic Mutations

In molecular terms, a gene is a sequence of DNA nucleotide bases and a genetic mutation is a change in this sequence. Mutations can be spontaneous or due to an environmental mutagen. Carcinogens are mutagens like radiation and organic chemicals that cause cancer.

Point mutations can range in effect depending on the particular codon change. Frameshift mutations result when a base is added or deleted, and the result is usually a nonfunctional protein. Nonfunctional proteins can affect the phenotype drastically as in albinism, which is due to a single faulty enzyme, and androgen insensitivity, which is due to a faulty receptor for testosterone.

Cancer is due to an accumulation of genetic mutations among regulatory genes that control a pathway that reaches from the plasma membrane to the nucleus. In particular, the cell cycle occurs inappropriately when proto-oncogenes become oncogenes and tumor-suppressor genes are no longer effective.

Reviewing the Chapter

1. Name and state the function of the three components of operons. 248
2. Explain the operation of the *trp* operon, and note why it is considered a repressible operon. 249
3. Explain the operation of the *lac* operon, and note why it is considered an inducible operon. 249–50
4. What are the four levels of genetic regulatory control in eukaryotes? 251
5. Relate euchromatin and heterochromatin to levels of chromosome organization. 252
6. With regard to transcriptional control in eukaryotes, explain how Barr bodies show that heterochromatin is genetically inactive. 253
7. Explain how lampbrush chromosomes in vertebrates and giant (polytene) chromosomal puffs in insects show that euchromatin is genetically active. 254
8. What do transcription factors do in eukaryotic cells? What are enhancers? 255
9. Give examples of posttranscriptional, translational, and posttranslational control in eukaryotes. 255–56
10. What are oncogenes and tumor-suppressor genes? What role do they play in a regulatory network that controls cell division and involves plasma membrane receptors and signaling proteins? 260

Testing Yourself

Choose the best answer for each question.

1. Which type of prokaryotic cell would be more successful as judged by its growth potential?
   a. One that is able to express all its genes all the time.
   b. One that is unable to express any of its genes any of the time.
   c. One that expresses some of its genes some of the time.
   d. One that limits its growth regardless of the richness of the culture medium.
   e. Both a and d are correct.
2. When lactose is present, and glucose is absent
   a. the repressor is able to bind to the operator.
   b. the repressor is unable to bind to the operator.
   c. transcription of structural genes occurs.
   d. transcription of the structural genes, operator, and promoter occur.
   e. Both b and c are correct.
3. When tryptophan is present
   a. the repressor is able to bind to the operator.
   b. the repressor is unable to bind to the operator.
   c. transcription of structural genes occurs.
   d. transcription of the structural genes, operator, and promoter occur.
   e. Both b and c are correct.
4. Which of these is mismatched?
   a. posttranslational control—nucleus
   b. transcriptional control—nucleus
   c. translational control—cytoplasm
   d. posttranscriptional control—nucleus
   e. Both b and c are mismatched.
5. RNA processing varies in different cells. This is an example of ______________ control of gene expression.
   a. transcriptional
   b. posttranscriptional
   c. translational
   d. posttranslational
   e. Both b and d are correct.
6. A scientist adds radioactive uridine (label for RNA) to a culture of cells and examines an autoradiograph. Which type of chromatin is apt to show the label?
   a. heterochromatin
   b. euchromatin
   c. the histones and not the DNA
   d. the DNA and not the histones
   e. Both a and d are correct.
7. Barr bodies are
   a. genetically active X chromosomes in males.
   b. genetically inactive X chromosomes in females.
   c. genetically active Y chromosomes in males.
   d. genetically inactive Y chromosomes in females.
   e. Both b and d are correct.
8. Which of these might cause a proto-oncogene to become an oncogene?
   a. exposure of cell to radiation
   b. exposure of cell to certain chemicals
   c. viral infection of the cell
   d. exposure of cell to pollutants
   e. All of these are correct.
9. A cell is cancerous. Where might you find an abnormality?
   a. only in the nucleus
   b. only in the plasma membrane receptors
   c. only in cytoplasmic reactions
   d. only in the cell cycle
   e. in any part of the cell concerned with growth and cell division
10. A tumor-suppressor gene
    a. inhibits cell division.
    b. opposes oncogenes.
    c. prevents cancer.
    d. is subject to mutations.
    e. All of these are correct.
11. Label this diagram of an operon.

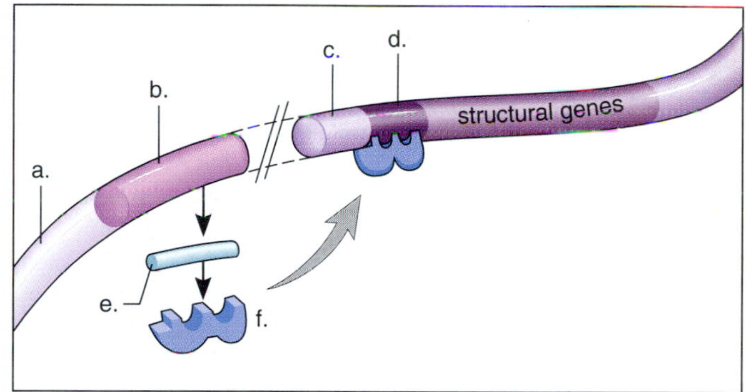

Thinking Scientifically

1. In patients with chronic myelogenous leukemia one sees an odd chromosome in all of the cancerous cells. A small piece of chromosome 9 is connected to chromosome 22. This 9:22 translocation has been termed the Philadelphia chromosome. How could a translocation cause genetic changes that result in cancer?
2. Interferon is a molecule that sets up an "alert" state in cells so that they can respond quickly to viral infection. This alert state actually causes the cell to self-destruct when a virus enters the cell. This is beneficial to the organism since the virus is prevented from producing progeny virus. For this system to work, the cell must be able to respond as rapidly as possible to the incoming virus. What hypothesis would explain how gene expression could produce the most rapid response? How could you use drugs that inhibit transcription to test the hypothesis?

Bioethical Issue

In the 1950s and 1960s, the U.S. government performed atomic bomb tests in Nevada. Radioactive iodine released into the air increased the risk of thyroid cancer for all U.S. citizens, and especially those who lived in Nevada, western states north of the test site, and states east of the site. People who were children then have an increased risk due to their small size at the time. Also, they probably drank milk from cows that fed on contaminated grass. It can take several decades for thyroid cancer to develop after exposure to radiation.

Physicians for Social Responsibility (PSR) are concerned because it took the government more than thirty years to decide that on the average people received 2 rads (radiation-absorbed doses) of radiation, and some children may have received levels above 120 rads. The Agency for Toxic Substances and Disease Registry sets a dose of 10 rads as the threshold for a need for a cancer monitoring program.

Still, the National Cancer Institute who released its report in two stages during 1997 and 1998, only after media pressure, does not believe that a monitoring program is necessary. Their argument is that there is no evidence that early detection of thyroid cancer will reduce mortality. Even so, the Physicians for Social Responsibility believe that monitoring should be offered, and further, a major research effort should begin on how to identify and treat people with thyroid cancer. A spokesperson for the group said, "Nuclear testing has exposed millions of American people to dangerous radioactivity without their knowledge or consent, and then the information about the exposure was withheld from them by federal officials. The Department of Health and Human Services should provide clear direction to citizens, physicians, and public health officials about [the] next steps [to be taken]." Should the U.S. government also be held responsible for cases of thyroid cancer that develop in the states mentioned? Why or why not?

Understanding the Terms

apoptosis 260
Barr body 253
cancer 258
carcinogen 258
corepressor 249
euchromatin 252
frameshift mutation 259
genetic mutation 258
genome 251
growth factor 260
heterochromatin 252
inducer 250
inducible operon 250
oncogene 260
operator 248
operon 248
point mutation 259
posttranscriptional control 251
posttranslational control 251
promoter 248
proto-oncogene 260
regulator gene 248
repressible operon 249
repressor 249
structural gene 248
transcription factor 254
transcriptional control 251
translational control 251
tumor 260
tumor-suppressor gene 260

Match the terms to these definitions:

a. ______________ Cancer-causing gene.
b. ______________ Abnormal growth derived from a single mutated cell that has repeatedly undergone cell division; may be benign or malignant.
c. ______________ Dark-staining body in the nuclei of female mammals that contains a condensed, inactive X chromosome.
d. ______________ Diffuse chromatin, which is being transcribed.
e. ______________ Environmental agent that causes mutations leading to the development of cancer.

Web Connections

Exploring the Internet

http://www.mhhe.com/biosci/genbio/mader
(click on *Biology 7/e*)

The *Biology 7/e* Online Learning Center provides many resources for studying the material in this chapter including links to the following sites:

The website of the National Institutes of Health contains an information index that will help you to find the institution responsible for the subject of your choice. The Health Information Page will also supply you with a link to that institute's home page.

http://www.nih.gov/health/

National Cancer Institute. Has a list of centers testing experimental therapies, as well as fact sheets and updates, plus news about ongoing clinical trials. Also provides access to PDQ (Physician Data Query), the largest resource on treatments and centers specializing in cancer care.

http://cancernet.nci.nih.gov/

American Cancer Society. The best site for information on diagnoses, treatments, and emotional support, as well as links to useful web pages screened for accuracy and reliability.

http://www.cancer.org/

Eukaryotic Cell Cycle and Cancer. This site includes a thorough discussion of the cell cycle, oncogenes, tumor-suppressor genes, how oncogenes cause cancer, and where these genes are located within the human genome.

http://www.cc.ndsu.nodak.edu/instruct/mcclean/plsc431/cellcycle/

Prokaryotic Genetics and Gene Expression. This MIT hypertextbook chapter covers gene expression, and includes practice problems.

http://esg-www.mit.edu:8001/esgbio/pge/pgedir.html

Biotechnology

chapter **17**

chapter concepts

17.1 Cloning of a Gene

- Using recombinant DNA technology, bacteria and viruses can be genetically engineered to clone a gene. 266
- A genomic library contains bacteria or viruses that carry fragments of the DNA of a particular organism. 267
- The polymerase chain reaction (PCR) makes multiple copies of DNA segments. Analysis of the DNA usually follows. 268

17.2 Biotechnology Products

- Bacteria, agricultural plants, and farm animals have been genetically engineered to produce commercially available products. 270
- Human tissues and farm animals are being genetically engineered to serve as a source of organs for human transplant patients. 271
- Agricultural plants and farm animals have been genetically engineered to improve their yield. 272
- It is now possible to clone animals, and cloning is used to produce multiple copies of farm animals that have been genetically engineered. 272

17.3 The Human Genome Project

- The Human Genome Project has two goals: to genetically map each chromosome and to sequence the DNA bases of each chromosome. 274

17.4 Gene Therapy

- Gene therapy is now being used to replace defective genes with healthy genes and to help cure various human ills. 275

The tomatoes in the foreground were genetically engineered to be insect resistant.

Since Mendel's work was rediscovered in 1900, geneticists have made startling advances which have led to a new era of DNA technology. Modern techniques enable genes to be removed from one organism and inserted into another in order to produce a desired substance, insulin for example. Not very long ago, people with insulin-dependent diabetes mellitus received the insulin they needed for survival from dead animals. Today, they receive human insulin, a product of biotechnology. Since the 1980s, biotechnology has produced drugs and vaccines to curb human illnesses, and nucleic acids for laboratory research.

Genetically engineered bacteria have been used to clean up environmental pollutants, increase the fertility of the soil, and kill insect pests. Biotechnology also extends beyond unicellular organisms; it is now possible to alter the genotype and subsequently the phenotype of plants and animals. Indeed, gene therapy in humans—attempting to repair a faulty gene—is already undergoing clinical trials. There are those who are opposed to manipulation of genes for any reason. Although there have been no ill effects as yet, they fear the possibility of health and ecological repercussions in the future.

17.1 Cloning of a Gene

The cloning of a gene produces many identical copies. Recombinant DNA technology is used when a very large quantity of the gene is required. The use of the polymerase chain reaction (PCR) creates a lesser number of copies within a laboratory test tube.

Recombinant DNA Technology

Recombinant DNA (rDNA) contains DNA from two different sources. To make rDNA, a technician often begins by selecting a **vector,** the means by which recombinant DNA is introduced into a host cell. One common type of vector is a plasmid. **Plasmids** are small accessory rings of DNA. The ring is not part of the bacterial chromosome and can be replicated independently. Plasmids were discovered by investigators studying the sex life of the intestinal bacterium *Escherichia coli.*

Two enzymes are needed to introduce foreign DNA into vector DNA (Fig. 17.1). The first enzyme, called a **restriction enzyme,** cleaves plasmid DNA, and the second, called **DNA ligase** [L. *ligo,* bind], seals foreign DNA into the opening created by the restriction enzyme.

Restriction enzymes occur naturally in bacteria, where they stop viral reproduction by cutting up viral DNA. They are called restriction enzymes because they *restrict* the growth of viruses. In 1970, Hamilton Smith, at Johns Hopkins University, isolated the first restriction enzyme; now hundreds of different restriction enzymes have been isolated and purified. Each one cuts DNA at a specific cleavage site. For example, the restriction enzyme called EcoRI always cuts double-stranded DNA when it has this sequence of bases at the cleavage site:

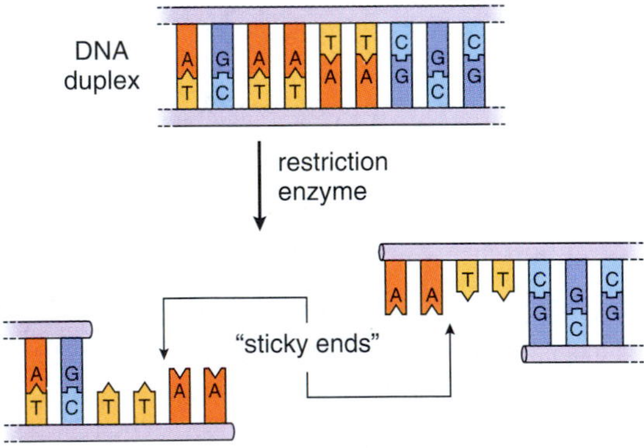

Notice there is now a gap into which a piece of foreign DNA can be placed if it ends in bases complementary to those exposed by the restriction enzyme. To assure this, it is only necessary to cleave the foreign DNA with the same type of restriction enzyme. The single-stranded but complementary ends of the two DNA molecules are called "sticky ends" because they can bind by complementary base pairing. They therefore facilitate the insertion of foreign DNA into vector DNA.

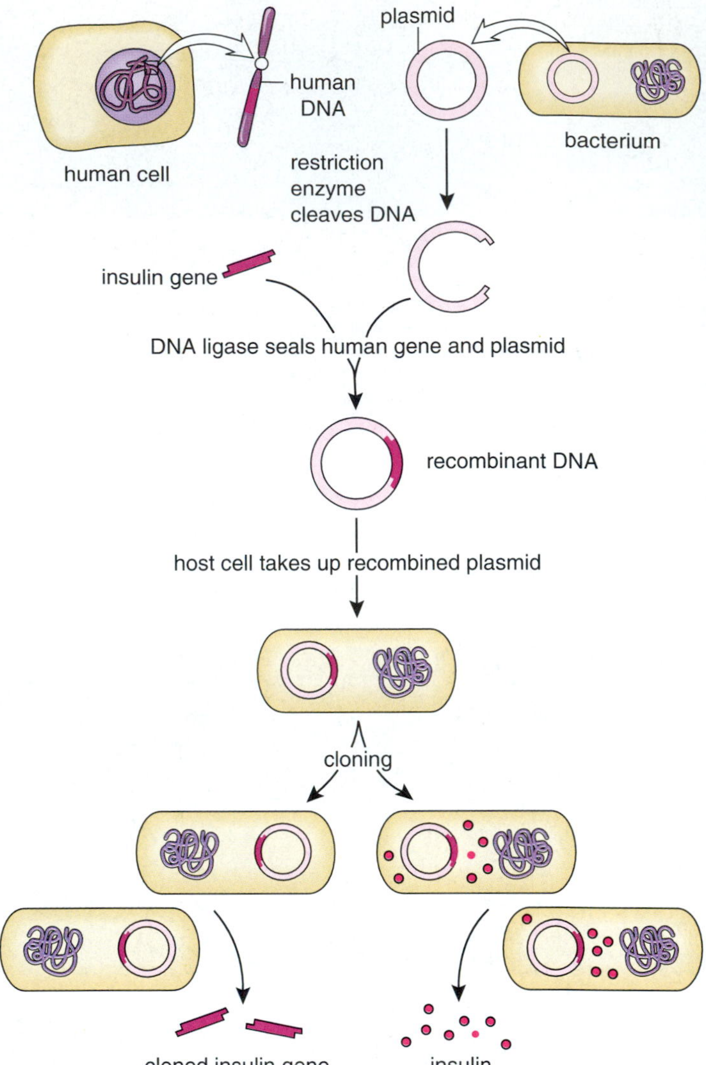

Figure 17.1 Cloning of a gene.
Human DNA and plasmid DNA are cleaved by the same type of restriction enzyme and spliced together by the enzyme DNA ligase. Gene cloning is achieved when a host cell takes up the recombined plasmid and the plasmid reproduces. Multiple copies of the gene are now available to an investigator. If the insulin gene functions normally as expected, the product (insulin) may also be retrieved.

The second enzyme needed for preparation of rDNA, DNA ligase, is a cellular enzyme that seals any breaks in a DNA molecule. Genetic engineers use this enzyme to seal the foreign piece of DNA into the vector. DNA splicing is now complete; an rDNA molecule has been prepared.

Plasmid Vector Compared to Viral Vector

A **clone** can be a large number of molecules (i.e., cloned genes) or cells (i.e., cloned bacteria) or organisms that are identical to an original specimen. Figure 17.2 compares the use of a plasmid and a virus to clone a gene. Bacterial cells take up recombined plasmids, especially if they are treated with calcium chloride to make them more permeable. Thereafter, as the host cell reproduces, a bacterial clone forms and each new cell contains at least one plasmid. Therefore, each of the bacteria contains the gene of interest which hopefully is expressing itself and producing a product. The investigator can recover either the cloned gene, or the protein product from this bacterial clone (see also Fig. 17.1).

Viruses that infect bacteria are called **bacteriophages.** In Figure 17.2*b*, the DNA of a bacteriophage called lambda is being used as a vector. After lambda attaches to a host bacterium, recombined DNA is released from the virus and enters the bacterium. Here, it will direct the reproduction of many more viruses. Each virus in the bacteriophage clone contains a copy of the gene being cloned.

Genomic Library

A **genome** is the full set of genes of an individual. A **genomic library** is a collection of bacterial or bacteriophage clones; each clone contains a particular segment of DNA from the source cell. When you make a genomic library, an organism's DNA is simply sliced up into pieces, and the pieces are put into vectors (i.e., plasmids or viruses) that are taken up by host bacteria. The entire collection of bacterial or bacteriophage clones that result contains all the genes of that organism.

In order for human gene expression to occur in a bacterium, the gene has to be accompanied by the proper regulatory regions. Also, the gene should not contain introns because bacterial cells do not have the necessary enzymes to process primary messenger RNA (mRNA). It's possible to make a human gene that lacks introns, however. The enzyme called reverse transcriptase can be used to make a DNA copy of all the mature mRNA molecules from a cell. This DNA molecule, called **complementary DNA (cDNA),** does not contain introns. Notice that a genomic library made from cDNA will contain only those genes that are being expressed in the source cell.

You can use a particular probe to search a genomic library for a certain gene. A **probe** is a single-stranded nucleotide sequence that will hybridize (pair) with a certain piece of DNA. Location of the probe is possible because the probe is either radioactive or fluorescent. Bacterial cells, each carrying a particular DNA fragment, can be plated onto agar in a petri dish. After the probe hybridizes with the gene of interest, the gene can be isolated from the fragment (Fig. 17.3). Now this particular fragment can be cloned further or even analyzed for its particular DNA sequence.

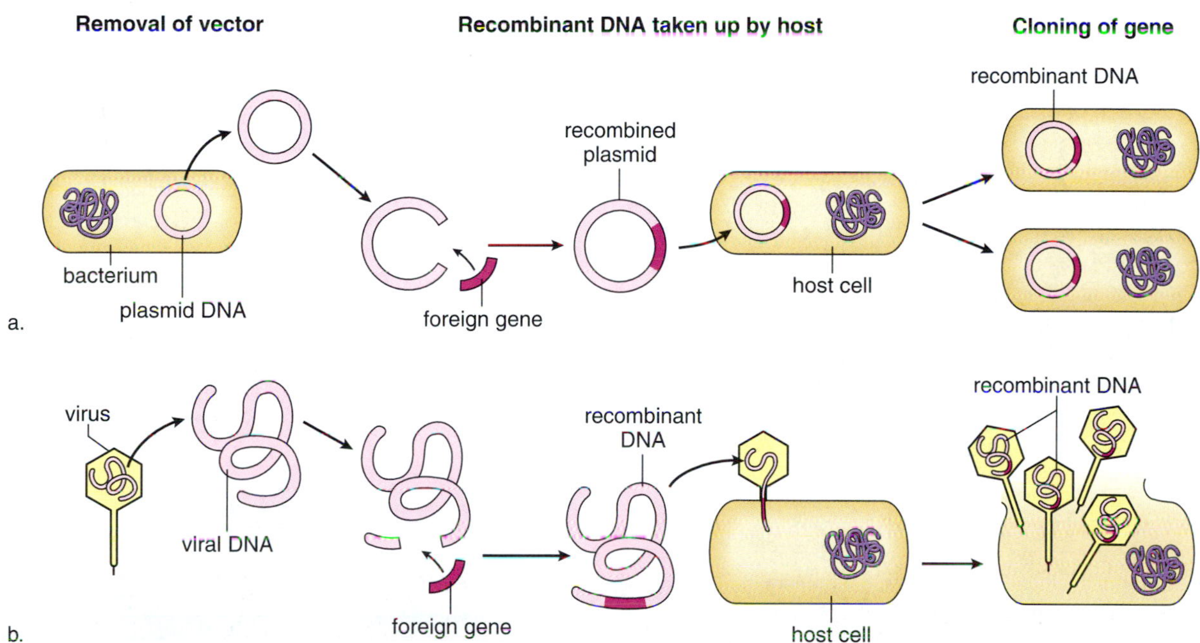

Figure 17.2 **Preparation of a genomic library.**
Each bacterial or viral clone in a genomic library contains a segment of DNA from a foreign cell. **a.** A plasmid is removed from a bacterium and is used to make recombinant DNA. After the recombined plasmid is taken up by a host cell, replication produces many copies. **b.** Viral DNA is removed from a bacteriophage such as lambda and is used to make recombinant DNA. The virus containing the recombinant DNA infects a host bacterium. Cloning is achieved when the virus reproduces and then leaves the host cell.

The Polymerase Chain Reaction

Kary B. Mullis developed the **polymerase chain reaction (PCR)** in 1983. Earlier methods of obtaining multiple copies of a specific sequence of DNA were time consuming and expensive. In contrast, PCR can create millions of copies of a single gene or any specific piece of DNA quickly in a test tube. PCR is very specific—the targeted DNA sequence can be less than one part in a million of the total DNA sample! This means that a single gene, or smaller piece of DNA, among all the human genes can be amplified (copied) using PCR.

PCR takes its name from DNA polymerase, the enzyme that carries out DNA replication in a cell. It is considered a chain reaction because DNA polymerase will carry out replication over and over again, until there are millions of copies of the desired DNA. PCR does not replace gene cloning, which is still used whenever a large quantity of gene or protein product is needed.

Before carrying out PCR, primers—sequences of about 20 bases that are complementary to the bases on either side of the "target DNA"—must be available. The primers are needed because DNA polymerase does not start the replication process; it only continues or extends the process. After the primers bind by complementary base pairing to the DNA strand, DNA polymerase copies the target DNA (Fig. 17.4).

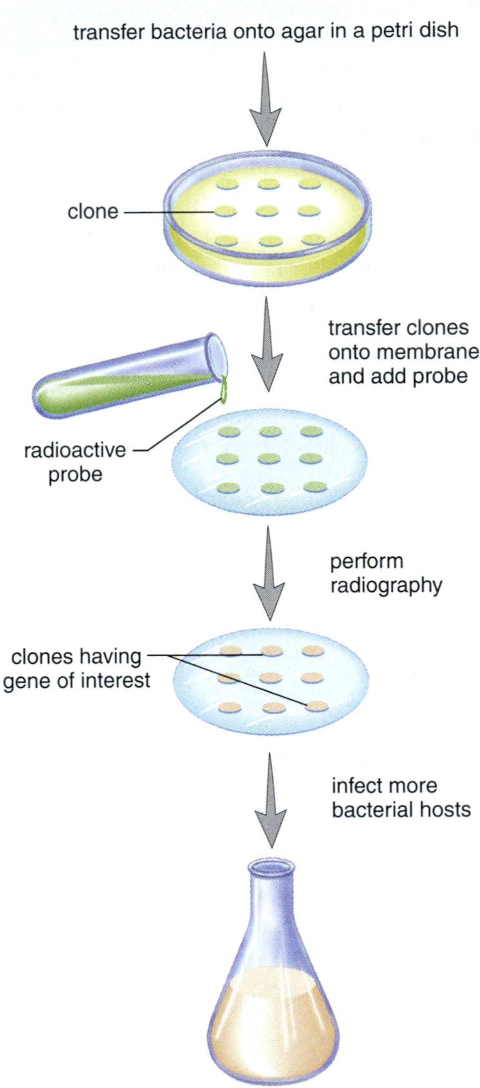

Figure 17.3 Identification of a cloned gene.
A probe, which may be single-stranded DNA or mRNA, hybridizes with the cloned DNA fragments containing the gene of interest. To see which clones (fragments) contain the gene, the probe has to be radioactive or fluorescent. Autoradiography is the use of a film that reveals the areas of radioactivity.

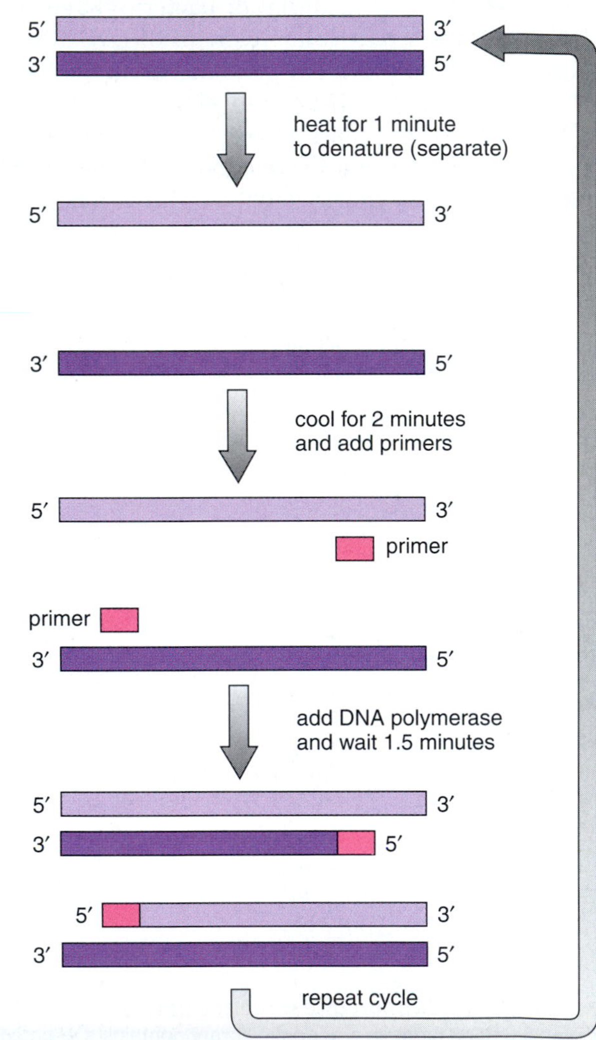

Figure 17.4 Polymerase chain reaction (PCR).
PCR is performed in laboratory test tubes. Primers (pink), which are DNA sequences complementary to the 3′ end of the targeted DNA, are necessary for DNA polymerase to make a copy of the DNA strand.

PCR has been in use for several years, and now almost every laboratory has automated PCR machines to carry out the procedure. Automation became possible after a temperature-insensitive (thermostable) DNA polymerase was extracted from the bacterium *Thermus aquaticus,* which lives in hot springs. This enzyme can withstand the high temperature used to separate double-stranded DNA; therefore, replication need not be interrupted by the need to add more enzyme.

Analyzing DNA

The entire genome of an individual can be subjected to **DNA fingerprinting,** a process described in Figure 17.5. The genome is treated with restriction enzymes, which results in a unique collection of different-sized fragments. Therefore, **restriction fragment length polymorphisms (RFLPs)** exist between individuals. During a process called gel electrophoresis, the fragments can be separated according to their lengths, and the result is a number of bands that are so close together they appear as a smear. However, the use of probes for genetic markers produces a distinctive pattern that can be recorded on X-ray film.

The DNA from a single sperm is enough to identify a suspected rapist. Since DNA is inherited, its fingerprint resembles that of one's parents. DNA fingerprinting successfully identified the remains of a teenager who had been murdered eight years before because the skeletal DNA was similar to that of the parents' DNA. DNA fingerprinting has also been helpful to evolutionists. For example, it was used to determine that the quagga, an extinct zebralike animal, was a zebra rather than a horse. The only remains of the quagga consisted of dried skin.

Following PCR, DNA segments, as opposed to the full genome, can also be cut by restriction enzymes and subjected to gel electrophoresis. A probe is not needed because the restriction fragments will appear as distinctive bands. PCR amplification and analysis can be used (1) to diagnose viral infections, genetic disorders, and cancer; (2) in forensic laboratories to identify criminals; and (3) to determine the evolutionary relationships of various organisms. Sequencing mitochondrial DNA segments helped determine the evolutionary history of human populations. It has even been possible to sequence DNA taken from a 76,000-year-old mummified human brain and from a 17- to 20-million-year-old plant fossil following PCR amplification.

Recombinant DNA technology and the polymerase chain reaction are two ways to clone a gene.

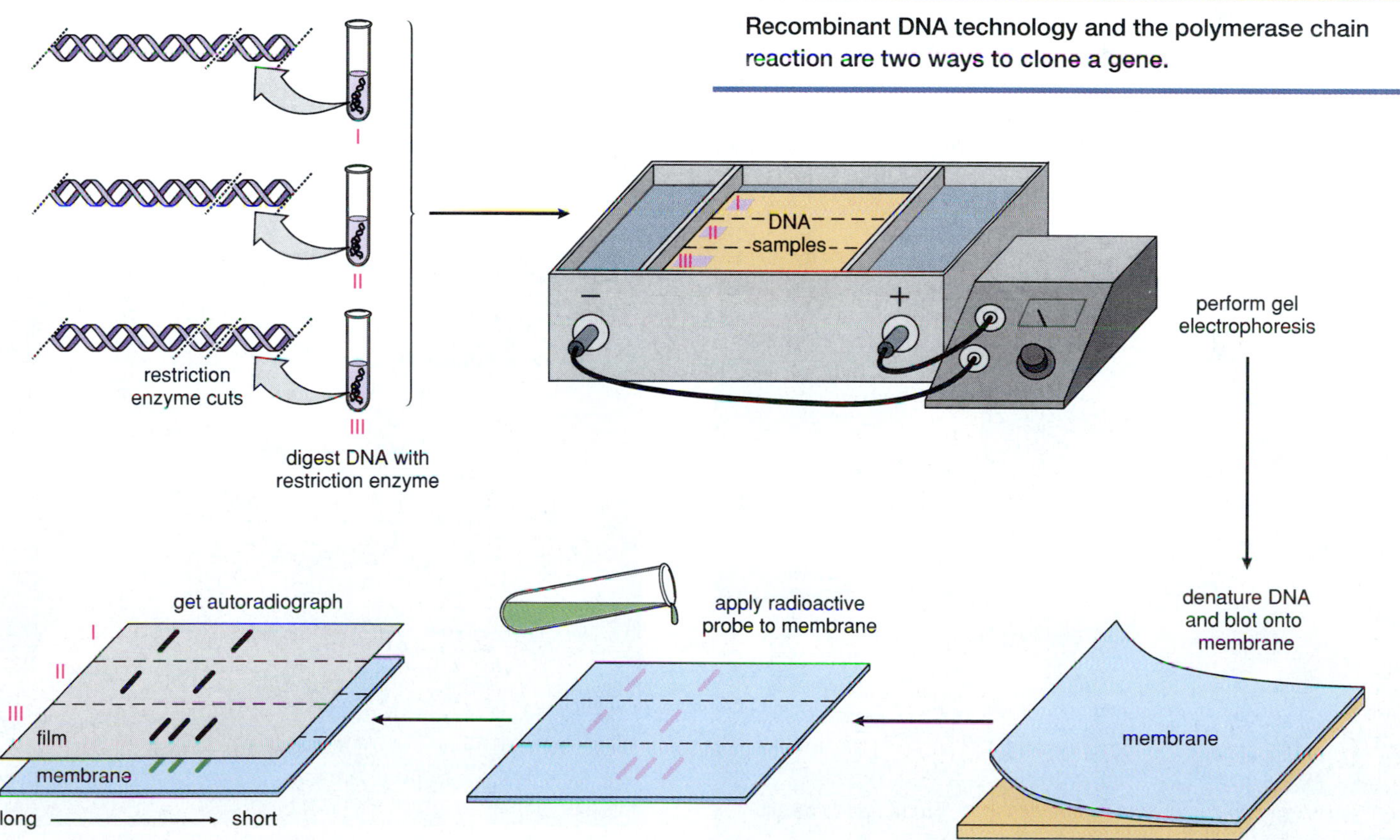

Figure 17.5 DNA fingerprinting.
DNA samples I and II are from the same individual. DNA sample III is from a different individual. Notice, therefore, that the restriction enzyme cuts are different for sample III. Gel electrophoresis separates the DNA fragments according to their length because shorter fragments migrate farther in an electrical field than do longer fragments. The fragments are denatured (separated) and transferred to a membrane where a radioactive probe can be applied. The resulting pattern (the DNA fingerprint) can then be detected by autoradiography. In a theoretical rape case, for example, sample I could be from the suspect's white blood cells, sample II could be from sperm in the victim's vagina, and sample III could be from the victim's white blood cells.

17.2 Biotechnology Products 

Today, bacteria, plants, and animals are **genetically engineered** to produce biotechnology products. Organisms that have had a foreign gene inserted into them are called **transgenic organisms.** Organs for transplant from nonhuman sources, as discussed in the reading on page 271, are also considered biotechnology products.

Transgenic Bacteria

Recombinant DNA technology is used to produce bacteria that reproduce in large vats called bioreactors. If the foreign gene is replicated and actively expressed, a large amount of protein product can be obtained. Biotechnology products produced by bacteria, such as insulin, human growth hormone, t-PA (tissue plasminogen activator), and hepatitis B vaccine, are now on the market (Fig. 17.6).

Transgenic bacteria have been produced to promote the health of plants. For example, bacteria that normally live on plants and encourage the formation of ice crystals have been changed from frost-plus to frost-minus bacteria. Also, a bacterium that normally colonizes the roots of corn plants has now been endowed with genes (from another bacterium) that code for an insect toxin. The toxin protects the roots from insects.

Bacteria can be selected for their ability to degrade a particular substance, and then this ability can be enhanced by genetic engineering. For instance, naturally occurring bacteria that eat oil can be genetically engineered to do an even better job of cleaning up beaches after oil spills (Fig. 17.7). Industry has found that bacteria can be used as biofilters to prevent airborne chemical pollutants from being vented into the air. They can also remove sulfur from coal before it is burned and help clean up toxic waste dumps. One such strain was given genes that allowed it to clean up levels of toxins that would have killed other strains. Further, these bacteria were given "suicide" genes that caused them to self-destruct when the job had been accomplished.

Organic chemicals are often synthesized by having catalysts act on precursor molecules or by using bacteria to carry out the synthesis. Today, it is possible to go one step further and to manipulate the genes that code for these enzymes. For instance, biochemists discovered a strain of bacteria that is especially good at producing phenylalanine, an organic chemical needed to make aspartame, the dipeptide sweetener better known as NutraSweet. They isolated, altered, and formed a vector for the appropriate genes so that various bacteria could be genetically engineered to produce phenylalanine.

Many major mining companies already use bacteria to obtain various metals. Genetic engineering may enhance the ability of bacteria to extract copper, uranium, and gold from low-grade sources. Some mining companies are testing genetically engineered organisms that have improved bioleaching capabilities.

Bacteria are being genetically altered to perform all sorts of tasks, not only in the factory, but also in the environment.

Figure 17.6 Biotechnology products.
Products like clotting factor VIII, which is administered to hemophiliacs, can be made by transgenic bacteria, plants, or animals. After being processed and packaged, it is sold as a commercial product.

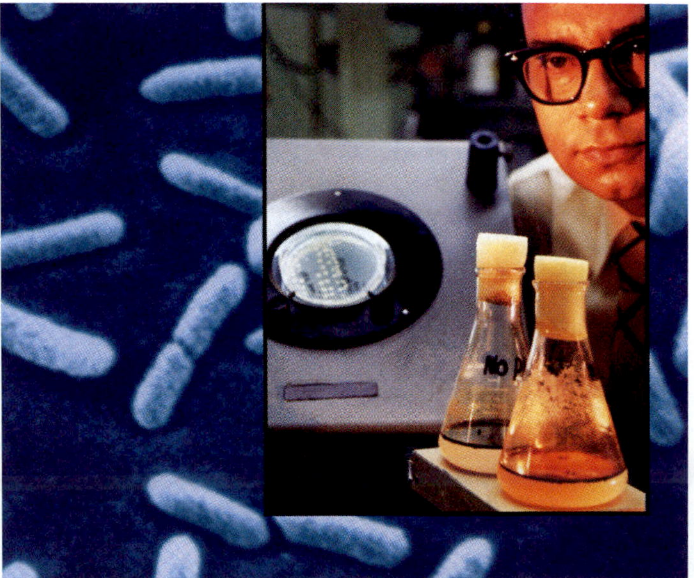

Figure 17.7 Bioremediation.
Bacteria capable of decomposing oil have been engineered and patented by the investigator Dr. Chakrabarty. In the inset, the flask toward the rear contains oil and no bacteria; the flask toward the front contains the bacteria and is almost clear of oil.

Organs for Transplant

We now have the ability to transplant human kidneys, heart, liver, pancreas, lung, and other organs for two reasons. First, solutions have been developed that preserve donor organs for several hours, and second, rejection of transplanted organs can be prevented by immunosuppressive drugs. Unfortunately, however, there are not enough human donors to go round. Thousands of patients die each year while waiting for an organ.

It's no wonder, then, that scientists are suggesting that we should get organs from a source other than another human. **Xenotransplantation** is the use of animal organs instead of human organs in transplant patients. You might think that apes, such as the chimpanzee or the baboon, might be a scientifically suitable species for this purpose. But apes are slow breeders and probably cannot be counted on to supply all the organs needed. Also, many people might object to using apes for this purpose. In contrast, animal husbandry has long included the raising of pigs as a meat source, and pigs are prolific. A female pig can become pregnant at six months and can have two litters a year, each averaging about ten offspring.

Ordinarily, humans would violently reject transplanted pig organs. Genetic engineering, however, can make these organs less antigenic. Scientists have produced a strain of pigs whose organs would most likely, even today, survive for a few months in humans. They could be used to keep a patient alive until a human organ was available. The ultimate goal is to make pig organs as widely accepted by humans as type O blood. A person with type O blood is called a universal donor because the red blood cells carry no A nor B antigens.

As xenotransplantation draws near, other concerns have been raised. Some experts fear that animals might be infected with viruses, akin to Ebola virus or the virus that causes "mad cow" disease. After infecting a transplant patient, these viruses might spread into the general populace and begin an epidemic. Scientists believe that HIV was spread to humans from monkeys when humans ate monkey meat. Those in favor of using pigs for xenotransplantation point out that pigs have been around humans for centuries without infecting them with any serious diseases.

Just a few years ago, scientists believed that transplant organs had to come from humans or other animals. Now, however, **tissue engineering** is demonstrating that it is possible to make some bioartificial organs—hybrids created from a combination of living cells and biodegradable polymers. Presently, only lab-grown hybrid tissues are on the market. A product composed of skin cells growing on a polymer is used to temporarily cover the wounds of burn patients. Similarly, damaged cartilage can be replaced with a hybrid tissue produced after chondrocytes are harvested from a patient. Another connective tissue product made from fibroblasts and collagen is available to help heal deep wounds without scarring. Soon to come are a host of other products including replacement corneas, heart valves, bladder valves, and breast tissue.

Tissue engineers have also created implants—cells producing a useful product encapsulated within a narrow plastic tube or a capsule the size of a dime or quarter (Fig. 17A). The pores of the container are large enough to allow the product to diffuse out but are too small for immune cells to enter and destroy the cells. An implant whose cells secrete natural pain killers will survive for months in the spinal cord and it can be easily withdrawn when desired. A "bridge to a liver transplant" is a bedside vascular apparatus. The patient's blood passes through porous tubes surrounded by pig liver cells. These cells convert toxins in the blood to nonpoisonous substances.

The goal of tissue engineering is to produce fully functioning organs for transplant. After nine years, a Harvard Medical School team headed by Anthony Atala has produced a working urinary bladder in the laboratory. The right culture conditions and combination of growth factors was needed to build the bladder wall from cells genetically engineered to be nonantigenic. After testing the bladders in laboratory animals, the Harvard group is ready to test them in humans whose own bladders have been damaged by accident or disease, or will not function properly due to a congenital birth defect. Another group of scientists has been able to grow arterial blood vessels in the laboratory. Still to come, tissue engineers are hopeful they can one day produce larger internal organs like the liver and kidney. Such organs will be more challenging because they are much more complex than hollow structures like bladders and arteries.

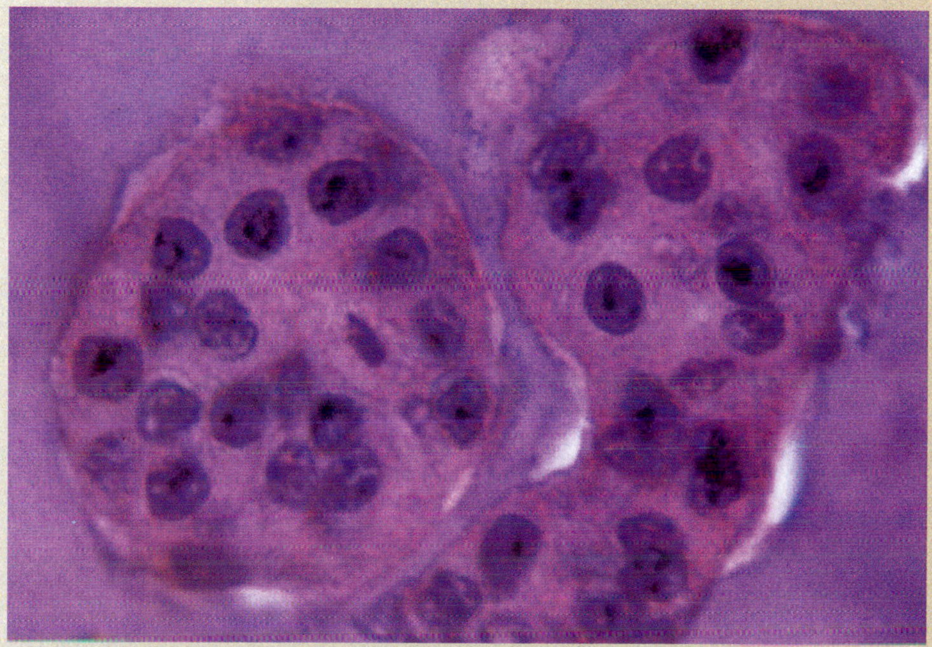

Figure 17A Microreactors.
Microreactors filled with insulin-producing pancreatic cells from pigs flourished for 10 weeks in a diabetic mouse without immune system-suppressing drugs.

Transgenic Plants

Techniques have been developed to introduce foreign genes into immature plant embryos, or into plant cells that have had the cell wall removed and are called protoplasts. It is possible to treat protoplasts with an electric current while they are suspended in a liquid containing foreign DNA. The electric current makes tiny, self-sealing holes in the plasma membrane through which genetic material can enter. Then a protoplast will develop into a complete plant.

Foreign genes transferred to cotton, corn, and potato strains have made these plants resistant to pests because their cells now produce an insect toxin. Similarly, soybeans have been made resistant to a common herbicide. Some corn and cotton plants are both pest and herbicide resistant. In 1999, these transgenic crops were planted on more than 70 million acres worldwide and the acreage is expected to triple in about five years. Improvements still to come are increased protein or starch content and modified oil or amino acid composition.

Agribusiness companies also are in the process of developing transgenic versions of wheat and rice in addition to corn. This is considered an absolute necessity if the 2020 global demand for rice, wheat, and corn is to be met. World grain harvests have continued to rise since the 1960s when special high-yield hybrid plants were developed during the so-called Green Revolution. But the per capita production has now flattened out because of continued population growth. The hope is that genetic engineering will allow farmers to surpass the yield barrier. Perhaps the stomates, the porelike openings in leaves, could be altered to boost carbon dioxide intake or cut down on water loss. Another possible goal is to increase the efficiency of the enzyme RuBP, which captures carbon dioxide in most plants. A team of Japanese scientists are attempting to introduce the C_4 cycle into rice. Plants that utilize the C_4 cycle avoid the inefficiency of carboxylase by using a different means of capturing carbon dioxide. Unlike the single gene transfers that have been done so far, these modifications would require a thorough re-engineering of plant cells.

Single gene transfers will cause plants to produce various products. A weed called mouse-eared cress has been engineered to produce a biodegradable plastic (polyhydroxybutyrate, or PHB) in cell granules. Plants are being engineered to produce human hormones, clotting factors, and antibodies, in their seeds. One type of antibody made by corn can deliver radioisotopes to tumor cells, and another made by soybeans can be used as treatment for genital herpes. Plant-made antibodies are inexpensive and there is little worry about contamination with pathogens that could infect people. Clinical trials have begun.

Farmers are now planting genetically altered crops. Medicines made by biotech plants will soon be used to treat cancer and other types of diseases.

Transgenic Animals

Techniques have been developed to insert genes into the eggs of animals. It is possible to microinject foreign genes into eggs by hand, but another method uses vortex mixing. The eggs are placed in an agitator with DNA and silicon-carbide needles, and the needles make tiny holes through which the DNA can enter. When these eggs are fertilized, the resulting offspring are transgenic animals. Using this technique, many types of animal eggs have taken up the gene for bovine growth hormone (bGH). The procedure has been used to produce larger fishes, cows, pigs, rabbits, and sheep. Genetically engineered fishes are now being kept in ponds that offer no escape to the wild because there is much concern that they will upset or destroy natural ecosystems.

Gene pharming, the use of transgenic farm animals to produce pharmaceuticals, is being pursued by a number of firms. Genes that code for therapeutic and diagnostic proteins are incorporated into the animal's DNA, and the proteins appear in the animal's milk. There are plans to produce drugs for the treatment of cystic fibrosis, cancer, blood diseases, and other disorders. Antithrombin III, for preventing blood clots during surgery, is currently being produced by a herd of goats, and clinical trials have begun. Figure 17.18*b* outlines the procedure for producing transgenic mammals: DNA containing the gene of interest is injected into donor eggs. Following in vitro fertilization, the zygotes are placed in host females where they develop. After female offspring mature, the product is secreted in their milk.

USDA scientists have been able to genetically engineer mice to produce human growth hormone in their urine instead of in milk. They expect to be able to use the same technique on larger animals. Urine is a preferable vehicle for a biotechnology product than milk because all animals in a herd urinate—only females produce milk; animals start to urinate at birth—females don't produce milk until maturity; and it's easier to extract proteins from urine than from milk.

Cloning of Transgenic Animals

Imagine that an animal has been genetically altered to produce a biotechnology product. What would be the best possible method of getting identical copies of the animal? Asexual reproduction through cloning the animal would be the preferred procedure to use. Cloning is a form of asexual reproduction because it requires only the genes of that one animal. For many years it was believed that adult vertebrate animals could not be cloned. Although each cell contains a copy of all the genes, certain genes are turned off in mature specialized cells. Different genes are expressed in muscle cells, which contract, compared to nerve cells, which conduct nerve impulses, and to glandular cells, which secrete. Cloning of an adult vertebrate requires that all genes of an adult cell be turned on again if development is to proceed normally. It had long been thought this would be impossible.

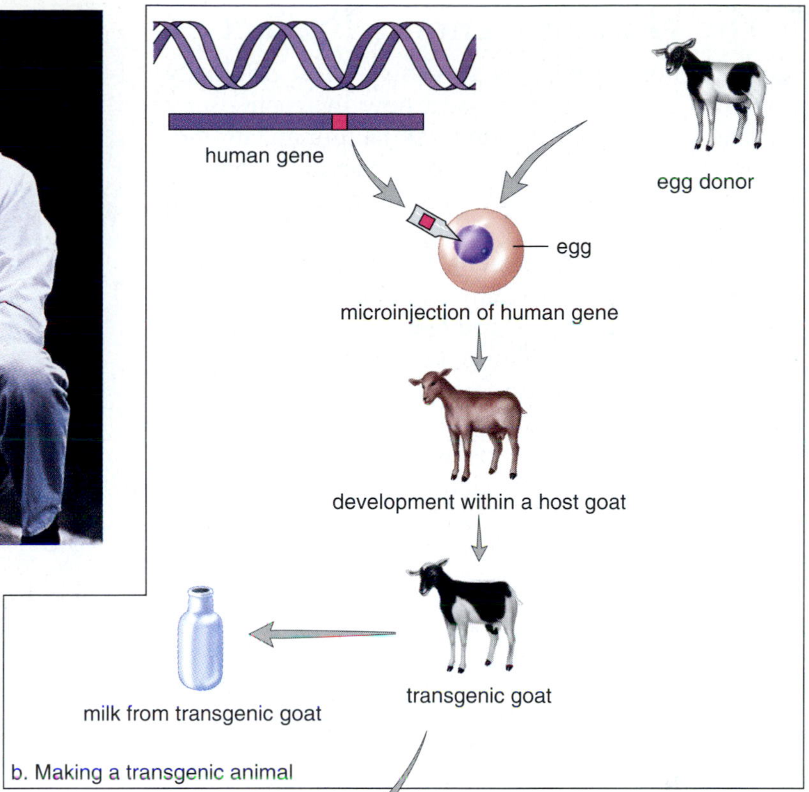

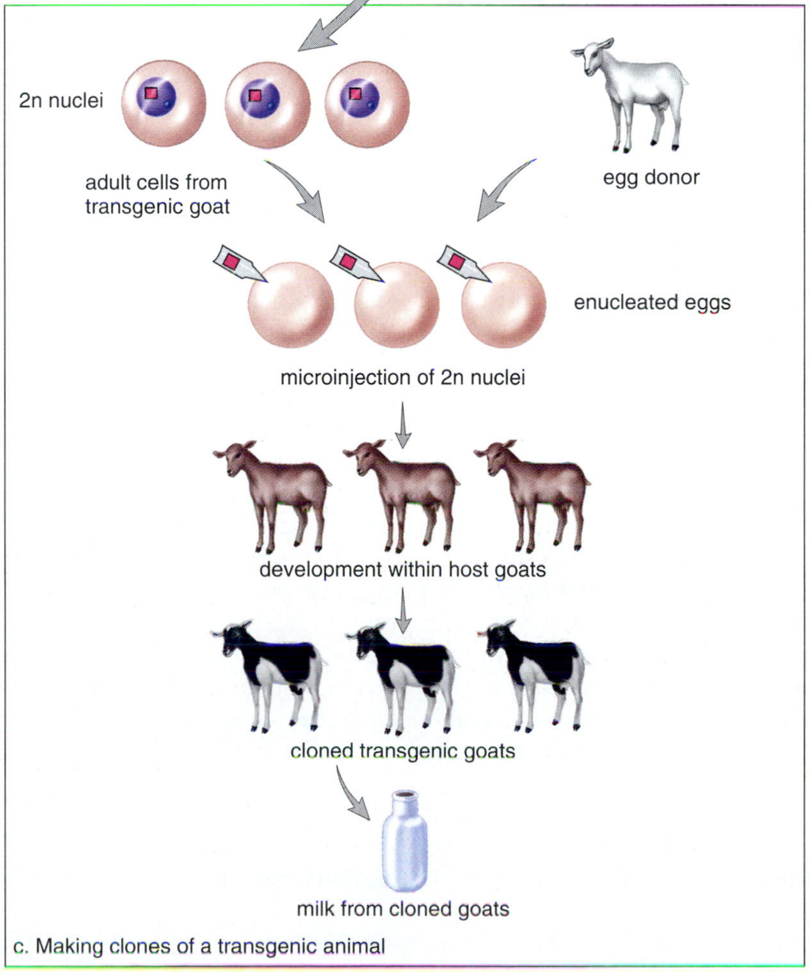

Figure 17.8 Genetically engineered animals. **a.** This goat is genetically engineered to produce antithrombin III, which is secreted in her milk. This researcher and many others are involved in the project. **b.** The procedure to produce a transgenic animal. **c.** The procedure to clone a transgenic animal.

In 1997, scientists at the Raslin Institute in Scotland announced that they achieved this feat and had produced a cloned sheep called Dolly. Since then calves and goats have also been cloned. Figure 17.8*c* shows that after enucleated eggs have been injected with 2n nuclei of adult cells, they can be coaxed to begin development. The offspring have the genotype and phenotype of the adult that donated the nuclei; therefore, the adult has been cloned. In the procedure that produced cloned mice, the 2n nuclei were taken from cumulus cells. Cumulus cells are those that cling to an egg after ovulation occurs. A specially prepared chemical bath was used to stimulate the eggs to divide and begin development. Now that scientists have a method to clone mammals, this procedure will undoubtedly be used routinely. In the United States, a presidential order prohibits the cloning of humans. But certain other countries are experimenting with the possibility.

Genetic engineering of animals has made much progress. Many firms are interested in gene pharming, the use of genetically engineered animals to produce pharmaceuticals in milk. Procedures have been developed to allow the cloning of these animals.

17.3 The Human Genome Project

The Human Genome Project is a massive effort originally funded by the U.S. government and now increasingly by U.S. pharmaceutical companies to map the human chromosomes. Many nonprofit and for profit biochemical laboratories about the world are now involved in the project which has two primary goals.

The first goal is to construct a genetic map of the human genome. The aim is to show the sequence of genes along the length of each type chromosome, such as depicted for the X chromosome in Figure 17.9. If the estimate of 1,000+ human genes is correct, each chromosome on average would contain about 50 alleles.

The map for each chromosome is presently incomplete, and in many instances scientists rely on the placement of RFLPs (see page 269). These sites eventually allow scientists to pinpoint disease-causing genes because a particular RFLP and a defective gene are often inherited together. For example, it is known that persons with Huntington disease have a unique site where a restriction enzyme cuts DNA. The test for Huntington disease relies on this difference from the normal.

The genetic map of a chromosome can be used not only to detect defective genes, but possibly also to tailor treatments to the individual. Only certain hypertension patients benefit from a low-salt diet, and it would be useful to know which patients these are. Myriad Genetics, a genome company, has developed a test for a mutant angiotensinogen gene because they want to see if patients with this mutation are the ones who benefit from a low-salt diet. Several other mutant genes have also been correlated with specific drug treatments (Table 17.1). One day the medicine you take might carry a label that it is effective only in persons with genotype #101!

The second goal is to construct a base sequence map. There are three billion base pairs in the human genome, and it's estimated it would take an encyclopedia of 200 volumes, each with 1,000 pages, to list all of these. Yet, this goal has been reached and all the chromosomes have been sequenced.

The methodology, thus far, has been to first chop up the genome into small pieces, each just 1,000 to 2,000 base pairs long. PCR instruments copy the pieces many times, and then an automatic DNA sequencer determines the order of the base pairs. You need many DNA copies because of the way the sequencer works. A computer program later strings the sequenced pieces together in the correct order by looking for base sequence overlaps between them. Instrumentation has gradually improved, and recently one scientist, J. Craig Venter, has founded a company which has now sequenced the entire genome.

Venter used what is called a whole-genome shotgun sequencing method. He worked with the entire human genome at once. Each overlapping fragment was about 5,000 bases long, and he sequenced the ends of each fragment using powerful new sequencing machines. Again, a computer program strung the fragments together by looking for overlapping regions. Why is Venter going off on his own, instead of participating in the worldwide effort by many laboratories and scientists to sequence the human genome? His backers expect to market the whole-genome database to subscribers, and to patent rare but pharmacologically interesting genes. Private enterprise is giving new impetus to the field now called genomics.

Knowing the base sequence of normal genes may make it possible one day to treat certain human ills by administering normal genes and/or their protein products to those who suffer from a genetic disease.

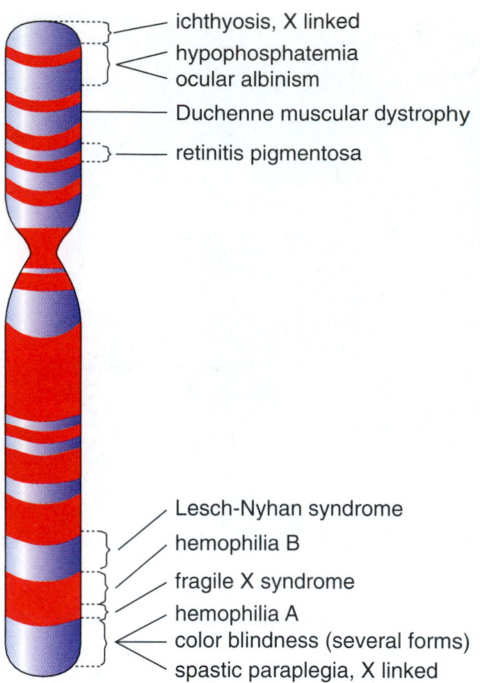

Figure 17.9 Genetic map of X chromosome.
The human X chromosome has been partially mapped, and this is the order of some of the genes now known to be on this chromosome.

Table 17.1

Customizing Drug Treatments

Mutant Gene for	Disease	Treatment
Apolipoprotein E	Alzheimer	Experimental Glaxo Wellcome drug
Cytochrome P-450	Cancer	Amonafide
Chloride gate	Cystic fibrosis	Pulmozyme
Dopamine receptor D4	Schizophrenia	Clozapine
Angiotensinogen	Hypertension	Low-salt diet

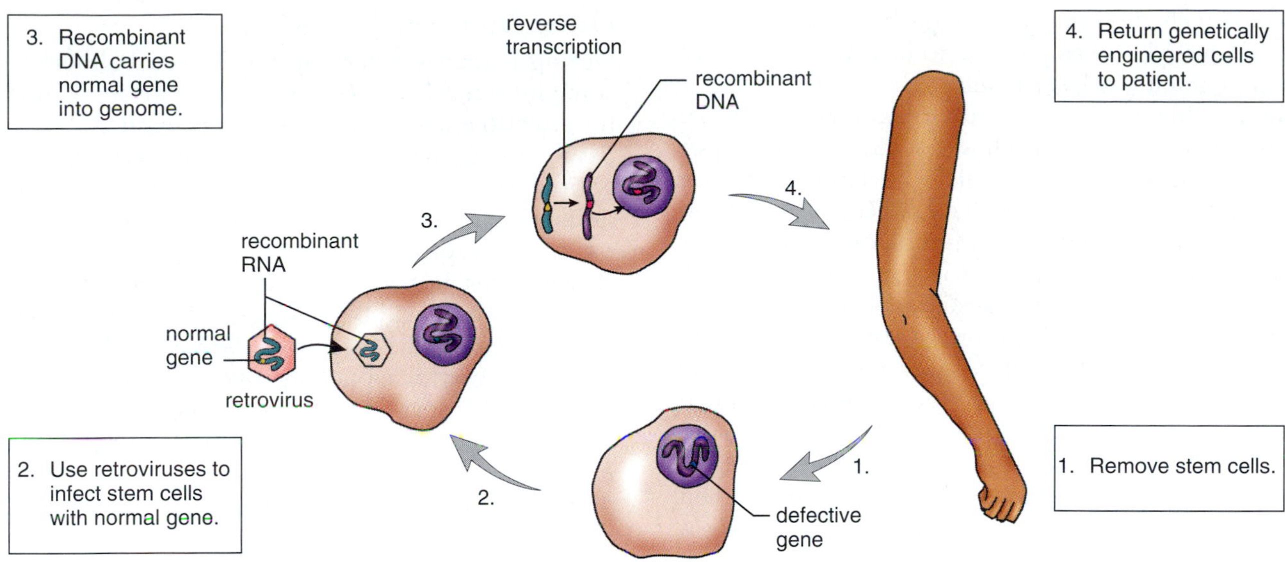
3. Recombinant DNA carries normal gene into genome.
reverse transcription
recombinant DNA
4. Return genetically engineered cells to patient.
4.
3.
recombinant RNA
normal gene
retrovirus
2. Use retroviruses to infect stem cells with normal gene.
2.
1.
defective gene
1. Remove stem cells.

Figure 17.10 **Ex vivo gene therapy in humans.**
Bone marrow stem cells are withdrawn from the body, a virus is used to insert a normal gene into them, and they are returned to the body.

17.4 Gene Therapy

Gene therapy is the insertion of genetic material into human cells for the treatment of a disorder. It includes procedures that give a patient healthy genes to make up for faulty genes and also includes the use of genes to treat various other human illnesses such as cancer and cardiovascular disease. Currently there are approximately 1,000 patients enrolled in nearly 200 approved gene therapy trials in the United States. There are ex vivo (outside the body) and in vivo (inside the body) methods of gene therapy.

Genetic Disorders

As an example of an ex vivo method of gene therapy, consider Figure 17.10, which describes a methodology for the treatment of children with severe combined immunodeficiency syndrome (SCID). These children lack an enzyme called adenosine deaminase (ADA) that is involved in the maturation of T and B cells, and therefore they are subject to life-threatening infections. Bone marrow stem cells are removed from the blood and infected with a retrovirus (RNA virus) that carries a normal gene for the enzyme. Then the cells are returned to the patient. Bone marrow stem cells are preferred for this procedure because they divide to produce more cells with the same genes. With an ex vivo method of genetic therapy, it is possible to test and make sure gene transfer has occurred before the cells are returned to the patient. Patients who have undergone this procedure do have a significant improvement in their immune function that is associated with a sustained rise in the level of ADA enzyme activity in the blood.

Among the many gene therapy trials, one is for the treatment of familial hypercholesterolemia, a condition that develops when liver cells lack a receptor for removing cholesterol from the blood. The high levels of blood cholesterol make the patient subject to fatal heart attacks at a young age. In a newly developed procedure, a small portion of the liver is surgically excised and infected with a retrovirus containing a normal gene for the receptor. Several patients have experienced a lowering of serum cholesterol levels following this procedure.

Cystic fibrosis patients lack a gene that codes for transmembrane carrier of the chloride ion. Patients often die due to numerous infections of the respiratory tract. An in vivo method of treatment is being tried. Liposomes—microscopic vesicles that spontaneously form when lipoproteins are put into a solution—have been coated with the gene needed to cure cystic fibrosis. Then the solution is sprayed into patients' nostrils. Due to limited gene transfer this methodology has not as yet been successful.

In a recent and surprising move, some researchers are investigating the possibility of directly correcting the base sequence of patients with a genetic disorder. The exact procedure has not yet been published.

Cancer

Chemotherapy in cancer patients often kills off healthy cells as well as cancer cells. In clinical trials, researchers have given genes to cancer patients that either make healthy cells more tolerant of chemotherapy or make tumors more vulnerable to it. In one ex vivo clinical trial, bone marrow stem cells from about 30 women with late-stage ovarian cancer were infected with a virus carrying a gene to make them more tolerant of chemotherapy. Once the bone marrow stem cells were protected, it was possible to increase the level of chemotherapy to kill the cancer cells.

In a recent in vivo study, a retrovirus carrying a normal *p53* gene was injected directly into the tumor of lung cancer patients. This gene, which helps regulate the cell cycle and also brings about apoptosis in cells with damaged DNA, is often mutated in tumor cells. No cures were reported but the tumors shrank in three patients and stopped growing in three others. Expression of *p53* is only needed for a short time, and an elevated amount in normal cells seems to do them no harm.

Some investigators prefer the use of adenoviruses rather than retroviruses. Adenoviruses can be produced in much larger quantities, and they are active even when they are not dividing. (Retroviruses integrate DNA into the host chromosome where it is inactive unless replication occurs.) When adenoviruses infect a cell, they normally produce a protein that binds to *p53* and inactivates it. In a cleverly designed procedure, researchers genetically engineered adenoviruses to lack the gene that produces this protein. Therefore, the virus will kill tumor cells that lack *p53* but not healthy cells that have a *p53* gene. Clinical trials are underway, and it is expected that the virus will spread through the cancer, killing the cancer cells, and stopping when it reaches normal cells.

Other Illnesses

During coronary artery angioplasty, a balloon catheter is sometimes used to open up a closed artery. Unfortunately, the artery has a tendency to close up again. But investigators have come up with a new procedure. The balloon is coated with a plasmid that contains a gene for VEGF (vascular endothelial growth factor). The expression of the gene, which promotes the proliferation of blood vessels to bypass the obstructed area, has been observed in at least one patient.

Perhaps it will be possible also to use in vivo therapy to cure hemophilia, diabetes, Parkinson disease, or AIDS. To treat hemophilia, patients could get regular doses of cells that contain normal clotting-factor genes. Or such cells could be placed in organoids, artificial organs that can be implanted in the abdominal cavity. To cure Parkinson disease, dopamine-producing cells could be grafted directly into the brain.

The Human Genome Project produces information useful to gene therapists. Researchers are envisioning all sorts of ways to cure human genetic disorders as well as many other types of illnesses.

Connecting Concepts

Basic research into the nature and organization of genes in various organisms allowed geneticists to produce recombinant DNA molecules. A knowledge of transcription and translation made it possible for scientists to manipulate the expression of genes in foreign organisms. These breakthroughs have spurred a biotechnology revolution. One result is that bacteria and eukaryotic cells are now used to produce vaccines, hormones, and growth factors for use in humans. Today, plants and animals are also engineered to produce a product or have characteristics desired by humans. In addition, biotechnology offers the promise of treating and even someday curing human genetic disorders such as muscular dystrophy, cystic fibrosis, hemophilia, and many others. Mapping the human genome relies on both traditional linkage data and biotechnology techniques to discover the locus of many other genetic disorders. This is the first step toward the gene therapy that may result in a cure of various genetic disorders.

Some people are concerned that transgenic organisms may spread out of control and wreak ecological havoc, much as plants and animals new to an area sometimes do. Or they worry that transgenic plants may pass on their traits, such as resistance to pesticides or herbicides, to weeds that will then prosper as never before. Thus far a "superweed" has not occurred but that does not mean it could not happen in the future. The alteration of organisms as a result of genetic changes is one of the definitions of evolution, a topic that will be discussed in detail in the next section of the text.

Summary

17.1 Cloning of a Gene

Two methods are currently available for making copies of DNA. Recombinant DNA contains DNA from two different sources. A restriction enzyme is used to cleave plasmid DNA and to cleave foreign DNA. The "sticky ends" produced facilitate the insertion of foreign DNA into vector DNA. The foreign gene is sealed into the vector DNA by DNA ligase. Both bacterial plasmids and viruses can be used as vectors to carry foreign genes into bacterial host cells. A genomic library can be used as a source of genes to be cloned. A radioactive or fluorescent probe is used to identify the location of a single gene among the cloned fragments of an organism's DNA.

If just a portion of the genome is of interest or available, PCR uses the enzyme DNA polymerase to make multiple copies of target DNA. Analysis of DNA segments following PCR can involve gel electrophoresis to identify the DNA as belonging to a particular organism or can involve determining the base sequence of the DNA segment. The entire genome is used for fingerprinting. First, restriction enzymes are used to fragment DNA. Then the use of probes is necessary in order to see a pattern following gel electrophoresis.

17.2 Biotechnology Products

Transgenic organisms are ones that have had a foreign gene inserted into them. Genetically engineered bacteria, agricultural plants, and farm animals now produce commercial products of interest to humans, such as hormones and vaccines. Bacteria secrete the product. The seeds of plants and the milk of animals contain the product.

Transgenic bacteria have been produced to promote the health of plants, perform bioremediation, extract minerals, and produce chemicals. Transgenic agricultural plants which have been engineered to resist herbicides and pests are commercially available. Transgenic animals have been given various genes, in particular the one for bovine growth hormone (bGH). Cloning of animals is now possible.

17.3 The Human Genome Project

The Human Genome Project has two goals. The first goal is to construct a genetic map which will show the sequence of the genes on each chromosome. Although the known location of genes is expanding, the genetic map still contains many RFLPs (restriction fragment length polymorphisms) instead as the first step toward finding more genes. The second goal of the project is to sequence the bases on all the chromosomes. To do this researchers rely on PCR and automatic sequencing instruments. As the sequencers have improved so has the speed with which DNA is sequenced. The hope is that knowing the location and sequence of bases in a gene will promote the possibility of gene therapy.

17.4 Gene Therapy

Gene therapy is used to correct the genotype of humans and to cure various human ills. There are ex vivo and in vivo methods of gene therapy. Gene therapy has apparently helped children with SCID to lead a normal life; the treatment of cystic fibrosis has been less successful. A number of imaginative therapies are being employed in the war against cancer and other human illness such as cardiovascular disease.

Reviewing the Chapter

1. What is the methodology for producing recombinant DNA to be used in gene cloning? 266
2. Bacteria can be used to clone a gene or produce a product. Explain. 266–67
3. What is a genomic library, and how do you locate a gene of interest in the library? 267
4. What is the polymerase chain reaction (PCR), and how is it carried out to produce multiple copies of a DNA segment? 268
5. What is DNA fingerprinting, a process that utilizes the entire genome? 269
6. What are some practical applications of DNA segments analysis following PCR? 268–69
7. For what purposes have bacteria, plants, and animals been genetically altered? 270–73
8. Explain how and why transgenic animals that secrete a product are often cloned. 272–73
9. Explain the two primary goals of the Human Genome Project. What are the possible benefits of the project? 274
10. Explain and give examples of ex vivo and in vivo gene therapies in humans. 275

Testing Yourself

Choose the best answer for each question.

1. Which of these is a true statement?
   a. Both plasmids and viruses can serve as vectors.
   b. Plasmids can carry recombinant DNA but viruses cannot.
   c. Vectors carry only the foreign gene into the host cell.
   d. Only gene therapy uses vectors.
   e. Both a and d are correct.
2. Which of these is a benefit to having insulin produced by biotechnology?
   a. It is just as effective.
   b. It can be mass-produced.
   c. It is nonallergenic.
   d. It is less expensive.
   e. All of these are correct.
3. Restriction fragment length polymorphisms (RFLPs)
   a. are achieved by using restriction enzymes.
   b. identify individuals genetically.
   c. are the basis for DNA fingerprints.
   d. can be subjected to gel electrophoresis.
   e. All of these are correct.
4. Which of these would you not expect to be a biotechnology product?
   a. vaccine
   b. modified enzyme
   c. DNA probe
   d. protein hormone
   e. steroid hormone
5. What is the benefit of using a retrovirus as a vector in gene therapy?
   a. It is not able to enter cells.
   b. It incorporates the foreign gene into the host chromosome.
   c. It eliminates a lot of unnecessary steps.
   d. It prevents infection by other viruses.
   e. Both b and c are correct.
6. Gel electrophoresis
   a. cannot be used on nucleotides.
   b. measures the size of plasmids.
   c. tells whether viruses are infectious.
   d. measures the charge and size of proteins and DNA fragments.
   e. All of these are correct.
7. Put these phrases in the correct order to form a plasmid carrying recombinant DNA.

   1 use restriction enzymes
   2 use DNA ligase
   3 remove plasmid from parent bacterium
   4 introduce plasmid into new host bacterium

   a. 1, 2, 3, 4
   b. 4, 3, 2, 1
   c. 3, 1, 2, 4
   d. 2, 3, 1, 4
8. Which of these is incorrectly matched?
   a. xenotransplantation—source of organs
   b. protoplast—plant cell engineering
   c. RFLPs—DNA fingerprinting
   d. DNA polymerase—PCR
   e. DNA ligase—mapping human chromosomes
9. The restriction enzyme called EcoRI has cut double-stranded DNA in this manner. The piece of foreign DNA to be inserted has what bases from left to the right?

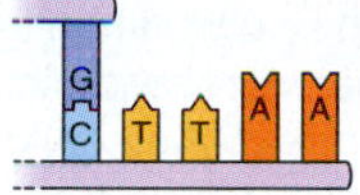

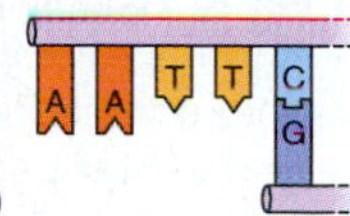

10. The following drawings pertain to gene therapy. Label the drawings, using these terms: retrovirus, recombinant RNA (twice), defective gene, recombinant DNA, reverse transcription, and human genome.

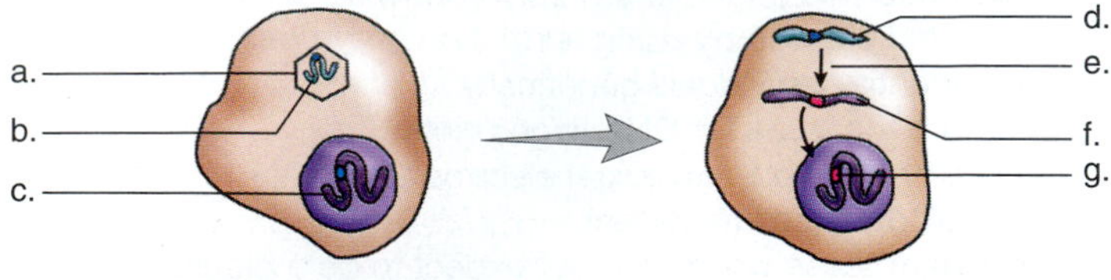

Thinking Scientifically

1. cDNA libraries contain only expressed DNA sequences. Therefore a cDNA library produced for a liver cell will contain genes unique to the cell. How could a CDNA library for a liver cell be used to acquire a complete copy of a liver-cell gene from a DNA library that contains the entire genome? A complete gene contains the promoter and introns.
2. There has been much popular interest in recreating extinct animals from DNA obtained from various types of fossils. However, such DNA is always badly degraded, consisting of extremely short pieces. Even if one hypothetically had 10 intact genes from a dinosaur, why might it still be impossible to create a dinosaur?

Bioethical Issue

Somatic gene therapy attempts to treat or prevent human illnesses. Some day, for example, it may be possible to give children who have cystic fibrosis, Huntington disease, or any other genetic disorder a normal gene to make up for the inheritance of a faulty gene. And gene therapy is even more likely for treatment of diseases like cancer, AIDS, and heart disease.

Germ-line gene therapy is the expression now being used to mean the use of gene therapy solely to improve the traits of an individual. It would be the genetic equivalent of procedures like body building, liposuction, or hair transplants. If genetic interference occurred early—that is, on the eggs, sperm, or embryo—it's possible that it would indeed affect the germ line—that is, all the future descendants of the individual.

How might germ-line gene therapy become routine? Consider this scenario. Presently, a gene for VEGF (vascular endothelial growth factor), a protein produced by cells to grow new blood vessels, is being used to treat atherosclerosis. Improved arterial circulation in the legs of some recent patients did away with the threat of possible amputation. This same gene is now being considered for the treatment of blocked coronary arteries. There may be instances, though, in which healthy people want to grow new blood vessels for enhancement purposes. Runners might want to improve their circulation in order to win races, and parents might think that increased circulation to the brain would increase the intelligence of their children. Bioethicist Eric Juengst of Case Western thinks that as a society, there is nothing we can now do to prevent us from crossing the line between the use of gene therapy for therapeutic purposes and its use for enhancement reasons. Do you have any concerns about gene therapy for enhancement reasons, such as its cost and general availability to everyone?

Understanding the Terms

bacteriophage 267
clone 266
complementary DNA (cDNA) 267
DNA fingerprinting 269
DNA ligase 266
gene therapy 275
genetic engineering 270
genome 267
genomic library 267
plasmid 266
polymerase chain reaction (PCR) 268
probe 267
recombinant DNA (rDNA) 266
restriction enzyme 266
restriction fragment length polymorphism (RFLP) 269
tissue engineering 271
transgenic organism 270
vector 266
xenotransplantation 271

Match the terms to these definitions:

a. ______________ Bacterial enzyme that stops viral reproduction by cleaving viral DNA; used to cut DNA at specific points during production of recombinant DNA.
b. ______________ Free-living organisms in the environment that have had a foreign gene inserted into them.
c. ______________ Known sequences of DNA that are used to find complementary DNA strands; can be used diagnostically to determine the presence of particular genes.
d. ______________ Production of identical copies; in genetic engineering, the production of many identical copies of a gene.
e. ______________ Self-duplicating ring of accessory DNA in the cytoplasm of bacteria.

Web Connections

Exploring the Internet

http://www.mhhe.com/biosci/genbio/mader
(click on *Biology 7/e*)

The *Biology 7/e* Online Learning Center provides many resources for studying the material in this chapter including links to the following sites:

The Howard Hughes Medical Institute's site, Blazing a Genetic Trail, focuses on the gene mutations that cause disease as they review recent genetic research and its applications.

http://www.hhmi.org/GeneticTrail/

The Human Genome Project describes the history of the project and progress made to date.

http://www.nhgri.nih.gov/HGP/

National Centre for Biotechnology Education. An overview of the origins and nature of biotechnology and enzymes used in food production can be found in Food Biotechnology: past, present, and future. This well-written, comprehensive website includes examples of current plant biotechnology, and plants that are currently undergoing field trials. Some of the concerns for environmental safety are explained.

http://www.ncbe.reading.ac.uk/

Biotechnology and Scientific Services Home Page. Information is found here on subjects ranging from GM (genetically modified) foods to regulation in the industry.

http://www.aphis.usda.gov/biotech/index.html

Recombinant DNA Chapter. This MIT hypertextbook chapter on immunology contains a wealth of information, particularly on the methodology involved in biotechnology.

http://esg-www.mit.edu:8001/esgbio/rdna/rdnadir.html

Further Readings for Part II

Ashley, M. January/February 1999. Molecular conservation genetics. *American Scientist* 87(1):28. Article discusses how knowing the structure of DNA in endangered species can help protect them.

Berns, M. W. April 1998. Laser scissors and tweezers. *Scientific American* 278(62):4. New laser techniques allow manipulation of chromosomes and other structures inside cells.

Borek, C. November/December 1997. Antioxidants and cancer. *Science & Medicine* 4(6):52. The importance of supplemental antioxidant vitamins depends on factors such as diet and lifestyle.

Curiel, T. September/October 1997. Gene therapy: AIDS-related malignancies. *Science & Medicine* 4(5):4. The field of AIDS-related gene therapies is advancing.

Galili, U. September/October 1998. Anti-Gal antibody prevents xenotransplantation. *Science & Medicine* 5(5):28. Prevention of interaction of the anti-Gal antibody with pig cells is necessary to the progress of xenotransplantation.

Garnick, M. B., and Fair, W. R. December 1998. Combating prostate cancer. *Scientific American* 279(6):74. Article details the recent developments in diagnosis and treatment of prostate cancer.

Glausiusz, J. May 1998. The great gene escape. *Discover* 19(5):90. Genes from genetically engineered plants can escape from crops into the wild, causing resistance in the wild plant.

Glausiusz, J. January 1999. The genes of 1998. *Discover* 20(1):33. Nine important human genes that were identified in 1998 through the Human Genome Project are examined.

Goldberg, J. April 1998. A head full of hope. *Discover* 19(4):70. A new gene therapy for killing brain cancer cells is presented.

Hawley, R. S., and Mori, C. A. 1999. *The human genome: A user's guide.* San Diego: Academic Press. This advanced text focuses on the genetics of human development, and covers other relevant genetic topics.

Jordan, V. C. October 1998. Designer estrogens. *Scientific American* 279(4):60. Selective estrogen receptor modulators may protect against breast and endometrial cancers, osteoporosis, and heart disease.

Kher, U. January 1998. A man-made chromosome. *Discover* 18(1):40. Researchers announce a promising new gene carrier, a human artificial chromosome, for use in gene therapy.

Miller, R. V. January 1998. Bacterial gene swapping in nature. *Scientific American* 278(1):66. The study of the process of DNA exchange between bacteria can help limit the risks of releasing genetically engineered microbes into the environment.

Moxon, E. R., and Wills, C. January 1999. DNA microsatellites: Agents of evolution? *Scientific American* 280(1):94. Repetitive DNA sequences may determine how an organism, such as a bacterium, adapts to its environment.

Nielson, P. E. September/October 1998. Peptide nucleic acids. *Science & Medicine* 5(5):48. Peptide nucleic acids mimic DNA and can substitute for DNA in gene therapy.

O'Brochta, D. A., and Atkinson, P. W. December 1998. Building a better bug. *Scientific American* 279(6):90. Transgenic insect technology could decrease pesticide use, and prevent certain infectious diseases. Article discusses the production of a transgenic insect.

Pennisi, E. 13 November 1998. Training viruses to attack cancers. *Science* 282(5392):1244. Certain viruses can replicate in and kill cancer cells, but leave normal tissue intact.

Plomerin, R., and DeFries, J. C. May 1998. The genetics of cognitive abilities and disabilities. *Scientific American* 278(5):62. The search is underway for the genes involved in cognitive abilities and disabilities, including dyslexia.

Pool, R. May 1998. Saviors. *Discover* 19(5):52. Genetic engineering may make animal organs compatible for human transplants.

Scientific American editors. 276(6):95. June 1997. Special report: Making gene therapy work. Obstacles must be overcome before gene therapy is ready for widespread use.

Scientific American 280(4):59–89. April 1999. The promise of tissue engineering. Much of the issue examines the hopes and challenges of tissue engineering for use in gene therapy and for the growth of new organs.

Scientific American Special Issue. 275(3). September 1996. What you need to know about cancer. The entire issue is devoted to the causes, prevention, and early detection of cancer, and cancer therapies—conventional and future.

Stix, G. October 1997. Growing a new field. *Scientific American* 277(4):15. Tissue engineers try to grow organs in the laboratory.

Stone, R. April 1999. Cloning the woolly mammoth. *Discover* 20(4):56. Researchers are searching for viable woolly mammoth sperm, with which they hope to impregnate an Asian elephant.

Van Noorden, C. J. F., et al. March/April 1998. Metastasis. *American Scientist* 86(2):130. The mechanisms by which cancer cells metastasize are discussed.

Velander, W. H., et al. January 1997. Transgenic livestock as drug factories. *Scientific American* 276(1):70. Farm animals can be bred to produce quantities of medicinal proteins in their milk.

Vogel, S. April 1999. Why we get fat. *Discover* 20(4):94. The genetic basis of obesity is examined.

Wills, C. January 1998. A sheep in sheep's clothing? *Discover* 18(1):22. Some pros and cons of cloning are discussed.

Wilmut, I. December 1998. Cloning for medicine. *Scientific American* 279(6):58. Cloning holds many benefits for the advancement of medical science and animal husbandry.

Winkonkal, N. M., and Brash, D. E. September/October 1998. Squamous cell carcinoma. *Science & Medicine* 5(5):18. Mutations of tumor-suppressor gene *p53* are commonly found in squamous cell carcinomas.

part iii

Evolution

Evolution refers to both descent with modification and adaptation to the environment. Descent from a common ancestor explains the unity of life—living things share a common chemistry and cellular structure because they are all descended from the same original source. Each type of living thing has a history that can be traced by way of the fossil record and discerned from a comparative study of other living things. Human evolution can also be understood by the application of these evolutionary principles.

Adaptation to a unique environment explains the diversity of life. Natural selection is a mechanism that results in adaptation to the environment. Individuals with variations that make them better adapted have more offspring than those who are not as well adapted. In that way certain characteristics become more common among a particular group of organisms. Each species has its own unique, evolved solutions to life's problems such as acquiring nutrients, finding a mate, and reproducing.

Darwin and Evolution

chapter **18**

chapter concepts

18.1 History of the Theory of Evolution

- Evolution has two aspects: descent from a common ancestor and adaptation to the environment. 283
- In the eighteenth century, scientists became especially interested in classifying and understanding the relationship among all the many forms of present and past life. 283
- Gradually, in the eighteenth century, scientists began to gather evidence and accept the idea that life-forms do change over time. Explanations for change varied. 284

18.2 Darwin's Theory of Evolution

- Charles Darwin's trip around the Southern Hemisphere aboard the HMS *Beagle* provided him with evidence that the earth is very old and that evolution does occur. 285
- Both Darwin and Alfred Wallace proposed natural selection as a mechanism by which adaptation to the environment takes place. This mechanism is consistent with our present-day knowledge of genetics. 288–89, 291

18.3 Evidence for Evolution

- The fossil record, biogeography, comparative anatomy, and comparative biochemistry support a hypothesis of common descent. 292

The modern horse, Equus, and its oligocene ancestor, *Mesohippus*.

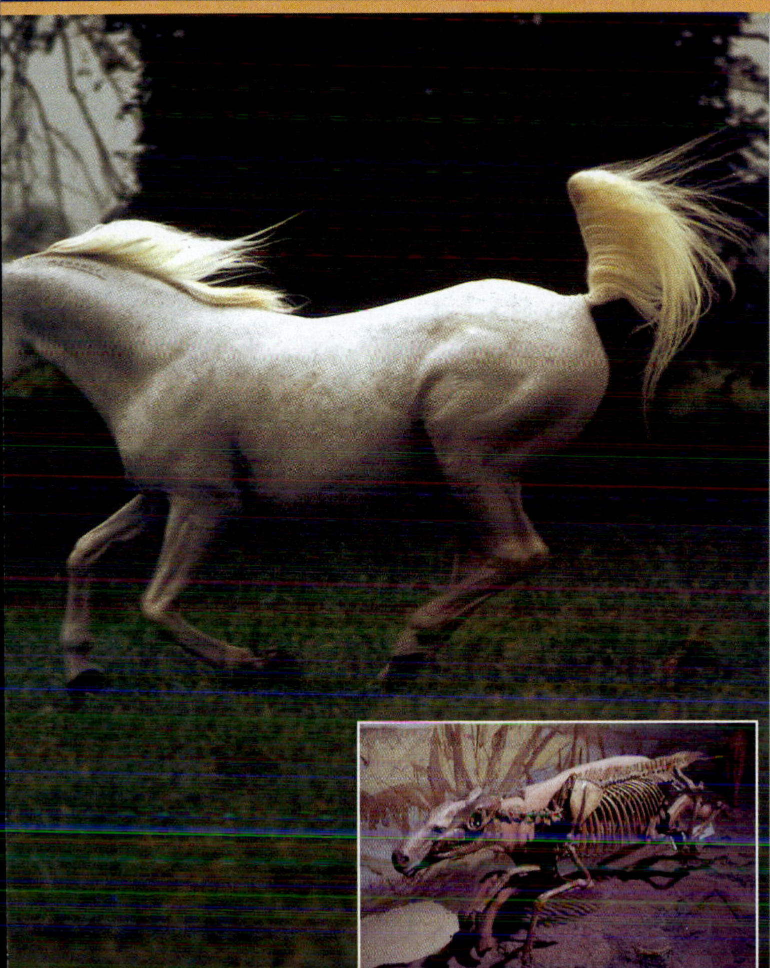

Modern geologists believe the earth is more than 4 billion years old and that life began about 3.5 billion years ago. From simple, unicellular organisms, new life-forms arose and changed in response to environmental pressures, producing the past and present biodiversity. Prior to the 1800s, however, most people thought the earth was only a few thousand years old. They also believed that species were specially created and fixed in time. Change was explained by the notion of global catastrophes—mass extinctions and repopulations of species—had occurred periodically throughout history.

Into this prevailing climate came Charles Darwin, Alfred Wallace, and other innovative minds who changed the field of biology forever. On a five-year ocean voyage, Darwin observed many diverse life-forms, and he eventually concluded that species evolve in response to their environment. Further, all living things share common characteristics because they have a common ancestry. Because the concept of evolution challenged the widely accepted biblical account of creation, acceptance required an intellectual revolution of great magnitude.

18.1 History of the Theory of Evolution

Charles Darwin was only 22 in 1831 when he accepted the position of naturalist aboard the HMS *Beagle,* a British naval ship about to sail around the world (Fig. 18.1). Darwin's major mission was to expand the navy's knowledge of natural resources (e.g., water and food) in foreign lands. The captain was also hopeful that Darwin would find evidence of the biblical account of creation. The results of Darwin's observations were just the opposite, however, as we shall examine later in the chapter.

Table 18.1 tells us that prior to Darwin, people had an entirely different way of looking at the world. Their mindset was determined by deep-seated beliefs held to be intractable truths. To turn from these beliefs and accept the

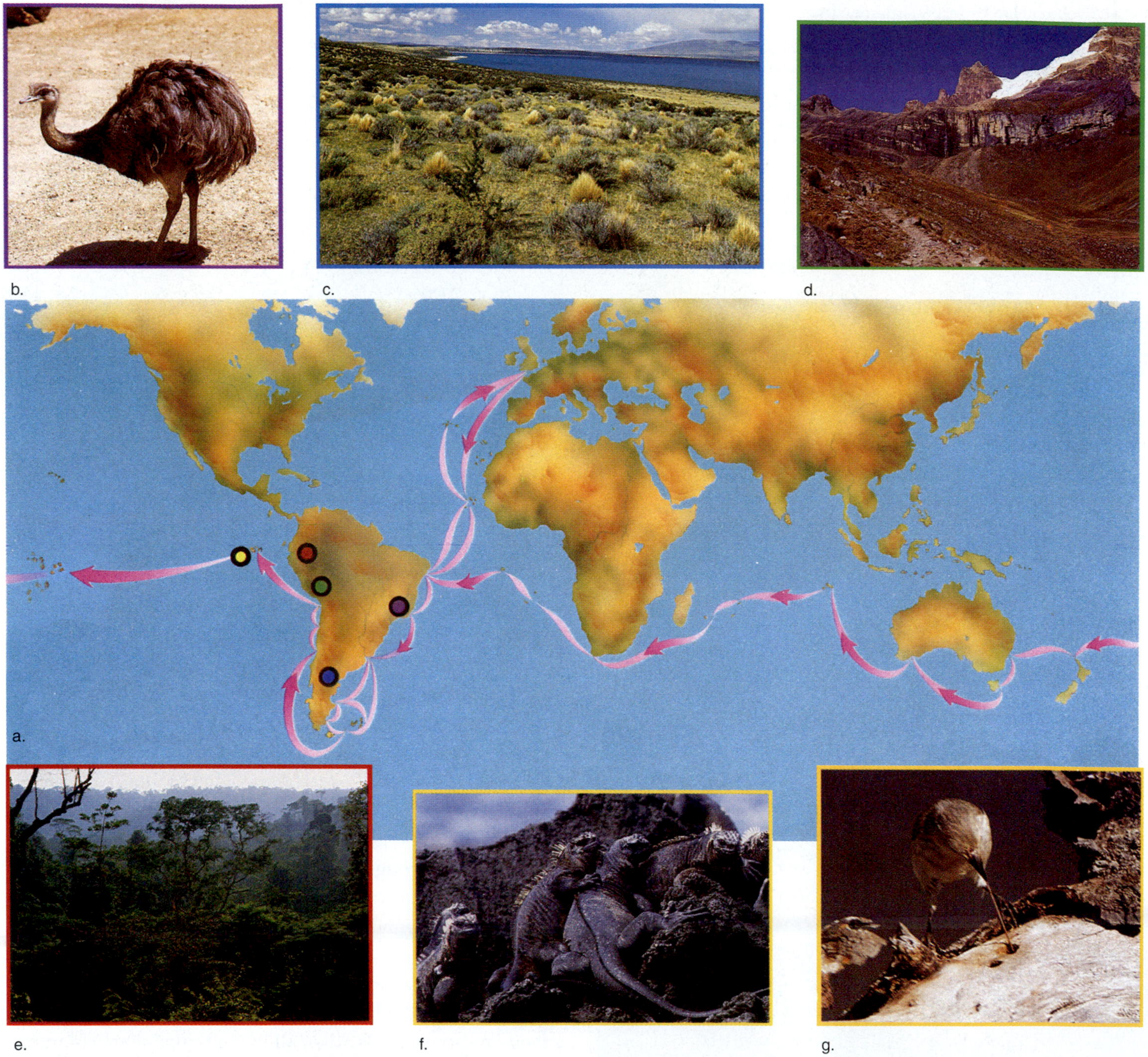

Figure 18.1 **Voyage of the HMS *Beagle.***
a. Map shows the journey of the HMS *Beagle* around the world. Notice the encircled colors are keyed to the frames of the photographs, which show us what Charles Darwin may have observed. **b.** As Darwin traveled along the east coast of South America, he noted that a bird called a rhea looked like the African ostrich. **c.** The sparse vegetation of the Patagonian Desert is in the southern part of the continent. **d.** The Andes Mountains of the west coast, with strata containing fossilized animals. **e.** The lush vegetation of a rain forest. **f.** On the Galápagos Islands, marine iguanas have large claws to help them cling to rocks and blunt snouts for eating seaweed. **g.** Galápagos finches are specialized to feed on various foods.

Table 18.1

Contrast of Worldviews

Pre-Darwinian View	Post-Darwinian View
1. Earth is relatively young—age is measured in thousands of years.	1. Earth is relatively old—age is measured in billions of years.
2. Each species is specially created; species don't change, and the number of species remains the same.	2. Species are related by descent—it is possible to piece together a history of life on earth.
3. Adaptation to the environment is the work of a creator, who decided the structure and function of each type of organism. Any variations are imperfections.	3. Adaptation to the environment is the interplay of random variations and environmental conditions.
4. Observations are supposed to substantiate the prevailing worldview.	4. Observation and experimentation are used to test hypotheses, including hypotheses about evolution.

Darwinian view of the world required an intellectual revolution of great magnitude. This revolution was fostered by changes in both the scientific and the social realms. Here, we will only touch on a few of the scientific contributions that helped bring about the new worldview.

Although it is often believed that Darwin forged this change in worldview by himself, biologists during the preceding century had slowly begun to accept the idea of **evolution** [L. *evolutio,* an unrolling]; that is, that species change with time. Evolution explains the unity and diversity of life. Living things share common characteristics because they have a common ancestry. Living things are diverse because each species is adapted to its habitat and way of life. We will see that the history of evolutionary thought is a history of ideas about descent and adaptation. Darwin himself used the expression "descent with modification," by which he meant that as descent occurs through time so does diversification. He saw the process of adaptation as the means by which diversification comes about.

Mid-Eighteenth-Century Contributions

Taxonomy, the science of classifying organisms, was an important endeavor during the mid-eighteenth century. Chief among the taxonomists was Carolus Linnaeus (1707–78), who gave us the binomial system of nomenclature (a two-part name for species, such as *Homo sapiens*) and who developed a system of classification for all known plants. Linnaeus, like other taxonomists of his time, believed in the fixity of species. Each species had an "ideal" structure and function and also a place in the *scala naturae,* a sequential ladder of life. The simplest and most material of beings was on the lowest rung of the ladder, and the most complex and most spiritual of beings was on the highest rung. Human beings occupied the last rung of the ladder.

These ideas, which were consistent with Judeo-Christian teachings of special creation, can be traced to the works of the famous Greek philosophers Plato and Aristotle. Plato said that every object on earth was an imperfect copy of an ideal form, which can be deduced upon reflection and study. To Plato, individual variations were imperfections that only distract the observer. Aristotle saw that organisms were diverse, and some were more complex than others. His belief that all organisms could be arranged in order of increasing complexity became the scala naturae just described.

Figure 18.2 Linnaeus.
For the most part, Linnaeus believed that each species was created separately and that classification should reveal God's divine plan. Variations were imperfections of no real consequence.

Linnaeus (Fig. 18.2) and other taxonomists wanted to describe the ideal characteristics of each species and also wanted to discover the proper place for each species in the scala naturae. Therefore, Linnaeus did not even consider the possibility of evolutionary change for most of his working life. There is evidence, however, that he did eventually perform hybridization experiments, which made him think that a species might change with time.

Georges-Louis Leclerc, better known by his title, Count Buffon (1707–88), was a French naturalist who devoted many years of his life to writing a 44-volume natural history that described all known plants and animals. He provided evidence of descent with modification, and he even speculated on various mechanisms such as environmental influences, migration, geographical isolation, and the struggle for existence. Buffon seemed to vacillate, however, as to whether or not he believed in evolutionary descent, and often he professed to believe in special creation and the fixity of species.

Erasmus Darwin (1731–1802), Charles Darwin's grandfather, was a physician and a naturalist. His writings on both botany and zoology contained many comments, although they were mostly in footnotes and asides, that suggested the possibility of common descent. He based his conclusions on changes undergone by animals during development, artificial selection by humans, and the presence of vestigial organs (organs that are believed to have been functional in an ancestor but are reduced and nonfunctional in a descendant). Like Buffon, Erasmus Darwin offered no mechanism by which evolutionary descent might occur.

Figure 18.3 **A mastodon.**
Although mastodons are known only from the fossil record, Cuvier reconstructed the structure of these animals when provided only with a single bone. He said these animals were shorter than modern elephants but had massive, pillarlike legs. The skull was lower and flatter and of generally simpler construction than in modern elephants.

Figure 18.4 **Inheritance of acquired characteristics.**
Jean-Baptiste de Lamarck is most famous for suggesting that the environment influences heritable traits. The long neck of the giraffe, he said, is due to continual stretching to reach food. Each generation of giraffes stretched a little farther, and this was passed to the next generation. With the advent of modern genetics, it became possible to explain why inheritance of acquired characteristics would not be possible.

Late Eighteenth-Century Contributions

Cuvier and Catastrophism

In addition to taxonomy, comparative anatomy was of interest to biologists prior to Darwin. Explorers and collectors traveled the world and brought back not only currently existing species to be classified but also fossils (remains of once-living organisms) to be studied. Georges Cuvier (1769–1832), a distinguished vertebrate (animals with backbones) zoologist, was the first to use comparative anatomy to develop a system of classifying animals. He also founded the science of **paleontology** [Gk. *palaios,* ancient, old, and *ontos,* having existed; -logy, study of, from *logikos,* rational, sensible], the study of fossils, and was quite skilled at using fossil bones to deduce the structure of an animal (Fig. 18.3).

Because Cuvier was a staunch advocate of special creation and fixity of species, he faced a real problem when a particular region showed a succession of life-forms in the earth's strata (layers). To explain these observations, he hypothesized that a series of local catastrophes or mass extinctions had occurred whenever a new stratum of that region showed a new mix of fossils. After each catastrophe, the region was repopulated by species from surrounding areas, and this accounted for the appearance of new fossils in the new stratum. The result of all these catastrophes was the appearance of change with time. Some of his followers even suggested that there had been worldwide catastrophes and that after each of these events, God created new sets of species. This explanation of the history of life came to be known as **catastrophism** [Gk. *katastrophe,* calamity, misfortune].

Lamarck's Theory of Evolution

Jean-Baptiste de Lamarck (1744–1829) was the first biologist to believe that evolution does occur and to link diversity with adaptation to the environment. Lamarck's ideas about descent were entirely different from those of Cuvier, perhaps because he was an invertebrate (animals without backbones) zoologist. Lamarck concluded, after studying the succession of life-forms in strata, that more complex organisms are descended from less complex organisms. He mistakenly said, however, that increasing complexity was the result of a natural force—a desire for perfection—that was inherent in all living things.

To explain the process of adaptation to the environment, Lamarck supported the idea of **inheritance of acquired characteristics**—that the environment can bring about inherited change. One example that he gave—and for which he is most famous—is that the long neck of a giraffe developed over time because animals stretched their necks to reach food high in trees and then passed on a long neck to their offspring (Fig. 18.4). The theory of the inheritance of acquired characteristics has never been substantiated by experimentation. The molecular mechanism of inheritance explains why. Phenotypic changes acquired during an organism's lifetime do not result in genetic changes that can be passed to subsequent generations.

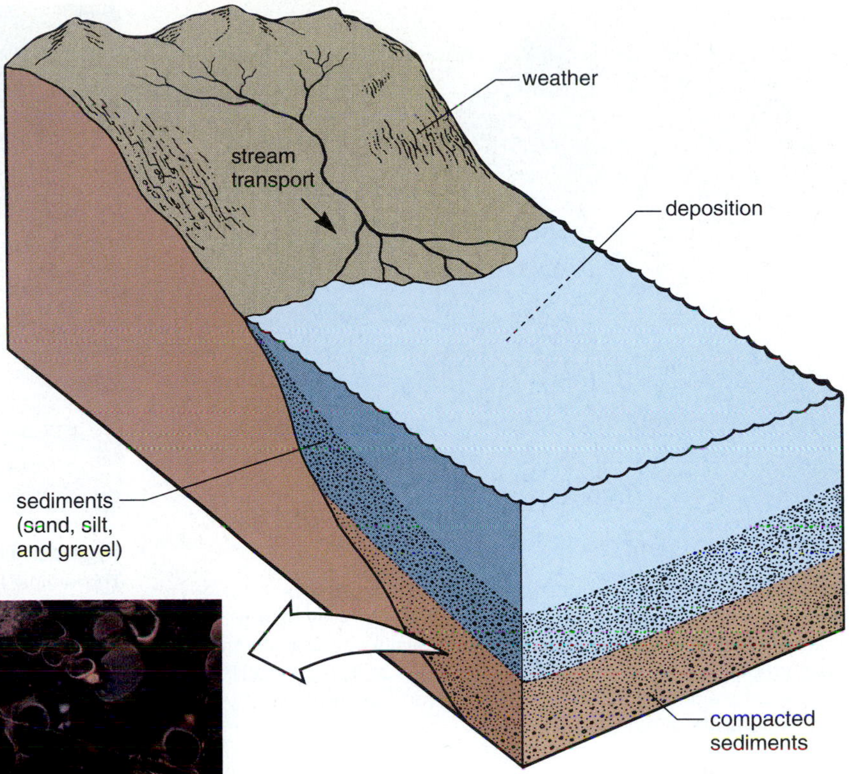

Figure 18.5 Formation of sedimentary rock.
a. This diagram shows how water brings sediments into the sea; the sediments then become compacted to form sedimentary rock. Fossils are often trapped in this rock, and as a result of a later geological upheaval, the rock may be located on land.
b. Fossil remains of freshwater snails, *Turritella,* in sedimentary rocks.

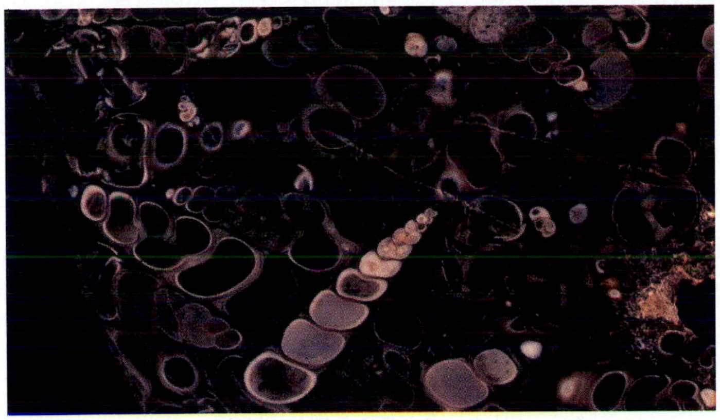

18.2 Darwin's Theory of Evolution

When Darwin signed on as naturalist aboard the HMS *Beagle,* he had a suitable background for the position. He was an ardent student of nature and had long been a collector of insects. His sensitive nature prevented him from studying medicine, and he went to divinity school at Cambridge instead. Even so, he attended many lectures in both biology and geology, and he was also tutored in these subjects by a friend and teacher, the Reverend John Henslow. As a result of arrangements made by Henslow, Darwin spent the summer of 1831 doing fieldwork with Adam Sedgwick, a geologist at Cambridge, and it was Henslow who recommended Darwin for the post aboard the HMS *Beagle*. The trip was to take five years, and the ship was to traverse the Southern Hemisphere (see Fig. 18.1), where we now know that life is most abundant and varied. Along the way, Darwin encountered forms of life very different from those in his native England.

Occurrence of Descent

Although it was not his original intent, Darwin began to gather evidence that organisms are related through common descent and that adaptation to various environments results in diversity.

Geology and Fossils

Darwin took Charles Lyell's *Principles of Geology* on the voyage. This book presented arguments to support a theory of geological change proposed by James Hutton. In contrast to the catastrophists, Hutton believed the earth was subject to slow but continuous cycles of erosion and uplift. Weather causes erosion; thereafter dirt and rock debris are washed into the rivers and transported to oceans. These loose sediments are deposited in thick layers, which are converted eventually into sedimentary rocks (Fig. 18.5). Then sedimentary rocks, which often contain fossils, are uplifted from below sea level to form land. Hutton concluded that extreme geological changes can be accounted for by slow, natural processes, given enough time. Lyell went on to propose a theory of **uniformitarianism,** that these slow changes occurred at a uniform rate. Hutton's general ideas about slow and continual geological change are still accepted today, although modern geologists realize that rates of change have not always been uniform. Darwin was not taken by the idea of uniform change, but he was convinced, as was Lyell, that the earth's massive geological changes are the result of slow processes and that the earth, therefore, must be very old.

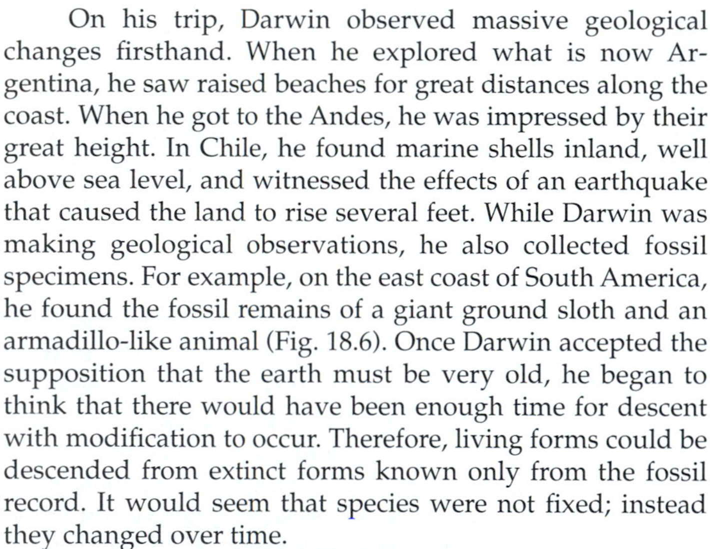

Figure 18.6 Glyptodont compared to an armadillo. **a.** A giant armadillo-like glyptodont, *Glyptodon,* known only by the study of its fossil remains. Darwin found such fossils and came to the conclusion that this extinct animal must be related to living armadillos. The glyptodont weighed 2,000 kilograms. **b.** A modern armadillo, *Dasypus,* weighs about 4.5 kilograms.

Figure 18.7 The Patagonian hare, *Dolichotis patagonium.* This animal has the face of a guinea pig and is native to South America, which has no native rabbits. The Patagonian hare has long legs and other adaptations similar to those of rabbits.

On his trip, Darwin observed massive geological changes firsthand. When he explored what is now Argentina, he saw raised beaches for great distances along the coast. When he got to the Andes, he was impressed by their great height. In Chile, he found marine shells inland, well above sea level, and witnessed the effects of an earthquake that caused the land to rise several feet. While Darwin was making geological observations, he also collected fossil specimens. For example, on the east coast of South America, he found the fossil remains of a giant ground sloth and an armadillo-like animal (Fig. 18.6). Once Darwin accepted the supposition that the earth must be very old, he began to think that there would have been enough time for descent with modification to occur. Therefore, living forms could be descended from extinct forms known only from the fossil record. It would seem that species were not fixed; instead they changed over time.

Darwin's geological observations were consistent with those of Hutton and Lyell. He began to think that the earth was very old, and there would have been enough time for descent with modification to occur.

Biogeography

Biogeography [Gk. *bios,* life, *geo,* earth, and *grapho,* writing] is the study of the geographic distribution of life-forms on earth. Darwin could not help but compare the animals of South America to those with which he was familiar. For example, instead of rabbits he found the Patagonian hare in the grasslands of South America. The Patagonian hare has long legs and ears but the face of a guinea pig, a rodent native to South America (Fig. 18.7). Did the Patagonian hare resemble a rabbit because the two types of animals were adapted to the same type of environment? Both animals ate grass, hid in bushes, and moved rapidly using long hind legs. Did the Patagonian hare have the face of a guinea pig because of common descent with guinea pigs?

As he sailed southward along the eastern coast of the continent of South America, Darwin saw how similar species replaced each other. For example, the greater rhea (an ostrich-like bird) found in the north was replaced by the lesser rhea in the south. Therefore, Darwin reasoned that related species could be modified according to the environment. When he got to the Galápagos Islands, he found further evidence of this. The Galápagos Islands are a small group of volcanic islands off the western coast of South America. The few types of plants and animals found there were slightly different from species Darwin had observed on the mainland, and even more important, they also varied from island to island.

Tortoises Each of the Galápagos Islands seemed to have its own type of tortoise, and Darwin began to wonder if this could be correlated with a difference in vegetation among islands (Fig. 18.8). Long-necked tortoises seemed to inhabit only dry areas, where food was scarce, most likely because the longer neck was helpful in reaching cacti. In moist regions with relatively abundant ground foliage, short-necked tortoises were found. Had an ancestral tortoise given rise to these different types, each adapted to a different environment?

a.

b.

Figure 18.8 **Galápagos tortoises, *Testudo.***
Darwin wondered if all of the tortoises of the various islands were descended from a common ancestor. **a.** The tortoises with dome shells and short necks feed at ground level and are from well-watered islands where grass is available. **b.** Those with shells that flare up in front have long necks and are able to feed on tall treelike cacti. They are from arid islands where prickly pear cactus is the main food source. Only on these islands are the cacti treelike.

a. *Geospiza magnirostris*

b. *Certhidea olivacea*

c. *Cactornis scandens*

Figure 18.9 **Galápagos finches.**
Each of the present-day thirteen species of finches has a bill adapted to a particular way of life. **a.** For example, the heavy beak of the large ground-dwelling finch is suited to a diet of seeds. **b.** The beak of the warbler-finch is suited to feeding on insects found among ground vegetation or caught in the air. **c.** The long, somewhat decurved beak and the split tongue of the cactus-finch is suited to probing cactus flowers for nectar.

Finches Although the finches on the Galápagos Islands seemed to Darwin like mainland finches, there are many more types (Fig. 18.9). Today, there are ground-dwelling finches with different-sized beaks, depending on the size of the seeds they feed on, and a cactus-eating finch with a more pointed beak. The beak size of the tree-dwelling finches also varies but according to the size of their insect prey. The most unusual of the finches is a woodpecker-type finch. This bird has a sharp beak to chisel through tree bark but lacks the woodpecker's long tongue which probes for insects. To make up for this, the bird carries a twig or cactus thorn in its beak and uses it to poke into crevices (see Fig. 18.1g). Once an insect emerges, the finch drops this tool and seizes the insect with its beak.

Later, Darwin speculated whether all the different species of finches he saw could have descended from a type of mainland finch. In other words, he wondered if a mainland finch was the common ancestor to all the types on the Galápagos Islands. Had speciation occurred because the islands allowed isolated populations of birds to evolve independently? Could the present-day species have resulted from accumulated changes occurring within each of these isolated populations?

Biogeography had a powerful influence on Darwin and made him think that adaptation to the environment accounts for diversification; one species can give rise to many species, each adapted differently.

Natural Selection and Adaptation

Once Darwin decided that adaptations develop over time (instead of being the instant work of a creator), he began to think about a mechanism by which adaptations might arise. Both Darwin and Alfred Russel Wallace, who is discussed in the reading on page 289, proposed **natural selection** as a mechanism for evolutionary change. Natural selection is a process in which preconditions (1–3) may result in certain consequences (4–5):

1. The members of a population have heritable variations.
2. In a population, many more individuals are produced each generation than the environment can support.
3. Some individuals have adaptive characteristics that enable them to survive and reproduce better than do other individuals.
4. An increasing proportion of individuals in succeeding generations have the adaptive characteristics.
5. The result of natural selection is a population adapted to its local environment.

Notice that because natural selection utilizes only variations that happen to be provided by genetic changes, it lacks any directedness or anticipation of future needs. Natural selection is an ongoing process because the environment of living things is constantly changing. Extinction (loss of a species) can occur when previous adaptations are no longer suitable to a changed environment.

Figure 18.10 Variation in a population.
For Darwin, variations, such as those seen in a human population, were highly significant and were required for natural selection to result in adaptation to the environment.

Organisms Have Variations

Darwin emphasized that the members of a population vary in their functional, physical, and behavioral characteristics (Fig. 18.10). Before Darwin (see Table 18.1), variations were imperfections that should be ignored since they were not important to the description of a species. For Darwin, variations were essential to the natural selection process. Darwin suspected—but did not have the evidence we have today—that the occurrence of variations is completely random; they arise by accident and for no particular purpose. New variations are just as likely to be harmful as helpful to the organism.

The variations that make adaptation to the environment possible are those that are passed on from generation to generation. The science of genetics was not yet well established, so Darwin was never able to determine the cause of variations or how they are passed on. Today, we realize that genes determine the phenotype of an organism, and that mutations and recombination of alleles during sexual reproduction can cause new variations to arise.

Organisms Struggle to Exist

In Darwin's time, a socioeconomist, Thomas Malthus, stressed the reproductive potential of human beings. He proposed that death and famine were inevitable because the human population tends to increase faster than the supply of food. Darwin applied this concept to all organisms and saw that the available resources were not sufficient for all members of a population to survive. He calculated the reproductive potential of elephants. Assuming a life span of about 100 years and a breeding span of from 30 to 90 years, a single female probably bears no fewer than six young. If all these young survive and continue to reproduce at the same rate, after only 750 years, the descendants of a single pair of elephants will number about 19 million!

Each generation has the same reproductive potential as the previous generation. Therefore, there is a constant struggle for existence, and only certain members of a population survive and reproduce each generation.

Organisms Differ in Fitness

Fitness is the reproductive success of an individual relative to other members of a population. The most fit individuals are the ones that capture a disproportionate amount of resources, and that convert these resources into a larger number of viable offspring. Since organisms vary anatomically and physiologically and the challenges of local environments vary, what determines fitness varies for different populations. For example, among western diamondback rattlesnakes *(Crotalus atrox)* living on lava flows, the most fit are those that are black in color. But among those living on desert soil, the most fit are those with the typical light and dark brown coloring. Background matching helps an animal both capture prey and avoid being captured; therefore, it is expected to lead to survival and increased reproduction.

Alfred Russel Wallace

Alfred Russel Wallace (1823–1913) is best known as the English naturalist who independently proposed natural selection as a process to explain the origin of species (Fig. 18A). Like Darwin, Wallace was a collector at home and abroad. Even at age 14, while learning the trade of surveying, he became interested in botany and started collecting plants. While he was a schoolteacher at Leicester in 1844–45, he met Henry Walter Bates, an entomologist, who interested him in insects. Together they went on a collecting trip to the Amazon, which lasted for several years. Wallace's knowledge of the world's extensive flora and fauna was much expanded by a tour he made of the Malay Archipelago from 1854–62. After studying the animals of every important island, he divided the islands into a western group, with animals like those of the Orient, and an eastern group, with animals like those of Australia. The dividing line between the islands of the archipelago is a narrow but deep strait that is now known as the Wallace Line (Fig. 18B).

Like Darwin, Wallace was a writer of articles and books. As a result of his trip to the Amazon, he wrote two books, entitled *Travels on the Amazon and Rio Negro* and *Palm Trees of the Amazon.* In 1855, during his trip to Malay, he wrote an essay called "On the Law Which Has Regulated the Introduction of New Species." In the essay, he said that "every species has come into existence coincident both in time and space with a preexisting closely allied species." It's clear, then, that by this date he believed in the origin of new species rather than the fixity of species. Later, he said that he had pondered for many years about a mechanism to explain the origin of species. He, too, had read Malthus's treatise on human population increases and, in 1858, while suffering an attack of malaria, the idea of "survival of the fittest" came upon him. He quickly completed an essay discussing a natural selection process, which he chose to send to Darwin for comment. Darwin was stunned upon its receipt. Here before him was the hypothesis he had formulated as early as 1844 but had never dared to publish. In 1856 he had begun to work on a book that would supply copious data to support natural selection as a mechanism for evolutionary change. He told his friend and colleague Charles Lyell that Wallace's ideas were so similar to his own that even Wallace's "terms now stand as heads of my chapters."

Figure 18A Alfred Russel Wallace.

Darwin suggested that Wallace's paper be published immediately, even though he as yet had nothing in print. Lyell and others who knew of Darwin's detailed work substantiating the process of natural selection suggested that a joint paper be read to the Linnean society. The title of Wallace's section was "On the Tendency of Varieties to Depart Indefinitely from the Original Type." Darwin presented an abstract of a paper he had written in 1844 and an abstract of his book *On the Origin of Species,* which was published in 1859.

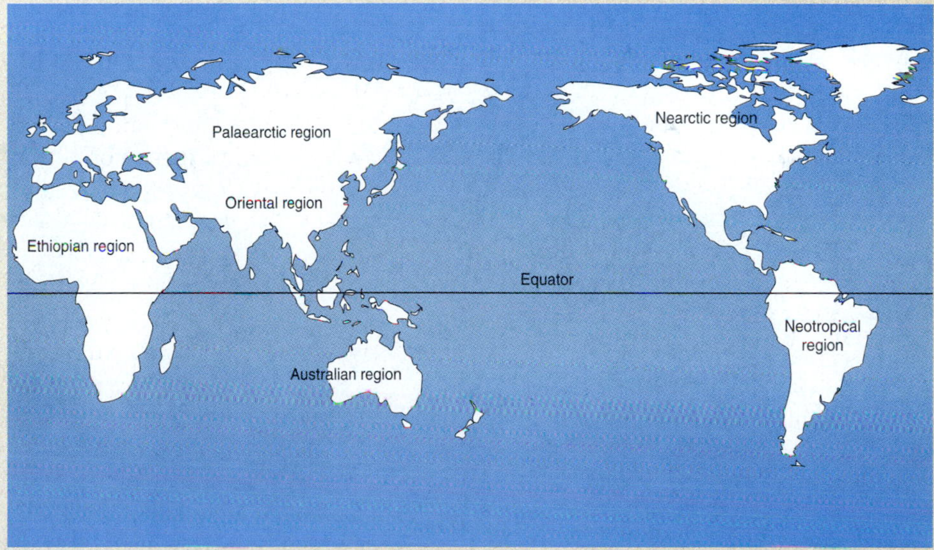

Figure 18B Biogeographical regions.
Aside from presenting a hypothesis that natural selection explains the origin of new species, Alfred Wallace is well known for another contribution. He said that the world can be divided into six biogeographical regions separated by impassable barriers. The deep waters between the Oriental and Australian regions are called the Wallace Line.

Figure 18.11 **Artificial selection of animals.**
All dogs, *Canis familiaris,* are descended from the wolf, *Canis lupus,* which began to be domesticated about 14,000 years ago. In evolutionary terms, the process of diversification has been exceptionally rapid. Several factors may have contributed: (1) the wolves under domestication were separated from other wolves because human settlements were separate, and (2) humans in each tribe selected for whatever traits appealed to them. Artificial selection of dogs continues even today.

Darwin noted that when humans help carry out *artificial selection,* the process by which a breeder chooses which traits to perpetuate, they select the animals that will reproduce. For example, prehistoric humans probably noted desirable variations among wolves and selected particular animals for breeding. Therefore, the desired traits increased in frequency in the next generation. This same process was repeated many times over. The result today is that there are many varieties of dogs all descended from the wolf (Fig. 18.11). In a similar way, several varieties of vegetables can be traced to a single ancestor. Chinese cabbage, brussels sprouts, and kohlrabi are all derived from a single species, *Brassica oleracea* (Fig. 18.12).

In nature, interactions with the environment determine which members of a population reproduce to a greater degree than other members. In contrast to artificial selection, the result of natural selection is not predesired. Natural selection occurs because certain members of a population happen to have a variation that allows them to survive and reproduce to a greater extent than other members. For example, any variation that increases the speed of a hoofed animal helps it escape predators and live longer; a variation that reduces water loss is beneficial to a desert plant; and one that increases the sense of smell helps a wild dog find its prey. Therefore, we expect organisms with these traits to have increased fitness.

Organisms Become Adapted

An **adaptation** [L. *ad,* toward, and *aptus,* fit, suitable] is a trait that helps an organism be more suited to its environment. We can especially recognize an adaptation when unrelated organisms, living in a particular environment, display similar characteristics. For example, manatees, penguins, and sea turtles all have flippers, which help them move through the water. Natural selection results in the adaptation of populations to their specific environments. Because of differential reproduction generation after generation, adaptive traits are increasingly represented in each succeeding generation. There are other processes of evolution aside from natural selection, but natural selection is the only process that results in adaptation to the environment.

On the Origin of Species by Darwin

After the HMS *Beagle* returned to England in 1836, Darwin waited more than 20 years to publish his ideas. During the intervening years, he used the scientific process to test his hypothesis that today's diverse life-forms arose by descent from a common ancestor and that natural selection is a mechanism by which species can change and even new species can arise. Darwin was prompted to publish his book after reading a similar hypothesis from Alfred Russel Wallace, as discussed in the reading on page 289.

Darwin became convinced that descent with modification explains the history of life. His theory of natural selection proposes a mechanism by which adaptation to the environment occurs.

a.

b.

c.

Figure 18.12 Artificial selection of plants.
All these vegetables are derived from a single species of *Brassica oleracea.* **a.** Chinese cabbage. **b.** Brussels sprouts. **c.** Kohlrabi. Darwin believed that artificial selection provided a model by which to understand natural selection. With natural selection, however, the environment provides the selective force.

18.3 Evidence for Evolution

Many different lines of evidence support the hypothesis that organisms are related through common descent. This is significant, because the more varied the evidence supporting a hypothesis, the more certain it becomes. Darwin cited much of the evidence we will discuss, except he had no knowledge, of course, of the biochemical data that became available after his time.

Fossil Evidence

The **fossil record** is the history of life recorded by remains from the past. Fossils are at least 10,000 years old and include such items as pieces of bone, impressions of plants pressed into shale, and even insects trapped in tree resin (which we know as amber). For the last two centuries, paleontologists have studied fossils in the earth's strata (layers) all over the world and have pieced together the story of past life.

The fossil record is rich in information. One of its most striking patterns is the succession of life-forms over time. Catastrophists offered an explanation for the extinction and subsequent replacement of one group of organisms by another group, but they never could explain successive changes that link groups of organisms historically. Particularly interesting are the fossils that serve as transitional links between groups. The amphibious fish *Eustheopteron,* the reptile-like amphibian *Seymouria,* and the mammal-like reptiles, or therapsids are well-known examples. Two newly found fossils from China, *Protarchaeopteryx* and *Caudipteryx* (Fig. 18.13), offer even more convincing evidence than the famous fossil, *Archaeopteryx,* that birds are descended from dinosaurs. Called by some, one of the most exciting discoveries of the century, if not the discovery of the century, these fossils are definitely recognized as dinosaurs that had feathers on their arms and tails. There is some evidence of body feathers as well. It is speculated that flapping their forelimbs while running may have given them added speed in chasing prey and escaping predators. Once lift-off occurred in a descendent, birds came into existence.

Sometimes the fossil record is complete enough to allow us to trace the history of an organism, such as the modern-day horse, *Equus* (Fig. 18.14). *Equus* evolved originally from *Hyracotherium,* which was about the size of a large dog (35 kg). This fossil animal had cusped, low-crowned molars, four toes on each front foot, and three toes on each hind foot. When grasslands replaced the forest home of *Hyracotherium,* the ancestors of *Equus* were subject to selective pressure for the development of strength, intelligence, speed, and durable grinding teeth. A larger size provided the strength needed for combat, a larger skull made room for a larger brain, elongated legs ending in hooves provided greater speed to escape enemies, and the durable grinding teeth enabled the animals to feed efficiently on grasses.

Figure 18.14 shows that the evolutionary history of *Equus* is like a tree, with multiple branchings and rebranchings from one source. A common ancestor is at each fork of the evolutionary tree, and we can trace the evolutionary history of *Equus* from common ancestor to common ances-

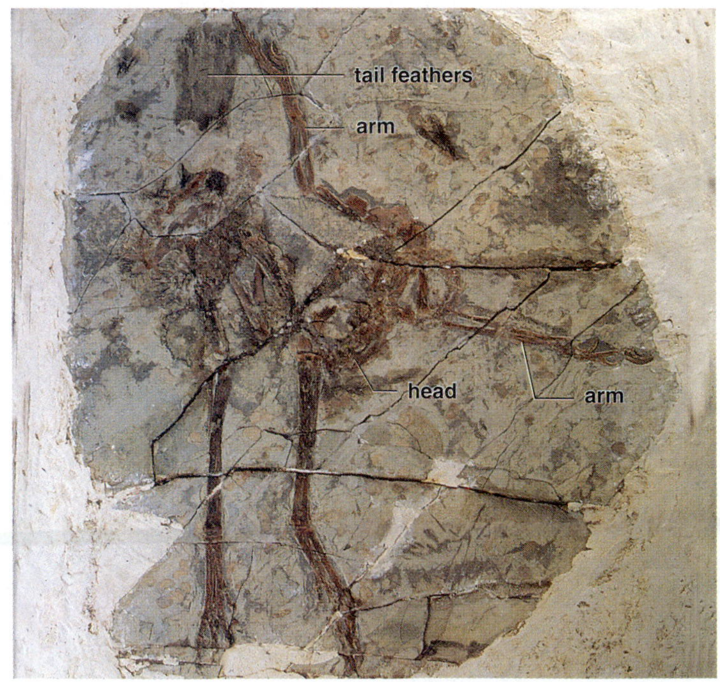

a.

b.

Figure 18.13 **Transitional link.**
a. *Caudipteryx,* discovered in 1998, is hailed by most paleontologists as unambiguous evidence of a dinosaur–bird link. Skeletal evidence that birds are descended from dinosaurs includes the similarity of their breastbones, three-toed feet, hollow bones, and swiveling wrist joints. **b.** Artist's representation of *Caudipteryx.*

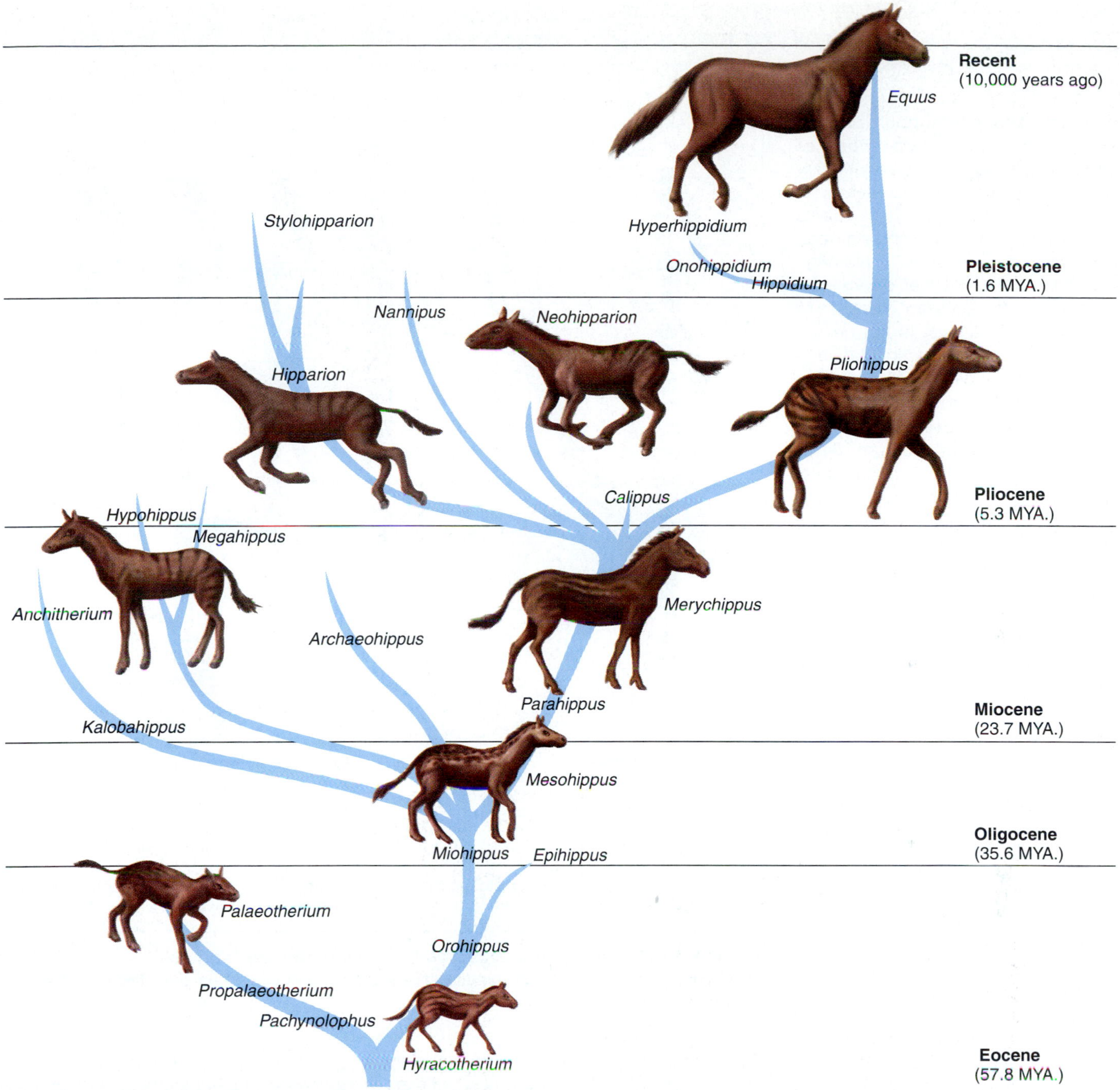

Recent
(10,000 years ago)
Equus
Stylohipparion
Hyperhippidium
Onohippidium
Hippidium
Pleistocene
(1.6 MYA.)
Nannipus
Neohipparion
Hipparion
Pliohippus
Calippus
Pliocene
(5.3 MYA.)
Hypohippus
Megahippus
Anchitherium
Archaeohippus
Merychippus
Parahippus
Kalobahippus
Miocene
(23.7 MYA.)
Mesohippus
Miohippus
Epihippus
Oligocene
(35.6 MYA.)
Palaeotherium
Orohippus
Propalaeotherium
Pachynolophus
Hyracotherium
Eocene
(57.8 MYA.)

Figure 18.14 Evolutionary history of *Equus*.
The evolutionary tree of *Equus* is known to have included many branchings, several of which came to dead ends. By ignoring these, it is possible to trace the history of *Equus* back to *Hyracotherium*.

tor through time. When we do so, there appears to have been a gradual change in form from *Hyracotherium* to *Equus*. We can also observe that many of the side branches became **extinct,** that is, died out. The paleontologist George Gaylord Simpson estimated that 99.9% of all species eventually become extinct. We can hypothesize that environments are constantly changing and that the ability to adapt to a changing environment is a requirement for long-term survival of the species.

Living organisms closely resemble the most recent fossils in their line of descent. Fossils can be linked over time because there is a similarity in form, despite observed changes; therefore, the fossil record supports common descent.

The fossil record broadly traces the history of life and more specifically allows us to study the history of particular groups.

The Pace of Evolution

Evolutionists who support phyletic gradualism [Gk. *phyle,* tribe], as did Darwin, suggest that evolutionary change is rather slow and steady. In other words, fossils of the same species designation can show a trend over time, say a change in plumage color (Fig. 18C*a*). Further divergence, when a common ancestor gives rise to two separate lineages, is not necessarily dependent upon speciation—that is, the origination of a new species. Indeed, the fossil record, even if complete, is unlikely to indicate when speciation has occurred. Since evolution occurs gradually, transitional links (see Fig. 18.13) are expected and most likely more will eventually be found in the fossil record.

Is the phyletic gradualism model applicable to the evolutionary history of *Equus* (Fig. 18.14)? Is it possible, for example, to show overall trends such as an increase in size, an increase in the grinding surface of the molar teeth, and a reduction in the number of toes? Most agree that it is applicable only if we pick and choose among the many fossils available. Closer examination reveals that as *Hyracotherium* evolved into *Equus,* the evolution of every character varied greatly, and there were even times of reversal. If one of the ancestral animals had lived on and *Equus* had become extinct, we no doubt would be discussing a different set of "trends."

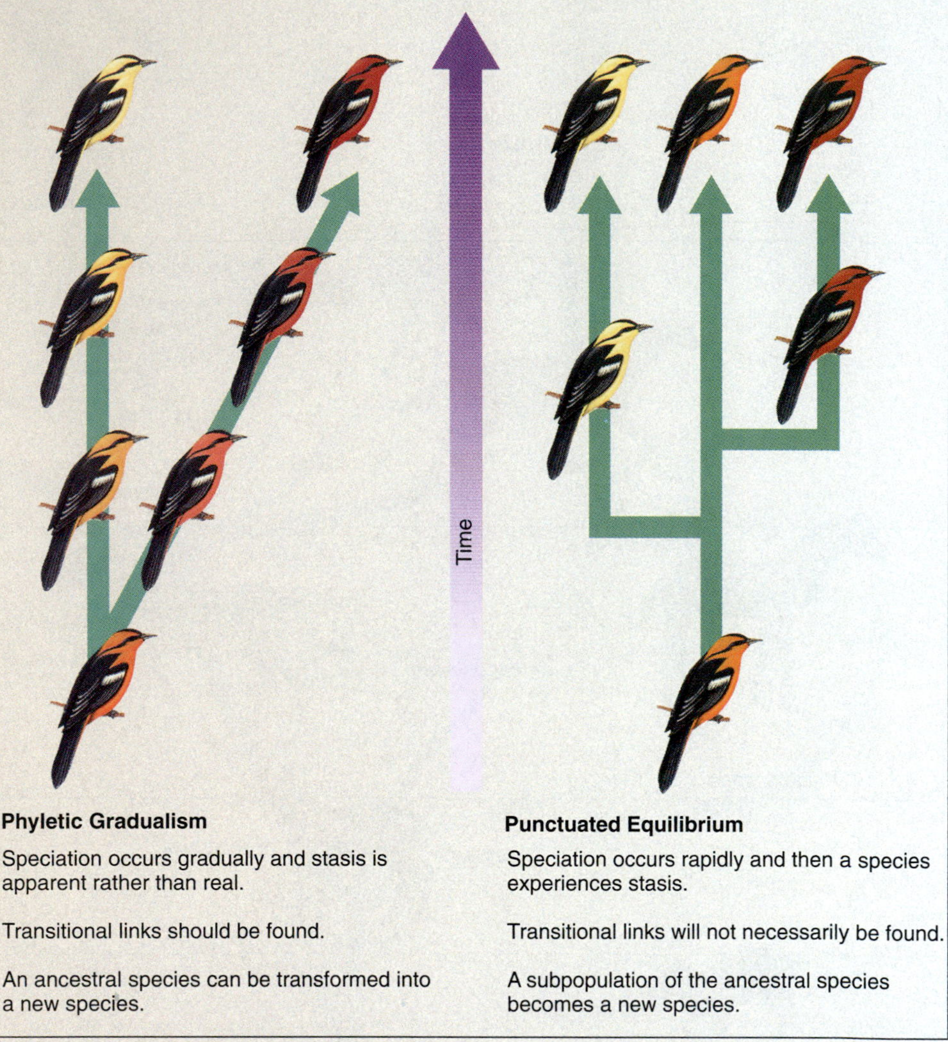

Time
Phyletic Gradualism
Speciation occurs gradually and stasis is apparent rather than real.
Transitional links should be found.
An ancestral species can be transformed into a new species.
Punctuated Equilibrium
Speciation occurs rapidly and then a species experiences stasis.
Transitional links will not necessarily be found.
A subpopulation of the ancestral species becomes a new species.

Figure 18C Phyletic gradualism versus punctuated equilibrium.
The differences between phyletic gradualism and punctuated equilibrium are reflected in these patterns of time versus speciation.

Considering such difficulties, other paleontologists—Stephen J. Gould, Nile Eldredge, and Steven Stanley in particular—have proposed a new model they call punctuated equilibrium (Fig. 18C*b*). They point out there are examples of organisms that are called *living fossils* because they are so similar to an ancestor known from the fossil record. A few years ago, investigators found exquisitely preserved specimens of cyanobacteria that have the same sizes, shapes, and organization as living forms. These findings suggest that the cyanobacteria of today have not changed at all in over 3 billion years. Among plants, the dawn redwood was thought to be extinct; then a living specimen was discovered in a small area of China. Horseshoe crabs, crocodiles, and coelacanth fish are animals that still resemble their earliest ancestors. A recently found scaly anteater fossil shows that these animals have changed minimally in 60 million years. A time of limited evolutionary change in a lineage is called *stasis.*

In most lineages, however, a period of equilibrium (stasis) is punctuated by evolutionary change—that is, speciation occurs. With reference to the length of the fossil record (about 3.5 billion years), speciation occurs relatively rapidly. Therefore, transitional links are less likely to become fossils, and to be found! Indeed, speciation most likely involves only an isolated population at one locale. Only when a new species evolves and displaces existing species, is it apt to show up in the fossil record.

Carlton Brett at the University of Rochester studied the sequence of fossil communities in the Devonian seas that covered the present state of New York 360 to 408 million years ago. He found that the mix of species in a community did not change as long as the sea level remained high. A low sea level disrupted the community and there was a change of species that maintained itself throughout the next interval of a high sea level. It would appear that gradual evolutionary processes are the norm but environmental disturbances promote rapid evolutionary change and the replacement of old species by new species.

Biogeographical Evidence

Biogeography is the study of the distribution of plants and animals throughout the world. Such distributions are consistent with the hypothesis that related forms evolved in one locale and then spread out into other accessible regions. As previously mentioned, Darwin noted that South America lacked rabbits, even though the environment was quite suitable to them. He concluded there are no rabbits in South America because rabbits originated somewhere else and they had no means to reach South America.

The islands of the world have many unique species of animals and plants found no place else, even when the soil and climate are the same as other places. Why are there so many species of finches on the Galápagos Islands when these same species are not on the mainland? The reasonable explanation is that the same ancestral finch originally inhabited the different islands. Geographic isolation allowed the ancestral finch to evolve into a different species on each island.

Physical factors, such as the location of continents, often determine where a population can spread. Both cacti and euphorbia are plants adapted similarly to a hot, dry environment—they both are succulent, spiny, flowering plants. Why do cacti grow in North American deserts and euphorbia grow in African deserts when each would do well on the other continent? It seems obvious that they just happened to evolve on their respective continents.

At one time in the history of the earth, South America, Antarctica, and Australia were all connected. Marsupials (pouched mammals) arose at this time, and today are found in both South America and Australia. When Australia separated and drifted away, the marsupials diversified into many different forms suited to various environments (Fig. 18.15). They were free to do so because there were few, if any, placental mammals in Australia. In other regions such as South America, where there are placental mammals, marsupials are not as diverse.

The distribution of organisms on the earth is explainable by assuming that related forms evolved in one locale. They then diversified as they spread out into other accessible areas.

Coarse-haired wombat, *Vombatus*, nocturnal and living in burrows

Sugar glider, *Petaurista*, a tree dweller

Kangaroo, *Macropus*, a herbivore of plains and forests

Australian native cat, *Dasyurus*, a carnivore of forests

Tasmanian wolf, *Thylacinus*, a nocturnal carnivore of deserts and plains

Figure 18.15 Biogeography.
Each type of marsupial in Australia is adapted to a different way of life. All of the marsupials in Australia presumably evolved from a common ancestor that entered Australia some 60 million years ago.

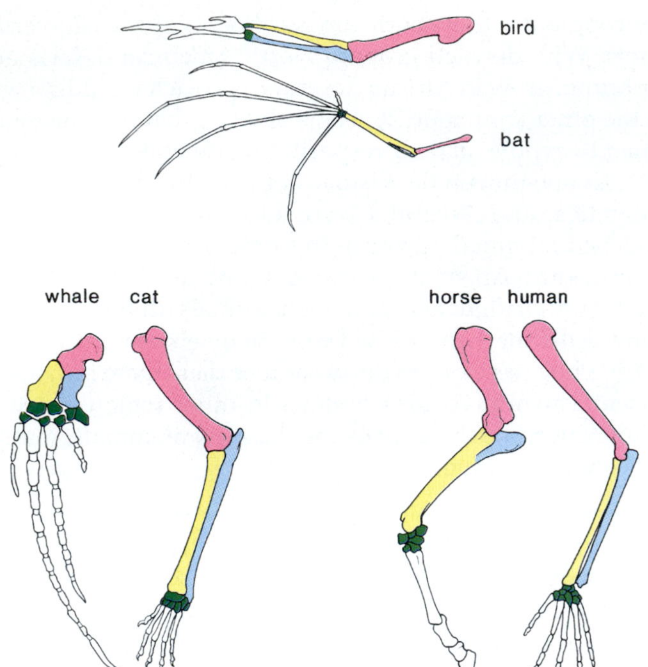

Figure 18.16 Significance of structural similarities. Although the specific design details of vertebrate forelimbs are different, the same bones are present (they are color-coded). This unity of plan is evidence of a common ancestor.

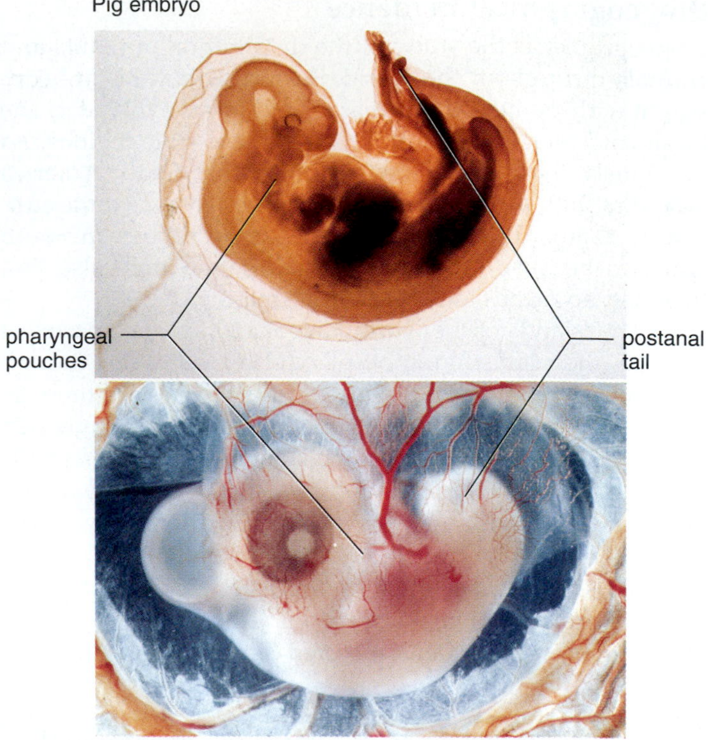

Figure 18.17 Significance of developmental similarities. At this comparable developmental stage, a chick embryo and a pig embryo have many features in common, which suggests they evolved from a common ancestor.

Anatomical Evidence

Darwin was able to show that a common descent hypothesis offers a plausible explanation for anatomical similarities among organisms. Vertebrate forelimbs are used for flight (birds and bats), orientation during swimming (whales and seals), running (horses), climbing (arboreal lizards), or swinging from tree branches (monkeys). Yet all vertebrate forelimbs contain the same sets of bones organized in similar ways, despite their dissimilar functions (Fig. 18.16). The most plausible explanation for this unity is that the basic forelimb plan belonged to a common ancestor, and then the plan was modified in the succeeding groups as each continued along its own evolutionary pathway. Structures that are anatomically similar because they are inherited from a common ancestor are called **homologous structures** [Gk. *homologos*, agreeing, corresponding]. In contrast, **analogous structures** serve the same function but they are not constructed similarly, nor do they share a common ancestry. The wings of birds and insects are analogous structures. The presence of homology, not analogy, is evidence that organisms are related.

Vestigial structures [L. *vestigium*, trace, footprint] are anatomical features that are fully developed in one group of organisms but are reduced and may have no function in similar groups. Most birds, for example, have well-developed wings used for flight. Some bird species (e.g., ostrich), however, have greatly reduced wings and do not fly. Similarly, snakes have no use for hindlimbs, and yet some have remnants of a pelvic girdle and legs. Humans have a tailbone but no tail. The presence of vestigial structures can be explained by the common descent hypothesis. Vestigial structures occur because organisms inherit their anatomy from their ancestors; they are traces of an organism's evolutionary history.

The homology shared by vertebrates extends to their embryological development (Fig. 18.17). At some time during development, all vertebrates have a postanal tail and exhibit paired pharyngeal pouches. In fishes and amphibian larvae, these pouches develop into functioning gills. In humans, the first pair of pouches becomes the cavity of the middle ear and the auditory tube. The second pair becomes the tonsils, while the third and fourth pairs become the thymus and parathyroid glands. Why should terrestrial vertebrates develop and then modify structures like pharyngeal pouches that have lost their original function? The most likely explanation is that fishes are ancestral to other vertebrate groups.

Organisms that share homologous structures are closely related and have a common ancestry. This is substantiated by comparative anatomy and embryological development.

Biochemical Evidence

Almost all living organisms use the same basic biochemical molecules, including DNA (deoxyribonucleic acid), ATP (adenosine triphosphate), and many identical or nearly identical enzymes. Further, organisms utilize the same DNA triplet code and the same 20 amino acids in their proteins. Organisms even share the same introns and type of repeats. There is obviously no functional reason why these elements need be so similar. But their similarity can be explained by descent from a common ancestor.

Also of interest, evo-devo researchers (evolutionary developmental biologists) have found that many developmental genes are shared in animals ranging from worms to humans. It appears that life's vast diversity has come about by only a slight difference in the same genes. The result has been widely divergent body plans. For example, a similar gene in arthropods and vertebrates determines the dorsal-ventral axis. But although the base sequences are similar, the genes have opposite effects. Therefore in arthropods, like fruit flies and crayfish, the neural tube is ventral, whereas in vertebrates, like chicks and humans, the neural tube is dorsal.

When the degree of similarity in DNA base sequences or the degree of similarity in amino acid sequences of proteins is examined, the data are as expected assuming common descent. Cytochrome *c* is a molecule that is used in the electron transport system of all the organisms appearing in Figure 18.18. Data regarding differences in the amino acid sequence of the cytochrome *c* show that in a human it differs from that in a monkey by only one amino acid, from that in a duck by 11, and from that in *Candida*, a yeast, by 51 amino acids. These data are consistent with other data regarding the anatomical similarities of these animals, and therefore their relatedness.

Darwin discovered that many lines of evidence support the hypothesis of common descent. Since his time, it has been found that biochemical evidence also supports the hypothesis. A hypothesis is strengthened when it is supported by many different lines of evidence.

Evolution is no longer considered a hypothesis. It is one of the great unifying theories of biology. In science, the word *theory* is reserved for those conceptual schemes that are supported by a large number of observations and have not yet been found lacking. The theory of evolution has the same status in biology that the germ theory of disease has in medicine.

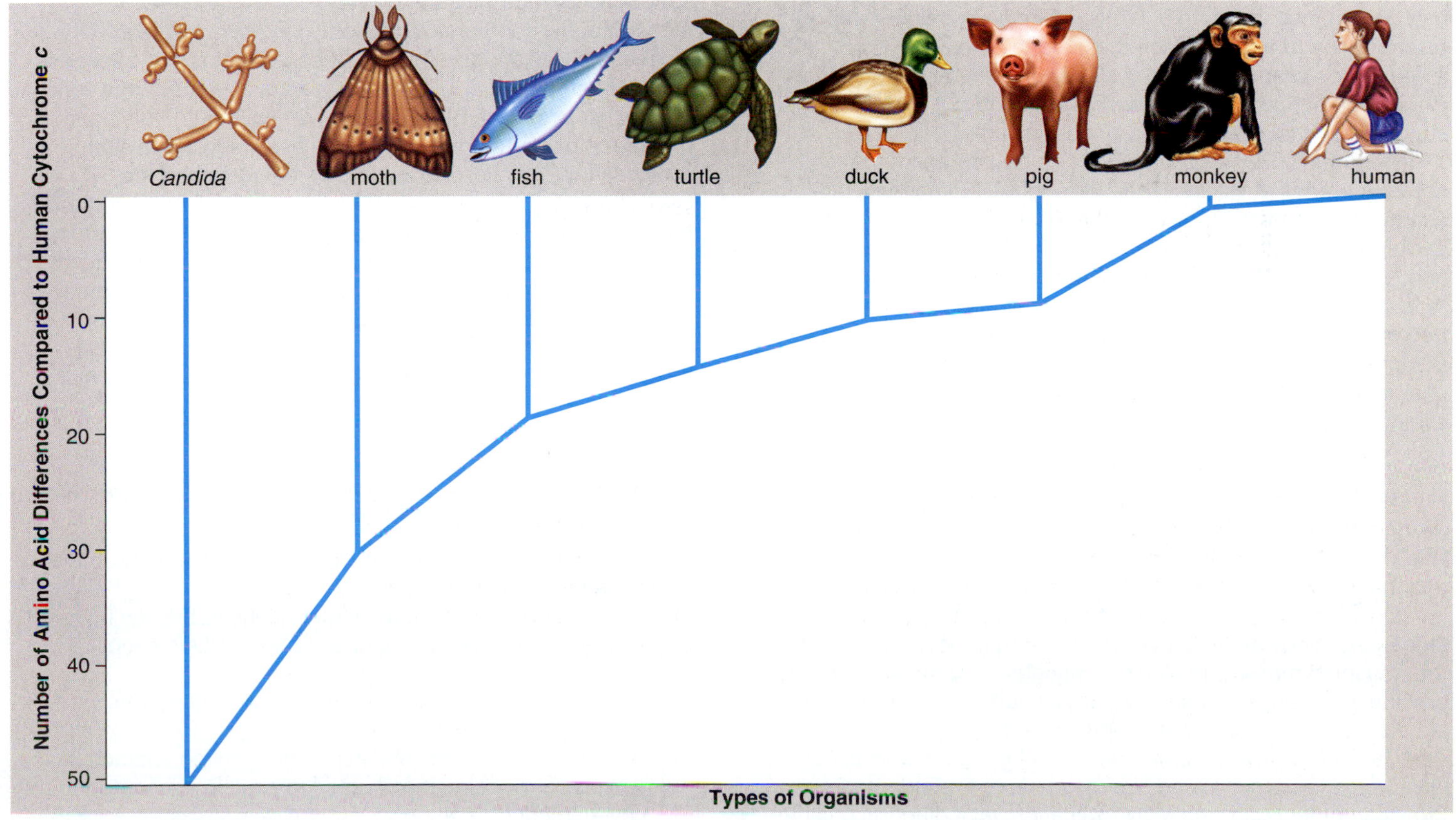

Figure 18.18 Significance of biochemical differences.
The branch points in this diagram tell the number of amino acids that differ between human cytochrome *c* and the organisms depicted. These biochemical data are consistent with that provided by a study of the fossil record and comparative anatomy.

Connecting Concepts

Before the 1800s, most people believed that the origin and diversity of life on earth was due to the work of a supernatural being who created each species at the beginning of the world, and modern organisms are essentially unchanged descendants of their ancestors. Scientists, however, seek natural, testable hypotheses to explain natural events rather than relying on religious dogma.

At the time Charles Darwin boarded the *Beagle,* he had studied writings of his grandfather Erasmus Darwin, James Hutton, Charles Lyell, Lamarck, Malthus, Linnaeus, and other original thinkers. He had ample time during his five-year voyage to reflect on the ideas of these authors, and collectively, they built a framework that helped support his theory of descent with modifications.

One aspect of scientific genius is the power of astute observation—to see what others miss or fail to appreciate. In this area, Darwin excelled. By the time he reached the Galápagos Islands, Darwin had already begun to hypothesize that species could be modified according to the environment. His observation of the finches on the isolated Galápagos Islands supported his hypothesis. He concluded that the finches on each island varied from each other and from mainland finches because each species had become adapted to a different habitat; therefore, one species had given rise to many. The fact that Alfred Wallace almost simultaneously proposed natural selection as an evolutionary mechanism suggests that the world was ready for a revised view of life on earth.

The theory of evolution has quite rightly been called the grand unifying theory (GUT) of biology. Fossils, comparative anatomy, biogeography, and biochemical data all indicate that living things share common ancestors. Evolutionary principles help us understand why organisms are different and alike, and why some species flourish and others die out. As the earth's habitats change over millions of years, those individuals with the traits best adapted to new environments survive and reproduce; thus populations change over time.

Summary

18.1 History of the Theory of Evolution

In general, the pre-Darwinian worldview was different from the post-Darwinian worldview (see Table 18.1). The scientific community, however, was ready for a new worldview, and it received widespread acceptance.

A century before Darwin's trip, classification of organisms had been a main concern of biology. Linnaeus thought that classification should describe the fixed features of species and reveal God's divine plan. Gradually, some naturalists, such as Count Buffon and Erasmus Darwin, began to put forth tentative suggestions that species do change over time.

Georges Cuvier and Jean-Baptiste de Lamarck, contemporaries of Darwin in the late eighteenth century, differed sharply on evolution. To explain the fossil record of a region, Cuvier proposed that a whole series of catastrophes (extinctions) and repopulations from other regions had occurred. Lamarck said that descent with modification does occur and that organisms do become adapted to their environments; however, he relied on commonly held beliefs (*scala naturae* and inheritance of acquired characteristics) to substantiate and provide a mechanism for evolutionary change.

18.2 Darwin's Theory of Evolution

Charles Darwin formulated hypotheses concerning evolution after taking a trip around the world as naturalist aboard the HMS *Beagle* (1831–36). His hypotheses were that common descent does occur and that natural selection results in adaptation to the environment.

Darwin's trip involved two primary types of observations. His study of geology and fossils caused him to concur with Lyell that the observed massive geological changes were caused by slow, continuous changes. Therefore, he concluded that the earth is old enough for descent with modification to occur.

Darwin's study of biogeography, including the animals of the Galápagos Islands, allowed him to conclude that adaptation to the environment can cause diversification, including the origin of new species.

Natural selection is the mechanism Darwin proposed for how adaptation comes about. Members of a population exhibit random but inherited variations. (In contrast to the previous worldview, variations are highly significant.) Relying on Malthus's ideas regarding overpopulation, Darwin stressed that there was a struggle for existence. The most fit organisms are those possessing characteristics that allow them to acquire more resources, survive, and reproduce more than the less fit. In this way, natural selection results in adaptation to a local environment.

18.3 Evidence for Evolution

The hypothesis that organisms share a common descent is supported by many lines of evidence. The fossil record, biogeography, comparative anatomy, and comparative biochemistry all support the hypothesis. The fossil record gives us the history of life in general and allows us to trace the descent of a particular group. Biogeography shows that the distribution of organisms on earth is explainable by assuming organisms evolved in one locale. Comparing the anatomy and the development of organisms reveals a unity of plan among those that are closely related. All organisms have certain biochemical molecules in common, and any differences indicate the degree of relatedness. A hypothesis is greatly strengthened when many different lines of evidence support it.

Today, the theory of evolution is one of the great unifying theories of biology because it has been supported by so many different lines of evidence.

Reviewing the Chapter

1. In general, contrast the pre-Darwinian worldview with the post-Darwinian worldview. 283
2. Cite naturalists who made contributions to biology in the mid-eighteenth century, and state their beliefs about evolutionary descent. 283
3. How did Cuvier explain the succession of life-forms in the earth's strata? 284
4. What is meant by the inheritance of acquired characteristics, a hypothesis that Lamarck used to explain adaptation to the environment? 284
5. What reading did Darwin do, and what observations did he make regarding geology? 285
6. What observations did Darwin make regarding biogeography? How did these influence his conclusions about the origin of new species? 286–87

7. What are the essential features of the process of natural selection as proposed by Darwin? 288
8. Distinguish between the concepts of fitness and adaptation to the environment. 288, 291
9. How do data from the fossil record support the concept that organisms are related through common descent? Explain why *Equus* is vastly different from its ancestor *Hyracotherium,* which lived in a forest. 292–93
10. How do data from biogeography support the concept of common descent? Explain why a diverse assemblage of marsupials evolved in Australia. 295
11. How do data from comparative anatomy support the concept of common descent? Explain why vertebrate forelimbs are similar despite different functions. 296
12. How do data from biochemical studies support the concept of common descent? Explain why the sequence of amino acids in cytochrome *c* differs between two organisms. 297

Testing Yourself

Choose the best answer for each question.

1. According to the theory of acquired inheritance,
   a. if a man loses his hand, then his children will also be missing a hand.
   b. changes in phenotype are passed on by way of the genotype to the next generation.
   c. organisms are able to bring about a change in their phenotype.
   d. evolution is striving toward particular traits.
   e. All of these are correct.
2. Why was it helpful to Darwin to learn that Lyell thought the earth was very old?
   a. An old earth has more fossils than a new earth.
   b. It meant there was enough time for evolution to have occurred slowly.
   c. There was enough time for the same species to spread out into all continents.
   d. Darwin said that artificial selection occurs slowly.
   e. All of these are correct.
3. All the finches on the Galápagos Islands
   a. are unrelated but descended from a common ancestor
   b. are descended from a common ancestor and therefore related.
   c. rarely compete for the same food source.
   d. Both a and c are correct.
   e. Both b and c are correct.
4. Organisms
   a. compete with other members of their species.
   b. differ in fitness.
   c. are adapted to their environment.
   d. are related by descent from common ancestors.
   e. All of these are correct.
5. DNA nucleotide differences between organisms
   a. indicate how closely related organisms are.
   b. indicate that evolution occurs.
   c. explain why there are phenotypic differences.
   d. are to be expected.
   e. All of these are correct.
6. If evolution occurs, we would expect different biogeographical regions with similar environments to
   a. all contain the same mix of plants and animals.
   b. each have its own specific mix of plants and animals.
   c. have plants and animals with similar adaptations.
   d. have plants and animals with different adaptations.
   e. Both b and c are correct.
7. The fossil record offers direct evidence for common descent because you can
   a. see that the types of fossils change over time.
   b. sometimes find common ancestors.
   c. trace the ancestry of a particular group.
   d. trace the biological history of living things.
   e. All of these are correct.
8. Organisms such as whales and sea turtles which are adapted to an aquatic way of life
   a. will probably have homologous structures.
   b. will have similar adaptations but not necessarily homologous structures.
   c. may very well have analogous structures.
   d. will have the same degree of fitness.
   e. Both b and c are correct.

For questions 9–12, offer an explanation for each of these observations based on information in the section indicated. Write out your answer.

9. Fossils can be dated according to the strata in which they are located. See Fossil Evidence (p. 292).
10. Cacti and euphorbia exist on different continents, but both have spiny, water-storing, leafless stems. See Biogeographical Evidence (p. 295).
11. Amphibians, reptiles, birds, and mammals all have pharyngeal pouches at some time during development. See Anatomical Evidence (p. 296).
12. The base sequence of DNA differs from species to species. See Biochemical Evidence (p. 297).

Thinking Scientifically

1. Because viruses are rapidly replicated, it is possible to observe hundreds of generations in a single host (infected organism). Some viruses (such as influenza and HIV) evolve rapidly, and others (such as rabies virus and poliovirus) are relatively stable. Two selective forces that influence the speed of viral evolution are (1) the strength of the host immune response which works to destroy the virus and (2) host behavior in assisting transmission of the virus. How could these two selective forces influence the speed of viral evolution?
2. DNA evidence shows that the closest living relatives of elephants are manatees, aquatic mammals found in the Atlantic Ocean. Two possibilities exist: manatees evolved from elephants (or their immediate ancestors), or elephants evolved from manatees (or their immediate ancestors). A study of embryonic and adult anatomy of manatees and elephants might reveal structures that would help to more firmly establish the evolutionary relationship between these two animals. Hypothesize what kind of structures these might be.

Bioethical Issue

Evolution is a scientific theory. So is the cell theory, which says that all organisms are composed of cells, and so is the atomic theory that says all matter is composed of atoms. Yet, no one argues that schools should teach alternatives to the cell theory or the subatomic theory. Confusion reigns over the use of the expression, "the theory of evolution." But the term *theory* in science is reserved for those ideas that scientists have found to be all-encompassing because they are based on data collected in a number of different fields.

No wonder most scientists in our country are dismayed when state legislatures or school boards rule that teachers must put forward a variety of "theories" on the origin of life, including one that runs contrary to the mass of data that supports the theory of evolution. An institute in California called the Institute for Creation Research advocates that students be taught an "intelligent-design theory" which says that DNA could never have arisen without the involvement of an "intelligent agent," and that gaps in the fossil record mean that species arose fully developed with no antecedents.

Since our country forbids the mingling of church and state—no purely religious ideas can be taught in the schools—the advocates for an intelligent-design theory are careful to never mention the Bible or any ideas like "God created the world in seven days." Still, teachers who have a good scientific background do not feel comfortable teaching an intelligent-design theory because it does not meet the test of a scientific theory. Science is based on hypotheses that have been tested by observation and/or experimentation. A scientific theory has stood the test of time—no hypotheses have been supported by observation and/or experimentation that runs counter to the theory. Indeed, the theory of evolution is supported by data collected in such wide-ranging fields as development, anatomy, geology, biochemistry, and so forth.

The polls consistently show that nearly half of all Americans prefer to believe the Old Testament account of how God created the world in seven days. That, of course, is their right, but should schools be required to teach an intelligent-design theory that traces its roots back to the Old Testament, and is not supported by observation and experimentation?

Understanding the Terms

adaptation 291
analogous structure 296
biogeography 286
catastrophism 284
evolution 283
extinct 293
fitness 288
fossil record 292
homologous structure 296
inheritance of acquired characteristics 284
natural selection 288
paleontology 284
uniformitarianism 285
vestigial structure 296

Match the terms to these definitions:

a. ______________ Study of the geographical distribution of organisms.

b. ______________ Study of fossils that results in knowledge about the history of life.

c. ______________ Poorly developed structure that was complete and functional in an ancestor but is no longer functional in a descendant.

d. ______________ Organism's modification in structure, function, or behavior suitable to the environment.

e. ______________ Lamarckian belief that organisms become adapted to their environment during their lifetime and pass on these adaptations to their offspring.

Web Connections

Exploring the Internet

http://www.mhhe.com/biosci/genbio/mader
(click on *Biology 7/e*)

The *Biology 7/e* Online Learning Center provides many resources for studying the material in this chapter including links to the following sites:

Darwin-Wallace 1858 Paper on Evolution. From this site you can search and read the seminal paper and related works.

http://www.inform.umd.edu/PBIO/darwin/darwindex.html

The Slow Death of Spontaneous Generation (1668–1859) describes the lengthy history of this belief and the major players who sought to refute this idea. Louis Pasteur finally laid this idea to rest, opening the door for ideas on evolution.

http://www.accessexcellence.org/AB/BC/Spontaneous_Generation.html

Enter Evolution: Theory and History is a well-written, thorough treatment of the founders of natural science, the great naturalists of the eighteenth century. The focus is on scientists who had ideas on evolution, and those who were proponents of natural selection.

http://www.ucmp.berkeley.edu/history/evolution.html

Evolution and Behavior. This site provides answers to FAQs related to many topics on evolution. It provides book reviews, information on evolutionary scientists, and perspectives on evolution and religion.

http://ccp.uchicago.edu/~jyin/evolution.html

The Talk Origins Archive. This site provides a discussion of Darwin's theory and traces the history of scientists' acceptance of evolution.

http://www.talkorigins.org/faqs/modern-synthesis.html

Process of Evolution

chapter 19

chapter concepts

19.1 Evolution in a Genetic Context

- The Hardy-Weinberg law defines evolution in terms of allele frequency changes in a population over time. 302
- Mutations, gene flow, nonrandom mating, genetic drift, and natural selection can cause allele frequency changes in a population. 304
- The raw material for evolutionary change is mutations. Recombination of genes is another source in sexually reproducing organisms. 304

19.2 Natural Selection

- Natural selection causes changes in allele frequencies in a population due to the differential ability of certain phenotypes to reproduce. 306
- Natural selection results in adaptation to the environment. The three types of natural selection are: directional selection, stabilizing selection, and disruptive selection. 306

19.3 Speciation

- New species come about when populations are reproductively isolated from other, similar populations. 310
- Adaptive radiation is the rapid development of several species from a single species; each species is adapted in a unique way. 313

Antibiotics are selective agents for evolution.

When your grandparents were young, infectious diseases such as tuberculosis, pneumonia, and syphilis killed thousands of people every year. Then in the 1940s, penicillin and other antibiotics were developed, and public health officials believed infectious diseases were a thing of the past. Today, however, tuberculosis, pneumonia, and many other ailments are back with a vengeance. What happened? Natural selection occurred.

As Darwin and Wallace emphasized, there is variation among individuals in all populations or groups of organisms. Just as the cats in your neighborhood differ from one another, bacteria that cause a disease differ from one another. Some of the tubercular bacteria, for example, just happened to be resistant to drug treatment. The antibiotic didn't cause this resistance; the bacteria were already resistant (preadapted). So the antibiotic-resistant bacteria survived and reproduced. Now, 50 years later, several strains of "superbugs" that cause various illnesses cannot be killed by antibiotics. One type of bacteria actually feeds on the antibiotic vancomycin!

19.1 Evolution in a Genetic Context

Darwin stressed that the members of a population vary but he did not know how variations come about and how they are transmitted. It was not until the 1930s that population geneticists were able to apply the principles of genetics to populations and thereafter develop a way to recognize when evolution has occurred. A **population** is all the members of a single species occupying a particular area at the same time. Evolution that occurs within a population is called **microevolution.**

Microevolution

In **population genetics,** the various alleles at all the gene loci in all individuals make up the **gene pool** of the population. It is customary to describe the gene pool of a population in terms of gene frequencies. Suppose that in a *Drosophila* population, one-fourth of the flies are homozygous dominant for long wings, one-half are heterozygous, and one-fourth are homozygous recessive for short wings. Therefore, in a population of 100 individuals, we have

25 *LL*, 50 *Ll*, and 25 *ll*

What is the number of the allele *L* and the allele *l* in the population?

Number of *L* alleles:			**Number of *l* alleles:**		
LL	(2 *L* × 25) =	50	*LL*	(0 *l*) =	0
Ll	(1 *L* × 50) =	50	*Ll*	(1 *l* × 50) =	50
ll	(0 *L*) =	0	*ll*	(2 *l* × 25) =	50
		100 *L*			100 *l*

To determine the frequency of each allele, calculate its percentage from the total number of alleles in the population; in each case, 100/200 = 50% = 0.5. The sperm and eggs produced by this population will also contain these alleles in these frequencies. Assuming random mating (all possible gametes have an equal chance to combine with any other), we can calculate the ratio of genotypes in the next generation by using a Punnett square.

There is an important difference between a Punnett square used for a cross between individuals and the one below. Below the sperm and eggs are those produced by the members of a population—not those produced by a single male and female. As you can see, the results of the Punnett square indicate that the frequency for each allele in the next generation is still 0.5.

		sperm	
		0.5 *L*	0.5 *l*
eggs	0.5 *L*	0.25 *LL*	0.25 *Ll*
	0.5 *l*	0.25 *Ll*	0.25 *ll*

Results:
0.25*LL* + 0.5*Ll* + 0.25*ll*
($\frac{1}{4}LL + \frac{1}{2}Ll + \frac{1}{4}ll$)

Therefore, sexual reproduction alone cannot bring about a change in allele frequencies. Also, the dominant allele need not increase from one generation to the next. Dominance does not cause an allele to become a common allele. The potential constancy, or equilibrium state, of gene pool frequencies was independently recognized in 1908 by G. H. Hardy, an English mathematician, and W. Weinberg, a German physician. They used the binomial expression ($p^2 + 2pq + q^2$) to calculate the genotypic and allele frequencies of a population. Figure 19.1 shows you how this is done.

The **Hardy-Weinberg law** states that an equilibrium of allele frequencies in a gene pool, calculated by using the expression $p^2 + 2pq + q^2$, will remain in effect in each succeeding generation of a sexually reproducing population as long as five conditions are met:

$$p^2 + 2pq + q^2$$

p^2 = % homozygous dominant individuals
p = frequency of dominant allele
q^2 = % homozygous recessive individuals
q = frequency of recessive allele
$2pq$ = % heterozygous individuals

Realize that $p + q = 1$ (there are only 2 alleles)
$p^2 + 2pq + q^2 = 1$ (these are the only genotypes)

Example

An investigator has determined by inspection that 16% of a human population has a continuous hairline (recessive trait). Using this information, we can complete all the genotypic and allele frequencies for the population, provided the conditions for Hardy-Weinberg equilibrium are met.

Given: $q^2 = 16\% = 0.16$ are homozygous recessive individuals

Therefore, $q = \sqrt{0.16} = 0.4$ = frequency of recessive allele
$p = 1.0 - 0.4 = 0.6$ = frequency of dominant allele
$p^2 = (0.6)(0.6) = 0.36 = 36\%$ are homozygous dominant individuals
$2pq = 2(0.6)(0.4) = 0.48 = 48\%$ are heterozygous individuals
or
$= 1.00 - 0.52 = 0.48$

} 84% have the dominant phenotype

Figure 19.1 Calculating gene pool frequencies using the Hardy-Weinberg equation.

1. No mutations: allelic changes do not occur, or changes in one direction are balanced by changes in the opposite direction.
2. No gene flow: migration of alleles into or out of the population does not occur.
3. Random mating: individuals pair by chance and not according to their genotypes or phenotypes.
4. No genetic drift: the population is very large, and changes in allele frequencies due to chance alone are insignificant.
5. No selection: no selective agent favors one genotype over another.

In real life, these conditions are rarely, if ever, met, and allele frequencies in the gene pool of a population do change from one generation to the next. Therefore, evolution has occurred. The significance of the Hardy-Weinberg law is that it tells us what factors cause evolution—those that violate the conditions listed. Evolution can be detected by noting any deviation from a Hardy-Weinberg equilibrium of allele frequencies in the gene pool of a population.

The accumulation of small changes in the gene pool over a relatively short period of two or more generations is called microevolution. Microevolution is involved in the origin of species to be discussed later in this chapter, as well as in the history of life recorded in the fossil record.

A change in allele frequencies results in a change in phenotype frequencies. Figure 19.2 illustrates a process called **industrial melanism.** Before soot is introduced into the air due to industry, the original peppered moth population has only 10% dark-colored moths. When dark-colored moths rest on light trunks, they are seen and eaten by predatory birds. With the advent of industry, the trunks of trees darken and it is the light-colored moths that stand out and are eaten. The birds are acting as a selective agent, and microevolution occurs; the last generation observed has 80% dark-colored moths.

A Hardy-Weinberg equilibrium provides a baseline by which to judge whether evolution has occurred. Any change of allele frequencies in the gene pool of a population signifies that evolution has occurred.

Practice Problems 1*

1. In a certain population, 21% are homozygous dominant, 49% are heterozygous, and 30% are homozygous recessive. What percentage of the next generation is predicted to be homozygous dominant, assuming a Hardy-Weinberg equilibrium?
2. Of the members of a population of pea plants, 9% are short (recessive). What are the frequencies of the recessive allele *t* and the dominant allele *T*? What are the genotypic frequencies in this population?

* *Answers to Practice Problems appear in Appendix A.*

Figure 19.2 Microevolution.
Microevolution has occurred when there is a change in gene pool frequencies—in this case, due to natural selection. On the far left, birds cannot see light moths on light tree trunks and, therefore, light moths are more frequent in the population. On the far right, birds cannot see dark moths on dark tree trunks, and dark moths are more frequent in the population. The percentage of the dark-colored phenotype has increased in the population because predatory birds can see light-colored moths against tree trunks that are now sooty due to pollution.

Causes of Microevolution

The list of conditions for allelic equilibrium implies that the opposite conditions can cause evolutionary change. The conditions that can cause a deviation from the Hardy-Weinberg equilibrium are mutation, gene flow, nonrandom mating, genetic drift, and natural selection. Only natural selection results in adaptation to the environment.

Genetic Mutations

Mutations are the raw material for evolutionary change; without mutations there could be no new variations among members of a population. Evidence of mutations in members of a *Drosophila pseudoobscura* population was gathered by R. C. Lewontin and J. L. Hubby in 1966. They extracted various enzymes and subjected them to electrophoresis, a process that separates proteins according to size and charge. These investigators concluded that a fly population has multiple alleles at no less than 30% of all its gene loci. Similar results have been found in studies on many species, demonstrating that high levels of allelic variation are the rule in natural populations.

Many mutations do not immediately affect the phenotype, and therefore they may not be detected. In a changing environment, even a seemingly harmful mutation can be a source of an adaptive variation. For example, the water flea *Daphnia* ordinarily thrives at temperatures around 20°C, but there is a mutation that requires *Daphnia* to live at temperatures between 25°C and 30°C. The adaptive value of this mutation is entirely dependent on environmental conditions.

Once alleles have mutated, certain combinations of several alleles might be more adaptive than others in a particular environment. The most favorable phenotype may not occur until just the right alleles are grouped by recombination.

Mutations cause many alleles in a gene pool to have multiple alleles. Recombination of these alleles increases the possibility for favorable phenotypes.

Gene Flow

Gene flow, also called gene migration, is the movement of alleles between populations by migration of breeding individuals. There can be constant gene flow between adjacent animal populations due to the migration of organisms (Fig. 19.3). Gene flow can increase the variation within a population by introducing novel alleles that were produced by mutation in another population. Continued gene flow makes gene pools similar and reduces the possibility of allele frequency differences among populations due to natural selection and genetic drift. Indeed, gene flow among populations can prevent speciation from occurring.

Gene flow tends to decrease the genetic diversity among populations, causing their gene pools to become more similar.

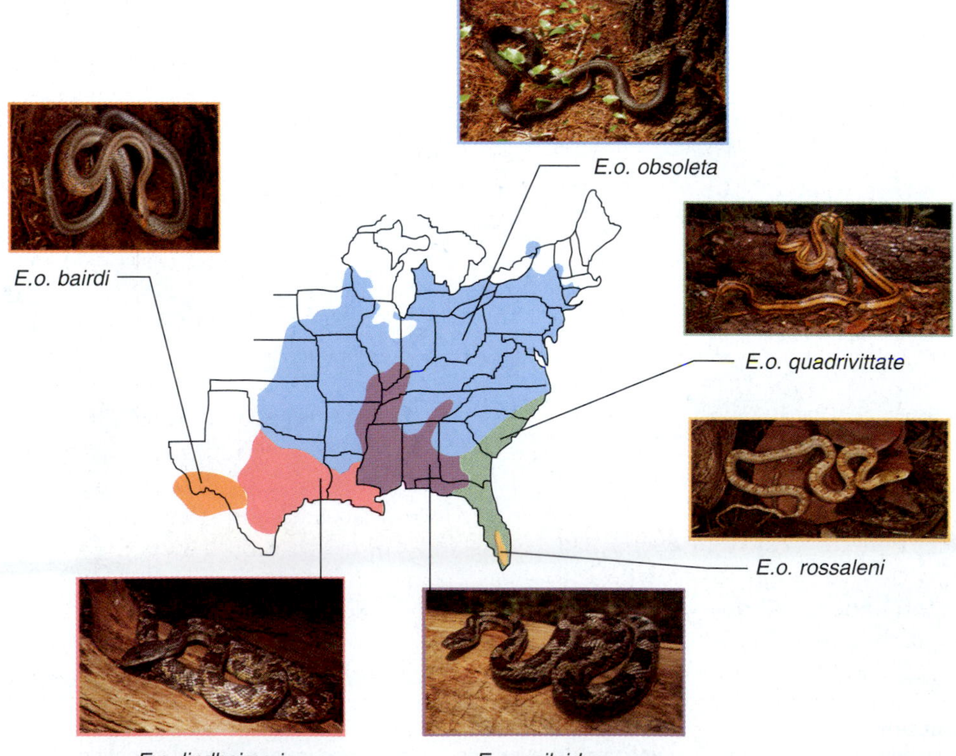

Figure 19.3 Gene flow.
Each rat snake represents a separate population of snakes. Because the populations are adjacent to one another, there is interbreeding and therefore gene flow among the populations. This keeps their gene pools somewhat similar, and each of these populations is a subspecies of the species *Elaphe obsoleta*. Therefore, each has a three-part name.

Nonrandom Mating

Random mating occurs when individuals pair by chance and not according to their genotypes or phenotypes. Inbreeding, or mating between relatives to a greater extent than by chance, is an example of **nonrandom mating.** Inbreeding does not change allele frequencies, but it does decrease the proportion of heterozygotes and increase the proportions of both homozygotes at all gene loci. In a human population, inbreeding increases the frequency of recessive abnormalities in the phenotype.

Assortative mating occurs when individuals tend to mate with those that have the same phenotype with respect to some characteristic. For example, in humans tall people seem to prefer to mate with each other. Assortative mating causes the population to subdivide into two phenotypic classes, between which there is reduced gene exchange. Homozygotes for the gene loci that control the trait in question increase in frequency, and heterozygotes for these loci decrease in frequency.

Sexual selection occurs when males compete for the right to reproduce and females choose to mate with males that have a particular phenotype. The elaborate tail of a peacock may have come about because peahens choose to mate with males with such tails.

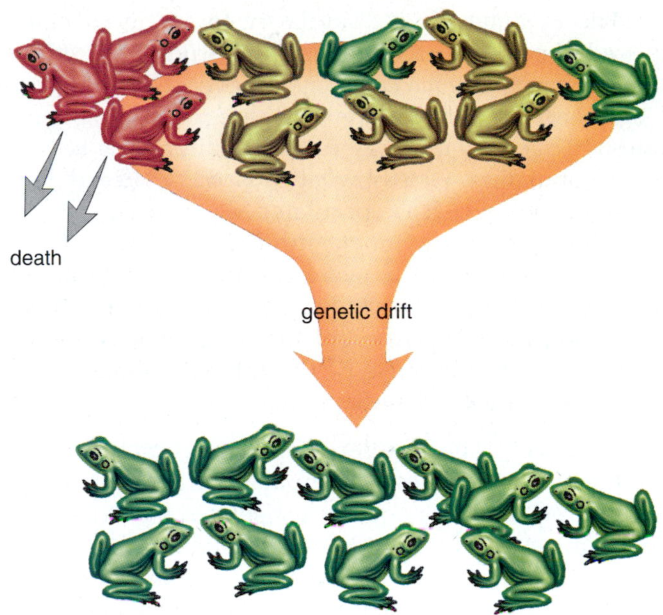

Figure 19.4 Genetic drift.
Genetic drift occurs when by chance only certain members of a population (in this case, green frogs) reproduce and pass on their genes to the next generation. The allele frequencies of the next generation's gene pool may be markedly different from that of the previous generation.

Nonrandom mating involves inbreeding and assortative mating. The former results in increased frequency of homozygotes at all loci, and the latter results in increased frequency of homozygotes at only certain loci.

Genetic Drift

Genetic drift refers to changes in allele frequencies of a gene pool due to chance. Although genetic drift occurs in both large and small populations, a larger population is expected to suffer less of a sampling error than a smaller population. Suppose you had a large bag containing 1,000 green balls and 1,000 blue balls, and you randomly drew 10%, or 200, of the balls. Because there is a large number of balls of each color in the bag, you can reasonably expect to draw 100 green balls and 100 blue balls or at least a ratio close to this. But suppose you had a bag containing only 10 green balls and 10 blue balls and you drew 10%, or only 2 balls. The chances of drawing one green ball and one blue ball with a single trial are now considerably less.

When a population is small, there is a greater chance that some rare genotype might not participate at all in the production of the next generation. Suppose there is a small population of frogs in which certain frogs for one reason or another do not pass on their traits. Certainly, the next generation will have a change in allele frequencies. When genetic drift leads to a loss of one or more alleles, other alleles over time become *fixed* in the population (Fig. 19.4).

In an experiment with 107 *Drosophila* populations, each population was in its own culture bottle. Every bottle contained eight heterozygous flies of each sex. There were no homozygous recessive or homozygous dominant flies. From the many offspring, the experimenter chose at random eight males and eight females. These were the parents for the next generation, and so forth for 19 generations. For the first few generations, most populations still contained many heterozygotes. But by the nineteenth generation, 25% of the populations contained only homozygous recessive flies and 25% contained only homozygous dominant flies for a brown-eyed allele.

Genetic drift is a random process, and therefore it is not likely to produce the same results in several populations. In California, there are a number of cypress groves, each a separate population. The phenotypes within each grove are more similar to one another than they are to the phenotypes in the other groves. Some groves have longitudinally shaped trees, and others have pyramidally shaped trees. The bark is rough in some colonies and smooth in others. The leaves are gray to bright green or bluish, and the cones are small or large. Because the environmental conditions are similar for all groves, and no correlation has been found between phenotype and environment across groves, it is hypothesized that these variations among populations are due to genetic drift.

Bottleneck Effect Sometimes a species is subjected to near extinction because of a natural disaster (e.g., earthquake or fire) or because of overharvesting and habitat loss. It is as if most of the population has stayed behind and only a few survivors have passed through the neck of a bottle. The **bottleneck effect** prevents the majority of genotype types from participating in the production of the next generation.

The extreme genetic similarity found in cheetahs is believed to be due to a bottleneck. In a study of 47 different enzymes, each of which can come in several different forms, all the cheetahs had exactly the same form. This demonstrates that genetic drift can cause certain alleles to be lost from a population. Exactly what caused the cheetah bottleneck is not known. It is speculated that perhaps cheetahs were slaughtered by nineteenth-century cattle farmers protecting their herds, or were captured by Egyptians as pets 4,000 years ago, or were decimated by a mass extinction tens of thousands of years ago. Today, cheetahs suffer from relative infertility because of the intense inbreeding that occurred after the bottleneck.

Founder Effect The **founder effect** is an example of genetic drift in which rare alleles, or combinations of alleles, occur at a higher frequency in a population isolated from the general population. After all, founding individuals contain only a fraction of the total genetic diversity of the original gene pool. Which particular alleles are carried by the founders is dictated by chance alone. The Amish of Lancaster County, Pennsylvania, are an isolated group that was begun by German founders. Today, as many as one in 14 individuals carries a recessive allele that causes an unusual form of dwarfism (affecting only lower arms and legs) and polydactylism (extra fingers) (Fig. 19.5). In the population at large, only one in 1,000 individuals has this allele.

Genetic drift is changes in gene pool frequencies due to chance.

Figure 19.5 Founder effect.
A member of the founding population of Amish in Pennsylvania had a recessive allele for a rare kind of dwarfism linked with polydactylism. The percentage of the Amish population now carrying this allele is much higher compared to that of the general population.

19.2 Natural Selection

Natural selection is the process that results in adaptation of a population to the biotic and abiotic environments. The biotic environment includes organisms that seek resources through competition, predation, and parasitism. The abiotic environment includes weather conditions dependent chiefly upon temperatures and precipitation. In the previous century, Charles Darwin, the father of evolution, became convinced that species evolve (change) with time and suggested natural selection as the mechanism for adaptation to the environment. Here, we restate Darwin's hypothesis of natural selection in the context of modern evolutionary theory.

Evolution by natural selection requires:

1. variation. The members of a population differ from one another.
2. inheritance. Many of these differences are heritable genetic differences.
3. differential adaptedness. Some of these differences affect how well an organism is adapted to its environment.
4. differential reproduction. Individuals that are better adapted to their environment are more likely to reproduce, and their fertile offspring will make up a greater proportion of the next generation.

Differential reproduction is the measure of an individual's fitness. Population geneticists speak of **relative fitness**, that is the fitness of one phenotype compared to another.

Types of Selection

Most of the traits on which natural selection acts are polygenic and controlled by more than one pair of alleles located at different gene loci. Such traits have a range of phenotypes, the frequency distribution of which usually resembles a bell-shaped curve.

Three types of natural selection have been described for any particular trait. They are directional selection, stabilizing selection, and disruptive selection.

Directional Selection

Directional selection occurs when an extreme phenotype is favored and the distribution curve shifts in that direction. Such a shift can occur when a population is adapting to a changing environment.

Industrial melanism, discussed earlier and depicted in Figure 19.2, is an example of directional selection. As you may know, indiscriminate use of antibiotics and pesticides results in a wide distribution of bacteria and insects that are resistant to these chemicals. When an antibiotic is administered, some bacteria may survive because they are genetically resistant to the antibiotic. These are the bacteria that are

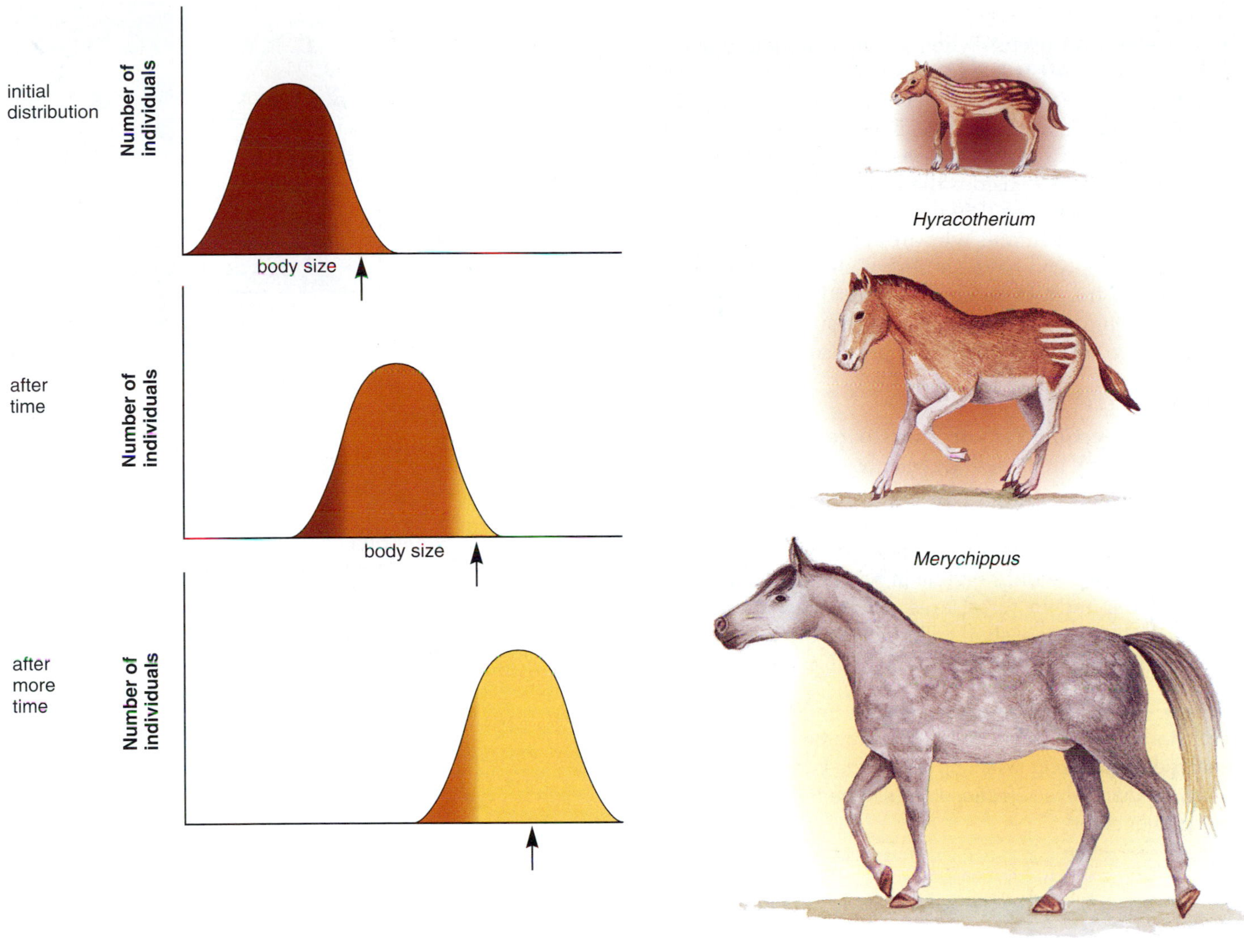

Figure 19.6 Directional selection.
Natural selection favors one extreme phenotype (see arrows), and there is a shift in the distribution curve. *Equus,* the modern-day horse, evolved from *Hyracotherium,* which was about the size of a dog. This small animal could have hidden among trees and had low-crowned teeth for browsing. When grasslands began to replace forest, the ancestors of *Equus* may have been subject to selective pressure for the development of strength, intelligence, speed, and durable grinding teeth. A larger size provided the strength needed for combat, a larger skull made room for a larger brain, elongated legs ending in hooves provided greater speed to escape enemies, and the durable grinding teeth enabled the animals to feed efficiently on grasses.

likely to pass on their genes to the next generation. As a result, the number of resistant bacteria ever increases. Drug-resistant strains of bacteria that cause tuberculosis have now become a serious threat to the health of people worldwide.

Another example of directional selection is the human struggle against malaria, a disease caused by an infection of the liver and the red blood cells. The *Anopheles* mosquito transmits the disease-causing protozoan *Plasmodium vivax* from person to person. In the early 1960s, international health authorities thought that malaria would soon be eradicated. A new drug, chloroquine, seemed effective against *Plasmodium,* and DDT (an insecticide) spraying had reduced the mosquito population. But in the mid-1960s, *Plasmodium* was showing signs of chloroquine resistance and, worse yet, mosquitoes were becoming resistant to DDT. A few drug-resistant parasites and a few DDT-resistant mosquitoes had survived and multiplied, making the fight against malaria more difficult than ever.

The gradual increase in the size of the modern horse, *Equus,* is an example of directional selection that can be correlated with a change in the environment from forestlike conditions to grassland conditions (Fig. 19.6). Even so, as discussed previously, the evolution of the horse should not be viewed as a straight line of descent because we know of many side branches that became extinct.

Stabilizing Selection

Stabilizing selection occurs when an intermediate phenotype is favored (Fig. 19.7). It can improve adaptation of the population to those aspects of the environment that remain constant. With stabilizing selection, extreme phenotypes are selected against, and individuals near the average are favored. As an example, consider the birth weight of human infants, which ranges from 0.89 to 4.9 kg (2 to 10.8 lb). The death rate is higher for infants who are at these extremes and is lowest for babies who have an intermediate birth weight (about 3.2 kg or 7 lb). Most babies have a birth weight within this range, which gives the best chance of survival. Similar results have been found in other animals, also.

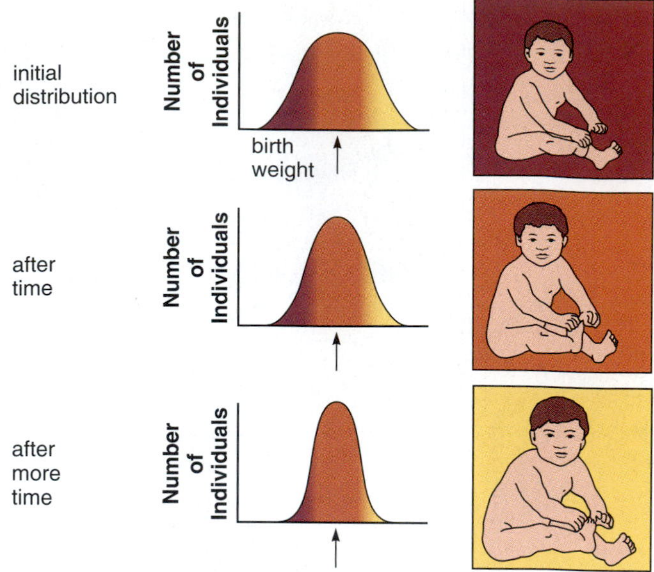

Figure 19.7 Stabilizing selection.
Natural selection favors the intermediate phenotype (see arrows) over the extremes. Today, it is observed that most human babies are of intermediate weight (about 3.2 kg, or 7 lb), and very few babies are either very small or very large.

Disruptive Selection

In **disruptive selection,** two or more extreme phenotypes are favored over any intermediate phenotype (Fig. 19.8). For example, British land snails *(Cepaea nemoralis)* have a wide habitat range that includes low-vegetation areas (grass fields and hedgerows) and forest areas. In low-vegetation areas, thrushes feed mainly on snails with dark shells that lack light bands, and in forest areas, they feed mainly on snails with light-banded shells. Therefore, these two distinctly different phenotypes are found in the population.

Directional selection favors one of the extreme phenotypes; stabilizing selection favors the intermediate phenotype; disruptive selection favors more than one extreme phenotype.

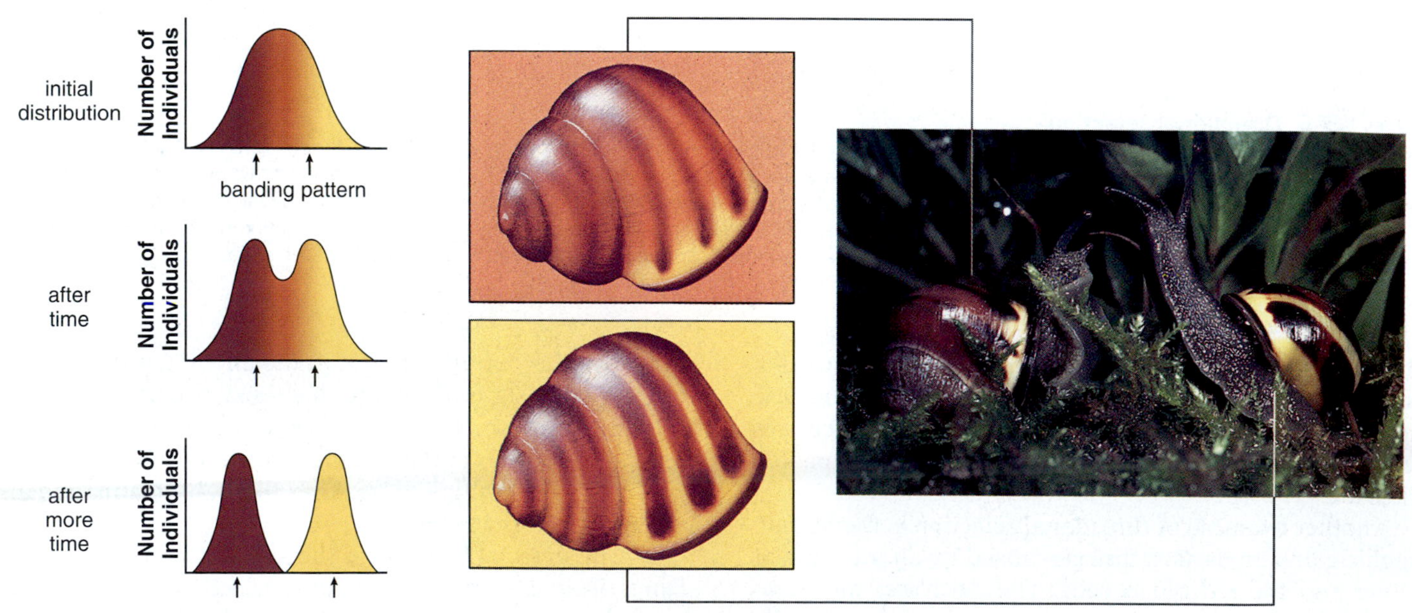

Figure 19.8 Disruptive selection.
Natural selection favors two extreme phenotypes (see arrows). Today, it is observed that British land snails comprise mainly two different phenotypes, each adapted to a different habitat.

Maintenance of Variations

A population always shows some genotypic variation. The maintenance of variation is beneficial because populations with limited variation may not be able to adapt to new conditions and may become extinct. How can variation be maintained in spite of selection constantly working to reduce it?

First, we must remember that the forces that promote variation are still at work: mutation still creates new alleles, and recombination still recombines these alleles during gametogenesis and fertilization. Second, gene flow might still occur. If the receiving population is small and is mostly homozygous, gene flow can be a significant source of new alleles. Finally, we have seen that natural selection reduces, but does not eliminate, the range of phenotypes. And disruptive selection even promotes polymorphism in a population. There are also other ways variation is maintained.

Diploidy and the Heterozygote

Only alleles that are exposed (cause a phenotypic difference) are subject to natural selection. In diploid organisms, this makes the heterozygote a potential protector of recessive alleles that otherwise would be weeded out of the gene pool. Consider these gene pool frequency distributions:

Frequency of Allele *a* in Gene Pool	Genotypic Frequencies		
	AA	*Aa*	*aa*
0.9	0.01	0.18	0.81
0.1	0.81	0.18	0.01

Notice that even when selection reduces the recessive allele from a frequency of 0.9 to 0.1, the frequency of the heterozygote remains the same. The heterozygote remains a source of the recessive allele for future generations. In a changing environment, the recessive phenotype may then be favored by natural selection.

Sickle-Cell Disease

In certain regions of Africa, the importance of the heterozygote in maintaining variation is exemplified by sickle-cell disease. The relative fitness of the heterozygote and the two homozygotes will determine their percentages in the population. When the ratio of two or more phenotypes remains the same in each generation, it is called *balanced polymorphism*. The optimum ratio will be the one that correlates with the survival of the most offspring.

Individuals with sickle-cell disease have the genotype Hb^SHb^S and tend to die at an early age due to hemorrhaging and organ destruction. Those who are heterozygous and have sickle-cell trait (Hb^AHb^S) are better off; their red blood cells usually become sickle shaped only when the oxygen content of the environment is low. Geneticists studying the distribution of sickle-cell disease in Africa have found that the recessive allele (Hb^S) has a higher frequency (0.2 to as high as 0.4 in a few areas) in regions with malaria (Fig. 19.9).

Malaria is caused by a parasite that lives in and destroys the red blood cells of the normal homozygote (Hb^AHb^A). The parasite is unable to live in the red blood cells of the heterozygote (Hb^AHb^S) because the infection causes the red blood cells to become sickle shape. Sickle-shaped red blood cells lose potassium and this causes the parasite to die. Therefore, the homozygote (Hb^SHb^S) is maintained at the same level in regions of Africa subject to malaria because the heterozygote is protected from both sickle-cell disease and from malaria.

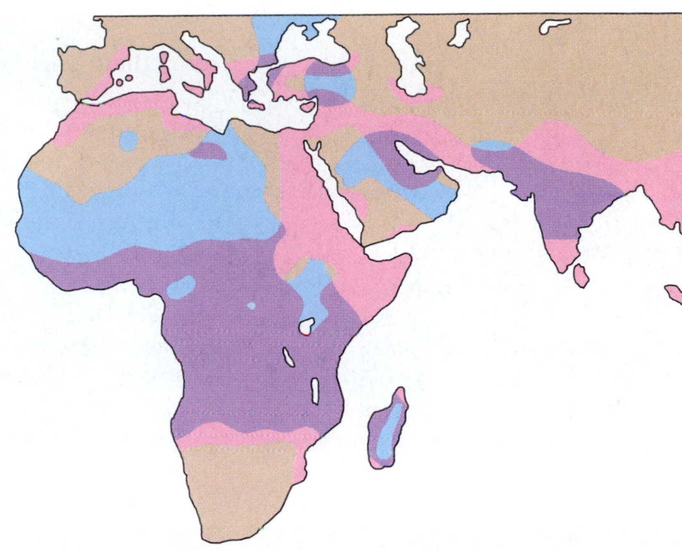

Figure 19.9 Distribution of sickle-cell disease and malaria. The pink color shows the areas where malaria was prevalent in Africa, the Middle East, and southern Europe in 1920, before eradication programs began; the blue color shows the areas where sickle-cell disease most often occurred. The overlap of these two distributions (purple) suggested that there might be a causal connection.

Genotype	Phenotype	Result
Hb^AHb^A	Normal	Dies due to malarial infection
Hb^AHb^S	Sickle-cell trait	Lives due to protection from both
Hb^SHb^S	Sickle-cell disease	Dies due to sickle-cell disease

Even after populations become adapted to their environments, variation is promoted and maintained by various mechanisms. In diploid organisms, the heterozygote genotype can help maintain variation.

least flycatcher, *Empidonax minimus*

Acadian flycatcher, *Empidonax virescens*

Traill's flycatcher, *Empidonax traillii*

Figure 19.10 Biological definition of a species.
Although the three species are nearly identical, we would still expect to find some structural feature that distinguishes them. We know they are separate species because they are reproductively isolated—the members of each species reproduce only with one another. Each species has a characteristic song and its own particular habitat during the mating season as well.

19.3 Speciation

Speciation is the splitting of one species into two or more species or the transformation of one species into a new species over time. Speciation is the final result of changes in gene pool allele and genotypic frequencies.

What Is a Species?

Sometimes it is very difficult to tell one **species** [L. *species*, a kind] from another. Before we consider the origin of species, then, it is first necessary to define a species. For Linnaeus, the father of taxonomy, one species was separated from another by morphology; that is, their physical traits differed. Darwin saw that similar species, such as the three flycatcher species in Figure 19.10, are related by common descent. The field of population genetics has produced the *biological definition of a species:* the members of one species interbreed and have a shared gene pool, and each species is reproductively isolated from every other species. The flycatchers in Figure 19.10 are members of separate species because they do not interbreed in nature.

Gene flow occurs between the populations of a species but not between populations of different species. For example, the human species has many populations that certainly differ in physical appearance. We know, however, that all humans belong to one species because the members of these populations can produce fertile offspring. On the other hand, the red maple and the sugar maple are separate species. Each species is found over a wide geographical range in the eastern half of the United States and is made up of many populations. The members of each species' populations, however, rarely hybridize in nature. Therefore, these two plants are related but separate species.

With the advent of biochemical genetics, a new way has arisen to help distinguish species. The phylogenetic species concept suggests that DNA/DNA comparison techniques can indicate the relatedness of groups of organisms.

Reproductive Isolating Mechanisms

For two species to be separate, they must be reproductively isolated; that is, gene flow must not occur between them. A *reproductive isolating mechanism* is any structural, functional, or behavioral characteristic that prevents successful reproduction from occurring. Table 19.1 lists the mechanisms by which reproductive isolation is maintained.

Prezygotic isolating mechanisms are those that prevent reproduction attempts and make it unlikely that fertilization will be successful if mating is attempted. Habitat isolation, temporal isolation, behavioral isolation, and mechanical isolation make it highly unlikely that particular genotypes will contribute to the gene pool of a population.

Table 19.1

Reproductive Isolating Mechanisms

Isolating Mechanism	Example
Prezygotic	
Habitat isolation	Species at same locale occupy different habitats.
Temporal isolation	Species reproduce at different seasons or different times of day.
Behavioral isolation	In animals, courtship behavior differs or they respond to different songs, calls, pheromones, or other signals.
Mechanical isolation	Genitalia unsuitable for one another.
Postzygotic	
Gamete isolation	Sperm cannot reach or fertilize egg.
Zygote mortality	Fertilization occurs, but zygote does not survive.
Hybrid sterility	Hybrid survives but is sterile and cannot reproduce.
F_2 fitness	Hybrid is fertile but F_2 hybrid has reduced fitness.

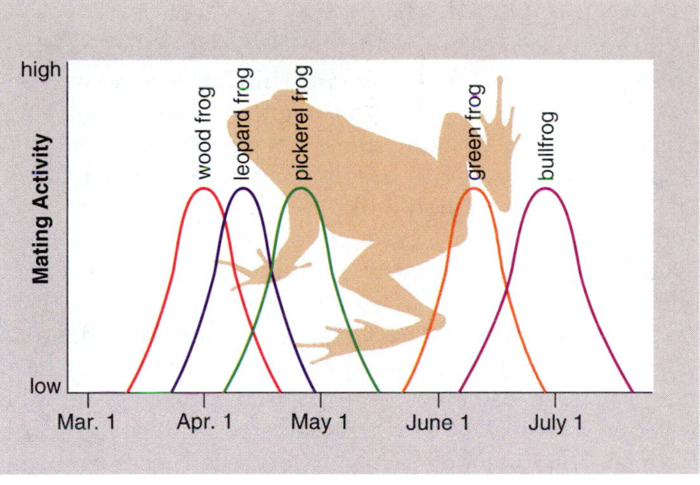

Figure 19.11 Temporal isolation.
Five species of frogs of the genus *Rana* are all found at Ithaca, New York. The species remain separate because the period of most active mating is different for each and because whenever there is an overlap, different breeding sites are used. For example, pickerel frogs are found in streams and ponds on high ground, leopard frogs in lowland swamps, and wood frogs in woodland ponds or shallow water.

Habitat isolation. When two species occupy different habitats, even within the same geographic range, they are less likely to meet and to attempt to reproduce. This is one of the reasons the flycatchers in Figure 19.10 do not mate, and the red maple and sugar maple already mentioned do not exchange pollen. In tropical rain forests, many animal species are restricted to a particular level of the forest canopy, and in this way they are isolated from similar species.

Temporal isolation. Two species can live in the same locale, but if each reproduces at a different time of year, they do not attempt to mate. For example, *Reticulitermes hageni* and *R. virginicus* are two species of termites. The former has mating flights in March through May, whereas the latter mates in the fall and winter months. Similarly, the frogs featured in Figure 19.11 have different periods of most active mating.

Behavioral isolation. Many animal species have courtship patterns that allow males and females to recognize one another. Male fireflies are recognized by females of their species by the pattern of their flashings; similarly, male crickets are recognized by females of their species by their chirping. Many males recognize females of their species by sensing chemical signals called pheromones. For example, female gypsy moths secrete chemicals from special abdominal glands. These chemicals are detected downwind by receptors on antennae of males.

Mechanical isolation. When animal genitalia or plant floral structures are incompatible, reproduction cannot occur. Inaccessibility of pollen to certain pollinators can prevent cross-fertilization in plants, and the sexes of many insect species have genitalia that do not match or other characteristics that make mating impossible. For example, male dragonflies have claspers that are suitable for holding only the females of their own species.

Postzygotic isolating mechanisms prevent hybrid offspring from developing or breeding, even if reproduction attempts have been successful. Gamete isolation, zygote mortality, hybrid sterility, and reduced F_2 fitness all make it unlikely that particular genotypes will contribute to the gene pool of a population.

Gamete isolation. Even if the gametes of two different species meet, they may not fuse to become a zygote. In animals, the sperm of one species may not be able to survive in the reproductive tract of another species, or the egg may have receptors only for sperm of its species. In plants, the stigma controls which pollen grains can successfully complete pollination.

Zygote mortality, hybrid sterility, and F_2 fitness. If by chance two of the frog species in Figure 19.11 do form hybrid zygotes, the zygotes fail to complete development or else the offspring are frail. As is well known, a cross between a horse and a donkey produces a mule, which is usually sterile—it cannot reproduce. In some cases, mules are fertile, but the F_2 generation is not. This has also been observed in both evening primrose and cotton plants.

The members of a biological species are able to breed and produce fertile offspring only among themselves. There are several mechanisms that keep species reproductively isolated from one another.

Modes of Speciation

Whenever reproductive isolation develops, speciation has occurred. Ernst Mayr, at Harvard University, proposed one model of speciation after observing that when a population is geographically isolated from other populations, gene flow stops. Variations due to different mutations, genetic drift, and directional selection build up, causing first postzygotic and then prezygotic reproductive isolation to occur. Mayr called this model **allopatric speciation** [Gk. *allo*, different, and *patri*, fatherland] (Fig. 19.12*a*).

With **sympatric speciation** [Gk. *sym*, together, and *patri*, fatherland], a population develops into two or more reproductively isolated groups without prior geographic isolation (Fig. 19.12*b*). The best evidence for this type of speciation is found among plants, where it can occur by means of polyploidy. In this case, there would be a multiplication of the chromosome number in certain plants of a single species. Sympatric speciation can also occur due to hybridization between two species, followed by a doubling of the chromosome number. Polyploid plants are reproductively isolated by a postzygotic mechanism; they can reproduce successfully only with other like polyploids, and backcrosses with diploid parents are sterile.

Adaptive Radiation

Adaptive radiation is the rapid development from a single ancestral species of many new species, which have spread out and become adapted to various ways of life. The 13 species of finches that live on the Galápagos Islands are believed to be descended from a single type of ancestral finch from the mainland. The populations on the various islands were subjected to the founder effect and the process of natural selection. Because of natural selection, each population became adapted to a particular habitat on its island. In time, the various populations became so genotypically different that now, when by chance they reside on the same island, they do not interbreed and are therefore separate species. There is evidence that the finches use beak shape to recognize members of the same species during courtship. Rejection of suitors with the wrong type of beak is a behavioral type of prezygotic isolating mechanism.

Similarly, on the Hawaiian Islands, there is a wide variety of honeycreepers that are descended from a common goldfinch-like ancestor that arrived from Asia or North America about 5 million years ago. Today, honeycreepers have a range of beak sizes and shapes for feeding on various food sources, including seeds, fruits, flowers, and insects (Fig. 19.13). Adaptive radiation has also been observed in plants on the Hawaiian Islands, as discussed in the reading on pages 314–315.

Allopatric speciation but not sympatric speciation is dependent on a geographic barrier. Adaptive radiation is a particular form of allopatric speciation.

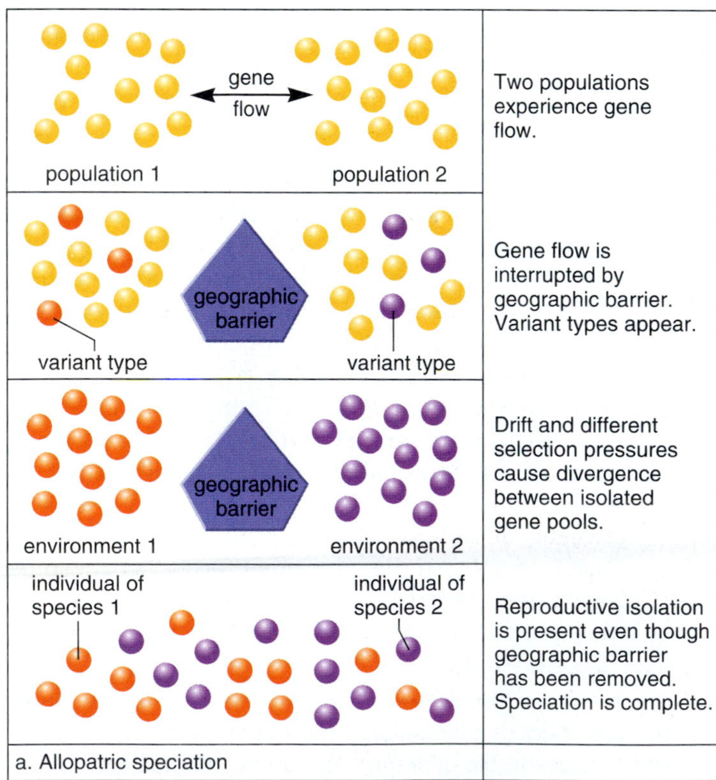

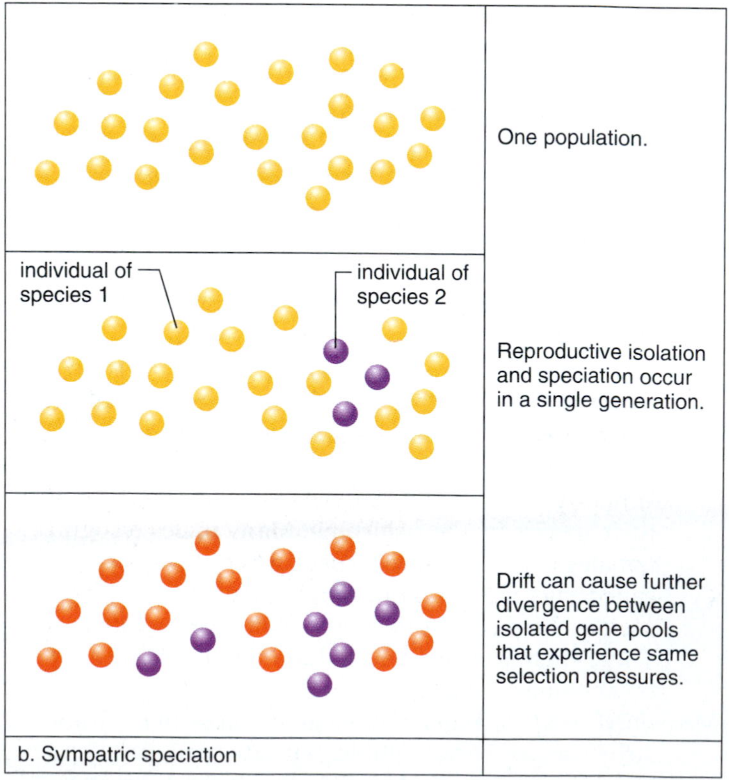

Figure 19.12 Allopatric versus sympatric speciation.
a. Allopatric speciation occurs after a geographic barrier prevents gene flow between populations that originally belonged to a single species. **b.** Sympatric speciation occurs when members of a population achieve immediate reproductive isolation without any prior geographic barrier.

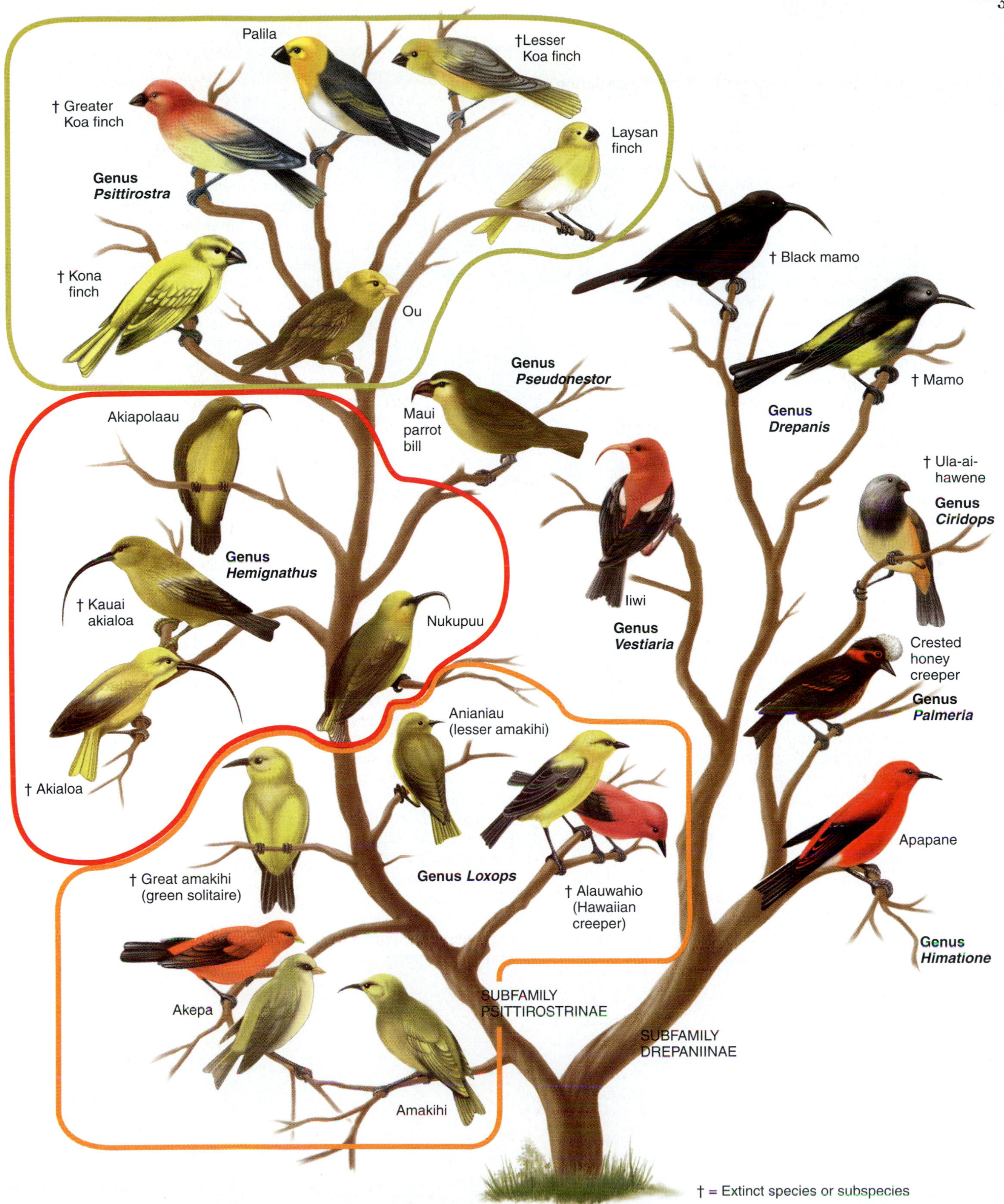

Figure 19.13 Adaptive radiation in Hawaiian honeycreepers.
More than 20 species evolved from a single species of a finch-like bird that colonized the Hawaiian Islands. The bills of honeycreepers are specialized for eating different types of foods.

Origin and Adaptive Radiation of the Hawaiian Silversword Alliance

When I was a graduate student at the University of California at Davis, I studied a genus of plants *(Calycadenia)* that belongs to a largely Californian group of plants called tarweeds. I knew that the tarweeds and plants belonging to the silversword alliance (an alliance is an assemblage of closely related species) are in the same large family of plants called Asteraceae (Compositae). But Sherwin Carlquist's anatomical research had suggested that the silversword plants and the tarweeds are even more closely related. The silverswords are found only in Hawaii, so when I took a faculty position at the University of Hawaii, I decided to gather evidence to possibly support Carlquist's interpretation.

There are 28 species of plants in the silversword alliance: five species of *Argyroxiphium,* two species of *Wilkesia,* and 21 species of *Dubautia.* This alliance constitutes one of the most spectacular examples of adaptive radiation among plants. Adaptive radiation characterizes many insular groups that have evolved in isolation away from related and nonrelated organisms. Members of the silversword alliance range in form from matlike subshrubs and rosette shrubs to large trees and climbing vines (lianas). They grow in habitats as diverse as exposed lava, dry scrub, dry woodland, moist forest, wet forest, and bogs. These habitats have a range in elevation from 75 to 3,800 meters (250– 12,500 feet), and in annual precipitation from 38 to 1,230 cm (15–485 inches).

Members of the alliance found in moist to wet forest habitats typically exhibit modifications that are adaptive in competition for light, such as increased height, vining habit, and thin leaves with a comparatively large surface area (Fig. 19A). Members of the alliance from open, more arid sites typically show features associated with conservation of water, such as decreased height, thickened leaves with comparatively low surface area, and compact internal tissues (Fig. 19B). Species of *Argyroxiphium* have more or less succulent leaves with compact tissue and channels filled with a water-binding gelatinous matrix of pectin (Fig. 19C). These adaptations help this species survive under the conditions of extreme water stress in its habitats—i.e., largely in oxygen-deficient, acid bogs or dry alpine cinder habitats.

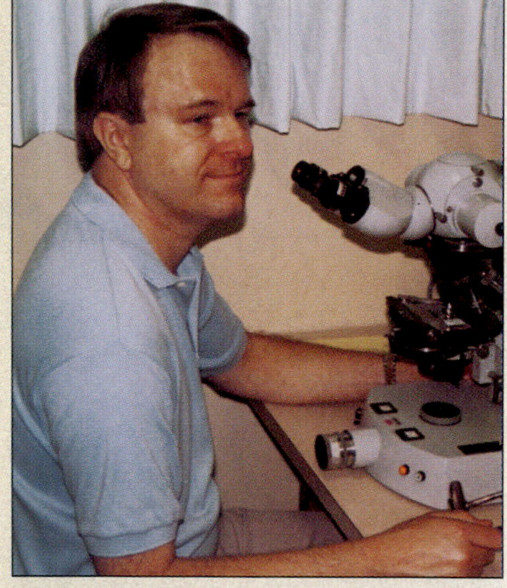

Gerald D. Carr
University of Hawaii at Manoa

I have done cytogenetic analyses of meiotic chromosome pairing in hybrids to determine relationships among the silversword plants and between these plants and the tarweeds. I have sought and found many natural interspecific and intergeneric hybrids among the diverse species of the silversword alliance. Many such hybrids also were produced artificially. Each newly discovered or created hybrid potentially provided information to fill in a missing piece of an intriguing and mysterious natural puzzle of evolution—or sometimes these findings complicated the picture further. Nearly 25 years later, the picture is still not complete. Perhaps the most exciting and personally gratifying result of my involvement with this group is the series of hybrids that have been produced between Hawaiian species and the tarweeds. These hybrids, and also a growing body of molecular data provided by one of my collaborators, Bruce Baldwin, clearly establish the origin of the Hawaiian silversword alliance from the tarweeds.

I like working with others who are innately curious about their natural surroundings and who do research for pure personal satisfaction. My scientific research helps provide insight into the mystery and meaning of the wonderful diversity of nature. An added element of excitement comes in sharing with others a discovery of one of nature's secrets that perhaps no other human has witnessed. Therefore, for me, teaching and research are naturally complementary and rewarding activities.

Readings and photographs courtesy of Gerald D. Carr, University of Hawaii at Manoa.

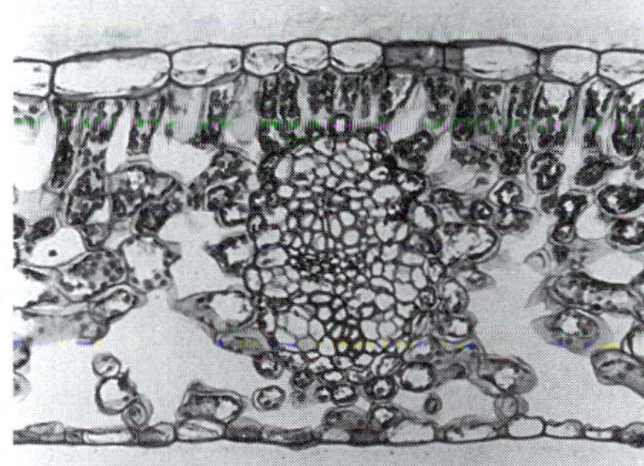

Figure 19A ***Dubautia knudsenii*** **is adapted to living in a moist habitat.**
In the cross section of the leaf, note the loose organization of tissue and thin cuticle on the upper and lower epidermis.

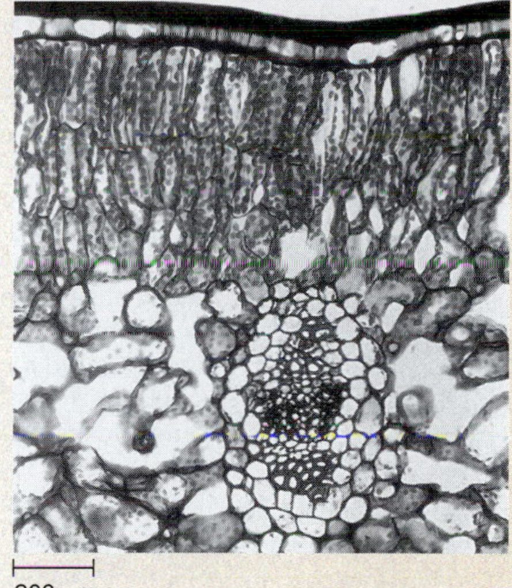

Figure 19B ***Dubautia menziesii*** **is adapted to living in open, more arid sites.**
In cross section, note the thick leaf with compact organization of tissue and very highly developed cuticle on upper and lower epidermis (lower epidermis not shown).

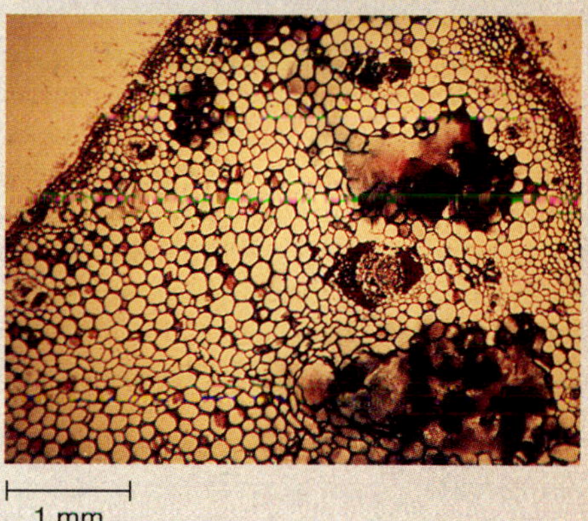

Figure 19C ***Argyroxiphium kauense*** **is adapted to living under conditions of water stress.**
In cross section, note the compact tissue and large channels of water-binding extracellular pectin that alternate with the major vascular bundles.

Connecting Concepts

We have seen that there are variations among the individuals in any population, whether the population is tuberculosis-causing bacteria in a city, the dandelions on a hill, or the squirrels in your neighborhood. Individuals vary because of the presence of mutations and, in sexually reproducing species, because of the recombination of alleles and chromosomes due to the process of meiosis and fertilization.

The field of population genetics, which utilizes the Hardy-Weinberg law, shows us how the study of evolution can be objective rather than subjective. A change in gene pool allele frequencies defines and signifies that evolution has occurred. There are various agents of evolutionary change, and natural selection is one of these.

Population genetics has also given us the biological definition of a species: the members of one species can successfully reproduce with each other but cannot successfully reproduce with members of another species. This criteria allows us to know when a new species has arisen. Species usually come into being after two populations have been geographically isolated. Suppose the present population of squirrels in your neighborhood was divided by a cavern following an earthquake. Then, there would be two genetically different populations of squirrels that are geographically isolated from each other. Suppose also that the area west of the cavern is slightly drier than that to the east. Over time, each population would become adapted to its particular environment. Eventually, the two populations may become so genetically different that even if members of each population came into contact, they would not be able to produce fertile offspring. Since gene flow between the two populations would not be possible, the squirrels would be considered separate species. By the same process, multiple species can repeatedly arise, as when a common ancestor led to thirteen species of Darwin's finches.

Summary

19.1 Evolution in a Genetic Context

All the various genes of a population make up its gene pool. The Hardy-Weinberg equilibrium is a constancy of gene pool allele frequencies that remains from generation to generation if certain conditions are met. The conditions are no mutations, no gene flow, random mating, no genetic drift, and no selection. Since these conditions are rarely met, a change in gene pool frequencies is likely. When gene pool frequencies change, evolution has occurred. Deviations from a Hardy-Weinberg equilibrium allow us to determine when evolution has taken place.

Mutations are the raw material for evolutionary change. Recombinations help bring about adaptive genotypes. Mutations, gene flow, nonrandom mating, genetic drift, and natural selection all cause deviations from a Hardy-Weinberg equilibrium. Certain genotypic variations may be of evolutionary significance only should the environment change. Gene flow occurs when a breeding individual (animals) migrates to another population or when gametes and seeds (plants) are carried into another population. Constant gene flow between two populations causes their gene pools to become similar. Nonrandom mating occurs when relatives mate (inbreeding) and assortative mating occurs. Both of these cause an increase in homozygotes. Genetic drift occurs when allele frequencies are altered by chance—that is, by sampling error. Genetic drift can cause the gene pools of two isolated populations to become dissimilar as some alleles are lost and others are fixed. Genetic drift is particularly evident after a bottleneck, when severe inbreeding occurs, or when founders start a new population.

19.2 Natural Selection

The process of natural selection can now be restated in terms of population genetics. A change in gene pool frequencies results in adaptation to the environment.

Most of the traits of evolutionary significance are polygenic; the many variations in population result in a bell-shaped curve. Three types of selection occur: (1) directional—the curve shifts in one direction as when dark-colored peppered moths become prevalent in polluted areas; (2) stabilizing—the peak of the curve increases, as when most human babies have a birth weight near the optimum for survival; and (3) disruptive—the curve has two peaks, as when *Cepaea* snails vary because a wide geographic range causes selection to vary.

Despite constant natural selection, variation is maintained. Mutations and recombination still occur; gene flow among small populations can introduce new alleles; and natural selection itself sometimes results in variation. In sexually reproducing diploid organisms, the heterozygote acts as a repository for recessive alleles whose frequency is low. In regard to sickle-cell disease, the heterozygote is more fit in areas with malaria and, therefore, both homozygotes are maintained in the population.

19.3 Speciation

The biological definition of a species recognizes that populations of the same species breed only among themselves and are reproductively isolated from other species. Reproductive isolating mechanisms prevent gene flow among species. Prezygotic isolating mechanisms (habitat, temporal, behavioral, and mechanical isolation) prevent mating from being attempted. Postzygotic isolating mechanisms (gamete isolation, zygote mortality, hybrid sterility, and F_2 fitness) prevent hybrid offspring from surviving and/or reproducing.

Allopatric speciation requires geographic isolation before reproductive isolation occurs. Sympatric speciation does not require geographic isolation for reproductive isolation to develop. The occurrence of polyploidy in plants is an example of this other type of speciation.

Adaptive radiation, as exemplified by the Hawaiian honeycreepers, is a form of allopatric speciation. It occurs because the opportunity exists for new species to adapt to new habitats.

Reviewing the Chapter

1. What is the Hardy-Weinberg law? 302–3
2. Name and discuss the five conditions of evolutionary change. 304–6
3. What is a bottleneck, and what is the founder effect? 305–6
4. State the steps required for adaptation by natural selection in modern terms. 306
5. Distinguish among directional, stabilizing, and disruptive selection by giving examples. 306–8
6. State ways in which variation is maintained in a population. 309
7. What is the biological definition of a species? 310
8. What is a reproductive isolating mechanism? Give examples of both prezygotic and postzygotic isolating mechanisms. 310–11
9. How does allopatric speciation occur? sympatric speciation? 312
10. What is adaptive radiation and how is it exemplified by the Galápagos finches and the Hawaiian honeycreepers? 313

Testing Yourself

Choose the best answer for each question.

1. Assuming a Hardy-Weinberg equilibrium, 21% of a population is homozygous dominant, 50% is heterozygous, and 29% is homozygous recessive. What percentage of the next generation is predicted to be homozygous recessive?
   a. 21%
   b. 50%
   c. 29%
   d. 42%
   e. 58%
2. A human population has a higher-than-usual percentage of individuals with a genetic disorder. The most likely explanation is
   a. mutations and gene flow.
   b. mutations and natural selection.
   c. nonrandom mating and genetic drift.
   d. nonrandom mating and gene flow.
   e. All of these are correct.
3. The offspring of better-adapted individuals are expected to make up a larger proportion of the next generation. The most likely explanation is
   a. mutations and nonrandom mating.
   b. gene flow and genetic drift.
   c. mutations and natural selection.
   d. mutations and genetic drift.
4. The continued occurrence of sickle-cell disease in parts of Africa with malaria is due to
   a. continual mutation.
   b. gene flow between populations.
   c. relative fitness of the heterozygote.
   d. disruptive selection.
   e. protozoan resistance to DDT.
5. Which of these is/are necessary to natural selection?
   a. variations
   b. differential reproduction
   c. inheritance of differences
   d. differential adaptiveness
   e. All of these are correct.
6. When a population is small, there is a greater chance of
   a. gene flow.
   b. genetic drift.
   c. natural selection.
   d. mutations occurring.
   e. sexual selection.
7. The biological definition of a species is simply the
   a. anatomical and developmental differences between two groups of organisms.
   b. geographic distribution of two groups of organisms.
   c. differences in the adaptations of two groups of organisms.
   d. reproductive isolation of two groups of organisms.
   e. difference in mutations between two groups of organisms.
8. Which of these is a prezygotic isolating mechanism?
   a. habitat isolation
   b. temporal isolation
   c. hybrid sterility
   d. zygote mortality
   e. Both a and b are correct.
9. Male moths recognize females of their species by sensing chemical signals called pheromones. This is an example of
   a. gamete isolation.
   b. habitat isolation.
   c. behavioral isolation.
   d. mechanical isolation.
   e. temporal isolation.
10. Allopatric but not sympatric speciation requires
    a. reproductive isolation.
    b. geographic isolation.
    c. prior hybridization.
    d. spontaneous differences in males and females.
    e. changes in gene pool frequencies.
11. The many species of Galápagos finches were each adapted to eating different foods. This is the result of
    a. gene flow.
    b. adaptive radiation.
    c. sympatric speciation.
    d. genetic drift.
    e. All of these are correct.

12. The following diagrams represent a distribution of phenotypes in a population. Superimpose another diagram on (a) to show that disruptive selection has occurred, on (b) to show that stabilizing selection has occurred, and on (c) to show that directional selection has occurred.

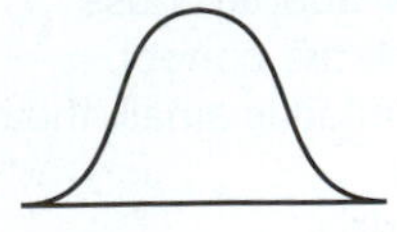
a. Disruptive selection

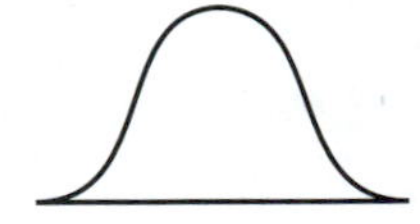
b. Stabilizing selection

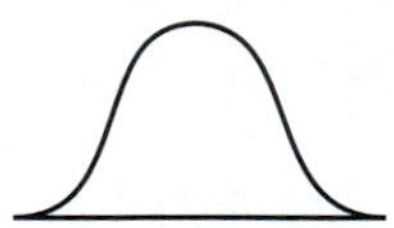
c. Directional selection

Additional Genetics Problems*

1. If $p^2 = 0.36$, what percentage of the population has the recessive phenotype, assuming a Hardy-Weinberg equilibrium?
2. If 1% of a human population has the recessive phenotype, what percentage has the dominant phenotype, assuming a Hardy-Weinberg equilibrium?
3. Four percent of the members of a population of pea plants are short (recessive characteristic). What are the frequencies of both the recessive allele and the dominant allele? What are the genotypic frequencies in this population, assuming a Hardy-Weinberg equilibrium?

Thinking Scientifically

1. You are observing a grouse population in which there are two feather phenotypes in males. One is relatively dark and blends into shadows well, and the other is relatively bright, and so is more obvious to predators. The females are uniformly dark-feathered. Observing the frequency of mating between females and the two types of males, you have recorded the following:

   matings with dark-feathered males: 13
   matings with bright-feathered males: 32

   Propose a hypothesis to explain why females apparently prefer bright-feathered males. What selective advantage might there be in choosing a male with alleles that make it more susceptible to predation? What data would help test the hypothesis?
2. A farmer uses a new pesticide. He applies the pesticide as directed by the manufacturer and loses about 15% of his crop to insects. A farmer in the next state learns of these results, uses three times as much pesticide and loses only 3% of her crop to insects. Each farmer follows this pattern for 5 years. At the end of 5 years, the first farmer is still losing about 15% of his crop to insects, but the second farmer is losing 40% of her crop to insects. How could these observations be interpreted on the basis of natural selection?

* Answers to Additional Genetics Problems appear in Appendix A.

Understanding the Terms

adaptive radiation 312
allopatric speciation 312
assortative mating 305
bottleneck effect 305
directional selection 306
disruptive selection 308
founder effect 306
gene flow 304
gene pool 302
genetic drift 305
Hardy-Weinberg law 302
industrial melanism 303
microevolution 302
natural selection 306
nonrandom mating 305
population genetics 302
postzygotic isolating mechanism 311
prezygotic isolating mechanism 310
relative fitness 306
sexual selection 305
speciation 310
species 310
stabilizing selection 308
sympatric speciation 312

Match the terms to these definitions:

a. ______________ Outcome of natural selection in which extreme phenotypes are eliminated and the average phenotype is conserved.
b. ______________ Anatomical or physiological difference between two species that prevents successful reproduction after mating has taken place.
c. ______________ Change in the genetic makeup of a population due to chance (random) events; important in small populations or when only a few individuals mate.
d. ______________ Evolution of a large number of species from a common ancestor.
e. ______________ Sharing of genes between two populations through interbreeding.

Web Connections

Exploring the Internet

http://www.mhhe.com/biosci/genbio/mader
(click on *Biology 7/e*)

The *Biology 7/e* Online Learning Center provides many resources for studying the material in this chapter including links to the following sites:

How convergent evolution led to flight in pterosaurs, birds, and bats is explored in Vertebrate Flight, and the pages linked to follow.

http://www.ucmp.berkeley.edu/vertebrates/flight/evolve.html

Punctuated Equilibrium. This site provides a history of the observations and ideas that led to the punctuated equilibrium model of evolution, as well as a detailed description of the model and the reasons why it is important, particularly to paleontologists.

http://www.talkorigins.org/faqs/punc-eq.html

Origin and History of Life

chapter 

chapter concepts

20.1 Origin of Life

20.2 History of Life

20.3 Factors That Influence Evolution

[1] BYA = billions of years ago

The earth was once devoid of life.

One of the most fascinating and frequently asked questions by laypeople and scientists alike is where did life come from? Although various answers have been proposed throughout history, today the most widely accepted hypothesis is that inorganic molecules in Earth's prebiotic oceans combined to produce organic molecules and eventually primitive cells arose. These earliest cells probably appeared around 3.5 billion years ago. About a billion years ago, oxygen-releasing photosynthesis began and an atmospheric ozone layer began to form. This shield protects the earth's surface from intense ultraviolet radiation, and consequently, life could move from water onto land. Shortly afterward (at least in geologic time), the number of multicellular life-forms increased dramatically.

If we could trace the lineages of all the millions of species ever to have evolved, the entire would resemble a dense bush. Some lines of descent are cut off close to the base; some continue in a straight line even to today; others have split, producing two or even several groups. The history of life on earth has many facets and twists and turns.

20.1 Origin of Life

Today we do not believe that life arises spontaneously from nonlife, and we say that "life comes only from life." But if this is so, how did the first form of life come about? Since it was the very first living thing, it had to come from nonliving chemicals. Could there have been an increase in the complexity of the chemicals—could a **chemical evolution** have produced the first cell(s) on the primitive earth?

The Primitive Earth

The sun and the planets, including Earth, probably formed over a 10-billion-year period from aggregates of dust particles and debris. At 4.6 billion years ago (BYA), the solar system was in place. Intense heat produced by gravitational energy and radioactivity caused the earth to become stratified into several layers. Heavier atoms of iron and nickel became the molten liquid core and dense silicate minerals became the semiliquid mantle. Upwellings of volcanic lava produced the first crust.

The mass of the earth is such that the gravitational field is strong enough to have an atmosphere. If the earth had less mass, atmospheric gases would escape into outer space. The earth's primitive atmosphere was not the same as today's atmosphere; it was produced primarily by outgassing from the interior, exemplified by volcanic eruptions. The primitive atmosphere most likely consisted mostly of water vapor (H_2O), nitrogen (N_2), and carbon dioxide (CO_2), with only small amounts of hydrogen (H_2) and carbon monoxide (CO). The primitive atmosphere, with little if any free oxygen, was a reducing atmosphere as opposed to the oxidizing atmosphere of today. This was fortuitous because oxygen (O_2) attaches to organic molecules, preventing them from joining to form larger molecules.

At first the earth was so hot that water was present only as a vapor that formed dense, thick clouds. Then as the earth cooled, water vapor condensed to liquid water, and rain began to fall. It rained in such enormous quantity over hundreds of millions of years that the oceans of the world were produced. The earth is an appropriate distance from the sun: any closer, water would have evaporated; any farther, water would have frozen.

It's also possible that the oceans were fed by celestial comets that entered Earth's gravitational field. In 1999, physicist Louis Frank presented images taken by cameras on NASA's Polar satellite to substantiate his claim that the earth is bombarded with 5 to 30 icy comets the size of a house every minute. The ice becomes water vapor that later comes down as rain, enough rain, says Frank, to raise the oceans' level by an inch in just 10,000 years.

Monomers Evolve

Other and larger comets, and pieces of them, called meteorites, frequently pelted the earth until about 3.8 BYA. Any one of these may have carried the first organic molecules to Earth. Others even suspect that bacterium-like cells evolved first on another planet and then were carried to Earth. A meteorite from Mars labeled ALH84001 landed on Earth some 13,000 years ago. When examined, experts found tiny rods similar in shape to fossilized bacteria.

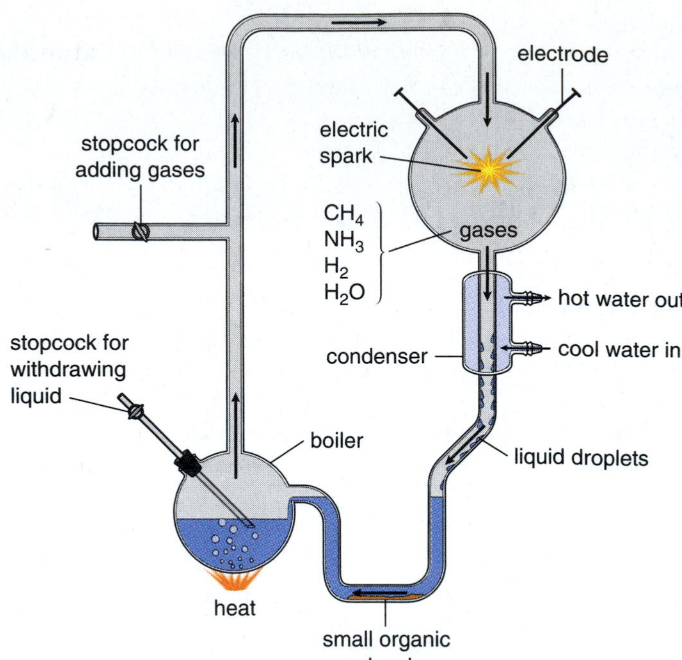

Figure 20.1 Chemical evolution in the atmosphere. Atmospheric gases were admitted to the apparatus, circulated past an energy source (electric spark), and cooled to produce a liquid that could be withdrawn. Upon chemical analysis, the liquid was found to contain various small organic molecules.

As early as 1938, Aleksandr Oparin, a Soviet biochemist, suggested that the first organic molecules could have been produced on Earth from primitive atmospheric gases in the presence of strong energy sources. The energy sources on the primitive earth included heat from volcanoes and meteorites, radioactivity from isotopes in the earth's crust, powerful electric discharges in lightning, and solar radiation, especially ultraviolet radiation.

In 1953, Stanley Miller provided support for Oparin's hypothesis through an ingenious experiment (Fig. 20.1). Miller placed a mixture resembling a strongly reducing atmosphere—methane (CH_4), ammonia (NH_3), hydrogen (H_2), and water (H_2O)—in a closed system, heated the mixture, and circulated it past an electric spark (simulating lightning). After a week's run, Miller discovered that a variety of amino acids and organic acids had been produced. Since that time, other investigators have achieved similar results by utilizing other, less-reducing combinations of gases dissolved in water.

These experiments apparently lend support to the hypothesis that the earth's first atmospheric gases could have reacted with one another to produce small organic compounds. If so, neither oxidation (there was no free oxygen) nor decay (there were no bacteria) would have destroyed these molecules, and they would have accumulated in the oceans for hundreds of millions of years. With the accumulation of small organic molecules, the oceans would have become a thick, warm organic soup.

Other investigators are concerned that Miller used ammonia as one of the atmospheric gases. They point out that whereas inert nitrogen gas (N_2) would have been abundant in the primitive atmosphere, ammonia (NH_3) would have been scarce. Where might ammonia have been abundant? A team of researchers at the Carnegie Institution in Washington, D.C. believe they have found the answer: hydrothermal vents on the ocean floor. These vents line huge **ocean ridges,** where molten magma wells up and adds material to the ocean floor (Fig. 20.2). Cool water seeping through the vents is heated to a temperature as high as 350°C, and when it spews back out it contains various mixed iron-nickel sulfides that can act as catalysts to change N_2 to NH_3. A laboratory test of this theory worked perfectly. Under ventlike conditions, 70% of various nitrogen sources were converted to ammonia within 15 minutes. German organic chemists Gunter Wachtershaüser and Claudia Huber have gone one more step. They have managed to show that organic molecules will react and amino acids will form peptides in the presence of iron-nickel sulfides under ventlike conditions.

Comets from outer space, atmospheric reactions, and hydrothermal vents could be responsible for the first organic molecules on Earth.

Figure 20.2 Chemical evolution at hydrothermal vents. Minerals that form at deep-sea hydrothermal vents like this one can catalyze the formation of ammonia and even organic molecules.

Polymers Evolve

In cells, monomers join to form polymers in the presence of enzymes which, of course, are proteins. How did the first organic polymers form if there were no proteins yet? As mentioned above, Gunter Wachtershaüser and Claudia Huber have managed to achieve the formation of peptides utilizing iron-nickel sulfides as inorganic catalysts under ventlike conditions of high temperature and pressure. These minerals have a charged surface that attracts amino acids and provides electrons so they can bond together.

Sidney Fox has shown that amino acids polymerize abiotically when exposed to dry heat. He suggests that once amino acids were present in the oceans, they could have collected in shallow puddles along the rocky shore. Then the heat of the sun could have caused them to form **proteinoids,** small polypeptides that have some catalytic properties. When he simulates this scenario in the lab and returns proteinoids to water, they form **microspheres** [Gk. *mikros,* small, little, and *sphaera,* ball], structures composed only of protein that have many properties of a cell. It's possible that even newly formed polypeptides had enzymatic properties, and some proved to be more capable than others. Those that led to the first cell or cells had a selective advantage. Fox's **protein-first hypothesis** assumes that DNA genes came after protein enzymes arose. After all, it is protein enzymes that are needed for DNA replication.

Another hypothesis is put forth by Graham Cairns-Smith. He believes that clay was especially helpful in causing polymerization of both proteins and nucleic acids at the same time. Clay also attracts small organic molecules and contains iron and zinc, which may have served as inorganic catalysts for polypeptide formation. In addition, clay has a tendency to collect energy from radioactive decay and to discharge it when the temperature and/or humidity changes. This could have been a source of energy for polymerization to take place. Cairns-Smith suggests that RNA nucleotides and amino acids became associated in such a way that polypeptides were ordered by and helped synthesize RNA. It is clear that this hypothesis suggests that both polypeptides and RNA arose at the same time.

There is still another hypothesis concerning this stage in the origin of life. The **RNA-first hypothesis** suggests that only the macromolecule RNA (ribonucleic acid) was needed to progress toward formation of the first cell or cells. Thomas Cech and Sidney Altman shared a Nobel Prize in 1989 because they discovered that RNA can be both a substrate and an enzyme. Some viruses today have RNA genes;

therefore, the first genes could have been RNA. It would seem, then, that RNA could have carried out the processes of life commonly associated today with DNA (deoxyribonucleic acid, the genetic material) and proteins (enzymes). Those who support this hypothesis say that it was an "RNA world" some 4 billion years ago.

Polymerization of monomers to produce proteins and nucleotides is the next step toward the first cell.

A Protocell Evolves

Before the first true cell arose, there would have been a **protocell** [Gk. *protos,* first], a structure that has a lipid-protein membrane and carries on energy metabolism (Fig. 20.3). Fox has shown that if lipids are made available to microspheres, lipids tend to become associated with microspheres producing a lipid-protein membrane.

Oparin, who was mentioned previously, showed that under appropriate conditions of temperature, ionic composition, and pH, concentrated mixtures of macromolecules tend to give rise to complex units called **coacervate droplets.** Coacervate droplets have a tendency to absorb and incorporate various substances from the surrounding solution. Eventually, a semipermeable-type boundary may form about the droplet. In a liquid environment, phospholipid molecules automatically form droplets called **liposomes** [Gk. *lipos,* fat, and *soma,* body]. Perhaps the first membrane formed in this manner, and the protocell contained only RNA, which functioned as both genetic material and enzymes.

The protocell would have had to carry on nutrition so that it could grow. If organic molecules formed in the atmosphere and were carried by rain into the ocean, nutrition would have been no problem because simple organic molecules could have served as food. This hypothesis suggests that the protocell was a heterotroph [Gk. *hetero,* different, and *trophe,* food], an organism that takes in preformed food. On the other hand, if the protocell evolved at hydrothermal vents, it may have carried out chemosynthesis. Chemosynthetic bacteria are autotrophs. They obtain energy for synthesizing organic molecules by oxidizing inorganic compounds like hydrogen sulfide (H_2S), a molecule that is abundant at the vents. When hydrothermal vents were first discovered in the 1970s, investigators were surprised to discover complex vent ecosystems supported by organic molecules formed by chemosynthesis, a process that does not require the energy of the sun.

At first, the protocell may have used preformed ATP (adenosine triphosphate), but as this supply dwindled, natural selection favored any cells that could extract energy from carbohydrates in order to transform ADP (adenosine diphosphate) to ATP. Glycolysis is a common metabolic pathway in living things, and this testifies to its early evolution in the history of life. Since there was no free oxygen, we can assume that the protocell carried on a form of fermentation. At first the protocell must have had limited ability to break down organic molecules and that it took millions of years for glycolysis to evolve completely. It is of interest that Fox has shown that microspheres from which protocells may have evolved have some catalytic ability and that Oparin found that coacervates do incorporate enzymes if they are available in the medium.

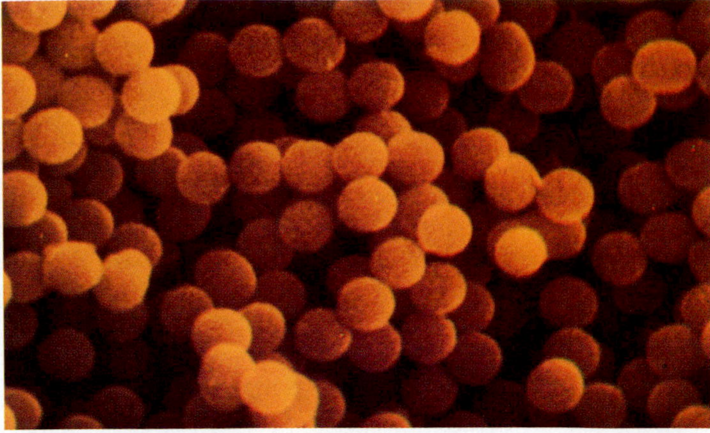

a.

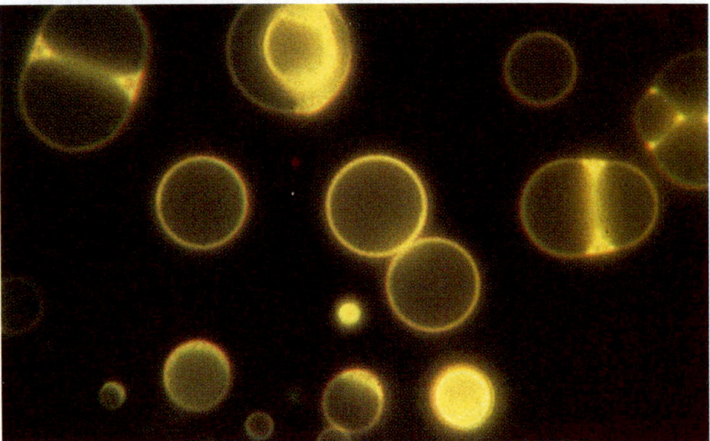

b.

Figure 20.3 Protocell anatomy.
a. Microspheres, which are composed only of protein, have a number of cellular characteristics and could have evolved into the protocell.
b. Liposomes form automatically when phospholipid molecules are put into water. Plasma membranes may have evolved similarly.

The protocell is hypothesized to have had a membrane boundary and to have been either a heterotroph or a chemoautotroph.

A Self-Replication System Evolves

Today's cell is able to carry on protein synthesis needed to produce the enzymes that allow DNA to replicate. The central dogma of genetics states that DNA directs protein synthesis and that there is a flow of information from DNA → RNA → protein. It is possible that this sequence developed in stages.

According to the RNA-first hypothesis, RNA would have been the first to evolve, and the first true cell would have had RNA genes. These genes would have directed and enzymatically carried out protein synthesis. As mentioned, ribozymes are composed of RNA and have enzymatic properties. Also, today we know there are viruses that have RNA genes. These viruses have a protein enzyme called reverse transcriptase that uses RNA as a template to form DNA. Perhaps with time, reverse transcription occurred within the protocell, and this is how DNA genes arose. If so, RNA was responsible for both DNA and protein formation. Once there were DNA genes, then protein synthesis would have been carried out in the manner dictated by the central dogma of genetics.

According to the protein-first hypothesis, proteins, or at least polypeptides, were the first of the three (i.e., DNA, RNA, and protein) to arise. Only after the protocell developed sophisticated enzymes did it have the ability to synthesize DNA and RNA from small molecules provided by the ocean. These researchers point out that a nucleic acid is a very complicated molecule, and they believe the likelihood that RNA arose *de novo* (on its own) is minimal. It seems more likely to them that enzymes were needed to guide the synthesis of nucleotides and then nucleic acids. Again, once there were DNA genes, protein synthesis would have been carried out in the manner dictated by the central dogma of genetics.

Cairns-Smith proposes that polypeptides and RNA evolved simultaneously. Therefore, the first true cell would have contained RNA genes that could have replicated because of the presence of proteins. This eliminates the baffling chicken-and-egg paradox: which came first, proteins or RNA? But it does mean, however, that two unlikely events would have to happen at the same time.

After DNA formed, the genetic code had to evolve before DNA could store genetic information. The present genetic code is subject to fewer errors than a million other possible codes. Also the present code is among the best at minimizing the effect of mutations. A single-base change in a present codon is likely to result in a substitution of a chemically similar amino acid, and therefore minimal changes in the final protein. This evidence suggests that the genetic code did undergo a natural selection process before finalizing into today's code.

Once there was a flow of information from DNA → RNA → protein, the protocell became a true cell, and biological evolution began.

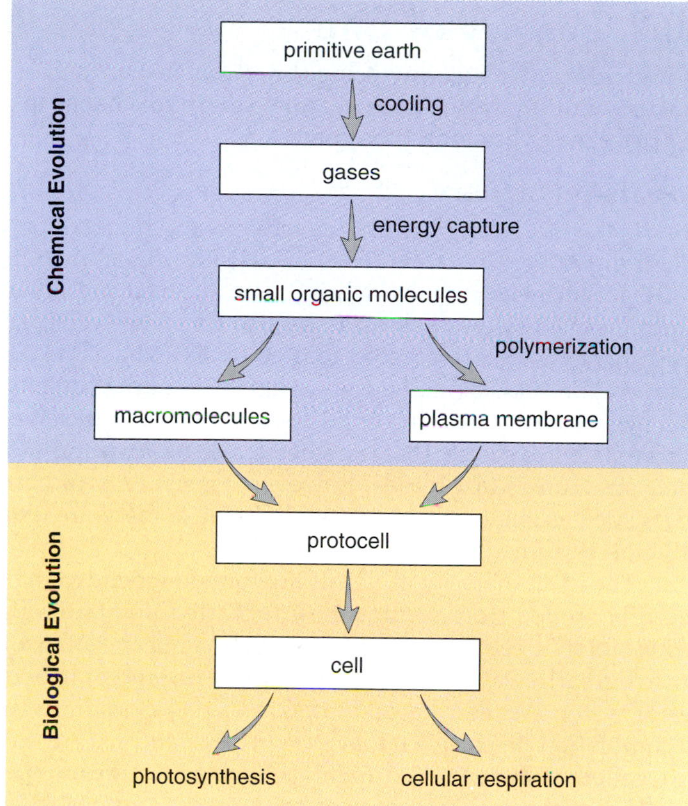

Figure 20.4 **Origin of the first cell(s).**
There was an increase in the complexity of macromolecules, leading to a self-replicating system (DNA → RNA → protein) enclosed by a plasma membrane. The protocell, a heterotrophic fermenter, underwent biological evolution, becoming a true cell, which then diversified.

Figure 20.4 reviews how most biologists believe life could have evolved on Earth:

1. There was an abiotic synthesis of small organic molecules such as amino acids and nucleotides either perhaps in the atmosphere or at hydrothermal vents.
2. These monomers joined together to form polymers either on land (warm seaside rocks or clay) or at the vents. The first polymers could have been proteins or RNA or they could have evolved together.
3. The aggregation of polymers inside a plasma membrane produced a protocell, which had some enzymatic properties such that it could grow. If the protocell developed in the ocean it was a heterotroph; if it developed at hydrothermal vents it was a chemoautotroph.
4. A true cell had evolved once the protocell contained DNA genes. The first genes may have been RNA molecules but later DNA became the information storage molecule of heredity. Biological evolution now began. The history of life began!

20.2 History of Life

In this chapter we will be tracing the events that encompass **macroevolution,** which is large-scale patterns of change taking place over very long time spans.

Fossils Tell a Story

Fossils [L. *fossilis,* dug up] are the remains and traces of past life or any other direct evidence of past life. Traces include trails, footprints, burrows, worm casts, or even preserved droppings. Usually when an organism dies, the soft parts are either consumed by scavengers or undergo bacterial decomposition. Occasionally, the organism is buried quickly and in such a way that decomposition is never completed or is completed so slowly that the soft parts leave an imprint of their structure. Most fossils, however, consist only of hard parts such as shells, bones, or teeth, because these are usually not consumed or destroyed.

The great majority of fossils are found embedded in or recently eroded from sedimentary rock. **Sedimentation** [L. *sedimentum,* a settling], a process that has been going on since the earth was formed, can take place on land or in bodies of water. Weathering and erosion of rocks produces an accumulation of particles that vary in size and nature and are called sediment. Sediment becomes a **stratum** (pl., strata), a recognizable layer in a stratigraphic sequence (Fig. 20.5). Any given stratum is older than the one above it and younger than the one immediately below it.

The fossils trapped in strata are the fossil record that tells us about the history of life (Fig. 20.6). **Paleontology** [Gk. *palaios,* ancient, old, and *ontos,* having existed; *-logy,* "study of" from *logikos,* rational, sensible] is the science of discovering and studying the fossil record and, from it, making decisions about the history of life.

Relative Dating of Fossils

In the early nineteenth century, even before the theory of evolution was formulated, geologists sought to correlate the strata worldwide. The problem was that strata change their character over great distances, and therefore a stratum in England might contain different sediments than one of the same age in Russia. Geologists discovered, however, that a stratum of the same age tended to contain the same fossil, and therefore the sequence of fossils comprises a **relative dating** method. For example, a particular species of fossil ammonite (an animal related to the chambered nautilus) has been found over a wide range and for a limited time period. Therefore, all strata around the world that contain this fossil must be of the same age.

This approach helped geologists determine the relative dates of the strata despite upheavals, but it was not particularly helpful to biologists who wanted to know the absolute age of fossils in years.

Absolute Dating of Fossils

An **absolute dating** method that relies on radioactive dating techniques assigns an actual date to a fossil. All radioactive isotopes have a particular half-life, the length of time it takes for half of the radioactive isotope to change into another stable element. If the fossil has organic matter, half of the carbon 14, ^{14}C, will have changed to nitrogen 14, ^{14}N, in 5,730 years. In order to know how much ^{14}C was in the organism to begin with, it is reasoned that organic matter always begins with the same amount of ^{14}C. (In reality, it is known that the ^{14}C levels in the air—and therefore the amount in organisms—can vary from time to time.) Now we need only compare the ^{14}C radioactivity of the fossil to that of a modern sample of organic matter. The amount of radiation left can be converted to the age of the fossil. After 50,000 years, however, the amount of ^{14}C radioactivity is so low it cannot be used to measure the age of a fossil accurately.

^{14}C is the only radioactive isotope contained within organic matter, but it is possible to use others to date rocks and from that infer the age of a fossil contained in the rock. For instance, the ratio of potassium 40 (^{40}K) to argon 40 trapped in rock is often used. If the ratio happens to be 1:1, then half of the ^{40}K has decayed and researchers know the rock is 1.3 billion years old.

Fossils, which can be dated relatively according to their location in strata and absolutely according to their content of radioactive isotopes, give us information about the history of life.

Figure 20.5 Strata.
The strata exposed by roadcuts is familiar to most travelers along major highways.

Figure 20.6 Fossils.
Fossils are the remains of past life. They can be impressions left in rocks, footprints, mineralized bones, shells, or any other evidences of life-forms that lived in the past.

The Precambrian

As a result of their study of fossils in strata, geologists have devised the **geological timescale** which divides the history of the earth into eras, then periods and epochs (Table 20.1). We will follow the biologist's tradition of first discussing the Precambrian, a period of time that encompasses the first two eras: the Archean era and the Proterozoic era.

The Precambrian is a very long period of time—it comprises about 87% of the geological timescale. It is during this period of time that life arose and the first cells came into existence. The first cells were probably prokaryotes. Prokaryotes do not have a nucleus or any membrane-bounded organelles. Of the living prokaryotes today, the archaea live in the most inhospitable of environments, such as hot springs,

Table 20.1

The Geological Timescale: Major Divisions of Geological Time with Some of the Major Evolutionary Events of Each Geological Period

Era	Period	Epoch	Millions of Years Ago	Plant Life	Animal Life
Cenozoic* (from the present to 66.4 million years ago)	Neogene	Holocene	0–0.01	Destruction of tropical rain forests by humans accelerates extinctions.	AGE OF HUMAN CIVILIZATION
		Significant Mammalian Extinction			
		Pleistocene	0.01–2	Herbaceous plants spread and diversify.	Modern humans appear.
		Pliocene	2–6	Herbaceous angiosperms flourish.	First hominids appear.
		Miocene	6–24	Grasslands spread as forests contract.	Apelike mammals and grazing mammals flourish; insects flourish.
	Paleogene	Oligocene	24–37	Many modern families of flowering plants evolve.	Browsing mammals and monkeylike primates appear.
		Eocene	37–58	Subtropical forests with heavy rainfall thrive.	All modern orders of mammals are represented.
		Paleocene	58–66	Angiosperms diversify.	Primitive primates, herbivores, carnivores, and insectivores appear.
		Mass Extinction: Dinosaurs and Most Reptiles			
Mesozoic (from 66.4 to 245 million years ago)	Cretaceous		66–144	Flowering plants spread; coniferous trees decline.	Placental mammals appear; modern insect groups appear.
	Jurassic		144–208	Cycads and other gymnosperms flourish.	Dinosaurs flourish; birds appear.
		Mass Extinction			
	Triassic		208–245	Cycads and ginkgoes appear; forests of gymnosperms and ferns dominate.	First mammals appear; first dinosaurs appear, corals and mollusks dominate seas.
		Mass Extinction			
Paleozoic (from 245 to 570 million years ago)	Permian		245–286	Conifers appear.	Reptiles diversify; amphibians decline.
	Carboniferous		286–360	Age of great coal-forming forests: club mosses, horsetails, and ferns flourish.	Amphibians diversify; first reptiles appear; first great radiation of insects.
		Mass Extinction			
	Devonian		360–408	First seed ferns appear.	Jawed fishes diversify and dominate the seas; first insects and first amphibians appear.
	Silurian		408–438	Low-lying vascular plants appear on land.	First jawed fishes appear.
		Mass Extinction			
	Ordovician		438–505	Marine algae flourish.	Invertebrates spread and diversify; jawless fishes, first vertebrates, appear.
	Cambrian		505–570	Marine algae flourish.	Invertebrates with skeletons are dominant.
Precambrian time (from 570 to 4,600 million years ago)			600 1,400–700 1,500 3,500 4,500	Oldest soft-bodied invertebrate fossils Protists evolve and diversify. Oldest eukaryotic fossils Oldest known fossils (prokaryotes) Earth forms.	

* *Many authorities divide the Cenozoic era into the Tertiary period (contains Paleocene, Eocene, Oligocene, Miocene, and Pliocene) and the Quaternary period (contains Pleistocene and Holocene).*

1.5 BYA	Oldest eukaryotic fossils
2.5 BYA	O_2 accumulates in atmosphere
3.5 BYA	Oldest known fossils
4.5 BYA	Formation of the earth

very salty lakes, and airless swamps—all of which may typify habitats on the primitive earth. The cell wall, plasma membrane, RNA polymerase, and ribosomes of archaea are more like those of eukaryotes than those of other bacteria.

The first identifiable fossils are those of complex prokaryotes. Sedimentary rocks from southwestern Greenland, dated at 3.8 BYA, contain chemical fingerprints of complex cells. And when paleobiologist J. William Schopf examined rock formations in northwestern Australia, he found 3.46 billion-year-old microfossils that resemble today's cyanobacteria, prokaryotes that carry on photosynthesis in the same manner as plants (Fig. 20.7*a*).

At this time only volcanic rocks jutted above the waves and there were as yet no continents. Strange looking boulders, called **stromatolites,** littered beaches and shallow waters (Fig. 20.7*b*). Living stromatolites can still be found today along Australia's southwestern coast. The outer surface of a stromatolite is alive with cyanobacteria that secrete a mucus. Grains of sand get caught in the mucus and bind with calcium carbonate from the water to form rock. To gain access to sunlight, the photosynthetic organisms move outward toward the surface before they are cemented in. They leave behind a menagerie of aerobic and then anaerobic bacteria caught in the layers of the rock.

The cyanobacteria in ancient stromatolites added oxygen to the atmosphere. By 2 BYA, the presence of oxygen was such that most environments were no longer suitable for anaerobic prokaryotes, and they began to decline in importance. Photosynthetic cyanobacteria and aerobic bacteria proliferated as new metabolic pathways evolved. Due to the presence of oxygen, the atmosphere became an oxidizing one instead of a reducing one. Oxygen in the upper atmosphere forms ozone (O_3), which filters out the ultraviolet (UV) rays of the sun. Before the formation of the **ozone shield,** the amount of ultraviolet radiation reaching the earth could have helped create organic molecules, but it would have destroyed any land-dwelling organisms. Once the ozone shield was in place, it meant that living things would be sufficiently protected and would be able to live on land.

The evolution of photosynthesizing organisms caused oxygen to enter the atmosphere.

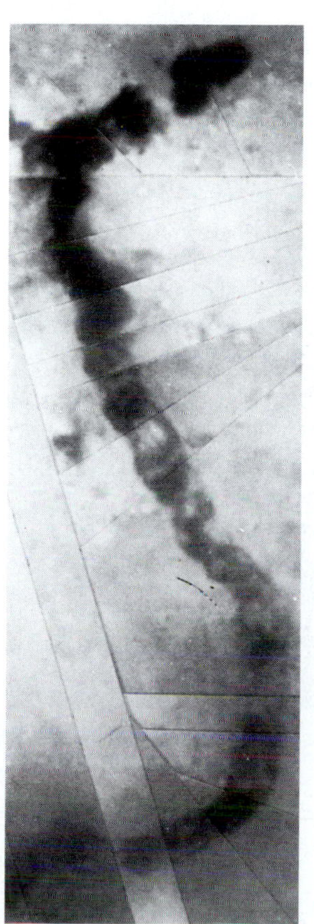

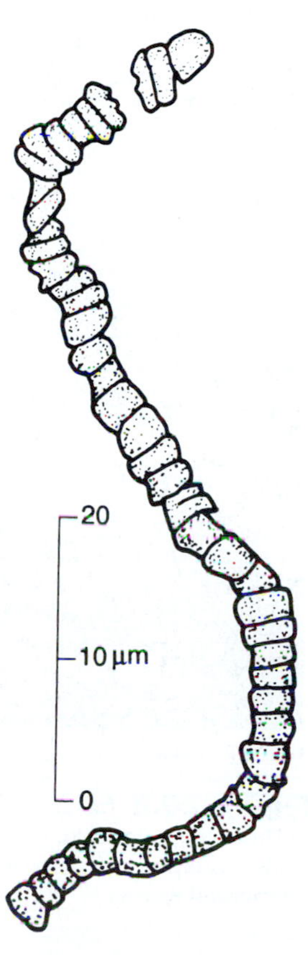

a.

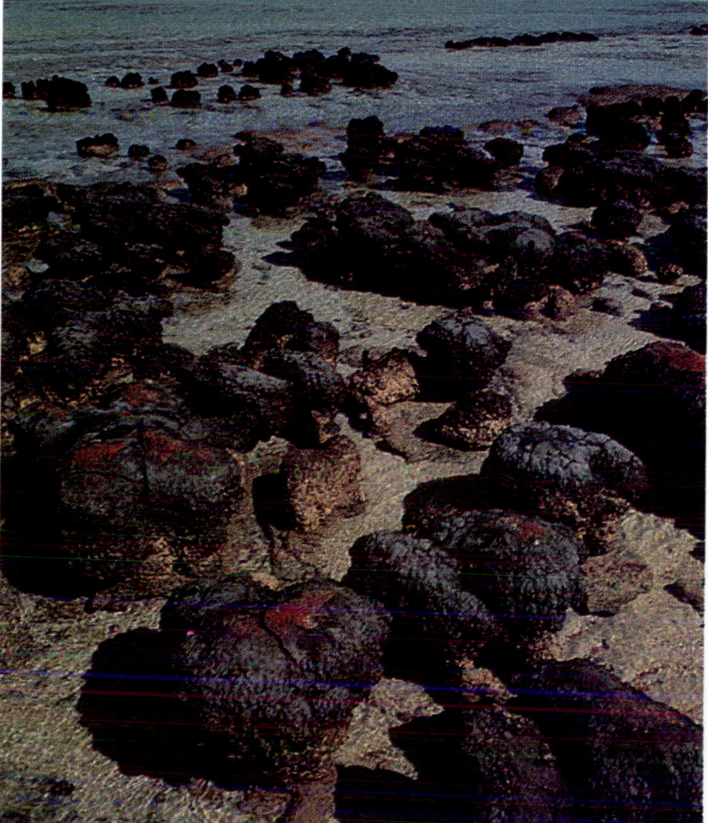

b.

Figure 20.7 Prokaryote fossil of the Precambrian.
a. The prokaryotic microorganism, *Primaevifilum* (with interpretive drawing), was found in rocks dated 3.46 BYA. **b.** Stromatolites, also date back to this time. Living stromatolites are located in shallow waters off the shores of western Australia and also in other tropical seas.

540 MYA	Cambrian animals
600 MYA	Ediacaran animals
1.40 BYA	Protists evolve and diversify
1.50 BYA	Oldest eukaryotic fossils

Eukaryotic Cells Arise

The eukaryotic cell, which originated about 1.5 BYA, is nearly always aerobic and contains a nucleus as well as other membranous organelles. Most likely the eukaryotic cell acquired its organelles gradually. It may be that the nucleus developed by an invagination of the plasma membrane. The mitochondria of the eukaryotic cell probably were once free-living aerobic prokaryotes, and the chloroplasts probably were free-living photosynthetic prokaryotes. The **endosymbiotic hypothesis** says that a nucleated cell engulfed these prokaryotes, which then became organelles. It's been suggested that flagella (and cilia) also arose by endosymbiosis. First, slender undulating prokaryotes could have attached themselves to a host cell in order to take advantage of food leaking from the host's outer membrane. Eventually, these prokaryotes adhered to the host cell and became the flagella and cilia we know today. The first eukaryotes were unicellular, as are prokaryotes.

Multicellularity Arises

It is not known when multicellularity began, but the very first multicellular forms were most likely microscopic. It's possible that the first multicellular organisms practiced sexual reproduction. Among protists (eukaryotes classified in the kingdom Protista) today, we find colonial forms in which some cells are specialized to produce gametes needed for sexual reproduction. Separation of germ cells, which produce gametes from somatic cells, may have been an important first step toward the development of complex macroscopic animals that appeared about 600 MYA.

In 1947, fossils of soft-bodied invertebrates dated 600 MYA were found in the Ediacara Hills in South Australia. Since then, similar fossils have been discovered on a number of other continents. They represent a community of animals that most likely lived on mudflats in shallow marine waters. Large immobile bizarre creatures that resembled spoked wheels, corrugated ribbons, and lettuce-like fronds had no internal organs (Fig. 20.8*a*). Perhaps they could have absorbed nutrients from the sea. Fewer in number were mobile organisms that bear some resemblance to animals of today (Fig. 20.8*b*). All were soft-bodied and their fossils are like footprints—impressions made in the sandy seafloor before their bodies decayed away. Soft-bodied animals could flourish because there were no predators to eat them.

For two billion years of the Precambrian era only prokaryotic cells existed. After eukaryotic cells arose it was nearly another million years before multicellularity occurred.

a. *Charniodiscus*

b. *Dickinsonia*

Figure 20.8 Ediacaran fossils.
a. These large frondlike organisms looked somewhat like soft corals, known today as sea pens. **b.** These fossils are interpreted to be segmented worms.

The Paleozoic Era

The Paleozoic era lasted over 300 million years. Many events occurred during this time, even though the era was quite short compared to the length of the Precambrian. For one thing, there were three major mass extinctions (see Table 20.1). An **extinction** is the total disappearance of all the members of a species or higher taxonomic group. **Mass extinctions,** which are the disappearance of a large number of species or higher taxonomic group within an interval of just a few million years, are discussed later in the chapter.

Cambrian Fossils

Figure 20.9 shows that the seas teemed with invertebrate life during the Cambrian period. Invertebrates are animals without a vertebral column. All of today's groups of animals can trace their ancestry to this time and perhaps earlier according to new molecular clock data. A **molecular clock** is based on the principle that DNA differences in certain parts of the genome occur at a fixed rate and are not tied to natural selection. The number of DNA base pair differences tells how long two species have been evolving separately.

Even if certain animals had evolved earlier no fossil evidence occurs until the Cambrian period. Why are fossils easier to find at this time? Because the animals had protective outer skeletons, and skeletons are capable of surviving the forces that are apt to destroy fossils. For example, Cambrian seafloors were dominated by now-extinct trilobites, which had thick, jointed armor covering them from head to tail. Trilobites are classified as arthropods, a major phylum of animals today. (Some Cambrian species, with most unusual eating and locomotion appendages, have been classified in phyla that no longer exist today.)

Paleontologists have sought an explanation for why animals had skeletons during the Cambrian period but not before. By this time, not only cyanobacteria but also various algae, which are floating photosynthetic organisms, were pumping oxygen into the atmosphere at a rapid rate. Perhaps the oxygen supply became great enough to permit aquatic animals to acquire oxygen even though they had outer skeletons. The presence of a skeleton reduces possible access to oxygen in seawater. Steven Stanley of Johns Hopkins University suggests that predation may have played a role. Skeletons may have evolved during the Cambrian period because skeletons help protect animals from predators.

The fossil record is rich during the Cambrian period but the animals may have evolved earlier. The richness of the Cambrian period may be due to the evolution of outer skeletons.

Figure 20.9 Sea life of the Cambrian period.
a. The animals depicted here are found as fossils in the Burgess Shale, a formation of the Rocky Mountains of British Columbia. Some lineages represented by these animals are still evolving today; others have become extinct. **b.** *Haplophrentis* has a tapering shell surrounded by an aperculum. The lateral appendages may have served as props.

Figure 20.10 Swamp forests of the Carboniferous period.
Vast swamp forests of treelike club mosses and horsetails dominated the land during the Carboniferous period (see Table 20.1). The air contained insects with wide wingspans, such as the dragonfly shown here, and amphibians lumbered from pool to pool.

Invasion of Land

Plants Sometime during the Paleozoic era, algae, which were common in the seas, most likely began to take up residence in bodies of fresh water; from there, they may have invaded damp areas on land. An association of plant roots with fungi called mycorrhizae is credited with allowing plants to live on bare rocks. The fungi are able to absorb minerals, which they pass to the plant, and the plant in turn passes carbohydrates, the product of photosynthesis, to the fungi.

Fossils of vascular plants with tissue for water transport date back to the Silurian period. They later flourished in the warm swamps of the Carboniferous period (Fig. 20.10). Club mosses, horsetails, and seed ferns were the trees of that time, and they grew to enormous size. A wide variety of smaller ferns and fernlike plants formed an underbrush.

Invertebrates The outer skeleton and jointed appendages of arthropods are adaptive for living on land. Various arthropods—spiders, centipedes, mites, and millipedes—all preceded the appearance of insects on land. Insects enter the fossil record in the Carboniferous period. The evolution of wings provided advantages that allowed insects to radiate into the most diverse and abundant group of animals today. Flying provides a way to escape enemies and find food.

Vertebrates Vertebrates are animals with a vertebral column. The vertebrate line of descent began in the early Ordovician period with the evolution of fishes. First there were jawless fishes and then fishes with jaws. Fishes are ectothermic (cold-blooded), aquatic vertebrates that have gills, scales, and fins. The cartilaginous and ray-finned fishes made their appearance in the Devonian period, which is called the Age of Fishes.

At this time the seas were filled with giant predatory fish covered with protective armor made of external bone. Sharks cruised up deep, wide rivers and the smaller lobe-finned fishes lived at the river's edge in waters too shallow for these large predators. Their fleshy fins helped them push aside debris or hold their place in strong currents. They may also have allowed these fishes to venture onto land and lay their eggs safe in inland pools. Lobe-finned fishes are believed to be ancestral to the amphibians.

Amphibians are thin-skinned vertebrates that are not fully adapted to life on land, particularly because they must return to water to reproduce. The Carboniferous swamp forests provided the water they needed and amphibians radiated into many different sizes and shapes. Some superficially resembled alligators and were covered with protective scales, others were small and snakelike, and a few were larger plant eaters. The largest measured 6 meters (20 feet) from snout to tail. The Carboniferous period is called the Age of the Amphibians.

There was a change of climate at the end of the Carboniferous period; cold and dry weather brought an end to the Age of the Amphibians and began the process that turned the great Carboniferous forests into the coal we use today to fuel our modern society.

Primitive vascular plants and amphibians were larger and more abundant during the Carboniferous period. Insects appeared and flourished to become the largest animal group today.

Figure 20.11 Dinosaurs of the Mesozoic era.
a. The dinosaurs of the Jurassic period included *Apatosaurus*, which fed on cycads and conifers. **b.** *Triceratops* (left) and *Tyrannosaurus rex* (right) were dinosaurs of the Cretaceous period, which ended with a mass extinction of dinosaurs.

a.

b.

The Mesozoic Era

Although there was a severe mass extinction at the end of the Paleozoic era, the evolution of certain types of plants and animals continued into the Triassic, the first period of the Mesozoic era. Nonflowering seed plants (collectively called gymnosperms), which had been present in the fossil record of the Permian period, became dominant. Among these largely cone-bearing plants were cycads and conifers. Cycads are short and stout with palmlike leaves; the female plant produces very large cones. Cycads were so prevalent during the Jurassic period that it is sometimes called the Age of the Cycads. Reptiles, too, can be traced back to the Permian period of the Paleozoic era. Unlike amphibians, reptiles can thrive in a dry climate because they have scaly skin and lay a shelled egg that hatches on land. Reptiles underwent an adaptive radiation during the Mesozoic era to produce forms that lived in the air, in the sea, and on the land. One group of reptiles, the therapsids, had several mammalian skeletal traits.

During the Jurassic period, large flying reptiles called pterosaurs ruled the air and giant marine reptiles with paddlelike limbs ate fishes in the sea, but on land it was the dinosaurs that prevented the evolving mammals from taking center stage. Some of the dinosaurs were of an enormous size. The gargantuan *Apatosaurus* (Fig. 20.11*a*) and the armored tractor-sized *Stegosaurus* fed on cycad seeds and conifer trees. One group of dinosaurs called theropods were bipedal and had an elongate, mobile S-shaped neck. The newly found fossil *Caudipteryx* (see Fig. 18.13) offers convincing evidence that birds are descended from dinosaurs. This fossil is a dinosaur with feathers on its arms and tail. The fossil record for birds begins with the famous fossil *Archaeopteryx*, which still retains some dinosaur-like features.

Up until 1999, Mesozoic mammal fossils largely consisted of teeth. This was changed by another fossil find from China, which has been dated at 120 MYA and named *Jeholodens*. The animal, identified as a mammal, apparently looked like a long-snouted rat. Surprisingly, *Jeholodens* still had the sprawling hindlimbs of a reptile but forelimbs under the belly like today's mammals. During the Cretaceous period, great herds of rhino-like dinosaurs, *Triceratops*, roamed the plains, as did the infamous *Tyrannosaurus rex*, which was carnivorous and played the same ecological role as lions do today (Fig. 20.11*b*). The size of a dinosaur such as *Apatosaurus* is hard for us to imagine. It was as tall as a four-story office building, and its weight was as much as that of a thousand people! How might dinosaurs have benefited from being so large? One theory is that being ectothermic (cold-blooded), the volume-to-surface ratio was favorable for retaining heat. Others believe dinosaurs were endothermic (warm-blooded) for reasons discussed in the reading on next page. At the end of the Cretaceous period, the dinosaurs became victims of a mass extinction, which will be discussed later.

During the Mesozoic era, the dinosaurs achieved an enormous size, while mammals remained small and insignificant.

Real Dinosaurs, Stand Up!

Today's paleontologists are setting the record straight about dinosaurs. Because dinosaurs are classified as reptiles, it is assumed that they must have had the characteristics of today's reptiles. They must have been ectothermic, slow-moving, and antisocial, right? Wrong!

First of all, not all dinosaurs were great lumbering beasts. Many dinosaurs were less than 1 meter (3 feet) long and their tracks indicate they moved "right along." These dinosaurs stood on two legs that were positioned directly under the body. Perhaps they were as agile as ostriches, which are famous for their great speed.

Dinosaurs may have been endothermic. Could they have competed successfully with the preevolving mammals otherwise? They must have been able to hunt prey and escape from predators as well as mammals, which are known to be active because of their high rate of metabolism. Some argue that, in contrast, ectothermic animals have little endurance and cannot keep up. They also believe that the bone structure of dinosaurs indicates they were endothermic.

Dinosaurs cared for their young much like birds do today. In Montana, paleontologist Jack Horner has studied fossilized nests complete with eggs, embryos, and nestlings (Fig. 20A). The nests are about 7.5 meters (24.6 feet) apart, the space needed for the length of an adult parent. About 20 eggs are laid in neatly arranged circles and may have been covered with decaying vegetation to keep them warm. Many contain the bones of juveniles as much as a meter long. It would seem then that baby dinosaurs remained in the nest to be fed by their parents. They must have obtained this size within a relatively short period of time, again indicating that dinosaurs were endothermic. Ectothermic animals grow slowly and take a long time to reach this size.

Dinosaurs were also social! An enormous herd of dinosaurs found by Horner and colleagues is estimated to have nearly 30 million bones, representing 10,000 animals in one area measuring about 1.6 square miles. Most likely, the herd kept on the move in order to be assured of an adequate food supply, which consisted of flowering plants that could be stripped one season and grow back the next season. The fossilized herd is covered by volcanic ash, suggesting that the dinosaurs died following a volcanic eruption.

Some dinosaurs, such as the duck-billed dinosaurs and horned dinosaurs, have a skull crest. How might it have functioned? Perhaps it was a resonating chamber, used when dinosaurs communicated with one another. Or, as with modern horned animals that live in large groups, the males could have used the skull crest in combat to establish dominance.

If dinosaurs were endothermic, fast-moving, and social animals, should they be classified as reptiles? Some say no!

a.

b.

Figure 20A **Behavior of dinosaurs.**
a. Nest of fossil dinosaur eggs found in Montana, dating from the Cretaceous period.
b. Bones of a hatchling (about 50 cm [20 inches]) found in the nest. These dinosaurs have been named *Maiosaura,* which means "good mother lizard" in Greek.

Figure 20.12 **Mammals of the Oligocene epoch.**
The artist's representation of these mammals and their habitat vegetation is based on fossil remains.

The Cenozoic Era

A new system divides the Cenozoic era into a Paleogene period and a Neogene period. We are living in the Neogene period.

Mammalian Diversification

At the end of the Mesozoic era, mammals began an adaptive radiation into the many habitats now left vacant by the demise of the dinosaurs. Mammals are endothermic and they have hair, which helps keep body heat from escaping. Their name refers to the presence of mammary glands, which produce milk to feed their young. At the start of the Paleocene epoch, mammals were small and resembled a rat. By the end of the Eocene epoch, mammals had diversified to the point that most of the modern orders were in existence. Bats are mammals that have conquered the air. Whales, dolphins, and other marine mammals live in the sea where vertebrates began their evolution in the first place. Hoofed mammals populate forests and grasslands and are fed upon by diverse carnivores. Many of the types of herbivores and carnivores of the late Paleogene period, however, are extinct today (Fig. 20.12).

Evolution of Primates

Flowering plants (collectively called angiosperms) evolved at the end of the Mesozoic era and began their radiation in the Cenozoic era. Primates are a type of mammal adapted to living in flowering trees where there is protection from predators and where food in the form of fruit is plentiful. The first primates were small squirrel-like animals, but from them evolved the first monkeys and then the apes. Apes diversified during the Miocene epoch and gave rise to the first hominids, a group that includes humans. Many of the skeletal differences between apes and humans relate to the fact that humans walk upright. Exactly what caused humans to adopt bipedalism is still being debated.

Figure 20.13 **Woolly mammoth of the Pleistocene epoch.**
Woolly mammoths, *Mammuthus,* were magnificent animals, which lived along the borders of continental glaciers.

The world's climate was progressively cooler during the Neogene period, so much so that the latter two epochs are known as the Ice Age. During periods of glaciation, snow and ice covered about one-third of the land surface of the Earth. The Pleistocene epoch was an age of not only humans, but also giant ground sloths, beavers, wolves, bison, woolly rhinoceroses, mastodons, and mammoths (Fig. 20.13). Humans have survived, but what happened to the oversized mammals just mentioned? Some think that humans became such skilled hunters they are at least partially responsible for the extinction of these awe-inspiring animals.

The Cenozoic era is the present era. Only during this time did mammals diversify and human evolution (the subject of the next chapter) begin.

20.3 Factors That Influence Evolution

It used to be thought that the earth's crust was immobile, that the continents had always been in their present positions, and that the ocean floors were only a catch basin for the debris that washed off the land. In 1920 Alfred Wegener, a German meteorologist, presented data from a number of disciplines to support his hypothesis of **continental drift.**

Continental Drift

Continental drift was finally confirmed in the 1960s; the continents are not fixed; instead, their positions and the position of the oceans have changed over time (Fig. 20.14). About 225 MYA, the continents joined to form one supercontinent that Wegener called Pangaea. First, Pangaea divided into two large subcontinents, called Gondwanaland and Laurasia, and then these also split to form the continents of today. Presently, the continents are still drifting in relation to one another.

Continental drift explains why the coastlines of several continents are mirror images of each other—the outline of the west coast of Africa matches that of the east coast of South America. The same geological structures are also found in many of the areas where the continents touched. A single mountain range runs through South America, Antarctica, and Australia. Continental drift also explains the unique distribution patterns of several fossils. Fossils of the same species of seed fern *(Glossopteris)* have been found on all the southern continents. No suitable explanation was possible previously, but now it seems plausible that the plant evolved on one continent and spread to the others when they were joined as one. Similarly, the fossil reptile *Cynognathus* is found in Africa and South America and *Lystrosaurus*, a mammal-like reptile, has now been discovered in Antarctica, far from Africa and southeast Asia, where it also occurs. With mammalian fossils, the situation is different: Australia, South America, and Africa all have their own distinctive mammals because mammals evolved after the continents separated. The mammalian biological diversity of today's world is the result of isolated evolution on separate continents.

The relationship of the continents to one another has affected the biogeography of the earth.

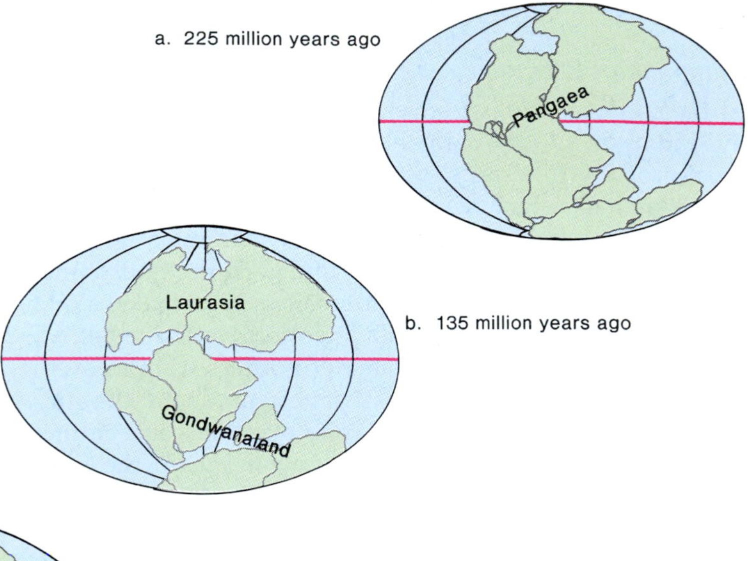

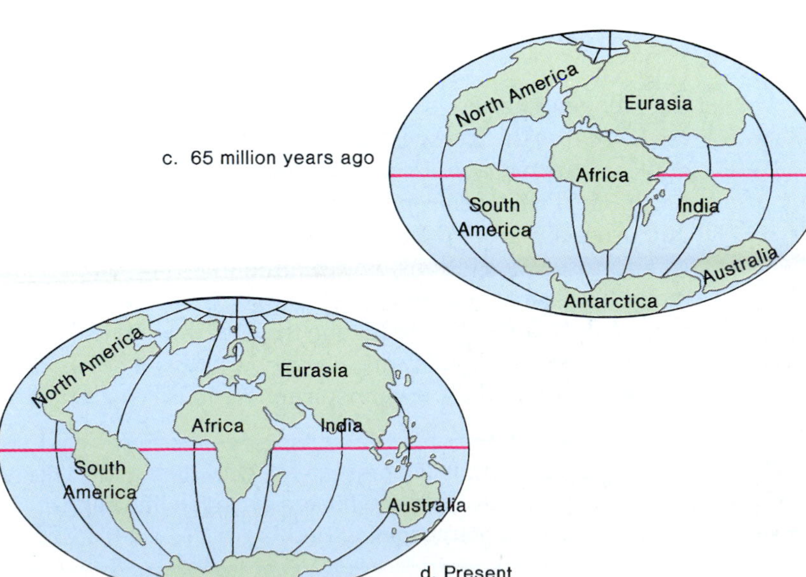

Figure 20.14 Continental drift.
a. About 225 million years ago, all the continents were joined into a supercontinent called Pangaea. **b.** When the joined continents of Pangaea first began moving apart, there were two large continents called Laurasia and Gondwanaland. **c.** By 65 million years ago, all the continents had begun to separate. This process is continuing today. **d.** North America and Europe are presently drifting apart at a rate of about 2 cm per year.

Why do the continents drift? According to a branch of geology known as **plate tectonics** [Gk. *tektos*, fluid, molten, able to flow] (tectonics refers to movements of the earth's crust), the earth's crust is fragmented into slablike plates that float on a lower hot mantle layer. The continents and the ocean basins are a part of these rigid plates, which move like conveyor belts. At **ocean ridges,** seafloor spreading occurs as molten mantle rock rises and material is added to the ocean floor (Fig. 20.15*a*). Seafloor spreading causes the continents to move a few centimeters a year on the average. At *subduction zones,* the forward edge of a moving plate sinks into the mantle and is destroyed. When an ocean floor is at the leading edge of a plate, a deep trench forms that is bordered by volcanoes or volcanic island chains. When two continents collide, the result is often a mountain range; for example, the Himalayas resulted when India collided with Eurasia. Two plates meet along a *transform boundary* where two plates scrape past one another (Fig. 20.15*b*). The San Andreas *fault* in southern California is at a transform boundary, and the movement of the two plates is responsible for the many earthquakes in that region.

The earth's crust is divided into plates that move because of seafloor spreading at ocean ridges.

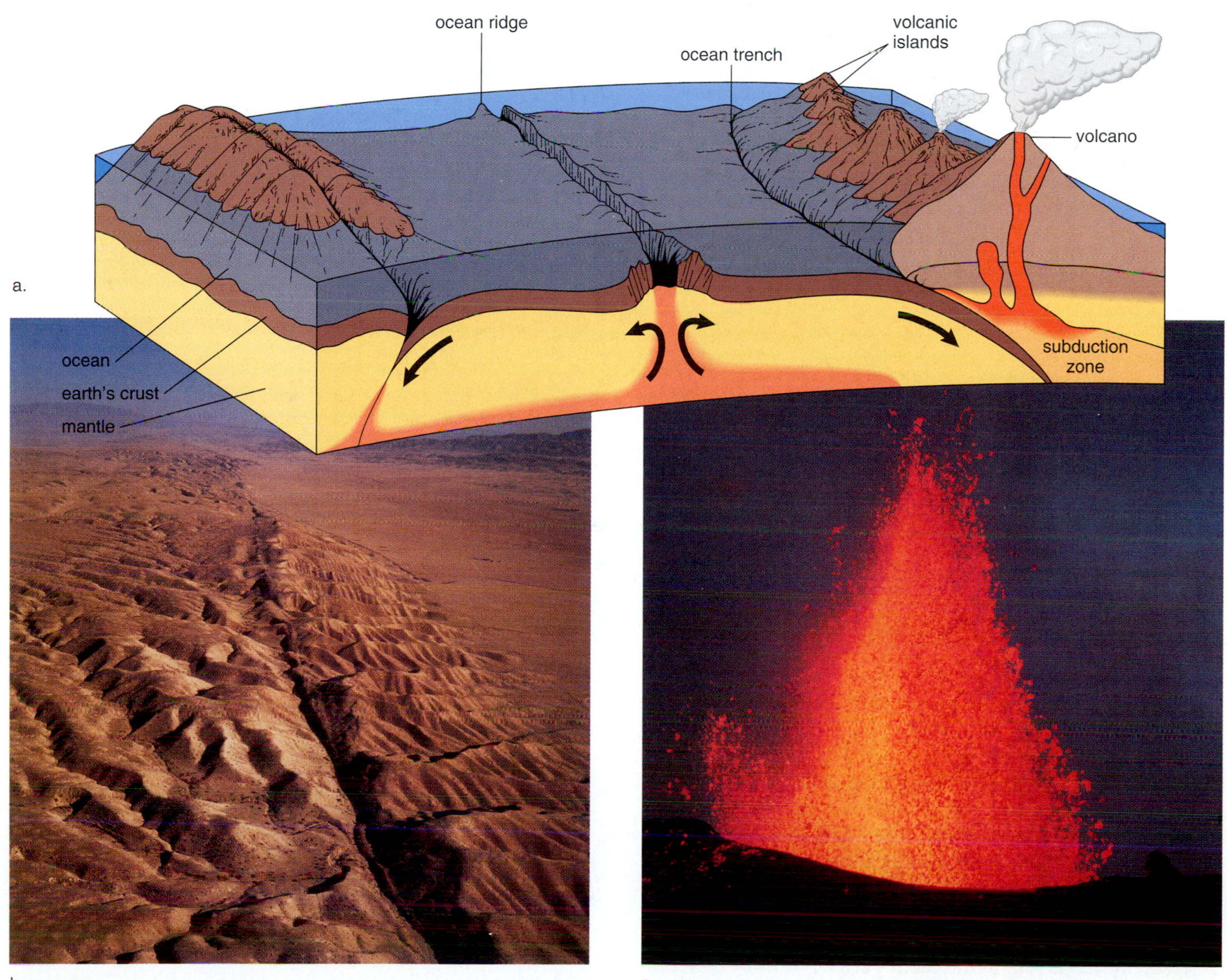

Figure 20.15 **Plate tectonics.**
a. Plates form and move away from ocean ridges toward subduction zones, where they are carried into the mantle and are destroyed. **b.** A transform boundary occurs where two plates scrape past each other. The San Andreas fault occurs at a transform boundary where earthquakes are apt to occur. **c.** Iceland is one of the few places in the world where an ocean ridge reaches the surface of the sea. The entire island is volcanic in origin.

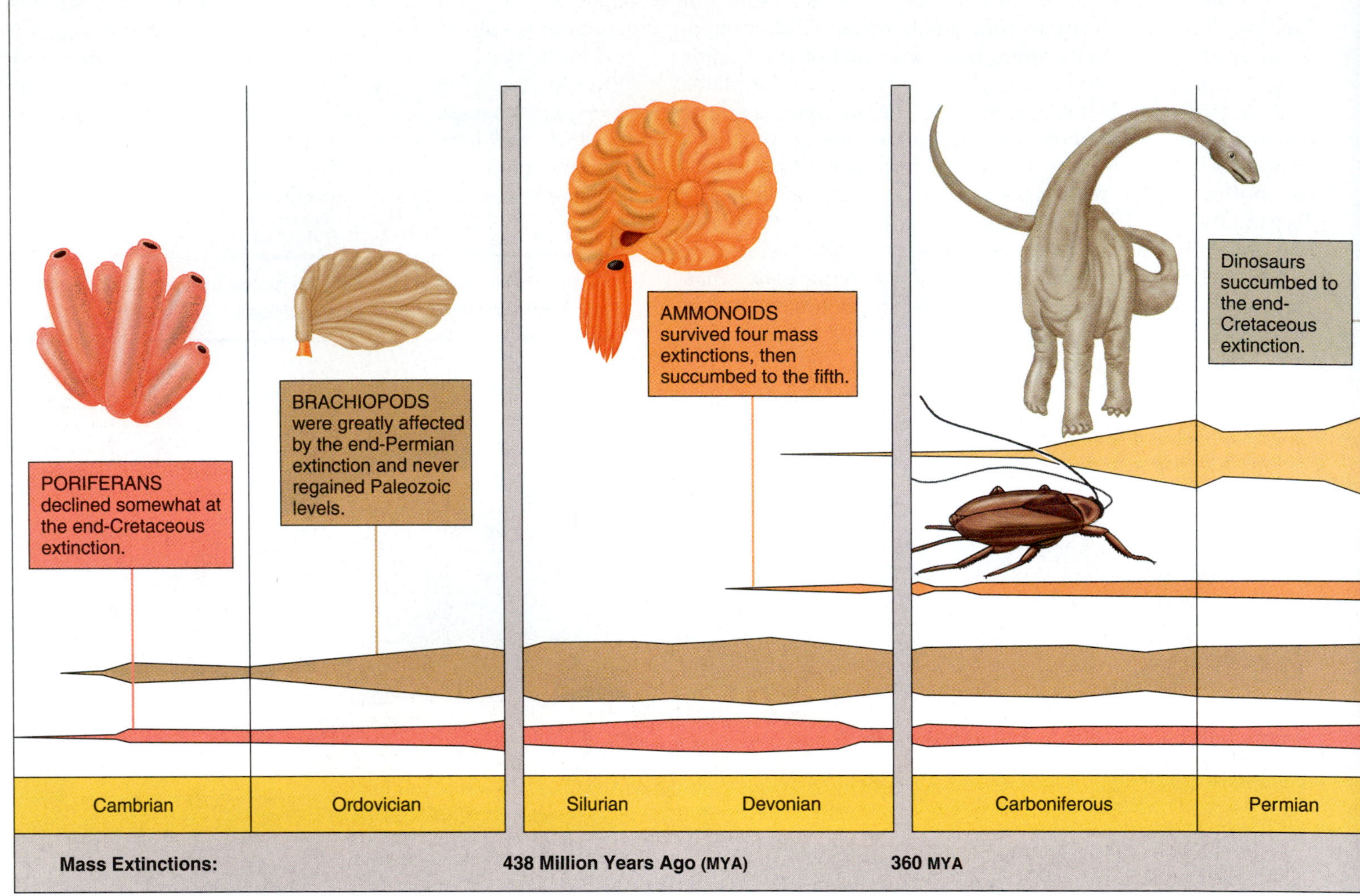
PORIFERANS declined somewhat at the end-Cretaceous extinction.
BRACHIOPODS were greatly affected by the end-Permian extinction and never regained Paleozoic levels.
AMMONOIDS survived four mass extinctions, then succumbed to the fifth.
Dinosaurs succumbed to the end-Cretaceous extinction.
Cambrian
Ordovician
Silurian
Devonian
Carboniferous
Permian
Mass Extinctions:
438 Million Years Ago (MYA)
360 MYA

Figure 20.16 Mass extinctions.
Five significant mass extinctions and their effects on the abundance of certain forms of marine and terrestrial life. The width of the horizontal bars indicates the varying abundance of each life-form considered.

Source: Data supplied and illustration reviewed by J. John Sepkoski, Jr., Professor of Paleontology, University of Chicago.

Mass Extinctions

There have been at least five mass extinctions throughout history: at the ends of the Ordovician, Devonian, Permian, Triassic, and Cretaceous periods (Fig. 20.16). Is a mass extinction due to some cataclysmic event, or is it a more gradual process brought on by environmental changes including tectonic, oceanic, and climatic fluctuations? This question was brought to the fore when Walter and Luis Alvarez proposed in 1977 that the Cretaceous extinction was due to a bolide. A bolide is an asteroid (minor planet) that explodes, producing meteorites that fall to earth. They found that Cretaceous clay contains an abnormally high level of iridium, an element that is rare in the earth's crust but more common in asteroids and meteorites. The result of a large meteorite striking the earth could have been similar to that from a worldwide atomic bomb explosion: a cloud of dust would have mushroomed into the atmosphere, blocking out the sun and causing plants to freeze and die. A layer of soot has been identified in the strata alongside the iridium, and a huge crater that could have been caused by a meteorite was found in the Caribbean–Gulf of Mexico region on the Yucatán peninsula.

In 1984, paleontologists David Raup and John Sepkoski suggested that the fossil record of marine animals shows that mass extinctions have occurred every 26 million years and, surprisingly, astronomers can offer an explanation. Our solar system is in a starry galaxy known as the Milky Way. Because of the vertical movement of our sun, our solar system approaches other members of the Milky Way every 26 to 33 million years, producing an unstable situation that could lead to the occurrence of a bolide.

Certainly continental drift contributed to the Ordovician extinction. This extinction occurred after Gondwanaland arrived at the south pole. Immense glaciers, which drew water from the oceans, chilled even the once-tropical land. Marine invertebrates and coral reefs, which were especially hard hit, didn't recover until Gondwanaland drifted away from the pole and warmth returned. The mass

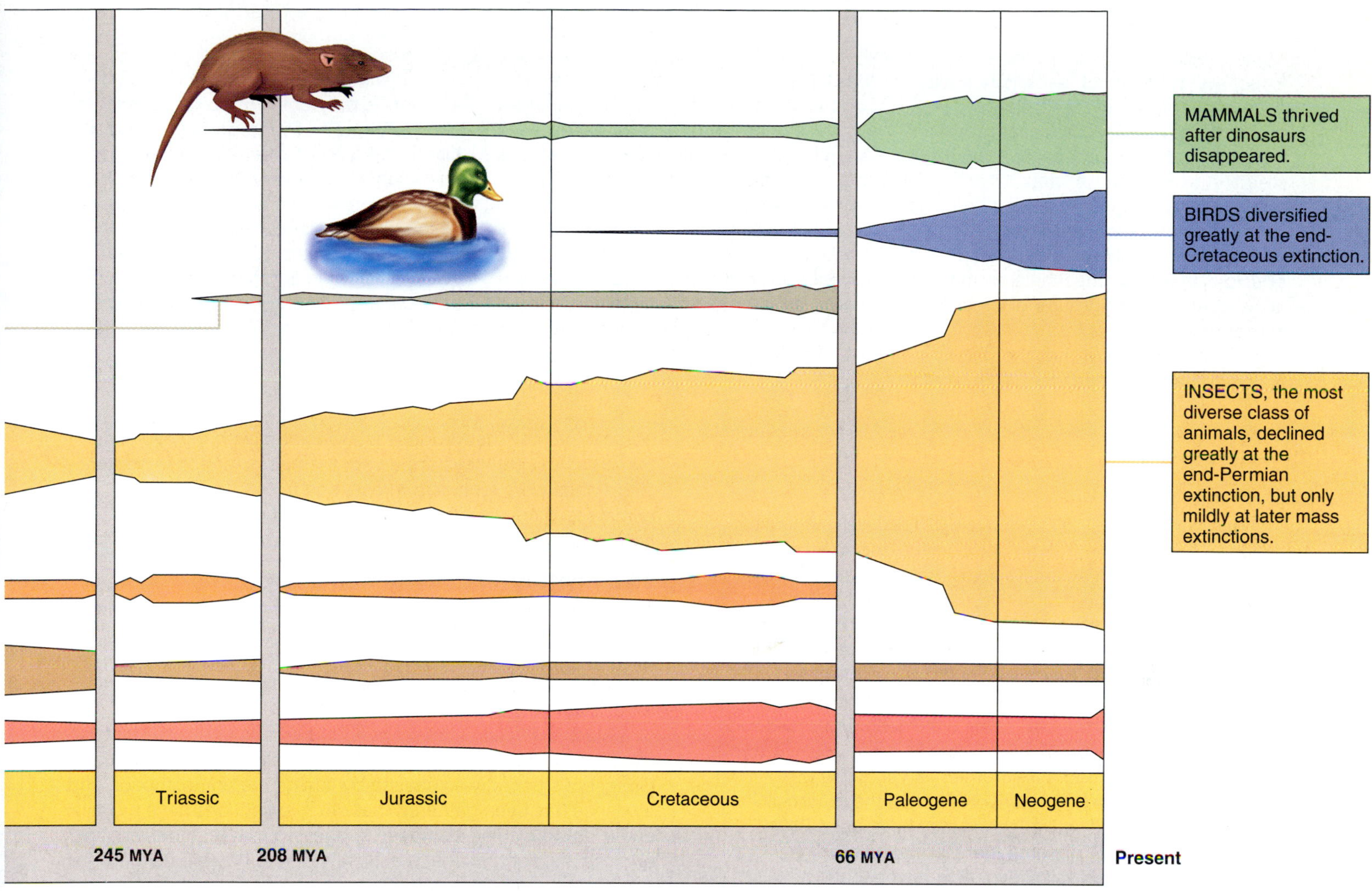
MAMMALS thrived after dinosaurs disappeared.
BIRDS diversified greatly at the end-Cretaceous extinction.
INSECTS, the most diverse class of animals, declined greatly at the end-Permian extinction, but only mildly at later mass extinctions.
Triassic
Jurassic
Cretaceous
Paleogene
Neogene
245 MYA
208 MYA
66 MYA
Present

extinction at the end of the Devonian period saw an end to 70% of marine invertebrates. Helmont Geldsetzer of Canada's Geological Survey notes that iridium has also been found in Devonian rocks in Australia, suggesting that a bolide event was involved in the Devonian extinction. Other scientists believe that this mass extinction could have been due to movement of Gondwanaland back to the south pole.

The extinction at the end of the Permian period was quite severe; 90% of species in the ocean and 70% on land disappeared. The latest hypothesis attributes the Permian extinction to excess carbon dioxide. When Pangaea formed, there were no polar ice caps to initiate ocean currents. The lack of ocean currents caused organic matter to stagnate at the bottom of the ocean. Then, as the continents drifted into a new configuration, ocean circulation switched back on. Now, the extra carbon on the seafloor was swept up to the surface where it became carbon dioxide, a deadly gas for sea life. The trilobites became extinct and the crinoids (sea lilies) barely hung on. Excess carbon dioxide on land led to a global warming which altered the pattern of vegetation. Areas that were wet and rainy became dry and warm and vice versa. Burrowing animals that could escape land surface changes seemed to have had the best chance of survival.

The extinction at the end of the Triassic period is another that has been attributed to the environmental effects of a meteorite collision with Earth. Central Quebec has a crater half the size of Connecticut that some believe is the impact site. The dinosaurs may have benefited from this event because this is the time when the first of the gigantic dinosaurs took charge of the land. The second wave occurred in the Cretaceous period.

Mass extinctions seem to be due to climatic changes that occur after a meteorite collision and/or after the continents drift into a new and different configuration.

Connecting Concepts

Does the process of evolution always have to be the same? The traditional view of the evolutionary process proposes that speciation occurs gradually and steadily over time. A new hypothesis suggests that long periods of little or no evolutionary change are punctuated by periods of relatively rapid speciation. Is it possible that both mechanisms may be at work in different groups of organisms and at different times?

Is it also possible that the history of life could have been different from what it was? The species alive today are the end product of the abiotic and biotic changes which occurred on Earth as life evolved. And what if the abiotic and biotic changes had been other than they were? For example, if the continents had not separated 65 million years ago, what types of mammals, if any, would be alive today? Given a different sequence of environments, another mix of plants and animals might very well have resulted than those with which we are familiar.

The history of life on earth, as we know it, is only one possible scenario. If we could rewind the "tape of life" and let history take its course anew, the end result might well be very different, depending on the geologic and biologic events which took place the second time around. As an analogy, consider that if you were born in another time period and in a different country, the end result of "you" might be very different than the "you" of today.

Summary

20.1 Origin of Life

The unique conditions of the primitive earth allowed a chemical evolution to occur. There was an abiotic synthesis of small organic molecules such as amino acids and nucleotides either perhaps in the atmosphere or at hydrothermal vents. These monomers joined together to form polymers either on land (warm seaside rocks or clay) or at the vents. The first polymers could have been proteins or RNA or they could have evolved together. The aggregation of polymers inside a plasma membrane produced a protocell which had some enzymatic properties such that it could grow. If the protocell developed in the ocean it was a heterotroph; if it developed at hydrothermal vents it was a chemoautotroph. A true cell had evolved once the protocell contained DNA genes. The first genes may have been RNA molecules but later DNA became the information storage molecule of heredity. Biological evolution now began.

20.2 History of Life

The fossil record allows us to trace the history of life. Prokaryotes evolved about 3.5 BYA during the Precambrian, and they were alone on earth for about 2 million years during which many metabolic pathways evolved. The oldest prokaryote fossils are cyanobacteria, which were the first organisms to add oxygen to the atmosphere. The eukaryotic cell evolved about 1.5 BYA but multicellularity not found until 600 MYA.

A rich anir..al fossil record starts at the Cambrian period of the Paleozoic era. The occurrence of external skeletons, which seems to explain the increased number of fossils at this time, may have been due to the presence of plentiful oxygen in the atmosphere, or perhaps it was due to predation. The fishes are the first vertebrates to diversify and become dominant. Amphibians are descended from lobe-finned fishes.

Plants, and then insects, invaded land during the Silurian period. The swamp forests of the Carboniferous period contained primitive vascular plants, insects, and amphibians. This period is sometimes called the Age of Amphibians.

The Mesozoic era was the Age of Cycads and Reptiles. Twice during this era, dinosaurs of enormous size evolved. By the end of the Cretaceous period the dinosaurs were extinct. New fossil finds show that birds are descended from dinosaurs and that mammals arose during the Cretaceous period from reptilian ancestors.

The Cenozoic era is divided into the Paleogene period and the Neogene period. The Paleogene is associated with the evolution of mammals and flowering plants that formed vast tropical forests. The Neogene is associated with the evolution of primates; first monkeys appeared, then apes, and then humans. Grasslands were replacing forests and this put pressure on primates, who were adapted to living in trees. The result may have been the evolution of humans—primates who left the trees.

20.3 Factors That Influence Evolution

The continents are on massive plates that move, carrying the continents with them. Plate tectonics is the study of the movement of the plates. Continental drift has affected biogeography and helps explain the distribution pattern of today's land organisms.

Mass extinctions have played a dramatic role in the history of life. It has been suggested that the extinction at the end of the Cretaceous period was caused by a large meteorite impact, and evidence has been gathered that other extinctions have a similar cause as well. It has also been suggested that tectonic, oceanic, and climatic fluctuations, particularly due to continental drift, can bring about mass extinctions.

Reviewing the Chapter

1. List and describe the various hypotheses concerning the chemical evolution that produced macromolecules. 320–22
2. Trace in general the steps by which the protocell may have evolved from macromolecules. 322
3. List and describe the various hypotheses concerning the origin of a self-replication system. 323
4. Explain how the fossil record develops and how fossils are dated relatively and absolutely. 324
5. When did the prokaryote arise and what are stromatolites? 326–27
6. When and how might the eukaryotic cell have arisen? 328
7. Describe the first multicellular animals found in the Ediacara Hills in South Australia. 328
8. Why might there be so many fossils from the Cambrian period? 329
9. Which plants, invertebrates, and vertebrates were present on land during the Carboniferous period? 330
10. Which type vertebrate was dominant during the Mesozoic era? Which types began evolving at this time? 331
11. Which type vertebrate underwent an adaptive radiation in the Cenozoic era? 333
12. What is continental drift, and how is it related to plate tectonics? Give examples to show how biogeography supports the occurrence of continental drift. 334–35

13. Identify five significant mass extinctions during the history of the earth. What may have been the cause and what types of organisms were most affected by each extinction? 336–37

Testing Yourself

Choose the best answer for each question.

1. Which of these did Stanley Miller place in his experimental system to show that organic molecules could have arisen from inorganic molecules on the primitive earth?
   a. microspheres
   b. purines and pyrimidines
   c. primitive gases
   d. only RNA
   e. All of these are correct.
2. Which of these is not a place where polymers may have arisen?
   a. hydrothermal vents
   b. on rocks beside the sea
   c. in clay
   d. in the atmosphere
   e. Both b and c are correct.
3. Which of these is the chief reason the protocell was probably a fermenter?
   a. The protocell didn't have any enzymes.
   b. The atmosphere didn't have any oxygen.
   c. Fermentation provides the most amount of energy.
   d. There was not ATP yet.
   e. All of these are correct.
4. The significance of liposomes, phospholipid droplets, is they show that
   a. the first plasma membrane contained protein.
   b. a plasma membrane could have easily evolved.
   c. a biological evolution produced the first cell.
   d. there was water on the primitive earth.
   e. the protocell had organelles.
5. Evolution of the DNA → RNA → protein system was a milestone because the protocell could now
   a. be a heterotrophic fermenter.
   b. pass on genetic information.
   c. use energy to grow.
   d. take in preformed molecules.
   e. All of these are correct.
6. Which of these events did not occur during the Precambrian era?
   a. evolution of the prokaryotic cell
   b. evolution of the eukaryotic cell
   c. evolution of multicellularity
   d. evolution of the first animals
   e. All of these occurred during the Precambrian era.
7. The organisms with the longest evolutionary history are
   a. prokaryotes which left no fossil record.
   b. eukaryotes which left a fossil record.
   c. prokaryotes which are still evolving today.
   d. animals who had a shell.
8. Which of these is mismatched?
   a. Cenozoic era—mammalian diversification
   b. Paleozoic era—invasion of land
   c. Carboniferous period—swamps, insects, and amphibians
   d. Mesozoic era—dinosaur diversity, evolution of birds and mammals
   e. Precambrian era—plants but not animals present
9. Continental drift helps explain the occurrence of
   a. mass extinctions.
   b. distribution of fossils on earth.
   c. geological upheavals like earthquakes.
   d. climatic changes.
   e. All of these are correct.
10. Which of these is mismatched?
    a. Mesozoic—cycads and dinosaurs
    b. Cenozoic—grasses and humans
    c. Paleozoic—prokaryotes and unicellular eukaryotes arise
    d. Cambrian—marine organisms with external skeletons
    e. Precambrian—origin of the cell at hydrothermal vents
11. Complete the following listings using these phrases: *oldest eukaryotic fossils, O_2 accumulates in atmosphere, Ediacaran animals, oldest known fossils, Cambrian animals, protists evolve and diversify*

1.5 BYA	a. ________	540 MYA	d. ________
2.5 BYA	b. ________	600 MYA	e. ________
3.5 BYA	c. ________	1.40 BYA	f. ________
4.5 BYA	formation of the earth	1.50 BYA	oldest eukaryotic fossils

Thinking Scientifically

1. From a scientific standpoint, trying to devise an experimental system that mimics the conditions of the early Earth is an inherently frustrating endeavor. While one might make some interesting hypotheses and experimental observations, there is no way to know for sure if experimental conditions are anything like those that really existed billions of years ago. Why do scientists continue in this quest?
2. Many environmentalists are concerned about global warming and ozone depletion. How would we know if current changes in climate are man-made or just part of natural, long-term cycles? Even if they aren't, if life has survived changes in the past why shouldn't it survive these changes now?

Bioethical Issue

Dr. H. M. Morris, Director of the Institute for Creation Research, lists these contradictions between evolution research and creation research:[2]

Evolution Research	Creation Research
Fishes evolved before fruit trees.	Fruit trees were created before fishes.
Insects evolved before birds.	Birds were created before insects.
Sun was present before land plants.	Land vegetation was created before the sun.
Reptiles evolved before birds.	Birds were created before reptiles.
Reptiles evolved before whales.	Whales were created before reptiles.
Rain was present before man.	Man was created before rain.
Evolution is still continuing.	Creation has been completed.

Do we have an obligation to accept one list over the other? Why or why not? On what basis?

---

[2] Montagu, A. (Ed). 1984. *Science and Creationism.* New York: Oxford University Press, p. 246.

Understanding the Terms

absolute dating (of fossils) 324
chemical evolution 320
coacervate droplets 322
continental drift 334
endosymbiotic hypothesis 328
extinction 329
fossil 324
geological timescale 326
liposome 322
macroevolution 324
mass extinction 329
microsphere 321
molecular clock 329
ocean ridge 321
ozone shield 327
paleontology 324
plate tectonics 335
protein-first hypothesis 321
proteinoid 321
protocell 322
relative dating (of fossils) 324
RNA-first hypothesis 321
sedimentation 324
stratum 324
stromatolite 327

Match the terms to these definitions:

a. ______________ Concept that the rate at which mutational changes accumulate in certain types of genes is constant over time.
b. ______________ Cell forerunner developed from cell-like microspheres.
c. ______________ Droplet of phospholipid molecules formed in a liquid environment.
d. ______________ A region where crust forms and from which it moves laterally in each direction.
e. ______________ Formed from oxygen in the upper atmosphere, it protects the earth from ultraviolet radiation.

Web Connections

Exploring the Internet

http://www.mhhe.com/biosci/genbio/mader
(click on *Biology 7/e*)

The *Biology 7/e* Online Learning Center provides many resources for studying the material in this chapter including links to the following sites:

Learn more about the geologic timescale and rock strata in Geology and Geologic Time.

http://www.ucmp.berkeley.edu/exhibit/geology.html

Geological Timescale. Text and nice graphics illustrate geological timescales on Earth.

http://www.geo.ucalgary.ca/~macrae/timescale/timescale.html

History of Evolutionary Thought. This is a timeline of the major players who have contributed to our understanding of evolutionary thought.

http://www.ucmp.berkeley.edu/history/evotmline.html

Evolutionary Relationships of Archosaurs. View images and read more about the Archosaurs, Pterosaurs, and Dinosaurs.

http://www.ucmp.berkeley.edu/diapsids/archosy.html

Dinosaurs in Cyberspace: Dinolinks. University of California at Berkeley Museum of Paleontology. A list of dinosaur-oriented websites, scientific and otherwise. This site has pages and pages of links!

http://www.ucmp.berkeley.edu/diapsids/dinolinks.html#dinosites

Human Evolution

chapter 21

chapter concepts

21.1 Humans Are Primates

- Humans *(Homo sapiens)* are in the order primates, the suborder Anthropoidea (monkeys, apes, and humans), the superfamily Hominoidea (apes and humans), and the family Hominidea (extinct and modern humans). 342
- Primate characteristics include an enlarged brain, opposable thumb, binocular vision, and an emphasis on learned behavior. 343
- The hominoid common ancestor lived during the Miocene epoch and the hominid common ancestor lived during the early Pliocene epoch. 345, 347

21.2 Evolution of Australopithecines

- About 5 MYA,[1] australopithecines (the first hominids) evolved in Africa. The australopithecines had a relatively small brain, but they could walk upright. 348

21.3 Evolution of Humans

- About 2 MYA, early *Homo* types evolved which had a larger brain than the australopithecines and were able to make primitive stone tools. 351
- An early form of *Homo ergaster* migrated out of Africa and evolved into *Homo erectus* in Asia. Later, *H. erectus* had a knowledge of fire and made more advanced tools. 352
- The out-of-Africa hypothesis says that modern humans evolved in Africa, and after migrating to Europe and Asia (about 100,000 years BP[2]), they replaced the archaic *Homo* species, including the Neanderthals. 353
- Cro-Magnon is the name given to modern humans who made sophisticated tools and definitely had culture. 354

[1] MYA = millions of years ago

[2] BP = before present

***Australopithecus africanus*, a hominid**

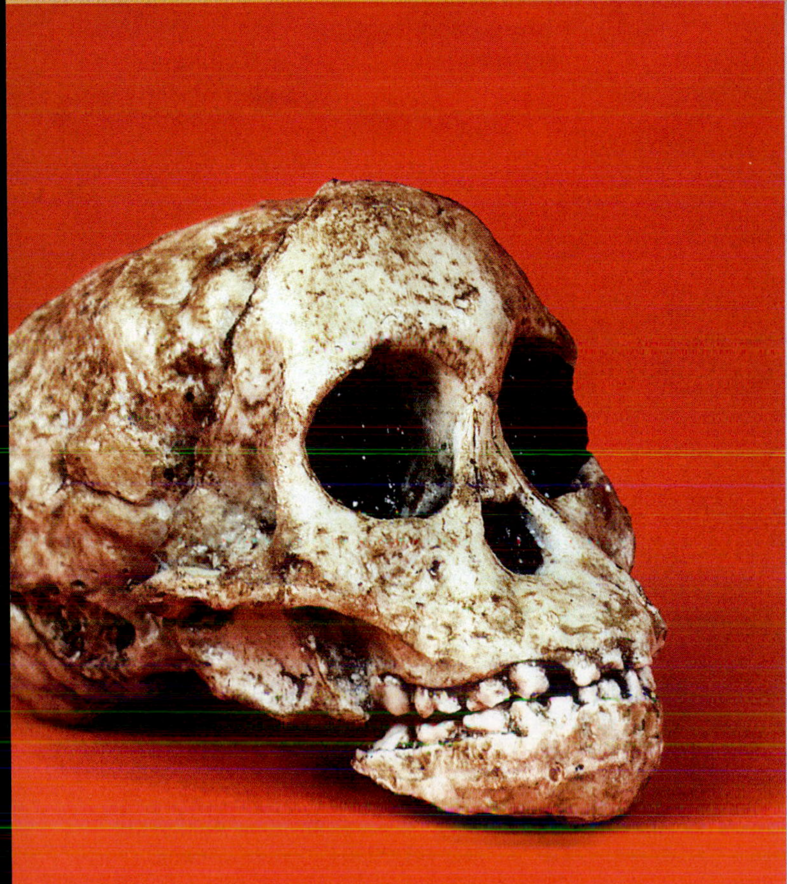

One of the most unfortunate misconceptions concerning human evolution is the belief that Darwin, Wallace, and others have suggested that humans evolved from apes. On the contrary, it is proposed that modern humans and apes evolved from a common apelike ancestor. Today's apes are our cousins, and we couldn't have evolved from our cousins because we are contemporaries—living on earth at the same time. Our relationship to apes is analogous to you and your first cousin being descended from your grandparents.

Some people feel that human evolution is somehow special or different from that of other species. Scientists, on the other hand, study human evolution in the same objective way they approach any other research topic. They have found that while human evolution follows the same patterns of evolutionary descent as other groups, it has more complexity. Various prehuman groups died out, migrated, and interbred with other groups in a short time; nevertheless, recently found fossils have brought knowledge of our evolutionary history closer to an apelike ancestor.

21.1 Humans Are Primates

Humans are a type of mammal called a primate. The classification table shows that there are two main groups of primates, the prosimians (lemurs, tarsiers, lorises) and the anthropoids (monkeys, apes, and humans). This listing also indicates the general order in which these groups diverged from the human line of descent; namely: prosimians, monkeys, apes, and humans. After divergence each group continued on its own evolutionary path. We must always remember that humans are not descended from apes; rather, they share a common ancestor with apes. This common ancestor was less specialized than any ape living today.

Primate Characteristics

Primates [L. *primus,* first] in contrast to other types of mammals are adapted for an **arboreal** life, that is, for living in the trees. The evolution of primates is characterized by trends toward mobile limbs, grasping hands, a flattened face, binocular vision, complex brain, and one birth at a time. These traits are particularly useful for living in trees.

Mobile Forelimbs and Hindlimbs

In primates, the limbs are mobile and the hands and feet both have five digits each. Flat nails replaced the claws of ancestral primates and sensitive pads on the underside of fingers and toes assist the ability to grasp objects. Many primates have both an opposable big toe and thumb; that is, the big toe or thumb can touch each of the other toes or fingers. Humans don't have an opposable big toe, but the thumb is opposable and this results in a grip that is both powerful and precise.

How are these features adaptive for a life in trees? Mobile limbs with clawless opposable digits allow primates to freely grasp and release tree limbs. They also allow a primate to easily reach out and bring food such as fruit to the mouth.

Binocular Vision

A foreshortened snout and a relatively flat face are evolutionary trends in primates. These may be associated with a general decline in the importance of smell and an increased reliance on vision, leading eventually to binocular vision. In most primates, the eyes are located in the front where they can focus on the same object from slightly different angles (Fig. 21.1). The stereoscopic (three-dimensional) vision and good depth perception that result permit primates to make accurate judgments about the distance and position of adjoining tree limbs.

Some primates, humans in particular, have color vision and greater visual acuity because the retina contains cone cells in addition to rod cells. Rod cells are activated in dim light but the blurry image is in shades of gray. Cone cells require bright light but the image is sharp and in color. The lens of the eye focuses light directly on the fovea, a region of the retina where cone cells are concentrated.

Large Complex Brain

Sense organs are only as beneficial as the brain that processes their input. The evolutionary trend among primates is toward a larger and more complex brain—the brain size is smallest in prosimians and largest in modern humans. The portion of the brain devoted to smell has gotten smaller and

CLASSIFICATION

Order Primates

Most species tree dwelling, enlarged, complex brain; five digits with nails on hands and feet adapted for grasping. Bodies, except humans, covered with hair.

Suborder Prosimii: lemurs, tarsiers, lorises

Suborder Anthropoidea: monkeys, humans, gibbons, great apes

- Superfamily Ceboidea: New World monkeys (e.g., howler, spider, squirrel monkeys)
- Superfamily Cercopithecoidea: Old World monkeys (e.g., baboons, macaques)
- Superfamily Hominoidea: apes, australopithecines,* humans
  - Family Hylobatidae: gibbons
  - Family Pongidae: Great apes (chimpanzee, orangutan, gorilla)
  - Family Hominidae: australopithecines,* humans
    - Genus: *Australopithecus afarensis,* Australopithecus boisei,* Australopithecus africanus,* Australopithecus robustus**
    - Genus: *Homo habilis,* Homo rudolfensis,* Homo ergaster,* Homo erectus,* Homo heidelbergensis,* Homo sapiens*

* extinct

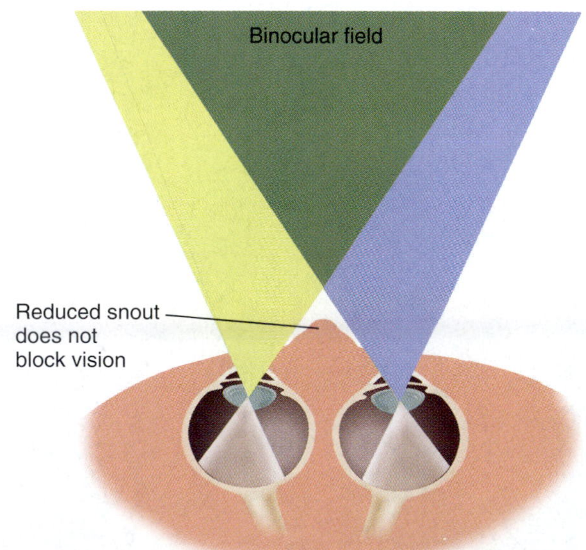

Figure 21.1 Binocular vision.
In primates, the snout is reduced and the eyes are at the front of the head. The result is a binocular field which aids depth perception.

the portions devoted to sight have increased in size and complexity. Also more of the brain is devoted to controlling and processing information received from the hands and the thumb. The result is good hand-eye coordination.

Reduced Reproductive Rate

Other trends in primate evolution are a general reduction in the rate of reproduction, associated with sexual maturity and extended life spans. Gestation is lengthy, allowing time for forebrain development. One birth at a time is the norm in primates; it is difficult to care for several offspring while moving from limb to limb. The juvenile period of dependency is extended, and there is an emphasis on learned behavior and complex social interactions.

These characteristics especially distinguish primates from other mammals:

Opposable thumb (and in some cases, big toe)	Binocular vision
Nails (not claws)	Expanded, complex brain
Single birth	Emphasis on learned behavior

Evolution of Major Groups

The primate order contains two suborders: prosimians and anthropoids.

Prosimians

Modern **prosimians** [L. *pro,* before, and *simia,* ape, monkey] have retained primitive characteristics and best resemble the first primates which may have taken to the trees in order to forage for insects. The largest and most diverse group of modern prosimians are the lemurs, which live on the island of Madagascar lying off the coast of Southeast Africa. The dwarf lemurs, featured in Figure 21.2*b*, are all nocturnal, arboreal, and either eat fruit or insects.

Unlike the lemurs, the lorises and bushbabies are found across Africa and India to Southeast Asia. Like the dwarf lemurs, however, these animals are all nocturnal and arboreal, feeding on either fruit or insects. Being nocturnal, they don't have to compete with monkeys and apes, which are active during the day.

Tarsiers (Fig. 21.2*c*), which are found in the Philippines and East Indies, are curious, mouse-sized creatures with enormous eyes suitable to their nocturnal way of life. In one species, a single eye weighs more than the brain! Tarsiers live in low forest and undergrowth and feed on insects.

a. *Tetonius* b. Dwarf lemur, *Cheirogaleus* c. Mindano tarsier, *Tarsius*

Figure 21.2 Prosimians.
a. The hypothesized Eocene primate *Tetonius* was small but agile and adapted to eating insects. b. There are over 40 species of lemurs on the island of Madagascar. The dwarf lemurs are particularly widespread. c. Tarsiers are vertical clingers and leapers. The enormous eyes allow the tarsier to judge a safe landing even at night.

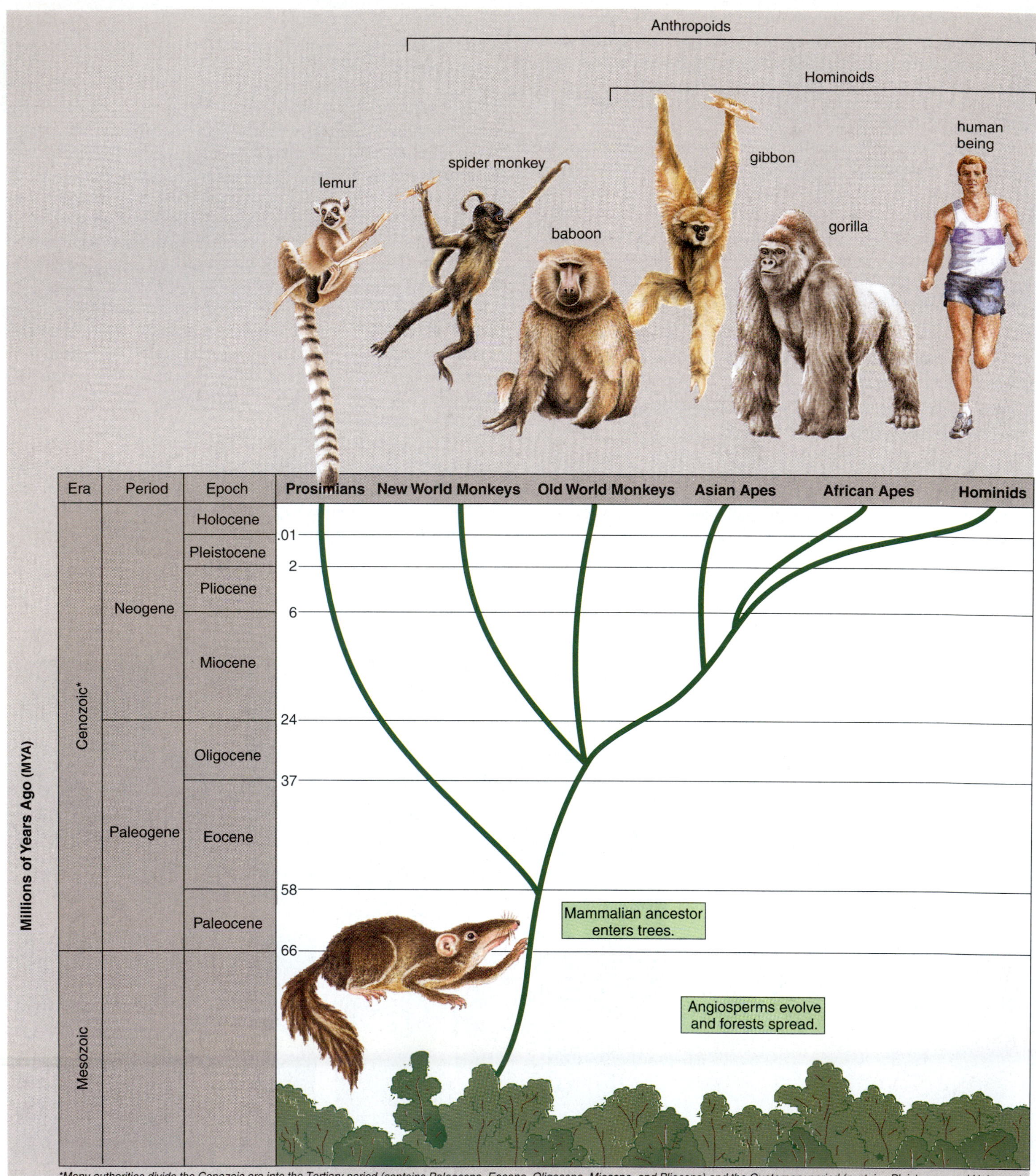

Figure 21.3 Primate phylogenetic tree.
A phylogenetic tree of the primate order began in the late Mesozoic era when a small insect-eating mammal climbed into the spreading angiosperm forests. The descendants of this mammal adapted to a new way of life and developed traits such as an enlarged brain, nails instead of claws, and binocular vision. Note the epoch in which each type of primate diverged and continued its evolution separate from the others.

Anthropoids

Figure 21.3 shows the sequence of anthropoid evolution during the Cenozoic era. The surviving **anthropoids** [Gk. *anthropos,* man, and *-eides,* like] are classified into three superfamilies: New World monkeys, Old World monkeys, and the hominoids. The New World monkeys often have long prehensile (grasping) tails and flat noses, and Old World monkeys, which lack such tails, have protruding noses. Two of the well-known New World monkeys are the spider monkey and the capuchin, the "organ grinder's" monkey. Some of the better-known Old World monkeys are now ground dwellers, such as the baboon and the rhesus monkey, which has been used in medical research.

Primate fossils similar to monkeys are first found in Africa, but not until the Oligocene epoch of the Cenozoic era (Fig. 21.3). At that time the Atlantic ocean would have been too expansive for some of them to have easily made their way to South America where the New World monkeys live today. It is hypothesized that a common ancestor to both the New World and Old World monkeys must have arisen much earlier when a narrower Atlantic would have made a crossing much more reasonable. The New World monkeys evolved in South America and the Old World monkeys evolved in Africa.

The **hominoids** [L. *homo,* man, and Gk. *-eides,* like] include all the apes and humans. During the Miocene epoch, there were dozens of ape species but the anatomy of *Proconsul* makes it a prime candidate to be the ancestral hominoid. *Proconsul* was about the size of a baboon, and the brain was a comparable size at about 165 cc. Notice that this hominoid didn't have a tail or the expansive pelvis of a monkey (Fig. 21.4). Although its elbow is similar to that of modern apes, its limb proportions suggested that it walked quadrupedally on top of tree limbs like the monkeys do. Although primarily a tree dweller, *Proconsul* may have also spent time exploring nearby grasslands for food.

At the end of the Miocene epoch, Africarabia (Africa plus the Arabian Peninsula) joined with Asia and the hominoids migrated into Europe and Asia. Two groups can be distinguished: dryomorphs and ramamorphs. While at one time it was believed that ramamorphs were ancestral to the later-appearing **hominids,** a group that contains only humans, ramamorphs are now classified as an ancestral orangutan group. In 1966, Spanish paleontologists announced the discovery of a specimen of *Dryopithecus* dated at 9.5 MYA near Barcelona. The anatomy of these bones clearly indicates that dryomorphs were tree dwellers and locomoted by swinging from branch to branch as orangutans do today. They did not walk along the top of tree limbs as *Proconsul* did.

The apelike *Proconsul,* which was prevalent in Africa during the Miocene epoch, is believed to be ancestral to today's hominoids—apes and humans.

monkey:
flat palms and soles
arched vertebral column
limbs equal in length
narrow rib cage
shoulder blades parallel
immobile shoulder joint

a. Monkey skeleton

b. *Proconsul* skeleton

c. *Proconsul* skull

Figure 21.4 Monkey skeleton compared to *Proconsul* skeleton.
Comparison of monkey skeleton (a) and *Proconsul* skeleton (b), a hominoid of the Miocene epoch. c. Skull of *Proconsul.*

a. White-handed gibbon, *Hylobates*

b. Orangutan, *Pongo*

c. Mountain gorilla, *Gorilla*

d. Common chimpanzee, *Pan*

Figure 21.5 Ape diversity.
a. Of the apes, gibbons are the most distantly related to humans. They dislike coming down from trees, even at watering holes. They will extend a long arm into the water and then drink collected moisture from the back of the hand. **b.** Orangutans are solitary except when a pair comes together to reproduce. Their name means "forest man"; early Malayans believed that they were intelligent and could speak but did not because they were afraid of being put to work. **c.** Gorillas are terrestrial and live in groups in which a silver-backed male, such as this one, is always dominant. **d.** Of the apes, chimpanzees sometimes seem the most humanlike.

Hominoids The close relationship between humans and apes is signified by their placement in the same superfamily, Hominoidea. Four types of apes have survived until today: the gibbon and the orangutan are found in Asia; the gorilla and the chimpanzee inhabit Africa (Fig. 21.5). The gibbon is the smallest of the apes, with a body weight of about 5.5 to 11 kilograms. Gibbons have extremely long arms that are specialized for swinging between tree limbs. The orangutan is large (75 kilograms) but nevertheless spends a great deal of time in trees. The gorilla, the largest of the apes (180 kilograms), spends most of its time on the ground. Chimpanzees, which are at home both in trees and on the ground, are the most humanlike of the apes in appearance and are frequently used in psychological experiments.

As yet, no fossil has been identified as a common ancestor for hominids and apes. Molecular biologists have studied our relationship to the apes. Recall that certain differences in DNA base pair sequences may occur at a constant rate and can be used as a molecular clock to judge when two groups have diverged from one another. A molecular clock study suggests that the orangutan diverged from the primate line of descent about 14 MYA, and African apes and humans did not split until around 8 MYA. Therefore, it is believed that the last common ancestor between the African apes and hominids lived during the early Pliocene epoch.

Hominids differ from modern African apes, such as the gorilla, in the ways noted in Figure 21.6. Anatomical differences of prime importance concern: (1) type of locomotion, which is dependent on skeletal features involving the spine, the pelvis, and the bones of the limbs; (2) jaw shape, which is related to the size and the shape of the teeth, and (3) brain size. Of these three, fossil evidence indicates that type of locomotion, namely **bipedalism,** is the first evolved feature that separated hominids from African apes.

Until recently, many scientists thought that hominids walk upright because of a dramatic change in climate that caused forests to be replaced by grassland. Now, some paleontologists suggest that the hominid ancestor began to develop bipedalism even while it lived in trees. Why? Because they cannot find evidence of a dramatic shift in vegetation 6 MYA. Even so, Africa as a whole was dryer than before. The first hominid's environment is now thought to have included some forest, some woodlands, and some grassland. While still living in trees, the first hominids may have walked upright on large branches as they collected fruit from overhead. Then, when they began to forage on the ground among bushes, it would have been easier to shuffle along on their hindlimbs. Bipedalism would also have prevented them from getting heat stroke: an upright stance exposes less of the body to the rays of the sun and more of the body to breezes. Bipedalism may have been an advantage in still another way. Males that acquired food far afield and could carry it back to females were more assured of sex than otherwise.

The evolution of bipedalism is believed to be the distinctive feature that separates the hominids from the apes. Scientists are still looking for reasons bipedalism evolved.

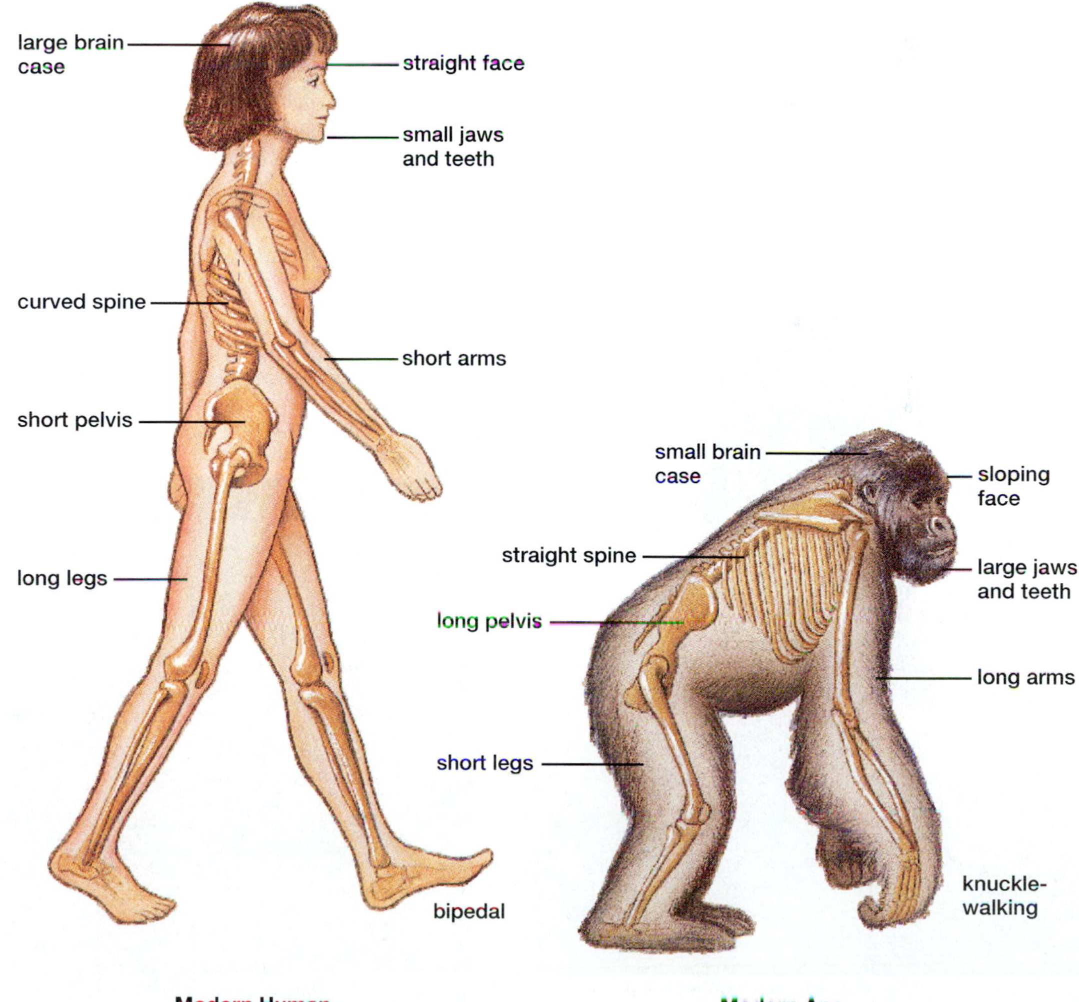

Figure 21.6 **Modern human skeletal features and those of a gorilla.**

21.2 Evolution of Australopithecines

The study of the fossil record during the past several years has shown that human evolution is not an orderly sequence of forms from one to the other. Rather, several hominids coexisted at the same time and the overall pattern resembles a bush instead of a straight line. Also, human characteristics did not evolve at the same rate. The term **mosaic evolution** is used because some body parts are more humanlike than others in early hominids. For example, in some fossils the legs have lengthened (humanlike) but forearms have remained long (apelike).

The hominid line of descent begins with the **australopithecines,** which evolved and diversified in Africa. Fossils have especially been discovered in southern Africa and eastern Africa. Some of these are slight of frame and termed gracile (slender) types. Some are robust (powerful) types that tend to have strong upper bodies, but especially massive jaws and chewing muscles anchored to a prominent bony crest along the top of the skull. The gracile types no doubt fed on soft fruits and leaves, while the robust types had a more fibrous diet that may have included chomping down on hard nuts.

Southern Africa

The first australopithecine to be discovered was unearthed in southern African by Raymond Dart in the 1920s. This hominid, named ***Australopithecus*** [L. *australis,* southern, *pithecus,* ape] ***africanus*** is a gracile type dated about 2.8 MYA. Abundant *A. africanus* fossils are now available that date from 2.4 to 3 MYA. ***A. robustus,*** dated from 2 to 1.5 MYA, is a robust type from southern Africa. Both *A. africanus* and *A. robustus* had a brain size of about 500 cc; their skull differences are essentially due to dental and facial adaptations to different diets as discussed.

Based on their skeletal remains, these hominids are believed to have walked upright. Nevertheless, the proportions of the limbs are apelike: the forelimbs are longer than the hindlimbs. Henry McHenry argues that since early *Homo* also had these limb proportions, *A. africanus,* with a relatively large brain size, is the best ancestral candidate for early *Homo*.

Eastern Africa

It wasn't until the 1960s that paleontologists, following the lead of the renowned couple Louis and Mary Leakey, began to concentrate their efforts in eastern as opposed to southern Africa. One 1994 find consisting of skull fragments, teeth, arm bones, and a part of a child's lower jaw has been dated at 4.4 MYA. At first it was called *Australopithecus ramidus*, but then the name was changed to *Ardipithecus ramidus* (*Ardipithecus,* ground ape, and *ramidus,* root). This fossil is close to the common ancestor for australopithecines and chimpanzees. In comparison to later australopithecines, the canine teeth are larger, and the skull is more like that of a chimpanzee. A determination is still being made whether *A. ramidus* walked upright or not. A 1995 discovery, *Australopithecus anamensis,* dated at just about 4 MYA, has jaws like that of an ape but legs like those of humans, so it is believed to have walked upright.

More than 20 years ago, a team led by Donald Johanson unearthed nearly 250 fossils of a hominid called ***A. afarensis.*** A now-famous female skeleton dated at 3.18 MYA is known worldwide by its field name, Lucy. (The name

a.

b.

Figure 21.7 ***Australopithecus afarensis.***
a. A reconstruction on display at the St. Louis Zoo. **b.** It's possible that *A. afarensis* made these fossilized footprints dated some 3.7 million years ago. The larger footprints are double—a smaller individual was stepping in the footprints of a larger individual. A youngster was walking to the side.

derives from the Beatles' song "Lucy in the Sky with Diamonds.") Although her brain was quite small (400 cc), the anatomy of her limbs indicate that Lucy did stand upright and walk bipedally (Fig. 21.7*a*). *A. afarensis* fossils are sexually dimorphic. Lucy was about four feet tall and weighed about 30 kilograms. In contrast, the males of the species were five feet tall and weighed up to 45 kilograms. It has long been thought that an *A. afarensis* made the famous footprints, discovered in Laetoli and dated at 3.7 MYA (Fig. 21.7*b*). Some experts now believe that the feet of *A. afarensis* are too primitive to have made these footprints. Perhaps another species still to be found made them.

A. afarensis, a gracile type, is believed to be ancestral to the robust types found in eastern Africa (Fig. 21.8). *A. aethiopicus* and *A. boisei* are both robust types; the former appears in the fossil record around 2.6 MYA. *A. boisei* had a powerful upper body and the largest molars of any hominid. Their diet consisted of foods that were more fibrous and hard to chew. Even though the robust types are dated later than Lucy, they are believed to be members of lineages that died out. Instead, the paleontologists working in eastern Africa have long supported the hypothesis that *A. afarensis* is ancestral to both *A. africanus* and early *Homo*.

Could it be, however, that some other hominid in eastern Africa might bridge the gap between *A. afarensis* and early *Homo*? Based on recent finds in eastern Africa, which consist of a skull, leg, and arm bones all dated at 2.5 MYA, Tim White and others have proposed that a hominid called *Australopithecus garhi* may be the long-sought ancestor to early *Homo*. The braincase is small (450 cc), but the shape of the premolars and size of the canine teeth to the molars does make the skull resemble that of humans, despite the fact the molars are as large as those of *A. robustus*. The upper leg bone is relatively long, but so is the forearm. Unfortunately, it is not possible to show beyond a doubt that the skull and the limb bones belong to the same individual. Only a meter away from the limb bones are those of horses, antelopes, and other animals with cut marks, suggesting these animals were butchered. One antelope bone was struck at both ends, apparently to get at the marrow inside. It has previously been theorized that the eating of bone marrow would have provided the nutrients necessary for the increase in brain size seen in early *Homo*.

Evolutionary Relationships Between Australopithecines

Each species of australopithecines existed for a range of years which overlap. As many as six or seven species of these hominids could have lived in Africa simultaneously. The evolutionary relationship between them is still being worked out. Previous to 1999, some phylogenetic trees show *A. africanus*, and others showed *A. afarensis* (Lucy), as the species that gave rise to early *Homo*. The other australopithecines belonged to lineages that eventually died out. Now the new species, *A. garhi*, is proposed to be a descendent of *A. afarensis* and a direct ancestor to early *Homo*.

Some investigators are interested in the relationship between the eastern and southern robust australopithecines. Although there are 50 or more skull characteristics shared by all the robust types, their facial traits could very well be developmental consequences of oversized molars and small front teeth. In other words, they very well could have evolved separately and still have faces that look alike. In fact, variations in tooth shape suggest that *A. boisei* (eastern fossil) and *A. robustus* (southern fossil) are not closely related.

Australopithecines, which arose in Africa, were the first hominids. They show that bipedalism was the first humanlike feature to evolve. The exact evolutionary relationship between australopithecines and early *Homo* is still not determined.

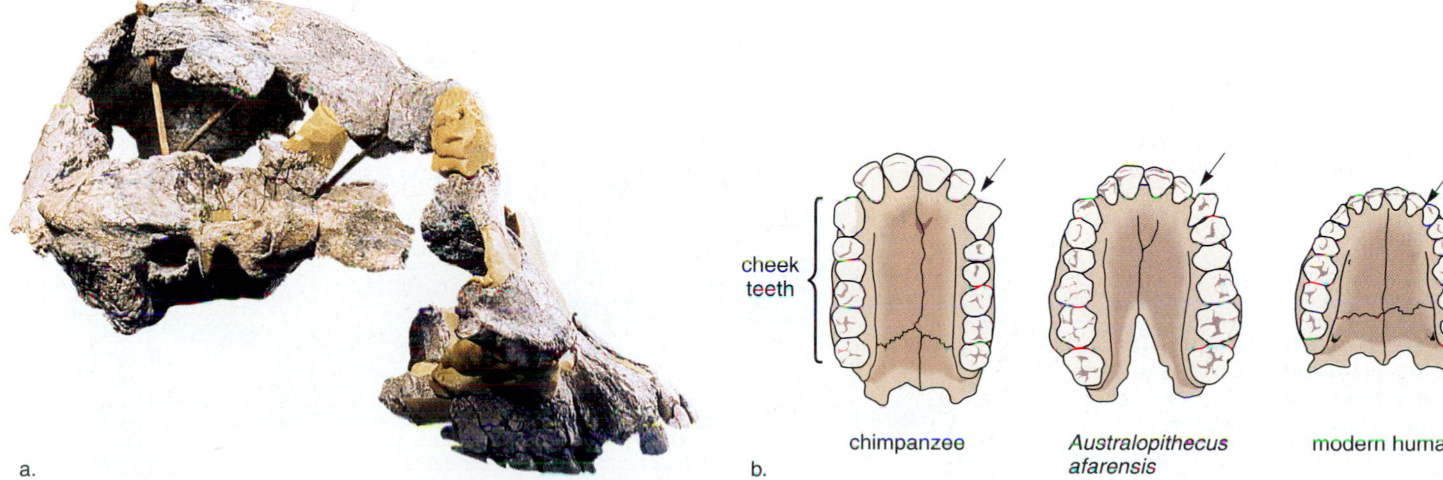

Figure 21.8 ***Australopithecus afarensis* skull and dentition.**
a. *A. afarensis* had an apelike face with prominent brow ridges and a projecting jaw. b. Dentition of an ape, *A. afarensis*, and a modern human. Note the size of the canine teeth, the gap between canine teeth and adjacent incisor (arrow), and the size of the cheek teeth.

Origin of the Genus *Homo*

Fossil evidence shows that the earliest *Homo* species evolved in Africa from the *Australopithecus* line of descent about 2.4 MYA. Remains of australopithecines indicate that they spent part of their time climbing trees and that they retained many apelike traits. Australopithecine arms, like those of an ape, were long, compared to the length of the legs. *A. afarensis* also had strong wrists and long, curved fingers and toes. These traits would have served well for climbing, and the australopithecines probably climbed trees for the same reason that chimpanzees do today: to gather fruits and nuts in trees and to sleep above ground at night so as to avoid predatory animals, such as lions and hyenas.

Whereas our brain is about the size of a grapefruit, that of the australopithecines was about the size of an orange. Their brain was only slightly larger than that of a chimpanzee. There is no evidence that the australopithecines manufactured stone tools; presumably, they were not smart enough to do so.

We know that the genus *Homo* evolved from the genus *Australopithecus,* but several years ago I [Steven Stanley] concluded that this could not have happened as long as the australopithecines climbed trees every day. The obstacle relates to the way in which we, members of *Homo,* develop our large brain. Unlike other primates, we retain the high rate of fetal brain growth through the first year after birth. (That is why a one-year-old child has a very large head.) The brain of other primates, including monkeys and apes, grows rapidly before birth, but immediately after birth their brain grows more slowly. An adult human brain is more than three times as large as that of an adult chimpanzee.

A continuation of the high rate of fetal brain growth eventually allowed the genus *Homo* to evolve from the genus *Australopithecus.* But there was a problem in that continued brain growth is linked to underdevelopment of the entire body. Although the human brain becomes much larger, human babies are remarkably weak and uncoordinated. Such helpless infants must be carried about and tended. Human babies are unable to cling to their mothers the way chimpanzee babies can (Fig. 21A).

How, then, did the evolutionary transition from *Australopithecus* to *Homo* take place? Fossilized pollen and other geological evidence reveal that at the time the *Homo* genus evolved, the climate was becoming cooler and drier. This caused forests to shrink and grasslands to expand across broad areas of Africa. The australopithecine line of descent became extinct, just as did many African antelope species that depended on forests for food and shelter. Before extinction occurred, however, humans—who walk erect, rarely climb trees, and have large brains—evolved.

The origin of the *Homo* genus entailed a great evolutionary compromise. Humans gained a large brain, but they were saddled with the largest interval of infantile helplessness in the entire class Mammalia. The positive value of a large brain must have outweighed the negative aspects of infantile helplessness, however, or natural selection would not have produced the *Homo* genus. Having a larger brain meant that humans were able to outsmart or ward off predators with weapons they were clever enough to manufacture.

Probably very few genetic changes were required to delay the maturation of *Australopithecus* and produce the large brain of *Homo*. The mutation of a regulatory gene that controls one or more other genes most likely could have delayed early maturation. As we learn more about the human genome, we will eventually uncover the particular gene or gene combinations that cause us to have a large brain, and this will be a very exciting discovery.

Courtesy of Steven Stanley, The Johns Hopkins University

Figure 21A **Human infant.** A human infant is often cradled and has no means to cling to its mother when she goes about her daily routine.

21.3 Evolution of Humans

It has become the practice to assign fossils to the species *Homo* if (1) the brain size is 600 cc or greater, (2) the jaw and teeth resemble that of humans, (3) the face is smaller and more flat as in humans, and (4) tool manufacture and use seems evident.

Early *Homo*

For many years, *Homo* fossils dated between 2.0 and 1.9 MYA have been categorized as *Homo habilis*, after fossils discovered in eastern Africa by Louis and Mary Leakey in the 1960s (Fig. 21.9). Recently, it's been suggested that these early *Homo* fossils are different enough to warrant the designation of two species: ***Homo habilis*** and ***Homo rudolfensis.*** *H. rudolfensis* is taller, has a broad face with flatter brow ridges, and a larger, rounder braincase than *H. habilis*.

The name *Homo habilis* means "handy man" and the remains of these fossils are often accompanied by stone tools. *H. rudolfensis* has a brain size as large as 775 cc, which is about 45% larger than that of *A. afarensis*. In addition, certain portions of the brain thought to be associated with speech areas are enlarged. The cheek teeth are smaller than even those of the gracile australopithecines. Cut marks on bones that could have been made by stone flakes have been found at many early *Homo* sites throughout eastern Africa. It seems clear that early humans made and used tools in order to strip meat off bones and, in keeping with the small size of their teeth, must have eaten meat to satisfy protein demands. As a scavenger, they may have depended simply on the kills of other animals; or as a predator, they may have killed small- to medium-sized prey. A predator's way of life became available to them when they had the ability to make and use tools intelligently.

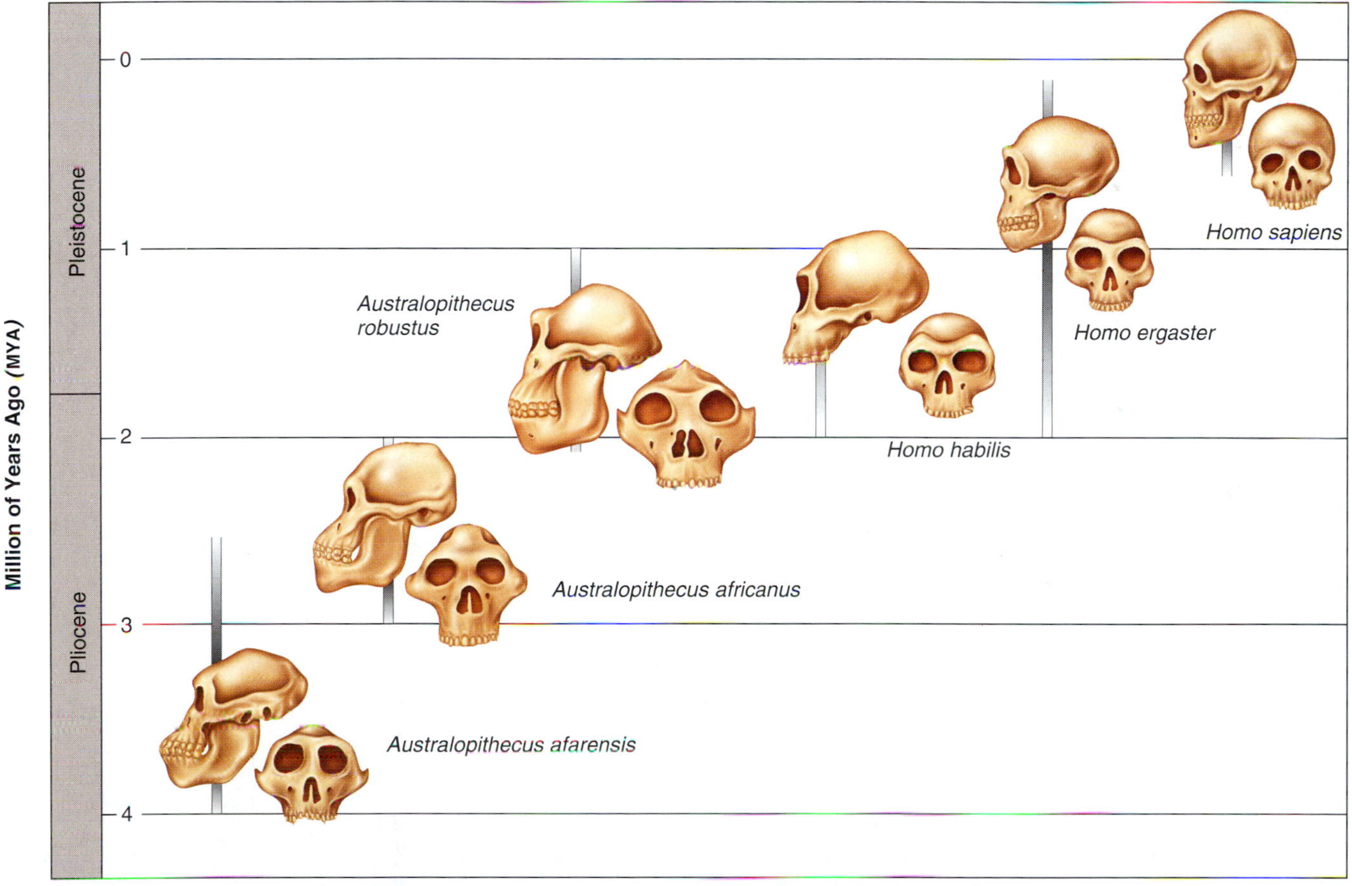

Figure 21.9 Human evolution.
The length of time each species existed is indicated by the vertical gray lines. Notice that there have been times when two or more hominids existed at the same time.

The stone tools made by *H. habilis* are called Oldowan tools because they were first identified as tools by the Leakeys in Olduvai Gorge (see Fig. 21.14). Oldowan tools are simple and look rather clumsy, but perhaps we are looking at the core that remains after flakes have been removed. The flakes would have been sharp and able to scrape away hide and cut tendons to easily remove meat from a carcass.

Notice in Figure 21.9 that *A. robustus*, *H. habilis*, and *H. ergaster* all coexisted at the same time. This supports the concept of a bushy pattern for human evolution rather than a linear pattern.

Homo ergaster and *Homo erectus*

A Dutch anatomist named Eugene Dubois was the first to unearth ***Homo erectus*** [L. *homo*, man, and *erectus*, upright] bones in Java in 1891. Since that time many fossils dated between 1.9 and 0.3 MYA were assigned to this species. Although they are similar in appearance, there is enough discrepancy to suggest that several different species have been included in this group. In particular, some experts believe that the African and Asian types are two different species. It is now customary to call the African type ***Homo ergaster*** (work man).

Compared to early *Homo*, *H. ergaster* had a larger brain (about 900 cc), more pronounced brow ridges, a flatter face, and a nose that projects like ours. This type of nose is adaptive for a hot, dry climate because it permits water to be removed before air leaves the body. The recovery of an almost complete skeleton of a ten-year-old boy (Fig. 21.10) indicates that *H. ergaster* was much taller than the hominids discussed thus far. As adults, males were 1.8 meters (about 6 feet) and females were 1.55 meters (approaching 5 feet) tall. Most likely, these hominids had a striding gait like ours. Also, the size of the birth canal indicates that infants were born in an immature state that required an extended period of care.

Previously, it was hypothesized that a migration out of Africa had occurred about 1 MYA, but recently *H. erectus* fossil remains in Java and the Republic of Georgia have been dated at 1.9 and 1.6 MYA, respectively. Therefore, it seems that an early form of *H. ergaster* may have migrated from Africa. Following this migration, *H. erectus* evolved in Asia and then spread to other areas.

In Africa, it is known that *H. ergaster* eventually had the knowledge of fire, and fashioned more advanced tools called Acheulean tools (see Fig. 21.14). Heavy teardrop-shaped axes and cleavers as well as flakes, which could be used for cutting and scraping, have been found. Some believe that *H. ergaster* was a systematic hunter and brought kills to the same site over and over again. In one location, paleontologists have found over 40,000 bones and 2,647 stones.

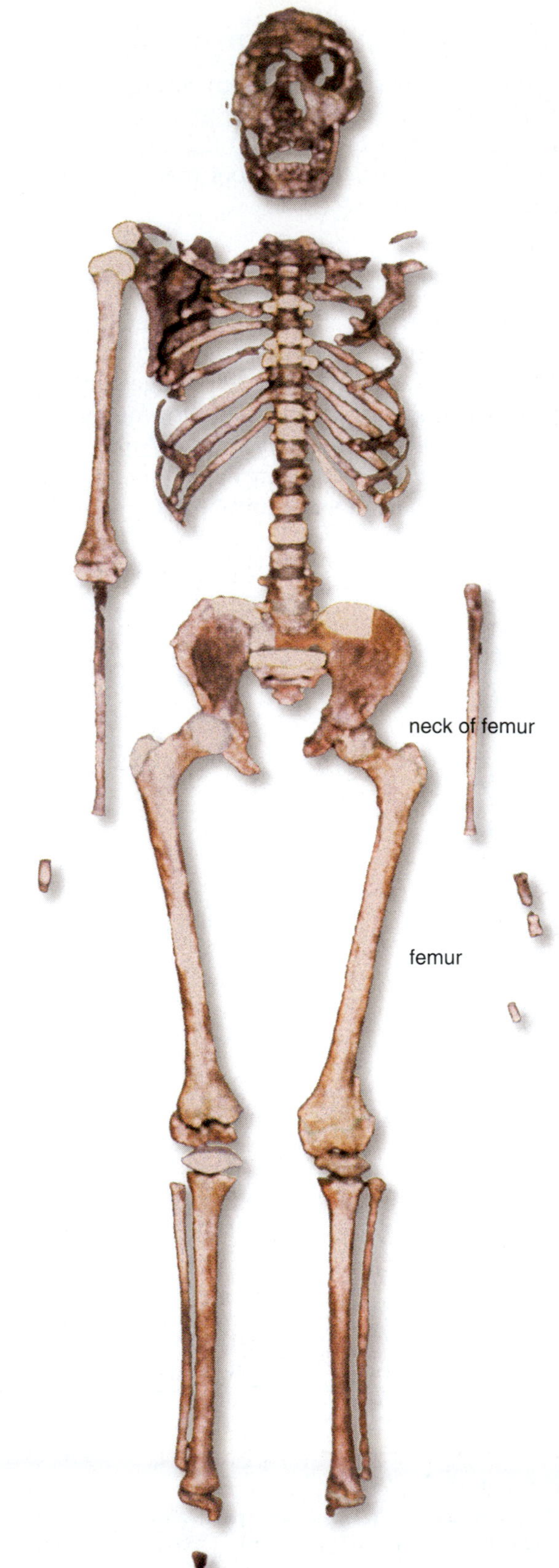

Figure 21.10 ***Homo ergaster.***
Skeleton of a ten-year-old boy who lived 1.6 MYA in eastern Africa. The femurs are angled and the necks are much longer than that of a modern human.

These sites could have been "home bases" where social interaction occurred and a prolonged childhood allowed time for much learning.

In contrast to the australopithecines, early *Homo* types had a larger brain, and probably manufactured tools. *H. ergaster* (in Africa) and *H. erectus* (in Asia) had a still larger brain, may have had a striding gait, and certainly *H. ergaster* in Africa manufactured more advanced tools.

Evolution of Modern Humans

Modern humans are placed in the species ***Homo sapiens.*** About 300,000 years BP (before present), so-called "archaic *Homo sapiens,*" now increasingly classified as *H. heidelbergensis,* were present in Europe, Asia, and Africa. Neanderthal is a well-known archaic *H. sapiens.* The hypothesis that each of these individual populations went on to evolve into modern humans is called the **multiregional continuity hypothesis** (Fig. 21.11*a*). This hypothesis requires that evolution be essentially similar in several different places. Each region would show a continuity of anatomical characteristics from about 2 MYA, when *H. ergaster* first arrived in Eurasia. Opponents argue that it seems highly unlikely that evolution would have produced essentially the same result in these different places. They suggest, instead, the **out-of-Africa hypothesis,** which proposes that *H. sapiens* arose only in Africa, and thereafter migrated to Europe and Asia about 100,000 years BP. If so, there would be no continuity of characteristics between earlier fossils and fossils of 100,000 years BP. Modern humans may have interbred to a degree with archaic populations but after coexisting with them for a while, they replaced them (Fig. 21.11*b*).

According to which hypothesis would modern humans be most genetically alike? With the multiregional hypothesis, human populations have been evolving separately for a long time and therefore genetic differences are expected. With the out-of-Africa hypothesis, we are all descended from a few individuals from about 100,000 years BP. Therefore, the out-of-Africa hypothesis suggests that we are more genetically similar. A few years ago, a study attempted to show that all the people of Europe (and the world for that matter) have essentially the same mitochondrial DNA. Called the mitochondrial Eve hypothesis by the press (note

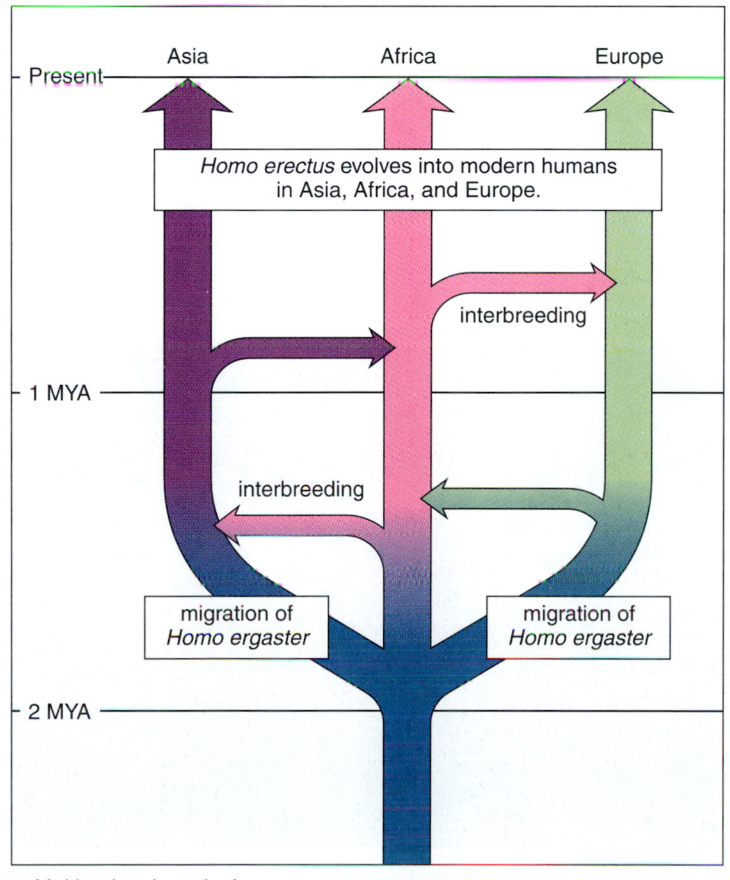

a. Multiregional continuity

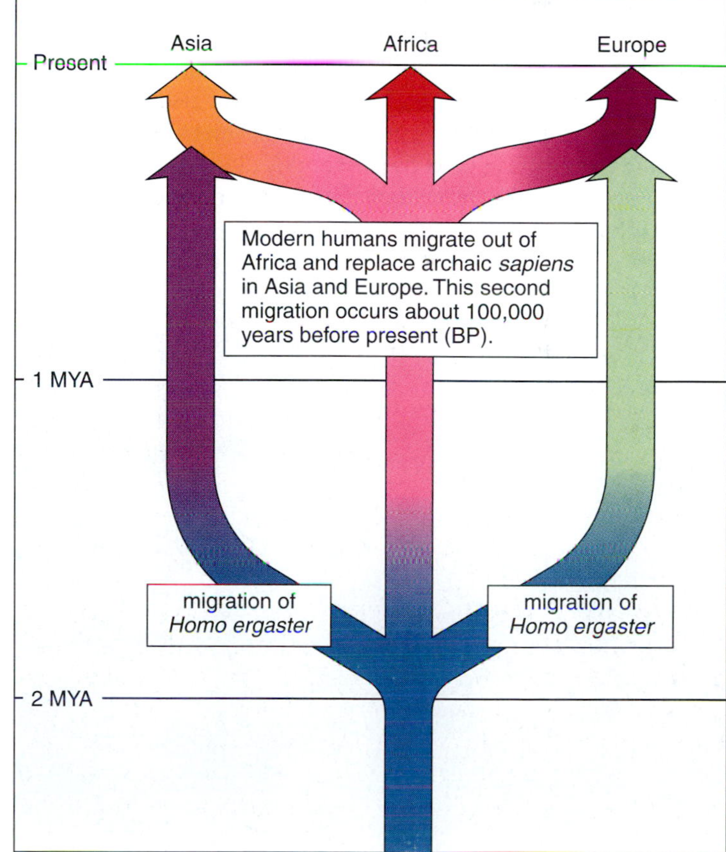

b. Out of Africa

Figure 21.11 Origin of modern humans.
a. The multiregional continuity hypothesis proposes that modern humans evolved separately in at least three different places: Asia, Africa, and Europe. Therefore, continuity of genotypes and phenotypes is expected in these regions. **b.** The out-of-Africa hypothesis proposes that modern humans originated only in Africa; then they migrated and supplanted populations of *Homo* in Asia and Europe about 100,000 years ago.

this is a misnomer because no single ancestor is proposed), the statistics that calculated the date of the African migration were found to be flawed. Still, the raw data—which indicate a close genetic relationship among all Europeans—support the out-of-Africa hypothesis.

A new nuclear DNA analysis, done in 1999, supports the multiregional hypothesis. DNA from a variety of ethnic groups around the globe was tested for different versions of a gene on the X chromosome called *PDHA1*. The investigators, Jody Hey and Eugene Harris, built a phylogenetic tree by assuming that the number of sequence differences between any two versions corresponds to the time since they diverged from one another. The tree showed that modern variants of the gene go back to two ancestral versions; one found in Africans today and one found in non-Africans. They suggested that more research is needed with other nuclear genes before either hypothesis can be rejected.

Investigators are currently testing two hypotheses: that modern humans (1) evolved separately in Europe, Africa, and Asia, and (2) evolved in Africa and then migrated.

Neanderthals

The **Neanderthals** *(H. neanderthalensis)* take their name from Germany's Neander Valley, where one of the first Neanderthal skeletons, dated some 50,000 years BP, was discovered. The Neanderthals had massive brow ridges, and the nose, jaws, and teeth protruded far forward. The forehead was low and sloping, and the lower jaw sloped back without a chin. New fossils show that the pubic bone is long, compared to ours.

At this time, the Neanderthals are thought to be a type of archaic *H. sapiens* and most likely not in the main line of *Homo* descent. Surprisingly, however, the Neanderthal brain was, on the average, slightly larger than that of modern humans (1,400 cc, compared to 1,360 cc in most modern humans). The Neanderthals were heavily muscled, especially in the shoulders and the neck (Fig. 21.12). The bones of the limbs were shorter and thicker than those of modern humans. It is hypothesized that a larger brain than that of modern humans was required to control the extra musculature. They lived in Eurasia during the last Ice Age, and their sturdy build could have helped conserve heat.

The Neanderthals give evidence of being culturally advanced. Most lived in caves, but those in the open may have built houses. They manufactured a variety of stone tools. They most likely successfully hunted bears, woolly mammoths, rhinoceroses, reindeer, and other contemporary animals. They used and could control fire, which probably helped in cooking frozen meat and in keeping warm. They even buried their dead with flowers and tools and may have had a religion.

Neanderthal fragments dated at 30,000 years BP still retain the classic features of the species, suggesting that they did not interbreed with humans. Modern humans most likely supplanted this archaic *H. sapiens* species. Why might that have been? One answer is that the Neanderthal way of life was hard. The teeth of Neanderthal children indicate they did not get enough to eat after they were weaned. And their bones show wear that would only occur if they were active; perhaps they had to keep up with the group rather than staying safely at home. The adult male fossils have multiple injuries. The stone points they made were meant to

Figure 21.12 Neanderthals.
This drawing shows that the nose and the mouth of these people protruded from the face, and their muscles were massive. They made stone tools and were most likely excellent hunters.

Figure 21.13 Cro-Magnon.
Cro-Magnon people are the first to be designated *Homo sapiens sapiens.* Their toolmaking ability and other cultural attributes, such as their artistic talents, are legendary.

be thrust at an animal rather than thrown. This presumably exposed them to the fury of their prey and many a nasty attack and broken bone.

Cro-Magnons

No one knows exactly when modern *Homo sapiens* evolved. But it is believed by many that *H. sapiens* coexisted for a time with the archaic versions before supplanting them.

The **Cro-Magnons,** named after a fossil location in France and found in Eurasia 100,000 years BP or even earlier, are classified as modern humans. Indeed, Cro-Magnons had a thoroughly modern appearance (Fig. 21.13). They made advanced stone tools called Aurignacian tools (Fig. 21.14). Included were compound tools, as when stone flakes were fitted to a wooden handle. They may have been the first to throw spears, enabling them to kill animals from a distance, or to have made knifelike blades. They were such accomplished hunters that some researchers believe they were responsible for the extinction of many larger mammals, such as the giant sloth, the mammoth, the saber-toothed tiger, and the giant ox during the late Pleistocene epoch.

Cro-Magnons hunted cooperatively, and perhaps they were the first to have had a language. They are believed to have lived in small groups, with the men hunting by day, while the women remained at home with the children. It's quite possible that a hunting way of life among prehistoric people influences our behavior even until today. The Cro-Magnon culture included art. They sculpted small figurines and even flutes out of reindeer bones and antlers. They also painted beautiful drawings of animals on cave walls in Spain and France (see Fig. 21.13). Most likely relations between members of a society reinforced creativity among them. Their technology was successful but they apparently also loved painting, music, and beauty.

It is hypothesized that modern humans lived side by side with archaic versions of our species and then supplanted them. Today there is only one type of *Homo sapiens* in existence.

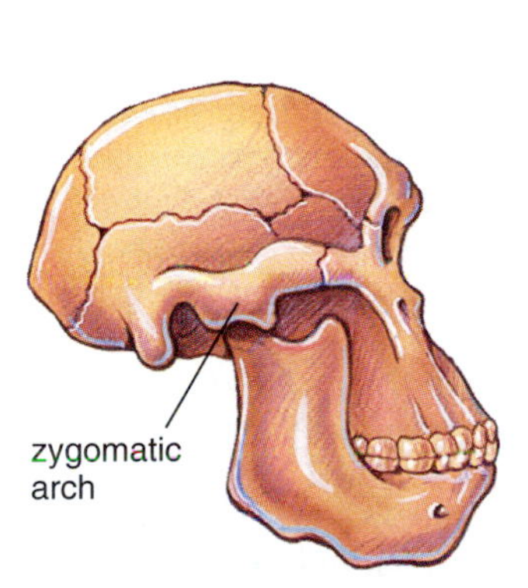

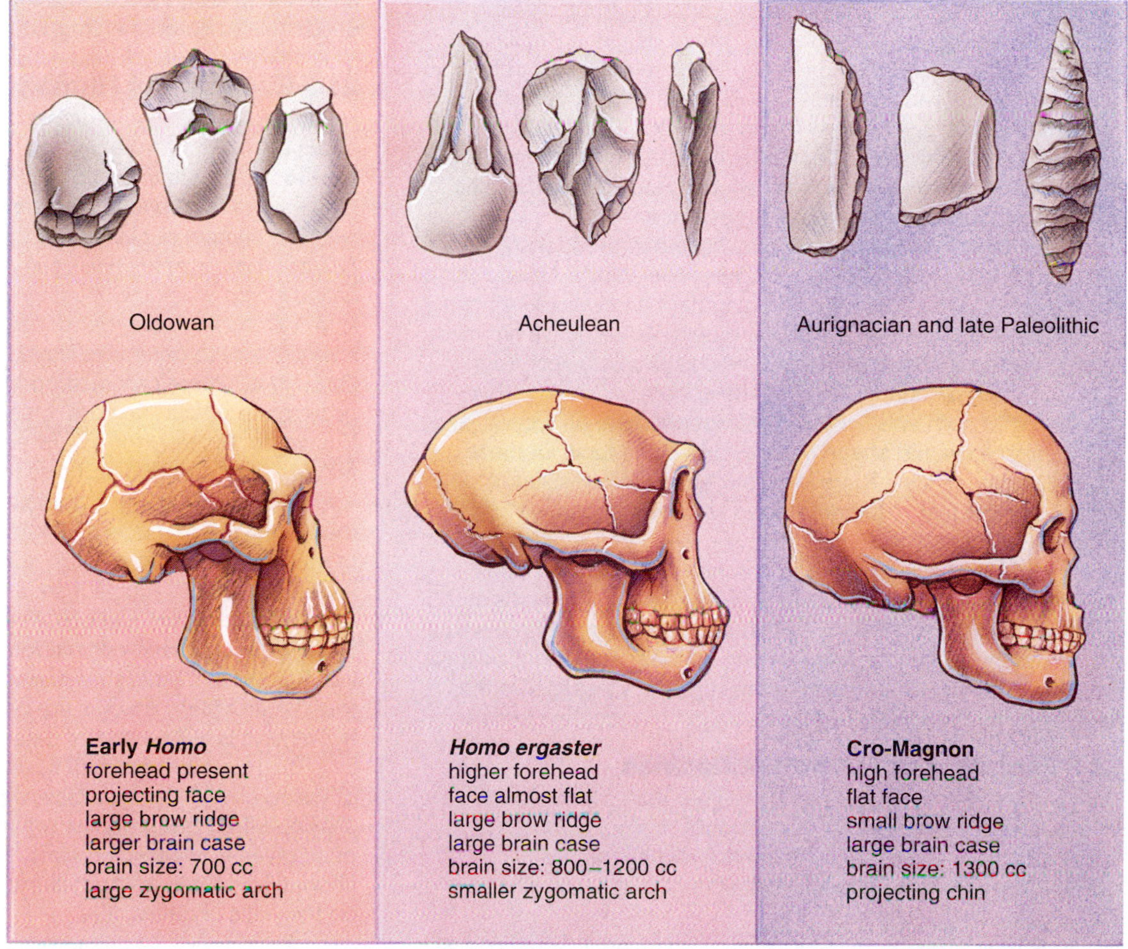

Figure 21.14 Hominid skulls and tools.
The listings compare hominid skulls. The Oldowan tools are crude. Acheulean tools are better made and more varied, and Aurignacian tools are well designed for specific purposes.

Connecting Concepts

Human beings are diverse, but even so, we are all classified as *H. sapiens.* This is consistent with the biological definition of species because it is possible for all types of humans to interbreed and to bear fertile offspring. While it may appear that there are various "races," molecular data show that the DNA base pair sequence varies as much between individuals of the same ethnicity as between individuals of different ethnicity.

It is generally accepted that the phenotype is adapted to the climate of a region. Although it might seem as if dark skin is a protection against the hot rays of the sun, it has been suggested that it is actually a protection against ultraviolet ray absorption. Dark-skinned persons living in southern regions and white-skinned persons living in northern regions absorb the same amount of radiation. (Some absorption is required for vitamin D production.) Other features that correlate with skin color, such as hair type and eye color, may simply be pleiotrophic effects of genes that control skin color.

Differences in body shape represent adaptations to temperature. A squat body with short limbs and a short nose retains more heat than an elongated body with long limbs and a long nose. Almond-shaped eyes, a flat nose and forehead, and broad cheeks are believed to be adaptations to the last Ice Age.

While it always has seemed to some that physical differences warrant assigning humans to different "races," this contention is not borne out by the molecular data mentioned previously.

Summary

21.1 Humans Are Primates

Primates, in contrast to other types of mammals, are adapted for an arboreal life. The evolution of primates is characterized by trends toward mobile limbs, grasping hands, a flattened face, binocular vision, large complex brain, and one birth at a time. These traits are particularly useful for living in trees.

During the evolution of primates, various groups diverged in a particular sequence from the main line of descent. Prosimians (tarsiers and lemurs), which diverged first, are most distantly related to us and most closely related to the original primate.

Anthropoids include New World monkeys, Old World monkeys, and hominoids. Soon after monkeys evolved in Africa, New World monkeys took up residence in South America while Old World monkeys continued living in Africa.

The first hominoid ancestor may have been *Proconsul,* whose apelike fossils date from the Miocene epoch. Among living apes, two are Asian (gibbon and orangutan) and two are African (gorilla and chimpanzee). Molecular biologists tell us we are most closely related to the African apes, whose ancestry split from ours about 8 MYA (during the Pliocene epoch).

We differ from apes in (1) type of locomotion, (2) type of jaw, and (3) brain size and shape of skull, brow, and face. Of these, bipedalism arose first. There is no evidence that grasslands replaced forest just at the time when hominids began walking upright. But there was an overall drying trend that reduced the amount of forest and introduced patches of woodlands and grasslands. Perhaps bipedalism developed when hominids stood on branches to reach fruit overhead and then they continued to use this stance when foraging among bushes. An upright posture reduces exposure of the body to the sun's rays and leaves the hands free to carry food, perhaps as a gift to receptive females.

21.2 Evolution of Australopithecines

Human evolution resembles a bush rather than a straight line from one species to another. Also, fossils exemplify mosaic evolution and have a mixture of apelike and humanlike characteristics. Hominid fossils have been found in southern Africa and eastern Africa.

In southern Africa, hominids classified as australopithecines include *Australopithecus africanus* (3 MYA), a gracile form, and *A. robustus* (2 MYA), a robust form. In eastern Africa, hominids classified as australopithecines include *Australopithecus ramidus* (4.4 MYA), *A. anamensis* (4 MYA), *A. afarensis* (Lucy) (3.18 MYA), *A. garhi* (2.5 MYA), and robust types (*A. aethiopicus* and *A. boisei*, dated at 2.6 MYA). These hominids walked upright and had a brain size of 400 to 500 cc.

Many of the australopithecines coexisted, and it is difficult to tell who is ancestral to whom. Debate is now going on as to whether *A. garhi, A. afarensis,* or *A. africanus* is directly ancestral to humans. *A. africanus* had a larger brain than *A. afarensis,* and the proportion of its limbs was more apelike, like that of early *Homo.* Perhaps *A. afarensis* is ancestral to *A. garhi*, which is ancestral to early *Homo.* There is some evidence that *A. garhi* used tools to butcher animal limbs and get at bone marrow.

21.3 Evolution of Humans

Homo habilis and *Homo rudolfensis* called early *Homo* are present about 2 MYA. *H. rudolfensis* had the larger brain, which at 775 cc was 45% larger than that of an australopithecine. Early *Homo* made stone tools and in fact *Homo habilis* means "handy man."

The custom of classifying fossils dated between 1.9 and 0.5 MYA as *Homo erectus* has been changed. Asian types are now called *Homo erectus* while the African types are now called *Homo ergaster.* A very good fossil find in Africa allowed investigators to determine that *H. ergaster* had a larger brain than early *Homo,* more pronounced brow ridges, a flatter face, and a nose that projects like ours. Most likely they had a striding gait like we do. An early form of *H. ergaster* is believed to have migrated out of Africa and given rise to *H. erectus* in Asia.

Two contradicting hypotheses have been suggested about the origination of modern humans. The multiregional continuity hypothesis says that modern humans originated separately in Asia, Europe, and Africa. If so, a difference in the genes is expected between locations. The out-of-Africa hypothesis says that modern humans originated only in Africa and, after migrating into Europe and Asia, they replaced the archaic *Homo* species found there. Many studies are being done to determine which hypothesis is supported by data.

The Neanderthals may have been an archaic *Homo* species of Europe. Their chinless face, squat frame, and heavy muscles are apparently adaptations to the cold. Cro-Magnon is a name often given to modern humans. Their tools were sophisticated, and they definitely had a culture, as witnessed by the paintings on the walls of caves.

Reviewing the Chapter

1. List and discuss various evolutionary trends among primates and state how they would be beneficial to animals with an arboreal life. 342

2. When did the prosimians and various anthropoids diverge from the human line of descent? 344
3. What are the two groups of living monkeys and how are they related? 345
4. What is the significance of the fossils classified as *Proconsul*? 345
5. What are the four groups of living apes and where do they live? 346–47
6. Assuming there was an overall drying trend in Africa, discuss the possible benefits of bipedalism in early hominids? 347
7. Name the various species of *Australopithecus,* and explain how they are believed to be related to one another and to humans. 348–49
8. Why are the early *Homos* classified as humans? If these hominids did make tools, what does this say about their possible way of life? 351
9. What role(s) might *H. ergaster* have played in the evolution of modern humans according to the multiregional continuity theory? the out-of-Africa theory? 353
10. Who were the Neanderthals and the Cro-Magnons, and what is their place in the evolution of humans according to the two hypotheses mentioned in question 9? 354–55

Testing Yourself

Choose the best answer for each question.

1. Which of these gives the correct order of divergence from the main line of descent leading to humans?
   a. prosimians, monkeys, gibbons, orangutans, African apes, humans
   b. gibbons, orangutans, prosimians, monkeys, African apes, humans
   c. monkeys, gibbons, prosimians, African apes, orangutans, humans
   d. African apes, gibbons, monkeys, orangutans, prosimians, humans
   e. *H. habilis, H. erectus, H. neanderthalensis,* Cro-Magnon
2. Lucy is a member of what species?
   a. *Homo erectus*
   b. *Australopithecus afarensis*
   c. *H. habilis*
   d. *A. robustus*
   e. *A. anamensis* and *A. afarensis* are alternate forms of Lucy.
3. What possibly may have influenced the evolution of bipedalism?
   a. A larger brain developed.
   b. Gathering of food was easier.
   c. The climate became drier.
   d. Both b and c are correct.
   e. Both a and c are correct.
4. Which of these is an incorrect association with robust types?
   a. massive chewing muscles attached to bony skull crest.
   b. *A. robustus* and *A. boisei*
   c. a fibrous diet
   d. lived during an Ice Age
   e. Both a and c are incorrect associations.
5. *H. ergaster* could have been the first to
   a. use and control fire.
   b. migrate out of Africa.
   c. make tools.
   d. have a brain of at least 850 cc.
   e. All of these are correct.
6. Which of these characteristics is not consistent with the others?
   a. brow ridges
   b. small cheek teeth (molars)
   c. high forehead
   d. projecting face
   e. binocular vision
7. Which of these is true? The last common ancestor for African apes and hominids
   a. has been found and it resembles a gibbon.
   b. has not been found but it is expected to be from the Pliocene epoch.
   c. has been found and it has been dated at 30 MYA.
   d. is not expected to be found because there was no such common ancestor.
   e. is now believed to have lived in Asia and not Africa.
8. Which of these is incorrectly matched?
   a. gibbon—hominoid
   b. *A. africanus*—hominid
   c. tarsier—anthropoid
   d. *H. erectus—Homo*
   e. Early *Homo—H. habilis*
9. If the multiregional continuity hypothesis is correct, then
   a. hominid fossils in China after 100,000 BP are not expected to resemble earlier fossils.
   b. hominid fossils in China after 100,000 BP are expected to resemble earlier fossils.
   c. the mitochondrial Eve study must be invalid.
   d. both a and c are correct.
   e. both b and c are correct.
10. Which of these is incorrectly matched?
    a. *H. ergaster*—made tools
    b. Neanderthal—good hunter
    c. *H. habilis*—controlled fire
    d. Cro-Magnon—good artist
    e. *A. robustus*—fibrous diet
11. Classify humans by filling in the missing lines.

Kingdom Animalia	
Phylum	a. ______________
Subphylum	b. ______________
c. ______________	Mammalia
d. ______________	Primates
Suborder	e. ______________
Superfamily	f. ______________
Family	g. ______________
h. ______________	*Homo*
Species	i. ______________

Thinking Scientifically

1. Bipedalism has many selective advantages. However, there is one particular disadvantage of walking on two feet: giving birth to offspring with a large head through the smaller pelvic opening that is necessitated by upright posture is very difficult. This situation results in a high percentage of deaths (of both mother and child) during birth compared to other primates. How do you explain the selection of a trait that is both positive and negative?
2. The molecular clock is an important tool in determining the number of years that have passed since two modern groups have diverged from a common ancestor. The only way this method of determining relationships can be accurate is if you know the rate at which DNA changes in the absence of selective pressure. How would you calibrate a molecular clock?

Bioethical Issue

Is it proper for humans to manipulate evolution? Since the dawn of civilization, humans have carried out artificial selection to develop plants and animals of interest to them. Some of our crops are bountiful but would be unable to survive without the fertilizer, pesticides, and hormones that are supplied to them. Many of our farm animals serve us well but would be unable to function in a natural environment.

With the advent of DNA technology we have entered a new era in which even greater control can be exerted over the evolutionary process. We can manipulate the genes and give organisms traits that they would not ordinarily possess. Some plants today produce human proteins that can be extracted from their seeds, and some animals grow larger because we have supplied them with an extra gene for growth hormone. Does this type of manipulation seem justifiable?

You might feel that it is all right to manipulate the evolution of plants and animals that have been under our care for hundreds of generations. Do you feel differently about wild plants and animals? In today's world the populations of plants and animals that can adjust best to the presence of humans are expanding, while the size of other populations that need an expansive space of their own are shrinking. In other words, humans are in the process of determining which organisms are fit and which are not fit in the evolutionary sense. Does this seem right to you?

What about the possibility that we are manipulating our own evolution? Should doctors increase the fitness of certain couples by providing them with a means to reproduce that they cannot achieve on their own? Is the use of alternate means of reproduction bioethically justifiable? And in the near future it may be possible for parents to choose the phenotypic traits of their offspring; it might be possible for humans to ensure that their offspring are stronger and brighter than you are. Does this seem ethical to you?

Understanding the Terms

arboreal 342
anthropoid 345
australopithecine 348
Australopithecus afarensis 348
Australopithecus africanus 348
bipedalism 347
Cro-Magnon 355
hominid 345
hominoid 345
Homo erectus 352
Homo ergaster 352
Homo habilis 351
Homo rudolfensis 351
Homo sapiens 353
mosaic evolution 348
multiregional continuity hypothesis 353
Neanderthal 354
out-of-Africa hypothesis 353
primate 342
prosimian 343

Match the terms to these definitions:

a. ______________ Group of primates that includes only monkeys, apes, and humans.
b. ______________ The common name for the first fossils generally accepted as being modern humans.
c. ______________ Hominid with a sturdy build who lived during the last Ice Age in Eurasia; hunted large game, and lived together in a kind of society.
d. ______________ Type of hominid which lived in Africa during the Pleistocene epoch; and had a striding gait similar to modern humans.
e. ______________ Member of a superfamily containing humans and the great apes.

Web Connections

Exploring the Internet

http://www.mhhe.com/biosci/genbio/mader
(click on *Biology 7/e*)

The *Biology 7/e* Online Learning Center provides many resources for studying the material in this chapter including links to the following sites:

A thorough treatment of what is known about human ancestry, along with fossil images, can be found in Fossil Hominids.

http://www.talkorigins.org/faqs/homs/

How scientists know that humans originated in Africa is explored in Human Origins and Evolution in Africa. Many links from this site.

http://www.indiana.edu/~origins/

Fossil Hominids. This site provides a summary of current thinking about human evolution and refutes creationist claims regarding human origins.

http://www.talkorigins.org/faqs/fossil-hominids.html

The Fossil Evidence for Human Evolution in China. This site documents early human evolution in China. It provides maps, photographs of fossil remains, full-length articles, and many links to other sites.

http://www.cruzio.com/~cscp/index.htm

Handprint: Ancestral Lines. This site describes the evolution of the genus Homo, *and related predecessors, with a very useful visual time line.*

http://www.handprint.com/LS/ANC/evol.html

Further Readings for Part iii

Agnew, N., and Demas, M. September 1998. Preserving the Laetoli footprints. *Scientific American* 279(3):44. This article recaps the discovery of hominid footprints in eastern Africa, and explains steps taken to preserve them.

Ben-Ari, E. T. February 1999. Molecular biographies. *BioScience* 49(2):98. Anthropological geneticists are using the genome to decode human evolution.

Berger, L. August 1998. The dawn of humans: Redrawing our family tree? *National Geographic* 194(2):90. Southern African fossils show that *A. africanus* was more apelike than Lucy.

Boroughs, D. March/April 1999. New life for a vanished zebra? *International Wildlife* 29(2):46. A program is now underway to try to breed back the extinct quagga.

Bower, B. April 24, 1999. African fossils flesh out humanity's past. *Science News* 155(17):262. Recent discoveries of a new fossil species (named *Australopithecus garhi*) in eastern Africa may be an ancestor of the *Homo* lineage.

Boyd, R., and Silk, J. B. 1997. *How humans evolved.* New York: W. W. Norton & Co., Inc. This introductory text integrates evolutionary theory, population genetics, and behavioral ecology with evidence from the hominid fossil record to emphasize the processes of human evolution.

Chiappe, L. M. September 1998. Wings over Spain. *Natural History* 107(7):30. The presence of a first digit on fossil remains of a 115-million-year-old bird shows advanced flying ability.

Clark, G. March 26, 1999. Highly visible, curiously intangible. *Science* 283(5410):2029. The two main theories of human evolution are subject to the interpretation.

Culotta, E. April 23, 1999. A new human ancestor? *Science* 284(5414):572. Ethiopian fossils reveal a new branch on the hominid family tree.

_____ May 14, 1999. Anthropologists probe bones, stones, and molecules. *Science* 284(5417):1109. Anthropologists explore brain shapes in hominids and the original Asian homeland of the first Americans.

Diamond, J. September 1998. Evolving backward. *Discover* 19(9):64. Studies of the blind mole rat shows how evolved traits are lost if they are not used.

Gore, R. September 1997. The dawn of humans. *National Geographic* 192(3):92. A footprint dated about 117,000 years ago was discovered in southern Africa.

Gould, S. J. February 1999. A division of worms. *Natural History* 108(1):18. Part One of a series about Jean-Baptiste Lamarck—who he was.

Gould, S. J. March 1999. Branching through a wormhole. *Natural History* 108(2):24. Part Two of a series about Jean-Baptiste Lamarck—why his system failed.

Leakey, M., and Walker, A. June 1997. Early hominid fossils from Africa. *Scientific American* 276(6):74. A bone unearthed in 1965 recently proved the existence of a new species of *Australopithecus,* showing ancestral humans existed 4 million years ago.

Levin, H. L. 1996. *The Earth through time.* Fort Worth, Texas: Saunders College Publishing. This introductory text provides background information on such topics as the geologic time scale, plate tectonics, the fossil record, and human origins.

Lewin, R. 1997. *Patterns in evolution: The new molecular view.* New York: Scientific American Library. This book explores how genetic information provides insights into evolutionary events.

Monastersky, R. March 1998. The rise of life on earth. *National Geographic* 193(3):54. Article discusses the origins of microbial life, stromatolite reefs, and Stanley Miller's model of the primitive atmosphere.

Morrell, V. April 9, 1999. Forming the robust Australopithecine face. *Science* 284(5412):230. The shape of teeth is important in determining modern human origins.

Nei, M., and Zhang, J. November 20, 1998. Molecular origin of species. *Science* 282(5393):1428. The genes involved in speciation are being identified and studied as to their effect on mating and development.

Padian, K., and Chiappe, L. M. February 1998. The origin of birds and their flight. *Scientific American* 278(2):38. Recent fossil discoveries confirm that birds descended from dinosaurs.

Pennisi, E. March 19, 1999. From a flatworm, new clues on animal origins. *Science* 283(5409):1823. Some flatworms seem to represent the earliest step in the evolution of bilaterally symmetrical organisms.

_____ March 19, 1999. Genetic study shakes up out-of-Africa theory. *Science* 283(5409):1828. Analysis and comparison of different versions of a gene on the X chromosome of early humans offers evidence of multiple ancient populations.

_____ March 26, 1999. Did cooked tubers spur the evolution of big brains? *Science* 283(5410):2004. Article discusses how diet may have affected the evolution of large brains, small teeth, and limb proportions.

Sandweiss, D., et al. January 22, 1999. Transitions in the mid-Holocene. *Science* 283(5401):499. Evidence for climatic and cultural change during the mid-Holocene points to a possible connection between climate and culture.

Seymour, R. S. March 1997. Plants that warm themselves. *Scientific American* 276(3):104. Some plants generate heat to keep blossoms at a constant temperature.

Shreeve, J. December 1997. Uncovering Patagonia's lost world. *National Geographic* 192(6):120. Recent fossil finds cause scientists to rethink the evolution of dinosaurs.

Stiassny, M. L. J., and Meyer, A. February 1999. Cichlids of the Rift lakes. *Scientific American* 280(2):64. The extraordinary diversity of cichlid fishes challenges entrenched ideas of how quickly new species can arise.

Swerdlow, J. February 1999. Biodiversity: Taking stock of life. *National Geographic* 195(2):2. The value of biodiversity on the face of increasing extinction is discussed.

Tattersall, I. April 1997. Out of Africa again . . . and again? *Scientific American* 276(4):60. Hominids may have migrated out of Africa several times, with each emigration sending a different species.

Thompson, K. May/June 1999. The coelocanth: Act three. *American Scientist* 213(87):3. Discusses this living fossil, and presents hypotheses to account for a living fossil lineage.

Tudge, C. 1996. *The time before history: 5 million years of human impact.* New York: Scribner. This book is a comprehensive record of changes in the Earth and its inhabitants during the period known as Plio/Pleistocene.

Vogel, G. January 15, 1999. Did early African hominids eat meat? *Science* 283(5400):303. Isotope testing of *Australopithecus africanus* has conclusively determined the diet of this hominid included meat and plants.

Webster, D. June 1999. A dinosaur named Sue. *National Geographic* 195(6):46. The largest *Tyrannosaurus rex* fossil skeleton ever found is being reconstructed.

Westenberg, K. May 1999. The rise of life on Earth. *National Geographic* 195(5):114. New discoveries suggest that legs for walking on land originated during life in the water, in the Devonian period.

Weuthrich, B. January 22, 1999. Stunning fossil shows breath of a dinosaur. *Science* 283(5401):468. Examination of 100-million-year-old dinosaur fossils with UV light supports the idea that dinosaurs were cold-blooded.

Zimmer, C. March 26, 1999. Fossil offers a glimpse into mammals' past. *Science* 283(5410):1989. This fossil find answers some questions about the history of Mesozoic mammals.

part iv

Behavior and Ecology

The behavior of an organism is regulated by the nervous and endocrine systems, whose development is under the control of inherited genes but also influenced by the environment. The genetic basis of behavior suggests that behavior is subject to natural selection and is adaptive. Organisms do not exist alone; rather, they are part of a population that interacts with both the abiotic (nonliving) and biotic (living) environment. A community is an assemblage of populations in a particular area; but ultimately, ecology considers the distribution and abundance of populations over the surface of the earth: in short, the biosphere. The biosphere performs services for us, such as regulating climate and the quality of the water we drink and the air we breathe. Unfortunately, human activities have reduced the size of natural ecosystems and strained the capacity of biogeochemical cycles to prevent pollution. It's possible that conservation, an active field of current research, will help preserve species and manage ecosystems for sustainable human welfare.

Animal Behavior

chapter 22

chapter concepts

Satin Bowerbirds, *Ptilonorhynchus violaceus*

At the start of the breeding season, male bowerbirds use small sticks and twigs to build elaborate display areas called bowers. They clear the space around the bower and decorate the area with fresh flowers, fruits, pebbles, shells, bits of glass, tinfoil, and any bright baubles they can find. Each species has its own preference in decorations. The Satin Bowerbird of eastern Australia prefers blue objects, a color that harmonizes with the male's glossy blue-black plumage.

After the bower is complete, a male spends most of his time near his bower, calling to females, renewing his decorations, and guarding his work against possible raids by other males. After inspecting many bowers and their owners, a female approaches one and the male begins a display. He faces her, fluffs up his feathers, and flaps his wings to the beat of a call. The female enters the bower, and if she crouches, the two mate.

Behavior is studied in the same objective manner as any other field of science. A behaviorist would determine how a male Satin Bowerbird is organized to perform this behavior and how the behavior helps him secure a mate.

22.1 Behavior Has a Genetic Basis

Two types of questions are central to the study of behavior. *Mechanistic questions* are answered by describing how an animal is biologically organized and equipped to carry out the behavior. *Survival value questions* are answered by describing how the behavior helps an animal exploit resources, avoid predators, or secure a mate. Both types of questions are grounded in the recognition that **behavior,** observable and coordinated responses to environmental stimuli, has at least a partial genetic basis.

Various types of experiments have been performed to determine if behavior has a genetic basis. Birds have the ability to return home, and some migrate great distances spring and fall. Birds that are caged and prevented from migrating exhibit increased activity that is called *migration restlessness.*

Peter Berthold and his colleagues noted that Blackcap Warblers from Germany migrate to Africa, while those that live in Cape Verde (islands off the west coast of Africa) do not migrate at all. He reasoned that if migration behavior is inherited, then hybrids might show intermediate behavior. Berthold captured birds from Germany and Cape Verde and mated them. He placed the hybrid offspring and Cape Verdean birds in separate cages that were equipped with perches that electronically recorded when birds landed on them. During the next migratory period, the hybrids jumped back and forth between perches every night for many weeks, but the Cape Verdean birds exhibited no migratory restlessness.

Andreas Helbig performed still another experiment with Blackcap Warblers. German blackcaps fly *southwest* from Germany to Spain and finally to southwest Africa. Austrian blackcaps fly *southeast*, from Austria to Israel and finally to southeast Africa. Helbig created hybrids and then placed the parents and the hybrids in a special funnel cage that allowed the birds to see the stars at night. (Migration studies have shown that birds navigate during migration by using the sun in the day and the stars at night as compasses to determine direction.) The floor and sides of the cages were covered with a special paper that recorded scratch marks as the birds jumped in their preferred direction during the next migratory period. When the choices of hybrid offspring were compared statistically with those of their parents, the hybrids were proved to be intermediate between them (Fig. 22.1). Both Berthold's and Helbig's studies with Blackcap Warblers support the hypothesis that behavior has at least a partial genetic basis.

Steven Arnold performed several experiments with the garter snake *Thamnophis elegans* after he observed a distinct difference between two different types of snake populations in California. Inland populations are aquatic and commonly feed underwater on frogs and fish. Coastal populations are terrestrial and feed mainly on slugs. In the laboratory, inland adult snakes refused to eat slugs, while coastal snakes readily did so. To test for possible genetic differences between the two populations, Arnold arranged matings

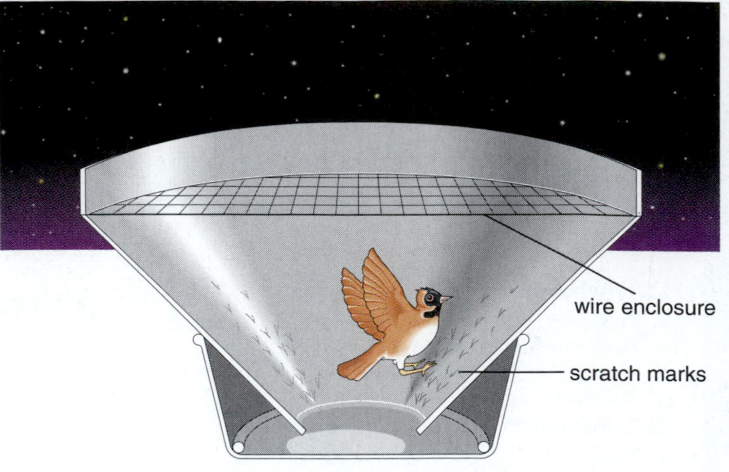

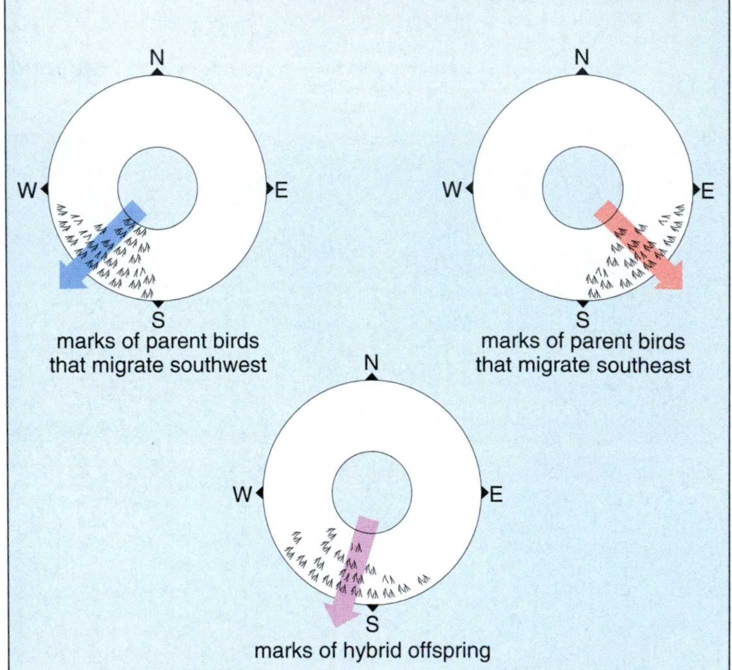

Figure 22.1 Inheritance of migratory behavior in Blackcap Warblers, *Sylvia*.
a. A funnel-shaped cage allowed birds to see the night sky because birds use the stars as a compass when migrating. **b.** Experimenters could determine the preferred direction of flight by scratch marks on the floor and sides of the cage. Parents preferred either to fly southeast or southwest; hybrids preferred an intermediate direction.

between inland and coastal individuals and found that isolated newborns show an overall intermediate incidence of slug acceptance.

The difference between slug acceptors and slug rejecters appears to be inherited, but what physiological difference have the genes brought about? Arnold devised a clever experiment to answer this causal question. When snakes eat, their tongues carry chemicals to an odor receptor in the roof of the mouth. They use tongue flicks to recognize

Inland garter snake does not eat slugs.

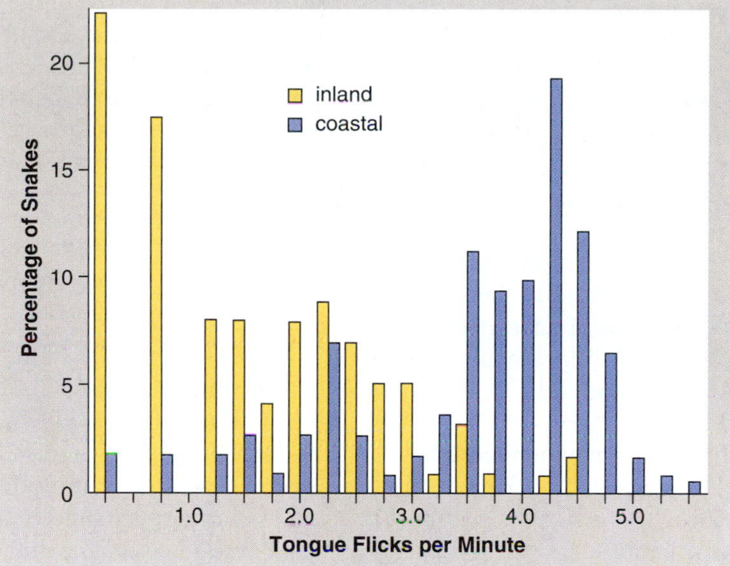

Coastal garter snake does eat slugs.

Figure 22.2 Feeding behavior of garter snakes, *Thamnophis elegans.*
The number of tongue flicks by inland and coastal garter snakes as a response to slug extract on cotton swabs. Coastal garter snakes do eat slugs; inland garter snakes do not eat slugs. Coastal snakes tongue-flicked more than inland snakes.

their prey! Even newborns will flick their tongues at cotton swabs dipped in fluids of their prey. Arnold dipped swabs in slug extract and counted the number of tongue flicks for newborn inland and coastal snakes. Coastal snakes had a higher number than inland snakes (Fig. 22.2). Although hybrids showed a great deal of variation in the number of tongue flicks, they were generally intermediate as predicted by the genetic hypothesis. These findings suggest that inland snakes do not eat slugs because they are not sensitive to their smell. It would seem that the genetic differences between the two populations of snakes has resulted in a physiological difference in their nervous systems.

Both nervous and endocrine systems are responsible for the integration of body systems. Is the endocrine system also involved in behavior, and is this system also regulated by the animal's genes? Various studies have been done to show that it is. For example, the egg-laying behavior in the sea slug *Aplysia* involves a set sequence of movements. Following copulation, the animal extrudes long strings of more than a million egg cases. It takes the egg case string in its mouth, covers it with mucus, and waves its head back and forth to wind the string into an irregular mass. This behavior attaches the mass to a solid object, like a rock. Several years ago, scientists isolated and analyzed an egg-laying hormone (ELH) that causes the snail to lay eggs even if it had not mated. ELH was found to be a small protein of 36 amino acids that diffuses into the circulatory system and excites the smooth muscle cells of the reproductive duct, causing them to contract and expel the egg string. Using recombinant DNA techniques, the investigators isolated the ELH gene. The gene's product turned out to be a protein with 271 amino acids. The protein can be cleaved into as many as 11 possible products, and ELH is one of these. ELH alone, or in conjunction with these other products, is thought to control all the components of egg-laying behavior in *Aplysia.*

Human Studies

The nature (inherited) versus nurture (environment) question has been asked for a long time. Twin studies have been employed to attempt to find the answer with regard to humans. It has been found that fraternal twins, even when raised in the same environment, are not remarkably similar in behavior, whereas identical twins raised separately are sometimes remarkably similar. Altogether the data seemed to show that about 50% of the differences in human personality traits are due to inheritance and about 50% are due to environmental influence.

The results of many types of studies support the hypothesis that behavior has at least a partial genetic basis, and that genes influence the development of neural and hormonal mechanisms that control behavior.

22.2 Behavior Undergoes Development

Given that all behaviors have at least a partial genetic basis, we can go on to ask if environmental experiences after hatching or birth also shape the behavior. Fixed action patterns (FAPs) are stereotyped behaviors—they are always performed the same way each time. FAPs are elicited by a sign stimulus, a cue that sets the behavior in motion. For example, human babies will smile when a flat but face-sized mask with two dark spots for eyes is brought near them. It's possible that some behaviors are FAPs, but increasingly investigators are finding that many behaviors, formerly thought to be FAPs, develop after practice.

One investigator, Jack Hailman, has exhaustively studied Laughing Gull chicks' begging behavior, which is always performed the same way in response to the parent's red beak. A chick directs a pecking motion toward the parent's beak, grasps it, and strokes it downward (Fig. 22.3). Hailman noticed that sometimes a parent stimulates the begging behavior by swinging its beak gently from side to side. After the chick responds, the parent regurgitates food onto the floor of the nest. If need be, the parent then encourages the chick to eat. This interaction between the chicks and their parents made Hailman think the begging behavior might involve learning. (**Learning** is defined as a durable change in behavior brought about by experience.) To test this hypothesis, Hailman painted diagrammatic pictures of gull heads on small cards and then collected eggs in the field. The eggs were hatched in a dark incubator to eliminate visual stimuli before the test. On the day of hatching, each chick was allowed to make about a dozen pecks at the model. The chicks were returned to the nest and were later retested. The tests showed that on the average, only one-third of the pecks by a newly hatched chick strike the model. But one day after hatching more than half of the pecks are accurate, and two days after hatching the accuracy reaches a level of more than 75%. His treatment and testing of other groups allowed him to conclude that improvement in motor skills, as well as visual experience, strongly effect development of chick begging behavior.

Behavior has a genetic basis, but the development of mechanisms that control behavior is subject to environmental influences, such as practice after birth.

Hailman went on to test how chicks recognize a parent. He found that newly hatched chicks peck equally at any model as long as it has a red beak. Chicks a week old, however, will peck only at models that closely resemble the parent. Hailman speculated that operant conditioning with a reward of food could account for this change in behavior.

Operant conditioning, which is one of many forms of learning, is often defined as the gradual strengthening of stimulus-response connections. In everyday life, most people know that animals can be taught tricks by giving rewards such as food or affection. The trainer presents the stimulus, say a hoop, and then gives a reward (food) for the proper response (jumping through the hoop). B. F. Skinner is well known for studying this type of learning in the laboratory. In the simplest type of experiment performed by Skinner, a caged rat happens to press a lever and is rewarded with sugar pellets, which it avidly consumes. Rewarding an animal for a particular behavior *reinforces* the behavior and makes it more likely to happen again. In this way the rat learns to regularly press the lever whenever it is hungry. In more sophisticated experiments, Skinner even taught pigeons to play Ping-Pong by reinforcing desired responses to stimuli.

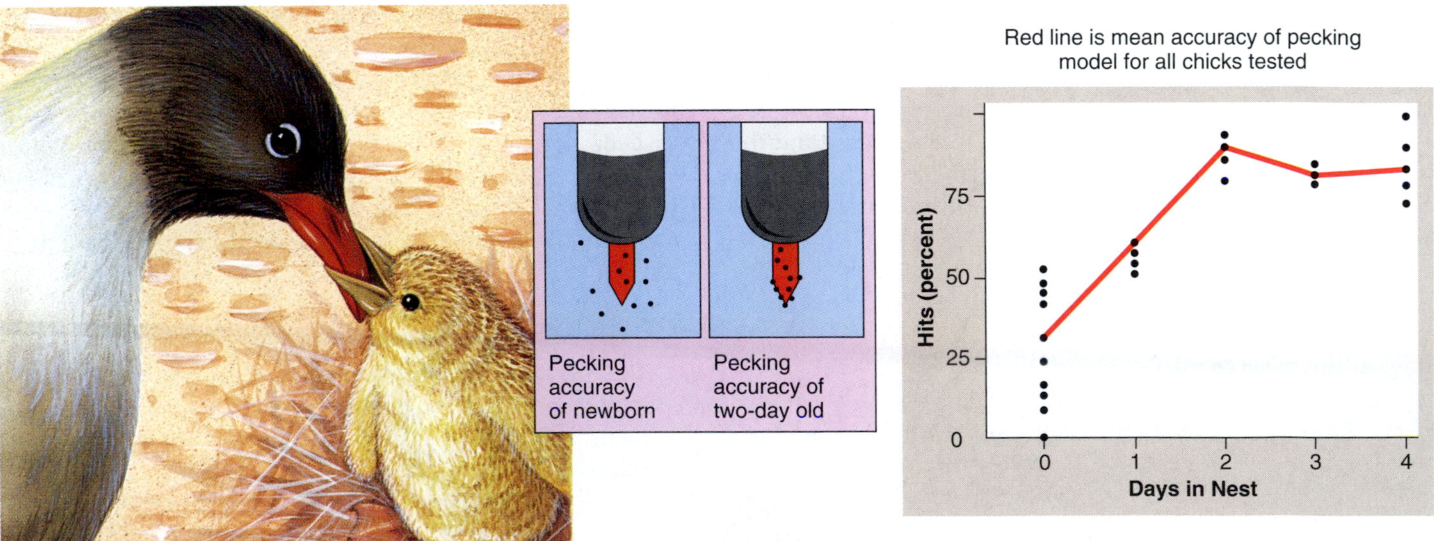

Figure 22.3 **Pecking behavior of Laughing Gull chicks, *Larus atricilla*.**
At about three days, Laughing Gull chicks grasp the beak of a parent, stroking it downward and the parent then regurgitates food.

Imprinting is another form of learning; chicks, ducklings, and goslings will follow the first moving object they see after hatching. This object is ordinarily their mother, but they seemingly can be imprinted on any object—a human or a red ball—if it is the first moving object they see during a sensitive period of two to three days after hatching. The term *sensitive period* means that the behavior only develops during this time. Although the Englishman Douglas Spalding first observed imprinting, the Austrian Konrad Lorenz is well known for investigating it. He found that imprinting not only served the useful purpose of keeping chicks near their mother, it also caused male birds to court a member of the correct species—someone who looks like mother! The goslings who had been imprinted on Lorenz courted human beings later in life.

In-depth studies on imprinting have shown that the process is more complicated than originally thought. Eckhard Hess found that mallard ducklings imprinted on humans in the laboratory would switch to a female mallard that had hatched a clutch of ducklings several hours before. Also, he found that vocalization before and after hatching was normally an important element in the imprinting process. Female mallards cluck during the entire time that imprinting is occurring. Do social interactions influence other forms of learning? Patterns of song learning in birds suggests that they can.

Song Learning in Birds

During the past several decades, an increasing number of investigators have studied song learning in birds. Although White-Crowned Sparrows have a species-specific song, males from different regions have different dialects. This caused Peter Marler to hypothesize that young birds learn how to sing from older birds. To test his hypothesis, Marler housed the birds in cages where he could completely control the songs they heard (Fig. 22.4). A group of birds that *heard no songs at all* sang a song, but it was not fully developed. Birds that *heard tapes of White-Crowned Sparrows singing* sang in that dialect, as long as the tapes were played during a sensitive period from about age ten to fifty days. White-Crowned Sparrows' dialects (or other species' songs) played before or after this sensitive period had no effect on their song. Apparently, their brain is especially primed to respond to acoustical stimuli during the sensitive period. Neurons that are critical for song production have been located, and they fire when a bird hears a recording of either its own song or its own dialect sung by another bird. Other investigators have shown that birds *given an adult tutor* will sing the song of even a different species—no matter when the tutoring begins! It would appear that social experience has a very strong influence over the development of singing.

Human Studies

Based on his studies, Jean Piaget, a famous Swiss psychologist, traced four stages in the development of children's thinking. The first, or sensorimotor stage, occurs in the first two years of life and is characterized by nonverbal hands-on learning. During a second stage, which lasts from two to seven years, objects are assigned words and the words are mentally manipulated. During a third stage, which lasts from 7 to 12 years, the child begins to classify objects by their similarities and differences. After 12 years, the individual begins to use formal logic and experiment with different modes of thinking. Educators have used Piaget's stages to develop teaching methods appropriate to the age of the child.

> **Animals have an ability to benefit from experience; learning occurs when a behavior changes with practice.**

Isolated bird sings but song is not developed.

Bird sings developed song played during a sensitive period.

Bird sings song of social tutor without regard to sensitive period.

Figure 22.4 Song learning by White-Crowned Sparrows, *Zonotrichia leucophrys*.
Three different experimental procedures are depicted and the results noted. These results suggest that there is both a genetic basis and an environmental basis for song learning in birds.

22.3 Behavior Is Adaptive

Since genes influence the development of behavior, it is reasonable to assume that behavioral traits (like other traits) can evolve. Our discussion will focus on reproductive behavior—specifically the manner in which animals secure a mate. But we will also touch on the other two survival issues—capturing resources and avoiding predators—because these help an animal survive, and without survival reproduction is impossible. Investigators studying survival value questions seek to test hypotheses that specify how a given trait might improve reproductive success.

Males can father many offspring because they produce sperm in great quantity. We would then expect competition among males to inseminate as many different females as possible. In contrast, females produce few eggs, so the choice of a mate becomes a prevailing consideration. Sexual selection can bring about evolutionary changes in the species. **Sexual selection** is changes in females and males, often due to female selectivity and male competition, leading to reproductive success.

Females Choice

Courtship displays are rituals that serve to prepare the sexes for mating. They help male and female recognize each other so that mating will be successful. They also play a role in a female's choice of a mate.

Gerald Borgia conducted a study of Satin Bowerbirds (see opening photo, p. 361) to test these two opposing models regarding female choice:

Good genes hypothesis: females choose mates on the basis of traits that are expected to improve the male's chances of survival.

Run-away hypothesis: females choose mates on the basis of traits that make them attractive to females. The term "run-away" pertains to the possibility that the trait will be exaggerated in the male until its reproductively favorable benefit is checked by the trait's unfavorable survival cost.

In both these models there is no paental investment on the part of males.

Borgia and his assistants watched bowerbirds at feeding stations and also monitored the bowers. They discovered that although males tend to steal blue feathers and/or actively destroy a neighbor's bower, more aggressive and vigorous males were able to keep their bowers in good condition. These were the males usually chosen as mates by females.

Borgia felt that the investigation did not clearly support either hypothesis. It could be that aggressiveness, if inherited, does improve the chances of survival, or it could be that females simply preferred bowers with the most blue feathers.

Bruce Beehler studied the behavior of the birds of paradise in New Guinea (Fig. 22.5). The Raggiana Bird of Paradise is remarkably dimorphic—the males are larger than females and have beautiful orange flank plumes. In contrast, the females are drab. Courting males gather and begin to call. (The gathering of courting males is called a lek.) If a female joins them, the males raise their orange display plumes, shake their wings and hop from side to side, while continuing to call. They then stop calling and lean upside down with the wings projected forward to show off their beautiful feathers.

Female choice can explain why male birds are so much more showy than females, even if it is not known which of Borgia's two hypotheses applies. It's possible that the remarkable plumes of Raggiana males do signify health and vigor. Or it's possible that females choose the flamboyant males on the basis that their sons will have an increased chance of being selected by females. Some investigators have hypothesized that extravagant male features could indicate that they are relatively parasite free. Anders Moller has tested this hypothesis using the barn swallow as his experimental material. He artificially shortened and lengthened the tails of male swallows and found that females chose those with the longest tails. Then he showed that males reared in nests sprayed with a mite killer had longer tails than otherwise.

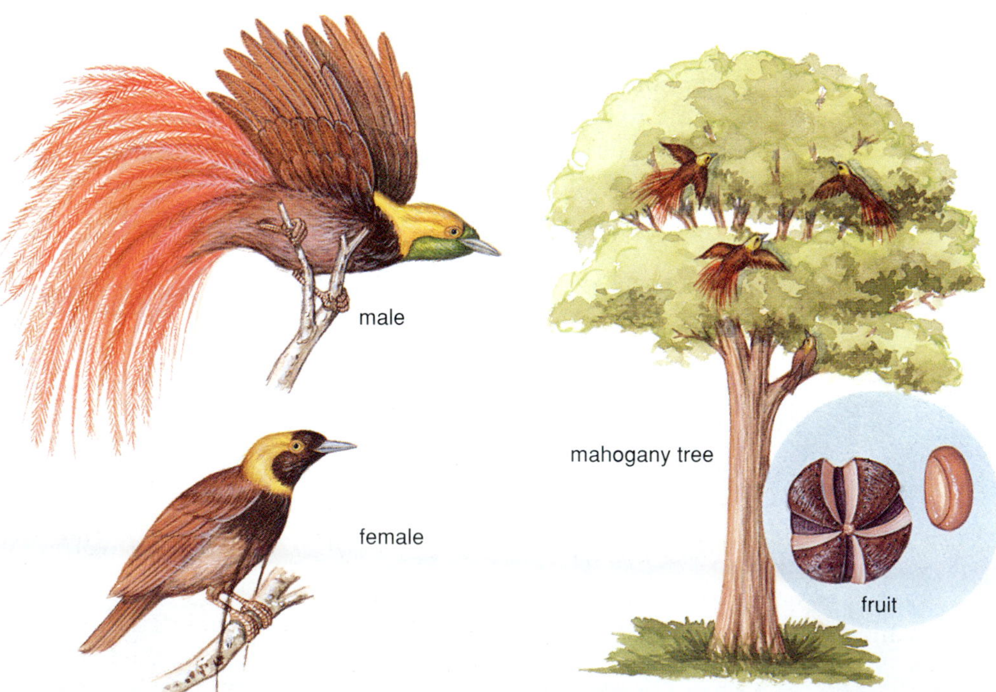

Figure 22.5 Mating behavior in Raggiana Birds of Paradise, *Paradisaea raggiana*. Mating behavior in Birds of Paradise is influenced by species' foraging habits. The Raggiana male has resplendent plumage brought about by sexual selection. The females are widely scattered, foraging for complex fruits; the males form leks that females visit to choose a mate.

Figure 22.6 A male olive baboon, *Papio anubis*, displaying full threat.
Males are larger than females and have enlarged canines. Competition between males establishes a dominance hierarchy for the distribution of resources.

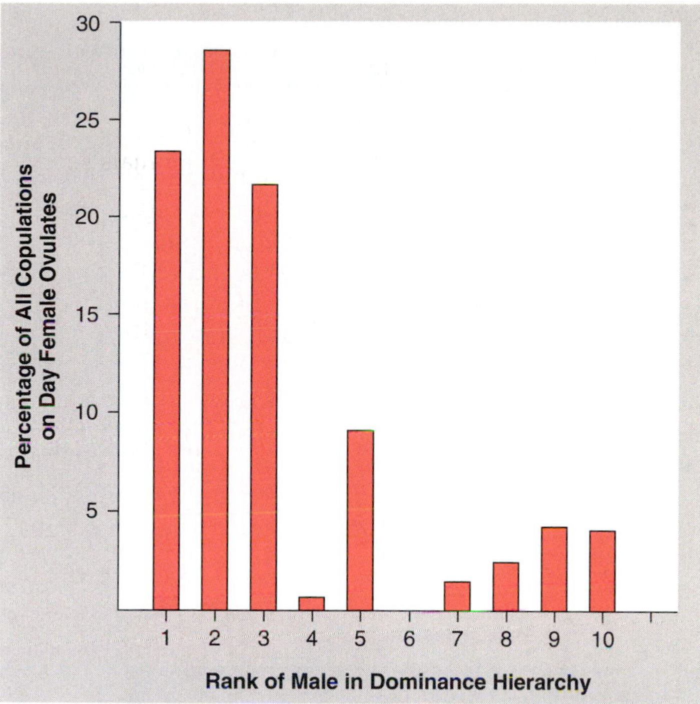

Figure 22.7 Female choice and male dominance among baboons.
Although it may appear that females mate indiscriminately, actually they mate more often with a dominant male when they are most fertile.

Male Competition

A trait that helps a male compete for mates can lead to a positive effect (e.g., more mates), but what are the possible negative effects (e.g., early death)? This type of cost-benefit analysis applies to fitness, traits that favor an individual's reproductive contribution to the subsequent generation. It can be reasoned that when costs are high, it is unlikely that a trait will evolve, but when benefits are high a trait is expected to evolve.

Dominance Hierarchy

Baboons, a type of Old World monkey, live together in a troop. Males and females have separate **dominance hierarchies** in which a higher ranking animal has greater access to resources than a lower ranking animal. Dominance is decided by confrontations, resulting in one animal giving way to the other.

Baboons are dimorphic; the males are larger than the females and they can threaten others with their long, sharp canines (Fig. 22.6). The baboons travel within a territory, hunting food each day, and sleeping in trees at night. The dominant males decide where and when the troop will move and, if the troop is threatened, they cover the troop as it retreats, and attack when necessary.

Females undergo a period known as estrus when they are willing to mate; however, she is fertile only the one day she ovulates. When first receptive, a female mates with subordinate males and older juveniles. Later, as ovulation nears, a female approaches a dominant male, and now he is very interested. They form a consort pair, which lasts only for an hour or several days. Dominant males are interested in and protective of young baboons, regardless of the true father.

However, the male baboon pays a cost for his dominant position. Being larger means that he needs more food, and being willing and able to fight predators means that he may get hurt, and so forth. Is there a reproductive benefit to his behavior? Glen Hausfater counted copulations of dominant males and found that they do indeed monopolize estrous females when he judged them to be most fertile (Fig. 22.7). Nevertheless, there are other avenues to fathering offspring. Some males act as helpers to particular females and her offspring; the next time she is in estrus she may mate preferentially with him instead of a dominant male. Or subordinate males may form a friendship group that can oppose a dominant male, making him give up a receptive female.

Territoriality

A territory is an area that is defended against competitors. **Territoriality** is protecting an area against other individuals. Vocalization and displays, rather than outright fighting, may be sufficient to defend a territory (Fig. 22.8). Male songbirds signify their territories by singing and in this way other males of the species know when an area is occupied.

T. H. Clutton-Brock has studied reproductive success among red deer (elk) on the Scottish island of Rhum. Stags (males) compete for a harem, a group of hinds (females) that mate only with one stag, called the harem master. The reproductive group occupies a territory that the harem master defends against other stags. Harem masters first attempt to repel challengers by roaring. If the challenger remains, the two lock antlers and push against one another. If the challenger now withdraws, the master pursues him for a short distance, roaring the whole time. If the challenger wins, he becomes the harem master.

After studying the red deer for more than twelve years, Clutton-Brock concluded that a harem master can father two dozen offspring at most, because he is at the peak of his fighting ability for only a short time. And what does it cost him to be able to father offspring? Stags must be large and powerful in order to fight; therefore, they grow faster and have less body fat. During bad times they are more likely to die of starvation and in general have shorter lives. The behavior of harem defense by stags will only persist in the population if its cost (reduction in the potential number of offspring because of a shorter life) is less than its benefit (increased number of offspring due to harem access).

Evolution by sexual selection can occur when females have the opportunity to select among potential mates and/or when males compete among themselves for access to reproductive females.

Figure 22.8 Competition between males among red deer, *Cervus elaphus.*
Male red deer compete for a harem within a particular territory. **a.** Roaring alone may frighten off a challenger. **b.** Outright fighting may be necessary, and the victor is most likely the stronger of the two animals.

Science Focus

Biological Rhythms

Certain behaviors in animals recur at regular intervals. Behaviors that occur on a daily basis are said to have a circadian (about a day) rhythm. For example, some animals, like humans, are usually active during the day and sleep at night. Others, such as bats, sleep during the day and hunt at night. There are also behaviors that occur on a yearly basis. For example, in the Northern Hemisphere, birds migrate south in the fall, and the young of many animals are born in the spring. Such behaviors have a circannual rhythm.

There were two hypotheses regarding the control of circadian rhythms: either their timing is controlled externally, or there is an internal timing device, often called a biological clock (Fig. 22A). These alternative hypotheses have been tested in crickets, which regularly call every night to attract females. When laboratory crickets are kept in a room under constant conditions with lights continually on or continually off, they continue to call every night, only calling starts as much as 26 hours later than it did on the day before. But exposure to night and day cycles will right the cycles; therefore, it was possible to conclude that there is a biological clock, but it does not keep perfect time and must be reset by environmental stimuli. Similar results have been obtained with all sorts of animals, from fiddler crabs to humans.

At a minimum, a circadian system must have three components (Fig. 22A). There must be a means to reset a pacemaker according to the current environmental light-dark cycle; a biological clock that keeps time; and the rhythmic behavior itself. By now, there is a general consensus that the mammalian biological clock is a collection of nerve cells in the hypothalamus of the brain known as the suprachiasmatic nucleus (SCN). Electrical and drug stimulation of the SCN upsets circadian rhythms and destruction of the nucleus does away with many rhythmic behaviors. Investigators have also discovered a number of genes whose protein products control the activity of the clock. Less is known about the other components of the circadian system. Whether the receptor that resets the clock is in the eye or not is still to be determined. In one recent study, investigators found that exposing the back of a knee to light reset the clock judged by body temperature and hormone levels. They concluded that any portion of the skin contains photoreceptor proteins that communicate with and reset the biological clock.

A number of investigators are interested in the observation that environmental light suppresses melatonin production by the pineal gland while we are awake. Conversely, as it gets dark, melatonin levels rise and we get sleepy. This knowledge causes some people to take melatonin for the symptoms of insomnia and "jet lag." When we travel by airplane from one part of the world to another it is difficult for our circadian system to adjust, and symptoms like insomnia, fatigue, headache, gastrointestinal distress, and moodiness occur. A medication for these symptoms would also be helpful to one out of every five people whose work shifts between day and night. Jet lag symptoms are reduced by exposure to daylight in the afternoon after westward flights and in the early morning after eastward flights. Similarly, an adjustment to a nighttime work shift is enhanced by exposure to bright light at night and to darkness during the day. Because light wakes us up and melatonin makes us sleepy, their use at different times of the day may be helpful in resetting the biological clock.

Many people are intrigued by the idea that some psychiatric conditions might be due to a disorder of the circadian system. The best example is seasonal affective disorder (SAD) which affects as much as 5% of the general population and is characterized by depressions during the fall and winter. The hypothesis that short winter days are responsible for SAD has led to successful therapies based on exposure to bright light. Circadian rhythms are clinically significant in other ways. The incidence of heart attacks, sudden cardiac death, and stroke peak in the late morning, and certain cancer drugs are more affective when given in the day or night. This is an exciting time for research concerning the human circadian system especially because such research is expected to have important applications that will help control many human ills, from those that are quite serious to those that are simply a nuisance.

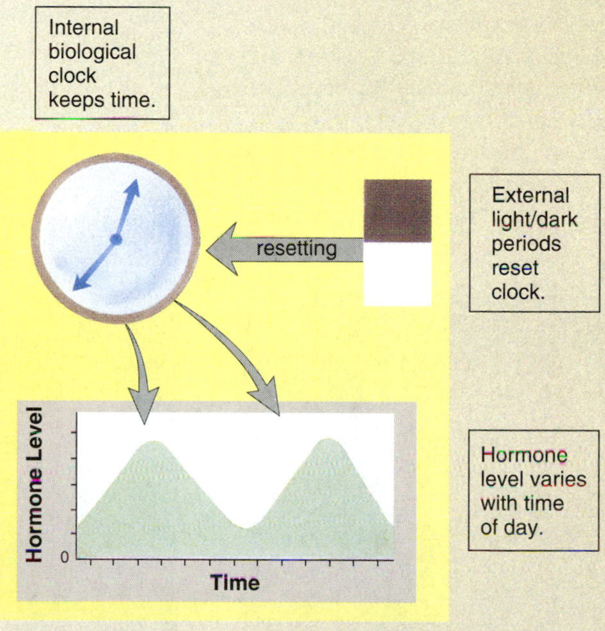
Internal biological clock keeps time.
resetting
External light/dark periods reset clock.
Hormone Level
0
Time
Hormone level varies with time of day.

Figure 22A **A circadian system.**
A circadian system has three components, as shown.

22.4 Animal Societies

Animals exhibit a wide range of degrees of sociality. Some animals are largely solitary and join with a member of the opposite sex only for the purpose of reproduction. Others pair, bond, and cooperate in raising offspring. Still others form a **society** in which members of species are organized in a cooperative manner, extending beyond sexual and parental behavior. We have already had occasion to mention the social groups of baboons and red deer. Social behavior in these and other animals requires that they communicate with one another.

Communicative Behavior

Communication is an action by a sender that influences the behavior of a receiver. The communication can be purposeful but does not have to be purposeful. Bats send out a series of sound pulses and listen for the corresponding echoes in order to find their way through dark caves and locate food at night. Some moths have an ability to hear these sound pulses, and they begin evasive tactics when they sense that a bat is near. Are the bats purposefully communicating with the moths? No, but the bat sounds are a *cue* to the moths that danger is near.

Chemical Communication

Chemical signals have the advantage of working both night and day. The term **pheromone** [Gk. *phero,* bear, carry, and *monos,* alone] is used to designate chemical signals that are passed between members of the same species. Female moths secrete chemicals from special abdominal glands, which are detected downwind by receptors on male antennae. The antennae are especially sensitive, and this assures that only male moths of the correct species (and not predators) will be able to detect them. Cheetahs and other cats mark their territories by depositing urine, feces, and anal gland secretions at the boundaries (Fig. 22.9). Klipspringers (small antelope) use secretions from a gland below the eye to mark twigs and grasses of their territory.

Human Studies Chemist George Preti and biologist Winnifred Cutler did a control study to see if humans are affected by pheromones. For three weeks, they collected underarm secretions from female volunteers and swabbed them on the upper lips of 10 female subjects. After three months, the subjects' menstrual cycles were roughly in synchrony with those of the women who donated the sweat. Women exposed to alcohol showed no change. Later these researchers found that the steroid androstenol peaked in female underarms just before ovulation. Perhaps this chemical is a pheromone causing menstrual synchrony.

Auditory Communication

Auditory *(sound)* communication has some advantages over other kinds of communication. It is faster than chemical communication, and it also is effective both night and day. Further, auditory communication can be modified not only by loudness but also by pattern, duration, and repetition. In an experiment with rats, a researcher discovered that an intruder can avoid attack by increasing the frequency with which it makes an appeasement sound.

Male crickets have calls, and male birds have songs for a number of different occasions. For example, birds may have one song for distress, another for courting, and still another for marking territories. Sailors have long heard the songs of humpback whales because they are transmitted through the hull of a ship. But only recently has it been shown that the song has six basic themes, each with its own phrases, that can vary in length and be interspersed with

Figure 22.9 Use of pheromone to mark a territory.
This male cheetah, *Acinonyx*, is spraying a pheromone onto a tree in order to mark its territory.

Figure 22.10 A chimpanzee with a researcher.
Chimpanzees are unable to speak but can learn to use a visual language consisting of symbols. Some believe chimps only mimic their teachers and never understand the cognitive use of a language. Here the experimenter shows Nim the sign for drink. Nim copies.

sundry cries and chirps. The purpose of the song is probably sexual and serves to advertise the availability of the singer. Language is the ultimate auditory communication, but only humans have the biological ability to produce a large number of different sounds and to put them together in many different ways. Nonhuman primates have at most only forty different vocalizations, each having a definite meaning, such as the one meaning "baby on the ground," which is uttered by a baboon when a baby baboon falls out of a tree. Although chimpanzees can be taught to use an artificial language, they never progress beyond the capability level of a two-year-old child (Fig. 22.10). It has also been difficult to prove that chimps understand the concept of grammar or can use their language to reason. It still seems as if humans possess a communication ability unparalleled by other animals.

Visual Communication

Visual signals are most often used by species that are active during the day. Contests between males make use of threat postures and possibly prevent outright fighting that might result in reduced fitness. A male baboon displaying full threat is an awesome sight that establishes his dominance and keeps peace within the baboon troop (see Fig. 22.6). Hippopotamuses have territorial displays that include opening the mouth.

The plumage of a male Raggiana Bird of Paradise allows him to put on a spectacular courtship dance to attract females and to give them a basis on which to select a suitable mate. Defense and courtship displays are exaggerated and are always performed in the same way so that their meaning is clear.

Tactile Communication

Tactile communication occurs when one animal touches another. For example, gull chicks peck at the parent's beak in order to induce the parent to feed them (see Fig. 22.3). A male leopard nuzzles the female's neck to calm her and to stimulate her willingness to mate. In primates, grooming—one animal cleaning the coat and skin of another—helps cement social bonds within a group.

Honeybees use a combination of tactile and auditory communication to impart information about the environment. Karl von Frisch, another famous behaviorist, did many detailed bee experiments in the 1940s. He discovered that when a foraging bee returns to the hive, it performs a **waggle dance** that indicates the distance and the direction of a food source (Fig. 22.11). As the bee moves between the two loops of a figure 8, it buzzes noisily and shakes its entire body in so-called waggles. Distance to the food source is believed to be indicated by the number of waggles and/or the amount of time taken to complete the straight run. Outside the hive, the dance is done on a horizontal surface and the straight run indicates the direction of the food. Inside the hive, the angle of the straight run to that of the direction of gravity is the same as the angle of the food source to the sun. In other words, a 40° angle to the left of vertical means that food is 40° to the left of the sun. Bees can use the sun as a compass to locate food because their biological clock allows them to compensate for the movement of the sun in the sky. Biological clocks are discussed in the reading on page 369.

Animals use a number of different ways to communicate, and communication sometimes facilitates cooperation.

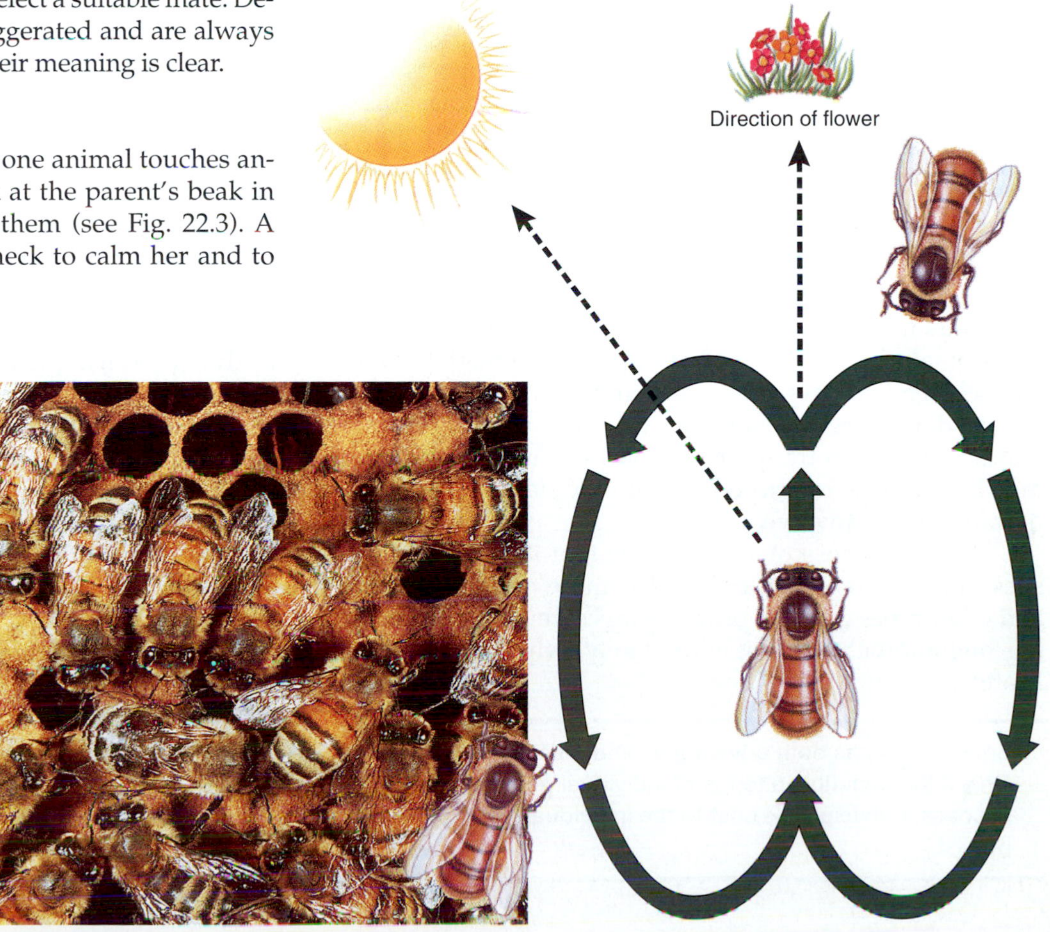

Figure 22.11 Communication among bees.
a. Honeybees, *Apis mellifera*, do a waggle dance to indicate the direction of food. b. If the dance is done outside the hive on a horizontal surface, the straight run of the dance will point to the food source.

22.5 Sociobiology and Animal Behavior

Sociobiology applies the principles of evolutionary biology to the study of social behavior in animals. Sociobiologists develop hypotheses about social living based on the assumption that a social individual derives more reproductive benefits than costs from living in a society. Then they may perform a cost-benefit analysis to see if their hypotheses are correct.

Group living does have benefits under certain circumstances. It can help an animal avoid predators, rear offspring, and find food. A group of impalas is more likely to hear an approaching predator than a solitary one. Many fish moving rapidly in many directions might distract a would-be predator.

Pair bonding of Trumpet Manucodes helps the birds raise their young. Due to their particular food source, the female cannot rear as many offspring alone as she can with the male's help. Weaver birds form giant colonies that help protect them from predators, but the birds may also share information about food sources. Primate members of the same troop signal to one another when they have found an especially bountiful fruit tree. Lions working together are able to capture large prey, such as zebra and buffalo.

Group living also has its disadvantages. When animals are crowded together into a small area, disputes can arise over access to the best feeding places and sleeping sites. Dominance hierarchies are one way to apportion resources, but this puts subordinates at a disadvantage. Clutton-Brock found that among red deer females, only dominant females can successfully rear sons; small subordinate females tend to rear daughters. From an evolutionary point of view, sons are preferable because, as a harem master, sons will result in a greater number of grandchildren. However, sons, which tend to be larger than daughters, need to be nursed more frequently and for a longer period of time. Subordinate females do not have access to enough food resources to adequately nurse sons and, therefore, they tend to rear daughters and not sons. Still, like the subordinate males in a baboon troop, the subordinate red deer females may be better off in fitness terms if they stay with a group, despite the cost involved.

Living in close quarters means that illness and parasites can pass from one to the other more rapidly. Baboons and other types of social primates invest much time grooming one another and this most likely helps them remain healthy.

Social living has both advantages and disadvantages. Only if the benefits, in terms of individual reproductive success, outweigh the cost to the individual will societies persist.

Altruism Versus Self-Interest

Altruism [L. *alter*, the other] is behavior that has the potential to decrease the lifetime reproductive success of the altruist while benefiting the reproductive success of another member of the group. In insect societies, especially, reproduction is limited to the queen and her mate. For example, among army ants the queen is inseminated only during her nuptial flight, and thereafter she spends her time reproducing. The society has three different sizes of sterile female workers. The smallest workers (3 mm), called the nurses, take care of the queen and larvae, feeding them and keeping them clean. The intermediately sized workers, constituting most of the population, go out on raids to collect food. The soldiers (14 mm), with huge heads and powerful jaws, run along the sides and rear of raiding parties where they can best attack any intruders.

Can the altruistic behavior of sterile workers be explained in terms of reproductive success? In 1964, the English biologist William Hamilton pointed out that a given gene can be passed from one generation to the next in two quite different ways. The first way is direct: a parent can pass the gene directly to an offspring. The second way is indirect: an animal can help a relative reproduce and thereby pass the gene to the next generation via this relative. Natural selection results in adaptation to the environment when the reproductive success of individuals differs. Natural selection can also result in adaptation to the environment when individuals promote the reproductive success of relatives. The **inclusive fitness** of an individual includes both personal reproduction and reproduction of relatives.

Among social bees, social wasps, and the ants, the queen is diploid but her mate is haploid. If the queen has had only one mate, sister workers are more closely related to each other (sharing on average 75% of their genes) than they are to their potential offspring (with which they share on average only 50% of their genes). Therefore, a worker can achieve a greater inclusive fitness benefit by aiding her mother (the queen) to produce additional sisters, than by directly reproducing herself. Under these circumstances, behavior that appears to be altruistic is more likely to evolve.

Indirect selection can also occur among animals whose offspring receive only a half set of genes from both parents. Consider that your brother or sister shares 50% of your genes, your niece or nephew shares 25%, and so forth. This means that the survival of two nieces (or nephews) is worth the survival of one sibling, assuming they both go on to reproduce.

Michael Ghiglieri, when studying chimpanzees in Africa, observed that a female in estrus frequently copulates with several members of the same group, and that the males make no attempt to interfere with each other's matings. How can they be acting in their own self-interest? Genetic relatedness of the males appears to underlie their apparent altruism; they all share genes in common because males never leave the territory in which they are born.

Figure 22.12 Inclusive fitness.
A meerkat, *Suricata*, is acting as a baby-sitter for its young sisters and brothers while their mother is away. Could this helpful behavior contribute to the baby-sitter's inclusive fitness?

Helpers at the Nest

In some bird species, offspring from one clutch of eggs may stay at the nest, helping parents rear the next batch of offspring. In a study of Florida scrub jays, the number of fledglings produced by an adult pair doubled when they had helpers. Mammalian offspring are also observed to help their parents (Fig. 22.12). Among jackals in Africa, pairs alone managed to rear an average of 1.4 pups, whereas pairs with helpers reared 3.6 pups.

Is the reproductive success of the helpers increased by their apparent altruistic behavior? It could be if the chance of their reproducing on their own is limited. David and Sandra Ligon studied the breeding behavior of Green Wood-hoopoes *(Phoeniculus purpurens)*, an insect-eating bird of Africa. A flock may have as many as sixteen members but only one breeding pair; the other sexually mature members help feed and protect the fledglings and protect the home territory from invasion by other Green Wood-hoopoes. The Ligons found that nest sites are rare, as acacia trees with cavities are relatively rare and the cavities are often occupied by other species. Moreover, predation by snakes within the cavities can be intense; even if a pair of birds acquire an appropriate cavity, they would be unable to protect their offspring by themselves. Therefore, the cost of trying to establish a territory is clearly very high.

What are the benefits of staying behind to help? First, a helper is contributing to the survival of its own kin. Therefore, the helper actually gains a fitness benefit (albeit a smaller benefit than it would achieve were it a breeder). Second, a helper is more likely than a nonhelper to inherit a parental territory—including other helpers. Helping, then, involves making a minimal, short-term reproductive sacrifice in order to maximize future reproductive potential. Once again, an apparently altruistic behavior turns out to be an adaptation.

Inclusive fitness is measured by the genes an individual contributes to the next generation, either directly by offspring or indirectly by way of relatives. Many of the behaviors once thought to be altruistic turn out, on closer examination, to be examples of kin selection, and are adaptive.

Application to Humans

Sociobiologists interpret human behavior according to these same principles. Human infants are born helpless and have a much better chance of developing properly if both parents contribute to the effort. Perhaps this explains why the human female has evolved to be continuously amenable to sexual intercourse. Under these circumstances, the male is more likely to remain and help care for offspring. In any case, parental love is clearly selfish in that it promotes the likelihood that an individual's genes will be present in the next generation's gene pool.

Studies of other human cultures lend themselves to sociobiological interpretations. Among African tribes, one man may have several wives. This is reproductively advantageous to the male, but is also advantageous to the woman. By this arrangement she has fewer children who are thereby assured a more nutritious diet. In Africa, sources of protein are scarce and early weaning poses a threat to the health of the child. In contrast, among the Bihari hill people of India, brothers have the same wife. Here the environment is hostile and it takes two men to provide the everyday necessities for one family. Since the men are brothers, they are actually helping each other look after common genes.

Some people object to an interpretation of human behavior based on evolutionary fitness because they want to stress that humans have the ability to control their behavior and it need not be determined by relatedness to others. And it is certainly possible to find examples that do not support the principles of sociobiology as when people adopt and care for children that are not related to them. Sociobiology also predicts that there would be less violence in small towns than large cities. Why? Because people are more likely to be related in small towns than in large cities. Yet, we know that parents sometimes neglect and even kill their own offspring. But according to evolutionary principles, these types of behavior would soon die out because of their extremely low fitness.

Connecting Concepts

Birds build nests, dogs bury bones, cats chase quick-moving objects, and snakes bask in the sun. All animals, including humans, behave—they respond to stimuli, both from the physical environment and from other individuals of the same or different species. Behaviorists are scientists who seek answers to two types of questions: what is the neurophysiological mechanism of the behavior, and how is the behavior beneficial to the organism? Using an evolutionary approach, behaviorists generate hypotheses that can be tested to better understand how the behavior increases individual fitness (i.e., the capacity to produce surviving offspring).

Both types of behavioral studies—those concerned with neurophysiology and those concerned with biological fitness—recognize that behavior has a genetic basis. Inheritance determines, for example, that hawks hunt by using vision rather than smell, and that cats, not dogs, climb trees.

Inheritance, too, produces the behavioral variations that are subject to natural selection, resulting in the most adaptive responses to stimuli. There is survival value, after all, in the ability of male baboons to react aggressively when the troop is under attack.

The study of behavior, however, does not settle the nature-nurture question because we now know that behavior can be modified by experience. Songbirds are born with the ability to sing, but which song or dialect they sing is strongly dependent on the songs they hear from their parents and siblings. A new chimpanzee mother naturally cares for her young, but she is a better mother if she observes other females in her troop raising young.

There is a cost to certain behaviors. Why should subordinate members of a baboon troop or subordinate females in a red deer harem remain in a situation that seemingly leads to reduced fitness? And why should older offspring help younger offspring? The evolutionary answer is that, in the end, there are benefits that outweigh the costs. Otherwise, the behavior would not continue. The evolutionary approach to studying behavior has proved fruitful in helping us understand why birds sing their melodious songs, dolphins frolic in groups, and wondrous Raggiana Birds of Paradise display to females.

Summary

22.1 Behavior Has a Genetic Basis

Behaviorists study how animals are organized to perform a particular behavior and how it benefits them to survive and reproduce. Hybrid studies with Blackcap Warblers produce results consistent with the hypothesis that behavior has at least a partial genetic basis. Garter snake experiments indicate that the nervous system controls behavior. *Aplysia* DNA studies indicate that the endocrine system also controls behavior.

22.2 Behavior Undergoes Development

The environment is influential in the development of behavioral responses, as exemplified by an improvement in Laughing Gull chick begging behavior and an increased ability of chicks to recognize parents. Modern studies suggest that most behaviors improve with experience. Even behaviors that were formerly thought to be fixed action patterns (FAPs) or otherwise thought to be inflexible sometimes can be modified.

Song learning in birds involves various elements—including the existence of a sensitive period when an animal is primed to learn—and the effect of social interactions.

22.3 Behavior Is Adaptive

Traits that promote reproductive success are expected to be conserved. Males who produce many sperm are expected to compete to inseminate females. Females who produce few eggs are expected to be selective about their mates. Experiments with Satin Bowerbirds and Raggiana Birds of Paradise support these bases for sexual selection.

Other aspects of reproductive behavior are also adaptive. The Raggiana Bird of Paradise males gather in a lek—most likely because females are widely scattered, as is their customary food source. The food source for a related species, the Trumpet Manucode, is readily available but less nutritious. These birds are monogamous—it takes two parents to rear the young.

A cost-benefit analysis can be applied to competition between males for mates in reference to a dominance hierarchy (e.g., baboons) and territoriality (e.g., red deer). If the behavior is conserved, the benefit most likely outweighs the cost.

22.4 Animal Societies

Animals that form social groups may communicate with one another. Communications such as chemical, auditory, visual, and tactile signals may foster cooperation that benefits both the sender and the receiver.

There are benefits and costs to living in a social group. If animals live in a social group, it is expected that the advantages (e.g., help to avoid predators, raise young, and find food) will outweigh the disadvantages (e.g., tension between members, spread of illness and parasites, and reduced reproductive potential). This expectation can sometimes be tested.

22.5 Sociobiology and Animal Behavior

In most instances, the individuals of a society act to increase their own reproductive success. Sometimes animals perform apparent altruistic acts, as when individuals help their parents rear siblings. There is a benefit to this behavior when one considers inclusive fitness, which involves both direct selection and indirect selection.

In social insects, apparent altruism is extreme but can be explained on the basis that they are helping a reproducing sibling survive. A study of Green Wood-hoopoes, an African bird, shows that younger siblings may help older siblings who reared them until the younger ones get a chance to reproduce themselves.

Reviewing the Chapter

1. State the two types of questions asked by behaviorists and explain how they differ. 362
2. Describe Helbig's experiment with Blackcap Warblers, and explain how it shows that behavior has a genetic basis. 362
3. What body system is involved in the behavior of garter snakes toward slugs according to Arnold's experiment? Explain the experiment and the results. 362–63
4. Studies of *Aplysia* DNA show that the endocrine system is also involved in behavior. Explain. 363
5. Some behaviors require practice before developing completely. How does Hailman's experiment with Laughing Gull chicks support this statement? 364

6. An argument can be made that social contact is an important element in learning. Explain with reference to imprinting in mallard ducks and song learning in White-Crowned Sparrows. 365
7. Why would you expect behavior to be subject to natural selection and be adaptive? 366
8. How do Borgia's studies of Satin Bowerbird behavior and Clutton-Brock's studies with red deer support the belief that reproductive behavior is related to female selectivity and male competition? 366, 368
9. Give examples of the different types of communication among members of a social group. 370–71
10. What is a cost-benefit analysis, and how does it apply to living in a social group? Give examples. 372
11. Give examples of behaviors that appear to be altruistic but actually increase the fitness of an individual. 372–73

Testing Yourself

Choose the best answer for each question.

1. Which of these questions is least likely to interest a behaviorist?
    a. How do genes control the development of the nervous system?
    b. Why do animals living in the tundra have white coats?
    c. Does aggression have a genetic basis?
    d. Why do some animals feed in groups and others feed singly?
    e. Why do some animals help tend their siblings?
2. Female Sage Grouse are widely scattered throughout the prairie. Which of these would you expect?
    a. A male will maintain a territory large enough to contain at least one female.
    b. Male and female birds will be monogamous, and both will help feed the young.
    c. Males will form a lek where females will choose a mate.
    d. Males will form a dominance hierarchy for the purpose of distributing resources.
    e. Female Sage Grouse have access to sources of extremely rich food.
3. White-Crowned Sparrows from two different areas sing with a different dialect. If the behavior is primarily genetic, newly hatched birds from each area will
    a. sing with their own dialect.
    b. need tutors in order to sing in their dialect.
    c. sing only when a female is nearby.
    d. sing a more complicated song than their parents.
    e. Both a and c are correct.
4. Orangutans are solitary but territorial. This would mean orangutans defend their territory's boundaries against
    a. other male orangutans.
    b. female orangutans.
    c. other types of primates.
    d. other types of animals.
    e. Both a and b are correct.
5. The resplendent plumes of a Raggiana Bird of Paradise are due to the fact that birds with the best display
    a. are dominant over other birds.
    b. have the best territories.
    c. are chosen by females as mates.
    d. live the longest.
    e. All of these are correct.
6. Subordinate females in a baboon troop do not produce offspring as often as dominant females. It is clear that
    a. the cost of being in the troop is too high.
    b. the dominant males do not mate with subordinate females.
    c. subordinate females must benefit in some way from being in the troop.
    d. dominant females have even fewer offspring than the subordinate females.
    e. Both a and b are correct.
7. German Blackcap Warblers migrate southwest to Africa, and Austrian Blackcap Warblers fly southeast to Africa. The fact that hybrids of these two are intermediate shows that
    a. the trait is controlled by the nervous system.
    b. nesting is controlled by hormones.
    c. the behavior is at least partially genetic.
    d. the behavior of German blackcaps is dominant over the behavior of Austrian blackcaps.
    e. Both a and c are correct.
8. At first Laughing Gull chicks peck at any model that looks like a red beak; later they will not peck at any model that does not look like a parent. This shows that the behavior
    a. is a fixed action pattern.
    b. undergoes development after birth.
    c. is controlled by the nervous system.
    d. is an example of auditory communication.
    e. All of these are correct.
9. Which of these could be an answer to a mechanistic question? Male red deer compete
    a. because they have the size and weapons with which to compete.
    b. because they produce many sperm for a long time.
    c. because the testes produce the hormone testosterone.
    d. by roaring and locking antlers.
    e. All of these are correct.
10. Which of these could be an answer to a survival value question? Females are choosy because
    a. they do not have the size and weapons with which to compete.
    b. they invest heavily in the offspring they produce.
    c. the ovaries produce the hormones estrogen and progesterone.
    d. the males tend to be more dressy than the females.
    e. All of these are correct.

Thinking Scientifically

1. Meerkats are said to exhibit altruistic behavior because certain members of a population act as sentries. These sentries stand on rocks or other high places and serve as lookouts while others feed. However, recent observations have shown that these sentries are the first ones to reach safety when a predator is spotted, and that sentries only serve after they have eaten. Their behavior is still beneficial to the group since without the sentries there would be no warning of approaching predators, but can it still be termed altruistic? How would you test the hypothesis that sentries are engaged in altruistic behavior?
2. You are testing the hypothesis that human infants instinctively respond to higher-pitched voices. Your design is to record head turns toward speakers placed on opposite sides of a month-old infant. The speakers would play voices

(all making the same noises) in different pitches and you would see if the infants turned toward some voices more often than others. When you do the experiment utilizing several different infants, your data support your hypothesis. However, prior learning by infants is still a serious criticism. What is the basis of this criticism?

Bioethical Issue

Is it ethical to keep animals in zoos where they are not free to behave as they do in the wild? If we keep animals in zoos are we depriving them of their freedom? Some point out that freedom is never absolute. Even an animal in the wild is restricted in various ways by its abiotic and biotic environment. The so-called five freedoms are to be free of starvation, cold, injury, and fear, as well as free to wander and express one's natural behavior. Perhaps it's worth giving up a bit of the last freedom to achieve the first four? Many modern zoos do keep animals in habitats that nearly match their natural one so that they do have some freedom to roam and behave naturally. Perhaps, too, we should consider the education and enjoyment of the many thousands of human visitors to a zoo compared to the loss to a much smaller number of animals kept in a zoo?

Today, reputable zoos rarely go out and capture animals in the wild—they usually get their animals from other zoos. Most people feel it is never a good idea to take animals from the wild except for very serious reasons. Certainly, zoos should not be involved in the commercial and often illegal trade of wild animals which still goes on today. When animals are captured it should be done by skilled biologists or naturalists who know how to care for and transport the animal.

Many zoos today are involved in the conservation of animals. They provide the best home possible while animals are recovering from injury or their numbers are increased until they can be released to the wild. Can we perhaps look at zoos favorably if they can show that their animals are being kept under good conditions, and that they are also involved in the preservation of animals?

Understanding the Terms

altruism 372
behavior 362
communication 370
dominance hierarchy 367
imprinting 365
inclusive fitness 372
learning 364
operant conditioning 364
pheromone 370
sexual selection 366
society 370
sociobiology 372
territoriality 368
waggle dance 371

Match the terms to these definitions:

a. ______________ Changes in males and females, often due to male competition and female selectivity, leading to reproductive success.
b. ______________ Chemical released by the body that causes a predictable reaction of another member of the same species.
c. ______________ Fitness that results from direct selection and indirect selection.
d. ______________ Form of learning that occurs early in the lives of animals; a close association is made that later influences sexual behavior.
e. ______________ Form of learning that results from rewarding or reinforcing a particular behavior.

Web Connections

Exploring the Internet

http://www.mhhe.com/biosci/genbio/mader
(click on *Biology 7/e*)

The *Biology 7/e* Online Learning Center provides many resources for studying the material in this chapter including links to the following sites:

Evolution and Behavior. This website provides information on evolution and links to the Animal Behavior Society and many other sites.

http://ccp.uchicago.edu/~jyin/evolution.html

Animal Behavior Society. Information at this website includes documents on the value of the study of animal behavior, lists of institutions with departments specializing in animal behavior, educational activities, the Animal Behavior Society newsletter, research reports, and interesting new information about animal behavior.

http://www.animalbehavior.org/ABS/

Center for Integrative Study of Animal Behavior. Programs in animal behavior, research articles, computer archives for the study of animal behavior, and articles from the Animal Behavior Bulletin are featured at this site.

http://www.cisab.indiana.edu/index.html

Animal Behavior Resources on the Internet. This site provides a wealth of information on animal behavior societies, resources, programs, mailing lists, and research results.

http://cricket.unl.edu/Internet.html

Ethology. University of Georgia Ethology and Animal Behavior resources. Includes electronic journals, directories, and many links.

http://scarlett.libs.uga.edu/science/ethology.html

Ecology of Populations

c h a p t e r

c h a p t e r   c o n c e p t s

Social unit of female elephants *(Loxodonta africana)*

Elephants have a large size, are social, live a long time, and produce few offspring. Females live in social family units, and the much larger males visit them only during a breeding season. Females give birth about every five years to a single calf that is well cared for and has a good chance of meeting the challenges of its lifestyle. Normally, an elephant population exists at the carrying capacity of the environment.

A population ecologist studies the distribution and abundance of organisms and relates the hard, cold statistics to the species life history in order to determine what causes population growth or decline. Elephants are threatened because of human population growth, and also because males, in preference to females, are killed for their tusks. Males don't tend to breed until they have reached their largest size—the size that makes them prized by humans. By now, even a moratorium on killing males will not help much. So few breeding males are left that elephant populations are expected to continue to decline for quite some time. The study of population ecology is necessary to preservation of species.

23.1 Scope of Ecology

In 1866, the German zoologist Ernst Haeckel coined the word ecology from two Greek roots [Gk. *oikos*, home, house, and *-logy*, "study of" from *logikos*, rational, sensible]. He said that **ecology** is the study of the interactions of organisms with other organisms and with the physical environment. And he pointed out that ecology and evolution are intertwined because ecological interactions are selection pressures that result in evolutionary change and evolutionary change affects ecological interactions, and so forth.

Ecology, like so many biological disciplines, is wide-ranging. At one of its lowest levels, ecologists study how the individual organism is adapted to its environment. For example, they study why fishes in a coral reef live only in warm tropical waters and how the fishes feed within that **habitat** (the place where an organism lives) (Fig. 23.1). Most organisms do not exist singly; rather, they are part of a population, a functional unit that interacts with the environment. A **population** is defined as all the organisms within an area belonging to the same species. At this level of study, ecologists are interested in factors that affect the growth and regulation of population size.

A **community** consists of all the various populations interacting at a locale. In a coral reef, there are numerous populations of fishes, crustacea, corals, and so forth. At this level ecologists want to know how interactions like predation and competition affect the organization of a community. An **ecosystem** contains a community of populations and also the abiotic environment. Energy flow and chemical cycling are significant aspects of understanding how an ecosystem functions. The **biosphere** is the zones of the earth's soil, water, and air in which living organisms are found.

Modern ecology is not just descriptive, it is predictive. It analyzes levels of organization and develops models and hypotheses that can be tested. A central goal of modern ecology is to develop models that explain and predict the distribution and abundance of organisms. Ultimately, ecology considers not one particular area, but the distribution and abundance of populations in the biosphere. What factors have brought about the mix of plants and animals in a tropical rain forest at one latitude and in a desert at another? While modern ecology is useful in and of itself, it also has almost unlimited application possibilities, such as the management of wildlife in order to prevent extinction, the maintenance of cultivated food sources, or even the ability to predict the course of an illness such as AIDS.

Ecology is the study of the interactions of organisms with other organisms and with the physical environment. These interactions determine the distribution and abundance of organisms at a particular locale and over the earth's surface.

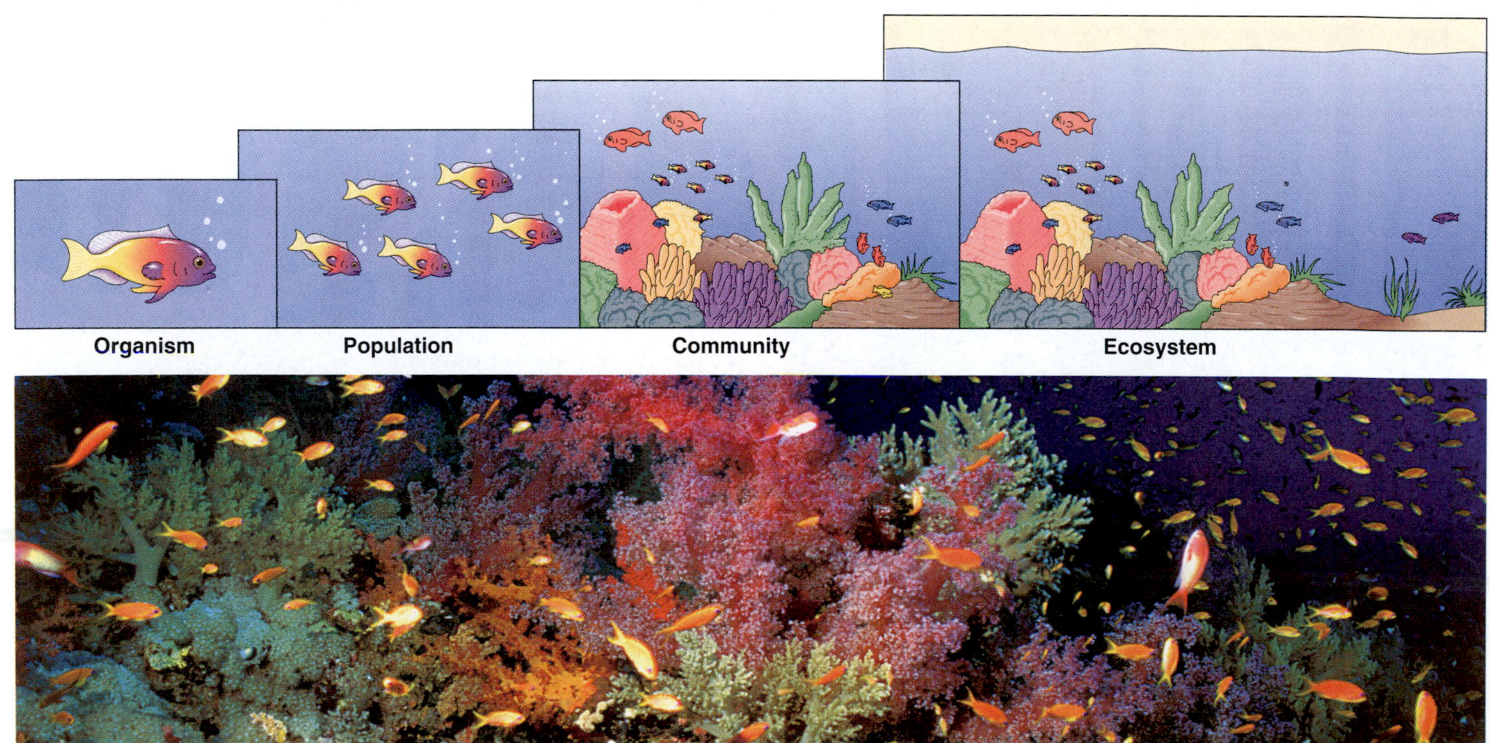

Figure 23.1 Ecological levels.
The study of ecology encompasses levels of organization from the individual organism to the population, community, and finally an ecosystem.

Density and Distribution of Populations

Population density is the number of individuals per unit area or volume, while **population distribution** is the pattern of dispersal of individuals within the area of interest. Population density figures make it seem as if individuals are uniformly distributed but, actually, the members of a population are not usually distributed uniformly (Fig. 23.2*a*). In some populations, individuals are randomly distributed as in Figure 23.2*b*, but most are clumped as illustrated in Figure 23.2*c*.

As an example of the relationship between density and distribution, consider that we can calculate the average density of people in the United States, but we know full well that most people live in cities where the number of people per unit area is dramatically higher than in the country. And even within a city more people live in particular neighborhoods than others, and such distributions can change over time. Therefore, basing ecological models solely on population density, as has often been done in the past, can be misleading.

Today ecologists want to analyze and discover what causes the spatial and temporal "patchiness" of organisms. For example, as discussed on page 392, a study of the distribution of hard clams in a bay on the south shore of Long Island, New York, showed that clam abundance is associated with sediment shell content. Indeed, it might be possible to use this information to transform areas that have few clams to high abundance areas.

As with the clams, the distribution of organisms can be due to *abiotic* (nonliving) factors. Physical factors like the type of precipitation, and not just the average temperature, but also the daily and seasonal variations in temperature, and the type of soil can affect where a particular organism lives. In fact, moisture, temperature, or a particular nutrient can be a limiting factor for the distribution of an organism. **Limiting factors** are those factors that particularly determine whether an organism lives in an area. Trout live only in cool mountain streams where there is a high oxygen content, but carp and catfish are found in rivers near the coast because they can tolerate warm waters, which have a low concentration of oxygen. The timberline is the limit of tree growth in mountainous regions or in high latitudes. Trees cannot grow above the high timberline because of low temperature and the fact that water remains frozen most of the year. The distribution of organisms can also be due to a *biotic* (living) factor. In Australia, the red kangaroo does not live outside arid inland areas because it is adapted to feeding on the grasses that grow there.

Ecology as a science includes a study of the distribution of organisms: where and why organisms are located in a particular place at a particular time.

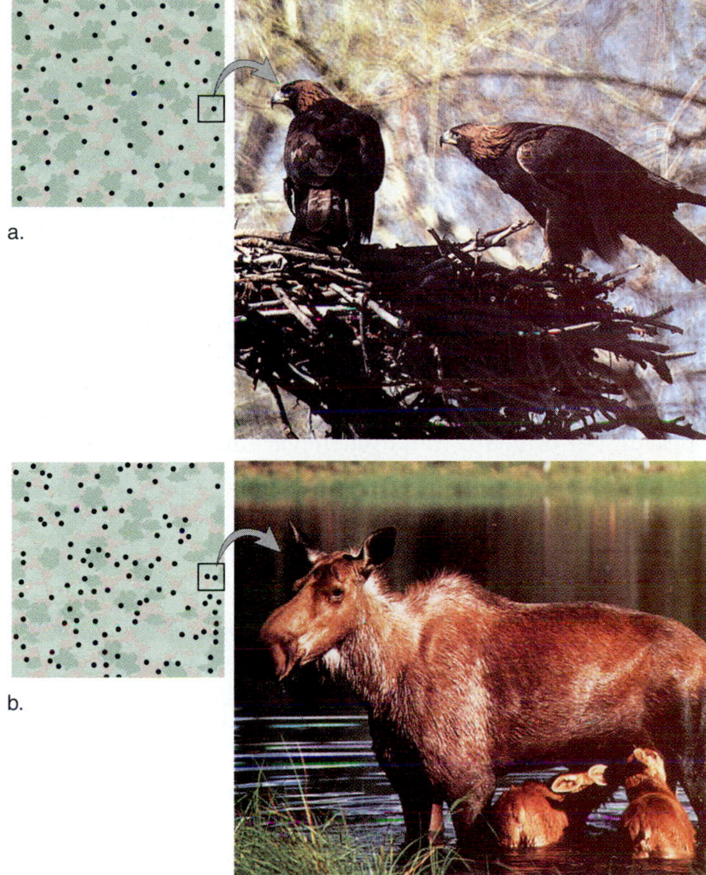

Figure 23.2 Patterns of dispersion within a population. Members of a population may be distributed uniformly, randomly, or usually in clumps. **a.** Golden eagle pair distribution is uniform over a suitable habitat area due to the territoriality of the birds. **b.** The distribution of moose aggregates is random over a suitable habitat. **c.** Cedar trees tend to be clumped near the parent plant because of poor distribution of seeds.

23.2 Characteristics of Populations

At any one point in time, populations have a certain size. **Population size** is the number of individuals contributing to the population's gene pool. Simply counting the number of individuals present in a population is not usually an option; instead it is necessary to estimate the present population size. Methods of estimating population size depend on the kind of species being studied and the validity of the estimate varies according to the method used.

Assuming that we can determine the present size, what factors would determine the future size of a population? Just as you might suspect, populations increase in size whenever natality (the number of births) exceeds mortality (the number of deaths) and whenever immigration exceeds emigration.

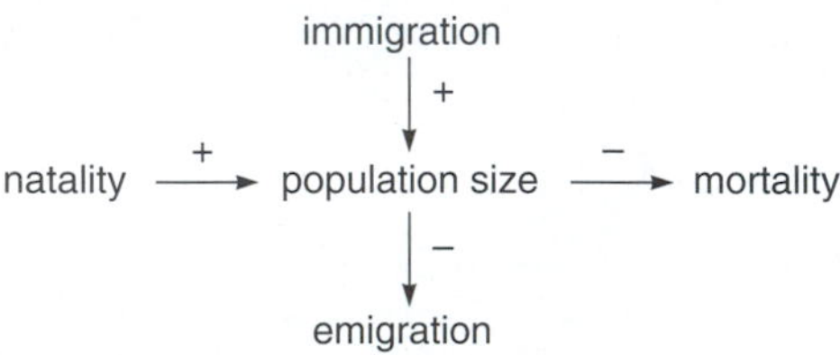

Usually it is possible to assume that immigration and emigration are about equal, and therefore it is only necessary to consider the birthrate and the death rate to arrive at the **intrinsic rate of natural increase,** or simply, r. Both birthrate and death rate are measured in terms of the individual; that is, per capita. For example, suppose a herd of elephants numbers 100. During the year, 10 births and 2 deaths occur; therefore, the birthrate is $10/100 = 0.10$ per elephant per year and the death rate is $2/100 = 0.02$ per elephant per year. We can combine both of these rates to arrive at $r = 0.08$ per capita per year. As we shall see, the intrinsic rate of natural increase allows us to calculate the growth and size of a population per any given unit of time.

Population Growth Models

We shall assume that there are two patterns of population growth. In the pattern called discrete breeding, the members of the population have only a single reproductive event in their lifetime. When this event draws near, the mature adults largely cease to grow, expend all their energy in reproduction, and then they die. Many insects and annual plants reproduce in this manner. They produce offspring or seeds that can survive dryness and/or cold and resume growth the next favorable season. In the pattern called continuous breeding, members experience many reproductive events throughout their lifetime. During their entire lives, they continue to invest energy in their future survival, increasing their chances of reproducing again. Most vertebrates, bushes, and trees have this pattern of reproduction.

Ecologists have developed mathematical models of population growth based on these two very different patterns of reproduction. Do these models have value even if the manner in which organisms reproduce does not always fit either of these two patterns, as exemplified in Figure 23.3? Although the mathematical models we will be describing are simplifications, they still may predict how best to control the distribution and abundance of organisms, or how to predict the responses of populations when their environment is altered in some way. Testing such predictions permit the development of new hypotheses that can then be tested.

a.

b.

Figure 23.3 Patterns of reproduction.
We shall assume that members of populations either have a single suicidal reproductive event or they reproduce repeatedly. But actually there are complications. **a.** Aphids reproduce repeatedly by asexual reproduction during the summer, and then reproduce sexually but once, right before winter comes on. Therefore, aphids utilize both patterns of reproduction. **b.** The offspring of annual plants can germinate several seasons later. Under these circumstances, population size could fluctuate according to environmental conditions.

Exponential Growth

As an example of discrete breeding, we will consider a population of insects in which females reproduce only once a year and then they die. Each female produces on the average 2.40 eggs per generation that will survive the winter and become offspring the next year. In the next generation, each female will again produce 2.40 eggs. In case of discrete breeding it is customary to replace r with R[1] = net reproductive rate. Why net reproductive rate? Because it is the observed reproductive rate after deaths have occurred.

Figure 23.4*a* shows how the population would grow year after year for ten years assuming that R stays constant from generation to generation. This growth is equal to the size of the population because all members of the previous generation have died. Figure 23.4*b* shows the growth curve for this population. This growth curve, which has a J shape, depicts exponential growth. With **exponential growth,** the number of individuals added each generation increases as the total number of females increases.

Notice that the curve has these phases:

lag phase: during this phase, growth is slow because the population is small.

exponential growth phase: during this phase, growth is accelerating.

Figure 23.4*c* gives the mathematical equation that allows you to calculate growth and size for any population that has discrete (nonoverlapping) generations. In other words, as discussed, all members of the previous generation die off before the new generation appears. In order to use this equation to determine future population size, it is necessary to know R, which is the net reproductive rate determined after gathering mathematical data regarding past population increases.

During exponential growth, a population is exhibiting its biotic potential. **Biotic potential** [Gk. *bios*, life] of a population is the maximum population growth that can possibly occur under ideal circumstances. These circumstances include plenty of room for each member of the population, unlimited resources, and no hindrances, such as predators or parasites. Resources include food, water, and a suitable place to exist. For exponential growth to occur, there must be no restrictions placed on growth. It is assumed that there is plenty of room, food, shelter, and any other requirements necessary to sustain unlimited growth. But in reality, exponential growth cannot continue for long because of environmental resistance. **Environmental resistance** is all those environmental conditions such as a limited supply of food, an accumulation of waste products, increased competition between members, or predation (if the population lives in the wild) that prevent populations from achieving their biotic potential.

Exponential growth produces a characteristic J-shaped curve. Because growth accelerates over time, the size of the population can increase dramatically.

Generation	Population size
0	10
1	24
2	57.6
3	138.2
4	331.7
5	796.1
6	1,910.6
7	4,585.4
8	11,005.0
9	26,412.0
10	63,388.8

a.

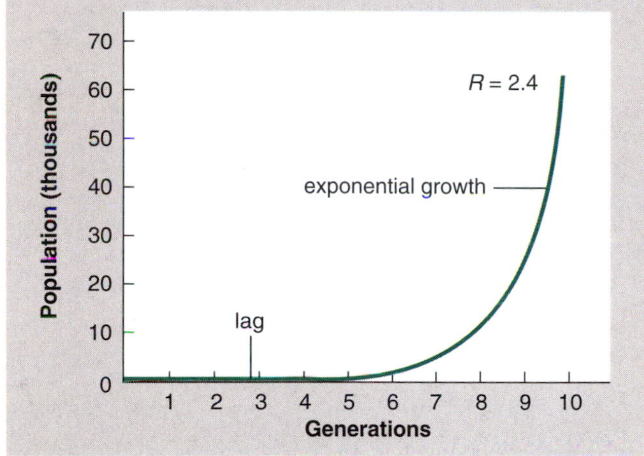

b.

To calculate population size from year to year, use this formula:

$$N_{t+1} = RN_t + N_1$$

N_t = number of females already present

R = net reproductive rate

N_{t+1} = population size the following year

c.

Figure 23.4 Model for exponential growth. When the data for discrete reproduction in **(a)** is plotted, the exponential growth curve in **(b)** results. **c.** This formula produces the same results as **(a)** and generates the same graph.

[1] The change of r to R is simply customary in discrete breeding calculations; both coefficients deal with the same thing (birth minus death).

Growth of Yeast Cells in Laboratory Culture		
Time (t) (hours)	Number of individuals (N)	Number of individuals added per 2-hour period $\left(\frac{\Delta N}{\Delta t}\right)$
0	9.6	0
2	29.0	19.4
4	71.1	42.1
6	174.6	103.5
8	350.7	176.1
10	513.3	162.6
12	594.4	81.1
14	640.8	46.4
16	655.9	15.1
18	661.8	5.9

a.

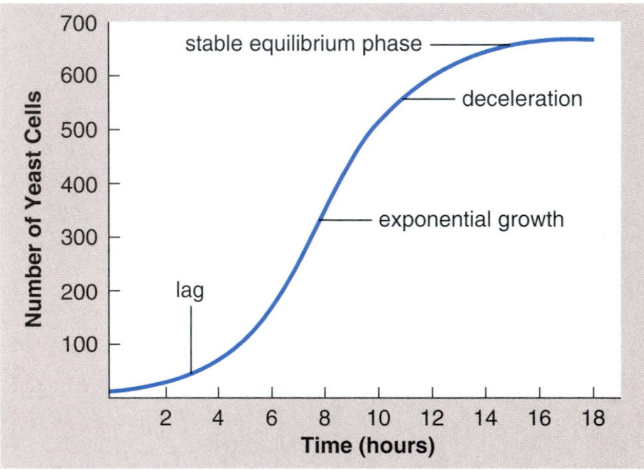

b.

To calculate population growth as time passes, use this formula:

$$\frac{dN}{dt} = rN\left(\frac{K-N}{K}\right)$$

N = population size

dN/dt = change in population size

r = intrinsic rate of natural increase

K = carrying capacity

$\frac{K-N}{K}$ = unutilized opportunity for population growth

c.

Figure 23.5 Model for logistic growth.
When the data for repeated reproduction (**a**) are plotted, the logistic growth curve in (**b**) results. **c.** This formula produces the same results as (**a**) and generates the same graph.

Logistic Growth

What type growth curve results when environmental resistance comes into play? In 1930 Raymond Pearl developed a method for estimating the number of yeast cells present every two hours in a laboratory culture vessel. His data are shown in Figure 23.5*a*. When the data are plotted, the growth curve has the appearance shown in Figure 23.5*b*. This type of growth curve is a sigmoidal (S) or S-shaped curve.

Notice that this so-called **logistic growth** has these phases:

lag phase: during this phase, growth is slow because the population is small.
exponential growth phase: during this phase, growth is accelerating.
deceleration phase: during this phase, the rate of population growth slows down.
stable equilibrium phase: during this phase, there is little if any growth because births and deaths are about equal.

Figure 23.5*c* gives you the mathematical equation that allows you to calculate logistic growth (so called because the exponential portion of the curve would produce a straight line if the log of N were plotted). The entire equation for logistic growth is:

$$\frac{dN}{dt} = rN\frac{(K-N)}{K}$$

but let's consider each portion of the equation separately.

Because the population has repeated reproductive events, we need to consider growth as a function of change in time (Δ):

$$\frac{\Delta N}{\Delta t} = rN$$

If the change in time is very small then we can turn to differential calculus, and the instantaneous population growth (d) is given by:

$$\frac{dN}{dt} = rN$$

This portion of the equation applies to the first two phases of growth—the lag phase and the exponential growth phase. During the exponential growth phase, the population is displaying its biotic potential. What would be the result if exponential growth was experienced by any population for any length of time? Apple trees produce many apples per season, and if each seed became an apple tree, we would soon see nothing but apple trees. Or, to take another example, it has been calculated that if a single female pig had her first litter at nine months and produced two litters a year, each of which contained an average of four females (which, in turn reproduced at the same rate), there would be 2,220 pigs by the end of three years (Fig. 23.6). Because of exponential growth, even species with a low biotic potential would soon

Figure 23.6 Biotic potential.
The ability of many populations like apple trees or pigs to reproduce exceeds by a wide margin the number necessary to replace those that die.

cover the face of the earth with their own kind. Even though elephants require 22 months to produce a single offspring, Charles Darwin calculated that a single pair of elephants could have over 19 million live descendants after 750 years.

Why did the growth curve level off for the yeast population and not for the insect population within the time span of the study? The yeast population, as you know, was grown in a vessel in which food could run short and waste products could accumulate. In other words, the pattern of growth was determined by the environmental resistance. As mentioned previously, environmental resistance opposes exponential growth when a population is displaying its biotic potential. Environmental resistance is all those factors such as a finite amount of resources and an accumulation of waste products that curtail unlimited population growth. Environmental resistance results in the *deceleration phase* and the *stable equilibrium phase* of the logistic growth curve (see Fig. 23.5). Now the population is at the carrying capacity of the environment.

Carrying Capacity

The **carrying capacity** of any environment is the maximum number of individuals of a given species the environment can support. The closer population size gets to the carrying capacity, the greater will be the environmental resistance—as resources become more scarce, the birthrate is expected to decline and the death rate is expected to increase. This will result in a decrease in population growth; eventually, the population stops growing and its size remains stable.

How does our mathematical model for logistic growth take this process into account? To our equation for growth under conditions of exponential growth we add the term

$$\frac{(K-N)}{K}$$

In this expression, K is the carrying capacity of the environment. The easiest way to understand the effects of this term is to consider two extreme possibilities. First, consider a time at which the population size is well below carrying capacity. Resources are relatively unlimited, and we expect rapid, nearly exponential growth to take place. Does the model predict this? Yes, it does. When N is very small relative to K, the term $(K-N)/K$ is very nearly $(K-0)/K$, or approximately 1. Therefore, $dN/dt =$ approximately rN.

Similarly, consider what happens when the population reaches carrying capacity. Here, we predict that growth will stop and the population will stabilize. What happens in the model? When $N=K$, the term $(K-N)/K$ declines from nearly 1 to 0, and the population growth slows to zero.

As mentioned, the model we have developed predicts that exponential growth will occur only when population size is much lower than the carrying capacity. So, as a practical matter, if we are using a fish population as a continuous food source, it would be best to maintain the population size in the exponential phase of the logistic growth curve. Biotic potential is having its full effect and the birthrate is the highest it can be during this phase. If we overfish, the population will sink into the lag phase, and it will be years before exponential growth recurs again. On the other hand, if we are trying to limit the growth of a pest, it is best, if possible, to reduce the carrying capacity rather than reduce the population size. Reducing the population size only encourages exponential growth to begin once again. Farmers can reduce the carrying capacity for a pest by alternating rows of different crops rather than growing one type of crop per the entire field.

Logistic growth curve produces a characteristic S-shaped curve. Population size stabilizes when the carrying capacity of the environment has been reached.

Table 23.1

A Life Table for a Bluegrass Cohort

Age (Months)	Number Observed Alive	Number Dying	Mortality Rate Per Capita	Avg. Number of Seeds/Individual
0–3	843	121	0.143	0
3–6	722	195	0.271	300
6–9	527	211	0.400	620
9–12	316	172	0.544	430
12–15	144	95	0.626	210
15–18	54	39	0.722	60
18–21	15	12	0.800	30
21–24	3	3	1.000	10
24	0	—	—	—

Mortality Patterns

Population growth patterns assume that populations are made up of identical individuals. Actually, the individuals are in different stages of their life span. Investigators studying population dynamics construct life tables that show how many members of an original group of individuals born at the same time, called a **cohort,** are still alive after certain intervals of time. For example, Table 23.1 is a life table for a bluegrass. The cohort contains 843 individuals. After three months, when 121 individuals have died, the mortality rate is 0.143 per capita. Another way to view this statistic, however, is to consider that 722 individuals are still alive—have survived—after three months. **Survivorship** is the probability of newborn individuals of a cohort surviving to particular ages. If we plot the number surviving, a survivorship curve is produced.

For the sake of discussion, three types of idealized survivorship curves are recognized (Fig. 23.7*a*). The type I curve is characteristic of a population in which most individuals survive well past the midpoint, and death does not come until near the end of the life span. On the other hand, the type III curve is typical of a population in which most individuals die very young. In the type II curve, survivorship decreases at a constant rate throughout the life span.

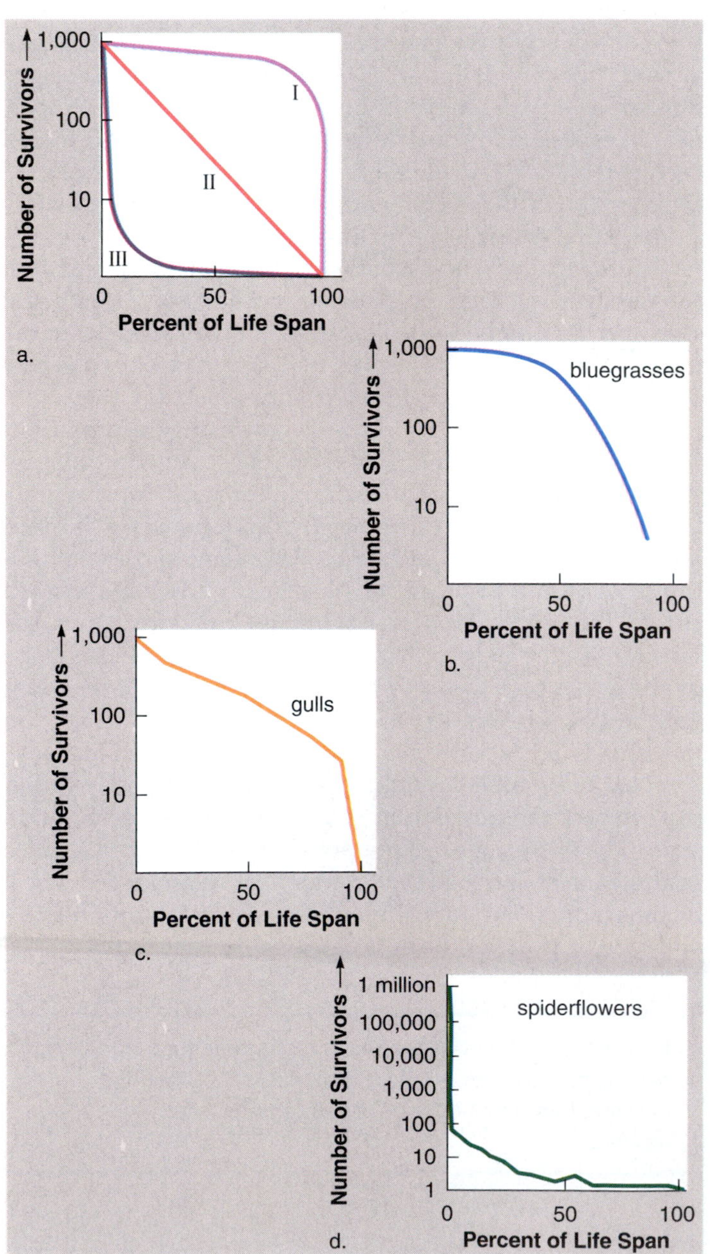

Figure 23.7 Survivorship curves. Survivorship curves show the number of individuals of a cohort that are still living over time. **a.** Three generalized survivorship curves. **b.** The survivorship curve for bluegrasses seems to be a combination of the type I and type II curves. **c.** Survivorship curve for gulls fits the type II curve somewhat. **d.** The survivorship curve for spiderflowers is a type III curve.

The survivorship curves of natural populations don't fit these three idealized curves exactly. In the bluegrass cohort, for example, most individuals survive till six to nine months, and then the chances of survivorship diminish at an increasing rate. Statistics for a gull cohort are close enough to classify the survivorship curve in the type II category, while a spiderflower cohort has a type III curve (Fig. 23.7*d*).

There is much that can be learned about the life history of a species by studying its life table and survivorship curve. Would you predict that most or few members of a population with a type III survivorship curve are contributing offspring to the next generation? Obviously since death comes early for most members, only a few are living long enough to reproduce. What about the other two types of survivorship curves? Look again at bluegrass life table. It tells us that per capita seed production increases as plants mature. How does that compare to the human situation?

Populations have a pattern of mortality/survivorship that becomes apparent from studying the life table and survivorship curve of a cohort.

Age Distribution

When the individuals in a population reproduce repeatedly, several generations may be alive at any given time. From the perspective of population growth, there are three major age groups in a population: prereproductive, reproductive, and postreproductive. Populations differ by what proportion of the population falls in each age group. At least three **age structure diagrams** are possible (Fig. 23.8).

When the prereproductive group is the largest of the three groups, the birthrate is higher than the death rate, and a pyramid-shaped diagram is expected. Under such conditions, even if the growth for that year was matched by the deaths for that year, the population would continue to grow in the following years. Why? Because there are more individuals entering than leaving the reproductive years. Eventually, as the size of the reproductive group equals the size of the prereproductive group, a bell-shaped diagram will result. The postreproductive group will still be the smallest, however, because of mortality. If the birthrate falls below the death rate, the prereproductive group will become smaller than the reproductive group. The age structure diagram will then be urn-shaped because the postreproductive group is now the largest.

The age distribution reflects the past and future history of a population. Because a postwar baby boom occurred in the United States between 1946 and 1964, the postreproductive group will soon be the largest group.

Age distributions contribute to our understanding of the past and future history of a population's growth.

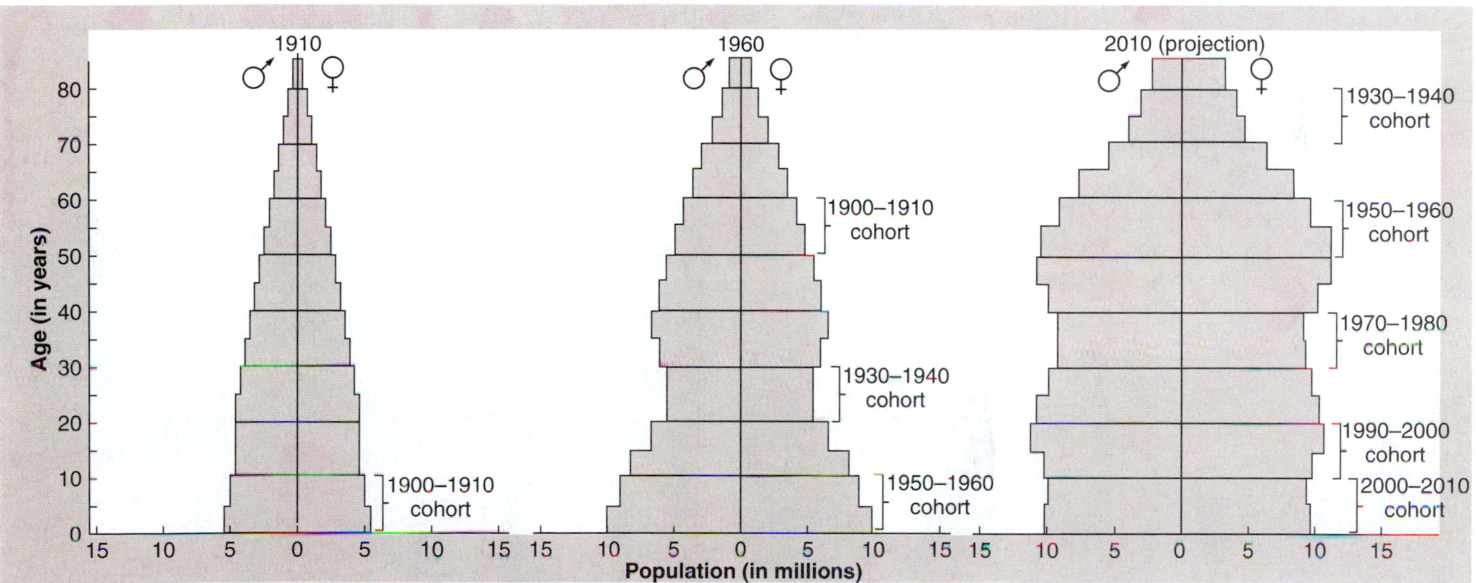

Figure 23.8 **U.S. age distributions, 1910, 1960, and 2010 (projected).**
In 1910, the distribution was shaped like a pyramid. In 1960, the distribution had shifted toward a bell shape, and in the year 201386 0, the age distribution is expected to be somewhat urn shaped. An unusually large number of offspring (called a baby-boom) were born following World War II. As the baby boomers grow older, there will be an unusually large number of persons in the postreproductive years.

23.3 Regulation of Population Size

We have developed two models of population growth: exponential growth with its J-shaped curve and logistic growth with its S-shaped curve. In a study of winter moth population dynamics, discussed in the reading on pages 388–389, it was discovered that a large proportion of eggs did not survive the winter and exponential growth never occurred. Perhaps the low number of individuals at the start of each season helps prevent the occurrence of exponential growth. This observation raises the question, "How well do the models for exponential and logistic growth predict population growth in natural populations?"

Is it possible, for example, that exponential growth may cause population size to rise above the carrying capacity of the environment, and as a consequence a population crash may occur? For example, in 1911, four male and 21 female reindeer *(Rangifer)* were released on St. Paul Island in the Bering Sea off Alaska. St. Paul Island had a completely undisturbed environment—there was little hunting pressure, and there were no predators. The herd grew exponentially to about 2,000 reindeer in 1938, overgrazed the habitat, and then abruptly declined to only eight animals in 1950 (Fig. 23.9).

Even though population growth may not follow our models exactly, we do find that environmental factors do regulate population size in natural environments. When ecologists first considered the question of environmental regulation, they emphasized that the environment contains abiotic (nonliving) and biotic (living) components. It seemed to them that both components could be involved in regulating population size. They suggested that abiotic factors, like weather and natural disasters, were **density-independent factors.** By this they meant that the number of organisms present did not influence the effect of the factor. The proportion of organisms killed by accidental fire, for example, is independent of density—fires don't necessarily kill a larger percentage of individuals in dense populations than they do in less dense populations (Fig. 23.10). On the other hand, biotic factors like parasitism, competition, and predation were designated as **density-dependent factors.** Predation increases when the prey population gets denser because it is easier for predators to find the prey they are looking for. Consider, for example, a population of crabs in which each crab needs a hole to hide in or else the crab is eaten by shorebirds. If there are only 100 holes, and 102 crabs inhabit the area, two will be without shelter. But since there are only two without shelter they may be hard to find. If neither crab is caught, then the predation rate is 0 captured/2 "available" = 0. However, if 200 crabs inhabit the area, 100 crabs will be without shelter. At this density, they are readily visible to birds and will probably attract a larger number of predators. If half of the exposed crabs are eaten, then the

a.

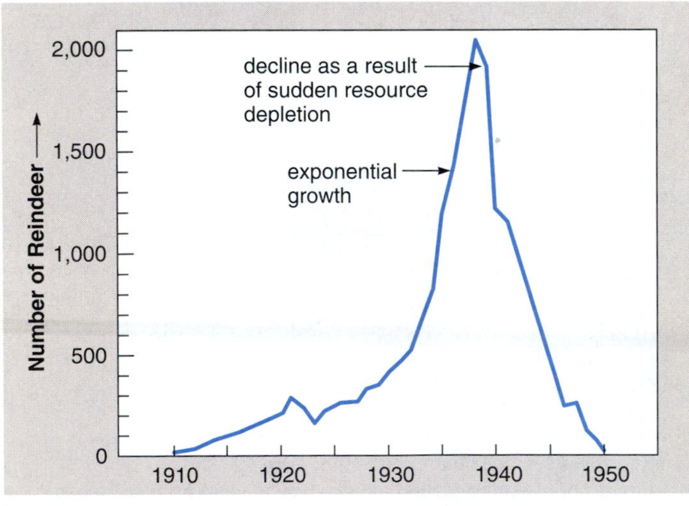

b.

Figure 23.9 Density-dependent effect.
a. A reindeer, *Rangifer.* b. On St. Paul Island, Alaska, reindeer grew exponentially for several seasons and then underwent a sharp decline as a result of overgrazing the available range.

Figure 23.10 Density-independent effect.
A fire can start and rage out of control regardless of how many organisms are present.

predation rate is 50 captured/100 "available" = 1/2. Thus, increasing the density has increased the proportion of individuals preyed upon.

Is it possible that other types of regulating factors sometimes keep the population size at just about the carrying capacity? Investigators have been collecting data on the population size of the bird called Great Tit *(Parus major)* for many years (Fig. 23.11*a*). Yearly counts during the breeding season for a population in Marley Wood, near Oxford, England, are given in Figure 23.11*b*. As you can see, the population size fluctuates above and below the carrying capacity. Clutch size (number of eggs produced per capita) does not seem to fluctuate from year to year, so other hypotheses are needed to explain why these fluctuations occur. Weather, resource abundance (food, shelter, etc.), and the presence and abundance of other animals (such as predators, pathogens, and parasites) are all clearly important—and all are extrinsic to the organism under investigation. Isn't it possible that intrinsic factors—those based on the anatomy, physiology, or behavior of the organisms—might have an effect on population size and growth rates?

Territoriality is apparent when spacing between members of a population is regular and more than seems necessary. Ninety to twenty meters are preferred by Great Tits, and the closer the nesting boxes the more likely some will go unoccupied. Therefore, territoriality is helping to regulate the Great Tit population size. Territoriality in birds and dominance hierarchies in mammals are behaviors that affect population size and growth rates. Recruitment, immigration, and emigration are other social means by which the population size of more complex organisms are regulated by intrinsic means.

Do populations ever show extreme fluctuations in size and growth rates in spite of extrinsic and intrinsic regulating mechanisms? Yes, they do. Outside of any density-dependent and density-independent factors, it could be that populations have an innate instability. Ecologists have developed models that predict complex, erratic changes in even simple systems. For example, a computer model of Dungeness crab populations assumed that adults produce many larvae and then they die. Most of the larvae do not survive, and those that do stay close to home. Under these circumstances, the model predicted wild fluctuations in population size that are now termed *chaos*.

Density-independent and density-dependent factors can often explain the population dynamics of natural populations. Both types of factors are extrinsic to the organism; perhaps intrinsic factors like territoriality in birds also play a role.

a.

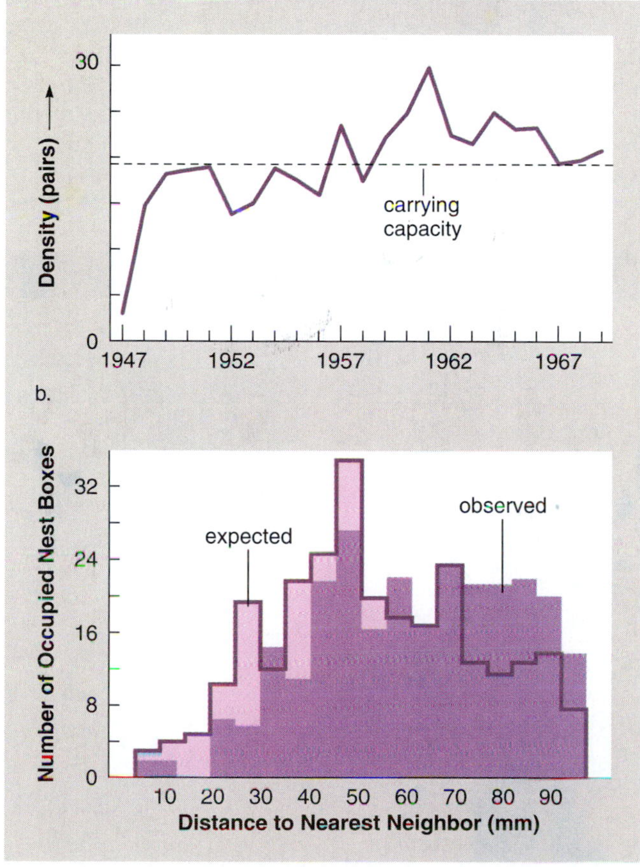

Figure 23.11 Great Tit, *Parus major.*
a. Photograph of Great Tit parents and offspring. b. Density dynamics of a breeding population in Marley Wood, near Oxford, England from 1947 to 1969. c. Great Tits are spaced out more than would be expected if distribution were random.

The Winter Moth: A Case Study

It would seem reasonable that many factors must be involved in regulating population size. When studies of population dynamics are done, investigators try to determine which are the key factors that determine mortality. For example, George Varley and George Gradwell studied the life cycle of the winter moth *Operophtera brumata* and several of its parasites and predators in order to determine which factor correlated best with observable fluctuations in population size.

Life Cycle of a Moth

Like other moths, winter moths have a life cycle that involves these stages: adult moths—larvae—pupae—adult moths. In the fall, adult females emerge from pupae in the soil litter and then climb (because they are wingless) the trunk of the nearest oak tree. After mating with a male (that has wings and can fly), each female lays about 150 eggs on twigs at the top of the tree. The eggs stay there all winter, and then in spring, they hatch when the buds on the tree open. The caterpillars feed on the growing leaves and then, when grown, they spin down to the ground on silk threads, crawl onto the litter, and pupate. In the fall the adults emerge once more (Fig. 23A).

The Parasites

The winter moth was accidently introduced from Europe into Nova Scotia sometime before 1949 and, within a few years, it was causing severe defoliation and beginning to spread westward. Two natural enemies of the moth were imported and became established. The adult female fly *Cyzenis albicans* deposits small resistant eggs on foliage, and these hatch only after they are eaten by the moth. Hatching occurs in the midgut of the moth, and the maggots bore from the gut into the blood cavity and then move to the salivary glands, where they feed slowly until the host pupates. After feeding some more, the maggots eventually form their puparia inside the moth's pupa. The adult flies emerge in the spring from these puparia.

The other parasite is a wasp, *Cratichneumon,* whose females lay an egg in each winter moth pupa they discover. The egg in turn hatches into a larva that consumes the winter moth pupa, and then pupates, later to emerge as an adult.

The Data

Varley and Gradwell counted the size of the population at each stage of the life cycle (Table 23A). They used sticky paper to capture and estimate the number of female moths climbing up the trees. They dissected several female moths to estimate the average number of eggs in each. They sampled the oak leaves to count the average number of caterpillars. They dissected larvae to see how many contained *Cyzenis* maggots. From this they knew that some larvae were killed by other parasites. They counted the number of moth pupae compared to how many *Cratichneumon* wasps emerged. From this they knew that some pupae were killed by predators.

The Conclusion

Varley and Gradwell used a mathematical formula that allowed them to assign a value (called a *k*-value in the table) to each environmental factor that caused moth mortality. Examine column three in Table 23A and note that most mortality was caused by winter disappearance due to inclement weather: females laid 658 eggs in the fall, but only 96.4 caterpillars (larvae) hatched

Table 23A

A Life Table for the Winter Moth

	Number Killed (per m^2)	Number Alive (per m^2)	*k*-value
Adult stage			
Females climbing trees, 1955	—	**4.39**	—
Egg stage			
Females × 150		658.0	—
Larval stage			
Winter disappearance	551.6	**96.4**	**0.84** = k_1
Attacked by *Cyzenis*	**6.2**	90.2	0.03 = k_2
Attacked by other parasites	**2.6**	87.6	0.01 = k_3
Infected by protozoa	**4.6**	83.0	0.02 = k_4
Pupal stage			
Killed by predators	54.6	28.4	**0.47** = k_5
Killed by *Cratichneumon*	**13.4**	15.0	0.27 = k_6
Adult stage			
Females climbing trees, 1956	—	7.5	—

Note: The figures in bold are those actually measured. The rest of the life table is derived from these.

1. winter moth adults
♂
♀
eggs
2. winter moth eggs
3. winter moth caterpillars feed on oak leaves and *Cyzenis* eggs
a. *Cyzenis* adult
b. *Cyzenis* develop inside winter moth caterpillars
winter moth adult trap
winter moth caterpillar trap
c. *Cyzenis* pupa
Cyzenis trap
4. winter moth pupa

October	November	December	January	February	March	April	May	June	July	August	September

Figure 23A **Life history of the winter moth *(Operophtera brumata)* and a parasitic fly *(Cyzenis)*.**
Winter moth: (light blue) 1. In fall, females emerge from pupae and climb trees and mate. 2. Eggs are laid. 3. Caterpillars hatch in spring to eat oak leaves. 4. Caterpillars pupate. *Cyzenis* (orange): a. Flies emerge from pupa in spring and lay eggs on leaves. b. When eggs are eaten by winter moth caterpillars, they develop inside the caterpillars. c. *Cyzenis* pupates inside of winter moth pupa. The thin black arrows in the diagram indicates stages at which collections were made.

in the spring. The *k*-value for winter disappearance was 0.84, and the *k*-values for all other factors are much less. Mortality due to predators such as shrews and beetles has the second largest *k*-value.

The *k*-values for winter disappearance correlate best with a sum of all mortality *k*-values as shown in Figure 23B. None of the other *k*-values, including pupal predation k_5, correlated well. Still, Varley and Gradwell developed a mathematical model that predicted moth population size from year to year based on density-dependent factors such as parasitism and predation. Their reasoning was that winter disappearance was due to the weather and weather can never be predicted from year to year.

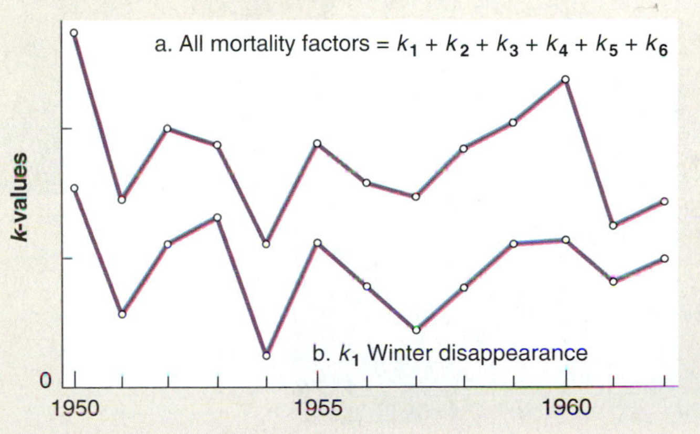

Figure 23B **Key factor analysis of the winter moth.**
Winter moth mortality, expressed as a *k*-value, due to (a) total mortality because of all factors, correlates well with winter moth mortality due to (b) winter disappearance. Therefore, winter disappearance largely accounts for population fluctuations in the adult population size.

23.4 Life History Patterns

We have already had an opportunity to point out that populations vary on such particulars as the number of births per reproduction, the age of reproduction, life span and the probability of living the entire life span. Such particulars are a part of a species' life history. Life histories contain characteristics that can be thought of as trade-offs. Each population is able to capture only so much of the available energy, and how this energy is distributed between its life span (short versus long), reproduction events (few versus many), care of offspring (little versus much), and so forth has evolved over the years. Natural selection shapes the final life history of individual species, and therefore it's not surprising that even related species may have different life history patterns, if they occupy different types of environments (Fig. 23.12).

The logistic population growth model has been used to suggest that members of some populations are subject to *r*-selection and some are subject to *K*-selection. In fluctuating and/or unpredictable environments, density-independent factors will keep populations in the lag or exponential phase of population growth. Population size is low relative to *K*, and ***r*-selection** favors *r*-strategists which produce large numbers of offspring when young. As a consequence of this pattern of energy allocation, small individuals that mature early and have a short life span are favored. They will tend to produce many relatively small offspring and to forego parental care in favor of a greater number of offspring. The more offspring, the more likely some of them will survive a population crash. Because of low population densities, most of the time density-dependent mechanisms such as predation and intraspecific competition are unlikely to play a major role in regulating population size and growth rates. Such organisms are often very good dispersers and colonizers of new habitats. Classic examples of such *opportunistic species* are many insects and annual plants (Fig. 23.13).

In contrast, we can imagine environments that are relatively stable and/or predictable, and in which populations tend to be near *K* with minimal fluctuations in population size. Resources such as food and shelter will be relatively scarce for these individuals, and those who are best able to compete will have the largest number of offspring.

strawberry arrow-poison frog, *Phyllobates lugubris*

wood frog, *Rana sylvatica*

Surinam toad, *Pipa pipa*

mouth-brooding frog, *Rhinoderma darwinii*

midwife toad, *Alyces obstetricans*

Figure 23.12 Parental care among frogs and toads.
Frogs and toads, which lay their eggs on land, as do wood frogs *(upper right)*, exhibit parental care that takes various forms. The midwife toad of Europe *(lower right)* carries strings of eggs entwined around his hind legs and takes them to water when they are ready to hatch. In mouth-brooding frogs of South America *(lower left)*, the male carries the larvae in a vocal pouch (brown area), which elongates the full length of his body before the froglets are released. In arrow-poison frogs of Costa Rica *(upper left)*, eggs are laid on land and after hatching, the tadpoles wiggle onto the parent's back and are then carried to water. Most amazing, in the Surinam toads of South America *(center)*, males fertilize the eggs during a somersaulting bout in which the eggs are placed on the female's back. Each egg develops in a separate pocket, where the tail of the tadpole acts as a placenta to take nourishment from the female's circulatory system.

K-selection favors *k*-strategists which allocate energy to their own growth and survival and to the growth and survival of their offspring. Therefore they are fairly large, are slow to mature, and have a fairly long life span. Because these organisms, termed *equilibrium species,* are strong competitors, they can become established and exclude opportunistic species. They are specialists rather than colonizers and tend to become extinct when their normal way of life is destroyed. The best possible examples of *K*-strategists are found among birds and mammals, such as bears (Fig. 23.13). The Florida panther is the largest of the animals in the Florida Everglades, requires a very large range, and produces few offspring that must be cared for. Currently, the Florida panther is unable to compensate for a reduction in its range, and is therefore on the verge of extinction.

Nature is actually more complex than these two possible life history patterns. It now appears that our description of *r*-strategist and *K*-strategist populations are at the ends of a continuum, and most populations lie somewhere in between these two extremes. For example, recall that plants have a two-generation life cycle, which includes a sporophyte and gametophyte generation. Ferns, which could be classified as *r*-strategists, distribute many spores and leave the gametophyte to fend for itself, but gymnosperms (e.g., pine trees) and angiosperms (e.g., oak trees), which could be classified as *K*-strategists, retain and protect the gametophyte. They produce seeds that contain the next sporophyte generation plus stored food. The added investment is significant, but these plants still release copious numbers of seeds.

A cod is a rather large fish weighing up to 25 pounds and measuring up to 6 feet in length—but the cod releases gametes in vast numbers, the zygotes form in the sea, and the parents make no further investment in developing offspring. Of the 6 to 7 million eggs released by a single female cod, only a few will become adult fish.

Differences in the environment result in different selection pressures and a range of life history characteristics.

Figure 23.13 Life history strategies. Are dandelions *r*-strategists with the characteristics noted, and are bears *K*-strategists with the characteristics noted? Most often the distinctions between these two possible life strategies are not as clear cut as they may seem.

Distribution of Hard Clams in the Great South Bay

The Great South Bay is a shallow bay located on the south shore of Long Island, New York, whose center is located about 1,010 kilometers east of New York City. The hard clam is a commercially harvested bivalve mollusc that is found throughout the Great South Bay. (It is eaten raw on the half shell, and is the key ingredient in baked clams as well as New England and Manhattan clam chowder.) The Great South Bay has often been referred to as a "hard clam factory" because in the 1970s, over half of the hard clams harvested in the United States came from its waters. For the past 20 years, I [Jeffrey Kassner] have been studying the distribution of hard clam abundance in the eastern third of the bay for the Town of Brookhaven which, by virtue of a 17th century grant from the King of England, owns 6,000 hectares of prime hard clam habitat.

Knowing the distribution and abundance of hard clams, as well as the responsible environmental factors, is important to Brookhaven because its ownership carries with it an obligation for managing a fishing industry for hard clams that has an annual value of $10 million and employs 300 fishermen. Of particular interest to Brookhaven is the possibility of using this information to develop projects, such as supplementing the natural hard clam production using aquaculture technology, that will increase hard clam abundance and hence the hard clam harvest and the economic benefits the hard clam resource provides to Brookhaven.

My study began with the censusing of the hard clam population. For several weeks during the summers, a barge-mounted crane with a 1 square meter clamshell bucket was used to take bottom grabs at 232 stations located throughout the study area. Each bottom sample was placed in a 1 square meter wire sieve and washed with a high pressure water hose to separate the hard clams from the sediment so that the hard clams could be counted and measured in order to calculate various demographic parameters of the hard clam population. The fieldwork was messy and physically hard, and by the end of the day, my field crew of students and biologists were tired, wet, and covered with mud. The results, however, have proven to be well worth the effort.

Working with Dr. Robert Cerrato of the State University of New York, I was able to draw up a composite census map showing the distribution of hard clam abundance. I was surprised to find that hard clams were not distributed uniformly throughout the study area, but occurred in distinct patches of high and low abundance. I termed the high abundance areas "beds" (a dense assemblage of clams is traditionally referred to as a "clam bed") and six such areas were identified.

When I overlaid the census map on a sediment map that I generated based on my field notes, nearly all of the beds coincided with areas of high shell content sediment associated with formerly productive oyster reefs, or what I call "relict" oyster reefs. This observation is of historical note as well as biological interest because up through the early years of this century, the Great South Bay was a major producer of oysters and was the source of the world famous "Blue Point Oyster." Although oysters are no longer found in the Great South Bay because of environmental changes, they left behind a legacy in the sediment that now supports high abundances of hard clams.

Having found that hard clam abundance was positively associated with relict oyster reefs, I wanted to find out what aspect of the sediment of the relict oyster reefs make them so productive, and conversely why low abundance areas have so few hard clams. This information would be useful to the Town of Brookhaven because it might lead to ways in which the sediment in low abundance areas can be transformed into high abundance areas. I therefore needed a sedimentary "portrait" of high and low abundance areas, and to develop one I borrowed techniques generally associated with other marine science disciplines: from shipwreck hunting, I used a side-scan sonar to map the topography of the bottom; from deep-sea research, an ROV (Remotely Operated Vehicle) to photograph the bottom; from commercial fishing, a fathometer to map the bathymetry; and from pollution studies, a sediment profile camera to photograph the sediment-water interface where hard clams live and feed.

I am now in the process of putting all these different information sources together on a single map (Fig. 23C). Because it will take time and perhaps more studies to develop the portrait, I am concurrently using my finding that hard clam abundance is strongly associated with sediment shell content to explore the feasibility of having shell placed on low abundance areas in order to create new relict oyster reefs. If this strategy works, it would be a tremendous boon to the shellfish industry because nearly three-quarters of the study area is low abundance.

My research project is not only scientifically interesting, but it is also personally rewarding. Over the years, I have become friends with many fishermen and I know that if I am successful, I will be helping them to continue in an occupation that in some cases goes back generations.

Figure 23C
Jeffrey Kassner is preparing a sedimentary "portrait" of high and low clam abundance areas in Great South Bay, Long Island, New York. This information will be used to increase the yield of clams for local fishermen.

23.5 Human Population Growth

The human population has an exponential pattern of growth and a J-shaped growth curve (Fig. 23.14). It is apparent from the position of 2000 on the growth curve in Figure 23.14 that growth is still quite rapid. The equivalent of a medium-sized city (200,000) is added to the world's population every day and 88 million (the equivalent of the combined populations of the United Kingdom, Norway, Ireland, Iceland, Finland, and Denmark) are added every year.

The present situation can be appreciated by considering the doubling time. The **doubling time**—the length of time it takes for the population size to double—is now estimated to be 47 years. Such an increase in population size will put extreme demands on our ability to produce and distribute resources. In 47 years, the world will need double the amount of food, jobs, water, energy, and so on just to maintain the present standard of living.

Many people are gravely concerned that the amount of time needed to add each additional billion persons to the world population has taken less and less time. The first billion didn't occur until 1800; the second billion arrived in 1930; the third billion in 1960, and today there are more than 6 billion. Only if the rate of natural increase declines can there be zero population growth when the birthrate equals the death rate and population size remains steady. The world's population may level off at 8, 10.5, or 14.2 billion, depending on the speed with which the net reproductive rate declines.

More-Developed Versus Less-Developed Countries

The countries of the world are divided into two groups. The **more-developed countries (MDCs),** typified by countries in North America and Europe, are those in which population growth is low and the people enjoy a good standard of living. The **less-developed countries (LDCs),** such as countries in Latin America, Africa, and Asia, are those in which population growth is expanding rapidly and the majority of people live in poverty. (Sometimes the term *third-world countries* is used to mean the less-developed countries. This term was introduced by those who thought of the United States and Europe as the first world and the former USSR as the second world.)

The more-developed countries (MDCs) doubled their populations between 1850 and 1950. This was largely due to a decline in the death rate, the development of modern medicine, and improved socioeconomic conditions. The decline in the death rate was followed shortly thereafter by a decline in the birthrate, so that populations in the MDCs experienced only modest growth between 1950 and 1975. This sequence of events (i.e., decreased death rate followed by decreased birthrate) is termed a **demographic transition.**

Yearly growth of the MDCs as a whole has now stabilized at about 0.1%. The populations of a few of the MDCs—Italy, Denmark, Hungary, Sweden—are not growing or are actually decreasing in size. In contrast, there is

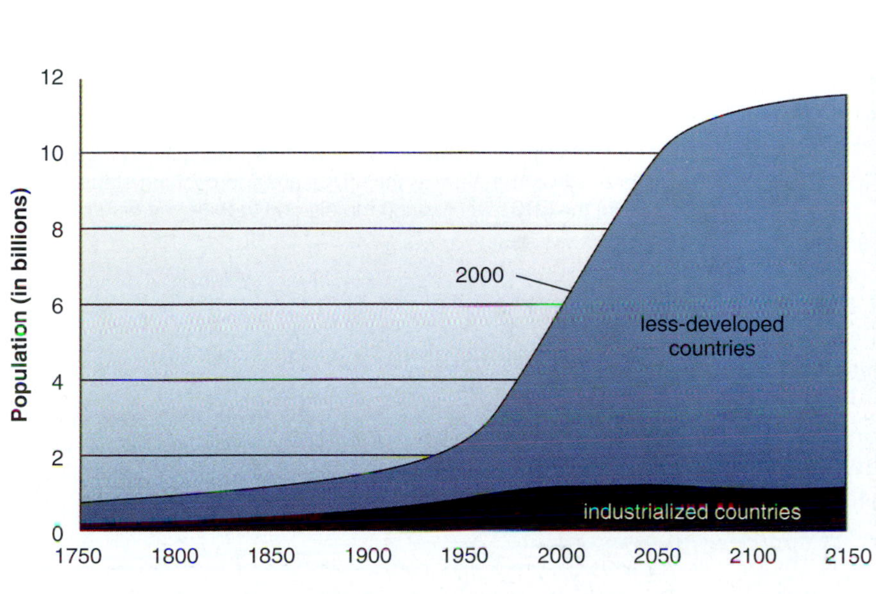

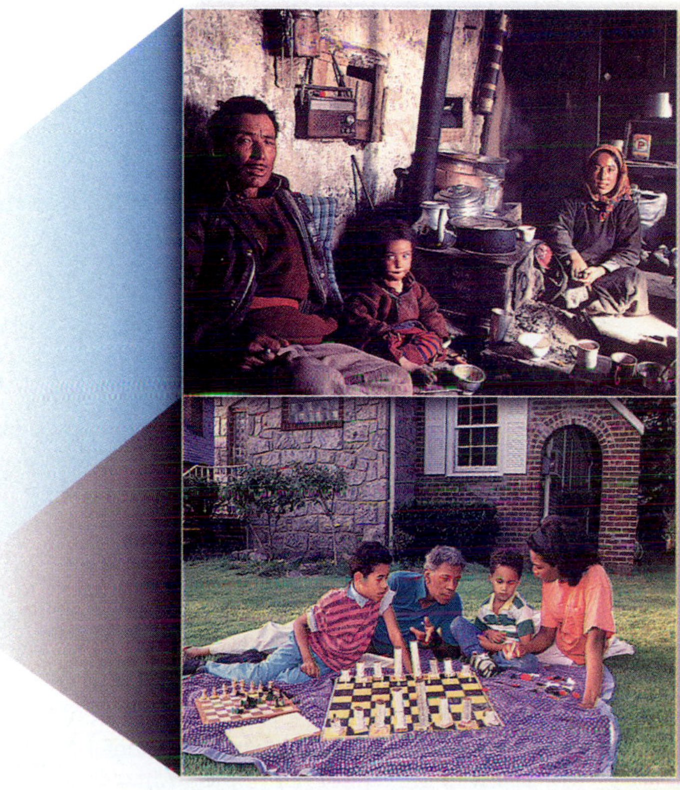

Figure 23.14 World population growth, 1750–2150.
The less-developed countries, with a minimal standard of living, will contribute most to world population growth. The more-developed or fully industrialized countries, with a high standard of living, will contribute least to world population growth.

no leveling off and no end in sight to U.S. population growth. Although yearly growth of the United States is only 0.6%, many people immigrate to the United States each year. In addition, there was an unusually large number of babies born between 1946 and 1964 (called a baby boom); therefore, a large number of women are still of reproductive age.

Although the death rate began to decline steeply in the LDCs following World War II with the importation of modern medicine from the MDCs, the birthrate remained high. The yearly growth of the LDCs peaked at 2.5% between 1960 and 1965. Since that time, a demographic transition has begun: the decline in the death rate has slowed and the birthrate has fallen. The yearly growth is now 1.7%. Still, the population of the LDCs may explode from 4.4 billion today to 10.2 billion in 2100 because of exponential growth. Most of this increase will occur in Africa, Asia, and Latin America. Ways to greatly reduce the expected increase have been suggested:

1. Establish and/or strengthen family planning programs. A decline in growth is seen in countries with good family planning programs supported by community leaders. Currently, 25% of women in sub-Saharan Africa say they would like to delay or stop childbearing, yet they are not practicing birth control; likewise, 15% of women in Asia and Latin America have an unmet need of birth control.
2. Use social progress to reduce the desire for large families. Many couples in the LDCs presently desire as many as four to six children. Providing education, raising the status of women, and reducing child mortality are desirable social improvements that seem to reduce the number of desired children.
3. Delay the onset of childbearing. A delay in the onset of childbearing and wider spacing of births could cause a temporary decline in the birthrate and reduce the present reproductive rate.

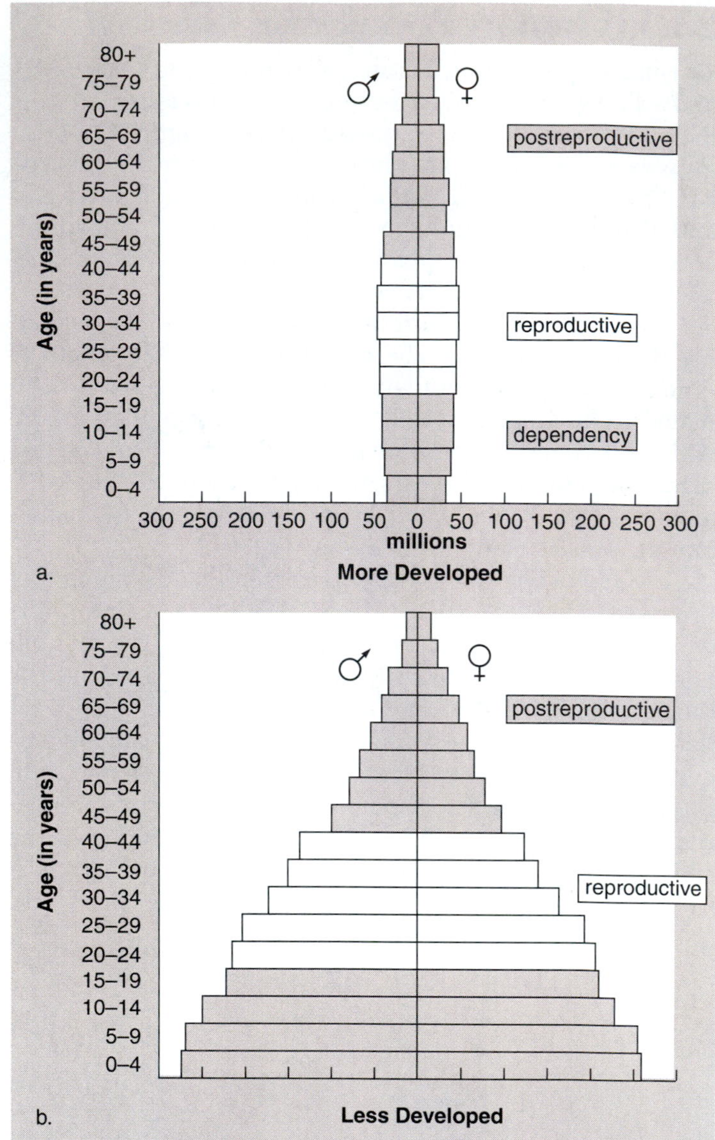

Figure 23.15 Age distributions for MDCs and LDCs (1989). The diagrams illustrate that (a) the MDCs are approaching stabilization, whereas (b) the LDCs will expand rapidly due to their age distributions.

Source: Data from *World Population Profiles,* WP-89.

Comparing Age Distributions

The age-structure diagrams of the MDCs and LDCs in Figure 23.15 divide the population into three age groups: dependency, reproductive, and postreproductive. The LDCs are experiencing a population momentum because they have more women entering the reproductive years than there are older women leaving them.

Laypeople are sometimes under the impression that if each couple has two children, **zero population growth** (no increase in population size) will take place immediately. However, **replacement reproduction,** as it is called, will still cause most countries today to continue growing due to the age structure of the population. If there are more young women entering the reproductive years than there are older women leaving them, then replacement reproduction will still result in growth of the population.

Many MDCs have a stabilized age structure, but most LDCs have a youthful profile—a large proportion of the population is younger than the age of 15. This means that their populations will still expand greatly, even after replacement reproduction is attained. The more quickly replacement reproduction is achieved, however, the sooner zero population growth will result.

Currently, the populations of Africa, Asia, and Latin America are expanding dramatically because of exponential growth.

A Sustainable World

While we are sometimes quick to realize that the growing populations of the LDCs are putting a strain on the environment, we should realize that the excessive resource consumption of the MDCs also stresses the environment. Environmental impact is measured not only in terms of population size, it is also measured in terms of resource consumption and the pollution caused by each person in the population. An average American family, in terms of per capita resource consumption and waste production, is the equivalent of 30 people in India (Fig. 23.16). It is time for us to realize that the carrying capacity of the earth is limited and only a certain number of people can be sustained at the standard of living in the developed countries.

Before the establishment of industrialized societies, people felt better connected to the plants and animals on which they depended, and they were better able to live in a sustainable way. After the Industrial Revolution, we especially began to think of ourselves as separate from nature and endowed with the right to exploit nature as much as possible. But our industrial society lives on *borrowed carrying capacity*—our cities not only borrow resources from the country, our entire population borrows from the past and future. The forests of the Carboniferous have become the fossil fuels that sustain our way of life today, and the environmental degradation we cause is going to be paid for by our children.

Overpopulation and overconsumption account for increased pollution and also for the mass extinction of wildlife that is going on. We are expected to lose one-third to two-thirds of the earth's species, any one of which could possibly have made a significant contribution to agriculture or medicine. It should never be said, "What use is this organism?" Aside from its contribution to the ecosystem in which it lives, one never knows how a particular organism might someday be useful to humans. Adult sea urchin skeletons are now used as molds for the production of small artificial blood vessels, and armadillos are used in leprosy research.

While it may seem like an either-or situation—either preservation of ecosystems or human survival—there is a growing recognition that this is not the case. Native peoples who harvest rubber from the trees of a tropical rain forest can have a sustainable income from the same trees year after year. One study calculated the market value of rubber and exotic produce that can be harvested continually from the Amazon rain forest. It concluded that selling these products would yield more than twice the income from either lumbering or cattle ranching, and the income would be sustainable.

It is clearly time for a new philosophy. We need to give up our desire for short-term personal gain and replace it with a reverence for the natural processes that make our lives possible. In a **sustainable world,** development will meet economic needs of all peoples while protecting the environment for future generations. Various organizations have singled out communities to serve as models of how to balance ecological and economic goals. For example, in Clinch Valley of southwest Virginia, the Nature Conservancy is helping to revive the traditional method of logging with draft horses. This technique, which allows the selective cutting of trees, preserves the forest and prevents soil erosion, which is so damaging to the environment. The United Nations has an established bioreserve system, a global network of sites that combine preservation with research on sustainable management for human welfare. More than 100 countries are now participants in the program. However, sustainability is more-than likely incompatible with the kinds of consumption/waste patterns currently practiced in more developed countries.

All peoples can benefit from a sustainable world where economic development and environmental preservation are considered complementary, rather than opposing, processes.

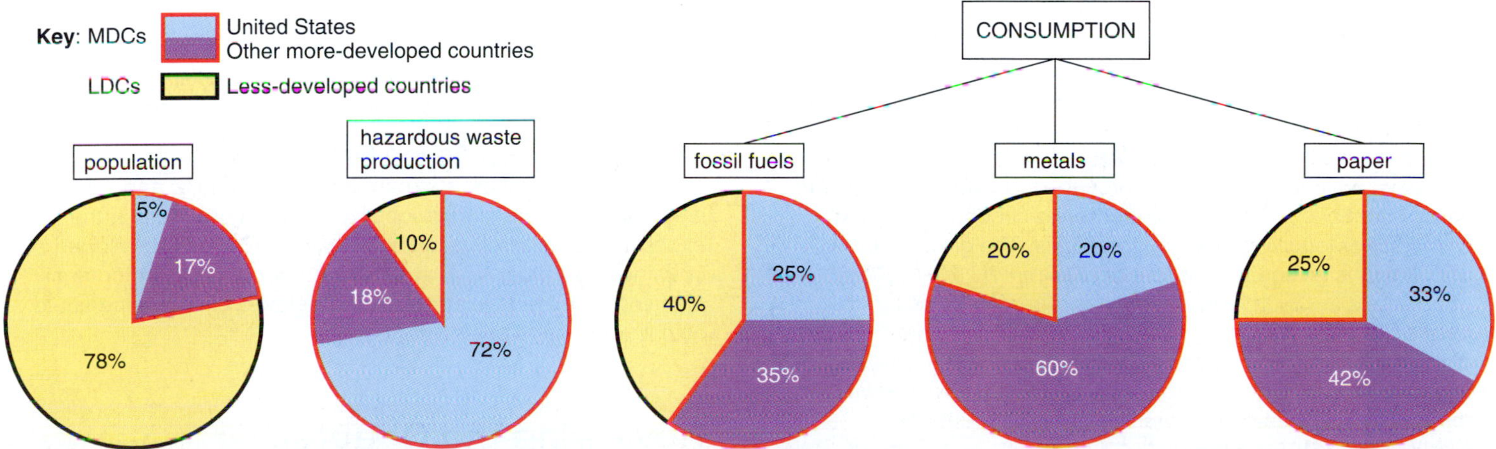

Figure 23.16 Resource consumption for MDCs and LDCs.
The populations of MDCs are smaller than LDCs, but MDCs produce most of the hazardous wastes because their consumption of fossil fuels, metals, and paper, for example, is much greater than the LDCs.

Connecting Concepts

Modern ecology began with descriptive studies by nineteenth century naturalists. In fact, an early definition of the field was "scientific natural history." However, modern ecology has grown to be much more than a simple descriptive field. Ecology is now very much an experimental, predictive science.

Much of the success in the development of ecology as a predictive science has come from studies of populations and the development of models that examine how populations change over time. The simplest models are based on population growth when there are unlimited resources. This results in exponential growth, a type of population growth that is only rarely seen in nature. Pest species may exhibit exponential growth until they run out of resources. Because so few natural populations exhibit exponential growth, population ecologists realized they must incorporate resource limitation in their models. The simplest models that account for limited resources result in sigmoidal or logistic growth. Populations which exhibit logistic growth will cease growth when they reach the environmental carrying capacity.

Many modern ecological studies are concerned with identifying the factors that place limits on population growth and that set the environmental carrying capacity. A combination of careful descriptive studies, experiments done in nature, and sophisticated models has allowed ecologists to make good predictions about which factors have the greatest influence on population growth. The example of the winter moth is a good case in point.

The next step in the development of modern ecology has been to try to understand how populations of different species affect each other. This is known as community ecology. Because each population in a community responds to environmental changes in slightly different ways, developing predictive models that explain how communities change has been challenging. However, ecologists are beginning to be able to predict how communities will change through time and to understand what factors influence community properties such as species number, abundance of individuals, and species interactions.

Summary

23.1 Scope of Ecology

Ecology is the study of the interactions of organisms with other organisms and with the physical environment. Ecology encompasses several levels of study: organism, population, community, ecosystem, and finally the biosphere. Ecologists are particularly interested in how interactions affect the distribution and abundance of organisms.

Population density is simply the number of individuals per unit area or volume. Distribution of these individuals can be uniform, random, or clumped. A population's distribution is often determined by limiting factors; that is, abiotic factors like water, temperature, and availability of nutrients.

23.2 Characteristics of Populations

Population size is dependent upon natality (number of births), mortality (number of deaths), immigration, and emigration. The number of births minus the number of deaths results in the net reproductive rate (symbolized as *r*) per capita per unit time.

One model for population growth assumes that the environment offers unlimited resources. In the example given, the members of the population have discrete reproductive events, and therefore the size of next year's population is given by the equation: $N_{t+1} = RN_t$. Under these conditions, exponential growth results in a J-shaped curve.

Most environments restrict growth, and exponential growth cannot continue indefinitely. Under these circumstances an S-shaped or logistic growth curve results. The growth of the population is given by the equation $dN/dt = rN\ (K-N)/K$ for populations in which individuals have repeated reproductive events. The term $(K-N)/K$ represents the unused portion of the carrying capacity (K). When the population reaches carrying capacity, the population stops growing because environmental resistance opposes biotic potential, the maximum net reproductive rate for a population.

Individuals are at different stages of their life span in a population. Mortality (deaths per capita) within a population can be recorded in a life table and illustrated by a survivorship curve. The pattern of population growth is reflected in the age distribution of a population, which consists of prereproductive, reproductive, and postreproductive segments. Populations that are growing exponentially have a pyramid-shaped age distribution pattern.

23.3 Regulation of Population Size

Population growth is limited by density-independent factors (e.g., weather) and density-dependent factors (predation, competition, and resource availability). Do some populations have an intrinsic means of regulating population growth as opposed to density-independent and density-dependent factors, which are extrinsic means? Territoriality is given as an example of a possible intrinsic means of regulation.

23.4 Life History Patterns

The logistic growth model has been used to suggest that the environment promotes either *r*-selection or *K*-selection. So-called *r*-selection occurs in unpredictable environments where density-independent factors affect population size. Energy is allocated to producing as many small offspring as possible. Adults remain small and do not invest in parental care of offspring. *K*-selection occurs in environments that remain relatively stable, where density-dependent factors affect population size. Energy is allocated to survival and repeated reproductive events. The adults are large and invest in parental care of offspring. Actual life histories contain trade-offs between these two patterns.

23.5 Human Population Growth

The human population is expanding exponentially, and it is unknown when the population size will level off. Most of the expected increase will occur in certain LDCs (less-developed countries) of Africa, Asia, and Latin America. Support for family planning, human development, and delayed childbearing could help prevent an expected increase.

Reviewing the Chapter

1. What are the various levels of ecological study? 378
2. What is density, and why is it sometimes misleading to do studies based on the average number of organisms per unit area? 379

3. What four population attributes determine population dynamics? 380
4. What type growth curve indicates that exponential growth is occurring? What are the environmental conditions for exponential growth? 381
5. What type growth curve indicates that biotic potential is being opposed by environmental resistance? What environmental conditions are involved in environmental resistance? 382–383
6. What is the carrying capacity of an area? 383
7. What is a life table and the three general types of survivorship curves? 384
8. What will the age distribution tell you about the population under study? 385
9. Does population growth of natural populations fit either the exponential growth model or the logistic growth model? Explain why or why not. 386–387
10. Give support to the belief that intrinsic factors might regulate population size in some populations. 387
11. Why would you expect the life histories of natural populations to vary and contain some characteristics that are so-called *r*-selected and some that are so-called *K*-selected? 390–391
12. What type of growth curve presently describes the population growth of the human populations? In what types of countries is most of this growth occurring, and how might it be curtailed? 393–394

Testing Yourself

Choose the best answer for each question.

1. Which of these levels of ecological study involves both abiotic and biotic components?
   a. organisms
   b. populations
   c. communities
   d. ecosystem
   e. All of these are correct.
2. When phosphorus is made available to an aquatic community, the algal populations suddenly bloom. This indicates that phosphorus is a
   a. density-dependent regulating factor.
   b. reproductive factor.
   c. limiting factor.
   d. *r*-selection factor.
   e. All of these are correct.
3. A J-shaped growth curve should be associated with
   a. exponential growth.
   b. biotic potential.
   c. no environmental resistance.
   d. rapid population growth.
   e. All of these are correct.
4. An S-shaped growth curve
   a. occurs when there is no environmental resistance.
   b. includes an exponential growth phase.
   c. occurs in natural populations but not laboratory ones.
   d. is subject to a sharp decline.
   e. All of these are correct.
5. If a population has a type I survivorship curve (most live the entire life span), which of these would you also expect?
   a. a single reproductive event per adult
   b. overlapping generations
   c. reproduction occurring near the end of the life span
   d. a very low birthrate
   e. None of these are correct.
6. A pyramid-shaped age distribution means that the
   a. prereproductive group is the largest group.
   b. population will grow for some time in the future.
   c. country is more likely an LDC rather than an MDC.
   d. fewer women are leaving the reproductive years than women entering them.
   e. All of these are correct.
7. Which of these is a population-independent regulating factor?
   a. competition
   b. predation
   c. weather
   d. resource availability
   e. the average age when child rearing begins
8. Fluctuations in population growth can correlate to changes in
   a. predation.
   b. weather.
   c. resource availability.
   d. parasitism.
   e. All of these are correct.
9. A species that has repeated reproductive events, lives a long time, but suffers a crash due to the weather is exemplifying
   a. *r*-selection.
   b. *K*-selection.
   c. a mixture of both.
   d. density-dependent and density-independent regulation.
   e. a pyramid-shaped age structure diagram.
10. The human population
    a. is undergoing exponential growth.
    b. is not subject to environmental resistance.
    c. fluctuates from year to year.
    d. only grows if emigration occurs.
    e. All of these are correct.

Thinking Scientifically

1. In the winter moth life cycle (p. 389) the two moth parasites were found to be a less important cause of mortality than mortality due to winter loss and predators. Give an evolutionary explanation for the inefficiency of parasites to control population size.
2. Many people enjoy backyard bird feeders and feel that they are helping preserve birds by providing them with food. Many more people unintentionally feed animals by leaving garbage cans and recycling bins where animals have access to the contents. This selects for what animal behaviors (and thus species with that behavior)? Which are selected against? How can feeding animals decrease diversity of local wildlife?

Bioethical Issue

The answer to how to curb the expected increase in the world's population lies in discovering how to curb the rapid population growth of the less-developed countries. In these countries, population experts have discovered what they call the "virtuous cycle." Family planning leads to healthier women, and healthier women have healthier children, and the cycle continues. Women no longer have to have many babies for a few to survive. More education is also helpful because better educated people are more interested in postponing childbearing and promoting women's rights. Women who have equal rights with men tend to have fewer children.

"There isn't any place where women have had the choice that they haven't chosen to have fewer children," says Beverly Winikoff at the Population Council in New York City. "Governments don't need to resort to force." Bangladesh is a case in point. Bangladesh is one of the densest and poorest countries in the world. In 1990 the birthrate was 4.9 children per woman and now it is 3.3. This achievement was due in part to the Dhaka-based Grameen Bank, which loans small amounts of money mostly to destitute women to start a business. The bank discovered that when women start making decisions about their lives, they also start making decisions about the size of their families. Family planning within Grameen families is twice as common as the national average; in fact, those women who get a loan promise to keep their families small! Also helpful has been the network of village clinics that counsel women who want to use contraceptives. The expression "contraceptives are the best contraceptives" refers to the fact that you don't have to wait for social changes to get people to use contraceptives—the two feed back on each other.

Recently, some of the less-developed countries, faced with economic crisis, have cut back on their family planning programs, and the more-developed countries have not taken up the slack. Indeed, some foreign donors have also cut back on aid—the U.S. by one-third. Are you in favor of foreign aid to help countries develop family planning programs. Why or why not?

Understanding the Terms

age structure diagram 385
biosphere 378
biotic potential 381
carrying capacity 383
cohort 384
community 378
demographic transition 393
density-dependent factor 386
density-independent factor 386
doubling time 393
ecology 378
ecosystem 378
environmental resistance 381
exponential growth 381
habitat 378
intrinsic rate of natural increase 380
K-selection 391
less-developed country (LDC) 393
limiting factor 379
logistic growth 382
more-developed country (MDC) 393
population 378
population density 379
population distribution 379
population size 380
replacement reproduction 394
r-selection 390
survivorship 384
sustainable world 395
zero population growth 394

Match the terms to these definitions:

a. ______________ Due to industrialization, a decline in the birthrate following a reduction in the death rate so that the population growth rate is lowered.
b. ______________ Group of organisms of the same species occupying a certain area and sharing a common gene pool.
c. ______________ Growth, particularly of a population, in which the increase occurs in the same manner as compound interest.
d. ______________ Largest number of organisms of a particular species that can be maintained indefinitely by a given environment.
e. ______________ Maximum population growth rate under ideal conditions.

Web Connections

Exploring the Internet

http://www.mhhe.com/biosci/genbio/mader
(click on *Biology 7/e*)

The *Biology 7/e* Online Learning Center provides many resources for studying the material in this chapter including links to the following sites:

Population Ecology. This site provides online data, information from lecture courses, and the names of organizations, people, and journals involved in population ecology.

http://www.ento.vt.edu/~sharov/popechome/refernce.html#lectures

The Ecological Society of America Homepage. This large organization has a homepage that will link the user to hundreds of useful sites.

http://www.sdsc.edu/~ESA/

Zero Population Growth. This is the homepage for ZPG. Were you aware that the human population on earth surpassed 6 billion in October of 1999? Links and much information here.

http://www.zpg.org/

Biological Control: A Guide to Natural Enemies in Nature. Information on using insects as biological control agents.

http://www.nysaes.cornell.edu:80/ent/biocontrol/index.html

Constructing Life Tables for Homo sapiens. *A fun, yet provocative laboratory exercise in the construction of a life table and survivorship curve for humans using data from cemeteries. This lab can also be done using obituaries from the newspaper or the web.*

http://www.auburn.edu/~nolanpm/Steve'sPage/GraveLab.html

and

Life Tables. Probably more math than you wanted to know, but this includes the equations used by population biologists.

http://www.ets.uidaho.edu/wlf448/LifeTables.htm

Community Ecology

chapter 

c h a p t e r   c o n c e p t s

24.1 Concept of the Community

- Communities are assemblages of interacting populations, which differ in composition and diversity. 400
- Environmental factors influence community composition and diversity. 401

24.2 Structure of the Community

- Community organization involves the interactions among species such as competition, predation, parasitism, and mutualism. 403
- The ecological niche is the role an organism plays in its community, including its habitat and its interactions with other organisms. 403
- Competition leads to resource partitioning, which reduces competition between species. 404
- Predation reduces prey population density but can lead to a reduction in predator population density also. 406
- There are a number of different kinds of prey defenses, including mimicry. 408
- Symbiotic relationships include parasitism, commensalism, and mutualism. 410

24.3 Community Development

- Ecological succession is a change in species composition and community structure and organization over time. 414

24.4 Community Biodiversity

- The intermediate disturbance hypothesis suggests that a moderate amount of environmental change leads to a diverse environment, and therefore more biodiversity. 416
- Predation and competition can help maintain biodiversity, and island biogeography suggests how to maintain species richness. 416

Coyotes *(Canis latrans)* and badgers *(Taxidea taxus)* interact within a prairie community

Coyotes and badgers sometimes improve their hunting of prairie dogs by joining forces. When coyotes are around, prairie dogs tend to stay in their burrows but badgers can flush them out to the waiting coyotes. For the coyote, it sure beats wasting time and energy stalking prairie dogs aboveground. Coyotes have even been seen to use various kinds of antics to encourage badgers to move on to more productive hunting grounds. The interaction of coyotes and badgers is a possible way for two different species living in the same community to interact.

There are numerous interactions between species in a community. One species may prey on or parasitize another species or may even assist another species. Often, interactions are not entirely black and white, as when coyotes sometimes kill young badgers. This chapter examines specific interactions and how they determine the distribution and abundance of species in a community. We will see that communities are subject to changes over time. Although an intermediate amount of disturbance seems to increase diversity, extreme disturbances can be damaging to the structure of a community.

24.1 Concept of the Community

Populations do not occur as single entities; they are part of a community. A **community** is an assemblage of populations interacting with one another within the same environment. Communities come in different sizes, and it is sometimes difficult to decide where one community ends and another begins. A fallen log can be considered a community because there are various populations in the log and these populations are interacting with one another. The fungi of decay break down the log and provide food for the various invertebrates living in the log that may feed on one another. Yet it is quite possible that bugs and worms living in the log could be eaten by a passing bird that flies about the entire forest. This interaction means that the log is a part of the forest community. And where does the forest end? Doesn't it gradually fade into the surrounding area? So the demarcation of any community turns out to be somewhat arbitrary.

This chapter examines the various types of community interactions and their importance to the structure of a community. Such interactions illustrate some of the most important selection pressures impinging on individuals. They also help us understand how biodiversity can be preserved.

Community Composition and Diversity

Two characteristics of communities—their composition and diversity—allow us to compare communities. This is a first step toward understanding how a community functions. The *composition* of a community is simply a listing of the various species in the community. The *diversity* includes both species richness (the number of species) and evenness (the relative abundance of individuals of different species).

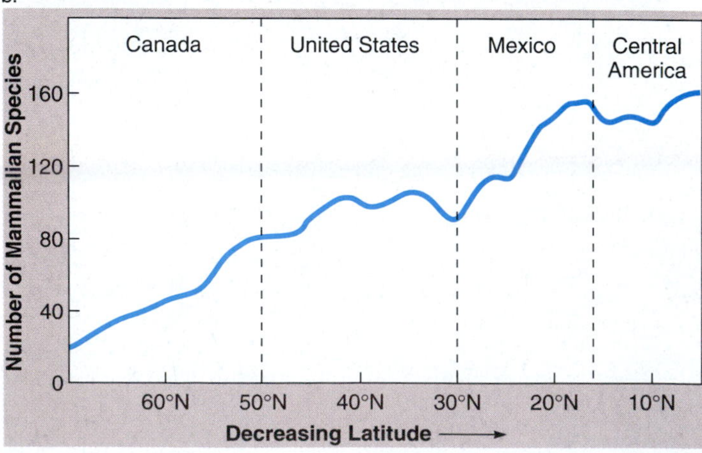

Figure 24.1 Community structure.
Communities differ in their composition, as witnessed by their predominant plants and animals. Diversity of communities is described by the richness of species and their relative abundance. **a.** A coniferous forest. Some mammals found here are listed to the left. **b.** A tropical rain forest. Some mammals found here are listed to the right. **c.** Richness increases with decreasing latitude.

Just glancing at Figure 24.1 makes it apparent that a coniferous forest has a different composition from a tropical rain forest. Pictorially we can see that narrow-leaved evergreen trees are present in a coniferous forest, and broad-leaved evergreen trees are numerous in a tropical rain forest. Mammals also differ between the two communities, as those listed demonstrate.

Diversity of a community goes beyond composition because it includes the number of species and also the abundance of each species. To take an extreme example: a forest in West Virginia has, among other species, 76 yellow poplar trees but only one American elm. If we were simply walking through this forest, we might miss seeing the American elm. If, instead, the forest had 36 poplar trees and 41 American elms, the forest would seem more diverse to us and indeed would be more diverse. The greater the diversity, the greater the number and the more even the distribution of species.

Models Regarding Composition and Diversity

What causes the populations within a community to assemble together in the same place at the same time? Figure 24.1*c* shows that the number of species in a community increases as we move from the northern latitudes to the equator. We know that at the equator the weather is warmer and there is more precipitation. Could it be that community composition is controlled by abiotic factors? A species' range is based on its tolerance for such abiotic factors in the environment as temperature, light, water availability, salinity, and so forth. In order to determine a species' tolerance range, we plot the species' ability to survive and reproduce under a particular environmental condition. The result is a bell-shaped curve that shows the species' range for that abiotic factor (Fig. 24.2*a*). It's possible that species may assemble because their tolerance ranges simply overlap (Fig. 24.2*b*).

This view, advanced by H. L. Gleason in the early 1900s, is called the *individualistic model* because it suggests that each population in a community is there because its own particular abiotic requirements are met by a particular habitat. The individualistic model predicts that species will have independent distributions, and that the boundaries between communities will not be distinct from one another.

For many years, most ecologists supported the *interactive model* of community structure suggested by F. E. Clements in the early 1900s. Clements saw a community as the highest level of organization arising from cell to tissue, to organism, to populations, and finally, to a community. Just as the parts of an organism are dependent on one another, so species are dependent on biotic interactions such as its food source. Further, like an organism, a community remains stable because of homeostatic mechanisms. This theory predicts that the same species will reoccur in communities whose boundaries are distinct from one another.

The data collected by modern ecologists lend support to the individualistic model. For example, F. H. Talbot and colleagues built artificial reefs of a constant size out of cement building blocks and put them in a warm tropical lagoon a few meters from one another. Altogether, 42 species of fish colonized the artificial reefs, but the similarity of species between the reefs was only 32%. Moreover, Talbot found that there was a very high turnover from month to month. From one month to the next, 20–40% of species had been replaced by other species. Therefore, species composition seemed to depend on chance migrations. However, we know that certain animals are more likely to occur where they find their usual food source. Koala bears feed on eucalyptus leaves and are found only in regions where these plants are found. Plants, in turn, are found where temperature, moisture, and soil are most favorable to them. Most likely community structure is dependent on both abiotic and biotic factors.

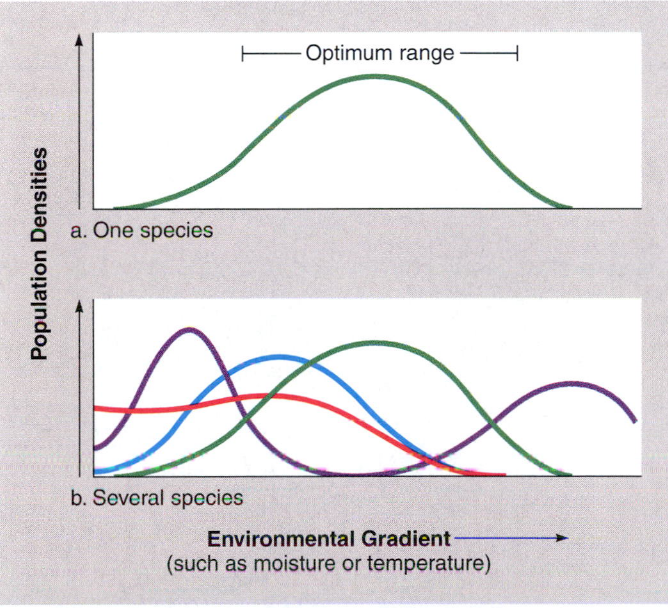

Figure 24.2 Species richness of communities. According to the individualistic hypothesis (a) each species is distributed along environmental gradients according to its own tolerance for abiotic factors, and (b) a community is believed by some to be an assemblage of species that happen to occupy the same area because of similar tolerances.

Community composition is probably dependent on both abiotic gradients (e.g., climate, inorganic nutrients) and biotic interactions (e.g., organic food source).

Island Biogeography

Would you expect larger coral reefs to have greater species richness (number of species) than smaller coral reefs? The area (space) occupied by a community can have a profound effect on its diversity. American ecologists Robert MacArthur and E. O. Wilson developed a general theory of *island biogeography* to explain and predict the effects of distance from the mainland and size of an island on community diversity.

Imagine two new islands that as yet contain no species at all. One of these islands is near the mainland, and one is far from the mainland. Which island will receive more immigrants from the mainland? Most likely the near one because it's easier for immigrants to get there (Fig. 24.3*a*).

Similarly, imagine two islands that differ in size. Which island will be able to support a greater number of species? The large one, because a greater amount of resources can support more populations (Fig. 24.3*b*). MacArthur and Wilson studied the diversity on many island chains including the West Indies, and discovered that species richness does correlate positively with island size. Conservationists note that the theory of island biogeography applies to their work because conserved land is like an island surrounded by farms, towns, and cities. The theory tells them that the larger the conserved area, the better the chance of preserving more species.

Is it possible to increase the amount of space without using more area? The *spatial heterogeneity model* holds that if the environment has *patches*, the greater the number of habitats, the greater the diversity. As gardeners, we are urged to create patches in our yards if we wish to attract more butterflies and birds! Stratification is one way to introduce patchiness. Just as a high-rise apartment building allows more human families to live in an area, so stratification within a community provides more and different types of living space for different species. Stratification of its canopy (treetops) is another reason why a tropical rain forest has many more species than a coniferous forest.

The fact that space is limited suggests that there must be an end point beyond which community richness cannot increase. Figure 24.3*c* suggests there will be an equilibrium point for four types of islands (i.e., near and large, near and small, far and large, far and small). Notice that the equilibrium point is highest for a large island that is near the mainland. An equilibrium is reached when the rate of species immigration matches the rate of species extinction. The equilibrium could be dynamic (new species keep on arriving, and new extinctions keep on occurring), or it could be that the composition of the community will remain steady unless disturbed. This is an idea that we will return to later in the chapter.

The theory of island biogeography says that distance from a population source and size affect species diversity. When immigration and extinction rates are equal an equilibrium in species diversity develops.

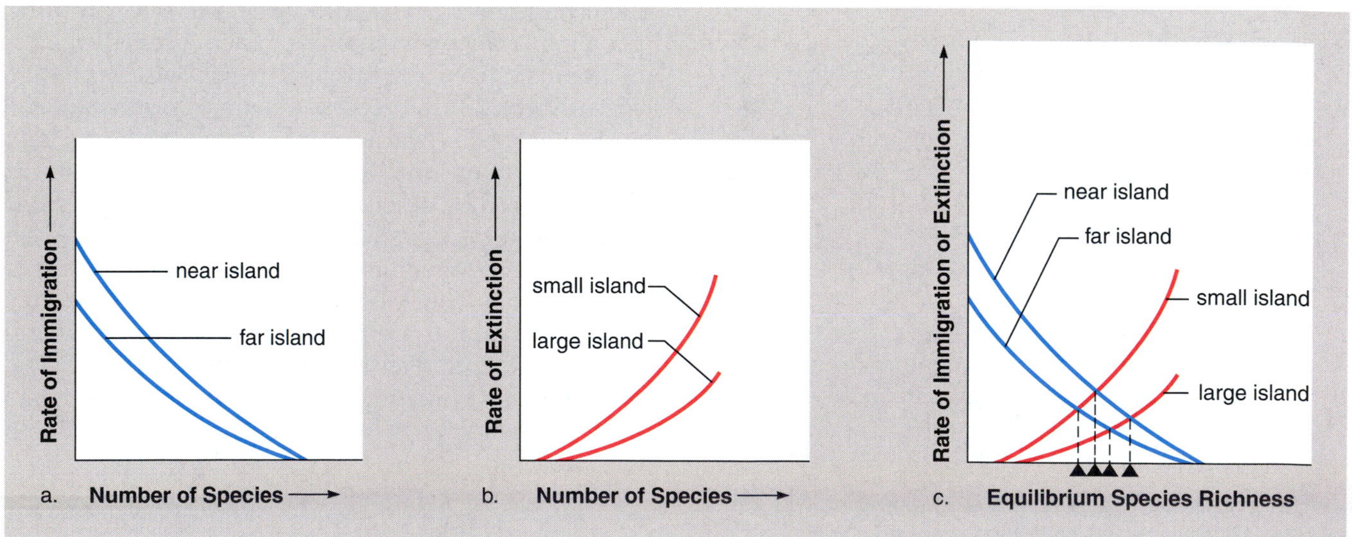

Figure 24.3 Theory of island biogeography.
a. An island that is near the mainland will have a higher immigration rate than an island that is far from the mainland. **b.** A small island will have a higher extinction rate than a large island. **c.** The balance between immigration and extinction for four possible types of islands. A large island that is near the mainland will have a higher equilibrium species richness than the other types of islands.

24.2 Structure of the Community

Competition for resources, predator-prey, parasite-host, and other types of interactions integrate species into a system of dynamic interacting populations (Table 24.1). Competition for limited resources between two species has a negative effect on the abundance of both species. In predation, one animal eats another called the prey, and in parasitism one individual obtains nutrients from another, called the host. Predation and parasitism are expected to increase the abundance of the predator and parasite at the expense of the prey and host abundance, respectively. In commensalism one species is benefited, but the other is not harmed. As an example, commensalism often occurs when one species provides a home or transportation for another. In mutualism, two species help one another and both species are benefited. In Table 24.1 a plus sign (+) is used whenever the fitness of the individual in the interaction is expected to increase, and a minus sign (−) is used whenever the fitness of the individual is expected to decrease. These effects are manifested by a change in the abundance of the species as indicated by the table.

Table 24.1

Species Interactions

Interaction	Expected Outcome
Predation (+ −)	Abundance of predator increases and abundance of prey decreases.
Symbiotic Relationships	
Parasitism (+ −)	Abundance of parasite increases and abundance of host decreases.
Competition (− −)	Abundance of both species decreases.
Commensalism (+ 0)	Abundance of one species increases, and the other is not affected.
Mutualism (+ +)	Abundance of both species increases.

Habitat and Ecological Niche

Each species occupies a particular position in the community, both in a spatial sense (where it lives) and in a functional sense (what role it plays). An organism lives and reproduces in its **habitat.** The habitat of an organism might be the forest floor, a swift stream, or the ocean's edge. The **ecological niche** of an organism is the role it plays in its community, including its habitat and its interactions with other organisms. The niche includes the resources an organism uses to meet its energy, nutrient, and survival demands. For a backswimmer, home is a pond or lake where it eats other insects. The pond must contain vegetation where the backswimmer can hide from its predators such as fish and birds. On the other hand, the water must be clear enough for the backswimmer to see its prey and warm enough for it to be in active pursuit. Since it's difficult to study the total niche of an organism, some observations focus on a certain aspect of an organism's niche, as with the birds featured in Figure 24.4.

Because an organism's niche is affected both by abiotic factors (such as climate and habitat) and biotic factors (such as competitors, parasites, and predators present in the habitat), ecologists distinguish between the fundamental and realized niches. An organism's *fundamental niche* comprises all conditions under which an organism can potentially survive and reproduce; the *realized niche* is the set of conditions under which it exists in nature.

Habitat is where an organism lives, and ecological niche is the role an organism plays in its community, including its habitat and its interactions with other organisms.

Flamingos feed on small mollusks, crustacea and vegetable matter strained from mud pumped through their bills by their powerful tongues.

Dabbling ducks feed by tipping, tail up, to reach aquatic plants, seeds, snails, and insects.

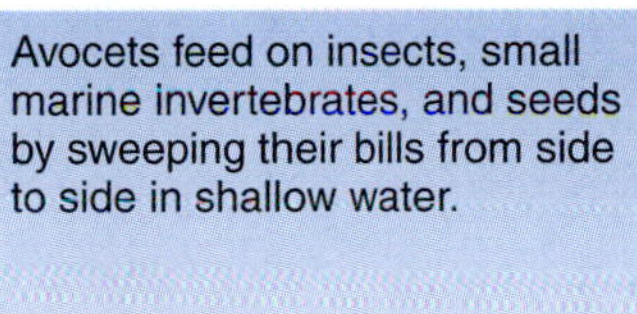

Avocets feed on insects, small marine invertebrates, and seeds by sweeping their bills from side to side in shallow water.

Oystercatchers pry open bivalve shells with their knifelike bills and probe sand for worms and crabs.

Plovers dart around on beaches and grasslands hunting for insects and small invertebrates.

Figure 24.4 Feeding niches for wading birds.
Flamingos feed in deeper water by filter feeding; dabbling ducks feed in shallower areas by upending; avocets feed by sifting. Oystercatchers and plovers are restricted to the shallows by their shorter legs.

Competition Between Populations

Interspecific competition occurs when members of different species try to utilize a resource (like light, space, or nutrients) that is in limited supply. If the resource is not in limited supply, there is no competition. In the 1920s, Lotka and Volterra developed mathematical formulas that predicted competition can lead to the predominance of one species and the virtual elimination of the other. Later, this was demonstrated in the laboratory by G. F. Gause, who grew two species of *Paramecium* in one test tube containing a fixed amount of bacterial food. Although each population survived when grown separately, only one survived when they were grown together (Fig. 24.5). The successful *Paramecium* population had a higher biotic potential than the unsuccessful population. After observing the outcome of similar other experiments in the laboratory, ecologists formulated the **competitive exclusion principle,** which states that no two species can occupy the same niche at the same time.

What does it take to have different ecological niches so that extinction of one species is avoided? In another laboratory experiment, Gause found that two species of *Paramecium* did continue to occupy the same tube when one species fed on bacteria at the bottom of the tube and the other fed on bacteria suspended in solution. Under these circumstances, **resource partitioning** decreased competition between the two species. What could have been one niche became two niches because of a divergence of behavior in this case.

It's possible to observe niche partitioning in nature. When three species of ground finches of the Galápagos Islands occur on separate islands, their bills tend to be the same intermediate size, enabling each to feed on a wider range of seeds. Where they co-occur, selection has favored divergence in beak size because the size of the beak affects the kinds of seeds that can be eaten. In other words, competition has led to resource partitioning. The tendency for characteristics to be more divergent when populations belong to the same community than when they are isolated is termed **character displacement.** Character displacement is often used as evidence that competition and resource partitioning have taken place (Fig. 24.6).

The niche partitioning we observe in nature can be very subtle. Five different species of warblers that occur in North American forests are all about the same size and all feed on budworms, a type of caterpillar found on spruce

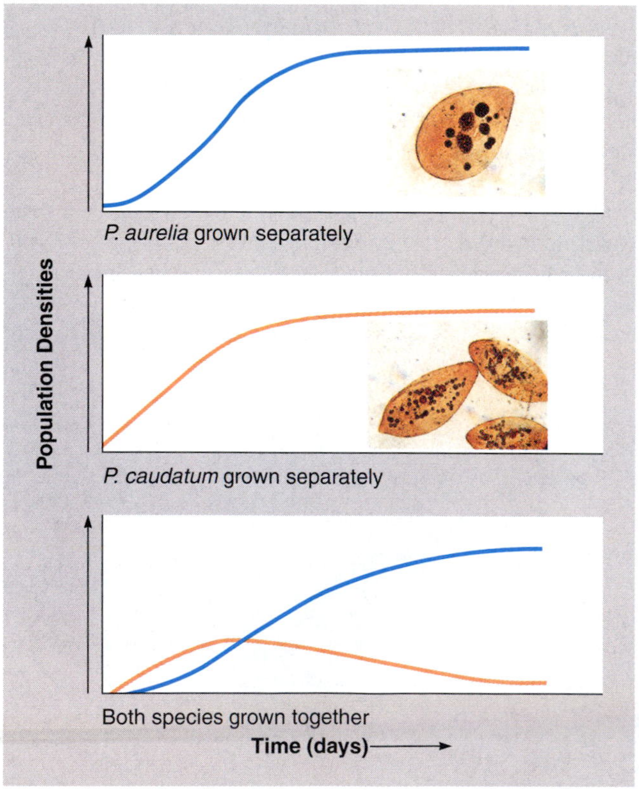

Population Densities

P. aurelia grown separately

P. caudatum grown separately

Both species grown together

Time (days)

Figure 24.5 Competition between two laboratory populations of *Paramecium.*
When grown alone in pure culture, *Paramecium caudatum* and *Paramecium aurelia* exhibit sigmoidal growth. When the two species are grown together in mixed culture, *P. aurelia* is the better competitor, and *P. caudatum* dies out.

Source: Data from G. F. Gause, *The Struggle for Existence,* 1934, Williams & Wilkins Company, Baltimore, MD.

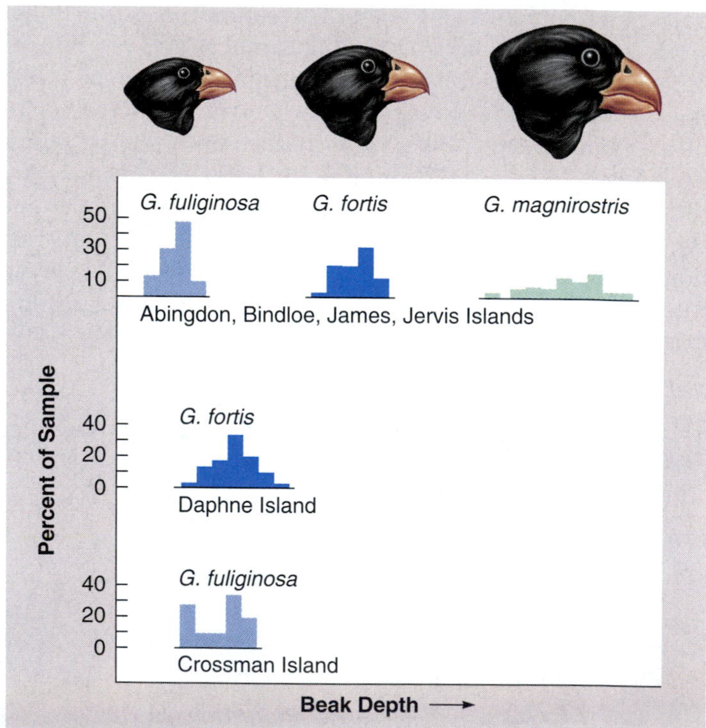

Figure 24.6 Character displacement in finches on Galápagos Islands.
When *Geospiza fuliginosa*, *G. fortis*, and *G. magnirostris* are on the same island, their beak sizes are appropriate to eating small-, medium-, and large-size seeds. When *G. fortis* and *G. fuliginosa* are on separate islands, their beaks have the same intermediate size, which allows them to eat seeds that vary in size.

Figure 24.7 Niche partitioning among five species of coexisting warblers.
The diagrams represent spruce trees. The time each species spent in various portions of the trees was determined; each species spent more than half its time in the blue regions.

trees. In a very famous study, Robert MacArthur recorded the length of time each of five species of warblers spent in different regions of spruce canopies to determine where each species did most of its feeding (Fig. 24.7). He discovered that each type of bird primarily used different parts of the tree canopy and in that way had a different niche.

As another example, consider that swallows, swifts, and martins all eat flying insects and parachuting spiders. These birds even frequently fly in mixed flocks. But each type of bird has different nesting sites and migrate at a slightly different time of year. Therefore they are not competing for the same food source when they are feeding their young.

In all these cases of niche partitioning, we have merely supposed that what we observe today is due to competition in the past. Some ecologists are fond of saying that in doing so we have invoked the "ghosts of competition past." Are there any instances in which competition has actually been observed? Perhaps we have one example. Joseph Connell has studied the distribution of barnacles on the Scottish coast, where a small barnacle (*Chthamalus stellatus*) lives on the high part of the intertidal zone, and a large barnacle (*Balanus balanoides*) lives on the lower part (Fig. 24.8). Free-swimming larvae of both species attach themselves to rocks at any point in the intertidal zone, where they develop into the sessile adult forms. In the lower zone, the large *Balanus* barnacles seem to either force the smaller *Chthamalus* individuals off the rocks or grow over them. To test his observation, Connell removed the larger barnacles and found that the smaller barnacles grew equally well on all parts of the rock. The entire intertidal zone is the fundamental niche for *Chthalamus*, but competition for space is restricting the range of *Chthalamus* on the rocks. *Chthalamus* is more resistant to drying out than is *Balanus*; therefore, it has an advantage that permits it to grow in the upper intertidal zone. The upper intertidal zone becomes the realized niche for *Chthalamus*.

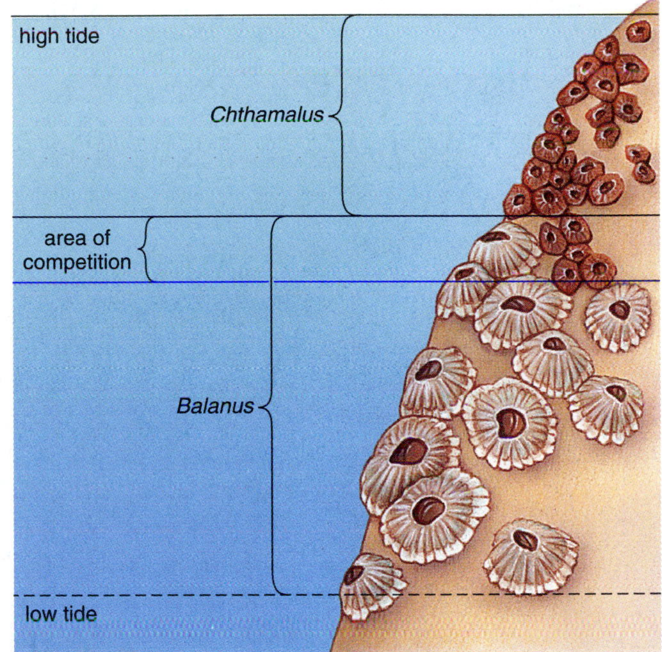

Figure 24.8 Competition between two species of barnacles.
Competition prevents two species of barnacles from occupying as much of the intertidal zone as possible. Both exist in the area of competition between *Chthamalus* and *Balanus*. Above this area only *Chthamalus* survives, and below it only *Balanus* survives.

Interspecific competition may result in niche partitioning. When similar species seem to be occupying the same ecological niche, it is usually possible to find differences that indicate niche partitioning has occurred.

Predator-Prey Interactions

Predation occurs when one living organism, called the **predator,** feeds on another, called the **prey.** In its broadest sense, predaceous consumers include not only animals like lions that kill zebra, but also filter-feeding blue whales that strain krill from ocean waters, parasitic ticks that suck blood from victims, and even herbivorous deer that browse on trees and bushes.

Predator-Prey Population Dynamics

Do predators reduce the population density of prey? On the face of it, you would probably hypothesize that they do, and your hypothesis would certainly be supported by a laboratory study done by Gause. When Gause reared the protozoan *Paramecium caudatum* (prey) and *Didinium nasutum* (predator) together in a culture medium, *Didinium* ate all the *Paramecium* and then died of starvation (Fig. 24.9). In nature, we can find a similar example. When a gardener brought prickly-pear cactus to Australia from South America, the cactus spread out of control until millions of acres were covered with nothing but cacti. The cacti were brought under control when a moth from South America, whose caterpillar feeds only on the cactus, was introduced. Now both cactus and moth are found at greatly reduced densities in Australia.

This raises an interesting point: the population density of the predator can be affected by the prevalence of the prey. In other words, the predator-prey relationship is actually a two-way street. In that context, consider that at first the biotic potential (maximum reproductive rate) of the prickley-pear cactus was maximized, but environmental resistance (factors that oppose biotic potential) increased after the moth was introduced. And the biotic potential of the moth was maximized when it was first introduced, but the carrying capacity decreased after its food supply was diminished.

In nature, presence of predators can decrease prey densities and vice versa.

There are mathematical formulas that predict a cycling of predator and prey populations instead of a steady state:

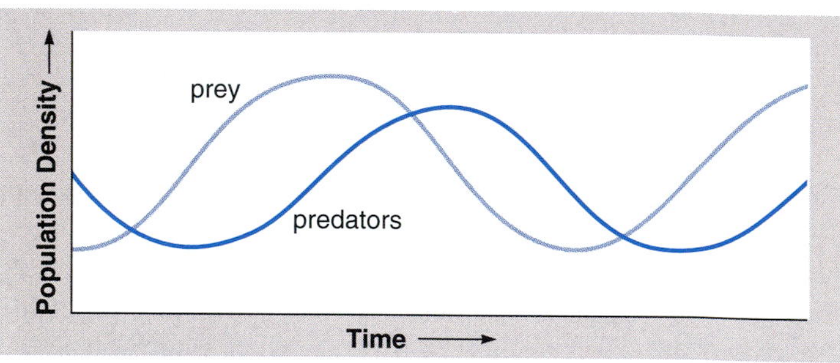

Notice that the predator population is smaller than the prey population, and that it lags behind the prey population. What might cause a cycling like this? We can certainly observe that as the prey population increases, the predator population also increases, most likely because more food has become available. At this point, there are at least two possibilities to account for the reduction in population densities that follow: (1) perhaps the biotic potential (reproductive rate) of the predator is so great, its increased numbers overconsume the prey, and then as the prey population declines, so does the predator population; or (2) perhaps the biotic potential of the prey is unable to keep pace, and the prey population overshoots the carrying capacity and suffers a crash. Now the predator population follows suit because of a lack of food. In either case (1 or 2), the result will be a series of peaks and valleys with the predator population lagging slightly behind the prey.

The famous case of predator/prey cycles occurs between the snowshoe hare and the Canadian lynx, a type of small cat (Fig. 24.10). The snowshoe hare is a common herbivore in the coniferous forests of North America, where it feeds on terminal twigs of various shrubs and small trees.

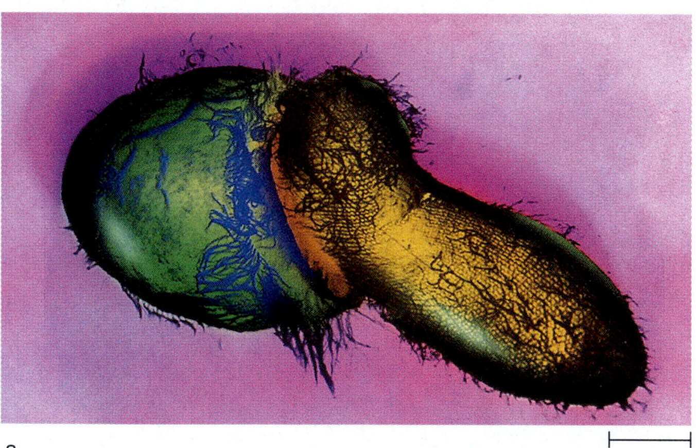

a.

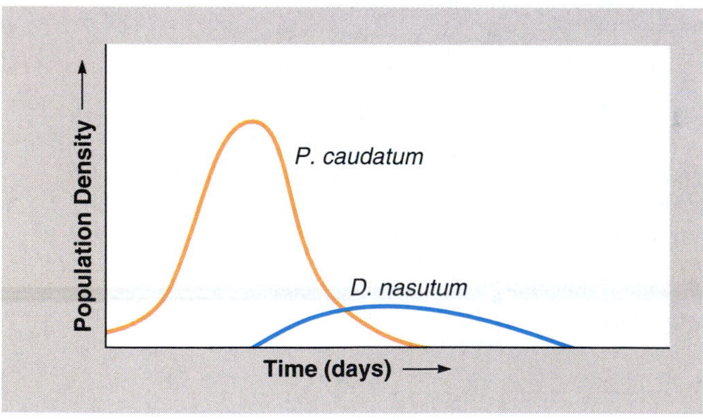

b.

Figure 24.9 Predator-prey interaction between *Paramecium caudatum* and *Didinium nasutum*.
a. Scanning electron micrograph of *Didinium* engulfing *Paramecium*.
b. In an oat medium, *Didinium* ate all the *Paramecium* and then died out.

Source: Data from G. F. Gause, *The Struggle for Existence*, 1934, Williams & Wilkins Company, Baltimore, MD.

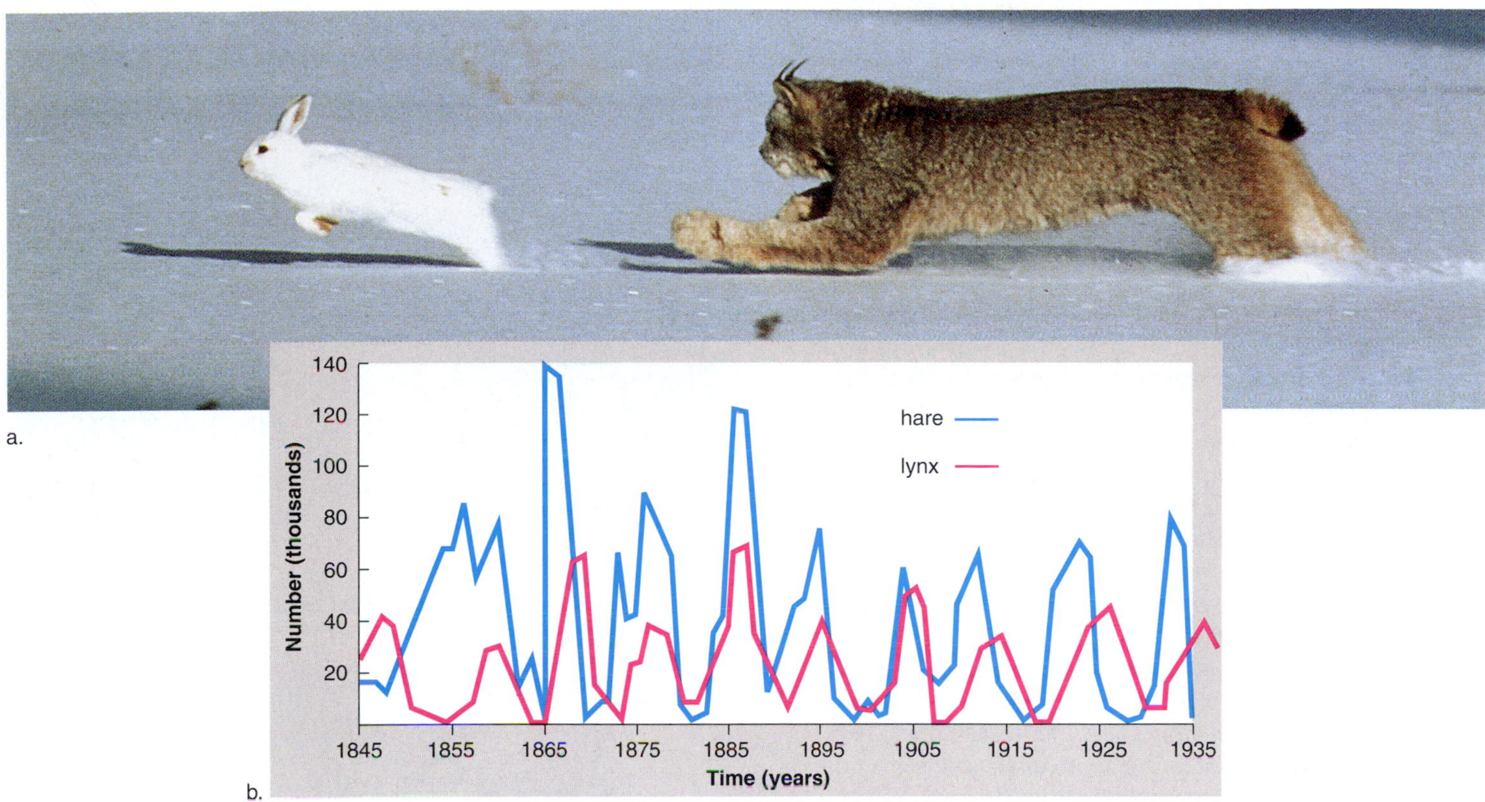

Figure 24.10 Predator-prey interaction between a lynx and a snowshoe hare.
a. A Canadian lynx (*Felis canadensis*) is a solitary predator. A long, strong forelimb with sharp claws grabs its main prey, the snowshoe hare (*Lepus americanus*). The lynx lives in northern forests. Its brownish-gray coat blends in well against tree trunks, and its long legs enable it to walk through deep snow. **b.** The number of pelts received yearly by the Hudson Bay Company for almost 100 years shows a pattern of ten-year cycles in population densities. The snowshoe hare population reaches a peak abundance before that of the lynx by a year or more.

The Canadian lynx feeds on snowshoe hares but also on ruffed grouse and spruce grouse, two types of birds. Investigators at first assumed that the lynx brings about a decline of the hare population and this accounts for the cycling seen in Fig. 24.10. But others noted that the decline in snowshoe hare abundance was accompanied by low growth and reproductive rates that could be signs of a food shortage.

To test whether the predator's (lynx) or the prey's (hare) limited food supply was causing the observed cycling between the two populations, three experiments were done:

1. To do away with both possibilities, a hare population was given a constant supply of food, and predators were excluded. Under these circumstances, the cycling of the hare population ceased.
2. To test the predator effect alone, a hare population was given a constant supply of food, but predators were not excluded. Under these circumstances, the cycling of both populations continued.
3. To test the food effect alone, predators were excluded, but no food was added to the environment of a hare population. Under these circumstances, the cycling of the hare population continued.

These results suggest that a hare-food cycle and a predator-hare cycle have combined to produce the cycling observed in Figure 24.10. It's interesting to note that the population densities of the grouse populations also cycle, perhaps because the lynx switches to this food source when the hare population declines. Our discussion points out that predators and prey do not normally exist as simple, two-species systems, and therefore abundance patterns should be viewed with the complete community in mind.

Interactions between predator/prey and between prey and its own food source can interact to produce complex cycles.

Prey Defenses and Other Interactions

Prey defenses are mechanisms that thwart the possibility of being eaten by a predator. Plants have defenses that discourage herbivores. The sharp spines of the cactus, the pointed leaves of the holly, and the tough, leathery leaves of oak trees make the plant less desirable as a food source. Plants even produce poisonous chemicals, some of which interfere with the normal metabolism of an adult insect, and others that act as hormone analogues that interfere with the development of insect larvae. Some insects can still inhabit trees that produce toxins because the insects have either evolved detoxifying enzymes or a method of holding the toxin within their bodies so it is not harmful to them.

Among animals we find a number of ways that predators fool their prey and prey avoid capture by predators. The anglerfish puts camouflage to work both as a prey defense and predator mechanism. The mottled skin, grotesque jaw fringe, and flat body allow it to blend into the ocean floor where it waits, while a fleshy wormlike appendage atop its head moves slowly back and forth (Fig. 24.11). When a fish draws near, the anglerfish opens its mouth and sucks in its prey. This anglerfish ranges from Newfoundland to North Carolina and appears in seafood stores and menus as "monkfish."

Many examples of protective **camouflage** are known: stick caterpillars look like twigs; katydids look like sprouting green leaves; and some moths resemble the barks of trees. Normally pepper moths are a light color. But in industrial areas where the tree trunks have turned dark due to

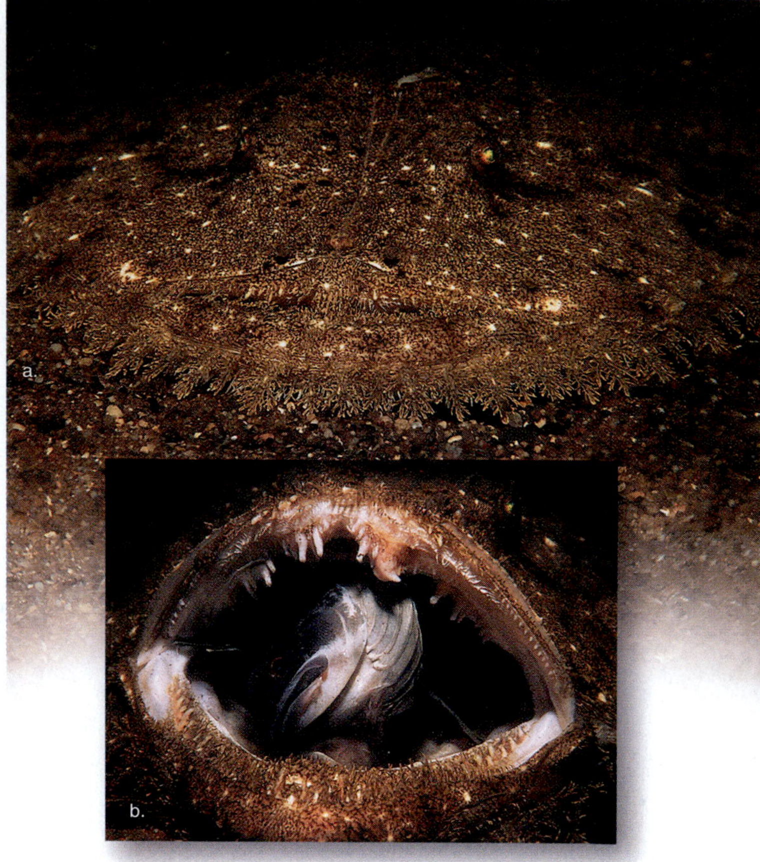

Figure 24.11 **Camouflage in the anglerfish.**
a. A wormlike appendage on the head of an anglerfish (*Lophius americanus*) lures possible prey. **b.** Then it opens its cavernous mouth and devours its prey like this black seabass.

Figure 24.12 Antipredator defenses.
a. The skin secretions of dart-poison frogs are so poisonous that they were used by natives to make their arrows instant lethal weapons. The coloration of these frogs warns others to beware. **b.** The South American lantern fly has a large false head that resembles that of an alligator. This may frighten a predator into thinking it is facing a dangerous animal.

pollution, the pepper moth populations have mostly dark-colored members. This change in population statistics is often used as an example of natural selection with predatory birds being the selective agent. The birds capture the more visible light-colored moths on dark tree trunks, and the dark-colored ones remain to contribute a greater number of offspring to the next generation.

Causing harm or fright is another possible prey defense. As a warning to possible predators, dart-poison frogs are brightly colored (Fig. 24.12*a*). The South American lantern fly has a large false head that resembles that of an alligator (Fig. 24.12*b*). Many moths have eyelike spots on their underwings that they can flash to startle a bird long enough to escape. The porcupine sends out arrowlike quills with barbs that dig into the predator's flesh and penetrate even deeper as the enemy struggles after being struck. In the meantime, the porcupine runs away.

Association with other prey may sometimes help avoid capture. Flocks of birds, schools of fish, and herds of mammals stick together as protection against predators. Baboons who detect predators visually, and antelopes who detect predators by smell, sometimes forage together, giving double protection against stealthy predators. The gazellelike springboks of southern Africa jump stiff-legged 8 to 12 feet into the air when alarmed. Such a jumble of shapes and motions might confuse an attacking lion, and the herd can escape.

When you think about it, many prey defenses involve not only structural but also behavioral adaptations. Those animals that use camouflage usually stay very still so as not to be detected; those that use fright must spring into action at the right time; and those that use other members as shields must stay together.

Mimicry

Warning coloration is a mechanism used by poisonous animals to prevent attack in the first place. The skin secretions of dart-poison frogs, such as the one shown in Figure 24.12*a*, are so poisonous that South American Indians used these secretions to turn their arrows into instant lethal weapons. **Mimicry** occurs when one species resembles another that possesses an overt antipredator defense. A mimic that lacks the defense of the organism it resembles is called a Batesian mimic (named for Henry Bates, who discovered it). In Figure 24.13, several insects are shown that resemble a wasp, but they lack the ability to sting. Once an animal experiences the defense of the model, it remembers the coloration and avoids all animals that look similar.

There are also examples of species that have the same defense and resemble each other. Among insects, many stinging insects—bees, wasps, and hornets—all have the familiar black and yellow color bands. Mimics that share the same protective defense are called Müllerian mimics after Fritz Müller who suggested that this, too, is a form of mimicry. Each of these Müllerian mimics reinforces the message that insects with this coloration can sting.

Just as with other prey defenses, behavior plays a role in mimicry. Mimicry works better if the mimic acts like the model. For example, beetles that resemble a wasp actively fly from place to place and spend most of their time in the same habitat as the wasp model. Their behavior makes them resemble a wasp to an even greater degree.

Prey escape predation by utilizing camouflage, fright, flocking together, warning coloration, and mimicry.

Figure 24.13 Mimicry among insects with yellow and black stripes. a. A yellow jacket (*Vespula*) and (b) a bumblebee (*Bombus*) are Müllerian mimics. They have a similar appearance, and they both use stinging as a defense. The other animals—(c) a flower fly (*Chrysotoxum*), (d) a longhorn beetle (*Strophiona*), and (e) a moth (*Aegeria*)—are Batesian mimics because they are incapable of stinging another animal, yet they have the same appearance as the yellow jacket.

a.

b.

c.

d.

e.

Symbiotic Relationships

Symbiosis refers to interactions in which there is a close relationship between members of two populations. Table 24.1 tells how these relationships affect the fitness of the individuals involved.

Parasitism

Parasitism is similar to predation in that an organism called a **parasite** derives nourishment from another called the **host** (just as the predator derives nourishment from its prey). Viruses, such as HIV, that reproduce inside human lymphocytes are always parasitic. Parasites occur in all kingdoms of life also. Bacteria (e.g., strep infection), protists (e.g., malaria), fungi (e.g., rusts and smuts), plants (e.g., mistletoe), and animals (e.g., leeches) all contain parasitic members. While small parasites can be endoparasites, larger ones are more likely to be ectoparasites that remain attached to the exterior of the body by means of specialized organs and appendages. The effects of parasites on the health of the host can range from a slight weakening effect to actually killing them over time. When host populations are at a high density, parasites readily spread from one host to the next, causing intense infestations and a subsequent decline in host density. Parasites that don't kill their host can still play a role in reducing the host's population density because an infected host is less fertile and becomes more susceptible to another cause of death.

In addition to nourishment, host organisms also provide their parasites with a place to live and reproduce, as well as a mechanism for dispersing offspring to new hosts. Many parasites have both a primary and secondary host. The secondary host may be a vector that transmits the parasite to the next primary host. Usually both hosts are required to complete the life cycle. The association between parasite and host is so intimate that parasites are usually specific and require only certain species as hosts.

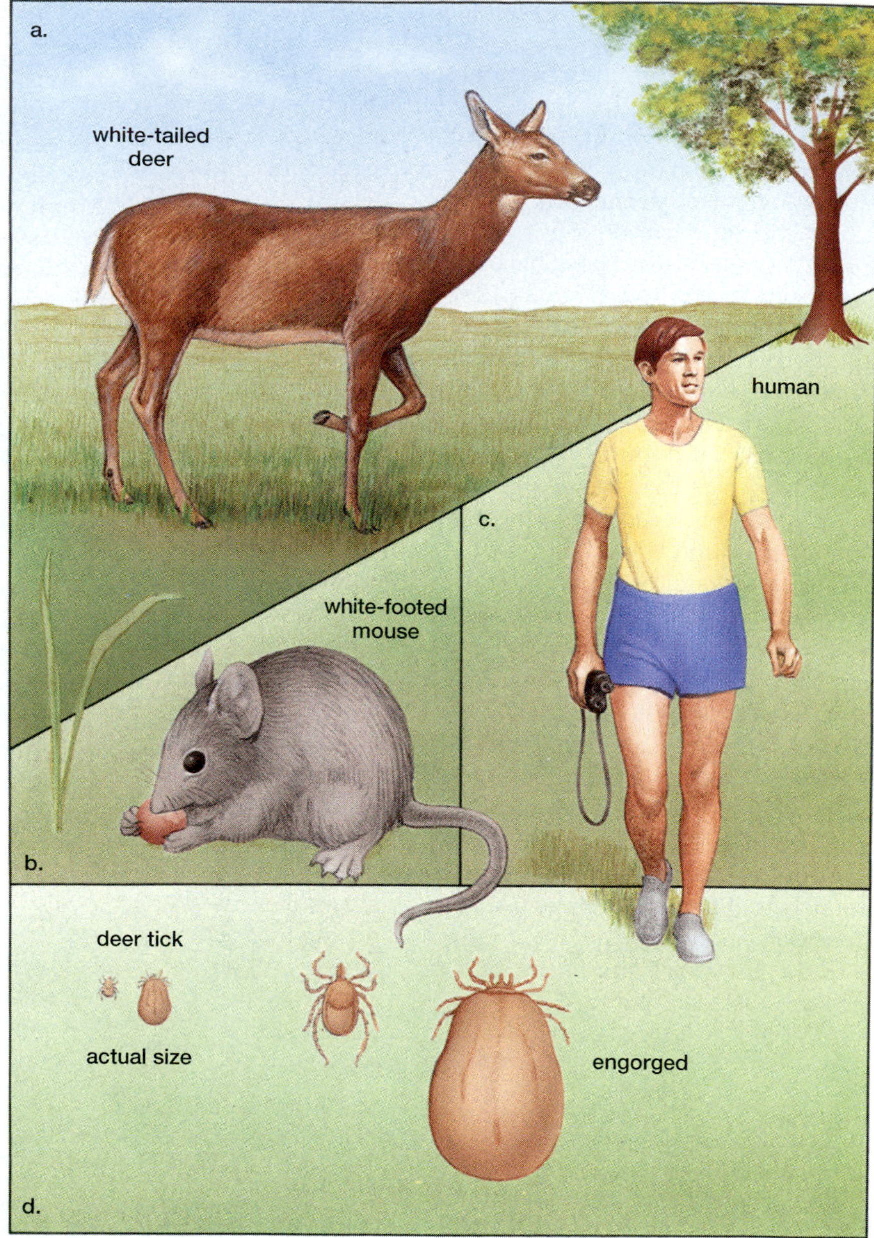

Figure 24.14 **The life cycle of a deer tick.**
a. Adult ticks feed and mate on white-tailed deer, which accounts for the common name of the ticks. Female adults lay their eggs in soil and then die. **b.** In the spring and summer, larvae feed mainly on white-footed mice which may be infected with the bacterium *Borrelia burgdorfei.* Larvae then overwinter. **c.** During the next summer, nymphs feed on white-footed mice or other animals, including humans. If a nymph is infected humans get Lyme disease. **d.** Deer tick before feeding and after feeding. Actual size is shown along with enlarged size.

As an example, we will consider the deer tick called *Ixodes dammini* in the east and *I. ricinus* in the west. Deer ticks are arthropods that go through a number of stages (egg, larva, nymph, adult). They are so named because adults feed and mate on white-tailed deer in the fall. The female lays her eggs on the ground, and when the eggs hatch in the spring they become larvae that feed primarily on white-footed mice. If a mouse is infected with the bacterium *Borrelia burgdorfei,* the larvae become infected also. The fed larvae overwinter and molt the next spring to become nymphs that can, by chance, take a blood meal from a human. It is at this time that the tick may pass the bacterium on to a human who subsequently comes down with Lyme disease characterized by arthritic-like symptoms. The fed nymphs develop into adults and the cycle begins again (Fig. 24.14).

Parasites are very different from predators. They take nourishment from their host but also use them as a habitat and a way to transmit themselves to the next host. Many complicated associations have evolved.

Interactions and Coevolution

Coevolution is present when two species adapt in response to selective pressure imposed by the other. Symbiosis (close association between two species), which includes parasitism, commensalism, and mutualism is especially prone to the process of coevolution. Flowers that have animal pollinators have features that attract them (see the reading on p. 704). As an example of this type of coevolution, a butterfly-pollinated flower is often a composite, containing many individual flowers; the broad expanse provides room for the butterfly to land, and the butterfly has a proboscis that it inserts into each flower in turn. In this case, coevolution is highly beneficial to the flower because the insect will carry pollen to other flowers of the same type.

Coevolution also occurs between predators and prey. For example, a predator like a cheetah sprints forward to catch its prey, and this behavior might be selective for those gazelles that are able to run away or jump high in the air (like a springbok) to avoid capture. The adaptation of the prey may very well put selective pressure on the predator for an adaptation to the prey's defense mechanism. Hence an "arms race" can develop. The process of coevolution has been studied in the cuckoo, a social parasite that reproduces at the expense of other birds by laying its eggs in their nests. It's a strange sight to see a small bird feeding a cuckoo nestling that is several times its size. How did this strange happening develop? Investigators discovered that in order to "trick" a host bird, the adult cuckoo has to (1) lay an egg that mimics the host's egg, (2) lay its egg very rapidly (only ten seconds are required) in the afternoon while the host is away from the nest, and (3) leave most of the eggs in the nest because hosts will desert a nest that has only one egg in it. (The cuckoo chick hatches first and is adapted to removing any other eggs in the nest [Fig. 24A].) It seems that the host birds might very well now evolve a way to distinguish the cuckoo from their own young.

Coevolution can take many forms. In the case of *Plasmodium,* a cause of malaria, the sexual portion of the life cycle occurs within mosquitoes (the vector), and the asexual portion occurs in humans. The human immune system uses surface proteins to detect pathogens, and *Plasmodium* has numerous genes for surface proteins. But it is capable of changing its surface proteins repeatedly, and in this way it stays one step ahead of the host's immune system. A similar capability of HIV has added to the difficulty of producing an AIDS vaccine.

The relationship between parasite and host can even include the ability of parasites to seemingly manipulate the behavior of their hosts in self-serving ways. Ants infected with the lance fluke (but not those uninfected) cling to blades of grass with their mouthparts. There the infested ants are eaten by grazing sheep, and the flukes are transmitted to the next host in their life cycle. Similarly, when snails of the genus *Succinea* are parasitized by worms of the genus *Leucochloridium,* they are eaten by birds. As the worms mature, they invade the snail's eyestalks, making them resemble edible caterpillars. Now the birds eat the snails, and the parasites release their eggs, which complete development inside the urinary tracts of birds.

It used to be thought that as host and parasite coevolved, each would become more tolerant of the other—if the opposite occurred, the parasite would soon run out of hosts. Parasites could first become commensal, or harmless to the host. Then, given enough time, the parasite and host might even become mutualists. The evolution of the eukaryotic cell by endosymbiosis is even predicated on the supposition that bacteria took up residence inside a larger cell, and then the parasite and cell became mutualists.

This argument is too teleological for some; after all, no organism is capable of "looking ahead" at its evolutionary fate. Rather, if an aggressive parasite can transmit more of itself in less time than a benign one, then aggressiveness would be favored by natural selection. On the other hand, other factors such as the life cycle of the host can determine whether aggressiveness is beneficial or not. For example, a benign parasite of newts will do better than an aggressive one. Why? Because newts take up solitary residence outside ponds in the forest for six years, and parasites have to wait that long before they are likely to meet up with another possible host.

Figure 24A Social parasitism by the Reed warbler-cuckoo. A cuckoo chick balances the eggs of its hosts on its back one by one, and heaves them from the nest. Its own egg (see inset) mimics and is accepted by the host as its own.

Commensalism

Commensalism is a symbiotic relationship between two species in which one species is benefited and the other is neither benefited nor harmed.

Instances are known in which one species provides a home and/or transportation for the other species. Barnacles that attach themselves to the backs of whales and the shells of horseshoe crabs are provided with both a home and transportation. Remoras are fishes that attach themselves to the bellies of sharks by means of a modified dorsal fin acting as a suction cup. The remoras obtain a free ride and also feed on the remains of the shark's meals. It's possible though that the movement of the host is impeded by the presence of the attached animals, and therefore some are reluctant to use these as instances of commensalism.

Epiphytes grow in the branches of trees, where they receive light, but they take no nourishment from the trees. Instead, their roots obtain nutrients and water from the air. Clownfishes live within the waving mass of tentacles of sea anemones (Fig. 24.15). Because most fishes avoid the poisonous tentacles of the anemones, clownfishes are protected from predators. Perhaps this relationship borders on mutualism, because the clownfishes actually may attract other fishes on which the anemone can feed.

Cases of commensalism may turn out, on closer examination, to be causes of either mutualism or parasitism. Cattle egrets benefit from grazing near cattle because the cattle flush insects and other animals from the vegetation as they graze. Under prey defenses, it was noted that baboons and antelopes normally forage together for added protection. Is this, then, a case of commensalism? To some it seems like wasted effort to try to classify symbiotic relationships into the three categories of parasitism, commensalism, and mutualism. The amount of harm or good two species seem to do one another is dependent on what the investigator chooses to measure.

Figure 24.15 Clownfish among sea anemone's tentacles. If the clownfish (*Premnas biaculeztus*) performs no service for the sea anemone, this association is a case of commensalism. If the clownfish lures other fish to be eaten by the sea anemone, this is a case of mutualism.

Mutualism

Mutualism is a symbiotic relationship in which both members of the association benefit. Mutualistic relationships often help organisms obtain food or avoid predation.

As with parasitism, it's possible to find examples of mutualism in all kingdoms of life. Bacteria that reside in the human intestinal tract are provided with food, but they also provide us with vitamins, molecules we are unable to synthesize for ourselves. Termites would not even be able to digest wood if it were not for the protozoa that inhabit their intestinal tract. These organisms digest cellulose, which termites cannot. Mycorrhizae are symbiotic associations between the roots of plants and fungal hyphae. It is now known that the roots of most plants form these relationships. The hyphae increase the solubility of minerals in the soil, improve the uptake of nutrients for the plant, protect the plant's roots against pathogens, and produce plant growth hormones. In return the fungus obtains carbohydrates from the plant.

Most mutualistic relations need not be equally beneficial to both species. The mutualistic relationship between plants and their animal pollinators has already been mentioned. We can imagine that the relationship began when herbivores feasted on pollen. The provision of nectar may have spared the pollen and at the same time allowed the animal to become an instrument of pollination. Lichens can grow on rocks because their fungal member conserves water and leaches minerals that are provided to the algal partner, which photosynthesizes and provides organic food for both populations. If this were a true mutualistic relationship, each partner would do poorly when grown alone. The algae seem to do fine without the fungus, which suffers when grown alone. For that reason, it's been suggested that the fungus is parasitic at least to a degree on the algae.

Ants form mutualistic relationships with both plants and insects. In tropical America, the bullhorn acacia tree is adapted to provide a home for ants of the species *Pseudomyrmex ferruginea* (Fig. 24.16). Unlike other acacias, this species has swollen thorns with a hollow interior, where ant larvae can grow and develop. In addition to housing the ants, the acacias provide them with food. The ants feed from nectaries at the base of the leaves and eat fat and protein-containing nodules called Beltian bodies, which are found at the tips of leaves. The ants constantly protect the plant from herbivores and other plants that might shade it because, unlike other ants, they are active 24 hours a day. Indeed, when the ants on experimental trees were poisoned, the trees died. Other types of trees, such as those in the genus *Croton,* also have nectaries for ants—the very same ants that form a mutualistic relationship with the butterfly *Thisbe irenea,* whose caterpillars feed on *Croton* saplings. *Thisbe* caterpillars have nectary organs that offer nourishment to ants, keeping them nearby. The caterpillar also exploits the social behavior of ants by releasing chemicals that normally cause ants to defend the ant colony against predators. The end result is that caterpillars are

a.

b.

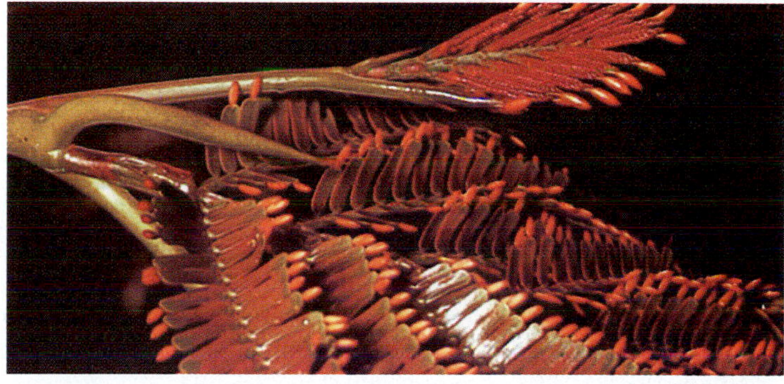

c.

Figure 24.16 Mutualism between the bullhorn acacia tree and ants.
The bullhorn acacia tree (*Acacia*) is adapted to provide nourishment for ants (*Pseudomyrmex ferruginea*). **a.** The thorns are hollow and the ants live inside. **b.** The base of leaves have nectaries (openings) where ants can feed. **c.** Leaves of the bullhorn acacia have Beltian bodies at the tips that ants harvest for larval food.

protected while they feast on the trees. In fact, the caterpillars, besides eating the leaves, feed from the ant nectaries on the trees!

Cleaning symbiosis is a symbiotic relationship in which the individual being cleaned is often a vertebrate (Fig. 24.17). Crustaceans, fish, and birds act as cleaners and are associated with a variety of vertebrate clients. Large fish in coral reefs line up at cleaning stations and wait their turn to be cleaned by small fish that even enter the mouths of the large fish. Whether cleaning symbiosis is an example of mutualism has been questioned because of the lack of experimental data. If clients respond to tactile stimuli by remaining immobile while cleaners pick at them, then cleaners may be exploiting this response by feeding on host tissues as well as on ectoparasites. Still, it could be that while stimulation may be the immediate cause of client behavior, cleaning could ultimately lead to net gains in client fitness.

> Symbiotic relationships do occur between species, but it may be too simplistic to divide them into parasitic, commensalistic, and mutualistic relationships.

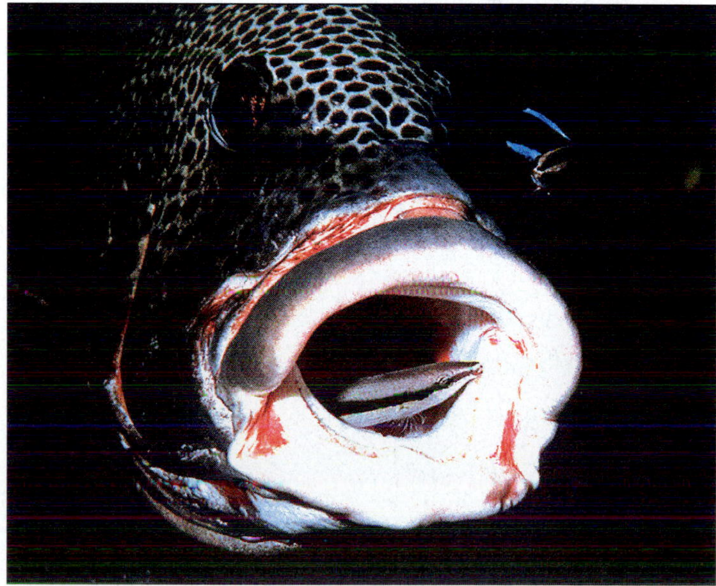

Figure 24.17 Cleaning symbiosis.
A cleaner wrasse (*Labroides dimidiatus*) in the mouth of a spotted sweetlip (*Plectorhincus chaetodontoides*) is feeding off parasites. Does this association improve the health of the sweetlip, or is the sweetlip being exploited? Investigation is under way.

24.3 Community Development

Each community has a history that can be surveyed over a short time period or even over geological time. We know that the distribution of life has been influenced by dynamic changes occurring during the history of earth. We have previously discussed how continental drifting contributed to various mass extinctions that have occurred in the past. When the continents joined to form the supercontinent Pangaea, many forms of marine life became extinct. Or when the continents drifted toward the poles, immense glaciers drew water from the oceans and even chilled once-tropical lands. During an Ice Age, glaciers moved southward and then, in between Ice Ages, glaciers retreated, changing the environment and allowing life to colonize the land once again. After time, complex communities came into being.

Many ecologists, however, try to observe changes as they occur during their own lifetime.

Ecological Succession

Communities are subject to disturbances that can range in severity from a storm blowing down a patch of trees to a beaver damming a pond to a volcanic eruption. We know from observation that following these disturbances, we'll see changes in the plant and animal communities over time; often, we'll wind up with the same kind of community with which we started.

Ecological succession is a change within a community following a disturbance. *Primary succession* occurs in areas where there is no soil formation such as following a volcanic eruption or a glacial retreat. *Secondary succession* begins in areas where soil is present, as when a cultivated field like a cornfield in New Jersey returns to a natural state (Fig. 24.18). Notice that we roughly observe a change from grasses to shrubs to a mixture of shrubs and trees.

a.

b.

c.

d.

e.

Figure 24.18 **Secondary succession.**
This is an example of secondary succession on the east coast of the United States, in New Jersey. **a.** During the first year, only the remains of corn plants are seen. **b.** During the second year, wild grasses have invaded the area. **c.** By the fifth year, the grasses look more mature and sedges have joined them. **d.** During the tenth year, there are goldenrod plants, shrubs (blackberry), and Eastern juniper trees. **e.** After twenty years, the juniper trees are mature and there are also birch and maple trees in addition to the blackberry shrubs.

The first species to begin secondary succession are called **pioneer species,** that is, plants that are invaders of disturbed areas, and then it progresses through a series of stages that are also described in Figure 24.19. Again we observe a series that begins with grasses and proceeds from shrub stages to a mixture of shrubs and trees, until finally there are only trees. Ecologists have tried to determine what the processes and mechanisms are by which the changes described in Figures 24.18 and 24.19 take place. And do these processes always have the same "end point" of community composition and diversity?

Theories About Succession

In 1916, F. E. Clements proposed the *climax-pattern model* of succession, and said that succession in a particular area will always lead to the same type of community, which he called a **climax community.** He believed that climate, in particular, determined whether a desert, a type of grassland, or a particular type of forest results. This is the reason, he said, that there is a coniferous forest in northern latitudes, a deciduous forest in temperate zones, and a tropical rain forest in the tropics. Secondarily, he believed that soil conditions might also affect the results. Shallow, dry soil might produce a grassland where a forest is expected, or the rich soil of a river bank might produce a woodland where a prairie is expected.

Further, Clements believed that each stage facilitated the invasion and replacement by organisms of the next stage. Shrubs can't grow on dunes until dune grass has caused soil to develop. Similarly, in the example given in Figure 24.19, shrubs can't arrive until grasses have made the soil suitable to them. Each successive community prepares the way for the next, so that grass-shrub-forest development occurs in a sequential way.

Aside from this *facilitation model*, there is also an *inhibition model*. The model predicts that colonists hold on to their space and inhibit the growth of other plants until the colonists die or are damaged. Still another model is called a tolerance model. The *tolerance model* predicts that different types of plants can colonize an area at the same time. Sheer chance determines which seeds arrive first, and successional stages may simply reflect the length of time it takes species to mature. This alone could account for the herb-shrub-forest development one often sees (Fig. 24.19). The length of time it takes for trees to develop might simply give the impression that there is a recognizable series of plant communities from the simple to the complex. In reality, the models we have mentioned are not mutually exclusive and succession is probably a multiple complex process.

Although it may not have been apparent to early ecologists, we now recognize that the most outstanding characteristic of natural communities is their dynamic nature. Also, it seems obvious to us now that the most complex communities most likely consist of habitat patches that are at various stages of succession. Each successional stage has its own mix of plants and animals, and if a sample of all stages is present, community diversity would be greatest. Further, we do not know if succession does continue to certain end points, because the process may not be complete anywhere on the face of the earth.

Ecological succession that occurs after a disturbance probably involves complex processes and the end result cannot be foretold.

Figure 24.19 Secondary succession in a forest.
In secondary succession in a large conifer plantation in central New York State, certain species are common to particular stages. However, the process of regrowth shows approximately the same stages as secondary succession from a cornfield (see Fig. 24.18).

24.4 Community Biodiversity

Community stability can be recognized in three ways: persistence through time, resistance to change, and recovery once a disturbance has occurred. Persistence occurs when, say, a forest remains just about the same year after year. A forest resists change when its trees are able to regrow their leaves after an insect infestation. And a chaparral community shows resilience when it quickly returns to its normal state after a fire.

The Intermediate Disturbance Hypothesis

The *intermediate disturbance hypothesis* states that a moderate amount of disturbance is required for a high degree of community diversity (Fig. 24.20). Fire, wind, moving water, and severe weather changes are possible abiotic factors that cause a disturbance. For example, fires promote understory plant diversity by preventing longer-lived shrub species from out-competing annuals. In a tropical rain forest, there are a great number of different types of trees, each with their own particular ecological niche requirements. This suggests that the environment is cut up into many small patches, each one slightly different from the other. As long as disturbances are small enough to affect one type of patch and not another, overall stability might still be maintained.

If widespread disturbances occur frequently, diversity is expected to be limited. The community will be dominated by species with a rapid growth rate, a short life span, and a strong ability to colonize disturbed areas. (Refer to Figure 23.13 and note that these attributes characterize *r*-strategists.) If disturbances are less widespread and less frequent, other types of species with slower growth rates and a longer life span will also be able to compete. (Refer to Figure 23.13 and note these are the attributes of *K*-strategists.) Under these circumstances, diversity will be the greatest.

It is possible for a disturbance to be so extensive, that the community never returns to its original state. Archaeological remains suggest that hundreds of square kilometers were cleared and cultivated by the Maya at Tikal from A.D. 300 to 900. This civilization collapsed for some unknown reason, and then the forest began to regrow. Even today, plant diversity in the community is low, and many of the common tree species are ones known to have been prized by the Maya for wood or for edible fruits or nuts. The animal wildlife is largely restricted to those animals that use these food resources. Even after 1,200 years, the site does not have the expected composition for a tropical rain forest in that area.

The intermediate disturbance hypothesis suggests that environmental heterogeneity increases species richness.

Predation, Competition, and Biodiversity

In certain communities, predation by a particular species reduces competition and increases diversity. Robert Paine was among the first to show this possible effect of predation. He

a.

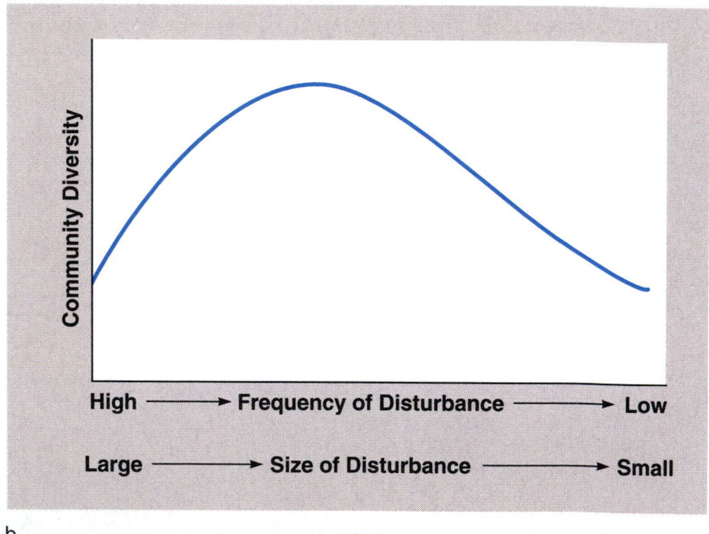

b.

Figure 24.20 The intermediate disturbance hypothesis. **a.** Disturbances like the Yellowstone fire in 1988 increase the diversity of a community by providing patches where annual/perennial plants can grow. **b.** This graph suggests that diversity is greatest when disturbances are intermediate in frequency and size.

removed the starfish *Pisaster* from test areas along the rocky intertidal zone on the west coast of North America but not from control areas. In the control areas, there was no change in the number of species, but in the test areas, the mussel *Mytilus* increased in number and excluded other invertebrates and algae from attachment sites on the rocks. The species richness declined drastically (Fig. 24.21). Predators that regulate competition and maintain the diversity of a community are now called **keystone predators.** The elephant is a possible keystone predator in Africa. Elephants feed on shrubs and small trees, causing woodland habitats to become open grassland. This is not beneficial to the elephant, which needs woody species in its diet, but it is beneficial to other ungulates that graze on grasses.

Island Biogeography and Biodiversity

Recall that MacArthur and Wilson have proposed a general theory of island biogeography, which states that a large island close to the mainland (the source of dispersing species) will have greater diversity than a small island that is distant from the mainland (see Fig. 24.3). Of late, humans have created many islands; some of these are not true islands surrounded by water—they are terrestrial patches surrounded by human cities, towns, or farms.

In Panama, Barro Colorado Island (BCI) was created in the 1910s when a river was dammed to form a lake. As predicted by the theory of island biogeography, BCI lost species because it was a small island now cut off from the mainland. Among those species that became extinct were the top predators on the island, namely the jaguar, puma, and ocelot. Thereafter, medium-sized terrestrial mammals such as the coatimundi increased in number. The coatimundi is an avid predator of bird eggs, nestlings, and other small vertebrates. Not surprisingly, there are now fewer bird species on BCI even though the island is large enough to support them. Therefore, it appears that the loss of certain predators from a community can sometimes result in chain reactions that lead to decreased diversity at various levels of community structure.

Such changes as this can also be expected in isolated terrestrial areas. We are finding that preserves need to be very large in order to conserve *K*-strategists, especially top predators.

Exotic Species and Biodiversity

These examples illustrate that unbridled competition and predation can reduce biodiversity. The introduction of alien (exotic) species into new areas also provides many good examples. African bees were introduced into Brazil to help increase honey production. Instead they displaced domestic honeybees and started moving northward. These so-called killer bees are now established as far north as Texas and New Mexico. The endemic bird populations of many Pacific islands have been devastated due to the accidental introduction of the brown tree snake which preys on the birds. The red fox was deliberately imported into Australia to prey on the introduced European rabbit, but it has now reduced the populations of native small mammals.

> **Predation and competition also influence biodiversity. *K*-strategists, which are sometimes top predators, are maintained best on large preserves. Exotic species outcompete and/or prey on native plants and animals.**

a.

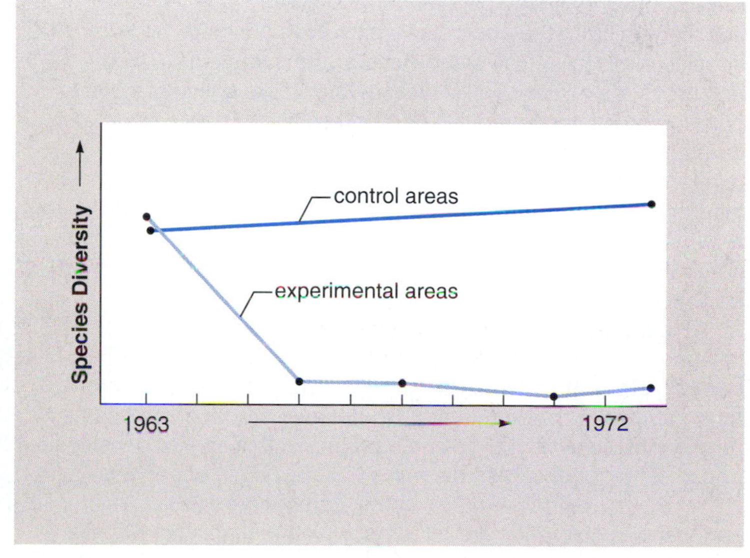

b.

Figure 24.21 Effect of a keystone predator.
a. The sea star *Pisaster ochracheous* is eating a mussel (*Mytilus*), its favorite food. **b.** Along the west coast, *Pisaster* was removed from experimental areas (tidepools) but not from control areas. Diversity decreased in experimental areas because the mussel population increased and crowded out other invertebrates and algae.

Connecting Concepts

Community ecology is concerned with how populations of different species interact with each other. Historically, some ecologists felt that communities functioned like "super-organisms," with each species a vital and integral part of the community. At the other extreme, other ecologists felt that each population simply occurred where conditions were best for it and communities were chance aggregations of species. Most likely, the truth lies somewhere in between the two extreme views. We know that a population of any species is distributed where conditions are best for it. However, communities do exhibit emergent properties that result from certain groupings of species co-occurring.

The number of individuals in many populations within ecological communities is determined by negative interactions such as interspecific competition, predation, and parasitism. The size of all populations are negatively affected by at least one of those interactions. Positive interactions such as mutualisms are fairly common in nature (especially for plants), but their effect on population size is not well understood yet.

Perhaps one of the most important recent discoveries about communities is that they are highly dynamic. The number of species, kinds of species, and size of populations within most communities is constantly changing due to disturbances and climatic variability.

Because the species composition of communities can be so variable, many ecologists study the movement of energy and nutrients through communities. The physical environment has a large influence on energy and nutrient flow, and thus the non-living world must be incorporated in our studies of communities as discussed in the next chapter.

Summary

24.1 Concept of the Community

A community is an assemblage of populations interacting with one another within the same environment. Communities differ in their composition (species found there) and their diversity (species richness and relative abundance). Abiotic factors such as latitude and environmental gradients seem to largely control community composition. As suggested by the theory of island biogeography, space limitations affect species diversity.

24.2 Structure of the Community

An organism's habitat is where it lives in the community. An ecological niche is defined by the role an organism plays in its community, including its habitat and how it interacts with other species in the community. Competition, predator-prey, parasite-host, commensalistic, and mutualistic relationships help organize populations into an intricate dynamic system.

According to the competitive exclusion principle, no two species can occupy the same niche at the same time. When resources are partitioned between two or more species, niche partitioning has occurred. Character displacement is a structural change that gives evidence of niche partitioning. But the difference between species can be more subtle as when warblers feed at different parts of the tree canopy. Usually we have to assume that we are seeing the results of competition. Barnacles competing on the Scottish coast may be an example of present ongoing competition.

Predator-prey interactions between two species are influenced by the biotic potential of the two populations, the carrying capacity of the environment, and the presence of other food sources in the community. Sometimes predation can cause prey populations to decline and remain at relatively low densities, or a cycling of population densities has also been observed.

Prey defenses take many forms: camouflage, use of fright, and warning coloration are three possible mechanisms. Batesian mimicry occurs when one species has the warning coloration but lacks the defense. Müllerian mimicry occurs when two species with the same warning coloration have the same defenses.

We would expect coevolution to occur within a community. The better the predator becomes at catching prey, the better the prey becomes at escaping the predator, for example. Flowers are adapted to attract specific pollinators, and hosts are adapted to cope with potential parasites.

Like predators, parasites take nourishment from their host. However, the host usually provides a home and a place where parasites reproduce. Parasites escape detection and manipulate their hosts in various ways. Whether they are aggressive (kill their host) or benign probably depends on which results in the highest fitness, considering the life cycles of both organisms in the relationship.

Perhaps symbiotic relationships are too complicated to try to classify them as commensalistic, parasitic, and mutualistic. Therefore, it is not surprising that certain relationships like clownfishes that live among sea anemone tentacles or lichens that contain fungal and algal members are questionable as to whether they are commensalistic or mutualistic, respectively. *Croton* saplings provide food for ants, but so do the caterpillars that feed on *Croton* saplings! In this case, the possible mutualistic relationship between tree and ant has been undermined by a predator.

24.3 Community Development

Directional patterns of community change are called ecological succession. Ecologists have recognized for some time that communities undergo succession both in terms of a geological timescale and present-day timescale. There is now much discussion about the cause of succession and whether it results in so-called climax communities or not. In the past, use of the term climax community has implied that a diverse community is stable in its composition.

24.4 Community Biodiversity

It is perfectly obvious now that heterogeneity or the presence of patches in different stages of succession would actually result in the most diversity. The intermediate disturbance hypothesis states that a moderate amount of disturbance is required for a high degree of community diversity.

Predation and competition are examples of interactions that affect biodiversity. Predation by keystone predators increases the diversity of a community. If a preserve is so small that top predators are eliminated, diversity will decrease because other predators are not held in check. Alien species often out-compete and/or prey on native species so that they are reduced in numbers.

Reviewing the Chapter

1. Contrast the individualistic model of community composition with the interactive model. 401
2. What do experiments with artificial reefs and the island biogeography hypothesis tell us about species richness? 401–2
3. Describe the habitat and ecological niche of some particular organism. 403
4. What is the competitive exclusion principle? How does the principle relate to character displacement and niche partitioning? 404–05
5. What is the special significance of the barnacle experiment off the coast of Scotland? 405
6. Would you expect all predator and prey interactions to produce similar results with regard to population densities? Why or why not? 406–07
7. What is mimicry, and why does it work as a prey defense? 409
8. Give an example of parasitism, commensalism, and mutualism, and show that it is not always easy to distinguish between these symbiotic relationships. 410, 412
9. Describe an example of coevolution between a predator and prey, between a parasite and host, and between mutualists. 411
10. What is ecological succession? What is the present controversy surrounding the concept? 414–15
11. What is the intermediate disturbance hypothesis, and how does it relate to community diversity? 416
12. What interactions affect biodiversity? In what way? 416–17

Testing Yourself

Choose the best answer for each question.

1. Six species of monkeys are found in a tropical forest. Most likely, they
   a. occupy the same ecological niche.
   b. eat different foods and occupy different ranges.
   c. spend much time fighting each other.
   d. are from different stages of succession.
   e. All of these are correct.
2. Leaf cutter ants keep fungal gardens. The ants provide food for the fungus but also feed on the fungus. This is an example of
   a. competition.
   b. predation.
   c. commensalism.
   d. parasitism.
   e. mutualism.
3. Clownfishes live among sea anemone tentacles, where they are protected. If the clownfish provides no service to the anemone, this is an example of
   a. competition.
   b. predation.
   c. parasitism.
   d. commensalism.
   e. mutualism.
4. Two species of barnacles vie for space in the intertidal zone. The one that remains is
   a. the better competitor.
   b. better adapted to the area.
   c. the better predator on the other.
   d. the better parasite on the other.
   e. Both a and b are correct.
5. A bullhorn acacia provides a home and nutrients for ants. Which statement is likely?
   a. The plant is under the control of pheromones produced by the ants.
   b. The ants protect the plant.
   c. They compete with one another.
   d. They have coevolved to occupy different ecological niches.
   e. All of these are correct.
6. The ecological niche of an organism
   a. is the same as its habitat.
   b. includes how it competes and acquires food.
   c. is specific to the organism.
   d. is usually occupied by another species.
   e. Both b and c are correct.
7. The frilled lizard of Australia suddenly opened its mouth wide and unfurled folds of skin around its neck. Most likely this was a way to
   a. conceal itself.
   b. warn that it was noxious to eat.
   c. scare a predator.
   d. scare its prey.
   e. All of these are correct.
8. When one species mimics another species, the mimic sometimes
   a. lacks the defense of the model.
   b. possesses the defense of the model.
   c. is brightly colored.
   d. competes with a mimic.
   e. All of these are correct.
9. Which of these most likely would help account for diversity in a coral reef?
   a. warm temperatures
   b. presence of predation
   c. intermediate disturbance
   d. prey defenses
   e. All of these are correct.
10. The presence of patches in an environment may be associated with
    a. spatial and temporal heterogeneity.
    b. a greater degree of diversity.
    c. past disturbances.
    d. community instability.
    e. All of these are correct.
11. Label this diagram, and tell how it applies to the individualistic concept of community composition.

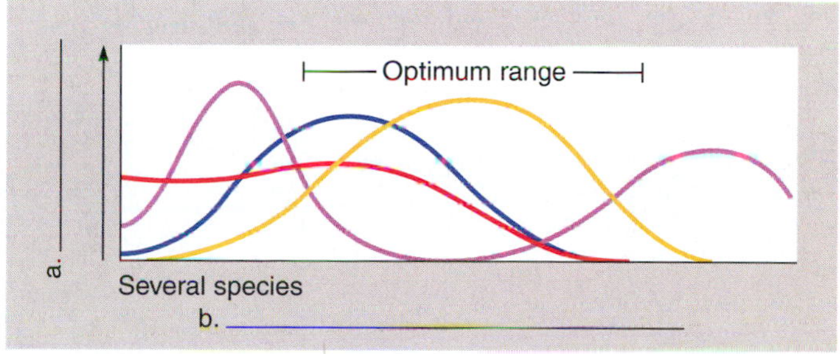

Thinking Scientifically

1. You suspect that a particular animal's coloration pattern is mimicking that of another, poisonous species. What data would support this hypothesis?
2. For a long time fire was considered destructive and was suppressed without regard to the role it might be playing in the environment. Now fire is recognized as a tool to help maintain healthy ecosystems. However, people living on the edge of lands susceptible to fires are nervous about allowing minimally controlled fires to burn out. Imagine you are a government ecologist in charge of determining which wild areas would benefit by allowing naturally occurring fires to burn with minimal control and which have no need of fire as a source of ecological disturbance. What criteria would help you make a decision?

Understanding the Terms

camouflage 408
character displacement 404
climax community 415
coevolution 411
commensalism 412
community 400
community stability 416
competitive exclusion principle 404
ecological niche 403
ecological succession 414
habitat 403
host 410
interspecific competition 404
keystone predator 417
mimicry 409
mutualism 412
parasite 410
parasitism 410
pioneer species 415
predation 406
predator 406
prey 406
resource partitioning 404
symbiosis 410

Match the terms to these definitions:

a. ______________ Assemblage of populations interacting with one another within the same environment.
b. ______________ Place where an organism lives and is able to survive and reproduce.
c. ______________ Directional pattern of change in which one community replaces another until a community typical of the area results.
d. ______________ Role an organism plays in its community, including its habitat and its interactions with other organisms.
e. ______________ Symbiotic relationship in which both species benefit, in terms of growth and reproduction.

Web Connections

Exploring the Internet

http://www.mhhe.com/biosci/genbio/mader/
(click on *Biology 7/e*)

The *Biology 7/e* Online Learning Center provides many resources for studying the material in this chapter including links to the following sites:

The Nature Conservancy. News about nature, ecology, and conservation.

http://www.tnc.org/

Outdoor Action Guide to Outdoor/Environmental Careers describes types of careers, graduate programs, and how to determine what job is right for you.

http://www.princeton.edu/~oa/jobs/careeroe.html

An in-class Laboratory Exercise in Competition. This illustrates the Lotka-Volterra competition model.

http://www.ets.uidaho.edu/wlf448/compex.htm

and

http://www.ets.uidaho.edu/wlf448/comprob.htm

Predation Laboratory. An exercise to illustrate some mathematical models which demonstrate predation.

http://www.ets.uidaho.edu/wlf448/pred1lab.htm#ProbSet

Ecosystems and Human Interferences

chapter 

chapter concepts

25.1 The Nature of Ecosystems

- The biosphere, where all living things exist, consists of ecosystems. 422
- In an ecosystem, organisms interact with one another and with the abiotic environment. 422
- Trophic (feeding) relationships are essential to the workings of an ecosystem. 423
- Autotrophs are self-feeders: photoautotrophs capture solar energy and produce organic nutrients. Heterotrophs, on the other hand, take in preformed organic nutrients. 423

25.2 Energy Flow and Nutrient Cycling

- Solar energy enters biotic communities via photosynthesis, and as organic nutrients pass from one trophic level to another, heat is returned to the atmosphere. 424
- Nutrients cycle within and between ecosystems in global biogeochemical cycles. 425

25.3 Global Biogeochemical Cycles

- Biogeochemical cycles are gaseous (carbon cycle, nitrogen cycle) or sedimentary (phosphorus cycle). 428
- The addition of carbon dioxide (and other gases) to the atmosphere is associated with global warming. 430
- The production of fertilizers from nitrogen gas is associated with acid deposition, photochemical smog, and temperature inversions. 432
- Fertilizer also contains mined phosphate; fertilizer runoff is associated with water pollution. 435

Deforestation due to clear-cutting of trees

An ecosystem is characterized by energy flow and chemical cycling. Both of these begin when algae and green plants capture a small percentage of the sun's energy and use it to transform inorganic chemicals like carbon dioxide and water to organic compounds that are used as food for themselves and all the other populations in an ecosystem. Eventually when decomposers break down organic matter, the inorganic chemicals are liberated once again, but the energy has dissipated as heat.

Ecosystems are not self-contained; they have inputs and outputs to the other ecosystems of the biosphere. Therefore, some ecologists think of the biosphere as a global ecosystem. Most persons now realize that humans are a part of the biosphere and that our activities affect all of its ecosystems. DDT is an insecticide that was first used during World War II in the Pacific to destroy mosquitoes that carry malaria. Today DDT is found in every body of water on earth and in the tissues of every human being. In the same way, our overuse of resources such as when we destroy forests may bring about changes that we find difficult to predict at this time.

25.1 The Nature of Ecosystems

Our planet is unique in many ways. Unlike the other planets in our solar system, Earth has water, an atmosphere, and abundant life. Perhaps our planet should have been called "water" instead of Earth. Water is present in the *hydrosphere* [Gk. *hydrias,* of water, and L. *sphaera,* ball], which covers over three-quarters of the earth's surface (Fig. 25.1). The oceans moderate the temperature of the earth; as surface temperatures rise, the oceans take up a great deal of heat, and then as temperatures cool, they return heat slowly to the atmosphere. This helps keep the temperature on Earth suitable to life.

The *atmosphere* [Gk. *atmos,* vapor, and L. *sphaera,* ball] is concentrated in the lowest 10 kilometers near Earth but extends out at least 1,000 kilometers. Among other gases, the atmosphere contains carbon dioxide, nitrogen, and oxygen, that are both used and released by living things. Carbon dioxide is necessary for photosynthesis. Oxygen is necessary for cellular respiration, and in the upper atmosphere it becomes ozone, a substance that shields Earth from damaging ultraviolet radiation and makes life on land possible.

A rocky substratum called the *lithosphere* [Gk. *lithos,* stone, and L. *sphaera,* ball] extends from Earth's surface to about 100 kilometers deep. The weathering of rocks supplies minerals to plants, which take root in weathered rocks and slowly form soil. Besides minerals, soil contains decaying organic material known as humus. The organisms of decay play a vital role in breaking down organic matter, returning inorganic nutrients to plants so that photosynthesis can continue.

The **biosphere** is that part of the atmosphere, hydrosphere, and lithosphere that contains living things. Taking the global view the entire biosphere is an **ecosystem,** a place where organisms interact among themselves and with the physical and chemical environment. These interactions help maintain ecosystems and in turn the biosphere. Human activities can alter the interactions between organisms and their environments in a way that reduces the abundance and diversity of life that environments can support. It is important to understand how ecosystems function so that we can repair past damages and predict how human activities might change current conditions.

Usually, ecologists study ecosystems on a smaller scale: a pond, a swamp, a wooded area, are also ecosystems. Regardless of their size, ecosystems are characterized by (1) one-way flow of energy through the biotic community of an ecosystem and (2) a cycling of materials from the abiotic environment through the biotic community and back to the abiotic environment.

The Biotic Community

The populations in the biotic community of an ecosystem are classified into **trophic** (feeding) **levels** according to the source of their nutrients (Fig. 25.2). Some populations are autotrophs and some are heterotrophs.

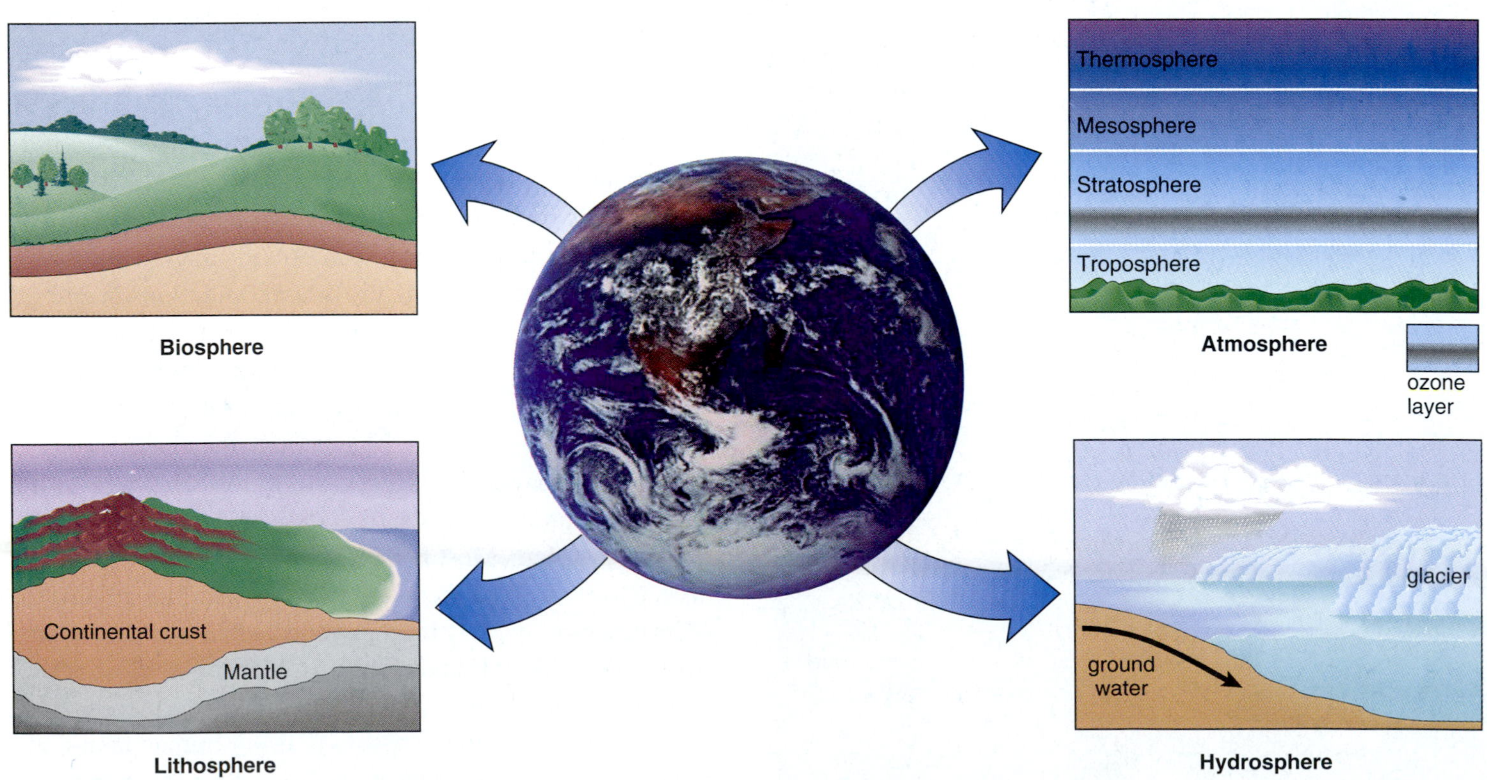

Figure 25.1 Biosphere.
The biosphere is that part of the lithosphere (crust), the atmosphere (gases) and the hydrosphere (water) that contains living things.

Autotrophs

Autotrophs require only inorganic nutrients and an outside energy source to produce sugars and other organic compounds for their own use and as a source of organic nutrients for other members of the community. Therefore, they are called **producers**—they produce food. Chemoautotrophs are bacteria that obtain energy by oxidizing inorganic compounds such as ammonia, nitrites, and sulfides, and they use this energy to synthesize organic compounds. Chemoautotrophs have been found to support some cave communities and also deep-sea communities at oceanic ridges where hot minerals spew out of the inner earth. The chemoautotrophs that function in the nitrogen cycle will be discussed in this chapter on page 432.

Photoautotrophs are photosynthesizers that produce most of the organic nutrients for the biosphere. Algae of all types possess chlorophyll and carry on photosynthesis in freshwater habitats and marine habitats. Algae make up the phytoplankton, which are photosynthesizing organisms suspended in water. Green plants are the dominant photosynthesizers on land.

Heterotrophs

Heterotrophs need a preformed source of organic nutrients. They are the **consumers**—they consume food. **Herbivores** are animals that graze directly on plants or algae. In terrestrial habitats, insects are small herbivores, while in aquatic habitats, zooplankton, such as protozoa, play that role. **Carnivores** feed on other animals; birds that feed on insects are carnivores, and so are hawks that feed on birds. This example allows us to mention that there are primary consumers (e.g., insects), secondary consumers (e.g., birds), and tertiary consumers (e.g., hawks). Sometimes tertiary consumers are called top predators. **Omnivores** are animals that feed both on plants and animals. As you most likely know, humans are omnivores.

Detritivores are organisms that feed on **detritus,** which is decomposing particles of organic matter. Fan worms feed off detritus that is floating in marine waters, while clams take it from the substratum. Earthworms and some beetles, termites, and maggots are all terrestrial detritivores. Nonphotosynthetic bacteria and fungi, including mushrooms carry out **decomposition,** the breakdown of dead organic matter, including animal wastes. They perform a very valuable service because they release inorganic nutrients that are taken up by plants once more. Otherwise, plants would have to wait for minerals to be released from rocks.

The populations in the biotic community of an ecosystem have a trophic (feeding) relationship. Autotrophs produce organic nutrients and heterotrophs consume organic nutrients.

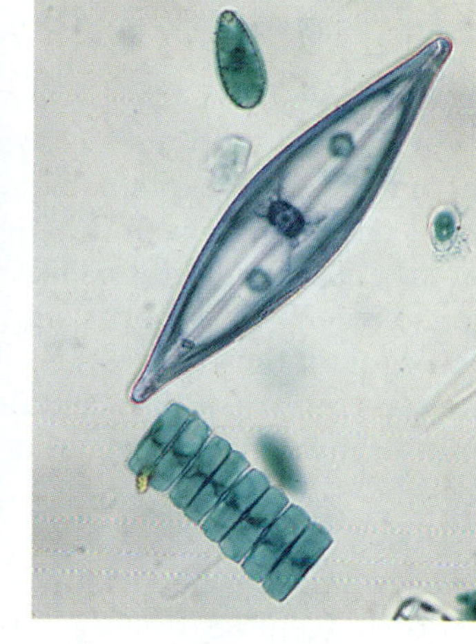

a. Producers

b. Herbivores

c. Carnivores

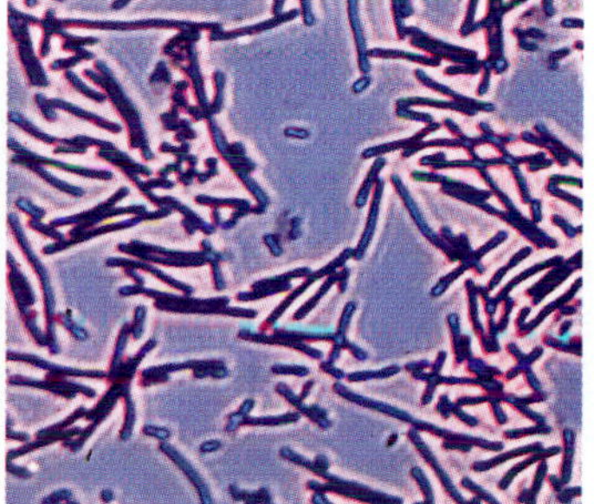

d. Decomposers

Figure 25.2 Biotic components.
Ecosystems include (a) autotrophic producers like green plants and certain protists, such as these diatoms; (b) heterotrophic herbivores like giraffes and caterpillars; (c) heterotrophic carnivores like the osprey and praying mantis; and (d) heterotrophic detritivores like mushrooms and soil-dwelling bacteria.

25.2 Energy Flow and Nutrient Cycling

With few exceptions ecosystems are dependent on an incoming supply of solar energy and nutrients that cycle between autotrophs and heterotrophs (Fig. 25.3). Nutrients include elements like CHNOPS (carbon, hydrogen, nitrogen, oxygen, phosphorus, and sulfur) as well as compounds that contain these and other elements. Carbon dioxide is an inorganic nutrient that contains carbon, while glucose is an organic nutrient that contains carbon.

Energy Flow

The laws of thermodynamics are consistent with the observation that energy flows through an ecosystem. The first law states that energy cannot be created (nor destroyed); it can only be transformed from one type of energy to another. The second law states that with every transformation some energy is degraded into a less available form such as heat. Eventually, then, all the solar energy converted to organic nutrients by autotrophs becomes dissipated as heat. Therefore, ecosystems need a continual supply of solar energy.

Energy flow begins when autotrophs produce organic nutrients after taking in inorganic nutrients and absorbing solar energy. In other words they transform solar energy to a form of chemical energy that is usable by all members of an ecosystem. Primary productivity is the rate at which an ecosystem's autotrophs capture and store energy within organic compounds over a certain length of time. Physical factors like climate (e.g., temperature, rainfall, and humidity) and the nature of the soil can all affect primary productivity. Only a portion of the total primary productivity, called gross primary productivity, is passed on to heterotrophs because plants use organic nutrients to fuel their own cellular respiration. The portion of gross primary productivity that is passed on as food for heterotrophs is called net primary productivity (Fig. 25.4).

Only about 55% of gross primary productivity is available to heterotrophs after autotrophs have fulfilled their energy needs. A much smaller percentage of organic nutrients taken in by heterotrophs is available to a higher level consumer because of all the processes indicated in this diagram:

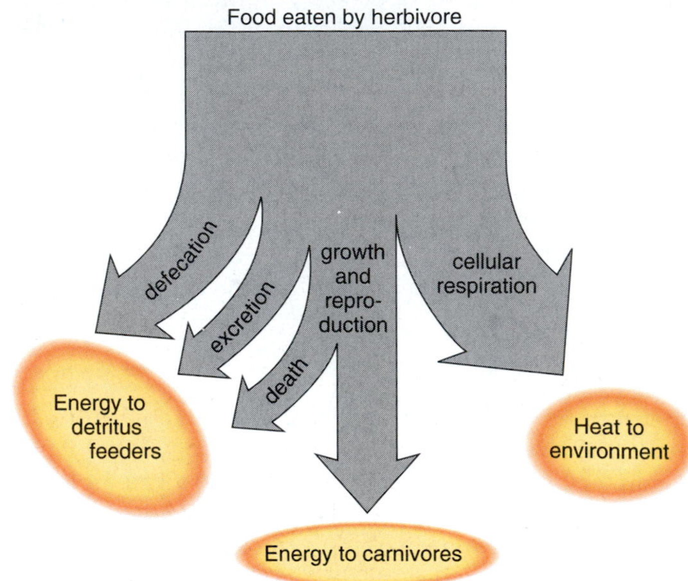

The diagram shows that a certain amount of energy obtained in organic nutrients (food) is egested in feces or

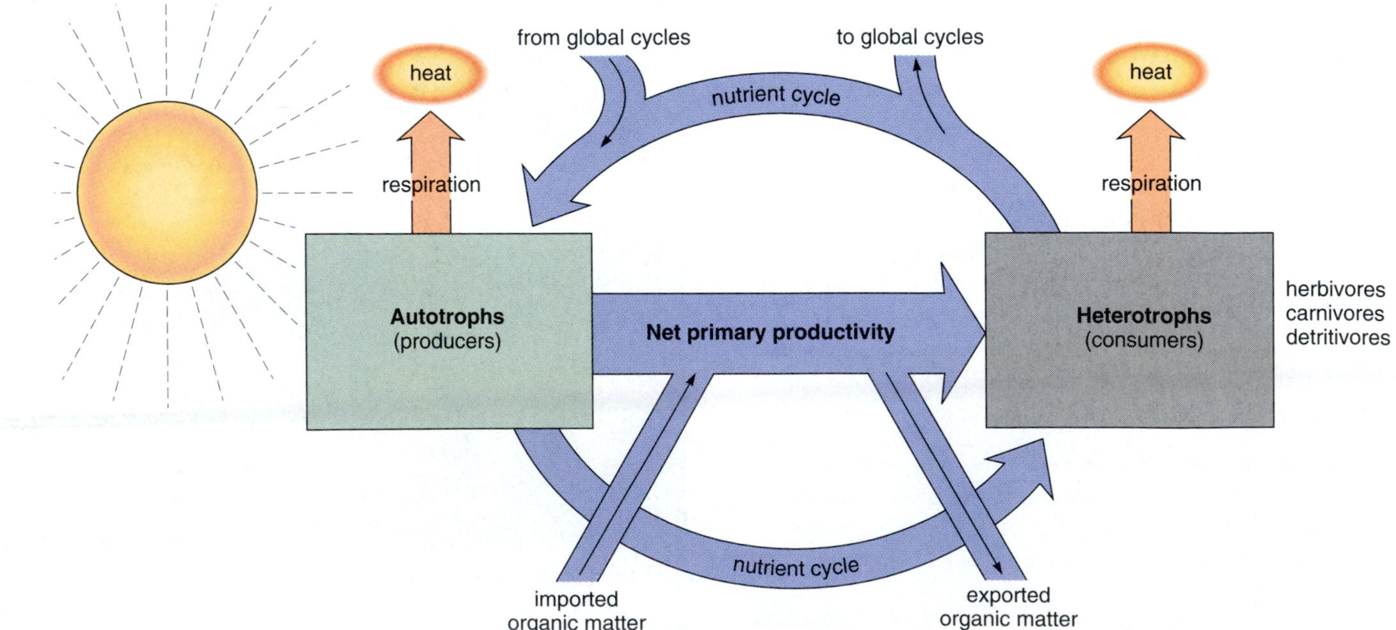

Figure 25.3 Ecosystems.
Energy flows from the sun to autotrophs and heterotrophs and to the environment in ecosystems. Nutrients cycle between autotrophs and heterotrophs within the ecosystem and in global cycles between ecosystems.

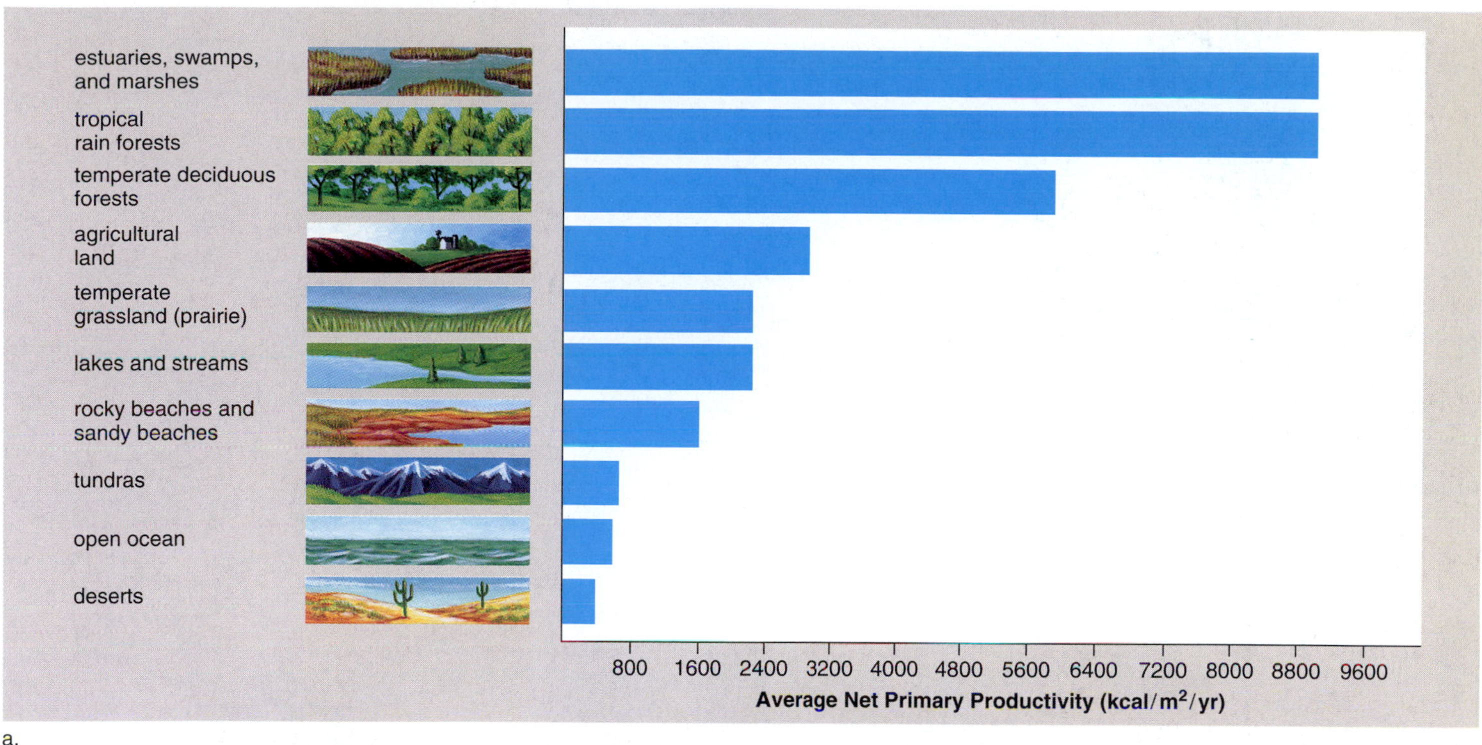

Gross primary productivity (GPP): Total rate of photosynthesis in a specified area. Autotrophs (most photosynthetic plants and algae) absorb solar energy and produce organic nutrients for all organisms in a community.

Net primary productivity (NPP): The rate at which plants produce organic nutrients for heterotrophs:

$$NPP = GPP - RS$$

Where RS is the energy used by autotrophs for respiration.

Plants use some of the organic nutrients they produce (GPP) for cellular respiration, and what is left (NPP) is available for heterotrophs. Usually between 50% and 90% of GPP remains as NPP.

b.

Figure 25.4 Gross and net primary productivity.
a. The net primary productivity (NPP) of ecosystems varies. Although the open ocean has a low productivity, per unit area it makes a significant contribution to the productivity of the biosphere because of its large size. Estuaries and tropical rain forests contribute significantly because of their high productivity. **b.** The mathematical relationship between gross and net primary productivity.

excreted in urine. Of the assimilated energy, a portion is utilized during cellular respiration and thereafter becomes heat. The remaining portion of energy (10–15%) is converted into increased body weight (or additional offspring). This latter use of energy is called *secondary productivity.* Only secondary productivity can be used as organic nutrients by an animal that feeds on a heterotroph. Of course, the feces and urine of a heterotroph provide energy to detritivores. Since detritivores are fed upon by other heterotrophs of an ecosystem, the situation can get a bit complicated. Still we can conceive that all the solar energy that enters an ecosystem eventually becomes heat as a higher level heterotroph feeds on the lower level heterotroph. In other words, energy flows through an ecosystem and does not cycle.

Nutrient Cycling

As Figure 25.3 indicates, nutrients do cycle between autotrophs and heterotrophs within ecosystems. When detritivores return inorganic nutrients to autotrophs the nutrient cycle begins again.

Even so, ecosystems are not self-contained with regard to inorganic nutrients either. Inorganic nutrients enter ecosystems by way of global cycles such as the carbon cycle and nitrogen cycle we will be discussing. They are also carried between ecosystems by rainfall and wind. Then, too, animals emigrate and immigrate, bringing nutrients into and taking nutrients out of ecosystems.

Energy flows through an ecosystem, while nutrients cycle within and among ecosystems.

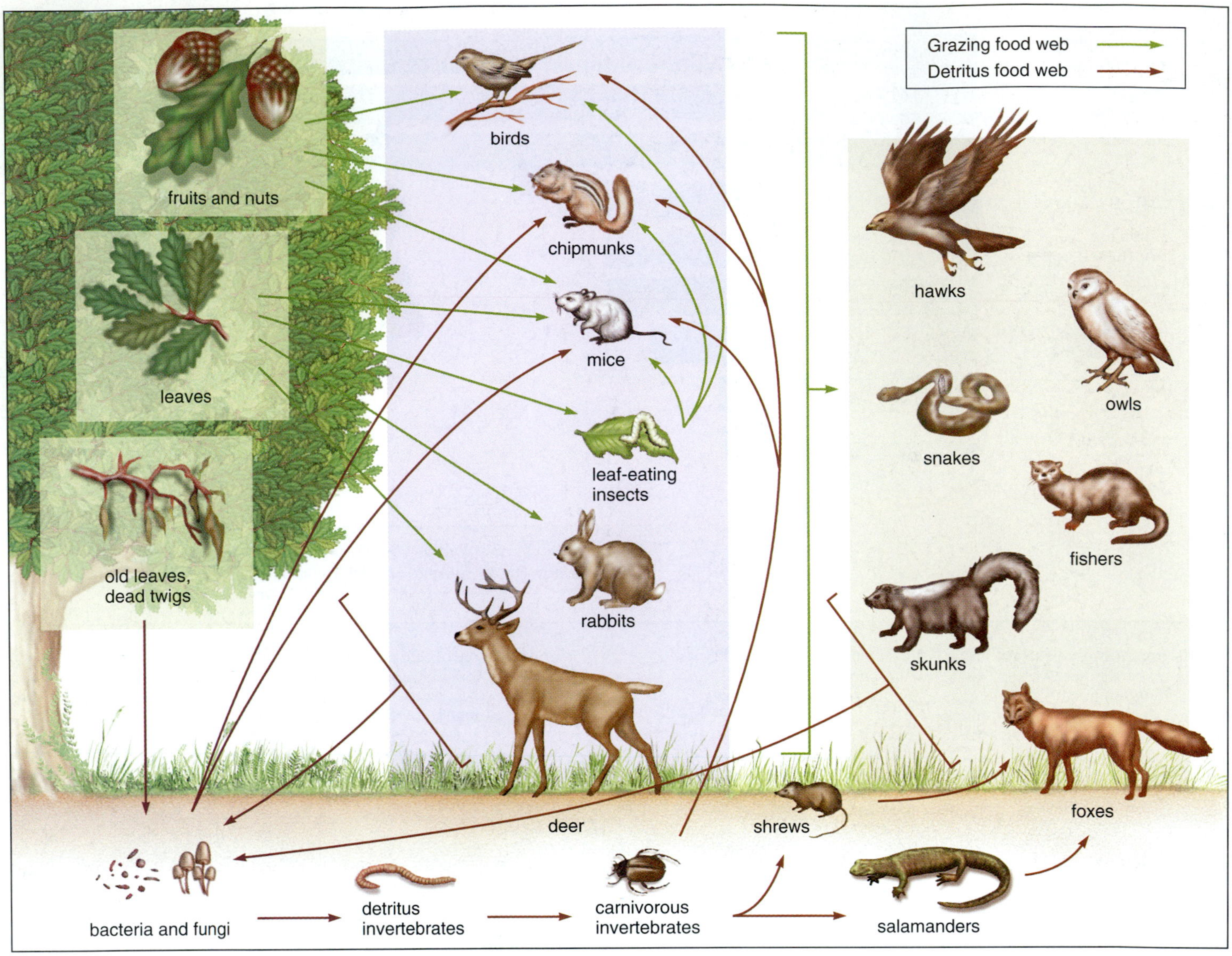

Grazing food web
Detritus food web
fruits and nuts
leaves
old leaves, dead twigs
birds
chipmunks
mice
leaf-eating insects
rabbits
deer
hawks
owls
snakes
fishers
skunks
foxes
shrews
bacteria and fungi
detritus invertebrates
carnivorous invertebrates
salamanders

Figure 25.5 Forest food webs.
Two linked food webs are shown for the Hubbard Brook Forest ecosystem: a grazing food web and a detrital food web.

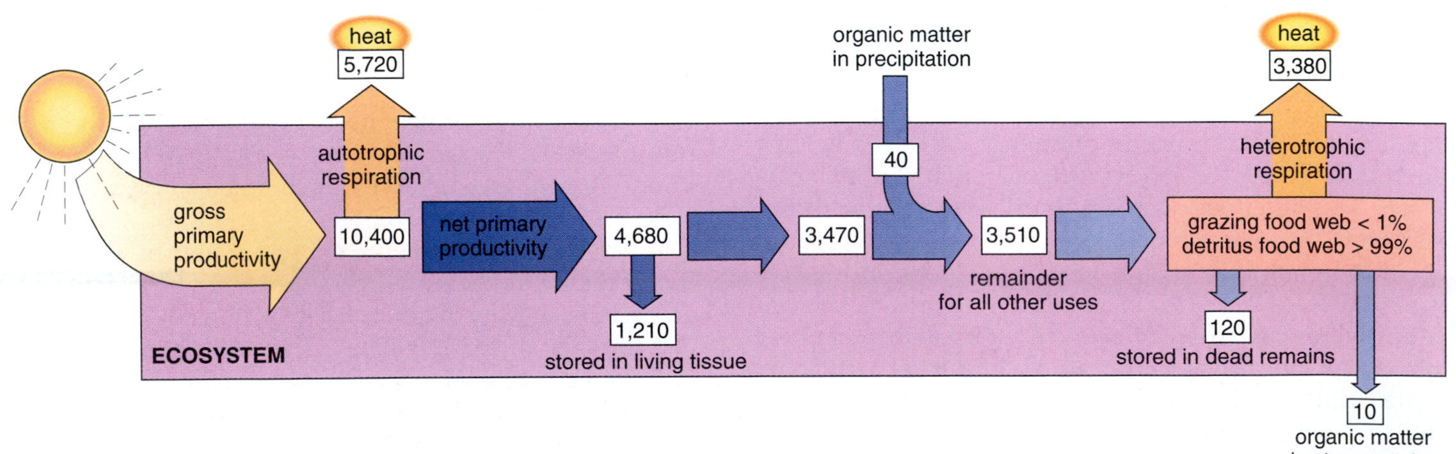

Figure 25.6 Flow of energy.
In this flow of energy diagram for the ecosystem pictured in Figure 25.5, gross primary productivity is 10,400 kcal/m^2/yr, of which 5,720 is used for respiration and becomes heat. Net primary productivity is 4,680 kcal/m^2/yr, of which 3,470 (increased by an input of 40 kcal/m^2/yr) is passed on to the grazing and detrital food webs. Less than 1% is utilized by the grazing food web, and the rest is funneled through the detrital food web. Of the 3,510 kcal/m^2/yr, 3,380 is used for respiration and becomes heat. A total of 10 kcal/m^2/yr is an output to other ecosystems. Notice that of the original 10,400 kcal/m^2/yr, 1,210 is stored in living plant tissue, and 120 is stored as dead remains in the soil.

Food Webs and Ecological Pyramids

Ecologists use a variety of ways to show the process and results of energy flow and nutrient cycling within ecosystems. A **food web** details the precise trophic relationships between specific populations of organisms. An **ecological pyramid** quantifies this information per trophic level making it easier to compare one ecosystem to another.

Food Web

In a forest of 132,300 square meters in New Hampshire, the producers include sugar maple, beech, and yellow birch trees (Fig. 25.5). Insects in the form of caterpillars feed on leaves, while mice, rabbits, and deer feed on leaf tissue at or near the ground. Birds, chipmunks, and mice feed on fruits and nuts, but they are in fact omnivores because they also feed on caterpillars. These herbivores and omnivores all provide nutrients for a number of different carnivores. This portion of the diagram is called a **grazing food web** because it begins with aboveground plant material.

The lower portion of Figure 25.5 is devoted to the **detrital food web.** The bacteria and fungi of decay decompose organic remains, making food available for other detritivores. Earthworms, shrews, and salamanders are examples of detritivores that become food for aboveground carnivores. In this way, the detrital and the grazing food webs are connected.

We naturally tend to think that aboveground vegetation like trees store the most organic matter and energy, but this is not necessarily the case. In this particular forest, the organic matter lying on the forest floor and mixed into the soil contains much more stored energy than does the leaf matter of living trees. The largest quantity of stored energy is in the soil, which contains over twice as much as the forest floor. Therefore, it's quite possible in a forest ecosystem for more energy to be funneling through the detrital food web than through the grazing food web. Figure 25.6 shows that less than 1% of net primary productivity moves through the grazing food web; 99% moves through the detrital food web.

Trophic Levels You can see that Figure 25.5 would allow us to link organisms one to another in a straight-line manner, according to who eats whom. For example, in the grazing food web we could link these organisms:

leaves → caterpillars → mice → hawks

And in the detrital food web we could designate these links:

dead organic matter → soil bacteria
→ earthworms → etc.

Diagrams like this that tell who eats whom are called **food chains.** All the organisms feed at a particular trophic level in a food chain. In the grazing food web, going from left to right, the trees are primary producers (first trophic level), the first series of animals are primary consumers (second trophic level), and the next group of animals are secondary consumers (third trophic level).

Ecological Pyramids

Ecological pyramids contain building blocks that designate either the number, biomass, or energy of the various trophic levels. A *pyramid of numbers* simply tells how many organisms there are at each trophic level. It's easy to see that a pyramid of numbers could be completely misleading. For example, in Figure 25.5 you would expect each tree to contain numerous caterpillars; therefore there would be more herbivores than autotrophs! The problem, of course, has to do with size. Autotrophs can be tiny, like microscopic algae, or they can be big like beech trees; similarly, herbivores can be small like caterpillars, or they can be large like elephants.

Pyramids of biomass eliminate size as a factor since biomass is the number of organisms multiplied by their dry weight and expressed as g/m^2. You would certainly expect the biomass of producers to be greater than the biomass of the herbivores, and that of the herbivores to be greater than the carnivores, as is shown in Figure 25.7, where the producers are not only algae but also eelgrass. In other aquatic ecosystems such as lakes and open seas, where algae are the only producers, the herbivores may have a greater biomass than the producers when you take their measurements. Why? The reason is that over time, the algae reproduce

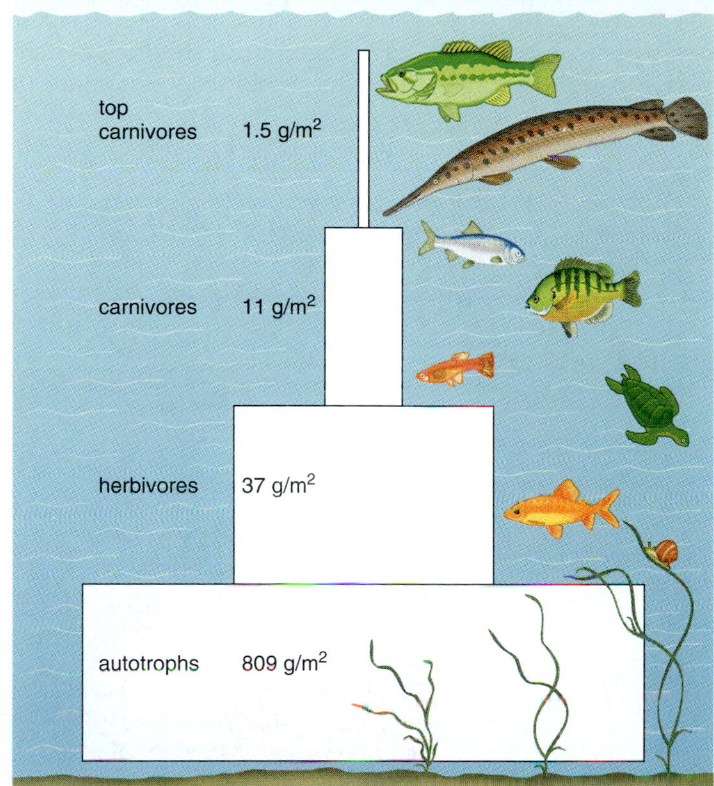

Figure 25.7 Ecological pyramid.
The biomass or dry weight (g/m^2) for trophic levels in a grazing food web in a bog at Silver Springs, Florida. There is a sharp drop in biomass between the producer level and herbivore level, which is consistent with the common knowledge that the detrital food web plays a significant role in bogs.

rapidly, but they are also consumed at a high rate. A pyramid, like this one, in which there are more herbivores than producers, is called an inverted pyramid:

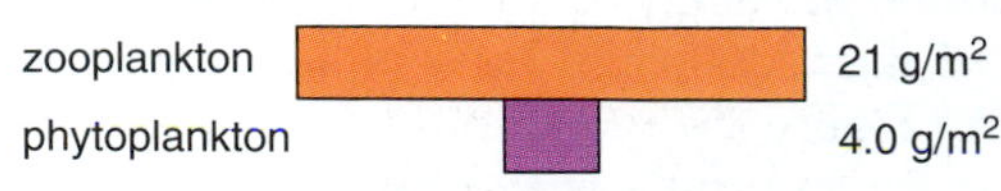

There are ecological *pyramids of energy* also, and they generally have the appearance of Figure 25.7. Ecologists are now beginning to rethink the usefulness of utilizing pyramids to describe energy relationships. One problem is what to do with the decomposers, which are rarely included in pyramids, and yet a large portion of energy becomes detritus in many ecosystems. It is sometimes difficult to decide where organisms belong in a pyramid. For example, chipmunks feed on fruits and nuts, but they also feed on leaf-eating insects.

There is a rule of 10% with regard to biomass (or energy) pyramids. It says that, in general, the amount of biomass (or energy) from one level to the next is reduced by a magnitude of 10. Thus, if an average of 1,000 kg of plant material is consumed by herbivores, about 100 kg is converted to herbivore tissue, 10 kg to first-level carnivores, and 1 kg to second-level carnivores. The rule of 10% suggests that few carnivores can be supported in a food web. This is consistent with the observation that each food chain has from three to four links, rarely five.

Food webs and ecological pyramids are two ways to represent nutrient cycling and energy flow within an ecosystem.

25.3 Global Biogeochemical Cycles

All organisms require a variety of organic and inorganic nutrients. Carbon dioxide and water are necessary for photosynthesis. Nitrogen is a component of all the structural and functional proteins and nucleic acids found in living tissues. Phosphorus is essential for ATP and nucleotide production. In contrast to energy, inorganic nutrients such as carbon dioxide, nitrate, and phosphate, are used over and over again by autotrophs.

Since the pathways by which nutrients circulate within and among ecosystems involve both living (biosphere) and nonliving (geological) components, they are known as **biogeochemical cycles.** For each element, chemical cycling may involve (1) a reservoir—a source normally unavailable to producers, such as fossilized remains, rocks, and deep-sea sediments; (2) an exchange pool—a source from which organisms do generally take chemicals, such as the atmosphere or soil; and (3) the biotic community—through which chemicals move along food chains, perhaps never entering a pool (Fig. 25.8).

There are two general categories of biogeochemical cycles. In a *gaseous cycle,* exemplified by the carbon and nitrogen cycles, the element returns to and is withdrawn from the atmosphere as a gas. In the *sedimentary cycle,* exemplified by the phosphorus cycle, the element is absorbed from the sediment by plant roots, passed to heterotrophs, and is eventually returned to the soil by decomposers, usually in the same general area.

The diagrams in the figures on the next few pages make it clear that nutrients can flow between terrestrial and aquatic ecosystems. In the nitrogen and phosphorus cycles, these nutrients run off from a terrestrial to an aquatic ecosystem and in that way enrich aquatic ecosystems. Decaying organic material in aquatic ecosystems can be a source of nutrients for intertidal inhabitants like fiddler crabs. Sea birds feed on fish but deposit guano (droppings) on land,

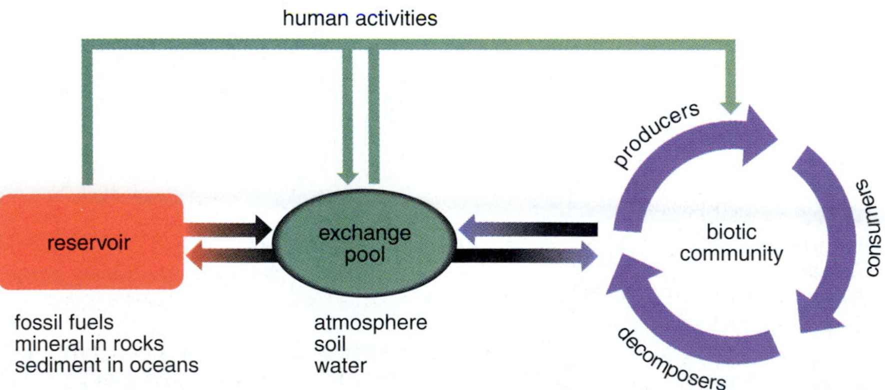

Figure 25.8 Model for chemical cycling.
Nutrients cycle between these components of ecosystems: Reservoirs such as fossil fuels, minerals in rocks, and sediments in oceans are normally relatively unavailable sources, but pools such as those in the atmosphere, soil, and water are available sources of chemicals for the biotic community. Human activities remove chemicals from reservoirs and pools and make them available to the biotic community, and the result can be pollution.

and in that way phosphorus from the water is deposited on land. It would seem that anything put into the environment in one ecosystem could find its way to another ecosystem. Scientists find the soot from urban areas and pesticides from agricultural fields in the snow and animals of the Arctic.

The Water Cycle

The **water (hydrologic) cycle** is described in Figure 25.9. Fresh water is distilled from salt water. The sun's rays cause fresh water to evaporate from seawater, and the salts are left behind. Vaporized fresh water rises into the atmosphere, cools, and falls as rain over the oceans and the land.

Water evaporates from land and from plants (evaporation from plants is called transpiration). It also evaporates from bodies of fresh water, but since land lies above sea level, gravity eventually returns all fresh water to the sea. In the meantime, water is contained within standing waters (lakes and ponds), flowing water (streams and rivers), and groundwater.

When rain falls, some of the water sinks or percolates into the ground and saturates the earth. The top of the saturation zone is called the groundwater table, or simply, the water table. Sometimes groundwater is also located in **aquifers,** rock layers that contain water and will release it in appreciable quantities to wells or springs. Aquifers are recharged when rainfall and melted snow percolate into the soil. In some parts of the country, especially arid areas and southern Florida, humans withdraw water from aquifers for agriculture and everyday uses. The amount withdrawn exceeds any possibility of recharge. This is called "groundwater mining." In these locations the groundwater is dropping, and residents may run out of groundwater, at least for irrigation purposes, within a few short years. Fresh water, which makes up only about 3% of the world's supply of water, is called a renewable resource because a new supply is always being produced. But it is possible to run out of fresh water when the available supply is not adequate and/or is polluted so that it is not usable.

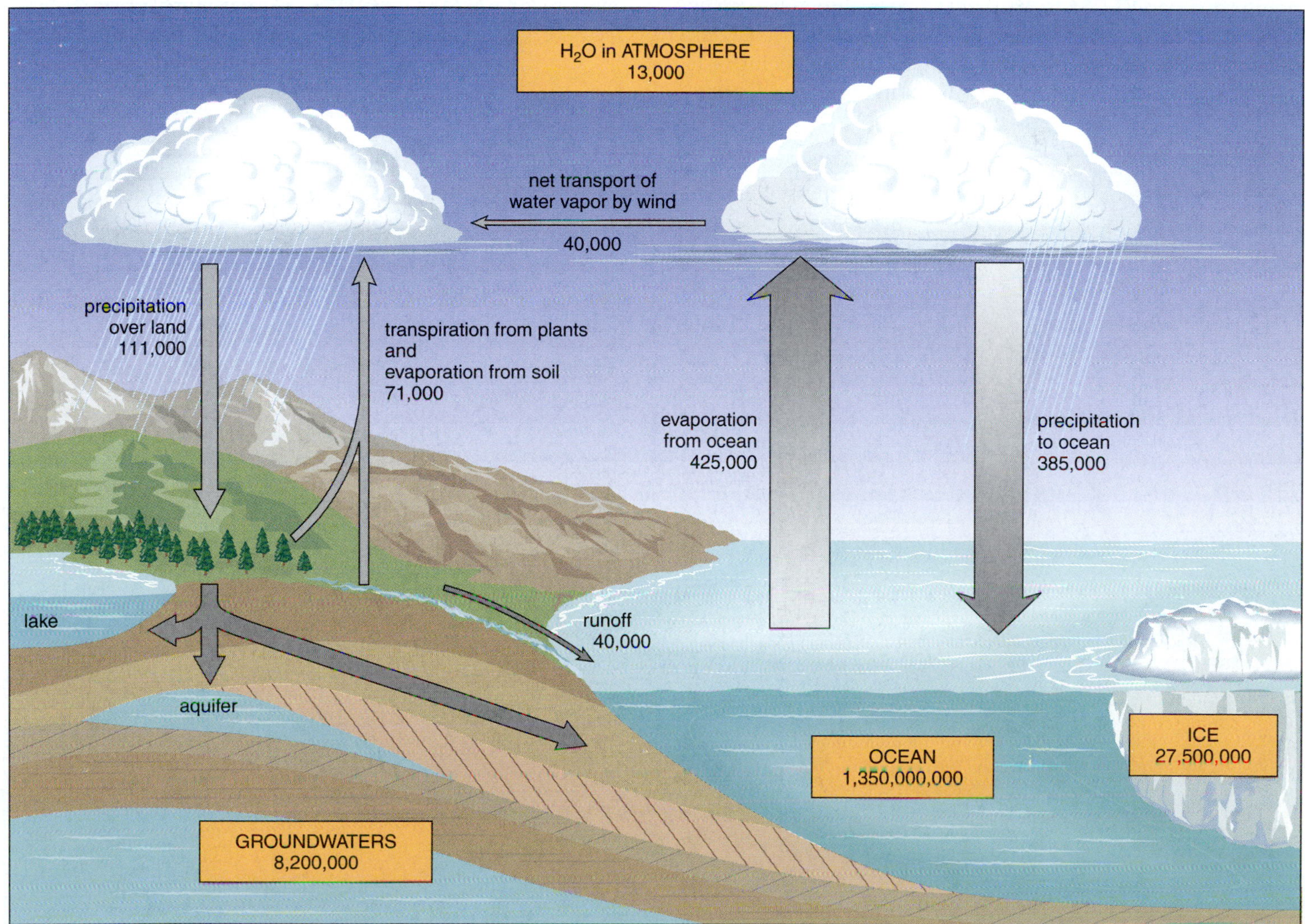

Figure 25.9 The hydrologic (water) cycle.
Evaporation from the ocean exceeds precipitation, so there is a net movement of water vapor onto land where precipitation results in surface water and groundwater that flow back to the sea. On land, transpiration by plants contributes to evaporation. The numbers in this diagram indicate water flow in km^3/yr.

The Carbon Cycle

In the **carbon cycle,** both terrestrial and aquatic organisms exchange carbon dioxide with the atmosphere (Fig. 25.10). On land, plants take up carbon dioxide from the air, and through photosynthesis they incorporate carbon into organic compounds (food) that is used for other living things. When organisms (e.g., plants, animals, and decomposers) respire, a portion of this carbon is returned to the atmosphere as carbon dioxide. In aquatic ecosystems, the exchange of carbon dioxide with the atmosphere is indirect. Carbon dioxide from the air combines with water to produce bicarbonate ion (HCO_3^-), a source of carbon for algae, which also produce food through photosynthesis. And when aquatic organisms respire, the carbon dioxide they give off becomes bicarbonate ions.

Living and dead organisms are reservoirs for carbon. If decomposition of dead remains fails to occur, they are subject to physical processes that transform them into coal, oil, and natural gas. We call these reservoirs for carbon the **fossil fuels.** Most of the fossil fuels were formed during the Carboniferous period, 286 to 360 million years ago, when an exceptionally large amount of organic matter was buried before decomposing. Another reservoir for carbon is calcium carbonate shells, which accumulate in ocean bottom sediments.

Carbon Dioxide and Global Warming

A **transfer rate** is defined as the amount of a nutrient that moves from one component of the environment to another within a specified period of time. The width of the arrows in Figure 25.10 indicates the transfer rate of carbon dioxide. The transfer rates due to photosynthesis and respiration, which includes decay, are just about even. However, there is now more carbon dioxide being deposited in the atmosphere than being removed. In 1850, atmospheric carbon dioxide was about 280 parts per million (ppm) and today it is about 350 ppm. This increase is largely due to the burning of fossil fuels and the destruction of forests to make way for farmland and pasture.

The emission of other gases due to human activities is also taking place. Altogether the following gases are expected to contribute significantly to an increase in atmospheric temperature called **global warming:**

Gas	From
Carbon dioxide (CO_2)	Fossil fuel and wood burning
Nitrous oxide (N_2O)	Fertilizer use and animal wastes
Methane (CH_4)	Biogas (bacterial decomposition, particularly in the guts of animals, in sediments, and in flooded rice paddies)

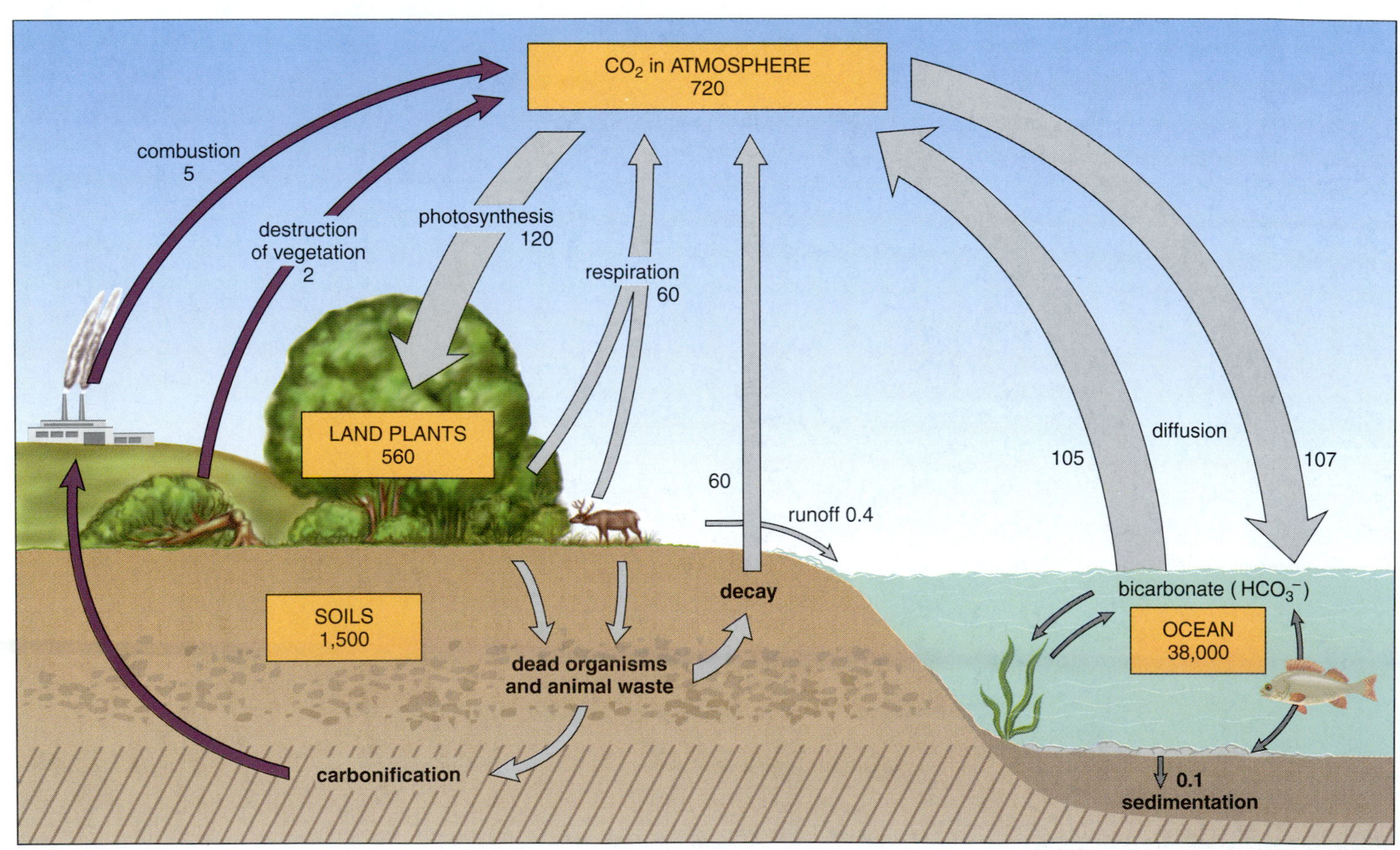

Figure 25.10 The carbon cycle.
The transfer rate of carbon into the atmosphere due to respiration just about matches the rate due to withdrawal by plants for photosynthesis. However, due to the burning of fossil fuels and destruction of vegetation by human activities (dark purple arrows), more carbon dioxide is added to the atmosphere than is withdrawn. The numbers in this diagram indicate 10^{15} g C/yr.

These gases contribute to the **greenhouse effect**—just like the panes of a greenhouse, they allow solar radiation to pass through but hinder the escape of infrared rays (heat) back into space. Figure 25.11 shows the earth's radiation balances. One thing to be learned from this diagram is that water vapor is a greenhouse gas: clouds also reradiate heat back to earth. If the earth's temperature rises due to the greenhouse effect, more water will evaporate, forming more clouds, setting up a positive feedback effect that could increase global warming.

Today, data collected around the world show a steady rise in the concentration of the various greenhouse gases. Methane, another significant greenhouse gas, given off by oil and gas wells, rice paddies, and organisms, is increasing by about 1% a year. Such data are used to generate computer models that predict the earth may warm to temperatures never before experienced by living things. The global climate has already warmed about 0.6°C since the Industrial Revolution. Computer models are unable to consider all possible variables, but the earth's temperature may rise 1.5° to 4.5°C by 2100 if greenhouse emissions continue at the current rates.

Global warming will bring about other effects, which computer models attempt to forecast. It is predicted that as the oceans warm, temperatures in the polar regions will rise to a greater degree than other regions. If so, glaciers would melt, and sea levels will rise, not only due to this melting but also because water expands as it warms. Water evaporation will increase, and most likely there will be increased rainfall along the coasts and dryer conditions inland. The occurrence of droughts will reduce agricultural yields and also cause trees to die off. Expansion of forests into Arctic areas might not offset the loss of forests in the temperate zones. Coastal agricultural lands such as the deltas of Bangladesh, India, and China would be inundated, and billions of dollars will have to be spent to keep coastal cities, like New York, Boston, Miami, and Galveston in the United States from disappearing into the sea.

The atmosphere is an exchange pool for carbon dioxide. Fossil fuel combustion in particular has increased the amount of carbon dioxide in the atmosphere. Global warming is predicted because carbon dioxide and other gases impede the escape of infrared radiation from the surface of the earth.

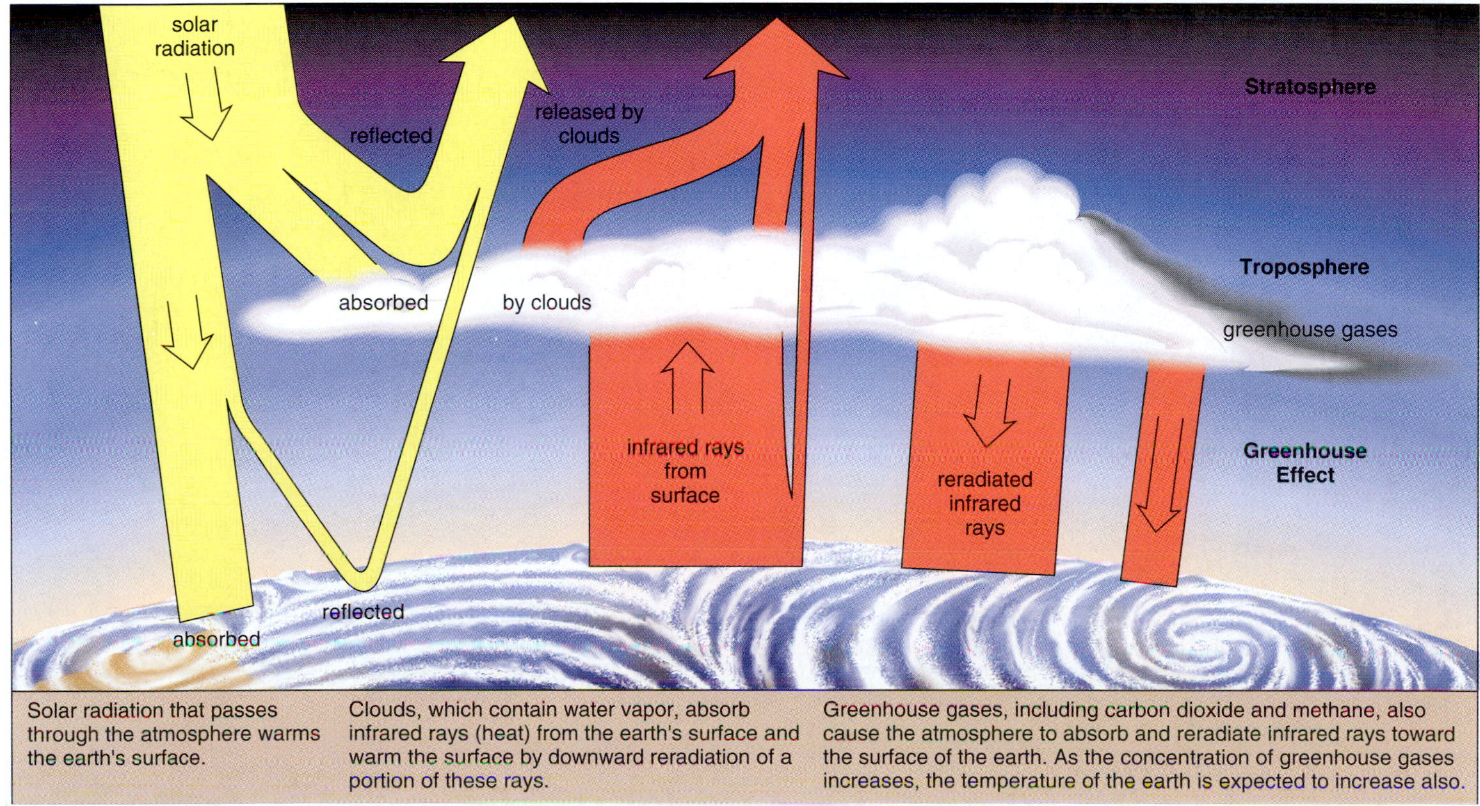
solar radiation
reflected
released by clouds
Stratosphere
Troposphere
greenhouse gases
absorbed
by clouds
infrared rays from surface
reradiated infrared rays
Greenhouse Effect
reflected
absorbed
Solar radiation that passes through the atmosphere warms the earth's surface.
Clouds, which contain water vapor, absorb infrared rays (heat) from the earth's surface and warm the surface by downward reradiation of a portion of these rays.
Greenhouse gases, including carbon dioxide and methane, also cause the atmosphere to absorb and reradiate infrared rays toward the surface of the earth. As the concentration of greenhouse gases increases, the temperature of the earth is expected to increase also.

Figure 25.11 Earth's radiation balances.
The contribution of greenhouse gases (far right) to the earth's surface is called global warming.

The Nitrogen Cycle

Nitrogen is an abundant element in the atmosphere. Nitrogen (N_2) makes up about 78% of the atmosphere by volume, yet nitrogen deficiency sometimes limits plant growth. Plants cannot incorporate nitrogen gas into organic compounds and therefore depend on various types of bacteria to make nitrogen available to them in a global cycle called the **nitrogen cycle** (Fig. 25.12).

Nitrogen fixation occurs when nitrogen (N_2) is converted to a form that plants can use. Some nitrogen-fixing bacteria live in nodules on the roots of legumes. They make nitrogen-containing organic compounds available to a host plant. Cyanobacteria in aquatic ecosystems and free-living bacteria in soil are able to fix nitrogen gas as ammonium (NH_4^+). Plants can use NH_4^+ and nitrate (NO_3^-) from the soil. After NO_3^- is taken up, it is enzymatically reduced to NH_4^+, which is used to produce amino acids and nucleic acids.

Nitrification is the production of nitrates. Nitrogen gas (N_2) is converted to nitrate (NO_3^-) in the atmosphere when cosmic radiation, meteor trails, and lightning provide the high energy needed for nitrogen to react with oxygen. Ammonium (NH_4^+) in the soil is converted to nitrate by chemoautotrophic soil bacteria in a two-step process. First, nitrite-producing bacteria convert ammonium to nitrite (NO_2^-), and then nitrate-producing bacteria convert nitrite to nitrate. Notice the subcycle in the nitrogen cycle that involves dead organisms and animal wastes, ammonium, nitrites, nitrates, and plants. This subcycle does not necessarily depend on nitrogen gas at all (Fig. 25.12).

Denitrification is the conversion of nitrate to nitrous oxide and nitrogen gas. There are denitrifying bacteria in both aquatic and terrestrial ecosystems. Denitrification balances nitrogen fixation, but not completely.

Nitrogen and Air Pollution

Human activities significantly alter transfer rates in the nitrogen cycle. Because we produce fertilizers, thereby converting N_2 to NO_3^-, and burn fossil fuels, the atmosphere contains three times the nitrogen oxides (NO_x) that it would otherwise. Fossil fuel combustion also pumps much sulfur

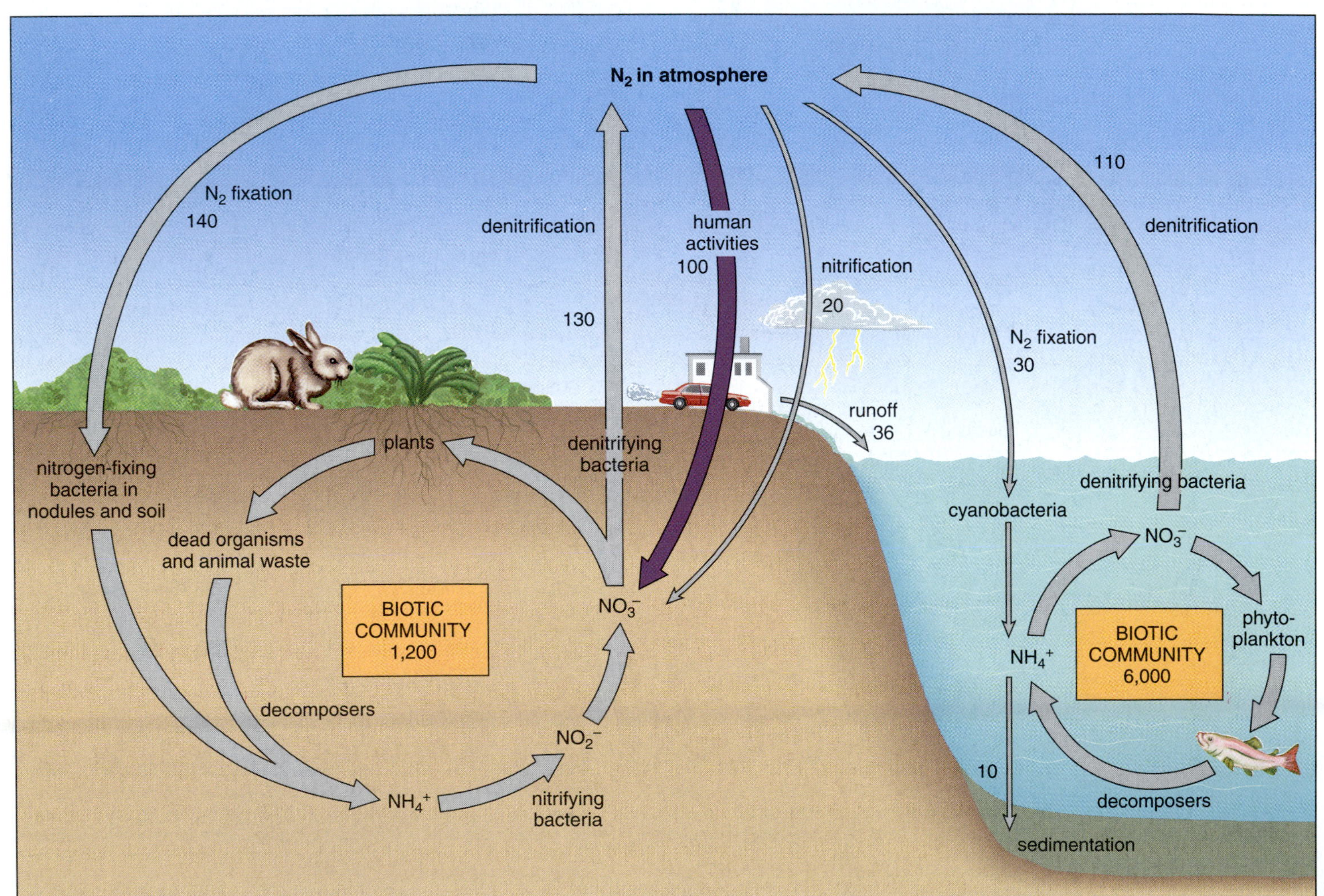

Figure 25.12 **The nitrogen cycle.**
Nitrogen is made available to biotic communities by internal cycling of the element. Without human activities the amount of nitrogen returned to the atmosphere (denitrification) exceeds withdrawal from the atmosphere (nitrogen fixation and nitrification). Human activities (dark purple arrow) result in an increased amount of NO_3^- in terrestrial communities with resultant runoff to aquatic biotic communities. The numbers in this diagram indicate 10^{12} g N/yr.

dioxide (SO_2) into the atmosphere. Both nitrogen oxides and sulfur dioxide are converted to acids during **acid deposition,** when they combine with water vapor in the atmosphere. These acids return to earth as either liquids (acid rain or snow) or solids (sulfate and nitrate salts).

Increased deposition of acids has drastically affected forests and lakes in northern Europe, Canada, and northeastern United States because their soils are naturally acidic and their surface waters are only mildly alkaline (basic) to begin with. The forests in these areas are dying (Fig. 25.13), and their waters cannot support normal fish populations. Acid deposition reduces agricultural yields and corrodes marble, metal, and stonework, an effect that is noticeable in cities.

Nitrogen oxides (NO_x) and hydrocarbons (HC) react with one another in the presence of sunlight to produce **photochemical smog,** which contains ozone (O_3) and PAN (peroxyacetylnitrate). Hydrocarbons come from fossil fuel combustion, but additional amounts come from various other sources as well, including paint solvents and pesticides. Breathing ozone affects the respiratory and nervous systems, resulting in respiratory distress, headache, and exhaustion. These symptoms are particularly apt to appear in young people. Ozone is especially damaging to plants, resulting in leaf mottling and reduced growth.

Normally, warm air near the ground is able to escape into the atmosphere. Sometimes, however, air pollutants, such as those in smog and soot, trap warm air near the earth. During a **thermal inversion** there is cold air at ground level beneath a layer of warm stagnant air above. Some areas surrounded by hills are particularly susceptible to the effects of a temperature inversion because the air tends to stagnate, and there is little turbulent mixing (Fig. 25.14).

Fertilizer use also results in the release of nitrous oxide (N_2O), a greenhouse gas and a contributor to ozone shield depletion in the stratosphere, a topic discussed in the reading on page 442.

Atmospheric N_2, a reservoir and exchange pool for nitrogen, must be fixed by bacteria in order to make nitrogen available to plants. Environmental problems are associated with the release of nitrous oxide (N_2O) and nitrogen oxides (NO_x) due to the action of bacteria on fertilizers and fossil fuel combustion, respectively.

Figure 25.13 **Acid deposition.**
a. Many forests in higher elevations of northeastern North America and Europe are dying due to acid deposition. **b.** Air pollution due to emissions from factories and fossil fuel burning is the major cause of acid deposition, which contains nitric acid (H_2NO_3) and sulfuric acid (H_2SO_4).

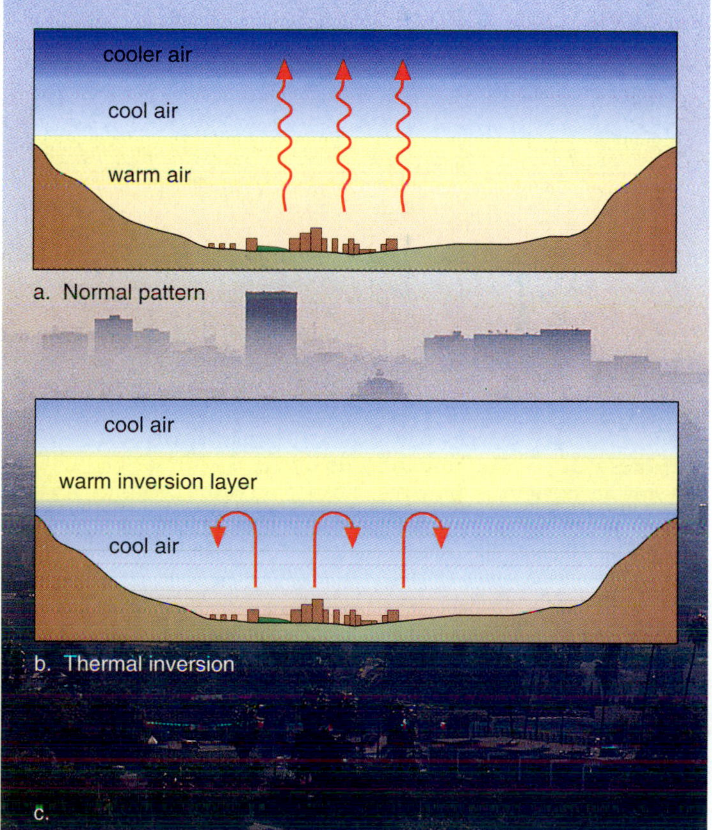

Figure 25.14 **Thermal inversion.**
a. Normally, pollutants escape into the atmosphere when warm air rises. **b.** During a thermal inversion, a layer of warm air (warm inversion layer) overlies and traps pollutants in cool air below. **c.** Los Angeles is particularly susceptible to thermal inversions, and this accounts for why this city is the "air pollution capital" of the United States.

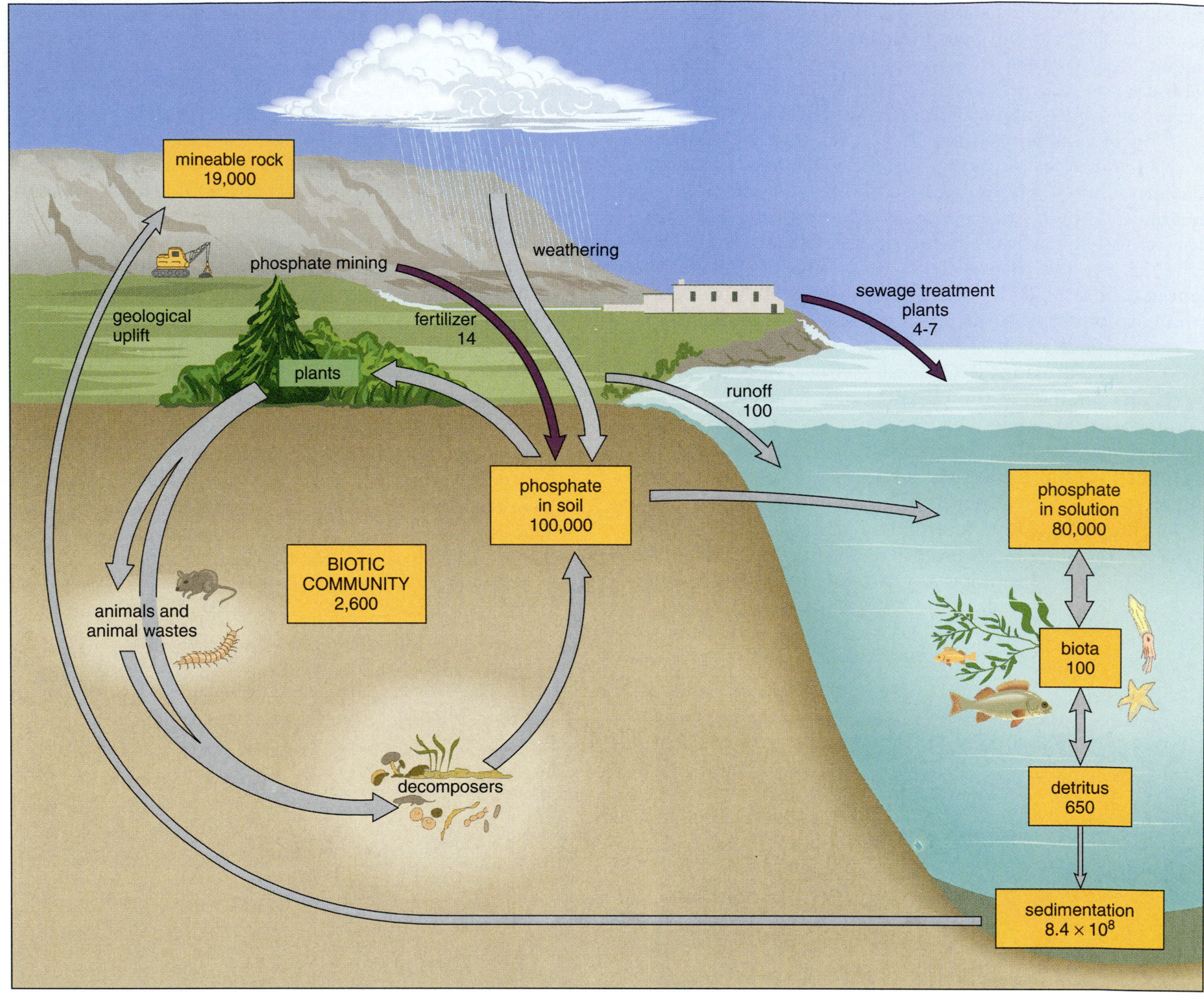

Figure 25.15 **The phosphorus cycle.**
The weathering of rocks provides phosphorus which cycles locally in both terrestrial and aquatic biota. When phosphorus becomes a part of oceanic sediments, it is lost to biotic communities for many years. The dark purple arrows show how humans alter transfer rates. Humans produce fertilizers which add to the amount of phosphorus available to biotic communities—eventually, fertilizers become a part of the runoff that enriches waters. Sewage treatment plants directly add phosphorus to local waters. The numbers in this diagram indicate 10^{12} g P/yr.

The Phosphorus Cycle

In the **phosphorus cycle,** phosphorus does not enter the atmosphere; therefore, this is a sedimentary cycle. On land, the weathering of rocks makes phosphate ions (PO_4^{3-} and HPO_4^{2-}) available to plants, which take up phosphate from the soil (Fig. 25.15). Some of this phosphate runs off into aquatic ecosystems where algae take phosphate up from the water before it becomes trapped in sediments. Phosphate in sediments only becomes available when a geological upheaval exposes sedimentary rocks to weathering once more. The phosphate taken up by producers is incorporated into a variety of molecules, including phospholipids and ATP or the nucleotides that become a part of DNA and RNA. Animals eat producers and incorporate some of the phosphate into teeth, bones, and shells that do not decompose for very long periods. Death and decay of all organisms, and also decomposition of animal wastes, do, however, make phosphate ions available to producers once again. Because available phosphate is generally taken up very quickly, it is often a limiting inorganic nutrient in most ecosystems. A limiting nutrient is one that regulates the growth of organisms because it is in shorter supply than other nutrients in the environment.

Phosphorus and Water Pollution

Human beings boost the supply of phosphate by mining phosphate ores for fertilizer and detergent production. Runoff of phosphate and nitrogen due to fertilizer use, animal wastes from livestock feedlots, as well as discharge from sewage treatment plants results in **eutrophication** (overenrichment). Eutrophication can lead to an algal bloom, apparent when green scum floats on the water. When the algae die off, decomposers use up all available oxygen during cellular respiration. The result is a massive fish kill.

Figure 25.16 lists the various sources of water pollution. Point sources are sources of pollution that are specific, and nonpoint sources are those caused by runoff from the land. Industrial wastes can include heavy metals and organochlorides, such as those in some pesticides. These materials are not degraded readily under natural conditions nor in conventional sewage treatment plants. They enter bodies of water and are subject to **biological magnification** because they remain in the body and are not excreted. Therefore, they become more concentrated as they pass along a food chain. Biological magnification occurs more readily in aquatic food chains because aquatic food chains have more links than terrestrial food chains. Humans are the final consumers in food chains, and in some areas, human milk contains detectable amounts of DDT and PCBs, which are organochlorides.

Coastal regions are the immediate receptors for local pollutants, and are the final receptors for pollutants carried by rivers that empty at a coast. Waste dumping occurs at sea, but ocean currents sometimes transport both trash and pollutants back to shore. Offshore mining and shipping add pollutants to the oceans. Some 5 million metric tons of oil a year—or more than one gram per 100 square meters of the oceans' surfaces—end up in the oceans. Large oil spills kill plankton, fish fry, and shellfishes, as well as birds and marine mammals. The largest tanker spill in U.S. territorial waters occurred on March 24, 1989, when the tanker Exxon *Valdez* struck a reef in Alaska's Prince William Sound and leaked 44 million liters of crude oil.

In the last 50 years, we have polluted the seas and exploited their resources to the point that many species are at the brink of extinction. Fisheries once rich and diverse, such as George's Bank off the coast of New England, are in severe decline. Haddock was once the most abundant species in this fishery, but now it accounts for less than 2% of the total catch. Cod and bluefin tuna have suffered a 90% reduction in population size. In warm, tropical regions, many areas of coral reefs are now overgrown with algae because the fish that normally keep the algae under control have been killed off.

Sedimentary rock is a reservoir for phosphorus; for the most part producers are dependent on decomposers to make phosphate available to them. Fertilizer production and other human activities add phosphate to aquatic ecosystems, contributing to water pollution.

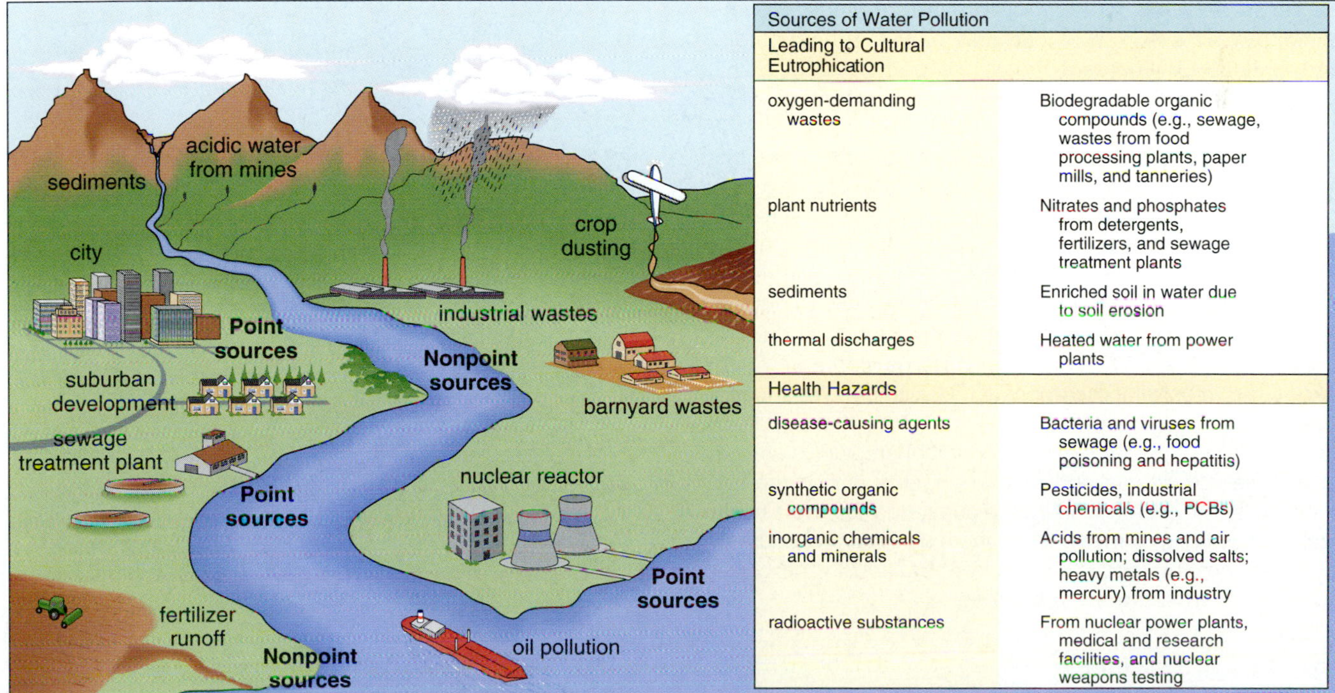

Sources of Water Pollution	
Leading to Cultural Eutrophication	
oxygen-demanding wastes	Biodegradable organic compounds (e.g., sewage, wastes from food processing plants, paper mills, and tanneries)
plant nutrients	Nitrates and phosphates from detergents, fertilizers, and sewage treatment plants
sediments	Enriched soil in water due to soil erosion
thermal discharges	Heated water from power plants
Health Hazards	
disease-causing agents	Bacteria and viruses from sewage (e.g., food poisoning and hepatitis)
synthetic organic compounds	Pesticides, industrial chemicals (e.g., PCBs)
inorganic chemicals and minerals	Acids from mines and air pollution; dissolved salts; heavy metals (e.g., mercury) from industry
radioactive substances	From nuclear power plants, medical and research facilities, and nuclear weapons testing

Figure 25.16 Sources of surface water pollution.
Many bodies of water are dying due to the introduction of pollutants from point sources, which are easily identifiable, and nonpoint sources, which cannot be specifically identified.

Stratospheric Ozone Depletion Threatens the Biosphere

The earth's atmosphere is divided into layers. The troposphere envelops us as we go about our day-to-day lives. When ozone (O_3) is present in the troposphere (called ground-level ozone), it is considered a pollutant because it adversely affects a plant's ability to grow and our ability to breathe oxygen (O_2). In the stratosphere, some 50 kilometers above the earth, ozone forms the **ozone shield** that absorbs much of the ultraviolet (UV) rays of the sun so that fewer rays strike the earth.

UV radiation causes mutations that can lead to skin cancer and can make the lens of the eyes develop cataracts. It also is believed to adversely affect the immune system and our ability to resist infectious diseases. Crop and tree growth is impaired, and UV radiation also kills off small plants (phytoplankton) and tiny shrimplike animals (krill) that sustain oceanic life. Without an adequate ozone shield, our health and food sources are threatened.

Depletion of the ozone layer within the stratosphere in recent years is, therefore, of serious concern. It became apparent in the 1980s that some worldwide depletion of ozone had occurred and that there was a severe depletion of some 40–50% above the Antarctic every spring. A vortex of cold wind (a whirlpool in the atmosphere) circles the pole during the winter months, creating ice crystals where chemical reactions occur that break down ozone. Severe depletions of the ozone layer are commonly called "ozone holes." Detection devices now tell us that the ozone hole above the Antarctic is about the size of the United States and growing. Of even greater concern, an ozone hole has now appeared above the Arctic as well, and ozone holes could also occur within northern and southern latitudes, where many people live. Whether or not these holes develop depends on prevailing winds, weather conditions, and the type of particles in the atmosphere. A United Nations Environment Program report predicts a 26% rise in cataracts and nonmelanoma skin cancers for every 10% drop in the ozone level. A 26% increase translates into 1.75 million additional cases of cataracts and 300,000 more skin cancers every year, worldwide.

The cause of ozone depletion can be traced to the release of chlorine atoms (Cl) into the stratosphere (Fig. 25A). Chlorine atoms combine with ozone and strip away the oxygen atoms, one by one. One atom of chlorine can destroy up to 100,000 molecules of ozone before settling to the earth's surface as chloride years later. These chlorine atoms come from the breakdown of chlorofluorocarbons (CFCs), chemicals much in use by humans. The best known CFC is Freon, a heat transfer agent found in refrigerators and air conditioners. CFCs are also used as cleaning agents and foaming agents during the production of styrofoam found in coffee cups, egg cartons, insulation, and paddings. Formerly, CFCs were used as propellants in spray cans, but this application is now banned in the United States and several European countries.

Most countries of the world have agreed to stop using CFCs by the year 2000. The United States halted production in 1995. Computer projections suggest that an 85% reduction in CFC emissions is needed to stabilize CFC levels in the atmosphere. Otherwise they keep on increasing. Scientists are now searching for CFC substitutes that will not release chlorine atoms (nor bromine atoms) to harm the ozone shield.

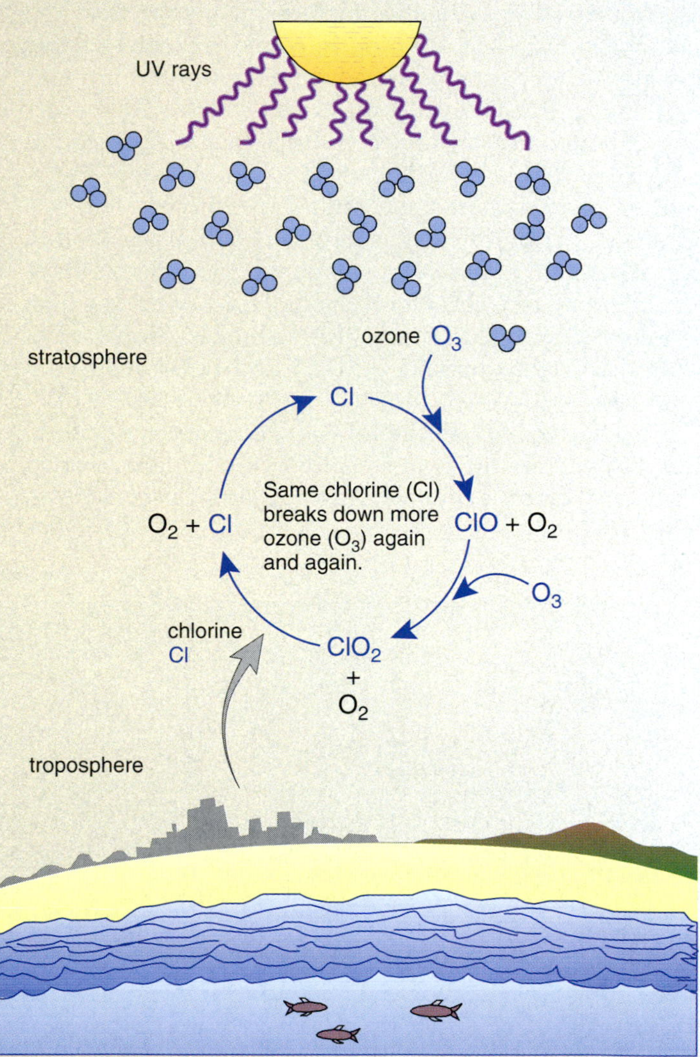

Figure 25A Ozone depletion.
CFCs (chlorofluorocarbons) release chlorine atoms that lead to the breakdown of ozone (O_3) and the buildup of oxygen (O_2) in the stratosphere. Oxygen does not absorb UV radiation and does not protect the earth.

Connecting Concepts

One of the most basic ways to study ecosystem function is to study energy flow within ecosystems. By measuring the amount of carbon entering an ecosystem via photosynthesis you can estimate how much solar energy is being fixed by primary producers. You can then determine how much carbon is moving from one trophic level to the next, and finally how much carbon is released through respiration to examine the entire path of energy flow. By determining which factors govern different levels of energy flow (factors such as precipitation, nutrient availability, and seasonality), you can develop a good model for each specific ecosystem.

Many ecologists feel that an understanding of energy flow is necessary to understanding nutrient cycling and availability. Nutrient availability does not just influence plant growth, it also determines distributions of animals. Many large grazing mammals preferentially consume plants with high nutrient content and thus tend to occur in areas with high nutrient availability.

Ecosystem studies have become increasingly sophisticated, and have moved beyond studying energy flow and nutrient cycling in general. Much current research is directed toward learning how particular species influence the function of an ecosystem. Thus we are now becoming able to predict which species are most important and to determine which species must be preserved in order to have natural ecosystem functioning.

Human activities are affecting the transfer rates of biogeochemical cycles and therefore the biosphere, the largest ecosystem of all. The pumping of water from aquifers is not normally a part of the water cycle. The burning of fossil fuels and trees is increasing the amount of carbon dioxide in the atmosphere and the end result may be global warming. Carbon dioxide and other greenhouse gases allow the sun's rays to pass through but they absorb and reradiate heat back to the earth. Transfer rates in both the phosphorus and nitrogen cycles are affected when we produce fertilizers and detergents. Nitrogen and phosphorus runoff in aquatic ecosystems cause eutrophication. Pollution can be defined as a change in transfer rates that lead directly or indirectly to a degradation of human health or a degradation of plant and animal life.

Summary

25.1 The Nature of Ecosystems

The biosphere, which contains living things, includes portions of the atmosphere, lithosphere, and hydrosphere. Ecosystems contain biotic (living) components and abiotic (physical) components where energy flows and nutrients cycle among different global chemical pools. The biotic components of ecosystems are either producers or consumers. Consumers may be herbivores, carnivores, omnivores, and detritivores.

25.2 Energy Flow and Nutrient Cycling

Energy flows through an ecosystem. When autotrophs photosynthesize, they transform solar energy into the chemical-bound energy of organic molecules, which become nutrients for themselves and all heterotrophs. As herbivores feed on plants (or algae), and carnivores feed on herbivores, some energy is converted to heat. Feces, urine, and dead bodies become food for detritivores. Eventually all the solar energy that enters an ecosystem is converted to heat, and thus ecosystems require a continual supply of solar energy. Thus ecosystems are open systems.

Nutrients are not lost from the biosphere as is heat. Nutrients recycle within and between ecosystems. Detritivores return some proportion of inorganic nutrients to autotrophs, and other portions are imported or exported between ecosystems in global cycles.

Ecosystems contain food webs, and a diagram of a food web shows how the various organisms are connected by eating relationships. Grazing food webs begin with vegetation that is fed on by herbivores, which become food for carnivores. In detrital food webs, decomposers act on organic material in the soil, and they are fed on by other detritivores, which become food for carnivores in the grazing food web. Thus the two food webs are connected.

It's possible to isolate food chains (straight-line diagrams of who eats whom) from food webs. A trophic level is all the organisms that feed at a particular link in a food chain. Ecological pyramids show trophic levels stacked one on the other like building blocks. Generally they show that biomass and energy content decrease from one trophic level to the next. Most pyramids pertain to grazing food webs and largely ignore the detrital food web portion of an ecosystem.

25.3 Global Biogeochemical Cycles

Biogeochemical cycles contain reservoirs, components of ecosystems like fossil fuels, sediments, and rocks that contain elements available on a limited basis to living things. Pools are components of ecosystems like the atmosphere, soil, and water—which are ready sources of nutrients for living things. Nutrients cycle among the members of the biotic component of an ecosystem.

In the water cycle, evaporation over the ocean is not compensated for by rainfall. Evaporation from terrestrial ecosystems includes transpiration from plants. Rainfall over land results in bodies of fresh water plus groundwater, including aquifers. Eventually all water returns to the oceans.

In the carbon cycle, organisms add as much carbon dioxide to the atmosphere as they remove. Shells in ocean sediments, organic compounds in living and dead organisms, and fossil fuels are reservoirs for carbon. Human activities such as the burning of fossil fuels and trees are adding carbon dioxide to the atmosphere. Like the panes of a greenhouse, carbon dioxide and other gases allow the sun's rays to pass through but impede the release of infrared wavelengths. It is predicted that a buildup of these "greenhouse gases" will lead to a global warming. The effects of global warming could be a rise in sea level and a change in climate patterns with disastrous effects.

In the nitrogen cycle, the biotic community, which includes several types of bacteria, keeps nitrogen recycling back to the producers. A few organisms (cyanobacteria in aquatic habitats and bacteria in soil and root nodules) can fix atmospheric nitrogen. Other bacteria return nitrogen to the atmosphere. Human activities convert atmospheric nitrogen to fertilizer which when broken down by soil bacteria adds nitrogen oxides to the atmosphere. Nitrogen oxides (NO_x) and sulfur dioxide (SO_2) react with water vapor to form acids that contribute to acid deposition. Nitrogen oxides and hydrocarbons (HC) react to form smog, which contains ozone and PAN (peroxyacetylnitrate).

In the phosphorus cycle, the biotic community recycles phosphorus back to the producers, and only limited quantities are made available by the weathering of rocks. Phosphates are mined for fertilizer production; when phosphates and nitrates enter lakes and ponds, overenrichment occurs. Many kinds of wastes enter rivers which flow to the oceans now degraded from added pollutants.

Reviewing the Chapter

1. Define the biosphere and tell where it is located. 422
2. Distinguish between autotrophs and heterotrophs, and describe four different types of heterotrophs found in natural ecosystems. 423
3. Discuss energy flow in an ecosystem and nutrient cycling within and between ecosystems. Why are ecosystems open systems? 424–425
4. Describe the two types of food webs typically found in ecosystems. Which of these typically moves more energy through an ecosystem? 426–427
5. With reference to food chains, what is a trophic level? what is an ecological pyramid? 427–428
6. Give examples of reservoirs and pools in biogeochemical cycles. Which of these is less accessible to living things? 428
7. Draw a diagram to illustrate the water cycle and the carbon cycle. 429–430
8. How and why is the global temperature expected to change, and what are the predicted consequences of this change? 431–432
9. Draw a diagram of the nitrogen cycle. What types of bacteria are involved in this cycle? 432
10. What causes acid deposition, and what are its effects? 433
11. How does photochemical smog develop, and what is a thermal inversion? 433
12. Draw a diagram of the phosphorus cycle. 434
13. What are several ways in which fresh water and marine waters can be polluted? What is biological magnification? 435
14. Of what benefit is the ozone shield? What pollutant in particular should be associated with ozone shield depletion, and what are the consequences of this depletion? 436

Testing Yourself

Choose the best answer for each question.

1. Of the total amount of energy that passes from one trophic level to another, about 10% is
   a. respired and becomes heat.
   b. passed out as feces or urine.
   c. stored as body tissue.
   d. recycled to autotrophs.
   e. All of these are correct.
2. Compare this food chain:
   algae → water fleas → fish → green herons
   to this food chain:
   trees → tent caterpillars → red-eyed vireos → hawks.
   Both water fleas and tent caterpillars are
   a. carnivores.
   b. primary consumers.
   c. detritus feeders.
   d. Both a and b are correct.
   e. present in both grazing and detrital food webs.
3. Which of the following contribute(s) to the carbon cycle?
   a. respiration
   b. photosynthesis
   c. fossil fuel combustion
   d. decomposition of dead organisms
   e. All of these are correct.
4. How do plants contribute to the carbon cycle?
   a. When plants respire, they release CO_2 into the atmosphere.
   b. When plants photosynthesize, they consume CO_2 from the atmosphere.
   c. When plants photosynthesize, they provide oxygen to heterotrophs.
   d. When plants emigrate they transport carbon molecules between ecosystems.
   e. Both a and b are correct.
5. How do nitrogen-fixing bacteria contribute to the nitrogen cycle?
   a. They return nitrogen (N_2) to the atmosphere.
   b. They change ammonium to nitrate.
   c. They change N_2 to ammonium.
   d. They withdraw nitrate from the soil.
   e. They decompose and return nitrogen to autotrophs.
6. In what way are detritivores like producers?
   a. Either may be the first member of both a grazing or a detrital food chain.
   b. Both produce oxygen for other forms of life.
   c. Both require a source of nutrient molecules and energy.
   d. Both are present only in the lithosphere.
   e. Both produce organic nutrients for other members of ecosystems.
7. Which statement is true concerning this food chain:
   grass → rabbits → snakes → hawks?
   a. Each predator population has a greater biomass than its prey population.
   b. Each prey population has a greater biomass than its predator population.
   c. Each population is omnivorous.
   d. Each returns inorganic nutrients and energy to the producer.
   e. Both a and c are correct.

For questions 8–11, match the terms with those in the key:

Key:

a. sulfate salts
b. ozone
c. carbon dioxide
d. chlorofluorocarbons (CFCs)
e. pesticides

8. acid deposition
9. ozone shield destruction
10. greenhouse effect
11. photochemical smog
12. Which of these is mismatched?
    a. fossil fuel burning—carbon dioxide given off
    b. nuclear power—radioactive wastes
    c. fertilizer use—phosphate in atmosphere
    d. aerosol sprays—chlorine in atmosphere
    e. pesticides—pollutants in water

13. Acid deposition causes
    a. lakes and forests to die.
    b. acid indigestion in humans.
    c. the greenhouse effect to lessen.
    d pests to increase decomposition.
    e. All of these are correct.
14. Water is a renewable resource, and
    a. there will always be a plentiful supply.
    b. the oceans can never become polluted.
    c. it is still subject to pollution.
    d. primary sewage treatment plants assure clean drinking water.
    e. Both a and c are correct.
15. Label the trophic levels using two of these terms for each level: producers, top carnivores, secondary consumers, autotrophs, primary consumers, tertiary consumers, carnivores, herbivores.
    e. Tell what the numbers on the left refer to.

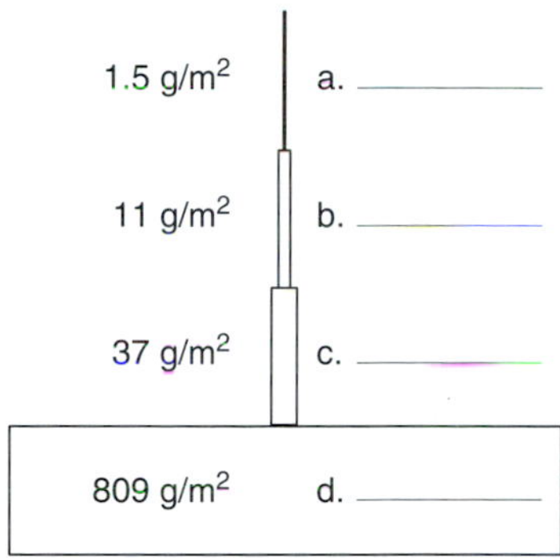

Thinking Scientifically

1. A large forest has been removed by clear-cutting and the land has not been replanted. After several years, humidity in the area seems to have decreased. The few plants that are present in the area are growing more slowly than would be expected based on the lack of competition. How can knowledge of the water cycle be used to interpret these observations?
2. In mountainous regions of western United States, there is interest in reintroducing large mammalian predators that were previously driven out of the area. For these projects to work, predator populations must have enough wild food or else domestic livestock may be preyed upon. What type data is needed concerning food webs and ecological pyramids to successfully reintroduce predators?

Bioethical Issue

Peter Jutro, a scientist working for the U.S. Environmental Protection Agency, wants to research native traditions for clues on how to preserve ecosystems. He has run into opposition from the indigenous groups because they mistrust conservationists. Take, as an example, the fact that the Kuna people in Panama refused to renew the lease for a Smithsonian Institution studying reef ecology for the past 21 years. Much of the trouble seems to have come from the failure of scientists to explain their program to local communities. In a meeting of the Kuna congress, the scientists were accused by the Kuna of "stealing their knowledge, stealing their reefs, stealing their sand." Local people find it hard to see a difference between a scientific study and commercial ventures which exploit their areas for minerals, timber, and other resources.

Laura Snook, a forester from Duke University, researches the growing habits of mahogany trees in Mexico. She came to the conclusion that because the trees grow slowly, logging shouldn't be done too fast. Most local foresters resented her findings, but she was able to establish a good relationship with two women foresters who were open to her ideas. Now local people are changing the way they replant mahogany trees so that the resource will be there for some time to come. The point is that conservationists working with people from different cultural backgrounds probably need to communicate their goals more clearly and involve local people in project planning. Perhaps they should learn to accept slower timescales and styles of decision making different from our own. Or do you think local groups should learn to think as we do?

Understanding the Terms

acid deposition	433	fossil fuel	430
aquifer	429	global warming	430
autotroph	423	grazing food web	427
biogeochemical cycle	428	greenhouse effect	431
biological magnification	435	herbivore	423
biosphere	422	heterotroph	423
carbon cycle	430	nitrification	432
carnivore	423	nitrogen cycle	432
consumer	423	nitrogen fixation	432
decomposition	423	omnivore	423
denitrification	432	ozone shield	436
detrital food web	427	phosphorus cycle	434
detritivore	423	photochemical smog	433
detritus	423	producer	423
ecological pyramid	427	thermal inversion	433
ecosystem	422	transfer rate	430
eutrophication	435	trophic level	422
food chain	427	water (hydrologic) cycle	429
food web	427		

Match the terms to these definitions:

a. ______________ Partially decomposed remains of plants and animals.
b. ______________ Formed from oxygen in the upper atmosphere, it protects the earth from ultraviolet radiation.
c. ______________ Remains of once living organisms that are burned to release energy, such as coal, oil, and natural gas.
d. ______________ Process by which atmospheric nitrogen gas is changed to forms that plants can use.
e. ______________ Complex pattern of interlocking and crisscrossing food chains.

Web Connections

Exploring the Internet

http://www.mhhe.com/biosci/genbio/mader
(click on *Biology 7/e*)

The *Biology 7/e* Online Learning Center provides many resources for studying the material in this chapter including links to the following sites:

Biodiversity and Biological Collections Web Server. This great site with lots of connections provides information on biodiversity studies from around the world.

http://biodiversity.uno.edu

Biodiversity. A fact sheet on biodiversity from the Ecological Society of America, with links to more information.

http://esa.sdsc.edu/biodiv2.htm

Virtual Library of Biodiversity, Ecology and the Environment. A clickable index, with lists of many endangered species, state issues, and legislation related to endangered species.

http://conbio.rice.edu/vl/browse/

National Wildlife *and* International Wildlife *Magazine Articles. Text of current and past articles from both magazines.*

http://www.nwf.org/natlwild/

Great Outdoor National Parks and Preserves: National Park Guides. Information and links to sites describing all national parks in all states, along with links to scenic areas, wilderness areas, beaches, and rivers.

http://www.gorp.com/gorp/resource/US_National_Park/main.htm

The Biosphere

chapter 

chapter concepts

26.1 Climate and the Biosphere

- Solar radiation provides the energy that drives climate differences in the biosphere. 442
- Global air circulation patterns and physical features help produce the various patterns of temperature and rainfall about the globe. 442

26.2 Biomes of the World

- Ecosystems can be divided into a few major types of terrestrial and aquatic biomes. 445
- Soil composition plays a major role in determining the plants and therefore the animals found in a particular region. 446

26.3 Terrestrial Biomes

- The Earth's major terrestrial biomes are forests (broad-leaf and coniferous), shrublands, grasslands (tropical savanna and temperate grasslands), tundra, and deserts. 447

26.4 Aquatic Biomes

- The Earth's major aquatic biomes are of two types: fresh water and salt water (usually marine). 455
- Ocean currents also affect the climate and the weather over the continents. 460

Satellite image of Earth

From outer space, the earth looks like an uninhabited, pristine aqua globe, hovering in space. Unfortunately, looks are deceiving—not only is Earth heavily populated, but there is nothing pristine about it. The farms, towns, and cities of human beings now dominate the biosphere, and most of the biomes of the world have been greatly affected by human activities. We have cleared vast areas of ecosystems after using the plants and killing the animals for our own purposes.

In this chapter, we see how ecosystems are distributed over the globe and how climate determines the characteristics of the major biomes, such as forest, deserts, grasslands, and oceans. Each biome has its own mix of species, which are adapted to living under particular environmental conditions. Through biogeochemical cycles, driven by solar energy, natural ecosystems transform the earth's crust, its waters, and the atmosphere into a life-supporting environment. It is critical for us to learn to value the services of ecosystems and to work toward preserving the remaining portions of the original biomes. In this way, we can help preserve species, including our own species.

26.1 Climate and the Biosphere

The distribution of biomes in the biosphere is dependent upon (1) variations in solar radiation reception due to a spherical earth, (2) the tilt of the earth's axis as it rotates about the sun, (3) distribution of landmasses and oceans, and (4) topography (landscape) features. Due to these factors, **climate,** particularly dictated by temperature and rainfall, differs throughout the biosphere.

Air Circulation

Because the earth is a sphere, the sun's rays are more direct at the equator and more spread out at polar regions (Fig. 26.1*a*). Therefore, the tropics are warmer than temperate regions. The tilt of the earth as it orbits around the sun causes one pole or the other to be closer to the sun (except at the spring and fall equinoxes), and this accounts for there being seasons in all parts of the earth except at the equator (Fig. 26.1*b*). When the Northern Hemisphere is having winter, the Southern Hemisphere is having summer and vice versa.

Atmospheric heat always passes from warm areas to colder areas. If the earth were standing still, and were a solid, uniform ball, all air movements—which we call winds—would be in two directions. Warm equatorial air would rise and move directly to the poles, creating a zone of lower pressure that would be filled by cold polar air moving equatorward.

Because the earth rotates, and because its surface consists of continents and oceans, the flows of warm and cold air are modified into three large circulation cells in each hemisphere (Fig. 26.2). At the equator, the sun heats the air and evaporates water. The warm moist air rises, cools, and loses most of its moisture as rain. The greatest amounts of rainfall on earth are near the equator. The rising air flows toward the poles, but at about 30° north and south latitude, it sinks toward the earth's surface and reheats. As the air descends and warms, it becomes very dry, creating zones of low rainfall. The great deserts of Africa, Australia, and the Americas occur at these latitudes. At the earth's surface, the air flows both poleward and equatorward. At about 60° north and south latitude, the air rises and cools, producing additional zones of high rainfall. This moisture supports the great forests of Temperate Zone. Part of this rising air flows equatorward, and part continues poleward, descending near the poles, which are zones of low precipitation.

The earth is not standing still; it rotates on its axis daily. The spinning of the earth affects the winds, so that the major global circulation systems flow toward the east or west rather than directly north or south (Fig. 26.2). Between about 30° north latitude and 30° south latitude, the winds blow from the east-southeast in the Southern Hemisphere and from the east-northeast in the Northern Hemisphere (the east coasts of continents at these latitudes are wet). These are called trade winds because sailors depended upon them to

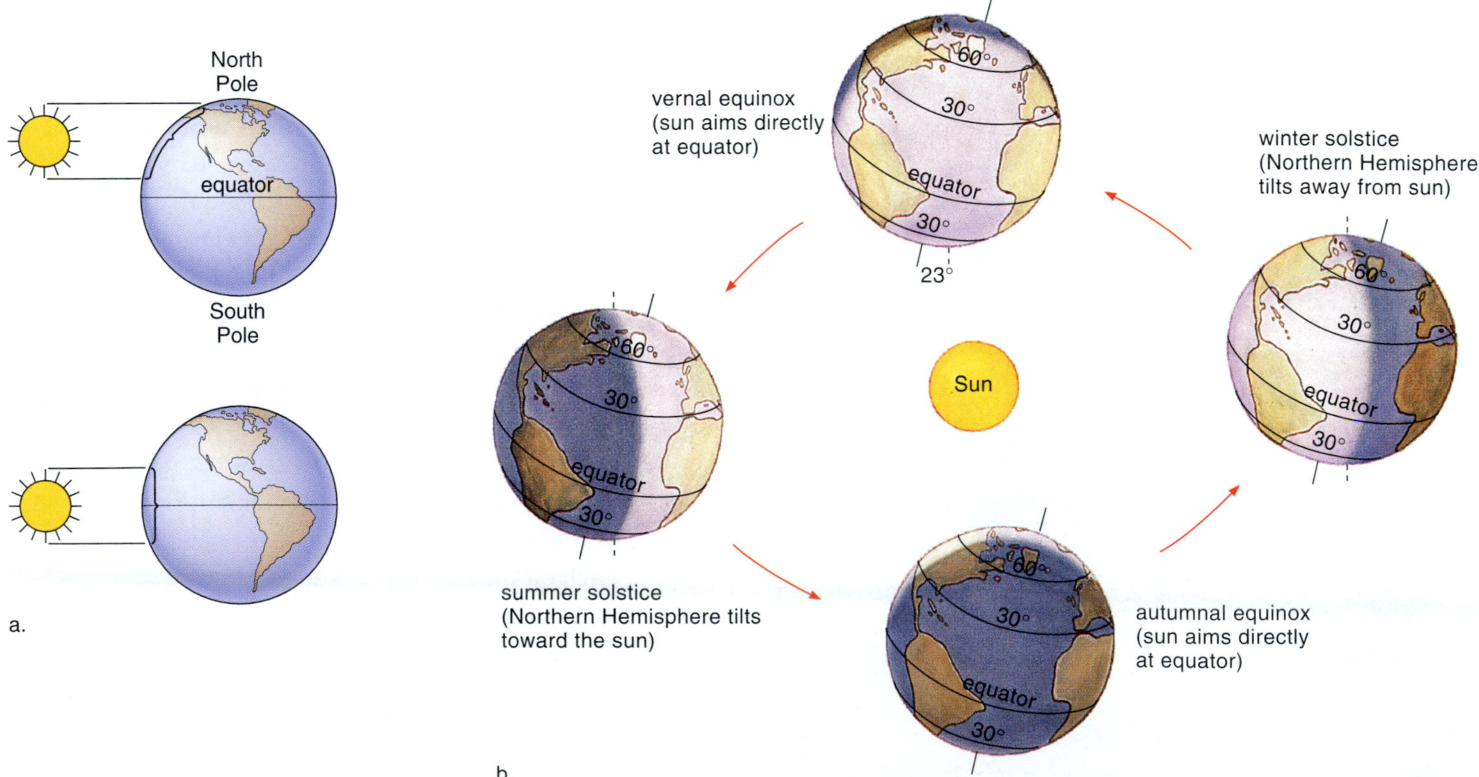

Figure 26.1 Distribution of solar energy.
a. Since the earth is a sphere, beams of solar energy striking the earth near one of the poles is spread over a wider area than similar beams striking the earth at the equator. **b.** The seasons of the Northern and Southern Hemispheres are due to the tilt of the earth on its axis as it rotates about the sun.

fill the sails of their trading ships. Between 30° and 60° north and south latitude, strong winds, called the prevailing westerlies, blow from west to east. The west coasts of the continents at these latitudes are wet, as is the Pacific Northwest where a massive evergreen forest is located. Weaker winds, called the polar easterlies, blow from east to west at still higher latitudes of their respective hemispheres.

The distribution of solar energy caused by a spherical earth, and the rotation and path of the earth about the sun, affect how the winds blow and the amount of rainfall various regions receive.

Effects of Topography

Topography means the physical features or "the lay" of the land. One physical feature that affects climate is the presence of mountains. As air blows up and over a mountain range, it rises and cools. This side of the mountain, called the windward side, receives more rainfall than the other side, called the leeward side. On the leeward side, the air descends, picks up moisture, and produces clear weather (Fig. 26.3). The difference between the windward side and the leeward side can be quite dramatic. In the Hawaiian Islands, for example, the windward side of the mountains receives more than 750 cm of rain a year, while the leeward side, which is in a **rain shadow,** gets on the average only 50 cm of rain and is generally sunny. In the United States, the western side of the Sierra Nevada Mountains is lush, while the eastern side is a semidesert.

The oceans are slower to change temperature—that is, to gain or lose their heat—than landmasses. This causes coasts to have a unique weather pattern that is not seen inland. During the day, the land warms more quickly than the ocean, and the air above the land rises. Then a cool sea breeze blows in from the ocean. At night the reverse happens; the breeze blows from the land to the sea.

India and some other countries in southern Asia have a **monsoon** climate, in which wet ocean winds blow onshore for almost half the year. The land heats more rapidly than the waters of the Indian Ocean during spring. The difference in temperature between the land and ocean causes a gigantic circulation of air: warm air rises over the land, and cooler air comes in off the ocean to replace it. As the warm air rises, it loses its moisture and the *monsoon season* begins. As just discussed, rainfall is particularly heavy on the windward side of hills. Cherrapunji in northern India receives an annual average of 1,090 cm of rain a year because of its high altitude. The weather pattern has reversed by November. The land is now cooler than the ocean; therefore, dry winds blow from the Asian continent across the Indian Ocean. In the winter, the air over the land is dry, the skies cloudless, and temperatures are pleasant. The chief crop of India is rice, which starts to grow when the monsoon rains begin.

In the United States, people often speak of the "lake effect," meaning that in the winter, Arctic winds blowing over the Great Lakes become warm and moisture laden. When these winds rise and lose their moisture, snow begins to fall. Places such as Buffalo, New York, get heavy snowfalls due to the lake effect, and there is snow on the ground for an average of 90 to 140 days every year.

Atmospheric circulations between the ocean and landmasses influence regional climate conditions and the distribution of biomes.

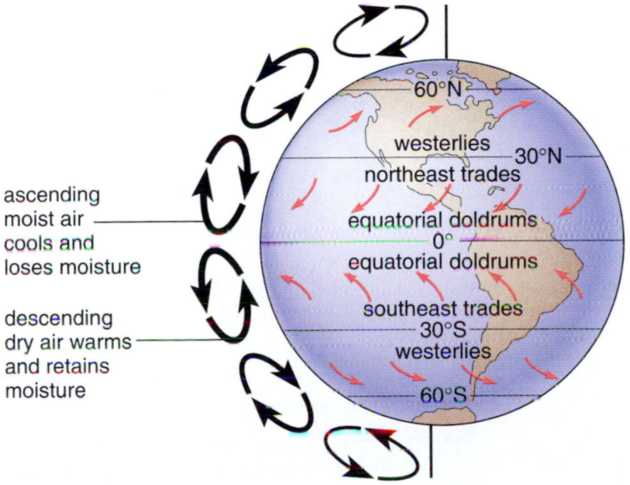

ascending moist air cools and loses moisture
descending dry air warms and retains moisture
60°N
westerlies
30°N
northeast trades
equatorial doldrums
0°
equatorial doldrums
southeast trades
30°S
westerlies
60°S

Figure 26.2 Global wind circulation.
At the equator, warm air rises and loses its moisture. At 30°, dry air descends; therefore, deserts occur at 30° latitude around the world. Because the earth is rotating on its axis, the trade winds move from the northeast to west in the Northern Hemisphere, and from the southeast to the west in the Southern Hemisphere. The westerlies move toward the east.

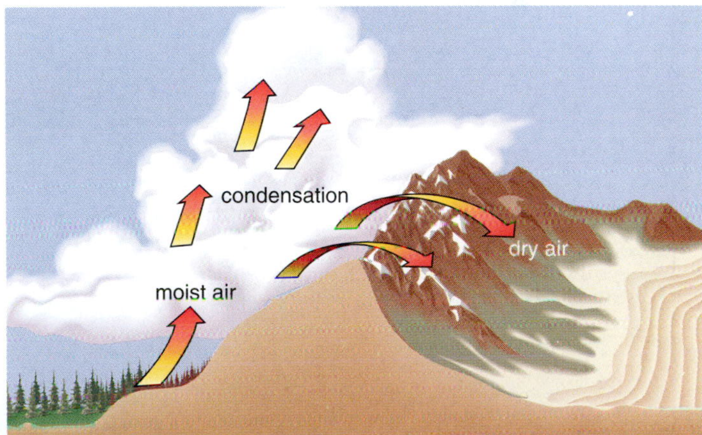

Figure 26.3 Formation of a rain shadow.
When winds from the sea cross a coastal mountain range, they rise and release their moisture as they cool this side of a mountain, which is called the windward side. The leeward side of mountains receives relatively little rain and is therefore said to lie in a "rain shadow."

Figure 26.4 Pattern of biome distribution.
a. Pattern of world biomes in relation to temperature and moisture. The dashed line encloses a wide range of environments in which either grasses or woody plants can dominate the area, depending on the soil type. **b.** Pattern of world biomes as shown on a global map.

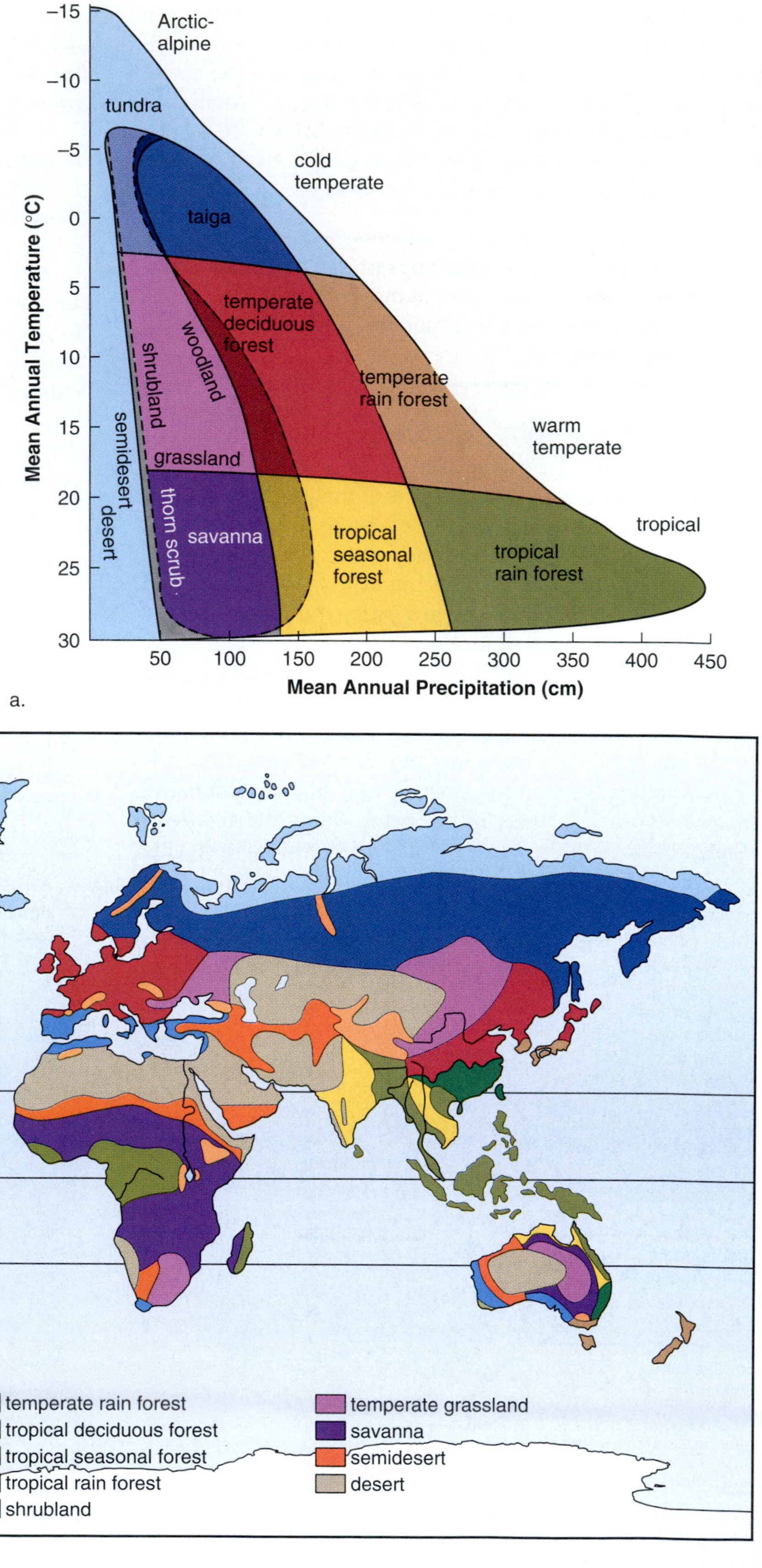

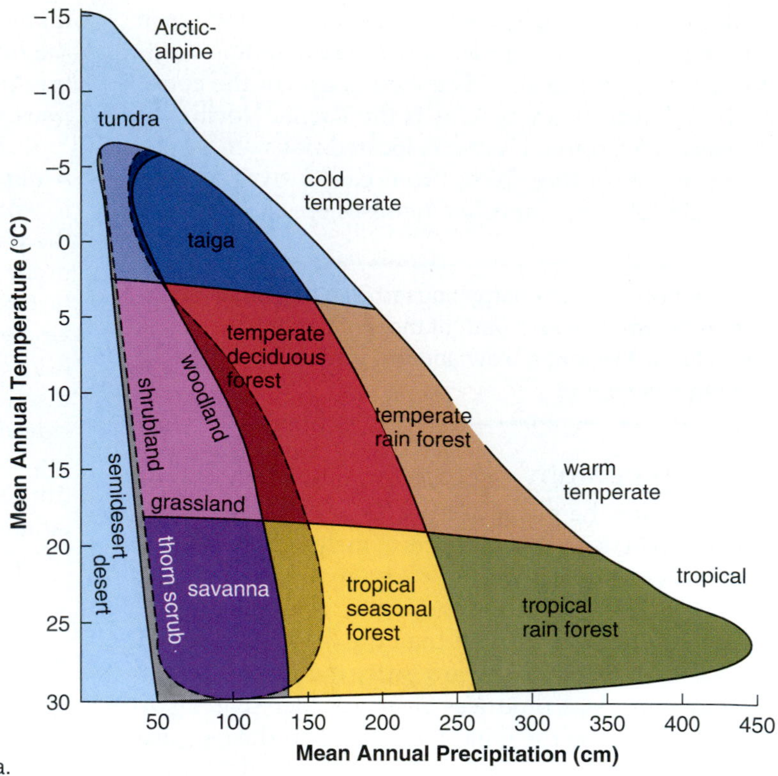

a.

polar ice
tundra
taiga
mountain zone
temperate deciduous forest
temperate rain forest
tropical deciduous forest
tropical seasonal forest
tropical rain forest
shrubland
temperate grassland
savanna
semidesert
desert

b.

26.2 Biomes of the World

A **biome** is the largest biogeographical unit of the biosphere. Although the term arose with reference only to terrestrial communities, we will be using it for both terrestrial and aquatic communities. A biome has a particular mix of plants and animals that are adapted to living under certain environmental conditions, of which climate has an overriding influence. For example, when terrestrial biomes are plotted according to their mean annual temperature and mean annual rainfall, a particular pattern results (Fig. 26.4*a*). The distribution of biomes is shown in Figure 26.4*b*. Even though Figure 26.4 shows definite demarcations, the biomes gradually change from one type to the other. Also, although we will be discussing each type of biome separately, we should remember that each biome has inputs from and outputs to all the other terrestrial and aquatic biomes of the biosphere.

The pattern of life on earth is determined principally by climate, which is influenced also by topographical features. The effect of a temperature gradient can be seen not only when we consider latitude but also when we consider altitude. If you travel from the equator to the North Pole, it is possible to observe first a tropical rain forest, followed by a temperate deciduous forest, a coniferous forest, and tundra, in that order, and this sequence is also seen when ascending a mountain (Fig. 26.5). The coniferous forest of a mountain is called a **montane coniferous forest,** and the tundra near the peak of a mountain is called an **alpine tundra.** When going from the equator to the South Pole, you would not reach a region corresponding to a coniferous forest and tundra of the Northern Hemisphere. Why not? Look at the distribution of the landmasses—they are shifted toward the north.

The distribution of biomes is determined by physical factors such as climate (principally temperature and rainfall), which varies according to latitude and altitude.

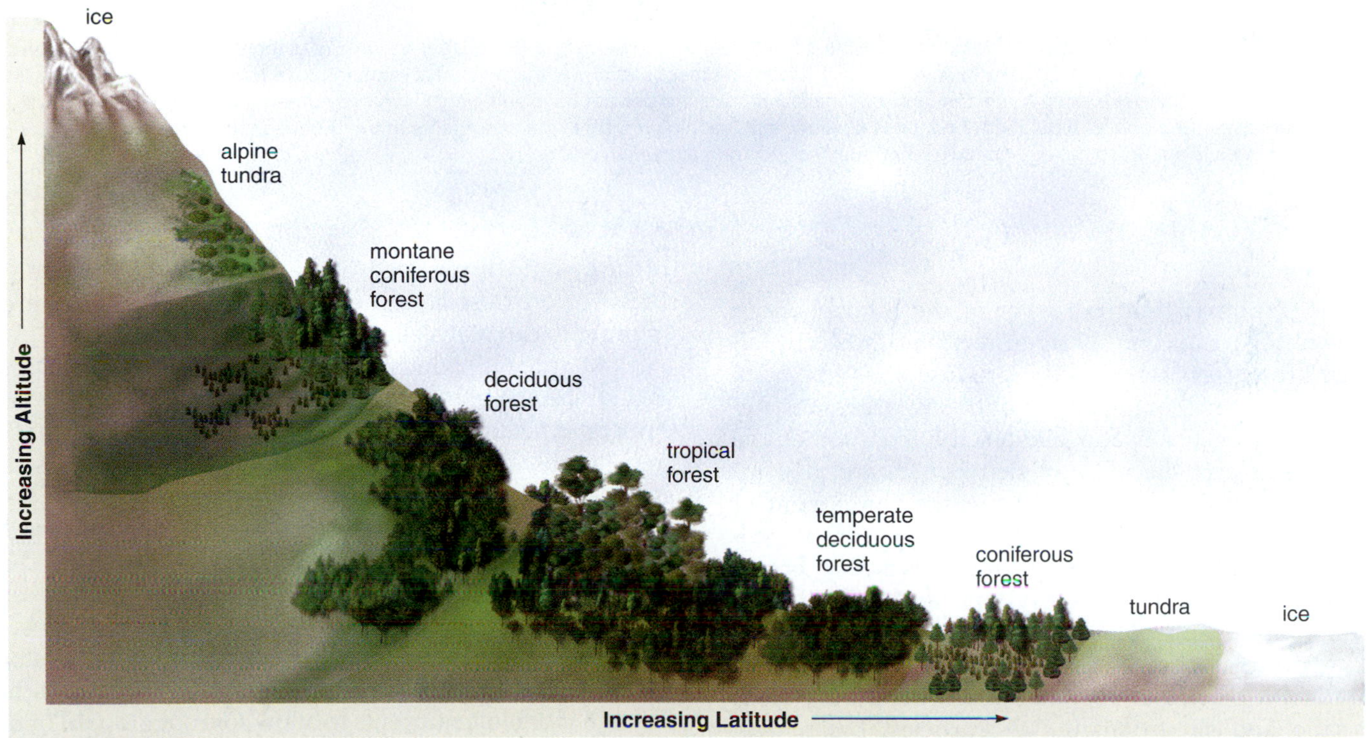

Figure 26.5 Climate and biomes.
Biomes change with altitude just as they do with latitude because vegetation is partly determined by temperature. Rainfall also plays a significant role, which is one reason why grasslands, instead of tropical or deciduous forests, are sometimes found at the base of mountains.

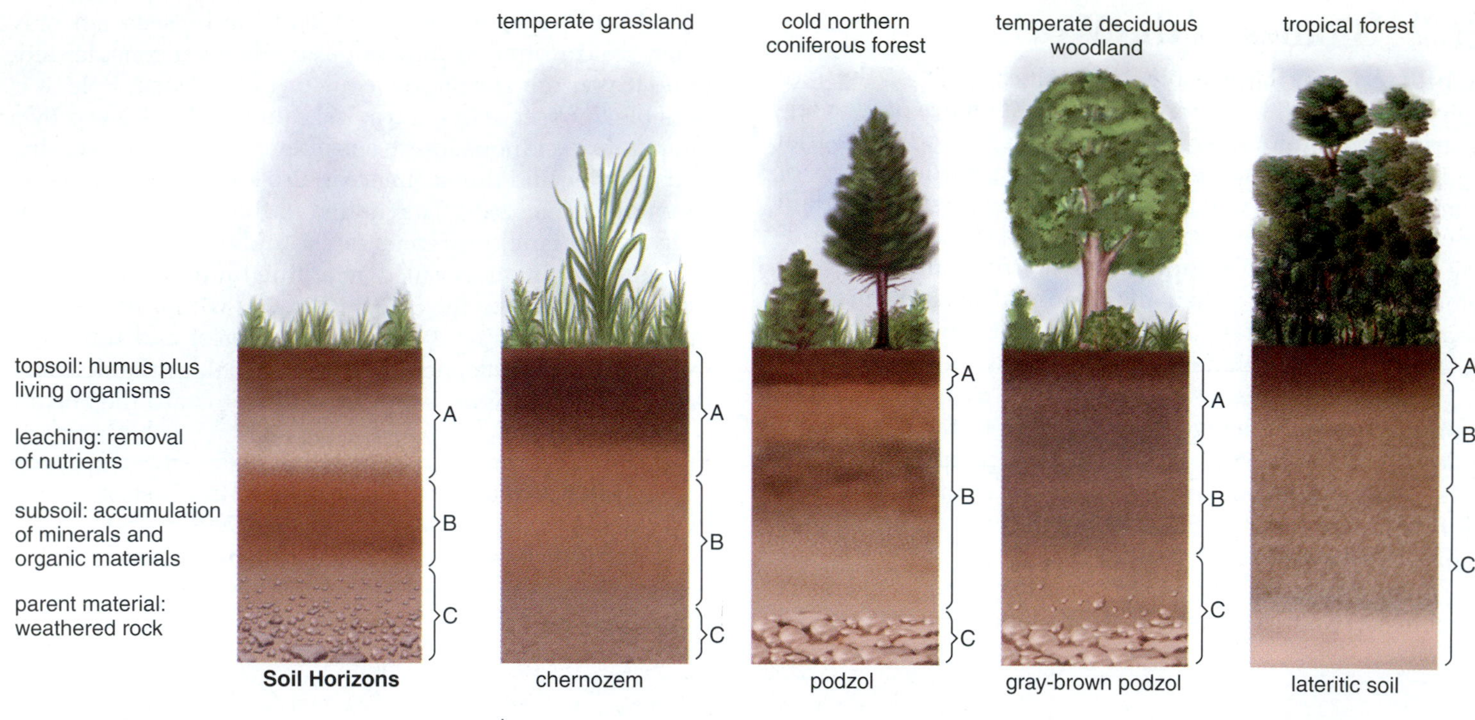

Figure 26.6 **Soil types.**
a. Soil horizons. The top layer (A horizon) contains most of the organic matter; the next layer (B horizon) accumulates inorganic nutrients as the water carries dissolved minerals downward; and the lowest layer (C horizon) is composed of weathered parent material. **b.** In temperate grassland soils (called chernozem soils), there is a deep A horizon, and the shallow B horizon does not have sufficient inorganic nutrients to support root growth. In temperate forest soils of the podzol type, the A horizon is shallow, and leaching results in many nutrients in the B horizon. In a tropical forest soil of the lateritic type, there is a reddish residue composed primarily of iron and aluminum oxides.

Soil Types

Soil is the uppermost surface layer of the lithosphere in which plants grow and on which, directly or indirectly, all terrestrial life depends. Soil is the product of the bedrock weathering and the reorganization of this weathered material by percolating water and biotic activities, such as the growth of trees and the work of detritivores and decomposers. *Humus* is the decomposed organic component of soil. It can take thousands, even millions, of years to produce rich topsoil, and therefore soil erosion is a very serious loss of a precious resource for ecosystems and agriculture.

Soil composition plays a major role in determining what kinds of plants, and therefore what kind of animals, are found in a particular region. Soil particles vary from coarse sand and silt to small clay particles. *Loam,* with roughly equal proportions of sand, silt, and clay, together with humus, is among the richest agricultural soils. Soils differ with regard to drainage. The speed with which gravity causes water to drain away correlates with the size of the soil pores, the spaces between the particles of the soil. Sandy soil has the largest pores, and therefore water drains rapidly away. Clay soil, which drains slowly, has the smallest pores because its particles are very tiny. You might think then that plants would grow well in clay, but this is not the case because pores also act as air spaces, and root cells require air in order to carry on cellular respiration.

Mature soil, which may reach to many meters in depth, is described in terms of three soil horizons (Fig. 26.6). The A horizon is the uppermost (or topsoil) layer that contains litter and humus, although most of the soluble chemicals have been leached (washed out). The B horizon has little or no organic matter but does contain the inorganic nutrients leached from A horizon. The C horizon is a layer of weathered and shattered rock. Soils formed in grasslands have a deep A horizon built up from decaying grasses over many years, but because of limited rain, there has been little leaching into the B horizon. In forest soils, both the A and B horizons have enough inorganic nutrients to allow for root growth. In tropical rain forests, the A horizon is more shallow than the generalized profile, and the B horizon is deeper, signifying that leaching is more extensive. Since the topsoil of rain forests lacks nutrients, it can only support crops for several years.

Soil type, determined in part by the amount of rainfall, influences the distribution of biomes.

26.3 Terrestrial Biomes

There are many major terrestrial biomes, and we will consider these: tundra, coniferous forests (taiga, temperate rain forest), temperate deciduous forest, tropical rain forest, savanna, temperate grasslands, shrubland, and desert.

Tundra

The **Arctic tundra** biome, which encircles the earth just south of ice-covered polar seas in the Northern Hemisphere, covers about 20% of the earth's land surface (Fig. 26.7). (A similar community, called the alpine tundra, occurs above the timberline on mountain ranges.) The Arctic tundra is cold and dark much of the year. Because rainfall amounts to only about 20 cm a year, the tundra could possibly be considered a desert, but melting snow creates a landscape of pools and mires in the summer, especially because so little evaporates. Only the topmost layer of earth thaws; the **permafrost** beneath this layer is always frozen, and therefore, drainage is minimal.

Trees are not found in the tundra because the growing season is too short, their roots cannot penetrate the permafrost, and they cannot become anchored in the boggy soil of summer. In the summer, the ground is covered with short grasses and sedges, but there also are numerous patches of lichens and mosses. Dwarf woody shrubs, such as dwarf birch, flower and seed quickly while there is plentiful sun for photosynthesis.

A few animals live in the tundra year-round. For example, the mouselike lemming stays beneath the snow; the ptarmigan, a grouse, burrows in the snow during storms; and the musk ox conserves heat because of its thick coat and short, squat body. In the summer, the tundra is alive with numerous insects and birds, particularly shorebirds and waterfowl that migrate inland. Caribou (reindeer) also migrate to and from the tundra, as do the wolves that prey upon them. Polar bears are common near the coast.

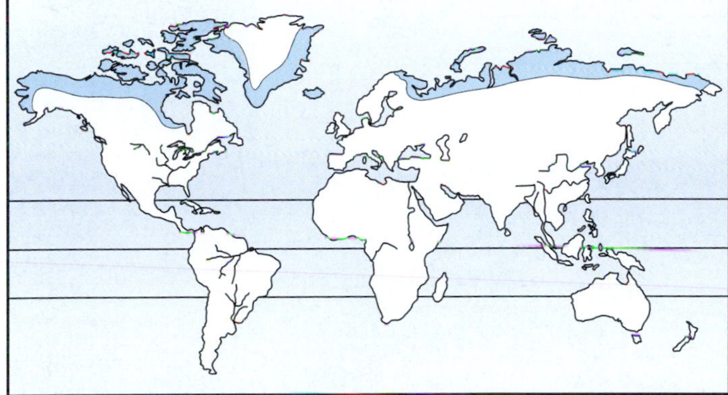

a.

b.

c. Caribou bull, *(Rangifer)*

Figure 26.7 The tundra.
a. In this biome, which is nearest the polar regions, the vegetation consists principally of lichens, mosses, grasses, and low-growing shrubs. **b.** Pools of water that do not evaporate nor drain into the permanently frozen ground attract many birds that feed on the plentiful insects in the summer. **c.** Caribou, more plentiful in the summer than the winter, feed on lichens, grasses, and shrubs.

Coniferous Forests

Coniferous forests are found in three locations: in the **taiga,** which extends around the world in the northern part of North America and Eurasia; near mountaintops (where it is called a montane coniferous forest), and also along the Pacific coast of North America, as far south as northern California.

The taiga (Fig. 26.8) typifies the coniferous forest with its cone-bearing trees, such as spruce, fir, and pine. These trees are well adapted to the cold because both the leaves and bark have thick coverings. Also, the needlelike leaves can withstand the weight of heavy snow. There is a limited understory of plants, but the floor is covered by low-lying mosses and lichens beneath the layer of needles. Birds harvest the seeds of the conifers, and bears, deer, moose, beavers, and muskrats live around the cool lakes and along the streams. Wolves prey on these larger mammals. A montane coniferous forest also harbors the wolverine and mountain lion.

The coniferous forest that runs along the west coast of Canada and the United States is sometimes called a **temperate rain forest.** The prevailing winds moving in off the Pacific Ocean lose their moisture when they meet the coastal mountain range. The plentiful rainfall along with a rich soil have produced some of the tallest conifer trees ever in existence, including the coastal redwoods. This forest is also called an old-growth forest because some trees are as old as 800 years. It truly is an evergreen forest because all trees are covered with mosses, ferns, and other plants that grow on their trunks. Whether the limited portion remaining should be preserved from logging has been quite a controversy. Unfortunately, the controversy has centered around the northern spotted owl, which is endemic to this area. The actual concern is conservation of this particular ecosystem.

Figure 26.8 The taiga.
The taiga, which means swampland, spans northern Europe, Asia, and North America. The appellation "spruce-moose" refers to the dominant presence of spruce trees and moose, which frequent the ponds.

Temperate Deciduous Forests

Temperate deciduous forests are found south of the taiga in eastern North America (Fig. 26.9), eastern Asia, and much of Europe. The climate in these areas is moderate, with relatively high rainfall (75–150 cm per year). The seasons are well defined, and the growing season ranges between 140 and 300 days. The trees, such as oak, beech, and maple, have broad leaves and are termed deciduous trees; they lose their leaves in the fall conserving water, and grow them in the spring.

The tallest trees form a canopy, an upper layer of leaves that are the first to receive sunlight. Even so, enough sunlight penetrates to provide energy for another layer of trees called understory trees. Beneath these trees are shrubs that may flower in the spring before the trees have put forth their leaves. Still another layer of plant growth—mosses, lichens, and ferns—resides beneath the shrub layer. This stratification provides a variety of habitats for insects and birds. Ground life is also plentiful. Squirrels, cottontail rabbits, shrews, skunks, woodchucks, and chipmunks are small herbivores. These and ground birds such as turkeys and grouse are preyed on by red foxes. The whitetail deer and black bears, which are herbivores, have increased in number of late. In contrast to the taiga, amphibians and reptiles are found in this biome because the winters are not as cold. Frogs and turtles prefer an aquatic existence, as do the beavers and muskrats, which are mammals.

Autumn fruits, nuts, and berries provide a supply of food for the winter, and the leaves, after turning brilliant colors and falling to the ground, contribute to the rich layer of humus. The minerals within the humus are washed far into the ground by the spring rains, but the deep tree roots capture these and bring them back up into the forest system again.

Figure 26.9 **Temperate deciduous forest.**
A temperate deciduous forest is home to many and varied plants and animals. Millipedes can be found among leaf litter; chipmunks feed on acorns; and bobcats prey on these and other small mammals.

Tropical Forests

In the **tropical rain forests** of South America, Africa, and the Indo-Malayan region near the equator, the weather is always warm (between 20° and 25°C), and rainfall is plentiful (with a minimum of 190 cm per year). This may be the richest biome, both in terms of number of different kinds of species and their abundance.

A tropical rain forest has a complex structure, with many levels of life (Fig. 26.10). Some of the broad-leaf evergreen trees grow from 15 to 50 meters or more. These tall trees often have trunks buttressed at ground level to prevent their toppling over. Lianas, or woody vines, that encircle the tree as it grows, also help to strengthen the trunk. The diversity of species is enormous—a 10-km^2 area of tropical rain forest may contain 750 species of trees and 1,500 species of flowering plants.

Although there is animal life on the ground (e.g., pacas, agoutis, peccaries, and armadillos), most animals live in the trees (Fig. 26.11). Insect life is so abundant that the majority of species have not been identified yet. Termites play a vital role in the decomposition of woody plant material, and ants are found everywhere, particularly in the trees. The various birds, such as hummingbirds, parakeets, parrots, and toucans, are often beautifully colored. Amphibians and reptiles are well represented by many types of frogs, snakes, and lizards. Lemurs, sloths, and monkeys are well-known primates that feed on the fruits of the trees. The largest carnivores are the big cats—the jaguars in South America and the leopards in Africa and Asia.

Many animals spend their entire life in the canopy, as do some plants. **Epiphytes** are air plants that grow on other plants but have roots of their own that absorb moisture and minerals leached from the canopy; others catch rain and debris in hollows produced by overlapping leaf bases. The most common epiphytes are related to pineapples, orchids, and ferns.

Figure 26.10 Tropical rain forest.
Levels of life in a tropical rain forest. Even the canopy (solid layer of leaves) has levels, and some organisms spend their entire life in one particular level. Long lianas (hanging vines) climb into the canopy, where they produce leaves. Epiphytes are air plants that grow on the trees but do not parasitize them.

While we usually think of tropical forests as being nonseasonal rain forests, there are tropical forests with wet and dry seasons in India, Southeast Asia, West Africa, South and Central America, the West Indies, and northern Australia. Here, there are deciduous trees, with many layers of growth beneath the trees. In addition to the animals just mentioned, certain of these forests also contain elephants, tigers, and hippopotamuses.

Whereas the soil of a temperate deciduous forest biome is rich enough for agricultural purposes, the soil of a tropical rain forest biome is not. Nutrients are cycled directly from the litter to the plants again. Productivity is high because of high temperatures, a yearlong growing season, and the rapid recycling of nutrients from the litter. (In humid tropical forests, iron and aluminum oxides occur at the surface, causing a reddish residue known as laterite. When the trees are cleared, laterite bakes in the hot sun to a bricklike consistency that will not support crops.) Swiden agriculture, often called slash-and-burn agriculture, is a destructive type of agriculture that is practiced in the tropics. Trees are felled and burned, and the ashes provide enough nutrients for several harvests. Thereafter, the forest must be allowed to regrow, and a new section must be cut and burned.

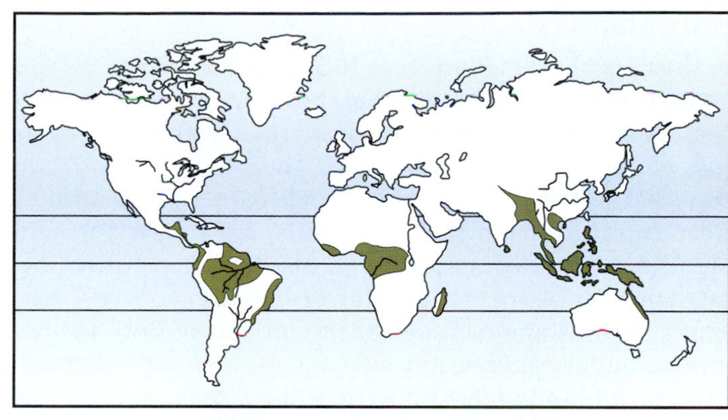

Figure 26.11 Animals of a tropical rain forest.

Shrublands

A shrub is shorter than trees (4.5–6 meters), and has multiple trunks. Shrubs have small but thick evergreen leaves that often are coated with a waxy material that prevents loss of moisture from the leaves. Their thick underground roots survive the dry summers and frequent fires and take deep moisture from the soil. Shrubs are adapted to withstand arid conditions and can also quickly sprout new growth after a fire. As a point of interest, you will recall that a shrub stage is a part of the process of both primary and secondary succession.

Shrublands tend to occur along coasts that have dry summers and receive most of their rainfall in the winter. A shrubland is found along the cape of South Africa, the western coast of North America, and the southwest and southern shores of Australia, around the Mediterranean Sea, and in central Chile. The dense shrubland that occurs in California is known as chaparral (Fig. 26.12). This type of shrubland, called the Mediterranean type, lacks an understory and ground litter, and is highly flammable. The seeds of many species require the heat and scarring action of fire to induce germination. Other shrubs sprout from the roots after a fire.

There is also a northern shrub area that lies west of the Rocky Mountains. This area is sometimes classified as a cold desert, but the region is dominated by sagebrush and other hardy plants. Some of the birds found here are dependent upon sagebrush for their existence.

Grasslands

Grasslands occur where rainfall is greater than 25 cm but is generally insufficient to support trees. Natural grasslands once covered more than 40% of the earth's land surface, but many areas that once were grasslands are now used for the cultivation of crops, such as wheat and corn. In temperate areas, where rainfall is between 10 and 30 inches a year, grasslands occur. Here, it is too dry for forests and too wet for deserts to form.

The grasses are well adapted to a changing environment and can tolerate a high degree of grazing, flooding, drought, and sometimes fire. Where rainfall is high, large tall grasses that reach more than 2 meters in height (e.g., pampas grass) can flourish. In drier areas, shorter grasses between 5 and 10 cm are dominant. Low-growing bunch grasses (e.g., grama grass) grow in the United States near deserts. Grasses also generally grow in different seasons; some grassland animals migrate, and ground squirrels hibernate, when there is little grass for them to eat.

The temperate grasslands include the Russian steppes, the South American pampas, and the North American **prairies** (Fig. 26.13). When traveling across the United States from east to west, the line between the temperate deciduous forest and a tall-grass prairie is roughly along the border between Illinois and Indiana. The tall-grass prairie requires more rainfall than does the short-grass prairie that occurs near deserts. Large herds of bison—estimated at hundreds of thousands—once roamed the prairies, as did herds of pronghorn antelope. Now, small mammals, such as mice, prairie dogs, and rabbits, typically live belowground, but usually feed aboveground. Hawks, snakes, badgers, coyotes, and foxes feed on these mammals. Virtually all of these grasslands, however, have been converted to agricultural lands.

Savannas, which are grasslands that contain some trees, occur in regions where a relatively cool, dry season is followed by a hot, rainy one (Fig. 26.14). One tree that can survive the severe dry season is the flat-topped acacia, which sheds its leaves during a drought. The African savanna supports the greatest variety and number of large herbivores of all the biomes. Elephants and giraffes are browsers that feed on tree vegetation. Antelopes, zebras, wildebeests, water buffalo, and rhinoceroses are grazers that feed on grasses. Any plant litter that is not consumed by grazers is attacked by a variety of small organisms, among them termites. Termites build towering nests in which they tend fungal gardens, their source of food. The herbivores support a large population of carnivores. Lions and hyenas hunt in packs, cheetahs hunt singly by day, and leopards hunt singly by night.

Figure 26.12 **Shrubland.**
Shrublands, such as chaparral in California, are subject to raging fires, but the shrubs are adapted to quickly regrow.

American bison,
Bison bison

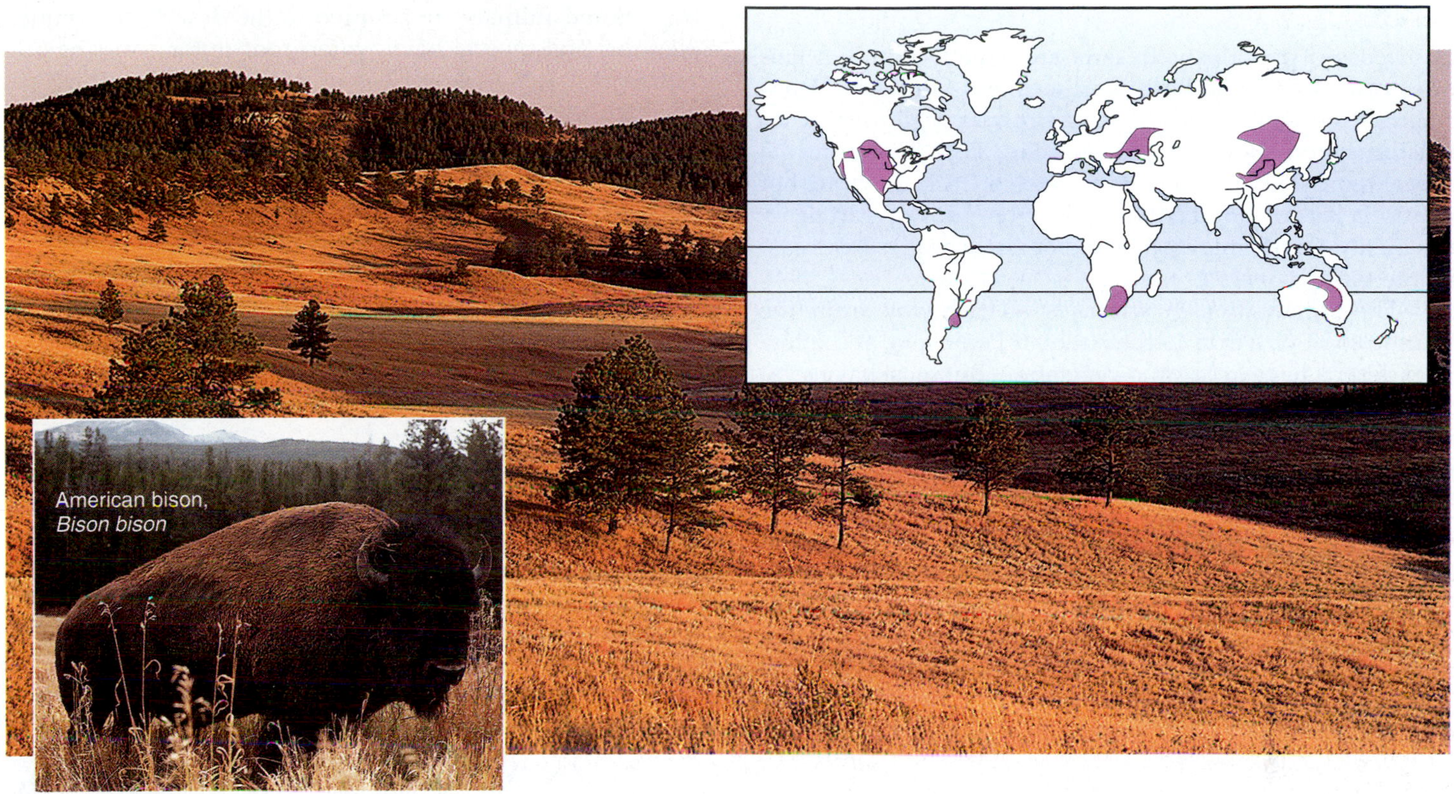

Figure 26.13 **The prairie.**
Tall-grass prairies are seas of grasses dotted by pines and junipers. Bison, once abundant, are now being reintroduced into certain areas.

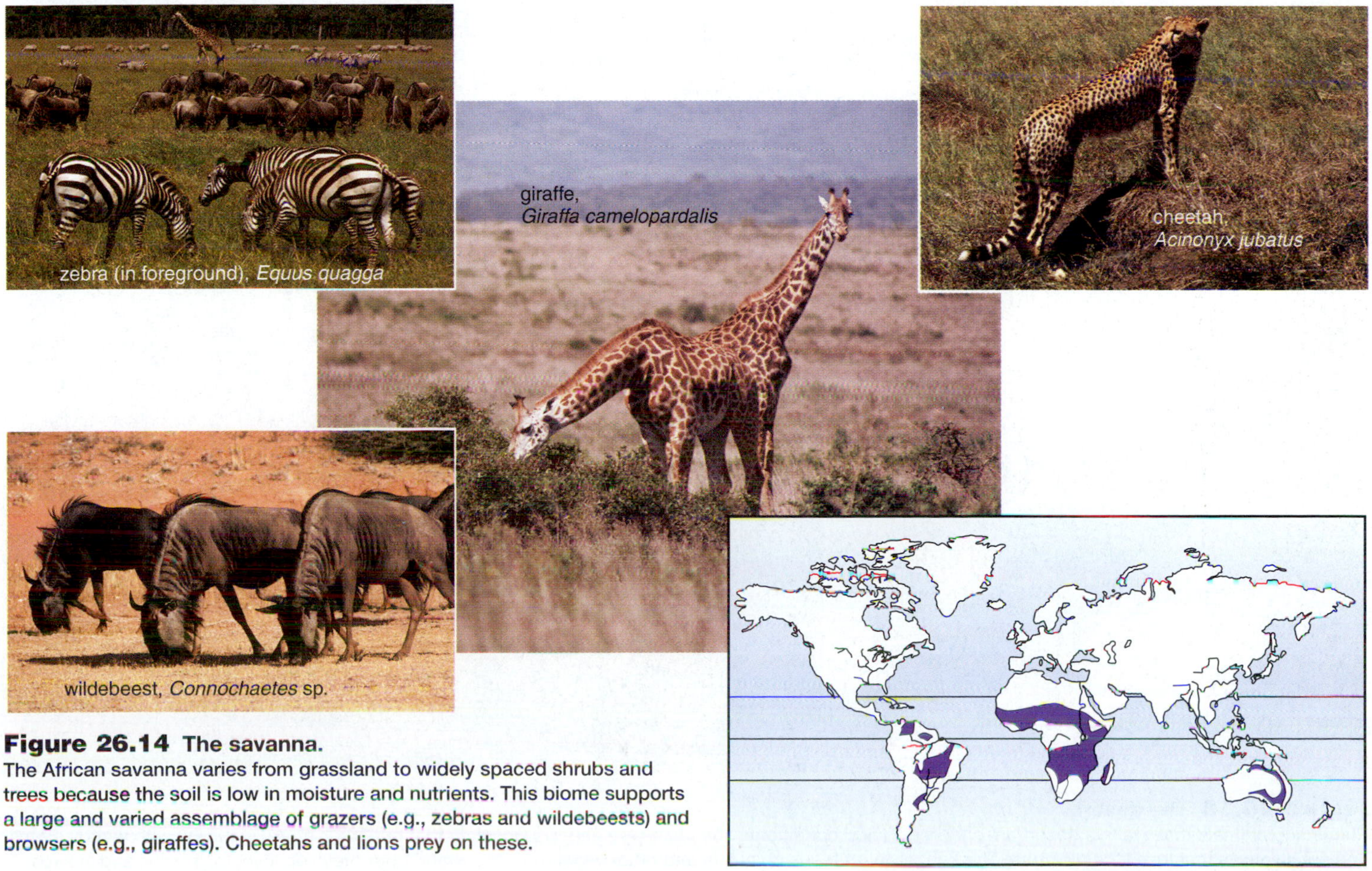

zebra (in foreground), Equus quagga
giraffe,
Giraffa camelopardalis
cheetah,
Acinonyx jubatus
wildebeest, Connochaetes sp.

Figure 26.14 **The savanna.**
The African savanna varies from grassland to widely spaced shrubs and trees because the soil is low in moisture and nutrients. This biome supports a large and varied assemblage of grazers (e.g., zebras and wildebeests) and browsers (e.g., giraffes). Cheetahs and lions prey on these.

Deserts

As discussed previously, **deserts** are usually found at latitudes of about 30°, in both Northern and Southern Hemispheres. The winds that descend in these regions pick up moisture before they rise again (see Fig. 26.2). Therefore, the annual rainfall is less than 25 cm (see Fig. 26.4). Days are hot because a lack of cloud cover allows the sun's rays to penetrate easily, but the nights are cold because heat escapes easily into the *atmosphere*.

The Sahara, which stretches all the way from the Atlantic coast of Africa to the Arabian Peninsula, and a few other deserts have little or no vegetation. But most have a variety of plants (Fig. 26.15). The best-known desert perennials in North America are the succulent, spiny-leafed cacti, which have stems that store water and carry on photosynthesis. Also common are nonsucculent shrubs, such as the many-branched sagebrush with silvery gray leaves and the spiny-branched ocotillo that produces leaves during wet periods and sheds them during dry periods.

Some animals are adapted to the desert environment. Reptiles and insects have waterproof outer coverings that conserve water. A desert has numerous insects, which pass through the stages of development from pupa to the next pupa again when there is rain. Reptiles, especially lizards and snakes, are perhaps the most characteristic group of vertebrates found in deserts, but running birds (e.g., the roadrunner) and rodents (e.g., the kangaroo rat) are also well known (Fig. 26.15). Larger mammals, like the coyote, prey on the rodents, as do the hawks.

bannertail kangaroo rat, *Dipodomys spectabilis*

greater roadrunner, *Geococcyx californianus*

Figure 26.15 **The desert.**
Plants and animals that live in a desert are adapted to arid conditions. The plants are either succulents that retain moisture or shrubs with woody stems and small leaves that lose little moisture. The kangaroo rat feeds on seeds and other vegetation; the roadrunner preys on insects, lizards, and snakes.

26.4 Aquatic Biomes

Aquatic biomes are classified as two types: fresh water (inland) or salt water (usually marine). Wetlands that lie near the sea have mixed fresh and salt water, called brackish water. Figure 26.16 shows how these communities are joined physically. They also interact by sharing nutrients and by the biogeochemical cycles. Consideration of the hydrologic cycle (p. 429) shows that fresh water is a distillate of salt water. As the sun's rays cause seawater to evaporate, the salts are left behind. The vaporized fresh water rises into the atmosphere, cools, and falls as rain either over the ocean or over the land. A lesser amount of water also evaporates from and returns to the land. Since land lies above sea level, gravity eventually returns all fresh water to the sea, but in the meantime, it is contained within standing waters (**lakes** and **ponds**), flowing waters (**streams** and **rivers**), and groundwater.

When rain falls, some of the water sinks or percolates into the ground and saturates the earth to a certain level. The top of the saturation zone is called the groundwater table, or simply the water table. Wherever the earth contains basins or channels, water will appear to the level of the water table. The water within basins is called lakes and ponds, and the water within channels is called streams or rivers. Sometimes groundwater is also located in underground rivers called aquifers.

Humans have the habit of channeling aboveground rivers and filling in wetlands (lands that are wet for at least part of the year). These activities degrade ecosystems and eventually cause seasonal flooding. Wetlands provide food and habitats for fish, waterfowl, and other wildlife. They also purify waters by filtering them and by diluting and breaking down toxic wastes and excess nutrients. Wetlands directly absorb storm waters and also absorb overflows from lakes and rivers. In this way they protect farms, cities, and towns from the devastating effects of floods. There are now federal and local laws for the protection of wetlands, but they are not always enforced.

Aquatic biomes can be classified as fresh water or salt water. However, it is important to realize that the two sets of communities interact and are joined by biogeochemical cycles.

stonefly larva, *Plecoptera* sp.

red banded trout, *Salmo gairdneri*

carp, *Cyprinus carpio*

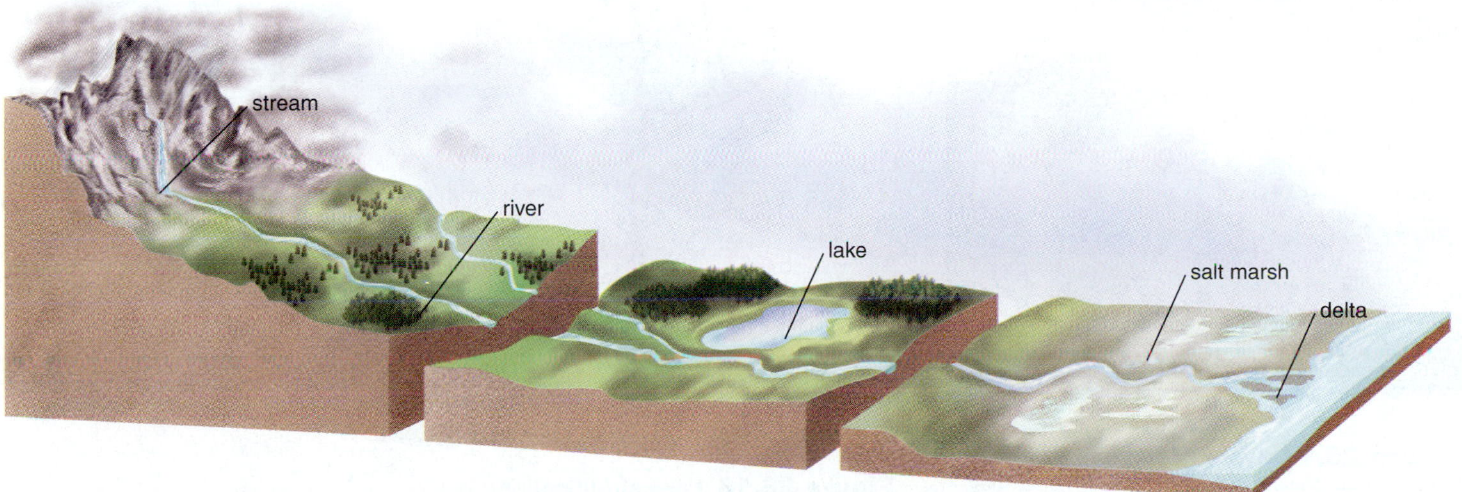

Figure 26.16 **Streams and rivers.**
Mountain streams have cold, clear water that flows over waterfalls and rapids. The feet of this long-legged stonefly insect larva are clawed, helping it to hold on to stones. Trout are found in occasional pools of the highly oxygenated water. As the streams merge, a river forms that gets increasingly wider and deeper until it meanders across broad, flat valleys. Carp are adapted to water that contains little oxygen and has much sediment. At its mouth, a river may divide into many channels where wetlands and estuaries are located.

Lakes

Lakes are bodies of fresh water often classified by their nutrient status. Oligotrophic (nutrient-poor) lakes are characterized by low organic matter and low productivity. Eutrophic (nutrient-rich) lakes are characterized by high organic matter and high productivity. Such lakes are usually situated in naturally nutrient-rich regions or are enriched by agricultural or urban and suburban runoffs. Oligotrophic lakes can become eutrophic through large inputs of nutrients (Fig. 26.17). This process is called **eutrophication.**

In the temperate zone, deep lakes are stratified in the summer and winter. In summer, lakes in the temperate zone have three layers of water that differ in temperature (Fig. 26.18). The surface layer, the epilimnion, is warm from solar radiation; the middle thermocline experiences an abrupt drop in temperature; and the hypolimnion is cold. These differences in temperature prevent mixing. The warmer, less dense water of the epilimnion "floats" on top of the colder, more dense water of the hypolimnion.

As the season progresses, the epilimnion becomes nutrient-poor, while the hypolimnion begins to be depleted of oxygen. The phytoplankton found in the sunlit epilimnion use up nutrients as they photosynthesize. Photosynthesis releases oxygen, giving this layer a ready supply. Detritus naturally falls by gravity to the bottom of the lake, and here oxygen is used up as decomposition occurs. Decomposition releases nutrients, however.

In the fall, as the epilimnion cools, and in the spring, as it warms, an overturn occurs. In the fall, the upper epilimnion waters become cooler than the hypolimnion waters. This causes the surface water to sink and the deep water to rise. The **fall overturn** continues until the temperature is uniform throughout the lake. At this point, wind aids in the circulation of water so that mixing occurs. Eventually, oxygen and nutrients become evenly distributed.

As winter approaches, the water cools. Ice formation begins at the top, and the ice remains there because ice is less dense than cool water. Ice has an insulating effect, preventing

a.

b.

Figure 26.17 Types of lakes.
Lakes can be classified according to whether they are (a) oligotrophic (nutrient-poor) or (b) eutrophic (nutrient-rich). Eutrophic lakes tend to have large populations of algae and rooted plants, resulting in a large population of decomposers that use up much of the oxygen, leaving little oxygen for fishes.

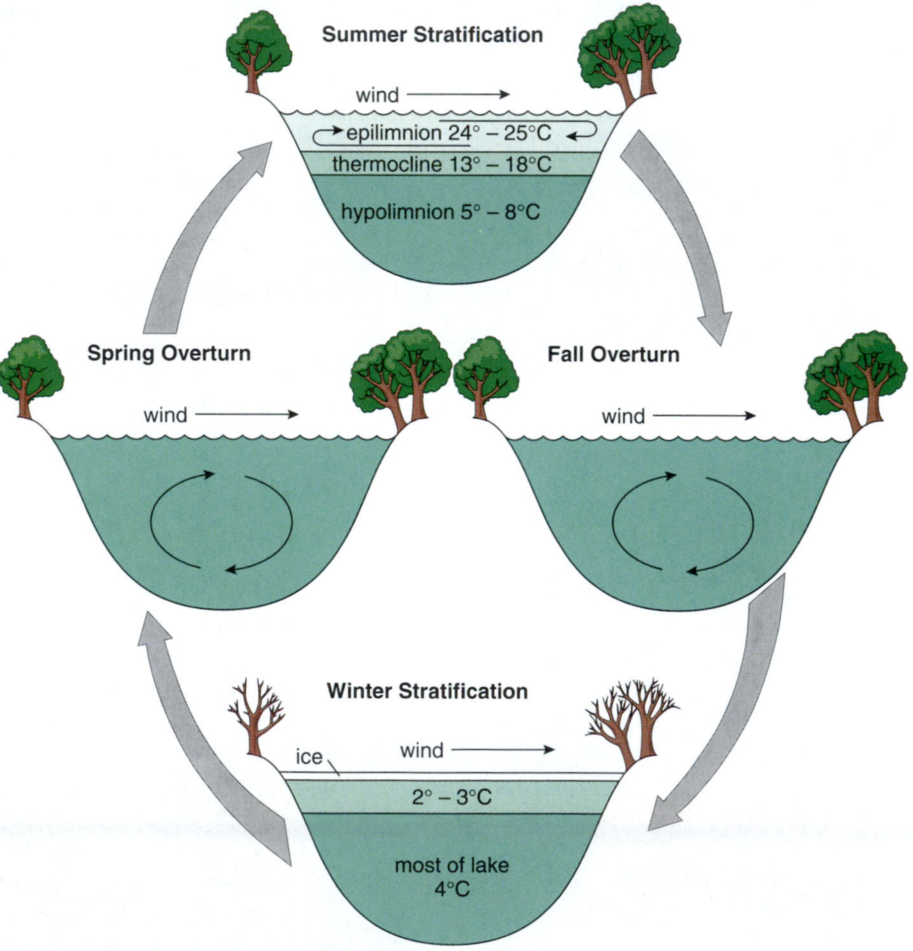

Figure 26.18 Lake stratification.
Temperature profiles of a large oligotrophic lake in a temperate region vary with the season. During spring and fall turnover, the deep waters receive oxygen from surface waters, and surface waters receive nutrients from deep waters.

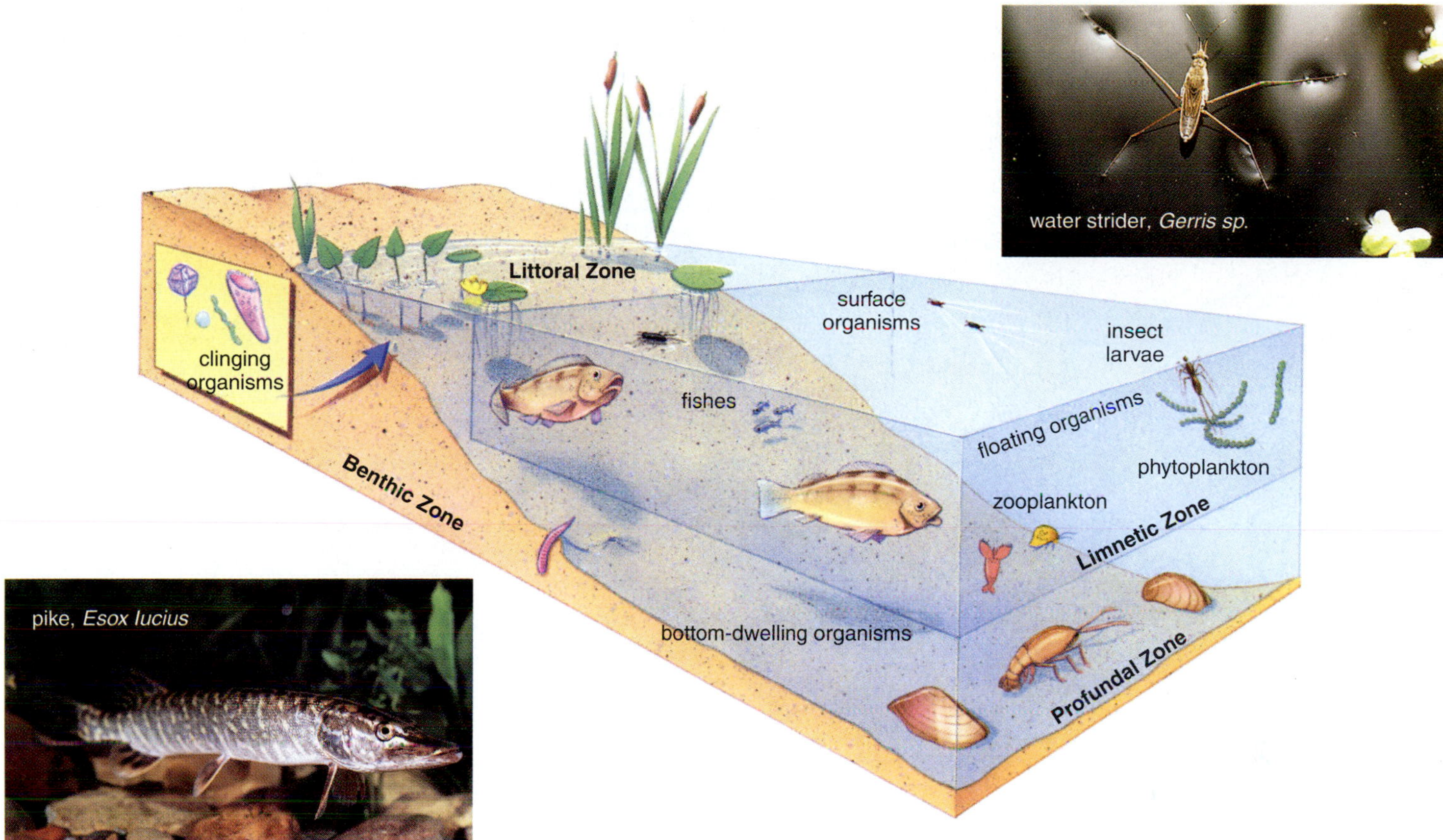

Figure 26.19 Zones of a lake.
Rooted plants and clinging organisms live in the littoral zone. Phytoplankton, zooplankton, and fishes are in the sunlit limnetic zone. Water striders stand on the surface film of water with water-repellent feet. Pike are top carnivores prized by fishermen. Bottom-dwelling organisms, like crayfishes and clams, are in the profundal zone.

further cooling of the water below. This permits aquatic organisms to live through the winter in the water beneath the surface of the ice.

In the spring, as the ice melts, the cooler water on top sinks below the warmer water on the bottom. The **spring overturn** continues until the temperature is uniform through the lake. At this point, wind aids in the circulation of water as before. When the surface waters absorb solar radiation, thermal stratification occurs once more.

This vertical stratification and seasonal change of temperatures in a lake basin influence the season distribution of fish and other aquatic life in the lake basin. For example, coldwater fish move to the deeper water in summer and inhabit the upper water in winter. In the fall and spring just after mixing occurs, phytoplankton growth at the surface is most abundant.

Life Zones

In both fresh and salt water, free-drifting microscopic organisms, called **plankton** [Gk. *planktos,* wandering], are important components of the community. **Phytoplankton** [Gk. *phyton,* plant, and *planktos,* wandering] are photosynthesizing plantlike algae that only become noticeable when they reproduce to the extent that a green scum or red tide appears on the water. **Zooplankton** [Gk. *zoon,* animal, *planktos,* wandering] are animals that feed on the phytoplankton. Lakes and ponds can be divided into several life zones. The *littoral zone* is closest to the shore, the *limnetic zone* forms the sunlit body of the lake, and the *profundal zone* is below the level of light penetration (Fig. 26.19). The *benthic zone* is at the soil-water interface. Aquatic plants are rooted in the shallow littoral zone of a lake, and various microscopic organisms cling to these plants and to rocks. Some organisms such as the water strider live at the water-air interface and can literally walk on water. In the limnetic zone, small fishes, such as minnows and killifish feed on plankton and also serve as food for large fishes. In the profundal zone, there are zooplankton and fishes such as whitefish that feed on debris that falls from above. Pike species are "lurking predators." They wait among vegetation around the margins of lakes and surge out to catch passing prey.

A few insect larvae are in the limnetic zone, but they are far more prominent in both the littoral and profundal zones. Midge larvae and ghost worms are common members of the benthos. The benthos are animals that live on the bottom in the benthic zone. In a lake, the benthos include crayfish, snails, clams, and various types of worms and insect larvae.

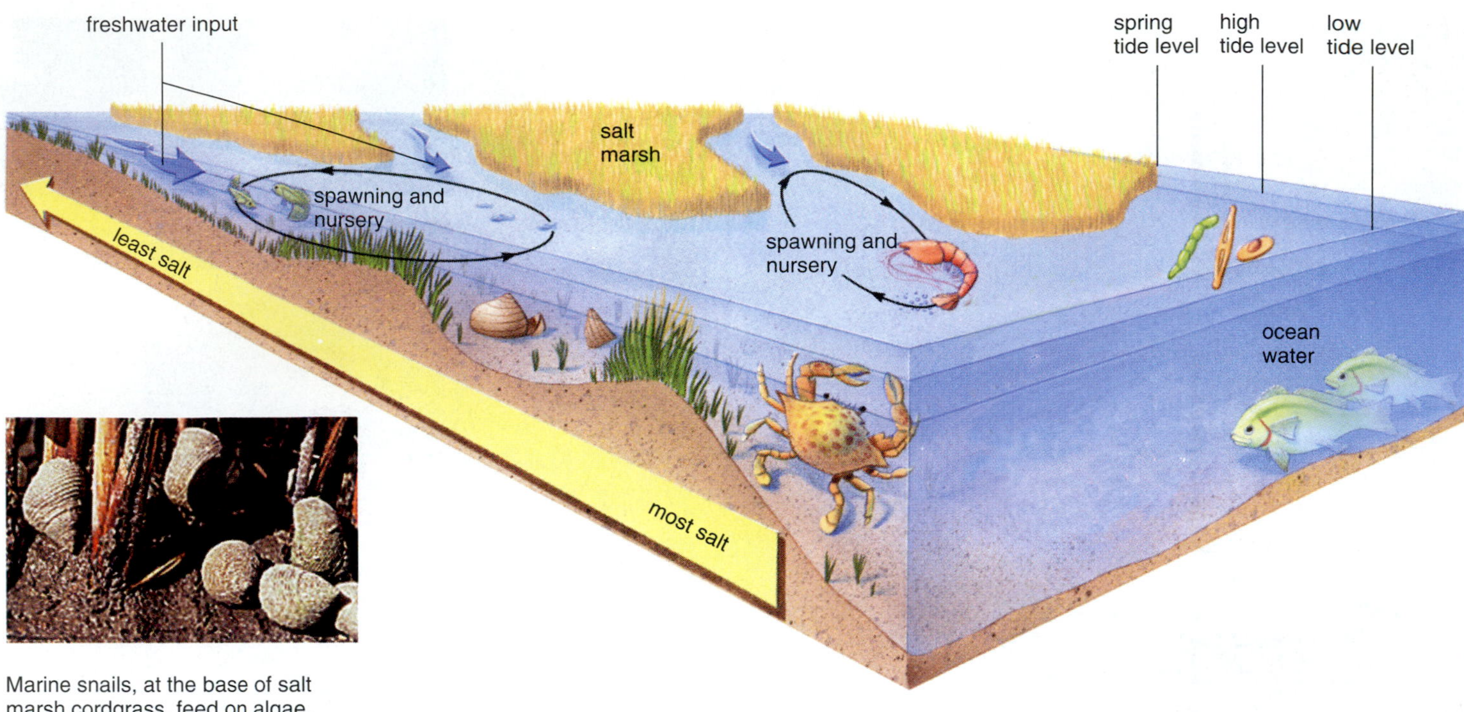

Marine snails, at the base of salt marsh cordgrass, feed on algae.

Figure 26.20 Estuary structure and function.
Since an estuary is located where a river flows into the ocean, it receives nutrients from land and also from the sea by way of the tides. Decaying grasses also provide nutrients. Estuaries serve as a nursery for the spawning and rearing of the young for many species of fishes, shrimp, and crustaceans, and mollusks such as the blue mussels, and their predator, dog whelks, depicted in the photograph.

Coastal Communities

Near the mouth of a river, a *salt marsh* in the temperate zone and a *mangrove swamp* in the subtropical and tropical zones are likely to develop. Also, the silt carried by a river may form mudflats. It is proper to think of seacoasts and mudflats, salt marshes, and mangrove swamps as belonging to one ecological system.

Estuaries

An **estuary** is a partially enclosed body of water where fresh water and seawater meet and mix (Fig. 26.20). A river brings fresh water into the estuary, and the sea, because of the tides, brings salt water. Coastal bays, tidal marshes, fjords (an inlet of water between high cliffs), some deltas (triangular-shaped areas of land at the mouths of rivers), and lagoons (a body of water separated from the sea by a narrow strip of land) are all examples of estuaries.

Organisms living in an estuary must be able to withstand constant mixing of waters and rapid changes in salinity. Not many organisms are suited to this environment, but for those that are suited, there is an abundance of nutrients. An estuary acts as a nutrient trap because the tides bring nutrients from the sea and at the same time prevent the seaward escape of nutrients brought by the river. Because of this, estuaries produce much organic food.

Although only a few small fish permanently reside in an estuary, many develop there so that there is always an abundance of larval and immature fish. It has been estimated that well over half of all marine fishes develop in the

a.

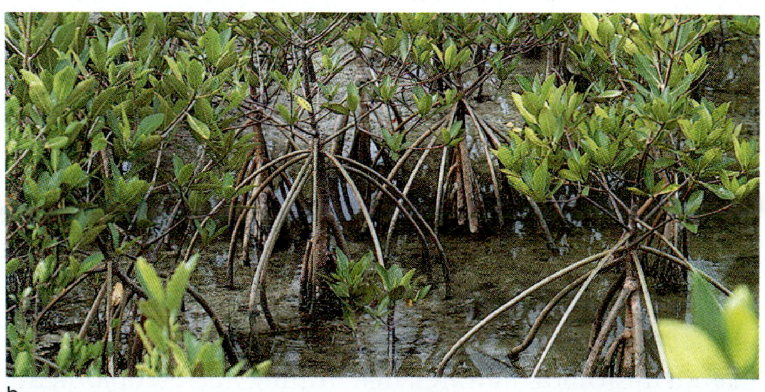

b.

Figure 26.21 Types of estuaries.
Many types of regions qualify as estuaries such as the salt marsh depicted in Figure 26.20 and (**a**) mudflats, which are frequented by migrant birds, and (**b**) mangrove swamps skirting the coastlines of many tropical and subtropical lands. The tangled roots of mangrove trees trap sediments and nutrients that sustain many immature forms of sea life.

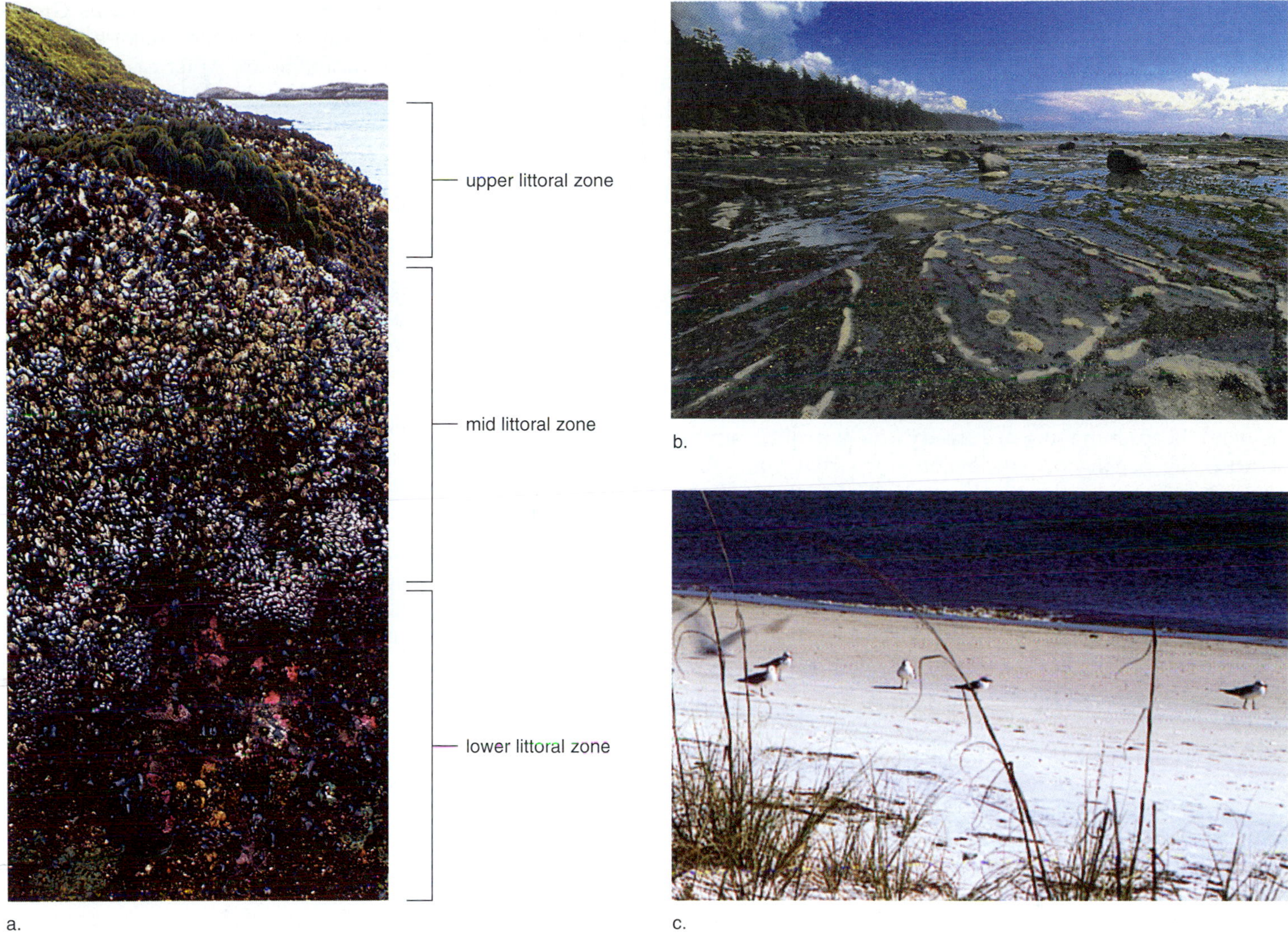

Figure 26.22 **Seacoasts.**
a. The littoral zone of a rocky coast, where the tide comes in and out, has different types of shelled and algal organisms at its upper, middle, and lower portions. **b.** Some organisms of a rocky coast live in tidal pools. **c.** A sandy shore looks devoid of life except for the birds that feed there. However, a number of invertebrate species burrow in the sand only during the day or all the time.

protective environment of an estuary, which explains why estuaries are called the *nurseries of the sea.* Estuaries are also feeding grounds for many birds, fish, and shellfish because they offer a ready supply of food.

Salt marshes dominated by salt marsh cordgrass are often associated with estuaries. Salt marshes contribute nutrients and detrital material to estuaries in addition to being an important wildlife habitat, a few of which are shown in Figure 26.21.

Seashores

Both rocky and sandy shores are constantly bombarded by the sea as the tides roll in and out (Fig. 26.22). A rocky seashore is divided into zones. The **littoral zone** of a rocky beach is divided into subzones. In the upper portion of the littoral zone, barnacles are glued so tightly to the stone by their own secretions that their calcareous outer plates remain in place even after the enclosed shrimplike animal dies. In the midportion of the littoral zone, brown algae known as rockweed may overlie the barnacles. In the lower portions of the littoral zone, oysters and mussels attach themselves to the rocks by filaments called byssal threads. Also present are snails called limpets and periwinkles. But periwinkles have a coiled shell and secure themselves by hiding in crevices or under seaweeds, while limpets press their single flattened cone tightly to a rock. Below the littoral zone, macroscopic seaweeds, which are the main photosynthesizers, anchor themselves to the rocks by holdfasts.

Organisms cannot attach themselves to shifting, unstable sands on a *sandy beach;* therefore, nearly all the permanent residents dwell underground. They either burrow during the day and surface to feed at night, or they remain permanently within their burrows and tubes. Ghost crabs and sandhoppers (amphipods) burrow themselves above the high tide mark and feed at night when the tide is out. Sandworms and sand (ghost) shrimp remain within their burrows in the littoral zone and feed on detritus whenever possible. Still lower in the beach, clams, cockles, and sand dollars are found.

Oceans

Climate is driven by the sun, but the oceans play a major role in redistributing heat in the biosphere. Water tends to be warm at the equator and much cooler at the poles because of the distribution of the sun's rays, as we have discussed before (see Fig. 26.1*a*). Air takes on the temperature of the water below, and warm air moves from the equator to the poles. In other words, the oceans make the winds blow. (The landmasses also play a role, but the oceans hold heat longer and remain cool longer during periods of changing temperature than do solid continents.)

When the wind blows strongly and steadily across a great expanse of ocean for a long time, friction from the moving air begins to drag the water along with it. Once the water has been set in motion, its momentum, aided by the wind, keeps it moving in a steady flow we call a current. Because the ocean currents eventually strike land, they move in a circular path—clockwise in the Northern Hemisphere and counterclockwise in the Southern Hemisphere (Fig. 26.23). As the currents flow, they take warm water from the equator to the poles. One such current, called the Gulf Stream, brings tropical Caribbean water to the east coast of North America and the higher latitudes of western Europe. Without the Gulf Stream, Great Britain, which has a relatively warm temperature, would be as cold as Greenland. In the Southern Hemisphere, another major ocean current warms the eastern coast of South America.

Also, in the Southern Hemisphere a current called the Humboldt Current flows toward the equator. The Humboldt Current carries phosphorus-rich cold water northward along the west coast of South America. During a process called **upwelling,** cold offshore winds cause cold nutrient-rich waters to rise and take the place of warm nutrient-poor waters. In South America, the enriched waters cause an abundance of marine life that supports the fisheries of Peru and northern Chile. Birds feeding on these organisms deposit their droppings on land, where it is mined as guano, a commercial source of phosphorus. When the Humboldt Current is not as cool as usual, upwelling does not occur, stagnation results, the fisheries decline, and climate patterns change globally. This phenomenon, which is discussed in the accompanying reading, is called **El Niño–Southern Oscillation.**

Major ocean currents move heat from the equator to cooler parts of the biosphere. The Gulf Stream warms the east coast of North America and parts of Europe.

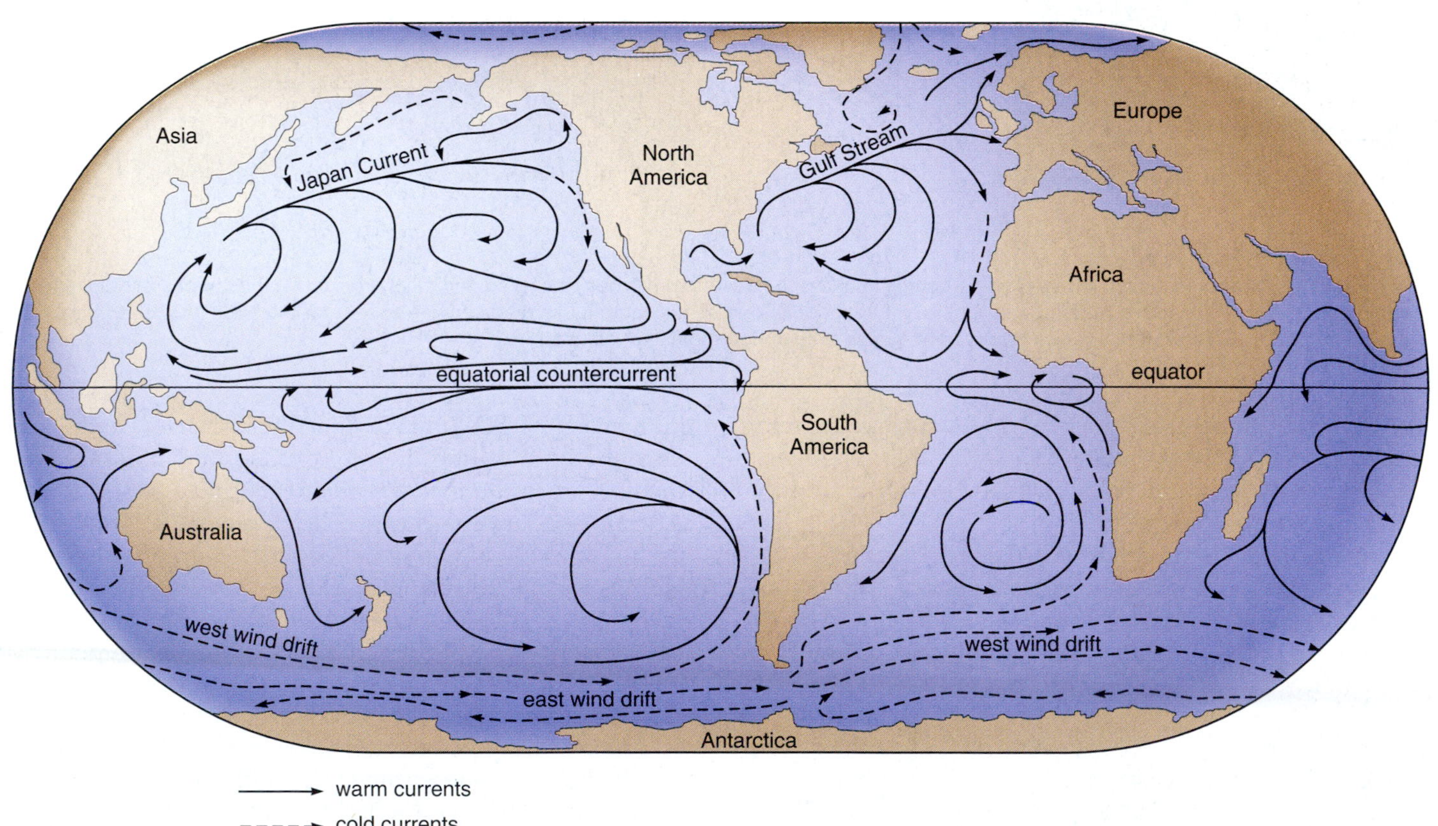

Figure 26.23 Ocean currents.
The arrows on this map indicate the locations and directions of the major ocean currents set in motion by the global wind circulation. By carrying warm water to cool latitudes (e.g., Gulf Stream) and cool water to warm latitudes (e.g., Humboldt Current), these currents have a major effect on the world's climates.

Ecology Focus

El Niño–Southern Oscillation

Climate largely determines the distribution of life on earth. Short-term variations in climate, which we call weather, also have a pronounced effect on living things. There is no better example than an El Niño. Originally, El Niño referred to a warming of the seas off the coast of Peru at Christmas time—hence its name El Niño, "the boy child" for the Christ child Jesus.

Now scientists prefer the term El Niño–Southern Oscillation (ENSO) for a severe weather change brought on by an interaction between the atmosphere and ocean currents. Ordinarily the southeast trade winds move along the coast of South America and turn west because of the earth's daily rotation on its axis. As the winds drag warm ocean waters from east to west, there is an upwelling of nutrient-rich cold water from the ocean's depths that results in a bountiful Peruvian harvest of anchovies. When the warm ocean waters reach their western destination, the monsoons bring rain to India and Indonesia. Scientists have noted that these events correlate with a difference in the barometric pressure over the Indian Ocean and the southeastern Pacific—the barometric pressure is low over the Indian Ocean and the barometric pressure is high over the southeastern Pacific. But when a "southern oscillation" occurs and the barometric pressures switch, an El Niño begins.

During an El Niño, both the northeast and the southeast trade winds slacken. Upwelling no longer occurs and the anchovy fishery off the coast of Peru plummets. During a severe El Niño, waters from the east never reach the west and the winds lose their moisture in the middle of the Pacific instead of over the Indian Ocean. The monsoons fail and there is drought in India, Indonesia, Africa, and Australia. Harvests decline, cattle must be slaughtered, and famine is likely in highly populated India and Africa, where funds to import replacement supplies of food are limited.

There can even be a backward movement of winds and ocean currents so that the waters can warm to more than 14° above normal along the west coast of the Americas. Now a severe El Niño has occurred, and the weather changes are dramatic in the Americas also. Southern California is hit by storms and even hurricanes, and the deserts of Peru and Chile receive so much rain that flooding occurs. A jet stream (strong wind currents) can carry moisture into Texas, Louisiana, and Florida, with flooding a near certainty. Or the winds can turn northward and deposit snow in the mountains along the west coast so that flooding occurs here in the spring. Some parts of the United States, however, benefit from an El Niño. The Northeast is warmer than usual, few if any hurricanes hit the east coast, and there is a lull in tornadoes throughout the Midwest. Altogether, a severe El Niño affects the weather over three-quarters of the globe.

Eventually an El Niño dies out and normal conditions return. The normal cold-water state off the coast of Peru is known as La Niña (the girl). Figure 26A contrasts the weather conditions of a La Niña with those of an El Niño. Since 1991, the sea surface has been almost continuously warm and there have been two record-breaking El Niños. What could be causing more of the El Niño state rather than the La Niña state? Some scientists are seeking data to relate this environmental change to global warming. Global warming is a rise in environmental temperature due to greenhouse gases, like carbon dioxide, in the atmosphere. Like the glass of a greenhouse, the gases allow the sun's rays to pass through, but trap the heat. Greenhouse gases are pollutants that human beings have been pumping into the atmosphere since the Industrial Revolution.

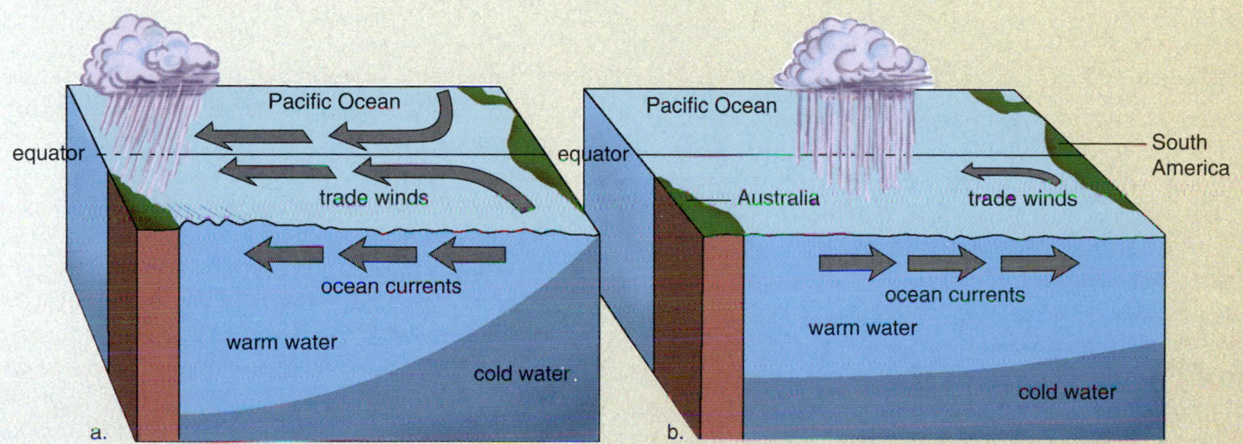

Figure 26A **a. La Niña**
- Upwelling off the west coast of South America brings cold waters to the surface.
- High barometric pressure over southeastern Pacific.
- Monsoons associated with Indian Ocean occur.
- Hurricanes off the east coast of the U.S.

b. El Niño
- Great ocean warming off the west coast of the Americas.
- Low barometric pressure over southeastern Pacific.
- Monsoons associated with Indian Ocean fail.
- Hurricanes off the west coast of the U.S.

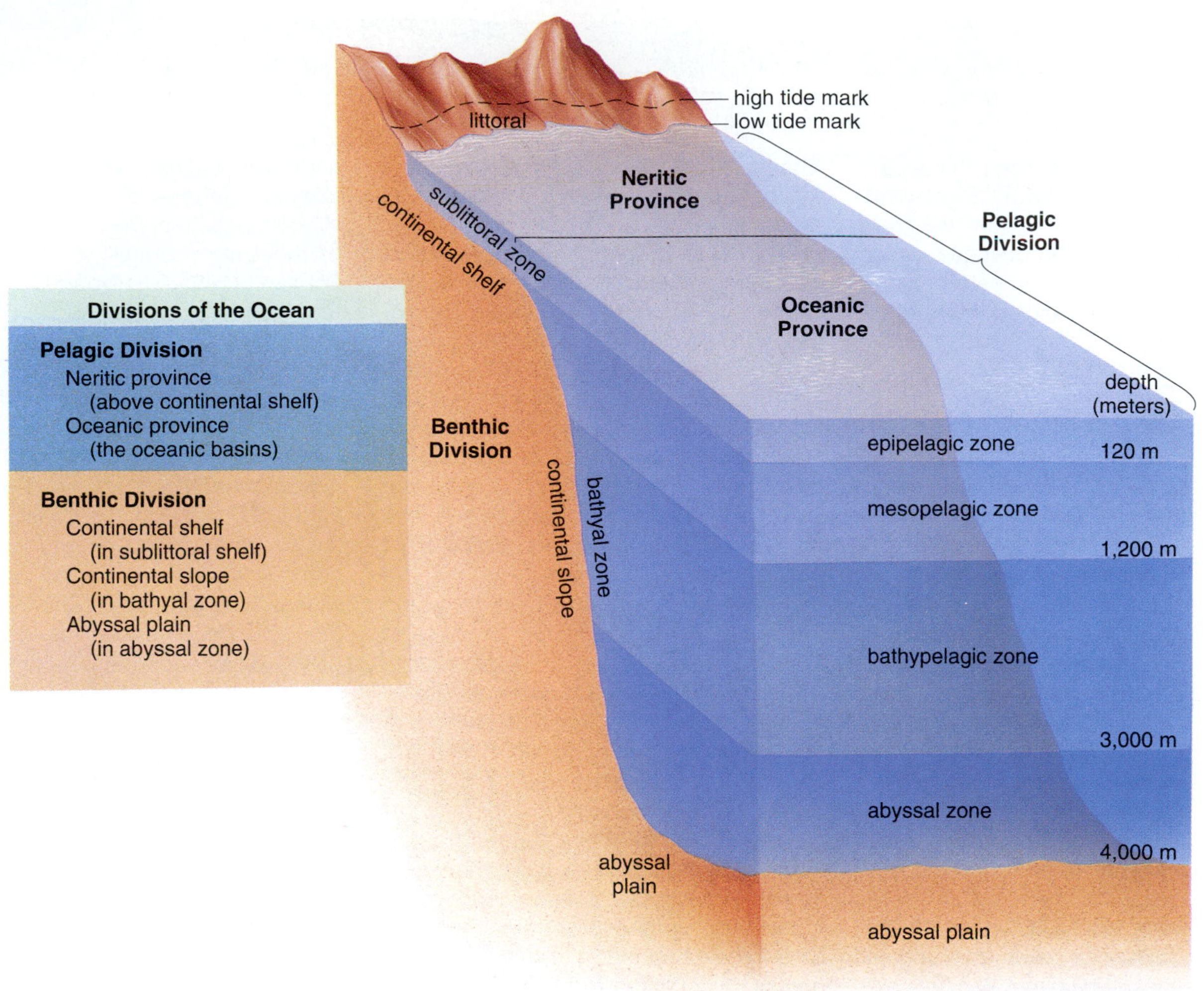

Figure 26.24 **Marine environment.**
Organisms reside in the pelagic division (blue), where waters are divided as shown. Organisms also reside in the benthic division (brown), which is divided as shown.

Pelagic Division

The oceans cover approximately three-quarters of our planet. The geographic areas and zones of an ocean are shown in Figure 26.24. It is customary to place the organisms of the oceans into either the pelagic division (open waters) or the benthic division (ocean floor).

The **pelagic division** [Gk. *pelagos,* the sea] includes the neritic province and the oceanic province. There is a greater concentration of organisms in the neritic province than in the oceanic province because sunlight penetrates the waters of the former, and this is where you find the most nutrients. Phytoplankton consisting of suspended algae is food not only for zooplankton but also for small fishes. These small fishes in turn are food for commercially valuable fishes—herring, cod, and flounder.

In the oceanic province, the epipelagic zone, although it is sunlit, does not have a high concentration of phytoplankton because it lacks nutrients. These photosynthesizers, however, still support a large assembly of zooplankton because the oceans are so large. The zooplankton are food for herrings and bluefishes, which in turn are eaten by larger mackerels, tunas, and sharks. Flying fishes, which glide above the surface, are preyed upon by dolphins (not to be confused with porpoises, which also are present). Whales are other mammals found in the epipelagic zone. Baleen whales strain krill (small crustacea) from the water, and the toothed sperm whales feed primarily on the common squid.

Animals in the mesopelagic zone are carnivores, are adapted to the absence of light, and tend to be translucent, red colored, or even luminescent. There are luminescent shrimps, squids, and fishes, such as lantern and hatchet fishes.

The bathypelagic zone is in complete darkness except for an occasional flash of bioluminescent light. Carnivores and scavengers are found in this zone. Strange-looking fishes with distensible mouths and abdomens and small, tubular eyes feed on infrequent prey.

Benthic Division

The **benthic division** [Gk. *benthos*, depths of the sea] includes organisms that live on or in the soil of the continental shelf, the continental slope, and the abyssal plain (Fig. 26.24). These are the organisms of the sublittoral, bathyal, and abyssal zones.

Seaweed grows on the continental shelf, and it can be found in batches on outcroppings as the water gets deeper. Nearly all benthic organisms, however, are dependent on the slow rain of plankton and detritus from the sunlit waters above. There is more diversity of life in the continental shelf and on the continental slope than in the abyssal zone. In these first two zones, clams, worms, and sea urchins are preyed upon by starfishes, lobsters, crabs, and brittle stars.

The abyssal zone is inhabited by animals that live at the soil-water interface on the dark abyssal plain (Fig. 26.25). It once was thought that few animals exist in this zone because of the intense pressure and the extreme cold. Yet many invertebrates live here by feeding on debris floating down from the mesopelagic zone. Sea lilies rise above the seafloor; sea cucumbers and sea urchins crawl around on the sea bottom; and tube worms burrow in the mud.

The flat abyssal plain is interrupted by enormous underwater mountain chains called oceanic ridges. Along the axes of the ridges, crustal plates spread apart and molten magma rises to fill the gap. At **hydrothermal vents,** seawater percolates through cracks and is heated to about 350°C, causing sulfate to react with water and form hydrogen sulfide (H_2S). The water temperature–hydrogen sulfide combination has been found to support communities that contain huge tube worms and clams. Chemosynthetic bacteria that obtain energy from oxidizing hydrogen sulfide live within these organisms. It was a surprise to find communities of organisms living so deep in the ocean, where light never penetrates. Unlike photosynthesis, chemosynthesis does not require light energy.

The neritic province and the epipelagic zone of the oceanic province receive sunlight and contain the organisms with which we are most familiar. The organisms of the benthic division are dependent upon debris that floats down from above.

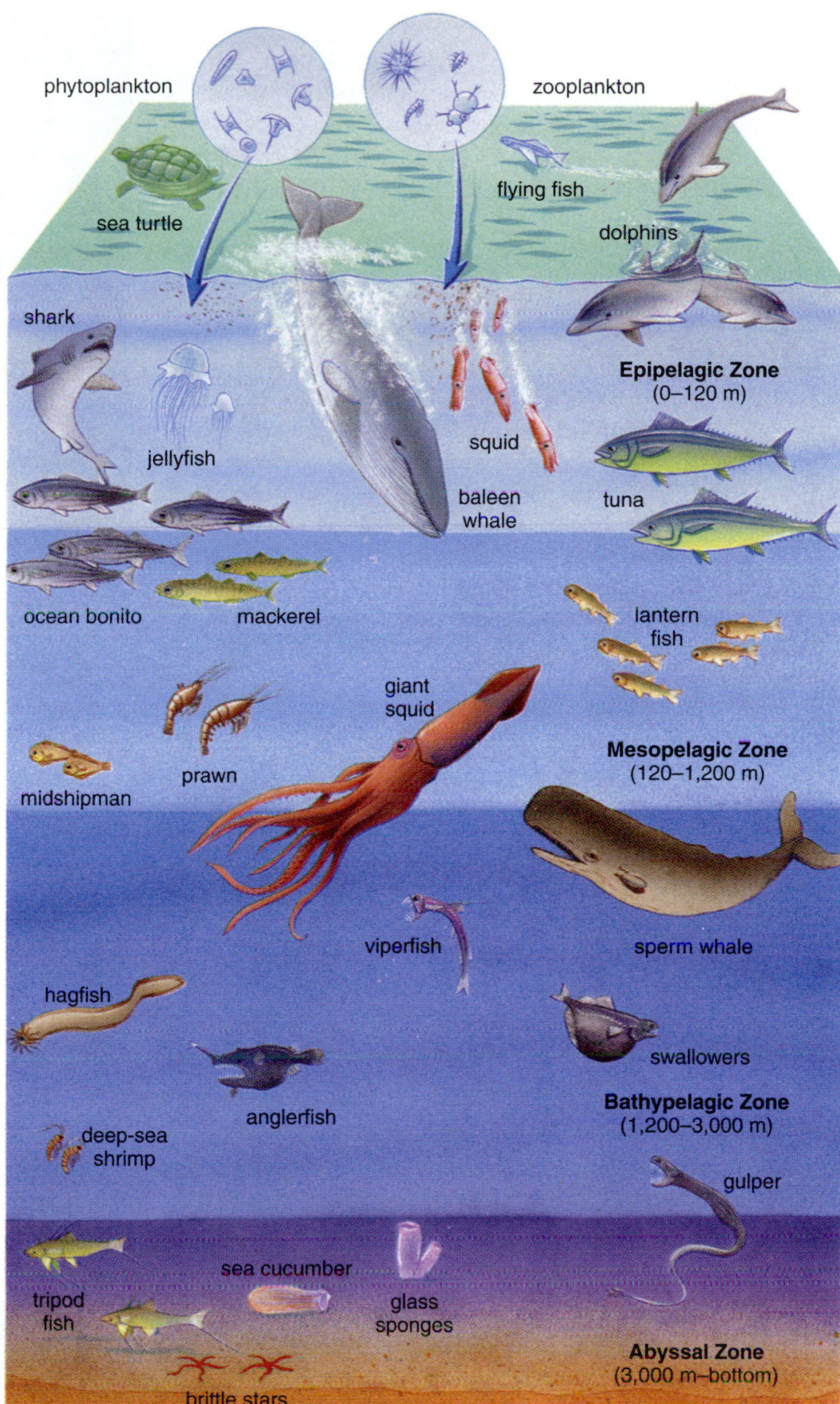

Figure 26.25 Pelagic division.
Organisms of the epipelagic, mesopelagic, and bathypelagic zones are shown. The abyssal zone is a part of the benthic division.

Coral Reefs

Coral reefs (Fig. 26.26) are areas of biological abundance found in shallow, warm, tropical waters just below the surface of the water. Their chief constituents are stony corals, animals that have a calcium carbonate (limestone) exoskeleton, and calcareous red and green algae. Corals do not usually occur individually; rather, they form colonies derived from an individual coral that has reproduced by means of budding. Corals provide a home for a microscopic alga called zooxanthellae. The corals, which feed at night, and the algae, which photosynthesize during the day, are mutualistic and share materials and nutrients. The close relationship of corals and zooxanthellae may be the reason why coral reefs form only in shallow sunlit water—the algae utilize sunlight for photosynthesis. Closely related to corals, sea anemones—the flowers of the sea with tentacles that bear some resemblance to petals—are also plentiful.

A reef is densely populated with life perhaps because reefs are areas of intermediate disturbance. The large number of crevices and caves provide shelter for filter feeders (sponges, sea squirts, and fan worms) and for scavengers (crabs and sea urchins). The barracuda, moray eel, and sharks are top predators in coral reefs (Fig. 26.26*b*). There are many types of small beautifully colored fishes. Parrot fishes feed directly on corals, and others feed on plankton or detritus. Small fishes become food for larger fishes like snappers that are caught for human consumption.

Coral reefs are subject the world over to overfishing and oceanic pollution. The crown-of-thorns starfish feeds on coral polyps, and the very existence of the Great Barrier Reef off the coast of Australia is threatened by a plague of these animals. The giant triton is a natural predator, and some believe that the crown-of-thorns is proliferating because humans have killed off the triton for its handsome spiral shell.

Coral reefs, which are areas of biological abundance, occur in tropical seas.

a.

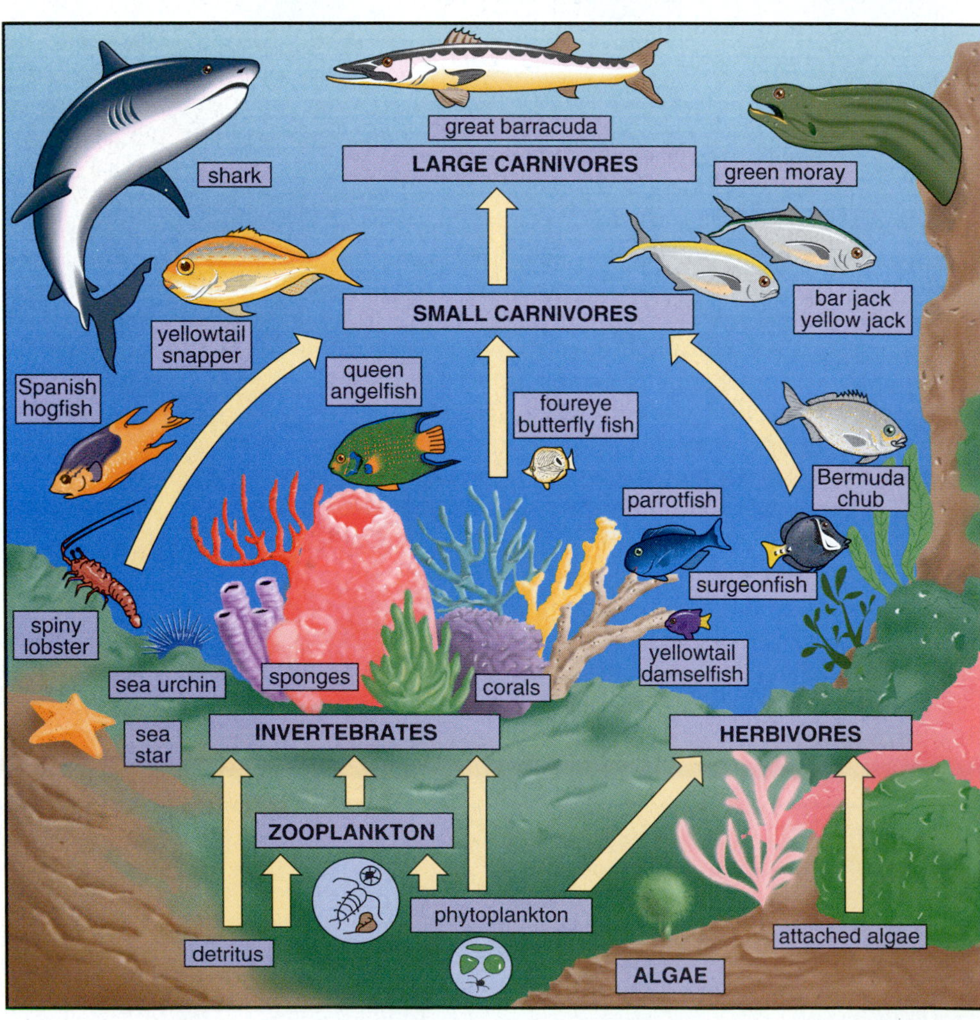

b.

Figure 26.26 Coral reefs.
A coral reef is (a) a unique community of marine organisms. b. The abundance of life results in a complex food web—a simplified representation is given here.

Connecting Concepts

The biosphere is the product of interaction of living organisms with the earth's physical and chemical environment over billions of years. Life has affected the atmosphere, modifying its composition and influencing global climate. Organisms have helped create the chemical and physical conditions of streams, lakes, and the oceans. The soils of terrestrial ecosystems and the sediments of aquatic ecosystems are structured largely by activities of organisms. Reef-building plants and animals have helped build the islands on which millions of humans live. The Earth's diverse biomes also result from interactions of the biota and the abiotic environment.

Over geological time, the biosphere has been changing constantly. The biomes of the age of dinosaurs were strikingly different from those of today. Astrophysical events have triggered some of this change. Changes in the sun's radiation output and in the tilt of the earth's axis have altered the pattern of solar energy reaching the earth's surface. Geological processes have also modified conditions for life. The drifting of continents has changed the arrangement of continents and oceans. Mountain ranges have been thrust up and eroded down. Through these changing conditions, life has evolved, and the structure of the earth's biomes has evolved as well. In the last few million years, humans appeared and learned to exploit the earth's biomes.

Humans have transformed vast areas of many of the terrestrial biomes into farmland, cities, highways, and other developments. Through our resource use and release of pollutants, we have now become an agent of global importance. Yet we still depend on the biodiversity that exists in the earth's biomes, and on the interactions of other organisms within the biosphere. These interactions still influence climate, patterns of nutrient cycling and waste processing, and basic biological productivity. The earth's biotic diversity also provides enjoyment and inspiration to millions of people, who spend billions of dollars to visit coral reefs, deserts, rain forests, and even the Arctic tundra. Unfortunately, as we shall see in the next chapter, many of the earth's biomes may not survive in their present form for our descendants' benefit. Although most of us value Earth's biodiversity, human activities are threatening many species with extinction.

Summary

26.1 Climate and the Biosphere

Because the earth is a sphere, the sun's rays at the poles are spread out over a larger area than the vertical rays at the equator. The temperature at the surface of the earth therefore decreases from the equator to each pole. The earth is tilted on its axis, and the seasons change as the earth rotates annually about the sun.

Warm air rises near the equator and loses its moisture and then descends at about 30° north and south latitude, and so forth to the poles. When the air descends, it is dry, and therefore the great deserts of the world are formed at 30° latitudes. Because the earth rotates on its axis daily, the winds blow in opposite directions above and below the equator.

Topography also plays a role in the distribution of moisture. Air rising over coastal ranges loses its moisture on the windward side, making the leeward side arid. The fact that the oceans hold heat to a greater degree than land causes the monsoon climate in southern Asia.

26.2 Biomes of the World

The term biome is used to refer to terrestrial and aquatic communities. Biomes are distributed according to climate; that is, temperature and rainfall influence the pattern of biomes about the world. The significance of temperature is also seen by observing that the same sequence of biomes is seen when traveling to northern latitudes as traveling up a mountain. Biomes blend one into the other, and there is no actual demarcation between them.

Soil types vary according to the biome. Grasslands have a deep A horizon, which contains organic nutrients, and forests have more of a B horizon, which contains leached inorganic nutrients.

26.3 Terrestrial Biomes

The tundra is the northernmost biome and consists largely of short grasses and sedges and dwarf woody plants. Because of the cold winters and short summers, most of the water in the soil is frozen the year round. This is called the permafrost.

The taiga, a coniferous forest, has less rainfall than other types of forests. The temperate deciduous forest has trees that gain and lose their leaves because of the alternating seasons of summer and winter. Tropical rain forests are the most complex and productive of all biomes.

Shrublands usually occur along coasts that have dry summers and receive most of their rainfall in the winter. Among grasslands, the savanna, a tropical grassland, supports the greatest number of different types of large herbivores. The prairie, found in the United States, has a limited variety of vegetation and animal life.

Deserts are characterized by a lack of water—they are usually found in places with less than 25 cm a year. Some plants, such as cacti, are succulents, and others are shrubs with thick leaves they often lose during dry periods.

26.4 Aquatic Biomes

Aquatic biomes are divided into freshwater communities (lakes, streams, rivers) and marine communities. Rainfalls in mountains, and streams develop that join to form a river that runs into the sea. Streams, rivers, and wetlands are different communities.

In deep lakes of the temperate zone, the temperature and the concentration of nutrients and gases in the water vary with depth. In the fall, the water on the surface cools and sinks to the bottom. In the spring, ice on the surface water melts, and this cool water sinks. The entire body of water is therefore cycled twice a year, distributing nutrients from the bottom layers. Lakes and ponds have three life zones. Rooted plants and clinging organisms live in the littoral zone, plankton and fishes are in the sunlit limnetic zone, and bottom-dwelling organisms like crayfishes and mollusks are in the profundal zone.

Estuaries (e.g., salt marshes, mudflats, mangrove forests) are near the mouth of a river. Estuaries are the nurseries of the sea. Marine communities are divided into coastal communities and the oceans. The coastal communities, especially estuaries, are more productive than the oceans.

The oceans play a role in distributing heat via ocean currents that move in a circular fashion from the equator. The Gulf Stream is an ocean current that brings warmth to northern Atlantic land masses. An ocean is divided into the pelagic division and the benthic division. The oceanic province of the pelagic division (open waters) has three zones. The epipelagic zone receives adequate sunlight and supports the most life. The mesopelagic zone contains organisms adapted to minimum and no light, respectively. The benthic division (ocean floor) includes the organisms that live on the continental shelf, the continental slope, and the abyssal zone.

Coral reefs are extremely productive communities found in shallow tropical waters. The intermediate disturbance theory is believed to account for the extreme diversity, which results in a complex food web.

Reviewing the Chapter

1. Tell how a spherical earth and the path of the earth about the sun affect climate. 442
2. Describe the air circulation about the Earth, and tell why deserts are apt to occur at 30° north and south of the equator. Why does the Pacific coast of California get plentiful rainfall? 442–443
3. How does a coastal range affect climate? What causes a monsoon climate? 443
4. Name the terrestrial biomes you would expect to find when going from the base to the top of a mountain. 445
5. What are the soil horizons, and how do they differ in forests and grasslands? 446
6. Describe the location, the climate, and the populations of the Arctic tundra, coniferous forests (both taiga and temperate rain forest), temperate deciduous forests, tropical forests, shrubland, grasslands (both prairie and savanna), and deserts. 447–454
7. Describe the overturn of a temperate lake and life zones of a lake, and the organisms you would expect to find in each zone. 456–457
8. Describe the coastal communities, and discuss the importance of estuaries to the productivity of the ocean. 458–459
9. Describe the ocean currents and how the Gulf Stream accounts for Great Britain having a mild temperature. 460
10. Describe the zones of the open ocean and the organisms you would expect to find in each zone. 462–463
11. Describe a coral reef including the varied organisms found there. 464

Testing Yourself

Choose the best answer for each question.

1. The seasons are best explained by
   a. the distribution of temperature and rainfall in biomes.
   b. the tilt of the earth on its axis as it rotates about the sun.
   c. the daily rotation of the earth on its axis.
   d. the fact that the equator is warm and the poles are cold.
   e. the monsoon and lake effect.
2. The location of deserts at 30° is best explained by which statement?
   a. Warm air rises and loses its moisture; cool air descends and becomes warmer and drier.
   b. Ocean currents give up heat as they flow from the equator to the poles.
   c. Cool ocean breezes cool the coast during the day.
   d. Terrestrial biomes cool off during the night.
   e. All of these are correct.
3. Which of these is mismatched?
   a. tundra—permafrost
   b. savanna—acacia trees
   c. prairie—epiphytes
   d. coniferous forest—evergreen trees
   e. mesopelagic zone—carnivorous animals
4. All of these phrases describe the tundra except
   a. low-lying vegetation.
   b. northernmost biome.
   c. short growing season.
   d. many different types of species.
   e. caribou migrations.
5. The forest with a multilevel understory is the
   a. tropical rain forest.
   b. coniferous forest.
   c. tundra.
   d. temperate deciduous forest.
   e. long-grass prairie.
6. All of these phrases describe a tropical rain forest except
   a. nutrient-rich soil.
   b. many arboreal plants and animals.
   c. canopy composed of many layers.
   d. broad-leaved evergreen trees.
   e. located at or near the equator.
7. Which type of animal would you be least likely to find in a grassland biome?
   a. hoofed herbivore
   b. active carnivore
   c. arboreal primate
   d. flying insect
   e. termites
8. Phytoplankton are more likely to be found in which life zone of a lake?
   a. limnetic zone
   b. profundal zone
   c. benthic zone
   d. mesopelagic zone
   e. All of these are correct.
9. An estuary acts as a nutrient trap because of the
   a. action of rivers and tides.
   b. depth at which photosynthesis can occur.
   c. amount of rainfall received.
   d. height of the water table.
   e. limited size.

10. The mild climate of Great Britain is best explained by
    a. the winds called the westerlies.
    b. the spinning of the earth on its axis.
    c. Great Britain being a mountainous country.
    d. the flow of ocean currents.
    e. a small landmass bringing more rain.
11. Which area of an ocean has the greatest concentration of nutrients?
    a. epipelagic zone and benthic zone
    b. epipelagic zone only
    c. benthic zone only
    d. neritic province
    e. Both c and d are correct.
12. Use plus signs (4+ is maximum) to indicate the degree of temperature and the amount of rainfall for these biomes.

	Temperature	*Rainfall*
a. tropical rain forest		
b. savanna		
c. taiga		
d. tundra		

Thinking Scientifically

1. Pharmaceutical companies are interested in "bioprospecting" in tropical rain forests. These companies are looking for naturally occurring compounds in plants or animals that can be used as drugs for a variety of diseases. The most promising compounds act as antibacterial or antifungal agents. Even discounting the fact that the higher density of species in tropical rain forests would produce a wider array of compounds than another biome, why would antibacterials and antifungal compounds be more likely to evolve in this particular biome?
2. Global warming is predicted by many scientists to lead to a partial melting of glaciers at the north and south poles and a rise in sea level. Among the many places that will be impacted by such a rise are the estuaries that now serve as breeding grounds for many birds and fish. While it is impossible to predict exactly how different estuaries would change in response to an increase in sea level, what are some possibilities that can be predicted based on the general structure of estuaries?

Bioethical Issue

Pollution of fresh water comes from two sources. Point sources, like paper mills or other industries that discharge pollutants directly into nearby rivers and bodies of water, are easily identifiable. Nonpoint sources, which are simply due to runoff from the land, are not easily identifiable. Although the public generally believes that big corporations are still causing most of the freshwater pollution, this is not the case. Instead, nonpoint sources are causing most of the problem. Agricultural fertilizers are the chief cause of nitrate contamination of drinking-water wells. Excessive nitrates in a baby's bloodstream can lead to a slow suffocation known as blue-baby syndrome. Agricultural herbicides are suspected carcinogens in the tap water of scattered communities coast to coast. A 1993 runoff from a dairy farm carried a parasite called cryptosporidium to the drinking water of Milwaukee, Wisconsin, sickening thousands of people, and killing a few. What can be done? Some farmers are already using irrigation methods that deliver water directly to plant roots, no-till agriculture that reduces the loss of topsoil and cuts back on herbicide use, and integrated pest management that relies heavily on good bugs to kill bad bugs. Perhaps more should do so. Encouraged—in some cases, compelled—by state and federal agents, dairy farmers have built sheds, concrete containments, and underground liquid storage tanks to hold the wastes when it rains. Later, the manure is trucked to fields and spread as fertilizer.

The manicuring of lawns, the use of motor vehicles, and the construction and use of roads and buildings also add contaminants to streams, lakes, and aquifers. Citizens around Grand Traverse Bay on the eastern shore of Lake Michigan have gotten the message, especially because they want to keep on enjoying water-dependent activities like boating, bathing, and fishing. James Haverman, a concerned member of the Traverse Bay Watershed Initiative says, "If we can't change the way people live their everyday lives, we are not going to be able to make a difference." Builders in Traverse County are already required to control soil erosion with filter fences, steer rainwater away from exposed soil, build sediment basins, and plant protective buffers. Presently homeowners must have a 25-foot setback from wetlands and a 50-foot setback from lakes and creeks. They are also encouraged to pump out their septic systems every two years. Do you approve of legislation that requires farmers and homeowners to protect freshwater supplies? Why or why not?

Understanding the Terms

alpine tundra 445
Arctic tundra 447
benthic division 463
biome 445
climate 442
coral reef 464
desert 454
El Niño–Southern Oscillation 460
epiphyte 450
estuary 458
eutrophication 456
fall overturn 456
grassland 452
hydrothermal vent 463
lake 455
littoral zone 459
monsoon 443
montane coniferous forest 445
pelagic division 462
permafrost 447
phytoplankton 457
plankton 457
pond 455
prairie 452
rain shadow 443
river 455
savanna 452
shrubland 452
soil 446
spring overturn 457
stream 455
taiga 448
temperate deciduous forest 449
temperate rain forest 448
tropical rain forest 450
upwelling 460
zooplankton 457

Match the terms to these definitions:

a. ______________ End of a river where fresh water and salt water mix as they meet.
b. ______________ Ocean floor, which supports a unique set of organisms in contrast to the pelagic division.
c. ______________ Terrestrial biome that is a coniferous forest extending in a broad belt across northern Eurasia and North America.
d. ______________ Terrestrial biome that is a grassland in Africa, characterized by few trees and a severe dry season.
e. ______________ Mixing process that occurs in spring in stratified lakes whereby the oxygen-rich top waters mix with nutrient-rich bottom waters.

Web Connections

Exploring the Internet

http://www.mhhe.com/biosci/genbio/mader
(click on *Biology 7/e*)

The *Biology 7/e* Online Learning Center provides many resources for studying the material in this chapter including links to the following sites:

The Reef Education Network will take you on an expedition of underwater communities, where you can meet the locals and ask personal questions about life undersea.

http://www.reef.edu.au/

Rainforest Action Network has a multitude of information about rain forests, areas that need help, campaigns, victories, and what you can do to help preserve these unique and biologically diverse ecosystems.

http://www.ran.org/ran/intro.html

ESA Aquatic Ecology Section. This site has many links to organizations and publications on the topic of aquatic ecology.

http://esa.sdsc.edu/aquatic/links.html

Ecological Society of America Panel on Vegetation Classification. A discussion of a new national standard for floristic levels/vegetation classifications common in the United States. These classifications would further divide biomes.

http://esa.sdsc.edu/vegwebpg.htm

Biomes. This site was apparently compiled by middle school students, but it is better than many other sites I've seen. Descriptions, pictures, and links to information about biomes.

http://ccwf.cc.utexas.edu/~bogler/ecology/biomes.html

Conservation Biology

chapter 27

chapter concepts

27.1 Conservation Biology and Biodiversity

- Conservation biology addresses a crisis—the loss of biodiversity. 470
- Conservation biology is an applied, goal-orientated, multidisciplinary field. 470
- Extinction rates have risen to many times their natural levels and many types of ecosystems are disappearing. 470
- Biodiversity includes species diversity, genetic diversity, community diversity and landscape diversity in marine, freshwater, and terrestrial habitats. 470

27.2 Value of Biodiversity

- Biodiversity has both direct and indirect value. 472, 474

27.3 Causes of Extinction

- Habitat loss, introduction of alien species, pollution, overexploitation and disease are now largely responsible for the loss of biodiversity. 476
- Global warming will shift the optimal range of many species northward and disrupt many coastal ecosystems. 478

27.4 Conservation Techniques

- Because of fragmented habitats, it is often necessary to conserve subdivided populations today. 480
- Identifying and conserving biodiversity hotspots and/or keystone species can save many other species. 480
- Computer analyses can be done to select areas for preservation and to determine the minimal population size needed for survival. 481
- Landscape preservation often involves restoration of habitats today. 482

Tusks of elephants confiscated from poachers are burned by Kenyan government.

The growth of the human population and our economic activities have brought us face to face with the limits set by the natural environment. The danger of permanent environmental damage has successively increased and the biosphere's biodiversity heritage is being challenged. The biodiversity crisis demands a twofold response. We must take quick action to protect biodiversity, but we must also address the human forces that are threatening this valuable resource.

The 1990s saw the emergence of conservation biology as a branch of environmental science directed at the challenge of managing our living heritage—biodiversity. Conservation biology is in part a basic science and in part an applied science. In the arena of basic science, for example, conservation biology examines the nature and evolutionary origin of biodiversity. It also considers how changing environmental conditions, especially those due to human activity, are reducing biodiversity. In the area of applied science, however, its ultimate role is goal-oriented: the preservation and management of ecosystems for human welfare.

27.1 Conservation Biology and Biodiversity

Conservation biology [L. *conservatio,* keep, save] is a new discipline which studies all aspects of biodiversity with the goal of conserving natural resources for this generation and all future generations. Conservation biology is unique in that it is concerned with both the development of scientific concepts and the application of these concepts to the everyday world. A primary goal is the management of biodiversity for sustainable use by humans. To achieve this goal, conservation biologists are interested in and come from many subfields of biology that only now have been brought together into a cohesive whole:

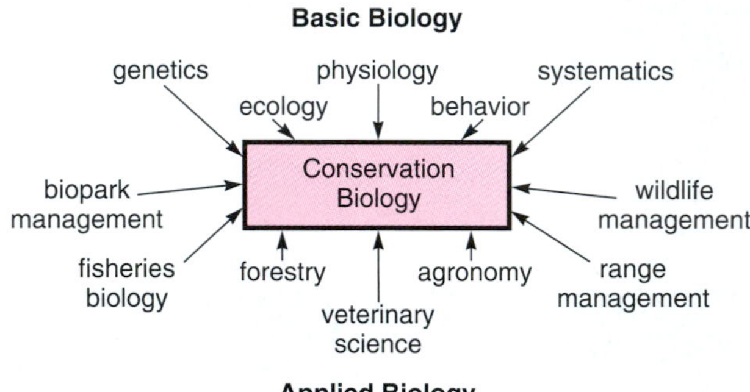

Like a physician, a conservation biologist must be aware of the latest findings both theoretical and practical, be able to use this knowledge to diagnose the source of the trouble, and suggest a suitable treatment. Often, it is necessary to work with government officials both at the local and federal levels.

Conservation biology is a unique science in another way. It takes a leap of faith and unabashedly supports these ethical principles: (1) biodiversity is desirable for the biosphere and therefore humans; (2) extinctions, due to human actions, are therefore undesirable; (3) the complex interactions in ecosystems supports biodiversity and are desirable; and (4) biodiversity brought about by evolutionary change has value in and of itself regardless of any practical benefit.

Conservation biology has emerged in response to a crisis—never before in the history of the earth are so many extinctions expected in such a short period of time. Estimates vary, but at least 10–20% of all species now living most likely will become extinct in the next 20 to 50 years unless immediate action is taken. It is urgently important, then, that all citizens understand the concept of biodiversity, the value of biodiversity, the likely causes of present-day extinctions, and what could possibly be done to prevent extinctions from occurring.

Conservation biology is a new discipline concerned with the study and sustainable management of biodiversity for the benefit of human beings.

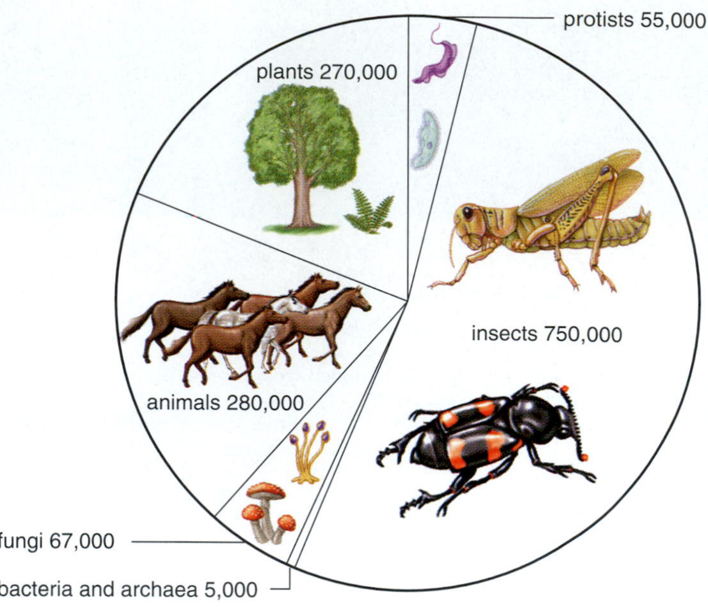

Figure 27.1 Number of described species.
There are only about 1.5 million described species; insects are far more prevalent than organisms in other groups. Undescribed species probably number far more than those species that have been described.

Biodiversity

At its simplest level, **biodiversity** [Gk. *bio,* life, and L. *diversus,* various] is simply the variety of life on earth. It is common practice to describe biodiversity in terms of the number of species among various groups of organisms. Figure 27.1 only accounts for the species that have so far been described and it's possible that many more are yet to be found. The majority of estimates state that there are probably 5 to 15 million species; if so, many species are still to be found and described.

You can well appreciate that simply giving the number of species on earth is not very meaningful and certainly ecologists describe biodiversity as an attribute of three other levels of biological organization: Genetic diversity refers to variations among the members of a population. Populations with high genetic diversity are more likely to have some individuals that can survive environmental changes such as a change in the structure of an ecosystem. The 1846 potato blight in Ireland; the 1922 wheat failure in the Soviet Union, and the 1984 outbreak of citrus canker in Florida were all made worse by limited genetic variation among these crops. If its populations are quite small and isolated, a species is more likely to eventually become extinct because of a loss of genetic diversity.

Community (and therefore ecosystem) diversity is dependent on the interactions of species at a particular locale. One community's species composition can be completely different from that of other communities. Community composition, therefore, increases the amount of biodiversity in the biosphere. Although in the past conservationists have frequently concentrated on saving particular species such as the California condor, the black-footed ferret, or the spotted owl, this is a shortsighted approach. Saving entire communities can save many species and the contrary is also true—

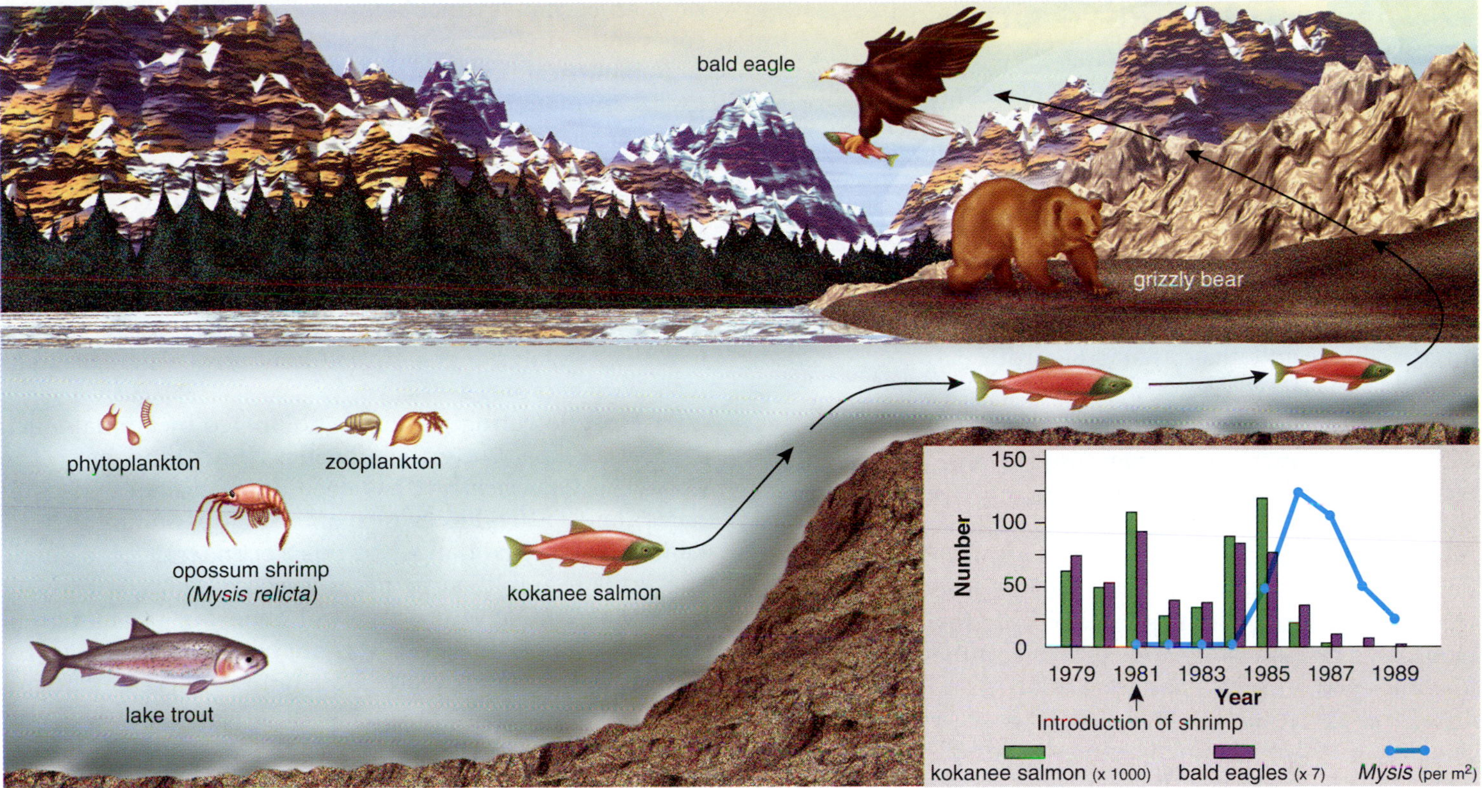

Figure 27.2 Eagles and bears feed on spawning salmon.
Humans introduced the opossum shrimp as prey for salmon. Instead the shrimp competed with salmon for zooplankton as a food source. The salmon, eagle, and bear populations subsequently declined.

disrupting a community threatens the existence of more than one species. Opossum shrimp *(Mytis relicta)* were introduced into Flathead Lake in Montana and its tributaries as food for salmon. The shrimp ate so much zooplankton there was in the end far less food for the fish and ultimately grizzly bears and bald eagles (Fig. 27.2).

Landscape diversity is a descriptive term more likely to be found in conservation literature. A **landscape** is a number of interacting ecosystems. In one landscape, for example, there may be plains, mountains, and rivers. Any of these ecosystems can be so fragmented that only patches (remnants) remain that must be connected by **habitat corridors**—strips of land that allow organisms to move from one patch to the other. Corridors might also help preserve animals that migrate between different ecosystems making up a landscape.

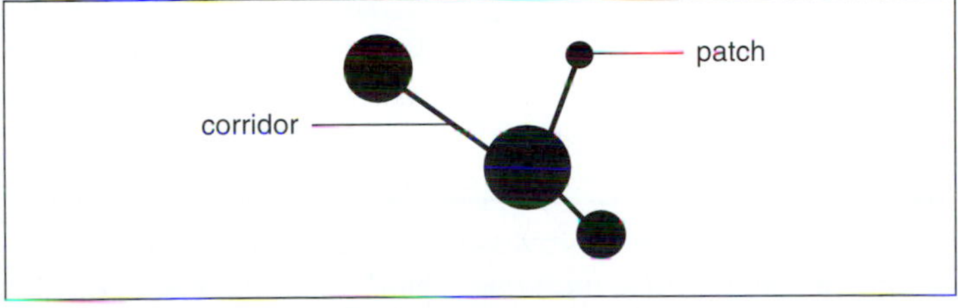

Distribution of Diversity

Biodiversity is not evenly distributed throughout the biosphere; therefore, protecting some areas will save more species than protecting other areas. Biodiversity is highest at the tropics and declines toward each pole on land, in fresh water, and in the ocean. Also, among coral reefs, there are more species in those of the Indonesian archipelago and the number declines in other coral reefs as one moves westward across the Pacific.

Some regions of the world are called **biodiversity hotspots** because they contain an unusually large concentrations of species. Hotspots contain about 20% of the earth's species but cover only about half of 1% of the earth's land area. The island of Madagascar, the Cape region of South Africa, the Great Barrier Reef of Australia are all biodiversity hotspots.

One surprise of late has been to discover that rain forest canopies and the deep sea benthos have many more species than formerly thought. Some conservationists refer to these two areas as biodiversity frontiers.

Biodiversity is often defined as variety of life at the species level, but biodiversity is also important at the genetic, community (ecosystem), and landscape levels.

27.2 Value of Biodiversity

Conservation biology strives to reverse the trend toward the possible extinction of thousands of plants and animals. To bring this about it is necessary to make all people aware that biodiversity is a resource of immense value.

Direct Value

Various individual species perform services for human beings and contribute greatly to the value we should place upon biodiversity. Only some of the most obvious values are discussed here and illustrated in Figure 27.3.

Medicinal Value

Most of the prescription drugs used in the United States were originally derived from living organisms. The rosy periwinkle from Madagascar is an excellent example of a tropical plant that has provided us with useful medicines. Potent chemicals from this plant are now used to treat two forms of cancer: leukemia and Hodgkin's disease. Hodgkin's disease is usually curable and the survival rate for childhood leukemia has gone from 10% to 90% because of these drugs. Although the value of saving a life cannot be calculated, still it is sometimes easier for us to appreciate the worth of a resource if it is explained in monetary terms. Researchers tell us that judging from the success rate in the past, there are 328 more types of drugs to be found in tropical rain forests and the value of this resource to society is probably $147 billion.

You may already know that the antibiotic penicillin is derived from a fungus and certain species of bacteria produce the antibiotics tetracycline and streptomycin. These drugs have proven to be indispensable in the treatment of diseases, including certain sexually transmitted diseases. Leprosy is among those diseases for which there is no cure yet. The bacterium that causes leprosy will not grow in the laboratory but scientists discovered that it grows naturally in the nine-banded armadillo. Having a source for the bacterium may make it possible to find a cure for leprosy. The blood of horseshoe crabs contains a substance called limulus amoebocyte lysate which is used to ensure that medical devices such as pacemakers, surgical implants, and prosthetic devices are free of bacteria. Blood is taken from 250,000 crabs a year and then they are returned to the sea unharmed.

Agricultural Value

Most crops such as wheat, corn, and rice are derived from wild plants that have been modified to be high producers. The same high-yield, genetically similar strains tend to be grown worldwide. When rice crops in Africa were being devastated by a virus, researchers grew wild rice plants from thousands of seed samples, until they found one that contained a gene for resistance to the virus. These wild plants were then used in a breeding program to transfer the gene into high-yield rice plants. If this variety of wild rice had become extinct before it could be discovered, rice cultivation in Africa may have collapsed.

Biological pest controls—natural predators and parasites—are often preferable to the use of chemical pesticides. When a rice pest called the brown planthopper became resistant to pesticides, farmers began to use natural brown planthopper enemies instead. The economic savings were calculated at well over $1 billion. Similarly, cotton growers in Cañete Valley, Peru found that pesticides were no longer working against the cotton aphid because of resistance. Research identified natural predators that are now being used to an ever greater degree by cotton farmers. Again, savings have been enormous.

Most flowering plants are pollinated by animals such as bees, wasps, butterflies, beetles, birds, and bats. The honeybee, *Apis mellifera*, has been domesticated and pollinates almost $10 billion worth of food crops annually in the United States. The danger of this dependency on a single species is exemplified in mites that have now wiped out more than 20% of the commercial honeybee population in the U.S. Where can we get resistant bees? From the wild, of course. The value of wild pollinators to the U.S. agricultural economy has been calculated to be worth $4.1 to 6.7 billion a year.

Consumptive Use Value

We have had much success cultivating crops, keeping cattle and sheep and other domesticated animals, growing trees in plantations, and so forth. But so far aquaculture, the growing of fish and shellfish for human consumption, has contributed only minimally to human welfare—most freshwater and marine harvests depend on the catching of wild animals, such as fishes (e.g., trout, cod, tuna, flounder) crustaceans (e.g., lobsters, shrimps, crabs) and mammals (e.g., whales). Obviously, this is a biodiversity resource that is invaluable.

The environment also provides all sorts of products that are sold in the marketplace worldwide, including wild fruits and vegetables, skins, fibers, beeswax, and seaweed. Also, some people obtain their meat directly from the environment. In one study, researchers calculated that the economic value of wild pig in the diet of native hunters in Sarawak, East Malaysia was approximately $40 million per year.

Similarly many trees are still felled in the natural environment for their wood. Researchers have calculated that a species-rich forest in the Peruvian Amazon is worth far more if the forest is used for fruit and rubber production rather than timber production. Fruit and latex, needed to produce rubber, can be brought to market for an unlimited number of years whereas once the trees are gone, no more timber can be harvested.

Wild species directly provide us with all sorts of goods and services whose monetary value can sometimes be calculated.

Wild species, like the lesser long-nosed bat, *Leptonycteris curasoae*, are pollinators of agricultural and other plants.
Wild species, like the rosy periwinkle, *Catharanthus roseus*, are sources of many medicines.
Wild species, like rubber trees, *Hevea*, can provide a product indefinitely if the forest is not destroyed.
Wild species, like many marine species, provide us with food.
Wild species, like lady bugs, *Coccinella*, play a role in biological control of agricultural pests.
Wild species, like the nine-banded armadillo, *Dasypus novemcinctus*, play a role in medical research.

Figure 27.3 Direct services of wildlife. The direct services of wild species benefits human beings immensely and it is sometimes possible to calculate the monetary value, which is always surprisingly large.

Indirect Value of Biodiversity

The wild species we have been discussing live in ecosystems (Fig. 27.4*a*). If we want to preserve them it is more economical to save ecosystems than individual species. Ecosystems perform many services for our modern society that increasingly lives in cities (Fig. 27.4*b*). These services are said to be indirect because they are pervasive and not easily discernible. Even so, our very survival depends on the functions that ecosystems perform for us.

Biogeochemical Cycles

You'll recall that ecosystems are characterized by energy flow and chemical cycling. The biodiversity within ecosystems contributes to the workings of the water, carbon, nitrogen, phosphorus cycles, and other biogeochemical cycles. We are dependent on these cycles for fresh water, removal of carbon dioxide from the atmosphere, uptake of excess soil nitrogen, and provision of phosphate. When human activities upset the usual workings of biogeochemical cycles, the dire environmental consequences include excess pollutants that are harmful to us. Technology is unable to artificially contribute to or create any of the biogeochemical cycles.

Waste Disposal

Decomposers break down dead organic matter and other types of wastes to nutrients that are used by the producers within ecosystems. This function aids humans immensely because we dump millions of tons of waste material into natural ecosystems each year. If it were not for decomposition, waste would soon cover the entire surface of our planet. We can build sewage treatment plants but they are expensive and few break down solid wastes completely to inorganic nutrients. It is less expensive and more efficient to water plants and trees with partially treated waste water and let soil bacteria cleanse it completely.

Biological communities are also capable of breaking down and immobilizing pollutants such as heavy metals and pesticides that humans release into the environment. A review of wetland functions in Canada assigned a value of $50,000 per hectare (100 acres or 10,000 square meters) per year to the ability of natural areas to purify water and take up pollutants.

Provision of Fresh Water

Few terrestrial organisms are adapted to living in a salty environment—they need fresh water. The water cycle continually supplies fresh water to terrestrial ecosystems. Humans use fresh water in innumerable ways, including drinking and irrigation of crops. Freshwater ecosystems, like rivers and lakes, also provide us with fish and other types of organisms for food.

Unlike other commodities there is no substitute for fresh water. We can remove salt from sea water to obtain fresh water, but the cost of desalination is about four to eight times the average cost of fresh water acquired via the water cycle.

Forests and other natural ecosystems exert a "sponge effect." They soak up water and then release it at a regular rate. When rain falls in a natural area, plant foliage and dead leaves lessen its impact and the soil slowly absorbs it, especially if the soil has been aerated by organisms. The water-holding capacity of forests reduces the possibility of flooding. The value of a marshland outside Boston, Massachusetts has been estimated at $72,000 per hectare per year solely on its ability to reduce floods. Forests release water slowly for days or weeks after the rains have ceased. Rivers flowing through forests in West Africa release twice as much water halfway through the dry season and between three to five times as much at the end of the dry season as do rivers from coffee plantations.

Prevention of Soil Erosion

Intact ecosystems naturally retain soil and prevent soil erosion. The importance of this ecosystem attribute is especially observed following deforestation. In Pakistan, the world's largest dam, the Tarbela Dam, is losing its storage capacity of 12 billion cubic meters many years sooner than expected because silt is building up behind the dam due to deforestation. At one time, the Philippines were exporting $100 million of oysters, mussels, clams, and cockles each year. Now silt carried down rivers following deforestation is smothering the mangrove ecosystem that serves as a nursery for the sea. Most coastal ecosystems are not as bountiful as they once were because of deforestation and a myriad of other assaults.

Regulation of Climate

At the local level, trees provide shade and reduce the need for fans and air conditioners during the summer.

Globally, tropical rain forests ameliorate the climate because they act as a sink for carbon dioxide. The leaves of tropical trees take up carbon dioxide when they photosynthesize and the body of the trees store carbon. When trees are cut and burned, carbon dioxide is released into the atmosphere. Carbon dioxide makes a significant contribution to global warming which is expected to be stressful for many plants and animals. Only a small percentage of wildlife will be able to move northward, where the weather will be suitable for them.

Ecotourism

Almost everyone prefers vacations in the natural beauty of an ecosystem. In the United States, nearly 100 million people enjoy vacationing in a natural setting. To do so, they spend $4 billion on fees, travel, lodging and food each year. Many tourists want to go sport fishing, whale watching, boat riding, hiking, birdwatching, and the like. Some want to merely immerse themselves in the beauty of a natural environment.

Figure 27.4 Indirect services of ecosystems.
Natural ecosystems (a) provide human impacted ecosystems (b) with many ecological services. c. Research results showing that the higher the biodiversity (measured by number of plant species) the greater the net primary productivity (NPP) of an experimental community.

Biodiversity and Natural Ecosystems

Massive changes in biodiversity, such as deforestation, have a significant impact on ecosystem services such as those we have been discussing. Researchers are interested in determining if a high degree of biodiversity also helps ecosystems function more efficiently. To test the benefits of biodiversity in a Minnesota grassland habitat, researchers sowed plots with seven levels of plant diversity. This study found that ecosystem performance improves with increasing species richness. The more diverse plots had lower concentrations of inorganic soil nitrogen, indicating a higher level of nitrate uptake. A similar study in California also showed greater overall resource use in more diverse plots.

Another group of experiments tested the effects of an increase in diversity at four levels: producers, herbivores, parasites, and decomposers. They found that net primary productivity (NPP) increased as diversity increased (Fig. 27.4*c*). A computer simulation has shown that the response of a deciduous forest to elevated carbon dioxide is a function of species diversity. The more complex community of nine tree species exhibited a 30% greater amount of photosynthesis than a community composed of a single species.

More studies are needed to test whether biodiversity maximizes resource acquisition and retention within an ecosystem. Also, are more diverse ecosystems better able to withstand environmental changes and invasions by other species, including pathogens? Then, too, how does fragmentation affect the distribution of organisms within an ecosystem and the functioning of an ecosystem?

Ecosystems perform services that most likely depend on a high degree of biodiversity.

27.3 Causes of Extinction

In order to stem the tide of extinction, it is first necessary to identify its causes. Researchers examined the records of 1,880 threatened and endangered wild species in the United States and found that habitat loss was involved in 85% of the cases (Fig. 27.5*a*). Alien species had a hand in nearly 50%, pollution was a factor in 24%, overexploitation in 17%, and disease in 3% of cases. The percentages add up to more than 100% because most of these species are imperiled for more than one reason. Macaws are a good example that a combination of factors lead to a species decline (Fig. 27.5*b*). Their habitat has been reduced by encroaching timber and mining companies. And macaws are hunted for food and collected for the pet trade.

Habitat Loss

Habitat loss has occurred in all ecosystems but concern has now centered upon tropical rain forests and coral reefs because they are particularly rich in species. A sequence of events in Brazil offers a fairly typical example of the manner in which rain forest is converted to noninhabitable land for wildlife. The construction of a major highway into the forest first provided a way to reach the interior of the forest (Fig. 27.5*c*). Small towns and industry sprang up along the highway, and roads branching off from the main highway gave rise to even tertiary ways into the forest. The end result was fragmentation of the once immense forest. The government offered subsidies to any that would take up residence in the forest and the people who came cut and burned trees in patches (Fig. 27.5*d*). Tropical soils contain limited nutrients but when the trees are burned, nutrients are released that support a lush growth for the grazing of cattle for about three years. Once the land is degraded (Fig. 27.5*e*), the farmer and his family move on to another portion of the forest.

Loss of habitat also affects freshwater and marine biodiversity. Coastal degradation is largely due to the large concentration of people that live there. Already 60% of coral reefs have been destroyed or are on the verge of destruction; it's possible that all coral reefs may disappear during the next 40 years. Mangrove forest destruction is also a problem; Indonesia, with the most mangrove acreage, has lost 45% of its mangroves and the percentage is even higher for other tropical countries. Wetland areas, estuaries, and seagrass beds are also being rapidly destroyed.

Figure 27.5 Habitat loss.
a. In a study examining records of imperiled U.S. plants and animals, habitat loss emerged as the greatest threat to wildlife. **b.** Macaws that reside in South American tropical rain forests are endangered for the reasons listed in the graph. **c.** The construction of roads opens up the rain forest and subjects it to fragmentation. **d.** The result is patches of forest and degraded land. **e.** Wildlife cannot live in destroyed portions of the forest.

a.

% Species Affected by Cause: Habitat loss, Alien species, Pollution, Over-exploitation, Disease

b.

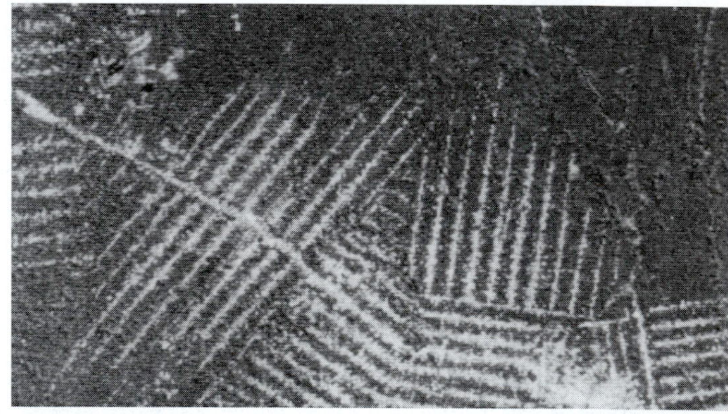

c.

d.

e.

Figure 27.6 Alien species.
Mongooses were introduced into Hawaii to control rats, but they also prey on native birds.

Alien Species

Alien species, sometimes called exotics, are nonnative species that migrate into a new ecosystem or are deliberately or accidently introduced by humans. Ecosystems about the globe are characterized by unique assemblages of organisms that have evolved together in one location. Migrating to a new location is not usually possible because of barriers such as oceans, deserts, mountains, and rivers. Humans, however, have introduced alien species into new ecosystems chiefly due to:

Colonization. Europeans, in particular, brought various familiar species with them when they colonized new places. For example, the pilgrims brought the dandelion to the United States as a familiar salad green.

Horticulture and agriculture. Some exotics now taking over vast tracts of land have escaped from cultivated areas. Kudzu is a vine from Japan that the Department of Agriculture thought would help prevent soil erosion. The plant now covers much landscape in the South including even walnut, magnolia, and sweet gum trees.

Accidental transport. Global trade and travel accidentally brings many new species from one country to another. Researchers found that the ballast water released from ships into Coos Bay, Oregon contained 367 marine species from Japan. Zebra mussel from the Caspian Sea was accidentally introduced into the Great Lakes in 1988. It now forms dense beds that squeeze out native mussels.

Alien species can disrupt food webs. When opossum shrimp were introduced into a lake in Montana, it introduced an additional trophic level that in the end meant less food for bald eagles and grizzly bears (see Fig. 27.2).

Exotics on Islands

Islands are particularly susceptible to environmental discord caused by the introduction of alien species. Islands have unique assemblages of native species that are closely adapted to one another and cannot compete well against exotics. Myrtle trees, *Myrica faya,* introduced into the Hawaiian Islands from the Canary Islands are symbiotic with a type of bacterium that is capable of nitrogen fixation. This feature allows the species to establish itself on nutrient-poor volcanic soil, a distinct advantage in Hawaii. Once established, myrtle trees call a halt to the normal succession of native plants on volcanic soil.

The brown tree snake has been introduced onto a number of islands in the Pacific Ocean. The snake eats eggs, nestlings, and adult birds. On Guam, it has reduced ten native bird species to the point of extinction. On the Galápagos Islands, black rats have reduced populations of giant tortoise; goats and feral pigs have changed the vegetation from highland forest to pampaslike grasslands and destroyed stands of cactus. Mongooses introduced into the Hawaiian Islands to control rats also prey on native birds (Fig. 27.6).

Pollution

In the present context, **pollution** can be defined as any environmental change that adversely affects the lives and health of living things. Pollution has been identified as the third main cause of extinctions. Pollution can also weaken organisms, and lead to disease, the fifth main cause of extinction. Biodiversity is particularly threatened by these environmental concerns:

Acid deposition. Both sulfur dioxide from power plants and nitrogen oxides in automobile exhaust are converted to acids when they combine with water vapor in the atmosphere. These acids return to earth as either wet deposition (acid rain or snow) or dry deposition (sulfate and nitrate salts). Sulfur dioxide and nitrogen oxides are emitted in one locale but deposition occurs across state and national boundaries. Acid deposition causes trees to weaken and increases their susceptibility to disease and insects. It also kills small invertebrates and decomposers so that the entire ecosystem is threatened. Many lakes in northern states are now lifeless because of the effects of acid deposition.

Eutrophication. Lakes are also under stress due to overenrichment. When lakes receive excess nutrients due to runoff from agricultural fields and waste water from sewage treatment, algae begin to grow in abundance. An algal bloom is apparent as a green scum or excessive mats of filamentous algae. Upon death, the decomposers break down the algae, but in so doing they use up oxygen. A decreased amount of oxygen is available to fish, leading sometimes to a massive fish kill.

Ozone depletion. The ozone shield is a layer of ozone (O_3) in the stratosphere, some 50 kilometers above the earth. The ozone shield absorbs most of the wavelengths of harmful ultraviolet (UV) radiation so that they do not strike the earth. The cause of ozone depletion can be traced to chlorine atoms (Cl^-) that come from the breakdown of chlorofluorocarbons (CFCs). The best known CFC is Freon, a heat transfer agent still found in refrigerators and air conditioners today. Severe ozone shield depletion will impair crop and tree growth and also kill plankton (microscopic plant and animal life) that sustain oceanic life. The immune system and the ability of all organisms to resist infectious diseases will most likely be weakened.

Organic chemicals. Our modern society makes use of organic chemicals in all sorts of ways. Nonylphenols are used in products from pesticides to dishwashing detergents, cosmetics, plastics, and spermicides. These particular organic chemicals mimic the effects of hormones, and in that way most likely harm wildlife. Salmon are born in fresh water but mature in salt water. Investigators exposed young fish to nonylphenol and found that 20–30% were unable to make the transition between fresh and salt water. Nonylphenols cause the pituitary to produce prolactin, a hormone that most likely prevents saltwater adaptation.

Global Warming

The expression, **global warming,** refers to an expected increase in average temperature during the twenty-first century. Carbon dioxide is a gas which comes from the burning of fossil fuels, and methane is a gas which comes from oil and gas wells, rice paddies, and animals. These gases are known as greenhouse gases because just like the panes of a greenhouse they allow solar radiation to pass through but hinder the escape of its heat back into space. Data collected around the world show a steady rise in the concentration of the various greenhouse gases. These data are used to generate computer models that predict the earth may warm to temperatures never before experienced by living things. The global climate has already warmed about 0.6°C since the Industrial Revolution and may rise as much as 1.5°–4.5°C by 2060 (Fig. 27.7*a*).

As the oceans warm, temperatures in the polar regions will rise to a greater degree than other regions. Therefore, glaciers will melt and the sea level will rise also because water expands as it warms. A one-meter rise in sea level in the next century could inundate 25–50% of United States coastal wetlands. This loss of habitat could be higher if wetlands cannot move inward because of coastal development and levees.

Global warming could very well cause many extinctions on land also. As temperatures rise, regions of suitable climate for various species will shift toward the poles and higher elevations. The present assemblages of species in ecosystems will be disrupted as some species migrate northward, leaving others behind. Plants migrate when seeds disperse and growth occurs in a new locale. To remain in a favorable habitat, it's been calculated that the rate of beech tree migration would have to be 40 times faster than has ever been observed. It seems unlikely that beech or any other types of tree could be able to meet the pace required. Then, too, many species of organisms are confined to relatively small habitat patches surrounded by agricultural or urban areas which they would not be able to cross. And even if they have the capacity to disperse to new sites, suitable habitat may not be available.

The tropics will also feel the effects of global warming. Coral growth is very dependent upon mutualistic algae which live in their walls. When the temperature rises by four degrees, corals expel their algae and are said to be "bleached" (Fig. 27.7*b*). Almost no growth and reproduction occurs until the algae return. Also, coral reefs prefer shallow waters and if sea levels rise they may "drown." Multiple assaults on coral are even now causing them to be stricken with various diseases.

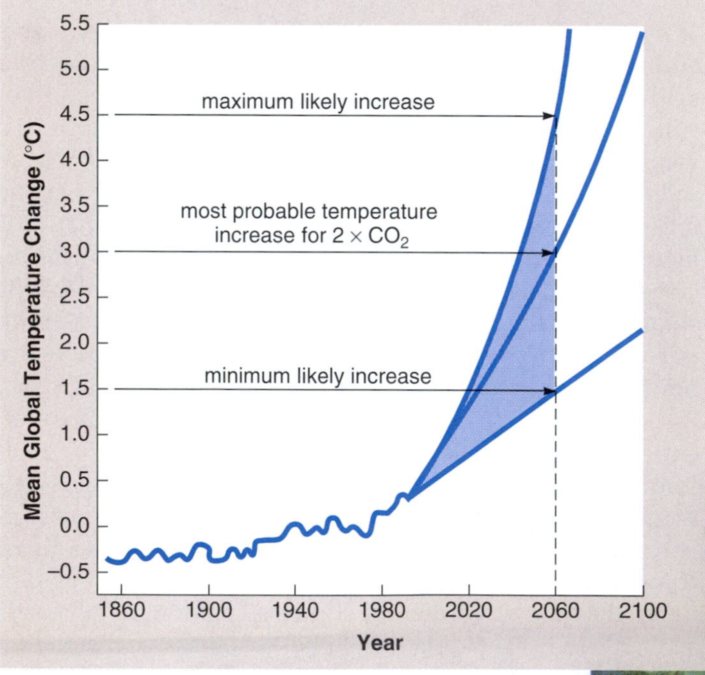

a.

b.

Figure 27.7 Global warming.
a. Mean global temperature change is expected to rise due to the introduction of greenhouse gases into the atmosphere. Global warming has the potential to significantly affect the world's biodiversity. **b.** A temperature rise of only a few degrees causes coral reefs to "bleach" and become lifeless.

Overexploitation

Overexploitation occurs when the number of individuals taken from a wild population is so great that the population goes into a severe reduction in numbers. The positive feedback cycle explains overexploitation: the smaller the population, the more valuable its members, and the greater the incentive to capture the few remaining organisms. Because of its rarity, a single Siberian tiger is now worth more than $500,000—its bones are pulverized and used as a medicinal powder.

A decorative plant and exotic pet market supports both a legal and illegal trade in wild species. Rustlers dig up rare cacti like the single crested saguaro and sell them to gardeners. Parakeets and macaws are among the birds taken from the wild for sale to pet owners. For every bird that is delivered alive, many more have died in the process. The same holds true for tropical fish, which often come from the coral reefs of Indonesia and the Philippines. Divers dynamite reefs or use plastic squeeze bottles of cyanide to stun them; in the process, many fish die.

Declining species of mammals, such as the Siberian tiger mentioned above, are still hunted for their hides, tusks, horns, or bones. The horns of rhinoceroses become ornate carved daggers and their bones are ground up to sell as a medicine. The ivory of an elephant's tusk is used to make art objects, jewelry, or piano keys. The fur of a Bengal tiger sells for as much as $100,000 in Tokyo.

The U.N. Food and Agricultural organization tells us that we have now overexploited 11 of 15 major oceanic fishing areas. Fish are a renewable resource if harvesting does not exceed the ability of the fish to reproduce. But our society uses larger and more efficient fishing fleets to decimate fishing stocks. Pelagic species, like tuna, are captured by purse-seine fishing. A very large net surrounds a school of fish and then the net is closed in the same manner as a drawstring purse. Dolphins that swim above schools of tuna are often captured and then killed in this type of net. Other fishing boats drag huge trawling nets, which can accommodate 12 jumbo jets, along the seafloor to capture bottom-dwelling fish (Fig. 27.8*a*). Only large fish are kept; undesirable small fish and sea turtles are discarded, dying, back into the ocean. Trawling has been called the marine equivalent of clear-cutting trees because after the net goes by, the sea bottom is devastated (Fig. 27.8*b*). Today's fishing practices don't allow fisheries to recover. Cod and haddock, once the most abundant bottom-dwelling fish along the northeast coast, are now often outnumbered by dogfish and skate.

A marine ecosystem can be disrupted by overfishing, as exemplified on the west coast. When sea otters began to decline in numbers, investigators found that they were being eaten by orcas (killer whales). Usually orcas prefer seals and sea lions to sea otters, but they began eating sea otters when few seals and sea lions could be found. What caused a decline in seals and sea lions? Their preferred food source—perch and herring—were no longer plentiful due to overfishing. Ordinarily, sea otters keep the population of sea urchins which feed on kelp under control. But with fewer sea otters around, the sea urchin population exploded and decimated the kelp beds. Thus, overfishing set in motion a chain of events that altered the food web of an ecosystem much to its detriment.

The five main causes of extinction are: habitat loss, introduction of alien species, pollution, over-exploitation, and disease.

a.

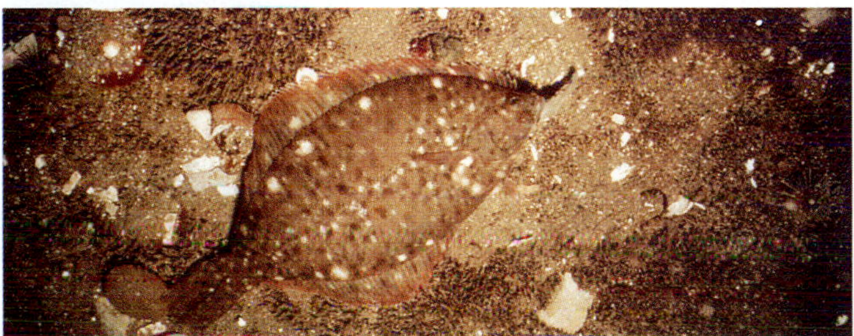

b.

Figure 27.8 Trawling.
a. These Alaskan pollock were caught by dragging net along the seafloor. **b.** *(top)* Appearance of the seabed before and *(bottom)* appearance after the net went by.

27.4 Conservation Techniques

Despite the value of biodiversity to our very survival, human activities are causing the extinction of thousands of species a year. Clearly, we need to reverse this trend and instead preserve as many species as possible. How should we go about it?

Habitat Preservation

Preservation of a species habitat is of primary concern, but first we must decide which species to conserve. As mentioned previously, the biosphere contains biodiversity hotspots, relatively small areas where there is a concentration of endemic (native) species not found any place else. In the tropical rain forests of Madagascar, 93% of the primate species, 99% of the frog species, and over 80% of the plant species are endemic to Madagascar. Preserving of these forests and other hotspots will save a wide variety of organisms.

Keystone species are species that influence the viability of a community more than you would suspect from their numbers. Because of this, the extinction of a keystone species can lead to other extinctions and a loss of biodiversity. Bats are designated a keystone species in tropical forests of the Old World. They are pollinators that also disperse the seeds of trees. When bats are killed off and their roosts destroyed, the trees fail to reproduce. The grizzly bear is a keystone species in northwestern United States and Canada (Fig. 27.9*a*). Bears disperse the seeds of berries; as many as 7,000 seeds can be in one dung pile. Grizzlies kill the young of many hoofed animals and thereby keep their populations under control. Grizzlies are also a principal mover of soil when they dig up roots and prey upon hibernating ground squirrels and marmots.

Metapopulations

The grizzly bear population is actually a **metapopulation,** [Gk. *meta,* between, and L. *populus,* people] that is, a population subdivided into several small and isolated populations due to habitat fragmentation. Originally there were probably 50,000 to 100,000 grizzlies south of Canada but this number has been reduced because communities have encroached on their home range and bears have been killed by frightened homeowners. Now there are six virtually isolated subpopulations totaling about 1,000 individuals. The Yellowstone National Park population numbers 200, but the others are even smaller.

Saving metapopulations sometimes requires determining which of the populations is a source and which are sinks. A **source population** is one that most likely lives in a favorable area and the birthrate is most likely higher than the death rate. Individuals from source populations move into **sink populations** where the environment is not as favorable and where at best birthrate equals death rate. When

a. Grizzly bear, *Ursus arctos horribilis*

b. Old-growth forest

c. Northern spotted owl, *Strix occidentalis caurina*

Figure 27.9 Habitat preservation. Other wildlife benefits when particular species are protected. **a.** The Greater Yellowstone Ecosystem has been delineated in an effort to save grizzly bears which need a very large habitat. **b.** Currently the remaining portions of old-growth forests in the Pacific Northwest are not being logged in order to save the **(c)** northern spotted owl.

trying to save the northern spotted owl, conservationists determined that it was best to avoid having owls move into sink habitats. The northern spotted owl reproduces successfully in old-growth rain forests of the Pacific Northwest (Fig. 27.9*b* and *c*) but not in nearby immature forests that are in the process of recovering from logging. Distinct boundaries that hindered the movement of owls into these sink habitats proved to be beneficial to maintaining source populations.

Landscape Dynamics

Grizzly bears make use of a number of different types of ecosystems including plains, mountains, and rivers. Saving any one of these particular types of ecosystems alone would not be sufficient to preserving grizzly bears. Instead, it is necessary to save diverse ecosystems that are at least connected by corridors. You will recall, a landscape encompasses different types of ecosystems. An area now called the Greater Yellowstone Ecosystem where bears will be free to roam has now been defined. It contains millions of acres in Yellowstone National Park, statelands in Montana, Idaho, and Wyoming, five different national forests, various wildlife refuges, and even private lands.

Landscape protection for one species is often beneficial for other wildlife that share the same space. The last of the contiguous 48 states' harlequin ducks, bull trout, westslope cutthroat trout, lynx, pine martens, wolverines, mountain caribou, and great gray owls are found in areas occupied by grizzlies. The recent return of gray wolves has occurred in this territory also. Then, too, grizzly range overlaps with 40% of Montana's vascular plants of special conservation concern.

The Edge Effect When preserving landscapes, it is necessary to consider the **edge effect.** An edge reduces the amount of habitat typical of an ecosystem because edges around a patch have a slightly different habitat than the interior of a patch. For example, forest edges are brighter, warmer, drier, windier, and have more vines, shrubs, and weeds than the interior of a forest. Also, Figure 27.10 shows that a small and a large patch of habitat have the same amount of edge; therefore the effective habitat shrinks as a patch gets smaller.

The edge effect can have a devastating effect on population size. Songbird populations west of the Mississippi have been declining of late, and ornithologists have noticed that the nesting success of songbirds is quite low at the edge of a forest. The cause turned out to be the brown-headed cowbird, a social parasite of songbirds. Adult cowbirds prefer to feed in open agricultural areas and they only briefly enter the forest when searching for a host nest in which to lay their eggs (Fig. 27.10*b*). Cowbirds are therefore benefited, while songbirds are disadvantaged, by the edge effect.

Computer Analyses

Two types of computer analyses, in particular, are now available to help conservationists plan how best to protect a species.

Gap analysis is the use of the computer to find gaps in preservation—places where biodiversity is high outside of preserved areas. First, a computerized map is drawn up which shows the topography, vegetation, hydrology, and land ownership of a region. Then, a computer map is done which shows the geographic distribution of a region's animal and plant species. These distribution maps are then superimposed onto the land-use maps. Then, it is obvious where preserved habitats still need to be and/or could be located.

A **population viability analysis** can help researchers determine how much habitat a species requires to maintain itself. First, it is necessary to calculate the minimum population size needed to prevent extinction. This size should protect the species from unforeseen events like natural catastrophes or chance swings in the birth and death rate. Another component to consider is the size needed to protect genetic diversity. This number varies according to the species. Analysis of red-cockaded woodpecker populations showed that an adult population of about 1,323 is needed to give a genetically effective population of 500 because of the breeding

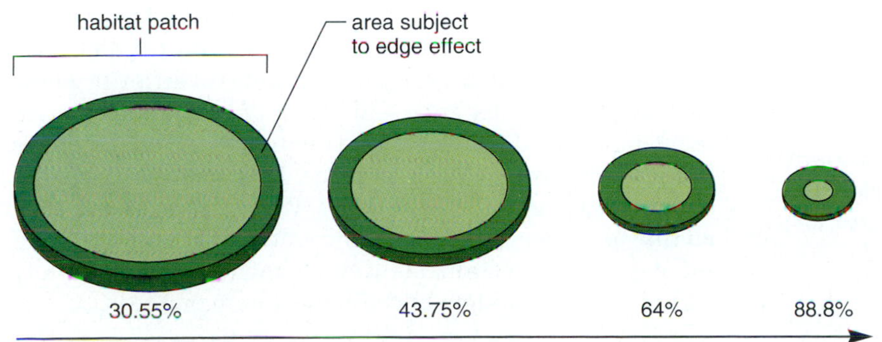

Figure 27.10 Edge effect.
a. The smaller the patch, the greater the proportion that is subject to the edge effect. **b.** Cowbirds lay their eggs in the nest of songbirds. A cowbird is bigger than a warbler nestling and will be able to acquire most of the food brought by the warbler parent.

system of the species. After you know the minimum population size, you can determine how much total acreage is needed for that size population.

All life history characteristics of organisms must be taken into account when a population viability analysis is done. For example, female grizzlies don't give birth until they are five or six and then they typically wait three years before reproducing again. After doing one of the first population viability analyses, Mark Shaffer of the Wilderness Society predicted that a total grizzly bear population of 70 to 90 individuals, each with a suitable home range of 6,000 miles, will have about a 95% chance of surviving for 100 years. But Fred Allendorf pointed out that because only a few dominant males breed, the population needs to be larger than this to protect genetic diversity. Also, to prevent inbreeding he recommended the introduction of one or two unrelated bears each decade into populations of 100 individuals. The bottom line is that there needs to be dispersal among subpopulations to prevent inbreeding and extinction. Sometimes outside assistance is needed. The Florida panther population is so severely inbred that eight female cougars from Texas were introduced into the Everglades in 1995. It was hoped they would mate with panthers and bolster genetic diversity; so far they have produced nine kittens, seven females and two males.

Habitat Restoration

Restoration ecology is a new subdiscipline of conservation biology that seeks scientific ways to return ecosystems to their former state. Three principles have so far emerged. First, it's best to begin as soon as possible before remaining fragments of the original habitat are lost. These fragments are sources of wildlife and seeds from which to restock the restored habitat. Second, once the natural history of the habitat is understood, then it is best to use biological techniques that mimic natural processes to bring about restoration. This might take the form of using controlled burns to bring back grassland habitats, biological pest controls to rid the area of aliens, or bioremediation techniques to clean up pollutants. Third, the goal is **sustainable development,** the ability of an ecosystem to maintain itself while providing services to human beings. We will use the Everglades ecosystem to illustrate these principles.

The Everglades

Originally, the Everglades encompassed the whole of southern Florida from Lake Okeechobee down to Florida Bay (Fig. 27.11). This ecosystem is a vast sawgrass prairie, interrupted occasionally by a cypress dome or hardwood tree island. Within these islands, both temperate and tropical evergreen trees grow amongst dense and tangled vegetation. Mangroves are found along sloughs (creeks) and at the shoreline. The prop roots of red mangroves protect over 40 different types of juvenile fishes as they grow to maturity. During a wet season, from May to November, animals disperse throughout the region, but in the dry season, from December to April, they congregate wherever pools of water are found. Alligators are famous for making "gator holes" where water collects, and fish, shrimps, crabs, birds, and a host of living things survive until the rains come again. The Everglades once supported millions of large and beautiful birds including herons, egrets, the roseate spoonbill, and the anhinga.

At the turn of the century, settlers began to drain the land just south of Lake Okeechobee to grow crops in the newly established Everglades Agricultural Area (EAA). A large dike now rings Lake Okeechobee and prevents water from overflowing its banks and moving slowly southward. To provide flood protection for urban development, water is shunted through the St. Lucie Canal to the Atlantic Ocean or through the canalized Caloosahatchee River to the Gulf of Mexico. In times of drought, water is contained not only in the lake but also in three so-called conservation areas established to the south of the lake. Water must be conserved for the irrigation of the farmland and to recharge the Biscayne aquifer (underground river), which supplies drinking water for the cities on the east coast of Florida. The Central and Southern Florida Flood Control Project (C&SF) included the construction of over 2,250 kilometers of canals, 125 water control stations, and 18 large pumping stations. Now the Everglades National Park receives water only when it is discharged artificially from a conservation area and the discharge is according to the convenience of the C&SF and not according to the natural wet/dry season of southern Florida. Largely because of this, the Everglades are now dying as witnessed by declining bird populations. The birds, which used to number in the millions, now number in the thousands.

Restoration Plan A restoration plan has been developed which will sustain the Everglades ecosystem while maintaining the services society requires. The U.S. Army Corps of Engineers is to redesign the C&SF so that the Everglades receive a more natural flow of water from Lake Okeechobee. This will require the flooding of the EAA and the growth of only crops like sugarcane and rice that can tolerate these conditions. This has the benefit of stopping the loss of topsoil and preventing possible residential development in the area. There will also be an extended east coast buffer zone between an expanded Everglades and the urban areas on the east coast. The buffer zone will contain a contiguous system of interconnected marsh areas, detention reservoirs, seepage barriers, and water treatment areas. This plan is expected to stop the decline of the Everglades, while still allowing agriculture to continue and providing water and flood control to the east coast. Sustainable development will indefinitely maintain the ecosystem and still meet human needs.

Today, landscape preservation is commonly needed to protect metapopulations. Often the preserved area must be restored before sustainable development is possible.

Roseate spoonbill wading

Florida panther kittens in their den

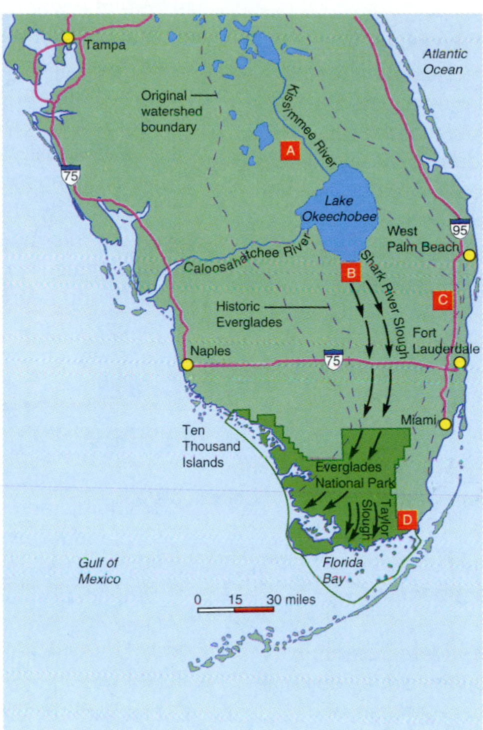

Alligator beside his "gater hole"

White ibis in breeding coloration

Figure 27.11 Restoration of the Everglades.
Restoration plans call for adding curves and habitat back to the Kissimmee River **(A)**; create large marshes and make Shark River Slough free-flowing again **(B)**; create a buffer zone of wetlands between urban development along Florida's eastern coast **(C)**; and reduce salinity by letting fresh water flow into and through Taylor Slough **(D)**.

Connecting Concepts

Our industrial societies are overusing the environment to the point of exhaustion. Forests throughout tropical, temperate, and subarctic regions are being harvested and cut for timber at unsustainable rates. Urban sprawl is completely replacing natural ecosystems in highly populated regions. Fresh waters are being diverted for agricultural and urban uses to the extent that river and lake beds are becoming dry in places. Dams are being constructed for hydropower and irrigation with little consideration of their impact on aquatic life. Alien species of plants, animals, and microbes are being released into new environments with little or no restraint. Marine fisheries are being exploited by major fishing nations at unsustainable levels.

All these actions, and others, are reducing biodiversity, which we now realize is a resource whose economic value is enormous. If properly managed, sustainable yields of food and fiber can be obtained from many natural lands and waters. Modern genetic engineering technologies make the genes of millions of wild species available for use for breeding improved crops, domestic animals, and biological control agents. Enjoyment of nature can also enrich human life enormously.

As natural forests, grasslands, streams, lakes, and seas are degraded, human society must expend greater amounts of nonrenewable energy and materials to substitute for benefits that biodiversity provides at no cost. Lost species, and ultimately, lost ecosystems cannot be replaced. Biodiversity is therefore a nonrenewable resource. The goal of conservation biology is to protect, restore, and use this resource wisely. To that end, the vision of conservation biology is:

> A world where leaders are committed to long-term environmental protection and to international leadership and cooperation in addressing the world's environmental problems.
>
> A world with an environmentally literate citizenry that has the knowledge, skills, and ethical values needed to achieve sustainable development.
>
> A world in which market prices and economic indicators reflect the full environmental and social costs of human activities.
>
> A world in which a new generation of technologies contributes to the conservation of resources and the protection of the environment.
>
> A world landscape that sustains natural systems, maximizes biological diversity, and uplifts the human spirit.
>
> A world in which human numbers are stabilized, all people enjoy a decent standard of living through sustainable development, and the global environment is protected for future generations.

Modified from *The Report of the National Commission on the Environment*, 1993.

Summary

27.1 Conservation Biology and Biodiversity

Conservation biology is the scientific study of biodiversity and its management for sustainable human welfare. The unequaled present rate of extinctions has drawn together scientists and environmentalists in basic and applied fields to address the problem.

Biodiversity is the variety of life on earth; the exact number of species is not known but there are many more insects than other types of organisms. Biodiversity must also be preserved at the genetic, community (ecosystem), and landscape levels of organization.

Conservationists have discovered that biodiversity is not evenly distributed in the biosphere, and therefore saving some particular areas may protect more species than saving other areas.

27.2 Value of Biodiversity

The direct value of biodiversity is the observable services of individual wild species. Wild species are our best source of new medicines to treat human ills. Wild species meet other medical needs: the bacteria that causes leprosy grows naturally in armadillos, and horseshoe crab blood contains a bacteria-fighting substance, for example.

Wild species have agricultural value. Domesticated plants and animals are derived from wild species; and they are a source of genes for the improvement of their phenotypes. Instead of pesticides, wild species can be used as biological controls, and most flowering plants make use of animal pollinators. Much of our food, particularly fish and shellfish, are still caught in the wild. Hardwood trees from natural forests supply us with lumber for various purposes, such as the making of furniture.

The indirect services provided by ecosystems is largely unseen but absolutely necessary to our well-being. These services include the workings of biogeochemical cycles, waste disposal, provision of fresh water, prevention of soil erosion, and regulation of climate. Many people enjoy vacationing in natural settings. Various studies show that more diverse ecosystems function better than less diverse systems.

27.3 Causes of Extinction

Researchers have identified the major causes of extinction. Habitat loss is the most frequent cause followed by introduction of alien species, pollution, overexploitation, and disease. Pollution often leads to disease and so these were discussed at the same time. Habitat loss has occurred in all parts of the biosphere, but concern has now centered upon tropical rain forests and coral reefs where biodiversity is especially high. Alien species have been introduced into foreign ecosystems because of colonization, horticulture and agriculture, and accidental transport. Among the various causes of pollution (acid rain, eutrophication, ozone depletion), global warming is expected to cause the most instances of extinction. Overexploitation is exemplified by commercial fishing which is so efficient that fisheries of the world are collapsing.

27.4 Conservation Techniques

To preserve species it is necessary to preserve their habitat. Some emphasize the need to preserve biodiversity hotspots because of their richness. Often today it is necessary to save metapopulations because of past habitat fragmentation. If so, it is best to determine the source populations and save those instead of sink populations. A keystone species like grizzly bears requires the preservation of a landscape consisting of several types of ecosystems over millions of acres of territory. Obviously, in the process many other species will also be preserved.

Conservation today is assisted by two types of computer analyses. A gap analysis tries for a fit between biodiversity concentrations and land still available to be preserved. A population viability analysis indicates the minimum size of a population needed to prevent extinction from happening.

Since many ecosystems have been degraded, habitat restoration may be necessary before sustainable development is possible. Three principles of restoration are: (1) start before sources of wildlife and seeds are lost; (2) use simple biological techniques that mimic natural processes; and (3) aim for sustainable development so that the ecosystem fulfills the needs of humans.

Reviewing the Chapter

1. Explain these attributes of conservation biology: (1) both academic and applied; (2) supports ethical principles; and (3) is responding to biodiversity crisis. 470
2. Discuss the conservation of biodiversity at the species, genetic, community, and landscape levels. 470–71
3. Describe the uneven distribution of diversity in the biosphere. What is the implication of uneven distribution for conservation biologists? 471
4. List various ways in which individual wild species provide us with valuable services. 472
5. List various ways in which ecosystems provide us with indispensable services. 474
6. List and discuss the five major causes of extinction, starting with the most frequent cause. 476–79
7. Introduction of alien species is usually due to what events? 477
8. List and discuss four major types of pollution that particularly affect biodiversity. Of these, why is global warming considered the most significant? 477–78
9. Use the positive feedback cycle to explain why overexploitation occurs. 479
10. Using the grizzly bear population as an example, explain keystone species, metapopulation, landscape preservation, and a population viability analysis. 480–82
11. Explain the three principles of habitat restoration with reference to restoration of the Everglades. 482

Testing Yourself

Choose the best answer for each question.

1. Which of these would not be within the realm of conservation biology?
   a. Helping to manage a national park.
   b. A government board charged with restoring an ecosystem.
   c. Writing textbooks and/or popular books on the value of biodiversity.
   d. Introducing endangered species back into the wild.
   e. All of these are concerns of conservation biology.
2. With reference to landscape biodiversity, habitat corridors
   a. promote interaction of communities.
   b. stifle genetic diversity.
   c. call for the use of fire to introduce biodiversity.
   d. increase the possibility of extinctions.
   e. Both a and c are correct.

3. Which of these is not a contrast in the number of species?
   a. temperate zone—tropical zone
   b. hotspots—cold spots
   c. rain forest canopy—rain forest floor
   d. pelagic zone—deep sea benthos
4. The value of wild pollinators to the U.S. agricultural economy has been calculated to be worth $4.1 to 6.7 billion a year. What is the implication?
   a. Society could easily replace the need for wild pollinators by domesticating various types of pollinators.
   b. Pollinators may be valuable but that doesn't mean that other species provide us with valuable services.
   c. If we did away with all natural ecosystems, we wouldn't be dependent on wild pollinators.
   d. Society doesn't always appreciate the services that wild species provide us naturally and without any fanfare.
   e. All of these statements are correct.
5. The services provided to us by ecosystems is unseen. This means
   a. they are not valuable.
   b. they are noticed particularly when the service is disrupted.
   c. biodiversity is not needed for ecosystems to keep on functioning as before.
   d. we should be knowledgeable about them and protect them.
   e. Both b and d are correct.
6. Which of these is a true statement?
   a. Habitat loss is the most frequent cause of extinctions today.
   b. Alien species are often introduced into ecosystems by accidental transport.
   c. Global warming is expected to cause many extinctions in the twenty-first century.
   d. Overexploitation of fisheries could very well lead to a complete collapse of the fishing industry.
   e. All of these statements are true.
7. Which of these is not expected because of global warming?
   a. the inability of species to migrate to cooler climates as environmental temperatures rise.
   b. the bleaching and drowning of coral reefs
   c. rise in sea levels and loss of wetlands
   d. preservation of species because cold weather causes hardships
   e. All of these are expected.
8. Why is a grizzly bear a keystone species existing as a metapopulation?
   a. Grizzly bears require many thousands of miles of preserved land because they are large animals.
   b. Grizzly bears have functions that increase biodiversity but presently the population is subdivided into isolated subpopulations.
   c. When grizzly bears are preserved so are many other types of species within a diverse landscape.
   d. Grizzly bears are a source population for many other types of organisms across several population types.
   e. All of these statements are correct.
9. Sustainable development of the Everglades will mean that
   a. the various populations that make up the Everglades will continue to exist indefinitely.
   b. human needs will also be met while successfully managing the ecosystem.
   c. the means used to maintain the Everglades will mimic the processes that naturally maintain the Everglades.
   d. the restoration plan is a workable plan.
   e. All of these statements are correct.
10. Which statement accepted by conservation biologists best shows that they support ethical principles?
    a. Biodiversity is the variety of life observed at various levels of biological organization.
    b. Wild species directly provide us with all sorts of goods and services.
    c. Reduction in the burning of fossil fuels would help reduce the effects of global warming.
    d. There are three principles of restoration biology that need to be adhered to in order to restore ecosystems.
    e. Biodiversity is desirable and has value in and of itself, regardless of any practical benefit.

Thinking Scientifically

1. The scale at which conservation biologists work often makes direct experimentation difficult. But computer models can assist in making predictions about the fate of populations or ecosystems. Some scientists feel that these models are inadequate because they cannot reproduce all the variables found in the real world. If you were trying to predict the impact on songbirds of clear-cutting a portion of a forest, what information would you need to develop a good model?
2. Bioprospecting is the search for medically useful molecules derived from living things. The desire for monetary gain from such discoveries provides an impetus to preserve endangered habitats. Bioprospecting protects ecosystems and in this way saves many species rather than individual species. What types of living things would bioprospectors be most interested in?

Bioethical Issue

In order to protect the state's declining bighorn sheep populations, the New Mexico Game Commission approved a plan that calls for killing scores of mountain lions over several years. The commission pointed out that the lions have killed 36 of 43 radio-collared bighorn sheep released into the wild since 1996 and have increasingly turned to killing sheep as the state's deer herd has declined.

Lisa Jennings, director of Animal Protection of New Mexico said that mountain lions and bighorns have long coexisted and the commission needs to consider a restoration management plan that would improve the ecosystem. She said that this was the best way to increase the bighorn sheep population. Then, too, the state recently engaged the Hornocker Wildlife Institute of Idaho to prepare a study of the situation. The report concluded that the number of lions does not necessarily affect the size of the sheep population. Rather, diseases from domestic livestock, grazing in the area, are responsible for more than 50% of the deaths in some bighorn sheep populations.

In reply, Bill Dunn, a state game department biologist who specializes in sheep, said that the plan to kill more lions was solidly based in science. He said that killing lions in selected areas will give the sheep population a chance to rebound because there are 2,000 lions and only 760 bighorn sheep in two separate populations. Do you approve of taking steps to protect bighorn sheep from lions? Why or why not? If you do approve, how would you proceed?

Understanding the Terms

alien species 477
biodiversity 470
biodiversity hotspot 471
conservation biology 470
edge effect 481
gap analysis 481
global warming 478
habitat corridor 471
keystone species 480
landscape 471
metapopulation 480
pollution 477
population viability analysis 481
restoration ecology 482
sink population 480
source population 480
sustainable development 482

Match the terms to these definitions:

a. ______________ A rather small area with an unusually large concentration of species.
b. ______________ A subdivided population in isolated patches of habitat.
c. ______________ A population that has a positive growth rate and net emigration rate to other locations.
d. ______________ Determination of whether a population of a given size can survive indefinitely in a particular region.
e. ______________ Regional assemblage of different, interacting ecosystem units, such as a forest, open fields, wetlands, and streams.

Web Connections

Exploring the Internet

http://www.mhhe.com/biosci/genbio/mader
(click on *Biology 7/e*)

The *Biology 7/e* Online Learning Center provides many resources for studying the material in this chapter including links to the following sites:

Want to know what legislation is in Congress regarding national parks, the ozone layer, or any other environmental issue? Use Thomas Legislative Information to search by specific bill number or topic.

http://thomas.loc.gov/

EE-Link. This site is a source of information on endangered species in the United States. It provides species lists sorted by region and taxonomic group, and information on natural history, laws, treaties, and periodicals.

http://www.nceet.snre.umich.edu/EndSpp/Endangered.html

and

http://www.fws.gov/r9endspp/endspp.html

Population Reference Bureau. A wealth of informational statistics concerning the human population.

http://www.prb.org/

National Wildlife Federation. This site, maintained by the National Wildlife Federation, provides an environmental hotline and presents information on pending legislation concerning environmental issues. It also provides activities and games for K–12 classrooms.

http://www.nwf.org/nwf/

Further Readings for Part IV

Alper, J. May 22, 1998. Ecosystem "engineers" shape habitats for other species. *Science* 280(5367):1195. Certain microbes create water pools, which give windblown seeds an opportunity to germinate, thus slowly changing the ecosystem.

Bavendam, F. July 1998. Lure of the frogfish. *National Geographic* 194(1):40. Use of camouflaging coloration to obtain prey.

Benchley, P. April 1999. Galápagos: Paradise in peril. *National Geographic* 195(4):2. The unique ecosystem of the Galápagos Islands is being threatened by tourism, development, and abnormal weather.

Chadwick, D. H. January 1999. Coral in peril. *National Geographic* 195(1):30. Coral reefs are at risk from overfishing, pollution, and disease.

Chapin, F., et al. January 1998. Ecosystem consequences of changing biodiversity. *BioScience* 48(1):45. Changes in biodiversity can have significant impacts on ecosystem and landscape processes.

Cohn, J. P. November 1999. Saving the California condor. *BioScience* 49(11):864. Condors are increasing in numbers; years of effort to save the species are paying off.

Cox, G. 1997. *Conservation ecology.* 2d ed. Dubuque, Iowa: Wm. C. Brown Publishers. Discusses the nature of the biosphere, the threats to its integrity, and ecologically sound responses.

Cunningham, W. P., and Saigo, B. W. 1997. *Environmental science: A global concern.* Dubuque, Iowa: Wm. C. Brown Publishers. Provides scientific principles plus insights into the social, political, and economic systems impacting the environment.

Dale, D. April 1999. Agent of ehrlichiosis, common in areas endemic for Lyme disease. *Scientific American Medicine* 22(4):3. Deer ticks also carry *Ixodes scapularis,* the agent of ehrlichiosis.

Doubilet, D. April 1999. Galápagos underwater. *National Geographic* 195(4):32. El Niño causes seas to warm, with dire consequences to penguins, iguanas, and sea lions.

Dugatkin, L. A., and Godin, J. J. April 1998. How females choose their mates. *Scientific American* 278(4):56. Female choice is studied in relation to a number of fish and bird species.

Duxbury, A. C., and Duxbury, A. B. 2000. *An introduction to the world's oceans.* Dubuque, Iowa: Wm. C. Brown Publishers. This introductory oceanographic text examines plate tectonics, climate, the geology of the sea, and the technologies and principles of oceanography.

Flather C., et al. May 1998. Threatened and endangered species geography. *BioScience* 48(5):365. Article suggests that conservation efforts should preserve an entire threatened ecosystem rather than aiding a single endangered species.

Goldfarb, T. 1997. *Taking sides: Clashing views on controversial environmental issues.* 7th ed. Guilford, Conn.: Dushkin/McGraw-Hill. This text examines the pros and cons and gives the history of 18 current environmental controversies.

Gorman, J. January 1998. They saw it coming. *Discover* 18(1):82. Article discusses how El Niño was forecast.

Harwell, M. A. September 1997. Ecosystem management of South Florida. *BioScience* 47(8):499. Article discusses the restoration of a sustainable natural ecosystem while maintaining the services that society requires.

Hornberg, G., et al. October 1998. Boreal swamp forests. *BioScience* 48(10):795. Swamp forests have gained status as biodiversity hotspots in conservation and forest management.

Kaiser, J. September 18, 1998. Impact of primate losses estimated. *Science* 281(5384):1780. An analysis of primate extinctions suggests that due to a ripple effect, extinctions of some species could damage the ecosystem more than expected.

Kerr, R. February 19, 1999. Big El Niños ride the back of slower climate change. *Science* 283(5404):1108. Observing the slower cycles of ocean warming and cooling could be a step toward forecasting an El Niño.

Kerr, R. A. July 16, 1999. Does a globe-girdling disturbance jigger El Niño? *Science* 285(5426):322. Unpredictable bursts of wind along the equator may increase Pacific warming, frustrating efforts to predict El Niño activity.

Knapp, A., et al. January 1999. The keystone role of bison in North American tallgrass prairie. *BioScience* 49(1):39. Bison alter plant communities and ecosystem processes; grazing activities are key to conserving and restoring the prairie.

Krajick, K., and Clark, R. December 1998. Nunataks: Icebound islands of life. *National Geographic* 194(6):60. This unique ecosystem is examined.

Levin, T. February/March 1999. To save a reef. *National Wildlife* 37(2):20. Article discusses the biology and ecology of the Florida Keys, and the threat to Florida coral reefs.

Moffett, M. May 1999. Ants and plants—friends and foes. *National Geographic* 195(5):100. Article explores symbiotic and hostile interactions between ants and plants.

Molles, M. C., Jr. 1999. *Ecology: Concepts and applications.* Boston: WCB/McGraw-Hill. An evolutionary perspective forms the foundation for this introductory ecology text.

National Geographic. February 1999. Millenium supplement: Biodiversity 195(2). The issue is devoted to biodiversity, including possible ways to preserve vanishing species.

National Geographic. October 1998. Millennium supplement: Population 4:2, 6, 36, 56. Article addresses issues such as birthrate, global food production, and migration.

Natural History Magazine. July/August 1998. 107(6):34–51. Preservation of Amazon rain forest diversity is discussed.

Nicol, S., and Allison, I. September/October 1997. The frozen skin of the Southern Ocean. *American Scientist* 85(5):426. Sea-ice and the organisms that occupy it interact with the ocean-atmosphere system in ways that may influence climate.

Normile, D. March 6, 1998. Habitat seen playing larger role in shaping behavior. *Science* 279(5356):1454. Studies suggest that environment influences a wide range of behaviors in nonhuman primates.

Ostfeld, R. S. July/August 1997. The ecology of Lyme-disease risk. *American Scientist* 85(4):338. History of Lyme disease, its symptoms and diagnosis, and deer tick life cycle are given.

Pitelka, L. F., et al. September/October 1997. Plant migration and climate change. *American Scientist* 85(5):464. There may be a relationship between plant migration and climate change, as evidenced by the fossil record and computer models.

Preston-Mafham, K. November 1998. Mating strategies of spiders. *Scientific American* 279(5):94. Article explores the behaviors used by spiders to woo cannibalistic mates.

Primack, R. B. 1998. *Essentials of conservation biology.* Sunderland, Mass.: Sinauer Associates, Inc. This text for ecology majors explores the expanding field of conservation biology.

Relman, D. April 1999. FDA approves vaccine against Lyme disease. *Scientific American Medicine* 22(4):4. LYMErix is the first vaccine for the prevention of Lyme disease.

Rice, R. E., et al. April 1997. Can sustainable management save tropical forests? *Scientific American* 276(4):44. The strategy of replacing harvested trees in rain forests often fails.

Rutowski, R. L. July 1998. Mating strategies in butterflies. *Scientific American* 279(1):64. Visual attributes (colorful wing patterns) and chemical signals (pheromones) play important roles in butterfly mating.

Safina, C. November 1995. The world's imperiled fish. *Scientific American* 273(5):46. Article discusses the decline of fish populations due to the commercial fishing industry.

Sandweiss, D., et al. January 22, 1999. Transitions in the mid-Holocene. *Science* 283(5401):499. Widespread evidence for climatic and cultural change during the mid-Holocene points to a possible connection between climate and culture.

Schmidt, M. J. January 1996. Working elephants. *Scientific American* 274(1):82. In Asia, teams of elephants serve as an alternative to destructive logging equipment.

Scientific American Quarterly. Fall 1998. The oceans. *Scientific American* 9(3). Articles discuss the origins of earth's water, polar ice cap melting, weather, pollution and legal issues, aquaculture, mineral mining, and marine diversity.

Simmons, L. M. August 1998. Indonesia's plague of fire. *National Geographic* 194(2):100. Slash-and-burn agricultural techniques result in air pollution and respiratory disease, as well as deforestation.

Smil, V. 1997. *Cycles of life: Civilization and the biosphere.* New York: Scientific American Library. This easy-to-understand text surveys the links and interactions among environments, populations, and economies.

Smith, R., et al. May 1999. Marine ecosystem sensitivity to climate change. *BioScience* 49(5):393. This century's rapid climate warming is occurring concurrently with a shift in population size and distribution of penguins.

Stone, R. June 5, 1998. Yellowstone rising again from ashes of devastating fires. *Science* 280(5369):1527. The post-fire ecosystem of Yellowstone National Park is essentially the same as before the fires.

Suplee, C. March 1999. El Niño/La Niña. *National Geographic* 195(3):72. Article discusses El Niño and La Niña, showing the evolution of peak storms and highs and lows of the cycle.

Suplee, C. May 1998. Unlocking the climate puzzle. *National Geographic* 193(5):38. The use of fossil fuels may be altering the earth's natural warming and cooling cycles.

Tangley, L. December/January 1999. How many species exist? *National Wildlife* 37(1):32. Plants and animals are vanishing before scientists can identify them.

Wilcove D., et al. August 1998. Quantifying threats to imperiled species in the United States. *BioScience* 48(8):607. Overexploitation, habitat destruction, and the introduction of alien species are threatening biodiversity.

Wilson, E. O. 1992. *The diversity of life.* Cambridge, Mass.: Belknap Press. The value of biodiversity, and the distribution and abundance of organisms within ecosystems is examined.

Wuethrick, B. December 4, 1998. Songbirds stressed in winter grounds. *Science* 282(5395):1791. This study shows that the quality of a migratory songbird's wintering grounds can affect its breeding success.

Zimmer, C. January 1999. The El Niño factor. *Discover* 20(1):98. Previous El Niño experiences help researchers predict coming El Niño effects.

part

V Diversity of Life

Using fossil, anatomical, molecular, and other types of evidence, biologists have determined the evolutionary relationships between major groups of organisms and know in general the history of life's diversity. Phylogenetic trees help us see these relationships, but it is important to realize that there are many more types of living things than depicted by any one tree. Each lineage is but a tiny twig on a sprawling bush that has no main trunk.

Adaptation to a particular way of life accounts for life's diversity. Unicellular protists are suited to living in a pond, mushrooms feed on organic matter in your lawn, and insects utilize flight as a way to escape enemies and seek food. These organisms are just as well adapted to their way of life as are the more complex flowering plants and the vertebrate animals to theirs. The study of biology is the study of the many ways that living organisms have evolved various solutions to the same fundamental problems. All living things acquire materials and energy, respond to stimuli, and reproduce.

Classification of Living Things

c h a p t e r

c h a p t e r c o n c e p t s

28.1 Taxonomy

- Each known species has been given a binomial name consisting of the genus and specific epithet. 490
- Species are distinguished on the basis of structure and reproductive isolation. This chapter stresses reproductive isolation. 492
- Classification usually involves the assignment of species to a genus, family, order, class, phylum, kingdom, and domain (the largest classification category). 492

28.2 Phylogenetic Trees

- Systematics encompasses both taxonomy (the naming of organisms) and classification (placing species in the proper categories). 494
- The fossil record, homology, and molecular data are used to decide the evolutionary relatedness of species. 495

28.3 Systematics Today

- There are three main schools of systematics: cladistic, phenetic, and traditional systematics. 498

28.4 Classification Systems

- The five-kingdom system contains these kingdoms: Monera, Protista, Plantae, Fungi, and Animalia. 501
- The three-domain system recognizes three domains: Bacteria, Archaea, and Eukarya. The domain Eukarya contains the kingdoms Protista, Fungi, Plantae, Animalia. 502

Golden toad, *Bufo periglenses*, Costa Rican cloud forest.

Faced with the enormous number of living things on earth, scientists realized long ago that we need a way to classify and name individual species. Although the ancient Greek philosopher Aristotle devised a primitive classification system over two thousand years ago, it wasn't until the 1700s that a Swedish biologist, Carolus Linnaeus, developed a systematic method of naming species that is still used today. A species' name consists of two Latin words, as in *Homo sapiens* for humans. No two species have the same scientific name.

An organism is generally classified on the basis of its evolutionary relationship to other species. Suppose, for example, a new toad were found in a rain forest. The animal's anatomy, genetics, and reproductive behavior would all be examined and compared to similar known toads. Once the new toad's relationship to other toads was determined, it would be possible to decide on its name and classification. There are various schools of classification and some of these are quite new. The classification of organisms is not a static field of biology; it changes over time as new discoveries and ideas are developed.

Figure 28.1 Classifying organisms.
How would you name and classify these organisms? After naming them would you assign them to a particular group? Based on what principles? An artificial system would not take into account how they might be related through evolution, as would a natural system.

28.1 Taxonomy

Suppose you were asked to classify the living organisms you know about (Fig. 28.1). Most likely you would begin by making a list, and naturally this would require that you give each organism a name. Then you would start assigning the organisms on your list to particular groups. But what criteria would you use—color, size, how the organisms relate to you? Deciding on the number, types, and arrangement of the groups would not be easy, and periodically you might change your mind or even start over. Biologists, too, have not had an easy time deciding how living things should be classified, and changes have been made throughout the history of this field. These changes are often brought about by an increase in fossil, anatomical, or molecular data. Classification is usually based on our understanding about how organisms are related to one another through evolution. A natural system of classification as opposed to a artificial system reflects the evolutionary history of organisms.

Taxonomy [Gk. *tasso,* arrange, classify, and *nomos,* usage, law], the branch of biology concerned with identifying, naming, and classifying organisms, began with the ancient Greeks and Romans. The famous ancient Greek philosopher Aristotle was interested in taxonomy, and he identified organisms as belonging to a particular group such as horses, birds, and oaks. In the middle ages these names were translated into Latin. The scientific names we use today are rendered in Latin. Much later, John Ray, a British naturalist of the seventeenth century, believed that each organism should have a set name. He said, "When men do not know the name and properties of natural objects—they cannot see and record accurately."

The Binomial System

The number of known types of organisms expanded greatly in the mid-eighteenth century due to European travel to distant parts of the world. It was during this time that Carolus Linnaeus developed the **binomial system** of naming species (Fig. 28.2). The name is a binomial because it has two parts. For example, *Lilium buibiferum* and *Lilium canadense* are two different species of lilies. The first word, *Lilium,* is the genus (pl., genera), a classification category

a.

Figure 28.2 Carolus Linnaeus.
a. Linnaeus was the father of taxonomy and gave us the binomial system of classifying organisms. He was particularly interested in classifying plants. Each of these two lilies (**b**) and (**c**) are species in the same genus, *Lilium.*

b. *Lilium buibiferum*

c. *Lilium canadense*

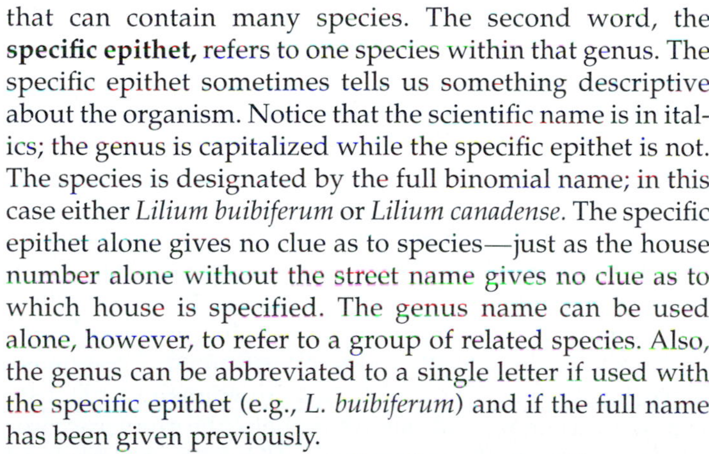

that can contain many species. The second word, the **specific epithet,** refers to one species within that genus. The specific epithet sometimes tells us something descriptive about the organism. Notice that the scientific name is in italics; the genus is capitalized while the specific epithet is not. The species is designated by the full binomial name; in this case either *Lilium buibiferum* or *Lilium canadense.* The specific epithet alone gives no clue as to species—just as the house number alone without the street name gives no clue as to which house is specified. The genus name can be used alone, however, to refer to a group of related species. Also, the genus can be abbreviated to a single letter if used with the specific epithet (e.g., *L. buibiferum*) and if the full name has been given previously.

Why do organisms need to have a scientific name in Latin? Why can't we just use common names for organisms? A common name will vary from country to country just because different countries use different languages. Hence the need for a universal language such as Latin, which not too long ago used to be well known by most scholars. Even those who speak the same language sometimes use different common names for the same organism. The Louisiana heron and the tricolored heron are the same bird found in southern United States. Between countries, the same common name is sometimes given to different organisms. A "robin" in England is very different from a "robin" in the United States, for example. When scientists use the same scientific name, they know they are speaking of the same organism.

The job of identifying and naming the species of the world is a daunting task. Of the estimated 3 to 30 million species now living on earth, we have named a million species of animals and a half million species of plants and microorganisms. We are further along on some groups than others; it's possible we have just about finished the birds, but there may yet be hundreds of thousands of unnamed insects. International associations of taxonomists govern the principles for the naming of organisms and rule on the appropriateness of new names. The same binomial name for each organism is used throughout the world.

The scientific name of an organism consists of its genus and a specific epithet. The complete binomial name indicates the species.

Figure 28.3 Members of a species.
Identifying the members of a species can be difficult—especially when the male and female members do not look alike, as in these mallards, *Anas platyrhynchos.*

Figure 28.4 Hybridization between species.
Zebroids are horse-zebra hybrids. Like mules, zebroids are generally infertile, due to differences in the chromosomes of their parents.

Identification of a Species

There are several ways to distinguish species, and each way has its advantages and disadvantages. For Linnaeus, every species has its own distinctive structural characteristics that are not shared by members of a similar species. In birds, the structural differences can involve the shape, size, and color of the body, feet, bill, or wings. We know very well, however, that variations do occur among members of a species. Differences between males and females or between juveniles and adults may even make it difficult to tell when an organism belongs to a particular species (Fig. 28.3).

The biological definition of a species rests on the recognition that distinctive characteristics are passed on from parents to offspring. This definition, which states that members of a species interbreed and share the same gene pool, applies only to sexually reproducing organisms and cannot apply to asexually reproducing organisms. Sexually reproducing organisms are not always as reproductively isolated as we would expect. When a species has a wide geographic range, there may be variant types that tend to interbreed where their populations overlap (see Fig. 19.3). This observation has led to calling these populations subspecies, designated by a three-part name. For example, *Elaphe obsoleta bairdi* and *Elaphe obsoleta obsoleta* are two subspecies within the same snake species *Elaphe obsoleta*. It could be that these subspecies are actually distinct species. Even species that seem to be obviously distinct interbreed on occasion (Fig. 28.4). Therefore, the presence or absence of hybridization may not be informative as to what constitutes a species.

In the context of this chapter which concerns classification, we defined species as a taxonomic category below the rank of genus. A **taxon** (pl., taxa) is a group of organisms that fill a particular category of classification; *Rosa* and *Felis* are taxa at the genus level. Species in the same genus share a more recent common ancestor than do species from different genera. A **common ancestor** is one that produced at least two lines of descent; there is one ancestor for all types of roses, for example.

Classification Categories

Classification, which begins when an organism is named, includes taxonomy, since genus and species are two classification categories. Individuals we have so far mentioned were taxonomists who contributed to classification. Aristotle divided living things into 14 groups—mammals, birds, fish, and so on. Then he went on to subdivide the groups according to the size of the organisms. Ray used a more natural system, since he grouped animals and plants according to how he thought they were related; but Linnaeus simply used flower part differences to assign plants to these categories: species, genus, order, and class. His studies were published in a book called *Systema Naturae.*

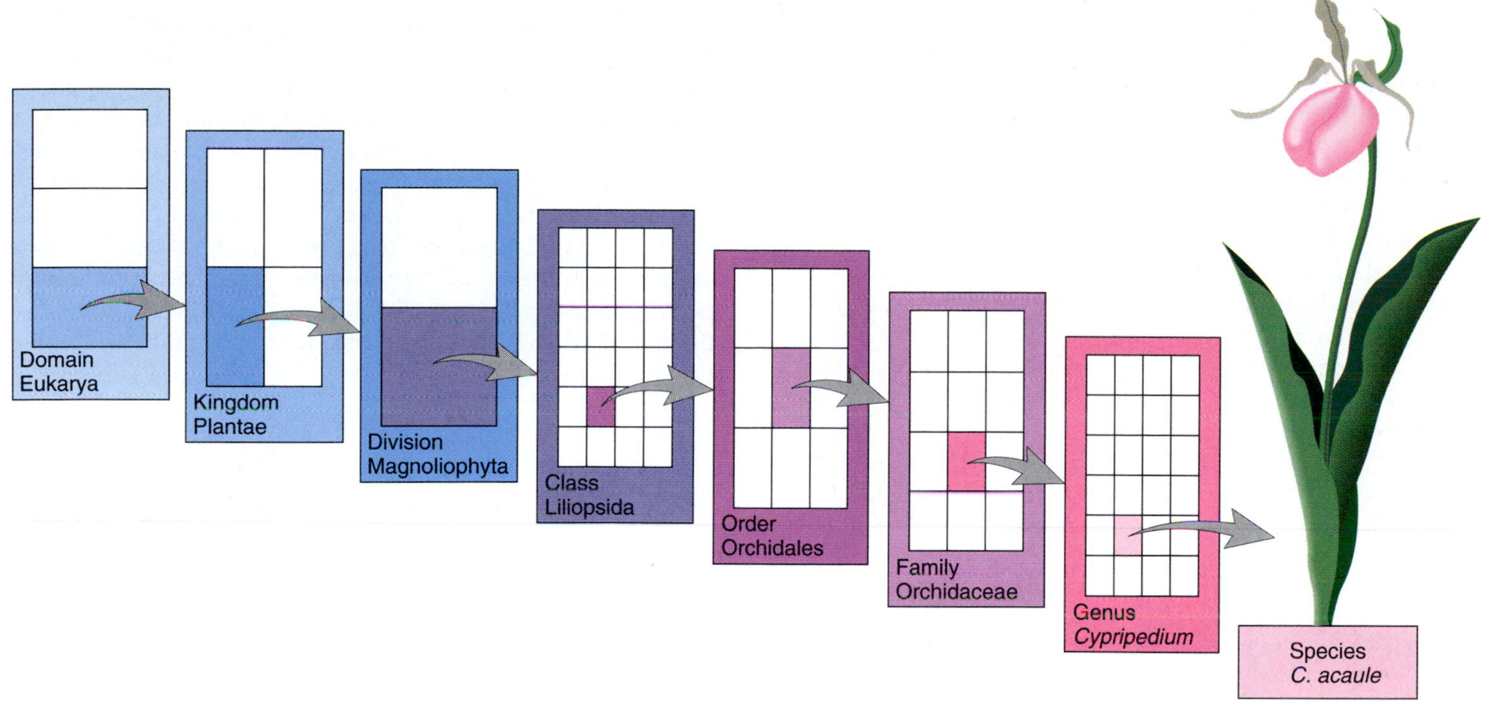

Figure 28.5 **Taxonomy hierarchy.**
A domain is the most inclusive of the classification categories. The plant kingdom is in the domain Eukarya. In the plant kingdom there are several divisions, each represented by a square in the first diagram. In the division Magnoliophyta, there are only two classes (the monocots and dicots). In the class Liliopsida, there are many orders. In the order Orchidales, there are many families, and in the family Orchidaceae, there are many genera, and in the genus *Cypripedium,* there are many species, for example, *Cypripedium acaule.*

Table 28.1

Hierarchy of the Taxa to Which Humans Are Assigned

Domain Eukarya	Organisms whose cells have a membrane-bounded nucleus
Kingdom Animalia	Usually motile, multicellular organisms, without cell walls or chlorophyll; usually, internal cavity for digestion of nutrients
Phylum Chordata	Organisms that at one time in their life history have a dorsal hollow nerve cord, a notochord, and pharyngeal pouches
Class Mammalia	Warm-blooded vertebrates possessing mammary glands; body more or less covered with hair; well-developed brain
Order Primates	Good brain development, opposable thumb and sometimes big toe; lacking claws, scales, horns, and hoofs
Family Hominidae	Limb anatomy suitable for upright stance and bipedal locomotion
Genus *Homo*	Maximum brain development, especially in regard to particular portions; hand anatomy suitable to the making of tools
Species *Homo sapiens**	Body proportions of modern humans; speech centers of brain well developed

* *To specify an organism, you must use the full name, such as* Homo sapiens.

Today, taxonomists make use of these categories of classification: **species, genus, family, order, class, phylum** and **kingdom.** (The plant kingdom uses the category *division* instead of phylum.) Recently, a higher taxonomic category, the **domain,** has been added to this list. There can be several species within a genus, several genera within a family, and so forth—the higher the category the more inclusive it is (Fig. 28.5). Therefore, there is a hierarchy of categories. The organisms that fill a particular classification category are distinguishable from other organisms by sharing a set of characteristics, or simply characters. A **character** is any structural, chromosomal, or molecular feature that distinguishes one group from another. Organisms in the same kingdom have general characters in common; those in the same species have quite specific characters in common. Table 28.1 lists some of the characters that help classify humans into major categories.

In most cases, categories of classification can be subdivided into three additional categories as in superorder, order, suborder, and infraorder. Considering this, there are more than 30 categories of classification.

28.2 Phylogenetic Trees

Taxonomy and classification are a part of the broader field of **systematics** [Gk. *systema*, an orderly arrangement], which is the study of the diversity of organisms at all levels of organization. One goal of systematics is to determine **phylogeny** [Gk. *phyle*, tribe, L. *genitus*, producing], or the evolutionary history of a group of organisms. Classification is a part of systematics because it lists the unique characters of each taxon and ideally is designed to reflect phylogeny. A species is most closely related to other species in the same genus, then to genera in the same family, and so forth, from order to class to phylum to kingdom. When we say that two species (or genera, families, etc.) are closely related, we mean they share a recent common ancestor.

Figure 28.6 shows how the classification of groups of organisms allows one to construct a **phylogenetic tree,** a diagram that indicates common ancestors and lines of descent (lineages). In order to classify organisms and to construct a phylogenetic tree, it is necessary to determine the characters of the various taxa. A **primitive character** is one that is present in the common ancestor and all members of a group. A **derived character** is one that is found only in a particular line of descent. Different lineages diverging from a common ancestor may have different derived characters. For example, all the animals in the family Cervidae have antlers, but they are highly branched in red deer and palmate (having the shape of a hand) in reindeer.

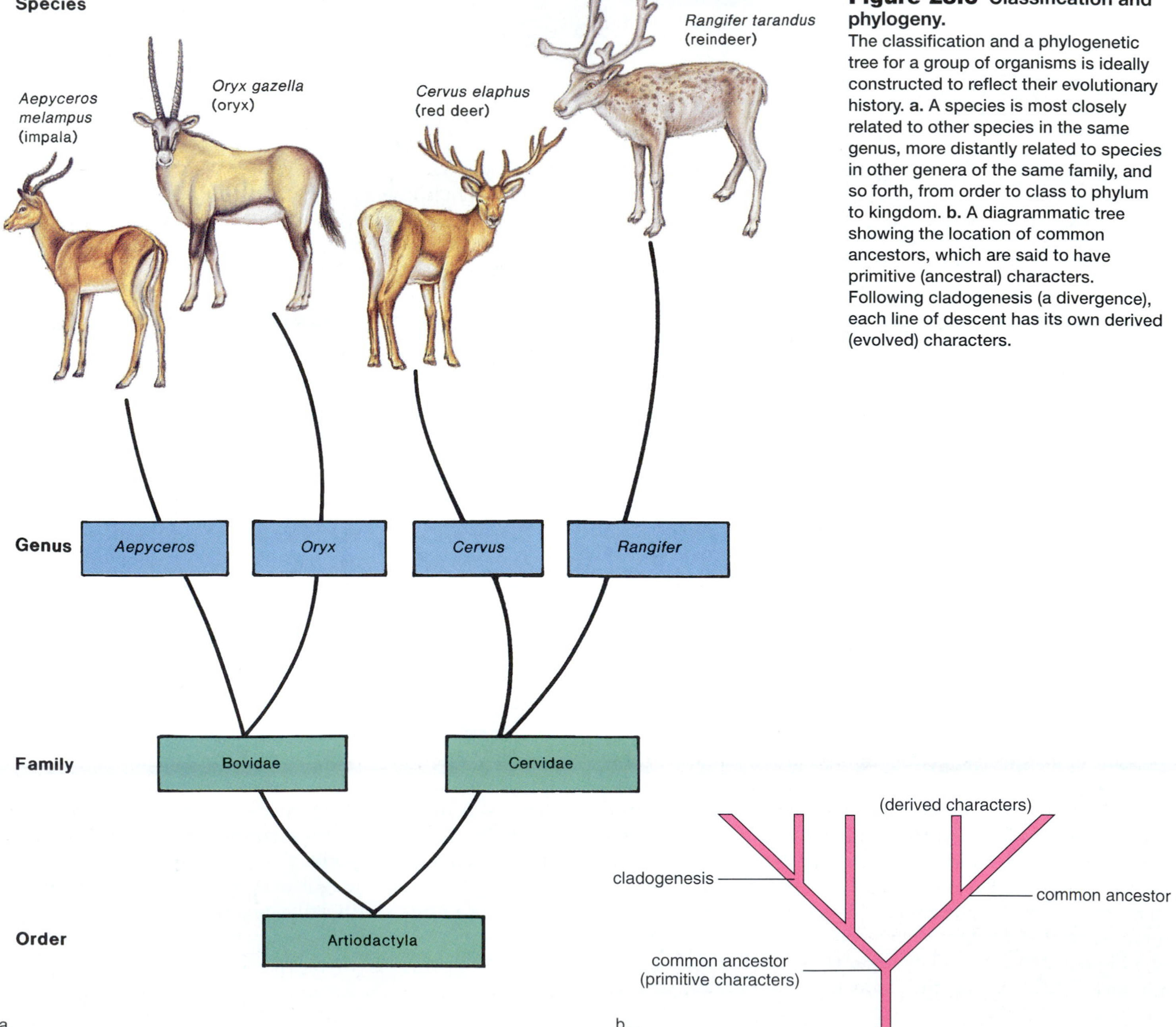

Figure 28.6 Classification and phylogeny.
The classification and a phylogenetic tree for a group of organisms is ideally constructed to reflect their evolutionary history. **a.** A species is most closely related to other species in the same genus, more distantly related to species in other genera of the same family, and so forth, from order to class to phylum to kingdom. **b.** A diagrammatic tree showing the location of common ancestors, which are said to have primitive (ancestral) characters. Following cladogenesis (a divergence), each line of descent has its own derived (evolved) characters.

Tracing Phylogeny

Systematists gather all sorts of data in order to discover the evolutionary relationship between species. They rely heavily on a combination of data from the fossil record, homology, and molecular data in order to determine the correct sequence of common ancestors in any particular group of organisms. If you can determine the common ancestors in the history of life then you know how evolution occurred and can classify organisms accordingly.

Fossil Record

As you know from a study of chapter 20, it is possible to use the fossil record to trace the history of life in broad terms, and sometimes to trace the history of lineages. One of the advantages of fossils is that they can be dated, but unfortunately it is not always possible to tell to which group, living or extinct, a fossil is related. For example, at the present, paleontologists are discussing whether fossil turtles indicate that turtles are distantly or closely related to crocodiles. On the basis of his interpretation of fossil turtles, Olivier C. Rieppel of the Field Museum of Natural History in Chicago is challenging the conventional interpretation that turtles are primitive (have traits seen in a common ancestor to all reptiles) and are not closely related to crocodiles, which evolved later. His interpretation is being supported by genetic analyses that shows turtles and crocodiles are closely related.

If the fossil record was more complete, there might be fewer controversies about the interpretation of fossils. One reason the fossil record is incomplete is that most fossils exist for only harder body parts, such as bones and teeth. Soft parts are usually eaten or decay before they have a chance to be buried. This may be one reason why it has been difficult to discover when angiosperms (flowering plants) first evolved. A Jurassic (see Geological Timescale, p. 326) fossil recently found, if accepted as an angiosperm by most botanists, may help pin down the date (Fig. 28.7).

Homology

Homology [Gk. *homologos*, agreeing, corresponding] is character similarity that stems from having a common ancestry. Comparative anatomy, including embryological evidence, provides information regarding homology. **Homologous structures** are related to each other through common descent, although they may now differ in their structure and function. The forelimbs of vertebrates are homologous because they contain the same bones organized in the same general way as in a common ancestor. For example, in a horse there is but a single digit and toe (the hoof) while in a bat, four lengthened digits provide support for the membranous wings.

Deciphering homology is sometimes difficult because of convergent evolution and parallel evolution. **Convergent evolution** is the acquisition of the same or similar characters in distantly related lines of descent. Similarity due to convergence is termed **analogy.** The wings of an insect and the wings of a bat are analogous. **Analogous structures** have the same function in different groups but do not have a common ancestry. Both spurges and cacti are adapted similarly to a hot, dry environment, and they both are succulent, spiny, flowering plants (Fig. 28.8). However, the details of their flower structure indicate that these plants are not closely related. **Parallel evolution** is the acquisition of the same or similar characters in two or more related lineages without it being present in a common ancestor. A similar banding pattern is found in several species of moths, for example. It is sometimes difficult to tell if features are primitive, derived, convergent, or parallel.

Figure 28.7 Ancestral angiosperm.
A newly found fossil, named *Archaefructus liaoningensis*, dated from the Jurassic period may be the earliest angiosperm to be discovered. Without knowing the anatomy of the first flowering plant, it has been difficult to work out the correct classification of these plants.

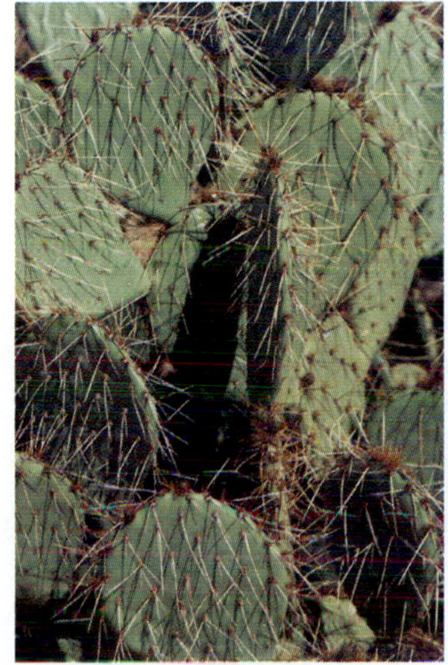

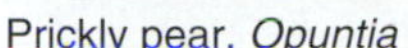

Prickly pear, *Opuntia*

Spurge, *Euphorbia*

Figure 28.8 Convergent evolution.
Cacti and spurges evolved on different continents, and yet they are both succulent flowering plants with spines. Cacti are adapted to living in American deserts, whereas spurges are adapted to living in tropical habitats of Africa. This is an example of convergent evolution.

Molecular Data

Speciation occurs when mutations bring about changes in the base pair sequences of DNA. Systematists, therefore, assume that the more closely species are related, the fewer changes there will be in DNA base pair sequences. Since DNA codes for amino acid sequences in proteins, it also follows that the more closely species are related, the fewer differences there will be in the amino acid sequences within their proteins.

Because molecular data is straightforward and numerical, it can sometimes sort out relationships obscured by inconsequently anatomical variations or convergence. Software breakthroughs have made it possible to analyze nucleotide sequences or amino acid sequences quickly and accurately using a computer. Also, these analyses are available to anyone doing comparative studies through the Internet so each investigator doesn't have to start from scratch. The combination of accuracy and availability of past data has made molecular systematics a standard way today to study the relatedness of groups of organisms.

Protein Comparisons Before amino acid sequencing became routine, immunological techniques were used to roughly judge the similarity of plasma membrane proteins. In one procedure, antibodies are produced by transfusing a rabbit with the cells of one species. Cells of the second species are exposed to these antibodies, and the degree of the reaction is observed. The stronger the reaction, the more similar the cells from the two species.

Later, it became customary to use amino acid sequencing to determine the number of amino acid differences in a particular protein. Cytochrome *c* is a protein that is found in all aerobic organisms and so its sequence has been determined for a number of different organisms. The amino acid difference in cytochrome *c* between chickens and ducks is only 3, but between chickens and humans there are 13 amino acid differences. From this data you can conclude that, as expected, chickens and ducks are more closely related than are chickens and humans. Since the number of proteins available for study in all living things at all times is limited, most new studies today study differences in RNA and DNA.

RNA and DNA Comparisons All cells have ribosomes because they are essential for protein synthesis. Further, the genes that code for ribosomal RNA (rRNA) have changed very slowly during evolution in comparison to other genes. Therefore, it is believed that comparative rRNA sequencing provides a reliable indicator of the similarity between organisms. Ribosomal RNA sequencing helped investigators conclude that all living things can be divided into the three domains that will be discussed later in this chapter.

It is possible to determine DNA similarities by **DNA-DNA hybridization.** The DNA double helix of each species is separated into single strands. Then strands from both species are allowed to combine. The more closely related the two species, the better the two strands of DNA will stick together. Some long-standing questions in systematics have been resolved by doing DNA-DNA hybridization (Fig. 28.9). The giant panda, which lives in China, was at one time considered to be a bear, but its bones and teeth resemble those of a raccoon. The giant panda eats only bamboo and has a false thumb by which it grasps bamboo stalks. The red panda, which lives in the same area and has the same raccoonlike features, also feeds on bamboo but lacks the false thumb. The results of DNA hybridization studies suggest that after raccoons and bears diverged from a common lineage 50 million years ago, the giant panda diverged from the bear line and the red panda diverged from the raccoon line. Therefore, it can be seen that some of the characters of the giant panda and the red panda are primitive (present in a common ancestor), and some are due to parallel evolution.

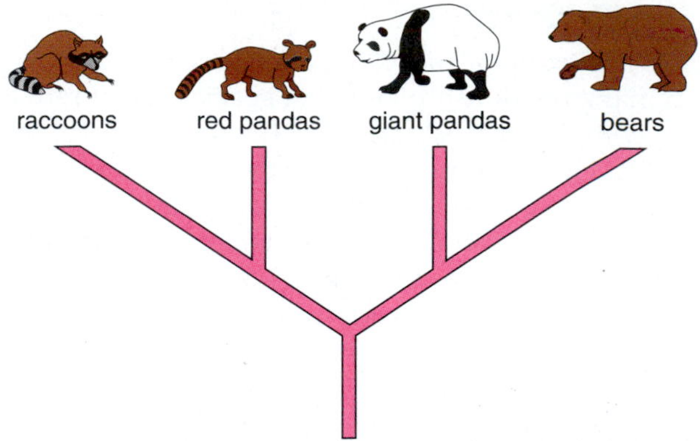

Figure 28.9 Ancestry of giant pandas.
DNA hybridization studies suggest that giant pandas are more closely related to bears than to raccoons.

Because hybridization studies do not provide numerical data, many researchers prefer to compare nucleotide sequences of a particular gene or genes. One study involving DNA differences produced the data shown in Figure 28.10. Although the data suggests that chimpanzees are more closely related to humans than they are to other apes, in most classifications, humans and chimpanzees are placed in different families; humans are in the family Hominidae and chimpanzees are in the family Pongidae. In contrast, the rhesus monkey and the green monkey, which have more numerous DNA differences, are placed in the same family (Cercopithecidae). To be consistent with the data, shouldn't humans and chimpanzees also be in the same family? Traditional systematists, in particular, believe that since humans are markedly different from chimpanzees because of adaptation to a different environment, it is justifiable to place humans in a separate family.

Mitochondrial DNA (mtDNA) changes ten times faster than nuclear DNA. Therefore, when determining the

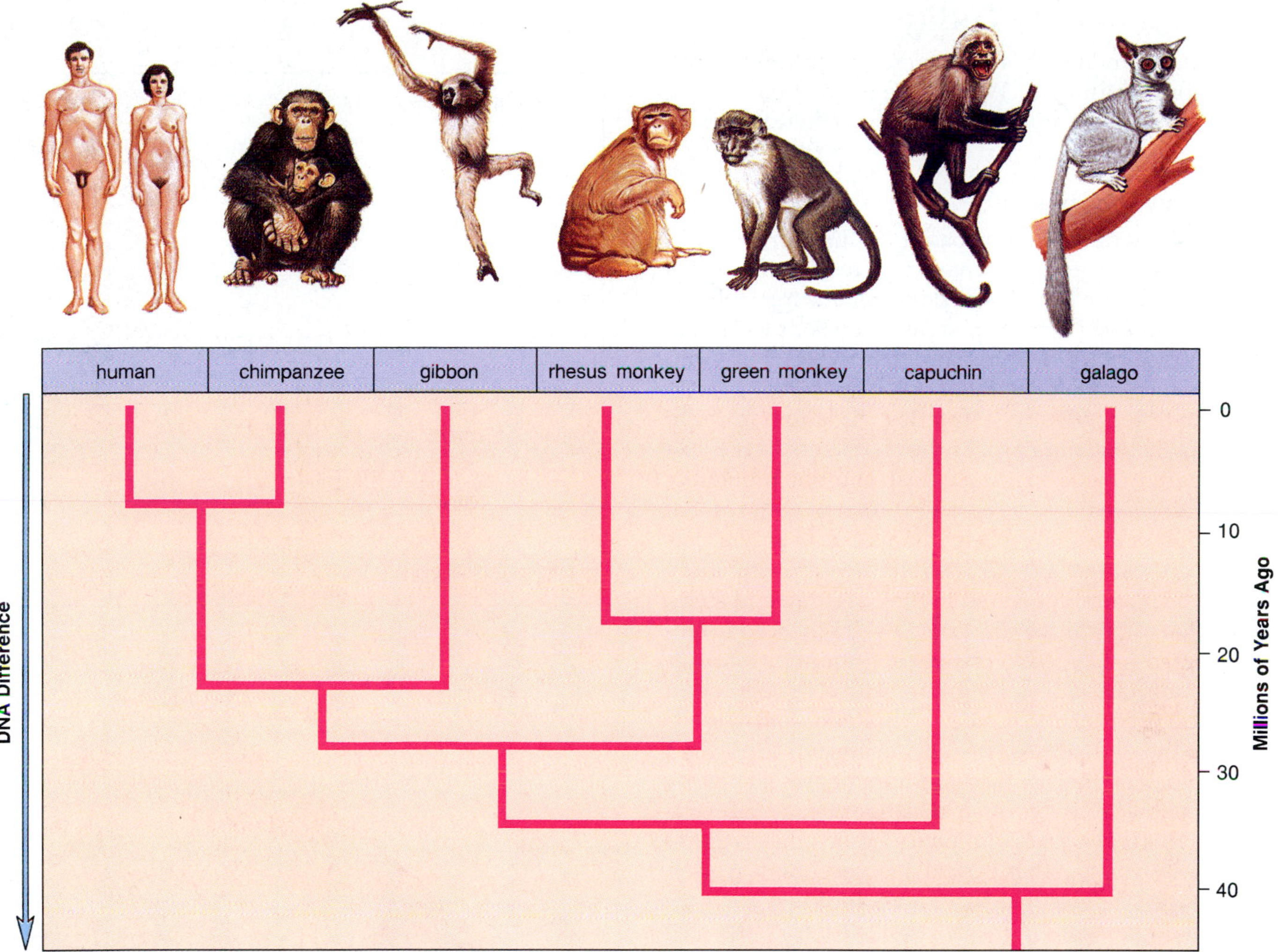

Figure 28.10 **Genetic data.**
The relationship of certain primate species based on a study of their genomes. The length of the branches indicates the relative number of nucleotide pair differences that were found between groups. This data along with knowledge of the fossil record for one divergence makes it possible to suggest a date for the other divergences in the tree.

phylogeny of closely related species, investigators often choose to sequence mtDNA instead of nuclear DNA. One such study concerned North American songbirds. It had long been suggested that these birds diverged into eastern and western subspecies due to retreating glaciers some 250,000 to 100,000 years ago. Sequencing of mtDNA allowed investigators to conclude that groups of North American songbirds diverged from one another an average of 2.5 MYA. Since the old hypothesis based on glaciation is flawed, a new hypothesis is required to explain why eastern and western subspecies arose among these songbirds.

Molecular Clocks When nucleic acid changes are neutral (not tied to adaptation) and accumulate at a fairly constant rate, these changes can be used as a kind of **molecular clock** to indicate relatedness and evolutionary time. The researchers doing comparative mtDNA sequencing used their data as a molecular clock when they equated a 5.1% nucleic acid difference among songbird subspecies to 2.5 MYA. In Figure 28.10 the researchers used their DNA sequence data to suggest how long the different types of primates have been separate. The fossil record was used to calibrate the clock: when the fossil record for one divergence is known, it tells you how long it probably takes for each nucleotide pair difference to occur. Even so, the tree drawn from molecular data is usually used as a hypothesis until it is confirmed by the fossil record. When the fossil record and molecular clock data agree, researchers have more confidence that the proposed phylogenetic tree is correct.

The fossil record, homology, and molecular data help systematists decipher phylogeny and construct phylogenetic trees.

28.3 Systematics Today

There are three main schools of systematics: cladistics, phenetics, and traditional. We will begin by considering cladistics and then compare the other methodologies to this school.

Cladistic Systematics

Cladistic systematics, which is based on the work of Willi Hennig, uses shared derived characters to classify organisms and arrange taxa in a type of phylogenetic tree, called a **cladogram** [Gk. *klados,* branch, stem, *gramma,* picture]. A cladogram traces the evolutionary history of the group being studied. Let's see how it works.

The first step when constructing a cladogram is to draw up a table that summarizes the characters of the taxa being compared (Fig. 28.11*a*). At least one but preferably several species are studied as an outgroup; taxon (taxa) that is (are) not part of the study group. In this example, lancelets are the outgroup and selected vertebrates are the study group. Any character found both in the outgroup and study group is a shared primitive character (e.g., notochord in embryo) presumed to have been present in a common ancestor to both the outgroup and study group. Any character found in one or scattered taxa (e.g., long cylindrical body) is excluded from the cladogram. The other characters are shared derived characters, that is, they are homologies shared by certain taxa of the study group. In a cladogram, a **clade** [Gk. *klados,* branch, stem] is an evolutionary branch that includes a common ancestor, together with all its descendant species. A clade includes the taxa that share homologies.

The cladogram in Figure 28.11*b* has three clades that differ in size because the first includes the other two, and so forth. Notice that the common ancestor at the root of the tree had one primitive character: notochord in embryo. There follows common ancestors which have vertebrae, lungs and three-chambered heart, and finally amniotic egg and internal fertilization. Therefore, this is the sequence in which these characters evolved during the evolutionary history of vertebrates. These are also the homologies that show the clade species are closely related to one another. All the taxa in the study group belong to the first clade because they all have vertebrae; newts, snakes, and lizards are in the clade that has lungs and a three-chambered heart; and only snakes and lizards have an amniote egg and internal fertilization.

A cladogram is objective because it lists the characters that were used to construct the cladogram. Cladists typically make use of many more characters than appear in our simplified cladogram. They also feel that a cladogram is a hypothesis that can be tested and either corroborated or refuted on the basis of additional data. These are the reasons that cladistics, a relatively young discipline, has now become a respected way to decipher evolutionary history. The terms that you need to master in order to understand cladistics are given in Table 28.2.

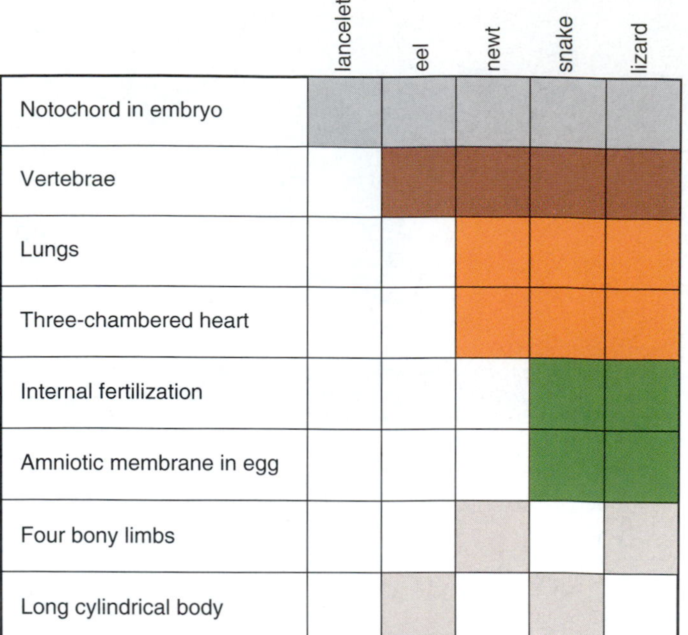

	lancelet	eel	newt	snake	lizard
Notochord in embryo	■	■	■	■	■
Vertebrae		■	■	■	■
Lungs			■	■	■
Three-chambered heart			■	■	■
Internal fertilization				■	■
Amniotic membrane in egg				■	■
Four bony limbs			■		■
Long cylindrical body		■		■	

a.

lancelet
eel
newt
snake
lizard
amniotic egg, internal fertilization
lungs, three-chambered heart
vertebrae

b.

Figure 28.11 Constructing a cladogram.
a. First, a table is drawn up which lists characters for all the taxa. An examination of the table shows which characters are primitive (notochord) and which are derived (brown, orange, and green). The shared derived characters distinguish the taxa. **b.** In a cladogram, the shared derived characters are sequenced in the order they evolved and are used to define clades. A clade contains a common ancestor and all the species that share the same derived characters (homologies). Four bony limbs and long cylindrical body were not used in constructing the cladogram because they are in scattered taxa.

Table 28.2

Terms Used in Cladistics

Term	Definition
Outgroup	Taxon (taxa) that define(s) the primitive characters of the study group
Study group	Taxa that will be placed into clades in a cladogram
Primitive characters	Structures that are present in the outgroup and also in study group
Clade	Evolutionary branch of a cladogram; a monophyletic taxon that contains a common ancestor and all its descendent species
Shared derived characters	Homologies that are present in a particular clade and not in the outgroup
Monophyletic taxon	Contains a single common ancestor and all its descendent species; no descendent species can be in any other taxon
Parsimony	Results in the simplest cladogram possible

Parsimony

Figure 28.12 shows a cladogram in which all three species represented by *X*, *Y*, and *Z* belong to the same monophyletic taxon, since they all trace their ancestry to same common ancestor which had the primitive characters designated by the first arrow. Species *Y* and *Z* are placed in the same clade because they share the derived characters designated by the second arrow. How do you know you have done the cladogram correctly, and that the other two patterns shown in Figure 28.12*b* and *c* are not likely? In the other two arrangements, the characters represented by the colored box would have had to evolve twice.

Cladists are always guided by the principle of *parsimony*—the minimum number of assumptions is the most logical. That is, they construct the cladogram that leaves the fewest number of shared derived characters unexplained or that minimizes the number of assumed evolutionary changes. However, cladists must be on the lookout for the possibility that convergent evolution has produced what appears to be common ancestry. Then, too, there is the realization that the reliability of a cladogram is dependent on the knowledge and skill of the particular investigator gathering the data and doing the character analysis.

Cladistics is based on the premise that shared derived characters (homologies) can be used to define monophyletic taxa and determine the sequence in which evolution occurred.

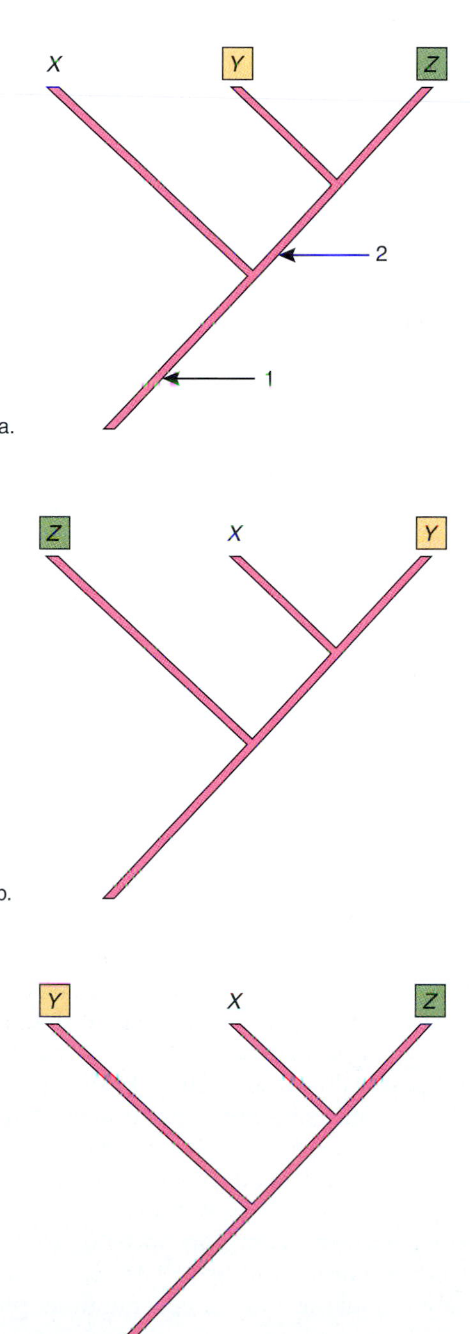

Figure 28.12 Alternate, simplified cladograms.
a. *X, Y, Z* share the same characters, designated by the first arrow, and are judged to form a monophyletic taxon. *Y* and *Z* are grouped together because they share the same derived character, designated by the second arrow and symbolized by the colored box. **b, c.** These cladograms are rejected because in each you would have to assume that the same character (colored boxes) evolved in different groups. Since this seems unlikely, the first branching pattern is chosen as the hypothesis.

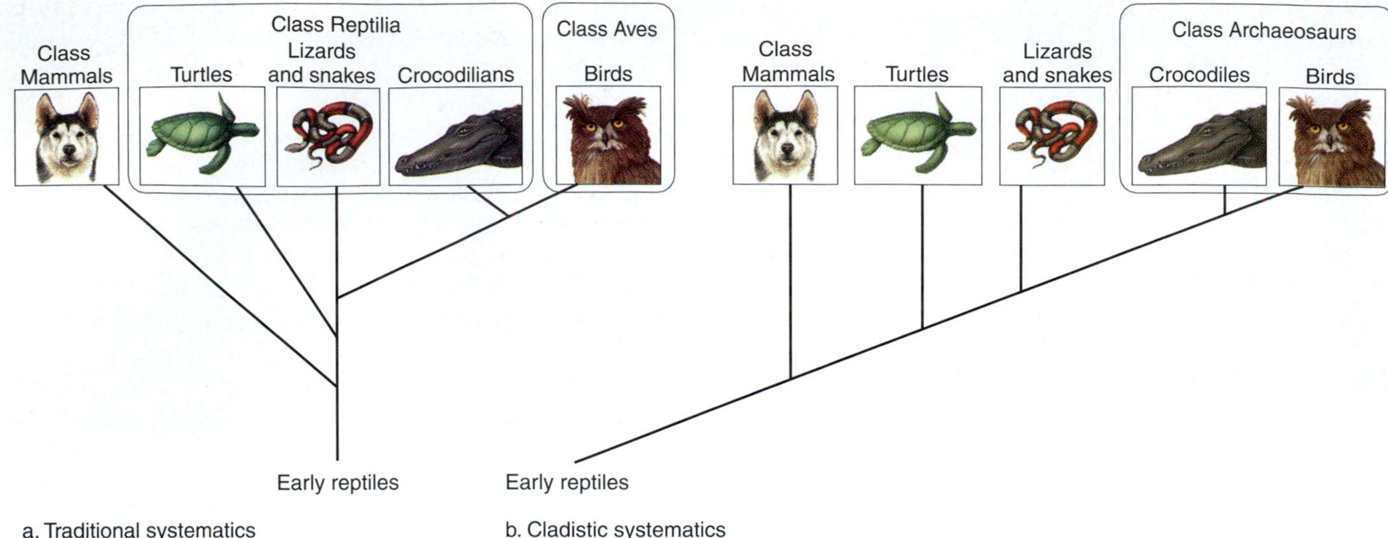

Figure 28.13 **Traditional versus cladistic view of reptilian phylogeny.**
a. According to traditionalists, mammals and birds are in separate classes. **b.** According to the cladists, crocodiles and birds all share a recent common ancestor and should be in the same subclass.

Phenetic Systematics

In **phenetic systematics,** species are classified according to the number of their similarities. Systematists of this school believe that since it is impossible to construct a classification that truly reflects phylogeny, it is better to rely on a method that does away with personal prejudices. They measure as many traits as possible, count the number the two species share, and then estimate the degree of relatedness. They simply ignore the possibility that some of the shared characters are probably the result of convergence or parallelism, or that some of the characters might depend on one another. For example, a large animal is bound to have larger parts. The results of their analysis are depicted in a phenogram. Figure 28.10 is an example of a phenogram that is based solely on the number of DNA differences among the species shown. (Phenograms have been known to vary for the same group of taxa, depending on how the data are collected and handled.)

Traditional Systematics

Soon after Darwin published his book, *The Origin of Species,* **traditional systematics** began and still continues today. These systematists mainly used anatomical data to classify organisms and construct phylogenetic trees based on evolutionary principles. Traditionalists differ from today's cladists largely by stressing both common ancestry *and* the degree of structural difference among divergent groups. Therefore, a group that has adapted to a new environment and shows a high degree of evolutionary change is not always classified with the common ancestor from which it evolved. In other words, traditionalists are not as strict as cladists are about making sure all taxa are monophyletic.

In the traditional phylogenetic tree shown in Figure 28.13*a,* birds and mammals are placed in different classes because it is quite obvious to the most casual observer that mammals (having hair and mammary glands) and birds (having feathers) are quite different in appearance from reptiles (having scaly skin). The traditionalist goes on to say that birds and mammals evolved from reptiles.

Cladists prefer the cladogram shown in Figure 28.13*b.* All the animals shown are in one clade because they all evolved from a common ancestor which laid eggs. Mammals can be placed in a class because they all have hair and mammary glands and three middle ear bones. Cladists doubt there should be a class Reptilia because the only thing that dinosaurs, crocodiles, snakes, lizards, and turtles have in common is that they are not birds or mammals. On the other hand, crocodiles and birds do share common derived characters not present in snakes and lizards. The fossil record indicates that snakes and lizards have a separate common ancestor from crocodiles and birds. Birds just seem different from crocodiles because each is adapted to a different way of life but actually they should be in a class called Archaeosaurs. Most biologists today are willing to admit to inconsistencies but still use the classes Aves and Reptilia for the sake of convenience.

Phenetics and traditionalists are not as strict as cladists about the use of only homologies and monophyletic groups to classify organisms and construct phylogenetic trees.

28.4 Classification Systems

From Aristotle's time to the middle of the twentieth century, biologists recognized only two kingdoms: kingdom Plantae (plants) and kingdom Animalia (animals). Plants were literally organisms that were planted and immobile, while animals were animated and moved about. After the light microscope was perfected in the late 1600s, unicellular organisms were revealed that didn't fit neatly into the plant or animal kingdoms. In the 1880s, a German scientist, Ernst Haeckel, proposed adding a third kingdom. The kingdom Protista (protists) included unicellular microscopic organisms and not multicellular, largely macroscopic, ones.

In 1969, R. H. Whittaker expanded the classification system to five kingdoms: Plantae, Animalia, Fungi, Protista, and Monera. Organisms were placed into these kingdoms based on type of cell (prokaryotic or eukaryotic), levels of organization (unicellular or multicellular), and type of nutrition.

The **five-kingdom system** of classification recognizes the fungi (yeast, mushrooms, and molds) as a separate kingdom. Fungi are eukaryotes that form spores, lack flagella, and have cell walls containing chitin. They also are saprotrophs, organisms that absorb nutrients from decaying organic matter. Whittaker pointed out that plants, animals, and fungi are all multicellular eukaryotes, but each has a distinctive nutritional mode: plants are autotrophic by photosynthesis, animals are heterotrophic by ingestion, and fungi are heterotrophic saprotrophs.

In Whittaker's system, the kingdom Protista contains a diverse group of organisms that are hard to classify and define. They are eukaryotes and mainly unicellular, but may be filaments, colonies, or multicellular sheets also. Protists do not have true tissues. They have ingestive, photosynthetic, or saprotrophic nutrition. There has been considerable debate over the classification of protists. In this text, the protists have been arranged in seven groups for the sake of discussion.

In the five-kingdom system, the Monera are distinguished by their structure—they are prokaryotic (lack a membrane-bounded nucleus)—whereas the organisms in the other kingdoms are eukaryotic (have a membrane-bounded nucleus). The only type of organism in kingdom Monera are called bacteria; therefore, all prokaryotes are called bacteria. As you can see in Figure 28.14, which depicts the five-kingdom system, monerans are at the base of the tree of life. It is suggested that protists evolved from the monerans, and the fungi, plants, and animals evolved from the protists via three separate lines of evolution.

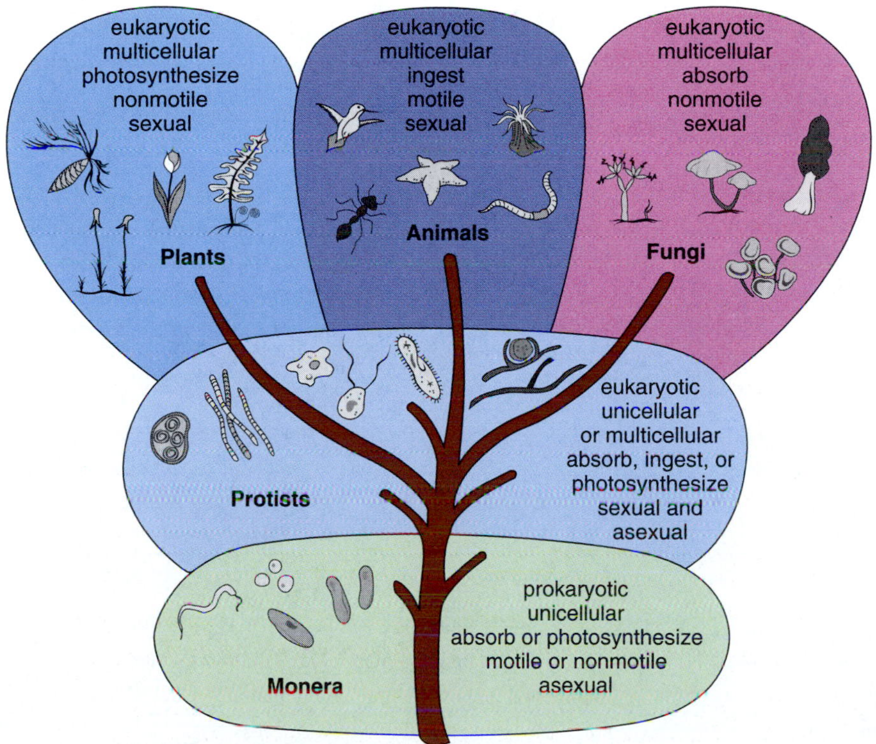

Figure 28.14 The traditional five-kingdom system of classification.
Representatives of each kingdom are depicted in the ovals, and the phylogenetic tree roughly indicates the lines of descent. In this system all prokaryotes are in the kingdom Monera. This text uses the three-domain system described in Figure 28.15.

Three-Domain System

Within the past ten years, new information has called into question the five-kingdom system of classification. Molecular data suggests that there are two groups of prokaryotes, the bacteria and archaea, and these groups are fundamentally different from each other—so different in fact that they should be assigned to separate domains, a category of classification that is higher than the kingdom category.

As mentioned previously, rRNA probably changes only slowly during evolution, and indeed may change only when there is a major evolutionary event. The sequencing of rRNA suggests that all organisms evolved from a common ancestor along three distinct lineages now called domain Bacteria, domain Archaea, and domain Eukarya. This is the **three-domain system** of classification (Table 28.3).

The bacteria diverged first in the tree of life, followed by the archaea and then the eukarya. The archaea and eukarya are more closely related to each other than either is to the bacteria:

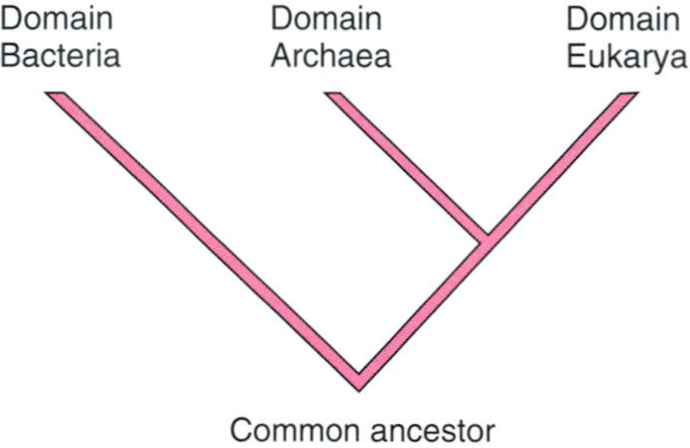

Domain Bacteria and **domain Archaea** contain prokaryotic unicellular organisms that reproduce asexually. The two domains are distinguishable by a difference in their rRNA base sequences and also by differences in their plasma membrane and cell wall chemistry. Systematists are in the process of sorting out what kingdoms belong within the domains Bacteria versus Archaea. The archaea live in extreme environments thought to be similar to those of the primitive earth. For example, the methanogens live in anaerobic environments, such as swamps and marshes; the halophiles are salt lovers living in bodies of water like the Great Salt Lake in Utah, and the thermoacidophiles are both temperature and acid loving. These archaea live in extremely hot acidic environments, such as hot springs and geysers. At least some of the differences between bacteria and archaea can be attributed to the adaptations of archaea to harsh environments. The branched nature of diverse lipids in the archaea plasma membrane is thought to make it resistant to extreme conditions. Such conditions would disrupt the phospholipid bilayer of bacterial and eukaryotic membranes. The chemical nature of the archaeal cell wall is diverse and never the same as that of the bacterial cell. Most prokaryotes are bacteria, a group of organisms that are so diversified and plentiful that they are found in large numbers everywhere on earth. Most of next chapter will be devoted to discussing the bacteria.

Domain Eukarya contains unicellular to multicellular organisms whose cells have a membrane-bounded nucleus. Sexual reproduction is common and all various types of life cycles are seen. We will be studying the individual kingdoms that occur within the domain Eukarya. These kingdoms are differentiated as in the five-kingdom system (Figure 28.15 and Table 28.4). The protists are not a monophyletic taxon, and some suggest that the kingdom Protista should be divided into many different kingdoms. The other kingdoms are thought to be monophyletic at this time.

Recently, it has been suggested that there are three evolutionary domains: Bacteria, Archaea, and Eukarya. The domain Eukarya contains four kingdoms.

Table 28.3

Major Distinctions Between the Three Domains of Life

	Bacteria	Archaea	Eukarya
Unicellularity	Yes	Yes	Some, many multicellular
Membrane lipids	Phospholipids, unbranched	Varied branched lipids	Phospholipids, unbranched
Cell wall	Yes (contains peptidoglycan)	Yes (no peptidoglycan)	Some yes, some no
Nuclear envelope	No	No	Yes
Membrane-bounded organelles	No	No	Yes
Ribosomes	Yes	Yes	Yes
Introns	No	Some	Yes

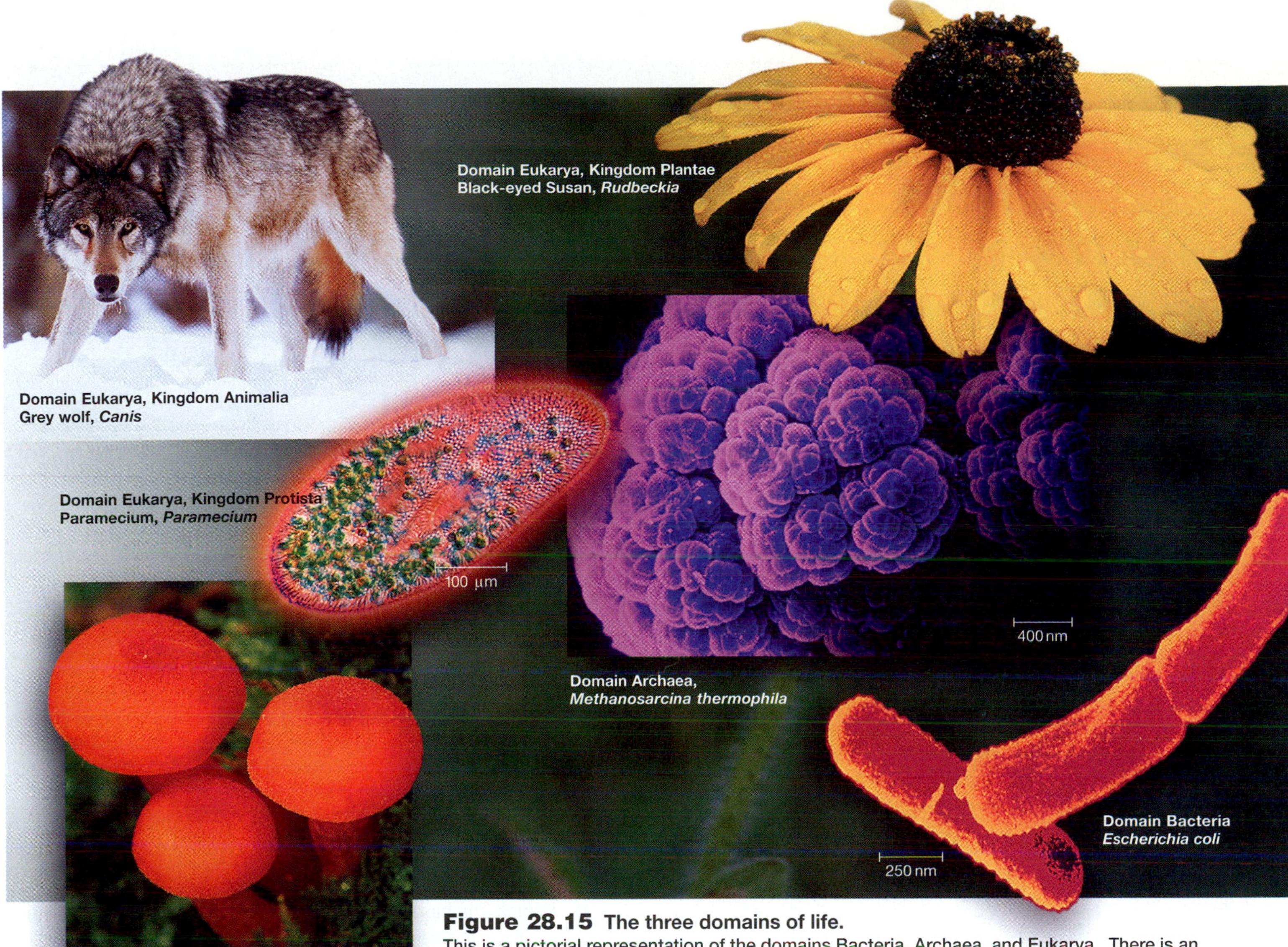

Figure 28.15 The three domains of life.
This is a pictorial representation of the domains Bacteria, Archaea, and Eukarya. There is an example for each of the four kingdoms in the domain Eukarya.

Table 28.4

Classification Criteria for Three Domains

	Domain Bacteria and Archaea	Domain Eukarya			
		Kingdom Protista	*Kingdom Fungi*	*Kingdom Plantae*	*Kingdom Animalia*
Type of cell	Prokaryotic	Eukaryotic	Eukaryotic	Eukaryotic	Eukaryotic
Complexity	Unicellular	Unicellular usual	Multicellular usual	Multicellular	Multicellular
Type of nutrition	Autotrophic or heterotrophic	Photosynthetic or heterotrophic by various	Heterotrophic saprotrophs	Photosynthetic	Heterotrophic by ingestion
Motility	Sometimes by flagella	Sometimes by flagella (or cilia)	Nonmotile	Nonmotile	Motile by contractile fibers
Life cycle*	Asexual usual	Various	Haplontic	Alternation of generations	Diplontic
Internal protection of zygote	No	No	No	Yes	Yes
Nervous system	None	Conduction of stimuli	None	None	Present

* See p. 532

Connecting Concepts

We have seen in this chapter that identifying, naming, and classifying living organisms is an ongoing process. Carolus Linnaeus' system of binomial nomenclature is still accepted by virtually all biologists—many species remain to be found and named (most in the rain forests). For several years now, Whittaker's five-kingdom concept of life on earth has been widely used. Now new findings suggest that there are three domains of life: Bacteria, Archaea, and Eukarya. The archaea are structurally similar to bacteria, but it has been found that their ribosomal subunits differ from those of bacteria and are instead similar to those of eukaryotes. Also, some archaeal genes are unique only to the archaea. Most likely, several kingdoms will eventually be recognized among the bacteria and archaea just as there are among the eukarya (protists, fungi, plants, and animals).

Most of today's systematists use evolutionary relationships among organisms for classification purposes. The traditional and cladistic schools differ as to how to determine such relationships. The traditionalist is willing to consider both structural similarities and obvious differences due to adaptations to new environments. The cladists believe that only similarities should be used to classify organisms. In the end, it may be the ability to quickly sequence genes that will do away with the need for any subjective analyses and make classification a purely objective science.

Summary

28.1 Taxonomy

Taxonomy deals with the naming of organisms; each species is given a binomial name consisting of the genus and specific epithet.

Distinguishing species on the basis of structure can be difficult because members of a species can vary in structure. Distinguishing species on the basis of reproductive isolation runs into problems because some species hybridize and because reproductive isolation is very difficult to observe. In this chapter, species is a taxon occurring below the level of genus. Species in the same genus share a more recent common ancestor than species in related genera, etc.

Classification involves the assignment of species to categories. When an organism is named, a species has been assigned to a particular genus. There are seven obligatory categories of classification: species, genus, family, order, class, phylum, and kingdom. Each higher category is more inclusive; species in the same kingdom share general characters, and species in the same genus share quite specific characters.

28.2 Phylogenetic Trees

Systematics, a very broad field, encompasses both taxonomy and classification. Classification should reflect phylogeny, and one goal of systematics is to create phylogenetic trees, based on primitive and derived characters.

The fossil record, homology, and molecular data are used to help decipher phylogenies. Because fossils can be dated, available fossils can establish the antiquity of a species. If the fossil record is complete enough, we can sometimes trace a lineage through time. Homology helps indicate when species belong to a monophyletic taxon (share a common ancestor); however, convergent evolution and parallel evolution sometimes make it difficult to distinguish homologous structures from analogous structures. Various molecular data are used to indicate relatedness, but DNA base sequence probably pertain more directly to phylogenetic characters.

28.3 Systematics Today

Today there are three main schools of systematics: the cladistic, the phenetic, and the traditional schools. The cladistic school analyzes primitive and derived characters and constructs cladograms on the basis of shared derived characters. A clade includes a common ancestor, and all the species derived from that common ancestor. Cladograms are diagrams based on homologies. The numerical phenetic school clusters species on the basis of the number of shared similarities regardless of whether they might be convergent, parallel, or depend on one another. The traditional school stresses common ancestry *and* the degree of structural difference among divergent groups in order to construct phylogenetic trees.

28.4 Classification Systems

The five-kingdom system of classification recognizes these kingdoms: Plantae, Animalia, Fungi, Protista, and Monera. On the basis of molecular data, it has been established that there are three evolutionary domains: Bacteria, Archaea, and Eukarya. The first two domains contain prokaryotes; the domain Eukarya contains the kingdoms Protista, Fungi, Plantae, and Animalia.

Reviewing the Chapter

1. Explain the binomial system of naming organisms. Why must species be designated by the complete name? 490–491
2. Why is it necessary to give organisms scientific names? 491
3. Discuss three ways to define a species. Which way relates to classification? 492
4. What are the seven obligatory classification categories? In what way are they a hierarchy? 493
5. How is it that taxonomy and classification are a part of systematics? What three types of data help systematists construct phylogenetic trees? 494–497
6. Discuss the principles of cladistics, and explain how to construct a cladogram. 498–499
7. In what ways do the cladistic school, the phenetic school, and the traditional school of systematics differ? 498–500
8. Compare the five-kingdom system of classification to the three-domain system. 501–502
9. Contrast the characteristics of archaea, the bacteria, and the eukarya. 502
10. Contrast the eukaryotic kingdoms: Protista, Fungi, Plantae, and Animalia. 503

Testing Yourself

Choose the best answer for each question.

1. Which is the scientific name of an organism?
   a. *Rosa rugosa*
   b. *Rosa*
   c. *rugosa*
   d. Rugosa rugosa
   e. Both a and d are correct.
2. Which of these best pertains to taxonomy? Species
   a. always have three-part names such as *Homo sapiens sapiens.*
   b. are always reproductively isolated from other species.
   c. always share the most recent common ancestor.
   d. always look exactly alike.
   e. Both c and d are correct.
3. The classification category below the level of family is
   a. class.
   b. species.
   c. phylum.
   d. genus.
   e. order.
4. Which kingdom is mismatched?
   a. Fungi—prokaryotic single cells
   b. Plantae—multicellular only
   c. Plantae—flowers and mosses
   d. Animalia—arthropods and humans
   e. Protista—unicellular eukaryotes
5. Which kingdom is mismatched?
   a. Fungi—usually saprotrophic
   b. Plantae—usually photosynthetic
   c. Animalia—rarely ingestive
   d. Protista—various modes of nutrition
   e. Both c and d are mismatched.
6. In a phylogenetic tree, which is incorrect?
   a. Dates of divergence are always given.
   b. Common ancestors occur at the notches.
   c. The more recently evolved are at the top of the tree.
   d. Ancestors have only primitive characters.
   e. All groups are the same level taxa.
7. Which is mismatched?
   a. homology—character similarity due to a common ancestor
   b. molecular data—DNA strands match
   c. fossil record—bones and teeth
   d. homology—functions always differ
   e. molecular data—molecular clock
8. One benefit of the fossil record is
   a. that hard parts are more likely to fossilize.
   b. fossils can be dated.
   c. its completeness.
   d. fossils congregate in one place.
   e. All of these are correct.
9. In cladistics
   a. a clade must contain the common ancestor plus all its descendants.
   b. derived characters help construct cladograms.
   c. data for the cladogram are presented.
   d. the species in a clade share homologous structures.
   e. All of these are correct.
10. In the traditional school of systematics, birds are assigned to a different group from reptiles because
    a. they evolved from reptiles and couldn't be a monophyletic taxon.
    b. they are adapted to a different way of life compared to reptiles.
    c. feathers came from scales and feet came before wings.
    d. all classes of vertebrates are only related by way of a common ancestor.
    e. All of these are correct.
11. Answer the following questions about this cladogram.
    a. This cladogram contains how many clades? How are they designated in the diagram?
    b. What character is shared by all taxa in the study group? What characters are shared by only snakes and lizards?
    c. Which taxa share a recent common ancestry? How do you know?

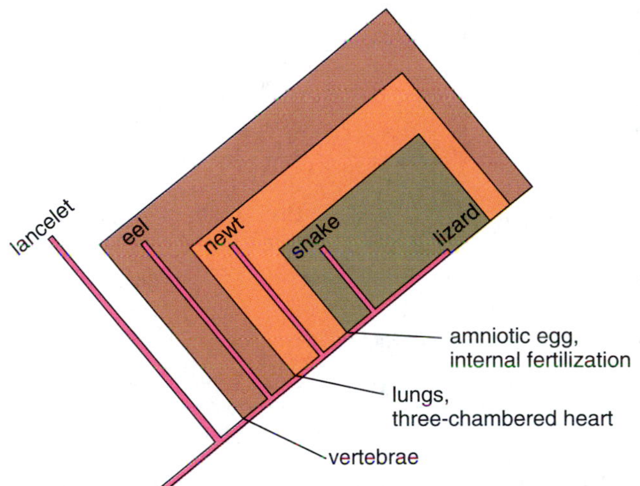

Thinking Scientifically

1. Recent DNA evidence suggests to some plant taxonomists that the traditional way of classifying flowering plants is not correct, and that flowering plants need to be completely reclassified. Other botanists disagree, saying it would be chaotic and unwise to disregard the historical classification groups. Argue for and against keeping traditional classification schemes.
2. Two populations of frogs apparently differ only in skin coloration. What data would you need to determine if both populations belong to the same species? If they are two different species, what data would you need to determine how closely related the two species are?

Understanding the Terms

analogous structure 495
analogy 495
binomial system 490
character 493
clade 498
cladogram 498
class 493
cladistic systematics 498
common ancestor 492
convergent evolution 495
derived character 494
DNA-DNA hybridization 496
domain 493
domain Archaea 502
domain Bacteria 502
domain Eukarya 502
family 493
five-kingdom system 501
genus 493
homologous structure 495
homology 495
kingdom 493
molecular clock 497
order 493
parallel evolution 495
phenetic systematics 500
phylogenetic tree 494
phylogeny 494
phylum 493
primitive character 494
species 493
specific epithet 491
systematics 494
taxon (pl., taxa) 492
taxonomy 490
three-domain system 502
traditional systematics 500

Match the terms to these definitions:

a. ______________ Branch of biology concerned with identifying, describing, and naming organisms.
b. ______________ Diagram that indicates common ancestors and lines of descent.
c. ______________ Group of organisms that fills a particular classification category.
d. ______________ School of systematics that determines the degree of relatedness by analyzing primitive and derived characters and constructing cladograms.
e. ______________ Similarity in structure due to having a common ancestor.

Web Connections

Exploring the Internet

http://www.mhhe.com/biosci/genbio/mader
(click on *Biology 7/e*)

The *Biology 7/e* Online Learning Center provides many resources for studying the material in this chapter including links to the following sites:

Journey into the World of Cladistics. This site discusses the introduction, methodology, implication, and the need for cladistics.

http://www.ucmp.berkeley.edu/clad/clad4.html

References About Phylogenetic Biology. A very long list of references (not links) on phylogenetic biology.

http://phylogeny.arizona.edu/tree/home.pages/references.html

BIOSIS. BIOSIS is a general but extensive Internet resource guide to animal taxa, animal diversity, various taxa, and mailing lists. Search for information by animal taxa or subject categories. This resource has an enormous number of links applicable for many zoological topics.

http://www.york.biosis.org/zrdocs/zoolinfo/grp_all.htm

NCBI (National Center for Biotechnology Information) Taxonomy Resources. This site provides information on systematics and molecular genetics.

http://www3.ncbi.nlm.nih.gov/Taxonomy/taxpage2.html

Taxonomic Resources and Expertise Directory (TRED). Find a new species in your backyard? This is the place to find information regarding how to classify it.

http://www.nbs.gov/cbi/programs/tred.html

Viruses, Bacteria, and Archaea

chapter

chapter concepts

Prokaryotes produce the nutrients for deep-sea vent communities.

Despite their simple structure, prokaryotes are metabolically diverse. They can live under conditions that are too hot, too salty, too acidic, or too cold for eukaryotes. They have been found hundreds of meters beneath ground level and hundreds of meters beneath Antarctic ice. Through their ability to oxidize sulfides that spew forth from deep-sea vents, and subsequently produce nutrients, they support communities of organisms where the sun never shines. Both on land and in the sea they routinely generate oxygen and recycle nutrients of dead plants and animals.

Prokaryotes include the archaea and the bacteria. Most people are familiar with bacteria because some cause deadly diseases. But others produce antibiotics that are used to cure such illnesses. Due to the ease with which they can be grown and manipulated in the laboratory, they are used to study basic life processes like protein synthesis. Humans also use bacteria to mine minerals, clean up oil spills, and produce industrial chemicals and medicines like vitamins and insulin. All in all, it is safe to say we could not live without the services of prokaryotes.

29.1 The Viruses *M*

Viruses [L. *virus,* poison] are nonliving particles with varied appearance, but even so they share certain common characteristics (Fig. 29.1). They are all infectious. In 1884, French chemist Louis Pasteur suggested that something smaller than a bacterium was the cause of rabies, and it was he who chose the word *virus* from a Latin word meaning poison. In 1892, Dimitri Ivanowsky, a Russian biologist, was studying a disease of tobacco leaves, called tobacco mosaic disease because of the leaves' mottled appearance. He noticed that an infective extract could be filtered through a fine-pore porcelain filter that retains bacteria, and it still caused disease. This substantiated Pasteur's belief because it meant that the disease-causing agent was smaller than any known bacterium. In the next century electron microscopy was born and viruses were seen for the first time. By the 1950s, virology was an active field of research; the study of viruses has contributed much to our understanding of disease, genetics, and even the characteristics of living things.

Viral Structure

The size of a virus is comparable to that of a large protein macromolecule; they are generally smaller than 200 nm in diameter. Many viruses can be purified and crystallized, and the crystals can be stored just as chemicals are stored. Still, viral crystals will become infectious when the viral particles they contain are given the opportunity to invade a host cell.

The following diagram summarizes viral structure:

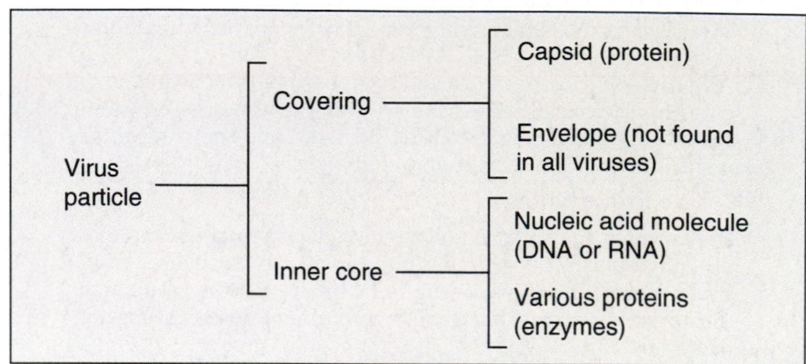

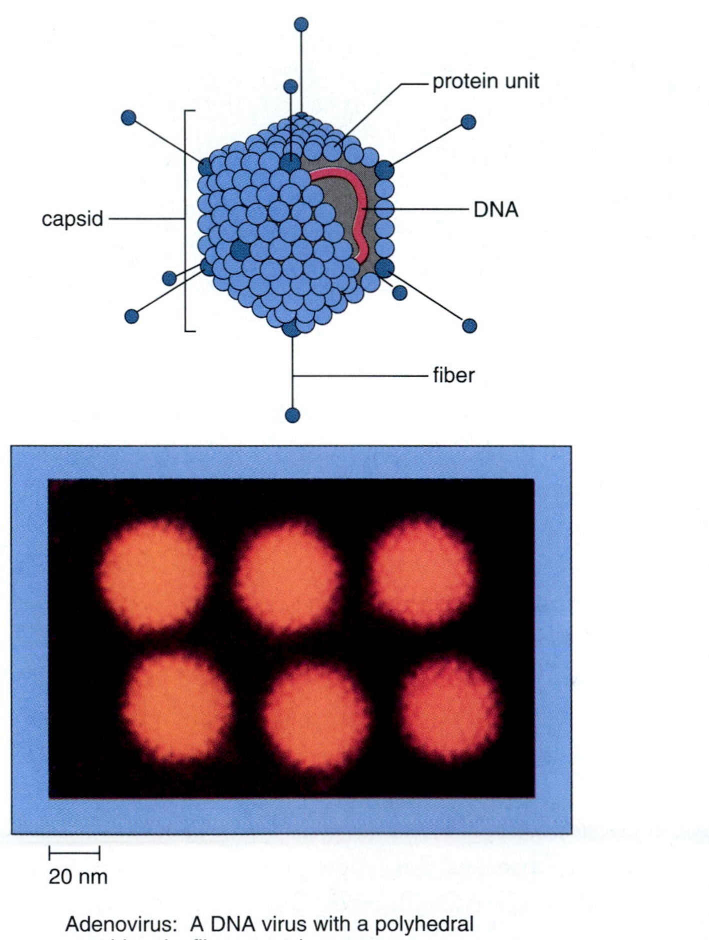

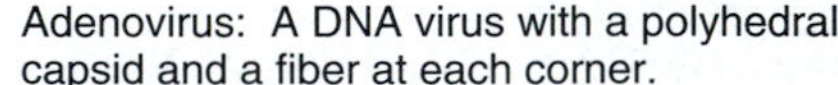

Adenovirus: A DNA virus with a polyhedral capsid and a fiber at each corner.

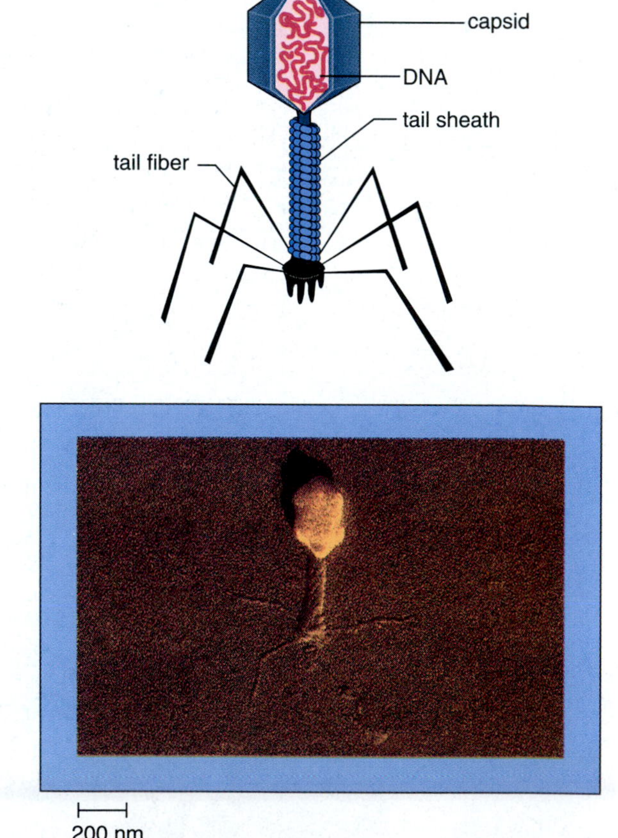

T-even bacteriophage: A DNA virus with a polyhedral head and a helical tail.

Figure 29.1 Viruses.
Despite their diversity, all viruses have an outer capsid composed of protein subunits and a nucleic acid core—either DNA or RNA, but not both. Some types of viruses also have a membranous envelope.

Each type of virus always has at least two parts: an outer *capsid* composed of protein subunits and an inner core of nucleic acid—either DNA (deoxyribonucleic acid) or RNA (ribonucleic acid), but not both. The viral genome at most has several hundred genes; a human cell contains thousands of genes. The capsid may be surrounded by an outer membranous envelope; if not, the virus is said to be naked. The envelope is actually a piece of the host's plasma membrane that also contains viral glycoprotein spikes. A viral particle may also contain various proteins, especially enzymes such as the polymerases, needed to produce viral DNA and/or RNA.

The classification of viruses is based on (1) their type of nucleic acid, including whether it is single stranded or double stranded, (2) their size and shape, and (3) the presence or absence of an outer envelope. Viruses differ from prokaryotes in the ways stated in Table 29.1. From this comparison you can see why viruses are considered nonliving and why they are not in the classification of organisms in Appendix B.

Table 29.1

Comparison of Viruses and Prokaryotes

Characteristic of Life	Viruses	Prokaryotes
Consist of cell	No	Yes
Metabolize	No	Yes
Respond to stimuli	No	Yes
Multiply	Yes (always inside living cell)	Yes (usually independently)
Evolve	Yes	Yes

Viruses are noncellular and have at least two parts: an outer capsid composed of protein subunits and an inner core of nucleic acid, either DNA or RNA but not both.

RNA

capsid

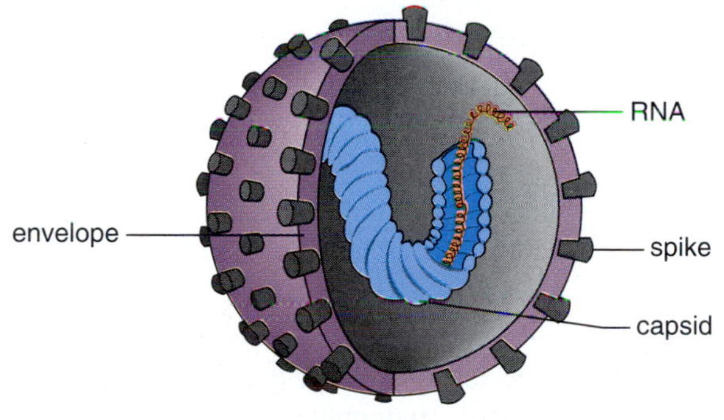

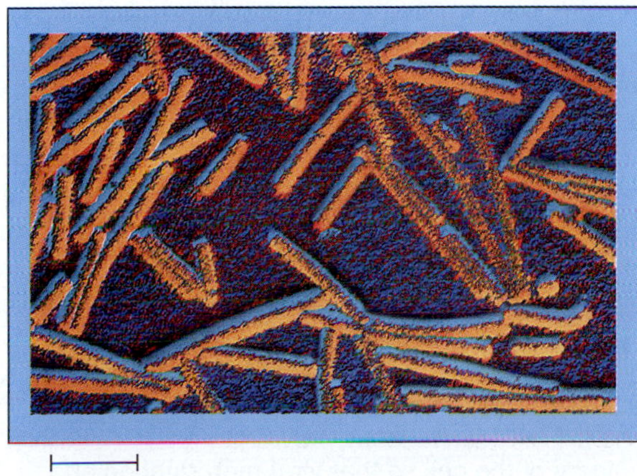

Tobacco mosaic virus: An RNA virus with a helical capsid.

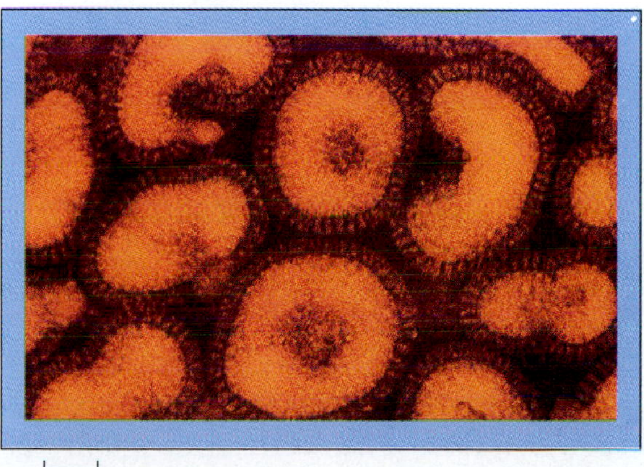

Influenza virus: An RNA virus with a helical capsid surrounded by an envelope with spikes.

Parasitic Nature

Viruses are *obligate intracellular parasites,* which means they cannot reproduce outside a living cell. To maintain animal viruses in the laboratory, they are sometimes injected into live chick embryos (Fig. 29.2). Today, host cells are often maintained in tissue (cell) culture by simply placing a few cells in a glass or plastic container with appropriate medium. The cells can then be infected with the animal virus to be studied. Viruses infect all sorts of cells—from bacterial cells to human cells—but they are very specific. Bacteriophages infect only bacteria, the tobacco mosaic virus infects only plants, and the rabies virus infects only mammals, for example. We know that some human viruses even specialize in a particular tissue. Human immunodeficiency virus (HIV) will enter only certain blood cells, the polio virus reproduces in spinal nerve cells, the hepatitis viruses infect only liver cells. What could cause this remarkable parasite-host cell correlation? It is now believed that viruses are derived from the very cell they infect; the nucleic acid of viruses came from their host cell genomes! Therefore, viruses must have evolved after cells came into existence, and new viruses are probably evolving even now.

Viruses can also mutate; therefore, it is correct to say that they evolve. Those that mutate often can be quite troublesome because a vaccine that is effective today may not be effective tomorrow. Flu viruses are well known for mutating, and this is why you have to have a flu shot every year—antibodies generated from last year's shot are not expected to be effective this year.

Viruses evolve and reproduce but they are obligate intracellular parasites. They only grow inside their specific host cells.

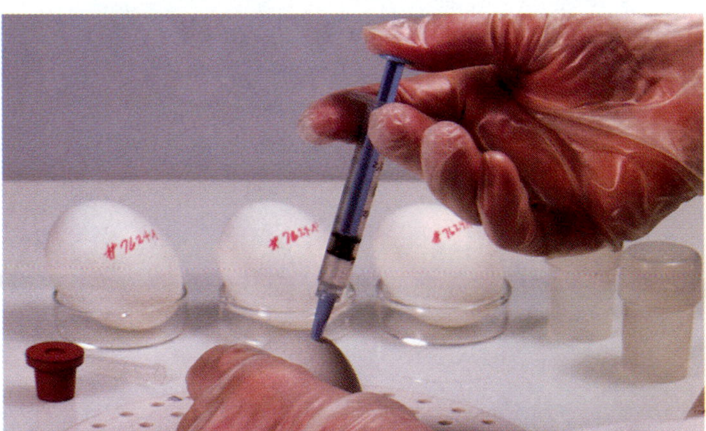

Figure 29.2 **Growing viruses.**
To grow a virus, you can inoculate live chick eggs with viral particles. A virus reproduces only inside a living cell, not because it uses the cell for nutrients, but rather because it takes over the machinery of the cell.

Viral Reproduction

Viruses gain entry into and are specific to a particular host cell because portions of the capsid (or the spikes of the envelope) adhere in a lock-and-key manner with a receptor on the host cell outer surface. The viral nucleic acid then enters the cell. Once inside, the nucleic acid codes for the protein units in the capsid. In addition, the virus may have genes for a few special enzymes needed for the virus to reproduce and exit from the host cell. In large measure, however, a virus relies on the host's enzymes, ribosomes, transfer RNA (tRNA), and ATP (adenosine triphosphate) for its own reproduction. In other words, a virus *takes over the metabolic machinery of the host cell* when it reproduces.

Reproduction of Bacteriophages

Bacteriophages [Gk. *bacterion,* rod, and *phagein,* to eat], or simply, phages, are viruses that parasitize bacteria; the bacterium in Figure 29.3 could be *Escherichia coli,* which lives in our intestines, for example. Two types of bacteriophage life cycles, termed the lytic cycle and the lysogenic cycle, have been carefully studied. In the lytic cycle, viral reproduction occurs and the host cell undergoes *lysis,* a breaking open of the cell to release viral particles. In the lysogenic cycle, viral reproduction does not immediately occur, but reproduction may take place sometime in the future. The life cycle of the bacteriophage, lambda, which can carry out either cycle is discussed below.

Lytic Cycle The **lytic cycle** [Gk. *lyo,* loose] may be divided into five stages: attachment, penetration, biosynthesis, maturation, and release. During *attachment,* portions of the capsid combine with a receptor on the rigid bacterial cell wall in a lock-and-key manner. During *penetration,* a viral enzyme digests away part of the cell wall, and viral DNA is injected into the bacterial cell. *Biosynthesis* of viral components begins after the virus brings about inactivation of host genes not necessary to viral replication. The virus takes over the machinery of the cell in order to carry out viral DNA replication and production of multiple copies of the capsid protein subunits. During *maturation,* viral DNA and capsids assemble to produce several hundred viral particles. Lysozyme, an enzyme coded for by a viral gene, is produced; this disrupts the cell wall, and the *release* of phage particles occurs. The bacterial cell dies as a result.

During the lytic cycle, a bacteriophage takes over the machinery of the cell so that viral reproduction and release occur.

Lysogenic Cycle With the **lysogenic cycle** [Gk. *lyo*, loose, break up, and *genitus*, producing], the infected bacteria does not immediately produce phage but may do so sometime in the future. In the meantime the phage is *latent*—not actively replicating. Following attachment and penetration, viral DNA becomes integrated into bacterial DNA with no destruction of host DNA. While latent, the viral DNA is called a *prophage*. The prophage is replicated along with the host DNA, and all subsequent cells, called lysogenic cells, carry a copy of the prophage. Certain environmental factors, such as ultraviolet radiation, can induce the prophage to enter the lytic stage of biosynthesis, followed by maturation and release.

During the lysogenic cycle, the phage becomes a prophage that is integrated into the host genome. At a later time, the phage may reenter the lytic cycle and reproduce itself.

Reproduction of Animal Viruses

Animal viruses reproduce in a manner similar to bacteriophages, but there are modifications. The viral genome covered by the capsid penetrate a host cell. Once inside, the virus is uncoated—that is, the capsid is removed. The viral genome, either DNA or RNA, is now free of its covering and biosynthesis proceeds. Also viral release occurs by budding. During budding, the virus picks up its envelope consisting of lipids, proteins, and carbohydrates either from the plasma membrane or the nuclear envelope of the host cell. Other envelope markers, such as the glycoproteins that allow the virus to enter a host cell, are coded for by viral genes.

After animal viruses enter the host cell, uncoating releases viral DNA or RNA, and reproduction occurs. If release is by budding, the viral particle acquires a membranous envelope.

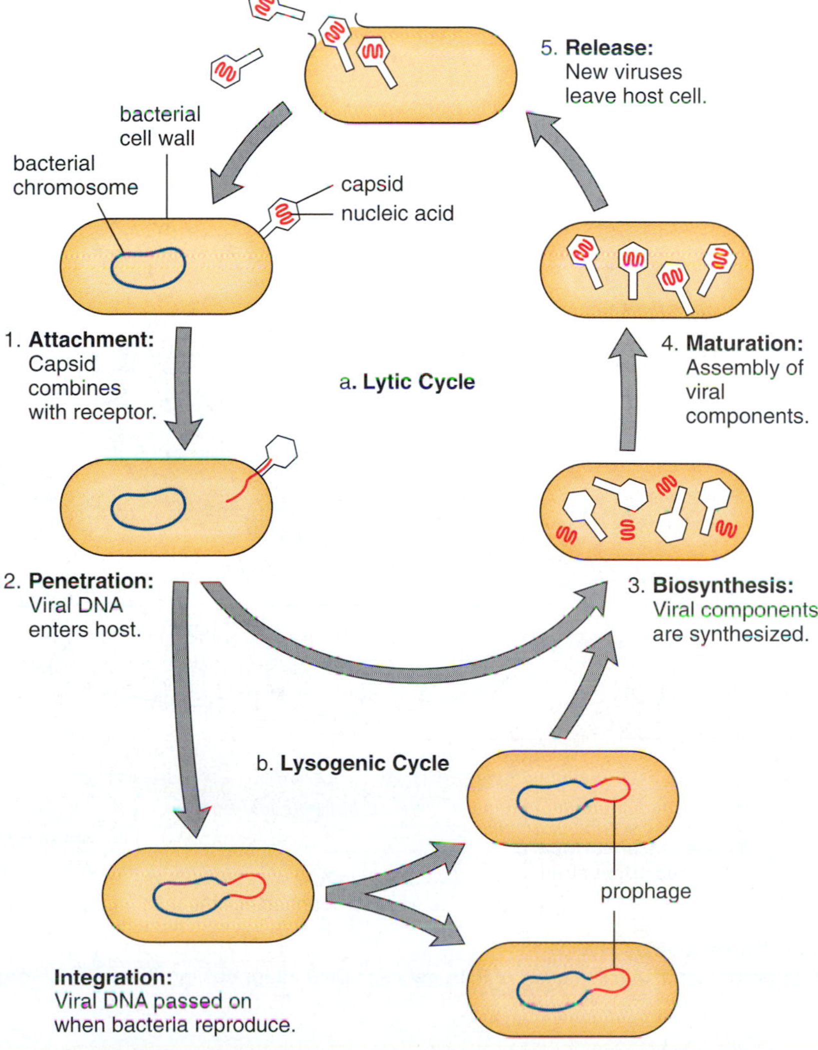

Figure 29.3 **Lytic and lysogenic cycles in prokaryotes.**
a. In the lytic cycle, viral particles escape when the cell is lysed (broken open). **b.** In the lysogenic cycle, viral DNA is integrated into host DNA. At some time in the future, the lysogenic cycle can be followed by the lytic cycle.

Retroviruses **Retroviruses** are RNA animal viruses that have a DNA stage. Figure 29.4 illustrates the reproduction of a retrovirus. A retrovirus [L. retro, backward, and virus, poison] contains a special enzyme called reverse transcriptase, which carries out RNA → cDNA transcription. The DNA is called cDNA because it is a DNA copy of the viral genome. Following replication of the single strand, the resulting double-stranded DNA is integrated into the host genome. The viral DNA remains in the host genome and is replicated when host DNA is replicated. When and if this DNA is transcribed, new viruses are produced by the steps we have already cited: biosynthesis, maturation, and release—not by destruction of the cell, but by budding.

Retroviruses are of interest because human immunodeficiency viruses (HIV), which cause AIDS, are retroviruses. Retroviruses also cause certain forms of cancer.

Viral Infections

Viruses are best known for causing infectious diseases in plants and animals, including humans. Of special concern are those viruses, such as the papillomaviruses, the herpes viruses, the hepatitis viruses, and the adenoviruses, which can cause infections that lead to specific types of cancer. These viruses undergo a period of latency, and their presence so alters the genome that the cells become cancerous. Some viruses are cancer-producing because they bring with them *oncogenes,* normal genes that have been transformed and that now can cause the cell to undergo repeatedly undergo the cell cycle.

The treatment of human viral diseases is discussed in the reading on page 521. Our discussion should allow you to see why antibiotics which are designed to interfere with bacterial metabolism have no effect on viral illnesses. Viruses lack most enzymes and instead utilize the enzymes of the host cell. Any drug that interferes with host cellular enzymes would kill off host cells.

At least a thousand different viruses cause diseases in plants. Since it is very difficult to distinguish a viral infection from a mineral deficiency and to treat a viral infection in plants, the policy is to propagate plants known to be healthy and to protect them from a viral infection. About a dozen crop diseases have been attributed not to viruses but to **viroids,** which are naked strands of RNA not covered by a capsid. Like viruses, though, viroids direct the cell to produce more viroids.

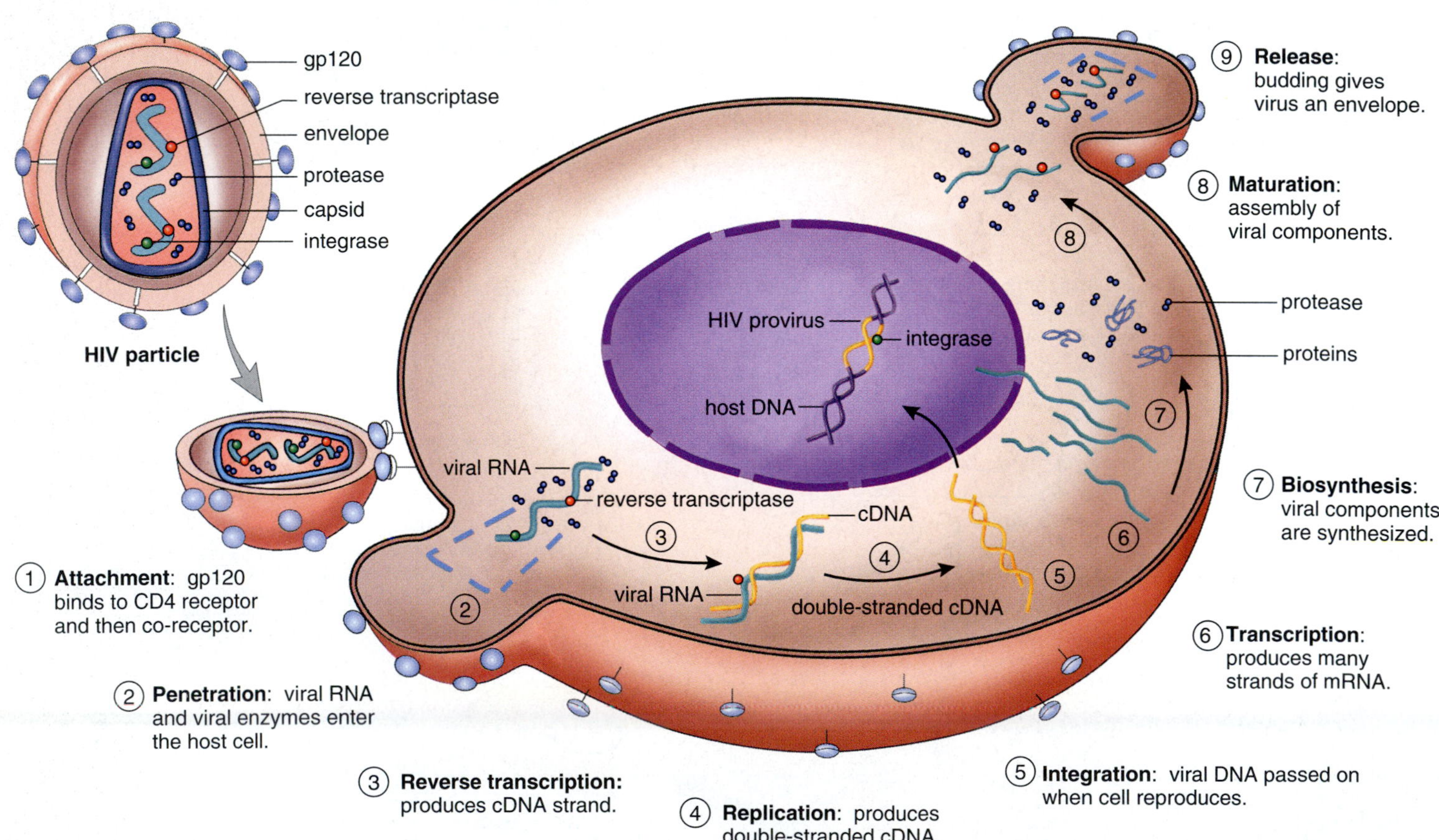

Figure 29.4 Reproduction of the retrovirus HIV-1.
HIV-1 utilizes reverse transcription to produce cDNA (DNA copy of RNA genes); cDNA integrates into the cell's chromosomes before it reproduces and buds from the cell.

Some diseases in humans have been attributed not to viruses but to **prions,** which are simply protein molecules. It seems that prions may have a misshapen tertiary structure and they can cause other proteins of their own kind to convert to this shape also. Prions are believed to cause Creutzfeldt-Jakob disease (CJD), a mental disorder that over a decade can lead to loss of vision and speech before paralysis and death occur. Also, prions have been linked to a serious outbreak in Great Britain of bovine spongiform encephalopathy (BSE), better known as mad cow disease. Cattle feed had been supplemented with the remains of sheep that had died of scrapie, another prion disease. In an effort to prevent an outbreak of BSE in this country, the importation of cattle feed, meat products, and live cattle from Great Britain and any other BSE-affected country has been banned since 1989.

Viruses, viroids, and prions are all known to cause diseases in humans.

29.2 The Prokaryotes

As previously mentioned, the **prokaryotes** include bacteria and archaea. The prokaryotes were not discovered until Dutch naturalist Antonie van Leeuwenhoek invented the first microscope in the seventeenth century and examined all sorts of specimens, including scrapings from his own teeth. Leeuwenhoek and others after him believed that the "little animals" he saw could arise spontaneously from inanimate matter. For about 200 years, scientists carried out various experiments to determine the origin of microorganisms in laboratory cultures. Finally, in about 1850, Louis Pasteur devised an experiment for the French Academy of Sciences that is described in Figure 29.5. It showed that a previously sterilized fluid medium cannot become cloudy with growth unless it is exposed directly to the air where bacteria are abundant. Today we know that bacteria are plentiful in air, water, and soil, and on most objects. A single spoonful of earth can contain 10^{10} prokaryotes. Clearly, the combined number of all prokaryotes exceeds that of any other type of organism on Earth.

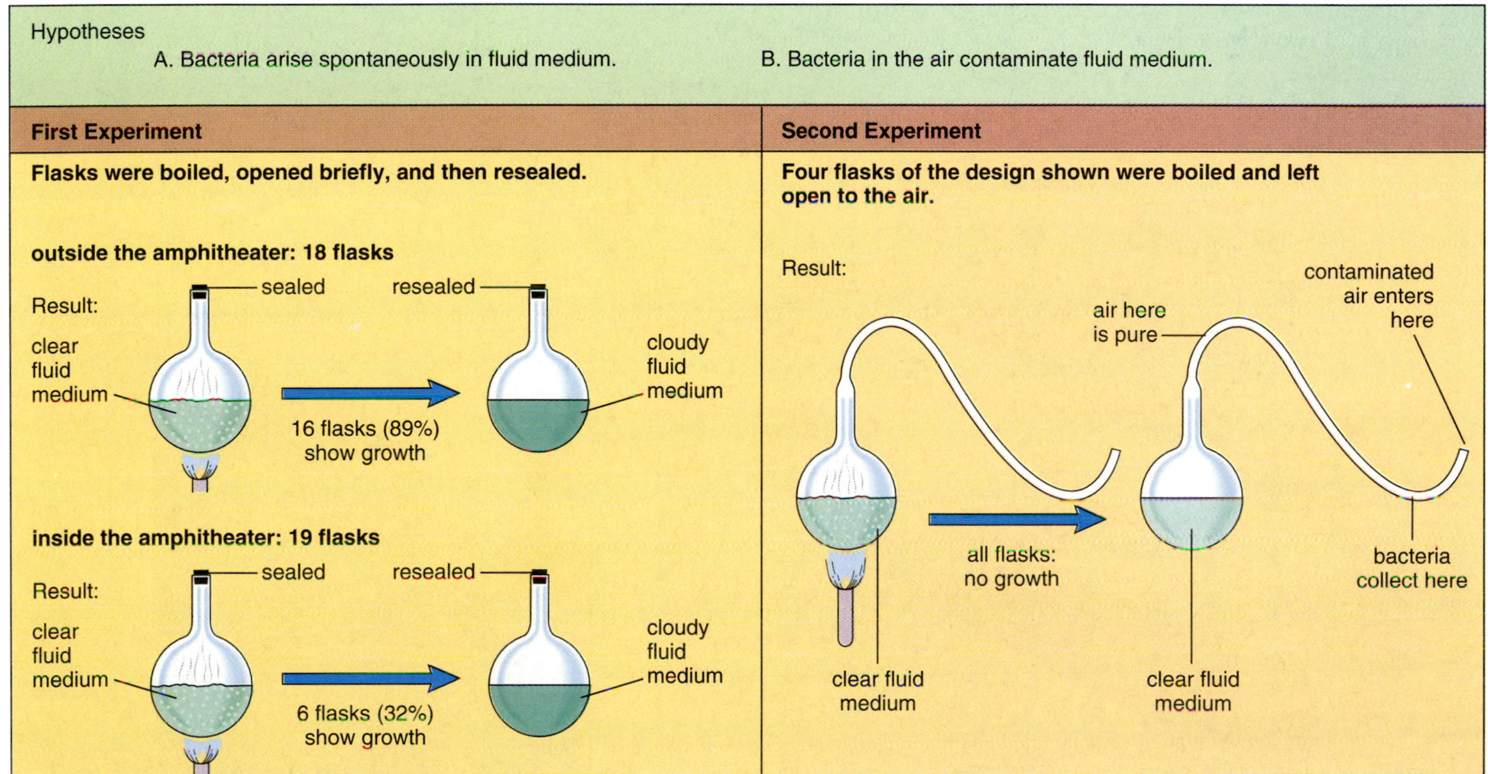

Conclusion:
With hypothesis A, all flasks would be cloudy. Relative concentrations of bacteria in the air explain the results; therefore, hypothesis B is supported.

Conclusion:
When air reaching the medium contains no bacteria, the medium remains free of growth; therefore, hypothesis B is further supported.

Figure 29.5 Pasteur's experiment.
Pasteur disproved the theory of spontaneous generation of microbes by performing experiments like this one.

Structure of Prokaryotes

Prokaryotes generally range in size from 1–10 µm in length and from 0.7–1.5 µm in width. The term prokaryote means "before a nucleus" and these organisms lack a eukaryotic nucleus (Fig. 29.6). There are prokaryotic fossils dated as long ago as 3.5 billion years, and the fossil record indicates that the prokaryotes were alone on Earth for at least 2 billion years. During that length of time, they became extremely diverse, not in structure but in metabolic capabilities. Prokaryotes are adapted to living in most environments because the various types differ in the ways they acquire and utilize energy.

Prokaryotes have an outer cell wall that prevents them from bursting or collapsing due to osmotic changes. The cell wall may be surrounded by an attached capsule and/or by a loose gelatinous sheath called a slime layer. In parasitic forms, these outer coverings protect the cell from host defenses.

Some prokaryotes move by means of **flagella.** The flagellum has a filament composed of three strands of the protein flagellin wound in a helix. The filament is inserted into a hook that is anchored by a basal body (Fig. 29.6*b*). The 360° rotation of the flagellum causes the cell to spin and move forward. Many prokaryotes adhere to surfaces by means of fimbriae, short hairlike filaments extending from the surface. The fimbriae of *Neisseria gonorrhoeae* allow it to attach to host cells and cause gonorrhea.

A prokaryotic cell lacks the membranous organelles of a eukaryotic cell, and various metabolic pathways are located on the plasma membrane. Although prokaryotes do not have a nucleus, they do have a dense area called a **nucleoid** where a single chromosome consisting largely of a circular strand of DNA is found. Many prokaryotes also have accessory rings of DNA called **plasmids.** Plasmids can be extracted and used as vectors to carry foreign DNA into host bacteria during genetic engineering processes. Protein synthesis in a prokaryotic cell is carried out by thousands of ribosomes, which are smaller than eukaryotic ribosomes. The following diagram summarizes prokaryotic cell structure:

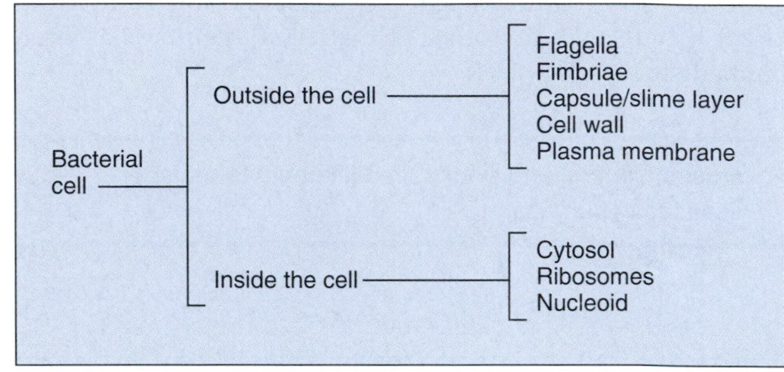

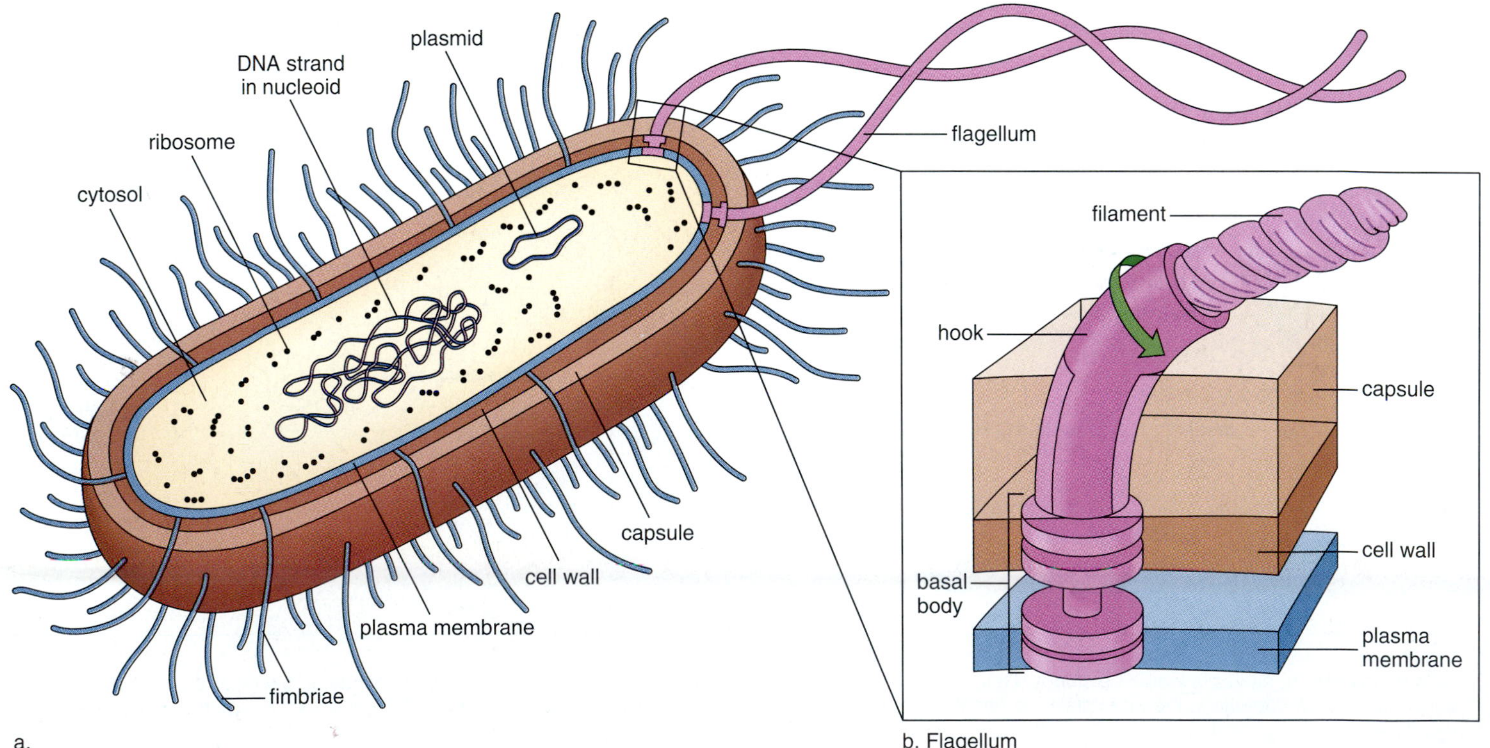

Figure 29.6 **Generalized structure of a prokaryote.**
a. The cell. **b.** The flagellum contains a basal body, a hook, and a filament. The arrow indicates that the hook (and filament) turn 360°.

Reproduction in Prokaryotes

Prokaryotes reproduce asexually by means of **binary fission** [L. *binarius,* of two, and *fissura,* cleft, break], a process that can be diagrammed as follows:

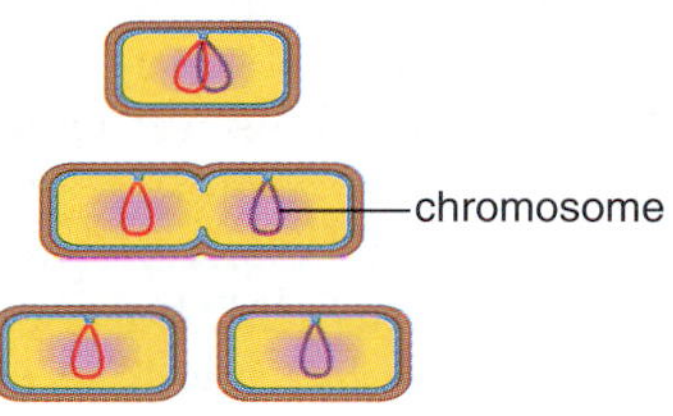

The single circular chromosome replicates, and then two copies separate as the cell enlarges. Newly formed plasma membrane and cell wall separate the cell into two cells. Mitosis, which requires the formation of a spindle apparatus, does not occur in prokaryotes.

Because prokaryotes have a generation time as short as 12 minutes under favorable conditions, mutations are generated and distributed throughout a population more quickly than in eukaryotes. Also, prokaryotes are haploid, and so mutations are immediately subjected to natural selection to assess any possible benefit.

In eukaryotes, genetic recombination occurs as a result of sexual reproduction. Sexual reproduction does not occur among prokaryotes, but three means of genetic recombination have been observed in bacteria. **Conjugation** occurs between bacteria when the donor cell passes DNA to the recipient cell by way of a sex pilus, which temporarily joins the two cells. Conjugation takes place only between bacteria in the same or closely related species. **Transformation** occurs when a bacterium picks up (from the surroundings) free pieces of DNA secreted by live prokaryotes or released by dead prokaryotes. During **transduction,** bacteriophages carry portions of bacterial DNA from one cell to another. Plasmids, which sometimes carry genes for resistance to antibiotics, can be transferred between infectious bacteria by any of these ways.

> Prokaryotes reproduce asexually by binary fission. Mutations are the chief means of achieving genetic variation.

When faced with unfavorable environmental conditions, some bacteria form **endospores** [Gk. *endon,* within, and *spora,* seed] (Fig. 29.7). A portion of the cytoplasm and a copy of the chromosome dehydrate and are then encased by three heavy, protective spore coats. The rest of the bacterial cell deteriorates and the endospore is released. Spores survive in the harshest of environments—desert heat and dehydration, boiling temperatures, polar ice, and extreme ultraviolet radiation. They also survive for very long periods. When anthrax spores 1,300 years old germinate, they can still cause a severe infection (usually seen in cattle and sheep). To germinate, the endospore absorbs water and grows out of the spore coat. In a few hours' time, it becomes a typical bacterial cell, capable of reproducing once again by binary fission. Humans particularly fear a deadly but uncommon type of food poisoning called botulism that is caused by the germination of endospores inside cans of food. Spore formation is not a means of reproduction, but it does allow survival and dispersal to new places.

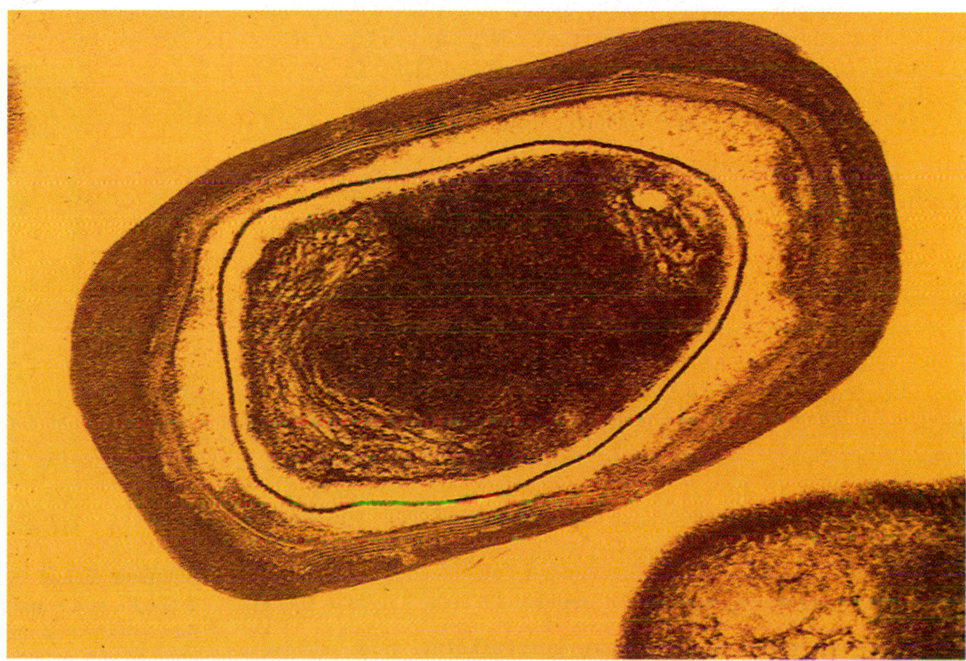

Figure 29.7 The endospore. An endospore is resistant to extreme environmental conditions. Sterilization, a process that kills all living organisms—even endospores—can be achieved by using an autoclave, a container that maintains steam under pressure. This bacterium, *Bacillus subtilis,* forms endospores. The large structure inside this cell is an endospore.

Prokaryotic Nutrition

With respect to nutrient requirements, prokaryotes are not much different from other organisms. One difference, however, concerns the need for oxygen. Some prokaryotes are **obligate anaerobes** and are unable to grow in the presence of free oxygen. A few serious illnesses—such as botulism, gas gangrene, and tetanus—are caused by anaerobic bacteria. Other prokaryotes, called **facultative anaerobes,** are able to grow in either the presence or the absence of gaseous oxygen. Most prokaryotes, however, are aerobic and, like animals, require a constant supply of oxygen to carry out cellular respiration.

Autotrophic Prokaryotes

Bacteria called **photoautotrophs** [Gk. *photos,* light, *auto,* self, and *trophe,* food] are photosynthetic. They use solar energy to reduce carbon dioxide to organic compounds. There are two types of photoautotrophs: bacteria that evolved first and do not give off oxygen (O_2) and bacteria that evolved later and do give off oxygen:

Do Not Give Off O_2	Do Give Off O_2
Photosystem I only	Photosystems I and II
Unique type of chlorophyll called bacteriochlorophyll	Type of chlorophyll found in plants

Green sulfur bacteria and some purple bacteria carry on the first type of photosynthesis. These bacteria do not give off oxygen because they do not use water as an electron donor; instead, they can, for example, use hydrogen sulfide (H_2S):

$$CO_2 + 2\,H_2S \longrightarrow (CH_2O)_n + 2\,S$$

These bacteria usually live in anaerobic conditions such as the muddy bottom of marshes and cannot photosynthesize in the presence of oxygen. In contrast, the cyanobacteria (see Fig. 29.10) contain chlorophyll *a* and carry on photosynthesis in the second way just as algae and plants do.

$$CO_2 + H_2O \longrightarrow (CH_2O)_n + O_2$$

Prokaryotes called **chemoautotrophs** [Gk. *chemo,* pertaining to chemicals, *auto,* self, and *trophe,* food] carry out chemosynthesis. They oxidize inorganic compounds such as hydrogen gas, hydrogen sulfide, and ammonia to obtain the necessary energy to produce their own organic compounds. The nitrifying bacteria oxidize ammonia (NH_3) to nitrites (NO_2^-) and nitrites to nitrates (NO_3). Their metabolic abilities keep nitrogen cycling through ecosystems. Other prokaryotes oxidize sulfur compounds at deep-sea vents 2.5 kilometers below sea level. The organic compounds they produce support the growth of a community of organisms found at vents. This discovery lends support to the suggestion that the first cells originated at deep-sea vents.

Heterotrophic Prokaryotes

Most prokaryotes, called **chemoheterotrophs** [Gk. *chemo,* pertaining to chemicals, *hetero,* different, and *trophe,* food], take in organic nutrients. They are aerobic **saprotrophs** that decompose most any large organic molecule to smaller ones that can be absorbed. There probably is no natural organic molecule that cannot be digested by at least one prokaryotic species. In ecosystems saprotrophic bacteria are called detritivores. They play a critical role in recycling matter and making inorganic molecules available to photosynthesizers.

The metabolic capabilities of chemoheterotrophic prokaryotes have long been exploited by human beings. Prokaryotes are used commercially to produce chemicals, such as ethyl alcohol, acetic acid, butyl alcohol, and acetones. Prokaryotic action is also involved in the production of butter, cheese, sauerkraut, rubber, cotton, silk, coffee, and cocoa. Even antibiotics, discussed later, are produced by some bacteria.

Chemoheterotrophs may be free-living or **symbiotic** [Gk. *sym,* together, and *bios,* life], meaning that they form mutualistic, commensalistic, or parasitic relationships. Some mutualistic bacteria are involved in nitrogen cycling. They live in the root nodules of soybean, clover, and alfalfa plants where they reduce atmospheric nitrogen (N_2) to ammonia, a process called nitrogen fixation (Fig. 29.8). Plants are unable to fix atmospheric nitrogen, and those without nodules take up nitrate and ammonia from the soil. Other mutualistic bacteria that live in human intestines release vitamins K and B_{12}, which we can use to help produce blood components. In the stomachs of cows and goats, special mutualistic prokaryotes digest cellulose, enabling these animals to feed on grass.

Commensalism often occurs when one population modifies the environment in such a way that a second population benefits. Obligate anaerobes can live in our intestines only because the bacterium *Escherichia coli* uses up the available oxygen. The parasitic bacteria causes disease including human diseases as discussed in the reading on page 521.

Figure 29.8 Nodules of a legume.
While there are some free-living bacteria that carry on nitrogen fixation, those of the genus *Rhizobium* invade the roots of legumes with the resultant formation of nodules. Here the bacteria convert atmospheric nitrogen to an organic nitrogen that the plant can use. These are nodules on the roots of a soybean plant, *Glycine.*

29.3 The Bacteria

Bacteria (domain Bacteria) are the more common type of prokaryote. Most bacteria cells are protected by a cell wall that contains a unique molecule, **peptidoglycan.** Groups of bacteria are commonly differentiated from one another by using the Gram stain procedure. This staining procedure was developed in the late 1880s by Hans Christian Gram, a Danish bacteriologist. Gram-positive bacteria retain a dye-iodine complex and appear purple under the light microscope, while Gram-negative bacteria do not retain the complex and appear pink. This difference is dependent on the construction of the cell wall; namely, the Gram-positive bacteria have a thick layer of peptidoglycan in their cell wall, whereas Gram-negative bacteria have only a thin layer.

Bacteria (and archaea) also are found in three basic shapes (Fig. 29.9): rod (bacillus, pl., bacilli); round or spherical (coccus, pl., cocci); and spiral or helical-shaped (spirillum, pl., spirilli). These three basic shapes may be augmented by particular arrangements or shapes of cells. For example, cocci may form clusters (staphylococci, diplococci) or chains (streptococci). Rod-shaped prokaryotes may appear as very short rods (coccobacilli) or as very long filaments (fusiform).

Historically, bacteria have been subdivided taxonomically into groups based on their cell wall type (Gram-positive or Gram-negative), presence of endospores, metabolism, growth and nutritional characteristics, physiological characteristics, and other criteria. For the past 75 years bacterial taxonomy has been compiled in *Bergey's Manual of Determinative Bacteriology.* The most recent edition of *Bergey's Manual* divided the prokaryotes into 19 major groups, which were further subdivided into orders, families, genera, and species. The names of the groups—for example, "nonmotile Gram-negative curved bacteria" or "nonsporeforming Gram-positive rods"—reflect the phenotypic bias used to group the bacteria. However, this edition was published just before the recognition of using ribosomal RNA analyses as the basis of phylogenetic grouping.

Since the 1980s, Carl Woese and other researchers have pioneered a new mode of bacterial taxonomy based on the phylogenetic comparisons of bacterial 16S ribosomal RNA sequences. Twelve groups are now recognized. Some of the characteristics and genera of these groups are listed in Table 29.2. Some of the new groups, such as the spirochetes, are essentially identical to early classification systems. However, other groups contain a diverse assortment of bacteria that appear to be physiologically distant, but nevertheless share common ribosomal RNA sequences. For example, the proteobacteria are a phenotypically diverse group of Gram-negative bacteria with many nutritional types. In spite of their obvious phenotypic differences, these seemingly diverse bacterial types are genetically related to one another. Further genetic studies may explain these phenotypic differences.

Cyanobacteria

Cyanobacteria [Gk. *kyanos,* blue, and *bacterion,* rod] are Gram-negative bacteria with a number of unusual traits. They photosynthesize in the same manner as plants and are believed to be responsible for first introducing oxygen into the primitive atmosphere. Formerly, the cyanobacteria were called blue-green algae and were classified with eukaryotic algae, but now we know that they are prokaryotes. They can have other pigments that mask the color of chlorophyll so that they appear, for example, not only blue-green but also red, yellow, brown, or black.

Cyanobacterial cells are rather large and range in size from 1–50 µm in width. They can be unicellular, colonial, or filamentous. Cyanobacteria lack any visible means of locomotion, although some glide when in contact with a solid surface and others oscillate (sway back and forth). Some cyanobacteria have a special advantage because they possess heterocysts, which are thick-walled cells without nuclei,

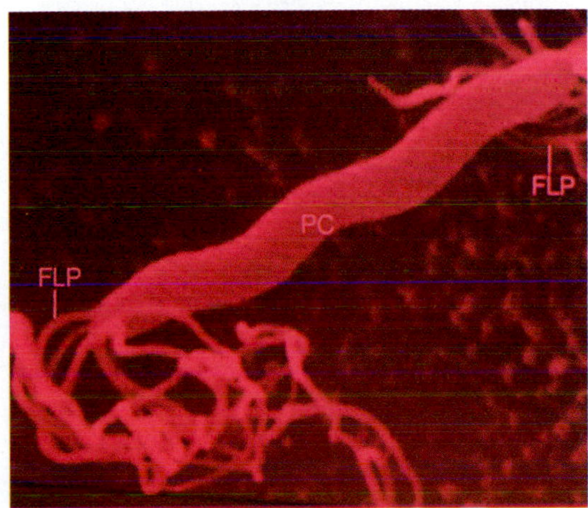

a. A spirillum with flagella

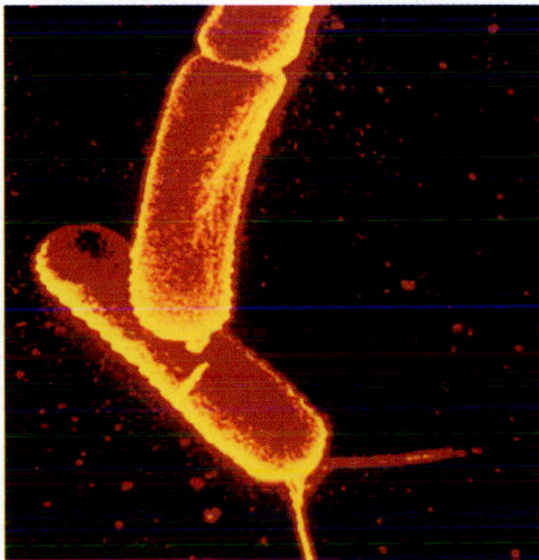
b. Bacilli in pairs

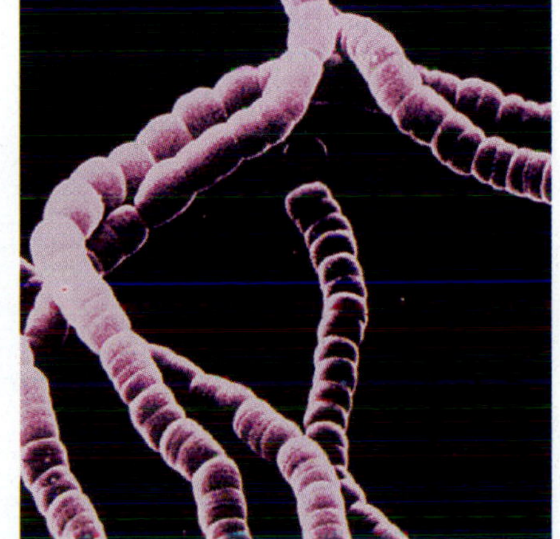
c. Cocci in chains

Figure 29.9 **Diversity of bacteria.**
a. Spirillum (spiral-shaped) bacterium. **b.** Bacillus (rod-shaped) bacterium. **c.** Coccus (round) bacterium.

Table 29.2

Major Phylogenetic Groups of Bacteria*

Group	Selected Characteristics of the Group	Representative Genera
Aquifex and relatives	Extremely thermophilic, chemoautotrophic bacteria. The oldest branch of the bacterial domain.	*Aquifex*
Deinococci	Gram-positive cocci that have atypical peptidoglycan cell walls (contain ornithine instead of diaminopimelic acid). Very resistant to radiation.	*Deinococcus*
Green nonsulfur bacteria	Can be either photosynthetic (*Chloroflexus*) or non-photosynthetic (*Herpetosiphon*). Some are thermophilic.	*Chloroflexus, Herpetosiphon*
Cyanobacteria	Photoautotrophic, gram-negative bacteria that have chlorophyll ***a*** and produce oxygen from water during photosynthesis.	*Prochloron, Oscillatoria, Anabaena*
Green sulfur bacteria	These photoautotrophic bacteria are anaerobic and use hydrogen sulfide, elemental sulfur, or hydrogen as an electron source in photosynthesis.	*Chlorobium, Pelodictyon*
Proteobacteria (purple bacteria)	A phenotypically diverse group of gram-negative bacteria with many nutritional types. Some are photoautotrophic, but do not produce oxygen during photosynthesis.Others are chemoautotrophs or chemoheterotrophs. A few genera fix atmospheric nitrogen. This very large, complex group is subdivided into alpha, beta, gamma, delta, and epsilon proteobacteria.	α: *Rhodospirillum, Rickettsia, Caulobacter, Rhizobium, Nitrobacter;* β: *Neisseria, Burkholderia;* γ: *Legionella, Pseudomonas, Vibrio, Salmonella, Shigella;* δ: *Bdellovibrio;* ϵ: *Campylobacter*
Low G + C Gram-positive bacteria	These gram-positive bacteria have DNA with a G + C (guanine + cytosine) content below 50%. Most are chemoheterotrophs and either rods or cocci. The mycoplasmas lack cell walls.	*Clostridium, Mycoplasma, Streptococcus, Enterococcus, Listeria,Staphylococcus*
High G + C Gram-positive Bacteria (Actinobacteria)	Chemoheterotrophic gram-positive bacteria with a G + C content above about 50–55%. Some are cocci, others are regular or irregular rods. The actinomycetes form complex, branching hyphae.	*Actinomyces, Corynebacterium, Mycobacterium, Nocardia*
Planctomyces, *Chlamydia,* and Relatives	These bacteria lack peptidoglycan in their walls. *Chlamydiae* are obligately intracellular parasites of mammals and birds.	*Planctomyces*
Spirochetes	Helically-shaped, motile, gram-negative bacteria with unique morphology and motility.	*Borrelia, Leptospira*
Bacteriodes	Gram-negative, anaerobic, chemoheterotrophic rods.	*Bacteroides, Porphyromonas*
Sphingobacteria, Flexibacteria, and *Cytophaga*	Gram-negative bacteria that often have sphingolipids in their cell walls. Many exhibit gliding motility.	*Sphingobacterium, Flexibacter, Cytophaga*

** These groups were constructed primarily using rRNA sequence comparisons.*

Table courtesy of Lansing M. Prescott

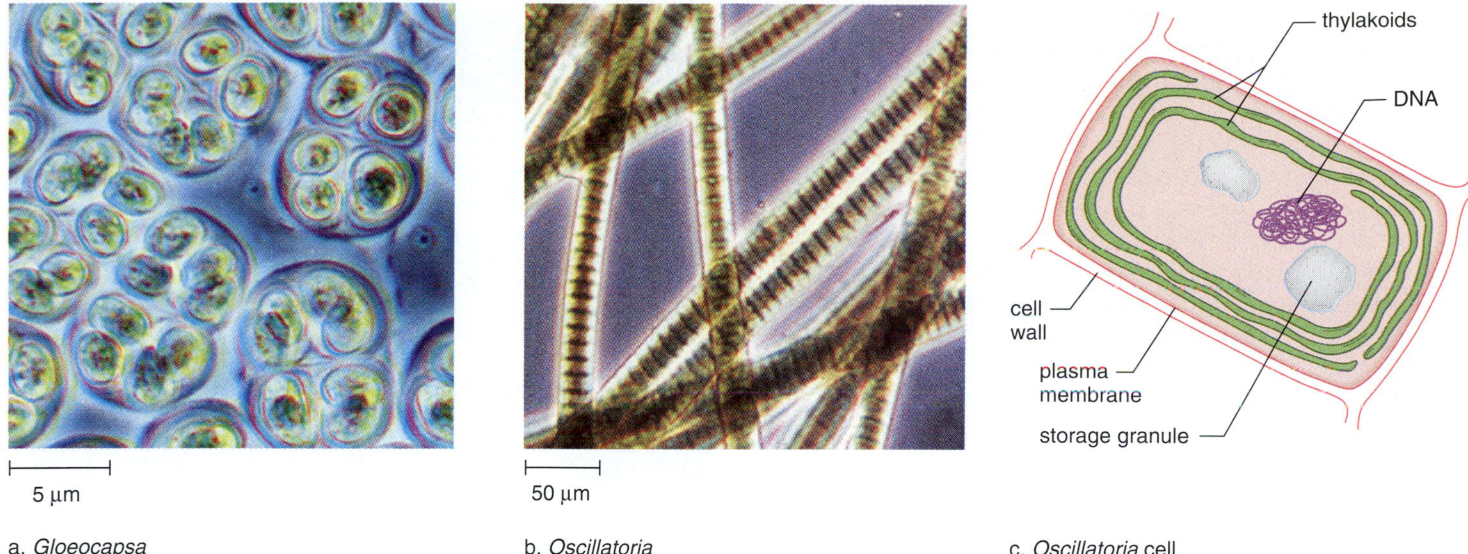

a. *Gloeocapsa* b. *Oscillatoria* c. *Oscillatoria* cell

Figure 29.10 Diversity among the cyanobacteria.
a. In *Gloeocapsa*, single cells are grouped in a common gelatinous sheath. **b.** Filaments of cells occur in *Oscillatoria*. **c.** One cell of *Oscillatoria* as it appears through the electron microscope.

where nitrogen fixation occurs. The ability to photosynthesize and also to fix atmospheric nitrogen (N_2) means that their nutritional requirements are minimal. They can serve as food for heterotrophs in ecosystems.

Cyanobacteria (Fig. 29.10) are common in fresh water, in soil, and on moist surfaces, but they are also found in harsh habitats, such as hot springs. They are symbiotic with a number of organisms, such as liverworts, ferns, and even at times invertebrates like corals. In association with fungi, they form **lichens** that can grow on rocks. A lichen is a symbiotic relationship in which the cyanobacterium provides organic nutrients to the fungus, while the fungus possibly protects and furnishes inorganic nutrients to its partner. It is also possible that the fungus is parasitic on the algal component. Lichens help transform rocks into soil; other forms of life then may follow. It is presumed that cyanobacteria were the first colonizers of land during the course of evolution.

Cyanobacteria are ecologically important in still another way. If care is not taken in the disposal of industrial, agricultural, and human wastes, phosphates drain into lakes and ponds, resulting in a "bloom" of these organisms. The surface of the water becomes turbid, and light cannot penetrate to lower levels. When a portion of the cyanobacteria die off, the decomposing prokaryotes use up the available oxygen, causing fishes to die from lack of oxygen.

Cyanobacteria are photosynthesizers that sometimes can also fix atmospheric nitrogen. In association with fungi, they form lichens, which contribute to soil formation.

29.4 The Archaea

Archaea (domain Archaea) are prokaryotes with biochemical characteristics that distinguish them from both bacteria and eukaryotes.

Relationship to Domain Bacteria and Domain Eukarya

The **archaea** used to be considered bacteria until Carl Woese discovered that their rRNA has a different sequence of bases than the rRNA of bacteria. He chose rRNA because of its involvement in protein synthesis—any changes in rRNA sequence probably occur in a slow-steady pace as evolution occurs. As discussed in the previous chapter (see pages 502–503), it is proposed that the tree of life contains three domains: Archaea, Bacteria, and Eukarya. Because archaea and some bacteria are both found in extreme environments (hot springs, thermal vents, salt basins), they may have diverged from a common ancestor relatively soon after life began. Then later, the eukarya are believed to have split off from the archaeal line of descent. In other words, the eukarya are believed to be more closely related to the archaea than to the bacteria. Archaea and eukarya share some of the same ribosomal proteins (not found in bacteria), initiate transcription in the same manner, and have similar types of tRNAs.

Structure and Function

The plasma membranes of archaea contain unusual lipids that allow them to function at high temperatures. Lipids of archaea contain glycerol linked to branched chain hydrocarbons in contrast to the lipids of prokaryotes that contain glycerol linked to fatty acids. The archaea also evolved diverse cell wall types, which facilitate their survival under extreme conditions. The cell walls of archaea do not contain peptidoglycan as do the cell walls of bacteria. In some archaea, the cell wall is largely composed of polysaccharides, and in others, the wall is pure protein. In a few, there is no cell wall.

Metabolically, the archaea have retained primitive and unique forms of metabolism. Methanogenesis, the ability to form methane, is one type of metabolism that is performed only by some methanogens.

Most archaea are chemoautotrophs (see page 516) and there are no photosynthetic archaea. This suggests that chemoautotrophy predated photoautotrophy during the evolution of prokaryotes.

Archaea are sometimes mutualistic or even commensalistic but there are no parasitic archaea—that is, they are not known to cause infectious diseases.

Types of Archaea

Archaea are often discussed in terms of their unique habitats. The **methanogens** (methane makers) are found in anaerobic environments in swamps, marshes, and the intestinal tracts of animals where they produce methane (CH_4) from hydrogen gas (H_2) and carbon dioxide coupled to the formation of ATP. This methane, which is also called biogas, is released into the atmosphere and contributes to the greenhouse effect and global warming. About 65% of the methane found in our atmosphere is produced by these methanogenic archaea.

The **halophiles** require high salt concentrations (usually 12–15%—the ocean is about 3.5% salt) for growth. They have been isolated from highly saline environments such as the Great Salt Lake in Utah, the Dead Sea, solar salt ponds, and hypersaline soils. These archaea have evolved a number of mechanisms to survive in environments that are high in salt. Their proteins have unique chloride pumps that use halorhodopsin (related to the rhodopsin pigment found in our own eyes) to pump chloride inside the cell and bacteriorhodopsin to synthesize ATP in the presence of light.

A third major type of archaea are the **thermoacidophiles** (Fig. 29.11). These archaea are isolated from extremely hot, acidic environments such as hot springs, geysers, submarine thermal vents, and around volcanoes. They reduce sulfides and survive best at temperatures above 80°C and some can even grow at 105°C (remember that water boils at 100°C)! Metabolism of sulfides results in acidic sulfates; these bacteria grow best at pH 1 to 2.

Archaea live in extremely stressful environments, perhaps like those of the primitive earth. Their biochemical properties suggest that they are more distantly related to bacteria than to eukaryotes.

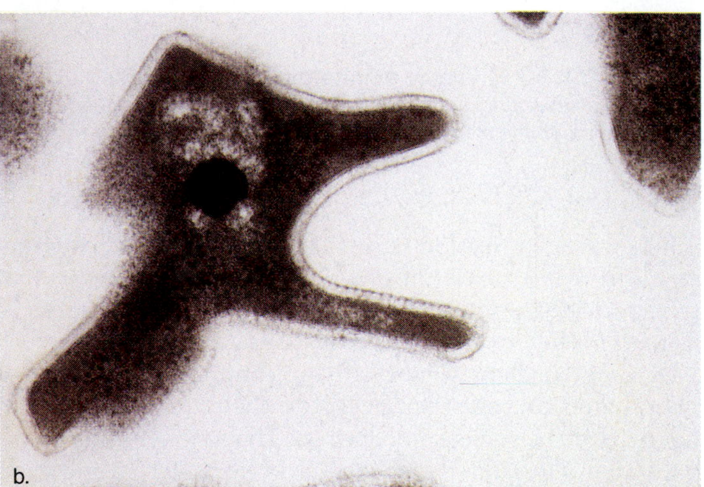

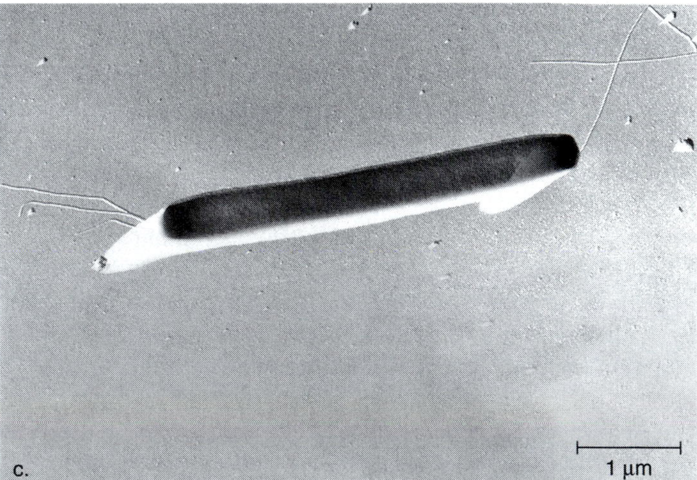

Figure 29.11 Archaea habitat and structure.
a. Boiling springs and geysers in Yellowstone National Park where thermoacidophiles live. **b.** Transmission electron micrograph of *Sulfolobus acidocaldarius*, a thermoacidophile. The dark central region has yet to be identified and could be an artifact of staining. **c.** Transmission electron micrograph of *Thermoproteus tenax*, also a thermoacidophile.

Health Focus

Treatment for Viral and Bacterial Infections

As shown in Table 29A and Table 29B, viruses and bacteria cause various diseases in humans. The development of drugs to treat viral infections has lagged far behind the development of those to treat bacterial infections. Viruses lack most enzymes and, instead, utilize the metabolic machinery of the host cell. Rarely has it been possible to find a drug that successfully interferes with viral reproduction without also interfering with host metabolism. One such drug, however, called vidarabine, was approved in 1978 for treatment of viral encephalitis, an infection of the nervous system. Acyclovir (ACV) seems to be helpful in treating genital herpes, and there are now various drugs (e.g., AZT) for the treatment of AIDS. Since viral drugs are difficult to develop, there is much concern about the possibility of other worldwide epidemics like the AIDS epidemic. One possibility is an Ebola virus epidemic. An infection with this virus begins with flulike symptoms and ends with vomiting and hemorrhaging. Spread by direct contact with a victim's blood or other body fluids, new cases are preventable only by strict hygienic and sanitary controls.

An antibiotic is a drug that selectively kills bacteria. Most antibiotics are produced naturally by soil microorganisms. Penicillin is made by the fungus *Penicillium,* and streptomycin, tetracycline, and erythromycin are all produced by the bacterium *Streptomyces.* Sulfa, a chemotherapeutic agent, is produced in the laboratory. Antibiotics inactivate bacterial enzymes without harming host enzymes. Penicillin blocks the synthesis of the bacterial cell wall; streptomycin, tetracycline, and erythromycin block protein synthesis; and sulfa prevents the production of a coenzyme. New antibiotics are being developed, but it will be some time before they are ready for general use.

There are problems associated with antibiotic therapy. Some patients are allergic to antibiotics, and the reaction can be fatal. Antibiotics not only kill off disease-causing bacteria, they also reduce the number of beneficial bacteria in the intestinal tract and other locations. These beneficial bacteria hold in check the growth of certain microbes that now begin to flourish. Diarrhea can result, as can a vaginal yeast infection. The use of antibiotics can also prevent natural immunity from occurring, leading to the need for recurring antibiotic therapy. Most important, perhaps, is the growing resistance of certain strains of bacteria to antibiotics. While penicillin used to be 100% effective against hospital strains of *Staphylococcus aureus,* today it is far less effective. Tetracycline and penicillin, long used to cure gonorrhea, now have a failure rate of more than 20% against certain strains of gonococcus. Pulmonary tuberculosis is on the rise, particularly among AIDS patients, the homeless, and the rural poor, and the new strains are resistant to the usual combined antibiotic therapy. A virulent streptococcal infection is now believed to be the cause of the much publicized "flesh-eating" condition more properly called necrotizing fasciitis. About 30% of the people in the United States who develop fasciitis usually die.

To keep antibiotics effective, most physicians believe that they should be administered only when absolutely necessary. Some believe that if antibiotic use is not strictly limited, resistant strains of bacteria will completely replace present strains and antibiotic therapy will no longer be effective. They are much opposed to the current practice of adding antibiotics to livestock feed as a preventive measure. Resistant bacteria are easily transferred from animals to humans. Antibiotics have been a boon to humans, but they should be used with care.

Table 29A

Viral Diseases in Humans

Category	Disease
Sexually transmitted diseases	AIDS (HIV), genital warts, genital herpes
Childhood diseases	Mumps, measles, chicken pox, German measles
Respiratory diseases	Common cold, influenza, acute respiratory infection
Skin diseases	Warts, fever blisters, shingles
Digestive tract diseases	Gastroenteritis, diarrhea
Nervous system diseases	Poliomyelitis, rabies, encephalitis
Other diseases	Cancer, hepatitis

Table 29B

Bacterial Diseases in Humans

Category	Disease
Sexually transmitted diseases	Syphilis, gonorrhea, chlamydia
Respiratory diseases	Strep throat, scarlet fever, tuberculosis, pneumonia, Legionnaires' disease, whooping cough
Skin diseases	Erysipelas, boils, carbuncles, impetigo, infections of surgical or accidental wounds and burns, acne
Digestive tract diseases	Gastroenteritis, food poisoning, dysentery, cholera
Nervous system diseases	Botulism, tetanus, spinal meningitis, leprosy
Systemic diseases	Plague, typhoid fever, diphtheria
Other diseases	Gas gangrene, puerperal fever, toxic shock syndrome, Lyme disease

Connecting Concepts

Microbiology began when Leeuwenhoek first used his microscope to observe microbes. Significant advances occurred with the discovery that bacteria and viruses cause disease, and again when microbes were first used in genetic studies. Although there are significant structural differences between prokaryotes and eukaryotes, many biochemical similarities exist between the two. Thus, the details of protein synthesis, first worked out in bacteria, are applicable to all cells including human cells. Today transgenic bacteria routinely make products and otherwise serve the needs of human beings.

Many prokaryotes can live in environments that may represent the kinds of habitats available when the earth first formed. We find prokaryotes in such hostile habitats as swamps, the Dead Sea, and hot sulfur springs. The fossil record suggests that the prokaryotes evolved before eukaryotes. Not only do all living things trace their ancestry to the prokaryotes, they are believed to have contributed to the evolution of the eukaryotic cell. The mitochondria and chloroplasts of the eukaryotic cell are probably derived from bacteria that took up residence inside a nucleated cell.

Cyanobacteria are believed to have introduced oxygen into the earth's primitive atmosphere and they may have been the first colonizers of the terrestrial environment. The decomposers include bacteria that recycle nutrients in both aquatic and terrestrial environments. Bacteria play significant roles in the carbon, nitrogen, and phosphorus cycles. Mutualistic bacteria also fix nitrogen in plant nodules, enabling herbivores to digest cellulose, and release certain vitamins in the human intestine. Clearly, humans are dependent on the past and present activities of prokaryotes.

Summary

29.1 The Viruses

Viruses are noncellular, while prokaryotes are fully functioning organisms. All viruses have at least two parts: an outer capsid composed of protein subunits and an inner core of nucleic acid, either DNA or RNA but not both. Some also have an outer membranous envelope.

Viruses are obligate intracellular parasites that can be maintained only inside living cells, such as those of a chick egg or those propagated in cell (tissue) culture.

The lytic cycle of a bacteriophage consists of attachment, penetration, biosynthesis, maturation, and release. In the lysogenic cycle of a bacteriophage, viral DNA is integrated into bacteria DNA for an indefinite period of time, but it can undergo the lytic cycle when stimulated.

The reproductive cycle differs for animal viruses. Uncoating is needed to free the genome from the capsid, and budding releases the viral particles from the cell. RNA retroviruses have an enzyme, reverse transcriptase, that carries out reverse transcription. This produces cDNA, which becomes integrated into host DNA. The AIDS virus is a retrovirus.

Viruses cause various diseases in plants and animals, including human beings. Viroids are naked strands of RNA not covered by a capsid which can cause disease. Prions are not viruses, they are protein molecules that have a misshapen tertiary structure. Prions cause diseases like CJD in humans and mad cow disease in cattle when they cause other proteins of their own kind to also become misshapen.

29.2 The Prokaryotes

The bacteria (domain Bacteria) and archaea (domain Archaea) are prokaryotes as discussed in the previous chapter. Prokaryotic cells lack a nucleus and most other cytoplasmic organelles found in eukaryotic cells. Prokaryotes reproduce asexually by binary fission. Their chief method for achieving genetic variation is mutation. Genetic recombination by means of conjugation, transformation, and transduction has been observed in bacteria. Some bacteria also form endospores, which are extremely resistant to destruction; the genetic material can thereby survive unfavorable conditions.

Prokaryotes differ in their need (and tolerance) for oxygen. There are obligate anaerobes, facultative anaerobes, and aerobic prokaryotes.

Some prokaryotes are autotrophic and are either photoautotrophs (photosynthetic) or chemoautotrophs (chemosynthetic). Some photosynthetic bacteria (cyanobacteria) give off oxygen and some (purple and green sulfur bacteria) do not. Chemoautotrophs oxidize inorganic compounds such as hydrogen gas, hydrogen sulfide, and ammonia to acquire energy to make their own food. Surprisingly, chemoautotrophs support communities at deep-sea vents.

Many prokaryotes are chemoheterotrophs (aerobic heterotrophs) and are saprotrophic decomposers that are absolutely essential to the cycling of nutrients in ecosystems. Their metabolic capabilities are so vast that they are used by humans both to dispose of and to produce substances. Many heterotrophic prokaryotes are symbiotic. The mutualistic nitrogen-fixing bacteria live in nodules on the roots of legumes. Some symbiotes, however, are parasitic and cause plant and animal, including human, diseases.

29.3 The Bacteria

Bacteria (domain Bacteria) are the more prevalent type of prokaryote. The classification of bacteria is still being developed. Of primary importance at this time is the shape of the cell and the structure of the cell wall, which affects Gram staining. There are three basic shapes: rod-shaped (bacillus), round (coccus), and spiral-shaped (spirillum). Of special interest are the cyanobacteria, which were the first organisms to photosynthesize in the same manner as plants. When cyanobacteria are symbionts with fungi, they form lichens.

29.4 The Archaea

The archaea (domain Archaea) are a second type of prokaryotes. On the basis of rRNA sequencing discussed on page 502, it is believed that there are three evolutionary domains: Bacteria, Archaea, and Eukarya. In addition, it appears that the archaea are more closely related to the eukarya than to bacteria. Archaea do not have peptidoglycan in their cell walls, as do the bacteria, and they share more biochemical characteristics with the eukarya than do bacteria.

The three types of archaea live under harsh conditions, such as anaerobic marshes (methanogens), salty lakes (halophiles), and hot sulfur springs (thermoacidophiles).

Reviewing the Chapter

1. Contrast viruses with prokaryotes as per the characteristics of life. 509
2. Describe the general structure of viruses, and describe both the lytic cycle and the lysogenic cycle of bacteriophages. 510–11
3. How do animal viruses differ in structure and reproductive cycle from bacteriophages? 511
4. How do retroviruses differ from other animal viruses? Describe the reproductive cycle of retroviruses in detail. 512
5. Explain Pasteur's experiment, which shows that bacteria do not arise spontaneously. 513
6. Describe the general structure of prokaryotes, and tell how they reproduce. 514–15
7. How do all prokaryotes introduce variations? How does genetic recombination occur in bacteria? 515
8. How do prokaryotes differ in their tolerance of and need for oxygen? 516
9. Compare photosynthesis between the green sulfur bacteria and the cyanobacteria. 516
10. What are chemoautotrophic prokaryotes, and where have they been found to support whole communities? 516
11. Discuss the importance of cyanobacteria in ecosystems and in the history of the earth. 518–19
12. How do the archaea differ from the bacteria? 518–20

Testing Yourself

Choose the best answer for each question.

1. Label this condensed version of bacteriophage reproductive cycles using these terms: bacterial chromosome, penetration, maturation, release, prophage, attachment, and integration.

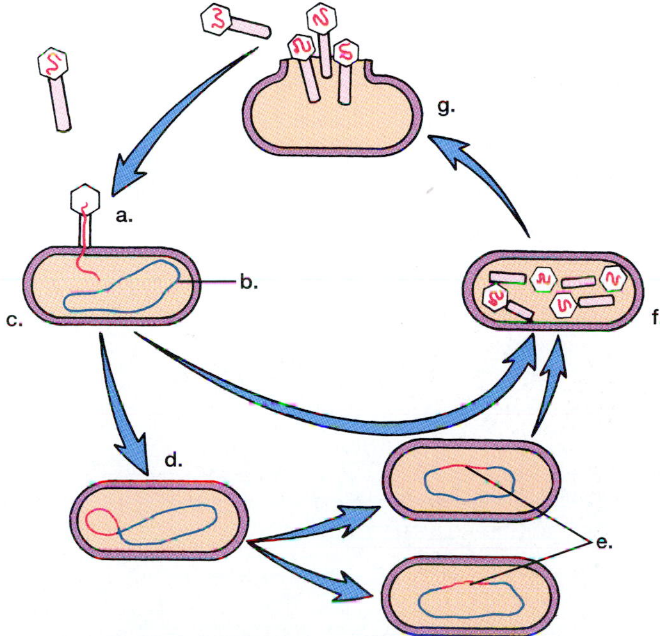

2. Viruses are considered nonliving because
   a. they do not locomote.
   b. they cannot reproduce independently.
   c. their nucleic acid does not code for protein.
   d. they are noncellular.
   e. All of these are correct.
3. Which of these are found in all viruses?
   a. envelope, nucleic acid, capsid
   b. DNA, RNA, and proteins
   c. proteins and a nucleic acid
   d. proteins, nucleic acids, carbohydrates, and lipids
   e. tail fibers, spikes, and rod-shaped
4. RNA retroviruses have a special enzyme that
   a. disintegrates host DNA.
   b. polymerizes host DNA.
   c. transcribes viral RNA to cDNA.
   d. translates host DNA.
   e. produces capsid proteins.
5. Which is not true of prokaryotes? They
   a. are living cells.
   b. lack a nucleus.
   c. all are parasitic.
   d. are both archaea and bacteria.
   e. evolved early in the history of life.
6. Facultative anaerobes
   a. require a constant supply of oxygen.
   b. are killed in an oxygenated environment.
   c. do not always need oxygen.
   d. are photosynthetic but do not give off oxygen.
   e. All of these are correct.
7. Cyanobacteria, unlike other types of bacteria that photosynthesize, do
   a. give off oxygen.
   b. not have chlorophyll.
   c. not have a cell wall.
   d. need a fungal partner.
8. Chemoautotrophic prokaryotes
   a. are chemosynthetic.
   b. use the rays of the sun to acquire energy.
   c. oxidize inorganic compounds to acquire energy.
   d. are always bacteria, not archaea.
   e. Both a and c are correct.
9. Archaea differ from bacteria in that they
   a. can form methanogen.
   b. have different rRNA sequences.
   c. do not have peptidoglycan in their cell walls.
   d. never photosynthesize.
   e. All of these are correct.
10. Which of these archaea would live at a deep-sea vent?
    a. thermoacidophile
    b. halophile
    c. methanogen
    d. parasitic forms
    e. All of these are correct.

Thinking Scientifically

1. While there are a few drugs that are effective against some viruses, they often impair the function of body cells and thereby have a number of side effects. Most antibiotics (antibacterial drugs) do not cause side effects. Why would antiviral medications be more likely to produce side effects?
2. *Escherichia coli* is a model organism for modern geneticists. What bacterial characteristics make *E. coli* particularly useful in genetic experiments?

Bioethical Issue

Carriers of disease are persons who do not appear to be ill but who can nonetheless pass on an infectious disease. The only way society can protect itself is to identify carriers and remove them from areas or activities where transmission of the pathogen is most likely. Sometimes it's difficult to identify all activities that might pass on a pathogen like HIV. A few people believe that they have acquired HIV from their dentists, and while this is generally believed to be unlikely, medical personnel are still required to identify themselves when they are carriers of HIV.

There are sports in which transmission of HIV is believed to be possible. The Centers for Disease Control did a statistical study to try to figure the odds of acquiring HIV from another football player. They figured that the odds were 1 in 85 million. The odds might be higher for boxing, a bloody sport. When two brothers, one of whom had AIDS, got into a vicious fight, the infected brother repeatedly bashed his head against his brother's. Both men bled profusely and soon after, the previously uninfected brother tested positive for the virus. The possibility of transmission of HIV in the boxing ring has caused several states to require boxers to undergo routine HIV testing. If they are HIV positive, they can't fight.

Should all people who are HIV positive always be required to identify themselves, no matter what the activity? Why or why not? By what method would they identify themselves at school, at work, and other places?

Understanding the Terms

archaea 519
bacteria 517
bacteriophage 510
binary fission 515
chemoautotroph 516
chemoheterotroph 516
conjugation 515
cyanobacteria 517
endospore 515
facultative anaerobe 516
flagella 514
halophile 520
lichen 519
lysogenic cycle 511
lytic cycle 510
methanogen 520
nucleoid 514
obligate anaerobe 516
peptidoglycan 517
photoautotroph 516
plasmid 514
prion 513
prokaryote 513
retrovirus 512
saprotroph 516
symbiotic 516
thermoacidophile 520
transduction 515
transformation 515
viroid 512
virus 508

Match the terms to these definitions:

a. ______________ Bacteriophage life cycle in which the virus incorporates its DNA into that of the bacteria; only later does it begin a lytic cycle, which ends with the destruction of the bacterium.

b. ______________ Organism that contains chlorophyll and uses solar energy to produce its own organic nutrients.

c. ______________ Organism that secretes digestive enzymes and absorbs the resulting nutrients back across the plasma membrane.

d. ______________ Relationship that could be mutualistic, commensalistic, or parasitic.

e. ______________ Type of prokaryote that is most closely related to eukarya.

Web Connections

Exploring the Internet

http://www.mhhe.com/biosci/genbio/mader
(click on *Biology 7/e*)

The *Biology 7/e* Online Learning Center provides many resources for studying the material in this chapter including links to the following sites:

The Microbe Zoo at Michigan State University's Communication Technology Laboratory. Visit the Habitat on Humanity in the Animal Pavilion to learn about the microbes that inhabit your body.

http://commtechlab.msu.edu/sites/dlc-me/zoo/zahmain.html

MSI- AIDS: The War Within. From Chicago's Museum of Science and Industry, latest news about AIDS and HIV.

http://www.msichicago.org/exhibit/AIDS/AIDShome.html

All the Virology on the WWW is a good resource on all facets of virology.

http://www.virology.net/garryfavweb.html

Learn more about viruses and see how different viruses appear in The Big Picture Book of Viruses.

http://www.tulane.edu/~dmsander/Big_Virology/BVHomePage.html

The Center for Complex Infectious Diseases site has information on current research, vaccine safety, and related links.

http://www.ccid.org

The Protists

chapter

chapter concepts

30.1 General Biology of the Protists

- Endosymbiosis played a role in the origin of the eukaryotic cell. 526
- The protists are largely unicellular but are quite varied in structure and life cycle. 526

30.2 Diversity of the Protists

- The green algae, the red algae, and the brown algae are protists. All of these have chlorophyll *a* and photosynthesize. 528
- Diatoms are unicellular with a two-piece construction of the cell wall. 533
- The flagellates include the dinoflagellates, the euglenoids, and the zooflagellates. 533
- The sarcodines include the amoeboids, the foraminiferans, and the actinopods which use pseudopods for locomotion. 536
- The protists called ciliates are a diverse group that use cilia for locomotion. 537
- The sporozoans are nonmotile spore-forming parasites. 538
- Slime molds include the plasmodial and the cellular slime molds. Water molds are also protists. 539

Spirogyra, a photosynthetic protist

In this light micrograph of *Spirogyra*, you can observe a nucleus and an obvious ribbonlike chloroplast, witnesses to the occurrence of membranous organelles in the cells of protists. Protists are biochemically diverse and every type of nutrition is seen. There are anaerobic, aerobic, and facultative aerobic heterotrophs.

Protists excel the prokaryotes in structural diversity. Some have cell walls, some not. Locomotion can be by pseudopods, flagella, or cilia. Most are unicellular. *Spirogyra* is filamentous—an end-to-end chain of cells form its threadlike body. But there are also multicellular protists. The multicellular giant kelp *Macrocystis* can be the length of a football field and its cells are specialized for particular purposes, a test of true multicellularity.

Two evolutionary events of monumental importance are first seen among the protists: the eukaryotic cell and the occurrence of multicellularity. These evolutionary occurrences led to organisms in the kingdoms we will study next. The fungi, plants, and animals are all multicellular eukaryotes and most are macroscopic. Unlike many protists, you can see them without the aid of a microscope.

30.1 General Biology of the Protists

The protists are classified in the domain Eukarya and the **kingdom Protista.** You will recall that the endosymbiotic hypothesis suggests how the eukaryotic cell arose. Mitochondria may have originated when a nucleated cell engulfed aerobic bacteria and chloroplasts may have originated when a nucleated cell engulfed cyanobacteria (Fig. 30.1*a*). It's also possible that the flagella of eukaryotes are derived from a spirochete bacterium. The protist *Giardia* has two nuclei but no mitochondria, suggesting that indeed the nucleated cell preceded the acquisition of mitochondria (Fig. 30.1*b*).

Most **protists** function at the cellular level but even so have achieved a high level of complexity. Traits occur in unique combinations. For example, euglenoids (see Fig. 30.10) are motile because they have flagella, but also are usually photosynthetic because they have chloroplasts. Euglenoids without chloroplasts are heterotrophic. Protists are also highly versatile in their life cycles. A plasmodial slime mold (see Fig. 30.15) is usually amoeboid, creeping along on the forest floor or agricultural field. But should a drought occur, this organism develops many sporangia where spores are produced. The life cycle of some protists is so complex that scientists have to study them in the laboratory in order to see all the stages.

Asexual reproduction is common among protists but sexual reproduction often occurs should the environment become stressful. Formation of spores helps a free-living species or a parasitic one survive in a hostile environment. Another common way for a protist to survive bad times is cyst formation. A **cyst** is a dormant cell with a resistant outer covering. A cyst can help a free-living protist overwinter and a parasite survive the digestive juices of a host. Some protists pass from host to host when food and water are contaminated by feces which contain cysts.

The complexity of protists is exemplified by the amoeboids and ciliates which have organelles not seen among the other eukaryotes. These organelles assist a unicellular organism in carrying out all the functions necessary to life. Food is digested in food vacuoles and excess water is expelled when contractile vacuoles discharge their contents.

Ecological Importance

While the protists have great medical importance because several cause diseases in humans, they also are of enormous ecological importance. Being aquatic, the photosynthesizers give off oxygen and are producers in both freshwater and saltwater ecosystems. They are a part of **plankton** [Gr. *plankt*, wandering], organisms that largely float near the surface and serve as food for heterotrophic protists and animals.

Protists enter symbiotic relationships ranging from parasitism to mutualism. Coral reef formation is greatly aided by the presence of a symbiotic photosynthetic protist that lives in the tissues of coral animals, for example.

Classification of Protists

The complexity and diversity of protists (Fig. 30.2) makes it difficult to classify them. None can be classified with the plants because even the multicellular photosynthesizers do not protect the gametes and zygote. None are animals because the heterotrophic ones do not undergo embryonic development. None are fungi because those that artificially resemble fungi have flagella and do not have chitin in their cell wall.

We can imagine that after the eukaryotic cell evolved, adaptive radiation occurred and many different evolutionary lineages began. The variety of protists is so great that it's been suggested that they should be split into more than a dozen kingdoms. Due to limited space, this text will discuss the artificially grouped phyla shown in the classification table on the next page.

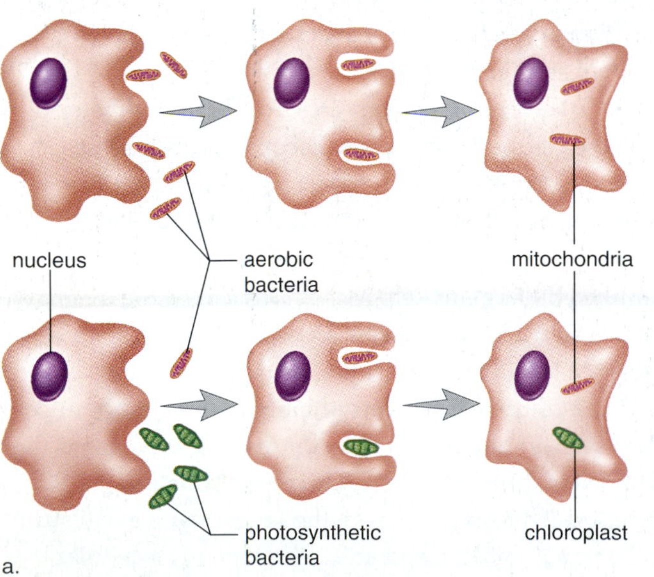

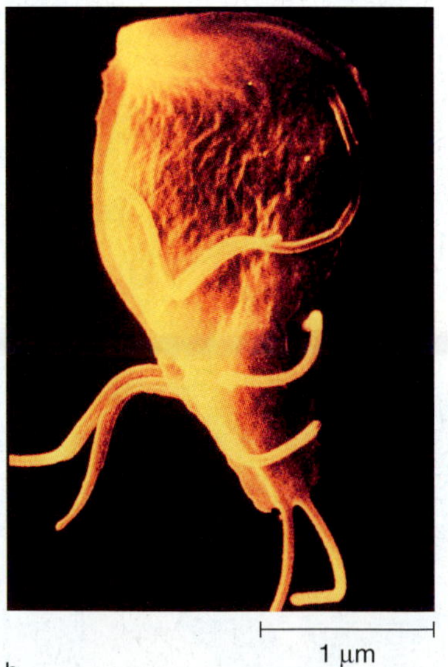

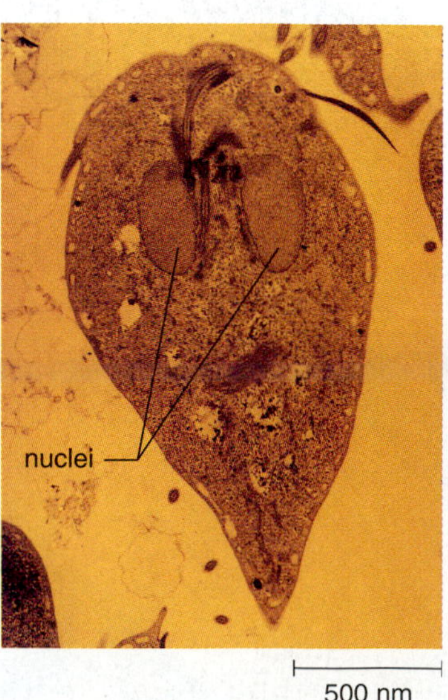

Figure 30.1 Origin of the eukaryotic cell.
a. According to the endosymbiotic hypothesis, mitochondria and chloroplasts were once free-living bacteria. **b.** *Giardia* is a protist that has two nuclei but lacks mitochondria.

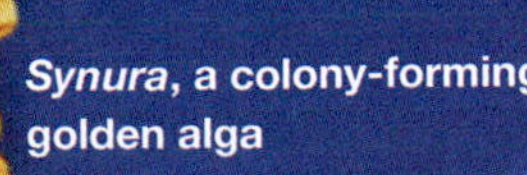

Synura, a colony-forming golden alga

CLASSIFICATION

Kingdom Protista

Eukaryotic, primarily unicellular. Metabolically diverse and structurally complex. Asexual reproduction usual; sexual reproduction diverse.

The Algae*
Phylum Chlorophyta: green algae
Phylum Rhodophyta: red algae
Phylum Phaeophyta: brown algae

The Diatoms*
Phylum Chrysophyta: diatoms, golden-brown algae

The Flagellates*
Phylum Pyrrophyta: dinoflagellates
Phylum Euglenophyta: euglenoids
Phylum Zoomastigophora: zooflagellates

The Sarcodines*
Phylum Rhizopoda: amoeboids
Phylum Foraminifera: foraminiferans
Phylum Actinopoda: radiolarians

The Ciliates*
Phylum Ciliophora: ciliates

The Sporozoans*
Phylum Apicomplexa: sporozoans

The Slime Molds and Water Molds*
Phylum Myxomycota: plasmodial slime molds
Phylum Acrasiomycota: cellular slime molds
Phylum Oomycota: water molds

* Not in the classification of organisms, but added here for clarity.

***Blepharisma*, a ciliate with visible vacuoles**

***Onychodromus*, a giant ciliate ingesting one of its own kind**

***Ceratium*, an armored dinoflagellate**

***Licmorpha*, a stalked diatom**

***Bossiella*, a coralline red alga**

***Acetabularia*, a single-celled green alga**

Figure 30.2
Protist diversity.

30.2 Diversity of the Protists

The Algae

The phyla names of three groups of protists are the green algae, the red algae, and the brown algae. Traditionally, the term **algae** means aquatic photosynthesizer. At one time botanists classified algae as plants because they have chlorophyll *a* and photosynthesize. Even so, algae are believed to have evolved from different ancestors and they should not be considered a monophyletic group.

Asexual reproduction that involves motile or nonmotile spores or vegetative growth is common among the algae. A sexual reproductive cycle can be dominated by a haploid phase or a diploid phase, or split evenly between the two as discussed in the reading on page 532.

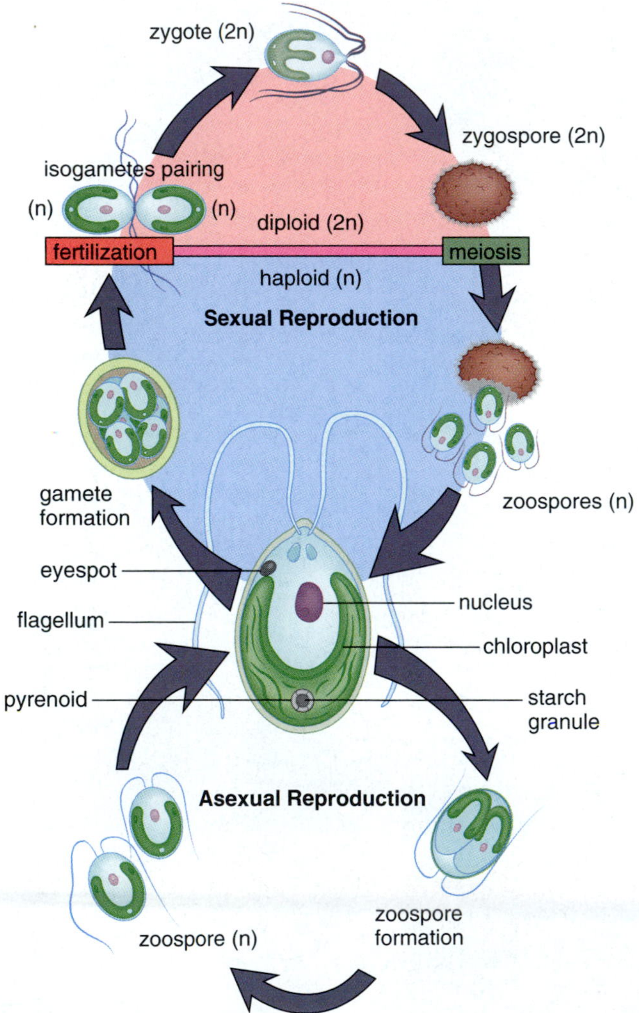

Figure 30.3 ***Chlamydomonas.***
Chlamydomonas is a motile green alga. During asexual reproduction, all structures are haploid; during sexual reproduction, meiosis follows the zygote stage, which is the only diploid part of the cycle.

The Green Algae

Green Algae **Green algae (phylum Chlorophyta)** live in the ocean but are more likely found in fresh water and can even be found on land, especially if moisture is available. The green algae also form symbiotic relationships with fungi, plants, and animals. As discussed on page 553, they associate with fungi in lichens. Some even have modifications that allow them to live on tree trunks, even in bright sun. Green algae are believed to be closely related to the first plants because both of these groups (1) have a cell wall that contains cellulose, (2) possess chlorophylls *a* and *b*, and (3) store reserve food as starch inside the chloroplast.

Green algae are not always green because some have pigments that give them an orange, red, or rust color. Body organization is also diverse: single cells, colonies, filaments, and multicellular forms will be studied in the section that follows.

Flagellated Green Algae *Chlamydomonas* is a unicellular green alga usually less than 25 μm long that has been studied in detail using the electron microscope (Fig. 30.3). It has a definite cell wall and a single, large, cup-shaped chloroplast that contains a *pyrenoid,* a dense body where starch is synthesized. The chloroplast also contains a red-pigmented eyespot (stigma), which is sensitive to light and helps bring the organism into the light, where photosynthesis can occur. Two long whiplike flagella project from the anterior end of this alga and curl backward to propel the cell freely toward the light.

When growth conditions are favorable, *Chlamydomonas* reproduces asexually. The adult divides, forming zoospores (flagellated spores) that resemble the parent cell. A **spore** is a haploid body that develops into a mature adult when conditions are favorable. When growth conditions are unfavorable, *Chlamydomonas* reproduces sexually. Gametes of two different mating types come into contact and join to form a zygote. A heavy wall forms around the zygote, and it becomes a resistant zygospore able to survive until conditions are favorable for germination. When a zygospore germinates, it produces four zoospores by meiosis. In most species, the gametes are identical, a condition known as **isogamy** [Gk. *isos,* equal, and *gamos,* marriage, union]. The gametes are called isogametes. In other species, there is a nonmotile egg specialized for storing food, and the motile sperm are specialized for seeking out an egg. This condition is known as **oogamy** [Gk. *oon,* egg, and *gamos,* marriage, union] and the gametes are called heterogametes.

Filamentous Green Algae **Filaments** [L. *filum,* thread] are end-to-end chains of cells that form after cell division occurs in only one plane. *Spirogyra,* a filamentous green alga, is found in green masses on the surfaces of ponds and streams. It has ribbonlike, spiralled chloroplasts (Fig. 30.4). **Conjugation** [L. *conjugalis,* pertaining to marriage], the temporary union of two individuals during which there is an exchange of genetic material, occurs during sexual reproduction. The two filaments line up parallel to each other, and the cell contents of one filament move into the cells of the other filament, forming diploid zygotes. These zygotes survive the winter, and in the spring they undergo meiosis to produce new haploid filaments.

Oogamy occurs among the filamentous green algae. The genus *Oedogonium* contains filamentous algae in which the cells are cylindrical with netlike chloroplasts; during sexual reproduction there is a definite egg and sperm.

Multicellular Green Algae Multicellular *Ulva* is commonly called sea lettuce because of its leafy appearance (Fig. 30.5). The thallus (body) is two cells thick and can be a meter long. *Ulva* has an alternation of generations like that of plants except that both generations look exactly alike, the gametes look alike (isogamy), and the spores are flagellated. In plants one generation is typically dominant over (lasts longer than) the other, egg and sperm are produced (oogamy), and the spores are not flagellated (see Fig. 30A*b*).

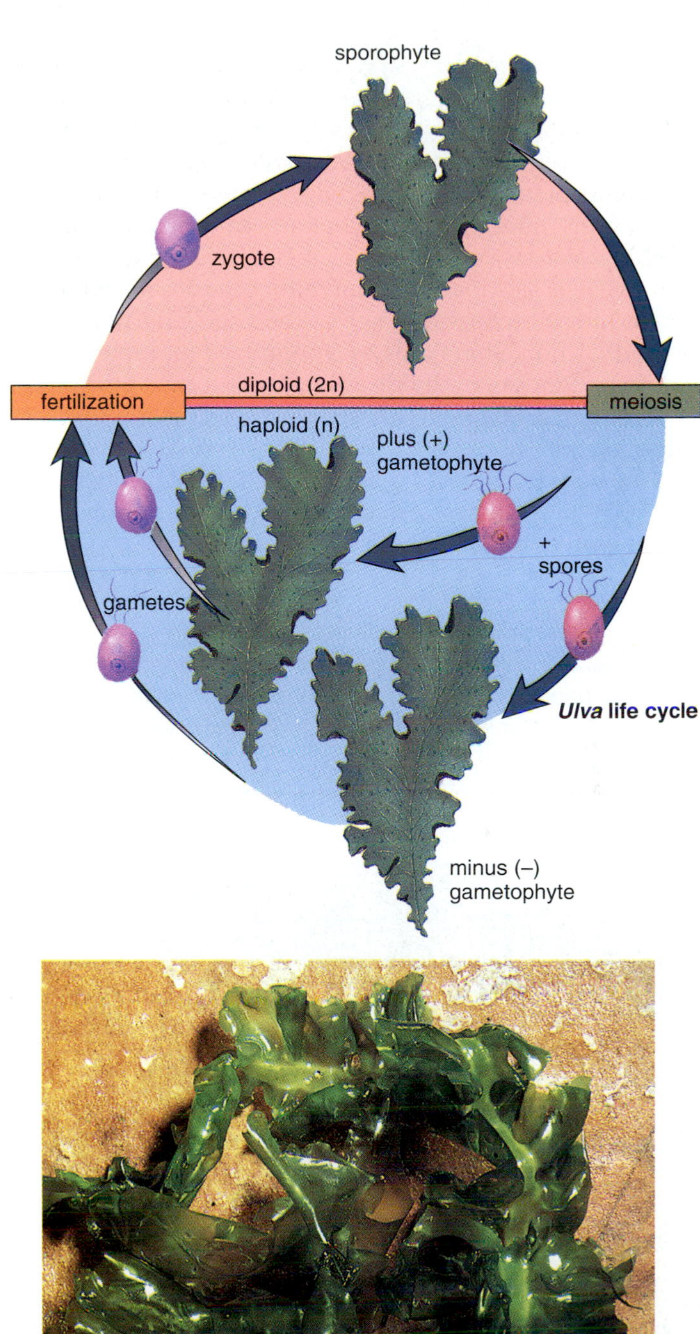

Figure 30.5 *Ulva.*
Ulva is a multicellular green alga. *Ulva* has an alternation of generations life cycle, as do plants. Notice that the sporophyte and gametophyte have the same appearance and that the gametophyte produces isogametes. These features are unlike plants.

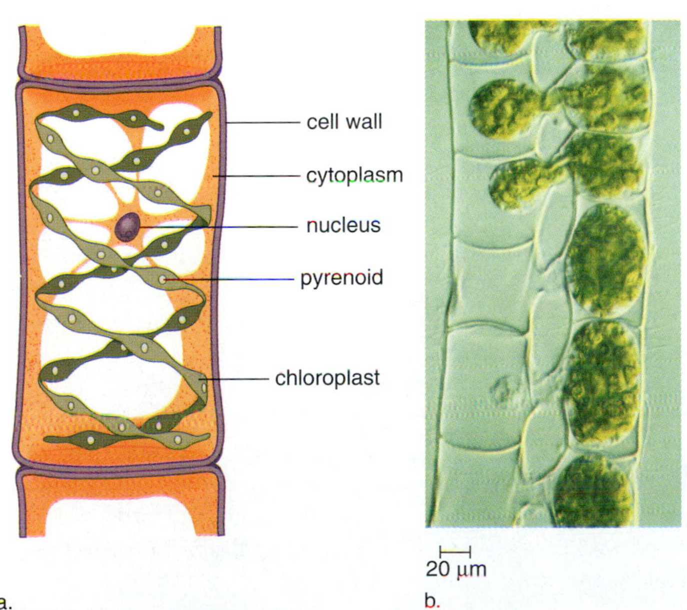

Figure 30.4 *Spirogyra.*
a. *Spirogyra* is a filamentous green alga, in which each cell has a ribbonlike chloroplast. **b.** During conjugation the cell contents of one filament enter the cells of another filament. Zygote formation follows.

Colonial Green Algae A number of *colonial* forms occur among the flagellated green algae. A **colony** is a loose association of independent cells in which there may be cells specialized for reproduction. A *Volvox* colony is a hollow sphere with thousands of cells arranged in a single layer surrounding a watery interior. Each cell of a *Volvox* colony resembles a *Chlamydomonas* cell—perhaps it is derived from daughter cells that fail to separate following zoospore formation. In *Volvox,* the cells cooperate in that the flagella beat in a coordinated fashion. Some cells are specialized for reproduction, and each of these can divide asexually to form a new daughter colony (Fig. 30.6). This daughter colony resides for a time within the parental colony, but then it leaves by releasing an enzyme that dissolves away a portion of the parental colony, allowing it to escape. Sexual reproduction among these algae involves oogamy.

Green algae occur in unicellular, colonial, and multicellular forms. During sexual reproduction, the zygote usually undergoes meiosis and the only adult is haploid. *Ulva* has an alternation of generations like plants do.

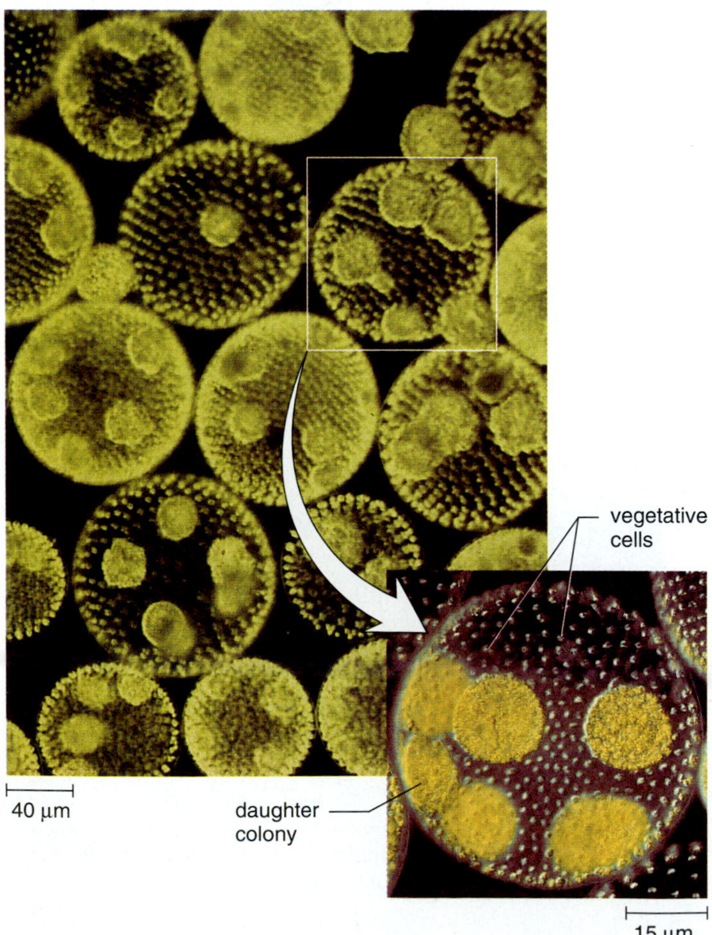

Figure 30.6 ***Volvox.***
Volvox is a colonial green alga. The adult *Volvox* colony often contains daughter colonies, which are asexually produced by special cells.

The Red Algae

Red algae (phylum Rhodophyta) are multicellular and they live chiefly in warmer seawater, growing in both shallow and deep waters. Red algae are usually much smaller and more delicate than brown algae, although they can be up to a meter long. Some forms of red algae are simple filaments, but more often they are complexly branched, with the branches having a feathery, flat, or expanded ribbonlike appearance (Fig. 30.7). Coralline algae are red algae that have cell walls impregnated with calcium carbonate. In some instances, they contribute as much to the growth of coral reefs as do coral animals.

Despite their macroscopic size, red algae show little tissue differentiation. Oogamy is seen during sexual reproduction, but the sperm are nonflagellated, an unusual feature in a water environment. Their chloroplasts resemble cyanobacteria in that they contain chlorophyll *a* and a type of pigment called phycobilin. The reserve food resembles glycogen and is called floridean starch.

Red algae are economically important. The mucilaginous material in the cell walls of certain genera of red algae is a source of agar used commercially to make capsules for vitamins and drugs, as a material for making dental impressions, and as a base for cosmetics. In the laboratory, agar is a solidifying agent for a bacterial culture medium. When purified, it becomes the gel for electrophoresis, a procedure that separates proteins or nucleotides. Agar is also used in food preparation—as an antidrying agent for baked goods and to make jellies and desserts set rapidly.

Many red algae have filamentous branches or are multicellular.

Figure 30.7 **Red alga.**
Red algae, represented by *Chondrus crispus,* are smaller and more delicate than brown algae.

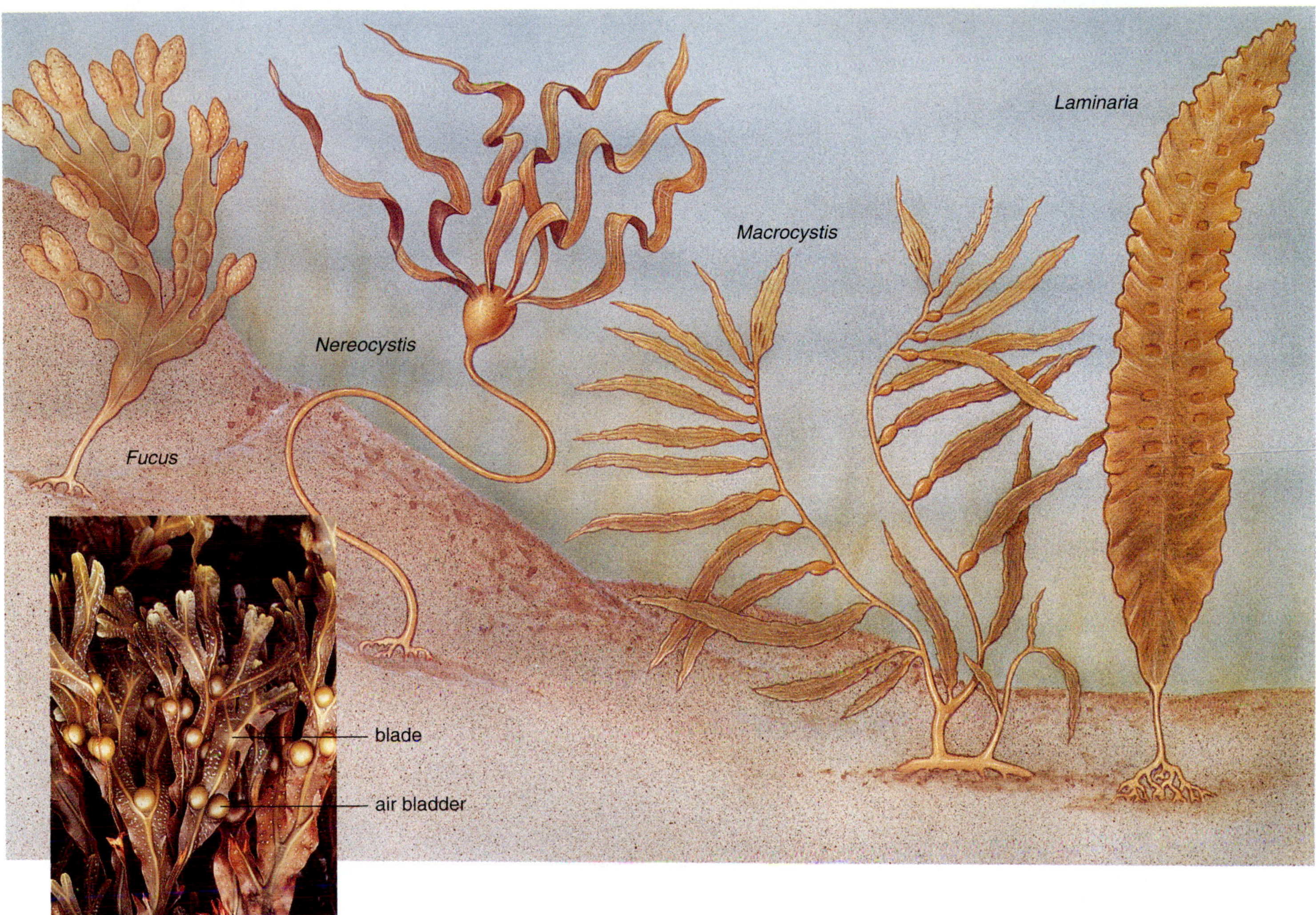

Rockweed, *Fucus*

Figure 30.8 **Brown algae.**
Laminaria and *Fucus* are seaweeds known as kelps. They live along rocky coasts of the north temperate zone. The other brown algae featured, *Nereocystis* and *Macrocystis*, form spectacular underwater "forests" at sea.

The Brown Algae

Brown algae (phylum Phaeophyta) range from small forms with simple filaments to large multicellular forms (50–100 m long) (Fig. 30.8). The brown algae have chlorophylls *a* and *c* in their chloroplasts and a type of carotenoid pigment (fucoxanthin) that gives them their color. The reserve food is a carbohydrate called laminarin.

The multicellular forms of green, red, and brown algae are called **seaweeds,** a common term for any large, complex alga. Brown algae are often observed along the rocky coasts in the north temperate zone, where they are pounded by waves as the tide comes in and are exposed to dry air as the tide goes out. They do not dry out, however, because their cell walls contain a mucilaginous, water-retaining material.

Both *Laminaria,* commonly called a *kelp,* and *Fucus,* known as rockweed, are examples of brown algae that grow along the shoreline. In deeper waters, the giant kelps (*Nereocystis* and *Macrocystis*) often grow extensively in vast beds. Individuals of the genus *Sargassum* sometimes break off from their holdfasts and form floating masses, where life-forms congregate in the ocean. Brown algae not only provide food and habitat for marine organisms, they are harvested for human food and for fertilizer in several parts of the world. They are also a source of algin, a pectinlike material that is added to ice cream, sherbet, cream cheese, and other products to give them a stable, smooth consistency.

Laminaria is unique among the protists because members of this genus show tissue differentiation—they transport organic nutrients by way of a tissue that resembles phloem in land plants. Most brown algae have the alternation of generations life cycle, but some species of *Fucus* are unique in that meiosis produces gametes and the adult is always diploid, as in animals.

The multicellular forms of brown algae include the largest of the algae, the giant kelps.

Life Cycles Among the Algae

Algae reproduce asexually and sexually. Asexual reproduction requires only one parent. The offspring are identical to this parent because the offspring receive a copy of only this parent's genes. Various modes of asexual reproduction occur. In any case, growth alone produces a new adult. Asexual reproduction is a frequent mode of reproduction among protists when the environment is favorable to growth. The new individuals, which are identical to the parent, are likely to survive and flourish.

Sexual reproduction, with its genetic recombination due in part to fertilization and independent assortment of chromosomes, is more likely to occur among protists when the environment is changing and is unfavorable to growth. Recombination of genes might produce individuals that are more likely to survive extremes in the environment—such as high or low temperatures, acidic or basic pH, or a lack of some particular nutrient.

Sexual reproduction requires two parents, each of which contributes chromosomes (genes) to the offspring by way of gametes. The gametes fuse to produce a diploid zygote. A reproductive cycle is isogamous when the gametes look alike—called isogametes—and the cycle is oogamous when the gametes are dissimilar—called heterogametes. Usually a small flagellated sperm fertilizes a large egg with plentiful cytoplasm.

Meiosis occurs during sexual reproduction—just *when* it occurs makes the sexual life cycles diagrammed below differ from one another. In these diagrams, the diploid phase is above and the haploid phase is below the line. In the haplontic cycle (Fig. 30A*a*), the zygote divides by meiosis to form haploid spores that develop into a haploid adult. In algae, the spores are typically zoospores. The zygote is only the diploid stage in this life cycle, and the haploid adult gives rise to gametes. This life cycle is seen in *Chlamydomonas* and a number of other algae.

In alternation of generations (sporic meiosis), the sporophyte (2n) produces haploid spores by meiosis (Fig. 30A*b*). A spore develops into a haploid gametophyte that produces gametes. The gametes fuse to form a diploid zygote, and the zygote develops into the sporophyte. This life cycle is characteristic of some algae (for example, *Ulva* and *Laminaria*) and all plants. In *Ulva,* the haploid and diploid generations have the same appearance. In plants, they are noticeably different from each other.

In the diplontic cycle, typical of animals, a diploid adult produces gametes by meiosis (Fig. 30A*c*). Gametes are the only haploid stage in this cycle. They fuse to form a zygote that develops into the diploid adult. This life cycle is rare in algae but does occur in a few species of the brown alga *Fucus.*

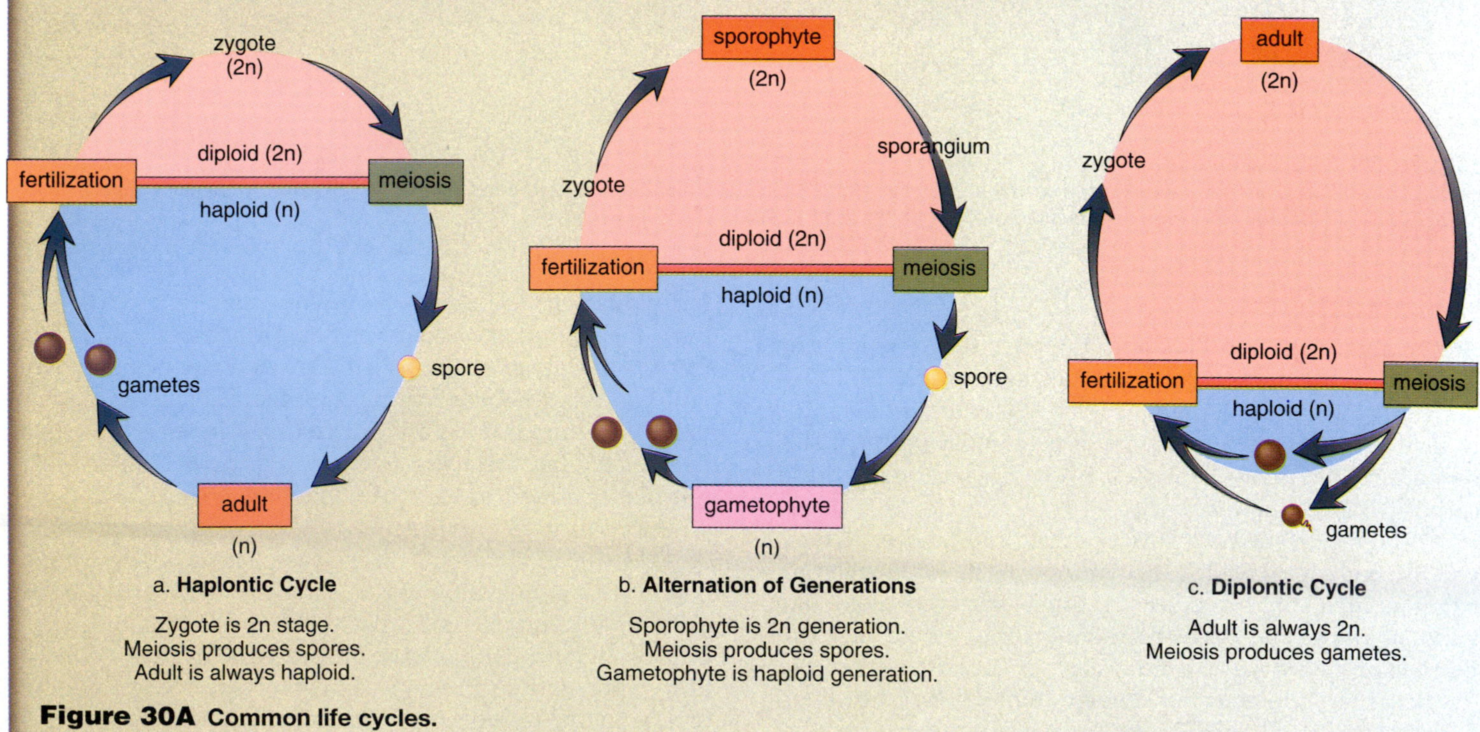

a. **Haplontic Cycle**
Zygote is 2n stage.
Meiosis produces spores.
Adult is always haploid.

b. **Alternation of Generations**
Sporophyte is 2n generation.
Meiosis produces spores.
Gametophyte is haploid generation.

c. **Diplontic Cycle**
Adult is always 2n.
Meiosis produces gametes.

Figure 30A Common life cycles.

The Diatoms

The diatoms and golden-brown algae **(phylum Chrysophyta)** are distinctive from one another, and some authorities place diatoms in their own phylum called phylum Bacillariophyta. **Diatoms** [Gk. *dia*, through, and *temno*, cut] are the most numerous unicellular algae in the oceans and they are plentiful in fresh water also. Because of their small size thousands can live in a single milliliter of water. Diatoms are a significant part of the **phytoplankton,** photosynthetic organisms that are suspended and float near the surface of the water. Phytoplankton are an important source of food and oxygen for heterotrophs in both freshwater and marine ecosystems.

The structure of a diatom is often compared to a box because the cell wall has two halves or valves, with the larger valve acting as a "lid" for the smaller valve (Fig. 30.9*a*). When diatoms reproduce asexually, each receives one old valve. The new valve fits inside the old one; therefore, new diatoms are smaller than the original ones. This continues until diatoms are about 30% of the original size. Then they reproduce sexually. The zygote becomes a structure that grows and then divides mitotically to produce diatoms of normal size.

The cell wall of a diatom has an outer layer of silica, a common ingredient of glass. The valves are covered with a great variety of striations and markings that form beautiful patterns when observed under the microscope. These are actually depressions or pores through which the organism makes contact with the outside environment. The remains of diatoms, called diatomaceous earth, accumulate on the ocean floor and are mined for use as filtering agents, soundproofing materials, and gentle abrasives.

The Flagellates

Thousands of species of protists have flagella but we have space only to discuss a few of these. Some flagellates are free-living but a variety are parasitic on animals, including humans. One group of zooflagellates, the Choanoflagellates, may be close relatives of animals. These protists resemble the collar cells of sponges (see Fig. 33.3).

Flagellates reproduce by a type of transverse binary fission that requires mitosis and cytokinesis.

The Dinoflagellates

Many **dinoflagellates** [Gk. *dinos*, whirling, and L. *flagello*, whip] **(phylum Pyrrophyta)** are bounded by protective cellulose plates (Fig. 30.9*b*). Most have two flagella; one lies in a longitudinal groove with its distal end free, and the other, which is flat and ribbonlike, lies in a transverse groove that encircles the organism. The longitudinal flagellum acts as a rudder, and the beating of the transverse one causes the cell to spin as it moves forward.

The chloroplasts of a dinoflagellate, which contain chlorophylls *a* and *c* as do those of the Chrysophyta, are probably derived from them following an endosymbiotic event (see Fig. 30.1). They vary in color from yellow-green to brown. Some species of dinoflagellates lack chloroplasts and are heterotrophic.

Like the diatoms, dinoflagellates are extremely numerous in the oceans; their density can equal 30,000 in a single milliliter. Under certain conditions, those in the genera *Gymnodinium* and *Gonyaulax* increase in number and cause a "red tide" in the ocean. At these times, they produce a

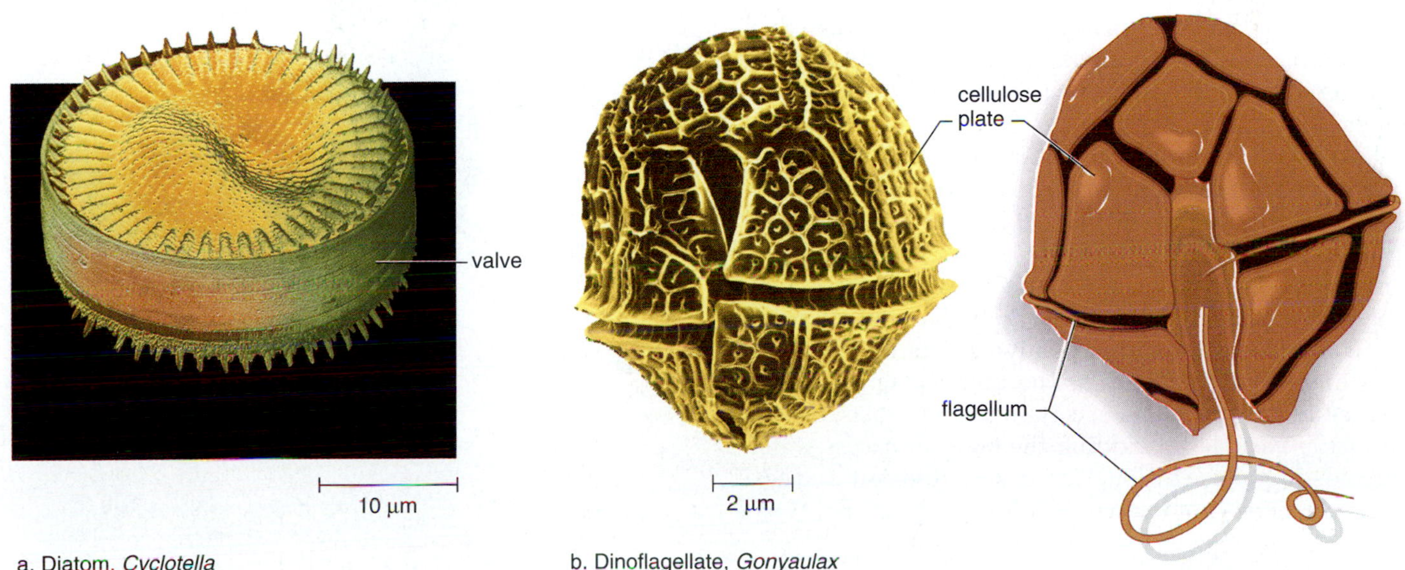

a. Diatom, *Cyclotella*

b. Dinoflagellate, *Gonyaulax*

Figure 30.9 Diatoms and dinoflagellates.
a. Diatoms may be variously colored, but even so their chloroplasts contain a unique golden-brown pigment (fucoxanthin), in addition to chlorophylls *a* and *c*. The beautiful pattern results from markings on the silica-embedded wall. **b.** Dinoflagellates have cellulose plates; these belong to *Gonyaulax*, the dinoflagellate that contains a red pigment and is responsible for occasional "red tides."

Table 30.1

Photosynthetic Protists

Name	Multicellular	Flagella	Examples
Green algae (Chlorophyta)	Yes, some forms	Yes, some forms	*Chlamydomonas, Volvox, Spirogyra, Ulva*
Red algae (Rhodophyta)	Yes, some forms	No	*Chondrus,* coralline algae
Brown algae (Phaeophyta)	Yes	On reproductive cells	*Laminaria, Macrocystis, Fucus*
Diatoms (Chrysophyta)	No	No	*Cyclotella, Diatoma*
Dinoflagellates (Pyrrophyta)	No	Yes	*Gonyaulax, Gymnodinium*
Euglenoids (Euglenophyta)	No	Yes	*Euglena*

neurotoxin that can kill fish and causes paralytic shellfish poisoning—humans who eat shellfish that have fed on these dinoflagellates suffer paralysis of the respiratory muscles.

Being a part of the phytoplankton, the dinoflagellates are an important source of food for small animals in the ocean. They also live within the bodies of some invertebrates as symbionts. For example, corals usually contain large numbers of these organisms, and this allows coral reefs to be one of the most productive ecosystems on earth.

The Euglenoids

Euglenoids (phylum Euglenophyta) are small (10–500 µm) freshwater unicellular organisms that typify the problem of classifying protists. One-third of all genera have chloroplasts; the rest do not. Those that lack chloroplasts ingest or absorb their food. This may not be surprising when one knows that their chloroplasts are like those of green algae and are probably derived from them through endosymbiosis. The chloroplasts are surrounded by three rather than two membranes. The pyrenoid is an organ outside the chloroplast that produces an unusual type of carbohydrate polymer (paramylon).

Euglenoids have two flagella, one of which typically is much longer than the other and projects out of an anterior vase-shaped invagination (Fig. 30.10). It is called a tinsel flagellum because it has hairs on it. Near the base of this flagellum is an eyespot, which shades a photoreceptor for detecting light. Because euglenoids are bounded by a flexible *pellicle* composed of protein strips lying side by side, they can assume different shapes as the underlying cytoplasm undulates and contracts. As in certain other protists, there is a contractile vacuole for ridding the body of excess water. Euglenoids reproduce by longitudinal cell division, and sexual reproduction is not known to occur.

The diatoms, dinoflagellates, and euglenoids are photosynthetic (Table 30.1). The diatoms and dinoflagellates, due to their abundance, play a significant role in the biosphere as a source of food and oxygen for heterotrophs.

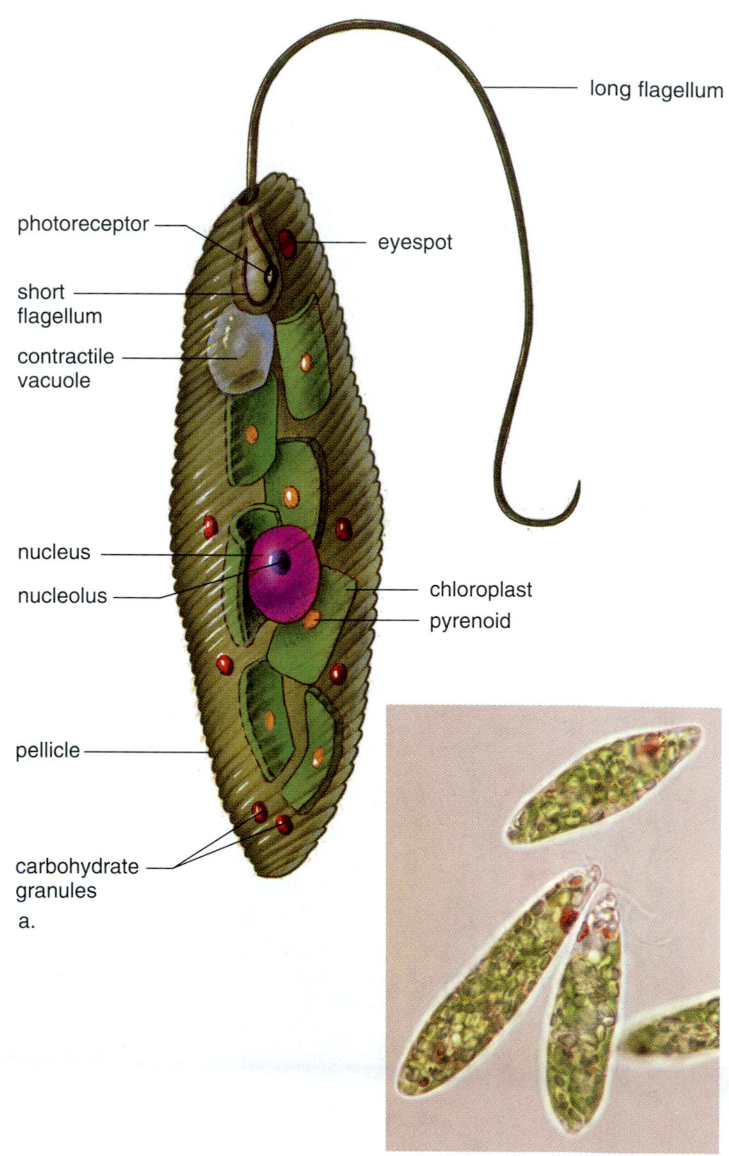

Figure 30.10 ***Euglena.***
a. In *Euglena*, a very long flagellum propels the body, which is enveloped by a flexible pellicle. A photoreceptor shaded by an eyespot allows *Euglena* to find light, after which photosynthesis can occur in the numerous chloroplasts. Pyrenoids synthesize a reserve carbohydrate, which is stored in the chloroplasts and also in the cytoplasm. b. Micrograph of several specimens.

The Zooflagellates

Zooflagellates (phylum Zoomastigophora) have one to thousands of flagella. They are covered by a pellicle that is often reinforced by underlying microtubules. Some consider the euglenoids a type of zooflagellate.

Many zooflagellates enter into symbiotic relationships (Fig. 30.11). *Trichonympha collaris* lives in the gut of termites; it contains a bacterium that enzymatically converts the cellulose of wood to soluble carbohydrates that are easily digested by the insect. *Giardia lamblia,* whose cysts are transmitted through contaminated water, attaches to the human intestinal wall and causes severe diarrhea. *Trichomonas vaginalis,* a sexually transmitted organism, infects the vagina and urethra of women and the prostate, seminal vesicles, and urethra of men. A **trypanosome,** *Trypanosoma brucei,* transmitted by the bite of the tsetse fly, is the cause of African sleeping sickness. The white blood cells in an infected human accumulate around the blood vessels leading to the brain and cut off circulation. The lethargy characteristic of the disease is caused by an inadequate supply of oxygen to the brain.

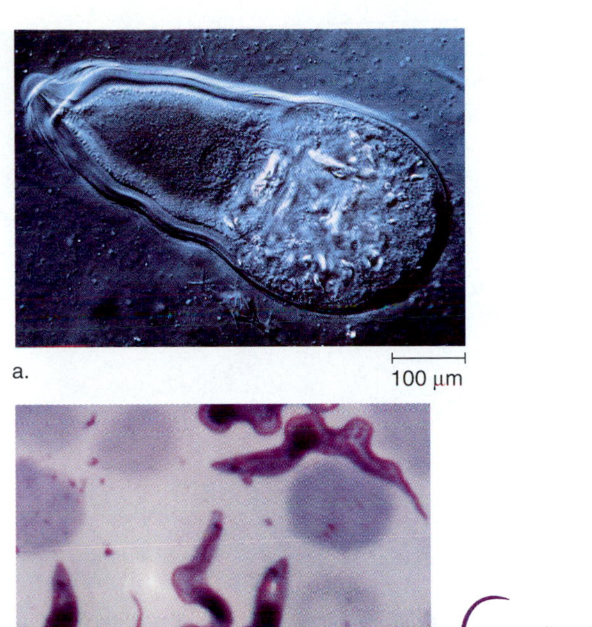

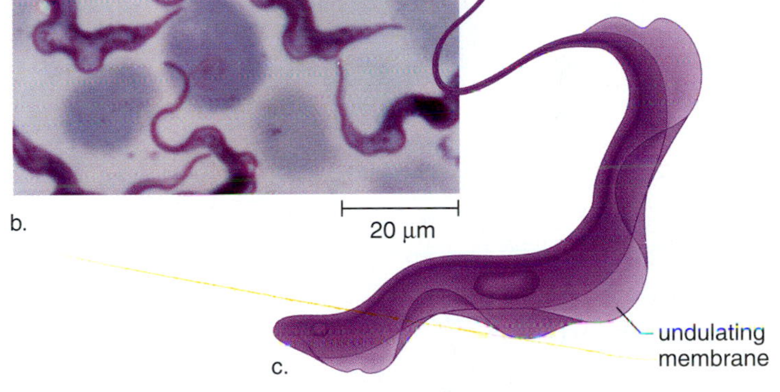

Figure 30.11 Zooflagellates.
a. *Trichonympha collaris* lives in the gut of termites and helps digest cellulose. **b.** Micrograph of *Trypanosoma brucei,* the cause of African sleeping sickness, among red blood cells. **c.** The drawing shows its general structure.

> The zooflagellates are well known for causing various diseases in humans ranging from those that are annoying to those that are extremely serious.

Some authorities include all the flagellates among the **protozoans,** a term that can simply mean a unicellular (or colonial) eukaryote. In that case, protozoans include both photosynthetic and heterotrophic organisms. Others prefer to restrict the term protozoan to unicellular heterotrophic organisms such as those listed in Table 30.2. Some of these heterotrophic forms ingest their food by endocytosis. Usually a protozoan has some form of locomotion, either by flagella, pseudopods, or cilia. At one time, zoologists classified protozoans as animals.

Table 30.2

Some Protozoans

Name	Unicellular	Locomotion	Examples
Zooflagellates	Yes	Flagella	*Trypanosoma, Euglena*
Amoeboids	Yes	Pseudopods	*Amoeba, Entamoeba*
Radiolarians	Yes	Pseudopods	—
Foraminiferans	Yes	Pseudopods	*Globigerina*
Ciliates	Yes	None	*Paramecium, Vorticella, Didinia, Stentor*
Sporozoans	Yes	Cilia	*Plasmodium*

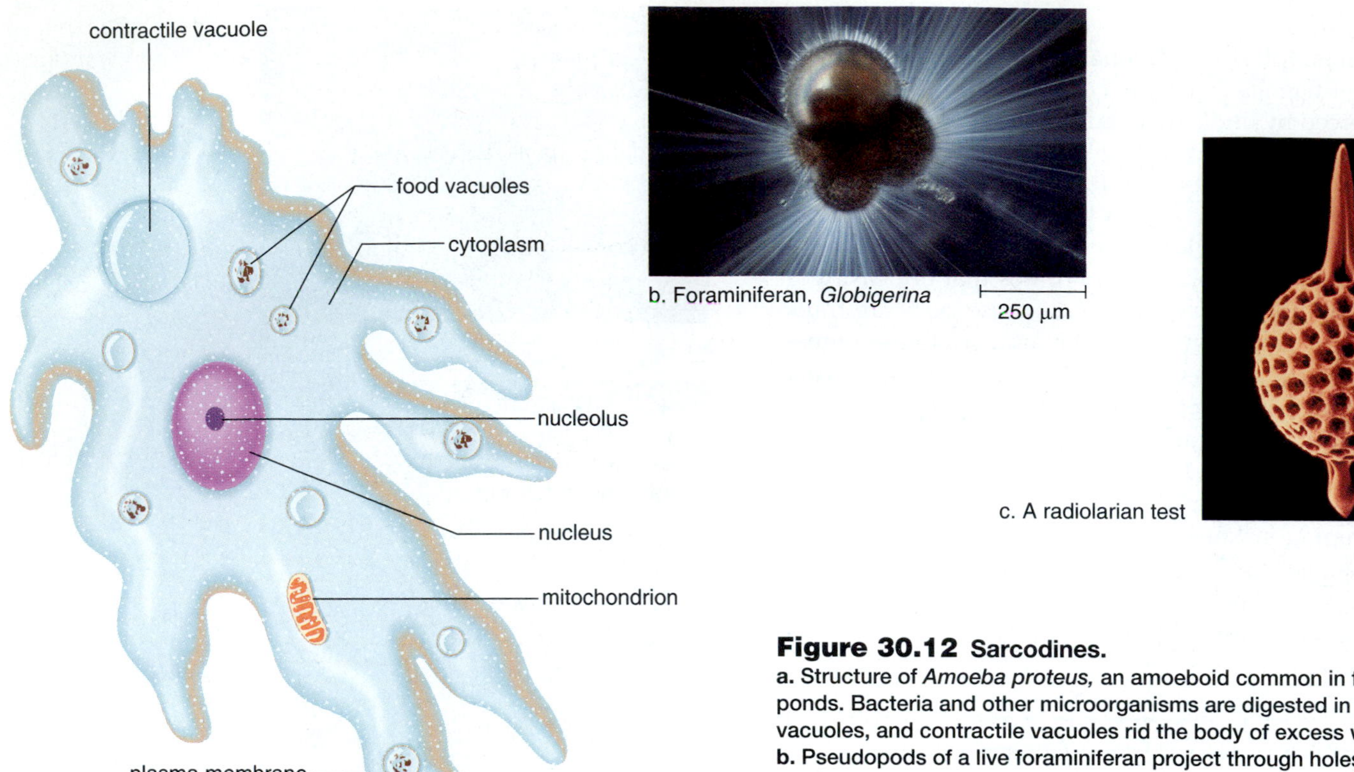

a. Amoeba, *Amoeba proteus*

b. Foraminiferan, *Globigerina*

c. A radiolarian test

Figure 30.12 **Sarcodines.**
a. Structure of *Amoeba proteus,* an amoeboid common in freshwater ponds. Bacteria and other microorganisms are digested in food vacuoles, and contractile vacuoles rid the body of excess water. **b.** Pseudopods of a live foraminiferan project through holes in the calcium carbonate shell. These shells were so numerous they became a large part of the White Cliffs of Dover when a geological upheaval occurred. **c.** Test of a radiolarian. In life pseudopods extend outward through the openings of the siliceous shell.

The Sarcodines

The sarcodines (move by pseudopods) and also the ciliates (move by cilia), discussed next, usually live in aquatic environments. In oceans and freshwater lakes and ponds, they are a part of the **zooplankton,** microscopic floating organisms that feed on other organisms.

The **amoeboids (phylum Rhizopoda)** are protists that move and also ingest their food with **pseudopods** [Gk. *pseudes,* false, and *podos,* foot] which are extensions of their body. Pseudopods form when the cytoplasm streams forward in a particular direction.

Amoeba proteus is a commonly studied freshwater member of this group (Fig. 30.12). When amoeboids feed, the pseudopods surround and **phagocytize** [Gk. *phagein,* eat, and *kytos,* cell] their prey, which may be algae, bacteria, or other protists. Digestion then occurs within a food vacuole. Freshwater amoeboids, including *Amoeba proteus,* have contractile vacuoles where excess water from the cytoplasm collects before the vacuole appears to "contract," releasing the water through a temporary opening in the plasma membrane.

Entamoeba histolytica is a parasitic amoeboid that lives in the human intestine and causes amoebic dysentery. Complications arise when this parasite invades the intestinal lining and reproduces there. If the parasites enter the body proper, liver and brain involvement can be fatal.

The **foraminiferans (phylum Foraminifera)** and the **radiolarians (phylum Actinopoda)** are sarcodines that have a skeleton called a **test.** In the foraminiferans the calcium carbonate test is often multichambered. The pseudopods extend through openings in the test which covers the plasma membrane. In the radiolarians, the silicon or strontium sulfate test is internal and usually has a radial arrangement of spines. The pseudopods called actinopods project from an external layer of cytoplasm and are supported by many rows of microtubules.

The tests of dead foraminiferans and the radiolarians form a deep layer (700–4,000 m) of sediment on the ocean floor. The radiolarians lie deeper than the foraminiferans because their glassy test is insoluble at greater pressures. The presence of either or both is used as an indicator of oil deposits on land and sea. Their fossils date back to even Precambrian times and are evidence of the antiquity of the protists. Because each geological period has a distinctive form of foraminiferan, they can be used to date sedimentary rock. Deposits for millions of years followed by geological upheaval formed the White Cliffs of Dover along the southern coast of England. Also, the great Egyptian pyramids are built of foraminiferan limestone.

The sarcodines have pseudopods for locomotion and feeding. *Amoeba proteus* lives in fresh water, but the shelled foraminiferans and radiolarians are very abundant in the ocean.

The Ciliates

The **ciliates (phylum Ciliophora)** such as those in the genus *Paramecium* are the most complex of the protozoans (Fig. 30.13). Hundreds of cilia, which beat in a coordinated rhythmic manner, project through tiny holes in a semirigid outer covering, or pellicle. Numerous oval capsules lying in the cytoplasm just beneath the pellicle contain **trichocysts.** Upon mechanical or chemical stimulation, trichocysts discharge long, barbed threads, useful for defense and for capturing prey. Toxicysts are similar, but they release a poison that paralyzes prey.

Most ciliates are **holozoic.** For example, when a paramecium feeds, food is swept down a gullet, below which food vacuoles form. Following digestion, the soluble nutrients are absorbed by the cytoplasm, and the indigestible residue is eliminated at the anal pore.

During asexual reproduction, ciliates divide by transverse binary fission. Ciliates have two types of nuclei: a large macronucleus and one or more small micronuclei. The macronucleus controls the normal metabolism of the cell, while the micronuclei are concerned with reproduction. Sexual reproduction involves conjugation. The macronucleus disintegrates and, after the micronuclei undergo meiosis, two ciliates exchange a haploid micronucleus. Then the micronuclei give rise to a new macronucleus, which contains copies of only certain housekeeping genes.

The diversity of ciliates is quite remarkable. The barrel-shaped didinia expand to consume paramecia much larger than themselves. *Suctoria* have an even more dramatic way of getting food. They rest quietly on a stalk until a hapless victim comes along. Then they promptly paralyze it and use their tentacles like straws to suck it dry. *Stentor* may be the prettiest ciliate. It resembles a giant blue vase decorated with stripes (see Fig. 30.13*a*).

Ciliates, which move by cilia, are diverse and very complex. The single cell has specialized regions to carry out various functions.

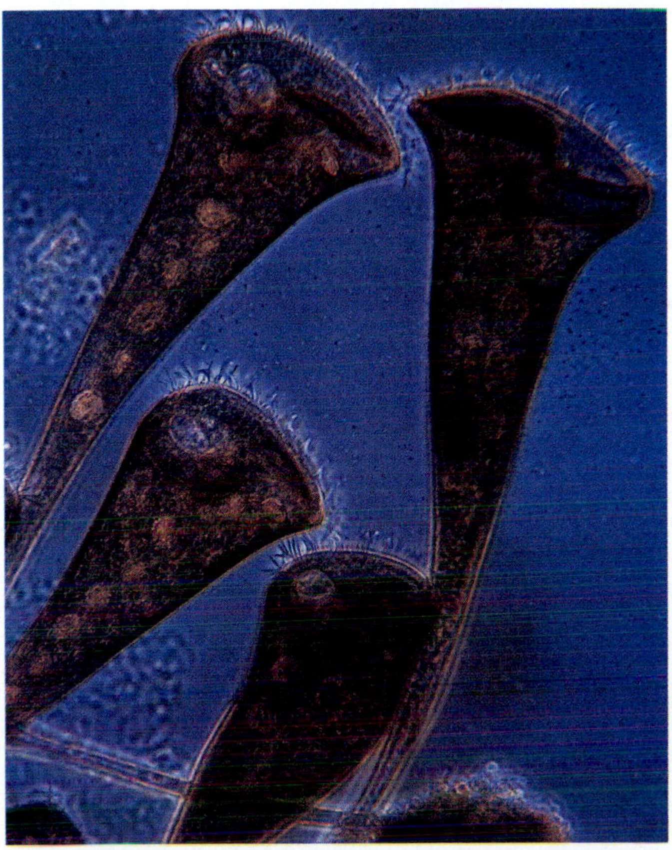

a. *Stentor* 200 µm

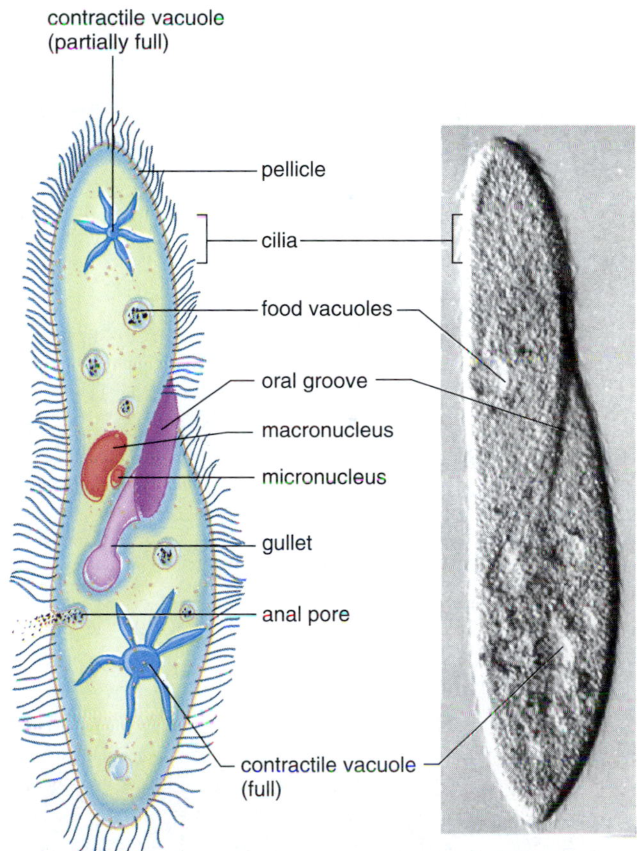

b. *Paramecium*

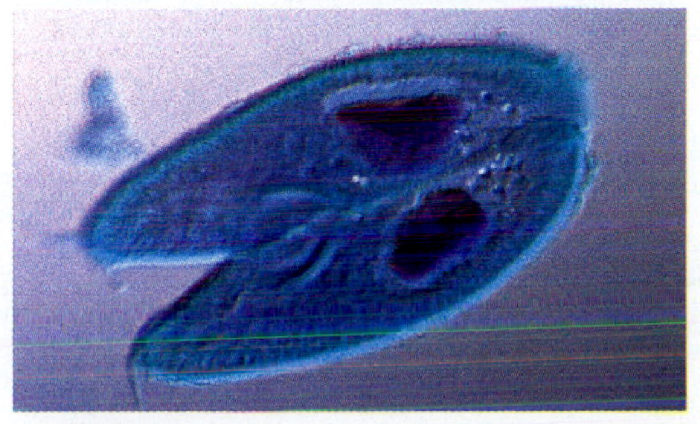

c. Conjugation

Figure 30.13 Ciliates.
a. *Stentor,* a large, vase-shaped, freshwater protist. **b.** Structure of *Paramecium,* adjacent to an electron micrograph. Ciliates are the most complex of the protists. Note the oral groove and the gullet and anal pore. **c.** A form of sexual reproduction called conjugation occurs every 700 generations or else the ciliate dies.

The Sporozoans

Sporozoans (phylum Apicomplexa) are nonmotile parasites. Their common name recognizes that these organisms form spores at some point in their life cycle. Their scientific name refers to a unique collection of organelles at one end of the infective motile stage.

Pneumocystis carinii causes the type of pneumonia seen primarily in AIDS patients. During sexual reproduction, thick-walled cysts form in the lining of pulmonary air sacs. The cysts contain spores that successively divide until the cyst bursts and the spores are released. Each spore becomes a new mature organism that can reproduce asexually but may also enter the sexual stage and form cysts.

The most widespread human parasite is *Plasmodium vivax,* the cause of one type of malaria. The life cycle alternates between a sexual and an asexual phase in different hosts. When a human is bitten by an infected female *Anopheles* mosquito, the parasite eventually invades the red blood cells. The chills and fever of malaria appear when the infected cells burst and release toxic substances into the blood (Fig. 30.14). Malaria is still a major killer of humans, despite extensive efforts to control it. A resurgence of the disease was caused primarily by the development of insecticide-resistant strains of mosquitoes and by parasites resistant to current antimalarial drugs.

Toxoplasma gondii, another sporozoan, causes toxoplasmosis, particularly in cats but also in people. In pregnant women the parasite can infect the fetus and cause birth defects; in AIDS patients it can infect the brain and cause neurological symptoms.

The malarial parasite *Plasmodium* is the best known of the sporozoans, but this phyla contains many examples of human parasites.

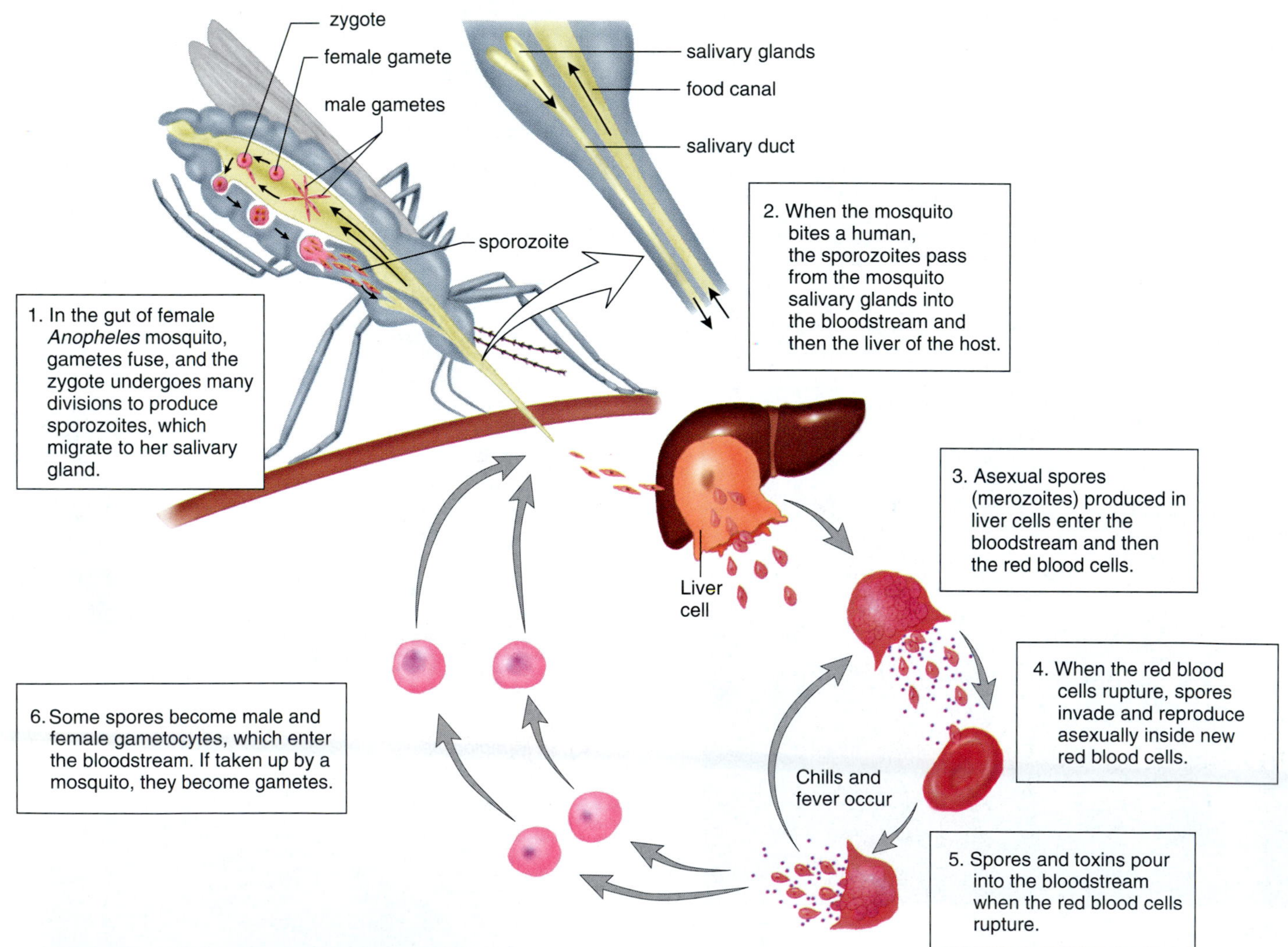
zygote
female gamete
male gametes
sporozoite
salivary glands
food canal
salivary duct
2. When the mosquito bites a human, the sporozoites pass from the mosquito salivary glands into the bloodstream and then the liver of the host.
1. In the gut of female Anopheles mosquito, gametes fuse, and the zygote undergoes many divisions to produce sporozoites, which migrate to her salivary gland.
3. Asexual spores (merozoites) produced in liver cells enter the bloodstream and then the red blood cells.
Liver cell
4. When the red blood cells rupture, spores invade and reproduce asexually inside new red blood cells.
Chills and fever occur
5. Spores and toxins pour into the bloodstream when the red blood cells rupture.
6. Some spores become male and female gametocytes, which enter the bloodstream. If taken up by a mosquito, they become gametes.

Figure 30.14 Life cycle of *Plasmodium vivax.*
Asexual reproduction occurs in humans, while sexual reproduction takes place within the *Anopheles* mosquito.

The Slime Molds and Water Molds

The organisms in this group were once classified as fungi but unlike fungi all have flagellated cells at some time during their life cycle. Only the water molds have a cell wall and it contains cellulose, not chitin as do the fungi. They all form spores; those of the water mold are 2n zoospores. Water molds are diploid and meiosis produces the gametes.

The evolutionary ties of these organisms are not strong and some may be more closely related to the amoeboids than to each other. Indeed the vegetative state of the slime molds is mobile and amoeboid! Like the amoeboids they ingest their food by phagocytosis.

The Plasmodial Slime Molds

Usually **plasmodial slime molds (phylum Myxomycota)** exist as a plasmodium, a diploid multinucleated cytoplasmic mass enveloped by a slime sheath that creeps along, phagocytizing decaying plant material in a forest or agricultural field (Fig. 30.15). The fan-shaped mass contains tubules formed by concentrated cytoplasm in which more liquefied cytoplasm streams. At times unfavorable to growth, such as during a drought, the plasmodium develops many sporangia. A **sporangium** [Gk. *spora*, seed, and *angeion* (dim. of *angos*), vessel] is a reproductive structure that produces spores. An aggregate of sporangia is called a fruiting body.

The spores produced by a sporangium can survive until moisture is sufficient for them to germinate. In plasmodial slime molds, spores release a haploid flagellated cell or an amoeboid cell. Eventually, two of them fuse to form a zygote that feeds and grows, producing a multinucleated plasmodium once again.

The Cellular Slime Molds

Cellular slime molds (phylum Acrasiomycota) are called such because they exist as individual amoeboid cells. They are common in soil where they feed on bacteria and yeast. Their small size prevents them from being seen.

As the food supply runs out, the cells release a chemical that causes them to aggregate into a pseudoplasmodium. The pseudoplasmodium stage is temporary and eventually gives rise to a fruiting body in which sporangia produce spores. When favorable conditions return, the spores germinate, releasing haploid amoeboid cells, and this asexual cycle begins again. A sexual cycle is known to occur under very moist conditions.

Slime molds, which produce nonmotile spores, are unlike fungi in that they have an amoeboid stage and are heterotrophic by ingestion.

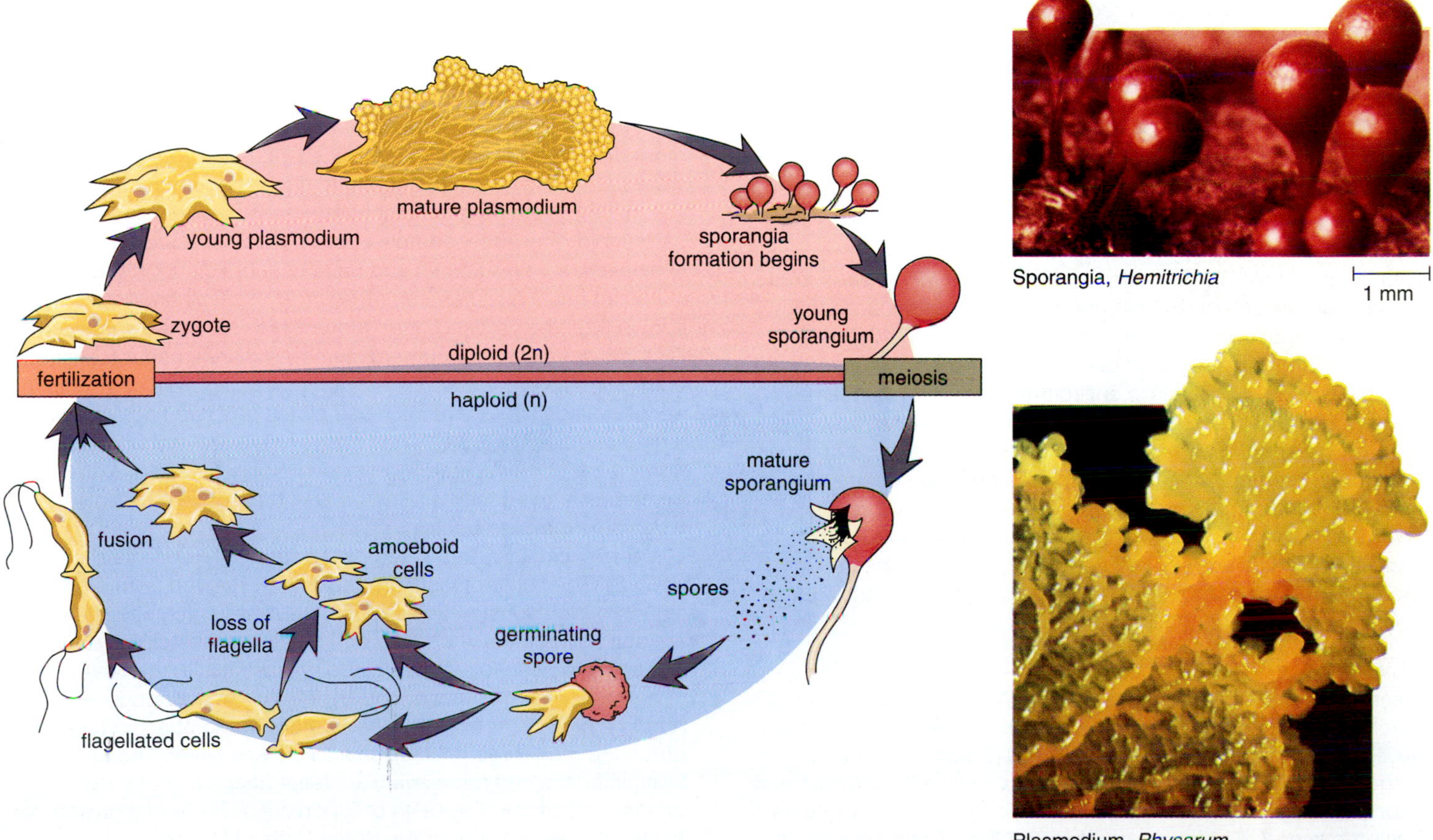

Sporangia, *Hemitrichia*

Plasmodium, *Physarum*

Figure 30.15 Plasmodial slime molds.
During sexual reproduction when conditions are unfavorable to growth, the diploid adult forms sporangia. Haploid spores germinate, releasing haploid amoeboid or flagellated cells that fuse.

The Water Molds

The **water molds (phylum Oomycota)** usually live in the water, where they form furry growths when they parasitize fishes and decompose remains. In spite of their common name, some water molds live on land and parasitize insects and plants. A water mold was responsible for the 1840s potato famine in Ireland. Most water molds are saprotrophic and live off dead organic matter, however.

Water molds have a filamentous body as do fungi, but their cell walls are largely composed of cellulose whereas fungi have cell walls of chitin. The life cycle of water molds differs from that of fungi. During asexual reproduction, water molds produce motile spores (2n zoospores), which are flagellated. The adult is diploid (not haploid as in the fungi), and meiosis produces gametes. Their phylum name refers to the enlarged tips (called oogonia) where eggs are produced.

Water molds have a filamentous body and are saprotrophic. The adult is diploid, and meiosis produces gametes; during asexual reproduction, there are motile zoospores.

Connecting Concepts

It's possible that the origin of sexual reproduction played a role in fostering the diversity of protists and their ability to inhabit water, soil, food, and even air. Unicellular *Chlamydomonas* usually reproduces asexually but when it does reproduce sexually, isogametes form. Most likely formation of isogametes represents an early stage in the development of gametogenesis and true syngamy—the formation of gametes and their union to form a 2n zygote. Some colonial forms of algae, like *Volvox,* and multicellular forms like the brown alga *Fucus* have a more specialized form of gametogenesis. They undergo oogamy: one gamete is larger and immobile, and the other is flagellated and motile. The green alga *Ulva,* and also the foraminiferans, have an alternation of generations in which only the haploid stage produces gametes.

Conjugation is another means to introduce genetic variation among offspring. *Spirogyra* undergoes conjugation but the ciliates have taken conjugation to another level. In ciliates, one haploid nucleus in each mating cell survives meiosis and divides again by mitosis. An exchange of haploid nuclei through a cytoplasmic channel restores diploidy and assures diversity of genes.

Reproduction in protists is adaptive in still another way. As a part of asexual and/or sexual reproduction, some protists form spores or cysts which can survive in a hostile environment. The cell emerging from a spore or cyst is often a reformed haploid or diploid organism depending on the protist. Spores and cysts can survive inclement conditions for years on end.

The diversity of reproduction among protists exemplifies why it is difficult to classify them and discover their evolutionary relationships to the fungi, plants, and animals. The protists we study today are distant cousins of these other eukaryotic groups, and therefore give only hints of what the ancestry of fungi, plants, and animals may have been.

Summary

30.1 General Biology of the Protists

Protists are in the domain Eukarya and the kingdom Protista. An endosymbiotic event may account for the presence of mitochondria, chloroplasts, and flagella in eukaryotic cells. Protists are generally unicellular but still are quite complex because they (1) have a combination of characteristics not seen in the other eukaryotic kingdoms, (2) have complicated life cycles that include the ability to withstand hostile environments, and (3) sometimes have organelles not seen in other types of eukaryotic cells.

Protists are of great ecological importance because in largely aquatic environments they are the producers that support communities of organisms. Protists also enter into various types of symbiotic relationships. Their great diversity makes it difficult to classify protists and, as yet, there is no general agreement about their classification. In this text, they have been arranged in groups convenient for discussion.

30.2 Diversity of the Protists

Green algae possess chlorophylls *a* and *b,* store reserve food as starch, and have cell walls of cellulose as do plants. Green algae are divided into those that are flagellated *(Chlamydomonas),* filamentous *(Spirogyra* and *Oedogonium),* multicellular *(Ulva),* and colonial *(Volvox).* The life cycle varies among the green algae. In most, the zygote undergoes meiosis and the adult is haploid. In this life cycle, both isogamy and oogamy are seen. *Ulva* has an alternation of generations like plants, but the sporophyte and gametophyte generations are similar in appearance; the gametes look alike, and the spores are flagellated.

Red algae are filamentous or multicellular seaweeds that are more delicate than brown algae, and are usually found in warmer waters. Like brown algae, red algae have economic importance.

Brown algae have chlorophylls *a* and *c* plus a brownish carotenoid. The large, complex brown algae, commonly called seaweeds, are well known and have economic importance.

Diatoms, which have an outer layer of silica, are better known than the golden-brown algae which are in the same phyla. Diatoms are extremely numerous in both marine and freshwater ecosystems.

Dinoflagellates have cellulose plates and two flagella, one at a right angle to the other. Their chloroplasts are probably derived from golden-brown algae by endosymbiosis. They are extremely numerous in the ocean and on occasion produce a neurotoxin when they form red tides.

Euglenoids are flagellated cells with a pellicle instead of a cell wall. Their chloroplasts are most likely derived from a green alga through endosymbiosis. Zooflagellates, which have flagella, are often symbiotic; *Trichonympha* helps termites digest wood, *Trypanosoma* causes African sleeping sickness in humans.

Amoeboids move and feed by forming pseudopods. In *Amoeba proteus,* food vacuoles form following phagocytosis of prey. Contractile vacuoles discharge excess water. The tests of foraminiferans and radiolarians cause a deep layer of sediment on the ocean floor. The tests of foraminiferans are responsible for the limestone deposits of the White Cliffs of Dover.

The ciliates move by their many cilia. They are remarkably diverse in form, and as exemplified by *Paramecium,* they show how complex a protist can be despite being a single cell.

The sporozoans are nonmotile parasites that form spores; *Plasmodium* causes malaria.

Slime molds, which produce nonmotile spores, are unlike fungi in that they have an amoeboid stage and are heterotrophic by ingestion. Water molds, which are filamentous and saprotrophic, are unlike fungi in that they produce flagellated 2n zoospores. Meiosis produces specialized gametes: sperm and egg.

Reviewing the Chapter

1. List and discuss three ways that protists are complex. 526
2. Describe the structure of *Chlamydomonas* and *Volvox;* contrast how they reproduce. 528, 530
3. Describe the structure of *Spirogyra* and *Oedogonium;* contrast how they reproduce. 529
4. Describe the structure of *Ulva,* and explain how its life cycle differs from that of plants. 529
5. Describe the structure of red algae and discuss their economic importance. 530
6. Describe the structure of brown algae and discuss their ecological and economic importance. 531
7. Describe the structure of diatoms and dinoflagellates. What is a red tide? 533
8. Describe the unique structure of euglenoids. 534
9. Name several illnesses caused by zooflagellates. 535
10. Define the term protozoan and list phyla that are sometimes included among the protozoans. 535
11. Distinguish between the sarcodines: amoeboids, foraminiferans, and radiolarians. 536
12. How are ciliates like and different from amoeboids? 537
13. Describe the life cycle of *Plasmodium vivax,* the causative agent of malaria. 538
14. What features distinguish slime molds and water molds from fungi? Describe the life cycle of a plasmodial slime mold. 539–40

Testing Yourself

Choose the best answer for each question.

1. Which of these is not a green alga?
   a. *Volvox*
   b. *Fucus*
   c. *Spirogyra*
   d. *Chlamydomonas*
   e. *Ulva*
2. Which is not a characteristic of brown algae?
   a. multicellular
   b. chlorophylls *a* and *b*
   c. lives along rocky coast
   d. harvested for commercial reasons
   e. contains a brown pigment
3. In *Chlamydomonas,*
   a. the adult is haploid.
   b. the zygospore survives times of stress.
   c. sexual reproduction occurs.
   d. asexual reproduction occurs.
   e. All of these are correct.
4. Which is found in *Ulva* but not in plants?
   a. alternation of generations
   b. generations look alike
   c. sperm and egg are produced
   d. protection of zygote
   e. Both b and c are correct.
5. Which of these algae are not flagellated?
   a. *Volvox*
   b. *Spirogyra*
   c. dinoflagellates
   d. *Chlamydomonas*
   e. trypanosome
6. Which is mismatched?
   a. diatoms—silica shell, boxlike, golden brown
   b. euglenoids—flagella, pellicle, eyespot
   c. *Fucus*—adult is diploid, seaweed, chlorophylls *a* and *c*
   d. *Paramecium*—cilia, calcium carbonate shell, gullet
   e. foraminiferan—test, pseudopod, digestive vacuole
7. Which is a false statement?
   a. Only heterotrophic and not photosynthetic protists are flagellated.
   b. Among protists, both flagellates and sporozoans are symbiotic.
   c. Among the protists, only green algae have a sexual life cycle.
   d. Ciliates exchange genetic material during conjugation.
   e. Slime molds have an amoeboid stage.
8. Which is mismatched?
   a. trypanosome—African sleeping sickness
   b. *Plasmodium vivax*—malaria
   c. amoeboid—severe diarrhea
   d. AIDS—*Giardia lamblia*
   e. dinoflagellates—coral
9. Which is found in slime molds but not fungi?
   a. nonmotile spores
   b. flagellated cells
   c. zygote formation
   d. photosynthesis
   e. chitin in cell walls
10. Label this diagram of the *Chlamydomonas* life cycle.

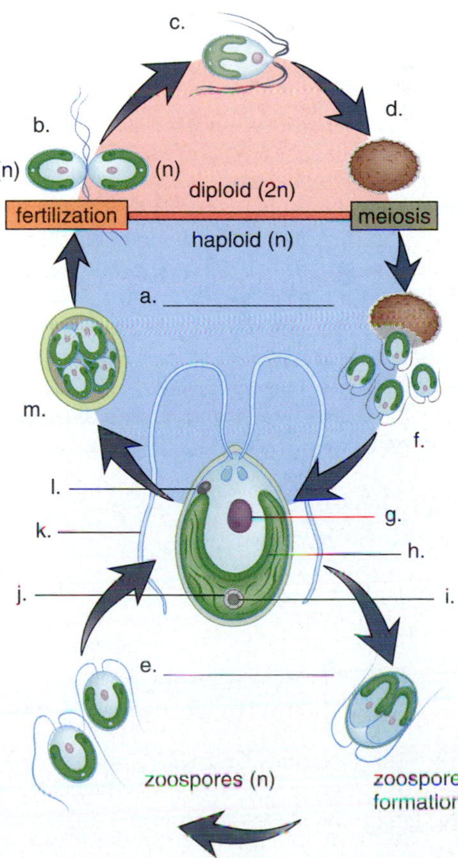

Thinking Scientifically

1. While studying a unicellular algae you discover a mutant in which the daughter cells do not separate after mitosis. This gives you an idea about how filamentous algae may have evolved. You hypothesize that the mutant alga is missing a protein or making a new form of a protein. How might each possibly lead to a filamentous appearance?
2. You are trying to develop a new antitermite chemical that will not harm environmentally beneficial insects. Since termites are adapted to eat only wood, they will starve if they cannot digest this food source. What are two possible non-insect targets for your chemical that would result in the death of termites?

Understanding the Terms

algae (sing., alga) 528
amoeboid 536
brown algae 531
cellular slime mold 539
ciliate 537
colony 530
conjugation 529
cyst 526
diatom 533
dinoflagellate 533
euglenoid 534
filament 529
foraminiferan 536
green algae 528
holozoic 537
isogamy 528
oogamy 528
phagocytize 536
phytoplankton 533
plankton 526
plasmodial slime mold 539
protist 526
protozoan 535
pseudopod 536
radiolarian 536
red algae 530
seaweed 531
sporangium 539
spore 528
sporozoan 538
test 536
trichocyst 537
trypanosome 535
water mold 540
zooflagellate 535
zooplankton 536

Match the terms to these definitions:

a. ______________ Cytoplasmic extension of amoeboid protists; used for locomotion and engulfing food.
b. ______________ Flexible freshwater unicellular organism that usually contains chloroplasts and is flagellated.
c. ______________ Freshwater or marine unicellular protist with a cell wall consisting of two silica-impregnated valves, which is extremely numerous in phytoplankton.
d. ______________ Causes severe diseases in human beings and domestic animals, including a condition called sleeping sickness.
e. ______________ Freshwater and marine organisms that are suspended on or near the surface of the water.

Web Connections

Exploring the Internet

http://www.mhhe.com/biosci/genbio/mader
(click on *Biology 7/e*)

The *Biology 7/e* Online Learning Center provides many resources for studying the material in this chapter including links to the following sites:

See the beauty and complexity of different types of algae in AAI's Phytoplankton Images Library.

http://www.paulsmiths.edu/aai/phyto.html

The World of Algae. Lots of links, and information on algae.

http://www.botany.uwc.ac.za/algae/

Society of Protozoologists. This site contains links to images of protozoans, databases, culture collections, and discussion groups.

http://www.uga.edu/~protozoa/

National Center for Infectious Diseases. This CDC site has many links to information on bacterial, viral, protozoan, and worm-related diseases (primarily affecting humans).

http://www.cdc.gov/ncidod/diseases/index.htm

Other specific links:

Trypanosomiasis.

http://www.cdc.gov/ncidod/dpd/list_try.htm

Malaria.

http://www.cdc.gov/ncidod/dpd/list_mal.htm

The Fungi

chapter

chapter concepts

31.1 Characteristics of Fungi

- Fungi are saprotrophic detritivores that aid the cycling of inorganic nutrients in ecosystems. 544
- The body of a fungus is multicellular; it is composed of thin filaments called hyphae. 544
- As an adaptation to life on land, fungi produce nonmotile and often windblown spores during asexual and sexual reproduction. 545

31.2 Classification of Fungi

- Fungi are classified according to aspects of their sexual life cycle. Zygospore fungi have a dormant stage consisting of a thick-walled zygospore. 546
- During sexual reproduction of sac fungi, saclike cells (asci) produce spores. Usually, asci are located in fruiting bodies. 549
- During sexual reproduction of club fungi, club-shaped structures (basidia) produce spores. Basidia are located in fruiting bodies. 551
- The imperfect fungi always reproduce asexually by conidiospores; sexual reproduction has not yet been observed in these organisms. 553

31.3 Symbiotic Relationships of Fungi

- Lichens, which may live in stressful environments, are an association between a fungus and a cyanobacterium or a green alga. The fungus may be somewhat parasitic on the alga. 553
- Mycorrhizae is an association between a fungus and the roots of a plant, such that the fungus helps the plant absorb minerals, and the plant supplies the fungus with carbohydrates. 554

Fly agaric mushroom, *Amanita muscaria,* Alaska

What do LSD, athlete's foot, and homemade bread have in common? They all involve species from the kingdom Fungi. You can find fungi growing on tree trunks and in the backyard after a rain. They resemble plants, but they lack chloroplasts and do not photosynthesize. Yet, fungi clearly aren't animals, nor do they resemble bacteria or protozoans. Based on their multicellular nature and mode of nutrition, Whittaker placed fungi in their own kingdom. You probably know that mushrooms are fungi, and you might also know that mildew, molds, and morels are fungi, too. From this litany you would think a fungus couldn't be very large. But researchers claim to have found one that spreads over 1,500 acres in Washington State. As fungi spread, they feed off the organic remains of plants and animals. Dead leaves, tree trunks, and even carcasses of animals are their daily fare. They, along with bacteria, are detritivores that enrich the immediate environs with inorganic nutrients and thereby keep chemicals cycling in ecosystems. Trees grow, grass turns green, and flowers appear—all because fungi, while digesting for themselves, have also digested for others.

31.1 Characteristics of Fungi

Kingdom Fungi contains the **fungi,** which are mostly multicellular eukaryotes of varied structure that share a common mode of nutrition. Like animals, they are heterotrophic and consume preformed organic matter. Animals, however, are heterotrophic by ingestion; fungi are heterotrophic by absorption. Their cells send out digestive enzymes into the immediate environment and then, when organic matter is broken down, the cells absorb the resulting nutrient molecules.

Most fungi are saprotrophic detritivores that decompose the waste products and dead remains of plants and animals. Some fungi are parasitic; they live off the tissues of living plants and animals. Fungi can enter leaves through the stomates, and this makes plants especially subject to fungal diseases. Fungal diseases account for millions of dollars in crop losses each year, and they have greatly reduced the numbers of various types of trees. They also cause human diseases such as ringworm, athlete's foot, and yeast infections.

Several types of fungi have a mutualistic relationship with the roots of seed plants; they acquire inorganic nutrients for plants, and in return they are given organic nutrients. Others form an association with a green alga or cyanobacterium within a lichen. Lichens can live on rocks and are important soil formers.

Fungi carry on external digestion and are heterotrophic by absorption. They are saprotrophic or parasitic and form symbiotic relationships.

Structure of Fungi

Fungi can be unicellular—yeast are the best known example. But the thallus (body) of most fungi is a multicellular structure known as a mycelium (Fig. 31.1). A **mycelium** [Gk. *mycelium,* fungus filaments] is a network of filaments called **hyphae** [Gk. *hyphe,* web]. Hyphae give the mycelium quite a large surface area per volume of cytoplasm, and this facilitates absorption of nutrients into the body of a fungus. Hyphae grow at their tips, and the mycelium absorbs and then passes nutrients on to the growing tips. When a fungus reproduces, a specific portion of the mycelium becomes a reproductive structure that is then nourished by the rest of the mycelium.

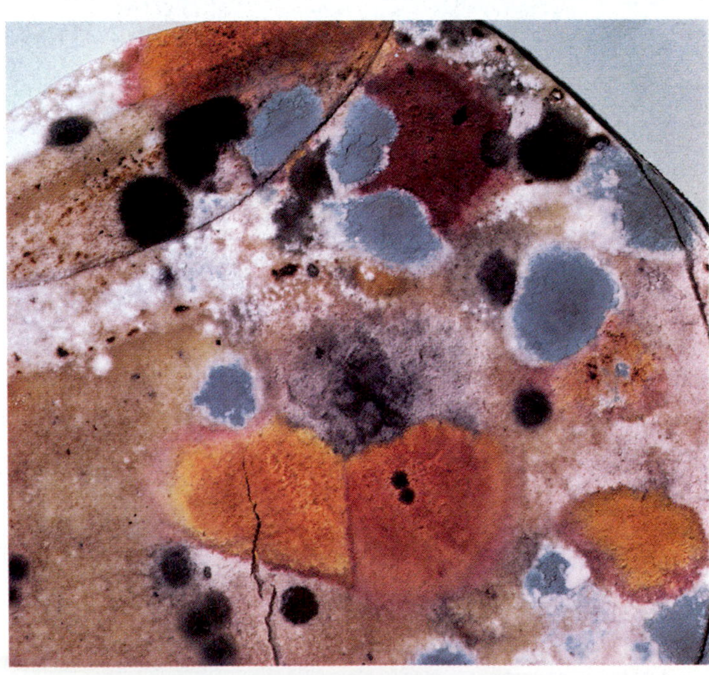

a. Fungal mycelium

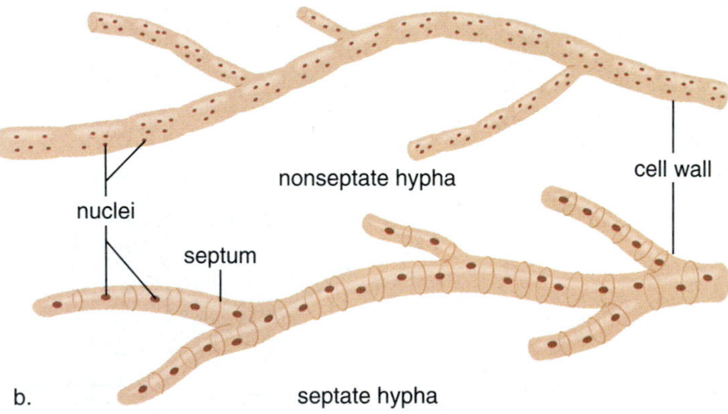

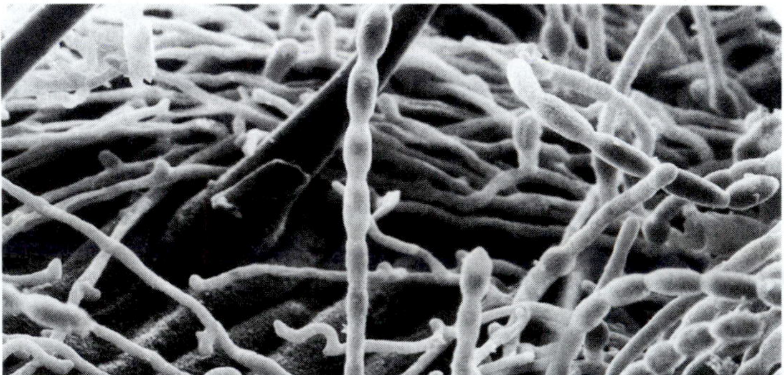

c. Fungal hyphae

Figure 31.1 Mycelium of fungi.
a. On a corn tortilla, each mycelium grown from a different spore is quite symmetrical. **b.** Hyphae are either nonseptate (do not have cross walls) or septate (have cross walls). **c.** Scanning electron micrograph showing fungal hyphae.

Fungal cells are quite different from plant cells not only by lacking chloroplasts but also by having a cell wall that contains chitin and not cellulose. Chitin, like cellulose, is a polymer of glucose organized into microfibrils. In chitin, however, each glucose molecule has a nitrogen-containing amino group attached to it. Chitin is also found in the external skeleton of insects and all arthropods. The energy reserve of fungi is not starch but glycogen, as in animals. Fungi are nonmotile; they lack basal bodies and do not have flagella at any stage in their life cycle. They move toward a food source by growing toward it. Hyphae can cover as much as a kilometer a day!

Some fungi have no cross walls in their hyphae. These hyphae are called **nonseptate**—septum (L. *septum,* fence, wall) is the term used for cross walls in fungi. Nonseptate fungi are multinucleated; they have many nuclei in the cytoplasm of a hypha. **Septate** fungi have cross walls, but actually the presence of septa makes little difference because pores allow cytoplasm and even sometimes organelles to pass freely from one cell to the other. The septa that separate reproductive cells, however, are complete in all fungal groups.

The nonmotile body of a fungus is made up of hyphae that form a mycelium.

Reproduction of Fungi

In general, fungal sexual reproduction involves these stages:

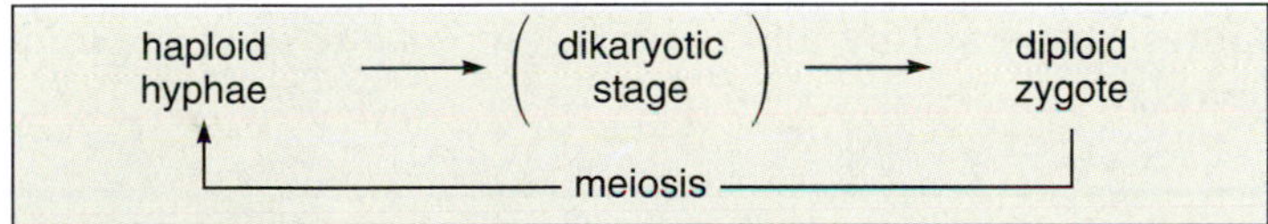

The relative length of time of each phase varies with the species.

During sexual reproduction, hyphae (or a portion thereof) from two different mating types make contact and fuse. You would expect the nuclei from the two mating types to also fuse immediately, and they do in some species. In other species the nuclei pair but do not fuse for days, months, or even years. The nuclei continue to divide in such a way that every cell (in septate hyphae) has at least one of each nucleus. A hypha that contains paired haploid nuclei is said to be n + n or **dikaryotic** [Gk. *dis,* two, and *karyon,* nucleus, kernel]. When the nuclei do eventually fuse, the zygote undergoes meiosis prior to spore formation. Fungal spores germinate directly into haploid hyphae without any noticeable embryological development.

How can a terrestrial and nonmotile organism ensure that the species will be dispersed to new locations? As an adaptation to life on land, fungi produce nonmotile, but usually windblown, spores during both sexual and asexual reproduction (Fig. 31.2). A **spore** is a reproductive cell that can grow directly into a new organism.

Asexual reproduction usually involves the production of spores by a single mycelium. Alternately, asexual reproduction can occur by fragmentation—a portion of a mycelium begins a life of its own. Also, unicellular yeast reproduce asexually by **budding;** a small cell forms and gets pinched off as it grows to full size (see Fig. 31.5).

a.

b.

Figure 31.2 Dispersal of spores.
a. Electron micrograph shows the external appearance of spores discharged by (**b**) earth star, *Geastrum,* which releases hordes of spores into the air.

Part *a.* From C. Y. Shih and R. G. Kessel *Living Images.* Science Books International, Boston, 1982.

31.2 Classification of Fungi

CLASSIFICATION

Kingdom Fungi

Multicellular eukaryote; heterotrophic by absorption; lack flagella; nonmotile spores form during both asexual and sexual reproduction

Division Zygomycota: zygospore fungi

Division Ascomycota: sac fungi

Division Basidiomycota: club fungi

Division Deuteromycota: imperfect fungi (i.e., means of sexual reproduction not known)

The ancestry of fungi, which evolved about 570 million years ago, has not been determined. It is possible that the organisms now classified in the **kingdom Fungi** do not share a recent common ancestor. In that case, they probably evolved separately from different protists. It's been suggested, though, that fungi evolved from red algae because both fungi and red algae lack flagella in all stages of the life cycle.

In the past, fungi have been classified in other kingdoms. Originally they were considered a part of the plant kingdom, and then they were placed in the kingdom Protista. But in 1969, R. H. Whittaker argued that their multicellular nature and mode of nutrition (extracellular digestion and absorption of nutrients) meant they should be in their own kingdom. Table 31.1 contrasts the features of fungi with the features of certain other organisms now included in kingdom Protista by this text. In the absence of knowledge of their evolutionary relations, fungal groups are classified according to differences in the life cycle and the type of structure that produces spores. Recently scientists have been using comparative DNA (deoxyribonucleic acid) data and analysis of enzymes and other proteins to decipher evolutionary relationships, but as yet this has not changed the classification of fungi.

Zygospore Fungi

The **zygospore fungi (division Zygomycota,** 665 species) are mainly saprotrophs living off plant and animal remains in the soil or bakery goods in your own pantry. Some are parasites of small soil protists or worms, and even insects such as the housefly.

The black bread mold, *Rhizopus stolonifer,* is commonly used as an example of this division. The body of this fungus, which is composed of nonseptate hyphae, demonstrates that although there is little cellular differentiation among fungi, the hyphae may be specialized for various purposes. In *Rhizopus,* stolons are horizontal hyphae that exist on the surface of the bread; rhizoids grow into the bread, anchor the mycelium and carry out digestion; and sporangiophores are stalks that bear sporangia. A **sporangium** is a capsule that produces spores called sporangiospores. During asexual reproduction all structures involved are haploid (Fig. 31.3).

The division name refers to the zygospore, which is seen during sexual reproduction. The hyphae of opposite mating types, termed plus (+) and minus (−), are chemically attracted, and they grow toward each other until they touch. The ends of the hyphae swell as nuclei enter; then cross walls develop a short distance behind each end, forming gametangia. The gametangia merge and the result is a large multinucleate cell in which nuclei of the two mating types pair and then fuse. A thick wall develops around the cell, which is now called a **zygospore.** The zygospore undergoes a period of dormancy before meiosis and germination takes place. One or more sporangiophores with sporangia at their tips develop, and many spores are released. The spores, dispersed by air currents, give rise to new haploid mycelia.

Zygospore fungi produce spores within sporangia. During sexual reproduction, a zygospore forms prior to meiosis and production of spores.

Table 31.1

Certain Protists Compared to Fungi

Feature	Fungi	Red Algae	Plasmodial Slime Molds	Water Molds
Body form	Filamentous	Filamentous	Multinucleate plasmodium	Filamentous
Mode of nutrition	Heterotrophic by absorption	Autotrophic by photosynthesis	Heterotrophic by absorption	Heterotrophic by absorption
Basal bodies/flagella	In no stages	In no stages	In one stage	Flagellated zoospores
Cell wall	Contains chitin	Contains cellulose	None	Contains cellulose
Life cycle	Zygotic meiosis (haplontic cycle)	Sporic meiosis (alternation of generations)	Unique	Gametic meiosis (diplontic cycle)

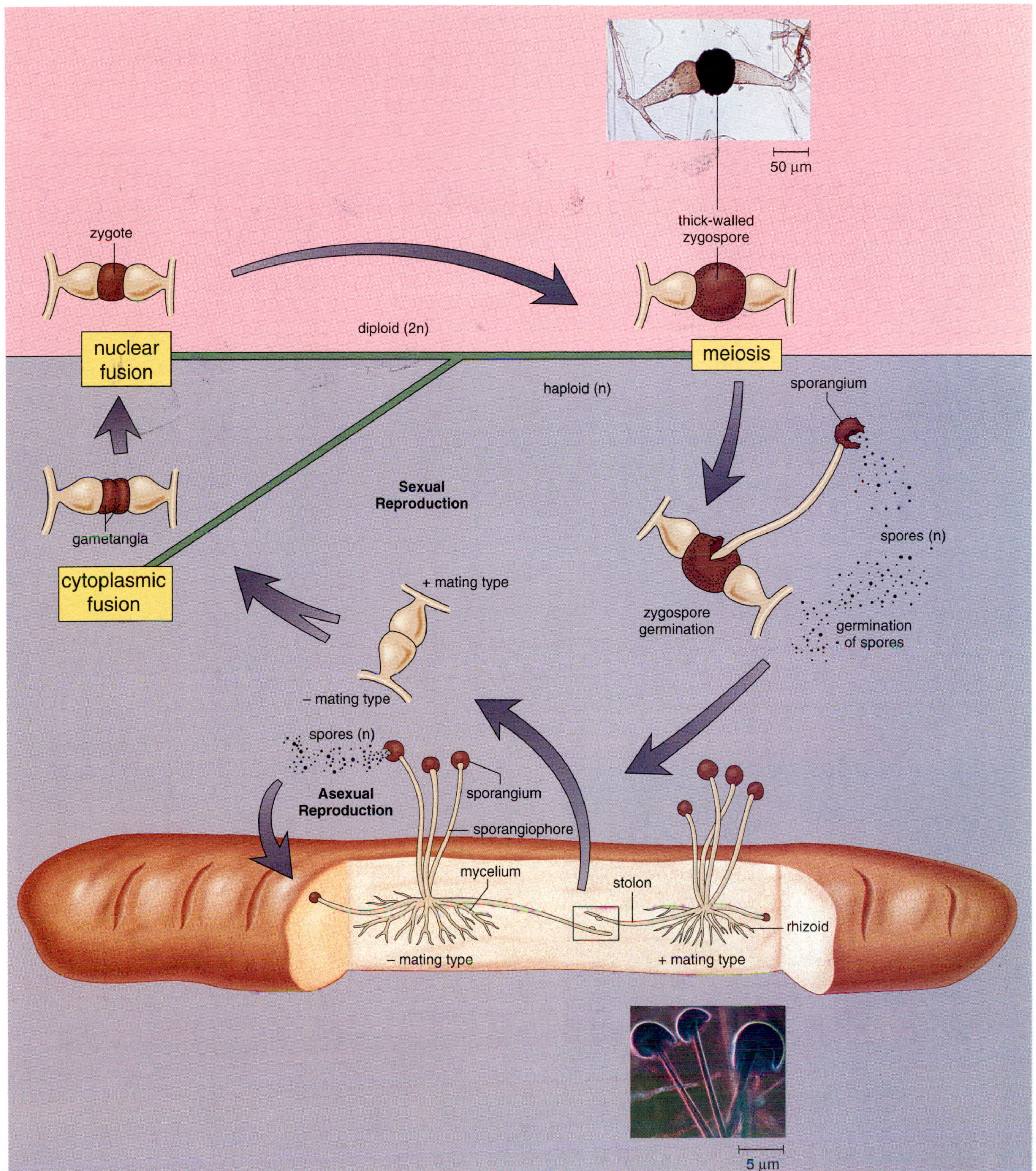

Figure 31.3 **Black bread mold, *Rhizopus stolonifer*.**
Asexual reproduction is the norm. As a result of sexual reproduction, the adult is haploid due to zygomeiosis, which occurs before or as the sporangiospores are produced. After two compatible mating types make contact, first gametangia fuse and then the nuclei fuse. The zygospore is a resting stage that can survive unfavorable growing conditions.

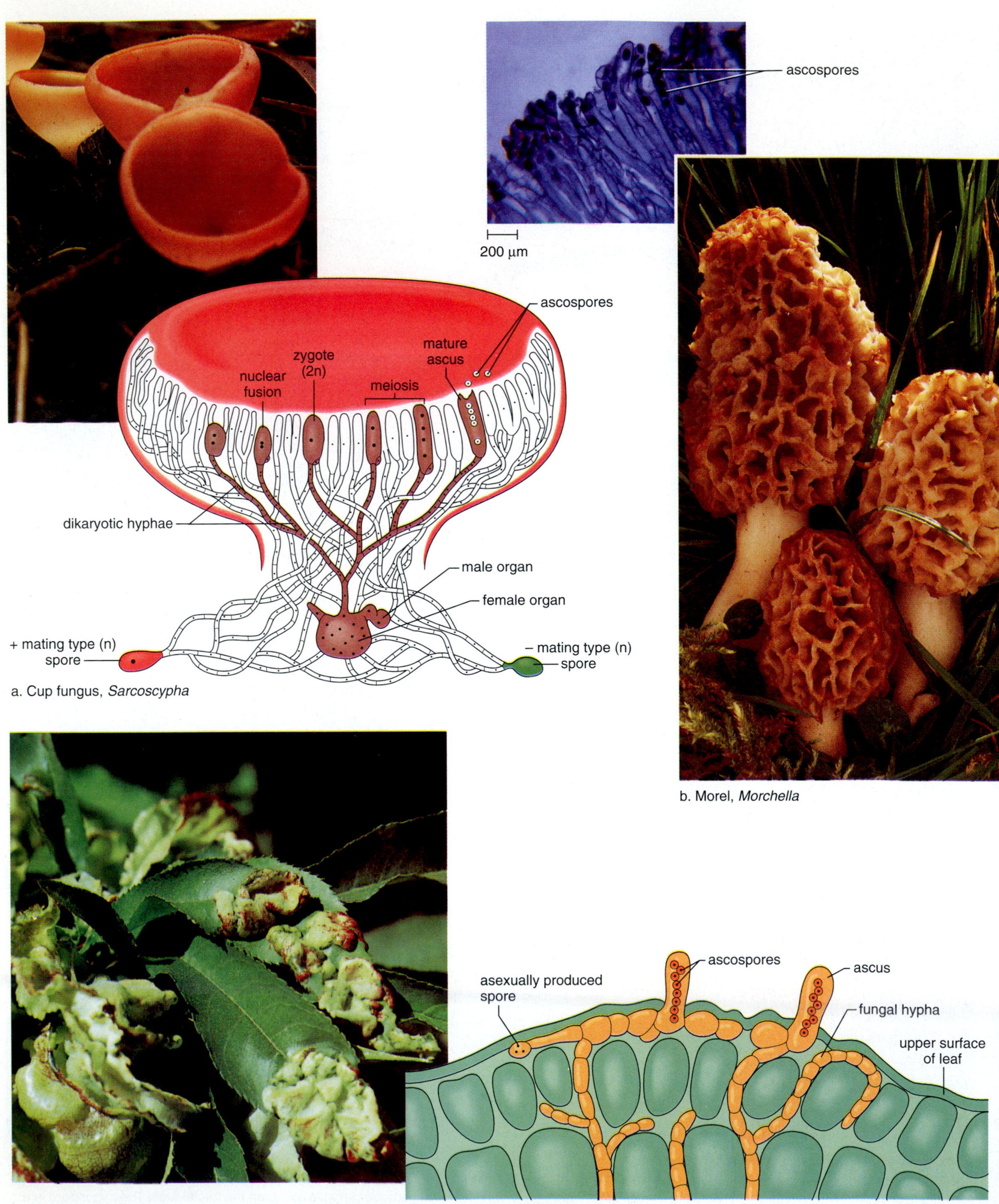

Figure 31.4 **Sac fungi.**
a. Cup fungi; photo and drawing of ascocarp section. Dikaryotic hyphae terminate in the ascocarp and develop asci where meiosis follows nuclear fusion and spore formation takes place. **b.** Morel; photo and light micrograph, ascocarp section. **c.** Peach leaf curl; photo and drawing.

Sac Fungi

Most **sac fungi (division Ascomycota,** 30,000 species) are saprotrophs that play an essential ecological role by digesting resistant (not easily decomposed) materials containing cellulose, lignin, or collagen. Red bread molds (e.g., *Neurospora*) are ascomycetes, as are cup fungi, morels, and truffles. Morels and truffles are highly prized as gourmet delicacies. A large number of ascomycetes are parasitic on plants; powdery mildews grow on leaves, as do leaf curl fungi; chestnut blight and Dutch elm disease destroy the trees named. Ergot, a parasitic sac fungus that infects rye and (less commonly) other grains, is discussed in the reading on page 552.

Yeasts are unicellular, but most ascomycetes are composed of septate hyphae. Their division name refers to the **ascus** [Gk. *askos*, bag, sac], a fingerlike sac that develops during sexual reproduction. Ascus-producing hyphae remain dikaryotic (Fig. 31.4) except in the walled-off portion that becomes the ascus where nuclear fusion, meiosis, and ascospore formation take place. Each ascus contains eight haploid nuclei and produces eight ascospores.

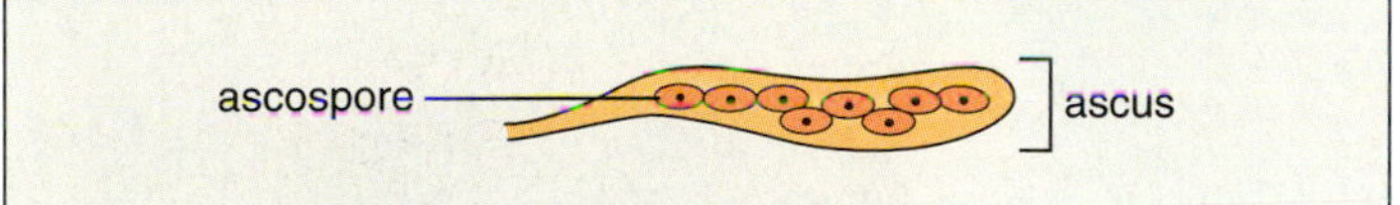

The asci are usually surrounded and protected by sterile hyphae within a fruiting body called an ascocarp. A **fruiting body** is a reproductive structure where spores are produced and released. Ascocarps can have different shapes; in cup fungi they are cup shaped, in molds they are flask shaped, and in morels they are stalked and crowned by bell-shaped convoluted tissue that bears the asci. In most ascomycetes, the asci become swollen as they mature and then they burst, expelling the ascospores. If released into the air, the spores are then windblown.

Asexual reproduction, which is the norm among ascomycetes, involves the production of spores called **conidiospores** [Gk. *konis*, dust, and *spora*, seed], or conidia, which vary in size and shape and may be multicellular. There are no sporangia in ascomycetes, and the conidiospores develop directly on the tips of modified aerial hyphae called conidiophores (see Fig. 31.8). When released, they are windblown.

Sac fungi usually produce asexual conidiospores. During sexual reproduction, asci within a fruiting body produce spores.

Yeasts

Yeasts are unicellular organisms that reproduce asexually either by mitosis or by budding. *Saccharomyces cerevisiae*, brewer's yeast, is representative of budding yeasts: a small cell forms and gets pinched off as it grows to full size (Fig. 31.5). Sexual reproduction, which occurs when the food supply runs out, results in the formation of asci and ascospores. Ascospores from two different mating types can fuse, and the result is a diploid cell that will reproduce asexually before meiosis occurs and ascospores are produced again. The haploid ascospores function directly as new yeast cells.

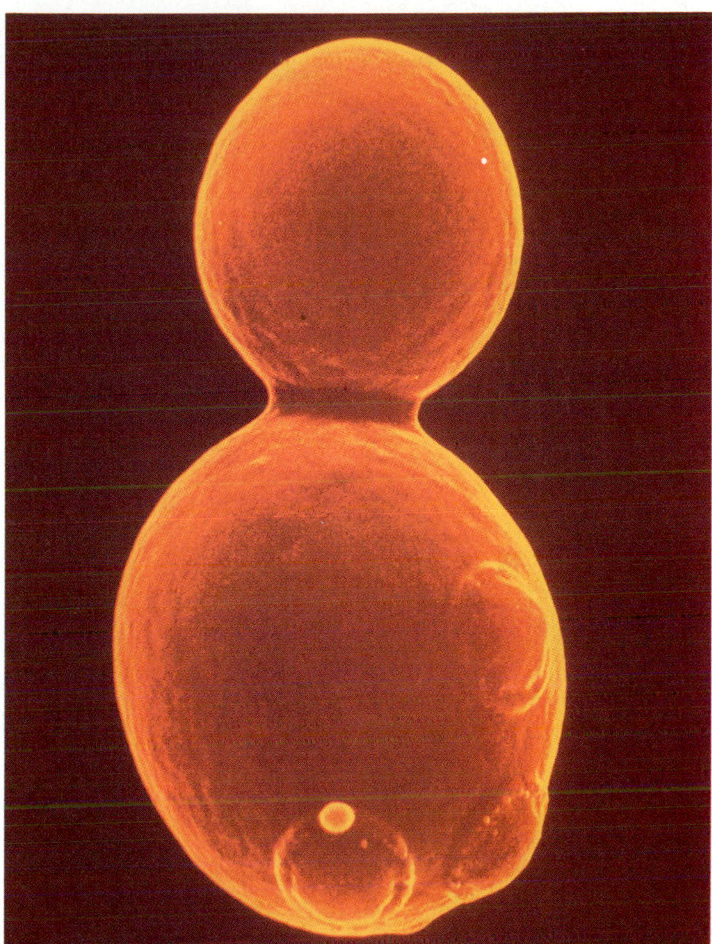

Figure 31.5 Yeast cells.
Saccharomyces cerevisiae is brewer's yeast, which reproduces asexually by budding.

When yeasts ferment, they produce ethanol and carbon dioxide. In the wild, yeasts grow on fruits, and historically the yeasts already present on grapes were used to produce wine. Today selected yeasts are added to relatively sterile grape juice in order to make wine. Also, yeasts are added to prepared grains to make beer. Both the ethanol and the carbon dioxide are retained for beers and sparkling wines; it is released for still wines. In baking, the carbon dioxide given off by yeast is the leavening agent that causes bread to rise.

Yeasts are serviceable to humans in another way. They have become the model of choice in genetic engineering experiments requiring a eukaryote. *Escherichia coli*, the usual model, is a prokaryote and does not function during protein synthesis as a eukaryote would.

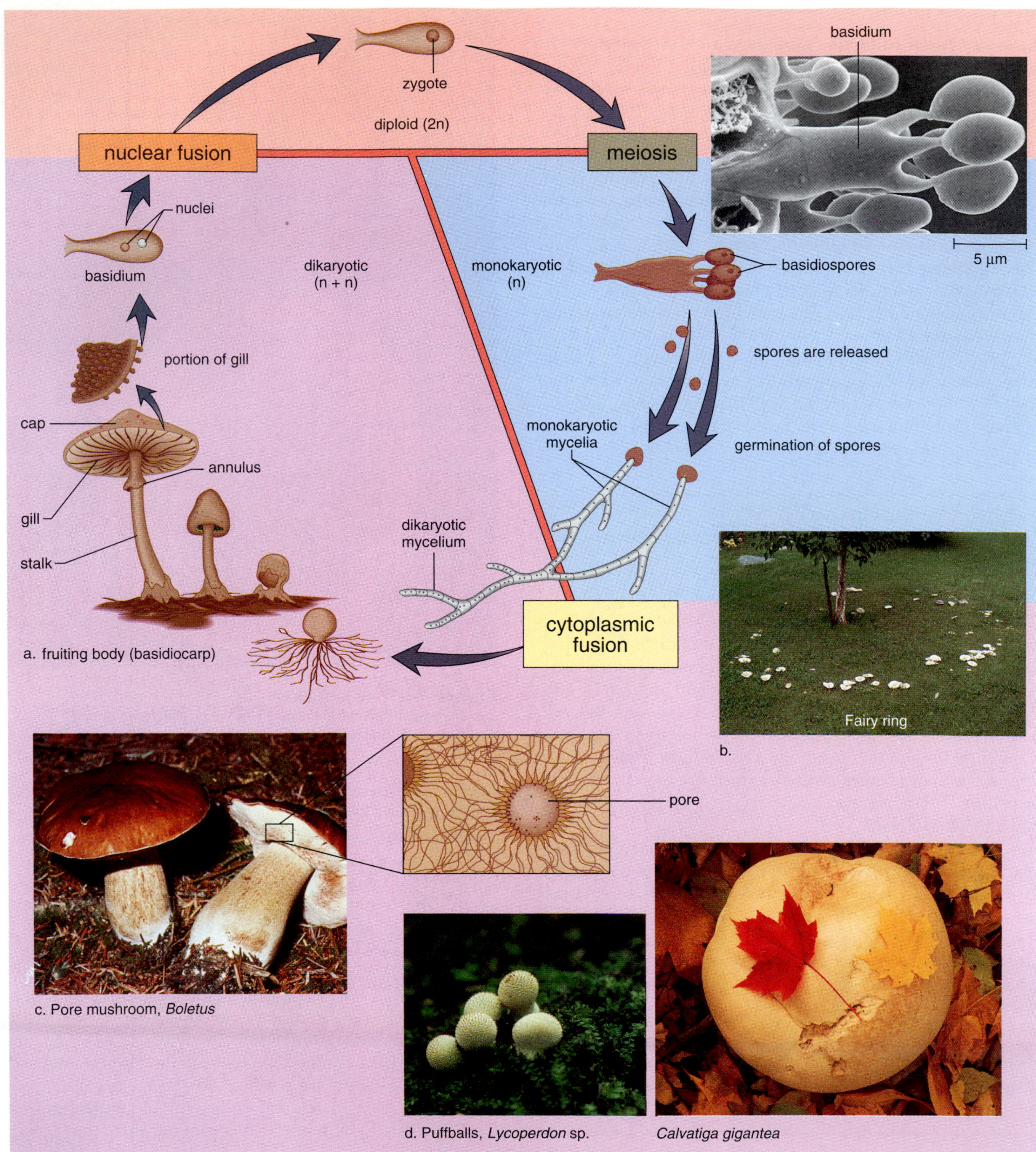

a. fruiting body (basidiocarp)

b.

c. Pore mushroom, *Boletus*

d. Puffballs, *Lycoperdon* sp.

Calvatiga gigantea

Figure 31.6 Club fungi.

a. Life cycle of a mushroom. Sexual reproduction is the norm. After hyphae from two compatible mating types fuse, the dikaryotic mycelium is long-lasting. Nuclear fusion results in a diploid nucleus within each basidium on the gills of the fruiting body shown. Zygomeiosis and production of basidiospores follow. Germination of a spore results in a haploid mycelium. **b.** Fairy ring. Mushrooms develop in a ring on the outer living fringes of a dikaryotic mycelium. The center has used up its nutrients and is no longer living. **c.** Fruiting bodies of *Boletus.* This mushroom is not gilled; instead it has basidia-lined tubes that open on the undersurface of the cap. **d.** In puffballs the spores develop inside an enclosed fruiting body. Giant puffballs are estimated to contain 7 trillion spores.

Club Fungi

Club fungi (**division Basidiomycota,** 16,000 species), which have septate hyphae, include the familiar mushrooms growing on lawns and the shelf or bracket fungi found on dead trees. Less well known are puffballs, bird's nest fungi, and stinkhorns. These structures are all fruiting bodies called basidiocarps. Basidiocarps contain the **basidia** [L. *basidium* (dim. originating from Gk. *basis*), pedestal], club-shaped structures that produce basidiospores and from which this division takes its name.

Although club fungi occasionally do produce conidiospores asexually, they usually reproduce sexually. When monokaryotic (n) hyphae of two different mating types meet, they fuse, and a dikaryotic (n + n) mycelium results (Fig. 31.6). The dikaryotic mycelium continues its existence year after year, even hundreds of years on occasion. In mushrooms, the dikaryotic mycelium often radiates out and produces basidiocarps in an ever larger so-called fairy ring. Basidiocarps are composed of nothing but tightly packed hyphae whose walled-off ends become the club-shaped basidia. In the gilled mushrooms, the hyphae terminate in radiating lamellae, and in pore mushrooms and shelf fungi the hyphae terminate in tubes. In any case, the extensive surface area is lined by basidia where nuclear fusion, meiosis, and spore production occur. A basidium has four projections into which cytoplasm and a haploid nucleus enter as the basidiospore forms.

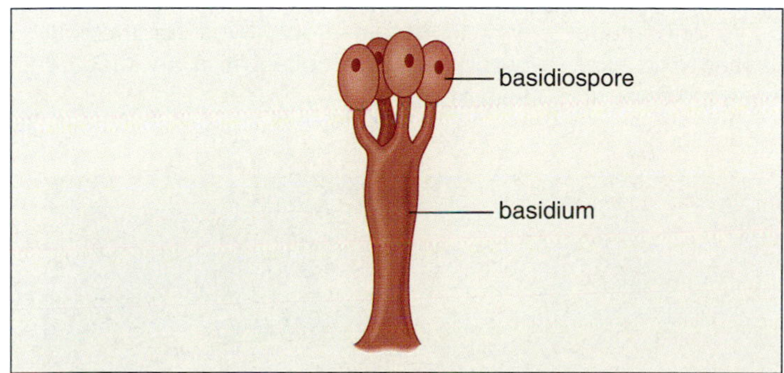

Basidiospores are often windblown; when they germinate, a new haploid mycelium forms.

In puffballs, spores are produced inside parchmentlike membranes, and the spores are released through a pore or when the membrane breaks down. In bird's nest fungi, falling raindrops provide the force that causes the nest's basidiospore-containing "eggs" to fly through the air and land on vegetation that may be eaten by an animal. If so, the spores pass unharmed through the digestive tract. Stinkhorns resemble a mushroom with a spongy stalk and a compact, slimy cap. The long stalk bears the elongated basidiocarp. Stinkhorns emit an incredibly disagreeable odor; flies are attracted by the odor, and when they linger to feed on a sweet jelly, the flies pick up spores that they later distribute.

Club fungi usually reproduce sexually. The dikaryotic stage is prolonged and periodically produces fruiting bodies where spores are produced in basidia.

a. Corn smut, *Ustilago*

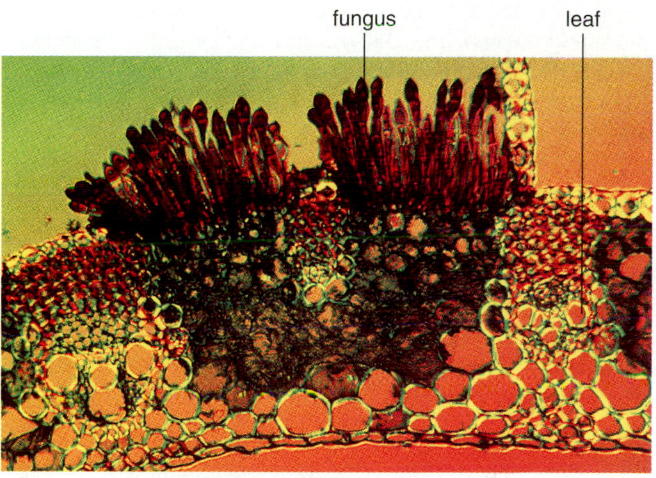

b. Wheat rust, *Puccinia*

Figure 31.7 Smuts and rusts.
a. Corn smut. **b.** Wheat rust.

Smuts and Rusts

Smuts and rusts are club fungi that parasitize cereal crops such as corn, wheat, oats, and rye. They are of great economic importance because of the crop losses they cause every year. Smuts and rusts don't form basidiocarps, and their spores are small and numerous, resembling soot. Some smuts enter seeds and exist inside the plant, becoming visible only near maturity. Other smuts externally infect plants. In corn smut, the mycelium grows between the corn kernels and secretes substances that cause the development of tumors on the ears of corn (Fig. 31.7*a*).

The life cycle of rusts, which may be particularly complex, often requires two different plant host species to complete the cycle. Black stem rust of wheat uses barberry bushes as an alternate host, and blister rust of white pine uses currant and gooseberry bushes. Campaigns to eradicate these bushes in areas where the alternate host grows help keep these rusts in check. Wheat rust (Fig. 31.7*b*) is also controlled by producing new and resistant strains of wheat. The process is continuous, because rust can mutate to cause infection once again.

Deadly Fungi

It is unwise for amateurs to collect mushrooms in the wild because certain mushroom species are poisonous. The red and yellow *Amanitas* are especially dangerous. This species is also known as fly agaric because it was thought to kill flies (it was gathered and then sprinkled with sugar to attract flies). Its toxins include muscarine and muscaridine, which produce symptoms similar to acute alcoholic intoxication. In one to six hours, the victim staggers, loses consciousness, and becomes delirious, sometimes suffering from hallucinations, manic conditions, and stupor. Luckily it also causes vomiting, which rids the system of the poison, so death occurs in less than 1% of cases. Death Angel (*Amanita phalloides,* Fig. 31A) causes 90% of the fatalities attributed to mushroom poisoning. When this mushroom is eaten, symptoms don't begin until ten to twelve hours later. Abdominal pain, vomiting, delirium, and hallucinations are not the real problem; rather, a poison interferes with RNA (ribonucleic acid) transcription by inhibiting RNA polymerase, and the victim dies from liver and kidney damage.

Some hallucinogenic mushrooms are used in religious ceremonies, particularly among Mexican Indians. *Psilocybe mexicana* contains a chemical called psilocybin that is a structural analogue of LSD and mescaline. It produces a dreamlike state in which visions of colorful patterns and objects seem to fill up space and dance past in endless succession. Other senses are also sharpened to produce a feeling of intense reality.

The only reliable way to tell a nonpoisonous mushroom from a poisonous one is to be able to correctly identify the species. Poisonous mushrooms cannot be identified with simple tests, such as whether they peel easily, have a bad odor, or blacken a silver coin during cooking.

In addition to club fungi, some sac fungi also contain chemicals that can be dangerous to people. *Claviceps purpurea*, the ergot fungus, infects rye and replaces the grain with ergot—hard, purple-black bodies consisting of tightly cemented hyphae containing ascospores (Fig. 31B). When ground with the rye and made into bread, the fungus releases toxic alkaloids that cause the disease ergotism. In humans, vomiting, feelings of intense heat or cold, muscle pain, a yellow face, and hand and feet lesions are accompanied by hysteria and hallucinations. We now know that ergot contains lysergic acid, from which LSD is easily synthesized. From recorded symptoms, historians believe that those who accused their neighbors of witchcraft in Salem, Massachusetts, in the seventeenth century could very well have been suffering from ergotism. As recently as 1951 an epidemic of ergotism occurred in Pont-Saint-Esprit, France. Over 150 persons became hysterical and four died.

Because the alkaloids that cause ergotism stimulate smooth muscle and selectively block the sympathetic nervous system, they can be used in medicine to cause uterine contractions and to treat certain circulatory disorders, including migraine headaches. Although the ergot fungus can be cultured in petri dishes, no one has succeeded in inducing it to form ergot in the laboratory. So far, the only way to obtain ergot, even for medical purposes, is to collect it in an infected field of rye.

Figure 31A Poisonous mushrooms, ***Amanita phalloides.***

Figure 31B Ergot infection of rye, caused by ***Claviceps purpurea.***

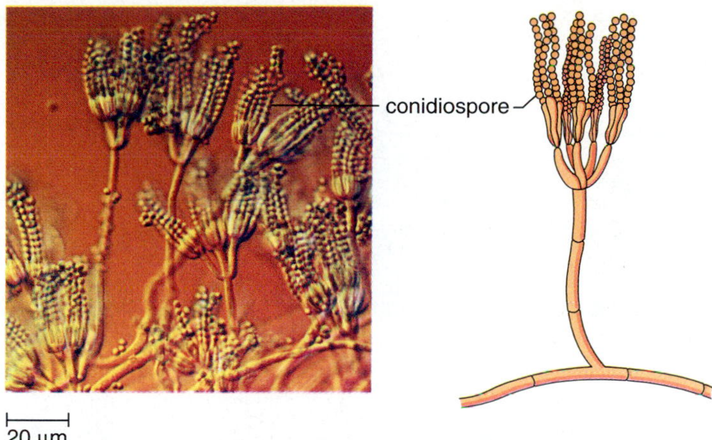

Figure 31.8 Conidiospores.
Sac fungi and imperfect fungi reproduce asexually by producing spores called conidiospores at the ends of certain hyphae. The organism shown here is *Penicillium*, an imperfect fungus.

Imperfect Fungi

The **imperfect fungi (division Deuteromycota,** 17,000 species) always reproduce asexually by forming conidiospores (Fig. 31.8). These fungi are "imperfect" in the sense that no sexual stage has yet been observed and may not exist. Without knowing the sexual stage, it is often difficult to classify a fungus as belonging to one of the other divisions. Usually, cellular morphology and biochemistry indicate these fungi are sac fungi that have lost the ability to reproduce sexually. Even so, genetic recombination does occur; hyphae fuse, and crossing-over between chromosomes results in haploid nuclei with varied genetic content.

Several imperfect fungi are serviceable to humans. Some species of the mold *Penicillium* are sources of the antibiotic penicillin, while other species give the characteristic flavor and aroma to cheeses such as Roquefort and Camembert. The bluish streaks in blue cheese are patches of conidiospores. The drug cyclosporine, which is administered to suppress the immune system following an organ transplant operation, is derived from an imperfect fungus found in soil. Another mold, *Aspergillus*, as well as other fungi, are used to produce soy sauce by fermentation of soybeans. In this country *Aspergillus* is used to produce citric and gallic acids, which serve as additives during the manufacture of different products—from inks to chewing gum.

Unfortunately, some imperfect fungi cause disease in humans. Certain dust-borne spores can cause infections of the respiratory tract, while athlete's foot and ringworm are spread by direct contact. *Candida albicans* is a yeastlike organism that causes infections of the vagina, especially in women on the birth control pill or certain types of antibiotic therapy. This organism also causes thrush, an inflammation of the mouth and throat.

> The imperfect fungi cannot be classified into one of the other divisions because their mode of sexual reproduction is unknown.

31.3 Symbiotic Relationships of Fungi

We have already mentioned several instances in which fungi are parasites of plants and animals. Two other associations are of interest.

Lichens

Lichens are an association between a fungus and a cyanobacterium or a green alga. The body of a lichen has three layers: the fungus forms a thin, tough upper layer and a loosely packed lower layer that shield the photosynthetic cells in the middle layer (Fig. 31.9). Specialized fungal hyphae, which penetrate or envelop the photosynthetic cells, transfer nutrients directly to the rest of the fungus. Lichens can reproduce asexually by releasing fragments that contain hyphae and an algal cell.

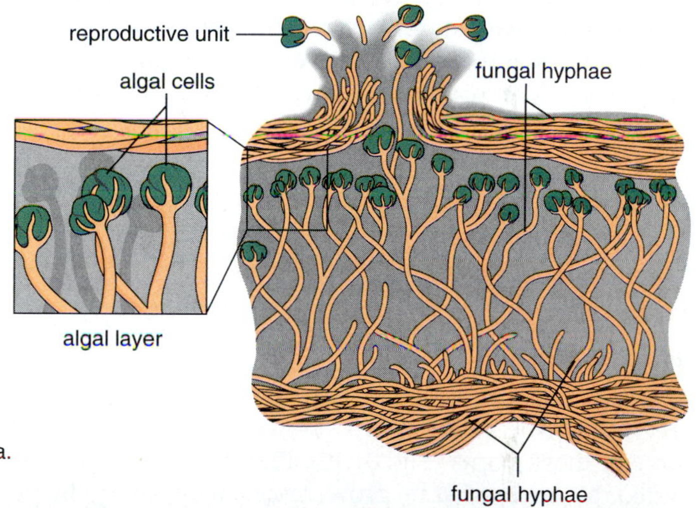

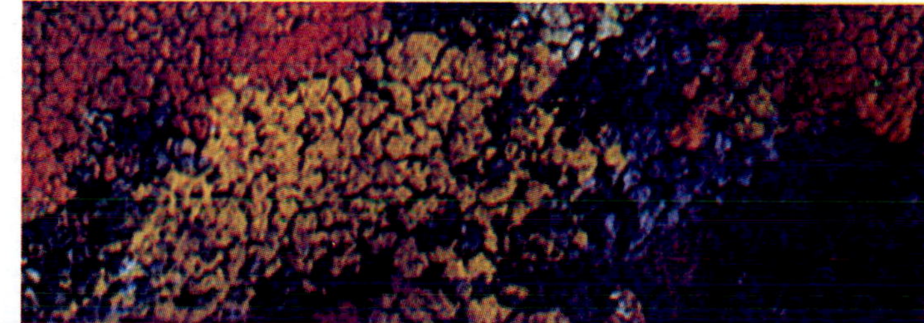

b. Mixture of crustose lichens

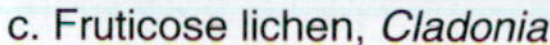

c. Fruticose lichen, *Cladonia*

d. Foliose lichen, *Xanthoparmelia*

Figure 31.9 Lichen morphology.
a. A section of a lichen shows the placement of the algal cells and the fungal hyphae, which encircle and penetrate the algal cells. **b.** Crustose lichens are compact. **c.** Fruticose lichens are shrublike. **d.** Foliose lichens are leaflike.

In the past, lichens were assumed to be mutualistic relationships in which the fungus received nutrients from the algal cells, and the algal cells were protected from desiccation by the fungus. Actually, lichens might involve a controlled form of parasitism of the algal cells by the fungus. Algae may not benefit at all from these associations. This is supported by experiments in which the fungal and algal components are removed and grown separately. The algae grow faster when they are alone rather than when they are part of a lichen. On the other hand, it is difficult to cultivate the fungus, which does not naturally grow alone. The different lichen species are identified according to the fungal partner.

Three types of lichens are recognized. Compact crustose lichens are often seen on bare rocks or on tree bark; foliose lichens are leaflike; and fruticose lichens are shrublike (Fig. 31.9). Lichens are efficient at acquiring nutrients and moisture, and therefore they can survive in areas of low moisture and low temperature as well as areas with poor or no soil. They produce and improve the soil, thus making it suitable for plants to invade the area. Unfortunately, lichens also take up pollutants and cannot survive where the air is polluted. Therefore, their presence is an indication that the air is healthy for humans to breathe.

Lichens, an association between fungal hyphae and algal cells, can live in areas of extreme conditions and contribute to the formation of soil.

Mycorrhizae

Mycorrhizae [Gk. *mykes*, fungus, and *rhizion*, dim. for root] are mutualistic relationships between soil fungi and the roots of most plants (Fig. 31.10). Plants whose roots are invaded by mycorrhizae grow more successfully in poor soils—particularly soils deficient in phosphates—than do plants without mycorrhizae. The fungal partner may enter the cortex of roots but does not enter the cytoplasm of plant cells. Ectomycorrhizae form a mantle that is exterior to the root, and they grow between cell walls. Endomycorrhizae penetrate only into the cell walls. In any case, the presence of the fungus gives the plant a greater absorptive surface for the intake of minerals. The fungus also benefits from the association by receiving carbohydrates from the plant.

Figure 31.10 Plant growth experiment.
Soybean plant, *Glycine*, without mycorrhizae grows poorly compared to two other plants infected with different strains of mycorrhizae.

It is of interest to know that the truffle, a gourmet delight whose ascocarp is somewhat prunelike in appearance, is a mycorrhizal fungus living in association with oak and beech tree roots. It used to be that the French used pigs to sniff out and dig up truffles, but now they have succeeded in cultivating truffles by inoculating the roots of seedlings with the proper mycelium.

Even the earliest fossil plants have mycorrhizae associated with them. It would appear, then, that mycorrhizae helped plants adapt to and flourish on land.

Mycorrhizae (fungus roots), a mutualistic association between a fungus and plant roots, help plants acquire mineral nutrients.

Connecting Concepts

Fungi clearly deserve their own kingdom. Like animals, they are heterotrophic, but they do not ingest food—they absorb nutrients. Like plants, they have cell walls, but their cell walls contain chitin instead of cellulose. Fungal cells store energy in the form of glycogen, as do animal cells. At one time fungi were considered a part of the plant kingdom, and then later, they were placed in the kingdom Protista. Whittaker argued that because of their multicellular nature and mode of nutrition, fungi should be in their own kingdom.

The range of species within the kingdom is broad. Some fungi are unicellular (such as yeast), some are multicellular parasites (such as those that cause athlete's foot), and some form mutualistic relationships with other species (such as those found in lichens). Most, however, are saprotrophic decomposers that play a vital role in all ecosystems. The decomposers work with bacteria to break down the waste products and dead remains of plants and animals so organic materials are recycled.

In addition to their various ways of life, fungi have been successful because they have diverse reproductive strategies. Asexual reproduction via spores, fragmentation, and budding is common. During sexual reproduction, the filaments of different mating types typically fuse. Asexual reproduction via spores, fragmentation, and budding is also common.

Fungi have been around for at least 570 million years. What is their evolutionary history? We don't know exactly. It is possible that the species now classified in the kingdom do not share a recent common ancestor, which means they probably evolved from different protists. It is also possible that fungi evolved from red algae because the two groups share similar morphological traits.

Summary

31.1 Characteristics of Fungi

Fungi are multicellular eukaryotes that are heterotrophic by absorption. After external digestion, they absorb the resulting nutrient molecules. Most fungi act as saprotrophic detritivores that aid the cycling of chemicals in ecosystems by decomposing dead remains. Some fungi are parasitic, especially on plants, and others are symbiotic with plant roots and algae.

The body of a fungus is composed of thin filaments called hyphae, which collectively are termed a mycelium. The cell wall contains chitin, and the energy reserve is glycogen. Fungi do not have flagella at any stage in their life cycle. Nonseptate hyphae have no cross walls; septate hyphae have cross walls, but there are pores that allow the cytoplasm and even organelles to pass through.

Fungi produce nonmotile and often windblown spores during both asexual and sexual reproduction. During sexual reproduction hyphae tips fuse so that dikaryotic ($n + n$) hyphae sometimes result, depending on the type of fungus. Following nuclear fusion, zygotic meiosis occurs during the production of the sexual spores.

31.2 Classification of Fungi

There are four divisions of fungi: division Zygomycota (zygospore fungi), division Ascomycota (sac fungi), division Basidiomycota (club fungi), and division Deuteromycota (imperfect fungi).

The zygospore fungi are nonseptate, and during sexual reproduction they have a dormant stage consisting of a thick-walled zygospore. When the zygospore germinates, sporangia produce windblown spores. Asexual reproduction occurs when nutrients are plentiful and again sporangia produce spores.

The sac fungi are septate, and during sexual reproduction saclike cells called asci produce spores. Asci are located in fruiting bodies called ascocarps. Asexual reproduction, which is dependent on the production of conidiospores, is more common.

The club fungi are septate, and during sexual reproduction club-shaped structures called basidia produce spores. Basidia are located in fruiting bodies called basidiocarps. Club fungi have a prolonged dikaryotic stage, and asexual reproduction by conidiospores is rare. A dikaryotic mycelium periodically produces fruiting bodies.

The imperfect fungi always reproduce asexually by conidiospores; they have not been observed to reproduce sexually. Therefore, they cannot be placed in one of the other divisions. Several imperfect fungi are of interest; for example, *Penicillium* is the source of penicillin, and *Candida* causes yeast infections.

31.3 Symbiotic Relationships of Fungi

Lichens are an association between a fungus and a cyanobacterium or a green alga. Traditionally, this association was considered to be mutualistic, but experimentation suggests a controlled parasitism by the fungus on the alga. Lichens may live in extreme environments and on bare rocks; they allow other organisms that will eventually form soil to establish. They are important in primary succession.

The term mycorrhizae refers to an association between a fungus and the roots of a plant. The fungus helps the plant absorb minerals, and the plant supplies the fungus with carbohydrates.

Reviewing the Chapter

1. Which characteristics best define fungi? Describe the body of fungi and how they reproduce. 544–545
2. On what basis are fungi classified? 546
3. Explain the term zygospore fungi. How does black bread mold reproduce asexually? sexually? 545–547
4. Explain the term sac fungi. How do sac fungi reproduce asexually? Describe the structure of an ascocarp. 549
5. Describe the structure of yeasts, and explain how they reproduce. How are yeasts useful to humans? 549
6. Explain the term club fungi. Draw and explain a diagram of the life cycle of a typical mushroom. 551
7. What is the economic importance of smuts and rusts? How can their numbers be controlled? 551
8. Explain the term imperfect fungi. Give examples of imperfect fungi that serve humans and examples of those that cause disease. 553
9. Describe the structure of a lichen, and name the three different types. What is the nature of this fungal association? 553–554
10. Describe the association known as mycorrhizae, and explain how each partner benefits. 554

Testing Yourself

Choose the best answer for each question.

1. An organism that decomposes remains is most likely to utilize which mode of nutrition?
   a. parasitic
   b. saprotrophic
   c. ingestion
   d. chemosynthesis
   e. Both a and b are correct.
2. Which feature is best associated with hyphae?
   a. strong impermeable walls
   b. rapid growth
   c. large surface area
   d. pigmented cells
   e. Both b and c are correct.
3. A fungal spore
   a. contains an embryonic organism.
   b. germinates directly into an organism.
   c. is always windblown.
   d. is most often diploid.
   e. Both b and c are correct.
4. The taxonomy of fungi is based on
   a. sexual reproductive structures.
   b. shape of the sporocarp.
   c. mode of nutrition.
   d. type of cell wall.
   e. level of organization.
5. In the life cycle of black bread mold, the zygospore
   a. undergoes meiosis and produces zoospores.
   b. produces spores as a part of asexual reproduction.
   c. is a thick-walled dormant stage.
   d. is equivalent to asci and basidia.
   e. All of these are correct.
6. In an ascocarp
   a. there are fertile and sterile hyphae.
   b. hyphae fuse, forming the dikaryotic stage.
   c. a sperm fertilizes an egg.
   d. hyphae do not have chitinous walls.
   e. conidiospores form.

7. In which fungus is the dikaryotic stage longer lasting?
   a. zygospore fungi
   b. imperfect fungi
   c. sac fungi
   d. club fungi
   e. fungal partner of a lichen
8. Conidiospores are formed
   a. asexually at the tips of special hyphae.
   b. during sexual reproduction.
   c. by all types of fungi except water molds.
   d. when it is windy and dry.
   e. as a way to survive a harsh environment.
9. Imperfect fungi are called imperfect because
   a. they have no zygospore.
   b. they cause diseases.
   c. they form conidiospores.
   d. sexual reproduction has not been observed.
   e. All of these are correct.
10. Lichens
   a. cannot reproduce.
   b. need a nitrogen source to live.
   c. are parasitic on trees.
   d. are able to live in extreme environments.
11. Label this diagram of the life cycle of a mushroom.

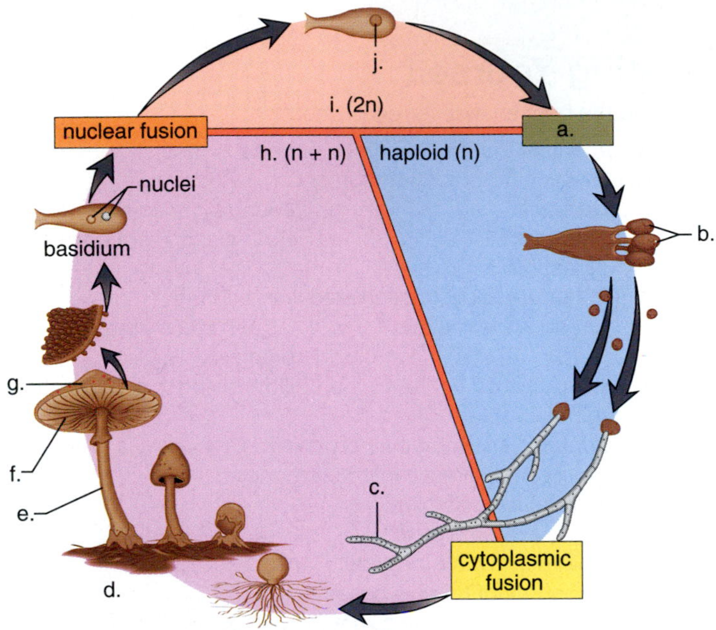

Thinking Scientifically

1. The very earliest bakers observed that dough left in the air would rise. Unknown to them, yeast from the air "contaminated" the bread, began to grow, and produced carbon dioxide. Carbon dioxide caused the bread to rise. Later, cooks began to save some soft dough (before much flour was added) from the previous loaf to use in the next loaf. The saved portion was called the mother. What is in the mother and why was it important to save it in a cool place?
2. There seems to be a fine line between symbiosis and parasitism when you examine the relationships between fungi and plants. What hypotheses could explain how different selective pressures might have caused particular fungal species to evolve one or the other relationship? Under what circumstances might a mutualistic relationship evolve between fungi and plants? a parasitic relationship evolve?

Understanding the Terms

ascus (pl., asci) 549
basidium (pl., basidia) 551
budding 545
conidiospore 549
dikaryotic 545
fruiting body 549
fungus (pl., fungi) 544
hypha (pl., hyphae) 544
lichen 553
mycelium 544
mycorrhiza (pl., mycorrhizae) 554
nonseptate 545
septate 545
sporangium (pl., sporangia) 546
spore 545
zygospore 546

Match the terms to these definitions:

a. ______________ Clublike structure in which nuclear fusion and meiosis occur during sexual reproduction of club fungi.
b. ______________ Tangled mass of hyphal filaments composing the vegetative body of a fungus.
c. ______________ Spore produced by sac and club fungi during asexual reproduction.
d. ______________ Spore-producing and spore-disseminating structure found in sac and club fungi.
e. ______________ Symbiotic relationship between fungal hyphae and roots of vascular plants. The fungus allows the plant to absorb more mineral ions and obtains carbohydrates from the plant.

Web Connections

Exploring the Internet

http://www.mhhe.com/biosci/genbio/mader
(click on *Biology 7/e*)

The *Biology 7/e* Online Learning Center provides many resources for studying the material in this chapter including links to the following sites:

Learn which mushrooms are nutritious and which are deadly in Edible and Poisonous Mushrooms. An extensive site with much text, and beautiful illustrations and photos.

http://www.conservation.state.mo.us/nathis/flora/mushroom/mushroom.html

Arizona's Tree of Life page on Basidiomycota. Much information on life cycles, references, and pictures.

http://phylogeny.arizona.edu/tree/eukaryotes/fungi/basidiomycota/basidiomycota.html

Arizona's Tree of Life page on the Ascomycota. Much information on life cycles, references, and pictures.

http://phylogeny.arizona.edu/tree/eukaryotes/fungi/ascomycota/ascomycota.html

Mushroom Heaven. An informative and fun site about collecting, identifying, eating, growing, and enjoying mushrooms.

http://www.aa.net/~reo/mushroom.html

Fungi Photo Index. As the name suggests, it is a compendium of photos of common mushrooms.

http://www.cinenet.net/users/velosa/photoalbum.html

The Plants

chapter

chapter concepts

32.1 Classification of Plants

- Plants are multicellular photosynthetic organisms adapted to living on land. Among various adaptations, plants generally have an alternation of generations life cycle. 558

32.2 Nonvascular Plants

- The nonvascular plants are low-growing and contain little or no vascular tissue. The gametophyte (haploid generation) is dominant, and produces windblown spores. 559

32.3 Vascular Plants

- In vascular plants, the sporophyte (diploid generation) is dominant and has transport (vascular) tissue. 558
- In seedless vascular plants, windblown spores disperse the species. In seed plants, seeds, transported by various means, disperse the species. 562

32.4 Ferns and Allies

- The seedless vascular plants were much larger and extremely abundant during the Carboniferous period. 563
- The seedless vascular plants include ferns and other species which do not produce seeds. 564
- The fern sporophyte is the familiar plant with large fronds that produces windblown spores; the independent gametophyte is a heart-shaped structure that produces flagellated sperm. 566

32.5 Gymnosperms

- The gymnosperms and angiosperms are the vascular plants that produce seeds. 568
- There are four divisions of gymnosperms; the familiar conifers and the little-known cycads, ginkgoes, and gnetophytes. 568
- The conifer sporophyte is usually a cone-bearing tree (e.g., pine tree). The pollen cones produce pollen, which is windblown to the seed cone where windblown seeds are produced. 568

32.6 Angiosperms

- Angiosperms are the very diversified flowering plants, which may depend on an animal pollinator to carry pollen to another flower. The seeds are enclosed within a fruit, which often assists dispersal of seeds. 572

Prairie biome, United States.

The terrestrial biomes of the world are defined by their vegetation: the tropical rain forests, the temperate deciduous forests, and the grasslands, for example. This is appropriate because plants dominate the landscape. They are the major producers on land and were the first organisms to invade the terrestrial environment. Animals cannot live on land without plants to sustain them. Plants store energy as starch and are at the base of most ecological pyramids, including those that sustain the world's human population. They also pull carbon dioxide out of the air and supply the oxygen that organisms require for aerobic respiration.

The plant kingdom includes over 270,000 species, and among them are the earth's largest species—some live oaks and redwoods tower a hundred feet tall and weigh thousands of pounds. The vascular plants have internal tissues that conduct water, nutrients, and minerals to all parts of their body. The seed plants do not depend on external water for the purpose of reproduction, and the flowering plants have developed unique mutualistic relationships with animals to carry pollen from flower to flower.

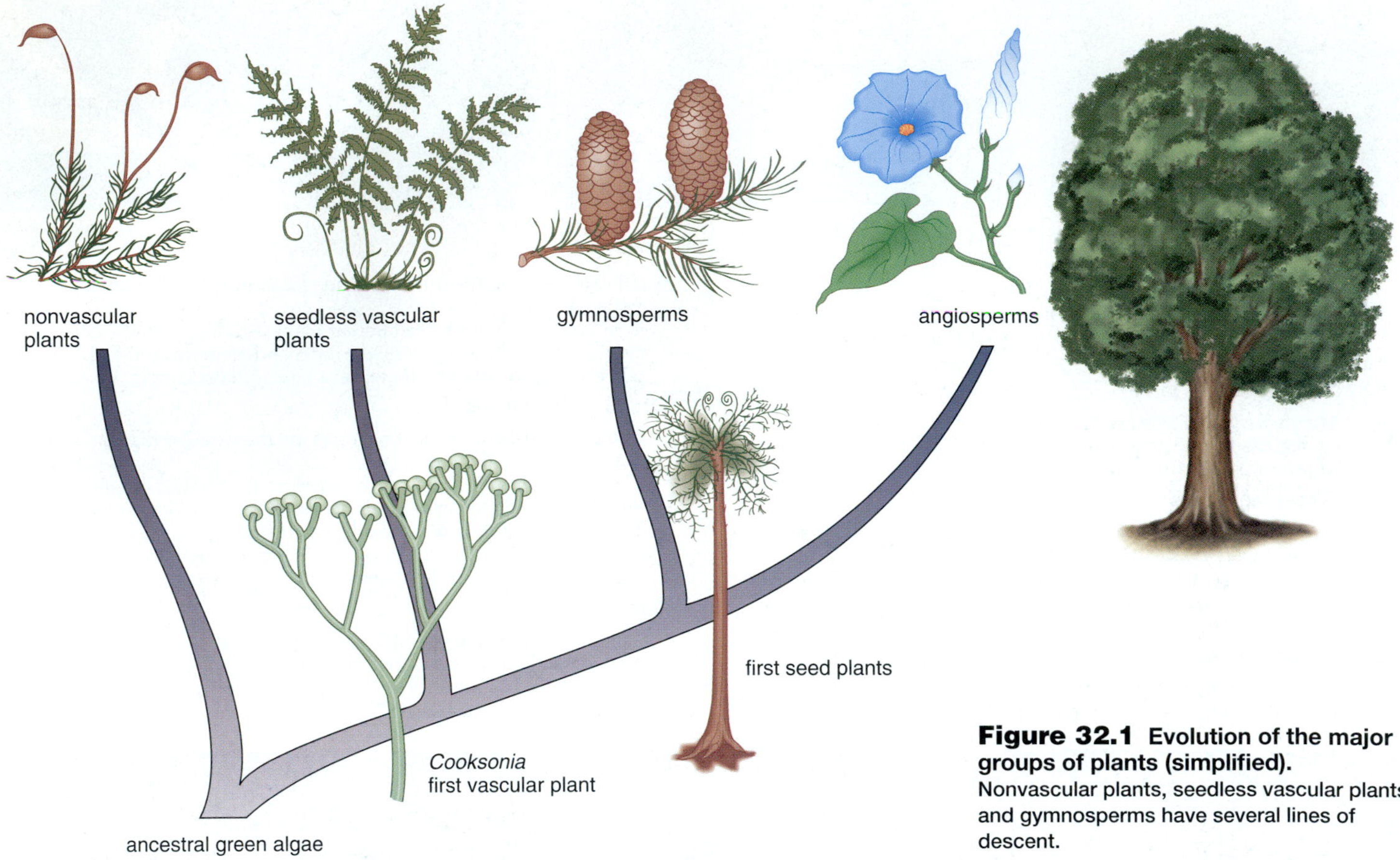

nonvascular plants
seedless vascular plants
gymnosperms
angiosperms
first seed plants
Cooksonia first vascular plant
ancestral green algae

Figure 32.1 Evolution of the major groups of plants (simplified). Nonvascular plants, seedless vascular plants, and gymnosperms have several lines of descent.

32.1 Classification of Plants

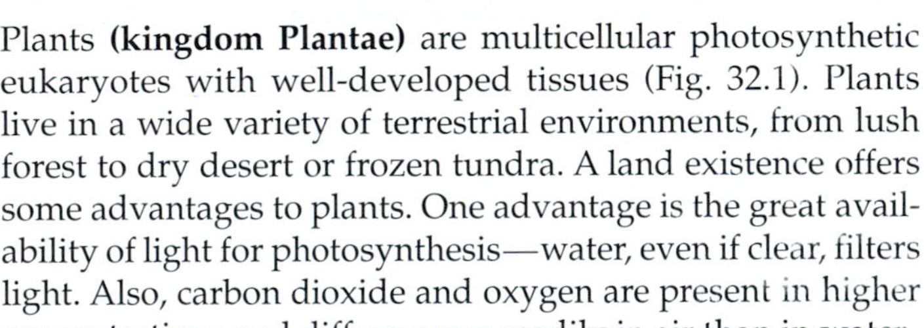

Plants **(kingdom Plantae)** are multicellular photosynthetic eukaryotes with well-developed tissues (Fig. 32.1). Plants live in a wide variety of terrestrial environments, from lush forest to dry desert or frozen tundra. A land existence offers some advantages to plants. One advantage is the great availability of light for photosynthesis—water, even if clear, filters light. Also, carbon dioxide and oxygen are present in higher concentrations and diffuse more readily in air than in water.

The land environment, however, requires adaptations to deal with the constant threat of desiccation (drying out). Most plants can obtain water from the substratum by means of roots. And to conserve water, leaves and stems are, at the very least, covered by a waxy **cuticle** that is impervious to water. The surface of leaves is interrupted by pores that can open and close to regulate gas exchange and water loss. Some plants have a vascular system that transports water up to and nutrients down from the leaves. In conjunction with this system, trees have a strong internal skeleton that opposes the force of gravity and lifts the leaves toward the sun. All plants enclose and protect their reproductive cells within a jacket of sterile cells. Seed plants protect the entire gametophyte generation; pollen grains that will produce sperm are transported by wind or motile animal to the egg. Then the subsequent embryo becomes a seed which has a seed coat.

> Plants are adapted to living on land. In general, they tend to have features that allow them to live and reproduce on land.

CLASSIFICATION

Kingdom Plantae

Multicellular; primarily terrestrial eukaryotes with well-developed tissues; autotrophic by photosynthesis; alternation of generations life cycle. Like green algae, plants contain chlorophylls *a* and *b* and carotenoids; store starch in chloroplasts; cell wall contains cellulose.

Nonvascular Plants*
Division Hepatophyta: liverworts
Division Bryophyta: mosses
Division Anthocerotophyta: hornworts

Seedless Vascular Plants*
Division Psilotophyta: whisk ferns
Division Lycopodophyta: club mosses
Division Equisetophyta: horsetails
Division Pteridophyta: ferns

Seed Vascular Plants*

Gymnosperms*
Division Pinophyta: conifers
Division Cycadophyta: cycads
Division Ginkgophyta: maidenhair tree
Division Gnetophyta: gnetophytes

Angiosperms*
Division Magnoliophyta: flowering plants
Class Magnoliopsida: dicots
Class Liliopsida: monocots

* Not in classification of organisms, but added here for clarity.

Relationship to Algae

Plants are believed to be closely related to green algae, which are classified in the kingdom Protista, because these two groups have several characteristics in common. Both utilize chlorophylls *a* and *b* as well as carotenoid pigments during photosynthesis. The primary food reserve is starch, which is stored inside the chloroplast. Their cell walls contain cellulose, and both types of organisms form a cell plate during cell division.

Why aren't multicellular green algae like *Ulva* considered plants? Plants, but not green algae, have multicellular **gametangia** (sex organs) and sporangia in which reproductive cells are protected by several layers of nonreproductive cells. Later, the zygote and then the embryo are protected within the body of the plant. In algae, fertilization is external and the embryo is not protected from the environment.

Plants and some algae have a two-generation life cycle called alternation of generations that involves sporic meiosis (Fig. 32.2):

1. The **sporophyte** [Gk. *spora*, seed, and *phyton*, plant], the diploid generation, produces haploid spores by meiosis. Spores develop into a haploid generation.
2. The **gametophyte** [Gk. *gamete*, wife, *gametes*, husband, and *phyton*, plant], the haploid generation, produces gametes that unite to form a diploid zygote.

The two generations are dissimilar, and one is dominant over the other. The dominant generation is larger and exists for a longer period of time. In addition, plants are oogamous; the gametes are eggs and sperm. The characteristics of plants are listed in Table 32.1.

Plants, unlike green algae, protect the embryo from drying out. During an alternation of generations life cycle, one generation is dominant over the other.

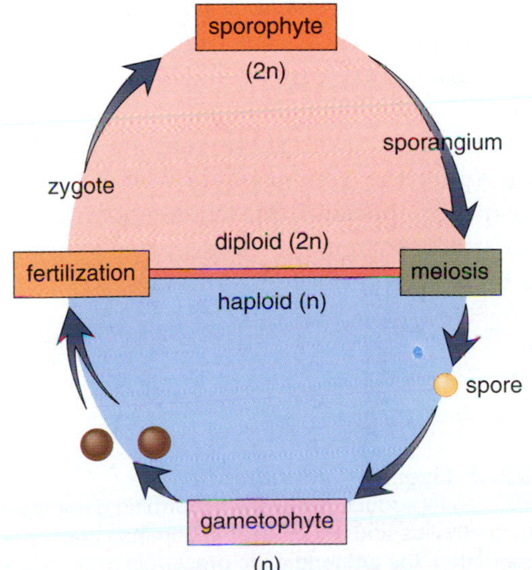

Figure 32.2 Alternation of generations.

32.2 Nonvascular Plants 

Plants are currently divided into two main groups: the nonvascular and the vascular plants. The **nonvascular plants** consist of three groups: hornworts (division Anthocerotophyta), liverworts (division Hepatophyta), and mosses (division Bryophyta). The nonvascular plants lack vascular plants' specialized means of transporting water and organic nutrients. Although they often have a "leafy" appearance, these plants do not have true roots, stems, and leaves—which by definition must contain true vascular tissue. Therefore, the nonvascular plants are said to have rootlike, stemlike, and leaflike structures.

The gametophyte is the dominant generation in the nonvascular plants—it is the generation we recognize as the plant. Flagellated sperm swim to the vicinity of the egg in a continuous film of water. The sporophyte, when present, is attached to and derives its nourishment from the photosynthetic gametophyte.

The nonvascular plants are quite small; the largest is no more than 20 cm tall. This characteristic is linked to the lack of an efficient means to transport water to any height. And because sexual reproduction involves flagellated sperm they are usually found in moist habitats. Nevertheless, mosses compete well in harsh environments because the gametophyte can reproduce asexually, allowing mosses to spread into stressful and even dry habitats. If need be, mosses can dry up and later, when water is available, begin to photosynthesize again.

It is not known how closely related the various nonvascular plants are. The current thinking is that these plants have individual lines of descent, so this text classifies them into three separate divisions. Some mosses have a rudimentary form of vascular tissue, and therefore mosses are believed to be more closely related to vascular plants than are liverworts and hornworts. We will be discussing the liverworts and mosses, the more familiar nonvascular plants.

In nonvascular plants, the dominant gametophyte is water dependent because it lacks vascular tissue and produces flagellated sperm.

Table 32.1

Characteristics of Plants

1. Are multicellular and have cells specialized to form tissues and organs
2. Contain chlorophylls *a* and *b* and carotenoids, store reserve food as starch, and have cellulose cell walls
3. Have gametangia (sex organs) with an outer layer of nonreproductive cells, which prevents desiccation of developing gametes
4. Protect the developing embryo from drying out by providing it with water and nutrients within the female reproductive structure
5. Have a life cycle that is described as alternation of generations

Liverworts

All **liverworts (division Hepatophyta,** 10,000 species) have either a flattened thallus or a leafy body, but those with a lobed thallus (body) are more familiar than those with "leaves." The name "liverwort" arose in the ninth century when it seemed to some that the plant had lobes like the liver. They thought the plant might therefore be beneficial in treating liver ailments, a hypothesis that has never been supported.

Marchantia is most often used as an example of this group of plants (Fig. 32.3). Each lobe of the thallus is perhaps a centimeter or so in length; the upper surface is smooth, and the lower surface bears numerous **rhizoids** [Gk. *rhizion,* dim. for root] (rootlike hairs) projecting into the soil. *Marchantia* reproduces both asexually and sexually. Gemma cups on the upper surface of the thallus contain *gemmae,* groups of cells that detach from the thallus and can start a new plant. Sexual reproduction depends on disk-headed stalks that bear **antheridia** [Gk. *anthos,* flower, and *-idion,* small], where flagellated sperm are produced, and on umbrella-headed stalks that bear **archegonia** [Gk. *arche-,* first, and *gone,* seed], where eggs are produced. Following fertilization a tiny sporophyte no more than a few millimeters in length is composed of a foot, a short stalk, and a capsule. Windblown spores are produced within the capsule.

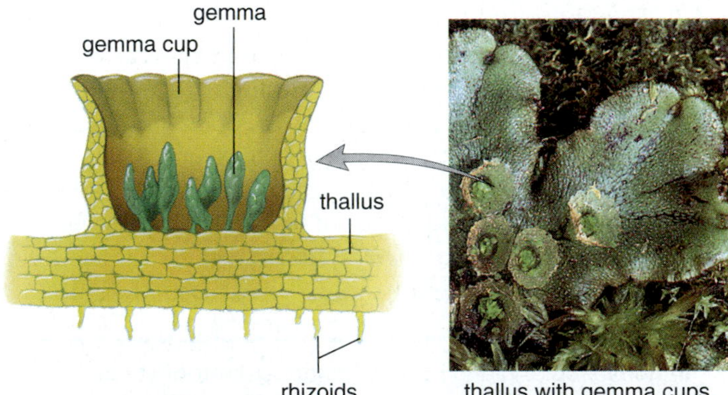

a. Gemma cup

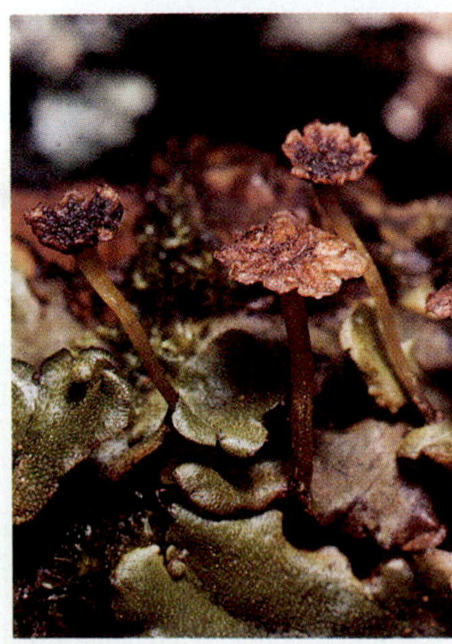

b. Structures that bear antheridia

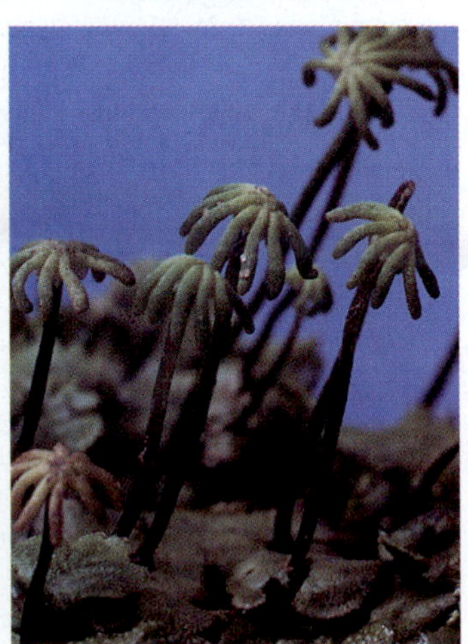

Structures that bear archegonia

Mosses

The body of a **moss (division Bryophyta,** 12,000 species) is usually a leafy shoot although some are secondarily flattened. Mosses can be found from the Arctic through the tropics to parts of the Antarctic. Although most prefer damp, shaded locations in the temperate zone, some survive in deserts and others inhabit bogs and streams. In forests, they frequently form a mat that covers the ground and rotting logs. Mosses can store large quantities of water in their cells, but if a dry spell continues for long, they become dormant. The whole plant shrivels, turns brown, and looks completely dead. As soon as it rains, however, the plant becomes green and resumes metabolic activity.

Some mosses are adapted to unusual conditions. The so-called copper mosses live only in the vicinity of copper and can serve as an indicator plant for copper deposits. Luminous moss, which glows with a golden green light, is found in caves, under the roots of trees, and other dimly lit places. Its cells, which are shaped like tiny lenses, focus what little light there is on the grana of chloroplasts.

The common name of several plants implies they are mosses when they are not. Irish moss is an edible alga that grows in leathery tufts along northern seacoasts. Reindeer moss, a lichen, is the dietary mainstay of reindeer and caribou in northern lands. Club mosses, discussed later in this chapter, are vascular plants; and Spanish moss, which hangs in grayish clusters from trees in the southeastern United States, is a flowering plant of the pineapple family.

Most mosses can reproduce asexually by fragmentation. Just about any part of the plant is able to grow and eventually produce leafy shoots. Figures 32.4 and 32B (see page 571) describe the life cycle of a typical temperate-zone moss. The gametophyte of mosses has two stages. First, there is the algalike protonema, a branching filament of cells. Then, after a period of favorable growing conditions, upright leafy shoots are seen at intervals along the protonema. Rhizoids anchor the shoots, which bear antheridia and archegonia. An antheridium consists of a short stalk, an outer layer of sterile cells, and an inner mass of cells that become the flagellated sperm. An archegonium, which looks like a vase with a long neck, has a single egg located at the base.

The dependent sporophyte consists of a foot, which grows down into the gametophyte tissue, a stalk, and an upper capsule, or **sporangium,** where spores are produced. At first the sporophyte is green and photosynthetic; at maturity it is brown and nonphotosynthetic.

Figure 32.3 **Liverwort, *Marchantia.***
a. *Marchantia* can reproduce asexually by forming gemmae, cells that detach from the thallus and are present in gemmae cups. **b.** During sexual reproduction, the antheridia are present in disk-shaped structures, and the archegonia are present in umbrella-shaped structures.

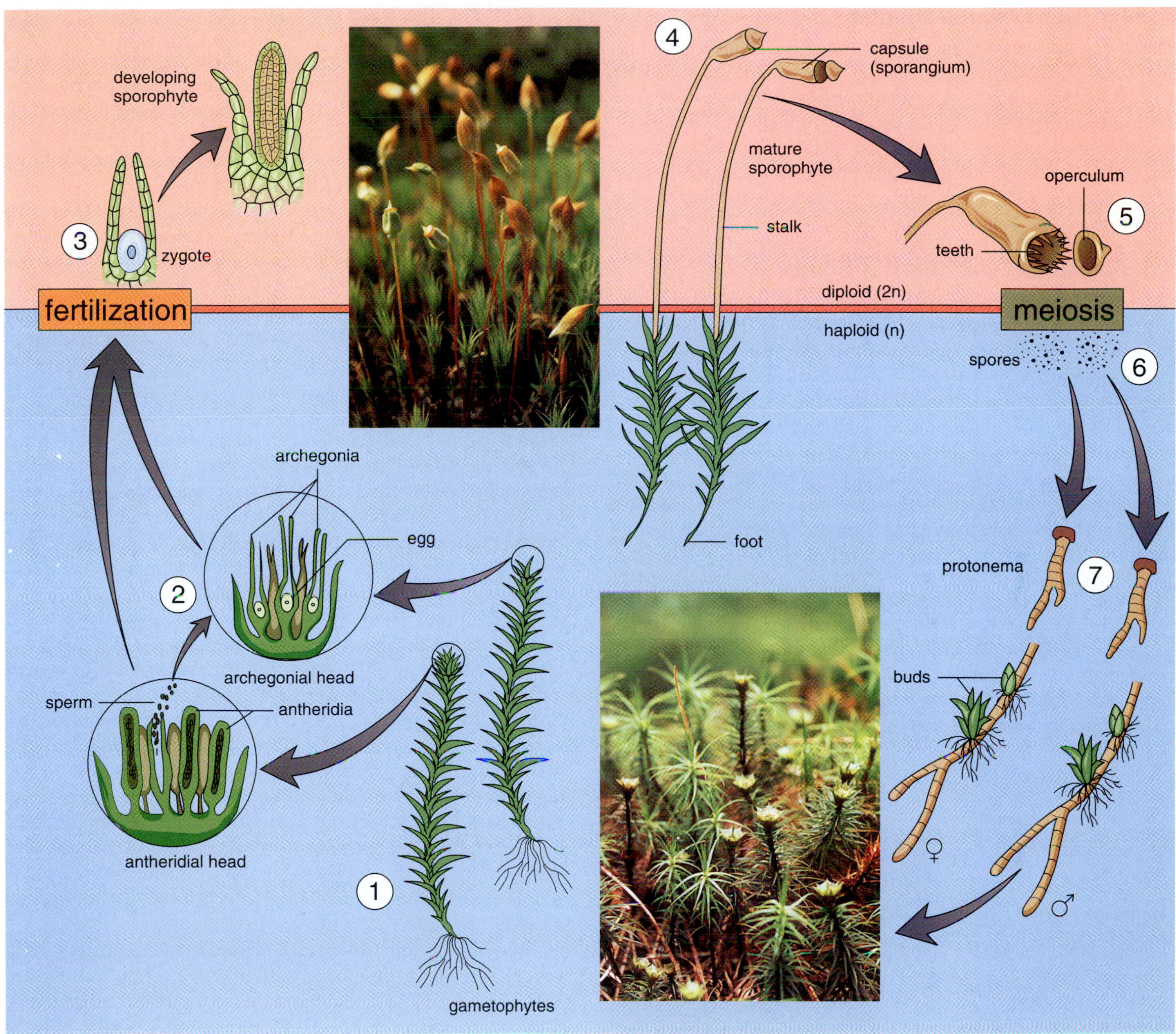

Figure 32.4 Moss life cycle.
The gametophyte is dominant in bryophytes, such as mosses. ① The leafy shoots bear separate antheridia and archegonia. ② Flagellated sperm are produced in antheridia, and these swim in external water to an archegonium that contains a single egg. ③ When the egg is fertilized, the zygote and developing sporophyte are retained within the archegonium. In some species of mosses, a hoodlike covering (calyptra) derived from the archegonium is carried upward by the growing sporophyte. ④ The mature sporophyte growing atop a gametophyte shoot consists of a foot that grows down into the gametophyte tissue, a stalk, and an upper capsule, or sporangium, where meiosis occurs and spores are produced. ⑤ When the covering and capsule lid (operculum) fall off, the spores are mature and are ready to escape. The release of spores is controlled by one or two rings of "teeth" that project inward from the margin of the capsule. The teeth close the opening when the weather is wet but curl up and free the spores when the weather is dry. ⑥ Spores are released at times when they are most likely to be dispersed by air currents. ⑦ When a spore lands on an appropriate site, it germinates into a protonema, the first stage of the gametophyte.

Uses of Nonvascular Plants

Sphagnum, also called bog or peat moss, has commercial importance. The cells of this moss have a tremendous ability to absorb water, which is why peat moss is often used in gardening to improve the water-holding capacity of the soil. In some areas where the ground is wet and acidic, such as bogs, dead mosses, especially sphagnum, accumulate and do not decay. This accumulated moss, called **peat,** can be used as fuel.

Adaptation of Nonvascular Plants

The nonvascular plants are small and simple plants generally found in moist habitats for two reasons: they lack vascular tissue, and their sexual reproduction involves flagellated sperm. Nonvascular plants do not have the complexity of, nor are they found in the variety of habitats occupied by, the vascular plants. Nevertheless, mosses are better than flowering plants at living on stone walls and on fences and even in the shady cracks of hot exposed rocks. They improve newly created soil so that it can be used for the growth of other organisms. For these particular microhabitats, there seems to be a selective advantage to being small and simple (Fig. 32.5).

The nonvascular plants (liverworts and mosses) lack specialized vascular tissue and have a dominant gametophyte. Fertilization requires an outside source of moisture, and they are often found in moist locations.

Figure 32.5 Moss growing on rocks.
The mosses that are adapted to growing on rocks contribute to the formation of soil so that other plants can enter the area.

32.3 Vascular Plants

Vascular plants [L. *vasculum,* dim. of *vas,* vessel] include ferns and their allies, gymnosperms, and angiosperms. Vascular tissue in these plants consists of **xylem** [Gk. *xylon,* wood], which conducts water and minerals up from the soil, and **phloem** [Gk. *phloios,* bark], which transports organic nutrients from one part of the plant to another. The vascular plants have true roots, stems, and leaves. The **roots** absorb water from the soil and the **stem** conducts water to the leaves. Xylem, with its strong-walled cells, supports the body of the plant against the pull of gravity. The leaves are fully covered by a waxy cuticle except where it is interrupted by **stomates,** little pores whose size can be regulated to control water loss.

The sporophyte generation is dominant in vascular plants. This is the generation that has vascular tissue and the other features just discussed. Another advantage of having a dominant sporophyte relates to its being diploid. If a faulty gene is present, it can be masked by a functional gene. Then, too, the greater the amount of genetic material, the greater the possibility of mutations that will lead to increased variety and complexity. Indeed, the vascular plants are complex, extremely varied, and widely distributed.

Ferns and their allies are seedless vascular plants. They disperse the species by producing windblown spores (see Fig. 32C, page 571). When the spores germinate, a small gametophyte is independent of the sporophyte for its nutrition. In these plants, flagellated sperm are released by antheridia and swim in a film of external water to the archegonia, where fertilization occurs. Because the water-dependent gametophyte is independent of the sporophyte, these plants cannot wholly benefit from the adaptations of the sporophyte to a terrestrial environment. As discussed in the reading on the following page, the seedless vascular plants formed the great swamp forest of the Carboniferous period.

In seed plants, there is a separate microgametophyte (male) and megagametophyte (female), as shown in Figure 32D (see page 571). The microgametophyte and megagametophyte are dependent on the sporophyte, which is fully adapted to a dry environment. The mature microgametophyte is the pollen grain. The megagametophyte, which is retained within the body of the sporophyte, becomes an **embryo sac.** Later, the embryo is enclosed within a seed. Seed dispersal occurs by various means. Some seeds are carried by wind and water or by animals to a new location.

In seedless vascular plants windblown spores disperse the species. In seed plants, seeds disperse the species.

Ecology Focus

The First Forests

Our industrial society runs on fossil fuels such as **coal.** The term "fossil fuel" might seem odd at first until one realizes that it refers to the remains of organic material from ancient times. During the Carboniferous period more than 300 million years ago, a great swamp forest (Fig. 32A) encompassed what is now the Appalachian Mountains, northern Europe, and Ukraine. The weather was warm and humid, and the trees grew very tall. These are not the trees we know today; instead they are related to today's seedless vascular plants: the club mosses, horsetails, and ferns! Club mosses today might stand as high as 30 cm; their ancient relatives were 30.5 meters (over 100 feet) tall and 1 meter wide. The spore-bearing cones were up to 30 cm long and some had leaves more than 1 meter long. Horsetails too—at 15 meters tall—were giants compared to today's specimens. The tree ferns were also taller than today's tree ferns found in the tropics, and there were two other types of trees: seed ferns and ancient gymnosperms. "Seed fern" is a misnomer because it has been shown that these plants, which only resemble ferns, were actually a type of gymnosperm.

The amount of biomass was enormous and occasionally the swampy water rose and the trees fell. Trees under water do not decompose well, and their partially decayed remains became covered by sediment that sometimes changed to sedimentary rock. Sedimentary rock applied pressure and the organic material then became coal, a fossil fuel. This process continued for millions of years, resulting in immense deposits of coal. Geological upheavals raised the deposits to the level they are today, where they can be mined.

With a change of climate, the trees of the Carboniferous period became extinct, and only their herbaceous relatives survived to our time. Without these ancient forests, our life today would be far different than it is because they helped bring about our industrialized society.

Figure 32A Swamp forest of Carboniferous period.

Table 32.2

The Geological Time Scale: Some Major Evolutionary Events of Plants

Era	Period	MYA*	Major Biological Events—Plants
Cenozoic	Neogene	2	Number of herbaceous plants increases
	Paleogene	66	Land dominated by angiosperms — **Age of Angiosperms**
Mesozoic	Cretaceous	144	Angiosperms spread
	Jurassic	208	First angiosperms appear
	Triassic	245	Land dominated by gymnosperms and ferns — **Age of Gymnosperms**
Paleozoic	Permian	286	Land covered by forests of seedless vascular plants
	Carboniferous	360	Age of great coal-forming forests, including club mosses, horsetails, and ferns — **Swamp Forest**
	Devonian	408	Expansion of seedless vascular plants on land
	Silurian	438	Seedless vascular plants appear on land
	Ordovician	505	First plant fossils
	Cambrian	570	Unicellular marine algae abundant — **Age of Algae**

* *MYA = millions of years ago*

32.4 Ferns and Allies

Because the seedless vascular plants are not closely related, each type is placed in its own division. The seedless vascular plants include whisk ferns (division Psilotophyta), club mosses (division Lycopodophyta), horsetails (division Equisetophyta), and ferns (division Pteridophyta).

Table 32.2 shows the evolutionary history of plants. Among the first vascular plants were the rhyniophytes, which were dominant from the mid-Silurian period until the mid-Devonian period. *Cooksonia* may have been the very first vascular plant, but *Rhynia,* which is also a rhyniophyte, is better known. In *Rhynia,* the erect stems fork repeatedly (are dichotomous). They are attached to a **rhizome,** a horizontal, underground stem that bears rhizoids below. The stem does have vascular tissue, but the plant has no true roots or leaves. Sporangia are at the ends of some of the branches. Ferns and their allies are believed to trace their ancestry back to the rhyniophytes or closely related forms.

Psilotophytes

The **psilotophytes (division Psilotophyta)** are represented by *Psilotum,* a plant that closely resembles *Rhynia* (Fig. 32.6). **Whisk ferns,** named for their resemblance to whisk brooms, are found in Arizona, Texas, Louisiana, and Florida, as well as Hawaii and Puerto Rico. The whisk ferns have no leaves or roots. A branched rhizome has rhizoids, and a mycorrhizal fungus helps gather nutrients. Aerial stems with tiny scales fork repeatedly and carry on photosynthesis. Sporangia are located at the ends of short branches. The independent gametophyte, which is found underground and is penetrated by a mycorrhizal fungus, produces flagellated sperm.

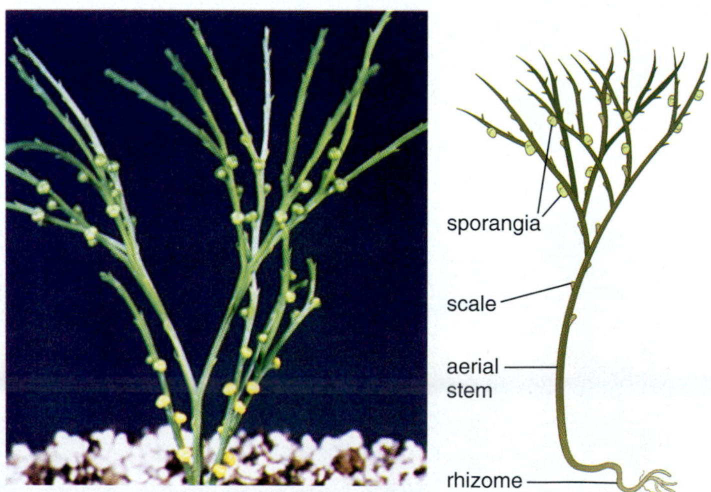

Figure 32.6 Whisk fern, *Psilotum*.
Whisk ferns have no roots or leaves—the branches carry on photosynthesis. The sporangia are yellow.

Figure 32.7 Club moss, *Lycopodium.*
Green photosynthetic stems are covered by scalelike leaves, and sporangia are found on leaves arranged as strobili.

Figure 32.8 Horsetail, *Equisetum.*
Whorls of branches and tiny leaves are at the joints of the stem. The sporangia are borne in strobili.

Club Mosses

The **club mosses (division Lycopodophyta,** 1,000 species) are common in moist woodlands of the temperate zone where they are known as ground pines. Typically, a branching rhizome sends up aerial stems less than 30 cm tall. Tightly packed, scalelike leaves cover the stem and branches of the plant, which has the appearance shown in Figure 32.7. The leaves are microphylls, so called because they have only one strand of vascular tissue. Such a leaf may have evolved from a scale:

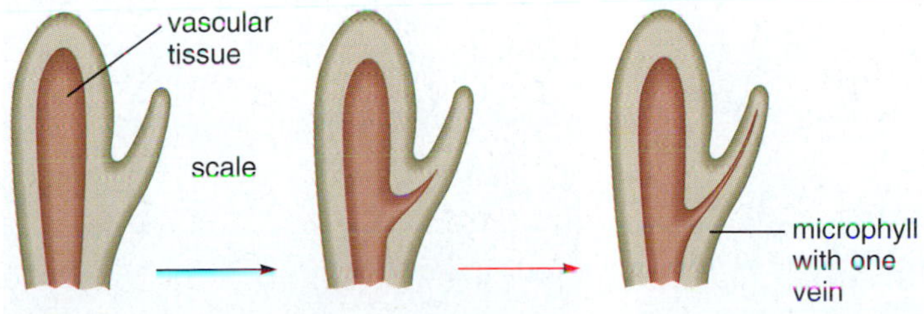

In club mosses, the sporangia are borne on terminal clusters of leaves, called strobili (sing., **strobilus**), which are club shaped. The spores germinate into inconspicuous and independent gametophytes. The majority of club mosses live in the tropics and subtropics where many of them are epiphytes—plants that live on, but are not parasitic on, trees. The closely related spike mosses (*Selaginella*) are extremely varied and include the resurrection plant, which curls up into a tight ball when dry but unfurls as if by magic when moistened.

Club mosses trace their ancestry to the Devonian period (Table 32.2). They are the only living plants to have microphylls, and it would appear that they are indirectly (rather than directly) related to the ferns and other allied groups.

Horsetails

Horsetails (division Equisetophyta, 15 species), which thrive in waste and wet places around the globe, are represented by *Equisetum*, the only genus in existence today (Fig. 32.8). A rhizome produces aerial stems that stand up to 1.3 meters. The whorls of slender green side branches at the joints (nodes) of the stem make the plant bear a fanciful resemblance to a horse's tail. The small, scalelike leaves also form whorls at the nodes. Although the leaves have only a single strand of vascular tissue, they are reduced megaphylls (with many strands of vascular tissue), not microphylls. Many horsetails have strobili at the tips of the stems; others send up special buff-colored stems that bear the strobili. The spores germinate into inconspicuous and independent gametophytes.

The stems are tough and rigid because of silica deposited in cell walls. Early Americans, in particular, used horsetails for scouring pots and called them "scouring rushes." Today they are still used as ingredients in a few abrasive powders.

Ferns

Ferns (division Pteridophyta, 12,000 species) are a widespread group of plants. They are most abundant in warm, moist tropical regions, but they are also found in northern regions and in dry, rocky places. They range in size from those that are low growing and resemble mosses to those that are tall trees. The fronds (leaves) in particular can vary. The royal fern has fronds that stand 6 feet tall; those of the Venus maidenhair fern are branched with broad leaflets. And those of the hart's tongue fern are straplike and leathery. In nearly all ferns, the leaves first appear in a curled-up form called a fiddlehead, which unrolls as it grows. Figure 32.9 shows the usual habitats of ferns and gives examples of their diversity. The life cycle of a typical temperate fern is shown in Figures 32.10 and 32C on page 571.

The fronds are megaphylls secondarily subdivided into leaflets. Megaphylls may have evolved in the following way:

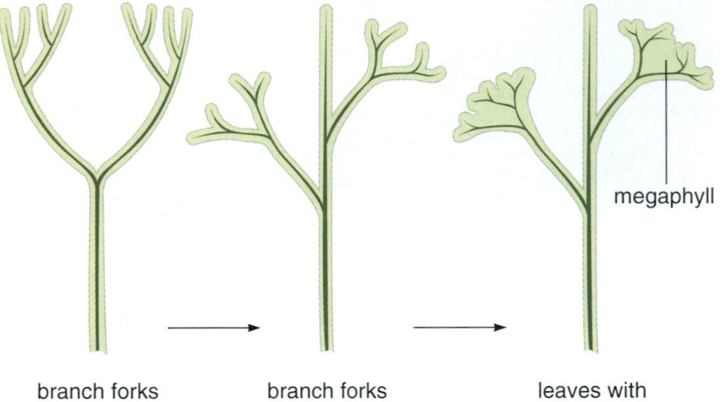

Uses of Ferns

At first it may seem that ferns do not have much economic value, but they are much used by florists in decorative bouquets and as ornamental plants in the home and garden. Wood from tropical tree ferns is often used as a building material because it resists decay, particularly by termites. Ferns have medicinal value; many Native Americans use ferns as an astringent during childbirth to stop bleeding, and the maidenhair fern is the source of an expectorant.

Adaptation of Ferns

Ferns have true roots, stems, and leaves. The well-developed leaves fan out, capture solar energy, and photosynthesize. The water-dependent gametophyte, which lacks vascular tissue, is separate from the sporophyte. Flagellated sperm require an outside source of water in which to swim to the eggs in the archegonia. Once established, some ferns, like the bracken fern *Pteridium aquilinum,* can spread into drier areas by means of vegetative (asexual) reproduction. Ferns also spread by means of the rhizomes growing horizontally in the soil, producing the fiddleheads that grow up as new fronds.

The seedless vascular plants include ferns and their allies. The sporophyte is dominant, and spores disperse the species. The independent, nonvascular gametophyte produces flagellated sperm.

Maidenhair fern, *Adiantum pedatum*

Royal fern, *Osmunda regalis*

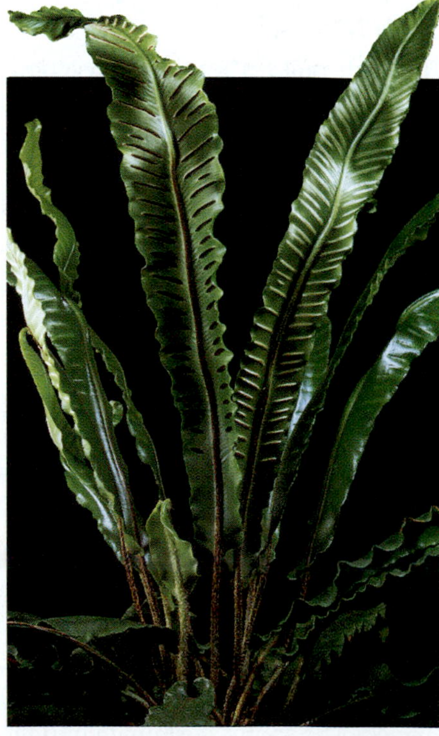

Hart's tongue fern, *Campyloneurum scolopendrium*

Figure 32.9 **Diversity of ferns.**

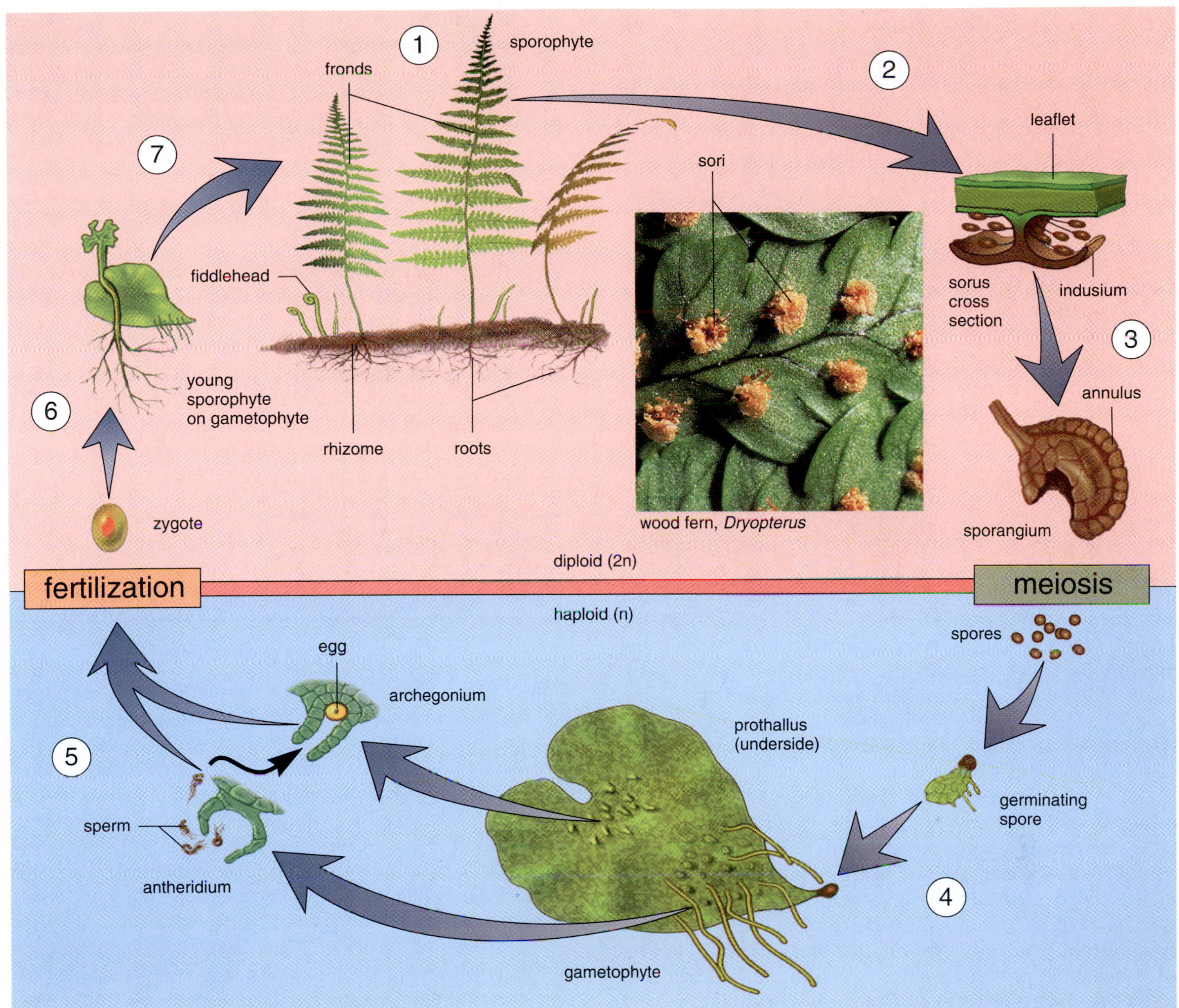

Figure 32.10 **Fern life cycle.**
① The sporophyte is dominant in ferns. ② In the fern shown here, the sori (sing., sorus), protected by an indusium, are on the underside of the leaflets. Within a sporangium, meiosis occurs and spores are produced. ③ As a band of thickened cells on the rim of a sporangium (the annulus) dries out, it moves backward, pulling the sporangium open, and the spores are released. ④ A spore germinates into a prothallus, which bears antheridia and archegonia on the underside. Typically, the archegonia are at the notch and antheridia are toward the tip, between the rhizoids. ⑤ Fertilization takes place when moisture is present, because the flagellated sperm must swim in a film of water from the antheridia to the egg within the archegonium. ⑥ The resulting zygote begins its development inside an archegonium, but the embryo soon outgrows the available space. As a distinctive first leaf appears above the prothallus, and as the roots develop below it, the sporophyte becomes visible. Often the sporophyte tissues and the gametophyte tissues are distinctly different shades of green. ⑦ The young sporophyte develops a root-bearing rhizome from which the fronds project.

32.5 Gymnosperms

There are four groups of **gymnosperms** [Gk. *gymnos,* naked, and *sperma,* seed]: conifers (division Pinophyta), cycads (division Cycadophyta), ginkgo (division Ginkgophyta), and gnetophytes (division Gnetophyta). Gymnosperms and angiosperms, discussed next, are the seed plants. A generalized life cycle for seed plants is shown in Figure 32D on page 571. Seed plants, like a few of their predecessors, produce heterospores called microspores and megaspores. There are also separate microgametophytes (male) and megagametophytes (female). A **microspore** develops into the immature microgametophyte contained within the **pollen grain** [L. *pollen,* fine dust]. After they are released, pollen grains develop into mature, sperm-bearing microgametophytes. **Pollination** is the transfer of pollen to the vicinity of the megagametophyte. The sperm is delivered to the egg through a pollen tube; therefore, no external water is required for fertilization. Independence from external water when reproducing is a major evolutionary trend in the plant kingdom.

A **megaspore** develops into an egg-bearing megagametophyte while still retained within an ovule. An **ovule** is the sporophyte structure that holds the megasporangium and then the megagametophyte. After fertilization, the zygote becomes an embryonic plant enclosed within the ovule, which then becomes the seed. Notice that megagametophytes and microgametophytes are dependent to a degree upon the diploid sporophyte. This means that the sporophyte can evolve into diverse forms without any corresponding changes in the gametophyte.

In seed plants, seeds are transported by various means to new locations. A **seed** is the mature ovule; it contains an embryonic sporophyte and stored food enclosed by a protective seed coat derived from the integuments (coverings) of the ovule. Seeds are resistant to adverse conditions, such as dryness or temperature extremes. They contain an embryo that is already partially developed and a food reserve that supports the emerging seedling until it can exist on its own. The survival value of seeds contributed greatly to the success of seed plants and their present dominance among plants.

Seed plants produce micro- and megagametophytes. This has led to two major evolutionary events: pollen grains that produce sperm, and seeds that disperse the sporophyte and the species.

In gymnosperms, the seeds are not covered. Instead, they are exposed on the surface of sporophylls, leaves that bear sporangia. (In angiosperms, seeds are enclosed within a fruit.) Reproductive organs are usually borne in **cones** on which the sporophylls are spirally arranged. Other than these features, the four divisions of gymnosperms have little in common.

The gymnosperms didn't begin to flourish until the Mesozoic era. When the supercontinent Pangaea formed during the later part of the former Paleozoic era, mountain ranges arose and deserts appeared on their leeward side. Land that had been swampy became much drier. A mass extinction occurred, and among those that all but vanished were the seedless vascular plants that had made up the great swamp forests of the Carboniferous period. Then the first seed plants, many of which are now extinct (such as the seed-bearing tree ferns), and also modern-day gymnosperms, with their many adaptations to life on land, came into their own.

The gymnosperms didn't begin to flourish until the Mesozoic era, when geological upheavals brought about a change in the climate that fostered their dominance.

Conifers

Conifers [L. *conus,* cone, and *fero,* carry] (**division Pinophyta,** 550 species) are cone-bearing trees such as pines, hemlocks, and spruces. Conifers, which usually have evergreen needlelike leaves, are well adapted to withstand extremes of temperature, humidity, and wind strength. The leaves have a thick cuticle, sunken stomates, and a reduced surface area. Conifers are found in windswept mountaintops, swampy lowlands, and semideserts—from the equator to the frigid far north. The taiga is a coniferous forest extending in a broad belt across northern Eurasia and North America.

Conifers set records for their size and longevity. A giant sequoia named the General Sherman tree, located in California's Sequoia National Park, is 84 meters tall, 10 meters in diameter, and weighs an estimated 1,385 tons. Douglas firs are quite tall and sometimes reach heights of 75 meters, but redwood trees commonly grow more than 90 meters high. Some redwood trees are 2,000 years old, but the gnarled and twisted bristlecone pines in the Nevada mountains are known to be more than 4,500 years old.

The life cycle of pines, outlined in Figure 32.11, demonstrates how fertilization does not require external water and the adaptation of the seed for dispersal by the wind.

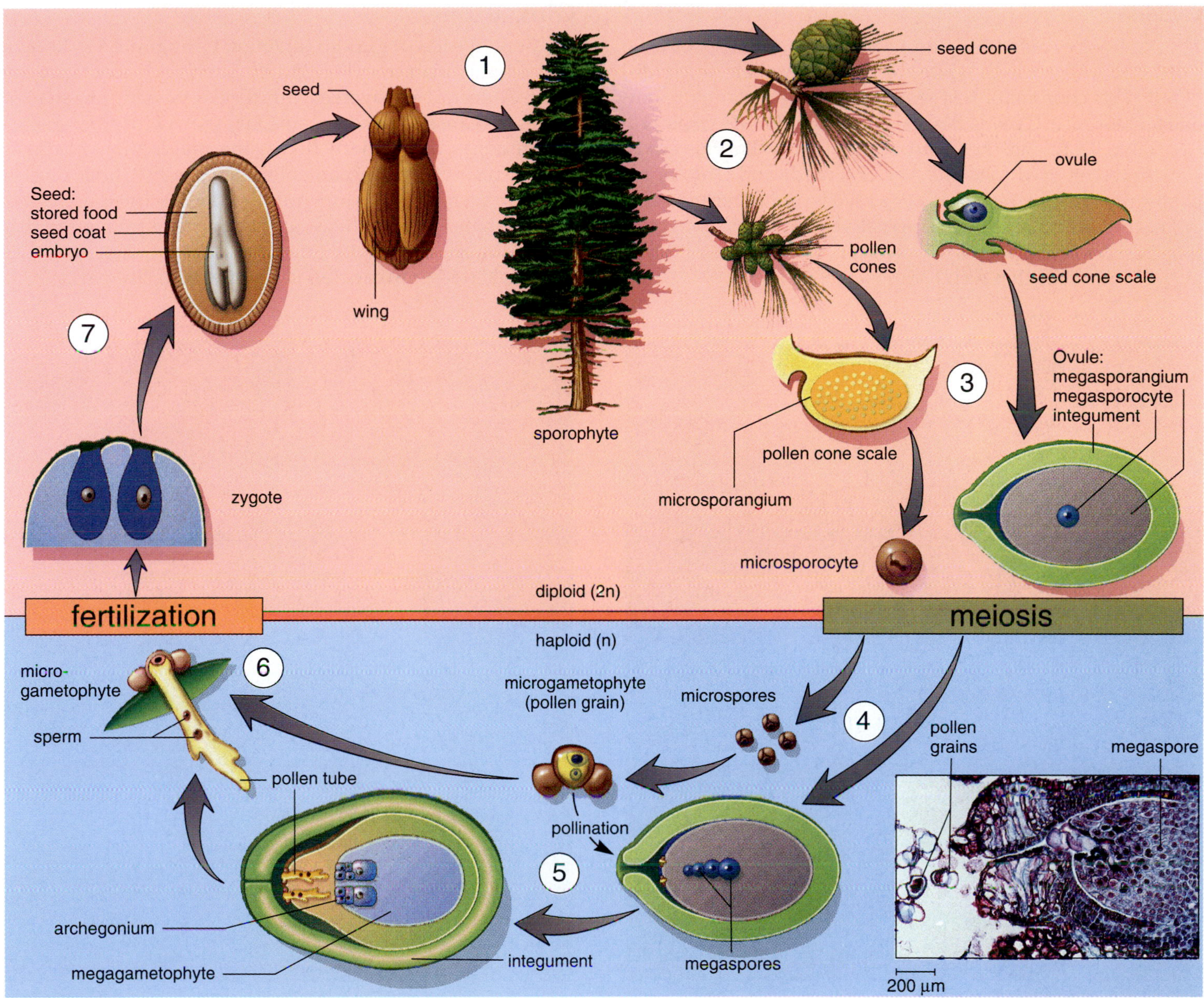

Figure 32.11 **Pine life cycle.**

① The sporophyte is dominant, and its sporangia are borne in cones. ② There are two types of cones: pollen cones and seed cones. Typically, the pollen cones are quite small and develop near the tips of lower branches. ③ Each scale of a pollen cone has two or more microsporangia on the underside. ④ Within these sporangia, each microsporocyte (microspore mother cell) undergoes meiosis and produces four microspores. ⑤ Each microspore develops into a microgametophyte, which is the pollen grain. The pollen grain has two wings and is carried by the wind to the seed cone during pollination. The seed cones are larger than the pollen cones and are located near the branch tips of higher branches. ③ Each scale of the seed cone has two ovules that lie on the upper surface. Each ovule is surrounded by a thick, layered integument, having an opening at one end. ④ The megasporangium is within the ovule, where a megasporocyte (megaspore mother cell) undergoes meiosis, producing four megaspores. ⑤ Only one of these spores develops into a megagametophyte, with two to six archegonia, each containing a single large egg lying near the ovule opening. ⑥ Once a pollen grain is enclosed within the seed cone, it develops a pollen tube that digests its way slowly toward a megagametophyte. The pollen tube discharges two nonflagellated sperm. One of these fertilizes an egg in an archegonium and the other degenerates. Fertilization, which takes place one year after pollination, is an entirely separate event from pollination. ⑦ After fertilization, the ovule matures and becomes the seed, composed of the embryo, the reserve food, and the seed coat. Finally, in the fall of the second season, the seed cone, by now woody and hard, opens to release winged seeds. When a seed germinates, the sporophyte embryo develops into a new pine tree, and the cycle is complete.

Cycads

The trunk of **cycads (division Cycadophyta,** 100 species) is stout and unbranched, while the large leaves are finely divided. This combination of features gives the plant a palm-like appearance. Cycads flourished during the Mesozoic era (see Table 32.2), and they were so numerous that this era is sometimes referred to as the Age of Cycads and Dinosaurs. Dinosaurs most likely fed on cycad seeds and foliage. Today, cycads are found mainly in tropical and subtropical regions (Fig. 32.12*a*). One species grows as tall as 20 meters and has meter-long cones that weigh almost 90 pounds.

Ginkgo Trees

Only one species of **ginkgo (division Ginkgophyta),** known as the maidenhair tree (Fig. 32.12*b*), survives today. Growing up to 70 feet, it is covered with forked-veined, fan-shaped leaves that are shed in autumn. Only the microsporangia occur in cones. In modern times, the maidenhair tree was largely restricted to ornamental gardens in China until it was discovered that it does quite well in polluted areas. Because female trees produce seeds with a fleshy covering that fall and make a foul-smelling mess, it is the custom to use only male trees, propagated vegetatively, in city parks.

a. Cycad, *Encephalartos humlis*

b. Maidenhair tree, *Ginkgo biloba*

c. Gnetophyte, *Welwitschia mirabilis*

Figure 32.12 **Lesser-known gymnosperms.**
a. Cycads resemble palm trees but are gymnosperms that produce cones. Cycads are dioecious; pollen cones and seed cones are on different plants. **b.** Ginkgoes exist only as a single species, the maidenhair tree. The plant in this photograph produces only seeds. Other plants have only pollen cones. **c.** There are three genera of gnetophytes—this is *Welwitschia mirabilis,* a very unusual African plant. The stem, which has two enormous straplike leaves, bears strobili.

A Closer Look

Comparing Plant Life Cycles

Three groups of plants are found in the fossil record. The nonvascular plants, represented today by the bryophytes, appear first. These are followed by the nonseed vascular plants and finally the seed plants. All three groups of plants have a life cycle described as alternation of generations: a diploid (2n) generation, the sporophyte, which produces haploid (n) spores by meiosis, alternates with a haploid generation, the gametophyte, which produces the gametes. Following fertilization, the diploid zygote develops into the sporophyte generation.

In mosses, a type of bryophyte, the haploid gametophyte is the long-lasting generation and the generation we recognize as the plant. Therefore, the gametophyte is dominant and more space is allotted to this generation in Figure 32B.

In bryophytes, the sperm are flagellated and swim in a film of water to the egg. The zygote is retained in the female gametangium, and the sporophyte is dependent upon the gametophyte—it is only temporarily present and photosynthesizes to a very limited extent. The sporophyte produces a single type of spore that is usually windblown and disperses the gametophyte generation.

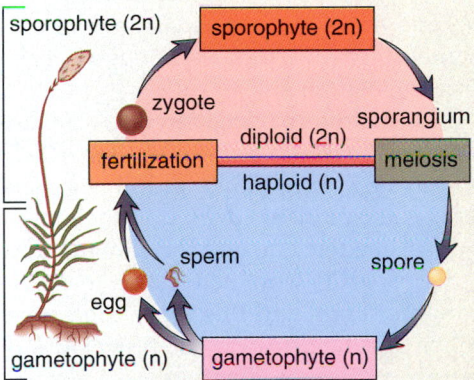

Figure 32B Nonvascular plants.
- Dominant gametophyte
- Flagellated sperm
- Dependent sporophyte
- Homospores disperse the species in mosses

In the ferns, a seedless vascular plant, both generations are independent but the sporophyte generation is the dominant one. How might this have happened? The sporophyte is attached to the almost microscopic gametophyte but, since it has vascular tissue, it becomes the larger and the longer lasting of the two generations. In Figure 32C more space is allotted to the sporophyte. The sporophyte generation produces usually windblown homospores following germination. The spore gives rise to a gametophyte, which produces both male and female gametes. Sperm are flagellated and swim to the egg in a film of water. These nonvascular plants were dominant and grew to enormous sizes during the Carboniferous period (see p. 559), when the weather was warm and wet. Many of these plants became extinct when the weather turned dryer and colder over much of the earth. Today's seedless vascular plants that live in the temperate zone use asexual propagation to spread into environments that are not favorable for a water-dependent gametophyte generation.

Figure 32C Seedless vascular plants.
- Dominant sporophyte
- Homospores disperse the species
- Independent gametophyte
- Flagellated sperm

Seed plants also have an alternation of generations life cycle, but it is much modified (Fig. 32D). The spores are retained within the sporophyte; there are heterosporangia and heterospores, termed microspores and megaspores. The gametophytes are so reduced that the megagametophyte is also retained within the structure (ovule) that held the megasporangium! Microspores become the windblown or animal-transported microgametophytes—the pollen grains. Pollen grains carry nonflagellated sperm to the egg-bearing megagametophyte. Following fertilization, the ovule becomes the seed. A seed contains a sporophyte embryo, and therefore seeds disperse the sporophyte generation. Fertilization no longer utilizes external water, and sexual reproduction is fully adapted to the terrestrial environment.

Figure 32D Seed plants.
- Dominant sporophyte
- Heterosporous
- Dependent micro-, megagametophytes
- Pollen grains
- Seeds disperse the species

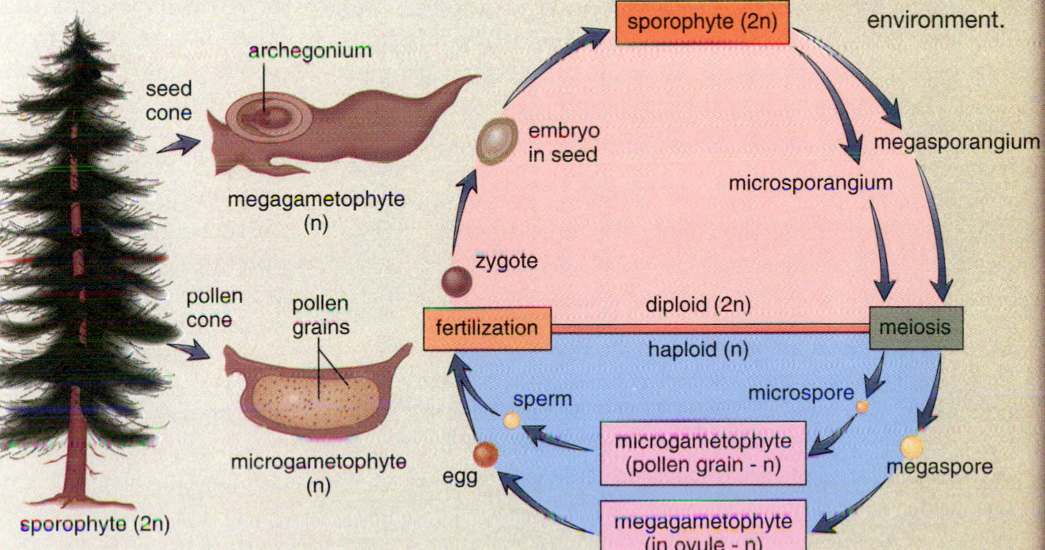

Gnetophytes

The **gnetophytes (division Gnetophyta,** 70 species), which also live mainly in the tropics, contain only three genera of plants. They look quite different from one another, though they share certain anatomical features (pertaining to their gametophytes and vascular tissue) that make them seem more related to angiosperms than other gymnosperms. Only one genus is found in the United States: *Ephedra* is a many-branched shrub with small, scalelike leaves that is found largely in desert regions. The leaves and stem of this plant are still brewed into Mormon tea. *Welwitschia* belongs to a genus of very unusual plants found in Africa. Most of the plant exists underground. Only a few centimeters of trunk grow aboveground, but the trunk can be over a meter wide! There are only two enormous, straplike leaves, which grow for hundreds of years. These may cover most of the exposed portion of the trunk (Fig. 32.12*c*).

Gymnosperms share a common characteristic: the seeds are naked (not covered). Conifers, with their ability to live under extreme conditions, are still a very prevalent and successful group.

Uses of Gymnosperms

Conifers grow on large areas of the earth's surface and are economically important. They supply much of the wood used for building construction and paper production. They also produce many valuable chemicals, such as those extracted from resin, a viscous liquid substance that protects the conifers from attack by fungi and insects.

Adaptation of Gymnosperms

Gymnosperms have well-developed roots and stems. Many are tall trees that can withstand heat, dryness, and cold. The reproductive pattern of conifers has several important innovations not found in the plants we have considered so far. There is no need of external water for fertilization to occur. Pollen grains are transferred by wind, and the growth of the pollen tube delivers a sperm to an egg. Enclosure of the dependent megagametophyte in an ovule protects it during its development and shelters the developing zygote as well. Finally, the embryo is protected within the seed and is provided with a store of nutrients that supports development for the first period of its growth following germination. All of these factors increase the chance for reproductive success on land.

Conifers are the most prevalent of the gymnosperms. In the conifer life cycle, windblown pollen grains make flagellated sperm unnecessary. Following fertilization, the seed develops from the ovule, a structure that is on a cone. The seeds are uncovered and dispersed by the wind.

32.6 Angiosperms

Angiosperms [Gk. *angion,* dim. of *angos,* vessel, and *sperma,* seed] **(division Magnoliophyta)** are the flowering plants; their seeds are enclosed by fruits. With 235,000 species, they are an exceptionally large and successful group of plants. This is six times the number of species of all other plant groups combined. Angiosperms live in all sorts of habitats, from fresh water to desert and from the frigid north to the torrid tropics. They range in size from the tiny, almost microscopic, duckweed to *Eucalyptus* trees that are over 100 meters tall. It would be impossible to exaggerate the importance of angiosperms in our everyday lives. They provide us with clothing, food, medicines, and commercially valuable products.

Origin and Evolution of Angiosperms

Angiosperms evolved from ancient gymnosperms. By the Jurassic period in the middle of the Mesozoic era, many of the gymnosperms had features similar to those we associate with angiosperms. Some had vascular tissue resembling that of angiosperms and some had sporophylls that were beginning to look like parts of flowers. The flowers were visited by beetles, which are among the earliest insects to evolve. Most of these transitional gymnosperms became extinct; only a few lines of descent are in existence today, notably the gnetophytes and the angiosperms. Among today's gymnosperms, the gnetophyte *Ephedra* is believed to be most closely related to angiosperms.

The oldest fossils definitely considered to be angiosperms come from the Cretaceous period (see Table 32.2) of the Mesozoic era, but flowering plants didn't diversify until the Cenozoic era. The continents drifted to their present locations during the Cenozoic era, and the climate grew colder. Perhaps the angiosperms were better able to diversify during these changing times, and this led to their present dominance. Perhaps also helpful was the prevalence of and diversity of insects to assist in pollination.

Rather than being tall trees, the first angiosperms may have been fast-growing woody shrubs that took on the appearance of weeds in open places. Both tall trees and herbaceous plants could have evolved from such an ancestor. Today many angiosperms are nonwoody **herbaceous plants** [L. *herba,* vegetation, plant] rather than being woody shrubs or trees. **Woody plants** have an internal skeleton composed of the xylem of previous seasons. Herbs are able to colonize disturbed places and exist in areas that woody plants cannot tolerate. Angiosperms that are adapted to the temperate, subarctic, and arctic zones are perennials or annuals. Perennials are plants that live two or more years; they die back seasonally in herbaceous plants and become dormant in woody plants. Annuals are plants that live for only one growing season.

Classification of Flowering Plants

Angiosperms are divided into two groups: dicotyledons (class Magnoliopsida) and monocotyledons (class Liliopsida). The **dicotyledons** [Gk, *dis,* two, and *cotyledon,* cup-shaped cavity] (or dicots) are either woody or herbaceous, and they have flower parts usually in fours and fives, net-veined leaves, vascular bundles arranged in a circle within the stem, and two cotyledons, or seed leaves. Dicot families include many familiar plant groups, such as the buttercup, mustard, maple, cactus, pea, and rose families. The rose family includes roses, apples, plums, pears, cherries, peaches, strawberries, raspberries, and a number of other shrubs. The **monocotyledons** (or monocots) are almost always herbaceous and have flower parts in threes, parallel-veined leaves, scattered vascular bundles in the stem, and one cotyledon, or seed leaf. Monocot families include the lily, palm, orchid, iris, and grass families. The grass family includes wheat, rice, corn (maize), and other agriculturally important plants.

The Flower

The flower consists of several kinds of highly modified leaves that are arranged in concentric rings and attached to a modified stem tip, the receptacle (Fig. 32.13). The sepals, which form the outermost ring, are frequently green and are quite similar to ordinary foliage leaves. They enclose the flower before it opens. Next are the petals, which are often large and colorful. Their color often helps attract pollinators. Within the petals, the stamens form a whorl around the pistil. **Stamens** are the pollen-producing portion of the flower. Each stamen has a slender filament with an anther at the tip. The **anthers** are modified sporophylls containing microsporangia where microspores which become pollen grains are produced. An anther could have evolved by the following stages involving a reduction in its leaflike portion:

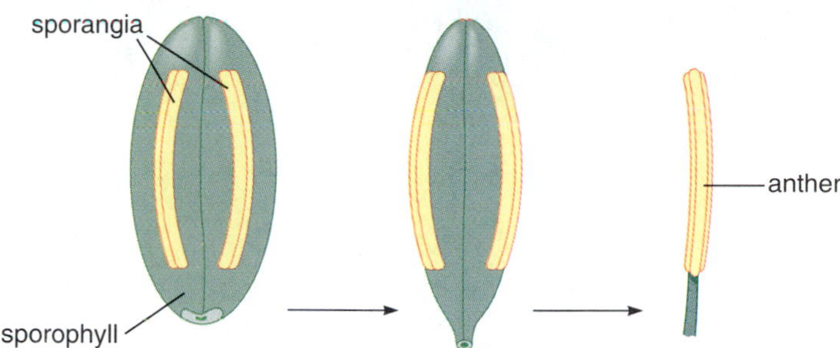

The pollen grains are either blown by the wind or carried by **pollinators** (most often insects) to the **pistil** which consists of **stigma, style,** and **ovary.** An ovary has one or more ovule-bearing units called a **carpel.** A simple ovary could have evolved by a folding of a single carpel:

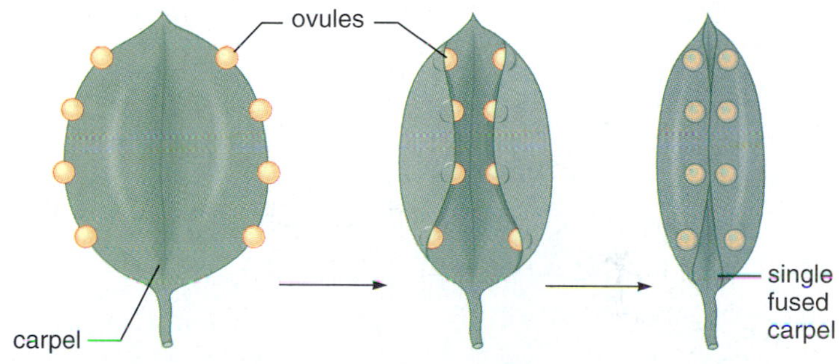

Compound ovaries formed by a fusing of separate folded carpels. Ultimately, the ovule becomes a seed and the ovary becomes a fruit; therefore, the seeds are enclosed by fruit. A **fruit** is a mature ovary and possible accessory part of a flower that provides a fleshy or dry covering for seeds. Fruits are sometimes specialized to aid in dispersal of seeds. Fleshy fruits may be eaten by animals, which transport the seeds to a new location and then deposit them when they defecate.

The life cycle of flowering plants is given in Figure 32.14.

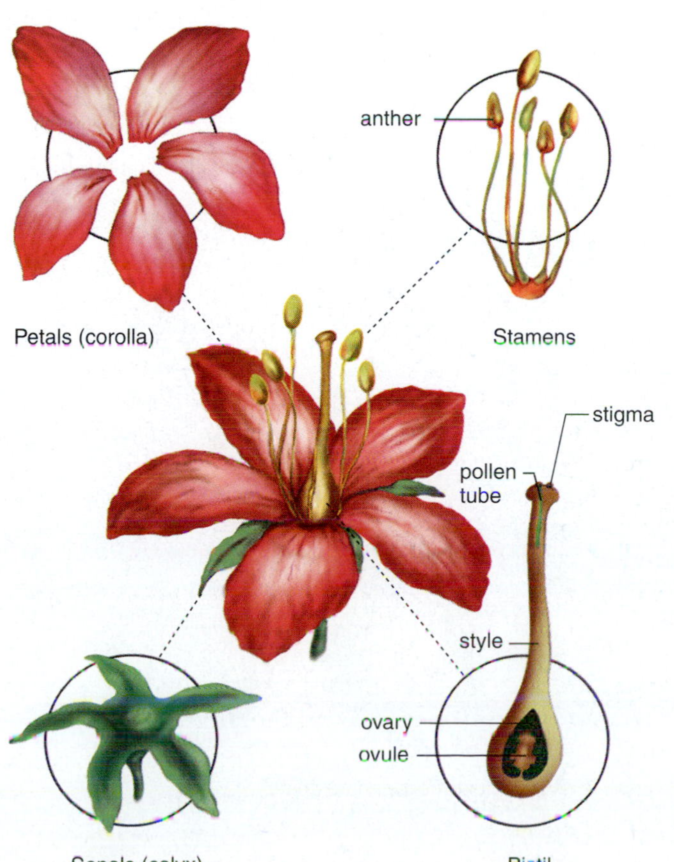

Figure 32.13 Generalized flower.
A flower has four main kinds of parts: sepals, petals, stamens, and pistils. A stamen has an anther and filament. A pistil has stigma, style, and ovary. An ovary contains ovules.

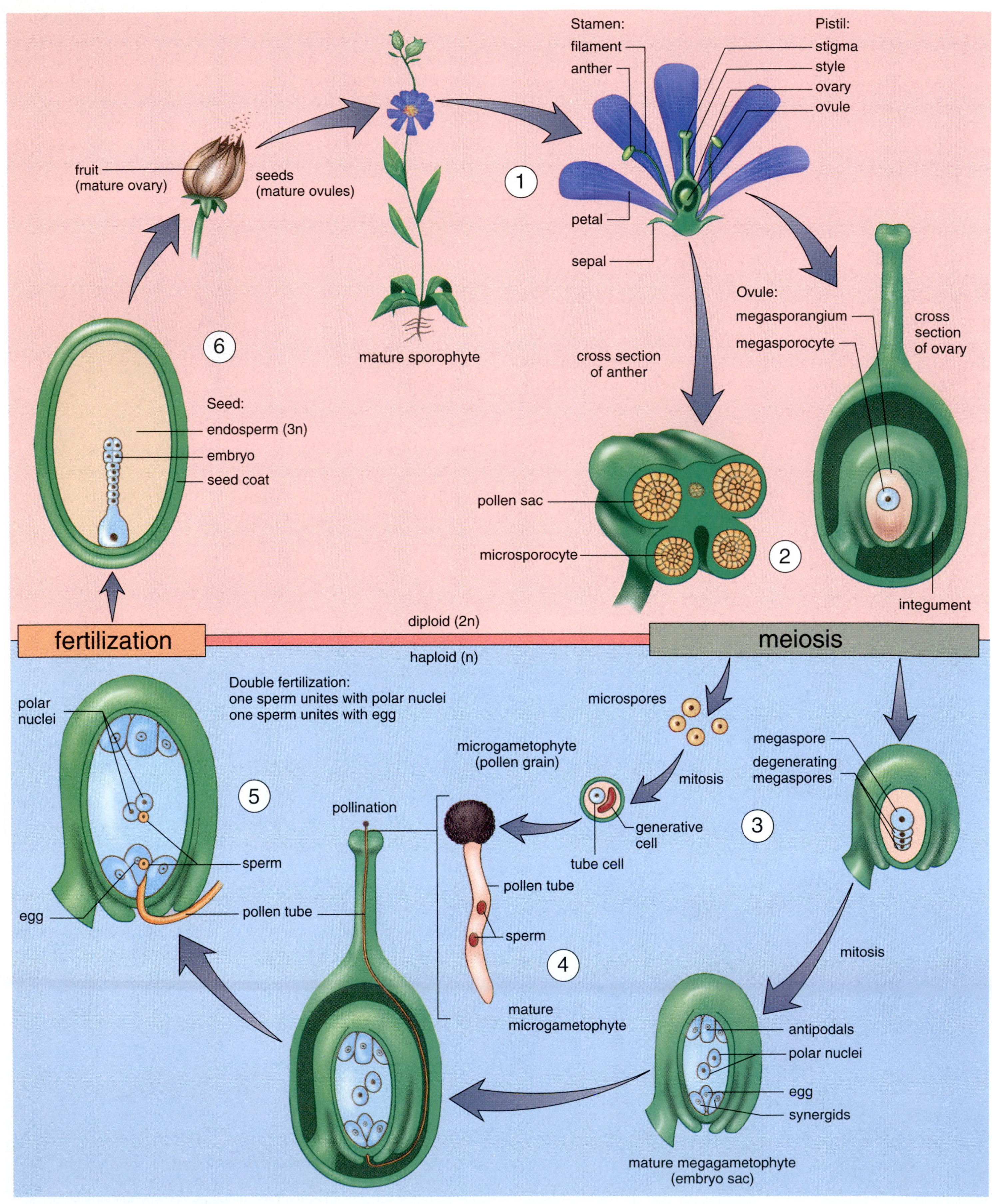
Stamen:
filament
anther
Pistil:
stigma
style
ovary
ovule
fruit (mature ovary)
seeds (mature ovules)
1
petal
sepal
mature sporophyte
Ovule:
megasporangium
megasporocyte
cross section of ovary
cross section of anther
6
Seed:
endosperm (3n)
embryo
seed coat
pollen sac
microsporocyte
2
integument
diploid (2n)
fertilization
meiosis
haploid (n)
Double fertilization:
one sperm unites with polar nuclei
one sperm unites with egg
polar nuclei
microspores
megaspore
degenerating megaspores
microgametophyte (pollen grain)
mitosis
5
pollination
generative cell
3
tube cell
sperm
pollen tube
egg
pollen tube
sperm
4
mitosis
mature microgametophyte
antipodals
polar nuclei
egg
synergids
mature megagametophyte (embryo sac)

Figure 32.14 Flowering plant life cycle.
① The parts of the flower involved in reproduction are the stamens and the pistil. Reproduction has been divided into development of the megagametophyte, development of the microgametophyte, double fertilization, and the seed. *Development of the megagametophyte.* The ovary at the base of the pistil contains one or more ovules. ② Within an ovule, a megasporocyte (megaspore mother cell) undergoes meiosis to produce four haploid megaspores. ③ Three of these megaspores disintegrate, leaving one functional megaspore, which divides mitotically. ④ The result is the megagametophyte, or embryo sac, which typically consists of eight haploid nuclei embedded in a mass of cytoplasm. The cytoplasm differentiates into cells, one of which is an egg and another of which is a diploid cell formed by the union of two polar nuclei. *Development of the microgametophyte.* ① The anther at the top of the stamen has pollen sacs, which contain numerous microsporocytes (microspore mother cells). ② Each microsporocyte undergoes meiosis to produce four haploid cells called microspores. When the microspores separate, each one becomes a microgametophyte or pollen grain. ③ At this point, the young microgametophyte contains two nuclei: the generative cell and the tube cell. Pollination occurs when pollen is windblown or carried by insects, birds, or bats to the stigma of the same type of plant. ④ Only then does a pollen grain germinate and produce a long pollen tube. This pollen tube grows within the style until it reaches an ovule in the ovary. Before fertilization occurs, the generative nucleus divides, producing two sperm, which have no flagella. This germinated pollen grain with its pollen tube and two sperm is the mature microgametophyte. *Double fertilization.* ⑤ On reaching the ovule, the pollen tube discharges the sperm. One of the two sperm migrates to and fertilizes the egg, forming a zygote; the other unites with the two polar nuclei, producing a 3n (triploid) endosperm nucleus. The endosperm nucleus divides to form endosperm, food for the developing plant. This so-called double fertilization is unique to angiosperms. *The seed.* ⑥ The ovule now develops into the seed, which contains an embryo and food enclosed by a protective seed coat. The wall of the ovary and sometimes adjacent parts develop into a fruit that surrounds the seeds. Therefore, angiosperms are said to have covered seeds.

Flowers and Diversification

We have seen that flowers are involved in the production and development of spores, gametophytes, gametes, and embryos enclosed within seeds. The successful completion of these processes requires the effective dispersal of pollen and then seeds. The various ways in which pollen and seeds can be dispersed has resulted in many different types of flowers.

Although some flowers disperse their pollen by wind, many are adapted to attract specific pollinators such as bees, wasps, flies, butterflies, moths, and even bats, which carry only a particular pollen from flower to flower. For example, bee-pollinated flowers are usually blue or yellow and have ultraviolet shadings that lead the pollinator to the location of nectar at the base of the flower. The mouthparts of bees are fused into a long tube that is able to obtain nectar from this location.

Not only do flowers lend themselves to efficient cross-pollination, they also aid in the dispersal of seeds by producing fruits. There are fruits that utilize wind, gravity, water, and animals for dispersal. Because animals live in particular habitats and/or have particular migration patterns, they are apt to deliver the fruit-enclosed seeds to a suitable location for seed germination (when the embryo begins to grow again) and development of the plant.

Flower diversification may be associated with the numerous means by which flowers are pollinated and fruits are dispersed.

Uses of Angiosperms

Angiosperms provide the food that sustains most animals on land, including humans. Humans also use plant fibers to produce cloth, to provide firewood, and to supply construction materials. Plant oils are not only used in cooking, they are also, for example, used in perfumes and as medicines. Spices and drugs are derived from different parts of various angiosperm plants. The uses of angiosperms are discussed further on pages 716–717.

Adaptation of Angiosperms

Angiosperms have true roots, stems, and leaves. The vascular tissue is well developed, and the leaves are generally broad. Angiosperms are found in all sorts of habitats; some have even returned to the water.

The reproductive organs are in the flowers, which often attract animal pollinators. Food is the reward that pollinators receive in exchange for their services. Some animals eat the pollen while many feed on the nectar secreted by special glands called nectaries that are hidden deep within the flower. Following fertilization the ovules located in ovaries develop into seeds and the ovaries develop into fruits. Therefore, angiosperms produce covered seeds. Fruits often help with dispersal of seeds.

Angiosperms are the flowering plants. The flower both attracts animals (e.g., insects) that aid in pollination and produces seeds enclosed by fruit, which aid dispersal.

Connecting Concepts

The first plants to invade land didn't have much competition for space and resources. Through the evolutionary process, they adapted to a dry environment and the threat of water loss (desiccation). Today, the nonvascular plants are generally small, and they are usually found in moist habitats. However, mosses can store large amounts of water and even become dormant during dry spells. The vascular plants, on the other hand, have specialized tissues to transport water and nutrients from one part of the plant to another. Therefore they can grow tall. And their leaves are covered by a waxy cuticle interrupted by small pores, which open and close to control water loss.

Reproductive strategies in plants are also adapted to a land environment. Mosses and ferns produce flagellated sperm that require external water, but then windblown spores disperse the species. In seed plants, pollen grains protect the sperm until they fertilize an egg. The sporophyte even retains the egg-producing megagametophyte and the resulting zygote. The seed protects the sporophyte embryo from drying out until it germinates in a new location. Dispersal of plants by spores or seeds reduces competition for resources.

Humans use plants for various purposes. Massive amounts of biomass were submerged by swamps and covered by sediment during the Carboniferous period. Due to extreme pressure, this organic material became the fossil fuel coal, which, along with the other fuels, makes our way of life today possible. Also, we must not forget that plants produce food and oxygen, two resources that keep our biosphere functioning today.

Summary

32.1 Classification of Plants

Plants are photosynthetic organisms adapted to a land existence. Among the various adaptations, all plants protect the developing embryo from desiccation. All plants have the alternation of generations life cycle. Some types of plants have a dominant gametophyte; others have a dominant sporophyte.

32.2 Nonvascular Plants

The nonvascular plants, which include the liverworts and mosses, are nonvascular plants and therefore lack true roots, stems, and leaves. In the moss life cycle, the gametophyte is dominant. Antheridia produce swimming sperm that need external water to reach the eggs in the archegonia. Following fertilization, the dependent sporophyte consists of a foot, a stalk, and a capsule within which windblown spores are produced by meiosis. Each spore germinates to produce a gametophyte.

32.3 Vascular Plants

Vascular plants arose during the Silurian period of the Paleozoic era. The extinct rhyniophytes may be ancestral vascular plants. These plants had photosynthetic stems (no leaves or roots) with sporangia at their tips. Most likely, the life cycle was similar to today's ferns. The sporophyte, which is diploid and has vascular tissue, is the dominant generation in ferns. The separate gametophyte produces flagellated sperm. The sporophyte of seed plants produces heterospores which develop into heterogametes. Every aspect of the life cycle is adapted to a dry environment.

32.4 Ferns and Allies

The seedless vascular plants include whisk ferns, club mosses, horsetails, and ferns. Lycopods, horsetails, and ferns were also trees during the Carboniferous period, although lycopods and horsetails are limited in diversity and rather small today. Seedless vascular plants have a life cycle like that of the ferns.

In ferns, the separate and water-dependent gametophyte (the heart-shaped prothallus) produces flagellated sperm in antheridia and eggs in archegonia. Following fertilization, the zygote develops into the sporophyte, which has large fronds. On the underside of the fronds are sori (sing., sorus), each containing several sporangia. Here meiosis produces windblown spores, each of which develops into a prothallus.

The Mesozoic era saw many geological changes as Pangaea formed and then broke apart. A mass extinction occurred that paved the way for the diversification of the seed plants. Those seed plants that are trees have especially well-developed roots and stems due to secondary growth of vascular tissue. Seed plants produce heterospores, microgametophytes, and megagametophytes. Microgametophytes develop into pollen grains that will produce sperm. Pollen grains which are often windblown make flagellated sperm unnecessary. The megagametophyte is retained within the ovule that develops into the seed.

32.5 Gymnosperms

There are four divisions of gymnosperms (seed plants that bear naked seeds): the familiar conifers and the little-known cycads, ginkgo, and gnetophytes. In conifers, pollen (male) and seed (female) cones are produced by the sporophyte plant. On the underside of a pollen cone scale, there are two microsporangia that produce microspores; each becomes a microgametophyte, or pollen grain. On the upper surface of a seed cone scale, there are two ovules, where meiosis produces one megaspore that develops into the megagametophyte. After windblown pollination, the pollen grain develops a tube through which sperm reach the egg. After fertilization, the ovule matures to be the seed.

32.6 Angiosperms

Angiosperms (seed plants that bear seeds protected by a fruit) are more diverse than the other types of plants. Their success may be associated with climatic changes in the Cenozoic era.

In a flower, the microsporangia develop within the anther portion of a stamen, and the megasporangia develop within ovules located in the ovary of the pistil. Pollination brings the mature microgametophyte (pollen grain) to the pistil, and the pollen tube brings the sperm to the ovule within the ovary. Angiosperms exhibit double fertilization: one sperm fertilizes the egg, and the other unites with the polar nuclei to form the endosperm, which is food for the embryo. The ovule develops into the seed, and the ovary becomes the fruit.

Angiosperms have complex vascular tissue and are found in various habitats. Their reproductive organs are found in flowers. Animal pollination increases the chance of appropriate fertilization, and fruit production often assists the dispersal of seeds.

Reviewing the Chapter

1. What are the characteristics that define plants? 558–559
2. What are the general characteristics of nonvascular plants, and what are the three main types? 559
3. Draw a diagram to describe the life cycle of the moss, pointing out significant features. 561
4. What are the human uses of nonvascular plants, and what are their adaptations? 562
5. What are the general characteristics of vascular plants, and how do features of the seedless plant life cycle differ from that of the seed plant life cycle? 564–565
6. What is the significance of rhyniophytes in the history of plants? What are the living seedless vascular plants, and describe the period of time in the earth's history when they were larger and more abundant than today. 554–565
7. Draw a diagram to describe the life cycle of the fern, pointing out significant features. What are the human uses of ferns, and what are their adaptations? 566–567
8. List and describe the four divisions of gymnosperms. 568, 570, 572
9. Use a diagram of the pine life cycle to point out significant features, including those that distinguish a seed plant's life cycle from that of a seedless vascular plant. 569
10. What are the human uses of gymnosperms, and what are their adaptations? 572
11. What role may have been played by beetles and insects in the evolution of flowering plants? 572
12. What are the parts of a flower? Use a diagram to explain and point out significant features of the flowering plant life cycle. 573–574
13. Offer an explanation as to why flowering plants are the dominant plants today. What are the human uses of angiosperms, and what are their adaptations? 575

Testing Yourself

Choose the best answer for each question.

1. Which of these are characteristics of plants?
   a. multicellular with specialized tissues and organs
   b. photosynthetic and contain chlorophylls *a* and *b*
   c. protect the developing embryo from desiccation
   d. have an alternation of generations life cycle
   e. All of these are correct.
2. In the moss life cycle, the sporophyte
   a. consists of leafy green shoots.
   b. is the heart-shaped prothallus.
   c. consists of a foot, a stalk, and a capsule.
   d. is the dominant generation.
   e. All of these are correct.
3. The rhyniophytes
   a. are a flourishing group of plants today.
   b. had large leaves like today's ferns.
   c. had sporangia at the tips of branches.
   d. have green stems.
   e. Both c and d are correct.
4. You are apt to find ferns in a moist location because they
   a. have a water-dependent sporophyte generation.
   b. have flagellated spores.
   c. are flagellated.
   d. are symbiotes that need water.
   e. All of these are correct.
5. Which of these is mismatched?
   a. pollen grain—microgametophyte
   b. ovule—megagametophyte
   c. seed—immature sporophyte
   d. pollen tube—spores
   e. tree—mature sporophyte
6. In the life cycle of the pine tree, the ovules are found on the
   a. needlelike leaves.
   b. seed cones.
   c. pollen cones.
   d. roots, stems, and leaves.
   e. All of these are correct.
7. Which of these is mismatched?
   a. anther—produces microsporangia
   b. pistil—produces pollen
   c. ovule—becomes seed
   d. ovary—becomes fruit
   e. flower—reproductive structure
8. Which of these plants contributed the most to our present-day supply of coal?
   a. nonvascular plants
   b. seedless vascular plants
   c. conifers
   d. angiosperms
   e. Both c and d are correct.
9. Which of these is found in seed plants?
   a. complex vascular tissue
   b. pollen grains that are not flagellated
   c. retention of megagametophyte within the ovule
   d. roots, stems, and leaves
   e. All of these are correct.
10. Label the following diagram of alternation of generations.

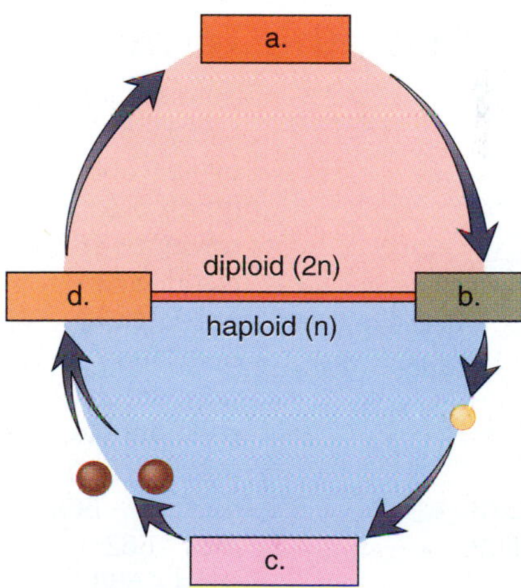

Thinking Scientifically

1. Design an experiment that tests the need for water in order for mosses to undergo alternation of generation instead of reproducing asexually by fragmentation.
2. Before the evolution of the vascular system the terrestrial landscape was essentially "two-dimensional." By allowing plants to rise above the ground, the vascular system created a three-dimensional world. How does being tall give plants an advantage?

Bioethical Issue

How far should you go to save an endangered species? As a society we seem to have answered that any amount of time, energy, and money is worth the saving of certain species. Think of all the effort to preserve the condor, a magnificent bird that now graces the California skies once more. Would we put so much effort into preserving any animal? any plant?

Largely because of human encroachment, dozens of American plant species are disappearing from their native habitat each year. Botanists estimate that some 3,000 of the 22,000 species of flowering plants in the U.S. may be facing extinction. Enter the Center for Plant Conservation, which has its headquarters at Harvard University's Arnold Arboretum. The center aims to preserve every kind of threatened plant in the U.S. through a network of 18 affiliated botanical gardens and horticultural research facilities in 14 states. Aside from growing the plants in greenhouses and other protected environments, the center stockpiles seeds of most species at the Department of Agriculture's Fort Collins, Colorado seed storage facility. In that way, despite any disaster that may wipe out a particular greenhouse or garden, the seeds would be available for propagation of new plants.

Should we spend money to save, say, the frostweed or the small whorled pogonia? Some say yes because of possible medical benefits. They point out, for example, that antitumor alkaloids found in the rosy periwinkle are now used in the treatment of childhood leukemia and Hodgkin's disease. Also, some of the plants resist disease or can survive droughts, characteristics that could be important to agriculture. Through biotechnology techniques, it may be possible to transfer these traits from wild plants to crop species some day. Are you in favor of preserving the habitats of plants if it means giving up jobs for humans?

Understanding the Terms

Match the terms to these definitions:

a. ______________ Diploid generation of the alternation of generations life cycle of a plant; meiosis produces haploid spores that develop into the gametophyte.
b. ______________ Flowering plant group; members have one embryonic leaf, parallel-veined leaves, scattered vascular bundles, and other characteristics.
c. ______________ Microgametophyte in seed plants.
d. ______________ Rootlike hair that anchors a nonvascular plant and absorbs minerals and water from the soil.
e. ______________ Vascular tissue that conducts organic solutes in plants; it contains sieve-tube elements and companion cells.

Web Connections

Exploring the Internet

http://www.mhhe.com/biosci/genbio/mader
(click on *Biology 7/e*)

The *Biology 7/e* Online Learning Center provides many resources for studying the material in this chapter including links to the following sites:

American Bryological and Lichenological Society. Many Internet resources on lichens and bryophytes.

http://www.unomaha.edu/~abls/

Botanical Society of America. Information on the American Journal of Botany, careers in botany, and many web links.

http://www.botany.org/

Careers in Botany, by the Botanical Society of America. Many associated links.

http://www.botany.org/bsa/careers/

Plant Nomenclature from the University of Maryland. Interesting, technical information on the taxonomy of plants.

http://www.inform.umd.edu:8080/EdRes/Colleges/LFSC/life_sciences/plant_biology/fam/revfam.html

Arizona's Tree of Life Page on the Green Plants. A description of the taxonomy of this group with links to some groups.

http://phylogeny.arizona.edu/tree/eukaryotes/green_plants/embryophytes/embryophytes.html

Animals: Introduction to Invertebrates

c h a p t e r

c h a p t e r c o n c e p t s

33.1 Classification of Animals

- Animals are multicellular heterotrophs that move about and ingest their food. They have the diplontic life cycle. 580
- Animals are classified according to certain criteria, including type of coelom, symmetry, body plan, development, and presence of segmentation. 580

33.2 Multicellularity

- Sponges have the cellular level of organization and lack tissues and symmetry. They depend on a flow of water through the body to acquire food. 582

33.3 Two Tissue Layers

- Cnidarians and comb jellies have a radially symmetrical saclike body consisting of two tissue layers derived from the germ layers ectoderm and endoderm. 584
- Cnidarians typically are either polyps (e.g., *Hydra*) or medusa (e.g., jellyfishes) or alternate between these two forms, the polyp being an asexual state and the medusa being a sexual state. 585

33.4 Bilateral Symmetry

- Ribbon worms and flatworms have tissues and organs derived also from mesoderm, the third germ layer. They have the organ level of organization and are bilaterally symmetrical. 588
- Planarians are free-living predators, but flukes and tapeworms are animals adapted to a parasitic way of life. 588

33.5 A Pseudocoelom

- Roundworms and rotifers have a coelom, a body cavity where organs are found and that can serve as a hydrostatic skeleton. Their coelom is a pseudocoelom because it is not completely lined by mesoderm. 593
- Roundworms take their name from a lack of segmentation; they are very diverse and include some well-known parasites. 593

Yellow coral polyps, *Parazoanthus gracilis*

Jeremy Jackson of the Smithsonian Tropical Research Institute wonders if he is doing enough to warn the public that we may lose 60% of all reefs by the year 2050. A coral reef is formed of limestone deposited by stony corals, an invertebrate animal. At a reef, many different types of aquatic protists and animals find a home and interact with one another in a complex ecosystem.

Reefs around the globe are in danger. Tons of soil from deforested tracts bring nutrients that stimulate the overgrowth of algae. This has caused a population explosion of the crown-of-thorn seastar that is ravaging the Great Barrier Reef of Australia. So-called coral bleaching occurs when pollutants and unusually warm seawater make corals expel their symbiotic dinoflagellates.

Overfishing of the reefs is commonplace. The methods used are sinister, including the use of dynamite to kill fish, cyanide to stun them, and satellite navigation systems to home in on areas where mature fish spawn. Yet, intact coral reefs are a storm barrier that protects the shoreline and provides a safe harbor for ships. And like a tropical rain forest they are most likely sources of medicines yet to be discovered.

33.1 Classification of Animals

Whereas plants are multicellular photosynthetic organisms, animals are multicellular eukaryotes that are heterotrophic by ingesting food (Fig. 33.1). Unlike the fungi that rely on external digestion, animals usually digest their food in a central cavity. Animals produce heterogametes (eggs and sperm), and they follow the diplontic life cycle in which the adult is always diploid. In this cycle, meiosis produces haploid gametes, which join to form a zygote that develops into an adult.

All animals are believed to have evolved from a protistan ancestor (Fig. 33.2). There are approximately 34 animal phyla, but we will consider in depth only the phyla illustrated in the tree. These are the ones recognized as the major animal phyla. All these phyla contain **invertebrates** [L. *in*, without, and *vertebra*, bones of backbone], which are animals without an endoskeleton of bone or cartilage. The phylum Chordata also contains the vertebrates, which are animals with an endoskeleton of bone or cartilage. Many invertebrates live in the sea, where early animal evolution occurred. All major animal phyla are represented by Cambrian fossils, but it has been very difficult to trace the complete evolutionary history because the fossil record is so much better for hard-shelled animals. In large part, the possible evolutionary relationship of living invertebrates has been worked out by using the anatomical criteria noted in the tree.

One criterion used to classify animals is type of symmetry. **Asymmetry** means that the animal has no particular symmetry. **Radial symmetry** means that the animal is organized circularly and, just as with a wheel, two identical halves are obtained no matter how the animal is sliced longitudinally. **Bilateral symmetry** means that the animal has definite right and left halves; only one longitudinal cut down the center of the animal will produce two equal halves. Radially symmetrical animals tend to be attached to a substrate; that is, they are **sessile.** This type of symmetry is useful to these animals since it allows them to reach out in all directions from one center. Bilaterally symmetrical animals tend to be active and to move forward at an anterior end.

One of the main events during the development of animals is the establishment of germ layers from which all other structures are derived. Although a total of three germ layers is seen in most animal embryos, some animals only have two germ layers: ectoderm and endoderm. Such animals have the tissue level of organization. Animals with three germ layers—ectoderm, mesoderm, and endoderm—have an organ level of organization.

The ribbon worms and flatworms lack a true **coelom** [Gk. *koiloma*, cavity], an internal body cavity completely lined by mesoderm, where internal organs are found. They are acoelomates because they have no coelom at all. The rotifers and roundworms are pseudocoelomates. They have a cavity, though it is incompletely lined with mesoderm. There is a layer of mesoderm beneath the body wall but not around the gut. The rest of the animal phyla are true coelomates—they have a coelom that is completely lined with mesoderm. Coelomates are either protostomes or deuterostomes. When the blastopore (the site of invagination of endoderm during development) is associated with the mouth, the animal is a protostome. When the blastopore is associated with the anus and a second opening becomes the mouth, the animal is a deuterostome (see Fig. 34.1).

Coelomates are also divided into those that are nonsegmented and those that are segmented. Segmented animals have repeating units. **Segmentation** leads to specialization of parts because the various segments can become differentiated for specific purposes.

Classification of animals is based on type of symmetry, number of tissue layers, type of coelom, and presence of segmentation.

American toad, *Bufo*, feeding on an earthworm, *Lumbricus*

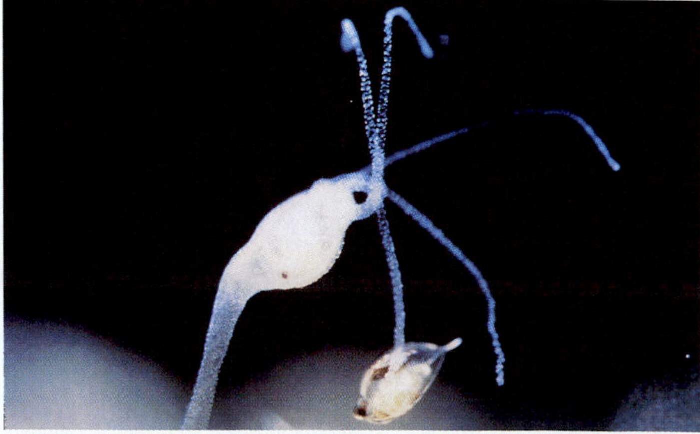

Hydra, *Hydra*, feeding on a water flea, *Daphnia*

Figure 33.1 Animals—multicellular, heterotrophic eukaryotes.
The toad is a vertebrate but the other animals depicted (hydra, water flea, and earthworm) are invertebrates. All animals are multicellular heterotrophic organisms that must take in preformed food.

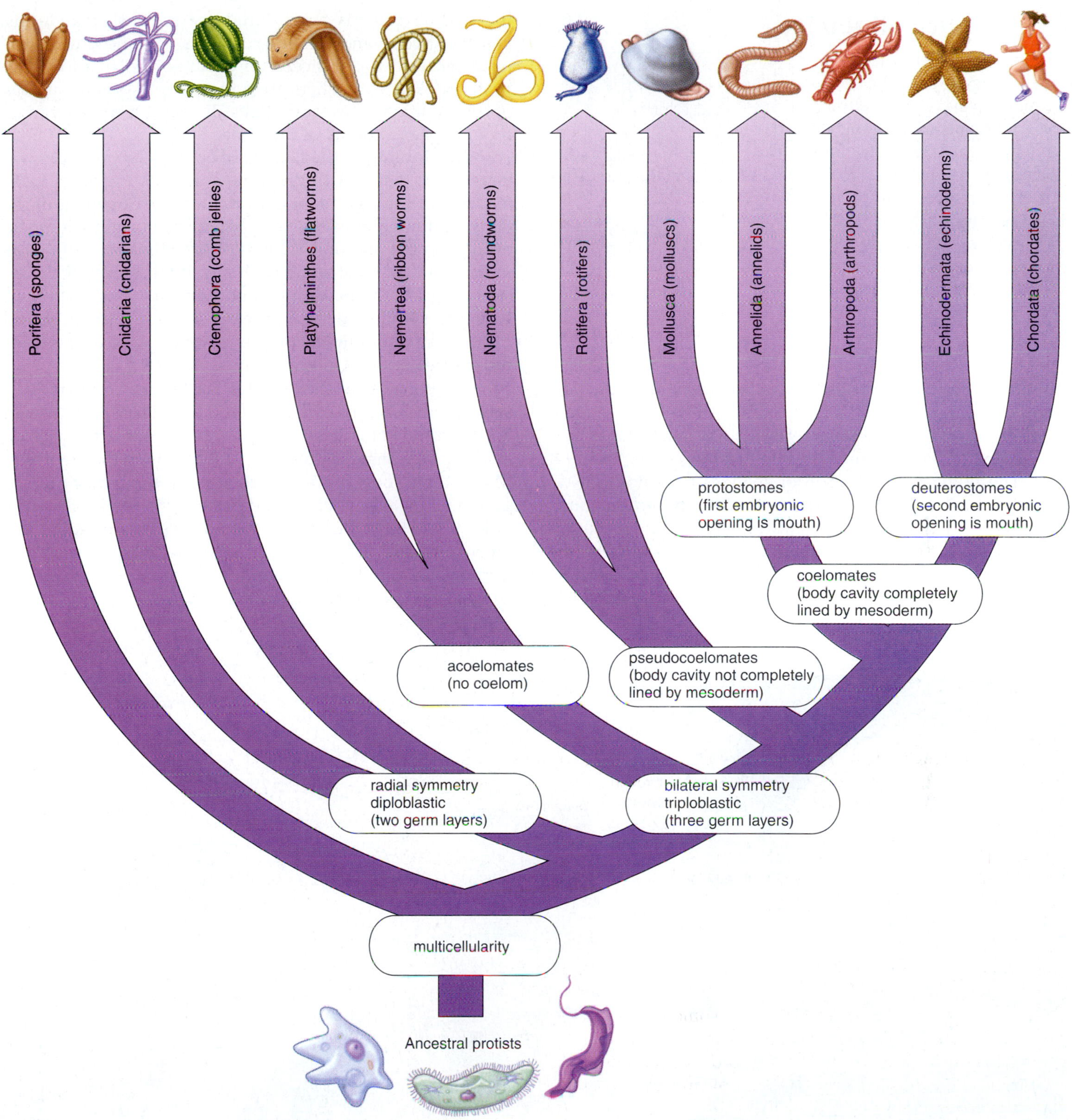

Figure 33.2 **Phylogenetic tree of the animal kingdom.**
All animals are believed to be descended from protists; however, the sponges may have evolved from protists separately from the rest of the animals.

33.2 Multicellularity

All animals are multicellular, but sponges are the only animals to have the cellular level of organization. Sponges, which have no symmetry and no tissues, are believed to be out of the mainstream of animal evolution. Most likely they evolved separately from protozoan ancestors and represent a dead-end branch of the evolutionary tree.

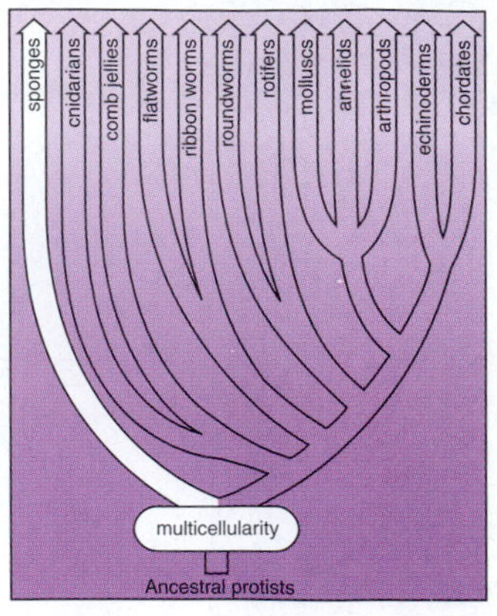

Sponges

Sponges (phylum Porifera, 5,000 species) are aquatic, largely marine animals, that vary greatly in size, shape, and color. Their saclike bodies are perforated by many pores; the phylum name, Porifera, means pore bearing.

The cellular organization of sponges is demonstrated by experimentation. After a sponge is broken into separate cells, the cells will exist individually until they spontaneously reorganize into a sponge once again. What types of cells are found in a sponge? The outer layer of the wall contains flattened epidermal cells, some of which have contractile fibers; the middle layer is a semifluid matrix with wandering amoeboid cells; and the inner layer is composed of flagellated cells called collar cells (or choanocytes) that look like protozoa (Fig. 33.3). There are no nerve cells or other means of coordination between the cells. To some, a sponge can be thought of as a colony of protozoa.

The beating of the flagella of collar cells produces water currents that flow through the pores into the central cavity and out through the osculum, the upper opening of the body. Although it may seem that sponges can't do much, even a simple one only 10 cm tall is estimated to filter as much as 100 liters of water each day. It takes this much water to supply the needs of the organism. Simple sponges have pores leading directly from the outside into the central cavity. Larger and more complex sponges have canals leading from external to internal pores.

A sponge is a **sessile filter feeder,** an organism that stays in one place and filters its food from the water. Oxygen is also supplied to the sponge, and waste products are taken away by the constant stream of water through the organism. Microscopic food particles brought by the water are engulfed by the collar cells and digested by them in food vacuoles or are passed to the amoeboid cells for digestion. The amoeboid cells also act as a circulatory device to transport nutrients from cell to cell, and they produce the sex cells (the egg and the sperm) and spicules.

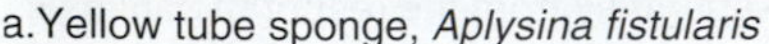
a. Yellow tube sponge, *Aplysina fistularis*

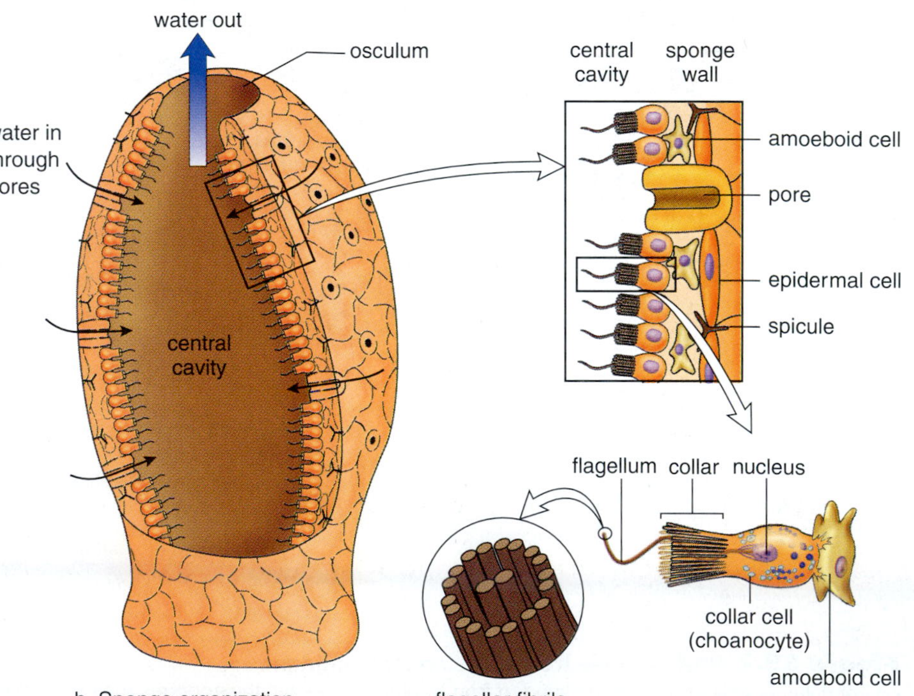

b. Sponge organization

Figure 33.3 Simple sponge anatomy.
a. Photograph of simple sponges. **b.** The wall contains two layers of cells: the outer epidermal cells and the inner collar cells. The collar cells (enlarged) have flagella that beat, moving the water through pores as indicated by the arrows. Food particles in the water are trapped by the collar cells and digested within their food vacuoles. Amoeboid cells transport nutrients from cell to cell; spicules compose an internal skeleton of some sponges.

CLASSIFICATION

Kingdom Animalia

Multicellular organisms with well-developed tissues; usually motile; heterotrophic by ingestion, generally in a digestive cavity; diplontic life cycle. Protostomes include phyla Mollusca, Annelida, and Arthropoda. Deuterostomes include phyla Echinodermata, Hemichordata, and Chordata.

Invertebrates*

Phylum Porifera: sponges
Phylum Cnidaria: jellyfishes, sea anemones, corals
Phylum Ctenophora: comb jellies, sea walnuts
Phylum Platyhelminthes: flatworms, e.g., planarians, flukes, tapeworms
Phylum Nemertea: ribbon worms
Phylum Nematoda: roundworms
Phylum Rotifera: rotifers
Phylum Mollusca: chitons, snails, slugs, clams, mussels, squids, octopuses
Phylum Annelida: segmented worms, e.g., clam worms, earthworms, leeches
Phylum Onychophora: walking worms
Phylum Arthropoda: spiders, scorpions, horseshoe crabs, lobsters, crayfish, shrimps, crabs, millipedes, centipedes, insects
Phylum Echinodermata: sea lilies, sea stars, brittle stars, sea urchins, sand dollars, sea cucumbers, sea daisies
Phylum Hemichordata: acorn worms
Phylum Chordata
- Subphylum Urochordata: tunicates
- Subphylum Cephalochordata: lancelets

Vertebrates*

- Subphylum Vertebrata
  - Superclass Agnatha: jawless fishes, e.g., lampreys, hagfishes
  - Superclass Gnathostomata: jawed fishes, all tetrapods
    - Class Chondrichthyes: cartilaginous fishes, e.g., sharks, skates, rays
    - Class Osteichthyes: bony fishes, e.g., herring, salmon, cod, eel, flounder
    - Class Amphibia: frogs, toads, salamanders, newts, caecilians
    - Class Reptilia: snakes, lizards, turtles, crocodiles
    - Class Aves: birds, e.g., sparrows, penguins, ostriches
    - Class Mammalia: mammals, e.g., cats, dogs, horses, rats, humans

* Not in the classification of organisms, but added here for clarity.

Sponges can reproduce asexually by fragmentation or by budding. During budding, a small protuberance appears and gradually increases in size until a complete organism forms. Budding produces colonies of sponges that can become quite large. During sexual reproduction, eggs and sperm are released into the central cavity. Fertilization results in a zygote that develops into a ciliated larva that may swim to a new location. Such a larva assures dispersal of the species for the sessile animal. Like all less specialized organisms, sponges are capable of regeneration, or growth of a whole from a small part. Thus, if a sponge is removed, chopped up, and returned to the water, each piece may grow into a complete sponge.

Sponges are classified on the basis of the type of skeleton they have. Some sponges also have an internal skeleton composed of **spicules** [L. *spicula*, dim. of *spika*, spear], little needle-shaped structures with one to six rays. Chalk sponges have spicules made of calcium carbonate; glass sponges have spicules that contain silica. Most sponges have fibers of spongin, a modified collagen. But some sponges contain only spongin fibers; a bath sponge is the dried spongin skeleton from which all living tissue has been removed. Today, however, commercial "sponges" are usually synthetic.

Sponges have a cellular level of organization and most likely evolved independently from protozoa. They are the only animals in which digestion occurs within cells.

33.3 Two Tissue Layers

As mentioned, during animal development there are a total of three possible germ layers: ectoderm, endoderm, and mesoderm. Animals in two phyla—comb jellies in phylum Ctenophora and cnidarians in phylum Cnidaria—develop only ectoderm and endoderm. Therefore, they are said to be diploblasts (Fig. 33.4). Animals in these phyla are radially symmetrical, meaning that any longitudinal cut, such as those shown, produces two identical halves. If an animal is bilaterally symmetrical, only the longitudinal cut shown yields two roughly identical halves.

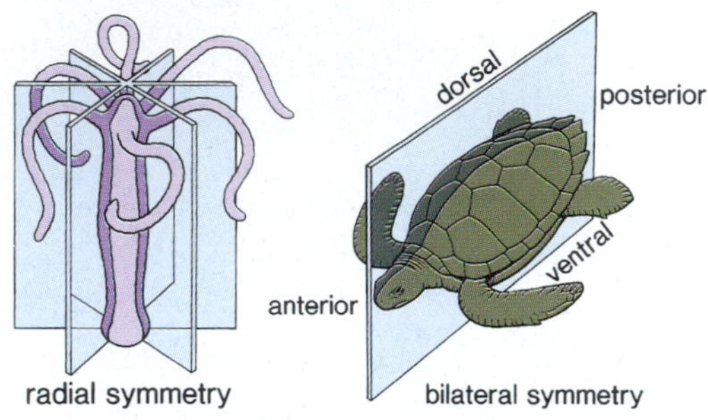

Comb jellies (**phylum Ctenophora,** 90 species) are small (a few cm), transparent, and often luminescent animals that take their name from their eight plates of fused cilia that resemble long combs. Most of their body is a jellylike packing material called **mesoglea** [Gk. *mesos,* middle, and *gloios,* glue]. They are the largest animals to be propelled by the beating of cilia. They capture prey either by means of long tentacles, covered with sticky filaments, or by using the entire body, which is covered with a sticky mucus.

Cnidarians

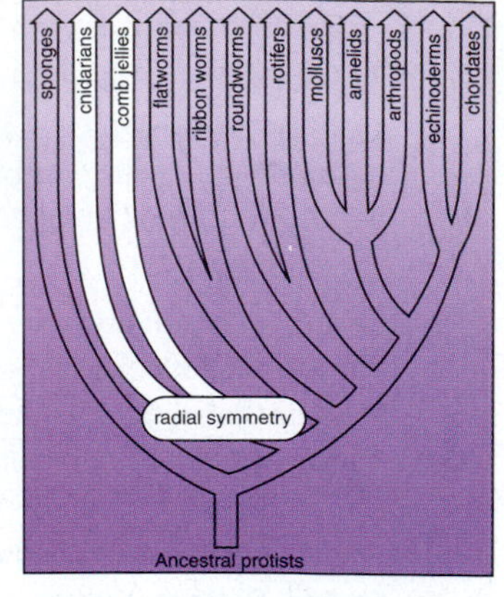

Cnidarians (**phylum Cnidaria,** 9,000 species) are tubular or bell-shaped animals that reside mainly in shallow coastal waters, except for the oceanic jellyfishes. Unique to cnidarians are specialized stinging cells, called cnidocytes, which give the phylum its name. Each cnidocyte has a fluid-filled capsule called a **nematocyst** [Gk. *nema,* thread, and *kystis,* bladder], which contains a long, spirally coiled hollow thread. When the trigger of the cnidocyte is touched, the nematocyst is discharged. Some threads merely trap a prey or predator; others have spines that penetrate and inject paralyzing toxins.

The body of a cnidarian is a two-layered sac. The outer tissue layer is a protective epidermis derived from ectoderm. The inner tissue layer, which is derived from endoderm, secretes digestive juices into the internal cavity, called the **gastrovascular cavity** [Gk. *gastros,* stomach, and L. *vasculum,* dim. of *vas,* vessel] because it serves for digestion of food and circulation of nutrients. The two tissue layers are separated by mesoglea. There are muscle fibers at the base of the epidermal and gastrodermal cells. Nerve cells located below the epidermis near the mesoglea interconnect and form a **nerve net** throughout the body. In contrast to highly organized nervous systems, the nerve net allows transmission of impulses in several directions at once—multiple firings of nematocysts in parts of the body not directly stimulated have been observed. Having both muscle fibers and nerve fibers, these animals are capable of directional movement; the body can contract or extend, and the tentacles that ring the mouth can reach out and grasp prey.

Two basic body forms are seen among cnidarians. The mouth of a polyp is directed upward, while the mouth of a jellyfish or medusa is directed downward. A medusa has more mesoglea than a polyp, and the tentacles are concentrated on the margin of the bell. At one time, both body forms may have been a part of the life cycle of all cnidarians. When both are present, the sessile polyp stage produces medusae and the motile medusan stage produces egg and sperm, thereby dispersing the species. In some cnidarians, one stage is dominant and the other is reduced; in other species one form is absent altogether.

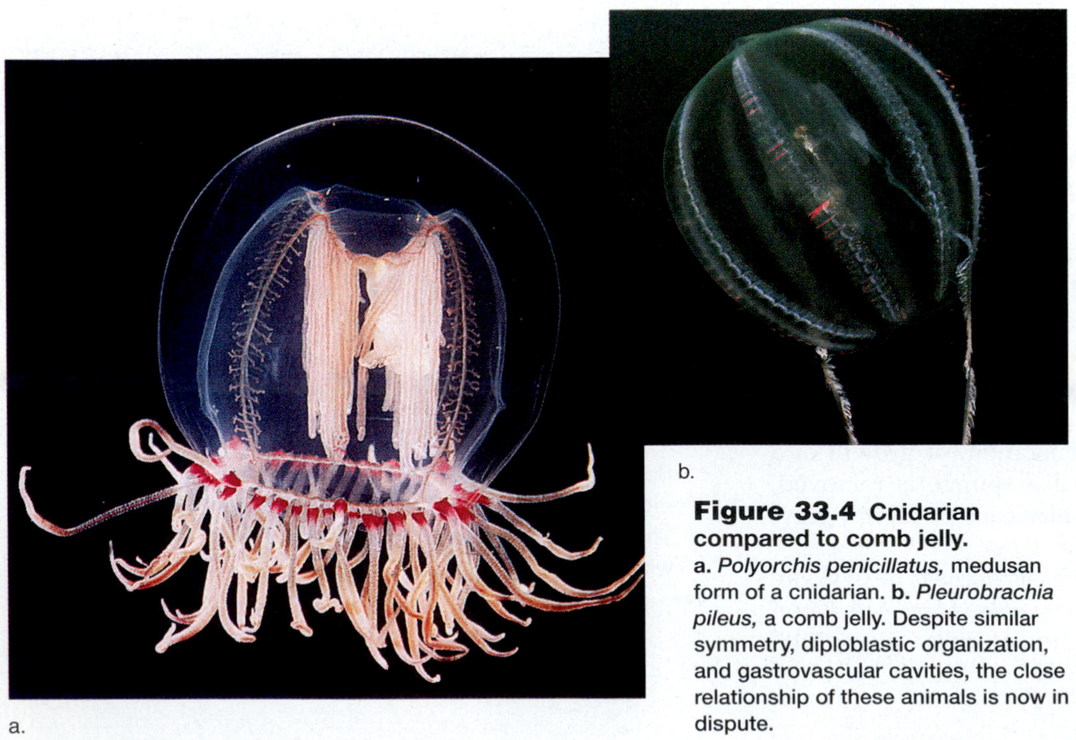

Figure 33.4 Cnidarian compared to comb jelly.
a. *Polyorchis penicillatus,* medusan form of a cnidarian. **b.** *Pleurobrachia pileus,* a comb jelly. Despite similar symmetry, diploblastic organization, and gastrovascular cavities, the close relationship of these animals is now in dispute.

Figure 33.5 Cnidarian diversity.
a. The life cycle of a cnidarian. In some cnidarians, there is both a polyp stage and a medusa stage; in others, one stage is dominant, or absent altogether. **b.** The anemone, which is sometimes called the flower of the sea, is a solitary polyp. **c.** Corals are colonial polyps residing in a calcium carbonate or proteinaceous skeleton. **d.** Portuguese man-of-war is a colony of modified polyps and medusae. **e.** True jellyfish undergo the complete life cycle; this is the medusa stage.

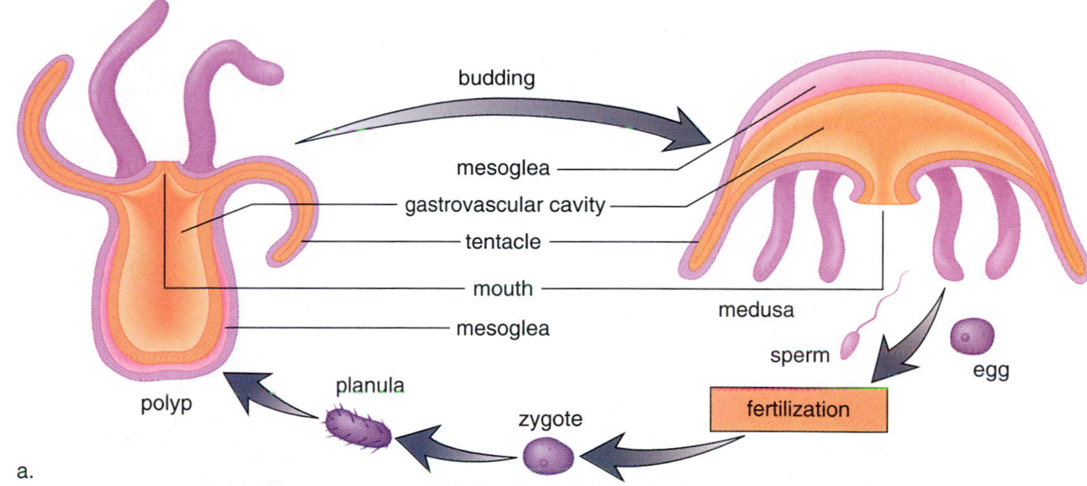

a.

b. Sea anemone, *Apitasia*

c. Cup coral, *Tubastrea*

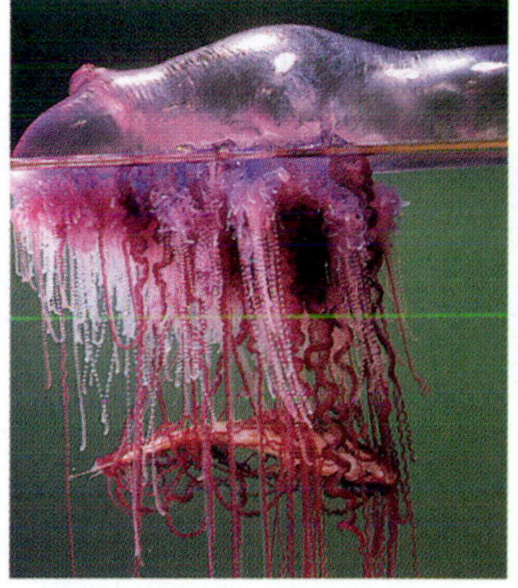
d. Portuguese man-of-war, *Physalia*

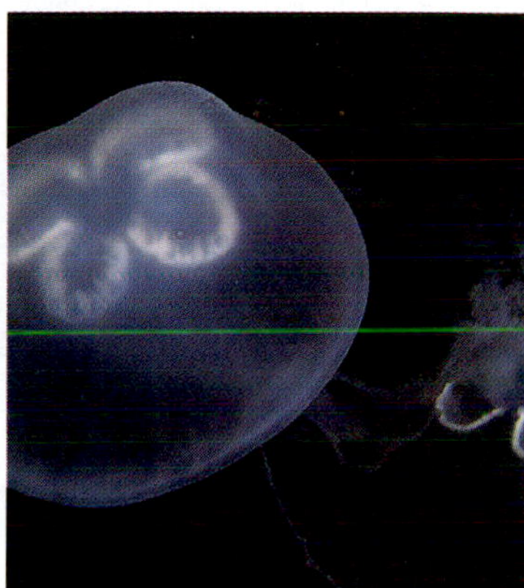
e. Jellyfish, *Aurelia*

Cnidarian Diversity

Among the cnidarian classes, class Anthozoa contains the sea anemones and corals (Fig. 33.5*b*, *c*). Sea anemones are solitary polyps that are large enough to be seen with the naked eye. Most sea anemones are 5–100 mm in height and 5–200 mm in diameter; some are much larger. They may be brightly colored and look like beautiful flowers.

The oral disk that bears a mouth of a sea anemone is surrounded by a large number of hollow tentacles. They feed on various invertebrates, and large species can capture fish. Sea anemones live attached to a submerged rock, timber, or shell. A number of species form mutualistic relationships (in which both species benefit) with hermit crabs and live attached to the shell of the crab. The anemone provides protection and camouflage for the crab, and the crab provides locomotion and perhaps some food for the sea anemone.

Corals, which are also anthozoans, resemble sea anemones in appearance. Some corals are solitary, but most are colonial with flat, rounded, or upright and branching colonies. Corals are usually found in shallow waters; this is particularly true of reef-building stony corals whose walls contain mutualistic algae. The slow accumulation of coral skeletal remains can result in massive structures such as the Great Barrier Reef along the eastern coast of Australia.

In class Hydrozoa, the polypoid stage is dominant. One of the most unusual hydrozoans is the Portuguese man-of-war, *Physalia*, which looks as if it might be an odd-shaped medusa (Fig. 33.5*d*) but actually is a colony of polyps. The original polyp becomes a gas-filled float that provides buoyancy—it keeps the colony afloat. Other polyps, which bud from this one, are specialized for feeding or for reproduction. A long single tentacle armed with numerous nematocysts arises from the base of each feeding polyp. Swimmers who accidently come upon a Portuguese man-of-war can receive painful, even serious, injuries from these stinging tentacles.

Class Scyphozoa includes the true jellyfishes, such as *Aurelia* (Fig. 33.5*e*). In jellyfishes, the medusa is the primary stage, and the polyp remains quite small and insignificant. Jellyfishes are a part of the zooplankton of the ocean and as such serve as food for larger animals.

Cnidarians have a tissue level of organization and are radially symmetrical. They have a sac body plan and exist as polyps or medusae.

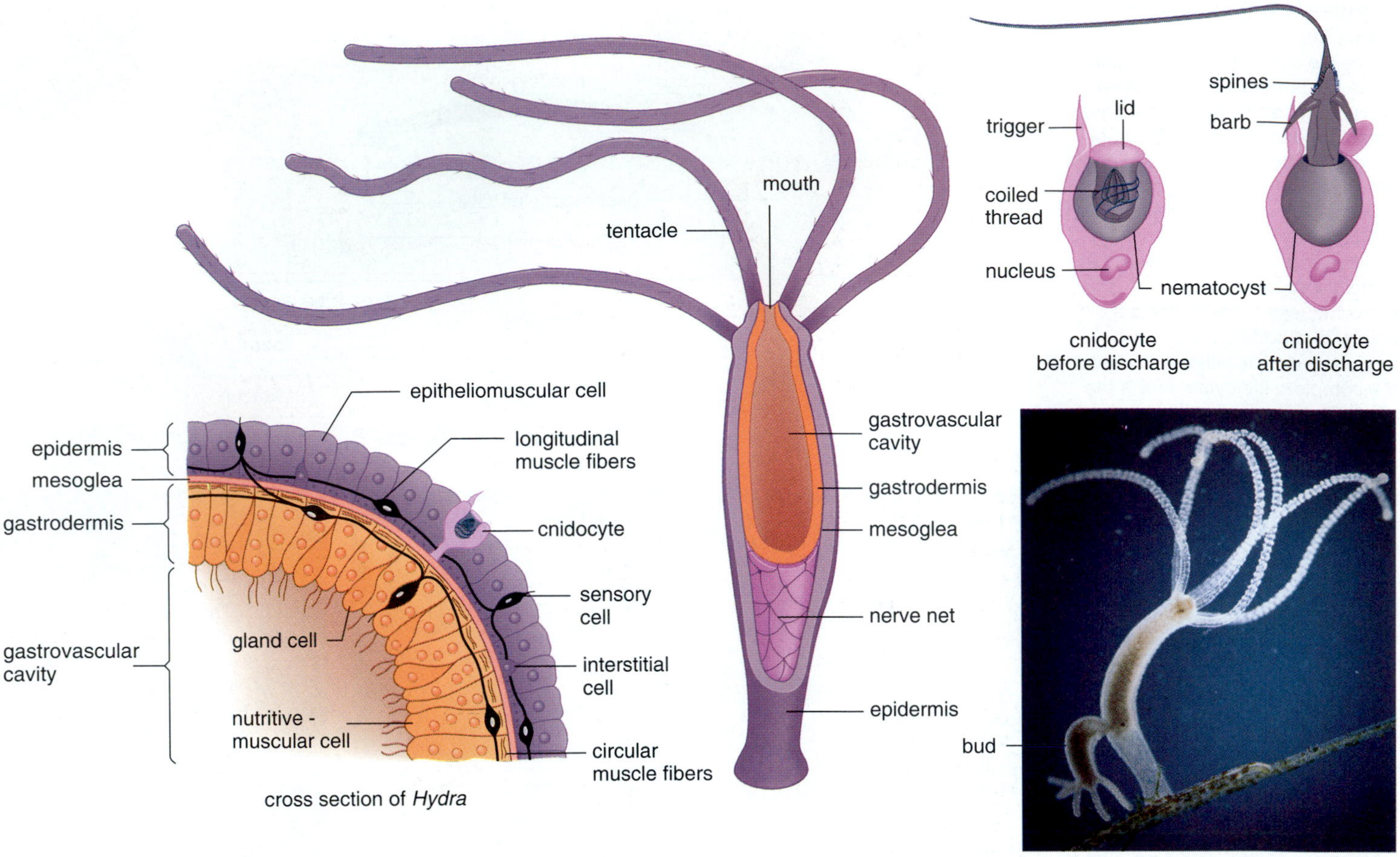

Figure 33.6 **Anatomy of *Hydra*.**
(center) The body of *Hydra* is a small tubular polyp whose body wall contains two tissue layers. *(left)* Various types of cells in the body wall. *(right, top)* Cnidocytes are cells that contain nematocysts. *(right, bottom) Hydra* reproduces asexually by forming outgrowths called buds which develop into a complete animal.

Hydra and *Obelia*

Hydra and *Obelia* in class Hydrozoa are two hydrozoans of particular interest. *Hydra* is a solitary polyp and *Obelia* is a colonial form, which has both a polyp and medusa stage in its life cycle.

Hydra **Hydras** [Gk. *hydra,* a many-headed serpent] are freshwater cnidarians. Hydras are likely to be found attached to underwater plants or rocks in most lakes and ponds. The body is a small tubular polyp about one-quarter inch in length. The only opening (the mouth) is in a raised area that is surrounded by four to six tentacles that contain a large number of nematocysts. The central cavity of the animal is the gastrovascular cavity.

Although a hydra usually remains in one location, it can glide along on its base or even move rapidly by means of somersaulting.

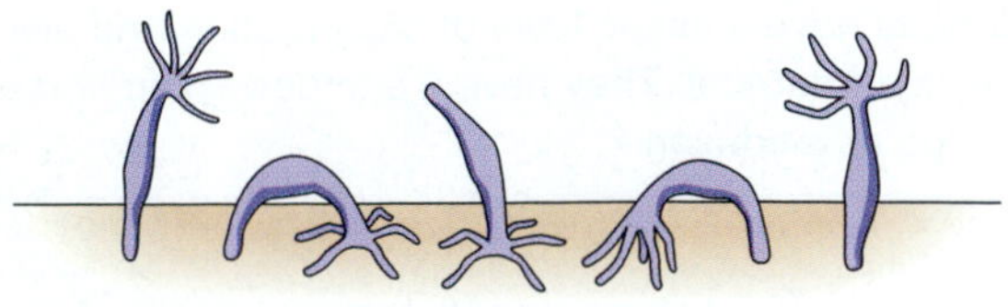

Hydras can respond to stimuli, and if a tentacle is touched with a needle, all the tentacles and the body contract, only to extend later. It is apparent, then, that like other animals capable of locomotion, hydras have both muscular and nerve fibers.

Figure 33.6 shows the microscopic anatomy of *Hydra.* The cells of the epidermis are termed epitheliomuscular cells because they contain muscle fibers. Also present in the epidermis are cnidocytes and sensory cells. The latter have long extensions that make contact with the nerve cells within the nerve net. The interstitial, or embryonic, cells also seen in this layer are capable of becoming other types of cells. For example, they can produce an ovary and/or a testis and probably also account for the animal's great regenerative powers. Like the sponges, cnidarians can grow an entire organism from a small piece.

Gland cells of the gastrodermis secrete digestive juices that pour into the gastrovascular cavity. Hydras feed on small prey that are captured when they trigger the release of nematocysts. The tentacles capture and stuff the prey into the gastrovascular cavity, which distends to accommodate the food. The enzymes released by the gland cells begin the digestive process, which is completed within food vacuoles

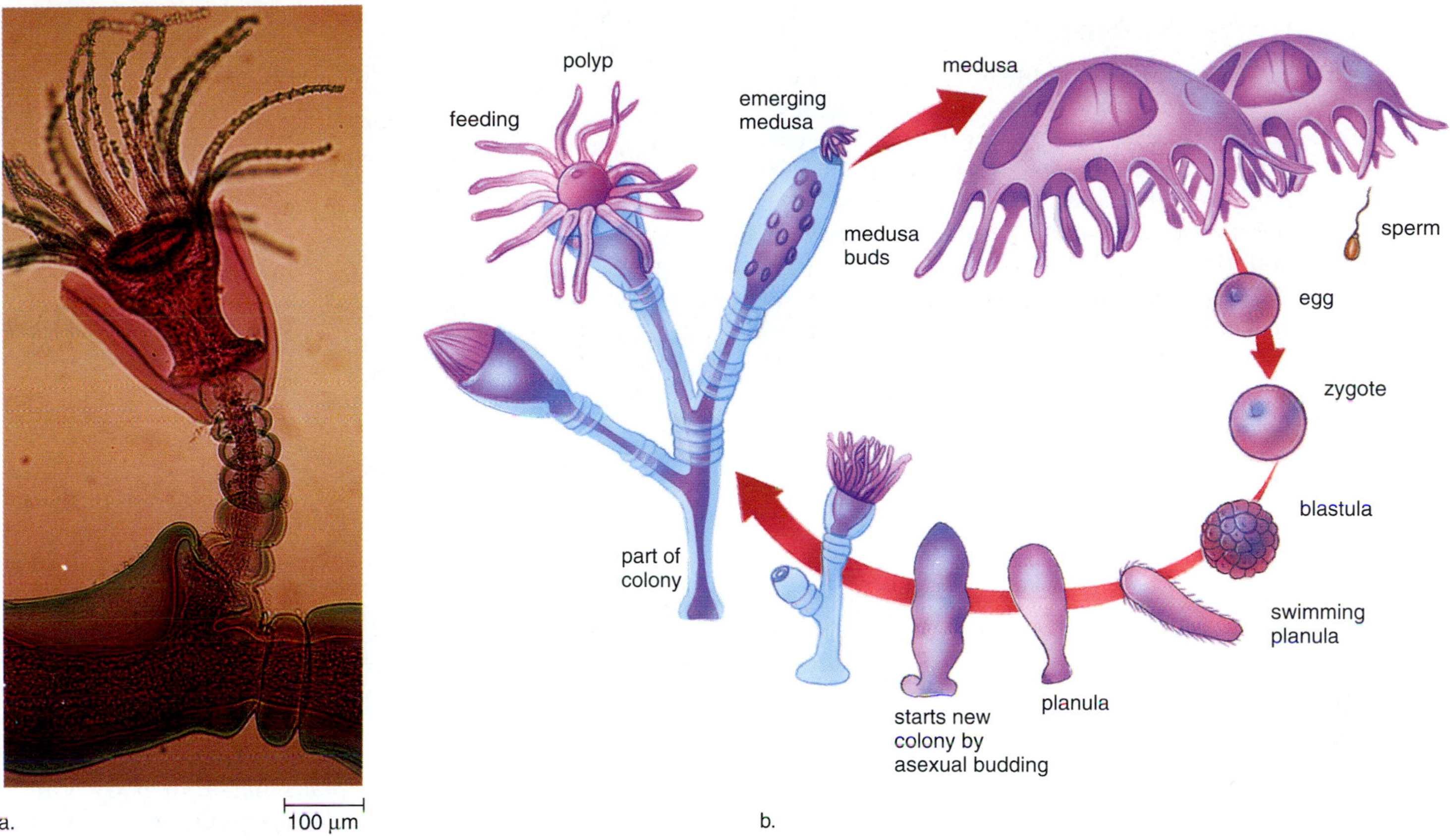

Figure 33.7 ***Obelia*** **structure and life cycle.**
Obelia, a colony of feeding polyps and reproductive polyps, undergoes an alternation of generations life cycle. **a.** Micrograph of feeding polyps. **b.** Life cycle.

of nutritive-muscular cells, the main type of gastrodermal cell. Nutrient molecules are passed by diffusion to the rest of the cells of the body. Nutritive-muscular cells also contain contractile fibers that run circularly about the body; when these contract, the animal lengthens.

Hydras can reproduce both asexually and sexually. They reproduce asexually by forming buds, small outgrowths that develop into a complete animal and then detach. When hydras reproduce sexually, sperm from a testis swim to an egg within an ovary. Fertilization and early development occur within the ovary, after which the embryo is encased within a hard, protective shell that allows it to survive until conditions are optimum for it to emerge and develop into a new polyp.

Obelia *Obelia* (Fig. 33.7) is a colony of polyps that is enclosed by a hard, chitinous covering. There are two types of polyps. The feeding polyps extend beyond the covering and can withdraw into it for protection. They have nematocyst-bearing tentacles that can capture and bring prey—such as tiny crustacea, worms, and larvae—into the gastrovascular cavity. The polyps are connected, and the partially digested food is distributed to the rest of the colony.

The colony increases in size asexually by the budding of new polyps. Sexual reproduction involves the production of medusae, which bud from the second type of polyp called reproductive polyps. Hydroid medusae tend to be smaller than those of the true jellyfishes. The tentacles attached to the bell margin have nematocysts, and they bring food into a gastrovascular cavity that extends even into the tentacles. The nerve net is concentrated into two nerve rings; the bell margin is supplied with sensory cells, such as statocysts, organs of equilibrium, and ocelli, light-sensitive organs.

While some species produce free-swimming medusae, in other species the medusae remain attached to the colony and shed only their gametes. The resulting zygote develops into a ciliated larva called a planula larva. The planula larva settles down and develops into a polyp colony.

Obelia is an example of a colonial hydroid consisting of feeding and reproductive polyps. Aside from this polyp stage, there is also a medusa stage in its life cycle.

The relationship of radially symmetrical cnidarians to the rest of the animal groups, which at some time during their life history are bilaterally symmetrical, has not been easy to determine. However, there are those who believe that a planuloid-type organism could have given rise to both the cnidarians and the flatworms, which are discussed next. The cnidarians have a two-tissue level of organization and radial symmetry. The flatworms have three germ layers and bilateral symmetry.

33.4 Bilateral Symmetry

All other animals to be studied are bilaterally symmetrical, at least in some stage of development. As embryos, they have three germ layers and are, therefore, called triploblasts. As adults, they have the organ level of organization. Flatworms in the phylum Platyhelminthes have these features, as do ribbon worms in the phylum Nemertea. Flatworms, however, like cnidarians, have a **sac body plan,** while ribbon worms have a **tube-within-a-tube body plan.** Animals with a sac body plan are said to have an incomplete digestive tract, while those with the tube-within-a-tube plan have a complete digestive tract.

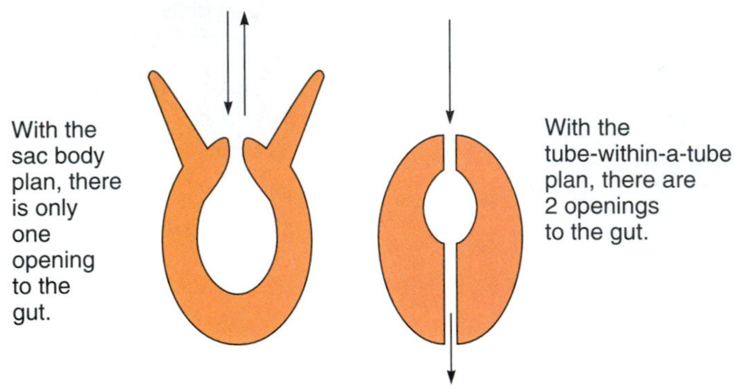

With the sac body plan, there is only one opening to the gut.
With the tube-within-a-tube plan, there are 2 openings to the gut.

In the tube-within-a-tube body plan, there is the possibility of specialization of parts along the length of the tube.

Ribbon Worms

Ribbon worms (phylum Nemertea, 650 species), which are mainly marine, have a distinctive proboscis apparatus—a long, hollow tube lying in a cavity called the rhynchocoel (Fig. 33.8). Contraction of the rhynchocoel wall causes the proboscis to evert and shoot outward through a pore located just above the mouth. The proboscis is used primarily for prey capture but also for defense, locomotion, and burrowing. Like flatworms, ribbon worms are acoelomates.

Figure 33.8 Ribbon worm, *Amphiporus.*
Ribbon worms, like flatworms, are bilaterally symmetrical, have three germ layers, and the organ level of organization. Unlike flatworms, ribbon worms have a complete digestive tract.

Flatworms

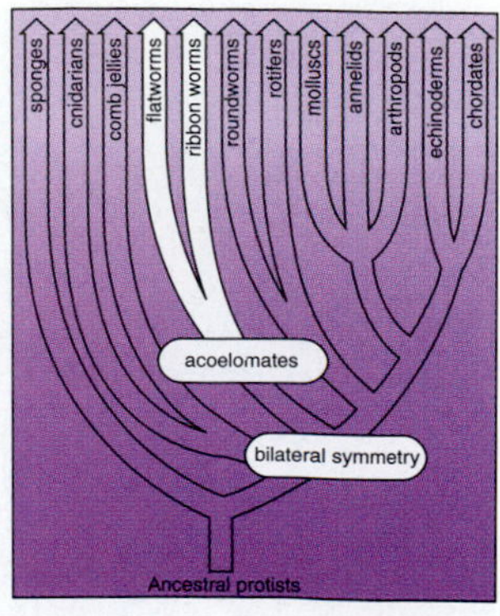

Flatworms (phylum Platyhelminthes, 13,000 species) can be either free-living or parasitic. Planarians and their relatives are freshwater animals in the class Turbellaria. The majority of flatworms are parasites. The flukes, which are either external or internal parasites, are in the class Trematoda. The tapeworms, which are intestinal parasites of vertebrates, are in the class Cestoda.

Flatworms are complex. In addition to an endodermis derived from endoderm and an epidermis derived from ectoderm, an embryonic mesoderm layer gives rise to muscles and reproductive organs. There is no coelom; these animals are, therefore, acoelomates.

Although flatworms have the organ level of organization, there are no specialized circulatory or respiratory structures. How are nutrients distributed about the body? The gastrovascular cavity, which is sometimes highly branched, serves this function. Gas exchange can occur by diffusion because of the animal's flat, thin body. Often, there is an excretory system that functions as an osmotic-regulating system.

Flatworms are bilaterally symmetrical, and free-living forms have undergone **cephalization** [Gk. *kaphale,* head]—the development of a head region. There is a ladder-type nervous system, so called because the two lateral nerve cords plus the connecting nerves look like a ladder. Paired ganglia (collection of nerve cells) function as a brain; sensory cells are located in the body wall, and the animal is able to respond to various stimuli.

Flatworms have a sac body plan but three germ layers and the organ level of organization. They are bilaterally symmetrical, and cephalization is present.

Planarians

The turbellarians include freshwater planarians such as *Dugesia,* which are small (several mm to several cm) literally flat worms, with brown or black pigmentation (Fig. 33.9). They live in lakes, ponds, and streams, where they feed on small, living or dead organisms.

The head is bluntly arrow shaped, with lateral extensions called auricles that function as sense organs to detect potential food sources and enemies. There are two light-sensitive eyespots whose pigmentation causes the worm to look cross-eyed. Inside, the brain is connected to a ladder-type nervous system. There are three kinds of muscle layers—an outer circular layer, an inner longitudinal layer, and a diagonal layer—that allow for quite varied movement. In larger forms, locomotion is accomplished by the

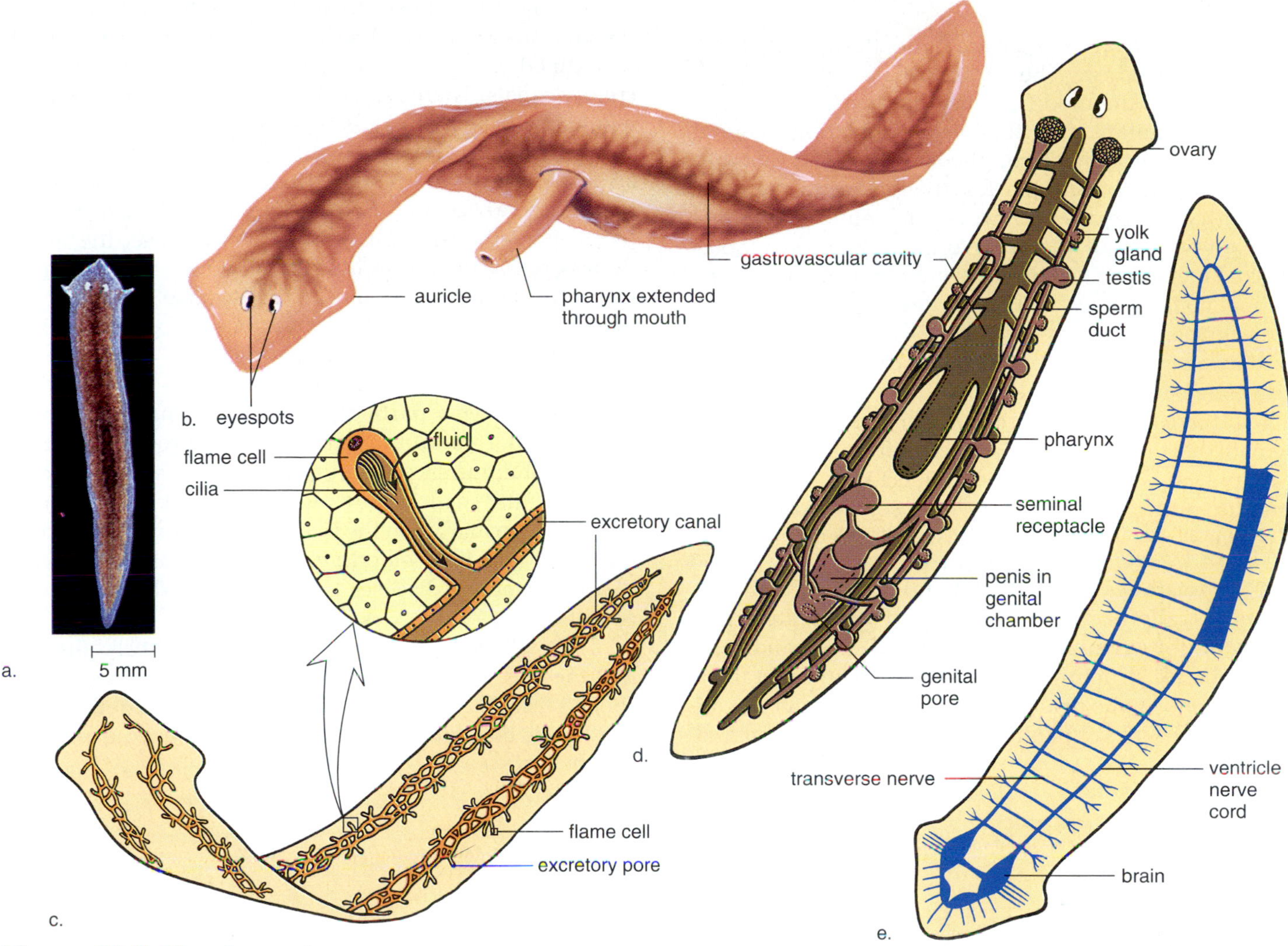

Figure 33.9 **Planarian anatomy.**
a. The photograph shows that flatworms, *Dugesia,* are bilaterally symmetrical and have a head region with eyespots. **b.** When the pharynx is extended as shown, food is sucked up into a gastrovascular cavity that branches throughout the body. **c.** The excretory system with flame cells is shown in detail. **d.** The reproductive system has both male and female organs, and the digestive system has a single opening. **e.** The nervous system has a ladderlike appearance.

movement of cilia on the ventral and lateral surfaces. Numerous gland cells secrete a mucous material upon which the animal moves.

The animal captures food by wrapping itself around the prey, entangling it in slime, and pinning it down. Then a muscular pharynx is extended, and by a sucking motion the food is torn up and swallowed. The pharynx leads into a three-branched gastrovascular cavity in which digestion is both extracellular and intracellular.

Why is it to be expected that *Dugesia,* which live in fresh water would have a well-developed excretory organ system? Their excretory organ functions in osmotic regulation as well as excreting water. The organ consists of a series of interconnecting canals that run the length of the body on each side. Bulblike structures containing cilia are at the ends of the side branches of the canals. The cilia move back and forth, bringing water into the canals that empty at pores. The beating of the cilia reminded some early investigator of the flickering of a flame, and so the excretory organ of the flatworm is called a flame-cell system.

Planarians can reproduce asexually. They constrict beneath the pharynx, and each part grows into a whole animal again. Many experiments in development have utilized planarians because of their marked ability to regenerate. Planarians also reproduce sexually. They are **hermaphroditic,** which means that they possess both male and female sex organs. The worms practice cross-fertilization when the penis of one is inserted into the genital pore of the other. The fertilized eggs are enclosed in a cocoon and hatch in two or three weeks as tiny worms.

Free-living planarians best exhibit the bilateral symmetry and organ development—including the nervous system and muscles—of a flatworm.

Parasitic Flatworms

Among the parasitic flatworms are flukes (trematodes) and tapeworms (cestodes). The structure of both of these worms illustrates the modifications that occur in parasitic animals. Concomitant with the loss of predation, there is an absence of cephalization; the anterior end notably carries hooks and/or suckers for attachment to the host. There is an extensive development of the reproductive system at the expense of the other organs. Well-developed nerves and a gastrovascular cavity are not needed because the animal no longer seeks out and digests prey; instead, it acquires nutrients from its host. Flukes and tapeworms are covered by a tegument, a specialized body wall resistant to host digestive juices.

Both flukes and tapeworms utilize a secondary, or intermediate, host to transport the species from primary host to primary host. The primary host is infected with the sexually mature adult; the secondary host contains the larval stage or stages.

Flukes **Trematodes** (class Trematoda) include the flukes, which are usually named for the type of vertebrate organ they inhabit; for example, there are blood, liver, and lung flukes. While the structure may vary slightly, in general the fluke body tends to be oval to elongate. At the anterior end surrounded by sensory papilla, there is an oral sucker and at least one other sucker for attachment to the host. Inside, the digestive system and nervous systems are reduced, and a modified excretory system has a reduced number of excretory canals. There is a well-developed reproductive system, and the adult fluke is usually hermaphroditic, as are planarians. An exception is the blood fluke, which causes **schistosomiasis** and is seen predominantly in the Middle East, Asia, Africa, and South America. About 800,000 infected persons die each year. In this disease, the female flukes deposit their eggs in small blood vessels close to the lumen of the intestine, and the eggs make their way into the digestive tract by a slow migratory process (Fig. 33.10). After the eggs pass out with the feces, they hatch into tiny larvae that swim about in the rice paddies and elsewhere until they enter a particular species of snail. Within the snail asexual reproduction occurs; sporocysts that are spore-containing sacs eventually produce new larval forms that leave the snail. If they penetrate the skin of a human, they begin to mature in the liver and implant themselves in the small intestinal blood vessels. Those infected usually die of secondary diseases brought on by their weakened condition.

The Chinese liver fluke requires two hosts: the snail and the fish. Humans become infected when they eat uncooked fish. The adults reside in the liver and deposit their eggs in the bile duct, which carries the eggs to the intestine.

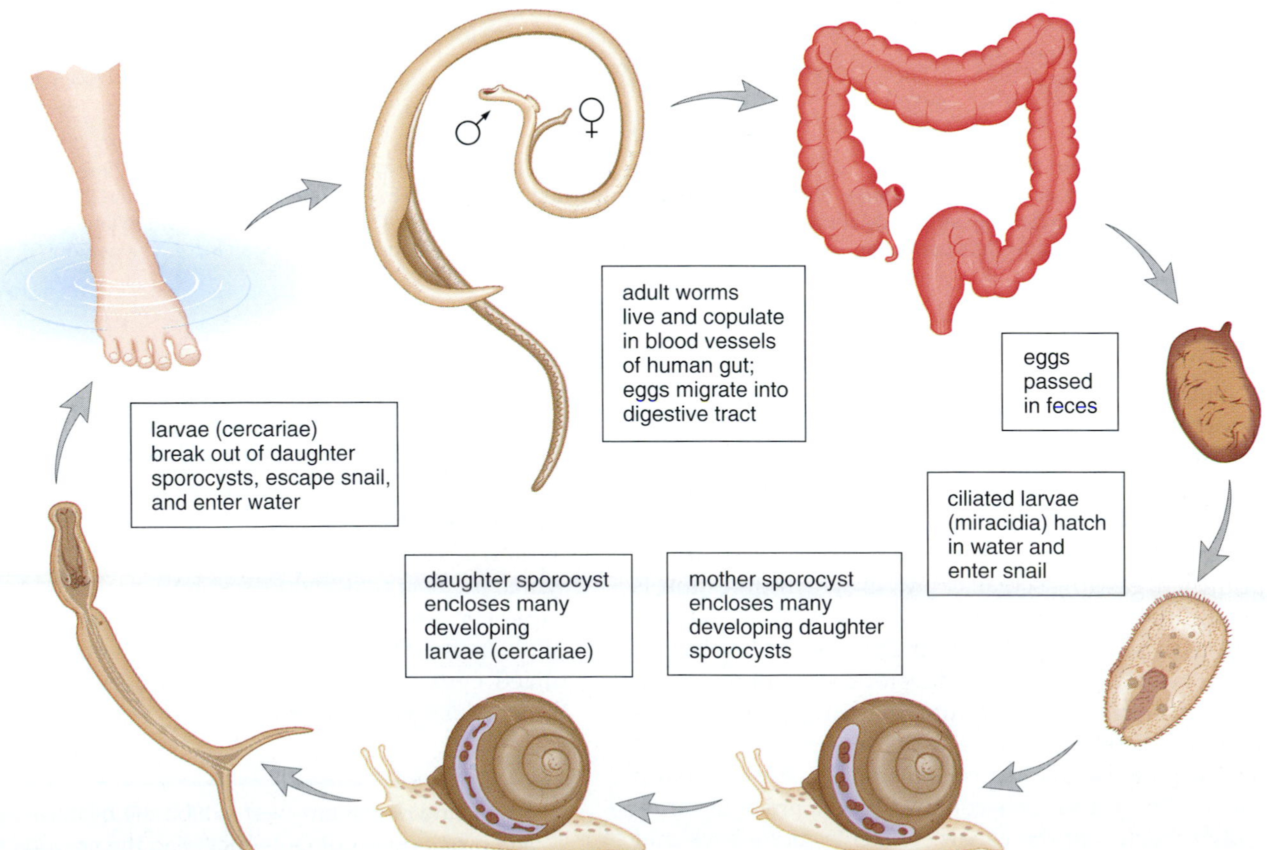
♂
♀
adult worms live and copulate in blood vessels of human gut; eggs migrate into digestive tract
eggs passed in feces
ciliated larvae (miracidia) hatch in water and enter snail
mother sporocyst encloses many developing daughter sporocysts
daughter sporocyst encloses many developing larvae (cercariae)
larvae (cercariae) break out of daughter sporocysts, escape snail, and enter water

Figure 33.10 **Schistosomiasis.**
This infection of humans, caused by blood flukes, *Schistosoma*, is an extremely prevalent disease in Egypt—especially since the building of the Aswan High Dam. Standing water in irrigation ditches, combined with unsanitary practices, has created the conditions for widespread infection.

Tapeworms **Cestodes** (class Cestoda) include the tapeworms. A tapeworm has an anterior region (Fig. 33.11) with modifications for attachment to the intestinal wall of the host. Behind the head region, called a **scolex,** there is a short neck and then a long series of proglottids. **Proglottids** are segments, each of which contains a full set of both male and female sex organs and little else. There are excretory canals but no digestive system and only the rudiments of nerves.

After fertilization, the organs disintegrate and the proglottids become nothing but a bag filled with maturing eggs. Gravid proglottids such as these break off and, as they pass out with the feces, the eggs are released.

If eggs are present in feces-contaminated food fed to pigs or cattle, larvae escape when the covering of the eggs is digested away. They burrow through the intestinal wall and travel in the bloodstream to finally lodge and encyst in muscle. Here a **cyst** means a small, hard-walled structure that contains a larva called a bladder worm. When humans eat infected meat that has not been thoroughly cooked, the bladder worms break out of the cyst, attach themselves to the intestinal wall, and grow to adulthood. Then the cycle begins again.

Flukes and tapeworms illustrate the modifications that occur when animals take up the parasitic way of life.

Table 33.1 contrasts features of planarians, tapeworms, and flukes to illustrate the anatomical changes to be associated with the parasitic way of life.

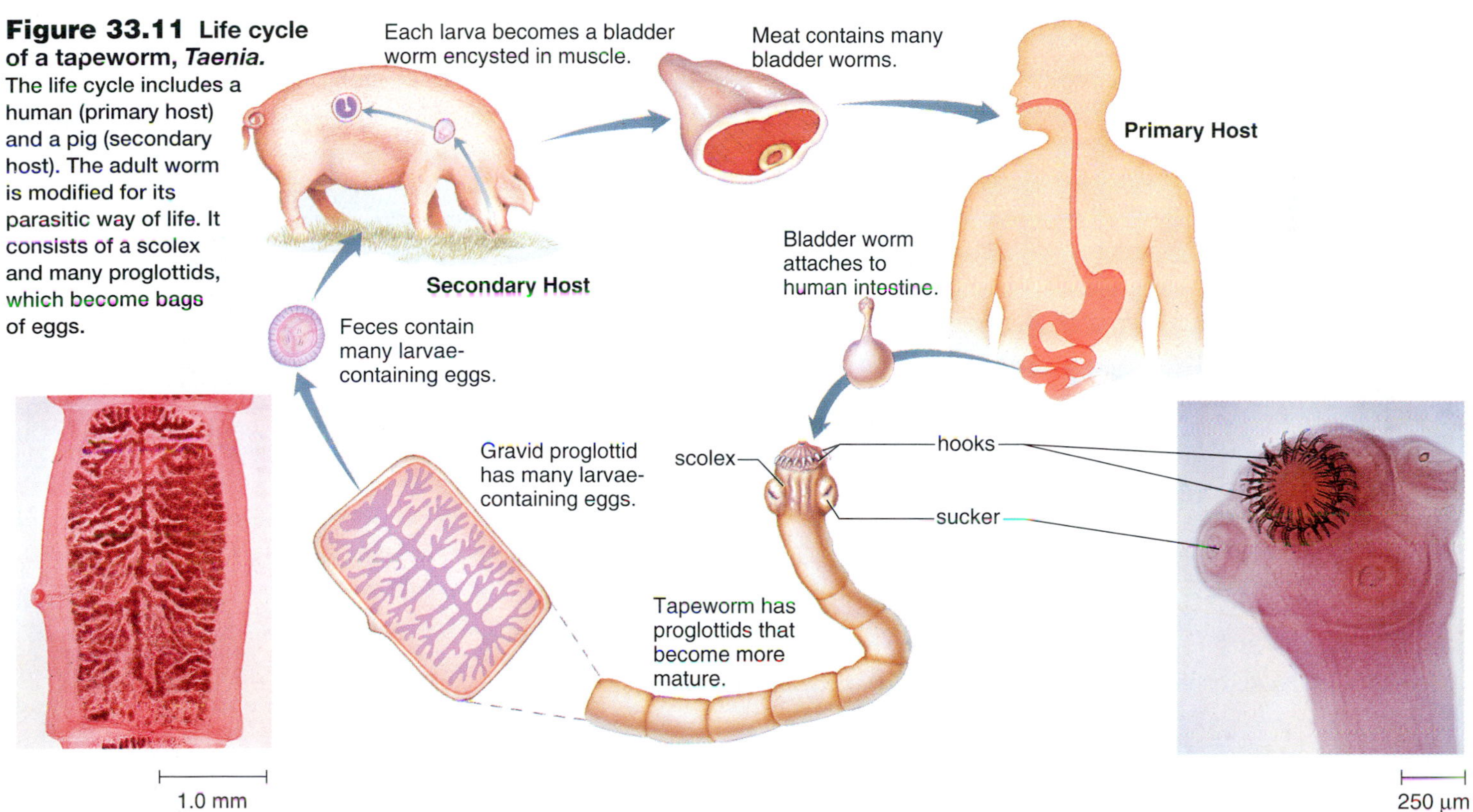
Each larva becomes a bladder worm encysted in muscle.
Meat contains many bladder worms.
Primary Host
Secondary Host
Bladder worm attaches to human intestine.
Feces contain many larvae-containing eggs.
Gravid proglottid has many larvae-containing eggs.
scolex
hooks
sucker
Tapeworm has proglottids that become more mature.
1.0 mm
250 µm

Figure 33.11 **Life cycle of a tapeworm, *Taenia*.** The life cycle includes a human (primary host) and a pig (secondary host). The adult worm is modified for its parasitic way of life. It consists of a scolex and many proglottids, which become bags of eggs.

Table 33.1

Free-Living Flatworms Versus Parasitic Flatworms

	Free Living	Parasitic	
	Planarians	*Flukes*	*Tapeworms*
Body wall	Ciliated epidermis	Tegument	Glycocalyx covers tegument
Cephalization	Yes—eyespots and auricles	No—oral suckers	No—scolex with hooks and suckers
Nervous organization	Nerves and brain	Reduced	Reduced
Gastrovascular cavity	Branched	Reduced	Absent
Reproductive organs	Hermaphroditic	Increased in volume	Extensively developed
Larva	Absent	Present	Present

Presence of a Coelom

A coelom is a body cavity surrounding the digestive organ or system. Flatworms are acoelomates (Fig. 33A*a*). They have no body cavity, and the mesoderm is packed solidly between the ectoderm and endoderm. The roundworms are pseudocoelomates (Fig. 33A*b*). They have a body cavity in which the organs lie loose. Apparently, this is disadvantageous to an increase in size, because most of these worms are small. In pseudocoelomates, a body cavity is incompletely lined by mesoderm. In coelomates, the coelom develops as a cavity within the mesoderm; therefore, it is completely lined with mesoderm (Fig. 33Ac). Such a coelom is often called a "true coelom." The organs in a true coelom are held in place by mesenteries, assuring a more stable arrangement with less crowding. Further, the gut is muscular (muscles are derived from mesoderm) and shows specialization of parts not seen in the pseudocoelomates. In the coelomates (all the other animals we will study), the internal organs are more complex.

The coelom also serves other functions. It allows the organs and the body wall to move independently. This means an animal can stretch and bend without putting a strain on the internal organs. The coelom is fluid filled, and this fluid protects and cushions the internal organs. In some animals, the fluid aids in the movement of materials, such as metabolic wastes; in others, this function is taken over by blood vessels. Not only metabolic wastes but also sex cells may be deposited into the coelomic cavity before they are transported away by ducts. The gastrovascular cavity of acoelomates (cnidarians and flatworms) and the fluid-filled coelom of soft-bodied coelomates can act as a hydrostatic skeleton. It offers some resistance to the contraction of muscles and yet permits flexibility, so that the animal can change shape and perform a variety of movements.

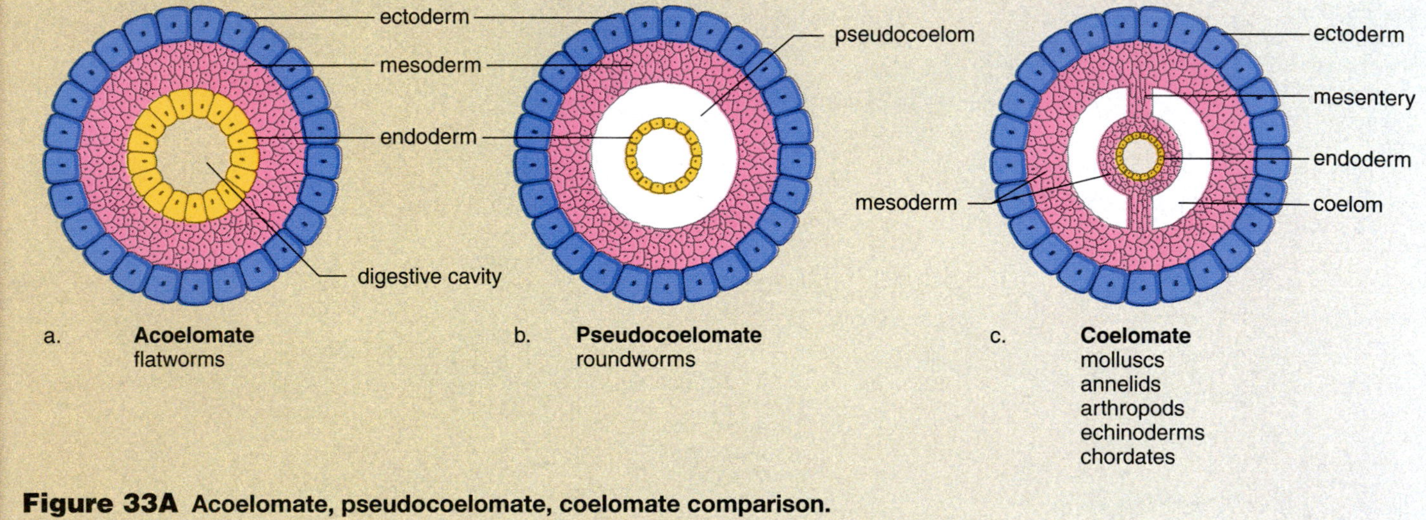

Figure 33A **Acoelomate, pseudocoelomate, coelomate comparison.**

33.5 A Pseudocoelom 

The accompanying reading compares animals on the basis of coelom type and describes how a coelom can serve as a *hydrostatic skeleton.* A **pseudocoelom** [Gk. *pseudes,* false, and *koiloma,* cavity] is a body cavity that is incompletely lined by mesoderm. In other words, mesoderm does not form a complete layer next to the body wall or around the gut. Two animal phyla consist of pseudocoelomates: roundworms (phylum Nematoda) and rotifers (phylum Rotifera). These animals also have the tube-within-a-tube body plan. The digestive tract is the inner tube within the rest of the animal, which is the outer tube.

Pseudocoelomate animals (e.g., roundworms and rotifers) have a coelom that is incompletely lined by mesoderm. A coelom provides a space for internal organs and can serve as a hydrostatic skeleton.

Roundworms

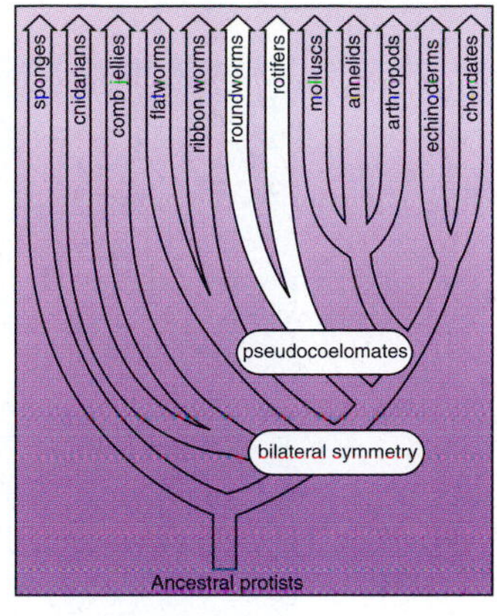

Roundworms (phylum Nematoda, 500,000 species), as their name implies, have a smooth outside wall, indicating that they are nonsegmented. These worms, which are generally colorless are found almost anywhere—in the sea, in fresh water, and in the soil—in such numbers and size that thousands of them can be found in a small area. Some are predators with teeth and other mouthparts, but many are scavengers or parasites. Among the latter, the pinworms, the hookworms, and *Trichinella* are small parasitic worms that are not easily examined. *Ascaris,* a large parasitic roundworm, is often studied as an example of this phylum.

Ascaris

Ascaris males (Fig. 33.12*a, b*) tend to be smaller (15–31 cm long) than females (20–49 cm long). In males, the posterior end is curved and comes to a point. Both sexes move by means of a characteristic whiplike motion because only longitudinal muscles lie next to the body wall.

The internal organs, including the tubular reproductive organs, lie within the pseudocoelom. Because mating produces eggs that mature in the soil, the parasite is limited to warmer environments. When these eggs are swallowed, larvae escape and burrow through the intestinal wall. Making their way through the organs of the host, they move from the intestine to the liver, the heart, and then the lungs. Within the lungs, molting takes place and, after about ten days, the larvae migrate up the windpipe to the throat where they are swallowed, once again reaching the intestine. Then the mature worms mate and the female deposits eggs that pass out with the feces. In this life cycle, as with that of other roundworms, feces must reach the mouth of the next host; therefore, proper sanitation is the best means to prevent infection with such worms as *Ascaris* and pinworms.

Other Roundworm Parasites

Trichinosis (Fig. 33.12*c*) is a serious infection that humans can contract when they eat rare pork containing encysted larvae. After maturation, the female adult burrows into the wall of the small intestine and produces living offspring that are carried by the bloodstream to the skeletal muscles, where they encyst.

Filarial worms, a type of roundworm, cause various diseases. *Dirofilaria,* the heartworm of dogs, is a common filarial worm of temperate zones. The cause of **elephantiasis,** a disease restricted to tropical areas of Africa, is caused by a filarial worm that utilizes the mosquito as a secondary host. Because the adult worms reside in lymphatic vessels,

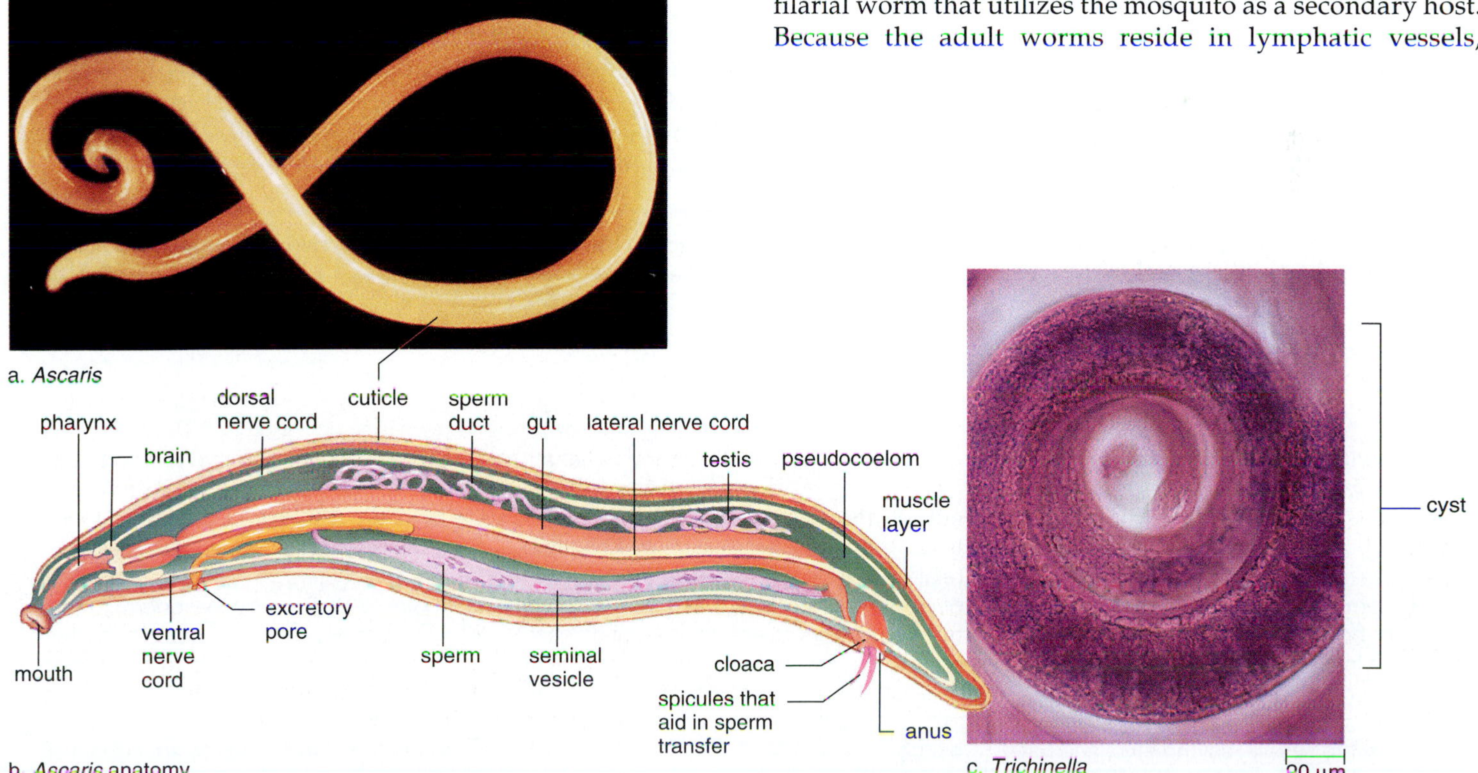

Figure 33.12 **Roundworm anatomy.**
a. The roundworm *Ascaris.* **b.** Roundworms such as *Ascaris* have a pseudocoelom and a complete digestive tract with a mouth and an anus. Therefore, roundworms have a tube-within-a-tube body plan. The sexes are separate; this is a male roundworm. **c.** The larvae of the roundworm, *Trichinella,* penetrate striated muscle fibers, where they coil in a sheath formed from the muscle fiber.

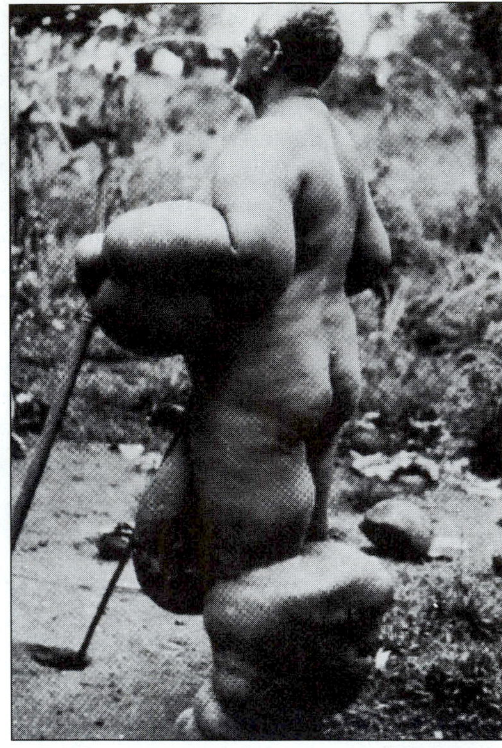

Figure 33.13 Filarial worm.
An infection from a filarial worm, *Wuchereria,* causes elephantiasis, a condition in which the individual experiences extreme swelling in regions where the worms have blocked the lymphatic vessels.

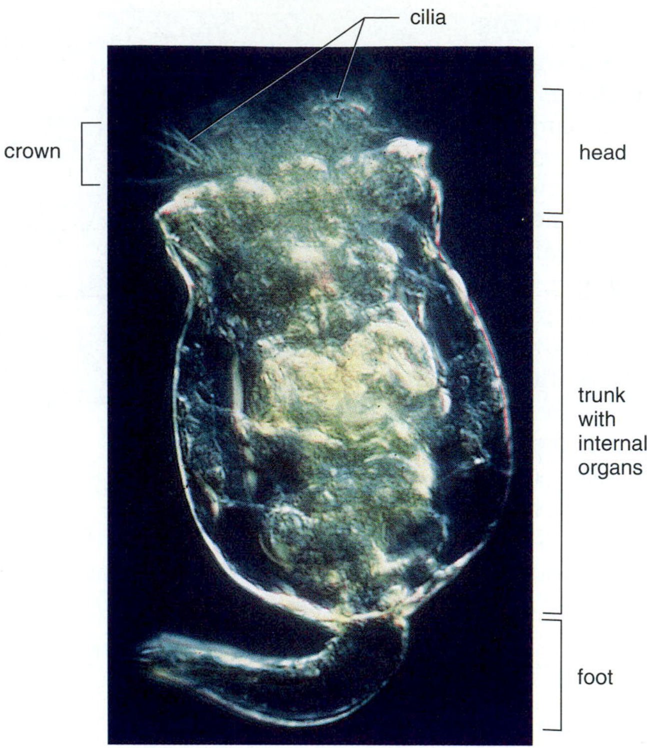

Figure 33.14 Rotifer.
Rotifers are microscopic animals only 0.1–3 mm in length. The beating of cilia on two lobes at the anterior end of the animal gives the impression of a pair of spinning wheels.

collection of fluid is impeded and the limbs of an infected person may swell to a monstrous size (Fig. 33.13). Elephantiasis is treatable in its early stages but not after scar tissue has blocked lymphatic vessels.

Rotifers

Students examining pond water are apt to see these tiny organisms and think that they are protozoans. Although microscopic, **rotifers (phylum Rotifera,** 2,000 species) are multicellular, with a pseudocoelom and internal organs (Fig. 33.14). They are named for a crown of cilia (corona) that resembles a rotating wheel and serves both as an organ of locomotion and as an aid in directing food to the mouth.

Both roundworms and rotifers have a pseudocoelom.

Connecting Concepts

Animals are a diverse group, but still they have certain features in common. In order to classify animals, zoologists examine and compare their symmetry, body plan, number of tissue layers derived from the germ layers ectoderm, mesoderm, and endoderm, and level of organization. The type of body cavity also helps categorize animals into certain groups—acoelomates, pseudocoelomates, and coelomates.

The sponges, the simplest multicellular animals, are sessile, asymmetric, and have several different kinds of cells but no true tissues. They are significantly different from other phyla, and apparently no other groups evolved from them. For this reason, many biologists consider sponges to be outside the mainstream of animal evolution. The sea anemones, jellyfish, and related species (cnidarians and ctenophorans) are radially symmetrical and have two germ layers as embryos. Most of these creatures also possess unique stinging cells containing nematocysts, which can put a serious damper on a day at the beach.

Animals more complex than the cnidarian group have bilateral symmetry at some stage in their development. They also have three germ layers as embryos and fairly well-developed organs. The flatworms and ribbon worms lack a body cavity; therefore, they are called acoelomates. Roundworms and rotifers have a body cavity that is incompletely lined by mesoderm; therefore, they are called pseudocoelomates. The remaining animal phyla to be studied in the next two chapters have a true body cavity and are called coelomates. Presence of a coelom has led to specialization of parts including a well-developed nervous and musculoskeletal system.

Summary

33.1 Classification of Animals

Animals are multicellular organisms that are heterotrophic and ingest their food. They have the diplontic life cycle. Typically, they have the power of motion by means of contracting fibers.

It's possible to construct a phylogenetic tree for animals, but this is largely based on a study of today's forms. Type of symmetry, number of tissue layers, type of coelom, and presence or absence of segmentation are criteria that are used in classification.

33.2 Multicellularity

Sponges may have evolved separately from other animals, since they have features that set them apart. They have the cellular level of organization and lack tissues and symmetry. Sponges are sessile and depend on a flow of water through the body to acquire food, which is digested in vacuoles within collar cells that line a central cavity.

33.3 Two Tissue Layers

Cnidarians and comb jellies have two tissue layers derived from the germ layers ectoderm and endoderm. They are radially symmetrical.

Cnidarians have a sac body plan. They exist as either polyps or medusae, or they can alternate between the two. Hydras and relatives—sea anemones and corals—are polyps; in jellyfishes the medusan stage is dominant. In *Hydra* and other cnidarians, an outer epidermis is separated from an inner gastrodermis by mesoglea. They possess tentacles to capture prey and nematocysts to stun it. A nerve net coordinates movements. Digestion of prey begins in the gastrovascular cavity and is finished within gastrodermal cells.

33.4 Bilateral Symmetry

Ribbon worms and flatworms have bilateral symmetry and the organ level of organization, including organs derived from mesoderm, a third germ layer, as do all the other phyla of animals to be studied.

Flatworms may be free living or parasitic. Freshwater planarians exemplify the features of flatworms in general and free-living forms in particular. They have muscles and a ladder-type nervous organization, and they show cephalization. They take in food through an extended pharynx leading to a gastrovascular cavity, which extends throughout the body. There is an osmotic-regulating organ that contains flame cells.

Flukes and tapeworms are parasitic. Flukes have two suckers by which they attach to and feed from their hosts. Tapeworms have a scolex with hooks and suckers for attaching to the host intestinal wall. The body of a tapeworm is made up of proglottids, which, when mature, contain thousands of eggs. If these eggs are taken up by pigs or cattle, larvae become encysted in their muscles. If humans eat this meat, they too may become infected with a tapeworm.

33.5 A Pseudocoelom

Roundworms and rotifers are pseudocoelomates. A coelom provides a space for internal organs and can serve as a hydrostatic skeleton. Roundworms are mostly small and very diverse; they are present almost everywhere in great numbers. The parasite *Ascaris* is representative of the group. Infections can also be caused by *Trichinella,* whose larval stage encysts in the muscles of humans. Elephantiasis is caused by a filarial worm that blocks lymphatic vessels.

Reviewing the Chapter

1. What are the characteristics that separate animals from plants? from fungi? 580
2. What does the phylogenetic tree (see Fig. 33.2) tell you about the evolution of the animals studied in this chapter? 581
3. List the types of cells found in a sponge, and describe their functions. 582–83
4. What features make sponges different from the other organisms placed in the animal kingdom? 583
5. What features do comb jellies and cnidarians have in common? How are they different? 584
6. What are the two body forms found in cnidarians? Explain how they function in the life cycle of various types of cnidarians. 584
7. Describe the anatomy of *Hydra,* pointing out those features that typify cnidarians. 586–87
8. What features do ribbon worms and flatworms have in common? How are they different? 588
9. Describe the anatomy of a free-living planarian, pointing out those features that typify nonparasitic flatworms. 588–89
10. Describe the parasitic flatworms, and give the life cycle of both the blood fluke that causes schistosomiasis and the pork tapeworm. 589–91
11. What is a pseudocoelom? What are the advantages of having a coelom? What two groups of animals have a pseudocoelom? 592
12. Describe the anatomy of *Ascaris,* pointing out those features that typify roundworms. 593

Testing Yourself

Choose the best answer for each question.

1. Which of these is not a characteristic of animals?
   a. heterotrophic
   b. diplontic life cycle
   c. have contracting fibers
   d. single cells or colonial
   e. lack of chlorophyll
2. The phylogenetic tree of animals shows that
   a. three germ layers evolved before a coelom.
   b. both molluscs and annelids are protostomes.
   c. some animals have radial symmetry.
   d. sponges were the first to evolve from ancestral protists.
   e. All of these are correct.
3. Which of these sponge characteristics is not typical of animals?
   a. They practice sexual reproduction.
   b. They have the cellular level of organization.
   c. They are asymmetrical.
   d. They have flagellated cells.
   e. Both b and c are not typical.
4. Which of these is mismatched?
   a. sponges—spicules
   b. tapeworms—proglottids
   c. cnidarians—nematocysts
   d. roundworms—cilia
   e. cnidarians—polyp and medusa
5. Flukes and tapeworms
   a. show cephalization.
   b. have well-developed reproductive systems.
   c. have well-developed nervous systems.
   d. are free living.
   e. have a tube-within-a-tube body plan.
6. The presence of mesoderm
   a. restricts the development of a coelom.
   b. is associated with the organ level of organization.

c. is associated with the development of muscles.
d. means the animal is diploblastic.
e. Both b and c are correct.

7. *Ascaris* is a parasitic
   a. roundworm.
   b. flatworm.
   c. hydra.
   d. sponge.
   e. comb jelly.

8. Label the following diagram of the cnidarian polyp:

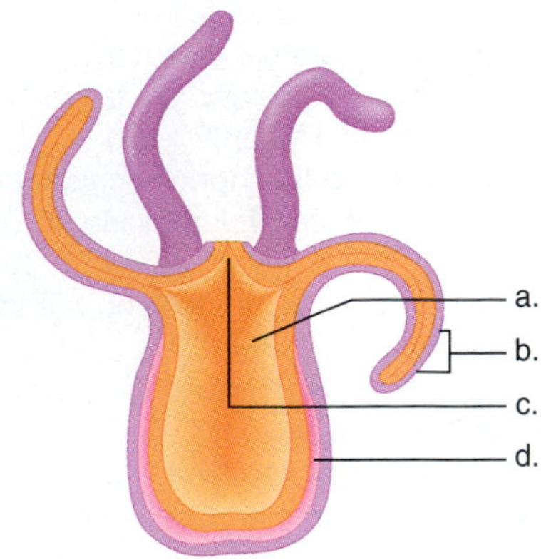

9. Write a correct phylum name beside each of these terms:
   a. radial symmetry:
   b. pseudocoelom:
   c. tissue level of organization:
   d. tube-within-a-tube body plan:
   e. cephalization:

10. Write the correct type animal beside each of these terms:
    a. proglottids:
    b. eyespots on head:
    c. collar cells:
    d. cnidocytes:
    e. crown of cilia:

Thinking Scientifically

1. Roundworms are tubular and have a complete digestive tract. What advantages can you associate with these anatomical features which might account for roundworms being more plentiful than flatworms?
2. You are taking a lab practical that covers just the phyla in this chapter. What criteria would you use to classify an unknown animal in one of these phyla?

Understanding the Terms

Match the terms to these definitions:

a. ______________ Blind digestive cavity that also serves a circulatory (transport) function in animals that lack a circulatory system.
b. ______________ Body cavity lying between the digestive tract and body wall that is completely lined by mesoderm.
c. ______________ Body cavity lying between the digestive tract and body wall that is incompletely lined by mesoderm.
d. ______________ Body plan having two corresponding or complementary halves.
e. ______________ Body with a digestive tract that has both a mouth and an anus.

Web Connections

Exploring the Internet

http://www.mhhe.com/biosci/genbio/mader
(click on *Biology 7/e*)

The *Biology 7/e* Online Learning Center provides many resources for studying the material in this chapter including links to the following sites:

Learn about the sponges and their fossil record in Introduction to Porifera.

http://www.ucmp.berkeley.edu/porifera/porifera.html

Cnidaria Home Page gives much information on this fascinating group of invertebrates, including images showing their diversity and explaining their evolution.

http://128.200.5.5/

Corals and Coral Reefs. Information from the education department of Sea World on coral reefs, destruction of coral, classification, many links.

http://www.seaworld.org/coral_reefs/introcr.html

Platyhelminthes. Arizona's Tree of Life Web Page. Pictures, characteristics, phylogenetic relationships, references on flatworms. A picture of a flatworm, and a link to trematodes (aspidogastreans).

http://phylogeny.arizona.edu/tree/eukaryotes/animals/platyhelminthes/platyhelminthes.html

Introduction to the Aschelminth Phyla. University of California at Berkeley Museum of Paleontology. This site contains many SEM images of microscopic nematodes, as well as links to other aschelminth phyla.

http://www.ucmp.berkeley.edu/aschelminthes/aschelminthes.html

Animals: The Protostomes

chapter

chapter concepts

34.1 A Coelom

- A coelom has many advantages: the digestive system can become more complex, coelomic fluid assists body processes and acts as a hydrostatic skeleton. 598
- In protostomes, the mouth appears at or near the blastopore, the first embryonic opening, and the coelom develops by a splitting of the mesoderm. 598

34.2 Molluscs

- The body of a mollusc typically contains a visceral mass, a mantle, and a foot. 600
- Clams are adapted to a sedentary coastal life, squids to an active life in the sea, and snails are adapted to a life on land. 600

34.3 Annelids

- Annelids are the segmented worms with a well-developed coelom, a closed circulatory system, a ventral solid nerve cord, and paired nephridia in each segment. 604
- Polychaetes include marine predators with a definite head region, and filter feeders with terminal tentacles to filter food from the water. 604
- Oligochaetes include the earthworms that burrow in the soil and use a moist body wall as a respiratory organ. 605

34.4 Arthropods

- Arthropods are segmented with specialized body regions and an external skeleton that includes jointed appendages. 606
- Among the many kinds of arthropods, crustaceans are adapted to a life at sea, and insects are adapted to a terrestrial existence. 608, 611

A destructive swarm of locusts, *Chortoicetes terminifera,* in Australia

In this chapter, we examine animals with which you may be more familiar—the arthropods, annelids, and molluscs. The arthropods are segmented, and they have a jointed exoskeleton that facilitates locomotion on land. The insects, which are arthropods, include more known species than any other group of animals, and most of these groups live on land. A swarm of locusts reminds us how plentiful insects can become when the environmental conditions promote increased reproduction.

Although annelids are a much smaller group than arthropods, they, too, are quite diversified. Annelids, exemplified by earthworms, are obviously segmented. The molluscs are a large diversified group with a unique body plan. Most molluscs are aquatic, but snails and slugs are adapted to living on land.

None of the types of animals studied in this chapter attain the size of the largest vertebrates but they still have biological and ecological importance. The same can be said for the sponges, jellyfish, corals, flatworms, and roundworms studied in the last chapter. An evolutionary history of millions of years testifies to the success of the invertebrates.

34.1 A Coelom

Protostomes are bilaterally symmetrical, have three germ layers, the organ level of organization, and the tube-within-a-tube body plan. In addition they have a true **coelom** [Gk. *koiloma,* cavity], a body cavity completely lined by mesoderm.

The presence of a coelom has many advantages. Body movements are freer because the outer wall can move independently of the enclosed organs. The ample space of a coelom allows complex organs and organ systems to develop. For example, the digestive tract can coil and provide a greater surface area for absorption of nutrients. And the coelomic cavity can serve as a storage area for eggs and sperm before they are released into the environment. Coelomic fluid protects internal organs against damage and against marked temperature changes. It can also assist in respiration and circulation by providing oxygen and nutrients to nearby cells. Metabolic wastes can accumulate in the cavity prior to being taken away by the excretory system. Also fluid within the cavity protects internal organs and can provide a hydrostatic skeleton—muscular contraction pushes against the fluid and allows the animal to move.

When a coelom is present, the digestive system and the body wall can move independently, and internal organs can become more complex. Coelomic fluid can assist respiration, circulation, and excretion. It also serves as a hydrostatic skeleton.

Phyla that have a coelom are divided into the **protostomes** [Gk. *protos,* first, and L. *stoma,* mouth] and **deuterostomes** [Gk. *deuteros,* second, and L. *stoma,* mouth]. There are three major differences in the embryological development of protostomes and deuterostomes (Fig. 34.1). Cleavage, the first event of development, is cell division without an increase in size of the cells. In protostomes, spiral cleavage occurs, and daughter cells sit in grooves formed by the previous cleavages. It can also be noted that the fate of these cells is fixed and determinate in protostomes; each can contribute to development in only one particular way. In deuterostomes, radial cleavage occurs, and the daughter cells sit right on top of the previous cells. The fate of these cells is indeterminate; that is, if they are separated from one another, each cell can go on to become a complete organism.

As development proceeds, a hollow sphere forms, and the indentation that follows produces an opening called the blastopore. In protostomes, the mouth appears at or near the blastopore, hence the origin of their name; in deuterostomes, the anus appears at or near the blastopore and only later does a new opening form the mouth, hence the origin of their name.

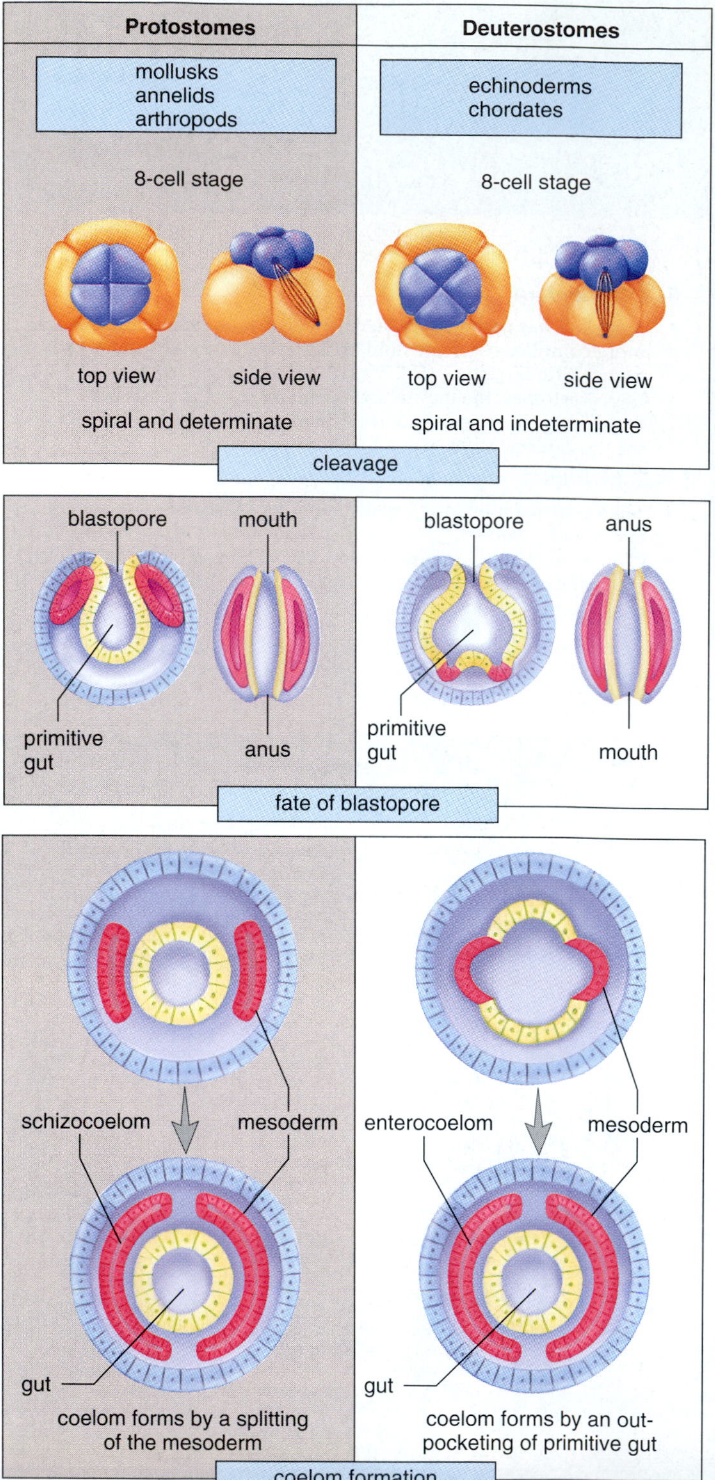

Figure 34.1 Protostomes versus deuterostomes.
In the embryo of protostomes (molluscs, annelids, arthropods), cleavage is spiral—new cells are at an angle to old cells—and each cell has limited potential and cannot develop into a complete embryo; the blastopore is associated with the mouth and the coelom is a schizocoelom. In deuterostomes (echinoderms and chordates), cleavage is radial—new cells sit on top of old cells—and each one can develop into a complete embryo; the blastopore is associated with the anus, and the coelom is an enterocoelom.

CLASSIFICATION

Kingdom Animalia

Multicellular organisms with well-developed tissues; usually motile; heterotrophic by ingestion, generally in a digestive cavity; diplontic life cycle. Protostomes include phyla Mollusca, Annelida, and Arthropoda. Deuterostomes include phyla Echinodermata, Hemichordata, and Chordata.

Invertebrates*

Phylum Porifera: sponges
Phylum Cnidaria: jellyfishes, sea anemones, corals
Phylum Ctenophora: comb jellies, sea walnuts
Phylum Platyhelminthes: flatworms, e.g., planarians, flukes, tapeworms
Phylum Nemertea: ribbon worms
Phylum Nematoda: roundworms
Phylum Rotifera: rotifers
Phylum Mollusca: chitons, snails, slugs, clams, mussels, squids, octopuses
Phylum Annelida: segmented worms, e.g., clam worms, earthworms, leeches
Phylum Onychophora: walking worms
Phylum Arthropoda: spiders, scorpions, horseshoe crabs, lobsters, crayfish, shrimps, crabs, millipedes, centipedes, insects
Phylum Echinodermata: sea lilies, sea stars, brittle stars, sea urchins, sand dollars, sea cucumbers, sea daisies
Phylum Hemichordata: acorn worms
Phylum Chordata
- Subphylum Urochordata: tunicates
- Subphylum Cephalochordata: lancelets

Vertebrates*

- Subphylum Vertebrata
  - Superclass Agnatha: jawless fishes, e.g., lampreys, hagfishes
  - Superclass Gnathostomata: jawed fishes, all tetrapods
    - Class Chondrichthyes: cartilaginous fishes, e.g., sharks, skates, rays
    - Class Osteichthyes: bony fishes, e.g., herring, salmon, cod, eel, flounder
    - Class Amphibia: frogs, toads, salamanders
    - Class Reptilia: snakes, lizards, turtles
    - Class Aves: birds, e.g., sparrows, penguins, ostriches
    - Class Mammalia: mammals, e.g., cats, dogs, horses, rats, humans

*Not in the classification of organisms, but added here for clarity.

The coelom develops differently in the two groups. In protostomes, the mesoderm arises from cells located near the embryonic blastopore, and a splitting occurs that produces the coelom, called a **schizocoelom.** In deuterostomes, the coelom arises as a pair of mesodermal pouches from the wall of the primitive gut. The pouches enlarge until they meet and fuse, forming an **enterocoelom.**

All complex animals have a true coelom. They are divided into the protostomes (spiral cleavage, blastopore associated with mouth, schizocoelom) and the deuterostomes (radial cleavage, blastopore associated with anus, enterocoelom).

This chapter will discuss the protostomes, and the next chapter will discuss the deuterostomes:

Protostomes	Deuterostomes
Molluscs	Echinoderms
Annelids	Chordates
Arthropods	

Protostomes, like other animals, evolved in the sea but some ventured onto land and became very successful there. In this chapter, we will have an opportunity to contrast animal adaptations suitable to living in water with adaptations suitable to living on land. Terrestrial existence requires breathing air, preventing desiccation, and having a means of locomotion and reproduction that are not dependent on external water. The excretory system may be modified for the excretion of a solid nitrogenous waste to help conserve water.

34.2 Molluscs

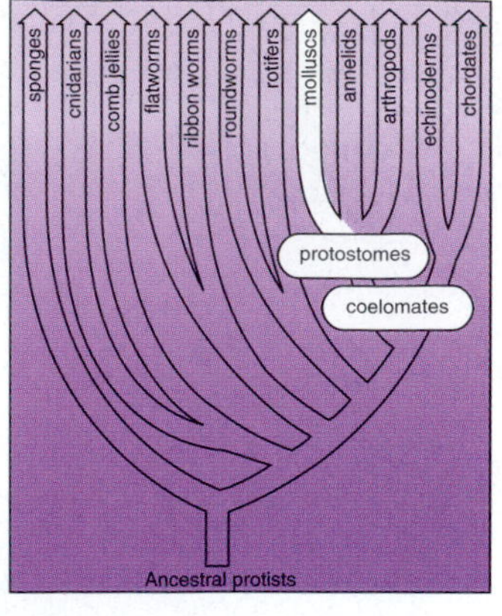

The **molluscs** [L. *mollusc*, soft] **(phylum Mollusca)** include over 110,000 living species—more than twice the number of vertebrate species! Most molluscs are marine, but there are also freshwater and terrestrial molluscs.

Molluscs have a three-part body plan—visceral mass, mantle, and foot—which distinguishes this phylum from others (Fig. 34.2*a*). The visceral mass contains the internal organs, including a highly specialized digestive tract, paired kidneys, and reproductive organs. The **mantle** is a covering that lies to either side of but does not completely enclose the visceral mass. It may secrete a shell and/or contribute to the development of gills or lungs. The space between the folds of the mantle is called the mantle cavity. The foot is a muscular organ that may be adapted for locomotion, attachment, food capture, or a combination of functions. Another feature often present is a **radula,** an organ that bears many rows of teeth and is used to obtain food (Fig. 34.2*b*).

The nervous system of a mollusc consists of several ganglia connected by nerve cords. The coelom is reduced and is largely limited to the region around the heart. Most molluscs have an open circulatory system. The heart pumps blood, more properly called hemolymph, through vessels into sinuses (cavities) collectively called a hemocoel. Blue hemocyanin, rather than red hemoglobin, is the respiratory pigment.

Some molluscs are slow moving and have no head; many others undergo marked cephalization, having both a head and sense organs, and are active predators. Chitons, class Polyplacophora, have a shell that consists of a row of eight overlapping plates. Their flat foot is used for creeping along or clinging to rocks. The chiton scrapes algae and other plant food from rocks with its well-developed radula.

Bivalves

Clams, oysters, mussels, and scallops are all **bivalves** (class Bivalvia) with a two-part shell that is hinged and closed by powerful muscles (Fig. 34.3). They have no head, no radula, and very little cephalization. Clams use their hatchet-shaped foot for burrowing in sandy or muddy soil, and mussels use their foot for production of threads, which attach them to nearby objects. Scallops both burrow and swim; rapid clapping of the valves releases water in spurts and causes the animal to move forward in jerklike fashion for a few feet.

In freshwater clams such as *Anodonta* (Fig. 34.3*c*), the shell, secreted by the mantle, is composed of protein and calcium carbonate with an inner layer of pearl. If a foreign body is placed between the mantle and the shell, pearls form as concentric layers of shell are deposited about the particle. The compressed muscular foot projects ventrally from the shell; by expanding the tip of the foot and pulling the body after it, the clam moves forward.

Within the mantle cavity, the ciliated gills hang down on either side of the visceral mass. The beating of the cilia causes water to enter the mantle cavity by way of the incurrent siphon and to exit by way of the excurrent siphon. The clam is a filter feeder; small particles in this constant stream of water adhere to the gills, and ciliary action sweeps them toward the mouth.

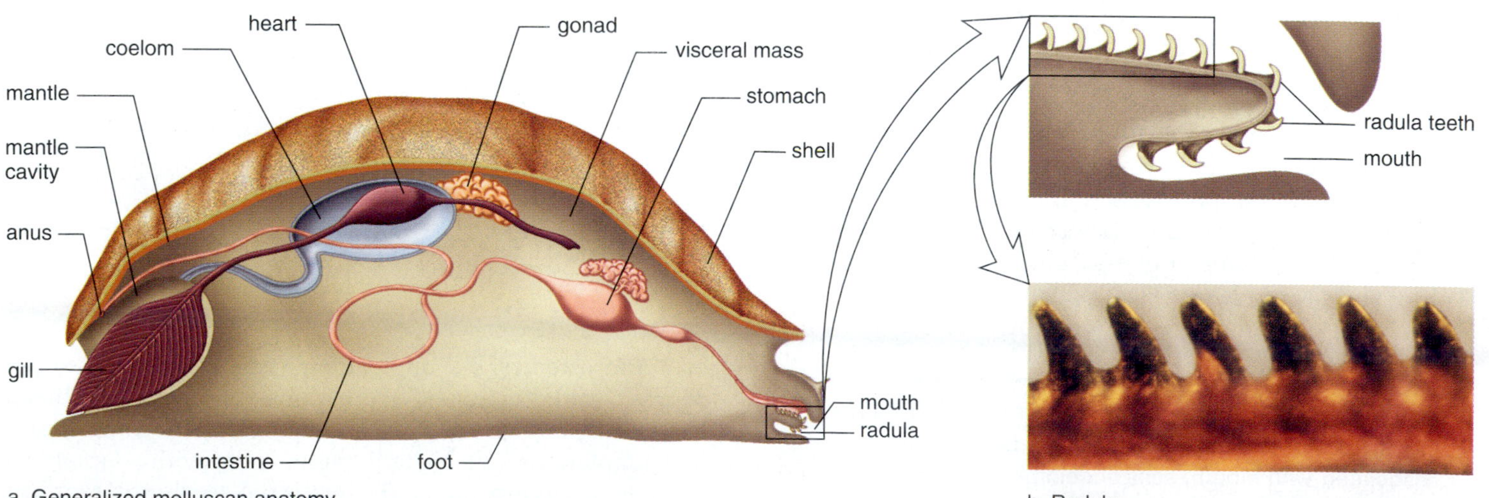

Figure 34.2 Body plan of molluscs.
a. Molluscs have a three-part body consisting of a ventral, muscular foot that is specialized for various means of locomotion; a visceral mass that includes the internal organs; and a mantle which covers the visceral mass and may secrete a shell. Ciliated gills may lie in the mantle cavity and direct food toward the mouth. **b.** In the mouth, the radula is a tonguelike organ that bears rows of tiny teeth that point backward, shown here in art and micrograph.

The mouth leads to a stomach and then to an intestine, which coils about in the visceral mass before going right through the heart and ending in an anus. The anus empties at the excurrent siphon. There is also an accessory organ of digestion called a digestive gland. The heart lies just below the hump of the shell within the pericardial cavity, the only remains of the coelom. The circulatory system is open; the heart pumps hemolymph into vessels that open into the hemocoel. The nervous system is composed of three pairs of ganglia (located anteriorly, posteriorly, and in the foot), which are connected by nerves.

There are two excretory kidneys, which lie just below the heart and remove waste from the pericardial cavity for excretion into the mantle cavity. The clam excretes ammonia (NH_3), a toxic substance that requires the concomitant excretion of water.

The sexes are separate. The gonad is located about the coils of the intestine. Certain clams and annelids have the same type of larva, and this indicates a possible evolutionary relationship between molluscs and annelids.

In clams, which are bivalves, the body is protected by a heavy shell. They are filter feeders with a hatchet foot that allows them to burrow slowly in sand and mud.

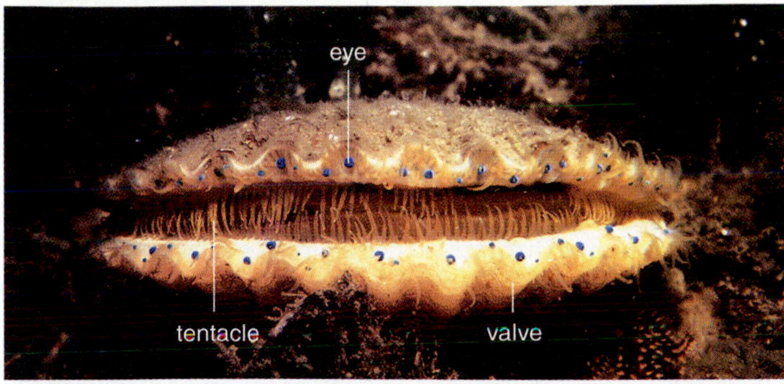

a. Scallop, *Pecten* sp.

b. Mussels, *Mytilus edulis*

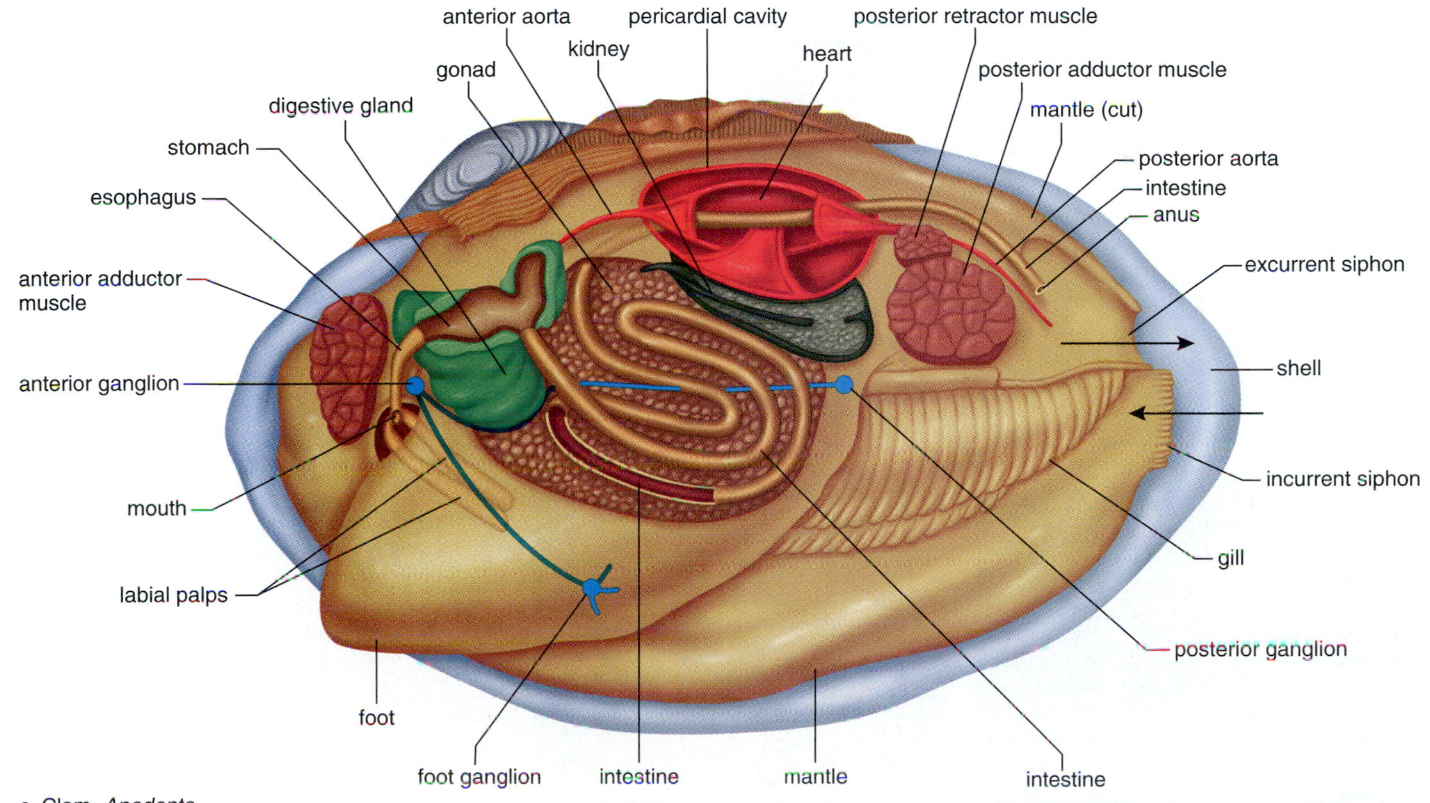

c. Clam, *Anodonta*

Figure 34.3 Bivalve diversity.
Bivalves have a two-part shell. **a.** Scallops clap their valves and swim by jet propulsion. Sensory organs consisting of blue eyes and tentacles along mantle edges occur in this scallop. **b.** Mussels form dense beds in the intertidal zone of northern shores. **c.** In this drawing of a clam, the mantle has been removed from one side. Trace the path of food from the incurrent siphon to the gills, to the mouth, the stomach, the intestine, the anus, and the excurrent siphon. Locate the three ganglia: anterior, foot, and posterior. The heart lies in the reduced coelom.

Cephalopods

Cephalopods [Gk. *kaphale,* head, and *podos,* foot] (class Cephalopoda) include squids, cuttlefish, octopuses, and nautiluses, all of which are fast-swimming predators in the open ocean (Fig. 34.4). Cephalopod means head footed; both squids and octopuses can squeeze their mantle cavity so that water is forced out through a funnel, propelling them by jet propulsion. Also, the tentacles and arms that circle the head capture prey by adhesive secretions or by suckers. A powerful, parrotlike beak is used to tear prey apart. They have well-developed sense organs, including eyes that are similar to those of vertebrates and focus like a camera. Cephalopods, particularly octopuses, have well-developed brains and show a remarkable capacity for learning. Nautiluses are enclosed in shells, but squids have a shell that is reduced and internal. Octopuses lack shells entirely. For protection, squids and octopuses possess ink sacs, from which they can squirt a cloud of brown or black ink. This action often leaves a potential predator completely confused.

In squids, such as *Loligo* (Fig. 34.4*c*), a tough, muscular mantle contains a vestigial skeleton called the pen that surrounds the visceral mass. In a squid the funnel can be directed anteriorly or posteriorly, resulting in either forward or backward movement. A squid has an effective means of seizing and eating food. For example, *Loligo* darts backward rapidly into a school of young mackerel, seizes a fish with its tentacles, quickly biting the neck and severing the nerve cord with its jaws.

Unlike other molluscs and in keeping with its active life, the squid has a closed circulatory system, meaning that blood is always enclosed within blood vessels or a heart. The squid has three hearts—one of which pumps blood to all the internal organs while the other two pump blood to the gills located in the mantle cavity. This efficient closed system effectively circulates oxygen and nutrients to body parts. The brain is formed from a fusion of the three molluscan ganglia. Nerves leave the brain and supply various parts of the body, including an especially large pair that control the rapid contraction of the mantle. The gonads take up a large part of the visceral mass, and the sexes are separate. Packets called spermatophores contain sperm, which the male passes to the female mantle cavity by means of its specialized tentacle. After the eggs are fertilized, they are attached to the substratum in elongated strings, each containing as many as 100 eggs.

Squids, which are cephalopods, have a closed circulatory system and a well-developed nervous system with cephalization. They are active predators in the deep ocean.

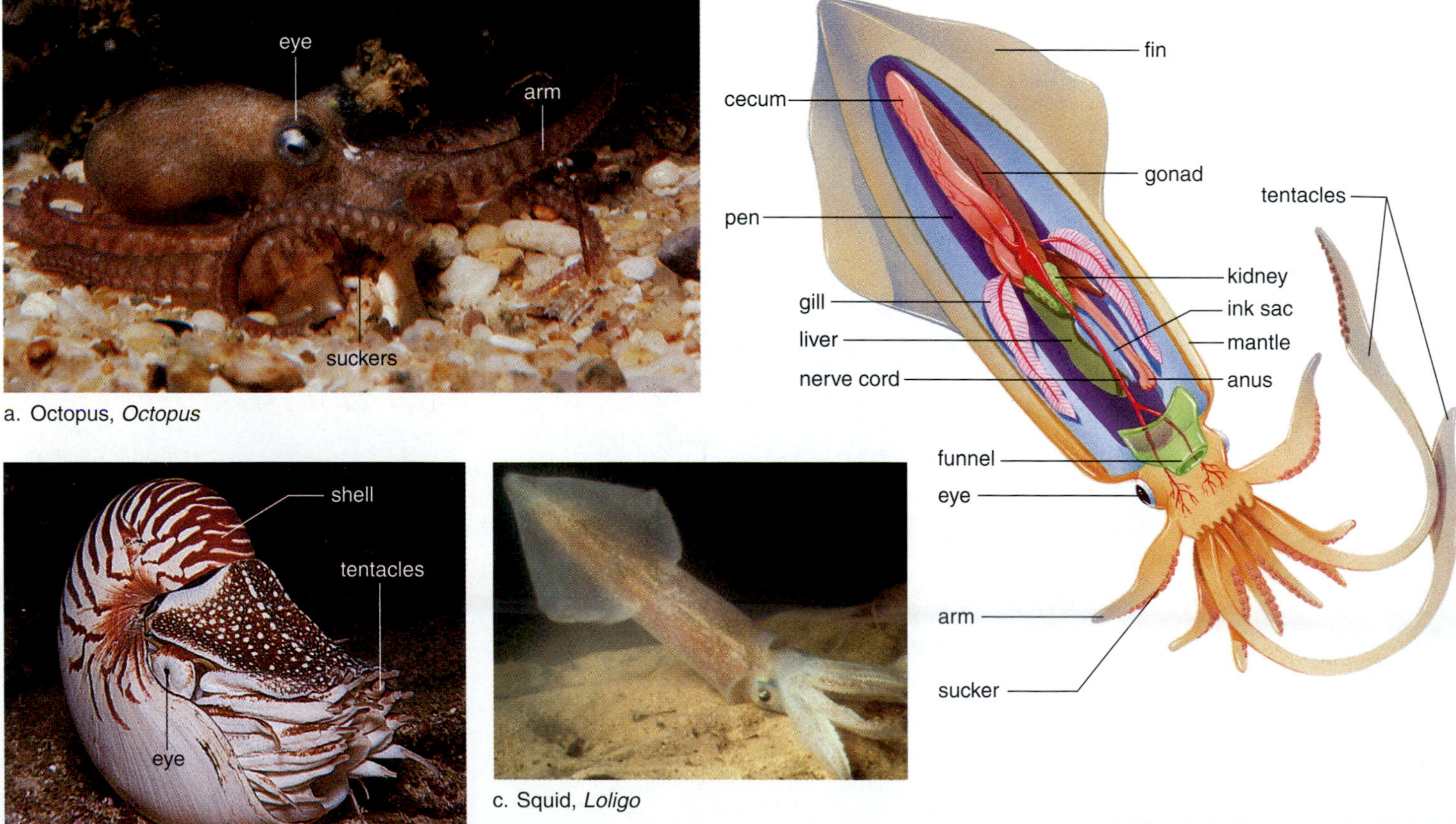

a. Octopus, *Octopus*

b. Chambered nautilus, *Nautilus*

c. Squid, *Loligo*

Figure 34.4 Cephalopod diversity.
Cephalopods have tentacles and/or arms. **a.** Octopuses use their eight arms for both swimming and crawling. **b.** When a nautilus is active, some 34 tentacles protrude from a coiled shell. **c.** Squids are torpedo-shaped, and adapted for fast swimming by jet propulsion.

Gastropods

Gastropods [Gk. *gastros,* stomach, and *podos,* foot] (class Gastropoda)—including snails, whelks, conchs, periwinkles, and sea slugs—are usually found in marine habitats, although they sometimes inhabit freshwater environments. In addition, garden snails and slugs are adapted to terrestrial habitats (Fig. 34.5). Many gastropods are herbivores that use their radula to scrape food from surfaces. Others are carnivores, using their radula to bore through surfaces such as bivalve shells to obtain food.

Gastropods have an elongated, flattened foot. In most, a well-developed head region with eyes and tentacles projects from a coiled shell that protects the visceral mass. Nudibranchs (sea slugs) and terrestrial slugs, however, lack a shell. During development, gastropods undergo a torsion, or twisting, that brings the anus and mantle cavity downward, then forward and around to a position above the head. Torsion positions the visceral mass squarely above the foot.

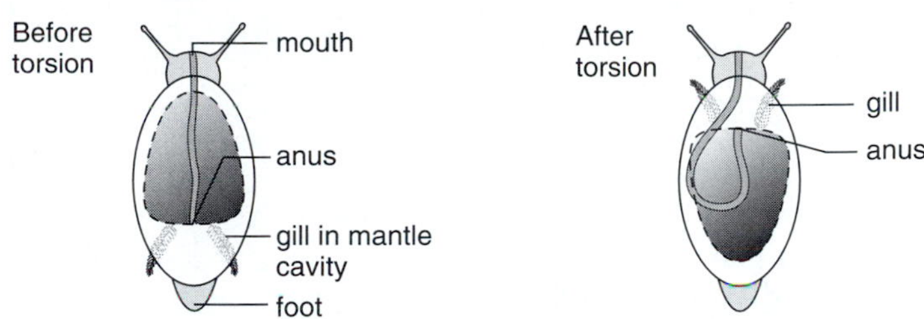

In aquatic gastropods, gills are found in the mantle cavity, but in those adapted to land, the mantle is richly supplied with blood vessels and functions as a lung when air is moved in and out through respiratory pores. As another adaptation to life out of water, terrestrial gastropod development does not include the swimming larval stage found in aquatic species.

Land snails, such as *Helix aspersa,* have three obvious divisions of the body: a head with two pairs of tentacles, one pair of which bears eyes at the tips; a flat, long, muscular foot; and a visceral mass surrounded by a shell (Fig. 34.5*c*). The shell not only offers protection but also prevents desiccation (drying out). Waves of contraction run from the anterior to the posterior portion of the foot, and a lubricating mucus is secreted to facilitate movement.

Land snails are hermaphroditic; when two snails meet, they shoot calcareous darts into each other's body wall as a part of premating behavior. Then each inserts a penis into the vagina of the other to provide sperm for the future fertilization of eggs that are deposited in the soil. Development proceeds directly without the formation of larvae.

The presence of a copulatory organ such as the penis, and even hermaphroditism, are adaptations to life on land. The penis allows easy transfer of sperm from one animal to another; hermaphroditism assures that any two animals can mate. This is especially useful in slow-moving animals that have large ranges.

Land snails have a coiled shell, a large, flat, muscular foot, and a head region. In addition, the mantle in a garden snail becomes a lung.

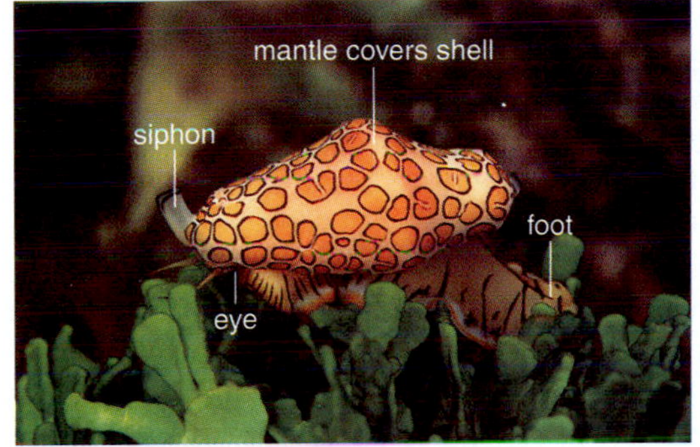

a. Flamingo tongue egg shell, *Cyphoma gibbosum*

b. Nudibranch, *Glossodoris macfarlandi*

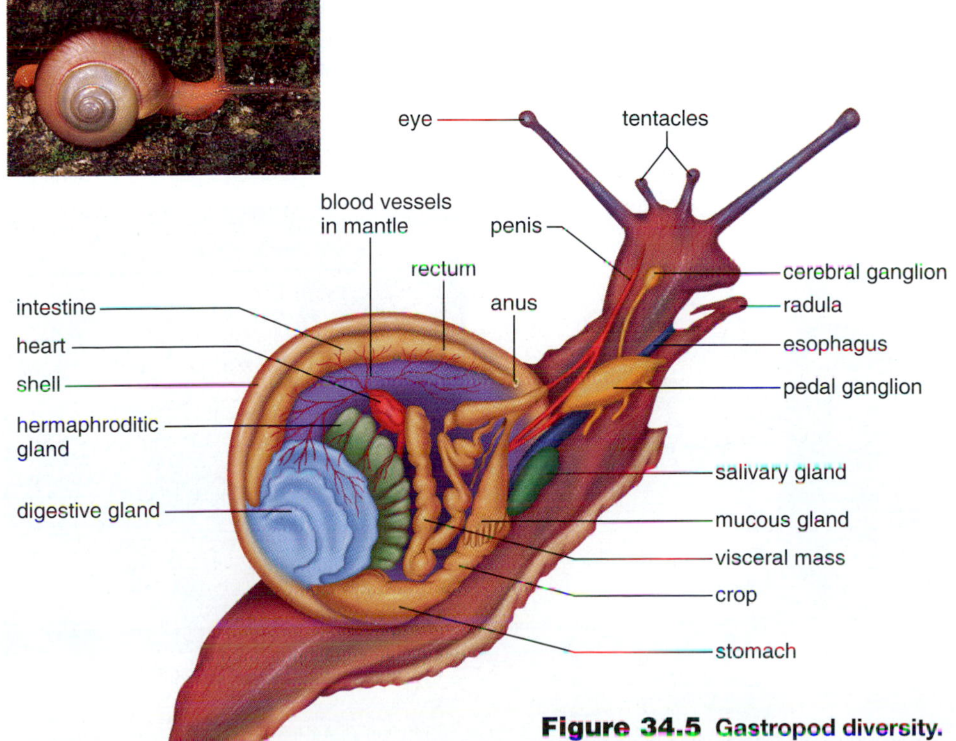

c. Rain forest snail, *Naninia glutinosa*

Figure 34.5 Gastropod diversity.
Gastropods have an elongated flat foot. **a.** Extensions of the mantle, having orange, brown, and white markings, cover the shell of this marine snail when it is active. **b.** Nudibranchs are also marine and, as their name suggests, have no shell. **c.** Drawing and photo of a land snail.

34.3 Annelids

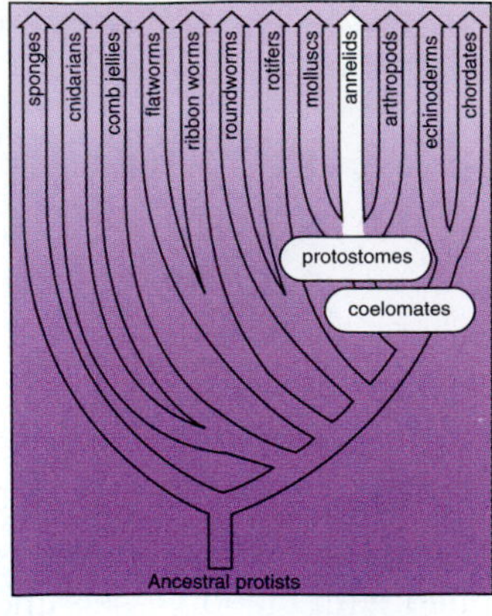

Annelids [L. *annelus,* dim. of *annulus,* ring] (**phylum Annelida,** 12,000 species) are *segmented,* as is externally evidenced by the rings that encircle the body. Internally, partitions called septa (sing., septum) divide the well-developed, fluid-filled coelom, which acts as a hydrostatic skeleton. A hydrostatic skeleton along with partitioning of the coelom permits each body segment independence of movement. Instead of just burrowing in the mud, an annelid can crawl on a surface.

Segmentation and the tube-within-a-tube body plan has led to specialization of the digestive tract in annelids. For example, the digestive system may include a pharynx, a stomach, and accessory glands. They have an extensive closed circulatory system with blood vessels that run the length of the body and branch to every segment. The nervous system consists of a brain connected to a ventral solid nerve cord, with a ganglion in each segment. The excretory system consists of paired **nephridia** [Gk. *nephros,* kidney], which are coiled tubules in each segment that collect waste material from the coelom and excrete it through openings in the body wall.

Annelids are segmented worms. The coelom is divided by septa, and organ systems have repeating parts. There is a closed circulatory system and a ventral solid nerve cord.

Marine Worms

Most annelids (class Polychaeta) are marine; their class name refers to the presence of many setae. **Setae** [L. *seta,* bristle] are bristles that anchor the worm or help it move. In polychaetes [Gk. *polys,* many, and *chaite,* bristle, long hair], the setae are in bundles on parapodia [Gk. *para,* beside, and *podos,* foot], which are paddlelike appendages found on most segments. These are used not only in swimming but also as respiratory organs where the expanded surface area allows for exchange of gases. Clam worms (Fig. 34.6*a*) such as *Nereis,* are predators. They prey on crustaceans and other small animals, which are captured by a pair of strong chitinous jaws that extend with a part of the pharynx when the animal is feeding. Associated with its way of life, *Nereis* undergoes cephalization and has a head region with eyes and other sense organs.

Other polychaetes are sedentary (sessile) tube worms, with tentacles that form a funnel-shaped fan. Water currents, created by the action of cilia, trap food particles that are directed toward the mouth. They are sessile filter feeders; a sorting mechanism rejects large particles, and only the smaller ones are accepted for consumption.

Polychaetes have breeding seasons, and only during these times do the worms have sex organs. In *Nereis,* many worms concurrently shed a portion of their bodies containing either eggs or sperm, and these float to the surface where fertilization takes place. The zygote rapidly develops into a type of larva that is similar to that in marine clams. The existence of this larva in both the annelids and molluscs shows that these two groups of animals are related.

Polychaetes are marine worms with bundles of setae attached to parapodia.

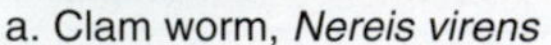
a. Clam worm, *Nereis virens*

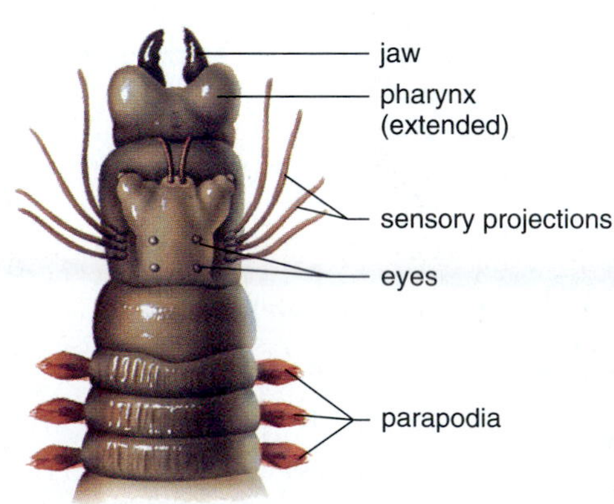

b. Fanworm, *Spirobranchus giganteus*

Figure 34.6 Polychaete diversity.
a. Clam worms are predaceous polychaetes that undergo cephalization. Note also the parapodia, which are used for swimming and as respiratory organs. **b.** Fanworms (a type of tube worm) are sessile filter feeders whose ciliated tentacles spiral in this example.

Earthworms

The **oligochaetes** [Gk. *oligos,* few, and *chaite,* bristle, long hair] (class Oligochaeta), which include earthworms, have few setae. Earthworms (e.g., *Lumbricus*) do not have a well-developed head or parapodia (Fig. 34.7). Their setae protrude in clusters directly from the surface of their body. Locomotion, which is accomplished section by section, utilizes muscle contraction and the setae. When longitudinal muscles contract, segments bulge and their setae protrude into the soil; then when circular muscles contract, the setae are withdrawn and these segments move forward.

Earthworms reside in soil where there is adequate moisture for the body wall to remain moist for gas exchange purposes. They are scavengers that feed on leaves or any other organic matter, living or dead, which can conveniently be taken into the mouth along with soil. Food drawn into the mouth by the action of the muscular pharynx is stored in a crop and ground up in a thick, muscular gizzard. Digestion and absorption occur in a long intestine whose dorsal surface is expanded by a **typhlosole** that allows additional surface for absorption.

Earthworm segmentation, which is so obvious externally, is also internally evidenced by septa. The long ventral solid nerve cord leading from the brain has ganglionic swellings and lateral nerves in each segment. The paired nephridia, or coiled tubules, in each segment have two openings: one is a ciliated funnel that collects coelomic fluid and the other is an exit in the body wall. Between the two openings is a convoluted region where waste material is removed from the blood vessels about the tubule. Red blood moves anteriorly in a dorsal blood vessel and then is pumped by five pairs of hearts into a ventral vessel. As the ventral vessel takes the blood toward the posterior regions of the worm's body, it gives off branches in every segment.

The worms are hermaphroditic; the male organs are the testes, the seminal vesicles, and the sperm ducts, and the female organs are the ovaries, the oviducts, and the seminal receptacles. Two worms lie parallel to each other facing in opposite directions. The fused midbody segment, called a clitellum, secretes mucus, protecting the sperm from drying out as they pass between the worms. After the worms separate, the clitellum of each produces a slime tube, which is moved along over the head by muscular contractions. As it passes, eggs and the sperm received earlier are deposited and fertilization occurs. The slime tube then forms a cocoon to protect the worms as they develop. There is no larval stage.

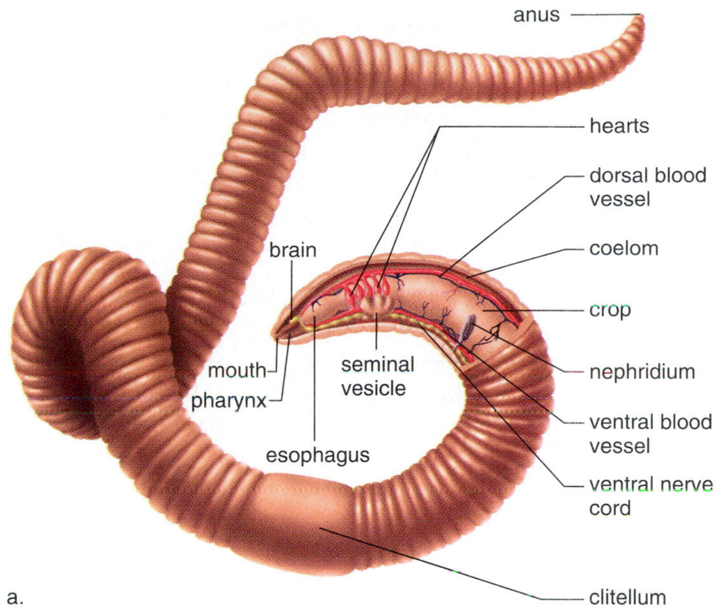

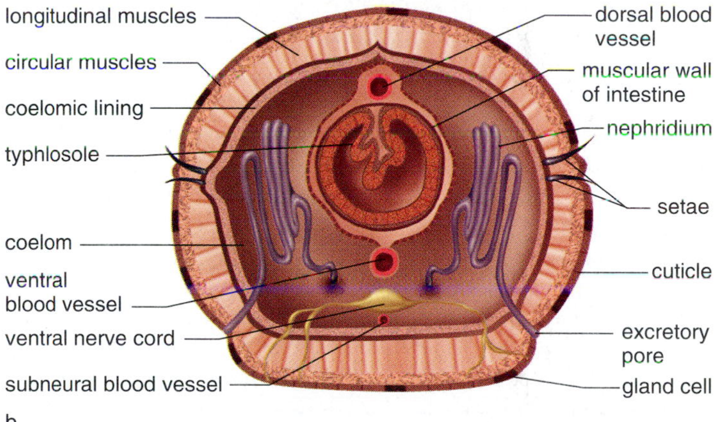

Figure 34.7 Earthworm, *Lumbricus*.
a. Internal anatomy of the anterior part of an earthworm. Notice that each body segment bears a pair of setae and that internal septa divide the coelom into compartments. **b.** Cross section of earthworm. **c.** When earthworms mate, they are held in place by a mucus secreted by the clitellum. The worms are hermaphroditic, and when mating, sperm pass from the seminal vesicles of each to the seminal receptacles of the other.

It is interesting to compare the anatomy of clam worms with earthworms because it highlights the manner in which earthworms are adapted to life on land. Jaws are not needed by the nonpredatory earthworms that extract organic remains from the soil they eat. The lack of parapodia helps reduce the possibility of water loss and facilitates burrowing in soil. The clam worm makes use of external water while the earthworm provides a mucous secretion to aid fertilization. It is the water form that has the swimming larva and not the land form.

Earthworms, which burrow in the soil, lack cephalization and parapodia. They are hermaphroditic, and there is no larval stage.

Leeches

Leeches (class Hirudinea) are usually found in fresh water, but some are marine or even terrestrial. They have the same body plan as other annelids but they have no setae, and each body ring has several transverse grooves. Most leeches are only 2–6 cm in length but some, including the medicinal leech, are as long as 20 cm.

Among their modifications are two suckers, a small oral one around the mouth and a large posterior one. Most leaches prey on small invertebrates or attach themselves to vertebrates, cut through the flesh with their proboscis, and feed on body fluids (Fig. 34.8). Leeches are able to keep blood flowing and prevent clotting by means of a substance in their saliva known as hirudin, a powerful anticoagulant. This has added to their potential usefulness in the field of medicine today.

Leeches are modified in a way that lends itself to the parasitic way of life. Some are external parasites known as bloodsuckers.

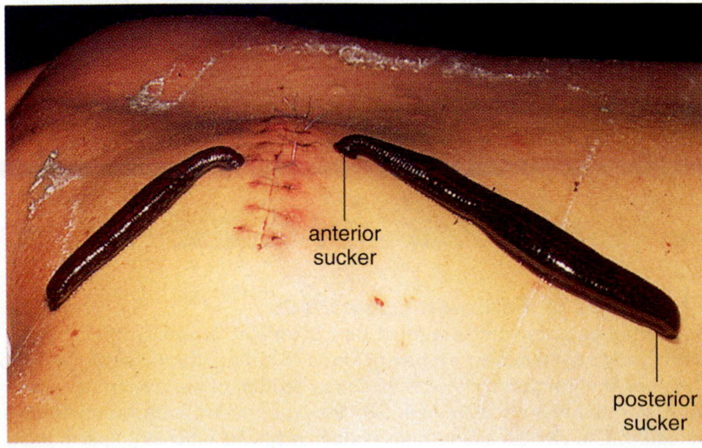

Figure 34.8 Medicinal leeches, *Hirudo.*
On occasion, leeches are used to remove blood from bruised skin. The anterior sucker of the medicinal leech has three jaws that saw through the skin, producing a Y-shaped incision through which blood is drawn up by the sucking action of the leech's muscular pharynx. The salivary glands secrete an anticoagulant called hirudin that keeps the blood flowing while the leech is feeding. Other salivary ingredients dilate the patient's blood vessels and act as an anesthetic.

34.4 Arthropods

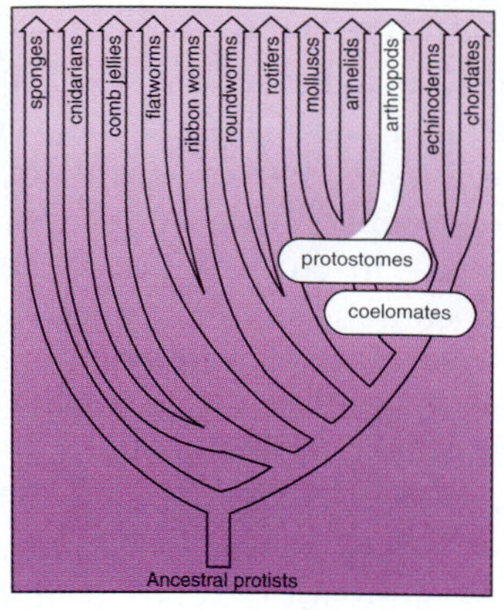

Arthropods [Gk. *arthron,* joint, and *podos,* foot] (**phylum Arthropoda,** over one million species) show such diversity and are adapted to so many different habitats that they are often said to be the most successful of all animals. Some claim the modern era should be called the age of arthropods, not the age of mammals! Arthropod literally means "jointed foot," but actually they have freely movable **jointed appendages.**

What characteristics account for the success of arthropods? Arthropods have a rigid, but jointed, **exoskeleton** (Fig. 34.9*a, b*). First, the exoskeleton of arthropods is composed primarily of **chitin** [Gk. *chiton,* tunic], a strong, flexible, nitrogenous polysaccharide. The exoskeleton serves many functions such as protection, attachment for muscles, locomotion, and prevention of desiccation. However, because it is hard and nonexpandable, arthropods must **molt,** or shed, the exoskeleton as they grow larger. Before molting, the body secretes a new, larger exoskeleton, which is soft and wrinkled, underneath the old one. After enzymes partially dissolve and weaken the old exoskeleton, the animal breaks it open and wriggles out. The new exoskeleton then quickly expands and hardens.

Second, arthropods are segmented but some segments are fused into regions, such as a head, a thorax, and an abdomen. In trilobites (subphylum Trilobitomorpha), which flourished during the Cambrian period, there was a pair of appendages on each body segment. In modern arthropods, appendages are specialized for such functions as walking, swimming, reproducing, eating, and sensory reception. These modifications account for much of the diversity of arthropods. Several arthropod groups, such as insects, arachnids, centipedes, and millipedes, contain species that are adapted to terrestrial life.

Third, arthropods have a well-developed nervous system. There is a brain and a ventral solid nerve cord. The head bears various types of sense organs, including eyes of two types—simple and compound (Fig. 34.9*c*). The compound eye is composed of many complete visual units, each of which operates independently. The lens of each visual unit focuses an image on the light-sensitive membranes of a small number of photoreceptors within that unit. The simple eye, like that of vertebrates, has a single lens that brings the image to focus onto many receptors, each of which receives only a portion of the image.

Fourth, arthropods have a variety of respiratory organs. Marine forms utilize gills, which are vascularized, highly convoluted, thin-walled tissue specialized for gas exchange. Terrestrial forms have book lungs (e.g., spiders) or air tubes called **tracheae** [L. *trachia*, windpipe]. Tracheae serve as a rapid way to transport oxygen directly to the cells.

Finally, the occurrence of metamorphosis has contributed to the success of arthropods. **Metamorphosis** [Gk. *meta*, implying change, and *morphe*, shape, form] is a drastic change in form and physiology that occurs as an immature stage, called a larva, becomes an adult. Among arthropods, the larva eats different food and lives in a different environment than the adult. This reduces competition and allows more members of a species to exist at one time. For example, larval crabs live among and feed on plankton, while adult crabs are bottom dwellers that catch live prey or scavenge dead organic matter. Among insects, such as butterflies, the caterpillar feeds on leafy vegetation while the adult feeds on nectar.

A phylum can be divided into subphyla, and we will consider three arthropod subphyla: the chelicerates (e.g., spiders), the crustaceans (e.g., crayfish), and the uniramians (e.g., millipedes, centipedes, and insects). There can also be superclasses, which include more than one class; the millipedes and centipedes are in a superclass and the insects are another superclass.

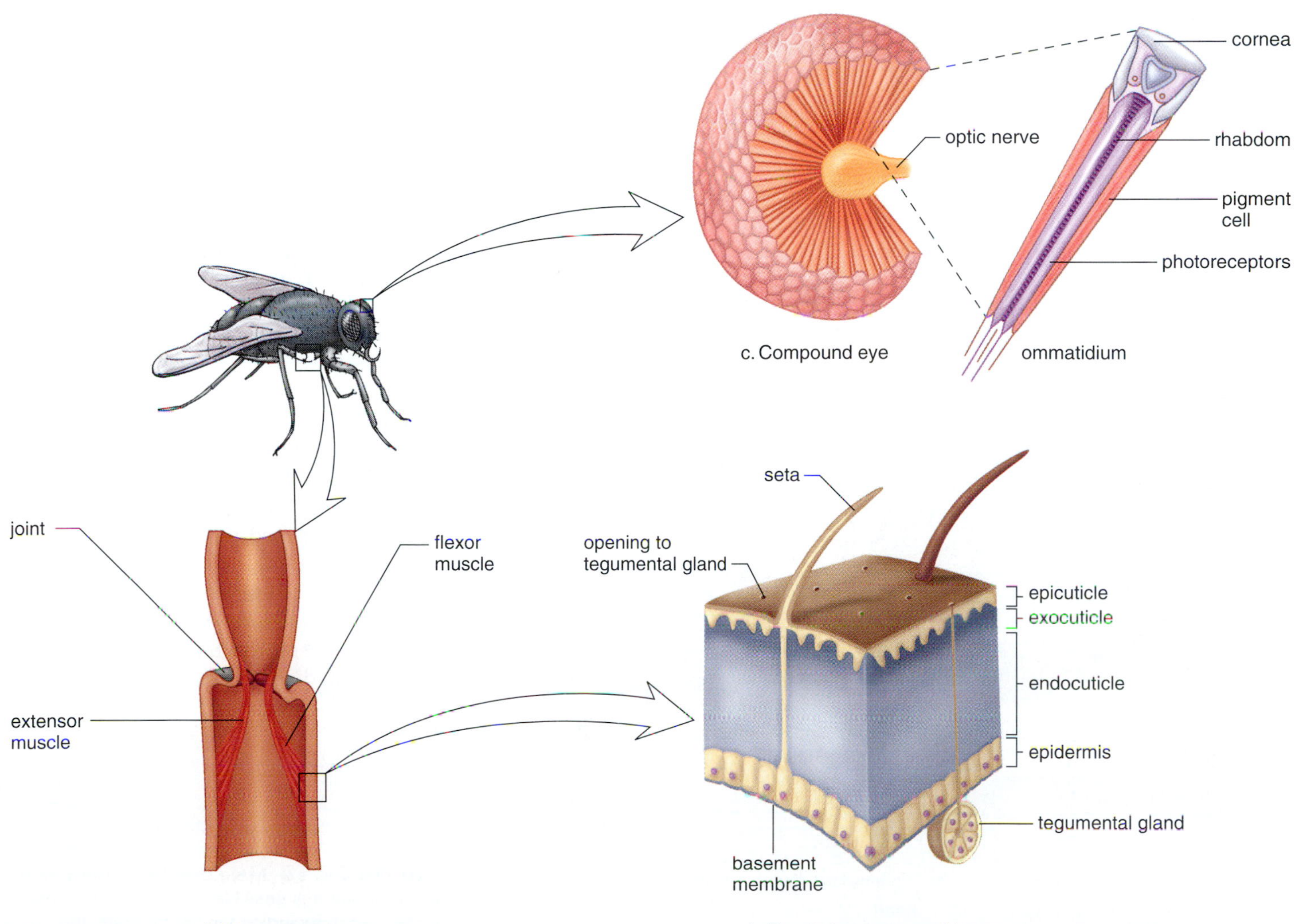

Figure 34.9 Arthropod skeleton and eye.
a. The joint in an arthropod skeleton is a region where the exocuticle is thinner and not as hard as the rest of the exocuticle. The direction of movement is toward the flexor muscle or the extensor muscle, whichever one has contracted. **b.** The exoskeleton is secreted by the epidermis and consists of the epicuticle, a waxy layer, an exocuticle, which is hardened by the deposition of calcium carbonate, and an endocuticle. Chitin makes up the bulk of the exo- and endocuticle. **c.** Arthropods have a compound eye which contains many individual units, each with its own lens and photoreceptors.

Crustaceans

Crustaceans (subphylum Crustacea, 40,000 species) are successful, largely marine, arthropods. Crustaceans are named for their hard shells; the exoskeleton is calcified. Although their anatomies are extremely diverse, the head usually bears a pair of compound eyes and five pairs of appendages. The first two pairs, called antennae and antennules, lie in front of the mouth and have sensory functions. The other three pairs (mandibles, first and second maxillae) lie behind the mouth and are usually used in feeding as mouthparts. Biramous [Gk. *bis,* two, and *ramus,* a branch] appendages on the thorax and abdomen are segmentally arranged; one branch is the gill branch and the other is the leg branch:

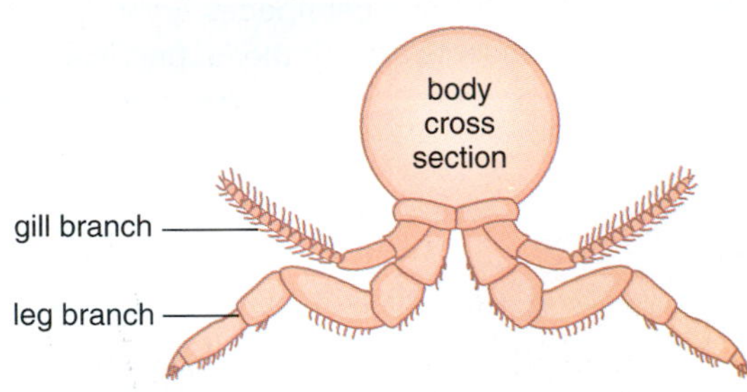

Copepods and krill are small crustaceans that live in the water, where they feed on algae. In the marine environment, they serve as food for fishes, sharks, and whales. They are so numerous that despite their small size, some believe that they are harvestable as food. Barnacles are also crustaceans but they have a thick, heavy shell as befits their inactive lifestyle. Stalked (gooseneck) barnacles are attached by a stalk, and stalkless (acorn) barnacles are attached directly by their shells. You see barnacles on wharf pilings, ship hulls, seaside rocks, and even the bodies of whales. They begin life as free-swimming larvae, but they undergo a metamorphosis that transforms their swimming appendages to cirri, feathery structures that are extended and allow them to filter feed when they are submerged.

Decapods are the most familiar and numerous of the crustaceans. They include shrimps, lobsters, crayfish, and crabs, in which the thorax bears five pairs of walking legs. The first pair may be modified as claws. Typically, there are gills situated above the walking legs. Figure 34.10*a* gives a view of the external anatomy of the crayfish. The head and thorax are fused into a **cephalothorax,** which is covered on the top and sides by a nonsegmented carapace. The abdominal segments are equipped with swimmerets, small paddle-like structures. The first two pairs of swimmerets in the male are quite strong and are used to pass sperm to the female.

The last two segments bear the uropods and the telson, which make up a fan-shaped tail. Ordinarily, a crayfish lies in wait for prey. It faces out from an enclosed spot with the claws extended and the antennae moving about. The claws seize any small animal, either dead ones or live ones that happen by, and carry them to the mouth. When a crayfish moves about, it generally crawls slowly but may swim rapidly by using its heavy abdominal muscles and tail.

The respiratory system consists of gills that lie above the walking legs protected by the carapace. As shown in Figure 34.10*b*, the digestive system includes a stomach, which is divided into two main regions: an anterior portion called the gastric mill, equipped with chitinous teeth to grind coarse food, and a posterior region, which acts as a filter to prevent coarse particles from entering the digestive glands where absorption takes place. Green glands lying in the head region, anterior to

cephalothorax
abdomen
first walking leg (modified as a pincerlike claw)
carapace
second walking leg
third walking leg
fourth walking leg
fifth walking leg
uropods
swimmerets
compound eye
telson
antennae
anus
a.
mouth
opening of sperm duct
claspers
excretory pore
gills

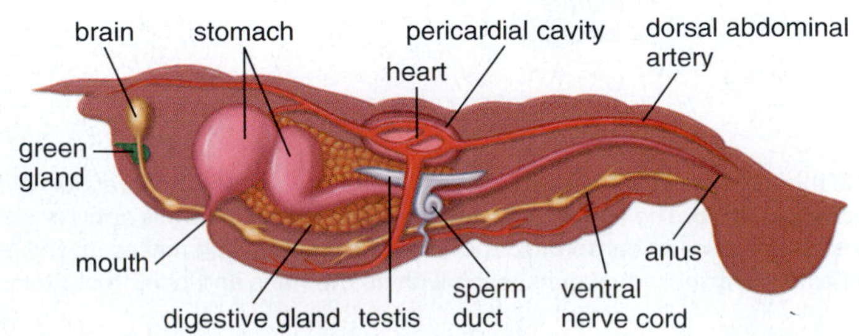

Figure 34.10 Male crayfish, *Cambarus*. **a.** Externally, it is possible to observe the jointed appendages, including the swimmerets, the walking legs, and the claws. These appendages, plus a portion of the carapace, have been removed from the right side so that the gills are visible. **b.** Internally, the parts of the digestive system are particularly visible. The circulatory system can also be clearly seen. Note the ventral solid nerve cord.

the esophagus, excrete metabolic wastes through a duct that opens externally at the base of the antennae. The coelom, which is so well developed in the annelids, is reduced in the arthropods and is composed chiefly of the space about the reproductive system. A heart pumps hemolymph containing the respiratory pigment hemocyanin into a **hemocoel** [Gk. *haima,* blood, and *koiloma,* cavity] consisting of sinuses (open spaces) where the hemolymph flows about the organs. (Whereas hemoglobin is a red pigment, hemocyanin is a blue pigment.) This is an open circulatory system because blood is not contained within blood vessels.

The nervous system is quite similar to that of the earthworm. There is a brain, as well as a ventral nerve cord that passes posteriorly. Along the length of the nerve cord, periodic ganglia give off lateral nerves.

The sexes are separate in the crayfish, and the gonads are located just ventral to the pericardial cavity. In the male, a coiled sperm duct opens to the outside at the base of the fifth walking leg. Sperm transfer is accomplished by the modified first two swimmerets of the abdomen. In the female, the ovaries open at the bases of the third walking legs. A stiff fold between the bases of the fourth and fifth pairs serves as a seminal receptacle. Following fertilization, the eggs are attached to the swimmerets of the female.

Crustaceans are diverse, as shown in Figure 34.11.

Crustaceans are mainly marine arthropods in which the head bears five pairs of appendages, including mandibles and maxillae. Typically, there are biramous appendages.

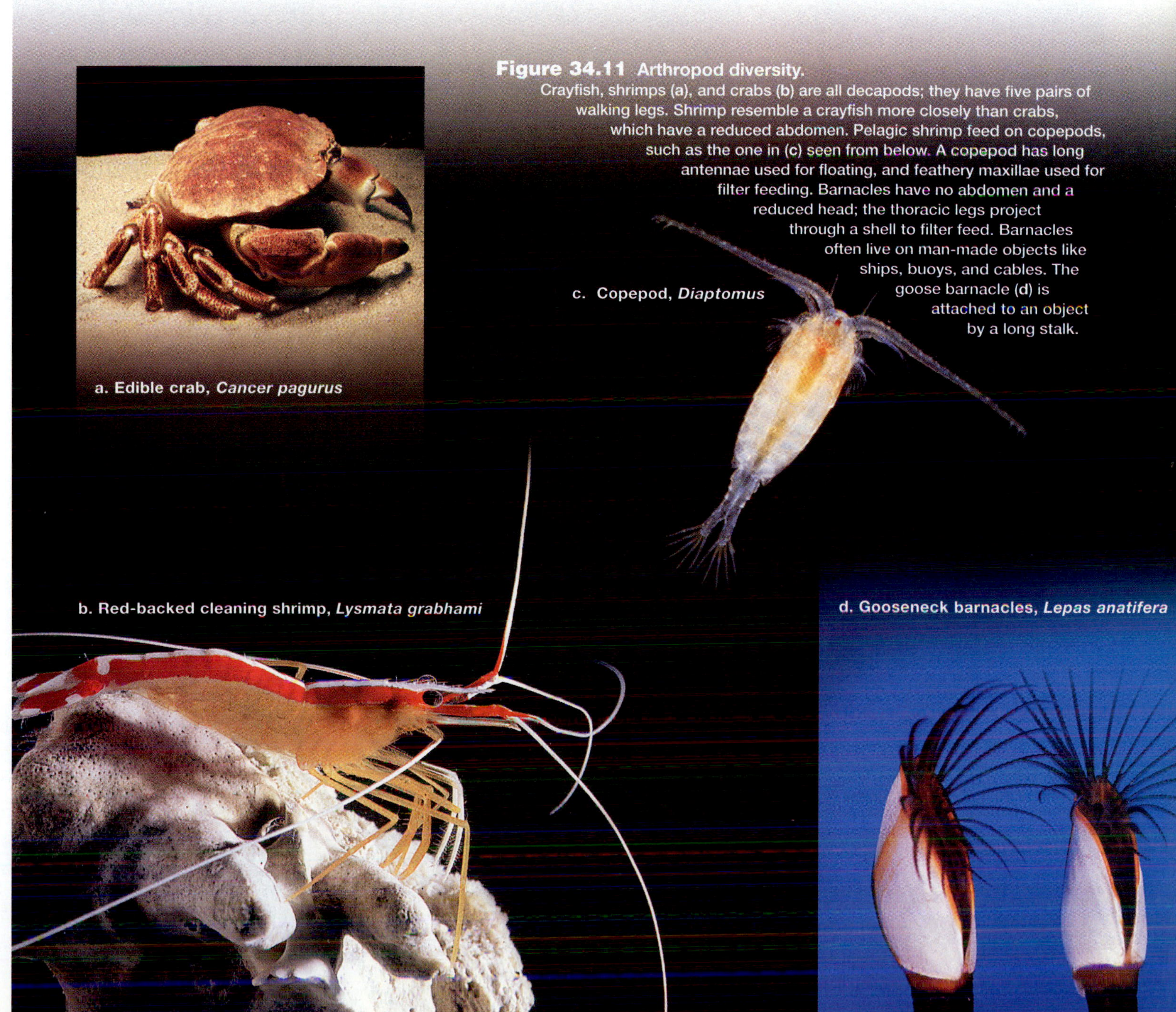

Figure 34.11 Arthropod diversity.
Crayfish, shrimps (a), and crabs (b) are all decapods; they have five pairs of walking legs. Shrimp resemble a crayfish more closely than crabs, which have a reduced abdomen. Pelagic shrimp feed on copepods, such as the one in (c) seen from below. A copepod has long antennae used for floating, and feathery maxillae used for filter feeding. Barnacles have no abdomen and a reduced head; the thoracic legs project through a shell to filter feed. Barnacles often live on man-made objects like ships, buoys, and cables. The goose barnacle (d) is attached to an object by a long stalk.

a. Edible crab, *Cancer pagurus*

b. Red-backed cleaning shrimp, *Lysmata grabhami*

c. Copepod, *Diaptomus*

d. Gooseneck barnacles, *Lepas anatifera*

White-tailed dragonfly, *Libellula*
Tortoise shell scale, *Lecanium*
Painted grasshopper, *Romalea*
Snout beetle, *Chloropholus*
Giant walking stick, *Diapheromera*
Green lacewing, *Chrysopa*

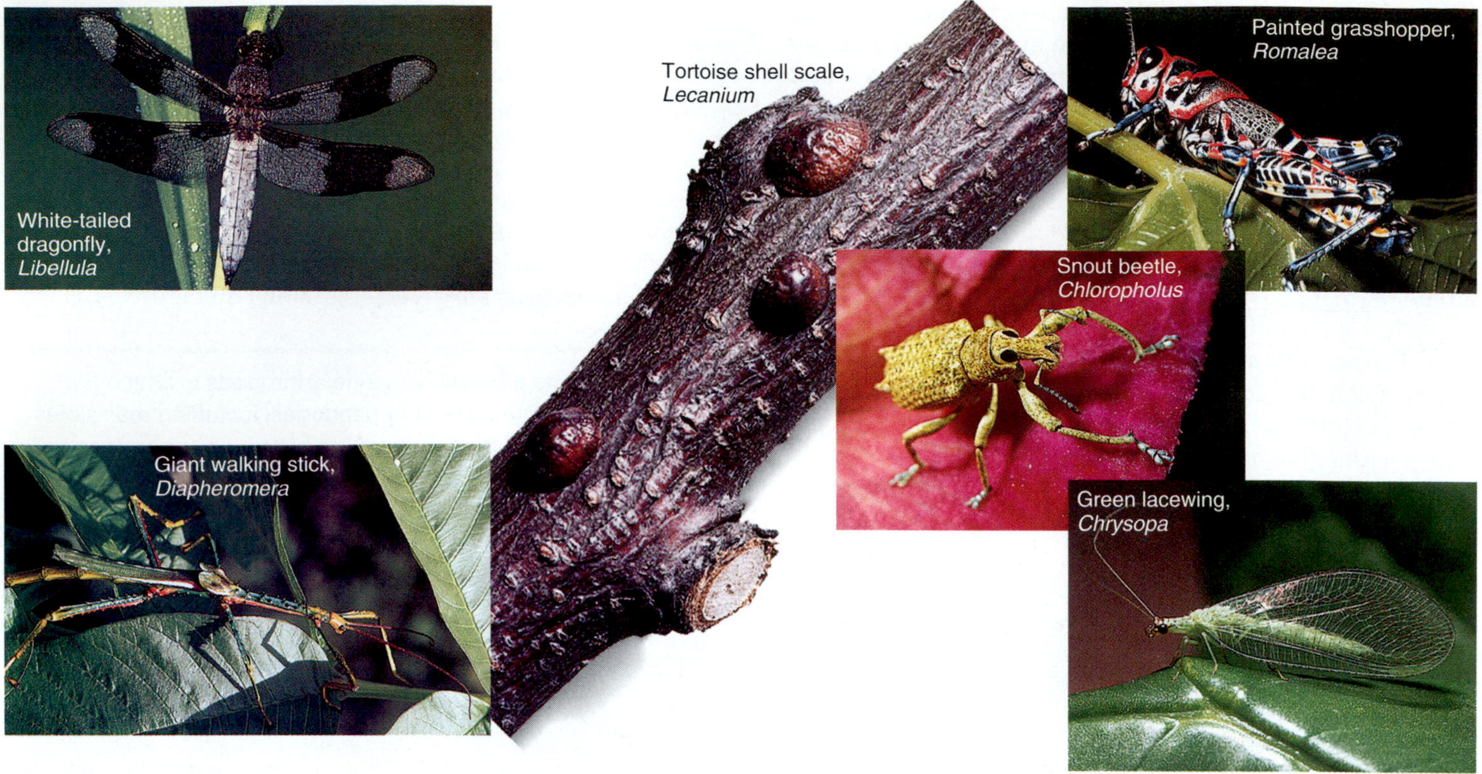

Figure 34.12 Insect (superclass Insecta) diversity.

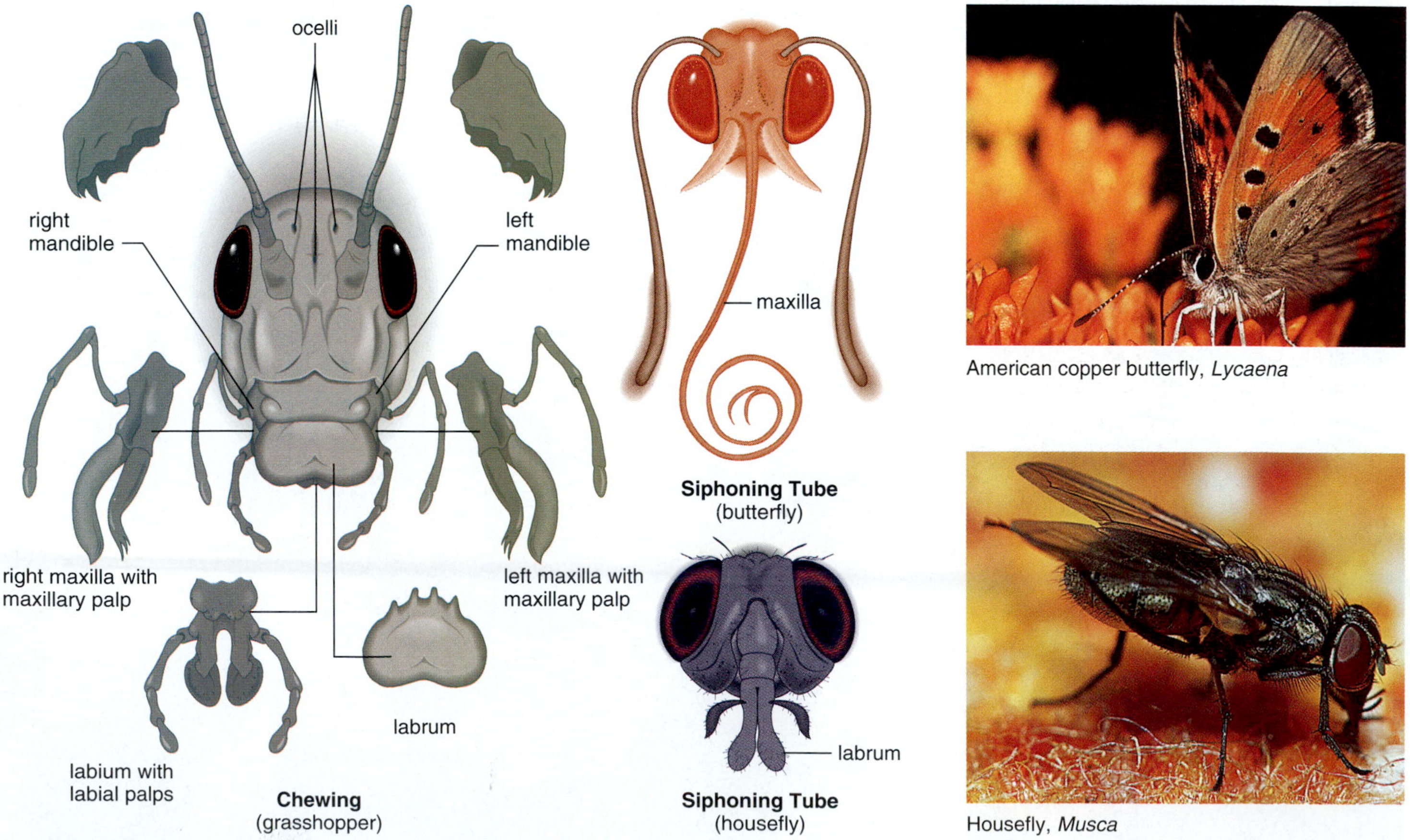

ocelli
right mandible
left mandible
right maxilla with maxillary palp
left maxilla with maxillary palp
labrum
labium with labial palps
Chewing (grasshopper)
maxilla
Siphoning Tube (butterfly)
labrum
Siphoning Tube (housefly)
American copper butterfly, *Lycaena*
Housefly, *Musca*

Figure 34.13 Three types of insect mouthparts.

Uniramians

Uniramians **(subphylum Uniramia)** include insects (superclass Insecta) and millipedes, centipedes (superclass Myriapoda). The uniramous appendages attached to the thorax and abdomen have only one branch—the leg branch. The head appendages include only one pair of antennae, one pair of mandibles, and one or two pairs of maxillae. The uniramians live on land and breathe by means of a system of air tubes called tracheae.

Insects

Insects (superclass Insecta, 900,000 species) include more known species than all other animal species combined. Only a few types can be represented in Figure 34.12. Insects are adapted for an active life on land, although some have secondarily invaded aquatic habitats. The body of an insect is divided into a head, a thorax, and an abdomen. The head bears the sense organs and mouthparts (Fig. 34.13); the thorax bears three pairs of legs and one or two pairs of wings; and the abdomen contains most of the internal organs. Wings enhance an insect's ability to survive by providing a way of escaping enemies, finding food, facilitating mating, and dispersing the species. The exoskeleton of an insect is lighter and contains less chitin than that of many other arthropods.

In the grasshopper (Fig. 34.14), the third pair of legs is suited to jumping. There are two pairs of wings. The forewings are tough and leathery, and when folded back at rest they protect the broad, thin hindwings. On the lateral surface, the first abdominal segment bears a large tympanum on each side for the reception of sound waves. The posterior region of the exoskeleton in the female has two pairs of projections that form an ovipositor, which is used to dig a hole in which eggs are laid.

The digestive system is suitable for a herbivorous diet. In the mouth, food is broken down mechanically by mouthparts and enzymatically by salivary secretions. Food is temporarily stored in the crop before passing into the gizzard, where it is finely ground. Digestion is completed in the stomach, and nutrients are absorbed into the hemocoel from outpockets called gastric ceca. A cecum is a cavity open at one end only. The excretory system consists of **Malpighian tubules,** which extend into a hemocoel and collect nitrogenous wastes that are concentrated and excreted into the digestive tract. The formation of a solid nitrogenous waste, namely uric acid, conserves water.

The respiratory system begins with openings in the exoskeleton called spiracles. From here, the air enters small tubules called tracheae (Fig. 34.14*a*). The tracheae branch and rebranch, finally ending in moist areas where the actual exchange of gases takes place. The movement of air through this complex of tubules is not a passive process; air is pumped through by a series of several bladderlike structures (air sacs), which are attached to the tracheae near the spiracles. Air enters the anterior four spiracles and exits by the posterior six spiracles. Breathing by tracheae may account for the small size of insects (most are less than 60 mm in length) since the tracheae are so tiny and fragile that they would be crushed by any amount of weight.

The circulatory system contains a slender, tubular heart that lies against the dorsal wall of the abdominal exoskeleton and pumps hemolymph into the hemocoel where it circulates before returning to the heart again. The hemolymph is colorless and lacks a respiratory pigment, and the tracheal system transports gases.

Reproduction is adapted to life on land. The male has a penis, which passes sperm to the female. Internal fertilization protects both gametes and zygotes from drying out. The female deposits the fertilized eggs in the ground with her ovipositor.

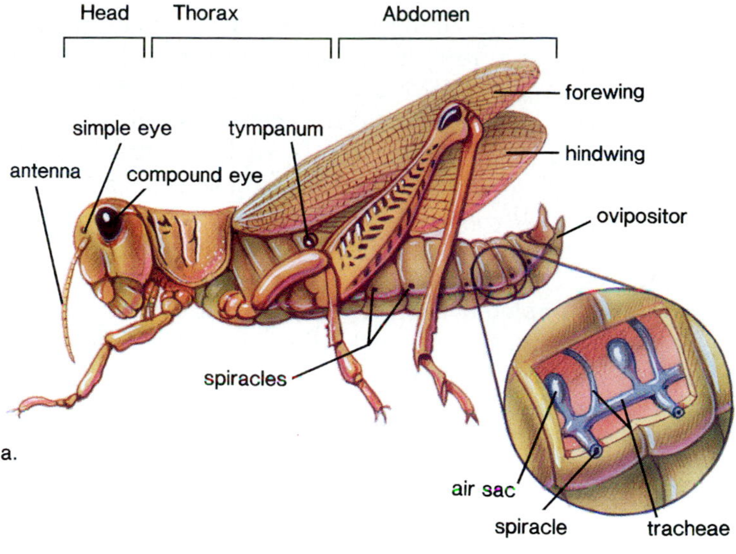

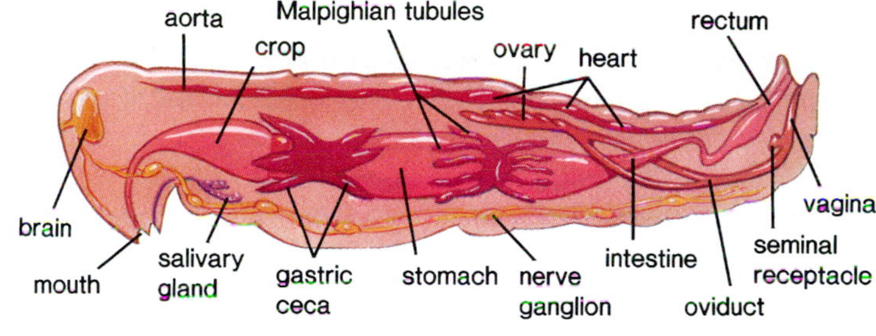

Figure 34.14 **Female grasshopper, *Romalea.***
a. Externally, the tympanum uses air waves for sound reception, and the hopping legs and the wings are for locomotion. **b.** Internally, the digestive system is specialized. The Malpighian tubules excrete a solid nitrogenous waste (uric acid). A seminal receptacle receives sperm from the male, which has a penis.

Grasshoppers undergo *incomplete metamorphosis*, a gradual change in form as the animal matures. The immature grasshopper, called a nymph, is recognizable as a grasshopper, even though it differs somewhat in shape and form from the adult. Other insects, such as butterflies, undergo *complete metamorphosis*, involving drastic changes in form. At first, the animal is a wormlike larva (caterpillar) with chewing mouthparts. It then forms a case, or cocoon, about itself and becomes a pupa. During this stage, the body parts are completely reorganized; the adult then emerges from the cocoon. This life cycle allows the larvae and adults to make use of different food sources.

Insects also show remarkable behavior adaptations, exemplified by the social systems of bees, ants, termites, and other colonial insects. Insects are so numerous and so diverse that the study of this one group is a major specialty in biology called entomology.

Uniramia—in which the legs have only one branch—include centipedes, millipedes, and insects. Insects, which are adapted to life on land, comprise more species than any other group of animals.

The grasshopper is adapted to a terrestrial environment while the crayfish is adapted to an aquatic environment. In crayfish, gills take up oxygen from water, while in the grasshopper, tracheae allow oxygen-laden air to enter the body. Appropriately, the crayfish has an oxygen-carrying pigment and a grasshopper has no such pigment in its blood. A liquid waste (ammonia) is excreted by a crayfish, while a solid waste (uric acid) is excreted by a grasshopper. Only in grasshoppers (1) is there a tympanum for the reception of sound waves and (2) do males have a penis and females an ovipositor. To swim, crayfish utilize tail (telson and uropods) and abdominal muscles; a grasshopper has legs for hopping and wings for flying.

Centipedes and Millipedes

Centipedes (2,500 species) have a body composed of a head and trunk. The body has many segments, and each segment has a pair of walking legs. They are carnivorous animals, and the head bears antennae and mouthparts with poison fangs.

Millipedes (10,000 species) have the same segmented organization as centipedes, but the body is cylindrical and some segments are fused. Therefore they appear to have two pairs of walking legs on each segment. Millipedes dwell in the soil, feeding on dead organic matter.

The term centipede means one hundred legs and millipede means one thousand legs. Although these arthropods do not have this many legs, a millipede does have more legs than a centipede (Fig. 34.15).

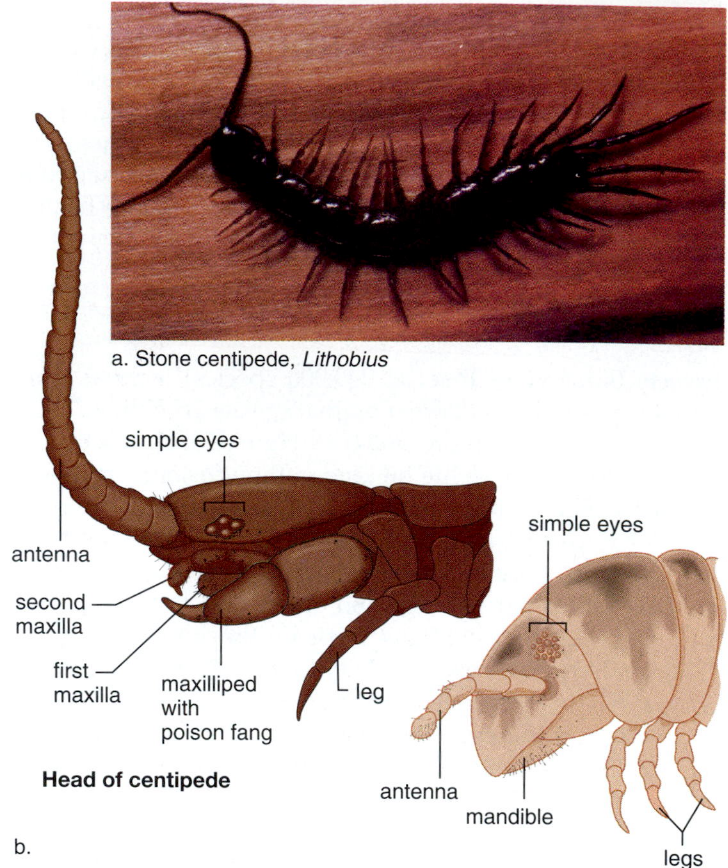

c. Flat-backed millipede, *Sigmoria*

Figure 34.15 Centipede and millipede.
a. A centipede has a pair of appendages on most every segment. **b.** Head of centipede compared to head of millipede. A centipede is a carnivorous animal and kills its prey with a poison fang. Most millipedes are herbivorous and usually feed on decayed plant matter. **c.** A millipede has two pairs of legs on most segments.

Chelicerates

The **chelicerates (subphylum Chelicerata,** 57,000 species) include terrestrial spiders, scorpions, ticks, mites, and the less familiar marine horseshoe crabs and sea spiders (Fig. 34.16). In this group, the first pair of appendages—the pincerlike chelicerae—are feeding organs. The second pair—pedipalps—are feeding or sensory in function. The pedipalps are followed by four pairs of walking legs. All of these appendages are attached to a **cephalothorax** [Gk. *kephale,* head, and *thorax,* breastplate] (fused head and thorax), which is followed by an abdomen that contains internal organs. There are no appendages found on the heads of these animals (antennae, mandibles, or maxillae), as in the other subphyla.

Horseshoe crabs of the genus *Limulus* are familiar along the east coast of North America. They scavenge sandy and muddy substrates for annelids, small molluscs, and other invertebrates. The body is covered by exoskeletal shields. The anterior shield is a horseshoe-shaped carapace, which bears two prominent compound eyes. A long, unsegmented telson projects to the rear. These marine animals have book gills, named for their resemblance to the pages of a closed book.

Scorpions, which are arachnids, are the oldest terrestrial arthropods. Today, they occur in the tropics, subtropics, and temperate regions of North America. They are nocturnal and spend most of the day hidden under a log or a rock. In these animals, the pedipalps are large pincers, and the long abdomen ends with a stinger that contains venom. Ticks and mites, with 25,000 species, may outnumber all other kinds of arachnids. They are often parasites. Ticks suck the blood of vertebrates and are sometimes transmitters of diseases, such as Rocky Mountain spotted fever or Lyme disease. Chiggers, the larvae of certain mites, feed on the skin of vertebrates.

Spiders, the most familiar arachnids, have a narrow waist that separates the cephalothorax from the abdomen. Spiders don't have compound eyes; instead, they have numerous simple eyes that perform a similar function. The chelicerae are modified as fangs, with ducts from poison glands, and the pedipalps are used to hold, taste, and chew food. The abdomen often contains silk glands, and they spin a web in which to trap their prey. Invaginations of the body wall form lamellae ("pages") of their so-called book lungs. Air flowing into the folded lamellae on one side exchanges gases with blood flowing in the opposite direction on the other side.

The chelicerates include horseshoe crabs, spiders, and scorpions. These animals have pincerlike appendages called chelicerae, which are modified as fangs in spiders.

a. Kenyan giant scorpion, *Pandinus*

book lung opening
reproductive opening
pedipalp
abdomen
spiracle
fang
anus
chelicera
spinnerets

c. Spider external anatomy, dorsal view

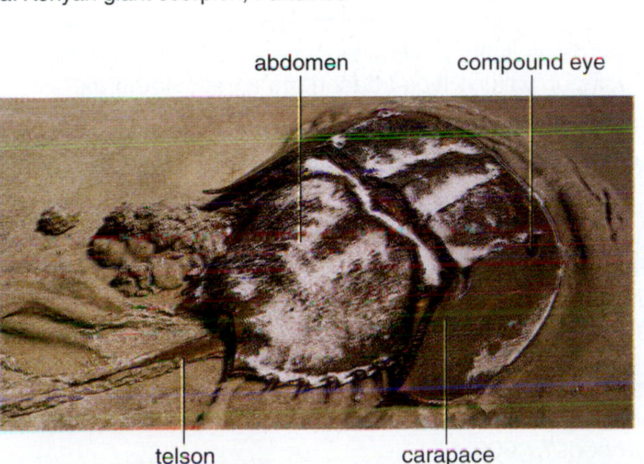

b. Horseshoe crab, *Limulus*

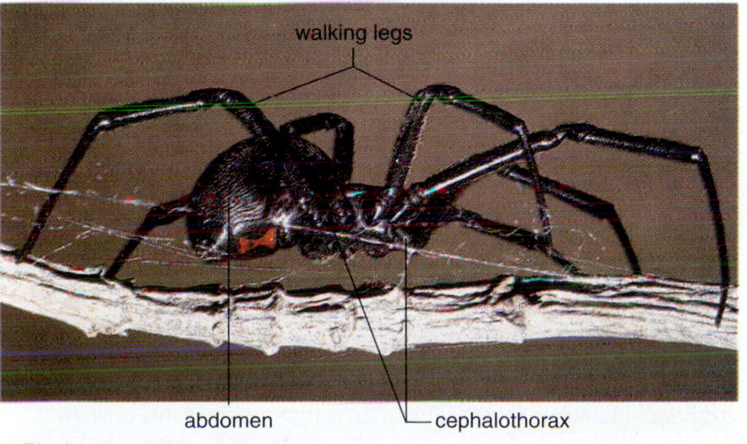

d. Black widow spider, *Latrodectus*

Figure 34.16 Chelicerate diversity.
a. Scorpions are more common in tropical areas. **b.** Horseshoe crabs are common along the east coast. **c.** Dorsal view of a spider. **d.** The black

Connecting Concepts

About ninety percent of all animal species have a coelom and are in the phyla Mollusca, Annelida, Arthropoda, and Chordata. (The chordates, which include the vertebrates, will be studied in the following chapter.) This would indicate that there are some adaptive advantages to having a coelom, or perhaps a combination of bilateral symmetry and the complexity made possible by the presence of a coelom. Most animals with a coelom have both a circulatory and respiratory system, which permit them to be active.

Animals with a coelom are divided into the protostomes (molluscs, annelids, and arthropods) and the deuterostomes (echinoderms and chordates) on the basis of embryological development. However, it is of interest that both lines of descent have representatives that are segmented. Segmentation of the annelids and many arthropods is obvious because the body has demarcations that can be seen externally. The repeating vertebrae of the backbone in vertebrates signal that they, too, are segmented. Was a coelom present before the evolution of segmentation? Perhaps it was. Later, a partitioned coelom may have provided a hydrostatic skeleton that enabled a worm to burrow more efficiently in the soil. In other words, there was a selective advantage to having a segmented coelom in organisms that lack limbs. The later evolution of limbs in jointed arthropods (e.g., insects) and vertebrates (e.g., frogs, lizards, horses) was even more adaptive for locomotion on land.

Animals that live on land must also have a means of reproduction that allows fertilization and development to take place without a dependence on external water.

Summary

34.1 A Coelom

Annelids, arthropods, and molluscs are all protostomial coelomates. As adults they all have bilateral symmetry, the tube-within-a-tube body plan, and organs derived from three germ layers.

The presence of a coelom has many advantages. The digestive system and body wall can move independently, and internal organs can become more complex. Coelomic fluids can assist respiration, circulation, and excretion, and can serve as a hydrostatic skeleton.

There are two groups of coelomate animals. In the protostomes, the mouth appears at or near the blastopore, and in the deuterostomes, the anus appears at or near the blastopore. Other differences between the two are spiral cleavage and a schizocoelom in protostomes versus radial cleavage and an enterocoelom in deuterostomes.

34.2 Molluscs

The body of a mollusc typically contains a visceral mass, a mantle, and a foot. Many also have a head and a radula. The nervous system consists of several ganglia connected by nerve cords. There is a reduced coelom and an open circulatory system. Clams (bivalves) are adapted to a sedentary coastal life, squids (cephalopods) to an active life in the sea, and snails (gastropods) to life on land.

Bivalves are filter feeders. Water enters and exits by siphons, and food trapped on the gills is swept toward the mouth. The nervous system has three major ganglia.

Predaceous squids undergo cephalization, move rapidly by jet propulsion, and have a closed circulatory system.

34.3 Annelids

Annelids are segmented worms. They have a well-developed coelom divided by septa, a closed circulatory system, a ventral solid nerve cord, and paired nephridia. Polychaetes ("many bristles") are marine worms that have parapodia. They may be predators, with a definite head region, or they may be filter feeders, with ciliated tentacles to filter food from the water. Earthworms are oligochaetes ("few bristles") that use the body wall for gas exchange. Leeches (class Hirudinea) are also part of this phylum.

34.4 Arthropods

Arthropods are the most varied and numerous of animals. Their success is largely attributable to a flexible exoskeleton, specialization of body regions, and jointed appendages. Also important are a high degree of cephalization, a variety of respiratory organs, and reduced competition through metamorphosis.

The arthropods are classified into four subphyla. Subphylum Trilobitomorpha includes animals that have been extinct for some 200 million years. The subphylum Crustacea (crayfish, lobsters, shrimps, copepods, krill, and barnacles) have a head that bears compound eyes, antennae, antennules, mandibles, and maxillae. There are biramous (two-branched) appendages on the thorax and abdomen. The crayfish illustrates other features such as an open circulatory system, respiration by gills, and a ventral solid nerve cord.

Arthropods in the subphylum Uniramia (insects, centipedes, and millipedes) have uniramous (one-branched) appendages on thorax and abdomen. The head has one pair of antennae, one pair of mandibles, and one or two pairs of maxillae. These terrestrial animals breathe by means of tracheae.

Insects include butterflies, grasshoppers, bees, and beetles. The anatomy of the grasshopper illustrates insect anatomy and ways they are adapted to life on land. Like other insects, grasshoppers have wings and three pairs of legs attached to the thorax. Grasshoppers have a tympanum for sound reception, a specialized digestive system for a grass diet, excretion of solid, nitrogenous waste by Malpighian tubules, tracheae for respiration, internal fertilization, and incomplete metamorphosis.

Arthropods in subphylum Chelicerata (horseshoe crabs, spiders, scorpions, ticks, and mites) have chelicerae, pedipalps, and four pairs of walking legs attached to a cephalothorax.

Reviewing the Chapter

1. List and discuss several advantages of the coelom. 598
2. Contrast the development of protostomes and deuterostomes in three ways. List the phyla which belong to each group. 598–99
3. Name at least three members of each phylum studied in this chapter. 599
4. What are the general characteristics of molluscs and the specific features of bivalves, cephalopods, and gastropods? 600–03
5. Contrast the anatomy of the clam, the squid, and the snail, indicating how each is adapted to its way of life. 601–03
6. What are the general characteristics of annelids and the specific features of the three major classes? 604–06

7. Contrast the anatomy of the clam worm and the earthworm, indicating how each is adapted to its way of life. 604–06
8. What are the general characteristics of arthropods and the specific features of the three major subphyla? 606–13
9. Describe the specific features of insects, using the grasshopper as an example. 611–12
10. Contrast the anatomy of the crayfish and the grasshopper, indicating how each is adapted to its way of life. 612
11. In what ways are spiders adapted to living on land? 613

Testing Yourself

Choose the best answer for each question.

1. Which of these does not pertain to a protostome?
   a. spiral cleavage
   b. blastopore is associated with the anus
   c. schizocoelom
   d. annelids, arthropods, and molluscs
   e. mouth is associated with first opening
2. Which of these best shows that annelids and arthropods are closely related? Both
   a. have a complete digestive tract.
   b. have a ventral solid nerve cord.
   c. are segmented.
   d. have a tube-within-a-tube body plan.
   e. All of these are correct.
3. Which of these best shows that snails are not closely related to crayfish?
   a. Snails are terrestrial, and crayfish are aquatic.
   b. Snails have a broad foot, and crayfish have jointed appendages.
   c. Snails are hermaphroditic, and crayfish have separate sexes.
   d. Snails are insects, but crayfish are fishes.
   e. Snails are bivalves and crayfish are chelicerates.
4. Which of these is mismatched?
   a. clam—gills
   b. lobster—gills
   c. grasshopper—book lungs
   d. polychaete—parapodia
   e. chelicerates—pedipalps
5. Which of these is an incorrect statement?
   a. Spiders are carnivores in the phylum Arthropoda.
   b. Clams are filter feeders in the phylum Mollusca.
   c. Earthworms are scavengers in the phylum Arthropoda.
   d. Squids are predators in the phylum Mollusca.
   e. Snails are gastropods in the phylum Mollusca.
6. A radula is a unique organ for feeding found in
   a. molluscs.
   b. annelids.
   c. arthropods.
   d. only insects.
   e. All of these are correct.
7. Which of these is mismatched?
   a. crayfish—walking legs
   b. clam—hatchet foot
   c. grasshopper—wings
   d. earthworm—many cilia
   e. squid—jet propulsion
8. Which of these is mismatched?
   a. molluscs—schizocoelom
   b. insects—hemocoel
   c. crayfish—coelom divided by septa
   d. clam worms—true coelom
   e. vertebrates—enterocoelom
9. Which of these is an adaptation to a terrestrial way of life in the snail?
   a. using the mantle instead of gills for respiration
   b. cephalization with antennae and eyes
   c. hatchet foot for crawling
   d. torsion so that anus is above the head
   e. incomplete metamorphosis
10. Associate each of these terms to annelids, to arthropods, and/or to molluscs.

Terms:

a. organ system level of organization
b. segmentation
c. true coelom
d. cephalization in some representatives
e. bilateral symmetry
f. complete gut
g. jointed appendages
h. three-part body plan

11. Associate each of these terms to clams and/or to earthworms.

Terms:

a. annelid
b. mollusc
c. three ganglia
d. gills
e. closed circulatory system
f. setae
g. open circulatory system
h. hatchet foot
i. hydrostatic skeleton
j. ventral solid nerve cord

12. Label this diagram.

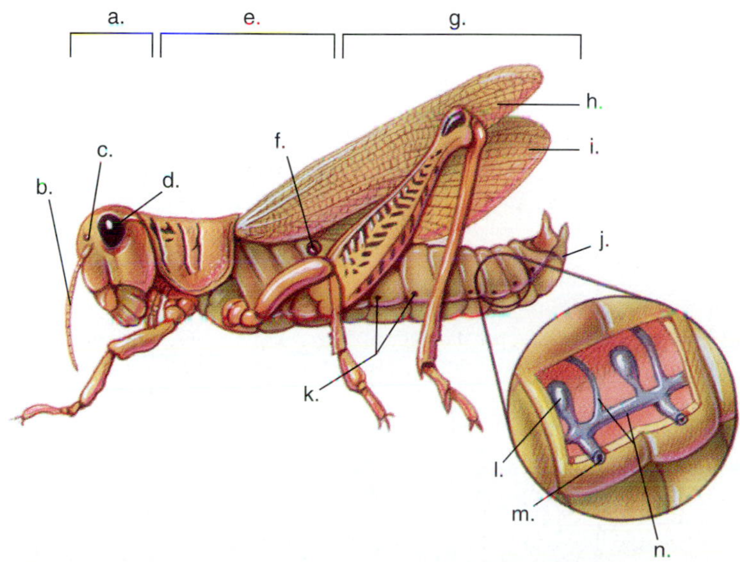

Thinking Scientifically

1. All life forms are believed to have evolved first in the ocean. What made water a better nursery than land for the evolution of new lineages?
2. A farmer sprays a field of soybeans with pesticide to kill an insect that has been eating his crop. Later in the season his plants are tall, but there are very few soybeans on the plants. What hypothesis could explain the connection between these two observations?

Understanding the Terms

annelid 604
arthropod 606
bivalve 600
centipede 612
cephalopod 602
cephalothorax 608, 613
chelicerate 613
chitin 606
coelom 598
crustacean 608
decapod 608
deuterostome 598
enterocoelom 599
exoskeleton 606
gastropod 603
hemocoel 609
insect 611
jointed appendages 606
leech 606
Malpighian tubule 611
mantle 600
metamorphosis 607
millipede 612
mollusc 600
molt 606
nephridium 604
oligochaete 605
protostome 598
radula 600
schizocoelom 599
seta (pl., setae) 604
trachea (pl., tracheae) 607
typhlosole 605

Match the terms to these definitions:

a. _______________ Air tube in insects that open at spiracles.
b. _______________ Change in shape and form that some animals, such as insects, undergo during development.
c. _______________ Repetition of body units as is seen in the earthworm.
d. _______________ Paired excretory tubules found in the earthworm and other invertebrates.
e. _______________ Strong but flexible nitrogenous polysaccharide found in the exoskeleton of arthropods.

Web Connections

Exploring the Internet

http://www.mhhe.com/biosci/genbio/mader
(click on *Biology 7/e*)

The *Biology 7/e* Online Learning Center provides many resources for studying the material in this chapter including links to the following sites:

Mollusca. Arizona's Tree of Life Web Page. Pictures, characteristics, phylogenetic relationships, references on molluscs. Pictures and references, and links to polyplacophorans, gastropods, bivalves, and cephalopods. Most sites have reference lists.

http://phylogeny.arizona.edu/tree/eukaryotes/animals/mollusca/mollusca.html

Introduction to the Annelida. University of California at Berkeley Museum of Paleontology. This site provides an introduction to the annelids. It contains information on annelids in the fossil record and annelid life histories, systematics, and morphology, as well as many links.

http://www.ucmp.berkeley.edu/annelida/annelida.html

BioPharm Leeches Homepage. This site provides interesting information on the use of leeches in anticoagulant research. In addition, it answers many frequently asked questions about leeches.

http://www.biopharm-leeches.com/index.html

Phylum Arthropoda. University of Michigan site describing arthropods. Links to chelicerates (merostomata, pycnogonida, arachnida), crustaceans (remipedians, cephalocarids, branchiopods, maxillopods, malacostracans), and uniramids (chilopodans, diplopodans, and insects). Each web link has pictures, descriptive material, links, references, and some have other web resources as well.

http://www.oit.itd.umich.edu/bio108/Arthropoda.html

Introduction to the Uniramia. University of California at Berkeley Museum of Paleontology. Links to the fossil record, life history and ecology, systematics, and more on morphology. Each site typically includes at least one picture of a member of the taxon.

http://www.ucmp.berkeley.edu/arthropoda/uniramia/uniramia.html

Animals: The Deuterostomes

chapter

chapter concepts

35.1 Echinoderms

- Echinoderms and chordates are both deuterostomes. 618
- Echinoderms have radial symmetry and a unique water vascular system for locomotion. 618

35.2 Chordates

- Lancelets are invertebrate chordates with the three chordate characteristics as adults: notochord, a dorsal tubular nerve cord, and pharyngeal pouches. 622

35.3 Vertebrates

- In vertebrates, the notochord is replaced by the vertebral column. Most vertebrates also have a head region, endoskeleton, and paired appendages. 622
- There are three groups of fishes: jawless (e.g., hagfishes and lampreys), cartilaginous (sharks and rays), and bony fishes (ray-finned and lobe-finned). Most modern-day fishes are ray-finned (e.g., trout, perch, etc.). 622
- Amphibians (e.g., frogs and salamanders), who were more numerous during the Carboniferous period, evolved from lobe-finned fishes. 625
- Modern-day reptiles (e.g., turtles, lizards, snakes) are the remnants of an ancient group that evolved from amphibians. 628
- The shelled egg of reptiles, which contains extraembryonic membranes, is an adaptation for reproduction on land. 628
- There is a close evolutionary relationship between birds and reptiles. However, birds maintain a constant body temperature. 632
- Birds have feathers and skeletal adaptations that enable them to fly. 632
- Mammals, which evolved from reptiles, were present when the dinosaurs existed. They did not diversify until the dinosaurs became extinct. 634
- Mammals are vertebrates with hair and mammary glands. The former helps them maintain a constant body temperature. 634
- Mammals are classified according to methods of reproduction: monotremes lay eggs, marsupials have a pouch in which newborn mature, and placental mammals retain the offspring in a uterus until birth. 634

Fisherman bat, *Noctilio leporinus*, in flight with fish in feet.

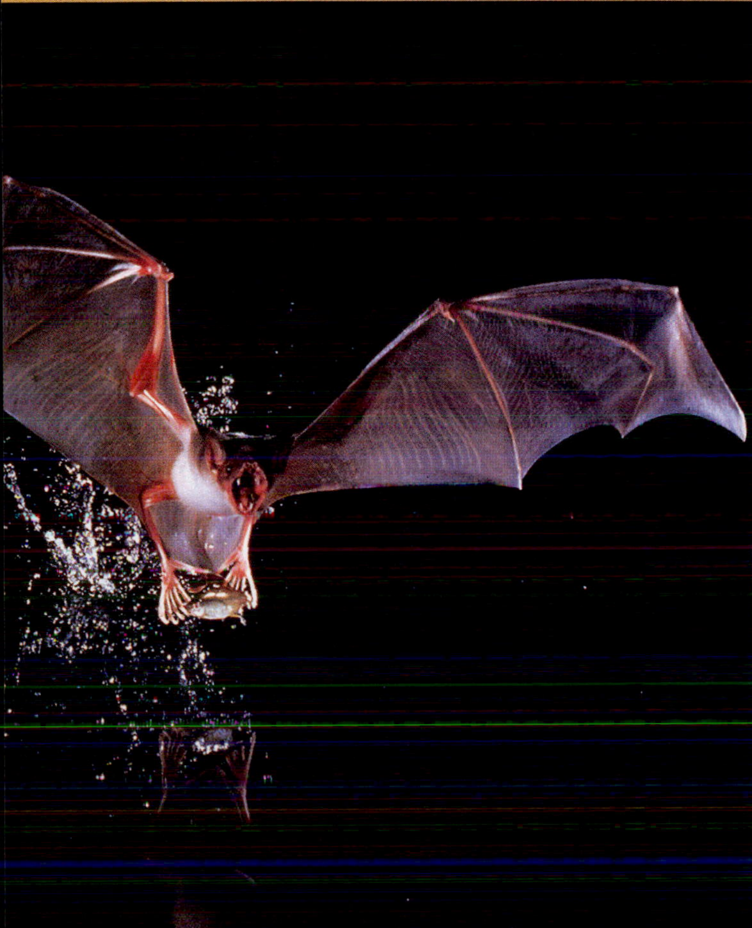

Human beings are vertebrate chordates, a phylum that includes a few invertebrate members. Without evidence to the contrary, who would think the invertebrate chordates are most closely related to the echinoderms—headless, brainless, and unsegmented creatures that spend their lives at sea? Sea stars, sea urchins, brittle stars, and sea lilies are all echinoderms, animals that are radially symmetrical as adults.

The fishes, which are aquatic vertebrates, far outnumber any of the other chordates. Still, there are more types of terrestrial animals among the chordates than in any other phylum. The reptiles were the first vertebrates to successfully reproduce on land by laying a shelled egg, as do birds. But within the egg, the embryo has a watery environment. Despite the live birth of placental mammals, such as ourselves, the embryo is still bathed by water while it develops. Today, birds and placental mammals are the most conspicuous vertebrates on land, because they underwent adaptive radiation after many reptiles, such as the dinosaurs, had become extinct. Bats are flying mammals that have diversified—some, such as the one pictured, even fish to obtain food.

35.1 Echinoderms

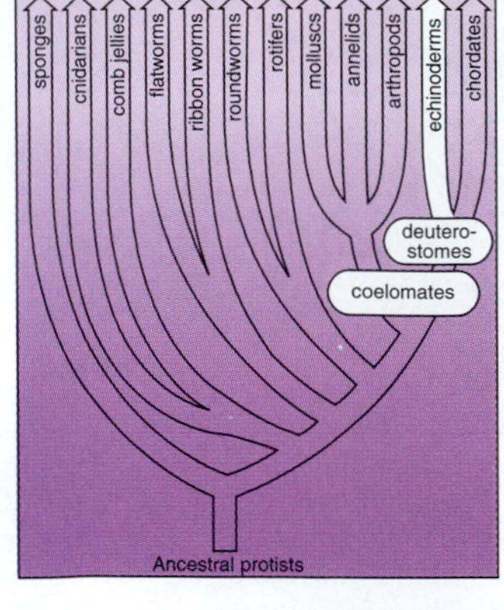

Phylum Echinodermata (6,000 species) and **phylum Chordata,** contain the **deuterostomes** [Gk. *deuteros*, second, and L. *stoma*, mouth], animals with embryonic characteristics described in Figure 34.1.

Echinoderms [Gk. *echinos*, spiny, and *derma*, skin] have an endoskeleton (internal skeleton) consisting of spine-bearing, calcium-rich plates. The spines, which stick out through the delicate skin, account for their name. Although echinoderms are radially symmetrical as adults, their larva is a free-swimming planktonic filter feeder with bilateral symmetry. Metamorphosis results in the radially symmetrical adult.

Class Crinoidea (600 species) includes the stalked sea lilies and the motile feather stars. Class Holothuroidea (1,500 species) are the sea cucumbers with a long leathery body that resembles a cucumber, except that there are feeding tentacles about the mouth. Class Echinoidea (950 species) includes sea urchins and sand dollars, both of which use their spines for locomotion, defense, and burrowing. Class Ophiuroidea (2,000 species) contains the brittle stars, which have a central disk from which long, flexible arms radiate. Class Asteroidea (1,500 species) consists of the sea stars, also called the starfishes.

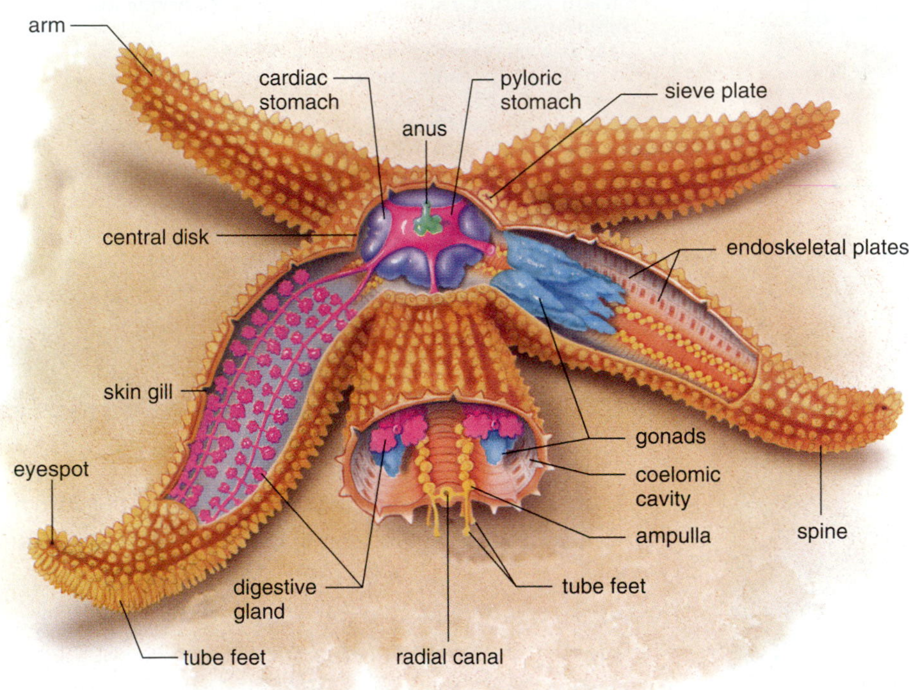

Figure 35.1 **Echinoderms.**
a. Sea star (starfish) anatomy and habitat. Like other echinoderms, sea stars have a water vascular system shown (light gold) which terminates in tube feet. The red sea star, *Mediastar,* uses the suction of its tube feet to open a clam, a primary source of food. **b.** Sea cucumber. **c.** Sea urchin.

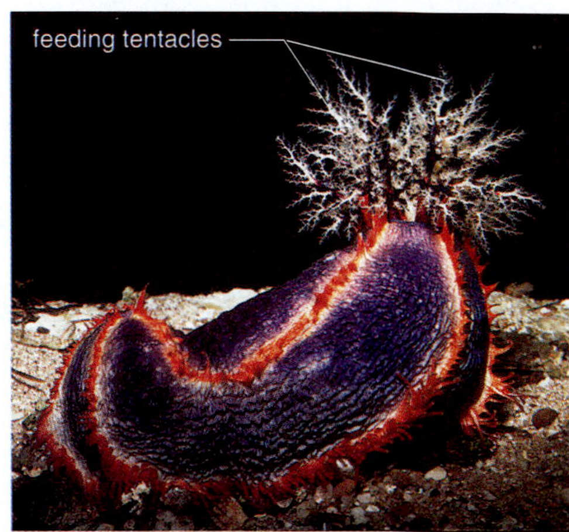

b. Sea cucumber, *Pseudocolochirus*

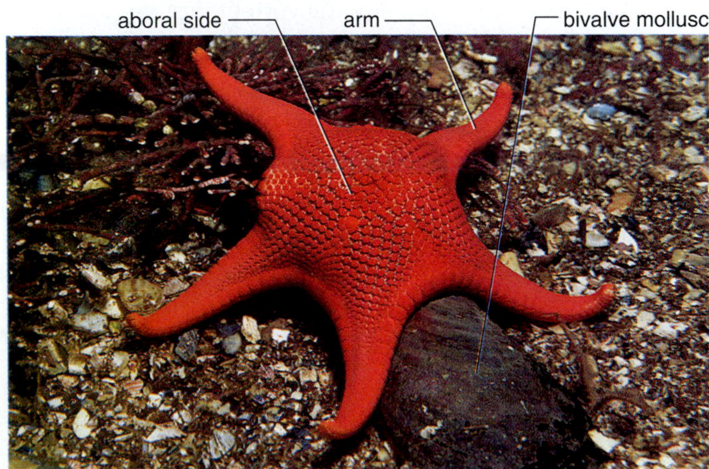

a. Red sea star, *Mediastar*

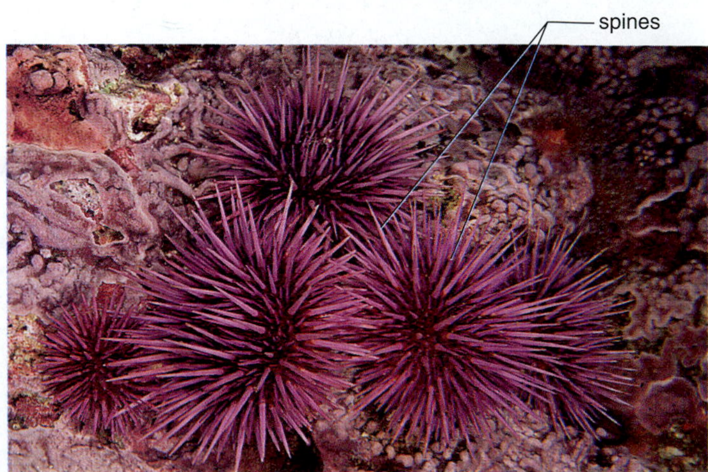

c. Purple sea urchin, *Strongylocentrotus*

Sea Stars

Sea stars are commonly found along rocky coasts where they feed on clams, oysters, and other bivalve molluscs. Various structures project through the body wall: (1) spines from the endoskeletal plates offer some protection; (2) pincerlike structures called pedicellarie keep the surface free of small particles; and (3) skin gills, tiny fingerlike extensions of the skin, are used for respiration. On the oral surface, each arm has a groove lined by little **tube feet** (Fig. 35.1).

To feed, a sea star positions itself over a bivalve and attaches some of its tube feet to each side of the shell. By working its tube feet in alternation, it pulls the shell open. A very small crack is enough for the sea star to evert its cardiac stomach and push it through the crack, so that it contacts the soft parts of the bivalve. The stomach secretes enzymes, and digestion begins even while the bivalve is attempting to close its shell. Later, partly digested food is taken into the sea star's body, where digestion continues in the pyloric stomach using enzymes from the digestive glands found in each arm. A short intestine opens at the anus on the aboral side.

In each arm the well-developed coelom contains a pair of digestive glands and gonads (either male or female) that open on the aboral surface by very small pores. The nervous system consists of a central nerve ring that gives off radial nerves in each arm. A light-sensitive eyespot is at the tip of each arm.

Locomotion depends on the **water vascular system.** Water enters this system through a structure on the aboral side called the sieve plate, or madreporite. From there it passes through a stone canal to a ring canal, which surrounds the mouth, and then to a radial canal in each arm. From the radial canals, many lateral canals extend into the tube feet, each of which has an ampulla. Contraction of an ampulla forces water into the tube foot, expanding it. When the foot touches a surface, the center is withdrawn, giving it suction so that it can adhere to the surface. By alternating the expansion and contraction of the tube feet, a starfish moves slowly along.

Echinoderms don't have a complex respiratory, excretory, or circulatory system. Fluids within the coelomic cavity and the water vascular system carry out many of these functions. For example, gas exchange occurs across the skin gills and the tube feet. Nitrogenous wastes diffuse through the coelomic fluid and the body wall. Cilia on the peritoneum lining the coelom keep the coelomic fluid moving.

Sea stars reproduce asexually and sexually. If the body is fragmented, each fragment can regenerate a whole animal. Sea stars spawn and release either eggs or sperm at the same time. The bilateral larva undergoes a metamorphosis to become the radially symmetrical adult.

Echinoderms have a well-developed coelom and internal organs despite being radially symmetrical. Spines project from their internal skeleton, and there is a unique water vascular system.

35.2 Chordates

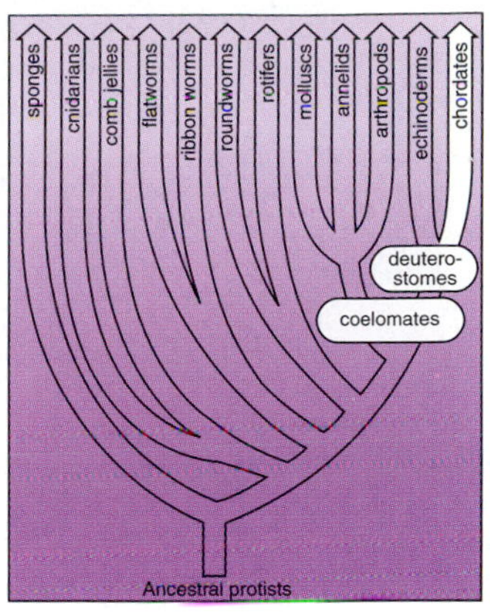

To be considered a **chordate** (**phylum Chordata,** 45,000 species), an animal must have the four basic characteristics listed below at some time during its life history:

1. A dorsal supporting rod called a **notochord** [Gk. *notos*, back, and *chorde*, string]. The notochord is located just below the **nerve cord.** Vertebrates have an embryonic notochord that is replaced by the vertebral column during development.
2. A dorsal tubular nerve cord. By tubular, it is meant that the cord contains a canal filled with fluid. In vertebrates, the nerve cord, more often called the spinal cord, is protected by the vertebrae.
3. Pharyngeal pouches. These are seen only during embryonic development in most vertebrates. In the invertebrate chordates, the fishes, and amphibian larvae, the pharyngeal pouches become functioning **gills** (respiratory organs of aquatic vertebrates). Water passing into the mouth and the pharynx goes through the gill slits, which are supported by gill arches. In terrestrial vertebrates, the pouches are modified for various purposes. In humans, the first pair of pouches become the auditory tubes. The second pair become the tonsils, while the third and fourth pairs become the thymus gland and the parathyroids.
4. A tail—as an embryo if not as an adult—that extends beyond the anus, is therefore called a postanal tail.

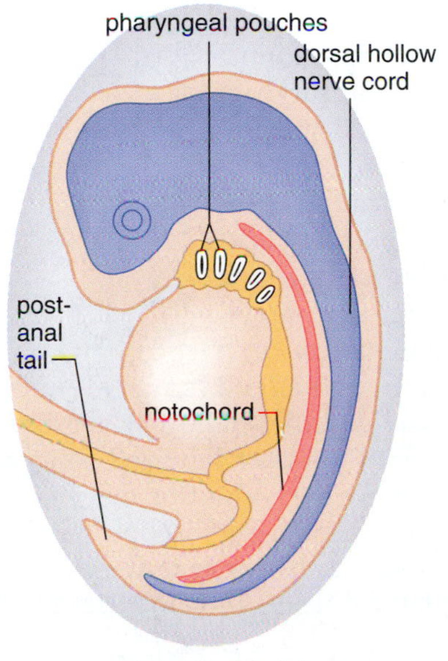

embryonic chordate

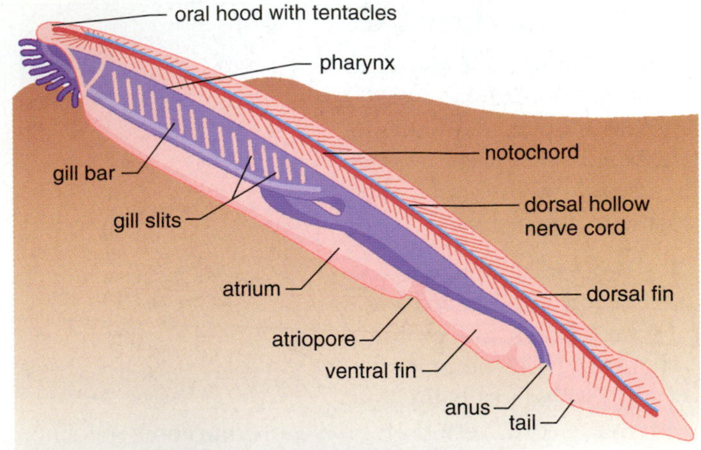

Figure 35.2 **Lancelet, *Branchiostoma.***
Lancelets are filter feeders. Water enters the mouth and exits at the atriopore after passing through the gill slits.

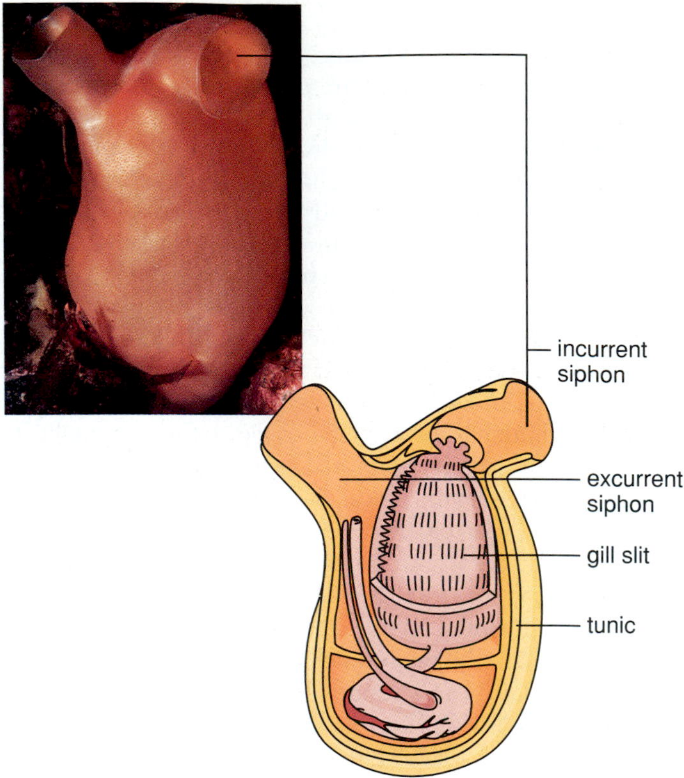

Figure 35.3 **Tunicate, *Halocynthia.***
Note that the only chordate characteristic remaining in the adult is gill slits.

Invertebrate Chordates

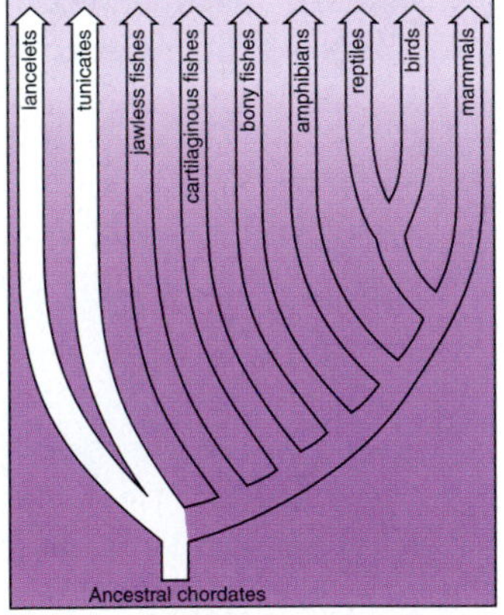

There are a few invertebrate chordates in which the notochord persists and is never replaced by a vertebral column.

Lancelets (**subphylum Cephalochordata,** 23 species) are in the genus *Branchiostoma,* formerly called Amphioxus. These marine chordates, which are only a few centimeters long, are named for their resemblance to a lancet—a small, two-edged surgical knife (Fig. 35.2). Lancelets are found in the shallow water along most coasts where they usually lie partly buried in sandy or muddy substrates with only their anterior mouth and gill apparatus exposed. They feed on microscopic particles filtered out of the constant stream of water that enters the mouth and passes through the gill slits into an atrium that opens at the atriopore.

Lancelets retain the three chordate characteristics as an adult. The notochord extends from the tail to the head and this accounts for their subphylum name, Cephalochordata. In addition, segmentation is present, as witnessed by the fact that the muscles are segmentally arranged and the dorsal tubular nerve cord has periodic branches.

Tunicates (**subphylum Urochordata,** 1,250 species) live on the ocean floor and take their name from a tunic that makes them look like thick-walled, squat sacs. They are also called sea squirts because they squirt out water from their excurrent siphon when disturbed. The tunicate larva is bilaterally symmetrical and has the three chordate characteristics. Metamorphosis produces the sessile adult with an incurrent and excurrent siphon (Fig. 35.3).

The pharynx is lined by numerous cilia whose beating creates a current of water that moves into the pharynx and out the numerous gill slits, the only chordate characteristic that remains in the adult. Microscopic particles adhere to a mucous secretion and are eaten.

Is it possible that the tunicates are directly related to the vertebrates? It has been suggested that a larva with the three chordate characteristics may have become sexually mature without developing the other adult tunicate characteristics. Then it may have evolved into a fishlike vertebrate. Figure 35.4 is a phylogenetic tree of chordates.

The invertebrate chordates include the tunicates and the lancelets. A lancelet has the three chordate characteristics as an adult.

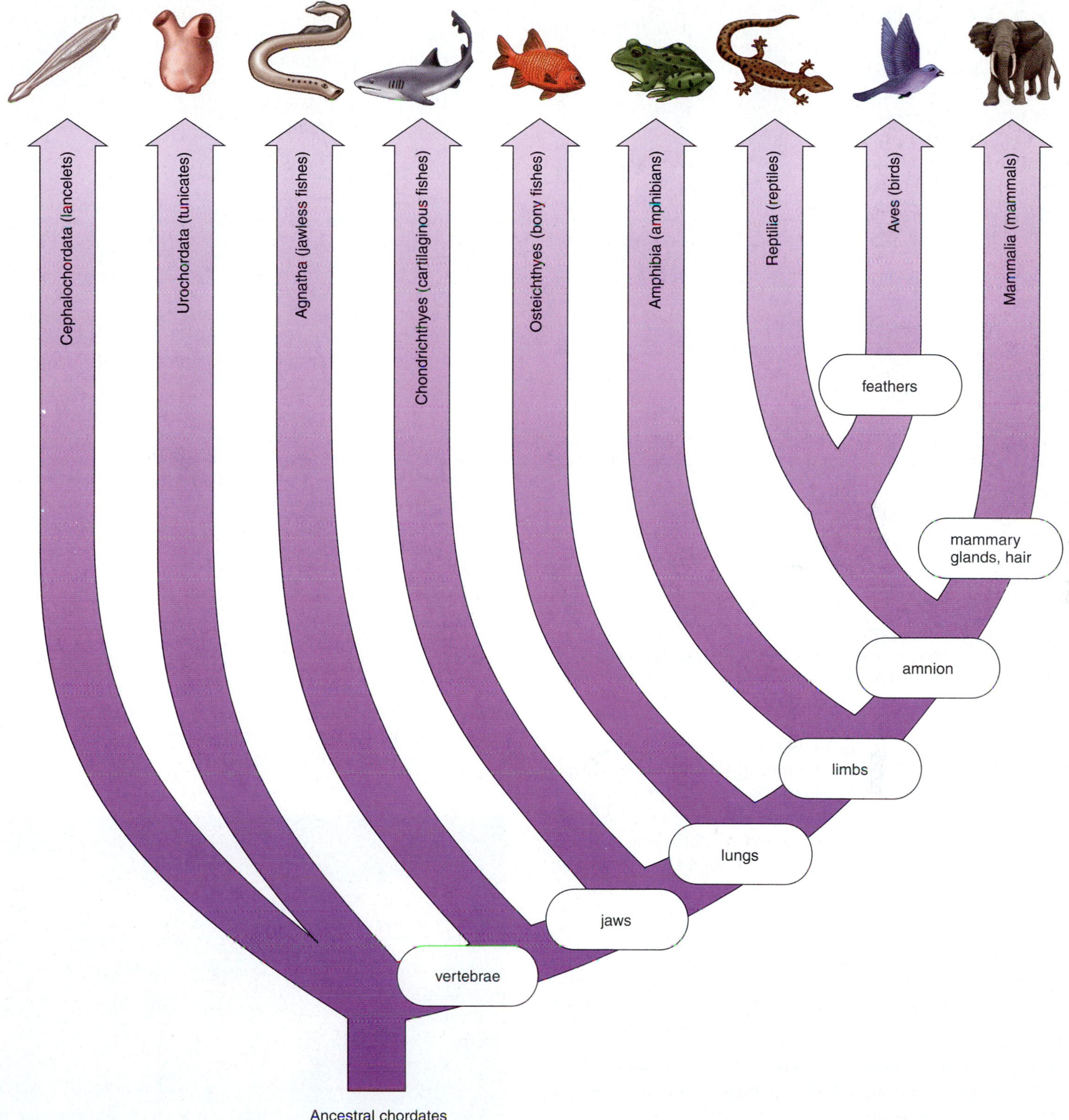

Figure 35.4 Phylogenetic tree of the chordates.
Each of the stated features is an evolved characteristic that is shared by the classes beyond this point.

35.3 Vertebrates

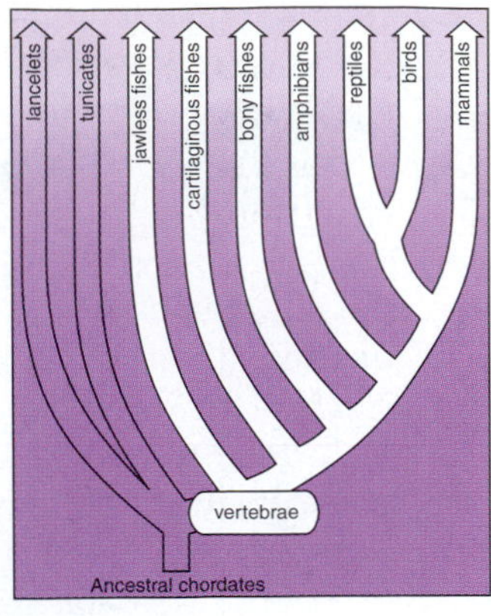

At some time in their life history, **vertebrates (subphylum Vertebrata,** 43,700 species) have all four chordate characteristics. The embryonic notochord, however, is generally replaced by a vertebral column composed of individual vertebrae. The vertebral column, which is a part of the flexible but strong endoskeleton, gives evidence that vertebrates are segmented. The vertebrate skeleton (either cartilage or bone) is a living tissue that grows with the animal. It also protects internal organs and serves as a place of attachment for muscles. Together the skeleton and muscles form a system that permits rapid and efficient movement. Two pairs of appendages are characteristic. The pectoral and pelvic fins of fish evolved into the jointed appendages that allowed vertebrates to move onto land.

The main axis of the internal jointed skeleton consists of not only the vertebral column, but also a skull that encloses and protects the brain. During vertebrate evolution the brain increased in complexity, and specialized regions developed to carry out specific functions. The high degree of cephalization is accompanied by complex sense organs. The eyes develop as outgrowths of the brain. The ears are primarily equilibrium devices in aquatic vertebrates, but they also function as sound-wave receivers in land vertebrates.

The evolution of jaws allowed some vertebrates to take up the predatory way of life. Vertebrates have a complete digestive tract and a large coelom. The circulatory system is closed (the blood is contained entirely within blood vessels). Vertebrates have an efficient means of obtaining oxygen from water or air, as appropriate. The kidneys are important excretory and water-regulating organs that conserve or rid the body of water as necessary. The sexes are generally separate, and reproduction is usually sexual. The evolution of the amnion allowed reproduction to take place on land. Reptiles, birds, and some mammals lay a shelled egg. In placental mammals development takes place within the uterus of the female.

Vertebrates are distinguished, in particular, by:

- living endoskeleton
- closed circulatory system
- paired appendages
- efficient respiration and excretion
- high degree of cephalization

In short, vertebrates are adapted to an active lifestyle.

Fishes

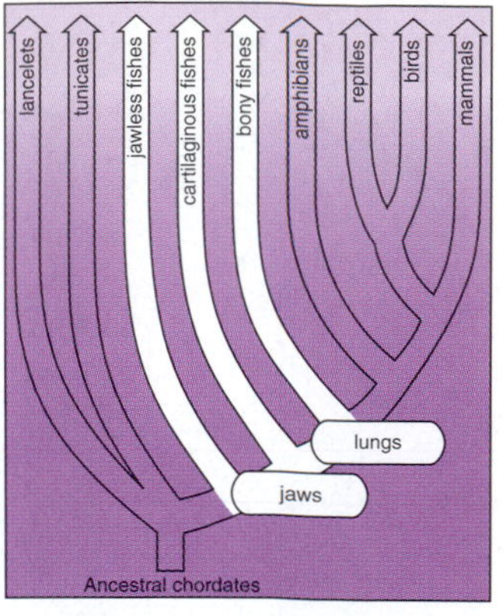

Fishes are aquatic, gill-breathing vertebrates that usually have fins and skin covered with **scales.** The small, jawless, and finless **ostracoderms** are the earliest vertebrate fossils. They were filter feeders, but most likely they were capable of moving water through their gills by muscular action. Ostracoderm fossils have been dated from the Cambrian and as late as the Devonian period, but then they apparently became extinct. Although none of the living jawless fishes has any external protection, large defensive head-shields were not uncommon in the early jawless fishes.

Jawless Fishes

All the **jawless fishes** are called **agnathans** [Gk. *a,* without, and *gnathos,* jaw] (superclass Agnatha, 63 species). Lampreys and hagfishes are modern-day jawless fishes that lack a bony skeleton. They are cylindrical, up to a meter long, and have smooth, nonscaly skin (Fig. 35.5). The hagfishes are scavengers feeding on soft-bodied invertebrates and dead fishes they suck into their mouths. Many lampreys are filter feeders like their ancestors. Parasitic lampreys have a round, muscular mouth to attach themselves to another fish and suck nutrients from the host's circulatory system. Marine parasitic lampreys gained entrance to the Great Lakes when a canal from the St. Lawrence River was deepened. The lamprey population grew quickly and caused extensive reduction in the trout population of the Great Lakes in the early 1950s.

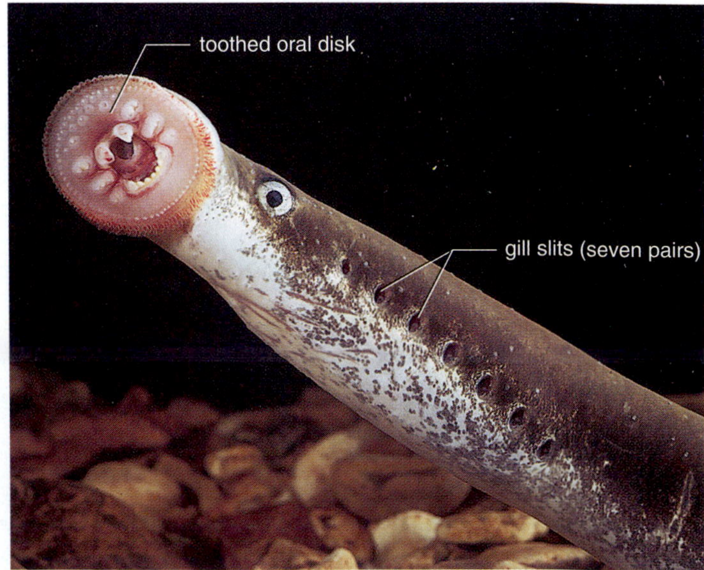

Figure 35.5 Lamprey, *Petromyzon.*
Note its toothed oral disk attached to aquarium glass. Lampreys, which are members of the vertebrate superclass Agnatha, have an elongated, rounded body and nonscaly skin.

Fishes with Jaws

All the other animals to be studied are **gnathostomates** (superclass Gnathostomata, 25,000 species); they have **jaws,** tooth-bearing bones of the head. Jaws are believed to have evolved from the first pair of gill arches of agnathans. The second pair of gill arches became support structures for the jaws.

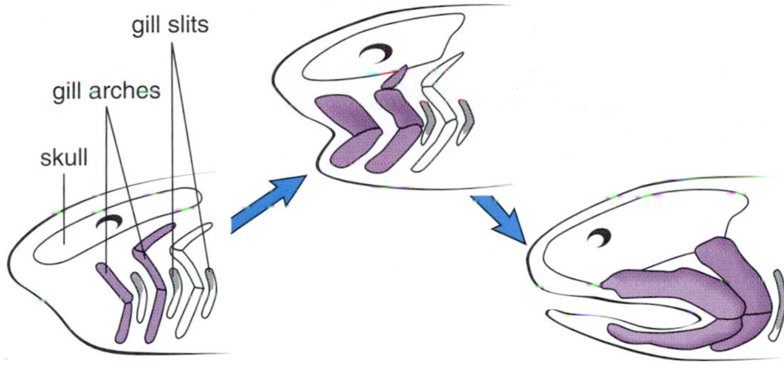

The **placoderms** are well-known extinct jawed fishes of the Devonian period. Placoderms were armored with heavy, bony plates and had strong jaws. Like modern-day fishes, they had paired pectoral and pelvic fins (see Fig. 35.7). Paired fins, which allow a fish to balance and to maneuver well in the water, are also an asset for predation.

Cartilaginous Fishes Sharks, rays, and skates (class Chondrichthyes, 850 species) are the largely marine **cartilaginous fishes;** they have a skeleton of cartilage instead of bone. There are five to seven gill slits on both sides of the pharynx, and they lack the gill cover of bony fishes. Their body is covered with epidermal placoid (toothlike) scales that project posteriorly, which is why a shark's skin feels like sandpaper. The menacing teeth of sharks and their relatives are simply larger, specialized versions of these scales. At any one time, a shark such as the great white shark may have up to 3,000 teeth in its mouth, arranged in six to twenty rows. Only the first row or two are actively used for feeding; the other rows are replacement teeth.

Three well-developed senses enable sharks and rays to detect their prey. They have the ability to sense electric currents in water—even those generated by the muscle movements of animals. They have a lateral line system, a series of pressure-sensitive cells that lie within canals along both sides of the body, which can sense pressure caused by a fish or other animal swimming nearby. They also have a very keen sense of smell; the part of the brain associated with this sense is very well developed. Sharks can detect about one drop of blood in 115 liters (25 gallons) of water.

The largest sharks are filter feeders, not predators. The basking sharks and whale sharks ingest tons of small crustacea, collectively called krill. Many sharks are fast-swimming predators in the open sea (Fig. 35.6*a*). The great white shark, about 7 meters (23 feet) in length, feeds regularly on dolphins, sea lions, and other seals. Humans are normally not attacked except when mistaken for sharks' usual prey. Tiger sharks, so named because the young have dark bands, reach 6 meters (18 feet) in length and are unquestionably one of the most dangerous sharks. As it swims through the water, a tiger shark will swallow anything it can—including rolls of tar paper, shoes, gasoline cans, paint cans, and even human parts.

In rays and skates (Fig. 35.6*b*), which live on the ocean floor, the pectoral fins are greatly enlarged into a pair of large, winglike **fins.** They usually swim slowly along the sea bottom and feed on animals that they dredge up. Members of the electric ray family are slow swimmers that feed on fishes they capture after stunning them with electric shocks. Their large electric organs, located at the bases of their pectoral fins, can discharge over 300 volts. Sawfish rays are named for their large, protruding anterior "saw." Swimming into a school of fishes, they rapidly move their saw back and forth, stunning or killing fish which they later eat.

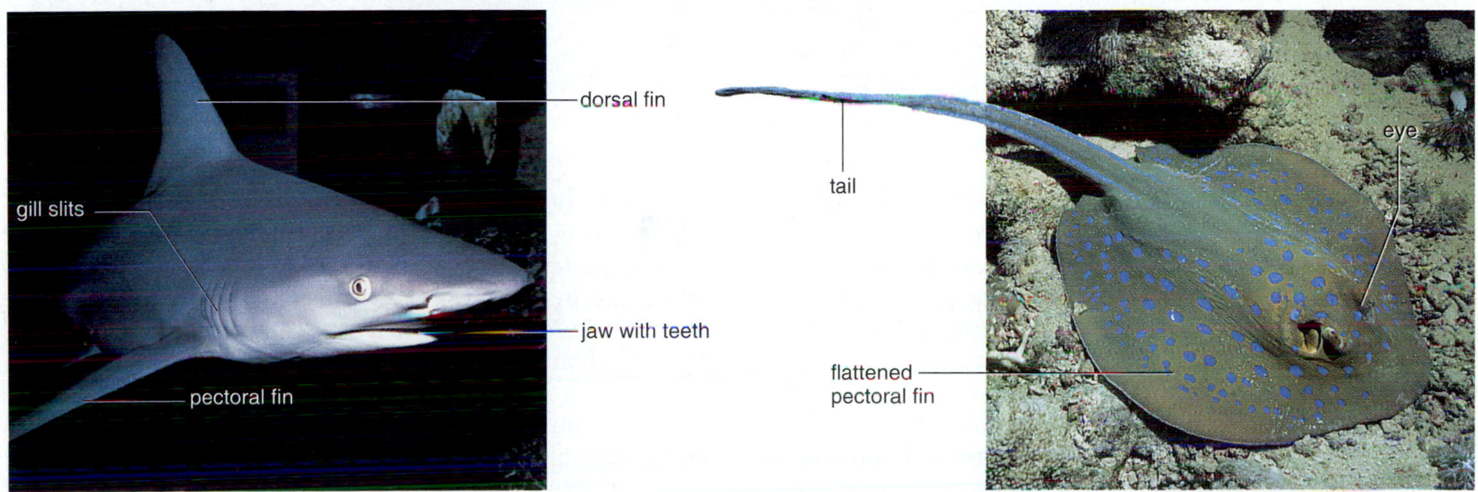

a. Bull shark, *Carcharhinus leucas*

b. Blue-spotted stingray, *Taeinura lymma*

Figure 35.6 Cartilaginous fishes.
a. Sharks are predators or scavengers that move gracefully through open ocean waters. b. Most stingrays grovel in the mud, feeding on bottom-dwelling invertebrates. This blue-spotted stingray is protected by its whiplike tail.

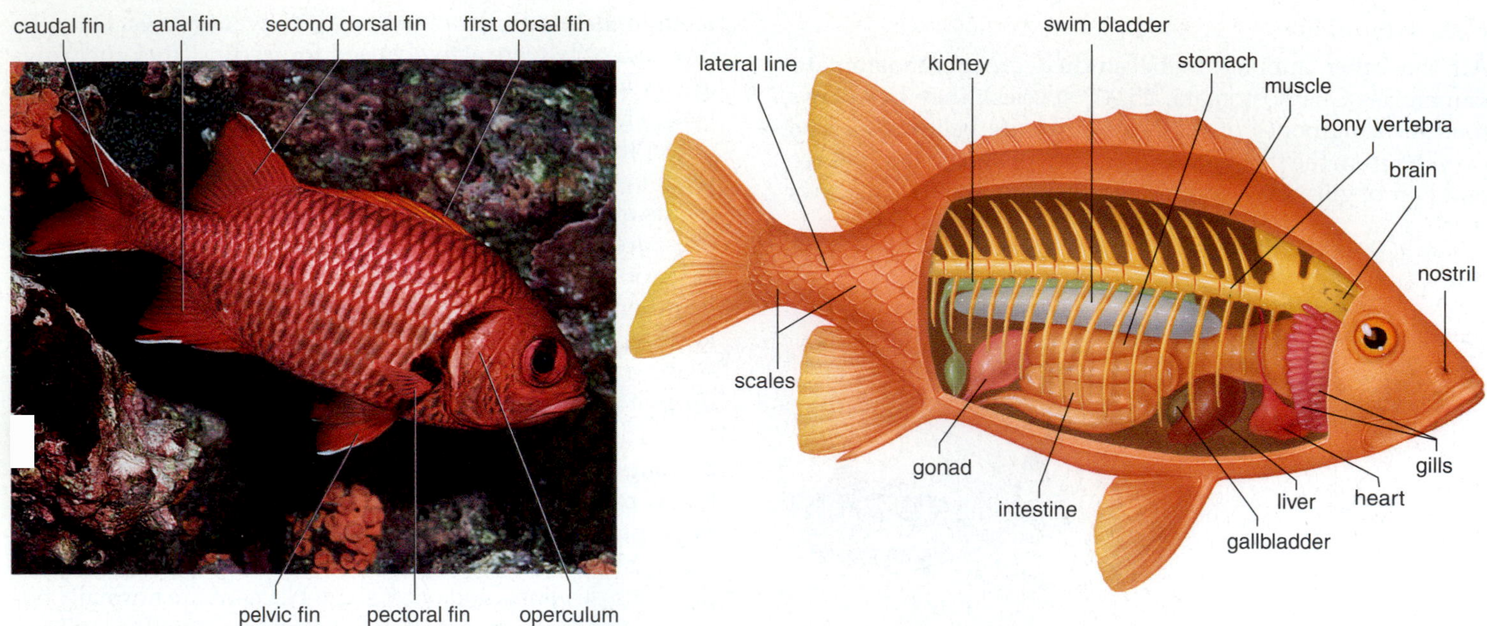

a. Soldierfish, *Myripristis jacobus*

b. Spotted lionfish, *Pterois antennata*

c. Seahorse, *Hippocampus kuda*

Figure 35.7 Ray-finned fishes. **a.** A soldierfish has the typical appearance and anatomy of a ray-finned fish. A lionfish **(b)** and a seahorse **(c)** show how diverse ray-finned fishes can be.

Bony Fishes **Bony fishes** (class Osteichthyes, 20,000 species) have a skeleton of bone. Most bony fishes are ray-finned fishes in which fan-shaped fins are supported by thin, bony rays. The lobe-finned fishes, a very small group of bony fishes, are important because lobe-finned fishes of the Devonian period are ancestral to amphibians.

The **ray-finned fishes** are the most successful and diverse of all the vertebrates. Some, like herrings, are filter feeders; others, like trout, are opportunists; and still others are predaceous carnivores, like piranhas and barracudas. Often the common names of fishes reflect their appearance. Zebra fish are striped, stone fish resemble stones, seahorses look like tiny upright horses, and porcupine fish (when inflated) are protected by lateral spikes.

Despite their diversity, bony fishes have features in common (Fig. 35.7). The skeleton is of bone and they are covered by scales formed of bone. The gills do not open separately and instead are covered by an operculum. They have a **swim bladder,** a gas-filled sac whose pressure can be altered to change buoyancy and, therefore, their depth in the water. Some fishes (trout, salmon, eels) can move from fresh water to salt water. When in fresh water, their kidneys excrete very dilute urine and their gills absorb salts from the water by active transport. Usually eggs and sperm are shed in the water or into a nest constructed by the parents. In almost all bony fishes, fertilization and embryo development occur outside the female's body.

In **lobe-finned fishes,** the fleshy fins are supported by central bones. They include six species of lungfishes with lungs and one species of coelacanth with especially noticeable lobes (Fig. 35.8*a*). Lungfishes live in Africa, South America, and Australia either in stagnant fresh water or in ponds that dry up annually. Coelacanths, in contrast, inhabit deep ocean environments. Because only Mesozoic fossils of these fishes had been found, it was once thought that coelacanths were extinct. Then one was captured off the eastern coast of South Africa in 1938, and now about 200 more have been captured. Comparisons of mitochondrial DNA suggest that lungfishes, and not coelacanths, are the closest living relatives of modern amphibians.

a. Coelacanth, *Latimeria*

Amphibians

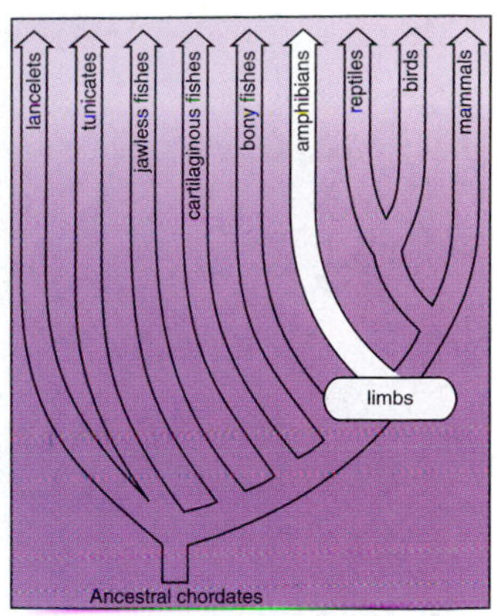

All the other animals to be studied are **tetrapods** [Gk. *tetra,* four, and *podos,* foot]—they have four limbs. The lobe-finned fishes of the Devonian period are ancestral to the amphibians, the first tetrapods. Figure 35.8*b* compares the limbs of a lobe-finned fish to those of an early amphibian, and you can see that the same bones are present in both animals. Animals that live on land use limbs to support the body especially since air is less buoyant than water. Their internal nares and lungs allowed lobe-finned fishes and early amphibians to breathe air.

Two hypotheses have been suggested to account for the evolution of **amphibians** [Gk. *amphibios,* living on both land and in water] (class Amphibia, 3,900 species) from lobe-finned fishes. Perhaps lobe-finned fishes, which were capable of using their limbs to move from pond to pond, had an advantage over those who could not do so. Or perhaps the supply of food on land in the form of plants and insects—and the absence of predators—promoted further adaptations to the land environment. In any case, the first amphibians diversified during the Carboniferous period, which is known as the Age of Amphibians.

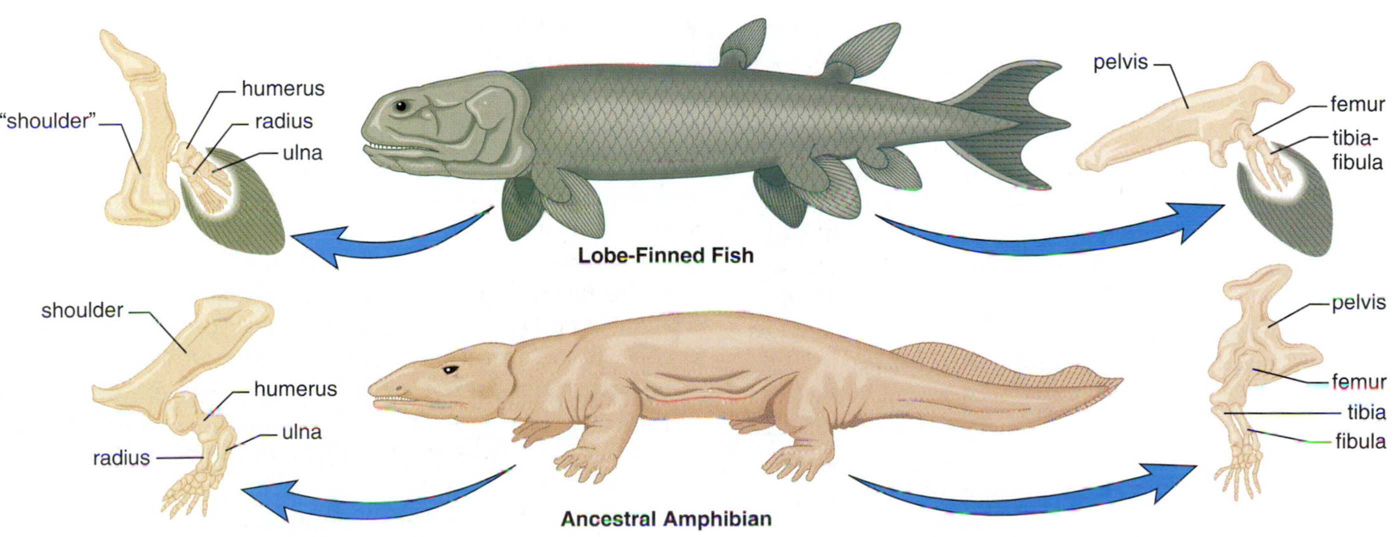

b. Comparison of limbs

Figure 35.8 Lobe-finned fish versus amphibian.
a. A coelacanth is a lobe-finned fish once thought to be extinct. **b.** These drawings compare the skeletal structure of the appendages of a lobe-finned fish and an ancestral amphibian. A shift in the position of the bones in the forelimbs and hindlimbs lifts and supports the body.

a. Barred tiger salamander, *Ambystoma tigrinum*

b. Tree frog, *Hyla andersoni*

c. Mexican caecilian, *Dermophis mexicanus*

Figure 35.9 Amphibians.
Living amphibians are divided into three orders: **a.** Salamanders and newts. Members of this order have a tail throughout their lives and, if present, unspecialized limbs. **b.** Frogs and toads. Like this frog, members of this order are tailless and have limbs specialized for jumping. **c.** The caecilians are wormlike burrowers.

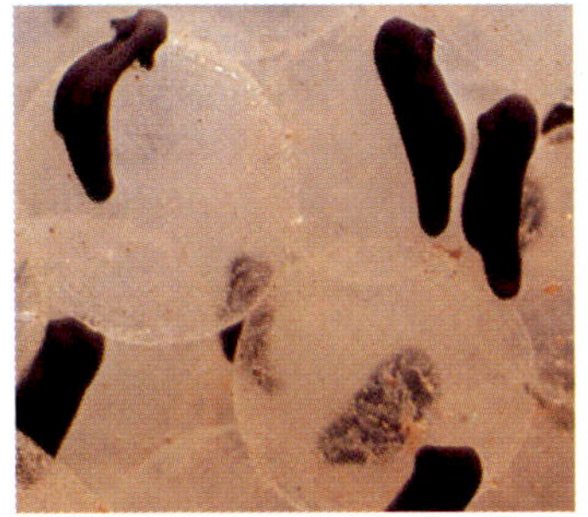

Hatching

Tadpole stage

Legs develop

Adult

Figure 35.10 Metamorphosis.
Salamanders and frogs undergo metamorphosis. During frog metamorphosis, the animal changes from a form specialized for an aquatic life to one that is specialized for a terrestrial one.

Diversity of Amphibians

The amphibians of today occur in three groups: frogs and toads; salamanders and newts; and caecilians. The salamanders and newts have an elongated body, with a long tail and usually two pairs of legs (Fig. 35.9*a*). Because their legs are set at right angles to the body, they resemble more closely the earliest fossil amphibians. They locomote like a fish, with side-to-side sinusoidal (S-shaped) movements:

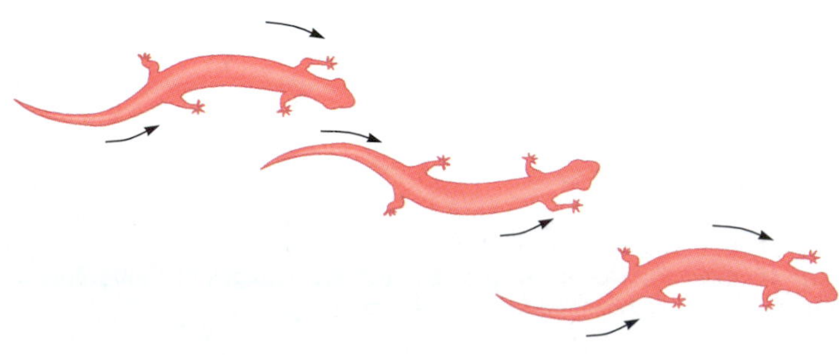

Both salamanders and newts are carnivorous, feeding on small invertebrates such as insects, slugs, snails, and worms. Salamanders practice internal fertilization; in most, males produce a sperm-containing spermatophore that females pick up with the **cloaca** (the common receptacle for the urinary, genital, and digestive canals). Then the eggs are laid in water or on land, depending on the species. Some amphibians, such as the mudpuppy of eastern North America, remain in the water and retain the gills of the larva.

Frogs and toads are the tailless amphibians (Fig. 35.9*b*). In these animals, the head and trunk are fused and the hindlimbs are specialized for jumping. Numerous muscles are located in the limbs as well as the trunk. Frogs have smooth skin and long legs, and they live in or near fresh water; toads have stout bodies and warty skin, and they live in dark, damp places away from the water.

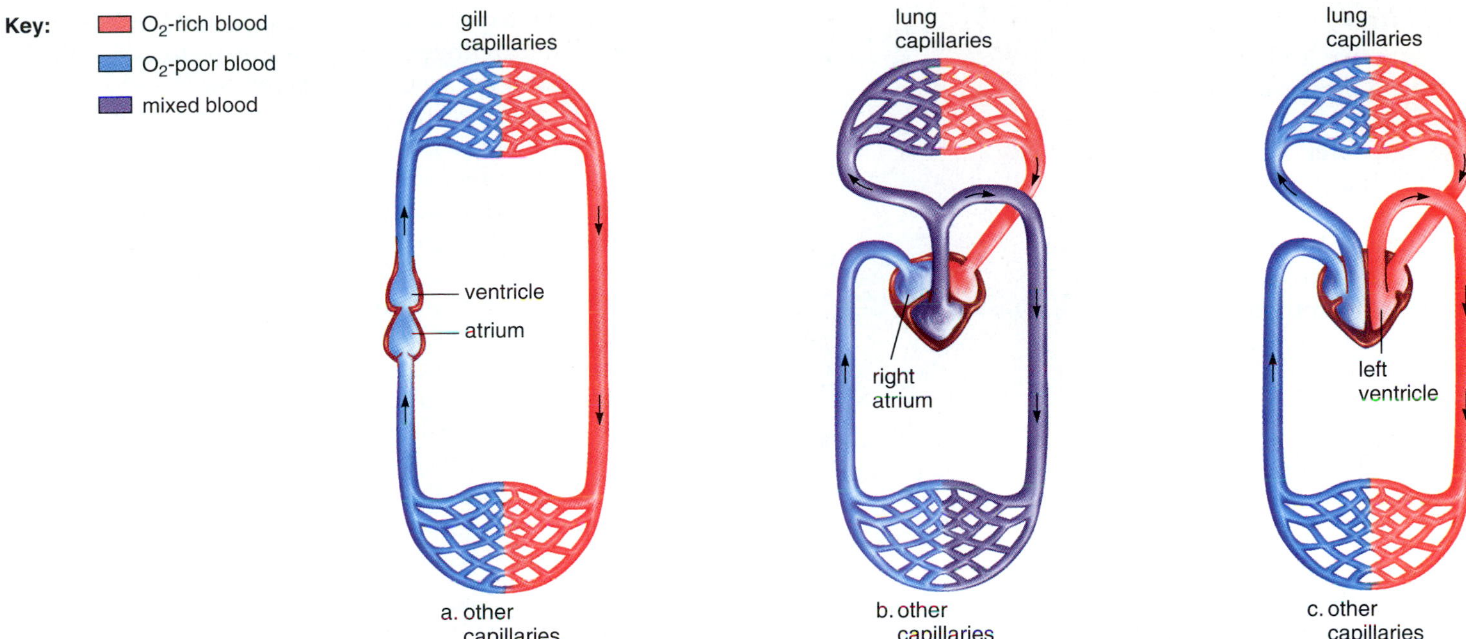

Figure 35.11 Vertebrate circulatory systems.
a. The single-loop system of fishes utilizes a two-chambered heart. **b.** The double-loop system of amphibians sends blood to the lungs and to the body. There is some mixture of O_2-rich and O_2-poor blood in the single ventricle. **c.** The four-chambered heart of birds and mammals sends only O_2-poor blood to the lungs and O_2-rich blood to the body.

Caecilians are legless, often sightless, worm-shaped amphibians that range in length from about 10 cm to more than 1 meter (Fig. 35.9*c*). Nearly all existing species burrow in moist soil, feeding on worms and other soil invertebrates. Some species have folds of skin that make them look like a segmented earthworm.

The class name, Amphibia, means "on both sides," a reference to the fact that most amphibians return to water for the purpose of reproduction. They shed their eggs and sperm into the water, where external fertilization takes place. Generally, the eggs are protected only by a jelly coat and not by a shell. When the young hatch, they are tadpoles (aquatic larvae with gills) that feed and grow in the water. After they undergo a metamorphosis, amphibians emerge from the water as adults that breathe air (Fig. 35.10). Some amphibians, however, have evolved mechanisms that allow them to reproduce on land.

Anatomy and Physiology of Amphibians

Aside from presence of limbs and appropriate modifications of the girdles, amphibians have other features not seen in bony fishes. Amphibians have a tongue for catching prey, eyelids for keeping eyes moist, ears adapted to picking up sound waves, and a voice-producing larynx. The brain is larger than that of a fish and the cerebral cortex is more developed.

Amphibians usually have **lungs,** respiratory organs of terrestrial vertebrates, but they are relatively small and respiration is supplemented by exchange of gases across the porous skin. The single-loop circulatory path of the fish has been replaced by a double loop (Fig. 35.11); but O_2-rich blood is partially mixed with O_2-poor blood in a single ventricle (lower chamber of the heart).

The smooth, nonscaly skin of an amphibian is kept moist by numerous mucous glands. Their skin plays an active role in water balance and respiration and can also help in temperature regulation when on land, through evaporative cooling. A thin, moist skin does mean, however, that most amphibians stay close to water or else they risk drying out. Glands in the skin secrete poisons that make the animal distasteful to eat. Some tropical species with brilliant fluorescent green and red coloration are particularly poisonous. Colombian Indians dip their darts in the deadly secretions of these frogs, aptly called poison-dart frogs.

Amphibians are **ectothermic** [Gk. *ekto*, outer, and *therme*, heat]; they depend on the environment to regulate their body temperature. During winters in the temperate zone, they become inactive and enter torpor. The European common frog can survive even if the temperature drops to −6°C.

These features, in particular, distinguish amphibians:

- usually tetrapods
- usually lungs in adult
- metamorphosis
- smooth and moist skin
- three-chambered heart

Reptiles

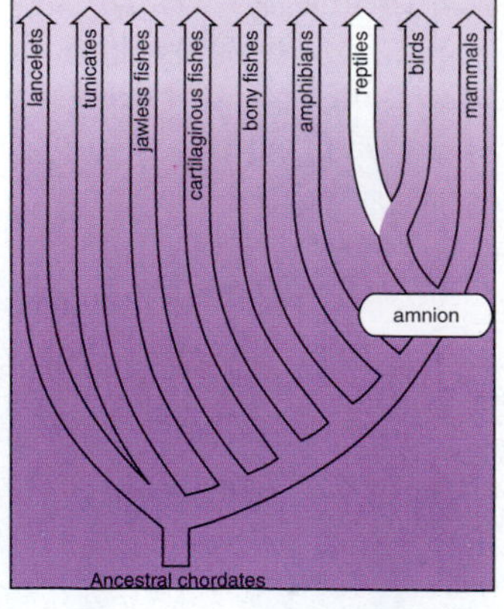

It is adaptive for land animals to have a means of reproduction that is not dependent on external water. Reptiles practice internal fertilization and lay eggs that are protected by a leathery shell (see Fig. 35.13). The **amniote egg** contains extraembryonic membranes, which protect the embryo, remove nitrogenous wastes, and provide the embryo with oxygen, food, and water. These membranes are not part of the embryo itself and are disposed of after development is complete. One of the membranes, the amnion, is a sac that fills with fluid and provides a "private pond," within which the embryo develops.

Reptiles [L. *reptile,* snake] (class Reptilia, 6,000 species) are believed to have evolved from amphibian ancestors by the Permian period (Fig. 35.12). The first reptiles, known as the stem reptiles, gave rise to several other lineages, each adapted to a different way of life. Of interest to us are the pelycosaurs, or sail lizards, so named because of a sail-like web of skin held above the body by slender spines. They are related to the **therapsids,** mammal-like reptiles that came later. Other lines of descent returned to the aquatic environment; one marine reptile (ichthyosaurs) of the Jurassic period was fishlike while another (plesiosaurs) had a long neck and large rowing paddles for limbs. The flying reptiles (pterosaurs) of the Jurassic period had a keel for the attachment of large flight muscles and air spaces in their bones to reduce weight. Their wings were membranous and supported by elongated bones of the fourth finger. The dinosaurs were varied in size and behavior but are well remembered for the great size of some. *Brachiosaurus,* a herbivore, was about 23 meters (75 feet) long and about 17 meters (56 feet) tall. *Tyrannosaurus rex,* a carnivore, was 5 meters (16 feet) tall when standing on its hind legs. A bipedal stance frees the forelimbs for seizing prey or fighting off predators. It is also preadaptive for the evolution of wings: birds are descended from dinosaurs—in fact some say birds are actually living dinosaurs.

Reptiles dominated the earth for about 170 million years during the Mesozoic era and then most died out. What could have caused the mass extinction that occurred at the end of the Cretaceous period? The answer is not known, but recently a layer of the mineral iridium, which is rare on earth but common in meteorites, has been found in rocks of that age. The impact of a large meteorite could have set off earthquakes and fires, raising enough dust to block out the sun. Death of most plants and animals would have followed. Such a scenario has been proposed by Luis and Walter Alvarez and several others who are still gathering evidence to support their hypothesis.

Diversity of Reptiles

Most of today's reptiles live in the tropics or subtropics. Among reptiles are alligators and crocodiles (Fig. 35.13*a*), and turtles (Fig. 35.13*b*) that live in water. Lizards (Fig. 35.13*c*) and snakes are reptiles that live on land. The tuataras are lizardlike reptiles that have remained almost identical to fossils almost 200 million years old (Fig. 35.13*d*).

Crocodiles and alligators lead a largely aquatic life, feeding on fishes, turtles, and terrestrial animals that

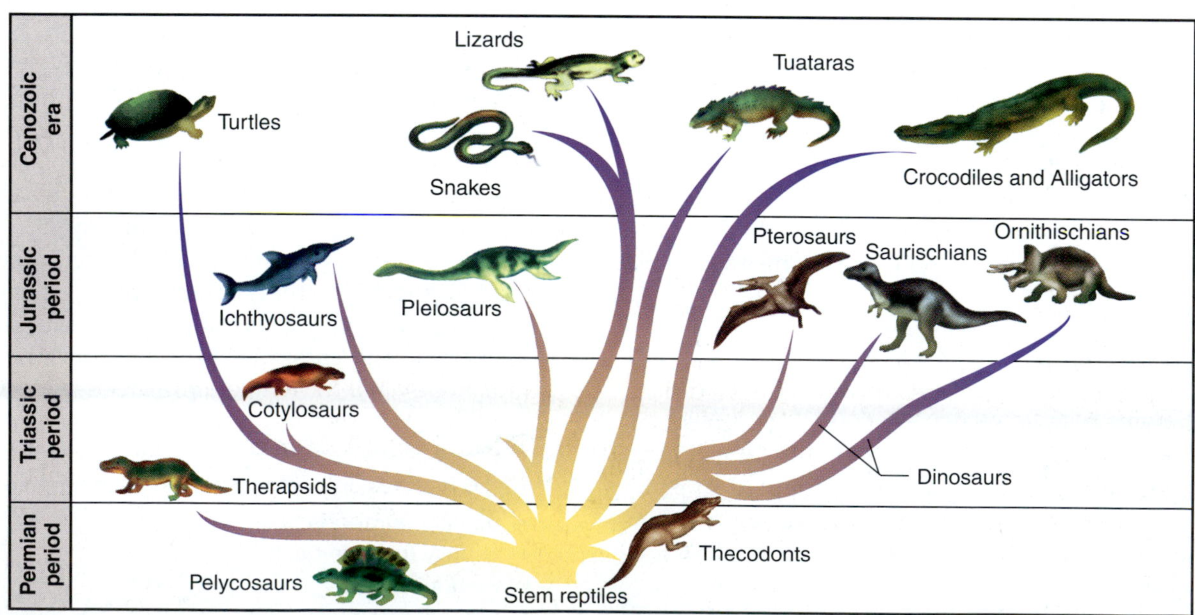

Figure 35.12 Phylogenetic tree.
This phylogenetic tree shows the presumed evolutionary relationship between major groups of reptiles starting with the stem reptiles in the Permian period.

venture close enough to be caught. They have long, powerful jaws with numerous teeth and a muscular tail that serves as both a weapon and a paddle. Although other reptiles are voiceless, male crocodiles and alligators bellow to attract mates. In some species, the male protects the eggs and cares for the young.

Turtles have a heavy shell to which the ribs and thoracic vertebrae are fused. They lack teeth but have a sharp beak. Most turtles spend some time in water. Sea turtles leave the ocean only to lay their eggs. Their legs are flattened and paddlelike, while tortoises, which are usually terrestrial, have strong legs for walking.

Lizards have four clawed feet and resemble their prehistoric ancestor in appearance. They are carnivorous and feed on insects and small animals, including other lizards. Marine iguanas of the Galápagos Islands are adapted to spend long periods of time at sea where they feed on sea lettuce and other marine vegetation. Chameleons are adapted to live in trees and have long, sticky tongues for catching insects some distance away. They can change color in order to blend in with their background. An Australian frilled lizard erects a collar of skin about its neck, which greatly increases its apparent size and makes it look frightening.

Snakes evolved from lizards and have also lost their legs as an adaptation to burrowing. They are carnivorous and have a jaw that is only loosely attached to the skull; therefore, they can eat prey, such as a mouse, that is much larger than their head size. When a snake flicks out its

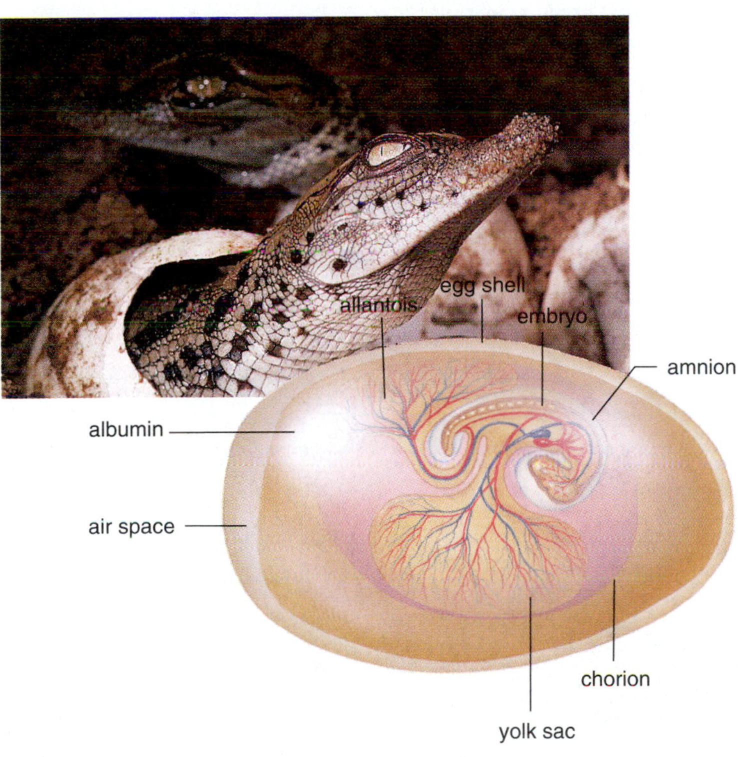

a. American crocodile, *Crocodylus acutus*

b. Green sea turtle, *Chelonia mydas*

c. Gila monster, *Heloderma suspectum*

d. Tuatara, *Sphenodon punctatus*

Figure 35.13 Reptilian diversity.
The living orders of reptiles include: **a.** Crocodiles and alligators. This is a photograph of a crocodile hatching from its egg. The shell is leathery and flexible, not brittle like birds' eggs. Inside the egg, the embryo is surrounded by membranes. The chorion aids gas exchange, the yolk sac provides nutrients, the allantois stores waste, and the amnion encloses a fluid that prevents drying out and provides protection. **b.** Turtles and tortoises. Turtles such as this green sea turtle migrate many miles to return to their nesting sites. **c.** Lizards and snakes. The Gila monster of southwestern United States is only one of two species of lizards that are poisonous. **d.** The tuataras. These animals come out of their burrows at dusk and dawn to feed on insects or small vertebrates.

tongue, it is collecting airborne molecules and transferring them to a Jacobson's organ for analysis. Jacobson's organ, which opens at the roof of the mouth, is an olfactory organ for the analysis of airborne chemicals. Though most snakes are not poisonous, the poisonous rattlesnakes, coral snakes, copperheads, and cobras have fangs for puncturing the skin and injecting venom, as discussed in the reading on page 631.

Anatomy and Physiology of Reptiles

Reptiles have a thick, scaly skin that is keratinized and impermeable to water. Keratin is the protein found in hair, fingernails, and feathers. The protective skin prevents water loss but requires several molts a year. The reptilian lung is more developed (Fig. 35.14), and air rhythmically moves in and out due to the presence of an expandable rib cage, except in turtles. The heart is nearly four chambered in most and is completely four chambered in crocodiles; therefore, O_2-poor blood is completely separate from O_2-rich blood. The well-developed kidneys excrete uric acid and, therefore, less water is required to rid the body of nitrogenous waste.

Reptiles, like amphibians, are ectothermic. This feature allows them to survive on a fraction of the food per body weight required by birds and mammals. Still, they are adapted behaviorally to maintain a warm body temperature by warming themselves in the sun.

These features, in particular, distinguish reptiles:

- usually tetrapods
- lungs with expandable rib cage
- shelled egg
- dry, scaly skin

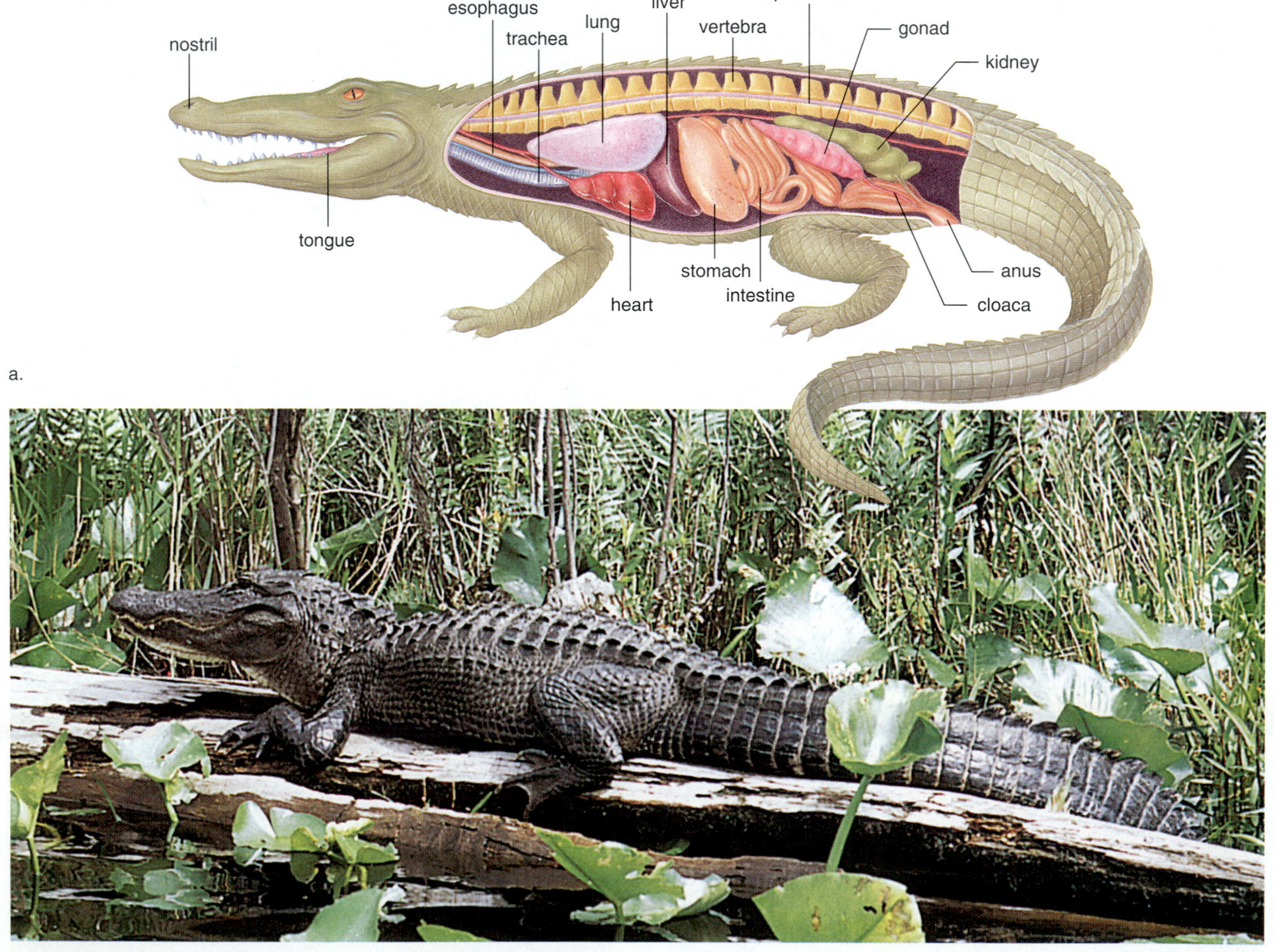

Figure 35.14 **Reptilian anatomy.**
Internal **(a)** and external anatomy **(b)** of the American alligator, *Alligator.*

Venomous Snakes Are Few

Modern snakes and lizards make up 95% of living reptiles. Snakes evolved from lizards during the Cretaceous period and became adapted to burrowing. They lack limbs, so their prey must be subdued and swallowed without the benefit of appendages for manipulating food. Most snakes, like the boas and pythons, are powerful constrictors, suffocating their struggling prey with strong coils. Smaller snakes, such as the familiar garter snakes and water snakes, frequently swallow their food while it is still alive. Still others use toxic saliva to subdue their prey, usually lizards. It is likely that snake venom evolved as a way to obtain food and is used only secondarily in defense.

There are two major groups of venomous snakes. Elapids are represented in the United States by the coral snakes, *Micrurus fulvius* and *Micruroides euryxanthus,* which inhabit the southern states and display bright bands of red, yellow (or white), and black that completely encircle the body (Fig. 35A). In these snakes, the fangs, which are modified teeth, are short and permanently erect. The venom is a powerful neurotoxin that usually paralyzes the nervous system. Actually, coral snakes are responsible for very few bites—probably because of their secretive nature, small size, and relatively mild manner.

Vipers, represented in the United States by pit vipers such as the copperhead and cottonmouth (both *Agkistrodon*) and about 15 species of rattlesnakes (*Sistrurus* and *Crotalus*), make up the remaining venomous snakes of the United States. They have a sophisticated venom delivery system terminating in two large, hollow, needle-like fangs that can be folded against the roof of the mouth when not in use. The venom destroys the victim's red blood cells and causes extensive local tissue damage. These snakes are readily identified by the combination of heat-sensing facial pits, elliptical pupils in the eyes, and a single row of scales on the underside of the tail (Fig. 35B). None of our harmless snakes has any combination of these characteristics.

First aid for snakebite is not advised if medical attention is less than a few hours away; application of a tourniquet, incising the wound to promote bleeding, and other radical treatments often cause more harm than good. The best method of treating snakebite is through the use of prescribed antivenin, a serum containing antibodies to the venom. A hospital stay is required because some people are allergic to the serum.

Most people are not aware that snakes are perhaps the greatest controllers of disease-carrying, crop-destroying rodents because they are well adapted to following such prey into their hiding places. Also, snakes are important food items in the diets of many other carnivores, particularly birds of prey such as hawks and owls. Their presence in an ecosystem demonstrates the overall health of the environment.

Figure 35A Coral snake, *Micrurus fulvius.*
This is a front-fanged snake, one of the major groups of venomous snakes.

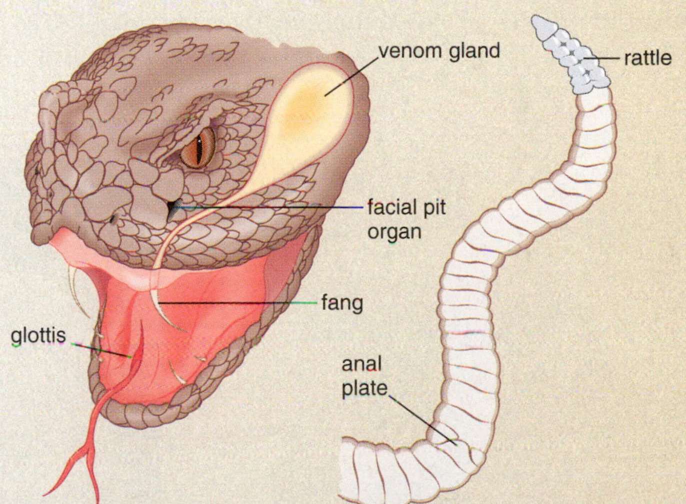

Figure 35B Viper characteristics.
Vipers, the other major group of venomous snakes, can be identified by these characteristics.

Birds

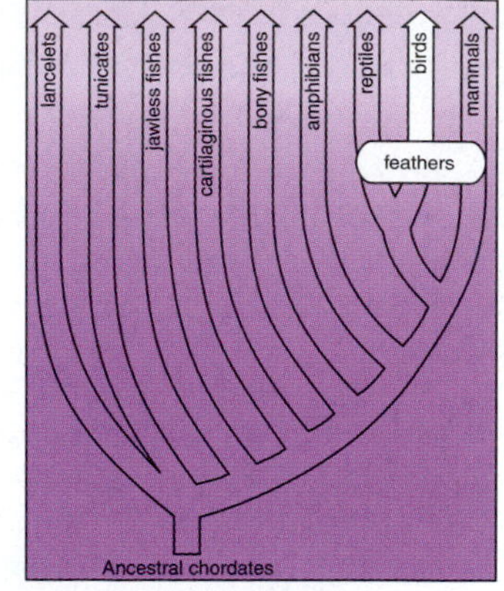

Today's **birds** (class Aves, 9,000 species) lack teeth and have only a vestigial tail, but they still retain many reptilian features such as the shape of the body, scales on their legs, claws on their toes, and a horny beak. They also lay amniotic eggs, although the shell is hard rather than leathery. The exact ancestry of birds is in dispute, but there are those who contend that birds are closely related to bipedal dinosaurs and that they should be classified as such.

Diversity of Birds

Classification of birds is particularly based on beak (Fig. 35.15) and foot types, and to some extent on habitat and behavior. The various orders include birds of prey with notched beaks and sharp talons; shorebirds with long, slender, probing bills and long, stiltlike legs; woodpeckers with sharp, chisel-like bills and grasping feet; waterfowl with webbed toes and broad bills; penguins with wings modified as paddles; and songbirds with perching feet.

Anatomy and Physiology of Birds

Birds are the only modern animals to have **feathers,** which are actually modified reptilian scales. Bird feathers are of two types. Contour feathers cover the body, and those attached to the wings (called flight feathers) overlap to produce a broad, flat surface beneficial for flight. Down feathers provide excellent insulation against body heat loss. This is important because birds are **homeothermic** [Gk. *homoios,* like, and *therme,* heat]; they maintain a constant, relatively high body temperature, which permits them to be continuously active even in cold weather. Perhaps feathers first provided insulation and then became adapted for flight.

Nearly every anatomical feature of birds can be related to their ability to fly (Fig. 35.16). Their forelimbs are modified as wings. They have hollow, very light bones laced with air cavities. A horny beak has replaced jaws equipped with teeth, and a slender neck connects the head to a rounded, compact torso. The breastbone is enlarged and has a keel, to which the strong muscles are attached for flying. Oxygen delivery to these muscles is efficient; there is no dead space in the lungs because air flows one way through the air sacs and the lungs, and the double-loop circulation is improved because the four-chambered heart keeps O_2-rich blood separated from O_2-poor blood.

Flight requires well-developed sense organs and nervous system. Birds have particularly acute vision and excellent muscle reflexes. Because birds can fly, they migrate and exploit food sources in widely separated habitats. Complex hormonal regulation and behavior responses are involved in bird behavior, which also includes caring for fledglings until they are independent.

These features, in particular, distinguish birds:

- feathers
- hard-shelled egg
- four-chambered heart
- usually wings for flying
- air sacs
- homeothermic

a. Cardinal, *Cardinalis cardinalis*

b. Bald eagle, *Haliaetus leucocephalus*

c. Flamingo, *Phoenicopterus ruber*

Figure 35.15 Bird beaks.
a. A cardinal's beak allows it to crack tough seeds. **b.** A bald eagle's beak allows it to tear prey apart. **c.** A flamingo's beak strains food from the water with bristles that fringe the mandibles.

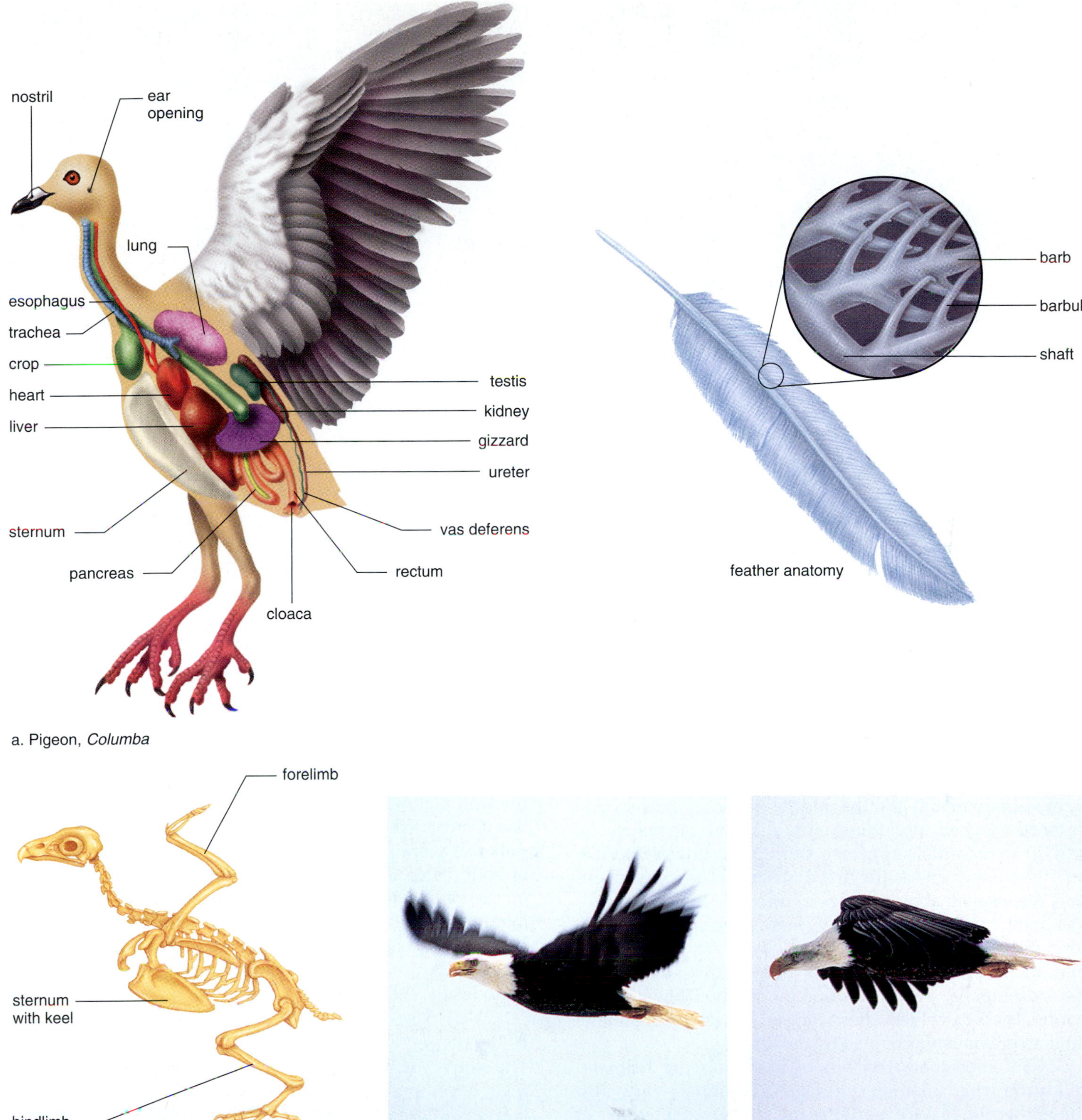

Figure 35.16 Bird anatomy and flight.
a. Bird anatomy. *(left)* Anatomy of a pigeon is representative of bird anatomy. *(right)* In feathers, a hollow central shaft gives off barbs and barbules, which interlock in a latticelike array. **b.** Bird flight. *(left)* The skeleton of an eagle shows that birds have a large, keeled sternum to which flight muscles attach. The fused bones of the forelimb help support the wings. *(right)* Birds fly by flapping their wings. Bird flight requires an airstream and a powerful wing downstroke for lift, a force at right angles to the airstream.

a. Duckbill platypus, *Ornithorhynchus anatinus*

b. Virginia opossum, *Didelphis virginianus*

c. Koala, *Phascolarctos cinereus*

Figure 35.17 Monotremes and marsupials.
a. The duckbill platypus is a monotreme that inhabits Australian streams. **b.** The opossum is the only marsupial in the Americas. The Virginia opossum is found in a variety of habitats. **c.** The koala is an Australian marsupial that lives in trees.

Mammals

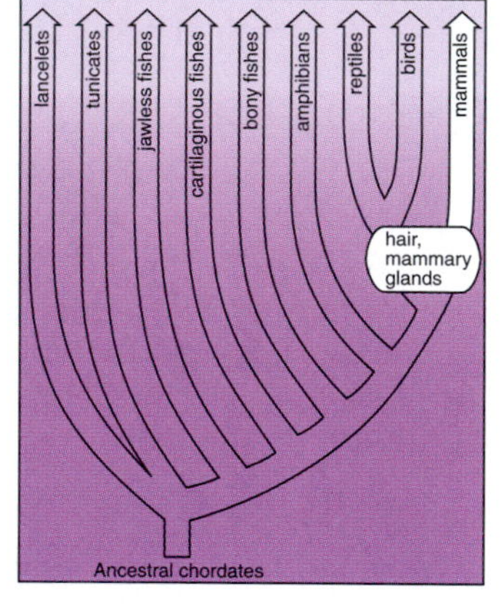

Mammals [L. *mamma,* breast, teat] (class Mammalia, 4,500 species) evolved during the Mesozoic era from therapsids, the mammal-like reptiles. The mammalian skull accommodates a larger brain relative to body size than does the reptilian skull; mammalian cheek teeth are differentiated as premolars and molars; their vertebrae are highly differentiated and the middle region of the vertebral column is arched, providing more effective movement on land. True mammals appeared during the Jurassic period, about the same time as the first dinosaurs. These first mammals were small, about the size of mice. All the time the dinosaurs flourished (165 million years), mammals were a minor group that changed little. Some of the earliest mammalian groups, represented today by the monotremes and marsupials, are not abundant today. The placental mammals that evolved later went on to occupy the many habitats previously occupied by the dinosaurs.

The chief characteristics of mammals are hair and milk-producing mammary glands. Mammals are also homeothermic, as are birds. Many of the adaptations of mammals are related to temperature control. Hair, for example, provides insulation against heat loss and allows mammals to be active even in cold weather. Like birds, mammals have efficient respiratory and circulatory systems, which assure a ready oxygen supply to muscles whose contraction produces body heat. Like birds, mammals have a double-loop circulation and a four-chambered heart.

Mammary glands enable females to feed (nurse) their young without deserting them to find food. Nursing also creates a bond between mother and offspring that helps ensure parental care while the young are helpless. In most mammals, the young are born alive after a period of development in the uterus, a part of the female reproductive tract. Internal development shelters the young and allows the female to move actively about while the young are maturing. Mammals are classified according to means of reproduction.

Mammals That Lay Eggs

Monotremes [Gk. *monos,* one, and *trema,* hole], mammals that have a cloaca and lay hard-shelled amniote eggs, are represented by the spiny anteater and the duckbill platypus, both of which are found in Australia (Fig. 35.17*a*). The female duckbill platypus lays her eggs in a burrow in the ground. She incubates the eggs and, after hatching, the young lick up milk that seeps from modified sweat glands on the abdomen of both males and females. The spiny anteater has a pouch on the belly side formed by swollen mammary glands and longitudinal muscle. The egg moves from the cloaca to this pouch where hatching takes place, and the young remain for about 53 days. Then they stay in a burrow, where the mother periodically visits and nurses them.

Mammals That Have Pouches

The young of **marsupials** [Gk. *marsupium,* pouch] begin their development inside the female's body, but they are born in a very immature condition. Newborns crawl up into a pouch on their mother's abdomen. Inside the pouch, they attach to nipples of mammary glands and continue to develop. Frequently, more are born than can be accommodated by the number of nipples, and it's "first come, first served."

Today, marsupial mammals are found mainly in Australia; only a few marsupials, such as the American opossums, are found outside that continent (Fig. 35.17*b*). In Australia, marsupials underwent adaptive radiation for several million years without competition from placental mammals,

a. White-tailed deer, *Odocoileus virginianus*

b. African lioness, *Panthera leo*

c. Golden tamarin monkey, *Leontopithecus rosalia*

d. Killer whale, *Orcinus orca*

Figure 35.18 Placental mammals.
Placental mammals have adapted to various ways of life. **a.** Deer are herbivores that live in forests. **b.** Lions are carnivores on the African plain. **c.** Monkeys inhabit tropical forests. **d.** Whales are sea-dwelling placental mammals.

which arrived there only recently. Among the herbivorous marsupials, koalas are tree-climbing browsers (Fig. 35.17*c*) and kangaroos are grazers. The Tasmanian wolf or tiger, thought to be extinct, was a carnivorous marsupial about the size of a collie dog.

Mammals That Have Placentas

Developing **placental mammals** are dependent on the **placenta,** an organ of exchange between maternal blood and fetal blood. Nutrients are supplied to the growing offspring, and wastes are passed to the mother for excretion. While the fetus is clearly parasitic on the female, in exchange, she is free to move about as she chooses while the fetus develops. The young are born at a relatively advanced stage of development.

Placental mammals lead an active life. The senses are acute and the brain is enlarged due to the expansion of the foremost part—the cerebral hemispheres. These have become convoluted and have expanded to such a degree that they hide many other parts of the brain from view. The brain is not fully developed for some time after birth, and there is a long period of dependency on the parents, during which the young learn to take care of themselves.

Placental mammals populate all continents except Antarctica. Most mammals live on land, but some (e.g., whales, dolphins, seals, sea lions, manatees) are secondarily adapted to live in water, and bats are able to fly. Classification is primarily based on mode of locomotion and methods of obtaining food. The following are some of the major orders of mammals:

The hoofed mammals include the orders Perissodactyla (e.g., horses, zebras, tapirs, rhinoceroses, 17 species) and the Artiodactyla (e.g., pigs, cattle, deer, hippopotami, buffaloes, giraffes, 185 species) whose elongated limbs are adapted for running often across open grasslands (Fig. 35.18*a*). Both groups of animals are herbivorous and have large grinding teeth.

Order Carnivora (270 species) includes dogs, cats, bears, raccoons, and skunks (Fig. 35.18*b*). All these animals have limbs adapted for running and have a well-developed sense of smell. The canines of meat eaters are large and conical. There are some aquatic carnivores—namely seals, sea lions, and walruses—which must return to land to reproduce.

Order Primates (180 species) includes lemurs, monkeys, gibbons, chimpanzees, gorillas, and humans (Fig. 35.18*c*). Typically, primates are tree-dwelling fruit eaters although some, like humans, are ground dwellers. They all have a freely movable head, they have five digits with nails (not claws), and the thumb in many (sometimes the large toe) is opposable. Primates, especially humans, are well known for their well-developed brains.

Order Cetacea (80 species) includes the whales and dolphins (Fig. 35.18*d*), which are mammals despite their lack of hair or fur. Blue whales, the largest animal ever to have lived on this planet, are baleen whales that feed by straining large quantities of water containing plankton. Toothed whales feed mainly on fish and squid.

Order Chiroptera (925 species) includes the nocturnal bats whose wings consist of two layers of skin and connective tissue stretched between the elongated bones of all fingers but the first. Many species use echolocation to locate their usual insect prey. But there are also bird-, fish-, frog-, and plant-eating bats.

Order Rodentia (1,760 species), the largest order, includes mice, rats, squirrels, beavers, and porcupines. Rodents have incisors that grow continuously. They usually feed on seeds but some are omnivorous and some eat mainly insects.

Order Proboscidea (2 species) includes the elephants, the largest living land mammals. The upper lip and nose have become elongated and muscularized to form a trunk.

Order Lagomorpha (65 species) includes the herbivorous rabbits, hares, and pikas—animals that superficially resemble rodents. They also have two pairs of continually growing incisors, and their hind legs are longer than their front legs.

These features, in particular, distinguish placental mammals:

- **body hair**
- **differentiated teeth**
- **well-developed brain**
- **infant dependency**
- **mammary glands**
- **homeothermic**
- **internal development**

Connecting Concepts

How do you measure success? By almost any criterion, vertebrates lose. Vertebrates are eukaryotes that have been assigned to one domain, while the prokaryotes now have two domains. The total number of prokaryotes is greater and there are possibly more types of prokaryotes than any other living form. The unseen world is much larger than the seen world! And prokaryotes are adapted to utilize most types of energy sources and live in most any type of environment.

If it were not for the myriad types of terrestrial insects, there would be more aquatic species than terrestrial ones. The adaptive radiation of mammals has taken place on land, and this might seem impressive to some, but actually the number of mammalian species (4,500) is small compared to, say, the molluscs (110,000 species), which radiated in the sea. Out of about three million species of organisms that have been discovered, at least 850,000 are insects. On land, vertebrate species are far outnumbered by the number of insect species. Being small has its advantages. To a large animal like a cow, grass is a more or less uniform carpet, but to small insects it is a highly varied habitat. Specific insects feed on the roots, stems, leaves, flowers, pollen, and seeds.

In terms of size, however, the vertebrates might win. A blue whale can weigh as much as 150 tons and have a length of 30 meters. But then again, redwood trees can be over 100 meters tall, and there are eucalyptus trees measuring as much as 140 meters. Metabolically speaking, plants are much more capable than animals. They can make their own food and change glucose into all the other organic molecules they require. An ecosystem only needs plants and decomposers to sustain itself.

The size and complexity of the brain is cited as a criterion by which vertebrates are more successful than other living things. However, it can be suggested that this characteristic is associated with others that make an animal prone to extinction. Which type animal is likely to become extinct? As we discussed earlier on page 391, animals that are large, have a long life span, are slow to mature, have few offspring, and expend much energy caring for their offspring (i.e., *K*-strategists) tend to become extinct when their normal way of life is destroyed. Vertebrates, in general, are more threatened than other types of organisms by our present biodiversity crisis—a crisis brought on by the activities of the vertebrate with the most complex brain of all, *Homo sapiens.*

Summary

35.1 Echinoderms

Echinoderms (e.g., sea stars, sea urchins, sea cucumbers, and sea lilies) have radial symmetry as adults (not as larvae) and internal calcium-rich plates with spines. Typical of echinoderms, sea stars have tiny skin gills, a central nerve ring with branches, and a water vascular system for locomotion. Each arm of a sea star contains branches from the nervous, digestive, and reproductive systems.

35.2 Chordates

Chordates (tunicates, lancelets, and vertebrates) have a notochord, a dorsal tubular nerve cord, and pharyngeal pouches at one time in their life history. Also, there is a postanal tail.

Lancelets and tunicates are the invertebrate chordates. Lancelets are the only chordate to have the three characteristics in the adult stage. Tunicates lack chordate characteristics (except gill slits) as adults, but they have a larva that could be ancestral for the vertebrates.

35.3 Vertebrates

Vertebrates are in the phylum Chordata, and have the four chordate characteristics as embryos. As adults, the notochord is replaced by the vertebral column. Vertebrates, which undergo cephalization, have an endoskeleton, paired appendages, and well-developed internal organs.

The first vertebrates lacked jaws and paired appendages. They are represented today by the hagfishes and lampreys. Ancestral bony fishes, which had jaws and paired appendages, gave rise during the Devonian period to two groups: today's cartilaginous fishes (skates, rays, and sharks) and the bony fishes, including the ray-finned fishes and the lobe-finned fishes. The ray-finned fishes became the most diverse group among the vertebrates. The lobe-finned fishes, represented today by three species of lungfishes and the coelacanth, gave rise to the amphibians.

Ancestral amphibians were tetrapods that diversified during the Carboniferous period. They are represented primarily today by frogs and salamanders, which usually return to the water to reproduce and then metamorphose into terrestrial adults.

Reptiles (today's crocodiles, turtles, lizards, and snakes) lay a shelled egg, which allows them to reproduce on land.

One main group of ancient reptiles, the stem reptiles that presumably evolved from amphibian ancestors, produced a line of descent that evolved into both dinosaurs and birds during the Mesozoic era. A different line of descent from stem reptiles evolved into mammals.

Birds are feathered, which helps them maintain a constant body temperature. They are adapted for flight: their bones are hollow, their shape is compact, their breastbone is keeled, and they have well-developed sense organs.

Mammals remained small and insignificant while the dinosaurs existed, but when the latter became extinct at the end of the Cretaceous period, mammals became dominant land organisms.

Mammals are vertebrates with hair and mammary glands. The former helps them maintain a constant body temperature, and the latter allows them to nurse their young. Monotremes lay eggs; marsupials have a pouch in which the newborn matures; and the placental mammals, which are far more varied and numerous, retain offspring inside the uterus until birth.

Reviewing the Chapter

1. What are the general characteristics of echinoderms? Explain how the water vascular system works in sea stars. 618–19
2. What four characteristics do all chordates have at some time in their life history? 619
3. Describe the two groups of invertebrate chordates, and explain how the tunicates might be ancestral to vertebrates. 620
4. What is the vertebrate body plan? Discuss the distinguishing characteristics of vertebrates. 622
5. Describe the jawless fishes, including ancient ostracoderms. 622
6. What is the significance of having jaws? Describe the ancient placoderms and today's cartilaginous and bony fishes. The amphibians evolved from what type of fish? 623–25
7. What is the significance of being a tetrapod? Discuss the characteristics of amphibians, stating which ones are especially adaptive to a land existence. Explain how their class name (Amphibia) characterizes these animals. 625–27
8. What is the significance of the amniote egg? What other characteristics make reptiles less dependent on a source of external water? 628
9. Draw a simplified phylogenetic tree for reptiles showing their relationship to birds and mammals. Your tree should also include stem reptiles and dinosaurs. 628
10. What is the significance of wings? In what other ways are birds adapted to flying? 632–33
11. What is the significance of homeothermy? What are the three subclasses of mammals, and what are their primary characteristics? 634–35

Testing Yourself

Choose the best answer for each question.

1. Which of these does not pertain to a deuterostome?
   a. Blastopore is associated with the anus.
   b. spiral cleavage
   c. enterocoelom
   d. echinoderms and chordates
   e. All of these do not pertain to deuterostomes.
2. The tube feet of echinoderms
   a. are their head.
   b. are a part of the water vascular system.
   c. are found in the coelom.
   d. help pass sperm to females during reproduction.
   e. All of these are correct.
3. Which of these is not a chordate characteristic?
   a. dorsal supporting rod, the notochord
   b. dorsal tubular nerve cord
   c. pharyngeal pouches
   d. postanal tail
   e. vertebral column
4. Adult tunicates
   a. do not have all three chordate characteristics.
   b. are known as sea squirts.
   c. are fishlike in appearance.
   d. are the first chordates to be terrestrial.
   e. All of these are correct.
5. Sharks and bony fishes are different in that only
   a. bony fishes have paired fins.
   b. bony fishes have a keen sense of smell.
   c. bony fishes have an operculum.
   d. sharks have a bony skeleton.
   e. sharks are predaceous.
6. Amphibians arose from
   a. tunicates and lancelets.
   b. cartilaginous fishes.
   c. jawless fishes.
   d. ray-finned fishes.
   e. bony fishes with lungs.
7. Which of these is not a feature of amphibians?
   a. dry skin that resists desiccation
   b. metamorphosis from a swimming form to a land form
   c. small lungs and a supplemental way of gas exchange
   d. reproduction in the water
   e. a single ventricle
8. How are salamanders different from snakes?
   a. The ancestors of snakes but not salamanders were tetrapods.
   b. Salamanders reproduce on land but snakes do not.
   c. Snakes but not salamanders lay a shelled egg.
   d. Salamanders are carnivores, and snakes are herbivores.
   e. Salamanders are sometimes aquatic but snakes never are.
9. Dinosaurs
   a. were dominant during the Mesozoic era.
   b. are closely related to the birds.
   c. lay shelled eggs.
   d. most likely cared for their young.
   e. All of these are correct.
10. Which of these is a true statement?
    a. In all mammals, offspring develop completely within the female.
    b. All mammals have hair and mammary glands.
    c. All mammals have one birth at a time.
    d. All mammals are land-dwelling forms.
    e. All of these are true.
11. Label the following diagram.

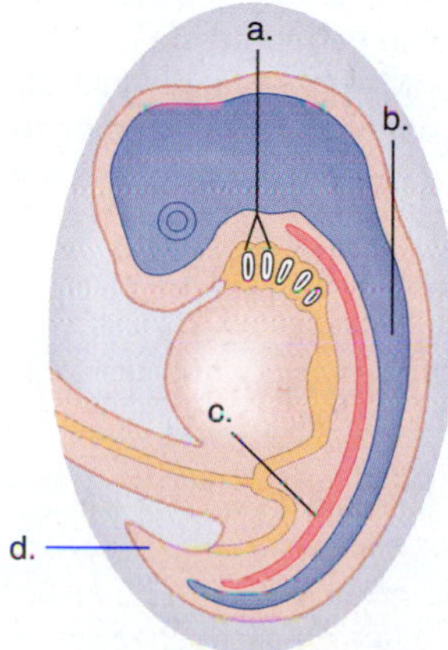

Thinking Scientifically

1. *Archaeopteryx* was a birdlike reptile that had a toothed beak. Modern birds have no teeth in their beaks. Even though modern birds are not thought to have evolved directly from *Archaeopteryx,* presumably teeth were present in the ancestors of birds. Give an evolutionary explanation for the elimination of teeth in a bird's beak.
2. While amphibians have rudimentary lungs, skin is also a respiratory organ. Because of this, they are much more sensitive to air pollution, especially particulate matter, than are higher vertebrates. Why would a thin skin be more sensitive to pollution than lungs?

Understanding the Terms

agnathan 622
amniote egg 628
amphibian 625
bird 632
bony fish 624
cartilaginous fish 623
chordate 619
cloaca 626
deuterostome 618
echinoderm 618
ectothermic 627
feather 632
fin 623
fish 622
gill 619
gnathostomate 623
homeothermic 632
jaw 623
jawless fish 622
lobe-finned fish 625
lung 627
mammal 634
marsupial 634
monotreme 634
nerve cord 619
notochord 619
ostracoderm 622
placenta 635
placental mammal 635
placoderm 623
ray-finned fish 624
reptile 628
scale 622
swim bladder 624
tetrapod 625
therapsid 628
tube foot 619
vertebrate 622
water vascular system 619

Match the terms to these definitions:

a. _______________ Animal (bird or mammal) that maintains a uniform body temperature independent of the environmental temperature.
b. _______________ Egg-laying mammal—for example, duckbill platypus and spiny anteater.
c. _______________ Member of a class of terrestrial vertebrates with internal fertilization, scaly skin, and an egg with a leathery shell; includes crocodiles, turtles, lizards, and snakes.
d. _______________ Dorsal supporting rod that exists in all chordates sometime in their life history; replaced by the vertebral column in vertebrates.
e. _______________ Member of a phylum of marine animals that includes sea stars, sea urchins, and sand dollars; characterized by radial symmetry and a water vascular system.

Web Connections

Exploring the Internet

http://www.mhhe.com/biosci/genbio/mader
(click on *Biology 7/e*)

The *Biology 7/e* Online Learning Center provides many resources for studying the material in this chapter including links to the following sites:

Echinodermata. Arizona's Tree of Life Web Page. Pictures, characteristics, phylogenetic relationships, references on echinoderms.

http://phylogeny.arizona.edu/tree/eukaryotes/animals/echinodermata/echinodermata.html

Chordata. Arizona's Tree of Life Web Page. An introduction, pictures, characteristics, phylogenetic relationships, and references on chordates.

http://phylogeny.arizona.edu/tree/eukaryotes/animals/chordata/chordata.html

Introduction to the Actinopterygii. University of California at Berkeley, Museum of Paleontology. Images, photos, systematics, more information, and links.

http://www.ucmp.berkeley.edu/vertebrates/actinopterygii/actinintro.html

Interactive Frog Dissection. This site is designed to prepare high school students for frog dissection. It is an interesting review for beginning zoology students or an introduction to frog anatomy for zoology students who did not dissect a frog in high school. The video may not work with all systems or platforms.

http://curry.edschool.Virginia.EDU/go/frog/

The Birds of North America. The Academy of Natural Sciences supports this site, where you can look up natural history information on almost all breeding birds in North America.

http://www.birdsofna.org/

Further Readings for Part V

Ben-Jacob, E., and Levine, H. October 1998. The artistry of microorganisms. *Scientific American* 279(4):82. Colonies of bacteria form geometric patterns, which reflect survival strategies.

Burkholder, J. M. August 1999. The lurking perils of *Pfiesteria. Scientific American* 281(2):42. Outbreaks of this dinoflagellate have killed fish by the millions in estuaries along the eastern United States.

Castro, P., and Huber, M. 1997. *Marine biology.* 2d ed. St. Louis: Mosby-Year Book, Inc. This introductory text is designed to provide a stimulating overview of marine biology.

Chadwick, D. H. March 1998. Planet of the beetles. *National Geographic* 193(3):100. With diverse sizes, forms, and functions, beetles make up one-third of the world's identified insects.

Crepet, W. L. November 27, 1998. The abominable mystery. *Science* 282(5394):1653. Researchers describe the first fossil evidence of a Jurassic angiosperm, which has important implications in determining the origin of flowering plants.

Finnell, R. B. May 1999. The flower issue. *Natural History* 108(4):36–68. This special issue includes articles examining flower fossils, evolution, pollination, and rare species.

Flannery, M. April 1999. Seeing plants a little more clearly. *American Biology Teacher* 61(4):303. Article examines the latest plant research.

Foster, K. R., et al. August 1998. The Philadelphia yellow fever epidemic of 1793. *Scientific American* 279(2):88. The history of this epidemic and the possibility of its recurrence are discussed.

Genthe, H. August 1998. The incredible sponge. *Smithsonian* 29(5):50. Sponge biology and the therapeutic uses of sponges in treating cancer are discussed.

Gwynne, D. T. August 1997. Glandular gifts. *Scientific American* 277(2):66. Male insects offer body parts and secretions as a strategy for fertilizing the female's eggs.

Huddler, G. W. 1998. *Magical mushrooms, mischievous molds.* Princeton, New Jersey: Princeton University Press. This introductory text discusses basic mycology and the impact of fungi on human life and history.

Jarrell, K. F., et al. July 1999. Recent excitement about Archaea. *BioScience* 49(7):530. Current topics on Archaea are presented.

Knols, B. G., and Meijerink, J. September/October 1997. Odors influence mosquito behavior. *Science & Medicine* 4(5):56. Definition of odors may lead to the control of insects that transmit infection diseases.

Laver, G., et al. January 1999. Disarming flu viruses. *Scientific American* 280(1):78. New flu treatments are examined.

Levy, S. B. March 1998. The challenge of antibiotic resistance. *Scientific American* 278(3):46. Misuse and overuse of antibiotics must end in order to preserve their effectiveness.

Lim, D. 1998. *Microbiology.* 2d ed. Dubuque, Iowa: WCB/McGraw-Hill. This introductory text shows how microorganisms relate to one another and to other organisms; includes new material in evolution and biodiversity.

Losick, R., and Kaiser, D. February 1997. Why and how bacteria communicate. *Scientific American* 276(2):68. Bacteria send and receive chemical messages and can organize into structures.

Luoma, J. R. March 1997. The magic of paper. *National Geographic* 191(3):88. The papermaking process is discussed in this article.

Madigan, M. T., and Marrs, B. L. April 1997. Extremophiles. *Scientific American* 276(4):86. Certain microorganisms can withstand extreme environments.

Margulis, L., et al. 1998. *Five kingdoms: An illustrated guide to the phyla of life on earth.* 3d ed. New York: W. H. Freeman & Company. Introduces the kingdoms of organisms.

Matthews, B. E. 1998. *An introduction to parasitology.* Cambridge, United Kingdom: Cambridge University Press. This introductory undergraduate text provides a concise overview of parasite biology.

Miller, S. A., and Harley, J. B. 1999. *Zoology.* 4th ed. Dubuque, Iowa: WCB/McGraw-Hill. This is an introductory general zoology text.

Moore-Landecker, E. 1996. *Fundamentals of the fungi.* 4th ed. Upper Saddle River, N.J.: Prentice-Hall. For intermediate students, this text presents a broad introduction to the field of mycology.

Moore, R., Clark, W. D., et al. 1998. *Botany.* 2d ed. Dubuque, Iowa: WCB/McGraw-Hill. This introductory botany text stresses the importance of plants and the process of science.

Murawski, D. A. March 1997. Moths come to light. *National Geographic* 191(3):40. Moths display a variety of disguises and survival techniques.

Nester, E. W., Roberts, C. E., et al. 1998. *Microbiology: A human perspective.* 2d ed. Dubuque, Iowa: WCB/McGraw-Hill. This introductory text relates basic microbiology to human health.

Parfit, M. October 1998. Antarctic desert. *National Geographic* (4):120. Microscopic organisms that survive in the Antarctic desert have been discovered.

Purnell, B. A., and Marx, J. May 21, 1999. Microbe management. *Science* 284(5418):1301. Special issue on microbes, immunity, and disease.

Rutowski, R. July 1998. Mating strategies in butterflies. *Scientific American* 279(1):64. Colorful wing patterns and pheromones play important roles in butterfly mating.

Seymour, R. S. March 1997. Plants that warm themselves. *Scientific American* 276(3):104. Some plants generate heat to keep blossoms at a constant temperature.

Sherman, P. W., and Billing, J. June 1999. Darwinian gastronomy: Why we use spices. *BioScience* 49(6):453. Some spices combat foodborne microorganisms with chemicals the plant uses for evolutionary survival.

Stern, K. 1999. *Introductory plant biology.* 8th ed. Dubuque, Iowa: WCB/McGraw-Hill. This text presents basic botany in a clear, informative manner.

Stokes, M. D., and Holland, N. D. November/December 1998. The lancelet. *American Scientist* 86(6):552. This creature, also known as "amphioxus," is a player in vertebrate phylogenetic history.

Sumich, J. L. 1998. *An introduction to the biology of marine life.* 7th ed. Dubuque, Iowa: Wm. C. Brown Publishers. This introductory text covers taxonomy, evolution, ecology, behavior, and physiology of selected groups of marine organisms.

Sze, P. 1998. *A biology of the algae.* 3d ed. Dubuque, Iowa: WCB/McGraw-Hill. This concise text introduces algae morphology, evolution, and ecology to the botany major.

Walters, D. R., and Keil, D. J. 1996. *Vascular plant taxonomy.* 4th ed. Dubuque, Iowa: Kendall/Hunt Publishing. Introduces plant families and experimental aspects of taxonomy.

Williams, N. September 4, 1998. Britain hunts down CJD epidemic in removed appendixes. *Science* 281(5382):1422. Cases of Creutzfeld-Jakob disease show prions in lymphoid tissue and in removed appendixes.

Wuethrich, B. April 16, 1999. Giant sulfur-eating microbe found. *Science* 284(5413):415. A giant new species of bacterium, with individual cells up to 0.75 millimeter in diameter, has been found in ocean sediments.

part VI

Plant Structure and Function

Flowering plants are terrestrial photosynthesizers. Their leaves, lifted skyward, capture solar energy and have pores that permit gas exchange. Their roots, anchored in the soil, absorb water and minerals. A vascular system transports water and minerals up the stem to the leaves and transports the products of photosynthesis to all body parts. Plant cells use sugars as the starting material for the production of all the metabolites they need.

A complex network of hormones controls plant growth and therefore their responses to the environment. Plants grow their entire lives and are ever capable of producing new body parts. Trees can have a much longer life span than animals: some live for thousands of years!

The adaptations of flowering plants to the land environment are especially evident when one considers the manner in which they carry on reproduction. Gametes, zygotes, and embryos are all protected from desiccation. And though flowering plants are nonmobile, many utilize the mobility of wind and animals to accomplish gamete and seed distribution.

Plant Structure

chapter 

Buttress roots of a tropical tree

There is a stunning array of plant life covering the earth, and over 80% of all living plants are flowering plants, or angiosperms. Therefore, it is fitting that we set aside one part of the text to primarily examine the structure and the function of flowering plants. The organization of a flowering plant is suitable to photosynthesizing on land. The elevated leaves have a shape that facilitates absorption of solar energy. Strong stems conduct water up to the leaves and organic food down to the extensive roots, which not only anchor the plant but also absorb water and minerals. It can be seen, then, that organs of flowering plants have a structure that suits their functions. Roots, stems, and leaves are adapted even further to suit particular environments.

Plants are essential for life on earth. Animals depend on them for food and oxygen; they also help regulate the water cycle and carbon cycle of the biosphere. Plants have many other uses for human beings. For example, they provide fibers for making clothes, wood for construction purposes, and chemicals for commercial and medicinal uses. Plants also give us much pleasure. We like to vacation in natural settings, and we use plants to beautify our parks, lawns, and homes.

36.1 Plant Organs

Flowering plants are extremely diverse because they are adapted to living in varied environments. There are even flowering plants that live in water! Despite their great diversity in size and shape, flowering plants usually have three vegetative organs. An **organ** is a structure which contains different types of tissues and performs one or more specific functions. The vegetative organs of a flowering plant—the root, the stem, and the leaf (Fig. 36.1)—allow a plant to live and grow. The flower, which functions during reproduction, contains a number of organs.

Roots

Although we are accustomed to speaking of the root, it is more appropriate to refer to the root system. The **root system** of a plant, like a tomato, has a main, or tap, root and many branch, or lateral, roots (Fig. 36.2*a*). As a rule of thumb, the root system is at least equivalent in size and extent to the **shoot system** (the part of the plant above ground). An apple tree, then, has a much larger root system than, say, a corn plant. A single corn plant may have roots as deep as 2.5 meters and spread out over 1.5 meters, but a mesquite tree that lives in the desert may have roots that penetrate to a depth of 20 meters. The extensive root system of a plant anchors it in the soil and gives it support.

The root system absorbs water and minerals from the soil, for the entire plant. The cylindrical shape of a root allows it to penetrate the soil as it grows and permits water to be absorbed from all sides. The absorptive capacity of a root is also increased by its many root hairs located in a special zone near the root tip. Root hairs, which are projections from root-hair cells, are especially responsible for the absorption of water and minerals. Root hairs are so numerous that they increase the absorptive surface of a root tremendously. It's been estimated that a single rye plant has about 14 billion hair cells and, if placed end to end, the root hairs would stretch 10,626 kilometers. Root hair cells are constantly being replaced. So this same rye plant most likely forms about 100 million new root hair cells every day. You are probably familiar with the fact that a plant yanked out of the soil will not fare well when transplanted; this is because the root hairs are torn off. Transplantation is more apt to be successful if you take a part of the surrounding soil along with the plant.

Roots have still other functions. Roots produce hormones that stimulate the growth of stems and coordinate their size with the size of the root. It is more efficient for a plant to have a root and stem size that is appropriate one to the other. Also, **perennial** plants, which die back and then regrow the next season, store the products of photosynthesis in their roots. Carrots and sweet potatoes come from the roots of such plants, for example.

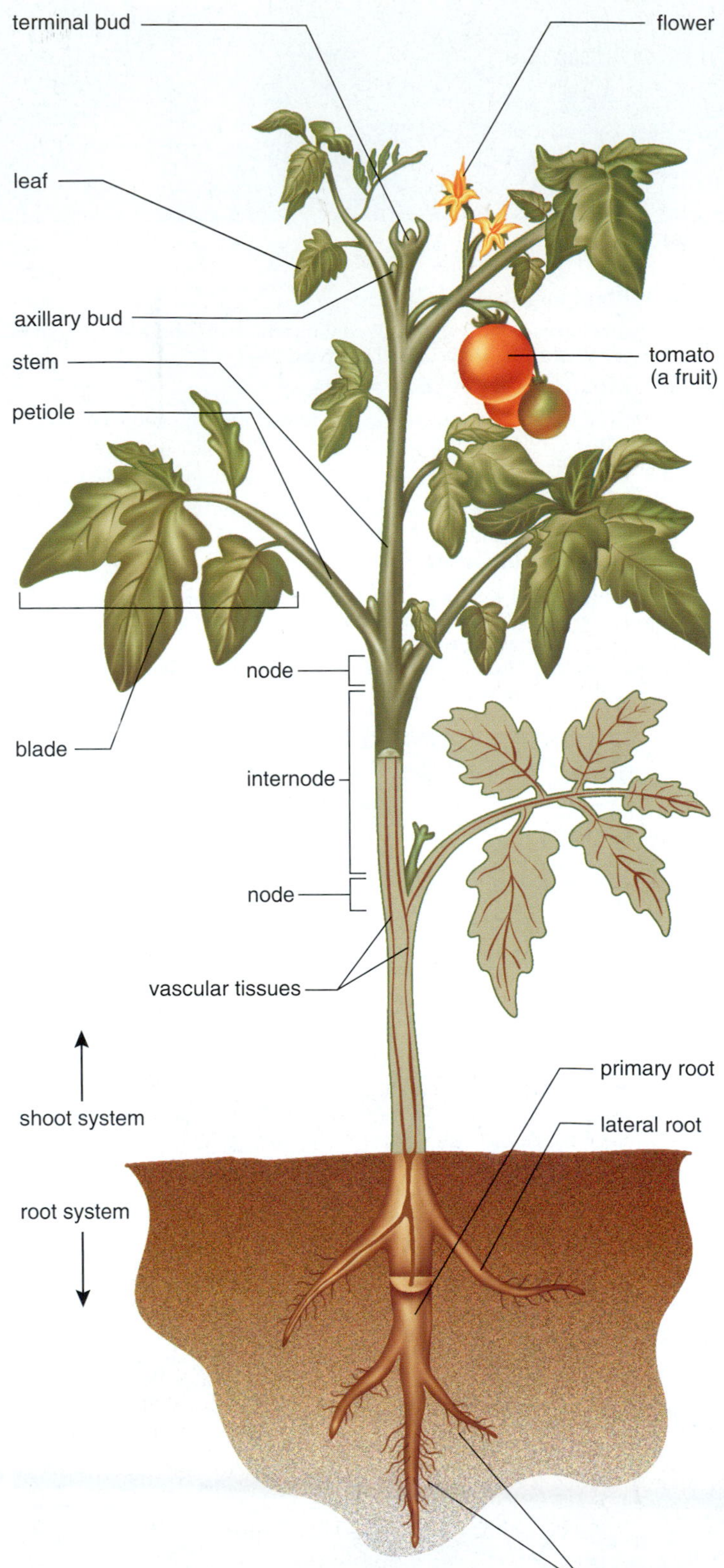

Figure 36.1 Organization of plant body.
The body of a plant consists of a root system and a shoot system. Roots are the only type of plant organ in the root system. The shoot system contains the stem and leaves, two other types of plant organs. Axillary buds can develop into branches of stems or flowers, the reproductive structures of a plant. The root system is connected to the shoot system by vascular tissue (brown) that extends from the roots to the leaves.

a. Root system

b. Shoot system

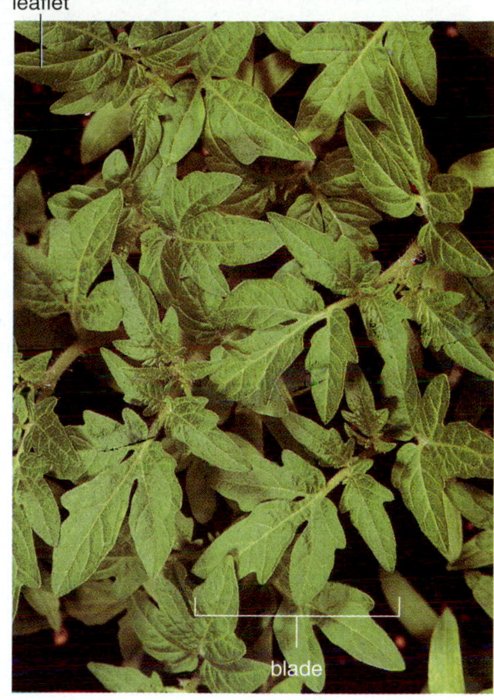

c. Leaves

Figure 36.2 Vegetative organs of the tomato, *Lycopersicon.*
a. The root system anchors the plant and absorbs water and minerals. **b.** The shoot system contains the stem and its branches which support the leaves and transport water and organic nutrients. **c.** The leaves, which are often broad and thin, carry on photosynthesis.

Stems

The shoot system of a plant contains both stems and leaves. A **stem** is the main axis of a plant along with its lateral branches (Fig. 36.2*b*). The stem of a flowering plant terminates in tissue that allows the stem to elongate and produce leaves. If upright, as most are, stems support leaves in such a way that each leaf is exposed to as much sunlight as possible. A **node** occurs where leaves are attached to the stem and an **internode** is the region between the nodes. The presence of nodes and internodes is used to identify a stem even if it happens to be an underground stem. In some plants the nodes of horizontal stems asexually produce new plants.

Aside from supporting the leaves, a stem has vascular tissue that transports water and minerals from the roots to the leaves and transports the products of photosynthesis, usually in the opposite direction. Nonliving cells form a continuous pipeline for water and mineral transport, while living cells join end to end for organic nutrient transport. A cylindrical stem can expand in girth as well as length. As trees grow taller each year they accumulate nonfunctional woody tissue that adds to the strength of their stems.

Some stems have functions other than transport. Some are specialized for storage. The stem stores water in cactuses and in other plants. Tubers are horizontal stems that store nutrients.

Leaves

A foliage **leaf** is that part of a plant that usually carries on photosynthesis, a process that requires water, carbon dioxide, and sunlight. Leaves receive water from the root system by way of the stem. In fact, as we shall see in the next chapter, stems and leaves function together to bring about water transport from the roots.

In contrast to the shape of stems, foliage leaves are usually broad and thin. This shape has the maximum surface area for the absorption of carbon dioxide and the collection of solar energy. Also unlike stems, leaves are almost never woody. All their cells are living, and the bulk of a leaf contains tissue specialized to carry on photosynthesis.

The wide portion of a foliage leaf is called the **blade.** A tomato plant has a compound blade with several leaflets (Fig. 36.2*c*). The **petiole** is a stalk that attaches the blade to the stem. The upper and acute angle between the petiole and stem is designated the leaf axil, and this is where an **axillary** (lateral) **bud,** which may become a branch or a flower, originates. Not all leaves are foliage leaves. Some are specialized to protect buds, attach to objects (tendrils), and store food (bulbs) or even capture insects.

A flowering plant has three vegetative organs: the root absorbs water and minerals, the stem supports and services leaves, and the leaf carries on photosynthesis.

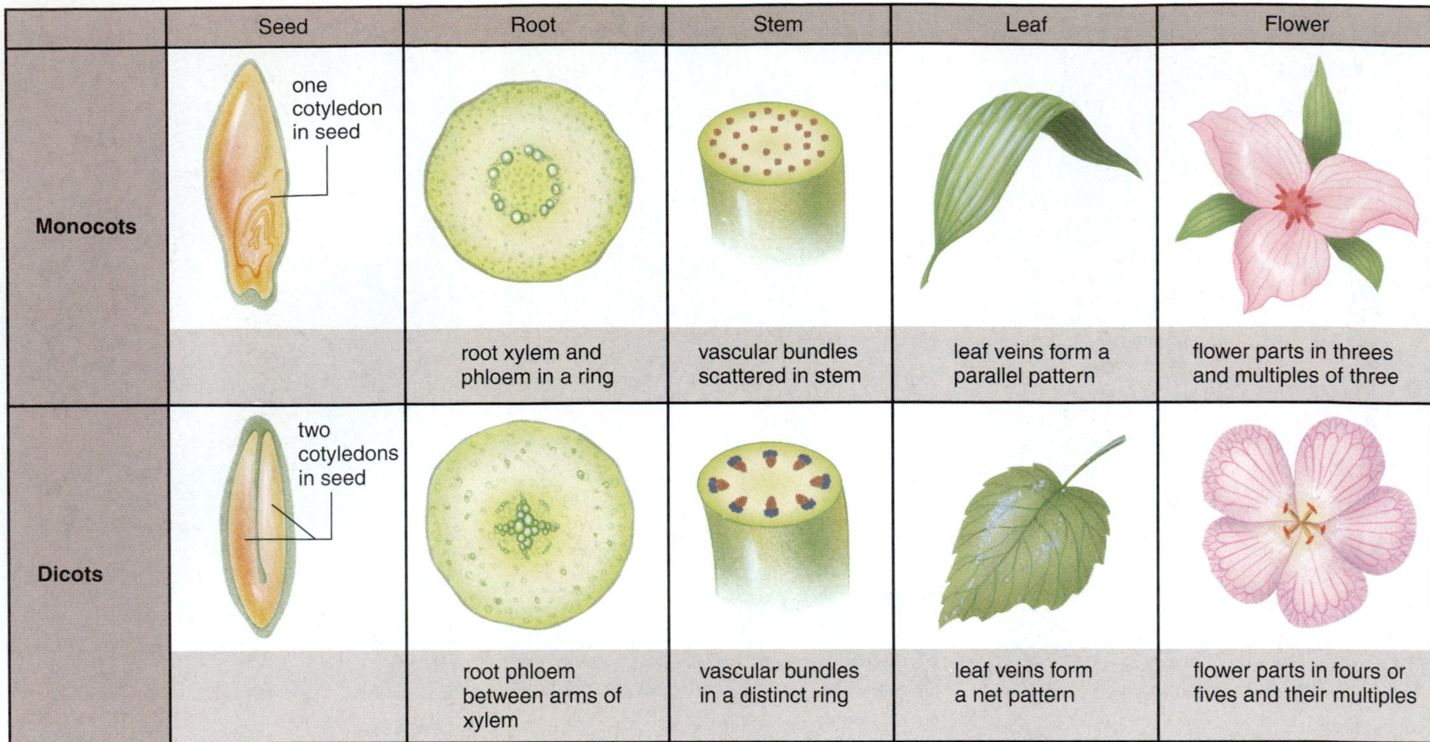

Seed
Root
Stem
Leaf
Flower
Monocots
one cotyledon in seed
root xylem and phloem in a ring
vascular bundles scattered in stem
leaf veins form a parallel pattern
flower parts in threes and multiples of three
Dicots
two cotyledons in seed
root phloem between arms of xylem
vascular bundles in a distinct ring
leaf veins form a net pattern
flower parts in fours or fives and their multiples

Figure 36.3 Flowering plants are either monocots or dicots.
Five features illustrated here are used to distinguish monocots from dicots: number of cotyledons; the arrangement of vascular tissue in roots, stems, and leaves; and the number of flower parts.

36.2 Monocot Versus Dicot Plants

Flowering plants are divided into two groups, depending on the number of **cotyledons** or seed leaves in the embryonic plant (Fig. 36.3). Cotyledons [Gk. *cotyledon*, cup-shaped cavity] provide nutrient molecules for seedlings before the true leaves begin photosynthesizing. Some plants have one cotyledon, and these plants are known as monocotyledons, or **monocots.** Other embryos have two cotyledons, and these plants are known as dicotyledons, or **dicots.**

The vascular (transport) tissue is organized differently in monocots and dicots. In the monocot root, vascular tissue occurs in a ring. In the dicot root, phloem (transports organic nutrients) is located between the arms of xylem (transports water and minerals) which has a star shape. In the monocot stem, the vascular bundles, which contain vascular tissue surrounded by a bundle sheath, are scattered. In a dicot stem, the vascular bundles occur in a ring. Figure 36.3 shows cross sections of stems and roots, but keep in mind that vascular tissue extends lengthwise from the roots to the leaves.

Leaf veins are vascular bundles within a leaf. Monocots exhibit parallel venation, and dicots exhibit netted venation, which may be either pinnate or palmate. Pinnate venation means that branch veins originate from points along the centrally placed main vein, and palmate venation means that the branch veins all originate at the point of attachment of the blade to the petiole:

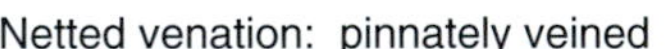

Adult monocots and dicots have other structural differences, such as differences in the usual number of flower parts and the number of apertures (thin areas in the wall) of pollen grains. Dicot pollen grains usually have three apertures, and monocot pollen grains usually have one aperture.

Although the division between monocots and dicots may seem of limited importance, it does in fact represent a separation that most likely dates back to the origin of flowering plants. The dicots are the larger group and include some of our most familiar flowering plants—from dandelions to oak trees. The monocots include grasses, lilies, orchids, and palm trees, among others. Some of our most significant food sources are monocots, including rice, wheat, and corn.

Flowering plants are divided into monocots and dicots on the basis of structural differences.

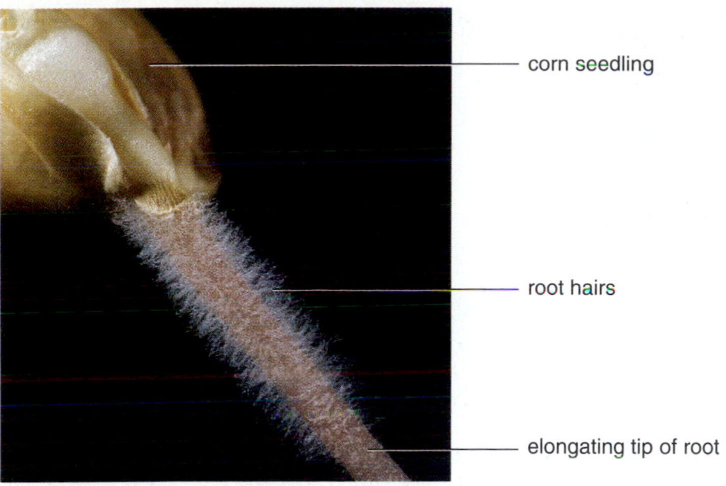

a. Root hairs

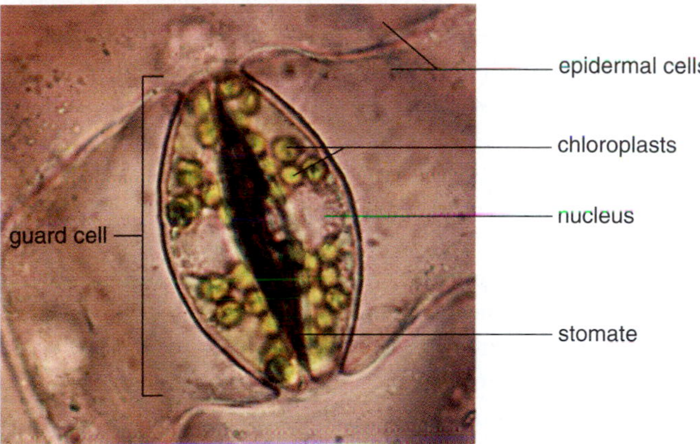

b. Stomate of leaf

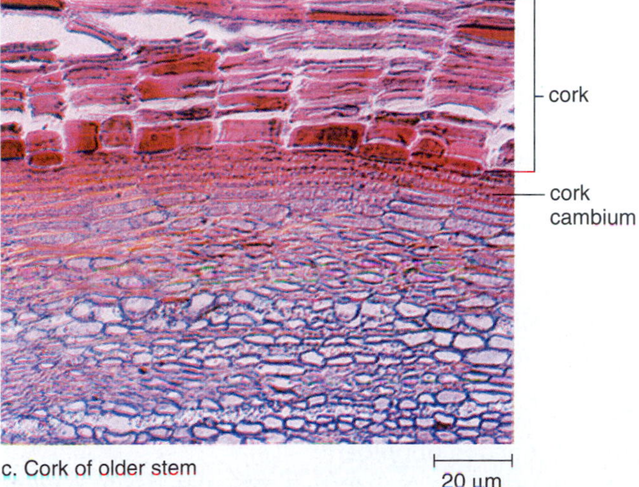

c. Cork of older stem

Figure 36.4 Modifications of epidermal tissue.
a. Root epidermis has root hairs to absorb water. **b.** Leaf epidermis contains stomates for gas exchange. **c.** Cork replaces epidermis in older, woody stems.

36.3 Plant Tissues

A plant grows its entire life because it has **meristem** (embryonic tissue) located in the stem and root tips (apexes). Three types of meristem continually produce the three types of specialized tissue in the body of a plant: protoderm, the outermost primary meristem, gives rise to epidermis; ground meristem produces ground tissue; and procambium produces vascular tissue.

We will be discussing these three specialized tissues:

1. **Epidermal tissue**—forms the outer protective covering of a plant.
2. **Ground tissue**—fills the interior of a plant.
3. **Vascular tissue**—transports water and nutrients in a plant and provides support.

Epidermal Tissue

The entire body of a nonwoody (herbaceous) and a young woody plant is covered by a layer of **epidermis** [Gk. *epi*, over, and *derma*, skin], which contains closely packed epidermal cells. The walls of epidermal cells that are exposed to air are covered with a waxy **cuticle** to minimize water loss. The cuticle also protects against bacteria and other organisms that might cause disease.

In roots, certain epidermal cells have long, slender projections called **root hairs** (Fig. 36.4*a*). As mentioned, the hairs increase the surface area of the root for absorption of water and minerals; they also help to anchor the plant firmly in place.

Protective hairs of a different nature are produced by epidermal cells of stems and leaves. Epidermal cells may also be modified as glands that secrete protective substances of various types.

In leaves, the lower epidermis in particular contains specialized cells, called guard cells (Fig. 36.4*b*). The guard cells, which unlike epidermal cells have chloroplasts, surround microscopic pores called **stomates** (also called stomata). When the stomates are open, gas exchange can occur. During the day, carbon dioxide diffuses in and oxygen diffuses out.

In older woody plants, the epidermis of the stem is replaced by cork tissue. **Cork,** the outer covering of the bark of trees, is made up of dead cork cells that may be sloughed off (Fig. 36.4*c*). New cork cells are made by a meristem called cork cambium. As the new cork cells mature, they increase slightly in volume and their walls become encrusted with suberin, a lipid material, so that they are waterproof and chemically inert. These nonliving cells protect the plant and make it resistant to attack by fungi, bacteria, and animals.

Epidermal tissue forms the outer protective covering of a herbaceous plant. It is modified in roots, stems, and leaves.

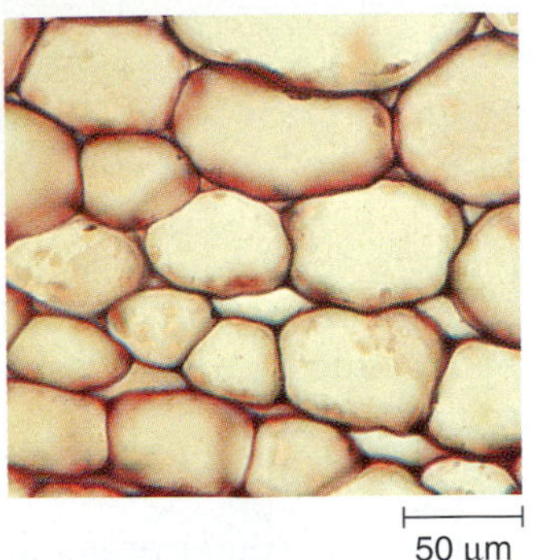

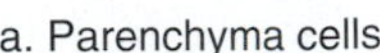

a. Parenchyma cells

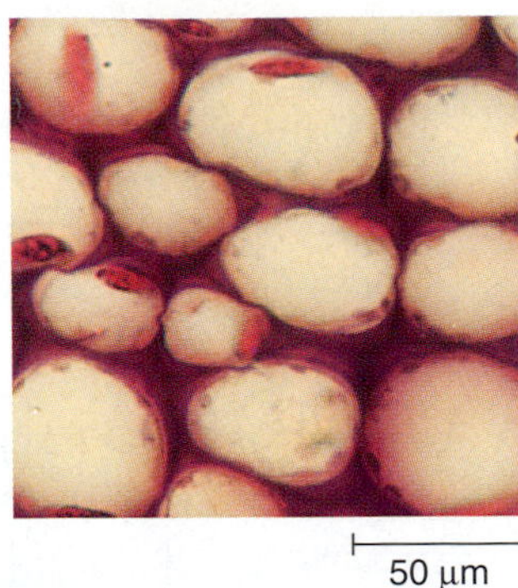

b. Collenchyma cells

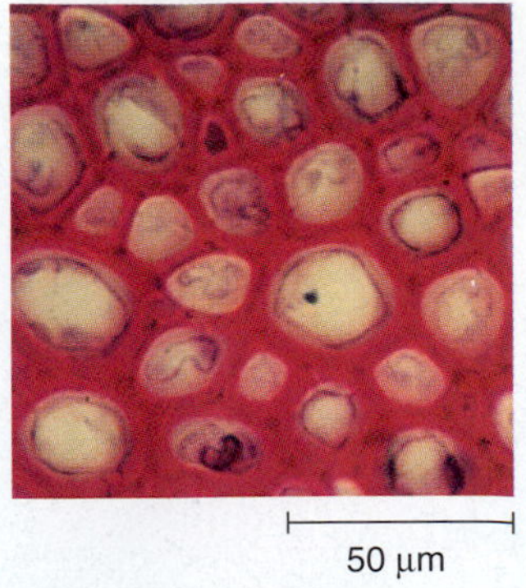

c. Sclerenchyma cells

Figure 36.5 **Ground tissue cells.**
a. Parenchyma cells are the least specialized of the plant cells. **b.** Collenchyma cells. Notice how much thicker the walls are compared to those of parenchyma cells. **c.** Sclerenchyma cells have very thick walls and are nonliving—their only function is to give strong support.

Ground Tissue

Ground tissue forms the bulk of a plant and contains parenchyma, collenchyma, and sclerenchyma cells (Fig. 36.5). **Parenchyma** [Gk. *para,* beside, and *enchyma,* infusion] cells correspond best to the typical plant cell. These are the least specialized of the cell types and are found in all the organs of a plant. They may contain chloroplasts and carry on photosynthesis, or they may contain colorless plastids that store the products of photosynthesis. Parenchyma cells can divide and give rise to more specialized cells, such as when roots develop from stem cuttings placed in water.

Collenchyma cells are like parenchyma cells except they have thicker primary walls. The thickness is uneven and usually involves the corners of the cell. Collenchyma cells often form bundles just beneath the epidermis and give flexible support to immature regions of a plant body. The familiar strands of celery stalks are composed mostly of collenchyma cells.

Sclerenchyma cells have thick secondary cell walls impregnated with lignin, which is a highly resistant organic substance that makes the walls tough and hard. Compared to re-enforced concrete, cellulose fibrils play the role of steel rods and lignin is analogous to cement. Most sclerenchyma cells are nonliving; their primary function is to support mature regions of a plant. Two types of sclerenchyma cells are fibers and sclereids. Although fibers are occasionally found in ground tissue, most are found in vascular tissue, which is discussed next. Fibers are long and slender and may be found in bundles that are sometimes commercially important. Hemp fibers can be used to make rope, and flax fibers can be woven into linen. Flax fibers, however, are not lignified, which is why linen is soft. Sclereids, which are shorter than fibers and more varied in shape, are found in seed coats and nutshells. They also give pears their characteristic gritty texture.

Vascular Tissue

There are two types of vascular (transport) tissue. **Xylem** transports water and minerals from the roots to the leaves, and **phloem** transports organic nutrients, usually from the leaves to the roots. Xylem contains two types of conducting cells: tracheids and vessel elements (Fig. 36.6). Both types of conducting cells are hollow and nonliving, but the vessel elements are larger, lack transverse end walls, and are arranged to form a continuous pipeline for water and mineral transport. The elongated tracheids, with tapered ends, form a less obvious means of transport, but water can move across the end walls and sidewalls because there are pits, or depressions, where the secondary wall does not form. In addition to vessel elements and tracheids, xylem contains parenchyma cells that store various substances. Vascular rays, which are flat ribbons or sheets of parenchyma cells located between rows of tracheids, conduct water and minerals laterally. Xylem also contains fibers, sclerenchyma cells that lend support.

The conducting cells of phloem are sieve-tube elements, each of which has a companion cell (Fig. 36.7). Sieve-tube elements contain cytoplasm but no nuclei. These elements have channels in their end walls that in cross section make them resemble a sieve. Plasmodesmata (sing., plasmodesma), which are strands of cytoplasm, extend from one cell to another through this so-called sieve plate. The smaller companion cells do have a nucleus in addition to cytoplasm. Companion cells are closely connected to sieve-tube elements by numerous plasmodesmata, and the nucleus of the companion cell may control and maintain the life of both cells. The companion cells are also believed to be involved in the transport function of phloem.

It is important to realize that vascular tissue (xylem and phloem) extends from the root to the leaves and vice versa (see Fig. 36.1). In the roots, the vascular tissue is located in the **vascular cylinder;** in the stem, it forms **vascular bundles;** and in the leaves, it is found in **leaf veins.**

The vascular tissues are xylem and phloem. Xylem transports water and minerals, and phloem transports organic nutrients.

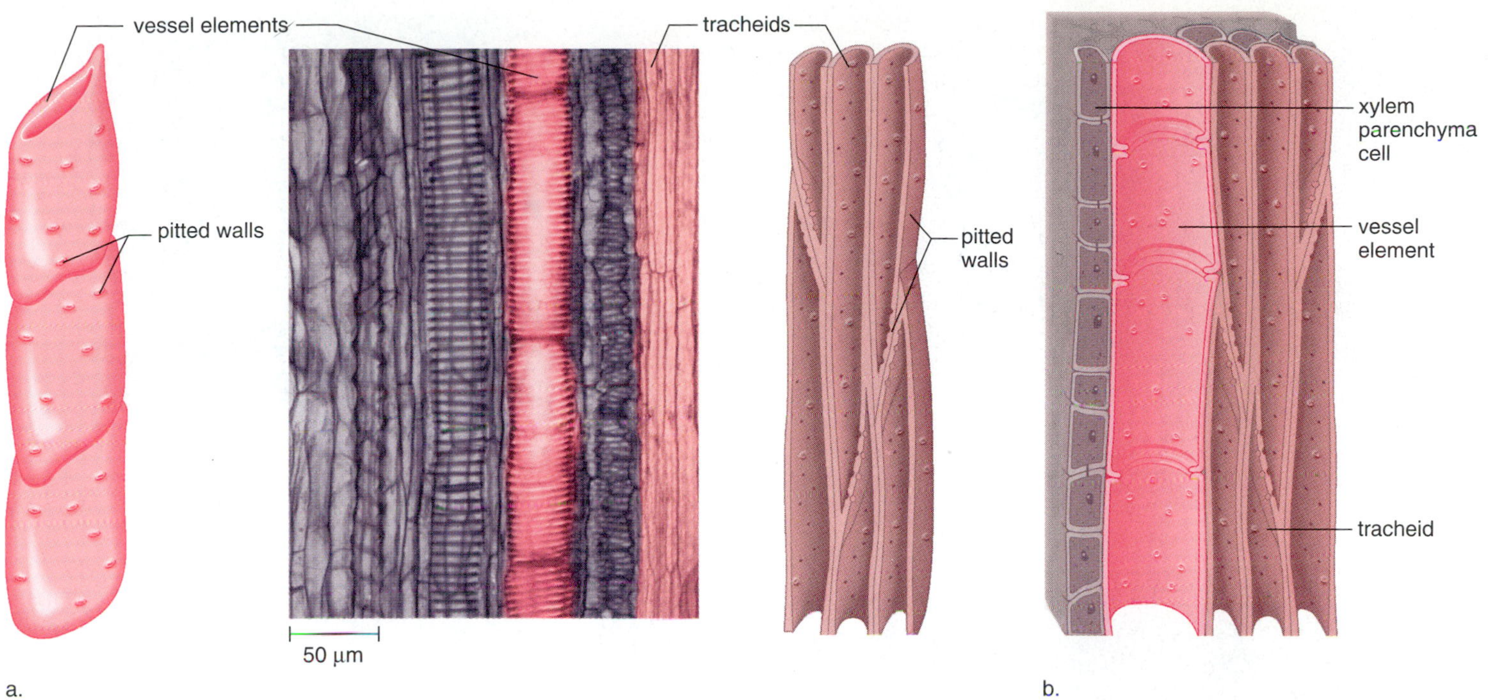

Figure 36.6 Xylem structure.
a. Photomicrograph of xylem vascular tissue with drawing of a vessel (composed of vessel elements) to left and tracheids to right. **b.** Drawing showing general organization of xylem tissue.

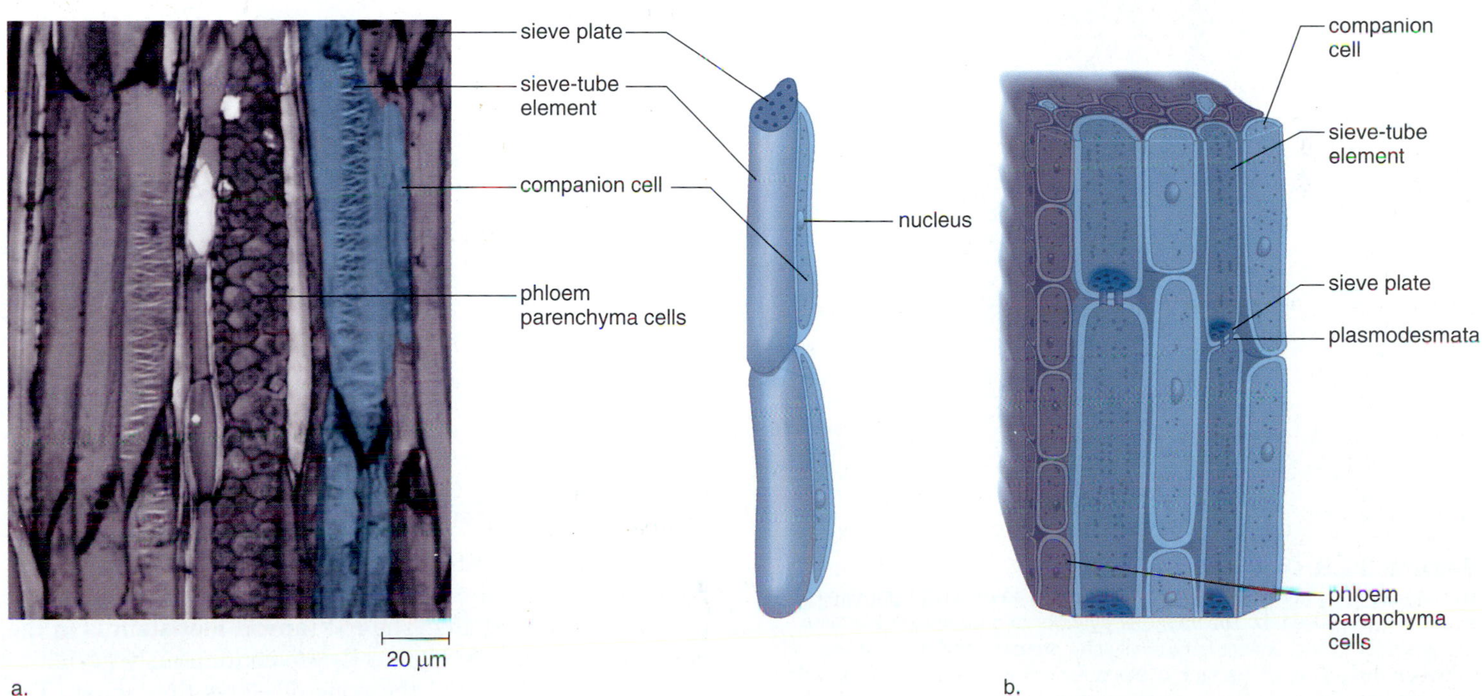

Figure 36.7 Phloem structure.
a. Photomicrograph of phloem vascular tissue with drawing of sieve tube (composed of sieve-tube elements) and companion cells to right.
b. Drawing showing general organization of phloem tissue.

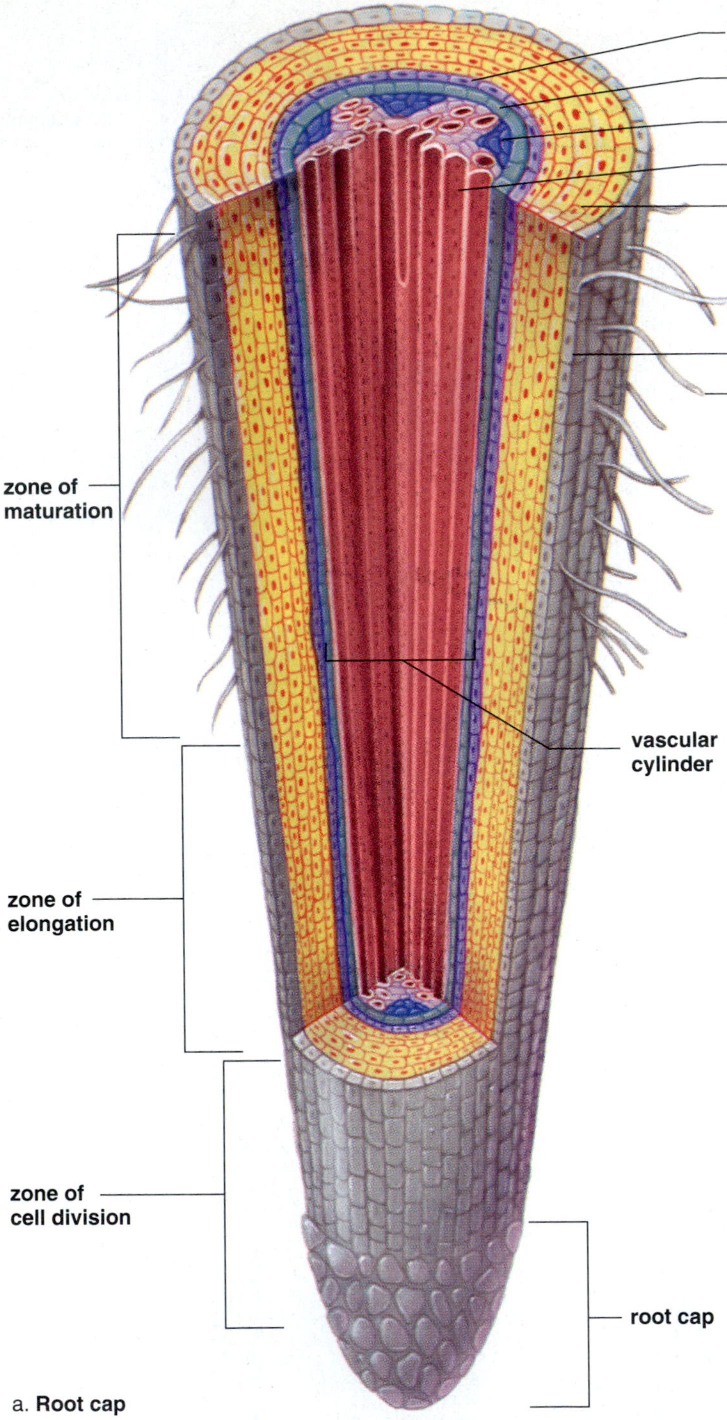

endodermis
pericycle
phloem
xylem
cortex
epidermis
root hair
zone of maturation
vascular cylinder
zone of elongation
zone of cell division
root cap

a. **Root cap**

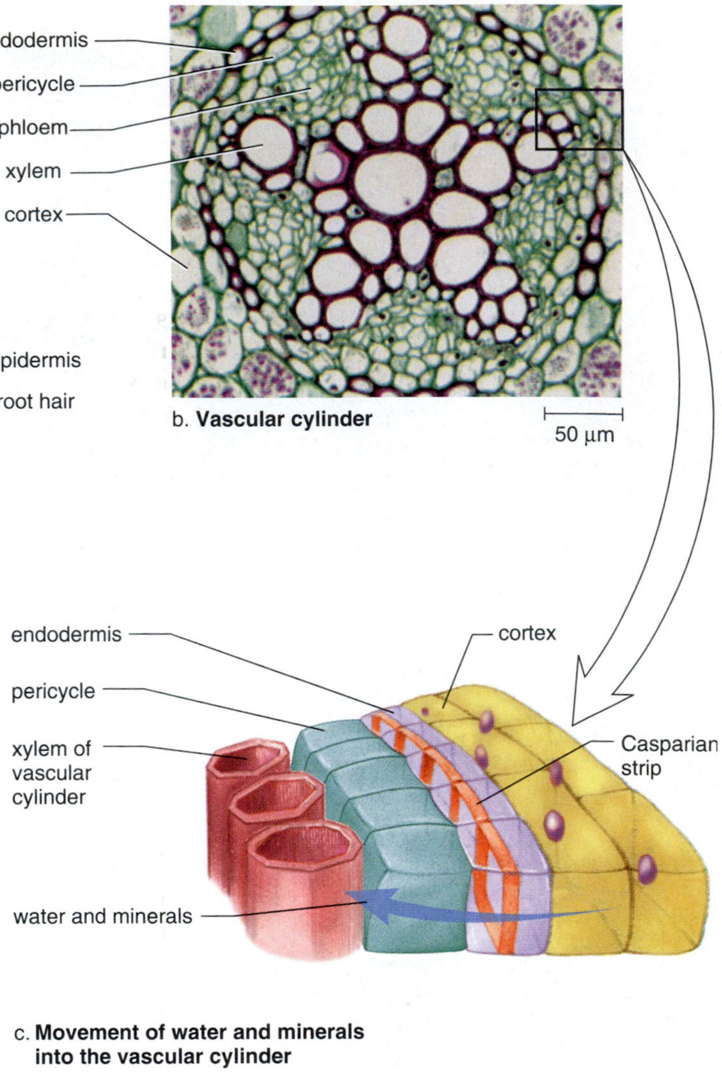

b. **Vascular cylinder**

50 μm

endodermis
cortex
pericycle
xylem of vascular cylinder
Casparian strip
water and minerals

c. **Movement of water and minerals into the vascular cylinder**

Figure 36.8 Dicot root tip.
a. The root tip is divided into four zones, best seen in a longitudinal section such as this. **b.** The vascular cylinder of a dicot root contains the vascular tissue. Xylem is typically star shaped, and phloem lies between the points of the star. **c.** Because of the Casparian strip, water and minerals must pass through the cytoplasm of endodermal cells in order to enter the vascular cylinder. In this way, endodermal cells regulate the passage of minerals into the vascular cylinder.

36.4 Organization of Roots

Figure 36.8*a*, a longitudinal section of a dicot root, reveals zones where cells are in various stages of differentiation as primary growth occurs. The root **apical meristem** is in the zone of cell division. Here cells are continuously added to the root cap below and the zone of elongation above. The root cap is a protective cover for the root tip. The cells in the root cap have to be replaced constantly, because they are ground off as the root pushes through rough soil particles. In the zone of elongation, the cells become longer as they become specialized. In the zone of maturation, the cells are mature and fully differentiated. This zone is recognizable even in a whole root because root hairs are borne by many of the epidermal cells.

Tissues of a Dicot Root

Figure 36.8*a* also shows a cross section of a root at the region of maturation. These specialized tissues are identifiable:

Epidermis The epidermis, which forms the outer layer of the root, consists of only a single layer of cells. The majority of epidermal cells are thin walled and rectangular, but in the zone of maturation, many epidermal cells have root hairs. These project as far as 5–8 mm into the soil particles.

Cortex Moving inward, next to the epidermis, large, thin-walled parenchyma cells make up the **cortex** of the root. These irregularly shaped cells are loosely packed, and it is possible for water and minerals to move through the cortex without entering the cells. The cells contain starch granules, and the cortex functions in food storage.

Endodermis The **endodermis** [Gk. *endon*, within, and *derma*, skin] is a single layer of rectangular endodermal cells that forms a boundary between the cortex and the inner vascular cylinder. The endodermal cells fit snugly together and are bordered on four sides (but not the two sides that contact the cortex and the vascular cylinder) by a layer of impermeable lignin and suberin known as the **Casparian strip** (Fig. 36.8*c*). This strip does not permit water and mineral ions to pass between adjacent cell walls. Therefore, the only access to the vascular cylinder is through the endodermal cells themselves, as shown by the arrow in Figure 36.8*c*. It is said that the endodermis regulates the entrance of minerals into the vascular cylinder.

Vascular Tissue The **pericycle,** the first layer of cells within the vascular cylinder, has retained its capacity to divide and can start the development of branch, or lateral, roots (Fig. 36.9). The main portion of the vascular cylinder, though, contains vascular tissue. The xylem appears star shaped in dicots because several arms of tissue radiate from a common center (Fig. 36.8*b*). The phloem is found in separate regions between the arms of the xylem.

Organization of Monocot Roots

Monocot roots have the same growth zones as a dicot root, but they do not undergo secondary growth as many dicot roots do. Also, the organization of their tissues is slightly different. In a monocot root's **pith,** which is centrally located, ground tissue is surrounded by a vascular ring composed of alternating xylem and phloem bundles (Fig. 36.10). They also have pericycle, endodermis, cortex, and epidermis.

The root system of a plant absorbs water and minerals. They cross the epidermis and cortex before entering the endodermis, the tissue that regulates their entrance into the vascular cylinder.

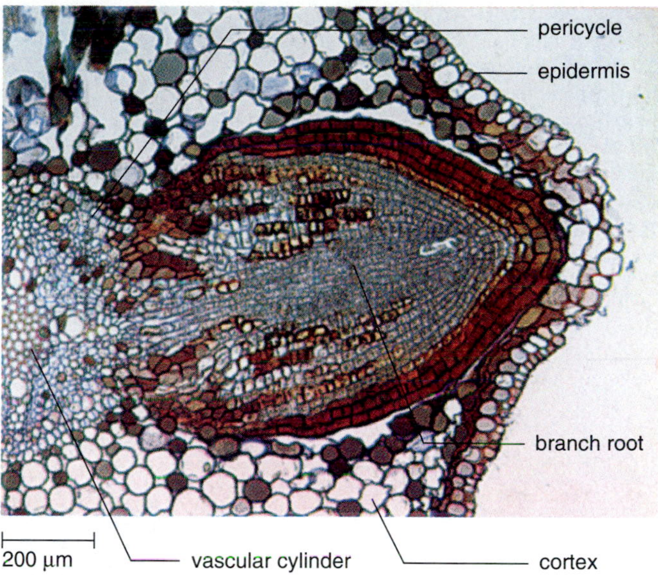

Figure 36.9 Branching of dicot root.
This cross section of a willow, *Salix*, shows the origination of a branch root from the pericycle.

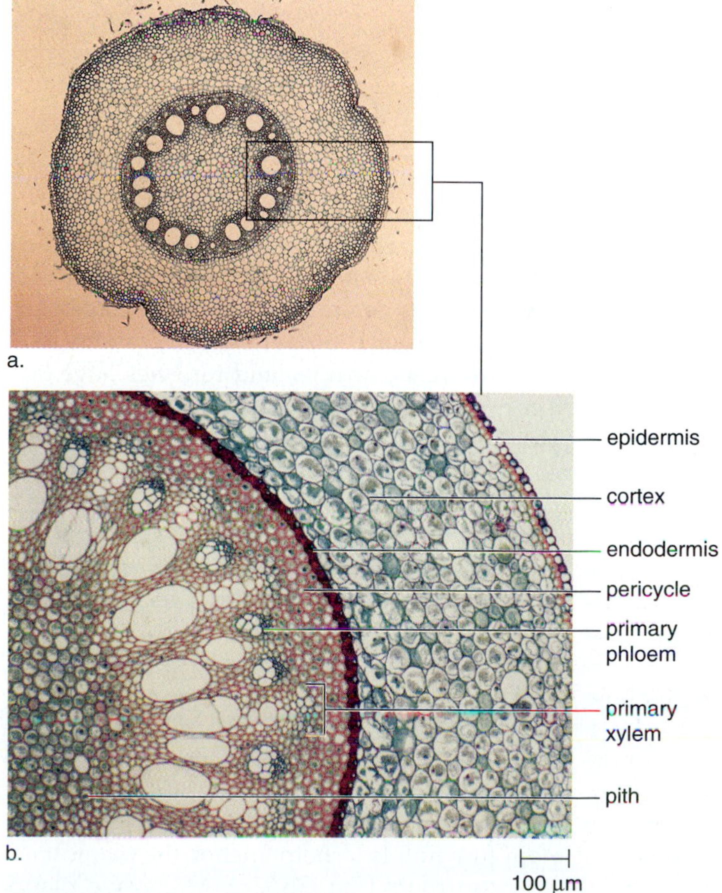

Figure 36.10 Monocot root.
a. In this overall cross section, it is possible to observe that a vascular ring surrounds a central pith. **b.** The enlargement shows the exact placement of various tissues.

Figure 36.11 Root diversity.
a. A taproot may have branch roots in addition to a main root. **b.** A fibrous root has many slender roots with no main root. **c.** Prop roots are specialized for support. **d.** *(right)* Dodder is a parasitic plant consisting mainly of orange-brown twining stems. (The green in the photograph is the host plant.) *(left)* Haustoria are rootlike projections of a stem that tap into the host's vascular system.

Root Diversity

Roots have various adaptations and associations to better perform their functions: anchorage, absorption of water and minerals, and storage of carbohydrates.

In some plants, notably dicots, the first or **primary root** grows straight down and remains the dominant root of the plant. This so-called **taproot** is often fleshy and stores food (Fig. 36.11*a*). Carrots, beets, turnips, and radishes have taproots that we consume as vegetables. Sweet potato plants don't have taproots but they do have roots with special storage areas for starch. We call these storage areas sweet potatoes.

In other plants, notably monocots, there is no single, main root; instead there are a large number of slender roots. These grow from the lower nodes of the stem when the first or primary root dies. These slender roots and their lateral branches make up a **fibrous root system** (Fig. 36.11*b*). Everyone has observed the fibrous root systems of grasses and has noted how these roots can hold the soil.

Since the fibrous roots of monocots develop from organs of the shoot system instead of the root system, they are known as **adventitious roots.** Some of these roots may emerge above the soil line, as they do in corn plants, in which their main function is to help anchor the plant. If so, they are called prop roots (Fig. 36.11*c*). Mangrove plants have large prop roots that spread away from the plant to help anchor it in marshy soil, where mangroves are typically found. Black mangroves grow in the water, and their roots have pneumatophores, which are root projections that rise above the surface of the water. In this way, the plants acquire oxygen from the air for cellular respiration. Other examples of adventitious roots are those found on underground stems (rhizomes), or the "holdfast" roots found along the internodes on the aerial shoots of ivy plants.

Some plants such as dodders and broomrapes are parasitic on other plants. Their stems have rootlike projections called haustoria (sing., haustorium) that grow into the host plant and make contact with vascular tissue from which they extract water and nutrients (Fig. 36.11*d*).

Mycorrhizae are fungus roots—an association between roots and fungi—which can extract water and minerals from the soil better than roots that lack a fungus partner. This is called a mutualistic relationship because the fungus receives sugars and amino acids from the plant, which receives water and minerals via the fungus.

Peas, beans, and other legumes have **root nodules** where nitrogen-fixing bacteria live. Plants cannot extract nitrogen from the air, but the bacteria within the nodules can take up and reduce atmospheric nitrogen. This means that the plant is no longer dependent upon a supply of nitrogen (i.e., nitrate or ammonium) in the soil, and indeed these plants are often planted just to bolster the nitrogen supply of the soil.

Roots have various adaptations and associations to enhance their ability to anchor a plant, absorb water and minerals, and store the products of photosynthesis.

Science Focus

Tree Rings Tell a Story

Each year a tree adds a new annual (growth) ring to its trunk. In the spring, growth is fast and the wood is light in color; in the summer, growth is slower and the wood is dark in color. Counting the dark rings tells you the age of the tree, but studying annual rings to determine their shape, thickness, color, and evenness lets you know what a tree has been through (Fig. 36A). Rings can reveal the years in which there were forest fires, smog damage, and leaf damage from caterpillars.

A great deal can also be learned from annual rings about past climatic conditions. For many years, A. E. Douglass, Harold C. Fritts, and others associated with the Laboratory of Tree-Ring Research at the University of Arizona studied ring widths in trees from arid sites. They discovered that very significant statistical relationships exist between growth of trees and climatic data. With the aid of computers and statistical analyses, they developed techniques that take into account subtle climatic and other environmental variables. They were able to reconstruct relatively precise histories of climatic fluctuations and changes dating back hundreds of years. Today, annual ring data are analyzed to try to determine climates dating back to prehistoric times.

Trees do not have to be cut down to examine the rings. A simple instrument called an *increment borer,* which consists primarily of a rigid metal cylinder, is driven into the trunk of a tree, and a core of wood is removed. The hole is then plugged to prevent disease-causing organisms from entering the tree, and the rings revealed by the core are then examined and analyzed.

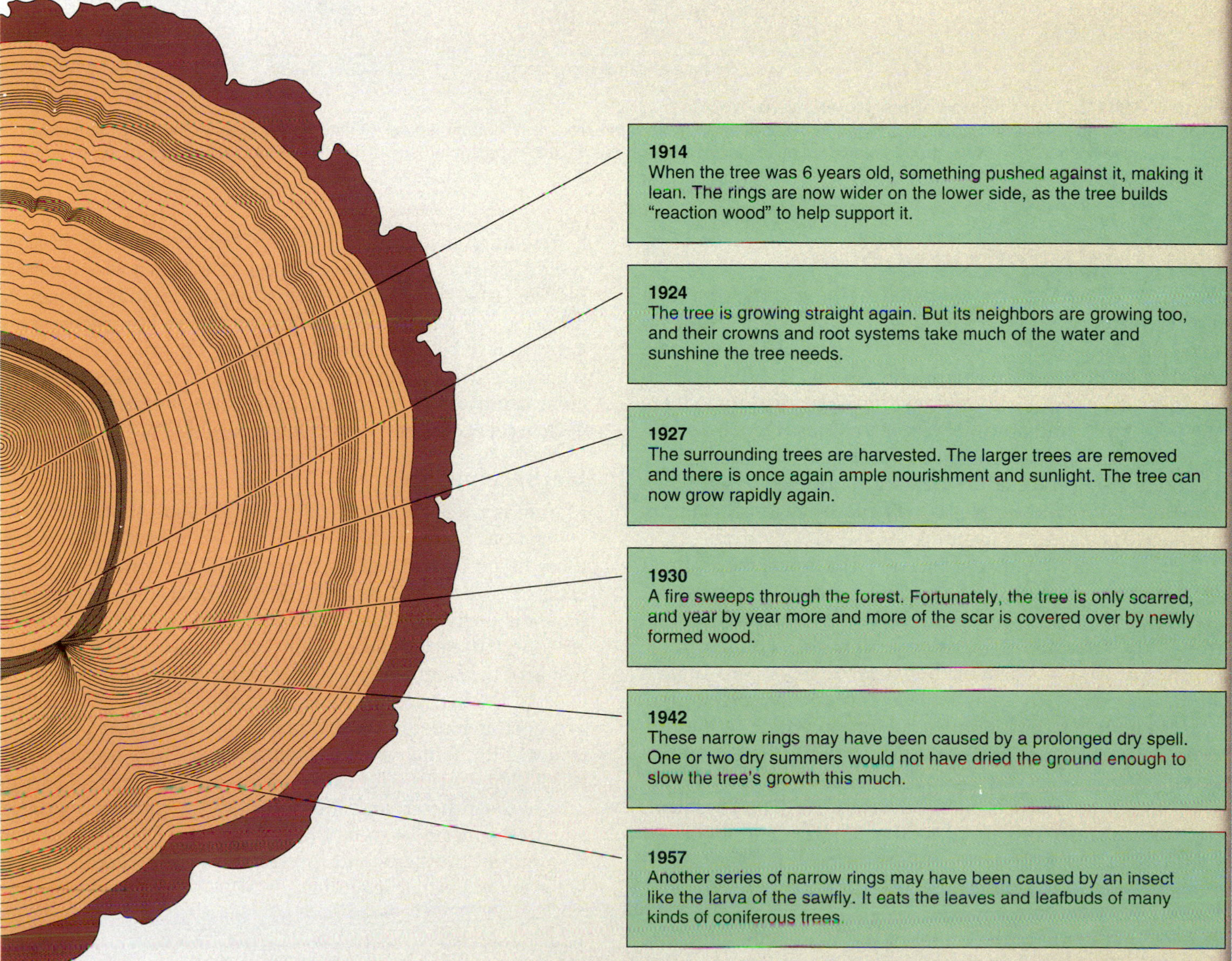

1914
When the tree was 6 years old, something pushed against it, making it lean. The rings are now wider on the lower side, as the tree builds "reaction wood" to help support it.
1924
The tree is growing straight again. But its neighbors are growing too, and their crowns and root systems take much of the water and sunshine the tree needs.
1927
The surrounding trees are harvested. The larger trees are removed and there is once again ample nourishment and sunlight. The tree can now grow rapidly again.
1930
A fire sweeps through the forest. Fortunately, the tree is only scarred, and year by year more and more of the scar is covered over by newly formed wood.
1942
These narrow rings may have been caused by a prolonged dry spell. One or two dry summers would not have dried the ground enough to slow the tree's growth this much.
1957
Another series of narrow rings may have been caused by an insect like the larva of the sawfly. It eats the leaves and leafbuds of many kinds of coniferous trees.

Figure 36A Life history of a tree.
This tree was planted in 1908 and cut down in 1970. The anatomy of the annual rings correlates to the events listed by date.

Source: Data from St. Regis Paper Company, New York, NY, 1966.

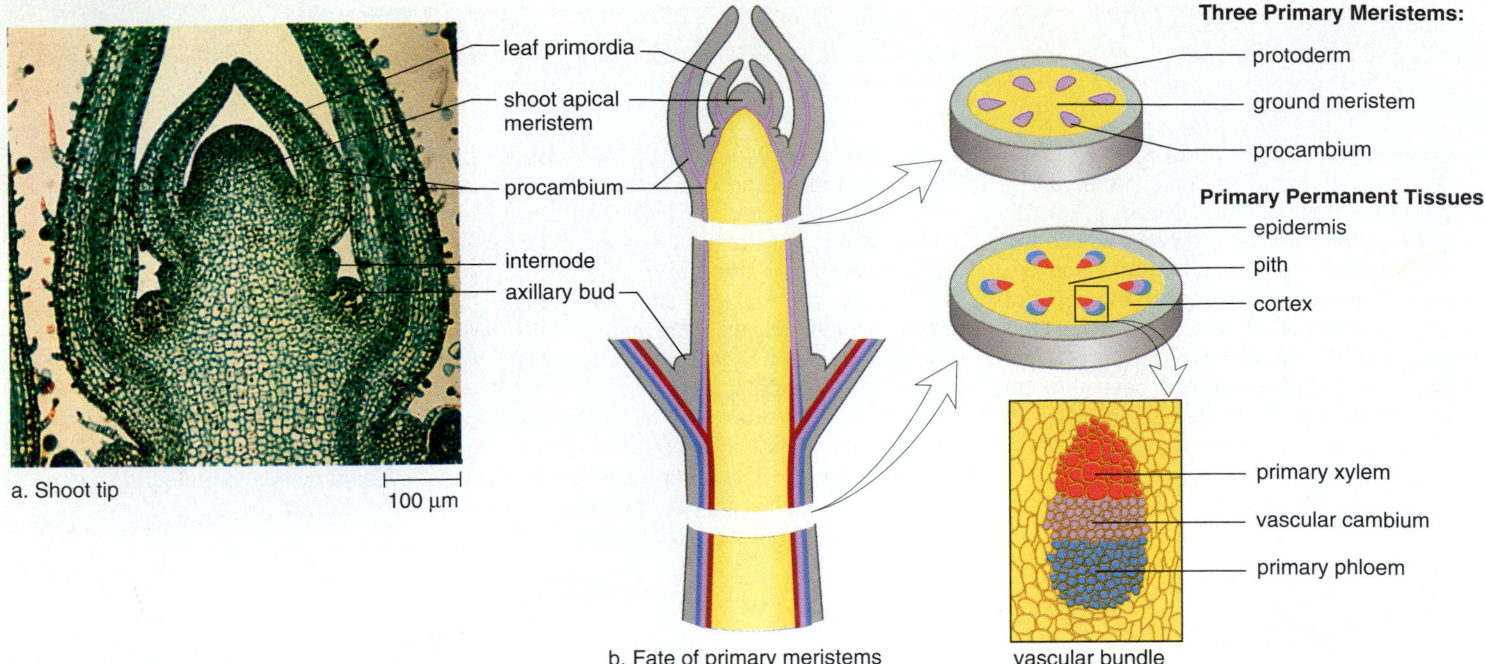

Figure 36.12 **Shoot tip and primary meristems.**
a. The shoot apical meristem within a terminal bud is surrounded by leaf primordia. **b.** The shoot apical meristem produces the primary meristems: protoderm gives rise to epidermis, ground meristem gives rise to pith and cortex, and procambium gives rise to vascular tissue including primary xylem, primary phloem, and vascular cambium.

36.5 Organization of Stems 

Growth of a stem can be compared to the growth of a root. During primary growth, apical meristem at the shoot tip called **shoot apical meristem** produces new cells that elongate and thereby increase the length of the stem. The shoot apical meristem, however, is protected within a **terminal bud** where leaf primordia (immature leaves) envelop it (Fig. 36.12*a*). In the temperate zone, a terminal bud stops growing in the winter and is then protected by bud scales. In the spring, when growth resumes, these scales fall off and leave a scar. You can tell the age of a stem by counting these bud scale scars.

Leaf primordia are produced by the apical meristem at regular intervals called nodes. The portion of a stem between two sequential nodes is called an internode. As a stem grows, the internodes increase in length. Axillary buds, which are usually dormant but may develop into branch shoots or flowers, are seen at the axes of the leaf primordia.

A shoot is a developing structure in which tissues are continually becoming differentiated. In addition to leaf primordia, three specialized types of primary meristem develop from shoot apical meristem (Fig. 36.12*b*). These primary meristems contribute to the length of a shoot. As mentioned, the protoderm, the outermost primary meristem, gives rise to epidermis. The ground meristem produces two tissues composed of parenchyma cells. The parenchyma tissue in the center of the stem is the pith, and the parenchyma tissue inside the epidermis is the cortex.

The procambium, seen as an obvious strand of tissue in Figure 36.12*a*, produces the first xylem cells, called primary xylem, and the first phloem cells, called primary phloem. Differentiation continues as certain cells become the first tracheids or vessel elements of the xylem within a vascular bundle. The first sieve-tube elements of the original phloem do not have companion cells and are short lived (some live only a day before being replaced). Mature phloem develops later, when all surrounding cells have stopped expanding and **vascular cambium** [L. *vasculum*, dim. of *vas*, vessel, and *cambio*, exchange] occurs between xylem and phloem.

Herbaceous Stems

Mature nonwoody stems, called **herbaceous stems** [L. *herba*, vegetation, plant], exhibit only primary growth. The outermost tissue of herbaceous stems is the epidermis, which is covered by a waxy cuticle to prevent water loss. These stems have distinctive vascular bundles, where xylem and phloem are found. In each bundle, xylem is typically found toward the inside of the stem and phloem is found toward the outside.

In the herbaceous dicot stem, the vascular bundles are arranged in a distinct ring that separates the cortex from the central pith, which stores water and products of photosynthesis (Fig. 36.13). The cortex is sometimes green and carries on photosynthesis, and the pith may function as a storage site. In the monocot stem, the vascular bundles are scattered throughout the stem, and there is no well-defined cortex or well-defined pith (Fig. 36.14).

As a stem grows, the shoot apical meristem produces new leaves and primary meristems. The primary meristems produce the other tissues found in herbaceous stems.

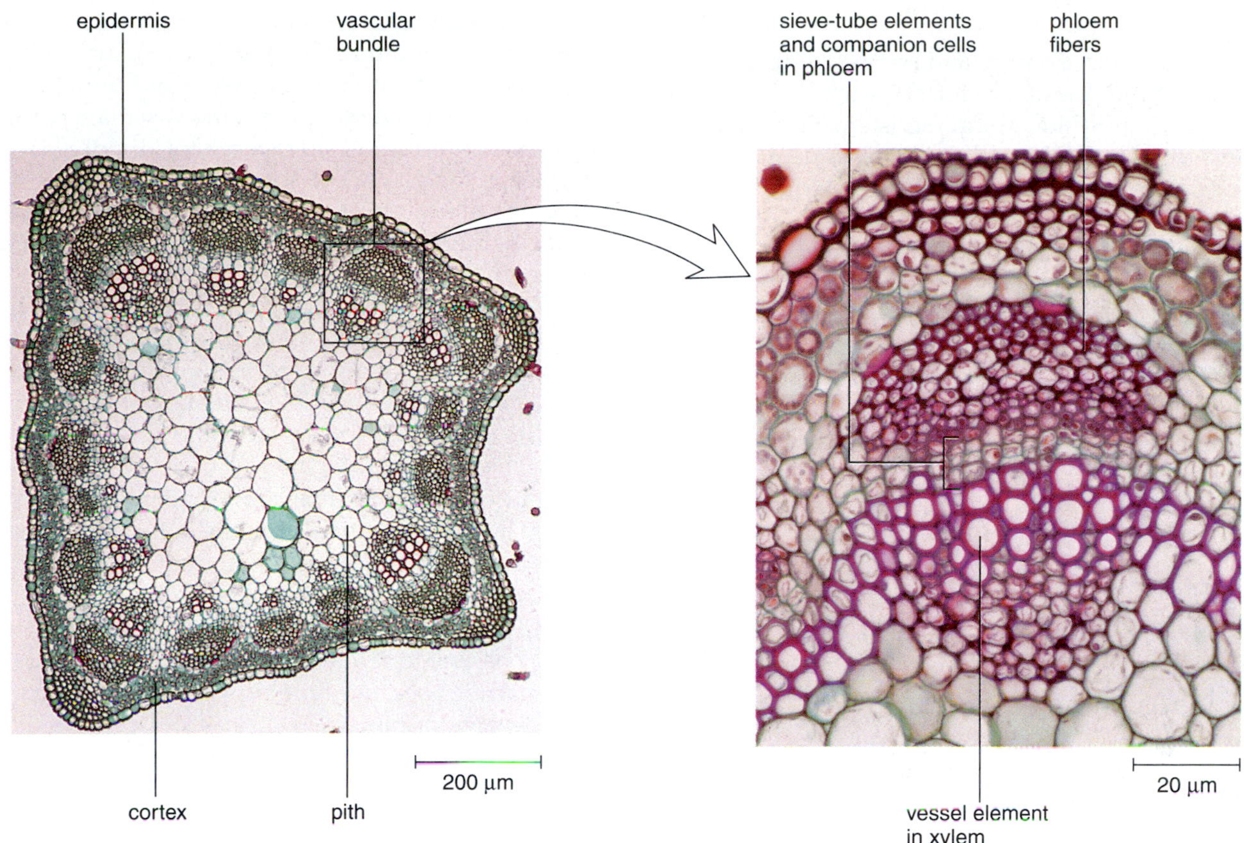

Figure 36.13 **Herbaceous dicot stem.**

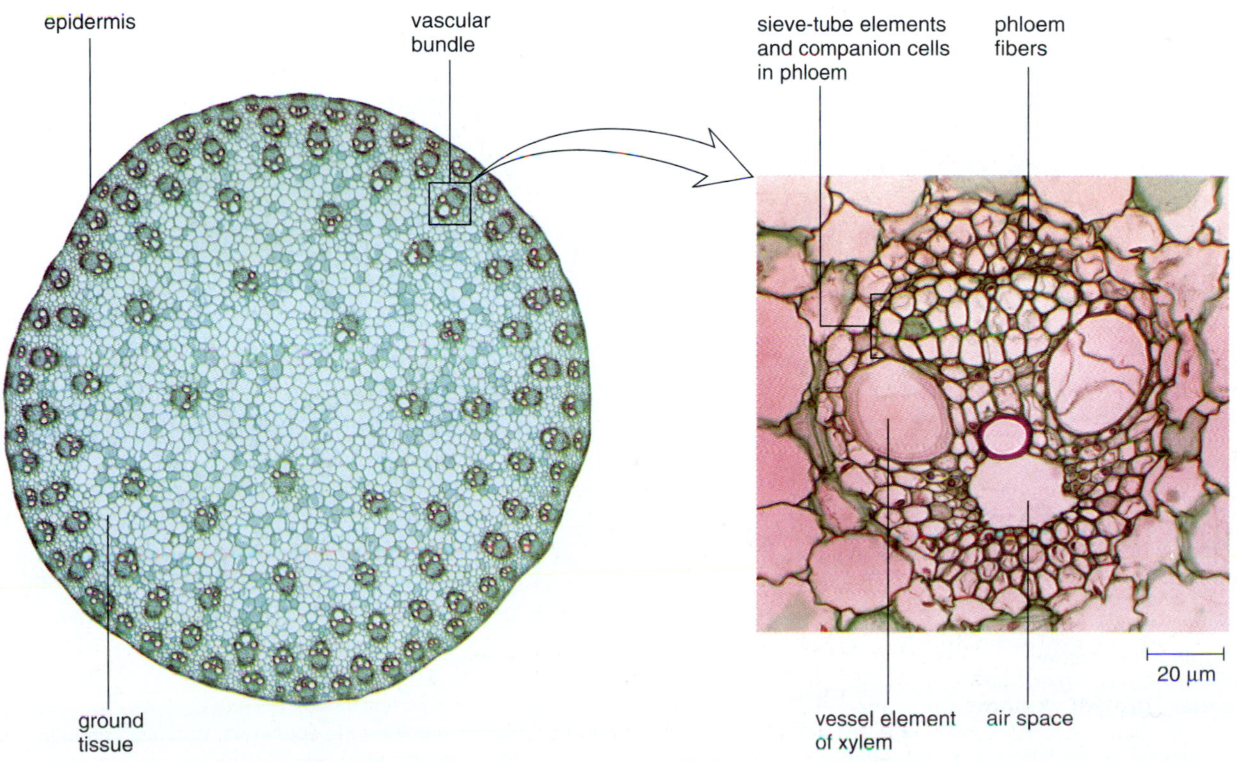

Figure 36.14 **Monocot stem.**

Woody Stems

A woody plant has both primary and secondary tissues. Primary tissues are those new tissues formed each year from primary meristems right behind apical meristem. Secondary tissues develop during the second and subsequent years of growth from lateral meristems: vascular cambium and cork cambium. Primary growth, which occurs in all plants, increases the length of a plant, and secondary growth, which occurs only in conifers and some dicots, increases its girth.

Trees undergo secondary growth because of a change in vascular cambium. Recall that vascular cambium begins as procambium between the xylem and phloem of each vascular bundle. In woody plants, the vascular cambium develops to form a ring of meristem that divides parallel to the surface of the plant. The secondary tissues produced by the vascular cambium, called secondary xylem and secondary phloem, therefore add to the girth of the stem instead of to its length (Fig. 36.15).

As a result of secondary growth, a woody dicot stem has an entirely different type of organization than that of a herbaceous dicot stem. After secondary growth has continued for a time, it is no longer possible to make out individual vascular bundles. Instead, a woody stem has three distinct areas: the pith, the wood, and the bark (Fig. 36.15*d*). Rays, such as pith rays, are of living cells that allow materials to move laterally.

The bark of a tree contains cork, cork cambium, and phloem [Gk. *phloios*, bark]. Although secondary phloem is produced each year by vascular cambium, phloem does not build up for many seasons. The phloem tissue is soft, making it possible to remove the bark of a tree; however, this is very harmful because without phloem there is no transport of organic nutrients.

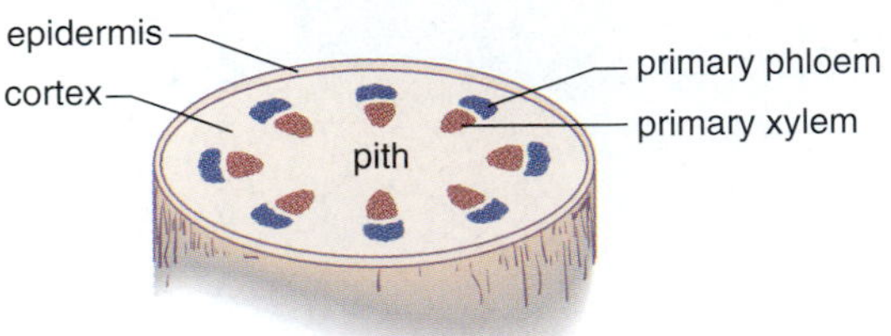

a. Dicot stem, no secondary growth

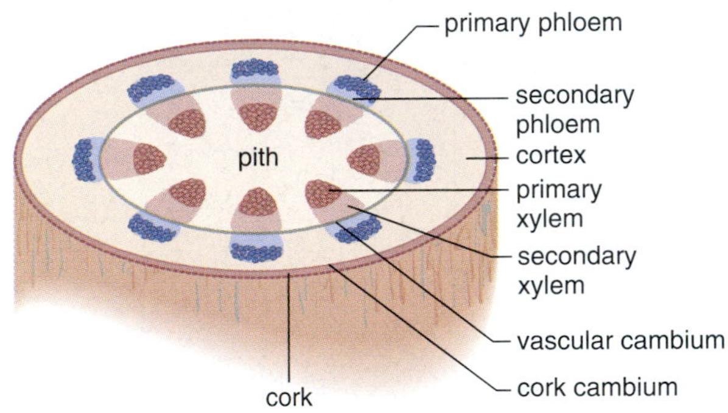

b. Dicot stem, some secondary growth

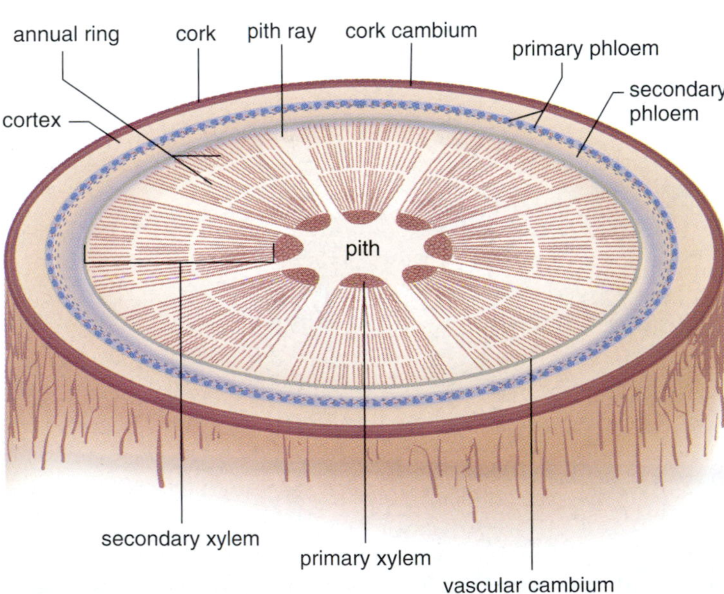

c. Dicot stem, secondary growth well underway

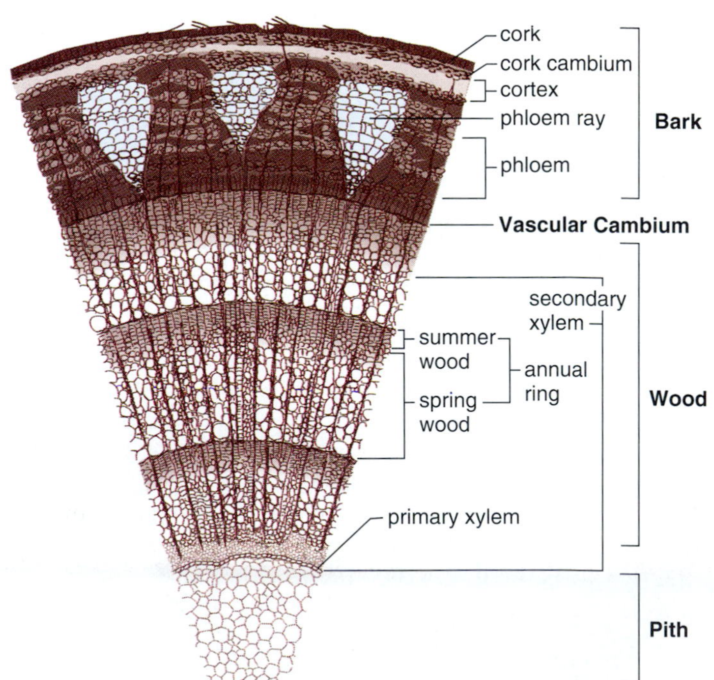

d. Section of woody stem

Figure 36.15 Secondary growth of stems.
a. A dicot herbaceous stem before secondary growth begins. **b.** Secondary growth has begun. Cork has replaced the epidermis. Vascular cambium produces secondary xylem and secondary phloem. **c.** A three-year-old stem in which cork cambium produces new cork. The primary phloem and cortex will eventually disappear, and only the secondary phloem (within the bark), produced by vascular cambium, will be active that year. The secondary xylem, also produced by vascular cambium, builds up to become the annual growth rings. **d.** Section of woody stem shows tissues at higher magnification.

a. Tree trunk, cross sectional view

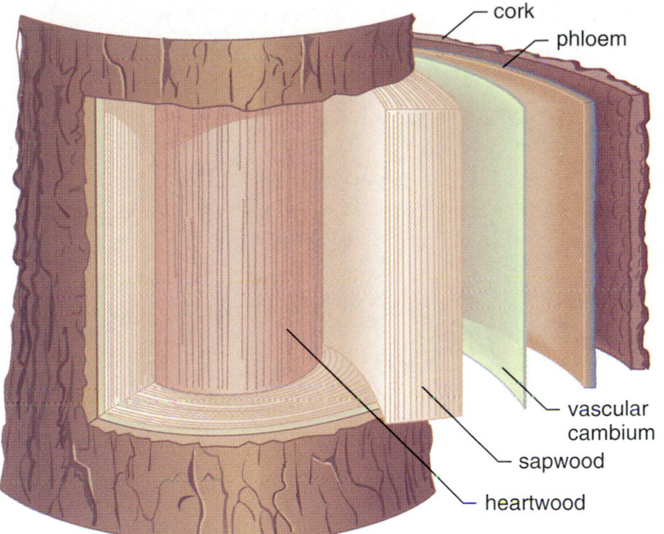

b. Tree trunk, longitudinal view

Figure 36.16 **Tree trunk.**
a. A cross section of a 39-year-old larch, *Larix decidua.* The xylem within the darker heartwood is inactive; the xylem within the lighter sapwood is active. **b.** The relationship of bark, vascular cambium, and wood is retained in a mature stem. The pith has disappeared.

Cork cambium [L. *cambio,* exchange] is meristem located beneath the epidermis. When cork cambium begins to divide, it produces tissue that disrupts the epidermis and replaces it with cork cells. Cork cells are impregnated with suberin, a lipid material that makes them waterproof but also causes them to die. This is protective because now the stem is less edible. But an impervious barrier means that gas exchange is impeded except at lenticels, which are pockets of loosely arranged cork cells not impregnated with suberin.

The first flowering plants to evolve may have been woody shrubs; herbaceous plants evolved later. Is it advantageous to be woody? If there is adequate rainfall, woody plants can grow taller and have more growth because they have adequate vascular tissue to support and service leaves. However, it takes energy to produce secondary growth and prepare the body for winter if the plant lives in the temperate zone. Also, there is a greater need for defense mechanisms because a long-lasting plant that stays in one spot is likely to be attacked by herbivores and parasites. Then, too, trees don't usually reproduce until they have grown several seasons, by which time they may have succumbed to an accident or disease. In certain habitats, it is more advantageous for a plant to put most of its energy into simply reproducing rather than being woody.

Annual Rings

In trees that have a growing season, vascular cambium is dormant during the winter. In the spring, when moisture is plentiful and leaves require much water for growth, the xylem [Gk. *xylon,* wood] contains wide vessels with thin walls. In this so-called spring wood, wide vessels transport sufficient water to the growing leaves. Later in the season, moisture is scarce and the wood at this time, called summer wood, has a lower proportion of vessels. Strength is required because the tree is growing larger and summer wood contains numerous fibers and thick-walled tracheids. At the end of the growing season, just before the cambium becomes dormant again, only heavy fibers with especially thick secondary walls may develop. When the trunk of a tree has spring wood followed by summer wood, the two together make up one year's growth, or **annual ring.** You can tell the age of a tree by counting the annual rings. The outer annual rings, where transport occurs, is called sapwood (Fig. 36.16).

In older trees, the inner annual rings, called the heartwood, no longer function in water transport. The cells become plugged with deposits, such as resins, gums, and other substances that inhibit the growth of bacteria and fungi. Heartwood may help support a tree, although some trees stand erect and live for many years after the heartwood has rotted away.

As described in the reading on page 651, the annual rings tell a story about the times of stress in the life of the tree.

Woody plants grow in girth due to the presence of vascular cambium and cork cambium. Their bodies have three main parts: bark (which contains cork, cork cambium, and phloem), wood (which contains xylem), and pith.

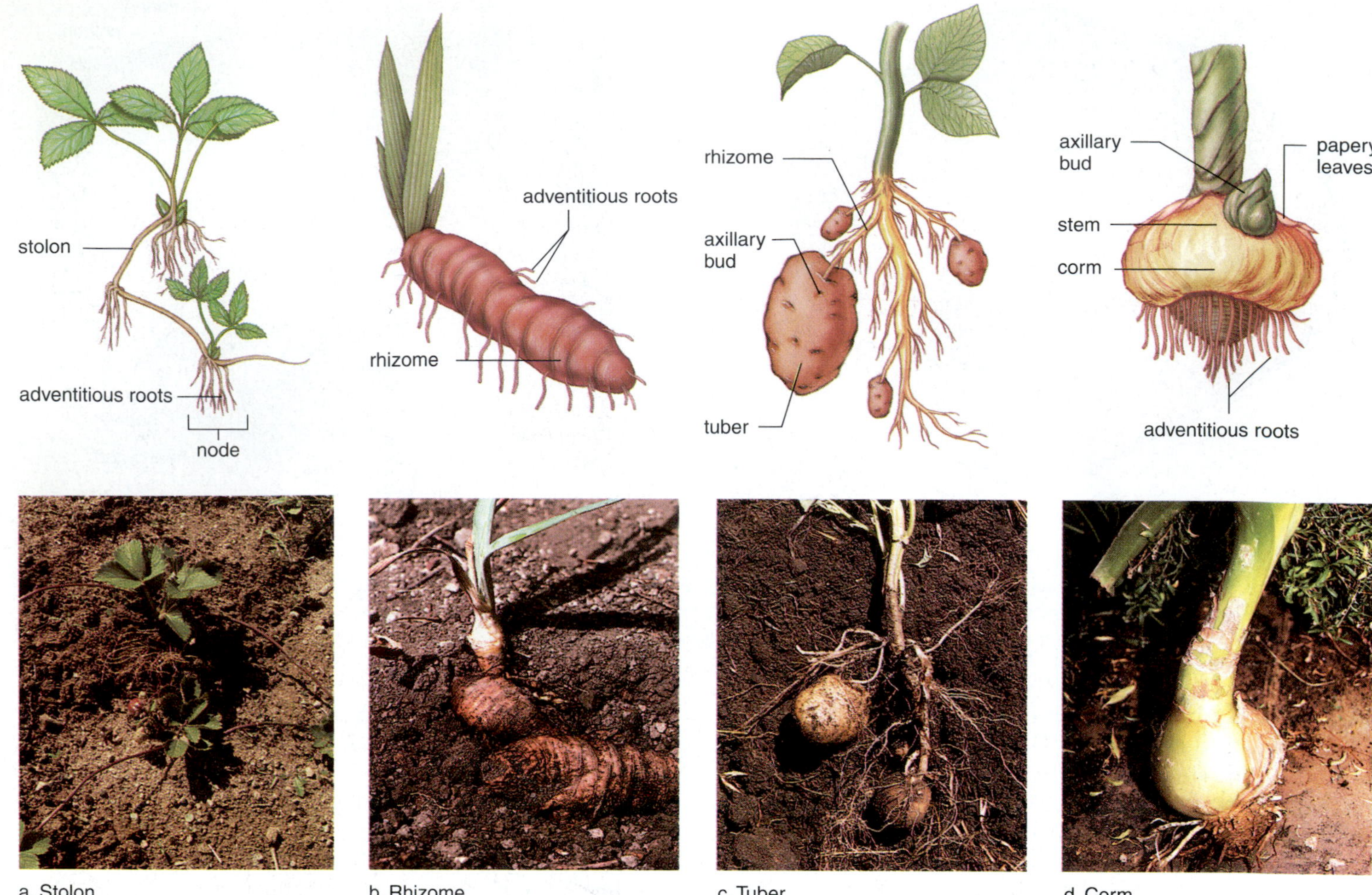

Figure 36.17 **Stem diversity.**
a. A strawberry plant has aboveground, horizontal stems called stolons. Every other node produces a new shoot system. **b.** The underground horizontal stem of an iris is a fleshy rhizome. **c.** The underground stem of a potato plant has enlargements called tubers. We call the tubers potatoes. **d.** The corm of a gladiolus is a stem covered by papery leaves.

Stem Diversity

Stem diversity is illustrated in Figure 36.17. Aboveground, horizontal stems, called **stolons** [L. *stolo,* shoot] or runners, produce new plants where nodes touch the ground. The strawberry plant is a common example of this type of stem which functions in vegetative reproduction.

Aboveground, vertical stems can also be modified. For example, cacti have succulent stems specialized for water storage, and the tendrils of grape plants (which are stem branches) allow them to climb. Morning glory and relatives have stems that twine around support structures. Tendrils and twining shoots help plants expose leaves to the sun.

Underground horizontal stems, **rhizomes** [Gk. *rhiza,* root], may be long and thin, as in sod-forming grasses, or thick and fleshy, as in iris. Rhizomes survive the winter and contribute to asexual reproduction because each node bears a bud. Some rhizomes have enlarged portions called tubers, which function in food storage. Potatoes are tubers, in which the eyes are buds that mark the nodes.

Corms are bulbous underground stems that lie dormant during the winter, just as rhizomes do. They also produce new plants the next growing season. Gladiolus corms are referred to as bulbs by laypersons, but the botanist reserves the term bulb for a structure composed of modified leaves attached to a short vertical stem. An onion is a bulb.

Humans make use of stems in many ways. The stem of the sugarcane plant is a primary source of table sugar. The spice cinnamon and the drug quinine are derived from the bark of different plants. And wood is necessary for the production of paper as discussed in the reading on the next page.

Plants use diverse stems for such functions as reproduction, climbing, survival, and food storage. Modified stems aid adaptation to different environments.

Paper Comes from Plants

The word *paper* takes its origin from papyrus, the plant Egyptians used to make the first form of paper. The Egyptians manually made sheets from the treated stems of papyrus grass and then strung them together into scrolls. From that beginning some 5,500 years ago, the production of paper is now a worldwide industry of major importance. The process is fairly simple. Plant material is ground up mechanically, and chemically treated to form a pulp that contains "fibers," which biologists know are the tracheids and vessel elements of a plant. The fibers automatically form a sheet when they are screened from the pulp. Today a revolving wire-screen belt is used to deliver a continuous wet sheet of paper to heavy rollers and heated cylinders which remove most of the remaining moisture and press the paper flat.

There are, of course, different types of paper dependent upon the plant material used and the way it is treated. Among the major plants used to make paper are:

Eucalyptus trees. In recent years Brazil has devoted huge areas in the Amazon to the growing of cloned eucalyptus seedlings, specially selected and engineered to be ready for harvest after about seven years.

Temperate hardwood trees. Plantation cultivation in Canada provides birch, beech, chestnut, poplar, and particularly aspen wood for paper making. Tropical hardwoods, usually coming from Southeast Asia, are also used.

Softwood trees. In the United States, several species of pine trees have been genetically improved to have a higher wood density and to be harvestable five years earlier than ordinary pines. Southern Africa, Chile, New Zealand, and Australia also devote thousands of acres to growing pines for paper pulp production.

Bamboo. Several Asian countries, especially India, provide vast quantities of bamboo pulp for the making of paper. Because bamboo is harvested without destroying the roots, and the growing cycle is favorable, this plant, which is actually a grass, is expected to be a significant source of paper pulp despite high processing costs to remove impurities.

Flax and cotton plants. Linen and cotton cloth from textile and garment mills are used to produce *rag paper* whose flexibility and durability are desirable in legal documents, high-grade bond paper, and high-grade stationery.

It has been known for some time that paper largely consists of the cellulose within plant cell walls. It seems reasonable to suppose, then, that paper could be made from synthetic polymers (e.g., rayon). Indeed, synthetic polymers produce a paper that has qualities superior to those of paper made from natural sources, but the cost thus far is prohibitive. Another consideration, however, is the ecological effects of making paper from trees. Plantations containing stands of uniform trees replace natural ecosystems and when the trees are clear-cut, the land is laid bare. Paper mill wastes, which include caustic chemicals, add significantly to the pollution of rivers and streams.

The use of paper for packaging and to make all sorts of products has increased dramatically in this century. Each person in the United States consumes about 318 kilograms (699 pounds) of paper products per year, and this compares to only 2.3 kilograms (5 pounds) of paper per person in India. It is clear, then, we should take the initiative in recycling paper. When newspaper, office paper, and photocopies are soaked in water, the fibers are released and they can be used to make a new batch of paper. It's estimated that recycling the Sunday newspapers alone would save an estimated number of 500,000 trees each week!

Figure 36B **Paper production.**
Machine No. 35 at Champion International's Courtland, Alabama, mill produces a 29-foot-wide roll of office paper every 60 minutes.

36.6 Organization of Leaves

Leaves are the organs of photosynthesis in vascular plants. As mentioned earlier, a leaf usually consists of a flattened blade and a petiole connecting the blade to the stem. The blade may be single or composed of several leaflets. Externally, it is possible to see the pattern of the leaf veins, which contain vascular tissue. Leaf veins have a net pattern in dicot leaves and a parallel pattern in monocot leaves (see Fig. 36.3).

Figure 36.18 shows a cross section of a typical dicot leaf of a temperate zone plant. At the top and bottom is a layer of epidermal tissue that often bears protective hairs and/or glands that produce irritating substances. These features may prevent the leaf from being eaten by insects. The epidermis characteristically has an outer, waxy cuticle [L. *cutis,* skin] that keeps the leaf from drying out. The cuticle also prevents gas exchange because it is not gas permeable. However, the epidermis, particularly the lower epidermis, contains stomates that allow gases to move into and out of the leaf. Each stomate has two guard cells that regulate its opening and closing.

The body of a leaf is composed of **mesophyll** [Gk. *mesos,* middle, and *phyllon,* leaf] tissue, which has two distinct regions: **palisade mesophyll,** containing elongated cells, and **spongy mesophyll,** containing irregular cells bounded by air spaces. The parenchyma cells of these layers have many chloroplasts and carry on most of the photosynthesis for the plant. The loosely packed arrangement of the cells in the spongy layer increases the amount of surface area for gas exchange.

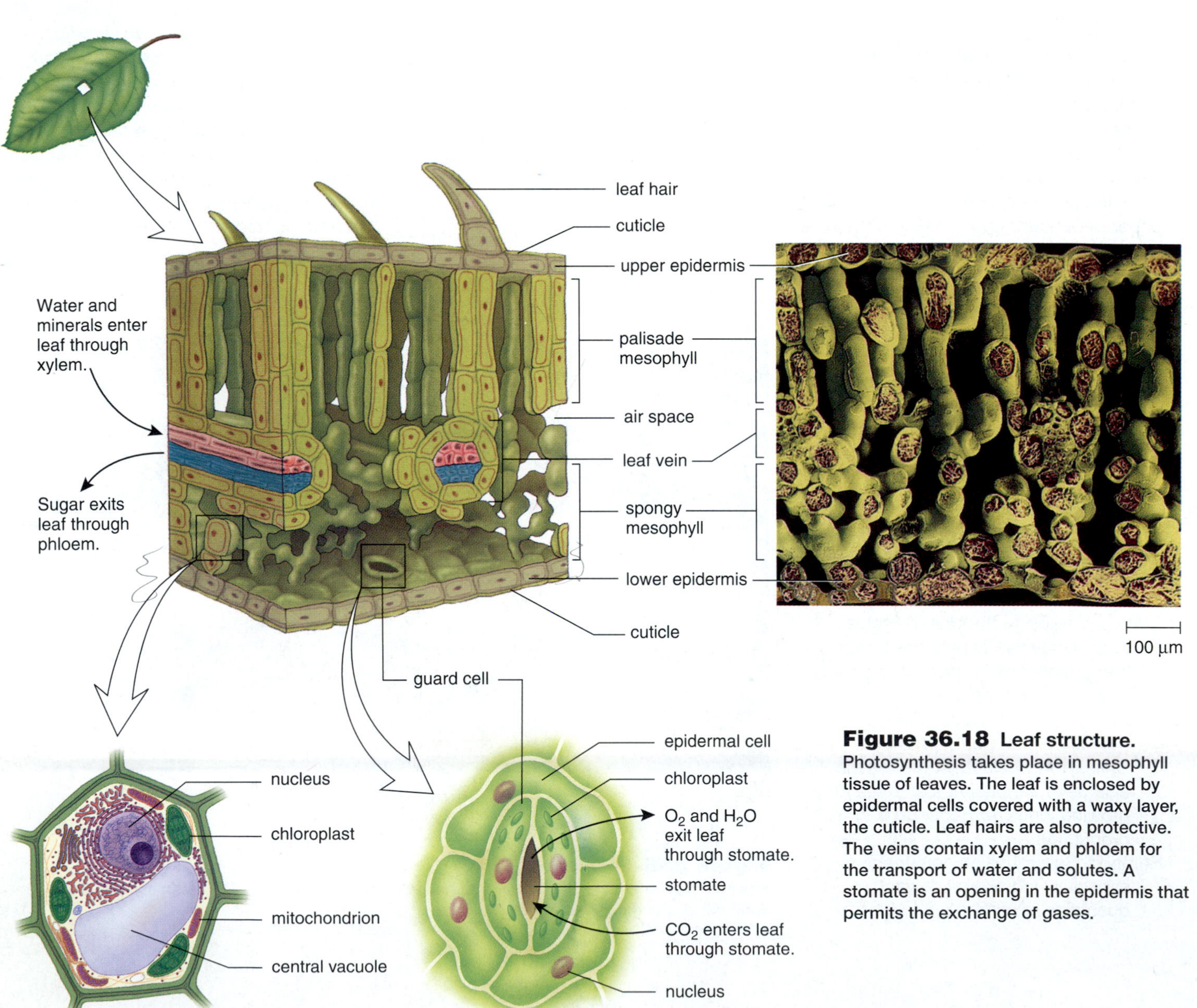

Figure 36.18 Leaf structure. Photosynthesis takes place in mesophyll tissue of leaves. The leaf is enclosed by epidermal cells covered with a waxy layer, the cuticle. Leaf hairs are also protective. The veins contain xylem and phloem for the transport of water and solutes. A stomate is an opening in the epidermis that permits the exchange of gases.

Leaf Diversity

The blade of a leaf can be simple or compound, with two or more separate leaflets making up the blade (Fig. 36.19). There are also various vascular arrangements in leaves and innumerable combinations of overall leaf shape, margin, and base modifications.

Leaves are adapted to environmental conditions. Shade plants tend to have broad, wide leaves, and desert plants tend to have reduced leaves with sunken stomates. The leaves of a cactus are the spines attached to the succulent stem (Fig. 36.20*a*). Other succulents have leaves adapted to hold moisture.

An onion bulb is made up of leaves surrounding a short stem. In a head of cabbage, large leaves overlap one another. The petiole of a leaf can be thick and fleshy, as in celery and rhubarb. Climbing leaves, such as those of peas and cucumbers, are modified into tendrils that can attach to nearby objects (Fig. 36.20*b*). The leaves of a few plants are specialized for catching insects. The leaves of a sundew have sticky epidermal hairs that trap insects and then secrete digestive enzymes. The Venus's-flytrap has hinged leaves that snap shut and interlock when an insect triggers sensitive hairs (Fig. 36.20*c*). The leaves of a pitcher plant resemble a pitcher and have downward-pointing hairs that lead insects into a pool of digestive enzymes. Insectivorous plants commonly grow in marshy regions, where the supply of soil nitrogen is severely limited. The digested insects provide the plants with a source of organic nitrogen.

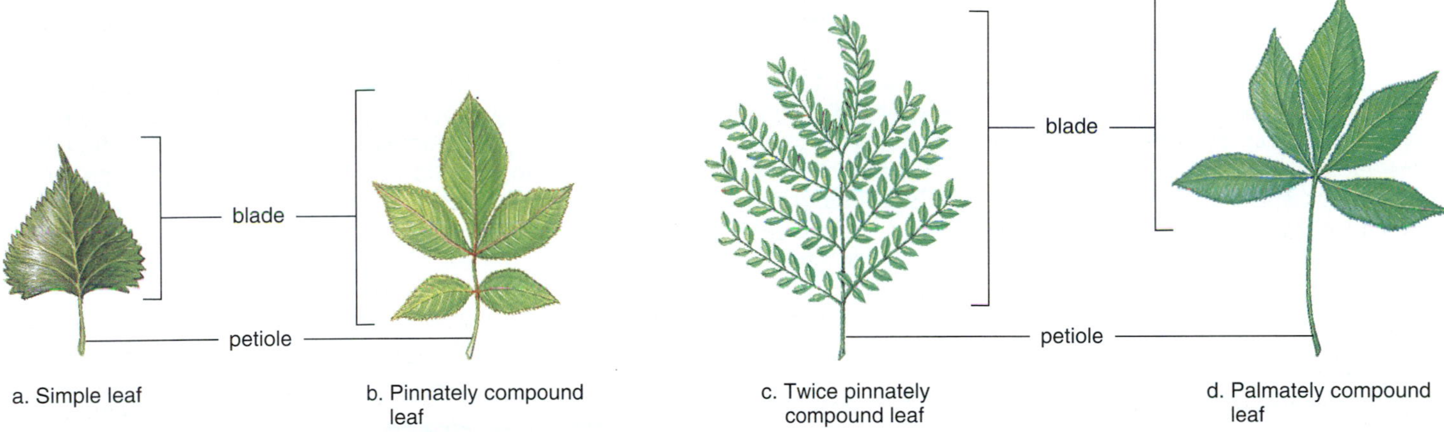

Figure 36.19 Classification of leaves.
a. The cottonwood tree has a simple leaf. **b.** The shagbark hickory has a pinnately compound leaf. **c.** The honey locust has a twice pinnately compound leaf. **d.** The buckeye has a palmately compound leaf.

a. Cactus, *Opuntia*

b. Cucumber, *Cucumis*

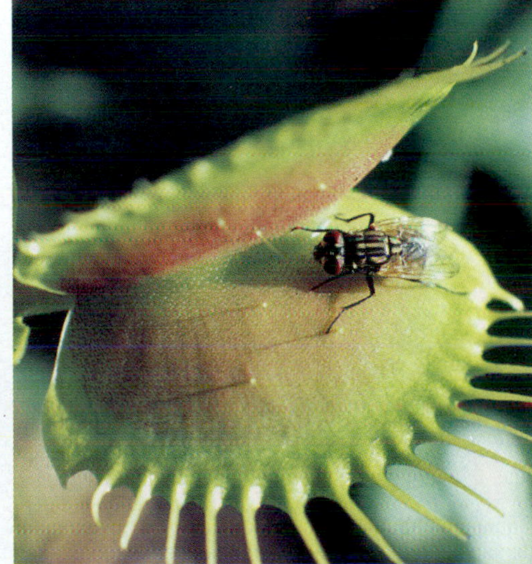

c. Venus's-flytrap, *Dionaea*

Figure 36.20 Leaf diversity.
a. The spines of a cactus plant are leaves modified to protect the fleshy stem from animal consumption. **b.** The tendrils of a cucumber are leaves modified to attach the plant to a physical support. **c.** The leaves of the Venus's-flytrap are modified to serve as a trap for insect prey. When triggered by an insect, the leaf snaps shut. Once shut, the leaf secretes digestive juices, which break down the soft parts of the insect's body.

Plants: Could We Do Without Them?

Plants define and are the producers in most ecosystems. Humans derive most of their sustenance from three flowering plants: wheat, corn, and rice. All three of these plants are of the grass family and are collectively, along with other species, called grains. Most of the earth's 5.6 billion people live a simple way of life, growing their food on family plots. The continued growth of these plants is essential to human existence. A virus or other disease could hit any one of these three plants and cause massive loss of life from starvation.

Corn, wheat, and rice originated and were first cultivated in different parts of the globe. Corn, or what is properly called maize, was first cultivated in Central America about 7,000 years ago. Disagreement still exists as to exactly what the wild plant looked like and where it originally grew. The most recent theory is that maize was developed from a plant called teosinte, which grows in the highlands of central Mexico. By the time Europeans were exploring Central America, over 300 varieties were already in existence—growing from Canada to Chile. We now commonly grow six major varieties of corn: sweet, pop, flour, dent, pod, and flint. Rice had its origin in southeastern Asia several thousand years ago, where it grew in swamps. Today we are familiar with white and brown rice, which differ in the extent of processing. Brown rice results when the seeds are thrashed to remove the hulls—the seed coat and complete embryo remain. If the seed coat and embryo are removed leaving only the starchy endosperm, white rice results. Unfortunately, it is the seed coat and embryo that contain the greatest nutritional value. Today rice is grown throughout the tropics and subtropics where water is abundant. It is also grown in some of our Western states by flooding diked fields with irrigation water. Wheat is commonly used in the United States to produce flour and bread. It was first cultivated in the Near East (Iran, Iraq, and neighboring countries) about 8,000 B.C.; hence, it is thought to be one of the earliest cultivated plants. Wheat was brought to North America in 1520 with early settlers; now the United States is one of the world's largest producers of wheat.

Many of us have an "addiction" to sugar. This nifty carbohydrate comes almost exclusively from two plants—sugarcane (grown in South America, Africa, Asia, and the Caribbean) and sugar beets (grown mostly in Europe). Each provides about 50% of the world's sugar.

Numerous other foods are bland or tasteless without spices. In the Middle Ages, wealthy Europeans spared no cost to obtain spices from the Near and Far East. In the fifteenth and sixteenth centuries, major expeditions were launched in an attempt to find better and cheaper routes for spice importation. The queen of Portugal became convinced that Columbus would actually find a shorter route to the Far East by traveling west by ocean rather than east by land. Columbus's idea was sound, but he encountered a little barrier, the New World. This later provided Europe with a wealth of new crops, including corn, potatoes, peppers, and tobacco.

Our most popular drinks—coffee, tea, and cola—also come from flowering plants. Coffee has its origin in Ethiopia, where it was first used (along with animal fat) during long trips for sustenance and to relieve fatigue. Coffee as a drink was not developed until the thirteenth century in Arabia and Turkey, and it did not catch on in Europe until the seventeenth century. Tea is thought to have been developed somewhere in central Asia. Its earlier uses were almost exclusively medicinal, especially among the Chinese, who still drink tea for medical reasons. The drink as we now know it was not developed until the fourth century. By the mid-seventeenth century it became popular in Europe. Cola is a common ingredient in tropical drinks and was used around the turn of the century, along

Wheat, *Triticum*

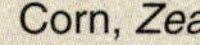
Corn, *Zea*

Rice, *Oryza*

Figure 36C Cereal grains.
These cereal grains are the principle source of calories and protein for our civilization.

with the drug coca (used to make cocaine), in the "original" Coca-Cola.

Until a few decades ago, cotton and other natural fibers were our only source of clothing. China is now the largest producer of cotton. Over thirty species of native cotton grow around the world. In the United States, cotton grows as a one-to-two meter annual, but in the tropics, woody shrubs up to six meters are not uncommon. The cotton fiber itself comes from filaments that grow on the seed. In sixteenth-century Europe, cotton was a little-understood fiber known only from stories brought back from Asia. Columbus and other explorers were amazed to see the elaborately woven cotton fabrics in the New World. But by 1800 Liverpool was the world's center of cotton trade. Interestingly, when Levi Strauss wanted to make a tough pair of jeans, he needed a stronger fiber than cotton, so he used hemp. No longer used for clothes, hemp (marijuana) is now known primarily as a hallucinogenic drug.

Rubber is another plant that has many uses today. The product had its origin in Brazil from the thick, white sap of the rubber tree. Once collected, the sap is placed in a large vat, where acid is added to coagulate the latex. When the water is pressed out, the product is formed into sheets or crumbled and placed into bales. Much stronger rubber, such as that in tires, was made by adding sulfur and heating in a process called vulcanization; this produces a flexible material less sensitive to temperature changes. Today, though, much rubber is synthetically produced.

Beyond these uses, plants have been used for centuries for a number of important household items, including the house itself. We are most familiar with lumber being used as the major structural portion in buildings. This wood comes mostly from a variety of conifers: pine, fir, and spruce, among others. In the tropics, trees and even herbs provide important components for houses. In rural parts of Central and South America, palm leaves are preferable to tin for roofs, since they last as long as ten years and are quieter during a rainstorm. In the Near East, numerous houses along rivers are made entirely of reeds.

An actively researched area of plant use today is that of medicinal plants. Currently about 50% of all pharmaceutical drugs have their origins from plants. The treatment of cancers appears to rest in the discovery of miracle plants. Indeed, the National Cancer Institute (NCI) and most pharmaceutical companies have spent millions (or, more likely, billions) of dollars to send botanists out to collect and test plant samples from around the world. Tribal medicine men, or shamans, of South America and Africa have already been of great importance in developing numerous drugs.

Over the centuries, malaria has caused far more human deaths than any other disease. After European scientists became aware that malaria can be cured by quinine, which comes from the bark of the cinchona tree, a synthetic form of the drug, chloroquine, was developed. But by the late 1960s, it was found that some of the malaria parasites, which live in red blood cells, had become resistant to the synthetically produced drug. Resistant parasites were first seen in Africa but now are showing up in Asia and the Amazon. Today the only 100%-effective drug for malaria treatment must come directly from the cinchona tree, common to northeastern South America.

Numerous plant extracts continue to be misused for their hallucinogenic or other effects on the human body; coca for cocaine and crack, opium poppy for morphine, and yam for steroids. In addition to all these uses of plants, we should not forget nor neglect the aesthetic value of plants. Flowers brighten any yard; ornamental plants accent landscaping, and trees provide cooling shade during the summer and break the wind of winter days. Plants also produce oxygen, which is so necessary for all plants and animals.

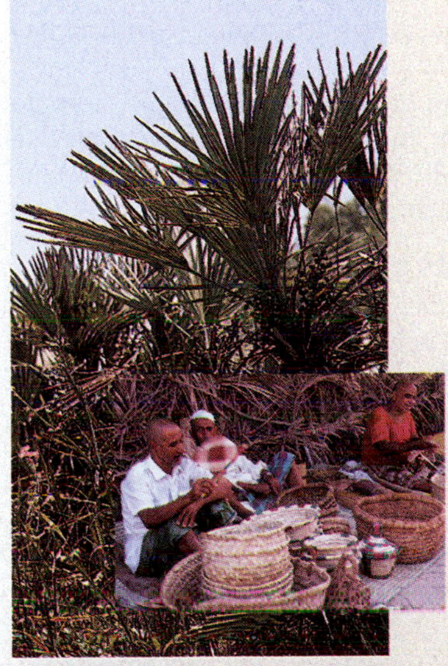

a. Dwarf fan palms, *Chamaerops*

b. Cotton, *Gossypium*

c. Rubber, *Hevea*

Figure 36D Uses of plants.
a. Dwarf fan palms can be used to make baskets. **b.** Cotton becomes clothing worn by all. **c.** A rubber plant provides latex for making tires.

Connecting Concepts

Perhaps a multicellular alga is ancestral to plants because multicellularity would have been adaptive for development of the specialized tissues found in plants—organisms that evolved on land. On land, as opposed to an aquatic environment, there is a danger of drying out. Even humid air is drier than a living cell, and the prevention of water loss is critical for plants. The epidermis and the cuticle it produces help prevent water loss and overheating in sunlight. (It is also protective against invasion by bacteria, fungi, and small insects.) The cork of woody plants is especially protective against water loss. Only the presence of lenticels still allows gas exchange to occur.

In an aquatic environment, water is available to all cells, but on land it is adaptive to have a means of water uptake and transport. In plants, the roots absorb water and have special extensions called root hairs that facilitate the uptake of water. Xylem transports water to all plant parts. The next chapter discusses how the drying effect of air is used by plants to help move water from the roots to the leaves. Roots buried in soil where they absorb water cannot photosynthesize, and the structure of phloem permits the transport of sugars down to the roots.

In an aquatic environment, water buoys up organisms and keeps them afloat, but on land it is adaptive to have a means to oppose the force of gravity. The stems of plants contain strong-walled sclerenchyma cells, tracheids, and vessel elements. The accumulation of annual rings in woody plants offers more support and allows a tree to grow ever taller.

In an aquatic environment, the surrounding water prevents the gametes and zygote from drying out. As we shall see in Chapter 39, flowering plants do not rely on external water for reproductive purposes and have evolved structures and a means of reproduction that prevent the gametes, zygote, and embryo from drying out.

Summary

36.1 Plant Organs

A flowering plant has three vegetative organs. A root anchors a plant, absorbs water and minerals, and stores the products of photosynthesis. Stems support leaves, conduct materials to and from roots and leaves, and help store plant products. Leaves take in carbon dioxide and carry on photosynthesis.

36.2 Monocot Versus Dicot Plants

Flowering plants are divided into the dicots and monocots according to the number of cotyledons in the seed, the arrangement of vascular tissue in roots, stems, and leaves, and the number of flower parts.

36.3 Plant Tissues

Three types of meristem continually divide and produce specialized tissues. Protoderm produces epidermis tissue, ground meristem produces ground tissue, and procambium produces primary vascular tissue.

Epidermal tissue contains the epidermis, which is modified in different organs of the plant. In the roots, epidermal cells bear root hairs; in the leaves, the epidermis contains guard cells. In a woody stem, epidermis is replaced by cork.

Ground tissue contains parenchyma cells, which are thin walled and capable of photosynthesis when they contain chloroplasts. Collenchyma cells have thicker walls for flexible support. Sclerenchyma cells are hollow, nonliving support cells with secondary walls.

Vascular tissue consists of xylem and phloem. Xylem contains vessels composed of vessel elements and tracheids, which are elongated and tapered with pitted end walls. Xylem transports water and minerals. Phloem contains sieve tubes composed of sieve-tube elements, each of which has a companion cell. Phloem transports organic nutrients.

36.4 Organization of Roots

A root tip shows three zones: the zone of cell division (apical meristem) protected by the root cap, the zone of elongation, and the zone of maturation.

A cross section of a herbaceous dicot root reveals the epidermis (protects), the cortex (stores food), the endodermis (regulates the movement of minerals), and the vascular cylinder (vascular tissue). In the vascular cylinder of a dicot, the xylem appears star shaped and the phloem is found in separate regions, between the arms of the xylem. In contrast, a monocot root has a ring of vascular tissue with alternating bundles of xylem and phloem surrounding pith.

Roots are diversified. Taproots are specialized to store the products of photosynthesis; a fibrous root system is composed of adventitious roots and so are prop roots, which are specialized to provide increased anchorage.

36.5 Organization of Stems

Primary growth of a stem is due to the activity of the shoot apical meristem, which is protected within a terminal bud. A terminal bud contains leaf primordia at nodes, and internodes between the nodes. When stems grow, the internodes lengthen.

In cross section, a nonwoody dicot has epidermis, cortex tissue, vascular bundles in a ring, and an inner pith. Monocot stems have scattered vascular bundles, and the cortex and pith are not well defined.

Secondary growth of a woody stem is due to vascular cambium, which produces new xylem and phloem every year, and cork cambium, which produces new cork cells when needed. Cork replaces epidermis in woody plants. The cross section of a woody stem shows bark, wood, and pith. The bark contains cork and phloem. Wood contains annual rings of xylem.

Stems are diverse. There are horizontal aboveground and underground stems. Corms and some tendrils are also modified stems.

36.6 Organization of Leaves

A cross section of a leaf shows the epidermis, with stomates mostly below. Vascular tissue is present within leaf veins.

Leaves are diverse. The spines of a cactus are leaves. Other succulents have fleshy leaves. An onion is a bulb with fleshy leaves, and the tendrils of peas are leaves. The Venus's-flytrap has leaves that trap and digest insects.

Reviewing the Chapter

1. Name and discuss the vegetative organs of a plant. 642–44
2. List five differences between monocots and dicots. 644
3. Epidermal cells are found in what type of plant tissue? Explain how epidermis is modified in various organs of a plant. Contrast an epidermal cell with a cork cell. 645
4. Contrast the structure and function of parenchyma, collenchyma, and sclerenchyma cells. These cells occur in what type of plant tissue? 646
5. Contrast the structure and function of xylem and phloem. Xylem and phloem occur in what type of plant tissue? 646–47
6. Name and discuss the zones of a root tip. Trace the path of water and minerals from the root hairs to xylem. Be sure to mention the Casparian strip. 648–49
7. Contrast a taproot with branch roots and a fibrous root system. What are adventitious roots? 650
8. Describe the primary and secondary growth of a stem. 652–54
9. Describe the cross section of a monocot, a herbaceous dicot, and a woody stem. 653–55
10. Discuss the diversity of stems by giving examples of several adaptations. 656
11. Describe the structure and organization of a typical dicot leaf. 658
12. Note diversity of leaves by giving examples of several adaptations. 658

Testing Yourself

Choose the best answer for each question.

1. Which of these is an incorrect contrast between monocots (stated first) and dicots (stated second)?
   a. one cotyledon—two cotyledons
   b. leaf veins parallel—net veined
   c. vascular bundles in a ring—vascular bundles scattered
   d. flower parts in threes—flower parts in fours or fives
   e. All of these are a correct contrast.
2. Which of these types of cells is most likely to divide?
   a. parenchyma
   b. meristem
   c. epidermis
   d. xylem
   e. sclerenchyma
3. Which of these cells in a plant is apt to be nonliving?
   a. parenchyma
   b. collenchyma
   c. sclerenchyma
   d. epidermal
   e. guard cells
4. Root hairs are found in the zone of
   a. cell division.
   b. elongation.
   c. maturation.
   d. apical meristem.
   e. All of these are correct.
5. Cortex is found in
   a. roots, stems, and leaves.
   b. roots and stems.
   c. roots and leaves.
   d. stems and leaves.
   e. roots only.
6. Between the bark and the wood in a woody stem, there is a layer of meristem called
   a. cork cambium.
   b. vascular cambium.
   c. apical meristem.
   d. the zone of cell division.
   e. procambium preceding bark.
7. Which part of a leaf carries on most of the photosynthesis of a plant?
   a. epidermis
   b. mesophyll
   c. epidermal layer
   d. guard cells
   e. Both a and b are correct.
8. Annual rings are the number of
   a. internodes in a stem.
   b. rings of vascular bundles in a monocot stem.
   c. layers of xylem in a stem.
   d. bark layers in a woody stem.
   e. Both b and c are correct.
9. The Casparian strip is found
   a. between all epidermal cells.
   b. between xylem and phloem cells.
   c. on four sides of endodermal cells.
   d. within the secondary wall of parenchyma cells.
   e. in both endodermis and pericycle.
10. Which of these is a stem?
    a. taproot of carrots
    b. stolon of strawberry plants
    c. spine of cacti
    d. prop roots
    e. Both b and c are correct.
11. Label this root using the terms endodermis, phloem, xylem, cortex, and epidermis:

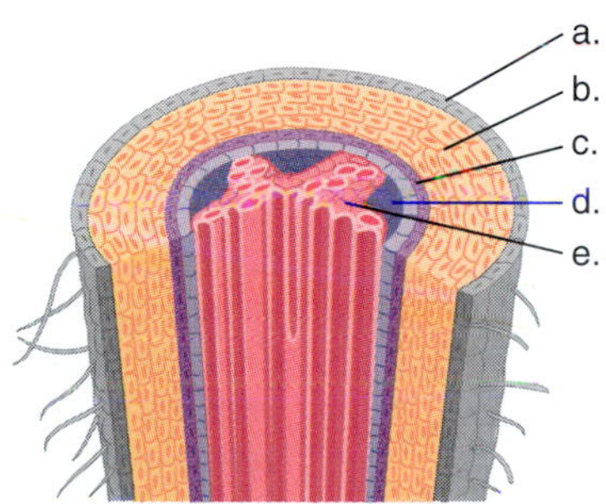

12. Label this leaf using the terms leaf vein, lower epidermis, palisade mesophyll, spongy mesophyll, and upper epidermis:

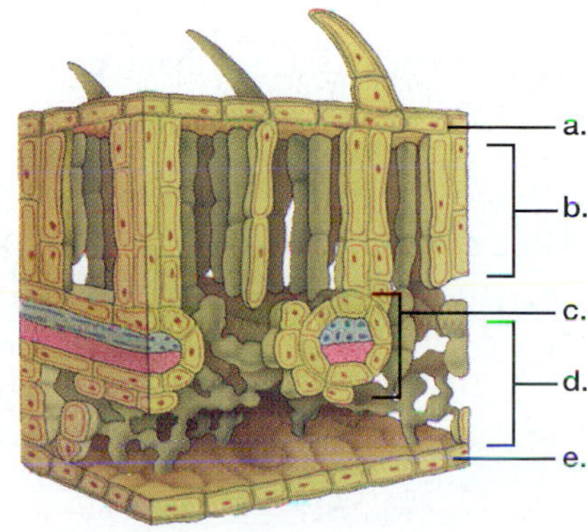

Thinking Scientifically

1. Carrots are taproots. It is somewhat strange that these roots are orange since they are belowground and cannot engage in photosynthesis. Hypothesize how such a pigment could have arisen and its possible advantages.
2. Design an experiment that tests the hypothesis that new plants arise at the nodes of a strawberry stolon according to environmental conditions pertaining to mechanical stimuli (whether nodes touch the ground or not), temperature, water, and sunlight.

Understanding the Terms

adventitious root 650
annual ring 655
apical meristem 648
axillary bud 643
blade 643
Casparian strip 649
collenchyma 646
cork 645
cork cambium 655
cortex 649
cotyledon 644
cuticle 645
dicot 644
endodermis 649
epidermal tissue 645
epidermis 645
fibrous root system 650
ground tissue 645
herbaceous stem 652
internode 643
leaf 643
leaf vein 646
meristem 645
mesophyll 658
monocot 644
node 643
organ 642
palisade mesophyll 658
parenchyma 646
perennial 642
pericycle 649
petiole 643
phloem 646
pith 649
primary root 650
rhizome 656
root hair 645
root nodule 650
root system 642
sclerenchyma 646
shoot apical meristem 652
shoot system 642
spongy mesophyll 658
stem 643
stolon 656
stomate 645
taproot 650
terminal bud 652
vascular bundle 646
vascular cambium 652
vascular cylinder 646
vascular tissue 645
xylem 646

Match the terms to these definitions:

a. ______________ Inner, thickest layer of a leaf consisting of palisade and spongy mesophyll; the site of most of photosynthesis.
b. ______________ Lateral meristem that produces secondary phloem and secondary xylem.
c. ______________ Seed leaf for embryonic plant, provides nutrient molecules for the developing plant before its mature leaves begin to photosynthesize.
d. ______________ Stem that grows horizontally along the ground and establishes plantlets periodically when it contacts the soil (e.g., the runners of a strawberry plant).
e. ______________ Vascular tissue that conducts organic solutes in plants; contains sieve-tube elements and companion cells.

Web Connections

Exploring the Internet

http://www.mhhe.com/biosci/genbio/mader
(click on *Biology 7/e*)

The *Biology 7/e* Online Learning Center provides many resources for studying the material in this chapter including links to the following sites:

Careers in Botany is a site produced by the Botanical Society of America and describes the nature of botany as a science as well as the diverse career choices a student could make in this interesting field.

http://www.botany.org/bsa/careers/

The Moore Botany page has links to further information on plants, and also, of use, a link to some very valuable information writing a research paper!

http://www.mhhe.com/biosci/pae/moore/index.mhtml

GardenWeb Glossary of Botanical Terms. A very user-friendly dictionary of botanical terms with definitions.

http://www.gardenweb.com/glossary/

California State University Bio. Web. A comprehensive list of sites of interest to botanists.

http://arnica.csustan.edu/site.asp

Missouri Botanical Garden Site's Other Sources of Information. An extensive list of other sites.

http://www.mobot.org/MOBOT/otherwww.html

Nutrition and Transport in Plants

chapter

chapter concepts

37.1 Plant Nutrition and Soil

- Certain inorganic nutrients (e.g., NO_3^-, K^+, Ca^{2+}) are essential to plants; others that are specific to a type of plant are termed beneficial. 666
- Soil is built up over time by the weathering of rock and the action of organisms. 668
- Soil particles, humus, and living organisms are components of soil that provide oxygen, water, and minerals to plants. 668
- Soil erosion is a serious threat to agriculture, worldwide. 669

37.2 Uptake of Water and Minerals

- The tissues of a root are organized so that water and minerals entering between or at the root hairs will eventually enter xylem. 670
- Mineral ions cross plasma membranes by a chemiosmotic mechanism. 670
- Plants have various adaptations that assist them in acquiring nutrients; e.g., symbiotic relationships are of special interest. 671

37.3 Transport Mechanisms in Plants

- The vascular system in plants is an adaptation to living on land. 672
- The vascular tissue, xylem, transports water and minerals; the vascular tissue, phloem, transports organic nutrients. 672
- Because water molecules are cohesive and adhere to xylem walls, the water column in xylem is continuous. 674
- Transpiration (evaporation) creates a tension that pulls water and minerals from the roots to the leaves in xylem. 674
- Stomates must be open for transpiration to occur. 676
- Active transport of sucrose into phloem creates a positive pressure that causes organic nutrients to flow in phloem from a source (where sucrose enters) to a sink (where sucrose exits). 679

Redwood, *Sequoia sempervirens*

Carbon, oxygen, hydrogen, nitrogen, potassium, calcium, and other elements are required in various amounts by virtually all plants. Gases such as carbon dioxide and oxygen are taken up by the leaves, but the roots absorb water and minerals in the form of NO_3^-, K^+, and Ca^{2+}. Plants, unlike most animals, have an amazing ability to concentrate minerals—that is, to take them up until they are many times more prevalent in the plant than in the soil. Some plants, such as those in the legume family, have roots infected with bacteria that make fixed nitrogen (e.g., NH_4^+) available to the plant for the production of proteins. This relationship allows legumes (e.g., beans) to be high in protein.

In humans, blood, which contains nutrients, salts, and other substances, is pumped throughout the body by the heart. In plants, there is no central pumping mechanism, yet materials move from the roots to the leaves and vice versa. How is this accomplished? The unique properties of water account for the movement of water and minerals in xylem, while osmosis plays a role in phloem transport of sugars. The same mechanisms account for transport in very tall redwood trees as in garden plants.

37.1 Plant Nutrition and Soil

The ancient Greeks believed that plants were "soil-eaters" and somehow converted soil into plant material. Apparently to test this hypothesis, a seventeenth-century Dutchman named Jean-Baptiste Van Helmont planted a willow tree weighing 5 pounds in a large pot containing 200 pounds of soil. He watered the tree regularly for five years and then reweighed both the tree and the soil. The tree weighed 170 pounds and the soil weighed only a few ounces less than the original 200 pounds. Van Helmont concluded that the increase in weight of the tree was due primarily to the addition of water.

Water is a vitally important nutrient for a plant, but Van Helmont was unaware that water and carbon dioxide (taken in at the leaves) combine in the presence of sunlight to produce carbohydrates, the chief organic matter of plants. In this chapter we will see that much of the water entering a plant evaporates at the leaves, and that roots carry on gas exchange, that is, oxygen enters root cells and carbon dioxide exits root cells (Fig. 37.1).

Essential Inorganic Nutrients

As we would suspect much (about 95%) of a plant's dry weight, that is, weight excluding free water, is carbon, hydrogen, and oxygen. Why? Because these are the elements that are found in most organic compounds, such as carbohydrates. Carbon dioxide (CO_2) is a source of carbon and oxygen, and water (H_2O) is a source of hydrogen and also oxygen found in the organic compounds of a plant.

Figure 37.1 Overview of plant nutrition.
Carbon dioxide, which enters leaves, and water, which enters roots, are combined during photosynthesis to form carbohydrates with the release of oxygen from the leaves. Roots carry on respiration—oxygen enters root cells and carbon dioxide exits. Aside from the elements carbon, hydrogen, and oxygen, plants require nutrients that are absorbed as minerals.

In addition to these nutrients, plants require certain others that are absorbed as minerals by the roots. A **mineral** is an inorganic substance usually containing two or more elements. Why are the nutrients absorbed as minerals from the soil needed by a plant? Nitrogen as you know is a part of nucleic acids and proteins; magnesium is a part of chlorophyll; and iron is part of the cytochrome molecules. The major functions of various **essential nutrients** for plants are listed in Table 37.1. A nutrient is essential if (1) it has an identifiable role, (2) no other nutrient can substitute and fulfill the same role, and (3) a deficiency of this nutrient causes the plant to die without completing its life cycle. Essential nutrients are divided into **macronutrients** and **micronutrients** according to their relative concentration in plant tissue.

Beneficial nutrients are another category of elements taken up by plants. Beneficial nutrients either are required or enhance the growth of a particular plant. Plants such as *Equisetum* (horsetails) require silicon as a mineral nutrient, and halophytes such as sugar beets show enhanced growth in the presence of sodium. Nickel is a beneficial mineral nutrient in soybeans when root nodules are present.

Determination of Essential Nutrients

If you burn a plant, its nitrogen component is given off as ammonia and other gases, but most other essential nutrients remain in the ash. The presence of a particular element in the ash, however, does not necessarily mean that the plant normally requires it. The preferred method for determining the mineral requirements of a plant was developed at the end of the nineteenth century. This method is called water culture, or **hydroponics** [Gk. *hydrias*, water, and *ponos*, hard work]. Hydroponics allows plants to grow well if they are supplied with all the nutrients they need. The investigator omits a particular mineral and observes the effect on plant growth. If growth suffers, it can be concluded that the omitted mineral is a required nutrient (Fig. 37.2). This method has been more successful for macronutrients than for micronutrients. For studies involving the latter, the water and the mineral salts used must be absolutely pure; purity is difficult to attain, because even instruments and glassware can introduce micronutrients. Then, too, the element in question may already be present in the seedling used in the experiment. These factors complicate the determination of essential plant micronutrients by means of hydroponics.

Certain elements are required by plants for good nutrition. Lack of an essential nutrient causes plants to die. Beneficial nutrients serve particular purposes in some plants.

Table 37.1

Some Essential Inorganic Nutrients in Plants

Elements	Symbol	Form	Major Functions
Macronutrients			
Carbon Hydrogen Oxygen	C H O	CO_2 H_2O O_2	Major component of organic molecules
Nitrogen	N	NO_3^- NH_4^+	Part of nucleic acids, proteins, chlorophyll, and coenzymes
Potassium	K	K^+	Cofactor for enzymes; water balance and opening of stomates
Calcium	Ca	Ca^{2+}	Regulates responses to stimuli and movement of substances through plasma membrane; involved in formation and stability of cell walls
Phosphorus	P	$H_2PO_4^-$ HPO_4^{2-}	Part of nucleic acids, ATP, and phospholipids
Magnesium	Mg	Mg^{2+}	Part of chlorophyll; activates a number of enzymes
Sulphur	S	SO_4^{2-}	Part of amino acids, some coenzymes
Micronutrients			
Chlorine	Cl	Cl^-	Role in water-splitting step of photosynthesis and water balance
Iron	Fe	Fe^{2+} Fe^{3+}	Part of cytochrome needed for cellular respiration; activates some enzymes
Manganese	Mn	Mn^{2+}	Required for photosynthesis; activates some enzymes such as those of the Krebs cycle
Zinc	Zn	Zn^{2+}	Role in chlorophyll formation; activates some enzymes
Boron	B	BO_3^{3-} $B_4O_7^{2-}$	Role in nucleic acid synthesis, hormone responses, and membrane function
Copper	Cu	Cu^{2+}	Part of certain enzymes, such as redox enzymes
Molybdenum	Mo	MoO_4^{2-}	Cofactor for enzyme used in nitrogen metabolism

a. Solution lacks nitrogen complete nutrient solution

b. Solution lacks phosphorus complete nutrient solution

c. Solution lacks calcium complete nutrient solution

Figure 37.2 Nutrient deficiencies.
The nutrient cause of poor plant growth is diagnosed when plants are grown in a series of complete nutrient solutions except for the elimination of just one nutrient at a time. Sunflower plants respond negatively to a deficiency of the following: (a) nitrogen, (b) phosphorus, (c) calcium.

Soil

Plants acquire carbon when carbon dioxide diffuses into leaves at the stomates. All the other essential nutrients are absorbed by roots from the soil. It would not be an exaggeration to say that terrestrial life is dependent on the quality of the soil and the ability of soil to provide plants with the nutrients they need.

Soil Formation

Soil formation begins with the weathering of rock, which is any mass of mineral matter found in the Earth's crust. Weathering gradually breaks down rock to rubble and then to soil particles. Some weathering mechanisms such as freeze-thaw cycle of ice or the grinding of rock on rock by the action of glaciers or river flow are purely mechanical. Other forces include a chemical effect as when acidic rain leaches (washes away) soluble components of rock or aerobic oxygen combines with the iron of rocks.

In addition to these forces, organisms also play a role in the formation of soil. Lichens and mosses grow on pure rock and trap particles that later allow grasses, herbs, and soil animals to follow. When these die, their remains are decomposed notably by bacteria and fungi. Decaying organic matter, called **humus,** begins to accumulate. Humus supplies nutrients to plants and its acidity also leaches minerals from rock.

Building soil takes a long time. Under ideal conditions, depending on the type of parent material (the original rock) and the various processes at work, a centimeter of soil may develop within 15 years.

Soil formation involves the interplay of physical, chemical, and biological processes that take place over an extended period of time.

The Nutritional Function of Soil

Soil is defined as a mixture of soil particles, decaying organic material, living organisms, air, and water that together support the growth of plants. In a good agricultural soil, the first three components will come together in such a way that there will be spaces for air and water. It's best if the soil contains particles of different sizes because only then will there be spaces for air. Notice that ideally water clings to particles by capillary action and does not fill the spaces. That's why you shouldn't overwater your house plants!

Soil Particles Soil particles vary in size: sand particles are the largest (0.05–2.0 mm in diameter), silt particles have an intermediate size (0.002–0.05 mm) and clay particles are the smallest (less than 0.002 mm). Soils are a mixture of these three types of particles. Because sandy soils have many large particles, they have large spaces and the water just drains right through them. In contrast to sandy soils, a soil composed mostly of clay particles has small spaces that fill completely with water. Most likely you have experienced the feel of sand and clay in your hand: sand having no moisture flows right through your fingers, while clay clumps together in one large mass because of its water content.

Clay particles have another benefit that sand particles do not have. Clay particles retain minerals because they are negatively charged. Look back at Table 37.1 and note that some minerals are negatively charged and others are positively charged. Clay retains positively charged minerals like calcium (Ca^{2+}) and potassium (K^+) preventing them from being washed away by leaching. Plants later exchange hydrogen ions for these minerals (Fig. 37.3). If rain is too acidic, the excess hydrogen ions displace the positive mineral ions and they drain away—one reason why acid rain kills trees. Also, note that clay particles are unable to retain negatively charged NO_3^-, and therefore the nitrogen content of soil is apt to be low. Legumes (see Fig. 37.6) are sometimes planted to replenish the soil nitrogen in preference to relying solely on the addition of fertilizer.

The type of soil called loam is composed of roughly one-third sand, silt, and clay particles. This combination sufficiently retains water and nutrients while still allowing the drainage necessary to provide air spaces. Roots take up oxygen from air spaces.

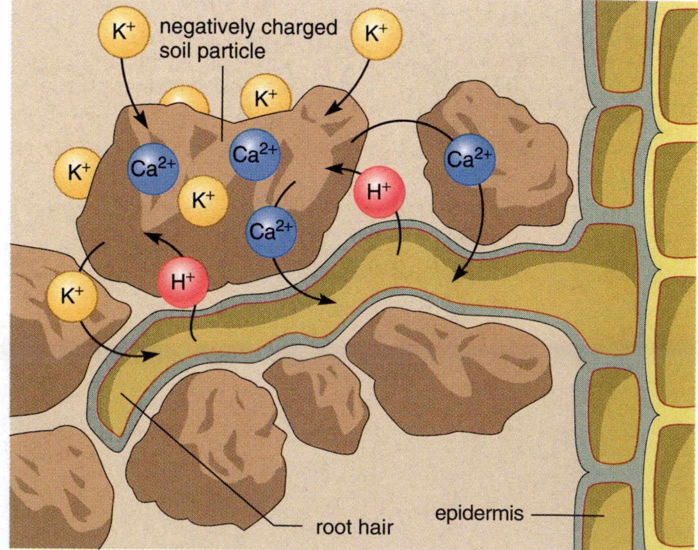

Figure 37.3 Absorbing minerals.
Negatively charged clay particles bind positively charged minerals such as Ca^{2+} and K^+. Plants extract these minerals by exchanging them for H^+. See also Figure 37.5.

Humus Humus, which mixes with the top layer of soil particles, increases the benefits of soil. Plants do well in soils that contain 10–20% humus.

Humus causes soil to have a loose, crumbly texture that allows water to soak in without doing away with air spaces. After a rain, the presence of humus decreases the chances of runoff. Humus swells when it absorbs water and shrinks as it dries. This action helps aerate soil.

Soil that contains humus is nutritious for plants. Humus is acidic; therefore, it retains positively charged minerals until plants take them up. When the organic matter in humus is broken down by bacteria and fungi, inorganic nutrients are returned to plants. Although soil particles are the original source of minerals in soil; recycling of nutrients, as you know, is a major characteristic of ecosystems.

Living Organisms Small plants play a major role in formation of soil from bare rock. Due to the process of succession (see Fig. 24.18), larger plants eventually become dominant in certain ecosystems. The roots of larger plants penetrate soil even to the cracks in hard rock. This action slowly opens up soil layers allowing water, air, and animals to follow.

There are many different types of soil animals. The largest of them, such as moles, badgers, and rabbits disturb and mix soil by burrowing. Smaller animals like earthworms ingest fine soil particles and deposit them on the surface as worm casts. A range of soil animals including mites, springtails, and millipedes help break down leaves and other plant remains by eating them.

The microorganisms in soil such as protozoans, fungi, algae, and bacteria are responsible for the final decomposition of organic remains in humus to inorganic nutrients. Recall that plants are unable to make use of atmospheric nitrogen (N_2) and that soil bacteria play an important nutrient role because they make nitrate available to plants.

Certain soil organisms, such as some roundworms and insects, are parasitic on plants. Insects improve the properties of soil but they are also major crop pests when they feed on plant roots.

Soil particles, humus, and living organisms, give soil the texture it needs to provide air, water, and minerals to plant roots.

Soil Profiles

A **soil profile** is a vertical section from the ground surface to the unaltered rock below. Usually, a soil profile has parallel layers known as **horizons.** Mature soil generally has three horizons (Fig. 37.4). The A horizon is the uppermost (or topsoil) layer that contains litter and humus, although most of the soluble chemicals may have been leached away. The B horizon has little or no organic matter but does contain the inorganic nutrients leached from the A horizon. The C horizon is a layer of weathered and shattered rock.

Because the parent material (rock) and climate (e.g., temperature and rainfall) differ in various parts of the biosphere, the soil profile varies according to the particular ecosystem (see Fig. 26.6). Soils formed in grasslands tend to have a deep A horizon built up from decaying grasses over many years, but because of limited rain, there has been little leaching into the B horizon. In forest soils, both the A and B horizons have enough inorganic nutrients to allow for root growth. In tropical rain forests, the A horizon is more shallow than the generalized profile, the B horizon is deeper, signifying that leaching is more extensive. Since the topsoil of a rain forest lacks nutrients, it can only support crops for a few years.

Soil Erosion

Soil erosion occurs when water or wind carry soil away to a new location. Erosion removes about 25 billion tons of topsoil yearly, worldwide. If this rate of loss continues, the earth will lose practically all of its topsoil by the middle of the next century. Deforestation (removal of trees) and desertification (increase in deserts due to overgrazing and overfarming marginal lands) contribute to the occurrence of erosion and so do poor farming practices.

In the United States, soil is eroding faster than it is being formed on about one-third of all cropland. Agricultural soils along with fertilizers and pesticides are being absorbed into groundwater and rivers and are threatening human health. To maintain productivity of eroding land, more fertilizers, more pesticides, and more energy is often applied. Instead, it is best to stop erosion before it occurs by following sound agricultural practices.

Erosion removes topsoil (A horizon) and seriously reduces the ability of soil to provide plants with the air, water, and minerals they need.

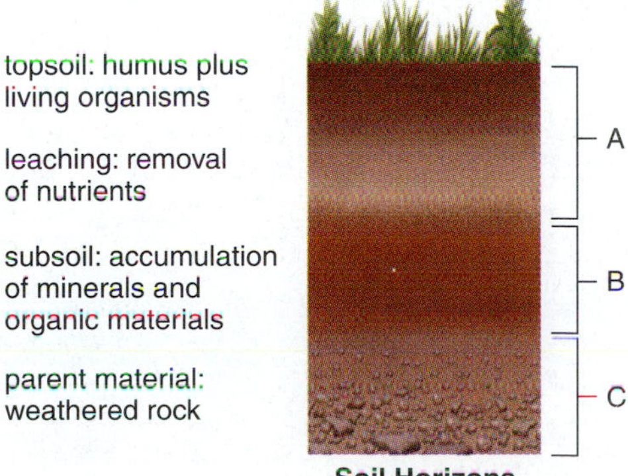

Figure 37.4 Generalized soil profile.
The top layer (A horizon) contains most of the humus; the next layer (B horizon) accumulates materials leached from A horizon; and the lowest layer (C horizon) is composed of weathered parent material. Erosion removes A horizon, a primary source of humus and minerals in soil.

37.2 Uptake of Water and Minerals

The pathway for water and mineral uptake and transport in a plant is the same. As Figure 37.5 shows, water along with minerals can enter the root of a flowering plant from the soil simply by diffusing between the cells. Eventually, however, the **Casparian strip,** a band of suberin and lignin bordering four sides of root endodermal cells, forces water to enter endodermal cells. Alternatively, water can enter epidermal cells at their **root hairs** and then progress through cells across the cortex and endodermis of a root. Regardless of the pathway, water enters root cells when they have a lower osmotic pressure than does soil solution.

In contrast to water, minerals are actively taken up by plant cells. Plants possess an astonishing ability to concentrate minerals—that is, to take up minerals until they are many times more concentrated in the plant than in the surrounding medium. The concentration of certain minerals in roots is as much as 10,000 times greater than in the surrounding soil. Following their uptake by root cells, minerals are secreted into xylem and are transported toward the leaves by the upward movement of water. Along the way, minerals can exit xylem and enter those cells that require them. Some eventually reach leaf cells. In any case, minerals must again cross a selectively permeable plasma membrane when they exit xylem and enter living cells. By what mechanism do minerals cross plasma membranes?

Recall that plant cells absorb minerals in the ionic form: nitrogen is absorbed as nitrate (NO_3^-), phosphorus as phosphate (HPO_4^{2-}), potassium as potassium ions (K^+), and so forth. Since the plasma membrane is charged, ions are likely to have difficulty crossing it. It has long been known that plant cells expend energy to actively take up and concentrate mineral ions. If roots are deprived of oxygen or are poisoned so that cellular respiration cannot occur, mineral ion uptake is diminished. The energy of ATP is required for mineral ion transport, but not directly (Fig. 37.5*b*). A plasma membrane pump, called a proton pump, hydrolyzes ATP and uses the energy released to transport hydrogen ions (H^+) out of the cell. This sets up an electrochemical gradient that drives positively charged ions like K^+ through a channel protein into the cell. Negatively charged mineral ions are transported, along with H^+, by carrier proteins. Since H^+ is moving down its concentration gradient, no energy is required. Notice that this model of mineral ion transport in plant cells is based on chemiosmosis, the establishment of an electrochemical gradient to perform work.

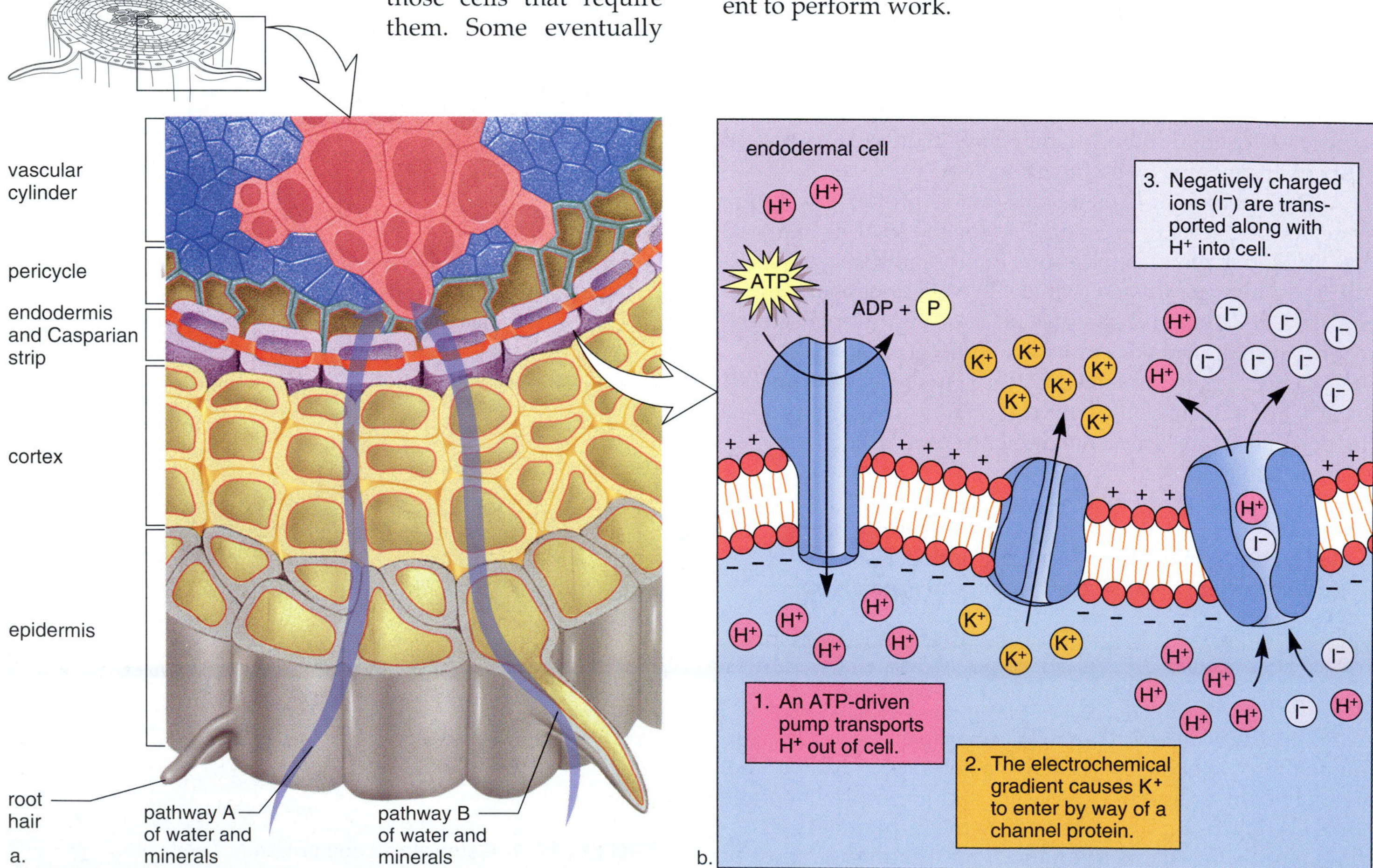

Figure 37.5 Water and mineral uptake.
a. Pathways of water and minerals. Water and minerals can travel via porous cell walls but then must enter endodermal cells because of the Casparian strip (pathway A). Alternatively, water and minerals can enter root hairs and move from cell to cell (pathway B). **b.** Transport of minerals across an endodermal plasma membrane. An ATP-driven pump removes hydrogen ions (H^+) from the cell. This establishes an electrochemical gradient that allows potassium (K^+) and other positively charged ions to cross the membrane via a channel protein. Negatively charged mineral ions (I^-) can cross the membrane by way of a carrier when they "hitch a ride" with hydrogen ions (H^+), which are diffusing down their concentration gradient.

a. Root nodule

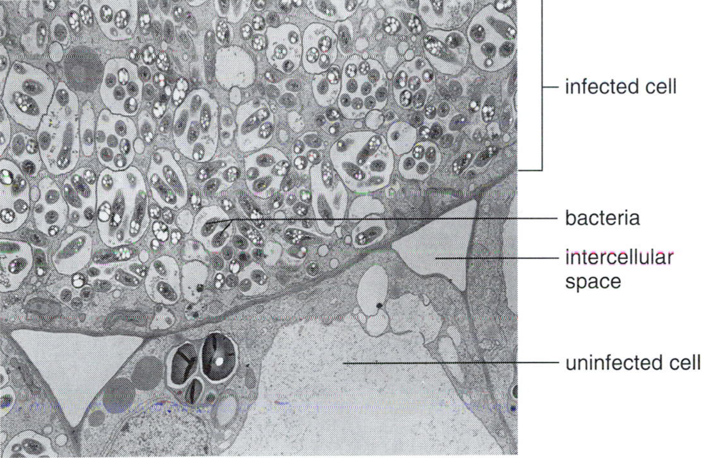

b. Cross section of nodule

Figure 37.6 Root nodules.
a. Nitrogen-fixing bacteria live in nodules on the roots of plants, particularly legumes. **b.** In infected cells bacteria fix atmospheric nitrogen and make reduced nitrogen available to a plant. The plant passes carbohydrates to the bacteria.

Adaptations of Roots for Mineral Uptake

Two symbiotic relationships are known to assist roots in taking up mineral nutrients. Some plants, such as legumes, soybeans, and alfalfa, have roots infected by *Rhizobium* bacteria, which can fix atmospheric nitrogen (N_2). They break the $N \equiv N$ bond and reduce nitrogen to NH_4^+ for incorporation into organic compounds. The bacteria live in **root nodules** [L. *nodulus,* dim. of *nodus,* knot] and are supplied with carbohydrates by the host plant (Fig. 37.6). The bacteria, in turn, furnish their host with nitrogen compounds.

The second type of symbiotic relationship, called a mycorrhizal association, involves fungi and almost all plant roots (Fig. 37.7). Only a small minority of plants do not have **mycorrhizae** [Gk. *mykes,* fungus, and *rhiza,* root], and these plants are most often limited as to the environment in which they can grow. Ectomycorrhizae form a mantle that is exterior to the root, and they grow between cell walls.

Figure 37.7 Mycorrhizae.
Experimental results show that rough lemon plants with mycorrhizae *(right)* grow much better than plants without mycorrhizae *(left).*

Endomycorrhizae can penetrate cell walls. In any case, the fungus increases the surface area available for mineral and water uptake and breaks down organic matter, releasing nutrients that the plant can use. In return, the root furnishes the fungus with sugars and amino acids. Plants are extremely dependent on mycorrhizae. Orchid seeds, which are quite small and contain limited nutrients, do not germinate until a mycorrhizal fungus has invaded their cells. Nonphotosynthetic plants, such as Indian pipe, use their mycorrhizae to extract nutrients from nearby trees.

Some plants have poorly developed roots or no roots at all because minerals and water are supplied by other mechanisms. **Epiphytes** [Gk. *epi,* over, and *phyton,* plant] are "air plants"; they do not grow in soil but on larger plants, which give them support. They do not receive nutrients from their host, however. Some epiphytes have roots that absorb moisture from the atmosphere, and many catch rain and minerals in special pockets at the base of their leaves. Parasitic plants such as dodders, broomrapes, and pinedrops send out rootlike projections called haustoria that tap into the xylem and phloem of the host stem (see Fig. 36.11*d*). Carnivorous plants such as the Venus's-flytrap and sundews obtain some nitrogen and minerals when their leaves capture and digest insects (see Fig. 36.20*c*).

Mineral uptake follows the same pathway as water uptake; however, mineral uptake takes an expenditure of energy. Most plants are assisted in acquiring minerals by a symbiotic relationship with microorganisms and/or fungi.

37.3 Transport Mechanisms in Plants

Flowering plants are well adapted to living in a terrestrial environment. Their leaves, which carry on photosynthesis, are positioned to catch the rays of the sun because they are held aloft by the stem (Fig. 37.8). Carbon dioxide enters leaves at the stomates but water, the other main requirement for photosynthesis, is absorbed by the roots. Water must be transported from the roots through the stem to the leaves. Vascular plants have a transport tissue, called **xylem,** that moves water and minerals from the roots to the leaves. Xylem contains two types of conducting cells: tracheids and vessel elements. **Tracheids** have end walls that are pitted, but the vessel elements have no end walls at all. The **vessel elements** placed end to end form a completely hollow pipeline from the roots to the leaves. Xylem, with its strong-walled nonliving cells gives trees much-needed internal support.

The process of photosynthesis results in sugars which are utilized as a source of energy and building blocks for other organic molecules throughout a plant. **Phloem** is the type of vascular tissue that transports organic nutrients to all parts of the plant. Roots buried in the soil cannot possibly carry on photosynthesis, but they still require a source of energy in order to carry on cellular metabolism. Vascular plants are able to transport the products of photosynthesis to regions that require them and/or where they will be stored for future use. The conducting cells in phloem are **sieve-tube elements,** each of which typically has a companion cell. **Companion cells** can provide proteins to sieve-tube elements, which contain cytoplasm but have no nucleus. The end walls of sieve-tube elements are called sieve plates because they contain numerous pores. The sieve-tube elements are aligned end to end, and strands of cytoplasm called plasmodesmata extend from one cell to the other through the sieve plates. In this way, sieve-tube elements form a continuous pathway for organic nutrient transport throughout the plant.

Knowing that vascular plants are structured in a way that allows materials to move from one part to another does not tell us the mechanisms by which they move. Plant physiologists have performed numerous experiments to determine how water and minerals rise to the top of very tall trees in xylem and how organic nutrients move in the opposite direction in phloem. We would expect these mechanisms to be mechanical in nature and based on the properties of water because water is a large part of both xylem sap and phloem sap. As you know, water molecules diffuse freely across plasma membranes from the area of higher concentration to the area of lower concentration. Botanists favor describing the movement of water in terms of water potential: water always flows passively from the area of higher water potential to the area of lower water potential. As you can see on the next page the concept of water potential has the benefit of considering water pressure in addition to osmotic pressure.

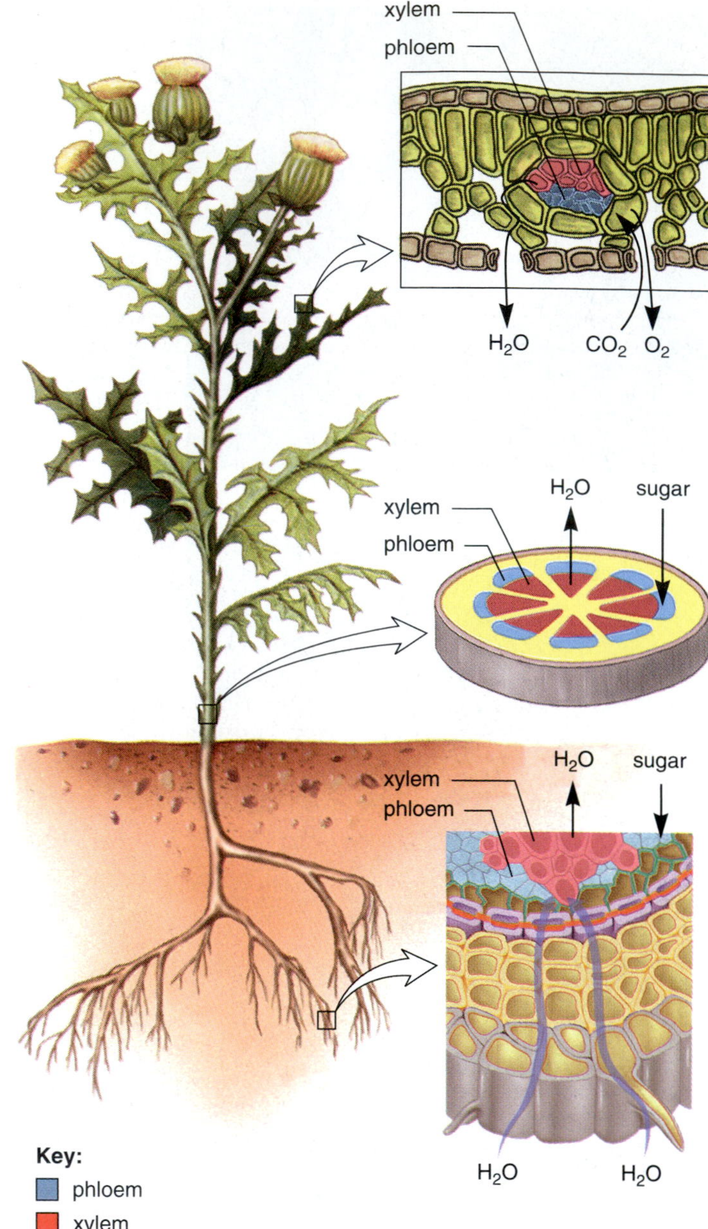

Figure 37.8 Plant transport system.
Vascular tissue in plants includes xylem, which transports water and minerals from the roots to the leaves, and phloem, which transports organic nutrients particularly in the opposite direction. Notice that the xylem and the phloem are in the roots, the stem, and the leaves, which are the vegetative organs of a plant.

Chemical properties of water are also important in movement of xylem sap. The polarity of water molecules, and the hydrogen bonding between water molecules allows water to fill xylem vessels.

Vascular plants have transport tissues; xylem transports water and minerals from the roots to the leaves, and phloem transports organic nutrients to various parts, particularly from the leaves to the roots.

A Closer Look

The Concept of Water Potential

As you may know, potential energy is stored energy due to the position of an object. A boulder placed at the top of a hill has potential energy. When pushed, the boulder moves down the hill as potential energy is converted into kinetic (motion) energy. The boulder at the bottom of the hill has lost much of its potential energy.

Let us define **water potential** as the energy of water. Just like the boulder, water at the top of a waterfall has a higher water potential than water at the bottom of the waterfall. As illustrated by this example, water moves from a region of higher water potential to a region of lower water potential.

In terms of cells, two factors usually determine water potential, which in turn determines the direction in which water will move across a plasma membrane. These factors concern differences in:

1. water pressure across a membrane
2. solute concentration across a membrane

Pressure potential is the effect that pressure has on water potential. With regard to pressure, it is obvious to us that water will move across a membrane from the area of higher pressure to the area of lower pressure. The higher the water pressure, the higher the water potential. The lower the water pressure, the lower the water potential and the more likely water will flow in that direction. Pressure potential is the concept that best explains the movement of sap in xylem and phloem.

To fully explain the movement of water into plant cells, the concept of osmotic potential is also required. Osmotic potential takes into account the effects of solutes on the movement of water. The presence of solutes restricts the movement of water because water tends to interact with solutes. Indeed, water tends to move across a membrane from the area of lower solute concentration to the area of higher solute concentration. The lower the concentration of solutes (osmotic potential) the higher the water potential. The higher the concentration of solutes, the lower the water potential and the more likely that water will flow in that direction.

Not surprisingly, increasing water pressure will counter the tendency of water to enter a cell because of the presence of solutes. Let us consider a common situation in regard to plant cells. As water enters a plant cell by osmosis, water pressure will increase inside the cell—a plant cell has a strong cell wall that allows water pressure to build up. When will water stop entering the cell? When the pressure potential inside the cell increases and balances the osmotic potential outside the cell.

Pressure potential that increases due to the process of osmosis is often called turgor pressure. Turgor pressure is critical, since plants depend on it to maintain the turgidity of their bodies (Fig. 37A). The cells of a wilted plant have insufficient turgor pressure and the plant droops as a result.

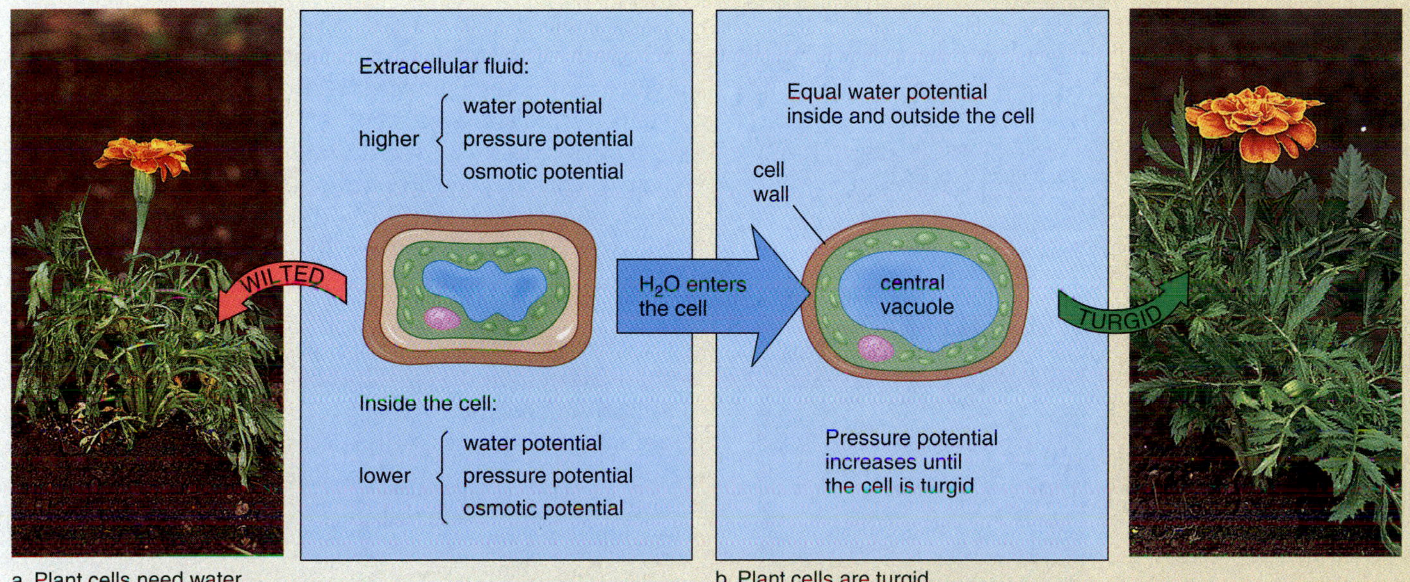

a. Plant cells need water.

b. Plant cells are turgid.

Figure 37A Water potential and turgor pressure.
a. Water flows from an area of higher water potential to an area of lower water potential. **b.** The cells of a wilted plant have a lower water potential; therefore, water enters the cells. Equilibrium is achieved when the water potential is equal inside and outside the cell. Cells are now turgid and the plant is no longer wilted.

Water Transport

Water and minerals enter the root of a plant, by the two pathways described in Figure 37.5. Eventually, water and minerals reach xylem which contains tracheids and vessel elements. The vessel elements are larger than the tracheids, and they are stacked one on top of the other to form vessels that stretch from the roots to the leaves. Vessels constitute an open pipeline because the vessel elements have no end walls separating one from the other. The tracheids, which are elongated with tapered ends, form a less obvious means of transport, but water can move across the end and side walls of tracheids because of pits, or depressions, where the secondary wall does not form (Fig. 37.9).

Water entering root cells creates a positive pressure called **root pressure.** Root pressure, which primarily occurs at night, tends to push xylem sap upward. Root pressure may be responsible for **guttation** (L. *gutta,* drops, spots] when drops of water are forced out of vein endings along the edges of leaves (Fig. 37.10). Although root pressure may contribute to the upward movement of water in some instances, it is not believed to be the mechanism by which water can rise to the tops of very tall trees.

Figure 37.10 Guttation.
Drops of guttation water on the edges of a strawberry leaf. Guttation, which occurs at night, may be due to root pressure. Root pressure is a positive pressure potential caused by the entrance of water into root cells.

Figure 37.9 Conducting cells of xylem.
a. Tracheids are long, hollow cells with tapered ends. Water can move into and out of tracheids through pits only. **b.** Vessel elements are wider and shorter than tracheids. Water can move from vessel element to vessel element through perforations (a large hole at each end). Vessel elements can also exchange water with tracheids through pits.

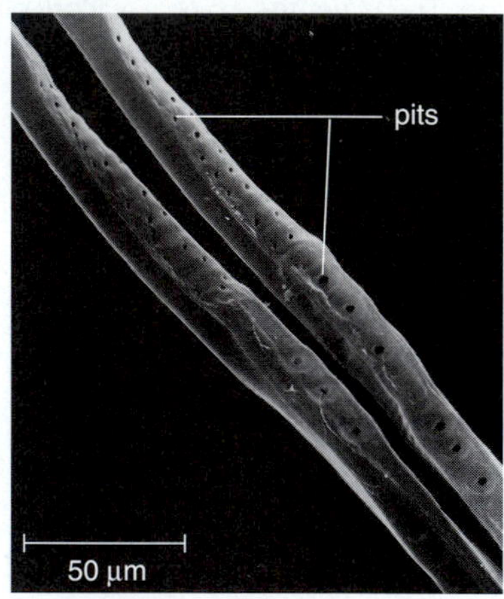

a. Tracheids

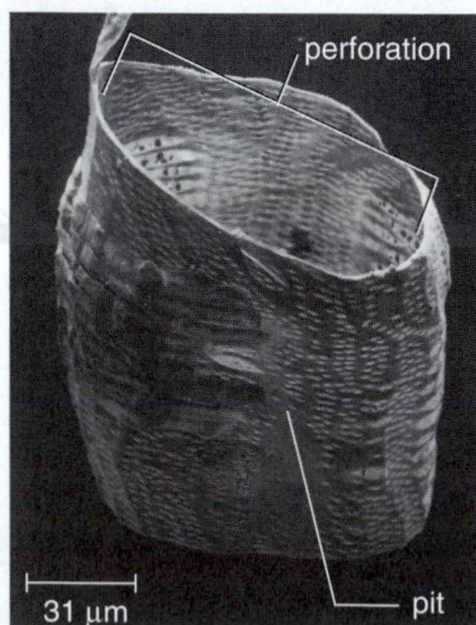

b. Vessel element

Cohesion-Tension Model of Xylem Transport

Once water enters xylem, it must be transported upward the entire height of a tree. This can be a daunting task, given that redwood trees can exceed 90 meters in height.

The **cohesion-tension model** of xylem transport outlined in Figure 37.11 suggests a mechanism for xylem transport. The term *cohesion* refers to the tendency of water molecules to cling together. Because of hydrogen bonding, water molecules interact with one another, and there is a continuous water column in xylem from the leaves to the roots that is not easily broken. Adhesion refers to the ability of water, a polar molecule, to interact with the molecules making up the walls of the vessels in xylem. Adhesion is a property of water that gives the water column extra strength and prevents it from slipping back.

Why does the continuous water column move upward? Consider the structure of a leaf. When the sun rises, the stomates open and carbon dioxide enters a leaf. Within the leaf the mesophyll cells—particularly the spongy layer—are exposed to the air, which can be quite dry. Water now evaporates from mesophyll cells. Evaporation of water from leaf cells is called **transpiration.** At least 90% of the water taken up by the roots is eventually lost by transpiration. This means that the total amount of water lost by a plant over a long period of time is surprisingly large. A single *Zea mays* (corn) plant loses somewhere between 135 and 200 liters of water through transpiration during a growing season.

The water molecules that evaporate are replaced by other water molecules from the leaf veins. In this way, transpiration exerts a driving force, that is, a *tension* which draws the water column up in vessels from the roots to leaves. As transpiration occurs, the water column is pulled upward, first within the leaf, then from the stem, and finally from the roots.

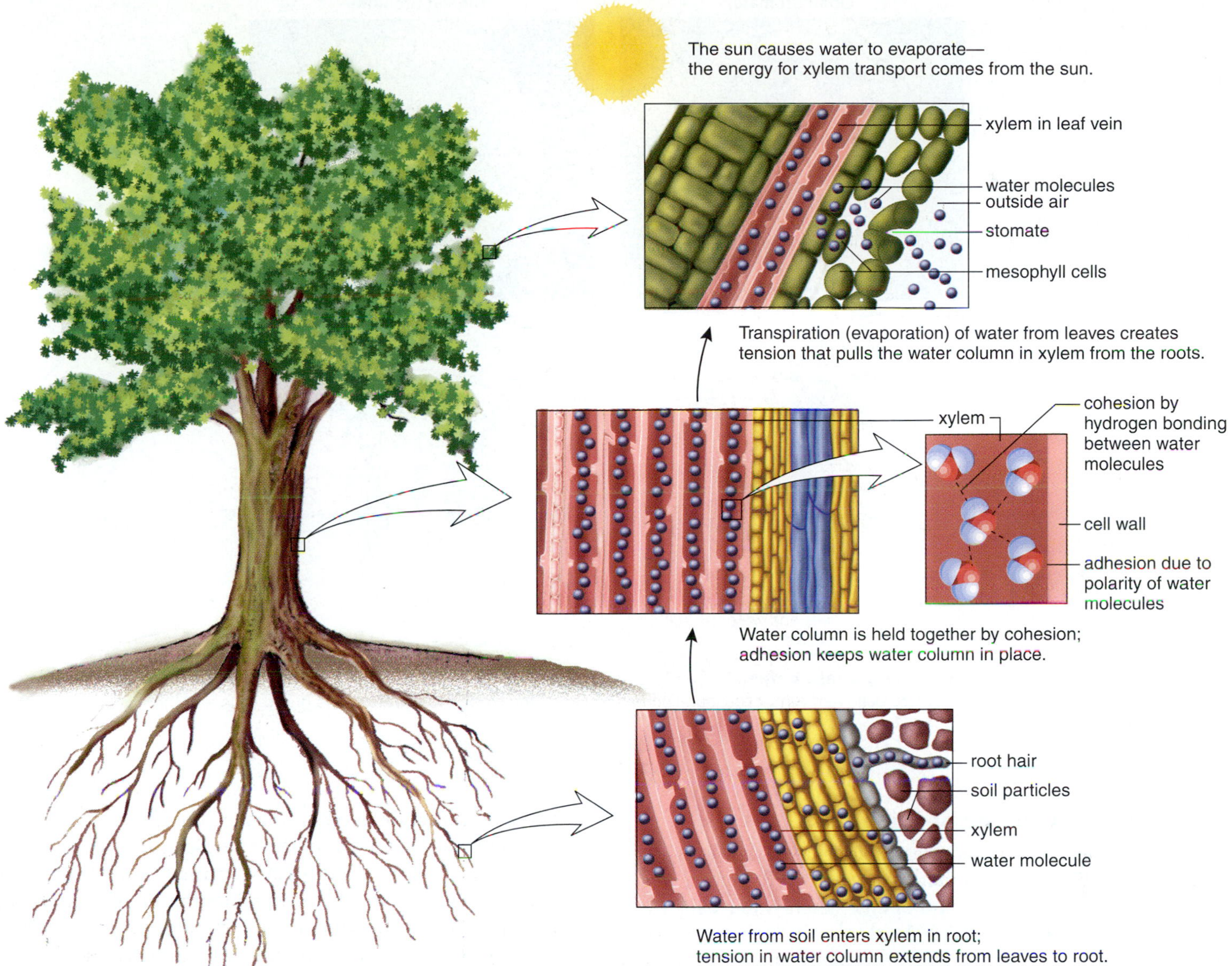
The sun causes water to evaporate—
the energy for xylem transport comes from the sun.
xylem in leaf vein
water molecules
outside air
stomate
mesophyll cells
Transpiration (evaporation) of water from leaves creates
tension that pulls the water column in xylem from the roots.
xylem
cohesion by
hydrogen bonding
between water
molecules
cell wall
adhesion due to
polarity of water
molecules
Water column is held together by cohesion;
adhesion keeps water column in place.
root hair
soil particles
xylem
water molecule
Water from soil enters xylem in root;
tension in water column extends from leaves to root.

Figure 37.11 **Cohesion-tension model of xylem transport.**
Tension created by evaporation (transpiration) at the leaves pulls water along the length of the xylem—from the root hairs to the leaves.

Tension can reach from the leaves to the root only if the water column is continuous. What happens if the water column within xylem is broken, as by cutting a stem? The water column "snaps back" down the xylem vessel away from the site of breakage, making it more difficult for conduction to occur. This is why it is best to maximize water conduction by cutting flower stems under water. This effect has also allowed investigators to measure the tension in stems. A device called the pressure bomb measures how much pressure it takes to push the xylem sap back to the cut surface of the stem.

There is an important consequence to the way water is transported in plants; when a plant is under water stress, the stomates close. Now the plant loses little water because the leaves are protected against water loss by the waxy **cuticle** of the upper and lower epidermis. When the stomates are closed, however, carbon dioxide cannot enter the leaves and plants are unable to photosynthesize. Photosynthesis, therefore, requires an abundant supply of water so that the stomates can remain open and allow carbon dioxide to enter.

Water is transported from the roots to the leaves in xylem. The water column remains intact as transpiration pulls water from the roots to the leaves.

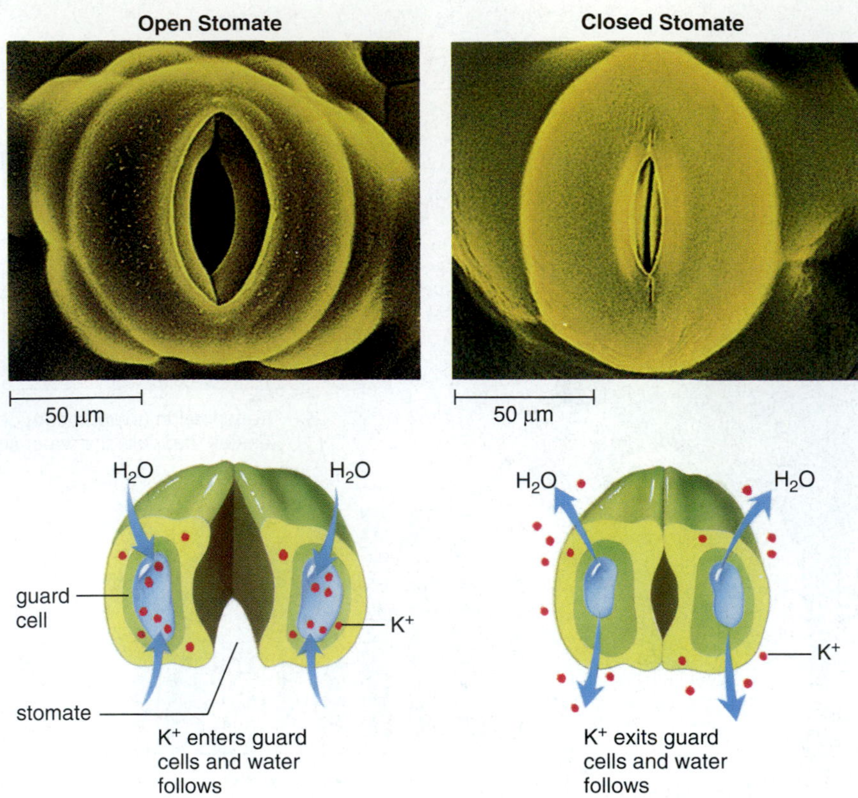

Figure 37.12 Opening and closing of stomates.
Stomates open when water enters guard cells and turgor pressure increases. Stomates close when water exits guard cells and there is a loss of turgor pressure.

Opening and Closing of Stomates

Each **stomate,** a small pore in leaf epidermis, is bordered by **guard cells.** When water enters the guard cells and turgor pressure increases, the stomate opens; when water exits the guard cells and turgor pressure decreases, the stomate closes. Notice in Figure 37.12 that the guard cells are attached to each other at their ends and that the inner walls are thicker than the outer walls. When water enters, a guard cell's radial expansion is restricted because of cellulose microfibrils in the walls, but lengthwise expansion of the outer walls is possible. When the guard cells expand lengthwise, they buckle out from the region of their attachment and the stomate opens.

Since about 1968 it has been clear that there is an accumulation of potassium ions (K^+) within guard cells when stomates open. In other words, active transport of K^+ into guard cells causes water to follow by osmosis and stomates to open. Also interesting is the observation that there is an accumulation of hydrogen ions (H^+) outside guard cells as K^+ moves into them. A proton pump run by the hydrolysis of ATP transports H^+ to the outside of the cell. This establishes an electrochemical gradient that allows K^+ to enter by way of a channel protein (see Fig. 37.5*b*).

What regulates the opening and closing of stomates? It appears that the blue-light component of sunlight is a signal that can cause stomates to open. There is evidence to suggest that a flavin pigment absorbs blue light, and then this pigment sets in motion the cytoplasmic response that leads to activation of the proton pump. Similarly, there could be a receptor in the plasma membrane of guard cells that brings about inactivation of the pump when carbon dioxide (CO_2) concentration rises, as might happen when photosynthesis ceases. Abscisic acid (ABA), which is produced by cells in wilting leaves, can also cause stomates to close. Although photosynthesis cannot occur, water is conserved.

If plants are kept in the dark, the stomates open and close on a 24-hour basis just as if they were responding to the presence of sunlight in the daytime and the absence of sunlight at night. This means that there must be some sort of internal biological clock that is keeping time. Circadian rhythms (a behavior that occurs every 24 hours) and biological clocks are areas of intense investigation at this time.

When stomates open, first K^+ and then water enters guard cells. Stomates open and close in response to environmental signals, and the exact mechanism is being investigated.

Competition for Resources and Biodiversity

From my earliest years in school, I have loved both mathematics and biology. In college, the biological issues that intrigued me most were those concerned with the relationship of species to their environment, including the effects of interactions such as competition or predation. It seemed to me that biodiversity might relate to these interactions and, if so, it might be possible to develop a theory, expressed mathematically, to explain biodiversity. The first step toward this goal is to establish that competition and/or predation do relate to biodiversity.

The Minnesota grasslands in which I work often have more than 100 plant species coexisting in an area the size of a few hectares. Long-term experiments have shown that all these plant species are held in check by (limited by) competition for the same resource, namely soil nitrogen. Mathematical models predict that the number of coexisting species can never be more than the number of resources that limit them. Is there an answer to this seeming paradox? There is if these species are competing with one another on some other basis besides soil nitrogen. To find out what this basis might be, we did a series of experiments. We found that the factor limiting these species was not insect or mammalian herbivores, light availability, or fire. However, the plant species did differ in their ability to disperse to new sites. We

Figure 37B
Little bluestem, *Schizachyrium scoparium*, is a North American bunchgrass that prefers sandy or rocky habitats. It is an excellent competitor for soil nitrogen because of its high allocation of roots.

G. David Tilman
University of Minnesota

discovered this by planting over 50 different plant species in several plots and finding how well they could germinate, grow, and reproduce there. The best competitors for nitrogen were native bunchgrasses that allocated 85% of their biomass to root but only 0.5% of their biomass to seed. Little bluestem *(Schizachyrium scoparium)* is an example of a bunchgrass (Fig. 37B). The best dispersers allocated 30% of their biomass to seed, but only 40% of their biomass to root. Bent grass *(Agrostis scabra)* is an example of a poor competitor for soil nitrogen but a good disperser (Fig. 37C).

A mathematical model showed that such allocation-based trade-offs could explain the stable coexistence of a whole range of plant species that differ according to their ability to compete for nitrogen and to disperse to new areas. Coexistence occurs because better competitors for soil nitrogen are poorer dispersers, and therefore they do not occupy all sites. Better dispersers are also better at finding and occupying all sites. Because their ability to compete and disperse vary, a number of species can coexist.

Another issue of interest to me is the relationship between biodiversity and the stability of an ecosystem. My colleagues and I were annually sampling 207 permanent plots when there was a drought in 1987–88. The 1988 drought was the third worst in the previous 150 years. We found that plots that contained only one to four species had their productivity (total mass of living plants) fall to between 1/8 and 1/16 of the pre-drought level. But plots that contained 16 to 26 species were able to maintain their productivity at about 1/2 the pre-drought level (Fig. 37D). This suggests that high biodiversity does buffer ecosystems against a disturbance and that it is wise to conserve the biodiversity of ecosystems in all areas—whether in Minnesota, New Jersey, Oregon, or the tropics.

Courtesy of G. David Tilman, University of Minnesota

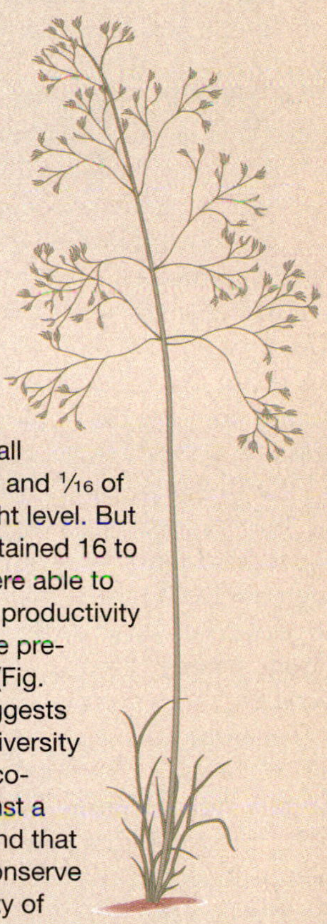

Figure 37C
Bent grass, *Agrostis scabra*, is a native North American prairie grass that is a poor competitor for soil nitrogen. It disperses rapidly into disturbed areas because of its high allocation of seed.

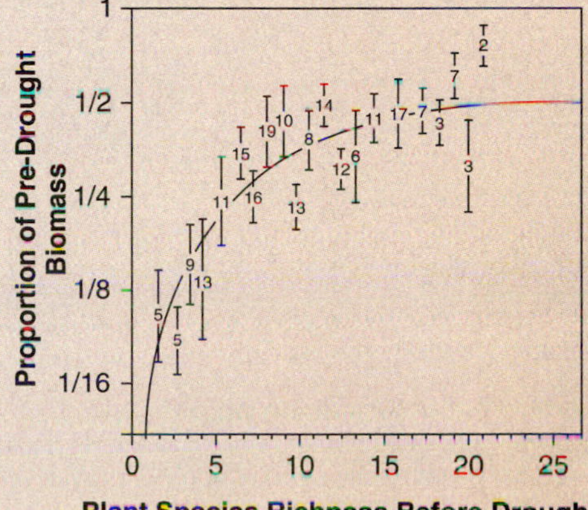

Figure 37D
During drought, plots with higher plant species richness attained plant biomasses that were a larger proportion of their pre-drought plant biomass. This graph compares biomass in 1988, a drought year, with that in 1986, two years before the drought. The numbers placed at the means indicate the number of plots used to calculate that mean. The high and low values for each biomass ratio are also indicated.

Organic Nutrient Transport

Not only do plants transport water and minerals from the roots to the leaves, they also transport organic nutrients to the parts of plants that have need of them. This includes young leaves that have not yet reached their full photosynthetic potential; flowers that are in the process of making seeds and fruits; and the roots whose location in the soil prohibits them from carrying on photosynthesis

Role of Phloem

As long ago as 1679, Marcello Malpighi suggested that bark is involved in translocating sugars from leaves to roots. He observed the results of removing a strip of bark from around a tree, a procedure called **girdling.** If a tree is girdled below the level of the majority of leaves, the bark swells just above the cut and sugar accumulates in the swollen tissue. We know today that when a tree is girdled, the phloem is removed but the xylem is left intact. Therefore, the results of girdling suggest that phloem is the tissue that transports sugars.

Radioactive tracer studies with ^{14}C have confirmed that phloem transports organic nutrients. When ^{14}C-labeled carbon dioxide (CO_2) is supplied to mature leaves, radioactively labeled sugar is soon found moving down the stem into the roots. It's difficult to get samples of sap from phloem without injuring the phloem, but this problem is solved by using aphids, small insects that are phloem feeders. The aphid drives its stylet, which is a sharp mouthpart that functions like a hypodermic needle between the epidermal cells and withdraws sap from a sieve-tube element (Fig. 37.13). If the aphid is anesthetized using ether, its body can be carefully cut away, leaving the stylet. Phloem can then be collected and analyzed by a researcher. By the use of radioactive tracers and aphids , it is known that the movement through phloem can be as fast as 60–100 cm per hour and possibly up to 300 cm per hour.

Pressure-Flow Model of Phloem Transport

The **pressure-flow model** is a current explanation for the movement or organic materials in phloem (Fig. 37.14). Consider the following experiment in which two bulbs are connected by a glass tube. The first bulb contains solute at a higher concentration than the second bulb. Each bulb is bounded by a differentially permeable membrane, and the entire apparatus is submerged in distilled water:

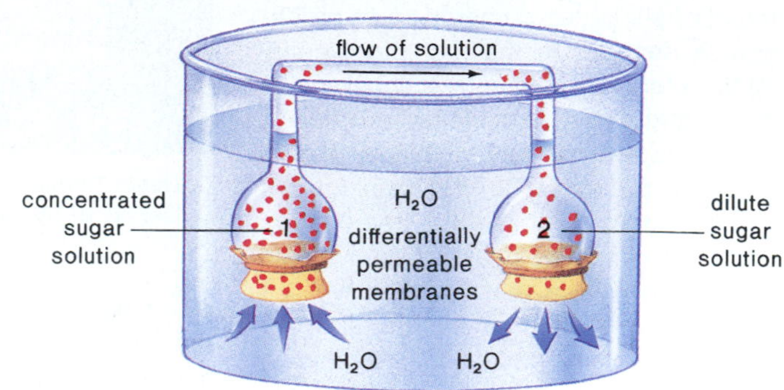

Distilled water flows into the first bulb because it has the higher solute concentration. The entrance of water creates a *pressure* and water *flows* toward the second bulb. This flow not only drives water toward the second bulb, it also provides enough force for water to move out through the membrane of the second bulb—even though the second bulb contains a higher concentration of solute than the distilled water.

In plants, sieve tubes are analogous to the glass tube that connects the two bulbs. Sieve tubes are composed of sieve-tube elements, each of which has a companion cell. It's possible the companion cells assist the sieve-tube elements in some way. The sieve-tube elements align end to end and strands of plasmodesmata (cytoplasm) extend through sieve plates from one sieve-tube element to the other. Sieve tubes,

a. An aphid feeding on a plant stem

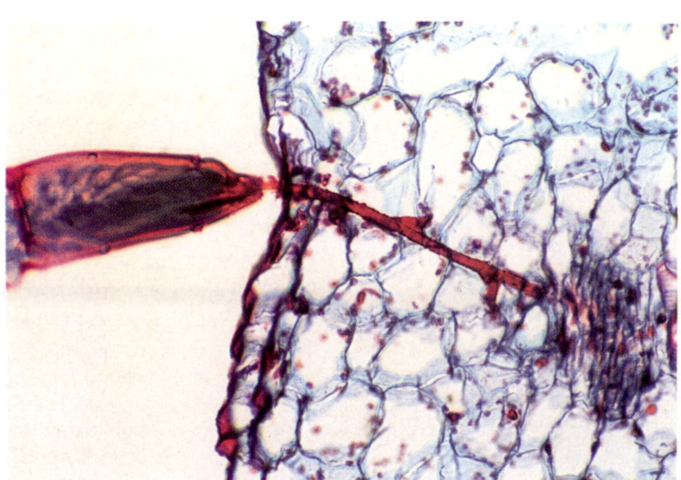

b. Aphid stylet in place

Figure 37.13 **Acquiring phloem sap.**
Aphids are small insects that remove nutrients from phloem by means of a needlelike mouthpart called a stylet. **a.** Excess phloem sap appears as a droplet after passing through the aphid's body. **b.** Micrograph of stylet in plant tissue. When an aphid is cut away from its stylet, phloem sap becomes available for collection and analysis.

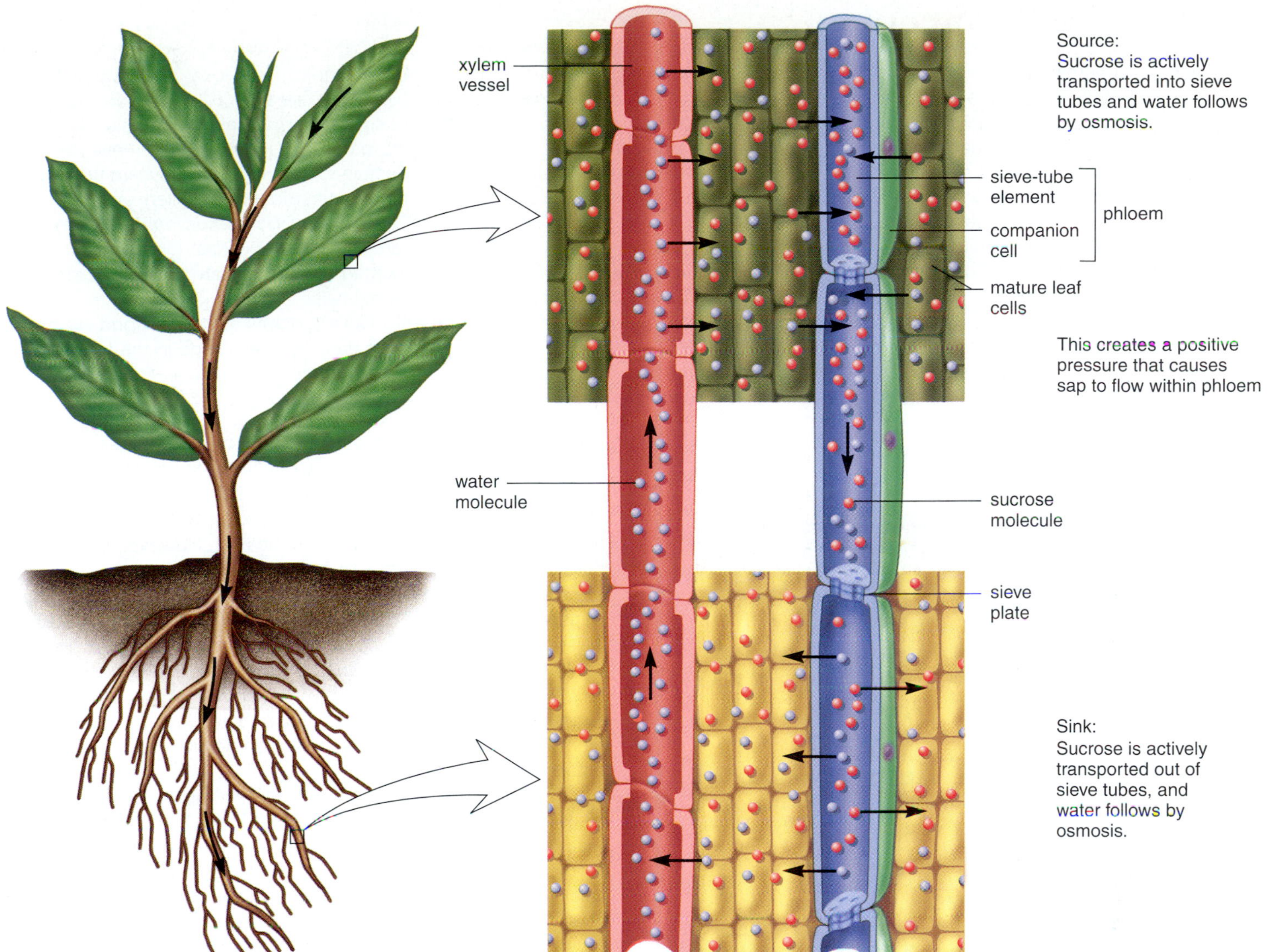
xylem vessel
Source:
Sucrose is actively transported into sieve tubes and water follows by osmosis.
sieve-tube element
companion cell
phloem
mature leaf cells
This creates a positive pressure that causes sap to flow within phloem
water molecule
sucrose molecule
sieve plate
Sink:
Sucrose is actively transported out of sieve tubes, and water follows by osmosis.

Figure 37.14 Pressure-flow model of phloem transport.
Sugar and then water enter sieve tubes at a source, in this case the leaves. This creates a positive pressure, which causes phloem contents to flow. Sieve tubes (composed of sieve-tube elements) form a continuous pipeline from a source to a sink, in this case a root, where sugar and then water exit sieve tubes.

therefore, form a continuous pathway for organic nutrient transport throughout a plant.

During the growing season, photosynthesizing leaves are producing sugar. Therefore, they are a source of sugar. This sugar is actively transported into phloem. Again, transport is dependent upon an electrochemical gradient established by a proton pump. Sucrose is carried across the membrane in conjunction with hydrogen ions (H^+), which are moving down their concentration gradient (see Fig. 37.5). After sugar enters sieve tubes, water follows passively by osmosis. The buildup of water within sieve tubes creates the positive pressure that starts a flow of phloem sap. The roots (and other growth areas) are a sink for sugar—the roots are removing sugar. After sugar is actively transported out of sieve tubes, water follows passively by osmosis. Now, phloem sap continues to flow from the leaves (source) to the roots (sink).

The pressure-flow model of phloem transport can account for any direction of flow in sieve tubes if we consider that the direction of flow is always from source to sink. For example, recently formed leaves can be a sink and they will receive sucrose until they begin to maximally photosynthesize.

The pressure-flow model of phloem transport suggests that phloem sap can move either up or down as appropriate for the plant at a particular time in its life cycle.

Connecting Concepts

The land environment offers many advantages for plants, such as greater availability of light and carbon dioxide for photosynthesis. (Water, even if clear, filters out light, and carbon dioxide concentration and rate of diffusion is less in water.) The evolution of a transport system was critical, however, for plants to make full use of these advantages. Only if a transport system is present can plants elevate the leaves so that they are better exposed to solar energy and carbon dioxide in the air. A transport system brings water, a raw material of photosynthesis, from the roots to the leaves, and also brings the products of photosynthesis down to the roots. Roots lie beneath the soil, and their cells depend on an input of organic food from the leaves in order to remain alive. An efficient transport system allows roots to penetrate deeply into the soil to absorb water and minerals.

The presence of a transport system also allows materials to be distributed to those parts of the body that are growing most rapidly. New leaves and flower buds would grow rather slowly if they had to depend on their own rate of photosynthesis, for example. Height in flowering plants, due to the presence of a transport system, has other benefits aside from elevation of leaves. It is also adaptive to have reproductive structures located where the wind can better distribute pollen and seeds. Once animal pollination came into existence, it was beneficial for flowers to be located where they are more easily seen by animals.

Clearly, plants with a transport system have a competitive edge in the terrestrial environment.

Summary

37.1 Plant Nutrition and Soil

Plants need both essential and beneficial inorganic nutrients. Carbon, hydrogen, and oxygen make up 96% of a plant's dry weight. The other necessary nutrients are taken up by the roots as mineral ions. Even nitrogen (N), which is present in the atmosphere, is most often taken up as NO_3^-.

You can determine mineral requirements by hydroponics, which is growing plants in a solution. The solution is varied by the omission of one mineral. If the plant grows poorly, then the missing mineral is required for growth.

Terrestrial life is dependent on soil which forms by the weathering of rock. Organisms contribute to the formation of humus and soil. Soil is a mixture of soil particles, humus, living organisms, air, and water. Soil particles are of three types from the largest to the smallest: sand, silt, and clay. Loam which contains about equal proportions of all three types retains water but still has air spaces. Humus contributes to the texture of soil and its ability to provide inorganic nutrients to plants. Topsoil (A horizon of a soil profile) contains humus and this is the layer that is lost by erosion; a worldwide problem.

37.2 Uptake of Water and Minerals

Water along with minerals can enter a root by moving between the cells until it reaches the Casparian strip, after which it passes through an endodermal cell before entering xylem. Water can also enter root hairs, then pass through the cells of the cortex and endodermis to reach xylem.

Mineral ions cross plasma membranes by a chemiosmotic mechanism. A proton pump transports H^+ out of the cell. This establishes an electrochemical gradient that causes positive ions to flow into the cells. Negative ions are carried across in conjunction with H^+, which is moving along its concentration gradient.

Plants have various adaptations that assist them in acquiring nutrients. Legumes have nodules infected with the bacterium *Rhizobium*, which makes nitrogen compounds available to these plants. Many other plants have mycorrhizae, or fungus roots. The fungus gathers nutrients from the soil, and the root provides the fungus with sugars and amino acids. Some plants have poorly developed roots. Most epiphytes live on, but do not parasitize, trees, whereas dodder and some other plants parasitize their host.

37.3 Transport Mechanisms in Plants

As an adaptation to life on land, plants have a vascular system that transports water and minerals from the roots to the leaves and must also transport the products of photosynthesis in the opposite direction. Vascular tissue includes xylem and phloem.

In xylem, vessels composed of vessel elements aligned end to end form an open pipeline from the roots to leaves. Particularly at night, root pressure can build in the root. However, this does not contribute significantly to xylem transport.

The cohesion-tension model of xylem transport states that transpiration creates a tension that pulls water upward in xylem. This means of transport works only because water molecules are cohesive with one another and adhesive with xylem walls.

Most of the water taken in by a plant is lost through stomates by transpiration. Only when there is plenty of water do stomates remain open, allowing carbon dioxide to enter the leaf and photosynthesis to occur.

Stomates open when guard cells take up water. They stretch lengthwise, and this causes them to buckle out. Water enters the guard cells after potassium ions (K^+) have entered. Light signals stomates to open, and high carbon dioxide (CO_2) signals stomates to close. Abscisic acid produced by wilting leaves also signals for closure.

In phloem, sieve tubes composed of sieve-tube elements aligned end to end form a continuous pipeline from the leaves to the roots. Sieve-tube elements have sieve plates through which plasmodesmata (strands of cytoplasm) extend from one to the other. The pressure-flow model of phloem transport proposes that a positive pressure drives sap in sieve tubes. Sucrose is actively transported into sieve tubes—also by a chemiosmotic mechanism—at a source, and water follows by osmosis. The resulting increase in pressure creates a flow that moves water and sucrose to a sink. A sink can be at the roots or any other part of the plant that requires organic nutrients.

Reviewing the Chapter

1. Name the elements that make up most of a plant's body. What are essential mineral nutrients and beneficial mineral nutrients? 666
2. Briefly describe the use of hydroponics to determine the mineral nutrients of a plant. 667
3. How is soil formed, and how do the components of soil join to provide nutrients to plants? Describe a generalized soil profile and how a profile is affected by erosion. 668
4. Give two pathways by which water and minerals can cross the epidermis and cortex of a root. What feature allows endodermal cells to regulate the entrance of molecules into the vascular cylinder? 670

5. Describe the chemiosmotic mechanism by which mineral ions cross plasma membranes. 670
6. Name two symbiotic relationships that assist plants in taking up minerals and two types of plants that tend not to take up minerals by roots from soil. 671
7. A vascular system is adaptive for a land existence. Explain. Describe the composition of a plant's vascular system. 672
8. What is root pressure, and why can't it account for transport of water in xylem? 674
9. Describe and give evidence for the cohesion-tension model of water transport. 674
10. Describe the structure of stomates and explain how they can open and close. By what mechanism do guard cells take up potassium (K^+) ions? 676
11. What data are available to show that phloem transports organic compounds? Explain the pressure-flow model of phloem transport. 678–79

Testing Yourself

Choose the best answer for each question.

1. Which of these is not an inorganic nutrient for plants?
   a. water
   b. carbon dioxide gas
   c. mineral ions
   d. nitrogen gas
   e. None of these are nutrients.
2. Which is a component of soil?
   a. soil particles
   b. humus
   c. organisms
   d. air and water
   e. All of these are correct.
3. The Casparian strip affects
   a. how water and minerals move into the vascular cylinder.
   b. vascular tissue composition.
   c. how soil particles function.
   d. how organic nutrients move into the vascular cylinder.
   e. Both a and d are correct.
4. Stomates are usually open
   a. at night, when the plant requires a supply of oxygen.
   b. during the day, when the plant requires a supply of carbon dioxide.
   c. day or night if there is excess water in the soil.
   d. during the day, when transpiration occurs.
   e. Both b and d are correct.
5. Which of these is not a mineral ion?
   a. NO_3^-
   b. Mg^+
   c. CO_2
   d. Al^{3+}
   e. All of these are correct.
6. What role do cohesion and adhesion play in xylem transport?
   a. Like transpiration, they create a tension.
   b. Like root pressure, they create a positive pressure.
   c. Like sugars, they cause water to enter xylem.
   d. They create a continuous water column in xylem.
   e. All of these are correct.
7. The pressure-flow model of phloem transport states that
   a. phloem sap always flows from the leaves to the root.
   b. phloem sap always flows from the root to the leaves.
   c. water flow brings sucrose from a source to a sink.
   d. water pressure creates a flow of water.
   e. Both c and d are correct.
8. Root hairs do not play a role in
   a. oxygen uptake.
   b. mineral uptake.
   c. water uptake.
   d. carbon dioxide uptake.
   e. the uptake of any of these.
9. Explain why this experiment supports the hypothesis that transpiration can cause water to rise to the top of tall trees.

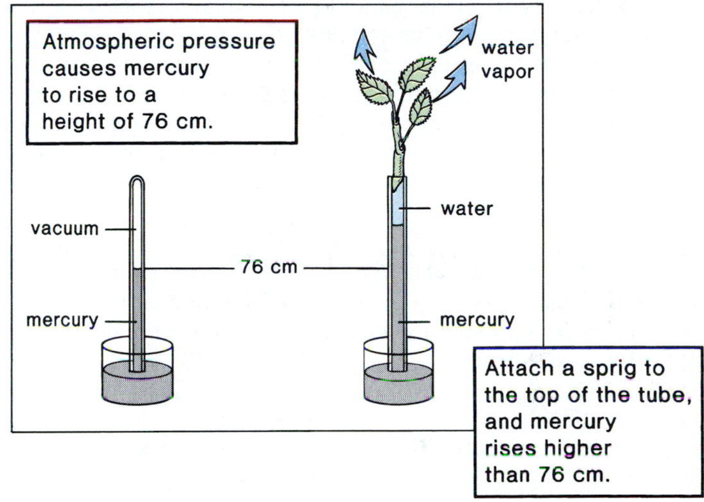

10. Label water (H_2O) and potassium (K^+) ions in these diagrams. What is the role of K^+ in the opening of stomates?

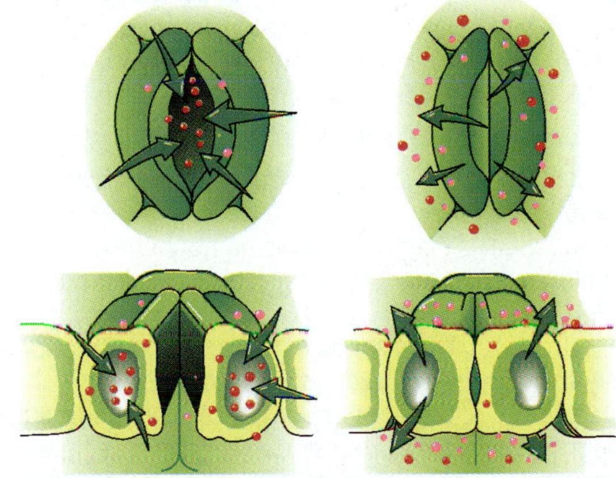

11. Explain why there is a flow of solution from the left bulb to the right bulb.

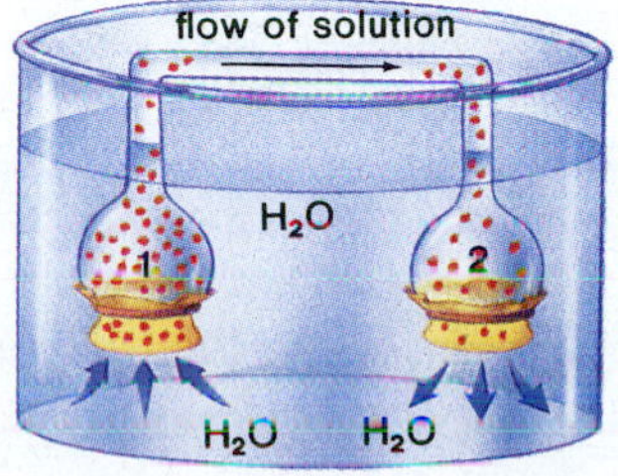

Thinking Scientifically

1. A field has been watered by irrigation for a number of years. The salts that were in the irrigation water have accumulated in the soil, making the solute concentration much higher than it was before. You notice that while water is still plentiful, plants in the field seem to be wilting. How would a change in the water's solute concentration affect its uptake by plants?
2. The force that pulls water up xylem is not unlike the force that pulls water up into a syringe. The wider the diameter of the syringe, the greater the force on the sides of the syringe. Vessel elements are larger in diameter than tracheids. Which would you expect to have stronger walls—vessel elements or tracheids? The xylem of gymnosperms, such as pine trees, contains only tracheids while the xylem in angiosperms, such as oak trees, contains both tracheids and vessel elements. Relate this information to the strength of the wood in pine trees and oak trees.

Understanding the Terms

Casparian strip 670
cohesion-tension model 674
companion cell 672
cuticle 675
epiphyte 671
essential nutrient 666
girdling 678
guard cell 676
guttation 674
horizons 669
humus 668
hydroponics 666
macronutrient 666
micronutrient 666
mineral 666
mycorrhiza 671
phloem 672
pressure-flow model 678
root hair 670
root nodule 671
root pressure 674
sieve-tube element 672
soil 668
soil erosion 669
soil profile 669
stomate 676
tracheid 672
transpiration 674
vessel element 672
water potential 673
xylem 672

Match the terms to these definitions:

a. ______________ Model explaining transport in sieve tubes of phloem.

b. ______________ Plant that takes its nourishment from the air because its attachment to other plants gives it an aerial position.

c. ______________ Plant's loss of water to the atmosphere, mainly through evaporation at leaf stomates.

d. ______________ Layer of impermeable lignin and suberin bordering four sides of root endodermal cells; causes water and minerals to enter endodermal cells before entering vascular tissue.

e. ______________ Type of plant cell that is found in pairs, with one on each side of a leaf stomate.

Web Connections

Exploring the Internet

http://www.mhhe.com/biosci/genbio/mader
(click on *Biology 7/e*)

The *Biology 7/e* Online Learning Center provides many resources for studying the material in this chapter including links to the following sites:

Essential Elements for Plant Growth describes the role of the 16 elements essential for plant growth. A large number of linked pages at the website provide more in-depth information.

http://bob.soils.wisc.edu/~barak/soilscience326/essentl.htm

and

http://bob.soils.wisc.edu/~barak/soilscience326/nitrogen.htm

http://bob.soils.wisc.edu/~barak/soilscience326/phosphorus.htm

http://bob.soils.wisc.edu/~barak/soilscience326/potassium.htm

http://bob.soils.wisc.edu/~barak/soilscience326/calcium.htm

http://bob.soils.wisc.edu/~barak/soilscience326/magnesium.htm

http://bob.soils.wisc.edu/~barak/soilscience326/sulfur.htm

Hydroponics. This site describes the principles of hydroponics, and answers FAQs.

http://www.interurban.com/hydroponics/index.html

Control of Plant Growth and Response

chapter

chapter concepts

38.1 Plant Responses

- Like animals, plants use a reception-transduction-response pathway when they respond to a stimulus. 684
- Tropisms are growth responses in plants toward or away from unidirectional stimuli such as light and gravity. 684
- Plants sometimes exhibit circadian rhythms (e.g., closing of stomates) that recur every 24 hours. 686

38.2 Plant Hormones

- Some plant hormones stimulate growth, and other plant hormones inhibit growth. 688
- The biochemical manner in which auxin and gibberellin function has been studied. 688
- It now appears that various plant hormones interact to bring about a response to a stimulus. 693

38.3 Photoperiodism

- Plant responses that are controlled by the length of daylight (photoperiod) involve the pigment phytochrome. 694

Live oak, *Quercus virginiana*

Plants respond to environmental stimuli like light, gravity, and seasonal changes. In the spring, seeds germinate and growth begins in the presence of sunlight if the soil is warm enough to contain liquid water. In the fall, when temperatures drop, shoot and root apical growth ceases.

Plants do not have eyes, ears, and an elaborate nervous system for quick responses to stimuli. Instead, hormones are involved in the perception of day length, in the shedding of leaves in the fall, in the breaking of seed dormancy, and a host of other phenomena.

Some plant responses are short-term. For example, plants will bend toward the light within a few days because a hormone produced by the growing tip has moved from the sunny side to the shady side of the stem. Even long-term plant responses to environmental conditions involve a change in the pattern of growth and development. When protected, oak trees tend to be tall with fewer branches, but when exposed to high winds, oak trees are strong and thickly branched like the one in the photograph. Plants could not meet such environmental challenges without a system for controlling growth and responses.

38.1 Plant Responses

Organisms are capable of responding to environmental stimuli, as when you withdraw your hand from a hot stove. It is adaptive for organisms to respond to environmental stimuli because it leads to their longevity and ultimately to the survival of the species. Animals often respond to external stimuli by an appropriate motion. Presented with a nipple, a newborn instinctively begins sucking. Sometimes plants, too, respond rapidly, as when the stomates (also called stomata) open at sunrise. While animals can change their location in response to a stimulus, plants, which are rooted in one place, change their growth pattern. This is why we can determine what a tree has been through—or even the history of the earth's climate—by studying tree rings!

Tropisms

Plant growth toward or away from a directional stimulus is called a **tropism** [Gk. *tropos*, turning]. Use of directional means that the stimulus is coming from only one direction instead of multiple directions. Growth toward a stimulus is called a positive tropism and growth away from a stimulus is called a negative tropism. Tropisms are due to differential growth—one side of an organ elongates faster than the other, and the result is a curving toward or away from the stimulus. Three well-known tropisms in plants are:

> Phototropism: a movement in response to a light stimulus
> Gravitropism: a movement in response to gravity
> Thigmotropism: a movement in response to touch

What mechanism permits plants to respond to stimuli? When humans respond to light, the stimulus is first received by a pigment in the retina at the back of the eyes, and then nerve impulses are generated that go to the brain. Thereafter, humans respond appropriately. In other words, the first step is *reception* (or perception) of the stimulus. The next step is *transduction,* meaning that the stimulus has been changed into a form that is meaningful to the organism. (In our example, the light stimulus was changed to nerve impulses.) Finally, there is the *response* by the organism. As discussed in the reading on page 691, investigators are beginning to study on a cellular basis how a signal is coupled to a response in plants.

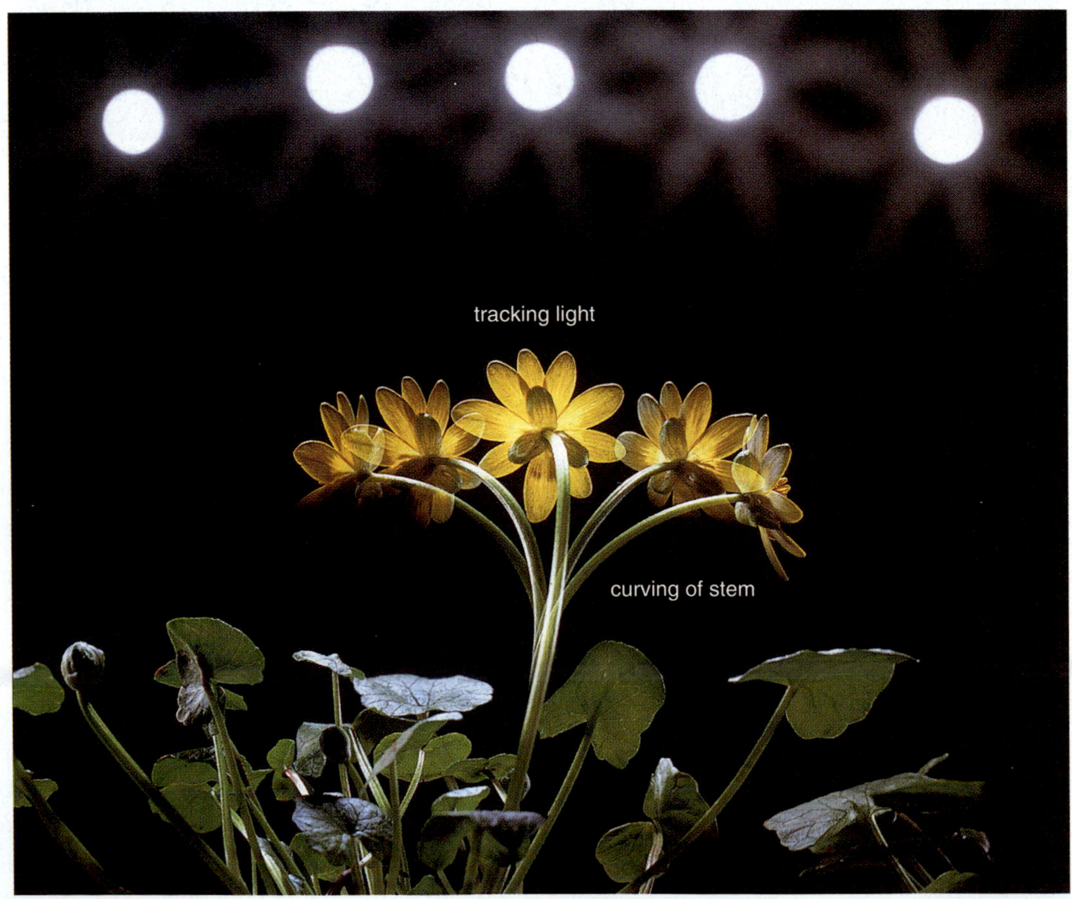

Figure 38.1 Phototropism.
Time-lapse photograph of a buttercup, *Ranunculus ficaria,* curving toward and tracking a source of light.

Phototropism

Early researchers, including Charles Darwin and his son Francis, observed that plants curve toward the light. The positive **phototropism** [Gk. *photos*, light, and *tropos*, turning] of stems occurs because the cells on the shady side of the stem elongate (Fig. 38.1). Curving away from light is called negative phototropism. Roots, depending on the species examined, are either insensitive to light or they exhibit negative phototropism.

It's believed that a pigment related to the vitamin riboflavin acts as a photoreceptor for light when phototropism occurs. Following reception, the plant hormone auxin migrates from the bright side to the shady side of a stem. The cells on that side elongate faster than those on the bright side, causing the stem to curve toward the light. It's not yet known how reception of the stimulus (light) is coupled to the production of auxin. That is, it is not known how transduction occurs.

As discussed in the next sections, auxin is also involved in gravitropism, **apical dominance,** root development, and seed development.

Gravitropism

When an upright plant is placed on its side, the stem displays negative **gravitropism** [L. *gravis*, heavy, and Gk. *tropos*, turning] because it grows upward, opposite the pull of gravity (Fig. 38.2*a*). Again, Charles Darwin and his son were among the first to say that roots, in contrast to stems, show positive gravitropism (Fig. 38.2*b*). Further, they discovered that if the root cap is removed, roots no longer respond to gravity. Later investigators came up with an explanation. Root cap cells contain sensors called **statoliths,** which are believed by some to be starch grains located within amyloplasts, a type of plastid. (Chloroplasts are a different type of plastid.) Due to gravity, the amyloplasts settle to the lowest part of the cell (Fig. 38.2*c*).

Again, the hormone auxin brings about the positive gravitropism of roots and the negative gravitropism of stems. The two types of tissues respond differently to auxin, which moves to the lower side of both stems and root after gravity has been perceived. Auxin inhibits the growth of root cells; therefore the cells of the upper surface elongate and the root curves downward. Auxin stimulates the growth of stem cells; therefore, the cells of the lower surface elongate and the stem curves upward.

a.

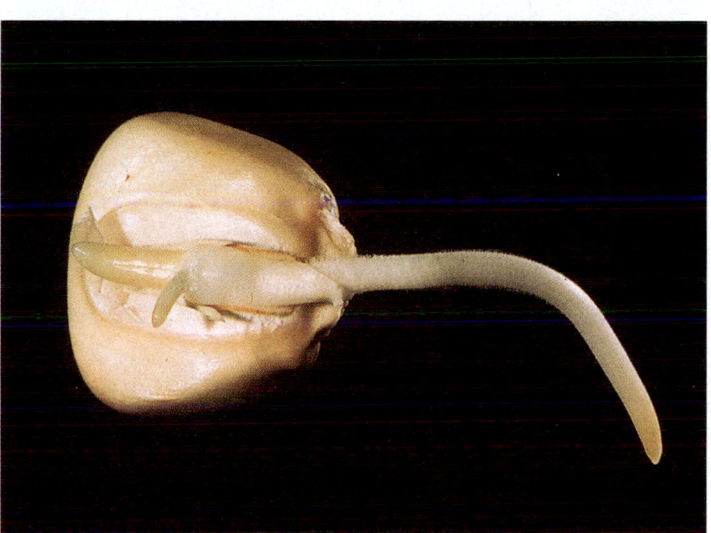

b.

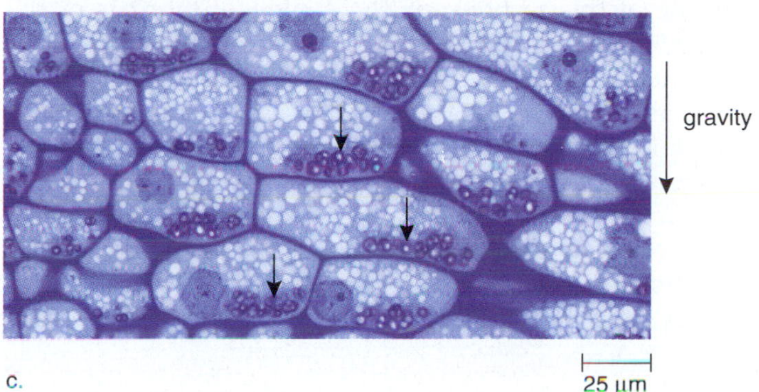

c.

Figure 38.2 Gravitropism.
a. Negative gravitropism of the stem of a *Coleus* plant 24 hours after the plant was placed on its side. **b.** Positive gravitropism of a root emerging from a corn kernel. **c.** Sedimentation of statoliths (see arrows), which are amyloplasts containing starch granules, is thought to explain how roots perceive gravity.

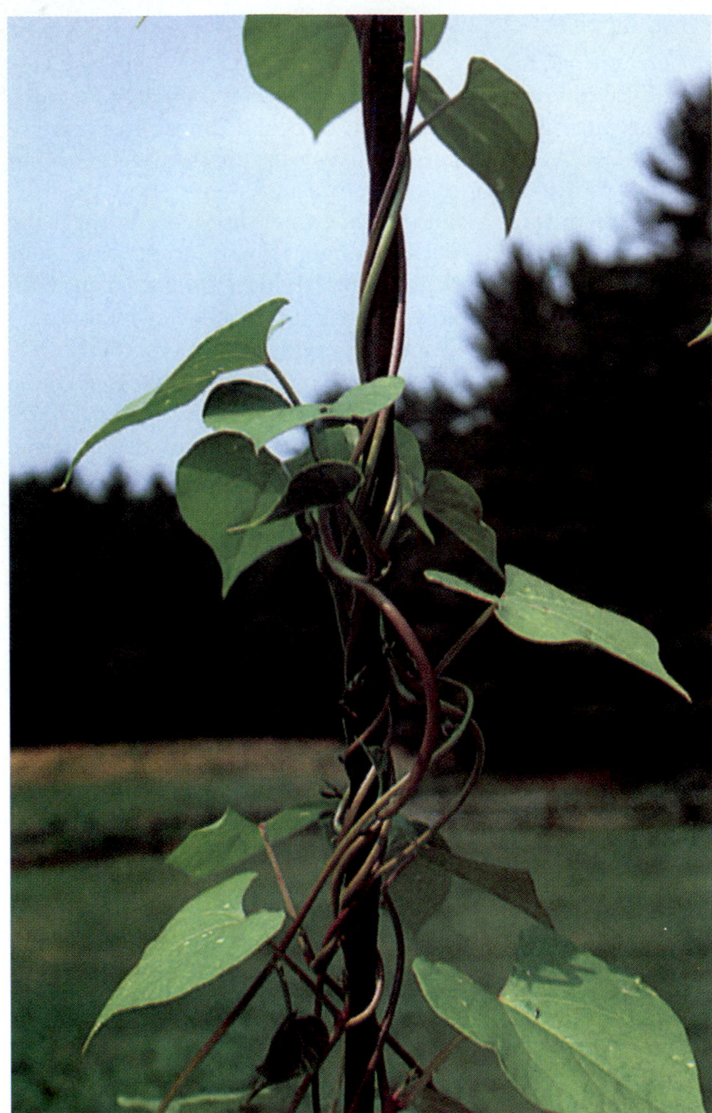

Figure 38.3 Coiling response.
The stem of a morning glory plant, *Ipomoea*, coiling around a pole.

Thigmotropism

Unequal growth due to contact with solid objects is called **thigmotropism** [Gk. *thigma*, touch, and *tropos*, turning]. An example of this response is the coiling of **tendrils** or the stems of plants such as morning glory (Fig. 38.3).

The plant grows straight until it touches something. Then the cells in contact with an object, such as a pole, grow less while those on the opposite side elongate. Thigmotropism can be quite rapid; a tendril has been observed to encircle an object within ten minutes. The response endures; a couple of minutes of stroking can bring about a response that lasts for several days. The response can also be delayed; tendrils touched in the dark will respond once they are illuminated. ATP (adenosine triphosphate) rather than light can cause the response; therefore, the need for light may simply be a need for ATP. Also, the hormones auxin and ethylene may be involved since they can induce curvature of tendrils even in the absence of touch.

Thigmomorphogenesis is a touch response related to thigmotropism. In this case, however, the entire plant responds to the presence of environmental stimuli like wind or rain. The same type of tree growing in a windy location often has a shorter, thicker trunk than one growing in a more protected location. Even simple mechanical stimulation like rubbing a plant with a stick can inhibit cellular elongation and produce a sturdier plant with increased amounts of support tissue.

Nastic Movements

In contrast to tropisms, nastic movements are independent of the direction of the stimulus. Seismonastic movements result from touch, shaking, or thermal stimulation. If you touch a *Mimosa pudica* leaf, the leaflets fold because the petiole droops (Fig. 38.4). This response, which takes only a second or two, is due to a loss of turgor pressure within cells located in a thickening, called a pulvinus, at the base of each leaflet. Investigation shows that potassium ions (K^+) move out of the cells and then water follows by osmosis. A single stimulus such as a hot needle is enough to cause all the leaves to respond. There must be some means of communication in order for this to occur, and in fact a nerve impulse-type stimulus has been recorded in these plants!

A Venus's-flytrap has three sensitive hairs at the base of the trap, and if these are touched by an insect, a nerve impulse-type stimulus brings about a closing of the trap. This, too, is due to turgor pressure changes in the leaf cells that form the trap.

Sleep Movements

A sleep movement is a nastic response that occurs daily in response to light and dark changes. One of the most common examples occurs in a houseplant called the prayer plant because at night the leaves fold upward into a shape resembling that of hands at prayer (Fig. 38.5). This movement is also due to changes in the turgor pressure of motor cells in a pulvinus located at the base of each leaf.

Circadian Rhythms Organisms exhibit periodic fluctuations that correspond to environmental changes. For example, your temperature and blood pressure tend to change with the time of day, and you become sleepy at a certain time of night. The prayer plant just mentioned displays rhythmic "sleep" behavior (Fig. 38.5). A biological rhythm with a 24-hour cycle is called a **circadian rhythm** [L. *circum*, about, and *dies*, day].

Figure 38.4 Seismonastic movement.
A leaf of the sensitive plant, *Mimosa pudica,* before and after it is touched.

Figure 38.5 Sleep movement.
Prayer plant, *Maranta leuconeura,* before dark and after dark when the leaves fold up.

Circadian rhythms tend to persist, even if the appropriate environmental cues are no longer present. For example, on a transcontinental flight, you will likely suffer jet lag, and it will take several days to adjust to the time change because your body will still be attuned to the day-night pattern of your previous environment. The internal mechanism by which a biological rhythm is maintained in the absence of appropriate environmental stimuli is termed a **biological clock.** Typically, if organisms are sheltered from environmental stimuli, their circadian rhythms continue, but the cycle extends. In prayer plants, for example, the sleep cycle changes to 26 hours. Therefore, it is believed that biological clocks are synchronized by external stimuli to 24-hour rhythms. The length of daylight compared to the length of darkness, called the photoperiod, sets the clock. Temperature has little or no effect. This is adaptive because the photoperiod indicates seasonal changes better than temperature changes. Spring and fall, in particular, can have both warm and cold days.

There are other examples of circadian rhythms in plants. Stomates and certain flowers usually open in the morning and close at night, and some plants secrete nectar at the same time of the day or night.

38.2 Plant Hormones

In order for plants to respond to stimuli, the activities of plant cells and structures have to be coordinated. Almost all communication in a plant is done by **hormones** [Gk. *hormao,* instigate], chemical messengers produced in very low concentrations and active in another part of the organism. A particular response is probably influenced by several hormones and most likely requires a specific ratio of two or more hormones. Hormones are synthesized or stored in one part of the plant, but will travel within phloem or from cell to cell after reception of the appropriate stimulus.

Each naturally occurring hormone has a specific chemical structure. Other chemicals, some of which differ only slightly from the natural hormones, also affect the growth of plants. These and the naturally occurring hormones are sometimes grouped together and called plant growth regulators.

Auxins

The most common naturally occurring **auxin** [Gk. *auximos,* promoting growth] is indoleacetic acid (IAA). It is produced in shoot apical meristem and is found in young leaves and in flowers and fruits. Therefore, you would expect that auxin would affect many aspects of plant growth and development. Apically produced auxin prevents the growth of axillary buds, a phenomenon called apical dominance. When a terminal bud is removed deliberately or accidentally, the nearest axillary buds begin to grow, and the plant branches. To achieve a fuller look, one generally prunes the top (apical meristem) of the plant. This removes apical dominance and causes more branching of the main body of the plant (Fig. 38.6).

The application of a weak solution of auxin to a woody cutting causes roots to develop (Fig. 38.7). Auxin production by seeds also promotes the growth of fruit. As long as auxin is concentrated in leaves or fruits rather than in the stem, leaves and fruits do not fall off. Therefore, trees can be sprayed with auxin to keep mature fruit from falling to the ground.

As discussed earlier, auxin is also involved in gravitropism and phototropism. After gravity has been perceived, auxin moves to the lower surface of roots and stems. Thereafter, roots curve downward and stems curve upward (see Fig. 38.2). The role of auxin in the positive phototropism of stems has been studied for quite some time. The experimental material of choice has been oat seedlings with coleoptile intact. A **coleoptile** is a protective sheath for the young leaves of the seedling. In 1881, the Darwins found that phototropism will not occur if the tip of the seedling is cut off or is covered by a black cap. They concluded that some influence that causes curvature is transmitted from the coleoptile tip to the rest of the shoot.

In 1926 Frits W. Went cut off the tips of coleoptiles and placed them on agar (a gelatinlike material). Then he placed an agar block to one side of a tipless coleoptile and found that the shoot would curve away from that side. The bending occurred even though the seedlings were not exposed to light (Fig. 38.8). Went concluded that the agar blocks contained a chemical that had been produced by the coleoptile tips. It was this chemical, he decided, that had caused the shoots to curve. He named the chemical substance auxin after the Greek word *aux,* which means to grow.

How Auxins Work

When a plant is exposed to unidirectional light, auxin moves to the shady side where it binds to receptors and activates the ATP-driven proton pump (Fig. 38.9). As hydrogen ions (H^+) are being pumped out of the cell, the cell wall becomes acidic, breaking hydrogen bonds. Cellulose fibrils are weakened, and activated enzymes further degrade the cell wall. The electrochemical gradient established by the proton pump causes solutes to enter the cell and water follows by osmosis. The turgid cell presses against the cell wall, stretching it so that elongation occurs. Auxin-mediated elongation is observed in younger, as opposed to more mature, cells. Perhaps older cells lack auxin receptors.

a.

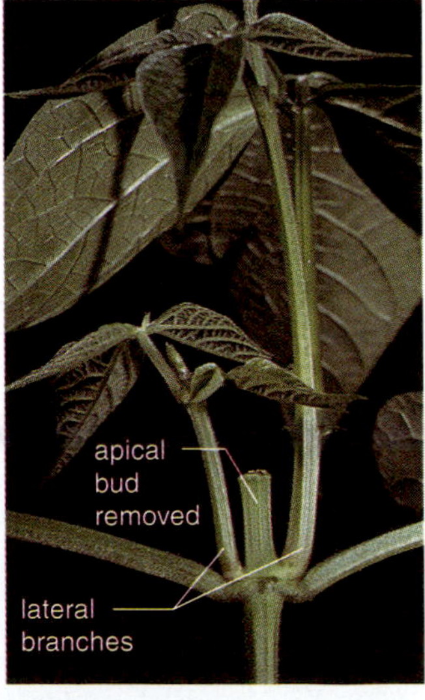

b.

Figure 38.6 Apical dominance.
a. Auxin in the apical bud inhibits lateral bud development and the plant exhibits apical dominance. **b.** When the apical bud is removed, lateral branches develop.

a.

b.

Figure 38.7
Adventitious roots.
Roots develop on an African violet stem cutting when it is placed in a solution containing IAA **(a)** but not when it is placed in pure water **(b)**.

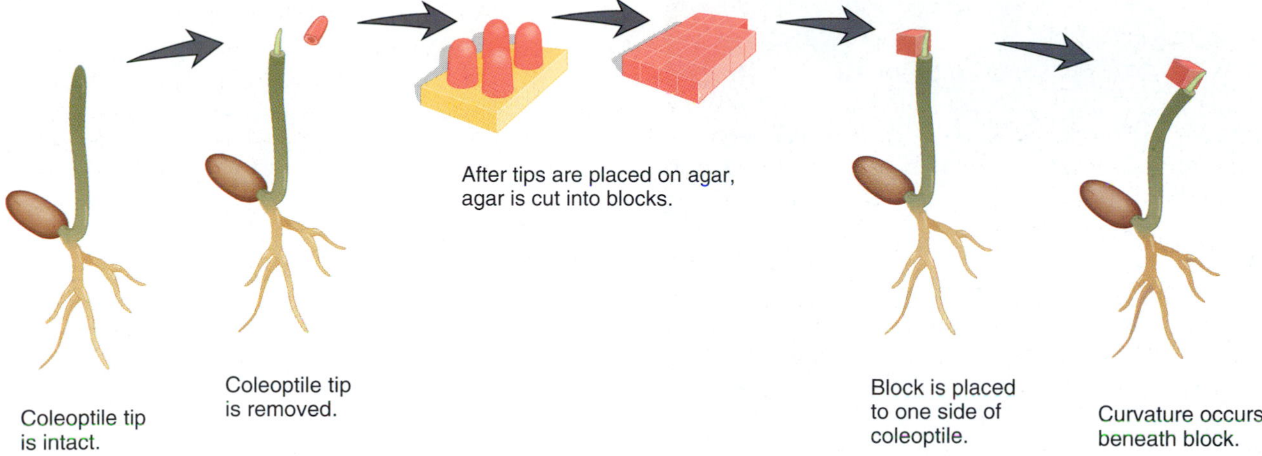
After tips are placed on agar, agar is cut into blocks.
Coleoptile tip is intact.
Coleoptile tip is removed.
Block is placed to one side of coleoptile.
Curvature occurs beneath block.

Figure 38.8 **Demonstrating phototropism.**
Oat seedlings are protected by a hollow sheath called a coleoptile. After a tip is removed and placed on agar, an agar block placed on one side of the coleoptile can cause it to curve.

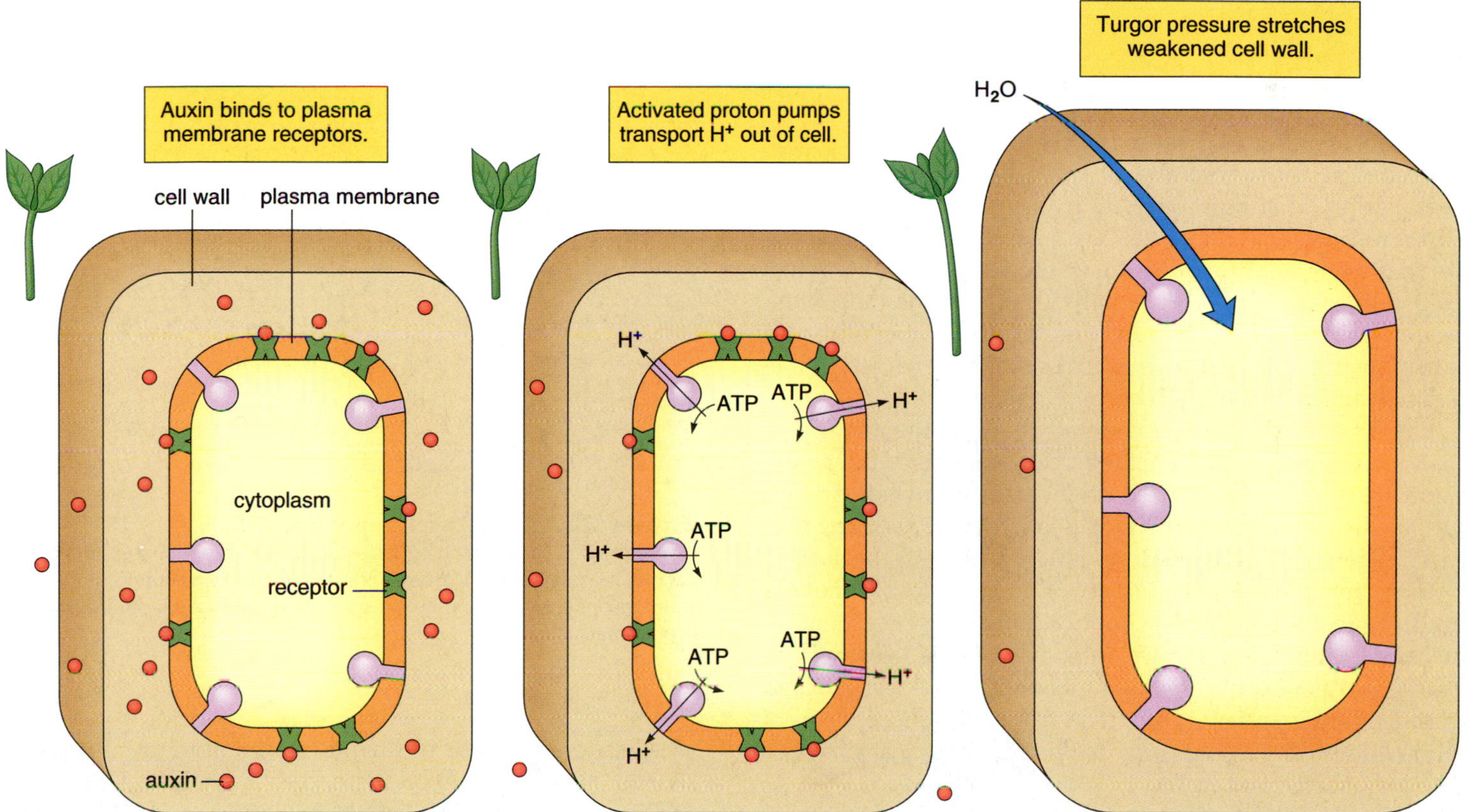
Auxin binds to plasma membrane receptors.
cell wall
plasma membrane
cytoplasm
receptor
auxin
Activated proton pumps transport H+ out of cell.
H+
ATP
Turgor pressure stretches weakened cell wall.
H_2O

Figure 38.9 **Auxin mode of action.**
After auxin binds to a receptor, the combination stimulates the proton pump so that hydrogen ions (H^+) are transported out of the cell. The resulting acidity causes the cell wall to weaken, and the electrochemical gradient causes solutes to enter the cell. Water follows by osmosis and the cell elongates.

Figure 38.10 Effect of gibberellins.
The *Cyclamen* plant on the right was treated with gibberellins; the plant on the left was not treated. Gibberellins are often used to promote stem elongation in economically important plants, but the exact mode of action still remains unclear.

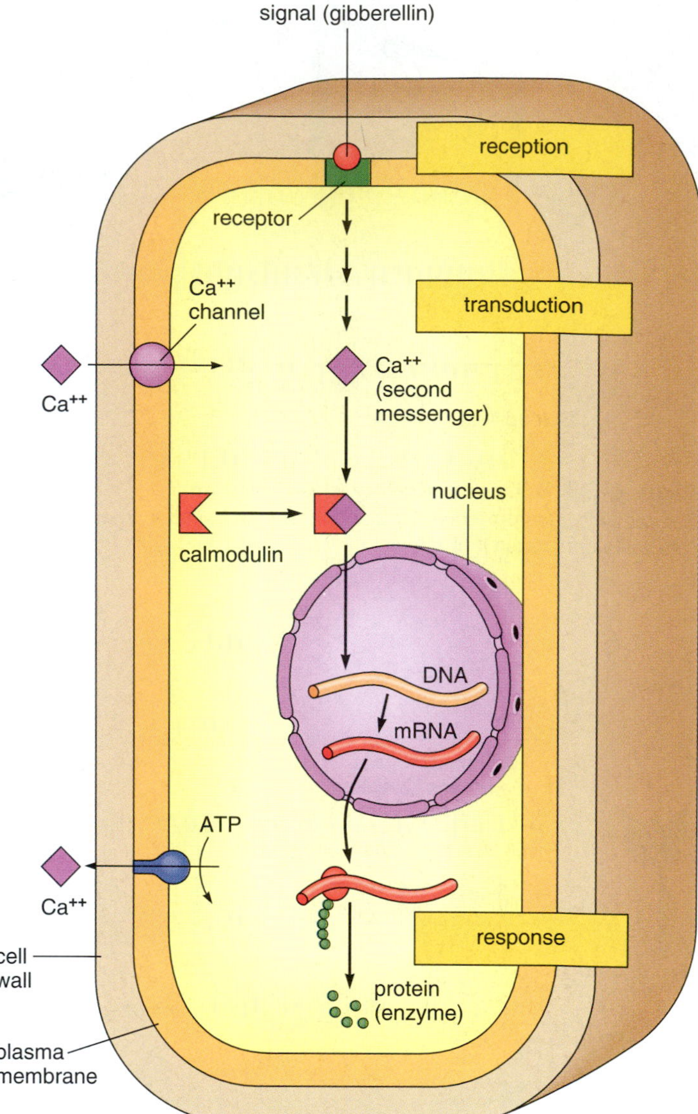

Figure 38.11 Gibberellin mode of action.
It appears that the hormone (gibberellin) binds to a receptor, and a second messenger (Ca^{2+}) inside the cell activates a protein (calmodulin) and the complex binds to DNA. Thereafter, an enzyme is produced.

Gibberellins

We know of about 70 gibberellins that chemically differ only slightly. The most common of these is GA_3 (the subscript designation distinguishes it from other gibberellins). **Gibberellins** [L. = dim. of *gibbus,* bent] are growth promoters that bring about elongation of the resulting cells. When gibberellins are applied externally to plants, the most obvious effect is stem elongation (Fig. 38.10). Gibberellins can cause dwarf plants to grow, cabbage plants to become 2 meters tall, and bush beans to become pole beans.

Gibberellins were discovered in 1926, the same year that Went performed his classic experiments with auxin. Ewiti Kurosawa, a Japanese scientist, was investigating a fungal disease of rice plants called "foolish seedling disease." The plants elongated too quickly, causing the stem to weaken and the plant to collapse. Kurosawa found that the fungus infecting the plants produced an excess of a chemical he called gibberellin, named after the fungus *Gibberella fujikuroi.* It wasn't until 1956 that gibberellic acid was isolated from a flowering plant rather than from a fungus. Sources of gibberellin in flowering plant parts are young leaves, roots, embryos, seeds, and fruits.

The **dormancy** of seeds and buds can be broken by applying gibberellins, and research with barley seeds has shown how GA_3 acts as a chemical messenger. Barley seeds have a large, starchy endosperm, which must be broken down into sugars to provide energy for growth. After the embryo produces gibberellins, amylase, an enzyme that breaks down the starch, appears in cells just inside the seed coat. It is hypothesized that GA_3 (the first chemical messenger) attaches to a receptor in the plasma membrane, and then a second messenger inside the cell, namely calcium ions (Ca^{2+}), combines with a protein called calmodulin. This complex is believed to activate the gene that codes for amylase (Fig. 38.11). Amylase then acts on starch to release sugars used as a source of energy by the growing embryo.

The action of hormones on the cellular level is another example of the reception-transduction-response pathway described at the beginning of this chapter. The following reading describes the work of two investigators in this area.

Science Focus

Husband and Wife Team Explores Signal Transduction in Plants

Donald Briskin and Margaret Gawienowski
University of Illinois at Urbana-Champaign

Plants perceive and react to a variety of environmental stimuli. Some examples include light intensity and quality, gravity, carbon dioxide levels, pathogen infection, drought, and touch. Plant responses may be short term, such as the rapid localized cell death that occurs at the site of infection by a pathogen. This rapid cell death, termed the hypersensitive response, serves to limit the further progression of infection. Another form of short-term response is the opening or closing of leaf stomates in response to light levels. High light causes the stomates to open, while low light or darkness causes them to close. In this way, the supply of carbon dioxide to the leaves via the stomates is coordinated with light-driven photosynthetic reactions occurring in leaf cells. On the other hand, plant responses to environmental stimuli may be more long term and involve actual plant growth. An example is the plant response to gravity (gravitropism), which results in the downward growth of the root and the upward growth of the stem. In each of these situations, the coupling of a stimulus to an appropriate response represents signal transduction.

Although we know that plants do not have a nervous system and a brain, some system or process must be present in these organisms to allow the perception of a stimulus and its conversion to an appropriate response. While we are both interested in attempting to understand this process, we approach this complex research area from different perspectives related to our training, expertise, and interests.

As a plant biochemist, Donald focuses on how membrane transport processes are involved in signal transduction. Here, research has shown that membrane transport of calcium ions (Ca^{2+}) can play a fundamental role in the signal transduction process. When a stimulus is perceived by a plant cell, a transient increase in cytoplasmic Ca^{2+} concentration occurs due to the rapid opening of Ca^{2+} channels at plant membranes, which operate not unlike Ca^{2+} channels present in animal membranes. This transient increase in cytoplasmic Ca^{2+} acts as a sort of message linking the stimulus to a response at the biochemical level. In particular, Donald studies how Ca^{2+}-transporting enzymes (called Ca^{2+}-ATPases) and Ca^{2+} channels are coordinated after a stimulus is received.

Just how a single type of message—such as a rapid change in cytoplasmic Ca^{2+}—can elicit so many different types of specific plant responses is Margaret's interest. As a plant molecular biologist, Margaret focuses on how proteins in the plant cytoplasm that bind Ca^{2+} might impart specificity to the response. Of particular interest to her is a Ca^{2+}-binding protein called calmodulin, which controls the activity of a number of other proteins once it binds to Ca^{2+}. Her work has shown that calmodulin in plants is encoded by multiple genes that produce variants of the protein called isoforms. The isoforms are present to a different extent in different plant tissues, have different properties (Ca^{2+} binding, target protein regulation), and may play a role in determining which responses are coupled to transient cytoplasmic Ca^{2+} changes. Together, we are conducting studies to determine how such signal transduction processes are involved in controlling the production of medicinal chemicals generated by medicinal plants such as Saint-John's-wort and Echinacea. We have found that the production of medicinal chemicals can be affected by environmental factors during growth of these plants. The environmental factors detected by the plant can include light levels, nitrogen supply, plant water status, and even the competition for these resources by other plants. Our work will lead to a better understanding of how the growth environment of medicinal plants can signal the plant to modify the quality and quantity of herbal medicines.

Although our approaches toward understanding plant signal transduction are very different, the work of each of us tends to complement the other's. We view each other's work from a slightly different perspective, and this helps generate ideas for new experiments and strategies. Furthermore, we work together in the same laboratory at the University of Illinois at Urbana-Champaign, and this leads to much discussion and exchange of opinions. For us, research is not only challenging and exciting, but it keeps us wondering what the next day will bring.

Courtesy of Donald Briskin and Margaret Gawienowski, University of Illinois at Urbana-Champaign.

a.

b.

c.

d.

Figure 38.12 **Interaction of hormones.**
Tissue culture experiments have revealed that auxin and cytokinin interact to affect differentiation during development. **a.** In tissue culture that has the usual amounts of these two hormones, tobacco strips develop into a callus of undifferentiated tissue. **b.** If the ratio of auxin to cytokinin is appropriate, the callus produces roots. **c.** Change the ratio, and vegetative shoots and leaves are produced. **d.** Yet another ratio causes floral shoots. It is now clear that each plant hormone rarely acts alone; it is the relative concentrations of hormones that produce an effect. The modern emphasis is to look for an interplay of hormones when a growth response is studied.

Cytokinins

The **cytokinins** [Gk. *kytos,* cell, and *kineo,* move] are a class of plant hormones that promote cell division; cytokinesis means cell division. These substances are derivatives of adenine, one of the purine bases in DNA and RNA. A naturally occurring cytokinin called zeatin has been extracted from corn kernels. Kinetin, a synthetic cytokinin, also promotes cell division.

The cytokinins were discovered as a result of attempts to grow plant tissue and organs in culture vessels in the 1940s (Fig. 38.12). It was found that cell division occurs when coconut milk (a liquid endosperm) and yeast extract are added to the culture medium. Although the effective agent or agents could not be isolated, they were collectively called cytokinins. Not until 1967 was the naturally occurring cytokinin called zeatin isolated from coconut milk. Cytokinins have been isolated from various seed plants, where they occur in the actively dividing tissues of roots and also in seeds and fruits.

Plant tissue culturing is now common practice, and researchers are well aware that the ratio of auxin to cytokinin and the acidity of the culture medium determine whether the plant tissue forms an undifferentiated mass, called a callus, or differentiates to form roots, vegetative shoots, leaves, or floral shoots (Fig. 38.12). Researchers have reported that chemicals they called oligosaccharins (chemical fragments released from the cell wall) are also effective in directing differentiation. They hypothesize that auxin and cytokinins are a part of a reception-transduction-response pathway (see Fig. 38.11), which leads to the activation of enzymes that release these fragments from the cell wall. Perhaps all plant hormones have a similar effect.

Senescence

When a plant organ, such as a leaf, loses its natural color, it is most likely undergoing an aging process called **senescence.** During senescence, large molecules within the leaf are broken down and transported to other parts of the plant. Senescence does not always affect the entire plant at once; for example, as some plants grow taller, they naturally lose their lower leaves. It has been found that senescence of leaves can be prevented by the application of cytokinins. Not only can cytokinins prevent death of leaves, they can also initiate leaf growth. Lateral buds begin to grow despite apical dominance when cytokinin is applied to them.

Cytokinins promote cell division, prevent senescence, and initiate growth. The interaction of hormones is well exemplified by the effect of varying ratios of auxin and cytokinins on differentiation of plant tissues.

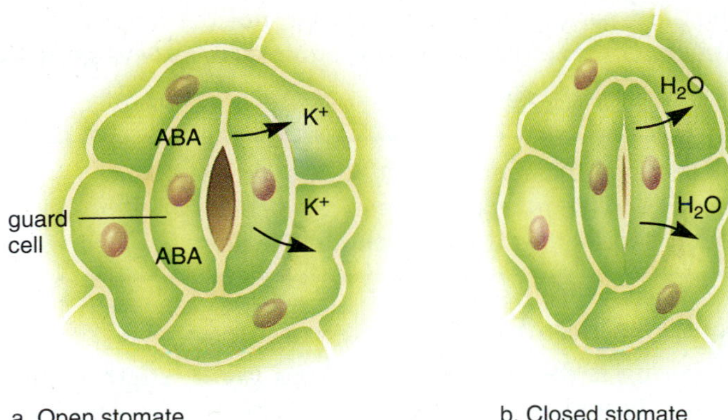

a. Open stomate b. Closed stomate

Figure 38.13 Control of stomate opening.
The hormone abscisic acid (ABA) is believed to bring about the closing of the stomates when water stress occurs. First, potassium ions (K^+) exit (**a**), and then water (H_2O) also leaves guard cells (**b**).

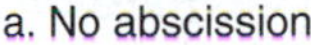

a. No abscission b. Abscission

Figure 38.14 Abscission.
a. Normally there is no abscission when a holly twig is placed under a glass jar for a week. **b.** When an ethylene-producing ripe apple is also under the jar, abscission of the holly leaves occurs.

Inhibitory Hormones

In contrast to the hormones discussed so far which stimulate growth, the plant hormones abscisic acid and ethylene inhibit growth. Therefore, they are the inhibitory hormones.

Abscisic Acid

Abscisic acid (ABA) is sometimes called the stress hormone because it initiates and maintains seed and bud dormancy and brings about the closure of stomates. Dormancy occurs when a plant organ readies itself for adverse conditions by stopping growth (even though conditions at the time are favorable for growth). For example, it is believed that abscisic acid moves from leaves to vegetative buds in the fall, and thereafter these buds are converted to winter buds. A winter bud is covered by thick and hardened scales. A reduction in the level of abscisic acid and an increase in the level of gibberellins are believed to break seed and bud dormancy. Then seeds germinate and buds send forth leaves.

Abscisic acid brings about the closing of stomates when a plant is under water stress (Fig. 38.13). In some unknown way, abscisic acid causes potassium ions (K^+) to leave guard cells. Thereafter, the guard cells lose water and the stomates close. It was once believed that abscisic acid functioned in **abscission** [L. *abscissus*, cut off], the dropping of leaves, fruits, and flowers from a plant. Although the external application of abscisic acid promotes abscission, this hormone is no longer believed to function naturally in this process. Instead, the hormone ethylene is believed to bring about abscission (Fig. 38.14). Abscisic acid is produced by any "green tissue" with chloroplasts, monocot endosperm, and roots.

Ethylene

In the early 1900s, it was common practice to prepare citrus fruits for market by placing them in a room with a kerosene stove; only later did researchers realize that an incomplete combustion product of kerosene, namely **ethylene,** ripens fruit. It does so by increasing the activity of enzymes that soften fruits. For example, it stimulates the production of cellulase, an enzyme that hydrolyzes cellulose in plant cell walls. Because it is a gas, ethylene moves freely through the air—a barrel of ripening apples can induce ripening of a bunch of bananas even some distance away. Ethylene is released at the site of a wound due to physical damage or infection (which is why one rotten apple spoils the whole bunch).

The presence of ethylene in the air inhibits the growth of plants in general; homeowners who use natural gas to heat their homes sometimes report difficulties in growing houseplants. Ethylene is also present in automobile exhaust, and it's possible that plant growth is inhibited by the exhaust that enters the atmosphere. It only takes one part of ethylene per 10 million parts of air to bring about inhibition of plant growth.

Ethylene is also involved in abscission. Low levels of auxin and perhaps gibberellin compared to the stem probably initiate abscission. But once the process of abscission has begun, ethylene stimulates such enzymes as cellulase, which cause leaf, fruit, or flower drop (Fig. 38.14).

Abscisic acid promotes dormancy of buds and seeds; it also brings about the closure of stomates. Ethylene ripens fruits and controls the abscission of leaves, flowers, and fruits.

38.3 Photoperiodism

Many physiological changes in plants are related to a seasonal change in day length. Such changes include seed **germination,** the breaking of bud dormancy, and the onset of senescence. A physiological response prompted by changes in the length of day or night is called **photoperiodism** [Gk. *photos*, light, and *periodus*, completed course]. In some plants, photoperiodism influences flowering: violets and tulips flower in the spring, and asters and goldenrods flower in the fall.

In the 1920s, when U.S. Department of Agriculture scientists began to study photoperiodism in more detail, they decided to grow plants in a greenhouse, where they could artificially alter the photoperiod. This work led them to conclude that plants can be divided into three groups:

1. **Short-day plants**—flower when the day length is shorter than a critical length (examples are cocklebur, poinsettia, and chrysanthemum).
2. **Long-day plants**—flower when the day length is longer than a critical length (examples are wheat, barley, clover, and spinach).
3. **Day-neutral plants**—flowering is not dependent on day length (examples are tomato and cucumber).

Further, we should note that both a long-day plant and a short-day plant can have the same critical length (Fig. 38.15). Spinach is a long-day plant that has a critical length of 14 hours; ragweed is a short-day plant with the same critical length. Spinach, however, flowers in the summer when the day length increases to 14 hours or more, and ragweed flowers in the fall, when the day length shortens to 14 hours or less. We now know that some plants may require a specific sequence of day lengths in order to flower.

In 1938, K. C. Hammer and J. Bonner began to experiment with artificial lengths of light and dark that did not necessarily correspond to a normal 24-hour day. They discovered that the cocklebur, a short-day plant, flowers as long as the dark period is continuous for 8.5 hours, regardless of the length of the light period. Further, if this dark period is interrupted by a brief flash of white light, the cocklebur does not flower. (Interrupting the light period with darkness has no effect.) Similar results have also been found for long-day plants. They require a dark period that is shorter than a critical length, regardless of the length of the light period. If a slightly longer-than-critical-length night is interrupted by a brief flash of light, however, long-day plants flower. We must conclude, then, that the length of the dark period controls flowering, not the length of the light period. Of course, in nature, short days always go with long nights and vice versa.

Short-day plants require a period of darkness that is longer, and long-day plants require a period of darkness that is shorter than a critical length to flower.

Phytochrome and Plant Flowering

If flowering is dependent on day and night length, plants must have some way to detect these periods. Many years of research by U.S. Department of Agriculture scientists led to the discovery of a plant pigment called phytochrome. **Phytochrome** [Gk. *phyton*, plant, and *chroma*, color] is a blue-green leaf pigment that alternately exists in two forms. As Figure 38.16 indicates:

P_r (phytochrome red) absorbs red light (of 660 nm wavelength) and is converted to P_{fr}.
P_{fr} (phytochrome far-red) absorbs far-red light (of 730 nm wavelength) and is converted to P_r.

Direct sunlight contains more red light than far-red light; therefore, P_{fr} is apt to be present in plant leaves during

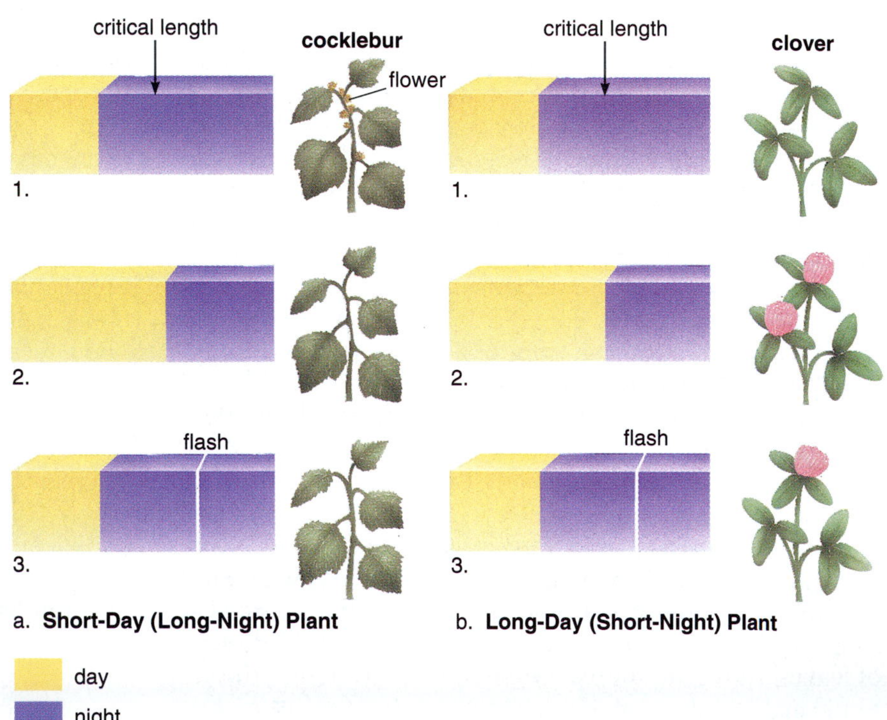

Figure 38.15 Photoperiodism and flowering.
a. Short-day plant. (1) When the day is shorter than a critical length, this type of plant flowers. (2) The plant does not flower when the day is longer than the critical length. (3) It also does not flower if the longer-than-critical-length night is interrupted by a flash of light. b. Long-day plant. (1) When the day is shorter than a critical length, this type plant does not flower. (2) The plant flowers when the day is longer than a critical length. (3) It also flowers if the slightly longer-than-critical-length night is interrupted by a flash of light.

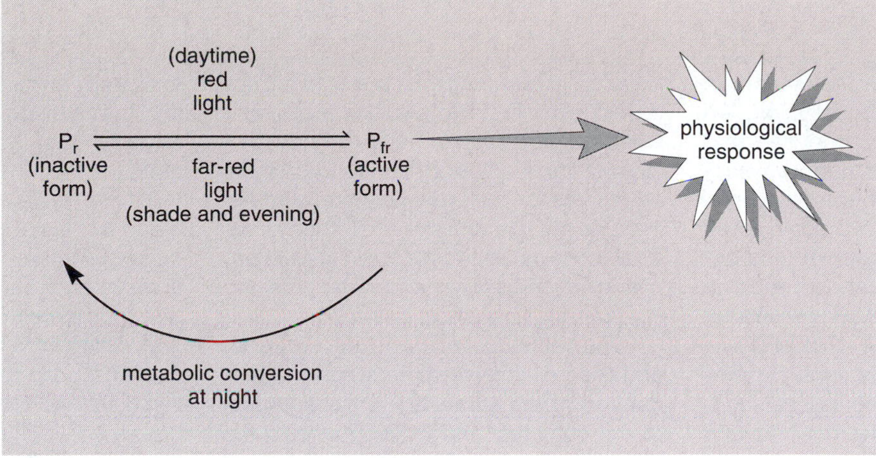

Figure 38.16 **Phytochrome conversion cycle.**
The inactive form P_r is prevalent during the night. At sunset or in the shade, when there is more far-red light, P_{fr} is converted to P_r. Also during the night, metabolic processes cause P_{fr} to be converted into P_r. P_{fr}, the active form of phytochrome, is prevalent during the day because at that time there is more red light than far-red light.

the day. In the shade and at sunset, there is more far-red light than red light; therefore, P_{fr} is converted to P_r as night approaches. There is also a slow metabolic replacement of P_{fr} by P_r during the night.

It is possible that phytochrome conversion is the first step in a reception-transduction-response pathway that results in flowering. At one time, researchers hypothesized that there was a special flowering hormone called florigen, but such a hormone has never been discovered.

Phytochrome alternates between two forms (P_{fr} during the day and P_r during the night), and this conversion allows a plant to detect photoperiod changes.

Other Functions of Phytochrome

The $P_r \rightarrow P_{fr}$ conversion cycle is now known to control other growth functions in plants. It promotes seed germination and inhibits stem elongation, for example. The presence of P_{fr} indicates to some seeds that sunlight is present and conditions are favorable for germination. This is why some seeds must be partly covered with soil when planted. Germination of other seeds is inhibited by light, so they must be planted deeper. Following germination, the presence of P_r indicates that stem elongation may be needed to reach sunlight. Seedlings that are grown in the dark etiolate; that is, the stem increases in length and the leaves remain small (Fig. 38.17). Once the seedling is exposed to sunlight and P_r is converted to P_{fr}, the seedling begins to grow normally—the leaves expand and the stem branches. It has now been shown that phytochrome in the P_{fr} form leads to the activation of one or more regulatory proteins in the cytosol. These proteins migrate to the nucleus where they bind to so-called "light-stimulated" genes that code for proteins found in chloroplasts.

a. Etiolation

b. Normal growth

Figure 38.17 **Phytochrome control of growth pattern.**
a. If far-red light is prevalent, as it is in the shade, etiolation occurs. **b.** If red light is prevalent, as it is in bright sunlight, normal growth occurs. These effects are due to phytochrome.

Connecting Concepts

Behavior in plants can be understood on three different levels of organization. On the species level, plant responses that promote survival and reproductive success have evolved through the process of natural selection. On the organismal level, hormones coordinate the growth and development of plant parts. And on the cellular level, hormones influence cellular metabolism. Let's take the example, why do plants bend toward the light? To provide an answer on the species level, consider that those plants that bend toward the light will be able to produce more organic food and be the ones to have more offspring. To provide an answer on the organismal level, consider that light causes certain plant parts to produce auxin, and when auxin moves from the lit side of a stem to the shady side, elongation occurs, and thereafter, the plant bends toward the light. To provide an answer on the cellular level, consider that after auxin is received by a plant cell, solutes and water enter the cell, causing its walls to stretch.

The response of both the organism and the cell involve three steps: (1) perception of the stimulus, (2) transduction of the stimulus, and (3) response to the stimulus. Can you designate these steps on the cellular level? Reception of auxin by plasma membrane receptors is the first step, entrance of solutes and water into the plant cell is the second step, and stretching of the cell wall is the third step.

If we were considering animals instead of plants, the same type of biological explanations would apply. Plants and animals and, indeed, all organisms share common ancestors, even back to the very first cell(s). You will recall that evolution explains both the unity and diversity of living things. Organisms are similar because they share common ancestors; they are different because they are adapted to different ways of life.

Summary

38.1 Plant Responses

Like animals, plants utilize a reception (or perception)-transduction-response pathway when they respond to a stimulus.

When plants respond to stimuli, growth and/or movement occurs. Tropisms are growth responses toward or away from unidirectional stimuli. The positive phototropism of stems results in a bending toward light, and the negative gravitropism of stems results in a bending away from the direction of gravity. Roots that bend toward the direction of gravity show positive gravitropism. Thigmotropism occurs when a plant part makes contact with an object, as when tendrils coil about a pole.

Nastic movements are not directional. Due to turgor pressure changes, some plants respond to touch and some perform sleep movements.

Plants exhibit circadian rhythms, which are believed to be controlled by a biological clock. The sleep movements of prayer plants, the closing of stomates, and the daily opening of certain flowers have a 24-hour cycle.

38.2 Plant Hormones

Both stimulatory and inhibitory hormones help control certain growth patterns of plants. There are hormones that stimulate growth (auxins, gibberellins, and cytokinins) and hormones that inhibit growth (abscisic acid and ethylene).

Auxin-controlled cell elongation is involved in phototropism and gravitropism. When a plant is exposed to light, auxin moves laterally from the bright to the shady side of a stem. There it activates the proton (H^+) pump, leading to the uptake of solutes and an increase in turgor pressure, and the result is elongation of cells on that side.

Gibberellin research indicates that after a hormone attaches to a plasma membrane receptor, a second chemical messenger inside the cell leads to transcription and protein synthesis.

Cytokinins cause cell division, the effects of which are especially obvious when plant tissues are grown in culture.

Abscisic acid (ABA) and ethylene are two plant growth inhibitors. ABA is well known for causing stomates to close, and ethylene is known for causing fruits to ripen.

38.3 Photoperiodism

Photoperiodism is seen in some plants. For example, short-day plants flower only when the days are shorter than a critical length, and long-day plants flower only when the days are longer than a critical length. Actually, research has shown that it is the length of darkness that is critical. Interrupting the dark period with a flash of white light prevents flowering in a short-day plant and induces flowering in a long-day plant.

Phytochrome is a pigment that responds to both red and far-red light and is involved in flowering. Daylight causes phytochrome to exist as P_{fr}, but during the night, it is reconverted to P_r by metabolic processes. Phytochrome in the P_{fr} form leads to activation of regulatory proteins that bind to genes.

In addition to being involved in the flowering process, P_{fr} promotes seed germination, leaf expansion, and stem branching. When P_r dominates, the stem elongates and grows toward sunlight.

Reviewing the Chapter

1. List and describe three tropisms, and indicate the involvement of any hormones in these responses. 684–86
2. What are nastic responses, and how do turgor pressure changes bring about movements of plants? 686
3. What is a biological clock, how does it function, and what is its primary usefulness in plants? 686–87
4. Name three types of hormones that promote growth processes, and give several specific functions for each. 688–93
5. Why does removing a terminal bud cause a plant to get bushier? 688
6. What experiments led to knowledge that a hormone is involved in phototropism? Explain the mechanism by which auxin brings about elongation of cells. 688–89

7. What is the hypothesized mode of hormone action according to gibberellin research? 690
8. Give experimental evidence to suggest that hormones interact when they bring about an effect. 693
9. Name two types of hormones that inhibit growth processes, and give several specific functions for each. 695
10. Define photoperiodism, and discuss its relationship to flowering in certain plants. 694
11. What is the phytochrome conversion cycle, and what are some possible functions of phytochrome in plants? 694–95

Testing Yourself

Choose the best answer for each question. For questions 1–5, match the items below with the hormones in the key.

Key:

a. auxin
b. gibberellin
c. cytokinin
d. ethylene
e. abscisic acid

1. One rotten apple can spoil the barrel.
2. Cabbage plants bolt (grow tall).
3. Stomates close when a plant is water stressed.
4. Sunflower plants all point toward the sun.
5. Coconut milk causes plant tissues to undergo cell division.
6. After unidirectional light is received, auxin moves to the shady side of a stem, and then the stem bends toward the light. Which step in this sequence represents transduction?
   a. sensitivity of a receptor to light
   b. auxin movement toward the shady side
   c. elongation of cells
   d. curving of stem toward light
   e. All of these are correct.
7. Which of these is a correct statement?
   a. Both stems and roots show positive gravitropism.
   b. Both stems and roots show negative gravitropism.
   c. Only stems show positive gravitropism.
   d. Only roots show positive gravitropism.
   e. Only roots show negative gravitropism.
8. The short-day plant
   a. is the same as the long-day plant.
   b. is apt to flower in the fall.
   c. photoperiod must be longer than a certain length.
   d. photoperiod must be shorter than a certain length.
   e. Both b and d are correct.
9. A plant requiring a dark period of at least 14 hours will
   a. flower if a 14-hour night is interrupted by a flash of light.
   b. not flower if a 14-hour night is interrupted by a flash of light.
   c. not flower if the days are 14 hours long.
   d. not flower if the nights are longer than 14 hours.
   e. Both b and c are correct.
10. Phytochrome
    a. is a plant pigment.
    b. is present as P_{fr} during the day.
    c. activates regulatory proteins.
    d. is a photoreceptor.
    e. All of these are correct.
11. Circadian rhythms
    a. require a biological clock.
    b. do not exist in plants.
    c. are involved in the tropisms.
    d. are involved in sleep movements.
    e. Both a and d are correct.
12. Label the following diagram. Explain what causes stomates to close when a plant is water stressed.

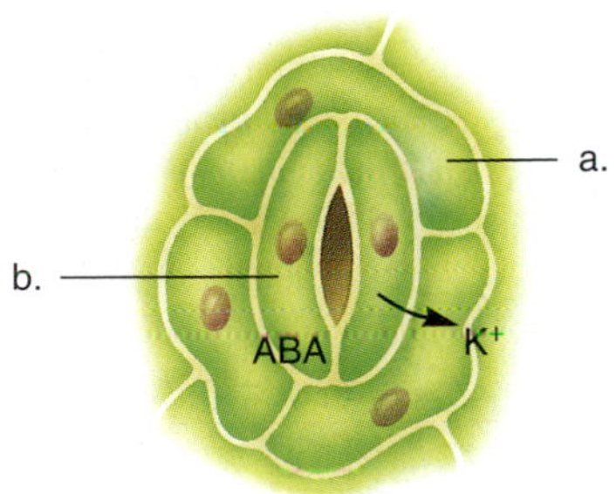

Thinking Scientifically

1. You wish to test the hypothesis that the gravitropic response is stronger than the phototropic response in stems. How could this hypothesis be tested and what results would be predicted?
2. Gibberellins are responsible for cell elongation between nodes of a stem. Miniature, or dwarf, varieties of plants have short internode cells. Remembering that a hormone only works after it binds to a receptor protein, propose at least two different hypotheses that could explain why a plant could fail to give a hormonal response.

Bioethical Issue

Marbled murrelets are endangered seabirds that nest in huge redwood trees along the coast of the Pacific northwest. Pacific Lumber was given the right to fell many acres in exchange for a few that would be preserved. The government paid the owner of Pacific Lumber some $500 million so that 3,500 acres would be preserved. Activists thought the deal was unfair—they wanted 60,000 acres preserved. Any less and there would not be enough habitat to keep the birds from sinking into oblivion. Not to mention that the trees themselves cannot regrow to their present size for hundreds upon hundreds of years.

Activists feel done in by their own representatives in the government. After all, the owner of Pacific Lumber is a big contributor to the reelection campaigns of elected officials who approved the deal. Under the circumstances, what would you do to save the trees? Some activists climbed the trees and refused to come down even when the trees were being cut. In September of 1998, David Chain, 24, lost his life when a tree fell on him and crushed his skull.

What is the proper action for activist groups when they feel their government is letting them down? Should they defy the law, and if so, what is the proper response of governmental officials?

Understanding the Terms

abscisic acid (ABA) 693
abscission 693
apical dominance 685
auxin 688
biological clock 687
circadian rhythm 686
coleoptile 688
cytokinin 692
day-neutral plant 694
dormancy 690
ethylene 693
germination 694
gibberellin 690
gravitropism 685
hormone 688
long-day plant 694
photoperiodism 694
phototropism 685
phytochrome 694
senescence 692
short-day plant 694
statolith 685
tendril 686
thigmotropism 686
tropism 684

Match the terms to these definitions:

a. ______________ Biological rhythm with a 24-hour cycle.
b. ______________ Directional growth of plants in response to the earth's gravity.
c. ______________ Dropping of leaves, fruits, or flowers from a plant.
d. ______________ Plant hormone producing increased stem growth; also involved in flowering and seed germination.
e. ______________ Relative lengths of daylight and darkness that affect the physiology and behavior of an organism.

Web Connections

Exploring the Internet

http://www.mhhe.com/biosci/genbio/mader
(click on *Biology 7/e*)

The *Biology 7/e* Online Learning Center provides many resources for studying the material in this chapter including links to the following sites:

Natural Vegetative Propagation is a site that describes the various vegetative means of plant propagation while describing the role of the hormones involved in each type.

http://koning.ecsu.ctstateu.edu/vegprop/vegpropn.html

Learn about the hormone (IAA) involved in plant growth in Apical Dominance, and see how it plays a role in bonsai, topiary, and other plant sculpture techniques.

http://koning.ecsu.ctstateu.edu/apical/apical.html

Plant Hormones. Information about plant hormones, and links to related sites.

http://www.plant-hormones.bbsrc.ac.uk/index.html

Plant Hormone Home Page. Information from a web page from Northern Illinois University on various plant hormones and other growth regulators.

http://www.plant-hormones.bbsrc.ac.uk/education/kenhp.htm

Educational Plant Hormone Resources. Information and links regarding plant hormones.

http://www.plant-hormones.bbsrc.ac.uk/educatio.html

Reproduction in Plants

c h a p t e r

c h a p t e r   c o n c e p t s

39.1 Life Cycle of Flowering Plants

- Flowering plants have an alternation of generation life cycle. The plant that bears flowers is the sporophyte. 700
- Flowering plants have two types of gametophytes: the microgametophyte (male) and the megagametophyte (female). 700
- Adaptation to a land environment includes pollen grains (microgametophytes) that carry sperm to the megagametophyte, which produces an egg. 703

39.2 Development of the Embryo

- The dicot embryo goes through a series of stages; once the cotyledons appear, it is possible to distinguish the shoot apex and the root apex. 707

39.3 Fruits and Seeds

- A seed has a seed coat, an embryo, and stored food. 708
- In flowering plants, seeds are enclosed by fruits, which develop from the ovary and accessory organs of a flower. 708
- Fruits aid the dispersal of seeds. 710

39.4 Asexual Reproduction in Plants

- Many flowering plants have an asexual means of propagation (i.e., from nodes of stems or from roots). 712
- Many plants can be regenerated in tissue of culture from meristem tissue and from individual cells. This has contributed to the genetic engineering of plants. 712

A flower of milkweed, *Asclepias*, is approached by a honeybee, *Apis*.

The flowering plants, or angiosperms, are the most diverse and widespread of all the plants, perhaps because they have a means of sexual reproduction that is well adapted to life on land. Pollen protects the sperm nucleus until fertilization takes place, and seeds protect the embryo until dispersal takes place. The structure of the flower allows angiosperms to produce seeds within fruits. The evolution of the flower on angiosperms also permits pollination to take place, not only by wind, but also by animals. Flowering plants that rely on animals for pollination have a mutualistic relationship with them. The flower provides nutrients for the pollinator such as a bee, a fly, a beetle, a bird, or even a bat. The animals in turn inadvertently carry pollen from one flower to another, allowing pollination to occur.

Many flowering plants can also reproduce asexually because their cells, especially meristem cells, are totipotent. Totipotency makes it possible for scientists to reproduce plants in laboratory cultures starting with individual cells or any organ of a plant. Plants grown from tissue cultures can be endowed with foreign genes that give them new and different characteristics of interest to human beings.

39.1 Life Cycle of Flowering Plants

A life cycle is the entire sequence of events from the time of fertilization and formation of the zygote to gamete formation once again. In contrast to animals, which have only one type of adult generation in their life cycle, plants have two types—a diploid generation and a haploid generation—that alternate with each other. The diploid generation is called the **sporophyte** because it produces spores. The spores are haploid, and they divide to become the haploid generation, which is called the **gametophyte** because it produces gametes. In flowering plants, the diploid sporophyte is said to be dominant. It is larger, longer lasting, and the generation we generally recognize as the plant. This is the generation that flowers (Fig. 39.1).

A flower produces two types of spores, microspores and megaspores. A **microspore** [Gk. *mikros,* small, little] develops into a microgametophyte; a **megaspore** [Gk. *megas,* great, large] develops into a megagametophyte. The microgametophyte is the pollen grain, which is either windblown or carried by an animal to the vicinity of the megagametophyte. When a pollen grain matures, it contains nonflagellated sperm, which travel by way of a pollen tube to the megagametophyte, which is called an embryo sac. Once a sperm fertilizes an egg in the embryo sac, the life cycle begins again.

Notice that the sporophyte is dominant in flowering plants. It is the generation that contains vascular tissue and has other adaptations suitable to living on land. The life cycle of a flowering plant is also adapted to a land existence. The gametophytes which produce the gametes are microscopic and dependent upon the sporophyte. Since the microgametophyte (pollen grain) is carried by wind or an animal, water is not needed for transport of pollen to the megagametophyte. The pollen tube provides passage for a sperm to reach an egg, and again no outside water is needed. (This may be compared to the manner in which fertilization is accomplished in human beings. At the time of sexual intercourse, the penis deposits sperm into the vagina of the female.) Following fertilization, the diploid zygote develops into an embryo protected within a seed enclosed by a fruit.

Unlike flowering plants, some plants produce only one kind of spore and this develops into a separate, but water-dependent, gametophyte. The evolution of two separate and dependent gametophytes in flowering plants led to the production of pollen and a means to carry on pollination and fertilization without need of external water.

Flowering plants undergo an alternation of generations life cycle that is modified in such a way that a sperm does not require an outside source of water to reach an egg.

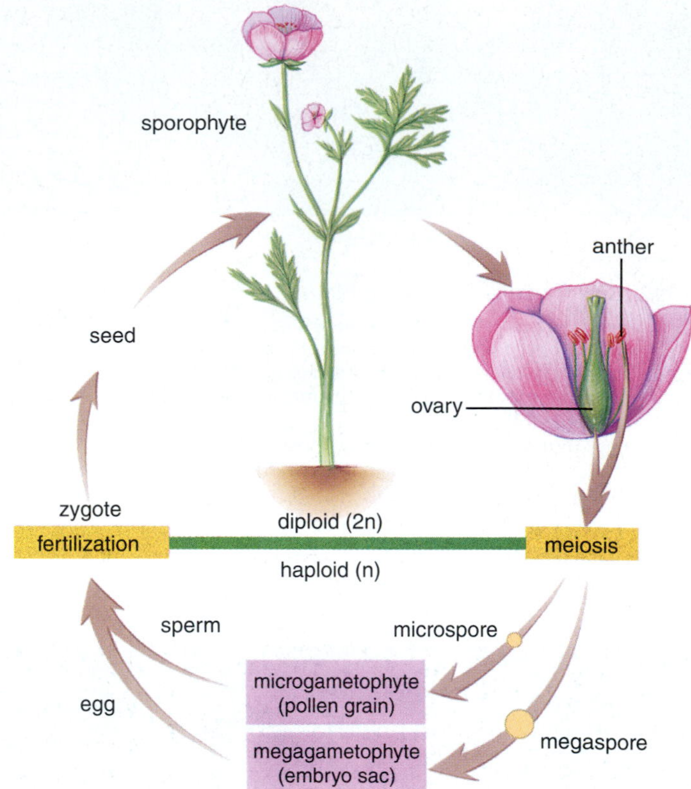

Figure 39.1 Alternation of generations overview. In flowering plants, the diploid zygote develops into an embryo enclosed within a seed. The embryo becomes the sporophyte, which bears flowers. The flower produces microspores and megaspores by meiosis. A megaspore becomes a megagametophyte, which produces an egg within an embryo sac and a microspore becomes a microgametophyte (pollen grain), which produces sperm. Fertilization results in a zygote.

The Sporophyte

In the life cycle of angiosperms, the sporophyte generation produces flowers. A **flower,** a reproductive structure, develops within a bud. In many plants, the same shoot apical meristem that previously formed leaves suddenly stops producing leaves and starts producing a flower. In other plants, axillary buds develop directly into flowers. Flower structures are modified leaves attached to a short stem tip called a receptacle (Fig. 39.2). In monocots, flower parts occur in threes and multiples of three. In dicots, flower parts are in fours or fives and multiples of four or five. The **sepals,** which are the most leaflike of all the flower parts, are usually green, and they protect the bud as the flower develops within. When the flower opens, there is an outer whorl of sepals and an inner whorl of **petals,** whose color accounts for the attractiveness of many flowers. The size, the shape, and the color of a flower are attractive to a specific pollinator. Wind-pollinated flowers often have no petals at all.

At the very center of a flower is the **pistil** [L. *pistillum,* club-shaped pounder, pestle], which is often vaselike in appearance. A pistil usually has three parts: the **stigma,** an enlarged sticky knob, the **style,** a slender stalk, and the **ovary,** an enlarged base. An ovary has one or more ovule-bearing units called a carpel [Gk. *karpos,* fruit]. A compound ovary has multiple carpels, which are often fused: A simple pistil contains a single **carpel.**

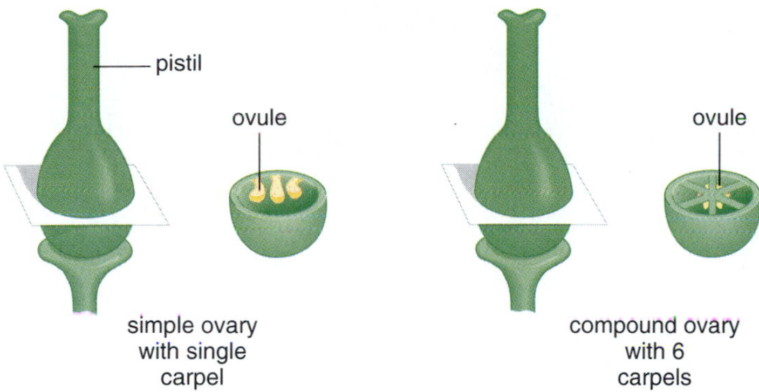

A carpel usually contains a number of **ovules** [L. *ovulum,* dim. of ovum: egg], which play a significant role in the production of megaspores and therefore megagametophytes. Grouped about the pistil are a number of **stamens,** each of which has two parts: the **anther,** a saclike container, and the **filament** [L. *filum,* thread], a slender stalk. Pollen grains develop from microspores in the anther.

Not all flowers have sepals, petals, stamens, and a pistil. Those that do are said to be complete and those that do not are said to be incomplete. Flowers that have both stamens and a pistil are called perfect flowers: those with only stamens are staminate flowers and those with only pistils are pistillate flowers. If staminate flowers and pistillate flowers are on one plant, as in corn, the plant is monoecious [Gk. *monos,* one, and *oikos,* home, house]. If staminate and pistillate flowers are on separate plants, the plant is dioecious. Holly trees are dioecious and, if red berries are a priority, it is necessary to acquire a plant with staminate flowers and another with pistillate flowers.

Sometimes the pistil is called the female part of the flower and the stamens are called the male part of the flower, but this is not strictly correct. The pistil and stamens do not produce gametes; they produce megaspores and microspores, respectively. A megaspore matures into a megagametophyte, which produces an egg, and a microspore matures into a microgametophyte (pollen grain), which produces sperm.

A flower contains four basic parts: sepals, petals, stamens, and a pistil.

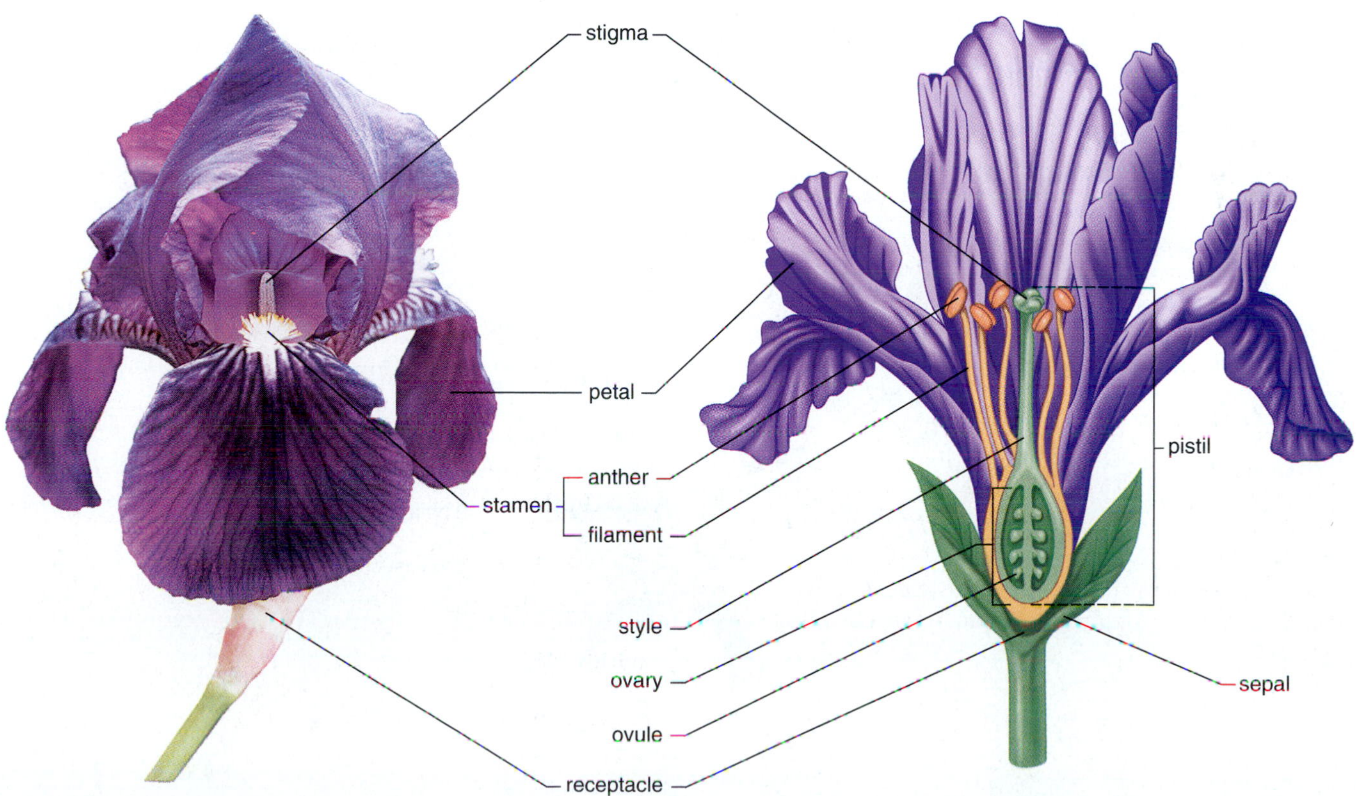

Figure 39.2 **Typical flower, *Iris.***
In angiosperms, the sporophyte produces flowers. A megagametophyte develops within an ovule and microgametophytes (pollen grains) develop within an anther.

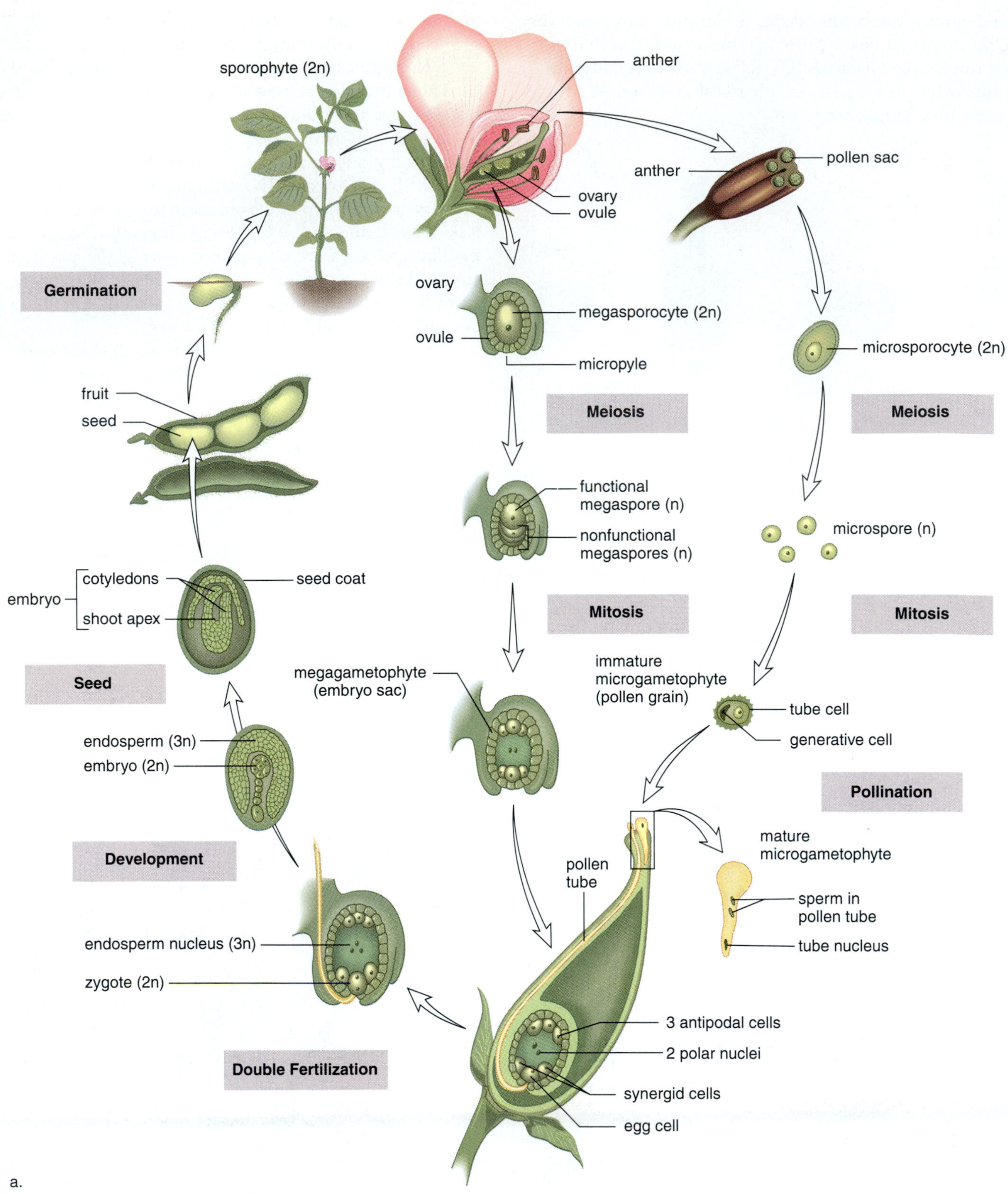
sporophyte (2n)
anther
anther
pollen sac
ovary
ovule
Germination
ovary
megasporocyte (2n)
ovule
micropyle
microsporocyte (2n)
fruit
seed
Meiosis
Meiosis
functional
megaspore (n)
nonfunctional
megaspores (n)
microspore (n)
embryo
cotyledons
shoot apex
seed coat
Mitosis
Mitosis
Seed
megagametophyte
(embryo sac)
immature
microgametophyte
(pollen grain)
tube cell
generative cell
endosperm (3n)
embryo (2n)
Pollination
mature
microgametophyte
Development
pollen
tube
sperm in
pollen tube
tube nucleus
endosperm nucleus (3n)
zygote (2n)
3 antipodal cells
2 polar nuclei
Double Fertilization
synergid cells
egg cell

a.

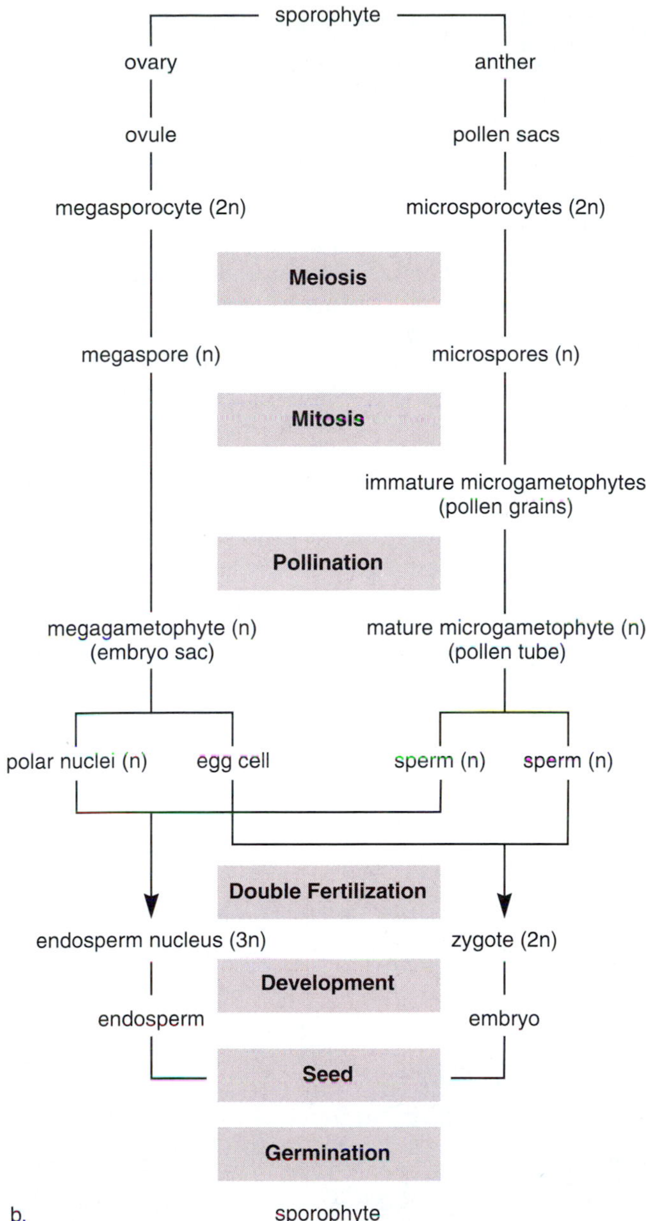

Figure 39.3 Life cycle of a flowering plant.
a. An ovary contains an ovule where a megasporocyte produces a megaspore by meiosis. A megaspore develops into an embryo sac containing seven cells, one of which is an egg. A pollen sac in the anther contains microsporocytes, which produce microspores by meiosis. A microspore develops into a pollen grain that contains sperm by the time it germinates. The sperm travel down a pollen tube; one sperm fertilizes the egg to form a diploid zygote, and the other fuses with the polar nuclei to form a triploid endosperm nucleus. A seed contains the developing embryo plus stored food. Germination and growth of a seed results in a new sporophyte plant. **b.** Flow diagram.

The Gametophytes

The ovary contains one or more ovules. An ovule has a central mass of parenchyma cells almost completely covered by integuments except where there is an opening, the micropyle (Fig. 39.3*a*). One parenchyma cell enlarges to become a **megasporocyte** (megaspore mother cell), which undergoes meiosis, producing four haploid megaspores. Three of these megaspores are nonfunctional and one is functional. The nucleus of the functional megaspores divides mitotically until there are eight nuclei of the **megagametophyte** [Gk. *megas*, great, large]. When cell walls form later, there are seven cells, one of which is binucleate. The megagametophyte, also called the **embryo sac,** consists of these seven cells:

one egg cell, associated with
two synergid cells;
one central cell, with two polar nuclei;
three antipodal cells

Microgametophytes are produced in the anther of the stamens (Fig. 39.3*a*, far right). An anther has four pollen sacs, each containing many **microsporocytes** (microspore mother cells). A microsporocyte undergoes meiosis to produce four haploid microspores. A microspore divides mitotically, forming two cells enclosed by a finely sculptured wall. This structure, called the **pollen grain,** is the immature **microgametophyte** and contains a tube cell and a generative cell. Either now or later, the generative cell divides mitotically to produce two sperm. The walls separating the pollen sacs in the anther break down when the pollen grains are ready to be released.

Look back at Figure 39.1 and realize that the sporophyte, the flowering diploid generation, has now produced a haploid megaspore and haploid microspores. The megaspore has developed into an embryo sac, the egg-producing megagametophyte generation. Each microspore has developed into a pollen grain, the sperm-producing microgametophyte generation. The haploid generation in flowering plants is represented by the megagametophyte and the microgametophyte, which are microscopic. The megagametophyte is retained within the body of the sporophyte. As discussed in the reading on pages 704–05, microgametophytes (pollen grains) are windblown or carried by various kinds of animals to the stigma of a pistil.

In flowering plants, the megagametophyte within the ovule produces an egg. The microgametophyte (pollen grain) produces two sperm.

Plants and Their Pollinators

A plant and its pollinator(s) are adapted to one another. They have a mutualistic relationship in which each benefits—the plant uses its pollinator to ensure that cross-pollination takes place, and the pollinator uses the plant as a source of food. This mutualistic relationship came about through the process of **coevolution;** that is, the codependency of the plant and the pollinator is the result of suitable changes in structure and function of each. The evidence for coevolution is observational. For example, floral coloring and odor are suited to the sense perceptions of the pollinator, the mouthparts of the pollinator are suited to the structure of the flower, the type of food provided is suited to the nutritional needs of the pollinator, and the pollinator forages at the time of day that specific plants are open. The following are examples of such coevolution.

Bee-Pollinated Flowers

There are now 20,000 different species of bees that pollinate flowers. The best-known pollinators are the honeybees (Fig. 39Aa). Bee eyes see a spectrum of light that is different from the spectrum seen by humans. The bee's visible spectrum is shifted so that they do not see red wavelengths but do see ultraviolet wavelengths. Bee-pollinated flowers are usually brightly colored and are predominantly blue or yellow; they are not entirely red. They may also have ultraviolet shadings called honey guides, which highlight the portion of the flower that contains the reproductive structures. The mouthparts of bees are fused into a long tube that contains a tongue. This tube is an adaptation for sucking up nectar provided by the plant, usually at the base of the flower.

Bee flowers are delicately sweet and fragrant to advertise that nectar is present. The honey guides often point to a narrow floral tube large enough for the bee's feeding apparatus but too small for other insects to reach the nectar. Bees also collect pollen as food for their larvae. Pollen clings to the hairy body of a bee, and the bees also gather it by means of bristles on their legs. They then store the pollen in pollen baskets on the third pair of legs. Bee-pollinated flowers are sturdy and irregular in shape because they often have a landing platform where the bee can alight. The landing platform requires the bee to brush up against the anther and stigma as it moves toward the floral tube to feed. One type of orchid, *Ophrys,* has evolved a unique adaptation. The flower resembles a female bee, and when the male of that species attempts to copulate with the flower, the bee receives pollen.

a.

b.

Figure 39A **Pollinators.**
a. A bee-pollinated flower is a color other than red (bees cannot detect this color) and has a landing platform where the reproductive structures of the flower brush up against the bee's body. **b.** A butterfly-pollinated flower is often a composite, containing many individual flowers. The broad expanse provides room for the butterfly to land, after which it lowers its proboscis into each flower in turn. **c.** Hummingbird-pollinated flowers are curved back, allowing the bird to insert its beak to reach the rich supply of nectar. While doing this, the bird's forehead and other body parts touch the reproductive structures. **d.** Bat-pollinated flowers are large, sturdy flowers that can take rough treatment. Here the head of the bat is positioned so that its bristly tongue can lap up nectar.

Moth- and Butterfly-Pollinated Flowers

Contrasting moth- and butterfly-pollinated flowers emphasizes the close adaptation between pollinator and flower. Both moths and butterflies have a long, thin, hollow proboscis, but they differ in other characteristics. Moths usually feed at night and have a well-developed sense of smell. The flowers they visit are visible at night because they are lightly shaded (white, pale yellow, or pink), and they have strong, sweet perfume, which helps attract moths. Moths hover when they feed, and their flowers have deep tubes with open margins that allow the hovering moths to reach the nectar with their long proboscis. Butterflies are active in the daytime and have good vision but a weak sense of smell. Their flowers have bright colors—even red because butterflies can see the color red—but the flowers tend to be odorless. Unable to hover, butterflies need a place to land. Flowers that are visited by butterflies often have flat landing platforms (Fig. 39A*b*). The flowers also tend to be composites, with many individual flowers clustered in a head. Each flower has a long, slender floral tube, accessible to the long, thin butterfly proboscis.

Bird- and Bat-Pollinated Flowers

In North America, the most well-known bird pollinators are the hummingbirds. These tiny animals have good eyesight but do not have a well-developed sense of smell. Like moths, they hover when they feed. Typical flowers pollinated by hummingbirds are red, with a slender floral tube and margins that are curved back and out of the way. And although they produce copious amounts of nectar, the flowers have little odor. As a hummingbird feeds on nectar with its long, thin beak, its head comes into contact with the stamens and pistil (Fig. 39A*c*).

Bats are adapted to gathering food in various ways, including feeding on the nectar and pollen of plants. Bats are nocturnal and have an acute sense of smell. Those that are pollinators also have keen vision and a long, extensible, bristly tongue. Typically, bat-pollinated flowers open only at night and are light-colored or white. They have a strong, musty smell similar to the odor that bats produce to attract one another. The flowers are generally large and sturdy and are able to hold up when a bat inserts part of its head to reach the nectar. While the bat is at the flower, its head is dusted with pollen (Fig. 39A*d*).

Coevolution

These examples are evidence of coevolution, but how did coevolution come about? Some 200 million years ago, when seed plants were just beginning to evolve and insects were not as diverse as they are today, wind alone was used to carry pollen. Wind pollination, however, is a hit-or-miss affair. Perhaps beetles feeding on vegetative leaves were the first insects to carry pollen directly from plant to plant by chance. This use of animal motility to achieve cross-fertilization no doubt resulted in the evolution of flowers, which have features, such as the production of nectar, to attract pollinators. Then, if beetles developed the habit of feeding on flowers, other features, such as the protection of ovules within ovaries, may have evolved.

As cross-fertilization continued, more and more flower variations likely developed, and pollinators became increasingly adapted to specific angiosperm species. Today, there are some 235,000 species of flowering plants and over 700,000 species of insects. This diversity suggests that the success of angiosperms has contributed to the success of insects, and vice versa.

c.

d.

Pollination and Fertilization

Pollination and fertilization are two separate events. **Pollination** is simply the transfer of pollen (Fig. 39.4) from the anther to the stigma of a pistil. **Fertilization** [L. *fertilis*, fruitful] is the fusion of nuclei, as when the sperm nucleus and the egg nucleus fuse.

Pollination

As mentioned previously, pollination is brought about by the wind or with the assistance of a particular pollinator. Self-pollination occurs if the pollen is from the same plant, and cross-pollination occurs if the pollen is from a different plant. Cross-pollination is preferred because it fosters genetic recombination resulting in new and varied plants.

Fertilization

When a pollen grain lands on the stigma of the same species, it germinates, forming a pollen tube (Fig. 39.5). The germinated pollen grain, containing a tube cell and two sperm, is the mature microgametophyte. As it grows, the pollen tube passes between the cells of the stigma and the style to reach the micropyle of the ovule. Now **double fertilization** occurs. One sperm nucleus unites with the egg nucleus, forming a 2n zygote, and the other sperm nucleus migrates and unites with the polar nuclei of the central cell, forming a 3n endosperm nucleus. The zygote divides mitotically to become the **embryo,** a young sporophyte, and the endosperm nucleus divides mitotically to become the endosperm. **Endosperm** [Gk. *endon*, within, and *sperma*, seed] is the tissue that will nourish the embryo and seedling as they undergo development.

Flowering plants practice double fertilization. One sperm nucleus unites with the egg nucleus, producing a zygote, and the other unites with the polar nuclei, forming a 3n endosperm cell.

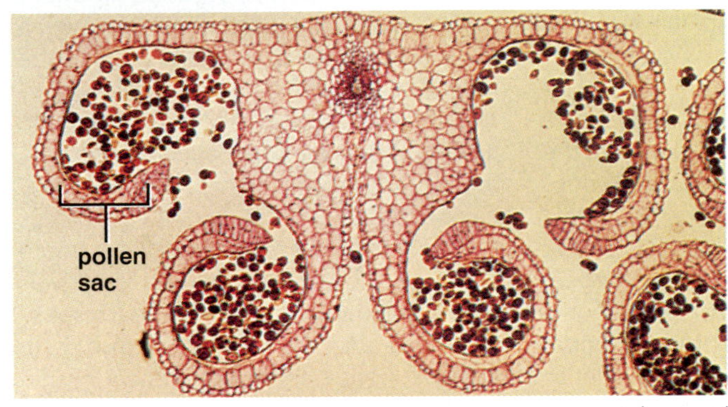

Figure 39.4 Development of pollen grains.
a. A mature anther showing that the walls between the pollen sacs have opened, allowing the pollen grains to be released. **b.** Germinating pollen grains of goosegrass, *Galium aparine*, are colored yellow in this false-colored scanning electron micrograph.

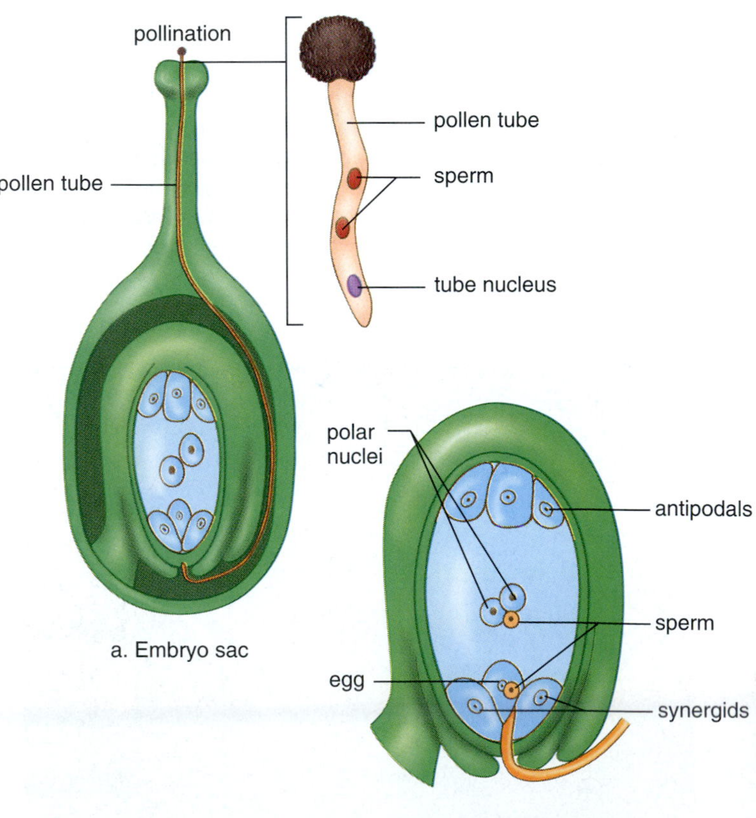

Figure 39.5 Fertilization.
a. The pollen tube grows down the style, carrying with it two sperm. The pollen tube grows through the micropyle and releases both sperm. **b.** One of these migrates to the egg; this sperm nucleus fuses with the egg nucleus, forming a zygote. The second sperm fuses with the polar nuclei, forming a large triploid endosperm nucleus.

39.2 Development of the Embryo

Stages in the development of a dicot embryo are shown in Figure 39.6. After double fertilization has taken place, the single-celled zygote lies beneath the endosperm nucleus. The endosperm nucleus divides to produce a mass of endosperm tissue surrounding the embryo. The zygote also divides, forming two parts: the upper part is the embryo, and the lower part is the suspensor, which anchors the embryo and transfers nutrients to it from the sporophyte plant. Soon the **cotyledons** [Gk. *cotyledon*, cup-shaped cavity], or seed leaves, can be seen. At this point, the dicot embryo is heart shaped. Later, when it becomes torpedo shaped, it is possible to distinguish the shoot tip and the root tip. These contain apical meristems, the tissues that bring about primary growth in a plant; the shoot apical meristem is responsible for aboveground growth, and the root apical meristem is responsible for underground growth.

Monocots, unlike dicots, have only one cotyledon. Another important difference between monocots and dicots is the manner in which nutrient molecules are stored in the seed. In a monocot, the cotyledon rarely stores food; rather, it absorbs food molecules from the endosperm and passes them to the embryo. During the development of a dicot embryo, the cotyledons usually store the nutrient molecules that the embryo uses. Therefore, in Figure 39.6 we can see that the endosperm seemingly disappears. Actually, it has been taken up by the two cotyledons. In a plant embryo, the epicotyl is the portion between the cotyledon(s) and the first leaves. This portion contributes to shoot development. The hypocotyl is that portion below the cotyledon(s) that contributes to stem development; and the radicle contributes to root development. The embryo plus stored food is now contained within a seed.

The plant embryo (which has gone through a set series of stages), plus its stored food, is contained within a seed.

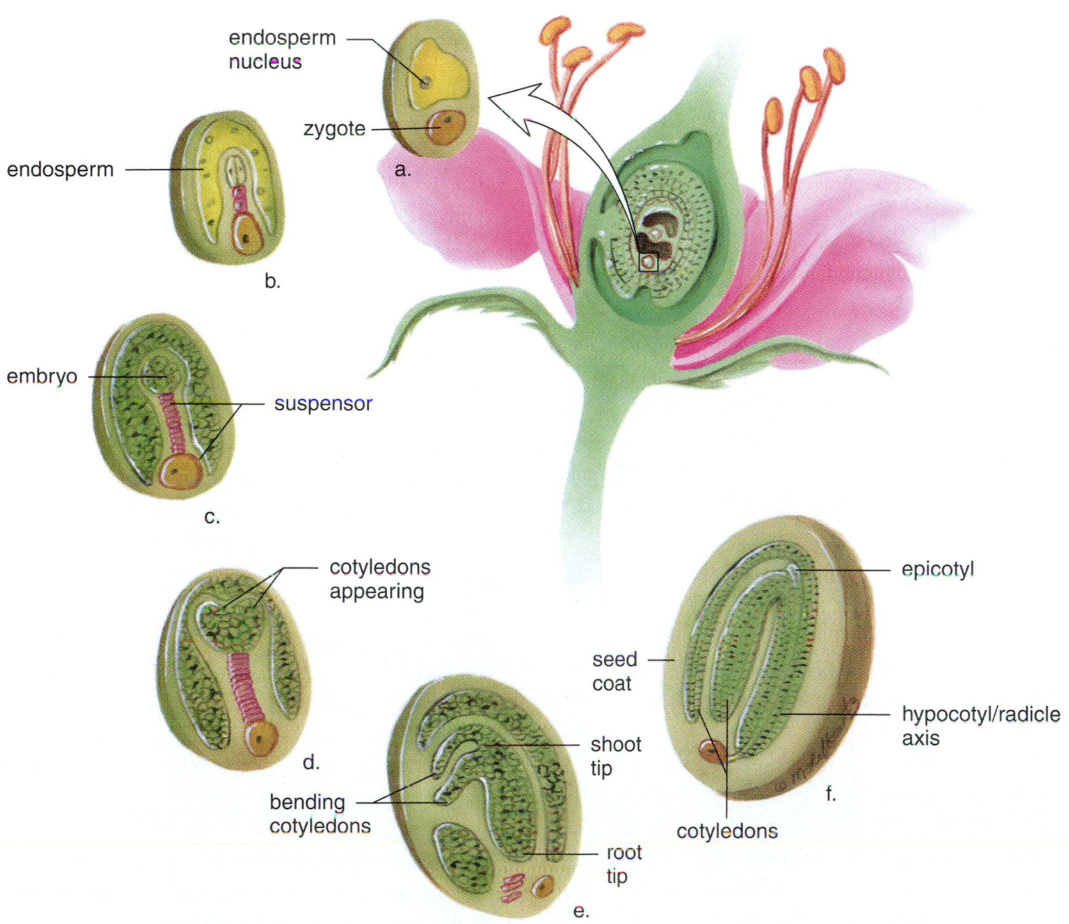

Figure 39.6 Development of a dicot embryo.
a. The single-celled zygote lies beneath the endosperm nucleus. **b, c.** The endosperm is a mass of tissue surrounding the embryo. The embryo is located above the suspensor. **d.** The embryo becomes heart shaped as the cotyledons begin to appear. **e.** There is progressively less endosperm as the embryo differentiates and enlarges. As the cotyledons bend, the embryo takes on a torpedo shape. **f.** The embryo consists of the epicotyl (represented here by the shoot apex), the hypocotyl, and the radicle (the latter of which contains the root apex).

a. Almond, *Prunus*

b. Tomato, *Lycopersicon*

Figure 39.7 Fruit diversity.
a. The almond fruit is fleshy, with a single seed enclosed by a hard covering. **b.** The tomato is derived from a compound ovary. **c.** Dry fruit of the pea plant develops from a simple ovary. **d.** Blackberry flowers contain several separate ovaries; each cluster is a berry from a single flower.

39.3 Fruits and Seeds

As the zygote develops into an embryo, the integuments of the ovule harden and become the seed coat. A **seed** is a structure formed by the maturation of the ovule; it contains a sporophyte embryo plus stored food. The ovary, and sometimes other floral parts, develops into a fruit. Although peas and beans in their shell, tomatoes, and cucumbers are commonly called vegetables, botanists categorize them as fruits. A **fruit** is a mature ovary that usually contains seeds.

Types of Fruits

As fruit develops from an ovary, the ovary wall thickens to become the pericarp. Most fruits are simple fruits; they are derived from an individual ovary, either simple or compound. In simple fleshy fruits, the pericarp is at least somewhat fleshy. Peaches and plums are good examples of simple fleshy fruits. In almonds, the fleshy part of the pericarp is a husk removed before marketing. We crack the remaining portion of the pericarp to obtain the seed (Fig. 39.7*a*). An apple develops from a compound ovary (one that has several sections), but much of the flesh comes from the receptacle, which grows around the ovary. It's more obvious that a tomato comes from a compound ovary, because in cross section, you can see several seed-filled cavities (Fig. 39.7*b*).

Dry fruits have a dry pericarp (Table 39.1). Legumes, such as peas (Fig. 39.7*c*) and beans, produce a fruit that splits along two sides, or seams. Not all dry fruits split at maturity. The pericarp of a grain is tightly fused to the seed and cannot be separated from it. A corn kernel is a grain, as are the fruits of wheat, rice, and barley plants.

Some fruits develop from several individual ovaries and therefore are compound fruits. A blackberry is an aggregate fruit in which each berry is derived from a separate ovary of a single flower (Fig. 39.7*d*). The strawberry is also an aggregate fruit, but each ovary becomes a one-seeded fruit called an achene. The flesh of a strawberry is from the receptacle. In contrast, a pineapple comes from the fruit of many individual flowers attached to the same fleshy stalk. As the ovaries mature, they fuse to form a large, multiple fruit.

In flowering plants, the seed develops from the ovule and the fruit develops from the ovary.

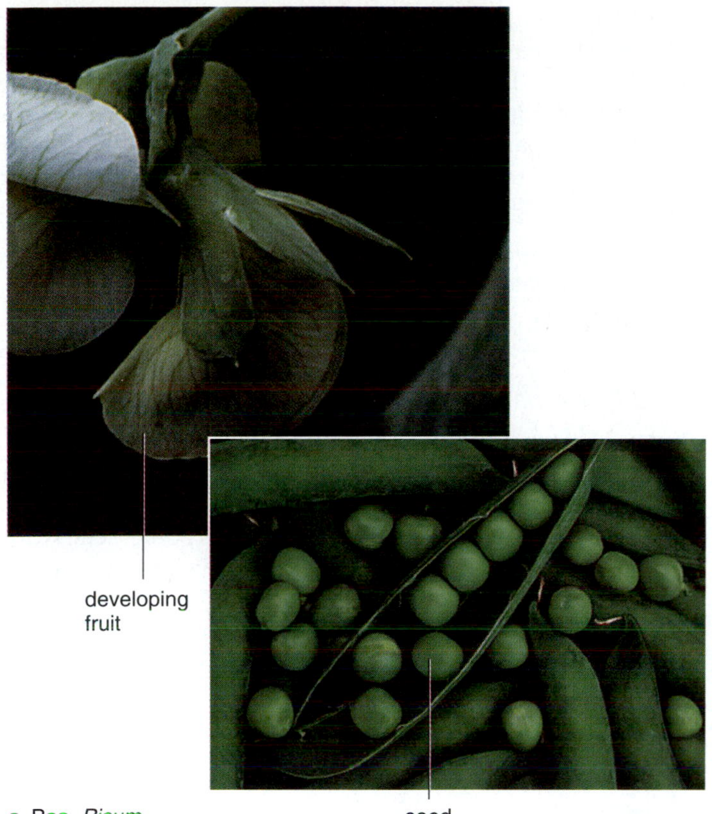

c. Pea, *Pisum*

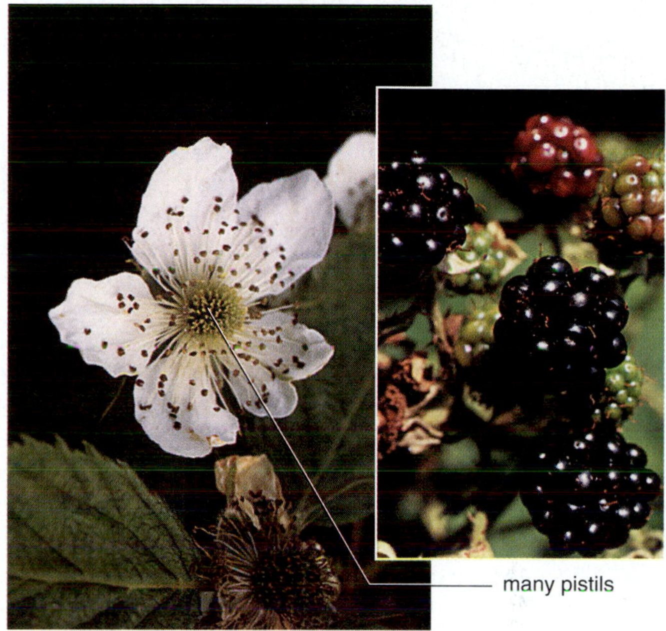

d. Blackberry, *Rubus*

Table 39.1

Kinds of Fruit

Name	Description	Example
Simple Fruits	***Develop from an Individual Ovary***	
Fleshy	Pericarp is usually fleshy.	
Drupe	From simple ovary with one seed (pit) and soft "skin"	Peach, plum, olive
Berry	From compound ovary with many seeds	Grape, tomato
Pome	From compound ovary; flesh is from accessory of flower parts.	Apple, pear
Dry	Pericarp is dry.	
Follicle	From simple ovary that splits open down one side	Milkweed, peony
Legume	From simple ovary that splits open on both sides	Pea, bean, lentil
Capsule	From compound ovary with capsules that split in various ways	Poppy
Achene	From simple ovary with one-seeded small fruit; pericarp easily removed	Sunflower, dandelion, strawberry
Nut	From simple ovary with one-seeded fruit; hard pericarp	Acorn, hickory nut, chestnut
Grain	From simple ovary with one-seeded small fruit; pericarp completely united with seed coat	Rice, oat, barley
Compound Fruits	***Develop from a Group of Individual Ovaries***	
Aggregate fruits	Ovaries are from a single flower.	Blackberry, raspberry
Multiple fruits	Ovaries are from separate flowers clustered together.	Pineapple

Seed Dispersal and Germination

For plants to be widely distributed, their seeds have to be dispersed—that is, distributed preferably long distances from the parent plant. Following dispersal, the seeds germinate: they begin to grow so that a seedling appears.

Seed Dispersal

Plants have various means to ensure that dispersal takes place. The hooks and spines of clover, bur, and cocklebur attach to the fur of animals and the clothing of humans. Birds and mammals sometimes eat fruits, including the seeds, which are then defecated (passed out of the digestive tract with the feces) some distance from the parent plant. Squirrels and other animals gather seeds and fruits, which they bury some distance away.

The fruit of the coconut palm, which can be dispersed by ocean currents, may land many hundreds of kilometers away from the parent plant. Some plants have fruits with trapped air or seeds with inflated sacs that help them float in water. Many seeds are dispersed by wind. Woolly hairs, plumes, and wings are all adaptations for this type of dispersal. The seeds of an orchid are so small and light that they need no special adaptation to carry them far away. The somewhat heavier dandelion fruit uses a tiny "parachute" for dispersal. The winged fruit of a maple tree, which contains two seeds, has been known to travel up to 10 kilometers from its parent. A touch-me-not plant has seed pods that swell as they mature. When the pods finally burst, the ripe seeds are hurled out.

Animals, water, and wind help plants disperse their seeds.

Seed Germination

Some seeds do not **germinate** until they have been dormant for a period of time. For seeds, dormancy is the time during which no growth occurs, even though conditions may be favorable for growth. In the temperate zone, seeds often have to be exposed to a period of cold weather before dormancy is broken. In deserts, germination does not occur until there is adequate moisture. This requirement helps ensure that seeds do not germinate until the most favorable growing season has arrived. Germination—an event that takes

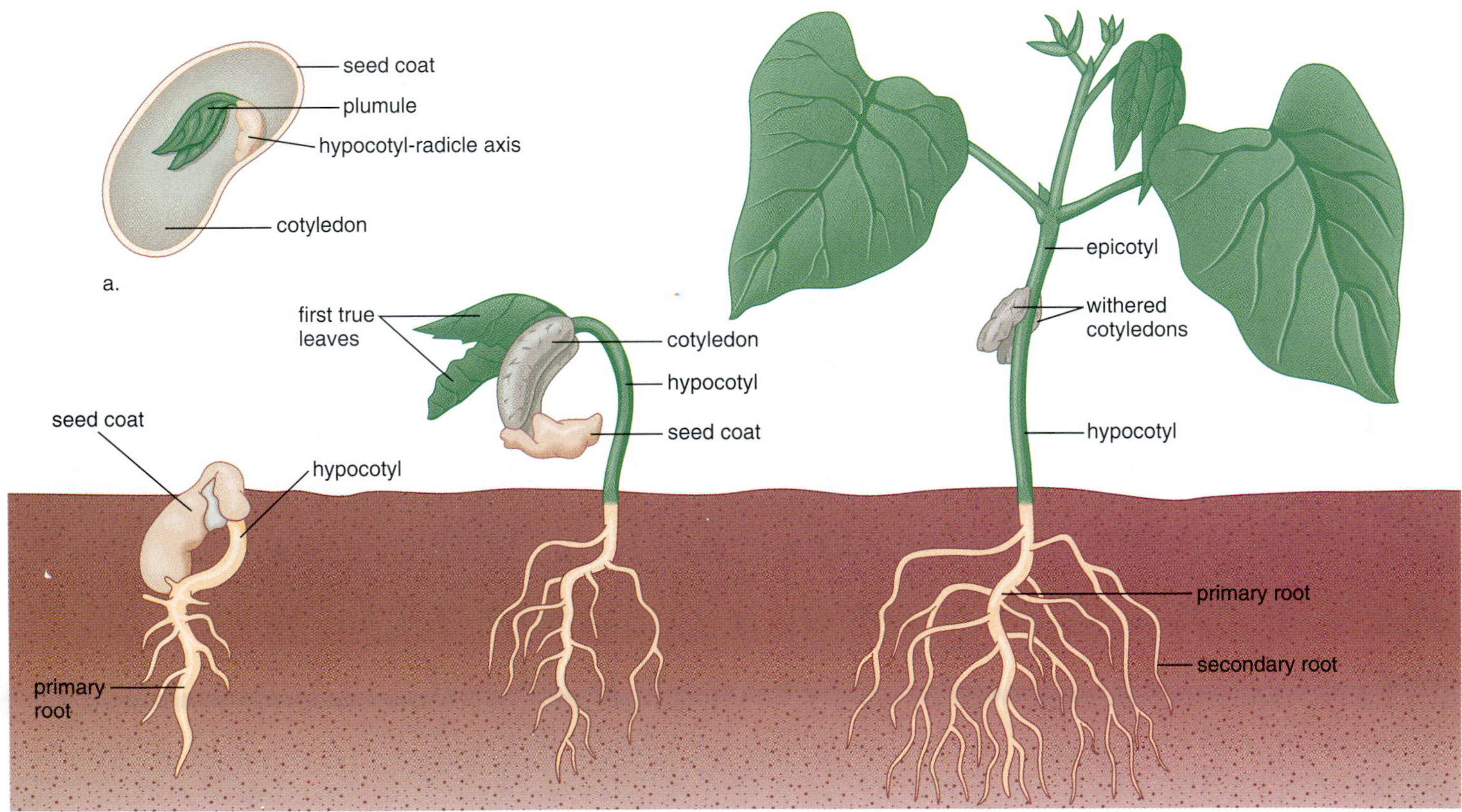

Figure 39.8 Common garden bean structure and germination. a. Seed structure. b. Germination and development of the seedling.

place if there is sufficient water, warmth, and oxygen to sustain growth—requires regulation, and both inhibitors and stimulators are known to exist. It is known that fleshy fruits (e.g., apples, pears, oranges, and tomatoes) contain inhibitors so that germination does not occur until the seeds are removed and washed. In contrast, stimulators are present in the seeds of some temperate zone woody plants. Mechanical action may also be required. Water, bacterial action, and even fire can act on the seed coat, allowing it to become permeable to water. The uptake of water causes the seed coat to burst.

Germination in Dicots and Monocots

As mentioned, the embryo of a dicot, such as a bean plant, has two seed leaves, called cotyledons. The cotyledons, which supply nutrients to the embryo and seedling, eventually shrivel and disappear. If the two cotyledons of a bean seed are parted, you can see a rudimentary plant (Fig. 39.8). The epicotyl bears young leaves and is called a **plumule** [L. *plumulla,* dim. of *pluma*: feather]. As the dicot seedling emerges from the soil, the shoot is hook shaped to protect the delicate plumule. The hypocotyl becomes the stem and the radicle develops into the roots. When a seed germinates in darkness, it etiolates—the stem is elongated, the roots and leaves are small, and the plant lacks color and appears spindly. Phytochrome, a pigment that is sensitive to red and far-red light, regulates this response and induces normal growth once proper lighting is available.

A corn plant is a monocot that contains a single cotyledon. Actually, the endosperm is the food-storage tissue in monocots, and the cotyledon does not have a storage role. A corn kernel is a type of fruit called a grain and the outer covering is the pericarp (Fig. 39.9). The plumule and radicle are enclosed in protective sheaths called the coleoptile and the coleorhiza, respectively. The plumule and the radicle burst through these coverings when germination occurs.

Germination is a complex event regulated by many factors. The embryo breaks out of the seed coat and becomes a seedling with leaves, stem, and roots.

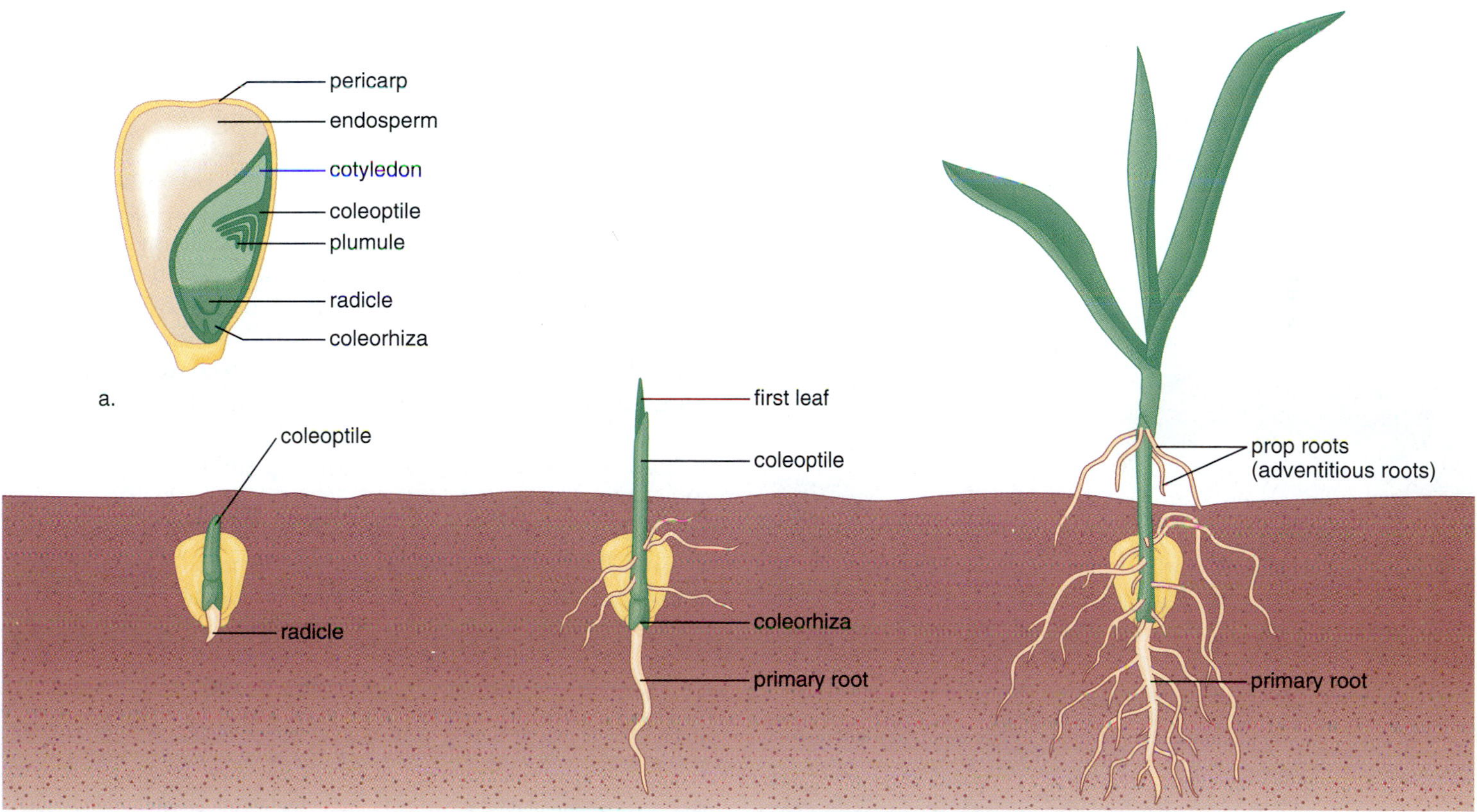

Figure 39.9 **Corn kernel structure and germination.**
a. Grain structure. b. Germination and development of the seedling.

39.4 Asexual Reproduction in Plants

Because plants contain nondifferentiated meristem tissue, they routinely reproduce asexually by vegetative propagation. In asexual reproduction there is only one parent, instead of two as in sexual reproduction. Violets will grow from the nodes of rhizomes (underground horizontal stems) and complete strawberry plants will grow from the nodes of stolons (aboveground horizontal stems) (Fig. 39.10). White potatoes are actually portions of underground stems, and each eye is a bud that will produce a new potato plant if it is planted with a portion of the swollen tuber. Sweet potatoes are modified roots; they can be propagated by planting sections of the root. You may have noticed that the roots of some fruit trees, such as cherry and apple trees, produce "suckers," small plants that can be used to grow new trees.

In addition to plants already mentioned, sugarcane, pineapple, cassava, and many ornamental plants have been propagated from stem cuttings for some time. In these plants, pieces of stem will automatically produce roots. The discovery that the plant hormone auxin will cause roots to develop has expanded the list of plants that can be propagated from stem cuttings.

Tissue Culture of Plants

Hydroponics, the growth of plants in aqueous solutions, had begun by the 1860s. This practice, along with the ability of plants to reproduce asexually, led the German botanist Gottlieb Haberlandt to speculate in 1902 that entire plants could be produced by tissue culture. **Tissue culture** is the growth of a *tissue* in an artificial liquid *culture* medium. Haberlandt said that plant cells are **totipotent** [L. *totus*, all, whole, and *potens*, powerful]—each cell has the full genetic potential of the organism—and therefore a single cell could become a complete plant. But it wasn't until 1958 that Cornell botanist F. C. Steward grew a complete carrot plant from a tiny piece of phloem. Like former investigators, he provided the cells with sugars, minerals, and vitamins, but he also added coconut milk. (Later, it was discovered that coconut milk contains the plant hormone cytokinin.) When the cultured cells began dividing, they produced a callus, an undifferentiated group of cells. Then the callus differentiated into shoot and roots and developed into complete plants.

Figure 39.10 Asexual reproduction in plants.
Meristem tissue at nodes can generate new plants, as when the "runners" of strawberry plants, *Fragaria*, give rise to new plants.

Tissue culture techniques have by now led to micropropagation, a commercial method of producing thousands, even millions, of identical seedlings in a limited amount of space. One favorite method to accomplish micropropagation is by meristem culture. If the correct proportions of auxin and cytokinin are added to a liquid medium, many new shoots will develop from a single shoot tip. When these are removed, more shoots form. Since the shoots are genetically identical, the adult plants that develop from them, called clonal plants, all have the same traits. Another advantage to meristem culture is that meristem, unlike other portions of a plant, is virus-free; therefore the plants produced are also virus-free. (The presence of plant viruses weakens plants and makes them less productive.)

Because plant cells are totipotent, it should be possible to grow an entire plant from a single cell. This, too, has been done. Enzymes are used to digest the cell walls of a small piece of tissue, usually mesophyll tissue, from a leaf, and the result is naked cells without walls, called **protoplasts** [Gk. *protos*, first, and *plastos*, formed, molded] (Fig. 39.11*a*). The protoplasts regenerate a new cell wall and begin to divide. These clumps of cells can be manipulated to produce **somatic** (asexually produced) **embryos** (Fig. 39.11*e*). Somatic embryos that are encapsulated in a protective hydrated gel (and sometimes called artificial seeds) can be shipped anywhere. It's possible to produce millions of somatic embryos at once in large tanks called bioreactors. This is done for certain vegetables like tomato, celery, and asparagus and for ornamental plants like lilies, begonias, and African violets. A mature plant develops from each somatic embryo (Fig. 39.11*f*). Plants generated from the somatic embryos vary somewhat because of mutations that arise during the production process. These so-called somaclonal variations are another way to produce new plants with desirable traits.

Anther culture is a technique in which mature anthers are cultured in a medium containing vitamins and growth regulators. The haploid tube cells within the pollen grains divide, producing proembryos consisting of as many as 20 to 40 cells. Finally the pollen grains rupture, releasing haploid embryos. The experimenter can now generate a haploid plant, or chemical agents can be added that encourage chromosomal doubling. After chromosomal doubling, the resulting plants are diploid but homozygous for all their alleles. Anther culture is a direct way to produce plants that express recessive alleles. If the recessive alleles govern desirable traits, the plants have these traits.

The culturing of plant tissues has led to a technique called cell suspension culture. Rapidly growing calluses are cut into small pieces and shaken in a liquid nutrient medium so that single cells or small clumps of cells break off and form a suspension. These cells will produce the same chemicals as the entire plant. For example, cell suspension cultures of *Cinchona ledgeriana* produce quinine and those of *Digitalis lanata* produce digitoxin. Scientists envision that it will also be possible to maintain cell suspension cultures in bioreactors for the purpose of producing chemicals used in the production of drugs, cosmetics, and agricultural chemicals. If so, it will no longer be necessary to farm plants simply for the purpose of acquiring the chemicals they produce.

Plant tissue culture is now well established. The starting material can be meristem tissue from most any part of a plant or it can be adult cells because plant cells are totipotent if provided with the correct hormonal/nutrient solution.

a. Protoplasts, naked cells

b. Cell division

c. Aggregates of cells

d. Callus, undifferentiated mass

e. Somatic embryo

f. Plantlet

Figure 39.11 **Tissue culture.**
a. When plant cell walls are removed by digestive enzyme action, the result is naked cells or protoplasts. **b.** Regeneration of cell walls and the beginning of cell division. **c.** Cell division produces an aggregate of cells. **d.** An undifferentiated mass, called a callus. **e.** Somatic cell embryos such as this one appear. **f.** The embryos develop into plantlets that can be transferred to soil for growth into adult plants.

Genetic Engineering of Plants

Traditionally **hybridization** [L. *hybrida,* mongrel], the crossing of different varieties of plants or even species, was used to produce plants with desirable traits. Hybridization, followed by vegetative propagation of the mature plants, generated a large number of identical plants with these traits. Today, it is possible to directly alter the genes of organisms. **Transgenic plants** carry a foreign gene that has been introduced into their cells so that they have new and different traits.

Tissue Culture and Genetic Engineering

Since a whole plant will grow from a protoplast, it is necessary only to place the foreign gene into a living protoplast. A foreign gene isolated from any type of organism is placed in the tissue culture medium. High-voltage electric pulses can then be used to create pores in the plasma membrane so that the DNA enters. In one of the first procedures carried out, a gene for the production of the firefly enzyme luciferase was inserted into tobacco protoplasts, and the adult plants glowed when sprayed with the substrate luciferin.

Unfortunately, the regeneration of cereal grains from protoplasts has been difficult. Corn and wheat protoplasts produce infertile plants. As a result, other methods are used to introduce DNA into plant cells with intact cell walls. In one technique, foreign DNA is inserted into the plasmid of the bacterium *Agrobacterium,* which normally infects plant cells. A plasmid, which is a circular fragment of DNA separated from the bacterial chromosome, can be used to produce recombinant DNA. Recombinant DNA contains genes from different sources, namely those of the plasmid and the foreign genes of interest. When the bacterium infects the plant, the recombinant plasmid is introduced into the plant cells. In 1987, John C. Sanford and Theodore M. Klein of Cornell University developed another method of introducing DNA into a plant tissue-culture callus. They constructed a device, called a particle gun, that bombards a callus with DNA-coated microscopic metal particles. Then, genetically altered somatic embryos develop into genetically adult plants. Many plants including corn and wheat varieties have been genetically engineered by this method.

Agricultural Plants with Improved Traits

Cotton, corn, potato, and soybean plants have been engineered to be resistant to either insect predation or herbicides that are judged to be environmentally safe (Fig. 39.12). Some corn and cotton plants have been produced that are both insect and herbicide resistant. In 1999, transgenic crops were planted on more than 70 million acres worldwide and the acreage is expected to triple in about five years. If crops are resistant to a broad-spectrum herbicide and weeds are not, then the herbicide can be used to kill the weeds. When herbicide-resistant plants were planted, weeds were easily controlled, less tillage was needed, and soil erosion was minimized.

One aim of genetic engineering is to produce crops that have the improved agricultural or food quality traits such as those listed in Figure 39.13*a*. To this end, investigators are putting much effort into sequencing the genomes of *Arabidopsis thaliana,* a dicot weed, and rice, a monocot grain. The reading on page 716 introduces you to *Arabidopsis thaliana,* which was literally unknown 20 years ago but is now much used in the laboratory for various avenues of research. The genomes of *Arabidopsis* and rice can be used as models because all flowering dicots and all flowering monocots are closely related. In other words, when scientists know the location and function of genes in *Arabidopsis* and rice, they will have a fair idea of their location and function in the genomes of crop plants like wheat, corn, sorghum, millet, and other cereals also.

a. Herbicide-resistant soybean plant

b. Nonresistant potato plant

c. Pest-resistant potato plant

Figure 39.12 **Genetically engineered plants.**

The production of a salt-tolerant *Arabidopsis* has already been achieved (Fig. 39.13*b*). First, scientists identified a gene coding for a channel protein that transports Na^+ along with H^+ across a vacuole membrane (see Fig. 37.5). Sequestering the Na^+ in a vacuole prevents it from interfering with plant metabolism. Then, the scientists cloned the gene and used it to genetically engineer plants that overproduce the channel protein. The modified plants thrived when watered with a salty solution. Irrigation, even with fresh water, inevitably leads to a salinization of soil that reduces crop yields. Today, crop production is limited by effects of salinization on about 50% of irrigated lands. The next step to solve this problem is to produce salt-tolerant crops. It's believed that the production not only of salt-, but also drought- and cold-tolerant crops will reduce the need for added farm acreage by increasing agricultural yields that will provide enough food for a world population that is expected to nearly double by 2050.

Some progress has also been made to increase the food quality of crops. Soybeans have been developed that mainly produce the monounsaturated fatty acid oleic acid, a change that may improve human health. These altered plants also produce vernolic acid and ricinoleic acids, derivatives of oleic acid that can be used as hardeners in paints and plastics. The necessary genes were derived from *Vernonia* and castor bean seeds and were transferred into the soybean genomes.

As mentioned, genetic engineering is expected to increase productivity. To that end, stomates might be altered to boost carbon dioxide intake or cut down on water loss. The efficiency of the enzyme RuBP carboxylase which captures carbon dioxide in plants could be improved. A team of Japanese scientists are working on introducing the C_4 photosynthetic cycle into rice. Unlike C_3 plants, C_4 plants do well in hot, dry weather. These modifications would require a more complete reengineering of plant cells than the single gene transfers that have been done so far.

Production of Products

Single-gene transfers have allowed plants to produce various products such as human hormones, clotting factors, and antibodies. One type of antibody made by corn can deliver radioisotopes to tumor cells, and another made by soybeans can be used as treatment for genital herpes. Clinical trials have begun.

Recently, a group of scientists from Biosource Technologies located in Vacaville, California, reported that they have been able to use the tobacco mosaic virus as a vector to introduce a human gene into adult tobacco plants in the field. (Note that this technology bypasses the need for tissue culture completely.) Tens of grams of α-galactosidase, an enzyme that can be used to treat a human lysosome storage disease, were harvested per acre of tobacco plants. And it only took thirty days to get tobacco plants to produce antigens to treat non-Hodgkin's lymphoma after being sprayed with a genetically engineered virus.

Genetic engineering of plants is now a reality. The next generation of transgenic crops is expected to have improved agricultural traits, improved food qualities, and result in a higher yield.

Figure 39.13 Transgenic crops of the future.
a. Transgenic crops of the future include those with improved agricultural or food quality traits such as those listed. **b.** A salt-tolerant *Arabidopsis* plant has been engineered. The plant to the left does poorly when watered with a salty solution, but the engineered plant to the right is tolerant of the solution.

Transgenic Crops of the Future	
Improved Agricultural Traits	
Herbicide resistant	Wheat, rice, sugar beets, canola
Salt tolerant	Cereals, rice, sugarcane
Drought tolerant	Cereals, rice, sugarcane
Cold tolerant	Cereals, rice, sugarcane
Improved yield	Cereals, rice, corn, cotton
Modified wood pulp	Trees
Improved Food Quality Traits	
Fatty acid/oil content	Corn, soybeans
Protein/starch content	Cereals, potatoes, soybeans, rice, corn
Amino acid content	Corn, soybeans
Disease protected	Wheat, corn, potatoes

a. Desirable traits

b. Salt intolerant Salt tolerant

Arabidopsis thaliana, the Valuable Weed

Plants are quite different from animals in many aspects of their development, cell biology, biochemistry, and environmental responses. Plants grow their entire lives and contain meristem tissue that allows them to continuously produce new cells that differentiate into specific tissues. Adult cells are totipotent and each one can easily give rise to a complete plant. Plant biochemistry involves metabolic pathways not seen in animals. Several pathways are needed for various types of photosynthesis and the synthesis of a wide range of chemicals; many of these are used defensively against herbivores and bacterial and fungal pathogens. Plants have their own hormones which are produced by many tissues and not by organs specialized for this function. Plant hormones, unlike those of animals, allow plants to respond in a unique way to environmental stimuli such as light and gravity. Because plants are unique, a classical and molecular study of their specific genetics is needed.

Until recently the study of plant genetics was hampered by the lack of a suitable experimental material. Like the fruit fly, *Drosophila,* a suitable material would not take up much laboratory space, but would produce many offspring within a short time. Enter *Arabidopsis thaliana,* a weed of no food or economic value, even though it is a member of the mustard family, as are cabbages and radishes. Unlike crop plants used formerly, *Arabidopsis* has the characteristics needed to promote the study of both plant classical and molecular genetics. *Arabidopsis* has a short generation time; its entire life cycle takes only about four to six weeks, and each adult plant produces 10,000 seeds (potential offspring). Dozens of *Arabidopsis* plants can be grown in a single pot because of its small size (Fig. 39B), and thousands can be grown on a lab bench under fluorescent, rather than sun, light. A total of 50,000 seeds will fit in a standard 1.5 ml tube. In contrast, crop plants such as corn have generation times of at least several months, and they require a great deal of field space for a large number to grow. In addition to thriving in soil, *Arabidopsis* will grow in a liquid media whose content can be biochemically controlled.

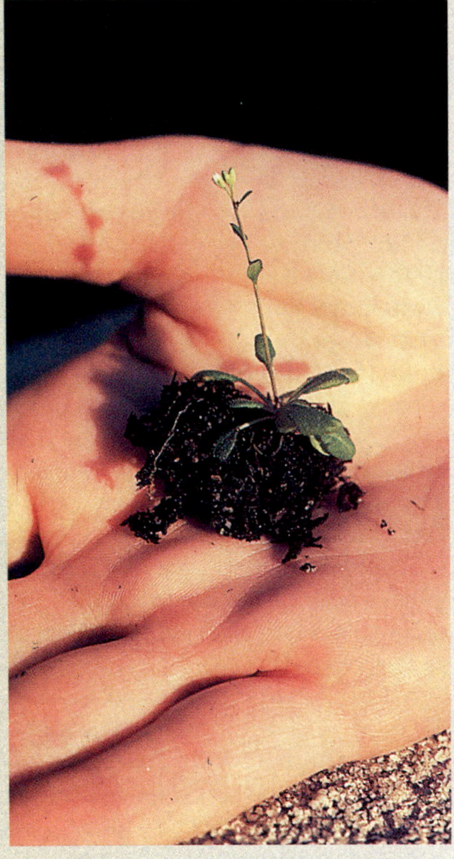

a.

There are many natural mutants of *Arabidopsis,* and much can be determined by studying them. In one instance, a dwarf plant was found to be deficient in the amount of gibberellin-producing enzymes. Since the plant showed a lack of internodal elongation but showed normal development of leaves and flowers, it was known that gibberellins were affecting only internodal elongation. Working with natural mutagens, researchers discovered that there are three classes of genes essential to normal floral pattern formation. These are homeotic genes because they cause either sepals, petals, stamens, or carpels to appear in place of one another. Triple mutants that lack all three types of genetic activities have flowers that consist entirely of leaves arranged in whorls (compare Fig. 39C and D). And a mutation of a regulatory gene results in flowers that have three whorls of petals (Fig. 39E). These floral organ identity genes appear to be regulated by transcription

b.

Figure 39B **Overall appearance of *Arabidopsis thaliana.*** Many investigators have turned to this weed as an experimental material to study the actions of genes, including those that control growth and development. **a.** Photograph of actual plant. **b.** Enlarged drawing.

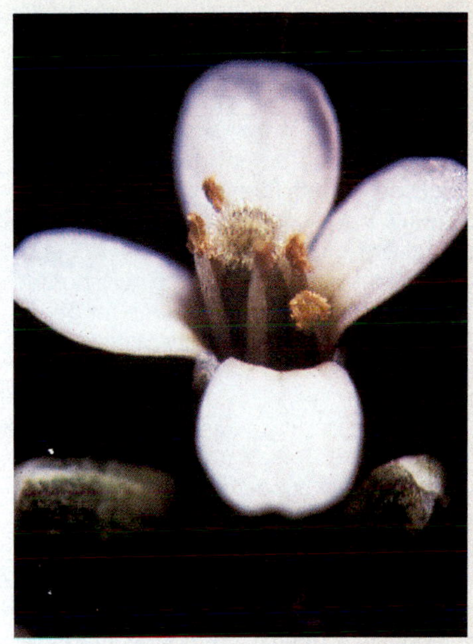

Figure 39C ***Arabidopsis thaliana*** **flower.**
The structure of the flower is determined in part by three classes of floral organ identity genes.

Figure 39D Mutated flower.
Mutations affecting all three classes of floral organ identity genes result in flowers that contain only leaves arranged in whorls.

Figure 39E Mutated flower.
Mutation of a regulatory gene controlling floral organ identity genes results in flowers that have three whorls of petals.

factors that are expressed and required for extended periods.

Artificial mutagenesis can also be easily accomplished in *Arabidopsis* when the seeds are exposed to chemical mutagens or radiation. *Arabidopsis* cells in tissue culture will take up the plasmid of *Agrobacterium tumefaciens,* the bacterium that can infect certain plants. On occasion the plasmid inserts itself into the middle of a normal gene, disrupting it and preventing it from functioning as it should. Or the plasmid can be engineered to carry a foreign gene into a host cell. Because the flowers of *Arabidopsis* self-fertilize, a significant number of plants in the next generation will be homozygous following mutagenesis. Many artificially produced mutants have been studied but we will consider just one example. Following mutagenesis, plants unable to grow in normal levels of CO_2 (but they could only grow in high levels of CO_2) were isolated. This classical genetics study became a molecular genetics study when it was discovered that this particular mutation was caused by a gene which codes for a protein regulating the activity of RuBP carboxylase. It was not known before that there was such a regulatory protein. By now the protein has been sequenced, and it is available for other plant cell studies.

A study of the *Arabidopsis* genome will undoubtedly promote plant molecular genetics in general. *Arabidopsis* has just five small chromosomes containing only 70,000 nucleotide base pairs. In contrast, tobacco has 15 times and corn has 30 times the number of nucleotide base pairs as *Arabidopsis.* However, crop plants have about the same number of functional genes as *Arabidopsis;* their excess DNA is noncoding repetitive DNA. Working with the *Arabidopsis* genome, but not the genome of crop plants, experimenters can easily find coding genes and determine what these genes do. Remarkably, it has been discovered that nearly all angiosperm plants contain approximately the same coding genes in the same sequence. Therefore, knowledge of the *Arabidopsis* genome can be used to locate specific genes in the genomes of other plants. Investigators expect to have completed sequencing the *Arabidopsis* genome in about one year's time. Genes of interest can be cloned from the *Arabidopsis* genome and then used as probes for the isolation of the homologous genes from plants of economic value. Also, cellular processes controlled by a family of genes in other plants requires only a single gene or fewer genes in *Arabidopsis.* This, too, facilitates molecular biological studies of the plant.

The application of *Arabidopsis* genetics to other plants has been shown. For example, one of the mutant genes that alters the development of flowers has been cloned and reintroduced into tobacco plants where, as expected, it caused sepals and stamens to appear where petals would ordinarily be. The investigators comment that the knowledge about the development of flowers in *Arabidopsis* can have far-ranging applications. It will undoubtedly lead someday to more productive crops.

Connecting Concepts

Life on earth would not be possible without vascular plants and specifically flowering plants which now dominate the biosphere. *Homo sapiens* evolved with flowering plants and therefore does not know a world without them. The earliest humans were mostly herbivores—they relied on foods they could gather for survival—fruits, nuts, seeds, tubers, roots, and so forth. Plants also provided protection from the environment, offering shelter from heavy rains and noonday sun. Later on, human civilizations could not have begun without the development of agriculture. The majority of the world's population still relies primarily on three flowering plants—corn, wheat, and rice—for the majority of its sustenance. Sugar, coffee, spices of all kinds, cotton, rubber, and tea are plants which have even promoted wars because of their importance to a country's economy. Although we now live in an industrialized society, we are still dependent on plants and have put them to even more uses. They produce substances needed to lubricate the engines of supersonic jets and to make cellulose acetate for films, to take a couple of examples. For millions of urban dwellers, plants are their major contact with the natural world. We grow them not only for food and shelter, but for their simple beauty.

Most people fail to appreciate the importance of plants, but plants may be even more critical to our lives today than they were to our early ancestors on the African plains. Currently, half of all pharmaceutical drugs have their origin in plants. The world's major drug companies are engaged in a frantic rush to collect and test plants in the rain forests for their drug-producing potential. Why the rush? Because the rain forests may be gone before all the possible cures for cancer, AIDS, and other killers have been found. Wild plants not only can help cure human ills, they also serve as a source for genes that can improve the quality of plants that support our way of life.

Summary

39.1 Life Cycle of Flowering Plants

Flowering plants produce flowers and often have a mutualistic relationship with an animal pollinator. They also produce seeds that are covered by fruits.

Flowering plants exhibit an alternation of generations life cycle that includes the microgametophyte and megagametophyte. The pollen grain is the microgametophyte. The megagametophyte is called the embryo sac, and it remains within the body of the sporophyte plant.

Typical parts of a flower are: sepals, which are usually green in color and form an outer whorl; petals, often colored, which form an inner whorl; the pistil, which is in the center and contains the carpels, each consisting of stigma, style, and ovary; and the stamens, each having a filament and anther, which are around the base of the carpels. The ovary contains ovules.

Each ovule within the ovary contains a megasporocyte, which divides meiotically to produce four haploid megaspores, only one of which survives. This megaspore divides mitotically to produce the megagametophyte (embryo sac), which usually has eight nuclei. The central cell contains two polar nuclei, and one of the three cells next to the micropyle is an egg cell.

The anthers contain microsporocytes, each of which divides meiotically to produce four haploid microspores. Each of these divides mitotically to produce a two-celled pollen grain. One cell is the tube cell, and the other is the generative cell. The generative cell later divides to produce two sperm cells. The pollen grain is the microgametophyte. After pollination, the pollen grain germinates, and as the pollen tube grows, the sperm cells travel to the embryo sac. Pollination is simply the transfer of pollen from anther to stigma.

Flowering plants practice double fertilization. One sperm nucleus unites with the egg nucleus, forming a 2n zygote, and the other unites with the polar nuclei of the central cell, forming a 3n endosperm cell.

After fertilization, the endosperm cell divides to form multicellular endosperm. The zygote becomes the sporophyte embryo. The ovule matures into the seed (its integuments become the seed coat). The ovary becomes the fruit.

39.2 Development of the Embryo

As the ovule is becoming a seed, the zygote is becoming an embryo. After the first several divisions, it is possible to discern the embryo and the suspensor. The suspensor attaches the embryo to the ovule and supplies it with nutrients. The dicot embryo becomes first heart shaped and then torpedo shaped. Once you can see the two cotyledons, it is possible to distinguish the shoot tip and the root tip, which contain the apical meristems. In dicot seeds, the cotyledons frequently take up the endosperm.

39.3 Fruits and Seeds

The seeds of flowering plants are enclosed by fruits. There are different types of fruits. Simple fruits are derived from a single ovary (which can be simple or compound). Some simple fruits are fleshy, such as a peach or an apple. Others are dry, such as peas, nuts, and grains. Aggregate fruits develop from a number of ovaries of a single flower, and compound fruits develop from multiple ovaries of separate flowers.

Flowering plants have several ways to disperse seeds. Seeds may be blown by the wind, attached to animals that carry them away, eaten by animals that defecate them some distance away, or adapted to water transport.

Prior to germination, you can distinguish a bean (dicot) seed's two cotyledons and plumule, which is the shoot that bears leaves. Also present are the epicotyl, the hypocotyl, and the radicle. In a corn kernel (monocot), the endosperm, the cotyledon, the plumule, and the radicle are visible.

39.4 Asexual Reproduction in Plants

Many flowering plants reproduce asexually, as when the nodes of stems (either aboveground or underground) give rise to entire plants, or when roots produce new shoots.

The practice of hydroponics—and the recognition that plant cells can be totipotent—led to plant tissue culture, a technique that now has many applications.

Micropropagation, the production of clonal plants as a result of meristem culture in particular, is now a commercial venture. Flower meristem culture results in somatic embryos that can be packaged in gel for worldwide distribution. Anther culture results in homozygous plants that express recessive genes. Leaf, stem, and root culture can result in cell suspensions that may eventually allow the production of plant chemicals in large tanks. Development of adult plants from protoplasts results in somaclonal variations, a new source of plant varieties.

Protoplasts in particular lend themselves to direct genetic engineering in tissue culture. Otherwise, the *Agrobacterium* technique or the particle-gun technique allow foreign genes to be introduced into plant cells, which then develop into adult plants with particular traits. Some crops (e.g., soybean, corn, and cotton) have been engineered to be herbicide and/or pest resistant. In the future, crops that have these and other improved agricultural traits, improved food quality, and higher productivity are expected.

Reviewing the Chapter

1. Name two unique features regarding the reproduction of flowering plants. 700
2. Draw a diagram of a flower and name the parts. 701
3. Draw a diagram that illustrates the life cycle of flowering plants. Why don't flowering plants require a source of outside water for pollination? 702–03
4. Describe the development of a megagametophyte, from the megasporocyte to the production of an egg. 702–03
5. Describe the development of the microgametophyte, from the microsporocyte to the production of sperm. 702–03
6. What is the difference between pollination and fertilization? 706
7. Describe the sequence of events as a dicot zygote becomes an embryo enclosed within a seed. 707
8. Distinguish between simple fleshy fruits and simple dry fruits. Give an example of each type. What is an aggregate fruit? A multiple fruit? 708–09
9. Name several mechanisms of seed and/or fruit dispersal. Contrast the germination of a bean seed with that of a corn kernel. 710–11
10. In what ways do plants ordinarily reproduce asexually? What is the importance of totipotency in regard to tissue culture? 712
11. How is plant tissue culture carried out, and what are the benefits of plant tissue culture? 712–13
12. How are transgenic plants produced? What types of plants have been produced, and for what purpose have they been genetically engineered? 714–15

Testing Yourself

Choose the best answer for each question.

1. In plants,
   a. gametes become a gametophyte.
   b. spores become a sporophyte.
   c. both sporophyte and gametophyte produce spores.
   d. only a sporophyte produces spores.
   e. Both a and b are correct.
2. The flower part that contains ovules is the
   a. carpel.
   b. stamen.
   c. sepal.
   d. petal.
   e. seed.
3. The megasporocyte and the microsporocyte
   a. both produce pollen grains.
   b. both divide meiotically.
   c. both divide mitotically.
   d. produce pollen grains and embryo sacs, respectively.
   e. All of these are correct.
4. A pollen grain is
   a. a haploid structure.
   b. a diploid structure.
   c. first a diploid and then a haploid structure.
   d. first a haploid and then a diploid structure.
   e. the mature gametophyte.
5. Which of these is mismatched?
   a. polar nuclei—plumule
   b. egg and sperm—zygote
   c. ovule—seed
   d. ovary—fruit
   e. stigma—pistil
6. Which of these is not a fruit?
   a. walnut
   b. pea
   c. green bean
   d. peach
   e. All of these are fruits.
7. Animals assist with
   a. pollination and seed dispersal.
   b. control of plant growth and response.
   c. translocation of organic nutrients.
   d. asexual propagation of plants.
   e. germination of seeds.
8. A seed contains
   a. a seed coat.
   b. an embryo.
   c. stored food.
   d. cotyledon(s).
   e. All of these are correct.
9. Which of these is mismatched?
   a. plumule—leaves
   b. cotyledon—seed leaf
   c. epicotyl—root
   d. pericarp—corn kernel
   e. carpel—ovule
10. Which of these is not a common procedure in the tissue culture of plants?
    a. shoot tip culture for the purpose of micropropagation
    b. flower meristem culture for the purpose of somatic embryos
    c. leaf, stem, and root culture for the purpose of cell suspension cultures
    d. protoplast culture for the purpose of genetic engineering of plants
    e. culture of hybridized mature plant cells
11. Label the following diagram of alternation of generations in flowering plants.

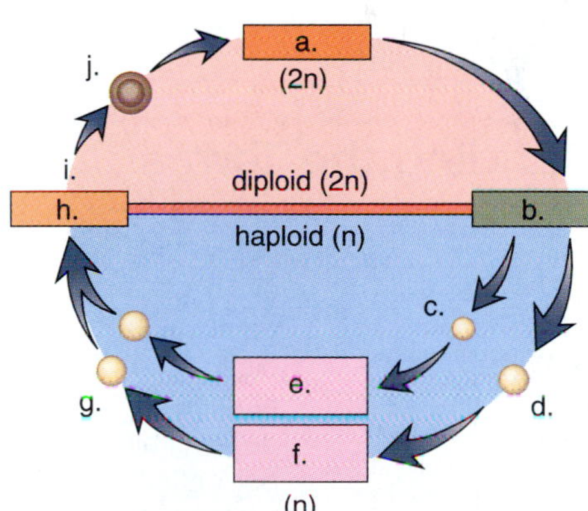

Thinking Scientifically

1. Peanuts are fruits that develop underground. The stems bend toward the ground and the flowers push into the soil like roots. Since this unusual gravitropic response only happens after flowering, it must be linked to reproduction. What hypothesis would explain bending? What steps in reproduction would you examine to test the hypothesis?
2. You have discovered an unusual lettuce variant with an orange leaf. Realizing the commercial potential of such a plant, you would like to propagate it. However, by the time of your discovery, the growing season is almost over and the

flowers are withered and the seeds gone. How would you propagate this plant? Why might it be beneficial to cross the propagated plant with green-leaf lettuce?

Bioethical Issue

We are in the midst of a pollination crisis due to a decline in the population of honeybees and many other insects, birds, and small mammals that transfer pollen from stamen to stigma. Pollinator populations have been decimated by pollution, pesticide use, and destruction or fragmentation of natural areas. Belatedly, we have come to realize that various types of bees are responsible for pollinating such cash crops as blueberries, cranberries, and squash, and are partly responsible for pollinating apple, almond, and cherry trees.

Why are we so shortsighted when it comes to protecting the environment and living creatures like pollinators? Because pollinators are a resource held in common. The term "commons" originally meant a piece of land where all members of a village were allowed to graze their cattle. The farmer who thought only of himself and grazed more cattle than his neighbor was better off. The difficulty is, of course, that eventually the resource is depleted and everyone loses.

So, when farmers or property owners use pesticides they are only thinking of their own field or lawn, and not the good of the whole community. The commons can only be protected if citizens have the foresight to enact rules and regulations by which all abide. DDT was outlawed in this country in part because it led to the decline of birds of prey. Similarly, we need legislation to protect pollinators from those factors that kill them off. As a society, we need legislation to protect pollinators because it helps protect the food supply for all of us. Are you willing to stop using pesticides that kill pollinators if it means your lawn will suffer? Why or why not? Are you willing to pressure your representatives for legislation to protect pollinators? Why or why not? Are you willing to turn your lawn and garden into a haven for pollinators? Why or why not?

Understanding the Terms

Match the terms to these definitions:

a. ______________ Flower structure consisting of an ovary, a style, and a stigma.
b. ______________ Flowering plant structure consisting of one or more ripened ovaries that usually contain seeds.
c. ______________ The gametophyte that produces an egg; an embryo sac in flowering plants.
d. ______________ Mature ovule that contains an embryo, with stored food enclosed in a protective coat.
e. ______________ Mature microgametophyte in seed plants.

Web Connections

Exploring the Internet

http://www.mhhe.com/biosci/genbio/mader
(click on *Biology 7/e*)

The *Biology 7/e* Online Learning Center provides many resources for studying the material in this chapter including links to the following sites:

Flowering Plant Reproduction: Flower Structure is a comprehensive site describing angiosperm life cycles, evolution of this group, and the anatomy of angiosperm flowers. Many links.

http://gened.emc.maricopa.edu/bio/bio181/BIOBK/BioBookflowers.html

Fruit facts, organized by common name. Description of the fruit, its growth form, cultivation, and references.

http://www.crfg.org/pubs/frtfacts.html

Seeds of Life. This site contains great diagrams and photographs of different types of seeds, fruits, etc.

http://versicolores.ca/seedsoflife/ehome.html

The Pollination Home Page. Much information as well as links regarding pollination, a process that is important to the fruit and vegetable industry.

http://users.aol.com/pollinator/polpage1.html

Plant Trivia Timeline. Despite the term "trivia" in the title of this site, there is nothing trivial about the content. In the nearly 50 pages of information contained in this site, there is a plethora of information about the economic importance of plants.

http://www.huntington.org/BotanicalDiv/Timeline.html

Further Readings for Part VI

Apse, M. P., et al. August 20, 1999. Salt tolerance conferred by overexpression of a vacuolar Na^+/H^+ antiport in *Arabidopsis*. *Science* 285(5431):1256. The mechanism by which genetically engineered *Arabidopsis* plants tolerate salinity is examined.

Arakawa, T., and Langridge, W. R. H. May 1998. Plants are not just passive creatures! *Nature Medicine* 4(5):550. Plants are being used to produce foreign proteins for human immunity.

Balick, M. J., and Cox, P. A. 1997. *Plants, people, and culture: The science of ethnobotany.* New York: Scientific American Library. This interesting, well-illustrated book discusses the medicinal and cultural uses of plants, and the importance of rain forest conservation.

Barinaga, M. March 19, 1999. Key molecular signals identified in plants. *Science* 283(5409):1825. A pathway that helps control growth at the apical meristem is found.

Busch, R. October/November 1998. The value of autumn leaves. *National Wildlife* 36(6):32. Article examines the physiology behind autumn leaf colors.

Canny, M. J. March/April 1998. Transporting water in plants. *American Scientist* 86(2):152. The cohesion-tension theory of water transport is discussed in detail.

Flannery, M. April 1999. Seeing plants a little more clearly. *American Biology Teacher* 61(4):303. Article discusses some of the latest research in the plant world.

Foyer, C., and Noctor, G. April 23, 1999. Leaves in the dark see the light. *Science* 284(5414):599. Excess light is potentially dangerous to plants because it can cause persistent decreases in the rates of photosynthesis.

Galen, C. August 1999. Why do flowers vary? *BioScience* 49(8):631. Article emphasizes interactions of plants with flower enemies, and resource costs associated with floral display.

Gibson, A. C. November 1998. Photosynthetic organs of desert plants. *BioScience* 48(11):911. The specialized leaf and stem adaptations in desert plants may be to maximize photosynthetic rates and energy production, rather than to conserve water.

Glenn, E. P., et al. August 1998. Irrigating crops with seawater. *Scientific American* 279(2):76. Certain useful salt-tolerant plants can flourish on seawater irrigation.

Grill, E., and Ziegler, H. October 9, 1998. A plant's dilemma. *Science* 282(5387):252. Article discusses the manipulation of stomates to close, which assists plants in conserving water.

Gura, T. August 27, 1999. New ways to glean medicines from plants. *Science* 285(5432):1347. Plant roots can be induced to secrete proteins to treat human diseases, and the tobacco mosaic virus can carry human genes into tobacco plants.

Jacoby, M. August 23, 1999. Botanists design plants with a taste for salty. *Chemical & Engineering News* 77(34):9. Genetically engineered *Arabidopsis* plants flourish when watered with concentrated salt solutions.

Lee, D. W. January/February 1997. Iridescent blue plants. *American Scientist* 85(1):56. Some tropical plants produce a blue color, as do some insects, by using layered filters.

Levetin, E., and McMahon, K. 1999. *Plants and Society.* 2d ed. Dubuque, Iowa: WCB/McGraw-Hill. Basic botany and the impact of plants on society are topics covered in this introductory text.

Luoma, J. R. March 1997. The magic of paper. *National Geographic* 191(3):88. The papermaking process is discussed.

Mauseth, J. 1998. *Botany: An Introduction to Plant Biology* (multimedia enhanced). 2d ed. Boston: Jones and Bartlett Publishers. Emphasizes evolution and diversity in botany and general principles of plant physiology and anatomy.

Milot, C. 17 July 1998. Plant biology in the genome era. *Science* 281(5375):331. Genomic data can be used to make plants that produce more vitamins and minerals, and can help explain physiological processes, such as flowering.

Moore, R., and Clark, W. D., et al. 1998. *Botany.* 2d ed. Dubuque, Iowa: WCB/McGraw-Hill. This introductory botany text stresses the process of science while presenting the evolution, anatomy, and physiology of plants.

Niklas, K. J. February 1996. How to build a tree. *Natural History* 105(2):48. Article discusses the properties of wood.

Normile, D. September 11, 1998. Multiplying knowledge of cell division, plant growth. *Science* 281(5383):1591. A new gene from *Arabidopsis thaliana* may provide insights into stem cell division and understanding plant development.

Northington, D., and Goodin, J. R. 1996. *The Botanical World.* 2d ed. St. Louis: Times-Mirror/Mosby College Publishing. This is an account of plant interactions and basic physiology.

Pitelka, L. F., et al. September/October 1997. Plant migration and climate change. *American Scientist* 85(5):464. A relationship between plant migration and climate change, as evidenced by the fossil record and computer models, is discussed.

Schmiedeskamp, M. December 1997. Pollution-purging poplars. *Scientific American* 277(6):46. Hybrid poplar trees may be used to break down toxic organic compounds in the soil.

Science 285(5426):367. July 16, 1999. This issue features several articles concerning plant biotechnology.

Seymour, R. S. March 1997. Plants that warm themselves. *Scientific American* 276(3):104. Some plants generate heat to keep blossoms at a constant temperature.

Stern, K. 2000. *Introductory Plant Biology.* 8th ed. Dubuque, Iowa: McGraw-Hill Higher Education. Presents basic botany in a clear, informative manner.

Strauss, E. January 1, 1999. RNA molecules may carry long-distance signals in plants. *Science* 283(5398):12. A carrier protein has been found that can truck large RNA molecules into the phloem.

———. July 24, 1998. How plants pick their mates. *Science* 281(5376):503. The stigma of plants are particular about what type of pollen grain they will accept.

———. October 2, 1998. When walls can talk, plant biologists listen. *Science* 282(5386):28. Plant cell walls are powerful signalers that can determine the fate of cells during development.

Stryer, L. 1995. *Biochemistry.* 5th ed. New York: W. H. Freeman and Company. Chapter 22 of this text presents an advanced but understandable treatment of photosynthesis.

Szalai, V. A., and Brudvig, G. W. November/December 1998. How plants produce oxygen. *American Scientist* 86(6):542. Oxygen production through photosynthesis involves the chemistry of a manganese-containing complex in the membranes of plant chloroplasts.

Sze, P. 1998. *A Biology of the Algae.* 3d ed. Dubuque, Iowa: WCB/McGraw-Hill. A text that introduces algae morphology, evolution, and ecology to the botany major.

Walsh, R. May 1997. The chocolate bug. *Natural History* 106(4):54. The cacao tree's survival depends upon a tiny insect, which dictates where cacao will grow.

part vii

Animal Structure and Function

In contrast to plants which are autotrophic and make their own organic food, animals are heterotrophic and feed on other organisms. Their mobility, which is dependent upon nerve fibers and muscle fibers, is essential in finding food. Then the food is digested and the nutrients are distributed to cells. Finally, wastes are expelled.

In complex animals, a distinct division of labor exists in that the body contains organ systems specialized to carry out specific functions. A circulatory system moves materials from one body part to another; a respiratory system carries out gas exchange; and an excretory system filters the blood. Coordination of the systems is accomplished by a nervous system and endocrine systems. The lymphatic system protects the body from infectious diseases.

Certain small, microscopic animals don't have these systems nor organs of any kind. In this part we trace the development of organ systems within the animal kingdom and contrast how they function in humans and other animals.

Animal Organization and Homeostasis

chapter 

chapter concepts

40.1 Types of Tissues

- Animals have these levels of organization: molecules—cells—tissues—organs—organ systems—organism. 724
- Animal tissues can be categorized into four major types: epithelial, connective, muscular, and nervous tissues. 724
- Epithelial tissues, which line body cavities and cover surfaces, are specialized in structure and function. 724
- Connective tissues, which protect, support, and bind other tissues, include cartilage and bone and also blood, the only liquid tissue. 726
- Muscular tissues, which contract, make body parts move. 728
- Nervous tissues coordinate the activities of the other tissues and body parts. 729

40.2 Organs and Organ Systems

- Organs usually contain several types of tissues. For example, although skin is composed primarily of epithelial tissue and connective tissue, it also contains muscle and nerve fibers. 730
- Organs are grouped into organ systems, each of which has specialized functions. 732
- The coelom, which arises during development, is later divided into various cavities where specific organs are located. 732

40.3 Homeostasis

- Homeostasis is the dynamic equilibrium of the internal environment. All organ systems contribute to homeostasis in animals. 734

Turtles, *Chrysemys,* sunning in the company of an alligator, *Alligator*

The organization of complex animals includes organ systems, such as the digestive and respiratory systems, that service the cells. Organ systems contain organs, which are composed of tissues, and each type of tissue has like cells. This chapter concerns these levels of organization and in particular examines the various types of tissues found in more complex animals, like the turtles in the photograph, and humans. Skin will be used as an example of an organ and we will also discuss the functions of various organ systems.

In order for organs and organ systems to function well, the internal environment needs to be relatively constant. The body should be warm enough for enzymes to speed reactions; the blood pressure high enough for blood to circulate; the oxygen content of mitochondria sufficient enough to produce ATP, to take just a few examples. The nervous and endocrine systems coordinate the other systems of the body and in that way help maintain homeostasis, a dynamic equilibrium of the internal environment.

40.1 Types of Tissues

All living things are organized, and animals are no exception. Animals begin life as a single cell, the fertilized egg or zygote. The zygote divides and soon there are many cells which in most animals go on to become the tissues we will be discussing. A **tissue** is a group of similar cells performing a similar function. Different types of tissues come together to form organs. An **organ** is a group of tissues that performs a specialized function. Several organs are found within an organ system, and the organ systems make up the organism. This sequence describes the levels of organization that make up the body of an animal: cells, tissues, organs, organ systems, body of the animal. Figure 40.1 takes an actual example. Specialized cells form the various tissues that are found within a kidney and the kidney is a part of the urinary system in humans.

The structure and functions of an organ system like the urinary system are dependent upon the specializations of the organs, tissues, and cell types contained therein. For instance, the intestine efficiently absorbs nutrients and the cells that line the lumen have microvilli, which increase surface area. Muscles contract when muscle cells shorten because they have intracellular components that move past one another. Nerve cells conduct impulses and they have long, slender projections that carry nerve impulses to distant body parts.

There are four major types of tissue in vertebrate animals: epithelial tissues cover body surfaces, line body cavities, and form glands; connective tissues bind and support body parts; muscular tissues cause body parts to move; and nervous tissue responds to stimuli and transmits impulses from one body part to another.

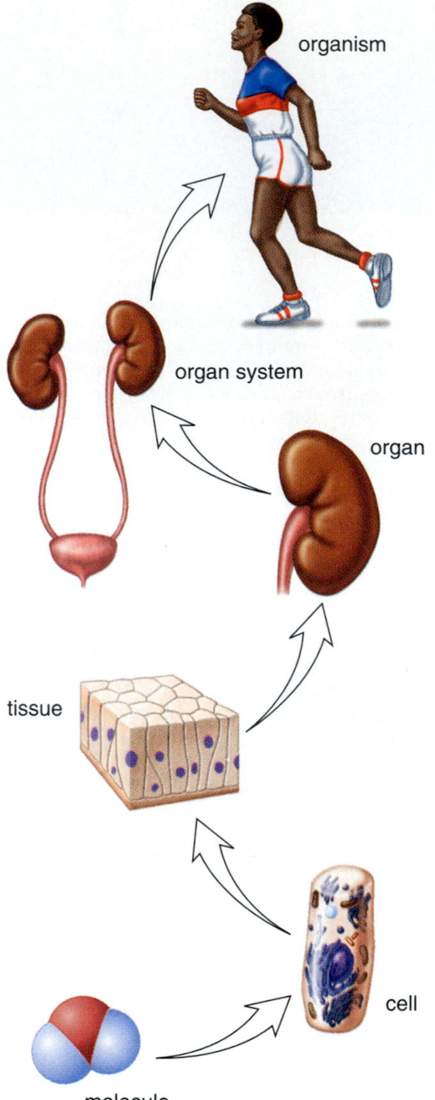

Figure 40.1 Levels of organization.
A cell is composed of molecules, a tissue is made up of cells; an organ is composed of tissues; and the organism contains organ systems. These are the levels of organization in the body of an animal.

Epithelial Tissues

Epithelial tissue [Gk. *epi*, over, and L. *theca*, case, container], also called epithelium, forms a continuous layer over body surfaces, lines inner cavities, and forms glands. One side of an epithelium is exposed at the skin surface or to a body cavity. The other side is usually attached to a basement membrane, which is a thin mat of specialized extracellular matrix (see Fig. 5.15).

Classified according to cell shape, there are three types of epithelial tissues. Squamous epithelium is composed of flat cells; cuboidal epithelium contains cube-shaped cells; and in columnar epithelium, the oblong cells resemble pillars or columns. Figure 40.2 describes the structure and function of epithelium (pl., epithelia) in vertebrates. Any epithelium can be simple or stratified. Simple means that the tissue has a single layer of cells, and stratified means that the tissue has layers piled one on top of the other. One type of epithelium is pseudostratified—it appears to be layered, but actually true layers do not exist because each cell touches the basement membrane, a thin, nonliving proteinaceous layer that anchors epithelium in place.

Epithelial tissues protect, but they also are specialized for specific functions. In vertebrates, simple squamous epithelium lines the air sacs of the lungs and forms the walls of the capillaries. The thin delicate nature of this epithelium facilitates the exchanges that take place in an organ like the lungs, where carbon dioxide is exchanged for oxygen. The columnar epithelium of the small intestine is specialized for the absorption of nutrient molecules. The pseudostratified epithelium of the respiratory tract has hairlike projections called cilia. The cilia sweep impurities toward the throat so they do not enter the lungs.

An epithelium sometimes secretes a product and is described as glandular. A gland can be a single epithelial cell, such as the mucus-secreting goblet cells in the lining of the human intestine, or a gland can contain numerous cells. **Exocrine glands** secrete their product into ducts or into cavities, and **endocrine glands** secrete their products directly into the bloodstream.

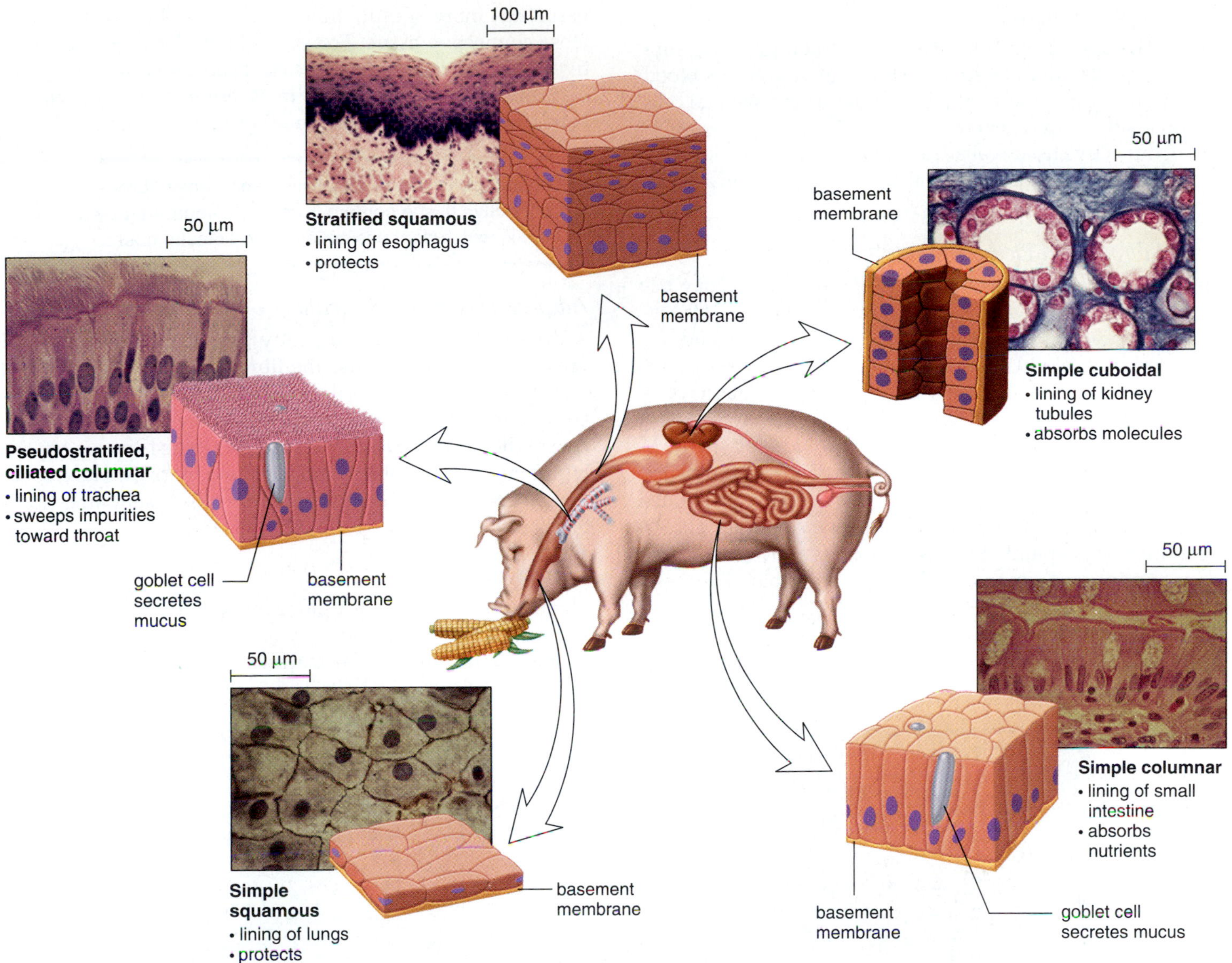

Figure 40.2 **Types of epithelial tissues in vertebrates.**
Epithelial tissues are classified according to shape of cell, whether they are simple or stratified, and whether they have cilia. A single location is given as an example for each type of tissue shown. Epithelial tissues have functions associated with protection, secretion, and absorption.

Epithelium forms the outer layer of skin of most animals. The skin of earthworms and snails is glandular and produces mucus that lubricates the body, helping to ease movement through a dry environment. In roundworms, annelids, and arthropods, an outer nonliving and protective cuticle is produced by epithelium. In terrestrial vertebrates, skin cells contain keratin, a substance that protects the skin from the possible loss of water.

Epithelial tissue cells are packed tightly and joined to one another in one of three ways (see Fig. 5.16). In tight junctions, plasma membrane proteins extending between neighboring cells bind them tightly. For example, they prevent digestive juices from passing between the epithelial cells lining the lumen (cavity). In adhesion junctions, cytoskeletal elements join internal plaques present in both cells. They allow the skin to withstand considerable stretching and mechanical stress. Gap junctions form when two identical plasma membrane channels join. They allow ions and small molecules to pass between the cells.

Epithelial tissue is classified according to cell shape. These tightly packed protective cells can occur in more than one layer, and the cells lining a cavity can be ciliated and/or glandular.

Connective Tissues

Connective tissue binds structures together, provides support and protection, fills spaces, stores fat, and forms blood cells. It provides the source cells for muscle and skeletal cells in animals that can regenerate lost parts.

Connective tissue cells are usually separated widely by a matrix, a noncellular material that varies in consistency from solid to semifluid to fluid.

Loose Fibrous and Dense Fibrous Tissues

The cells of loose fibrous and dense fibrous connective tissues, called **fibroblasts** [L. *fibra*, thread, and Gk. *blastos*, bud], are located some distance from one another and are separated by a jellylike matrix that contains white collagen fibers and yellow elastic fibers. Collagen fibers contain collagen, a protein that gives them flexibility and strength. Elastic fibers contain elastin, a protein that provides elasticity.

Loose fibrous connective tissue supports epithelium and also many internal vertebrate organs (Fig. 40.3*a*). Its presence in lungs, arteries, and the urinary bladder allows these organs to expand. It forms a protective covering encasing many internal organs, such as muscles, blood vessels, and nerves.

Dense fibrous connective tissue contains many collagenous fibers that are packed closely together. This type of tissue has more specific functions in vertebrates than does loose connective tissue. For example, dense fibrous connective tissue is found in **tendons** [L. *tendo*, stretch], which connect muscles to bones, and in **ligaments** [L. *ligamentum*, band], which connect bones to other bones at joints.

Loose fibrous connective tissue and dense fibrous connective tissue contain fibroblasts separated by a matrix, which contains collagen and elastic fibers.

Adipose Tissue and Reticular Connective Tissue

Adipose tissue [L. *adipalis*, fatty], insulates the body and provides padding because the fibroblasts enlarge and store fat (Fig. 40.3*b*). In mammals, adipose tissue is found particularly beneath the skin, around the kidneys, and on the surface of the heart. Reticular connective tissue is present in the lymph nodes, the spleen, and the bone marrow. Here, reticular fibers, associated with reticular cells resembling fibroblasts, support many free blood cells.

Cartilage and Bone

Cartilage and bone are rigid connective tissues in which structural proteins (cartilage) or calcium salts (bone) are deposited in the intercellular matrix.

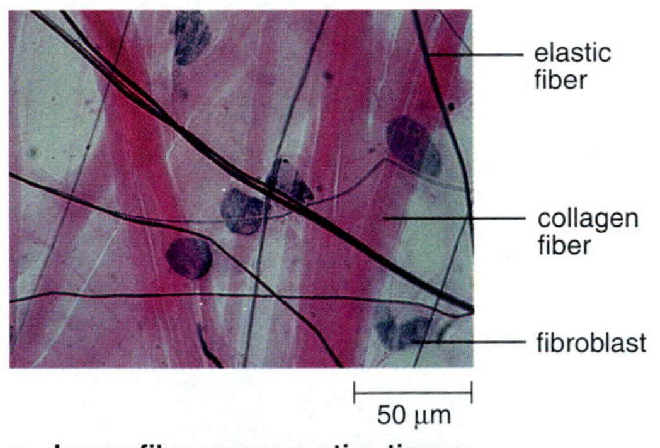

a. **Loose fibrous connective tissue**
- has space between components.
- occurs beneath skin and most epithelial layers.
- functions in support and binds organs.

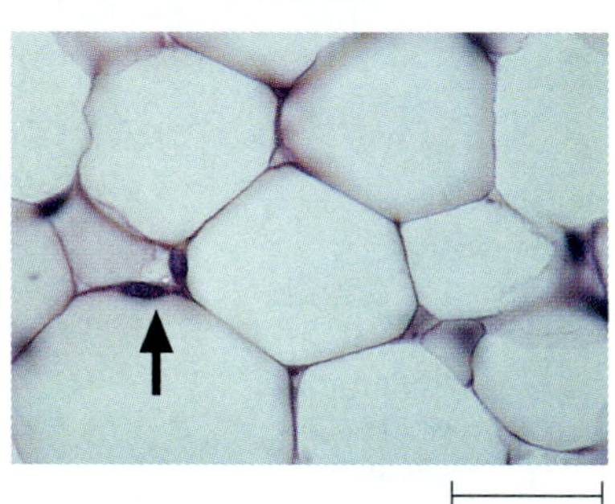

b. **Adipose tissue**
- cells are filled with fat.
- occurs beneath skin, around organs and heart.
- functions in insulation, stores fat.

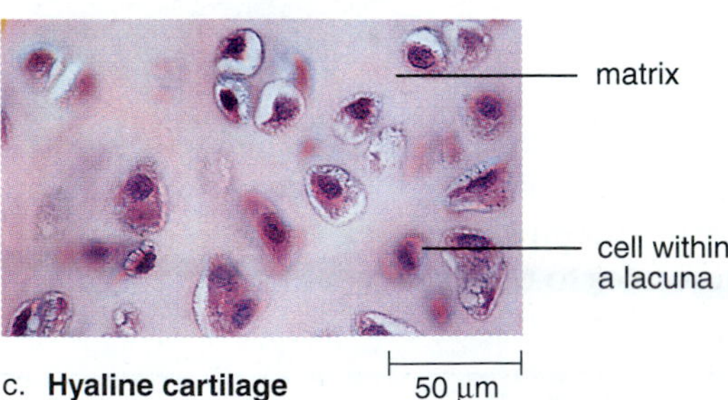

c. **Hyaline cartilage**
- has cells in lacunae.
- occurs in nose and walls of respiratory passages; at ends of bones including ribs.
- functions in support and protection.

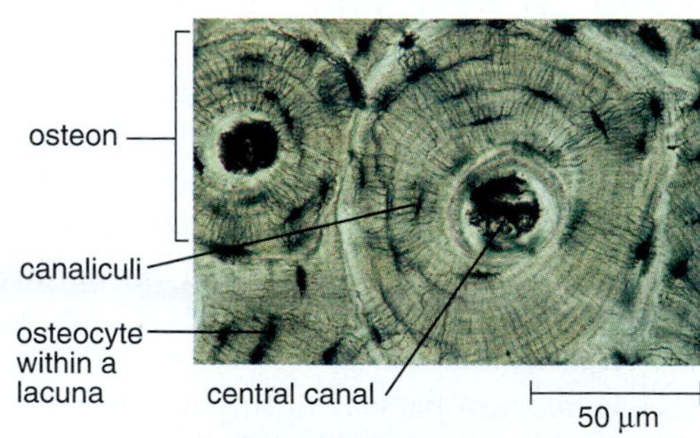

d. **Compact bone**
- has cells in concentric rings.
- occurs in bones of skeleton.
- functions in support and protection.

Figure 40.3 Connective tissue examples.
a. In loose connective tissue, fibroblasts are separated by a matrix that is jellylike but contains fibers. b. Adipose tissue cells are filled with fat and the nuclei (arrow) are pushed to one side. c. In hyaline cartilage, a flexible matrix has a translucent appearance. d. In compact bone, the hard matrix contains concentric rings of osteocytes in elongated cylinders called osteons. The central canal contains blood vessels and nerve fibers.

In **cartilage** [L. *cartilago,* gristle], cells, called chondrocytes, lie in small chambers called **lacunae** (sing., lacuna), separated by a matrix that is strong yet flexible (Fig. 40.3*c*). There are various types of cartilage, which are classified according to type of collagen and elastic fiber found in the matrix. In some vertebrates, notably sharks and rays, the entire skeleton is made of cartilage. In humans, the fetal skeleton is cartilage, but it is later replaced by bone. Cartilage is retained at the ends of long bones, at the end of the nose, in the framework of the ear, in the walls of respiratory ducts, and within intervertebral disks.

In bone, the matrix of inorganic, chiefly calcium, salts is deposited around protein fibers, especially collagen fibers. The minerals give bone rigidity, and the protein fibers provide elasticity and strength, much as steel rods do in reinforced concrete.

In **compact bone,** bone cells, called osteocytes, are located in lacunae that are arranged in concentric circles within osteons (Haversian systems) around tiny tubes called central canals (Fig. 40.3*d*). Nerve fibers and blood vessels are in these canals. The latter bring the nutrients that allow bone to renew itself. The nutrients can reach all of the cells because there are minute canals (canaliculi) containing thin processes of the osteocytes that connect them with one another and with the central canals.

The ends of a long bone contain spongy bone, which has an entirely different structure. **Spongy bone** contains numerous bony bars and plates separated by irregular spaces that may contain red bone marrow. Although lighter than compact bone, spongy bone still is designed for strength. Just as braces are used for support in buildings, the solid portions of spongy bone follow lines of stress.

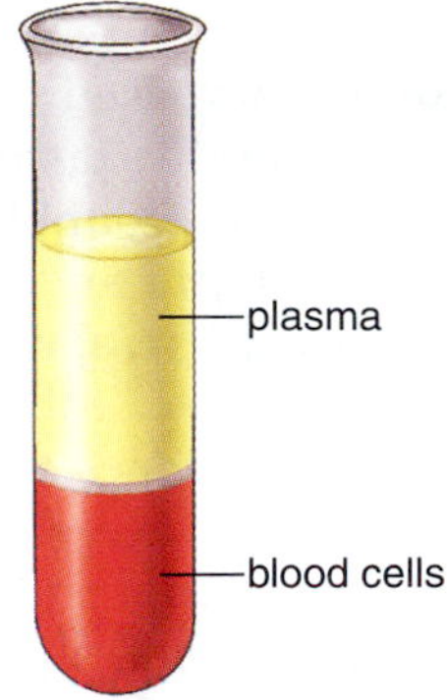

a. Blood sample

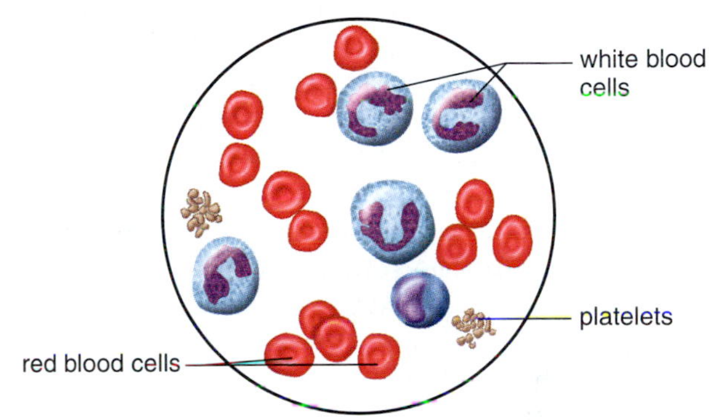

b. Blood smear

Figure 40.4 Blood, a liquid tissue.
a. Blood is classified as connective tissue because the cells are separated by a matrix—plasma. Plasma, the liquid portion of blood, usually contains several types of cells. This shows the cells in human blood (red blood cells, white blood cells, and platelets, which are actually fragments of a larger cell). **b.** Drawing of the components of blood.

Cartilage and bone are support tissues. Cartilage is more flexible than bone because the matrix is rich in protein; bone is rich in calcium salts.

Blood

Blood has transporting, regulating, and protective functions. It transports nutrients and oxygen to cells and removes carbon dioxide and other wastes. It helps distribute heat and also plays a role in fluid, ion, and pH balance. Blood is a connective tissue in which the cells are separated by a liquid called plasma (Table 40.1). In vertebrates, blood cells are primarily two types: red blood cells (erythrocytes), which carry oxygen, and white blood cells (leukocytes), which aid in fighting infection (Fig. 40.4). Also present in plasma are platelets, which are important in blood clotting. Platelets are not complete cells; rather, they are fragments of giant cells found in the bone marrow.

Blood is unlike other types of connective tissue in that the intercellular matrix (i.e., plasma) is not made by the cells. Plasma is a mixture of different types of molecules that enter the blood at various locations.

Blood is a connective tissue in which the matrix is plasma.

Table 40.1

Blood Plasma	
Water (92% of Total)	
Solutes (8% of Total)	
Inorganic ions (salts)	Na^+, Ca^{2+}, K^+, Mg^{2+}, Cl^-, HCO_3^-, HPO_4^{2+}, SO_4^{2+}
Gases	O_2, CO_2
Plasma proteins	Albumin, globulins, fibrinogen
Organic nutrients	Glucose, fats, phospholipids, amino acids, etc.
Nitrogenous waste products	Urea, ammonia, uric acid
Regulatory substances	Hormones, enzymes

Muscular Tissue

Muscular (contractile) tissue is composed of cells called muscle fibers. Muscle fibers contain actin filaments and myosin filaments, whose interactions account for the movements we associate with animals. There are three types of vertebrate muscle tissue: skeletal, cardiac, and smooth.

Skeletal muscle (Fig. 40.5*a*), also called voluntary muscle, is attached by tendons to the bones of the skeleton, and when it contracts, body parts move. Contraction of skeletal muscle is under voluntary control and occurs faster than in all the other muscle types. Skeletal muscle fibers are cylindrical and quite long—sometimes they run the length of the muscle. They arise during development when several cells fuse, producing one muscle fiber with multiple nuclei. The nuclei are located at the periphery of the fiber, just inside the plasma membrane. The fibers have alternating light and dark bands that give them a **striated** [L. *stria*, lined] appearance. These bands are due to the placement of actin filaments and myosin filaments in the cell.

Smooth (visceral) muscle is so named because its fibers lack striations. The spindle-shaped fibers form layers in which the thick middle portion of one fiber is opposite the thin ends of adjacent fibers. Consequently, the nuclei form an irregular pattern in the tissue (Fig. 40.5*b*). Smooth muscle is not under voluntary control and therefore is said to be involuntary. Smooth muscle, found in walls of viscera (intestine, stomach, and other internal organs) and blood vessels, contracts more slowly than skeletal muscle but can remain contracted for a longer time. When the smooth muscle of the intestine contracts, food moves along inside the lumen (central cavity). When the smooth muscle of the blood vessels contracts, it constricts the blood vessels, helping to raise the blood pressure.

Cardiac muscle (Fig. 40.5*c*) is found only in the walls of the heart. Its contraction pumps blood and accounts for the heartbeat. Cardiac muscle combines features of both smooth muscle and skeletal muscle. It has striations like skeletal muscle, but the contraction of the heart is involuntary for the most part. Cardiac muscle fibers also differ from skeletal muscle fibers in that they have a single, centrally placed nucleus. These fibers are branched and seemingly fused one with the other, and the heart appears to be composed of one large interconnecting mass of muscle fibers. Actually, cardiac muscle fibers are separate and individual, but they are bound end to end at intercalated disks, areas where folded plasma membranes between two fibers contain adhesion junctions and gap junctions. Intercalated disks allow impulses to move from fiber to fiber so that the beat of the heart is coordinated.

All muscle tissue contains actin and myosin filaments; these overlap, forming striations in skeletal and cardiac muscle but not in smooth muscle.

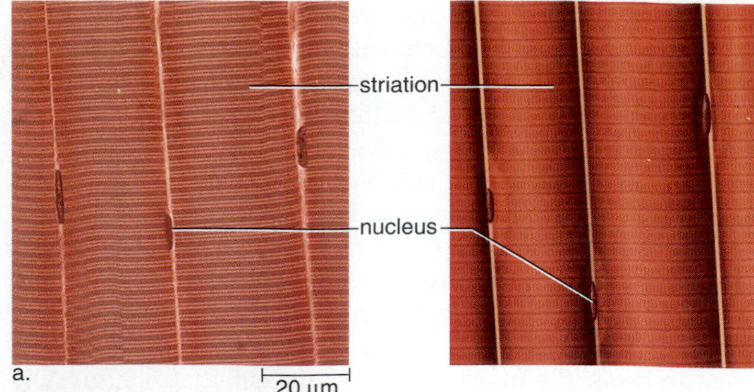

Skeletal muscle
- has striated cells with multiple nuclei.
- usually attached to skeleton.
- functions in voluntary movement.

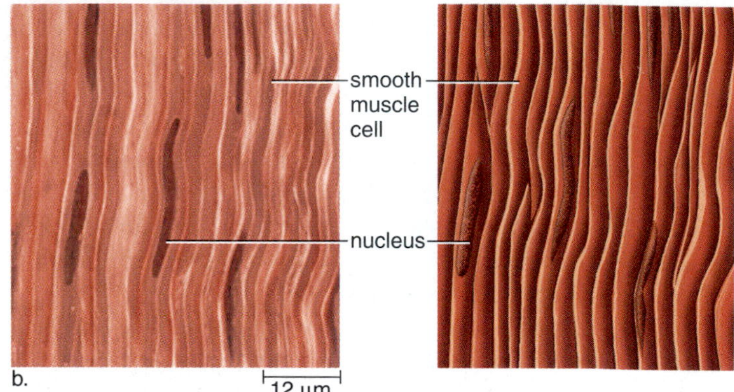

Smooth muscle
- has spindle-shaped cells, each with a single nucleus.
- occurs in walls of hollow internal organs.
- functions in movement of substances in lumens of body.
- no cross striations, involuntary.

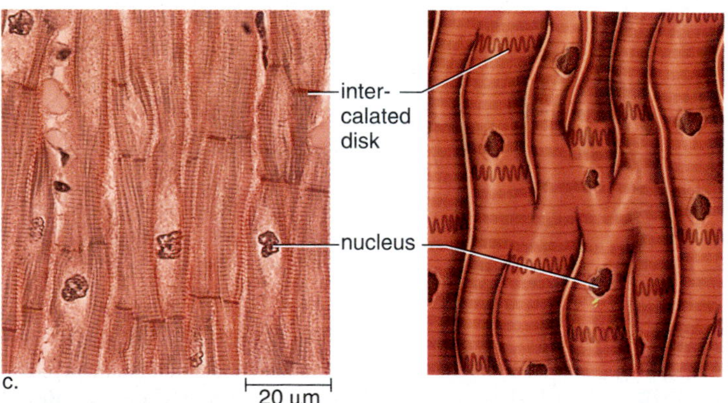

Cardiac muscle
- has branching striated cells, each with a single nucleus.
- occurs in the wall of the heart.
- functions in the pumping of blood.
- involuntary.

Figure 40.5 Muscular tissue.
a. Skeletal muscle is voluntary and striated. **b.** Smooth muscle is involuntary and nonstriated. **c.** Cardiac muscle is involuntary and striated. The plasma membranes of the branching cells fit together at intercalated disks.

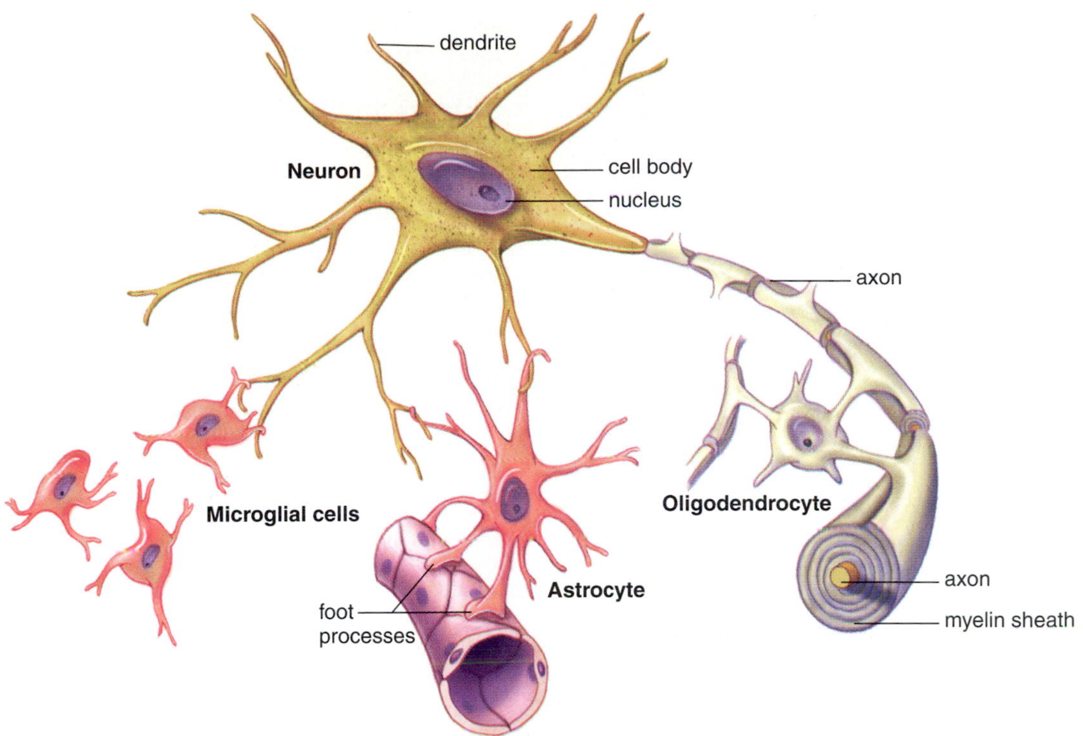

Figure 40.6 Neurons and neuroglia.
Neurons conduct nerve impulses. Neuroglia consists of cells which support and service neurons and have various functions: microglial cells are phagocytes that clean up debris. Astrocytes lie between neurons and a capillary; therefore, substances entering neurons from the blood must first pass through astrocytes. Oligodendrocytes form the myelin sheaths around fibers in the brain and spinal cord.

Nervous Tissue

Nervous tissue, which contains nerve cells called **neurons** [Gk. *neuron,* nerve] (Fig. 40.6), is present in the brain, spinal cord, and nerves. A neuron is a specialized cell that has three parts: dendrites, cell body, and an axon. A dendrite is a process that typically conduct signals toward the cell body. The cell body contains the major concentration of the cytoplasm and the nucleus of the neuron. An axon is a process that typically conducts nerve impulses away from the cell body. Axons can be quite long, and outside the brain and the spinal cord are bound by connective tissue to form the nerves.

The nervous system has just three functions: sensory input, integration of data, and motor output. Nerves conduct impulses from sensory receptors to the spinal cord, and the brain where integration occurs. The phenomenon called sensation occurs only in the brain, however. Nerves also conduct nerve impulses away from the spinal cord and brain to the muscles and glands, causing them to contract and secrete, respectively. In this way, a response to the stimulus is achieved. In addition to neurons, nervous tissue contains neuroglia.

Neuroglia

There are several different types of cells collectively called **neuroglia** in the central nervous system (Fig. 40.6). Much research is currently being conducted to determine how much these cells contribute to the functioning of the central nervous system. Neuroglia outnumber neurons nine to one and take up more than half the volume of the brain, but until recently, they were thought to merely support and nourish neurons. Three types of neuroglia cells are oligodendrocytes, microglial cells, and astrocytes. Oligodendrocytes form the myelin sheath around an axon. Microglial cells, in addition to supporting neurons, phagocytize bacterial and cellular debris. Astrocytes provide nutrients to neurons and produce a hormone, glial-derived growth factor (GDGF), that someday might be used as a cure for Parkinson disease and other diseases caused by neuron degeneration. Neuroglia don't have long processes, but even so, researchers are now beginning to gather evidence that they do communicate among themselves and with neurons!

Nerve cells, called neurons, have fibers (processes) called axons and dendrites. Axons are found in nerves. Neuroglia support and service neurons.

40.2 Organs and Organ Systems

We tend to associate particular tissues with particular organs. For example, we associate muscular tissue with muscles and nervous tissue with the brain. In actuality, however, an **organ** is a structure that is composed of two or more types of tissues working together to perform particular functions. An **organ system** contains many different organs that cooperate to carry out a process such as digestion of food. We are going to examine human skin as an example of an organ. Some authorities even call skin the integumentary system (especially since it cannot be placed in one of the other systems). They maintain that the hair follicles, the oil and sweat glands, the sensory receptors, and the skin are separate organs, and these organs work together to perform various functions.

Skin as an Organ

Human skin (Fig. 40.7) covers the body, protecting underlying parts from physical trauma, pathogen invasion, and water loss. Skin cells manufacture a precursor molecule that is converted to vitamin D in the body after it is exposed to ultraviolet (UV) light. Only a small amount of UV radiation is needed to change the precursor to vitamin D. The skin also helps regulate body temperature, and because it contains sensory receptors, the skin helps us to be aware of our surroundings and to know when to communicate with others.

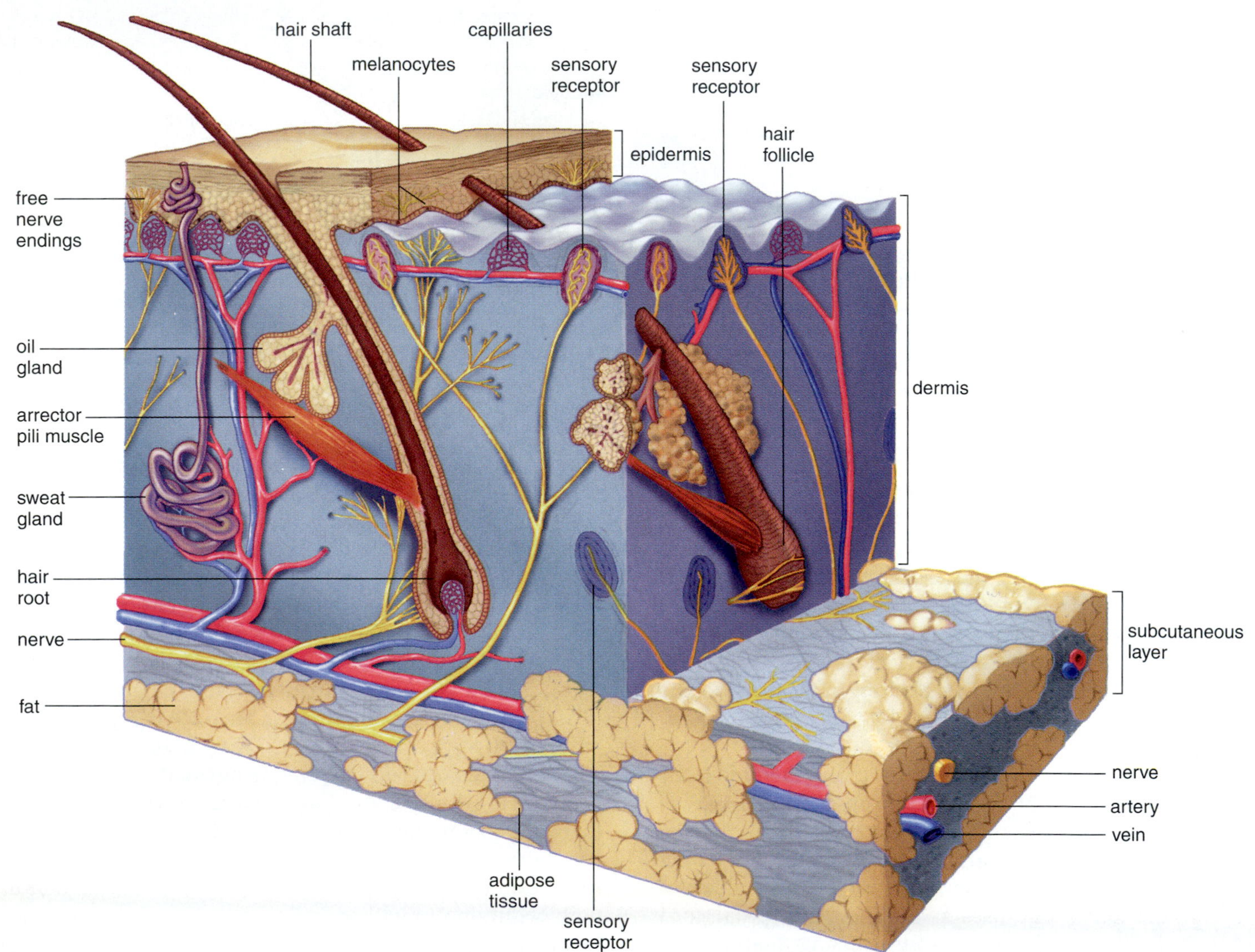

Figure 40.7 **Human skin anatomy.**
Skin consists of two regions, the epidermis and the dermis. A subcutaneous layer lies below the dermis.

Regions of Skin

The skin has two regions: the epidermis and the dermis. A subcutaneous layer found beneath the dermis binds the skin to underlying organs.

The **epidermis** [Gk. *epi,* over, and *derma,* skin] is the outer, thinner region of the skin. It is a stratified squamous epithelium, whose cells are derived from the basal cells, which undergo continuous cell division. As newly formed cells are pushed to the surface away from their blood supply, they gradually flatten and harden. Eventually, they die and are sloughed off. Hardening is caused by cellular production of a waterproof protein called keratin. Dandruff occurs when the rate of keratinization is two or three times the normal rate. Over much of the body, keratinization is minimal, but the palm of the hand and the sole of the foot have a particularly thick outer layer of dead keratinized cells.

Specialized cells in the dividing layer of epidermis, called melanocytes, produce melanin, the pigment responsible for skin color in dark-skinned persons. When you sunbathe, the melanocytes become more active, producing melanin which protects the skin from the damaging effects of the ultraviolet (UV) radiation in sunlight.

Nails grow from special epidermal cells at the base of the nail in the region called the nail root. These cells become keratinized as they grow out over the nail bed. The visible portion of the nail is called the nail body. The pink color of nails is due to the vascularized dermal tissue beneath the nail. The whitish color of the half-moon–shaped base results from the thicker germinal layer in this area. Ordinarily, nails grow only about one millimeter a week.

The epidermis of skin is made up of stratified squamous epithelium. In this layer, new cells are pushed outward, become keratinized, die, and are sloughed off.

The **dermis** [Gk. *derma,* skin] is a region of fibrous connective tissue that is deeper and thicker than the epidermis. It contains elastic fibers and collagen fibers. The collagen fibers form bundles that interlace and run, for the most part, parallel to the skin surface. If a surgeon's cut runs with the collagen fibers the resulting scar will most likely be quite thin; if the cut is against the grain, a thick scar will form.

There are several types of structures in the dermis. A hair, except for the root, is formed of dead, hardened epidermal cells; the root is alive and resides in a hair follicle found in the dermis. Each follicle has one or more oil (sebaceous) glands that secrete sebum, an oily substance that lubricates the hair and the skin. If sebaceous glands fail to discharge, the secretions collect and form "whiteheads," or "blackheads." The color of blackheads is due to oxidized sebum. A smooth muscle called the arrector pili muscle is attached to the hair follicle in such a way that when contracted, the muscle causes the hair to stand on end. When you are frightened or cold, goose bumps appear due to a mounding up of the skin from the contraction of these muscles.

Sweat (sudoriferous) glands are present in most regions of the skin. A sweat gland begins as a coiled tubule within the dermis, but then it straightens out near its opening. Some sweat glands open into hair follicles, but most open onto the surface of the skin.

Small receptors are present in the dermis. There are different receptors for pressure, touch, temperature, and pain. Pressure receptors are in onion-shaped sense organs called Pacinian corpuscles that lie deep inside the dermis and around joints and tendons. They are also believed to provide instant information about how and where we are moving. In cats, Pacinian corpuscles are concentrated on the paws, the leg joints, and the connective tissue of the abdomen. Those close to the ground may provide information about the location of prey. Closely related sensors on the tongues of woodpeckers help them find insects in tree bark.

Touch receptors, which are flat and oval shaped, are concentrated in the fingertips, the palms, the lips, the tongue, the nipples, the penis, and the clitoris. Their prevalence is thought to provide these regions with special sensitivity. Heat and cold sense organs are encapsulated by sheaths of connective tissue and contain lacy networks of nerve fibers. Nerve fibers branch out through all skin, and free nerve endings are believed to be the receptors for pain.

The dermis also contains blood vessels. When blood rushes into these vessels, a person blushes, and when blood volume is reduced in them, a person turns ashen or white.

The dermis, composed of fibrous connective tissue, lies beneath the epidermis. It contains hair follicles, sebaceous glands, and sweat glands. It also contains receptors, blood vessels, and nerve fibers.

The subcutaneous layer, which lies below the dermis, is composed of loose connective tissue, including adipose tissue. Adipose tissue helps to insulate the body by minimizing both heat gain and heat loss. A well-developed subcutaneous layer gives a rounded appearance to the body. Excessive development of this layer accompanies obesity.

Skin Cancer

In recent years, there has been a great increase in the number of persons with skin cancer, and physicians believe this is due to sunbathing or even to the use of tanning machines. Protecting your skin from the sun's damaging rays is discussed in the reading on page 733.

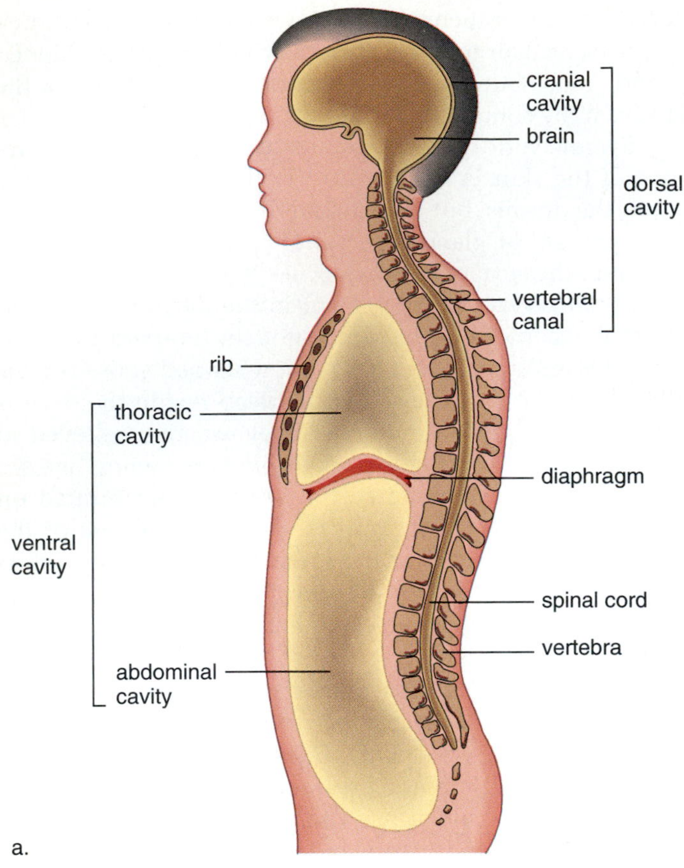

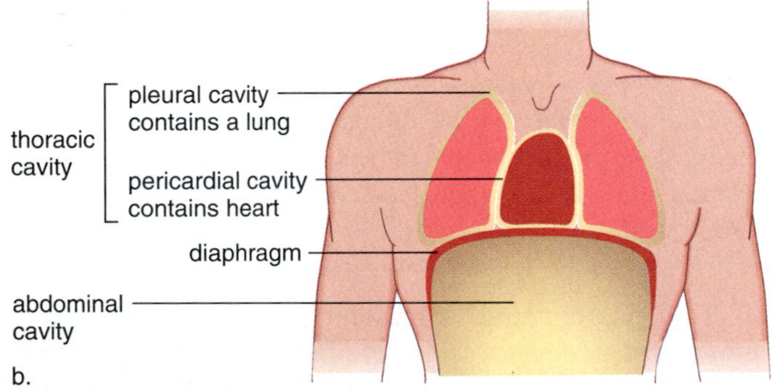

Figure 40.8 Mammalian body cavities.
a. Side view. There is a dorsal cavity, which contains the cranial cavity and the vertebral canal. The brain is in the cranial cavity, and the spinal cord is in the vertebral canal. There is a well-developed ventral cavity, which is divided by the diaphragm into the thoracic cavity and the abdominal cavity. The heart and lungs are in the thoracic cavity, and most other internal organs are in the abdominal cavity. **b.** Anterior view of the thoracic cavity.

Organ Systems

In most animals, individual organs function as part of an organ system, the next higher level of animal organization (see Fig. 40.1). These same systems are found in all vertebrate animals. The organ systems of vertebrates carry out the life processes that are common to all animals, and indeed to all organisms:

Life Processes	Human Systems
Coordinate body activities	Nervous system Endocrine system
Acquire materials and energy (food)	Skeletal system Muscular system Digestive system
Maintain body shape	Skeletal system Muscular system
Exchange gases	Respiratory system
Transport materials	Cardiovascular system
Excrete wastes	Urinary system
Protect the body from disease	Lymphatic system Immune system
Produce offspring	Reproductive system

Body Cavities

Each organ system has a particular distribution within the human body. There are two main body cavities: the smaller dorsal body cavity and the larger ventral body cavity (Fig. 40.8). The brain and the spinal cord are in the dorsal body cavity.

During development, the ventral body cavity develops from the coelom. In humans and other mammals, the coelom is divided by a muscular diaphragm that assists breathing. The heart (a pump for the cardiovascular system) and the lungs are located in the upper (thoracic or chest) cavity. The major portions of the digestive system, the excretory system, and much of the reproductive system are located in the lower (abdominal) cavity. The major organs of the urinary system are the paired kidneys. The digestive system has accessory organs such as the liver and pancreas. Each sex has characteristic sex organs.

The animal body is organized; the organs have a specific structure, function, and location. The performance of the organs of each system are coordinated.

Health Focus

Skin Cancer on the Rise

In Victorian days, Caucasian women carried parasols to keep their skin fair. But early in this century, some fair-skinned people began to prefer the golden-brown look, and they took up sunbathing as a way to achieve a tan. A few hours after exposure to the sun, pain and redness due to dilation of blood vessels occur. Tanning occurs when melanin granules increase in keratinized cells at the surface of the skin as a way to prevent any further damage by ultraviolet (UV) rays. The sun gives off two types of UV rays: UV-A rays and UV-B rays. UV-A rays penetrate the skin deeply, affect connective tissue, and cause the skin to sag and wrinkle. UV-A rays are also believed to increase the effects of the UV-B rays, which are the cancer-causing rays. UV-B rays are more prevalent at midday.

Skin cancer is categorized as either nonmelanoma or melanoma. Nonmelanoma cancers are of two types. Basal cell carcinoma, the most common type, begins when UV radiation causes epidermal basal cells to form a tumor, while at the same time suppressing the immune system's ability to detect the tumor. The signs of a tumor are varied. They include an open sore that will not heal, a recurring reddish patch, a smooth, circular growth with a raised edge, a shiny bump, or a pale mark (Fig. 40A). About 95% of patients are easily cured, but reoccurrence is common.

Squamous cell carcinoma begins in the epidermis proper. Squamous cell carcinoma is five times less common than basal cell carcinoma, but it is more likely to spread to nearby organs, and the death rate is about 1% of cases. The signs of squamous cell carcinoma are the same as that for basal cell carcinoma, except that it may also show itself as a wart that bleeds and scabs.

Melanoma that starts in the melanocytes has the appearance of an unusual mole. Unlike a dark mole that is circular and confined, melanoma moles look like spilled ink spots. A variety of shades can be seen in the same mole, and they can itch, hurt, or feel numb. The skin around the mole turns grey, white, or red. Melanoma is most apt to appear in persons who have fair skin, particularly if they have suffered occasional severe burns as children. The chance of melanoma increases with the number of moles a person has. Most moles appear before the age of 14, and their appearance is linked to sun exposure. Melanoma rates have risen since the turn of the century, but the incidence has doubled in the last decade. Now about 32,000 cases of melanoma are diagnosed each year, and one in five persons diagnosed dies within five years.

Since the incidence of skin cancer is related to UV exposure, scientists have developed a UV index to determine how powerful the solar rays are in different U.S. cities. The index assigns a baseline level of 100 to Anchorage, Alaska. In general, the more southern the city, the higher the UV index, and the greater the risk of skin cancer.

Looking ahead, we should note that for every 10% decrease in the ozone layer, the UV index per city rises by 13–20%, and the chance of skin cancer rises as well. Even if you live in Anchorage, you should take the following steps to protect yourself from the sun.

To prevent the possible occurrence of skin cancer, observe the following:

- Use a broad-spectrum sunscreen, which protects you from both UV-A and UV-B radiation, with an SPF (sun protection factor) of at least 15. (This means, for example, that if you usually burn after a 20-minute exposure, it will take 15 times that long before you will burn.)
- Wear protective clothing. Choose fabrics with a tight weave and wear a wide-brimmed hat. A baseball cap does not protect the rims of the ears.
- Stay out of the sun altogether between the hours of 10 A.M. and 3 P.M. This will reduce your annual exposure by as much as 60%.
- Wear sunglasses that have been treated to absorb both UV-A and UV-B radiation. Otherwise, sunglasses can expose your eyes to more damage than usual because pupils dilate in the shade.
- Avoid tanning machines. Although most tanning devices use high levels of only UV-A, the deep layers of the skin become more vulnerable to UV-B radiation when you are later exposed to the sun.

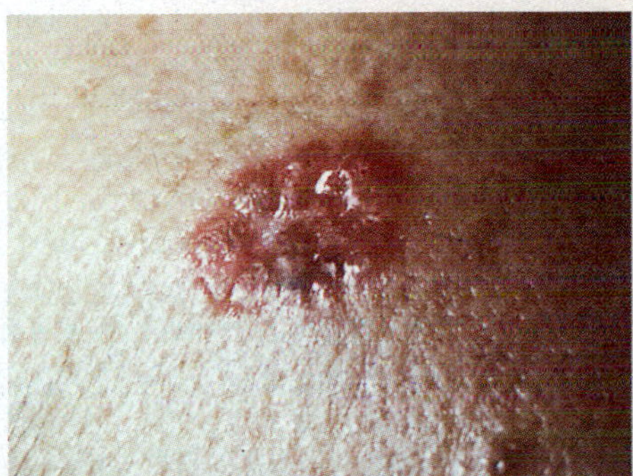

a. Basal cell carcinoma

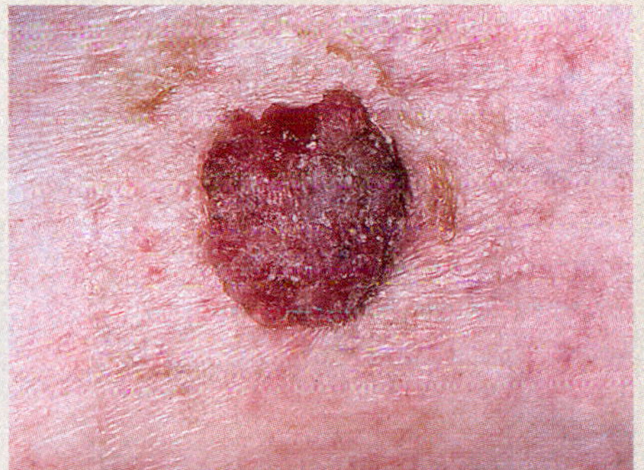

b. Squamous cell carcinoma

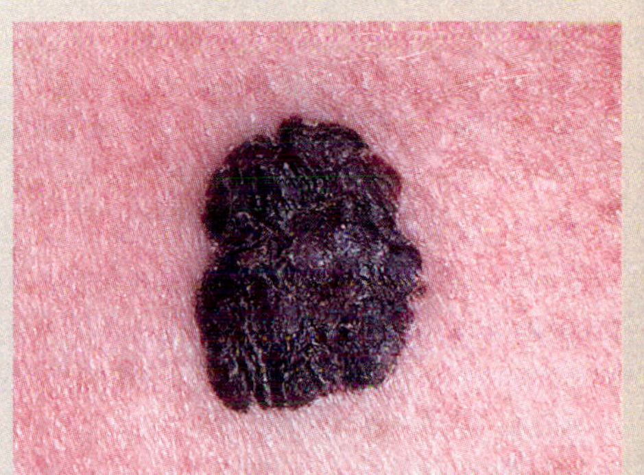

c. Melanoma

Figure 40A Skin cancer.
a. Basal cell carcinoma occurs when basal cells proliferate abnormally. b. Squamous cell carcinoma arises in epithelial cells derived from basal cells. c. Malignant melanoma is due to a proliferation of pigmented cells. About one-third develop from pigmented moles.

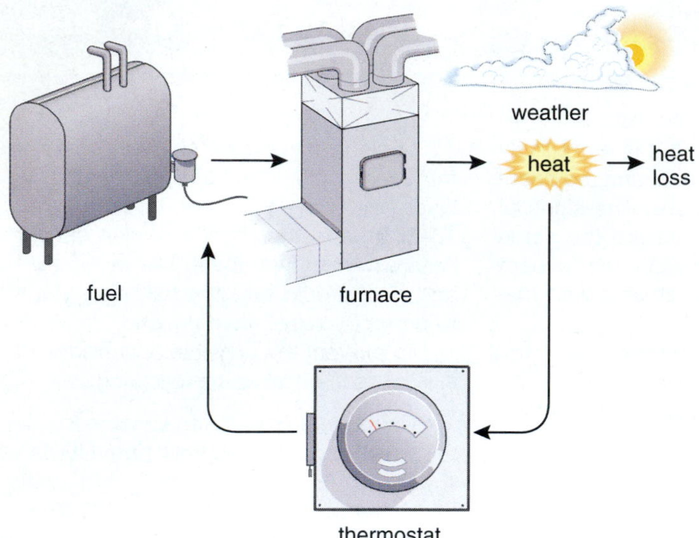

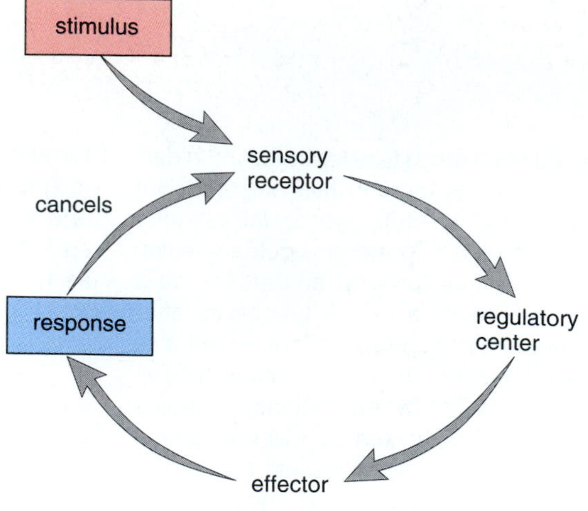

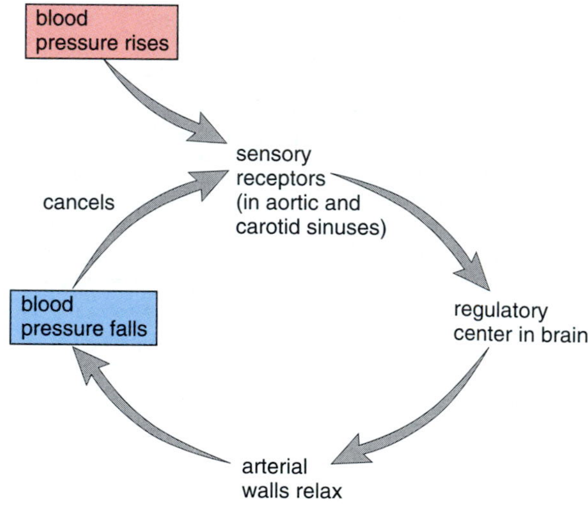

Figure 40.9 Negative feedback control.
a. Mechanical model. A thermostat that senses the room temperature can signal the furnace to turn on to help maintain a relatively stable room temperature. **b.** Biological model. A stimulus causes a receptor to signal a regulatory center in the brain. The regulatory center signals effectors to respond, and a response cancels the stimulus. **c.** Biological example. When blood pressure rises, special receptors in blood vessels signal a particular center in the brain. The brain signals the arteries to relax, and blood pressure declines.

40.3 Homeostasis

Claude Bernard, a famous French physiologist, suggested in 1859 that while an animal lives in an external environment, the cells of the body lie within an internal environment. The internal environment is largely a fluid, called **tissue fluid,** that bathes the cells of the body. Bernard concluded that the relative stability of the internal environment allows animals to live in an external environment that can vary considerably. Later, Walter Cannon, an American physiologist, introduced the term **homeostasis** [Gk. *homoios,* like, resembling, and *stasis,* standing]. He said there is a dynamic interplay between events that tend to change the internal environment and those that act against this possibility. To achieve homeostasis, the composition of blood and tissue fluid, the body temperature, and the blood pressure, for example, must stay within a normal range.

The internal environment of an animal's body consists of tissue fluid, which bathes the cells.

Most organ systems of the human body contribute to homeostasis. The digestive system takes in and digests food, providing nutrient molecules that enter the blood and replace the nutrients that are constantly being used by the body cells. The respiratory system adds oxygen to the blood and removes carbon dioxide. The amount of oxygen taken in and carbon dioxide given off can be increased to meet body needs. The liver and the kidneys contribute greatly to homeostasis. For example, immediately after glucose enters the blood, it can be removed by the liver and stored as glycogen. Later, glycogen is broken down to replace the glucose used by the body cells; in this way, the glucose composition of the blood remains constant. The hormone insulin, secreted by the pancreas, regulates glycogen storage. The kidneys are also under hormonal control as they excrete wastes and salts, substances that can affect the pH level of the blood.

Although homeostasis is, to a degree, controlled by hormones, it is ultimately controlled by the nervous system. In humans, the brain contains centers that regulate such factors as blood pressure. Maintaining blood pressure levels requires a receptor that detects unacceptable levels and signals a regulatory center. If a correction is required, the center then directs an adaptive response. Once normalcy is obtained, the receptor is no longer stimulated. This is called control by **negative feedback** because the regulatory centers bring about a response that is opposite to present conditions. A

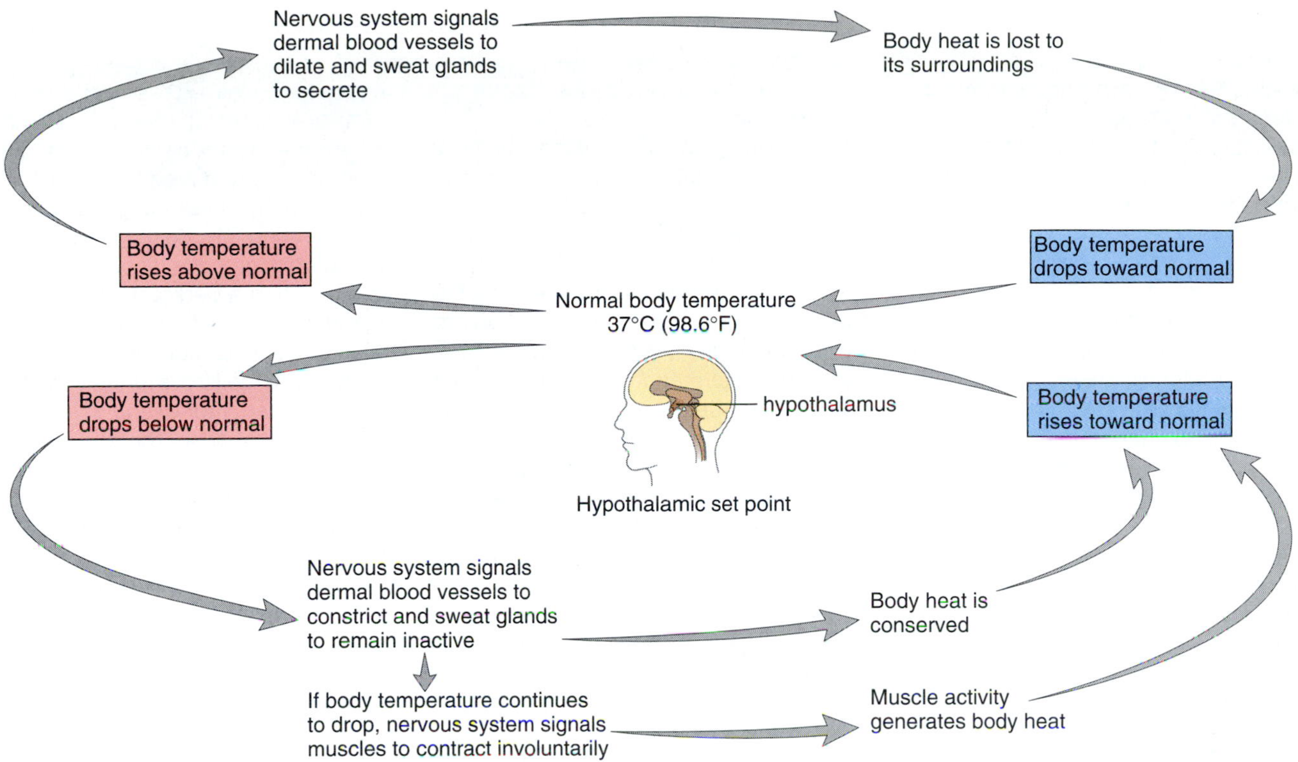

Figure 40.10 Homeostasis and temperature control.
A negative feedback mechanism controls body temperature in the manner described.

useful analogy that helps illustrate how such a negative feedback mechanism works is the control of the heating system of a house by a thermostat. Through negative feedback control, indoor temperature can be maintained within a relatively narrow range with a slight fluctuation above and below a mean (Fig. 40.9*a*).

Figure 40.9*b* shows that in the body there are sensory receptors that fulfill the role of sensing devices. When a receptor is stimulated, it signals a regulatory center that then turns on an effector. The effector brings about a response that negates the original conditions that stimulated the receptor. In the absence of suitable stimulation, the receptor no longer signals the regulatory center.

Figure 40.9*c* gives an example involving the nervous system. When blood pressure rises, receptors signal a regulatory center, which then sends out nerve impulses to the arterial walls, causing them to relax, and the blood pressure now falls. Therefore, the sensory receptors are no longer stimulated, and the system shuts down. Notice that negative feedback control results in a fluctuation above and below an average. Thus, there is a dynamic equilibrium of the internal environment.

Positive feedback also occurs on occasion. In these instances, certain events increase the likelihood of a particular response. For example, once the childbirth process begins, each succeeding event makes it more likely that the process will continue until completion.

Regulation of Body Temperature

Although the skin senses a change in external temperature, the regulatory center for body temperature, located in the hypothalamus, is sensitive only to the temperature of the blood. When the body temperature falls below normal, the regulatory center directs (via nerve impulses) the blood vessels of the skin to constrict (Fig. 40.10). This conserves heat. Also, in furry animals the arrector pili muscles pull hairs erect, and a layer of insulating air is trapped next to the skin. If body temperature falls even lower, the regulatory center sends nerve impulses to the skeletal muscles, and shivering occurs. Shivering generates heat, and gradually body temperature rises to 37°C. If the body temperature is normal, the regulatory center is not active.

When the body temperature is higher than normal, the regulatory center is activated and then directs the blood vessels of the skin to dilate. This allows more blood to flow near the surface of the body, where heat can be lost to the environment. In addition, the nervous system activates the sweat glands, and the evaporation of sweat also helps lower body temperature. Gradually, body temperature decreases to 37°C.

The temperature of the human body is maintained at about 37°C due to the action of a regulatory center in the hypothalamus.

Connecting Concepts

In this chapter we have concentrated on complex coelomate animals but homeostasis occurs even in the simplest of animals. Homeostasis in unicellular organisms and thin acoelomate animals occurs only at the cellular level, and each cell must carry out its own exchanges with the external environment to maintain a relative constancy of cytoplasm. In complex animals, there are localized boundaries where materials are exchanged with the external environment. In terrestrial animals, gas exchange usually occurs within lungs, food is digested within a digestive tract, and kidneys collect and excrete metabolic wastes. Exchange boundaries are an effective way to regulate the internal environment if there is a transport system to carry materials from one body part to another. Circulation in invertebrates and vertebrates carries out this function.

Regulating mechanisms occur at all levels of organization. At the cellular level, the actions of enzymes are often controlled by feedback mechanisms. However, in animals with organ systems, the nervous and endocrine systems regulate the actions of organs. In addition, an animal's nervous system gathers and processes information about the external environment. Sense receptors act as specialized boundaries through which external stimuli are received and converted into a form that can be processed by the nervous system. The information received and processed by the nervous system may then influence the organism's behavior in a way that contributes to homeostasis.

Homeostasis is so critical that without it the organism dies. Let us take a familiar example in humans. After eating, when the hormone insulin is present, glucose is removed from blood and stored in the liver as glycogen. In between eating, glycogen breakdown keeps the blood glucose level at just about 0.1%. When a person has diabetes type I, the pancreas fails to secrete insulin and glucose is not stored in the liver. Worse yet, cells are unable to take up glucose even after eating when there is a plentiful supply in the blood. Lacking glucose for cellular respiration, cells begin to break down fats with the result that acids are released in cells and enter the bloodstream. Now the person has acidosis—a low pH that may hinder enzymatic activity to the point that cellular metabolism falters and the person dies.

Summary

40.1 Types of Tissues

During development, the zygote divides to produce cells that go on to become tissues, like cells specialized for a particular function. Tissues make up organs and organ systems make up the organism. This sequence describes the levels of organization within an organism.

Tissues are categorized into four groups. Epithelial tissue covers the body and lines cavities. There is squamous, cuboidal, and columnar epithelium. Each type can be simple or stratified; it can also be glandular, or have modifications like cilia. Epithelial tissue protects, absorbs, secretes, and excretes.

Connective tissue has a matrix between cells. Loose fibrous connective tissue and dense fibrous connective tissue contain fibroblasts and fibers. Loose fibrous connective tissue has both collagen and elastic fibers. Dense fibrous connective tissue, like that of tendons and ligaments, contains closely packed collagen fibers. In adipose tissue, the cells enlarge and store fat.

Both cartilage and bone have cells within lacunae, but the matrix for cartilage is more flexible than that for bone, which contains calcium salts. In bone, the lacunae lie in concentric circles within an osteon (or Haversian system) about a central canal. Blood is a connective tissue in which the matrix is a liquid called plasma.

Muscular (contractile) tissue can be smooth or striated (skeletal and cardiac); involuntary (smooth and cardiac) or voluntary (skeletal). In humans, skeletal muscle is attached to bone, smooth muscle is in the wall of internal organs, and cardiac muscle makes up the heart.

Nervous tissue has one main type of conducting cell, the neuron, and several types of neuroglial cells. Each neuron has dendrites, a cell body, and an axon. The brain and spinal cord contain complete neurons, while the nerves contain only neuron fibers. Axons are specialized to conduct nerve impulses.

40.2 Organs and Organ Systems

Organs contain various tissues. Skin is an organ that has two regions. Epidermis (stratified squamous epithelium) overlies the dermis (fibrous connective tissue containing sensory receptors, hair follicles, blood vessels, and nerves). A subcutaneous layer is composed of loose connective tissue.

Organ systems contain several organs. Organ systems of humans have specific functions and carry out the life processes that are common to all organisms.

The human body contains two main cavities. The dorsal cavity contains the brain and spinal cord. The ventral cavity is divided into the thoracic cavity (heart and lungs) and the abdominal cavity (most other internal organs).

40.3 Homeostasis

Homeostasis is the dynamic equilibrium of the internal environment that allows the blood and tissue constituents and values stay within a normal range. All organ systems contribute to homeostasis but special contributions are made by the liver, which keeps the blood glucose constant, and the kidneys, which regulate the pH. The nervous and hormonal systems regulate the other body systems. Both of these are controlled by negative feedback mechanisms, which result in slight fluctuations above and below desired levels. Body temperature is regulated by a center in the hypothalamus.

Reviewing the Chapter

1. Name the four major types of tissues. 724
2. Describe the structure and the functions of three types of epithelial tissue. 724–25
3. Describe the structure and the functions of six types of connective tissue. 726–27
4. Describe the structure and the functions of three types of muscular tissue. 728

5. Nervous tissue contains what types of cells? 729
6. Describe the structure of skin, and state at least two functions of this organ. 730–31
7. In general terms, describe the location of the human organ systems. 732
8. Tell how the various systems of the body contribute to homeostasis. 734
9. What is the function of receptors, the regulatory center, and effectors in a negative feedback mechanism? Why is it called negative feedback? 734

Testing Yourself

Choose the best answer for each question.

1. Which of these is mismatched?
   a. tissues—like cells
   b. epithelial tissue—protection and absorption
   c. muscular tissue—contraction and conduction
   d. connective tissue—binding and support
   e. nervous tissue—conduction and message sending
2. Which of these are not epithelial tissue?
   a. simple cuboidal and stratified columnar
   b. bone and cartilage
   c. stratified squamous and simple squamous
   d. pseudostratified and ciliated
   e. All of these are epithelial tissue.
3. Which tissue is more apt to line a lumen?
   a. epithelial tissue
   b. connective tissue
   c. nervous tissue
   d. muscular tissue
   e. only smooth muscle
4. Tendons and ligaments
   a. are connective tissue.
   b. are associated with the bones.
   c. are found in vertebrates.
   d. contain collagen.
   e. All of these are correct.
5. Which tissue has cells in lacunae?
   a. epithelial tissue
   b. cartilage
   c. bone
   d. smooth muscle
   e. Both b and c are correct.
6. Cardiac muscle is
   a. striated.
   b. involuntary.
   c. smooth.
   d. many fibers fused together.
   e. Both a and b are correct.
7. Which of these components of blood fights infection?
   a. red blood cells
   b. white blood cells
   c. platelets
   d. hydrogen ions
   e. All of these are correct.
8. Which of these body systems contribute to homeostasis?
   a. digestive and excretory systems
   b. respiratory and nervous systems
   c. nervous and endocrine systems
   d. immune and cardiovascular systems
   e. All of these are correct.
9. In a negative feedback mechanism
   a. the output cancels the input.
   b. there is a fluctuation above and below the average.
   c. there is self-regulation.
   d. a regulatory center communicates with other body parts.
   e. All of these are correct.
10. When a human being is cold, the superficial blood vessels
    a. dilate, and the sweat glands are inactive.
    b. dilate, and the sweat glands are active.
    c. constrict, and the sweat glands are inactive.
    d. constrict, and the sweat glands are active.
    e. contract so that shivering occurs.
11. Give the name, the location, and the function for each of these tissues in the human body.
    a. Type of epithelial tissue
    b. Type of muscular tissue
    c. Type of connective tissue

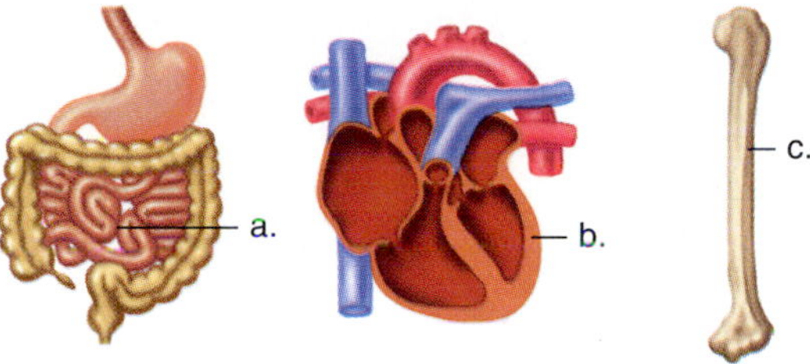

Thinking Scientifically

1. Many cancers develop from epithelial tissue. These include lung, colon, and skin cancers. What are two attributes of this tissue type that makes cancer more likely to develop?
2. When infected with certain bacteria or viruses, our bodies produce a febrile, or fever, response. Fevers occur when the hypothalamus changes its temperature set point. Signaling of the hypothalamus could be direct (from the infectious agent itself) or indirect (from the immune system). Which of these would enable the hypothalamus to respond to the greatest variety of infectious agents? Is there any disadvantage to such a signaling system?

Bioethical Issue

Transplantation of the kidney, heart, liver, pancreas, lung, and other organs is now possible due to two major breakthroughs. First, solutions have been developed that preserve donor organs for several hours. This made it possible for one young boy to undergo surgery for 16 hours, during which time he received five different organs. Second, rejection of transplanted organs is now prevented by immunosuppressive drugs; therefore, organs can be donated by unrelated individuals, living or dead. After death, it is possible to give the "gift of life" to someone else—over 25 organs and tissues from one cadaver can be used for transplants. Survival rate after a transplant operation is good. So many heart recipients are now alive and healthy, they have formed basketball and softball teams, demonstrating the normalcy of their lives after surgery.

One problem persists however, and that is the limited availability of organs for transplantation. At any one time, at least 27,000 Americans are waiting for a donated organ. Keen competition for organs can lead to various bioethical inequities. When the governor of Pennsylvania received a heart and lungs within a relatively short period of time, it appeared that his social status may have played a role. When Mickey Mantle received a liver transplant, people asked if it was right to give an organ to an older man who had a diseased liver due to the consumption of alcohol. If a father gives a kidney to a child, he has to undergo a major surgical operation that leaves him vulnerable to possible serious consequences in the future. If organs are taken from those who have just died, who guarantees that the individual is indeed dead? And is it right to genetically alter animals to serve as a source of organs for humans? Such organs will most likely be for sale, and does this make the wealthy more likely to receive a transplant than those who cannot pay? How can we be certain that the distribution of organs for transplant is equitable?

Understanding the Terms

adipose tissue 726
blood 727
cardiac muscle 728
cartilage 727
compact bone 727
connective tissue 726
dense fibrous connective tissue 726
dermis 731
endocrine gland 724
epidermis 731
epithelial tissue 724
exocrine gland 724
fibroblast 726
homeostasis 734
lacuna 727
ligament 726
loose fibrous connective tissue 726
muscular (contractile) tissue 728
negative feedback 734
nervous tissue 728
neuroglia 728
neuron 728
organ 724, 730
organ system 730
positive feedback 735
skeletal muscle 728
smooth (visceral) muscle 728
spongy bone 727
striated 728
tendon 726
tissue 724
tissue fluid 734

Match the terms to these definitions:

a. ______________ Fibrous connective tissue that joins bone to bone at a joint.
b. ______________ Outer region of the skin composed of stratified squamous epithelium.
c. ______________ Having bands such as in cardiac and skeletal muscle.
d. ______________ Self-regulatory state in which imbalances result in a fluctuation above and below a mean.
e. ______________ Porous bone found at the ends of long bones where blood cells are formed.

Web Connections

Exploring the Internet

http://www.mhhe.com/biosci/genbio/mader
(click on *Biology 7/e*)

The *Biology 7/e* Online Learning Center provides many resources for studying the material in this chapter including links to the following sites:

Guided Tour of the Visible Human. Sections through the body, with clickable buttons for further information.

http://www.madsci.org/~lynn/VH/annotated.html

U. Pennsylvania's Basic Embryology Review Program. Winner of two "Best of the Web" directories.

http://www.med.upenn.edu/meded/public/berp/

Cancer: Prevention and Treatment of Cancer. The Mayo Clinic informational site on cancer. Reference articles, information, references, links, and quizzes.

http://www.mayohealth.org/mayo/common/htm/canhpage.htm

National Skin Cancer Prevention Program. Information on skin cancer, prevention, and treatment from the CDC.

http://www.cdc.gov/nccdphp/dcpc/nscpep/skin.htm

Consumer Health Information: Skin. Information from the National Institutes of Health on acne, psoriasis, rosacea, and other skin problems.

http://www.nih.gov/health/consumer/conkey.htm#S

Circulation

chapter

chapter concepts

Scanning electron micrograph of mammalian blood cells

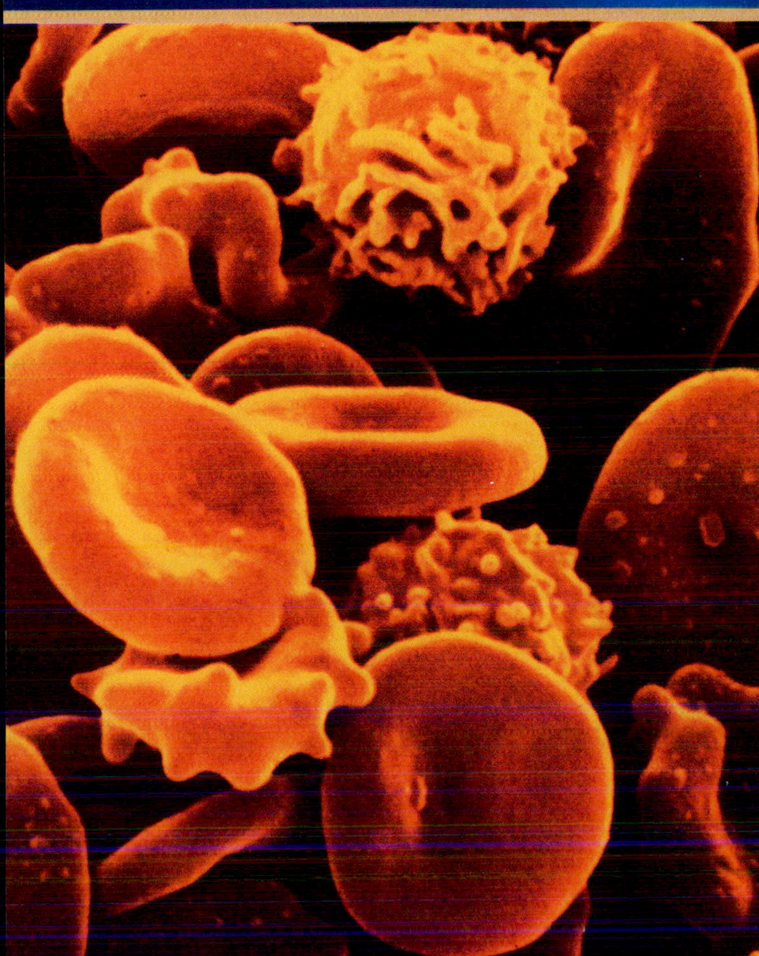

All animal cells acquire nutrients and oxygen from the environment and give off carbon dioxide and other wastes to the environment. In small aquatic animals, each cell directly exchanges materials with the external environment by utilizing diffusion and plasma membrane transport mechanisms. These animals have no need of a circulatory system.

Larger, more active, animals have a circulatory system in which the heart pumps a fluid about the body to all organs. In humans, an eleven ounce heart keeps blood flowing into a system of vessels 60,000 miles long. The pumping of the heart allows the brain to think, the lungs to breathe, and the muscles to move. The heart is one of the first organs to form during development and when it stops, death occurs.

The blood transports nutrients, gases, and metabolic wastes to or from organs with exchange boundaries to all the cells of the body. Red blood cells assist the transport of gases while nutrients and wastes are carried within the liquid portion of blood. White blood cells are also active in homeostasis by helping the body fight infections.

41.1 Transport in Invertebrates

Unicellular protists with a high surface area to volume ratio rely on diffusion for gas, nutrients, and waste exchanges with the external environment that surrounds them. Even some small multicellular animals do not have an internal transport system because their cells can be serviced without one (Fig. 41.1). Larger invertebrates usually have a circulatory system. Some of these have an open—and others have a closed—circulatory system.

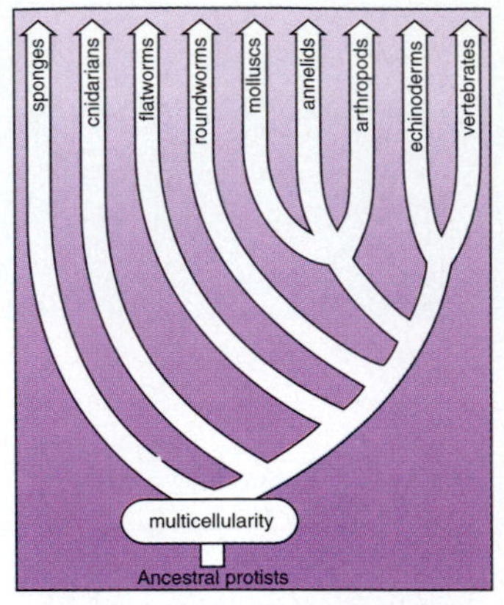

Invertebrates without a Circulatory System

Sea anemones, which are cnidarians, and planarians, which are flatworms, have a sac body plan (Fig. 41.1*b*, *c*). This body plan makes a circulatory system unnecessary. In a sea anemone, cells are either part of an external layer or they line the gastrovascular cavity. In either case, each cell is exposed to water and can independently exchange gases and rid itself of wastes. The cells that line the gastrovascular cavity are specialized to carry out digestion. They pass nutrient molecules to other cells by diffusion. In a planarian, a trilobed gastrovascular cavity ramifies throughout the small and flattened body. No cell is very far from one of the three digestive branches, so nutrient molecules can diffuse from cell to cell. Similarly, diffusion meets the respiratory and excretory needs of the cells.

Pseudocoelomate invertebrates, such as nematodes, use the coelomic fluid of their body cavity for transport purposes. Echinoderms also rely on movement of coelomic fluid within a body cavity as a circulatory system.

Figure 41.1 Aquatic organisms without a circulatory system.
a. A paramecium is a unicellular organism that carries on gas exchange across its cell surface. Food particles that flow into a specialized region called a gullet are enclosed within food vacuoles (green), where digestion occurs. Molecules leave these vacuoles as they are distributed about the cell by a movement of the cytoplasm. **b.** In a sea anemone, a cnidarian, digestion takes place inside the gastrovascular cavity, so named because it (like a vascular system) makes digested material available to the cells that line the cavity. These cells can also acquire oxygen from the watery contents of the cavity and discharge their wastes there. **c.** In a planarian, a flatworm, the gastrovascular cavity ramifies throughout the body, bringing nutrients to body cells. Diffusion is sufficient to pass molecules to every cell from either the cavity or the exterior surface.

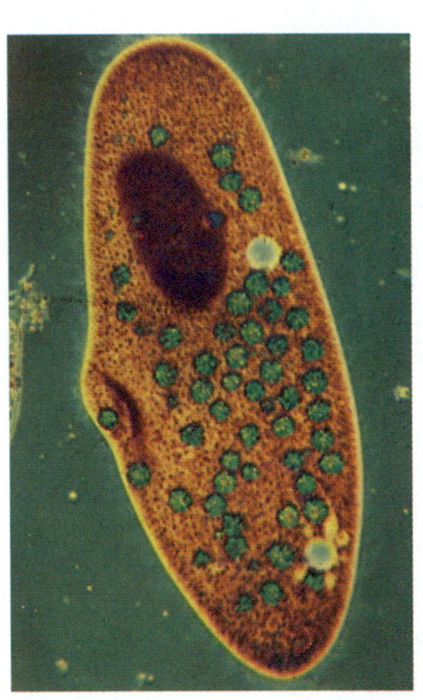

a. Paramecium, *Paramecium*

20 µm

b. Sea anemone, *Apitasia*

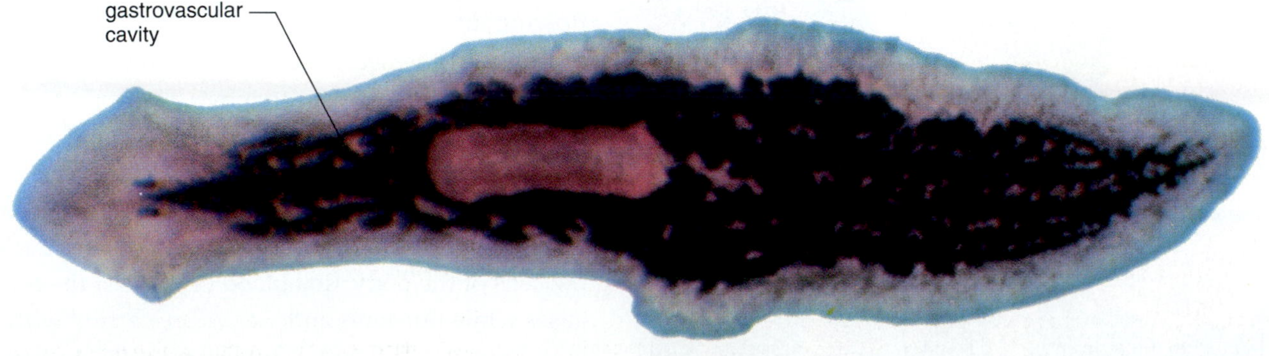

c. Flatworm, *Dugesia*

200 µm

Invertebrates with an Open or a Closed Circulatory System

All other animals have a **circulatory system** in which a pumping heart moves a fluid into blood vessels. There are two types of circulatory fluids: **blood,** which is always contained within blood vessels, and **hemolymph,** which flows into a body cavity called a hemocoel. Hemolymph is a mixture of blood and interstitial fluid.

Hemolymph is seen in animals that have an **open circulatory system.** In most molluscs and arthropods, the heart pumps hemolymph via vessels into tissue spaces that are sometimes enlarged into saclike sinuses (Fig. 41.2*a*). Eventually hemolymph drains back to the heart. In the grasshopper, an arthropod, the dorsal heart pumps hemolymph into a dorsal aorta, which empties into the hemocoel. When the heart contracts, openings called ostia (sing., ostium) are closed; when the heart relaxes, the hemolymph is sucked back into the heart by way of the ostia. The hemolymph of a grasshopper is colorless because it does not contain hemoglobin or any other respiratory pigment. It carries nutrients but no oxygen. Oxygen is taken to cells and carbon dioxide is removed from them by way of air tubes, called tracheae, which are found throughout the body. The tracheae provide an efficient transport and delivery of respiratory gases while at the same time restricting water loss.

Some invertebrates (like earthworms—annelids; squids and octopuses—molluscs), have a **closed circulatory system.** Blood, which usually consists of cells and plasma, is pumped by the heart into a system of blood vessels (Fig. 41.2*b*). There are valves that prevent the backward flow of blood. In the segmented earthworm, five pairs of anterior lateral vessels pump blood into the ventral blood vessel, which has a branch in every segment of the worm's body. Blood moves through these branches into capillaries, where exchanges with tissue fluid take place. Blood then moves into veins that return it to the dorsal blood vessel. This dorsal blood vessel returns blood to the heart for repumping.

The earthworm has red blood that contains the respiratory pigment hemoglobin. Hemoglobin is dissolved in the blood and is not contained within cells. The earthworm has no specialized boundary (e.g., lungs) for gas exchange with the external environment. Gas exchange takes place across the body wall, which must always remain moist for this purpose.

Animals with a gastrovascular cavity use this cavity for transport purposes. Other animals have an open or a closed circulatory system.

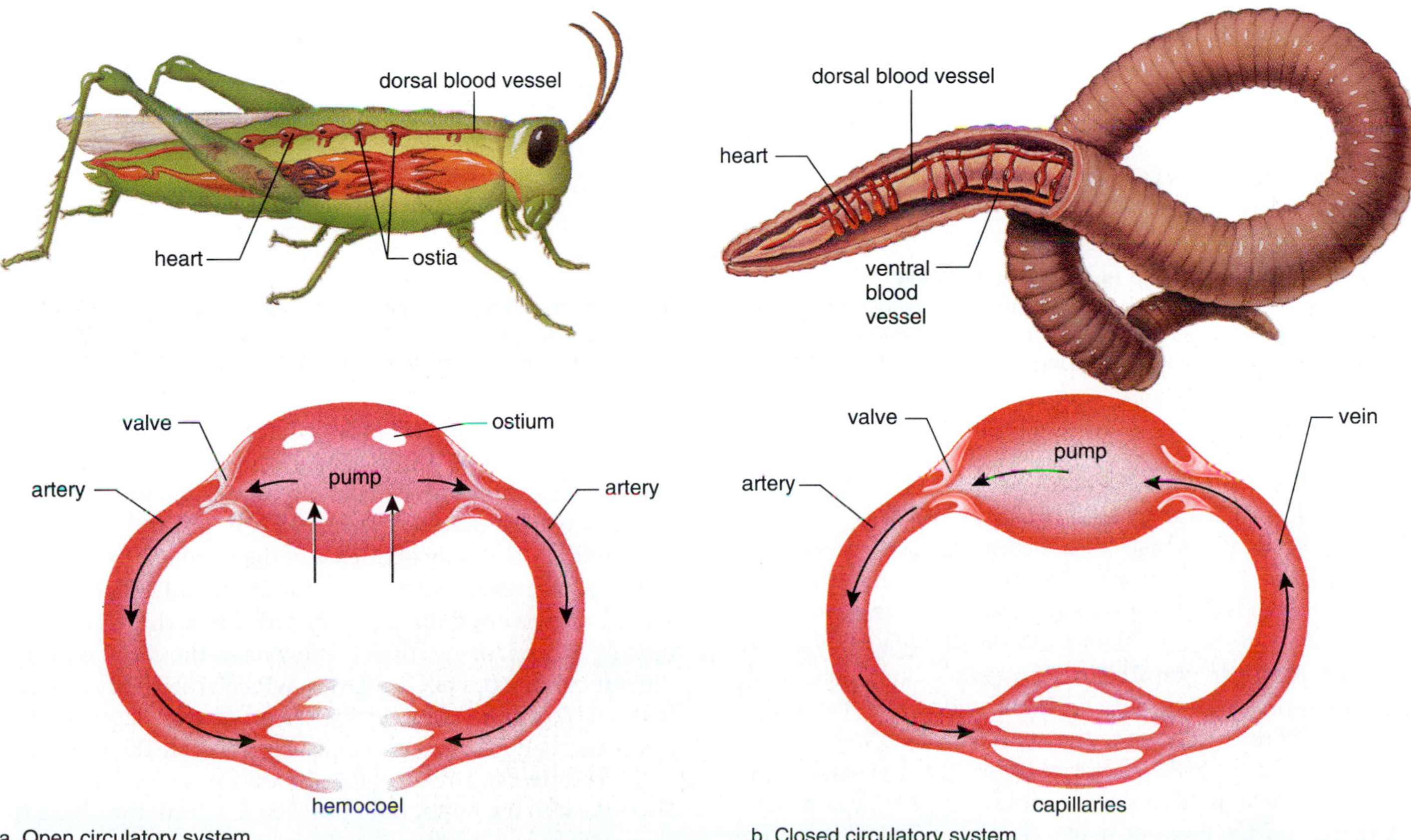

Figure 41.2 Open versus closed circulatory system.
a. The grasshopper, an arthropod, has an open circulatory system. A hemocoel is a body cavity filled with hemolymph, which freely bathes the internal organs. The heart keeps the hemolymph moving, but this open system probably could not supply oxygen to wing muscles rapidly enough. These muscles receive oxygen directly from tracheae (air tubes). **b.** The earthworm, an annelid, has a closed circulatory system. The dorsal and ventral blood vessels are joined by five pairs of anterior hearts and by branch vessels in the rest of the worm.

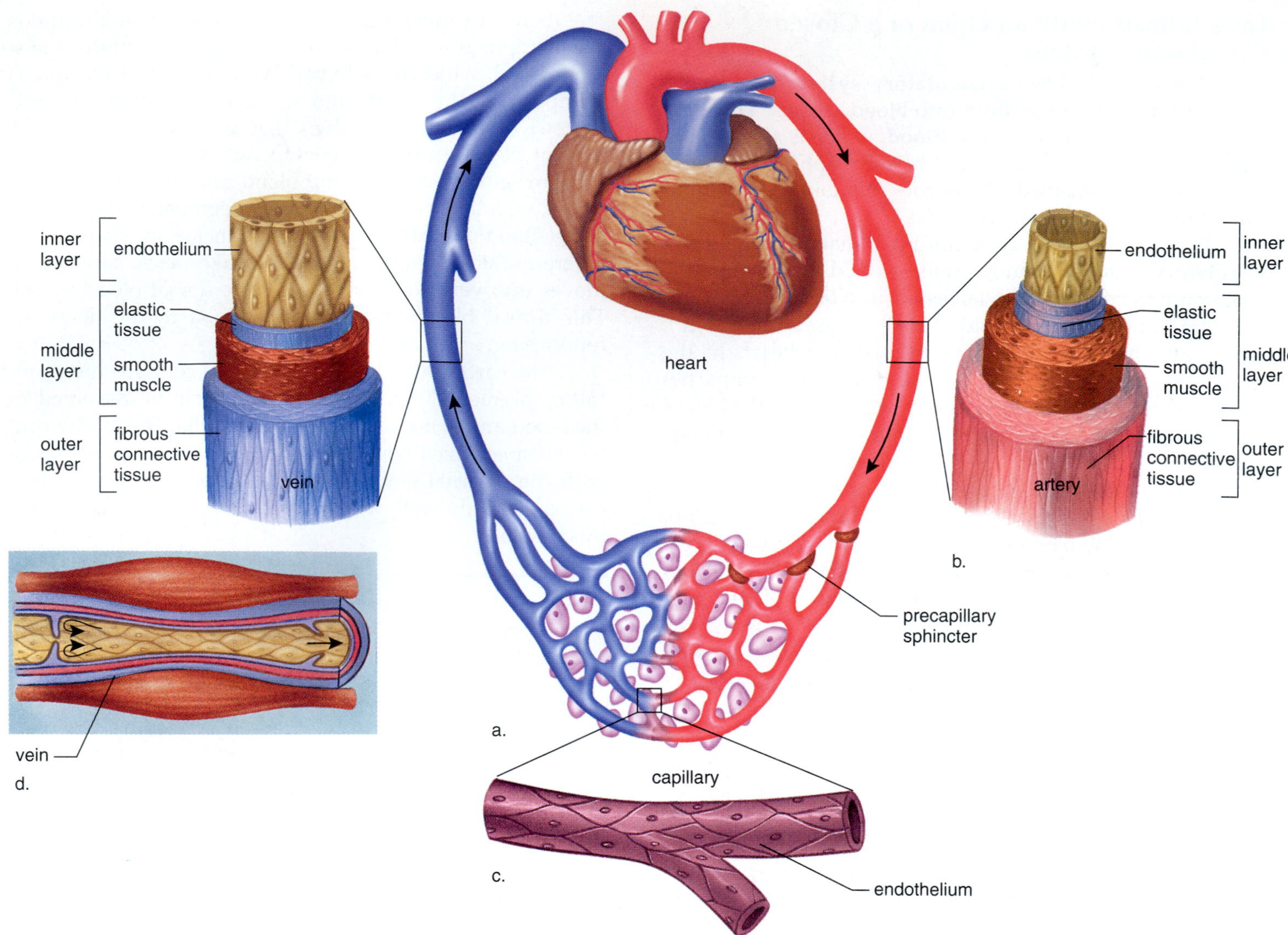

Figure 41.3 **Transport in birds and mammals.**
a. Blood leaving the heart moves from an artery to arterioles to capillaries to venules and then returns to the heart by way of a vein. **b.** Arteries have well-developed walls with a thick middle layer of elastic tissue and smooth muscle. **c.** Capillary walls are only one cell thick. **d.** Veins have flabby walls, particularly because the middle layer is not as thick as in arteries. Veins have valves, which point toward the heart.

41.2 Transport in Vertebrates

All vertebrate animals have a closed circulatory system, which is called a **cardiovascular system** [Gk. *kardia*, heart, L. *vascular*, vessel]. It consists of a strong, muscular heart, in which the atria (sing., atrium) receive blood and the muscular ventricles pump blood out through the blood vessels. There are three kinds of blood vessels: **arteries,** which carry blood away from the heart; **capillaries** [L. *capillus*, hair], which exchange materials with tissue fluid; and **veins** [L. *vena*, blood vessel], which return blood to the heart (Fig. 41.3).

Arteries have thick walls, and those attached to the heart are resilient, meaning that they are able to expand and accommodate the sudden increase in blood volume that results after each heartbeat. **Arterioles** are small arteries whose diameter can be regulated by the nervous system. Arteriole constriction and dilation affect blood pressure in general. The greater the number of vessels dilated, the lower the blood pressure.

Arterioles branch into capillaries, which are extremely narrow, microscopic tubes with a wall composed of only one layer of cells. Capillary beds (many capillaries interconnected) are so prevalent that in humans, all cells are within 60–80 μm of a capillary. But only about 5% of the capillary beds are open at the same time. After an animal has eaten, precapillary sphincters relax and the capillary beds in the digestive tract are usually open. During muscular exercise, the capillary beds of the muscles are open. Capillaries, which are usually so narrow that red blood cells pass through in single file, allow exchange of nutrient and waste molecules across their thin walls.

Venules and veins collect blood from the capillary beds and take it to the heart. First the venules drain the blood from the capillaries, and then they join to form a vein. The wall of a vein is much thinner than that of an artery and this may be associated with a lower blood pressure in the veins. Valves within the veins point, or open, toward the heart, preventing a backflow of blood when they close (Fig. 41.3*d*).

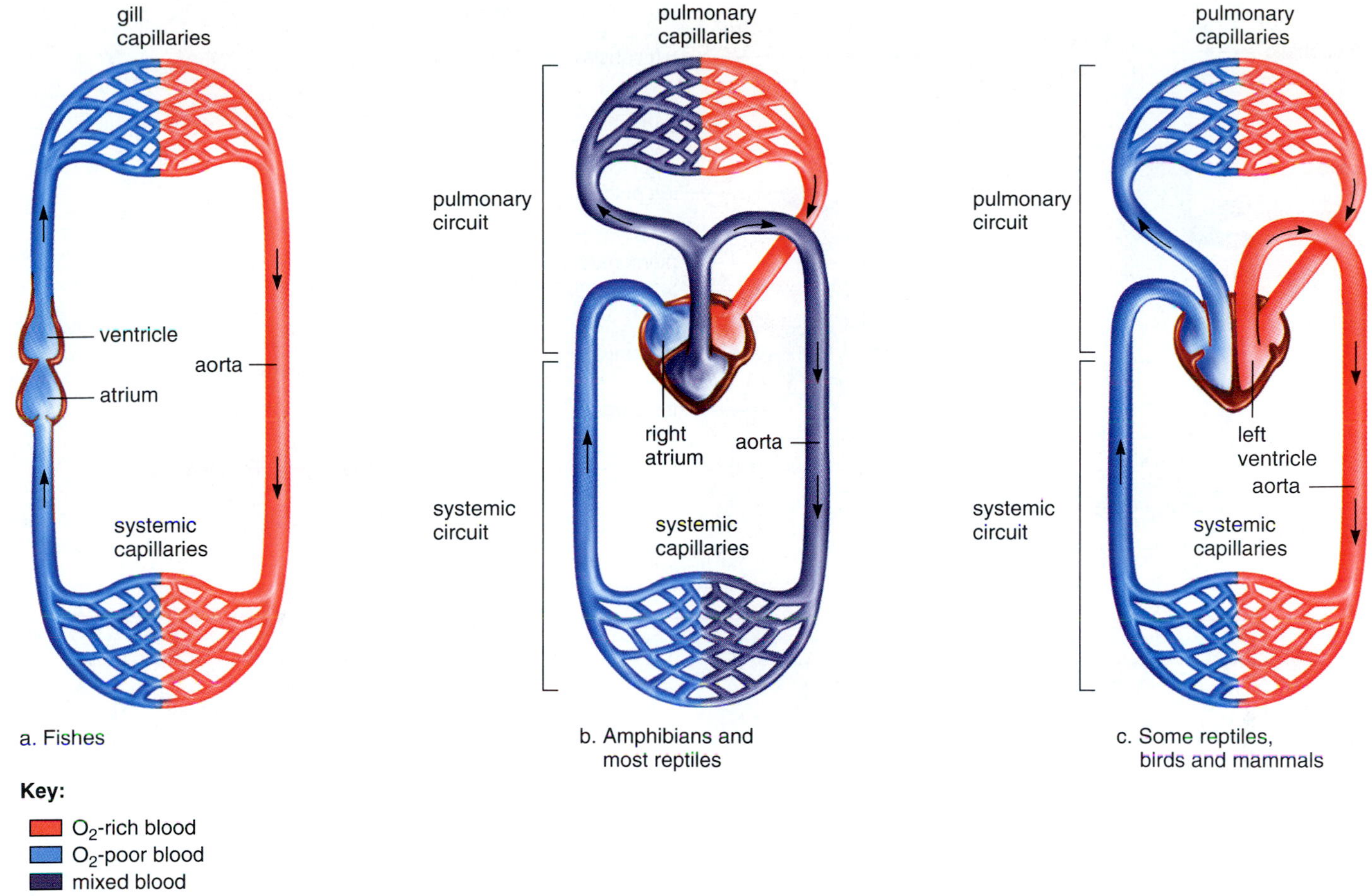

Figure 41.4 Comparison of circulatory circuits in vertebrates.
a. In a fish, the blood moves in a single circuit. The heart has a single atrium and ventricle and pumps the blood into the gill region, where gas exchange takes place. Blood pressure created by the pumping of the heart is dissipated after the blood passes through the gill capillaries. This is a disadvantage of this one-circuit system. **b.** Amphibians and most reptiles have a two-circuit system in which the heart pumps blood to both the lungs and the body itself. Although there is a single ventricle, there is little mixing of O_2-rich and O_2-poor blood. **c.** The pulmonary and systemic circuits are completely separate in crocodiles (a reptile) and in birds and mammals, because the heart is divided by a septum into a right and left half. The right side pumps blood to the lungs, and the left side pumps blood to the body proper.

Comparison of Circulatory Pathways

Among vertebrate animals, there are three different types of circulatory pathways. In fishes, blood follows a one-circuit (single-loop circulatory) pathway through the body. The heart has a single atrium and a single ventricle (Fig. 41.4*a*). The pumping action of the ventricle sends blood under pressure to the gills, where gas exchange occurs. After passing through the gills, blood is under reduced pressure and flow. However, this single circulatory loop has advantages in that the gill capillaries receive O_2-poor blood and the systemic capillaries receive fully O_2-rich blood.

As a result of evolutionary changes, the other vertebrates have a two-circuit (double-loop circulatory) pathway. The heart pumps blood to the tissues, called a systemic circuit, and also pumps blood to the lungs, called a pulmonary circuit. This double pumping action is an adaptation to breathing air on land.

In amphibians, the heart has two atria, but there is only a single ventricle (Fig. 41.4*b*). The same holds true for most reptiles, except that the ventricle has a partial septum. The hearts of crocodiles, which are reptiles, and all birds and mammals are divided into right and left halves (Fig. 41.4*c*). The right ventricle pumps blood to the lungs, and the left ventricle, which is larger than the right ventricle, pumps blood to the rest of the body. This arrangement provides adequate blood pressure for both the pulmonary and systemic circuits.

> In mammals, birds, and crocodiles, the heart is divided into a right side and a left side; this ensures adequate blood pressure for both the pulmonary and systemic circuits.

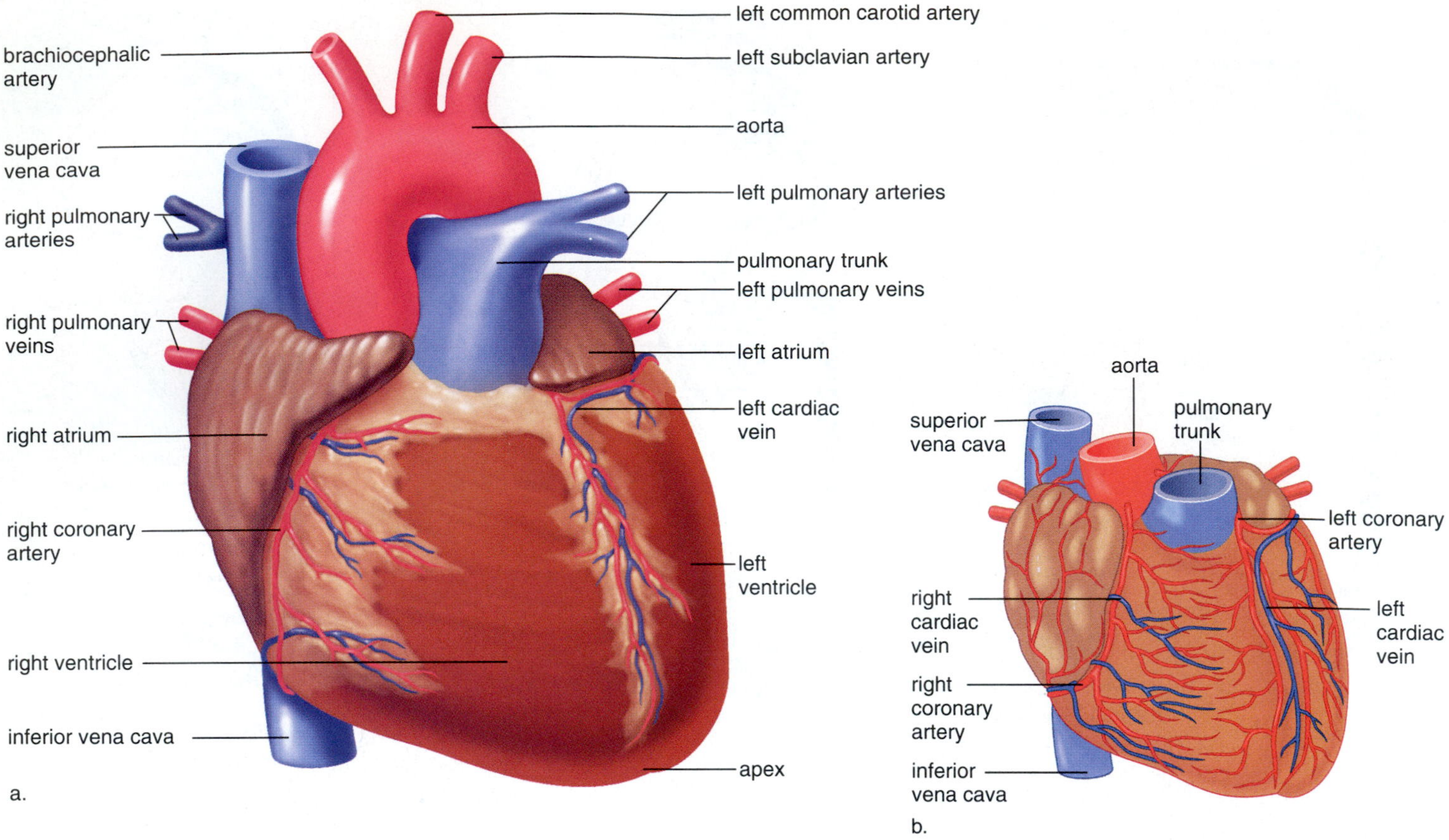

brachiocephalic artery
superior vena cava
right pulmonary arteries
right pulmonary veins
right atrium
right coronary artery
right ventricle
inferior vena cava
left common carotid artery
left subclavian artery
aorta
left pulmonary arteries
pulmonary trunk
left pulmonary veins
left atrium
left cardiac vein
left ventricle
apex
a.

aorta
superior vena cava
pulmonary trunk
left coronary artery
right cardiac vein
left cardiac vein
right coronary artery
inferior vena cava
b.

Figure 41.5 **External heart anatomy.**
a. The venae cavae and the pulmonary trunk are attached to the right side of the heart. The aorta and the pulmonary veins are attached to the left side of the heart. **b.** The coronary arteries and cardiac veins pervade cardiac muscle. They bring oxygen and nutrients to cardiac cells, which derive no benefit from blood coursing through the heart.

41.3 Transport in Humans

In the cardiovascular system of humans, the pumping of the heart keeps blood moving in the arteries. Skeletal muscle contraction pressing against veins is primarily responsible for the movement of blood in the veins.

The Human Heart

The **heart** is a cone-shaped, muscular organ about the size of a fist (Fig. 41.5). It is located between the lungs directly behind the sternum (breastbone) and is tilted so that the apex (the pointed end) is oriented to the left. The major portion of the heart, called the myocardium, consists largely of cardiac muscle tissue. The muscle fibers of the myocardium are branched and tightly joined to one another. The heart lies within the pericardium, a thick, membranous sac that secretes a small quantity of lubricating liquid. The inner surface of the heart is lined with endocardium, a membrane composed of connective tissue and endothelial tissue.

Internally, a wall called the **septum** separates the heart into a right side and a left side (Fig. 41.6). The heart has four chambers. The two upper, thin-walled atria (sing., **atrium**) have wrinkled protruding appendages called auricles. The two lower chambers are the thick-walled **ventricles,** which pump the blood.

The heart also has four valves, which direct the flow of blood and prevent its backward movement. The two valves that lie between the atria and the ventricles are called the **atrioventricular valves.** These valves are supported by strong fibrous strings called chordae tendineae. The chordae, which are attached to muscular projections of the ventricular walls, support the valves and prevent them from inverting when the heart contracts. The atrioventricular valve on the right side is called the tricuspid valve because it has three flaps, or cusps. The valve on the left side is called the bicuspid (or the mitral) because it has two flaps. The remaining two valves are the **semilunar valves,** whose flaps resemble half-moons, between the ventricles and their attached vessels. The pulmonary semilunar valve lies between the right ventricle and the pulmonary trunk. The aortic semilunar valve lies between the left ventricle and the aorta.

Humans have a four-chambered heart (two atria and two ventricles). A septum separates the right side from the left side.

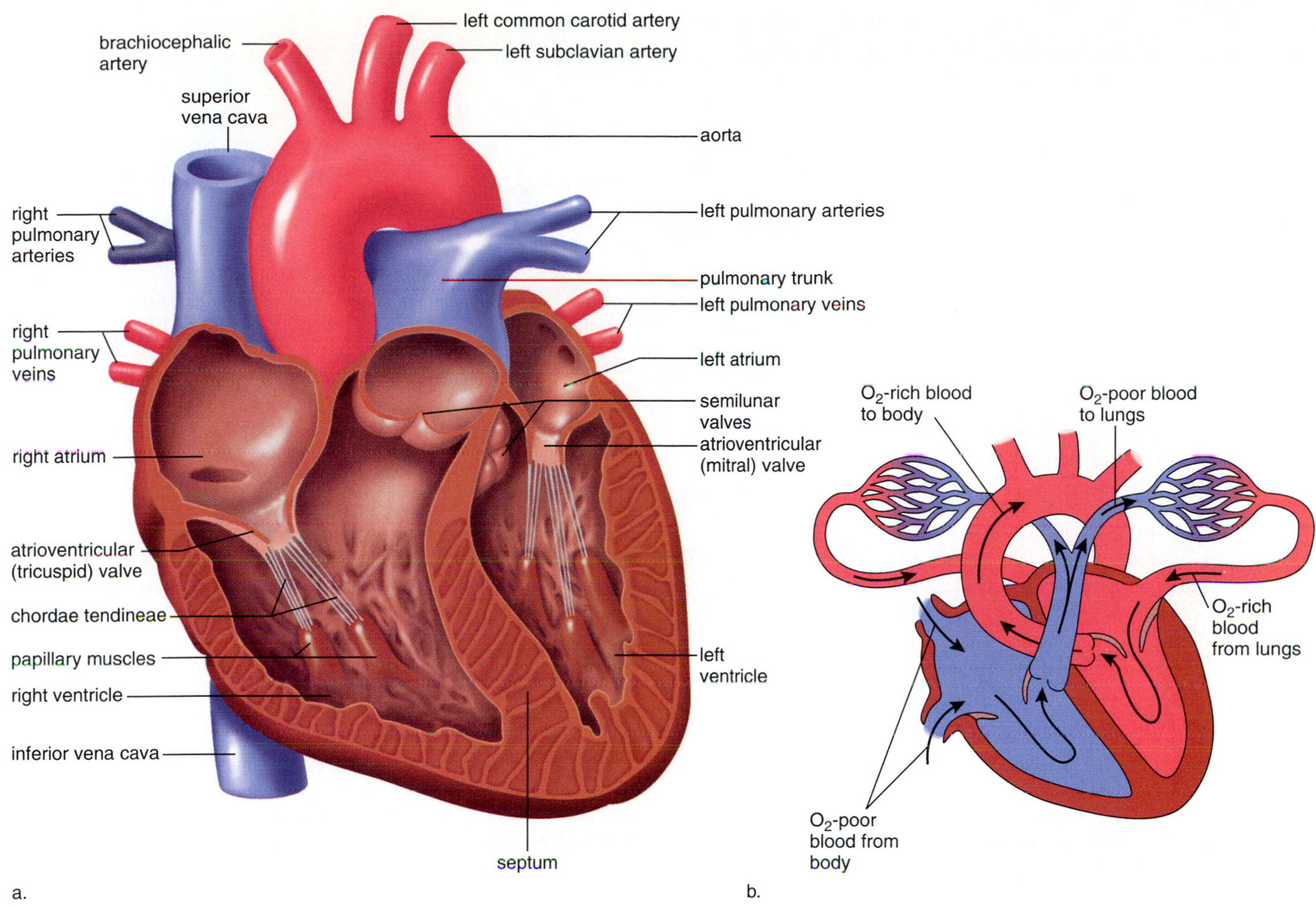

Figure 41.6 **Internal view of the heart.**
a. The heart has four valves. The atrioventricular valves allow blood to pass from the atria to the ventricles, and the semilunar valves allow blood to pass out of the heart. **b.** This diagrammatic representation of the heart allows you to trace the path of the blood through the heart.

Path of Blood Through the Heart

We can trace the path of blood through the heart (Fig. 41.6) in the following manner:

The superior vena cava and the inferior **vena cava,** which carry O_2-poor blood that is relatively high in carbon dioxide, enter the right atrium.

The right atrium sends blood through an atrioventricular valve (the tricuspid valve) to the right ventricle.

The right ventricle sends blood through the pulmonary semilunar valve into the pulmonary trunk and the two pulmonary arteries to the lungs.

Four pulmonary veins, which carry O_2-rich blood that is relatively low in carbon dioxide, enter the left atrium.

The left atrium sends blood through an atrioventricular valve (the bicuspid or mitral valve) to the left ventricle.

The left ventricle sends blood through the aortic semilunar valve into the aorta to the body proper.

From this description, you can see that O_2-poor blood never mixes with O_2-rich blood and that blood must go through the lungs in order to pass from the right side to the left side of the heart. In fact, the heart is a double pump because the right ventricle of the heart sends blood into the pulmonary circuit, and the left ventricle sends blood into the systemic circuit. Since the left ventricle has the harder job of pumping blood to the entire body, its walls are thicker than those of the right ventricle, which pumps blood a relatively short distance to the lungs.

The right side of the heart pumps blood into the pulmonary circuit, and the left side of the heart pumps blood into the systemic circuit.

William Harvey and Circulation of the Blood

William Harvey (1578–1657) was the first to offer data that the blood circulates in the body of humans and other animals. Harvey was an English scientist of the seventeenth century, a time of renewed interest in the collection of facts, the use of the hypothesis, the use of experimentation, and the respect for mathematics. This time period is known as the scientific revolution.

After many years of research and study, Harvey hypothesized that the heart is a pump for the entire circulatory system and that blood flows in a circuit. In contrast to former anatomists, Harvey dissected not only dead but also live organisms and observed that when the heart beats, it contracts, forcing blood into the aorta. Had this blood come from the right side of the heart? To do away with the complication of the lungs (pulmonary circulation), Harvey turned to fishes and noticed that the heart first receives and then pumps the blood forward. He observed that blood in the mammalian fetus passes directly from the right side of the heart through the septum to the left side. He felt confident that in mature higher organisms, all blood moves from the right to the left side of the heart by way of the lungs.

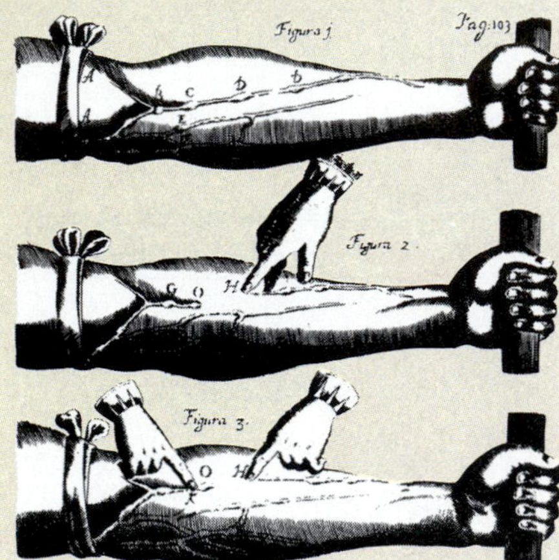

Figure 41A Harvey's drawings showing how he concluded that the veins return blood to the heart. Harvey tied a ligature (figure 1) above the elbow to observe the accumulation of blood in the veins. Blood can be forced past a valve from H to O (figure 2), but not in the opposite direction (figure 3). Therefore, it can be deduced that venous blood ordinarily moves toward the heart.

Harvey then wanted to show an intimate connection between the arteries and veins in the tissues of the body. Again using live organisms, he demonstrated that if an artery is slit, the whole blood system empties, including arteries and veins. He measured the capacity of the left ventricle in humans and found it to be 2 oz. Since the heart beats 70 times a minute, in one hour the left ventricle forces into the aorta no less than $70 \times 60 \times 2 = 8{,}400$ oz = 525 lb, or about three times a man's weight! Could so much blood be created and consumed every hour? The same blood must return again and again to the heart.

Harvey also studied the valves in the veins and suggested their true purpose. By the use of ligatures, he demonstrated that a tight ligature on the arm causes the artery to swell on the side of the heart, and a slack ligature causes the vein to swell on the opposite side (Fig. 41A). He said, "This is an obvious indication the blood passes from the arteries into the veins . . . and there is an anastomosis of the two orders of vessels."

Harvey's methods showed how fruitful research might be done. He established that physical and mechanical evidence could provide data for a theory of circulation. He erred, however, when he speculated that the heart heated the blood, and the lung served to cool it or control the degree of heat. His basic method, however, contributed to the scientific revolution and set an example for others to follow.

The Heartbeat

The heart contracts, or beats, about seventy times a minute, and each heartbeat lasts about 0.85 second. The term **systole** [Gk. *systole,* contraction] refers to contraction of heart chambers, and the word **diastole** [Gk. *diastole,* dilation, spreading] refers to relaxation of these chambers. Each heartbeat, or **cardiac cycle,** consists of the following phases:

Cardiac Cycle		
Time	**Atria**	**Ventricles**
0.15 sec	Systole	Diastole
0.30 sec	Diastole	Systole
0.40 sec	Diastole	Diastole

First the atria contract (while the ventricles relax), then the ventricles contract (while the atria relax), and then all chambers rest. Note that the heart is in diastole about 50% of the time. The short systole of the atria is appropriate since the atria send blood only into the ventricles. It is the muscular ventricles that actually pump blood out into the circulatory system proper. When the word systole is used alone, it usually refers to the left ventricular systole.

When the heart beats, the familiar lub-dub sound is heard as the valves of the heart close. The longer and lower-pitched *lub* is caused by vibrations of the heart when the atrioventricular valves close due to ventricular contraction. The shorter and sharper *dub* is heard when the semilunar valves close due to back pressure of blood in the arteries. A heart murmur, or a slight slush sound after the lub, is often due to ineffective valves, which allow blood to pass back into the atria after the atrioventricular valves have closed.

The **pulse** is a wave effect that passes down the walls of the arterial blood vessels when the aorta expands and then recoils following ventricle systole. Because there is one arterial pulse per ventricular systole, the arterial pulse rate can be used to determine the heart rate.

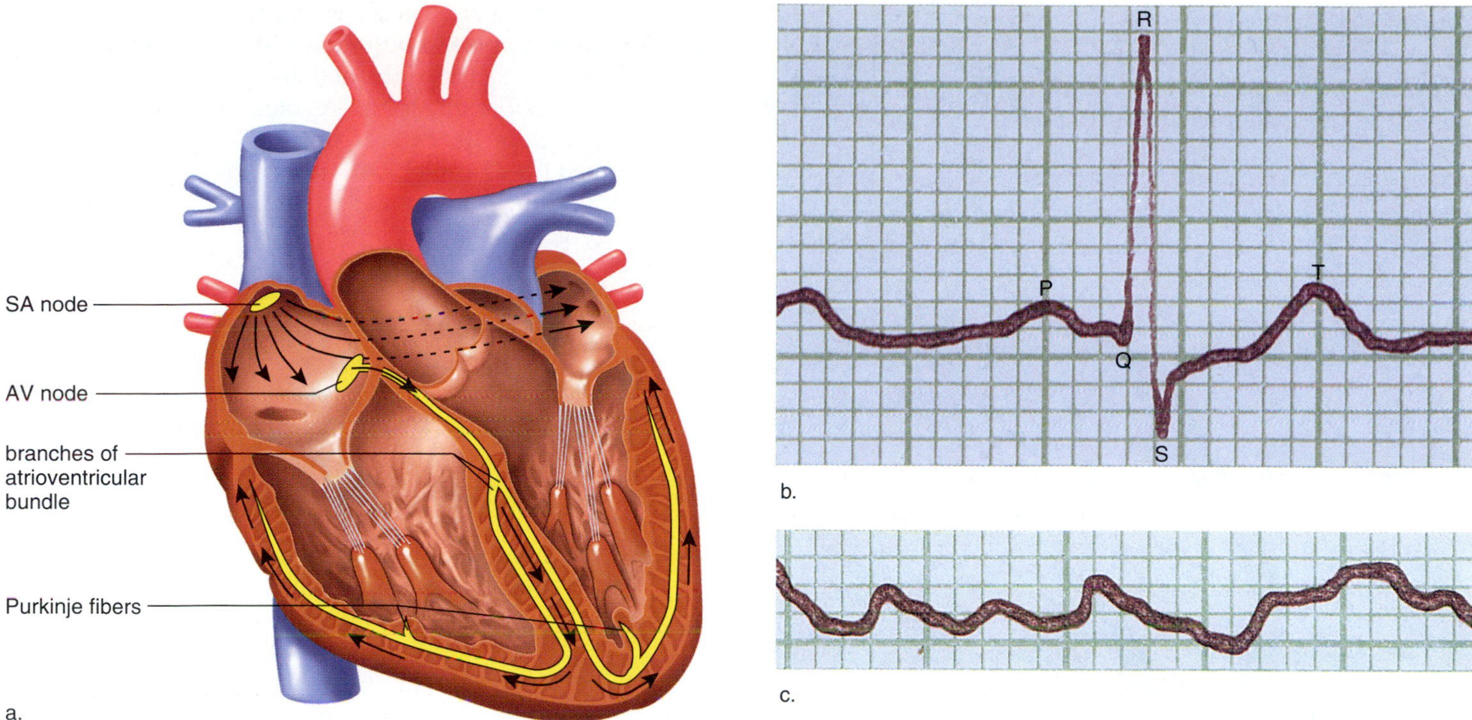

Figure 41.7 **Cardiac conduction system.**
a. The SA (sinoatrial) node sends out a stimulus (black arrows), which causes the atria to contract. When this stimulus (black arrows) reaches the AV (atrioventricular) node, it signals the ventricles to contract. Impulses pass down the two branches of the atrioventricular bundle to the Purkinje fibers and thereafter the ventricles contract. **b.** A normal electrocardiogram (ECG) indicates that the heart is functioning properly. The P wave occurs just prior to atrial contraction; the QRS complex occurs just prior to ventricular contraction and the T wave occurs when the ventricles are recovering from contraction. **c.** Ventricular fibrillation produces an irregular electrocardiogram due to irregular stimulation of the ventricles.

The rhythmical contraction of the heart is due to the **cardiac conduction system.** Nodal tissue, which has both muscular and nervous characteristics, is a unique type of cardiac muscle located in two regions of the heart. The SA (sinoatrial) node, is found in the upper dorsal wall of the right atrium; the other, the AV (atrioventricular) node, is found in the base of the right atrium very near the septum (Fig. 41.7*a*). The SA node initiates the heartbeat and every 0.85 second automatically sends out an excitation impulse, which causes the atria to contract. Therefore the SA node is called the **cardiac pacemaker** because it usually keeps the heartbeat regular. When the impulse reaches the AV node, the AV node signals the ventricles to contract by way of large fibers terminating in the more numerous and smaller Purkinje fibers. Although the beat of the heart is intrinsic, it is regulated by the nervous system, which can increase or decrease the heartbeat rate.

An **electrocardiogram (ECG)** is a recording of the electrical changes that occur in myocardium during a cardiac cycle. Body fluids contain ions that conduct electrical currents, and therefore the electrical changes in myocardium can be detected on the skin's surface. When an electrocardiogram is being taken, electrodes placed on the skin are connected by wires to an instrument that detects the myocardium's electrical changes. Thereafter a pen rises or falls on a moving strip of paper. Figure 41.7*b* depicts the pen's movements during a normal cardiac cycle.

When the SA node triggers an impulse, the atrial fibers produce an electrical change that is called the P wave. The P wave indicates that the atria are about to contract. After that, the QRS complex signals that the ventricles are about to contract. The electrical changes that occur as the ventricular muscle fibers recover produces the T wave.

Various types of abnormalities can be detected by an electrocardiogram. One of these, called ventricular fibrillation, is caused by uncoordinated contraction of the ventricles (Fig. 41.7*c*). Ventricular fibrillation is of special interest because it can be caused by an injury or drug overdose. It is the most common cause of sudden cardiac death in a seemingly healthy person. Once the ventricles are fibrillating, they have to be defibrillated by applying a strong electric current for a short period of time. Then the SA node may be able to reestablish a coordinated beat.

The conduction system of the heart includes the SA node, the AV node, and the Purkinje fibers. An ECG helps to determine if the conduction system, and therefore the heartbeat, is regular.

Vascular Pathways

The human cardiovascular system includes two major circular pathways, the **pulmonary circuit** [L. *pulmonarius*, of the lungs] and the **systemic circuit** (Fig. 41.8).

The Pulmonary Circuit

In the pulmonary circuit, the path of blood can be traced as follows. O_2-poor blood from all regions of the body collects in the right atrium and then passes into the right ventricle, which pumps it into the pulmonary trunk. The pulmonary trunk divides into the right and left pulmonary arteries, which carry blood to the lungs. As blood passes through the pulmonary capillaries, carbon dioxide is given off and oxygen is picked up. O_2-rich blood returns to the left atrium of the heart, through pulmonary venules that join to form pulmonary veins.

The Systemic Circuit

The **aorta** [L. *aorte*, great artery] and the **venae cavae** (sing., vena cava) [L. *vena*, blood vessel, and *cavus*, hollow] are the major blood vessels in the systemic circuit. To trace the path of blood to any organ in the body, you need only mention the aorta, the proper branch of the aorta, the organ, and the vein returning blood to the vena cava. In the systemic circuit, arteries contain O_2-rich blood and have a bright red color, but veins contain O_2-poor blood and appear dull red or, when viewed through the skin, blue.

The coronary arteries are extremely important because they serve the heart muscle itself (see Fig. 41.5). (The heart is not nourished by the blood in its chambers.) The coronary arteries arise from the aorta just above the aortic semilunar valve. They lie on the exterior surface of the heart, where they branch into arterioles and then capillaries. The capillary beds enter venules, which join to form the cardiac veins, and these empty into the right atrium.

A **portal system** [L. *porto*, carry, transport] is one that begins and ends in capillaries. The hepatic portal system takes blood from the intestines to the liver. The liver, an organ of homeostasis, modifies substances absorbed by the intestines and monitors the normal composition of the blood. The hepatic vein leaves the liver and enters the inferior vena cava.

The pulmonary circuit takes O_2-poor blood to the lungs and returns O_2-rich blood to the heart. The systemic circuit takes blood throughout the body from the aorta to the venae cavae.

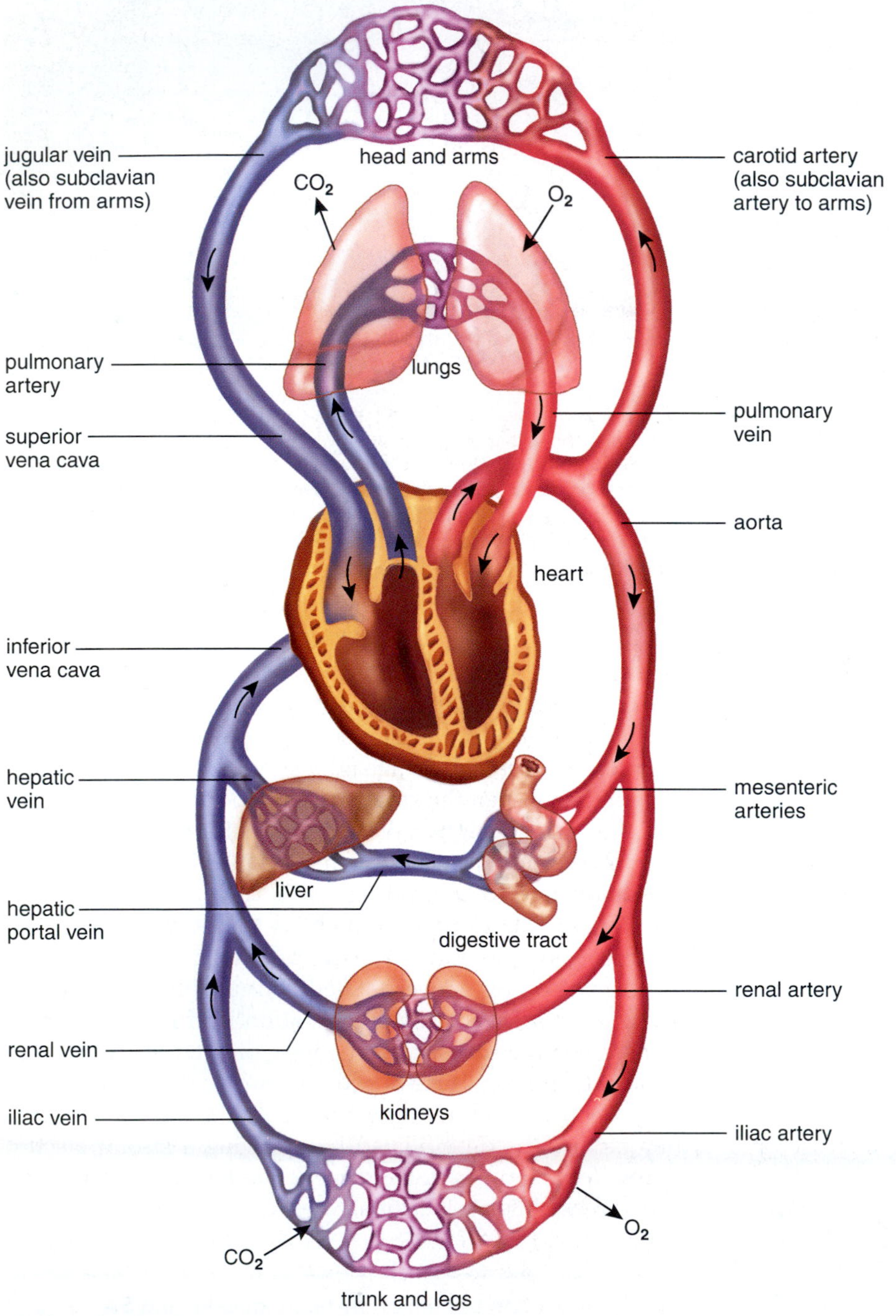

Figure 41.8 **Path of blood.**
When tracing blood from the right to the left side of the heart, you must mention the pulmonary vessels. When tracing blood from the digestive tract to the right atrium, you must mention the hepatic portal vein, the hepatic vein, and the inferior vena cava. The blue-colored vessels carry O_2-poor blood, and the red-colored vessels carry O_2-rich blood; the arrows indicate the flow of blood.

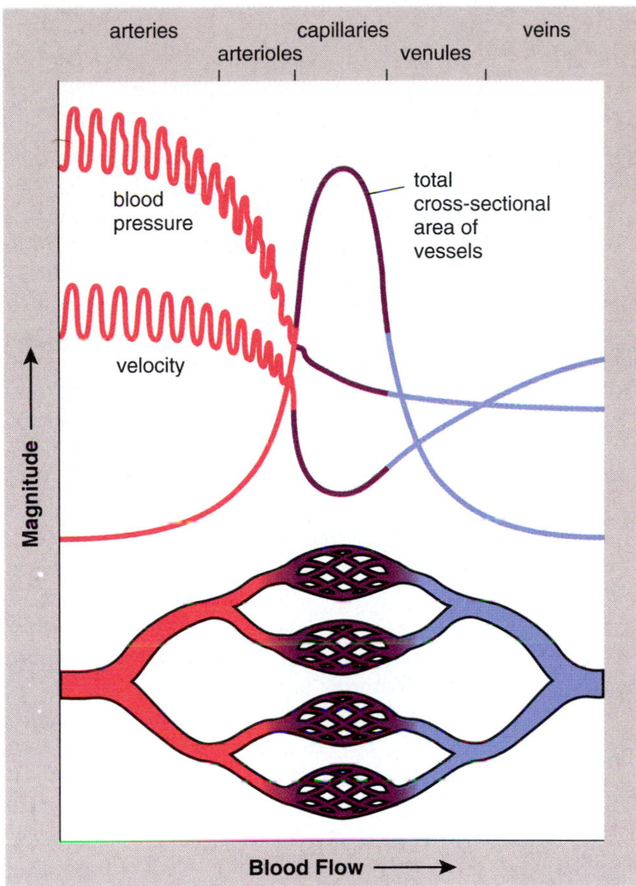

Figure 41.9 Velocity and blood pressure related to vascular cross-sectional area.
In capillaries, blood is under minimal pressure and has the least velocity. Blood pressure and velocity drop off because capillaries have a greater cross-sectional area than arterioles.

Blood Pressure

When the left ventricle contracts, blood is forced into the systemic arteries under pressure. Systolic pressure results from blood being forced into the arteries during ventricular systole, and diastolic pressure is the pressure in the arteries during ventricular diastole. Human **blood pressure** is measured with a sphygmomanometer, which has a pressure cuff that measures the amount of pressure required to stop the flow of blood through an artery. Blood pressure which is normally measured on the brachial artery, an artery in the upper arm, is stated in millimeters of mercury (mm Hg). A blood pressure reading consists of two numbers, for example, 120/80—which represents systolic and diastolic pressures, respectively.

As blood flows from the aorta into the various arteries and arterioles, blood pressure falls. Also, the difference between systolic and diastolic pressure gradually diminishes. In the capillaries, there is a slow, fairly even, flow of blood.

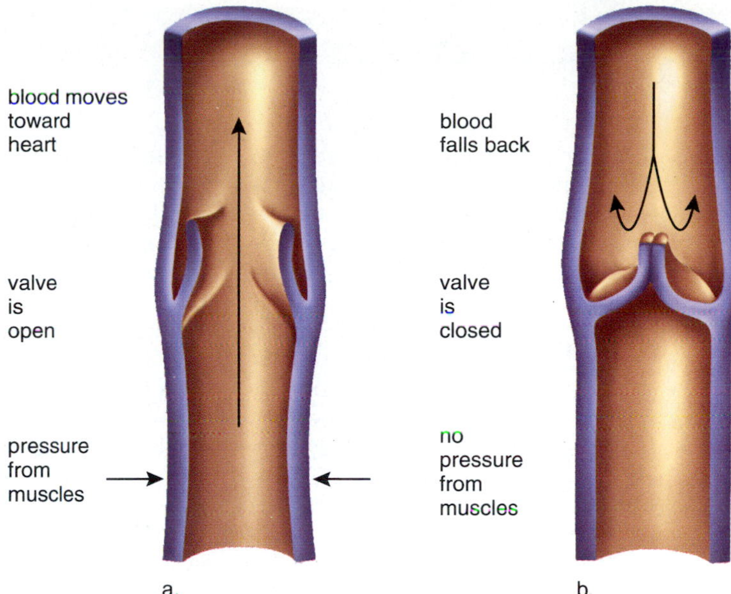

Figure 41.10 Cross section of a valve in a vein.
a. Pressure on the walls of a vein, exerted by skeletal muscles, increases blood pressure within the vein and forces a valve open. **b.** When external pressure is no longer applied to the vein, blood pressure decreases, and back pressure forces the valve closed. Closure of the valves prevents the blood from flowing in the opposite direction.

This may be related to the very high total cross-sectional area of the capillaries (Fig. 41.9). It's been calculated that if all the blood vessels in a human were connected end to end, the total distance would reach around the earth two times at the equator. A large portion of this distance would be due to the quantity of capillaries.

Blood pressure in the veins is low and cannot move blood back to the heart, especially from the limbs of the body. When skeletal muscles near veins contract, they put pressure on the collapsible walls of the veins and on blood contained in these vessels. Veins, however, have valves that prevent the backward flow of blood, and therefore pressure from muscle contraction is sufficient to move blood through veins toward the heart (Fig. 41.10). Varicose veins, abnormal dilations in superficial veins, develop when the valves of the veins become weak and ineffective due to a backward pressure of the blood.

The beat of the heart supplies the pressure that keeps blood moving in the arteries, and skeletal muscle contraction pushes blood in the veins toward the heart. Veins have valves which prevent the backward flow of blood.

Prevention of Cardiovascular Disease

All of us can take steps to prevent the occurrence of cardiovascular disease, the most frequent cause of death in the United States. There are genetic factors that predispose an individual to cardiovascular disease, such as family history of heart attack under age 55, male gender, and ethnicity (African Americans are at greater risk). Those with one or more of these risk factors need not despair, however. It means only that they should pay particular attention to these guidelines for a heart-healthy lifestyle.

The Don'ts

Smoking

Hypertension is well recognized as a major contributor to cardiovascular disease. When a person smokes, the drug nicotine, present in cigarette smoke, enters the bloodstream. Nicotine causes the arterioles to constrict and the blood pressure to rise. Restricted blood flow and cold hands are associated with smoking by most people. More serious is the need for the heart to pump harder to propel the blood through the lungs at a time when the oxygen-carrying capacity of the blood is reduced.

Drug Abuse

Stimulants, such as cocaine and amphetamines, can cause an irregular heartbeat and lead to heart attacks and strokes in people who are using drugs even for the first time. Intravenous drug use may result in a cerebral embolism.

Too much alcohol can destroy just about every organ in the body, the heart included. But investigators have discovered that people who take an occasional drink have a 20% lower risk of heart disease than do teetotalers. Two to four drinks a week is the recommended limit for men, one to three drinks for women.

Weight Gain

Hypertension is prevalent in persons who are more than 20% above recommended weight for their height. More tissues require servicing, and the heart sends the extra blood out under greater pressure in those who are overweight. It may be harder to lose weight once it is gained, and therefore it is recommended that weight control be a lifelong endeavor. Even a slight decrease in weight can bring with it a reduction in hypertension. A 4.5-kilogram weight loss doubles the chance that blood pressure can be normalized without drugs.

The Do's

Healthy Diet

It was once thought that a low-salt diet was protective against cardiovascular disease, and it still may be in certain persons. Theoretically, hypertension occurs because the saltier the blood, the greater the osmotic pressure and the higher the water content.

In recent years, the emphasis has switched to a diet low in saturated fats and cholesterol as protective against cardiovascular disease. Cholesterol is ferried in the blood by two types of plasma proteins called LDL (low-density lipoprotein) and HDL (high-density lipoprotein). LDL (called "bad" lipoprotein) takes cholesterol from the liver to the tissues, and HDL (called "good" lipoprotein) transports cholesterol out of the tissues to the liver. When the LDL level in blood is abnormally high or the HDL level is abnormally low, cholesterol accumulates in the cells. When cholesterol-laden cells line the arteries, plaque develops, which interferes with circulation (Fig. 41B).

It is recommended that everyone know his or her blood cholesterol level. Individuals with a high blood cholesterol level (240 mg/100 ml) should be further tested to determine their LDL-cholesterol level. The LDL-cholesterol level together with other risk factors such as age, family history, general health, and whether the patient smokes will determine who needs dietary therapy to lower their LDL. Eating foods high in saturated fat (red meat,

41.4 Cardiovascular Disorders

Cardiovascular disease (CVD) is the leading cause of untimely death in the Western countries. Modern research efforts have resulted in improved diagnosis, treatment, and prevention. This section discusses the range of advances that have been made in these areas. The reading on this page emphasizes how to prevent CVD from developing in the first place.

Hypertension

It is estimated that about 20% of all Americans suffer from hypertension, which is high blood pressure. Women of any age are considered to have hypertension if their blood pressure reading is 160/95 or above. For a man under age 45, a reading above 130/90 is hypertensive, and beyond age 45, a reading above 140/95 is considered hypertensive. While both systolic and diastolic pressures are considered important, it is the diastolic pressure that is emphasized when medical treatment is being considered.

Hypertension is sometimes called a silent killer because it may not be detected until a stroke or heart attack occurs. It has long been thought that a certain genetic makeup might account for the development of hypertension. Now researchers have discovered two genes that may be involved in some individuals. One gene codes for angiotensinogen, a plasma protein that is converted to a powerful vasoconstrictor in part by the product of the second gene. Persons with hypertension due to overactivity of these genes might one day be cured by gene therapy.

Atherosclerosis

Hypertension also is seen in individuals who have atherosclerosis (formerly called arteriosclerosis), an accumulation of soft masses of fatty materials, particularly cholesterol, beneath the inner linings of arteries (Fig. 41B). Such deposits cream, and butter) and foods in so-called trans-fats (margarine, commercially baked goods, and deep-fried foods) raises the LDL-cholesterol level. Replacement of these harmful fats with healthier ones like monounsaturated fats (olive and canola oil) and polyunsaturated ones (corn safflower, soybean oil) is recommended.

Evidence is mounting to suggest a role for antioxidant vitamins (A, E, and C) in the prevention of cardiovascular disease. Antioxidants protect the body from free radicals that may damage HDL cholesterol through oxidation or damage the lining of an artery, leading to a blood clot that can block the vessel. Nutritionists believe that the consumption of at least five servings of fruit and vegetables a day may be protective against cardiovascular disease.

Exercise

Those who exercise are less apt to have cardiovascular disease. One study found that moderately active men who spent an average of 48 minutes a day on a leisure-time activity such as gardening, bowling, or dancing had one-third fewer heart attacks than peers who spent an average of only 16 minutes each day. Exercise helps to keep weight under control, may help minimize stress, and reduces hypertension. The heart beats faster when exercising, but exercise slowly increases its capacity. This means that the heart can beat slower when we are at rest and still do the same amount of work. One physician recommends that his cardiovascular patients walk for one hour, three times a week, and in addition they are to practice meditation and yoga-like stretching and breathing exercises to reduce stress.

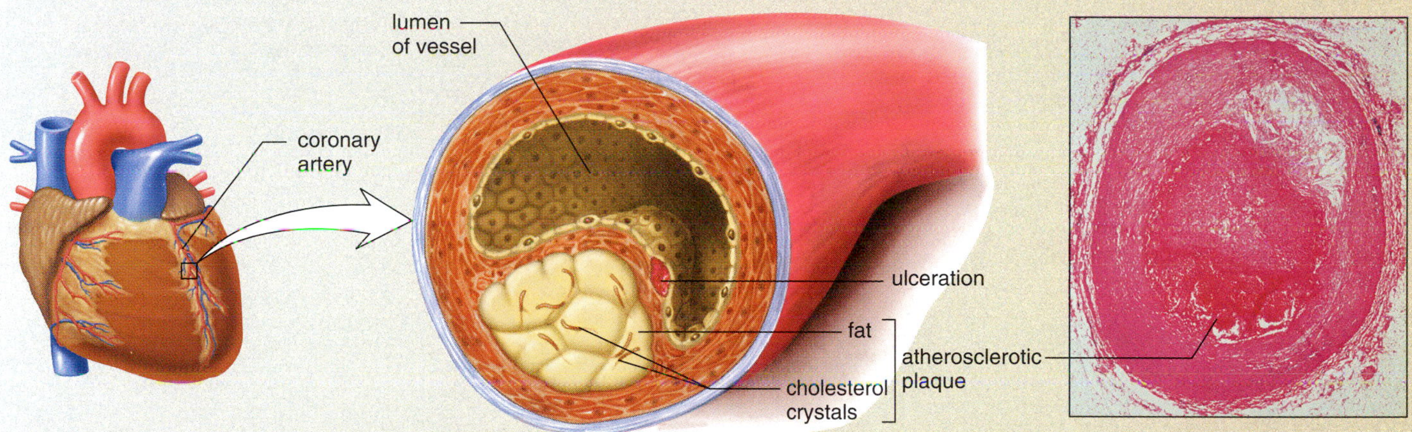

Figure 41B Coronary arteries and plaque.
Atherosclerotic plaque is an irregular accumulation of cholesterol and fat. When plaque (yellow) is present in a coronary artery, a heart attack is more likely to occur because of restricted blood flow.

are called plaque. As deposits occur, plaque tends to protrude into the lumen of the vessel, interfering with the flow of blood. Atherosclerosis begins in early adulthood and develops progressively through middle age, but symptoms may not appear until an individual is 50 or older. To prevent its onset and development, the American Heart Association and other organizations recommend a diet low in saturated fat and cholesterol and rich in fruits and vegetables.

Plaque can cause a clot to form on the irregular arterial wall. As long as the clot remains stationary, it is called a thrombus, but when and if it dislodges and moves along with the blood, it is called an embolus. If thromboembolism is not treated, complications, which are mentioned in the following section, can arise.

In certain families, atherosclerosis is due to an inherited condition such as familial hypercholesterolemia. The presence of the disease-associated mutation can be detected, and this information is helpful if measures are taken to prevent the occurrence of the disease.

Stroke and Heart Attack

Stroke, heart attack, and aneurysm are associated with hypertension and atherosclerosis. A cardiovascular accident (CVA), also called a stroke, often results when a small cranial arteriole bursts or is blocked by an embolus. A lack of oxygen causes a portion of the brain to die, and paralysis or death can result. A person sometimes is forewarned of a stroke by a feeling of numbness in the hands or the face, difficulty in speaking, or temporary blindness in one eye.

A myocardial infarction (MI), also called a heart attack, occurs when a portion of the heart muscle dies due to a lack of oxygen. If a coronary artery becomes partially blocked, the individual may then suffer from angina pectoris, characterized by chest pains and/or a radiating pain in the left arm. Nitroglycerin or related drugs dilate blood vessels and help relieve the pain. When a coronary artery is completely blocked, perhaps because of thromboembolism, a heart attack occurs.

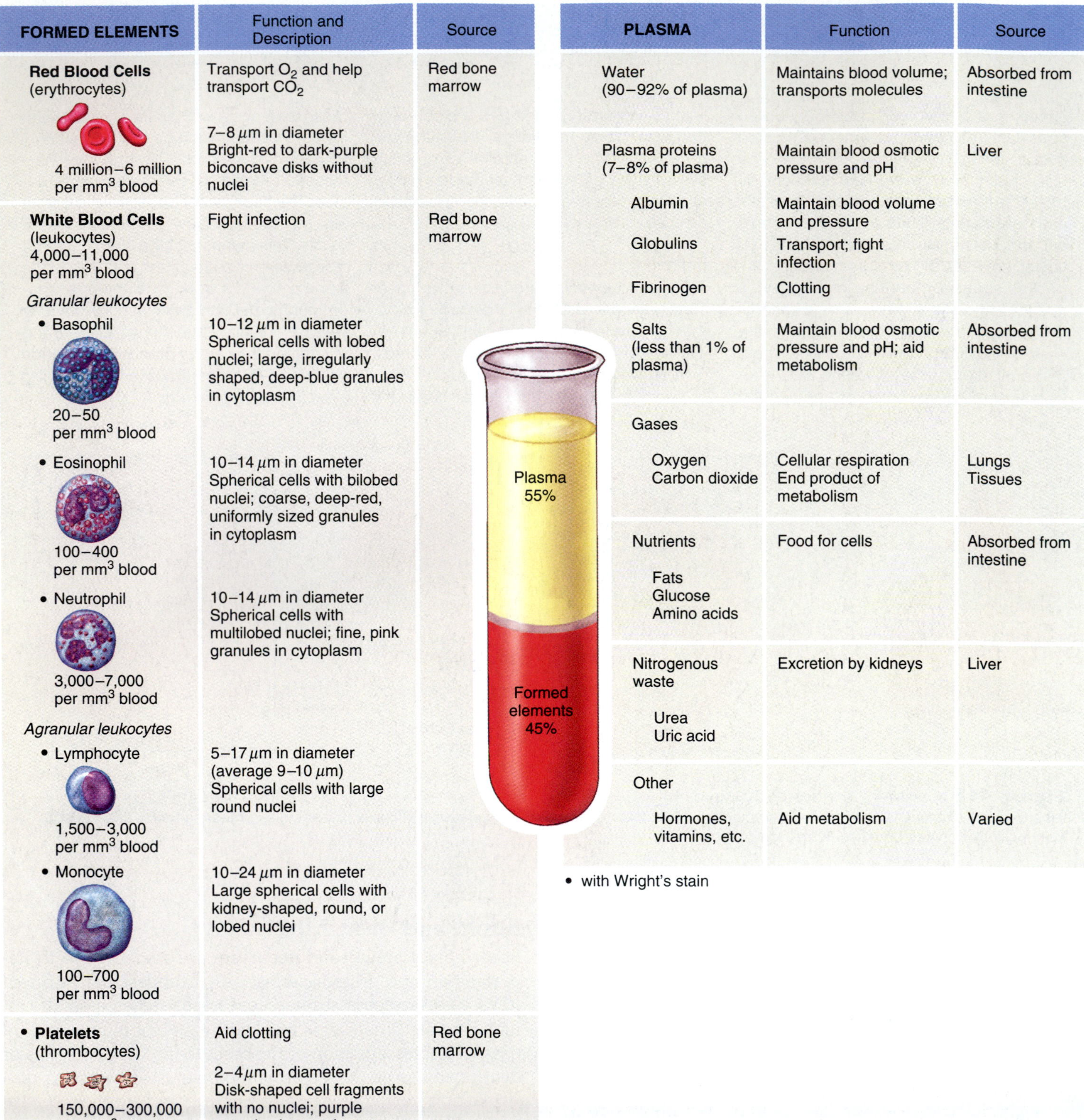

FORMED ELEMENTS	Function and Description	Source
Red Blood Cells (erythrocytes) 4 million–6 million per mm^3 blood	Transport O_2 and help transport CO_2 7–8 μm in diameter Bright-red to dark-purple biconcave disks without nuclei	Red bone marrow
White Blood Cells (leukocytes) 4,000–11,000 per mm^3 blood	Fight infection	Red bone marrow
Granular leukocytes		
• Basophil 20–50 per mm^3 blood	10–12 μm in diameter Spherical cells with lobed nuclei; large, irregularly shaped, deep-blue granules in cytoplasm	
• Eosinophil 100–400 per mm^3 blood	10–14 μm in diameter Spherical cells with bilobed nuclei; coarse, deep-red, uniformly sized granules in cytoplasm	
• Neutrophil 3,000–7,000 per mm^3 blood	10–14 μm in diameter Spherical cells with multilobed nuclei; fine, pink granules in cytoplasm	
Agranular leukocytes		
• Lymphocyte 1,500–3,000 per mm^3 blood	5–17 μm in diameter (average 9–10 μm) Spherical cells with large round nuclei	
• Monocyte 100–700 per mm^3 blood	10–24 μm in diameter Large spherical cells with kidney-shaped, round, or lobed nuclei	
• **Platelets** (thrombocytes) 150,000–300,000 per mm^3 blood	Aid clotting 2–4 μm in diameter Disk-shaped cell fragments with no nuclei; purple granules in cytoplasm	Red bone marrow

PLASMA	Function	Source
Water (90–92% of plasma)	Maintains blood volume; transports molecules	Absorbed from intestine
Plasma proteins (7–8% of plasma)	Maintain blood osmotic pressure and pH	Liver
Albumin	Maintain blood volume and pressure	
Globulins	Transport; fight infection	
Fibrinogen	Clotting	
Salts (less than 1% of plasma)	Maintain blood osmotic pressure and pH; aid metabolism	Absorbed from intestine
Gases		
Oxygen	Cellular respiration	Lungs
Carbon dioxide	End product of metabolism	Tissues
Nutrients Fats Glucose Amino acids	Food for cells	Absorbed from intestine
Nitrogenous waste Urea Uric acid	Excretion by kidneys	Liver
Other		
Hormones, vitamins, etc.	Aid metabolism	Varied

• with Wright's stain

Figure 41.11 Composition of blood.
When blood is transferred to a test tube and is prevented from clotting, it forms two layers. The transparent, yellow, top layer is plasma, the liquid portion of blood. The formed elements are in the bottom layer. The tables describe these components in detail.

41.5 Blood, a Transport Medium

The blood of mammals has numerous functions that help maintain homeostasis. Blood (1) transports substances to and from the capillaries, where exchanges with tissue fluid take place; (2) helps guard against invasion by pathogens, such as bacteria and viruses; (3) helps regulate body temperature; and (4) clots, preventing a potentially life-threatening loss of blood.

In humans, blood has two main portions: the liquid portion, called plasma, and the formed elements consisting of various cells and platelets (Fig. 41.11). **Plasma** [Gk. *plasma*, something molded] contains many types of molecules, including nutrients, wastes, salts, and proteins. The salts and proteins are involved in buffering the blood, effectively keeping the pH near 7.4. They also maintain blood's osmotic pressure so that water has an automatic tendency to enter blood capillaries. Several plasma proteins are involved in blood clotting and others transport large organic molecules in the blood. Albumin, the most plentiful of the plasma proteins transport bilirubin, a breakdown product of hemoglobin. Globulins have various functions and among them are the lipoproteins that transport cholesterol.

Formed Elements

The formed elements are of three types: red blood cells (RBCs), or erythrocytes [Gk. *erythros*, red, *kytos*, cell]; white blood cells (WBCs), or leukocytes [Gk. *leukos*, white, *kytos*, cell]; and platelets, or thrombocytes [Gk. *thrombos*, blood clot, *kytos*, cell].

Red Blood Cells

Red blood cells are small biconcave disks that at maturity lack a nucleus and contain the respiratory pigment hemoglobin. There are 6 million red blood cells per mm^3 of whole blood, and each one of these cells contains about 250 million hemoglobin molecules. **Hemoglobin** [Gk. *haima*, blood, and L. *globus*, ball] contains four globin protein chains, each associated with heme, an iron-containing group. Iron combines loosely with oxygen, and in this way oxygen is carried in the blood. If there is an insufficient number of red blood cells, or if the cells do not have enough hemoglobin, the individual suffers from anemia and has a tired, rundown feeling.

Red blood cells are manufactured continuously in the red bone marrow of the skull, the ribs, the vertebrae, and the ends of the long bones. The hormone erythropoietin is produced by the kidneys when it acts on a precursor made by the liver. Erythropoietin stimulates the production of red blood cells. Now available as a drug, erythropoietin is helpful to persons with anemia and is also sometimes abused by athletes who want to increase performance.

Before they are released from the bone marrow into blood, red blood cells lose their nucleus and synthesize hemoglobin. After living about 120 days, they are destroyed chiefly in the liver and the spleen, where they are engulfed by large phagocytic cells. When red blood cells are destroyed, hemoglobin is released. The iron is recovered and is returned to the red bone marrow for reuse. The heme portions of the molecules undergo chemical degradation and are excreted by the liver as bile pigments in the bile. The bile pigments are primarily responsible for the color of feces.

White Blood Cells

White blood cells differ from red blood cells in that they are usually larger and have a nucleus, they lack hemoglobin, and, without staining, they appear translucent. With staining, white blood cells appear light blue unless they have granules that bind with certain stains. The following are granular leukocytes with a lobed nucleus: neutrophils, which have granules that stain slightly pink; eosinophils, which have granules that take up the red dye eosin; and basophils, which have granules that take up a basic dye, staining them a deep blue. The agranular leukocytes with no granules and a circular or indented nucleus are the larger monocytes and the smaller lymphocytes. The newly discovered stem cell growth factor (SGF) can be used to increase the production of all white blood cells, and there are also various specific stimulating factors that can be used to stimulate the production of specific stem cells. These growth factors are helpful to patients with low immunity, such as AIDS patients.

When microorganisms enter the body due to an injury, the response is called an inflammatory reaction because there is swelling and reddening at the injured site. The damaged tissue has released kinins, which cause vasodilation, and histamines, which cause increased capillary permeability. **Neutrophils** [Gk. *neuter*, neither, and *phileo*, love], which are amoeboid, squeeze through the capillary wall and enter the tissue fluid, where they phagocytize foreign material. Monocytes appear and are transformed into **macrophages** [Gk. *makros*, long, and *phagein*, to eat], which are large phagocytizing cells that release white blood cell growth factors. Soon there is an explosive increase in the number of leukocytes. The thick, yellowish fluid called pus contains a larger proportion of dead white blood cells that have fought the infection.

Lymphocytes [L. *lympha*, clear water, and Gk. *kytos*, cell] also play an important role in fighting infection. Certain lymphocytes called T cells attack infected cells that contain viruses. Other lymphocytes called B cells produce antibodies. Each B cell produces just one type of antibody, which is specific for one type of antigen. An **antigen** [Gk. *anti*, against, and L. *genitus*, forming, causing], which is most often a protein but is sometimes a polysaccharide, causes the body to produce an antibody because the antigen doesn't belong to the body. Antigens are found in the outer covering of parasites or are present in their toxins. When **antibodies** [Gk. *anti*, against] combine with antigens, the complex is often phagocytized by a macrophage. An individual is actively immune when a large number of B cells are all producing the specific antibody needed for a particular infection.

Platelets

Platelets (thrombocytes) result from fragmentation of certain large cells, called megakaryocytes, in the red bone marrow. Platelets are produced at a rate of 200 billion a day, and the blood contains 150,000–300,000 per mm^3. These formed elements are involved in the process of blood clotting, or coagulation.

There are at least 12 clotting factors in the blood that participate in the formation of a blood clot. Hemophilia is an inherited clotting disorder in which the liver is unable to produce one of the clotting factors. The slightest bump can cause the affected person to bleed into the joints and this leads to degeneration of the joints. Bleeding into muscles can lead to nerve damage and muscular atrophy. The most frequent cause of death due to hemophilia is bleeding into the brain.

We will discuss the roles played by platelets, prothrombin, and fibrinogen in blood clotting. Fibrinogen and prothrombin are proteins manufactured and deposited in blood by the liver. Vitamin K, found in green vegetables and also formed by intestinal bacteria, is necessary for the production of prothrombin, and if by chance this vitamin is missing from the diet, hemorrhagic disorders can also develop.

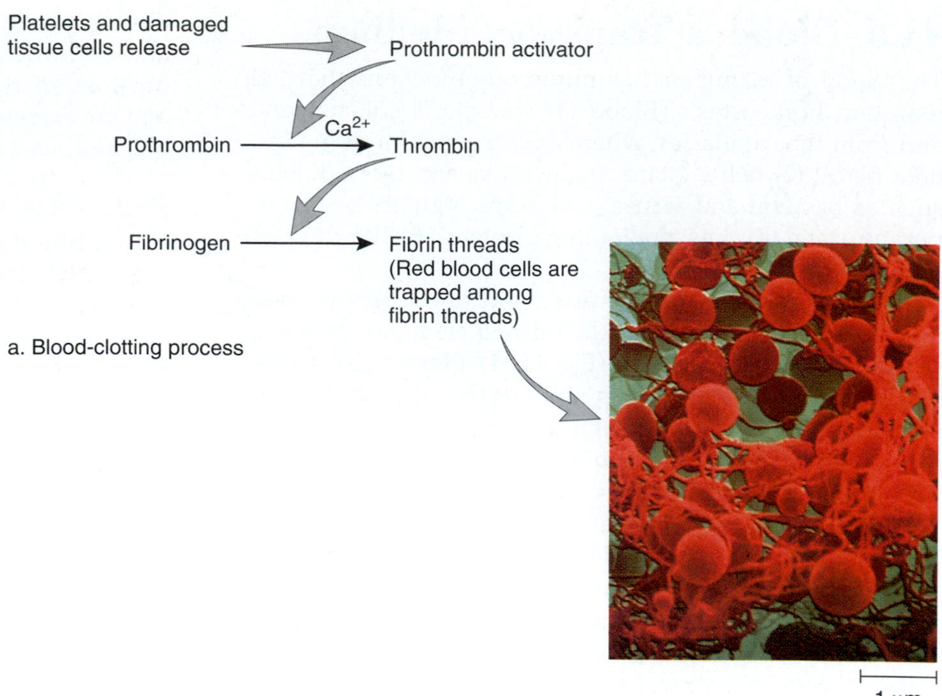

Figure 41.12 Blood clotting.
a. Platelets and damaged tissue cells release prothrombin activator, which acts on prothrombin in the presence of Ca^{2+} (calcium ions) to produce thrombin. Thrombin acts on fibrinogen in the presence of Ca^{2+} to form fibrin threads. **b.** A scanning electron micrograph of a blood clot shows red blood cells caught in the fibrin threads.

Blood Clotting When a blood vessel in the body is damaged, platelets clump at the site of the puncture and partially seal the leak. They and the injured tissues release a clotting factor called prothrombin activator that converts prothrombin to thrombin. This reaction requires calcium ions (Ca^{2+}). **Thrombin,** in turn, acts as an enzyme that severs two short amino acid chains from each fibrinogen molecule. These activated fragments then join end to end, forming long threads of fibrin. Fibrin threads wind around the platelet plug in the damaged area of the blood vessel and provide the framework for the clot. Red blood cells also are trapped within the fibrin threads; these cells make a clot appear red (Fig. 41.12). A fibrin clot is present only temporarily. As soon as blood vessel repair is initiated, an enzyme called plasmin destroys the fibrin network and restores the fluidity of plasma.

If blood is allowed to clot in a test tube, a yellowish fluid develops above the clotted material. This fluid is called serum, and it contains all the components of plasma except fibrinogen. Table 41.1 lists the many different terms used to refer to various body fluids related to blood.

A blood clot consists of platelets and red blood cells entangled within fibrin threads.

Capillary Exchange

Two forces primarily control movement of fluid through the capillary wall: osmotic pressure, which tends to cause water to move from tissue fluid to blood, and blood pressure, which tends to cause water to move in the opposite direction. At the arterial end of a capillary, blood pressure (30 mm Hg) is higher than the osmotic pressure of blood (21 mm Hg) (Fig. 41.13). Osmotic pressure is created by the presence of salts and the plasma proteins. Because blood pressure is higher than osmotic pressure at the arterial end of a capillary, water exits a capillary at this end.

Midway along the capillary, where blood pressure is lower, the two forces essentially cancel each other, and there

Table 41.1
Body Fluids

Name	Composition
Blood	Formed elements and plasma
Plasma	Liquid portion of blood
Serum	Plasma minus fibrinogen
Tissue fluid	Plasma minus most proteins
Lymph	Tissue fluid within lymphatic vessels

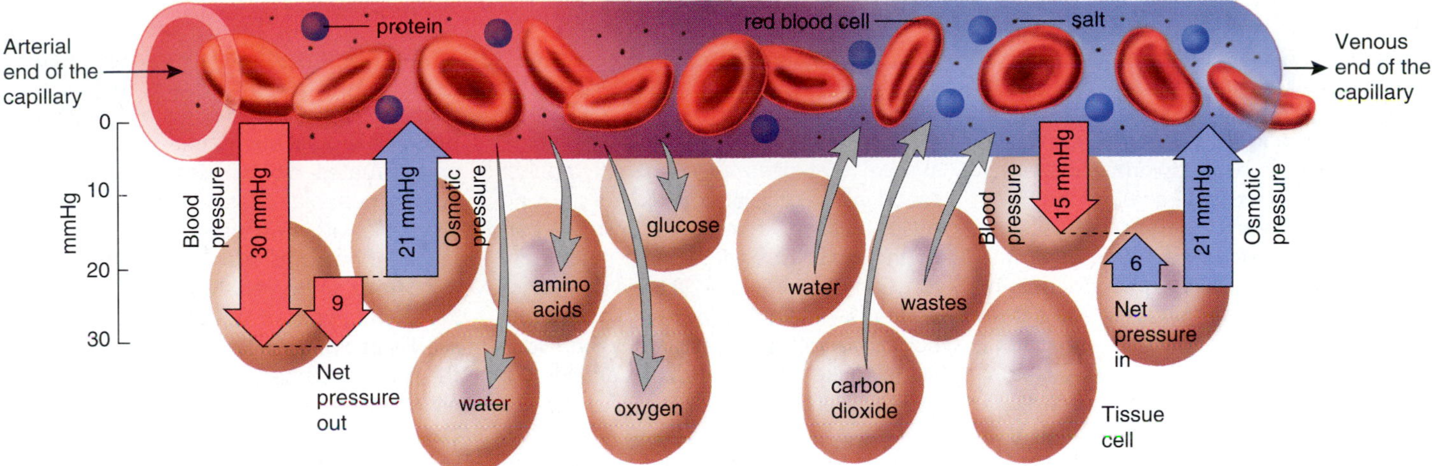

Figure 41.13 Capillary exchange.
A capillary, illustrating the exchanges that take place and the forces that aid the process. At the arterial end of a capillary, the blood pressure is higher than the osmotic pressure; therefore, water (H_2O) tends to leave the bloodstream. In the midsection, molecules including oxygen (O_2) and carbon dioxide (CO_2) follow their concentration gradients. At the venous end of a capillary, the osmotic pressure is higher than the blood pressure; therefore, water tends to enter the bloodstream. Notice that the red blood cells and the plasma proteins are too large to exit a capillary.

is no net movement of water. Solutes now diffuse according to their concentration gradient—oxygen and nutrients (glucose and amino acids) diffuse out of the capillary; carbon dioxide and wastes diffuse into the capillary. Red blood cells and almost all plasma proteins remain in the capillaries. The substances that leave a capillary contribute to **tissue fluid,** the fluid between the body's cells. Since plasma proteins are too large to readily pass out of the capillary, tissue fluid tends to contain all components of plasma except much lesser amounts of protein.

At the venule end of a capillary, where blood pressure has fallen even more, osmotic pressure is greater than blood pressure, and water tends to move into the capillary. Almost the same amount of fluid that left the capillary returns to it, although there is always some excess tissue fluid collected by the lymphatic capillaries (Fig. 41.14). Tissue fluid contained within lymphatic vessels is called **lymph.** Lymph is returned to the systemic venous blood when the major lymphatic vessels enter the subclavian veins in the shoulder region.

Not all capillary beds are open at the same time. When the precapillary sphincters (circular muscles) shown in Figure 41.14 are relaxed, the capillary bed is open and blood flows through the capillaries. When precapillary sphincters are contracted, blood flows through a shunt that carries blood directly from an arteriole to a venule. In addition to nutrients and wastes, the blood distributes heat to body parts. When you are warm, many capillaries that serve the skin are open and your face is flushed. This helps rid the body of excess heat. When you are cold skin capillaries close, conserving heat and the skin turns blue.

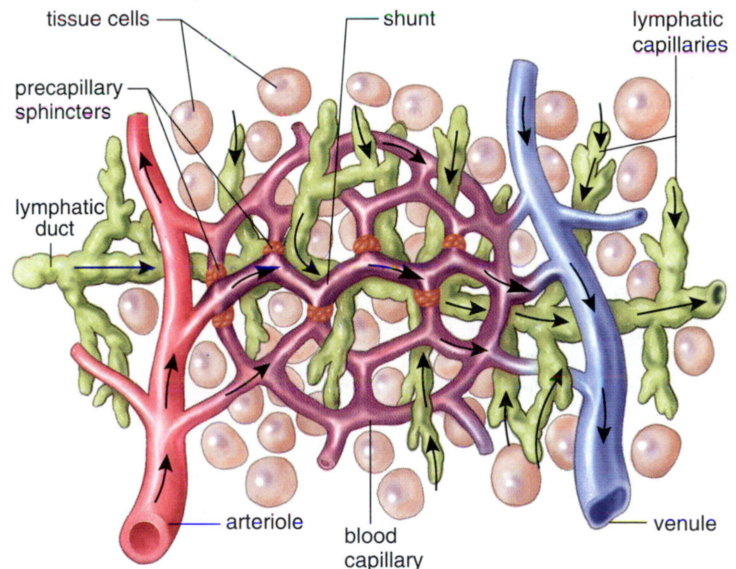

Figure 41.14 Capillary bed.
A lymphatic capillary bed lies near a blood capillary bed. When lymphatic capillaries take up excess tissue fluid, it becomes lymph. Precapillary sphincters can shut down a blood capillary and blood then flows through the shunt.

Oxygen and nutrients exit a capillary near the arterial end; carbon dioxide and waste molecules enter a capillary near the venous end.

Connecting Concepts

Small aquatic animals with no circulatory system may rely on the passage of external water through a gastrovascular cavity to service cells. Roundworms and other pseudocoelomates use a fluid-filled body cavity as a means of transporting substances. A fluid-filled cavity, whether it is a gastrovascular cavity, pseudocoelom, or a coelom as in earthworms, can act as a hydrostatic skeleton (see also page 592). Unlike a hard endo- or exoskeleton, a hydrostatic skeleton doesn't take much energy to carry around. Even animals that have a rigid skeleton may still rely on body fluids for the purpose of locomotion. Bivalves, for example, pump hemolymph into the foot, thereby expanding it for digging into the mud.

Most animals do have a circulatory system that uses either hemolymph or blood as a circulatory fluid for moving gases, nutrients, and wastes from one part of the body to another. The respiratory function of a circulatory fluid is often augmented by the presence of a pigment that combines with and carries oxygen. Body fluids make ideal culture media for the growth of infectious parasites, and these fluids often have some way to ward off an invasion. In mammals, the white blood cells perform this function by phagocytizing microbes and producing antigens.

In birds and mammals, which are homeothermic, blood plays a role in temperature regulation. In animals that live in cold climates, blood vessels are arranged so that heat can pass from the arteries to the veins within exposed limbs. This conserves heat and also keeps the limbs from freezing. In vertebrates that dive under water, most of the blood pumped by the heart flows through the capillaries of the brain and heart. Closure of other possible pathways keeps these organs active despite the cessation of breathing. These extreme examples shows how the circulatory system can be modified as an adaptation to the environment.

Summary

41.1 Transport in Invertebrates

Some invertebrates do not have a transport system. The presence of a gastrovascular cavity allows diffusion alone to supply the needs of cells in cnidarians and flatworms. Roundworms make use of their pseudocoelom.

Other invertebrates do have a transport system. Insects have an open circulatory system, and earthworms have a closed one.

41.2 Transport in Vertebrates

Vertebrates have a closed system in which arteries carry blood away from the heart to capillaries, where exchange takes place, and veins carry blood to the heart.

Fishes have a single-loop circulatory pathway because the heart, with the single atrium and ventricle, pumps blood only to the gills. The other vertebrates have both pulmonary and systemic circuits. Amphibians have two atria but a single ventricle. Birds and mammals, including humans, have a heart with two atria and two ventricles, in which O_2-rich blood is always separate from O_2-poor blood.

41.3 Transport in Humans

The heartbeat in humans begins when the SA node (pacemaker) causes the two atria to contract and blood moves through the atrioventricular valves to the two ventricles. The SA node also stimulates the AV (atrioventricular) node, which in turn causes the two ventricles to contract. Ventricular contraction sends blood through the semilunar valves to the pulmonary trunk and the aorta. Now all chambers rest. The heart sounds, lub-dub, are caused by the closing of the valves.

In the pulmonary circuit, blood can be traced to and from the lungs. In the systemic circuit, the aorta divides into blood vessels that serve the body's cells. The venae cavae return O_2-poor blood to the heart.

Blood pressure created by the beat of the heart accounts for the flow of blood in the arteries, but skeletal muscle contraction is largely responsible for the flow of blood in the veins, which have valves preventing a backward flow.

41.4 Cardiovascular Disorders

Hypertension and atherosclerosis are two circulatory disorders that lead to heart attack and to stroke. A heart-healthy diet, getting regular exercise, maintaining a proper weight, and not smoking cigarettes are protective against the development of these conditions.

41.5 Blood, a Transport Medium

Blood has two main parts: plasma and formed elements. Plasma contains mostly water (90–92%) and proteins (7–8%) but also nutrients and wastes.

The red blood cells contain hemoglobin and function in oxygen transport. Defense against disease depends on the various types of white blood cells. Neutrophils and monocytes are phagocytic and are especially responsible for the inflammatory reaction. Lymphocytes are involved in the development of immunity to disease.

The platelets and two plasma proteins, prothrombin and fibrinogen, function in blood clotting, an enzymatic process that results in fibrin threads. Blood clotting is a complex process that includes three major events: platelets and injured tissue release prothrombin activator, which enzymatically changes prothrombin to thrombin; thrombin is an enzyme that causes fibrinogen to be converted to fibrin threads.

When blood reaches a capillary, water moves out at the arterial end, due to blood pressure. At the venule end, water moves in, due to osmotic pressure. In between, nutrients diffuse out and wastes diffuse into a capillary.

Reviewing the Chapter

1. Describe transport in those invertebrates that have no circulatory system; in those that have an open circulatory system; and in those that have a closed circulatory system. 740–41
2. Compare the circulatory systems of a fish, an amphibian, and a mammal. 743
3. Trace the path of blood in humans from the right ventricle to the left atrium; from the left ventricle to the kidneys and to the right atrium; from the left ventricle to the small intestine and to the right atrium. 745

4. Describe the mechanism of a heartbeat, mentioning all the factors that account for this repetitive process. Describe how the heartbeat affects blood flow. What other factors are involved in blood flow? 746
5. Define these terms: pulmonary circuit, systemic circuit, and portal system. 748
6. Discuss the life cycle and function of red blood cells. 753
7. How are white blood cells classified? What are the functions of neutrophils, monocytes, and lymphocytes? 753
8. Name the steps that take place when blood clots. Which substances are present in blood at all times, and which appear during the clotting process? 754
9. What forces facilitate exchange of molecules across the capillary wall? 754–55

Testing Yourself

Choose the best answer for each question.

1. Which one of these would you expect to be part of a closed, but not an open, circulatory system?
   a. ostia
   b. capillary beds
   c. hemocoel
   d. heart
   e. All of these are correct.
2. In a one-circuit circulatory system, blood pressure
   a. is constant throughout the system.
   b. drops significantly after gas exchange has taken place.
   c. is higher at the intestinal capillaries than at the gill capillaries.
   d. brings O_2-rich blood directly to the heart.
   e. does not occur in the animal kingdom.
3. In which animal does aortic blood have less oxygen than blood in the pulmonary vein?
   a. frog
   b. chicken
   c. monkey
   d. fish
   e. All of these are correct.
4. Which of these factors has little effect on blood flow in arteries?
   a. heartbeat
   b. blood pressure
   c. total cross-sectional area of vessels
   d. skeletal muscle contraction
   e. the amount of blood leaving the heart.
5. In humans, blood returning to the heart from the lungs returns to
   a. the right ventricle.
   b. the right atrium.
   c. the left ventricle.
   d. the left atrium.
   e. both the right and left sides of the heart.
6. Systole refers to the contraction of the
   a. major arteries.
   b. SA node.
   c. atria and ventricles.
   d. major veins.
   e. All of these are correct.
7. Which of these associations is incorrect?
   a. white blood cells—infection fighting
   b. red blood cells—blood clotting
   c. plasma—water, nutrients, and wastes
   d. red blood cells—hemoglobin
   e. platelets—blood clotting
8. Water enters capillaries on the venule side as a result of
   a. active transport from tissue fluid.
   b. an osmotic pressure gradient.
   c. higher blood pressure on this side.
   d. higher blood pressure on the arterial side.
   e. higher red blood cell concentration on this side.
9. The last step in blood clotting
   a. is the only step that requires calcium ions.
   b. occurs outside the bloodstream.
   c. is the same as the first step.
   d. converts prothrombin to thrombin.
   e. converts fibrinogen to fibrin.
10. Label arrows as either blood pressure or osmotic pressure.

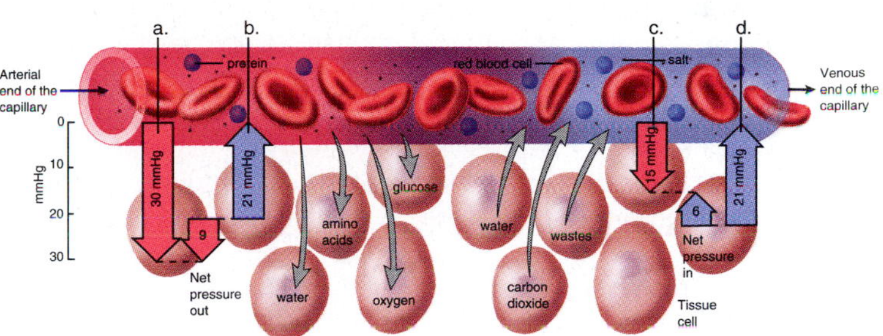

11. Label this diagram of the heart.

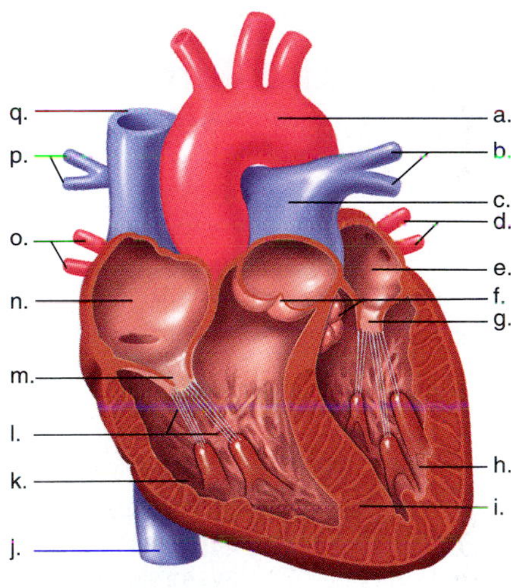

Thinking Scientifically

1. For several years, researchers have attempted to produce an artificial blood for transfusions. Artificial blood would most likely be safer and more readily available than human blood. While artificial blood may not have all the characteristics of human blood it would be useful on the battlefield and in emergency situations. Which characteristics must artificial blood have to be useful and which would probably be too difficult to reproduce?
2. You have to stand in front of the class to give a report and you are nervous. Your heart is pounding. How would your ECG appear?

Bioethical Issue

According to a 1993 study, about one million deaths a year in the United States could be prevented if people adopted the healthy lifestyle described in the reading on page 750. Tobacco, lack of exercise, and a high-fat diet probably cost the nation about $200 billion per year in health-care costs. To what lengths should we go to prevent these deaths and reduce health-care costs?

E. A. Miller, a meatpacking entity of ConAgra in Hyrum, Utah, charges extra for medical coverage of employees who smoke. Eric Falk, Miller's director of human resources says, "We want to teach employees to be responsible for their behavior." Anthem Blue Cross–Blue Shield of Cincinnati, Ohio takes a more positive approach. They give insurance plan participants $240 a year in extra benefits, like additional vacation days, if they get good scores in five out of seven health-related categories. The University of Alabama, Birmingham, School of Nursing has a health-and-wellness program that counsels employees about how to get into shape in order to keep their insurance coverage. Audrey Brantley is in the program and says she has mixed feelings. She says, "It seems like they are trying to control us, but then, on the other hand, I know of folks who found out they had high blood pressure or were borderline diabetics and didn't know it."

Does it really work is another question. Turner Broadcasting System in Atlanta has a policy that affects all employees hired after 1986. They will be fired if caught smoking—whether at work or at home—but some admit they still manage to sneak a smoke. What steps do you think are ethical to encourage people to adopt a healthy lifestyle?

Understanding the Terms

antibody 753
antigen 753
aorta 748
arteriole 742
artery 742
atrioventricular valve 744
atrium (pl., atria) 744
blood 741
blood pressure 749
capillary 742
cardiac conduction system 747
cardiac cycle 746
cardiac pacemaker 747
circulatory (or cardiovascular) system 741, 742
closed circulatory system 741
diastole 746
electrocardiogram (ECG) 747
heart 744
hemoglobin 763
hemolymph 741
lymph 755
lymphocyte 753
macrophage 753
neutrophil 753
open circulatory system 741
plasma 753
platelet 754
portal system 748
pulmonary circuit 748
pulse 746
red blood cell 753
semilunar valve 744
septum 744
systemic circuit 748
systole 746
thrombin 754
tissue fluid 755
vein 742
vena cava 748
ventricle 744
venule 742
white blood cell 753

Match the terms to these definitions:

a. ______________ Blood vessel that transports blood away from the heart.

b. ______________ Cell fragment that is necessary to blood clotting.

c. ______________ The liquid portion of blood; contains nutrients, wastes, salts, and proteins.

d. ______________ The major systemic veins that take blood to the heart from the tissues.

e. ______________ Iron-containing respiratory pigment occurring in vertebrate red blood cells and in blood plasma of many invertebrates.

Web Connections

Exploring the Internet

http://www.mhhe.com/biosci/genbio/mader
(click on *Biology 7/e*)

The *Biology 7/e* Online Learning Center provides many resources for studying the material in this chapter including links to the following sites:

Healthfinder is a gateway website developed by the Department of Health and Human Services. It offers information on circulatory diseases as well as a multitude of other issues in health care indexed A-Z.

http://www.healthfinder.gov/

National Heart, Lung, and Blood Institute Home Page. Information and many links to resources on the circulatory and respiratory systems.

http://www.nhlbi.nih.gov/

Heart Disease, Hypertension, Cholesterol, Treatment and Prevention. The Mayo Clinic informational site on heart health. Reference articles, information, references, links, and quizzes.

http://www.mayohealth.org/mayo/common/htm/heartpg.htm

Cholesterol Counts for Everyone. More information on cholesterol and heart disease from the National Heart, Lung, and Blood Institute.

http://rover.nhlbi.nih.gov/chd/

The Heart: An Online Exploration. This site covers a variety of heart-related topics.

http://sln.fi.edu/biosci/heart.html

Lymph Transport and Immunity

c h a p t e r

c h a p t e r   c o n c e p t s

HIV (red) buds from a cell of the immune system.

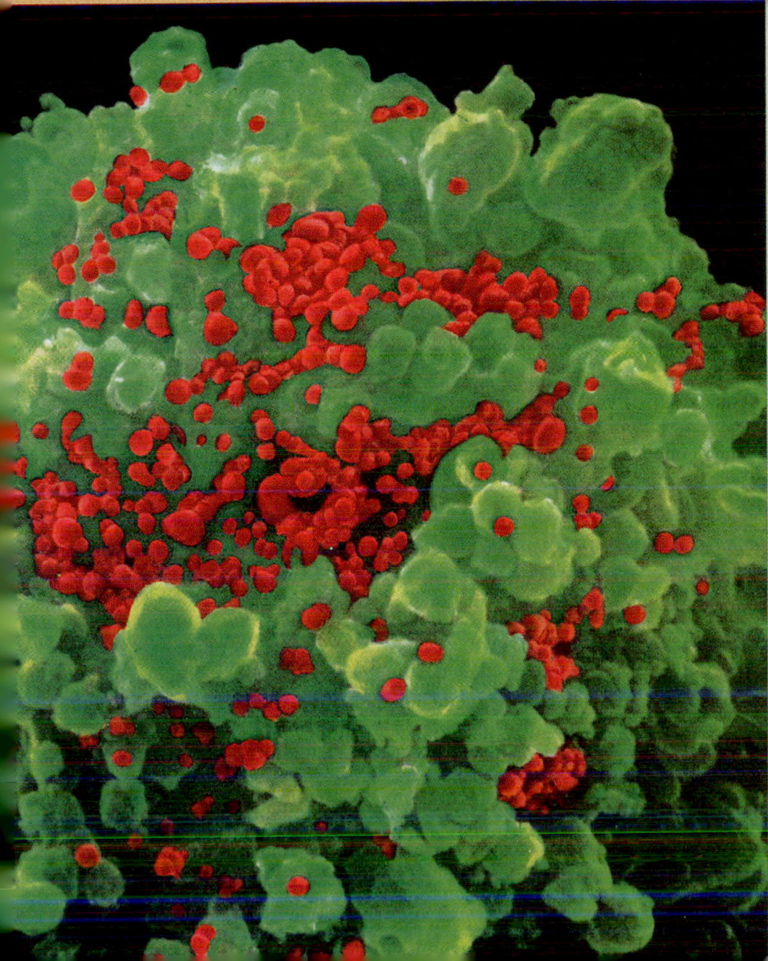

Once HIV came on the scene in the 1970s, the general public began to learn more about the immune system than past generations ever dreamed of knowing. HIV—human immunodeficiency virus—is a group of viruses transmitted by the exchange of internal body fluids. HIV does not kill directly; it kills because it destroys the cells of the host's immune system, thereby enabling pathogens to invade the body. In contrast, individuals with an intact immune system can usually defend themselves against these opportunistic diseases. HIV sufferers cannot.

This illustrates that homeostasis is impossible without an efficient immune system. Immunity includes all the possible means to defend ourselves, including assistance from the lymphatic system. The mammalian lymphatic system consists of one-way lymph vessels, lymph nodes, lymphatic fluid, and several other organs. These organs produce and store the cells of the immune system. In addition to its role in defense, the lymphatic system has several other vital functions such as returning excess tissue fluid back to the bloodstream. It also absorbs fats in the small intestine and transports them to the cardiovascular system.

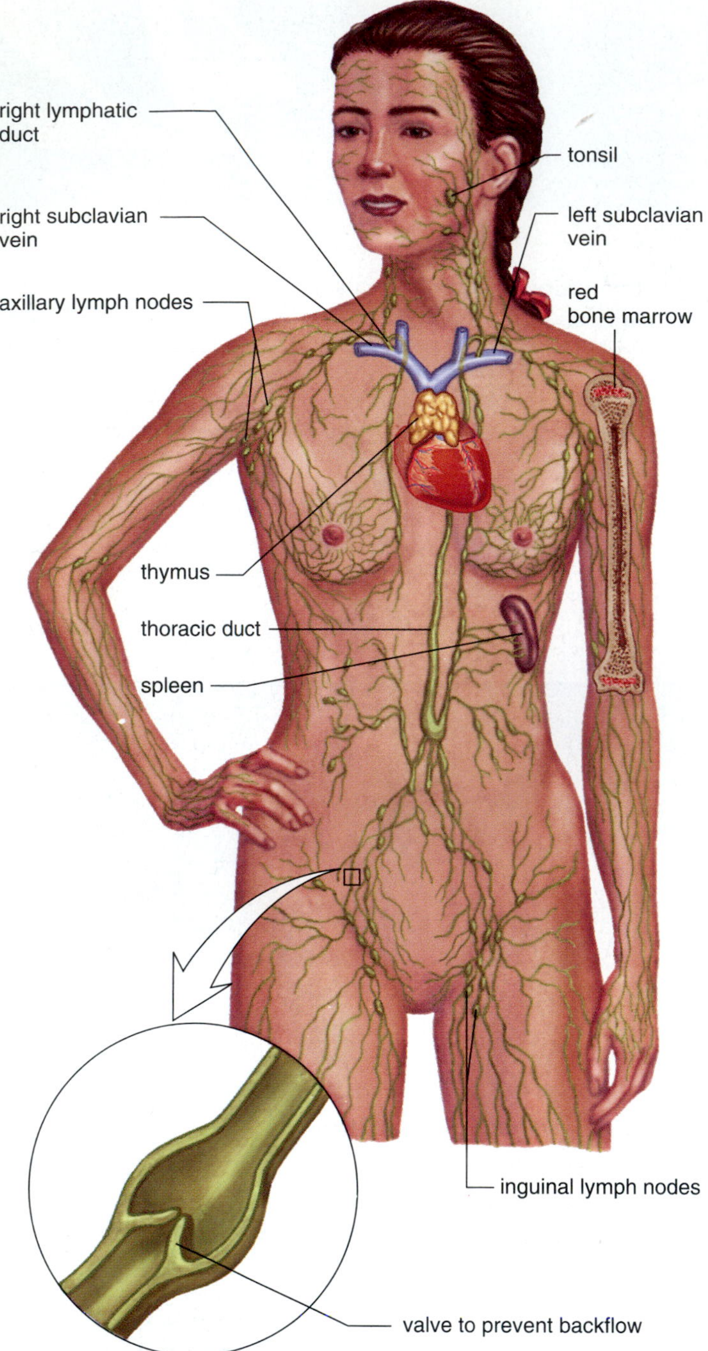

Figure 42.1 **Lymphatic system.**
Lymphatic vessels drain excess fluid from the tissues and return it to the cardiovascular system. The enlargement shows that lymphatic vessels like cardiovascular veins have valves to prevent backward flow. The tonsils, spleen, thymus gland, and red bone marrow are lymphoid organs that assist immunity.

42.1 Lymphatic System

The mammalian **lymphatic system** [L. *lympha,* clear water] consists of **lymphatic vessels** and **lymphoid organs.** This system, which is closely associated with the cardiovascular system, has three main functions: (1) lymphatic capillaries take up excess tissue fluid and return it to the bloodstream; (2) lymphatic capillaries absorb fats at the intestinal villi and transport them to the bloodstream; and (3) the lymphatic system helps to defend the body against disease.

Lymphatic Vessels

Lymphatic vessels are quite extensive; most regions of the body are richly supplied with lymphatic capillaries (Fig. 42.1). The construction of the larger lymphatic vessels is similar to that of cardiovascular veins, including the presence of valves. Also, the movement of lymph within these vessels is dependent upon skeletal muscle contraction. When the muscles contract, the lymph is squeezed past a valve that closes, preventing the lymph from flowing backwards.

The lymphatic system is a one-way system that begins with lymphatic capillaries. These capillaries take up fluid that has diffused from and has not been reabsorbed by the blood capillaries. **Edema** [Gk. *oidema,* swelling with fluid] is localized swelling caused by the accumulation of tissue fluid. This can happen if too much tissue fluid is made and/or not enough of it is drained away. Once tissue fluid enters the lymphatic vessels, it is called **lymph.** The lymphatic capillaries join to form lymphatic vessels that merge before entering one of two ducts: the thoracic duct or the right lymphatic duct. The thoracic duct is much larger than the right lymphatic duct. It serves the lower extremities, the abdomen, the left arm, and the left side of both the head and the neck. The right lymphatic duct serves the right arm, the right side of both the head and the neck, and the right thoracic area. The lymphatic ducts enter the subclavian veins, which are cardiovascular veins, in the shoulder region.

Lymph flows one way from a capillary to ever-larger lymphatic vessels and finally to a lymphatic duct, which enters a subclavian vein.

Lymphoid Organs

The lymphatic organs of special interest are the lymph nodes, the tonsils, the spleen, the thymus gland, and the bone marrow (Fig. 42.2). They help the body overcome **pathogens** which are disease-causing agents like viruses and some bacteria.

Lymph nodes, which are small (about 1–25 mm) ovoid or round structures, are found at certain points along lymphatic vessels. Lymph nodes are named for their location. Inguinal nodes are in the groin and axillary nodes are in the

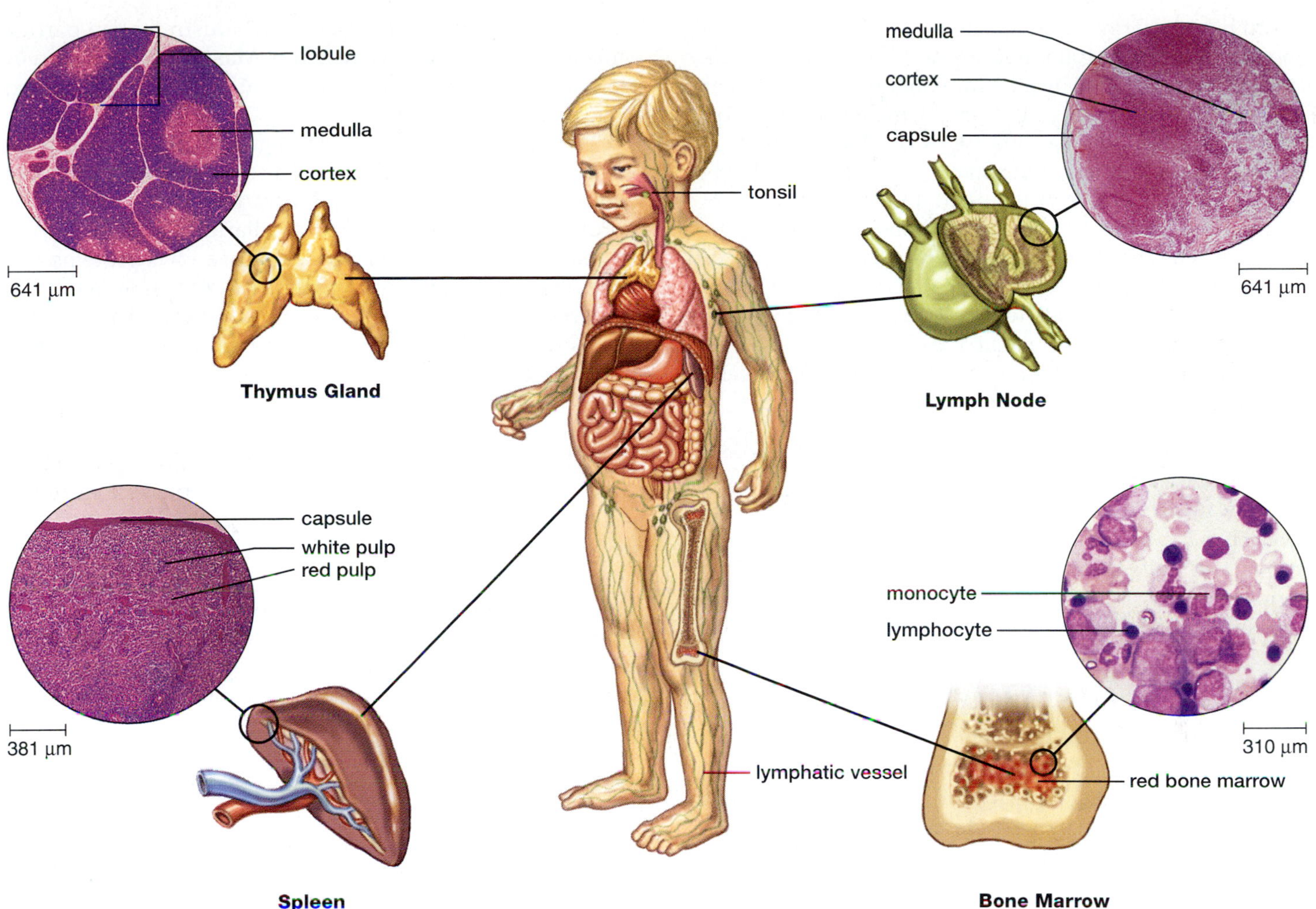

Figure 42.2 The lymphoid organs.
The lymphoid organs include the lymph nodes, the tonsils, the spleen, the thymus gland, and the red bone marrow, which all contain lymphocytes.

armpits. Physicians often feel for the presence of swollen, tender lymph nodes in the neck as evidence that the body is fighting an infection.

A capsule surrounds the cortex and medulla of a lymph node. Lymphocytes congregate and multiply in the medulla. Macrophages cleanse lymph as it courses through the node. Any pathogens present cause macrophages and lymphocytes to mount a specific immune response against them.

The **tonsils** are partially encapsulated lymphatic tissue located in a ring about the pharynx. The well-known pharyngeal tonsils are also called adenoids, while the larger palatine tonsils located on either side of the posterior oral cavity are most apt to be infected. The tonsils perform the same functions as lymph nodes. Because of their location, tonsils are the first to encounter pathogens that enter the body by way of the nose and mouth.

The **spleen** is located in the upper left region of the abdominal cavity just beneath the diaphragm. It is much larger than a lymph node, about the size of a fist. Whereas the lymph nodes cleanse lymph, the spleen cleanses blood. A capsule surrounds tissue known as white pulp and red pulp. White pulp contains lymphocytes and performs the immune functions of the spleen. The red pulp contains red blood cells and plentiful macrophages. The red pulp helps to purify blood that passes through the spleen by removing pathogens and worn-out or damaged red blood cells.

The spleen's outer capsule is relatively thin, and an infection and/or a blow can cause the spleen to burst. Although its functions are replaced by other organs, a person without a spleen is often slightly more susceptible to infections and may have to receive antibiotic therapy indefinitely.

The **thymus gland** is located along the trachea behind the sternum in the upper thoracic cavity. This gland varies in size, but it is larger in children than in adults and may disappear completely in old age. The thymus is divided into lobules by connective tissue. T lymphocytes mature in the lobules of a spleen. The interior (medulla) of a lobule, which consists mostly of epithelial cells, stains lighter than the cortex. It produces thymic hormones, such as thymosin, that are thought to aid in maturation of T lymphocytes. Thymosin may also have other functions in immunity.

Red bone marrow is the site of origination for all types of blood cells, including the five types of white blood cells pictured in Figure 41.11. The marrow contains stem cells that are ever capable of dividing and producing cells that go on to differentiate into the various types of blood cells. In a child, most bones have red bone marrow, but in an adult it is present only in the bones of the skull, the sternum (breastbone), the ribs, the clavicle, the pelvic bones, and the vertebral column. The red bone marrow consists of a network of connective tissue fibers, called reticular fibers, which are produced by cells called reticular cells. These and the stem cells and their progeny are packed around thin-walled sinuses filled with venous blood. Differentiated blood cells enter the bloodstream at these sinuses.

The lymphoid organs have specific functions that assist immunity. Lymph is cleansed in lymph nodes; blood is cleansed in the spleen; T lymphocytes mature in the thymus; and white blood cells are made in the bone marrow.

42.2 Nonspecific Defenses

Immunity is the ability of the body to defend itself against infectious agents, foreign cells, and even abnormal body cells, such as cancer cells. Thereby, the internal environment has a better chance of remaining stable. Immunity includes nonspecific and specific defenses. The four types of nonspecific defenses—barriers to entry, the inflammatory reaction, natural killer cells, and protective proteins—are effective against many types of pathogens.

Barring Entry

Skin and the mucous membranes lining the respiratory, digestive, and urinary tracts serve as mechanical barriers to entry by pathogens. Oil gland secretions contain chemicals that weaken or kill certain bacteria on skin. The upper respiratory tract is lined by ciliated cells that sweep mucus and trapped particles up into the throat, where they can be swallowed, or expectorated (coughed out). The stomach has an acidic pH, which inhibits the growth of or kills many types of bacteria. The various bacteria that normally reside in the intestine and other areas, such as the vagina, prevent pathogens from taking up residence.

Inflammatory Reaction

Whenever the skin is broken due to a minor injury, a series of events occurs that is known as the **inflammatory reaction.** The inflamed area has four outward signs: redness, heat, swelling, and pain. Figure 42.3 illustrates the participants in the inflammatory reaction. **Mast cells,** which occur in tissues, resemble **basophils,** one of the white cells found in the blood.

When an injury occurs, damaged tissue cells and mast cells release inflammatory chemicals, such as **histamine** and **kinins,** which cause the capillaries to dilate and become more permeable. The enlarged capillaries cause the skin to redden, and the increased permeability allows proteins and fluids to escape and swelling results. A rise in temperature increases phagocytosis by white blood cells. The swollen area as well as kinins stimulate free nerve endings, causing the sensation of pain.

Neutrophils and monocytes migrate to the site of injury. They are amoeboid and can change shape to squeeze through capillary walls to enter tissue fluid. Neutrophils, and also mast cells, can phagocytize bacteria. The engulfed bacteria are destroyed by hydrolytic enzymes when the endocytic vesicle combines with a lysosome, one of the cellular organelles.

Monocytes differentiate into **macrophages,** large phagocytic cells that are able to devour a hundred bacteria or viruses and still survive. Some tissues, particularly connective tissue, have resident macrophages, which routinely act as scavengers, devouring old blood cells, bits of dead tissue, and other debris. Macrophages can also bring about an explosive increase in the number of leukocytes by liberating colony-stimulating hormones, which pass by way of blood to the red bone marrow, where they stimulate the production and the release of white blood cells, primarily neutrophils.

When a blood vessel ruptures, the blood clots to seal the break. The inflammatory chemicals, mentioned earlier, and antigens move through the tissue fluid and lymph to the lymph nodes. Now lymphocytes can also be activated to react to the threat of an infection. As the infection is being overcome, some neutrophils may die. These—along with dead tissue, cells, bacteria, and living white blood cells—form pus, a whitish material. Pus indicates that the body is trying to overcome the infection.

Sometimes inflammation persists and the result is chronic inflammation that is often treated by the administration of anti-inflammatory agents such as aspirin, ibuprofen, or cortisone. They act against the inflammatory chemicals released by the white blood cells in the area.

The inflammatory reaction is a "call to arms"—it marshals phagocytic white blood cells to the site of bacterial invasion and stimulates the immune system to react against a possible infection.

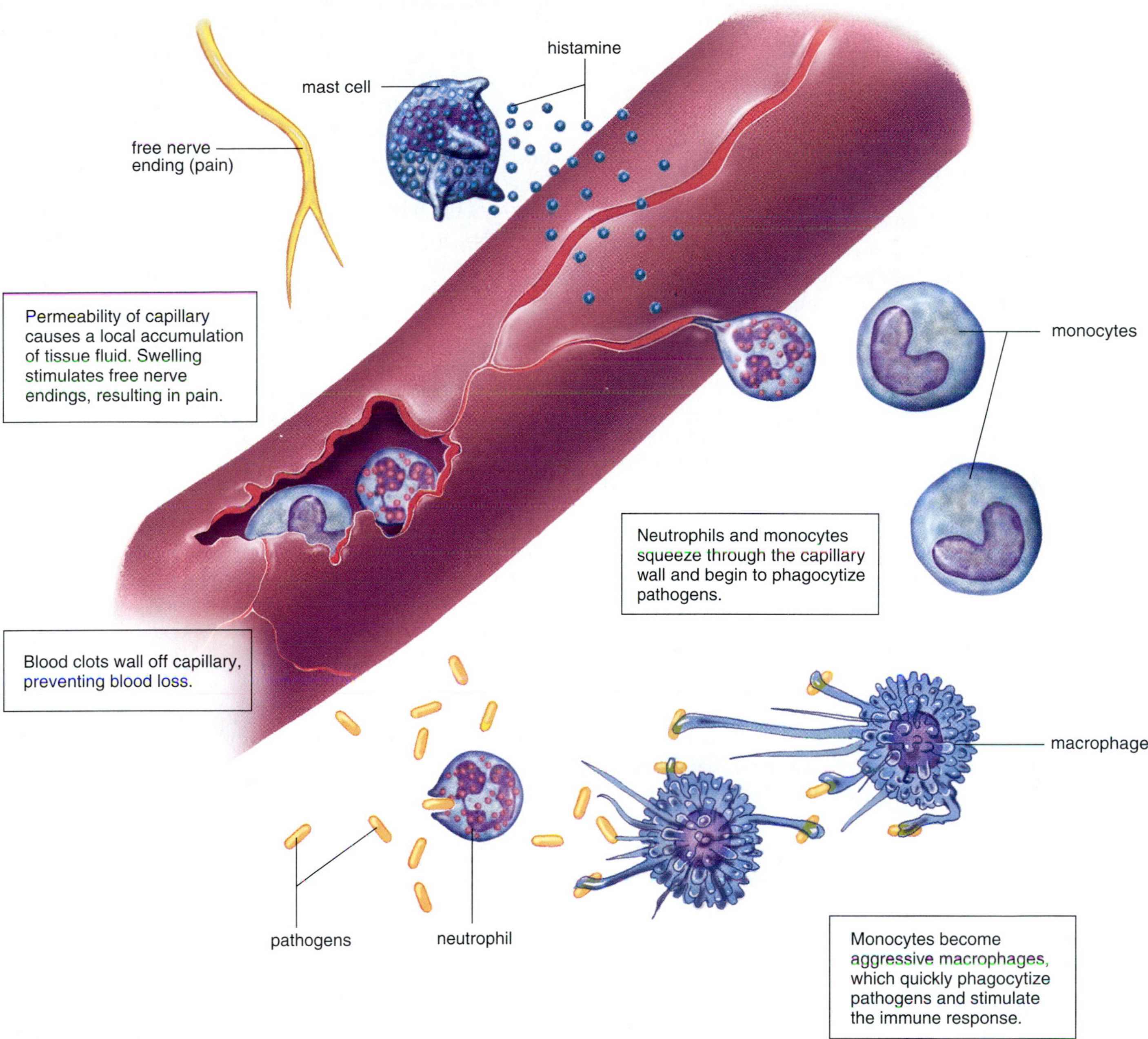
Injured tissue cells and mast cells release inflammatory chemicals (e.g., histamine) that dilate capillaries, bringing blood to the scene. Redness and heat result.
histamine
mast cell
free nerve ending (pain)
Permeability of capillary causes a local accumulation of tissue fluid. Swelling stimulates free nerve endings, resulting in pain.
monocytes
Neutrophils and monocytes squeeze through the capillary wall and begin to phagocytize pathogens.
Blood clots wall off capillary, preventing blood loss.
macrophage
pathogens
neutrophil
Monocytes become aggressive macrophages, which quickly phagocytize pathogens and stimulate the immune response.

Figure 42.3 Inflammatory reaction.
Mast cells, which are related to basophils, a type of white blood cell, are involved in the inflammatory reaction. When a blood vessel is injured, mast cells release substances like histamine. Histamine dilates blood vessels and increases permeability so that tissue fluid leaks from the vessel. Swelling in the area stimulates pain receptors (free nerve endings). Neutrophils and monocytes which become macrophages squeeze through the capillary wall. These white blood cells begin to phagocytize pathogens (bacteria and viruses), especially those combined with antibodies. Blood clotting seals off the capillary, preventing blood loss.

Natural Killer Cells

Natural killer (NK) cells kill virus-infected cells and tumor cells by cell-to-cell contact. They are large granular lymphocytes. They have no specificity and no memory. Their number is not increased by immunization.

Protective Proteins

The **complement system,** often simply called complement, is a number of plasma proteins designated by the letter C and a subscript. A limited amount of activated complement protein is needed because a domino effect occurs: each activated protein in a series is capable of activating many other proteins.

Complement is activated when pathogens enter the body. It "complements" certain immune responses, which accounts for its name. For example, it is involved in and amplifies the inflammatory response because complement proteins attract phagocytes to the scene. Some complement proteins bind to the surface of pathogens already coated with antibodies, which ensures that the pathogens will be phagocytized by a neutrophil or macrophage.

Certain other complement proteins join to form a membrane attack complex that produces holes in bacterial cell walls and plasma membranes of bacteria. Fluids and salts then enter the bacterial cell to the point that it bursts (Fig. 42.4).

Interferon is a protein produced by virus-infected cells. Interferon binds to receptors of noninfected cells, causing them to prepare for possible attack by producing substances that interfere with viral replication. Interferon is specific to the species; therefore, only human interferon can be used in humans.

Immunity includes these nonspecific defenses: barriers to entry, the inflammatory reaction, natural killer cells, and protective proteins.

42.3 Specific Defenses

When nonspecific defenses have failed to prevent an infection, specific defenses come into play. An **antigen** is any foreign substance (often a protein or polysaccharide) that stimulates the **immune system** to react to it. Pathogens have antigens, but antigens can also be part of a foreign cell or a cancer cell. Because we do not ordinarily become immune to our own cells, it is said that the immune system is able to distinguish self from nonself.

Immunity usually lasts for some time. For example, once we recover from the measles, we usually do not get the illness a second time. Immunity is primarily the result of the action of the **B lymphocytes** and the **T lymphocytes.** B lymphocytes[1] mature in the *b*one marrow, and T lymphocytes mature in the *t*hymus gland. B lymphocytes, also called B cells, give rise to plasma cells, which produce **antibodies,** proteins that are capable of combining with and neutralizing antigens. These antibodies are secreted into the blood, lymph, and other body fluids. In contrast, T lymphocytes, also called T cells, do not produce antibodies. Instead, certain T cells directly attack cells that bear antigens. Other T cells regulate the immune response.

Lymphocytes are capable of recognizing an antigen because they have receptor molecules on their surface. The shape of the receptors on any particular lymphocyte is complementary to a specific antigen. It is often said that the receptor and the antigen fit together like a lock and a key. It is estimated that during our lifetime, we encounter a million different antigens, so we need a great diversity of lymphocytes to protect us against antigens. It is remarkable that during the maturation process diversification occurs to such an extent that there is a lymphocyte type for any possible antigen. Just how this occurs is discussed in the reading on page 767.

[1] Historically, the *B* stands for bursa of Fabricius, an organ in the chicken where these cells were first identified. As it turns out, however, the *B* can conveniently be thought of as referring to bone marrow.

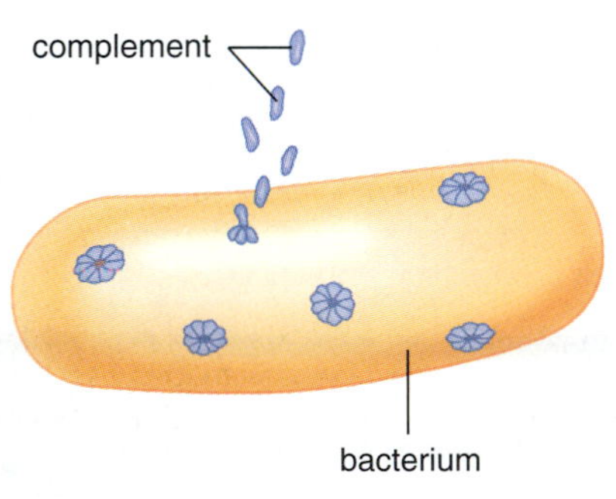

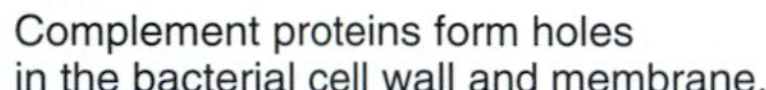

Complement proteins form holes in the bacterial cell wall and membrane.

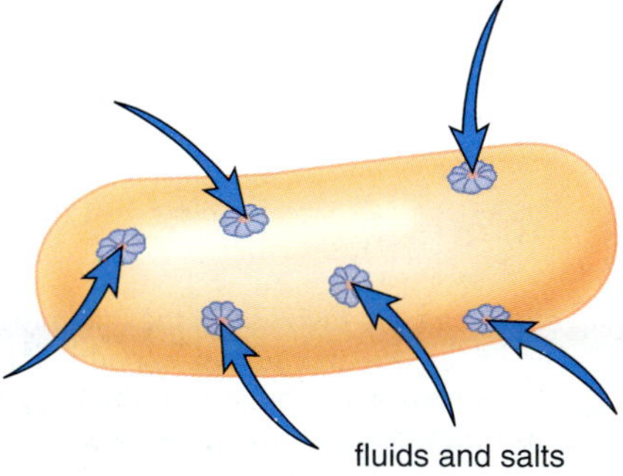

Holes allow fluids and salts to enter the bacterium.

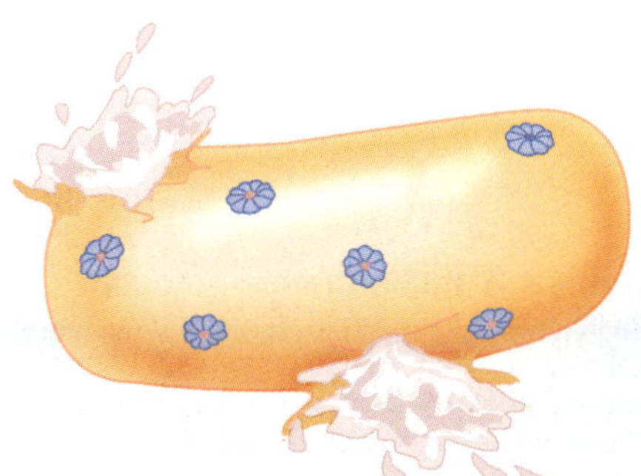

Bacterium expands until it bursts.

Figure 42.4 Action of the complement system against a bacterium.
When complement proteins in the plasma are activated by an immune reaction, they form holes in bacterial cell walls and plasma membranes, allowing fluids and salts to enter until the cell eventually bursts.

B Cells and Antibody-Mediated Immunity

Each type of B cell carries its specific antibody, as a membrane-bound receptor, on its surface. When a B cell in a lymph node or the spleen encounters a bacterial cell or a toxin bearing an appropriate antigen, it becomes activated to divide many times. Most of the resulting cells are plasma cells, which secrete antibodies against this antigen. A **plasma cell** is a mature B cell that mass-produces antibodies in lymph nodes and in the spleen.

The **clonal selection theory** states that the antigen selects which lymphocyte will undergo clonal expansion and produce more lymphocytes bearing the same type of receptor (Fig. 42.5). Notice that a B cell does not divide until its antigen is present and binds to its receptors. B cells are then stimulated to divide and become plasma cells by helper T-cell secretions, as is discussed in the next section. Some members of the clone become memory B cells which are the means by which long-term immunity is possible. If the same antigen enters the system again, memory B cells quickly divide and give rise to plasma cells capable of producing antibodies.

Once the threat of an infection has passed, the development of new plasma cells ceases and those present undergo apoptosis. **Apoptosis** is a process of programmed cell death (PCD) involving a cascade of specific cellular events leading to the death and destruction of the cell. The methodology of PCD is still being worked out, but we know it is an essential physiological mechanism regulating the cell population within an organ system. PCD normally plays a central role in maintaining tissue homeostasis.

Defense by B cells is called **antibody-mediated immunity** because plasma cells produce antibodies. It is also called humoral immunity because these antibodies are present in blood and lymph. A humor is any fluid normally occurring in the body.

Characteristics of B cells:

- Antibody-mediated immunity
- Produced and mature in bone marrow
- Reside in spleen and lymph nodes, circulate in blood and lymph
- Directly recognize antigen and then undergo clonal selection
- Clonal expansion produces antibody-secreting plasma cells as well as memory B cells.

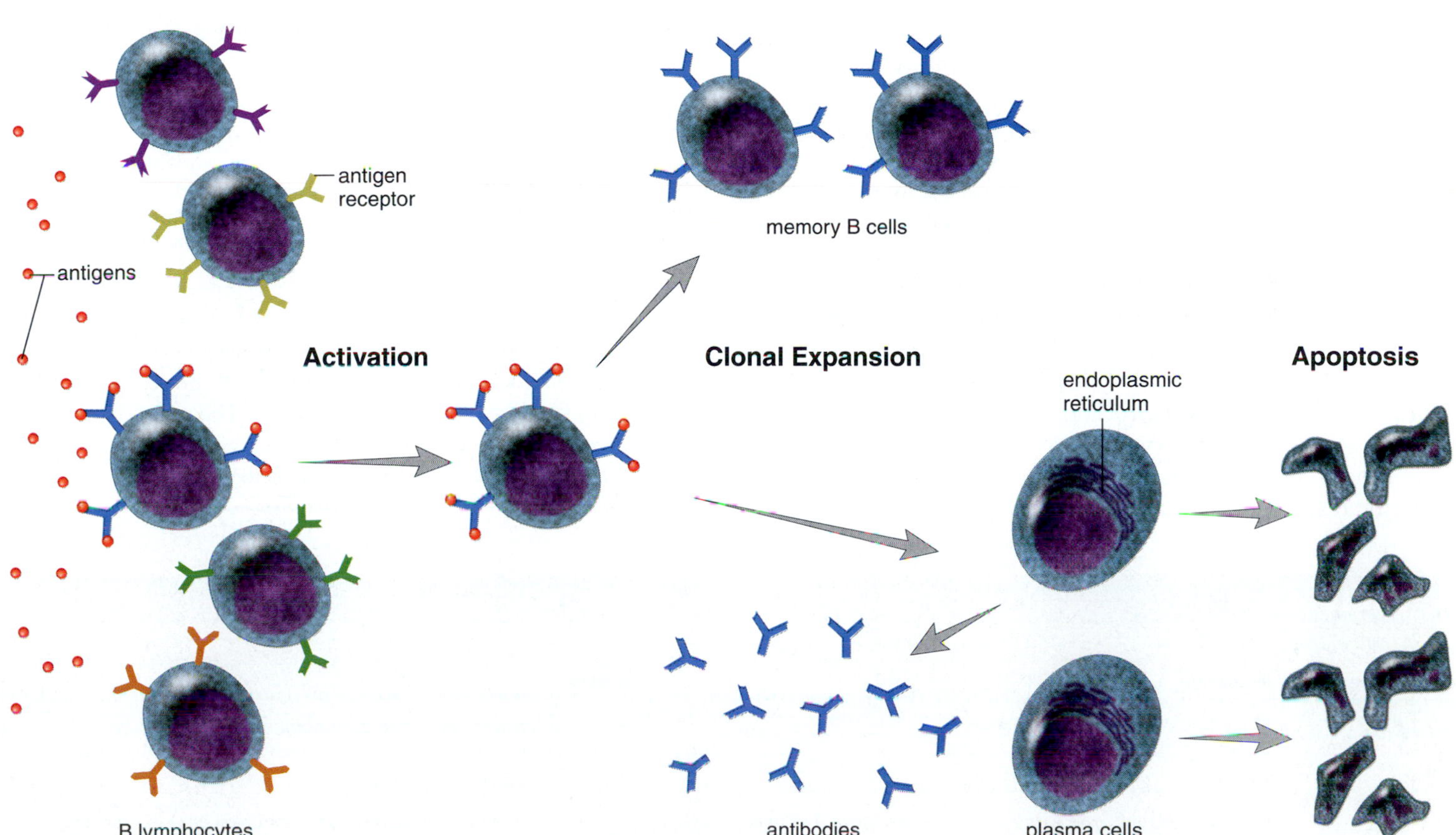

Figure 42.5 Clonal selection theory as it applies to B cells.
An antigen activates only the B cell whose receptors can combine with the antigen. This B cell then undergoes clonal expansion. During the process many plasma cells and memory B cells are produced. Plasma cells produce specific antibodies against this antigen. After the infection passes, plasma cells undergo apoptosis. Memory B cells, which retain the ability to recognize this antigen, are retained in the body.

Structure of IgG

The most common type of antibody (IgG) is a Y-shaped protein molecule with two arms. Each arm has a "heavy" (long) polypeptide chain and a "light" (short) polypeptide chain. These chains have constant regions, where the sequence of amino acids is set, and variable regions, where the sequence of amino acids varies between antibodies (Fig. 42.6). The constant regions are not identical among all the antibodies. Instead, they are almost the same within different classes of antibodies. The variable regions form an antigen-binding site, and their shape is specific to a particular antigen, as discussed in the reading on the next page. The antigen combines with the antibody at the antigen-binding site in a lock-and-key manner.

The antigen-antibody reaction can take several forms, but quite often the reaction produces complexes of antigens combined with antibodies. Such antigen-antibody complexes, sometimes called immune complexes, mark the antigens for destruction. For example, an antigen-antibody complex may be engulfed by neutrophils or macrophages, or it may activate complement. Complement makes pathogens more susceptible to phagocytosis, as discussed previously.

Other Types of Antibodies

There are five different classes of circulating antibody proteins or **immunoglobulins (Igs)** (Table 42.1). IgG antibodies are the major type in blood, and lesser amounts are also found in lymph and tissue fluid. IgG antibodies bind to pathogens and their toxins. A toxin is a specific chemical, produced by bacteria, that is poisonous to other living things. IgM antibodies are pentamers, meaning that they contain five of the Y-shaped structures shown in Figure 42.6*a*. These antibodies appear in blood soon after an infection begins and disappear before it is over. They are good activators of the complement system. IgA antibodies are monomers, dimers, or larger molecules containing two Y-shaped structures. They are the main type of antibody found in bodily secretions. They bind to pathogens before they reach the bloodstream. The main function of IgD antibodies seems to be to serve as receptors for antigens on mature B cells. IgE antibodies, which are involved in immediate allergic responses, are discussed on page 776.

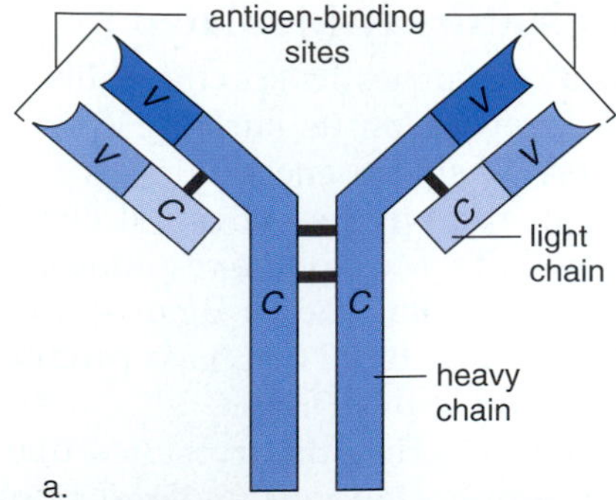

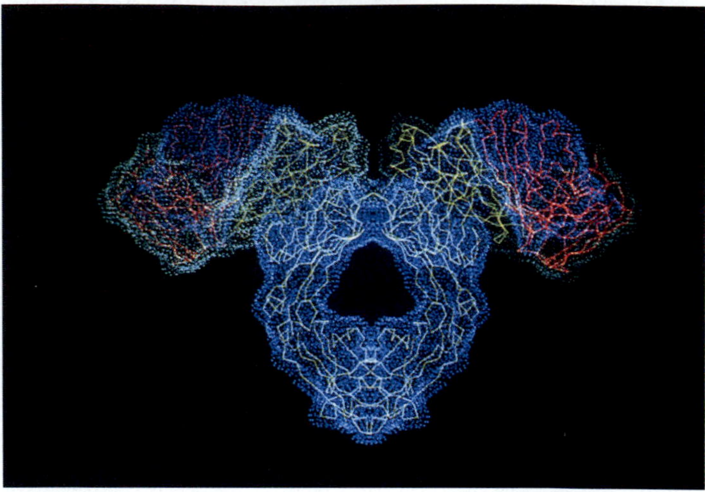

Figure 42.6 Structure of the most common antibody (IgG). **a.** An IgG antibody contains two heavy (long) polypeptide chains and two light (short) chains arranged so there are two variable regions, where a particular antigen is capable of binding with the antibody (*V* = variable region, *C* = constant region). **b.** Computer model of an antibody molecule. The antigen combines with the two side branches.

An antigen combines with an antibody at the antigen-binding site in a lock-and-key manner. The reaction can produce antigen-antibody complexes, which contain several molecules of antibody and antigen.

Table 42.1

Antibodies

Classes	Presence	Function
IgG	Main antibody type in circulation	Binds to pathogens, activates complement, and enhances phagocytosis
IgM	Antibody type found in circulation; largest antibody	Activates complement; clumps cells
IgA	Main antibody type in secretions such as saliva and milk	Prevents pathogens from attaching to epithelial cells in digestive and respiratory tract
IgD	Antibody type found in circulation in extremely low quantity	Presence signifies maturity of B cell
IgE	Antibody type found as membrane-bound receptor on basophils in blood and on mast cells in tissues	Responsible for immediate allergic response and protection against certain parasitic worms

Science Focus

Susumu Tonegawa and Antibody Diversity

In 1987, Susumu Tonegawa became the first Japanese scientist to win the Nobel Prize in Physiology or Medicine. He had dedicated himself to finding the solution to an engrossing puzzle. Immunologists and geneticists knew that each B cell makes an antibody especially equipped to recognize the specific shape of a particular antigen. But they did not know how the human genome contained enough genetic information to permit the production of up to a million different antibody types needed to combat all of the pathogens we are likely to encounter during the course of our lives.

An antibody is composed of two light and two heavy polypeptide chains, which are divided into constant and variable regions. The constant region determines the antibody class and the variable region determines the specificity of the antibody, because this is where an antigen binds to a specific antibody (see Fig. 42.6). Each B cell must have a genetic way to code for the variable regions of both the light and heavy chains.

Tonegawa's colleagues say that he is a creative genius who intuitively knows how to design experiments to answer specific questions. In this instance, he examined the DNA sequences of lymphoblasts and compared them to mature B cells. He found that the DNA segments coding for the variable and constant regions were scattered throughout the genome in B lymphocyte stem cells and that only certain of these segments appeared in each mature antibody-secreting B cell where they randomly came together and coded for a specific variable region. Later, the variable and constant regions are joined to give a specific antibody (Fig. 42A*b*). As an analogy, consider that each person entering a supermarket chooses various items for purchase, and that the possible combination of items in any particular grocery bag is astronomical. Tonegawa also found that mutations occur as the variable segments are undergoing rearrangements. Such mutations are another source of antibody diversity.

Invariably some B cells with receptors that could bind to the body's own cell surface molecules arise. It is believed that these cells undergo apoptosis, or programmed cell death.

Tonegawa received his B.S. in chemistry in 1963 at Kyoto University and earned his Ph.D. in biology from the University of California at San Diego (UCSD) in 1969. After that he worked as a research fellow at UCSD and the Salk Institute. In 1971, he moved to the Basel Institute for Immunology and began the experiments that eventually led to his Nobel Prize-winning discovery. Tonegawa also contributed to the effort to decipher the receptors of T cells. This was an even more challenging area of research than the diversity of antibodies produced by B cells. Since 1981, he has been a full professor at the Massachusetts Institute of Technology (MIT), where he has a reputation for being an "aggressive, determined researcher."

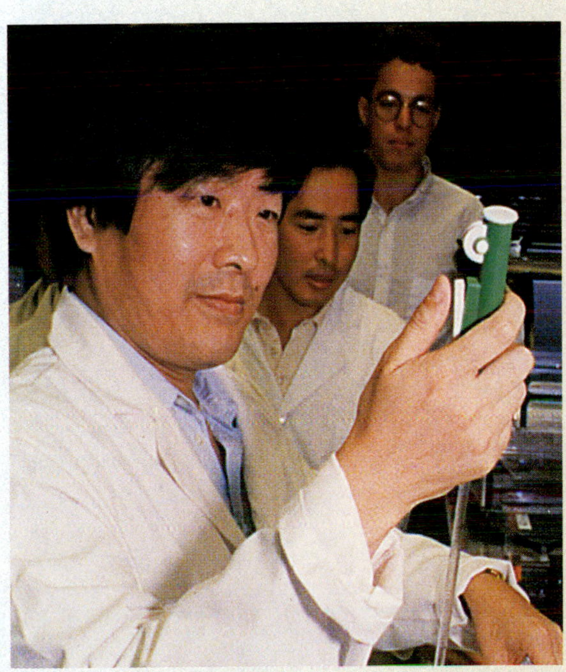

a.

genes for variable regions

Z Y X W V U

Z Y X V U

amino acids in variable region of one B cell antigen receptor (heavy or light chain)

Z Y X W

amino acids in variable region of another B cell antigen receptor (heavy or light chain)

b.

Figure 42A Antibody diversity.
a. Susumu Tonegawa received a Nobel Prize for his findings regarding antibody diversity. b. Different genes for the variable regions of heavy and light chains are brought together during the production of B lymphocytes so that their antigen receptors can combine with only a particular antigen.

T cells and Cell-Mediated Immunity

The two main types of T cells are cytotoxic T cells and helper T cells. **Cytotoxic T cells** are the type of T cell responsible for **cell-mediated immunity,** so called because T cells bring about the destruction of antigen-bearing cells, such as virus-infected or cancer cells. Cytotoxic T cells have storage vacuoles containing perforin molecules. When released, **perforin** molecules perforate a plasma membrane, forming a pore that allows water and salts to enter. The cell then swells and eventually bursts (Fig. 42.7).

Helper T cells regulate immunity by enhancing the response of other immune cells. When exposed to an antigen, they enlarge and secrete cytokines, stimulatory molecules that cause helper T cells to divide and other immune cells to perform their functions. For example, cytokines stimulate macrophages to phagocytize and stimulate B cells to become antibody-producing plasma cells. Because HIV, which causes AIDS, infects helper T cells and certain other cells of the immune system, it inactivates the immune response.

As we shall see, there are also memory T cells that remain in the body and can jump-start an immune reaction to an antigen previously present in the body.

Activation of T Cells

When T cells leave the thymus they have unique receptors just as B cells do. Unlike B cells, however, cytotoxic T cells and helper T cells are unable to recognize an antigen present in lymph, blood, or the tissues without help. The antigen must be presented to them by an **antigen-presenting cell (APC).** When an APC, usually a macrophage, engulfs a pathogen, the pathogen is broken down to fragments within an endocytic vesicle. These fragments are antigenic; that is, they have the properties of an antigen. After the fragments are linked to a major histocompatibility complex (MHC) protein in the plasma membrane, they can be presented to a T cell.

Human MHC proteins are self proteins called **HLA (human leukocyte associated) antigens.** They mark the cell

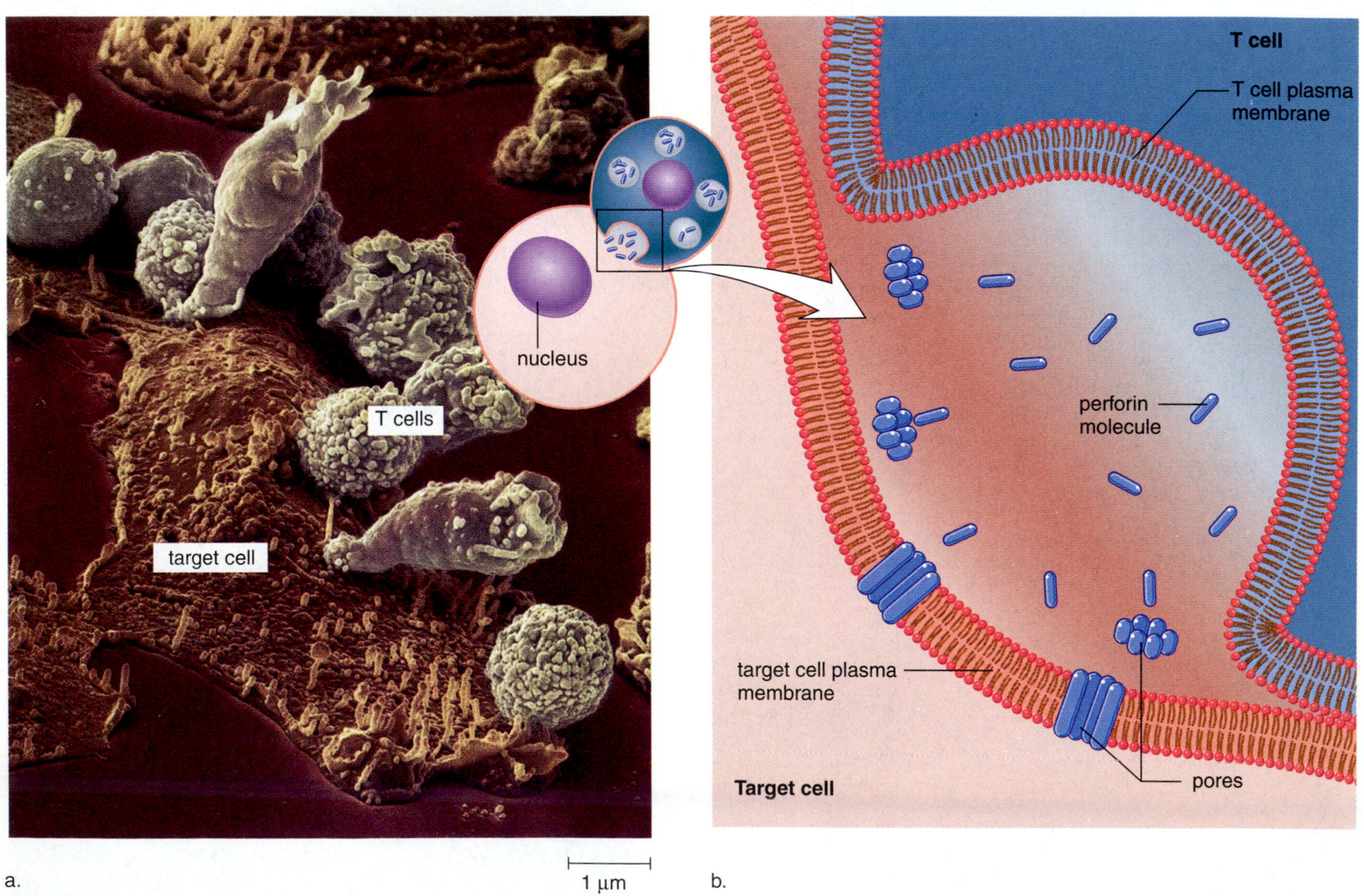

Figure 42.7 Cell-mediated immunity.
a. The scanning electron microscope shows cytotoxic T cells attacking and destroying a cancer cell. **b.** During the killing process, the vacuoles in a cytotoxic T cell release perforin molecules. These molecules combine to form pores in the target cell plasma membrane. Thereafter, fluid and salts enter so that the target cell eventually bursts.

as belonging to a particular individual. The importance of self proteins was first recognized when it was discovered that they make it difficult to transplant tissue from one human (or animal) to another. In other words, when the donor and the recipient are histo (tissue)-compatible (the same or nearly so), a transplant is more likely to be successful.

Figure 42.8 shows a macrophage presenting an antigen to a T cell. Once a helper T cell is activated in this manner, it undergoes clonal expansion and produces signaling molecules called **cytokines** that stimulate immune cells to remain active. Once a cytotoxic T cell is activated in this manner, it undergoes clonal expansion and destroys any cell infected with the same virus. (The infected cell bears the antigen linked to an HLA.) As the infection disappears, the immune reaction wanes and fewer cytokines are produced. Now, the activated T cells become susceptible to apoptosis. As mentioned previously, apoptosis is a programmed cell death that contributes to homeostasis by regulating the number of cells that are present in an organ, or in this case the immune system. A few of the clonally expanded T cells do not undergo apoptosis. The survivors are memory T cells—T cells that can rapidly respond should the same antigen be present at a later time.

Apoptosis also occurs in the thymus as T cells are maturing. A T cell that bears a receptor with the potential to recognize a self antigen undergoes suicide. When apoptosis does not occur as it should, T-cell cancers (i.e., lymphomas and leukemias) can result.

Characteristics of T cells:

- Cell-mediated immunity
- Produced in bone marrow, mature in thymus
- Antigen must be presented in groove of an HLA molecule.
- Cytotoxic T cells destroy antigen-bearing cells.
- Helper T cells secrete cytokines that control the immune response.

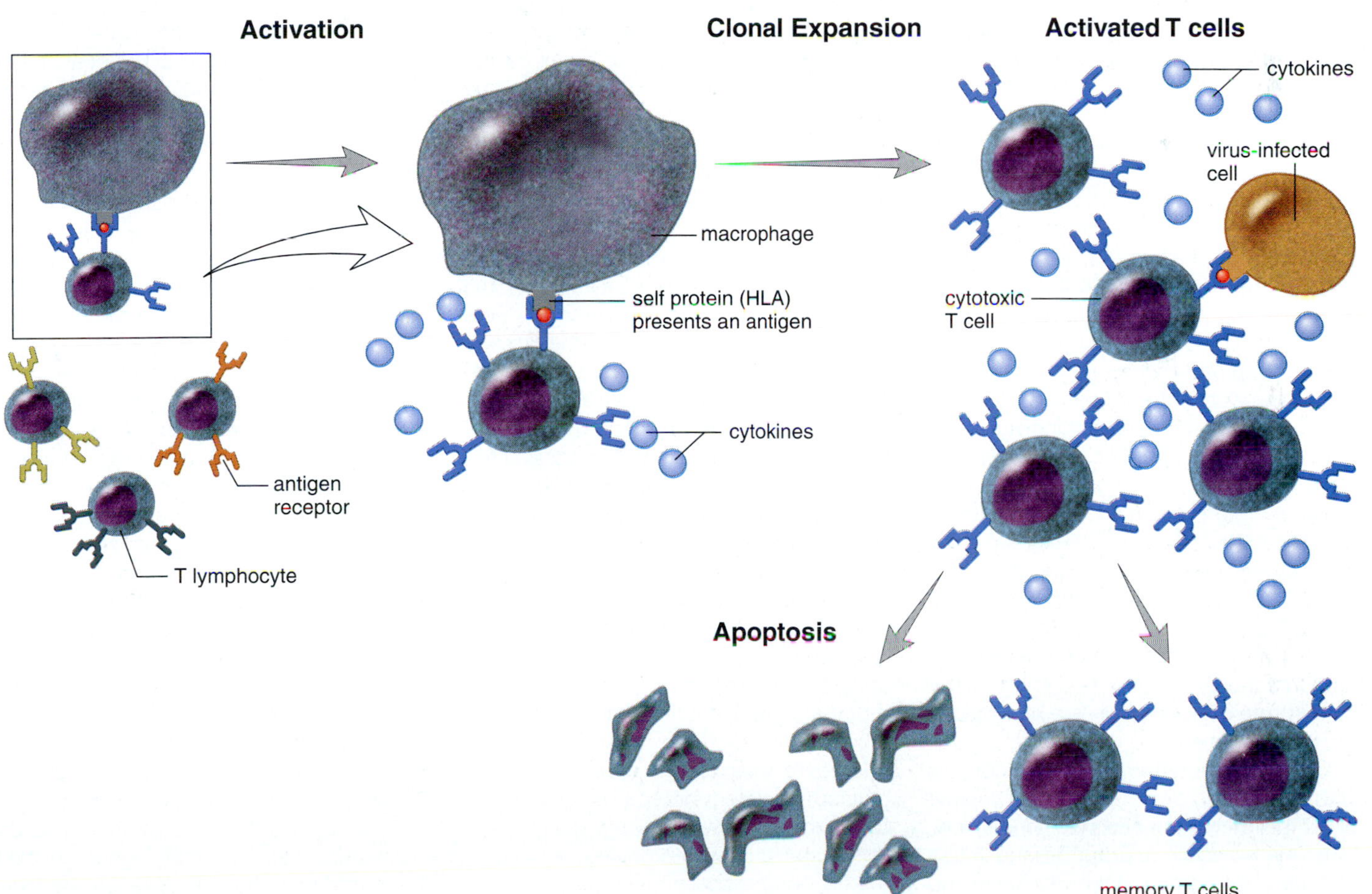

Figure 42.8 Clonal selection theory as it applies to T cells.
Each type of T cell bears a specific antigen receptor. When a macrophage presents an antigen in the groove of an HLA molecule to the correct T cell, it undergoes clonal expansion. After the immune response has been successful, the majority of T cells undergo apoptosis while a small number may become memory T cells. Memory T cells provide protection should the same antigen enter the body again at a future time.

AIDS

Acquired immunodeficiency syndrome (AIDS) is caused by a group of related retroviruses known as HIV (human immunodeficiency viruses). In the United States, AIDS is usually caused by HIV-1, which enters a host by attaching itself to a plasma protein called a CD4 receptor. HIV-1 infects helper T cells, the type of lymphocyte which stimulates B cells to produce antibodies and cytotoxic T cells to destroy virus-infected cells. Macrophages, which present antigens to helper T cells and thereby stimulate them, are also under attack.

HIV is a retrovirus, meaning that its genetic material consists of RNA instead of DNA. Once inside the host cell, HIV uses a special enzyme called reverse transcriptase to make a DNA copy (called cDNA) of its genetic material. Now cDNA integrates into a host chromosome, where it directs the production of more viral RNA. Each strand of viral RNA brings about synthesis of an outer protein coat called a capsid. The viral enzyme protease is necessary to the formation of capsids. Capsids assemble with RNA strands to form viruses which bud from the host cell (see Figure 29.4 page 512).

Transmission of HIV

Infection spreads when infected cells in body secretions (e.g., semen), and in blood are passed to another individual. To date, as many as 50 million people worldwide may have contracted HIV and almost 14 million have died. A new infection is believed to occur every 15 seconds, the majority in heterosexuals. HIV infections are not distributed equally throughout the world. Most infected people live in Africa (66%) where the infection first began, but new infections are now occurring at the fastest rate in Southeast Asia and the Indian subcontinent.

HIV is transmitted by sexual contact with an infected person, including vaginal or rectal intercourse and oral/genital contact. Also, needle-sharing among intravenous drug users is high-risk behavior. Babies born to HIV-infected women may become infected before or during birth, or through breast-feeding after birth.

HIV first spread through the homosexual community, and male-to-male sexual contact still accounts for the largest percentage of new AIDS cases in the United States. But the largest increases of HIV infections are occurring through heterosexual contact or by intravenous drug use. Now, women account for 20% of all newly diagnosed cases of AIDS. The rise in the incidence of AIDS among women of reproductive age is paralleled by a rise in the incidence of AIDS in children younger than 13.

Stages of an HIV Infection

The Centers for Disease Control and Prevention recognize three stages of an HIV-1 infection called category A, B, and C (Fig. 42B). During a category A stage, the helper T-lymphocyte count is 500 per mm^3 or greater. For a period of time after the initial infection with HIV, people don't usually have any symptoms at all. A few (1–2%) do have mononucleosis-like symptoms that may include fever, chills, aches, swollen lymph nodes, and an itchy rash. These symptoms disappear, however, and there are no other symptoms for quite some time. Although there are no symptoms, the person is highly infectious. Although there is a large number of viruses in the plasma, the HIV blood test is not yet positive because it tests for the presence of antibodies and not for the presence of HIV itself. This means that HIV can still be transmitted before the HIV blood test is positive.

Several months to several years after a nontreated infection, the individual will probably progress to category B in which the helper T-lymphocyte count is 200 to 499 per mm^3. During this stage there will be swollen lymph nodes in the neck, armpits, or groin that persist for three months or more. Other symptoms that indicate category B are severe fatigue not related to exercise or drug use; unexplained persistent or recurrent fevers, often with night sweats; persistent cough not associated with smoking, a cold, or the flu; and persistent diarrhea.

When the individual develops nonlife-threatening but recurrent infections, it is a signal that the disease is progressing. One possible infection is thrush, a fungal infection that is identified by the presence of white spots and ulcers on the tongue and inside the mouth. The fungus may also spread to the vagina, resulting in a chronic infection there. Another frequent infection is herpes simplex, with painful and persistent sores on the skin surrounding the anus, the genital area, and/or the mouth.

Previously, the majority of infected persons proceeded to category C, in which the helper T-lymphocyte count is 200 per mm^3 and the lymph nodes degenerate. The patient, who is now suffering from AIDS, characterized by severe weight loss and weakness due to persistent diarrhea and coughing, will most likely contract an opportunistic infection. An opportunistic infection is one that only has the opportunity to occur because the immune system is severely weakened. Persons with AIDS die from one or more opportunistic diseases and not from the HIV infection itself. The opportunistic diseases include: *Pneumocystis carinii* pneumonia—the lungs become useless as they fill with fluid and debris due to an infection with this organism. *Mycobacterium tuberculosis*—a bacterial infection, usually of the lungs that is seen more often as an infection of lymph nodes and other organs in patients with AIDS. Of special concern, tuberculosis is spreading into the general populace and is multidrug resistant. Toxoplasmic encephalitis—caused by a one-cell parasite that lives in cats and other animals as well as humans. Many persons harbor a latent infection in the brain or muscle, but in AIDS patients the infection leads to loss of brain cells, seizures, weakness, or decreased sensation on one side of the body. Kaposi's sarcoma—an unusual cancer of blood vessels, which gives rise to reddish purple, coin-size spots and lesions on the skin. Invasive cervical cancer—a cancer of the cervix that spreads to nearby tissues. This condition has been added to the list because the incidence of AIDS has now increased in women.

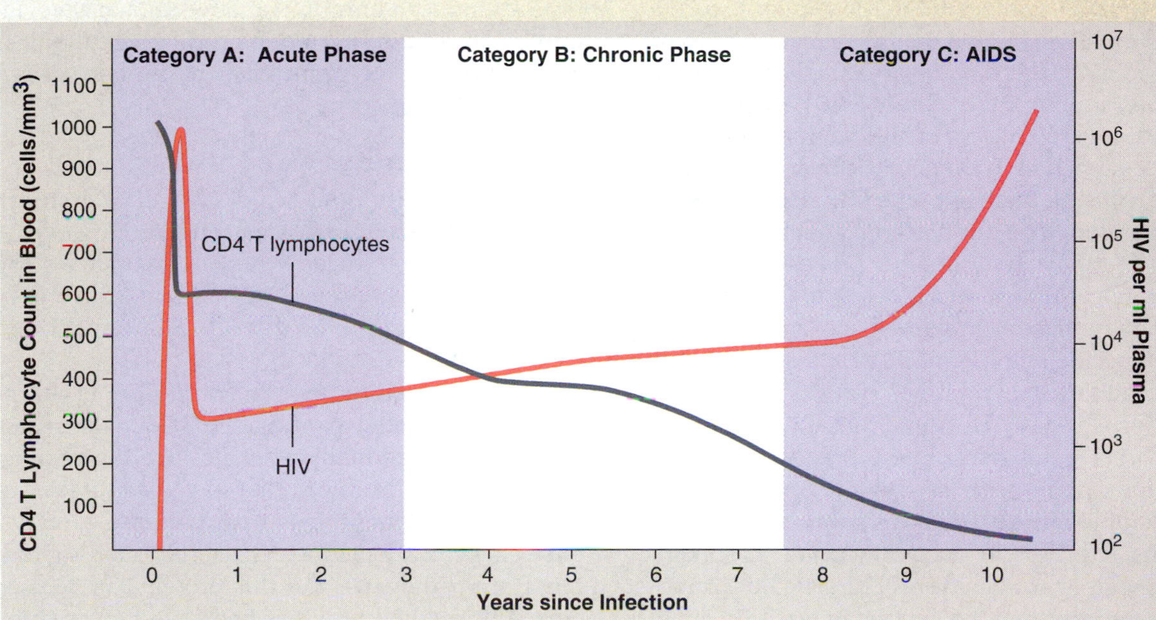

Figure 42B **Stages of an HIV infection.**
In category A individuals, the number of HIV in plasma rises upon infection and then falls. The number of CD4 T lymphocytes falls, but stays above 400 per mm^3. In category B individuals, the number of HIV in plasma is slowly rising and the number of T lymphocytes is decreasing. In category C individuals, the number of HIV in plasma rises dramatically as the number of T lymphocytes falls below 200 per mm^3.

Treatment for AIDS

Therapy usually consists of combining two drugs that inhibit reverse transcriptase with another that inhibits protease, an enzyme needed for formation of a viral capsid. This multidrug therapy, when taken according to the manner prescribed, seems to usually prevent mutation of the virus to a resistant strain. The sooner drug therapy begins after infection, the better the chances that the immune system will not be destroyed by HIV. And medication must be continued indefinitely. Unfortunately, an HIV strain resistant to all known drugs has been reported—persons who become infected with this strain have no drug therapy available to them.

The likelihood of transmission from mother to child at birth can be lessened if the mother takes an inhibitor of reserve transcriptase called AZT and the child is delivered by cesarean section.

Many investigators are working on a vaccine for AIDS. Some are trying to develop a vaccine in the traditional way. Others are working on subunit vaccines that utilize just a single HIV protein as the vaccine. For example, a protein from the viral envelope has been produced by genetic engineering of bacteria. This protein is undergoing clinical trial as a vaccine in Thailand. Also, investigators have found that viral cDNA for this protein can act as a vaccine because it causes cells to produce and display the protein at their plasma membrane.

To Stop the Spread of AIDS:

- Abstain from sexual intercourse, or develop a long-term monogamous (always the same partner) sexual relationship with a partner who is free of HIV and is not an intravenous drug user.
- Practice safe sex. If you do not know for certain that your partner has been free of HIV for the past five years, always use a latex condom during sexual intercourse. Be sure to follow the directions supplied by the manufacturer. Use of a spermicide containing nonoxynol-9 in addition to the condom can offer further protection because nonoxynol-9 immobilizes the virus and virus-infected lymphocytes.
- Avoid fellatio (kissing and insertion of the penis into a partner's mouth) and cunnilingus (kissing and insertion of the tongue into the vagina) because they may be a means of transmission. The mouth and gums often have cuts and sores that facilitate the entrance of infected T lymphocytes.
- Stop, if necessary, or do not start the habit of injecting drugs into veins. If you are a drug user and cannot stop your behavior, then always use a new sterile needle for injection or one that has been cleaned in bleach.
- Aside from intravenous drug use, do not use alcohol or any drugs in a way that may prevent you from being able to control your behavior.

42.4 Immunity in Other Animals

The abundance of invertebrates suggests that they must have some way to defend themselves against pathogens. As far back as 1882, the Russian zoologist Elie Metchnikoff stuck a rose thorn into a starfish and noted that phagocytes gathered in an attempt to remove the foreign object. Chemowarfare has also been found among invertebrates. In 1979, Swedish Hans G. Boman discovered a class of silk moth antibacterial peptides, which he named cecropins. Like complement, they perforate bacteria, causing them to burst. (These peptides are currently being developed as antibacterial agents for use in humans.) Similarly, many invertebrates possess molecules related to vertebrate cytokines. In the sea star, cells that resemble macrophages release a cytokine-like chemical.

Apparently, specific defense mechanisms evolved only among the vertebrates. Gary W. Litman has studied specific immunity in the horned shark. He found that antibody diversity has a genetic basis in this shark, just as it does in humans. But the horned shark relies more heavily on "inherited" diversity. This shark more quickly produces antibodies directed against pathogens it is apt to encounter than antibodies against new and different antigens. This characteristic of antibody-mediated immunity may be an advantage if the environment remains relatively constant. Investigators have hypothesized that cell-mediated immunity based on the presence of T cells predates antibody-mediated immunity based on B cells. Therefore, Litman decided to test the horned shark for evidence of T-cell activity. He used the polymerase chain reaction (PCR) to replicate human T-cell receptor genes prior to sequencing them. Thereafter, he found evidence of corresponding proteins in the horned shark.

There is evidence of nonspecific defenses in invertebrates, but specific defenses seem to be unique to vertebrates.

42.5 Induced Immunity

Immunity occurs naturally through infection or is brought about artificially by medical intervention. There are two types of induced immunity: active and passive. In active immunity, the individual alone produces antibodies against an antigen; in passive immunity, the individual is given prepared antibodies.

Active Immunity

Active immunity sometimes develops naturally after a person is infected with a pathogen. However, active immunity is often induced when a person is well so that possible future infection will not take place. To prevent infections, people can be artificially immunized against them (Fig. 42.9).

Immunization involves the use of **vaccines,** substances that contain an antigen to which the immune system responds. Traditionally, vaccines are the pathogens themselves, or their products, that have been treated so they are no longer virulent (able to cause disease). Today, it is possible to genetically engineer bacteria to mass-produce a protein from pathogens, and this protein can be used as a vaccine. This method now has been used to produce a vaccine against hepatitis B, a viral disease, and is being used to prepare a vaccine against malaria, a protozoan disease.

After a vaccine is given, it is possible to follow an immune response by determining the amount of antibody present in a sample of plasma—this is called the antibody titer. After the first exposure to a vaccine, a primary response occurs. For a period of several days, no antibodies are present; then, there is a slow rise in the titer, followed by first a plateau and then a gradual decline as the antibodies bind to the antigen or simply break down (Fig. 42.9). After a second exposure, a secondary response is expected. The titer rises rapidly to a plateau level much greater than before. The second exposure is called a "booster" because it boosts the

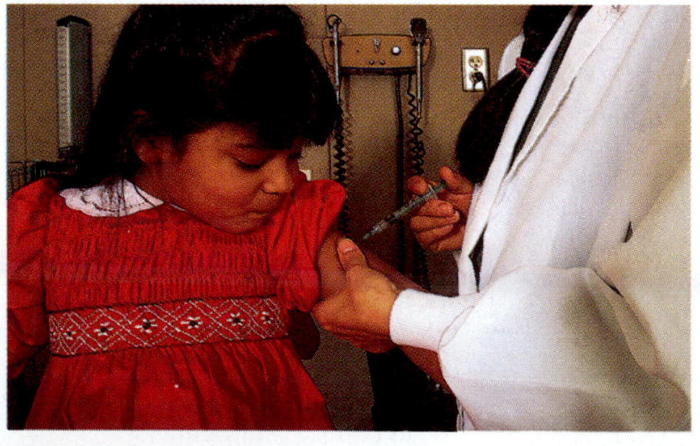

a.

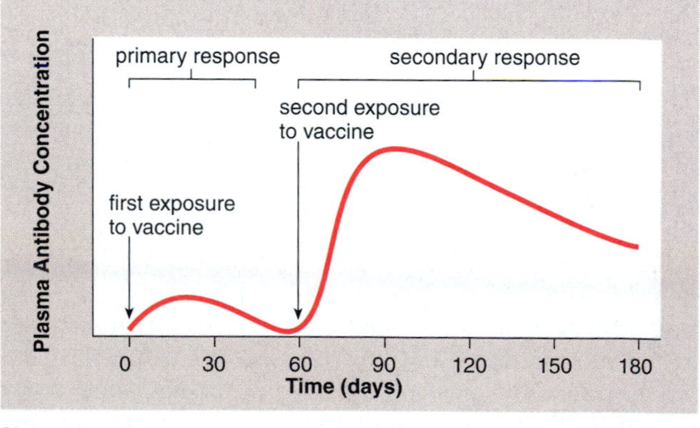

ßb.

Figure 42.9 Active immunity due to immunizations.
a. Vaccines are used to immunize children against various childhood diseases. **b.** The primary response, after the first exposure to a vaccine, is minimal, but the secondary response, which may occur after the second exposure, shows a dramatic rise in the amount of antibody present in plasma.

antibody titer to a high level. The high antibody titer now is expected to help prevent disease symptoms even if the individual is exposed to the disease-causing antigen.

Active immunity is dependent upon the presence of memory B cells and memory T cells which are capable of responding to lower doses of antigen. Active immunity is usually long-lived, although a booster may be required every so many years.

Passive Immunity

Passive immunity occurs when an individual is given prepared antibodies (immunoglobulins) to combat a disease. Since these antibodies are not produced by the individual's B cells, passive immunity is short-lived. For example, newborn infants are passively immune to some diseases because antibodies have crossed the placenta from the mother's blood. These antibodies soon disappear, however, so that within a few months, infants become more susceptible to infections. Breast-feeding prolongs the natural passive immunity an infant receives from the mother because antibodies are present in the mother's milk (Fig. 42.10).

Even though passive immunity does not last, it sometimes is used to prevent illness in a patient who has been unexpectedly exposed to an infectious disease. Usually, the patient receives a gamma globulin injection (serum that contains antibodies), perhaps taken from individuals who have recovered from the illness. In the past, horses were immunized, and serum was taken from them to provide the needed antibodies against such diseases as diphtheria, botulism, and tetanus. In the past, a patient who received these antibodies became ill about 50% of the time because the serum contained proteins that the individual's immune system recognized as foreign. This was called serum sickness. But problems can still occur with products produced in other ways. An immunoglobulin intravenous product called Gammagard was withdrawn from the market because of possible implication in the transmission of hepatitis.

Active (long-lived) immunity, which is dependent upon the presence of memory B cells and memory T cells, can be induced by the use of vaccines. Passive immunity is short-lived because there are no memory cells.

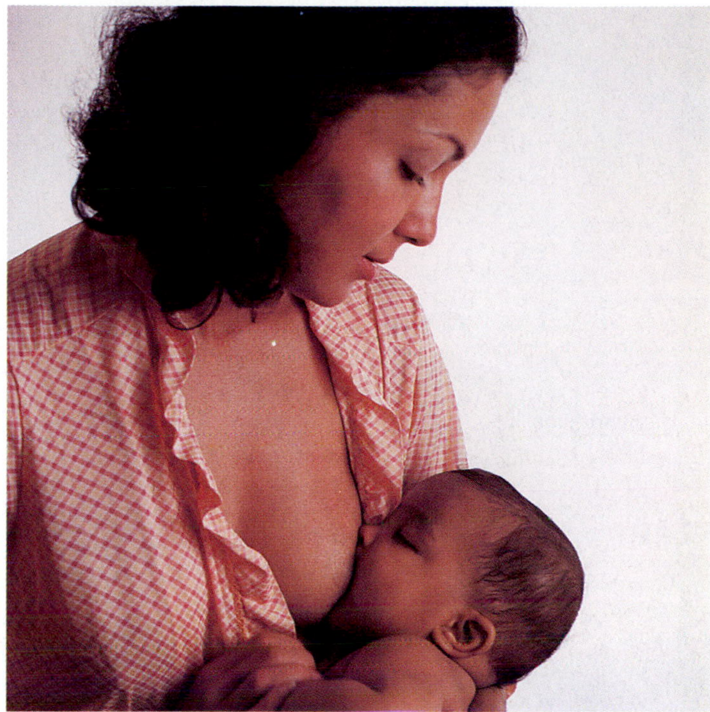

Figure 42.10 Passive immunity.
Breast-feeding is believed to prolong the passive immunity an infant receives from the mother because antibodies are present in the mother's milk.

Cytokines and Immunity

Cytokines are signaling molecules produced by lymphocytes, monocytes, or other cells. Because cytokines regulate white blood cell formation and/or function, they are being investigated as possible adjunct therapy for cancer and AIDS. Both interferon and **interleukins,** which are cytokines produced by various white blood cells, have been used as immunotherapeutic drugs, particularly to enhance the ability of the individual's own T cells (and possibly B cells) to fight cancer.

Interferon, discussed previously on page 764, is a substance produced by leukocytes, fibroblasts, and probably most cells in response to a viral infection. Interferon still is being investigated as a possible cancer drug, but so far it has proven to be effective only in certain patients, and the exact reasons for this as yet cannot be discerned.

When and if cancer cells carry an altered protein on their cell surface, they should be attacked and destroyed by cytotoxic T cells. Whenever cancer does develop, it is possible that the cytotoxic T cells have not been activated. In that case, cytokines might awaken the immune system and lead to the destruction of the cancer. In one technique being investigated, researchers first withdraw T cells from the patient and activate the cells by culturing them in the presence of an interleukin. The cells then are reinjected into the patient, who is given doses of interleukin to maintain the killer activity of the T cells.

Those who are actively engaged in interleukin research believe that interleukins soon will be used as adjuncts for vaccines, for the treatment of chronic infectious diseases, and perhaps for the treatment of cancer. Interleukin antagonists also may prove helpful in preventing skin and organ rejection, autoimmune diseases, and allergies.

Various cytokines show some promise of potentiating the individual's own immune system.

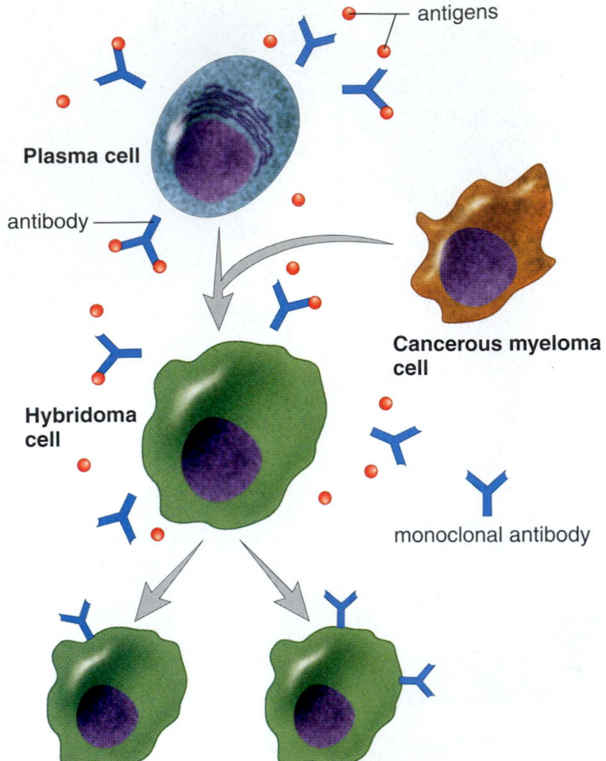

Figure 42.11 **Production of monoclonal antibodies.** Plasma cells (derived from immunized mice) are fused with myeloma (cancerous) cells, producing hybridoma cells that are "immortal." Hybridoma cells divide and continue to produce the same type of antibody, called monoclonal antibodies.

Monoclonal Antibodies

Every plasma cell derived from the same B cell secretes antibodies against a specific antigen. These are **monoclonal antibodies** because all of them are the same type and because they are produced by plasma cells derived from the same B cell. One method of producing monoclonal antibodies in vitro (outside the body in a laboratory) is depicted in Figure 42.11. B lymphocytes are removed from an animal (today, usually mice are used) and are exposed to a particular antigen. The resulting plasma cells are fused with myeloma cells (malignant plasma cells that live and divide indefinitely). The fused cells are called hybridomas; *hybrid* because they result from the fusion of two different cells, and *oma* because one of the cells is a cancer cell.

At present, monoclonal antibodies are being used for quick and certain diagnosis of various conditions. For example, a particular hormone is present in the urine of a pregnant woman. A monoclonal antibody can be used to detect this hormone; if it is present, the woman knows she is pregnant. Monoclonal antibodies also are used to identify infections. And because they can distinguish between cancer and normal tissue cells, they are used to carry radioactive isotopes or toxic drugs to tumors so that cancer cells can be selectively destroyed.

42.6 Immunity Side Effects

The immune system usually protects us from disease because it can distinguish self from nonself. Sometimes, however, it responds in a manner that does harm to the body, as when individuals receive the wrong blood type, suffer tissue rejection, have an autoimmune response, or develop allergies.

Blood Types

When blood transfusions were first attempted, illness and even death sometimes resulted. Eventually, it was discovered that only certain types of blood are compatible because red blood cell plasma membranes carry proteins or sugar residues that are antigens to blood recipients. The ABO system of typing blood is based on this principle.

ABO System

Blood typing in the ABO system is based on two antigens known as antigen A and antigen B. There are four blood types: O, A, B, and AB. Type O has neither the A antigen nor the B antigen on red blood cells; the other types of blood have antigen A, B, or both A and B, respectively (Table 42.2).

Within plasma, there are naturally occurring antibodies to the antigens not present on the person's red blood cells. This is reasonable, because if the same antigen and antibody are present in blood, **agglutination,** or clumping of red blood cells, occurs. Agglutination causes blood to stop circulating and red blood cells to burst.

Figure 42.12 shows a way to use the antibodies derived from plasma to determine the blood type. If agglutination occurs after a sample of blood is mixed with a particular antibody, the person has that type of blood.

Rh System

Another important antigen in matching blood types is the Rh factor. Persons with the Rh factor on their red blood cells are Rh positive (Rh^+); those without it are Rh negative (Rh^-). Rh-negative individuals normally do not have antibodies to the Rh factor, but they may make them when exposed to the Rh factor during pregnancy or blood transfusion.

If a mother is Rh negative and a father is Rh positive, a child may be Rh positive. The Rh-positive red blood cells of the child may begin leaking across the placenta into the mother's circulatory system, as placental tissues normally break down before and at birth. This sometimes causes the mother to produce anti-Rh antibodies. In this or a subsequent pregnancy with another Rh-positive child, anti-Rh antibodies may cross the placenta and destroy the child's red blood cells. This condition is called hemolytic disease of the newborn (HDN).

The Rh problem has been solved by giving Rh-negative women an Rh-immunoglobulin injection (often a RhoGam injection) either midway through the first pregnancy or no later than 72 hours after giving birth to any Rh-positive child. This injection contains anti-Rh antibodies, which

Table 42.2

The ABO System

Blood Type	Antigen on Red Blood Cells	Antibody in Plasma	% U.S. African American	% U.S. Caucasian	% U.S. Asian	% North American Indians	% Americans of Chinese Descent
A	A	Anti-B	27	41	28	8	25
B	B	Anti-A	20	9	27	1	35
AB	A, B	None	4	3	5	0	10
O	None	Anti-A and anti-B	49	47	40	91	30

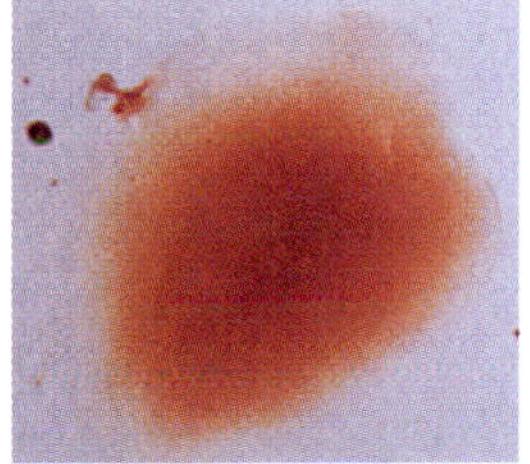

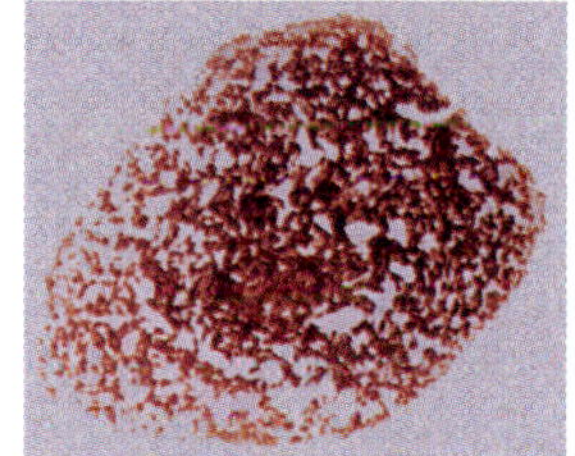

a. No agglutination Agglutination

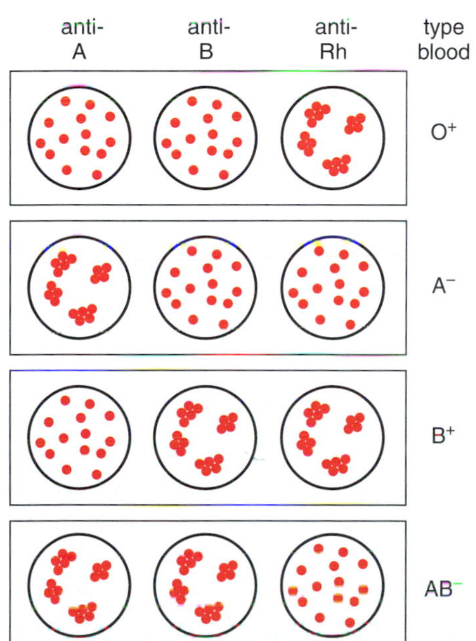

b.

Figure 42.12 **Blood typing.**
The standard test to determine ABO and Rh blood type consists of putting a drop of anti-A antibodies, anti-B antibodies, and anti-Rh antibodies on a slide. To each of these, a drop of the person's blood is added. **a.** If agglutination occurs, as seen in the photo on the right, the person has this antigen on red blood cells. **b.** Several possible results.

attack any of the child's red blood cells in the mother's blood before these cells can stimulate her immune system to produce her own antibodies. This injection is not beneficial if the woman has already begun to produce antibodies; therefore, the timing of the injection is most important.

Blood is often typed according to the ABO system combined with the Rh system. The possibility of hemolytic disease of the newborn exists when the mother is Rh negative and the father is Rh positive.

Tissue Rejection

The body will reject blood cells and also other tissues that carry foreign antigens. Certain organs, such as skin, the heart, and the kidneys, could be transplanted easily from one person to another if the body did not attempt to reject them. Rejection occurs because cytotoxic T cells bring about destruction of foreign tissue in the body.

Organ rejection can be controlled by careful selection of the organ to be transplanted and the administration of immunosuppressive drugs. It is best if the transplanted organ has the same type of self proteins (HLA antigens) as those of the recipient, because cytotoxic T cells recognize foreign HLA antigens. The immunosuppressive drug cyclosporine has been used for many years. The drug tacrolimus (formerly known as FK-506) shows some promise, especially in liver transplant patients. However, both drugs, which act by inhibiting the response of T cells to cytokines, are known to adversely affect the kidneys.

When an organ is rejected, the immune system has recognized and destroyed cells that bear HLA antigens different from those of the individual.

Autoimmune Diseases

When T cells or antibodies mistakenly attack the body's own cells as if they bore foreign antigens, the resulting condition is known as an **autoimmune disease.** Exactly what causes autoimmune diseases is not known. However, sometimes they occur after an individual has recovered from an infection.

In the autoimmune disease myasthenia gravis, transmission at neuromuscular junctions is impaired and muscular weakness results. In multiple sclerosis, the myelin sheath of nerve fibers breaks down, and this causes various neuromuscular disorders. A person with systemic lupus erythematosus has various symptoms prior to death due to kidney damage. In rheumatoid arthritis, the joints are affected. Researchers suggest that heart damage following rheumatic fever and type I diabetes are also autoimmune illnesses. As yet there are no cures for autoimmune diseases, but they can be controlled with drugs.

Autoimmune diseases occur when antibodies and cytotoxic T cells recognize and destroy the body's own cells.

Allergies

Allergies are hypersensitivities to substances such as pollen or animal hair that ordinarily would do no harm to the body (Fig. 42.13). The response to these antigens, called **allergens,** usually includes some degree of tissue damage. There are four types of allergic responses, but we will consider only two of these: immediate allergic responses and delayed allergic responses.

Immediate Allergic Response

An **immediate allergic response** occurs within seconds of contact with the antigen. Coldlike symptoms are common. Anaphylactic shock is a severe reaction characterized by a sudden and life-threatening drop in blood pressure.

One type of immediate allergic response is caused by antibodies known as IgE (see Table 42.1). IgE antibodies are attached to the plasma membrane of mast cells in the tissues and basophils in the blood. When an allergen attaches to the IgE antibodies on these cells, they release histamine and other substances that bring about the coldlike symptoms or, rarely, anaphylactic shock.

Allergy shots sometimes prevent the onset of an allergic response. It's been suggested that injections of the allergen may cause the body to build up high quantities of IgG antibodies, and these combine with allergens received from the environment before they have a chance to reach the IgE antibodies located in the membrane of mast cells and basophils.

Delayed Allergic Response

Delayed allergic responses are initiated by sensitized T cells at the site of allergen in the body. A sensitized T cell is one that is ready to respond to the antigen because the antigen has been in the body before. T cells initiate the response by recruiting the help of macrophages, which are able to phagocytize offending viral particles or infectious cells. The overall response is regulated by the cytokines secreted by both the T cells and macrophages.

A classic example of a delayed allergic response is the tuberculin skin test. When the result of the test is positive, there is a reddening and hardening of tissue where the antigen was injected. This shows that there was prior exposure to tubercle bacilli which cause TB. Contact dermatitis, such as occurs when one is allergic to poison ivy, jewelry, cosmetics, and so forth, is also an example of a delayed allergic response.

Allergies occur when the immune system reacts to environmental substances that would do the body no harm.

Figure 42.13 Protection against allergies.
The allergic reaction known as hay fever, and asthma attacks, can have many triggers, one of which is the pollen of a variety of plants. A dramatic solution to the problem has been found by these people.

Connecting Concepts

The role of the lymphatic system in homeostasis cannot be overemphasized. The internal environment of cells consists of tissue fluid and lymph. If the composition of these fluids stays relatively constant, homeostasis is maintained. The lymphatic system is also intimately involved in immunity. White blood cells are made in the red bone marrow, where B cells also mature. T cells mature in the thymus. The spleen filters the blood directly. Clonal expansion of lymphocytes occurs in the lymph nodes, which also filter the lymph.

The defense systems of humans have been extensively studied, but little is known about these same systems in other animals. In humans, the levels of defense against invasion of the body by pathogens can be compared to how we protect our homes. Homes usually have external defenses such as a fence, a dog, or locked doors. Similarly, the body has barriers, like the skin, that prevent pathogens from entering the blood and lymph. Like a home alarm system, if invasion does occur, a signal goes on. First nonspecific defense mechanisms like the complement system and phagocytosis by white blood cells come into play. Finally, specific defense, which is dependent on the activities of B and T cells, occurs.

In humans, a strong connection exists between the immune, nervous, and endocrine systems. Lymphocytes have receptors for a wide variety of hormones, and the thymus gland produces hormones which influence the immune response. Cytokines help the body recover from disease by affecting the brain's temperature control center. A fever is thought to create an unfavorable environment for foreign invaders. Also, cytokines bring about a feeling of sluggishness, sleepiness, and loss of appetite. These behaviors tend to make us take care of ourselves until we feel better. A close connection between the immune and hormonal systems is illustrated by the ability of cortisone to mollify the inflammatory reaction in joints.

Summary

42.1 Lymphatic System

The lymphatic system consists of lymphatic vessels and lymphoid organs. The lymphatic vessels collect fat molecules at intestinal villi and excess tissue fluid at blood capillaries, and carry these to the bloodstream.

Lymphocytes are produced and accumulate in the lymphoid organs (red bone marrow, lymph nodes, spleen, and thymus gland). Lymph is cleansed of pathogens and/or their toxins in lymph nodes, and blood is cleansed of pathogens and/or their toxins in the spleen. T lymphocytes mature in the thymus, while B lymphocytes mature in the red bone marrow where all blood cells are produced. White blood cells are necessary for nonspecific and specific defenses.

42.2 Nonspecific Defenses

Immunity involves nonspecific and specific defenses. Nonspecific defenses include barriers to entry, the inflammatory reaction, natural killer cells, and protective proteins.

42.3 Specific Defenses

Specific defenses require lymphocytes, which are produced in the bone marrow. B cells mature in the bone marrow. They undergo clonal selection with production of plasma cells and memory B cells after their specific plasma membrane receptors directly combine with a particular antigen. Plasma cells secrete antibodies and eventually undergo apoptosis. Plasma cells are responsible for antibody-mediated immunity. IgG antibody is a Y-shaped molecule that has two binding sites for a specific antigen. Memory B cells remain in the body and produce antibodies if the same antigen enters the body at a later date.

T cells, which are responsible for cell-mediated immunity, mature in the thymus. The two main types of T cells are cytotoxic T cells and helper T cells. Cytotoxic T cells kill infected cells that bear a foreign antigen on contact; helper T cells stimulate other immune cells and produce cytokines. Like B cells, each T cell bears a specific receptor. However, for a T cell to recognize an antigen, the antigen must be presented by an antigen-presenting cell (APC), usually a macrophage. The antigen is linked to a self protein [HLA (human lymphocyte-associated) antigen]. Thereafter, the activated T cell undergoes clonal expansion until the infection has been stemmed. Then most of the activated T cells undergo apoptosis. A few cells remain, however, as memory T cells.

42.4 Immunity in Other Animals

There is evidence of only nonspecific defenses in invertebrates. Other vertebrates besides humans apparently have both nonspecific and specific defenses. Investigative work has particularly been done in echinoderms and sharks.

42.5 Induced Immunity

Immunity can be induced in various ways. Vaccines are available to induce long-lived active immunity, and antibodies sometimes are available to provide an individual with short-lived passive immunity.

Cytokines, including interferon, are used in an attempt to promote the body's ability to recover from cancer and to treat AIDS.

42.6 Immunity Side Effects

In the ABO blood system there are four types of blood: A, B, AB, and O, depending on the type of antigen present on red blood cells. If the red blood cells have a particular antigen, they clump when exposed to the corresponding antibody. Table 42.2 tells which types of antibodies are present in the plasma for the various ABO blood groups.

Allergic responses occur when the immune system reacts vigorously to substances not normally recognized as foreign. Immediate allergic responses, usually consisting of coldlike symptoms, are due to the activity of antibodies. Delayed allergic responses, such as contact dermatitis, are due to the activity of T cells.

The immune system sometimes responds in a way that does harm to the body, as when a person receives the wrong blood type, suffers tissue rejection, or has an autoimmune disease or allergies.

Reviewing the Chapter

1. What is the lymphatic system, and what are its three functions? 760
2. Describe the structure and the function of lymph nodes, the spleen, the thymus, and red bone marrow. 761–62
3. What are the body's nonspecific defense mechanisms? 762–64
4. Describe the inflammatory reaction, and give a role for each type of cell and molecule that participates in the reaction. 762–63
5. Describe the clonal selection theory as it applies to B cells. B cells are responsible for which type of immunity? 765
6. Describe the structure of an antibody, and define the terms variable regions and constant regions. 766
7. Name the two main types of T cells, and state their functions. 768
8. Describe the clonal selection theory as it applies to T cells. 768–69
9. Describe the evidence for immunity in other animals. 772
10. How is active immunity artificially achieved? How is passive immunity achieved? 772–73
11. What are cytokines, and how are they used in immunotherapy? 773
12. How are monoclonal antibodies produced, and what are their applications? 774
13. Discuss blood typing, tissue rejection, autoimmune diseases, and allergies as they relate to the immune system. 775–76

Testing Yourself

Choose the best answer for each question.

1. Complement
   a. is a general defense mechanism.
   b. probably occurs in invertebrates.
   c. is a series of proteins present in the plasma.
   d. plays a role in destroying bacteria.
   e. All of these are correct.
2. Which of these pertain(s) to T cells?
   a. have specific receptors
   b. are more than one type
   c. are responsible for cell-mediated immunity
   d. help stimulate antibody production by B cells
   e. All of these are correct.
3. Which one of these does not pertain to B cells?
   a. have passed through the thymus
   b. have specific receptors
   c. are responsible for antibody-mediated immunity
   d. undergoes clonal expansion
   e. become plasma cells which liberate antibodies
4. The clonal selection theory says that
   a. an antigen selects certain B cells and suppresses them.
   b. an antigen stimulates the multiplication of B cells that produce antibodies against it.
   c. T cells select those B cells that should produce antibodies, regardless of antigens present.
   d. T cells suppress all B cells except the ones that should multiply and divide.
   e. Both b and c are correct.
5. Plasma cells are
   a. the same as memory cells.
   b. formed from blood plasma.
   c. B cells that are actively secreting antibody.
   d. inactive T cells carried in the plasma.
   e. a type of red blood cell.
6. For a T cell to recognize an antigen, it usually interacts with
   a. complement.
   b. a macrophage.
   c. a B cell.
   d. a thymus cell.
   e. All of these are correct.
7. Antibodies combine with antigens
   a. at variable regions.
   b. at constant regions.
   c. only if macrophages are present.
   d. only if T cells are present.
   e. Both a and c are correct.
8. Which one of these is mismatched?
   a. helper T cells—help complement react
   b. cytotoxic T cells—active in tissue rejection
   c. macrophage—activate T cells
   d. memory T cells—long-living line of T cells
   e. T cells—mature in thymus
9. Vaccines are
   a. the same as monoclonal antibodies.
   b. treated bacteria or viruses, or one of their proteins.
   c. short-lived.
   d. MHC proteins.
   e. All of these are correct.
10. The theory behind the use of cytokines in cancer therapy is that
    a. if cancer develops, the immune system has been ineffective.
    b. cytokines stimulate the immune system.
    c. cancer cells bear antigens that should be recognizable by cytotoxic T cells.
    d. cytokines can be isolated from the blood.
    e. All of these are correct.
11. During blood typing, agglutination indicates that the
    a. plasma contains certain antibodies.
    b. red blood cells carry certain antigens.
    c. plasma contains certain antigens.
    d. red blood cells carry certain antibodies.
    e. white blood cells fight infection.
12. Use these terms to label this IgG molecule: antigen-binding sites, light chain, heavy chain.

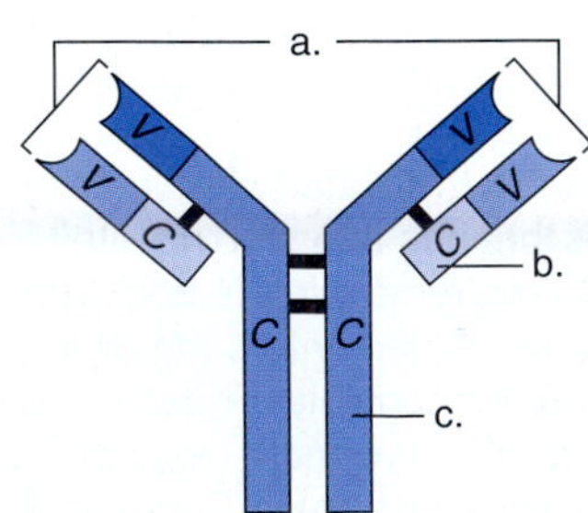

    d. What does *V* stand for in the diagram?
    e. What does *C* stand for in the diagram?

Thinking Scientifically

1. You are feeling very tired and lethargic. You believe you might have mononucleosis. The results of your blood test show that you have IgG antibodies against EBV, the virus that causes mononucleosis, but no IgM. Is mononucleosis likely to be the cause of these symptoms?
2. Laboratory mice are immunized with a measles vaccine. When the mice are challenged with measles virus to test the strength of their immunity, the memory cells do not completely prevent replication of the measles virus. The virus undergoes a few rounds of replication before the immune response is observed. You have developed a strain of mice with a much faster response to a viral challenge, but these mice often develop an autoimmune disease. Speculate on the connection between speed of response and an autoimmune disease.

Bioethical Issue

The United Nations estimates that at least 16,000 people become newly infected with a human immunodeficiency virus (HIV) each day, or 5.8 million per year. Ninety percent of these infections occur in the less-developed countries[2] where infected persons do not have access to antiviral therapy. In Uganda, for example, there is only one physician per 100,000 people, and only $6.00 is spent annually on health care, per person. In contrast, in the United States $12,000 to $15,000 is sometimes spent on treating an HIV-infected person per year.

The only methodology presently available to prevent the spread of HIV is counseling against behaviors that increase the risk of infection. Clearly an effective vaccine would be most beneficial to all countries, and especially less-developed countries. Several HIV vaccines are in various stages of development, and all need to be clinically tested in order to see if they are effective. It seems reasonable to carry out such trials in less-developed countries, but there are many ethical questions.

A possible way to carry out the trial is this: vaccinate the uninfected sexual partners of HIV-infected individuals. If the uninfected partner remains free of the disease, then the vaccine is effective. But is it ethical to allow a partner identified as having an HIV infection to remain untreated for the sake of the trial?

And should there be a placebo group—a group that does not get the vaccine? If a greater number of persons in the placebo group become infected than those in the vaccine group, then the vaccine is effective. But if members of the placebo group become infected, shouldn't they be given effective treatment? For that matter, even participants in the vaccine group might become infected. Shouldn't any participant of the trial be given proper treatment if they become infected? Who would pay for such treatment when the trial could involve thousands of persons?

[2] Country in which population growth is expanding rapidly and the majority of people live in poverty.

Understanding the Terms

Match the terms to these definitions:

a. ______________ Antigens prepared in such a way that they can promote active immunity without causing disease.
b. ______________ Fluid, derived from tissue fluid, that is carried in lymphatic vessels.
c. ______________ Foreign substance, usually a protein or a polysaccharide, that stimulates the immune system to react, such as to produce antibodies.
d. ______________ Process of programmed cell death involving a cascade of specific cellular events leading to the death and destruction of the cell.
e. ______________ Lymphocyte that matures in the thymus and exists in three varieties, one of which kills antigen-bearing cells outright.

Web Connections

Exploring the Internet

http://www.mhhe.com/biosci/genbio/mader
(click on *Biology 7/e*)

The *Biology 7/e* Online Learning Center provides many resources for studying the material in this chapter including links to the following sites:

The website of the National Institute of Health contains an information index that will help you to find the institution responsible for the subject of your choice. The Health Information Page will also supply you with a link to that institute's home page.

http://www.nih.gov/health/

The AIDS Virtual Library contains volumes of information on health care topics, drug trials, social services, and even safer sex as it relates to AIDS.

http://planetq.com/aidsvl/index.html

HIV InSite. Gateway to AIDS knowledge. Links to a variety of peer-reviewed articles on HIV infection, many statistics.

http://HIVInSite.ucsf.edu/

The American Academy of Allergy, Asthma, and Immunology Home Page. Offers helpful information and current events concerning the treatment of allergic diseases.

http://www.aaaai.org/

Immune Explorer. This interactive site explores the biology of the immune system.

http://www-wilson.ucsd.edu/ve/index_noframes.html

Digestion and Nutrition

chapter

c h a p t e r   c o n c e p t s

43.1 Digestive Tracts

- An incomplete digestive tract with only one opening has little specialization of parts; a complete digestive tract with two openings does have specialization of parts. 782
- Discontinuous feeders, rather than continuous feeders, have a storage area for food. 783
- The dentition of herbivores, carnivores, and omnivores is adapted to the type of food they eat. 784

43.2 Human Digestive Tract

- The human digestive tract has many specialized parts and several accessory organs of digestion, which contribute in their own way to the digestion of food. 785
- The digestive enzymes are specific and have an optimum temperature and pH at which they function best. 788
- The products of digestion are small molecules, such as amino acids and glucose, that can cross plasma membranes. 789

43.3 Nutrition

- Proper nutrition supplies the body with energy and nutrients, including all vitamins and minerals. 792

Long-nosed tree snake, *Ahaetulla prasinus*, feeding on prey.

Animals are heterotrophic organisms that must take in organic molecules. The variety of diets found in the animal kingdom is astounding, from beetles which feed on rotting material to killer whales which go after live prey. Most species have a complete digestive tract with one opening that serves as an entrance and another that serves as an exit. With this type of tract, it is possible to have a greater range of adaptations to manipulate and digest food.

Some animals like plankton-eating baleen whales are continuous feeders while others like snakes are discontinuous eaters. Few snakes eat more than once a week and many eat but once a month. Snakes can eat large prey because their jaws and their entire digestive tract are expandable. A snake's curved teeth hold onto the prey animal and pull it back into the esophagus.

In a study of mantled howling monkeys, it was concluded that howlers choose foods that give them the greatest nutrient return. In contrast, adult humans often must be taught good nutrition. Eating correctly may very well help prevent major killer diseases like cardiovascular disease and cancer.

43.1 Digestive Tracts

Most animals have some sort of gut or digestive tract where food is digested to small nutrient molecules that can cross plasma membranes. Digestion contributes to homeostasis by providing the body with the nutrients needed to sustain the life of cells. A digestive tract (1) ingests food, (2) breaks food down into small molecules that can cross plasma membranes, (3) absorbs these nutrient molecules, and (4) eliminates nondigestible remains.

Incomplete Versus Complete Tracts

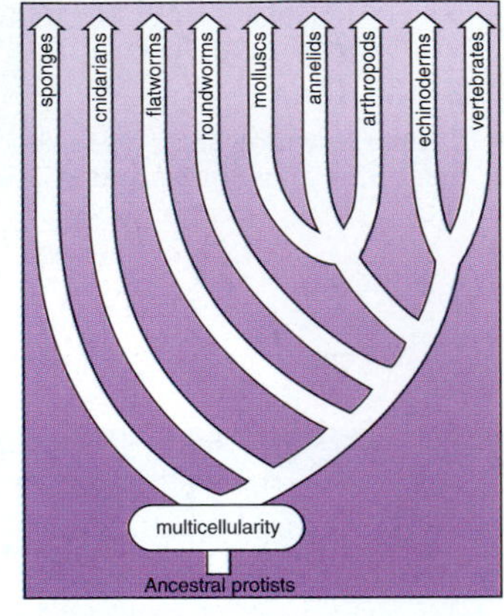

An **incomplete gut** has a single opening, usually called a mouth. Planarians, which are flatworms have an incomplete gut (Fig. 43.1). The digestive tract, a gastrovascular cavity, which branches throughout the body begins with a mouth and muscular pharynx. Planarians are carnivorous and feed largely on smaller aquatic animals. When a worm is feeding, the pharynx actually extends beyond the mouth. The worm wraps its body about the prey and uses its pharynx to suck up minute quantities at a time. Digestive enzymes present in the tract allow some extracellular ("outside the cells") digestion to occur. Digestion is finished intracellularly by the cells that line the tract. No cell in the body is far from the digestive tract; therefore diffusion alone is sufficient to distribute nutrient molecules.

The digestive tract of a planarian is notable for its lack of specialized parts. It is saclike because the pharynx serves not only as an entrance for food but also as an exit for nondigestible material. Specialization of parts does not occur under these circumstances.

Planarians have some modified parasitic relatives. Tapeworms, for example, have no digestive tract at all—nutrient molecules are simply absorbed from intestinal juices that surround the body. The body wall is highly modified for this purpose: it has millions of microscopic fingerlike projections that increase the surface area for absorption.

In contrast to planarians, earthworms, which are annelids, have a **complete gut,** meaning that the digestive tract has a mouth and an anus (Fig. 43.2). Earthworms feed mainly on decayed organic matter in soil. The muscular pharynx draws in food with a sucking action. The crop is a storage area that has thin expansive walls. The gizzard has thick muscular walls for churning and grinding the food. Digestion is extracellular—in the intestine. The surface area of digestive tracts is often increased for absorption of nutrient molecules, and in earthworms, this is accomplished by an intestinal fold called the typhlosole. Undigested remains pass out of the body at the anus. Specialization of parts is obvious in the earthworm because the pharynx, the crop, the gizzard, and the intestine each has a particular function in the digestive process.

In contrast to the incomplete, saclike gut, the complete gut, with both a mouth and an anus, can have many specialized parts depending on the way of life of the animal.

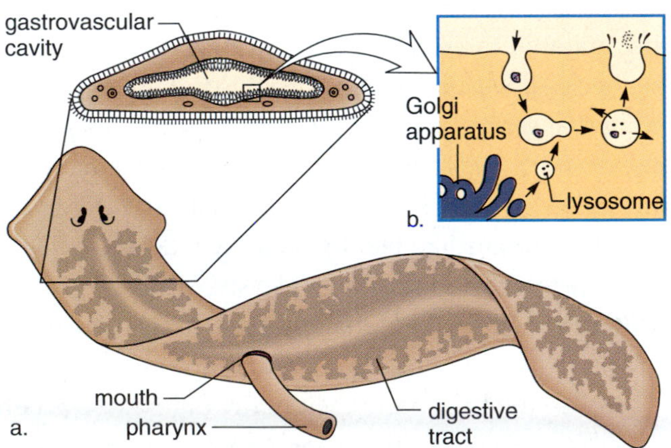

Figure 43.1 Incomplete digestive tract of a planarian.
a. Planarians, which are flatworms, have a gastrovascular cavity with a single opening that acts as both an entrance and an exit. Planarians rely on intracellular digestion to complete the digestive process.
b. Phagocytosis produces a vacuole, which joins with an enzyme-containing lysosome. The digested products pass from the vacuole into the cytoplasm before any nondigestible material is eliminated at the plasma membrane.

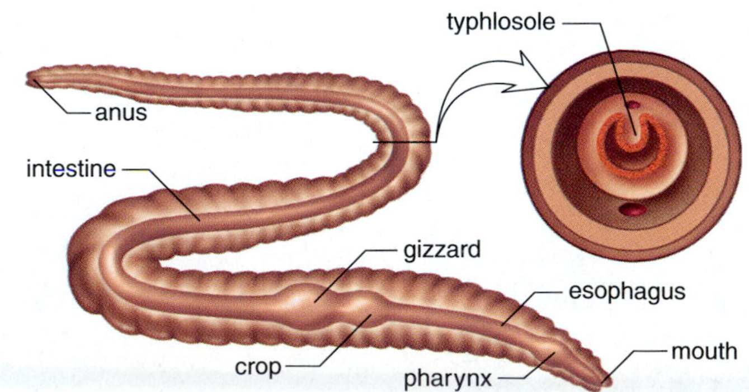

Figure 43.2 Complete digestive tract of an earthworm.
Complete digestive tracts have both a mouth and an anus and can have many specialized parts as those labeled in this drawing. Also in earthworms, which are annelids, the absorptive surface of the intestine is increased by an internal fold called the typhlosole.

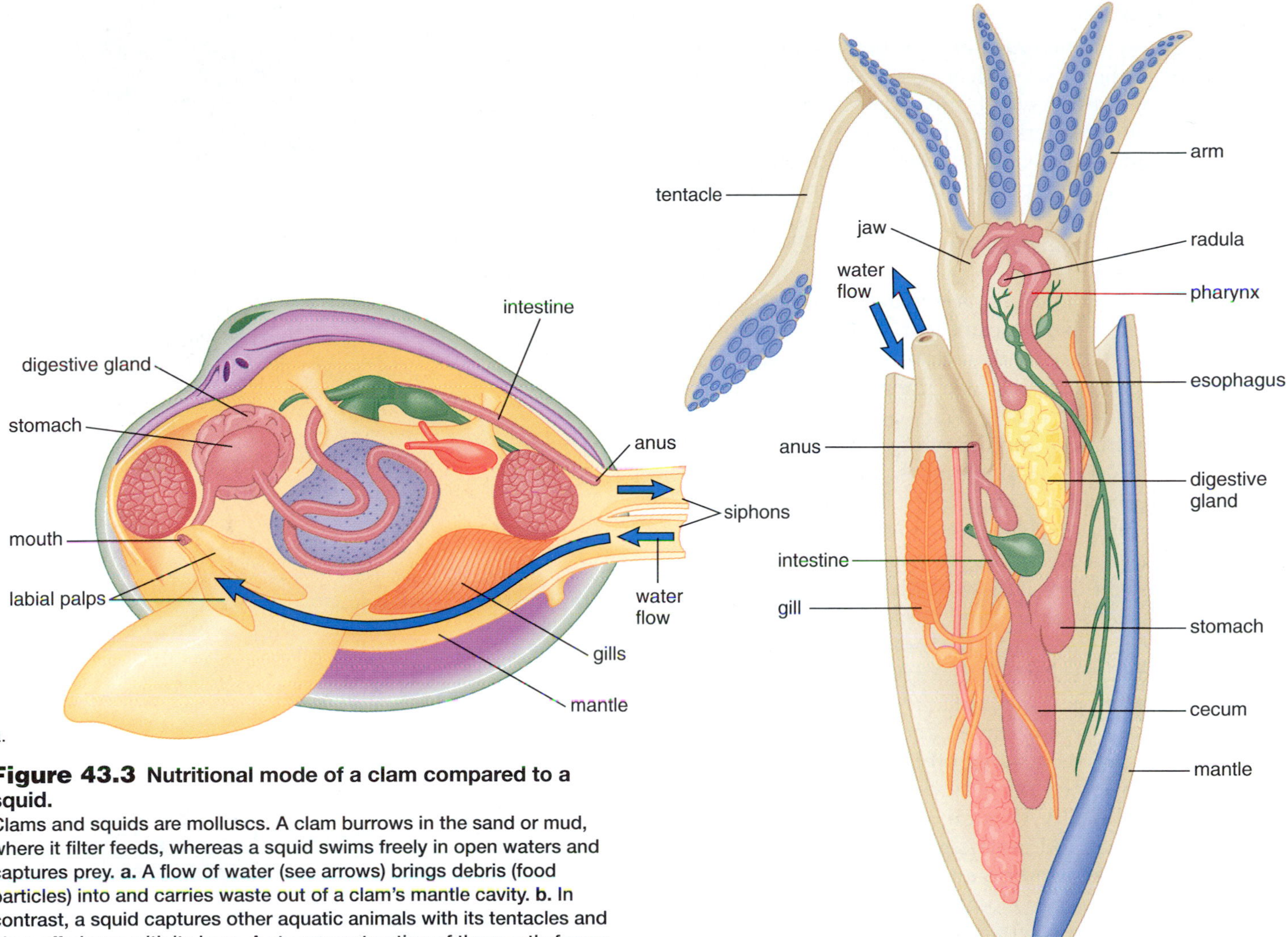

Figure 43.3 Nutritional mode of a clam compared to a squid.
Clams and squids are molluscs. A clam burrows in the sand or mud, where it filter feeds, whereas a squid swims freely in open waters and captures prey. **a.** A flow of water (see arrows) brings debris (food particles) into and carries waste out of a clam's mantle cavity. **b.** In contrast, a squid captures other aquatic animals with its tentacles and bites off pieces with its jaws. A strong contraction of the mantle forces water (see arrows), which has entered the mantle cavity, out the funnel, resulting in a type of "jet propulsion."

Continuous Versus Discontinuous Feeders

Clams, which are molluscs, are continuous feeders, called filter feeders (Fig. 43.3*a*). Water is always moving into the mantle cavity by way of the incurrent siphon (slitlike opening) and depositing particles on the gills. The size of the incurrent siphon permits the entrance of only small particles, which adhere to the gills. Ciliary action moves suitably sized particles to the labial palps, which force them through the mouth into the stomach. Digestive enzymes are secreted by a large digestive gland, but amoeboid cells present throughout the tract are believed to complete the digestive process by intracellular digestion.

Marine fanworms, which are annelids, are filter feeders. The feathery tentacles of these worms are specialized for gathering fine particles and microscopic plankton from the water. They allow only small particles to enter their digestive tract. Larger particles are rejected. A baleen whale is an active filter feeder. Baleen, a curtainlike fringe, hangs from the roof of the mouth and filters small shrimp called krill from the water. A baleen whale filters up to a ton of krill every few minutes.

Squids, which are molluscs, are discontinuous feeders (Fig. 43.3*b*). The body of a squid is streamlined, and the animal moves rapidly through the water using jet propulsion (forceful expulsion of water from a tubular funnel). The head of a squid is surrounded by ten arms, two of which have developed into long, slender tentacles whose suckers have toothed, horny rings. These tentacles seize prey (fishes, shrimps, and worms) and bring it to the squid's beaklike jaws, which bite off pieces pulled into the mouth by the action of a radula, a toothy tongue. An esophagus leads to a stomach and a cecum (blind sac) where digestion occurs. The stomach, supplemented by the cecum, retains food until digestion is complete. Discontinuous feeders, whether they are carnivores or herbivores, require such a storage area.

Continuous feeders, such as sessile filter feeders, do not need a storage region for food; discontinuous feeders, do need a storage area for food.

Adaptation to Diet

Some animals are omnivores; they eat both plants and animals. Others are herbivores; they feed only on plants. Still others are carnivores; they eat only other animals. Among invertebrates, filter feeders like clams and tube worms are omnivores. Land snails, which are terrestrial molluscs, and some insects like grasshoppers and locusts, are herbivores. Spiders (arthropods) are carnivores, as are sea stars (echinoderms) that feed on clams. A sea star positions itself above a clam and uses its tube feet to pull the valves of the shell apart. Then, it everts a part of its stomach to start the digestive process, even while the clam is trying to close its shell. Some invertebrates are cannibalistic. A female praying mantis (an insect), if starved, will feed upon her mate as the reproductive act is taking place!

Among mammals, the dentition differs according to mode of nutrition (Fig. 43.4). Humans, as well as raccoons, rats, and brown bears, are omnivores. Therefore, the dentition has a variety of specializations to accommodate both a vegetable diet and a meat diet. An adult human has thirty-two teeth. One-half of each jaw has teeth of four different types: two chisel-shaped incisors for shearing; one pointed canine for tearing; two fairly flat premolars for grinding; and three molars, well-flattened for crushing.

Among herbivores, the koala of Australia is famous for its diet of only eucalyptus leaves, and likewise many other mammals are browsers, feeding off bushes and trees. Grazers, like the horse, feed off grasses. The horse has sharp, even incisors for neatly clipping off blades of grass and large, flat premolars and molars for grinding and crushing the grass. Extensive grinding and crushing disrupts plant cell walls, allowing bacteria located in a part of the digestive tract called the cecum to get at and digest cellulose. Other mammalian grazers, like the cow and deer, are ruminants. In contrast to horses, they graze quickly and swallow partially chewed grasses into a special part of the stomach called a rumen. Here, microorganisms start the digestive process and the result, called cud, is regurgitated at a later time when the animal is no longer feeding. The cud is chewed again before being swallowed for complete digestion.

Many mammals, including dogs, toothed whales, and polar bears, are carnivores. A carnivore, like a lion, uses pointed incisors and enlarged canine teeth to shear off pieces small enough to be quickly swallowed. Meat is rich in protein and fat and is easier to digest than plant material. Therefore, the digestive system of carnivores is shorter and doesn't have the specialization seen in herbivores.

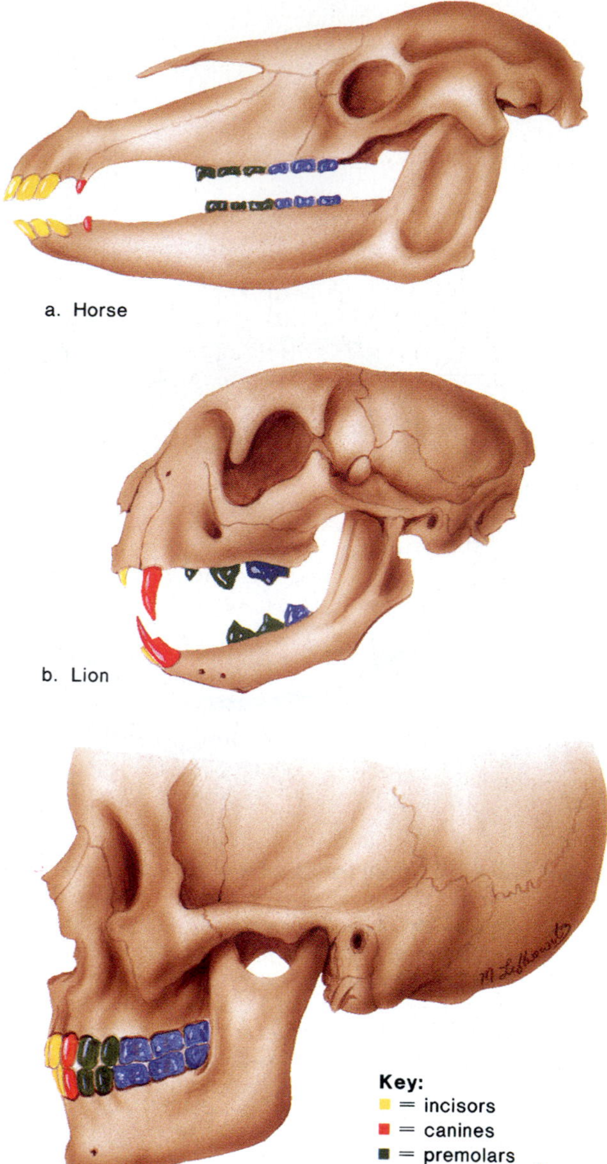

Figure 43.4 Dentition among vertebrates.
a. Horses are herbivores that graze on grasses. Note the sharp incisors, reduced canines, and large, flat premolars and molars. b. Lions are carnivores that prey on other animals. Note the pointed incisors, enlarged canines, and jagged premolars and molars. c. Humans are omnivores that have nonspecialized teeth.

Omnivores have nonspecialized teeth. Herbivores have large, flat molars that grind food; carnivores have incisors and canines to tear off chunks of meat.

43.2 Human Digestive Tract

Humans have the tube-within-a-tube body plan, and therefore the human digestive tract is complete—there is both a mouth and an anus. Table 43.1 lists the parts and functions of the human digestive system, and Figure 43.5 depicts these parts.

Digestion of food in humans is an extracellular process, and digestive enzymes are secreted by the digestive tract or by glands that lie nearby. Food is never found within these accessory glands, only within the tract itself. Digestion requires a cooperative effort between different parts of the body. The production of hormones and the performance of the nervous system achieve the cooperation of body parts.

Mouth

We have already mentioned that human dentition has a variety of specializations because humans are omnivores (see Fig. 43.4*c*). Food is chewed in the mouth, where it is mixed with saliva. There are three major pairs of **salivary glands** that send their juices by way of ducts to the mouth. Saliva contains the enzyme **salivary amylase,** which begins the process of starch digestion. The disaccharide maltose is a typical end product of salivary amylase digestion:

$$\text{starch} + H_2O \xrightarrow[\text{amylase}]{\text{salivary}} \text{maltose}$$

While in the **mouth,** food is manipulated by a muscular tongue, which has touch and pressure receptors similar to those in the skin. Taste buds, sensory receptors that are stimulated by the chemical composition of food, are also found primarily on the tongue as well as on the surface of the mouth. The tongue, which is composed of striated muscle and an outer layer of mucous membrane, mixes the chewed food with saliva. It then forms this mixture into a mass called a bolus in preparation for swallowing.

The salivary glands send saliva into the mouth where food is chewed and formed into a bolus for swallowing.

Table 43.1

Path of Food

Organ	Special Features	Function
Mouth	Teeth, tongue	Chewing of food; digestion of starch
Esophagus		Movement of food by peristalsis
Stomach	Gastric glands	Storage of food; acidity kills some bacteria; digestion of protein
Small intestine	Villi	Digestion of all foods; absorption of nutrients
Large intestine		Absorption of water; storage of nondigestible remains
Anus		Defecation

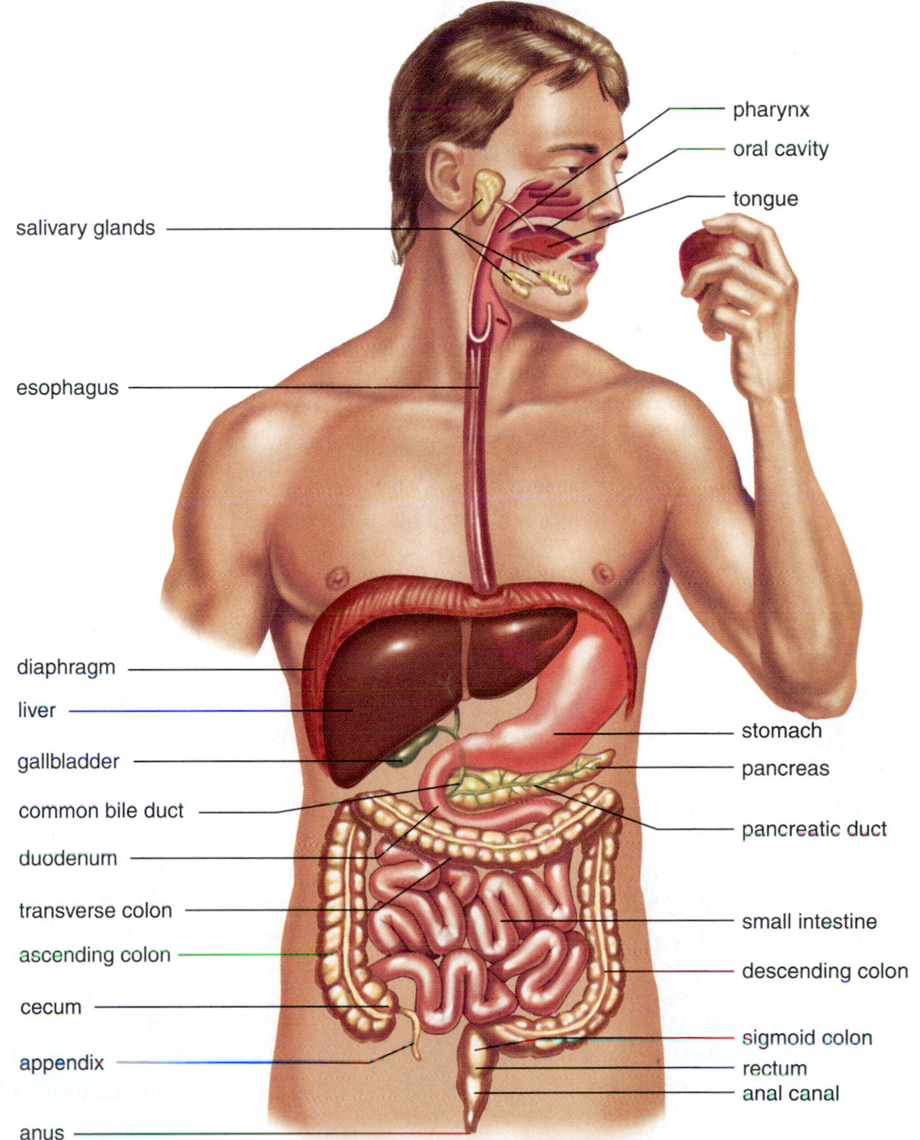

Figure 43.5 The human digestive tract.
Trace the path of food from the mouth to the anus. Note the placement of two accessory organs of digestion—the liver and the pancreas.

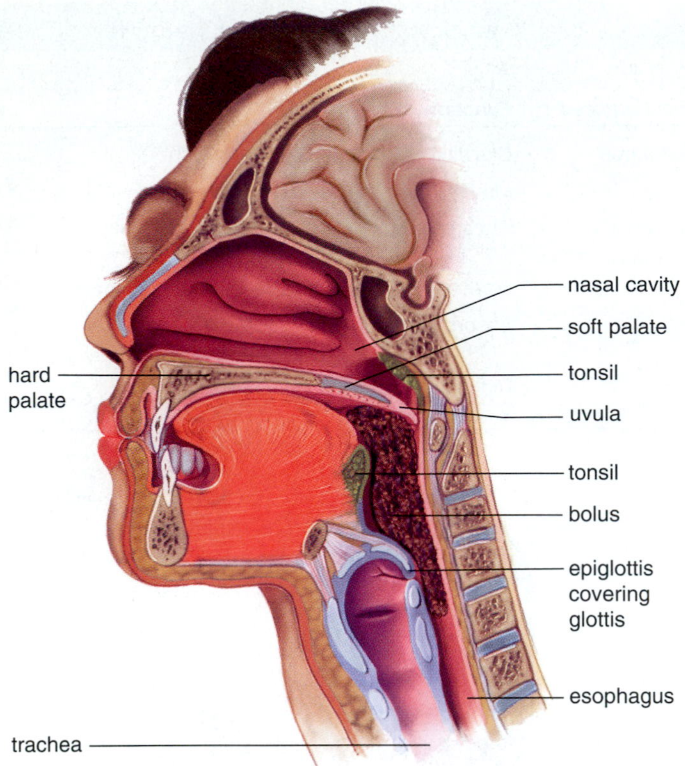

Figure 43.6 Swallowing.
Respiratory and digestive passages converge and diverge in the pharynx. During swallowing, the soft palate closes off the nasal cavity, and the epiglottis covers the glottis, forcing the bolus to pass down the esophagus. Therefore, you do not breathe when swallowing.

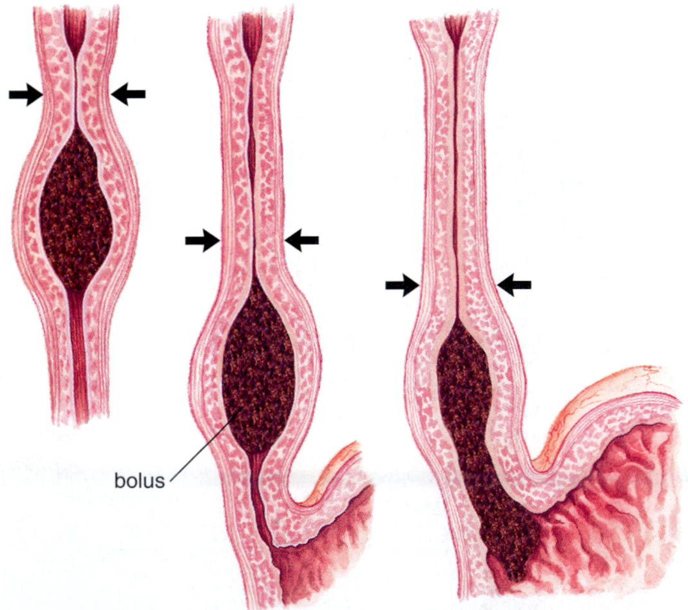

Figure 43.7 Peristalsis in the digestive tract.
These three drawings show how a peristaltic wave moves through a single section of the esophagus over time. The arrows point to areas of contraction.

The Pharynx and the Esophagus

The digestive and respiratory passages come together in the **pharynx** and then separate (Fig. 43.6). When food is swallowed, the soft palate, the rear portion of the mouth's roof, moves back to close off the nasal cavities. A flap of tissue called the epiglottis covers the glottis, an opening into the trachea. Now the bolus must move through the pharynx into the esophagus because the air passages are blocked.

The **esophagus** [Gk. *eso,* within, and *phagein,* eat] is a tubular structure that takes food to the stomach. When food enters the esophagus, peristalsis begins (Fig. 43.7). **Peristalsis** [Gk. *peri,* around, and *stalsis,* compression] is a rhythmical contraction that serves to move the contents along in tubular organs, such as the digestive tract.

During swallowing the air passages are blocked and the bolus enters the esophagus which conducts it to the stomach.

Stomach

The **stomach** (Fig. 43.8) is a thick-walled, J-shaped organ that lies on the left side of the body beneath the diaphragm. The wall of the stomach has deep folds which disappear as the stomach fills to an approximate capacity of one liter. Therefore, humans can periodically eat relatively large meals and spend the rest of their time at other activities.

But the stomach is much more than a mere storage organ, as was discovered by William Beaumont in the mid-nineteenth century. Beaumont, an American doctor, had a French Canadian patient, Alexis St. Martin. St. Martin had been shot in the stomach, and when the wound healed, he was left with a fistula, or opening, that allowed Beaumont to look inside the stomach and to collect gastric (stomach) juices produced by gastric glands. Beaumont was able to determine that the muscular walls of the stomach contract vigorously and mix food with juices that are secreted whenever food enters the stomach. He found that gastric juice contains hydrochloric acid (HCl) and a substance, now called pepsin, which is active in digestion. He also found that the gastric juices are produced independently of the protective mucus secretions of the stomach. Beaumont's work, which was very carefully and painstakingly done, pioneered the study of the physiology of digestion.

We now know that the epithelial lining of the stomach has millions of gastric pits, which lead into gastric glands. The gastric glands produce gastric juice. So much hydrochloric acid is secreted by the gastric glands that the stomach routinely has a pH of about 2. Such a high acidity usually is sufficient to kill bacteria and other microorganisms that might be in food. This low pH also stops the activity of salivary amylase, which functions optimally at the near-neutral pH of saliva, but it promotes the activity of pepsin. **Pepsin** is a hydrolytic enzyme that acts on protein to produce peptides:

$$\text{protein} + H_2O \xrightarrow{\text{pepsin}} \text{peptides}$$

As with the rest of the digestive tract, a thick layer of mucus protects the wall of the stomach from enzymatic action. Still, an ulcer, which is an open sore in the wall caused by the gradual destruction of tissues, does occur in some individuals. It is now believed that ulcers are due to an infection by an acid-resistant bacterium, *Helicobacter pylori,* which is able to attach to the epithelial lining. Wherever the bacterium attaches, the lining stops producing mucus and the area becomes exposed to digestive action. Now an ulcer develops.

Eventually, the stomach contents become **chyme,** which has a thick, soupy consistency. At the base of the stomach is a narrow opening controlled by a sphincter. A sphincter is a muscle that surrounds a tube and closes or opens the tube by contracting and relaxing. Whenever the sphincter relaxes, a small quantity of chyme passes through the opening into the small intestine. When chyme enters the small intestine, it sets off a neural reflex that causes the muscles of the sphincter to contract vigorously and to close the opening temporarily. Then the sphincter relaxes again and allows more chyme to enter. The slow manner in which chyme enters the small intestine allows for thorough digestion.

The stomach can expand to accommodate large amounts of food. When food is present, the stomach churns, mixing food with acidic gastric juice.

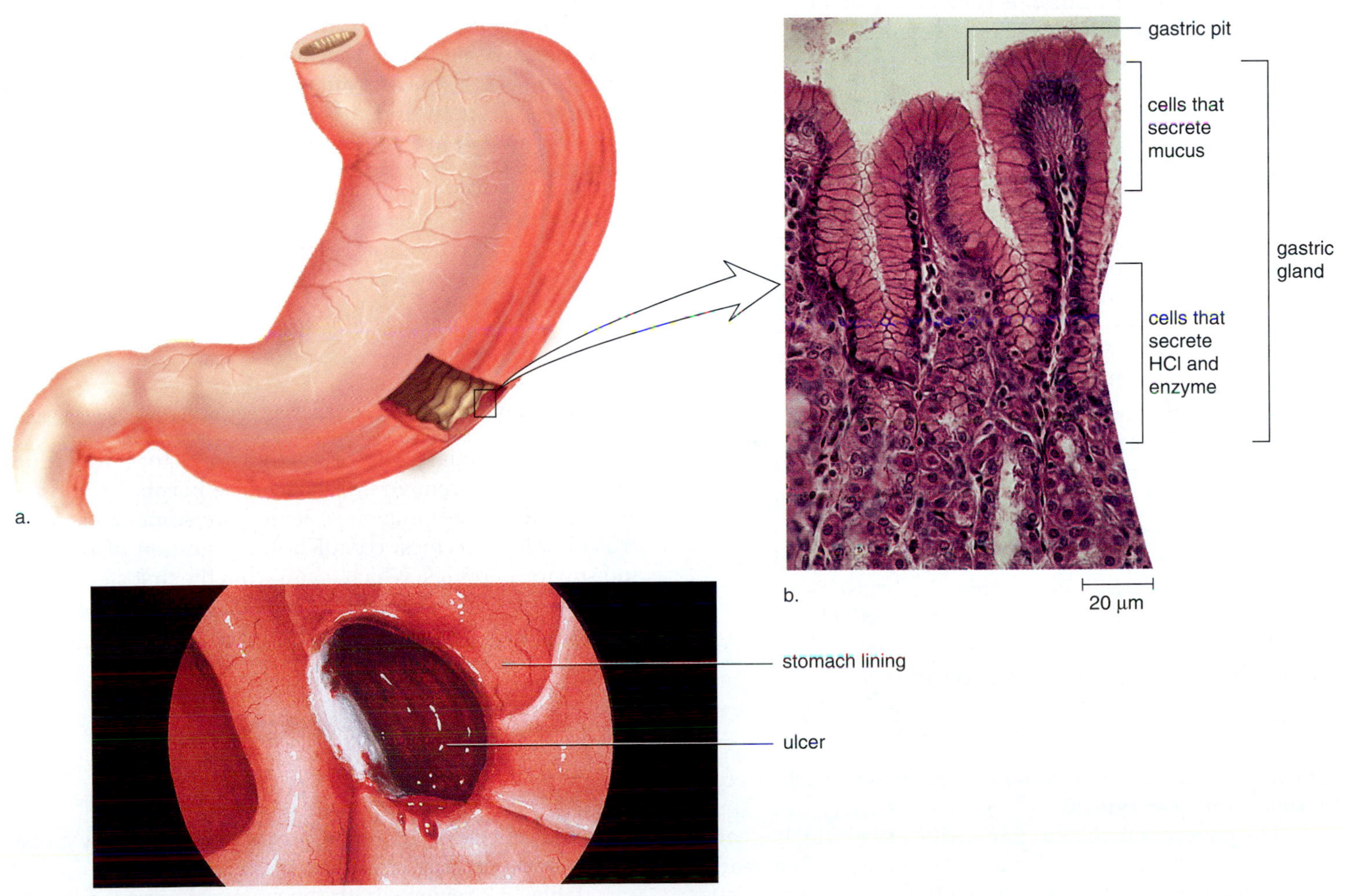

Figure 43.8 Anatomy of the stomach.
a. The stomach, which has thick walls, expands as it fills with food. **b.** The mucous membrane layer of its walls secretes mucus and contains gastric glands, which secrete a gastric juice active in the digestion of protein. **c.** View of a bleeding ulcer by using an endoscope (a tubular instrument bearing a tiny lens and a light source) that can be inserted into the abdominal cavity.

Table 43.2

Major Digestive Enzymes

Food	Digestion	Enzyme	Optimum pH	Produced by	Site of Action
Starch	Starch + $H_2O \rightarrow$ maltose	Salivary amylase Pancreatic amylase	Neutral Basic	Salivary glands Pancreas	Mouth Small intestine
	Maltose + $H_2O \rightarrow$ glucose	Maltase	Basic	Small intestine	Small intestine
Protein	Protein + $H_2O \rightarrow$ peptides	Pepsin Trypsin	Acidic Basic	Gastric glands Pancreas	Stomach Small intestine
	Peptide + $H_2O \rightarrow$ amino acids	Peptidases	Basic	Small intestine	Small intestine
Nucleic acid	RNA and DNA + $H_2O \rightarrow$ nucleotides	Nuclease	Basic	Pancreas	Small intestine
	Nucleotides + $H_2O \rightarrow$ base + sugars + phosphate	Nucleosidases phosphate	Basic	Small intestine	Small intestine
Fat	Fat droplet + $H_2O \rightarrow$ glycerol + fatty acids	Lipase	Basic	Pancreas	Small intestine

Small Intestine

The human **small intestine,** a coiled tube about 3 meters long, functions in the digestion of chyme and in the absorption of nutrient molecules.

Digestion by Enzymes

Chyme from the stomach enters the **duodenum,** the first part of the small intestine. Proteins and carbohydrates in chyme are only partly digested, and fat digestion still needs to be carried out. Considerably more digestive activity is required before these nutrients can be absorbed through the intestinal wall. Two important accessory glands, the **liver,** the largest organ in the body, and the **pancreas,** located behind the stomach, send secretions to the duodenum (see Fig. 43.5).

The liver produces **bile,** which is stored in the **gallbladder,** an organ that acts on bile so that it is a thick, mucous-like material. Bile is sent to the duodenum by way of a duct. Bile looks green because it contains pigments that are products of hemoglobin breakdown. This green color is familiar to anyone who has observed how bruised tissue changes color. Hemoglobin within the bruise is breaking down into the same types of pigments found in bile.

Bile also contains bile salts, which break up fat into fat droplets by a process called **emulsification.**

$$\text{fat} \xrightarrow{\text{bile salts}} \text{fat droplets}$$

Fat droplets mix with the water and have more surface area for digestion by enzymes.

The pancreas produces pancreatic juice, which is conducted to the duodenum, also by way of a duct. Pancreatic juice contains sodium bicarbonate ($NaHCO_3$), which neutralizes the chyme and makes the pH of the small intestine slightly basic. A higher pH helps prevent autodigestion of the intestinal lining by pepsin and optimizes the pH for intestinal and pancreatic enzymes. Pancreatic juice also contains digestive enzymes that act on every major component of food (Table 43.2). **Pancreatic amylase** digests starch to maltose; **trypsin** and other enzymes digest protein to peptides; and **lipase** digests fat droplets to glycerol and fatty acids:

$$\text{starch} + H_2O \xrightarrow{\text{pancreatic amylase}} \text{maltose}$$

$$\text{protein} + H_2O \xrightarrow{\text{trypsin}} \text{peptides}$$

$$\text{fat droplet} + H_2O \xrightarrow{\text{lipase}} \text{glycerol} + \text{fatty acids}$$

The epithelial cells of the villi produce **intestinal enzymes,** which remain attached to the plasma membrane of microvilli. These enzymes, which are sometimes called brush-border enzymes, complete the digestion of peptides and sugars. Peptides, which result from the first step in protein digestion, are digested by peptidases to amino acids. Maltose, which results from the first step in starch digestion, is digested by maltase to glucose:

$$\text{peptide} + H_2O \xrightarrow{\text{peptidases}} \text{amino acids}$$

$$\text{maltose} + H_2O \xrightarrow{\text{maltase}} \text{glucose} + \text{glucose}$$

Other disaccharides, each of which is acted upon by a specific enzyme, are digested in the small intestine.

Table 43.2 reviews both the reactions required for the digestion of food and the various digestive enzymes

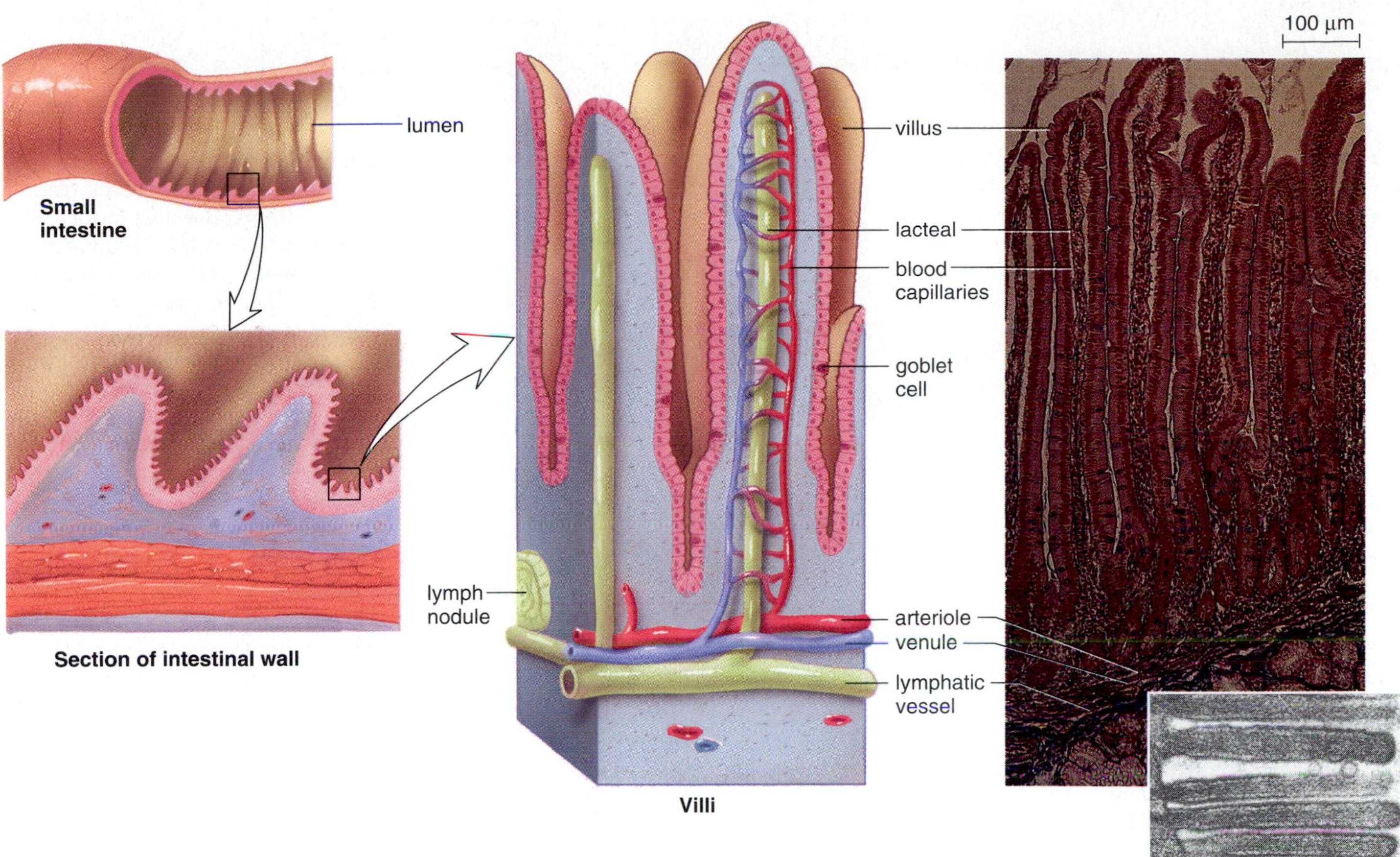

Figure 43.9 **Anatomy of intestinal lining.**
The products of digestion are absorbed by villi, fingerlike projections of the intestinal wall that contain blood vessels and a lacteal. Each villus has many microscopic extensions called microvilli.

discussed in this chapter. Bile is not listed because it contains an emulsifier and not an enzyme.

Absorption by Villi

The mucous membrane layer of the small intestine has ridges and furrows that give it an almost corrugated appearance (Fig. 43.9). On the surface of these ridges and furrows are small, fingerlike projections called **villi** [L. *villus*, shaggy hair]. Cells on the surfaces of the villi have minute projections called **microvilli.**

Small molecules are absorbed by the huge number of villi that line the intestinal wall. Each villus has microscopic extensions called microvilli, which greatly increase its absorptive surface. Collectively in electron micrographs, microvilli give the villi a fuzzy border known as a "brush border."

Carbohydrates, proteins, and fats have now been broken down by digestive enzymes to small molecules that can be absorbed by intestinal villi and their microvilli. If the small intestine were simply a smooth tube, it would have to be 500 to 600 meters long to have a comparable surface area for absorption.

Each villus contains blood vessels and a lymphatic vessel called a **lacteal.** Sugars and amino acids enter villi cells and then are absorbed into the bloodstream. In contrast, glycerol and fatty acids enter villi cells and are reassembled into fat molecules, which move into the lacteals.

Absorption continues until almost all products of digestion have been absorbed. Absorption involves both diffusion (passive transport) and active transport. Active transport requires an expenditure of cellular energy.

The wall of the small intestine is lined by villi. Sugars and amino acids enter the blood vessels, and re-formed fats enter the lacteals of the villi.

Control of Digestive Juices

The study of the control of digestive gland secretion began in the late 1800s. At that time, Ivan Pavlov showed that dogs would begin to salivate at the ringing of a bell because they had learned to associate the sound of the bell with being fed. Pavlov's experiments demonstrated that even the thought of food can cause the nervous system to order the secretion of digestive juices. If food is present in the mouth, the stomach, and the small intestine, digestive secretion occurs because of simple reflex action. The presence of food sets off nerve impulses that travel to the brain. Thereafter, the brain stimulates the digestive glands to secrete.

In this century, investigators discovered that specific control of digestive secretions is achieved by hormones. Hormones are chemical messengers often transported in the bloodstream from one set of cells to another. When investigators cut the nerves leading to the digestive tract, the stomach and pancreas continued to secrete their enzymes. The gallbladder continued to secrete bile. They concluded that hormones, not nerves, controlled these secretions.

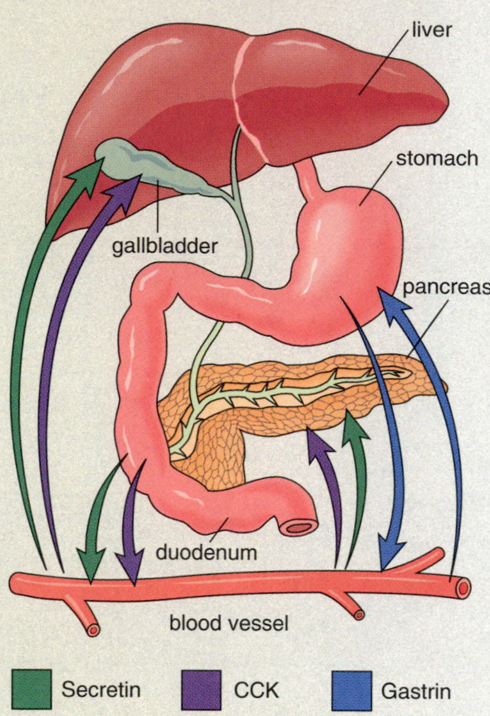

Figure 43A Hormonal control of digestive gland secretions. The arrows to the blood vessel are secretion into the bloodstream, and the away arrows are hormone reception by target organs.

When a person has eaten a meal particularly rich in protein, the gastric glands of the lower stomach wall produce the hormone gastrin (Fig. 43A). Gastrin enters the bloodstream, and soon stomach churning and the secretory activity of gastric glands increase.

Cells of the duodenal wall also produce hormones, two of which are of particular interest—secretin and CCK (cholecystokinin). Acid, especially hydrochloric acid (HCl) present in chyme, stimulates the release of secretin, while partially digested protein and fat stimulate the release of CCK. Soon after these hormones enter the bloodstream, the pancreas increases its output of pancreatic juice and the liver increases its output of bile. The gallbladder contracts to release bile.

Another hormone produced by the duodenal wall, GIP (gastric inhibitory peptide), works opposite to gastrin—it inhibits gastric gland secretion and stomach motility. This is not surprising, because the body's hormones often have opposite effects.

Accessory Organs

The pancreas and the liver are accessory organs of digestion along with the teeth, salivary glands, and gallbladder.

Pancreas

The pancreas lies deep in the abdominal cavity, resting on the posterior abdominal wall. It is an elongated and somewhat flattened organ that functions as both an endocrine gland and exocrine gland. It is an endocrine gland when it produces and secretes insulin and glucagon into the bloodstream. It is an exocrine gland when it produces and secretes pancreatic juice into the duodenum.

Most pancreatic cells produce pancreatic juice, which contains sodium bicarbonate ($NaHCO_3$) and digestive enzymes for all types of food. Pancreatic enzymes function best at a slightly basic pH and sodium bicarbonate neutralizes chyme which is acid. Pancreatic juice travels by way of ducts to the duodenum of the small intestine where they function in carbohydrate, protein, and fat digestion. Regulation of pancreatic secretion is discussed in the reading on this page.

Liver

The liver has numerous functions, including the following: (1) detoxifies the blood by removing and metabolizing poisonous substances; (2) produces the plasma proteins such as albumin and fibrinogen; (3) destroys old red blood cells and converts hemoglobin to the breakdown products in bile (bilirubin and biliverdin); (4) produces bile, which is stored in the gallbladder before entering the small intestine, where it emulsifies fats; (5) stores glucose as glycogen and breaks down glycogen to glucose between meals to maintain a constant glucose concentration in the blood; and (6) produces urea from amino groups and ammonia.

Blood vessels from the large and small intestines merge to form the hepatic portal vein, which leads to the liver (Fig. 43.10). The liver helps maintain the glucose concentration in blood at about 0.1% by removing excess glucose from the hepatic portal vein and storing it as glycogen. Between meals, glycogen is broken down and glucose enters the hepatic vein. Glycogen is sometimes called animal starch because both starch and glycogen are made up of glucose molecules. If by chance the supply of glycogen

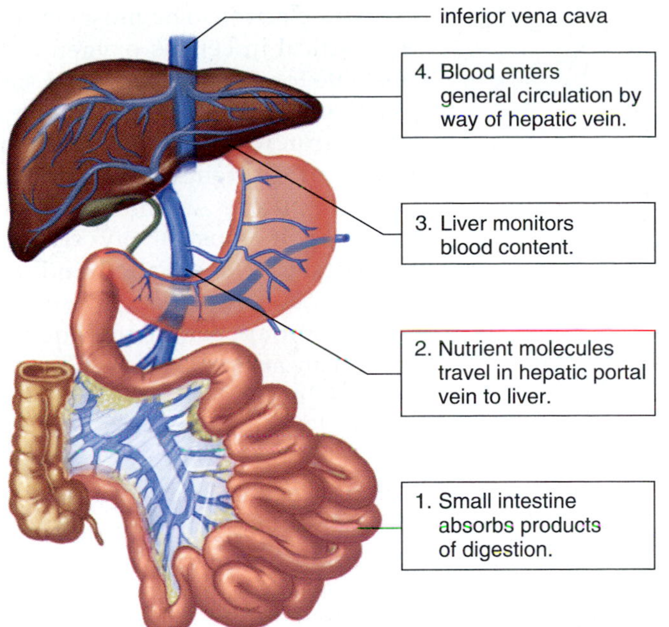
inferior vena cava
4. Blood enters general circulation by way of hepatic vein.
3. Liver monitors blood content.
2. Nutrient molecules travel in hepatic portal vein to liver.
1. Small intestine absorbs products of digestion.

Figure 43.10 Hepatic portal system.
The hepatic portal vein takes the products of digestion from the digestive system to the liver, where they are processed before entering the hepatic vein.

and glucose runs short, the liver converts amino acids to glucose molecules.

Amino acids contain nitrogen in the form of amino groups, whereas glucose contains only carbon, oxygen, and hydrogen. Therefore, before amino acids can be converted to glucose molecules, deamination, the removal of amino groups from amino acids, must occur. By a complex metabolic pathway, the liver converts the amino groups to urea, the most common nitrogenous waste product of humans. After urea is formed in the liver, it is transported by the bloodstream to the kidneys, and eventually it is excreted.

Liver Disorders When a person is jaundiced, there is a yellowish tint to the skin due to an abnormally large amount of bile pigments in the blood. In hemolytic jaundice, red blood cells are broken down in abnormally large amounts; in obstructive jaundice, there is an obstruction of the bile duct or damage to the liver cells. Obstructive jaundice often occurs when crystals of cholesterol precipitate out of bile and form gallstones.

Jaundice can also result from viral hepatitis. Hepatitis A is most often caused by eating contaminated food. Hepatitis B and C are commonly spread by blood transfusions, kidney dialysis, and injection with unsterilized needles. All three types of hepatitis can be spread by sexual contact.

Cirrhosis is a chronic liver disease in which the organ first becomes fatty. Liver tissue is then replaced by inactive fibrous scar tissue. Alcoholics often get cirrhosis, a condition most likely is caused at least in part by the excessive amounts of alcohol the liver is forced to break down.

Large Intestine

The **large intestine,** which includes the cecum, the colon, the rectum, and the anal canal, is larger in diameter but shorter than the small intestine. The large intestine absorbs water, salts, and some vitamins. It also stores indigestible material until it is eliminated at the anus.

The cecum, which lies below the junction with the small intestine, is the blind end of the large intestine. The cecum has a small projection called the vermiform **appendix** [L. *verm*, worm, and *form*, shape, and *append*, an addition]. In humans, the appendix may play a role in fighting infections. In the case of appendicitis, the appendix becomes infected and so filled with fluid that it may burst. If an infected appendix bursts before it can be removed, it can lead to a serious, generalized infection of the abdominal lining, called peritonitis.

The **colon** is subdivided into the ascending, transverse, descending, and sigmoid colon (see Fig. 43.5). The sigmoid colon enters the rectum, the last 20 cm of the large intestine. About 1.5 liters of water enter the digestive tract daily as a result of eating and drinking. An additional 8.5 liters enter the digestive tract each day carrying the various substances secreted by the digestive glands. About 95% of this water is absorbed by the small intestine, and much of the remaining portion is absorbed into cells of the colon. If this water is not reabsorbed, **diarrhea** can lead to serious dehydration and ion loss, especially in children.

The large intestine has a large population of bacteria, notably *Escherichia coli.* The bacteria break down indigestible material and they produce some vitamins, such as vitamin K. Vitamin K is necessary to blood clotting. Digestive wastes (feces) eventually leave the body through the **anus,** the opening of the anal canal. Feces are about 75% water and 25% solid matter. Almost one-third of this solid matter is made up of intestinal bacteria. The remainder is undigested plant material, fats, waste products (such as bile pigments), inorganic material, mucus, and dead cells from the intestinal lining.

The colon is subject to the development of **polyps,** which are small growths arising from the epithelial lining. Polyps, whether they are benign or cancerous, can be removed surgically. Some investigators believe that dietary fat increases the likelihood of colon cancer. Dietary fat causes an increase in bile secretion and it could be that intestinal bacteria convert bile salts to substances that promote the development of colon cancer. Dietary fibers absorbs water and adds bulk, thereby diluting the concentration of bile salts and facilitating the movement of substances through the intestine. Regular elimination reduces the time that the colon wall is exposed to any cancer-promoting agents in feces.

Digestion is facilitated by the production of pancreatic juice by the pancreas and bile by the liver. The large intestine does not produce digestive enzymes; it absorbs water, salts, and some vitamins.

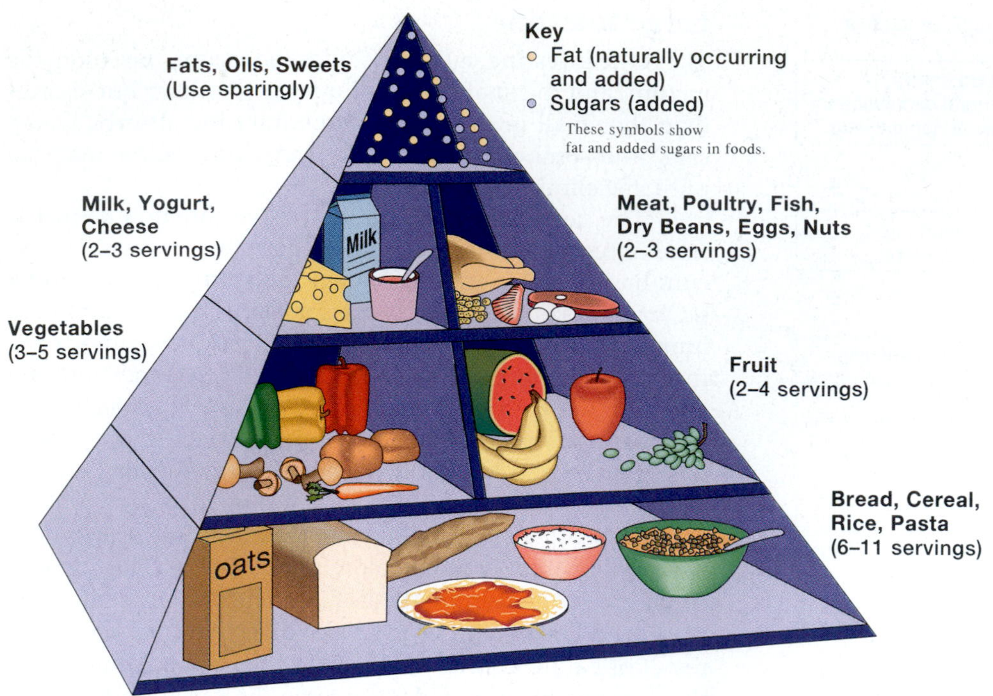

Figure 43.11 Ideal American diet.
The U.S. Department of Agriculture uses a pyramid to show the ideal diet because it emphasizes the importance of including grains, fruits, and vegetables in the diet. Meats and dairy products are needed in limited amounts; fats, oils, and sweets should be used sparingly.

Source: U.S. Department of Agriculture.

43.3 Nutrition

To be sure that the essential nutrients are included in the diet, it is necessary to eat a balanced diet. A balanced diet includes a variety of foods proportioned as shown in Figure 43.11.

Vitamins

Vitamins are organic compounds (other than carbohydrate, fat, and protein) that the body is unable to produce but uses for metabolic purposes. Many vitamins are portions of coenzymes, which are enzyme helpers. For example, niacin is part of the coenzyme NAD^+ and riboflavin is part of FAD. Coenzymes are needed in only small amounts because each can be used over and over again. Not all vitamins are coenzymes; vitamin A, for example, is a precursor for the visual pigment that prevents night blindness. It has been known for several years that if vitamins are lacking in the diet, various symptoms develop. Altogether there are 13 vitamins, which are divided into those that are fat soluble and those that are water soluble (Table 43.3).

Antioxidants

Over the past 20 years, numerous statistical studies have been done to determine whether a diet rich in fruits and vegetables is protective against cancer. Cellular metabolism generates free radicals, unstable molecules that carry an extra electron. The most common free radical in cells is oxygen in the unstable form, O_3^-. In order to stabilize themselves, free radicals donate an electron to another molecule like DNA, or proteins, including enzymes, plasma membranes, and lipids. This most likely damages these cellular molecules and may lead to disorders, perhaps even cancer. Also, it appears that plaque formation in arteries may begin when arterial linings are injured by oxidized LDL cholesterol.

Vitamins C, E, and A are believed to defend the body against free radicals and, therefore, they are termed antioxidants. These vitamins are especially abundant in fruits and vegetables. The new dietary guidelines shown in Figure 43.11 suggest that we eat several servings of fruits and vegetables a day. It is not difficult to achieve this goal because a serving can be salad greens, raw or cooked vegetables, dried fruit, and fruit juice in addition to traditional apples and oranges and such.

Dietary supplements may provide a potential safeguard against cancer and cardiovascular disease, but it is important to realize that it is not appropriate to take supplements instead of improving intake of fruits and vegetables. There are many beneficial compounds in fruits that cannot be obtained from a vitamin pill. These substances enhance each other's absorption or action and also perform independent biological functions.

Vitamin D

Skin cells contain a precursor cholesterol molecule that is converted to vitamin D after UV exposure. Since only a small amount of UV radiation is needed to change the precursor molecule to vitamin D, the skin should not be exposed unnecessarily. Vitamin D leaves the skin and is modified first in the kidneys and then in the liver until finally it is activated into a form called calcitriol. Calcitriol circulates throughout the body, regulating calcium uptake and metabolism. It promotes the absorption of calcium by the intestines, and the absence of vitamin D leads to rickets in children. Rickets, characterized by a bowing of the legs, is caused by defective mineralization of the skeleton. Most milk today is fortified with vitamin D, which helps prevent the occurrence of rickets.

Vitamins are essential to cellular metabolism; many are protective against identifiable illnesses and conditions.

Table 43.3

Vitamins: Their Role in the Body and Their Food Sources

Vitamin	Major Role in Body	Good Food Sources
Fat Soluble		
Vitamin A	Vision, health of skin, hair, bones, and sex organs	Deep green or yellow vegetables, dairy products
Vitamin D	Health of bones and teeth	Dairy products, tuna, eggs
Vitamin E	Strengthening of red blood cell membrane	Green leafy vegetables, whole grains
Vitamin K	Clotting of blood, bone metabolism	Green leafy vegetables, cabbage, cauliflower
Water Soluble		
Thiamine (B_1)	Carbohydrate metabolism	Pork, whole grains
Riboflavin (B_2)	Energy metabolism	Whole grains, milk, green vegetables
Niacin (B_3)	Energy metabolism	Organ meats, whole grains
Pyridoxine (B_6)	Amino acid metabolism	Meats, fish, whole grains
Vitamin B_{12}	Red blood cell formation	Meats, dairy foods
Biotin	Carbohydrate metabolism	Eggs, most foods
Folic acid	Formation of red blood cells, DNA and RNA	Green leafy vegetables, nuts, whole grains
Pantothenic acid	Energy metabolism	Most foods
Vitamin C	Collagen formation	Citrus fruits, tomatoes

Table 43.4

Minerals: Their Role in the Body and Their Food Sources

Mineral	Major Role in Body	Good Food Sources
Macrominerals		
Calcium (Ca)	Strong bones and teeth, nerve conduction, muscle contraction	Dairy products, green leafy vegetables
Phosphorus (P)	Strong bones and teeth	Meat, dairy products, whole grains
Potassium (K)	Nerve conduction, muscle contraction	Many fruits and vegetables
Sodium (Na)	Nerve conduction, pH balance	Table salt
Chlorine (Cl)	Water balance	Table salt
Magnesium (Mg)	Protein synthesis	Whole grains, green leafy vegetables
Microminerals		
Zinc (Zn)	Wound healing, tissue growth	Whole grains, legumes, meats
Iron (Fe)	Hemoglobin synthesis	Whole grains, legumes, eggs, green leafy vegetables
Fluorine (F)	Strong bones and teeth	Fluoridated drinking water, tea
Copper (Cu)	Hemoglobin synthesis	Seafood, whole grains, legumes
Iodine (I)	Thyroid hormone synthesis	Iodized table salt, seafood

Minerals

In addition to vitamins, various **minerals** are also required by the body (Table 43.4). Some minerals (calcium, phosphorus, potassium, sulfur, sodium, chlorine, and magnesium) are recommended in amounts more than 100 mg per day. These macrominerals serve as constituents of cells and body fluids, and as structural components of tissues. Calcium is needed for the construction of bones and teeth and also for nerve conduction and muscle contraction.

Other minerals (iron, manganese, copper, iodine, cobalt, and zinc) are recommended in amounts of less than 20 mg per day. These microminerals are more likely to have very specific functions. Iron is needed for the production of hemoglobin, and iodine is used in the production of thyroxine, a hormone produced by the thyroid glands. As research continues, more and more elements have been added to the list of those considered essential. During the past three decades, very small amounts of molybdenum, selenium, chromium, nickel, vanadium, silicon, and even arsenic have been found to be essential to good health.

Some individuals do not receive enough iron (especially women), calcium, magnesium, or zinc in their diets.

Adult females need more iron in their diet than males (18 mg compared to 10 mg) because they lose hemoglobin each month during menstruation. Stress can bring on a magnesium deficiency, and a vegetarian diet may lack zinc, which is usually obtained from meat. A varied and complete diet, however, usually supplies the recommended daily allowances for minerals.

Calcium

There is much interest in calcium supplements to counteract the development of osteoporosis [Gk. *osteon,* bone, and *poros,* hole, passage], a degenerative bone disease that afflicts an estimated one-fourth of older men and one-half of older women in the United States. Osteoporosis develops because bone-eating cells, called osteoclasts, are more active than bone-forming cells, called osteoblasts. Therefore, the bones are porous, and they break easily because they lack sufficient calcium. Due to recent studies that show consuming more calcium does slow bone loss in elderly people, the guidelines have been revised. A calcium intake of 1,000–1,500 mg a day is recommended. To achieve this amount, supplemental calcium is most likely necessary.

Exercise is also effective in building bone mass. There are also medications that slow bone loss while increasing skeletal mass but these should be taken only when under a physician's care.

Sodium

The recommended amount of sodium intake per day is 400–3,300 mg, and the average American intake is 4,000–4,700 mg, mostly as salt (sodium chloride). In recent years, this imbalance has caused concern because high-sodium intake has been linked to hypertension in some people. About one-third of the sodium we consume occurs naturally in foods; another third is added during commercial processing; and we add the last third either during home cooking or at the table in the form of table salt.

Clearly, it is possible for us to cut down on the amount of sodium in the diet by reducing the amount of salt we eat.

Excess sodium in the diet can lead to hypertension; therefore, we should reduce sodium intake.

Connecting Concepts

In humans, digestive enzymes are produced by the salivary glands, gastric glands, and intestinal glands. Two other organs (the pancreas and the liver) also contribute secretions that help break down food. The liver produces bile (stored by the gallbladder) which emulsifies fat. The pancreas produces enzymes for the digestion of carbohydrates, proteins, and fat. Secretions from these glands, which are sent by ducts into the small intestine, are regulated by hormones such as secretin produced by the digestive tract.

Assigning any organ to a particular body system seems arbitrary. Granted, a good argument can be made for declaring that the mouth, esophagus, stomach, and the intestines are in the digestive system. But actually, even these organs assist other body systems. The stomach and small intestine contribute to the endocrine system, for example, by producing hormones (see page 790). And how could muscles and nerves function without a supply of calcium from the digestive tract? Or for that matter, any other system in the body without a supply of nutrients absorbed by the digestive tract and distributed by the cardiovascular system?

The liver has so many functions it really belongs to many systems of the body. Doesn't it belong to the cardiovascular system because it produces plasma proteins and the urinary system because it produces urea, as well as the digestive system because it produces bile? The liver is a vital organ, meaning that we cannot live without it. Among its many functions, it detoxifies blood by removing and metabolizing poisonous substances. The liver has amazing regenerative powers and in some instances can recover if the rate of regeneration exceeds the rate of damage. Otherwise liver transplantation is usually the preferred treatment, but artificial livers have been developed and tried in a few cases. One type of artificial liver consists of a cartridge that contains liver cells. The patient's blood passes through a cellulose acetate tube of the cartridge and is serviced in the same manner as with a normal liver. In the meantime, the patient's liver has a chance to recover.

Summary

43.1 Digestive Tracts

Some animals (e.g., planarians) have an incomplete digestive tract, which shows little specialization of parts. Other animals (e.g., earthworms) have a complete digestive tract, which does show specialization of parts.

Some animals are continuous feeders (e.g., clams, which are sessile filter feeders); others are discontinuous feeders (e.g., squid). Discontinuous feeders need a storage area for food.

Most mammals have teeth. Herbivores need teeth that can clip off plant material and grind it up. Also, the herbivore's stomach contains bacteria that can digest cellulose.

Carnivores need teeth that can tear and rip meat into pieces. Meat is easier to assimilate, so the digestive system of carnivores has less specialization of parts and a shorter intestine than that of herbivores.

43.2 Human Digestive Tract

In the human digestive tract, food is chewed and manipulated in the mouth, where salivary glands secrete saliva. Saliva contains salivary amylase, which begins carbohydrate digestion.

Food then passes to the pharynx and down the esophagus by peristalsis to the stomach. The stomach stores and mixes food with mucus and gastric juice to produce chyme. Pepsin begins protein digestion here.

Chyme gradually enters the duodenum where bile, pancreatic juice, and the intestinal secretions are found. Enzymes in the small intestine hydrolyze all of the organic nutrients. Table 43.2 summarizes the enzymes involved in digesting food.

There are three hormones that regulate digestive tract secretions: gastrin, which stimulates acid- and enzyme-secreting cells in the stomach; secretin, which stimulates the pancreas to release sodium bicarbonate; and CCK (cholecystokinin), which stimulates the gallbladder to release bile and the pancreas to release digestive enzymes.

The pancreas produces and sends digestive enzymes to the small intestine for every major component of food.

The liver produces bile, which is stored in the gallbladder before entering the small intestine. The liver is also involved in the processing of absorbed nutrient molecules and in maintaining the blood concentration of nutrient molecules, such as glucose. The liver converts ammonia to urea and breaks down toxins.

Nutrient absorption takes place in the small intestine, but some water and minerals are absorbed in the colon. Digestive wastes leave the colon by way of the anus.

43.3 Nutrition

A balanced diet is required for good health. Food should provide us with all the necessary vitamins, minerals, amino acids, fatty acids, and an adequate amount of energy.

Reviewing the Chapter

1. Contrast the incomplete gut with the complete gut, using the planarian and earthworm as examples. 782
2. Contrast a continuous feeder with a discontinuous feeder, using the clam and squid as examples. 783
3. Contrast the dentition of the mammalian herbivore with that of the mammalian carnivore, using the horse and lion as examples. 784
4. List the parts of the human digestive tract, anatomically describe them, and state the contribution of each to the digestive process. 785–89
5. Assume that you have just eaten a ham sandwich. Discuss the digestion of the contents of the sandwich. Mention all necessary enzymes. 785–89
6. Discuss the absorption of the products of digestion into the lymphatic and cardiovascular systems. 789
7. What are gastrin, secretin, and CCK (cholecystokinin)? Where are they produced, and what are their functions? 790
8. State the location and describe the functions of both the pancreas and the liver. 790–91
9. Name and discuss three serious illnesses of the liver. 791
10. Describe the structure and function of the large intestine. Name two medical conditions associated with the large intestine. 791
11. Explain why vitamins and minerals are necessary to good nutrition. Give examples. 792–93

Testing Yourself

Choose the best answer for each question.

1. Animals that feed discontinuously
   a. have digestive tracts that permit storage.
   b. are always filter feeders.
   c. exhibit extremely rapid digestion.
   d. have a nonspecialized digestive tract.
   e. usually eat only meat.
2. In which of the following types of animals would you expect the digestive tract to be more complex?
   a. those with a single opening for the entrance of food and exit of wastes
   b. those with two openings, one serving as an entrance and the other as an exit
   c. only those complex animals that also have a respiratory system
   d. those with two openings that use the digestive tract to help fight infections
   e. All but a are correct.
3. The typhlosole within the gut of an earthworm compares best to which of these organs in humans?
   a. teeth in the mouth
   b. esophagus in the thoracic cavity
   c. folds in the stomach
   d. villi in the small intestine
   e. the large intestine because it absorbs water
4. Which of these animals is a continuous feeder with a complete gut?
   a. planarian
   b. clam
   c. squid
   d. lion
   e. humans
5. The products of digestion are
   a. large macromolecules needed by the body.
   b. enzymes needed to digest food.
   c. small nutrient molecules that can be absorbed.
   d. regulatory hormones of various kinds.
   e. the food we eat.
6. Which of these could be absorbed directly without need of digestion?
   a. glucose
   b. fat
   c. protein
   d. nucleic acid
   e. All of these are correct.
7. Which association is incorrect?
   a. protein—trypsin
   b. fat—lipase
   c. maltose—pepsin
   d. starch—amylase
   e. protein—pepsin
8. Most of the absorption of the products of digestion takes place in humans across the
   a. squamous epithelium of the esophagus.
   b. convoluted walls of the stomach.
   c. fingerlike villi of the small intestine.
   d. smooth wall of the large intestine.
   e. lacteals of the lymphatic system.
9. The hepatic portal vein is located between
   a. the hepatic vein and the vena cava.
   b. two capillary beds.
   c. the pancreas and the small intestine.
   d. small intestine and the liver.
   e. Both b and d are correct.
10. Bile in humans
    a. is an important enzyme for the digestion of fats.
    b. is made by the gallbladder.
    c. emulsifies fat.
    d. must be activated during the first few weeks of life.
    e. All of these are correct.
11. Which of these is not a function of the liver in adults?
    a. produce bile
    b. store glucose
    c. produce urea
    d. make red blood cells
    e. produce proteins needed for blood clotting
12. The large intestine in humans
    a. digests all types of food.
    b. is the longest part of the intestinal tract.
    c. absorbs water.
    d. is connected to the stomach.
    e. All of these are correct.

13. Predict and explain the expected digestive results per test tube for this experiment.

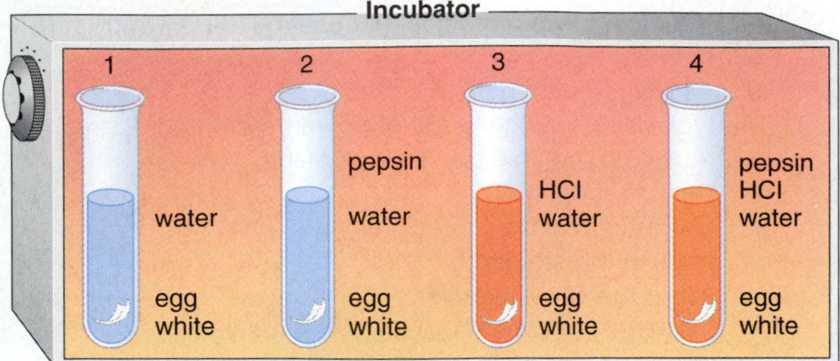

Thinking Scientifically

1. A drug for leukemia is not broken down in the stomach and is well absorbed by the intestine. However, the molecular form of the drug collected from the blood is not the same as the form that was swallowed by the patient. What explanation is most likely?
2. Snakes often swallow whole animals, a process that takes a long time. Then snakes spend some time digesting their food. What structural modifications would allow slow swallowing and storage of a whole animal to occur ? What chemical modifications would be necessary to digest a whole animal?

Understanding the Terms

anus 791
appendix 791
bile 788
chyme 787
colon 791
complete gut 782
diarrhea 791
duodenum 788
emulsification 788
esophagus 786
gallbladder 788
incomplete gut 782
intestinal enzyme 788
lacteal 789
large intestine 791
lipase 788
liver 788
microvillus 788
mineral 793
mouth 785
pancreas 788
pancreatic amylase 788
pepsin 787
peristalsis 786
pharynx 786
polyp 791
salivary amylase 785
salivary gland 785
small intestine 788
stomach 786
trypsin 788
villus (pl., villi) 788
vitamin 792

Match the terms to these definitions:

a. ______________ Essential requirement in the diet, needed in small amounts. They are often part of coenzymes.

b. ______________ Fat-digesting enzyme secreted by the pancreas.

c. ______________ Lymphatic vessel in an intestinal villus, it aids in the absorption of fats.

d. ______________ Muscular tube for moving swallowed food from the pharynx to the stomach.

e. ______________ Organ attached to the liver that serves to store and concentrate bile.

Web Connections

Exploring the Internet

http://www.mhhe.com/biosci/genbio/mader
(click on *Biology 7/e*)

The *Biology 7/e* Online Learning Center provides many resources for studying the material in this chapter including links to the following sites:

Visit the Cyberkitchen at Shapeup America where you can develop your own customized meal plan to help you balance your food intake with your level of physical activity.

http://www.shapeup.org/

Human Anatomy Online. This site provides much information on the human body, including a detailed look at the digestive system.

http://www.innerbody.com/image/digeov.html

National Digestive Diseases Information Clearinghouse. Extensive information on Crohn's Disease.

http://www.niddk.nih.gov/health/digest/pubs/crohns/crohns.htm

and

NIDDK home page which will lead to sites on other digestive disorders, ranging from gas to porphyria.

http://www.niddk.nih.gov/health/digest/digest.htm

CDC page on Giardiasis. Information on this parasitic infection which can lead to chronic gastrointestinal upsets.

http://www.cdc.gov/ncidod/dpd/list_gir.htm

Respiration

chapter

chapter concepts

44.1 Gas Exchange Surfaces

- Respiration comprises breathing, external and internal respiration, and cellular respiration. 798
- Most animals have a special, localized gas-exchange area such as the gills in aquatic animals and lungs in terrestrial animals. 798
- Amphibians use positive pressure, but other vertebrates use negative pressure, to ventilate the lungs. 801
- Birds have a complete ventilation system in that there is a one-way flow of air over the gas-exchange area. 801

44.2 Human Respiratory System

- There is a continuous pathway for air from the nose (or mouth) to the lungs in humans. 802
- The breathing rate is regulated and increases when there is a greater demand of exchange of gases in the lungs. 803
- The respiratory pigment hemoglobin transports oxygen from the lungs to the tissues and aids the transport of carbon dioxide from the tissues to the lungs. 805

44.3 Respiration and Health

- The respiratory tract is especially subject to infections. Cigarette smoking contributes to two major lung disorders—emphysema and cancer. 807

The bottle-nosed dolphin, *Tursiops*, and boy, *Homo*, are mammals that breathe air.

Many animals have a storage area for food that allows them to eat first and digest later. When it comes to gases, animals have no such storage area; oxygen and carbon dioxide are continually exchanged with the environment.

In many small aquatic invertebrates, gas exchange occurs directly between the animal's cells and the environment. Dissolved oxygen in water diffuses into cells, while carbon dioxide diffuses from the cells into the water. In these animals, no specialized respiratory structure is present. Other aquatic invertebrates have highly branched gills filled with blood vessels. Water moving over the gills supplies oxygen to and removes carbon dioxide from the blood.

Among vertebrate animals, fishes and some amphibians also rely on gills for respiration. Most adult amphibians have lungs, but also use their thin skin for gas exchange. Reptiles, birds, and mammals have efficient vascularized lungs for respiration. Air tubes bring oxygen into the lungs where gas exchange occurs with the blood. Diffusion is the basic mechanism for gas exchange in all animals, whether the animal has no respiratory organ, gills, or lungs.

44.1 Gas Exchange Surfaces

Respiration is the sequence of events that results in gas exchange between the body's cells and the environment. Breathing is only the first step of respiration, which includes the following steps in terrestrial vertebrates:

- Breathing: includes inspiration (entrance of air into lungs) and expiration (exit of air from the lungs).
- External respiration: gas exchange between air and blood within the lungs. Blood transports gases from the lungs about the body.
- Internal respiration: gas exchange between blood and tissue fluid. The body's cells exchange gases with tissue fluid.

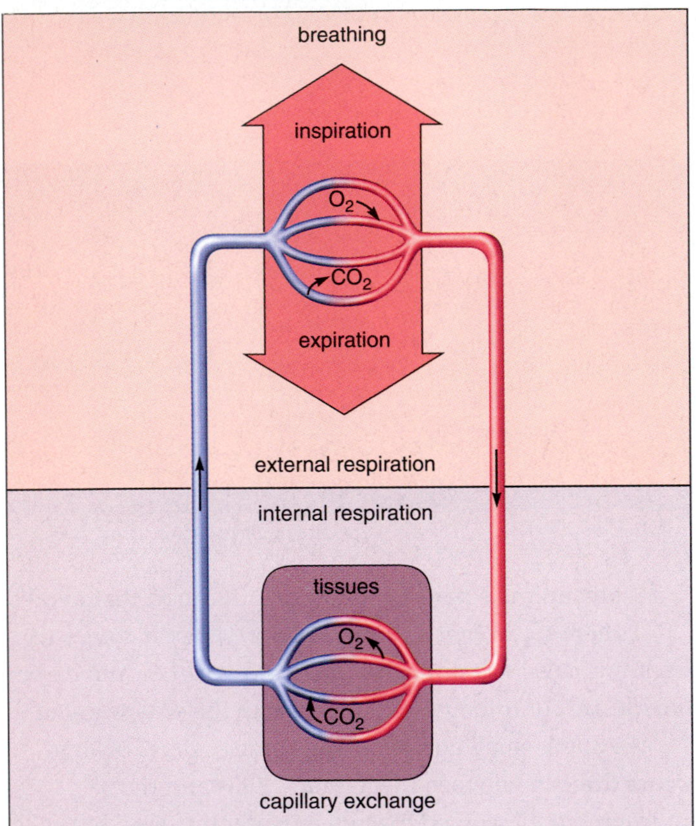

Gas exchange takes place by the physical process of diffusion. For diffusion to be effective, the gas-exchange region must be (1) moist, (2) thin, and (3) large in relation to the size of the body. Some animals are small and shaped in a way that allows the surface of the animal to be the gas-exchange surface. Other animals are complex and have specialized gas-exchange surfaces. The effectiveness of diffusion is enhanced by vascularization, and delivery to cells is promoted when the blood contains a respiratory pigment such as hemoglobin.

Regardless of the particular gas exchange surface and the manner in which gases are delivered to the cells, in the end oxygen enters mitochondria where cellular respiration takes place. As glucose is oxidized to carbon dioxide and water, ATP molecules are formed. Without the delivery of oxygen to the body's cells, ATP production does not take place and life ceases.

Water Environments

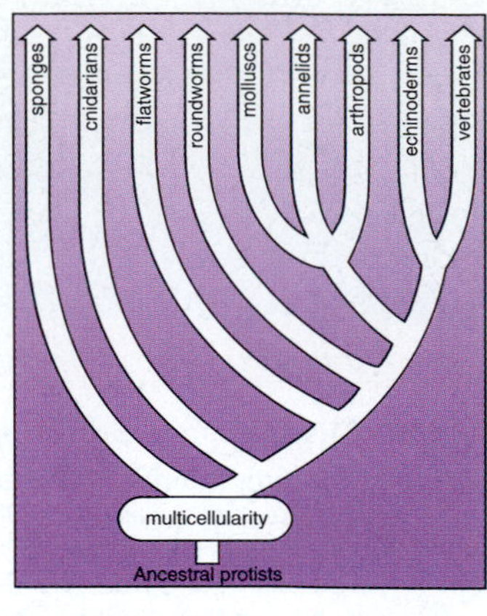

It is more difficult for animals to obtain oxygen from water than from air. Water fully saturated with air contains only a fraction of the amount of oxygen in the same volume of air. Also, water is more dense than air. Therefore, aquatic animals expend more energy to breathe than do terrestrial animals. Fishes use up to 25% of their energy output to breathe, while terrestrial mammals use only 1–2% of their energy output.

Hydras, which are cnidarians, and planarians, which are flatworms, have a large surface area in comparison to their size (Fig. 44.1). This makes it possible for most of their cells to exchange gases directly with the environment. In hydras, the outer layer of cells is in contact with the external environment, and the inner layer can exchange gases with the water in the gastrovascular cavity. In planarians, the flattened body permits cells to exchange gases with the external environment.

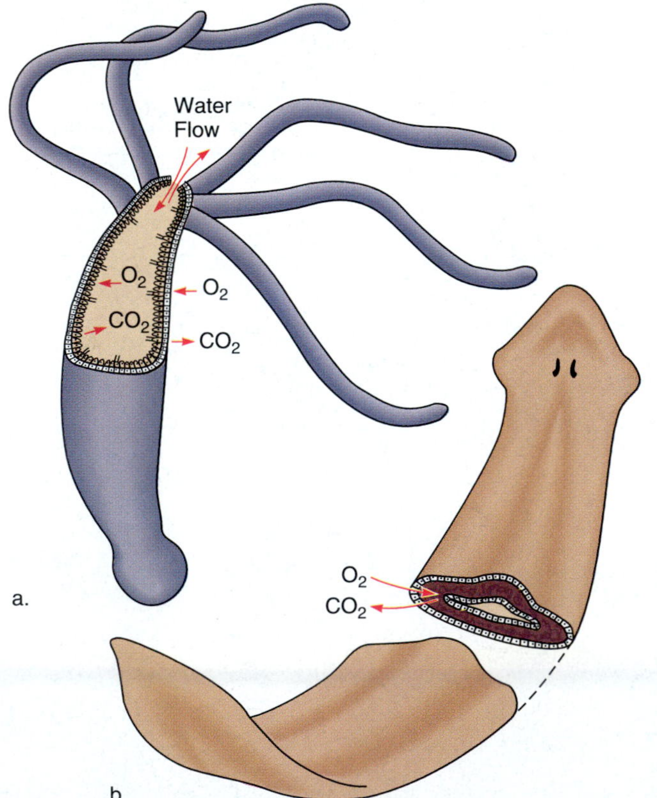

Figure 44.1 Animal shapes and gas exchange. Some small aquatic animals use the body surface for gas exchange. This works because the body surface is large compared to the size of the animal. **a.** In hydras, which are cnidarians, every cell is near a source of oxygen. **b.** The flattened body of planarians, which are flatworms, allows most cells to carry out gas exchange with the external environment.

The tubular shape of annelids also provides a surface area adequate for gas-exchange purposes. In addition to a tubular shape, polychaete worms have extensions of the body wall called parapodia, which are vascularized. Quite often, aquatic animals have **gills,** which are finely divided and vascularized outgrowths of either the outer or the inner body surface. Among molluscs, such as a clam or squid, water is drawn into the mantle cavity, where it passes through the gills. In decapod crustacea, which are arthropods, the gills are located in brachial chambers covered by the exoskeleton. Water is kept moving by the action of specialized appendages located near the mouth.

Among aquatic vertebrates, the gills of bony fishes are outward extensions of the pharynx (Fig. 44.2). Ventilation is brought about by the combined action of the mouth and gill covers, or opercula (sing., operculum). When the mouth is open, the opercula are closed and water is drawn in. Then the mouth closes and the opercula open, drawing the water from the pharynx through the gill slits located between the gill arches. On the outside of the gill arches, the gills are composed of filaments that are folded into platelike lamellae.

In the capillaries of each lamella, the blood flows in a direction opposite to the movement of water across the gills. With a countercurrent flow as blood gains oxygen, it always encounters water having an even higher oxygen content and no equilibrium point is reached. If the flow were concurrent (oxygen-rich water and oxygen-poor blood would flow in the same direction) an equilibrium point would occur and only about half the oxygen in the water would be captured. With a countercurrent mechanism an equilibrium point is not reached and about 80–90% of the initial dissolved oxygen in water is extracted.

Small aquatic animals sometimes use the body surface for gas exchange, but many larger ones have localized gas-exchange surfaces known as gills.

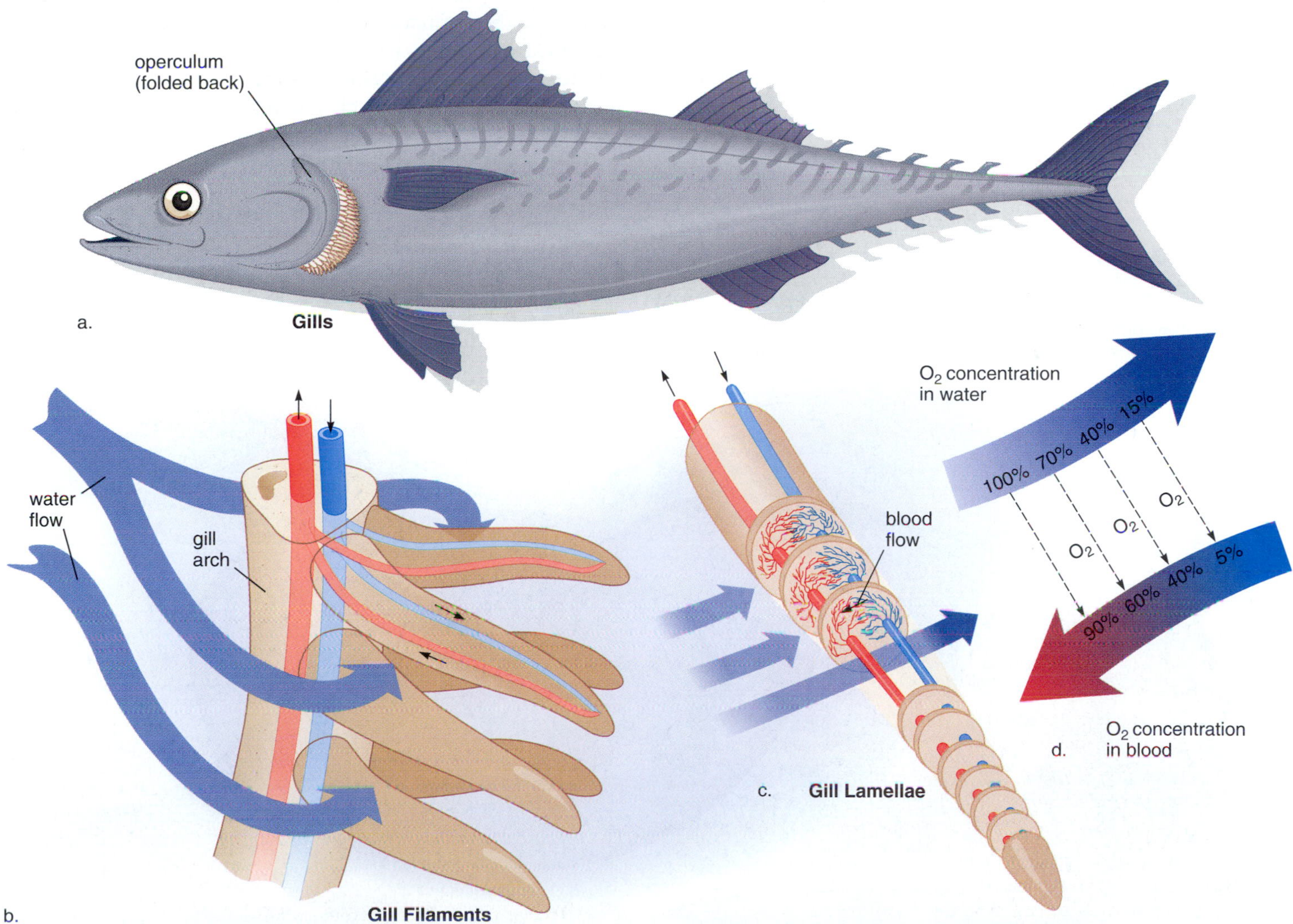

Figure 44.2 Anatomy of gills in bony fishes.
a. The operculum (folded back) covers and protects several layers of delicate gills. b. Each gill layer has two rows of gill filaments. c. Each filament has many thin, platelike lamellae. Gases are exchanged between the capillaries inside the lamellae and the water that flows between the lamellae. d. Blood in the capillaries flows in the direction opposite to that of the water. Blood takes up almost all of the oxygen in the water as a result of this countercurrent flow.

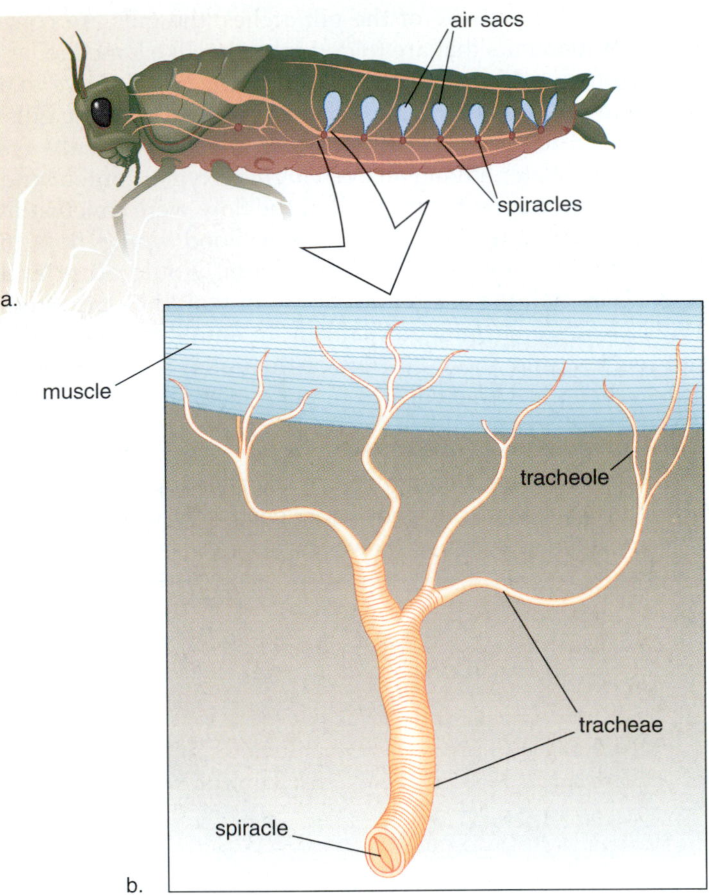

Figure 44.3 Tracheal system of insects.
a. A system of air tubes extends throughout the body of an insect and they, rather than blood, carry oxygen to the cells. **b.** Air enters the tracheae at openings called spiracles. From here, the air moves to the smaller tracheoles, which take it to the cells, where gas exchange takes place.

Land Environments

Air is a rich source of oxygen compared to water; however, it does have a drying effect on respiratory surfaces. A human loses about 350 ml of water per day when the air has a relative humidity of only 50%.

The earthworm, which is an annelid, is an example of an invertebrate terrestrial animal that uses its body surface for respiration. An earthworm expends much energy to keep its body surface moist by secreting mucus and by releasing fluids from excretory pores. Further, the worm is behaviorally adapted to remain in damp soil during the day, when the air is driest.

Insects and certain other terrestrial arthropods have a respiratory system known as a tracheal system (Fig. 44.3). Oxygen enters the tracheae at spiracles, which are valvelike openings on each side of the body. The tracheae branch and then rebranch, ending in tiny channels, the tracheoles, that are in direct contact with the body cells. Larger insects have a ventilation system to keep the air moving in and out of the trachea. Many have air sacs located near major muscles; contraction of these muscles causes the air sacs to empty, and relaxation causes the air sacs to expand and draw air in. The tracheal system of insects is very effective in delivering oxygen to the cells so that the circulatory system plays no role in gas transport.

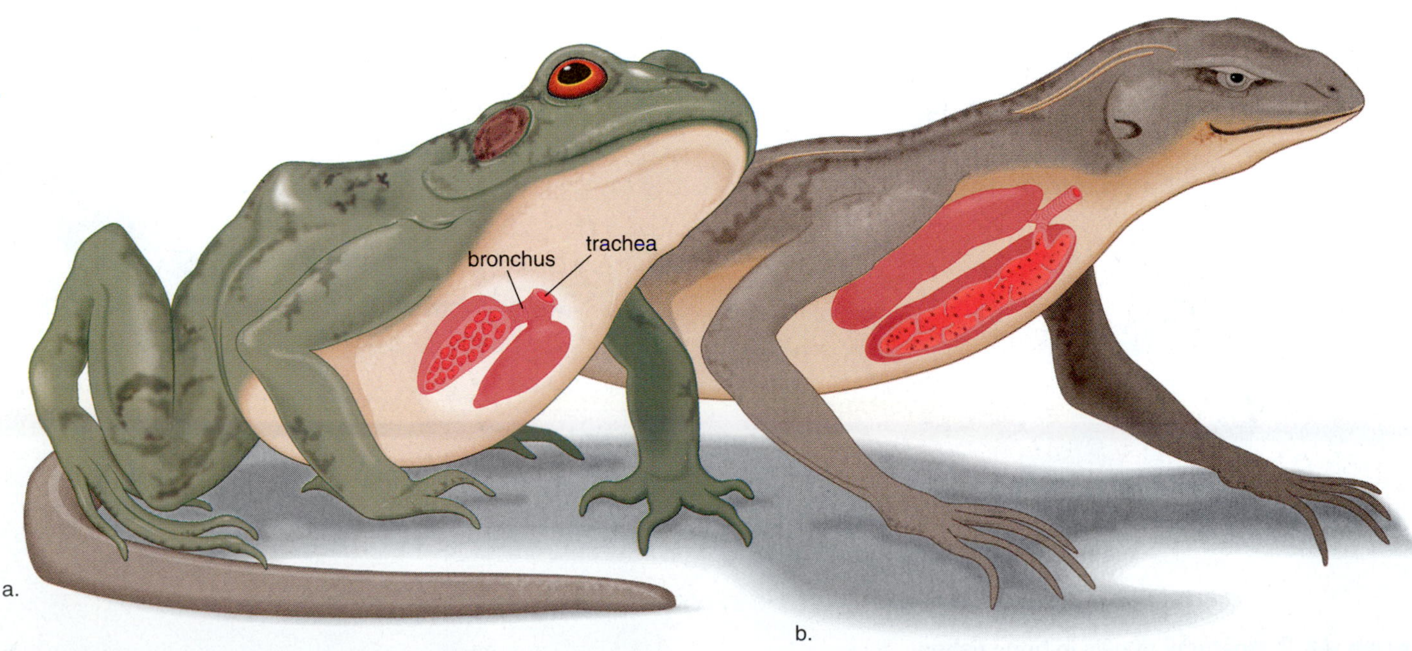

Figure 44.4 Respiration in amphibians compared to reptiles.
a. Amphibians use positive pressure to push air into small and saclike lungs. The moist, thin skin of amphibians is often used as an auxiliary gas-exchange surface. **b.** Reptiles use negative pressure to draw air into lungs, which are more convoluted. The tough, scaly skin protects the animal from drying out and is not used for respiration.

Terrestrial vertebrates have evolved **lungs,** which are vascularized outgrowths from the lower pharyngeal region. The lungs of amphibians are simple, saclike structures (Fig. 44.4*a*). Many amphibians possess a short trachea, which divides into two bronchi that open into the lungs. Most amphibians breathe to some extent through the skin, which is kept moist by the presence of mucus produced by numerous glands on the surface of the body. During the winter in temperate climates, amphibians burrow in the mud, and all gas exchange occurs by way of the skin.

The inner lining of the lungs is more finely divided in reptiles than in amphibians (Fig. 44.4*b*). The lungs of birds and mammals are elaborately subdivided into small passageways and spaces. It has been estimated that human lungs have a total surface area that is at least 50 times the skin's surface area. To keep the lungs from drying out, air is moistened as it moves through passageways leading to the lungs.

Terrestrial vertebrates ventilate the lungs by moving air into and out of the respiratory tract. Frogs use positive pressure to force air into the respiratory tract. With the nostrils firmly shut, the floor of the mouth rises and pushes the air into the lungs. Reptiles, birds, and mammals use negative pressure to move air into the lungs. Reptiles have jointed ribs that can be raised to expand the lungs. Mammals have a rib cage that is lifted up and out and a muscular **diaphragm** [Gk. *diaphragma,* partition wall] that is flattened. As the thoracic cavity (chest cavity) expands and lung volume increases, air will flow into the lungs due to differences in air pressure. Following **inspiration** (or inhalation), **expiration** (or exhalation) occurs. In reptiles, lowering the ribs exerts a pressure that forces air out. In mammals, when the rib cage is lowered and the diaphragm rises, the thoracic pressure increases, forcing air out of the lungs.

The lungs of amphibians, reptiles, and mammals are not completely emptied and refilled during each breathing cycle. Because of this incomplete ventilation method, the air entering mixes with used air remaining in the lungs. While this does help conserve water, it also decreases gas-exchange efficiency. The high oxygen requirement of flying birds, however, requires a method of complete ventilation (Fig. 44.5). Incoming air is carried past the lungs by a trachea that takes it to a set of posterior air sacs. The air then passes forward through the lungs into a set of anterior air sacs. From here, it is finally expelled. Fresh, oxygen-rich air passes through the lungs in one direction only and does not mix with used air.

Most terrestrial animals have a specialized gas-exchange region. Insects have a tracheal system, while terrestrial vertebrates depend on lungs, which are ventilated variously.

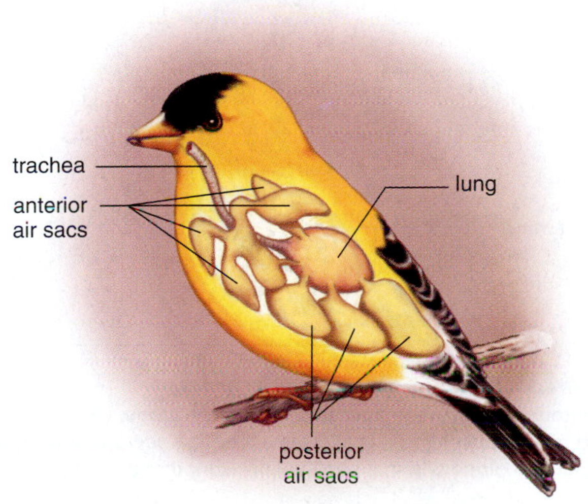

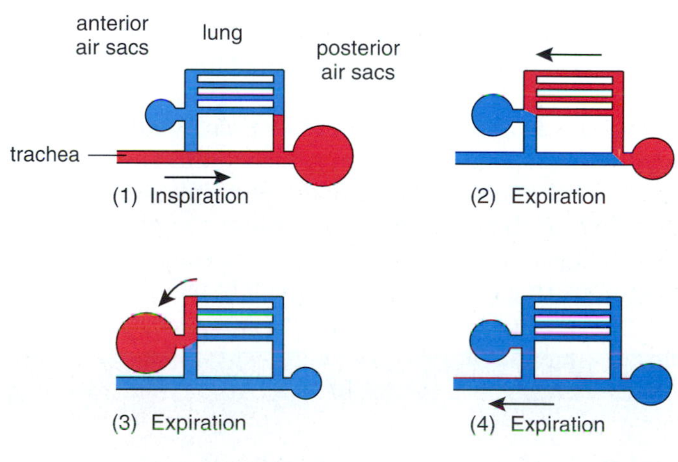

Figure 44.5 Respiratory system in birds.
a. Birds have a system of air sacs in their bones. **b.** When a bird inhales, air enters the posterior air sacs, and when a bird exhales, air moves through the lungs to the anterior air sacs before exiting the trachea. This one-way flow of air through the lungs is efficient and allows more oxygen to be removed from one breath of air.

44.2 Human Respiratory System

The human respiratory system includes all structures that conduct air to and from the lungs (Fig. 44.6 and Table 44.1). The lungs lie deep within the thoracic cavity, where they are protected from drying out. As air moves through the nose, the pharynx, the trachea, and the bronchi to the lungs, it is filtered so that it is free of debris, warmed, and humidified. By the time the air reaches the lungs, it is at body temperature and is saturated with water. In the nose, hairs and cilia act as a screening device. In the trachea and the bronchi, cilia beat upward, carrying mucus, dust, and occasional bits of food that "went down the wrong way" into the throat, where the accumulation may be swallowed or expectorated.

The hard and soft palates separate the nasal cavities from the mouth, but the air and food passages cross in the **pharynx** [Gk. *pharynx,* throat]. This may seem inefficient, and there is danger of choking if food accidentally enters the trachea, but this arrangement does have the advantage of letting you breathe through your mouth in case your nose is plugged up. In addition, it permits greater intake of air during heavy exercise, when greater gas exchange is required.

Air passes from the pharynx through the **glottis** [Gk. *glotti,* tongue], an opening into the **larynx** [Gk. *larynx,* gullet], or voice box. At the edges of the glottis, embedded in mucous membrane, are the **vocal cords.** These flexible and pliable bands of connective tissue vibrate and produce sound when air is expelled past them through the glottis from the larynx.

The larynx and the **trachea** [L. *trachia,* windpipe] are permanently held open to receive air. The larynx is held open by the complex of cartilages; among them is the cartilage of the Adam's apple. The trachea is held open by a series of C-shaped, cartilaginous rings that do not completely meet in

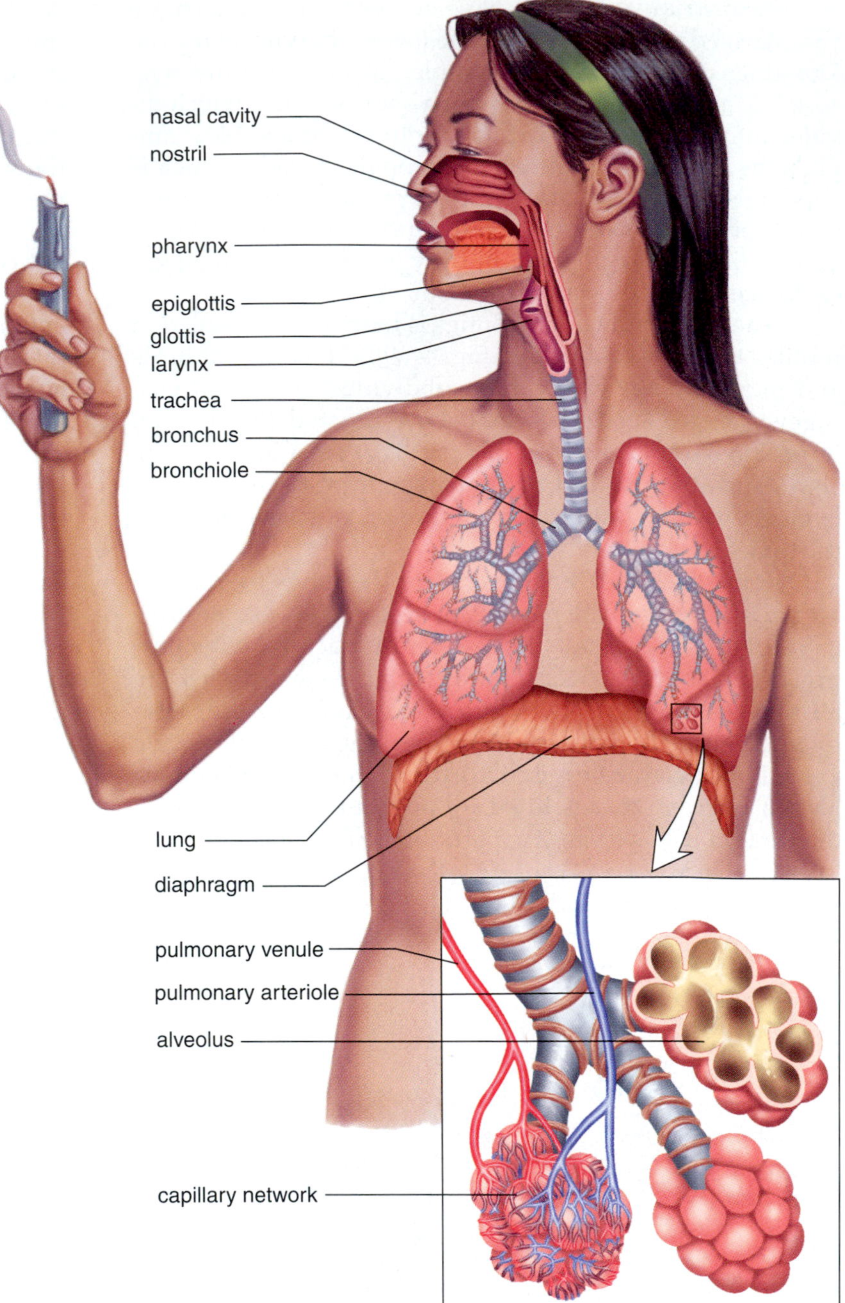

Figure 44.6 The human respiratory tract.
The respiratory tract extends from the nose to the lungs, which are composed of air sacs called alveoli. Gas exchange occurs between air in the alveoli and blood within a capillary network that surrounds the alveoli.

Table 44.1

Path of Air

Structure	Function
Nasal cavities	Filter, warm, and moisten
Pharynx (throat)	Connection to larynx
Glottis	Permits passage of air
Larynx (voice box)	Sound production
Trachea (windpipe)	Passage of air to bronchi
Bronchi	Passage of air to lung
Bronchioles	Passage of air to alveoli
Alveoli	Air sacs for gas exchange

the rear. When food is being swallowed, the larynx rises, and the glottis is closed by a flap of tissue called the **epiglottis** [Gk. *epi,* over, and *glotta,* tongue]. A backward movement of the soft palate covers the entrance of the nasal passages into the pharynx. The food then enters the esophagus, which lies behind the larynx.

The trachea divides into two primary **bronchi** [Gk. *bronchos,* windpipe], which enter the right and left lungs. Branching continues until there are a great number of smaller passages called **bronchioles** [Gk. dim. of *bronchos,* windpipe]. The two bronchi resemble the trachea in structure, but as the bronchial tubes divide and subdivide, their walls become thinner, and rings of cartilage are no longer present. Each bronchiole terminates in an elongated space enclosed by a multitude of air pockets, or sacs, called **alveoli** [L. *alveolus,* dim. of *alveus,* cavity], which make up the lungs.

Breathing

Humans breathe using the same mechanism employed by all other mammals. The volume of the thoracic cavity and lungs is increased by muscle contractions that lower the diaphragm and raise the ribs (Fig. 44.7). These movements create a negative pressure in the thoracic cavity and lungs, and air then flows into the lungs. When rib and diaphragm muscles relax, air is exhaled as a result of increased pressure in the thoracic cavity and lungs.

Increased hydrogen ion (H^+) and carbon dioxide (CO_2) concentrations in the blood are the primary stimuli that increase breathing rate. The chemical content of the blood is monitored by chemoreceptors called the aortic and carotid bodies, which are specialized structures located in the walls of the aorta and the carotid arteries. These receptors are very sensitive to changes in hydrogen ion and carbon dioxide concentrations, but they are only minimally sensitive to a lower oxygen (O_2) concentration. Information from the chemoreceptors goes to the respiratory center in the medulla oblongata of the brain, which then increases the breathing rate when concentrations of hydrogen ions and carbon dioxide rise. The respiratory center is itself sensitive to the chemical content of the blood reaching the brain.

Air moves through passages that repeatedly divide until the smallest passages terminate at the alveoli of the lungs. As the lungs expand, inspiration occurs; as the lungs recoil, expiration occurs

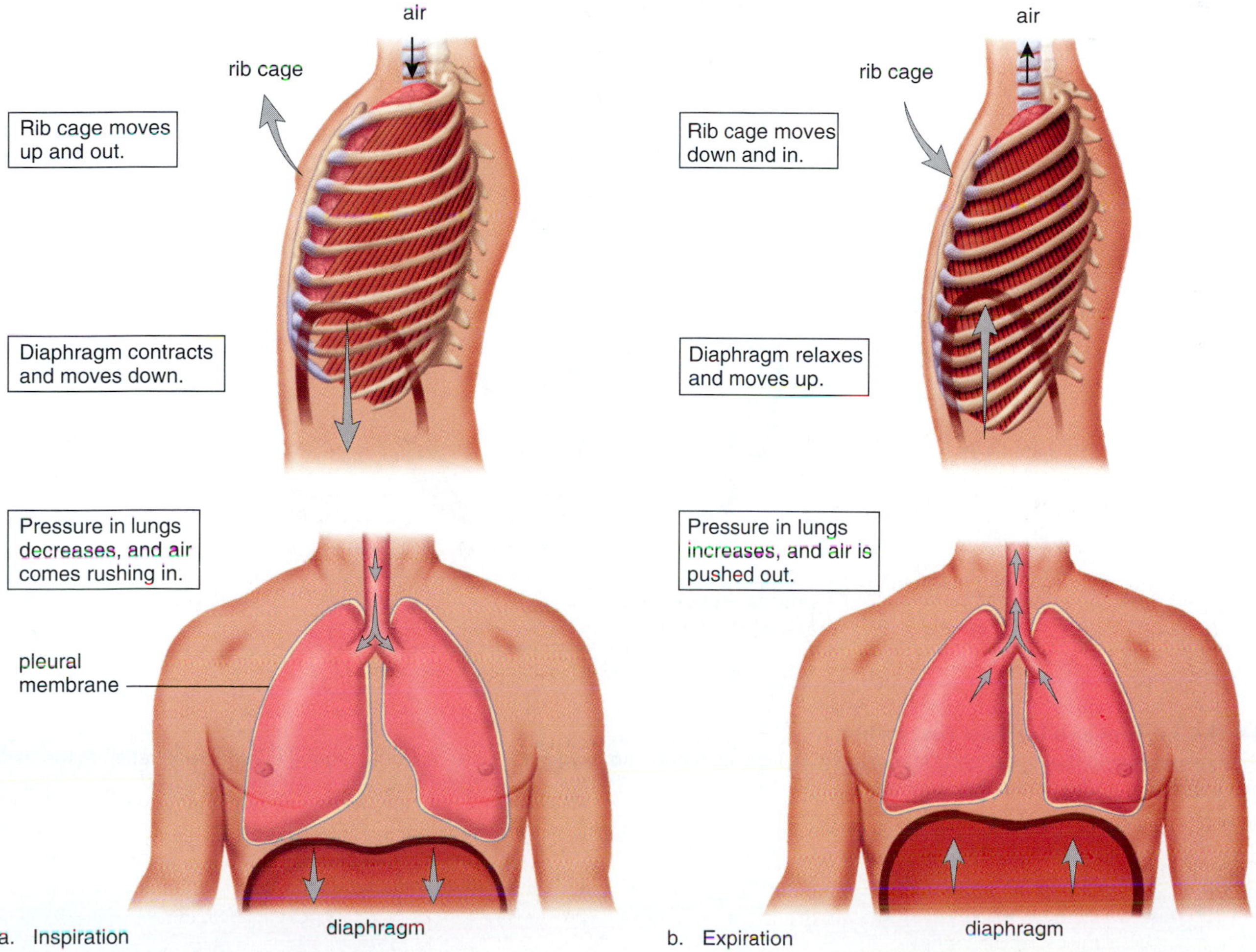

Figure 44.7 Inspiration versus expiration.
a. During inspiration, the thoracic cavity and lungs expand so that air is drawn in. **b.** During expiration, the thoracic cavity and lungs resume their original positions and pressures. Now, air is forced out.

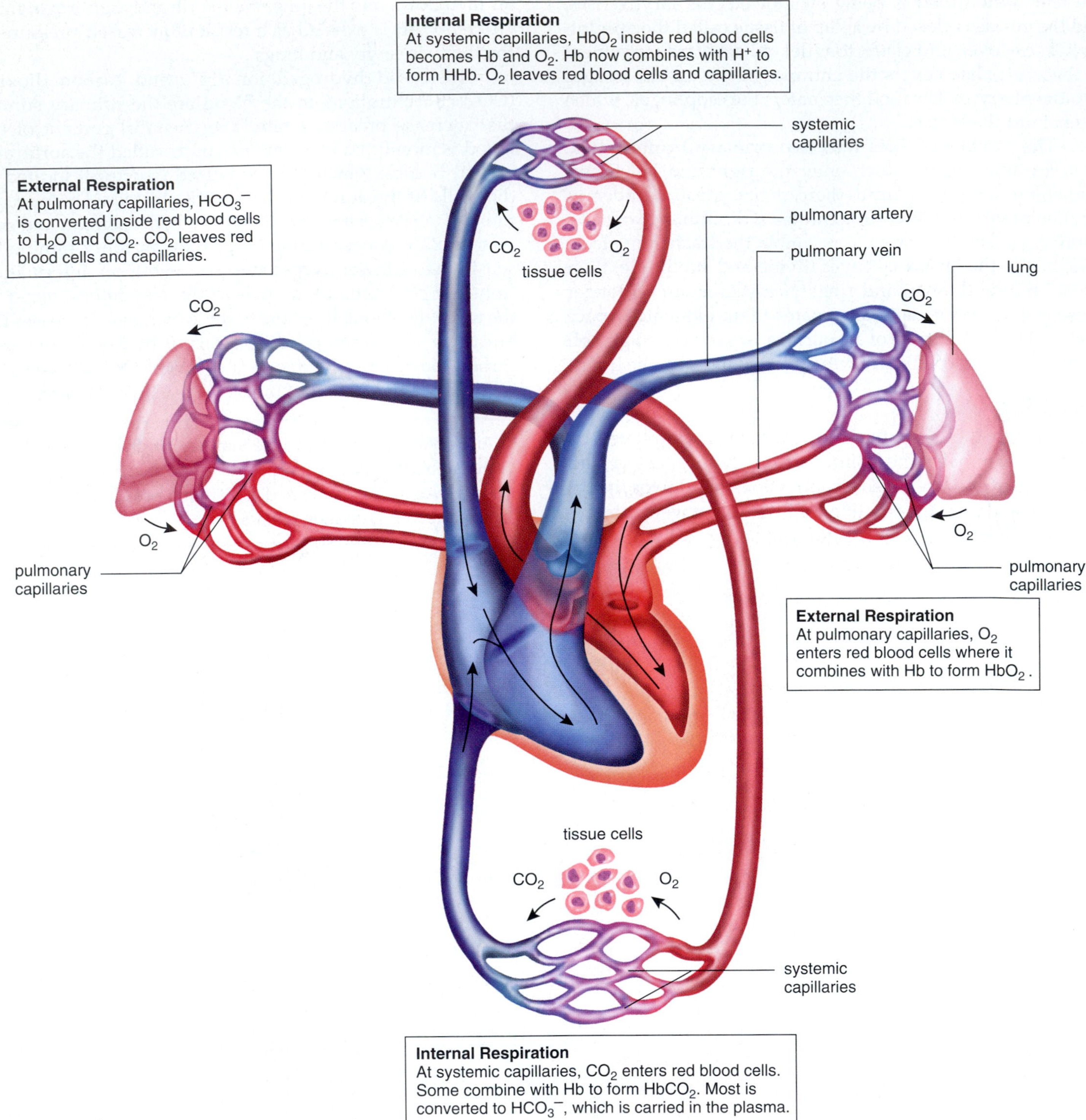
Internal Respiration
At systemic capillaries, HbO_2 inside red blood cells becomes Hb and O_2. Hb now combines with H^+ to form HHb. O_2 leaves red blood cells and capillaries.
External Respiration
At pulmonary capillaries, HCO_3^- is converted inside red blood cells to H_2O and CO_2. CO_2 leaves red blood cells and capillaries.
systemic capillaries
CO_2
O_2
tissue cells
pulmonary artery
pulmonary vein
lung
CO_2
CO_2
O_2
O_2
pulmonary capillaries
pulmonary capillaries
External Respiration
At pulmonary capillaries, O_2 enters red blood cells where it combines with Hb to form HbO_2.
tissue cells
CO_2
O_2
systemic capillaries
Internal Respiration
At systemic capillaries, CO_2 enters red blood cells. Some combine with Hb to form $HbCO_2$. Most is converted to HCO_3^-, which is carried in the plasma.

Figure 44.8 External and internal respiration.
During external respiration in the lungs, carbon dioxide (CO_2) leaves blood and oxygen (O_2) enters blood. During internal respiration in the tissues, oxygen leaves blood and carbon dioxide enters blood.

Gas Exchange and Transport

Diffusion primarily accounts for the exchange of gases between the air in the alveoli and the blood in the pulmonary capillaries (Fig. 44.8). Blood flowing into the pulmonary capillaries has a higher carbon dioxide concentration than alveolar air. Therefore, carbon dioxide diffuses out of the blood and into the alveoli. The pattern is the reverse for oxygen. Blood coming into the pulmonary capillaries has a lower concentration of oxygen than alveolar air. Therefore, oxygen diffuses from the alveoli into the capillaries.

Transport of Oxygen and Carbon Dioxide

Most oxygen entering the pulmonary capillaries from the alveoli combines with **hemoglobin** (Hb) [Gk. *haima*, blood, and L. *globus*, ball] in red blood cells to form **oxyhemoglobin:**

$$\underset{\text{deoxyhemoglobin}}{Hb} + \underset{\text{oxygen}}{O_2} \rightarrow \underset{\text{oxyhemoglobin}}{HbO_2}$$

Since there are about 250 million hemoglobin molecules in each red blood cell, each red blood cell is capable of carrying more than one billion molecules of oxygen. Each hemoglobin molecule contains four polypeptide chains, and each chain is folded around an iron-containing group called **heme.** It is actually the iron that forms a loose association with oxygen.

Figure 44.9 shows the percent saturation of hemoglobin at various partial pressures, depending on the temperature and acidity. A partial pressure of a gas is simply the amount of pressure exerted by that gas among all the gases present. At the normal partial pressure of oxygen in the lungs, hemoglobin is practically saturated with oxygen, but at the partial pressure in the tissues, oxyhemoglobin quickly gives up much of its oxygen:

$$HbO_2 \rightarrow Hb + O_2$$

The acid pH and warmer temperature of the tissues also promote the release of oxygen by hemoglobin. In tissues, oxygen leaves the blood and enters tissue fluid where it is taken up by cells.

Carbon dioxide enters blood in the tissues. Some carbon dioxide combines with hemoglobin to form **carbaminohemoglobin.** Most of the carbon dioxide, however, is transported in the form of the **bicarbonate ion** (HCO_3^-). First carbon dioxide combines with water, forming carbonic acid, and then this dissociates to a hydrogen ion (H^+) and HCO_3^-:

$$\underset{\text{carbon dioxide}}{CO_2} + \underset{\text{water}}{H_2O} \rightarrow \underset{\text{carbonic acid}}{H_2CO_3} \rightarrow \underset{\text{hydrogen ion}}{H^+} + \underset{\text{bicarbonate ion}}{HCO_3^-}$$

Carbonic anhydrase [Gk. *an*, without, and *hydrias*, water], an enzyme in red blood cells, speeds this reaction. The release of H^+, could drastically change the pH of the blood. However, the H^+ is absorbed by the globin portions of hemoglobin, and the HCO_3^- diffuse out of the red blood cells to be carried in the plasma. Hemoglobin that combines with a H^+ is called reduced hemoglobin and can be symbolized as HHb. HHb plays a vital role in maintaining the normal pH of the blood.

As blood enters the pulmonary capillaries, most of the carbon dioxide is present in plasma as HCO_3^-. Reduced hemoglobin (HHb) gives up the H^+ it has been carrying, as carbonic anhydrase speeds this reaction:

$$H^+ + HCO_3^- \rightarrow H_2CO_3 \rightarrow H_2O + CO_2$$

Now free carbon dioxide diffuses out of blood into the alveoli of the lungs.

Oxygen is transported in the blood by hemoglobin, and carbon dioxide is largely carried in plasma as the bicarbonate ion.

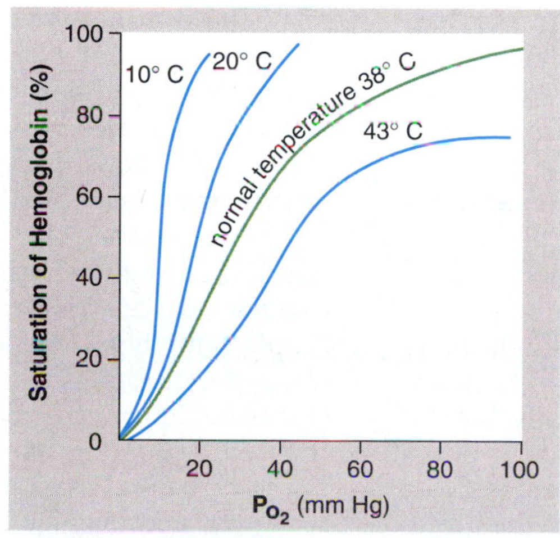

a. Saturation of Hb relative to temperature

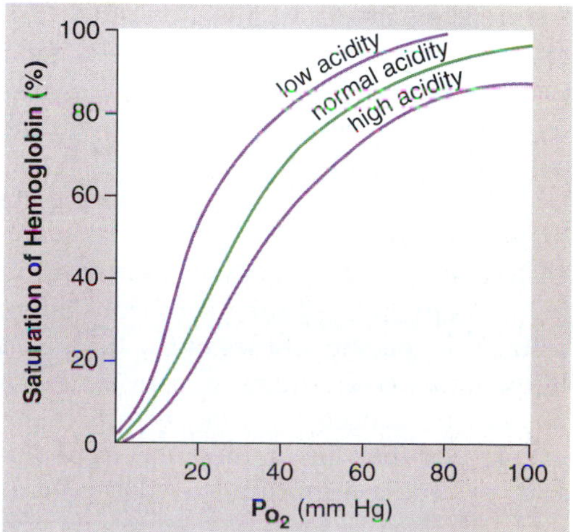

b. Saturation of Hb relative to pH

Figure 44.9 Environmental conditions and hemoglobin saturation.
The partial pressure of oxygen (P_{O_2}) in pulmonary capillaries is about 98–100 mm Hg, but only about 40 mm Hg in tissue capillaries. Hemoglobin is about 98% saturated in the lungs because of P_{O_2}, and also because (a) the temperature is cooler and (b) the pH is higher in lungs. On the other hand, hemoglobin is only about 60% saturated in the tissues because of the P_{O_2} and also because (a) the temperature is warmer and (b) the pH is lower in the tissues.

44.3 Respiration and Health

We have seen that the entire respiratory tract has a warm, wet, mucous membrane lining, which is constantly exposed to environmental air. The quality of this air, determined by the pollutants and the pathogens therein, can affect our health.

Upper Respiratory Tract Infections

The upper respiratory tract consists of the nose, the pharynx, and the larynx. Upper respiratory infections (URI) can spread from the nasal cavities to the sinuses, to the middle ears, and to the larynx. Viral infections sometimes lead to secondary bacterial infections. What we call "strep throat" is a primary bacterial infection caused by *Streptococcus pyogenes* that can lead to a generalized upper respiratory infection and even a systemic (affecting the body as a whole) infection.

Sinusitis is an infection of the sinuses, cavities within the facial skeleton that drain into the nasal cavities. Only about 1–3% of upper respiratory infections are accompanied by sinusitis. Sinusitis develops when nasal congestion blocks the tiny openings leading to the sinuses. Tonsillitis occurs when tonsils become inflamed and enlarged. **Tonsils** are masses of lymphatic tissue that occur in the pharynx. The tonsils in the dorsal wall of the nasopharynx are often called adenoids. The tonsils remove many of the pathogens that enter the pharynx; therefore, they are a first line of defense against invasion of the body. Laryngitis is an infection of the larynx with an accompanying hoarseness leading to the inability to talk in an audible voice. Usually laryngitis disappears with treatment of the upper respiratory infection. Persistent hoarseness without the presence of an upper respiratory infection is one of the warning signs of cancer and therefore should be looked into by a physician.

Lower Respiratory Tract Disorders

Lower respiratory tract disorders are illustrated in Figure 44.10.

Lower Respiratory Infections

Acute bronchitis is an infection of the primary and secondary bronchi. Usually it is preceded by a viral URI that has led to a secondary bacterial infection. Most likely, a nonproductive cough has become a deep cough that expectorates mucus and perhaps pus.

Pneumonia is a viral or bacterial infection of the lungs in which bronchi and alveoli fill with thick fluid (Fig. 44.10). Most often it is preceded by influenza. Rather than being a generalized lung infection, pneumonia may be localized in specific lobules of the lungs. Obviously the more lobules involved, the more serious the infection. Pneumonia can be caused by a bacterium that is usually held in check, but that has gained the upper hand due to stress and/or reduced immunity. AIDS patients are subject to a particularly rare form of pneumonia caused by the protozoan *Pneumocystis carinii*. Pneumonia of this type is almost never seen in individuals with a healthy immune system. High fever and chills with headache and chest pain are symptoms of pneumonia.

Pulmonary tuberculosis is caused by the tubercle bacillus, a type of bacterium. It is possible to tell if a person has ever been exposed to tuberculosis with a skin test in which a highly diluted extract of the bacillus is injected into the skin of the patient. A person who has never been in contact with tubercle bacillus shows no reaction, but one who has developed immunity to the organism shows an area of inflammation that peaks in about 48 hours. When tubercle bacilli invade the lung tissue, the cells build a protective capsule about the foreigners, isolating them from the rest of the body. This tiny capsule is called a tubercle. If the resistance of the body is high, the imprisoned organisms die, but if the resistance is low, the organisms eventually can be liberated. If a chest X ray detects active tubercles, the individual is put on appropriate drug therapy to ensure the localization of the disease and the eventual destruction of any live bacterial organisms.

Tuberculosis was a major killer in the United States before the middle of this century, after which antibiotic therapy brought it largely under control. In recent years, however, the incidence of tuberculosis is on the rise, particularly among AIDS patients, the homeless, and the rural poor. Worse, the new strains are resistant to the usual antibiotic therapy.

Pulmonary Disorders

Inhaling particles such as silica (sand), coal dust, asbestos, and, now it seems, fiberglass can lead to pulmonary fibrosis, a condition in which fibrous connective tissue builds up in the lungs. The lungs cannot inflate properly and are always tending toward deflation. Breathing asbestos is also associated with the development of cancer. Since asbestos has been used so widely as a fireproofing and insulating agent, unwarranted exposure has occurred. It is projected that two million deaths could be caused by asbestos exposure—mostly in the workplace—between 1990 and 2020.

In chronic bronchitis, the airways are inflamed and filled with mucus. A cough that brings up mucus is common. The bronchi have undergone degenerative changes including the loss of cilia and their normal cleansing action. Under these conditions an infection is more likely to occur. Smoking cigarettes and cigars is the most frequent cause of chronic bronchitis. Exposure to other pollutants can also cause chronic bronchitis.

Emphysema is a chronic and incurable disorder in which the alveoli are distended and their walls damaged so that the surface area available for gas exchange is reduced. Emphysema is often preceded by chronic bronchitis. Air trapped in the lungs leads to alveolar damage and a noticeable ballooning of the chest. The elastic recoil of the lungs is reduced, so not only are the airways narrowed but the driving force behind expiration is also reduced. The victim is breathless and may have a cough. Because the surface area for gas exchange is reduced, oxygen reaching the heart and the brain is reduced. Even so, the heart works furiously to force more blood through the lungs, and an increased workload on the heart can result. Lack of oxygen to the brain can make the person feel depressed, sluggish, and irritable. Exercise, drug therapy, and supplemental oxygen, along with giving up smoking, may relieve the symptoms and possibly slow the progression of emphysema.

Asthma is a disease of the bronchi and bronchioles that is marked by wheezing, breathlessness, and sometimes cough and expectoration of mucus. The airways are unusually sensitive to specific irritants, which can include a wide range of allergens such as pollen, animal dander, dust, cigarette smoke, and industrial fumes. Even cold air, however, can be an irritant. When exposed to the irritant, the smooth muscle in the bronchioles undergoes spasms. It now appears that chemical mediators given off by immune cells in the bronchioles result in the spasms. Most asthma patients have some degree of bronchial inflammation that reduces the diameter of the airways and contributes to the seriousness of an attack. Asthma is not curable but is treatable. There are inhalers that control the inflammation and hopefully prevent an attack, but there are also inhalers that stop the muscle spasms should an attack occur.

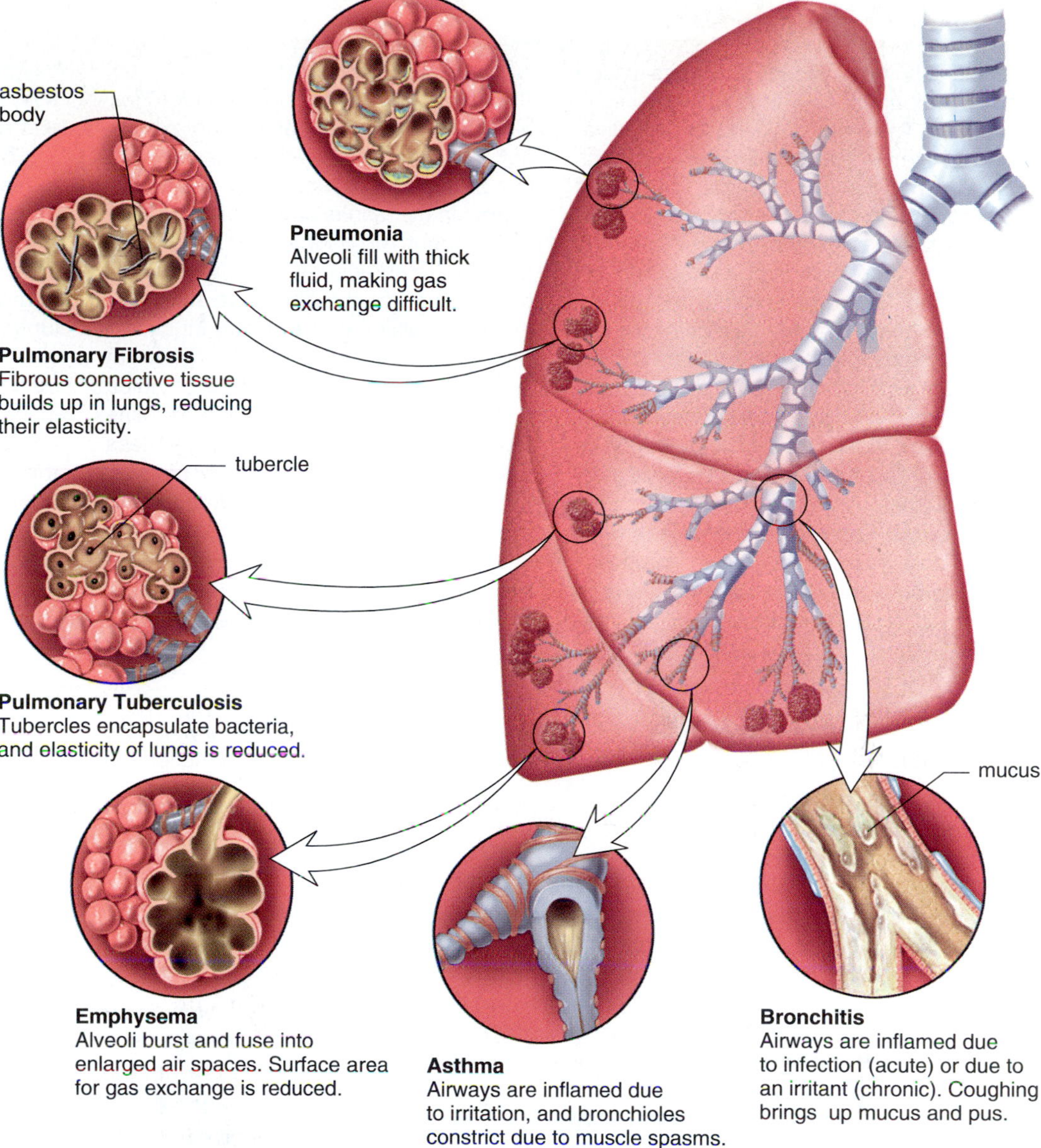
asbestos body
Pulmonary Fibrosis
Fibrous connective tissue builds up in lungs, reducing their elasticity.
Pneumonia
Alveoli fill with thick fluid, making gas exchange difficult.
tubercle
Pulmonary Tuberculosis
Tubercles encapsulate bacteria, and elasticity of lungs is reduced.
mucus
Emphysema
Alveoli burst and fuse into enlarged air spaces. Surface area for gas exchange is reduced.
Asthma
Airways are inflamed due to irritation, and bronchioles constrict due to muscle spasms.
Bronchitis
Airways are inflamed due to infection (acute) or due to an irritant (chronic). Coughing brings up mucus and pus.

Figure 44.10 **Common bronchial and pulmonary diseases.**
Exposure to infectious pathogens and/or polluted air, including tobacco smoke, causes the diseases and disorders shown here.

Lung Cancer

Lung cancer used to be more prevalent in men than in women, but recently it has surpassed breast cancer as a cause of death in women. This can be linked to an increase in the number of women who smoke today. Autopsies on smokers have revealed the progressive steps by which the most common form of lung cancer develops. The first event appears to be thickening and callusing of the cells lining the airways. (Callusing occurs whenever cells are exposed to irritants.) Then there is a loss of cilia so that it is impossible to prevent dust and dirt from settling in the lungs. Following this, cells with atypical nuclei appear in the callused lining. A tumor consisting of disordered cells with atypical nuclei is considered to be cancer in situ (at one location). A final step occurs when some of these cells break loose and penetrate other tissues, a process called metastasis. Now the cancer has spread. The original tumor may grow until a bronchus is blocked, cutting off the supply of air to that lung. The entire lung then collapses, the secretions trapped in the lung spaces become infected, and pneumonia or a lung abscess (localized area of pus) results. The only treatment that offers a possibility of cure is to remove a lobe or the lung completely before metastasis has had time to occur. This operation is called a **pneumonectomy.**

Lung infections and disorders occur at a higher incidence among smokers than nonsmokers.

Connecting Concepts

Both the human intestine and lungs are exchange boundaries with the external environment. The lungs are composed of alveoli that number about 300 million, and each alveolus may have as many as 1,800 blood capillary contacts. The lungs are adapted to facilitate gas exchange, freely allowing oxygen to enter and carbon dioxide to exit the blood. The characteristics that enhance the function of the lungs for gas exchange also facilitate the entrance of substances and pathogens that could be harmful to the body. Defense mechanisms are in place that limit possible pathogen invasion of the body along the respiratory tract. Numerous lymphoid nodules located in the bronchial lining are a source of white blood cells that protect against infection.

The body, however, has no means by which to inhibit the entrance of dangerous chemicals across the alveolar wall. Breathing air pollutants can result in respiratory distress, headache, and exhaustion. On occasion, when smog gets trapped above a city by a blanket of warm air, the results can be disastrous. In 1966, about 168 people died in New York City due to an accumulation of air pollutants over several days. We know the manner in which one air pollutant causes death. Carbon monoxide (CO) is an air pollutant that comes from the incomplete combustion of natural gas and gasoline. Because carbon monoxide is a colorless, odorless gas, people can be unaware that it has entered their system by way of the lungs. But once CO is in the bloodstream, it combines with the iron of hemoglobin 200 times more tightly than oxygen. The result is that the delivery of oxygen to mitochondria is impaired and so is the functioning of mitochondria. The end result can be death.

Those who smoke cigars and cigarettes should be aware that CO is in tobacco smoke. The soot in cigarette smoke impairs the ability of the lungs to breathe, and the CO impairs the ability of hemoglobin to transport oxygen to the lungs. Cells that lack sufficient oxygen turn to fermentation as a way to keep producing ATP. Fermentation, as you know, is inefficient and results in the end product lactate which increases the acidity of the blood. As Figure 44.9*b* shows, a higher acidity also reduces the ability of hemoglobin to carry oxygen. Clearly, smoking makes it more difficult for blood concentrations to remain normal, and threatens homeostasis.

Summary

44.1 Gas Exchange Surfaces

Some aquatic animals like hydras and planarians use their entire body surface for gas exchange.

Most animals have a special, localized gas-exchange area. Most large aquatic animals pass water over gills. On land, insects utilize tracheal systems and vertebrates have lungs.

Lungs are found inside the body, where water loss is reduced. To ventilate the lungs, some vertebrates use positive pressure, but most inhale, using muscular contraction to produce a negative pressure that causes air to rush into the lungs. When the breathing muscles relax, air is exhaled.

Birds have a series of air sacs that allow a one-way flow of air over the gas-exchange area. This is called a complete ventilation system.

44.2 Human Respiratory System

Table 44.1 lists the structures found in the human respiratory system.

Humans breathe by negative pressure, as do other mammals. During inspiration, the rib cage goes up and out, and the diaphragm lowers. The lungs expand and air comes rushing in. During expiration the rib cage goes down and in, and the diaphragm rises. Therefore, air rushes out.

The rate of breathing is dependent upon the amount of carbon dioxide in the blood, as detected by chemoreceptors such as the aortic and carotid bodies.

Gas exchange in the lungs and tissues is brought about by diffusion. Hemoglobin transports oxygen in the blood; carbon dioxide is mainly transported in plasma as the bicarbonate ion. The enzyme carbonic anhydrase found in red blood cells speeds the formation of the bicarbonate ion.

44.3 Respiration and Health

The respiratory tract is subject to infections. Two disorders of the lungs, emphysema and cancer, are usually due to cigarette smoking.

Reviewing the Chapter

1. Compare the respiratory organs of aquatic animals to those of terrestrial animals. 798–801
2. How does the countercurrent flow of blood within gill capillaries and water passing across the gills assist respiration in fishes? 799
3. Why don't insects require circulatory system involvement in air transport? Why is it beneficial for the body wall of earthworms to be moist? 800
4. Explain the phrase, "breathing by using negative pressure." 801
5. Contrast incomplete ventilation in humans with complete ventilation in birds. 801
6. Name the parts of the human respiratory system, and list a function for each part. 802
7. The concentration of what substances in blood controls the breathing rate in humans? Explain. 803–05
8. How are oxygen and carbon dioxide transported in blood? What does carbonic anhydrase do? 805
9. Which conditions depicted in Figure 44.10 are due to infection? Which are due to behavioral or environmental factors? Explain. 806–07

Testing Yourself

Choose the best answer for each question.

1. One problem faced by terrestrial animals with lungs, but not by freshwater aquatic animals with gills, is that
   a. gas exchange involves water loss.
   b. breathing requires considerable energy.
   c. oxygen diffuses very slowly in air.
   d. the concentration of oxygen in water is greater than that in air.
   e. All of these are correct.
2. In which animal is the circulatory system not involved in gas transport?
   a. mouse
   b. dragonfly
   c. trout
   d. sparrow
   e. humans
3. Birds have a more efficient lung than humans because the flow of air
   a. is the same during both inspiration and expiration.
   b. travels in only one direction through the lungs.
   c. never backs up like in human lungs.
   d. is not hindered by a larynx.
   e. enters their bones.
4. Which animal breathes by positive pressure?
   a. fish
   b. human
   c. bird
   d. frog
   e. planarians
5. Which of these is a true statement?
   a. In lung capillaries, carbon dioxide combines with water to produce carbonic acid.
   b. In tissue capillaries, carbonic acid breaks down to carbon dioxide and water.
   c. In lung capillaries, carbonic acid breaks down to carbon dioxide and water.
   d. In tissue capillaries, carbonic acid combines with hydrogen ions to form the carbonate ion.
   e. All of these statements are true.
6. Air enters the human lungs because
   a. atmospheric pressure is less than the pressure inside the lungs.
   b. atmospheric pressure is greater than the pressure inside the lungs.
   c. although the pressures are the same inside and outside, the partial pressure of oxygen is lower within the lungs.
   d. the residual air in the lungs causes the partial pressure of oxygen to be less than it is outside.
   e. the process of breathing pushes air into the lungs.
7. If the digestive and respiratory tracts were completely separate in humans, there would be no need for
   a. swallowing.
   b. a nose.
   c. an epiglottis.
   d. a diaphragm.
   e. Any of these are correct.
8. To trace the path of air in humans you would place the trachea
   a. directly after the nose.
   b. directly before the bronchi.
   c. before the pharynx.
   d. directly before the lungs.
   e. Both a and c are correct.
9. In humans, the respiratory center
   a. is stimulated by carbon dioxide.
   b. is located in the medulla oblongata.
   c. controls the rate of breathing.
   d. is stimulated by hydrogen ion concentration.
   e. All of these are correct.
10. Carbon dioxide is carried in the plasma
    a. in combination with hemoglobin.
    b. as the bicarbonate ion.
    c. combined with carbonic anhydrase.
    d. only as a part of tissue fluid.
    e. All of these are correct.
11. Label this diagram of the human respiratory system:

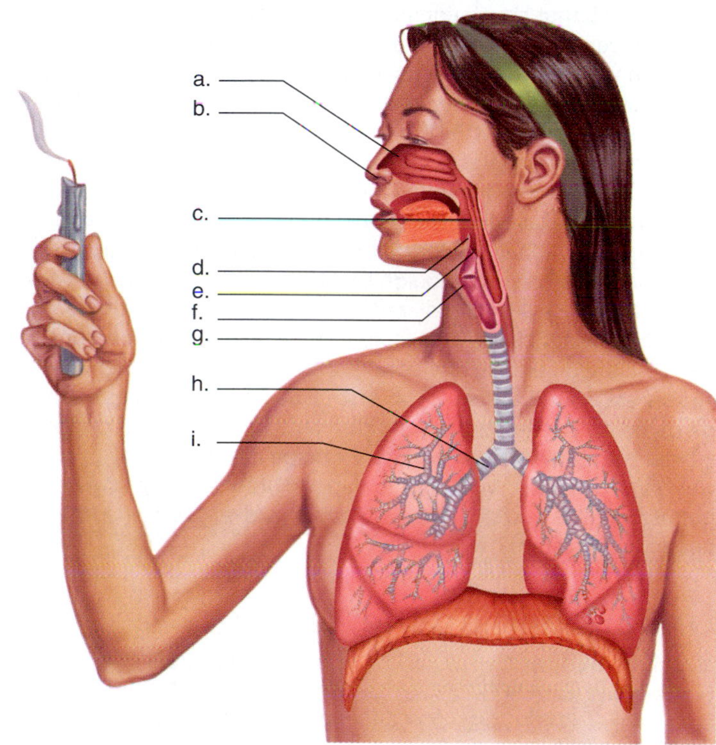

Thinking Scientifically

1. A roadside museum advertises that it has fossils of meter-long insects. Explain why the respiratory system of insects would not support a large size and therefore the existence of such fossils is highly unlikely.
2. Fetal hemoglobin picks up oxygen from the maternal blood. If the oxygen-binding characteristics of hemoglobin in the fetus were identical to the hemoglobin of the mother, oxygen could never be transferred at the placenta to fetal circulation. What hypothesis about the oxygen-binding characteristics of fetal hemoglobin would explain how fetuses get the oxygen they need?

Bioethical Issue

Antibiotics cure respiratory infections, but there are problems associated with antibiotic therapy. Aside from a possible allergic reaction, antibiotics not only kill off disease-causing bacteria, they also reduce the number of beneficial bacteria in the intestinal tract and other locations. These beneficial bacteria hold in check the growth of other pathogens that now begin to flourish. Diarrhea can result, as can a vaginal yeast infection. The use of antibiotics can also prevent natural immunity from occurring, leading to the need for recurring antibiotic therapy. Especially alarming at this time is the occurrence of resistance. Resistance takes place when vulnerable bacteria are killed off by an antibiotic, and this allows resistant bacteria to become prevalent. The bacteria that cause ear, nose, and throat infections, and scarlet fever and pneumonia are becoming widely resistant because we have not been using antibiotics properly. Tuberculosis is on the rise, and the new strains are resistant to the usual combined antibiotic therapy

Every citizen needs to be aware of our present crisis situation. Stuart Levy, a Tufts University School of Medicine microbiologist, says that we should do what is ethical for society and ourselves. What is needed? Antibiotics kill bacteria, not viruses—therefore, we shouldn't take antibiotics unless we know for sure we have a bacterial infection. And we shouldn't take them prophylactically—that is, just in case we might need one. If antibiotics are taken in low dosages and intermittently, resistant strains are bound to take over. Animal and agricultural use should be pared down, and household disinfectants should no longer be spiked with antibacterial agents. Perhaps then, Levy says, vulnerable bacteria will begin to supplant the resistant ones in the population. Are you doing all you can to prevent bacteria from becoming resistant?

Understanding the Terms

alveolus 803
bicarbonate ion 805
bronchiole 803
bronchus 803
carbaminohemoglobin 805
carbonic anhydrase 805
diaphragm 801
epiglottis 803
expiration 801
gill 799
glottis 802
heme 805
hemoglobin 805
inspiration 801
larynx 802
lung 801
oxyhemoglobin 805
pharynx 802
pneumonectomy 807
respiration 798
tonsils 806
trachea 802
vocal cord 802

Match the terms to these definitions:

a. ______________ Common passageway for both food intake and air movement, located between the mouth and the esophagus.

b. ______________ Dome-shaped muscularized sheet separating the thoracic cavity from the abdominal cavity in mammals.

c. ______________ Fold of tissue within the larynx; creates vocal sounds when it vibrates.

d. ______________ Respiratory organ in most aquatic animals; in fish an outward extension of the pharynx.

e. ______________ Stage during breathing when air is pushed out of the lungs.

Web Connections

Exploring the Internet

http://www.mhhe.com/biosci/genbio/mader
(click on *Biology 7/e*)

The *Biology 7/e* Online Learning Center provides many resources for studying the material in this chapter including links to the following sites:

Asthma. A fact sheet from the CDC about asthma.

http://www.cdc.gov/od/oc/media/fact/asthma.htm

Asthma Center: Information on Asthma and Allergies. The Mayo Clinic informational site on allergies and asthma. Reference articles, information, references, links, and quizzes.

http://www.mayohealth.org/mayo/common/htm/allergy.htm

Tobacco Information. A CDC-supported site containing information on tobacco use and prevention. Links to many other sites.

http://www.cdc.gov/tobacco/

CDC site on Tuberculosis. Much information concerning TB, the current rise in occurrence, and the multidrug resistant strains.

http://www.cdc.gov/nchstp/tb/faqs/qa.htm

CDC site on Pneumocystis carinii *Pneumonia (PCP). This condition is particularly common in persons who are HIV infected, or otherwise immune compromised.*

http://www.cdc.gov/ncidod/dpd/list_pcp.htm

Body Fluid Regulation and Excretion

chapter 45

c h a p t e r   c o n c e p t s

Hydras **are microscopic aquatic animals.**

Some aquatic animals, like the hydra pictured here, don't have an excretory organ. The body wall has only two layers of cells, and water enters and exits a large, central, fluid-filled cavity. Each cell excretes metabolic wastes directly into the cavity and water washes them away. In more complex animals, nitrogenous wastes and excess fluids or salts are discharged into the external environment by an excretory organ.

An abnormal osmolarity of internal fluids can cause an animal to lose or gain fluids to or from the external environment. Excretory organs play a major role in maintaining homeostasis because they control the salt balance and in that way regulate the water balance. As in humans, excretory organs also play a role in maintaining the normal pH of body fluids.

Homeostasis includes the distribution of oxygen and nutrients to cells but it also includes getting rid of their wastes. **Excretion** is the elimination of molecules that have taken part in metabolic reactions and should not be confused with defecation, which is the elimination of nondigested material from the digestive tract.

45.1 Body Fluid Regulation

The excretory system is involved in regulating body fluid concentrations. It does this by retaining or eliminating certain ions and water. The regulation of body fluids is dependent upon the concentration of mineral ions such as sodium (Na^+), chloride (Cl^-) potassium (K^+), and the bicarbonate ion (HCO_3^-). Body fluids gain mineral ions as a result of eating foods and drinking fluids. Excretion is the primary way that the body loses ions.

In many animals, water can enter the body through metabolism (e.g., cellular respiration produces water); by eating foods that contain water; and by drinking water. Water is lost from the body through evaporation (e.g., from skin and lungs), feces formation, and excretion (Fig. 45.1). To be in fluid balance, the amount of water exiting the body must equal the amount that enters.

When there are differences in osmolarity between two regions, water tends to move into the region with the higher amount of ions. A marine environment, which is high in salts, tends to promote the osmotic loss of water and the gain of ions by such means as drinking water. Fresh water tends to promote a gain of water by osmosis and a loss of ions as excess water is excreted. Terrestrial animals tend to lose both water and ions to the environment.

Aquatic Animals

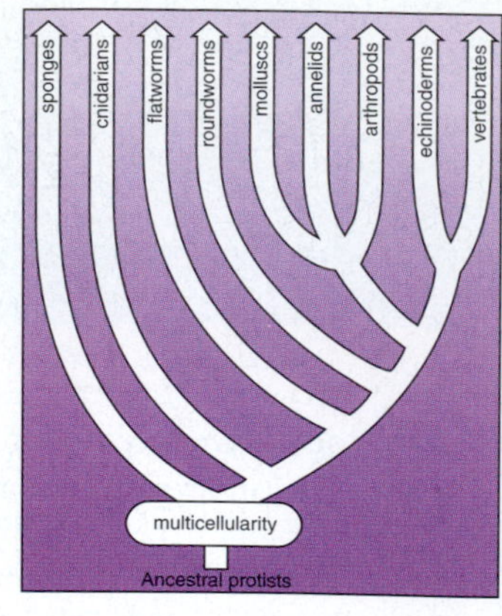

Among aquatic animals, only marine invertebrates, such as molluscs and arthropods; and cartilaginous fishes, such as sharks and rays, have body fluids that are nearly isotonic to seawater. These organisms have little difficulty maintaining their normal salt and water balance. It is surprising, though, that while they are isotonic, the body fluids of cartilaginous fishes do not contain the same amount of mineral ions as seawater. The answer to this paradox is that their blood contains a concentration of urea high enough to match the tonicity of the sea! For some unknown reason, this amount of urea is not toxic to cartilaginous fishes.

Bony Fishes

The body fluids of all bony fishes normally have only a moderate amount of salt. Apparently, their common ancestor evolved in fresh water, and only later did some groups in-

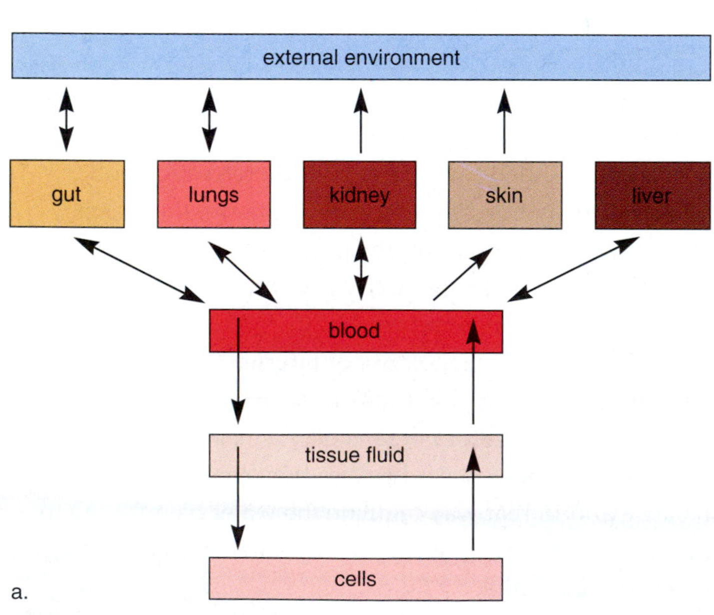

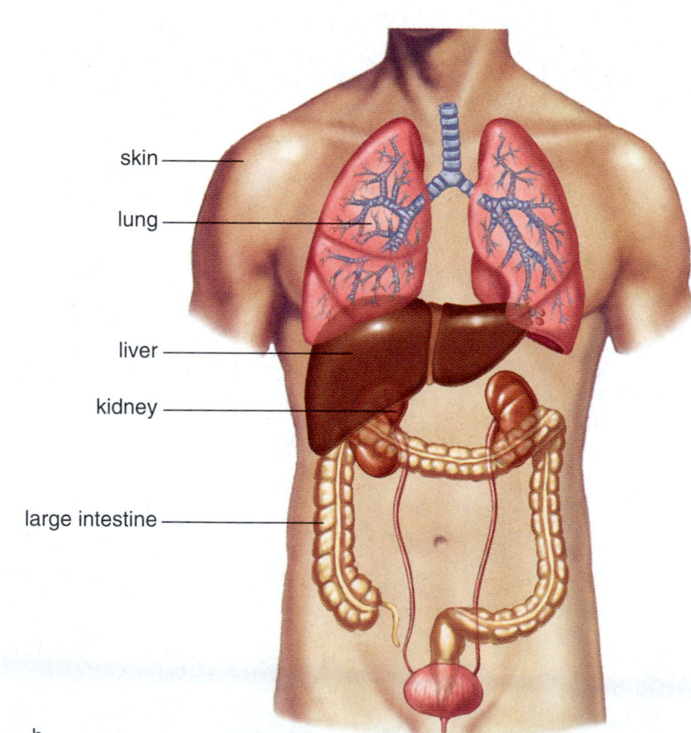

Figure 45.1 Excretory functions.
a. The internal environment of cells (blood and tissue fluid) in humans and other animals stays relatively constant because the blood is continually exchanging substances with the environment. Fluids and solutes enter blood and then tissue fluid by way of the digestive tract; water and mineral ions are lost by way of the lungs, skin, and kidneys. **b.** Excretion is ridding the body of metabolic wastes, as when the lungs excrete carbon dioxide and the kidneys excrete urea produced by the liver. The skin also excretes urea.

vade the sea. Marine bony fishes (Fig. 45.2*a*) are therefore prone to water loss and could become dehydrated. To counteract this, they drink seawater almost constantly. On the average, marine bony fishes swallow an amount of water estimated to be equal to 1% of their body weight every hour. This is equivalent to a human drinking about 700 ml of water every hour around the clock. While they get water by drinking, this habit also causes these fishes to acquire salt. To rid the body of excess salt, they actively transport sodium (Na^+) and chloride (Cl^-) ions into the surrounding seawater at the gills. This causes a passive loss of water through gills.

The osmotic problems of freshwater bony fishes are exactly opposite to those of marine bony fishes (Fig. 45.2*b*). The body fluids of freshwater bony fishes are hypertonic to fresh water, and they are prone to passively gain water. These fishes never drink water, but instead eliminate excess water through production of large quantities of dilute (hypotonic) urine. They discharge a quantity of urine equal to one-third their body weight each day. Because this causes them to lose salts, they actively transport salts into the blood across the membranes of their gills.

The difference in adaptation between marine and freshwater bony fishes makes it remarkable that some fishes actually can move between the two environments during their life cycle. Salmon, for example, begin their lives in freshwater streams and rivers, move to the ocean for a period of time, and finally return to fresh water to breed. These fishes alter their behavior and their gill and kidney functions in response to the osmotic changes they encounter when moving from one environment to the other.

Terrestrial Animals

Most terrestrial animals need to drink water occasionally to make up for the water lost by excretion, respiration, in sweat and feces. Like marine bony fishes, birds and reptiles that live near the sea are able to drink seawater despite its high osmolarity because they have a nasal salt gland that can excrete large volumes of concentrated salt solution.

To prevent water loss, some animals excrete a nitrogenous waste that is rather insoluble. An impermeable outer covering also helps. Compare the moist, thin, permeable skin of a frog to the dry, horny, thick skin of a lizard and you know immediately which one is adapted to a dry terrestrial environment. Aside from these measures to prevent water loss, unique adaptations abound. To prevent loss of water during the process of breathing, certain desert animals, like the camel and kangaroo rat, have a nasal passage that has a highly convoluted mucous membrane surface. This surface allows them to capture moisture from exhaled air, which they use to humidify air that is being inhaled. In contrast, the air we exhale is full of moisture, which is why you can see it on cold winter mornings—the moisture in exhaled air is condensing. As we shall see, humans mainly conserve water by producing a hypertonic urine. The kangaroo rat also forms a very concentrated urine, and its fecal material is almost completely dry. This animal is so adapted to conserving water that it can survive using metabolic water derived from cellular respiration.

To maintain body fluids, animals regulate salts and water excretion.

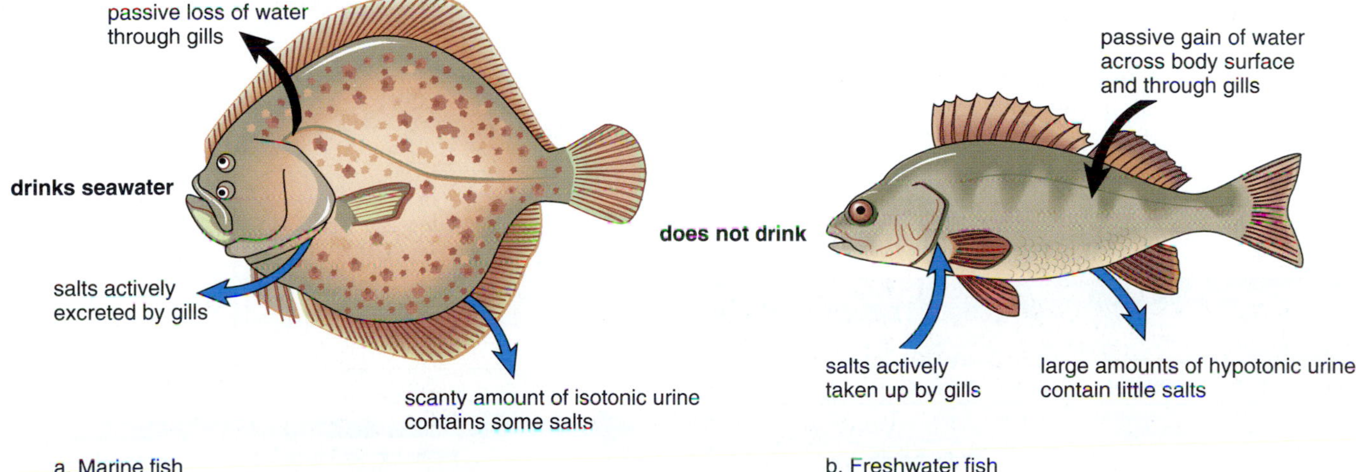

Figure 45.2 Body fluid regulation in bony fishes.
The black arrows represent passive transport from the environment, and the blue arrows represent active transport by the fishes to counteract environmental pressures. **a.** Marine bony fish. **b.** Freshwater bony fish.

45.2 Nitrogenous Waste Products

The breakdown of various molecules, including nucleic acids and amino acids, results in nitrogenous wastes. For simplicity's sake, however, we will limit our discussion to amino acid metabolism. Amino acids not used for protein synthesis are broken down by the body to generate energy, or are converted to fats or carbohydrates that can be stored. In either case, the amino groups (—NH_2) must be removed because only the carbon chains are utilized when amino acids are broken down or converted to forms of stored energy. Once the amino groups have been removed from amino acids, they may be excreted from the body in the form of ammonia, urea, or uric acid, depending on the species (Table 45.1). Removal of amino groups from amino acids requires a fairly set amount of energy. The energy requirement for the conversion of amino groups to ammonia, urea, or uric acid, however, differs, as indicated in Figure 45.3.

Excretion of Nitrogenous Wastes

Amino groups removed from amino acids immediately form **ammonia** (NH_3) by addition of a third hydrogen ion. Little or no energy is required to convert an amino group to ammonia by the addition of a hydrogen ion. Ammonia is quite toxic and can be used as a nitrogenous excretory product only if a good deal of water is available to wash it from the body. The high solubility of ammonia permits this means of excretion in bony fishes, aquatic invertebrates, and amphibians, whose gills and skin surfaces are in direct contact with the water of the environment.

Terrestrial amphibians and mammals usually excrete **urea** [Gk. *urina,* urine] as their main nitrogenous waste. Urea is much less toxic than ammonia and can be excreted in a moderately concentrated solution. This allows body water to be conserved, an important advantage for terrestrial animals with limited access to water. Production of urea, however, requires the expenditure of energy. Urea is produced in the liver by a set of energy-requiring enzymatic reactions known as the urea cycle. In the cycle, carrier molecules take up carbon dioxide and two molecules of ammonia, finally releasing urea.

Table 45.1

Nitrogenous Waste Excretion

Product	Habitat	Animals
Ammonia	Water	Aquatic invertebrates Bony fishes Amphibian larvae
Urea	Land	Adult amphibians Mammals
Uric acid	Land	Insects Birds Reptiles

Uric acid is routinely excreted by insects, reptiles, and birds. Uric acid is not very toxic, and is poorly soluble in water. Poor solubility is an advantage if water conservation is needed, because uric acid can be concentrated even more readily than can urea. In reptiles and birds, a dilute solution of uric acid passes from the kidneys to the *cloaca,* a common reservoir for the products of the digestive, urinary, and reproductive systems. The cloacal contents are refluxed into the large intestine, where water is reabsorbed. Embryos of reptiles and birds develop inside completely enclosed, shelled eggs. The production of insoluble, relatively nontoxic uric acid is advantageous for shelled embryos because all nitrogenous wastes are stored inside the shell until hatching takes place.

Uric acid is synthesized by a long, complex series of enzymatic reactions that requires expenditure of even more ATP than does urea synthesis. Here again, there seems to be a trade-off between the advantage of water conservation and the disadvantage of energy expenditure for synthesis of an excretory molecule.

Animals excrete nitrogenous wastes as ammonia, urea, or uric acid. Ammonia requires the most water to excrete; uric acid requires the most energy to produce.

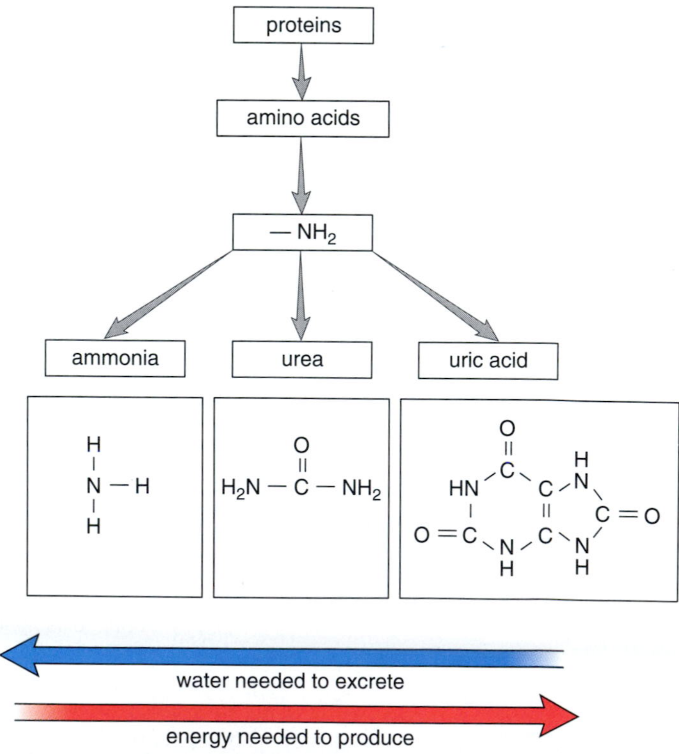

Figure 45.3 Nitrogenous wastes.
Proteins are hydrolyzed to amino acids, whose breakdown results in carbon chains and amino groups (—NH_2). The carbon chains can be used as an energy source, but the amino groups must be excreted as ammonia, urea, or uric acid.

45.3 Organs of Excretion

Most animals have tubular excretory organs that regulate the salt-water balance and excrete metabolic wastes into the environment.

Flame Cells in Planarians

Planarians, which are flatworms, live in fresh water, and have two strands of branching excretory tubules that open to the outside of the body through excretory pores (Fig. 45.4*a*). Located along the tubules are bulblike **flame cells,** each of which contains a cluster of beating cilia that looks like a flickering flame under the microscope. The beating of flame-cell cilia propels fluid through the excretory tubules and out of the body. The system is believed to function in ridding the body of excess water and also in excreting wastes.

Nephridia in Earthworms

The body of an earthworm, an annelid, is divided into segments, and nearly every body segment has a pair of excretory structures called **nephridia** [Gk. *nephros,* kidney]. Each nephridium is a tubule with a ciliated opening (the nephridiostome) and an excretory pore (the nephridiopore) (Fig. 45.4*b*). As fluid from the coelom is propelled through the tubule by beating cilia, nutrient substances are reabsorbed and carried away by a network of capillaries surrounding the tubule. The urine of an earthworm contains only metabolic wastes, salts, and water.

Each day, an earthworm excretes a lot of water and may produce a volume of urine equal to 60% of its body weight. The excretion of ammonia is consistent with these data.

Malpighian Tubules in Insects

Insects, which are arthropods, have a unique excretory system consisting of long, thin tubules, **Malpighian tubules,** attached to the gut. Uric acid simply flows from the surrounding hemolymph into these tubules, and water follows a salt gradient established by active transport of K^+. Water and other useful substances are reabsorbed at the rectum but the uric acid leaves the body at the anus. Insects that live in water or that eat large quantities of moist food reabsorb little water. But insects in dry environments reabsorb most of the water and excrete a dry, semisolid mass of uric acid.

Most animals have a primary excretory organ. Planarians have flame cells, earthworms have nephridia, and insects have Malpighian tubules.

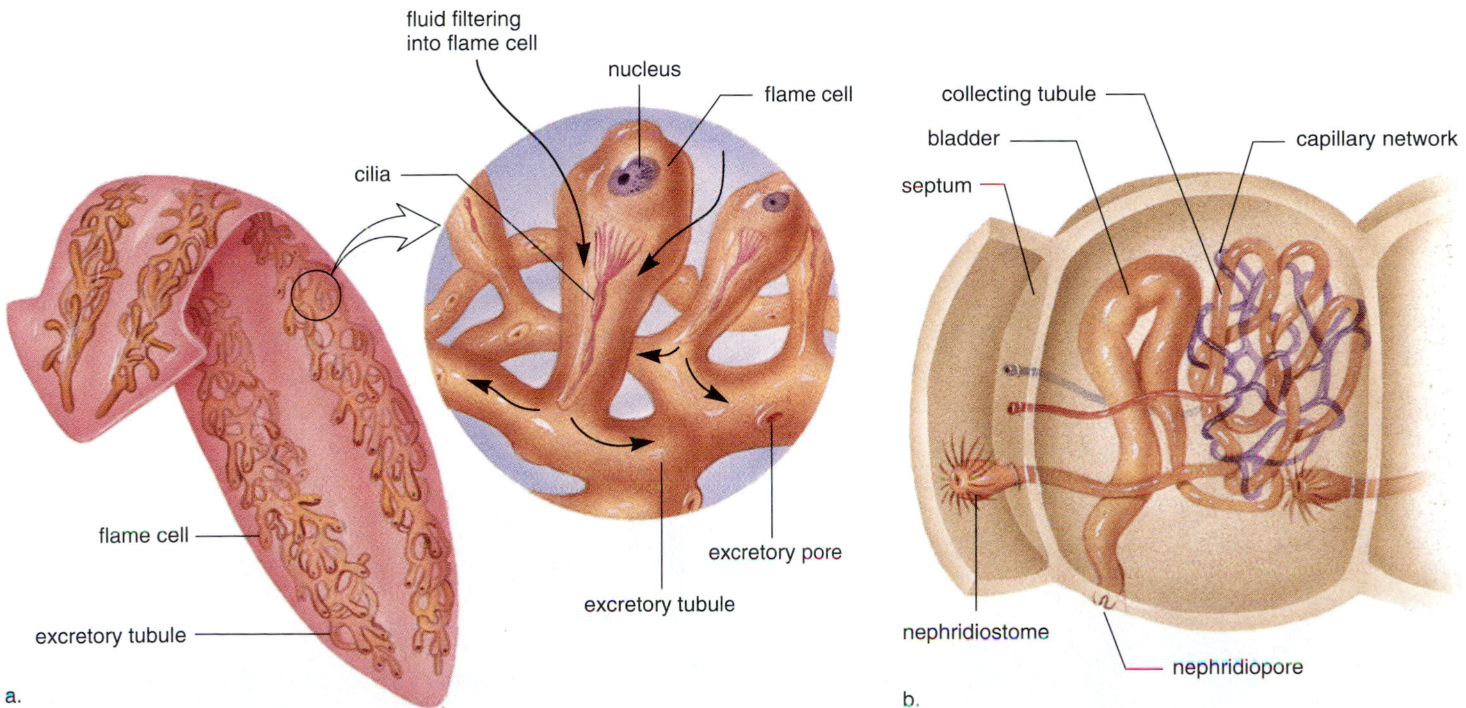

Figure 45.4 Excretory organs in animals.
a. The flame cell excretory system in planarians. Two or more tracts of branching tubules run the length of the body and open to the outside by pores. At the ends of side branches there are small flame cells, bulblike cells whose beating cilia cause fluid to enter the tubules that remove excess water from the body. b. The earthworm nephridium. The nephridium has a ciliated opening, the nephridiostome, that leads to a coiled tubule surrounded by a capillary network. Urine can be temporarily stored in the bladder before being released to the outside via a pore termed a nephridiopore. Most segments contain a pair of nephridia, one on each side.

45.4 Urinary System in Humans

The human urinary system consists of the organs shown in Figure 45.5. The human **kidneys** are bean-shaped, reddish brown organs, each about the size of a fist. They are located on either side of the vertebral column just below the diaphragm, where they are partially protected by the lower rib cage. **Urine** [Gk. *urina*, urine] made by the kidneys is conducted from the body by the other organs in the urinary system. Each kidney is connected to a **ureter,** a duct that takes urine from the kidney to the **urinary bladder,** where it is stored until it is voided from the body through the single **urethra.** In males, the urethra passes through the penis, and in females, it opens ventral to the opening of the vagina. There is no connection between the genital (reproductive) and urinary systems in females, but there is a connection in males. In males, the urethra also carries sperm during ejaculation.

Kidneys

If a kidney is sectioned longitudinally, three major parts can be distinguished (Fig. 45.6). The outer region is the **renal cortex,** which has a somewhat granular appearance. The **renal medulla** consists of renal pyramids that lie on the inner side of the renal cortex. The innermost part of the kidney is a hollow chamber called the **renal pelvis,** which is the flared end of the ureter inside the kidney. Urine formed in the kidney collects in the renal pelvis before entering the ureter. Microscopically, each kidney is composed of about one million tiny tubules called **nephrons** [Gk. *nephros*, kidney].

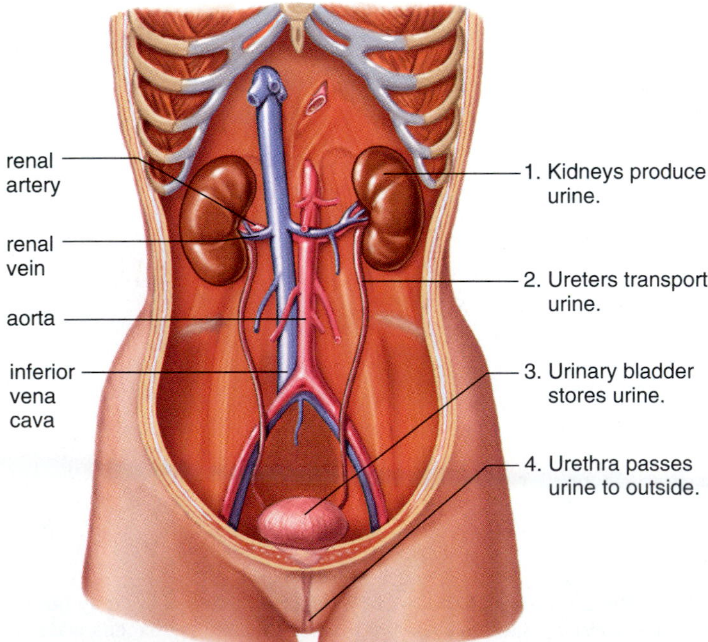

Figure 45.5 The human urinary system.
Urine is found only within the kidneys, the ureters, the urinary bladder, and the urethra.

Nephrons

Some nephrons are located primarily in the renal cortex, but others dip down into the renal medulla, as shown in Figure 45.6*b*. Each nephron is made of several parts (Fig. 45.7). The blind end of a nephron is pushed in on itself to form a cuplike structure called the **glomerular capsule** [L. *glomeris*, ball]. The outer layer of the glomerular capsule is composed of squamous epithelial cells; the inner layer is composed of specialized cells that allow easy passage of molecules. Leading from the glomerular capsule is a portion of the nephron known as the **proximal convoluted tubule** [L. *proximus*, nearest], which is lined by cells with many mitochondria and tightly packed microvilli. Then simple squamous epithelium appears in the **loop of the nephron,** which has a descending limb and an ascending limb. This is followed by the **distal convoluted tubule** [L. *distantia*, far]. Several distal convoluted tubules enter one **collecting duct.** The collecting duct transports urine down through the renal medulla and delivers it to the renal pelvis. The loop of the nephron and

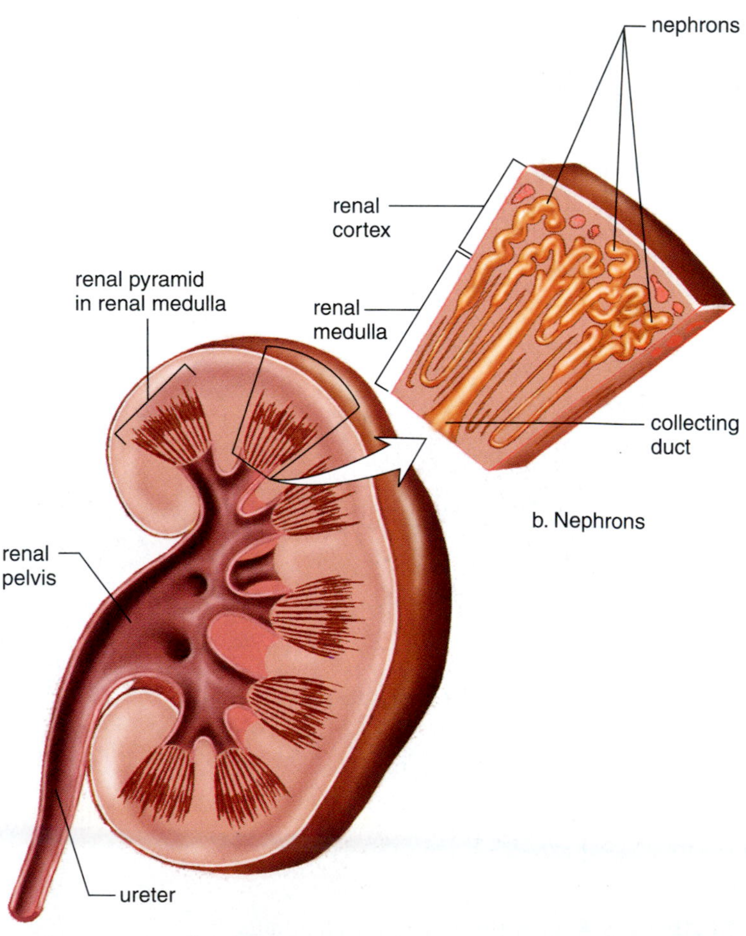

Figure 45.6 Macroscopic and microscopic anatomy of the kidney.
a. Longitudinal section of kidney showing the location of the renal cortex, the renal medulla, and the renal pelvis. **b.** An enlargement of one renal lobe showing the placement of nephrons.

the collecting duct give the pyramids of the renal medulla their striped appearance (Fig. 45.6).

Each nephron has its own blood supply (Fig. 45.7). The renal artery branches into numerous small arteries, which branch into arterioles, one for each nephron. Each arteriole, called an afferent arteriole, divides to form a capillary bed, the **glomerulus** [L. *glomeris*, ball], which is surrounded by the glomerular capsule. The glomerulus drains into an efferent arteriole, which subsequently branches into a second capillary bed around the tubular parts of the nephron. These capillaries, called peritubular capillaries, lead to venules that join to form veins leading to the renal vein, a vessel that enters the inferior vena cava.

Microscopically, the human kidney is composed of nephrons, tubules with specific parts that are richly supplied with capillaries.

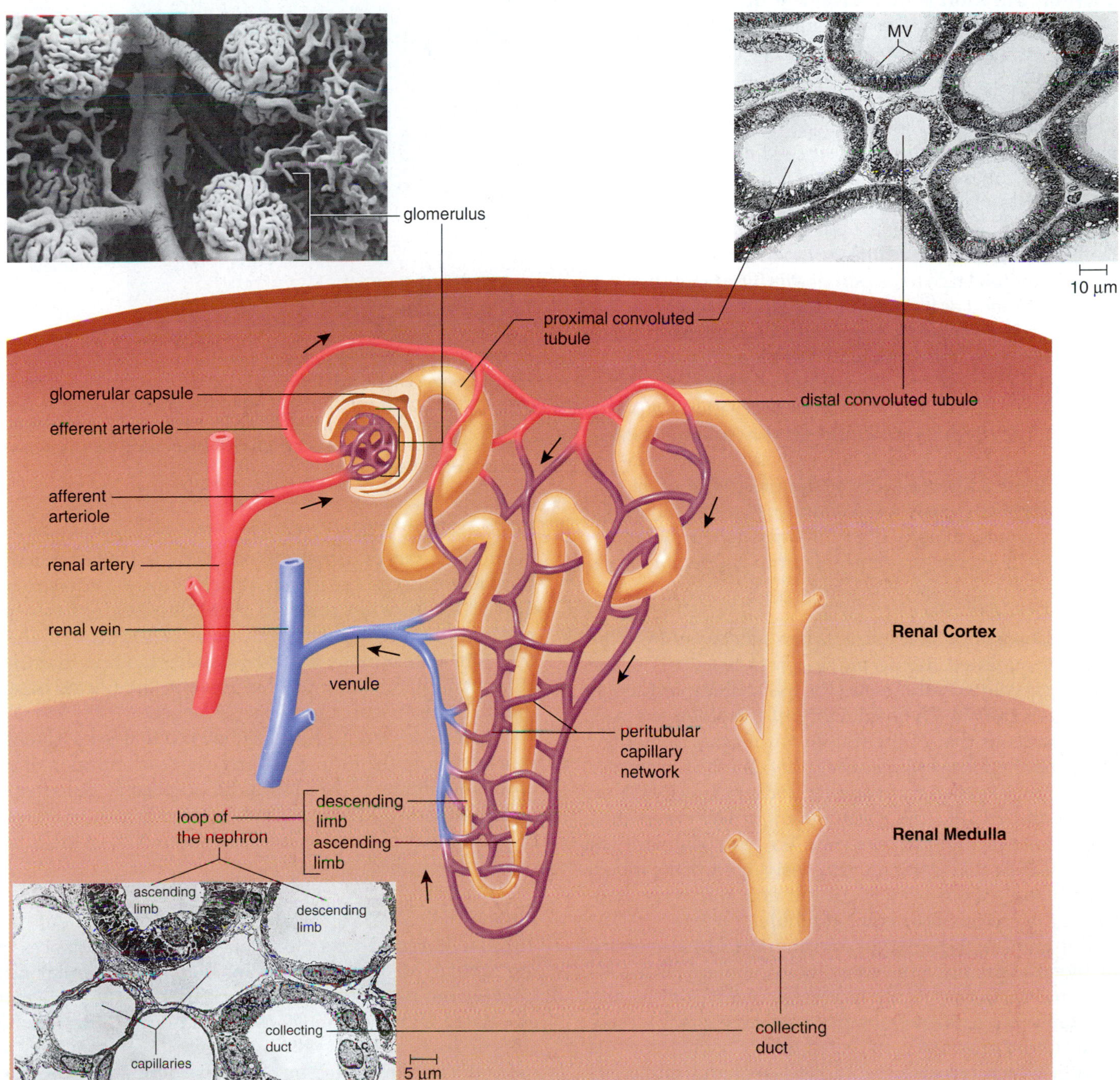

Figure 45.7 Nephron anatomy.
A nephron is made up of a glomerular capsule, the proximal convoluted tubule, the loop of the nephron, the distal convoluted tubule, and the collecting duct. The micrographs show these structures in cross section; MV = microvilli. You can trace the path of blood about the nephron by following the arrows.

Urine Formation

Urine production requires three distinct processes:

1. Glomerular filtration at the glomerular capsule;
2. Tubular reabsorption at the proximal convoluted tubule, in particular; and
3. Tubular secretion at the distal convoluted tubule in particular.

Glomerular Filtration

Glomerular filtration is the movement of small molecules across the glomerular wall into the glomerular capsule as a result of blood pressure. When blood enters the glomerulus, blood pressure is sufficient to cause small molecules, such as water, nutrients, salts, and wastes, to move from the glomerulus to the inside of the glomerular capsule, especially since the glomerular walls are 100 times more permeable than the walls of most capillaries elsewhere in the body. The molecules that leave the blood and enter the glomerular capsule are called the glomerular filtrate. Blood proteins and blood cells are too large to be part of this filtrate, so they remain in the blood as it flows into the efferent arteriole.

Glomerular filtrate is essentially protein free but otherwise has the same composition as blood plasma. If this composition were not altered in other parts of the nephron, death from loss of nutrients (starvation), and loss of water (dehydration) would quickly follow. The total blood volume averages about 5 liters and this amount of fluid is filtered every 40 minutes. Most of the filtered water is obviously quickly returned to the blood or a person would actually die from urination. Tubular reabsorption, however, prevents this from happening.

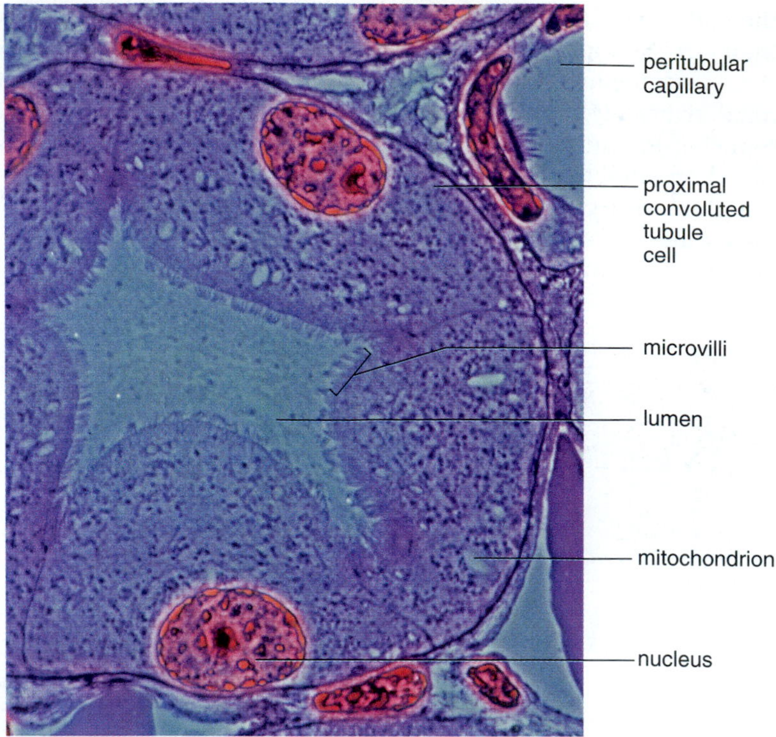

Figure 45.8 **Proximal convoluted tubule.**
This photomicrograph shows that the cells lining the proximal convoluted tubule have a brush border composed of microvilli, which greatly increases the surface area exposed to the lumen. The peritubular capillary adjoins the cells. There are also numerous mitochondria, which provide the energy for active transport of nutrients.

Tubular Reabsorption

Tubular reabsorption takes place when substances from the proximal convoluted tubule move across the walls of the proximal convoluted tubule into the associated peritubular capillaries (Fig. 45.8). The osmolarity of the blood is essentially the same as that of the filtrate within the glomerular capsule, and therefore osmosis of water from the filtrate into the blood cannot yet occur. However, sodium ions (Na^+) are actively pumped into the peritubular capillary and then chloride ions (Cl^-) follow passively. Now the osmolarity of the blood is such that water moves passively from the tubule into the blood. About 60–70% of salt and water are reabsorbed at the proximal convoluted tubule.

Nutrients, such as glucose and amino acids, also return to the blood at the proximal convoluted tubule. This is a selective process, because only molecules recognized by carrier proteins in plasma membranes are actively reabsorbed. The cells of the proximal convoluted tubule have numerous microvilli, which increase the surface area, and numerous mitochondria, which supply the energy needed for active transport (Fig. 45.8). Glucose is an example of a molecule that ordinarily is reabsorbed completely because there is a plentiful supply of carrier molecules for it. However, if there is more glucose in the filtrate than there are carriers to handle it, glucose will exceed its renal threshold, or transport maximum. When this happens, the excess glucose in the filtrate will appear in the urine. In diabetes mellitus, there is an abnormally large amount of glucose in the filtrate because the liver fails to store glucose as glycogen.

Urea is an example of a substance that is passively reabsorbed from the filtrate. At first, the concentration of urea within the filtrate is the same as blood plasma. But after water is reabsorbed, the urea concentration is greater than that of peritubular plasma. In the end, about 50% of the filtered urea is reabsorbed.

Tubular Secretion

Tubular secretion refers to the transport of substances into the tubular lumen by means other than glomerular filtration. For our purposes, tubular secretion may be particularly associated with the distal convoluted tubule (Fig. 45.9). Substances such as uric acid, hydrogen ions, ammonia, and penicillin are eliminated by tubular secretion. The process of tubular secretion may be viewed as helping to rid the body of potentially harmful compounds that were not filtered into the glomerulus.

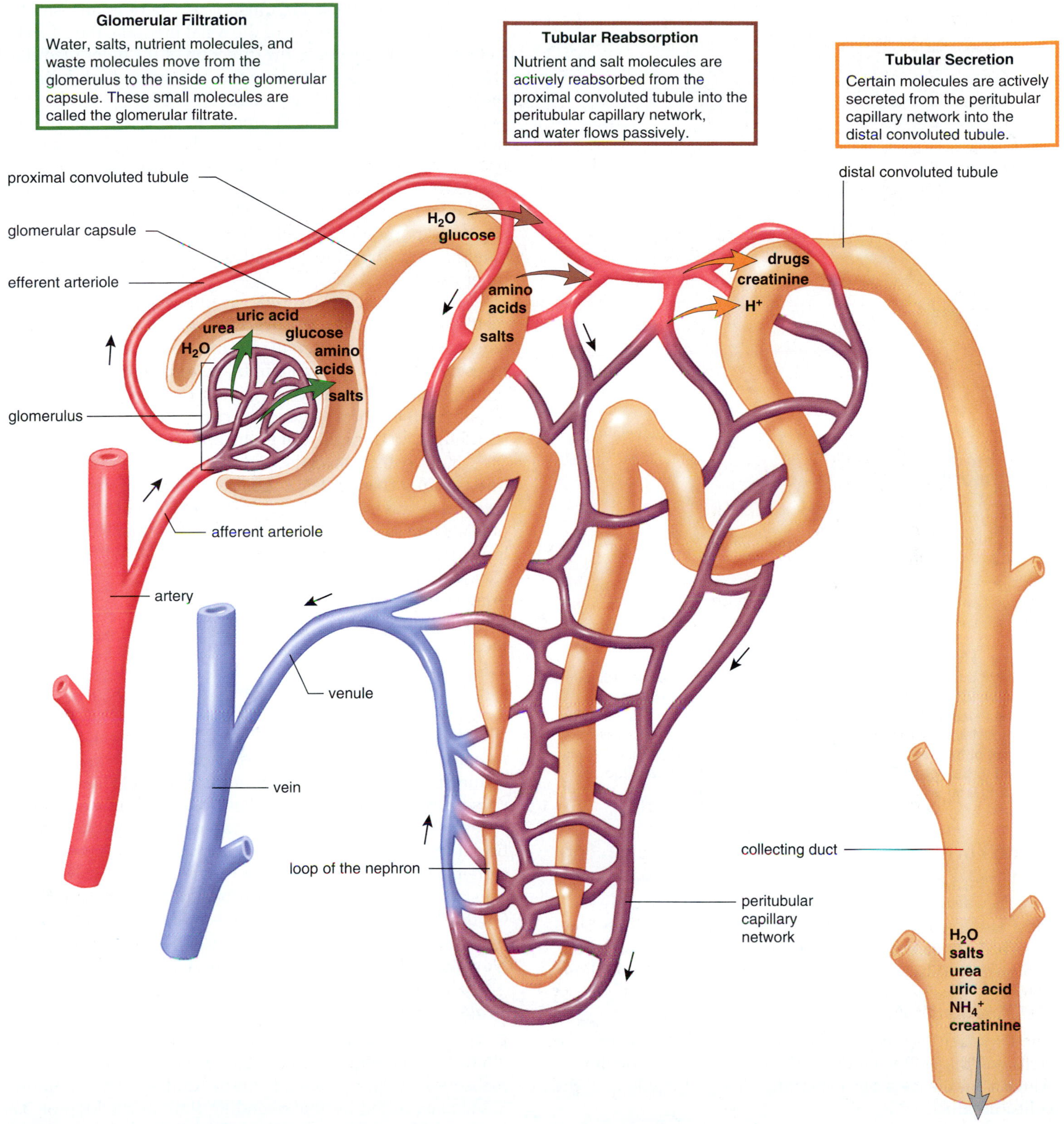

Figure 45.9 Steps in urine formation.
The three main steps in urine formation are color coded to arrows that show the movement of molecules into or out of the nephron at specific locations. In the end, urine is composed of the substances within the collecting duct (see gray arrow).

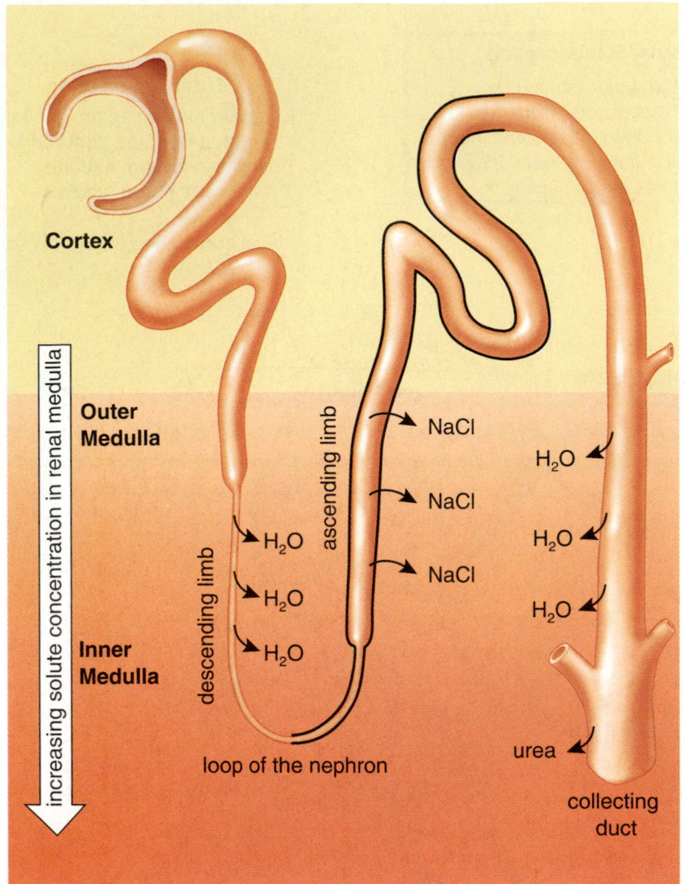

Figure 45.10 **Countercurrent mechanism.**
Salt (NaCl) diffuses and is actively transported out of the ascending limb of the loop of the nephron into the renal medulla; also, urea is believed to leak from the collecting duct and to enter the tissues of the renal medulla. This creates a hypertonic environment, which draws water out of the descending limb and the collecting duct. This water is returned to the cardiovascular system. (The thick line means the ascending limb is impermeable to water.)

Urine Formation and Homeostasis

The kidneys regulate the water balance of the blood. In this way, they also maintain the blood volume and blood pressure. Most of the water and salt (NaCl) present in the filtrate is reabsorbed across the wall of the proximal convoluted tubule. Reabsorption also occurs along the remainder of the nephron.

Reabsorption of Water

The excretion of a hypertonic urine (one that is more concentrated than blood) is dependent upon the reabsorption of water from the loop of the nephron (loop of Henle) and the collecting duct.

A long loop of the nephron, which typically penetrates deep into the renal medulla, is made up of a descending (going down) limb and an ascending (going up) limb. Salt (NaCl) passively diffuses out of the lower portion of the ascending limb, but the upper, thick portion of the limb actively extrudes salt out into the tissue of the outer renal medulla (Fig. 45.10). Less and less salt is available for transport as fluid moves up the thick portion of the ascending limb. Because of these circumstances, there is an osmotic gradient within the tissues of the renal medulla: the concentration of salt is greater in the direction of the inner medulla. (Note that water cannot leave the ascending limb because the limb is impermeable to water.)

The white arrow in Figure 45.10 indicates that the innermost portion of the inner medulla has the highest concentration of solutes. This cannot be due to salt because active transport of salt does not start until fluid reaches the thick portion of the ascending limb. Urea is believed to leak from the lower portion of the collecting duct, and it is this molecule that contributes to the high solute concentration of the inner medulla.

Because of the osmotic gradient within the renal medulla, water leaves the descending limb along its entire length. This is a countercurrent mechanism: as water diffuses out of the descending limb, the remaining fluid within the limb encounters an even greater osmotic concentration of solute; therefore, water will continue to leave the descending limb from the top to the bottom.

Fluid enters the collecting duct from the distal convoluted tubule. This fluid is isotonic to the cells of the renal cortex. This means that to this point, the net effect of reabsorption of water and salt is the production of a fluid that has the same tonicity as blood plasma. However, the filtrate within the collecting duct also encounters the same osmotic gradient mentioned earlier (Fig. 45.10). Therefore, water diffuses out of the collecting duct into the renal medulla, and the urine within the collecting duct becomes hypertonic to blood plasma.

Antidiuretic hormone (ADH) [Gk. *anti*, against, and L. *ouresis*, urination] released by the posterior lobe of the pituitary plays a role in water reabsorption at the collecting duct. In order to understand the action of this hormone, consider its name. Diuresis means increased amount of urine, and antidiuresis means decreased amount of urine. When ADH is present, more water is reabsorbed (blood volume and pressure rise), and a decreased amount of urine results. In practical terms, if an individual does not drink much water on a certain day, the posterior lobe of the pituitary releases ADH, causing more water to be reabsorbed and less urine to form. On the other hand, if an individual drinks a large amount of water and does not perspire much, ADH is not released. Now more water is excreted, and more urine forms.

Reabsorption of Salt

Usually, more than 99% of sodium (Na^+) filtered at the glomerulus is returned to the blood. Most sodium (67%) is reabsorbed at the proximal tubule, and a sizable amount (25%) is extruded by the ascending limb of the loop of the nephron. The rest is reabsorbed from the distal convoluted tubule and collecting duct.

Hormones regulate the reabsorption of sodium at the distal convoluted tubule. **Aldosterone** is a hormone secreted by the adrenal cortex, the outer portion of the adrenal glands, which lie atop the kidneys. Aldosterone promotes the excretion of potassium ions (K^+) and the reabsorption of sodium ions (Na^+).

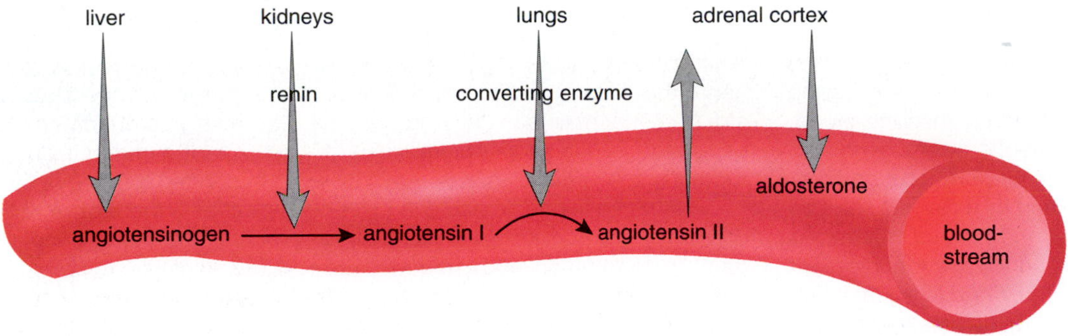

Figure 45.11 The renin-angiotensin-aldosterone system.
The liver secretes angiotensinogen into the bloodstream. Due to the action of renin from the kidneys and converting enzyme in lung capillaries, angiotensin II results. Angiotensin II acts on the adrenal cortex to secrete aldosterone which causes reabsorption of sodium ions and a subsequent rise in blood pressure.

When blood volume, and therefore blood pressure, is not sufficient to promote glomerular filtration, the kidneys secrete renin. **Renin** is an enzyme that changes angiotensinogen (a large plasma protein produced by the liver) into angiotensin I. Later, angiotensin I is converted to angiotensin II, a powerful vasoconstrictor that also stimulates the adrenal cortex to release aldosterone (Fig. 45.11). The reabsorption of sodium ions is followed by the reabsorption of water. Therefore, blood volume and blood pressure increase.

Atrial natriuretic hormone (ANH) is a hormone secreted by the atria of the heart when cardiac cells are stretched due to increased blood volume. ANH inhibits the secretion of renin by the juxtaglomerular apparatus and the secretion of aldosterone by the adrenal cortex. Its effect, therefore, is to promote the excretion of Na^+, that is, natriuresis. When Na^+ is excreted, so is water, and therefore blood volume and blood pressure decrease.

These examples show that the body regulates the water balance in blood by controlling the excretion and the reabsorption of various ions. Sodium (Na^+) is an important ion in plasma that must be regulated, but the kidneys also excrete or reabsorb other ions, such as potassium ions (K^+), bicarbonate ions (HCO_3^-), and magnesium ions (Mg^{2+}), as needed.

Adjustment of Blood pH

The bicarbonate (HCO_3^-) buffer system and breathing work together to maintain the pH of the blood. Central to the mechanism is this reaction, which you have seen before:

$$H^+ + HCO_3^- \rightleftharpoons H_2CO_3 \rightleftharpoons H_2O + CO_2$$

The excretion of carbon dioxide (CO_2) by the lungs helps keep the pH within normal limits, because when carbon dioxide is exhaled this reaction is pushed to the right and hydrogen ions are tied up in water. Indeed, when blood pH decreases, chemoreceptors in the carotid bodies (located in the carotid arteries) and in aortic bodies (located in the aorta) stimulate the respiratory center, and the rate and depth of breathing increases. On the other hand, when blood pH begins to rise, the respiratory center is depressed and the bicarbonate ion increases in the blood.

As powerful as this system is, only the kidneys can rid the body of a wide range of acidic and basic substances. The kidneys are slower acting than the buffer/breathing mechanism, but they have a more powerful effect on pH. For the sake of simplicity, we can think of the kidneys as reabsorbing bicarbonate ions and excreting hydrogen ions as needed to maintain the normal pH of the blood.

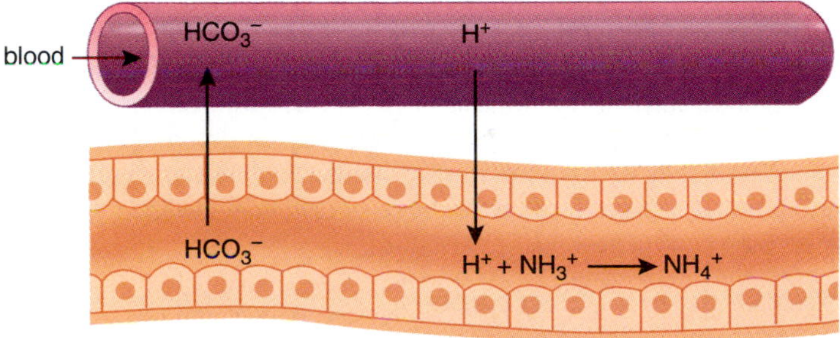

If the blood is acidic, hydrogen ions are excreted and bicarbonate ions are reabsorbed. If the blood is basic, hydrogen ions are not excreted and bicarbonate ions are not reabsorbed. Since the urine is usually acidic, it shows that usually an excess of hydrogen ions are excreted. Ammonia (NH_3) provides a means for buffering these hydrogen ions in urine: ($NH_3 + H^+ \rightarrow NH_4^+$). Ammonia (whose presence is quite obvious in the diaper pail or kitty litter box) is produced in tubule cells by the deamination of amino acids. Phosphate provides another means of buffering hydrogen ions in urine.

The acid-base balance of the blood is adjusted by the reabsorption of the bicarbonate ions (HCO_3^-) and the secretion of hydrogen ions (H^+) as appropriate.

Connecting Concepts

An artificial kidney employs the principle of dialysis in order to cleanse the blood. Blood from a patient's artery is passed through a semimembranous tubing, which is surrounded by a dialysis solution that contains salt and various other small molecules. The concentration gradient between blood and the dialysis solution is such that waste molecules and excess salts diffuse through the membrane from the blood into the dialysis solution, while blood cells and plasma proteins stay behind. Because the volume of dialysis solution is large, wastes continually diffuse out of the blood. In the course of a 6-hour treatment, 50 to 250 grams of urea can be removed from a patient, an amount that greatly exceeds the urea clearance rate of normal kidneys. On the other hand, the dialysis solution contains salts at a level that maintains their proper concentration in the blood before it is returned to the patient.

How does a kidney machine function differently from an animal's excretory organ? In the human body, blood is also passing through a tube, namely the glomerulus. The solution that forms on the other side of the glomerulus (the glomerular filtrate) is free of proteins, but otherwise has the same composition as blood plasma. Why? Because blood pressure forces water and small molecules to cross a highly permeable membrane. There is no opposing solution that maintains the normal nutrient concentration and osmolarity of the blood, nor an extreme concentration gradient that draws urea out of the blood.

In the human kidney, energy must be expended to retrieve the water, salts, and nutrients that are on their way out of the body. Following simple diffusion, pumps are employed to actively return Na^+ to the blood, and then water follows passively. Nutrients, such as glucose and amino acids, are reabsorbed by carrier-mediated active transport processes. Therefore, we see that in contrast to an artificial kidney, an animal's kidney is selective about what small molecules are returned to the blood, rather than what small molecules leave the blood. Finally, hormones regulate the osmolarity of the blood and fine-tune just how much water and salts are excreted.

Summary

45.1 Body Fluid Regulation

Osmotic regulation is important to animals. Most have to balance their water and salt intake and excretion to maintain normal solute and water concentration in body fluids. Marine fishes constantly drink water, excrete salts at the gills, and pass an isotonic urine. Freshwater fishes never drink water; they take in salts at the gills and excrete a hypotonic urine.

45.2 Nitrogenous Waste Products

Animals excrete nitrogenous wastes. The amount of water and energy required to excrete nitrogenous wastes differs. Aquatic animals usually excrete ammonia (needs much water to excrete), and land animals excrete either urea or uric acid (needs much energy to produce).

45.3 Organs of Excretion

Animals often have an excretory organ. The flame cells of planarians rid the body of excess water. Earthworm nephridia exchange molecules with the blood in a manner similar to vertebrate kidneys. Malpighian tubules in insects take up metabolic wastes and water from the hemolymph. Later, the water is absorbed by the gut.

45.4 Urinary System in Humans

Kidneys are a part of the human urinary system. Microscopically, each kidney is made up of nephrons, each of which has several parts and its own blood supply.

Urine formation requires three steps: glomerular filtration, when nutrients, water, and wastes enter the glomerular capsule; tubular reabsorption, when nutrients and most water is reabsorbed into the proximal convoluted tubule; and tubular secretion, when additional wastes are added to the tubule (e.g., the distal convoluted tubule).

Humans excrete a hypertonic urine. The ascending limb of the loop of the nephron actively extrudes salt so that the renal medulla is increasingly hypertonic relative to the contents of the descending limb and the collecting duct. Since urea leaks from the lower end of the collecting duct, the inner renal medulla has the highest concentration of solute. Therefore, a countercurrent mechanism assures that water will diffuse out of the descending limb and the collecting duct.

Three hormones are involved in maintaining the water content of the blood. The hormone ADH (antidiuretic hormone), which makes the collecting duct more permeable, is secreted by the posterior pituitary in response to an increase in the osmotic pressure of the blood.

The hormone aldosterone is secreted by the adrenal cortex after low blood pressure has caused the kidneys to release renin. The presence of renin leads to the formation of angiotensin II, which causes the adrenal cortex to release aldosterone. Aldosterone acts on the kidneys to retain Na^+; therefore, water is reabsorbed and blood pressure rises. The atrial natriuretic hormone prevents the secretion of ADH and aldosterone.

The kidneys keep blood pH within normal limits. They reabsorb HCO_3^- and excrete H^+ as needed to maintain the pH at about 7.4.

Reviewing the Chapter

1. Contrast the osmotic regulation of a marine bony fish with that of a freshwater bony fish. 812–13
2. Give examples of how other types of animals regulate their water and salt balance. 812–13
3. Relate the three primary nitrogenous wastes to the habitat of animals. 814
4. Describe how the excretory organs of the earthworm and the insect function. 815
5. Describe the path of urine, and give a function for each structure mentioned. 816
6. Describe the macroscopic anatomy of a human kidney, and relate it to the placement of nephrons. 816
7. List the parts of a nephron, and give a function for each structure mentioned. 816–17
8. Describe how urine is made by outlining what happens at each part of the nephron. 818–19

9. Describe the reabsorption of water and salt along the length of the nephron. Include the contribution of the loop of the nephron. 820
10. Name and describe the action of antidiuretic hormone (ADH), the renin-aldosterone connection, and the atrial natriuretic hormone (ANH). 820–21
11. How does the nephron regulate the pH of the blood? 821

Testing Yourself

Choose the best answer for each question.

1. Which of these is mismatched?
   a. insects—excrete uric acid
   b. humans—excrete urea
   c. fishes—excrete ammonia
   d. birds—excrete ammonia
   e. All of these are correct.
2. One advantage of urea excretion over uric acid excretion is that urea
   a. requires less energy than uric acid to form.
   b. can be concentrated to a greater extent.
   c. is not a toxic substance.
   d. requires no water to excrete.
   e. is a larger molecule.
3. Freshwater bony fishes maintain water balance by
   a. excreting salt across their gills.
   b. periodically drinking small amounts of water.
   c. excreting a hypotonic urine.
   d. excreting wastes in the form of uric acid.
   e. Both a and c are correct.
4. Animals with which of these are most likely to excrete a semisolid nitrogenous waste?
   a. nephridia
   b. Malpighian tubules
   c. human kidneys
   d. flame cells
   e. All of these are correct.
5. In which of these human structures are you least apt to find urine?
   a. large intestine
   b. urethra
   c. collecting duct
   d. bladder
   e. Both a and c are correct.
6. Excretion of a hypertonic urine in humans is associated best with the
   a. glomerular capsule.
   b. proximal convoluted tubule.
   c. loop of the nephron.
   d. collecting duct.
   e. Both c and d are correct.
7. The presence of ADH (antidiuretic hormone) causes an individual to excrete
   a. less salt.
   b. less water.
   c. more water.
   d. more salt.
   e. Both a and c are correct.
8. In humans, water is
   a. found in the glomerular filtrate.
   b. reabsorbed from the nephron.
   c. in the urine.
   d. reabsorbed from the collecting duct.
   e. All of these are correct.
9. Glomerular filtration is associated with the
   a. glomerular capsule.
   b. proximal convoluted tubule.
   c. collecting duct.
   d. distal convoluted tubule.
   e. All of these are correct.
10. Normally in humans, glucose
    a. is always in the filtrate and urine.
    b. is always in the filtrate with little or none in urine.
    c. undergoes tubular secretion and is in urine.
    d. undergoes tubular secretion and is not in urine.
    e. is not in the filtrate and is not in the urine.
11. Label this diagram of a nephron.

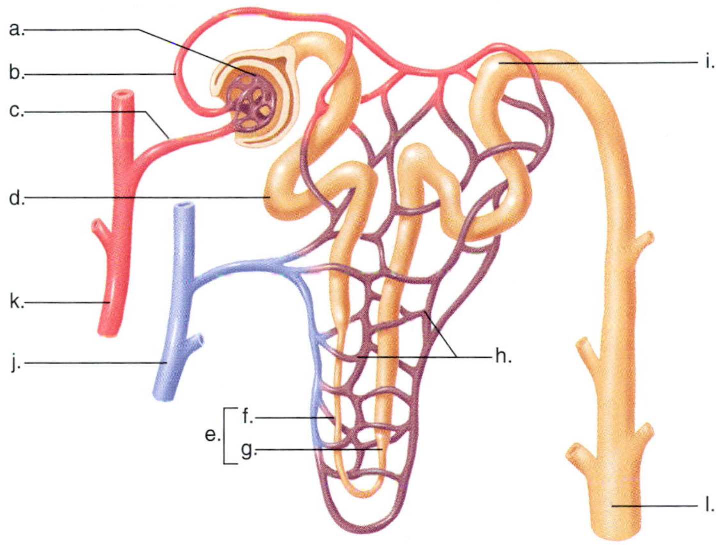

Thinking Scientifically

1. High blood pressure often is accompanied by kidney damage. In some the kidney damage is subsequent to the high blood pressure but in others the kidney damage is what caused the high blood pressure. Would a low salt diet enable you to determine whether the high blood pressure or the kidney damage came first?
2. The renin-angiotensin-aldosterone system can be inhibited in order to reduce high blood pressure. Usually, the angiotensin-converting enzyme is inhibited by drug therapy. Why would this enzyme be the most effective point to disrupt the system?

Bioethical Issue

As a society we are accustomed to thinking that as we grow older, diseases like urinary disorders will begin to occur. Almost everyone is aware that most males are subject to enlargement of the prostate as they age, and that cancer of the prostate is not uncommon among elderly men. However, like many illnesses associated with aging, medical science now knows how to treat or even cure prostate problems. Because of these successes, medical science has lengthened our life span. A child born in the

United States in 1900 lived to, say, 47 years. If that same child were born today, it would probably live to at least 76. Even more exciting is the probability that scientists will improve the life span. People could live beyond 100 years and have the same vigor and vitality they had when they were young.

Most people are appreciative of living longer, especially if they can expect to be free of the illnesses and inconveniences associated with aging. But have we examined how we feel about longevity as a society? Whereas we are accustomed to considering that if the birthrate increases so does the size of a population, what about the death rate? If the birthrate stays constant and the death rate decreases, obviously population size also increases. Most experts agree that population growth depletes resources and increases environmental degradation. An older population can also put a strain on the economy if they are unable to meet their financial, including medical, needs without governmental assistance.

What is the ethical solution to this problem? Should we just allow the population to increase due to older people living longer? Should we decrease the birthrate? Should we reduce governmental assistance to older people so they realize that they must be able to take care of themselves? Should we call a halt to increasing the life span through advancements in medical science?

Understanding the Terms

aldosterone 820
ammonia 814
antidiuretic hormone (ADH) 820
atrial natriuretic hormone (ANH) 821
collecting duct 816
distal convoluted tubule 816
excretion 811
flame cell 815
glomerular capsule 816
glomerular filtration 818
glomerulus 817
kidney 816
loop of the nephron 816
Malpighian tubule 815
nephridium 815
nephron 816
proximal convoluted tubule 816
renal cortex 816
renal medulla 816
renal pelvis 816
renin 821
tubular reabsorption 818
tubular secretion 818
urea 814
ureter 816
urethra 816
uric acid 814
urinary bladder 816
urine 816

Match the terms to these definitions:

a. _______________ Blind, threadlike excretory tubule near the anterior end of an insect hindgut.
b. _______________ Cuplike structure that is the initial portion of a nephron; where glomerular filtration occurs.
c. _______________ Final portion of a nephron that joins with a collecting duct; associated with tubular secretion.
d. _______________ Hormone secreted by the adrenal cortex that regulates the sodium and potassium ion balance of the blood.
e. _______________ Main nitrogenous waste of insects, reptiles, and birds.

Web Connections

Exploring the Internet

http://www.mhhe.com/biosci/genbio/mader
(click on *Biology 7/e*)

The *Biology 7/e* Online Learning Center provides many resources for studying the material in this chapter including links to the following sites:

Patient Information Documents on Kidney Diseases. Describes end-stage renal disease, kidney diseases of diabetics, kidney stones, and polycystic kidney disease.

http://www.niddk.nih.gov/health/kidney/kidney.htm

Normal Anatomy and Physiology. Learn more about the normal male and female urinary tracts and normal bladder anatomy and physiology at this site.

http://www.life-tech.com/uro/urolib/normant.htm

The Basics of the Kidney. A good introduction to the basic function of the kidney is presented at this site.

http://www.clark.net/pub/nhp/med/kidney/basics.html

Urinary System. The urinary system removes waste from the blood and helps regulate body fluids. These functions are covered at this site.

http://www.aps.uoguel.ph.ca/swatland/html/ch1.4.html

Animal Survival. Information on animals (amphibians, reptiles, birds, and mammals) adapted to desert environments. Links to relevant organizations.

http://www.desertusa.com/survive.html

Neurons and Nervous Systems

c h a p t e r

c h a p t e r   c o n c e p t s

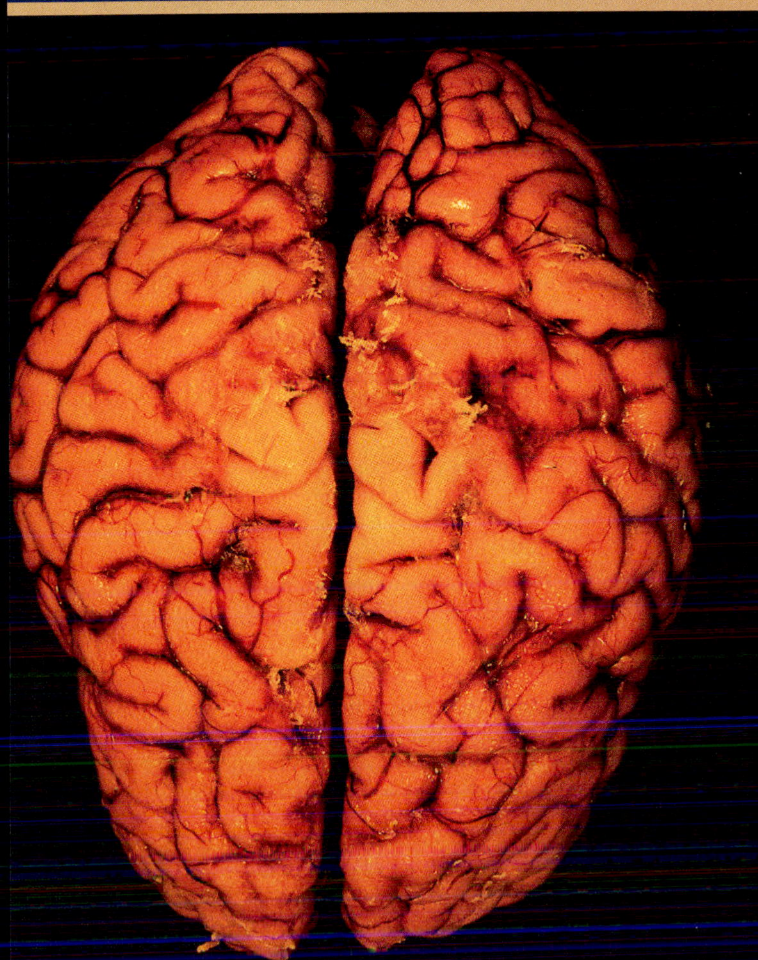

Superior view of human brain

The human nervous system consists of the brain, spinal cord, and nerves. **Sensory receptors** detect changes in stimuli, and nerve impulses race through sensory fibers to the brain and spinal cord. The brain and spinal cord sum up the data before sending impulses via motor fibers to **effectors** (muscles and glands) so that a response to stimuli is possible. The principles of operation are simple, but the human nervous system is intricately complex. The human brain carries out all sorts of higher mental functions, and slight anatomical differences may benefit its performance. Einstein's brain has less of a fissure than normal in the portion dealing with mathematical relationships, and this may account for his genius.

In complex animals, coordination of internal systems is especially important to achieve homeostasis that is a relatively stable internal environment. Digestion of food, breathing, and transport of nutrients are all regulated by the nervous system, with the help of the endocrine system. The survival of all animals, even the minuscule rotifer or flatworm, depends on sensing the environment and responding to changes appropriately.

46.1 Evolution of the Nervous System

A comparative study of animal nervous organization indicates the evolutionary trends in nervous system organization that may have led to the nervous system of vertebrates.

Invertebrate Nervous Organization

Even sponges, which have the cellular level of organization, can respond to stimuli; the most common observable response is closure of the osculum (central opening). Hydras, which are cnidarians with the tissue level of organization, can contract and extend their bodies, move their

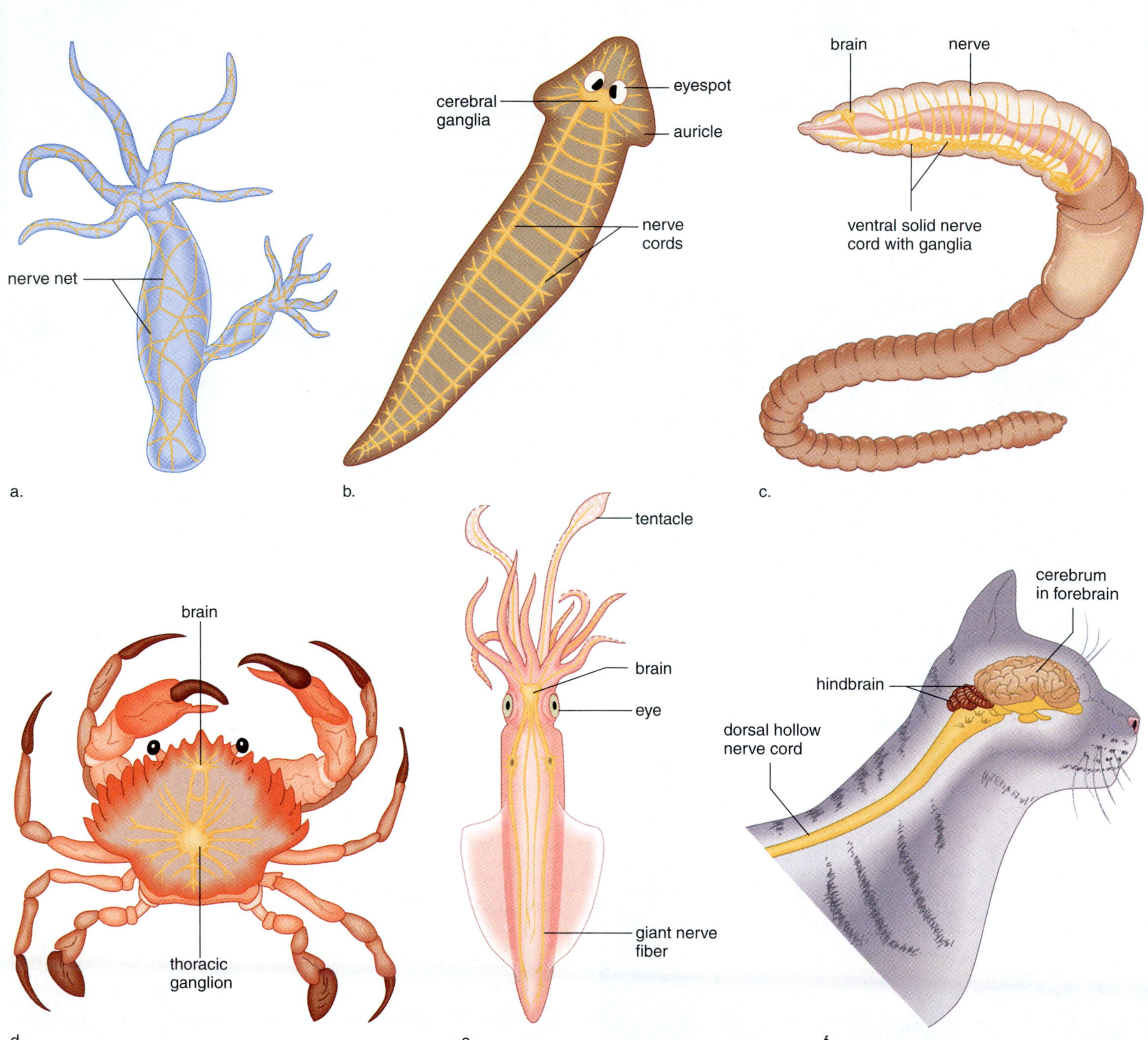

Figure 46.1 Evolution of the nervous system.
a. The nerve net of a hydra, a cnidarian. **b.** In planarians, a flatworm, the paired nerve cords with transverse nerves have the appearance of a ladder. **c.** The earthworm, an annelid, has a central nervous system consisting of a brain and a ventral solid nerve cord. It also has a peripheral nervous system consisting of nerves. **d.** The crab, an arthropod, has a nervous system that resembles that of annelids, but the ganglia are larger. **e.** The squid, a mollusc, has a definite brain with well-developed giant nerve fibers that produce rapid muscle contraction so the squid can move quickly. **f.** A cat, like other vertebrates, has a dorsal tubular nerve cord in the central nervous system.

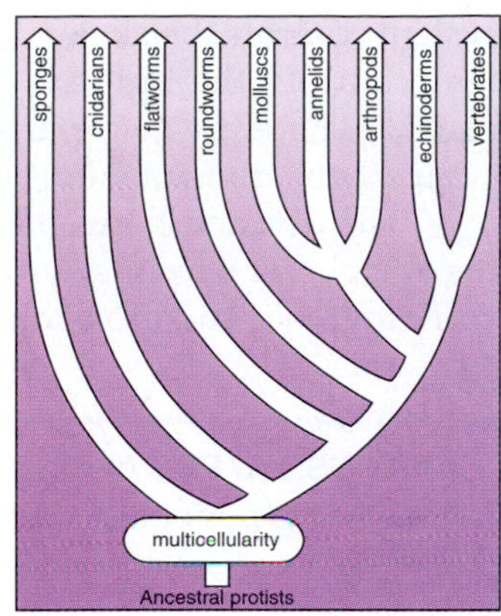

tentacles to capture prey, and even turn somersaults. They have a **nerve net** that is composed of neurons in contact with one another and with contractile epitheliomuscular cells (Fig. 46.1*a*). Sea anemones and jellyfishes, which are also cnidarians, seem to have two nerve nets. A fast-acting one allows major responses, particularly in times of danger, and the other coordinates slower and more delicate movements.

Planarians, which are flatworms, have a nervous organization in keeping with their bilateral symmetry. They have two lateral nerve cords (bundles of nerves) and transverse nerves that join them. The entire arrangement is a **ladderlike nervous system. Cephalization** has occurred because there is a concentration of ganglia and sensory receptors in a head region. Anterior cerebral ganglia receive sensory information from photoreceptors in the eyespots and sensory cells in the auricles (Fig. 46.1*b*). The two longitudinal nerve cords allow a rapid transfer of information from the cerebral ganglia to the posterior end, and the transverse nerves between the nerve cords keep the movement of the two sides coordinated. Bilateral symmetry plus cephalization are two significant trends in the development of a nervous organization that is adaptive for an active way of life. Also, the nervous organization in planarians is a foreshadowing of a central nervous system (cerebral ganglia and nerve cords) and a peripheral nervous system (transverse nerves) seen in vertebrates.

Annelids (e.g., earthworm, Fig. 46.1*c*), arthropods (e.g., crab, Fig. 46.1*d*), and molluscs (e.g., squid, Fig. 46.1*e*), are complex animals with true nervous systems. The annelids and arthropods have what is usually considered the typical invertebrate nervous system. There is a brain and a ventral solid nerve cord having a ganglion in each segment. The brain, which normally receives sensory information, controls the activity of the ganglia and assorted nerves so that the muscle activity of the entire animal is coordinated. The crab and squid show marked cephalization—the anterior end has a well-defined brain, and there are well-developed sense organs. The presence of a brain and other ganglia in the body of all these animals indicate an increase in the number of neurons (nerve cells) among invertebrates.

Bilateral symmetry, cephalization, and an increase in the number of neurons are evolutionary trends that are observable among the invertebrates.

Vertebrate Nervous Organization

In vertebrates (e.g., cat, Fig. 46.1*f*) the central nervous system, consisting of a brain and spinal cord, develops from an embryonic dorsal tubular nerve cord. Cephalization, coupled with bilateral symmetry, results in several types of paired sensory receptors such as the paired eyes, ears, and olfactory structures that allow the animal to gather information from the environment. There are paired cranial and spinal nerves that contain numerous nerve fibers. In vertebrates, there is a vast increase in the number of neurons. For example, an insect's entire nervous system may contain a total of about one million neurons, while a vertebrate nervous system may contain many thousand to several billion times that number.

The Vertebrate Brain

The vertebrate **brain** is the enlarged anterior end of the dorsal tubular nerve cord. It is customary to divide the brain into the hindbrain, midbrain, and forebrain. Nearly all vertebrates have a well-developed hindbrain that regulates organs below the level of consciousness. In humans, for example, the lungs and heart function even when we are sleeping. The hindbrain also functions in the coordination of motor activity associated with limb movement, posture, and balance.

The optic lobes are a part of the midbrain, which was originally a center for coordinating reflex responses to visual input. The forebrain receives sensory input from the midbrain and the hindbrain and regulates their output. The cerebrum is the foremost part of the brain in vertebrates. The cerebrum, which is highly developed in mammals, integrates sensory and motor input and is particularly associated with higher mental capabilities. In humans, the outer layer of the cerebrum, called the cerebral cortex, is particularly large and complex.

In vertebrates, the brain is organized into three areas: the hindbrain, the midbrain, and the forebrain. The forebrain is highly developed in mammals, particularly humans.

The Human Nervous System

The nervous system has three specific functions: (1) it receives sensory input—sensory receptors present in skin and other organs respond to external and internal stimuli by generating nerve impulses that travel to the central nervous system (CNS); (2) it performs integration—the CNS sums up the input it receives from all over the body; and (3) it generates motor output—nerve impulses from the CNS go to the muscles and glands. Muscle contractions and gland secretions are therefore responses to stimuli received by sensory receptors. As an example, consider the events that occur as a person raises a glass to the lips. Continual sensory input to the CNS from the eyes and hand inform the CNS of the position of the glass, and the CNS continually sums up the incoming data before commanding the hand to proceed. At any time, integration with other sensory data might cause the CNS to command a different motion instead.

In humans, the **central nervous system (CNS)** includes the brain and spinal cord, which have a central location—they lie in the midline of the body (Fig. 46.2). The **peripheral nervous system (PNS)** [Gk. *periphereia*, circumference] lies outside the central nervous system and contains both cranial nerves and spinal nerves. The paired cranial nerves connect to the brain, and the paired spinal nerves lie on either side of the spinal cord. In the PNS, the somatic system controls the skeletal muscles and the autonomic system controls the smooth muscles, cardiac muscles, and glands. There are two parts to the autonomic system: the sympathetic division and the parasympathetic division.

The division between the central nervous system and the peripheral nervous system is arbitrary; the two systems work together and are connected to one another.

The CNS and PNS of the human nervous system work together to perform the functions of the nervous system.

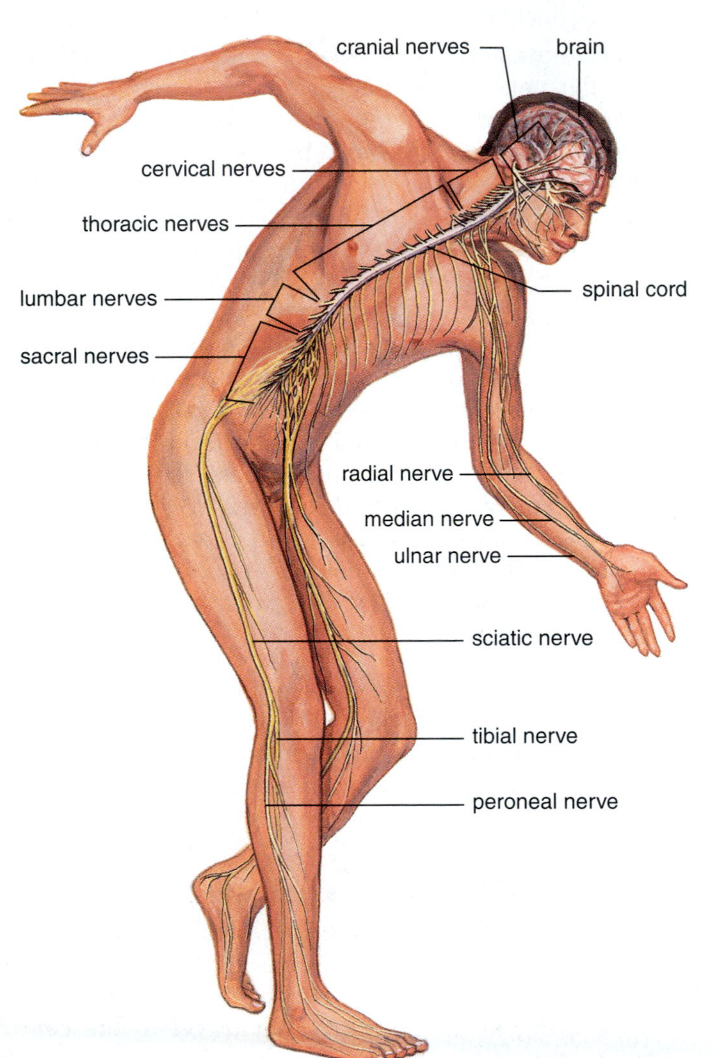

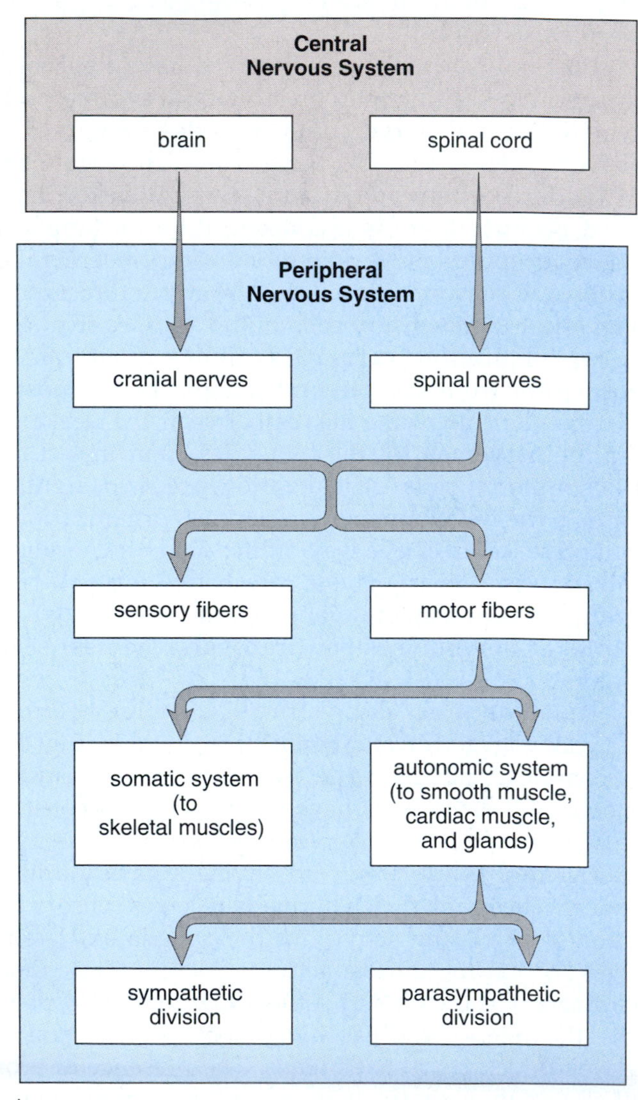

Figure 46.2 Organization of the nervous system in humans.
a. Pictorial representation. The central nervous system (CNS, composed of brain and spinal cord) and some of the nerves of the peripheral nervous system (PNS) are shown. **b.** The CNS communicates with the PNS, which contains nerves. In the somatic system, nerves conduct impulses from sensory receptors to the CNS and motor impulses from the CNS to the skeletal muscles. In the autonomic system, consisting of the sympathetic and parasympathetic divisions, motor impulses travel to smooth muscle, cardiac muscle, and the glands.

46.2 Nervous Tissue

Although exceedingly complex, nervous tissue is made up of just two principal types of cells: (1) **neurons,** also called nerve cells, which transmit nerve impulses; and (2) **neuroglia,** which supports and nourishes neurons.

Neurons

Neurons [Gk. *neuron,* nerve] are cells that vary in size and shape, but they all have three parts: the dendrites, an axon, and the cell body. In motor neurons (Fig. 46.3*a*), the **dendrites** [Gk. *dendron,* tree] are short processes that receive information from other neurons and conduct signals *toward* the cell body. The **axon** [Gk. *axon,* axis], on the other hand, conducts nerve impulses *away* from the cell body. The **cell body** contains the nucleus and other organelles typically found in cells. One of the main functions of the cell body is to manufacture neurotransmitters, which are chemicals stored in secretory vesicles at the ends of axons. When neurotransmitters are released, they influence the excitability of nearby neurons.

Any long axon is also called a **nerve fiber.** Long axons are covered by a white **myelin sheath** [Gk. *myelos,* spinal cord] formed from the membranes of tightly spiraled neuroglial cells. In the PNS, a neuroglial cell called a **neurolemmocyte** (Schwann cell) performs this function, leaving gaps called neurofibril nodes (nodes of Ranvier). There is another type of neuroglial cell that performs a similar function in the CNS.

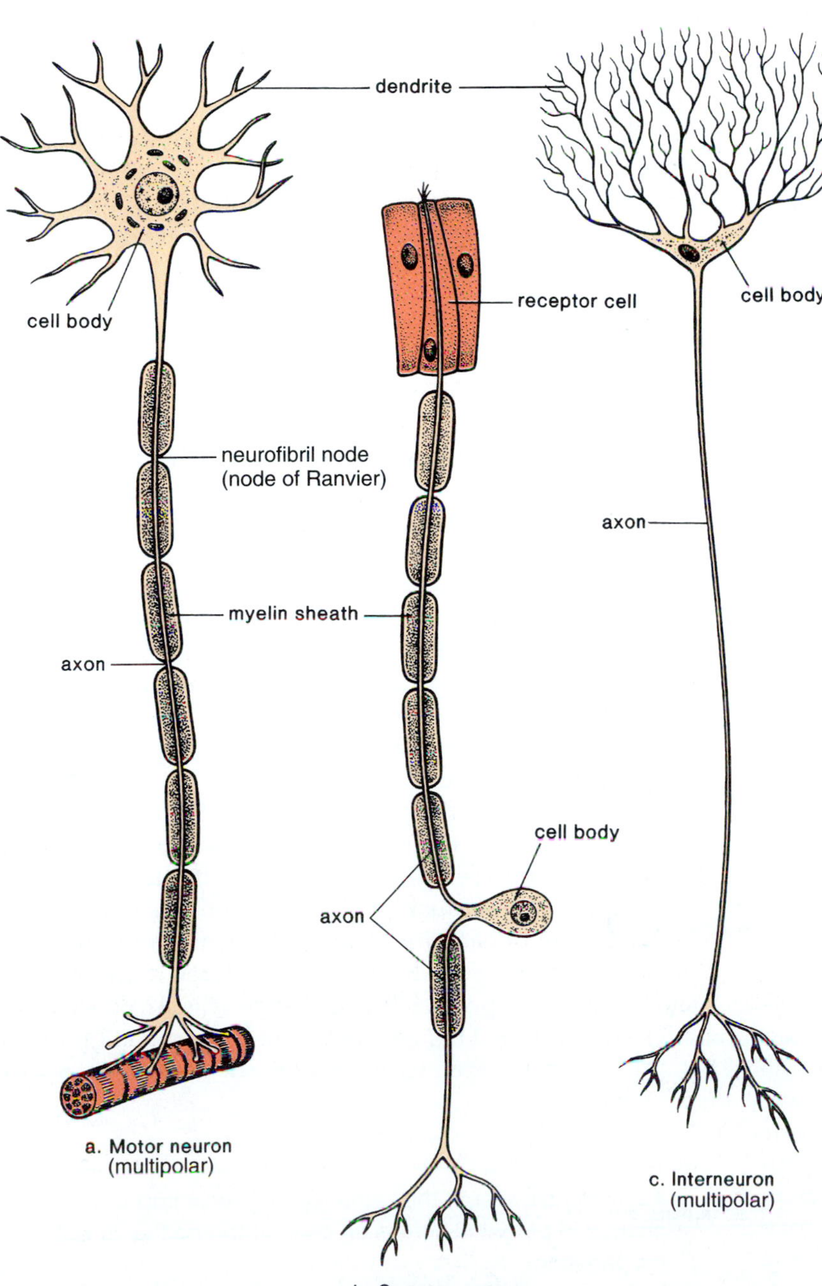

Types of Neurons

Neurons can be classified according to their function and shape. **Motor neurons** take messages from the CNS to muscle or glands. Motor neurons are said to be multipolar because they have many dendrites and a single axon (Fig. 46.3*a*). Motor neurons cause muscle fibers to contract or glands to secrete, and therefore they are said to innervate these structures.

Sensory neurons take messages from sensory receptors to the CNS. The sensory receptor, which is the distal end of the long axon of a sensory neuron, may be as simple as a naked nerve ending (a pain receptor), or may be built into a highly complex organ, such as the eye or ear. Almost all sensory neurons have a structure that is termed unipolar (Fig. 46.3*b*). In unipolar neurons, the process that extends from the cell body divides into a branch that extends to the periphery and another that extends to the CNS. Since both of these extensions are long and myelinated and transmit nerve impulses, it is now generally accepted to refer to them as an axon.

Interneurons [L. *inter,* between, and Gk. *neuron,* nerve], also known as association neurons, occur entirely within the CNS. Interneurons, which are typically multipolar (Fig. 46.3*c*), convey messages between various parts of the CNS. Some lie between sensory neurons and motor neurons, and some take messages from one side of the spinal cord to the other or from the brain to the cord, and vice versa. They also form complex pathways in the brain where processes accounting for thinking, memory, and language occur.

Figure 46.3 Neuron anatomy.
a. Motor neuron. Note the branched dendrites and the single, long axon, which branches only near its tip.
b. Sensory neuron with dendrite-like structures projecting from the peripheral end of the axon. c. Interneuron (from the cortex of the cerebellum) with very highly branched dendrites.

Transmission of the Nerve Impulses

Italian investigator Luigi Galvani discovered in 1786 that a nerve can be stimulated by an electric current. But it was realized later that the speed of the nerve impulse is too slow to be simply an electric current traveling within an axon. In the early 1900s, Julius Bernstein, at the University of Halle, Germany, suggested that the nerve impulse is an electrochemical phenomenon involving the movement of unequally distributed ions on either side of an axomembrane, the plasma membrane of an axon. It was not until later, however, that the investigators developed a technique that enabled them to support this hypothesis. A. L. Hodgkin and A. F. Huxley, English neurophysiologists, received the Nobel Prize in 1963 for their work in this field. They and a group of researchers, headed by K. S. Cole and J. J. Curtis, at Woods Hole, Massachusetts, managed to insert a very tiny electrode into the giant axon of the squid *Loligo*. This internal electrode was then connected to a voltmeter and an oscilloscope, an instrument with a screen that shows a trace or pattern indicating a change in voltage with time (Fig. 46.4). Voltage is a measure of the electrical potential difference between two points, which in this case is the difference between two electrodes—one placed inside and another placed outside the axon. (An electrical potential difference across a membrane is called the membrane potential.) When a membrane potential exists, we can say that a plus pole and a minus pole exist; therefore, an oscilloscope indicates the existence of polarity and records polarity changes.

Resting Potential

When the axon is not conducting an impulse, the oscilloscope records a membrane potential equal to about −65 mV (millivolts), indicating that the inside of the neuron is more negative than the outside (Fig. 46.4*a*). This is called the **resting potential** because the axon is not conducting an impulse.

The existence of this polarity can be correlated with a difference in ion distribution on either side of the axomembrane. As Figure 46.4*a* shows, there is a higher concentration of sodium ions (Na^+) outside the axon and a higher concentration of potassium ions (K^+) inside the axon. The unequal distribution of these ions is in part due to the action of the sodium-potassium pump. This pump is an active transport system in the plasma membrane that pumps three sodium ions out of and two potassium ions into the axon. The pump is always working because the membrane is somewhat permeable to these ions and they tend to diffuse toward their lesser concentration. Since the membrane is more permeable to potassium ions than to sodium ions, there are always more positive ions outside the membrane than inside; this accounts for some of the polarity recorded by the oscilloscope. There are also large, negatively charged proteins in the cytoplasm of the axon; altogether, then, the oscilloscope records that the cytoplasm is −65 mV compared to tissue fluid. This is the resting potential.

Action Potential

The occurrence of an **action potential** is obvious when a rapidly moving pattern, indicating rapid changes in membrane potential, appears on the oscilloscope screen (Fig. 46.4*b*). An action potential requires two types of gated channel proteins in the axomembrane. There is a gated channel protein that allows sodium (Na^+) to pass through the membrane and another that allows potassium (K^+) to pass through the membrane.

The action potential is generated only after the occurrence of a threshold value. Threshold is the minimum change in polarity across the axomembrane that is required to generate an action potential. During a depolarization, the inside of a neuron at a particular location becomes positive because of the sudden entrance of sodium ions. If threshold depolarization occurs, many more sodium channels open, and the action potential begins. As sodium ions rapidly move across the membrane to the inside of the axon, the action potential swings up from −65 mV to +40 mV. This reversal in polarity causes the sodium channels to close and the potassium channels to open. Now potassium ions move from inside the axon to the outside of the axon. As potassium ions leave, the action potential swings down from +40 mV to −65 mV. In other words, a repolarization occurs (Fig. 46.4*c*).

Propagation of Action Potentials

If an axon is unmyelinated, an action potential at one locale stimulates an adjacent part of the axomembrane to produce an action potential. In myelinated fibers, an action potential at one neurofibril node causes an action potential at the next node. This type of conduction, which is called **saltatory conduction** [L. *saltator*, hopper], is much faster than otherwise. In thin, unmyelinated axons, the action potential travels about 1.0 meter/second, and in thick, myelinated axons, the rate is more than 100 meters/second. In any case, action potentials are self-propagating; each action potential generates another along the length of an axon.

The conduction of a nerve impulse (action potential) is an all-or-none event; that is, either a fiber conducts a nerve impulse or it does not. Intensity of a message is determined by how many nerve impulses are generated within a given time span. A fiber can conduct a volley of nerve impulses because only a small number of ions are exchanged with each impulse. As soon as an impulse has passed by each successive portion of a fiber, it undergoes a short refractory period during which it is unable to conduct an impulse. This ensures a one-way direction of the impulse. During a refractory period, the sodium gates cannot yet open.

All neurons transmit the same type of nerve impulse—a change in potential across an axomembrane that is self-propagating.

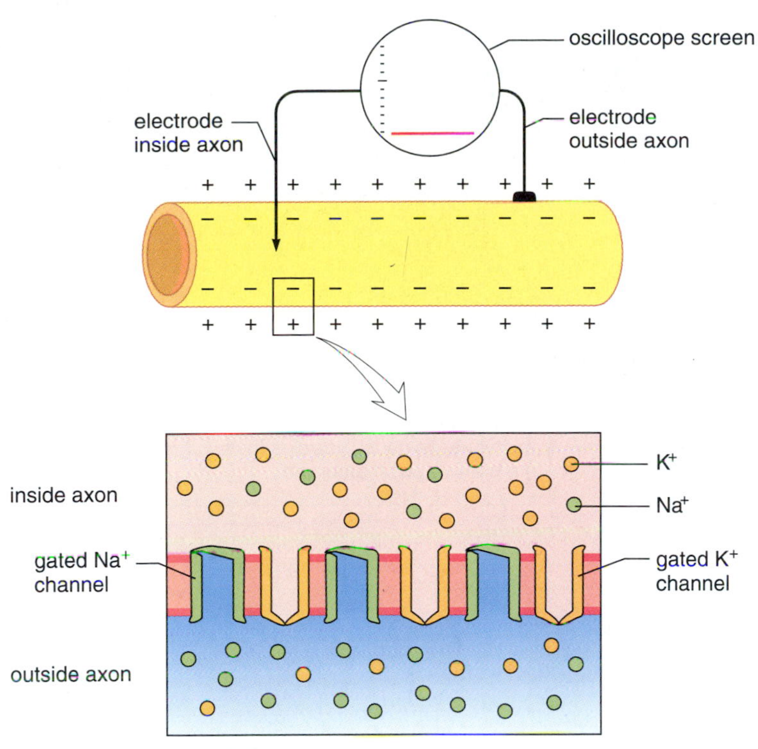

a. Resting Potential

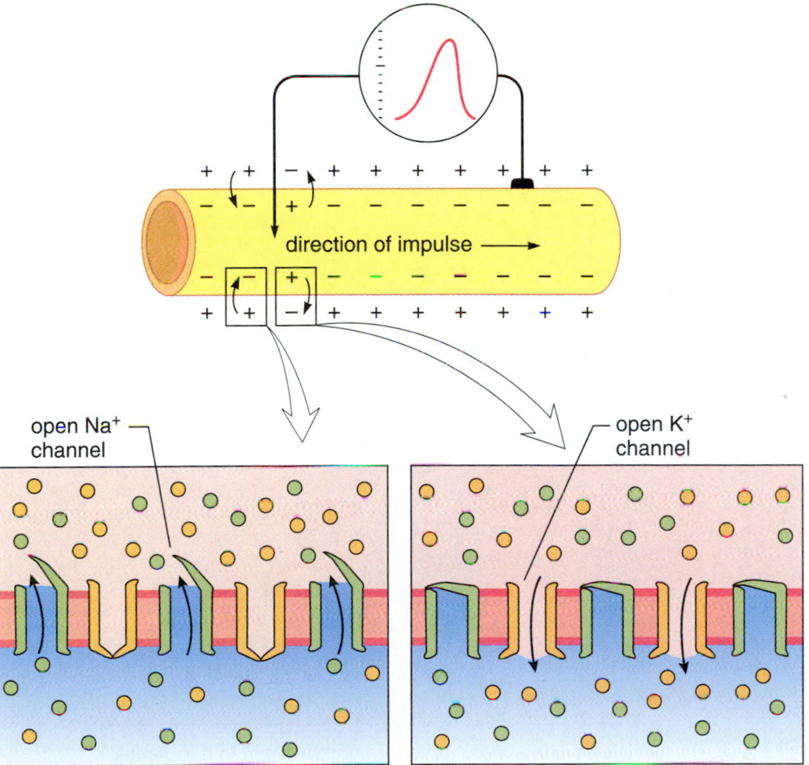

b. Action Potential

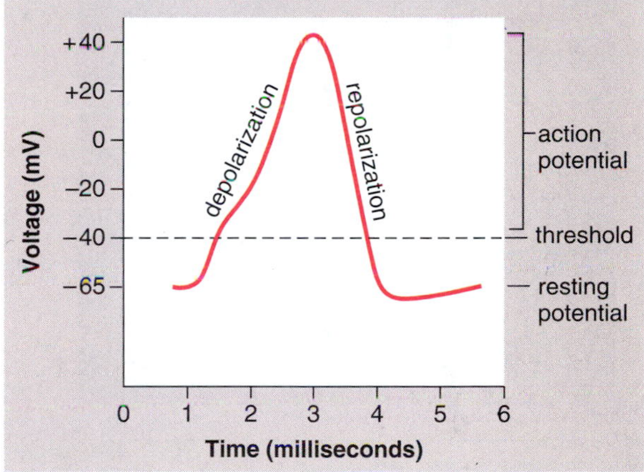

c. Enlargement of Action Potential

Figure 46.4 Resting and action potential.
a. Resting potential. An oscilloscope records a resting potential of −65 mV. There is a preponderance of Na^+ outside the axon and a preponderance of K^+ inside the axon. The permeability of the membrane to K^+ compared to Na^+ causes the inside to be negative compared to the outside. **b.** Action potential. A depolarization occurs when Na^+ gates open and Na^+ moves to inside the axon, and a repolarization occurs when K^+ gates open and K^+ moves to outside the axon. **c.** Enlargement of the action potential, which is seen by an experimenter using an oscilloscope, an instrument that records voltage changes.

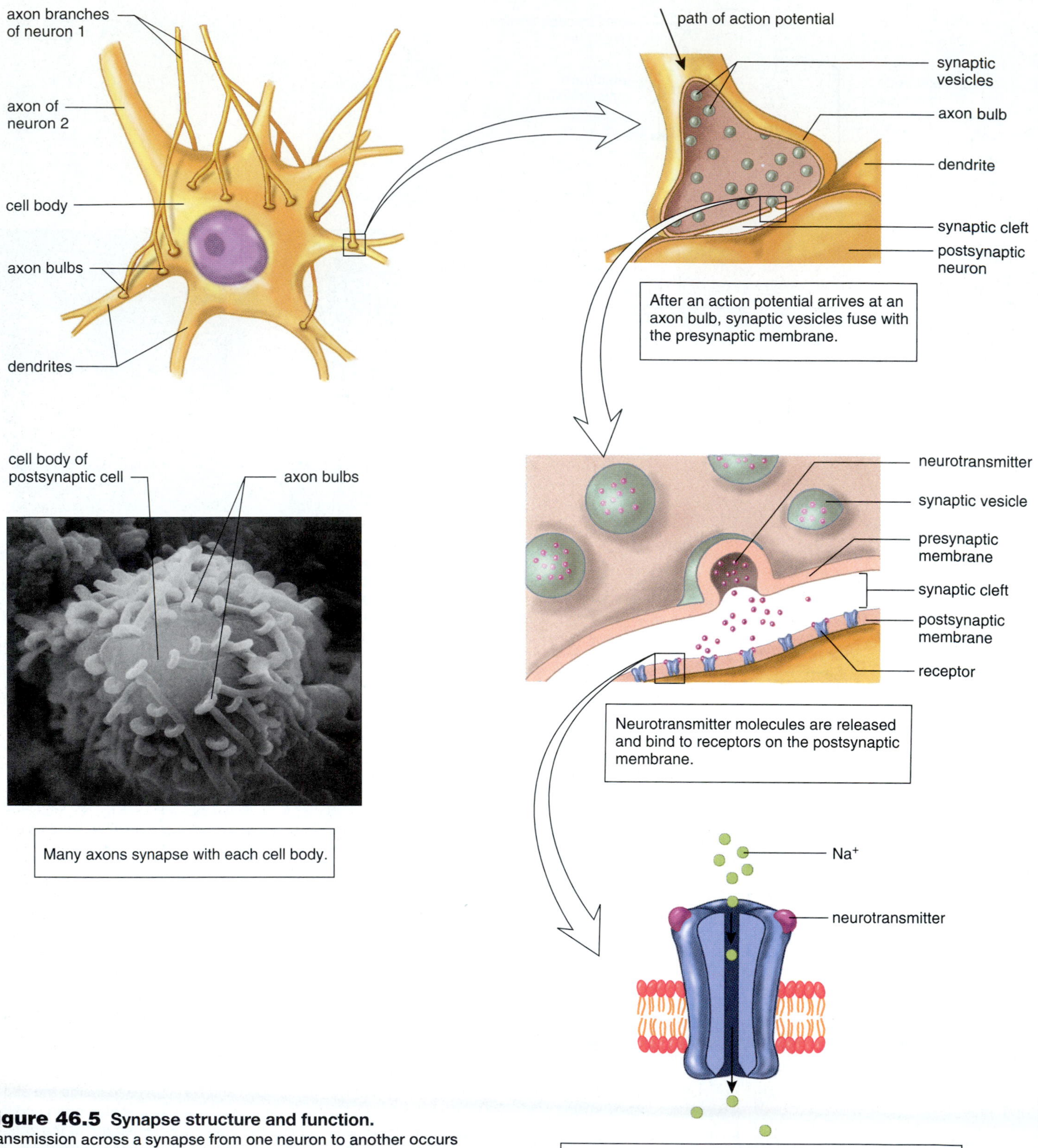

Figure 46.5 **Synapse structure and function.**
Transmission across a synapse from one neuron to another occurs when a neurotransmitter is released at the presynaptic membrane and diffuses across a synaptic cleft and binds to a receptor in the postsynaptic membrane.

Transmission Across a Synapse

Every axon branches into many fine endings, each tipped by a small swelling, called an axon bulb (Fig. 46.5). Each bulb lies very close to the dendrite (or the cell body) of another neuron. This region of close proximity is called a **synapse.** At a synapse, the membrane of the first neuron is called the *pre*synaptic membrane, and the membrane of the next neuron is called the *post*synaptic membrane. The small gap between is the **synaptic cleft.**

Transmission across a synapse is carried out by molecules called **neurotransmitters,** which are stored in synaptic vesicles. When nerve impulses traveling along an axon reach an axon bulb, gated channels for calcium ions (Ca^{2+}) open and calcium enters the bulb. This sudden rise in Ca^{2+} stimulates synaptic vesicles to merge with the presynaptic membrane, and neurotransmitter molecules are released into the synaptic cleft. They diffuse across the cleft to the postsynaptic membrane, where they bind with specific receptor proteins.

Depending on the type of neurotransmitter and/or the type of receptor, the response of the postsynaptic neuron can be toward excitation or toward inhibition. Excitatory neurotransmitters that utilize gated ion channels are fast acting. Other neurotransmitters affect the metabolism of the postsynaptic cell and therefore are slower acting.

Neurotransmitter Molecules

At least 25 different neurotransmitters have been identified, but two very well-known neurotransmitters are **acetylcholine (ACh)** and **norepinephrine (NE).**

Once a neurotransmitter has been released into a synaptic cleft and has initiated a response, it is removed from the cleft. In some synapses, the postsynaptic membrane contains enzymes that rapidly inactivate the neurotransmitter. For example, the enzyme **acetylcholinesterase (AChE)** breaks down acetylcholine. In other synapses, the presynaptic membrane rapidly reabsorbs the neurotransmitter, possibly for repackaging in synaptic vesicles or for molecular breakdown. The short existence of neurotransmitters at a synapse prevents continuous stimulation (or inhibition) of postsynaptic membranes.

It is of interest to note here that many drugs that affect the nervous system act by interfering with or potentiating the action of neurotransmitters. As described in the reading on pages 842–43, drugs can enhance or block the release of a neurotransmitter, mimic the action of a neurotransmitter or block the receptor, or interfere with the removal of a neurotransmitter from a synaptic cleft.

> Transmission across a synapse is dependent on the release of neurotransmitters, which diffuse across the synaptic cleft from one neuron to the next.

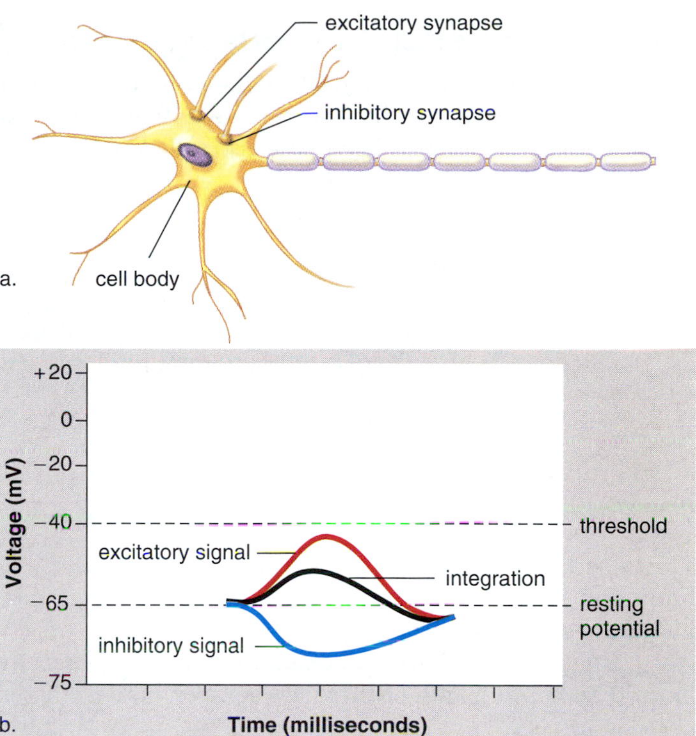

Figure 46.6 Integration.
a. Inhibitory signals and excitatory signals are summed up in the dendrite and cell body of the postsynaptic neuron. Only if the combined signals cause the membrane potential to rise above threshold does an action potential occur. **b.** In this example, threshold was not reached.

Synaptic Integration

A single neuron has many dendrites plus the cell body and both can have synapses with many other neurons. One thousand to ten thousand synapses per a single neuron is not uncommon. Therefore, a neuron is on the receiving end of many excitatory and inhibitory signals. An excitatory neurotransmitter produces a potential change called a signal that drives the neuron closer to an action potential, and an inhibitory neurotransmitter produces a signal that drives the neuron further from an action potential. Excitatory signals have a depolarizing effect, and inhibitory signals have a hyperpolarizing effect (Fig. 46.6).

Neurons integrate these incoming signals. Integration is the summing up of excitatory and inhibitory signals. If a neuron receives many excitatory signals (either from different synapses or at a rapid rate from one synapse), the chances are the axon will transmit a nerve impulse. On the other hand, if a neuron receives both inhibitory and excitatory signals, the summing up of these signals may prohibit the axon from firing.

> Integration is the summing up of inhibitory and excitatory signals received by a postsynaptic neuron.

46.3 Peripheral Nervous System

The peripheral nervous system (PNS) lies outside the central nervous system and contains **nerves** which are bundles of axons. Axons that occur in nerves are also called nerve fibers.

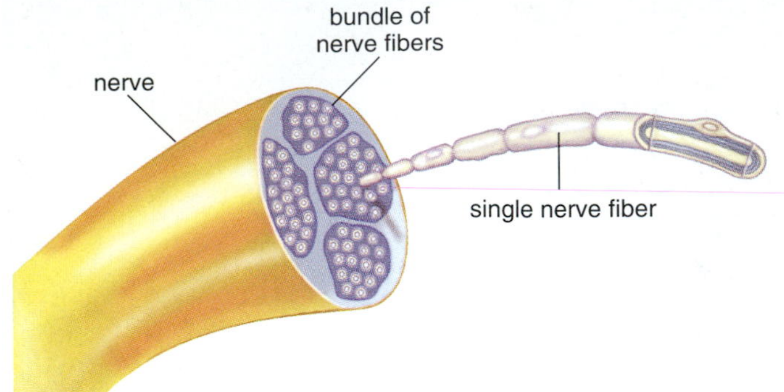

The cell bodies of neurons are found in the CNS—that is, the brain and spinal cord—or in ganglia. Ganglia (sing., **ganglion**) are collections of cell bodies within the PNS.

Humans have 12 pairs of **cranial nerves** attached to the brain (Fig. 46.7*a*). Some of these are sensory nerves; that is, they contain only sensory nerve fibers. Some are motor nerves that contain only motor fibers, and others are mixed nerves that contain both sensory and motor fibers. Cranial nerves are largely concerned with the head, neck, and facial regions of the body. However, the vagus nerve has branches not only to the pharynx and larynx, but also to most of the internal organs.

Humans have 31 pairs of **spinal nerves** (Figs. 46.7*b* and 46.8). The paired spinal nerves emerge from the spinal cord by two short branches, or roots. The dorsal root contains the axons of sensory neurons, which are conducting impulses to the spinal cord from sensory receptors. The cell body of a sensory neuron is in the **dorsal-root ganglion.** The ventral root contains the axons of motor neurons, which are conducting impulses away from the cord to effectors. These two roots join to form a spinal nerve. All spinal nerves are mixed nerves that contain many sensory and motor fibers. Each spinal nerve serves the particular region of the body in which it is located.

In the PNS, cranial nerves take impulses to and/or from the brain, and spinal nerves take impulses to and from the spinal cord.

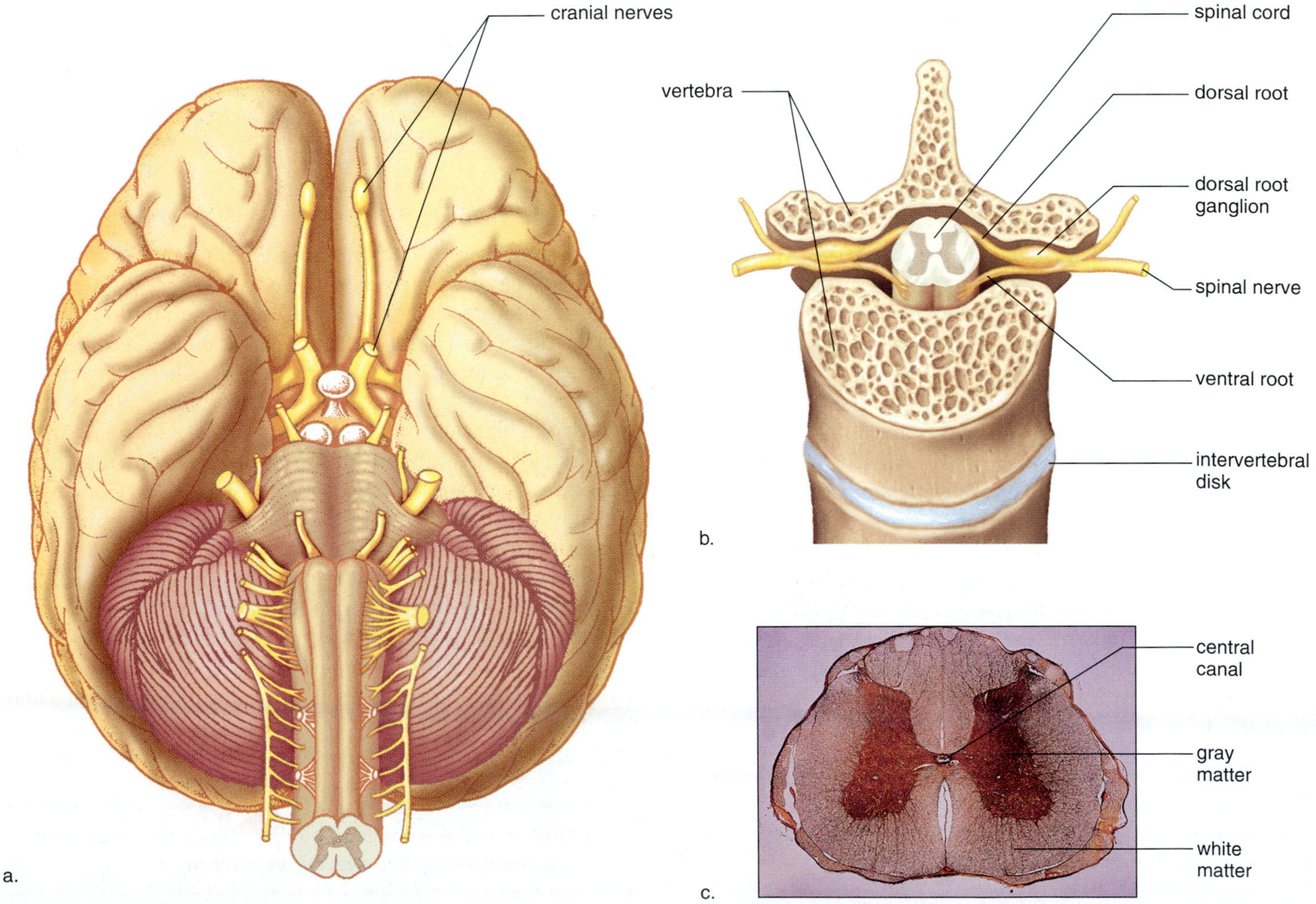

Figure 46.7 Cranial and spinal nerves.
a. Ventral surface of the brain showing the attachment of the cranial nerves. **b.** Cross section of the spinal cord, showing a spinal nerve. Each spinal nerve has a dorsal root and a ventral root attached to the spinal cord. **c.** Photomicrograph of spinal cord cross section.

Somatic System

The **somatic system** includes the nerves that take sensory information from external sensory receptors to the CNS and motor commands away from the CNS to skeletal muscles. Voluntary control of skeletal muscles always originates in the brain. Involuntary responses to stimuli, called **reflexes,** can involve either the brain or just the spinal cord. Flying objects cause eyes to blink and sharp pins cause hands to jerk away even without us having to think about it. The neurotransmitter acetylcholine is active in the somatic system (see Table 46.1).

The Reflex Arc

Figure 46.8 illustrates the path of a reflex that involves only the spinal cord. If your hand touches a sharp pin, sensory receptors in the skin generate nerve impulses that move along sensory axons toward the spinal cord. Sensory neurons which enter the cord dorsally pass signals on to many interneurons. Some of these interneurons synapse with motor neurons. The short dendrites and the cell bodies of motor neurons are in the spinal cord but their axons leave the cord ventrally. Nerve impulses travel along motor axons to an effector, which brings about a response to the stimulus. In this case a muscle contracts so that you withdraw your hand from the pin. Various other reactions are possible—you will most likely look at the pin, wince, and cry out in pain. This whole series of responses is explained by the fact some of the interneurons involved carry nerve impulses to the brain. The brain makes you aware of the stimulus and directs these other reactions to it.

In the somatic system, nerves take information from external sensory receptors to the CNS and motor commands to skeletal muscles. Involuntary reflexes allow us to respond rapidly to external stimuli.

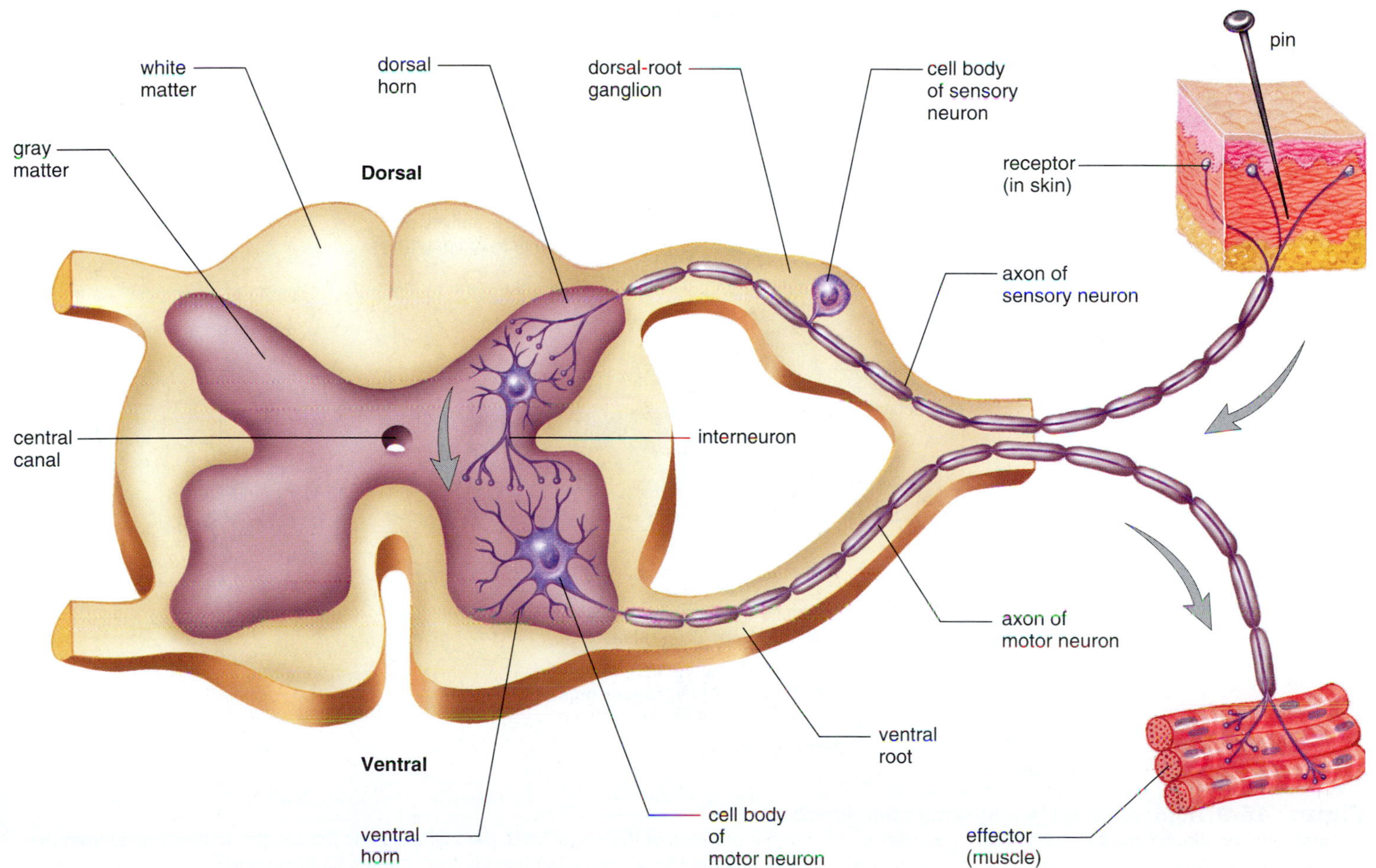

Figure 46.8 A reflex arc showing the path of a spinal reflex.
A stimulus (e.g., sharp pin) causes sensory receptors in the skin to generate nerve impulses that travel in sensory axons to the spinal cord. Interneurons integrate data from sensory neurons and then relay signals to motor axons. Motor axons convey nerve impulses from the spinal cord to a skeletal muscle which contracts. Movement of the hand away from the pin is the response to the stimulus.

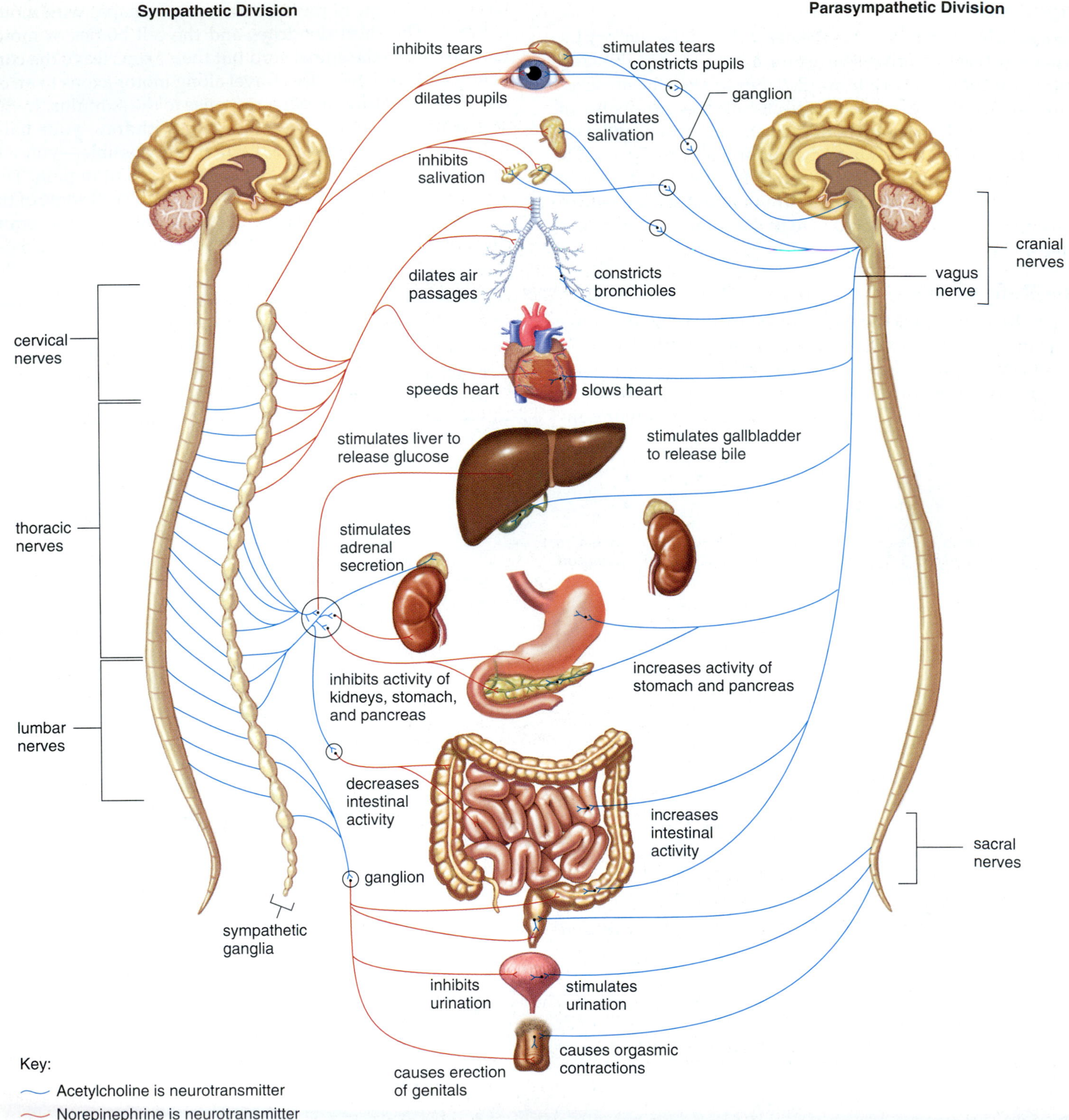

Figure 46.9 Autonomic system structure and function.
Sympathetic preganglionic fibers arise from the thoracic and lumbar portions of the spinal cord; parasympathetic preganglionic fibers arise from the brain and the sacral portion of the spinal cord. Each system innervates the same organs but has contrary effects as described.

Autonomic System

The **autonomic system** of the PNS regulates the activity of cardiac and smooth muscle and glands. The system is divided into the sympathetic and parasympathetic divisions (Fig. 46.9 and Table 46.1). Both of these divisions (1) function automatically and usually in an involuntary manner; (2) innervate all internal organs; and (3) utilize two neurons and one ganglion for each impulse. The first neuron has a cell body within the CNS and a preganglionic fiber. The second neuron has a cell body within the ganglion and a postganglionic fiber.

Reflex actions, such as those that regulate the blood pressure and breathing rate, are especially important to the maintenance of homeostasis. These reflexes begin when the sensory neurons in contact with internal organs send information to the CNS. They are completed by motor neurons within the autonomic system.

Sympathetic Division

Most preganglionic fibers of the **sympathetic division** arise from the middle, or thoracic-lumbar, portion of the spinal cord and almost immediately terminate in ganglia that lie near the cord. Therefore, in this division, the preganglionic fiber is short, but the postganglionic fiber that makes contact with an organ is long.

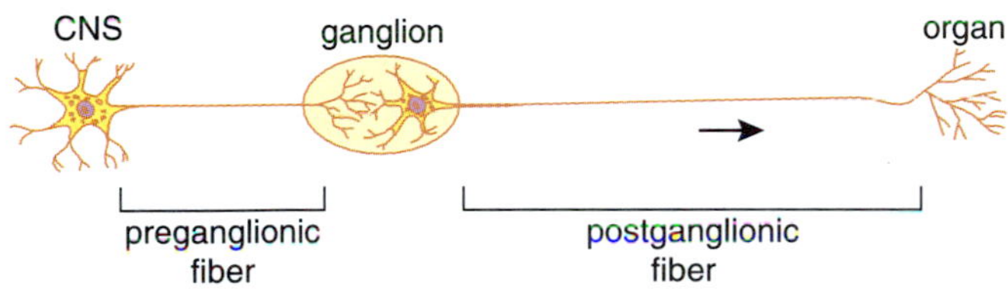

The sympathetic division is especially important during emergency situations and is associated with "fight or flight." If you need to fend off a foe or flee from danger, active muscles require a ready supply of glucose and oxygen. The sympathetic division accelerates the heartbeat and dilates the bronchi. On the other hand, the sympathetic division inhibits the digestive tract—digestion is not an immediate necessity if you are under attack. The neurotransmitter released by the postganglionic axon is primarily norepinephrine (NE). The structure of NE is like that of epinephrine (adrenaline), an adrenal medulla hormone that usually increases heart rate and contractility.

The sympathetic division brings about those responses we associate with "fight or flight."

Parasympathetic Division

The **parasympathetic division** includes a few cranial nerves (e.g., the vagus nerve) and also fibers that arise from the sacral (bottom) portion of the spinal cord. Therefore, this division often is referred to as the craniosacral portion of the autonomic system. In the parasympathetic division, the preganglionic fiber is long, and the postganglionic fiber is short because the ganglia lie near or within the organ.

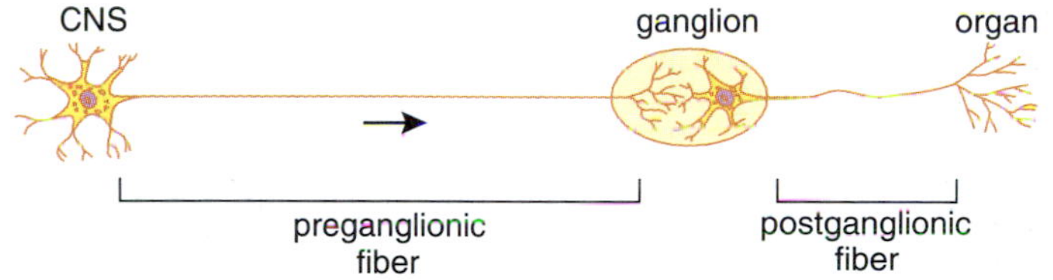

The parasympathetic division, sometimes called the "housekeeper division," promotes all the internal responses we associate with a relaxed state; for example, it causes the pupil of the eye to contract, promotes digestion of food, and retards the heartbeat. The neurotransmitter utilized by the parasympathetic division is acetylcholine (ACh).

The parasympathetic division brings about the responses we associate with a relaxed state.

Table 46.1

Comparison of Somatic Motor and Autonomic Motor Pathways

Items	Somatic Motor Pathway	Autonomic Motor Pathways	
		Sympathetic	**Parasympathetic**
Type of control	Voluntary/involuntary	Involuntary	Involuntary
Number of neurons per message	One	Two (preganglionic shorter than postganglionic)	Two (preganglionic longer than postganglionic)
Location of motor fiber	Most cranial nerves and all spinal nerves	Thoracolumbar spinal nerves	Cranial (e.g., vagus) and sacral spinal nerves
Neurotransmitter	Acetylcholine	Norepinephrine	Acetylcholine
Effectors	Skeletal muscles	Smooth and cardiac muscle, glands	Smooth and cardiac muscle, glands

46.4 Central Nervous System: Brain and Spinal Cord

The central nervous system (CNS) consists of the spinal cord and the brain, where sensory information is received and motor control is initiated. The spinal cord and the brain are both protected by bone; the spinal cord is surrounded by vertebrae and the brain is enclosed by the skull (Figs. 46.7*b* and 46.10). Both the spinal cord and brain are wrapped in three protective membranes known as **meninges** [Gk. *meninga,* membranes covering the brain]. The spaces between the meninges are filled with **cerebrospinal fluid** [L. *cerebrum,* brain, and *spina,* backbone], which cushions and protects the central nervous system. Cerebrospinal fluid is contained in the central canal of the spinal cord and within the **ventricles** of the brain, which are interconnecting spaces that produce and serve as reservoirs for cerebrospinal fluid. A small amount of cerebral spinal fluid is withdrawn from around the cord for laboratory testing when a spinal tap is performed.

The Spinal Cord

The **spinal cord** has two main functions: (1) it is the center for many reflex actions, which are discussed on page 835; and (2) it provides a means of communication between the brain and the spinal nerves, which leave the spinal cord.

The spinal cord has white matter and gray matter (see Fig. 46.8). Cell bodies and short unmyelinated fibers give the **gray matter** its color. In cross section, the gray matter looks like a butterfly or the letter *H.* Portions of sensory neurons and motor neurons are found here, as are short interneurons that connect these two types of neurons.

Myelinated long fibers of interneurons that run together in bundles called **tracts** give **white matter** its color. These tracts connect the spinal cord to the brain. Dorsally, there are primarily ascending tracts taking information to the brain, and ventrally, there are primarily descending tracts carrying information from the brain. Because the tracts at one point cross over, the left side of the brain controls the right side of the body, and the right side of the brain controls the left side of the body.

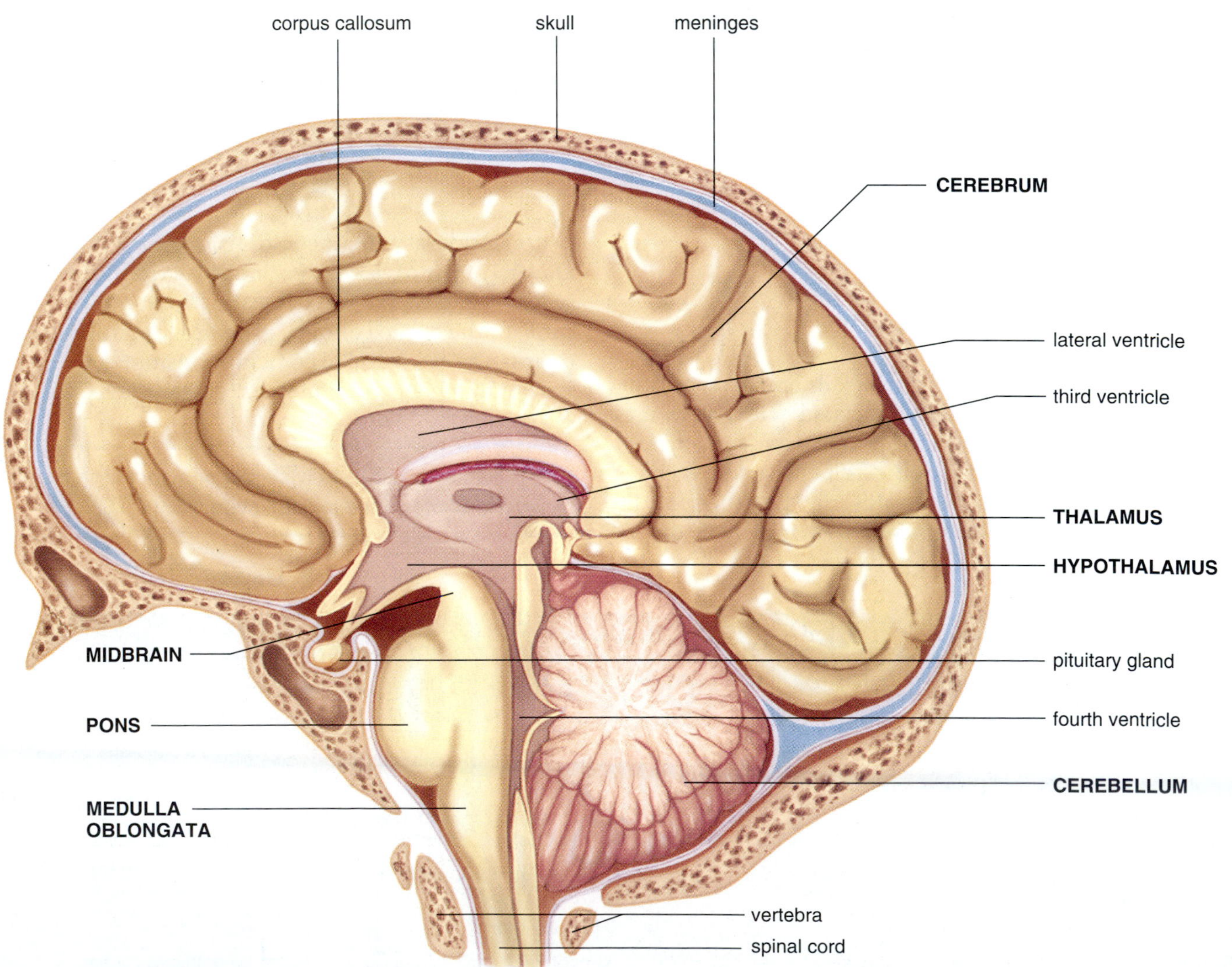

Figure 46.10 The human brain.
The cerebrum is divided into the right and left cerebral hemispheres. The right cerebral hemisphere is shown here. The hemispheres are connected by the corpus callosum.

The Brain

The human brain has been called the last great frontier of biology. The ventricles of the brain are four in number: two lateral ventricles, the third ventricle and the fourth ventricle (Fig. 46.10). It may be helpful to you to associate the brain stem with the fourth ventricle, the diencephalon with the third ventricle, and the cerebrum with the two lateral ventricles.

The Brain Stem

The medulla oblongata, the pons, and the midbrain lie in a portion of the brain known as the **brain stem** where the fourth ventricle is located. The **medulla oblongata** [L. *medulla,* marrow, innermost part, and *oblongus,* longer than broad] lies between the spinal cord and the pons and is anterior to the cerebellum. It contains a number of vital centers for regulating heartbeat, breathing, and vasoconstriction (blood pressure). It also contains the reflex centers for vomiting, coughing, sneezing, hiccuping, and swallowing. The medulla contains tracts that ascend or descend between the spinal cord and the brain's higher centers.

The **pons** [L. *pontis,* bridge] contains bundles of axons traveling between the cerebellum and the rest of the central nervous system. In addition, the pons functions with the medulla to regulate the breathing rate and has reflex centers concerned with head movements in response to visual and auditory stimuli. Aside from acting as a relay station for tracts passing between the cerebrum and the spinal cord or cerebellum, the **midbrain** has reflex centers for visual, auditory, and tactile responses.

The Diencephalon

The hypothalamus and thalamus are in a portion of the brain known as the **diencephalon** [Gk. *dien,* between, and *cephal,* brain] where the third ventricle is located. The hypothalamus forms the floor of the third ventricle. The **hypothalamus** [Gk. *hypo,* under] maintains homeostasis and contains centers for regulating hunger, sleep, thirst, body temperature, water balance, and blood pressure. The hypothalamus controls the pituitary gland and thereby serves as a link between the nervous and endocrine systems.

The **thalamus** consists of two masses of gray matter located in the sides and roof of the third ventricle. The thalamus integrates sensory information and serves as a central relay station for sensory impulses traveling upward from other parts of the brain to the cerebrum. Only one type sensation does not travel through the thalamus and it is our oldest sense—the sense of smell. The thalamus is also involved in arousal and higher mental functions such as memory and emotion.

The Cerebellum

The **cerebellum,** which lies below the posterior portion of the cerebrum, is separated from the brain stem by the fourth ventricle. The cerebellum functions in muscle coordination, integrating impulses received from higher centers to ensure that all of the skeletal muscles work together to produce smooth and graceful motions. The cerebellum is also responsible for maintaining normal muscle tone and transmitting impulses to muscles that maintain posture. It receives information from the inner ear indicating position of the body and then sends impulses to the muscles, whose contraction maintains or restores balance.

The Cerebrum

The **cerebrum,** the foremost part of the brain, is the largest part of the brain in humans (Fig. 46.10). The cerebrum is the highest center to receive sensory input and carry out integration before commanding voluntary motor responses. It is in communication with and coordinates the activities of the other parts of the brain. The cerebrum carries out higher thought processes required for learning and memory and for language and speech.

The **reticular formation** is a complex network of nuclei (masses of cell bodies) and fibers that extend the length of the brain stem (Fig. 46.11). One portion of the reticular formation called the reticular activating system (RAS) arouses the cerebrum via the thalamus and causes a person to be alert. It is believed to filter out unnecessary sensory stimuli and this may account for why you can study with the TV on. An inactive reticular formation results in sleep; a severe injury to the RAS can cause a person to be comatose.

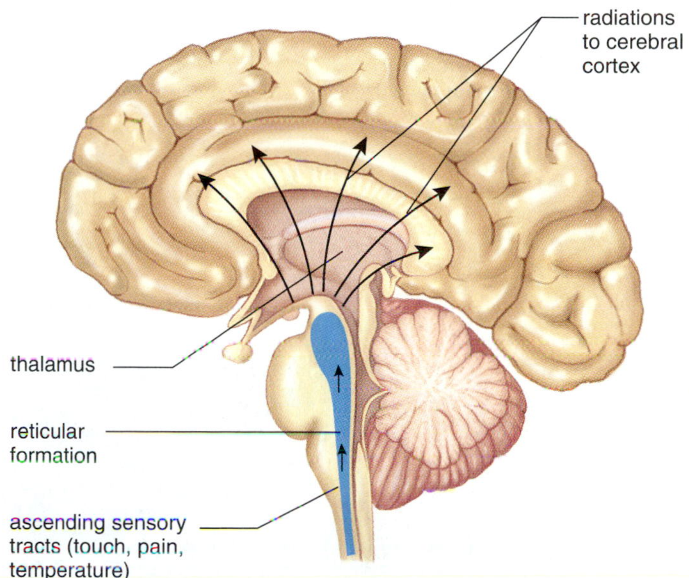

Figure 46.11 The reticular activating system.
The reticular formation receives and sends on motor and sensory information to various parts of the CNS. One portion, the reticular activating system (arrows), arouses the cerebrum and in this way controls alertness versus sleep.

The Cerebral Hemispheres

The cerebrum consists of the left and right **cerebral hemispheres** which are connected by a bridge of nerve fibers called the corpus callosum. The outer portion of the cerebral hemispheres, the cerebral cortex, is highly convoluted and gray in color because it contains cell bodies and short unmyelinated fibers. The cerebral cortex of each hemisphere contains four surface lobes: frontal, parietal, temporal, and occipital (Fig. 46.12).

A comparative study of vertebrates indicates a progressive increase in the relative size of the cerebrum from fishes to humans, and the cerebral cortex is more convoluted in humans than other vertebrates. The function of the cerebrum varies in different groups of animals. In fishes and amphibians, the cerebrum largely has an olfactory function, but in reptiles, birds, and mammals, the cerebrum receives information from other parts of the brain and coordinates sensory data and motor functions.

Voluntary commands begin in the primary motor area of the frontal lobe. Each part of the body is sequentially represented. Our versatile hand takes up an especially large area of the primary motor area. Ventral to the primary motor area is a premotor area. The premotor area organizes motor functions for skilled motor activities before the primary motor area sends signals to the cerebellum which integrates them. The unique ability of humans to speak, is partially dependent upon a motor speech area located in the left frontal lobe. Signals originating here pass to the premotor area before reaching the primary motor area.

Sensory information from the skin and skeletal muscles arrives at the primary somatosensory area, where each part of the body is sequentially represented. A primary visual area in the occipital lobe receives information from our eyes, and a primary auditory area in the temporal lobe receives information from our ears. A primary taste area in the parietal lobe accounts for taste sensations.

Association areas are places where cell bodies integrate information. The somatosensory association area processes and analyzes sensory information from the skin and muscles. The visual association area in the occipital lobe associates new visual information with old information. It might "decide," for example, if we have seen this face or tool or whatever before. The auditory association area in the temporal lobe performs the same functions with regard to sounds. A general interpretation area receives information from all the sensory association areas and allows us to quickly integrate incoming signals and send them on to the prefrontal area in the frontal lobe so an immediate response is possible.

The prefrontal area in the frontal lobe receives input from the other association areas and uses this information to reason and plan our actions. Integration in this area accounts for our most cherished human abilities to think critically and to formulate appropriate behaviors.

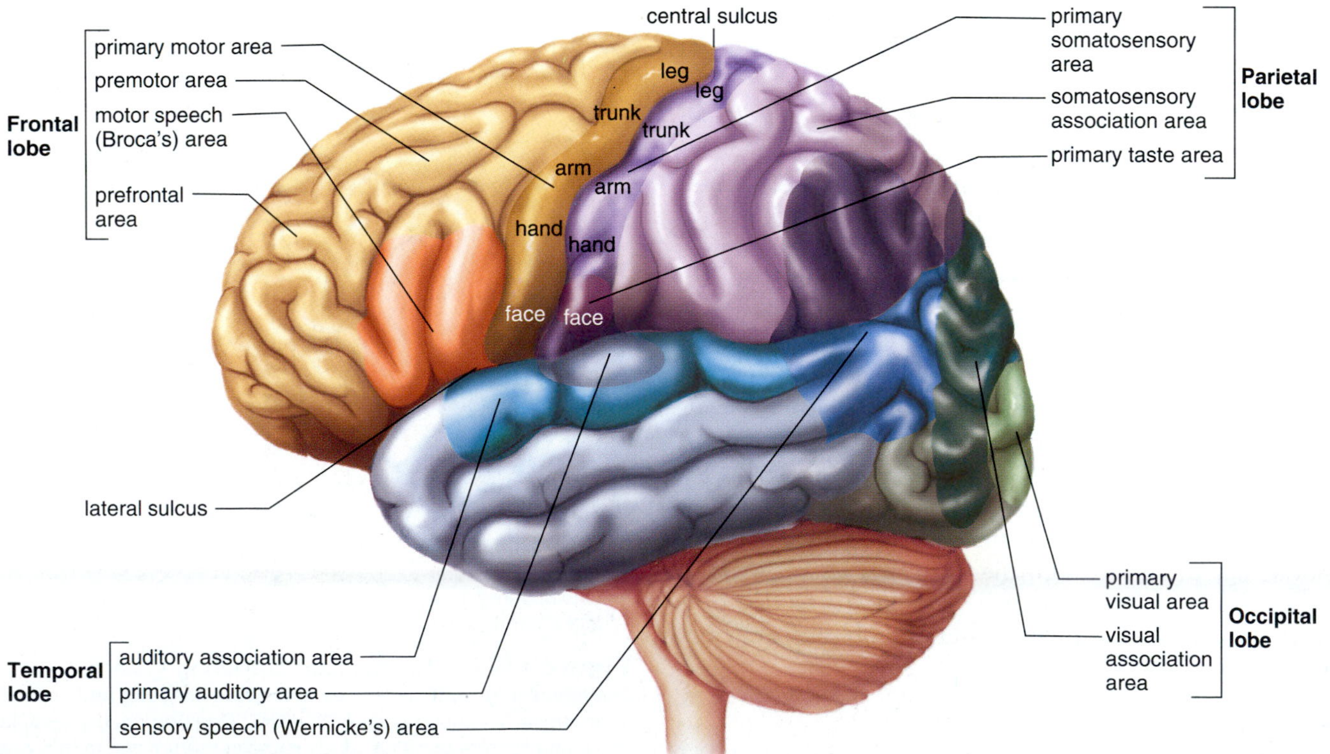

Figure 46.12 The cerebral cortex.
The convoluted cerebral cortex is divided into four surface lobes: frontal, parietal, temporal, and occipital. The frontal lobe has motor areas and an association area called the prefrontal area. The other lobes have sensory areas and also association areas.

The Limbic System

The **limbic system** is a complex network of tracts and nuclei that incorporates medial portions of the cerebral lobes, subcortical nuclei, and the diencephalon (Fig. 46.13). The limbic system blends higher mental functions and primitive emotions into a united whole. It accounts for why activities like sexual behavior and eating seem pleasurable and also why say, mental stress can cause high blood pressure.

Two significant structures within the limbic system are the hippocampus and the amygdala, which are essential for learning and memory. The hippocampus, a seahorse-shaped structure that lies deep in the temporal lobe, is well situated in the brain to make the prefrontal area aware of past experiences stored in sensory association areas. The amygdala, in particular, can cause these experiences to have emotional overtones. The inclusion of the frontal lobe in the limbic system means that reason can keep us from acting out strong feelings.

Learning and Memory **Memory** is the ability to hold a thought in mind or recall events from the past, ranging from a word we learned only yesterday to an early emotional experience that has shaped our lives. Learning takes place when we retain and utilize past memories.

The prefrontal area in the frontal lobe is active during short-term memory as when we temporarily recall a telephone number. Some telephone numbers go into long-term memory. Think of a telephone number you know by heart and see if you can bring it to mind without also thinking about the place or person associated with that number. Most likely you cannot because typically long-term memory is a mixture of what is called semantic memory (numbers, words, etc.) and episodic memory (persons, events, etc.). Skill memory is a type of memory that can exist independent of episodic memory. Skill memory is being able to perform motor activities like riding a bike or playing ice hockey.

What parts of the brain are functioning when you remember something from long ago? As mentioned, our long-term memories are stored in bits and pieces throughout the sensory association areas of the cerebral cortex. The hippocampus gathers this information together for use by the prefrontal area of the frontal lobe when we remember Uncle Frank or our summer holiday. Why are some memories so emotionally charged? The amygdala is responsible for fear conditioning and associating danger with sensory information received from the thalamus and the cortical sensory areas.

Long-term potentiation (LTP) is an enhanced response at synapses within the hippocampus. LTP is most likely essential to memory storage but, unfortunately, it sometimes causes a postsynaptic neuron to become so excited, it undergoes apoptosis, a form of cell death. This phenomenon called excitotoxicity is due to the action of glutamate, a neurotransmitter. When glutamate binds to the postsynaptic membrane, calcium may rush in too fast due to a receptor that is malformed due to a mutation. A gradual extinction of brain cells in the hippocampus appears to be the underlying cause of Alzheimer disease (AD), a condition characterized by a gradual loss of memory.

AD neurons have neurofibrillary tangles (bundles of fibrous protein) surrounding the nucleus and protein-rich accumulations called amyloid plaques, enveloping the axon branches. Although it is not yet known how excitotoxicity is related to structural abnormalities of AD neurons, some researchers are trying to develop neuroprotective drugs that can possibly guard brain cells against damage due to glutamate.

The cerebral cortex is divided into lobes with specific functions. The limbic system is a unique combination of various portions of the brain which unify brain functions and sensations.

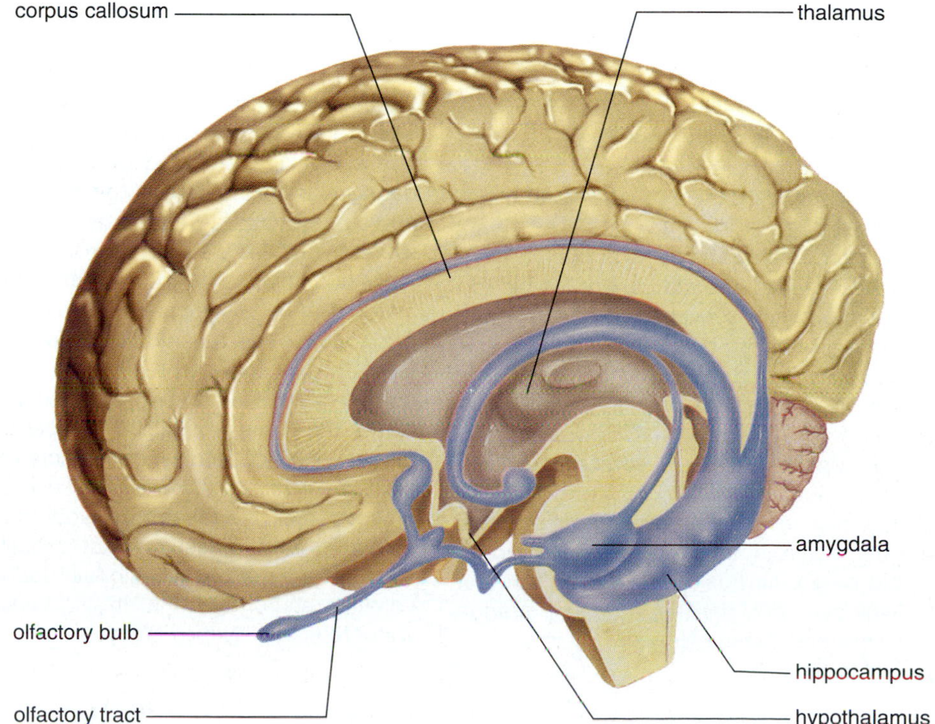

Figure 46.13 **The limbic system.**
Structures deep within the cerebral hemispheres and surrounding the diencephalon join higher mental functions like reasoning with more primitive feelings like fear and pleasure.

Five Drugs of Abuse

Alcohol

Drugs that people take to alter the mood and/or emotional state affect normal body functions, often by interfering with neurotransmitter release or uptake in the brain. It is possible to drink alcoholic beverages in moderation, but alcohol is often abused. Alcohol use becomes "abuse," or an illness, when alcohol ingestion impairs an individual's social relationships, health, job efficiency, or judgment. While it is general knowledge that alcoholics are prone to drink until they become intoxicated, there is much debate as to what causes alcoholism. Some believe that alcoholism is due to an underlying psychological disorder, while others maintain that the condition is due to an inherited physiological disorder.

Alcohol is primarily metabolized in the liver, where it disrupts the normal workings of glycolysis and the Krebs cycle. The liver contains dehydrogenase enzymes, which carry out these reactions, reducing NAD in the process.

$$\begin{array}{ccc} \text{NAD} & \longrightarrow & \text{NADH} \\ \text{alcohol} \longrightarrow & & \longrightarrow \text{acetyl-CoA} \end{array}$$

The supply of NAD in liver cells is used up by these reactions, and there is not enough free NAD left to keep glycolysis and the Krebs cycle running. The cell begins to ferment and lactic acid builds up. The pH of the blood decreases and becomes acidic.

Since the Krebs cycle is not working, excess active acetate cannot be broken down and it is converted to fat—the liver turns fatty. Fat accumulation, the first stage in liver deterioration, begins after only a single night of heavy drinking. If heavy drinking continues, fibrous scar tissue appears during a second stage of deterioration. If heavy drinking stops, the liver can still recover and become normal once again. If not, the final and irrevocable stage, cirrhosis of the liver, occurs: liver cells die, harden, and turn orange ("cirrhosis" means "orange").

The surgeon general recommends that pregnant women drink no alcohol at all. Alcohol crosses the placenta freely and can cause fetal alcohol syndrome, which is characterized by mental retardation and various physical defects.

Another problem is that heavy drinking interferes with good nutrition. Alcohol is energy intensive—the NADH molecules that result from its breakdown can be used to produce ATP molecules. However, these calories are empty because they do not supply any amino acids, vitamins, and minerals as other energy sources do. Without adequate vitamins, red and white blood cells cannot be formed in the bone marrow. The immune system becomes depressed, and the chances of stomach, liver, lung, pancreas, colon, and tongue cancer increase. Protein digestion and amino acid metabolism are so upset that even adequate protein intake will not prevent amino acid deficiencies. Muscles atrophy and weakness results. Fat deposits accumulate in the heart wall and hypertension develops. There is an increased risk of cardiac arrhythmias and stroke.

Nicotine

Nicotine, an alkaloid derived from tobacco, is a widely used neurological agent. When smoking a cigarette, nicotine is quickly distributed to all body organs including the central and peripheral nervous systems. In the central nervous system, nicotine causes neurons to release dopamine, a neurotransmitter associated with behavioral states. The excess of dopamine has a reinforcing effect that leads to dependence on the drug. In the peripheral nervous system, nicotine stimulates the same postsynaptic receptors as acetylcholine and leads to increased skeletal muscular activity. It also increases the heartbeat rate and blood pressure, and digestive tract mobility. Nicotine may even occasionally induce vomiting and/or diarrhea. It also causes water retention by the kidneys.

Many cigarette smokers find it difficult to give up the habit because nicotine induces both physiological and psychological dependence. Withdrawal symptoms include headache, stomach pain, irritability, and insomnia. Tobacco not only contains nicotine, it also contains many other harmful substances. Cigarette smoking contributes to early death from cancer, including not only lung cancer but also cancer of the larynx, mouth, throat, pancreas, and urinary bladder. Chronic diseases like bronchitis and emphysema are likely to develop and there is increased risk of heart attack due to cardiovascular disease.

Now that women are as apt to smoke as men, lung cancer has surpassed breast cancer as a cause of death. Cigarette smoking in young women who are sexually active is most unfortunate because nicotine, like other psychoactive drugs, adversely affects a developing embryo and fetus.

Marijuana

The dried flowering tops, leaves, and stems of the Indian hemp plant *Cannabis sativa* contain and are covered by a resin that is rich in THC (tetrahydrocannabinol). The names *Cannabis* and marijuana apply to either the plant or THC.

The effects of marijuana differ depending upon the strength and the amount consumed, the expertise of the user, and the setting in which it is taken. Usually, the user reports experiencing a mild euphoria along with alterations in vision and judgment, which result in distortions of space and time. Motor incoordination occurs, as well as the inability to concentrate and to speak coherently.

Intermittent use of low-potency marijuana generally is not associated with obvious symptoms of toxicity, but heavy use can produce chronic intoxication. Intoxication is recognized by the presence of hallucinations, anxiety, depression, rapid flow of ideas, body image distortions, paranoid reactions, and similar psychotic symptoms. The terms cannabis psychosis and cannabis delirium refer to such reactions.

Marijuana is classified as a hallucinogen. It is possible that, like LSD (lysergic acid diethylamide), it has an effect on the action of serotonin, an excitatory neurotransmitter in the brain.

Marijuana use does not seem to produce physical dependence, but a psychological dependence on the euphoric and sedative effects can develop. Craving can also occur as a part of regular heavy use.

Usually marijuana is smoked in a cigarette form called a joint. Since this allows toxic substances, including carcinogens, to enter the lungs, chronic respiratory disease and lung cancer are considered dangers of long-term, heavy use. Some researchers claim that marijuana use leads to long-term brain impairment as well. Others report that males and females suffer reproductive dysfunctions. Fetal cannabis syndrome, which resembles fetal alcohol syndrome, has also been reported. In addition, marijuana has been called a gateway drug because adolescents who have used marijuana also tend to try other drugs. For example, in a study of 100 cocaine abusers, 60% had smoked marijuana for more than ten years.

Some psychologists are very concerned about the use of marijuana among adolescents. Marijuana can be used to avoid dealing with the personal problems that often develop during this maturational phase.

Cocaine

Cocaine is an alkaloid derived from the shrub *Erythroxylon coca*. Cocaine is sold in powder form and as crack, a more potent extract. Users often describe the feeling of euphoria that follows intake of the drug as a rush. Snorting (inhaling) produces this effect in a few minutes, injection, within 30 seconds, and smoking, in less than 10 seconds. Persons dependent upon the drug are, therefore, most likely to smoke cocaine. The rush lasts only a few seconds and then is replaced by a state of arousal, which lasts from 5 to 30 minutes. Then the user begins to feel restless, irritable, and depressed. To overcome these symptoms, the user is apt to take more of the drug, repeating the cycle again and again. A binge of this sort can go on for days, after which the individual suffers a crash. During the binge period, the user is hyperactive and has little desire for food or sleep but has an increased sex drive. During the crash period, the user is fatigued, depressed, and irritable, has memory and concentration problems, and displays no interest in sex. Indeed, men are often impotent. Other drugs, such as marijuana, alcohol, or heroin, often are taken to ease the symptoms of the crash.

Cocaine affects the concentration of dopamine, a neurotransmitter associated with behavioral states. After release into a synapse, dopamine ordinarily is withdrawn into the presynaptic cell for recycling and reuse. Cocaine prevents the reuptake of dopamine by the presynaptic membrane; this causes an excess of dopamine in the synaptic cleft so that the user experiences the sensation of a rush. The epinephrine-like effects of dopamine account for the state of arousal that lasts for some minutes after the rush experience.

With continued cocaine use, the body begins to make less dopamine to compensate for a seemingly excess supply. The user then experiences tolerance, withdrawal symptoms, and an intense craving for the drug. Cocaine, then, is extremely addictive.

The number of deaths from cocaine and the number of emergency-room admissions for drug reactions involving cocaine have increased greatly. High doses can cause seizures and cardiac and respiratory arrest.

Individuals who snort the drug can suffer damage to the nasal tissues and even perforation of the septum between the nostrils. Whether or not long-term cocaine abuse causes brain damage is not yet known. It is known, however, that babies born to addicts suffer withdrawal symptoms and may suffer neurological and developmental problems.

Heroin

Heroin is derived from morphine, an alkaloid of opium. Heroin usually is injected. After intravenous injection, the onset of action is noticeable within one minute and reaches its peak in about five minutes. There is a feeling of euphoria along with relief of pain. Side effects can include nausea, vomiting, dysphoria, and respiratory and circulatory depression leading to death.

Heroin binds to receptors meant for the endorphins, the special neurotransmitters that kill pain and produce a feeling of tranquility. They are believed to alleviate pain by preventing the release of a neurotransmitter termed substance P from certain sensory neurons in the region of the spinal cord. When substance P is released, pain is felt, and when substance P is not released, pain is not felt. Endorphins and heroin also bind to receptors on neurons that travel from the spinal cord to the limbic system. Stimulation of these can cause a feeling of pleasure.

Individuals who inject heroin become physically dependent on the drug. With time, the body's production of endorphins decreases. Tolerance develops so that the user needs to take more of the drug just to prevent withdrawal symptoms. The euphoria originally experienced upon injection is no longer felt.

Heroin withdrawal symptoms include perspiration, dilation of pupils, tremors, restlessness, abdominal cramps, gooseflesh, defecation, vomiting, and increase in systolic pressure and respiratory rate. Those who are excessively dependent may experience convulsions, respiratory failure, and death. Infants born to women who are physically dependent also experience these withdrawal symptoms.

Connecting Concepts

Like the wiring of a modern office building, the peripheral nervous system of humans contains nerves that reach to all parts of the body. There is a division of labor among the nerves. The cranial nerves serve the face, teeth, and mouth; below the head there is only one cranial nerve, the vagus nerve. All bodily movements are controlled by spinal nerves, and this is why paralysis may follow a spinal injury. Except for the vagus nerve, only spinal nerves make up the autonomic system, which controls the internal organs. Like most other animals, much of the work of the nervous system in humans is below the level of consciousness. The nervous system has just three functions: sensory input, integration, and motor output. Sensory input would be impossible without sensory receptors, which are sensitive to external and internal stimuli. You might even argue that sense organs like the eyes and ears should be considered a part of the nervous system, since there would be no sensory nerve impulses without their ability to generate them. Nerve impulses are the same in all neurons, so how is it that stimulation of eyes causes us to see, and stimulation of ears causes us to hear? Essentially, the central nervous system carries out the function of integrating incoming data. The brain allows us to perceive our environment, reason, and remember. After sensory data have been processed by the CNS, motor output occurs. Muscles and glands are the effectors that allow us to respond to the original stimuli. Without the musculoskeletal system we would never be able to respond to a danger detected by our eyes and ears.

Summary

46.1 Evolution of the Nervous System

A comparative study of the invertebrates shows a gradual increase in the complexity of the nervous system. The vertebrate nervous system, like that of the earthworm, is divided into the central and peripheral nervous systems.

46.2 Nervous Tissue

The anatomical unit of the nervous system is the neuron, of which there are three types: sensory, motor, and interneuron. Each of these is made up of a cell body, an axon, and dendrites.

When an axon is not conducting an action potential (nerve impulse), the resting potential indicates that the inside of the fiber is negative compared to the outside. The sodium-potassium pump helps maintain a concentration of Na^+ ions outside the fiber and K^+ ions inside the fiber. When the axon is conducting a nerve impulse, an action potential (i.e., a change in membrane potential) travels along the fiber. Depolarization (inside becomes positive) due to the movement of Na^+ to the inside, and then repolarization (inside becomes negative again) due to the movement of K^+ to the outside of the fiber, occurs.

Transmission of the nerve impulse from one neuron to another takes place across a synapse. In humans, synaptic vesicles release a chemical, known as a neurotransmitter, into the synaptic cleft. The binding of neurotransmitters to receptors in the postsynaptic membrane can either increase the chance of an action potential (stimulation) or decrease the chance of an action potential (inhibition) in the next neuron. A neuron usually transmits several nerve impulses, one after the other.

46.3 Peripheral Nervous System

The peripheral nervous system contains the somatic system and the autonomic system. Reflexes are automatic, and some do not require involvement of the brain. A simple reflex requires the use of neurons that make up a reflex arc. In the somatic system, a sensory neuron conducts nerve impulses from a sensory receptor to an interneuron, which in turn transmits impulses to a motor neuron, which stimulates an effector to react.

While the motor portion of the somatic system of the PNS controls skeletal muscle, the motor portion of the autonomic system controls smooth muscle of the internal organs and glands. The sympathetic division, which is often associated with those reactions that occur during times of stress, and the parasympathetic division, which is often associated with those activities that occur during times of relaxation, are both parts of the autonomic system.

46.4 Central Nervous System: Brain and Spinal Cord

The CNS consists of the spinal cord and brain. The gray matter of the cord contains cell bodies; the white matter contains tracts that consist of the long axons of interneurons. These run from all parts of the cord, even up to the cerebrum.

The brain integrates all nervous system activity and commands all voluntary activities. In the brain stem, the medulla oblongata and pons have centers for visceral functions. The cerebellum coordinates muscle contractions. In the diencephalon, the hypothalamus in particular controls homeostasis, and the thalamus specializes in sense reception.

The cerebrum has two cerebral hemispheres. The cerebral cortex of each hemisphere is divided into a frontal, parietal, occipital and temporal lobe. Each lobe has specific functions. Voluntary motor control, sensation, reasoning, learning and memory, and also language and speech take place in the cerebrum.

The limbic system is a complex network of cortical and subcortical nuclei (masses of cell bodies) within the brain that is known to generate primitive emotions and function in higher mental functions, like memory. The hippocampus acts as a conduit for sending information to long-term memory and retrieving it once again. The amygdala adds emotional overtones to memories. On the cellular level, long-term potentiation seems to be required for long-term memory. Unfortunately, long-term potentiation can go awry when neurons become overexcited and die. Neuroprotective drugs are being developed in the hope that they will prevent disorders like Alzheimer disease.

Reviewing the Chapter

1. Trace the evolution of the nervous system by contrasting the organization of the nervous system in hydras, planarians, earthworms, and humans. 826–27
2. Describe the structure of a neuron, and give a function for each part mentioned. 829
3. Name three types of neurons, and give a function for each. 829
4. What are the major events of an action potential, and what ion changes are associated with each event? 830–31

5. Describe the mode of action of a neurotransmitter at a synapse, including how it is stored and how it is destroyed. 832–33
6. Contrast the structure and function of the peripheral and central nervous systems. 828, 834, 838
7. Trace the path of a spinal reflex. 836–37
8. Contrast the sympathetic and parasympathetic divisions of the autonomic system. 837
9. Name the major parts of the human brain, and give a principal function for each part. 838–40
10. Describe the limbic system, and discuss its possible involvement in learning and memory. 841

Testing Yourself

Choose the best answer for each question.

1. Which is the most complete list of animals that have a central nervous system (CNS) and a peripheral nervous system (PNS)?
   a. hydra, planarian, earthworm, rabbit, human
   b. planarian, earthworm, rabbit, human
   c. earthworm, rabbit, human
   d. rabbit, human
   e. All of these are correct.
2. Which of these are the first and last elements in a spinal reflex?
   a. axon and dendrite
   b. sense organ and muscle effector
   c. ventral horn and dorsal horn
   d. motor neuron and sensory neuron
   e. sensory receptor and the brain
3. Which term does not belong with the others?
   a. cerebrum
   b. cerebral cortex
   c. cerebral hemispheres
   d. association areas
   e. cerebellum
4. A spinal nerve takes nerve impulses
   a. to the CNS.
   b. away from the CNS.
   c. both to and away from the CNS.
   d. only inside the CNS.
   e. only from the cerebrum.
5. Which of these correctly describes the distribution of ions on either side of an axon when it is not conducting a nerve impulse?
   a. more sodium ions (Na^+) outside and less potassium ions (K^+) inside
   b. K^+ outside and Na^+ inside
   c. charged protein outside; Na^+ and K^+ inside
   d. Na^+ and K^+ outside and water only inside
   e. Ca^{2+} inside and outside
6. When the action potential begins, sodium gates open, allowing Na^+ to cross the membrane. Now the polarity changes to
   a. negative outside and positive inside.
   b. positive outside and negative inside.
   c. There is no difference in charge between outside and inside.
   d. Any one of these could be correct.
   e. None of these are correct.
7. Transmission of the nerve impulse across a synapse is accomplished by the
   a. release of Na^+ at the presynaptic membrane.
   b. release of neurotransmitters at the postsynaptic membrane.
   c. reception of neurotransmitters at the postsynaptic membrane.
   d. Only a and c are correct.
   e. All of these are correct.
8. The autonomic system has two divisions called the
   a. CNS and PNS.
   b. somatic and skeletal systems.
   c. efferent and afferent systems.
   d. sympathetic and parasympathetic divisions.
   e. somatic and skeletal systems.
9. Synaptic vesicles are
   a. at the ends of dendrites and axons.
   b. at the ends of axons only.
   c. along the length of all long fibers.
   d. at the ends of interneurons only.
   e. Both b and d are correct.
10. Which of these is mismatched?
    a. cerebrum—thinking and memory
    b. thalamus—motor and sensory centers
    c. hypothalamus—internal environment regulator
    d. cerebellum—motor coordination
    e. medulla oblongata—fourth ventricle
11. Label this diagram of a reflex arc.

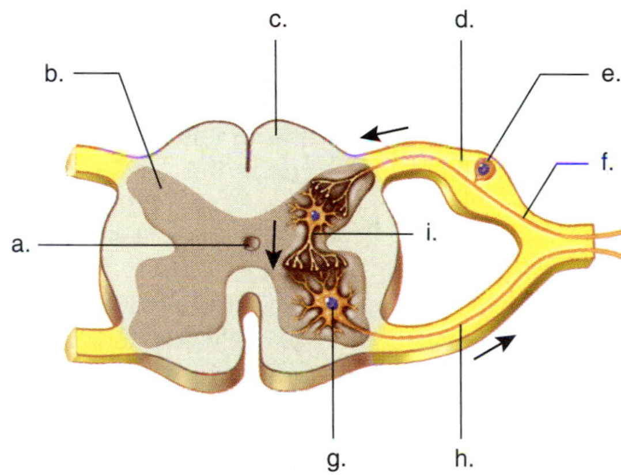

Thinking Scientifically

1. In individuals with panic disorder, the fight-or-flight response is activated by inappropriate stimuli. How might it be possible to directly control this response in order to treat panic disorder? Why is such control often impractical?
2. A man who lost his leg several years ago continues to experience pain as though it were coming from the missing limb. What hypothesis could explain the neurological basis of this pain?

Bioethical Issue

To control their weight, many people in the United States have turned to a new diet pill called Redux, which stimulates the production and availability of the neurotransmitter serotonin in the brain. This very same neurotransmitter is released when we eat a high carbohydrate-rich meal. Redux also prevents serotonin from being reabsorbed at presynaptic membranes. The result is spirits are lifted and appetite is squelched. It's not uncommon for patients to lose 20 or more pounds a week, simply because they have lost all interest in eating!

Unfortunately, Redux has side effects. Some, like fatigue, diarrhea, vivid dreams, and a dry mouth can be tolerated. Others, however, are low in incidence but very serious. Studies suggest that the drug increases the incidence of primary pulmonary hypertension from 1 to 2 per one million to as much as 46 per one million. Primary pulmonary hypertension destroys blood vessels in the lungs and heart, and can lead to death. Also, the drug causes significant, and possibly permanent, brain damage in lab animals. An abundance of serotonin makes the neurons that ordinarily produce the neurotransmitter swell, wither, and then die according to Dr. Mark Molliver, a Johns Hopkins neurologist.

Are you acting recklessly, and therefore unethically, if you take this drug? To decide, a risk-benefit analysis may be appropriate. Severe obesity puts people at risk for hypertension, heart attacks, diabetes, and some cancers. And these illnesses contribute to 300,000 deaths a year in the United States. The slim risk of serious side effects may be worth it for the obese, but the same conclusion does not hold for those who are merely overweight. Even so, most doctors will probably prescribe the drug for anyone who asks for it. Just three months after the introduction of Redux, doctors were writing 85,000 prescriptions a week!

Understanding the Terms

acetylcholine (ACh) 833
acetylcholinesterase (AChE) 833
action potential 830
autonomic system 837
axon 829
brain 827
brain stem 839
cell body 829
central nervous system (CNS) 828
cephalization 827
cerebellum 839
cerebral hemisphere 840
cerebrospinal fluid 838
cerebrum 839
cranial nerve 834
dendrite 829
diencephalon 839
dorsal-root ganglion 834
ganglion 834
gray matter 838
hypothalamus 839
interneuron 829
ladderlike nervous system 827
limbic system 841
medulla oblongata 839
memory 841
meninges 838
midbrain 839
motor neuron 829
myelin sheath 829
nerve 834
nerve fiber 829
nerve net 827
neuroglia 829
neurolemmocyte 829
neuron 829
neurotransmitter 833
norepinephrine (NE) 833
parasympathetic division 837
peripheral nervous system (PNS) 828
pons 839
reflex 835
resting potential 830
reticular formation 839
saltatory conduction 830
sensory neuron 829
somatic system 835
spinal cord 838
spinal nerve 834
sympathetic division 837
synapse 833
synaptic cleft 833
thalamus 839
tracts 838
ventricle 838
white matter 838

Match the terms to these definitions:

a. ______________ Automatic, involuntary response of an organism to a stimulus.
b. ______________ Chemical stored at the ends of axons that is responsible for transmission across a synapse.
c. ______________ Division of the peripheral nervous system that regulates internal organs.
d. ______________ Collection of neuron cell bodies usually outside the central nervous system.
e. ______________ Neurotransmitter active in the somatic system of the peripheral nervous system.

Web Connections

Exploring the Internet

http://www.mhhe.com/biosci/genbio/mader
(click on *Biology 7/e*)

The *Biology 7/e* Online Learning Center provides many resources for studying the material in this chapter including links to the following sites:

Comparative Mammalian Brain Collection. This site compares the anatomy of many different mammalian brains.

http://www.neurophys.wisc.edu/brain/

Neurosciences on the Internet. An index of neuroscience resources: neurology, neurosurgery, psychiatry, psychology, and neurological diseases of humans.

http://www.neuroguide.com/

Shuffle Brain. This unique site features how the brain and mind work.

http://www.indiana.edu/~pietsch/home.html

Virtual Hospital: The Human Brain. Images, diagrams, and information on structures of the brain, cranial vessels, and the spinal cord.

http://www.vh.org/Providers/Textbooks/BrainAnatomy/BrainAnatomy.html

Neurological Disease Information. Human neurological diseases may be accessed via a clickable list.

http://neuroguide.com/neurodis.html

Sense Organs

c h a p t e r

c h a p t e r   c o n c e p t s

47.1 Chemical Senses

- Chemoreceptors are almost universally found in animals for sensing chemical substances in food, liquids, and air. 848
- Human taste buds and olfactory cells are chemoreceptors that respond to chemicals in food and the air respectively. 848

47.2 Sense of Vision

- The eye of arthropods is a compound eye made up of many individual units; the human eye is a camera-type eye with a single lens. 850
- Photoreceptors contain visual pigments that respond to light rays. 853
- In the human eye, the rods work in minimal light and detect motion; the cones require bright light and detect color. 853
- A great deal of integration occurs in the retina of the human eye before nerve impulses are sent to the brain. 854

47.3 Sense of Hearing and Balance

- The inner ear of humans contains mechanoreceptors for a sense of balance and for hearing. 856
- The mechanoreceptors for hearing are hair cells in the cochlea of the inner ear, which respond to pressure waves. 857
- The mechanoreceptors for balance are hair cells in the vestibule and semicircular canals of the inner ear, which respond to the tilt of the head and the movement of the body, respectively. 858–59

Smelling fireweed flowers, *Epilobium angustifolium*

When you smell a flower, molecules in the air bind to olfactory cells, causing them to generate nerve impulses that travel to the brain. Interpretation of these impulses is the function of the brain, which has a specific region for receiving information from each of the sense organs. Impulses arriving at a particular sensory area of the brain can be interpreted in only one way; for example, those arriving at the olfactory area result in smell sensation, and those arriving at the visual area result in sight sensation. The brain integrates data from various sensory receptors in order to perceive, for example, a flower that led to the sensations.

Our sensory receptors form an exchange area with the external environment, just as our digestive tract and lungs are exchange areas. They gather the information that allows the brain to make decisions about finding prey, escaping a predator, and any number of other adaptive behaviors. Therefore, sensory receptors play a significant role in maintaining homeostasis. Learning also helps survival and sensations can rekindle memories of meaningful experiences that occurred years before.

47.1 Chemical Senses

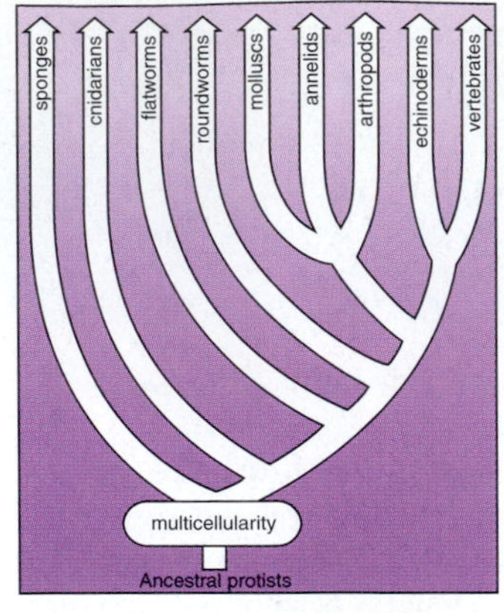

The sensory receptors responsible for taste and smell are termed **chemoreceptors** [Gk. *chemo,* pertaining to chemicals, and L. *receptor,* receiver] because they are sensitive to certain chemical substances in food, liquids, and air. Chemoreception is found almost universally in animals and is therefore believed to be the most primitive sense. Chemoreceptors are present all over the body of planarians, which are flatworms, but they are concentrated on the auricles at the sides of the head. Insects and crustaceans are arthropods. In insects, such as the housefly, chemoreceptors are found largely on the feet—flies taste with their feet instead of their mouth. Insects also detect airborne pheromones, which are chemical messages passed between individuals. In crustaceans (e.g., lobsters and crabs), chemoreceptors are widely distributed over all the appendages and antennae. In amphibians, which are vertebrates, chemoreceptors are located in the nose, in the mouth, and over the entire skin. They are used to locate mates, detect harmful chemicals, and find food. In mammals, the receptors for taste are located in the mouth, and the receptors for smell are located in the nose.

Sense of Taste

In humans, taste buds are located primarily on the tongue (Fig. 47.1). Many lie along the walls of the papillae, the small elevations on the tongue that are visible to the naked eye. Isolated ones are also present on the hard palate, the pharynx, and the epiglottis.

Taste buds open at a taste pore. Taste buds have supporting cells and a number of elongated taste cells that end in microvilli. The microvilli, which project into the taste pore, bear receptor proteins for certain molecules. When molecules bind to receptor proteins, nerve impulses are generated in associated sensory nerve fibers. These nerve impulses go to the brain, including cortical areas which interpret them as tastes.

There are four primary types of tastes (bitter, sour, salty, sweet) and taste buds for each are concentrated on the tongue in particular regions as shown in Figure 47.1*a*. A particular food can stimulate more than one of these types of taste buds. In this way, the response of taste buds can result in a range of sweet, sour, salty, and bitter tastes. The brain appears to survey the overall pattern of incoming sensory impulses and to take a "weighted average" of their taste messages as the perceived taste.

The taste cells of humans are located in taste buds. The microvilli of taste cells have receptor proteins for molecules that cause the brain to distinguish between sweet, sour, salty, and bitter tastes.

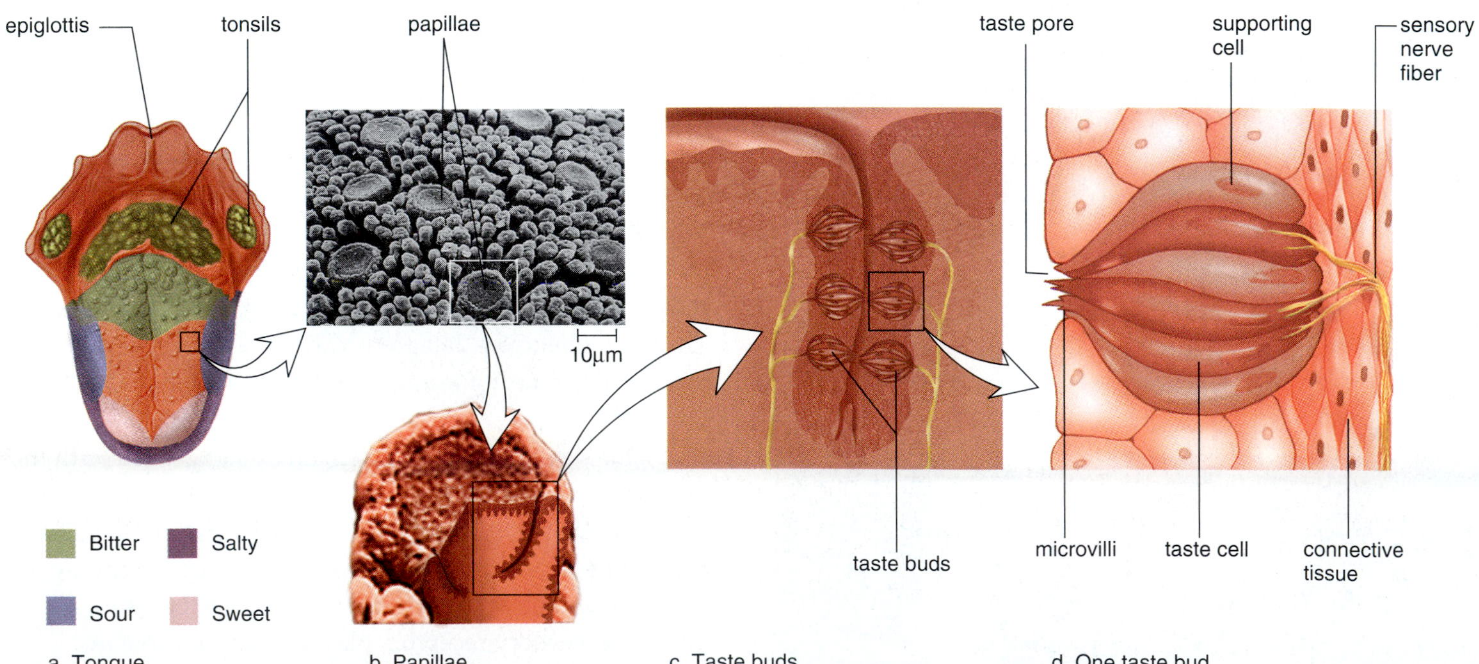

Figure 47.1 Taste buds in humans.
a. Papillae on the tongue containing taste buds that are sensitive to sweet, sour, salt, and bitter are indicated. **b.** Enlargement of papillae. **c.** Taste buds occur along the walls of the papillae. **d.** Taste cells end in microvilli that bear receptor proteins for certain molecules. When molecules bind to the receptor proteins, nerve impulses are generated that go to the brain where the sensation of taste occurs.

Sense of Smell

Our sense of smell is dependent on **olfactory cells** located high in the roof of the nasal cavity (Fig. 47.2). Olfactory cells are modified neurons. Each cell ends in a tuft of about five olfactory cilia, which bear receptor proteins for odor molecules. Each olfactory receptor cell has only one out of 1,000 different types of receptor proteins. Nerve fibers from like olfactory cells lead to the same interneuron in the olfactory bulb, an extension of the brain. An odor contains many odor molecules which activate a characteristic combination of receptor proteins. A rose might stimulate olfactory cells designated by purple and green, while a daffodil might stimulate a different combination. An odor's signature in the olfactory bulb is determined by which corresponding interneurons are stimulated. When the interneurons communicate this information via the olfactory tract to the olfactory areas of the cerebral cortex, we know we have smelled a rose or a daffodil.

Have you ever noticed that an aroma will bring to mind a vivid memory of a person or place? A person's perfume may remind you of someone else, or the smell of boxwood may remind you of your grandfather's farm. Look again at Figure 46.13 and notice that the olfactory bulbs have direct connections with the limbic system and its centers of emotions and memory. One investigator showed that when subjects smelled an orange when viewing a painting, they not only remembered the painting, they had many deep feelings about it.

The sense of taste and the sense of smell supplement each other, creating a combined effect when interpreted by the cerebral cortex. For example, when you have a cold, you think that food has lost its taste, but actually you have lost the ability to sense its smell. This method works in reverse also. When you smell something, some of the molecules move from the nose down into the mouth region and stimulate the taste buds there. Therefore, part of what we refer to as smell may actually be taste.

The olfactory epithelium is the sense organ that contains olfactory cells. The cilia of olfactory cells have receptor proteins for odor molecules that cause the brain to distinguish odors.

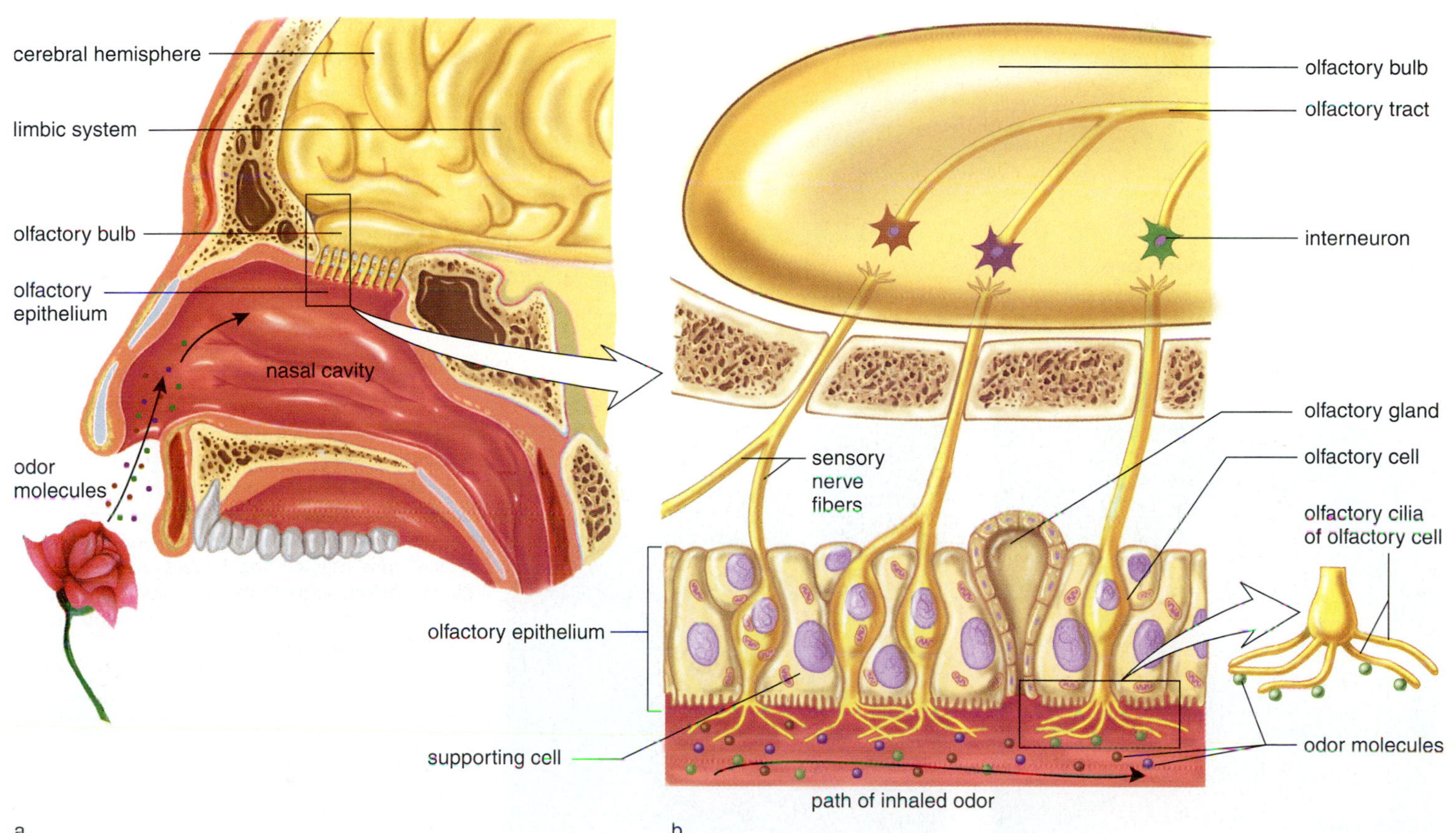

Figure 47.2 **Olfactory cells in humans.**
a. The olfactory epithelium in humans is located high in the nasal cavity. b. Olfactory cells end in cilia that bear receptor proteins for specific odor molecules. The cilia of each olfactory cell can bind to only one type of odor molecule signified here by color. If a rose causes olfactory cells sensitive to purple and green odor molecules to be stimulated, then interneurons designated by purple and green in the olfactory bulb are activated. The primary olfactory area of the cerebral cortex interprets the pattern of interneurons stimulated as the scent of a rose.

47.2 Sense of Vision

Photoreceptors [Gk. *photos*, light, and L. *receptor*, receiver] are sensory receptors for light rays. Some animals lack photoreceptors and depend on senses like smelling and hearing instead; other animals have photoreceptors but live in environments that do not require them. For example, moles live underground and utilize their sense of smell and touch rather than eyesight.

Not all photoreceptors form images. The "eyespots" of planarians allow these animals to determine the direction of light. Image-forming eyes are found among four invertebrate groups: cnidarians, annelids, molluscs, and arthropods. Arthropods have **compound eyes** composed of many independent visual units called ommatidia [Gk. *ommation*, dim. of *omma*, eye], each possessing all the elements needed for light reception (Fig. 47.3). Both the cornea and crystalline cone function as lenses to direct light rays toward the photoreceptors. The photoreceptors generate nerve impulses, which pass to the brain by way of optic nerve fibers. The outer pigment cells absorb stray light rays so that the rays do not pass from one visual unit to the other. The image, which results from all the stimulated visual units, is crude because the small size of compound eyes limits the number of visual units that still might number as many as 28,000. How arthropod brains integrate images from the compound eye to perceive objects is not known.

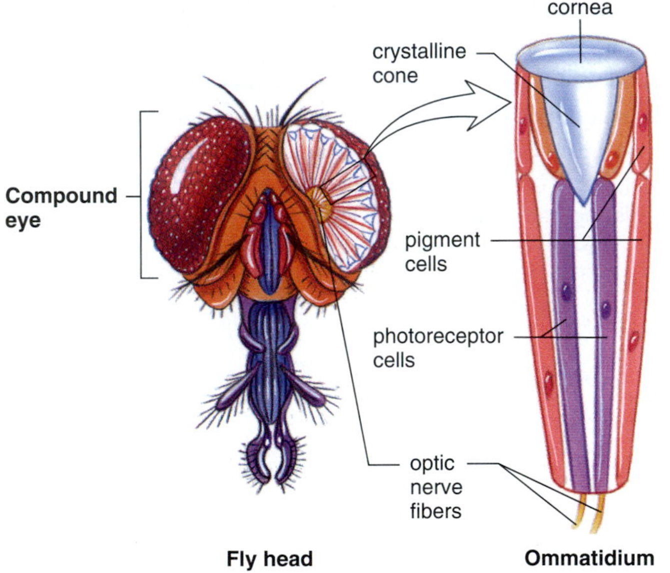

Figure 47.3 Compound eye.
Each visual unit of a compound eye has a cornea and a lens that focus light onto photoreceptors. The photoreceptors generate nerve impulses that are transmitted to the brain, where interpretation produces a mosaic image.

Insects have color vision, but they make use of a slightly shorter range of the electromagnetic spectrum compared to humans. However, they can see the longest of the ultraviolet rays, and this enables them to be especially sensitive to the reproductive parts of flowers, which have particular ultraviolet patterns (Fig. 47.4). Some fishes, reptiles, and most birds are believed to have color vision, but among mammals, only humans and other primates have color vision. It would seem, then, that this trait was adaptive for a diurnal habit (active during the day), which accounts for its retention in a few mammals.

Vertebrates and certain molluscs, like the squid and the octopus, have a **camera-type eye.** Since molluscs and vertebrates are not closely related, this similarity is an example of convergent evolution. A single lens focuses an image of the visual field on photoreceptors, which are closely packed together. In vertebrates the lens changes shape to aid focusing, but in molluscs the lens moves back and forth. All of the photoreceptors taken together can be compared to a piece of film in a camera. The human eye is more complex than a camera, however, as we shall see.

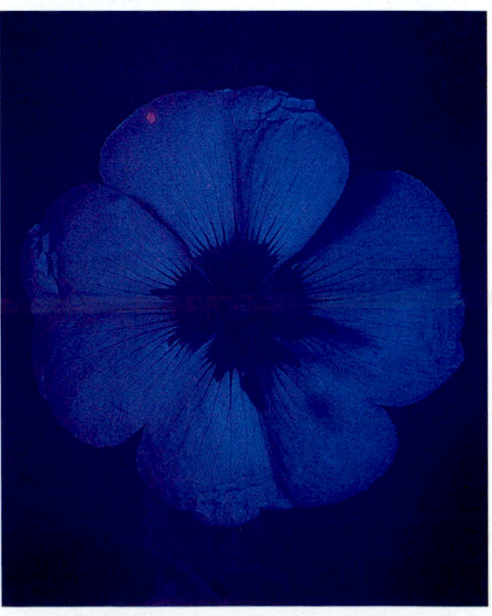

Figure 47.4 Nectar guides.
Evening primrose, *Oenothera*, as seen by humans *(left)* and insects *(right)*. Humans see no markings, but insects see distinct blotches because their eyes respond to ultraviolet rays. These types of markings, known as nectar guides, often highlight the reproductive parts of flowers where insects feed on nectar and pick up pollen at the same time.

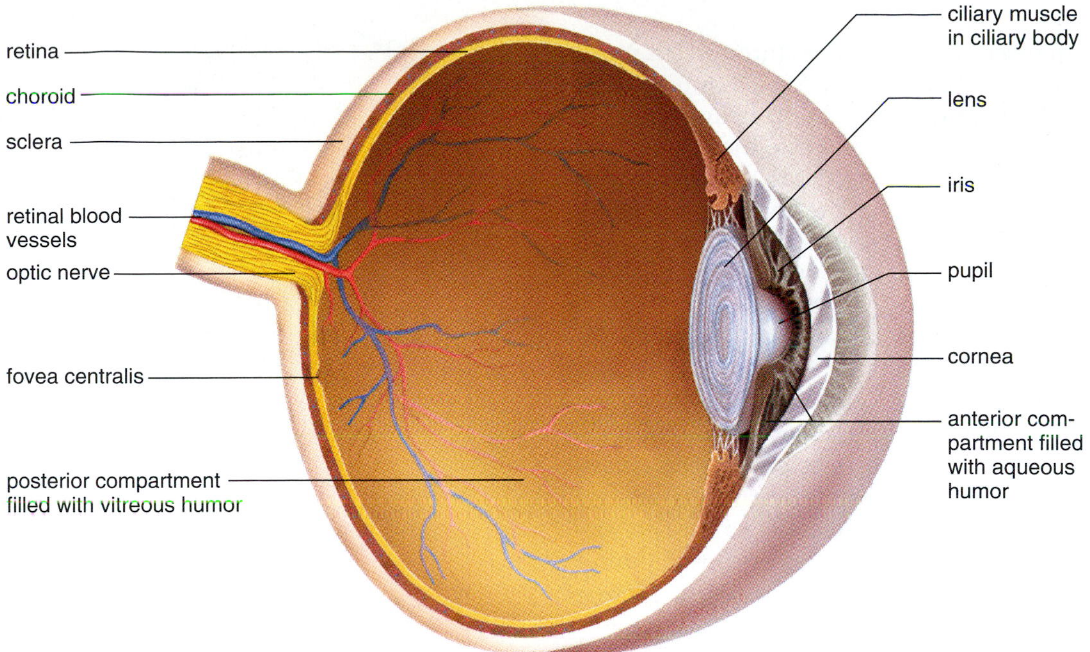

Figure 47.5 **Anatomy of the human eye.**
Notice that the sclera, the outer layer of the eye, becomes the cornea and that the choroid, the middle layer, is continuous with the ciliary body and the iris. The retina, the inner layer, contains the photoreceptors for vision. The fovea centralis is the region where vision is most acute.

The Human Eye

The most important parts of the human eye and their functions are listed in Table 47.1. The human eye, which is an elongated sphere about 2.5 cm in diameter, has three layers, or coats: the sclera, the choroid, and the retina (Fig. 47.5). The outer layer, the **sclera** [Gk. *skleros,* hard], is an opaque, white, fibrous layer that covers most of the eye; in front of the eye the sclera becomes the transparent cornea, the window of the eye. The middle, thin, dark-brown layer, the **choroid** [Gk. *chorion,* membrane], contains many blood vessels and pigment that absorb stray light rays. Toward the front of the eye, the choroid thickens and forms the ring-shaped ciliary body and a thin, circular, muscular diaphragm, the iris. The iris regulates the size of an opening called the pupil. The lens, which is attached to the ciliary body by ligaments, divides the cavity of the eye into two portions. A basic, watery solution called aqueous humor fills the anterior compartment between the cornea and the lens. A viscous, gelatinous material, the vitreous humor, fills the large posterior compartment behind the lens.

The inner layer of the eye, the **retina** [L. *rete,* net], is located in the posterior compartment. The retina contains photoreceptors called rod cells and cone cells. The rods are very sensitive to light but they do not see color; therefore, at night or in a darkened room we see only shades of gray. The cones, which require bright light, are sensitive to different wavelengths of light and, therefore, we have the ability to distinguish colors. The retina has a very special region called the fovea centralis, where cone cells are densely packed. Light is normally focused on the fovea when we look directly at an object. This is helpful because vision is most acute in the fovea centralis. Sensory fibers form the optic nerve which takes nerve impulses to the brain.

The human eye has three layers: the outer sclera, the middle choroid, and the retina. Only the retina contains receptors for sight.

Table 47.1

Function of Parts of the Eye

Part	Function
Lens	Refracts and focuses light rays
Iris	Regulates light entrance
Pupil	Admits light
Choroid	Blood supply and absorbs stray light
Sclera	Protects and supports eyeball
Cornea	Refracts light rays
Humors	Refract light rays
Ciliary body	Holds lens in place, accommodation
Retina	Contains receptors for sight
Rod cells	Make black-and-white vision possible
Cone cells	Make color vision possible
Optic nerve	Transmits impulses to brain
Fovea centralis	Makes acute vision possible

Focusing of the Eye

When we look at an object, light rays pass through the **pupil** and are focused on the retina (Fig. 47.6*a*). The image produced is much smaller than the object because light rays are bent (refracted) when they are brought into focus. Focusing starts at the **cornea** and continues as the rays pass through the **lens** and the humors. Notice that the image on the retina is inverted (it is upside down) and reversed from left to right.

The lens provides additional focusing power as visual accommodation occurs for close vision. The shape of the lens is controlled by the **ciliary muscle** within the ciliary body. When we view a distant object, the ciliary muscle is relaxed, causing the suspensory ligaments attached to the ciliary body to be taut; therefore, the lens remains relatively flat (Fig. 47.6*b*). When we view a near object, the ciliary muscle contracts, releasing the tension on the suspensory ligaments, and the lens rounds up due to its natural elasticity (Fig. 47.6*c*). Because close work requires contraction of the ciliary muscle, it very often causes muscle fatigue known as eyestrain.

The lens, assisted by the cornea and the humors, focuses images on the retina.

With normal aging, the lens loses its ability to accommodate for near objects (Fig. 47.6*c*); therefore, persons frequently need reading glasses once they reach middle age.

Aging, or possibly exposure to the sun, also makes the lens subject to cataracts; the lens can become opaque and therefore incapable of transmitting light rays. Currently surgery is the only viable treatment for cataracts. First, a surgeon opens the eye near the rim of the cornea. The enzyme zonulysin may be used to digest away the ligaments holding the lens in place. Most surgeons then use a cryoprobe, which freezes the lens for easy removal. An intraocular lens attached to the iris can then be implanted so that the patient does not need to wear thick glasses or contact lenses.

Distance Vision Persons who can see a near object but have trouble seeing what is designated as a size 20 letter 20 feet away are said to be nearsighted. These individuals often have an elongated eyeball, and when they attempt to look at a distant object, the image is brought to focus in front of the retina. Usually these people must wear concave lenses, which diverge the light rays so that the image can be focused on the retina. There is a new treatment for nearsightedness called radial keratotomy, or radial K. From four to eight cuts are made in the cornea so that they radiate out from the center like spokes on a wheel. When the cuts heal, the cornea is flattened. Although many patients are satisfied with the result, others complain of glare and varying visual acuity.

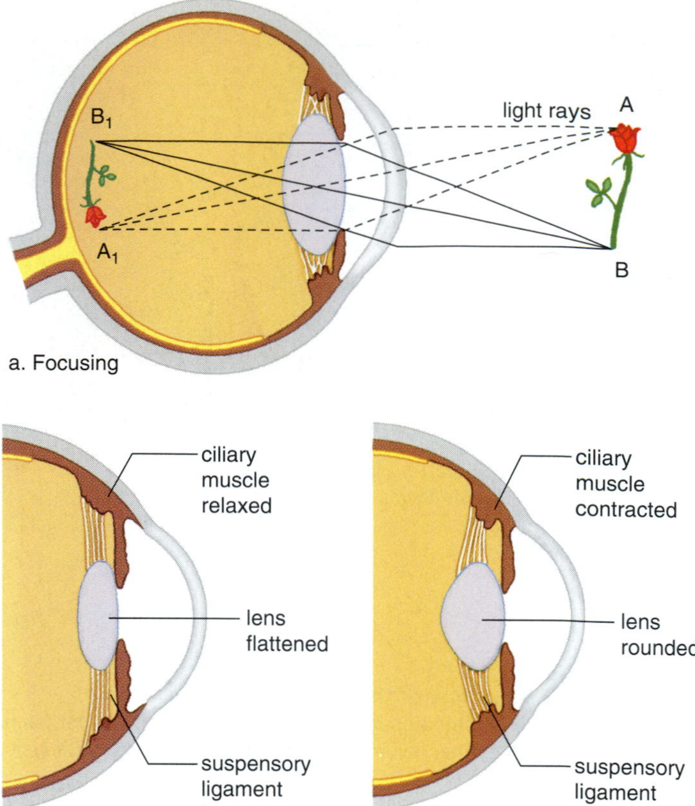

b. Focusing on distant object c. Focusing on near object

Figure 47.6 Focusing of the human eye.
a. Light rays from each point on an object are bent by the cornea and the lens in such a way that an inverted and reversed image of the object forms on the retina. **b.** When focusing on a distant object, the lens is flat because the ciliary muscle is relaxed and the suspensory ligament is taut. **c.** When focusing on a near object, the lens accommodates: it becomes rounded because the ciliary muscle contracts, causing the suspensory ligament to relax.

Persons who can easily see the optometrist's chart but cannot easily see near objects are farsighted. They often have a shortened eyeball, and when they try to see near objects, the image is focused behind the retina. These persons must wear a convex lens to increase the bending of light rays so that the image can be focused on the retina. When the cornea or lens is uneven, the image is fuzzy. This condition, called astigmatism, can be corrected by an unevenly ground lens to compensate for the uneven cornea.

The inability of the lens to accommodate as we age requires corrective lenses for close vision. The shape of the eyeball determines the need for corrective lenses for distance vision.

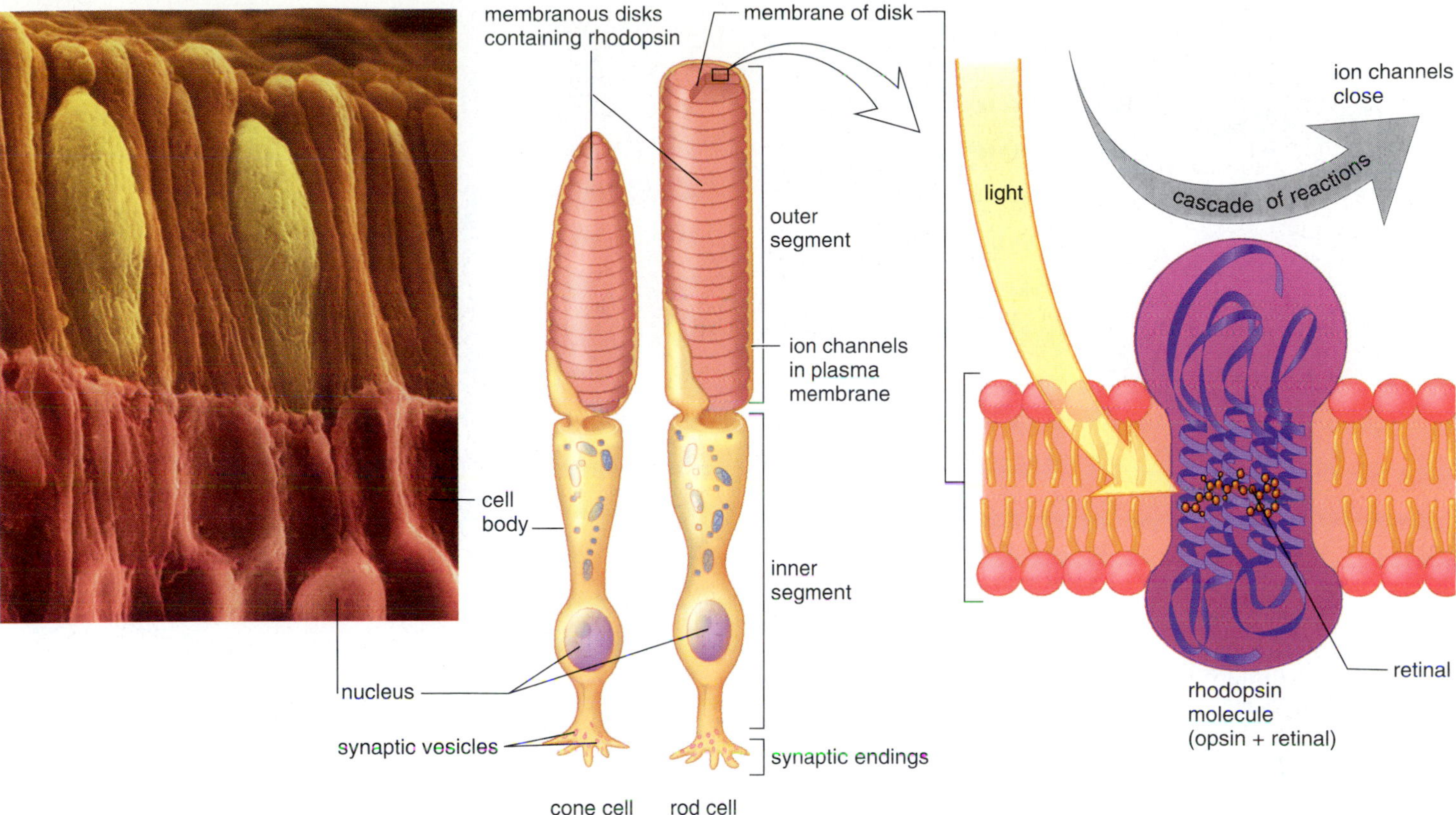

Figure 47.7 Photoreceptors in the human eye.
The outer segment of rods and cones has stacks of membranous disks, which contain visual pigments. In rods, the membrane of each disk contains rhodopsin, a complex molecule containing the protein opsin and the pigment retinal. When rhodopsin absorbs light energy, it splits releasing opsin, which sets in motion a cascade of reactions that ends when ion channels in the plasma membrane close. Thereafter, nerve impulses traveling in the optic nerve to the brain result in vision.

Photoreceptors of the Eye

Vision begins once light has been focused on the photoreceptors in the retina. Figure 47.7 illustrates the structure of our photoreceptors called **rod cells** and **cone cells.** Both rods and cones have an outer segment joined to an inner segment. The outer segment contains many membranous disks. And many pigment molecules are embedded in the membrane of these disks. Synaptic vesicles are located at the synaptic endings of the inner segment.

The visual pigment in rods is a deep purple pigment called rhodopsin. **Rhodopsin** is a complex molecule made up of the protein opsin and a light-absorbing molecule called retinal, which is a derivative of vitamin A. When a rod absorbs light, rhodopsin splits into opsin and retinal, leading to a cascade of reactions and the closure of ion channels in the rod cell's plasma membrane. This stops the release of inhibitory transmitter molecules from the rod's synaptic vesicles and starts the signals that result in nerve impulses going to the brain. Rods are very sensitive to light and therefore are suited to night vision. (Since carrots are rich in vitamin A, it is true that eating carrots can improve your night vision.) Rod cells are plentiful throughout the entire retina; therefore, they also provide us with peripheral vision and a perception of motion.

The cones, on the other hand, are located primarily in the **fovea centralis** and are activated by bright light. They allow us to detect the fine detail and the color of an object. Color vision depends on three different kinds of cones, which contain a blue, green, or red pigment. Each pigment is made up of retinal and opsin, but there is a slight difference in the opsin structure of each, which accounts for their individual absorption patterns. Various combinations of cones are believed to be stimulated by in-between shades of color. Normally, an individual is able to distinguish about 17,000 different hues.

The photoreceptors for sight are the rods and the cones. The rods permit vision in dim light at night, and the cones permit vision in bright light needed for color vision.

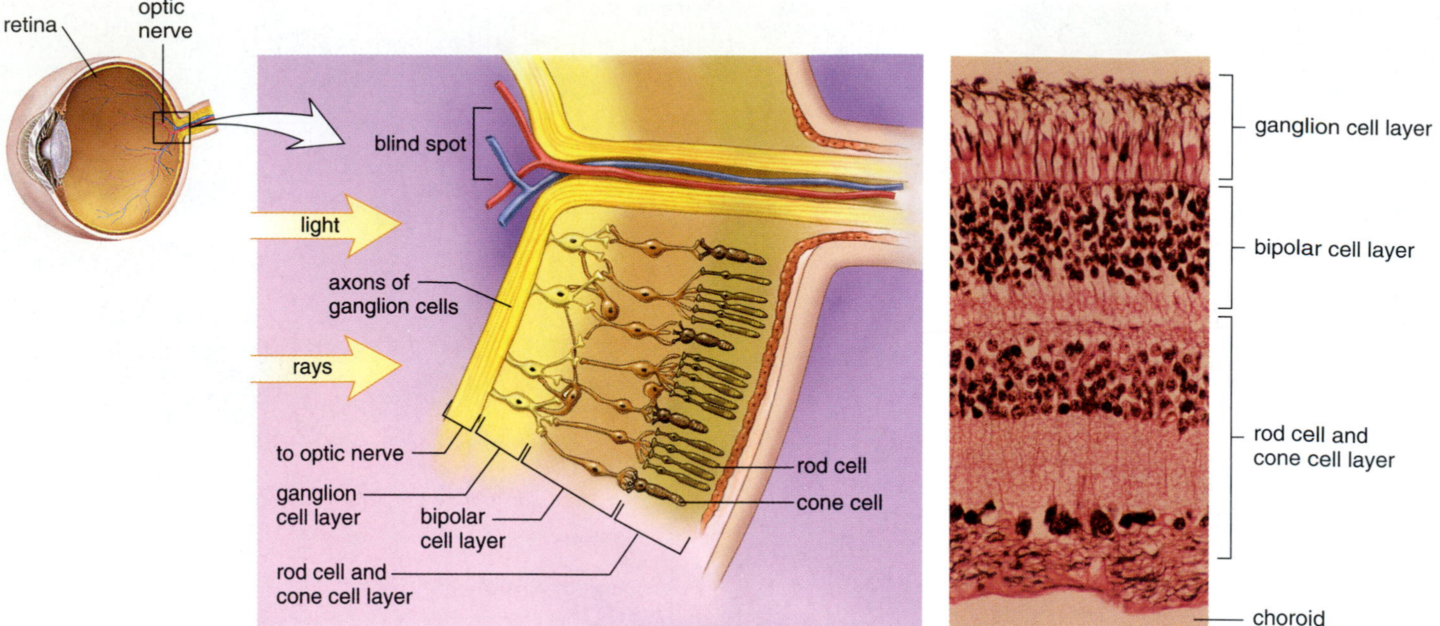

Figure 47.8 **Integration of visual signals.**
Rods and cones, located at the back of the retina, synapse with bipolar cells which synapse with ganglion cells. Integration of signals occurs at these synapses; therefore much processing occurs in bipolar and ganglion cells. Further, notice that many rod cells share a bipolar cell, but each cone cell synapses with only one ganglion cell. Cone cells, therefore, distinguish more detail than do rod cells.

Integration of Visual Signals in the Retina

The retina has three layers of neurons (Fig. 47.8). The layer closest to the choroid contains the rod cells and cone cells; the middle layer contains bipolar cells; and the innermost layer contains ganglion cells, whose sensory fibers become the optic nerve. Only the rod cells and the cone cells are sensitive to light, and therefore light must penetrate to the back of the retina before they are stimulated.

The rod cells and the cone cells synapse with the bipolar cells, which in turn synapse with ganglion cells that initiate nerve impulses. Notice in Figure 47.8 that there are many more rod cells and cone cells than ganglion cells. In fact, the retina has as many as 150 million rod cells and 6 billion cone cells but only one million ganglion cells. The sensitivity of cones versus rods is mirrored by how directly they connect to ganglion cells. As many as 100 rods may synapse with the same ganglion cell. No wonder that stimulation of rods results in vision that is only blurred and indistinct. In contrast, each cone, particularly in the fovea centralis, synapses with only one ganglion cell. This accounts for the reason cones provide us with a sharper, more detailed image of an object.

As signals pass to bipolar cells and ganglion cells, integration occurs. Each ganglion cell receives signals from rod cells covering about one square millimeter of retina (about the size of a thumbtack hole). This region is called the ganglion cell's receptive field. Some time ago, scientists discovered that a ganglion cell is stimulated only when light hits the center of its receptive field, and is inhibited when light hits the area of the receptive field surrounding the center. If all the rod cells in the receptive field are stimulated, the ganglion cell responds in a neutral way—it reacts only weakly or perhaps not at all. This supports the hypothesis that considerable processing occurs in the retina before nerve impulses are sent to the brain. Integration also occurs in the visual areas of the cerebral cortex.

Synaptic integration and processing begins in the retina before nerve impulses are sent to the brain.

Blind Spot

Figure 47.8 provides an opportunity to point out that there are no rods and cones where the optic nerve exits the retina. Therefore, no vision is possible in this area. You can prove this to yourself by putting a dot to the right of center on a piece of paper. Use the right hand to move the paper slowly toward the right eye while you look straight ahead. The dot will disappear at one point—this is your **"blind spot."**

Health Focus

Protecting Vision and Hearing

Age can be accompanied by a serious loss of vision and hearing. The time to start preventive measures for such problems, however, is when we are younger.

Preventing a Loss of Vision

The eye is subject to both injuries and disorders. Although flying objects sometimes penetrate the cornea and damage the iris, lens, or retina, careless use of contact lenses is the most common cause of injuries to the eye. Injuries cause only 4% of all cases of blindness; the most frequent causes are retinal disorders, glaucoma, and cataracts, in that order. Retinal disorders are varied. In diabetic retinopathy, which blinds many people between the ages of 20 and 74, capillaries to the retina burst and blood spills into the vitreous fluid. Careful regulation of blood glucose levels in these patients may be protective. In macular degeneration, the cones are destroyed because thickened choroid vessels no longer function as they should. Glaucoma occurs when the drainage system of the eyes fails, so that fluid builds up and destroys nerve fibers responsible for peripheral vision. Eye doctors always check for glaucoma, but it is advisable to be aware of the disorder in case it comes on quickly. Those who have experienced acute glaucoma report that the eyeball feels as heavy as a stone. In cataracts, cloudy spots on the lens of the eye eventually pervade the whole lens. The milky yellow-white lens scatters incoming light and blocks vision.

There are preventive measures that we can take to reduce the chance of defective vision as we age. Accumulating evidence suggests that both macular degeneration and cataracts, which tend to occur in the elderly, are caused by long-term exposure to the ultraviolet rays of the sun. It is recommended, therefore, that everyone, especially those who live in sunny climates or work outdoors, wear glass and not plastic sunglasses to absorb ultraviolet light. Large lenses worn close to the eyes offer further protection. Special purpose lenses that block at least 99% of UV-B and 60% UV-A, and 20–97% of visible light are good for bright sun combined with sand, snow, or water. Health-care providers have found an increased incidence of cataracts in heavy cigarette smokers. In men, smoking 20 cigarettes or more a day, and in women, smoking more than 35 cigarettes a day doubles the risk of cataracts. It is possible that smoking reduces the delivery of blood and therefore nutrients to the lens.

Preventing a Loss of Hearing

Especially when we are young, the middle ear is subject to infections that can lead to hearing impairments if they are not treated promptly by a physician. The mobility of ossicles decreases with age, and in otosclerosis, new filamentous bone grows over the stirrup impeding its movement. Surgical treatment is the only remedy for this type of conduction deafness. However, age-associated nerve deafness due to stereocilia damage from exposure to loud noises is preventable. Hospitals are now aware that even the ears of the newborn need to be protected from noise and are taking steps to make sure neonatal intensive care units and nurseries are as quiet as possible.

In today's society, exposure to excessive noise is a possibility. Noise is measured in decibels, and any noise above a level of 80 decibels could result in damage to the hair cells of the spiral organ (organ of Corti). Eventually, the stereocilia and then the hair cells disappear completely (Fig. 47A). If listening to city traffic for extended periods can damage hearing, it stands to reason that frequent attendance at rock concerts, constantly playing a stereo loudly, or using earphones at high volume is also damaging to hearing. The first hint of danger could be temporary hearing loss, a "full" feeling in the ears, muffled hearing, or tinnitus (e.g., ringing in the ears). If you have any of these symptoms, modify your listening habits immediately to prevent further damage. If exposure to noise is unavoidable, specially designed noise-reduction earmuffs are available, and it is also possible to purchase earplugs made from a compressible, spongelike material at the drugstore or sporting-goods store. These earplugs are not the same as those worn for swimming, and they should not be used interchangeably.

Aside from loud music, noisy indoor or outdoor equipment, such as a rug-cleaning machine or a chain saw, is also troublesome. Even motorcycles and recreational vehicles such as snowmobiles and motocross bikes can contribute to a gradual loss of hearing. Exposure to intense sounds of short duration, such as a burst of gunfire, can result in an immediate hearing loss. Hunters may have a significant hearing reduction in the ear opposite to the shoulder where the gun is carried. The butt of the rifle offers some protection to the ear nearest the gun when it is shot.

Finally, people need to be aware that some medicines are ototoxic. Anticancer drugs, most notably cisplatin, and certain antibiotics (e.g., streptomycin, kanamycin, gentamicin) make ears especially susceptible to a hearing loss. Anyone taking such medications needs to be especially careful to protect the ears from any loud noises.

Figure 47A Hearing loss.
Damaged hair cells in the spiral organ of a guinea pig. This damage occurred after 24-hour exposure to a noise level typical of rock concerts.

47.3 Sense of Hearing and Balance

The ear has two sensory functions: hearing and balance (equilibrium). The sensory receptors for both of these are located in the inner ear, and each consists of hair cells with stereocilia that respond to mechanical stimulation. Therefore they are **mechanoreceptors.**

Anatomy of the Ear

Figure 47.9 shows that the ear has three divisions: outer, middle, and inner. The outer ear consists of the pinna (external flap) and the auditory canal. The opening of the auditory canal is lined with fine hairs and sweat glands. Modified sweat glands are located in the upper wall of the canal; they secrete earwax, a substance that helps to guard the ear against the entrance of foreign materials, such as air pollutants.

The middle ear begins at the **tympanic membrane** (eardrum) and ends at a bony wall containing two small openings covered by membranes. These openings are called the oval window and the round window. Three small bones are found between the tympanic membrane and the oval window. Collectively called the **ossicles,** individually they are the malleus (hammer), the incus (anvil), and the stapes (stirrup) because their shapes resemble these objects. The malleus adheres to the tympanic membrane, and the stapes touches the oval window. An auditory (eustachian) tube, which extends from each middle ear to the nasopharynx, permits equalization of air pressure. Chewing gum, yawning, and swallowing in elevators and airplanes help to move air through the auditory tubes upon ascent and descent. As this occurs we often hear the ears "pop."

Whereas the outer ear and the middle ear contain air, the inner ear is filled with fluid. The inner ear, anatomically speaking, has three areas: the semicircular canals and the vestibule are concerned with equilibrium; the **cochlea** is concerned with hearing. The cochlea resembles the shell of a snail because it spirals.

Process of Hearing

The process of hearing begins when sound waves enter the auditory canal (Fig. 47.10). Just as ripples travel across the surface of a pond, sound waves travel by the successive vibrations of molecules. Ordinarily, sound waves do not carry much energy, but when a large number of waves strike the tympanic membrane, it moves back and forth (vibrates) ever so slightly. The malleus then takes the pressure from the inner surface of the tympanic membrane and passes it by means of the incus to the stapes in such a way that the pressure is multiplied about 20 times as it moves from the tympanic membrane to the stapes. The stapes strikes the membrane of the oval window, causing it to vibrate, and in this way, the pressure is passed to the fluid within the cochlea.

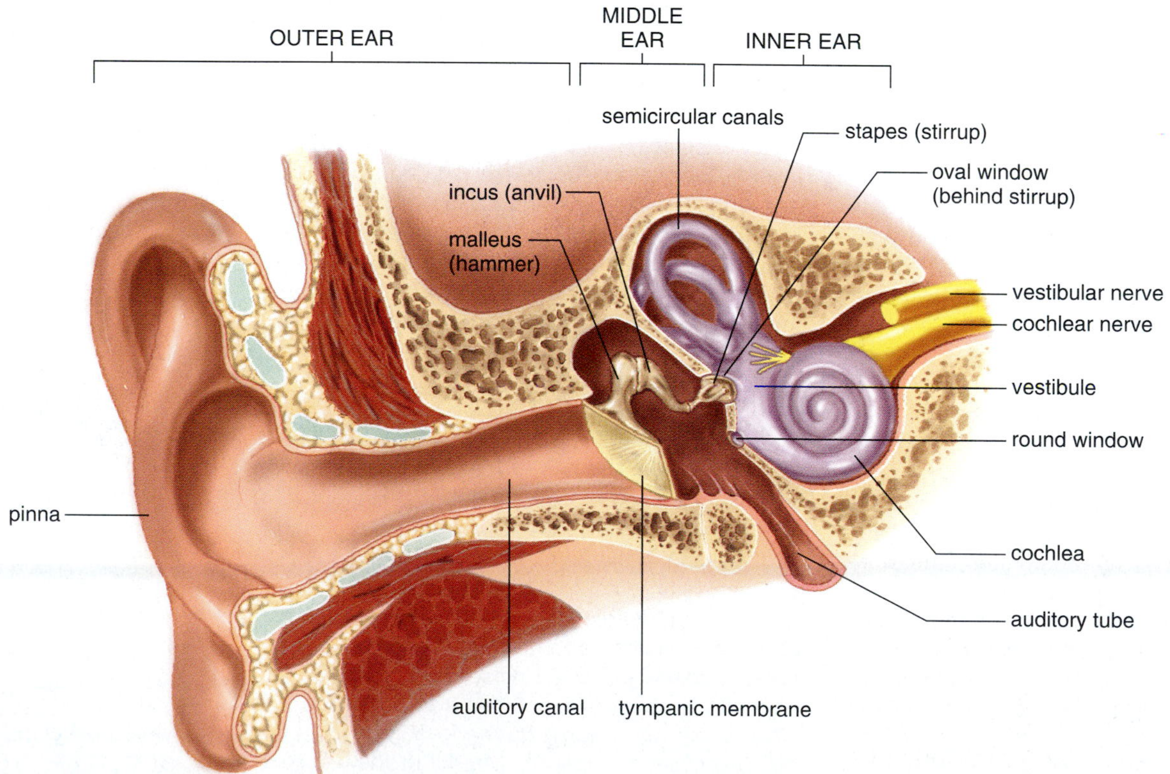

Figure 47.9 Anatomy of the human ear.
In the middle ear, the malleus (hammer), the incus (anvil), and the stapes (stirrup) amplify sound waves. The inner ear contains the sensory receptors for balance located in the semicircular canals and the vestibule, and the sensory receptors for hearing are located in the cochlea.

If the cochlea is unwound and examined in cross section (Fig. 47.10) you can see that it has three canals: the vestibular canal, the cochlear canal, and the tympanic canal. The vestibular canal connects with the tympanic canal, which leads to the round window membrane. Along the length of the basilar membrane, which forms the lower wall of the cochlear canal, are little hair cells whose stereocilia are embedded within a gelatinous material called the tectorial membrane. The hair cells of the cochlear canal, called the **spiral organ** (organ of Corti), synapse with nerve fibers of the cochlear (auditory) nerve.

When the stapes strikes the membrane of the oval window, pressure waves move from the vestibular canal to the tympanic canal and across the basilar membrane, and the round window bulges. As the basilar membrane moves up and down, the stereocilia of the hair cells embedded in the tectorial membrane bend. Then, nerve impulses begin in the cochlear nerve and travel to the brain stem. When they reach the auditory areas of the cerebral cortex they are interpreted as a sound.

Each part of the spiral organ is sensitive to different wave frequencies, or pitch. Near the tip, the spiral organ responds to low pitches, such as a tuba, and near the base, it responds to higher pitches, such as a bell or a whistle. The nerve fibers from each region along the length of the spiral organ lead to slightly different areas in the brain. The pitch sensation we experience depends upon which region of the basilar membrane vibrates and which area of the brain is stimulated.

Volume is a function of the amplitude of sound waves. Loud noises cause the fluid of the cochlea to vibrate to a greater degree, and this, in turn, causes the basilar membrane to move up and down to a greater extent. The resulting increased stimulation is interpreted by the brain as volume. It is believed that the tone of a sound is an interpretation of the brain based on the distribution of hair cells stimulated.

The mechanoreceptors for sound are hair cells on the basilar membrane (the spiral organ). When the basilar membrane vibrates, the stereocilia of the hair cells bend, and nerve impulses are transmitted to the brain.

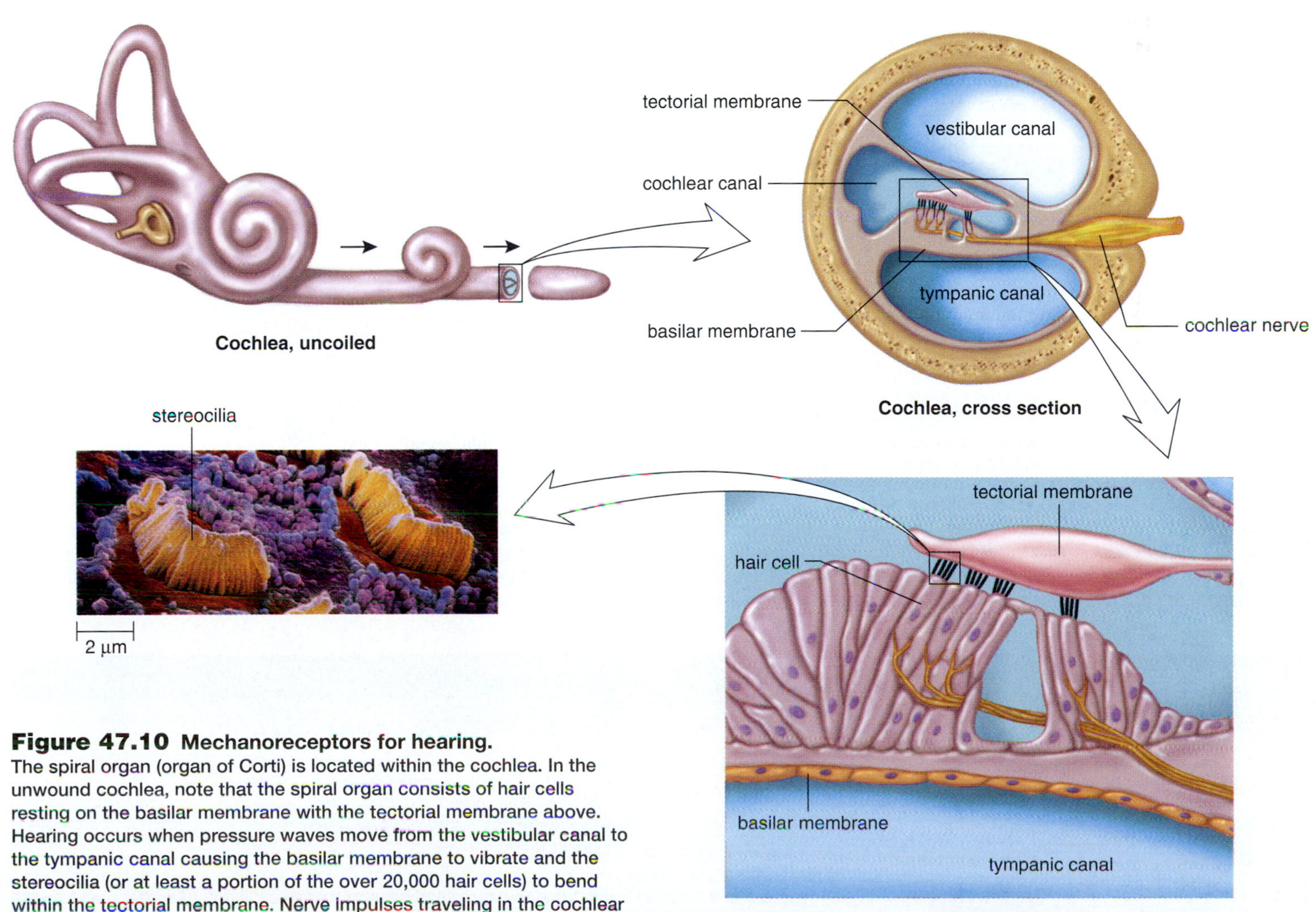

Figure 47.10 Mechanoreceptors for hearing.
The spiral organ (organ of Corti) is located within the cochlea. In the unwound cochlea, note that the spiral organ consists of hair cells resting on the basilar membrane with the tectorial membrane above. Hearing occurs when pressure waves move from the vestibular canal to the tympanic canal causing the basilar membrane to vibrate and the stereocilia (or at least a portion of the over 20,000 hair cells) to bend within the tectorial membrane. Nerve impulses traveling in the cochlear nerve to the brain result in hearing.

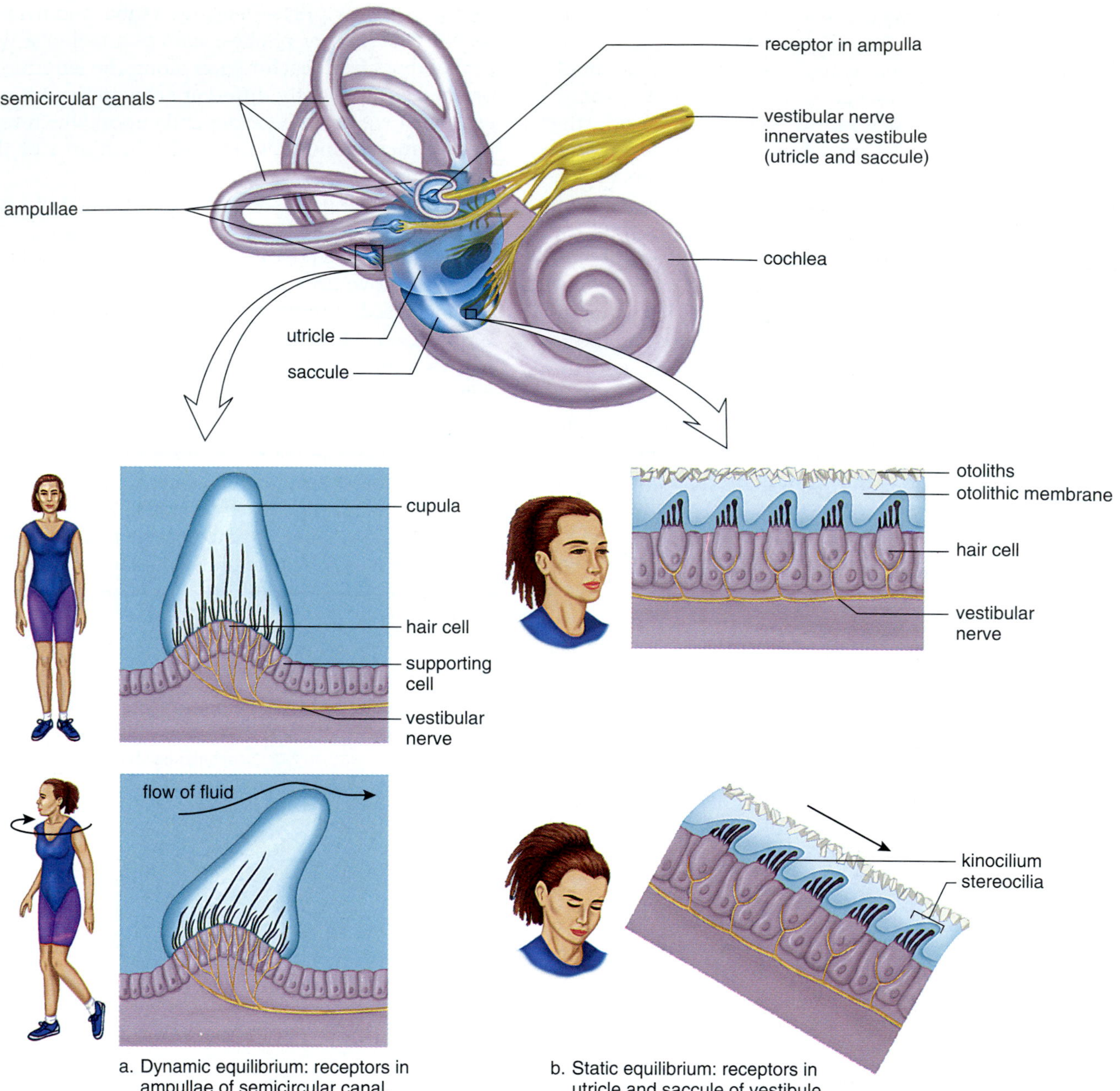

a. Dynamic equilibrium: receptors in ampullae of semicircular canal.

b. Static equilibrium: receptors in utricle and saccule of vestibule.

Figure 47.11 Mechanoreceptors for balance.
a. Dynamic equilibrium. The ampullae of the semicircular canals contain hair cells with stereocilia embedded in a gelatinous cupula. When the head rotates, the cupula is displaced, bending the stereocilia. Thereafter, nerve impulses travel in the vestibular nerve to the brain. **b.** Static equilibrium. The utricle and the saccule contain hair cells with stereocilia embedded in a gelatinous otolithic membrane. When the head bends, otoliths are displaced, causing the membrane to sag and the stereocilia to bend. The rapidity of nerve impulses in the vestibular nerve tells the brain how much the head has moved.

Table 47.2

Function of Parts of the Ear

	Part of Ear and Function
Outer Ear	Pinna collects sound waves; auditory canal filters air.
Middle Ear	Tympanic membrane and ossicles amplify sound waves; auditory tube equalizes air pressure.
Inner Ear	Semicircular canals function in balance (dynamic equilibrium). Utricle and saccule function in balance (static equilibrium). Cochlea transmits pressure waves that cause the spiral organ to generate nerve impulses, resulting in hearing.

Sense of Balance

The sense of balance has been subdivided into two senses: dynamic equilibrium, involving angular and/or rotational movement of the head, and static equilibrium, involving movement of the head in one plane, either vertical or horizontal.

Dynamic Equilibrium

Dynamic equilibrium utilizes the **semicircular canals** which are arranged so that there is one in each dimension of space (Fig. 47.11*a*). The base of each of the three canals, called the ampulla, is slightly enlarged. Little hair cells whose stereocilia are embedded within a gelatinous material called a cupula are found within the ampullae. Because there are three semicircular canals, each ampulla responds to head rotation in a different plane of space. As fluid within a semicircular canal flows over and displaces a cupula, the stereocilia of the hair cells bend, and the pattern of impulses carried by the vestibular nerve to the brain changes. Continuous movement of fluid in the semicircular canals causes one form of motion sickness.

Vertigo is dizziness and a sensation of rotation. It is possible to simulate a feeling of vertigo by spinning rapidly and stopping suddenly. When the eyes are rapidly jerked back to a midline position, the person feels like the room is spinning. This shows that the eyes are also involved in our sense of balance.

Static Equilibrium

Static equilibrium depends on the **utricle** and **saccule,** two membranous sacs located in the vestibule (Fig. 47.11*b*). Both of these sacs contain little hair cells, whose stereocilia are embedded within a gelatinous material called an otolithic membrane. Calcium carbonate ($CaCO_3$) granules, or **otoliths,** rest on this membrane. The utricle is especially sensitive to horizontal movements and the bending of the head, while the saccule responds best to vertical (up-down) movements. When the body is still, the otoliths in the utricle and the saccule rest on the otolithic membrane above the hair cells. When the head bends or the body moves in the horizontal and vertical planes, the otoliths are displaced and the otolithic membrane sags, bending the stereocilia of the hair cells beneath. If the stereocilia move toward the kinocilium, the largest stereocilium, nerve impulses in the vestibular nerve increase. If the stereocilia move away from the kinocilium, nerve impulses in the vestibular nerve decrease. These data tell the brain the direction of the movement of the head.

Movement of a cupula within the semicircular canals contributes to the sense of dynamic equilibrium. Movement of the otolithic membrane within the utricle and the saccule accounts for static equilibrium.

Invertebrate Sensory Receptors

The **lateral line** system of fishes and amphibians, which allows them to detect water currents and pressure waves from nearby objects, utilizes hair cells in the same manner as the sensory receptors in the human ear. In bony fishes, the sensory receptors are located within a canal that has openings to the outside. A lateral line receptor is a collection of hair cells with cilia embedded in a gelatinous cupula.

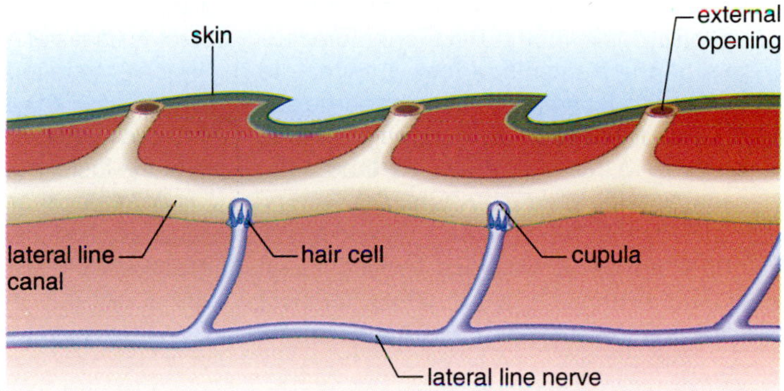

When the cupula bends due to pressure waves, the hair cells initiate nerve impulses.

Static equilibrium organs, called **statocysts,** are found in cnidarians, molluscs, and crustaceans which are arthropods. These organs give information only about the position of the head; they are not involved in the sensation of movement (Fig. 47.12).

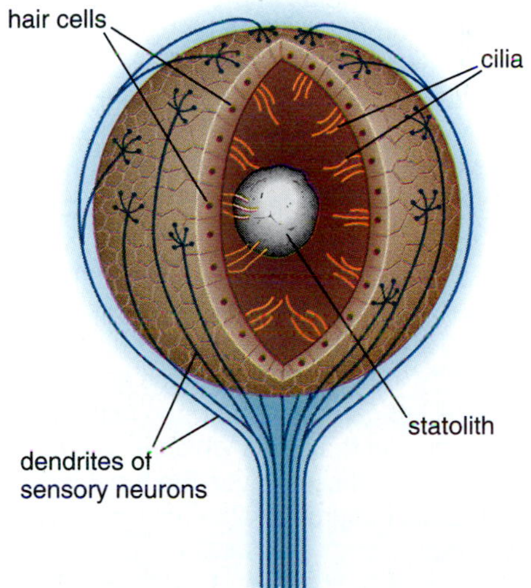

Figure 47.12 Statocysts.
In molluscs and crustaceans, a small particle, the statolith, moves in response to a change in the animal's position. When the statolith stops moving, it stimulates the cilia of the closest hair cells. These cilia generate impulses, indicating the position of the head.

Connecting Concepts

An animal's information exchange with the internal and external environment is dependent upon just a few types of sensory receptors. We have examined chemoreceptors, such as taste cells and olfactory cells, photoreceptors, such as eyes, and mechanoreceptors, such as the hair cells for balance and hearing. The senses are not equally developed in all animals. Male moths have chemoreceptors on the filaments of their antennae to detect minute amounts of an airborne sex attractant released by a female. This is certainly a more efficient method than searching for a mate by sight.

Birds that live in forested areas signal that a territory is occupied by singing—it is difficult to see a bird in a tree, as most birders know. On the other hand, hawks have such a keen sense of sight that they are able to locate a small mouse far below them. Insectivorous bats have an unusual adaptation for finding prey in the dark. They send out a series of sound pulses and listen for the echoes that come back. The time it takes for an echo to return indicates the location of an insect. A unique adaptation is found among the so-called electric fishes of Africa and Australia. They have electroreceptors that can detect disturbances in an electrical current they emit into the water. These disturbances indicate the location of obstacles and prey.

Animals that migrate use various senses to find their way. Salmon hatch in a freshwater stream, but drift to the ocean as larvae. By the end of the third or fourth year, they migrate back to where they were hatched. Like other migrating animals, they apparently can use the sun as a compass to find their way back to the vicinity of their home river, but then salmon switch to a sense of smell to find the exact location of their hatching. Perhaps the odor of the plants and soil in this stream was imprinted on the nervous system of the larval fish.

Through the evolutionary process, animals tend to rely on those stimuli and senses that are adaptive to their particular environment and way of life. In all cases, sensory receptors generate nerve impulses that travel to the brain where sensation occurs. In mammals, and particularly human beings, integration of the data received from various sensory receptors results in a perception of events occurring in the external environment.

Summary

47.1 Chemical Senses

Chemoreception is found universally in animals and is, therefore, believed to be the most primitive sense.

Human olfactory cells and taste buds are chemoreceptors. They are sensitive to chemicals in water and air.

47.2 Sense of Vision

Vision is dependent on the eye, the optic nerves, and the visual areas of the cerebral cortex. The eye has three layers. The outer layer, the sclera, can be seen as the white of the eye; it also becomes the transparent bulge in the front of the eye called the cornea. The middle pigmented layer, called the choroid, absorbs stray light rays. The rod cells (sensory receptors for dim light) and the cone cells (sensory receptors for bright light and color) are located in the retina, the inner layer of the eyeball. The cornea, the humors, and especially the lens bring the light rays to focus on the retina. To see a close object, accommodation occurs as the lens rounds up.

When light strikes rhodopsin within the membranous disks of rod cells, rhodopsin splits into opsin and retinal. A cascade of reactions leads to the closing of ion channels in a rod cell's plasma membrane. Inhibitory transmitter molecules are no longer released and nerve impulses are carried in the optic nerve to the brain.

Integration occurs in the retina which is composed of three layers of cells: the rod and cone layer, the bipolar cell layer, and the ganglion cell layer. Integration also occurs in the visual areas of the cerebral cortex.

47.3 Sense of Hearing and Balance

Hearing in humans is dependent on the ear, the cochlear nerve, and the auditory areas of the cerebral cortex. The ear is divided into three parts. The outer ear consists of the pinna and the auditory canal, which direct sound waves to the middle ear. The middle ear begins with the tympanic membrane and contains the ossicles (malleus, incus, stapes). The malleus is attached to the tympanic membrane, and the stapes is attached to the oval window, which is covered by membrane. The inner ear contains the cochlea and the semicircular canals, plus the utricle and saccule.

Hearing begins when the outer and middle portions of the ear convey and amplify the sound waves that strike the oval window. Its vibrations set up pressure waves within the cochlea, which contains the spiral organ, consisting of hair cells whose stereocilia are embedded within the tectorial membrane. When the stereocilia of the hair cells bend, nerve impulses begin in the cochlear nerve and are carried to the brain.

The ear also contains receptors for our sense of balance. Dynamic equilibrium is dependent on the stimulation of hair cells within the ampullae of the semicircular canals. Static equilibrium relies on the stimulation of hair cells within the utricle and the saccule.

Reviewing the Chapter

1. Discuss the structure and the function of human chemoreceptors. 848–49
2. In general, how does the eye in arthropods differ from that in humans? 850
3. What types of animals have eyes that are constructed like the human eye? 850
4. Name the parts of the human eye, and give a function for each part. 851
5. Describe the anatomy of the eye, and explain focusing and accommodation. 851–52
6. Contrast the location and the function of rod cells to those of cone cells. 853–54
7. Explain the process of integration in the retina and the brain. 854
8. Describe the structure of the human ear. 856–57
9. Describe how we hear. 856–57
10. Describe the role of the semicircular canals, utricle, and the saccule in balance. 858–59

Testing Yourself

Choose the best answer for each question.

1. A sensory receptor
   a. is the first portion of a reflex arc.
   b. initiates nerve impulses.
   c. responds to only one type of stimulus.
   d. is associated with a sensory neuron.
   e. All of these are correct.
2. Which of these gives the correct path for light rays entering the human eye?
   a. sclera, retina, choroid, lens, cornea
   b. fovea centralis, pupil, aqueous humor, lens
   c. cornea, pupil, lens, vitreous humor, retina
   d. optic nerve, sclera, choroid, retina, humors
   e. All of these are correct.
3. Which gives an incorrect function for the structure?
   a. lens—focusing
   b. iris—regulation of amount of light
   c. choroid—location of cones
   d. sclera—protection
   e. fovea centralis—acute vision
4. Retinal is
   a. a derivative of vitamin A.
   b. sensitive to light energy.
   c. a part of rhodopsin.
   d. found in rod cells.
   e. All of these are correct.
5. Which association is incorrect?
   a. taste buds—humans
   b. compound eye—arthropods
   c. camera-type eye—squid
   d. statocysts—sea stars
   e. chemoreceptors—planarians
6. Which one of these wouldn't you mention if you were tracing the path of sound vibrations?
   a. auditory canal
   b. tympanic membrane
   c. semicircular canals
   d. cochlea
   e. ossicles
7. Which one of these correctly describes the location of the spiral organ?
   a. between the tympanic membrane and the oval window in the inner ear
   b. in the utricle and saccule within the vestibule
   c. between the tectorial membrane and the basilar membrane in the cochlear canal
   d. between the outer and inner ear within the semicircular canals
8. Which of these is mismatched?
   a. semicircular canals—inner ear
   b. utricle and saccule—outer ear
   c. auditory canal—outer ear
   d. ossicles—middle ear
   e. cochlear nerve—inner ear
9. Both olfactory receptors and sound receptors have cilia, and they both
   a. are chemoreceptors.
   b. have receptor proteins.
   c. initiate nerve impulses.
   d. are located near balance receptors.
   e. All of these are correct.
10. Label this diagram of the human eye. State a function for each structure labeled:

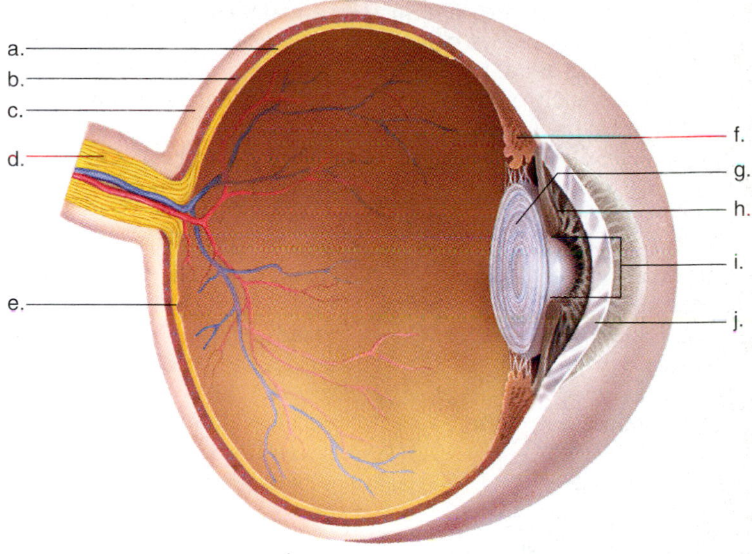

Thinking Scientifically

1. The density of taste buds on the tongue can vary. Some obese individuals have a lower density of taste buds than those who are not obese. Assume that taste perception is related to taste bud density. If so, what hypothesis would you test to see if there is a relationship between taste bud density and obesity?
2. A man who has spent many years serving on submarines complains of hearing loss, particularly the inability to hear high tones. When a submarine submerses, the inside air pressure intensifies. What hypotheses might explain hearing loss in this individual?

Bioethical Issue

A cataract is a cloudiness of the lens that occurs in 50% of people between the ages 65 and 74, and in 70% of those age 75 or older. The extent of visual impairment depends on the size, density of the cataract, and where it is located in the lens. A dense centrally placed cataract causes severe blurring of vision.

Are cataracts preventable? For most people the answer is yes, they are preventable. These factors have been identified as contributing to the chances of having a cataract:

- Smoking 20 or more cigarettes a day doubles the risk of cataracts in men—women have to smoke more than 30 cigarettes a day to increase their chances of a cataract.
- Exposure to the ultraviolet radiation in sunlight can more than double the risk of a cataract. In addition, this relationship is dose-dependent—the more sunlight, the higher the risk.
- Also, as many as one-third of cataracts may be caused by being overweight. Diet, rather than exercise, seems to reduce cataract formation, perhaps through lower blood sugar levels or improved antioxidant properties of the blood.

It's clear, then, that our own behavior contributes to the occurrence of cataracts for most of us. Should we be responsible in our actions and take all possible steps to prevent developing a cataract, such as not smoking, wearing sunglasses and a wide-brim hat, and watching our weight? Or should we simply rely on medical science to restore our eyesight? Most cataract operations today are performed on an outpatient basis with minimal postoperative discomfort and with a high expectation of restoration of sight. Perhaps it's better, though, to take all possible steps to prevent the occurrence of cataracts, just in case our experience is atypical.

Understanding the Terms

blind spot 854
camera-type eye 850
chemoreceptor 848
choroid 851
ciliary muscle 852
cochlea 856
compound eye 850
cone cell 853
cornea 852
fovea centralis 853
lateral line 859
lens 852
mechanoreceptor 856
olfactory receptor cell 848
ossicle 856
otolith 859
photoreceptor 850
pupil 852
retina 851
rhodopsin 853
rod cell 853
saccule 859
sclera 851
semicircular canal 859
spiral organ 857
statocyst 859
taste bud 848
tympanic membrane 856
utricle 859

Match the terms to these definitions:

a. ______________ Structure that receives sensory stimuli and is a part of a sensory neuron or transmits signals to a sensory neuron.

b. ______________ Inner layer of the eyeball containing the photoreceptors—rod cells and cone cells.

c. ______________ Outer, white, fibrous layer of the eye that surrounds the eye except for transparent cornea.

d. ______________ Receptor that is sensitive to chemical stimulation—for example, receptors for taste and smell.

e. ______________ Specialized region of the cochlea containing the hair cells for sound detection and discrimination.

Web Connections

Exploring the Internet

http://www.mhhe.com/biosci/genbio/mader
(click on *Biology 7/e*)

The *Biology 7/e* Online Learning Center provides many resources for studying the material in this chapter including links to the following sites:

Anatomy and Physiology of the Ear. Information on the structure and function of the ear.

http://psych.hanover.edu/classes/hfnotes/sld041.html

The EyeCare Reports—The Aging Eye. Describes what can be expected from our vision as we age. The treatment of presbyopia, cataracts, macular degeneration, and retinal detachment are described here.

http://www.west.net/~eyecare/the_aging_eye_report.html

Seeing, Hearing, and Smelling the World. A full text reprint of a Howard Hughes Medical Institute report on "making sense of our senses."

http://www.hhmi.org/senses/

CDC site on River Blindness (Onchocerciasis). Causes blindness, affects millions of people in lesser developed, tropical areas.

http://www.cdc.gov/ncidod/dpd/list_rb.htm

Laser Eye Surgery. Information, FAQs, and much more at this site which describes Lasik (laser-assisted in situ keratomileusis), a relatively new surgical alternative which corrects myopia (nearsightedness) or astigmatism. This is just one of hundreds of commercial sites describing this procedure. (To see others, simply search on the web using the keyword "Lasik.")

http://www.vision-institute.com/

Support Systems and Locomotion

chapter

chapter concepts

48.1 Diversity of Skeletons

- Animals have one of three types of skeletons: a hydrostatic skeleton, an exoskeleton, or an endoskeleton. 864
- The strong but flexible skeleton of arthropods and vertebrates is adaptive for living on land. 865

48.2 The Human Skeletal System

- The cartilaginous skeleton of the fetus is converted to a skeleton of bone which continually undergoes remodeling. 866
- There are two types of bone tissue called compact bone and spongy bone that differ in structure and function. 866
- The human skeleton is divided into these portions: the axial skeleton consists of the skull, the ribs, the sternum, and the vertebrae, and the appendicular skeleton contains the girdles and the limbs. 868
- The human skeleton is jointed; the joints differ in movability. 871

48.3 The Human Muscular System

- Macroscopically, human skeletal muscles work in antagonistic pairs and exhibit tone. 873
- Microscopically, muscle fiber contraction is dependent on filaments of both actin and myosin, and a ready supply of calcium ions (Ca^{2+}) and ATP. 875
- Motor nerve fibers release ACh at a neuromuscular junction and thereafter a muscle fiber contracts. 876

Horses, *Equus*, crossing water

Not all animals have muscles and bones but they all use contractile fibers for locomotion. In vertebrates, the muscular system and the skeletal system work together to provide movement, whether it be a horse running, a fish swimming, or an eagle flying. Most animals also have a nervous system, which integrates data received from sensory receptors and coordinates muscular activity so that animal movement achieves goals like feeding, escaping enemies, reproducing, or simply playing.

In order to create movement, muscle contraction must be directed against some sort of medium. In planarians, hydras, and earthworms, muscles push against body fluids inside a gastrovascular cavity or coelom. In vertebrates, the muscles are attached to a bony endoskeleton. Both the skeletal system and the muscular system contribute to homeostasis. Aside from giving the body shape, and protecting internal organs, the skeleton serves as a storage area for inorganic calcium and produces blood cells. The skeleton also supports the body and its organs against the pull of gravity. Aside from moving the body and body parts, the skeletal muscles give off heat which warms the body.

48.1 Diversity of Skeletons

Different types of skeletons occur in the animal kingdom. A hydrostatic skeleton is seen in cnidarians, flatworms, roundworms, and annelids. Molluscs and arthropods have an **exoskeleton** (external skeleton); the molluscan skeleton is composed of calcium carbonate while the arthropod skeleton contains chitin. Echinoderms and vertebrates have a bony **endoskeleton** (internal skeleton).

Hydrostatic Skeleton

In animals that lack a hard skeleton, a fluid-filled gastrovascular cavity or coelom can act as a hydrostatic skeleton. A **hydrostatic skeleton** [Gk. *hydrias*, water, and *stasis*, standing] offers support and resistance to the contraction of muscles so that mobility results. As analogies, consider that a garden hose stiffens when filled with water, and that a water-filled balloon changes shape when squeezed at one end. Similarly, an animal with a hydrostatic skeleton can change shape and perform a variety of movements.

Hydras, which are cnidarians, and planarians, which are flatworms, use their fluid-filled gastrovascular cavity as a hydrostatic skeleton. When muscle fibers at the base of epidermal cells in a hydra contract, the body or tentacles shorten rapidly. Planarians usually glide over a substrate with the help of muscular contractions that control the body wall and cilia. Roundworms have a fluid-filled pseudocoelom and move in a whiplike manner when their longitudinal muscles contract. Annelids, such as earthworms, are segmented and have septa that divide their coelom into compartments (Fig. 48.1). Each segment has its own set of longitudinal and circular muscles and its

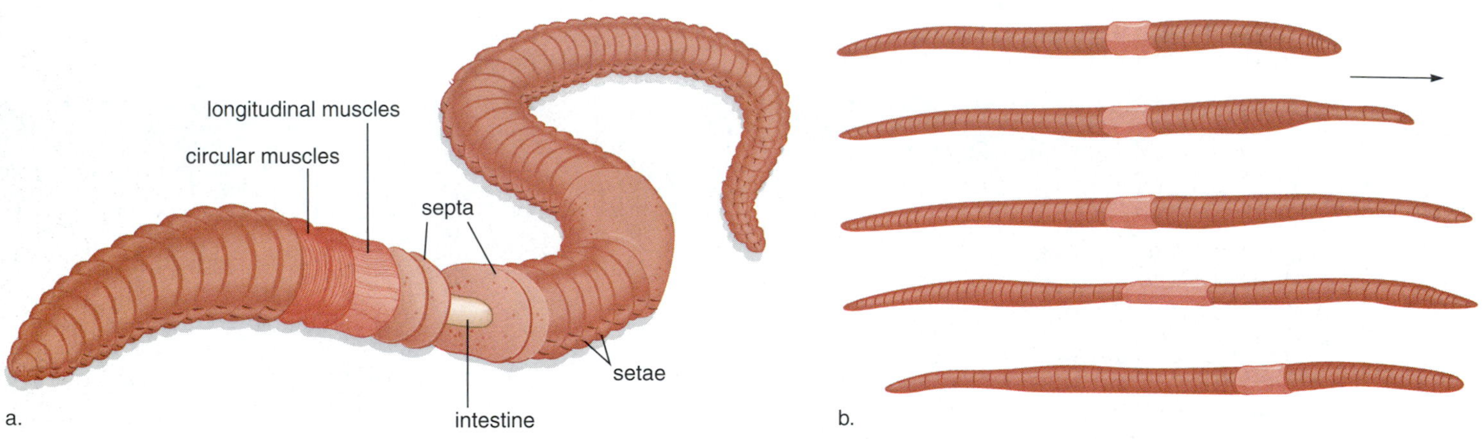

Figure 48.1 **Locomotion in an earthworm.**
a. The coelom is divided by septa, and each body segment is a separate locomotor unit. There are both circular and longitudinal muscles.
b. As circular muscles contract, a few segments extend. The worm is held in place by setae, needlelike chitinous structures on each segment of the body. Then, as longitudinal muscles contract, a portion of the body is brought forward. This series of events occurs down the length of the worm.

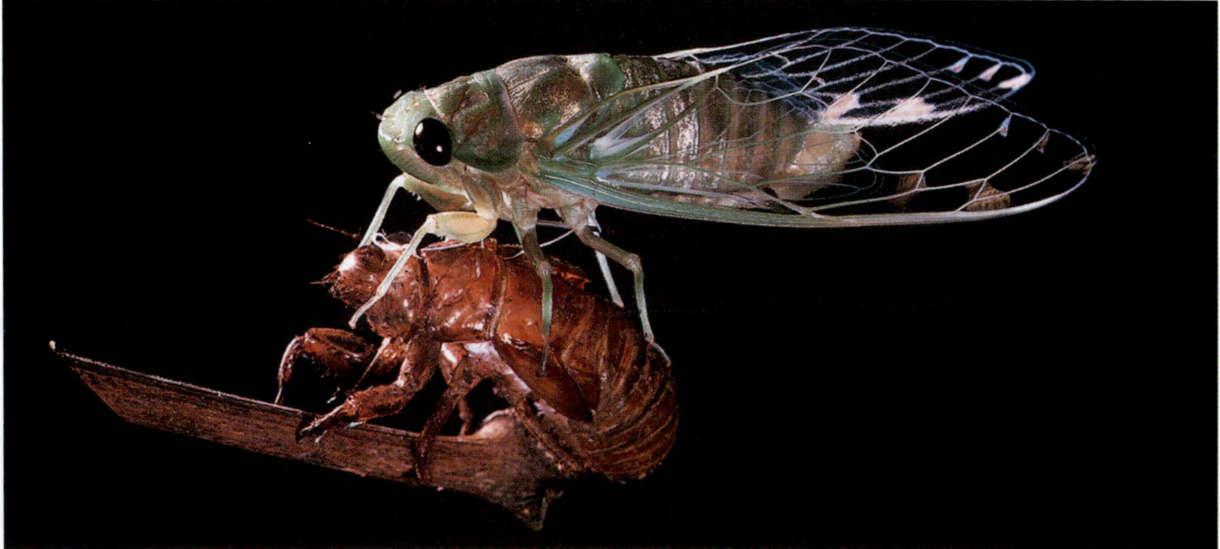

Figure 48.2 **Exoskeleton.**
Exoskeletons support muscle contraction and prevent drying out. The chitinous exoskeleton of arthropods is shed as the animal molts; until the new skeleton dries and hardens, the animal is vulnerable to predators, and muscle contractions may not translate into body movements. In this photo a dog-day cicada, *Tibicen*, has just finished molting.

own nerve supply, so each segment or group of segments may function independently. When circular muscles contract, the segments become thinner and elongate. When longitudinal muscles contract, the segments become thicker and shorten. By alternating circular muscle contraction and longitudinal muscle contraction, the animal moves forward.

Exoskeletons and Endoskeletons

Calcium carbonate forms the exoskeleton of molluscs, such as clams and snails, but arthropods, such as insects and crustaceans, have a chitinous exoskeleton. Chitin is a strong, flexible, nitrogenous polysaccharide. Besides providing protection against wear and tear and against enemies, an exoskeleton also prevents drying out. This is an important feature for animals that live on land. Although the stiffness of an exoskeleton provides superior support for muscle contractions, an exoskeleton is not as strong as an endoskeleton. Strength can be achieved by increasing the thickness of an exoskeleton, but this increases its weight and leaves less room for internal organs.

In molluscs the exoskeleton grows as the animal grows. The thick and nonmobile calcium carbonate exoskeleton is largely for protection. The chitinous exoskeleton of arthropods, which is jointed and movable, is suitable for life on land. Arthropods, however, molt to rid themselves of an exoskeleton that has become too small, and molting makes an animal vulnerable to predators (Fig. 48.2).

Vertebrates have an endoskeleton, composed of bone and cartilage, which grows with the animal. Endoskeletons do not limit the space available for internal organs, and they support greater weight. The soft tissues that surround an endoskeleton protect it, and injuries to soft tissues are apt to be easier to repair than is a broken hard skeleton. Even so, endoskeletons do usually protect vital internal organs (Fig. 48.3).

The exoskeleton of arthropods and the endoskeleton of vertebrates not only offer support; they are also jointed. A strong but flexible skeleton helped the arthropods and vertebrates successfully colonize the terrestrial environment.

All types of skeletons assist movement; endoskeletons and exoskeletons are protective of vital organs.

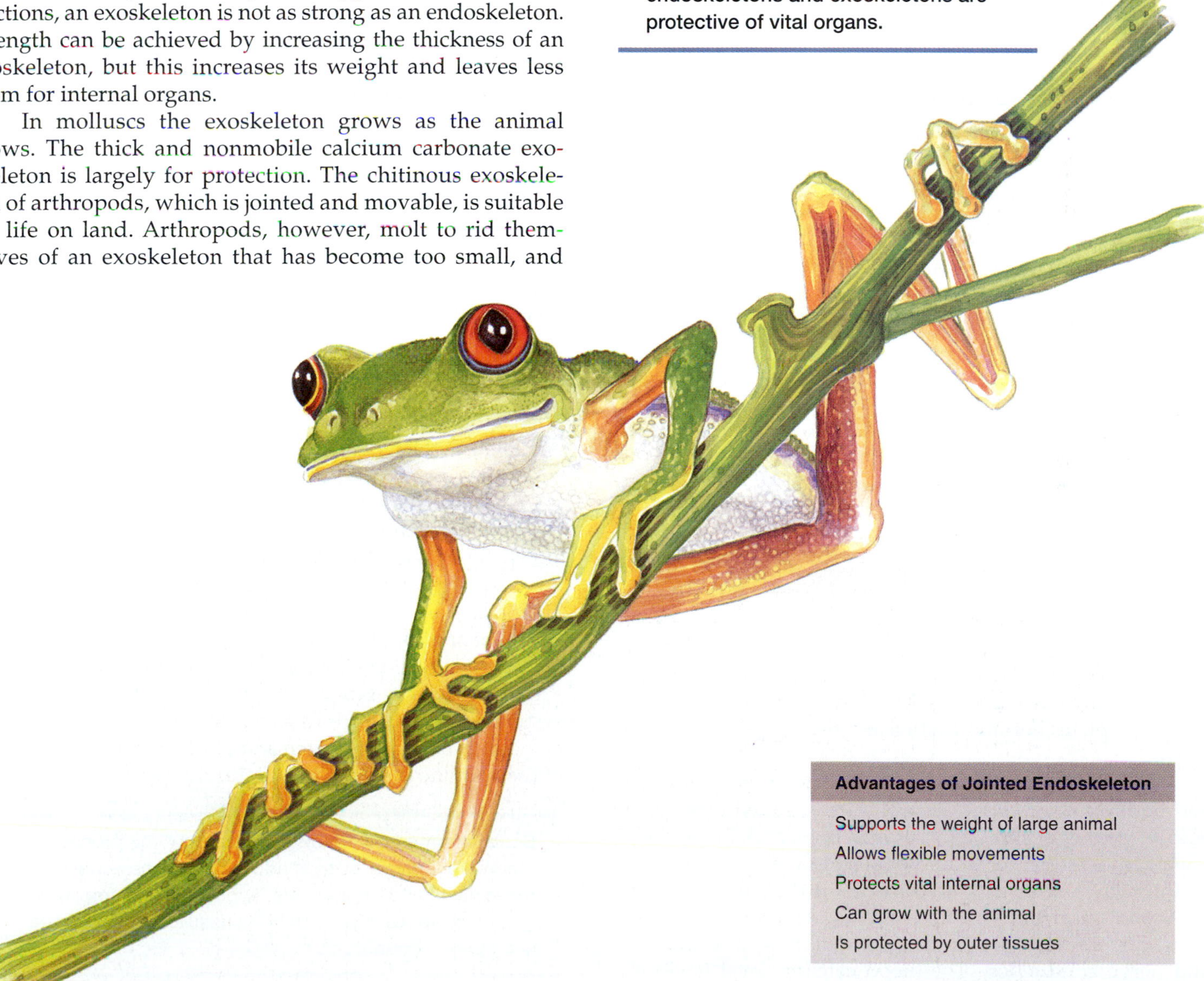

Advantages of Jointed Endoskeleton
Supports the weight of large animal
Allows flexible movements
Protects vital internal organs
Can grow with the animal
Is protected by outer tissues

Figure 48.3 **The vertebrate endoskeleton.**
The vertebrate jointed endoskeleton has the advantages listed. In addition, an endoskeleton lends itself to adaptation to the environment. Vertebrates move in various ways (e.g., jumping, flying, swimming, running).

48.2 The Human Skeletal System

The human skeletal system has many functions that contribute to homeostasis. It protects vital internal organs. For example, the skull forms a protective encasement for the brain, as does the rib cage for the heart and the lungs. The bones serve as sites for muscle attachment and particularly those of the arms and legs permit flexible body movement. The large, heavy bones of the legs support the body against the pull of gravity.

Flat bones—such as those of the skull, the ribs, and the breastbone—produce red blood cells. All bones are storage areas for inorganic calcium and phosphate ions.

The skeleton supports and protects the body while at the same time permitting flexible movement. All bones serve as a storehouse for calcium and phosphate ions and certain ones produce red blood cells.

Bone Growth and Renewal

Most of the bones of the human skeleton are cartilaginous during prenatal development. Since the cartilaginous structures are shaped like the future bones, they provide "models" of these bones. The cartilaginous models are converted to bones when calcium salts are deposited in the matrix, first by the cartilaginous cells and later by bone-forming cells called **osteoblasts** [Gk. *osteon*, bone, and *blastos*, bud]. The conversion of cartilaginous models to bones is called endochondral ossification.

There are also examples of ossification that have no previous cartilaginous model. Facial bones and certain other bones of the skull are formed by direct ossification. During intramembranous ossification, fibrous connective tissue membranes gives support as ossification begins.

During endochondral ossification of a long bone, there is at first only a primary ossification center in the middle of the cartilaginous model. Later, secondary ossification centers form at the ends of the model. A cartilaginous growth plate remains between the primary ossification center and each secondary center. As long as these plates remain, growth is possible. The rate of growth is controlled by hormones, particularly growth hormone (GH) and the sex hormones. Eventually, the plates become ossified and the growth of the bone stops.

In the adult, bone is continually being broken down and built up again. Bone-absorbing cells, called **osteoclasts** [Gk. *osteon*, bone, and *klastos*, broken in pieces], break down bone, remove worn cells, and deposit calcium in the blood. In this way, osteoclasts help maintain the blood calcium level and contribute to homeostasis. Among other functions, calcium ions play a major role in muscle contraction and nerve conduction. The blood calcium level is closely regulated by the antagonistic hormones parathyroid hormone (PTH) and calcitonin. PTH promotes the activity of osteoclasts and calcitonin inhibits their activity to keep the blood calcium level within normal limits.

Assuming that the blood calcium level is normal, bone destruction caused by the work of osteoclasts is repaired by osteoblasts. As they form bone, osteoblasts take calcium from the blood. Eventually some of these cells get caught in the matrix they secrete and are converted to **osteocytes** [Gk. *osteon*, bone, and *kytos*, cell], the cells found within the lacunae of osteons. Strange as it may seem, adults are thought to require more calcium in the diet than do children in order to promote the work of osteoblasts.

Through this process of remodeling, old bone tissue is replaced by new bone tissue. Therefore, the thickness of bones can change depending on exercise and hormone balances. As discussed in the reading on page 872, a thinning of the bones called osteoporosis can occur as we age if the proper precautions are not taken.

Anatomy of a Long Bone

A long bone, such as the humerus, illustrates principles of bone anatomy. When the bone is split open, as in Figure 48.4, the longitudinal section shows that it is not solid but has a cavity called the medullary cavity bounded at the sides by compact bone and at the ends by spongy bone. Beyond the spongy bone, there is a thin shell of compact bone and finally a layer of hyaline cartilage. The cavity of a long bone usually contains yellow bone marrow, which is a fat-storage tissue.

Compact bone contains many osteons (Haversian systems) where osteocytes lie in tiny chambers called lacunae. The lacunae are arranged in concentric circles around central canals, which contain blood vessels and nerves. The lacunae are separated by a matrix of collagen fibers and mineral deposits, primarily calcium and phosphorous salts.

Spongy bone has numerous bony bars and plates separated by irregular spaces. Although lighter than compact bone, spongy bone is still designed for strength. Just as braces are used for support in buildings, the solid portions of spongy bone follow lines of stress. The spaces in spongy bone are often filled with **red bone marrow,** a specialized tissue that produces blood cells. This function of the skeletal system is of assistance to homeostasis. As you know, red blood cells transport oxygen and white blood cells are a part of the immune system which fights infection.

Bone is formed during development but the process of renewal continues even in adults due to the action of osteoclasts and osteoblasts. A long bone exemplifies that there are two types of bone tissue: compact bone and spongy bone.

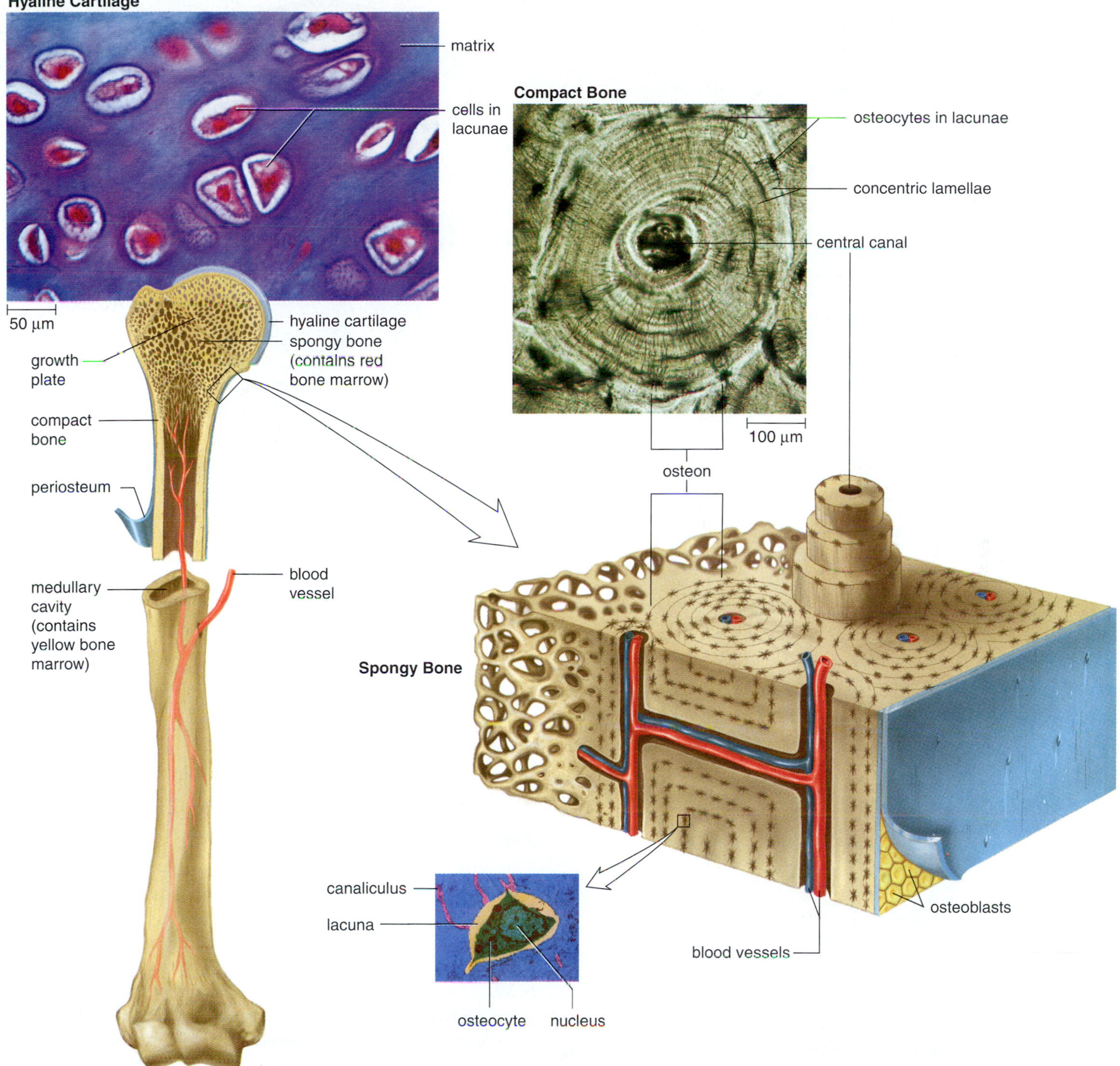

Figure 48.4 **Anatomy of a long bone.**
A long bone is encased by fibrous membrane except where it is covered at the ends by hyaline cartilage (see micrograph). Spongy bone located beneath the cartilage may contain red bone marrow. The central shaft contains yellow bone marrow and is bordered by compact bone which is shown in the enlargement and micrograph.

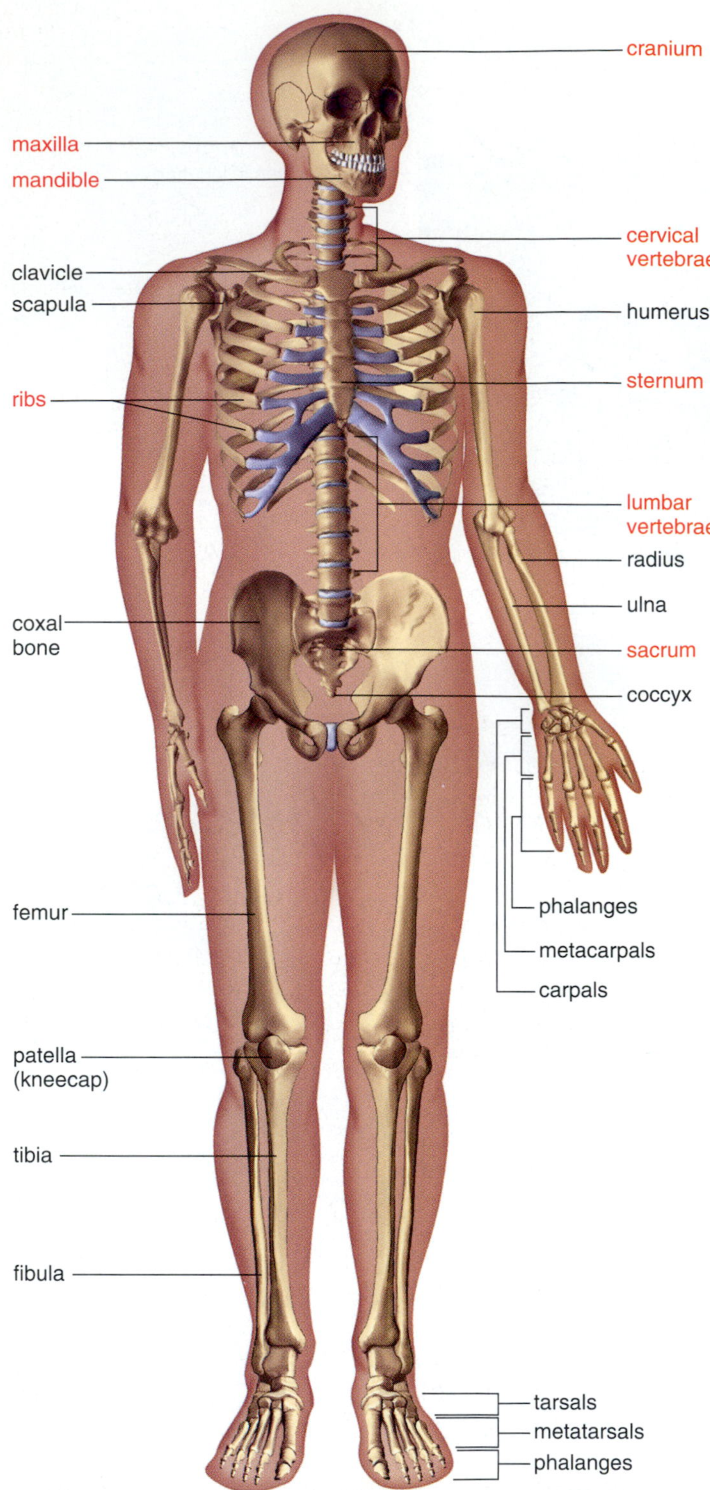

Figure 48.5 **The human skeleton.**
In the human skeleton, the axial skeleton is composed of the skull, the vertebral column, the sternum, and the ribs (red labels). The rest of the bones belong to the appendicular skeleton (black labels).

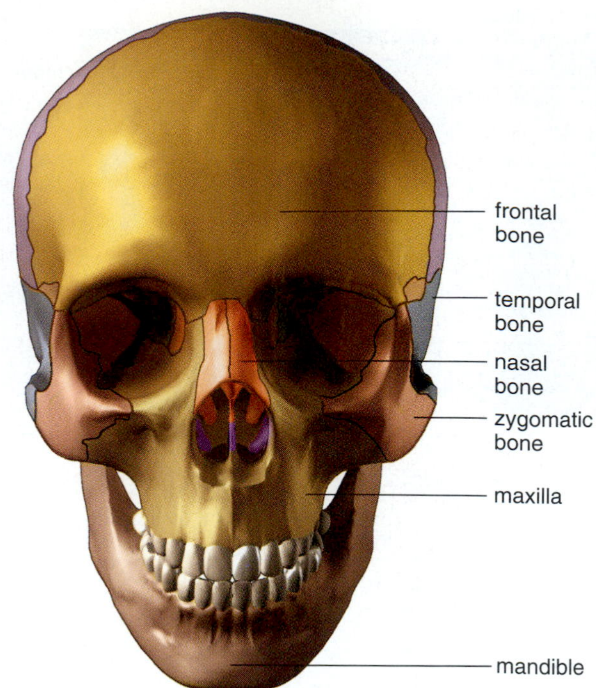

Figure 48.6 **The skull.**
The skull consists of the cranium and the facial bones. The frontal bone is the forehead; the zygomatic bones form the cheekbones, and the maxillae form the upper jaw. The mandible has a projection we call the chin.

The Axial Skeleton

The **axial skeleton** [L. *axis,* axis, hinge, and Gk. *skeleton,* dried body] lies in the midline of the body and consists of the skull, the vertebral column, the sternum, and the ribs (red labels in Fig. 48.5).

The Skull

The skull which protects the brain is formed by the cranium and the facial bones (Fig. 48.6). In newborns, certain bones of the cranium are joined by membranous regions called **fontanels,** all of which usually close by the age of two years. The bones of the cranium contain the sinuses [L. *sinus,* hollow], air spaces lined by mucous membrane that reduce the weight of the skull and give a resonant sound to the voice. Two sinuses called the mastoid sinuses drain into the middle ear. Mastoiditis, a condition that can lead to deafness, is an inflammation of these sinuses.

The major bones of the cranium have the same names as the lobes of the brain. On the top of the cranium, the frontal bone forms the forehead, the parietal bones extend to the sides. Below the much larger parietal bones, each temporal bone has an opening that leads to the middle ear. In the rear of the skull, the occipital bone (not shown) curves to form the base of the skull. At the base of the skull, the spinal cord passes through a large opening called the **foramen magnum** [L. *foramen,* hole, and *magnus,* great, large], and becomes the brain stem.

The temporal and frontal bones are cranial bones that contribute to the face. The temporal bones account for the flattened areas on each side of the forehead, which we call the temples. The frontal bone not only forms the forehead it also has supraorbital ridges where the eyebrows are located. Glasses sit where the frontal bone joins the nasal bones.

The most prominent of the facial bones are the mandible [L. *mandibula,* jaw], the maxillae, the zygomatic bones, and the nasal bones. The mandible, or lower jaw, is the only freely movable portion of the skull (Fig. 48.6), and its action permits us to chew our food. It also forms the "chin." Tooth sockets are located on the mandible and on the maxillae, which form the upper jaw and a portion of the hard palate. The zygomatic bones are the cheekbone prominences, and the nasal bones form the bridge of the nose. Other bones make up the nasal septum which divides the nose cavity into two regions.

While the ears are formed only by elastic cartilage and not by bone, the nose is a mixture of bones, cartilages and fibrous connective tissue. The lips and cheeks have a core of skeletal muscle.

The skull is formed by the cranium, whose major bones are named after the lobes of the brain, and the facial bones. The most prominent facial bones are the mandible, maxillae, the zygomatic bones, and the nasal bones.

The Vertebral Column and Rib Cage

The **vertebral column** [L. *vertebra,* bones of backbone] supports the head and trunk and protects the spinal cord and roots of spinal nerves. It is a longitudinal axis that serves either directly or indirectly as an anchor for all the other bones of the skeleton.

The vertebrae make up the vertebral column (see Fig. 48.5). The cervical vertebrae are located in the neck and the thoracic vertebrae are in the thorax. The lumbar vertebrae are found in the small of the back. The five sacral vertebrae are fused to form the sacrum. The coccyx, or tailbone, is also composed of fused vertebrae. Normally, the vertebral column has four normal curvatures that provide more resilience and strength for an upright posture than a straight column could provide. Scoliosis is an abnormal lateral (sideways) curvature of the spine. Another well-known abnormal curvature results in a hunchback and still another results in a swayback.

Intervertebral disks, composed of fibrocartilage between the vertebrae act as a kind of padding. They prevent the vertebrae from grinding against one another and absorb shock caused by movements such as running, jumping, and even walking. The presence of the disks allows vertebrae to move as we bend forward, backward, and from side to side. Unfortunately, these disks become weakened with age and can slip or even rupture. Pain results when the damaged disk presses against the spinal cord and/or spinal nerves. The body may heal itself, or the disk can be removed surgically. If so, the vertebrae can be fused together, but this limits the flexibility of the body.

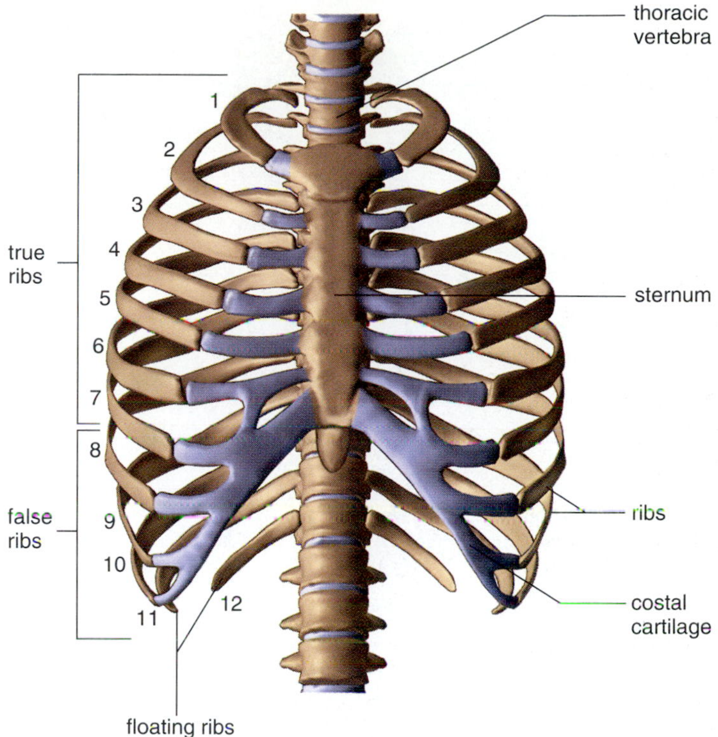

Figure 48.7 The rib cage.
The rib cage consists of the thoracic vertebrae, the twelve pairs of ribs, the costal cartilages, and the sternum or breastbone.

Rib Cage

The thoracic vertebrae are a part of the rib cage. The rib cage also contains the ribs and their associated cartilages, and the sternum or breastbone (Fig. 48.7).

There are twelve pairs of ribs. The upper seven pairs of ribs are "true ribs" because they attach to the sternum. The lower five pairs do not connect directly to the sternum and are called the "false ribs." Three pairs of false ribs attach by means of a common cartilage and two pairs are "floating ribs" because they do not attach to the sternum at all.

The rib cage demonstrates how the skeleton is protective but also flexible. The rib cage protects the heart and lungs; yet it swings outward and upward upon inspiration and then downward and inward upon expiration.

The vertebral column contains the vertebrae and forms the longitudinal axis of the body. The rib cage protects the heart and lungs and assists breathing.

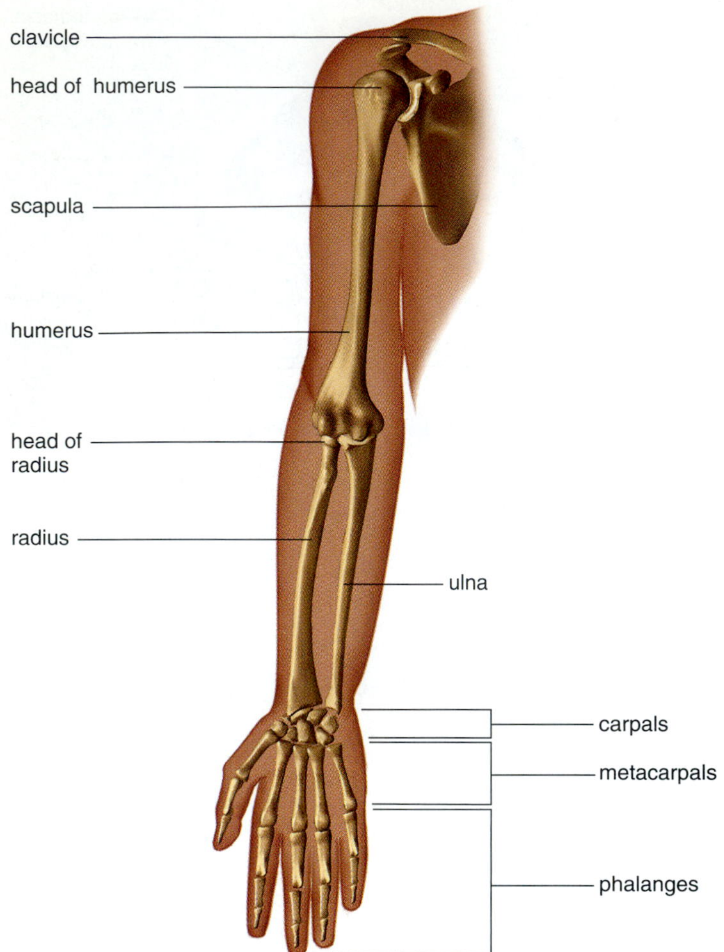

Figure 48.8 Bones of the pectoral girdle, the arm, and the hand.
The humerus is known as the "funny bone" of the elbow. The sensation upon bumping it is due to the activation of a nerve that passes across its end.

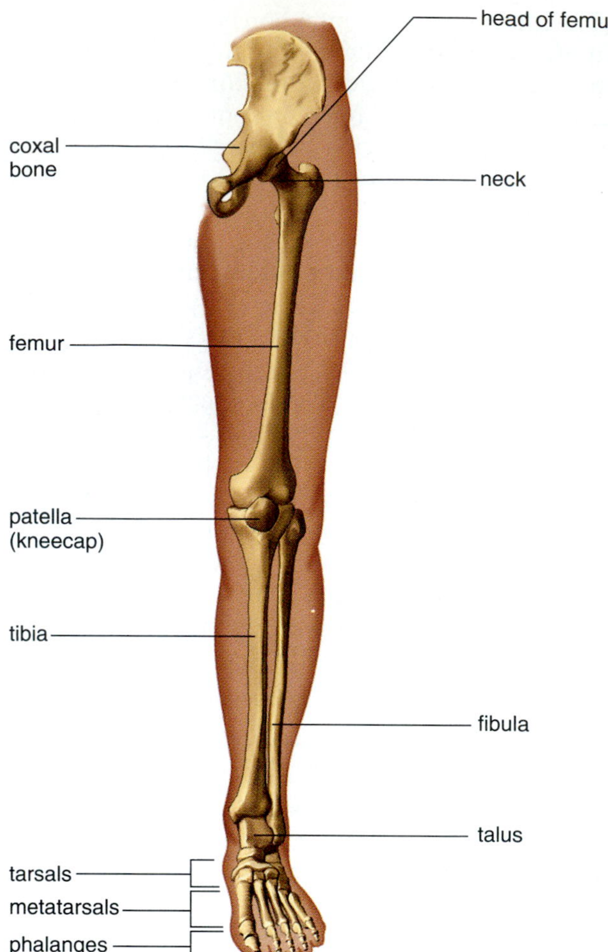

Figure 48.9 Bones of the pelvic girdle, the leg, and the foot.
The femur is our strongest bone—it withstands a pressure of 540 kilograms per 2.5 cm^3 when we walk.

The Appendicular Skeleton

The **appendicular skeleton** [L. *appendicula*, dim. of *appendix*, appendage] consists of the bones within the pectoral and pelvic girdles and the attached limbs (see Fig. 48.5). The pectoral (shoulder) girdle and arms are specialized for flexibility, but the pelvic girdle (hipbones) and legs are specialized for strength.

The Pectoral Girdle and Arm

The components of the **pectoral girdle** [Gk. *pechys*, forearm] are only loosely linked together by ligaments (Fig. 48.8). Each clavicle (collarbone) connects with the sternum in front and the scapula (shoulder blade) behind, but the scapula is freely movable and held in place only by muscles. This allows it to follow freely the movements of the arm. The single long bone in the upper arm, the humerus, has a smoothly rounded head that fits into a socket of the scapula. The socket, however, is very shallow and much smaller than the head. Although this means that the arm can move in almost any direction, there is little stability. Therefore, this is the joint that is most apt to dislocate. The opposite end of the humerus meets the two bones of the lower arm, the ulna and the radius, at the elbow. (The prominent bone in the elbow is the topmost part of the ulna.) When the arm is held so that the palm is turned frontward, the radius and ulna are about parallel to one another. When the arm is turned so that the palm is next to the body, the radius crosses in front of the ulna, a feature that contributes to the easy twisting motion of the forearm.

The many bones of the hand increase its flexibility. The wrist has eight carpal bones, which look like small pebbles. From these, five metacarpal bones fan out to form a framework for the palm. The metacarpal bone that leads to the thumb is placed in such a way that the thumb can reach out and touch the other digits. (Digits is a term that refers to either fingers or toes.) Beyond the metacarpals are the phalanges, the bones of the fingers and the thumb. The phalanges of the hand are long, slender, and lightweight.

The Pelvic Girdle and Leg

The **pelvic girdle** [L. *pelvis*, basin] (Fig. 48.9) consists of two heavy, large coxal bones (hipbones). The coxal bones are anchored to the sacrum, and together these bones form a hollow cavity that is wider in females (to accommodate childbearing) than in males. The weight of the body is transmitted through the pelvis to the legs and then onto the ground. The largest bone in the body is the femur, or thighbone.

In the lower leg, the larger of the two bones, the tibia, has a ridge we call the shin. Both of the bones of the lower leg have a prominence that contributes to the ankle—the tibia on the inside of the ankle and the fibula on the outside of the ankle. Although there are seven tarsal bones in the ankle, only one receives the weight and passes it on to the heel and the ball of the foot. If you wear high-heeled shoes, the weight is thrown toward the front of your foot. The metatarsal bones participate in forming the arches of the foot. There is a longitudinal arch from the heel to the toes and a transverse arch across the foot. These provide a stable, springy base for the body. If the tissues that bind the metatarsals together become weakened, flat feet are apt to result. The bones of the toes are called phalanges, just like those of the fingers, but in the foot the phalanges are stout and extremely sturdy.

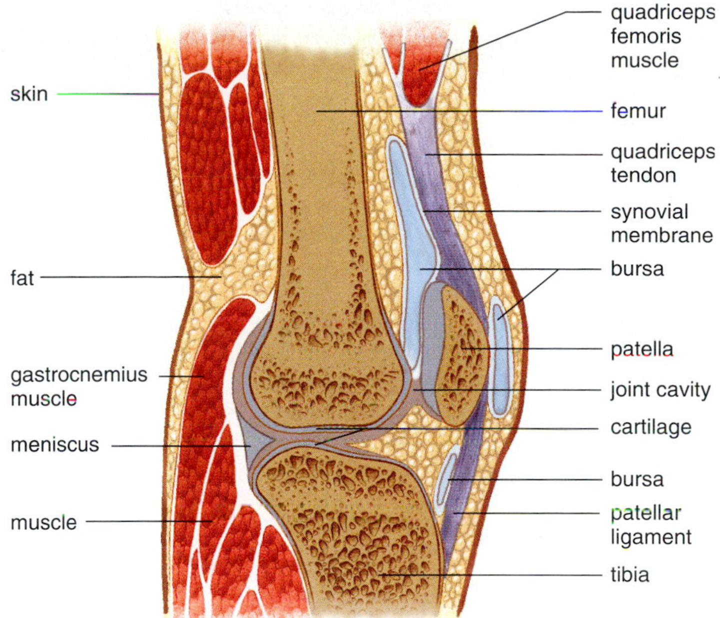

Figure 48.10 Knee joint.
The knee joint is an example of a synovial joint. The cavity between the bones is encased by ligaments and lined by synovial membrane. The patella (kneecap) serves to guide the quadriceps tendon over the joint when flexion or extension occurs.

The appendicular skeleton contains the bones of the girdles and the limbs.

Classification of Joints

Bones are joined at the **joints,** which are classified as fibrous, cartilaginous, and synovial. Fibrous joints such as those between the cranial bones are immovable. Cartilaginous joints such as those between the vertebrae are slightly movable. The vertebrae are also separated by disks, which increase their flexibility. The two hipbones are slightly movable because they are ventrally joined by cartilage. Owing to hormonal changes, this joint becomes more flexible during late pregnancy, allowing the pelvis to expand during childbirth.

In freely movable **synovial joints,** the two bones are separated by a cavity. **Ligaments,** composed of fibrous connective tissue, bind the two bones to each other, holding them in place as they form a capsule. In a "double-jointed" individual, the ligaments are unusually loose. The joint capsule is lined by synovial membrane, which produces synovial fluid, a lubricant for the joint.

The knee is an example of a synovial joint (Fig. 48.10). In the knee, as in other freely movable joints, the bones are capped by cartilage, although there are also crescent-shaped pieces of cartilage between the bones called menisci [Gk. *meniscus*, crescent]. These give added stability, helping to support the weight placed on the knee joint. Unfortunately, athletes often suffer injury of the meniscus, known as torn cartilage. Thirteen fluid-filled sacs called bursae (sing., **bursa**) occur around the joint. Bursae ease friction between tendons and ligaments and between tendons and bones. Inflammation of the bursae is called bursitis. Tennis elbow is a form of bursitis.

There are different types of movable joints. The knee and elbow joints are hinge joints because, like a hinged door, they largely permit movement in one direction only. More movable are the ball-and-socket joints; for example, the ball of the femur fits into a socket on the hipbone. Ball-and-socket joints allow movement in all planes and even a rotational movement.

Synovial Joints

Synovial joints are subject to arthritis. In rheumatoid arthritis, the synovial membrane becomes inflamed and thickened. Degenerative changes take place that make the joint almost immovable and painful to use. There is evidence that these effects are brought on by an autoimmune reaction. In old-age arthritis, or osteoarthritis, the cartilage at the ends of the bones disintegrates so that the two bones become rough and irregular. This type of arthritis is apt to affect the joints that have received the greatest use over the years.

Bones are joined at joints. Synovial joints are freely movable and allow particular types of movements.

You Can Avoid Osteoporosis

Osteoporosis is a condition in which the bones are weakened due to a decrease in the bone mass that makes up the skeleton. Throughout life, bones are continuously remodeled. While a child is growing, the rate of bone formation is greater than the rate of bone breakdown. The skeletal mass continues to increase until ages 20 to 30. After that, there is an equal rate of formation and breakdown of bone mass until ages 40 to 50. Then, reabsorption begins to exceed formation, and the total bone mass slowly decreases.

Over time, men are apt to lose 25% and women lose 35% of their bone mass. But we have to consider that men tend to have denser bones than women anyway, and their testosterone (male sex hormone) level generally does not begin to decline significantly until after age 65. In contrast, the estrogen (female sex hormone) level in women begins to decline at about age 45. Since sex hormones play an important role in maintaining bone strength, this difference means that women are more likely than men to suffer a higher incidence of fractures, involving especially the hip vertebrae, long bones, and pelvis. Although osteoporosis may at times be the result of various disease processes, it is essentially a disease of aging.

There are measures that everyone can take to avoid osteoporosis when they get older. Adequate dietary calcium throughout life is an important protection against osteoporosis. The U.S. National Institutes of Health recommend a calcium intake of 1,200–1,500 mg per day during puberty. Males and females require 1,000 mg per day until the age of 65 and 1,500 mg per day after age 65. In older women not receiving estrogen replacement therapy, 1,500 milligrams per day is especially desirable.

A small daily amount of vitamin D is also necessary in order to use calcium correctly. Exposure to sunlight is required to allow skin to synthesize a precursor to vitamin D. If you reside on or north of a "line" drawn from Boston to Milwaukee, to Minneapolis, to Boise, chances are you're not getting enough vitamin D during the winter months. Therefore, you should avail yourself of vitamin D present in fortified foods such as low-fat milk and cereal.

Very inactive people, such as those confined to bed, lose bone mass 25 times faster than people who are moderately active. On the other hand, a regular, moderate, weight-bearing exercise like walking or jogging is another good way to maintain bone strength (Fig. 48A).

Postmenopausal women with any of these risk factors should have an evaluation of their bone density:

- white or Asian race
- thin body type
- family history of osteoporosis
- early menopause (before age 45)
- smoking
- a diet low in calcium, or excessive alcohol consumption and caffeine intake
- sedentary lifestyle

Presently bone density is measured by a method called dual energy X-ray absorptiometry (DEXA). This test measures bone density based on the absorption of photons generated by an X-ray tube. Soon there may be a blood and urine test to detect the biochemical markers of bone loss. Then it may be possible for physicians to screen all older women and at-risk men for osteoporosis.

If the bones are thin, it is worthwhile to take other measures to gain bone density because even a slight increase can significantly reduce fracture risk. Usually, experts believe that estrogen therapy in women is the treatment of choice, but there are other options. Evidence shows that estrogen replacement therapy will delay accelerated bone loss in postmenopausal women. Hormone replacement is most effective when begun at the start of menopause and continued over the long term. Some benefit, however, may still be obtained when treatment is begun later. A combination of hormone replacement and exercise apparently yields the best results. But there are other options, also. Calcitonin is a naturally occurring hormone whose main site of action is the skeleton where it inhibits the action of osteoclasts, the cells that break down bone. Also, alendronate is a drug that acts similarly to calcitonin. After three years of alendronate therapy, an increase in spinal density by about 8% and hip density by about 7% is obtained. Promising new drugs include slow-release fluoride therapy and certain growth hormones. These medications stimulate the formation of new bone.

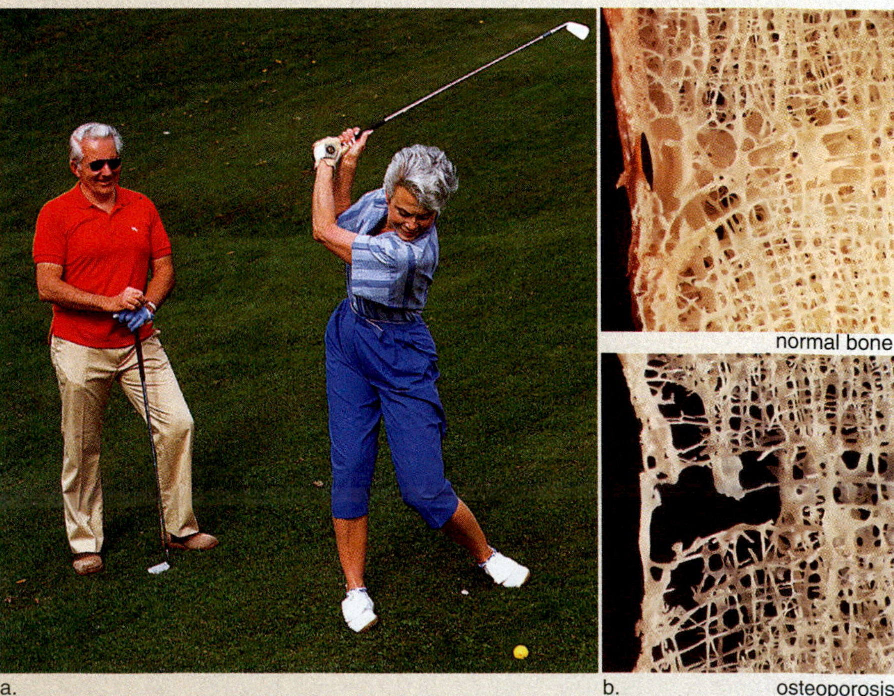

Figure 48A Osteoporosis.
a. Exercise can help prevent osteoporosis, but when playing golf you should carry your own clubs and walk instead of using a golf cart. b. Normal bone compared to bone from a person with osteoporosis.

48.3 The Human Muscular System

Figure 48.11 gives an anterior view of some of the major superficial skeletal muscles. Aside from supporting the body and allowing the body and body parts to move, skeletal muscle contraction assists homeostasis by helping maintain a constant body temperature. Skeletal muscle contraction causes ATP breakdown, releasing heat that is distributed about the body.

Macroscopic Anatomy and Physiology

Skeletal muscles are attached to the skeleton by bands of fibrous connective tissue called **tendons** [L. *tendo,* stretch]. When muscles contract, they shorten. Therefore, muscles can only pull; they cannot push. Because of this, skeletal muscles must work in antagonistic pairs. If one muscle of an antagonistic pair flexes the joint and brings the limb toward the body, the other one extends the joint and straightens the limb. Figure 48.12 illustrates this principle and compares the actions of muscles in the limb of an arthropod, in which the muscles are attached to an exoskeleton, with those of a human, in whom the muscles are attached to an endoskeleton.

In the laboratory, if a muscle is given a rapid series of threshold stimuli, it can respond to the next stimulus without relaxing completely. In this way, muscle contraction summates until maximal sustained contraction, called **tetanus,** is achieved. Tetanic contractions ordinarily occur in the body's muscles whenever skeletal muscles are actively used. Even when muscles appear to be at rest, they exhibit **tone,** in which some of their fibers are contracting. Muscle tone is particularly important in maintaining posture. If all the fibers within the muscles of the neck, trunk, and legs were to suddenly relax, the body would collapse.

Muscles work in antagonistic pairs. A muscle at rest exhibits tone dependent on tetanic contractions.

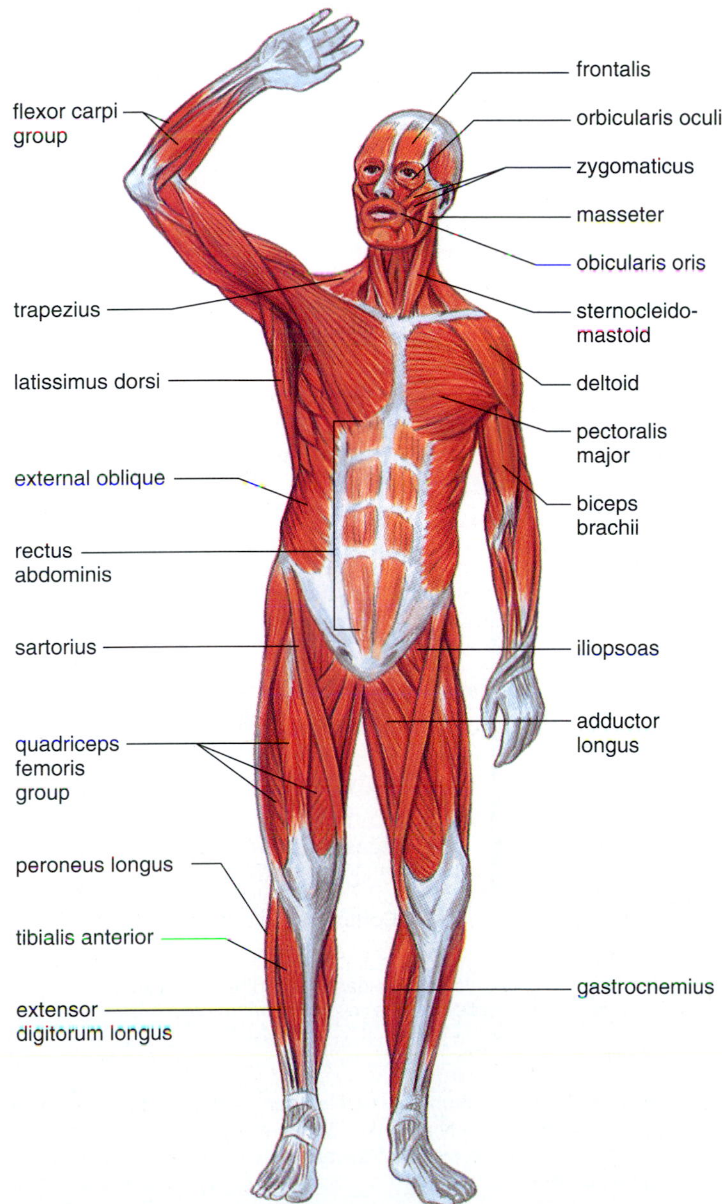

Figure 48.11 Human musculature.
Anterior view of some of the major superficial skeletal muscles.

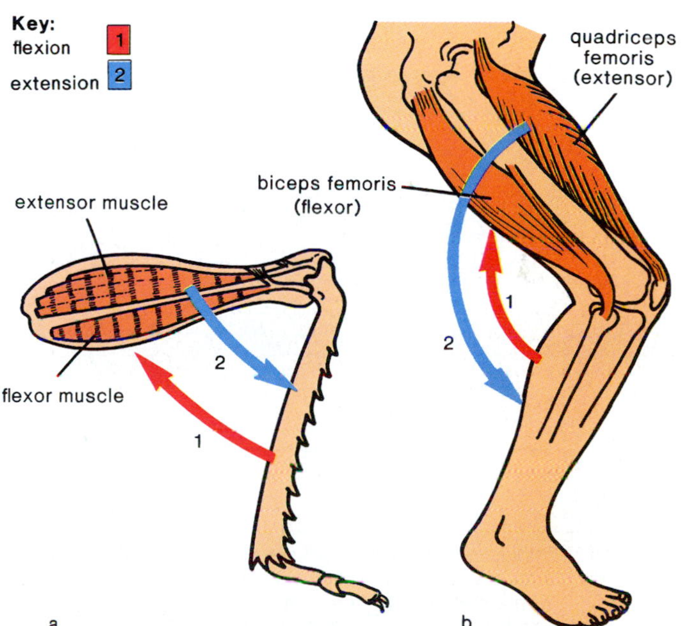

Figure 48.12 Antagonistic muscle pairs.
Muscles can exert force only by shortening. Movable joints are supplied with double sets of muscles, which work in opposite directions. As indicated by the arrows, flexor muscles decrease the angle at a joint, and extensors increase the angle at a joint. **a.** Muscles in an insect's leg as an example of antagonistic muscles attached to the inside of an arthropod exoskeleton. **b.** Muscles in the human leg as an example of antagonistic muscles attached to a vertebrate endoskeleton.

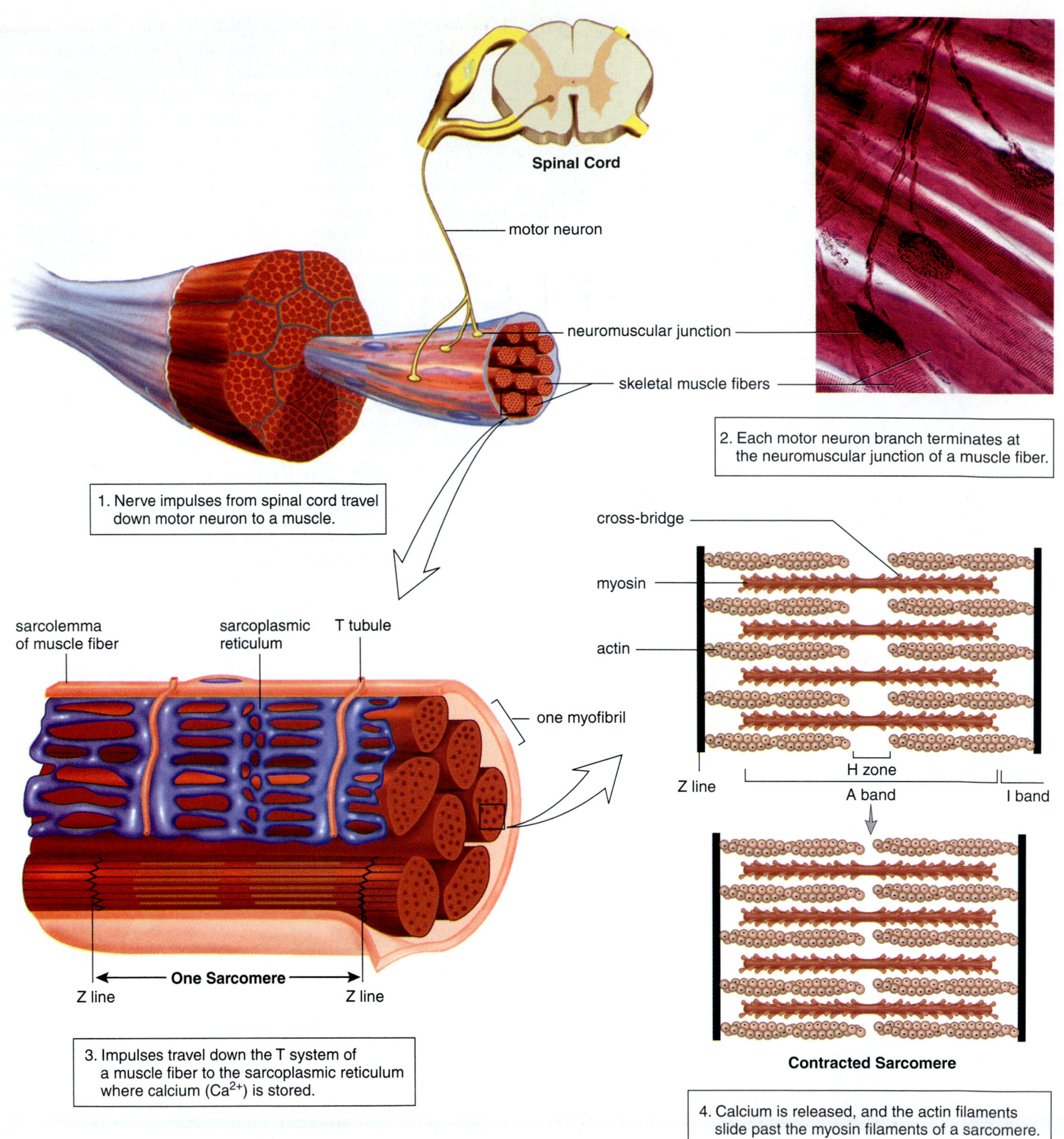

Figure 48.13 **Contraction of a muscle.**
A whole muscle contains bundles of skeletal muscle fibers. A motor neuron ends in neuromuscular junctions at muscle fibers. A muscle fiber contains many myofibrils. Each myofibril is divided into sarcomeres, where actin and myosin filaments are precisely arranged. When motor impulses reach a neuromuscular junction, transmission across a synapse occurs. Impulses travel down the T system of a muscle fiber and calcium is released from the sarcoplasmic reticulum. Then, actin filaments slide past myosin filaments and sarcomeres contract.

Microscopic Anatomy and Physiology

A whole skeletal muscle is composed of a number of muscle fibers in bundles. Each muscle fiber is a cell containing the usual cellular components but some components have special features (Fig. 48.13). The **sarcolemma** or plasma membrane [Gk. *sarkos,* flesh, and *lemma,* sheath], forms a T (for transverse) system. The T tubules penetrate, or dip down, into the cell so that they come into contact—but do not fuse—with expanded portions of modified endoplasmic reticulum called the **sarcoplasmic reticulum.** These expanded portions serve as storage sites for calcium ions (Ca^{2+}), which are essential for muscle contraction. The sarcoplasmic reticulum encases hundreds and sometimes even thousands of **myofibrils** [Gk. *myos,* muscle, and L. *fibra,* thread], which are the contractile portions of muscle fiber.

Myofibrils are cylindrical in shape and run the length of the muscle fiber. The light microscope shows that a myofibril has light and dark bands called striations. It is these bands that cause skeletal muscle to appear striated. The electron microscope shows that the striations of myofibrils are formed by the placement of protein filaments within contractile units called **sarcomeres** [Gk. *sarkos,* flesh, and *meros,* part]. A sarcomere extends between two dark lines called the Z lines. There are two types of protein filaments: the thick filaments are made up of **myosin,** and the thin filaments are made up of **actin.** The I band is light colored because it contains only actin filaments attached to a Z line. The dark regions of the A band contain overlapping actin and myosin filaments, and its H zone has only myosin filaments.

Sliding Filament Model

As a muscle fiber contracts, the sarcomeres within the myofibrils shorten. When a sarcomere shortens, the actin (thin) filaments slide past the myosin (thick) filaments and approach one another. This causes the I band to shorten and the H zone to nearly or completely disappear. The movement of actin filaments in relation to myosin filaments is called the **sliding filament model** of muscle contraction. During the sliding process, the sarcomere shortens even though the filaments themselves remain the same length.

The participants in muscle contraction have the functions listed in Table 48.1. ATP supplies the energy for muscle contraction. Although the actin filaments slide past the myosin filaments, it is the myosin filaments that do the work. Myosin filaments break down ATP and form crossbridges that attach to and pull the actin filaments toward the center of the sarcomere.

The sliding filament model states that actin filaments slide past myosin filaments because myosin has crossbridges, which pull the actin filaments inward.

Table 48.1

Muscle Contraction

Name	Function
Actin filaments	Slide past myosin, causing contraction
Ca^{2+}	Needed for myosin to bind to actin
Myosin filaments	Pull actin filaments by means of cross-bridges; are enzymatic and split ATP
ATP	Supplies energy for muscle contraction

ATP ATP provides the energy for muscle contraction. Although muscle cells contain **myoglobin,** a molecule that stores oxygen, cellular respiration does not immediately supply all the ATP that is needed. In the meantime, muscle fibers rely on **creatine phosphate** (phosphocreatine), a storage form of high-energy phosphate. Creatine phosphate cannot directly participate in muscle contraction. Instead, it anaerobically regenerates ATP by the following reaction:

$$\text{creatine—P} + \text{ADP} \longrightarrow \text{ATP} + \text{creatine}$$

This reaction occurs in the midst of sliding filaments, and therefore, this method of supplying ATP is the speediest energy source available to muscles.

When all of the creatine phosphate is depleted, mitochondria may by then be producing enough ATP for muscle contraction to continue. If not, fermentation is a second way muscles have of supplying ATP without consuming oxygen. Fermentation, which is apt to occur when strenuous exercise first begins, can supply ATP for only a short time because of lactate buildup. The buildup is noticeable when it produces muscle aches and fatigue upon exercising.

We all have had the experience of needing to continue deep breathing following strenuous exercise. This continued intake of oxygen required to complete the metabolism of lactate and restore cells to their original energy state represents an **oxygen debt.** The lactate is transported to the liver, where 20% of it is completely broken down to carbon dioxide (CO_2) and water (H_2O). The ATP gained by this respiration then is used to reconvert 80% of the lactate to glucose. In persons who train, the number of mitochondria increases, and there is a greater reliance on them rather than fermentation to produce ATP. Less lactate is produced, and there is less oxygen debt.

Working muscles require a supply of ATP. Anaerobic creatine phosphate breakdown and fermentation quickly generate ATP. Sustained exercise requires aerobic respiration.

Muscle Innervation

Muscles are stimulated to contract by motor nerve fibers. Nerve fibers have several branches, each of which ends in an axon bulb that lies in close proximity to the sarcolemma of a muscle fiber. A small gap, called a synaptic cleft, separates the axon bulb from the sarcolemma. This entire region is called a **neuromuscular junction** (Fig. 48.14).

Axon bulbs contain synaptic vesicles that are filled with the neurotransmitter acetylcholine (ACh). When nerve impulses traveling down a motor neuron arrive at an axon bulb, the synaptic vesicles release ACh into the synaptic cleft. ACh quickly diffuses across the cleft and binds to receptors in the sarcolemma. Now the sarcolemma generates impulses that spread over the sarcolemma and down T tubules to the sarcoplasmic reticulum. The release of calcium from the sarcoplasmic reticulum causes the filaments within sarcomeres to slide past one another. Sarcomere contraction results in myofibril contraction, which in turn results in muscle fiber and finally muscle contraction.

At a neuromuscular junction, nerve impulses bring about the release of a neurotransmitter that signals a muscle fiber to contract.

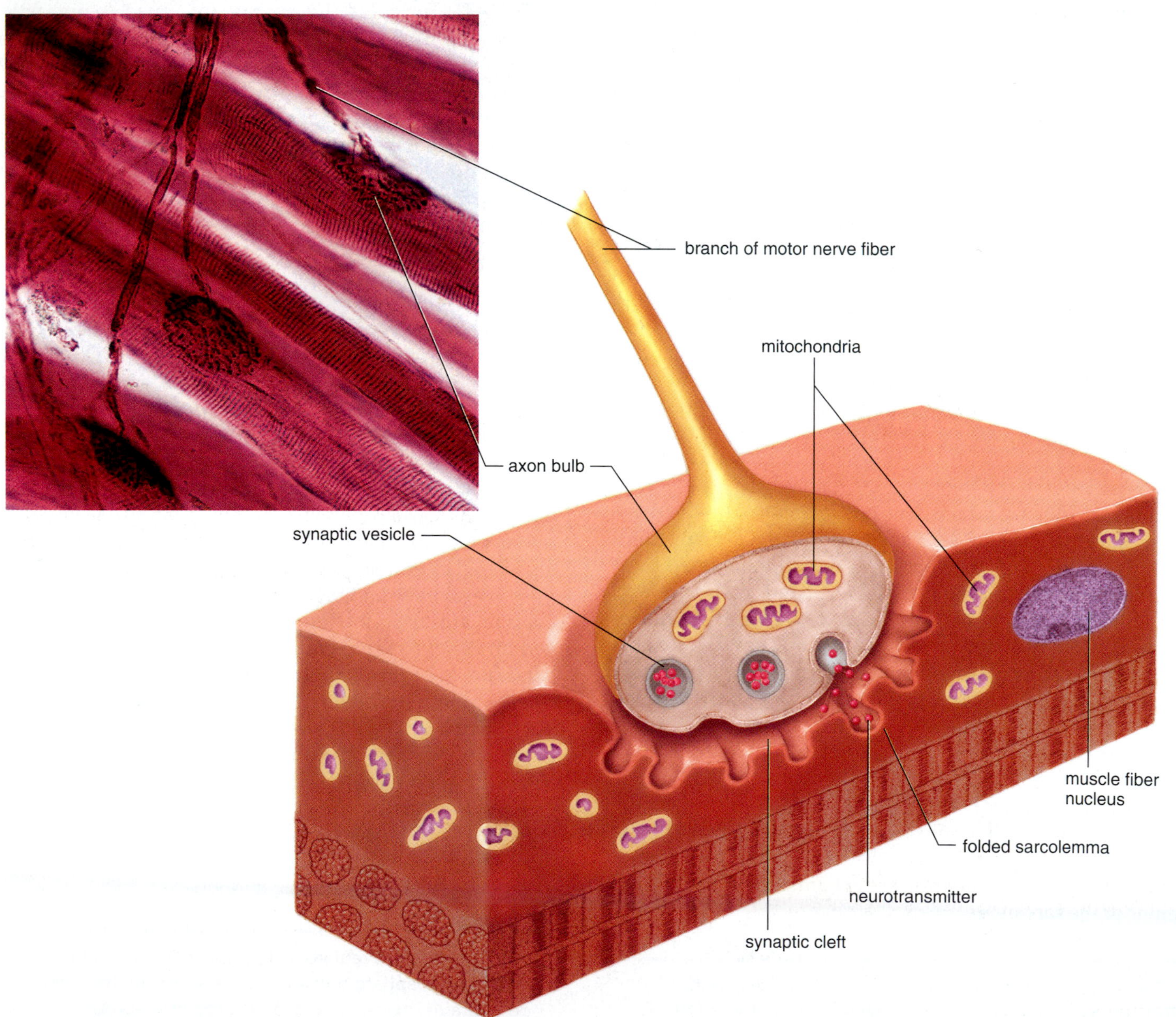

Figure 48.14 Neuromuscular junction.
The branch of a motor nerve fiber terminates in an axon bulb that meets but does not touch a muscle fiber. A synaptic cleft separates the axon bulb from the sarcolemma of the muscle fiber. Nerve impulses traveling down a motor fiber cause synaptic vesicles to discharge a neurotransmitter, which diffuses across the synaptic cleft. When the neurotransmitter is received by the sarcolemma of a muscle fiber, impulses begin which lead to muscle fiber contraction.

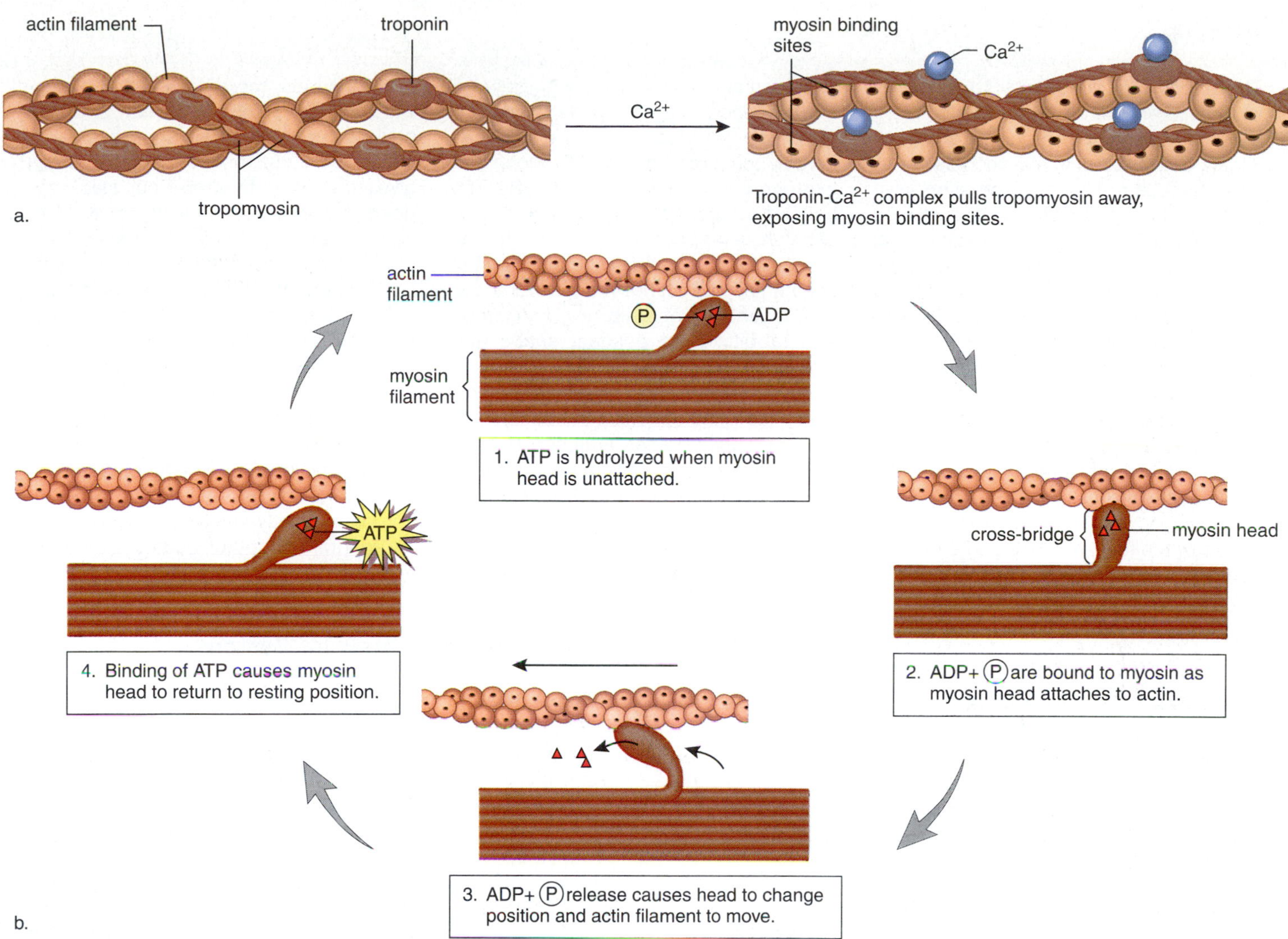

Figure 48.15 The role of calcium and myosin in muscle contraction.
a. Upon release, calcium binds to troponin, exposing myosin binding sites. b. After breaking down ATP (1), myosin heads bind to an actin filament (2), and later, a power stroke causes the actin filament to move (3). When another ATP binds to myosin, the head detaches from actin (4), and the cycle begins again. Although only one myosin head is shown, many heads are active at the same time.

Figure 48.15 shows the placement of two other proteins associated with a thin filament, which is composed of a double row of twisted actin molecules. Threads of tropomyosin wind about an actin filament, and troponin occurs at intervals along the threads. Calcium ions (Ca^{2+}) that have been released from the sarcoplasmic reticulum combine with troponin. After binding occurs, the tropomyosin threads shift their position, and myosin binding sites are exposed.

The thick filament is actually a bundle of myosin molecules, each having a double globular head with an ATP binding site. The heads function as ATPase enzymes, splitting ATP into ADP and Ⓟ. This reaction activates the heads so that they bind to actin. The ADP and Ⓟ remain on the myosin heads until the heads attach to actin, forming cross-bridges. Now, ADP and Ⓟ are released, and this causes the cross-bridges to change their positions. This is the power stroke that pulls the thin filaments toward the middle of the sarcomere. When another ATP molecule binds to a myosin head, the cross-bridge is broken as the head detaches from actin. The cycle begins again; the actin filaments move nearer the center of the sarcomere each time the cycle is repeated.

Contraction continues until nerve impulses cease and calcium ions are returned to their storage sites. The membranes of the sarcoplasmic reticulum contain active transport proteins that pump calcium ions back into the calcium storage sites.

Myosin filament heads break down ATP and then attach to an actin filament, forming cross-bridges that pull the actin filament to the center of a sarcomere.

Connecting Concepts

The adage that structure fits function is evident when one observes how different animals locomote. Among a group of widely diversified animals, such as mammals, various modes of locomotion have evolved. Humans are bipedal and walk on the soles of their feet formed by the tarsal bones. This form of locomotion allows the hands to be free and may have evolved from the habit of monkeys and apes to use only forelimbs as they swing through the branches of trees. Dexterity of hands and feet are actually the ancestral mammalian condition. In humans and apes, the bones of the hands and feet are not fused and the wrist and ankle can rotate in three dimensions.

In carnivores, like dogs and cats, the bones of the feet are fused and these animals walk on their toes. This is an obvious adaptation to running—we also raise the heel and engage the toes in order to run faster. Hoofed mammals, such as horses and deer, have greatly elongated legs and run on the tips of their digits. The hoof of a horse is its third digit only. A cheetah can sprint faster than a horse, but a horse will eventually outdistance the cheetah because it has more endurance.

Mammals that jump, like kangaroos and rabbits, have a squat shape and elongated hindlimbs that propel them forward. Several groups of arboreal mammals, like flying squirrels and sugar gliders, have membranes attached to their bodies that permit them to glide through the air from tree to tree. Among mammals, only bats truly fly. Their membranous wings are stretched between greatly elongated forelimbs and fingers. In both birds and bats, the wing is moved downward and forward in one motion, and then backward and upward in another.

In terrestrial animals, the skeleton gives the body its shape but also supports it against the pull of gravity. Because water is buoyant, gravity isn't much of a problem for aquatic animals, but a shape that reduces friction is quite helpful. Seals, sea lions, whales, and dolphins have a streamlined, torpedo shape that facilitates movement through water. A whale has few protruding parts; a male's penis is completely hidden within muscular folds, and the teats of the female lie behind slits on either side of the genital area. Thus, we see that while locomotion chiefly involves musculoskeletal adaptations, other body systems are involved as well.

Summary

48.1 Diversity of Skeletons

Three types of skeletons are found in the animal kingdom: hydrostatic skeleton (cnidarians, flatworms, and segmented worms); exoskeleton (certain molluscs and arthropods); and endoskeleton (vertebrates). The rigid but jointed skeleton of arthropods and vertebrates helped them colonize the terrestrial environment.

48.2 The Human Skeletal System

The human skeleton gives support to the body, helps protect internal organs, provides sites for muscle attachment, and is a storage area for calcium and phosphorous salts, as well as a site for blood cell formation.

Most bones are cartilaginous in the fetus but are converted to bone during development. A long bone undergoes endochondral ossification in which a cartilaginous growth plate remains between the primary ossification center in the middle and the secondary centers at the ends of the bones. Growth of the bone is possible as long as the growth plates are present, but eventually they, too, are converted to bone. Bone is constantly being renewed; osteoclasts break down bone, and osteoblasts build new bone. Osteocytes are in the lacunae of osteon; long bone has a shaft of compact bone and two ends that contain spongy body. The shaft contains a medullary cavity with yellow marrow, and the ends contain red marrow.

The human skeleton is divided into two parts: (1) the axial skeleton, which is made up of the skull, the ribs, the sternum, and the vertebrae; and (2) the appendicular skeleton, which is composed of the girdles and their appendages.

Joints are classified as immovable, like those of the cranium; slightly movable, like those between the vertebrae; and freely movable or synovial joints, like those in the knee and hip. In synovial joints, ligaments bind the two bones together, forming a capsule in which there is synovial fluid.

48.3 The Human Muscular System

Whole skeletal muscles can only shorten when they contract; therefore, they work in antagonistic pairs. For example, if one muscle flexes the joint and brings the limb toward the body, the other one extends the joint and straightens the limb. A muscle at rest exhibits tone dependent on tetanic contractions.

A whole skeletal muscle is composed of muscle fibers. Each muscle fiber is a cell that contains myofibrils in addition to the usual cellular components. Longitudinally, myofibrils are divided into sarcomeres, which display the arrangement of actin and myosin filaments.

The sliding filament model of muscle contraction says that myosin filaments have cross-bridges, which attach to and detach from actin filaments, causing actin filaments to slide and the sarcomere to shorten. (The H zone disappears as actin filaments approach one another.) Myosin is an ATPase—ATP is needed for muscle contraction. Anaerobic creatine phosphate breakdown and fermentation quickly generate ATP. Sustained exercise requires cellular respiration for the generation of ATP.

Nerves innervate muscles. Nerve impulses traveling down motor neurons to neuromuscular junctions cause the release of ACh, which binds to receptors on the sarcolemma (plasma membrane of a muscle fiber). Impulses begin and move down T tubules that approach the sarcoplasmic reticulum (endoplasmic reticulum of muscle fibers) where calcium is stored. Thereafter, calcium ions are released and bind to troponin. The troponin-Ca^{2+} complex causes tropomyosin threads winding around actin filaments to shift their position, revealing myosin binding sites. The myosin filament is composed of many myosin molecules, each containing a head with an ATP binding site. When a myosin head breaks down ATP, it is ready to attach to actin. The release of ATP + Ⓟ causes the head to change its position. This is the power stroke that causes the actin filament to slide toward the center of a sarcomere. When another ATP binds to myosin, the head detaches from actin and the cycle begins again.

Reviewing the Chapter

1. What are the three types of skeletons found in the animal kingdom and how do they differ? Cite animals that have these types of skeletons. 864–65
2. Give several functions of the skeletal system in humans. How does the skeletal system contribute to homeostasis? 866
3. Contrast compact bone with spongy bone. Explain how bone grows and is renewed. 866–67
4. Distinguish between the axial and appendicular skeletons. 868, 870
5. List the bones that form the pectoral girdle and arm; the pelvic girdle and leg. 870–71
6. How are joints classified? Describe the anatomy of a freely movable joint. 871
7. Give several functions of the muscular system in humans. How does the muscular system contribute to homeostasis? 873
8. Describe how muscles are attached to bones. What is accomplished by muscles acting in antagonistic pairs? 873
9. Discuss the microscopic structural features of a muscle fiber and a sarcomere. What is the sliding filament model? 875
10. Discuss the availability and the specific role of ATP during muscle contraction. What is oxygen debt, and how is it repaid? 875
11. Describe the structure and function of a neuromuscular junction. 876
12. Describe the cyclical events as myosin pulls actin toward the center of a sarcomere. 877

Testing Yourself

Choose the best answer for each question. For questions 1–4, match the following items with the correct locations given in the key.

Key:

a. upper arm
b. lower arm
c. pectoral girdle
d. pelvic girdle
e. upper leg
f. lower leg

1. ulna
2. tibia
3. clavicle
4. femur
5. Spongy bone
   a. contains osteons.
   b. contains red bone marrow where blood cells are formed.
   c. lends no strength to bones.
   d contributes to homeostasis.
   e. Both b and d are correct.
6. Which of these is mismatched?
   a. slightly movable joint—vertebrae
   b. hinge joint—hip
   c. synovial joint—elbow
   d. immovable joint—sutures in cranium
   e. ball and socket joint—hip
7. In a muscle fiber
   a. the sarcolemma is connective tissue holding the myofibrils together.
   b. the sarcoplasmic reticulum stores calcium.
   c. both myosin and actin filaments have cross-bridges.
   d. there is a T system but no endoplasmic reticulum.
   e. All of these are correct.
8. When muscles contract
   a. sarcomeres shorten.
   b. myosin breaks down ATP.
   c. actin slides past myosin.
   d. the H zone disappears.
   e. All of these are correct.
9. Which of these is the direct source of energy for muscle contraction?
   a. ATP
   b. creatine phosphate
   c. lactic acid
   d. glycogen
   e. Both a and b are correct.
10. Nervous stimulation of muscles
    a. occurs at a neuromuscular junction.
    b. involves the release of ACh.
    c. results in impulses that travel down the T system.
    d. causes calcium to be released from the sarcoplasmic reticulum.
    e. All of these are correct.
11. Label this diagram of muscle fiber, using these terms: myofibril, Z line, T tubule, sarcomere, sarcolemma, sarcoplasmic reticulum.

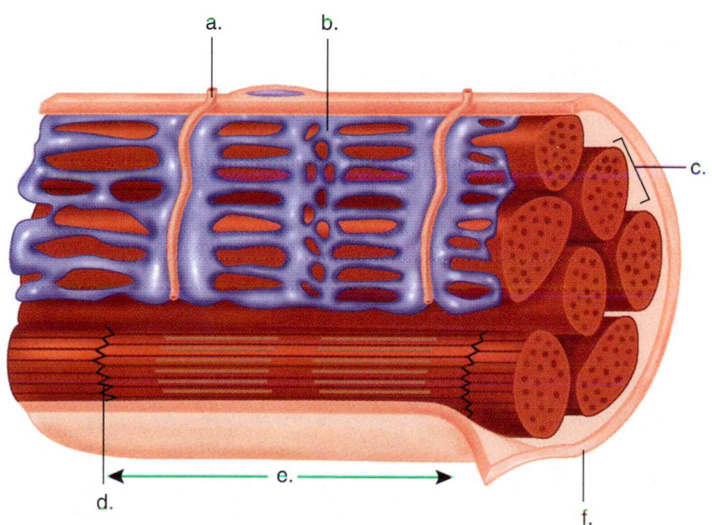

Thinking Scientifically

1. It is observed that some motor neurons innervate only a few muscle fibers in the biceps brachii. Other motor neurons each innervate many muscle fibers. How might this observation correlate with our ability to pick up a pencil or a two-liter soda bottle? On what basis would the brain bring about the correct level of contraction?
2. Some athletes believe that if they take oral creatine it will increase their endurance because it will increase the amount of phosphate available to their muscles for ATP synthesis. This statement can be regarded as two hypotheses: (1) oral creatine will increase endurance; (2) oral creatine increases the amount of creatine available in muscles for ATP synthesis. How could these two hypotheses be tested?

Bioethical Issue

A natural advantage does not bar an athlete from participating in and winning a medal in a particular sport at the Olympic Games. Nor are athletes restricted to a certain amount of practice or required to eliminate certain foods from their diet.

Athletes are, however, prevented from participating in the Olympic Games if they have taken certain performance-enhancing drugs. There is no doubt that regular use of drugs like anabolic steroids leads to kidney disease, liver dysfunction, hypertension, and a myriad of other undesirable side effects. Even so, shouldn't the individual be allowed to take these drugs if they want to? Anabolic steroids are synthetic forms of the male sex hormone testosterone. Large doses along with strength training leads to much larger muscles than otherwise. Extra strength and endurance can give an athlete an advantage in certain sports such as racing, swimming, and weight lifting.

Should the Olympic committee outlaw the taking of anabolic steroids and, if so, on what basis? The basis can't be an unfair advantage, because some athletes naturally have an unfair advantage over other athletes. Should these drugs be outlawed on the basis of health reasons? Excessive practice alone and a purposeful decrease or increase in weight to better perform in a sport can also be injurious to one's health. In other words, how can you justify allowing some behaviors that enhance performance and not others?

Understanding the Terms

actin 875
appendicular skeleton 870
axial skeleton 868
bursa 871
compact bone 866
creatine phosphate 875
endoskeleton 864
exoskeleton 864
fontanel 868
foramen magnum 868
hydrostatic skeleton 864
joint 871
ligament 871
myofibril 875
myoglobin 875
myosin 875
neuromuscular junction 876
osteoblast 866
osteoclast 866
osteocyte 866
oxygen debt 875
pectoral girdle 870
pelvic girdle 871
red bone marrow 866
sarcolemma 875
sarcomere 875
sarcoplasmic reticulum 875
sliding filament model 875
spongy bone 866
synovial joint 871
tendon 873
tetanus 873
tone 873
vertebral column 869

Match the terms to these definitions:

a. ______________ Bone-forming cell.
b. ______________ Movement of actin filaments in relation to myosin filaments, which accounts for muscle contraction.
c. ______________ Muscle protein making up the thin filaments in a sarcomere; its movement shortens the sarcomere, yielding muscle contraction.
d. ______________ Part of the skeleton that consists of the shoulder and pelvic girdles and the bones of the arms and legs.
e. ______________ Portion of the skeleton that provides support and attachment for the arms.

Web Connections

Exploring the Internet

http://www.mhhe.com/biosci/genbio/mader
(click on *Biology 7/e*)

The *Biology 7/e* Online Learning Center provides many resources for studying the material in this chapter including links to the following sites:

Shapeup America's site. Determine your body mass index and answers to commonly asked fitness questions at this site.

http://www.shapeup.org/

The Arthritis Foundation's site Arthritis Research. Describes the advances in treatment and medication that ameliorate symptoms of arthritis.

http://www.arthritis.org/research/arthritis_research.asp

NIH Osteoporosis and Related Bone Diseases. Information, a bibliography, newsletters and more at this comprehensive site on osteoporosis and related disorders.

http://www.osteo.org/

Osteoporosis—Doctor's Guide to the Internet. Much information, including full-text articles on current research on treatments and prevention relating to osteoporosis.

http://www.pslgroup.com/OSTEOPOROSIS.HTM

101 Skeleton Jokes. For a lighter approach to the skeleton, check out these terrible jokes!

http://www.laffnow.com/humor/hallskel.htm

Hormones and Endocrine Systems

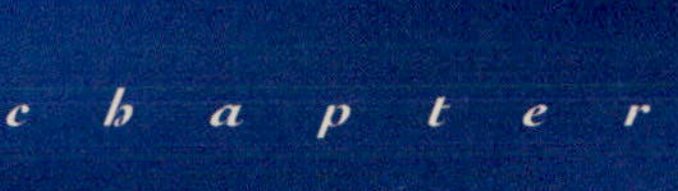

chapter concepts

49.1 Chemical Signals

- In general, there are three types of chemical signals, and hormones are one of these types. 882
- Hormones influence the metabolism of their target cells. 883

49.2 Human Endocrine System

- The hypothalamus controls the function of the pituitary gland, which in turn controls several other glands. 886
- The parathyroid glands are embedded in the thyroid but have an entirely separate structure and function. 889
- The thyroid produces two hormones that speed metabolism and another that lowers the blood calcium level. The parathyroid glands produce a hormone that raises the blood calcium level. 889–90
- The adrenal medulla and the adrenal cortex are separate parts of the adrenal glands that have functions in relation to stress. 891
- The pancreas secretes hormones that help control the blood glucose level. 894
- Diabetes mellitus occurs when cells are unable to take up glucose and it spills over into the urine. 895
- The gonads produce the sex hormones that control secondary sexual characteristics. 896

Cheetah, *Acinonyx jubatus*, hunting Thomson's gazelle, *Gazella thomsoni*

The "fight-or-flight" reaction illustrates that it is difficult to separate the nervous system from the endocrine system. Sympathetic nerve fibers control the release of epinephrine and norepinephrine from the adrenal medulla (an endocrine gland) when a gazelle is trying to escape a cheetah.

The nervous system is well known for bringing about an immediate response to environmental stimuli, as when you move your hand away from a flame. The nervous system integrates sensory information and controls the action of effectors—muscles and glands—by using chemical signals called neurotransmitters.

The **endocrine system** coordinates body parts by using chemical signals called hormones. It is usually slower acting than the nervous system and regulates processes that occur over days or even months. Most often, hormones secreted into the bloodstream control whole body processes like growth, reproduction, and complex behaviors, including courtship and migration. Once hormones arrive at those cells with appropriate receptor proteins, they influence the metabolism of the cell. It takes a while to metabolically change a cell and the effect is longer lasting.

49.1 Chemical Signals

Chemical signals are a general phenomenon in the animal kingdom. It is apparent to us that the nervous system uses chemical signals called neurotransmitter molecules and the endocrine system uses chemical signals called hormones. But chemical signals are used at various levels of communication: between individuals, between body parts, and between cells.

Figure 49.1*a* illustrates that chemical signals sometimes act at a distance between individuals. Female moths release a sex attractant that is received by male moth antennae even several miles away. The receptors on the antennae of the male silkworm moth are so sensitive that only 40 out of 40,000 receptor proteins need to be activated in order for the male to respond and find the female. Such extreme sensitivity is believed to ensure that only male moths of the correct species and none others will respond. Many other types of organisms also release chemical signals, called **pheromones,** as a way to communicate with other members of their species. Ants lay down a pheromone trail which directs other ants to food. Mammals also release pheromones. Mammals as varied as dogs and antelopes use their urine, feces, and anal gland secretions to mark their territories. Studies are being conducted to determine if humans also have pheromones.

Figure 49.1*b* illustrates that chemical signals sometimes act at a distance between body parts. This category includes hormones produced by the endocrine glands. A **hormone** can be defined as an organic chemical produced by one set of cells that affects a different set. A hormone, such as insulin produced by the pancreas, travels through the bloodstream to cells that have receptor proteins to receive it. The receptors combine with insulin in a lock-and-key manner. Cells respond to one hormone and not another, depending on their specific receptors.

This category also includes the secretions of neurosecretory cells in the hypothalamus. These secretions travel in a capillary network that runs between the hypothalamus and the pituitary gland. Some of these secretions stimulate and others prevent the pituitary from secreting its hormones.

Figure 49.1*c* illustrates that chemical signals sometimes act locally between adjacent cells. Neurotransmitters released by neurons belong in this category. Prostaglandins are substances produced by cells that can affect their neighbors, sometimes promoting pain and inflammation. Aspirin seems to offset the effect of prostaglandins. Growth factors are local hormones which promote cell division and mitosis. Some growth factors are now being used to treat various human conditions. Antagonists are being developed against a growth factor called tumor angiogenesis factor, which is released by tumor cells and promotes the formation of capillary networks.

There are chemical signals that work at a distance between individuals, or at a distance between body parts, or locally between adjacent cells.

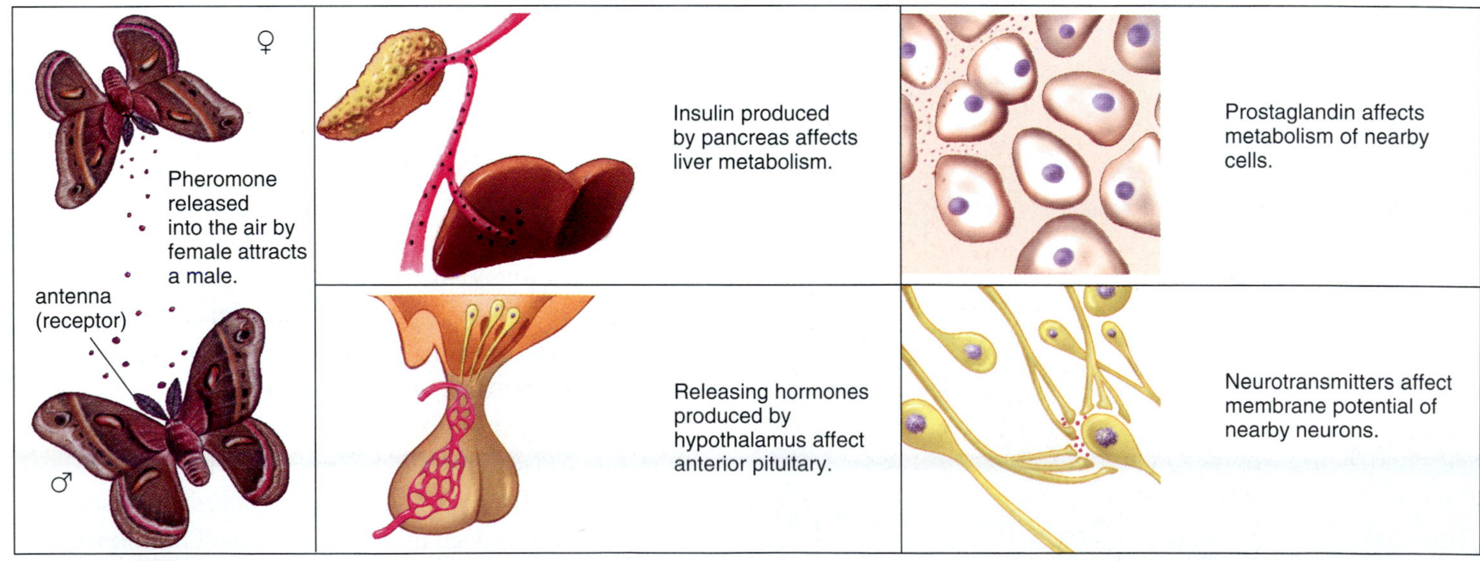

Figure 49.1 Chemical signals.
There are three categories of chemical signals (red dots). **a.** Pheromones are chemical signals that act at a distance between individuals. **b.** Endocrine hormones and neurosecretions typically are carried in the bloodstream and act at a distance within the body of a single organism. **c.** Some chemical signals have local effects only; they pass between cells that are adjacent to one another. Neurotransmitters belong to this category as do such local hormones including prostaglandins.

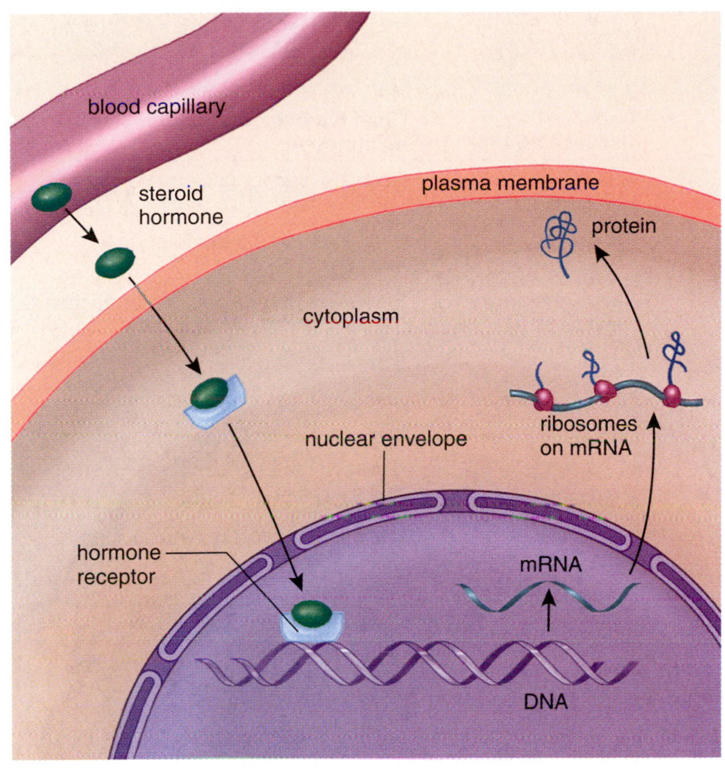

a. Action of steroid hormone

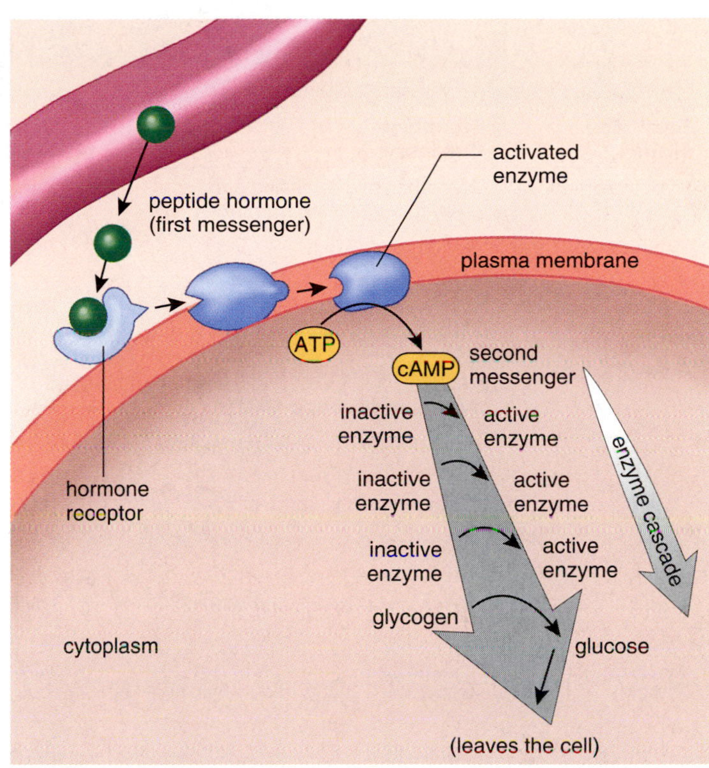

b. Action of peptide hormone

Figure 49.2 Cellular activity of hormones.
a. After passing through the plasma membrane, a steroid hormone binds to a receptor protein inside the cytoplasm or the nucleus. The hormone-receptor complex binds to DNA, and this leads to activation of certain genes and protein synthesis. **b.** Peptide hormones, called first messengers, bind to a receptor protein in the plasma membrane. A protein relay in the membrane ends when an enzyme converts ATP to cAMP, the second messenger, which activates an enzyme cascade.

The Action of Hormones

Hormones fall into two basic chemical classes: (1) in the class **steroid hormone** the hormone always has the same complex of four carbon rings, but each one has different side chains; (2) the class **peptide hormone** includes those hormones that are peptides, proteins, glycoproteins or modified amino acids. Because their effect is amplified through cellular mechanisms, both classes of hormones can function at extremely low concentrations.

A hormone does not seek out a particular organ; rather the organ is awaiting the arrival of the hormone. Cells that can react to a hormone have receptor proteins that combine with the hormone in a lock-and-key manner. Steroid hormones are lipids, and therefore they cross cell membranes (Fig. 49.2*a*). Only after they are inside the cytoplasm or the nucleus do steroid hormones, such as estrogen and progesterone, bind to receptor proteins. The hormone-receptor complex then binds to DNA, activating particular genes. Activation leads to production of a cellular enzyme in multiple quantities.

Most peptide hormones cannot pass through cell membranes, and they bind to a receptor protein in the plasma membrane (Fig. 49.2*b*). After epinephrine binds to a receptor protein, a relay system leads to the conversion of ATP to cyclic AMP (cyclic adenosine monophosphate). Cyclic AMP (cAMP) contains only one phosphate group attached to the adenosine portion of the molecule at two spots. Therefore, the molecule is cyclic. The peptide hormone is called the **first messenger** and cAMP is called the **second messenger.** Calcium is also a common second messenger, and this helps explain why calcium regulation in the body is so important.

The second messenger sets in motion an enzyme cascade. In muscle cells, the reception of epinephrine leads to the breakdown of glycogen to glucose (Fig. 49.2*b*). An enzyme cascade is so called because each enzyme in turn activates another. Because enzymes work over and over, every step in an enzyme cascade leads to more reactions—the binding of a single peptide hormone molecule can result even in a thousandfold response.

Hormones are chemical signals that influence the metabolism of the cell either indirectly by regulating the production of a particular protein (steroid hormone) or directly by activating an enzyme cascade (peptide hormone).

Table 49.1

Principal Endocrine Glands and Hormones

Endocrine Gland	Hormone Released	Chemical Class	Target Tissues/Organs	Chief Function(s) of Hormone
Hypothalamus	Hypothalamic-releasing and release-inhibiting hormones	Peptide	Anterior pituitary	Regulate anterior pituitary hormones
Posterior pituitary	Antidiuretic (ADH)	Peptide	Kidneys	Stimulates water reabsorption by kidneys
	Oxytocin	Peptide	Uterus, mammary glands	Stimulates uterine muscle contraction release of milk by mammary glands
Anterior pituitary	Thyroid-stimulating (TSH)	Glycoprotein	Thyroid	Stimulates thyroid
	Adrenocorticotropic (ACTH)	Peptide	Adrenal cortex	Stimulates adrenal cortex
	Gonadotropic [follicle-stimulating (FSH), luteinizing (LH)]	Glycoprotein	Gonads	Egg and sperm production, and sex hormone production
	Prolactin (PRL)	Protein	Mammary glands	Milk production
	Growth (GH)	Protein	Soft tissues, bones	Cell division, protein synthesis, and bone growth
	Melanocyte-stimulating (MSH)	Peptide	Melanocytes in skin	Unknown function in humans; regulates skin color in lower vertebrates
Thyroid	Thyroxine (T_4) and triiodothyronine (T_3)	Iodinated amino acid	All tissues	Increases metabolic rate; regulates growth and development
	Calcitonin	Peptide	Bones, kidneys, intestine	Lowers blood calcium level
Parathyroids	Parathyroid (PTH)	Peptide	Bones, kidneys, intestine	Raises blood calcium level
Adrenal cortex	Glucocorticoids (cortisol)	Steroid	All tissues	Raise blood glucose level; stimulate breakdown of protein
	Mineralocorticoids (aldosterone)	Steroid	Kidneys	Reabsorb sodium and excrete potassium
	Sex hormones	Steroid	Gonads, skin, muscles, bones	Stimulate reproductive organs and and bring about sex characteristics
Adrenal medulla	Epinephrine and norepinephrine	Modified amino acid	Cardiac and other muscles	Emergency situations; raise blood glucose level
Pancreas	Insulin	Protein	Liver, muscles, adipose tissue	Lowers blood glucose level; promotes formation of glycogen
	Glucagon	Protein	Liver, muscles, adipose tissue	Raises blood glucose level
Gonads				
Testes	Androgens (testosterone)	Steroid	Gonads, skin, muscles, bones	Stimulate secondary male sex characteristics
Ovaries	Estrogens and progesterone	Steroid	Gonads, skin, muscles, bones	Stimulate female sex characteristics
Thymus	Thymosins	Peptide	T lymphocytes	Production and maturation of T lymphocytes
Pineal gland	Melatonin	Modified amino acid	Brain	Circadian and circannual rhythms; possibly involved in maturation of sex organs

49.2 Human Endocrine System

Endocrine glands can be contrasted with exocrine glands. Exocrine glands have ducts and secrete their products into these ducts for transport into body cavities. For example, the salivary glands send saliva into the mouth by way of the salivary ducts. Endocrine glands are ductless; they secrete their hormones directly into the bloodstream for distribution throughout the body.

The human endocrine system can be used to exemplify the vertebrate endocrine system. Even so, there are marked differences in the vertebrate hormones. Prolactin in humans stimulates female breasts to secrete milk, but in pigeons it stimulates the secretion of crop milk, a product of the gut. Some hormones have entirely different functions in different vertebrates. Thyroxine in humans stimulates metabolism, but in frogs it induces metamorphosis (the transformation of tadpole to adult).

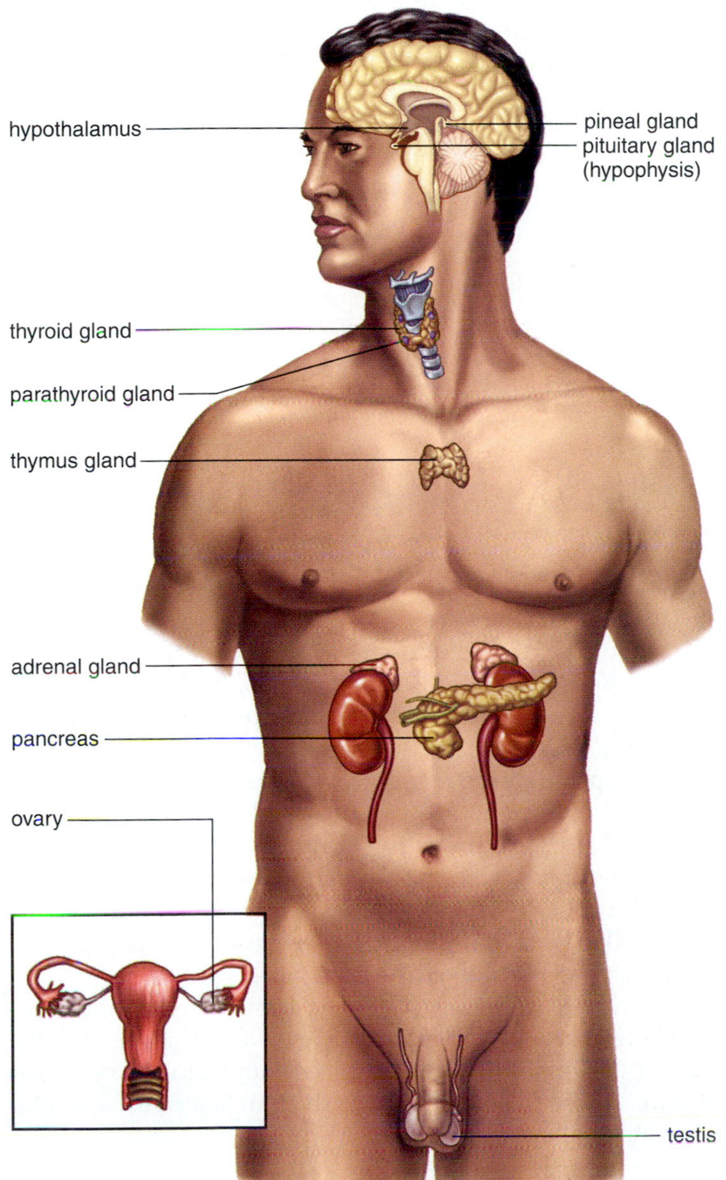

Figure 49.3 The human endocrine system. Anatomical location of major endocrine glands in the body.

Table 49.1 lists the hormones released by the principal human endocrine glands, which are depicted in Figure 49.3. The hypothalamus and pituitary gland are located in the brain, the thyroid and parathyroids are located in the neck, while the adrenal glands and pancreas are located in the pelvic cavity. The gonads include the ovaries, located in the pelvic cavity, and the testes, located outside this cavity in the scrotum. Also listed in Table 49.1 are the pineal gland, located in the brain, and the thymus, which lies ventral to the thorax.

Like the nervous system, the endocrine system is especially involved in homeostasis; that is, a relatively stable internal environment. The internal environment is the blood and tissue fluid that surrounds the body's cells. Notice that several hormones directly affect the blood glucose, calcium, and sodium levels. Other hormones are involved in the maturation and function of the reproductive organs. In fact, many people are most familiar with the effect of hormones on sexual functions.

Quite often a **negative feedback** mechanism controls the secretion of hormones involved in maintaining homeostasis. An endocrine gland can be sensitive to either the condition it is regulating or to the blood level of the hormone it is producing. For example, when the blood glucose level rises, the pancreas produces insulin, which causes the liver to store glucose. The stimulus for the production of insulin has thereby been dampened, and therefore, the pancreas stops producing insulin. The effect of insulin, however, can also be offset by the production of glucagon by the pancreas. In other words, two or more hormones can have antagonistic actions and this regulates the effect of each.

Notice there are other examples of antagonistic hormonal actions in Table 49.1. The thyroid lowers the blood calcium level, but the parathyroids raise the blood calcium level. We will also have the opportunity to point out other instances in which hormones work opposite to one another and thereby bring about the regulation of a substance in the blood.

The effect of hormones is usually controlled in two ways: (1) negative feedback opposes their release, and (2) antagonistic hormones oppose each other's actions. The end result is regulation of a bodily substance or function within normal limits.

Hypothalamus and Pituitary Gland

The **hypothalamus** regulates the internal environment through the autonomic system. For example, it helps control heartbeat, body temperature, and water balance. The hypothalamus also controls the glandular secretions of the **pituitary gland.** The pituitary, a small gland about 1 cm in diameter, is connected to the hypothalamus by a stalklike structure. The pituitary has two portions: the anterior pituitary and the posterior pituitary (Fig. 49.4, left).

Posterior Pituitary

There are neurons in the hypothalamus called neurosecretory cells which produce the hormones antidiuretic hormone (ADH) [Gk. *anti*, against, and L. *ouresis*, urination] and oxytocin. These hormones pass through axons into the posterior pituitary where they are stored in axon endings.

ADH promotes the reabsorption of water from the collecting ducts attached to nephrons within the kidneys. Certain neurons in the hypothalamus are sensitive to the osmolarity of the blood. When they determine that the blood is too concentrated, ADH is released into the bloodstream from the axon endings in the posterior pituitary. Upon reaching the kidneys, ADH causes water to be reabsorbed. As the blood becomes dilute, ADH is no longer released. This is an example of control by negative feedback because the hormonal effect (to dilute blood) acts to shut down the release of the hormone. Negative feedback maintains stable conditions and homeostasis.

Inability to produce ADH causes diabetes insipidus (watery urine), in which a person produces copious amounts of urine with a resultant loss of ions from the blood. The condition can be corrected by the administration of ADH.

Oxytocin [Gk. *oxys*, quick, and *tokos*, birth] is the other hormone that is made in the hypothalamus and stored in the posterior pituitary. Oxytocin causes uterine contraction during childbirth and milk letdown when a baby is nursing. When labor begins, pressure receptors in the uterine wall send nerve impulses to the hypothalamus and thereafter oxytocin is released. The more the uterus contracts, the more likely oxytocin is released. Similarly, the more a baby suckles, the more likely nerve impulses reach the hypothalamus and oxytocin is released. In both instances, the release of oxytocin from the posterior pituitary is controlled by **positive feedback**—the stimulus continues to bring about an effect that ever increases in intensity. Positive feedback is not a way to maintain stable conditions and homeostasis.

The posterior pituitary releases ADH and oxytocin, both of which are produced by the hypothalamus.

Anterior Pituitary

A portal system, consisting of two capillary systems connected by a vein, lies between the hypothalamus and the anterior pituitary (Fig. 49.4, right). The hypothalamus controls the anterior pituitary by producing hypothalamic-releasing hormones and hypothalamic-inhibiting hormones. For example, there is a thyroid-releasing hormone (TRH) and a thyroid-inhibiting hormone (TIH). TRH stimulates the anterior pituitary to secrete thyroid-stimulating hormone, and TIH inhibits the pituitary from secreting thyroid-stimulating hormone.

Three of the six hormones produced by the anterior pituitary (hypophysis) have an effect on other glands: (1) thyroid-stimulating hormone (TSH) stimulates the thyroid to produce thyroid hormones; (2) adrenocorticotropic hormone (ACTH) stimulates the adrenal cortex to produce cortisol; and (3) gonadotropic hormones (FSH and LH), stimulate the gonads—the testes in males and the ovaries in females—to produce gametes and sex hormones. In each instance, the blood level of the last hormone in the sequence exerts negative feedback control over the secretion of the first two hormones:

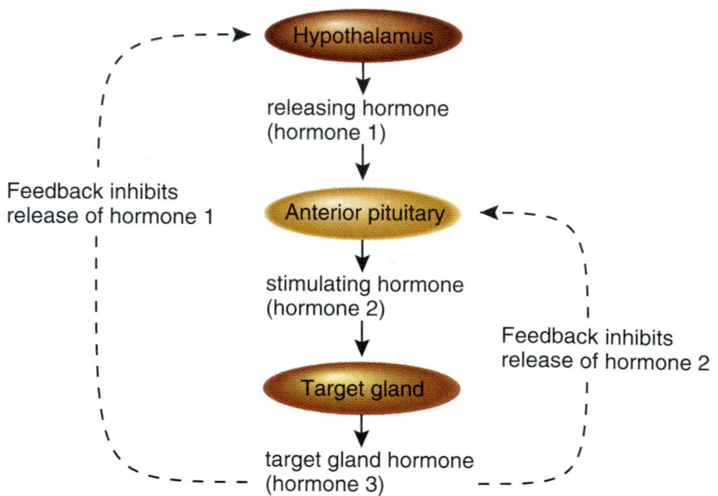

The other three hormones produced by the anterior pituitary do not affect other endocrine glands. Prolactin (PRL) [L. *pro*, before, and *lactis*, milk] is produced in quantity only after childbirth. It causes the mammary glands in the breasts to develop and produce milk. It also plays a role in carbohydrate and fat metabolism.

Melanocyte-stimulating hormone (MSH) [Gk. *melanos*, black, and *kytos*, cell] causes skin-color changes in many fishes, amphibians, and reptiles that have melanophores, special skin cells that produce color variations. The concentration of this hormone in humans is very low.

Growth hormone (GH), or somatotropic hormone, promotes skeletal and muscular growth. It stimulates the rate at which amino acids enter cells and protein synthesis occurs. It also promotes fat metabolism as opposed to glucose metabolism.

The hypothalamus, the anterior pituitary, and other glands controlled by the anterior pituitary are all involved in self-regulating negative feedback mechanisms which maintain stable conditions.

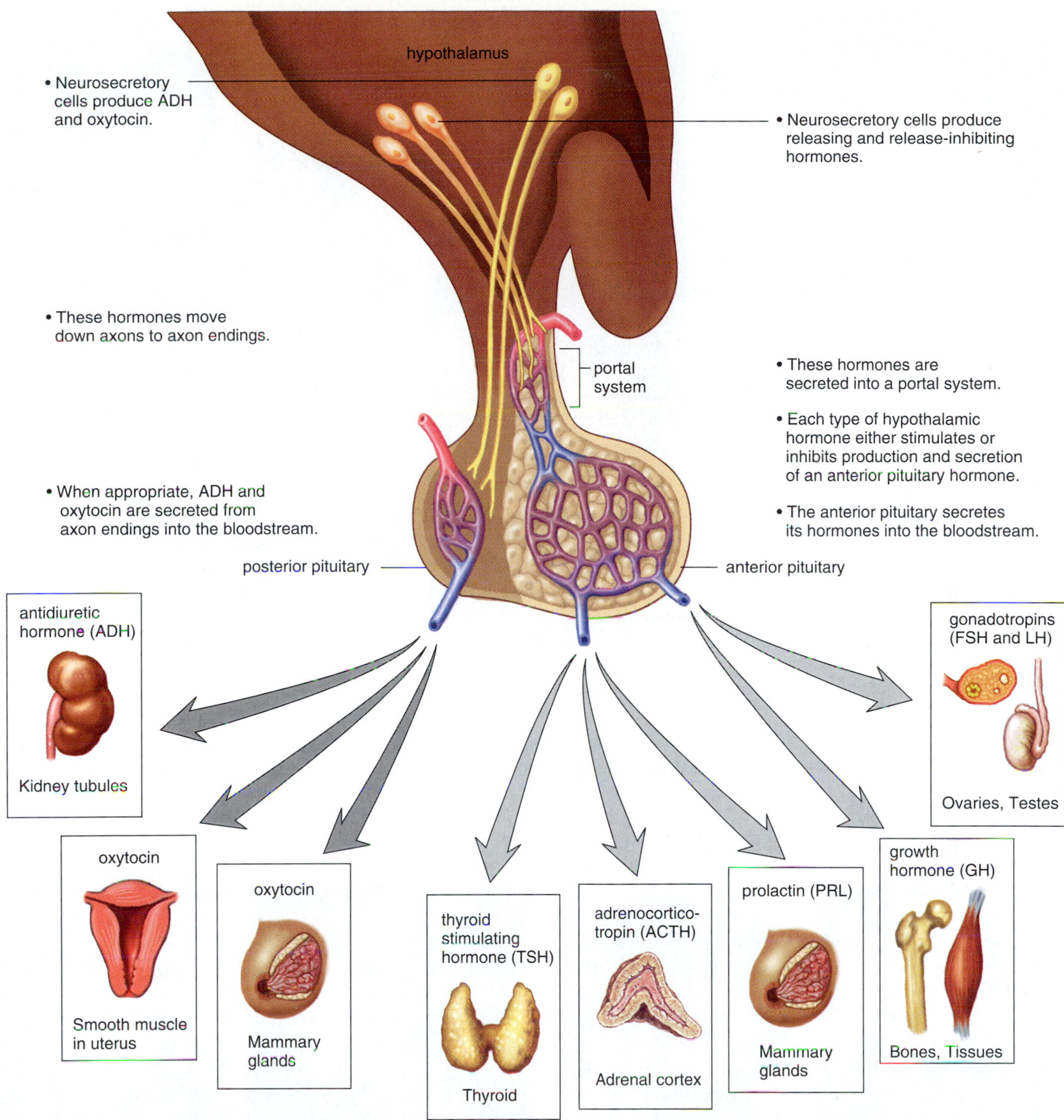
hypothalamus
• Neurosecretory cells produce ADH and oxytocin.
• Neurosecretory cells produce releasing and release-inhibiting hormones.
• These hormones move down axons to axon endings.
portal system
• These hormones are secreted into a portal system.
• Each type of hypothalamic hormone either stimulates or inhibits production and secretion of an anterior pituitary hormone.
• When appropriate, ADH and oxytocin are secreted from axon endings into the bloodstream.
• The anterior pituitary secretes its hormones into the bloodstream.
posterior pituitary
anterior pituitary
antidiuretic hormone (ADH)
Kidney tubules
gonadotropins (FSH and LH)
Ovaries, Testes
oxytocin
Smooth muscle in uterus
oxytocin
Mammary glands
thyroid stimulating hormone (TSH)
Thyroid
adrenocortico-tropin (ACTH)
Adrenal cortex
prolactin (PRL)
Mammary glands
growth hormone (GH)
Bones, Tissues

Figure 49.4 Hypothalamus and the pituitary.
(Left) The hypothalamus produces two hormones, ADH and oxytocin, which are stored and secreted by the posterior pituitary. *(Right)* The hypothalamus controls the secretions of the anterior pituitary, and the anterior pituitary controls the secretions of the thyroid, adrenal cortex, and gonads, which are also endocrine glands.

Effects of Growth Hormone GH is produced by the anterior pituitary. The quantity is greatest during childhood and adolescence, when most body growth is occurring. If too little GH is produced during childhood, the individual becomes a pituitary dwarf, characterized by perfect proportions but small stature. If too much GH is secreted, a person can become a giant (Fig. 49.5). Giants usually have poor health, primarily because GH has a secondary effect on the blood glucose level, promoting an illness called diabetes mellitus.

On occasion, there is overproduction of growth hormone in the adult, and a condition called acromegaly results. Long bone growth is no longer possible in adults. Only the feet, hands, and face (particularly the chin, nose, and eyebrow ridges) can respond, and these portions of the body become overly large (Fig. 49.6).

The amount of growth hormone during childhood and adulthood affects the height of an individual.

Figure 49.5 **Effect of growth hormone.**
The amount of growth hormone production by the anterior pituitary during childhood affects the height of an individual. Plentiful growth hormone produces very tall basketball players *(left)* and even giants *(right)*. Little growth hormone results in limited stature and even pituitary dwarfism *(right)*.

Age 9

Age 16

Age 33

Age 52

Figure 49.6 **Acromegaly.**
Acromegaly is caused by overproduction of GH in the adult. It is characterized by an enlargement of the bones in the face, the fingers, and the toes of an adult.

Thyroid and Parathyroid Glands

The **thyroid gland** [Gk. *thyreos*, large, door-shaped shield] is a large gland located in the neck, where it is attached to the trachea just below the larynx (see Fig. 49.3). The parathyroid glands are imbedded in the posterior surface of the thyroid gland.

Thyroid Gland

The thyroid gland is composed of a large number of follicles, each a small spherical structure made of thyroid cells filled with thyroxine (T_4), which contains four iodine atoms and triiodothyronine (T_3), which contains three iodine atoms.

Thyroid Hormones To produce thyroxine and triiodothyronine, the thyroid gland actively acquires iodine. The concentration of iodine in the thyroid gland can increase to as much as 25 times that of blood. If iodine is lacking in the diet, the thyroid gland is unable to produce the thyroid hormones. In response to constant stimulation by the anterior pituitary, it enlarges, resulting in a simple goiter (Fig. 49.7). Some years ago it was discovered that the use of iodized salt allows the thyroid to produce the thyroid hormones, and therefore helps prevent simple goiter.

Thyroid hormones increase the metabolic rate. They do not have one target organ; instead, they stimulate all organs of the body to metabolize at a faster rate. More glucose is broken down and more energy is utilized.

If the thyroid fails to develop properly, a condition called cretinism results (Fig. 49.8). Individuals with this condition are short and stocky and have had extreme hypothyroidism since infancy or childhood. Thyroid hormone therapy can initiate growth, but unless treatment is begun within the first two months, mental retardation results. The occurrence of hypothyroidism in adults produces the condition known as myxedema, which is characterized by lethargy, weight gain, loss of hair, slower pulse rate, lowered body temperature, and thickness and puffiness of the skin. The administration of adequate doses of thyroid hormones restores normal function and appearance.

In the case of hyperthyroidism, or Graves' disease, the thyroid gland is overactive and a goiter forms. The eyes protrude because of edema in eye socket tissues and swelling of muscles that move the eyes. This type of goiter is called exophthalmic goiter. The patient usually becomes hyperactive, nervous, irritable, and suffers from insomnia. Removal or destruction of a portion of the thyroid by means of radioactive

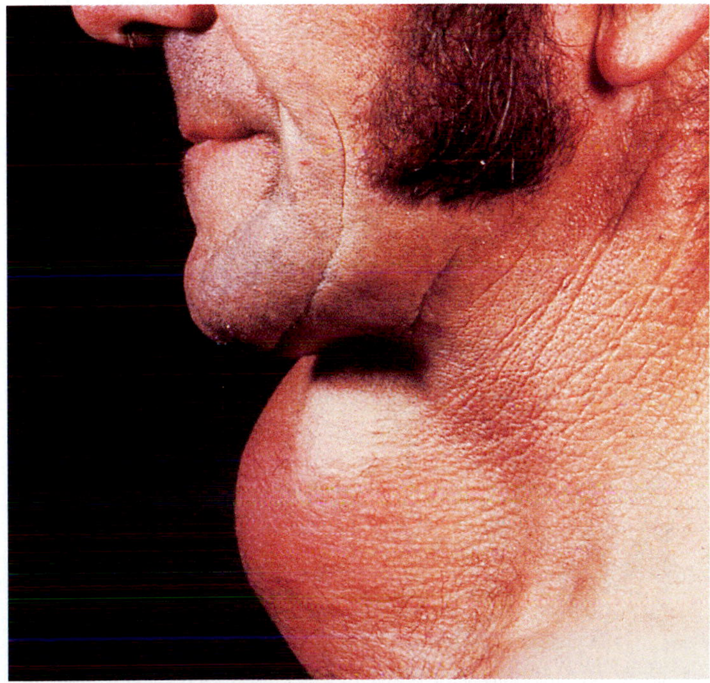

Figure 49.7 Simple goiter.
An enlarged thyroid gland often is caused by a lack of iodine in the diet. Without iodine, the thyroid is unable to produce thyroid hormones, and continued anterior pituitary stimulation causes the gland to enlarge.

Figure 49.8 Cretinism.
Individuals who have hypothyroidism since infancy or childhood do not grow and develop as others do. Unless medical treatment is begun, the body is short and stocky; mental retardation is also likely.

iodine is sometimes effective in curing the condition. Hyperthyroidism can also be caused by a thyroid tumor, which is usually detected as a lump during physical examination. Again, the treatment is surgery in combination with administration of radioactive iodine. The prognosis for most patients is excellent.

Calcitonin Calcium (Ca^{2+}) plays a significant role in both neural conduction and muscle contraction. It is also necessary for blood clotting. The blood calcium level is regulated in part by calcitonin, a hormone secreted by the thyroid gland when the blood calcium level rises (Fig. 49.9). The primary effect of calcitonin is to bring about the deposit of calcium in the bones. It does this by temporarily reducing the activity and number of osteoclasts. When the blood calcium is normal, the release of calcitonin by the thyroid is inhibited; a low level stimulates the release of parathyroid hormone (PTH) by the parathyroid glands.

Parathyroid Glands

Many years ago, the four parathyroid glands were sometimes mistakenly removed during thyroid surgery because they are so small. Parathyroid hormone (PTH), the hormone produced by the **parathyroid glands,** causes the blood phosphate (HPO_4^{2-}) level to decrease and the blood calcium level to increase. A low blood calcium level stimulates the release of PTH. The hormone has several effects on the body. PTH promotes the activity of osteoclasts and the release of calcium from the bones. PTH also promotes the reabsorption of calcium by the kidneys where it also activates vitamin D. Vitamin D, in turn, stimulates the absorption of calcium from the intestine. These effects bring the blood calcium level back to the normal range so that the parathyroid glands no longer secrete PTH.

When an insufficient PTH production leads to a dramatic drop in the blood calcium level, tetany results. In tetany, the body shakes from continuous muscle contraction. The effect is brought about by increased excitability of the nerves, which initiate nerve impulses spontaneously and without rest.

The antagonistic actions of calcitonin, from the thyroid gland, and parathyroid hormone, from the parathyroid glands, maintain the blood calcium level within normal limits.

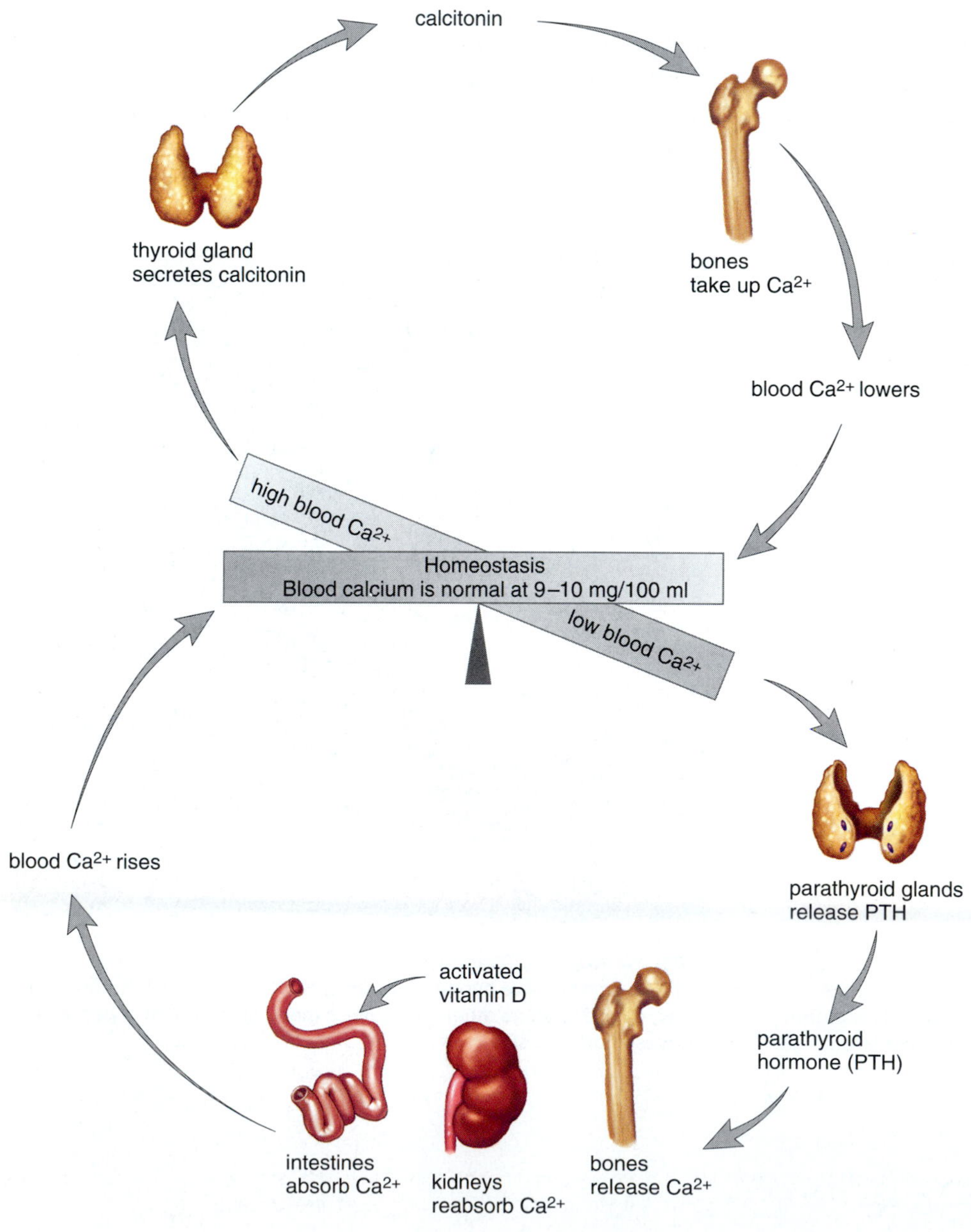

Figure 49.9 Regulation of blood calcium level.
(Top) When the blood calcium (Ca^{2+}) level is high, the thyroid gland secretes calcitonin. Calcitonin promotes the uptake of Ca^{2+} by the bones, and therefore the blood calcium level returns to normal. *(Bottom)* When the blood calcium level is low, the parathyroid glands release parathyroid hormone (PTH). PTH causes the bones to release Ca^{2+}, the kidneys to reabsorb Ca^{2+}, and the intestines to absorb Ca^{2+}; therefore, the blood calcium level returns to normal.

Adrenal Glands

We have two adrenal glands [L. *ad*, toward, and *renis*, kidney] that sit atop the kidneys (see Fig. 49.3). Each adrenal gland consists of an inner portion called the **adrenal medulla** and an outer portion called the **adrenal cortex.** These portions, like the anterior pituitary and the posterior pituitary, have no physiological connection with one another.

The hypothalamus exerts control over the activity of both portions of the adrenal glands. It can initiate nerve impulses that travel by way of the brain stem, spinal cord, and sympathetic nerve fibers to the adrenal medulla, which then secretes its hormones. The hypothalamus, by means of ACTH-releasing hormone, controls the anterior pituitary's secretion of ACTH, which, in turn, stimulates the adrenal cortex. Stress of all types, including both emotional and physical trauma, prompts the hypothalamus to stimulate the adrenal glands.

There are several adrenal hormones. Epinephrine (adrenaline) and norepinephrine (noradrenaline) produced by the adrenal medulla rapidly bring about all the bodily changes that occur when an individual reacts to an emergency situation as listed in Figure 49.10. In contrast, the hormones produced by the adrenal cortex provide a sustained response to stress. The two major types of hormones produced by the adrenal cortex are the mineralocorticoids and the glucocorticoids. The mineralocorticoids regulate salt and water balance leading to increase in blood volume and blood pressure. The glucocorticoids regulate carbohydrate, protein, and fat metabolism, leading to an increase in blood glucose level. Cortisone, the medication that is often administered for inflammation of joints, is a glucocorticoid.

The adrenal cortex also secretes a small amount of male sex hormones and a small amount of female sex hormones in both sexes—that is, in the male, both male and female sex hormones are produced by the adrenal cortex, and in the female, both male and female sex hormones are also produced by the adrenal cortex.

The adrenal medulla is under nervous control, and the adrenal cortex is under the control of ACTH, an anterior pituitary hormone. The adrenal hormones help us respond to stress.

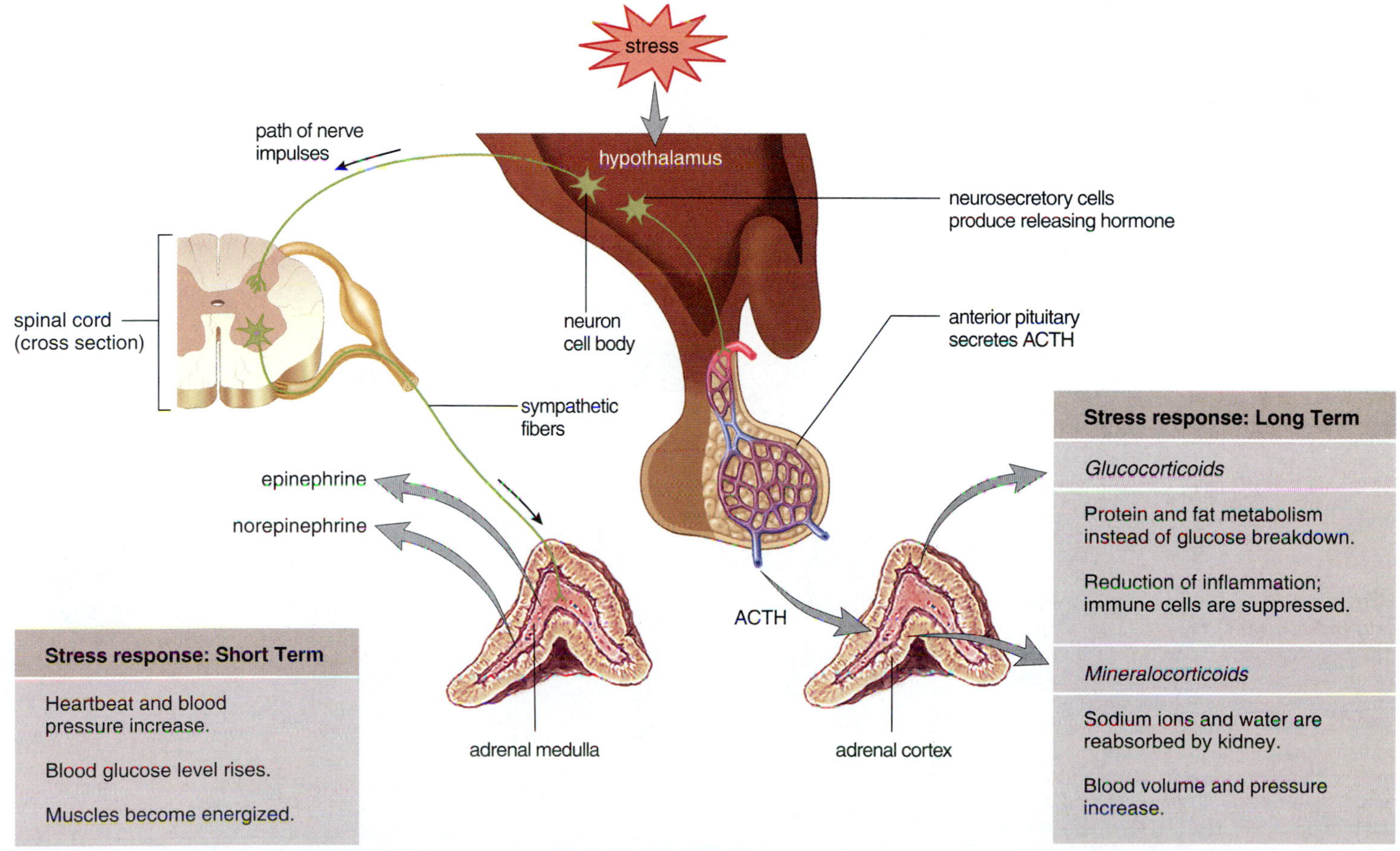

Figure 49.10 Adrenal glands.
Both the adrenal medulla and the adrenal cortex are under the control of the hypothalamus when they help us respond to stress. *(Left)* The adrenal medulla provides a rapid, but short-term stress response. *(Right)* The adrenal cortex provides a slower but long-term stress response.

Glucocorticoids

Cortisol is a biologically significant glucocorticoid produced by the adrenal cortex. Cortisol raises the blood glucose level in at least two ways. It promotes the breakdown of muscle proteins to amino acids which are taken up by the liver from the bloodstream. The liver breaks down these excess amino acids to glucose which enters the blood. Cortisol also promotes the metabolism of fatty acids rather than carbohydrates and this spares glucose.

Cortisol counteracts the inflammatory response that leads to the pain and the swelling of joints in arthritis and bursitis. The administration of cortisol aids these conditions because it reduces inflammation. Very high levels of glucocorticoids in the blood can suppress the body's defense system, including the inflammatory response that occurs at infection sites. Cortisone and other glucocorticoids can relieve swelling and pain from inflammation, but by suppressing pain and immunity they can also make a person highly susceptible to injury and infection.

Mineralocorticoids

Aldosterone is the most important of the mineralocorticoids. The primary target organ of aldosterone is the kidney, where it promotes renal absorption of sodium (Na^+) and renal excretion of potassium (K^+).

The secretion of mineralocorticoids is not under the control of the anterior pituitary. When the blood sodium level is low and, therefore, the blood pressure is low, the kidneys secrete renin (Fig. 49.11). Renin is an enzyme that converts the plasma protein angiotensinogen to angiotensin I, which is changed to angiotensin II by a converting enzyme found in lung capillaries. Angiotensin II stimulates the adrenal cortex to release aldosterone. The effect of this system, called the renin-angiotensin-aldosterone system, is to raise blood pressure in two ways. Angiotensin II constricts the arterioles, which increases blood pressure, and aldosterone causes the kidneys to reabsorb sodium. When the blood sodium level rises, water is also reabsorbed by the kidneys especially because the hypothalamus secretes ADH (see page 886). Then blood volume and pressure increases to normal.

There is an antagonistic hormone to aldosterone, as you might suspect. When the atria of the heart are stretched due to increased blood volume, cardiac cells release a hormone called atrial natriuretic hormone (ANH), which inhibits the secretion of aldosterone from the adrenal cortex. The effect of this hormone is, therefore, to cause the excretion of sodium, that is, natriuresis. When sodium is excreted, so is water, and therefore blood pressure lowers to normal.

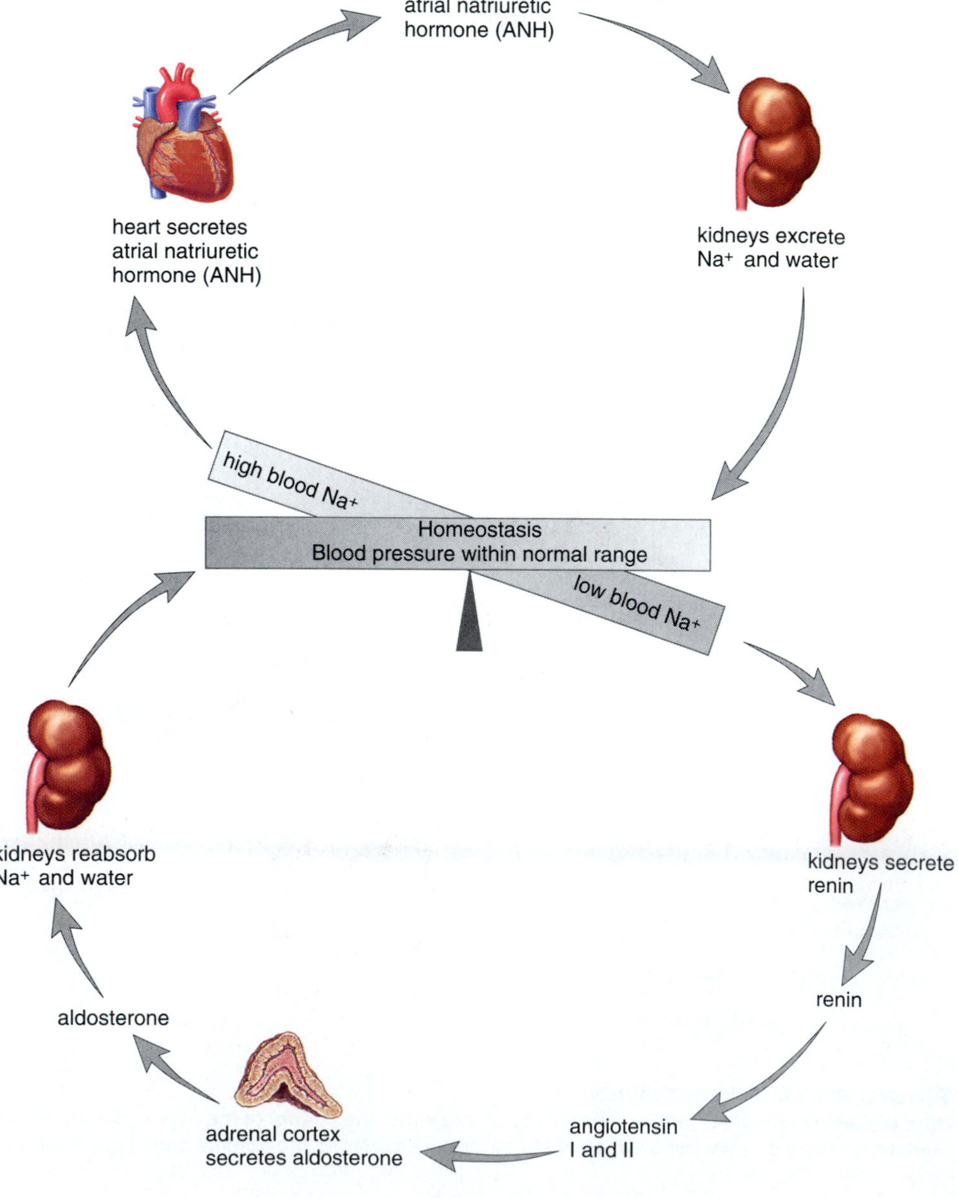

Figure 49.11 Blood sodium level and blood pressure.
(Bottom) When the blood Na^+ is low, a low blood pressure causes the kidneys to secrete renin. Renin leads to the secretion of aldosterone from the adrenal cortex. Aldosterone causes the kidneys to reabsorb Na^+ and water follows so that blood volume and pressure return to normal. *(Top)* When the blood sodium (Na^+) level is high, a high blood volume causes the heart to secrete atrial natriuretic hormone (ANH). ANH causes the kidneys to secrete Na^+ and water follows. The blood volume and pressure return to normal.

Malfunction of the Adrenal Cortex

When there is a low level of adrenal cortex hormones due to hyposecretion, a person develops Addison disease. The presence of ACTH, which is in excess but ineffective, causes a bronzing of the skin because ACTH, like MSH, can lead to a buildup of melanin (Fig. 49.12). The lack of cortisol results in an inability to replenish the blood glucose level when a stressful situation arises. Even a mild infection can lead to death. The lack of aldosterone results in a loss of sodium and water, and the development of low blood pressure and possibly severe dehydration. Left untreated, Addison disease can be fatal.

When there is a high level of adrenal cortex hormones due to hypersecretion, a person develops Cushing syndrome (Fig. 49.13). The excess of cortisol results in a tendency toward diabetes mellitus as muscle protein is metabolized and subcutaneous fat is deposited in the midsection. The trunk is obese while the arms and legs remain a normal size. An excess of aldosterone and reabsorption of sodium and water by the kidneys leads to a basic blood pH and hypertension. The face is moon-shaped due to edema. Masculinization may occur in women because of excess adrenal male sex hormones.

The adrenal cortex hormones are essential to homeostasis. Addison disease is due to adrenal cortex hyposecretion, and Cushing syndrome is due to adrenal cortex hypersecretion.

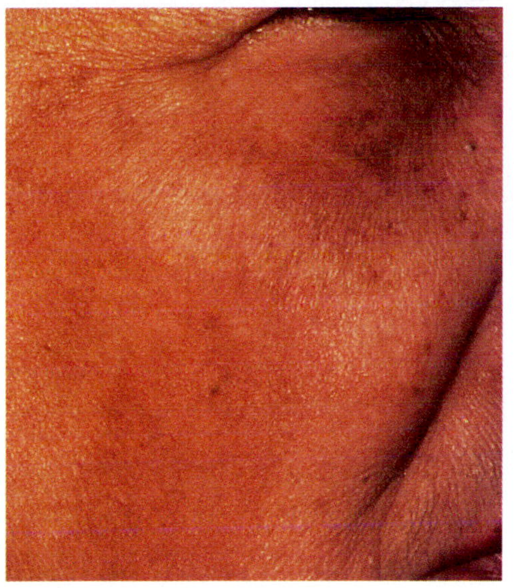

a.

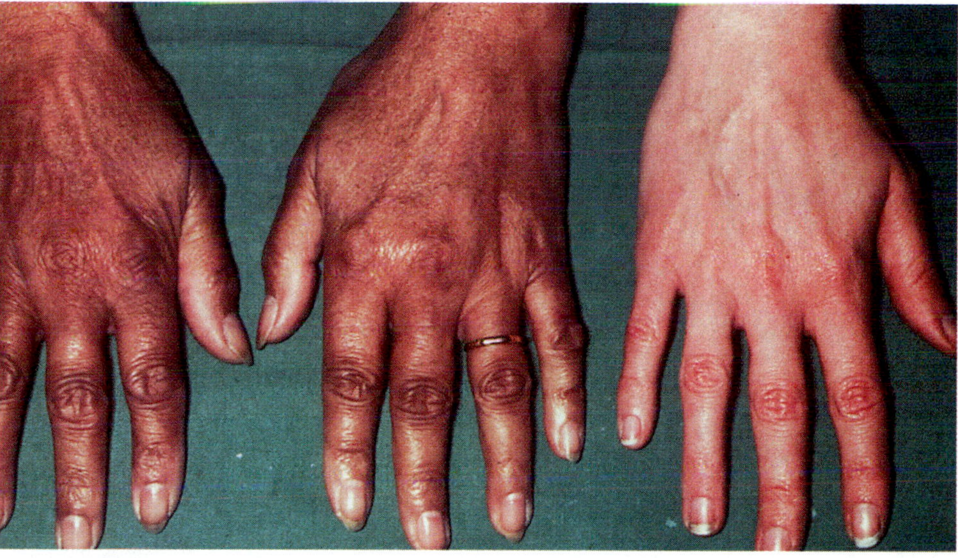

b.

Figure 49.12 Addison disease.
Addison disease is characterized by a peculiar bronzing of the skin, particularly noticeable in light-skinned individuals. Note the color of (a) the face and (b) the hands compared to the hand of an individual without the disease.

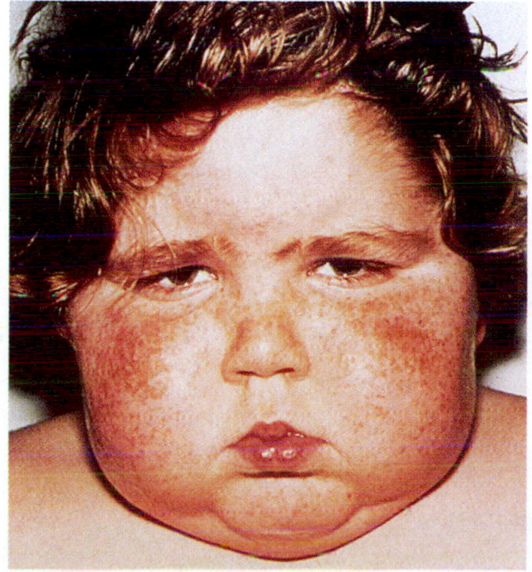

a.

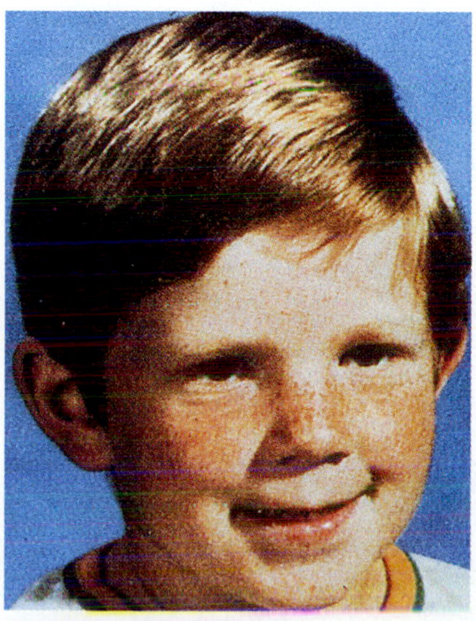

b.

Figure 49.13 Cushing syndrome.
Cushing syndrome results from hypersecretion due to an adrenal cortex tumor. **a.** First diagnosed with Cushing syndrome. **b.** Four months later, after therapy.

Pancreas

The **pancreas** is a long organ that lies transversely in the abdomen between the kidneys and near the duodenum of the small intestine. It is composed of two types of tissue. Exocrine tissue produces and secretes digestive juices that go by way of ducts to the small intestine. Endocrine tissue, called the **pancreatic islets** (islets of Langerhans), produces and secretes the hormones insulin and glucagon directly into the blood (Fig. 49.14).

Insulin is secreted when there is a high blood glucose level, which usually occurs just after eating. Insulin stimulates the uptake of glucose by cells, especially liver cells, muscle cells, and adipose tissue cells. In liver and muscle cells, glucose is then stored as glycogen. In muscle cells the breakdown of glucose supplies energy for protein metabolism, and in fat cells the breakdown of glucose supplies glycerol for the formation of fat. In these various ways insulin lowers the blood glucose level.

Glucagon is secreted from the pancreas, usually in between eating, when there is a low blood glucose level. The major target tissues of glucagon are the liver and adipose tissue. Glucagon stimulates the liver to break down glycogen to glucose and to use fat and protein in preference to glucose as energy sources. Adipose tissue cells break down fat to glycerol and fatty acids. The liver takes these up and uses them as substrates for glucose formation. In these various ways glucagon raises the blood glucose level.

The two contrary hormones insulin and glucagon, both produced by the pancreas, maintain the normal level of glucose in the blood.

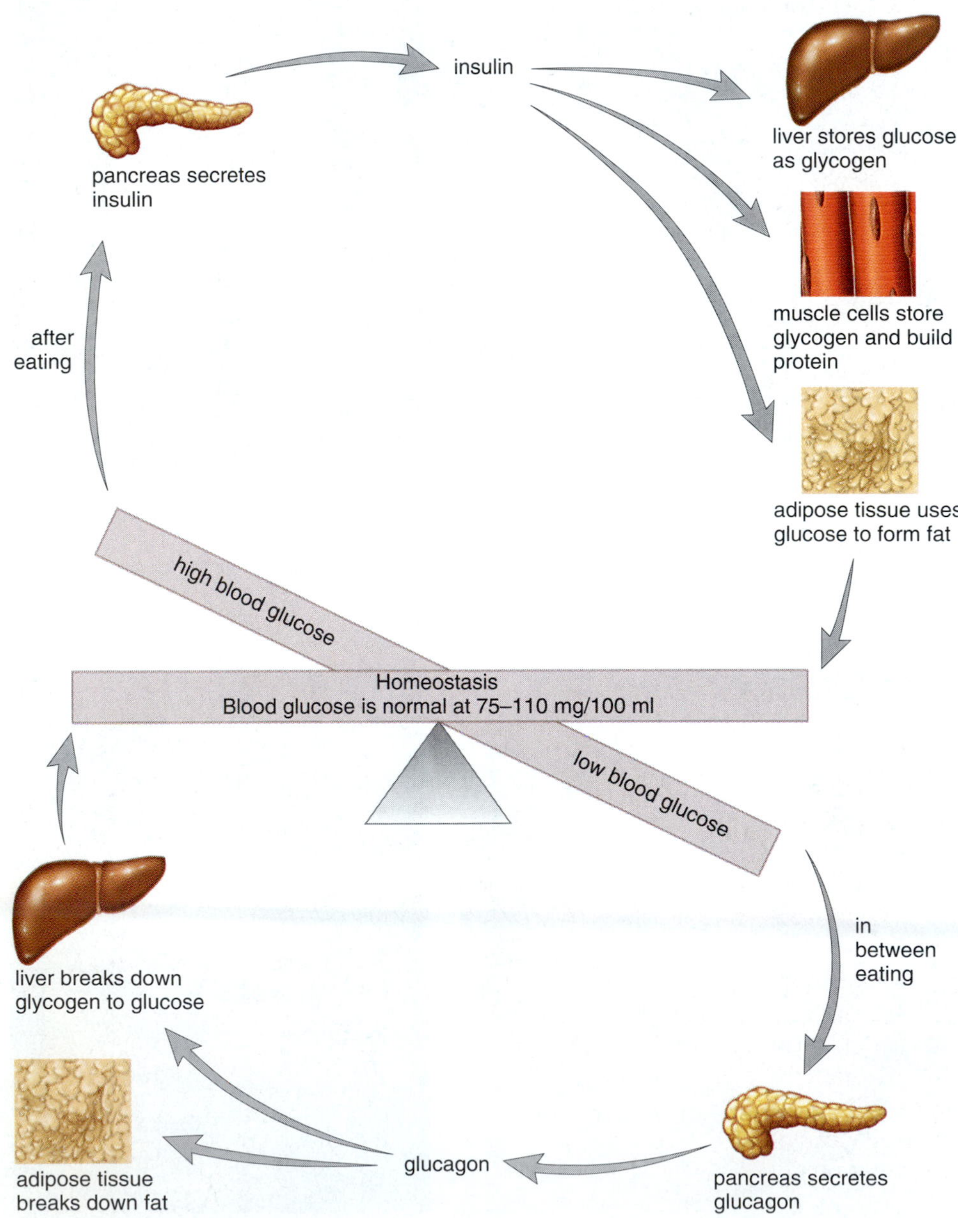

Figure 49.14 **Regulation of blood glucose level.**
(Top) When the blood glucose level is high, the pancreas secretes insulin. Insulin promotes the storage of glucose as glycogen and the synthesis of proteins and fats (as opposed to their use as energy sources). Therefore, insulin lowers the blood glucose level to normal. *(Bottom)* When the blood glucose level is low, the pancreas secretes glucagon. Glucagon acts opposite to insulin; therefore, glucagon raises the blood glucose level to normal.

Science Focus

Isolation of Insulin

The pancreas is both an exocrine gland and an endocrine gland. It sends digestive juices to the duodenum by way of the pancreatic duct, and it secretes the hormones insulin and glucagon into the bloodstream. In 1920, physician Frederick Banting decided to try to isolate insulin. Previous investigators had been unable to do this because the enzymes in the digestive juices destroyed insulin (a protein) during the isolation procedure. Banting hit upon the idea of tying off the pancreatic duct, which he knew from previous research would lead to the degeneration only of the cells that produce digestive juices and not of the pancreatic islets (of Langerhans), where insulin is made. J. J. Macleod made a laboratory available to him at the University of Toronto and also assigned a graduate student, Charles Best, to assist him. Banting and Best had limited funds and spent that summer working, sleeping, and eating in the lab. By the end of the summer, they had obtained pancreatic extracts that did lower the blood glucose level in diabetic dogs. Macleod then brought in biochemists, who purified the extract. Insulin therapy for the first human patient began in 1922, and large-scale production of purified insulin from pigs and cattle followed. Banting and Macleod received a Nobel Prize for their work in 1923. The amino acid sequence of insulin was determined in 1953. Insulin is presently synthesized using recombinant DNA technology. Banting and Best followed the required steps given in the chart to identify a chemical messenger.

Steps	Example
Identify the source of the chemical	Pancreatic islets are source
Identify the effect to be studied	Presence of pancreas in body lowers blood sugar
Isolate the chemical	Insulin isolated from pancreatic secretions
Show that the chemical alone has the effect	Insulin alone lowers blood sugar

Diabetes Mellitus

Diabetes mellitus is a fairly common hormonal disease in which liver, and indeed all body cells, do not take up and/or metabolize glucose. Therefore cellular famine exists in the midst of plenty. As the blood glucose level rises, glucose, along with water, is excreted in the urine. The loss of water in this way causes the diabetic to be extremely thirsty. Since glucose is not being metabolized, the body turns to the breakdown of protein and fat for energy. The metabolism of fat leads to the excessive presence of ketones in the blood and acidosis (acid blood) that can eventually cause coma and death.

There are two types of diabetes mellitus. In type I (insulin-dependent) diabetes, the pancreas is not producing insulin. The condition is believed to be brought on by exposure to an environmental agent, most likely a virus, whose presence causes cytotoxic T cells to destroy the pancreatic islets. As a result, the individual must have daily insulin injections. These injections control the diabetic symptoms but still can cause inconveniences, since either an overdose of insulin or the absence of regular eating can bring on the symptoms of hypoglycemia (low blood sugar). These symptoms include perspiration, pale skin, shallow breathing, and anxiety. Because the brain requires a constant supply of glucose, unconsciousness can result. The cure is quite simple: immediate ingestion of a sugar cube or fruit juice can very quickly counteract hypoglycemia.

Of the 16 million people who now have diabetes in the United States, most have type II (noninsulin-dependent) diabetes. This type of diabetes mellitus usually occurs in people of any age who are obese and inactive. The pancreas produces insulin, but the liver and muscle cells do not respond to it in the usual manner. They may increasingly lack the receptor proteins that bind to insulin. If type II diabetes is untreated, the results can be as serious as type I diabetes. (Diabetics are prone to blindness, kidney disease, and circulatory disorders. Pregnancy carries an increased risk of diabetic coma, and the child of a diabetic is somewhat more likely to be stillborn or to die shortly after birth.) It is possible to prevent or at least control type II diabetes by adhering to a low-fat diet and exercising regularly. If this fails, oral drugs that stimulate the pancreas to secrete more insulin and enhance the metabolism of glucose in the liver and muscle cells are available.

Diabetes mellitus is caused by the lack of insulin or the insensitivity of cells to insulin. It is important to discover which of these is causing diabetes and follow an appropriate therapeutic regimen.

Testes and Ovaries

The **gonads** are the testes in the male and the ovaries in the female. The testes are located in the scrotum, and the ovaries are located in the pelvic cavity. The testes produce the androgens [Gk. *andros*, male, and L. *genitus*, producing] (e.g., testosterone), which are the male sex hormones, and the ovaries produce estrogens and progesterone, the female sex hormones.

Puberty is the time of life in humans when the sexual organs mature and the secondary sexual organs appear. Testosterone secretion at the time of puberty stimulates maturation of the testes and other sexual organs. It also brings about the secondary sexual characteristics. Testosterone causes growth of a beard, axillary (underarm) hair, and pubic hair. It prompts the larynx and the vocal cords to enlarge, causing the voice to change. Testosterone is believed to be largely responsible for the sex drive and may even contribute to the supposed aggressiveness of males. It is responsible for the muscular strength of males, and this is the reason some athletes take supplemental amounts of anabolic steroids, which are either testosterone or related chemicals. (The contraindications of taking anabolic steroids are listed in Fig. 49.15.) Testosterone is largely responsible for acne and body odor, and also baldness. Genes for baldness probably are inherited by both sexes, but baldness is seen more often in males because of the presence of testosterone.

The female sex hormones, estrogens and progesterone, have many effects on the body. In particular, estrogens secreted at the time of puberty stimulate the maturation of the ovaries and other sexual organs. Estrogen [Gk. *oistros*, sexual heat, and L. *genitus*, producing] is necessary for development of oocytes (egg production) and is largely responsible for the secondary sexual characteristics in females. It is responsible for female body hair and fat distribution. In general, females have a more rounded appearance than males because of a greater accumulation of fat beneath the skin. Also, the pelvic girdle enlarges in females, facilitating the process of giving birth. Both estrogen and progesterone are required for breast development and regulation of the uterine cycle, which includes monthly menstruation (discharge of blood and mucosal tissues from the uterus).

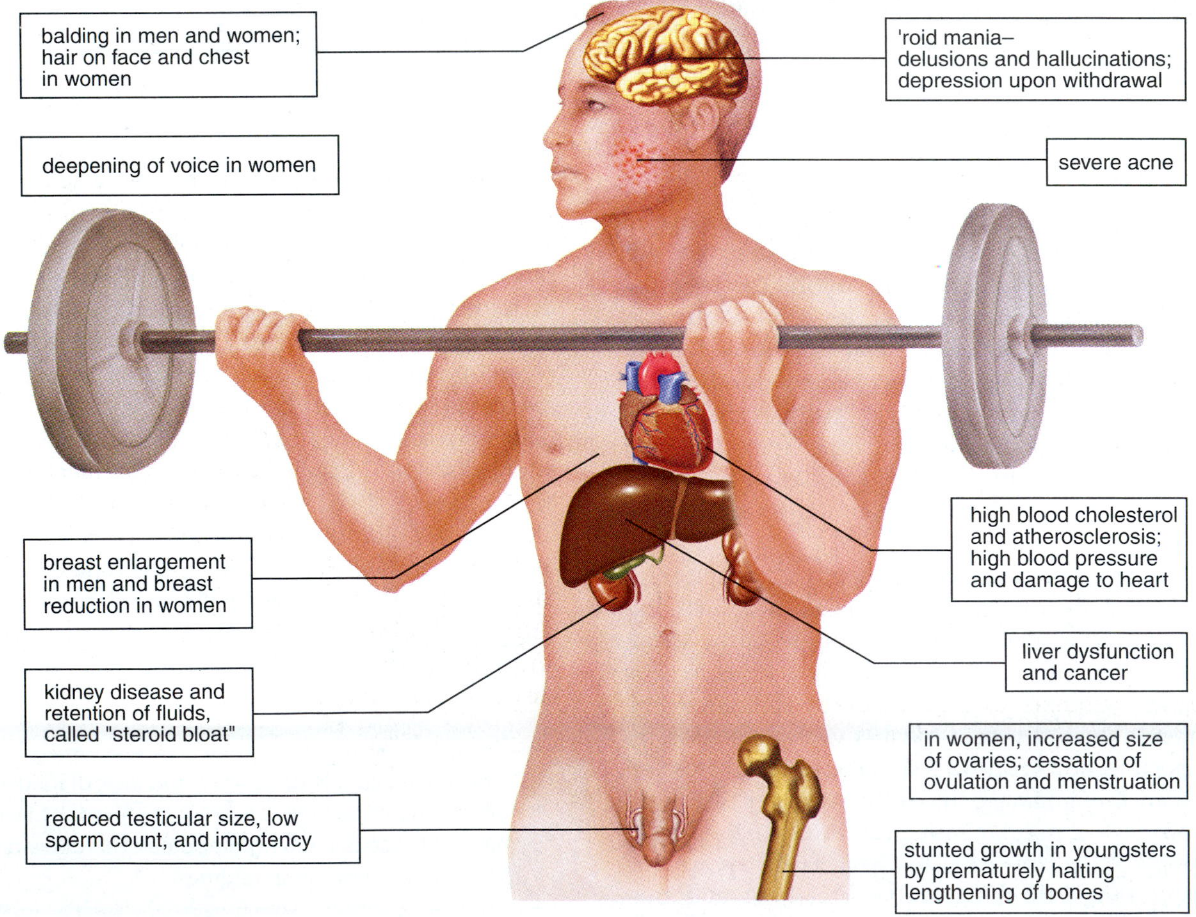

Figure 49.15 The effects of anabolic steroid use.

Thymus Gland

The **thymus gland** is a lobular organ that lies in the upper thoracic cavity (see Fig. 49.3). This organ reaches its largest size and is most active during childhood. With aging, the organ gets smaller and becomes fatty. Certain lymphocytes that originate in the bone marrow and then pass through the thymus are transformed into T cells. The thymus produces various hormones called thymosins, which aid the differentiation of T cells and may stimulate immune cells in general. There is hope that these hormones can be used in conjunction with lymphokine therapy to restore or to stimulate T cell function in patients suffering from AIDS or cancer.

Pineal Gland

The **pineal gland** produces the hormone called melatonin; this occurs primarily at night. In fishes and amphibians, the pineal gland is located near the surface of the body and is a "third eye," which receives light rays directly. In mammals, the pineal gland is located in the third ventricle of the brain and cannot receive direct light signals (see Fig. 49.3). However, it does receive nerve impulses from the eyes by way of the optic tract. Again, we see that the nervous and endocrine systems are often connected.

The pineal gland and melatonin are involved in daily cycles called **circadian rhythms** [L. *circum*, around, and *dies*, day]. Normally we grow sleepy at night when melatonin levels are high and we awaken once daylight returns and melatonin levels are low. Shift work is usually troublesome because it upsets this normal daily rhythm. Similarly, international travel often results in jet lag because the body is still producing melatonin according to the old schedule. In seasonal affective disorder (SAD), sufferers become depressed and have an uncontrollable desire to sleep with the onset of winter. Giving melatonin makes their symptoms worse, but exposure to a bright light improves them.

Many animals go through a yearly cycle that includes enlargement of reproductive organs during the summer when melatonin levels are low. Mating occurs in the fall and young are born in the spring. It is of interest that children with a brain tumor that destroys the pineal gland experience early puberty. Therefore, it's possible that the pineal gland is also involved in human sexual development.

The gonads, thymus, and pineal gland are also endocrine organs. The gonads secrete the sex hormones, the thymus secretes thymosins, and the pineal gland secretes melatonin.

Connecting Concepts

The nervous system and the endocrine system are structurally and functionally related. The hypothalamus, a portion of the brain, controls the pituitary, an endocrine gland. The hypothalamus even produces the hormones that are released by the posterior pituitary. Neurosecretory cells in the hypothalamus produce chemical signals that control the activity of the anterior pituitary. Such signals fit the definition of a hormone because they are carried in a portal system from hypothalamus to the anterior pituitary.

A survey of the animal kingdom also shows an overlap of the two systems. In the snail *Aplysia*, a hormone called egg-laying hormone (ELH) stimulates the laying of long strings of eggs. ELH is a excitatory neurotransmitter that also diffuses into the circulatory system and excites the smooth muscle cells of the reproductive duct, causing them to contract and expel strings of eggs. Similarly in mammals, norepinephrine is both a neurotransmitter in the sympathetic division of the autonomic system and a hormone released by the adrenal medulla. Sympathetic nerve endings stimulate the adrenal medulla to release norepinephrine which promotes the "fight-or-flight" reaction.

Molting in insects is controlled by the balance of three different hormones, one of which is called brain hormone because it is produced by neurosecretory cells. Brain hormone controls a gland that releases ecdysone, which promotes molting. As long as another gland is producing a hormone called juvenile hormone, molting produces a more mature larva. With time, the level of juvenile hormone falls off and molting produces a pupa in which metamorphosis produces the adult insect.

The close connection between the nervous and endocrine systems suggests that the categorization of organs in any one system is somewhat arbitrary. Recall that the digestive tract produces hormones that control the secretion of digestive juices (page 790); that the kidneys produce renin that leads to the release of aldosterone; and that the heart produces atrial natriuretic hormone that opposes the action of aldosterone (page 892). Local hormones that affect only their neighbors are produced by most cells. It has even been discovered that brain cells produce insulin which is used locally to influence the metabolism of adjacent cells.

Summary

49.1 Chemical Signals

There are three categories of chemical signals: those that act at a distance between individuals (pheromones); those that act at a distance within the individual (traditional endocrine hormones and secretions of neurosecretory cells); and local hormones (such as prostaglandins, growth factors, and neurotransmitters).

Steroid hormones combine with a receptor protein in the cell, and the complex attaches to and activates DNA. Transcription and translation lead to protein synthesis. Peptide hormones are usually received by a receptor protein located in the plasma membrane. Most often their reception leads to activation of an enzyme that changes ATP to cyclic AMP (cAMP). cAMP then activates an enzyme cascade. Hormones work in small quantities because their effect is amplified.

49.2 Human Endocrine System

Neurosecretory cells in the hypothalamus produce antidiuretic hormone (ADH) and oxytocin, which are stored in axon endings in the posterior pituitary until they are released.

The hypothalamus produces hypothalamic-releasing and hypothalamic-inhibiting hormones, which pass to the anterior pituitary by way of a portal system. The anterior pituitary produces several types of hormones, and some of these stimulate other hormonal glands to secrete hormones.

The thyroid gland requires iodine to produce thyroxine and triiodothyronine, which increase the metabolic rate. If iodine is available in limited quantities, a simple goiter develops; if the thyroid is overactive, an exophthalmic goiter develops. The thyroid gland also produces calcitonin, which helps lower the blood calcium level. The parathyroid glands secrete parathyroid hormone which raises the blood calcium and decreases the blood phosphate level.

The adrenal glands respond to stress. The adrenal medulla secretes epinephrine and norepinephrine, which bring about a rapid response in an emergency situation. On a long-term basis, the adrenal cortex produces the glucocorticoids (e.g., cortisol) and the mineralocorticoids (e.g., aldosterone). Cortisol stimulates hydrolysis of proteins to amino acids that are converted to glucose; in this way, it raises the blood glucose level. Aldosterone causes the kidneys to reabsorb sodium ions (Na^+) and excrete potassium ions (K^+). Addison disease develops when the adrenal cortex is underactive, and Cushing syndrome develops when the adrenal cortex is overactive.

The pancreatic islets secrete insulin, which lowers the blood glucose level, and glucagon, which has the opposite effect. The most common illness due to hormonal imbalance is diabetes mellitus, which is due to the failure of the pancreas to produce insulin or the cells to take it up.

The gonads produce the sex hormones; the pineal gland produces melatonin, which may be involved in circadian rhythms and the development of the reproductive organs; and the thymus secretes thymosins, which stimulate T lymphocyte production and maturation.

Reviewing the Chapter

1. Categorize chemical signals into three groups based upon the distance between site of secretion and site of reception, and give examples for each group. 882
2. Explain how steroid hormones and peptide hormones affect the metabolism of the cell. 883
3. Explain the relationship of the hypothalamus to the posterior pituitary gland and to the anterior pituitary gland. List the hormones secreted by the posterior and anterior pituitary. 886
4. Give an example of the negative feedback relationship among the hypothalamus, the anterior pituitary, and other endocrine glands. 886
5. Discuss the effect of growth hormone on the body as a result of there being too much or too little growth hormone when a young person is growing. What is the result if the anterior pituitary produces growth hormone in an adult? 888
6. What types of goiters are associated with a malfunctioning thyroid? Explain each type. 889
7. How do the thyroid and the parathyroid work together to control the blood calcium level? 890
8. How do the adrenal glands respond to stress? What hormones are secreted by the adrenal medulla, and what effects do these hormones have? 891
9. Name the most significant glucocorticoid and mineralocorticoid, and discuss their functions. Explain the symptoms of Addison disease and Cushing syndrome. 893
10. Draw a diagram to explain how insulin and glucagon maintain the blood glucose level. Use your diagram to explain three major symptoms of type I diabetes mellitus. 894–95
11. Name the other endocrine glands discussed in this chapter, and discuss the function of the hormones they secrete. 897

Testing Yourself

Choose the best answer for each question.

1. One of the chief differences between pheromones and local hormones is
   a. the distance over which they act.
   b. that one is a chemical signal and the other is not.
   c. that one is made by only invertebrates and the other is made by only vertebrates.
   d. that one is a steroid and the other a peptide.
   e. All of these are correct.
2. Peptide hormones
   a. are received by a receptor located in the plasma membrane.
   b. are received by a receptor located in the cytoplasm.
   c. bring about the transcription of DNA.
   d. activate a second messenger.
   e. Both a and d are correct.

Match the hormone in questions 3–7 to the correct gland in the key.

Key:

a. pancreas
b. anterior pituitary
c. posterior pituitary
d. thyroid
e. adrenal medulla
f. adrenal cortex

3. cortisol
4. growth hormone (GH)
5. oxytocin storage
6. insulin
7. epinephrine
8. The blood cortisol level controls the secretion of
   a. a releasing hormone from the hypothalamus.
   b. adrenocorticotropic hormone (ACTH) from the anterior pituitary.
   c. cortisol from the adrenal cortex.
   d. all these hormones, but not mineralocorticoids.
   e. All of these statements are correct.
9. The anterior pituitary controls the secretion(s) of
   a. both the adrenal medulla and the adrenal cortex.
   b. both cortisol and aldosterone.
   c. thyroxine.
   d. the thymus.
   e. All of these are correct.
10. Aldosterone
   a. is opposed by atrial natriuretic hormone (ANH).
   b. causes the kidneys to reabsorb sodium.
   c. causes the blood volume to increase.
   d. causes the blood pressure to rise.
   e. All of these are correct.
11. Diabetes mellitus is associated with
   a. too much insulin in the blood.
   b. too high a blood glucose level.
   c. blood that is too dilute.
   d. high metabolic rate.
   e. All of these are correct.
12. Which of these is not a pair of antagonistic hormones?
   a. insulin—glucagon
   b. calcitonin—parathyroid hormone
   c. cortisol—epinephrine
   d. aldosterone—atrial natriuretic hormone (ANH)
   e. thyroxine—growth hormone
13. Which hormone and condition is mismatched?
   a. growth hormone—acromegaly
   b. thyroxine—goiter
   c. parathyroid hormone—tetany
   d. epinephrine—Addison disease
   e. insulin—diabetes
14. Fill in this diagram to explain the negative feedback relationship between the hypothalamus, the anterior pituitary, and the target gland.

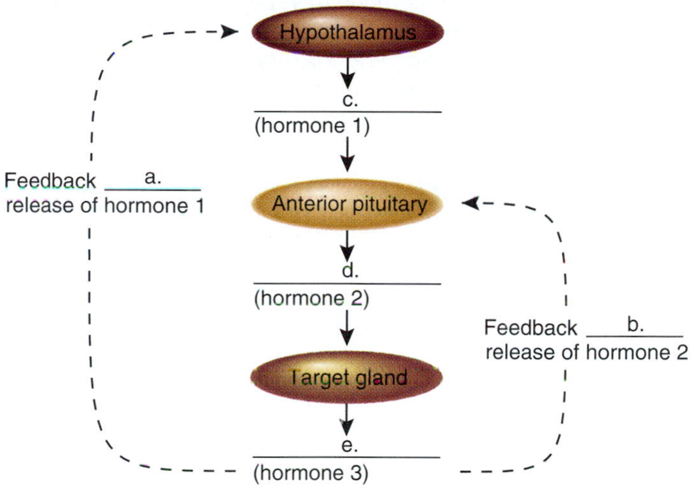

Thinking Scientifically

1. Caffeine inhibits the breakdown of cAMP in the cell. Referring to Figure 49.2, how would this influence a stress response brought about by epinephrine?
2. Both males and females can develop secondary sex characteristics of the opposite sex if they take enough of the appropriate sex hormone. Hypothesize the pattern of gene inheritance for these characteristics and explain why certain of these genes and not others are expressed.

Bioethical Issue

Hormone therapy saves lives. Diabetics and those with Addison disease would soon succumb if hormone therapy was not available for their conditions. Other hormone therapies are a matter of choice and raise ethical questions. If you were the parent of a child who was short for his or her age group, on what basis might you decide to have the child undergo hormone therapy?

Researchers have found that human growth hormone doesn't work. It causes puberty to arrive earlier than usual, and in this way does away with any gain in height that occurred before puberty. However, studies are being conducted with a synthetic growth hormone called somatotropin that seems to work better. Researchers in England conducted a survey of about 14,000 girls, and decided to treat seven girls who were short for their age. They compared their progress to a group of untreated girls. All girls started puberty at about the same age—13.5 years. The girls in the treated group received a daily injection of somatotropin between the ages of 8 and 14. They grew to 5′1″ by age 16, when most girls stop growing. That was 2.4 to 3.0 inches taller than the girls in the untreated group.

The researchers were interested in whether height affected psychological outlook. They found that girls who were 4′10″ appeared to be just as happy and well balanced as those who were taller. Is psychological outlook a good criterion to use when deciding whether to use hormone therapy to increase height? Is this criterion more appropriately applied to boys than to girls? When does short height become a disability in our society? Are tall people likely to have more successful careers than short people? Should short people receive hormone therapy?

Understanding the Terms

adrenal cortex 891
adrenal medulla 891
circadian rhythm 897
endocrine system 881
first messenger 883
gonads 896
hormone 882
hypothalamus 886
negative feedback 885
pancreas 894
pancreatic islet 894
parathyroid gland 890
peptide hormone 883
pheromone 882
pineal gland 897
pituitary gland 886
positive feedback 886
second messenger 883
steroid hormone 883
thymus gland 897
thyroid gland 889

Match the terms to these definitions:

a. _______________ Organ that is in the neck and secretes several important hormones, including thyroxine and calcitonin.
b. _______________ Chemical signal produced in one part of the body that controls the activity of other parts.
c. _______________ Organ—either at the skin surface (fish, amphibians) or in the third ventricle of the brain (mammals)—that produces melatonin.
d. _______________ Type of hormone that binds to a plasma membrane receptor and results in activation of an enzyme cascade.
e. _______________ Chemical substance secreted by one organism that influences the behavior of another.

Web Connections

Exploring the Internet

http://www.mhhe.com/biosci/genbio/mader
(click on *Biology 7/e*)

The *Biology 7/e* Online Learning Center provides many resources for studying the material in this chapter including links to the following sites:

Drug-Free Resource Net. What anabolic steroids really do to a human body is explained in this site. How steroids increase body muscle yet lead to dangerous, even fatal, side effects is the focus of this site.

http://www.drugfreeamerica.org/steroids.html

Reproductive Endocrinology. This site, maintained by Pennsylvania State University, provides information on reproductive endocrinology.

http://www.hmc.psu.edu/depts/obgyn/repend.htm

Endocrines and Reproduction. This site includes valuable information on the endocrine system and hormones vital in reproduction.

http://www.umds.ac.uk/elsewhere/physiology/banks/endorep.html

Human Anatomy Online: Endocrine System. This site provides an abundance of information on the human endocrine system.

http://www.innerbody.com/image/endoov.html

Facts About Diabetes. A CDC fact sheet on diabetes.

http://www.cdc.gov/od/oc/media/fact/diabetes.htm

Reproduction

chapter 50

chapter concepts

50.1 How Animals Reproduce

- Among animals, there are two patterns of reproduction: asexual reproduction and sexual reproduction. 902
- Sexually reproducing animals have gonads for the production of gametes, and many have accessory organs for the storage and passage of gametes into or from the body. 902
- Animals have various means of assuring fertilization of gametes and protecting immature stages. 903

50.2 Male Reproductive System

- The human male reproductive system continuously produces a large number of sperm that are transported within a fluid medium. 904
- Hormones control the production of sperm and maintain the primary and secondary sexual characteristics of males. 907

50.3 Female Reproductive System

- The female reproductive system produces one egg monthly, and prepares the uterus to house the developing fetus. 908
- Hormones control the monthly reproductive cycle in females and play a significant role in maintaining pregnancy, should it occur. 910

50.4 Control of Reproduction

- There are alternative methods of reproduction today, including in vitro fertilization followed by introduction of the zygote to the uterus. 912
- Birth control measures vary in effectiveness from those that are very effective to those that are minimally effective. 913

Hippopotamus and calf, *Hippopotamus amphibius*, Kenya

The life processes we have discussed so far, like digesting food, exchanging gases, maintaining salt and water balance, and coordinating body systems, are necessary to the survival of the individual. Reproduction is different because it pertains not to the survival of the individual, but to the survival of an individual's genes in future generations of a species.

The many different ways that animals reproduce can be categorized as asexual or sexual. Asexual reproduction does not involve the use of sex cells like sperm and egg—the offspring have exactly the same traits as the single parent. The parent's adaptations for survival are passed on unchanged to the offspring, and this can be an advantage if the environment is not changing.

Sexual reproduction requires two parents: the egg of one parent is fertilized by the sperm of another parent. Sexual reproduction has the advantage of producing offspring that are not exactly like either parent. The introduction of genetic variation among the offspring may very well help to ensure the survival of the species if environmental conditions are changing.

50.1 How Animals Reproduce

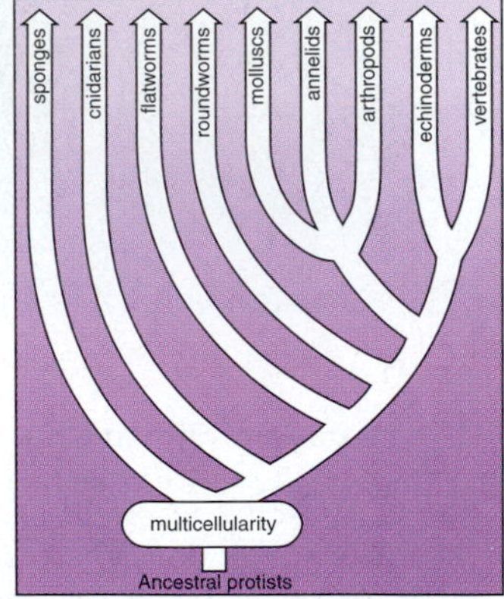

Asexual and sexual reproduction are the two fundamental patterns of reproduction in organisms including animals. In asexual reproduction, there is only one parent, and in sexual reproduction there are two parents.

Asexual Reproduction

Some animals usually reproduce asexually but still on occasion practice sexual reproduction. Hydras, which are cnidarians, can reproduce by budding (Fig. 50.1). A new individual arises as an outgrowth (bud) of the parent. In other cnidarians there are two diploid generations. The polypoid stage of *Obelia,* which is a colony made up of many hydralike polyps, is sessile and produces diploid medusae by budding. The medusa stage, which looks like a jellyfish, is motile and produces haploid eggs and sperm. The motile stage disperses the species. Many flatworms can constrict into two halves; each half regenerates to become a new individual. Fragmentation followed by regeneration is also seen among sponges and echinoderms. Chopping up a sea star does not kill it; instead, each fragment grows into another animal.

Several types of flatworms, roundworms, crustaceans, annelids, and insects, fishes, and lizards have the ability to reproduce parthenogenetically. **Parthenogenesis** [Gk. *parthenos,* virgin, and *genitus,* producing] is a modification of sexual reproduction in which an unfertilized egg develops into a complete individual. In honeybees, the queen bee can fertilize or allow eggs to pass unfertilized as she lays them. The fertilized eggs become diploid females called workers, and the unfertilized eggs become haploid males called drones.

Sexual Reproduction

Usually during sexual reproduction, the egg of one parent is fertilized by the sperm of another. Even among earthworms, which are hermaphroditic—each worm has both male and female sex organs—cross-fertilization occurs. Sequential hermaphroditism or sex reversal also occurs. In coral reef fishes called wrasses, a male has a harem of several females. If the male dies, the largest female becomes a male.

Animals usually produce gametes in specialized organs called **gonads** [Gk. *gone,* seed]. Sponges are an exception to this rule because in sponges the collar cells lining the central cavity give rise to sperm and eggs. Hydras and other cnidarians produce only temporary gonads in the fall when sexual reproduction occurs (Fig. 50.1). Animals in other phyla have permanent reproductive organs. The gonads are **testes,** which produce sperm and **ovaries** [L. *ovaris,* eggkeeper], which produce eggs. Eggs or sperm are derived from germ cells, which become specialized for this purpose during early development. Other cells in a gonad support and nourish the developing gametes or produce hormones necessary to the reproductive process. There are also accessory organs—ducts and storage areas that aid in bringing the gametes together.

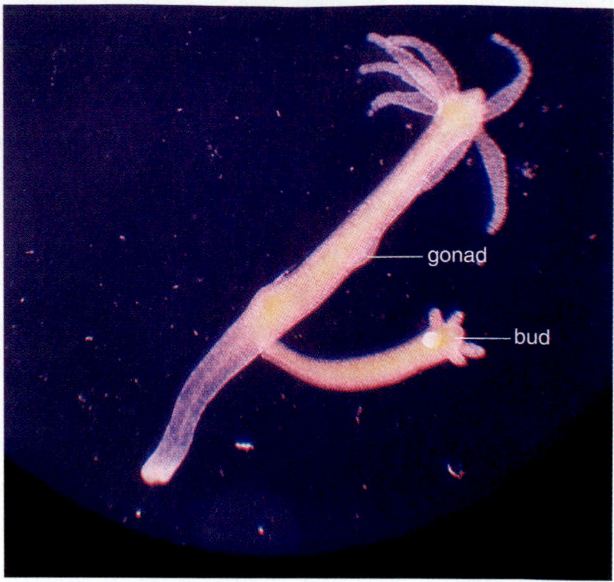

Figure 50.1 **Reproduction in *Hydra.***
Hydras reproduce asexually and sexually. During asexual reproduction, a new polyp buds from the parental polyp. During sexual reproduction, temporary gonads develop in the body wall.

Sexually reproducing animals have all sorts of ways for making sure that the gametes find each other. Aquatic animals that practice external fertilization are programmed to release their eggs in the water only at certain times. One environmental signal that seems to work is the lunar cycle. Each month the moon moves closer to the earth and the tides become somewhat higher. Aquatic animals able to sense this change can release their gametes at the same time. Hundreds of thousands of palolo worms rise to the surface of the sea and release their eggs during a two- to four-hour period on two or three specific successive days of the year. Most likely they are under the control of a biological clock that can sense the passage of time so that their reproductive behavior is synchronized.

Copulation [L. *copulatus,* join] is sexual union to facilitate the reception of sperm by a partner, usually a female. In terrestrial animals, males typically have a penis for depositing sperm into the vagina of females. Aquatic animals also have other types of copulatory organs. Lobsters and crayfish, which are arthropods, have modified swimmerets. Cuttlefish and octopuses, which are molluscs, use an arm; and sharks, which are vertebrates, have a modified pelvic fin that passes packets of sperm to the female. Among terrestrial animals, most birds lack a penis and vagina. They have a cloaca, a chamber that receives products from the digestive, urinary, and reproductive tracts. A male transfers sperm to a female after placing his cloacal opening adjacent to hers.

Life History Strategies

Many aquatic animals practice external fertilization; that is, eggs and sperm join outside the body in the water. Terrestrial animals tend to practice internal fertilization—egg and sperm join inside the female's body. Both types of animals are usually oviparous, meaning that they deposit eggs in the external environment. In insects, the eggs are produced in the ovaries, and as they mature they increase in size because yolk has been added to them. **Yolk** is stored food to be used by the developing embryo. Then, to prevent the eggs from drying out, they are covered by a shell consisting of several layers of protein- and wax-containing material. Small holes are left at one end of the egg for the entry of sperm. Some insects have a special internal organ for storing sperm for some time after copulation so that the eggs can be fertilized internally before they are deposited in the environment.

A **larva** is an independent form that is often quite different in appearance and way of life from the adult. A larva is able to seek its own food and to sustain itself until it becomes an adult. In some terrestrial insects, there are several larval stages and then the animal pupates. A pupa is enclosed by a hardened cuticle, often within a cocoon. Here **metamorphosis** [Gk. *meta*, implying change, and *morphe*, shape, form], which is a dramatic change in shape, takes place. Then the adult insect emerges and flies off to find a mate and reproduce. Other insects (e.g., grasshoppers) undergo incomplete metamorphosis; pupation does not occur and there are a number of nymph stages, each one looking more like the adult.

Many aquatic forms also have a larval stage. Since the larva has a different lifestyle, it is able to make use of a different food source than the adult. In sea stars, which are echinoderms, the bilaterally symmetrical larva simply attaches itself to a substratum and undergoes metamorphosis to become a radially symmetrical juvenile. Among barnacles, which are arthropods, the free-swimming larva metamorphoses into the sessile adult with calcareous plates. Crayfish, on the other hand, do not have a larval stage; the egg hatches into a tiny juvenile with the same form as the adult.

Reptiles and particularly birds provide their eggs with plentiful yolk; there is no larval stage. Complete development takes place within a shelled egg containing **extraembryonic membranes** [L. *extra*, on the outside] to serve the needs of the embryo. The outermost membrane, the chorion, lies next to the shell and functions in gas exchange. The amnion forms a water-filled sac around the embryo, ensuring that it will not dry out. The yolk sac holds the yolk, which nourishes the embryo, and the allantois holds nitrogen waste products (see Fig. 51.11). The shelled egg frees these animals from the need to reproduce in the water and is a significant adaptation to the terrestrial environment.

Birds in particular tend their eggs, and newly hatched birds usually have to be fed before they are able to fly away and seek food for themselves. Complex hormones and neural regulation are involved in reproductive behavior of parental birds (Fig. 50.2). Some animals take a different tactic. They do not deposit and tend their eggs; instead they are ovoviviparous, meaning that their eggs are retained in the body until they hatch. Then fully developed offspring, which have a way of life like the parent, are released. Oysters, which are molluscs, retain their eggs in the mantle cavity, and male sea horses which are vertebrates, have a special brood pouch in which the eggs develop. Garter snakes, water snakes, and pit vipers retain their eggs in their bodies until they hatch and give birth to living young.

Figure 50.2 Parenting in birds.
Birds, such as the American goldfinch, *Carduelis tristis*, are oviparous and lay hard-shelled eggs. They are well known for incubating their eggs and caring for their offspring after they hatch.

Finally, mammals in particular are viviparous and produce living young. After offspring are born, the nutrients needed for further growth are supplied by the mother. Viviparity represents the ultimate in caring for the zygote and embryo. How did viviparity among certain mammals come about? Some mammals, such as the duckbill platypus and the spiny anteater, are egg-laying mammals. Marsupial offspring are born in a very immature state; they finish their development within a pouch, where they are nourished on milk. In marsupials, the embryos develop within a duplex uterus, having two chambers. The uterus of most placental mammals has two so-called horns where several embryos can attach and develop in sequence. Only among primates, including humans, is there a simplex uterus where usually a single embryo develops. The **placenta** is a complex structure derived in part from the chorion, which first appears in a shelled egg. The evolution of the placenta allowed the developing offspring to exchange materials with the mother and made the shelled egg unnecessary.

Most animals have gonads in which gametes are produced. The details concerning gamete union, care of the eggs and young vary greatly.

50.2 Male Reproductive System

The human male reproductive system includes the organs pictured in Figure 50.3 and listed in Table 50.1. The male gonads are paired testes, which are suspended within the scrotal sacs of the scrotum. The testes begin their development inside the abdominal cavity, but they descend into the scrotal sacs as development proceeds. If the testes do not descend—and the male does not receive hormone therapy or undergo surgery to place the testes in the scrotum—sterility (the inability to produce offspring) results. Sterility occurs because normal sperm production is inhibited at body temperature; a slightly cooler temperature is required.

Sperm produced by the testes mature within the epididymides (sing., epididymis), which are tightly coiled tubules lying just outside the testes. Maturation seems to be required for the sperm to swim to the egg. Once the sperm have matured, they are propelled into the vasa deferentia (sing., vas deferens) by muscular contractions. Sperm are stored in both the epididymides and the vasa deferentia. When a male becomes sexually aroused, sperm enter the urethra, part of which is located within the penis.

The **penis** is a cylindrical organ that usually hangs in front of the scrotum. Three cylindrical columns of spongy, erectile tissue containing distensible blood spaces extends through the shaft of the penis (Fig. 50.4). During sexual arousal, nervous reflexes cause an increase in arterial blood

Table 50.1

Male Reproductive System

Organ	Function
Testes	Produce sperm and sex hormones
Epididymides	Maturation and some storage of sperm
Vasa deferentia	Conduct and store sperm
Seminal vesicles	Contribute fluid to semen
Prostate gland	Contributes fluid to semen
Urethra	Conducts sperm (and urine)
Bulbourethral glands	Contribute fluid to semen
Penis	Organ of copulation

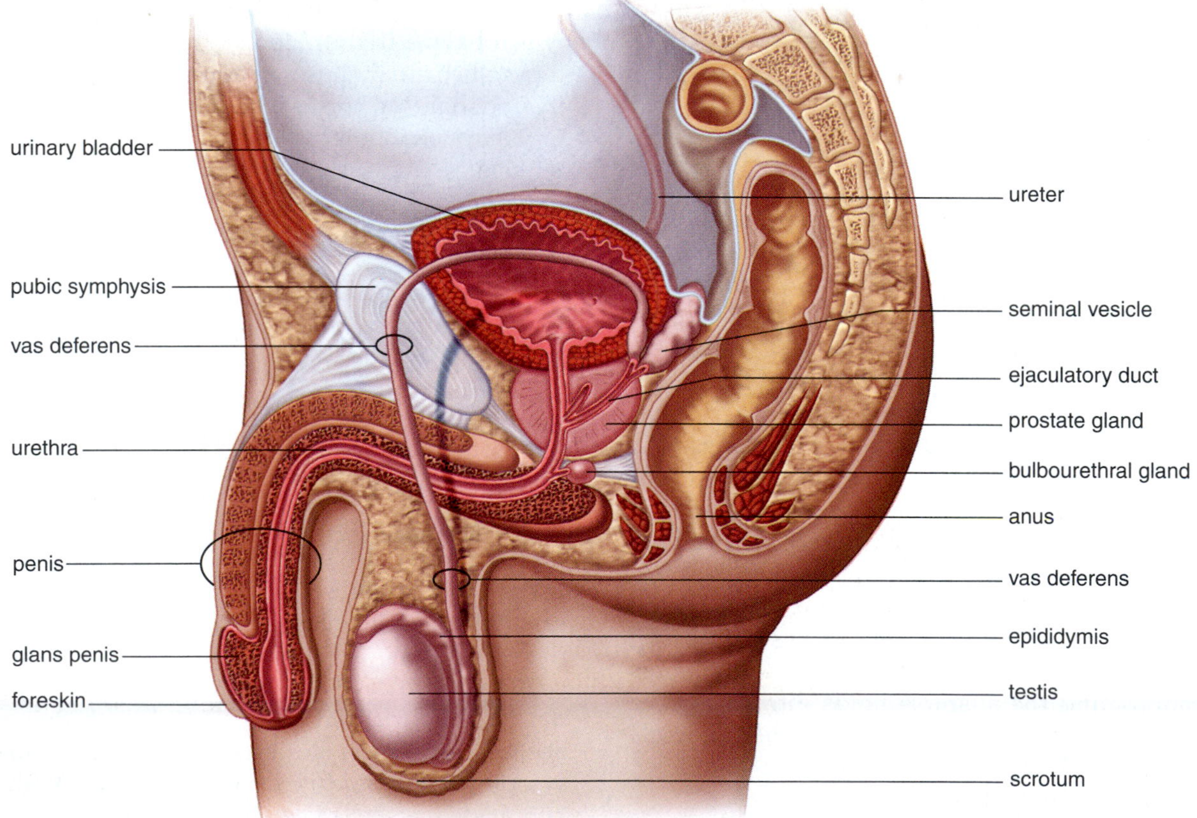

Figure 50.3 The male reproductive system.
The testes produce sperm. The seminal vesicles, the prostate gland, and the bulbourethral glands provide a fluid medium for the sperm. Circumcision is the removal of the foreskin. Notice that the penis in this drawing is not circumcised because the foreskin is present.

flow to the penis. This increased blood flow fills the blood space in the erectile tissue, and the penis, which is normally limp (flaccid), stiffens and increases in size. These changes are called erection. If the penis fails to become erect, the condition is called impotency.

Semen (seminal fluid) [L. *semen*, seed] is a thick, whitish fluid that contains sperm and secretions from three glands (Table 50.1). The seminal vesicles lie at the base of the bladder. Each joins a vas deferens to form an ejaculatory duct that enters the urethra. As sperm pass from the vasa deferentia into the ejaculatory duct, these vesicles secrete a thick, viscous fluid containing nutrients for possible use by the sperm. Just below the bladder is the prostate gland, which secretes a milky alkaline fluid believed to activate or increase the motility of the sperm. In older men, the prostate gland may become enlarged, thereby constricting the urethra and making urination difficult. Also, prostate cancer is the most common form of cancer in men. Slightly below the prostate gland, on either side of the urethra, is a pair of small glands called bulbourethral glands, which have mucous secretions with a lubricating effect. Notice from Figure 50.3 that the urethra also carries urine from the bladder during urination.

Sperm produced by the testes mature in the epididymides and pass from the vasa deferentia to the urethra, where certain glands add fluid to semen.

Ejaculation

If sexual arousal reaches its peak, ejaculation follows an erection. The first phase of ejaculation is called emission. During emission, the spinal cord sends nerve impulses via appropriate nerve fibers to the epididymides and vasa deferentia. Their subsequent motility causes sperm to enter the ejaculatory duct, whereupon the seminal vesicles, prostate gland, and the bulbourethral glands release their secretions. Secretions from the bulbourethral glands are the first to enter the urethra and they function to cleanse the urethra of acidic residue from urine. This fluid does not normally contain sperm, but considering that the young human adult male produces up to one billion sperm a day, it may. It is therefore possible though not probable for fertilization to occur in the female even though ejaculation has not occurred.

During the second phase of ejaculation, called expulsion, rhythmical contractions of muscles at the base of the penis and within the urethral wall expel semen in spurts from the opening of the urethra. These rhythmical contractions are an example of release from myotonia, or muscle tenseness. Myotonia is another important sexual response. An erection lasts for only a limited amount of time. The penis now returns to its normal flaccid state. Following ejaculation, a male may typically experience a period of time, called refractory period, during which stimulation does not bring about an erection. The contractions that expel semen from the penis are a part of male **orgasm** [Gk. *orgasmos*, sexual excitement], the physiological and psychological sensations that occur at the climax of sexual stimulation.

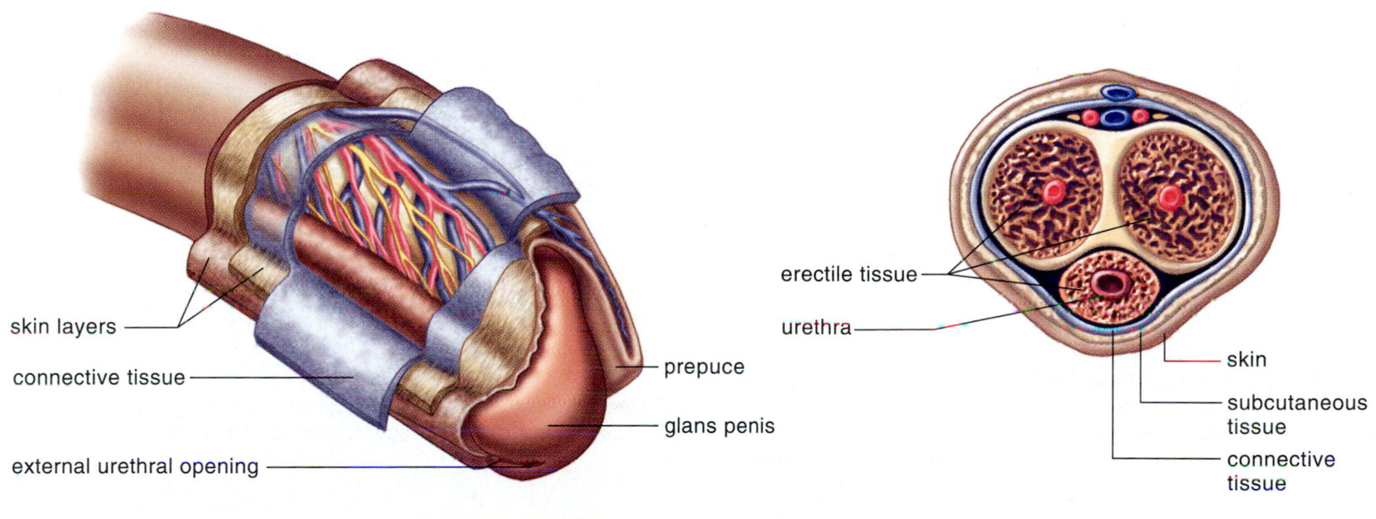

Figure 50.4 Penis anatomy.
a. Beneath the skin and the connective tissue lies the urethra, surrounded by erectile tissue. This tissue expands to form the glans penis, which in uncircumcised males is partially covered by the prepuce (foreskin). b. Two other columns of erectile tissue in the penis are dorsally located.

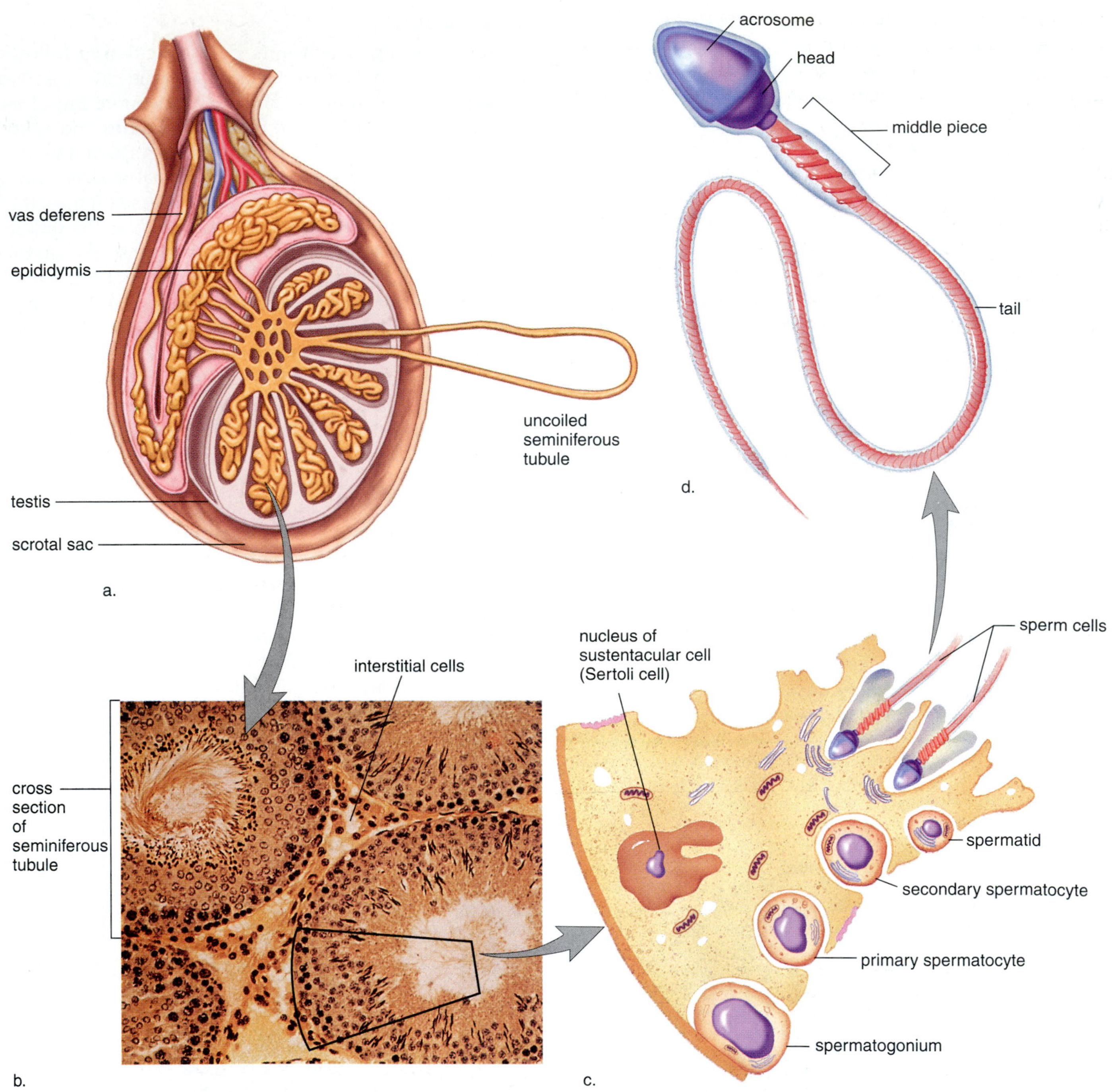

Figure 50.5 **Testis and sperm.**
a. The lobules of a testis contain seminiferous tubules. **b.** Photomicrograph of seminiferous tubules in cross section where spermatogenesis occurs. **c.** Diagrammatic representation of spermatogenesis, which occurs in the wall of the tubules. **d.** Mature sperm consist of a head, a middle piece, and a tail. The nucleus is in the head, capped by an enzyme-containing acrosome.

The Testes

A longitudinal section of a testis shows that it is composed of compartments called lobules, each of which contains one to three tightly coiled **seminiferous tubules** (Fig. 50.5*a*). Altogether, these tubules have a combined length of approximately 250 meters. A microscopic cross section of a seminiferous tubule shows that it is packed with cells undergoing spermatogenesis (Fig. 50.5*b*), a process that involves meiosis. Also present are sustentacular (Sertoli) cells, which support, nourish, and regulate the spermatogenic cells (Fig. 50.5*c*).

Mature **sperm,** or spermatozoa, have three distinct parts: a head, a middle piece, and a tail (Fig. 50.5*d*). The middle piece and the tail contain microtubules, in the characteristic 9 + 2 pattern of cilia and flagella. In the middle piece, mitochondria are wrapped around the microtubules and provide the energy for movement. The head contains a nucleus covered by a cap called the acrosome [Gk. *akros*, at the tip, and *soma*, body], which stores enzymes needed to penetrate the egg. The human egg is surrounded by several layers of cells and a thick membrane—the acrosomal enzymes play a role in allowing a sperm to reach the surface of the egg. The ejaculated semen of a normal human male contains several hundred million sperm, assuring an adequate number for fertilization to take place. Fewer than 100 ever reach the vicinity of the egg, however, and only one sperm normally enters an egg.

Hormonal Regulation in Males

The hypothalamus has ultimate control of the testes' sexual function because it secretes a hormone called gonadotropin-releasing hormone, or GnRH, that stimulates the anterior pituitary to produce the gonadotropic hormones. There are two gonadotropic hormones—follicle-stimulating hormone (FSH) and luteinizing hormone (LH)—in both males and females. In males, FSH promotes spermatogenesis in the seminiferous tubules.

LH in males is sometimes given the name interstitial cell-stimulating hormone (ICSH) because it controls the production of the androgen testosterone by the interstitial cells, scattered in the spaces between the seminiferous tubules (Fig. 50.5*b*). All these hormones, including inhibin, a hormone released by seminiferous tubules, are involved in a negative feedback relationship that maintains the fairly constant production of sperm and testosterone (Fig. 50.6).

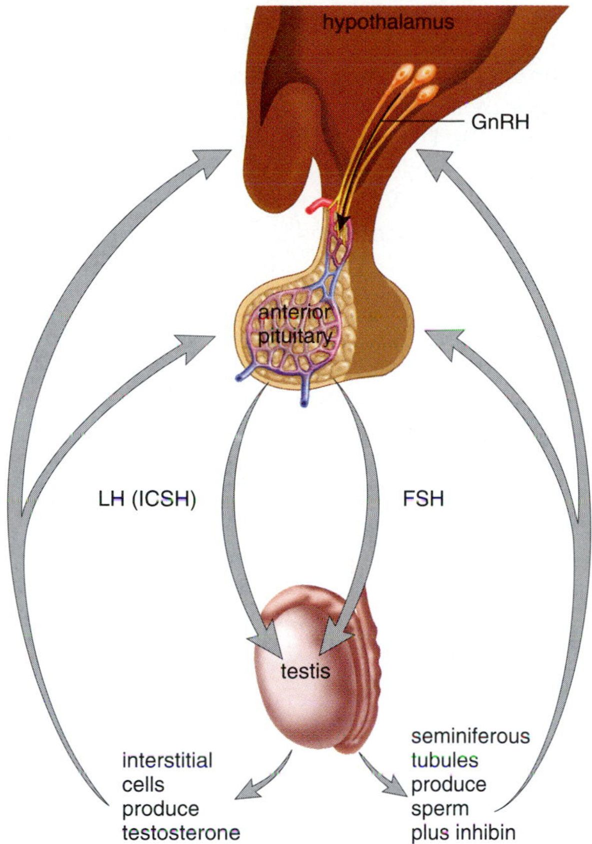

Figure 50.6 Hormonal control of testes.
GnRH (gonadotropin-releasing hormone) stimulates the anterior pituitary to secrete the gonadotropic hormones FSH and LH. In the male, LH is sometimes appropriately called interstitial cell-stimulating hormone (ICSH). FSH stimulates the testes to produce sperm, and LH stimulates the testes to produce testosterone. Testosterone and inhibin exert negative feedback control over the hypothalamus and the anterior pituitary, and this ultimately regulates the level of testosterone in the blood.

Functions of Testosterone

Testosterone is the main sex hormone in males. It is essential for the normal development and functioning of the organs listed in Table 50.1. Testosterone is also necessary for the maturation of sperm.

Testosterone also brings about and maintains the male **secondary sexual characteristics** that develop at the time of **puberty.** Males are generally taller than females and have broader shoulders and longer legs relative to trunk length. The deeper voice of males compared to females is due to the fact they have a larger larynx with longer vocal cords. Since the so-called Adam's apple is a part of the larynx, it is usually more prominent in males than in females.

Testosterone is responsible for the greater muscular development in males. Knowing this, males and females sometimes take anabolic steroids (either the natural or synthetic form of testosterone) to build up their muscles (see page 896). Testosterone causes males to develop noticeable hair on the face, chest, and occasionally on other regions of the body such as the back. Testosterone also leads to the receding hairline and pattern baldness that occurs in males.

The gonads in males are the testes, which produce sperm and testosterone, the most significant male sex hormone.

50.3 Female Reproductive System

The human female reproductive system includes the ovaries, the oviducts, the uterus, and the vagina (Fig. 50.7 and Table 50.2). The ovaries, which produce a secondary **oocyte** each month, lie in shallow depressions, one on each side of the upper pelvic cavity. The oviducts [L. *ovum*, egg, and *duco*, lead out], also called uterine or fallopian tubes, extend from the ovaries to the uterus; however, the oviducts are not attached to the ovaries. Instead, they have fingerlike projections called fimbriae (sing., fimbria) that sweep over the ovaries. When an oocyte bursts from an ovary during ovulation, it usually is swept into an oviduct by the combined action of the fimbriae and the beating of cilia that line the oviducts. Fertilization, if it occurs, normally takes place in an oviduct, and the developing embryo is propelled slowly by ciliary movement and tubular muscle contraction to the uterus. The **uterus** [L. *uterus*, womb] is a thick-walled muscular organ about the size and shape of an inverted pear. The narrow end of the uterus is called the cervix. The embryo completes its development after embedding itself in the uterine lining, called the **endometrium** [Gk. *endon*, within, and *metra*, womb]. A small opening at the cervix leads to the vaginal canal. The vagina [L. *vagina*, sheath] is a tube at a 45° angle with the small of the back. The mucosal lining of the vagina lies in folds and can extend. This is especially important when the vagina serves as the birth canal, and it also can facilitate intercourse, when the vagina receives the penis during copulation.

The external genital organs of the female are known collectively as the vulva (Fig. 50.7*b*). The mons pubis and two folds of skin called labia minora and labia majora are on either side of the urethral and vaginal openings. At the juncture of the labia minora is the clitoris, which is homologous to the penis in males. The clitoris has a shaft of erectile tissue and is capped by a pea-shaped glans. The many sense receptors of the clitoris allow it to function as a sexually sensitive organ. Orgasm in the female is a release of neuromuscular tension in the muscles of the genital area, vagina, and uterus.

Table 50.2

Female Reproductive Organs

Organ	Function
Ovaries	Produce egg and sex hormones
Oviducts (fallopian tubes)	Conduct egg; location of fertilization
Uterus (womb)	Houses developing embryo and fetus
Cervix	Contains opening to uterus
Vagina	Receives penis during copulation and serves as birth canal

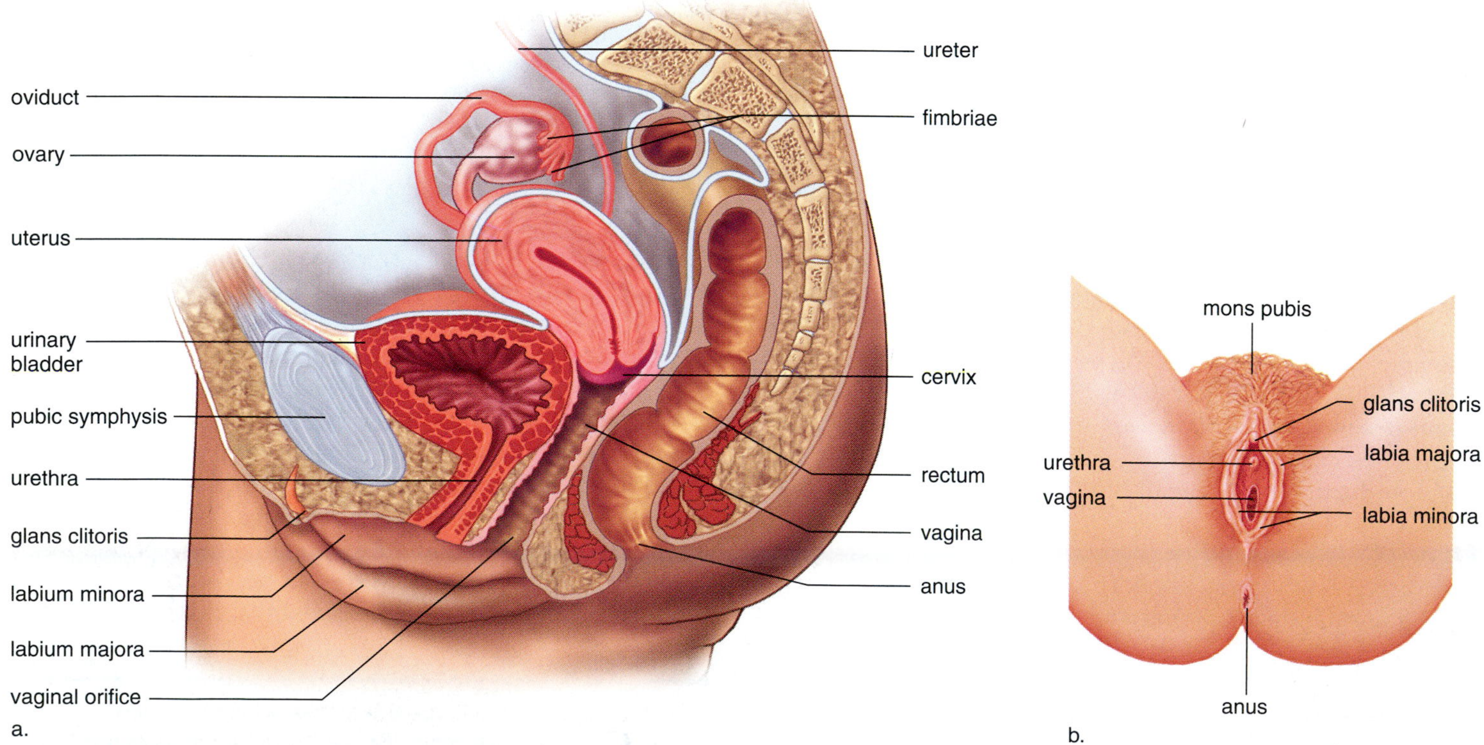

Figure 50.7 **Female reproductive system.**
a. The ovaries produce one oocyte (egg) per month. Fertilization occurs in the oviduct, and development occurs in the uterus. The vagina is the birth canal and organ of sexual intercourse. **b.** Vulva. At birth, the opening of the vagina is partially occluded by a membrane called the hymen. Physical activities and sexual intercourse disrupt the hymen.

The Ovaries

The ovaries alternate in producing one oocyte each month. For the sake of convenience, the released oocyte is often called an **ovum,** or egg. The ovaries also produce the female sex hormones, **estrogens** and **progesterone**, during the ovarian cycle.

The Ovarian Cycle

The **ovarian cycle** is described in Figure 50.8, a longitudinal section through an ovary. The ovary contains many cellular **follicles** [L. dim. of *folliculus*, bag], each containing an oocyte. At birth a female has as many as 2 million follicles, but the number is reduced to 300,000–400,000 by the time of puberty. Only a small number of follicles (about 400) ever mature, and produce an oocyte.

As the follicle undergoes maturation, it develops from a primary follicle to a secondary follicle to a vesicular (Graafian) follicle. Oogenesis has begun, and a secondary follicle contains a secondary oocyte with a reduced number of chromosomes. In a secondary follicle, the secondary oocyte is pushed to one side in a fluid-filled cavity. In a vesicular follicle, the fluid-filled cavity increases to the point that the follicle wall balloons out on the surface of the ovary and bursts, releasing the secondary oocyte surrounded by glycoprotein and follicular cells. The release of a secondary oocyte from a vesicular follicle is termed **ovulation.**

Oogenesis is completed when and if the secondary oocyte is fertilized by a sperm. In the meantime, the follicle is developing into the **corpus luteum** [L. *corpus*, body, and *luteus*, yellow]. If fertilization and pregnancy do not occur, the corpus luteum begins to degenerate after about ten days.

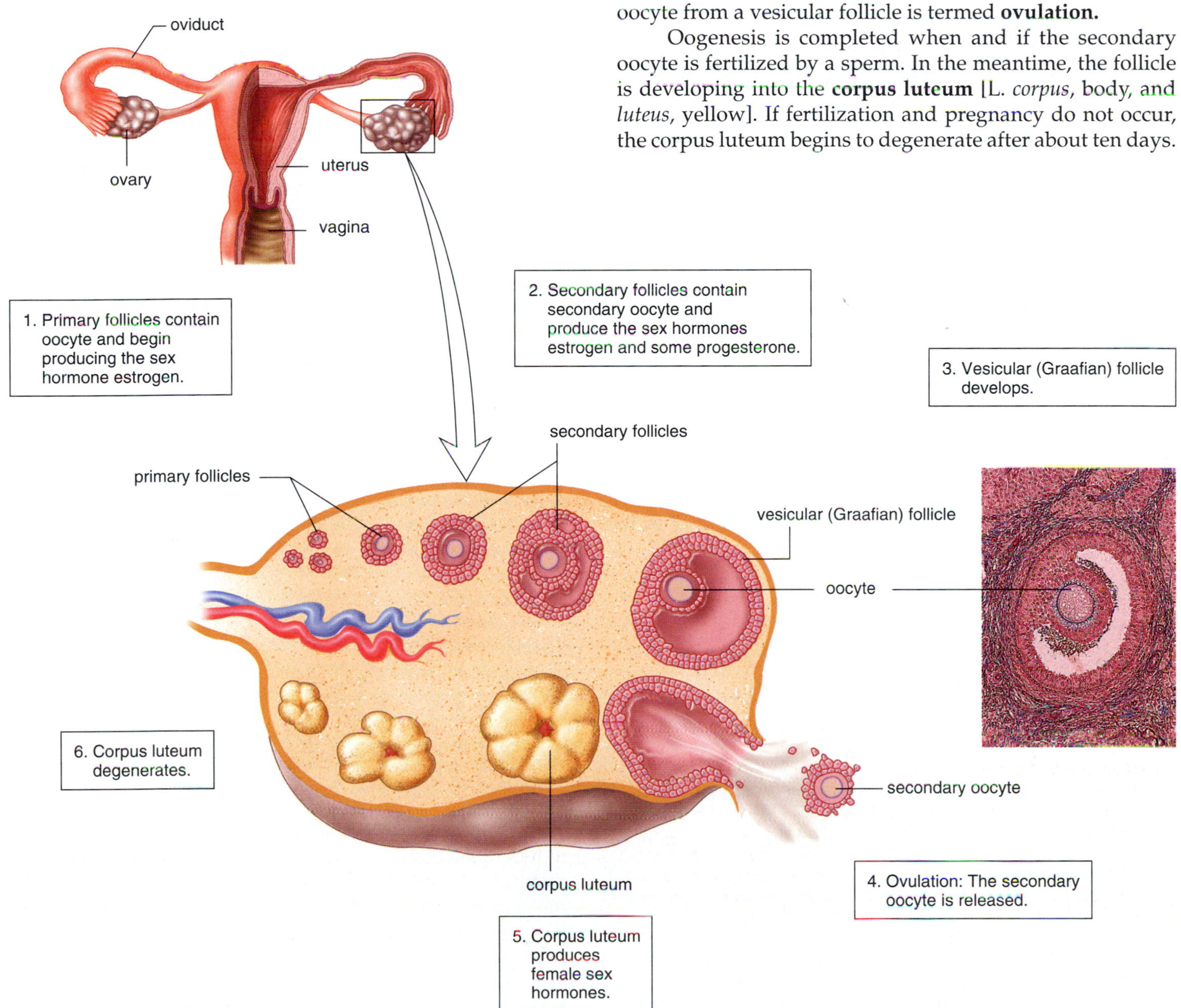

Figure 50.8 Ovarian cycle.
As a follicle matures, the oocyte enlarges and is surrounded by layers of follicular cells and fluid. Eventually, ovulation occurs, the mature follicle ruptures, and the secondary oocyte is released. A single follicle actually goes through all stages in one place within the ovary.

Table 50.3

Ovarian and Uterine Cycles (Simplified)

Ovarian Cycle	Events	Uterine Cycle	Events
Follicular phase—Days 1–13	FSH Follicle maturation Estrogens	Menstruation—Days 1–5 Proliferative phase—Days 6–13	Endometrium breaks down Endometrium rebuilds
Ovulation—Day 14*	LH spike		
Luteal phase—Days 15–28	LH Corpus luteum Progesterone	Secretory phase—Days 15–28	Endometrium thickens and glands are secretory

* *Assuming a 28-day cycle*

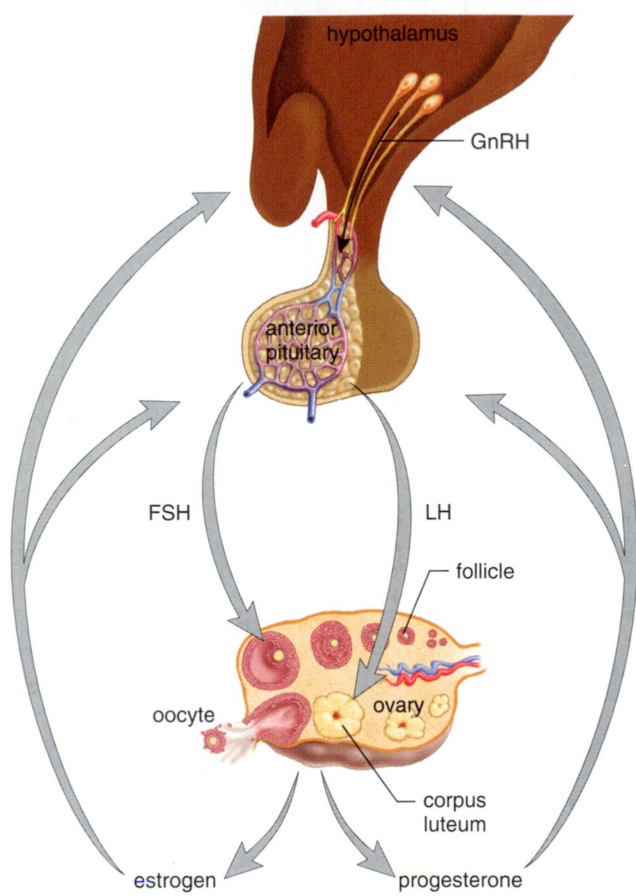

Figure 50.9 Hormonal control of ovaries.
The hypothalamus produces GnRH (gonadotropin-releasing hormone). GnRH stimulates the anterior pituitary to produce FSH and LH. FSH stimulates a follicle to produce estrogen, and LH stimulates the corpus luteum to produce progesterone. Estrogen and progesterone maintain the sex organs (e.g., uterus) and the secondary sexual characteristics, and exert feedback control over the hypothalamus and the anterior pituitary.

The ovarian cycle is under the control of the gonadotropic hormones, follicle-stimulating hormone (FSH) and luteinizing hormone (LH) (Fig. 50.9 and Table 50.3). The gonadotropic hormones are not present in constant amounts and instead are secreted at different rates during the cycle. For simplicity's sake, it is convenient to emphasize that during the first half, or **follicular phase,** of the cycle, FSH promotes the development of a follicle, which secretes estrogens. As the estrogen level in the blood rises, it exerts feedback control over the anterior pituitary secretion of FSH so that the follicular phase comes to an end.

Presumably, the high level of estrogens in the blood also causes the hypothalamus suddenly to secrete a large amount of GnRH. This leads to a surge of LH production by the anterior pituitary and to ovulation at about the fourteenth day of a 28-day cycle (Fig. 50.10).

During the second half, or **luteal phase,** of the ovarian cycle, it is convenient to emphasize that LH promotes the development of the corpus luteum, which secretes progesterone. As the blood level of progesterone rises, it exerts feedback control over anterior pituitary secretion of LH so that the corpus luteum begins to degenerate. As the luteal phase comes to an end, menstruation occurs.

One ovarian follicle per month produces a secondary oocyte. Following ovulation, the follicle develops into the corpus luteum.

The Uterine Cycle

The female sex hormones, estrogens and progesterone, have numerous functions. The effects of these hormones on the endometrium of the uterus causes the uterus to undergo a cyclical series of events known as the **uterine cycle** (Table 50.3 and Fig. 50.10). Twenty-eight-day cycles are divided as follows.

During *days 1–5,* there is a low level of female sex hormones in the body, causing the endometrium to disintegrate and its blood vessels to rupture. A flow of blood, known as the **menses,** passes out of the vagina during **menstruation** [L. *menstrualis,* happening monthly], also known as the menstrual period.

During *days 6–13,* increased production of estrogens by an ovarian follicle causes the endometrium to thicken and to become vascular and glandular. This is called the proliferative phase of the uterine cycle.

Ovulation usually occurs on the fourteenth day of the 28-day cycle.

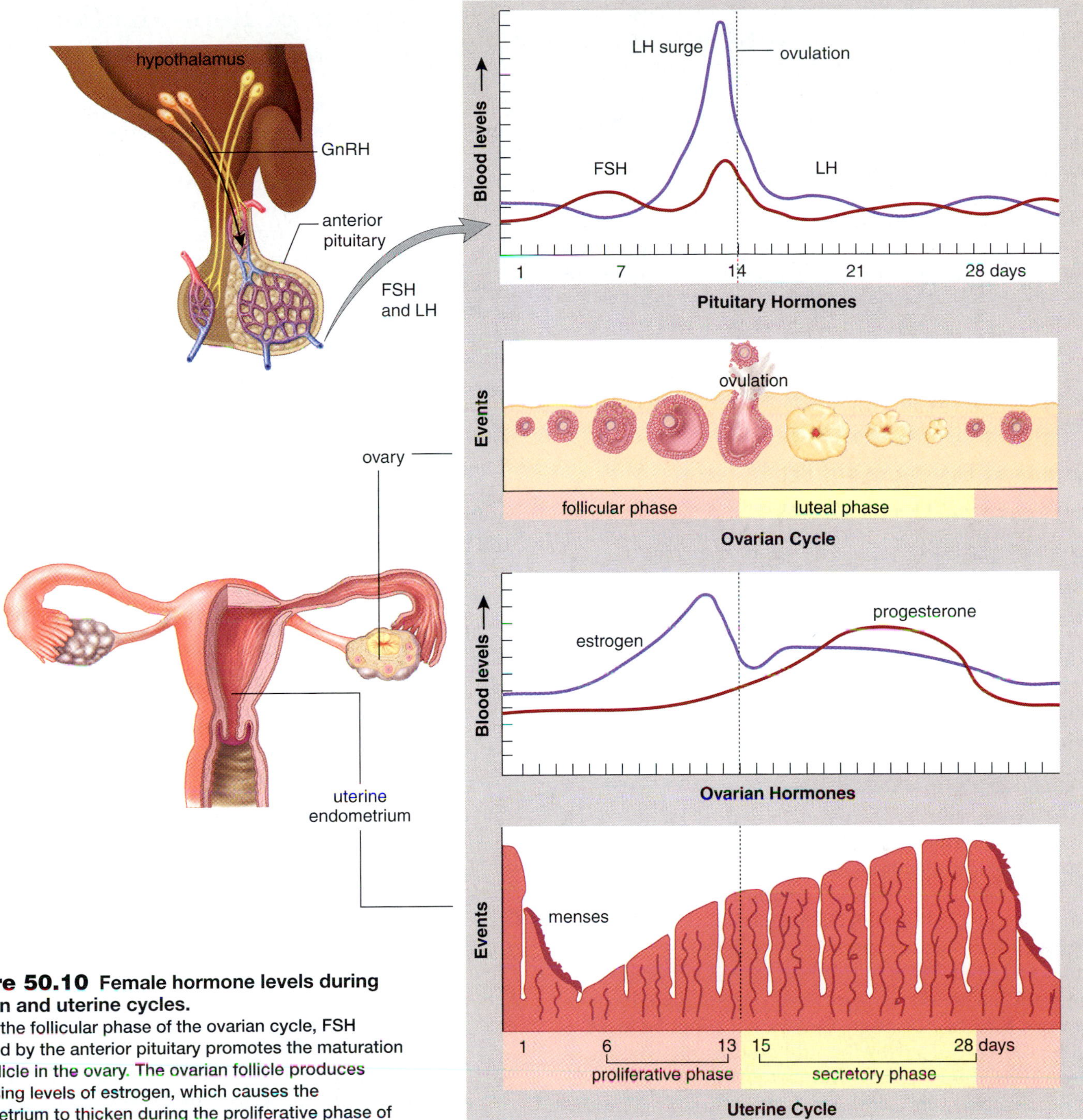

Figure 50.10 Female hormone levels during ovarian and uterine cycles.
During the follicular phase of the ovarian cycle, FSH released by the anterior pituitary promotes the maturation of a follicle in the ovary. The ovarian follicle produces increasing levels of estrogen, which causes the endometrium to thicken during the proliferative phase of the uterine cycle. After ovulation and during the luteal phase of the ovarian cycle, LH promotes the development of the corpus luteum. This structure produces increasing levels of progesterone, which causes the endometrium to become secretory. Menstruation begins when progesterone production declines to a low level.

During *days 15–28*, increased production of progesterone by the corpus luteum causes the endometrium to double in thickness and the uterine glands to mature, producing a thick mucoid secretion. This is called the secretory phase of the uterine cycle. The endometrium now is prepared to receive the developing embryo. If pregnancy does not occur, the corpus luteum degenerates and the low level of sex hormones in the female body causes the endometrium to break down. Menses begins and this is day one of the next cycle. Even while menstruation is occurring, the anterior pituitary begins to increase its production of FSH and a new follicle begins to mature. Table 50.3 indicates how the ovarian cycle controls the uterine cycle.

Events Following Fertilization

If fertilization does occur, an embryo begins development even as it travels down the oviduct to the uterus. The endometrium is now prepared to receive the developing embryo, which becomes embedded in the lining several days following fertilization. The placenta originates from both maternal and embryonic tissues. It is shaped like a large, thick pancake and is the site of exchange of gases and nutrients between fetal and maternal blood although there is rarely any mixing of the two. At first the placenta produces human chorionic gonadotropin (HCG), which maintains the corpus luteum until the placenta begins its own production of progesterone and estrogens. Progesterone and estrogens have two effects. They shut down the anterior pituitary so that no new follicles mature, and they maintain the lining of the uterus so that the corpus luteum is not needed. There is no menstruation during pregnancy.

Functions of Estrogen and Progesterone

Estrogens in particular are essential for the normal development and functioning of the organs listed in Table 50.2. Estrogens are also largely responsible for the secondary sexual characteristics in females, including body hair and fat distribution. In general, females have a more rounded appearance than males because of a greater accumulation of fat beneath the skin. Also, the pelvic girdle enlarges so that females have wider hips than males and the thighs converge at a greater angle toward the knees. Both estrogen and progesterone are also required for breast development.

The Female Breast

A female breast contains between 15 and 24 lobules, each with its own mammary duct (Fig. 50.11). This duct begins at the nipple and divides into numerous other ducts, which end in blind sacs called alveoli. **Lactation** is the production of milk by the cells of the alveoli. Milk is not produced during pregnancy. Prolactin causes the alveoli to begin producing milk, and production of this hormone is suppressed by the feedback inhibition estrogens and progesterone have on the anterior pituitary during pregnancy. It takes a couple of days after delivery of a baby for milk production to begin. In the meantime, the breasts produce a watery, yellowish-white fluid called colostrum, which has a similar composition to milk but contains more protein and less fat. Colostrum (and later milk) is rich in IgA antibodies that may provide some degree of immunity to the newborn.

Breast cancer is the most common form of cancer in females. Women should regularly check their breasts for lumps and have mammograms (X-ray photographs) taken as recommended by their physician.

50.4 Control of Reproduction

The two major causes of infertility in females are blocked oviducts, possibly due to a sexually transmitted disease, and failure to ovulate due to low body weight. Endometriosis, the spread of uterine tissue beyond the uterus, is also a cause. If no obstruction is apparent and body weight is normal, it is possible to give females HCG extracted from the urine of pregnant women, along with HMG (human menopausal gonadotropin) extracted from the urine of postmenopausal women. This treatment causes multiple ovulations and sometimes multiple pregnancies. The most frequent causes of sterility and infertility in males are low sperm count and/or a large proportion of abnormal sperm. Disease, radiation, chemical mutagens, too much heat near the testes, and the use of psychoactive drugs can contribute to this condition.

Birth Control Methods

Sometimes couples wish to prevent a possible pregnancy. The most reliable method of birth control is abstinence; that is, the absence of sexual intercourse. This form of birth control has the added advantage of preventing transmission of a sexually transmitted disease. Other, perhaps more common, means of birth control used in the United States are listed in Table 50.4. The effectiveness of the method refers to

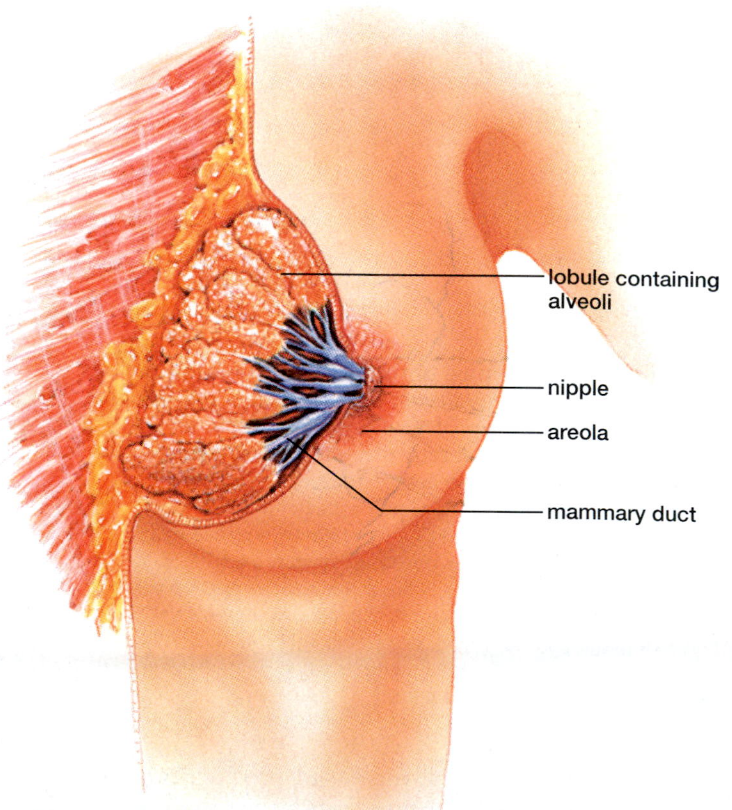

Figure 50.11 Anatomy of breast.
The female breast contains lobules consisting of ducts and alveoli. The alveoli are lined by milk-producing cells in the lactating (milk-producing) breast.

the number of women per year who will not get pregnant even though they are regularly engaging in sexual intercourse. The male and female condom also offer some protection against sexually transmitted diseases.

Investigators have long searched for a "male pill." Analogues of gonadotropic-releasing hormone prevent the hypothalamus from stimulating the anterior pituitary. Feminization due to low testosterone blood level has thus far made these possible birth control methods undesirable. Inhibin has also been shown to inhibit spermatogenesis in males, but this hormone must be administered by injection.

Morning-After Pills

The expression "morning-after pill" refers to a medication that will prevent pregnancy after unprotected intercourse. The expression is a misnomer in that medication can begin one to several days after unprotected intercourse.

A kit, now called Preven, is made up of four synthetic progesterone pills; two are taken up to 72 hours after unprotected intercourse, and two more are taken 12 hours later. The medication upsets the normal uterine cycle, making it difficult for the embryo to implant itself in the endometrium. In a recent study, it was estimated that the medication was 85% effective in preventing unintended pregnancies.

Mifepristone, better known as RU-486, is a pill that is presently used to cause the loss of an implanted embryo by blocking the progesterone receptors of endometrial cells. Without functioning receptors for progesterone, the endometrium sloughs off, carrying the embryo with it. When taken in conjunction with a prostaglandin to induce uterine contractions, RU-486 is 95% effective. It is possible that some day this medication will also be a "morning-after pill," taken when menstruation is late without evidence that pregnancy has occurred.

There are numerous well-known birth control methods and devices available to those who wish to prevent pregnancy. Their effectiveness varies. In addition, new methods are expected to be developed.

Table 50.4

Common Birth Control Methods

Name	Procedure	Methodology	Effectiveness	Risk
Abstinence	Refrain from sexual intercourse	No sperm in vagina	100%	None
Vasectomy	Vasa deferentia cut and tied	No sperm in seminal fluid	Almost 100%	Irreversible sterility
Tubal ligation	Oviducts cut and tied	No eggs in oviduct	Almost 100%	Irreversible sterility
Oral contraception	Hormone medication is taken daily	Anterior pituitary does not release FSH and LH	Almost 100%	Thromboembolism, especially in smokers
Depo-Provera injection	Four injections of progesterone-like steroid given per year	Anterior pituitary does not release FSH and LH	About 99%	Breast cancer? Osteoporosis?
Contraceptive implants	Tubes of progestin (form of progesterone) implanted under skin	Anterior pituitary does not release FSH and LH	More than 90%	Presently none known
Intrauterine device (IUD)	Plastic coil inserted into uterus by physician	Prevents implantation	More than 90%	Infection (pelvic inflammatory disease, PID)
Diaphragm	Latex cup inserted into vagina to cover cervix before intercourse	Blocks entrance of sperm to uterus	With jelly, about 90%	Presently none known
Cervical cap over cervix	Latex cap held by suction	Delivers spermicide near cervix	Almost 85%	Cancer of cervix
Male condom	Latex sheath fitted over erect penis	Traps sperm and prevents STDs	About 85%	Presently none known
Female condom	Polyurethane liner fitted inside vagina	Blocks entrance of sperm to uterus and prevents STDs	About 85%	Presently none known
Coitus interruptus	Penis withdrawn before ejaculation	Prevents sperm from entering vagina	75%	Presently none known
Jellies, creams, foams	These spermicidal products inserted before intercourse	Kills a large number of sperm	About 75%	Presently none known
Natural family planning	Day of ovulation determined by record keeping; various methods of testing	Intercourse avoided on certain days of the month	About 70%	Presently none known
Douche	Vagina and uterus cleansed after intercourse	Washes out sperm	Less than 70%	Presently none known

Sexually Transmitted Diseases

Sexually transmitted diseases (STDs) are caused by organisms ranging from viruses to arthropods; however, we will discuss only certain STDs caused by viruses and bacteria. Unfortunately, for unknown reasons, humans cannot develop good immunity to any STDs. Therefore, prompt medical treatment should be received when exposed to an STD. To prevent the spread of STDs, a latex condom can be used; the concomitant use of a spermicide containing nonoxynol 9 gives added protection.

It is difficult to cure the STDs caused by viruses (e.g., AIDS, genital herpes, and genital warts), but treatment is available for AIDS and genital herpes. Those STDs caused by bacteria (e.g., gonorrhea, chlamydia, and syphilis) are treatable with antibiotics.

AIDS

The organism that causes acquired immunodeficiency syndrome (AIDS) is a virus called **human immunodeficiency virus (HIV).** HIV attacks the type of lymphocyte known as helper T cells. Helper T cells, you will recall, stimulate the activities of B lymphocytes, which produce antibodies. After an HIV infection sets in, helper T cells begin to decline in number, and the person becomes more susceptible to other types of infections.

Symptoms

AIDS has three stages of infection called category A, B, and C. During category A stage, which may last about a year, the individual is an asymptomatic carrier. There may be no symptoms, but the individual can pass on the infection. Immediately after infection, and before the blood test becomes positive, there is a large number of infectious viruses in the blood that could be passed on to another person. Even after the blood test becomes positive, the person remains well as long as the body produces sufficient helper T lymphocytes to keep the count higher than 500 per mm^3. During the category B stage, which may last six to eight years, the lymph nodes swell and there may also be weight loss, night sweats, fatigue, fever, and diarrhea. Infections like thrush (white sores on the tongue and in the mouth) and herpes reoccur. Finally, the person may progress to category C, which is full-blown AIDS characterized by nervous disorders and by the development of an opportunistic disease such as an unusual type of pneumonia or skin cancer. Opportunistic diseases are ones that occur only in individuals who have little or no capability of fighting an infection. Without intensive medical treatment, the AIDS patient dies about seven to nine years after infection. Now, with a combination therapy of several drugs, AIDS patients are beginning to live longer in the United States.

Transmission

AIDS is transmitted by sexual contact with an infected person, including vaginal or rectal intercourse and oral/genital contact. Also, needle-sharing among intravenous drug users is high-risk behavior. A less common mode of transmission (and now rarely in countries where donated blood is screened for HIV) is through transfusions of infected blood or blood-clotting factors.

HIV first spread through the homosexual community, and male-to-male sexual contact still accounts for the largest percentage of new AIDS cases in the United States. But the largest increases of HIV infections are occurring through heterosexual contact or by intravenous drug use. Now, women account for 19% of all newly diagnosed cases of AIDS. The rise in the incidence of AIDS among women of reproductive age is paralleled by a rise in the incidence of AIDS in children younger than 13. Babies born to HIV-infected women may become infected before or during birth, or through breast-feeding after birth.

Genital Warts

Genital warts are caused by the human papillomaviruses (HPVs). Many times, carriers do not have any sign of warts, or merely flat lesions may be present. When present, the warts commonly are seen on the penis and foreskin of men and near the vaginal opening in women. A newborn can become infected by passage through the birth canal.

Presently, there is no cure for an HPV infection, but it can be treated effectively by surgery, freezing, application of an acid, or laser burning, depending on severity. If visible warts are removed, they may recur. Genital warts are associated with cancer of the cervix, as well as tumors of the vulva, the vagina, the anus, and the penis. Some researchers believe that the viruses are involved in 90–95% of all cases of cancer of the cervix.

Genital Herpes

Genital herpes is caused by herpes simplex virus—type 1 usually causes cold sores and fever blisters, while type 2 more often causes genital herpes.

Persons usually get infected with herpes simplex virus type 2 when they are adults. In some people there are no symptoms. Or there may be a tingling or itching sensation before blisters appear on the genitals. Once the blisters rupture, they leave painful ulcers that may take as long as three weeks or as little as five days to heal. The blisters may be accompanied by fever, pain on urination, swollen lymph nodes in the groin, and in women, a copious discharge. At this time, the individual has an increased risk of acquiring an AIDS infection.

After the ulcers heal, the disease is only latent, and blisters can recur, although usually at less frequent intervals and with milder symptoms. Fever, stress, sunlight, and menstruation are associated with reoccurrence of symptoms. Exposure to herpes in the birth canal can cause an infection in the newborn, which leads to neurological disorders and even death. Birth by cesarean section prevents this possibility.

Hepatitis

There are several types of hepatitis. The type of hepatitis and the virus that causes it is designated by the same letter. Hepatitis A is usually acquired from sewage-contaminated drinking water, but this infection can also be sexually transmitted through oral/anal contact. Hepatitis B, which is spread in the same manner as AIDS, is even more infectious. Fortunately, a vaccine is now available for hepatitis B. Hepatitis C is called post-transfusion hepatitis. Hepatitis infections cause infection of the liver and can lead to liver failure, liver cancer, and death.

Chlamydia

Chlamydia is named for the tiny bacterium that causes it *(Chlamydia trachomatis).* The incidence of new chlamydial infections has steadily increased since 1984 (Fig. 50A).

Chlamydial infections of the lower reproductive tract usually are mild or asymptomatic, especially in women. About 8 to 21 days after infection, men may experience a mild burning sensation on urination and a mucoid discharge. Women may have a vaginal discharge along with the symptoms of a urinary tract infection. Chlamydia also causes cervical ulcerations, which increase the risk of acquiring AIDS.

If the infection is misdiagnosed, or the person does not seek medical help, there is a particular risk of the infection spreading from the cervix to the oviducts so that pelvic inflammatory disease (PID) results. This very painful condition can result in a blockage of the oviducts with the possibility of sterility and infertility. If a baby comes in contact with chlamydia during birth, inflammation of the eyes or pneumonia can result.

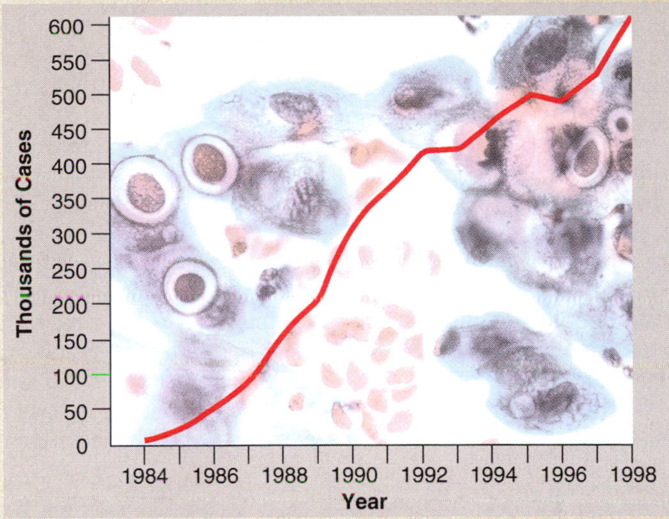

Figure 50A **Chlamydial infection.**
A graph depicting the incidence of reported cases of chlamydia in the United States from 1984 to 1998 is superimposed on a micrograph of a cell containing different stages of the organism.

Source: Sexually Transmitted Disease Surveillance, 1997 and 1998. Atlanta: Centers for Disease Control and Prevention.

Gonorrhea

Gonorrhea is caused by the bacterium *Neisseria gonorrhoeae.* Diagnosis in the male is not difficult, since typical symptoms are pain upon urination and a thick, greenish-yellow urethral discharge. In males and females, a latent infection leads to pelvic inflammatory disease (PID), in which the vasa deferentia or oviducts are affected. As the inflamed tubes heal, they may become partially or completely blocked by scar tissue, resulting in sterility or infertility. If a baby is exposed during birth, an eye infection leading to blindness can result. All newborns are given eyedrops to prevent this possibility.

Gonorrhea proctitis, an infection of the anus, with symptoms including anal pain and blood or pus in the feces, also occurs in patients. Oral/genital contact can cause infection of the mouth, throat, and tonsils. Gonorrhea can spread to internal parts of the body, causing heart damage or arthritis. If, by chance, the person touches infected genitals and then touches his or her eyes, a severe eye infection can result. Up to now, gonorrhea was curable by antibiotic therapy, but resistance to antibiotic therapy is becoming more and more common, and 40% of all strains are now known to be resistant to therapy.

Syphilis

Syphilis, which is caused by the bacterium *Treponema pallidum,* has three stages, which are typically separated by latent periods. In the primary stage, a hard chancre (ulcerated sore with hard edges) appears. In the secondary stage, a rash appears all over the body—even on the palms of the hands and the soles of the feet. During the tertiary stage, syphilis may affect the cardiovascular and/or nervous system. An infected person may become mentally retarded, become blind, walk with a shuffle, or show signs of insanity. Gummas, which are large destructive ulcers, may develop on the skin or within the internal organs. Syphilitic bacteria can cross the placenta, causing birth defects or a stillbirth. Unlike the other STDs discussed, there is a blood test to diagnose syphilis.

Syphilis is a very devastating disease. Control depends on prompt and adequate treatment of all new cases; therefore, it is very important for all sexual contacts to be traced so they can be treated with antibiotic therapy.

Vaginitis

Women are subject to vaginitis, an infection that is caused by the flagellated protozoan *Trichomonas vaginalis* or by the yeast *Candida albicans*. The protozoan infection causes a frothy white or yellow foul-smelling discharge accompanied by itching, and the yeast infection causes a thick, white, curdy discharge, also accompanied by itching. *Trichomonas* is most often acquired through sexual intercourse, and an asymptomatic partner is usually the reservoir of infection. *Candida albicans,* however, is a normal organism found in the vagina; its growth simply increases beyond normal under certain circumstances. Women taking the birth control pill or who have been on antibiotic therapy are sometimes prone to yeast infections.

Connecting Concepts

When reproduction does not occur in the usual manner, couples often seek alternative reproductive methods, which include surrogate motherhood (a woman has another couple's child), artificial insemination (sperm are placed in the vagina by a physician), and in vitro fertilization (fertilization takes place in laboratory glassware). In vitro fertilization has progressed to the ability to freeze eggs or sperm or even embryos for future use. Older women who never had the opportunity to freeze their eggs can still have children if they use donated eggs—perhaps today harvested from a fetus.

The dizzying array of reproductive technologies has resulted in many legal complications. Questions abound about which mother has first claim to the child—the surrogate mother, the woman who donated the egg, or the primary care giver—to which partner has first claim to frozen embryos following a divorce. Legal decisions about who has the right to use what techniques have rarely been discussed, much less decided upon. Some clinics will help anyone, male or female, no questions asked, as long as they have the ability to pay. And most clinics are heading toward doing any type of procedure, including guaranteeing the sex of the child, and making sure the child will be free from some particular genetic disorder. It would not be surprising if, in the future, zygotes could be engineered to have any particular trait desired by the parents.

Even today eugenic (good gene) goals are evidenced by the fact that reproductive clinics advertise for egg and sperm donors, primarily in elite college newspapers. Is it too late for us as a society to make ethical decisions about reproductive issues? Should we come to a consensus about what techniques should be allowed and who should be able to use them? We all want to avoid, if possible, what happened to Jonathan Alan Austin. Jonathan, who was born to a surrogate mother, later died from injuries inflicted by his father. Should background checks be legally required? Should surrogate mothers only make themselves available to individuals or couples who fulfill certain psychological characteristics?

Summary

50.1 How Animals Reproduce

Ordinarily, asexual reproduction may quickly produce a large number of offspring genetically identical to the parent. Sexual reproduction involves gametes and produces offspring that are genetically slightly different from the parents. The gonads are the primary sex organs, but there are also accessory organs. The accessory organs consist of storage areas for sperm and ducts that conduct the gametes. They also contribute to formation of the semen.

Animals typically protect their eggs and embryos. Those that are oviparous provide them with yolk, and if the animal is terrestrial there is typically a shell to prevent drying out. The amount of yolk is dependent on whether there is a larval stage.

Reptiles and birds have extraembryonic membranes that allow them to develop on land; these same membranes are modified for internal development in mammals. Ovoviviparous animals retain their eggs until the offspring have hatched, and viviparous animals retain the embryo. Placental mammals exemplify viviparous animals.

50.2 Male Reproductive System

In human males, sperm are produced in the testes, mature in the epididymides, and may be stored in the vasa deferentia before entering the urethra, along with seminal fluid (produced by seminal vesicles, prostate gland, and bulbourethral glands). Sperm are ejaculated during male orgasm, when the penis is erect.

Spermatogenesis occurs in the seminiferous tubules of the testes, which also produce testosterone in interstitial cells. Testosterone brings about the maturation of the primary sex organs during puberty and promotes the secondary sexual characteristics of males, such as low voice, facial hair, and increased muscle strength.

Follicle-stimulating hormone (FSH) from the anterior pituitary stimulates spermatogenesis, and luteinizing hormone (LH, also called ICSH) stimulates testosterone production. A hypothalamic-releasing hormone, gonadotropic-releasing hormone (GnRH), controls anterior pituitary production and FSH and LH release. The level of testosterone in the blood controls the secretion of GnRH and the anterior pituitary hormones by a negative feedback system.

50.3 Female Reproductive System

In females, an oocyte produced by an ovary enters an oviduct, which leads to the uterus. The uterus opens into the vagina. The external genital area of women includes the vaginal opening, the clitoris, the labia minora, and the labia majora.

In either ovary, one follicle a month matures, produces a secondary oocyte, and becomes a corpus luteum. This is called the ovarian cycle. The follicle and the corpus luteum produce estrogens and progesterone, the female sex hormones.

The uterine cycle occurs concurrently with the ovarian cycle. In the first half of these cycles (days 1–13, before ovulation), the anterior pituitary produces FSH and the follicle produces estrogens. Estrogens cause the endometrium to increase in thickness. In the second half of these cycles (days 15–28, after ovulation), the anterior pituitary produces LH and the follicle produces progesterone. Progesterone causes the endometrium to become secretory. Feedback control of the hypothalamus and anterior pituitary causes the levels of estrogens and progesterone to fluctuate. When they are at a low level, menstruation begins.

If fertilization occurs, a zygote is formed and development begins. The resulting embryo travels down the oviduct and implants itself in the prepared endometrium. A placenta, which is the region of exchange between the fetal blood and mother's blood, forms. At first, the placenta produces HCG, which maintains the corpus luteum; later, it produces progesterone and estrogens.

Estrogens and progesterone are the female sex hormones. Primarily estrogens bring about the maturation of the primary sex organs during puberty and promotes the secondary sexual characteristics of females, including less body hair than males, a wider pelvic girdle, a more rounded appearance, and development of breasts.

50.4 Control of Reproduction

Infertile couples are increasingly resorting to alternative methods of reproduction. Numerous birth control methods and devices are available for those who wish to prevent pregnancy.

Reviewing the Chapter

1. Contrast asexual reproduction with sexual reproduction, primary sex organs with accessory organs, reproduction in water with reproduction on land, the life history of an insect with that of a bird. 902–03
2. Trace the path of sperm in a human male. What glands contribute fluids to semen? 904–05
3. Discuss the anatomy and physiology of the testes. Describe the structure of sperm. 906–07
4. Name the endocrine glands involved in maintaining the sexual characteristics of males and the hormones produced by each. 907
5. Trace the path of an oocyte in a human female. Where do fertilization and implantation occur? Name two functions of the vagina. 908–09
6. Describe the external genital organs in females. 908
7. Discuss the anatomy and physiology of the ovaries. Describe the ovarian cycle and ovulation. 909–10
8. Describe the uterine cycle and relate it to the ovarian cycle. In what way is menstruation prevented if pregnancy occurs? 909–11
9. Name three functions of the female sex hormones. 912
10. Describe the anatomy and physiology of the breast. 912
11. Which means of birth control require surgery, utilize hormones, use barrier methods, are dependent on none of these? 912–13

Testing Yourself

1. Label this diagram of the male reproductive system and trace the path of sperm:

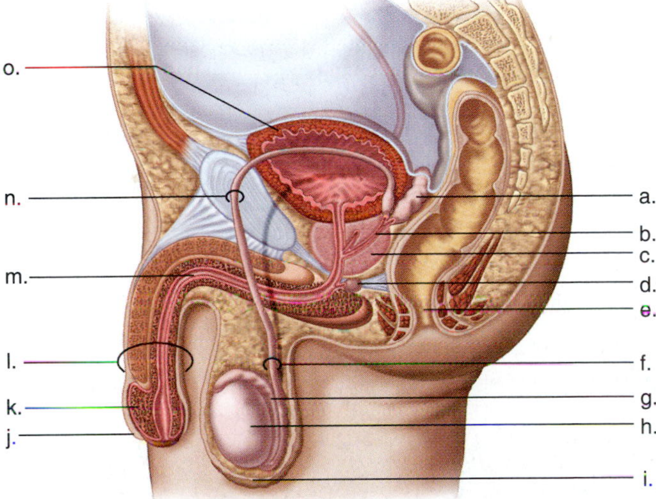

Choose the best answer for each question.

2. Which of these is a requirement for sexual reproduction?
   a. male and female parents
   b. production of gametes
   c. optimal environmental conditions
   d. aquatic habitat
   e. All of these are correct.
3. Internal fertilization
   a. can prevent the drying out of gametes and zygotes.
   b. must take place on land.
   c. is practiced by humans.
   d. requires that males have a penis.
   e. Both a and c are correct.
4. Which of these is mismatched?
   a. interstitial cells—testosterone
   b. seminiferous tubules—sperm production
   c. vasa deferentia—seminal fluid production
   d. urethra—conducts sperm
   e. Both c and d are mismatched.
5. Follicle-stimulating hormone (FSH)
   a. is secreted by females but not males.
   b. stimulates the seminiferous tubules to produce sperm.
   c. secretion is controlled by gonadotropic-releasing hormone (GnRH).
   d. is the same as luteinizing hormone.
   e. Both b and c are correct.
6. Which of these combinations is most likely to be present before ovulation occurs?
   a. FSH, corpus luteum, estrogen, secretory uterine lining
   b. luteinizing hormone (LH), follicle, progesterone, thick endometrium
   c. FSH, follicle, estrogen, endometrium becoming thick
   d. LH, corpus luteum, progesterone, secretory endometrium
   e. Both c and d are correct.
7. In tracing the path of sperm, you would mention vasa deferentia before
   a. testes.
   b. epididymides.
   c. urethra.
   d. uterus.
   e. All of these are correct.
8. An oocyte is fertilized in the
   a. vagina.
   b. uterus.
   c. oviduct.
   d. ovary.
   e. All of these are correct.
9. During pregnancy
   a. the ovarian and uterine cycles occur more quickly than before.
   b. GnRH is produced at a higher level than before.
   c. the ovarian and uterine cycles do not occur.
   d. the female secondary sexual characteristics are not maintained.
   e. Both b and c are correct.
10. Which of the following means of birth control is most effective in preventing sexually transmitted diseases?
    a. condom
    b. pill
    c. diaphragm
    d. spermicidal jelly
    e. vasectomy

Thinking Scientifically

1. Female athletes who train intensively often stop menstruating. The important factor appears to be the reduction of body fat below a certain level. Give a possible evolutionary explanation for a relationship between body fat in females and reproductive cycles.
2. The average sperm count in males is now lower than it was several decades ago. The reasons for the lower sperm count usually seen today are not known. What data might be helpful in order to formulate a testable hypothesis?

Understanding the Terms

Match the terms to these definitions:

a. ______________ Release of an oocyte from the ovary.
b. ______________ Development of an egg into a whole organism without fertilization.
c. ______________ Female sex hormone that causes the endometrium of the uterus to become secretory during the uterine cycle; along with estrogen, it maintains secondary sexual characteristics in females.
d. ______________ Thick, whitish fluid consisting of sperm and secretions from several glands of the male reproductive tract.
e. ______________ Organ that produces gametes; the ovary, which produces eggs, and the testis, which produces sperm.

Web Connections

Exploring the Internet

http://www.mhhe.com/biosci/genbio/mader
(click on *Biology 7/e*)

The *Biology 7/e* Online Learning Center provides many resources for studying the material in this chapter including links to the following sites:

The website of the National Institute of Health contains an information index that will help you to find the institution responsible for the subject of your choice. The Health Information Page will also supply you with a link to that institute's home page.

http://www.nih.gov/health/

Human Anatomy Online; Reproductive Systems. These sites provide images of the male and female reproductive tracts.

http://www.innerbody.com/image/repmov.html

http://www.innerbody.com/image/repfov.html

Breast Cancer Network. The American Cancer Society's home page with information about breast cancer survivorship, news, advocacy, and resources.

http://www2.cancer.org/bcn/index.html

Genetic Testing for Breast Cancer Risk. Information from the National Cancer Institute about the genetic causes for breast and ovarian cancers, testing, and factors which may affect a woman's choice for testing.

http://cancernet.nci.nih.gov/clinpdq/risk/Genetic_Testing_for_Breast_Cancer_Risk:__It's_Your_Choice.html

Reproduction: A Last Hope for Some Endangered Species. This is a page from the National Zoological Park. It explains the importance of reproductive technologies for some rare animals and the importance of a large gene pool for a population.

http://www.si.edu/natzoo/zooview/newsserv/artrepro.htm

Development

chapter

chapter concepts

51.1 Early Developmental Stages

- Development begins when a sperm fertilizes an egg. 920
- The first stages of embryonic development in animals lead to the establishment of the embryonic germ layers. 921
- The presence of yolk affects the manner in which animal embryos go through the early developmental stages. 922
- In vertebrates, the nervous system develops above the notochord after formation of a neural tube. 923

51.2 Developmental Processes

- Cytoplasmic segregation and induction help bring out cellular differentiation and morphogenesis. 924
- Developmental genetics has benefited from research into the development of *Caenorhabditis elegans*, a roundworm, and *Drosophila melanogaster*, a fruit fly. 926, 928
- Homeotic genes are involved in shaping the outward appearance of animals. 928

51.3 Human Embryonic and Fetal Development

- Humans, like chicks, are dependent upon extraembryonic membranes that perform various services that contribute to development. 930
- During the embryonic period of human development, all systems appear. 931
- Humans are placental mammals; the placenta is a unique organ where exchange between fetal blood and mother's blood takes place. 934
- During fetal development, the fetus grows large enough to live on its own. Birth is a multistage process that includes delivery of the child and the extraembryonic membranes. 935

As development occurs, complexity increases.

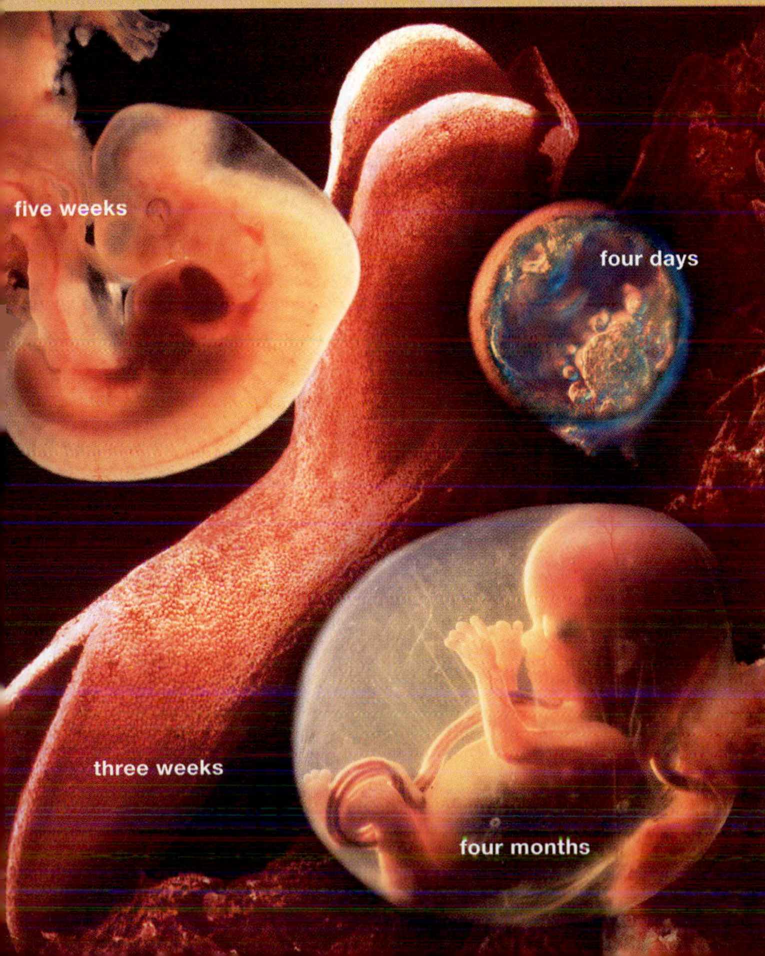

As development proceeds, specialization of cells becomes apparent. How does this occur, since all cells contain the same genes? Scientists do not have all the answers to this question, but they do know that specific molecules signal specific genes to become active at different times. In the end, nerve cells have axons and dendrites, not myofibrils like muscle cells, and bone cells deposit calcium salts, not skin pigment. And a baby has eyes and ears, fingers and toes in the right number and place.

The same processes observed during embryological development are also seen as a newborn matures, as lost parts regenerate, as a wound heals, and even as organisms age. Therefore, it has become increasingly clear that the study of development encompasses not only embryology but these other events as well.

The goal of developmental genetics is to discover which signaling molecules turn on which genes to make specialization come about. It's not possible to perform experiments on human embryos, but luckily scientists have discovered that concepts gained from studying frog, roundworm, and fly embryos apply to humans too.

51.1 Early Developmental Stages

Fertilization [L. *fertilis,* fruitful], which results in a zygote, requires that the sperm and egg interact. Figure 51.1 shows the manner in which an egg is fertilized by a sperm in sea stars. The sperm has three distinct parts: a head, a middle piece, and a tail. The tail is a flagellum, which allows the sperm to swim toward the egg, and the middle piece contains ATP-producing mitochondria. The head contains a haploid nucleus capped by a membrane-bounded acrosome containing enzymes that allow the sperm to penetrate the egg.

Several mechanisms have evolved to assure that fertilization takes place and in a species-specific manner (Fig. 51.1). A male releases so many sperm that the egg is literally covered by them. The sea star egg has a plasma membrane, a glycoprotein layer called the vitelline envelope, and a jelly coat. The acrosome enzymes digest away the jelly coat as the acrosome extrudes a filament that attaches to a receptor located in the vitelline envelope. This is a lock-and-key reaction that is species-specific. Then, the egg plasma membrane and sperm nuclear membrane fuse, allowing the sperm nucleus to enter. A zygote is present when the sperm nucleus fuses with the egg nucleus. Following fusion, the egg plasma membrane and the vitelline envelope undergo changes that prevent the entrance of any other sperm. The vitelline envelope is now called the fertilization envelope.

Embryonic Development

Development is all the changes that occur during the life cycle of an organism. During the first stages of development an organism is called an **embryo.** Most animals go through the same embryonic stages of zygote, morula, blastula, early gastrula, and late gastrula. As an example, we will consider the lancelet, an invertebrate chordate (see Fig. 35.2) whose egg has little yolk. **Yolk** is dense nutrient material.

Following fertilization, the zygote undergoes **cleavage,** which is cell division without growth (Fig. 51.2). DNA replication and mitotic cell division occur repeatedly, and the cells get smaller with each division. Because the lancelet has

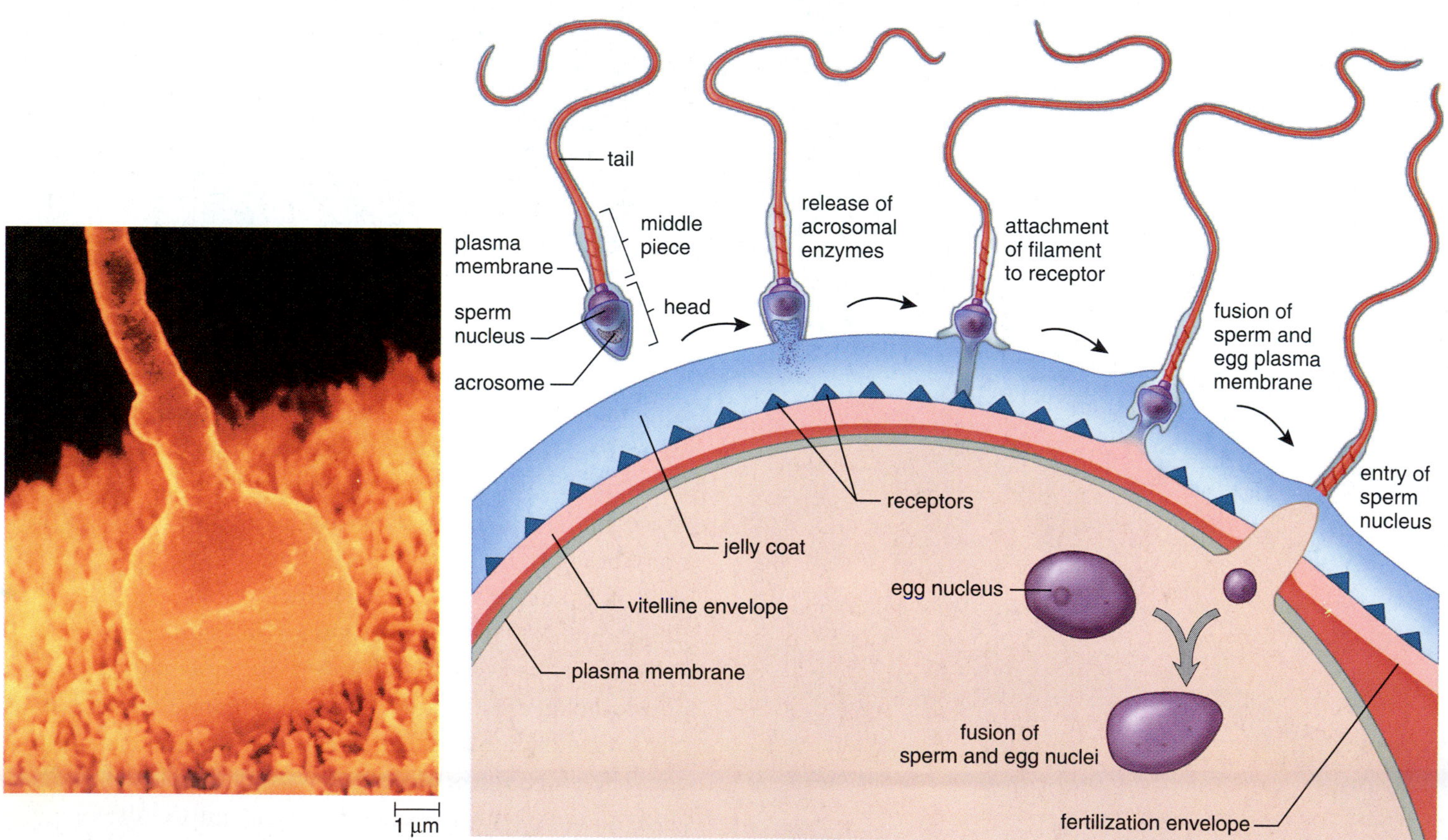

Figure 51.1 Fertilization of a sea star egg.
A head of a sperm has a membrane-bounded acrosome filled with enzymes. When released, these enzymes digest away the jelly coat around the egg, and the acrosome extrudes a filament that attaches to a receptor on the vitelline envelope. Now the sperm nucleus enters and fuses with the egg nucleus, and the resulting zygote begins to divide. The vitelline envelope becomes the fertilization envelope, which prohibits any more sperm from entering the egg.

little yolk, the cell divisions are equal, and the cells are of uniform size in the resulting **morula** [L. dim. of *morus*, mulberry]. Then a cavity called the **blastocoel** [Gk. *blastos*, bud, and *koiloma*, cavity] develops and a hollow ball of cells known as the **blastula** [Gk. dim. of *blastos*, bud, and L. *ula*, little] forms.

The **gastrula** [Gk. dim. of *gastros*, stomach] stage is evident in a lancelet when certain cells begin to push, or invaginate, into the blastocoel, creating a double layer of cells. The outer layer of cells is called the **ectoderm,** and the inner layer is called the **endoderm.** The space created by invagination will become the gut, but at this point it is termed either the archenteron or the primitive gut. The pore, or hole, created by invagination is the blastopore, and in a lancelet the blastopore eventually becomes the anus.

Gastrulation is not complete until three layers of cells are present. The third, or middle, layer of cells is called the **mesoderm.** In a lancelet, this layer begins as outpocketings from the primitive gut. These outpocketings grow in size until they meet and fuse. In effect then, two layers of mesoderm are formed, and the space between them is the coelom. The coelom is a body cavity that contains internal organs.

Ectoderm, mesoderm, and endoderm are called the embryonic **germ layers.** No matter how gastrulation takes place, the end result is the same: three germ layers are formed. It is possible to relate the development of future organs to these germ layers:

Embryonic Germ Layer	Vertebrate Adult Structures
Ectoderm (outer layer)	Epidermis of skin; epithelial lining of oral cavity and rectum; nervous system
Mesoderm (middle layer)	Skeleton; muscular system; dermis of skin; cardiovascular system; excretory system; reproductive system—including most epithelial linings; outer layers of respiratory and digestive systems
Endoderm (inner layer)	Epithelial lining of digestive tract and respiratory tract; associated glands of these systems; epithelial lining of urinary bladder

Karl E. Von Baer, the nineteenth-century embryologist, first related development to the formation of germ layers. This is called the germ layer theory.

The three embryonic germ layers arise during gastrulation, when cells invaginate into the blastocoel. The development of organs can be related to the three germ layers: ectoderm, mesoderm, and endoderm.

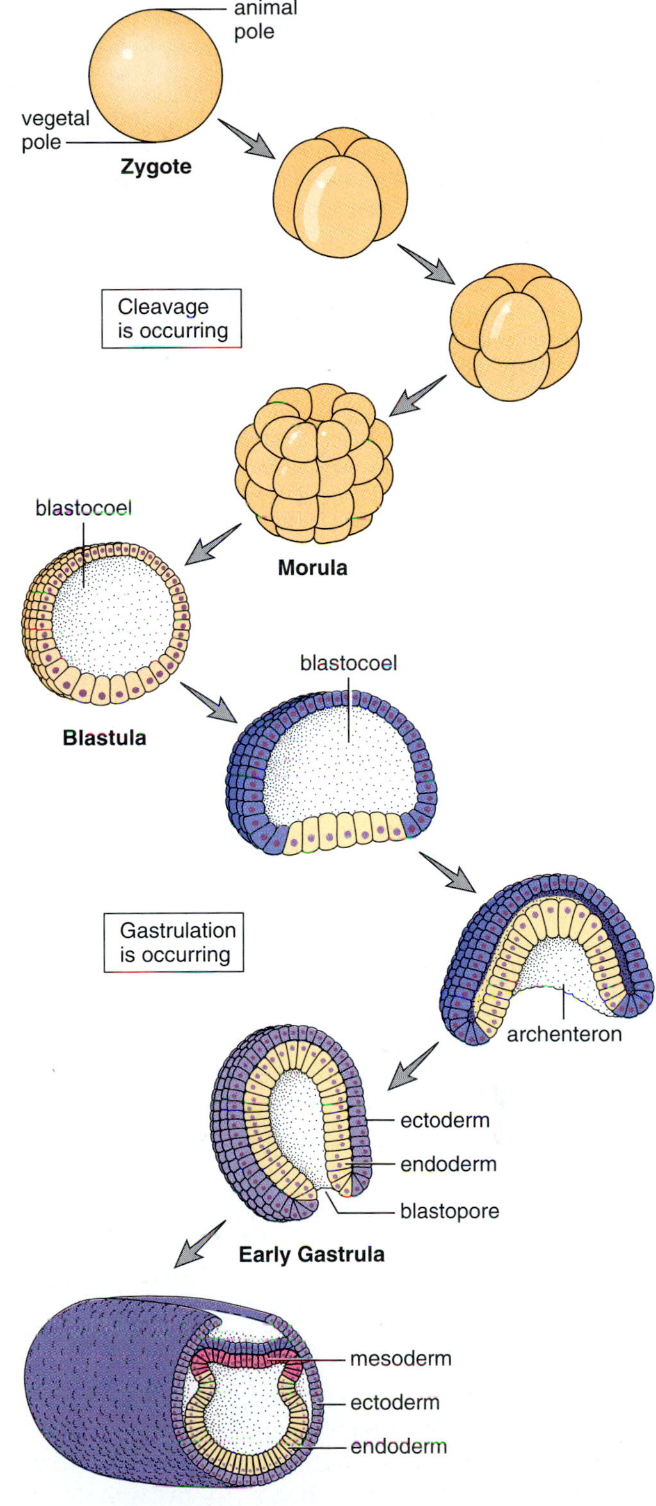

Figure 51.2 Lancelet early development.
A lancelet has little yolk, as an embryo, and it can be used to exemplify the early stages of development in such animals. Cleavage produces a number of cells that form a cavity. Invagination during gastrulation produces the germ layers ectoderm and endoderm. Mesoderm arises from pouches that pinch off from the endoderm.

The Effect of Yolk

Table 51.1 indicates the amount of yolk in four types of embryos and relates the amount of yolk to the environment in which the animal develops. The lancelet and frog develop in water, and they have less yolk than the chick because their development proceeds quickly to a swimming larval stage that can feed itself. The chick is representative of vertebrate animals that develop on land and lay a hard-shelled egg that contains plentiful yolk. Development continues in the shell until there is an offspring capable of land existence.

Early stages of human development resemble those of the chick embryo, yet this resemblance cannot be related to the amount of yolk because the human egg contains little yolk. But the evolutionary history of these two animals can provide an answer for this similarity. Both birds and mammals are related to reptiles. This explains why all three groups develop similarly, despite a difference in the amount of yolk in the eggs.

Figure 51.3 compares the appearance of early developmental stages in the lancelet, the frog, and the chick. In the frog embryo, cells at the animal pole have little yolk while those at the vegetal pole contain more yolk. The presence of yolk causes cells to cleave more slowly, and you can see that the cells of the animal pole are smaller than those of the vegetal pole. In the chick, cleavage is incomplete—only those cells lying on top of the yolk cleave. This means that although cleavage in the lancelet and the frog results in a morula, no such ball of cells is seen in the chick. Instead, during the morula stage the cells spread out on a portion of the yolk.

In the frog, the blastocoel is formed at the animal pole only. The heavily laden yolk cells of the vegetal pole do not participate in this step. In a chick, the blastocoel is created when the cells lift up from the yolk and leave a space between the cells and the yolk.

In the frog, the cells containing yolk do not participate in gastrulation and therefore they do not invaginate. Instead, a slitlike blastopore is formed when the animal pole cells begin to invaginate from above. Following this, other animal pole cells move down over the yolk, and the blastopore becomes rounded when these cells also invaginate from below. At this stage, there are some yolk cells temporarily left in the region of the blastopore; these are called the yolk plug. In the chick, there is so much yolk that endoderm formation does not occur by invagination. Instead, an upper layer of cells differentiates into ectoderm, and a lower layer differentiates into endoderm.

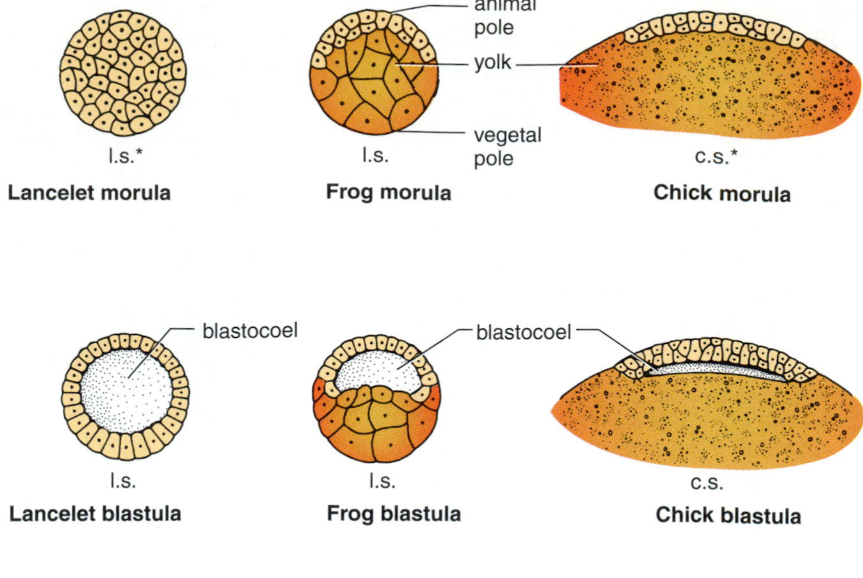

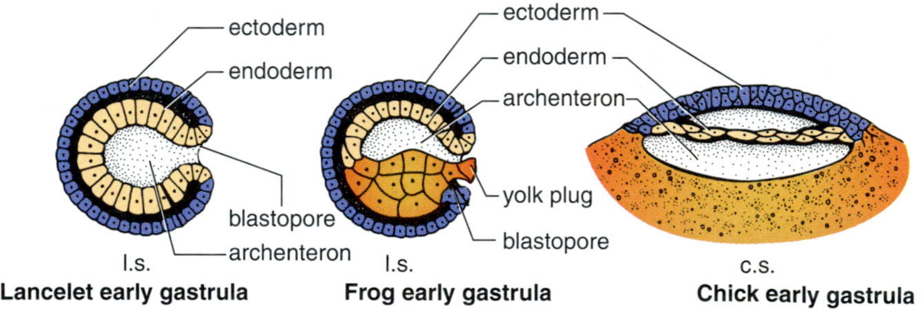

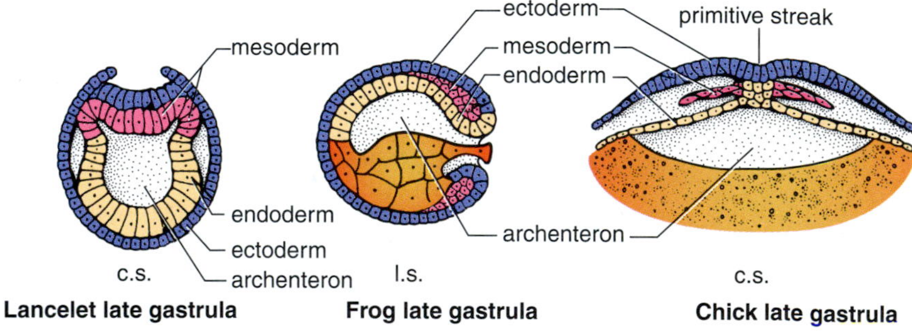

*l.s. = longitudinal section; c.s. = cross section

Figure 51.3 **Comparative animal development.**

Table 51.1

Amount of Yolk in Eggs Versus Location of Development

Animal	Yolk	Location of Development
Lancelet	Little	External in water
Frog	Some	External in water
Chick	Much	Within hard shell
Human	Little	Inside mother

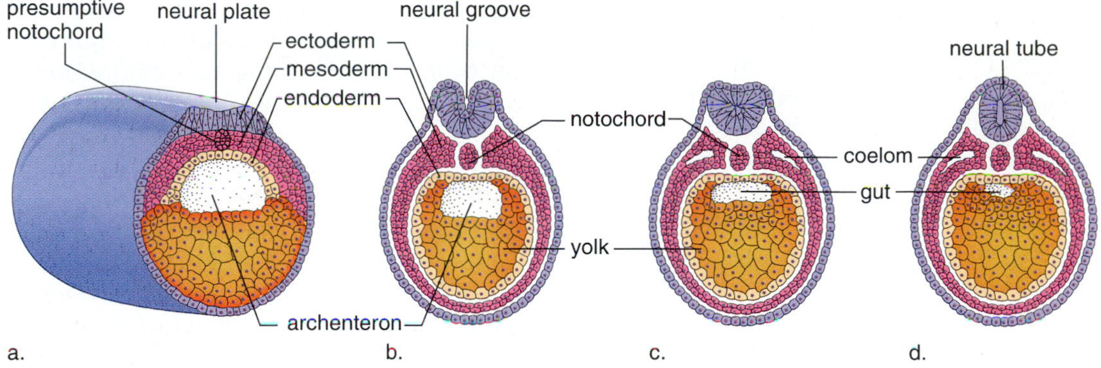

Figure 51.4 Development of neural tube and coelom in a frog embryo.
a. Ectoderm cells that lie above the future notochord (called presumptive notochord) thicken to form a neural plate. **b.** The neural groove and folds are noticeable as the neural tube begins to form. **c.** A splitting of the mesoderm produces a coelom, which is completely lined by mesoderm. **d.** A neural tube and a coelom have now developed.

In the frog, cells from the dorsal lip of the blastopore migrate between the ectoderm and endoderm, forming the mesoderm. Later, a splitting of the mesoderm creates the coelom. In the chick, the mesoderm layer arises by an invagination of cells along the edges of a longitudinal furrow in the midline of the embryo. Because of its appearance, this furrow is called the primitive streak. Later, the newly formed mesoderm will split to produce a coelomic cavity.

The amount of yolk typically affects the manner in which animals complete the first three stages of development.

Neurulation and the Nervous System

The newly formed mesoderm cells that lie along the main longitudinal axis of the animal coalesce to form a dorsal supporting rod called the **notochord** [Gk. *noto,* back, and *chord,* string]. The notochord persists in lancelets but in frogs, chicks, and humans, it is later replaced by the vertebral column. Therefore they are called vertebrates.

The nervous system develops from midline ectoderm located just above the notochord. At first, a thickening of cells called the **neural plate** is seen along the dorsal surface of the embryo. Then, neural folds develop on either side of a neural groove, which becomes the **neural tube** when these folds fuse. Figure 51.4 shows cross sections of frog development to illustrate the formation of the neural tube. At this point, the embryo is called a neurula. Later, the anterior end of the neural tube develops into the brain and the rest becomes the spinal cord.

Midline mesoderm cells that did not contribute to the formation of the notochord now become two longitudinal masses of tissue. These two masses become blocked off into the somites, which give rise to segmental muscles in all chordates. In vertebrates, the somites also produce the vertebral bones.

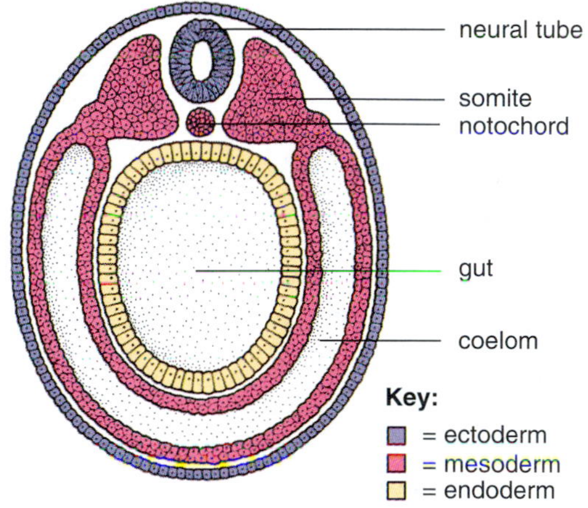

Figure 51.5 Chordate embryo, cross section.
At the neurula stage, each of the germ layers, indicated by color (see key), can be associated with the later development of particular parts. The somites give rise to the muscles of each segment and to the vertebrae, which replace the notochord in vertebrates.

The chordate embryo in Figure 51.5 shows the location of various parts. This figure and the chart on page 921 will help you relate the formation of chordate structures and organs to the three embryonic layers of cells: the ectoderm, the mesoderm, and the endoderm.

During neurulation, the neural tube develops just above the notochord. At the neurula stage of development, a cross section of all chordate embryos is similar in appearance.

51.2 Developmental Processes

Development requires (1) growth, (2) cellular differentiation, and (3) morphogenesis. **Cellular differentiation** occurs when cells become specialized in structure and function; a muscle cell looks different and acts differently than a nerve cell. **Morphogenesis** produces the shape and form of the body. One of the earliest indications of morphogenesis is cell movement as discussed in the reading on page 927. Later, morphogenesis includes **pattern formation** when tissues are organized into specific structures.

At one time investigators mistakenly believed that irreversible genetic changes must account for differentiation and morphogenesis. Perhaps the genes are parcelled out as development occurs and that's why cells of the body have a different structure and function. Our ability today to clone mammals such as sheep, goats, and mice from specialized adult cells shows that every cell in an organism's body contains a full complement of genes. It's said that cells in the adult body are **totipotent;** each one contains all the instructions needed by any other specialized cell in the body.

The answer to this puzzle becomes clear when we consider that only muscle cells produce the proteins myosin and actin; only red blood cells produce hemoglobin; and only skin cells produce keratin. In other words, we now know that specialization is not due to a parceling out of genes rather it is due to differential gene expression. Certain genes and not others are turned on in differentiated cells. In recent years, investigators have turned their attention to discovering the mechanisms that lead to differential gene expression. Two mechanisms—cytoplasmic segregation and induction—seem to be especially important.

Cytoplasmic Segregation

Differentiation must begin long before we can recognize specialized types of cells. Ectodermal, endoderm, and mesodermal cells in the gastrula look quite similar, but yet they must be different because they develop into different organs. The egg is now known to contain substances called maternal determinants because they influence the course of development. Cytoplasmic segregation is the parceling out of maternal determinants as mitosis occurs:

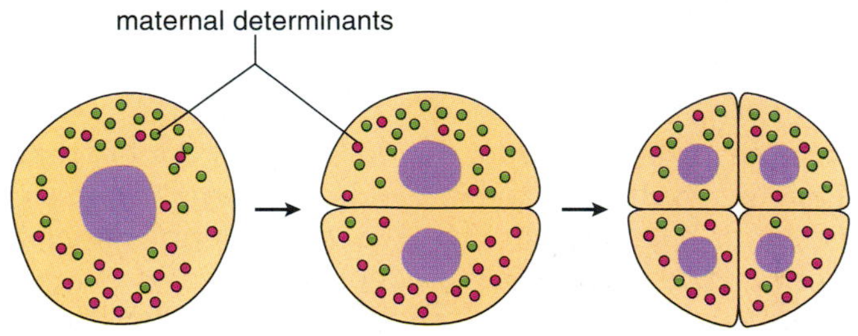

Ooplasmic Segregation

Cytoplasmic segregation helps determine how the various cells of the morula will develop.

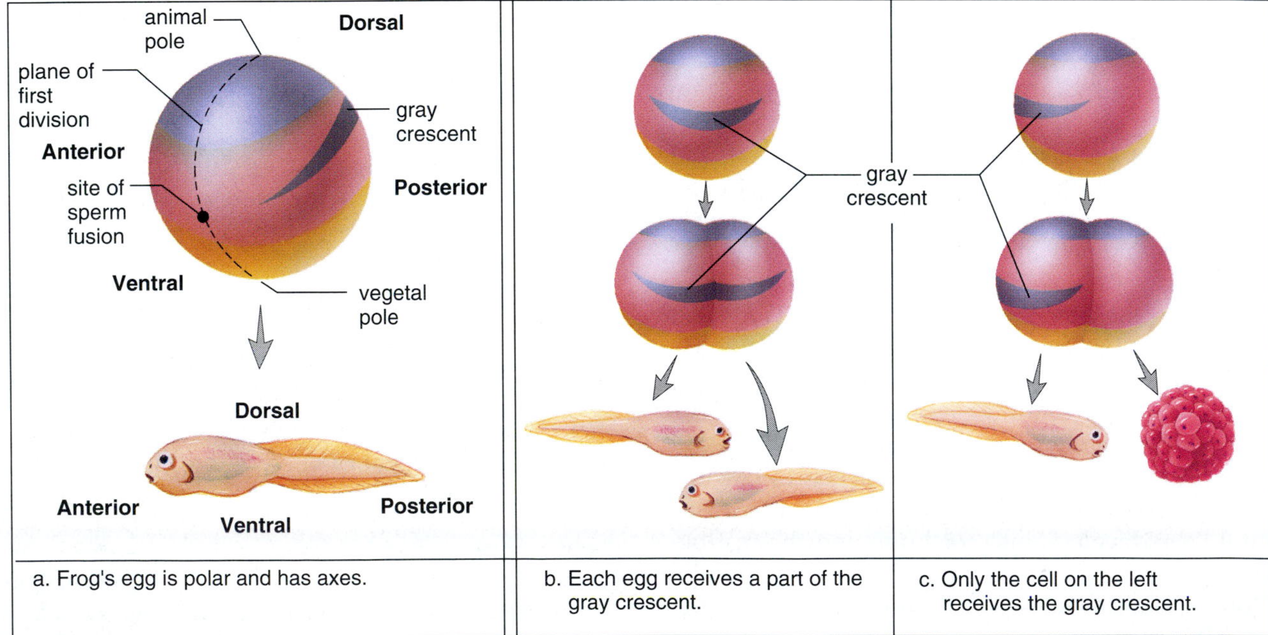

Figure 51.6 Cytoplasmic influence on development.
a. A frog's egg has anterior/posterior and dorsal/ventral axes that correlate with the position of the gray crescent. **b.** The first cleavage normally divides the gray crescent in half, and each daughter cell is capable of developing into a complete tadpole. **c.** But if only one daughter cell receives the gray crescent, then only that cell can become a complete embryo. This shows that maternal determinants are present in the cytoplasm of a frog's egg.

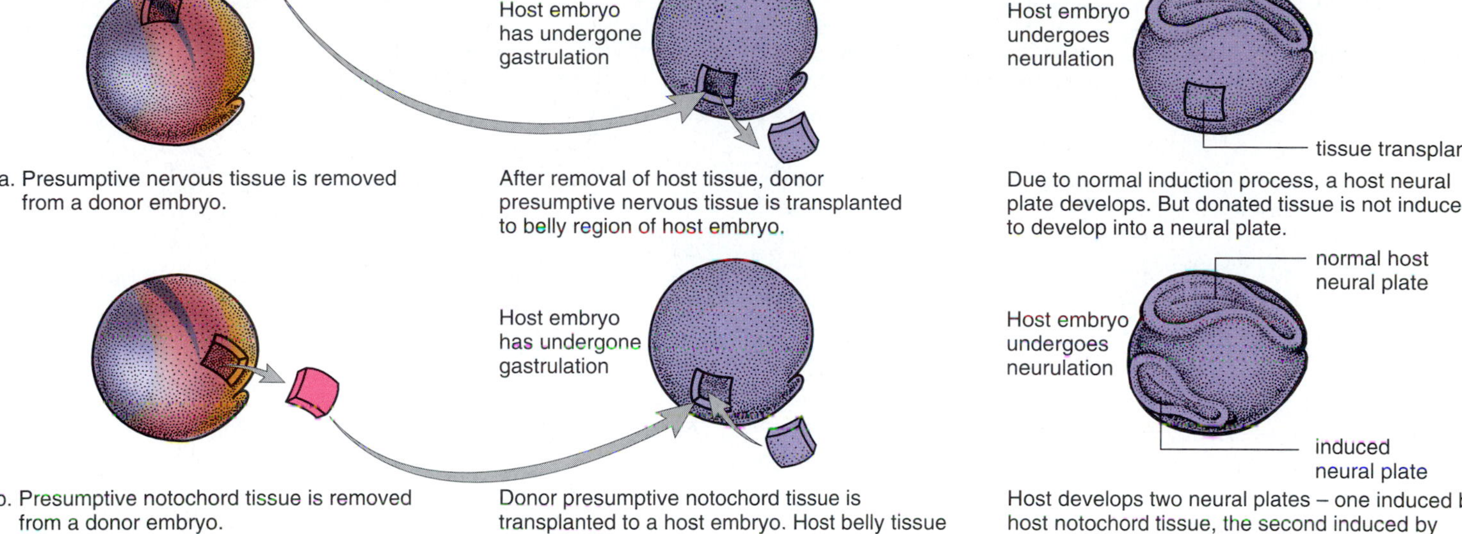

a. Presumptive nervous tissue is removed from a donor embryo.

After removal of host tissue, donor presumptive nervous tissue is transplanted to belly region of host embryo.

Due to normal induction process, a host neural plate develops. But donated tissue is not induced to develop into a neural plate.

b. Presumptive notochord tissue is removed from a donor embryo.

Donor presumptive notochord tissue is transplanted to a host embryo. Host belly tissue (which was removed) is returned to the host.

Host develops two neural plates – one induced by host notochord tissue, the second induced by transplanted notochord tissue.

Figure 51.7 Control of nervous system development.
a. In this experiment, the presumptive nervous system (blue) does not develop into the neural plate if moved from its normal location. **b.** In this experiment, the presumptive notochord (red) can cause even belly ectoderm to develop into the neural plate (blue). This shows that the notochord induces ectoderm to become a neural plate, most likely by sending out chemical signals.

An early experiment showed that the cytoplasm of a frog's egg is not uniform. It is polar and has both an anterior/posterior axis and a dorsal/ventral axis, which can be correlated with the **gray crescent,** a gray area that appears after the sperm fertilizes the egg (Fig. 51.6*a*). Hans Spemann, who received a Nobel Prize in 1935 for his extensive work in embryology, showed that if the gray crescent is divided equally by the first cleavage, each experimentally separated daughter cell develops into a complete embryo (Fig. 51.6*b*). If the egg divides so that only one daughter cell receives the gray crescent, however, only that cell becomes a complete embryo (Fig. 51.6*c*). This experiment allows us to speculate that the gray crescent must contain particular chemical signals that are needed for development to proceed normally.

Induction

As development proceeds, specialization of cells and formation of organs is influenced not only by maternal determinants but also by signals given off by neighboring cells. **Induction** [L. *in,* into, and *duco,* lead] is the ability of one embryonic tissue to influence the development of another tissue.

Spemann showed that a frog embryo's gray crescent becomes the dorsal lip of the blastopore, where gastrulation begins. Since this region is necessary for complete development, he called the dorsal lip of the blastopore the primary organizer. The cells closest to the Spemann's primary organizer become endoderm, those farther away become mesoderm, and those farthest away become ectoderm. This suggests that there may be a molecular concentration gradient that acts as a chemical signal to induce germ layer differentiation.

The gray crescent of a frog's egg marks the dorsal side of the embryo where the mesoderm becomes notochord and ectoderm becomes nervous system. In a classic experiment Spemann and his colleague Hilde Mangold showed that presumptive (potential) notochord tissue induces the formation of the nervous system (Fig. 51.7). If presumptive nervous system tissue, located just above the presumptive notochord, is cut out and transplanted to the belly region of the embryo, it does not form a neural tube. On the other hand, if presumptive notochord tissue is cut out and transplanted beneath what would be belly ectoderm, this ectoderm differentiates into neural tissue. Still other examples of induction are now known. In 1905, Warren Lewis studied the formation of the eye in frog embryos. He found that an optic vesicle, which is a lateral outgrowth of developing brain tissue, induces overlying ectoderm to thicken and become a lens. The developing lens in turn induces an optic vesicle to form an optic cup, where the retina develops.

Cytoplasmic segregation and induction are two mechanisms that help explain the developmental processes of differentiation and morphogenesis.

Figure 51.8 Development of *C. elegans*, a nematode.
a. A fate map of the worm showing that as cells arise by cell division they are destined to become particular structures.
b. Apoptosis. When a cell-death signal is received, a master protein is inactive and no longer inhibits a cell-death cascade that ends with proteases and nucleases that destroy the cell.

a. Fate map

b. Apoptosis

Model Organisms

Developmental genetics has benefited from research utilizing the roundworm, *Caenorhabditis elegans*, the fruit fly, *Drosophila melanogaster*, and the mouse, *Mus musculus*. These organisms are referred to as model organisms because the study of their development produced concepts that help us understand development in general.

Caenorhabditis elegans

This tiny worm is only one millimeter long and vast numbers can be raised in the laboratory either in petri dishes or a liquid medium. The worm is hermaphroditic and self-fertilization is the rule. Therefore, even though induced mutations may be recessive, the next generation will yield individuals which are homozygous recessive and will show the mutation. Many modern genetic studies have been done. The entire genome has been sequenced. Individual genes have been altered, cloned, and their products injected into cells or extracellular fluid.

What have we learned? Development of *C. elegans* takes only three days and the adult worm contains only 959 cells. It's been possible for investigators to watch the process from beginning to end, especially since the worm is transparent. **Fate maps** have been developed that show the destiny of each cell as it arises following successive cell divisions (Fig. 51.8*a*). Some investigators have studied in detail the development of the vulva, a pore through which eggs are laid. A cell called the anchor cell induces the vulva to form. The cell closest to the anchor cell receives the most inducer and becomes the inner vulva. This cell in turn produces another inducer which acts on its two neighboring cells and they become the outer vulva. The inducers are growthlike factors that alter the metabolism of the receiving cell and activate particular genes. Work with *C. elegans* has shown that induction requires the transcriptional regulation of genes in a particular sequence. This diagram shows how induction can occur sequentially:

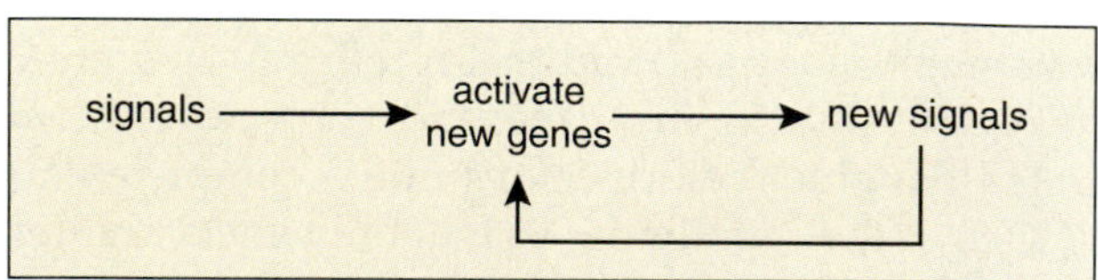

Apoptosis We have already discussed the importance of **apoptosis** (programmed cell death) in the normal day-to-day operation of the immune system and in preventing the occurrence of cancer. Apoptosis is also an important part of development in all organisms. In humans we know that apoptosis is necessary to the shaping of the hands and feet; if it does not occur, the child is born with webbing between fingers and toes.

The fate maps of *C. elegans* indicate that apoptosis occurs in 131 cells as development takes place. Researchers have been able to study the molecular basis for apoptosis in these cells. Certain proteins, called ced (for *c*ell *d*eath) proteins, are part of a cell-death pathway always present in the cell. The cell-death pathway is ordinarily inhibited by a master protein present in the plasma membrane. But when a receptor protein binds to a cell-death signal, the master protein becomes inactive. Then, the cell-death cascade produces proteases and nucleases that slice up proteins and DNA in the cell (Fig. 51.8*b*).

Research with a Cellular Slime Mold Fruiting Body

We do not often think how remarkable it is that most of our cells and tissues are in the correct places and in the correct proportion to other tissues. Further, we do not always realize that there are mechanisms controlling these events—until one of these mechanisms fails. Only then is it obvious that control mechanisms play such important roles in our development. Some of the most exciting questions in developmental biology involve these fundamental processes of positioning and proportioning of different cell types. In large complex organisms such as ourselves, the controls are likely to be more complex than in more simple species. Thus, research into these questions often centers on comparatively simple organisms such as the cellular slime mold, *Dictyostelium.*

Much of my research has involved study of the proportions and movements of cells in *Dictyostelium.* I first became interested in these questions as a graduate student in John Bonner's laboratory at Princeton University. Although I certainly found *Dictyostelium's* apparently simple development intriguing, I think what initially attracted me (a wide-eyed, eager graduate student) was my first visit to Dr. Bonner's lab. After introducing me around the lab, he spent a few minutes talking with his technician about a particular experiment. Dr. Bonner concluded that the experiment had not been fruitful, yet he was reluctant to drop it because it had been "fun with slime molds." I had known that Dr. Bonner was an excellent scientist, and now I saw that he found just doing experiments to be terrifically enjoyable. From that moment, I knew I wanted to join his lab; and since that time, I have also spent many pleasant years working with slime molds.

Dictyostelium exhibits a fascinating life cycle that includes a unicellular stage and several multicellular stages. In the unicellular stage, individual amoeboid cells feed on bacteria in the soil and grow and divide. When the food supply is diminished, the cells aggregate into a tiny multicellular mound. The mound forms a slug about a millimeter long that subsequently develops into a fruiting body. The fruiting body stands erect on the substrate and consists of a stalk holding aloft a small mass of spores. Each spore can germinate to produce a single amoeba to begin the life cycle again.

Recently my research has focused on one aspect of fruiting body formation. Stalk cells stack on top of one another as a fruiting body grows upward. Even so, the spores stay near the tip. How can they do that when they have lost their ability to move by amoeboid action? My hypothesis was that the cells nearest the spores, called anterior-like cells are involved in raising the spores to the tip of the fruiting body. To test the hypothesis I did experiments in which I removed the anterior-like cells. I found that when I removed the upper group of anterior-like cells, the spores were left along the middle of the fruiting body; they were not elevated to the tip. If undifferentiated stalk cells were transplanted in place of the upper anterior-like cells, the spores also remained along the middle of the fruiting body. This suggests that not just any motile cells can replace the function of the anterior-like cells. However, when I replaced the anterior-like cells with more anterior-like cells, the spores were again (although not 100% of the time) elevated to the tip. Altogether, these experiments indicate that the upper layer of anterior-like cells act as "elevator" cells, moving the spores to the top of the fruiting body (Fig. 51A).

When we learn about events and mechanisms in simple organisms, such as *Dictyostelium,* we are able to turn to more complex organisms, including ourselves, and ask if similar processes occur during our development. The type of research that undergraduates in my lab and I do holds a fascination for me and many others in biology. For me personally, I get enjoyment in performing the experiments and in the discovery of new facts about *Dictyostelium.* In addition, an investigator knows that while basic research may not be directed toward solving some immediate problem of humankind, it may eventually contribute to the betterment of our species, either directly or indirectly.

John Sternfeld

John Sternfeld
State University of New York College at Cortland

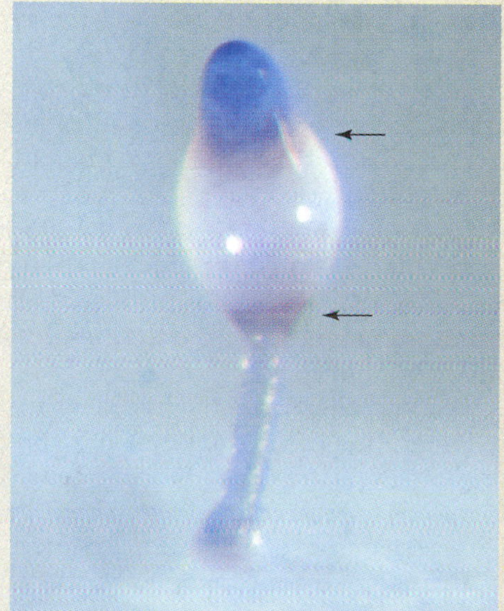

Figure 51A Cellular slime mold fruiting body.
The cells stained red (see arrows) are "elevator cells" which raise the spores (in the white area) to the top of the 0.6 mm high fruiting body.

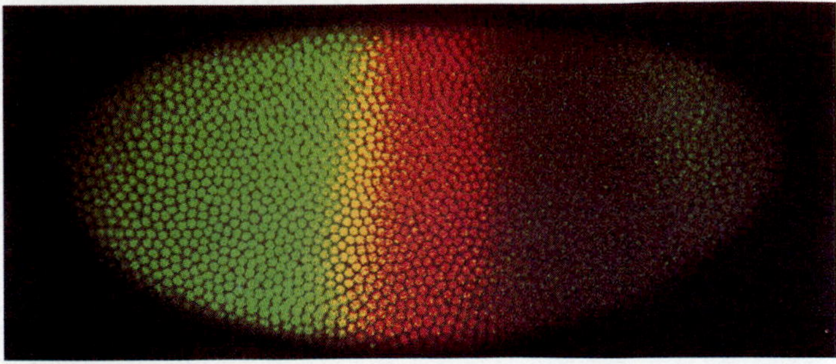

a. Protein products of *gap* genes

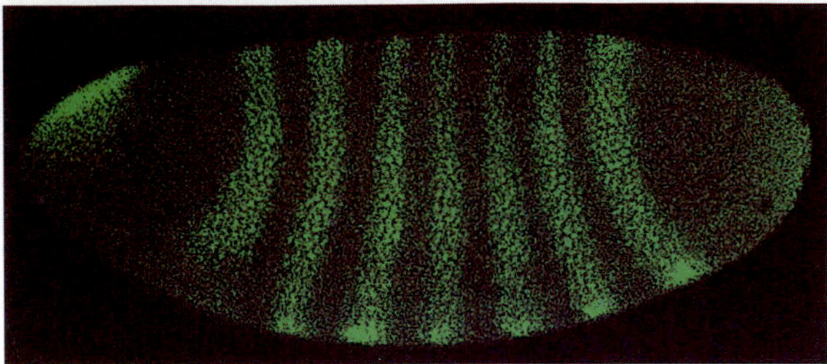

b. Protein product of *pair-rule* gene

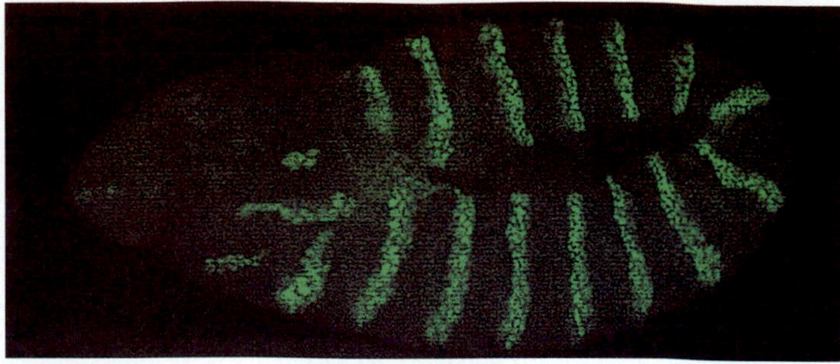

c. Protein product of a *segment-polarity* gene

Figure 51.9 Development in *Drosophila*, a fruit fly.
a. The different colors show that two different *gap* gene proteins are present from the anterior to the posterior end of an egg. **b.** The green stripes show that a *pair-rule* gene is being expressed as segmentation of the fly occurs. **c.** Now *segment-polarity* genes help bring about division of each segment into an anterior and posterior end.

Drosophilia melanogaster

Investigators studying morphogenesis in *Drosophila* (fruit fly) have discovered that there are some genes that determine the animal's anterior/posterior and dorsal/ventral axes, others that determine the fly's segmentation pattern, and still others, called homeotic genes, that determine the body parts on each segment.

The Anterior/Posterior Axes One of the first events toward successful development is the establishment of the body axes: location of the head versus the tail and location of the back versus the belly. In the *Drosophila* egg there is a greater concentration of a protein called bicoid at one end. This end becomes the anterior region where the head develops. (*Bicoid* means "two tailed" and a *bicoid* gene mutation can cause the embryo to have only two posterior ends.) Researchers have cloned the *bicoid* gene and used it as a probe to establish that mRNA for the bicoid protein is present in a concentration gradient from the anterior end to the posterior end of the embryo.

Many other protein gradients in the *Drosophila* egg help determine the axes of the body. Proteins that influence morphogenesis are called **morphogens.** The bicoid gradient is a morphogen gradient. The bicoid gradient switches on the expression of segmentation genes in *Drosophila*. What is the advantage of a morphogen gradient? A morphogen gradient can have a range of effects depending on its concentration in a particular portion of the animal.

The Segmentation Pattern The next event in the development of *Drosophila* is the establishment of its segments. The bicoid gradient initiates a cascade in which a series of segmentation gene sets are turned on, one after the other.

Christiane Nusslein-Vollard and Eric Wieschaus received a Nobel Prize for their work in discovering the segmentation genes of *Drosophila*. They exposed the flies to mutagenic chemicals and then performed innumerable crosses to map the mutated genes that caused segmental abnormalities. They went on to clone many of these genes. The first set of segmental genes to be activated are called *gap* genes (Fig. 51.9*a*). (If one of these genes mutates there are *gaps,* that is, large blocks of segments are missing.) Then the *pair-rule* genes become active and the embryo has precisely 14 segments (Fig. 51.9*b*). If one of these mutates, the animal has half the number of segments. Next the *segment-polarity* genes are expressed and each segment has an anterior and posterior half (Fig. 51.9*c*).

Work with *Drosophila* has suggested how morphogenesis comes about. Sequential sets of master genes code for morphogen gradients that activate the next set of master genes in turn. How do morphogen gradients turn on genes? They are transcription factors which regulate which genes are active in which parts of the embryo in what order.

Homeotic Genes **Homeotic genes** control pattern formation, the organization of differentiated cells into specific three-dimensional structures. During normal development of *Drosophila* the homeotic genes are activated after the segmentation genes. If a mutant is missing the set of homeotic genes called the *bithorax complex,* the fly still has the normal number of posterior segments.

a.

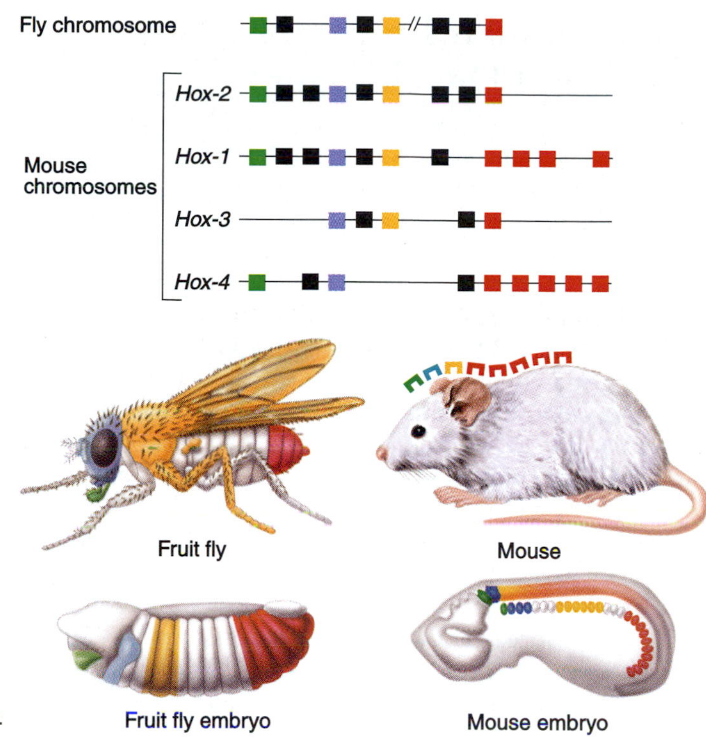

b.

Figure 51.10 Pattern formation in *Drosophila*. Homeotic genes control pattern formation, an aspect of morphogenesis. **a.** If homeotic genes are activated at inappropriate times, abnormalities such as a fly with four wings occurs. **b.** The green, blue, yellow, and red colors show that homologous homeotic genes occur on a fly chromosome and four mouse chromosomes in the same order. These genes are color coded to the region of the embryo and, therefore, the adult, where they regulate pattern formation. The black boxes are homeotic genes that are not identical between the two animals. In mammals, homeotic genes are called *Hox* genes.

As early as the 1940s, Edward B. Lewis had discovered the homeotic genes in *Drosophila*. He was able to determine that certain genes controlled whether a particular segment would bear antennae, legs, or wings. A homeotic mutation caused these appendages to be misplaced—a mutant fly could have extra legs where antennae should be or two pairs of wings (Fig. 51.10*a*).

Homeotic genes have now been found in many other organisms and, surprisingly, they all contain the same particular sequence of nucleotides, called a **homeobox.** (Because homeotic genes contain a homeobox in mammals they are called *Hox* genes.) The homeobox codes for a particular sequence of 60 amino acids called a homeodomain:

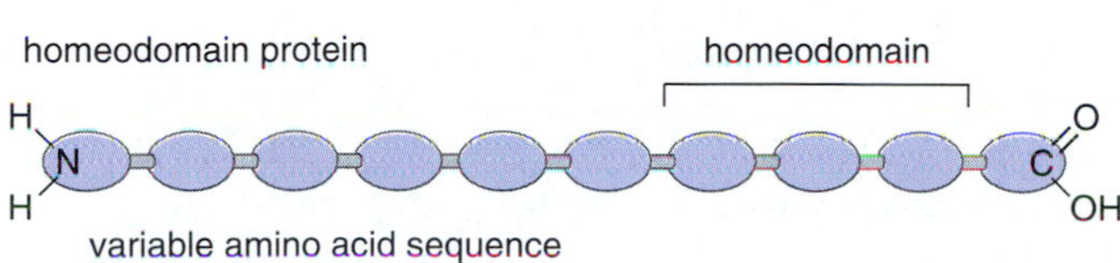

Homeotic genes, like many other developmental genes, code for transcription factors. The homeodomain protein is that part of a transcription factor that binds to DNA but the other more variable sequences of a transcription factor determine which particular genes are turned on. Researchers envision that a homeodomain protein produced by one homeotic gene binds to and turns on the next homeotic gene and so forth. This orderly process in the end determines the morphology of particular segments.

Mice and humans have the same four clusters of homeotic genes located on four different chromosomes. In *Drosophila*, homeotic genes are located on a single chromosome. In all three types of animals, homeotic genes are expressed from anterior to posterior in the same order. The first clusters determine the final development of anterior segments of the animal, while those later in the sequence determine the final development of posterior segments of the embryo.

Since the homeotic genes of so many different organisms contain the same homeodomain, we know that this nucleotide sequence arose early in the history of life that has been largely conserved as evolution occurred. In general it has been very surprising to learn how similar developmental genetics is in organisms from yeasts to plants to a wide variety of animals. Certainly, the genetic mechanisms of development appear to be quite similar in all animals.

There is a hierarchy of gene activity that causes development to be orderly. Morphogen gradients contain transcription factors that cause cells to produce yet other morphogen gradients until finally specific structures are formed.

51.3 Human Embryonic and Fetal Development

In humans, the length of the time from conception (fertilization followed by **implantation**) to birth (parturition) is approximately nine months. It is customary to calculate the time of birth by adding 280 days to the start of the last menstruation, because this date is usually known, whereas the day of fertilization is usually unknown. Because the time of birth is influenced by so many variables, only about 5% of babies actually arrive on the forecasted date.

Human development is often divided into embryonic development (months 1 and 2) and fetal development (months 3–9). The **embryonic period** consists of early formation of the major organs, and fetal development is the refinement of these structures.

Before we consider human development chronologically, we must understand the placement of **extraembryonic membranes** [L. *extra,* on the outside]. Extraembryonic membranes are best understood by considering their function in reptiles and birds. In reptiles, these membranes made development on land first possible. If an embryo develops in the water, the water supplies oxygen for the embryo and takes away waste products. The surrounding water prevents desiccation, or drying out, and provides a protective cushion. For an embryo that develops on land, all these functions are performed by the extraembryonic membranes.

In the chick, the extraembryonic membranes develop from extensions of the germ layers, which spread out over the yolk. Figure 51.11 shows the chick surrounded by the membranes. The **chorion** [Gk. *chorion,* membrane] lies next to the shell and carries on gas exchange. The **amnion** [Gk. *amnion,* membrane around fetus] contains the protective amniotic fluid, which bathes the developing embryo. The **allantois** [Gk. *allantos,* sausage] collects nitrogenous wastes, and the **yolk sac** surrounds the remaining yolk, which provides nourishment.

Humans (and other mammals) also have these extraembryonic membranes. The chorion develops into the fetal half of the placenta; the yolk sac, which lacks yolk, is the first site of blood cell formation; the allantoic blood vessels become the umbilical blood vessels; and the amnion contains fluid to cushion and protect the embryo, which develops into a fetus. Therefore, the function of the membranes in humans has been modified to suit internal development, but their very presence indicates our relationship to birds and to reptiles. It is interesting to note that all chordate animals develop in water, either in bodies of water or within amniotic fluid.

The presence of extraembryonic membranes in reptiles made development on land possible. Humans also have these membranes, but their function has been modified for internal development.

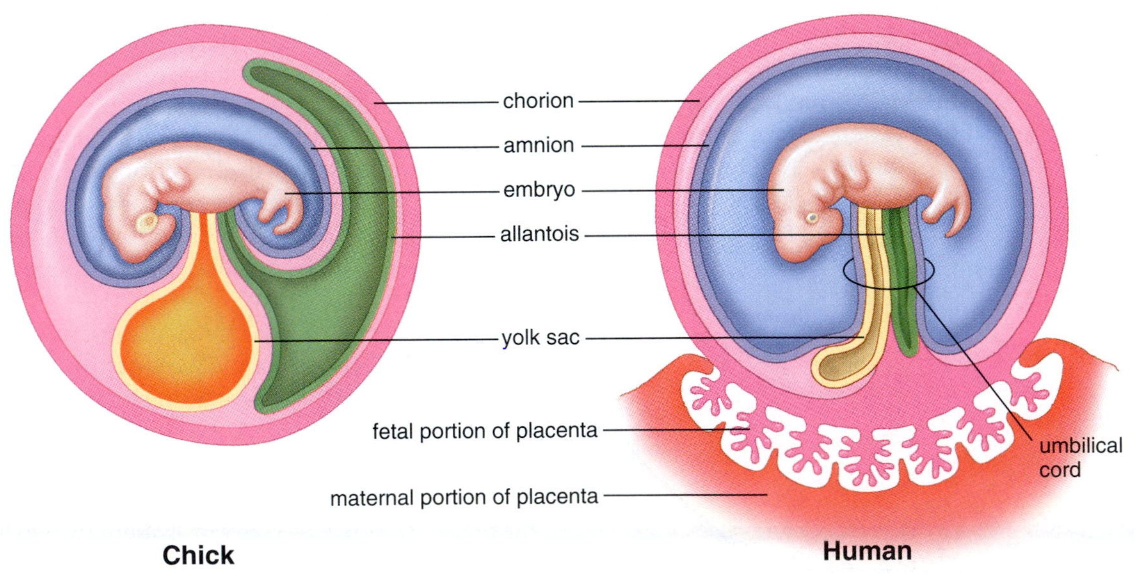

Figure 51.11 Extraembryonic membranes.
Extraembryonic membranes, which are not part of the embryo, are found during the development of chicks and humans, where each has a specific function.

Embryonic Development

Embryonic development includes the first two months of development.

The First Week

Fertilization occurs in the upper third of an oviduct (Fig. 51.12), and cleavage begins even as the embryo passes down this duct to the uterus. By the time the embryo reaches the uterus on the third day, it is a morula. The morula is not much larger than the zygote because, even though multiple cell divisions have occurred, there has been no growth of these newly formed cells. By about the fifth day, the morula is transformed into the blastocyst. The **blastocyst** has a fluid-filled cavity, a single layer of outer cells called the **trophoblast** [Gk. *trophe*, food, and *blastos*, bud], and an inner cell mass. Later, the trophoblast, reinforced by a layer of mesoderm, gives rise to the chorion, one of the extraembryonic membranes (Fig. 51.11). The inner cell mass eventually becomes the embryo, which develops into a fetus.

The Second Week

At the end of the first week, the embryo begins the process of implanting in the wall of the uterus. The trophoblast secretes enzymes to digest away some of the tissue and blood vessels of the endometrium of the uterus (Fig. 51.12). The embryo is now about the size of the period at the end

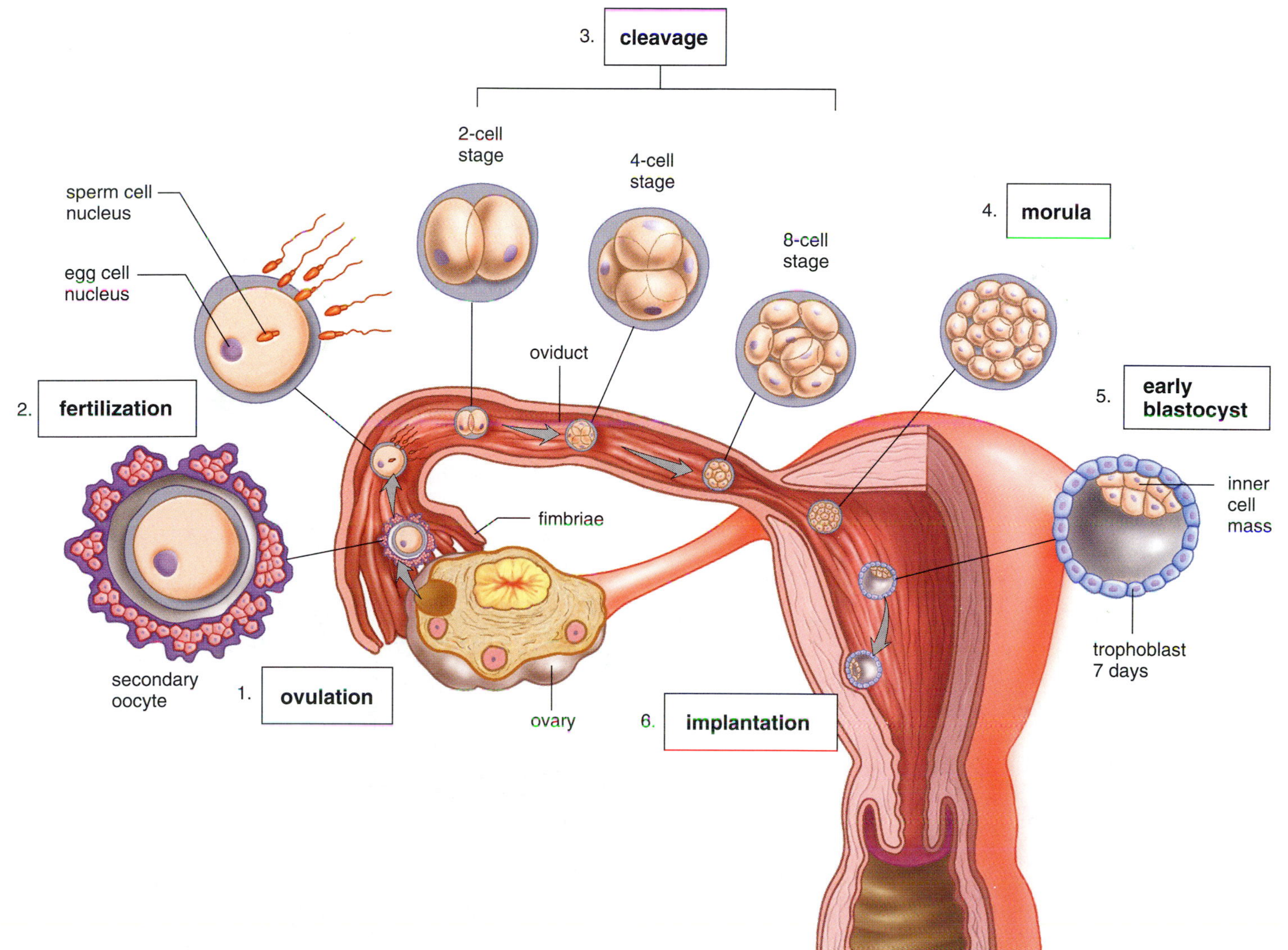

Figure 51.12 Human development before implantation.
Structures and events proceed counterclockwise. At ovulation (1), the secondary oocyte leaves the ovary. A single sperm nucleus enters the egg, and fertilization (2) occurs in the oviduct. As the zygote moves along the oviduct, it undergoes cleavage (3) to produce a morula (4). The blastocyst forms (5) and implants itself in the uterine lining (6).

of this sentence. The trophoblast begins to secrete **human chorionic gonadotropin (HCG),** the hormone that is the basis for the pregnancy test and that serves to maintain the corpus luteum past the time it normally disintegrates. Because of this, the endometrium is maintained and menstruation does not occur.

As the week progresses, the inner cell mass detaches itself from the trophoblast, and two more extraembryonic membranes form (Fig. 51.13*a*). The yolk sac, which forms below the embryonic disk, has no nutritive function as in chicks, but it is the first site of blood cell formation. However, the amnion and its cavity are where the embryo (and then the fetus) develops. In humans, amniotic fluid acts as an insulator against cold and heat and also absorbs shock, such as that caused by the mother exercising.

Gastrulation occurs during the second week. The inner cell mass now has flattened into the **embryonic disk,** composed of two layers of cells: ectoderm above and endoderm below. Once the embryonic disk elongates to form the primitive streak, the third germ layer, mesoderm, forms by invagination of cells along the streak. The trophoblast is reinforced by mesoderm and becomes the chorion (Fig. 51.13*b*).

It is possible to relate the development of future organs to these germ layers (see page 921).

The Third Week

Two important organ systems make their appearance during the third week. The nervous system is the first organ system to be visually evident. At first, a thickening appears along the entire dorsal length of the embryo, and then

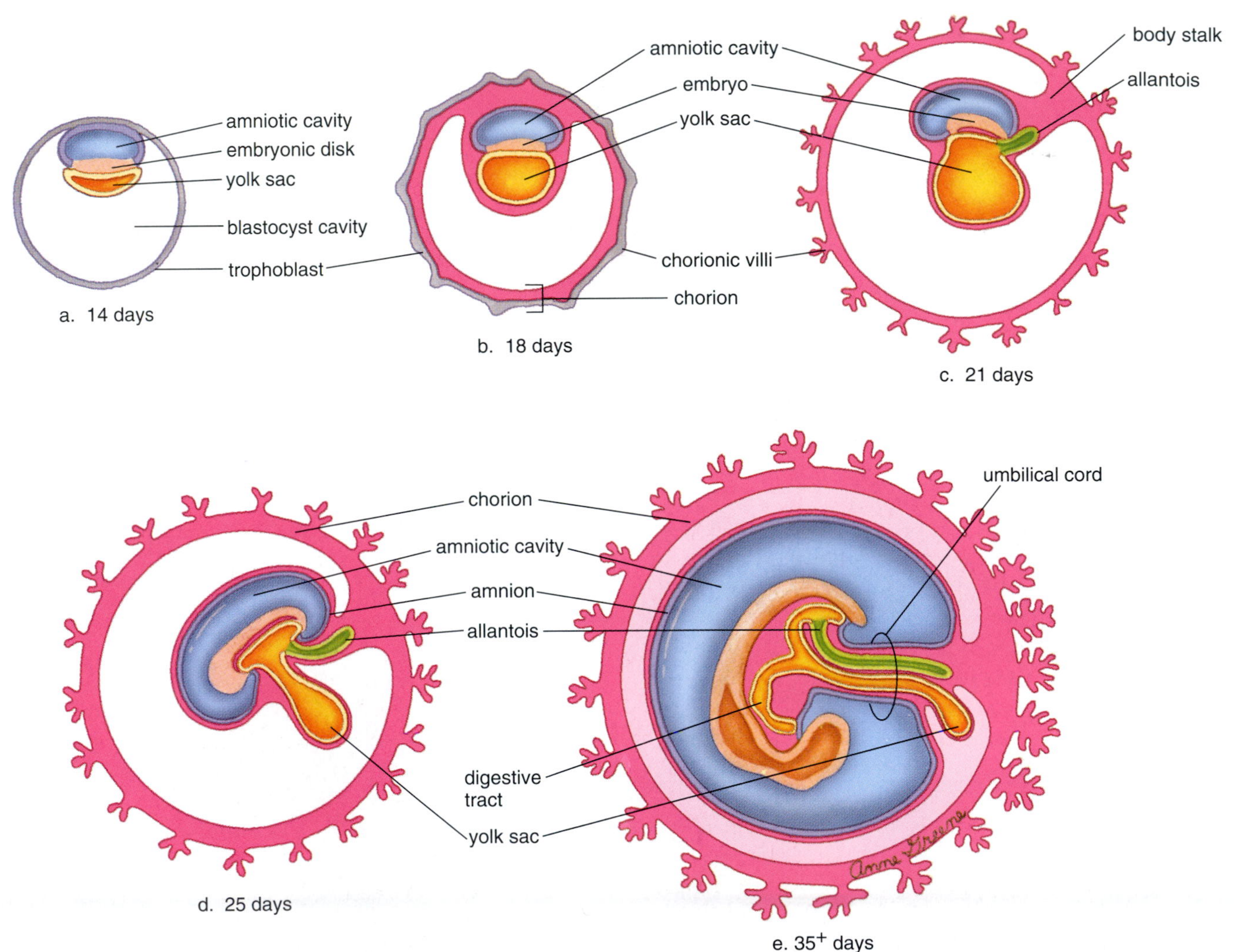

Figure 51.13 Human embryonic development.
a. At first there are no organs present in the embryo, only tissues. The amniotic cavity is above the embryo, and the yolk sac is below. **b.** The chorion is developing villi, so important to exchange between mother and child. **c.** The allantois and yolk sac are two more extraembryonic membranes. **d.** These extraembryonic membranes are positioned inside the body stalk as it becomes the umbilical cord. **e.** At 35+ days, the embryo has a head region and a tail region. The umbilical cord takes blood vessels between the embryo and the chorion (placenta).

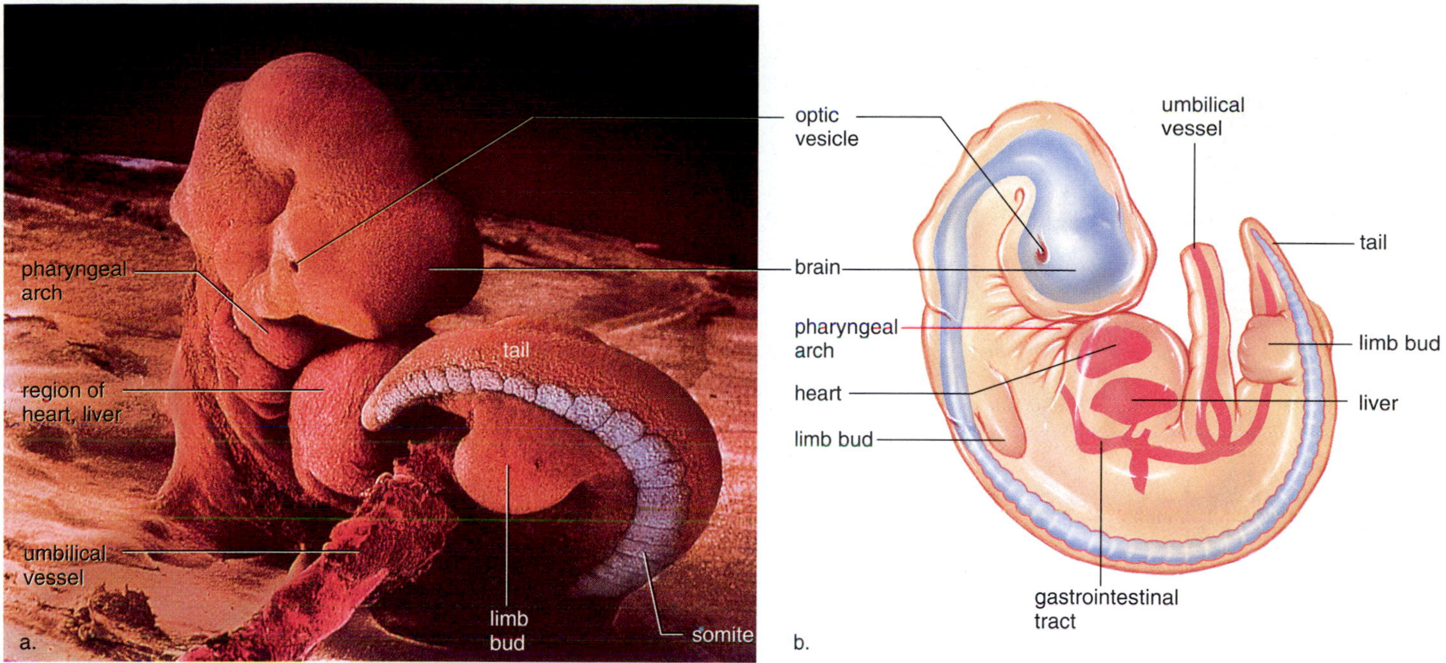

Figure 51.14 Human embryo at beginning of fifth week.
a. Scanning electron micrograph. **b.** The embryo is curled so that the head touches the heart, the two organs whose development is further along than the rest of the body. The organs of the gastrointestinal tract are forming, and the arms and the legs develop from the bulges that are called limb buds. The tail is an evolutionary remnant; its bones regress and become those of the coccyx (tailbone). The pharyngeal arches become functioning gills only in fishes and amphibian larvae; in humans, the first pair of pharyngeal pouches becomes the auditory tubes. The second pair becomes the tonsils, while the third and fourth become the thymus gland and the parathyroid glands.

invagination occurs as neural folds appear. When the neural folds meet at the midline, the neural tube, which later develops into the brain and the nerve cord, is formed (see Fig. 51.4). After the notochord is replaced by the vertebral column, the nerve cord is called the spinal cord.

Development of the heart begins in the third week and continues into the fourth week. At first, there are right and left heart tubes; when these fuse, the heart begins pumping blood, even though the chambers of the heart are not fully formed. The veins enter posteriorly and the arteries exit anteriorly from this largely tubular heart, but later the heart twists so that all major blood vessels are located anteriorly.

The Fourth and Fifth Weeks

At four weeks, the embryo is barely larger than the height of this print. A bridge of mesoderm called the body stalk connects the caudal (tail) end of the embryo with the chorion, which has treelike projections called **chorionic villi** [Gk. *chorion,* membrane, and L. *villus,* shaggy hair] (Fig. 51.13*c* and *d*). The fourth extraembryonic membrane, the allantois, is contained within this stalk, and its blood vessels become the umbilical blood vessels. The head and the tail then lift up, and the body stalk moves anteriorly by constriction. Once this process is complete, the **umbilical cord** [L. *umbilicus,* navel], which connects the developing embryo to the placenta, is fully formed (Fig. 51.13*e*).

Little flippers called limb buds appear (Fig. 51.14); later, the arms and the legs develop from the limb buds, and even the hands and the feet become apparent. At the same time—during the fifth week—the head enlarges and the sense organs become more prominent. It is possible to make out the developing eyes, ears, and even nose.

The Sixth Through Eighth Weeks

There is a remarkable change in external appearance during the sixth through eighth weeks of development—a form that is difficult to recognize as a human becomes easily recognized as human. Concurrent with brain development, the head achieves its normal relationship with the body as a neck region develops. The nervous system is developed well enough to permit reflex actions, such as a startle response to touch. At the end of this period, the embryo is about 38 mm (1.5 inches) long and weighs no more than an aspirin tablet, even though all organ systems are established.

During the embryonic period of development, the extraembryonic membranes appear and serve important functions; the embryo acquires organ systems.

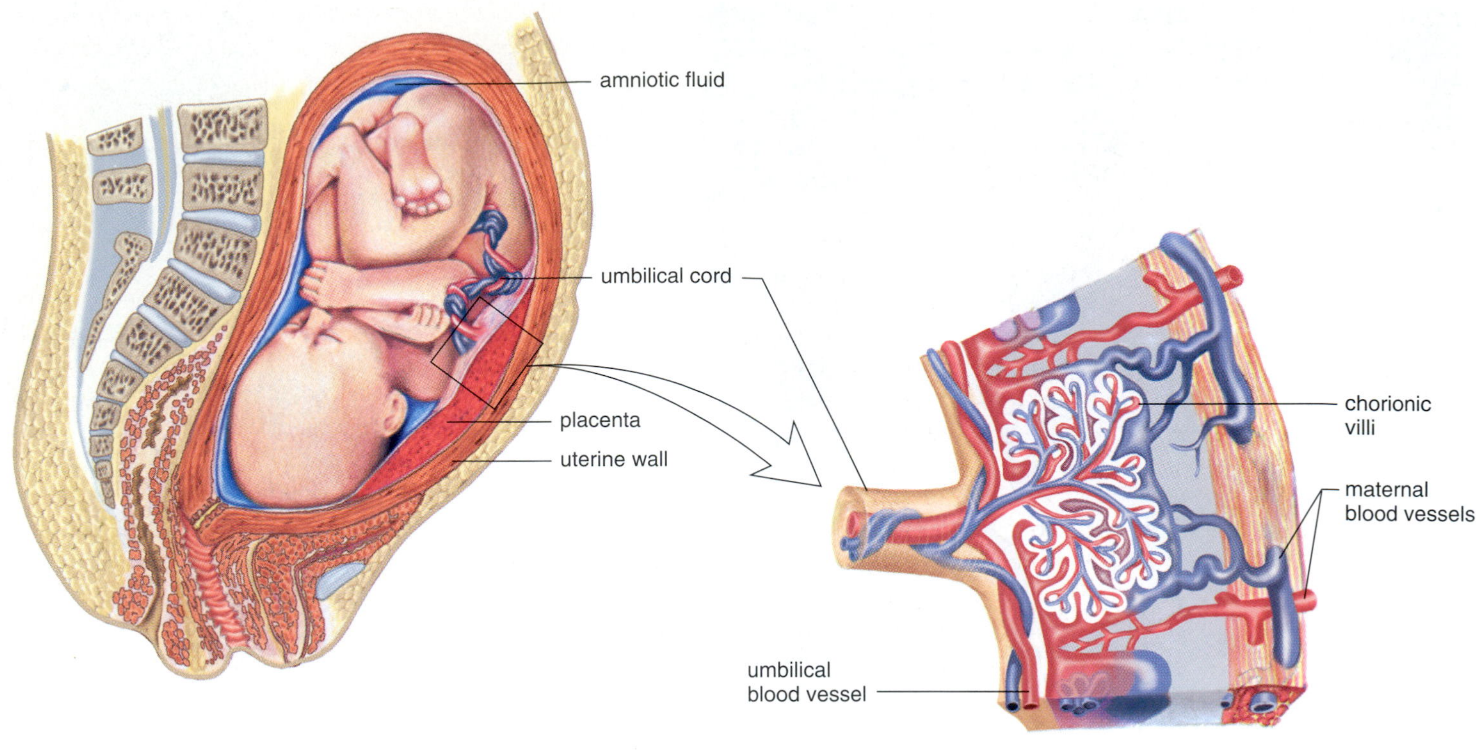

Figure 51.15 **Anatomy of the placenta in a fetus at six to seven months.**
The placenta is composed of both fetal and maternal tissues. Chorionic villi penetrate the uterine lining and are surrounded by maternal blood. Exchange of molecules between fetal and maternal blood takes place across the walls of the chorionic villi.

The Structure and Function of the Placenta

The **placenta** is a mammalian structure that functions in gas, nutrient, and waste exchange between embryonic (later fetal) and maternal circulatory systems. The placenta begins formation once the embryo is fully implanted. At first the entire chorion has chorionic villi that project into endometrium. Later, these disappear in all areas except where the placenta develops. By the tenth week, the placenta (Fig. 51.15) is fully formed and is producing progesterone and estrogen. These hormones have two effects: due to their negative feedback control of the hypothalamus and the anterior pituitary, they prevent any new follicles from maturing, and they maintain the lining of the uterus—now the corpus luteum is not needed. There is no menstruation during pregnancy.

The placenta has a fetal side contributed by the chorion and a maternal side consisting of uterine tissues. Notice in Figure 51.15 how the chorionic villi are surrounded by maternal blood; yet maternal and fetal blood never mix, since exchange always takes place across plasma membranes. Carbon dioxide and other wastes move from the fetal side to the maternal side, and nutrients and oxygen move from the maternal side to the fetal side of the placenta. The umbilical cord stretches between the placenta and the fetus. Although it may seem that the umbilical cord travels from the placenta to the intestine, actually the umbilical cord is simply taking fetal blood to and from the placenta. The umbilical cord is the lifeline of the fetus because it contains the umbilical arteries and vein, which transport waste molecules (carbon dioxide and urea) to the placenta for disposal and take oxygen and nutrient molecules from the placenta to the rest of the fetal circulatory system.

Harmful chemicals can also cross the placenta. This is of particular concern during the embryonic period, when various structures are first forming. Each organ or part seems to have a sensitive period during which a substance can alter its normal development. For example, if a woman takes the drug thalidomide, a tranquilizer, between days 27 and 40 of her pregnancy, the infant is likely to be born with deformed limbs. After day 40, however, the infant is born with normal limbs.

During mammalian development, the embryo and later the fetus are dependent on the placenta for gas exchange and also for acquiring nutrients and for ridding the body of wastes.

Fetal Development and Birth

Fetal development (months 3–9) is marked by an extreme increase in size. Weight multiplies 600 times, going from less than 28 grams to 3 kilograms. During this time, too, the fetus grows to about 50 cm in length. The genitalia appear in the third month, so it is possible to tell if the fetus is male or female.

Soon, hair, eyebrows, and eyelashes add finishing touches to the face and head. In the same way, fingernails and toenails complete the hands and feet. A fine, downy hair (lanugo) covers the limbs and trunk, only to later disappear. The fetus looks very old because the skin is growing so fast that it wrinkles. A waxy, almost cheeselike substance (vernix caseosa) [L. *vernix*, varnish, and *caseus*, cheese] protects the wrinkly skin from the watery amniotic fluid.

The fetus at first only flexes its limbs and nods its head, but later it can move its limbs vigorously to avoid discomfort. The mother feels these movements from about the fourth month on. The other systems of the body also begin to function. After 16 weeks, the fetal heartbeat is heard through a stethoscope. A fetus born at 24 weeks has a chance of surviving, although the lungs are still immature and often cannot capture oxygen adequately. Weight gain during the last couple of months increases the likelihood of survival.

The Stages of Birth

The latest findings suggest that when the fetal brain is sufficiently mature, the hypothalamus causes the pituitary to stimulate the adrenal cortex so that androgens are released into the bloodstream. The placenta utilizes androgens as a precursor for estrogens, hormones that stimulate the production of prostaglandin (a molecule produced by many cells that acts as a local hormone) and oxytocin. All three of these molecules cause the uterus to contract and expel the fetus.

The process of birth (parturition) includes three stages. During the first stage, the cervix dilates to allow passage of the baby's head and body. The amnion usually bursts along about this time. During the second stage, the baby is born and the umbilical cord is cut. During the third stage, the placenta is delivered (Fig. 51.16).

During the fetal period there is a refinement of organ systems and the fetus gains weight. Finally birth occurs.

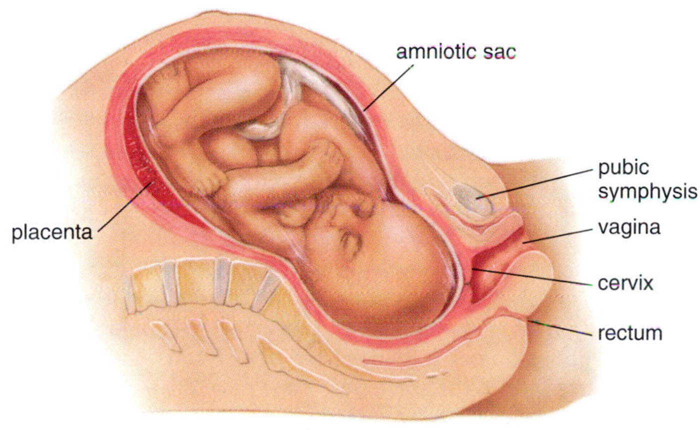

a. 9-month-old fetus

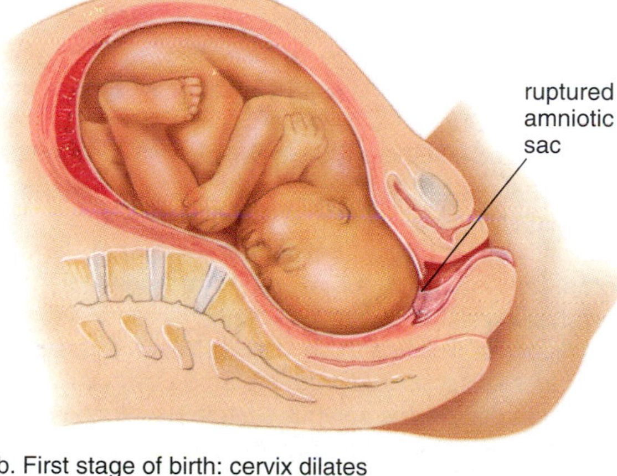

b. First stage of birth: cervix dilates

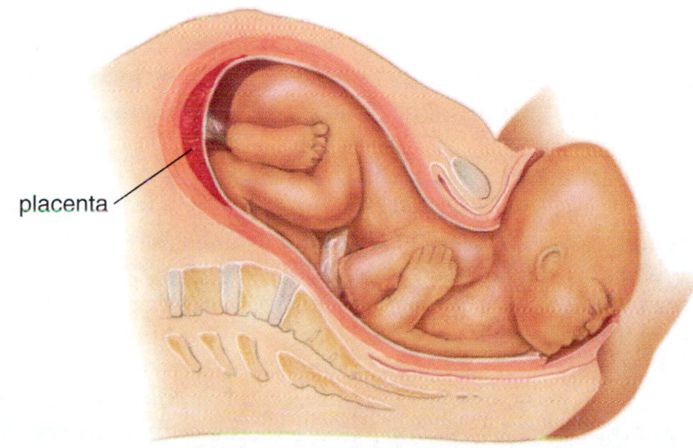

c. Second stage of birth: baby emerges

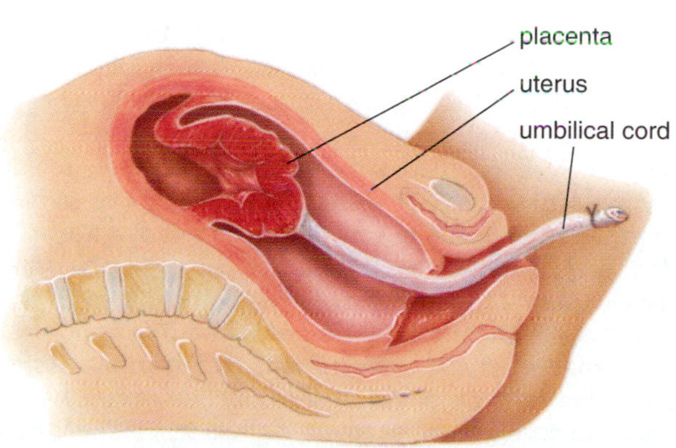

d. Third stage of birth: expelling afterbirth

Figure 51.16 Three stages of parturition.
a. Position of fetus just before birth begins. b. Dilation of cervix. c. Birth of baby. d. Expulsion of afterbirth.

Connecting Concepts

We have come full circle. We began our study of biology by considering the structure of the cell and its genetic machinery, including how the expression of genes is regulated. In this chapter, we have observed that animals go through the same early embryonic stages of morula, blastula, gastrula, and so forth. The set sequence of these stages is due to the expression of genes that bring about cellular changes. Therefore, once again, we are called upon to study organisms at the cellular level of organization.

We have seen that hormones are signals that affect cellular metabolism. Steroid hormones in animals, like gibberellins in plants, turn on the expression of genes. When a steroid hormone binds to a specific hormone receptor, a gene is transcribed, and then translation produces the corresponding protein. This same type of signal transduction pathway occurs during development. Transduction means that the signal has been transformed into an event that has an effect on the organism.

A set sequence of signaling molecules are produced as development occurs. Each new signal in the sequence turns on a specific gene, or more likely, a sequence of genes. Gene expression cascades are common during development. Homeotic genes are arranged in sets on the chromosomes, and a protein product from one set of genes acts as a transcription factor to turn on another set, and so forth. During the development of flies, it is possible to observe that first one region of a chromosome and then another puffs out, indicating that genes are transcribed in sequence as development occurs.

Developmental biology is now making a significant contribution to the field of evolution. Homeotic genes with the same homeoboxes (sequence of about 180 base pairs) have been discovered in many different types of organisms. This suggests that homeotic genes arose early in the history of life, and mutations in these genes could possibly account for macroevolution—the appearance of new species or even higher taxons.

Summary

51.1 Early Developmental Stages

Development occurs after fertilization. The acrosome of a sperm releases enzymes that digest away the jelly coat around the egg; the acrosome extrudes a filament that attaches to a receptor on the vitelline membrane. The sperm nucleus enters the egg and fuses with the egg nucleus. The resulting zygote begins to divide.

The early developmental stages in animals include the following events. During cleavage, division occurs, but there is no overall growth. The result is a morula, which becomes the blastula when an internal cavity (the blastocoel) appears. During the gastrula stage, invagination of cells into the blastocoel results in formation of the germ layers: ectoderm, mesoderm, and endoderm. Later development of organs can be related to these layers.

The development of three types of animals (lancelet, frog, and chick) is compared. The first three stages (cleavage, blastulation, and gastrulation) differ according to the amount of yolk in the egg.

During neurulation, the nervous system develops from midline ectoderm, just above the notochord. At this point, it is possible to draw a typical cross section of a chordate embryo (see Fig. 51.5).

51.2 Developmental Processes

Two important mechanisms—cytoplasmic segregation and induction—bring about cellular differentiation and morphogenesis as development occurs. The egg contains chemical signals called maternal determinants which are parcelled out during cell division. After the first cleavage of a frog embryo, only a daughter cell that receives a portion of the gray crescent is able to develop into a complete embryo. This illustrates the importance of cytoplasmic segregation to early development of a frog.

Induction is the ability of one embryonic tissue to influence the development of another tissue. The notochord induces the formation of the neural tube in frog embryos. The reciprocal induction that occurs between the lens and the optic vesicle is another good example of induction. Induction occurs because the inducing cells give off chemical signals that influence their neighbors.

C. elegans and *Drosophila* are two model organisms that have contributed to our knowledge of developmental genetics. Fate maps and the development of vulva in *C. elegans* have shown that induction is an ongoing process in which one tissue after the other regulates the development of another through chemical signals coded for by particular genes. Apoptosis is necessary to development and the process has been studied on the cellular level in *C. elegans.*

Work with *Drosophila* has allowed researchers to identify genes that determine the axes of the body and segmentation genes that regulate the development of segments. An important concept has emerged: during development sequential sets of master genes code for morphogen gradients that activate the next set of master genes in turn. Morphogens are transcription factors that bind to DNA.

Homeotic genes control pattern formation such as the presence of antennae, wings, and limbs on the segments of *Drosophila*. Homeotic genes code for proteins that contain a homeodomain, a particular sequence of 60 amino acids. Homeotic genes are also transcription factors and the homeodomain is the portion of the protein that binds to DNA. Homologous homeotic genes have been found in a wide variety of organisms and therefore they must have arisen early in the history of life and they have been conserved.

51.3 Human Embryonic and Fetal Development

Human development can be divided into embryonic development (months 1 and 2) and fetal development (months 3–9). The early stages in human development resemble those of the chick. The similarities are probably due to their evolutionary relationship, not the amount of yolk the eggs contain, because the human egg has little yolk.

The extraembryonic membranes appear early in human development. The trophoblast of the blastocyst is the first sign of the chorion, which goes on to become the fetal part of the placenta. Exchange occurs between fetal and maternal blood at the placenta. The amnion contains amniotic fluid, which cushions and protects the embryo. The yolk sac and allantois are also present.

Fertilization occurs in the oviduct, and cleavage occurs as the embryo moves toward the uterus. The morula becomes the blastocyst before implanting in the endometrium of the uterus. Organ development begins with neural tube and heart formation. There follows a steady progression of organ formation during embryonic development. During fetal development, refinement of features occurs, and the fetus adds weight. Birth occurs about 280 days after the start of the mother's last menstruation.

Reviewing the Chapter

1. Describe how fertilization of a sea star egg occurs. 920
2. State the germ layer theory, and tell which organs are derived from each of the germ layers. 921
3. Compare the process of cleavage and the formation of the blastula and gastrula in lancelets, frogs, and chicks. 922–23
4. Draw a cross section of a typical chordate embryo at the neurula stage, and label your drawing. 923
5. Describe two mechanisms that are known to be involved in the processes of cellular differentiation and morphogenesis. 924–25
6. Describe an experiment performed by Spemann suggesting that the notochord induces formation of the neural tube. Give another well-known example of induction between tissues. 925
7. With regard to *C. elegans,* what is a fate map? how does induction occur? and what causes apoptosis? 926
8. With regard to *Drosophila,* what is a morphogen gradient and what does such a gradient do? 928
9. What is the function of homeotic genes, and what is the significance of the homeobox within these genes? 929
10. List the human extraembryonic membranes, give a function for each, and compare their functions to those in the chick. 930
11. Tell where fertilization, cleavage, the morula stage, and the blastocyst stage occur in humans. What happens to the embryo in the uterus? 931
12. Describe the structure and the function of the placenta in humans. 932
13. List and describe the stages of birth. 935

Testing Yourself

Choose the best answer for each question.

1. Which of these stages is the first one out of sequence?
   a. cleavage
   b. blastula
   c. morula
   d. gastrula
   e. neurula
2. Which of these stages is mismatched?
   a. cleavage—cell division
   b. blastula—gut formation
   c. gastrula—three germ layers
   d. neurula—nervous system
   e. Both b and c are mismatched.
3. Which of the germ layers is best associated with development of the heart?
   a. ectoderm
   b. mesoderm
   c. endoderm
   d. neurula
   e. All of these are correct.
4. In many embryos, differentiation begins at what stage?
   a. cleavage
   b. blastula
   c. gastrula
   d. neurula
   e. after the completion of these stages
5. Morphogenesis is associated with
   a. protein gradients.
   b. induction.
   c. transcription factors.
   d. homeotic genes.
   e. All of these are correct.
6. In humans, the placenta develops from the chorion. This indicates that human development
   a. resembles that of the chick.
   b. is associated with extraembryonic membranes.
   c. cannot be compared to lower animals.
   d. begins only upon implantation.
   e. Both a and b are correct.
7. In humans, the fetus
   a. is surrounded by four extraembryonic membranes.
   b. has developed organs and is recognizably human.
   c. is dependent upon the placenta for excretion of wastes and acquisition of nutrients.
   d. is embedded in the endometrium of the uterus.
   e. Both b and c are correct.
8. Developmental changes
   a. require growth, differentiation, and morphogenesis.
   b. stop occurring when one is grown.
   c. are dependent upon a parceling out of genes into daughter cells.
   d. are dependent upon activation of master genes in an orderly sequence.
   e. Both a and d are correct.
9. Which of these is mismatched?
   a. brain—ectoderm
   b. gut—endoderm
   c. bone—mesoderm
   d. lens—endoderm
   e. heart—mesoderm
10. Label this diagram illustrating the placement of the extraembryonic membranes, and give a function for each membrane in humans:

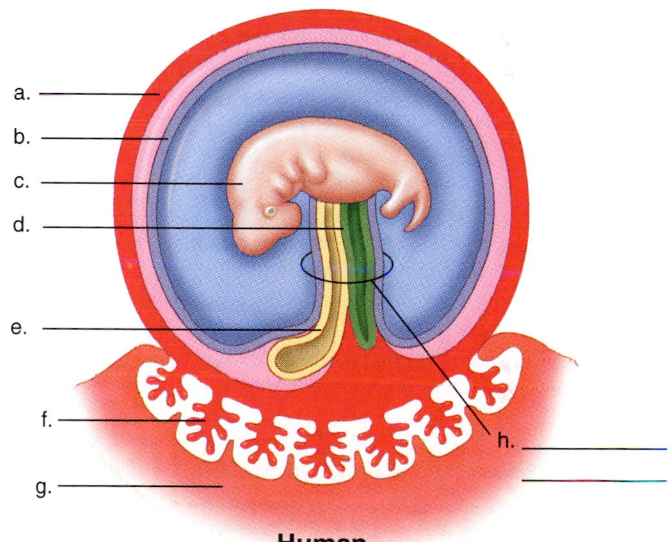

Thinking Scientifically

1. A mutant gene is known to disrupt the earliest stages of development in sea urchins. Individuals with two copies of the mutant gene seem to develop normally. When normal males are crossed with mutant (normal-appearing) females, however, none of the offspring develop at all. How could this pattern of expression be explained?
2. Babies that were malnourished in utero are at a higher risk for many adult-onset diseases than those that were well-nourished. To determine how well babies were nourished in utero, physicians often determine the ratio of head circumference to abdominal circumference. If malnourished, the head circumference to abdominal circumference will be larger than usual because blood in the malnourished fetuses is preferentially directed toward the brain. Why is it better to use the circumference data rather than birth weight data to detect malnourished babies?

Bioethical Issue

The fetus is subject to harm by maternal use of medicines and drugs of abuse, including nicotine and alcohol. Also, various sexually transmitted diseases, notably an HIV infection, can be passed on to the fetus by way of the placenta. Women need to be aware of the need to protect their unborn child from harm. Indeed, their behavior should be protective if they are sexually active, even if they are using a recognized form of birth control. Harm can occur before a woman realizes she is pregnant!

Because we are now aware of the need for maternal responsibility before a child is born, there has been a growing acceptance of prosecuting women when a newborn has a condition such as fetal alcohol syndrome, a condition that can only be caused by the drinking habits of the mother. Employers have also become aware that they might be subject to prosecution. To protect themselves, Johnson Controls, a U.S. battery manufacturer, developed a fetal protection policy. No woman who could bear a child was offered a job that might expose her to toxins that could negatively affect the development of her baby. To get such a job, a woman had to show that she had been sterilized or was otherwise incapable of having children. In 1991, the U.S. Supreme Court declared this policy unconstitutional on the basis of sexual discrimination. The decision was hailed as a victory for women, but was it? The decision was written in such a way that women alone, and not an employer, are responsible for any harm done to the fetus by workplace toxins.

Some have noted that prosecuting women for causing prenatal harm can itself have a detrimental effect. The women may tend to avoid prenatal treatment, thereby increasing the risk to their children. Or they may opt for an abortion in order to avoid the possibility of prosecution. The women feel they are in a no-win situation. If they have a child that has been harmed due to their behavior, they feel they are bad mothers or if they abort, they feel they are also bad mothers. Should sexually active women who can bear a child be expected to avoid substances or situations that could possibly harm an unborn even if they are using birth control? Why or why not?

Understanding the Terms

Match the terms to these definitions:

a. ______________ Ability of a chemical or a tissue to influence the development of another tissue.
b. ______________ Primary tissue layer of a vertebrate embryo; namely ectoderm, mesoderm, or endoderm.
c. ______________ Extraembryonic membrane of birds, reptiles, and mammals that forms an enclosing, fluid-filled sac.
d. ______________ A 180-nucleotide sequence located in nearly all homeotic genes.
e. ______________ Stage of early animal development during which the germ layers form, at least in part, by invagination.

Web Connections

Exploring the Internet

http://www.mhhe.com/biosci/genbio/mader
(click on *Biology 7/e*)

The *Biology 7/e* Online Learning Center provides many resources for studying the material in this chapter including links to the following sites:

The Virtual Embryo. A guide to animal development.

http://www.ucalgary.ca/UofC/eduweb/virtualembryo/

The Visible Embryo. This site provides the user with a detailed look at the first few stages of embryonic development.

http://visembryo.ucsf.edu/

Embryo Development Overview. This site lets you view human development from conception to week 38.

http://www.med.upenn.edu/embryo_project/overview/overview.html

Society for Developmental Biology. This site includes many valuable websites of the members of the society.

http://sdb.bio.purdue.edu/index.html

Bill Wasserman's Developmental Biology Page. Many links are found in "web resources."

http://www.luc.edu/depts/biology/dev.htm

Further Readings for Part vii

Arakawa, T., and Langridge, W. R. H. May 1998. Plants are not just passive creatures! *Nature Medicine* 4(5):550. Plants are being used to produce foreign proteins for human immunity.

Barinaga, M. July 23, 1999. Mapping smells in the brain. *Science* 285(5427):508. An optical imaging technique detects neuron patterns that occur in response to particular odors—each smell activates a unique pattern, or code, for that smell.

Barkley, R. A. September 1998. Attention-deficit hyperactivity disorder. *Scientific American* 279(3):66. ADHD may result from neurological abnormalities with a genetic basis.

Beardsley, T. August 1997. The machinery of thought. *Scientific American* 277(2):78. Researchers have identified the area of the brain responsible for memory.

Belmonte, J. C. I. June 1999. How the body tells left from right. *Scientific American* 280(6):46. The precise orientation of our internal organs is controlled in part by proteins that are produced on one side of an embryo.

Blumenthal, M., et al. March 1999. Discoveries in allergy and asthma. *Discover* 20(3):S-1. This supplement examines advances in understanding and treating allergy and asthma.

Crooks, R., and Baur, K. 1998. *Our sexuality.* 7th ed. Redwood City, Calif.: Benjamin/Cummings Publishing. Introduction to the biological, psychosocial, behavioral, and cultural aspects of sexuality.

Crowley, L. 1997. *Introduction to human disease.* Boston, Mass.: Jones & Bartlett Publishers. This well-illustrated text for study in the allied health fields describes diseases, and their symptoms, diagnoses, and treatments.

Debinski, W. May/June 1998. Anti-brain tumor cytotoxins. *Science & Medicine* 5(3):36. Delivery of bacterial toxins to tumor cells is a new therapy for brain tumor treatment.

Deyo, R. A. August 1998. Low-back pain. *Scientific American* 279(2):48. Treatment options for low-back pain which don't involve bed rest or surgery are improving.

Dickman, S. July 1997. Mysteries of the heart. *Discover* 18(7):117. Article discusses why coronary arteries may still become blocked after treatment for atherosclerosis.

Gazzaniga, M. S. July 1998. The split brain revisited. *Scientific American* 279(1):50. Recent research on split brains has led to new insights into brain organization and consciousness.

Glausiusz, J. October 1997. The good bugs on our tongues. *Discover* 18(10):32. Without friendly bacteria that live on our tongues, we would be vulnerable to harmful bacteria such as *Salmonella.*

Glausiusz, J. September 1998. Infected hearts. *Discover* 19(9):30. Infectious bacteria may play a role in heart disease; antibiotics could prevent the need for heart surgery.

Halstead, L. S. April 1998. Post-polio syndrome. *Scientific American* 278(4):42. Recovered polio victims are experiencing fatigue, pain, and weakness, resulting from degeneration of motor neurons.

Hanson, L. A. November/December 1997. Breast-feeding stimulates the infant immune system. *Science & Medicine* 4(6):12. Long-lasting protection against some infectious diseases has been reported in breast-fed infants.

Harvard Health Letter. April 1998. A special report: Parkinson's disease. This overview presents the symptoms and diagnosis, and treatment of Parkinson's disease.

Jordan, V. C. October 1998. Designer estrogens. *Scientific American* 279(4):60. Selective estrogen receptor modulators may protect against breast and endometrial cancers, osteoporosis, and heart disease.

Julien, R. M. 1997. *A primer of drug action.* 8th ed. New York: W. H. Freeman and Company. A concise, nontechnical guide to the actions, uses, and side effects of psychoactive drugs.

Kempermann, G., and Gage, F. May 1999. New nerve cells for the adult brain. *Scientific American* 280(5):48. The knowledge that the human brain can produce new nerve cells in adulthood could lead to better treatments for neurological diseases.

Kooyman, G. L., and Ponganis, P. J. November/December 1997. The challenges of diving to depth. *American Scientist* 85(6):530. Marine animals do not experience problems associated with pressure at depth.

Kunzig, R. February 1999. What's a pinna for? *Discover* 20(2):24. Article examines the function of the folds of the outer ear.

MacDonald, P. C., and Casey, M. L. March/April 1996. Preterm birth. *Scientific American Science & Medicine* 3(2):42. Article discusses the role of oxytocin, prostaglandins, and infections in the initiation of human labor.

Mader, S. S. 1990. *Human reproductive biology.* 2d ed. Dubuque, Iowa: Wm. C. Brown Publishers. An introductory text covering human reproduction in a clear, easily understood manner.

Mader, S. S. 1998. *Human biology.* 6th ed. Dubuque, Iowa: WCB/McGraw-Hill, Inc. A student-friendly text that covers the principles of biology with emphasis on human anatomy and physiology.

Mader, S. S. 2000. *Understanding anatomy and physiology.* 4th ed. Dubuque, Iowa: Wm. C. Brown Publishers. A text that emphasizes the basics for beginning allied health students.

Mattson, M. P. March/April 1998. Experimental models of Alzheimer's disease. *Science & Medicine* 5(2):16. In Alzheimer's disease, mutations accelerate changes that occur during normal aging.

McDonald, J. W., et al. September 1999. Repairing the damaged spinal cord. *Scientific American* 281(3):64. New treatments being studied may minimize or reverse damage to the spinal cord, and give hope for some spinal cord restoration.

Moore, K., and Persaud, T. 1998. *Before we are born: Essentials of embryology and birth defects.* 5th ed. Philadelphia: W. B. Saunders and Co. For medical and associated health students, this text presents the essentials of normal and abnormal human embryological development.

Nature Medicine Vaccine Supplement. May 1998. Vol. 4, no. 5. The entire issue is devoted to the topic of vaccines, including history, recent developments and research in malaria, cancer, and HIV vaccines.

Newman, J. April 2000. How old is too old to have a baby? *Discover* 21(4):60. Article discusses advancements in reproductive technology that may make it possible for couples in their late forties to sixties to have children.

Nolte, J. 1998. *The human brain.* 4th ed. St. Louis: Mosby-Year Book, Inc. Beginners are guided through the basic aspects of brain structure and function.

Nucci, M. L., and Abuchowski, A. February 1998. The search for blood substitutes. *Scientific American* 278(2):72. Artificial blood substitutes based on hemoglobin are being developed from synthetic chemicals.

Osorio, D. July/August 1997. The evolution of arthropod nervous systems. *American Scientist* 85(3):244. Nervous systems of insects and crustaceans have common features.

Packer, C. July/August 1998. Why menopause? *Natural History* 107(6):24. Article addresses reasons why menopause occurs so early in life, compared to other aging processes.

Powledge, T. M. July 1999. Addiction and the brain. *BioScience* 49(7):513. Drug use changes the biochemistry and anatomy of neurons and alter the way they work.

Appendix A

ANSWER KEY

Chapter 1

Testing Yourself

1. d; **2.** c; **3.** b; **4.** e; **5.** c; **6.** b; **7.** c; **8.** b; **9.** c; **10.** d; **11. a.** After dye is spilled on culture plate, investigator notices that bacteria live despite exposure to sunlight; **b.** Dye protects bacteria against death by UV light; **c.** Expose all culture plates to UV light: one set of plates contains bacteria and dye, and the other set contains only bacteria. The bacteria in both sets die; **d.** Dye does not protect bacteria against death by UV light.

Understanding the Terms

a. metabolism; **b.** evolution; **c.** experimental variable; **d.** photosynthesis; **e.** control group

Chapter 2

Testing Yourself

1. c; **2.** e; **3.** e; **4.** d; **5.** e; **6.** d; **7.** b; **8.** c; **9.** a; **10.** 7p and 7n in nucleus; two electrons in first shell, and five electrons in outer shell. This means that nitrogen needs three more electrons in the outer shell to be stable; because each hydrogen contributes one electron, the formula for ammonia is NH_3.

Understanding the Terms

a. polar covalent bond; **b.** ion; **c.** acid; **d.** molecule; **e.** pH scale

Chapter 3

Testing Yourself

1. c; **2.** a; **3.** e; **4.** b; **5.** c; **6.** b; **7.** c; **8.** c; **9.** e; **10.** c; **11.** e; **12.** c; **13. a.** monomer; **b.** condensation; **c.** polymer; **d.** hydrolysis. The diagram shows the manner in which polymers are synthesized and degraded in cells.

Understanding the Terms

a. carbohydrate; **b.** lipid; **c.** polymer; **d.** isomer; **e.** peptide

Chapter 4

Testing Yourself

1. c; **2.** c; **3.** d; **4.** c; **5.** a; **6.** c; **7.** c; **8.** a; **9.** d; **10.** b; **11.** e; **12. a.** example; **b.** Mitochondria and chloroplasts are a pair because they are both membranous structures involved in energy metabolism; **c.** Centrioles and flagella are a pair because they both contain microtubules; centrioles give rise to the basal bodies of flagella; **d.** ER and ribosomes are a pair because together they are rough ER, which produces proteins. **13. a.** chromatin—DNA specifies the order of amino acids in proteins; **b.** nucleolus—forms ribosomal RNA, which participates in protein synthesis; **c.** smooth ER—forms transport vesicles; **d.** rough ER—produces proteins; **e.** Golgi apparatus—processes and packages proteins for distribution.

Understanding the Terms

a. Golgi apparatus; **b.** chromatin; **c.** nucleolus; **d.** cytoskeleton; **e.** lysosome

Chapter 5

Testing Yourself

1. a. hypertonic—cell shrinks due to loss of water; **b.** hypotonic—central vacuole expands due to gain of water; **2.** b; **3.** b; **4.** b; **5.** c; **6.** c; **7.** e; **8.** e; **9.** b; **10.** b

Understanding the Terms

a. differentially permeable; **b.** osmosis; **c.** hypertonic solution; **d.** osmotic pressure; **e.** cholesterol

Chapter 6

Testing Yourself

1. e; **2.** e; **3.** d; **4.** d; **5.** b; **6.** e; **7.** e; **8.** c; **9. a.** active site; **b.** substrates; **c.** product; **d.** enzyme; **e.** enzyme-substrate complex; **f.** enzyme. The shape of an enzyme is important to its activity because it allows an enzyme-substrate complex to form. **10. a.** high H^+ concentration; **b.** low H^+ concentration; **c.** H^+ pump in electron transport system; **d.** H^+; **e.** ATP synthase complex; **f.** ADP; **g.** P; **h.** ATP; **i.** energy from electron transfers. See also Figure 6.10, page 107.

Understanding the Terms

a. metabolism; **b.** potential energy; **c.** vitamin; **d.** entropy; **e.** coenzyme

Chapter 7

Testing Yourself

1. e; **2.** d; **3.** a; **4.** e; **5.** c; **6.** e; **7.** c; **8.** e; **9. a.** outer membrane; **b.** inner membrane; **c.** stroma; **d.** granum; **e.** thylakoid; **f.** thylakoid; **g.** stroma; **10. a.** water; **b.** oxygen; **c.** carbon dioxide; **d.** carbohydrate; **e.** ADP + Ⓟ → ATP; **f.** $NADP^+ \rightarrow NADPH$

Understanding the Terms

a. light-dependent reaction; **b.** photosystem; **c.** electron transport system; **d.** photosynthesis; **e.** Calvin cycle

Chapter 8

Testing Yourself

1. b; **2.** c; **3.** a; **4.** c; **5.** c; **6.** c; **7.** a; **8.** b; **9.** e; **10.** c; **11.** a; **12.** c; **13.** b; **14. a.** cristae, contains electron transport system and ATP synthase complex; **b.** matrix, location of transition reaction and Krebs cycle; **c.** outer membrane, defines the boundary of the mitochondrion; **d.** intermembrane space, accumulation of H^+; **e.** inner membrane, partitions the mitochondrion into the intermembrane space and the matrix.

Understanding the Terms

a. fermentation; **b.** Krebs cycle; **c.** metabolic pool; **d.** acetyl-CoA; **e.** electron transport system

Chapter 9

Testing Yourself

1. c; **2.** b; **3.** e; **4.** d; **5.** a; **6.** c; **7.** b; **8.** d; **9.** b; **10.** b; **11.** a; **12. a.** chromatid of chromosome; **b.** centriole; **c.** spindle fiber or aster; **d.** nuclear envelope (fragment)

Understanding the Terms

a. centrosome; **b.** centromere; **c.** spindle; **d.** sister chromatid; **e.** apoptosis

Chapter 10

Testing Yourself

1. b; **2.** d; **3.** e; **4.** b; **5.** a; **6.** d; **7.** d; **8.** e; **9.** c; **10.** The cell on the right represents metaphase I because bivalents are present at the metaphase plate.

Understanding the Terms

a. spermatogenesis; **b.** bivalent; **c.** polar body; **d.** secondary oocyte; **e.** homologue

Chapter 11

Practice Problems 1

1. a. all *W*; **b.** 1/2 *W*, 1/2 *w*; **c.** 1/2 *T*, 1/2 *t*; **d.** all *T*; **2. a.** gamete; **b.** genotype; **c.** gamete

Practice Problems 2

1. *bb*; **2.** 75% yellow; 25% green; **3.** 75%; **4.** *Tt* × *tt*, *tt*

Practice Problems 3

1. Testcross black-coated horse × brown-coated horse; **2.** *Ll* × *ll*; **3.** 200 have wrinkled seeds

Practice Problems 4

1. a. 1/2 *TG*, 1/2 *tG*; **b.** 1/4 *TG*, 1/4 *Tg*, 1/4 *tG*, 1/4 *tg*; **c.** 1/2 *TG*, 1/2 *Tg*; **2. a.** gamete; **b.** genotype; **c.** gamete

Practice Problems 5

1. *BbTt* only; **2. a.** *LlGg* × *llgg*; **b.** *LlGg* × *LlGg*; **3.** 9/16

Testing Yourself

1. b; **2.** a; **3.** c; **4.** d; **5.** c; **6.** d; **7.** a; **8.** b; **9.** b; **10.** b; **11.** c; **12.** d

Additional Genetics Problems

1. 100% chance for widow's peak and 0% chance for continuous hairline; **2.** *ee* (recessive); **3.** 50%; **4.** 210 gray bodies and 70 black bodies; 140 = heterozygous; cross fly with recessive (black body); **5.** $BBHH = 2^0 = 1$ possibility: *BH*; $BbHh = 2^2 = 4$ possibilities: *BH*, *Bh*, *bH*, *bh*; $BBHh = 2^1 = 2$ possibilities: *BH* and *Bh*; **6.** F_1 = all black with short hair; F_2 = 9 black, short: 3 black, long: 3 brown, short: 1 brown, long; offspring would be 1 brown long: 1 brown short : 1 black long : 1 black short; **7.** *Bbtt* × *bbTt* and *bbtt*; **8.** *GGLl*; **9.** 25%

Understanding the Terms

a. recessive allele; **b.** allele; **c.** dominant allele; **d.** testcross; **e.** genotype

Chapter 12

Practice Problems 1

1. *MM'*, *Mm* (both melanic), *M'm* (insularia), *mm* (typical); **2.** 1 pink: 1 white; or 50% pink, 50% white; **3.** 7; **4.** pleiotropy accounts for albinism and epistasis accounts for albinism and crossed eyes occurring together.

Practice Problems 2

1. a. X^BX^B, X^BX^b (bar-eyed females), X^bX^b (normal-eyed female); **b.** 2 possible genotypes for males are: X^BY (bar-eyed male has the gametes X^B, Y) and X^bY (normal-eyed male has the gametes X^b, Y); **2. a. (2)** b, ratio 1:1; **3. a.** 100%; **b.** none; **c.** 100%

Practice Problems 3

1. a. 9:3:3:1; **b.** linkage; **2.** 12 map units; **3.** bar eye, scalloped wings, garnet eye

Testing Yourself

1. b; **2.** c; **3.** e; **4.** a; **5.** b; **6.** b; **7.** d; **8.** b; **9.** c; **10.** d

Additional Genetics Problems

1. pleiotropy; **2.** 50%; **3.** 6, yes; **4.** linkage, 10 map units; **5.** Both males and females are 1:1; **6.** $X^BX^B \times X^bY$ = all males and females are bar-eyed, $X^bX^b \times X^BY$ = females are bar-eyed, males are all normal; **7.** X^bX^bWw

Understanding the Terms

a. autosome; **b.** polyploid; **c.** X-linked gene; **d.** polygenic inheritance; **e.** polyploid

Chapter 13

Testing Yourself

1. c; **2.** a; **3.** b; **4.** c; **5.** d; **6.** c; **7.** b; **8.** d; **9.** c; **10.** a; **11.** autosomal dominant condition

Additional Genetics Problems

1. 0%, 0%, 100%; **2.** 25%; **3.** AB, yes, A, B, AB, or O; **4.** X^cYww, X^CX^CWw, X^CYWw, both sexes have normal vision with widow's peak; **5.** sex-linked recessive, X^AX^a

Understanding the Terms

a. autosome; **b.** sex-influenced trait; **c.** karyotype; **d.** nondisjunction; **e.** carrier

Chapter 14

Testing Yourself

1. b; **2.** d; **3.** a; **4.** c; **5.** c; **6.** c; **7.** a; **8.** b; **9.** e; **10.** The parental helix is heavy-heavy, and each daughter helix is heavy-light. This shows that each daughter helix is composed of one old strand and one new strand, which is consistent with semiconservative replication.

Understanding the Terms

a. genetic mutation; **b.** complementary base pairing; **c.** DNA polymerase; **d.** purine; **e.** bacteriophage

Chapter 15

Testing Yourself

1. b; **2.** a; **3.** c; **4.** a; **5.** a; **6.** c; **7.** b; **8.** c; **9.** b; **10. a.** ACU CCU GAA UGC AAA; **b.** UGA GGA CUU ACG UUU; **c.** threonine-proline-glutamate-cysteine-lysine

Understanding the Terms

a. RNA polymerase; **b.** intron; **c.** translation; **d.** polyribosome; **e.** codon

Chapter 16

Testing Yourself

1. c; **2.** e; **3.** a; **4.** a; **5.** b; **6.** b; **7.** b; **8.** e; **9.** e; **10.** e; **11. a.** DNA; **b.** regulator gene; **c.** promoter; **d.** operator; **e.** mRNA; **f.** active repressor

Understanding the Terms

a. oncogene; **b.** tumor; **c.** Barr body; **d.** euchromatin; **e.** carcinogen

Chapter 17

Testing Yourself

1. a; **2.** e; **3.** e; **4.** e; **5.** b; **6.** d; **7.** c; **8.** e; **9.** AATTTTAA; **10. a.** retrovirus; **b.** recombinant RNA; **c.** human genome; **d.** recombinant RNA; **e.** reverse transcription; **f.** recombinant DNA; **g.** defective gene. See also Figure 17.10, page 275.

Understanding the Terms

a. restriction enzyme; **b.** transgenic organism; **c.** probe; **d.** clone; **e.** plasmid

Chapter 18

Testing Yourself

1. e; **2.** b; **3.** e; **4.** e; **5.** e; **6.** e; **7.** e; **8.** e; **9.** Life has a history, and it's possible to trace the history of individual organisms. **10.** Two different continents can have similar environments, and therefore unrelated organisms that are similarly adapted. **11.** All vertebrates share a common ancestor, who had pharyngeal pouches during development. **12.** Genetic differences account for speciation, and therefore evolution.

Understanding the Terms

a. biogeography; **b.** paleontology; **c.** vestigial structure; **d.** adaptation; **e.** inheritance of acquired characteristics

Chapter 19

Practice Problems 1

1. 21%; **2.** $t = q = 0.3$, $T = p = 0.7$; homozygous recessive $= q^2 = 9\%$; homozygous dominant $= p^2 = 49\%$; heterozygous $= 2pq = 42\%$

Testing Yourself

1. c; **2.** c; **3.** c; **4.** c; **5.** e; **6.** b; **7.** d; **8.** e; **9.** c; **10.** b; **11.** b; **12. a.** See Figure 19.8, page 308; **b.** See Figure 19.7, page 308; **c.** See Figure 19.6, page 307.

Additional Genetics Problems

1. 16%; **2.** 81%; **3.** recessive allele = 0.2, dominant allele = 0.8; homozygous recessive = 0.04, homozygous dominant = 0.64, heterozygous = 0.32

Understanding the Terms

a. stabilizing selection; **b.** postzygotic isolating mechanism; **c.** genetic drift; **d.** adaptive radiation; **e.** gene flow

Chapter 20

Testing Yourself

1. c; **2.** d; **3.** b; **4.** b; **5.** b; **6.** e; **7.** c; **8.** e; **9.** e; **10.** c; **11. a.** oldest eukaryotic fossils; **b.** O_2 accumulates; **c.** oldest known fossils; **d.** Cambrian animals; **e.** Ediacaran animals; **f.** protists evolve and diversify

Understanding the Terms

a. molecular clock; **b.** protocell; **c.** liposome; **d.** ocean ridge; **e.** ozone shield

Chapter 21

Testing Yourself

1. a; **2.** b; **3.** d; **4.** d; **5.** e; **6.** d; **7.** b; **8.** c; **9.** e; **10.** c; **11. a.** Chordata; **b.** Vertebrata; **c.** Class; **d.** Order; **e.** Anthropoidea; **f.** Hominoidea; **g.** Hominidae; **h.** Genus; **i.** *Homo sapiens*

Understanding the Terms

a. anthropoid; **b.** Cro-Magnon; **c.** Neanderthal; **d.** *Homo ergaster*; **e.** hominoid

Chapter 22

Testing Yourself

1. b; **2.** c; **3.** a; **4.** e; **5.** c; **6.** c; **7.** c; **8.** b; **9.** c; **10.** b

Understanding the Terms

a. sexual selection; **b.** pheromone; **c.** inclusive fitness; **d.** imprinting; **e.** operant conditioning

Chapter 23

Testing Yourself

1. d; **2.** c; **3.** e; **4.** b; **5.** b; **6.** e; **7.** c; **8.** e; **9.** c; **10.** a

Understanding the Terms

a. demographic transition; **b.** population; **c.** exponential growth; **d.** carrying capacity; **e.** biotic potential

Chapter 24

Testing Yourself

1. b; **2.** e; **3.** d; **4.** e; **5.** b; **6.** e; **7.** c; **8.** e; **9.** e; **10.** e; **11. a.** population densities; **b.** environmental gradient. A community contains species with overlapping tolerances to environmental factors (see also Fig. 24.2, page 401).

Understanding the Terms

a. community; **b.** habitat; **c.** ecological succession; **d.** ecological niche; **e.** mutualism

Chapter 25

Testing Yourself

1. c; **2.** b; **3.** e; **4.** e; **5.** c; **6.** c; **7.** b; **8.** a; **9.** d; **10.** c; **11.** b; **12.** c; **13.** a; **14.** c; **15. a.** top carnivore, tertiary consumers; **b.** carnivore, secondary consumers; **c.** herbivore, primary consumers; **d.** producer, autotrophs; **e.** The numbers are the dry biomass weights of the organisms at each level.

Understanding the Terms

a. detritus; **b.** ozone shield; **c.** fossil fuel; **d.** nitrogen fixation; **e.** food web

Chapter 26

Testing Yourself

1. b; **2.** a; **3.** c; **4.** d; **5.** d; **6.** a; **7.** c; **8.** a; **9.** a; **10.** d; **11.** d; **12. a.** 4 + for both; **b.** 3 + for both; **c.** 2 + for both; **d.** 1 + for both

Understanding the Terms

a. estuary; **b.** benthic division; **c.** taiga; **d.** savanna; **e.** spring overturn

Chapter 27

Testing Yourself

1. e; **2.** a; **3.** b; **4.** d; **5.** e; **6.** e; **7.** d; **8.** b; **9.** e; **10.** e

Understanding the Terms

a. biodiversity hotspot; **b.** metapopulation; **c.** source population; **d.** population viability analysis; **e.** landscape

Chapter 28

Testing Yourself

1. a; **2.** c; **3.** d; **4.** a; **5.** c; **6.** b; **7.** d; **8.** b; **9.** e; **10.** b; **11. a.** three by color; **b.** all taxa: vertebra; only snakes and lizards: amniote egg and internal fertilization; **c.** the taxa in a clade; they share the same derived characters.

Understanding the Terms

a. taxonomy; **b.** phylogenetic tree; **c.** taxon; **d.** cladistics; **e.** homology

Chapter 29

Testing Yourself

1. a. attachment; **b.** bacterial chromosome; **c.** penetration; **d.** integration; **e.** prophage; **f.** maturation; **g.** release; See also Figure 29.3, page 511; **2.** d; **3.** c; **4.** c; **5.** c; **6.** c; **7.** a; **8.** e; **9.** e; **10.** a

Understanding the Terms

a. lysogenic; **b.** photoautotroph; **c.** saprotroph; **d.** symbiotic; **e.** archaea

Chapter 30

Testing Yourself

1. b; **2.** b; **3.** e; **4.** b; **5.** b; **6.** d; **7.** c; **8.** d; **9.** b; **10. a.** sexual reproduction; **b.** isogametes pairing; **c.** zygote (2n); **d.** zygospore (2n); **e.** asexual reproduction; **f.** zoospores (n); **g.** nucleus; **h.** chloroplast; **i.** starch granule; **j.** pyrenoid; **k.** flagellum; **l.** eyespot; **m.** gamete formation. See also Figure 30.3, page 526.

Understanding the Terms

a. pseudopod; **b.** euglenoid; **c.** diatom; **d.** trypanosome; **e.** plankton

Chapter 31

Testing Yourself

1. b; **2.** c; **3.** e; **4.** a; **5.** c; **6.** a; **7.** d; **8.** a; **9.** d; **10.** d; **11. a.** meiosis; **b.** basidiospores; **c.** dikaryotic mycelium; **d.** fruiting body (basidiocarp); **e.** stalk; **f.** gill; **g.** cap; **h.** dikaryotic (n + n); **i.** diploid (2n); **j.** zygote. See also Figure 31.6, page 550.

Understanding the Terms

a. basidium; **b.** mycelium; **c.** conidiospore; **d.** fruiting body; **e.** mycorrhizae

Chapter 32

Testing Yourself

1. e; **2.** c; **3.** e; **4.** c; **5.** d; **6.** b; **7.** b; **8.** b; **9.** e; **10. a.** sporophyte (2n); **b.** meiosis; **c.** gametophyte (n); **d.** fertilization. See also Figure 32.2, page 559.

Understanding the Terms

a. sporophyte; **b.** monocotyledon; **c.** pollen grain; **d.** rhizoid; **e.** phloem

Chapter 33

Testing Yourself

1. d; **2.** e; **3.** e; **4.** d; **5.** b; **6.** e; **7.** a; **8. a.** gastrovascular cavity; **b.** tentacle; **c.** mouth; **d.** mesoglea; **9. a.** Cnidaria; **b.** Nematoda; **c.** Cnidaria; **d.** Nematoda; **e.** Platyhelminthes; **10. a.** tapeworm; **b.** planarian; **c.** sponge; **d.** *Hydra*; **e.** rotifer

Understanding the Terms

a. gastrovascular cavity; **b.** coelom; **c.** pseudocoelom; **d.** bilateral symmetry; **e.** tube-within-a-tube body plan

Chapter 34

Testing Yourself

1. b; **2.** c; **3.** b; **4.** c; **5.** c; **6.** a; **7.** d; **8.** c; **9.** a; **10. a.** all three; **b.** annelids, arthropods; **c.** all three; **d.** all three; **e.** all three; **f.** all three; **g.** arthropods; **h.** molluscs; **11. a.** earthworms; **b.** clams; **c.** clams; **d.** clams; **e.** earthworms; **f.** earthworms; **g.** clams; **h.** clams; **i.** earthworms; **j.** earthworms; **12. a.** head; **b.** antenna; **c.** simple eye; **d.** compound eye; **e.** thorax; **f.** tympanum; **g.** abdomen; **h.** forewing; **i.** hindwing; **j.** ovipositor; **k.** spiracles; **l.** air sac; **m.** spiracle; **n.** tracheae. See also Figure 34.14*a*, page 611.

Understanding the Terms

a. trachea; **b.** metamorphosis; **c.** segmentation; **d.** nephridia; **e.** chitin

Chapter 35

Testing Yourself

1. b; **2.** b; **3.** e; **4.** b; **5.** c; **6.** e; **7.** a; **8.** c; **9.** e; **10.** b; **11. a.** pharyngeal pouches; **b.** dorsal tubular nerve cord; **c.** notochord; **d.** postanal tail

Understanding the Terms

a. homeothermic; **b.** monotreme; **c.** reptile; **d.** notochord; **e.** echinoderm

Chapter 36

Testing Yourself

1. c; **2.** b; **3.** c; **4.** c; **5.** b; **6.** b; **7.** b; **8.** c; **9.** c; **10.** b; **11. a.** epidermis; **b.** cortex; **c.** endodermis; **d.** phloem; **e.** xylem. See also Figure 36.8, page 648; **12. a.** upper epidermis; **b.** palisade mesophyll; **c.** leaf vein; **d.** spongy mesophyll; **e.** lower epidermis. See also Figure 36.18, page 658.

Understanding the Terms

a. mesophyll; **b.** vascular cambium; **c.** cotyledon; **d.** stolon; **e.** phloem

Chapter 37

Testing Yourself

1. d; **2.** e; **3.** a; **4.** e; **5.** c; **6.** d; **7.** e; **8.** d; **9.** The diagram shows that air pressure pushing down on mercury in the pan can raise a column of mercury only to 76 cm. When water above the column is transpired, it pulls on the mercury and raises it higher. This suggests that transpiration would be able to raise water to the top of trees. **10.** See Figure 37.12, page 676. After K^+ enters guard cells, water follows by osmosis and the stomate opens. **11.** There is more solute in the left bulb than in the right; therefore water enters the left bulb. This creates a positive pressure that causes water, along with solute, to flow toward the right bulb. See also illustration, page 678.

Understanding the Terms

a. pressure-flow model; **b.** epiphyte; **c.** transpiration; **d.** Casparian strip; **e.** guard cell

Chapter 38

Testing Yourself

1. d; **2.** b; **3.** e; **4.** a; **5.** c; **6.** b; **7.** d; **8.** e; **9.** e; **10.** e; **11.** e; **12. a.** epidermal cell; **b.** guard cell. The hormone ABA in some unknown manner causes K^+ to leave the guard cell. Water follows the movement of K^+ and the guard cell closes.

Understanding the Terms

a. circadian rhythm; **b.** gravitropism; **c.** abscission; **d.** gibberellin; **e.** photoperiodism

Chapter 39

Testing Yourself

1. d; **2.** a; **3.** b; **4.** a; **5.** a; **6.** e; **7.** a; **8.** e; **9.** c; **10.** e; **11. a.** sporophyte; **b.** meiosis; **c.** microspore; **d.** megaspore; **e.** microgametophyte (pollen grain); **f.** megagametophyte (embryo sac); **g.** egg and sperm; **h.** fertilization; **i.** zygote; **j.** seed. See also Figure 39.1, page 700.

Understanding the Terms

a. pistil; **b.** fruit; **c.** megagametophyte; **d.** seed; **e.** pollen grain

Chapter 40

Testing Yourself

1. c; **2.** b; **3.** a; **4.** e; **5.** e; **6.** e; **7.** b; **8.** e; **9.** e; **10.** c; **11. a.** columnar epithelium, lining of intestine (digestive tract), protection and absorption; **b.** cardiac muscle, wall of heart, pumps blood; **c.** compact bone, skeleton, support and protection.

Understanding the Terms

a. ligament; **b.** epidermis; **c.** striated; **d.** homeostasis; **e.** spongy bone

Chapter 41

Testing Yourself

1. b; **2.** b; **3.** a; **4.** d; **5.** d; **6.** c; **7.** b; **8.** b; **9.** e; **10. a.** blood pressure; **b.** osmotic pressure; **c.** blood pressure; **d.** osmotic pressure; **11. a.** aorta; **b.** left pulmonary arteries; **c.** pulmonary trunk; **d.** left pulmonary veins; **e.** left atrium; **f.** semilunar valves; **g.** atrioventricular (mitral) valve; **h.** left ventricle; **i.** septum; **j.** inferior vena cava; **k.** right ventricle; **l.** chordae tendineae; **m.** atrioventricular (tricuspid) valve; **n.** right atrium; **o.** right pulmonary veins; **p.** right pulmonary arteries; **q.** superior vena cava. See also Figure 41.6, page 745.

Understanding the Terms

a. artery; **b.** platelet; **c.** plasma; **d.** vena cavae; **e.** hemoglobin

Chapter 42

Testing Yourself

1. e; **2.** e; **3.** a; **4.** b; **5.** c; **6.** b; **7.** a; **8.** a; **9.** b; **10.** e; **11.** b; **12. a.** antigen-binding sites; **b.** light chain; **c.** heavy chain; **d.** variable region; **e.** constant region

Understanding the Terms

a. vaccine; **b.** lymph; **c.** antigen; **d.** apoptosis; **e.** T lymphocyte

Chapter 43

Testing Yourself

1. a; **2.** b; **3.** d; **4.** b; **5.** c; **6.** a; **7.** c; **8.** c; **9.** e; **10.** c; **11.** d; **12.** c; **13.** Test tube 1: no digestion—no enzyme and no HCl; Test tube 2: some digestion—no HCl; Test tube 3: no digestion—no enzyme; Test tube 4: digestion—both enzyme and HCl are present

Understanding the Terms

a. vitamins; **b.** lipase; **c.** lacteal; **d.** esophagus; **e.** gallbladder

Chapter 44

Testing Yourself

1. a; **2.** b; **3.** b; **4.** d; **5.** c; **6.** b; **7.** c; **8.** b; **9.** e; **10.** b; **11. a.** nasal cavity; **b.** nostril; **c.** pharynx; **d.** epiglottis; **e.** glottis; **f.** larynx; **g.** trachea; **h.** bronchus; **i.** bronchiole. See also Figure 44.6, page 802.

Understanding the Terms

a. pharynx; **b.** diaphragm; **c.** vocal cord; **d.** gill; **e.** expiration

Chapter 45

Testing Yourself

1. d; **2.** a; **3.** c; **4.** b; **5.** a; **6.** e; **7.** b; **8.** e; **9.** a; **10.** b; **11. a.** glomerulus; **b.** efferent arteriole; **c.** afferent arteriole; **d.** proximal convoluted tubule; **e.** loop of the nephron; **f.** descending limb; **g.** ascending limb; **h.** peritubular capillary network; **i.** distal convoluted tubule; **j.** renal vein; **k.** renal artery; **l.** collecting duct

Understanding the Terms

a. Malpighian tubule; **b.** glomerular capsule; **c.** distal convoluted tubule; **d.** aldosterone; **e.** uric acid

Chapter 46

Testing Yourself

1. b; **2.** b; **3.** e; **4.** c; **5.** a; **6.** a; **7.** c; **8.** d; **9.** b; **10.** b; **11. a.** central canal; **b.** gray matter; **c.** white matter; **d.** dorsal root ganglion; **e.** cell body of sensory neuron; **f.** axon of sensory neuron; **g.** cell body of motor neuron; **h.** axon of motor neuron; **i.** interneuron

Understanding the Terms

a. reflex; **b.** neurotransmitter; **c.** autonomic system; **d.** ganglion; **e.** acetylcholine

Chapter 47

Testing Yourself

1. e; **2.** c; **3.** c; **4.** e; **5.** d; **6.** c; **7.** c; **8.** b; **9.** c; **10. a.** retina—contains sensory receptors; **b.** choroid—absorbs stray light; **c.** sclera—protects and supports eyeball; **d.** optic nerve—transmits impulses to brain; **e.** fovea centralis—makes acute vision possible; **f.** muscle in ciliary body—holds lens in place, accommodation; **g.** lens—refracts and focuses light rays; **h.** iris—regulates light entrance; **i.** pupil—admits light; **j.** cornea—refracts light rays

Understanding the Terms

a. sensory receptor; **b.** retina; **c.** sclera; **d.** chemoreceptor; **e.** spiral organ

Chapter 48

Testing Yourself

1. b; **2.** f; **3.** c; **4.** e; **5.** e; **6.** b; **7.** b; **8.** e; **9.** a; **10.** e; **11. a.** T tubule; **b.** sarcoplasmic reticulum; **c.** myofibril; **d.** Z line; **e.** sarcomere; **f.** sarcolemma of muscle fiber

Understanding the Terms

a. osteoblast; **b.** sliding filament model; **c.** actin; **d.** appendicular skeleton; **e.** pectoral girdle

Chapter 49

Testing Yourself

1. a; **2.** e; **3.** f; **4.** b; **5.** c; **6.** a; **7.** e; **8.** e; **9.** c; **10.** e; **11.** b; **12.** e; **13.** d; **14. a.** inhibits; **b.** inhibits; **c.** releasing hormone; **d.** stimulating hormone; **e.** target gland hormone

Understanding the Terms

a. thyroid gland; **b.** hormone; **c.** pineal gland; **d.** peptide hormone; **e.** pheromone

Chapter 50

Testing Yourself

1. a. seminal vesicle; **b.** ejaculatory duct; **c.** prostate gland; **d.** bulbourethral gland; **e.** anus; **f.** vas deferens; **g.** epididymis; **h.** testis; **i.** scrotum; **j.** foreskin; **k.** glans penis; **l.** penis; **m.** urethra; **n.** vas deferens; **o.** urinary bladder. Path of sperm: testis, epididymis, vas deferens, urethra (in penis). See also Figure 50.3, page 904. **2.** b; **3.** e; **4.** c; **5.** e; **6.** c; **7.** c; **8.** c; **9.** c; **10.** a

Understanding the Terms

a. ovulation; **b.** parthenogenesis; **c.** progesterone; **d.** semen; **e.** gonad

Chapter 51

Testing Yourself

1. b; **2.** b; **3.** b; **4.** a; **5.** e; **6.** b; **7.** e; **8.** e; **9.** d; **10. a.** chorion (contributes to forming placenta where wastes are exchanged for nutrients and oxygen); **b.** amnion (protects and prevents desiccation); **c.** embryo; **d.** allantois (blood vessels become umbilical blood vessels); **e.** yolk sac (first site of blood cell formation); **f.** chorionic villi (embryonic portion of placenta); **g.** maternal portion of placenta; **h.** umbilical cord (connects developing embryo to the placenta). See also Figure 51.11, page 930.

Understanding the Terms

a. induction; **b.** germ layer; **c.** amnion; **d.** homeobox; **e.** gastrula

Appendix B

CLASSIFICATION OF ORGANISMS

Domain Archaea

Prokaryotic, unicellular organisms which lack a membrane-bounded nucleus and reproduce asexually. Many are autotrophic by chemosynthesis; some are heterotrophic by absorption. Most live in extreme or anaerobic environments. Archaea are distinguishable from bacteria by their unique rRNA base sequence; and their distinctive plasma membrane and cell wall chemistry. (519)

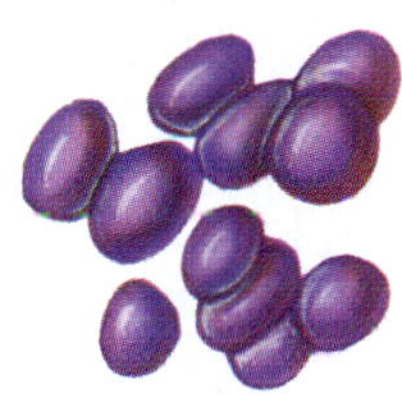

Domain Bacteria

Prokaryotic, unicellular organisms which lack a membrane-bounded nucleus and reproduce asexually. Metabolically diverse being heterotrophic by absorption; autotrophic by chemosynthesis or by photosynthesis. Motile forms move by flagella consisting of a single filament. (517)

Domain Eukarya

Eukaryotic, unicellular to multicellular organisms which have a membrane-bounded nucleus containing several chromosomes. Sexual reproduction is common. Phenotypes and nutrition are diverse; each kingdom has specializations that distinguish it from the other kingdoms. Flagella, if present, have a 9 + 2 organization.

Kingdom Protista

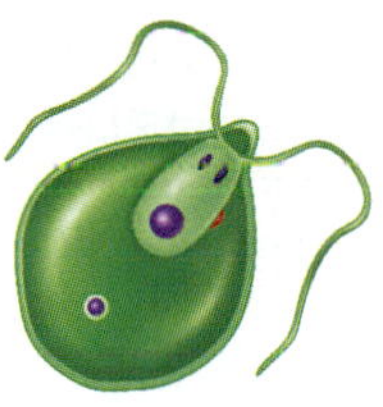

Eukaryotic, unicellular organisms and their immediate multicellular descendants. Asexual reproduction is common, but sexual reproduction as a part of various life cycles does occur. Phenotypically and nutritionally diverse, being either photosynthetic or heterotrophic by various means. Locomotion, if present, utilizes flagella, cilia, or pseudopods.

Algae*

Phylum Chlorophyta: green algae (528)
Phylum Rhodophyta: red algae (530)
Phylum Phaeophyta: brown algae (531)

Diatoms

Phylum Chrysophyta: golden-brown algae (533)

Flagellates

Phylum Pyrrophyta: dinoflagellates (533)
Phylum Euglenophyta: euglenoids (534)
Phylum Zoomastigophora: zooflagellates (535)

Sarcodines

Phylum Rhizopoda: amoeboids (536)
Phylum Foraminifera: foraminiferans (536)
Phylum Actinopoda: radiolarians (536)

Ciliates

Phylum Ciliophora: ciliates (537)

Sporozoans

Phylum Apicomplexa: sporozoans (538)

Slime Molds and Water Molds

Phylum Myxomycota: plasmodial slime molds (539)
Phylum Acrasiomycota: cellular slime molds (539)
Phylum Oomycota: water molds (540)

* *Not a classification category, but added for clarity*

Kingdom Fungi

Multicellular eukaryotes which form nonmotile spores during both asexual and sexual reproduction as a part of the haplontic life cycle. The only multicellular forms of life to be heterotrophic by absorption. They lack flagella in all life cycle stages. Approximately 77,000 species

- Division Zygomycota: zygospore fungi (546)
- Division Ascomycota: sac fungi (549)
- Division Basidiomycota: club fungi (551)
- Division Deuteromycota: imperfect fungi (means of sexual reproduction is not known) (553)

Kingdom Plantae

Multicellular, primarily terrestrial, eukaryotes with well-developed tissues. Plants have an alternation of generations life cycle and are usually photosynthetic. Like green algae, they contain chlorophylls *a* and *b*, carotenoids; store starch in chloroplasts; and have a cell wall which contains cellulose. Approximately 270,000 species.

Nonvascular Plants

- Division Hepatophyta: liverworts (560)
- Division Bryophyta: mosses (560)
- Division Anthocerotophyta: hornworts (560)

Seedless Vascular Plants

- Division Psilotophyta: whisk ferns (564)
- Division Lycopodophyta: club mosses, spike mosses, quillworts (565)
- Division Equisetophyta: horsetails (565)
- Division Pteridophyta: ferns (566)

Gymnosperms

- Division Pinophyta: conifers, such as pines, firs, yews, redwoods, spruces (568)
- Division Cycadophyta: cycads (570)
- Division Ginkgophyta: maidenhair tree (570)
- Division Gnetophyta: gnetophytes (572)

Angiosperms

- Division Magnoliophyta: flowering plants (572)
    - Class Liliopsida: monocots
    - Class Magnoliopsida: dicots

Kingdom Animalia

Multicellular organisms with well-developed tissues that have the diplontic life cycle. Animals tend to be mobile and are heterotrophic by ingestion, generally in a digestive cavity. Complexity varies; the more complex forms have well-developed organ systems. More than a million species have been described.

Invertebrates

- Phylum Porifera: sponges (582)
- Phylum Cnidaria: hydras, jellyfishes (584)
    - Class Anthozoa: sea anemones, corals (585)
    - Class Hydrozoa: *Hydra, Obelia* (585–87)
    - Class Scyphozoa: *Aurelia* (585)
- Phylum Ctenophora: comb jellies, sea walnuts (584)
- Phylum Nemertea: ribbon worms (588)
- Phylum Platyhelminthes: flatworms (588)
    - Class Turbellaria: planarians (588)
    - Class Trematoda: flukes (590)
    - Class Cestoda: tapeworms (591)
- Phylum Nematoda: roundworms: *Ascaris* (593)
- Phylum Rotifera: rotifers (594)
- Phylum Mollusca: molluscs (600)
    - Class Polyplacophora: chitons (600)
    - Class Bivalvia: clams, scallops, oysters, mussels (600)
    - Class Cephalopoda: squids, chambered nautilus, octopus (602)
    - Class Gastropoda: snails, slugs, nudibranchs (603)
- Phylum Annelida: annelids (604)
    - Class Polychaeta: clam worms, tube worms (604)
    - Class Oligochaeta: earthworms (605)
    - Class Hirudinea: leeches (606)
- Phylum Arthropoda: arthropods (606)
    - Subphylum Trilobitomorpha: trilobites (606)
    - Subphylum Crustacea: crustaceans (608)
    - Subphylum Uniramia: millipedes, centipedes, insects (611)
    - Subphylum Chelicerata: spiders, scorpions, horseshoe crabs (613)
- Phylum Echinodermata: echinoderms (618)
    - Class Crinoidea: sea lilies, feather stars (618)
    - Class Asteroidea: sea stars (618)
    - Class Ophiuroidea: brittle stars (618)
    - Class Echinoidea: sea urchins, sand dollars (618)
    - Class Holothuroidea: sea cucumbers (618)
- Phylum Chordata: chordates (619)
    - Subphylum Cephalochordata: lancelets (620)
    - Subphylum Urochordata: tunicates (620)

Vertebrates

- Subphylum Vertebrata (622)
    - Superclass Agnatha: jawless fishes (622)
    - Superclass Gnathostomata: jawed fishes, tetrapods (623)
        - Class Chondrichthyes: cartilaginous fishes (623)
        - Class Osteichthyes: bony fishes (624)
        - Class Amphibia: amphibians (625)
        - Class Reptilia: reptiles (628)
        - Class Aves: birds (632)
        - Class Mammalia
            - Monotremes (634)
            - Marsupials (634)
            - Placental mammals (635)

* Extinct

Glossary

a

abscisic acid (ABA) Plant hormone that causes stomates to close and that initiates and maintains dormancy. 693

abscission Dropping of leaves, fruits, or flowers from a plant. 693

acetyl-CoA Molecule made up of a two-carbon acetyl group attached to coenzyme A. During cellular respiration, the acetyl group enters the Krebs cycle for further breakdown. 132

acetylcholine (ACh) Neurotransmitter active in both the peripheral and central nervous systems. 833

acetylcholinesterase (AChE) Enzyme that breaks down acetylcholine within a synapse. 833

acid Molecules tending to raise the hydrogen ion concentration in a solution and to lower its pH numerically. 28

acid deposition Return to earth as rain or snow of the sulfate or nitrate salts of acids produced by commercial and industrial activities. 433

actin Muscle protein making up the thin filaments in a sarcomere; its movement shortens the sarcomere, yielding muscle contraction. 875

action potential Electrochemical changes that take place across the axomembrane; the nerve impulse. 830

active site Region on the surface of an enzyme where the substrate binds and where the reaction occurs. 103

active transport Use of a plasma membrane carrier protein to move a molecule or ion from a region of lower to higher concentration; it opposes equilibrium and requires energy. 88

adaptation Organism's modification in structure, function, or behavior suitable to the environment. 4, 291

adaptive radiation Evolution of several species from a common ancestor into new ecological or geographical zones. 312

adenine (A) One of four nitrogen-containing bases in nucleotides composing the structure of DNA and RNA. 223

adenosine Portion of ATP and ADP that is composed of the base adenine and the sugar ribose. 50

adhesion junction Junction between cells in which the adjacent plasma membranes do not touch but are held together by intercellular filaments attached to buttonlike thickenings. 93

adipose tissue Connective tissue in which fat is stored. 726

ADP (adenosine diphosphate) Nucleotide with two phosphate groups that can accept another phosphate group and become ATP. 50, 101

adrenal cortex Outer portion of the adrenal gland; secretes hormones such as mineralocorticoid aldosterone and glucocorticoid cortisol. 891

adrenal medulla Inner portion of the adrenal gland; secretes the hormones epinephrine and norepinephrine. 891

adventitious root Fibrous roots which develop from stems or leaves, such as prop roots of corn or holdfast roots of ivy. 650

agglutination Clumping of red blood cells due to a reaction between antigens on red blood cell plasma membranes and antibodies in the plasma. 774

agnathan Vertebrates that lack jaws; includes only the jawless fishes; i.e., the lampreys and hagfishes. 622

aldosterone Hormone secreted by the adrenal cortex that regulates the sodium and potassium ion balance of the blood. 820

algae (sing., alga) Type of protist that carries on photosynthesis; unicellular forms are a part of phytoplankton and multicellular forms are called seaweed. 528

alien species Nonnative species that migrate or are introduced by humans into a new ecosystem; also called exotics. 477

allantois Extraembryonic membrane that accumulates nitrogenous wastes in birds and reptiles and contributes to the formation of umbilical blood vessels in mammals. 930

allele Alternative form of a gene—alleles occur at the same locus on homologous chromosomes. 177

allergen Foreign substance capable of stimulating an allergic response. 776

allopatric speciation Origin of new species between populations that are separated geographically. 312

allosteric site Regulatory binding site on an enzyme that controls the activity of that enzyme. 105

altruism Social interaction that has the potential to decrease the lifetime reproductive success of the member exhibiting the behavior. 372

alveolus In humans, terminal, microscopic, grapelike air sac found in lungs. 803

amino acid Organic molecule having an amino group and an acid group, that covalently bonds to produce peptide molecules. 45

amniocentesis Procedure for removing amniotic fluid surrounding the developing fetus for the testing of the fluid or cells within the fluid. 203

amnion Extraembryonic membrane of birds, reptiles, and mammals that forms an enclosing, fluid-filled sac. 930

amoeboid Cell that moves and engulfs debris with pseudopods. 536

amphibian Member of a class of vertebrates that includes frogs, toads, and salamanders; they are still tied to a watery environment for reproduction. 625

anabolism Metabolic process by which larger molecules are synthesized from smaller ones; anabolic metabolism. 139

analogous structure Structure that has a similar function in separate lineages but differs in anatomy and ancestry. 296, 495

angiosperm Flowering plant; the seeds are borne within a fruit. 572

annual ring Layer of wood (secondary xylem) usually produced during one growing season. 655

anther In flowering plants, pollen-bearing portion of stamen. 573, 701

antheridium Sperm-producing structure, as in the moss life cycle. 560

anthropoid Group of primates that includes monkeys, apes, and humans. 345

antibody Protein produced in response to the presence of an antigen; each antibody combines with a specific antigen. 753, 764

antibody-mediated immunity Specific mechanism of defense in which plasma cells derived from B cells produce antibodies that combine with antigens. 765

anticodon Three-base sequence in a transfer RNA molecule base that pairs with a complementary codon in mRNA. 240

antidiuretic hormone (ADH) Hormone secreted by the posterior pituitary that increases the permeability of the collecting ducts in a kidney. 820

antigen Foreign substance, usually a protein or a polysaccharide, that stimulates the immune system to react, such as to produce antibodies. 753, 764

anus Outlet of the digestive tube. 791

aorta In humans, the major systemic artery that takes blood from the heart to the tissues. 748

apical dominance Influence of a terminal bud in suppressing the growth of lateral buds. 685

apoptosis Programmed cell death involving a cascade of specific cellular events leading to death and destruction of the cell. 69, 152, 260, 765, 926

appendicular skeleton Part of the vertebrate skeleton forming the appendages, shoulder girdle, and hip girdle. 870

appendix In humans, small, tubular appendage that extends outward from the cecum of the large intestine. 791

aquifer Rock layers that contain water and will release it in appreciable quantities to wells or springs. 429

arboreal Living in trees. 342

Archaea One of the three domains of life containing prokaryotic cells that often live in extreme habitats and which have unique genetic, biochemical, and physiological characteristics. 502, 519

archegonium Egg-producing structure as in the moss life cycle. 560

arteriole Vessel that takes blood from an artery to capillaries. 742

artery Blood vessel that transports blood away from the heart. 742

ascus (pl., asci Fingerlike sac in which nuclear fusion, meiosis, and ascospore production occur during sexual reproduction of sac fungi. 549

asexual reproduction Reproduction that requires only one parent and does not involve gametes. 144

aster Short, radiating fibers produced by the centrosomes in animal cells. 146

asymmetry Body plan having no particular symmetry. 580

atom Smallest particle of an element that displays the properties of the element. 18

atomic mass Mass of an atom equal to the number of protons plus the number of neutrons within the nucleus. 19

atomic number Number of protons within the nucleus of an atom. 19

ATP (adenosine triphosphate) Nucleotide with three phosphate groups. The breakdown of ATP into ADP + Ⓟ makes energy available for energy-requiring processes in cells. 48, 50, 101

ATP synthase Enzyme that is part of an ATP synthase complex and functions in the production of ATP in chloroplasts and mitochondria. 107, 118

atrial natriuretic hormone (ANH) Hormone secreted by the heart that increases sodium excretion. 821

atrioventricular valve Heart valve located between an atrium and a ventricle. 744

atrium (pl., atria) Chamber; particularly an upper chamber of the heart lying above a ventricle. 744

australopithecine Any of the first evolved hominids; classified into several species of *Australopithecus*. 348

autoimmune disease Disease that results when the immune system mistakenly attacks the body's own tissues. 776

autonomic system Portion of the peripheral nervous system that regulates internal organs. 831

autosome Any chromosome other than the sex-determining pair. 192, 202

autotroph Organism that can capture energy and synthesize organic molecules from inorganic nutrients. 423

auxin Plant hormone regulating growth, particularly cell elongation; also called indoleacetic acid (IAA). 688

axial skeleton Part of the vertebrate skeleton forming the vertical support or axis, including the skull, the rib cage, and the vertebral column. 868

axillary bud Bud located in the axil of a leaf. 643

axon Elongated portion of a neuron that conducts nerve impulses typically from the cell body to the synapse. 829

b

B lymphocyte Lymphocyte that matures in the bone marrow and, when stimulated by the presence of a specific antigen, gives rise to antibody-producing plasma cells. 764

Bacteria One of the three domains of life containing prokaryotic cells that differ from archaea because they have their own unique genetic, biochemical, and physiological characteristics. 60, 502, 517

bacteriophage Virus that infects bacteria. 221, 267, 510

Barr body Dark-staining body (discovered by M. Barr) in the nuclei of female mammals that contains a condensed, inactive X chromosome. 253

base Molecules tending to lower the hydrogen ion concentration in a solution and raise the pH numerically. 28

basidium (pl., basidia) Clublike structure in which nuclear fusion, meiosis, and basidiospore production occur during sexual reproduction of club fungi. 551

basophil Leukocyte with a granular cytoplasm and that is able to be stained with a basic dye. 762

behavior Observable, coordinated responses to environmental stimuli. 362

benthic division Ocean or lake floor from the high-tide mark to the deepest depths; which supports a unique set of organisms. 463

bicarbonate ion Ion that participates in buffering the blood, and the form in which carbon dioxide is transported in the bloodstream. 805

bilateral symmetry Body plan having two corresponding or complementary halves. 580

bile Secretion of the liver that is temporarily stored and concentrated in the gallbladder before being released into the small intestine, where it emulsifies fat. 788

binary fission Splitting of a parent cell into two daughter cells; serves as an asexual form of reproduction in bacteria. 144, 515

binomial system Assignment of two names to each organism, the first of which designates the genus and second of which is the specific epithet. 490

biodiversity Total number of species, the variability of their genes, and the communities in which they live. 7, 470

biodiversity hotspots Regions of the world that contain unusually large concentrations of species. 471

biogeochemical cycle Circulating pathway of elements such as carbon and nitrogen involving exchange pools, storage areas, and biotic communities. 428

biogeography Study of the geographical distribution of organisms. 286

biological clock Internal mechanism that maintains a biological rhythm in the absence of environmental stimuli. 687

biological magnification Process by which substances become more concentrated in organisms in the higher trophic levels of a food web. 435

biome One of the biosphere's major communities characterized in particular by certain climatic conditions and particular types of plants. 445

biosphere Zone of air, land, and water at the surface of the earth in which living organisms are found. 5, 378, 422

biotic potential Maximum population growth rate under ideal conditions. 381

bipedalism Walking erect on two feet. 347

bird Homeothermic vertebrate characterized by the presence of feathers, having wings often adapted for flight and laying hard-shelled eggs. 632

bivalent Homologous chromosomes, each having sister chromatids that are joined by a nucleoprotein lattice during meiosis; also called tetrad. 161

bivalve Type of mollusc with a shell composed of two valves; includes clams, oysters, and scallops. 600

blade Broad expanded portion of a plant leaf that may be single or compound leaflets. 643

blastocoel Fluid-filled cavity of a blastula. 921

blastocyst Early stage of human embryonic development that consists of a hollow fluid-filled ball of cells. 931

blastula Hollow, fluid-filled ball of cells occurring during animal development prior to gastrula formation. 921

blind spot Region of the retina lacking rods or cones and where the optic nerve leaves the eye. 854

blood Fluid circulated by the heart through a closed system of vessels. 727, 741

blood pressure Force of blood pushing against the inside wall of blood vessels. 749

bony fish Member of a class of vertebrates (class Osteichthyes) containing numerous, diverse fishes, with a bony rather than cartilaginous skeleton. 624

bottleneck effect Cause of genetic drift; occurs when a majority of genotypes is prevented from participating in the production of the next generation as a result of a natural disaster or human interference. 305

brain Ganglionic mass at the anterior end of the nerve cord; in vertebrates the brain is located in cranial cavity of the skull. 827

brain stem In mammals; portion of the brain consisting of the medulla oblongata, pons, and midbrain. 839

bronchiole In terrestrial vertebrates, small tube that conducts air from a bronchus to the alveoli. 803

bronchus (pl., bronchi) In terrestrial vertebrates, branch of the trachea that leads to lungs. 803

brown alga Marine photosynthetic protist with a notable abundance of xanthophyll pigments; this group includes well-known seaweeds of northern rocky shores. 531

budding Asexual form of reproduction whereby a new organism develops as an outgrowth of the body of the parent. 545

buffer Substance or group of substances that tend to resist pH changes of a solution, thus stabilizing its relative acidity and basicity. 29

bursa Saclike, fluid-filled structure, lined with synovial membrane, that occurs near a synovial joint. 871

C

C_3 plant Plant that fixes carbon dioxide via the Calvin cycle; the first stable product of C_3 photosynthesis is a three-carbon compound. 122

C_4 plant Plant that fixes carbon dioxide to produce a C_4 molecule that releases carbon dioxide to the Calvin cycle. 122

Calvin cycle Primary pathway of the light-independent reactions of photosynthesis; converts carbon dioxide to carbohydrate. 119

CAM plant Plant that fixes carbon dioxide at night to produce a C_4 molecule that releases carbon dioxide to the Calvin cycle during the day; CAM stands for crassulacean-acid metabolism. 122

camera-type eye Type of eye found in vertebrates and certain molluscs; a single lens focuses an image on closely-packed photoreceptors. 850

camouflage Method of hiding from predators in which the organism's behavior, form, and pattern of coloration allow it to blend into the background and prevent detection. 408

cancer Malignant tumor whose nondifferentiated cells exhibit loss of contact inhibition, uncontrolled growth, and the ability to invade tissue and metastasize. 152, 258

capillary Microscopic blood vessel; gases and other substances are exchanged across the walls of a capillary between blood and tissue fluid. 742

capsule Gelatinous layer surrounding the cells of blue-green algae and certain bacteria. 60

carbaminohemoglobin Hemoglobin carrying carbon dioxide. 805

carbohydrate Class of organic compounds that includes monosaccharides, disaccharides, and polysaccharides. 37

carbon cycle Continuous process by which carbon circulates in the air, water, and organisms of the biosphere. 430

carbonic anhydrase Enzyme in red blood cells that speeds the formation of carbonic acid from water and carbon dioxide. 805

carcinogen Environmental agent that causes mutations leading to the development of cancer. 258

cardiac conduction system System of specialized cardiac muscle fibers that conducts impulses from the SA node to the chambers of the heart causing them to contract. 747

cardiac cycle One complete cycle of systole and diastole for all heart chambers. 746

cardiac muscle Striated, involuntary muscle tissue found only in the heart. 728

cardiac pacemaker Mass of specialized cardiac muscle tissue that controls the rhythm of the heartbeat; the SA node. 747

carnivore Consumer in a food chain that eats other animals. 423

carpel Ovule-bearing unit that is a part of a pistil. 573, 701

carrier Heterozygous individual who has no apparent abnormality but can pass on an allele for a recessively inherited genetic disorder. 207

carrier protein Protein that combines with and transports a molecule or ion across the plasma membrane. 83, 88

carrying capacity Largest number of organisms of a particular species that can be maintained indefinitely by a given environment. 383

cartilage Connective tissue in which the cells lie within lacunae embedded in a flexible proteinaceous matrix. 727

cartilaginous fish Member of a class of vertebrates (class Chondrichthyes) with a cartilaginous rather than bony skeleton, including sharks, rays, and skates. 623

Casparian strip Layer of impermeable lignin and suberin bordering four sides of root endodermal cells; prevents water and solute transport between adjacent cells. 649, 670

catabolism Metabolic process that breaks down large molecules into smaller ones; catabolic metabolism. 139

catastrophism Belief espoused by Georges Cuvier that periods of catastrophic extinctions occurred, after which repopulation of surviving species took place, giving the appearance of change through time. 284

cell Smallest unit that displays the properties of life; composed of cytoplasm surrounded by a plasma membrane. 56

cell cycle Repeating sequence of events in eukaryotes that involves cell growth and nuclear division; consists of the stages G_1, S, G_2, and M. 150

cell plate Structure across a dividing plant cell that signals the location of new plasma membranes and cell walls. 148

cell theory One of the major theories of biology which states that all organisms are made up of cells; cells are capable of self-reproduction and they come only from pre-existing cells. 56

cell wall Structure that surrounds a plant, protistan, fungal, or bacterial cell and maintains the cell's shape and rigidity. 60, 61, 92

cell-mediated immunity Specific mechanism of defense in which T cells destroy antigen-bearing cells. 768

cellular differentiation Process and developmental stages by which a cell becomes specialized for a particular function. 924

cellular respiration Metabolic reactions that use the energy from carbohydrate or fatty acid or amino acid breakdown to produce ATP molecules. 128

cellular slime mold Free-living amoeboid cells that feed on bacteria by phagocytosis and aggregate to form a plasmodium that produces spores. 539

cellulose Polysaccharide that is the major complex carbohydrate in plant cell walls. 38

centriole Cell organelle, existing in pairs, that occurs in the centrosome and may help organize a mitotic spindle for chromosome movement during animal cell division. 74, 146

centromere Constriction where sister chromatids of a chromosome are held together. 145

centrosome Central microtubule organizing center of cells. In animal cells, it contains two centrioles. 72, 144, 146

cephalization Having a well-recognized anterior head with a brain and sensory receptors. 588, 827

cephalopod Type of mollusc in which a modified foot develops into the head region; includes squids, cuttlefish, octopuses, and nautiluses. 602

cephalothorax Fused head and thorax found in decapods (shrimps, lobsters, crayfish, and crabs). 608, 613

cerebellum In terrestrial vertebrates, portion of the brain that coordinates skeletal muscles to produce smooth, graceful motions. 839

cerebral hemisphere Either of the two lobes of the cerebrum in vertebrates. 840

cerebrospinal fluid Fluid found in the ventricles of the brain, in the central canal of the spinal cord, and in association with the meninges. 839

cerebrum Largest part of the brain in mammals, the cortex receives and responds to sensory information and carries out mental processes. 839

channel protein Protein that forms a channel to allow a particular molecule or ion to cross the plasma membrane. 83

character Any structural, chromosomal, or molecular feature that distinguishes one group from another. 493

character displacement Tendency for characteristics to be more divergent when similar species belong to the same community than when they are isolated from one another. 404

chemical energy Energy associated with the interaction of atoms in a molecule. 98

chemical evolution Increase in the complexity of chemicals over time that could have led to the first cells. 320

chemiosmosis Ability of certain membranes to use a hydrogen ion gradient to drive ATP formation. 107, 118, 135

chemoautotroph Organism able to synthesize organic molecules by using carbon dioxide as the carbon source and the oxidation of an inorganic substance (such as hydrogen sulfide) as the energy source. 516

chemoheterotroph Organism that is unable to produce its own organic molecules and therefore requires organic nutrients in its diet. 516

chemoreceptor Sensory receptor that is sensitive to chemical stimulation—for example, receptors for taste and smell. 848

chitin Strong but flexible nitrogenous polysaccharide found in the exoskeleton of arthropods. 39, 606

chlorophyll Green pigment that absorbs solar energy and is important in algal and plant photosynthesis; occurs as chlorophyll *a* and chlorophyll *b*. 114

chloroplast Membrane-bounded organelle in algae and plants with chlorophyll-containing membranous thylakoids; where photosynthesis takes place. 70, 114

cholesterol One of the major lipids found in animal plasma membranes; makes the membrane impermeable to many molecules. 81

chordate Animals in the phylum Chordata that have a dorsal tubular nerve cord, a notochord, pharyngeal gill pouches and past anal tail at some point in their life cycle. 619

chorion Extraembryonic membrane functioning for respiratory exchange in birds and reptiles; contributes to placenta formation in mammals. 930

chorionic villi sampling Prenatal test in which a sample of chorionic villi cells are removed for diagnostic purposes. 212

chorionic villus Treelike extension of the chorion of the embryo, projecting into the maternal tissues at the placenta. 933

chromatin Network of fibrils consisting of DNA and associated proteins observed within a nucleus that is not dividing. 65, 145

chromosomal mutation Alteration in chromosome structure or number typical of the species. 197

chromosome Structure consisting of DNA complexed with proteins that transmits genetic information from the previous generation of cells and organisms to the next generation. 65, 144

chyme Thick, semiliquid food material that passes from the stomach to the small intestine. 787

ciliary muscle Within the ciliary body of the vertebrate eye, the ciliary muscle controls the shape of the lens. 852

ciliate Complex unicellular protist that moves by means of cilia and digests food in food vacuoles. 537

cilium Short, hairlike projection from the plasma membrane, occurring usually in larger numbers. 74

circadian rhythm Biological rhythm with a 24-hour cycle. 686, 897

cladistic systematics (cladistics) School of systematics that uses derived characters to determine monophyletic groups and construct cladograms. 498

cladogram In cladistics, a branching diagram that shows the relationship among species in regard to their shared, derived characters. 498

class One of the categories, or taxa, used by taxonomists to group species; taxon above the order level. 493

cleavage Cell division without cytoplasmic addition or enlargement; occurs during the first stage of animal development. 920

climate Weather condition of an area including especially prevailing temperature and average daily/yearly rainfall. 442

climax community In ecology, community that results when succession has come to an end. 415

cloaca Posterior portion of the digestive tract in certain vertebrates that receives feces and urogenital products. 626

clonal selection theory States that the antigen selects which lymphocyte will undergo clonal expansion and produce more lymphocytes bearing the same type of receptor. 765

clone Production of identical copies; in organisms, the production of organisms with same genes, in genetic engineering, the production of many identical copies of a gene. 266

club moss Type of seedless vascular plant that are also called ground pine because they appear to be miniature pine trees. 575

cnidarian Invertebrate in the phylum Cnidaria existing as either a polyp or medusa with two tissue layers and radial symmetry. 584

coal Fossil fuel formed millions of years ago from plant material that did not decay. 563

cochlea Spiral-shaped structure of the vertebrate inner ear containing the sensory receptors for hearing. 856

codominance Inheritance pattern in which both alleles of a gene are equally expressed. 188

codon Three-base sequence in messenger RNA that causes the insertion of a particular amino acid into a protein or termination of translation. 237

coelom Body cavity lying between the digestive tract and body wall that is completely lined by mesoderm. 580, 598

coenzyme Nonprotein organic molecule that aids the action of the enzyme to which it is loosely bound. 105

coevolution Joint evolution in which one species exerts selective pressure on the other species. 411, 704

cofactor Nonprotein adjunct required by an enzyme in order to function; many cofactors are metal ions, others are coenzymes. 105

cohesion-tension model Explanation for upward transport of water in xylem based upon transpiration-created tension and the cohesive properties of water molecules. 674

cohort Group of individuals having a statistical factor in common, such as year of birth, in a population study. 384

coleoptile Protective sheath that covers the young leaves of a seedling. 688

collecting duct Duct within the kidney that receives fluid from several nephrons; the reabsorption of water occurs here. 816

collenchyma Plant tissue composed of cells with unevenly thickened walls; supports growth of stems and petioles. 646

colon In humans, the major portion of the large intestine. 791

colony Loose association of cells that remain independent for most functions. 530

commensalism Symbiotic relationship in which one species is benefited, and the other is neither harmed nor benefited. 412

common ancestor Ancestor held in common by at least two lines of descent. 492

community Assemblage of populations interacting with one another within the same environment. 5, 378, 400

compact bone Type of bone that contains osteons consisting of concentric layers of matrix and osteocytes in lacunae. 727, 866

companion cell Cell associated with sieve-tube elements in phloem of vascular plants. 672

competitive exclusion principle Theory that no two species can occupy the same niche. 404

complement system Series of proteins in plasma that form a nonspecific defense mechanism against a microbe invasion; it complements the antigen-antibody reaction. 764

complementary base pairing Hydrogen bonding between particular purines and pyrimidines in DNA. 49, 224

complementary DNA (cDNA) DNA that has been synthesized from mRNA by the action of reverse transcriptase. 267

complete gut Digestive tract that has both a mouth and an anus. 782

compound Substance having two or more different elements united chemically in fixed ratio. 22

compound eye Type of eye found in arthropods; it is composed of many independent visual units. 850

concentration gradient Gradual change in chemical concentration from one point to another. 84

conclusion Statement made following an experiment as to whether the results support or falsify the hypothesis. 10

condensation synthesis Chemical reaction resulting in a covalent bond with the accompanying loss of a water molecule. 36

cone Structure comprised of scales which bear sporangia; pollen cones bear microsporangia and seed cones bear megasporangia. 568

cone cell Photoreceptor in vertebrate eyes that responds to bright light and allows color vision. 853

conidiospore Spore produced by sac and club fungi during asexual reproduction. 549

conjugation Transfer of genetic material from one cell to another. 515, 529

connective tissue Type of animal tissue that binds structures together, provides support and protection, fills spaces, stores fat, and forms blood cells; adipose tissue, cartilage, bone, and blood are types of connective tissue. 726

consumer Organism that feeds on another organism in a food chain; primary consumers eat plants, and secondary consumers eat animals. 423

continental drift Movement of continents with respect to one another over the earth's surfaces. 334

control group Sample that goes through all the steps of an experiment but does not contain the variable being tested; a standard against which results of an experiment are checked. 10

convergent evolution Similarity in structure in distantly related groups due to adaptation to the environment. 495
copulation Sexual union between a male and a female. 902
coral reef Area of biological abundance found in shallow, warm tropical waters on and around coral formations. 464
corepressor Molecule that binds to a repressor, allowing the repressor to bind to an operator in a repressible operon. 249
cork Outer covering of bark of trees; made of dead cells that may be sloughed off. 645
cork cambium Lateral meristem that produces cork. 655
cornea Transparent, anterior portion of the outer layer of the eyeball. 852
corpus luteum Follicle that has released an egg and increases its secretion of progesterone. 909
cortex In plants, ground tissue bounded by the epidermis and vascular tissue in stems and roots; in animals, outer layer of an organ such as the cortex of the kidney or adrenal gland. 649
cotyledon Seed leaf for embryo of a flowering plant; provides nutrient molecules for the developing plant before photosynthesis begins. 644
coupled reactions Reactions that occur simultaneously; one is an exergonic reaction that releases energy and the other is an endergonic reaction that requires an input of energy in order to occur. 100
covalent bond Chemical bond in which atoms share one pair of electrons. 23
cranial nerve Nerve that arises from the brain. 834
creatine phosphate High energy phosphate molecule found in vertebrate muscles used to generate ATP molecules for muscle contraction. 875
crenation In animal cells, shriveling of the cell due to water leaving the cell when the environment is hypertonic. 87
cristae (sing., crista) Short, fingerlike projections formed by the folding of the inner membrane of mitochondria. 71
crossing-over Exchange of segments between nonsister chromatids of a bivalent during meiosis. 162
cuticle Waxy layer covering the epidermis of plants that protects the plant against water loss and disease-causing organisms. 558, 675
cyanobacterium Photosynthetic bacterium that contains chlorophyll and releases oxygen; formerly called a blue-green alga. 517
cycad Type of gymnosperm with palmate leaves and massive cones; cycads are most often found in the tropics and subtropics. 570
cyclic electron pathway Portion of the light-dependent reaction that involves only photosystem I and generates ATP. 116
cyclin Protein that cycles in quantity as the cell cycle progresses; combines with and activates the kinases that function to promote the events of the cycle. 150
cyst In protists and invertebrates, resting stage that contains reproductive bodies or embryos. 526, 591
cytokine Type of protein secreted by a T lymphocyte that attacks viruses, virally infected cells, and cancer cells. 768
cytokinesis Division of the cytoplasm following mitosis and meiosis. 145
cytokinin Plant hormone that promotes cell division; often works in combination with auxin during organ development in plant embryos. 692
cytoplasm Contents of a cell between the nucleus (nucleoid) region of bacteria and the plasma membrane. 60
cytosine (C) One of four nitrogen-containing bases in nucleotides composing the structure of DNA and RNA; pairs with guanine. 223
cytoskeleton Internal framework of the cell, consisting of microtubules, actin filaments, and intermediate filaments. 72
cytosol Semifluid medium between the nucleus and the plasma membrane. 60
cytotoxic T cell T lymphocyte that attacks and kills antigen-bearing cells. 768

d

datum (pl., data) Fact or piece of information collected through observation and/or experimentation. 10
day-neutral plant Plant whose flowering is not dependent on day length, i.e., tomato and cucumber. 694
deamination Removal of an amino group ($—NH_2$) from an amino acid or other organic compound. 139
decapod Type of crustacean in which the thorax bears five pairs of walking legs; includes shrimps, lobsters, crayfish, and crabs. 608
decomposition The breakdown of organic matter accompanied by the release of inorganic compounds; a means by which nutrients cycle in ecosystems. 423
deductive reasoning Process of logic and reasoning, using "if . . . then" statements. 9
delayed allergic response Allergic response initiated at the site of the allergen by sensitized T cells, involving macrophages and regulated by cytokines. 776
demographic transition Due to industrialization, a decline in the birthrate following a reduction in the death rate so that the population growth rate is lowered. 393
denatured Loss of an enzyme's normal shape so that it no longer functions; caused by a less than optimal pH and temperature. 47, 104
dendrite Part of a neuron that sends signals toward the cell body. 829
denitrification Conversion of nitrate or nitrite to nitrogen gas by bacteria in soil. 432
dense fibrous connective tissue Type of connective tissue containing many collagen fibers packed together; found in tendons and ligaments, for example. 726
density-dependent factor Biotic factor, such as disease and competition, that affects population size according to the population's density. 386
density-independent factor Abiotic factors, such as fire and flood, that affects population size independent of the population's density. 386
deoxyribose Pentose sugar found in DNA that has one less hydroxyl group than ribose. 37
dependent variable Result or change that occurs when an experimental variable is utilized in an experiment. 10
derived character Structural, physiological, or behavioral trait that is present in a specific lineage and is not present in the common ancestor for several lineages. 494
dermis In mammals, thick layer of the skin underlying the epidermis. 731
detrital food web Complex pattern of interlocking and crisscrossing food chains that begins with a population of detritivores. 427
detritivore Organism that feeds on freshly dead or partially decomposed organic matter. 423
detritus Organic matter produced by decomposition of substances such as tissues and animal wastes. 423
deuterostome Group of coelomate animals in which the second embryonic opening is associated with the mouth; the first embryonic opening, the blastopore, is associated with the anus. 598, 618
diaphragm In mammals, dome-shaped muscularized sheet separating the thoracic cavity from the abdominal cavity. 801
diarrhea Excessively frequent bowel movements. 791
diastole Relaxation period of a heart chamber during the cardiac cycle. 746
diatom Golden brown alga with a cell wall in two parts, or valves; significant part of phytoplankton. 533
dicotyledon Flowering plant group; members have two embryonic leaves, (cotyledons) net-veined leaves, cylindrical arrangement of vascular bundles, flower parts in fours or fives, and other characteristics. 573
diencephalon In vertebrates, portion of the brain in the region of the third ventricle that includes the thalamus and hypothalamus. 839
differentially permeable Ability of plasma membranes to regulate the passage of substances into and out of the cell, allowing some to pass through and preventing the passage of others. 84
diffusion Movement of molecules or ions from a region of higher to lower concentration; it requires no energy and tends to lead to an equal distribution. 85
dikaryotic Having two haploid nuclei that stem from different parent cells; during sexual reproduction sac and club fungi have dikaryotic cells. 545
dinoflagellate Photosynthetic unicellular protist, with two flagella, one whiplash and the other located within a groove between protective cellulose plates; significant part of phytoplankton. 533
diploid (2n) number Cell condition in which two of each type of chromosome are present. 145, 160

directional selection Outcome of natural selection in which an extreme phenotype is favored, usually in a changing environment. 306

disaccharide Sugar that contains two units of a monosaccharide; e.g., maltose. 37

disruptive selection Outcome of natural selection in which the two extreme phenotypes are favored over the average phenotype, leading to more than one distinct form. 308

distal convoluted tubule Final portion of a nephron that joins with a collecting duct; associated with tubular secretion. 816

DNA (deoxyribonucleic acid) Nucleic acid polymer produced from covalent bonding of nucleotide monomers that contain the sugar deoxyribose; the genetic material of nearly all organisms. 48, 220

DNA fingerprinting Using DNA fragment lengths, resulting from restriction enzyme cleavage, to identify particular individuals. 269

DNA ligase Enzyme that links DNA fragments; used during production of recombinant DNA to join foreign DNA to vector DNA. 266

DNA polymerase During replication, an enzyme that joins the nucleotides complementary to a DNA template. 226

DNA repair enzyme One of several enzymes that restore the original base sequence in an altered DNA strand. 229

DNA replication Synthesis of a new DNA double helix prior to mitosis and meiosis in eukaryotic cells and during prokaryotic fission in prokaryotic cells. 226

DNA-DNA hybridization Method to determine relatedness by allowing single DNA strands from two different species to join and thereafter observing how well they joined. 496

domain Largest of the categories, or taxa, used by taxonomists to group species; the three domains are Archaea, Bacteria, and Eukarya. 493

dominance hierarchy Organization of animals in a group that determines the order in which the animals have access to resources. 367

dominant allele Allele that exerts its phenotypic effect in the heterozygote; it masks the expression of the recessive allele. 177

dormancy In plants, a cessation of growth under conditions that seem appropriate for growth. 690

dorsal-root ganglion Mass of sensory neuron cell bodies located in the dorsal root of a spinal nerve. 834

double fertilization In flowering plants, one sperm fuses with the egg and a second sperm fuses with the polar nucleus of an embryo sac. 706

doubling time Number of years it takes for a population to double in size. 393

duodenum First part of the small intestine where chyme enters from the stomach. 788

e

ecological niche Role an organism plays in its community, including its habitat and its interactions with other organisms. 403

ecological pyramid Pictorial graph based on the biomass, number of organisms, or energy content of various trophic levels in a food web—from the producer to the final consumer populations. 427

ecological succession The gradual replacement of communities in an area following a disturbance (secondary succession) or the creation of new soil (primary succession). 414

ecology Study of the interactions of organisms with other organisms and with the physical and chemical environment. 378

ecosystem Biological community together with the associated abiotic environment; characterized by a flow of energy and a cycling of inorganic nutrients. 5, 378, 422

ectoderm Outermost primary tissue layer of an animal embryo that gives rise to the nervous system and the outer layer of the integument. 921

ectothermic Having a body temperature that varies according to the environmental temperature. 627

edema Swelling due to tissue fluid accumulation in the intercellular spaces. 760

edge effect Edges around a landscape patch have a slightly different habitat than favorable habitat in the interior of the patch. 481

effector Muscle or gland that receives signals from motor fibers and thereby allows an organism to respond to environmental stimuli. 825

electrocardiogram (ECG) Recording of the electrical activity associated with the heartbeat. 747

electromagnetic spectrum Solar radiation divided on the basis of wavelength, with gamma rays having the shortest wavelength and radio waves having the longest wavelength. 112

electron Negative subatomic particle, moving about in an energy level around the nucleus of an atom. 18

electron shells Concentric energy levels in which electrons orbit. 20

electron transport system Passage of electrons along a series of membrane-bounded carrier molecules from a higher to lower energy level; the energy released is used for the synthesis of ATP. 106, 116, 129

electronegativity The ability of an atom to attract electrons toward itself in a chemical bond. 24

element Substance that cannot be broken down into substances with different properties; composed of only one type atom. 18

elephantiasis A swelling of the limbs due to blockage of lymphatic vessels by parasitic filarial roundworms. 593

embryo Stage of a multicellular organism that develops from a zygote before it becomes free living; in seed plants the embryo is part of the seed. 706, 920

embryo sac Megagametophyte of flowering plants. 562, 706

embryonic disk During human development, flattened area during gastrulation from which the embryo arises. 932

embryonic period From approximately the second to the eighth week of human development, during which the major organ systems are organized. 930

emulsification Breaking up of fat globules into smaller droplets by the action of bile salts or any other emulsifier. 788

endergonic reaction Chemical reaction that requires an input of energy; opposite of exergonic reaction. 100

endocrine gland Ductless organ that secretes (a) hormone(s) into the bloodstream. 724

endocytosis Process by which substances are moved into the cell from the environment by phagocytosis (cellular eating) or pinocytosis (cellular drinking; includes receptor-mediated endocytosis). 90

endoderm Innermost primary tissue layer of an animal embryo that gives rise to the linings of the digestive tract and associated structures. 921

endodermis Internal plant root tissue forming a boundary between the cortex and the vascular cylinder. 649

endometrium Mucous membrane lining the interior surface of the uterus. 908

endoplasmic reticulum (ER) System of membranous saccules and channels in the cytoplasm, often with attached ribosomes. 67

endoskeleton Protective, internal skeleton, as in vertebrates. 864

endosperm In flowering plants, nutritive storage tissue that is derived from the fusion of a sperm cell and polar nuclei in the embryo sac. 706

endospore Spore formed within a cell; certain bacteria form endospores. 515

endosymbiotic hypothesis Possible explanation of the evolution of eukaryotic organelles by phagocytosis of prokaryotes. 64, 328

energy Capacity to do work and bring about change; occurs in a variety of forms. 3, 20, 98

energy of activation Energy that must be added in order for molecules to react with one another. 102

enterocoelom Body cavity that forms by the fusion of a pair of mesodermal pouches from the wall of the primitive gut. 597

entropy Measure of disorder or randomness. 98

environmental resistance Sum total of factors in the environment that limit the numerical increase of a population in a particular region. 381

enzymatic protein Protein that catalyzes a specific reaction. 83

enzyme Organic catalyst, usually a protein, that speeds up a reaction in cells due to its particular shape. 102

enzyme inhibition Means by which cells regulate enzyme activity; there is competitive and noncompetitive inhibition. 105

epidermis In mammals, the outer, protective layer of the skin; in plants, tissue that covers roots and leaves and stems of nonwoody organisms. 645, 731

epiglottis Structure that covers the glottis, the air-tract opening, during the process of swallowing. 803

epiphyte Plant that takes its nourishment from the air because its placement in other plants gives it an aerial position. 450, 671

epistasis Inheritance pattern in which one gene masks the expression of another gene that is at a different locus and is independently inherited. 189

epithelial tissue Tissue that lines hollow organs and covers surfaces. 724

esophagus Muscular tube for moving swallowed food from the pharynx to the stomach. 786

estrogen Female sex hormone that helps maintain sexual organs and secondary sexual characteristics. 909

estuary Portion of ocean located where a river enters and fresh water mixes with salt water. 458

ethylene Plant hormone that causes ripening of fruit and is also involved in abscission. 693

euchromatin Chromatin that is extended and accessible for transcription. 252

euglenoid Flagellated and flexible freshwater unicellular protists that usually contains chloroplasts and has a semirigid cell wall. 534

Eukarya One of the three domains of life consisting of organisms with eukaryotic cells and further classified into the kingdoms Protista, Fungi, Plantae, and Animalia. 502

eukaryotic cell Type of cell that has a membrane-bounded nucleus and membranous organelles; found in organisms within the domain Eukarya. 60

eutrophication Enrichment of water by inorganic nutrients used by phytoplankton; often overenrichment caused by human activities leading to excessive bacterial growth and oxygen depletion. 435, 456

evolution Descent of organisms from common ancestors with the development of genetic and phenotypic changes over time that make them more suited to the environment. 4, 283

excretion Elimination of metabolic wastes by an organism at exchange boundaries such as the plasma membrane of unicellular organisms and excretory tubules of multicellular animals. 811

exergonic reaction Chemical reaction that releases energy; opposite of endergonic reaction. 100

exocrine gland Gland that secretes its product to an epithelial surface directly or through ducts. 724

exon Coding segment of an eukaryotic gene that is expressed; exons are separated by introns. 239

exoskeleton Protective external skeleton, as in arthropods. 864

experiment Artificial situation devised to test a hypothesis. 9

experimental variable Component that is tested in an experiment by manipulating it and observing the results. 10

expiration Act of expelling air from the lungs; exhalation. 801

exponential growth Growth, particularly of a population, in which the increase occurs in the same manner as compound interest. 381

extinction Total disappearance of a species or higher group. 293, 329

extraembryonic membrane Membrane that is not a part of the embryo but is necessary to the continued existence and health of the embryo. 903, 930

f

facilitated transport Passive transfer of a substance into or out of a cell along a concentration gradient by a process that requires a carrier. 88

facultative anaerobe Prokaryote that is able to grow in either the presence or the absence of gaseous oxygen. 516

FAD Flavin adenine dinucleotide; a coenzyme of oxidation-reduction that becomes $FADH_2$ as oxidation of substrates occurs and then delivers electrons to the electron transport system in mitochondria during cellular respiration. 128

fall overturn Mixing process that occurs in fall in stratified lakes whereby the oxygen-rich top waters mix with nutrient-rich bottom waters. 456

falsify To show a hypothesis to be untrue. 10

family One of the categories, or taxa, used by taxonomists to group species; taxon above the genus level. 493

fat Organic molecule that contains glycerol and fatty acids and is found in adipose tissue of vertebrates. 40

fate map Diagram that traces the differentiation of cells during development from their origin to their final structure and function. 926

fatty acid Molecule that contains a hydrocarbon chain and ends with an acid group. 40

feather One of the light, horny epidermal outgrowths that form the external covering of the body of birds and the greater part of the surface of their wings. 632

fermentation Anaerobic breakdown of glucose that results in a gain of two ATP and end products such as alcohol and lactate. 129, 137

fertilization Fusion of sperm and egg nuclei producing a zygote which develops into a new individual. 706, 920

fibroblast Connective tissue cell that synthesizes fibers and ground substance. 726

fibrous root system In most monocots, a mass of similarly sized roots that cling to the soil. 650

filament End to end chains of cells that form as cell division occurs in only one plane; in plants, the elongated stalk of a stamen. 529, 701

fin In fish and other aquatic animals, membranous, winglike or paddlelike process used to propel, balance, or guide the body. 623

first messenger Chemical signal such as a peptide hormone that binds to a plasma membrane receptor and alters the metabolism of a cell because a second messenger is activated. 883

fitness Ability of an organism to reproduce and pass its genes to the next fertile generation; measured against the ability of other organisms to reproduce in the same environment. 288

five-kingdom system System of classification that contains the kingdoms Monera, Protista, Plantae, Animalia, and Fungi. 501

flagellum (pl., flagella) Slender, long extension used for locomotion by some bacteria, protozoans, and sperm. 60, 74, 514

flame cell Found along excretory tubules of planarians; functions in propulsion of fluid through the excretory canals and out of the body. 815

flower Reproductive organ of a flowering plant, consisting of several kinds of modified leaves arranged in concentric rings and attached to a modified stem called the receptacle. 700

fluid-mosaic model Model for the plasma membrane based on the changing location and pattern of protein molecules in a fluid phospholipid bilayer. 80

follicle Structure in the ovary of animals that contains an oocyte; site of oocyte production. 909

follicular phase First half of the ovarian cycle, during which the follicle matures and much estrogen (and some progesterone) is produced. 910

fontanel Membranous region located between certain cranial bones in the skull of a vertebrate fetus or infant. 868

food chain The order in which one population feeds on another in an ecosystem, thereby showing the flow of energy from a detritivore (detrital food chain) or producer (grazing food chain) to final consumer. 427

food web In ecosystems, complex pattern of interlocking and crisscrossing food chains. 427

foramen magnum Opening in the occipital bone of the vertebrate skull through which the spinal cord passes. 868

fossil Any past evidence of an organism that has been preserved in the earth's crust. 324

fossil fuel Fuels such as oil, coal, and natural gas that are the result of partial decomposition of plants and animals coupled with exposure to heat and pressure for millions of years. 430

founder effect Cause of genetic drift due to colonization by a limited number of individuals who, by chance, have different gene frequencies than the parent population. 306

fovea centralis Region of the retina consisting of densely packed cones that is responsible for the greatest visual acuity. 853

frameshift mutation Insertion or deletion of at least one base so that reading frame of the corresponding mRNA changes. 259

free energy Useful energy in a system that is capable of performing work. 100

fructose Hexose monosaccharide frequently found in fruits. 37

fruit Flowering plant structure consisting of one or more ripened ovaries that usually contain seeds. 573, 708

fruiting body Spore-producing and spore-disseminating structure found in sac and club fungi. 549

functional group A specific cluster of atoms attached to the carbon skeleton of organic molecules that enters into reactions and behaves in a predictable way. 35

fungus Saprotrophic decomposer; the body is made up of filaments called hyphae that form a mass called a mycelium. 544

g

gallbladder Organ attached to the liver that serves to store and concentrate bile. 788

gametangia Multicellular sex organs; plants, but not green algae, have gametangia. 559

gamete Haploid sex cell; e.g., egg and sperm. 160

gametophyte Haploid generation of the alternation of generations life cycle of a plant; produces gametes that unite to form a diploid zygote. 559, 700

ganglion Collection or bundle of neuron cell bodies usually outside the central nervous system. 834

gap analysis Use of computers to discover places where biodiversity is high outside of preserved areas. 481

gap junction Junction between cells formed by the joining of two adjacent plasma membranes; it lends strength and allows ions, sugars, and small molecules to pass between cells. 93

gastrovascular cavity Blind digestive cavity in animals that have a sac body plan. 584

gastrula Stage of animal development during which the germ layers form, at least in part, by invagination. 921

gene Unit of heredity existing as alleles on the chromosomes; in diploid organisms typically two alleles are inherited—one from each parent. 4

gene flow Sharing of genes between two populations through interbreeding. 304

gene locus Specific location of a particular gene on homologous chromosomes. 177

gene pool Total of all the genes of all the individuals in a population. 302

gene therapy Correction of a detrimental mutation by the addition of new DNA and its insertion in a genome. 275

genetic disorder An abnormal phenotype often resulting in a medical condition. 207

genetic drift Mechanism of evolution due to random changes in the allelic frequencies of a population; more likely to occur in small populations or when only a few individuals of a large population reproduce. 305

genetic engineering Alteration of genomes for medical or industrial purposes. 229, 258, 270

genetic mutation Altered gene whose sequence of bases differs from the previous sequence. 229

genome Full set of genetic information within an organism or a virus. 251, 267

genomic imprinting Inheritance pattern that differs depending upon the sex of the parent passing it on. 208

genomic library Collection of engineered bacteriophages or viruses that together carry all of the genes of the species. 267

genotype Genes of an organism for a particular trait or traits; often designated by letters, for example, *BB* or *Aa*. 177

genus One of the categories, or taxa, used by taxonomists to group species; contains those species that are most closely related through evolution. 493

geological timescale History of the earth based on correlations between rocks (or the fossils contained in them) and time periods of the past. 326

germ layer Primary tissue layer of a vertebrate embryo; namely, ectoderm, mesoderm, or endoderm. 921

germination Beginning of growth of a seed, spore, or zygote, especially after a period of dormancy. 694

gibberellin Plant hormone promoting increased stem growth; also involved in flowering and seed germination. 690

gill Respiratory organ in most aquatic animals; in fish, an outward extension of the pharynx. 619, 799

girdling Removing a strip of bark from around a tree. 678

global warming Predicted increase in the earth's temperature, due to human activities that promote the greenhouse effect. 430, 478

glomerular capsule Cuplike structure that is the initial portion of a nephron. 816

glomerular filtration Movement of small molecules from the glomerulus into the glomerular capsule due to the action of blood pressure. 818

glomerulus Capillary network within a glomerular capsule of a nephron. 817

glottis Opening for airflow in the larynx. 802

glucose Six-carbon sugar that organisms degrade as a source of energy during cellular respiration. 37

glycerol Three-carbon carbohydrate with three hydroxyl groups attached that is a component of fats and oils. 40

glycogen Storage polysaccharide, found in animals, that is composed of glucose molecules joined in a linear fashion but having numerous branches. 38

glycolipid Lipid in plasma membranes that bears a carbohydrate chain attached to a hydrophobic tail. 81

glycolysis Anaerobic breakdown of glucose that results in a gain of two ATP and the end product pyruvate. 129, 130

glycoprotein Protein in plasma membranes that bears a carbohydrate chain. 82

gnathostomate Jawed vertebrates; includes jawed fishes and all tetrapods. 623

Golgi apparatus Organelle consisting of saccules and vesicles that processes, packages, and distributes molecules about or from the cell. 68

gonad Organ that produces gametes; the ovary produces eggs, and the testis produces sperm. 896, 902

granum Stack of chlorophyll-containing thylakoids in a chloroplast. 71, 114

gravitropism Growth response of roots and stems of plants to the earth's gravity; roots demonstrate positive gravitropism, and stems demonstrate negative gravitropism. 685

gray crescent Gray area that appears in an amphibian egg after being fertilized by the sperm, thought to contain chemical signals that turn on the genes that control development. 925

gray matter Nonmyelinated axons and cell bodies in the central nervous system. 838

grazing food web Complex pattern of interlocking and crisscrossing food chains that begins with a population of photosynthesizers serving as a producer. 427

green alga Member of a diverse group of photosynthetic protists that contains chlorophylls *a* and *b* and has other biochemical characteristics like those of plants. 528

greenhouse effect Reradiation of solar heat toward the earth, caused by gases such as carbon dioxide, methane, nitrous oxide, water vapor, ozone, and nitrous oxide in the atmosphere. 431

ground tissue Tissue that constitutes most of the body of a plant; consists of parenchyma, collenchyma, and sclerenchyma cells which function in storage, basic metabolism, and support. 645

growth factor Chemical messenger that promotes the growth of development of particular types of cells. 260

guanine (G) One of four nitrogen-containing bases in nucleotides composing the structure of DNA and RNA; pairs with cytosine. 223

guard cell One of two cells that surround a leaf stomate; changes in the turgor pressure of these cells cause the stomate to open or close. 676

guttation Liberation of water droplets from the edges and tips of leaves. 674

gymnosperm Type of woody seed plant in which the seeds are not enclosed by fruit and are usually borne in cones such as those of the conifers. 568

h

habitat Place where an organism lives and is able to survive and reproduce. 378, 403

habitat corridor Strips of land that allow organisms to move from one fragment of an ecosystem to another or one ecosystem to another. 471

halophile Type of archaea that lives in extremely salty habitats. 520

haploid (n) number Cell condition in which only one of each type of chromosome is present. 145, 160

Hardy-Weinberg law Law stating that the gene frequencies in a population remain stable if evolution does not occur due to nonrandom mating, selection, migration, and genetic drift. 302

heart Muscular organ whose contraction causes blood to circulate in the body of an animal. 744

helper T cell Secretes lymphokines, which stimulate all kinds of immune cells. 768

heme Iron-containing group found in hemoglobin. 805

hemocoel Residual coelom found in arthropods that is filled with hemolymph. 609

hemoglobin Iron-containing respiratory pigment occurring in vertebrate red blood cells and in blood plasma of some invertebrates. 763, 805

hemolymph Circulatory fluid which is a mixture of blood and interstitial fluid; seen in animals that have an open circulatory system, such as molluscs and arthropods. 741

herbaceous plant Plant that lacks persistent woody tissue. 572, 652

herbivore Primary consumer in a grazing food chain; a plant eater. 423

hermaphroditic Type of animal that has both male and female sex organs. 589

heterochromatin Highly compacted chromatin that is not accessible for transcription. 252

heterotroph Organism that cannot synthesize organic compounds from inorganic substances and therefore must take in organic food. 423

heterozygous Possessing unlike alleles for a particular trait. 177

hexose Six-carbon sugar. 37

histamine Substance, produced by basophils in blood and mast cells in connective tissue, that causes capillaries to dilate. 762

HLA (human leukocyte associated) antigen Protein in a plasma membrane that identifies the cell as belonging to a particular individual and acts as an antigen in other organisms. 768

holozoic Obtaining nourishment by ingesting solid food particles. 537

homeobox 180 nucleotide sequence located in all homeotic genes. 929

homeostasis Maintenance of normal internal conditions in a cell or an organism by means of self-regulating mechanisms. 3, 734

homeothermic Maintenance of a uniform body temperature independent of the environmental temperature. 632

homeotic genes Genes that control the overall body plan by controlling the fate of groups of cells during development. 928

Homo sapiens Modern humans. 353

homologous chromosome Member of a pair of chromosomes that are alike and come together in synapsis during prophase of the first meiotic division; a homologue. 160

homologous structure In evolution, a structure that is similar in different types of organisms because these organisms are derived from a common ancestor. 296, 495

homologue Member of a homologous pair of chromosomes. 160

homozygous Possessing two identical alleles for a particular trait. 177

horizons A major layer of soil visible in vertical profile; topsoil is the A horizon. 669

hormone Chemical messenger produced in one part of the body that controls the activity of other parts. 688, 882

host Organism that provides nourishment and/or shelter for a parasite. 410

human chorionic gonadotropin (HCG) Gonadotropin hormone produced by the chorion that functions to maintain the uterine lining. 932

human immunodeficiency virus (HIV) Virus responsible for AIDS. 914

humus Decomposing organic matter in the soil. 668

hybridization Crossing of different species. 714

hydrogen bond Weak bond that arises between a slightly positive hydrogen atom of one molecule and a slightly negative atom of another molecule or between parts of the same molecule. 24

hydrogen ion Hydrogen atom that has lost its electron and therefore bears a positive charge (H^+). 28

hydrolysis Splitting of a compound by the addition of water, with the H^+ being incorporated in one fragment and the OH^- in the other. 36

hydrophilic Type of molecule that interacts with water by dissolving in water and/or by forming hydrogen bonds with water molecules. 26, 35

hydrophobic Type of molecule that does not interact with water because it is nonpolar. 35

hydroponics Technique for growing plants by suspending them with their roots in a nutrient solution. 666

hydrostatic skeleton Fluid-filled body compartment which provides support for muscle contraction resulting in movement; seen in cnidarians, flatworms, roundworms, and segmented worms. 864

hydrothermal vent Hot springs in the sea floor along ocean ridges where heated seawater and sulfate react to produce hydrogen sulfide; here chemosynthetic bacteria support a community of varied organisms. 463

hydroxide ion One of two ions that results when a water molecule dissociates; it has gained an electron and therefore bears a negative charge (OH^-). 28

hypertonic solution Higher solute concentration (less water) than the cytosol of a cell; causes cell to lose water by osmosis. 87

hypha (pl., hyphae) Filament of the vegetative body of a fungus. 544

hypothalamus In vertebrates, part of the brain that helps regulate the internal environment of the body; for example, involved in control of heart rate, body temperature, water balance. 839, 886

hypothesis Supposition that is established by reasoning after consideration of available evidence; it can be tested by obtaining more data, often by experimentation. 9

hypotonic solution Lower solute (more water) concentration than the cytosol of a cell; causes cell to gain water by osmosis. 86

i

immediate allergic response Allergic response that occurs within seconds of contact with an allergen, caused by the attachment of the allergen to IgE antibodies. 776

immunity Ability of the body to protect itself from foreign substances and cells, including disease-causing agents. 762

immunization Strategy for achieving immunity to the effects of specific disease-causing agents. 772

immunoglobulin (Ig) Globular plasma protein that functions as an antibody. 766

implantation In placental mammals, the embedding of an embryo at the blastocyst stage into the endometrium of the uterus. 930

imprinting Learning to make a particular response to only one type of animal or object. 365

inclusive fitness Fitness that results from personal reproduction and from helping nondescendant relatives reproduce. 372

incomplete dominance Inheritance pattern in which the offspring has an intermediate phenotype as when a red-flowered plant and a white-flowered plant produce pink-flowered offspring. 188

incomplete gut Digestive tract that has a single opening, usually called a mouth. 782

independent assortment Alleles of unlinked genes segregate independently of each other during meiosis so that the gametes contain all possible combinations of alleles. 163

induced-fit model Change in the shape of an enzyme's active site which enhances the fit between the enzyme's active site and its substrate(s). 103

inducer Molecule that brings about activity of an operon by joining with a repressor and preventing it from binding to the operator. 250

induction Ability of a chemical or a tissue to influence the development of another tissue. 925

inductive reasoning Using specific observations and the process of logic and reasoning to arrive at a hypothesis. 9

industrial melanism Increased frequency of darkly pigmented (melanic) forms in a population when soot and pollution make lightly pigmented forms easier for predators to see against a pigmented background. 303

inflammatory reaction Tissue response to injury that is characterized by redness, swelling, pain, and heat. 762

inheritance of acquired characteristics Lamarckian belief that characteristics acquired during the lifetime of an organism can be passed on to offspring. 284

inorganic molecule Type of molecule that is not an organic molecule; not derived from a living organism. 34

insect Type of arthropod, the head has antennae, compound eyes, and simple eyes; a thorax has three pairs of legs and often wings; and an abdomen has internal organs. 611

inspiration Act of taking air into the lungs; inhalation. 801

interferon Antiviral agent produced by an infected cell that blocks the infection of another cell. 764

interleukin Cytohine produced by macrophages and T lymphocytes that functions as a metabolic regulator of the immune response. 773

interneuron Neuron, located within the central nervous system, conveying messages between parts of the central nervous system. 829

internode In vascular plants, the region of a stem between two successive nodes. 643

interphase Stages of the cell cycle (G_1, S, G_2) during which growth and DNA synthesis occur when the nucleus is not actively dividing. 150

interspecific competition Similar species trying to occupy the same niche in an ecosystem compete with one another for a share of resources and in this way the number of niches increases. 404

intestinal enzymes Enzymes, produced by the epithelial cells on the surface of villi, which function in the digestion of small organic molecules. 788

intrinsic rate of natural increase Maximum per capita rate of population increase under ideal conditions. 380

intron Noncoding region of an eukaryotic gene; introns are transcribed but removed from mRNA before mRNA leaves the nucleus. 239

invertebrate Referring to an animal without a serial arrangement of vertebrae. 580

ion Charged particle that carries a negative or positive charge. 22

ionic bond Chemical bond in which ions are attracted to one another by opposite charges. 22

isogamy Sexual reproduction by means of gametes that look alike; there is no definite egg and sperm. 528

isomer Molecules with the same molecular formula but different structure, and therefore shape. 35

isotonic solution Solution that is equal in solute concentration to that of the cytosol of a cell; causes cell to neither lose nor gain water by osmosis. 86

isotopes Atoms having the same atomic number but a different atomic mass due to the number of neutrons. 19

j

jaw Tooth-bearing bone of the head. 623

jawless fish Type of fish that has no jaws, includes today's hagfishes and lampreys. 622

joint Articulation between two bones of a skeleton. 871

jointed appendages Freely moving appendages of arthropods. 606

k

***k*-selection** A favorable life-history strategy under stable environmental conditions characterized by the production of a few offspring with much attention given to offspring survival. 391

karyotype Chromosomes arranged by pairs according to their size, shape, and general appearance in mitotic metaphase. 202

keystone species Species whose activities have a significant role in determining community structure. 480

kidney One of the paired organs of the vertebrate urinary system that regulates the chemical composition of the blood and produces a waste product called urine. 816

kinase Any one of several enzymes that phosphorylate their substrates. 150

kinetic energy Energy associated with motion. 98

kingdom One of the categories, or taxa, used by taxonomists to group species; taxon above phylum (in animals) and division (in plants). 493

kinin Chemical mediator, released by damaged tissue cells and mast cells, which causes the capillaries to dilate and become more permeable. 762

Krebs cycle Cycle of reactions in mitochondria that begins with citric acid; it breaks down an acetyl group and produces CO_2, ATP, NADH, and $FADH_2$; also called the citric acid cycle. 129

l

lactation Secretion of milk by mammary glands usually for the nourishment of an infant. 912

lacteal Lymphatic vessel in an intestinal villus, it aids in the absorption of fats. 789

lactose Disaccharide that contains galactose and glucose; found in milk. 37

lacuna Small pit or hollow cavity, as in bone or cartilage, where a cell or cells are located. 727

ladderlike nervous system In planarians, two lateral nerve cords joined by transverse nerves. 827

landscape A number of interacting ecosystems. 471

large intestine In vertebrates, portion of the digestive tract that follows the small intestine; in humans, consists of the cecum, colon, rectum, and anal canal. 791

larva Immature form in the life cycle of some animals; it sometimes undergoes metamorphosis to become the adult form. 903

larynx Cartilaginous organ located between the pharynx and the trachea; in humans, contains the vocal cords; voice box. 802

lateral line Canal system containing sensory receptors that allow fishes and amphibians to detect water currents and pressure waves from nearby objects. 859

law Theory that is generally accepted by an overwhelming number of scientists. 12

leaf Lateral appendage of a stem, highly variable in structure, often containing cells that carry out photosynthesis. 643

leaf vein Vascular tissue within a leaf. 646

learning Relatively permanent change in an animal's behavior that results from practice and experience. 364

leech Blood-sucking annelid, usually found in fresh water, with a sucker at each end of a segmented body. 606

lens Clear membranelike structure found in the vertebrate eye behind the iris; brings objects into focus. 852

less-developed country (LDC) Country which is becoming industrialized; typically population growth is expanding rapidly and the majority of people live in poverty. 393

lichen Symbiotic relationship between certain fungi and algae, in which the fungi possibly provide inorganic food or water and the algae provide organic food. 519, 553

life cycle Recurring pattern of genetically programmed events by which individuals grow, develop, maintain themselves, and reproduce. 168

ligament Tough cord or band of dense fibrous tissue that binds bone to bone at a joint. 726, 871

light-dependent reactions Portion of photosynthesis that captures solar energy and takes place in thylakoid membranes of chloroplasts; it produces ATP and NADPH. 114

light-independent reactions Portion of photosynthesis that takes place in the stroma of chloroplasts and can occur in the dark; it uses the products of the light-dependent reactions to reduce carbon dioxide to a carbohydrate. 114

limbic system In humans, functional association of various brain centers including the amygdala and hippocampus; governs learning and memory and various emotions such as pleasure, fear, and happiness. 841

limiting factor Resource or environmental condition that restricts the abundance and distribution of an organism. 379

linkage group Alleles of different genes that are located on the same chromosome and tend to be inherited together. 194

lipase Fat-digesting enzyme secreted by the pancreas. 788

lipid Class of organic compounds that tends to be soluble in nonpolar solvents such as alcohol; includes fats and oils. 40

liposome Droplet of phospholipid molecules formed in a liquid environment. 322

littoral zone Shore zone between high-tide mark and low-tide mark; also shallow water of a lake where light penetrates to the bottom. 459

liver Large dark red internal organ that produces urea and bile, detoxifies the blood, stores glycogen, and produces the plasma proteins among other functions. 788

logistic growth Population increase that results in an S-shaped curve; growth is slow at first, steepens, and then levels off due to environmental resistance. 382

long-day plant Plant which flowers when day length is longer than a critical length, i.e., wheat, barley, clover, spinach. 694

loop of the nephron Portion of a nephron between the proximal and distal convoluted tubules; function in water reabsorption. 816

loose fibrous connective tissue Tissue composed mainly of fibroblasts widely separated by a matrix containing collagen and elastic fibers. 726

lung Internal respiratory organ containing moist surfaces for gas exchange. 627, 801

luteal phase Second half of the ovarian cycle, during which the corpus luteum develops and much progesterone (and some estrogen) is produced. 910

lymph Fluid, derived from tissue fluid, that is carried in lymphatic vessels. 755, 760

lymph nodes Mass of lymphoid tissue located along the course of a lymphatic vessel. 760

lymphatic vessel Vessel that carries lymph. 760

lymphocyte Specialized white blood cell that functions in specific defense; occurs in two forms—T lymphocyte and B lymphocyte. 753

lymphoid organ Organ other than a lymphatic vessel that is part of the lymphatic system; includes lymph nodes, tonsils, spleen, thymus gland, and bone marrow. 760

lysogenic cycle Bacteriophage life cycle in which the virus incorporates its DNA into that of a bacterium; occurs preliminary to the lytic cycle. 511

lysosome Membrane-bounded vesicle that contains hydrolytic enzymes for digesting macromolecules. 69

lytic cycle Bacteriophage life cycles in which the virus takes over the operation of the bacterium immediately upon entering it and subsequently destroys the bacterium. 510

m

macroevolution Large-scale evolutionary change; e.g., the formation of new species. 324

macronutrient Essential element needed in large amounts for plant growth such as nitrogen, calcium, sulfur. 666

macrophage In vertebrates, large phagocytic cell derived from a monocyte that ingests microbes and debris. 753, 762

Malpighian tubule Blind, threadlike excretory tubule near the anterior end of an insect hindgut. 611, 815

maltose Disaccharide composed of two glucose molecules; found in the human digestive tract as a result of starch digestion. 37

mammal Homeothermic vertebrate characterized especially by the presence of hair and mammary glands. 634

mantle In molluscs an extension of the body wall which may secrete a shell. 600

mass extinction Episode of large-scale extinction in which large numbers of species disappear in a few million years or less. 329

mast cell A connective tissue cell that releases histamine in allergic reactions. 762

matrix Unstructured semifluid substance that fills the space between cells in connective tissues or inside organelles. 71

matter Anything that takes up space and has mass. 18

mechanoreceptor Sensory receptor that responds to mechanical stimuli, such as that from pressure, sound waves, and gravity. 856

medulla oblongata In vertebrates, part of the brain stem that is continuous with the spinal cord; controls heartbeat, blood pressure, breathing, and other vital functions. 839

megagametophyte In seed plants, the gametophyte that produces an egg; in flowering plants, an embryo sac. 703

megaspore One of the two types of spores produced by seed plants; develops into a megagametophyte (embryo sac). 568, 700

megasporocyte Megaspore mother cell; produces megaspores by meiosis; only one megaspore persists. 703

meiosis Type of nuclear division that occurs as part of sexual reproduction in which the daughter cells receive the haploid number of chromosomes in varied combinations. 160

memory Capacity of the brain to store and retrieve information about past sensations and perceptions; essential to learning. 841

meninges Protective membranous coverings about the central nervous system. 838

menses Flow of blood during menstruation. 910

menstruation Periodic shedding of tissue and blood from the inner lining of the uterus in primates. 910

meristem Undifferentiated embryonic tissue in the active growth regions of plants. 712

mesoderm Middle primary tissue layer of an animal embryo that gives rise to muscle, several internal organs, and connective tissue layers. 645, 921

mesoglea Jellylike layer between the epidermis and the gastrodermis of a cnidarian. 584

mesophyll Inner, thickest layer of a leaf consisting of palisade and spongy mesophyll; the site of most of photosynthesis. 658

messenger RNA (mRNA) Type of RNA formed from a DNA template and bearing coded information for the amino acid sequence of a polypeptide. 236

metabolic pathway Series of linked reactions, beginning with a particular reactant and terminating with an end product. 102

metabolic pool Metabolites that are the products of and/or the substrates for key reactions in cells, allowing one type of molecule to be changed into another type, such as the conversion of carbohydrates to fats. 139

metabolism All of the chemical reactions that occur in a cell during growth and repair. 3, 100

metamorphosis Change in shape and form that some animals, such as insects, undergo during development. 607, 903

metapopulation Population subdivided into several small and isolated populations due to habitat fragmentation. 480

metastasis Spread of cancer from the place of origin throughout the body; caused by the ability of cancer cells to migrate and invade tissues. 153

methanogen Type of archaea that lives in oxygen-free habitats, such as swamps, and releases methane gas. 520

microevolution Change in gene frequencies between populations of a species over time. 302

microgametophyte In seed plants, the gametophyte that produces sperm; a pollen grain. 703

micronutrient Essential element needed in small amounts for plant growth such as boron, copper, and zinc. 666

microsphere Formed from proteinoids exposed to water; has properties similar to today's cells. 321

microspore One of the two types of spores produced by seed plants; develops into a microgametophyte (pollen grain). 568, 700

microsporocyte Microspore mother cell; produces microspores by meiosis. 703

microtubule Small cylindrical organelle composed of tubulin protein about an empty central core; present in the cytoplasm, centrioles, cilia, and flagella. 72

microvillus Cylindrical process that extends from an epithelial cell of a villus and serves to increase the surface area of the cell. 788

midbrain In mammals, part of the brain located below the thalamus and above the pons. 839

mimicry Superficial resemblance of two or more species; a mechanism that avoids predation by appearing to be noxious. 409

mineral Naturally occurring inorganic substance containing two or more elements; certain minerals are needed in the diet. 666, 793

mitochondrion Membrane-bounded organelle in which ATP molecules are produced during the process of cellular respiration. 70, 132

mitosis Process in which a parent nucleus produces two daughter nuclei, each having the same number and kinds of chromosomes as the parent nucleus. 145

molecular clock Idea that the rate at which mutational changes accumulate in certain genes is constant over time and is not involved in adaptation to the environment. 329, 497

molecule Union of two or more atoms of the same element; also the smallest part of a compound that retains the properties of the compound. 22

molt Periodic shedding of the exoskeleton in arthropods. 606

monoclonal antibody One of many antibodies produced by a clone of hybridoma cells which all bind to the same antigen. 774

monocotyledon Flowering plant group; members have one embryonic leaf, parallel-veined leaves, scattered vascular bundles, and other characteristics. 573

monomer Small molecule that is a subunit of a polymer; e.g., glucose is a monomer of starch. 36
monosaccharide Simple sugar; a carbohydrate that cannot be decomposed by hydrolysis; e.g., glucose. 37
monotreme Egg-laying mammal—for example, duckbill platypus and spiny anteater. 634
monsoon Climate in India and southern Asia caused by wet ocean winds that blow onshore for almost half the year. 443
more-developed country (MDC) Country that is industrialized; typically population growth is low and the people enjoy a good standard of living. 393
morphogen Protein that is part of a gradient which influences morphogenesis. 928
morphogenesis Emergence of shape in tissues, organs, or entire embryo during development. 924
morula Spherical mass of cells resulting from cleavage during animal development prior to the blastula stage. 921
mosaic evolution Concept that human characteristics did not evolve at the same rate; e.g., some body parts are more humanlike than others in early hominids. 348
motor molecule Protein that moves along either actin filaments or microtubules and translocates organelles. 72
motor neuron Nerve cell that conducts nerve impulses away from the central nervous system and innervates effectors (muscle and glands). 829
mouth In humans, organ of the digestive tract where food is chewed and mixed with saliva. 785
multiple allele Inheritance pattern in which there are more than two alleles for a particular trait; each individual has only two of all possible alleles. 190, 210
multiregional continuity hypothesis Proposal that modern humans evolved separately in at least three different places: Asia, Africa, and Europe. 353
muscular (contractile) tissue Type of animal tissue composed of fibers that shorten and lengthen to produce movements. 728
mutation Alteration in chromosome structure or number and also an alteration in a gene due to a change in DNA composition. 196
mutualism Symbiotic relationship in which both species benefit in terms of growth and reproduction. 412
mycelium Tangled mass of hyphal filaments composing the vegetative body of a fungus. 544
mycorrhiza (pl., mycorrhizae) Mutualistic relationship between fungal hyphae and roots of vascular plants. 554, 671
myelin sheath White, fatty material—derived from the membrane of neurolemmocytes—that forms a covering for nerve fibers. 829
myofibril Specific muscle cell organelle containing a linear arrangement of sarcomeres, which shorten to produce muscle contraction. 875
myoglobin Pigmented molecule in muscle tissue that stores oxygen. 875
myosin Muscle protein making up the thick filaments in a sarcomere; it pulls actin to shorten the sarcomere, yielding muscle contraction. 875

n

NAD$^+$ (nicotinamide adenine dinucleotide) Coenzyme of oxidation-reduction that accepts electrons and hydrogen ions to become NADH + H$^+$ as oxidation of substrates occurs; during cellular respiration, NADH carries electrons to the electron transport system in mitochondria. 106, 128
NADP$^+$ (nicotinamide adenine dinucleotide phosphate) Coenzyme of oxidation-reduction that accepts electrons and hydrogen ions to become NADPH + H$^+$. During photosynthesis NADPH participates in the reduction of carbon dioxide to glucose. 106
natural killer (NK) cell Lymphocyte that causes an infected or cancerous cell to burst. 764
natural selection Mechanism of evolution caused by environmental selection of organisms most fit to reproduce; results in adaptation to the environment. 4, 288, 306
negative feedback Mechanism of homeostatic response by which the output of a system suppresses or inhibits activity of the system. 734, 885
nematocyst In cnidarians, a capsule that contains a threadlike fiber whose release aids in the capture of prey. 584
nephridium Segmentally arranged, paired excretory tubules of many invertebrates, as in the earthworm. 604, 815
nephron Microscopic kidney unit that regulates blood composition by glomerular filtration, tubular reabsorption, and tubular secretion. 816
nerve Bundle of long axons outside the central nervous system. 834
nerve cord In many complex animals, a centrally placed cord of nervous tissue that receives sensory information and exercises motor control. 619
nerve net Diffuse noncentralized arrangement of nerve cells in cnidarians. 584, 827
nervous tissue Tissue that contains nerve cells (neurons), which conduct impulses, and neuroglial cells, which support, protect, and provide nutrients to neurons. 728
neural plate Region of the dorsal surface of the chordate embryo that marks the future location of the neural tube. 923
neural tube Tube formed by closure of the neural groove during development. In vertebrates, the neural tube develops into the spinal cord and brain. 923
neuroglia Nonconducting nerve cells that are intimately associated with neurons and function in a supportive capacity. 728, 829
neurolemmocyte Type of neuroglial cell that forms a myelin sheath around axons; also called Schwann cell. 829
neuromuscular junction Region where an axon bulb approaches a muscle fiber; contains a presynaptic membrane, a synaptic cleft, and a postsynaptic membrane. 876
neuron Nerve cell that characteristically has three parts: dendrites, cell body, and axon. 728, 829
neurotransmitter Chemical stored at the ends of axons that is responsible for transmission across a synapse. 833
neutron Neutral subatomic particle, located in the nucleus and having a weight of approximately one atomic mass unit. 18
neutrophil Granular leukocyte that is the most abundant of the white blood cells; first to respond to infection. 753
nitrification Process by which nitrogen in ammonia and organic compounds is oxidized to nitrites and nitrates by soil bacteria. 432
nitrogen cycle Continuous process by which nitrogen circulates in the air, soil, water, and organisms of the biosphere. 432
nitrogen fixation Process whereby free atmospheric nitrogen is converted into compounds, such as ammonium and nitrates, usually by bacteria. 432
node In plants, the place where one or more leaves attach to a stem. 643
noncyclic electron pathway Portion of the light-dependent reaction of photosynthesis that involves both photosystem I and photosystem II. It generates both ATP and NADPH. 117
nondisjunction Failure of homologous chromosomes or daughter chromosomes to separate during meiosis I and meiosis II respectively. 203
nonpolar covalent bond Bond in which the sharing of electrons between atoms is fairly equal. 24
nonrandom mating Mating among individuals on the basis of their phenotypic similarities or differences rather than mating on a random basis. 305
nonseptate Lacking cell walls; some fungal species have hyphae that are nonseptate. 545
nonvascular plant Bryophytes such as mosses and liverworts that have no vascular tissue and either occur in moist locations or have special adaptations for living in dry locations. 559
norepinephrine (NE) Neurotransmitter of the postganglionic fibers in the sympathetic division of the autonomic system; also a hormone produced by the adrenal medulla. 833
notochord Cartilaginous-like supportive dorsal rod in all chordates sometime in their life cycle; is replaced by vertebrae in vertebrates. 619, 923
nuclear envelope Double membrane that surrounds the nucleus in eukaryotic cells and is connected to the endoplasmic reticulum; has pores that allow substances to pass between the nucleus and the cytoplasm. 65

nuclear pore Opening in the nuclear envelope which permits the passage of proteins into the nucleus and ribosomal subunits out of the nucleus. 65

nucleic acid Polymer of nucleotides; both DNA and RNA are nucleic acids. 48, 220

nucleoid Region of prokaryotic cells where DNA is located; it is not bounded by a nuclear envelope. 60, 144, 514

nucleolus Dark-staining, spherical body in the nucleus that produces ribosomal subunits. 65

nucleoplasm Semifluid medium of the nucleus, containing chromatin. 65

nucleotide Monomer of DNA and RNA consisting of a five-carbon sugar bonded to a nitrogenous base and a phosphate group. 48, 220

nucleus Membrane-bounded organelle within a eukaryotic cell that contains chromosomes and controls the structure and function of the cell. 65

o

obligate anaerobe Prokaryote unable to grow in the presence of free oxygen. 516

ocean ridge Ridge on the ocean floor where oceanic crust forms and from which it moves laterally in each direction. 321

octet rule An atom other than hydrogen tends to form bonds until it has eight electrons in its outer shell; an atom that already has eight electrons in its outer shell does not react and is inert. 21

olfactory cell Modified neuron that is a sensory receptor for the sense of smell. 848

omnivore Organism in a food chain that feeds on both plants and animals. 423

oncogene Cancer-causing gene. 260

oocyte Immature egg that is undergoing meiosis; upon completion of meiosis the oocyte becomes an egg. 908

oogamy Sexual reproduction in which the gametes are dissimilar; the egg is large and nonmotile and the sperm is small and motile. 528

oogenesis Production of eggs in females by the process of meiosis and maturation. 168

operant conditioning Learning that results from rewarding or reinforcing a particular behavior. 364

operator In an operon the sequence of DNA that binds tightly to a repressor and thereby regulates the expression of structural genes. 248

operon Group of structural and regulating genes that functions as a single unit. 248

orbital Volume of space around a nucleus where electrons can be found most of the time. 21

order One of the categories, or taxa, used by taxonomists to group species; taxon above the family level. 493

organ Combination of two or more different tissues performing a common function. 642, 724, 730

organ system Group of related organs working together. 730

organelle Small, often membranous structure in the cytoplasm having a specific structure and function. 61

organic molecule Type of molecule that contains carbon and hydrogen; it may also have oxygen attached to the carbon(s). 34

orgasm Physiological and psychological sensations that occur at the climax of sexual stimulation. 905

osmosis Diffusion of water through a differentially permeable membrane. 86

osmotic pressure Measure of the tendency of water to move across a differentially permeable membrane; visible as an increase in liquid on the side of the membrane with higher solute concentration. 86

ossicle One of the small bones of the vertebrate middle ear—malleus, incus, stapes. 856

osteoblast Bone-forming cell. 866

osteoclast Cell that causes erosion of bone. 866

osteocyte Mature bone cell located within the lacunae of bone. 866

ostracoderm Earliest vertebrate fossils of the Cambrian and Devonian periods, these fishes were small, jawless, and finless. 622

otolith Calcium carbonate granule associated with sensory receptors for detecting movement of the head; in vertebrates, located in the utricle and saccule. 859

out-of-Africa hypothesis Proposal that modern humans originated only in Africa; then they migrated and supplanted populations of *Homo* in Asia and Europe about 100,000 years ago. 353

ovarian cycle Monthly changes occurring in the ovary that determine the level of sex hormones in the blood. 909

ovary Female gonad in animals that produces an egg and female sex hormones; in flowering plants, the enlarged ovule-bearing portion of the pistil which develops into a fruit. 573, 701, 902

ovulation Bursting of a follicle when a secondary oocyte is released from the ovary; if fertilization occurs the secondary oocyte becomes an egg. 909

ovule In seed plants, a structure that contains the megagametophyte and has the potential to develop into a seed. 560, 701

ovum Haploid egg cell which is usually fertilized by a sperm to form a diploid zygote. 909

oxidation Loss of one or more electrons from an atom or molecule; in biological systems, generally the loss of hydrogen atoms. 106

oxidative phosphorylation Process by which ATP production is tied to an electron transport system that uses oxygen as the final acceptor; occurs in mitochondria. 134

oxygen debt Amount of oxygen required to oxidize lactic acid produced anaerobically during strenuous muscle activity. 137, 875

oxyhemoglobin Compound formed when oxygen combines with hemoglobin. 805

ozone shield Accumulation of O_3, formed from oxygen in the upper atmosphere; a filtering layer that protects the earth from ultraviolet radiation. 327, 436

p

***p53* gene** Involved in cell division control, the *p53* gene halts the cell cycle when DNA mutates and is in need of repair. 152

paleontology Study of fossils that results in knowledge about the history of life. 284, 324

palisade mesophyll Layer of tissue in a plant leaf containing elongated cells with many chloroplasts. 658

pancreas Internal organ that produces digestive enzymes and the hormones insulin and glucagon. 788, 894

pancreatic amylase Enzyme that digests starch to maltose. 788

pancreatic islet Masses of cells that constitute the endocrine portion of the pancreas. 894

parallel evolution Similarity in structure in related groups that cannot be traced to a common ancestor. 495

parasite Organism that lives off of and takes nourishment from an organism called the host. 410

parasitism Symbiotic relationship in which one species (parasite) benefits in terms of growth and reproduction to the harm of the other species (host). 410

parasympathetic division Division of the autonomic system that is active under normal conditions; uses acetylcholine as a neurotransmitter. 837

parathyroid gland Gland embedded in the posterior surface of the thyroid gland; it produces parathyroid hormone. 890

parenchyma Plant tissue composed of least specialized of all plant cells; found in all organs of a plant. 646

parthenogenesis Development of an egg cell into a whole organism without fertilization. 902

pathogen Disease-causing agent. 760

pattern formation Positioning of cells during development that determines the final shape of an organism. 924

peat Organic fuel that consists of the partially decomposed remains of peat mosses which accumulate in bogs. 562

pectoral girdle Portion of the vertebrate skeleton that provides support and attachment for the upper (fore) limbs; consists of the paired scapula and clavicle. 870

pedigree chart A family tree that uses standardized genetic symbols and shows the inheritance pattern for a specific phenotypic characteristic. 207

pelagic division Open portion of the sea. 462

pelvic girdle Portion of the vertebrate skeleton to which the lower (hind) limbs are attached; consists of the coxal bones. 871

penis Male copulatory organ; in humans, the male organ of sexual intercourse. 904

pentose Five-carbon sugar; deoxyribose is the pentose sugar found in DNA; ribose is a pentose sugar found in RNA. 37

pepsin Enzyme secreted by gastric glands that digests proteins to peptides. 787

peptide Two or more amino acids joined together by covalent bonding. 45

peptide bond Type of covalent bond that joins two amino acids. 45

peptide hormone Type of hormone that is a protein, a peptide, or is derived from an amino acid. 883

peptidoglycan Unique molecule found in bacterial cell walls. 517

perennial Flowering plant that lives more than one growing season because the underground parts regrow each season. 642

perforin Molecule secreted by a cytotoxic T cell that perforates the plasma membrane of the target cell so that water and salts enter, causing the cell to swell and burst. 768

pericycle Layer of cells surrounding the vascular tissue of roots; produces branch roots. 649

peristalsis Wavelike contractions that propel substances along a tubular structure like the esophagus. 786

permafrost Permanently frozen ground usually occurring in the tundra, a biome of Arctic regions. 447

peroxisome Enzyme-filled vesicle in which fatty acids and amino acids are metabolized to hydrogen peroxide that is broken down to harmless products. 69

petal A flower part that occurs just inside the sepals; often conspicuously colored to attract pollinators. 700

petiole Part of a plant leaf that connects the blade to the stem. 643

pH scale Measurement scale for the hydrogen ion concentration. 28

phagocytize To ingest extracellular particles by ingesting them as do amoeboid-type cells. 536

phagocytosis Process by which amoeboid-type cells engulf large substances, forming an intracellular vacuole. 90

pharynx In vertebrates, common passageway for both food intake and air movement, located between the mouth and the esophagus. 786, 802

phenetic systematics School of systematics that determines the degree of relatedness between species by counting the number of their similarities. 500

phenomenon Observable event. 9

phenotype Visible expression of a genotype—for example, brown eyes or attached earlobes. 177

pheromone Chemical messenger that works at a distance and alters the behavior of another member of the same species. 370, 882

phloem Vascular tissue that conducts organic solutes in plants; contains sieve-tube elements and companion cells. 562, 646

phospholipid Molecule that forms the bilayer of cell membranes; has a polar, hydrophilic head bonded to two nonpolar hydrophobic tails. 42, 81

phosphorus cycle Continuous process by which phosphorus circulates in the soil, water, and organisms of the biosphere. 434

photoautotroph Organism able to synthesize organic molecules by using carbon dioxide as the carbon source and sunlight as the energy source. 516

photochemical smog Air pollution that contains nitrogen oxides and hydrocarbons which react to produce ozone and PAN (peroxyacetylnitrate). 433

photon Discrete packet of solar energy; the amount of energy in a photon is inversely related to the wavelength of the photon. 112

photoperiodism Relative lengths of daylight and darkness that affect the physiology and behavior of an organism. 694

photoreceptor Sensory receptor that responds to light stimuli. 850

photorespiration Series of reactions that occurs in plants when carbon dioxide levels are depleted but oxygen continues to accumulate and the enzyme RuBP carboxylase fixes oxygen instead of carbon dioxide. 123

photosynthesis Process occurring usually within chloroplasts whereby chlorophyll-containing organelles trap solar energy to reduce carbon dioxide to carbohydrate. 3, 112

photosystem Photosynthetic unit where solar energy is absorbed and high energy electrons are generated; contains a pigment complex and an electron acceptor; occurs as PS (photosystem) I and PS II. 116

phototropism Growth response of plant stems to light; stems demonstrate positive phototropism. 685

phylogenetic tree Diagram that indicates common ancestors and lines of descent among a group of organisms. 494

phylogeny Evolutionary history of a group of organisms. 494

phylum One of the categories, or taxa, used by taxonomists to group species; taxon above the class level. 493

phytochrome Photoreversible plant pigment that is involved in photoperiodism and other responses of plants such as etiolation. 694

phytoplankton Part of plankton containing organisms that photosynthesize releasing oxygen to the atmosphere and serving as food producers in aquatic ecosystems. 437, 533

pineal gland Gland—either at the skin surface (fish, amphibians) or in the third ventricle of the brain (mammals)—that produces melatonin. 897

pinocytosis Process by which vesicle formation brings macromolecules into the cell. 90

pioneer species Early colonizer of barren or disturbed habitats that usually has a rapid growth and high dispersal rate. 415

pistil Reproductive flower structure composed of one or more carpels; consisting of an ovary, style, and stigma. 573, 700

pith Parenchyma tissue in the center of some stems and roots. 649

pituitary gland Small gland that lies just inferior to the hypothalamus; consists of the anterior and posterior pituitary, both of which produce hormones. 886

placenta Organ formed during the development of placental mammals from the chorion and the uterine wall that allows the embryo, and then the fetus, to acquire nutrients and rid itself of wastes; produces hormones that regulate pregnancy. 635, 903, 934

placental mammal Member of mammalian subclass characterized by the presence of a placenta during the development of an offspring. 635

placoderm First jawed vertebrates; heavily armored fishes of the Devonian period. 623

plankton Freshwater and marine organisms that are suspended on or near the surface of the water; includes phytoplankton and zooplankton. 457, 526

plasma In vertebrates, the liquid portion of blood; contains nutrients, wastes, salts, and proteins. 753

plasma cell Cell derived from a B-cell lymphocyte that is specialized to mass-produce antibodies. 765

plasma membrane Membrane surrounding the cytoplasm that consists of a phospholipid bilayer with embedded proteins; functions to regulate the entrance and exit of molecules from cell. 60

plasmid Self-duplicating ring of accessory DNA in the cytoplasm of bacteria. 60, 266, 514

plasmodesmata In plants, cytoplasmic strands that extend through pores in the cell wall and connect the cytoplasm of two adjacent cells. 92

plasmodial slime mold Free-living mass of cytoplasm that moves by pseudopods on a forest floor or in a field feeding on decaying plant material by phagocytosis; reproduces by spore formation. 539

plasmolysis Contraction of the cell contents due to the loss of water. 87

plate tectonics Concept that the earth's crust is divided into a number of fairly rigid plates whose movements account for continental drift. 335

platelet Component of blood that is necessary to blood clotting. 754

pleiotropy Inheritance pattern in which one gene affects many phenotypic characteristics of the individual. 189

plumule In flowering plants, the embryonic plant shoot that bears young leaves. 711

pneumonectomy Surgical removal of all or part of a lung. 807

point mutation Alternation of one base only in the sequence of bases in a gene. 259

polar body In oogenesis, a nonfunctional product; two to three meiotic products are of this type. 169

polar covalent bond Bond in which the sharing of electrons between atoms is unequal. 24

pollen grain In seed plants, structure that is derived from a microspore and develops into a microgametophyte. 568, 703

pollination In gymnosperms, the transfer of pollen from pollen cone to seed cone; in angiosperms, the transfer of pollen from anther to stigma. 568, 706

pollinator Animal; e.g., bee, that inadvertently transfers pollen from anther to stigma. 573
pollution Any environmental change that adversely affects the lives and health of living things. 477
polygenic inheritance Inheritance pattern in which a trait is controlled by several allelic pairs; each dominant allele contributes to the phenotype in an additive and like manner. 190
polymer Macromolecule consisting of covalently bonded monomers; for example, a polypeptide is a polymer of monomers called amino acids. 36
polymerase chain reaction (PCR) Technique that uses the enzyme DNA polymerase to produce millions of copies of a particular piece of DNA. 268
polyp Small, abnormal growth that arises from the epithelial lining. 791
polypeptide Polymer of many amino acids linked by peptide bonds. 45
polyploid (polyploidy) Having a chromosome number that is a multiple greater than twice that of the monoploid number. 196
polyribosome String of ribosomes simultaneously translating regions of the same mRNA strand during protein synthesis. 66, 241
polysaccharide Polymer made from sugar monomers; the polysaccharides starch and glycogen are polymers of glucose monomers. 38
pond Standing water within a basin. 455
population Group of organisms of the same species occupying a certain area and sharing a common gene pool. 5, 378
population genetics Study of gene frequencies and their changes within a population. 302
population viability analysis Calculation of the minimum population size needed to prevent extinction. 481
portal system Pathway of blood flow that begins and ends in capillaries, such as the portal system located between the small intestine and liver. 748
positive feedback Mechanism of homeostatic response in which the output of the system intensifies and increases the activity of the system. 735, 886
posttranscriptional control Gene expression following translation regulated by the way mRNA transcripts are processed. 251
posttranslational control Gene expression following translation regulated by the activity of the newly synthesized protein. 251
postzygotic isolating mechanism Anatomical or physiological difference between two species that prevents successful reproduction after mating has taken place. 311
potential energy Stored energy as a result of location or spatial arrangement. 98
prairie Terrestrial biome that is a temperate grassland. When travelling east to west in the midwest of the United States, a tall-grass prairie changes to a short-grass prairie. 452
predation Interaction in which one organism uses another, called the prey, as a food source. 406
predator Organism that practices predation. 406
pressure-flow model Explanation for phloem transport; osmotic pressure following active transport of sugar into phloem brings a flow of sap from a source to a sink. 678
prey Organism that provides nourishment for a predator. 406
prezygotic isolating mechanism Anatomical or behavioral difference between two species that prevents the possibility of mating. 310
primary root Original root that grows straight down and remains the dominant root of the plant; contrast fibrous root system. 650
primitive character Structural, physiological, or behavioral trait that is present in a common ancestor and all members of a group. 494
principle Theory that is generally accepted by an overwhelming number of scientists. Also called law. 12
prion Infectious particle consisting of protein only and no nucleic acid. 513
probe Known sequence of DNA that is used to find complementary DNA strands; can be used diagnostically to determine the presence of particular genes. 267
producer Photosynthetic organism at the start of a grazing food chain that makes its own food (e.g., green plants on land and algae in water). 423
product Substance that forms as a result of a reaction. 100
progesterone Female sex hormone that helps maintain sexual organs and secondary sexual characteristics. 909
proglottid Segment of a tapeworm that contains both male and female sex organs and becomes a bag of eggs. 591
prokaryote Organism that lacks the membrane-bounded nucleus and membranous organelles typical of eukaryotes; bacteria and archaea are prokaryotes. 513
prokaryotic cell Lacking a membrane-bounded nucleus and organelles; the cell type within the domain Bacteria and Archaea. 60
promoter In an operon, a sequence of DNA where RNA polymerase binds prior to transcription. 238, 248
proofreading Process used to check the accuracy of DNA replication as it occurs, and to replace a mispaired base with the right one. 229
protein Molecule consisting of one or more polypeptides. 44
protein-first hypothesis In chemical evolution, the proposal that protein originated before other macromolecules and allowed the formation of protocells. 321
protist Member of the kingdom Protista. 526
proto-oncogene Normal gene that can become an oncogene through mutation. 260
protocell In biological evolution, a possible cell forerunner that became a cell once it acquired genes. 322
proton Positive subatomic particle, located in the nucleus and having a weight of approximately one atomic mass unit. 18
protoplast Plant cell from which the cell wall has been removed. 712
protostome Group of coelomate animals in which the first embryonic opening (the blastopore) is associated with the mouth. 598
protozoan Heterotrophic unicellular protist that moves by flagella, cilia, pseudopodia, or are immobile. 535
proximal convoluted tubule Portion of a nephron following the glomerular capsule where tubular reabsorption of filtrate occurs. 816
pseudocoelom Body cavity lying between the digestive tract and body wall that is incompletely lined by mesoderm. 591
pseudopod Cytoplasmic extension of amoeboid protists; used for locomotion and engulfing food. 536
puberty Period of time in life when secondary sex changes occur in humans, marked by the onset of menses in females and sperm production in males. 907
pulmonary circuit Circulatory pathway between the lungs and the heart. 748
pulse Vibration felt in arterial walls due to expansion of the aorta following ventricle contraction. 746
pupil Opening in the center of the iris of the vertebrate eye. 852
purine Type of nitrogen-containing base, such as adenine and guanine, having a double-ring structure. 223
pyrimidine Type of nitrogen-containing base, such as cytosine, thymine, and uracil, having a single-ring structure. 223
pyruvate End product of glycolysis; its further fate, involving fermentation or entry into a mitochondrion, depends on oxygen availability. 129

r

***r*-selection** A favorable life history strategy under certain environmental conditions; characterized by a high reproductive rate with little or no attention given to offspring survival. 390
radial symmetry Body plan in which similar parts are arranged around a central axis, like spokes of a wheel. 580
radioactive isotope Unstable form of an atom that spontaneously emits radiation in the form of radioactive particles or radiant energy. 19
radula Tonguelike organ found in molluscs that bears rows of tiny teeth that point backward; used to obtain food. 600
rain shadow Lee side (side sheltered from the wind) of a mountainous barrier, which receives much less precipitation than the windward side. 443
reactant Substance that participates in a reaction. 100

receptor protein Protein located in the plasma membrane or within the cell that binds to a substance that alters some metabolic aspect of the cell. 83

receptor-mediated endocytosis Selective uptake of molecules into a cell by vacuole formation after they bind to specific receptor proteins in the plasma membrane. 91

recessive allele Allele that exerts its phenotypic effect only in the homozygote; its expression is masked by a dominant allele. 177

recombinant DNA (rDNA) DNA that contains genes from more than one source. 266

red algae Marine photosynthetic protist with a notable abundance of phycobilin pigments; includes coralline algae of coral reefs. 530

red blood cell Erythrocyte; contains hemoglobin and carries oxygen from the lungs or gills to the tissues in vertebrates. 753

red bone marrow Vascularized modified connective tissue that is sometimes found in the cavities of spongy bone; site of blood cell formation. 762, 866

reduction Gain of electrons by an atom or molecule with a concurrent storage of energy; in biological systems, the electrons are accompanied by hydrogen ions. 106

reflex Automatic, involuntary response of an organism to a stimulus. 835

regulator gene In an operon, a gene that codes for a protein that regulates the expression of other genes. 248

relative dating (of fossils) Determining the age of fossils by noting their sequential relationships in strata. 324

renal cortex Outer portion of the kidney that appears granular. 816

renal medulla Inner portion of the kidney that consists of renal pyramids. 816

renal pelvis Hollow chamber in the kidney that lies inside the renal medulla and receives freshly prepared urine from the collecting ducts. 816

renin Enzyme released by kidneys that leads to the secretion of aldosterone and a rise in blood pressure. 821

replacement reproduction Population in which each person is replaced by only one child. 394

replication fork In eukaryotes, the point where the two parental DNA strands separate to allow replication. 229

repressible operon Operon that is normally active because the repressor is normally inactive. 249

repressor In an operon, protein molecule that binds to an operator, preventing transcription of structural genes. 249

reproduce To produce a new individual of the same kind. 4

resource partitioning Mechanism that increases the number of niches by apportioning the supply of a resource such as food and living space between species. 404

respiration Sequence of events that results in gas exchange between the cells of the body and the environment. 798

resting potential Membrane potential of an inactive neuron. 830

restoration ecology Subdiscipline of conservation biology that seeks ways to return ecosystems to their former state. 482

restriction enzyme Bacterial enzyme that stops viral reproduction by cleaving viral DNA; used to cut DNA at specific points during production of recombinant DNA. 266

restriction fragment length polymorphism (RFLP) Differences in DNA sequence between individuals that result in different patterns when DNA is cleaned by a restriction enzyme. 269

reticular formation Complex network of nerve fibers within the central nervous system that arouses the cerebrum. 839

retina Innermost layer of the vertebrate eyeball containing the photoreceptors—rod cells and cone cells. 851

retrovirus RNA virus containing the enzyme reverse transcriptase that carries out RNA/DNA transcription. 512

rhizoid Rootlike hair that anchors a plant and absorbs minerals and water from the soil. 560

rhizome Rootlike, underground stem. 564, 656

rhodopsin Light-absorbing molecule in rod cells and cone cells that contains a pigment and the protein opsin. 853

ribbon worm Marine invertebrate of the phylum Nemertea having a distinctive proboscis apparatus. 588

ribose Pentose sugar found in RNA. 37

ribosomal RNA (rRNA) Type of RNA found in ribosomes that translate messenger RNAs to produce proteins. 236

ribosome RNA and protein in two subunits; site of protein synthesis in the cytoplasm. 60, 66

RNA (ribonucleic acid) Nucleic acid produced from covalent bonding of nucleotide monomers that contain the sugar ribose; occurs in three forms: messenger RNA, ribosomal RNA, and transfer RNA. 48, 220, 236

RNA polymerase During transcription, an enzyme that joins nucleotides complementary to a DNA template. 238

RNA transcript mRNA molecule formed during transcript that has a sequence of bases complementary to a gene. 238

RNA-first hypothesis In chemical evolution, the proposal that RNA originated before other macromolecules and allowed the formation of the first cell(s). 321

rod cell Photoreceptor in vertebrate eyes that responds to dim light. 853

root In vascular plants, the underground organ that anchors the plant in the soil, absorbs water and minerals, and stores the products of photosynthesis. 562

root hair Extension of a root epidermal cell that collectively increases the surface area for the absorption of water and minerals. 562, 645, 670

root nodule Structure on plant root that contains nitrogen-fixing bacteria. 650

root pressure Osmotic pressure, caused by active movement of mineral into root cells; serves to elevate water in xylem for a short distance. 674

root system Includes the main root and any and all of its lateral (side) branches. 642

rough ER (endoplasmic reticulum) Membranous system of tubules, vesicles, and sacs in cells; with attached ribosomes. 67

roundworm Member of the phylum Nematoda with a cylindrical body that has a complete digestive tract and a pseudocoelom; some forms free-living in water and soil and many are parasitic. 593

RuBP (ribulose bisphosphate) Five-carbon compound that combines with and fixes carbon dioxide during the Calvin cycle and is later regenerated by the same cycle. 120

S

sac body plan Body with a digestive cavity that has only one opening, as in cnidarians and flatworms. 588

saccule Saclike cavity in the vestibule of the vertebrate inner ear; contains sensory receptors for static equilibrium. 859

salivary amylase In humans, enzyme in saliva that digests starch to maltose. 785

salivary gland In humans, gland associated with the mouth that secretes saliva. 785

saltatory conduction Movement of nerve impulses from one neurolemmal node to another along a myelinated axon. 830

saprotroph Organism that secretes digestive enzymes and absorbs the resulting nutrients back across the plasma membrane. 516

sarcolemma Plasma membrane of a muscle fiber; also forms the tubules of the T system involved in muscular contraction. 875

sarcomere One of many units, arranged linearly within a myofibril, whose contraction produces muscle contraction. 875

sarcoplasmic reticulum Smooth endoplasmic reticulum of skeletal muscle cells; surrounds the myofibrils and stores calcium ions. 875

saturated Fatty acid molecule that lacks double bonds between the carbons of its hydrocarbon chain. The chain bears the maximum number of hydrogens possible. 40

savanna Terrestrial biome that is a grassland in Africa, characterized by few trees and a severe dry season. 452

scale In fishes and reptiles, a thin flake, many of which cover the body and offer protection. 622

schistosomiasis Disease caused by the blood fluke, a parasitic flatworm of the phylum Platyhelminthes. 590

schizocoelom In protostomes, coelom formed by splitting of the embryonic mesoderm. 599

science Human endeavor that considers only what is observable by the senses or by instruments that extend the ability of the senses. 9

scientific method Process by which scientists formulate a hypothesis, gather data by observation and experimentation, and come to a conclusion. 12

scientific theory Concept supported by a broad range of observations, experiments, and data. 12

sclerenchyma Plant tissue composed of cells with heavily lignified cell walls and functions in support. 646

scolex Tapeworm head region; contains hooks and suckers for attachment to host. 591

seaweed Multicellular forms of red, green, and brown algae found in marine habitats. 531

second messenger Chemical signal such as cyclic AMP that causes the cell to respond to the first messenger—a hormone bound to plasma membrane receptor. 883

secondary oocyte In oogenesis the functional product of meiosis I; becomes the egg. 169

secondary sexual characteristic Trait that is sometimes helpful but not absolutely necessary for reproduction and is maintained by the sex hormones in males and females. 907

sedimentation Process by which particulate material accumulates and forms a stratum. 324

seed Mature ovule that contains an embryo, with stored food enclosed in a protective coat. 568, 708

segmentation Repetition of body units as seen in the earthworm. 580

semen (seminal fluid) Thick, whitish fluid consisting of sperm and secretions from several glands of the male reproductive tract. 905

semicircular canal One of three half-circle-shaped canals of the vertebrate inner ear that contains sensory receptors for dynamic equilibrium. 859

semiconservative replication Duplication of DNA resulting in two double helix molecules each having one parental and one new strand. 226

semilunar valve Valve resembling a half moon located between the ventricles and their attached vessels. 744

seminiferous tubule Long, coiled structure contained within chambers of the testis where sperm are produced. 907

senescence Sum of processes involving aging, decline, and eventual death of a plant or plant part. 692

sensory neuron Nerve cell that transmits nerve impulses to the central nervous system after a sensory receptor has been stimulated. 829

sensory receptor Structure that receives either external or internal environmental stimuli and is a part of a sensory neuron or transmits signals to a sensory neuron. 825

sepal Outermost, sterile, leaflike covering of the flower; usually green in color. 700

septate Having cell walls; some fungal species have hyphae that are septate. 545

septum Partition or wall that divides two areas; the septum in the heart separates the right half from the left half. 744

sessile filter feeder Animal that stays in one place and filters small food particles from the water. 582

seta (pl., setae) A needlelike chitinous bristle of annelids, arthropods, and others. 604

sex chromosome Chromosome that determines the sex of an individual; in humans, females have two X chromosomes and males have an X and Y chromosome. 192, 202

sex-influenced trait Autosomal phenotype controlled by an allele that is expressed differently in the two sexes; for example, the possibility of pattern baldness is increased by the presence of testosterone in human males. 215

sex-linked Allele that occurs on the sex chromosomes but may control a trait that has nothing to do with sexual characteristics of an animal. 214

sexual reproduction Reproduction involving meiosis, gamete formation, and fertilization; produces offspring with chromosomes inherited from each parent with a unique combination of genes. 160

sexual selection Changes in males and females, often due to male competition and female selectivity leading to increased fitness. 305, 366

shoot apical meristem Group of actively dividing embryonic cells at the tips of plant shoots. 652

shoot system Aboveground portion of a plant consisting of the stem, leaves, and flowers. 642

short-day plant Plant which flowers when day length is shorter than a critical length, i.e., cocklebur, poinsettia, and chrysanthemum. 694

sieve-tube element Member that joins with others in the phloem tissue of plants as a means of transport for nutrient sap. 672

sink population Population that is found in an unfavorable area where at best the birthrate equals the death rate; sink populations receive new members from source populations. 480

sister chromatid One of two genetically identical chromosomal units that are the result of DNA replication and are attached to each other at the centromere. 145

skeletal muscle Striated, voluntary muscle tissue that comprises skeletal muscles; also called striated muscle. 728

sliding filament model An explanation for muscle contraction based on the movement of actin filaments in relation to myosin filaments. 875

slime layer Gelatinous sheath surrounding the cell wall of certain bacteria. 60

small intestine In vertebrates, the portion of the digestive tract that precedes the large intestine; in humans, consists of duodenum, jejunum, and ileum. 788

smooth (visceral) muscle Nonstriated, involuntary muscles found in the walls of internal organs. 728

smooth ER (endoplasmic reticulum) Membranous system of tubules, vesicles, and sacs in eukaryotic cells; without attached ribosomes. 67

society Group in which members of species are organized in a cooperative manner, extending beyond sexual and parental behavior. 370

sociobiology Application of evolutionary principles to the study of social behavior of animals, including humans. 372

sodium-potassium pump Carrier protein in the plasma membrane that moves sodium ions out of and potassium into animal cells; important in nerve and muscle cells. 88

soil Accumulation of inorganic rock material and organic matter that is capable of supporting the growth of vegetation. 446, 668

soil erosion Movement of topsoil to a new location due to the action of running water or wind. 669

soil profile Vertical section of soil from the ground surface to the unaltered rock below. 669

solute Substance that is dissolved in a solvent, forming a solution. 85

solution Fluid (the solvent) that contains a dissolved solid (the solute). 85

solvent Liquid portion of a solution that serves to dissolve a solute. 85

somatic embryo Plant cell embryo which is asexually produced through tissue culture techniques. 712

source population Population that can provide members to other populations of the species because it lives in a favorable area, and the birthrate is most likely higher than the death rate. 480

speciation Origin of new species due to the evolutionary process of descent with modification. 310

species Group of similarly constructed organisms capable of interbreeding and producing fertile offspring; organisms that share a common gene pool; taxon at the lowest level of classification. 310, 493

specific epithet In the binomial system of taxonomy, the second part of an organism's name, which may be descriptive. 491

sperm Male gamete having a haploid number of chromosomes and the ability to fertilize an egg, the female gamete. 907

spermatogenesis Production of sperm in males by the process of meiosis and maturation. 168

spicule Skeletal structure of sponges composed of calcium carbonate or silicate. 583

spinal cord In vertebrates, the nerve cord that is continuous with the base of the brain and housed within the vertebral column. 838

spinal nerve Nerve that arises from the spinal cord. 834

spindle Microtubule structure that brings about chromosomal movement during nuclear division. 146

spiral organ Structure in the vertebrate inner ear that contains auditory receptors (also called organ of Corti). 857

spleen Large, glandular organ located in the upper left region of the abdomen that stores and purifies blood. 761

sponge Invertebrate animal of the phylum Porifera; pore-bearing filter feeder whose inner body wall is lined by collar cells. 582

spongy bone Type of bone that has an irregular meshlike arrangement of thin plates of bone. 727, 866

spongy mesophyll Layer of tissue in a plant leaf containing loosely packed cells increasing the amount of surface area for gas exchange. 658

sporangium (pl., sporangia) Structure that produces spores. 539, 546, 560

spore Asexual reproductive or resting cell capable of developing into a new organism without fusion with another cell, in contrast to a gamete. 545

sporophyte Diploid generation of the alternation of generations life cycle of a plant; produces haploid spores that develop into the haploid generation. 559, 700

sporozoan Spore-forming protist that has no means of locomotion and is typically a parasite with a complex life cycle having both sexual and asexual phases. 538

spring overturn Mixing process that occurs in spring in stratified lakes whereby the oxygen-rich top waters mix with nutrient-rich bottom waters. 457

stabilizing selection Outcome of natural selection in which extreme phenotypes are eliminated and the average phenotype is conserved. 308

stamen In flowering plants, the portion of the flower that consists of a filament and an anther containing pollen sacs where pollen is produced. 573, 701

starch Storage polysaccharide found in plants that is composed of glucose molecules joined in a linear fashion with few side chains. 38

statocyst Static equilibrium organs of cnidarians, molluscs, and crustaceans; gives information about the position of the head. 859

statolith Sensors found in root cap cells which cause a plant to demonstrate gravitropism. 685

stem Usually the upright, vertical portion of a plant, which transports substances to and from the leaves. 643

steroid Type of lipid molecule having a complex of four carbon rings; examples are cholesterol, progesterone, and testosterone. 42

steroid hormone Type of hormone that has a complex of four carbon rings but different side chains from other steroid hormones. 883

stigma In flowering plants, portion of the pistil where pollen grains adhere and germinate before fertilization can occur. 573, 700

stolon Stem that grows horizontally along the ground and may give rise to new plants where it contacts the soil (e.g., the runners of a strawberry plant). 656

stomach In vertebrates, muscular sac that mixes food with gastric juices to form chyme, which enters the small intestine. 786

stomate Small opening between two guard cells on the underside of leaf epidermis through which gases pass. 562, 645, 676

stratum Ancient layer of sedimentary rock; results from slow deposition of silt, volcanic ash, and other materials. 324

striated Having bands; in cardiac and skeletal muscle, alternating light and dark crossbands produced by the distribution of contractile proteins. 728

strobilus In club mosses, terminal clusters of leaves which bear sporangia. 565

stroma Fluid within a chloroplast that contains enzymes involved in the synthesis of carbohydrates during photosynthesis. 71, 114

stromatolite Domed structures found in shallow seas consisting of cyanobacteria bound to calcium carbonate. 327

structural gene Gene that codes for an enzyme in a metabolic pathway. 248

style Elongated, central portion of the pistil between the ovary and stigma. 573, 701

substrate Reactant in a reaction controlled by an enzyme. 102

substrate-level phosphorylation Process in which ATP is formed by transferring a phosphate from a metabolic substrate to ADP. 130

sucrose Disaccharide that contains glucose and fructose; found in plants and is refined as table sugar. 37

survivorship Probability of newborn individuals of a cohort surviving to particular ages. 384

sustainable development Management of an ecosystem so that it maintains itself while providing services to human beings. 282

swim bladder In fishes, a gas-filled sac whose pressure can be altered to change buoyancy. 624

symbiosis Relationship that occurs when two different species live together in a unique way; it may be beneficial, neutral, or detrimental to one and/or the other species. 410, 516

sympathetic division Division of the autonomic system that is active when an organism is under stress; uses norepinephrine as a neurotransmitter. 837

sympatric speciation Origin of new species in populations that overlap geographically. 312

synapse Junction between neurons consisting of the presynaptic (axon) membrane, the synaptic cleft, and the postsynaptic (usually dendrite) membrane. 833

synapsis Pairing of homologous chromosomes during meiosis I. 161

synaptic cleft Small gap between presynaptic and postsynaptic membranes of a synapse. 833

synovial joint Freely moving joint in which two bones are separated by a cavity. 871

systematics Study of the diversity of organisms to determine evolutionary relationships and classify organisms. 494

systemic circuit Circulatory pathway of blood flow between the tissues and the heart. 748

systole Contraction period of a heart during the cardiac cycle. 746

t

T lymphocyte Lymphocyte that matures in the thymus and exists in four varieties, one of which kills antigen-bearing cells outright. 764

taiga Terrestrial biome that is a coniferous forest extending in a broad belt across northern Eurasia and North America. 448

taproot Main axis of a root that penetrates deeply and is used by certain plants such as carrots for food storage. 650

taste bud Structure in the vertebrate mouth containing sensory receptors for taste; in humans most taste buds are on the tongue. 848

taxon (pl., taxa) Group of organisms that fills a particular classification category. 492

taxonomy Branch of biology concerned with identifying, describing, and naming organisms. 8, 490

template Parental strand of DNA that serves as a guide for the complementary daughter strand produced during DNA replication. 226

tendon Strap of fibrous connective tissue that connects skeletal muscle to bone. 726, 873

tendril Threadlike extension of a modified leaf or stem used for grasping objects; a means of mechanical support. 686

terminal bud Bud that develops at the apex of a shoot. 652

terminator Specific DNA sequence that signals transcription to terminate. 238

territoriality Marking and/or defending a particular area against invasion by another species member; area often used for the purpose of feeding, mating, and caring for young. 368

testcross Cross between an individual with the dominant phenotype and an individual with the recessive phenotype. The resulting phenotypic ratio indicates whether the dominant phenotype is homozygous or heterozygous. 180

testis Male gonad which produces sperm and the male sex hormones. 902

tetanus Sustained muscle contraction without relaxation. 873

tetrapod Four-footed vertebrates; includes amphibians, reptiles, birds, and mammals. 625

thalamus In vertebrates, the portion of the diencephalon that passes on selected sensory information to the cerebrum. 839

therapsid Mammallike reptiles appearing in the middle Permian period; ancestral to mammals. 628

thermal inversion Temperature inversion that traps cold air and its pollutants near the earth with the warm air above it. 433

thermoacidophile Type of archaea that lives in hot and acidic aquatic habitats, such as hot springs or near hydrothermal vents. 520

thigmotropism In plants, unequal growth due to contact with solid objects, as the coiling of tendrils around a pole. 686

three-domain system System of classification that recognizes three domains: Bacteria, Archaea, and Eukarya. 502

thrombin Enzyme that converts fibrinogen to fibrin threads during blood clotting. 754

thylakoid Flattened sac within a granum whose membrane contains chlorophyll and where the light-dependent reactions of photosynthesis occur. 60, 71, 114

thymine (T) One of four nitrogen-containing bases in nucleotides composing the structure of DNA; pairs with adenine. 223

thymus gland Lymphoid organ involved in the development and functioning of the immune system; T lymphocytes mature in the thymus gland. 761, 897

thyroid gland Large gland in the neck that produces several important hormones, including thyroxine, triiodothyronine and calcitonin. 889

tight junction Junction between cells when adjacent plasma membrane proteins join to form an impermeable barrier. 93

tissue Group of similar cells combined to perform a common function. 724

tissue culture Process of growing tissue artificially in usually a liquid medium in laboratory glassware. 712

tissue engineering Biotechnology which creates products from a combination of living cells and biodegradable polymers resulting in bioartificial organs and implants. 271

tissue fluid Fluid that surrounds the body's cells consists of dissolved substances that leave the blood capillaries by filtration and diffusion. 734, 755

tone Continuous, partial contraction of muscle. 873

tonicity Osmolarity of a solution compared to that of a cell; if the solution is isotonic to the cell, there is no net movement of water; if the solution is hypotonic, the cell gains water; and if the solution is hypertonic, the cell loses water. 86

tonsils Partially encapsulated lymph nodules located in the pharynx. 761, 806

totipotent Cell that has the full genetic potential of the organism and has the potential to develop into a complete organism. 712, 924

trachea (pl., tracheae) Air tube in insects that is located between the spiracles and the tracheoles. 802

trachea Air tube (windpipe) in tetrapod vertebrates that runs between the larynx and the bronchi. 607

tracheid In vascular plants, type of cell in xylem that has tapered ends and pits through which water and minerals flow. 672

tract Bundle of myelinated axons in the central nervous system. 838

traditional systematics School of systematics that takes into consideration the degree of difference between derived characters to construct phylogenetic trees. 500

transcription Process whereby a DNA strand serves as a template for the formation of mRNA. 236

transcription factor In eukaryotes, protein required for the initiation of transcription by RNA polymerase. 254

transcriptional control Control of gene expression during the transcriptional phase determined by mechanisms that control whether transcription occurs or the rate at which it occurs. 251

transduction Exchange of DNA between bacteria by means of a bacteriophage. 515

transfer rate Amount of a substance that moves from one component of the environment to another within a specified period of time. 430

transfer RNA (tRNA) Type of RNA that transfers a particular amino acid to a ribosome during protein synthesis; at one end it binds to the amino acid and at the other end it has an anticodon that binds to an mRNA codon. 236

transformation Taking up of extraneous genetic material from the environment by bacteria. 515

transgenic organism Free-living organisms in the environment that have had a foreign gene inserted into them. 270, 714

transition reaction Reaction that oxidizes pyruvate with the release of carbon dioxide; results in the acetyl-CoA and connects glycolysis to the Krebs cycle. 129

translation Process whereby ribosomes use the sequence of codons in mRNA to produce a polypeptide with a particular sequence of amino acids. 236

translational control Gene expression regulated by the activity of mRNA transcripts. 251

transpiration Plant's loss of water to the atmosphere, mainly through evaporation at leaf stomates. 674

trichinosis Serious infection caused by parasitic roundworm of the phylum Nematoda whose larvae encyst in muscles. 593

trichocyst Found in ciliates, trichocysts contain long, barbed threads useful for defense and capturing prey. 537

triglyceride Neutral fat composed of glycerol and three fatty acids. 40

triplet code During gene expression, each sequence of three nucleotide bases stands for a particular amino acid. 237

trophic level Feeding level of one or more populations in a food web. 422

trophoblast Outer membrane surrounding the embryo in mammals, when thickened by a layer of mesoderm, it becomes the chorion, an extraembryonic membrane. 931

tropism In plants, a growth response toward or away from a directional stimulus. 684

trypanosome Parasitic zooflagellate that causes severe disease in human beings and domestic animals, including a condition called sleeping sickness. 535

trypsin Protein-digesting enzyme secreted by the pancreas. 788

tube foot Part of the water vascular system in sea stars, located on the oral surface of each arm; functions in locomotion. 619

tube-within-a-tube body plan Body with a digestive tract that has both a mouth and an anus. 588

tubular reabsorption Movement of primarily nutrient molecules and water from the contents of the nephron into blood at the proximal convoluted tubule. 818

tubular secretion Movement of certain molecules from blood into the distal convoluted tubule of a nephron so that they are added to urine. 818

tumor Cells derived from a single mutated cell that has repeatedly undergone cell division; benign tumors remain at the site of origin, and malignant tumors metastasize. 152, 260

tumor-suppressor gene Gene that codes for a protein that ordinarily suppresses cell division; inactivity can lead to a tumor. 260

turgor pressure Pressure of the cell contents against the cell wall, in plant cells, determined by the water content of the vacuole; gives internal support to the plant cell. 87

tympanic membrane Membranous region that receives air vibrations in an auditory organ; in humans, the eardrum. 856

typhlosole Expanded dorsal surface of long intestine of earthworms, allowing additional surface for absorption. 605

u

umbilical cord Cord connecting the fetus to the placenta through which blood vessels pass. 933

uniformitarianism Belief espoused by James Hutton that geological forces act at a continuous, uniform rate. 285

unsaturated Fatty acid molecule that has one or more double bonds between the carbons of its hydrocarbon chain. The chain bears fewer hydrogens than the maximum number possible. 40

upwelling Upward movement of deep, nutrient-rich water along coasts; it replaces surface waters that move away from shore when the direction of prevailing wind shifts. 460

uracil The pyrimidine base that replaces thymine in RNA; pairs with adenine. 236

urea Main nitrogenous waste of terrestrial amphibians and most mammals. 814

ureter Tubular structure conducting urine from the kidney to the urinary bladder. 816

urethra Tubular structure that receives urine from the bladder and carries it to the outside of the body. 816

uric acid Main nitrogenous waste of insects, reptiles and birds. 814

urinary bladder Organ where urine is stored. 816

urine Liquid waste product made by the nephrons of the vertebrate kidney through the processes of glomerular filtration, tubular reabsorption, and tubular secretion. 816

uterine cycle Cycle that runs concurrently with the ovarian cycle; it prepares the uterus to receive a developing zygote. 910

uterus In mammals, expanded portion of the female reproductive tract through which eggs pass to the environment or in which an embryo develops and is nourished before birth. 908

utricle Saclike cavity in the vestibule of the vertebrate inner ear; contains sensory receptors for static equilibrium. 859

v

vaccine Antigens prepared in such a way that they can promote active immunity without causing disease. 772
vacuole Fluid-filled membrane-bounded sac that fills much of the interior of a plant cell and contains a variety of substances. 69
variable Factor that can cause an observable change during the progress of the experiment. 10
vascular bundle In plants, primary phloem and primary xylem enclosed by a bundle sheath. 646
vascular cambium In plants, lateral meristem that produces secondary phloem and secondary xylem. 652
vascular cylinder In dicots, the tissues in the middle of a root consisting of the pericycle and vascular tissues. 646
vascular plant Type of plant that contains vascular tissue; e.g., ferns, gymnosperms, and angiosperms. 562
vascular tissue Transport tissue in plants consisting of xylem and phloem. 645
vector In genetic engineering, a means to transfer foreign genetic material into a cell; e.g., a plasmid. 266
vein Blood vessel that arises from venules and transports blood toward the heart. 742
vena cava Large systemic vein that returns blood to the right atrium of the heart in tetrapods; either the superior or inferior vena cava. 748
ventricle Cavity in an organ, such as a lower chamber of the heart; or the ventricles of the brain. 744, 838
venule Vessel that takes blood from capillaries to a vein. 742
vertebral column Portion of the vertebrate endoskeleton that houses the spinal cord; consists of many vertebrae separated by intervertebral disks. 869
vertebrate Chordate in which the notochord is replaced by a vertebral column. 622
vesicle Small, membrane-bounded sac that stores substances within a cell. 66
vessel element Cell which joins with others to form a major conducting tube found in xylem. 672
vestigial structure Remains of a structure that was functional in some ancestor but is no longer functional in the organism in question. 296
villus (pl., villi) Small, fingerlike projection of the inner small intestinal wall. 788
viroid Infectious strand of RNA devoid of a capsid and much smaller than a virus. 512
virus Noncellular parasitic agent consisting of an outer capsid and an inner core of nucleic acid. 508
visible light Portion of the electromagnetic spectrum that is visible to the human eye. 112
vitamin Essential requirement in the diet, needed in small amounts. They are often part of coenzymes. 105, 792
vocal cord In humans, fold of tissue within the larynx; creates vocal sounds when it vibrates. 802

w

waggle dance Performed by honeybees, this action indicates the distance and direction of a food source. 371
water (hydrologic) cycle Interdependent and continuous circulation of water from the ocean, to the atmosphere, to the land, and back to the ocean. 429
water mold Unicellular or filamentous organisms having cell walls made of cellulose; typically decomposers of dead freshwater organisms but some are parasites of aquatic or terrestrial organisms. 540
water potential Potential energy of water; it is a measure of the capability to release or take up water relative to another substance. 673
water vascular system Series of canals that takes water to the tube feet of an echinoderm, allowing them to expand. 619
wax Sticky, solid, waterproof lipid consisting of many long-chain fatty acids usually linked to long-chain alcohols. 40
whisk fern Common name for seedless vascular plant that consists only of stems and has no leaves or roots. 564
white blood cell Leukocyte, of which there are several types, each having a specific function in protecting the body from invasion by foreign substances and organisms. 753
white matter Myelinated axons in the central nervous system. 838
woody plant Plant that contains wood; usually trees such as evergreen trees (gymnosperms) and flowering trees (angiosperms). Alternative is a herbaceous plant. 542

x

X chromosome Female sex chromosome that carries genes involved in sex determination; see Y chromosome. 202
X-linked Allele located on an X chromosome but may control a trait that has nothing to do with the sexual characteristics of an animal. 192, 214
xenotransplantation Use of animal organs, instead of human organs, in human transplant patients. 271
xylem Vascular tissue that transports water and mineral solutes upward through the plant body; it contains vessel elements and tracheids. 562, 646, 672

y

Y chromosome Male sex chromosome that carries genes involved in sex determination; see X chromosome. 202
yolk Dense nutrient material that is present in the egg of a bird or reptile. 903, 920
yolk sac One of the extraembryonic membranes which in shelled vertebrates contains yolk for the nourishment of the embryo and in placental mammals is the first site for blood cell formation. 930

z

zero population growth No growth in population size. 394
zooflagellate Nonphotosynthetic protist that moves by flagella; typically zooflagellates enter into symbiotic relationships and some are parasitic. 535
zooplankton Part of plankton containing protozoa and other types of microscopic animals. 457, 536
zygospore Thick-walled, resting cell formed during sexual reproduction of zygospore fungi. 546
zygote Diploid cell formed by the union of two gametes; the product of fertilization. 160

Credits

Line Art and Text

Chapter 2

Ecology Focus: Data from G. Tyler Miller, *Living in the Environment*, 1993, Wadsworth Publishing Company, Belmont, CA; and Lester R. Brown, *State of the World*, 1992, W.W. Norton & Company, Inc., New York, NY. p. 29.

Chapter 3

3.9B, C: From R.R. Seeley & T.D. Stephens, *Anatomy & Physiology*, 3rd edition. Copyright 1995 McGraw-Hill Companies, Inc. All Rights Reserved. Reprinted by Permission; **3.19A:** From Peter H. Raven & George B. Johnson, *Biology*, 5th edition. Copyright 1996 McGraw-Hill Companies, Inc. All Rights Reserved. Reprinted by Permission. p. 41.

Chapter 6

6.6: From Peter H. Raven & George B. Johnson, *Biology*, 5th edition. Copyright 1996 McGraw-Hill Companies, Inc. All Rights Reserved. Reprinted by Permission. p. 103.

Chapter 8

Health Focus: Scott K. Powers and Edward T. Howley, *Exercise Physiology*, 2d edition. Copyright 1994 McGraw-Hill Companies, Inc., Dubuque, Iowa. All Rights Reserved. Reprinted by Permission. p. 138.

Chapter 9

9.5: From Peter H. Raven and George B. Johnson, *Biology*, 5th edition. Copyright 1996 McGraw-Hill Companies, Inc. All Rights Reserved. Reprinted by Permission. p. 149; **9.6:** R.G. Kessel and C.Y. Shih, *Scanning Electron Microscopy in Biology: A Student's Atlas on Biological Organization* (1974). Reprinted with Permission of Springer-Verlag, New York. p. 149.

Chapter 11

11.8: From Burton S. Guttman, *Biology*. Copyright 1999 McGraw-Hill Companies, Inc. All Rights Reserved. Reprinted by Permission. p. 182; **11.9:** From Burton S. Guttman, *Biology*. Copyright 1999 McGraw-Hill Companies, Inc. All Rights Reserved. Reprinted by Permission. p. 182.

Chapter 12

12.5: From Burton S. Guttman, *Biology*. Copyright 1999 McGraw-Hill Companies, Inc. All Rights Reserved. Reprinted by Permission. p. 193.

Chapter 13

13.2B: From Robert F. Weaver and Philip W. Hedrick, *Genetics*, 2nd edition. Copyright 1994 McGraw-Hill Companies, Inc. All Rights Reserved. Reprinted by Permission. p. 203.

Chapter 16

Science Focus: Courtesy of Joyce Haines. p. 257.

Chapter 18

18.14: From Burton S. Guttman, *Biology*. Copyright 1999 McGraw-Hill Companies, Inc. All Rights Reserved. Reprinted by Permission. p. 293; **18C:** From Peter H. Raven & George B. Johnson, *Biology*, 5th edition. Copyright 1996 McGraw-Hill Companies, Inc. All Rights Reserved. Reprinted by Permission. p. 294.

Chapter 19

19.13: From Burton S. Guttman, *Biology*. Copyright 1999 McGraw-Hill Companies, Inc. All Rights Reserved. Reprinted by Permission. p. 313; **Science Focus:** Courtesy of Gerald D. Carr, University of Hawaii at Manoa. p. 314.

Chapter 20

20.16: Data supplied and illustration reviewed by J. John Sepkoski, Jr., Professor of Paleontology, University of Chicago. p. 336.

Chapter 21

21.1: From Burton S. Guttman, *Biology*. Copyright 1999 McGraw-Hill Companies, Inc. All Rights Reserved. Reprinted by Permission. p. 342; **21.2:** Redrawn from *Vertebrates, Phylogeny, and Philosophy*, 1986. Contributions to Geology, University of Wyoming, Special Paper 3 (frontispiece and article) by Rose and Bown; and Thomas M. Bown and Kenneth D. Rose, "Patterns of Dental Evolution in Early Eocene Anaptomorphine Primates (Omomyidae) from Bighorn Basin, Wyoming" in Paleontological Society Memoir 23 (*Journal of Paleontology*, Vol. 61, supplement to no. 5). Courtesy of Kenneth D. Rose of The Johns Hopkins University, Baltimore, MD. p. 343; **21.4C:** Reprinted by Permission from Alan Walker; **Science Focus:** Courtesy of Steven Stanley, The Johns Hopkins University. p. 345; **21.9:** From C. Hickman, *Integrated Principles of Zoology*, 9th edition, 1997. Reproduced with Permission of The McGraw-Hill Companies, Inc. All Rights Reserved. Reprinted by Permission. p. 351.

Chapter 22

22.2: Data from S.J. Arnold, "The Microevolution of Feeding Behavior" in *Foraging Behavior: Ecology, Ethological, and Psychological Approaches*, edited by A. Kamil and T. Sargent, 1980, Garland Publishing Company, New York, NY. p. 363; **22.7:** Data from G. Hausfater, "Dominance and Reproduction in Baboons (Papio cynocephalus): A Quantitive Analysis," *Contributions in Primatology*, 7:1–150, 1975. p. 367.

Chapter 23

23.5: From Raymond Pearl, *The Biology of Population Growth*. Copyright 1925 McGraw-Hill Companies, Inc. All Rights Reserved. Reprinted by Permission. p. 382; **23.7B:** Data from W.K. Purves, et al., *Life: The Science of Biology*, 4th edition, Sinauer & Associates. p. 384; **23.7C:** Data from E.J. Kormondy, 1984, *Concepts of Ecology*, 3rd edition, Prentice-Hall, Inc., Figure 4.6, page 107; **23.7D:** Data from A.K. Hegazy, 1990, "Population Ecology & Implications for Conservation of Cleome Droserifolia: A Threatened Xerophyte," *Journal of Arid Environments*, 19:269–82. p. 384; **23.9B:** Data from Charles J. Krebs, *Ecology*, 3rd edition, 1984, Harper & Row; after Scheffer, 1951; **23.11B:** Data from Charles J. Krebs, *Ecology*, 3rd edition, 1984, Harper & Row; after Lack 1966 and J. Krebs, personal communication. p. 387; **23.11C:** Data from Michael Begon, et al., *Population Ecology*, 3rd edition, 1996, Blackwell Science; **23A table:** Data from Charles J. Krebs, *Ecology*, 3rd edition, 1984, Harper & Row. p. 388; **23A:** Data from Karen Arms & Pamela Camp, *Biology*, 4th edition, 1995, Saunders College Publishing. p. 389; **23B:** Data from Charles J. Krebs, *Ecology*, 3rd edition, 1984, Harper & Row.

p. 389; **Science Focus and 23C:** Reading and Photograph Courtesy of Jeffrey Kassner. p. 392; **23.14:** Data from *Population Today* Vol. 24, No. 4, April 1996, Population Reference Bureau, Inc., Washington, D.C. p. 393; **23.15:** Data from *World Population Profiles*, WP-89; **23.16:** Data from *Natural Defense Council*, 1993. p. 395.

Chapter 24

24.1C: Data from G.G. Simpson, "Species Density of North America Recent Mammals" in *Systematic Zoology*, Vol. 13:57–73, 1964. p. 400; **24.2:** Data from Charles J. Krebs, *Ecology*, 3rd edition, 1984, Harper & Row. p. 401; **24.5:** Data from G.F. Gause, *The Struggle for Existence*, 1934, Williams & Wilkins Company, Baltimore, MD. p. 404; **24.6:** From Peter H. Raven and George B. Johnson, *Biology*, 5th edition. Copyright 1996 McGraw-Hill Companies, Inc. All Rights Reserved. Reprinted by Permission. p. 404; **24.8:** From Peter H. Raven and George B. Johnson, *Biology*, 4th edition. Copyright 1996, McGraw-Hill Company, Inc., Dubuque, Iowa. All Rights Reserved. Reprinted by Permission. p. 405; **24.9:** Data from G.F. Gause, *The Struggle for Existence*, 1934, Williams & Wilkins Company, Baltimore, MD. p. 406; **24.10:** Data from D.A. MacLulich, *Fluctuations in the Numbers of the Varying Hare (Lepus americanus)*, University of Toronto Press, Toronto, 1937, reprinted 1974. p. 407.

Chapter 26

26.1B: From Peter H. Raven and George B. Johnson, *Biology*, 5th edition. Copyright 1996 McGraw-Hill Companies, Inc. All Rights Reserved. Reprinted by Permission. p. 442; **26.2:** From Peter H. Raven and George B. Johnson, *Biology*, 5th edition. Copyright 1996 McGraw-Hill Companies, Inc. All Rights Reserved. Reprinted by Permission. p. 443; **26.4B:** From Peter H. Raven and George B. Johnson, *Biology*, 5th edition. Copyright 1996 McGraw-Hill Companies, Inc. All Rights Reserved. Reprinted by Permission; **26.7:** From Peter H. Raven and George B. Johnson, *Biology*, 5th edition. Copyright 1996 McGraw-Hill Companies, Inc. All Rights Reserved. Reprinted by Permission. p. 444; **26.8:** From Peter H. Raven and George B. Johnson, *Biology*, 5th edition. Copyright 1996 McGraw-Hill Companies, Inc. All Rights Reserved. Reprinted by Permission. p. 448; **26.9:** From Peter H. Raven and George B. Johnson, *Biology*, 5th edition. Copyright 1996 McGraw-Hill Companies, Inc. All Rights Reserved. Reprinted by Permission. p. 449; **26.11:** From Peter H. Raven and George B. Johnson, *Biology*, 5th edition. Copyright 1996 McGraw-Hill Companies, Inc. All Rights Reserved. Reprinted by Permission. p. 453; **26.13:** From Peter H. Raven and George B. Johnson, *Biology*, 5th edition. Copyright 1996 McGraw-Hill Companies, Inc. All Rights Reserved. Reprinted by Permission. p. 453; **26.14:** From Peter H. Raven and George B. Johnson, *Biology*, 5th edition. Copyright 1996 McGraw-Hill Companies, Inc. All Rights Reserved. Reprinted by Permission. p. 453; **26.15:** From Peter H. Raven and George B. Johnson, *Biology*, 5th edition. Copyright 1996 McGraw-Hill Companies, Inc. All Rights Reserved. Reprinted by Permission. p. 454.

Chapter 27

27.2: Redrawn from "Shrimp Stocking, Salmon Collapse, and Eagle Displacement: Cascading Interactions in the Food Web of a Large Aquatic Ecosystem," by Spencer, C.N., B.R. McClelland, and J.A. Stanford, *BioScience*, 41(1):14–21. © 1991 American Institute of Biological Sciences. p. 471; **27.4C:** Reprinted by permission from *Nature* 368:734–737. From "Declining Biodiversity Can Alter the Performance of Ecosystems," by Thompson, Lawler, Lawton, & Woodfin. Copyright 1994 Macmillan Magazines Ltd. p. 475; **27.5B:** From "Qualifying Threats to Imperiled Species in the United States," by Wilcove, Rothstein, Dubow, Phillips & Losos, *BioScience*, 48(8) August 1998, page 609. © 1998 American Institute of Biological Sciences. p. 476; **27.7A:** Data from David M. Gates, *Climate Change and Its Biological Consequences*, 1993, Sinauer Associates, Inc., Sunderland, MA. p. 478.

Chapter 28

28.13: From P.H. Raven & G.B. Johnson, *Biology*, 5th edition. Copyright 1996 McGraw-Hill Companies, Inc. All Rights Reserved. Reprinted by Permission. p. 500.

Chapter 29

29.2 table: Courtesy of Lansing M. Prescott. p. 518; **29.2 table frame 2, 3, 6, 7, 8:** From Peter H. Raven and George B. Johnson, *Biology*, 5th edition. Copyright 1999 McGraw-Hill Companies, Inc. All Rights Reserved. Reprinted by Permission. p. 518.

Chapter 32

32.13: From Moore, Clark & Vodopich, *Botany*, 2nd edition. Copyright 1995 McGraw-Hill Companies, Inc. All Rights Reserved. Reprinted by Permission. p. 573.

Chapter 33

33.7: From Stephen Miller and John Harley, *Zoology*, 3rd edition. Copyright 1996 McGraw-Hill Companies, Inc., Dubuque, Iowa. All Rights Reserved. Reprinted by Permission. p. 587. **33.9C:** Source: Population Reference Bureau. p. 393.

Chapter 35

35.2: From Stephen Miller and John Harley, *Zoology*, 3rd edition. Copyright 1996 McGraw-Hill Companies, Inc. All Rights Reserved. Reprinted by Permission. p. 620; **35.12:** From Burton S. Guttman, *Biology*. Copyright 1999 McGraw-Hill Companies, Inc. All Rights Reserved. Reprinted by Permission. p. 628; **A Closer Look:** A Closer Look Courtesy of Gregory J. McConnell. p. 631; **35.16A:** From Burton S. Guttman, *Biology*. Copyright 1999 McGraw-Hill Companies, Inc. All Rights Reserved. Reprinted by Permission. p. 631.

Chapter 36

36A: Data from St. Regis Paper Company, New York, NY, 1966. p. 651; **36.18:** From Burton S. Guttman, *Biology*. Copyright 1999 McGraw-Hill Companies, Inc. All Rights Reserved. Reprinted by Permission. p. 658; **Ecology Focus:** Courtesy of Charles Horn. p. 657.

Chapter 37

37.3: From Moore, Clark & Vodopich, *Botany*, 2nd edition. Copyright 1995 McGraw-Hill Companies, Inc. All Rights Reserved. Reprinted by Permission. p. 668; **37.8:** From Peter H. Raven & George B. Johnson, *Biology*, 5th edition. Copyright 1996 McGraw-Hill Companies, Inc. All Rights Reserved. Reprinted by Permission. p. 672; **Science Focus:** Courtesy of G. David Tilman, University of Minnesota. p. 677.

Chapter 38

Science Focus: Courtesy of Donald Briskin and Margaret Gawienowski, University of Illinois at Urbana–Champaign. p. 691.

Chapter 39

39.3: From Moore, Clark & Vodopich, *Botany*, 2nd edition. Copyright 1995 McGraw-Hill Companies, Inc. All Rights Reserved. Reprinted by Permission. p. 703; **39.8:** Kingsley R. Stern, *Introductory Plant Biology*, 6th edition. Copyright 1994 McGraw-Hill Companies, Inc., Dubuque, Iowa. All Rights Reserved. Reprinted by Permission. p. 710; **39.9:** Kingsley R. Stern, *Introductory Plant Biology*, 6th edition.

Copyright 1994 McGraw-Hill Companies, Inc., Dubuque, Iowa. All Rights Reserved. Reprinted by Permission. p. 711.

Chapter 40

40.4: From John W. Hole, Jr., *Human Anatomy & Physiology*, 6th edition. Copyright 1993 McGraw-Hill Companies, Inc., Dubuque, Iowa. All Rights Reserved. Reprinted by Permission. p. 727; **40.10:** From D. Shier, et al., *Hole's Human Anatomy & Physiology*, 8th edition. Copyright 1996 McGraw-Hill Companies, Inc. All Rights Reserved. Reprinted by Permission. p. 735.

Chapter 43

43.7: From Kent M. Van De Graaff and Stuart Ira Fox, *Concepts of Human Anatomy and Physiology*, 4th edition. Copyright 1995 McGraw-Hill Companies, Inc., Dubuque, Iowa. All Rights Reserved. Reprinted by Permission. p. 786; **43.11:** U.S. Department of Agriculture. p. 792.

Chapter 49

49.10: From K.S. Saladin, *Anatomy & Physiology*. Copyright 1999 McGraw-Hill Companies. Reprinted with Permission of The McGraw-Hill Companies. p. 891.

Chapter 50

50.11: From Kent M. Van De Graaff and Stuart Ira Fox, *Concepts of Human Anatomy and Physiology*, 4th edition. Copyright 1995 McGraw-Hill Companies, Inc., Dubuque, Iowa. All Rights Reserved. Reprinted by Permission. p. 912; **50A:** Sexually Transmitted Disease Surveillance, 1997 and 1998. Atlanta: Centers for Disease Control and Prevention. p. 915; **50.2 text art:** From Ruth Bernstein and Stephen Bernstein, *Biology*. Copyright 1996 McGraw-Hill Companies, Inc., Dubuque, Iowa. All Rights Reserved. Reprinted by Permission. p. 917.

Chapter 51

51.8: From Ricki Lewis, *Life*, 3rd edition. Copyright 1997 McGraw-Hill Companies, Inc. All Rights Reserved. Reprinted by Permission. p. 926; **Science Focus:** Courtesy of John Sternfeld, State University of New York College at Cortland. p. 927; **51.10B:** From Burton S. Guttman, *Biology*. Copyright 1999 McGraw-Hill Companies, Inc. All Rights Reserved. Reprinted by Permission. p. 929; **51.16:** Redrawn from Kent M. Van De Graaff and Stuart Ira Fox, *Concepts of Human Anatomy and Physiology*, 4th edition. Copyright 1995, McGraw-Hill Companies, Inc., Dubuque, Iowa. All Rights Reserved. Reprinted by Permission. p. 935.

Photos

History of Biology

Leeuwenhoek, Darwin, Pasteur, Koch, Pavlov, Lorenz, Pauling: © Bettmann/Corbis; McClintock: © AP/Wide World; Franklin: © Cold Springs Harbor Laboratory.

Chapter 1

Opener: © Joe McDonald/Animals Animals/Earth Scenes; **1.2a:** © Mitch Reardon/Photo Researchers, Inc.; **1.2b:** © David C. Fritts/Animals Animals/Earth Scenes; **1.3:** © Francisco Erize/Bruce Coleman, Inc.; **1.5a:** Rainforest: © Barbara Von Hoffman/Tom Stack & Associates; **1.5b:** Toucan: © Ed Reschke/Peter Arnold, Inc.; **1.5 (Orchid):** © Max and Bea Hunn/Visuals Unlimited; **1.5 (Butterfly):** © Kjell Sandved/Butterfly Alphabet; **1.5 (Jaguar):** © BIOS (Seitre)/Peter Arnold, Inc.; **1.5 (Frog):** © Kevin Schafer/Martha Hill/Tom Stack & Assoc.; **1.6a:** © T.J. Beveridge/Visuals Unlimited; **1.6 (Nostoc):** © Eric Grave/Photo Researchers, Inc.; **1.6 (Euglena):** © John D. Cunningham/ Visuals Unlimited; **1.6d (Mushroom):** © Rod Planck/Tom Stack & Assoc.; **1.6 (Rosa):** © Farell Grehan/Photo Researchers, Inc.; **1.6 (Lynx):** © Leonard L. Rue.

Chapter 2

Opener: © Dani/Jeske/Animals Animals/Earth Scenes; **2.1:** © The McGraw-Hill Companies, Inc./Carlyn Iverson, photographer; **2.5b:** © Charles M. Falco/Photo Researchers, Inc.; **2.8a:** © Grant Taylor/Tony Stone Images, Inc.; **2.10a:** © Di Maggio/Peter Arnold, Inc.; **p. 29:** © Frederica Georgia/Photo Researchers, Inc.

Chapter 3

Opener: © Dr. Gopal Murti/SPL/Photo Researchers, Inc.; **3.1a:** © John Gerlach/Tom Stack & Assoc.; **3.1b:** © Leonard Lee Rue/ Photo Researchers, Inc.; **3.1c:** © H. Pol/ CNRI/SPL/Photo Researchers, Inc.; **3.7a:** © Jeremy Burgess/SPL/Photo Researchers, Inc.; **3.7b:** © CNRI/SPL/Photo Researchers, Inc.; **3.8:** © BioPhoto Assoc./Photo Researchers, Inc.; **3.18:** Courtesy Ealing Corporation.

Chapter 4

Opener: © Biophoto Associates/Photo Researchers, Inc.; **4.1a:** © Tony Stone Images; **4.1b:** © Jim Solliday/Biological Photo Service; **4.1c:** © Barbara J. Miller/Biological Photo Service; **4.1d:** © Ed Reschke; **4Aa:** © Robert Brons/Tony Stone Images; **4Ab:** © M. Schliwa/Visuals Unlimited; **4Ac:** © Kessel/Shih/Peter Arnold, Inc.; **4Ba:** © Ed Reschke; **4Bb:** © Biophoto Associates/Photo Researchers, Inc.; **4Bc:** © David M. Phillips/Visuals Unlimited; **4Bd, 4Be:** © David M. Phillips/Visuals Unlimited; **4.3a:** © David M. Phillips/Visuals Unlimited; **4.3b:** © Biophoto Assoc./Photo Researchers, Inc.; **4.4b:** © Alfred Pasieka/Photo Researchers, Inc.; **4.5b:** © Newcomb/Wergin/BPS/Tony Stone Images; **4.7b:** Courtesy Ron Milligan/Scripps Research Institute; **4.7c:** Courtesy E.G. Pollock; **4.8a:** © Warren Rosenberg/Biological Photo Service; **4.10b:** © Richard Rodewald/Biological Photo Service; **4.10c:** © K.G. Murti/Visuals Unlimited; **4.11b:** © S.E. Frederick & E.H. Newcomb/Biological Photo Service; **4.12a:** Courtesy Herbert W. Israel, Cornell University; **4.13a:** Courtesy Dr. Keith Porter; **4.14 (top):** © M. Schliwa/Visuals Unlimited; **4.14 (middle):** © K.G. Murti/Visuals Unlimited; **4.14 (bottom):** © K.G. Murti/Visuals Unlimited; **4.15b:** Courtesy Kent McDonald, University of Colorado Boulder; **4.15c:** From Manley McGill, D.P. Highfield, T.M. Monahan, and B.R. Brinkley, *Journal of Ultrastructure Research* 57, 43–53 pp. 48, 6 (1976) Academic Press; **4.16a:** Sperm: © David M. Phillips/Photo Researchers, Inc.; **4.16b,c:** Flagellum and Basal Body: © William L. Dentler/Biological Photo Service.

Chapter 5

Opener: © Hank Morgan/Photo Researchers, Inc.; **5.1a:** © Warren Rosenberg/
Biological Photo Service; **5.1d:** © Don Fawcett/Photo Researchers, Inc.; **5.13b:** Courtesy Mark Bretscher; **5.16a:** Courtesy Camillo Peracchia; **5.16b:** © David M. Phillips/Visuals Unlimited; **5.16c:** From Douglas E. Kelly, J. Cell Biol. 28(1966):51. Reproduced by copyright permission of The Rockefeller University Press.

Chapter 6

Opener: © Patti Murray/Animals Animals/Earth Scenes; **6.1:** © Gregg Adams/Tony Stone Images.

Chapter 7

Opener: © Keren SU/Tony Stone Images; **7.2b:** © Newcomb & Wergin/BPS/Tony Stone Images; **7.2c:** Courtesy Herbert W. Israel, Cornell University; **7.6a:** Courtesy Melvin P. Calvin; **7.7:** © The McGraw-Hill Companies, Inc./Bob Coyle, photographer; **7.9a:** © Jim Steinberg/Photo Researchers, Inc.; **7.9b:** © Charlie Waite/Tony Stone Images; **7.9c:** © Beverly Factor/Tony Stone Images.

Chapter 8

Opener: © Jess Stock/Tony Stone Images; **8.5:** Courtesy Dr. Keith Porter; p. 138: © Tim Davis/Photo Researchers, Inc.

Chapter 9

Opener: © CNRI/Photo Researchers, Inc.; **9.1b:** © Stanley C. Holt/Biological Photo Service; **9.1c:** © Stanley C. Holt/Biological Photo Service; **9.1d:** © Stanley C. Holt/Biological Photo Service; **9.2a:** © Biophoto Assoc./Photo Researchers, Inc.; **9.3 (Nondividing cell, Prophase, Metaphase, Anaphase, Telophase):** © Ed Reschke; **9.3 (Prometaphase):** © Michael Abbey/Photo Researchers, Inc.; **9.4a–d:** © Ed Reschke; **9.5a:** © B.A. Palevitz and E.H. Newcomb BPS/Tom Stack & Associates; **9.6a,c:** © R.G. Kessel and C.Y. Shih (1974) *Scanning Electron Microscopy in Biology: A Student Atlas on Biological Organization* Springer-Verlag, New York; **9.9a:** Courtesy Klaus Hahn, Cecelia Subauste, and Gary Bokoch. From *Science,* vol. 280, April 3, 1998, p. 32.

Chapter 10

Opener: © Yorgos Nikis/Tony Stone Images; **10.1a:** © Department of Clinical Cytogenetics, Addenbrookes Hosp./SPL/Photo Researchers, Inc.; **10.3a:** Courtesy of Dr. D. Von Wettstein; **10.4:** Courtesy of Dr. Bernard John.

Chapter 11

Opener: © Darrell Gulin/Tony Stone Images; **11.1:** © Bettmann/Corbis.

Chapter 12

Opener: © Gunther F. Bahr/AFIP/Tony Stone Images; **12.4a:** © Jane Burton/Bruce Coleman, Inc.; **12.8a–c:** © The McGraw-Hill Companies, Inc./Bob Coyle, photographer.

Chapter 13

Opener: © Karl Weatherly/Tony Stone Images; **13.1(4):** © CNRI/SPL/Photo Researchers, Inc.; **13.1(5):** © CNRI/SPL/Photo Researchers, Inc.; **13.3a:** © Jill Cannefax/EKM-Nepenthe; **13.4a,b:** From Kampheier, R., *Physical Examination of Health Diseases* © 1958 F.A. Davis Company; **13.5a:** © David M. Phillips/Visuals Unlimited; **13.5b,c:** From R. Simensen and R. Curtis Rogers, "Fragile X Syndrome," *American Family Physician* 39(5):186, May 1989. © American Academy of Family Physicians; **13.8a,b:** © Steve Uzzell; **13.9:** Courtesy PathoGenesis; **13.10b:** © Bill Longcore/Photo Researchers, Inc.

Chapter 14

Opener: © Tony Hutchings/Tony Stone Images; **14.2b:** © Lee Simon/Photo Researchers, Inc.; **14.5b:** Courtesy of Biophysics Dept., Kings College, London; **14.6a:** © Nelson Max/Peter Arnold, Inc.

Chapter 15

Opener: © C.W. Perkins/Animals Animals/Earth Scenes; **15.2 (both):** © Bill Longcore/Photo Researchers, Inc.; **15.7a:** © Oscar L. Miller/Photo Researchers, Inc.; **15.9b:** Courtesy University of California Lawrence Livermore National Library and the U.S. Department of Energy; **15.10d:** Courtesy Alexander Rich.

Chapter 16

Opener: © David M. Phillips/Visuals Unlimited; **16.5:** Courtesy Stephen Wolfe; **16.8a:** Courtesy *Nature* 163:676 (1949); **16.9a:** From M.B. Roth and J.G. Gall *Journal of Cell Biology,* 105:1047–1054, 1987. The Rockefeller University Press; **16.10:** Courtesy Jose Mariano Amabis, University of Sao Paulo; **16A:** Courtesy Cold Spring Harbor Archive; **16B:** © John N.A. Lott/Biological Photo Service; **16.14:** Courtesy James E. Cleaver, University of California, San Francisco; **16.16:** Courtesy Dr. Howard Jones, Eastern VA Medical School.

Chapter 17

Opener: © James Balog/Tony Stone Images; **17.6:** © Will & Deni McIntyre/Photo Researchers, Inc.; **17.7 (both):** Courtesy General Electric Research & Development; **17.A:** Courtesy of Robert P. Lanza; **17.8a:** Courtesy Genzyme Corporation and Tufts University School of Medicine.

Chapter 18

Opener (inset): © Tom McHugh/Photo Researchers, Inc.; **Opener:** © C. Seghers/Photo Researchers, Inc.; **18.1 (Rhea):** © Tom Stack/Tom Stack & Associates; **18.1 (Desert):** © C. Luiz Claudio Marigo/Peter Arnold, Inc.; **18.1 (Mountains):** © Gary J. James/ Biological Photo Service; **18.1 (Rain forest):** © C. Luiz Claudio Mariog/Peter Arnold, Inc.; **18.1 (Iguanas):** © Ken Lucas; **18.1 (Finch):** © Miguel Castro/Photo Researchers, Inc.; **18.2:** © Bettmann/Corbis; **18.3:** © John D. Cunningham/Visuals Unlimited; **18.4:** © Malcolm Boulton/Photo Researchers, Inc.; **18.5b:** © J. & L. Weber/Peter Arnold, Inc.; **18.7:** © Juan & Carmecita Munoz/Photo Researchers, Inc.; **18.8a:** © Walt Anderson/Visuals Unlimited; **18.8b:** © Michael Dick/Animals Animals/Earth Scenes; **18.9a:** © Adrienne T. Gibson/Animals Animals/Earth Scenes; **18.9b:** © Joe McDonald/Animals Animals/Earth Scenes; **18.9c:** © Leonard Lee Rue/Animals Animals/Earth Scenes; **18.10:** © Wendell Metzen/Bruce Coleman, Inc.; **18A:** © Stock Montage; **18.11 (Timber wolf):** © Gary Milburn/Tom Stack & Assoc.; **18.11 (Red chow):** © Jeanne White/Photo Researchers; **18.11 (Bloodhound):** © Mary Bloom/Peter Arnold, Inc.; **18.11 (French bulldog):** © Salana/Jacana/Photo Researchers, Inc.; **18.11 (Boston terrier):** © Mary Bloom/Peter Arnold, Inc.; **18.11 (Shetland):** © Carolyn McKeone/Photo Researchers, Inc.; **18.11 (Beagle):** © Mary Bloom/Peter Arnold, Inc.; **18.11 (Dalmatian):** © Alexander Lowry/Photo Researchers, Inc.; **18.11 (Chihuahua):** © Mary Eleanor Browning/Photo Researchers, Inc.; **18.12a–c:** Courtesy W. Atlee Burpee Company; **18.13a,b:** © O. Louis Mazzatenta/National Geographic; **18.15a (Wombat):** © Adrienne Gibson/Animals Animals/Earth Scenes; **18.15b (Sugar glider):** © John Sundance/Jacana/Photo Researchers, Inc.; **18.15c (Kangaroo):** © George Holton/Photo Researchers, Inc.; **18.15d (Dasyurus):** © Tom McHugh/Photo Researchers, Inc.; **18.15e (Tasmanian wolf):** © Tom McHugh/Photo Researchers, Inc.; **18.17a:** © Carolina Biological Supply/Phototake NYC; **18.17b:** © Carolina Biological Supply/Phototake NYC.

Chapter 19

Opener: © Oscar Burriel/SPL/Photo Researchers, Inc.; **19.2a:** © Breck Kent/Animals Animals/Earth Scenes; **19.2d:** © Michael Tweedie/Photo Researchers, Inc.; **19.3b:** © Visuals Unlimited; **19.3c:** © Zig Leszczynski/Animals Animals/Earth Scenes; **19.3d:** © Joseph Collins/Photo Researchers, Inc.; **19.3e:** © Dale Jackson/Visuals Unlimited; **19.3f:** © Zig Leszczynski/Animals Animals/Earth Scenes; **19.3g:** © William Weber/Visuals Unlimited; **19.5:** Courtesy Victor McKusick; **19.8b:** © Bob Evans/Peter Arnold, Inc.; **19.10a (Least):** © Stanley Maslowski/Visuals Unlimited; **19.10b (Acadian):** © Karl Maslowski/Visuals Unlimited; **19.10c (Traill's):** © Ralph Reinhold/Animals Animals/Earth Scenes, p. 314; **19A,B,C:** Courtesy Gerald D. Carr.

Chapter 20

Opener: © Nada Pecnik/Visuals Unlimited; **20.2:** © Dudley Foster/Woods Hole Oceanographic Institution; **20.3a:** © Science VU/Visuals Unlimited; **20.3b:** Courtesy Dr. David Deamer; **20.5:** © Henry W. Robinson/ Visuals Unlimited; **20.6 (Trees):** © Doug Sherman/Geofile; **20.6 (Ammonites):** © Sinclair Stammers/SPL/

Photo Researchers, Inc.; **20.6 (Tracks):** © Scott Berner/Visuals Unlimited; **20.6 (Midge):** © Alfred Pasieka/SPL/Photo Researchers, Inc.; **20.6 (Fern):** © Carolina Biological Supply/Phototake; **20.6 (Insectivore):** © Gary Retherford/Photo Researchers, Inc.; **20.7a:** Courtesy J. William Schopf; **20.7b:** © Francois Gohier/Photo Researchers, Inc.; **20.8a:** © G.R. Roberts; **20.8b:** Courtesy Dr. Bruce N. Runnegar; **20.9a:** #1096 Courtesy of American Museum of Natural History; **20.9b:** Courtesy of the National Museum of Natural History, Smithsonian Institution, photograph by Chip Clark; **20.10, 20.11a, 20.11b:** Courtesy of The Field Museum; **20Aa, 20Ab:** Courtesy Museum of the Rockies; **20.12, 20.13:** Courtesy of The Field Museum; **20.15b:** © David Parker/Photo Researchers, Inc.; **20.15c:** © Matthew Shipp/Photo Researchers, Inc.

Chapter 21

Opener: © CABISCO/Visuals Unlimited; **21.2b:** © Howard Uible/Photo Researchers, Inc.; **21.2c:** © Ron Austing/Photo Researchers, Inc.; **21.4c:** © National Museum of Kenya; **21.5a (Gibbon):** © Martha Reeves/Photo Researchers, Inc.; **21.5b (Orangutan):** © BIOS/Peter Arnold, Inc.; **21.5c (Gorilla):** © George Holton/Photo Researchers, Inc.; **21.5d (Chimpanzee):** © Tom McHugh/Photo Researchers, Inc.; **21.7a:** © Dan Dreyfus and Associates; **21.7b:** © John Reader/Photo Researchers, Inc.; **21.8a:** © Institute of Human Origins; **21A:** © Margaret Miller/Photo Researchers, Inc.; **21.10:** © National Museum of Kenya; **21.12:** Courtesy of The Field Museum; **21.13:** Transp. #608 Courtesy Dept. of Library Services, American Museum of Natural History.

Chapter 22

Opener: © T & P Gardner/Bruce Coleman, Inc.; **22.2a:** © R. Andrew Odum/Peter Arnold, Inc.; **22.2b:** © R. Andrew Odum/Peter Arnold, Inc.; **22.6:** © Frans Lanting/Minden Pictures; **22.8a:** © Y. Arthus-Bertrand/Peter Arnold, Inc.; **22.8b:** © FPG International; **22.9:** © Jonathan Scott/Planet Earth; **22.10:** © Susan Kuklin/Photo Researchers, Inc.; **22.11a:** © OSF/Animals Animals/Earth Scenes; **22.12:** © J & B Photo/Animals Animals/Earth Scenes.

Chapter 23

Opener: © Nigel Dennis/Photo Researchers, Inc.; **23.1b:** © David Hall/Photo Researchers; **23.2a:** © C. Palck/Animals Animals/Earth Scenes; **23.2b:** © S.J. Krasemann/Peter Arnold, Inc.; **23.2c:** © Peter Arnold/Peter Arnold, Inc.; **23.3a:** © Breck P. Kent/Animals Animals/Earth Scenes; **23.3b:** © Doug Sokell/Visuals Unlimited; **23.6a,b:** © The McGraw-Hill Companies, Inc./Bob Coyle, photographer; **23.9a:** © Paul Janosi/Valan Photos; **23.10:** © Kent & Donna Dennen/Photo Researchers, Inc.; **23.11a:** © Neil Bromhall/OSF/Animals Animals/Earth Scenes; **23.12a (Strawberry arrow poison):** © Michael Fogden/Animals Animals/Earth Scenes; **23.12b (Darwin's frog):** © Michael Fogden/Animals Animals/Earth Scenes; **23.12c (Surinam toad):** © Tom McHugh/Photo Researchers, Inc.; **23.12d (Wood frog pair):** © Matt Meadows/Peter Arnold, Inc.; **23.12e (Midwife toad):** © Mike Linley/OSF/Animals Animals/Earth Scenes; **23.13a:** © Ted Levin/Animals Animals/Earth Scenes; **23.13b:** © Michio Hoshino/Minden Pictures; **23C:** Courtesy Jeffrey Kassner; **23.14a:** © Ben Osborne/OSF/Animals Animals/Earth Scenes; **23.14b:** © Jeff Greenberg/Peter Arnold, Inc.

Chapter 24

Opener: © Michael H. Francis; **24.1a:** © Charlie Ott/Photo Researchers, Inc.; **24.1b:** © Michael Graybill and Jan Hodder/Biological Photo Service; **24.5:** © Ken Wagner/Phototake; **24.9a:** © Biophoto Assoc./Photo Researchers, Inc.; **24.10a:** © Alan Carey/Photo Researchers, Inc.; **24.11a,b:** © Herb Segars; **24.12a:** © Zig Leszczynski/Animals Animals/Earth Scenes; **24.12b:** © National Audubon Society/A. Cosmos Blank/Photo Researchers, Inc.; **24.13a–e:** © Edward Ross; **24A:** Courtesy Dr. Ian Wyllie; **24.15:** © Dave B. Fleetham/Visuals Unlimited; **24.16a,b,c:** Courtesy Daniel Janzen; **24.17:** © Bill Wood/Bruce Coleman, Inc.; **24.18a,b,c,d,e:** © Breck P. Kent/Animals Animals/Earth Scenes; **24.20a:** © Jeff Foott/Bruce Coleman, Inc.; **24.21a:** © William E. Townsend, Jr./Photo Researchers, Inc.

Chapter 25

Opener: © Andras Dancs/Tony Stone Images; **25.1:** Courtesy NASA; **25.2 (Tree):** © Hermann Eisenbeiss/Photo Researchers, Inc.; **25.2 (Diatom):** © Ed Reschke/Peter Arnold, Inc.; **25.2 (Giraffe):** © George W. Cox; **25.2 (Caterpillar):** © Muridsany et Perennoy/Photo Researchers, Inc.; **25.2 (Osprey):** © Joe McDonald/Visuals Unlimited; **25.2 (Mantis):** © Scott Camazine; **25.2 (Mushroom):** © Michael Beug; **25.2 (Acetobacter):** © David M. Phillips/Visuals Unlimited; **25.13a:** © John Shaw/Tom Stack & Assoc.; **25.13b:** © Thomas Kitchin/Tom Stack & Assoc.; **25.14c:** © Bill Aron/Photo Edit.

Chapter 26

Opener: © NASA/TSADO/Tom Stack & Associates; **26.7a:** © John Shaw/Tom Stack & Assoc.; **26.7b:** © John Eastcott/Animals Animals/Earth Scenes; **26.7c:** © John Shaw/Bruce Coleman, Inc.; **26.8b:** © Mack Henly/Visuals Unlimited; **26.8c:** © Bill Silliker, Jr./Animals Animals/Earth Scenes; **26.9 (Forest):** © E.R. Degginger/Animals Animals/Earth Scenes; **26.9 (Chipmunk):** © Zig Lesczynski/Animals Animals/Earth Scenes; **26.9 (Millipede):** © OSF/Animals Animals/Earth Scenes; **26.9 (Bobcat):** © Tom McHugh/Photo Researchers, Inc.; **26.9 (Marigolds):** © Virginia Neefus/Animals Animals/Earth Scenes; **26.11 (Ocelot):** © Martin Wendler/Peter Arnold, Inc.; **26.11 (Katydid, Poison dart frog):** © James Castner; **26.11 (Macaw, Butterfly, Arboreal lizard):** © Kjell Sandved/Butterfly Alphabet; **26.11 (Lemur):** © Erwin & Peggy Bauer/Bruce Coleman, Inc.; **26.12:** © Bruce Iverson; **26.12 (Inset):** © Kathy Merrifield/Photo Researchers, Inc.; **26.13a (Prairie):** © Jim Steinberg/Photo Researchers, Inc.; **26.13b (Bison):** © Steven Fuller/Animals Animals/Earth Scenes; **26.14a,b,e:** © Darla G. Cox; **26.14c:** © George W. Cox; **26.15 (Kangaroo rat):** © Bob Calhoun/Bruce Coleman, Inc.; **26.15 (Roadrunner):** © Jack Wilburn/Animals Animals/Earth Scenes; **26.15c:** © John Shaw/Bruce Coleman; **26.16 (Stonefly):** © Kim Taylor/Bruce Coleman, Inc.; **26.16 (Trout):** © William H. Mullins/Photo Researchers, Inc.; **26.16 (Carp):** © Robert Maier/Animals Animals/Earth Scenes; **26.17a:** © Roger Evans/Photo Researchers, Inc.; **26.17b:** © Michael Gadomski/Animals Animals/Earth Scenes; **26.19 (Pike):** © Robert Maier/Animals Animals/Earth Scenes; **26.19 (Pond skater):** © G.I. Bernard/Animals Animals/Earth Scenes; **26.20b:** © Heather Angel; **26.21a:** © John Eastcott/Yva Momatiuk/Animals Animals/Earth Scenes; **26.21b:** © James Castner; **26.22a:** © Ann Wertheim/Animals Animals/Earth Scenes; **26.22b:** © Carlyn Iverson/Absolute Images; **26.22c:** © Jeff Greenburg/Photo Researchers, Inc.; **26.26a:** © Mike Bacon/Tom Stack & Assoc.

Chapter 27

Opener: © Tom Stoddart/Katz Pix/Woodfin Camp; **27.3a:** © Kevin Schaefer/Peter Arnold, Inc.; **27.3b:** © John Cancalosi/Peter Arnold, Inc.; **27.3c:** © Anthony Mercieca/Photo Researchers, Inc.; **27.3d:** © Merlin D. Tuttle/Bat Conservation International; **27.3e:** © Herve Donnezan/Photo Researchers, Inc.; **27.3f:** © Bryn Campbell/Tony Stone Images; **27.4a:** © The McGraw-Hill Companies, Inc./Carlyn Iverson, photographer; **27.4b:** © Don and Pat

Valenti/Tony Stone Images; **27.5b:** © Gunter Ziesler/Peter Arnold, Inc.; **27.5c:** Courtesy Woods Hole Research Center; **27.5d:** Courtesy R.O. Bierregaard; **27.5e:** Courtesy Thomas G. Stone, Woods Hole Research Center; **27.6:** © Chris Johns/ National Geographic Society Image Collection; **27.7b:** Courtesy Walter C. Jaap/ Florida Fish & Wildlife Conservation Commission; **27.8a:** © Shane Moore/ Animals Animals/Earth Scenes; **27.8b (both):** © Peter Auster/University of Connecticut; **27.9a:** © Gerard Lacz/Peter Arnold, Inc.; **27.9b:** © Art Wolfe; **27.9c:** © Pat & Tom Leeson/Photo Researchers, Inc.; **27.10b:** © Jeff Foott Productions; **27.11 (top left):** © Kim Heacox/Peter Arnold, Inc.; **27.11 (lower left):** Courtesy Darrell Land/Florida Game and Freshwater Fish Commission; **27.11 (top right):** © Bruce S. Cushing/Visuals Unlimited; **27.11 (lower right):** © Steven J. Maka.

Chapter 28

Opener: © Michael Fogden/Bruce Coleman, Inc.; **28.2a:** Courtesy Uppsala University Library, Sweden; **28.2b (Bubil lily):** © Arthur Gurmankin/Visuals Unlimited; **28.2c (Canada lily):** © Dick Poe/ Visuals Unlimited; **28.3:** © Tim Davis/ Photo Researchers, Inc.; **28.4:** © Jen & Des Bartlett/Bruce Coleman, Inc.; **28.8a:** © John D. Cunningham/Visuals Unlimited; **28.8b:** © John Shaw/Tom Stack & Assoc.; **28.7c:** Courtesy Dr. David Dilcher, Florida Museum of Natural History, University of Florida; **28.15 (E. coli):** © David M. Phillips/ Visuals Unlimited; **28.15 (Archaea):** © Ralph Robinson/Visuals Unlimited; **28.15 (Flower):** © Ed Reschke/Peter Arnold, Inc.; **28.15 (Paramecium):** © M. Abbey/Visuals Unlimited; **28.15 (Mushroom):** © Rod Planck/Tom Stack & Assoc.; **28.15 (Wolf):** © Art Wolfe/Tony Stone Images.

Chapter 29

Opener: © WHOI/Dudley Foster/Visuals Unlimited; **29.1a:** © Robert Caughey/ Visuals Unlimited; **29.1b:** © Michael Wurz/ Biozentrum, Univ. of Basel/Photo Researchers, Inc.; **29.1c:** © Science Source/ Photo Researchers, Inc.; **29.1d:** © K.G. Murti/Visuals Unlimited; **29.2:** © Ed Degginger/Color Pic Inc.; **29.7:** © Dr. Tony Brain/SPL/Photo Researchers, Inc.; **29.8:** Courtesy Nitragin Company, Inc.; **29.9a:** © R.G. Kessel - C.Y. Shih/Visuals Unlimited; **29.9b,c Bacillus, Coccus:** © David M. Phillips/Visuals Unlimited; **29.10a:** © R. Knauft/Biology Media/Photo Researchers, Inc.; **29.10b:** © Eric Grave/ Photo Researchers, Inc.; **29.11 (top):** © Jeff Lepore/Photo Researchers, Inc.; **29.11 (center):** Courtesy Dennis W. Grogan, Univ. of Cincinnati; **29.11 (lower):** Courtesy Prof. Dr. Karl O. Stetter, Univ. Regensburg, Germany.

Chapter 30

Opener: © Triarch/Visuals Unlimited; **30.1b (left):** © Hal Beral/Visuals Unlimited; **30.1b (right):** © E. White/Visuals Unlimited; **30.2 (Blepharisma):** © Eric Grave/Photo Researchers, Inc.; **30.2b (Onychodromus):** Courtesy Dr. Barry Wicklow; **30.2 (Ceratium):** © D.P. Wilson/ Photo Researchers, Inc.; **30.2 (Bossiella):** © Daniel V. Gotschall/Visuals Unlimited; **30.2 (Acetabularia):** © L.L. Sims/ Visuals Unlimited; **30.2 (Naegleria):** © Science VU/David John/Visuals Unlimited; **30.2g (Synura):** Courtesy Dr. Ronald W. Hoham; **30.2h (Licmophorta):** © Biophoto Assoc./Photo Researchers, Inc.; **30.4b:** © M.I. Walker/Science Source/Photo Researchers, Inc.; **30.5b:** © William E. Ferguson; **30.6a:** © John D. Cunningham/ Visuals Unlimited; **30.6b:** © CABISCO/ Visuals Unlimited; **30.7:** © Walter Hodge/ Peter Arnold, Inc.; **30.8b:** © D.P. Wilson/ Eric & David Hosking/Photo Researchers, Inc.; **30.9a:** © Dr. Ann Smith/Photo Researchers, Inc.; **30.9b:** © Biophoto Assoc./Photo Researchers, Inc.; **30.10, 30.11a:** © Michael Abbey/Visuals Unlimited; **30.11b:** © Ed Reschke/Peter Arnold, Inc.; **30.12b:** © Manfred Kage/Peter Arnold, Inc.; **30.12c:** © G. Shih, R. Kessel/ Visuals Unlimited; **30.13a:** © Eric Grave/ Photo Researchers, Inc.; **30.13b:** © CABISCO/Phototake; **30.13c:** © Manfred Kage/Peter Arnold, Inc.; **30.15 (top):** © V. Duran/Visuals Unlimited; **30.15 (bottom):** © CABISCO/Visuals Unlimited.

Chapter 31

Opener: © Phil Dotson/Photo Researchers, Inc.; **31.1a:** © Gary R. Robinson/Visuals Unlimited; **31.1c:** © Gary T. Cole/Biological Photo Service; **31.2a:** From C.Y. Shih and R.G. Kessel, *Living Images* Science Books International, Boston, 1982; **31.2b:** © Jeffrey Lepore/Photo Researchers, Inc.; **31.3a:** © James W. Richardson/Visuals Unlimited; **31.3c:** © David M. Phillips/Visuals Unlimited; **31.4a:** © Walter H. Hodge/Peter Arnold, Inc.; **31.4b (top):** © James Richardson/Visuals Unlimited; **31.4b (bottom):** © Michael Viard/Peter Arnold, Inc.; **31.4c:** © Kingsley Stern; **31.5:** © J. Forsdyke/GeneCox/SPL/Photo Researchers, Inc.; **31.6a (right):** © Biophoto Associates; **31.6b:** © Glenn Oliver/Visuals Unlimited; **31.6c:** © M. Eichelberger/Visuals Unlimited; **31.6d (left):** © Dick Poe/Visuals Unlimited; **31.6d (right):** © L. West/Photo Researchers, Inc.; **31.7a:** © Leonard L. Rue/ Photo Researchers, Inc.; **31.7b:** © Arthur M. Siegelman/Visuals Unlimited; **31A:** © G. Tomsich/Photo Researchers, Inc.; **31B:** © R. Calentine/Visuals Unlimited; **31.8a:** Courtesy G.L. Barron/University of Guelph; **31.9b:** © Stephen Krasemann/Peter Arnold, Inc.; **31.9c:** © John Shaw/Tom Stack & Assoc.; **31.9d:** © Kerry T. Givens/ Tom Stack & Assoc.; **31.10:** © R. Roncadori/ Visuals Unlimited.

Chapter 32

Opener: © The McGraw-Hill Companies, Inc./Carlyn Iverson, photographer; **32.3a:** © Ed Reschke/Peter Arnold, Inc.; **32.3b (left):** © J.M. Conrarder/Nat'l Audubon Society/ Photo Researchers, Inc.; **32.3b (right):** © R. Calentine/Visuals Unlimited; **32.4 (top):** © Heather Angel/Biofotos; **32.4 (bottom):** © Bruce Iverson; **32.5:** © Doug Sokell/Tom Stack & Assoc.; **32.6a:** © CABISCO/Phototake; **32.7a:** © Steve Solum/Bruce Coleman, Inc.; **32.8a:** © Robert P. Carr/Bruce Coleman, Inc.; **32.9 (left):** © John Gerlach/Visuals Unlimited; **32.9 (center):** © Walter H. Hodge/Peter Arnold, Inc.; **32.9 (right):** © Forest W. Buchanan/Visuals Unlimited; **32.10b:** © Matt Meadows/Peter Arnold, Inc.; **32.11b:** © CABISCO/Phototake; **32.12a:** © David Hosking/Photo Researchers, Inc.; **32.12b:** © Kingsley Stern; **32.12c:** © Edward Ross; **32.13:** © Dwight Kuhn.

Chapter 33

Opener: © James Castner; **33.1 (top):** © Joe McDonald/Visuals Unlimited; **33.1 (lower):** © CABISCO/Visuals Unlimited; **33.3a:** © Andrew J. Martinez/Photo Researchers, Inc.; **33.4a:** © J. Mcullagh/Visuals Unlimited; **33.4b:** © Jeff Rotman; **33.5b:** © CABISCO/Phototake; **33.5c:** © Ron Taylor/Bruce Coleman; **33.5d:** © Runk/ Schoenberger/Grant Heilman Photography; **33.5e:** © Gregory Ochocki/Photo Researchers, Inc.; **33.6b:** © CABISCO/ Visuals Unlimited; **33.7a:** © Runk/ Schoenberger/Grant Heilman Photography; **33.8:** © Fred Bavendam/Peter Arnold, Inc.; **33.9a:** © Tom E. Adams/Peter Arnold, Inc.; **33.11a:** © John D. Cunningham/ Visuals Unlimited; **33.11b:** © James Webb/ Phototake NYC; **33.12a:** © Arthur Siegelman/Visuals Unlimited; **33.12c:** © James Solliday/Biological Photo Service; **33.13:** From E.K. Markell and M. Voge *Medical Parasitology, 7/e.* 1992 W.B. Saunders Co.; **33.14:** © Peter Parks/OSF/Animals , Animals/Earth Scenes.

Chapter 34

Opener: © Andrew Henley/Biofotos; **34.2b:** © Kjell Sandved/Butterfly Alphabet; **34.3a:** Courtesy of Larry S. Roberts; **34.3b:** © Rick Harbo; **34.4a:** © Alex Kerstitch/Bruce Coleman, Inc.; **34.4b:** © Douglas Faulkner/Photo Researchers, Inc.; **34.4c:** © Michael DiSpezio; **34.5a:** © M. Gibbs/OSF/Animals Animals/Earth Scenes; **34.5b:** © Kenneth W. Fink/Bruce Coleman, Inc.; **34.5c:** © James H. Carmichael; **34.6a:** © Heather Angel; **34.6b:** © James H. Carmichael; **34.7c:** © Roger K. Burnard/Biological Photo Service; **34.8:** © St. Bartholomew's Hospital/SPL/Photo Researchers, Inc.; **34.11a:** © James H. Carmichael; **34.11b:** © Natural History Photographic Agency; **34.11c:** © Kim Taylor/Bruce Coleman, Inc.; **34.11d:** © Kjell Sandved/Butterfly Alphabet; **34.12 (Dragonfly):** © John Gerlach/Tony Stone Images; **34.12 (Walking stick):** © Art Wolfe/Tony Stone Images; **34.12 (Scale):** © Science VU/Visuals Unlimited; **34.12 (Grasshopper):** © Alex Kerstitch/Visuals Unlimited; **34.12 (Beetle):** © Kjell Sandved/Bruce Coleman, Inc.; **34.12 (Lacewing):** © Glenn Oliver/Visuals Unlimited; **34.13 (top):** © Bill Beatty/Visuals Unlimited; **34.13 (lower):** © L. West/Bruce Coleman, Inc.; **34.15a:** © Dwight Kuhn; **34.15c:** © John MacGregor/Peter Arnold, Inc.; **34.16 (Scorpion):** © Tom McHugh/Photo Researchers, Inc.; **34.16 (Crab):** © Zig Leszczynski/Animals Animals/Earth Scenes; **34.16 (Spider):** © Ken Lucas/Planet Earth Pictures Limited.

Chapter 35

Opener: © Merlin D. Tuttle/Bat Conservation International; **35.1a:** © Randy Morse/Tom Stack & Assoc.; **35.1b:** © Alex Kerstitch/Visuals Unlimited; **35.1c:** © Randy Morse/Animals Animals/Earth Scenes; **35.2b:** © Heather Angel; **35.3a:** © Rick Harbo; **35.5:** © Heather Angel; **35.6a:** © Norbert Wu; **35.6b:** © Fred Bavendam/Minden; **35.7a:** © Ron & Valarie Taylor/Bruce Coleman, Inc.; **35.7b:** © Hal Beral/Visuals Unlimited; **35.7c:** © Jane Burton/Bruce Coleman, Inc.; **35.8a:** © Estate of Dr. Jerome Metzner/Peter Arnold, Inc.; **35.9a:** © Suzanne L. Collins & Joseph T. Collins/Photo Researchers, Inc.; **35.9b:** © Joe McDonald/Visuals Unlimited; **35.9c:** Courtesy Dr. Marvalee H. Wake; **35.10a–d:** © Jane Burton/Bruce Coleman, Inc.; **35.13a:** © Bruce Davidson/Animals Animals/Earth Scenes; **35.13b:** © H. Hall/OSF/Animals Animals/Earth Scenes; **35.13c:** © Joe McDonald/Visuals Unlimited; **35.13d:** © Nathan W. Cohen/Visuals Unlimited; **35.14b:** © R.F. Ashley/Visuals Unlimited; **35A:** © William Weber/Visuals Unlimited; **35.15a:** © Kirtley Perkins/Visuals Unlimited; **35.15b:** © Thomas Kitchin/Tom Stack & Associates; **35.15c:** © Brian Parker/Tom Stack & Associates; **35.16c,d:** © Daniel J. Cox/Tony Stone Images; **35.17a:** © Tom McHugh/Photo Researchers, Inc.; **35.17b:** © Leonard Lee Rue/Photo Researchers, Inc.; **35.17c:** © Tony Stone Images; **35.18a:** © Leonard Lee Rue; **35.18b:** © Stephen J. Krasemann/DRK Photo; **35.18c:** © Denise Tackett/Tom Stack & Associates; **35.18d:** © Mike Bacon/Tom Stack & Associates.

Chapter 36

Opener: © Richard Thom/Visuals Unlimited; **36.2a:** © Michael Gadomski/Photo Researchers, Inc.; **36.2b:** © Norman Owen Tomalin/Bruce Coleman, Inc.; **36.2c:** © Brian Stablyk/Tony Stone Images; **36.4a:** © Runk Schoenberger/Grant Heilman Photography; **36.4b:** © J.R. Waaland/Biological Photo Service; **36.4c:** © Biophoto Assoc./Photo Researchers, Inc.; **36.5a,b,c:** © Biophoto Associates/Photo Researchers, Inc.; **36.6a:** © J. Robert Waaland/Biological Photo Service; **36.7a:** © George Wilder/Visuals Unlimited; **36.8b:** © CABISCO/Phototake; **36.9:** © Dwight Kuhn; **36.10a:** © John D. Cunningham/Visuals Unlimited; **36.10b:** Courtesy of George Ellmore, Tufts University; **36.11a:** © G.R. Roberts; **36.11b:** © Ed Degginger/Color Pic; **36.11c:** © David Newman/Visuals Unlimited; **36.11d (left):** © Biophot; **36.11d (right):** © Barbara J. Miller/Biological Photo Service; **36.12a:** © J. Robert Waaland/Biological Photo Service; **36.13a:** © Ed Reschke; **36.13b:** © Dwight Kuhn; **36.14a:** © CABISCO/Phototake; **36.14b:** © Runk/Schoenberger/Grant Heilman Photography; **36.16a:** © Ardea London Limited; **36.17a:** © John D. Cunningham/Visuals Unlimited; **36.17b:** © William E. Ferguson; **36.17c,d:** © The McGraw-Hill Companies, Inc./Carlyn Iverson, photographer; **36B:** © James Schnepf Photography, Inc.; **36.18b:** © Jeremy Burgess/SPL/Photo Researchers, Inc.; **36.20 (Cactus):** © Patti Murray/Animals Animals/Earth Scenes; **36.20 (Cucumber):** © Michael Gadomski/Photo Researchers, Inc.; **36.20 (Flytrap):** © CABISCO/Phototake; **36C (Wheat plants):** © C.P. Hickman/Visuals Unlimited; **36C (Wheat grains):** © Philip Hayson/Photo Researchers, Inc.; **36C (Corn plants):** © Adam Hart-Davis/SPL/Photo Researchers, Inc.; **36C (Corn grains):** © Mark S. Skalny/Visuals Unlimited; **36C (Rice plants):** © Scott Camazine; **36C (Rice grains):** © John Tiszler/Peter Arnold, Inc.; **36.Da (both):** © Heather Angel; **36Db (left):** © Dale Jackson/Visuals Unlimited; **36Db (right):** © Bob Daemmrich/The Image Works, Inc.; **36Dc:** © Steven King/Peter Arnold, Inc.; **36Dc (inset):** © Will & Deni McIntyre/Photo Researchers, Inc.

Chapter 37

Opener: © Lawrence Migdale/Photo Researchers, Inc.; **37.1:** © Dwight Kuhn; **37.2a:** Courtesy Mary E. Doohan; **37.2b,c:** Courtesy Mary E. Doohan; **37.6a:** © Dwight Kuhn; **37.6b:** © E.H. Newcomb & S.R. Tardon/Biological Photo Service; **37.7:** © Runk Schoenberger/Grant Heilman; **37Aa, Ab:** © Dwight Kuhn; **37.9a,b:** Courtesy W.A. Cote, Jr., N.C. Brown Center for Ultrastructure Studies, SUNY-ESF; **37.10:** © Ed Reschke/Peter Arnold, Inc.; **37.12a:** © Jeremy Burgess/SPL/Photo Researchers, Inc.; **37.12b:** © Jeremy Burgess/SPL/Photo Researchers, Inc.; Science Focus p. 677: Courtesy G. David Tilman; **37.13a:** From M.H. Zimmerman. "Movement of Organic Substances in Trees. SCIENCE 133 (13 Jan. 1961), p. 667 © AAAS; **37.13b:** © Bruce Iverson/SPL/Photo Researchers, Inc.

Chapter 38

Opener: © Bates Littlehales/Animals Animals/Earth Scenes; **38.1:** © Kim Taylor/Bruce Coleman, Inc.; **38.2a:** © Kingsley Stern; **38.2b:** Courtesy Malcom Wilkins, Botany Department, Glascow University; **38.2c:** © Biophot; **38.3:** © John D. Cunningham/Visuals Unlimited; **38.4a,b:** © John Kaprielian/Photo Researchers, Inc.; **38.5a,b:** © Tom McHugh/Photo Researchers, Inc.; **38.6a,b:** Courtesy Prof. Malcolm B. Wilkins; **38.7a,b:** © Runk Schoenberger/Grant Heilman Photography; **38.10:** © Robert E. Lyons/Visuals Unlimited; p. 691: Courtesy Donald Briskin; **38.12a–d:** Courtesy Alan Darvill and Stefan Eberhard, Complex Carbohydrate Research Center, University of Georgia; **38.14a,b:** © Kingsley Stern; **38.17a,b:** © Grant Heilman Photography.

Chapter 39

Opener: © Dwight Kuhn; **39.2:** © Michael Viard/Peter Arnold, Inc.; **39Aa:** © Comstock; **39Ab:** © H. Eisenbeiss/Photo Researchers, Inc.; **39Ac:** © Anthony Mercieca/Photo Researchers, Inc.; **39Ad:** © Merlin D. Tuttle/Bat Conservation International; **39.4a:** © Biophoto Assoc./Photo Researchers, Inc.; **39.4b:** © Jeremy Burgess/SPL/Photo Researchers, Inc.; **39.7a (left):** © Joe Munroe/Photo Researchers, Inc.; **39.7a (inset):** © Kingsley Stern; **39.7b (left):** © Ralph Reinhold/Animals Animals/Earth Scenes; **39.7b (right):** © W. Ormerod/Visuals Unlimited; **39.7c (both):** © Dwight Kuhn; **39.7d (left):** © Christopher Lobban/Biological Photo Service; **39.7d (right):** © John N.A. Lott/Biological Photo Service; **39.10:** © G.I. Bernard/Animals Animals/Earth Scenes; **39.11a–f:** Courtesy Prof. Dr. Hans-Ulrich

Koop, from *Plant Cell Reports,* 17:601–604; **39.12a–c:** Courtesy Monsanto; **39.13b:** Courtesy Eduardo Blumwald; **39B,C,D,E:** Courtesy Elliot Meyerowitz/California Institute of Technology.

Chapter 40

Opener: © Zig Leszczynski/Animals Animals/Earth Scenes; **40.2 (Stratified squamous, Pseudostratified, Simple squamous, Simple cubodial):** © Ed Reschke; **40.2 (Simple columnar):** © Ed Reschke/Peter Arnold, Inc.; **40.3a–d:** © Ed Reschke; **40.5a–c:** © Ed Reschke; **40Aa:** © Ken Greer/Visuals Unlimited; **40Ab:** © Dr. P. Marazzi/ SPL/Photo Researchers, Inc.; **40Ac:** © James Stevenson/SPL/Photo Researchers, Inc.

Chapter 41

Opener: © David M. Phillips/Visuals Unlimited; **41.1a:** © Eric Grave/Photo Researchers, Inc.; **41.1b:** © CABISCO/ Phototake; **41.1c:** © Michael DiSpezio; **41A:** © Bettmann/Corbis; **41B:** © Biophoto Associates/Photo Researchers, Inc.; **41.12b:** © Manfred Kage/Peter Arnold, Inc.

Chapter 42

Opener: © NIBSC/SPL/Photo Researchers, Inc.; **42.2 (Thymus):** © Ed Reschke/Peter Arnold, Inc.; **42.2 (Spleen):** © Ed Reschke/ Peter Arnold, Inc.; **42.2 (Bone marrow):** © R. Calentine/Visuals Unlimited; **42.2 (Lymph node):** © Fred E. Hossler/ Visuals Unlimited; **42.6b:** Courtesy Dr. Arthur J. Olson, Scripps Institute; **42Aa:** © AP/Wide World Photo; **42.7a:** © Bohringer Ingelheim International, photo by Lennart Nilsson; **42.9a:** © Matt Meadows/Peter Arnold, Inc.; **42.10:** © Estate of Ed Lettau/ Peter Arnold, Inc.; **42.12 (both):** Courtesy of Stuart I. Fox; **42.13:** © Martha Cooper/Peter Arnold, Inc.

Chapter 43

Opener: © Matt Meadows/Peter Arnold, Inc.; **43.8b:** © Ed Reschke/Peter Arnold, Inc.; **43.8c:** © St. Bartholomew's Hospital/SPL/ Photo Researchers, Inc.; **43.9c:** © Manfred Kage/Peter Arnold, Inc.; **43.9d:** Photo by Susumu Ito, from Charles Flickinger *Medical Cellular Biology* W.B. Saunders, 1979.

Chapter 44

Opener: © Bates Littlehales/Animals Animals/Earth Scenes.

Chapter 45

Opener: © Biophoto Associates/Photo Researchers, Inc.; **45.7 (top left):** © R.G. Kessel and R.H. Kardon, *Tissues and Organs: A Text-Atlas of Scanning Electron Microscopy,* 1979; **45.7 (top right and bottom):** © 1966 Academic Press, from A.B. Maunsbach, J. Ultrastruct. Res. 15:242–282; **45.8:** © J. Gennaro/Photo Researchers, Inc.

Chapter 46

Opener: © Dr. Colin Chumbley/SPL/ Photo Researchers, Inc.; **46.4:** © Linda Bartlett; **46.5a:** Courtesy Dr. E.R. Lewis, University of California Berkeley; **46.7c:** © Manfred Kage/Peter Arnold, Inc.

Chapter 47

Opener: © Johnny Johnson/Animals Animals/Earth Scenes; **47.1:** © Omikron/ SPL/Photo Researchers, Inc.; **47.4a:** © Heather Angel; **47.4b:** © Heather Angel; **47.7:** © Lennart Nilsson, from *The Incredible Machine;* **47.8:** © Biophoto Associates/ Photo Researchers, Inc.; **47A:** Robert S. Preston, courtesy Prof. J.E. Hawkins, Kresge Hearing Research Institute, Univ. of Michigan Medical School; **47.10a:** © P. Motta SPL/Photo Researchers, Inc.

Chapter 48

Opener: © Robert Maier/Animals Animals/Earth Scenes; **48.2:** © Michael Fogden/OSF/Animals Animals/Earth Scenes; **48.4 (Hyaline cartilage, Compact bone):** © Ed Reschke; **48.4 (Osteocyte):** © Biophoto Associates/Photo Researchers, Inc.; **48Aa:** © Royce Bair/Unicorn Stock Photo; **48Ab-1,2:** © Michael Klein/Peter Arnold, Inc.; **48.13b:** © Ed Reschke/Peter Arnold, Inc.; **48.14:** © Victor B. Eichler.

Chapter 49

Opener: © A.M. Shah/Animals Animals/ Earth Scenes; **49.5a:** © Bob Daemmrich/Stock Boston; **49.5b:** © Ewing Galloway, Inc.; **49.6a–d:** From Clinical Pathological Conference, "Acromegaly, Diabetes, Hypermetabolism, Proteinura and Heart Failure," *American Journal of Medicine,* 20 (1956): 133. Reprinted with permission from Excerpta Medica Inc.; **49.7:** © Biophoto Associates/Photo Researchers, Inc.; **49.8:** © John Paul Kay/Peter Arnold, Inc.; **49.12a:** © Custom Medical Stock Photos; **49.12b:** © NMSB/Custom Medical Stock Photos; **49.13a,b:** "Atlas of Pediatric Physical Diagnosis," Second Edition by Zitelli & Davis, 1992. Mosby-Wolfe Europe Limited, London, UK.

Chapter 50

Opener: © Stephen Krasemann/Tony Stone Images; **50.1:** © Runk/Schoenberger/Grant Heilman Photography; **50.2:** © Anthony Mercieca/Photo Researchers, Inc.; **50.5b:** © Biophoto Assoc./Photo Researchers, Inc.; **50.8b:** © Ed Reschke/Peter Arnold, Inc.; **50A:** © G.W. Willis/BPS/Tony Stone Images.

Chapter 51

Opener (four days): Lennart Nilsson, "A Child is Born," 1990, Delacorte Press, p. 63; **Opener (five weeks):** Lennart Nilsson, "A Child is Born," 1990, Delacorte Press, p. 81; **Opener (three weeks):** Lennart Nilsson, "A Child is Born," 1990, Delacorte Press, p. 76; **Opener (four months):** Lennart Nilsson, "A Child is Born," 1990, Delacorte Press, p. 111 top frame; **51.1:** © David M. Phillips/Visuals Unlimited; **51A–1, A–2:** Courtesy John Sternfeld; **51.9a–c:** Courtesy Steve Paddock, Howard Hughes Medical Research Institute; **51.10a:** Courtesy E.B. Lewis; **51.14a:** © Lennart Nilsson *A Child is Born* Dell Publishing.

Index

Note: Page numbers followed by f refer to figures; page numbers followed by t refer to tables.

a

∂

e

g

j

k

l

m

n

O

q

r

t

u

v